Delmar's Agriscience Dictionary

with
Searchable CD–ROM

Ray V. Herren & Roy L. Donahue

Online Services

Delmar Online
To access a wide variety of Delmar products and services on the World Wide Web, point your browser to:
http://www.delmar.com
or email: info@delmar.com

A service of I(T)P®

Delmar's
Agriscience Dictionary

with
Searchable CD–ROM

Ray V. Herren & Roy L. Donahue

Delmar Publishers

 an International Thomson Publishing company

Albany • Bonn • Boston • Cincinnati • Detroit • London • Madrid
Melbourne • Mexico City • New York • Pacific Grove • Paris • San Francisco
Singapore • Tokyo • Toronto • Washington

NOTICE TO THE READER

Publisher does not warrant or guarantee any of the products described herein or perform any independent analysis in connection with any of the product information contained herein. Publisher does not assume, and expressly disclaims, any obligation to obtain and include information other than that provided to it by the manufacturer.

The reader is expressly warned to consider and adopt all safety precautions that might be indicated by the activities herein and to avoid all potential hazards. By following the instructions contained herein, the reader willingly assumes all risks in connection with such instructions.

The publisher makes no representation or warranties of any kind, including but not limited to, the warranties of fitness for particular purpose or merchantability, nor are any such representations implied with respect to the material set forth herein, and the publisher takes no responsibility with respect to such material. The publisher shall not be liable for any special, consequential, or exemplary damages resulting, in whole or part, from the readers' use of, or reliance upon, this material.

Cover design: Carolyn Miller

Delmar Staff:

Publisher: Susan Simpfenderfer
Acquisitions Editor: Jeff Burnham
Developmental Editor: Andrea Edwards Myers

Production Manager: Wendy Troeger
Production Editor: Carolyn Miller
Marketing Manager: Katherine M. Hans

COPYRIGHT© 2000
By Delmar Publishers

an International Thomson Publishing company I(T)P®

The ITP logo is a trademark under license
Printed in the United States of America

For more information contact:

Delmar Publishers
3 Columbia Circle, Box 15015
Albany, New York 12212-5015

International Thomson Publishing Europe
Berkshire House
168–173 High Holborn
London, WC1V7AA
United Kingdom

Nelson ITP, Australia
102 Dodds Street
South Melbourne,
Victoria, 3205 Australia

Nelson Canada
1120 Birchmont Road
Scarborough, Ontario
M1K5G4, Canada

International Thomson Publishing France
Tour Maine-Montparnasse
33 Avenue du Maine
75755 Paris Cedex 15, France

International Thomson Editores
Seneca 53
Colonia Polanco
11560 Mexico D. F. Mexico

International Thomson Publishing GmbH
Königswinterer Straße 418
53227 Bonn
Germany

International Thomson Publishing Asia
60 Albert Street
#15-01 Albert Complex
Singapore 189969

International Thomson Publishing Japan
Hirakawa-cho Kyowa Building, 3F
2-2-1 Hirakawa-cho, Chiyoda-ku,
Tokyo 102, Japan

ITE Spain/Paraninfo
Calle Magallanes, 25
28015-Madrid, Espana

All rights reserved. No part of this work covered by the copyright hereon may be reproduced or used in any form or by any means—graphic, electronic, or mechanical, including photocopying, recording, taping, or information storage and retrieval systems—without the written permission of the publisher.

1 2 3 4 5 6 7 8 9 10 XXX 03 02 01 00 99 98

Library of Congress Cataloging-in-Publication Data

Herren, Ray V.
    Delmar's agriscience dictionary with searchable CD-ROM / Ray
V. Herren & Roy L. Donahue.
       p. cm.
    Summary: Provides definitions for more than 15,000 terms in the fields of Animal Science,
Business and Mechanical Technology, Forestry, and Natural Resources Management.
    ISBN 0-7668-1146-8
    1. Agriculture--Dictionaries, Juvenile. [1. Agriculture--Dictionaries.] I. Donahue, Roy Luther, 1908-   .
II. Title. III. Title: Agriscience dictionary with searchable CD-ROM.
S411.H476 1999
630'.3--dc21
                                        98-45647
                                            CIP
                                             AC

Contents

Preface........vi

Part 1
Animal Science........AS-1

Part 2
Business and Mechanical Technology........BT-1

Part 3
Forestry and Natural Resources Management........NR-1

Part 4
Plant Science........PS-1

Preface

Agriculture today is becoming increasingly more technical and all encompassing. Modern agriculture is a broad, diverse science and has, like any other discipline, a language of its own. Keeping abreast of the terminology of an ever-changing industry is a complicated task. *Delmar's Agriscience Dictionary with Searchable CD–ROM* is a unique product that can be used in the classroom as well as for individual studies and it will help to bridge the gap across all branches of agriculture. It is an up-to-date, authoritative source of over 15,000 dictionary terms and definitions. These terms and definitions are divided into four sections to correspond with the four agriscience topic areas of Animal Science, Business and Mechanical Technology, Forestry and Natural Resources Management, and Plant Science. The terms in each of these sections are presented in A–Z order for easy access, and the terms that have different definitions for the same word or process will be repeated in the sections where the definition would be most appropriate based on definition content.

The searchable CD–ROM adds to the uniqueness of this dictionary. All of the terms and definitions in the print dictionary are available on the CD–ROM and are also divided into the four agriscience topic areas. The CD–ROM also allows the user to search within the four agriscience sections for a term and the definition of that term will automatically appear once the term is found. Terms are available alphabetically and in a scrollable format per letter of the alphabet by section. A navigation bar and agriscience icons will help the user to reach the main screens and each of the four sections quickly and easily enhancing the user-friendliness of this added value technology component.

Part 1

Animal Science

A.U.–(Abbr.) Animal unit.

A-type Merino–A type variation in the Merino breed of sheep which has very wrinkled skin covered with a fine-wool fleece and many body folds. See Merino.

Abacterial–Free from bacteria.

Abattoir–A slaughterhouse.

Abderdeen Angus–A breed of beef cattle originally from Scotland, which are black, hornless, and hardy. Also called Angus, Black Angus, doddie.

Abdomen–In mammals, that part of the body which contains the digestive organs; in lower orders of animals and insects that part of the body which roughly corresponds to the abdomen; belly.

Aberration–In genetics, an irregularity in chromosome distribution during cell division that results in deviation from normal.

Abiotic Disease–A disease caused by an inanimate agent; a nonparasitic disease. Also called physiogenic disease; e.g., a mineral deficiency in plants or animals.

Ablactate–To wean.

Abnormal–Deviating from that which is typical.

Abomasum–The fourth, or true, stomach division of a ruminant, or cud-chewing animal, such as a cow or sheep. The compound stomach consists of: (1) rumen or paunch, (2) reticulum, (3) omasum, (4) abomasum. Herbage is swallowed nearly unchewed, and passes into the rumen or reticulum, from which it is regurgitated and masticated in the mouth while the animal is resting. It is then swallowed again and passed through the reticulum ad omasum into the abomasum, where it is acted on by the gastric juices.

Abort–(1) In animals, to expel the embryo or fetus from the womb prematurely. See Brucellosis, Leptospirosis. (2) In plants, to produce few or no seeds, or to drop fruit prematurely.

Abortion–The presentation of a dead, undeveloped fetus.

Abortive–Defective or barren.

Abscess–Localized collection of pus in a cavity formed by the disintegration of tissue.

Absconding Swarm–A bee colony which abandons a hive because of disease, wax moth, or other causes; unusual in modern apiaries.

Absolute Age–Age in years rather than developmental age.

Absorb–(1) To assimilate or incorporate one substance into another, as a blotter absorbs ink. (2) In botany, to assimilate water and nutrients.

Absorption–(1) The processes by which water enters the earth. (2) The passage of digested food from the alimentary tract into the circulatory system. (3) The intake of water, gases, nutrients or other substances by plants.

Absorption Rate–The rate at which a chemical enters the body.

Abundance–The number of plants or animals per unit area.

Acariasis–The condition of plants or animals infested with mites or ticks.

Acaricide–A substance, solution, or paste which kills mites or ticks.

Acarid–Any mite or tick of the family Acaridae or order Acarina.

Acarine–Any mite or tick.

Acarine Disease–A disease of adult bees caused by a mite, *Acarapis woodi*, infesting the tracheae; not known to be present in North America.

Acarology–The branch of science that deals with mites and ticks.

Acceptable Daily Intake (ADI)–The daily dosage of a drug or a chemical residue that appears to present no appreciable risk to health during the entire lifetime of a human being or animal.

Acclimate–To become conditioned to a new climate or different growing environment, usually applied to plants. (See Hardening-off.) May also be applied to animals adjusting to a new environment.

Accredited Herd–Dairy cattle certified to be free from tuberculosis as a result of two successive tests given under the direction of the United States Department of Agriculture. The term is sometimes erroneously applied to a brucellosis-free herd.

Accumulated Grazing–Forage reserved for later use by livestock.

Accumulating Pesticide–A chemical that can build up in animals or in the environment.

Accuracy (of selection)–The correlation between an animal's unknown actual breeding value and a calculated estimated breeding value.

Acetic Acid–An organic acid, CH_3COOH, which is the chief active component of vinegar; also important as a product in lactic acid fermentation and therefore, an important constituent of flavor in many milk products.

Acetic Bacteria–The bacteria that produce acetic acid (vinegar acid) from alcohol.

Acetic Fermentation–The process by means of which acetic acid is formed from ethyl alcohol in weak solution through the action of bacterial ferments; e.g., the formation of vinegar from hard cider or from wine.

Acetobacter–A bacterium, family Nitobacteriaceae, which grows in beer and wine.

Acetone–CH_3COCH_3, a ketone; a clear, rapidly evaporating liquid which: (1) in nature is obtained from fermentation of sugar and starch and is present in diabetic urine, breath, blood. Its presence in the breath of a lactating animal (cattle, sheep, milk goats) indicates her failure to oxidize the fatty material of her feed, making her deficient in carbohydrates; (2) is commercially obtained from distillation of wood and has wide application as a solvent.

Acetonemia–A disease characterized by the presence of excessive amounts of acetone bodies in the blood. Also called ketosis; false milk fever; chronic milk fever; acidosis; acetonuria; pregnancy disease in ewes.

Achondroplasia–Skeletal malformation during prenatal development caused by a genetic factor, e.g., a bulldog calf whose legs are very short, the head short and thick, the upper lip split, the lower jaw protruding, the nose flattened, and the tongue swollen. In some instances the fetus is carried to the full term of pregnancy, in which case the calf dies within a few days.

Acid–(1) A substance containing hydrogen that dissociates to form hydrogen ions when dissolved in water (or which reacts with water to form hydronium ions). (2) A substance capable of donating protons to other substances. (3) A term applied to igneous rocks having a higher percentage of silica (66%) than orthoclase. (4) Term applied to any substance with a pH less than 7.0. See Acetic Acid.

Acid Board–A sheet of metal fastened to a wooden frame which is the same size as the top of the beehive. The underside of the sheet is covered with cloth, and the topside is painted black. The cloth is saturated with carbolic acid, the fumes of which drive the bees away from the honey in the super so that a person can work with the hive.

Acid Casein–The precipitate or curd of milk produced by the addition of acid or by developed acidity. Used for food, for making cold water paints and glue, wallpaper and fabrics, coating paper, and plastics.

Acid Cleaner–Cleaning compound made up of a combination of mild acids and wetting agents used to clean milk stone from milking equipment. See Milk Stone.

Acid Food–Food with a pH of 4.6 or below. An acid food can be safely processed for canning in a boiling-water bath for specified times. Includes most fruits, some tomatoes, and pickled vegetables. See Low-Acid Food.

Acid Hydrolysis–Decomposition or alteration of a chemical substance by acid.

Acid-fast–Property of not being readily decolorized by acids.

Acidified Silage–Silage preserved by the addition of acid, such as commercial phosphoric acid, sulfuric acid, or hydrochloric acid, or a combination of the latter two.

Acidity–The measure of how many hydrogen ions a solution contains.

Acidophilus–Refers to organisms that grow well or exclusively in an acidic soil or medium.

Acidophilus Milk–The food which results from souring milk with *Lactobacillus acidophilus*. Used for therapeutic reasons, since this bacterium, established in the intestines with fermentable carbohydrates, causes a mild acid condition; often prescribed for cases of intestinal autointoxication; also called acidophilus cultured buttermilk.

Acquired Character–A change in character of a plant or animal, morphological or physiological, due to the environment, which is not passed on to the next generation; not a genetic change.

Acquired Immunity–Ability to resist a disease to which the individual would ordinarily be susceptible; opposed to natural immunity. It may be a result of antibodies built up through exposure to, or inoculation with, pathogens (active immunity), or it may result from the individual who has become immune (passive immunity).

Acre Per Animal Unit Month–The estimated number of acres necessary to provide forage for one animal unit for one month under proper use.

Acrosome–The thin covering or cap over the head of a sperm cell.

ACTH–An abbreviation for the adrenocorticotrophic hormone secreted by the front lobe of the pituitary gland. the hormone stimulates the outer part or cortex of the adrenal gland to secrete its hormones into the blood.

Actinobacillosis–A chronic, infectious, rarely fatal disease of cattle, sheep, and swine caused by *Actinobacillus lignieresi*. In cattle, it is characterized by movable swellings up to the size of an egg, involving the soft tissues of the head and neck region. The lymph nodes of the neck may become enlarged. Sometimes the tongue is affected, and there may be ulcerations of the mucous membranes of the throat. It may spread to various other internal organs. In sows, it is responsible for udder abscesses. In sheep, it is usually a secondary infection. Also called lumpy jaw, big jaw, wooden tongue. (Sometimes it is mistaken for actinomycosis.) See Actinomycosis.

Actinomycosis–A chronic, infectious, rarely fatal disease caused by *Actinomyces bovis*, which affects the jawbones of cattle and swine and is characterized by difficulty in breathing, enlarged spongy bones with cavities filled with pus, swollen gums and loosened teeth, bony growths through the tissues and skin, resulting in an abscess with a foul odor. It is sometimes seen in the udders of swine, where it causes a tumorlike enlargement. Humans are susceptible to a similar disease caused by *Actinomyces hominis*. Also called lumpy jaw, big jaw, clams, clyers, wooden tongue. (Sometimes it is mistaken for actinobacillosis.) See Actinobacillosis.

Active Agent (Active Ingredient)–That part of an insecticide, herbicide, or fungicide formula that has toxic properties to target species.

Active Immunity–The type of immunity in animals where the animal plays a part in the development of that immunity.

Actual Analysis–The composition of a material based on a laboratory chemical analysis rather than a generalized guaranteed analysis.

Actual Use–The use made of an area by livestock and/or game animals without reference to a recommended utilization; usually expressed in terms of animal units or animal unit months. See Use.

Acute–(1) Usually refers to a disease that runs a short, severe course. (2) Having the shape of an acute angle.

Acute Toxicity–The potential of a substance to cause injury or illness when given in a single dose or in multiple doses over a period of 24 hours or less. In aquatic studies, exposure to a given concentration for 96 hours or less.

Ad Lib Feeding—A system of feeding livestock in which no limit is placed on feed intake.

Ad Libitum—The total voluntary intake when feed is available to animals at all times without restriction.

Adaptability—The capability of an organism to make changes that make it more fit to its environmental conditions.

Adaptation—(1) A measure of physiological fitness of the organism to one or several conditions of its environment. (2) The structures or activities of an organism, or of one or more of its parts, which tend to fit it better for life in its environment or for particular functions. See Adjustments.

Adder—(1) The harmless North American milk snake, *Lampropeltis triangulum*, so called because it is sometimes found in dairy buildings, where it goes in search of mice rather than milk. (2) The harmless North American puffing adder, *Heterodan contortix*; also called blowing adder, flat-headed adder, hognose snake, spreading adder, sand viper. (3) The common venomous European snake *Vipera berus*. (4) Any species of the venomous European snake family Viperidae. (5) The Australian venomous death adder, *Acanthophis antarctius*. (6) The Indian banded adder or cobralike krait, *Bunganis coeruleus*. (7) The African venomous puff adder, *Bitus arietans* or *Clotho arietans*. (8) The venomous Near Eastern horned adder, *Cerastes cornutes*. (9) The venomous Indian asp or serpent *Naje Haje*. (10) The sea adder, *Spinachia spinachia*, or pipe-fish of the genus *Nerophis*, family Syngnathidae.

Additivies—Materials added to food to help manufacture and preserve it and to improve nutritive value, palatability, and eye appeal. May be classified as emulsifiers, flavors, thickeners, curing agents, humectants, colors, nutrients, or as mold, yeast, or bacterial inhibitors. Amounts used in food are regulated by law.

Addled Egg—An egg in which the yolk has become mixed with the white; sometimes a rotten egg, hence considered inedible.

Adherence—The act of sticking to a surface.

Adhesion—Molecular attraction which holds the surfaces of two unlike substances in contact, such as water and soil particles. See Cohesion.

Adhesions—The adherence of separate tissues or organs of the body to each other as the result of an inflammation.

Adipose—Refers to the fat-filled cells of connective tissue.

Adjusted Weaning Weight (Adj.205-day Wt.)—Weight of a calf at weaning, adjusted to a standard 205 days of age and adjusted for the age of the dam.

Adjusted Yearling Weight (Adj. 365-day Wt.)—Weight of a calf as a yearling, adjusted to a standard 365 days of age and adjusted for the age of the dam.

Adjustments—(1) Range management:changes in animal numbers, seasons of use, kinds or classes of animals, or management practices as warranted by specific conditions. (2) Ecological: the processes by which an organism becomes better fitted to its environment; functional, never structural. See Adaptation.

Adjuvant—(1) Any solid or liquid added to a substance, such as a pesticide or a fertilizer, to increase its effectiveness, e.g., solvents, diluents, carriers, emulsifiers, stickers, spreaders, or sometimes a pesticide to another pesticide or a fertilizer to another fertilizer. (2) A carrier for a biological that releases the biological into the bloodstream over an extended period, thus serving the function of a series of booster shots; consequently, the adjuvant helps lengthen the period of immunity provided by the biological.

Adrenal Glands—A pair of ductless glands, each located on or adjacent to a kidney, which secrete adrenalin and cortin, whence cortisone; suprarenal glands.

Adrenalin—One of the hormones produced by the medulla of the adrenal glands; also called epinephrine; helps in preparing the body for emergency actions.

Adrenocorticotrophic Hormone—See ACTH.

Adulterant—An impurity not allowed by law in a food, plant, animal, fertilizer, or pesticide formulation.

Adulterate—To lower the quality of a product by mixing in another substance; as to adulterate milk by adding water to it. The federal government and most states legislate against the adulteration of food, drugs, fertilizers, and pesticides.

Aegagropila—See Hair Ball, Bezoar.

Aerobe—Bacteria or other organisms which live only in free oxygen.

Aerobic—Pertaining to organisms that grow only in the presence of oxygen, as bacteria in a properly prepared compost.

Aerobiology—The study of microorganisms carried in the air.

Aerogens—Gas-producing bacteria.

Aerosol—Finely atomized spray or smoke with particles ranging in size between 0.1 and 50 microns. The particles are produced by blasts of heated air, or exhaust gases, or rapid volatilization of a liquefied gas containing a nonvolatile chemical solution. Aerosols include insecticides, antibiotics, germicides, and deodorants.

Aestival—Pertaining to summer.

Aestival Pond—A pond which contains some water throughout the open season but freezes to the bottom in winter.

Afebrile—Without fever.

Aflatoxin—The substance produced by some strains of the fungus *Aspergillus Flavus*, the most potent natural carcinogen yet discovered; a persistent contaminant of corn, other grains, and peanuts.

African Horse Sickness (AHS)—A highly fatal, insect-borne, febrile viral disease of the family Equidae. Symptoms are a swelling of the heart and internal bleeding.

African Swine Fever—A highly contagious, often acute, viral disease of domestic swine characterized by fever, marked cyanosis of skin areas, and pronounced hemorrhages of the internal organs, particularly the lymph nodes, kidney, and gastrointestinal mucosa. Mortality frequently approaches 100 percent in initial epizootics.

Afterbirth–The matter which surrounds and attaches the fetus to the uterus. It is expelled after the fetus.

Aftermath–The regrowth of range or artificial pasture forage after grazing or mowing. More commonly used to refer to grazing of forage or cropped areas after harvest where there may not always be regrowth but just crop residues.

Aftosa Fever–See Foot-and-Mouth Disease.

Agalactia–A failure to secrete milk following the birth of offspring.

Age Class–A descriptive term to indicate the relative age grouping of plants or animals.

Aged Horse–Correctly speaking, a horse 8 years of age or over; but the term is often used to indicate a horse that is smooth-mouth—that is, 12 years of age or older. Since one year of a horse's life corresponds to approximately three of a human's, it follows that at age 7 a horse "comes of age," or attains maturity.

Agent–One who represents another from whom he has derived authority.

Agglutinins–(1) Antibodies produced in the animal body in response to an infection or to the injection of microorganisms. They cause agglutination of the organisms responsible for their formation. (2) Any antibody capable of clumping the organism which stimulated its production in the animal body.

Aging–A term applied to the ripening process of meat. The action of enzymes on the connective tissue increases the tenderness of the meat.

Agister–A person who takes charge of animals, properly cares for them, and charges for his or her service.

Agitation–The process of stirring or mixing in a sprayer.

Agitator–Revolving paddles which keep a liquid, powder, or gas in motion to maintain a proper mixture within a tank. Used in dairying, spraying, fertilizer drill, etc.

Agonistic–A type of animal behavior that involves offensive or defensive activities.

Agoraphobia–A fear of open places; especially that occurring in animals kept in stalls for too long a period. Among horses, also called picket-line-bound, barn rat.

Agostadero–A pasture used only in the summer (southwestern United States).

Agrarian–(1) Pertaining to agriculture. (2) Pertaining to political action or movements for the benefit of farmers. See Rural Population, Urban.

Agrarian Zone–Of or pertaining to the cultivated portion of an area.

Agribusiness–A term referring to the full scale of operations related to the business of agriculture. It connotes the interrelationships of farming, farm services, soil science, agronomy, land grant universities, county extension services, state and federal experiment stations, soil and water conservation services, plant and animal nutrition, plant and animal protection, transportation, finance, and marketing.

Agrichemicals–A term used to designate chemical materials used in agriculture, such as herbicides, insecticides, fungicides, and fertilizers.

Agricola–(Latin) Farmer.

Agricultural Chemistry–The branch of the science of chemistry which is concerned with the composition and transformation of the plants and animals on which the economy of the farmer rests.

Agricultural College–An educational institution devoted to study, research, and the dissemination of knowledge in agronomy, horticulture, animal husbandry, agricultural economics, etc. The U.S.D.A. works in conjunction with most agricultural colleges (sometimes called a school or college of agriculture within a university).

Agricultural Land–All the land devoted to raising crops and livestock, including farmstead, roadways, drainage and irrigation ditches, ponds, water supply, cropland, and grazing land of every kind. (The term is not strictly synonymous with land in farms, cropland, pasture land, land suitable for crops or land suitable for farming)

Agricultural Pollution–The liquid and solid wastes from all types of farming, including runoff from pesticides, fertilizers, and feedlots; erosion and dust from plowing; animal manure and carcasses; and crop residues and debris. Old cars and trucks are a part of aesthetic agricultural pollution.

Agriculture–The broad industry engaged in the production of plants and animals for food and fiber, the provision of agricultural supplies and services, and the processing, marketing, and distribution of agricultural products.

Agriculturist–A person engaged in the production of food and/or fiber; also ancillaries such as teachers of agriculture, farm editors, researchers, etc.

AI–Artificial insemination.

Aids–The legs, hands, weight, and voice, as used in controlling a horse being ridden.

Air Sac Mite–*Cytodites (=Cytoleichus) nudus,* order Acarina; a mite that invades the lungs, windpipe, air sacs, and air-containing cavities of the bone in fowls. Also called pulmonary mite.

Aitchbone–(1) The bone of the rump of an animal; the floor of the pelvis. (2) The retail cut of meat that contains the Aitchbone, also called edge bone.

Alate–Winged; having wings.

Alba, Albidus, Albus–Terms meaning white.

Albino–(1) A human or animal with pink eyes and white skin and hair, caused by a congenital deficiency of pigment in the skin. (2) In botany, plants with a chlorophyll deficiency.

Albumin–Proteins consisting primarily of amino acids. Albumin contains about 2 peercent sulfur. It is soluble in water and coagulated by heat; found in milk, blood, egg white, muscle, and vegetables.

Albuminuria–The presence of albumin in the urine.

Alcaligenes–A bacterial genus of the family Bacteriaceae found in human and animal intestines. They usually do not cause disease, but *Alcaligenes viscosum* does make milk glutinous or ropy.

Alcohol–The family name of a group of organic chemical compounds composed of carbon, hydrogen, and oxygen; a series of molecules that vary in chain length and are composed of a hydrocarbon plus a hydroxyl group, $CH_2\text{-}(CH_2)N\text{-}OH$; includes methanol, ethanol, isopropyl alcohol, and others.

Alcohol Test–A test in which equal parts of milk and ethyl alcohol are mixed to discover abnormal milk or milk with unusual salt balances. Normal milk breaks clear of the test tube, while the curd of abnormal milk clings to the glass. The test also detects milk which is apt to coagulate during sterilization of condensed milk.

Alcohol-Alizarin Test–A test in which alcohol and alizarin are added to milk. A color reaction of lilac-red indicates normal milk, yellow-brown indicates sour milk, and violet indicates mastitis milk.

Alcoholic Fermentation–The transformation of simple hexose sugars, especially glucose, into alcohol and carbon dioxide. Useful as a method of preservation.

Alderney–A breed of cattle identical to the Guernsey from the Channel Island of Alderney.

Alemtejo–A soft, ripened cheese made from ewe's milk, sometimes with goat's milk added.

Aleukia–(1) Absence of leukocytes in the blood. (2) Alimentary toxic aleukia. Mycotoxicosis in people and some animals caused by ingesting one or more of the following genera of fungi that develop on damp grain crops: *Actinomyces, Alternaria, Cladosporium, Fusarium, Penicillium, Piptocephalis, Rhizopus, Thammidium, Tricoderma, Trichothecium,* and *Verticillium.*

Alevin–(1) Newly hatched salmon or related fish (usually) with a yolk sac attached, before it emerges from the spawning gravel to begin swimming freely. (2) Newly hatched, incompletely developed fishes (usually salmonids) still in nest or inactive on the bottom, living off stored yolk. (3) The stage from hatching to end of dependence on yolk sac as primary source of nutrition (usually salmonids).

Alfalfa Leaf Meal–A stock feed consisting chiefly of ground, leafy materials separated from alfalfa hay or meal. It is reasonably free from other crop plants and weeds, and does not contain more than 18 percent of crude fiber.

Alfalfa Meal–A stock feed of ground alfalfa hay. It is reasonably free from other crop plants and weeds, and does not contain more than 33 percent of crude fiber.

Alfalfa-Molasses Feed–A stock feed of alfalfa meal and cane or beet molasses, in which the molasses makes up from 20 to 40 percent of the mixture. Used as replacement to supply proteins and vitamins in the rations of dairy cows, beef cattle, and sheep.

Algae–(singular, alga) Comparatively simple plants containing photosynthetic pigments. A majority are aquatic, and many are microscopic. They grow in sunlit waters. They are food for fish and small aquatic animals and, like all green plants, put oxygen in the water when carrying on photosynthesis. Although some forms of algae are necessary and desirable, excessive concentrations tend to discolor water and cause objectionable tastes and odors, severely limiting all uses of the water.

Algal "Bloom"–A term for a rapid increase in numbers of algae in a body of water, often caused by a sudden increase in available nitrogen, phosphorus, temperature, and/or sunshine. A density of 500 or more individual algae per milliliter of water is often considered a "bloom." Excess algae are not desirable because they physically reduce the use of bodies of water for swimming and boating; at night they deplete oxygen for fish and other aquatic organisms, and upon decomposition have a foul odor of hydrogen sulfides, H_2S.

Algicide–A chemical highly toxic to algae and satisfactory for application to water, commonly copper sulfate. See Copper Sulfate.

Alighting Board–The extended entrance of a beehive on which incoming bees land.

Alimentary Canal or Tract–The passage in an animal's body through which food passes from mouth to anus.

Alkali–(1) A chemical compound of oxygen and hydrogen with one element, such as sodium, potassium, calcium, magnesium, or the ammonium radical, capable of neutralizing acids. (2) A general term denoting salts of sodium, calcium, potassium, and magnesium which injure plant growth. (3) A general term for caustic soda, sodium hydroxide, caustic potash, sodium carbonate, etc., all cleansing agents in food processing plants.

Alkali Disease–Selenium poisoning of animals caused by grazing on plants containing excessive amounts of selenium. Characterized by emaciation, loss of hair (mane and tail among horses), deformed hoofs, and blind staggers.

Alkaline–A chemical term referring to basic reaction where the pH reading is above 7, as distinguished from acidic reaction where the pH reading is below 7.

Allantois–One of the three fetal membranes; located between the amnion and the chorion.

Allegheny Mound Ant–*Formica exsectoides;* a mound-building ant, family Formicidae; common in the northeastern United States. It kills trees, shrubs, etc., within a radius of 20 to 30 feet., that threaten to shade its mound.

Allele–(1) The alternative forms of genes having the same place in homologous chromosomes which influence the development of alternative traits or characters. (2) A pair of Mendelian genes at the same locus as a pair of homologous chromosomes. Also called allelomorph.

Allergy–An exaggerated susceptibility to a substance harmless to most members of the same species.

Alliaceous–Having the smell or taste of onion or garlic.

Alligator Farming–The commercial raising of alligators. Under state approval, alligators are raised in captivity for their hides. The leather is used to make shoes, handbags, suitcases, etc. The meat is sold for human consumption.

Allite–Soils characterized by a silica-alumina ratio less than 2:1 in the hydrochloric acid extract. A soil from which silica has been removed during formation, and in whose clay fractions aluminum is in the form of AL_2O_3, as in bauxite and laterite.

Allogamy–Cross-fertilization.

Allopolyploid–A polyploid containing genetically different chromosome sets, e.g., from two or more species.

Allosomes–Chromosomes distinguished by peculiarities of behavior or sometimes by a difference in size or shape.

Allotment–(1) The number of acres, etc., a producer is allowed to grow of a particular crop under a government program. (2) An area designated for the use of a prescribed number of cattle or sheep, or for common use of both.

Alopecia–Loss of hair.

Alpaca–(1) *Lama pacos;* a llama domesticated in Peru and adjoining countries. (2) The long wool of the alpaca which is woven into a cloth.

Alter–(1) To neuter by castrating or spaying. Also called cut, geld, emasculate. (2) To prune. (3) To remove the comb from a beehive.

Alternate Grazing–Changing pastures or ranges so that the forage grows back before it is grazed again. Also called rotational grazing.

Alternate Host–A plant or animal upon which a disease organism exists for only a part of its life cycle; e.g., cedar and apple for rust.

Alternation of Generations–Reproduction in which common characteristics are found only in every second generation, e.g., one generation reproduces sexually and the next asexually.

Alveolus–(Plural, alveoli) A small cavity or sac. (1) A tiny, thin-walled air sac of the lung, in the walls of which gases are exchanged between the air and the blood. (2) An acinus or terminal lobule of a racemose gland. (3) One of the honeycomb pits in the wall of the stomach. (4) A tooth socket. (5) A cell of the honeycomb. (6) The tiny, balloonlike structure of the udder lined with epithelial cells that are active in the secretion of milk. (7) A pore or pitted perithecium in a fungus.

Amasesis–Incapable of masticating (chewing).

Ambidexterous–Able to use both hands well.

Ambler–A horse whose best gait is pacing; a pacer.

Amblotic–Likely to cause abortion.

Ambosexous–Bisexual.

Ameba–A protozoan organism with no specific shape that moves in a flowing manner by the extension of a false foot or pseudopod.

American Cockroach–*Periplaneta americana,* family Blattidae; house-dwelling insect, active at night, feeding on book bindings, paper, garbage, and food. Also inhabits greenhouses where it feeds on seedlings, flowers, and girdle plants.

American Dog Tick–*Dermacentor variabilis,* family Ixodidae; a blood-sucking acarid. It carries Rocky Mountain spotted fever, tularemia, bovine anaplasmosis. Also called wood tick.

American Foulbrood–An infectious disease of the honeybee larvae, caused by *Bacillus larvae;* characterized by sunken, discolored, perforated brood cappings. If widespread, the hive ceases to function. See European Foulbrood.

American Grades of Wool–Twelve classifications established by the U.S.D.A. They are designated by numbers, according to the diameter of wool fibers.

American Grasshopper–*Schistocera americana,* family Acrididae; an insect pest in the southeastern United States.

American Horse Council, Inc.–Represents all sectors of the United States horse industry; formed in 1969. It is dedicated to the development of the American equine industry.

American Jack–A breed of ass (*Equus asinus*) developed largely from the Catalonian, black with white points, weighting about 1,300 pounds, standing about 15 hands high (60 inches or 52.4 centimeters). Also called mammoth jack.

American Landrace–See Landrace.

American Quarter Horse–A type of horse developed for work; popular as a cow horse in the southwestern United States. It weighs 1,100 pounds or more and has a short back and good feet. It is active, quick, and given to sudden bursts of speed; popular in quarter-mile racing.

American Saddle Horse–A light breed of horse developed from the Narragansett pacers, the Canadian pacers, the Morgans, and Thoroughbred breeds. Popular for pleasure riding and in horse shows because of its beauty, style, high action, and ability to perform three to five gaits.

American Standard of Perfection–A book published by the American Poultry Association, Inc., that gives a complete description of all recognized varieties of fowls.

American Thoroughbred–A branch of the Thoroughbred horse which differs from the English Thoroughbred only because in 1913 the English restricted the Thoroughbreds to descendants of certain mares in England.

AMF–See Anhydrous Milkfat.

Amino Acids–Organic substances from which organisms build proteins, or the end product of protein decomposition.

Aminosis–An abnormal condition of the body caused by excessive ingestion of amino acids.

Amitosis–Cell division without formation and splitting of chromosomes. See Mitosis.

Ammonia–NH_3; a chemical compound composed of 82.25 percent nitrogen and 17.75 percent hydrogen. At ordinary temperatures, it is a colorless, pungent gas about one-half as heavy as air. Liquid ammonia is used as a fertilizer. Sometimes added to livestock feed to increase nutritional value.

Amnion–The fetal membrane located nearest the fetus; it is filled with amniotic fluid to protect the fetus from shock.

Amoeba–A genus of unicellular protozoan organisms of microscopic size, existing in nature in large numbers; many live as parasites; some species are pathogenic to humans.

Amoebiasis–A disease caused by single-celled, or protozoan, parasites belonging to the amoeba group, phylum Protozoa; also called amebiasis, amebiosis.

Amorphus–A heartless, limbless, and headless fetus.

Amotus–Denoting the hind toe of certain birds, which does not touch the ground.

Amphibian–A cold-blooded animal with legs, feet, and lungs that breathes by means of gills in the early stages and by means of lungs in the later stages of life.

Amphigean–Native of both the Old and New Worlds.

Amphivorous–Omnivorous.

Amphoteric–Capable of reacting as either an acid or a base, as casein.

Ampule–A unit of packaging for frozen semen; semen is now shipped in straws instead of glass ampules.

Ampulla–A funnel-shaped structure located at or near the end of a duct; the part of the Fallopian tube between the infundibulum and the isthmus; a part of the vas deferens where it joins the urethra in some species.

Amylase–An enzyme secreted by the pancreas and delivered to the small intestine that aids in the digestion of starch.

Amylodextrins–See Dextrins.

Amylopsin–A pancreatic enzyme which breaks down starch to sugars; the chief carbohydrate of pancreatic juice.

Anabiosis–Resuscitation or revival of an organism after apparent death.

Anabolism–Constructive process by which simple substances are converted by living cells into more-complex compounds. See Catabolism.

Anadipsia–Excessive thirst.

Anadromous–Pertaining to fishes that leave seas and ascend freshwater streams to spawn; e.g., salmon.

Anaerobe–Organisms (usually bacteria) which live and multiply without free oxygen.

Anaerobic–(1) Living or active in the absence of free oxygen. (2) Pertaining to or induced by organisms that can live in the absence of free oxygen.

Anaerobic Bacteria–Bacteria not requiring the presence of free or dissolved oxygen for metabolism. Strict anaerobes are hindered or completely blocked by the presence of dissolved oxygen and sometimes by the presence of highly oxidized substances, such as sodium nitrates, nitrites, and, perhaps, sulfates. End-product gases include methane and hydrogen sulfide.

Anaerobic Decomposition–Reduction of the net energy level and change in chemical composition of organic matter caused by microorganisms in an anaerobic environment.

Anal–Pertaining to the last abdominal segment, which bears the anus.

Analogous–Refers to an organ of one organism that corresponds in function to an organ of another animal or plant but which is not homologous. See Homologous.

Analysis–The percentage composition of fertilizers, feeds, etc., as determined by chemical analysis, expressed in terms specified by law.

Anaphylactic Shock–Violent attack of symptoms produced by an injection of a serum or protein given to a sensitive animal or person.

Anaphylaxis–State of increased sensitivity in an animal following an injection of foreign matter.

Anaplasmosis–A blood disease of cattle caused by the parasite *Anaplasma marginale*, phylum Protozoa, located in the red blood cells. Symptoms are: anemia, milk flow ceases, high fever, almost total collapse; in a day or two the hairless areas become pale yellow with jaundice; death may occur in two or three days. Also called yellow teat disease or gall sickness. In Africa, a much milder form of the disease is caused by *Anaplasma centrale* or *Anaplasma marginale* var. *centrale*.

Anatomy–The branch of biology that deals with the structure of organisms.

Ancestor–An individual from whom an animal or person is descended.

Ancestor Merit–An unbiased estimate of the production of future daughters of a bull based on an accurate evaluation of the bull's sire and maternal grandsire.

Ancona–A breed of chickens originating in Italy, having black feathers with white tips; excellent layers of white eggs.

Andalusian–A Spanish breed of horse of medium size, 15 hands or less in height, gray with well-shaped legs and large bones.

Androgenesis–Development of offspring with the paternal chromosomes only.

Androgynized Cow–A cow that has received synthetic testosterone (male sex hormone). A bell-shaped marker is attached to the cow's neck to mark cows that are in estrus (heat) as she mounts them.

Anemia–(1) A deficiency of hemoglobin, iron, or red blood cells. (2) Condition in which the blood is deficient either in quality or quantity. (3) Anemia in suckling pigs; low blood pressure in young pigs caused by a nutritional deficiency of iron or copper; produced by housing the pigs indoors on wood or concrete floors during their first few weeks; characterized by decrease of vigor, and paleness of the mouth. When on dirt floors, pigs get enough iron and copper form the soil.

Anesthesia–Loss of feeling,e specially loss of tactile and pain sensations, with or without loss of consciousness.

Anesthetics–Drugs which produce insensibility to pain.

Anestrus–The nonbreeding season; the period of time when a female is not cycling.

Aneuploid–An organism or cell which has a chromosome number other than an exact multiple of the monoploid or basic number; i.e., Hyperploids=higher; hypoploid=lower.

Anezeh–A horse bred by that Arab tribe in the Syrian desert.

Angora Goat–A breed of domestic goat from Asia Minor; bred especially for mohair.

Angora Rabbit–A species whose mohairlike fur is valued. See Rabbit.

Angus–See Aberdeen Angus.

Anhidrosis–A deficiency of sweat in any animal; also spelled anidrosis.

Anhydride–An oxide that will react with water to form the corresponding acid or base; e.g., P_2O_5 is the anhydride of H_3PO_4; CaO is the anhydride of $Ca(OH)_2$.

Anhydrite–($CaSO_4$) Anhydrous calcium sulfate; used for the same purposes as land plaster or gypsum, i.e., as a soil amendment; also used in India and the United Kingdom for making sulfuric acid.

Anhydrous Milkfat (AMF)–A product similar in composition to butteroil. However, it is made directly from cream and is used most commonly in milk-deficit countries to combine with nonfat dry milk to make fluid milk, ice cream, and other dairy products.

Animal Husbandry–Agriculture which deals with livestock production, usually limited to meat animals and horses, although it may include poultry and dairy cattle.

Animal Industry–Agriculture which deals with livestock and the processing of the products derived therefrom.

Animal Kingdom–All animals collectively. In the kingdom Animalia there are twelve branches or phyla ranging from I, Protozoa, to XII, Chordata, which includes the class Mammalia in the subphylum Vertebrata.

Animal Month–A month's tenure upon range by one animal.

Animal Protein Factor (APF)–The term which formerly designated unknown growth factors in feeds of animal and marine origin, such as tankage, meat scraps, fish meals, milk products, liver meals, etc.

Animal Rights–A philosophy that animals have the same rights as humans and that they should not be used for human consumption. See Animal Welfare.

Animal Science–See Animal Industry.

Animal Sign–The evidence of an animal's presence in an area.

Animal Therapy–The use of domestic animals to provide therapy for humans; to relieve lonesomeness, make them feel needed, and soothe the emotional self.

Animal Type–The combination of characteristics of an animal appropriate to a special kind of use or desirability, such as beef type.

Animal Unit–Measurement based on the amount of feed eaten and manure produced by an average mature horse, cow, or the equivalent; as two heifers over one year, four calves under one year, seven ewes or bucks, two and one-half brood sows or boars, five hogs raised to 200 pounds each, or 100 hens.

Animal Unit Conversion Factor–As used in the United States, a numerical figure which allows conversion from one kind or class of animal to another. A conversion factor is satisfactory in respect to the amount of forage required to maintain an animal, but may have no application in determining stocking rates for range use for particular kinds or classes. In the United States, the generally accepted per-month conversion factors are: mature cow with calf, 1.0; mature bull, 1.12; weaned calf, 0.6; yearling over 12 months and under 17 months, 0.7; yearling from 17 to 22 months, 0.75; two-year-old from 22 to 32 months, 0.9; elk, 0.7; white-tail deer, 0.14; mule deer, 0.2; mature ewe with lamb, 0.2; weaned wether, 0.17; doe goat with kid, 0.17; buck, 0.17; weaned wether, 0.14; and grown horse, 1.25.

Animal Unit Months (AUMs)–Amount of grazing required by a 1,000-pound (454 kg) cow or equivalent weight of other domestic animal for one month.

Animal Welfare–A philosophy that animals should be treated in a kind and caring manner.

Animalcule–Minute animals such as those belonging to the Protozoa phylum.

Ankle Hobble–A restraining device which prevents a horse or other animal from straying, consisting of a short piece of rope or chain with straps at each end fastened to the legs of the animal just above the hoofs.

Annulus–The annual mark or zone on the scales, vertebrae, or other hard portion of a fish, which is formed once each year.

Anorexia–Lack or loss of appetite; a symptom of many diseases.

Anoxia–A deficiency of oxygen.

Antagonism–The loss of activity of a chemical when exposed to another chemical.

Ante Partum Paralysis–An abnormal condition of the cow (seldom of sheep or goats and rare in mares), in which the hindquarters of the pregnant animal become paralyzed about a week to three weeks before parturition (giving birth); a symptom of many diseases.

Antefebrile–Before the onset of fever.

Antennae–Slender jointed feelers, which bear certain sense organs, on head of insects.

Anterior–In four-legged animals, toward the head; the opposite of posterior.

Anterior Pituitary–The front part, or lobe, of the pituitary gland, located at the base of the brain. It produces and secretes specific hormones, regulates growth of the body, and regulates and controls the thyroid, adrenal cortex, ovaries, and testicles.

Anterior Presentation–Normal birth position of mammals; front feet and head presented first.

Anthelmintic–Any drug used for expelling or killing stomach or intestinal worms. See Vermifuge.

Anthony Pig–The smallest pig or runt in a litter.

Anthrax–Malignant anthrax; a fatal infectious disease of cattle and sheep caused by a bacterium, ***Bacillus anthracis***, and characterized by hard ulcers at point of inoculation and by symptoms of collapse; also occurs in humans.

Antibacterial–Any substance that has the ability, even in dilute solutions, to destroy or inhibit the growth or reproduction of bacteria and other microorganisms; used especially in the treatment of infectious diseases of people, animals, and plants.

Antibiosis–An association between two organisms in which one harms the other.

Antibiotic–Germ-killing substance produced by a bacterium or mold.

Antibiotic Feeds–Animal feeds that contain antibiotics. The use of antibiotics in animal feeds is controversial because of the perceived potential human health hazard associated with the practice. The animal science industry contends that meat from animals that consume these feeds are perfectly safe.

Antibody–The very specific biological substance that the body itself manufactures to combat specific diseases following an attack of a disease or following a vaccination.

Anticarcinogen–A substance that inhibits or eliminates the activity of a carcinogen (cancer-producing substance).

Antidiuresis–A decrease in the production and excretion of urine. See Diuresis.

Antidote–A remedy for counteracting a poison.

Antidote Statement–A required statement on the containers of chemicals explaining methods that may be used to counteract the effects of the chemical.

Antigen–Any substance that stimulates the formation of antibodies when it is introduced into the body. Pure cultures of organisms or their products may be employed as antigens in various biological testing methods to measure the antibody content of blood serum and other body fluids.

Antimutagen–A substance that inhibits or eliminates the activity of a mutagen. See Mutagen.

Antioxidant–A compound which prevents oxidation. Used in mixed feeds to prevent rancidity or loss of vitamin potency.

Antisepsis–Prevention of infection by the exclusion, inhibition, or destruction of the causative organisms.

Antiseptic–An agent that destroys or severely inhibits microorganisms that cause disease, decomposition, or fermentation.

Antiserum–A serum containing antibodies to give temporary protection against certain specific infectious diseases; obtained from the blood of animals that possess a high degree of immunity against certain infectious diseases.

Antispasmodic–Any drug that prevents or counteracts spasms (muscular, intestinal, bronchial, etc.).

Antitoxin–An antibody that neutralizes toxin of a bacterium.

Antrum–The cavity in a hollow organ or structure, such as the cavity inside a developing follicle.

Anus–The posterior opening of the digestive tract.

APF–See Animal Protein Factor.

APHIS–Animal and Plant Health Inspection Service.

Aphtha–A vesicle or sac containing a thin fluid on the udder, inside the mouth, or sometimes between the toes of a cloven-footed animal. The condition is characteristic of foot-and-mouth disease. See Foot-and-Mouth Disease.

Aphthous Fever–See Foot-and-Mouth Disease.

Apiarist–A beekeeper.

Apiary–Bee colonies, hives, and other honeybee equipment assembled in one location; also called bee yard.

Apiculture–The science and art of studying and using honeybees for human benefit.

Apidae–The bee family, including the subfamilies of the common bee (Apinae) and the stingless bee (Meliponinae).

Apiology–The scientific study of bees, especially honeybees.

Apis–The genus to which the honeybee belongs.

Apis cerana–Scientific name of the Eastern honeybee, the honey producer of South Asia; also called *Apis indica*.

Apis dorsata–Scientific name for the large honeybee of Asia, which builds open-air nests of single comb suspended from tree branches, rocky ledges, etc.

Apis florea–Scientific name for the small honeybee of Asia.

Apis mellifera–Scientific name of the Western honeybee.

Apnea–Cessation of breathing.

Apogamous–Developed without fertilization; parthenogenetic.

Apoplectic Anthrax–A form of anthrax from which death may occur in a few minutes, although in the case of sheep, cattle, and goats, death may not occur for three hours. See Anthrax.

Apoplectiform Septicemia–An infectious, lethal disease of fowls, caused by the streptococci *Streptococcus capsulatus gallinarum* and *S. zymogenes*; characterized by listlessness, lack of appetite, staggering gait, drawn-back head, and ruffled feathers. There may be a discharge from the eyes; also called streptococcus infection, avian streptococcosis.

Appaloosa–A strain of western stock horse characterized by color spots, particularly across the croup; from 14 to 15 hands high, having a short back; also called rain drop, dollar spot, leopard spot, Colorado ranger, buttermilk horse.

Apparel Wools–Include any wools that are manufactured into cloth for use as clothing.

Appendage–An attachment to something larger; a leg or limb.

Apterium–A featherless portion on the body of a bird.

Apterous–Wingless.

Aquaculture–Underwater agriculture, commonly called fish farming, that includes ordinarily the raising of water animals such as fish and shrimp, but which also includes the growing of water vegetation such as kelp. Mariculture is a broad category within the field of aquaculture, but is limited to the raising of marine animals in saltwater rather than in fresh water. See Pisciculture.

Aquatic–Living in water.

Aquatic animal–An animal that lives in water.

Aqueous–(1) Pertaining to water. (2) Pertaining to sediment deposited by water. (3) Ammonia in water; also called ammonium hydroxide (NH_4OH) solution. A low-pressure solution, about 20 to 21 percent N, used as a fertilizer.

Arabian–A type of saddle horse from 14 to 15 hands high; from 350 to 1,000 pounds; predominant colors are gray, bay, and chestnut. It possesses endurance and docility and belongs to perhaps the oldest known pure breed.

Arachidonic Acid–A substance essential to body tissues built by the animal body from the simpler fatty acids that are derived from the food fats.

Arachnid–A group of arthropods having four pairs of legs and one or two body segments; mites, ticks, and spiders are examples.

Arachnoid–Cobwebby; of slender entangled hairs.

Aranose–Like a spider web.

Arch–(1) A curve. (2) The anterior-posterior curvature of the backs of hogs. (3) A log-supporting device used with a tractor in skidding logs. (4) The curved portion of a dam, bridge, or other structure.

Archetype–In biology, the antecedent of a group of plants or animals from which certain typical characteristics have been inherited; a progenitor.

Arena–(1) Sand. (2) Sphere of action, such as a livestock arena.

Areole–(1) A small area, especially the open space between anastomosing veins. (2) A spine-bearing sunken or raised spot on the stem in cacti.

Argas–A genus of ticks, family Argasidae, which in warm climates infest poultry.

Ark–A small house with ridged roof, often used for a small number of poultry or rabbits.

Armed Tapeworm–*Taenia solium*; a dangerous parasite transmitted to humans through pork infected with larvae. See Tapeworm.

Arsenic Poisoning–A condition which results from ingestion or absorption through the skin of an arsenical; acute cases are characterized by trembling, weakness, prostration, and diarrhea.

Arteriosclerosis–A disease involving the thickening and hardening of the walls of the arteries.

Artery–A blood vessel that functions in carrying blood from the heart. Veins carry blood back to the heart.

Arthritis–Inflammation of the joints of mammals characterized by stiffness, swelling, and soreness.

Arthropod–A phylum or division of the animal kingdom; includes insects, spiders, and Crustacea; characterized by a coating which serves as an external skeleton and by legs with distinct movable segments or joints.

Articular–Pertaining to the joints of plants and animals.

Artificial Hormone–A manufactured substance that is used in the place of a naturally produced hormone.

Artificial Insemination–The deposition of spermatozoa in the female genitalia by artificial rather than by natural means.

Artificial Vagina–In artificial insemination, a device used to collect semen; consisting of a heavy rubber or metal cylinder, an inner liner of soft rubber, and a tube.

Artificially Acquired Immunity–Immunity that comes about as a result of a vaccination.

Artiodactyla–The zoological order of mammals that includes hoofed animals with an even number of toes such as cattle, sheep, and swine.

As-Fed Basis–Data on feed composition or nutrient requirements calculated on the basis of the average amount of moisture found in the feed as it is used on the farm; sometimes referred to as air-dry.

Ascariasis–A chicken disease caused by a parasitic roundworm, family Ascarideae; the symptoms are diarrhetic.

Ascarid–Family Ascaridae; a parasitic roundworm that infests the intestines of humans and animals.

Ascorbic Acid (Vitamin C)–A chemical compound, $C_6H_8O_6$, which occurs in fruits and vegetables and prevents scurvy in mammals.

Aseptic–Being free from infectious microorganisms.

Asiago Cheese–An Italian hard cheese made from cows' milk in the province of Vicenza.

Asiatic Class–Domesticated chickens including Brahmas, Cochins, and Langshans which are heavy and have feathered legs. They lay brownshelled eggs.

Asiatic Urial–A breed of wild sheep believed to be ancestors of some present-day domestic breeds.

Asper–Rough.

Aspergillosis–A respiratory disease of chickens and turkeys caused by the fungus *Aspergillus fumigatus*, family Moniliaceae, in the lungs and air sacs. Characterized by listlessness, loss of appetite, a darkening of the comb, ruffled feathers, high temperature, thirst, diarrhea, gasping for breath; high mortality among chicks and poults; also called brooder pneumonia, pneumomycosis.

Asperous–Rough or harsh to the touch.

Asphyxiation–(1) Suffocation or death from the lack of oxygen. (2) A systemic condition of plants, etc., brought on by oxygen deficiency; characterized by seeds not germinating, blighting of plants, and breakdown of fruits, tubers, etc.

Aspirate–To draw by suction.

Ass–(1) *Equus asinus*; the donkey; a domesticated beast of burden used in the breeding of mules by crossing an ass with a horse. (2) Any of several species of *Equus*; e.g., wild asses of Asia and North Africa.

Assimilation–The process of converting food nutrients to body tissue.

Asthenia–Debility, or the lack of body strength.

Astomous–Mouthless.

Astrakhan–The lambskin of Karakul or Bokhara sheep, which is less lustrous, has longer hair and more open curl than Persian lambskin.

Astringent–A drug, such as tannic acid, alum, and zinc oxide or zinc sulphate, that causes contraction of tissues.

Astrologer–A poetical name sometimes given to the cock as the harbinger of dawn.

Asymmetrical–Without proper proportion of parts; unsymmetrical.

At the End of the Halter–An animal that is sold with no guarantee except title.

Atavism–Reversion; reappearance of a characteristic or disease after a lapse of one or more generations.

Ataxia–Lack of muscular coordination.

Atom–The smallest unit of an element to retain the chemical characteristics of that element. It consists of negatively charged particles called electrons orbiting around the nucleus and, within the small mass of the nucleus, other particles—protons (which are positively charged and balance the extranuclear electrons) and neutrons (which have no charge). It is the number and arrangement of the atom's electrons and protons that make one element differ from another; the number of neutrons distinguishes one isotope from another of the same element.

Atomic Energy–Energy released in nuclear reactions. Of particular interest is the energy released when a neutron splits an atom's nucleus into smaller pieces (fission) or when two nuclei are joined together at hundreds of millions of degrees of heat (fusion). Atomic energy is really a popular misnomer; it is more correctly called nuclear energy.

Atomic Weight–The average relative weight of an atom of an element as compared with another element (usually oxygen) that is taken as a standard. Isotopes of the same element have different atomic weights because the number of neutrons differs from isotope to isotope; a quoted atomic weight is the average weight unless a specific isotope is named.

Atomization–The process of breaking a liquid into a fine spray.

Atomize–To reduce a liquid into very fine particles in a sprayer.

Atrium–One of the upper chambers of the heart.

Atrophy–(1) The wasting away or decreasing in size of muscle, fat, or any tissue or organ. May result from weakening disease or from disease. (2) Following the use of the elastrator for castration, the scrotum and testicles will decrease in size and finally wither away because the blood supply is cut off. See Elastrator.

Atropine–A poisonous, crystalline alkaloid used in medicine; a specific antidote for poisoning by organic phosphate insecticides.

Atrovirens–Dark green.

Attaint–A wound on a horse's leg caused by its own hoofs.

Attenuate–To dilute, thin down, enfeeble or reduce. Specifically, to reduce the virulence of an organism; to destroy the power of disease-producing bacteria by means of chemical action, or of viruses by passing them through unnatural hosts, etc. This principle is used to develop certain bacterins and vaccines.

Atypical–Disagreeing with the form, state, or situation usually found under similar circumstances; not typical.

Aubin–An irregular gait of a horse; slower than a gallop but resembling it.

Aureomycin–An antibiotic derived from the organism *Streptomyces aurefaciens*; used in livestock feeds as a growth stimulant and for treatment of certain diseases. This antibiotic was discovered on a very poor soil in research plots at the University of Missouri in Columbia.

Auto Gate–See Cattle Guard.

Autocidal Control–A technique that is used primarily for insect control. It involves the rearing and release of insects that are sterile or altered genetically in order to suppress members of their own species that are causing pest problems. A type of autocidal control is the sterile male method whereby large numbers of male insects are artificially sterilized by irradiation or chemical sterilants. This method is most popularly exemplified in the eradication of the screwworm fly in the Southwest and parts of Mexico.

Autoclave–Vessel in which high temperatures can be reached by using high pressure; e.g., the domestic pressure cooker.

Autogenous Vaccine–A vaccine made from the patient's own bacteria instead of from stock cultures.

Autoimmunity–Production by an animal of an allergic reaction to its own tissues, which may produce clinical disease.

Autointoxication–Literally, self-poisoning; the formation and absorption of a poison within one's body.

Autolysis–Self-digestion. The natural softening process of fruits or vegetables after picking, or meat after slaughtering.

Autoparasitism–A parasite growing on a parasite; also called secondary parasitism or superparasitism.

Autopass–A device constructed to allow automobile traffic but to prevent cattle or sheep from getting out of a pasture; also known as cattle gap, cattle guard, auto gate. See Cattle Guard.

Autopsy–Examination, including dissection, of a carcass to learn the cause and nature of a disease or the cause of death; also called post mortem.

Autosomes–All the chromosomes except the sex chromosomes.

Autumn Lag–A period during the autumn and early winter when rabbits and pigeons slow down or stop reproducing.

Available Energy–See Metabolizable Energy.

Available Forage–Forage that is accessible for animal consumption.

Aveoli–Small grapelike structures in the mammary system that take raw materials from the bloodstream and synthesize them into milk.

Average Daily Gain (ADG)–The calculation of an animal's postweaning gain figured by dividing the weight gain by the days on feed.

Avermectin–*Streptomyces avermitilis*; a soil fungus used to spray on wool to protect it from moth and beetle damage.

Avian–Refers to birds; the class Aves.

Avian Diphtheria–See Fowl Pox.

Avian Encephalomyelitis–A viral disease of chickens which affects the nervous system; found only in chicks from one to three weeks old; characterized by tremor of the head, lack of coordination, sitting on haunches.

Avian Leukosis Complex–A widespread, viral, transmissible disease-complex of chickens that is more destructive than all other poultry diseases combined.

Avian Pasteurellosis–See Fowl Cholera.

Avian Plague–A lethal, infectious, viral disease of fowls; characterized by dullness, drooping wings and tails, closed eyes, running nose and eyes, thirst, rapid breathing, high temperature, brownish combs and wattles, and sudden death; the whole range of symptoms occurs within 24 to 36 hours; prevalent in Europe and the Middle East; also called bird pest, fowl plague, brunswick bird plague, fowl pest.

Avian Pneumoencephalitis–See Newcastle Disease.

Avian Tuberculosis–A disease of fowls, particularly of chickens; caused by the acid-fast microorganism *Mycobacterium avium*; characterized by loss of weight, listlessness, lameness, good appetite, diarrhea; the comb, wattle, and skin become pale; may result in death.

Avian Typhoid–A widespread, infectious disease of fowls caused by the microorganism *Salmonella gallinarum*; characterized by droopiness, listlessness, head drawn in, sagging wings, ruffled feathers, pale comb and wattles, yellowish or greenish diarrhea; may result in death.

Aviary–A place for keeping birds; pigeon loft, etc.

Avicide–A substance used to kill birds.

Aviculture–The caring for and raising of birds.

Avidin–A protein material that can combine with the B vitamin biotin, causing the vitamin to be unavailable to the body. Cooking renders avidin inactive.

Avirulent–Without the ability to produce disease.

Avitaminosis–A condition or disease of an animal resulting from a vitamin deficiency.

Avives–See Strangles.

Avoidance–A response whereby organisms prolong their dormancy, thereby achieving lesser vulnerability to environmental stresses.

Axial Feather–The short feather in the middle of the wing of a bird, which separates the primary from the secondary feathers.

Axon–The long fiber of a nerve cell that carries nerve impulses away from the cell body.

Axunge–Rich fat from the center of the kidneys of pigs and geese.

Ayrshire–An important breed of dairy cattle originating in Scotland. The animals vary in color from red to brown or mahogany interspersed with the white spots; between the Jersey and Holstein in size; the bulls weigh from 1,400 to 2,000 pounds.

Azoturia–An ailment common to draft horses kept on full feed when idle; characterized by excessive sweating, stiff movements, lameness, reluctance to move, sitting down or lying on one side, stiffness of muscles, passing red or brown urine; also called azotemia, blackwater, holiday disease, lumbago, Monday morning disease, myochaemoglubinuria, paralytic haemoglobinaemia.

B. T.–See Bacillus thuringiensis.

B.T.V. (Boynton's Tissue Vaccine)–A vaccine used to prevent hog cholera.

Babcock Test–A test of the fat in milk that involves centrifugation with sulfuric acid. It was developed by Dr. S. M. Babcock at the University of Wisconsin in 1890.

Babesia–A genus of single-celled organisms, class Sporozoa, which parasitize cattle, sheep, and horses, causing diseases of the bloodstream.

Baby Beef–The market classification of a beef carcass that grades prime, is between 8 and 18 months of age, and came from an animal that weighed between 600 and 1,100 pounds on foot.

Baby Combing Wool–Short, fine wool which is usually manufactured on the French system of worsted manufacture; synonymous with French combing wool.

Bacillus–A genus of single-celled, rod-shaped organisms, family Bacillaceae. Although most species are harmless, they do cause chemical changes in animal and vegetable matter.

Bacillus Larvae–A bacterial organism causing a disease called American Foulbrood in bees. See American Foulbrood.

Bacitracin–An antibiotic derived from the organism *Bacillus subtilin*.

Backcross Method of Mating–The breeding of F1 females to a male of the same breed as the F1's parents.

Backfat–The amount of fat covering on the back of a live animal or a carcass. The measurement is usually taken over the ribs and is used to determine yield grade.

Backfat Probing–Taking a measurement of the fat thickness with a probing knife over the last rib of a pig.

Backgrounding–The growing and feeding of calves from weaning until they are ready to enter the feedlot.

Bacon–The cured or smoked back and sides of a pig carcass; formerly used to describe any part of a hog carcass.

Bacon-type Hog–A term referring to the type of hog bred to have a large proportion of lean meat in relation to fat. The term was used as producers began raising hogs for meat rather than for lard.

Bacteremia–A disease caused by the presence of bacteria in the blood.

Bacteria–Single-celled microorganisms; some cause human, animal, or plant diseases; others are beneficial.

Bacteria, Facultative–Bacteria capable of adapting to an aerobic or anaerobic environment.

Bactericide–Anything which destroys bacteria; a germicide.

Bacterin–A vaccine made from killed or inactivated bacteria; injected into animals to increase their resistance to infection.

Bacteriolysis–The destruction or dissolution of bacteria inside or outside the animal body.

Bacteriophage–A viruslike, bacteria destroying agent that can propagate itself only in the presence of young, active, susceptible bacteria.

Bacteriostat–A compound which inhibits growth and reproduction of or kills certain bacteria.

Bad Quarter–See Mastitis.

Badger–A nocturnal carnivorous mammal; *Taxidea taxus*, family Mustelidae; has a black and gray coat, with a white streak running from the muzzle to a point behind the ears, short legs, a heavy body, and a round back; native to North America.

Bag–The udder; the scrotum; a sack; the total kill in a hunter's possession.

Bail–A half-ring or half-circle such as the handle of a bucket; a doubletree; a portable or mobile dairy house.

Bait Insecticide–An appetizing mixture of poison and food which, when eaten by an insect, results in its death.

Bait Shyness–The tendency for rodents, birds, or other pests to avoid a poisoned bait.

Balance–(1) The skeletal and muscular makeup of an animal, which gives the animal visual appeal. A well-balanced animal's body parts appear to fit together and blend harmoniously and symmetrically. (2) A device for determining weights. (3) To put in proper proportions, such as balancing a feed ration. (4) A term used to express the ratio between the resources of land, labor, capital, and management that attains the optimum use of each resource in the production of crops and livestock to maximize financial returns and to maintain or improve soil productivity.

Balanced Operation–A range livestock enterprise which provides sufficient feed and forage resources at each season to sustain continuous satisfactory maintenance of its livestock and game throughout the year. Also, an enterprise in which the gross income either equals or exceeds, by a suitable margin for profit, the cost of production.

Balanced Ration–A daily allowance of livestock or fowl feed; mixed to contain suitable nutrients required to promote normal development, maintenance, lactation, gestation, etc.

Balanced Seat–That position of the mounted rider that requires the minimum of muscular effort to remain in the saddle and which interferes least with the horse's movements and equilibrium.

Balantidium coli–A parasitic species which attacks the intestines of mammals, especially pigs; can also cause severe dysentery in humans.

Bald Face–A white face or a white streak on the face of an animal. On a horse the streak extends out and around the eyes and down to the nostrils.

Baldy–An animal that is naturally polled or hornless.

Balk–(1) A ridge of untilled land missed in plowing. (2) The act of an animal stopping and refusing to move.

Balling Forceps–A pair of tongs for administering medicine, especially pills, capsules, etc., to animals; used to place medicine on the back of the tongue.

Balling Gun–A dispenser with a plunger mechanism in the barrel used to place medicine in the form of pills or capsules in the back of an animal's mouth or down its throat so that the animal cannot spit the medicine out.

Balling Iron–An open paddle-shaped instrument inserted into the mouth of an animal to permit placing pills or capsules on the back of the tongue so that the animal is less likely to spit out the medicine.

Band Day–Range forage required by a band of sheep of a given size and class during one day.

Bandage–(1) A cloth used for dressing wounds. (2) A cloth used to cover and compress cheese.

Banding–(1) The marking of animals or birds by fastening bands to their legs or ears or by painting their fathers for identification. (2) The placing of a collar or band around the stem of a plant or the trunk of a tree to poison, repel, or trap insects. (3) A method of castration in which a tight rubber band is placed around the scrotum; circulation is stopped and the testicles atrophy. (4) The placement of fertilizer or herbicide in a band along a row.

Bang's Disease–See Brucellosis.

Bangtail–Slang term for a racehorse, as in the old days running horses usually had banged tails, often banged close to the dock, or docked and banged; also, a wild horse.

Bantam–A naturally occurring, small-sized fowl. There has been so much selective breeding to scale down the standard-sized birds, the term is generally taken to mean any small-sized fowl.

Bar–(1) The mouth piece of a bridle bit. (2) The landside of a plow. (3) The straight line of a cattle brand. (4) The part of a horse's hoof that is bent inward, extending toward the center of the hoof. (5) Ridges in the roof of a horse's mouth. (6) The space in front of a horse's molars where the bit is placed. (7) A pole used to close a fence gap. (8) 14.50 pounds per square inch of air pressure.

Bar Bit–A specific type of horse bit consisting of a solid metal piece that has no lever action.

Bar Pad–A leather covering for a horse's hoof to prevent slipping. It extends to each heel over the bars of the hoof and is fitted as a part of a short shoe having a soft pad to cover the frog. See Bar Shoe.

Bar Shoe–A horseshoe which has a bar plate consisting of a flat piece of metal across the heel opening to protect the frog of the hoof. See Bar Pad.

Barb–(1) A breed of horses related to the Arab and native to Barbary; probably introduced into Spain by the Moors and probably smaller and coarser than the Arab; its strain is evident in all known present breeds. (2) A hairlike side branch of a feather. (3) Mucous membrane projections for the openings of submaxillary glands under the tongue of horses and cattle. (4) A pointed projection on a fence wire. (5) In botany, a hooked hair or bristle. (6) The teeth or spines on the awns of grasses, especially barley.

Barbed Wire—Fencing made of twisted wires to form one strand and to which wire barbs are attached at short intervals.

Bareback—(1) A poultry term denoting a chicken which is not fully feathered out on the back. (2) To ride bareback is to ride a horse without a saddle.

Barium Carbonate—A solid material used as a rat poison and as a water purifier; a mineral known as witherite in its natural state.

Barium Fluosilicate—$BaSiF_6$; a compound used as an insecticide and a rodent poison.

Barium Poisoning—Livestock poisoning caused by excessive ingestion of preparations containing barium; characterized by profuse salivation, sweating, convulsions, palpitations, and general paralysis.

Barking Squirrel—A prairie dog.

Barn—A farm building used for storage of hay, grain, farm implements, etc., or for housing of domestic animals.

Barn Book—The record book in which breeding and management data on livestock is kept. Commonly included are breeding dates, birth dates, health data, and weight data.

Barnes Dehorner—A device with two handles joined by a hinge that has curved blades at the end. When the handles are folded together the blades are fit over the horn of a calf; when the handles are snapped open, the horn is removed.

Barny—A term used in the judging of dairy products to designate an off flavor characterized by cow-stable odor and its persistence on the palate.

Barnyard—Land immediately adjacent to a barn, fenced in to enclose livestock or fowls; also called barn lot.

Barnyard Fowl—Any commonly domesticated fowls; chickens, ducks, geese, turkeys, etc.

Barred—A trait in chickens involving the alternating white and dark stripes on the feathers.

Barred Plymouth Rock—A breed of chicken reared for both egg production and meat. The cock weighs about 9.5 pounds and the hen weighs about 7.5 pounds. The feathers are striped horizontally with black or almost black alternating bars; also called barred rock.

Barrel—(1) The trunk of a domesticated animal. (2) A measure of corn on the ear: one barrel equals one bushel of shelled corn. (3) In the earlier period of flour milling in the United States, the wooden barrel, net weight 196 pounds of flour, was a common container for shipment. (4) A cylindrical container. (5) A liquid measure of 42 United States gallons.

Barren—(1) Denoting an area unable to support vegetation. (2) Incapable of producing offspring, seed or fruit.

Barrenness—The condition of a female animal that renders her sterile.

Barrow—(1) A male pig castrated before sexual maturity. (2) In New Zealand, to shear or partially shear a sheep. (3) A small cart (wheelbarrow). (4) A hill or mound.

Barton—(1) A poultry yard or farmyard. (2) Farm buildings attached to a farmhouse; common in New England. (3) Land belonging to an estate.

Barzona—A breed of cattle originating in Arizona; genetics include Africaner, Hereford, Shorthorn, Brahman, and Angus.

Basal Feed—A feed used primarily for its energy content.

Basal Metabolism—The energy necessary to sustain the vital functions of cellular change, respiration, and circulation when an organism is at rest.

Base Exchange Capacity—(1) In soil science, the cations that a soil can absorb; expressed in milli-equivalent per 100 grams of soil. (2) In medical science, the maximum capacity of the blood to exchange base (alkali) for acid in maintaining a physiologic balance.

Base Exchange Solid—Nonfat milk solid from which most of the calcium has been removed by ion exchange; used in making ice cream.

Base Narrow—A term used to describe a horse that stands with front or rear feet close together, yet standing with legs vertical.

Base Wide—A term used to describe a horse that stands with front or rear feet wide apart, yet with legs vertical.

Basic—On the basic side of neutral pH (above 7.0).

Battery—In poultry husbandry, a series of cages used for raising chickens without a hen, for fattening chickens and broilers for market, and for increasing egg production in laying hens.

Bay—(1) A horse with a body color ranging from tan to red to reddish brown. The mane and tail are black, and the lower legs are usually black. (2) The laurel or its berries. (3) A shallow swamp supporting dense tree and shrub vegetation, and containing peat or muck soil. See Pocosin. (4) A coastal inlet. (5) A compartment in a barn for storage or for a special use, as a horse barn.

Bay Roan—A coat color of a horse which is brown with gray.

Beak—(1) The upper and lower mandibles or nibs of a fowl. (2) The awn on the outer chaff of wheat. (3) Any part of a plant resembling a bird's beak.

Bean Pork—Fresh trimmed hog jowls usually used to season beans.

Bean-Shooter—A horse that throws its front feet violently forward at the trot, with little flexion, "landing" about 12 inches above the ground; a very undesirable trait.

Bear Trap Bridle—A homemade, bitless bridle that has a light chain across the horse's nose.

Beard—(1) Hairy appendages on the face of a man or animal, such as the whiskers of a goat. (2) A group of feathers hanging from the throat of certain breeds of chickens. (3) A tuft of bristly hair projecting from the upper part of a turkey's breast. (4) A long awn or bristlelike hair, as in the inflorescence of some grasses. (5) A tuft, line, or zone of pubescence, as on the falls of the bearded irises. (6) The part of a horse's lower jaw which carries the curb of the bridle.

Bearing Rein–A rein pushed against the neck of a horse toward the direction of turn; also called neck rein.

Beast–A four-footed animal.

Beast of Burden–A domesticated animal used to carry, haul, or pull, etc., as a horse, ox, ass, mule, camel, etc.

Bed–(1) A small plot of soil used for growing seedlings, vegetables, flowers or shrubs, and often raised above the level of the surrounding soil. (2) A place for animals to sleep. (3) The hauling platform of a wagon or truck.

Bed Bug–*Cimex lectularius*, family Cimicidae; an insect that feeds by sucking blood from humans and animals.

Bedding–(1) Plowing or otherwise elevating surface soil into a series of crowned parallel beds or lands with shallow surface drains separating them. (2) A method of surface drainage consisting of narrow plowed lands in which dead furrows run parallel to the prevailing slope; the land between adjacent dead furrows is called a bed. (3) Straw, leaves, sawdust, sand, peat moss, etc., to make a bed for an animal. It is used where an animal may lie to absorb its urine. (4) Flowers that are appropriate for growing in flower beds for a massed decoration effect.

Bedding Out–A system of handling sheep wherein the band is bedded on the same spot, usually for not more than three successive nights.

Bee–(1) Any insect of the families Prospidae, Colletidae, Megachillidae, Xylocopidae, Ceratinidae, Nomadidae, Andrenidae, Anthophoridae, Bomidae, or Apidae. They are generally beneficial, being insect parasites, predators, pollenizers, or producers of honey and beeswax. (2) See Honeybee, Drone, Worker, Queen Bee, Bumblebee. (3) A social gathering of neighbors, especially in rural areas, to husk corn, to raise a barn, or to quilt in the home, etc.

Bee Blower–A portable machine that produces large volumes of rapidly moving air to blow bees from their combs.

Bee Bread–Pollen stored in the cells of honeycomb; used as food for bees.

Bee Brush–A soft-bristled brush used for removing bees from combs.

Bee Dance–Movement of bees on a comb as a means of communication; best known to indicate the direction and distance of a source of nectar or pollen.

Bee Escape–A device to let bees pass in one direction only; usually inserted between honey supers and brood chambers for removal of bees from honey chambers.

Bee Farming–The raising of bees for the honey and beeswax they produce, and for their pollinating of plants.

Bee Gum–Usually a hollow log used as a hive. The term is also used to refer to any beehive.

Bee Louse–*Braula coeca*; a relatively harmless insect that gets on honeybees, but the larvae can damage honeycomb.

Bee Metamorphosis–The transformation of the bee from egg to larva to pupa and finally to the adult stage.

Bee Moth–See Wax Moth.

Bee Paralysis–An adult bee disease of chronic and acute type caused by different viruses.

Bee Pasture–Vegetation, attractive to bees, which is within flying distance of a hive; also called bee pasturage.

Bee Space–A space ($\frac{1}{4}$ to $\frac{5}{16}$ inch) big enough to permit the free passage for a bee but too small to encourage comb building. Leaving bee space between parallel beeswax combs and between the outer comb and the hive walls is the basic principle of hive construction.

Bee Veil–A wire screen or cloth enclosure worn over the head and neck to protect from bee stings.

Bee Venom–Poison injected by bee stings.

Bee Wolf–A solitary wasp, *Philanthus triangulum*, which captures honeybees and stores its nest with them as food for its larvae. Where plentiful, it can cause beekeepers considerable trouble. Found in Europe; related species have been observed in the United States preying on native bees but not on honeybees.

Bee Yard–See Apiary.

Beef–(1) Any cow, bull, or steer which is fattened for slaughter for food. (2) The meat derived from cattle nearly one year old or older.

Beef Breed–Any breed of cattle which is especially developed for use as a source of meat, as Hereford, Angus, Shorthorn, etc. Also called beef cattle.

Beef Carcass Data Service–A program whereby producers, for a fee, can receive carcass evaluation data on their cattle by using a special "carcass data" eartag for their slaughter animals.

Beef Improvement Federation (BIF)–A federation of organizations, businesses, and individuals interested or involved in performance evaluation of beef cattle. The purposes of BIF are to bring about uniformity of procedures, development of programs, cooperation among interested entities, education of its members and the ultimate consumers of performance evaluation methods, and to build confidence of the beef industry in the principles and potentials of performance testing.

Beefalo–A cross between a beef animal and a buffalo or a bison.

Beehive–A domicile prepared for a colony of honeybees.

Beeswax–Wax secreted from glands on the underside of a bee's abdomen; molded by bees to form honeycomb; used in the manufacture of cosmetics, candles, salves, comb foundation, polishes, adhesives, etc.

Beget–To procreate, as a sire.

Belgian–(1) A breed of horses originating in Belgium from the Flemish horse. They are strong, brown draft horses; a stallion may weigh over a ton. (2) A breed of rabbits raised for meat production.

Bell Boots–Rubber protective boots that are bell-shaped, fitting over the coronet bands and down on the hoof of a horse.

Bellwether–A wether (a male sheep castrated before sexual maturity) with a bell tied around his neck who serves as a flock leader.

Belly–The abdomen of a human or an animal.

Belly Wool–The wool that grows on the belly of the sheep and occasionally extends up the side in irregular patches; usually uneven, different in grade (generally grades one grade or finer) from the body of the fleece; shorter and less desirable.

Belt–(1) A geographical term denoting an area of similar soil and climate particularly suited to specific crops or animals, such as the wheat belt. (2) A band of hair or skin of another color, often white, around an animal's body, as in Dutch belted cattle. (3) In machinery, a broad flexible band passed around two wheels to transmit motion from one to the other.

Belted–(1) Any fruit which has a retarded growth area extending completely around its middle; often due to frost injury when the fruit is very young; band is usually shrunken; also called banded fruit. (2) Any animal having a band of different color around its body.

Benign–In reference to disease or disease processes, mild or nonmalignant, such as a benign tumor or growth, in contrast to malignant or cancerous tumor or growth.

Benthic Region–The bottom of a body of water.

Benzaldehyde–A liquid used to drive bees from honeycombs; a component of oil of bitter almond. It has a smell that is pleasant to humans.

Berkshire–A popular breed of hogs originating from Berkshire, England. It is black with white feet, nose, and tail, and has a very short, dished face with erect ears.

Bezoar–Stones, hair balls, wool balls, etc., found in the alimentary canals of ruminants and sometimes other animals. At one time these were thought to have antipoison properties. See Hair Ball.

Bicolor–Having two colors.

Bidenate–Having two teeth.

BIF–See Beef Improvement Federation.

Bilateral–Having two sides.

Bilateral Symmetry–Similarity of form, one side with the other.

Bilharziasis–A disease caused by blood flukes (parasitic trematodes) belonging to the genus *Schistosoma* in humans and animals; more commonly known as schistosomiasis.

Billygoat–A male goat.

Binomial–By international agreement, all plants and animals have two Latin names: genus and species.

Bioaccumulation–The process by which plants and animals accumulate substances, especially pollutants, that may not be injurious to that organism but may injure other organisms that eat them. For example, nitrates may accumulate in corn and oats and be injurious to animals feeding on them but not to the plants themselves. In a like manner, fish may accumulate DDT or PCB in their fat, which may be toxic to an animal eating the fish but not to the fish themselves.

Biochemical Oxygen Demand (BOD)–The amount of oxygen required by the biological population of a water sample to oxidize the organic matter in that water. It is usually determined over a five-day period under standardized laboratory conditions and hence may not represent actual field conditions. Also known as Biological Oxygen Demand.

Biodegradable–Substances capable of being degraded into their constituent elements. This term is used especially in reference to toxins, such as pesticides, being degraded to nontoxins.

Biogenesis–(1) Formation by the action of organisms. (2) The doctrine that all life has been derived from previously living organisms.

Biological–Products derived from a living process or living matter, such as sera, vaccines, bacterins, and antitoxins, etc. Also called biologics.

Biological Control–A method of pest control by the use of predatory insects, fungi, or viruses; as contrasted to control by chemical pesticides. See Biophage.

Biologist–A person who studies living organisms as a career.

Biology–The field of study dealing with living organisms. It may be divided into the study of plants (botany) and of animals (zoology).

Biophages–Organisms that obtain nourishment from other organisms, e.g., predators, parasites, and pathogens. See Biological Control.

Biopsy–The microscopic or chemical analysis of tissue removed from a living body, usually to discover the cause of illness.

Biostress–Difficulties that plants and animals have in obtaining the necessities of life: food, water, and living space.

Biotechnology–Technology concerning the application of biological and engineering techniques to microorganisms, plants, and animals, sometimes used in the narrower sense of genetic engineering.

Biotic–Pertaining to life; biological.

Biotic Influence–Biological influences on plant and animal life as contrasted to climatic influences.

Biotic Potential–(1) The maximum reproduction power or ability. The inherent ability of an organism to reproduce and survive in greater numbers. (2) The ability of an organism to reproduce in an optimum, unrestricted, and noncompetitive environment.

Biotype–Groups of plants or animals primarily distinguishable on the basis of interaction with relatively genetically stable varieties or clones of host plants; a strain of plant or animal species.

Birth–See Parturition.

Birth Weight–The weight of a calf taken within 24 hours after birth. Heavy birth weights tend to be correlated with calving problems, but the conformation of the calf and the cow are contributing factors.

Bisexual–An animal or plant that produces both eggs and sperm, or a flower that bears both stamens and pistils.

Bishop–Specifically, to drill small holes in a horse's teeth and to grind the surface of the front teeth so as to make the horse appear to be younger than it is. The term is derived from a horse dealer named Bishop.

Bit–(1) The part of a horse bridle that is placed in the mouth; usually made of steel. (2) An earmark for cattle. (3) The blade of an axe or similar tool. (4) A tool used to drill a hole.

Bitch–A female dog, fox, wolf, or other canine.

Bitting–Teaching a horse to give (yield) to the bit before it is used for riding.

Black Baldy–A term used to describe a calf or older animal that is a cross between a Hereford and an Angus; characterized by a black body with a white face and white markings.

Black Blowfly–*Phormia regina*, family Calliphoridae; breeds primarily on animal carcasses; also attacks soiled sheep's wool, in which the maggots feed on the wool around the rump; in attacking horns, it causes the adjacent skin to fester and the wool to loosen and become putrid.

Black Disease–A highly infectious disease of cattle and sheep caused by the anaerobic bacillus *Clostridiumoedematiens*. Symptoms are similar to those of blackleg and malignant edema. Prevention is by vaccination; sometimes called false blackleg.

Black Fly–An insect, family Simuliidae, which frequently bites with very irritating results. Some species are vectors of several diseases of humans and animals. The larvae live in streams.

Black Roan–Used to describe a horse whose coat is black and white, rather uniformly mixed. Also called blue.

Black Top Wool–Wool containing a large amount of wool grease combined at the tip of wool staples with dirt, usually obtained from Merino sheep. This wool is usually fine in quality, of good character, and desirable in type, but the shrinkage is high.

Black Water–See Azoturia, Cattle Tick Fever.

Black Wool–A term reserved for any wool containing black fibers. A fleece having only a few black fibers is rejected by a grader and goes into black wool because there is no way of separating the few black fibers in the manufacturing process. Black wool is usually run in lots that are to be dyed black.

Black-faced Highland Sheep–A small, black-faced hardy breed originating in the Highlands of Scotland; noted for its long wool and yielding good mutton.

Blackhead–A chronic disease of poultry, chiefly of turkeys, caused by a protozoan parasite, *Histomonas melegridis*, in which the principal lesions are confined to the liver and the spleen. The disease is also characterized by listlessness, diarrhea with droppings from light-green to brown and darkening of the skin on the head. Chickens may harbor the parasite, although they are seldom seriously affected, whereas with turkeys, the disease is commonly fatal. Also called enterohepatitis.

Blackleg–(1) A highly infectious, rapidly fatal disease of cattle and sheep caused by the anaerobic bacillus *Clostridium chauvaei*. It has been found in almost all range-grazing areas. Symptoms are so similar to those of malignant edema that it is almost impossible to distinguish between the two; also called black quarter, quarter ill. (2) A disease of cabbage and other crucifers caused by the fungus *Phoma lingam*, family Sphaeriodaceae. The disease persists in the soil and is carried by the seed. It is characterized by lesions on the stem near the ground or on the petioles and leaves, which soon become sunken and dark. (3) A potato disease caused by a complex of soft rot bacteria, including *Erwinia* and *E. atroseptica*. Bacteria cause decay from the seed to the stem, producing a black moist soft rot on immature plants. (4) A disease of geranium caused by *Botrytis* spp., *Rhizoctonia* spp., or a bacterial leaf spot; characterized by stems rotting and blackening at the base.

Blacksmith–A mechanic who welds and fashions iron and other metals, sharpens plowpoints, shoes horses, etc.

Blacktongue–(1) A disease of dogs caused by a deficiency of one of the B vitamins, nicotinic acid (niacin); characterized by ulcers in the mouth and inflammation of the alimentary canal and severe nervousness. Identical with pellagra in humans. (2) A variation of anthrax disease that attacks cattle and horses; characterized by ulcerations on the tongue.

Bladderworm–The newly hatched embryos of certain tapeworms.

Blade–(1) The expanded portion of a leaf. (2) A cutting tool, as the blade of an axe or the blade of a bulldozer. (3) A retail cut of meat from the forequarter or shoulder. (4) The hind part of a fowl's single comb.

Bland–Mild-flavored, not stimulating to the taste; smooth, soft-textured.

Bland Lard–A deodorized pork product to which hydrogenated lard is added, as well as a stabilizing agent to retard rancidity.

Blanket Algae–A mass of filamentous algae floating as a visible mat on the surface of the water.

Blast–(1) A disease of sheep characterized by flatulence (gas in the digestive tract). (2) A disease in which a plant dries up or shrivels because of electricity or some other agent in the atmosphere.

Blast Freezing–Freezing meat with a continuous blast of cold air moving around the meat.

Blastula–A mass of cells with a cavity that occurs from the dividing of a fertilized egg. From this stage the cells begin to differentiate.

Blaze–(1) A white mark, such as a star, on the face of an animal. (2) A cut or other mark (such as paint) on a tree which indicates a trail or timber to be felled.

Bleed–(1) The process of removing air from a hydraulic or fuel system of a tractor or other machine. (2) To remove the flood from an animal during the slaughter process.

Bleeding Pit–A pit in which the blood from a hanging slaughtered animal is collected; also called bleeding trough. The blood is used as a high-protein feed or a high-nitrogen organic fertilizer.

Blend Price–Price paid a milk producer based on the proportion of milk utilized in each price class, such as fluid milk and manufacturing milk.

Bliinkers–See Blinders.

Blind Gut–The cecum; also called bung.

Blind Quarter–A quarter of an udder that will not secrete milk during lactation or one that has an obstruction in the teat; may be a common hereditary defect in swine and first-calf heifers; a serious defect; also called blind teat, blind nipple, inverted nipples.

Blind Staggers–A disease of horses, cattle, and swine caused by an excessive intake of the mineral selenium; characterized by the loss of hair from the mane and tail in horses, from the tail in cattle, and the general loss of hair in swine; the hoofs slough off, lameness occurs, food consumption decreases even to the point of starvation, and vision becomes gradually impaired. The disease is prevalent on the plains of the United States, particularly in South Dakota. See Alkali Disease.

Blind Teat (Blind Nipple)–A teat on a sow that is not connected with the milk gland and will not produce milk.

Blinders–Small pieces of material, usually leather, fastened to a bridle to keep a horse from seeing anything not straight ahead; also called blinkers.

Blindness–Failure of vision in humans and animals.

Blinky–Sour or slightly sour milk.

Blissom–A ewe in heat.

Blizzard–A severe storm of high wind, low temperatures, and heavy snow.

Bloat–(1) A severe distension of the abdomen by gas; usually in ruminant animals. Caused by eating watery food or by eating too quickly. In horses, the condition is called colic. (2) Distention of one end of a can because of gas pressure.

Block–(1) A piece of wood used as a divider between two bales of hay. (2) A pulley used to increase pulling or hoisting power. (3) In logging, an administrative division of a forest. (4) About 5 pounds of cotton hanks. (5) A portion of hay thrown into a baler. (6) To thin out plants, especially sugar beets, into smaller bunches, which will be again thinned into single plants. (7) To trim fleece to enhance the appearance of a sheep. (8) See Cutting Block.

Block Salt–A cube of about 10 inches of compressed, sometimes medicated salt for consumption by domesticated animals.

Blocky Foot–A horse with a noticeably more upright slope to the hoof than is normal.

Blond D'Aquitaine–A large breed of beef cattle that originated in France. They are horned and fawn in color.

Blood–(1) The fluid that carries nourishment and oxygen to all parts of a body and carries away waste products. (2) In plant or animal breeding, descended from parents of registered breed; relationship by blood. (3) A market term referring to the fineness of wool, as in three-eighths blood, quarter blood, low quarter blood, half-blood, blood classification. (4) A market classification for a type of Spanish orange.

Blood Flour–Dried blood reduced to fine powder. Principally fed to dairy calves on a minimum milk diet. See Blood Meal.

Blood Horse–A purebred horse.

Blood Meal–Dried, ground animal blood which is heated to coagulation, drained, and then pressed to remove remaining water. The meal is 80 percent hard-to-digest protein, but it is mixed into calf starter feed. See Blood Flour.

Blood Poisoning–See Hemorrhagic Septicemia.

Blood Spavin–A varicose vein enlargement on a horse's leg which appears on the inside of the hock but immediately above the location of bog spavin.

Blood Spot–An egg defect that occurs most commonly at the beginning of the laying year; usually detected during candling.

Blood Sugar Level–The amount of sugar, usually glucose, in blood. Abnormally high or low blood sugar levels may indicate the presence of a disease.

Bloodline–A family or family lines of breeding.

Bloodstopper–A veterinary product that contains a coagulant to cause the clotting of blood and a chemical astringent to shrink blood vessels or tissue when applied to animal wounds.

Bloodsucking Conenoses–The insect *Triatoma sanguisuga*, family Pseudococcidae, whose intensely painful bite causes swelling over much of the body, faintness, and vomiting, which may last for weeks or months; also called Mexican bed bug.

Bloodsucking Insect–Any insect (e.g., mosquito, fly, tick) which draws the blood of its host.

Bloodworms–Cylindrical, elongated midge larvae (family Chironomideae) with pairs of prolegs on both the first thoracic and last abdominal segments. Although many species are blood-red in color, some are pale yellowish, yellowish-red, brownish, pale greenish-yellow, or green. Most feed on diatoms, algae, tissues of aquatic plants, decaying organic matter, and plankton. Some are associated with rich organic deposits. Midge larvae are important as food for fish.

Bloody Milk–Milk from an animal suffering from ruptured blood vessels and injured inner gland tissues. Such milk should always be discarded.

Bloody Murrain–See Cattle Tick Fever.

Bloody Scours–A form of diarrhea that attacks small calves and pits; characterized by bloody feces. It is highly contagious and often fatal. See Calf Scours.

Bloody Whites–See Blood Spot.

Blow-out–(1) An excavation by the wind of loose soil, usually sand. (2) A hole in soil above a tile line caused by excessive pressure in the drain. (3) To walk or exercise a horse either to loosen its muscles for further exercise, or to prevent chilling and stiffening after a hard workout.

Blowfly–Any fly that breeds in decaying flesh, or that infests wounds. Many brightly colored species are known as bluebottle and green bottle flies.

Blue Cheese–Any cheese ripened by a blue mold, thereby resembling Roquefort. Made from cow's milk (in France from ewe's milk) and cured for two or three months. Also called blue-mold cheese, blue-veined cheese.

Blue Eye–In marketing, an apparent defect in the eye of a horse or mule.

Blue John–Colloquial term for skimmed or watered, slightly sour milk having a bluish tint.

Blue Louse–An insect (*Solenopotes capillatus*) that infests cattle.

Blue Milk–See Blue John.

Blue Ribbon–A prize or award for the best livestock or other agricultural commodity in a class; made of inscribed blue cloth.

Blue Roan–A horse with black hair interspersed with white, giving an impression of bluish or grayish color.

Blue Skin–Bluish spots on the skin of purebred sheep; a fault or disqualification.

Blue-gray–(1) A calf of a white Shorthorn bull and an Aberdeen Angus or Galloway cow. (2) A horse having steely-blue hairs scattered through a gray coat.

Blue-green Algae–The group Myxophyceae; characterized by simplicity of structure and reproduction, with cells in a slimy matrix and containing no starch, nucleus, or plastids and with a blue pigment present in addition to the green chlorophyll. It is capable of fixing atmospheric nitrogen.

Bluebag–Gangrenous form of mastitis that causes a bluish discoloration of the udder; most frequently seen in cattle, sheep, and goats.

Bluetongue–An infectious but noncontagious viral disease of ruminants transmitted by insects and characterized by congestion, edema, and hemorrhage in the affected animal. In sheep the disease is characterized by fever, emaciation, oral lesions, lameness, and a substantial death rate, frequently with the heaviest losses in lambs. Cattle usually have milder disease with a low mortality rate; sometimes the disease in cattle may be diagnosed only by inoculating sheep with bovine materials. In Africa bluetongue has been confused with foot-and-mouth disease.

Boar–A male pig which has not been castrated.

Boar Pig–Male swine under one year of age.

Board–(1) Lumber less than 2 inches thick and more than 8 inches wide. (2) Lumber 1½ inches thick, 6 or more inches wide, and 8 or more feet long (British). (3) Lumber of all widths 1 inch thick; widths of less than 6 inches are sometimes called strips. (4) Plank floor on which sheep are sheared (New Zealand).

Boardman Feeder–See Entrance Feeder.

Bob–(1) Veal from calves slaughtered when less than four to six weeks old; also called slunk. (2) The docked tail of a horse. (3) In transporting lumber, a single pair of runners on which the forward ends of the logs are placed. (4) To transport logs on a bob.

Bob Wire–See Barbed Wire.

Bobbed Wire–See Barbed Wire.

Bobtail–An animal with a short tail.

BOD–See Biochemical Oxygen demand.

Body–(1) The physique of an animal. (2) The consistency of a substance, such as honey, butter, cheese, etc. (3) The fullness of flavor, especially of a beverage.

Bog Spavin–In horses, a distention of the joint caused by excess fluid on the inner surface of the hock. It makes a horse absolutely unsound; also called boggy in the hocks.

Boiled Ham–A retail pork product. The outside fat and bone is removed from cured, unsmoked ham. The meat is then compressed, tied with cord, and simmered 30 minutes per pound.

Boiled Milk–Denoting an off-taste that milk acquires when heated to 180°F during pasteurization.

Boiling Point–The temperature at which the vapor pressure of a liquid equals the atmospheric pressure. At the boiling point, bubbles of vapor rise continually and break on the surface. The boiling temperature of pure water at sea level (barometric pressure of 30 inches of mercury) is 212°F (100°C). At high altitudes, the boiling point of water is lower because the atmospheric pressure is lower. At 5,000 feet above sea level, for example, the boiling point of water is 203°F (95°C); at 10,000 feet it is 194°F (90°C).

Bold–(1) Designating the assured gait of a horse. (2) Designating assurance of character in an animal.

Bologna Bull–Colloquially, a medium-sized bull, the meat of which, because it carries little fat and absorbs water, is used in making bologna.

Bolster–(1) The padded undersection of a saddle. (2) In growing plants, the excrescence at the leafstalk where it joins the axis. (3) A cross member of a vehicle on which the bed, rack, or load rests.

Bolus–(1) A large pill for dosing animals. (2) Regurgitated food that has been chewed and is ready to be swallowed; a cud.

Bolus Gun–See Balling Gun.

Bone–(1) A piece of the skeleton of a vertebrate. (2) To remove bones from fish, meat, or poultry. This is best done by fish or meat dealer. A special, short, sharp-pointed boning knife is used. (3) The measurement of the circumference around the cannon bone of a horse about halfway between the knee and fetlock joints. Eight inches of bone is average for the Thoroughbred. "Flat bone" indicate that the cannon and the back tendon are parallel, with the tendon cleancut and standing well away form the cannon bone. The word flat refers to the appearance of the cannon, which is wide and flat when viewed from the side although narrow from the front, and does not mean that the bone itself is flat.

Bone Ash–The highly calcic and phosphatic residue of bones burned in air; of limited use as a fertilizer. Also used in making assay cups and in cleaning jewelry.

Bone Chewing–The eating, by animals, usually cattle and horses, of the bones of dead animals; indicative of a mineral deficiency, usually of phosphorus and calcium. Also called pica, depraved appetite, licking disease.

Bone Meal–(1) The product of drying and grinding animal bones not previously steamed under pressure. The composition is nitrogen 3.3 to 4.1 percent, bone phosphate 43 to 50 percent or phosphorus 10 to 12 percent, and calcium 21 to 24 percent. Used as a fertilizer and stock feed. (2) (Steamed) A product of grinding animal bones previously steamed under pressure. The composition is nitrogen 1.65 to 2.5 percent and calcium 24 to 33 percent. Used as a fertilizer and stock feed.

Bone Phosphate–The calcium phosphate obtained from bones; also in commerce, applied to calcium phosphate obtained from phosphatic rocks, e.g., those of Florida.

Bone Products–The skeleton of vertebrates (20 to 30 percent total P_2O_5); earliest source of fertilizer phosphorous, now used largely in animal feeds and, to some extent, in specialty fertilizers such as for growing roses.

Bone Spavin–A bony growth in the hock which lames a horse. Also called jack spavin, spavin, spavined hock. See Spavin.

Boner–A utility-grade beef carcass that is devoid of marbling. The meat is usually "boned out" and ground or processed.

Boning–The process of making the hair stand out on a beef animal's leg. The purpose is to make the leg and/or hind quarter appear thicker in the show ring.

Boot–(1) The hollow metal casting on a planter or a drill through which the seed passes to be planted. (2) Profuse feathering on the shank and toes of fowls. (3) The sheathlike leaf structure on the upper end of grain or grass plants that encloses the inflorescence prior to its emergence. During inflorescence the plant is said to be in the boot stage or in boot. (4) In a grain elevator, the box which contains the lower pulley or sprocket.

Booted–Fowls that are feathered on shanks and toes and having vulture hocks are said to be booted, as in booted and sultans.

Bos Indicus–Scientific name for domestic humped cattle common to the tropical countries; Zebu or Brahman.

Bos Taurus–Scientific name for domestic cattle common to the temperate zones; Hereford, shorthorn, Angus, etc.

Bosal Hackamore–A bitless bridle with a braided noseband (bosal) that controls a horse with leverage and pressure on the nose and jaw.

Boss–(1) A lumpy formation on an animal or plant. (2) Metal stud on either side of the bit of a horse's bridle. (3) The dominant individual among a group of animals. See Pecking Order.

Bossy–An affectionate name for a cow.

Boston Shoulder–The top part of the shoulder of a pork carcass; sometimes called the Boston butt.

Botany Wools–In Great Britain, all fine Australian wools are known as Botany. This designation is also extensively used in the United States. Also a general term for all classes of fine wool.

Bots–The larval or maggot stage of the bot fly or nit fly that migrates from the tongue or mouth to the stomach in horses (*Gastrophilus intestinalis* and *G. nasalis*), or from the nostrils to the sinus cavities of sheep, goats, and deer (*Oestrus ovis*), where they may grow to be over ½-inch long; hard to control.

Bottom–(1) (Often plural) Low-lying land adjacent to a river, usually rich in alluvial deposits, e.g., Mississippi River bottoms. (2) A section of a plow consisting of the moldboard, share, frame, and landside. (3) Stamina in a horse.

Bottom Board–The floor of a beehive.

Bottom Chuck–A boned cut from a side of beef below the blade bone.

Bottom Round–A cut of beef composed of the two tender muscles of the leg; used for either steaks or roasts.

Botulism–A food poisoning caused by the bacterium ***Clostridium botulinum*** in preserved foods or feed for animals. The poison can be fatal to humans or animals. Called limberneck in poultry. Also called food poisoning, allantiasis.

Bougie–An instrument used for insertion into natural passages of a body for dilation or medication, e.g., a teat bougie; usually constructed of steel, rubber, or plastic.

Bouillon–(1) An abnormal growth on a horse's frog. (2) A beef extract or clear soup.

Bourbon Red–A breed of turkey with dark, reddish-brown feathers, except for the white tail and primary and secondary wing feathers.

Bovine–An animal of the family Bovideae; a cow, bull, steer, calf, or ox.

Bovine Farcy–A slow, chronic, and fatal disease of cattle, caused by ***Actinomyces farcinicus***; characterized by small swellings on the legs, especially above the hocks.

Bovine Genital Trichomoniasis–A venereal disease of cattle caused by a flagellated protozoan, ***Trichomonas fetus***; characterized by difficulty in breeding, temporary sterility, and an accumulation of pus in the uterus without a fever. Also called tric, bovine venereal trichomoniasis.

Bovine Papular Stomatitis–A widely distributed viral disease that produces papular and occasionally erosive lesions on the muzzle and buccal mucous membranes of young cattle.

Bovine Somatotropin (BST)–A naturally occurring hormone that aids in stimulating the production of milk in cows.

Bovine Viral Diarrhea (Mucosal Disease)–An infectious viral disease of cattle, manifested clinically by an acute erosive stomatitis, gastroenteritis, and diarrhea.

Bowl–(1) A vat or tank in which wool is scoured. (2) A basin for holding liquids.

Box Stall–An enclosure in which an animal can be confined without being tied.

Boxed Beef–Beef that is sold by the meatpacker in primal, subprimal, or final retail cuts rather than as a carcass; this beef is usually frozen and boxed for shipping convenience.

Brace–(1) Any device designed to strengthen or support, such as a corner-post brace, wall brace, etc. (2) Specifically, a rod connected to the beam, which reinforces the landside of a plow. (3) A curved tool for holding and rotating bits. (4) A pair, as a brace of pheasants.

Brace Bandages–Resilient bandages on the legs of horses worn in some cases in an effort to support lame legs, and worn in other cases to protect a horse from cutting and skinning its legs while racing.

Brace Comb–A section of honeycomb built between and attached to other combs.

Bracken Sickness–A livestock disorder caused by eating eastern or western bracken. See Eastern Bracken, Western Bracken.

Bradford System–The system of combing and spinning wool commonly used in the United States. The longer wools are used for this system of manufacture. Bradford system yarns are used to make worsted fabrics.

Bradycardia–Abnormally slow heartbeat.

Braford–A cross between Hereford and Zebu breeds of cattle.

Brahma–Large or bantam meat chickens of the Asiatic class, of which there are the light (Columbian pattern), dark (silver-penciled), and buff varieties. The Brahma has a pea comb, yellow skin, heavily feathered shanks and toes, and the eggs are brown.

Brahman–A breed of hump-backed cattle originating in India. The proper name, however, is Zebu (*Bos indicus*). Most cattle in the United States originated in Europe and are *Bos taurus*.

Braid–The coarsest grade of wool.

Braining–(1) Fowl slaughter, in which a knife is inserted through the roof of the mouth to the rear of the skull, piercing the medulla oblongata. (2) Removing the brain from a carcass.

Brains–(1) The nerve tissues enclosed in the skull or cranium of vertebrates. (2) Calf brains, sold for human consumption.

Branch–(1) A lateral stem arising from the main stem of a plant, bough, or limb. (2) Part of a horseshoe from the first nailhole to the end of the heel. (3) A tributary stream. (4) An ancillary pipe attached to a main pipe.

Branch of Frog–The soft, rear part of a horse's foot, which branches into two sections that are divided by the middle cleft of the frog.

Brand–(1) A mark of identification burned into the skin of cattle or horses. See Tag. (2) The steel or iron die used to mark the animal. (3) The design of the mark of identification commonly used to identify foods and feedstuffs. (4) A German hard cheese made from sour milk curd and butter. Beer is used to moisten the cheese as it ripens in kegs.

Brand Artist–(1) A person who brands cattle. (2) A person skilled in forging brands.

Brand Book–An official record of ranch brands used to mark cattle.

Brand Inspector–A government official who inspects the brands of cattle at licensed auction yard sales, etc., to authenticate ownership.

Branding Bee–A festive, annual gathering at which animals born within the past year are branded.

Branding Iron–An iron handle on the end of which is welded a unique, registered mark (brand); it is heated and burned through the hair and shallow into the flesh of cattle and horses. Some more recent branding irons are electric. In many places, large, colored plastic ear tags are replacing the branding iron.

Branding Paint–A paint, indelible or removable, used for marking animals, particularly sheep, for identification.

Brangus–A breed of beef cattle originating in the United States; developed from the Brahman and Angus breeds. They combine the hardiness of the Brahman with the carcass qualities of the Angus.

Brassiness–A term descriptive of a light yellowish metallic cast commonly found in the plumage of White and Parti-White varieties of chickens, and to a lesser degree in several other varieties, particularly in the hackle, wing-bow, and saddle of the male; a serious defect; may be hereditary or affected by exposure to sun rays and certain items in the diet.

Brawn–(1) Muscles, especially of the arm and leg. (2) Boar's flesh, especially the boiled and pickled meat. (3) A pig fattened for its meat. (4) Hardened skin.

Braxy–An acute infection of Norwegian, Scottish, and Faroe Island sheep; characterized by hemorrhagic abomasitis. Apparently caused by *Clostridium septicum*, the bacterial organism which causes malignant edema in United States livestock. Also called bradsot, bradshot.

Bray–The cry of a donkey or mule.

Breachy–Describing an animal who passes over, or through fences.

Bread Madness–See Ergotism.

Break In–(1) To discipline; to train, as in a horse. (2) The process of wearing in a desirable fit between the surfaces of two new or reconditioned parts of a machine, as in the pistons and cylinders of an engine.

Break Joint–In grading sheep carcasses, the metacarpal (forearm bone) is struck with a heavy object. If the resulting break occurs at the center of the joint, it is referred to as a break joint, and the carcass is classified as a lamb carcass. If the break occurs above the joint and the joint does not separate cleanly, it is termed a spool joint. The carcass is then classified as a mutton carcass.

Breaker–(1) A wave meeting a shore, reef, sandbar, or rock and collapsing. (2) For watering: a widened fan-shaped device screwed on the end of a hose to decrease (break) the velocity of water; used in watering plants. (3) For plowing: an extended moldboard on a turning plow to more completely turn heavy sod upside down. (4) For electrical overloading: an automatic flip switch on an electrical circuit that breaks (shuts off) the current when overloaded. (5) Utility-grade slaughtered cattle in which some degree of marbling is in the meat. The loins and rounds are "broken out" and sold as steaks.

Breast–(1) The front of the body between the neck and the abdomen. (2) The breastbone and flesh surrounding it, especially in fowl, lamb, or veal. (3) The forepart of a moldboard on a plow. (4) The part of a cotton gin consisting of a frame, which contains the huller ribs, the picker rollers, and the roll box. (5) That part of a bulk milk tank which connects the inside lining to the outer wall.

Breast Band–A strap, used in place of a collar, which extends over the forechest of a horse; also called breast collar. See Breast Collar.

Breast Collar–An alternative to a neck collar for a horse. The breast collar is a wide horizontal strap around the lower part of the neck to which traces are attached for traction. See Breast Band.

Breastbone–Breastbone

Bred Heifer–A pregnant, young cow which has not previously borne a calf. See Heifer, Open Heifer.

Breech–(1) Presentation of a fetus at birth with hindquarters foremost. (2) That part of an animal from the base of the tail to the hocks. (3) Coarse wool from the breech and hind legs of a sheep.

Breech Presentation–A birth that is not normal, that is, hind feet and rump presented first.

Breeching–A part of a harness, usually of leather, which is fastened around the hind quarters. It allows a horse to stop or back a wagon.

Breed–(1) Animals having a common origin and distinguishing characteristics. (2) To improve, through control, characteristics in plants and animals.

Breed Association–A group organized to safeguard the purity of a livestock or poultry breed and to promote the breed. It sets the standards and maintains registers for the breed.

Breed Character–Details of conformation or color, e.g., shape of horns in cattle, size and shape of ear in swine, etc., which distinguish a breed of animals.

Breed Out–To eliminate undesirable characteristics by selective breeding.

Breed Trademark–The symbol of a distinguishing characteristic.

Breed True–The ability to uniformly transmit a characteristic to an offspring.

Breed Type–Distinctive features in which one breed differs from another.

Breeder–(1) The owner of the dam when she was bred. The definition holds true in registering all classes of livestock. (2) A specialist in breeding. (3) An animal or plant used for breeding. (4) Also known as a converter; a nuclear reactor that converts nonfissionable atoms of one element into fissionable atoms of the same or a different element. An advanced breeder is a converter that produces more fissionable fuel than it consumes.

Breeder Hatchery–An establishment which promotes a particular strain of poultry.

Breeder House–See Breeding House.

Breeder Tom–A male turkey used for breeding, which is marketed between April and August.

Breeder's Young Herd–A show group consisting of one bull less than two years of age, two heifers less than one year of age, and two heifers between one and two years of age.

Breeder-Feeder Litter–A show term for a group of swine; a barrow, a boar, or a gilt from the same litter.

Breeding Chute (Rack) (Crate)–A structure built to confine a cow, mare, sow, or other female domestic animal at the time of mating to aid in the mating act.

Breeding Class–Designated animals used for breeding purposes.

Breeding Herd–The livestock retained to provide for the perpetuation of the herd or band.

Breeding Hobbles–A fetter fastened to the feet of a mare to prevent her from kicking the stallion during mating.

Breeding House–(1) A building, divided into pens about 6 × 10 feet in size, used for breeding poultry. One male and approximately a dozen females are placed in each pen. (2) Any structure used for breeding.

Breeding Program Goals–The objective or "direction" of breeder's selection programs. Goals are basic decisions breeders must make to give direction to their breeding program. Goals should vary among breeders because of relative genetic merit of their animals, their resources, and their markets.

Breeding Season–The period of the year when animals normally reproduce.

Breeding Unit Index–A measure of a breeding herd, including the total number of female animals capable of giving birth, weighted by the production per head in a base period.

Breeding Unsoundness–Malformation of the genitals, cryptorchidism, scrotal rupture, and other abnormalities which disqualify animals for procreation though not for other purposes. See Hereditary Unsoundness, Serviceable Unsoundness, Temporary Unsoundness.

Breeding Value–The ability of an animal to transmit the genetic capability to produce meat, milk, eggs, or other economically important products; the value of an animal as a parent. The working definition is twice the difference between a very large number of progeny and the population average when individuals are mated at random within the population and all progeny are managed alike. The difference is doubled because only a sample half (one gene of each pair) is transmitted from a parent to each progeny. Breeding value exists for each trait and is dependent on the population in which the animal is evaluated. For a given trait, an individual can be an above-average producer in one herd and a below-average producer in another herd.

Breedy–Cows with a high degree of femininity; bulls with strong masculine features.

Brewers' Dried grains–The dried residue left after wort has been extracted form barley malt; used in mixed feeds and equals corn gluten in content of digestible protein; it is unpalatable; also called dried brewers' grains, beer grains.

Brewers' Dried Yeast–A source of B complex vitamins from yeast filtered from fermented beer. It has roughly 50 percent protein and 1.5 percent phosphorus.

Bridle–(1) The part of a harness that includes the bit, reins, and headstall. (2) A short rope with hooks on either end. (3) On logging sleds, a chain placed around the rear runners to control the speed.

Bridle Wise–Describing a horse accustomed to a bridle.

Bright Bay–A horse with a light, glossy, tan coat.

Bright Wool–Light-colored farm wool, as compared with semibright wool, which is darker because of soiled conditions. Both types scour out to a good white color.

Brindle–(1) An animal which is tawny, gray, or brown, having irregular, dark streaks or spots. (2) Tobacco mosaic.

Brine Poisoning–Excessive consumption of salt by animals, resulting in severe vomiting, thirst, salivation, diarrhea, flatulence, dilation of the pupils, and hypersensitivity to touch. Convulsions and blindness occur in severe attacks. Also called salt poisoning.

Bring Forth–To produce; to give birth.

Brisket–(1) In quadrupeds, the breast or lower chest between the forelegs. (2) The meat from these parts.

Bristle–Stiff, sharp hair or hairlike parts which grow on animals and plants.

Britch Wool–Wool from the thigh and twist region of the sheep; the coarsest and poorest wool on the entire fleece; usually manure-encrusted and urine-stained.

British Breeds–Breeds of cattle such as Angus, Hereford, and shorthorn originating in Great Britain.

British Red-water Disease–A parasitic disease of cattle, caused by the protozoan *Babesia bovis*; characterized by fever, emaciation, diarrhea, anemia; occurs in Europe, Africa, South America, and the East Indies; also called bovine hemoglobinuria.

Brittle–Describing crumbly fat particles in butter or meat which do not stick together.

Broad-breated Bronze–A variety of the bronze turkeys noted for its unusually meaty breast, thighs, and drumsticks.

Broad-spectrum Antibiotic–An antibiotic that attacks both Gram-positive and Gram-negative bacteria, and which may also show activity against other diseases.

Broad-spectrum Pesticide–A pesticide that kills a wide variety of insects. It may kill many beneficial insects, fish, birds, and mammals as well as target pests.

Broadtail–A type of skin obtained from lambs of the Karakul breed.

Brod–A blacksmith's nail.

Broiler–A chicken from 8 to 12 weeks old, weighing 2½ or more pounds, sufficiently tender to be broiled.

Broke–Describing a gentled horse that can be harnessed or ridden.

Broken Coat–In rabbits, a condition in which the guard hair is broken or missing in spots, exposing the undercoat.

Broken Crest–A heavy neck in a horse, which breaks over and falls to one side.

Broken Ear–A distinct break in the cartilage of a rabbit's ear which prevents erect ear carriage; also called lop ear.

Broken Mouth–A condition occurring in sheep and goats, usually at about the age of five or six years, whereby some of the permanent teeth are missing. See Full Mouth, Gummer.

Broken Wind–A heredity, incurable respiratory disease of old horses, in which the air cells of the lungs are ruptured or dilated. The symptom is forced, double exhaling. Probably caused by hard exercise or consumption of dusty feed. Also called heaves.

Bromelin–Proteolytic enzyme in fresh pineapple juice; used for tenderizing meat and sausage casings.

Bronc–A wild or unbroken horse. Also called bronco.

Bronc Buster–(1) A trainer of wild horses. (2) In fairs and rodeos, a rider of a bucking horse.

Bronc Spur–A pointed implement, secured to a cowboy's boot heel by strong straps; used to urge or goad a bucking horse, usually in rodeo arenas. Its short, dull rowels are bent to the inside of the boot heel, rather than projecting straight out, as in the ordinary spur.

Bronchiole–One of many subdivisions of the bronchial tubes within a lung; small branch of the bronchus.

Bronchitis–Inflammation of mucous membranes in the bronchia, which are the tubes connecting the lungs to the windpipe. It affects horses, sheep, pigs, cattle, and domestic animals. The symptoms are hard, dry coughing, vomiting, rising temperature, nasal discharge, and difficulty of respiration. In horses, it sometimes results in broken wind. Also called infectious bronchitis, chick bronchitis, gaping disease.

Broncho-pneumonia–Inflammation of an animal's lungs, which usually occurs after another illness has become established. It is characterized by a bronchitis attack with severe rise in temperature; increased difficulty in breathing; short, moist cough; thick white catarrhal discharge from the nostrils; fast pulse; and general impression of severe illness. The appetite is lost, the head is extended, and depth of breathing is increased. In sheep and cattle, rumination ceases. In dogs, the lips are blown out and sucked in with each breath. The disease may be fatal within three or four days, but with proper treatment, most animals recover within two weeks.

Bronchus–A major airway of the respiratory system.

Bronze–(1) A popular breed of turkey with brilliant reddish-bronze plumage. It is the heaviest breed of turkeys; the adult tom may weigh more than 36 pounds, and the adult hen as much as 20 pounds. (2) The metallic bronze-colored cast sometimes found in the black plumage of black or part black chickens; a defect.

Brood–(1) The young of any animal. (2) A single generation of insects. (3) To incubate eggs or protect young from the weather. (4) Of or pertaining to an individual kept for breeding, as a brood mare, brood sow, brood cow. (5) A hen inclined to set on eggs, as a brood hen. (6) Immature or developing stages of bees; includes eggs, larvae (unsealed brood), and pupae (sealed brood).

Brood Capping–A porous substance with which honeybees seal the brood cells in a beehive.

Brood Cell–A cell in which the larvae of bees are reared.

Brood Chamber–That part of a beehive in which the brood is reared. It includes one or more hive bodies and the combs within.

Brood Comb–The wax comb from brood chamber of a beehive containing brood.

Brood Foundation–A base used to control the size of bee combs, consisting of a sheet of pure beeswax embossed with the shape of cells. Bees follow this pattern to build up combs.

Brood Frame–Wooden frame containing the brood comb.

Brood Hen–A hen used to hatch eggs and rear young chicks.

Brood Nest–The area within the combs in which young bees are reared; may include only part of one comb or many combs.

Brood Sow–A sow reserved for breeding.

Brooder–The enclosure in which young animals are kept, or are allowed free access to, which furnishes heat until they adapt to outside temperatures.

Brooder House–A structure in which chicks or baby animals are raised without a mother; also called brooder.

Brooder Ring–A small wall (usually of cardboard) placed immediately around the brooder house during the first couple of weeks of the birds' life to keep them from wandering off and becoming chilled.

Broodiness–Describing a hen which attempts constantly to sit on eggs. Also used to describe female animals (especially swine) that have characteristics that indicate reproduction efficiency.

Brooding–Raising newly hatched chicks in a protected environment.

Broodmare–A female horse used for breeding.

Broody–(1) A livestock-judging term used to describe an animal, especially a gilt or sow, that possesses characteristics that indicate she will be reproductively efficient. (2) A bird manifesting a desire to sit on a clutch of eggs.

Broody Coop–An enclosure used to confine a broody hen and hasten her return to normal egg production.

Broomtail–(1) Wild horse or a horse raised on the open range. (2) A pony with a short bushy tail (western United States).

Brothers in Blood–A horse-breeding term for colts by the same sire out of full sisters, or by full brothers of the same mare.

Brow Band–The part of the bridle that goes around the brow of the horse's head.

Brown Bee–The honeybee native to Europe; reputedly the first honeybee introduced into the United States in 1638.

Brown Dog Tick–*Rhipicephalus sanguineus*, family Ixodidae; a hard-backed tick that infests dogs and ruminants; a carrier of piroplasmosis; also called European dog tick, European brown tick.

Brown Grease–A grease obtained form inedible hog fat. Its oil is used in heavy lubricants, leather and illuminating oils, stearic acid, and in candle making.

Brown Rat–*Rattus norvegicus*, family Muridae; the common rat; also called Norway rat, house rat, wharf rat, sewer rat.

Brown Swiss–A breed of large dairy cattle that originated in Switzerland as a milk, meat, and draft animal but which is used for dairy purposes in the United States. It has the oldest of the purebred dairy breeds. The calves make good vealers.

Brown-ticked Gray–The color of a horse with very small spots of brown scattered over a background of gray.

Browning of Milk–Discoloration that occurs when milk is heated above 212°F (100°C).

Browntail Moth–*Nygmia phaeorrhoea*, family Lymantriidae; this moth is destructive to foliage of many trees. Its hair carries a poison which irritates the skin of humans.

Browse–Leaves, small twigs, and shoots of shrubs, seedling, and sapling trees, and vines available for forage for livestock and wildlife.

Brucellosis–A highly contagious disease of cattle, goats, hogs, etc., caused by ***Brucella abortus, B. melitensis,*** and ***B. suis.*** The animal appears to be normal, but the disease brings about abortion, infertility, and low milk production. Easily communicable to humans. Also called contagious abortion, Bang's disease of cattle, infectious abortion, epizootic abortion.

Brush–(1) Shrub vegetation and trees that do not produce timber. (2) A revolving cylinder in a cotton gin, with bristles on its outer surface to remove cotton lint from the gin saw. (3) The lower part of the tail of the cow or horse. (4) The panicle and flower clusters of broomcorn. (5) The tuft of hair on the tip of a wheat kernel. (6) A device with set-in bristles for grooming the coat of an animal; used with a curry comb.

Brushing–The habit of a horse striking the fetlock with the other hoof, which may result in either roughing the fetlock hair or an actual injury.

BST–See Bovine Somatotropin.

Buck–(1) A male goat, sheep, rabbit, deer, or antelope. (2) Wool from rams. See Bucks. (3) To saw felled trees into logs. (4) The common beech or its nuts. (5) Of a horse, to quickly leap with back arched and head held low.

Buck Fat–Thte fat of a goat.

Buck Herd–A group of rams.

Buckaroo–A cowboy (southwestern United States).

Buckbrush Ceanothus–*Ceanothus cuneatus*, a shrub, family Rhamnaceae; a forage for livestock on the western ranges of the United States. Also called wedgeleaf ceanothus.

Bucked Knees–Used to describe a horse or cow with knees bent out.

Bucket Tree–A sugar maple producing sufficient sap to warrant hanging a bucket on it.

Buckeye Poisoning–Poisoning caused in livestock by eating plants belonging to the genus ***Aesculus***. It causes inflammation of the mucous membranes, vomiting, depression, stupor, lack of coordination, twitching, and paralysis.

Buckeyed–Describing a horse's eye with a convex cornea, which appears to protrude beyond the eyelids. The horse is shortsighted and shies easily.

Bucking Bronco–Usually an unbroken horse ridden at rodeos.

Bucking Chute–(1) In logging, a chute at a landing, in which logs are cut into equal lengths before loading. (2) An enclosure where broncos are saddled.

Bucking Range–In certain localities, range selected for placing rams with ewes.

Buckling–Male goat between one and two years of age.

Bucks (Buck Fleece)–The wool obtained from rams, distinguishable from ewe or wether wool by its odor and usually heavier shrinkage. It often sells at a lower price because of its shrinkage and coarseness.

Buckskin–(1) A disease of citrus, usually grapefruit and oranges, caused by the mite ***Phyllocoptes oleivorous***; characterized by a crusted, leathery condition of the rind. (2) A term for a grayish-brown horse. (3) The skin of a buck deer. (4) A disease of sweet cherry, the symptoms of which vary according to the strain of virus and the stock. In general, the fruit loses its luster and drops or fails to ripen. (5) A horse that is light brown and has a black mane and tail.

Buckstrap–A saddle strap used for a hand-hold on a rearing horse.

Buff Orpington–A variety of Orpington fowl. It is buff-colored with red comb, face, and wattles. Raised for brown egg production and for meat. The cock may weigh 10 pounds and the hen 8 pounds.

Buffalo–The American buffalo, which is better known as the American bison. The bison has been crossed with cattle to produce cattalo or beefalo.

Buffalo Cholera–A disease of buffalo, which may spread as an epidemic among domestic animals.

Buffalo Gnat–A small insect of the genus *Simulium*, found in the lower Mississippi Valley. Many bites may kill a domestic animal.

Buffer–(1) A substance in soil, such as organic matter, clay, carbonates, or phosphates, which resists changes of soil pH. (2) Animals which serve as food for predators, thus reducing danger to game. (3) A tool used to cut clinches from horseshoe nails before removing the shoe; also called clinch cutter.

Buffer Solution–A solution to which large amounts of acid or base may be added with only a very small resultant change in the hydrogen-ion concentration.

Buffering Action–Resistance to change in acidity or alkalinity. Also called buffering.

Bug–(1) Any insect, specifically of the order Hemiptera. (2) A flaw in the construction and operation of machinery.

Buggy–In the United States, a light, four-wheeled, horse-drawn carriage. In the United Kingdom, the buggy usually has two wheels.

Bulb–(1) The subterranean bud of some plants, which has a short stem bearing overlapping, membranelike leaf bases, as in onions and tulips. It stores food for reproduction and represents the inactive stage of the plant. (2) Any plant or flower shaped like a bulb. (3) The upper part of a horse's heel.

Bulgarian Buttermilk–A fermented milk produced by growing a culture of *Lactobacillus bulgaricus* (Bulgarian bacteria) in sterile milk at a comparatively high temperature. It is generally thick with a distinctly acid taste. Also called Bulgarian milk, Bulgarian sour milk, Bulgarian milk, yogurt.

Bulimia–An abnormal increase in the appetite. See Cynorexia, Polyphagia, Sitomania.

Bulk Bin–A large storage bin used for storing feed. The feed is put in through the top using an auger from a feed truck or a feed mill. The feed is usually discharged by an auger from the bottom and is moved to the animal's trough or feeder by an auger or conveyer.

Bulk Blending–The practice of mixing dry, individual, granular fertilizer materials. The practice adheres to a marketing system intended to decrease the costs of transportation, chemical processing, and bagging. It usually brings the raw materials directly from the primary producer to the bulk-blending plant located in the market area. The blender often applies the fertilizer to the land according to the specifications of a soil test.

Bulk Comb Foundation–A thick base for beehives, $4\frac{1}{2} \times 16\frac{1}{2}$ inches, which has heavier walls than an ordinary comb foundation; used for bulk production of honey. Also called cut comb foundation.

Bull–(1) An uncastrated male bovine. (2) A market term for a male bovine at sexual maturity. (3) The male elk, moose, and other ruminants. (4) The central bar of a harrow.

Bull Beef–A federal meat grading for beef from a bull's carcass.

Bull Frogs–*Rana catesbeiana*, large frogs that are grown commercially for food. The legs are considered to be a delicacy.

Bull Nurse–In the southwestern United States, a man who tends cattle during shipment.

Bull Pen–(1) A small wood or metal enclosure for confining bulls. (2) The sale ring at a livestock auction. (3) Living quarters for a group of men. (4) Storage room for machinery.

Bulldog–(1) A steel clip fitted into the nostrils of a bull, used to control him; also called bull holder. (2) To throw a steer or cow to the ground by grasping its horns and twisting its neck (United States).

Buller–A cow in heat for a long time; a cow difficult to get with calf.

Bulling–Presenting signs of being ready for mating (heifers and cows).

Bullock–A castrated bull. See Ox.

Bumblefoot–Hard cores or large swellings in the foot or foot pad of poultry and other birds that cause birds to walk clumsily or to stumble. Generally believed to be caused by bacterial invasion of the foot through bruises or injuries to the foot.

Bunch–(1) A cluster of plants or fruits, as a bunch of grapes. (2) In pigs, an improperly healed wound following castration. (3) To skid logs together for hauling. (4) To pile harvested crops in the field.

Bunker Silo–A silo for storing silage, consisting of a wide trench constructed in the side of a hill from which surface water has been diverted, such as by a diversion terrace. The vertical walls of the horizontal silo may be form 6 to 8 feet and the length from about 50 to 100 feet. When feeding the silage, movable self-feeders can be constructed in either or both ends.

Burdizzo Forceps–A pincers-like instrument with wide jaws, used for bloodless castration of animals with a pendant scrotum. Also used for bloodless killing of poultry and for docking lamb tails.

Burglar–A horse, good in appearance, but with a slight defect that can be remedied temporarily when offered for sale. Also called robber.

Burr Comb–Honeycomb built out of place, between movable frames or between the hive bodies.

Burrel–(1) A variety of pear with soft, tasty flesh. (2) Blue-colored wild sheep found in the Himalayas.

Burro–A donkey; ass.

Burry–Describing wool with excessive quantities of burrs, birdseeds, chaff, etc., in the fibers.

Bursa– A sac or pouch in connective tissue; specifically in birds, the bursa of fabricious, which is located just inside and above the vent.

Bursattae– A skin disease of horses caused by larvae of the large stomach worms ***Draschia megastoma***. The larvae migrate into pre-existing wounds or sores on the head and front of the body, making them large and filled with pus. The afflicted areas become chronic and difficult to heal. Endemic in wet season in the East Indies. Also called summer sores, jack sores.

Bush– (1) A grove of sugar maple trees. See Sugar Bush. (2) A forest wilderness or vast scrubland (Canada, Alaska, New Zealand, and Australia; formerly used in the northern United States also). (3) A shrub. (4) A method of training plants and fruit trees to assume a desired shape by means of stakes, ties, wires, and other supports. (5) The brush of an animal; the tail. (6) To force a seller to accept a lower price for a horse than was bid in the auction ring.

Bush Sickness– In New Zealand, a cobalt deficiency or nutritional anemia of ruminants. See Cobalt Deficiency, Nutritional Anemia, Salt Sick.

Buster– (1) One who trains animals for saddle or work. (2) Middlebuster (a lister) that moves soil to the right and left simultaneously.

Butcher– (1) A person who slaughters animals or dresses meat. (2) One who sells meat. (3) Animals which are to be killed for their meat; heifers, cows, steers, or bulls suitable to be sold in carcass as block beef. (4) To slaughter or dress animals.

Butcher Run– Unsorted and ungraded rabbits purchased for slaughter.

Butcher Tie– A method of trussing fowl. A string is drawn across the shoulders of the carcass, over the wings and drumsticks and tied at the base of the tail.

Butcher's Round– A wholesale cut of meat; about 20 percent of the hindquarter.

Butt– (1) The thick base or lower end of a tree or plant. (2) A stump, especially a walnut stump. (3) In butchering, the upper half of a ham or shoulder. (4) In Australia, a package of wool containing 196 pounds greasy or 112 poundsscoured, with a tare weight of more than 11 pounds. (5) To strike with the head or horns, as do cattle, goats, or rams.

Butt End Sirloin– The first cut of steak taken from the rump end of the hip loin.

Butter– Solidified milk or cream fat to which salt and artificial coloring may be added. In commerce, nonfat constituents are limited to 20 percent by weight.

Butter Oil– Melted, clarified butterfat used in making process butter. It contains some water and nonfat solids and can be shipped without refrigeration. In India known as gee.

Butterfat– The fat in milk ; also called milk fat.

Buttermilk– The milk that remains after butter has been obtained from the churning process; also made from skim milk by adding certain kinds of bacteria that produce the characteristic buttermilk flavors.

Buttermilk Horse– A roan with a coat color between red and blue.

Button– (1) An irregularly shaped berry. (2) A bud. (3) A round seed vessel. (4) Any stunted or immature fruit. (5) A round, firm, cheesy curd of condensed milk; a defect of the body of the milk. (6) A nipple, especially of a hog. (7) A partially dismantled queen bee cell in a beehive, which resembles a small acorn cup. (8) Onion set. (9) A stunted or immature horn growth, as on a calf. (10) A metal clasp used to connect sections of a check row wire. (11) Cartilage on the chine bone of cattle (12) Any shell-like bone construction of the body; also called concha. (13) A leather ring for adjusting a horse's bridle.

Butyric Anhydride– A liquid used to drive bees from honeycombs. It has an odor unpleasant for humans similar to that found in rancid butter and perspiration.

By– Begotten by a male; sired.

By-product– A product of significantly less value than the major product. In beef cattle, the major product is meat; by-products include the hide and other items.

C-type Merino– The largest type of Merino sheep that is free from folds of skin on the body and that has only two or three small folds on the neck. These sheep produce more mutton, although the wool is less dense on the legs and head than on other types. Also called Delaine Merino, Delaine type. See A-type Merino, B-type Merino.

Caballada– A herd of horses or mules (southwestern United States).

Caballo– (Spanish) Horse.

Cabresto– A lariat or halter made of horse hair.

Cabrito– The name given by Mexicans to the meat of a young Spanish goat.

Cadence– The rhythm of a horse's gait.

Cadmium– (Greek; ***kadmia***, earth) A bivalent metal similar to tin. Its atomic number is 48 and atomic weight 112.40. A metallic "heavy metal" used in the production of copper, lead, silver, and aluminum alloys, in photography, ceramics, and in insecticides. Cadmium in sewage sludge is of grave concern when sludge is applied to soils used for the production of food and feed crops. Cadmium is a hazardous pollutant to people, domestic animals, and shellfish. It also is an experimental carcinogen.

Caecum– In humans and animals, the closed sac which is the beginning of the large intestine. Also spelled cecum.

Cage Battery– System of keeping poultry and rabbits in cages in a controlled environment throughout their life.

Cake–A residue from the pressed kernels of cottonseed, linseed, or soybean, etc., from which the oil has been removed. It is pressed into hard cakelike masses and is used as a feed.

Caked Udder–(1) Physiological edema of the udder most often associated with recent freshening. (2) A doughy, swollen udder.

Calciferous Glands–Glands that produce or secrete carbonate of lime (calcium carbonate, $CaCO_3$). Such glands are found near the esophagus of earthworms and in some other lower animals.

Calcium Deficiency–A physiological condition of plants or animals which results from insufficient calcium in the body. In plants, new fibrous roots die within four weeks, then the terminal bud, and ultimately the plant. A deficiency of vitamin D in birds and mammals makes it impossible for them to utilize the calcium in their diets. In such cases, hens may lay normal eggs at the expense of calcium needed for bone development, resulting in crooked breastbones. In animals, it produces rickets, abnormal and fragile bones, and low blood calcium.

Calculi (singular, Calculus)–Stones of calcium carbonate or calcium oxalate which may form inside the bodies of people or animals due to stagnant fluids and other secretions.

Calendar Year–A period from January 1 through December 31 of the same year.

Calf–(1) The young of certain large mammals. Of the bovine animals, specifically those more than three months old but sexually immature. (2) The hide of such a calf.

Calf Crop–The number of calves produced by a given number of cows, usually expressed in percent of calves weaned of cows bred.

Calf Diphtheria–A herd infection of half-grown calves caused by the necrosis germ *Spherophorus necrophorus* (formerly *Actinomyces necrophorus*); characterized by difficulty in eating, drinking, and breathing. The mouth and throat are affected with ulcers and diphtheritic membranes to a varying degree. Also called calf diphtheroid.

Calf Pen–A small cubicle in which a calf is raised. The pens are raised off the floor or are movable. The purpose is to lessen the chance of the calf's contracting a disease such as scours.

Calf Pneumonia–An infectious respiratory disease of calves, caused by faulty sanitation, inadequate ventilation, overcrowding, or debility from other ailments, such as diarrhea. Its symptoms are high temperature, rough coat, poor appetite, weakness, and coughing when the chest is tapped or compressed by the hand.

Calf Puller–A device used to assist a cow in delivering a calf. A chain or cord is attached to the calf's front foot, and the calf is drawn from its mother.

Calf Roping–A rodeo sport in which a rider lassoes a running calf, dismounts, throws the calf to the ground, and ties its feet together. The sport derives from old round-up practices.

Calf Scours–An acute, infectious disease, cyclic in occurrence, which reaches panzootic proportions for calves less than ten days old. There is marked depression accompanied by severe diarrhea. The most lethal form appears at birth or 6 to 72 hours after birth. In these cases there may be no diarrhea; the calf is found in a cold, weak, and dying condition. Also called white scours, infectious diarrhea, calf septicemia, three-day calf disease.

Calfskin–A subclass of packer hides and country hides. Used in making gloves, shoes, and in bookbinding, etc.

Calico–(1) (a) Any cotton cloth from India. (b) Coarse printed cotton cloth from the United States. (c) Plain white cotton cloth from the United Kingdom. (2) A spotted horse. (3) A viral disease of the potato, characterized by irregular blotches of various shades on the leaflets. It is one of the potato mosaics. (4) Tobacco mosaic.

California Flu–See Newcastle Disease.

Calk–A downward-pointed part of a horse's shoe, which prevents slipping. Also called caltrop.

Calking–Injury to the coronary band by the shoe of the horse. Usually incurred by horses whose shoes have calks or by horses that are "roughshod" for walking on ice.

Callus–(1) In plants, protective covering that forms over a wounded surface. (2) Undifferentiated or unorganized tissue that grows from a plant cell or a piece of leaf when it is placed on media containing certain hormones; an early event in the regeneration of a whole plant from engineered cells. (3) In animals, an area of skin which becomes horny and thickens. (4) A hard exudate which forms at the end of a broken bone and eventually becomes part of it.

Calves–Cattle of either sex under one year of age.

Calving–(1) Parturition; a cow having a calf. (2) Breaking off and floating away, as icebergs of large masses of a glacier that reach the sea.

Calving Difficulty (Dystocia)–In cattle, abnormal or difficult labor, causing difficulty in delivering the fetus and/or placenta.

Calving Ease Index–A rating for calving ease that combines cleaving ease scores for heifers, second calf and older cows and birth weight as reported to breed associations and published in their annual sire summaries. Some of them label it Ease of Calving Index. The index is used to select replacement heifers.

Calving Season–The time of the year when the calves of a herd are born; occurs at different times in different parts of the country.

Camp Tender–In the United States, one who transports supplies to a sheep herder and moves his supply camp from place to place on the range.

Camp Unit–A subdivision of a sheep allotment on federally owned land.

Canadian Bacon–Smoked pork cut from the back and loin. Also called English bacon.

Canary-stained Wool–In wool, the occurrence of a yellowish coloration which cannot be removed by the ordinary scouring methods. Such stain has been observed in the wools obtained from New Zealand in Corriedale and Romney sheep, West Australian and Queensland Merino sheep, and in the Cape Merino wools.

Cancer–(1) A malignant tumor of humans and animals. (2) In plants, a growth similar to that caused by crown gall.

Cancer Eye–A noninfectious, sporadic, malignant ailment of the eyeball and the contiguous tissues in cattle (particularly Hereford) and occasionally in horses.

Candle–To examine an egg in front of a light to observe internal characteristics associated with edible quality or hatchability, such as air cell size, yolk shadow position, presence of blood or meat spots, and presence or lack of germ development.

Canine–A member of the dog family; includes dogs, wolves, foxes, jackals, etc.

Canine Madness–Rabies.

Canine Tooth–In mammals, one of the four sharp teeth on both jaws between the incisors and bicuspids.

Canker–(1) A diseased lesion of the bark and underlying tissue in woody plants. (2) An ulcer in the mouth. (3) A serious disease of the horse's hoof. The frog frequently discharges a stinking fluid, and ultimately the sole and the frog rot.

Canner–(1) A marketing classification for low-grade beef. (2) A wild horse, from the western range of the United States, slaughtered for its meat. (3) Berries, etc., too soft for shipment but fresh enough for immediate table use or canning.

Cannibalism–Habit of some birds in a poultry flock of repeatedly pecking and clawing other birds in the flock, often drawing blood and occasionally causing death. Weaker or smaller birds are usually the victims of cannibalism. Feather picking often leads to cannibalism, but overcrowding is probably the main cause.

Canning Factory Silage–Husks, cobs, and undesirable kernels of sweet corn, pea vines, pear wastes, etc., which are unfit for canning, but usable as a stock feed.

Cannon–(1) That portion of an animal's leg, ankle, or pastern from the knee to the fetlock, in the front legs, or the hock to the fetlock, ankle, or pastern in the rear legs. (2) A horse's bit. (3) To lift and swing a log crosswise to a load of logs.

Cannula–A metal, rubber, or glass tube inserted into a body cavity to allow the escape of fluids or gas and through which liquids may be introduced into the body.

Canter–The gait of a horse in which the feet strike the ground one at a time. An abbreviation of Canterbury gallop, describing the gait of horses ridden by pilgrims to Canterbury. Also called slow gallop.

Canvas Jacket–An apron made of canvas or other heavy material which is placed on the back of a female turkey to prevent injury during breeding. Also called saddle, apron.

Cap–(1) A piece of metal placed over the end of a log to make it skid over obstacles. (2) The top member of a trestle or similar support. (3) Milk tooth of an animal. (4) The crown of a kernel of corn. (5) The calyx of the strawberry. (6) To remove the calyx of the berry after it is picked. (7) To place a cover over a stack of hay, bound grain, etc. (8) To seal a cell in the honeycomb with wax, as by a bee.

Capacity, Carrying–In wildlife or livestock management, the optimum density of animals that a given environment or range is capable of sustaining permanently.

Cape–The short feathers underneath the hackle of chickens.

Cape Wool or Capes–Wool from South Africa.

Capillaria Worms–Internal parasites of the genus *Capillaria*, which infest the crop, esophagus, small intestines, and ceca of chickens, ducks, turkeys, and some wild birds. Symptoms are unthriftiness, weakness, diarrhea, loss of weight, ruffled feathers, soiled vent, and a tendency to sit on the ground. Also called hair worms, threadworms.

Capillary–Extremely narrow, microscopic blood vessel. The presence of blood in capillaries produces the pink color of the skin.

Capitate–(1) Headlike, formed like a head. (2) In heads, aggregated into a very dense or compact cluster.

Capon–A cockerel castrated to fatten it for the market. It is usually seven to ten months old and weighs 6 pounds.

Caponette–A male chicken that is neutered before sexual maturity by the implantation of female hormones. See Capon.

Capped Brood–In honeybees, brood (either last larval stage or pupal stage) that has been capped over in its cell.

Capped Hock–In cattle or horses, an enlargement of the point of the hock; usually caused by bruising.

Capped Honey–Cells full of honey, closed or capped with beeswax.

Cappings–Beeswax covering of cells of honey, which are removed before extracting the honey.

Capric Acid–A fatty acid found in butter, fusel oil, etc., which gives a characteristic buttery flavor to cheese. Occurs more in goat's than in cow's milk. Also called caprinic acid, decanoic acid.

Capsule–(1) A fruit (dry at maturity) of more than one carpel but not necessarily of more than one locule, which dehisces at maturity, permitting the escape of the enclosed seeds. Typical capsules are found in the iris, lily, poppy, cotton. (2) Gelatine covering for medicines.

Capsule Breeding–An older term for artificial insemination. See Artificial Insemination.

Car Lot–The quantity of produce or number of animals which can be transported in a railroad car.

Caracul–A type of pelt produced by young lambs of Karakul breeding. See Karakul.

Caraway–(1) *Carum carvi*, family Umbelliferae; an annual or biennial herb whose aromatic seeds are used as a condiment. (2) Any cheese spiced with caraway seeds.

Carbohydrate–Any of certain organic chemical compounds of carbon, hydrogen, and oxygen, which include sugars and starches. Formed in plants by photosynthesis, carbohydrates make up a large part of animal feed.

Carbon–An essential chemical element component in plants and animals. It is present in soils, in humus, plant residues, charcoal, and particles of coal, carbonaceous shale, etc.

Carbon Dioxide–CO_2; a colorless, odorless gas constituting 0.03 percent of unpolluted air. It is absorbed by green plants through the leaf stomata and is used as the source of carbon for manufacturing sugars, starches, proteins, and fats. The burning of fossil fuels has put so much CO_2 in congested cities that the concentration may reach more than three times the background (normal) level. This is causing a "greenhouse effect," resulting in a slow warming of the earth's surface.

Carbon Dioxide Evolution–The liberation of gaseous carbon dioxide from soil by biological processes.

Carbon Disulfide–See Carbon Bisulfide.

Carbon Tetrachloride–CCl_4, a chemical that is an excellent solvent for grease. It is not inflammable and can be used as a fire extinguisher. Frequently used as a fumigant, though not in large-scale operations; toxic when inhaled.

Carbon-14 Dating–The use of radioactive carbon, which has an atomic mass of 14 and an approximate half-life of 5,000 years, for determining approximately the age of organic materials in soils, buried materials such as wood, and other organic materials.

Carbonaceous–(1) Pertaining to, or composed largely of, carbon. (2) The carbonaceous sediments include original organic tissues and subsequently produced derivatives of which the composition is chemically organic.

Carbonate Cider–Aerated apple cider.

Carbonizing–A process for extracting vegetable matter from wool by treating it with acid or some other chemical. It carbonizes the vegetable matter without harming the wool fiber. Wool thus treated is known as carbonized wool.

Carcass–The major portion of a meat animal remaining after slaughter. Varies among animals, but usually the head and internal organs have been removed. Skin and shanks are removed from cattle and sheep.

Carcass Evaluation–Techniques of measuring components of meat quality and quantity in carcasses.

Carcass Merit–Desirability of a carcass relative to quantity of components (muscle, fat, and bone), USDA quality grade, plus potential eating qualities.

Carcass Quality Grade–An estimate of the palatability of meat within a beef carcass. Quality grade is based primarily on marbling and maturity. See Marbling, Maturity.

Carcass Quantity–Amount of salable meat (muscle) a carcass will yield. Cutability is an estimate of this.

Carcass Weight–The weight of an animal carcass after the hide, head, feet, and entrails have been removed.

Carcinogen–A chemical, physical, or biological agent that increases the incidence of cancer.

Carcinoma–Malignant cancerous growth.

Card–A machine which is used to separate the wool fibers by opening the locks or tufts of wool. The machine contains multiple rolls with teeth. Hand cards are used chiefly in the fitting of show sheep.

Cardia–Juncture of the esophagus and stomach.

Cardiac–(1) That which is related to the heart. (2) A heart stimulant.

Carding–One of the first steps in the preparation of scoured wool; fibers are separated from other fibers in the locks or bunches of wool.

Carding Wools–Synonymous with clothing wools—wools that are too short to be manufactured by either the Bradford or French systems, and so must be manufactured by the woolen system.

Carminative–A chemical which expels gas from the alimentary canal to relieve colic, etc.

Carnivore–A member of a large order of mammals which customarily eat flesh; e.g., dogs, cats, bears, and seals.

Carotenase–An enzyme in the human body capable of converting carotene into vitamin A. See Carotene.

Carpet Wool–Wool that is coarse, harsh, strong, and more suitable for carpets than fabrics. Very little wool of this type is produced in the United States. Some of the choicer carpet wools are used to make tweeds or other rough sport clothing.

Carriage–(1) A frame on which a log is held. (2) Manner or bearing of an animal; its physical control and behavior.

Carriage Horse–Breeds of light harness horses, weighing up to 1,300 pounds, formerly used primarily for pulling carriages, but now bred mostly for the show ring. See Cleveland Bay, French Coach, German Coach, hackney, Russian Orloff, Yorkshire Coach.

Carrier–(1) A genetic term that refers to an animal that expresses the dominant trait but is heterozygous for the recessive gene. (2) Animal or person in apparently good health who harbors pathogenic microorganisms. (3) The liquid or solid material added to a chemical compound to facilitate uniformity of application in the field.

Carrion–Putrefying carcass.

Carry–(1) To keep an animal on a maintenance ration without obtaining any weight gain or any products from it. (2) To bear, as a pregnant cow carries a calf. (3) To sustain, as a farm carries a debt, or as a range carries stock. (4) To convey during transportation and marketing, as a container carries satisfactorily or a product carries well.

Carrying Capacity–In its true sense, the maximum number of individual animals that can survive the greatest period of stress each year on a given land area. It does not refer to sustained production. In range management, the term has become erroneously synonymous with grazing capacity.

Cartilage–A firm but pliant type of tissue forming portions of the skeleton. Some of the cartilage becomes calcified, changing into bone. Hence in young animals the proportion of cartilage in the skeleton is greater than in mature animals. In certain fishes, sharks, and rays, it is permanent, and never becomes ossified. Also called gristle.

Caruncles–(1) Specialized points of attachment in the uteri of ruminants for fetal membranes. (2) The fleshy protuberances on the naked portions of the head, face, and neck of the turkey and Muscovy duck.

Casein–(1) The chief protein of milk; a phosphoprotein which occurs in the milk of different animals. Also called casseinogen.

Cashmere Goat–A domestic goat raised in the Himalayas for its soft fleece, which is used in making rich shawls and cloth.

Casing–(1) A large pipe sunk into a well to prevent the walls form caving in and within which the pipe for pumping liquids is placed. (2)

Cleaned sections of hog, cattle, or sheep intestine used as sausage skins.

Cast–(1) The number of young produced. (2) A caterpillar which has had its entire larval body filled and displaced with a fungus which replaces any normal host structures, as by the fungus ***Cordyceps***. (3) Device for immobilization of fractured bones. (4) A sheep which is unable to rise (New Zealand). (5) Wool which has fallen off a sheep before shearing (New Zealand). (6) Angleworm excreta. (7) To throw off a horseshoe, as by a horse when the nails become worn or loosened, etc. (9) To force to the ground, as to cast an animal to the ground to control it. (10) To broadcast seed or fertilizer.

Castes–The different forms of adult female bees in a colony; workers and queens. See Honeybee.

Castings–(1) Excrement of earthworms, an important factor in the enrichment of soils. (2) Fecal pellets of the animal kingdom.

Castor–A callus on a horse's leg.

Castrate–(1) To remove the testicles or to destroy their use; to geld. See also Spay. (2) To remove the stamens from a flower.

Cat Flea–*Ctenocephalides felis*, family Pulicidae; a biting insect which infests humans, dogs, and cats.

Cat-hammed–A term used to describe a horse having long, relatively thin thighs and legs.

Catabolism–Process of destroying or breaking down tissues and cells of the body from complex to simpler compounds. See Anabolism.

Catalase–Enzyme in plants and animals; splits hydrogen peroxide into water and gaseous oxygen.

Catalonian–A quality breed of Spanish ass that has contributed to the American jack in the breeding of mules. They are black or brown with light spots.

Catalyst–An agent that promotes interaction between or among other chemical substances without itself being changed.

Cataract–(1) An opaque or cloudy condition of the lens of the eye, or of the capsule for the lens, which obstructs vision. (2) A waterfall. See Cascade.

Catarrh–An inflammatory condition of any mucous membrane, in which the discharge is thin and watery.

Catch–(1) A sheep whose fleece is cut after the day's work has stopped (New Zealand). (2) A successful establishment from a seeding, e.g., a catch of grass or grain. (3) To germinate, sprout, and become established as a crop after sowing. (4) To conceive after breeding, as a mare.

Catchment–The area draining into a stream, lake, or discrete sewer. Identical to a watershed.

Catfish Farming–A branch of aquaculture in which catfish are raised in constructed ponds and fed a balanced diet. See Aquaculture.

Catgut–The tough string manufactured from the intestines of sheep and other animals, used in surgical sutures, musical instruments, etc.

Cathartics–Drugs that cause the evacuation of the bowels.

Catheter–A slender, tubelike, metal, rubber, plastic, or glass instrument inserted into body passages for the insertion or removal of fluids, e.g., removal of urine form the bladder or insertion of semen into the cervix in artificial insemination.

Cattalo–A cross between the bison (buffalo) and domestic cattle.

Cattle–(1) Collectively, mature bovine animals. Calves and yearlings may sometimes be included. In a broad sense, the term may also include all domesticated quadrupeds. (2) Denoting beef carcasses in the meat trade.

Cattle Country–(1) The grazing area of other western plains and mountains in the United States where cattle are one of the principal sources of income. (2) A country whose principal product is cattle, e.g., Argentina.

Cattle Drive–Moving cattle on foot from one place to another.

Cattle Feeding–The finishing of cattle for market by feeding them mainly on corn and other feed concentrates for fattening. Cattle feeding is most extensive in the corn belt states of the United States.

Cattle Grub (Heel Fly)–The common cattle grub, ***Hypoderma lineatum***, and the northern grub, ***H. bovis***; a serious pest of cattle. The mature female heel fly lays eggs on the animal's body, usually on or around the heel. When the eggs hatch, the larvae enter the body through the skin and eventually migrate to the back, where they open a hole through the skin on the back of the animal.

Cattle Guard–A device or structure at points where roads or railroads cross a fence line, so designed that vehicular travel is unimpeded but crossing by all kinds of livestock is prevented. Syn., auto gate. See Cattle Pass.

Cattle Leader–A rod, or rod and rope, attached to a ring fastened in the nostril of a cow, bull, or steer for restraining or leading the animal.

Cattle Marker–A metal or composition tag with identifying numbers or characters, fastened to the ear, neck, or horns. Also called cattle tag.

Cattle Pass–A culvert constructed under a road for cattle to pass from one pasture to another without obstructing traffic. See Cattle Guard.

Cattle Prod–A device that delivers a harmless electrical current to the end of a rod; used to prod cattle or other animals into moving. Also called stock prod.

Cattle Ranch–Usually a very large area for the purpose of breeding and/or raising cattle.

Cattle Range–Particular sections of a country devoted to the grazing of cattle, e.g., certain areas of the western states.

Cattle Run–(1) The number of cattle sent to market within a stated period, e.g., a day, week, etc. (2) An alley or passageway for cattle.

Cattle Scabies–Psoroptic cattle scabies is caused by tiny parasitic mites that puncture the skin of cattle and feed on the body fluids released from the wounds. These fluids dry and form scabs. As mites increase in number, an infested animal's hair will fall out, and the lesions that are formed can eventually cover much of the body with thick, crusty scabs—hence the name, scabies. Heavy infestations can, in some cases, cause death. Scabies does not affect the wholesomeness of meat from infested cattle, but the intense itching caused by the mites produces loss of appetite and lowered gains and feed efficiency. Economic losses, particularly in feedlots, can be severe.

Cattle Tag–See Cattle Marker.

Cattle Tick–*Boophilus annulatus*, family Ixodidae; an insect that sucks the blood of cattle, reducing their resistance to disease, and causing general unthriftiness. It is the vector of cattle tick fever and sometimes of anaplasmosis. See Anaplasmosis.

Cattle Tick Fever–An infectious disease of cattle caused by the protozoan parasite *Babesia bigemina (Piroplasma bigemina)*, whose vector is the cattle tick. The protozoan enters the blood and destroys red corpuscles, causing high fever, an enlarged spleen, engorged liver, thick flaky bile, red urine, jaundice, and emaciation. It is fatal in about 90 percent of acute cases. Also called Texas fever, red water, black water, southern cattle fever, acclimation fever, murrain, bloody murrain, Mexican fever, Spanish fever, splenetic fever, hemoglobinuria, bovine piroplasmosis, bovine malaria, cattle malaria.

Cattle Yak–A crossbreed of the yak and domesticated cattle.

Cattle Yearlong–The forage or feed required to maintain an animal unit for one year. See Animal Unit, Animal Unit Month.

Cattle-proof–Designating a fence which will successfully restrain cattle.

Caucasian Bee–A dark honeybee race originating in the Caucasus Mountains.

Caudal–A directional term meaning toward the tail; the opposite of cranial.

Caudal Fold–Fold of skin at the junction of tail and body.

Caul–(1) The fat around the stomach of cattle, sheep, and swine. (2) In embryology, a part of the thin membrane or web surrounding the fetus.

Causal Organism (Causative Agent)–The organism (pathogen) that produces a specific disease.

Cauterize–To burn the flesh or skin of an animal with a drug or heated metal to induce healing.

Cavesson–Head stall with a noseband (often quite large) used for exercising and training horses.

Cavity–(1) A hollow space within a body, e.g., the buccal cavity. (2) Any hole in the trunk or branches of a tree.

Cavy–(1) An American Indian pony. (2) One or a group of stray animals (United States). (3) Any South American rodent of the family Caviidae of which the capybara and guinea pig are the main species.

Cayuga–A breed of duck with lustrous greenish-black plumage. The drake weighs up to 8 pounds, and the duck up to 7 pounds.

Cayuse–A riding horse or pony of no particular breed (western United States). Also called fuzz tail. See Broom Tail.

Cecum–A large pouch that is the forward part of the large intestine of a horse.

Cell–(1) A hexagonal unit compartment of a honeycomb. (2) The ultimate functional unit of an organic structure, plant, or animal. It consists of a microscopic mass of protoplasm which includes a nucleus surrounded by a membrane. In most plants it is surrounded by a cell wall. (3) A single element of an electric battery, either primary or secondary, generally consisting of a jar filled with a liquid or a pasty electrolyte, into which the electrodes are inserted or connected. (4) A very small, enclosed compartment.

Cell Cup–(1) An artificial cell made of beeswax for rearing queen bees. (2) Initially constructed base of queen cell; also made artificially for queen rearing.

Cell Grazing Management–A system of pasture rotation whereby pastures are divided into equal-sized segments with respect to carrying capacity and are arranged in a pie-shaped design around a central core area for supplemental feeding, watering, and handling pens. Under this system, livestock are rotated among the pastures using high stocking densities and little time on pasture for any one grazing period.

Celtic Ox–*Bos taurus longifrons*, a fairly small, Neolithic ox with a high forehead; probably the progenitor of the Brown Swiss and Jersey breeds of cattle.

Center Cut–The tender parts near the center of a piece of meat, e.g., pork chops.

Central Nervous System–The brain and spinal cord in vertebrates, which receive sensory impulses and from which motor impulses issue.

Central Test–A location where animals are assembled from several herds to evaluate differences in certain performances traits under uniform management conditions.

Centrals–The two central incisors of a horse. Also called pincers.

Centrifugal Separation–The separation of substances of different densities by means of centrifugal force, as the separation of fat from milk.

Centrifuge–A machine, such as a Babcock milk tester, used to separate the fat from the acid-milk mixture.

Centriole–A small organelle located near the nuclear membrane of cells that divides during mitosis and forms the centers toward which the chromosomes move upon division of the cell.

Centromere–A small structure located on a chromosome that appears to form an attachment to the spindle fibers during cell division.

Centrosome–A minute protoplasmic body sometimes held to be the dynamic center of mitotic activity.

Certified Boar–A boar that has sired several animals meeting the requirements for certification as established by a breed association.

Certified Mating–The mating of a certified boar to a certified sow.

Certified Meat Litter–A litter of pigs meeting the production requirement as established by a breed association.

Certified Meat Sire–A boar whose progeny in a specified number of litters has met certain requirements regarding carcass excellence, growth rate, and size, etc.

Certified Sow–A sow that has met the production requirements as established by a breed association.

Cervix–A part of the reproductive tract of female mammals that forms a seal or doorway between the uterus and the vagina.

Cesarean Section (Cesarean Birth)–Technique used in some S.P.F. (Specific Pathogen-Free) laboratories to remove unborn pigs from the sow. Pigs are removed through the wall of the uterus, leaving the sow capable of further reproduction.

Chaffy Wool–Wool containing a considerable amount of chaff.

Chalaza–(1) Either of two cordlike, opalescent, albuminous strands in the white of a fowl's egg, which are attached to the yolk and prolonged toward the ends of the egg. The chalazas aid in keeping the yolk in proper position. (2) The place in an ovule or seed at which the integuments diverge from the nucellus.

Chalkbrood–A disease of brood combs in a beehive caused by the fungus *Pericystis apis*, family Pericystaceae, which usually attacks only the drone brood. Uncommon in the United States.

Chamois–(1) *Rupicapra rupicapra*, a small antelope that resembles a goat. Found in the mountains of Europe and New Zealand. (2) Soft, pliable leather made by rubbing oil into the skins of the chamois, goat, sheep, etc. Also called shammy.

Champ–To chew noisily, as a horse champs at the bit.

Champignon–(1) A mushroom, especially *Marasmius oreades*, the fairy ring mushroom. (2) Inflammation of a horse's spermatic cord.

Chantecler–A Canadian breed of fowls with good winter laying qualities; capable of resisting a rigorous climate. Characterized by small cushion-shaped combs, small wattles, tight feathering, and yellow skins; the eggs are brown. There are two varieties, White and Partridge.

Chaps–Leather coverings from the waist to the feet, worn by cowboys to protect their legs from cacti, trees, or brush. Also called chaparajos.

Character–(1) One of the details of structure, form, substance, or function which make up a species and distinguish it from others. (2) A judging term for desirable, necessary, or attractive qualities of an animal or product. (3) The evenness and distinctness of crimp in wool fibers.

Charalo–A breed of cattle obtained by the systematic breeding of Charolais cattle and bison.

Charbray–A breed of beef cattle, about three-fourths Charolais and one-fourth Zebu.

Charlier Shoe–A narrow, small horseshoe without a toe-clip which is nailed into a prepared groove in the hoof. Sometimes used when the horse is put out to pasture for a period of time.

Charlock–*Brassica kaber*, family Cruciferae; a weed, bearing yellow flowers, found in grainfields and among other cereal crops. Feeds containing large amounts of its seed may result in chronic enteritis, hemorrhagic diarrhea, colic, abortion, nephritis, apathy, etc., in animals. Native to Eurasia. Also called wild mustard, corn mustard.

Charolais–A breed of French beef cattle now also raised in the United States. The cattle are large-framed and white.

Charqui–"Jerked" beef; naturally dried strips of meat.

Check–(1) In irrigation (a) A basin into which the flow of water is regulated by levees and dykes. (b) An adjustable gate on a canal to regulate the flow of water. (c) Concrete blocks or wooden ties placed on a channel bottom to reduce erosion. (d) A crack which appears in drying soil. (2) In lumbering, a lengthwise separation in the grain, caused by strains during seasoning, which extends across the annual rings. (3) A short crack within the body of a cheese. (4) A narrow crack in a rice grain that may cause it to break during milling. (5) An egg which has a cracked shell, with the inner membrane intact. When the crack is naturally mended, it is called a blind crack. Also called crack, dent. (6) To retard growth.

Check Bit–A small bit attached to the checkrein of a bridle. The check rein is attached to the backpad of the harness, thus holding up the horse's head. Also called checkrein and bearing rein. See Checkrein.

Check Dam–A small, low dam constructed in a shallow watercourse to decrease the velocity of stream flow and to promote the deposition of eroded material.

Check Strap–A part of a harness that passes between a horse's forelegs. It is fastened to the collar on one side and bellyband on the other. It prevents the collar form rising when the horse is backing up or stopping.

Checkrein–A leather strap fastened to the backpad of a harness to prevent a horse from lowering its head. Also called bearing rein. See Check Bit.

Cheddar–A popular English cheese, now made in many countries. It is made from cow's whole or skimmed milk, which is then cured for about six months at 40° to 50°F. The finished product is hard and may be white or yellow. The term *cheddar cheese* includes cheeses of the granular or stirred curd types.

Cheek–(1) The fleshy side of the face in animals. (2) The part of the ventral side of a wheat kernel next to the crease.

Cheek Teeth–The premolar and molar teeth; the large teeth at the side of the mouth.

Cheeking–Grasping of the cheekpiece of a bridle to pull the head of the horse toward the rider.

Cheese–(1) Food made from consolidated curds which have been separated from milk by a coagulating agent. Cheese may be made from the whole milk, skim milk, or cream from any mammal. (2) The beeswax refuse, usually wrapped in burlap, from which beeswax is forced when the mass is submerged in a hot water press.

Cheese Rennet–A solution of the enzyme rennin used in setting milk for cheese making. Also called cheeselip.

Cheesecloth–A thin, loosely woven cloth made of cotton, used in the pressing of cheese curds.

Chemical–(1) A substance obtained by or used in a chemical process. (2) In dairying, an uncommon flavor defect which occurs when cheese has been made from contaminated milk.

Chemical Additives–Substances added to foods to improve their flavor, color, texture, or keeping quality.

Chemical Bond–See Valence.

Chemical Caponizing– The injection or implantation of a hormone, diethylstilbestrol (stilbestrol) under the neck skin of a male fowl, including old roosters, to cause it to have feminine characteristics. After six weeks it results in loss of fighting instinct, paleness of the undeveloped comb and wattles, paleness of shanks and skin, and an increase in weight.

Chemical Castration– The process of injecting a chemical into the testicles of an animal. The chemical causes the testicles to atrophy.

Chemical Dehorning– The application of chemicals to the horn buttons of young calves to stop horn growth.

Chemical Energy– The energy contained in the chemical bond between atoms; it can be released into the environment by a chemical reaction, e.g., combustion.

Chemical Oxygen Demand (COD)– The quantity of oxygen required to oxidize organic matter in a sample of waste under specific conditions of oxidizing agent, temperature, and time. Potassium dichromate dissolved in 50% sulfuric acid is the oxidizing agent. Silver sulfate is added as a catalyst, and mercuric sulfite is added to remove interference of chlorides. Excess dichromate is titrated with standard ferrous ammonium sulfate, using an orthophenanthroline-ferrous complex as the indicator.

Chemistry– (Greek, chemeia) Science of compounds, elements, and atoms.

Chemolysis– Decomposition of organic matter brought about by chemical agents.

Chemosterilant– A chemical that can prevent reproduction. See Chemical Castration, Chemical Caponizing.

Chemotherapy– Use of chemicals to treat infectious diseases.

Cheshire– (1) A breed of white swine which originated in New York State from crossing the Yorkshire with White Suffolk hogs. (2) A popular old English cheese; less firm and compact than cheddar, which has white or yellow curd. It is made of cow's milk, and annatto added for coloring. It is cured for three weeks (early ripening), for two months (medium ripening), or ten weeks to ten months (late ripening).

Chestnut Roan– A horse on which white hairs are mixed with the basic chestnut color. Sometimes called strawberry roan.

Cheviot– A hardy, hornless breed of English sheep; the rams weight up to 200 lbs. and ewes up to 100 lbs. These sheep are small, thick-set, and well proportioned, and are able to withstand severe climatic conditions.

Chevon– Goat meat.

Chewing Louse– Any wingless, parasitic insect which feeds on birds and sometimes on mammals.

Chianina– A large, white breed of beef cattle originating in Italy.

Chicago Style– A method of cutting a beef carcass into standard wholesale cuts (ribs, chucks, rounds, etc.); in general use in the Midwestern and western United States.

Chick– A young chicken.

Chick Bronchitis– See Infectious Bronchitis.

Chick Dermatosis– A deficiency of pantothenic acid in the ration of chicks, characterized by sores and incrustations on the corners of the eyes, the mouth, bottom of the feet and the joints of the toes. It also upsets the feather development and the general growth.

Chick Feathers– The feathers from chickens, a by-product of the poultry industry, gathered for use in the millinery trade.

Chick Starter– A balanced feed for the quick growth of baby chicks, consisting of ground grains, meat scraps, leaf meal, soybean meal, dried milk, limestone, iodized salt, vitamins, antibiotics, and other items.

Chicken– *Gallus domesticus*, family Phasianidae; the domestic fowl, whose ancestry, while obscured by antiquity, is attributed to *Gallus gallus* and other species of jungle fowl of India and the Malay Peninsula. The chicken is one of the most widely distributed and most commonly kept of all farm animals.

Chicken Body Louse– *Menacanthus stamineus*, family Menoponidae; a biting pest that lives on the skin of chickens or turkeys, especially around the vent and under the wings. It feeds by nibbling on dried skin, scales, and feathers, resulting in irritation to the skin of the infested bird.

Chicken Cholera– See Fowl Cholera.

Chicken Corn– (1) *Sorghum vulgare* var. *drummondi*, family Gramineae; an herbaceous plant, which has escaped from cultivation and is a troublesome weed in some places. Native to North America. (2) An old United States southern term for sorgo (sweet sorghum).

Chicken Fat– The raw or rendered fat from chickens; an edible by-proudct of poultry; particularly in demand by the Jewish bakery trade.

Chicken Feet– A by-product of poultry processing, especially rich in gelatin. By removing the outer skin and boiling the inner flesh, a rich stock is obtained that gels readily.

Chicken Head– A poultry by-product, a table delicacy. Some consider the edible portion of the head, that is, the brains, eyes, and tongue served in the jelly of the feet, as a delicacy.

Chicken Head Louse– *Cuclotogaster heterographus*, family Philopteridae; a biting pest which infests young chickens and turkeys. See Chicken Body Louse.

Chicken Horse– A wild horse of poor quality in the western United States, used for chicken and dog feed.

Chicken Mite– *Dermanyssus gallinae*, family Dermanyssidae; a blood-sucking pest that attacks fowl. It lives in the cracks about the roosts, nests, and walls of the poultry house in the daytime and on nesting fowls at night. In severe cases it causes unthriftiness, low egg production, and at times even proves fatal. Also called bird mite.

Chicken Pox– See Fowl Pox.

Chicken Septicemia– See Fowl Cholera.

Chicken Sticker– A poultry knife with a short, thin blade, used for killing poultry. The blade is forced through the roof of the mouth into the brain.

Chicken Stomach Worm–*Tetrameres americana*; an internal parasite which enters the gland of the proventriculus. Infested birds do not always show symptoms, but young chickens appear emaciated and sometimes die.

Chilblain–A painful swelling on feet or hands due to exposure to cold, sometimes resulting in sores.

Chill–To reduce the temperature of a product, such as a meat carcass or milk, to cure it, improve the flavor, prevent spoilage, and increase market life.

Chilled Brood–Honeybee brood that has died because of chilling.

Chimera–A plant or part of it, composed of tissues of two or more genetically different types. If tissue of one genetic type is external to and surrounds another genetic type, it is a periclinal chimera. If the tissue extends from the surface into and is deeply seated in another genetic type, it is a sectorial chimera. When the tissues are greatly intermingled, it is hyperchimera.

Chimerism–The quality of being a chimera; i.e., in genetics, the presence in an individual plant or animal of cells from a different species, a different genotype, or with different antigens, caused by radiation, grafting, or mutation; e.g., a fruit, half orange and half lemon; an apple with sweet and sour flesh; peaches with half fuzzy and half smooth skin.

Chin Ball Marker–A device worn beneath the chin of a surgically altered bull or androgenized cow used to detect and mark cows in heat. The cows are marked by the device when they are mounted.

Chine–The backbone of an animal carcass.

Chit–(1) Second- or third-grade rice. (2) A young animal. (3) A plant shoot.

Chitterlings–The boiled intestine or gut of the pig, but also of cattle. Chitterlings also means sausages in some areas of the United States.

Chlorinator–A device for adding chlorine gas to sewage to kill infectious organisms.

Choice–A market designation of high quality, usually second from highest; e.g., choice beef as opposed to prime beef.

Choke–(1) An acute condition in livestock brought about by a food mass lodged in the esophagus. (2) Hairy or filamentous undeveloped scales at the base of glove artichoke heads, which are removed before eating. (3) In engine carburetion, to increase the ratio of gasoline to air in the fuel mixture when starting a cold engine. (4) In plants, to kill, dwarf, or stunt through excessive competition for space and nutrients, e.g., the choking of plants by dodder or bindweed. (5) In plowing, the gathering of wet straw, weeds, etc., about the plowshare, which reduces its efficiency.

Choke Down–To break in or control an animal by putting a rope around its neck so that it chokes if it struggles.

Choker–A rope with a noose for pulling logs.

Cholera–(Greek; *chole*, bile) A highly infectious and dehydrating disease (***Vibrio cholerae***), often transmitted by contaminated drinking water. Most commonly endemic and epidemic in densely populated Asia, with a high mortality rate. See Fowl Cholera, Hog Cholera.

Cholesterol–A fat-soluble substance found in the fat, liver, nervous system, and other areas of the body; it plays an important role in the synthesis of bile, sex hormones, and vitamin D.

Choline–An organic compound present in animals and plants; essential for growth.

Cholinesterase–A chemical catalyst (enzyme) found in animals that helps regulate the activity of nerve impulses.

Cholinesterase Testing–A pesticide-monitoring program that monitors a person's blood at various intervals during periods that he/she is using pesticides to determine the degree to which the person has been exposed to the pesticide.

Chop–(1) Animal feed of coarsely crushed or finely ground cereal grains. (2) A small cut of meat which usually includes a rib, e.g., mutton chop. (3) To cut hay into small portions for easy storage without baling. (4) To hoe a row crop, especially cotton; often the first hoeing is called chopping. (5) To crush grain. (6) Jaw; generally in the plural to denote the jaws forming the mouth. Also called chap. See Green Chop.

Choppers–Aged ewes in medium flesh not good enough to grade as fat.

Chorion–(1) The outermost membrane that encloses the unborn fetus in mammals. (2) The tender, fleshy substance of the original nucleus of a plant seed.

Chorizo–Spiced pork or pork and beef sausage.

Chroma–(Greek) Color. See Mundell Color Standard.

Chromatid–One strand of a doubled chromosome seen in the prophase and metaphase of mitosis.

Chromoseres–The smallest particles identifiable by characteristic size and position in the chromosome thread. They are minute subdivisions of chromatin arranged in a linear, beadlike manner on the chromosome.

Chromosome–A microscopic, dark-staining body, visible in the nucleus of the cell at the time of nuclear division, which carries the genes, arranged in linear order. Its number in any species is usually constant, and it serves as the bridge of inheritance, i.e., the sole connecting link between two succeeding generations.

Chronic–Refers to a disease that is marked by long duration and frequent recurrence.

Chronic Alveolar Emphysema–See Broken Wind, Heaves.

Chronic Molybdenum Poisoning (Molybdenosis)–A condition which develops when animals constantly take more molybdenum with their feed than they can absorb. It causes profuse diarrhea, abnormal bone formation, rapid emaciation, swollen genitals, marked anemia, general weakness, stiffness, and fading of coat color. Molybdenum is a trace element in soils, a lack of which limits plant growth.

Chronological Age–The actual age of an animal in days, weeks, months, or years.

Chuck–A cut of meat from the neck including the part around the shoulder blades and the upper three ribs.

Chuck Rib Roast–A roast cut from the five upper ribs of a beef carcass.

Chuck Steak–A steak cut from the upper five ribs of a beef carcass.

Chuck Wagon–A wagon that carried food and cooking and eating utensils for cattlemen tending livestock in the United States.

Chunk Honey–A jar of honey containing both liquid (extracted) honey and a piece of comb with honey.

Churn–A vessel in which cream is agitated vigorously to obtain butter.

Churned Buttermilk–Cultured buttermilk to which churned cream, containing small lumps of butter, is added.

Churning–The beating, shaking, or stirring of whole milk or cream to make butter.

Churnmilk–See Buttermilk.

Chute–(1) A narrow passage through which animals are moved for branding, spraying, or loading, or through which grain slides to a lower level. (2) A trough constructed of round timbers in which logs are slid up or down a grade. (3) A stampede of animals (western United States). (4) A high-velocity conduit for carrying water. (5) An inclined drop or fall.

Chyme–The partly digested food passed from the stomach into the duodenum.

Cide–A suffix which indicates a killer; e.g., amoebacide, that which kills amoeba.

Cierra–(Spanish) Corral.

Cilia–(1) Minute hairlike formations on the cells of many animals. Constantly vibrating, they serve as means of locomotion in aquatic unicellular animals and help remove mucus and fluid residue in higher forms. (2) Tender, sensitive hair forming a fringe along the edges of leaves.

Ciliate–Bearing cilia, fringed with hairs, bearing hairs on the margin.

Cinch–A band of leather or heavy canvas attached to the saddle. It is tied around the horse's barrel to hold the saddle in place.

Circa–(Latin) About, approximate.

Circle Rider–A horseback rider who rounds up cattle for branding, etc. (western United States).

Circling Disease–An infectious disease of sheep, swine, cattle, and goats caused by the bacterium *Listeria monocytogenes*. So named because it affects the central nervous system, causing the animal to move in circles. Also called listeriosis.

Circulation–(1) The pulsatory movement of blood in the body. (2) The flow of cytoplasm in the cells of a plant.

Cirrhosis–A diffuse fibrosis (hardening) of the liver and some other organs.

Citric Acid–An acidic white, crystalline substance present in lemons, limes, currants, gooseberries, raspberries, etc. It is usually obtained from citrate of lime by decomposition and filtration. Used in artificial lemonade, medicines, and dyeing.

Clabber–Curdled milk in which whey has not separated from curd. See Curdled Milk.

Claiming Pen–A pen or small enclosure within a pen in which a ewe and her newborn lamb are placed until the ewe accepts the lamb.

Clamp–To castrate an animal by severing the sperm cords by pinching them with a clamp.

Clarified Milk–Milk that has been cleared of solid impurities by passing through a centrifugal separator.

Clarify–To remove undesirable, solid substances from a liquid, such as milk or fruit juice, by ordinary or by centrifugal filtration.

Class–A division of the plant or animal kingdom lower than a phylum and higher than an order; e.g., the class Insecta.

Class of Animal–Age and/or sex groups of a kind of animal; e.g., calves, cows, does, ewes, fawns, yearlings, etc.

Class of Chickens–Refers to the area where the breed or variety of chicken was developed.

Classification–The forming, sorting, apportioning, grouping, or dividing of objects into classes to form an ordered arrangement of items having a defined range of characteristics. Classification systems may be taxonomic, mathematical, or other types, depending on the purpose to be served.

Classify–(1) To systematically categorize plants or animals according to a set scheme. (2) To sort individuals together into groups having common characteristics or attributes.

Claw–(1) The sharp nail on the toe of an animal or bird. (2) The slender, extended lower part of the petal, as in the iris, lily, etc. (3) A device on a milking machine to which the stanchion tubes are connected.

Claybark Dun–A coat color of animals, especially horses. It is dun imposed on a sorrel chestnut background. Also called copper dun.

Clean Wool–This term usually refers to scoured wool but occasionally it is used to describe grease wool that has a minimum amount of vegetable matter.

Cleansing Flight–A flight of honeybees during which they are able to void their feces, following a period of confinement to the hive, caused by inclement weather.

Clear Egg–An infertile egg.

Clearly Visible Germ Development–The development of a germ spot in the yolk of a fertile egg which is plainly visible as a definite, circular area or spot with no blood in evidence.

Cleavage–(1) In animal or plant reproduction, the splitting of one cell into two identical parts. Each resulting daughter cell matures and may divide again. (2) Tendency of certain minerals or woods to split along particular planes or angles. (3) The weight required to cause splitting in a standard piece of wood three inches long, expressed in pounds per inch of width.

Cleaver–A heavy, wide butcher's knife for cutting bones.

Cleft-footed–An animal having a divided hoof or foot, as a cow or hog. Also called cloven-footed.

Clefts–In blacksmithing, cracks in the heels of horses.

Clinch—Part of a nail bent over to keep it from being pulled out, as in horseshoeing, box manufacture, etc.

Clinical—Concerning the investigation of decease on a living subject by observation, as contrasted to a controlled experiment.

Clinical Evidence—Any symptom of disease that can be determined by direct observation, such as fever, lack of appetite, swellings, paralysis, etc.

Clip—(1) A semicircular metal piece extending from the outer surface of the horseshoe at the toe or side to prevent the shoe from shifting on the hoof. (2) Shears. (3) (a) An inclusive term for shorn wool. (b) The process of shearing wool. (c) The year's production of wool. (4) (a) To shear the hair of an animal close to the skin, as to clip a dog, or a cow's flank and udder. (b) To remove fleece from a sheep or goat. Also called shear. (c) To cut the feathers from a fowl's wing to prevent it from flying. (5) (a) To trim a plant. (b) To cut off the tops of a crop, as clover or alfalfa, at an early stage of growth when the crop is to be harvested for seed. Also called preclipping.

Clipped Queen—Queen bee whose wing (or wings) has been clipped for identification purposes.

Clitoris—A small organ located at the ventral part of the female vulva; it is homologous with the male penis.

Cloaca—The common area or chamber that serves as the terminal part of the urinary, digestive, and reproductive tracts of birds, reptiles, and amphibians.

Cloacitis—A foul-smelling inflammative condition of the cloaca and vent caused by a noninfectious unknown organism. Also called vent gleet, vent disease of fowl.

Clog—(1) A heavy, wooden block attached to a horse's hind pasterns by a strap to prevent it from kicking when put out to pasture. (2) An impediment, encumbrance, or restraint. (3) A wooden shoe. (4) To stop a machine, as a harvester, by feeding material in too rapidly; to plug up; to choke up.

Close Breeding—A form of livestock inbreeding that involves very close relatives.

Close Cropping—The grazing by animals of grasses and other forage dangerously close to the crown of the plant, which can ruin grazing lands. Also called close grazing.

Close-feathered—Said of a fowl in which the feathers are held closely to the body, i.e., at no perceptible angle to the body.

Closed Breed—A breed of livestock where registration is restricted to progeny of animals themselves registered in that breed.

Closed Formula—The statement of ingredients on a tag attached to a sack of feed which gives the guaranteed percentages of protein, fat, and fiber. It does not give the amount of each ingredient.

Closed Herd—A herd in which no outside blood is introduced.

Closed Loop System—A system of tanks where fish are raised. Oxygen is provided and the water is filtered.

Closed Pedigree—Hybrid combinations of unrevealed inbreds which have been put together by privately owned organizations in the seed business. Also called unpublished pedigree.

Closed Side—The right side of a beef carcass.

Closed-faced—A sheep that has considerable wool covering about the face and eyes. This often leads to a condition known as wool blindness.

Clostridial Organism—Bacteria of the genus *Clostridium*, which cause diseases such as tetanus, malignant edema, and blackleg.

Clot—(1) A coagulum; a semisolid mass, as a clot of cream or a blood clot. (2) To coagulate, as blood.

Clothing Wool—Short staple wool, too short to be combed, suitable for woolen clothing. Also called carding wool.

Clotted Cream—Devonshire cream; an English dish made from rich milk allowed to separate and then scalded. The thickened cream is then skimmed off and served for dessert as an accompaniment to strawberries or eaten on bread.

Clouding—An undesirable translucent condition of clear liquids.

Cloudy Wool—Wool that is off-color; may be caused by wool becoming wet while in a pile.

Cloven-footed—See Cleft-footed.

Clover-sick—Designating soils on which clovers fail because of infestation by parasitic fungi or low soil fertility.

Club Foot—A deformity in an animal's foot in which the foot turns inward to resemble a club. See Clubroot of Cabbage.

Club Steak—A cut of beef from the loin.

Cluck—The call of a brooding hen to her chicks.

Cluster—(1) The form in which bees cling together in the hive after swarming or during winter. Also called clustering. (2) A large number of plants grouped, arranged, or growing in close proximity to one another. (3) An inflorescence.

Clutch—(1) (a) A nest of eggs in a hatchery. (b) A brood of chickens. Also called cletch. (2) A device to engage or disengage the power from various working parts of machinery.

Clydesdale—A quality breed of Scottish draft horse, distinguished by an abundance of fetlock hair. Usually bay-colored with white points, sometimes roans and browns are also found among them. A stallion weighs up to 2,000 pounds and a mare up to 1,800 pounds.

Coach Breed—Any of several breeds of horses adapted for drawing coaches, as the Cleveland Bay, the Yorkshire Coach, and German Coach Horse. They are usually heavier than riding horses.

Coagulant—(1) A substance which acts on a liquid to coagulate it, as rennin. (2) An agent which aids the clotting of blood.

Coagulation—The clumping together of solids to make them settle faster out of sewage. Coagulation of solids is brought about with the use of certain chemicals such as lime, alum, or iron salts.

Coal-tar Dip—An emulsified coal-tar disinfectant used in a dipping vat to rid animals of external parasites. It consists of coal-tar oils, tar acids, and soap mixed with water to form a milky emulsion.

Coarse-wool–Referring to a breed of sheep which produces a coarse wool, as the Cotswold, Leicester, Lincoln, Romney Marsh. See Fine Wool, Medium Wools.

Coaster–The long-horned cattle found along the Texas coast of the United States.

Coat–The outer covering of an animal, i.e., the fur of a beaver, the feathers of a fowl.

Cob–(1) The chaffy axis upon which the kernels of corn grow. Also called corn cob. (2) A stocky, short-legged horse used for pulling light carts. (3) Male swan.

Cobalt–Co; a trace element that occurs in soils in quantities up to 15 parts per million. It is constituent of vitamin B_{12}, and is more significant in animal than in plant nutrition.

Cobalt Deficiency–Chiefly a vitamin B_{12} deficiency in ruminant animals due to a reduced synthesis of this vitamin, which results in anemia, and a thin, emaciated appearance in the animal.

Cobalt Sulfate (Cobaltous Sulfate)–$CoSO_4 \times 7H_2O$; a pinkish crystalline salt, soluble in water; used in animal feeds and sometimes added in small quantities in mixed fertilizers for use on pastures where the forage is deficient in cobalt. Cobalt is essential for animal nutrition and there is slight evidence that it is beneficial for some plants.

Coccidioidomycosis–An infection causing swelling of the lymph nodes in humans and animals. It is produced by the fungus *Coccidioides immitis*, probably related to the yeasts. The characteristics may be vague unless the affected lymph node ruptures, in which case septic infection follows. The disease is sometimes confused with actinomycosis.

Coccidiosis–A disease caused by one or more of the different kinds of protozoans belonging to the order Coccidia. In most domestic animals, coccidiosis is a disease of the digestive tract and causes diarrhea.

Coccidiostat–Any of a group of chemical agents mixed in feed or drinking water to control coccidiosis, a growth-retarding and occasionally fatal intestinal disease of poultry caused by infestation of protozoans of the order Coccidia. See Coccidiosis.

Cocculiferous–Referring to plants that bear berries.

Coccus–A spherical bacterium, as of the family Coccaceae.

Coccygeal–Of, or pertaining to, vertebrae of the tail.

Cochin–An Asiatic breed of very large, egg-producing chickens with profuse, soft plumage and fluff, and full-feathered shanks and feet. The cock weighs up to 11 lbs., and the hen up to 8½ pounds. Varieties are Buff Cochin, Partridge Cochin, White Cochin, Black Cochin. Also called Chinese Shanghai fowl, Cochin China.

Cock–(1) The adult male fowl. Also called rooster, stag. (2) A small conical pile of hay, manure, etc. (3) A device for regulating the flow of fluids through a pipe. it consists of a conical plug working in a shell of iron or brass bored out to receive it, which is lodged in the pipe. The passage of fluids is controlled by rotation of the plug.

Cocked Ankle–A dislocation of the fetlock in a horse caused by contraction of the tendon. Also called knuckling over.

Cockerel–A male chicken less than one year old.

Coconut Oil Cake–A stock feed, a by-product of dried coconut meal after coconut oil has been extracted. It averages 21.3% protein; it equals corn gluten in feed value. Also called copra oil cake.

Cod–(1) The part of the scrotum that remains after castration. (2) A species of edible fish.

Cod Liver Oil–Oil obtained from the codfish, *Gadus morrhuae*, or from other species of *Gadus*. It is sometimes mixed with other fish liver oils or synthetic vitamins to increase its food value.

Codominance–A kind of gene action where one allele does not exhibit complete dominance over the other.

Codominant–(1) A tree which grows close enough to other trees as to receive very little sunlight from the sides. (2) Any plant or animal species that shares dominance with other species in an area.

Codominant Genes–Genes that are neither dominant nor recessive.

Coefficient of Digestibility–The amount of a particular nutrient digested and absorbed by an animal; expressed as a percentage of the amount that was in the animal's feed.

Coenzyme–A partner needed by some enzymes to accomplish a biochemical change.

Coffee Cream–Cream that contains 18 to 22 percent milk fat. Also called table cream.

Coffin Bone–The bone of the foot of a horse, enclosed within the hoof.

Coggins Test–A test for diagnosing equine infectious anemia. It was developed by Dr. Leroy Coggins, Cornell University.

Coincidence–In genetics, a term used in estimating the distance between two genes. It is the ratio between the actual percentage of double crossing over and the percentage expected on the assumption that each crossing over is an independent event.

Coitus–Sexual intercourse; copulation; service; mating.

Colanut–See Sudan Colanut.

Colby–An open-textured cheese made similarly to cheddar, except that the curd is kept from mating by stirring.

Cold Carcass Weight–The weight of a carcass after the carcass has cooled and shrunk. See Hot Carcass Weight, Shrinkage.

Cold Collar–(1) A freshly harnessed horse (southwestern United States). (2) A balky horse.

Cold Manure–Farmyard manure which does not ferment and heat unduly while in storage, such as cattle manure. See Hot Manure.

Cold Sterilization–(1) The use of cathode ray or electron beam gun in food processing to kill bacterial or insect life. (2) Chemical sterilization of instruments.

Cold Test–(1) A test for viability in which sample seeds are placed on wet soil, sand, cloth, or blotter at a constant temperature of 40°F for three days or longer. The sample is then placed in a warm place and the percentage of germination recorded. (2) Synonymous with cold shrinkage. After wool scouring, the hot wool is allowed to stand in a cool room until it has taken on a normal amount of moisture from the air. The wool is then weighed, and the shrinkage figured to give the cold shrinkage, or cold test.

Cold Water Fish–A fish that will not thrive in water temperatures above 70 degrees Fahrenheit.

Cold-backed–Describes a horse that humps his back and does not settle down until the saddle has been on a few minutes. Some cold-backed horses will merely tuck their tails and arch their backs when first mounted, but others will take a few crow hops until warmed up.

Cold-blooded–(1) Referring to a horse of unknown breeding or one which does not have an ancestor of pure blood. (2) Referring to certain animals whose blood temperature depends on the environment, e.g., fish and snakes. See Hot-blooded.

Cold-jawed–Describing a horse difficult to control with the bit and bridle. Also called hard-mouthed.

Cold-storage Eggs–Eggs held in storage at a temperature of 45°F (7.2°C) or less for more than thirty days.

Colic–(1) Pertaining to the colon. (2) A pain in the abdomen caused by irregular muscular contractions, obstruction, spasm, or distension of the viscera.

Coliform Bacteria–A group of bacteria predominantly inhabiting the intestines of humans or animals, but also occasionally found elsewhere. Fecal coliform bacteria are those organisms associated with the intestine of warm-blooded animals that are used commonly to indicate the presence of fecal material and the potential presence of organisms capable of causing disease in humans.

Colitis–Inflammation of the mucous membrane of the colon.

Collagen–A protein that forms the chief constituent of the connective tissue, cartilage, tendon, bone, and skin. Collagen is changed to gelatin by the action of water and heat.

Collar–(1) The padded and reinforced part of a harness which girdles the neck. Straps for pulling are attached to it by means of hames. (2) A closely bound retail cut of rolled meat. (3) The part of a tree or other higher plant at the line of union of the roots and the trunk.

Collateral Relative–Animals related not by being ancestors or descendants, but by having one or more common ancestors.

Collected–(1) A collected horse has full control over its limbs at all gaits and is ready and able to respond to the signals or aids of its rider. (2) Plants dug from the wild and offered for sale.

Collection–(1) The balance of a horse and rider which makes for instinctive coordinated movement. (2) The accumulation of sperm for use in artificial insemination. (3) In flower shows, an exhibit of several varieties of plants, flowers, fruits, or vegetables. A collection differs from a group in that the number of varieties is an important factor. Artistic arrangement is not necessarily mandatory.

Collector Tube–A tube fastened to the artificial vagina to collect semen from the male animal for use in artificial insemination.

Collop–A unit of grazing area which can support a full-grown horse or cow for one year.

Colon–That part of the large intestine extending from the cecum to the rectum.

Colony–(1) An aggregate of worker bees, drones, and a queen bee, living together in a hive as a unit. (2) A cluster of bacteria or fungi grown on a culture medium, usually originating from a single bacterium, spore, or inoculation transfer. (3) A group of people bound by communal or religious ideals that occupies tracts of land, often wild, for permanent settlement, especially in the early development of the United States and Canada.

Colony Morale–The zeal of bees in rearing a brood and accumulating stores; determined by the resistance of the colonies to disturbances by animals or people.

Color Defect–Any color that is not removable in wool scouring, such as urine stain, dung stain, canary yellow stain, and black fibers.

Colorados–Sheep that were bred in Colorado, United States.

Colostrum–The milk secreted by the udder for 48 hours after birth of the young. It contains a concentration of antibodies which are passed to the young and act as a valuable protection against a number of infectious conditions. The young animal can only absorb these antibodies for a matter of hours (possibly only 24 hours), so it is crucial that it should suckle colostrum from its mother frequently during that short period.

Colpindach–A young heifer.

Colt–A young male horse, ass, or mule under four years of age. See Filly.

Columbia–A breed of sheep developed in the United States from the Lincoln and Rambouillet breeds. It is large, vigorous, heavy-boned and long-legged, with the rams averaging 275 pounds and ewes up to 135 pounds. It yields long-stapled fleece of quarter-blood wool.

Columbian–A plumage color pattern in chickens. The main body plumage is white, but certain feathers in the neck, tail, and wing sections are variously tinged with black. It is the plumage pattern of the Light Brahma, Columbian Plymouth Rock, and the Columbian Wyandotte.

Colyone–The secretion of one of the ductless glands in animals that inhibits metabolism. Also called chalone.

Coma–(1) A tuft of soft hairs on a seed, as in milkweeds (*Asclepias*). (2) A tuft of leaves or bracts at the apex of an inflorescence, as in pineapple (*Ananas*). (3) A leafy crown or head, as in many palms. (4) Insensibility of an animal or person caused by poison, injury, or disease.

Comatose–Unconscious; in a coma.

Comb–(1) A fleshy crest on the head of fowls. (2) A back-to-back arrangement of two series of hexagonal cells in a beehive. Worker cells are approximately five cells to a linear inch and drone cells about four cells. (3) An instrument consisting principally of metal, bone, plastic, etc., in which long, narrow, thin teeth are cut for the purpose of arranging and cleaning hair on an animal. Also to clean and arrange the hair of an animal with such an instrument. (4) A device for cleaning and removing buds, stems, etc., from produce. See Currycomb.

Comb Foundation–(1) A thin sheet of beeswax impressed by a mill to form bases of cells; some foundation also is made of plastic or metal. (2) A sheet of beeswax embossed on each side with the cell pattern.

Comb Honey–Honey in the sealed comb in which it was produced; also called section comb honey when produced in thin wooden frames called sections, and bulk comb honey when produced in shallow frames.

Comb Honey Section–A square or rectangular wooden container in which section comb honey is produced; usually made of sanded basswood ⅛-inches thick, which is grooved for folding and dovetailed on the ends for fitting together. Plastic sections are also used.

Comb Honey Super–One of several types of hive body in which section comb honey is produced.

Combing–An operation in the manufacture of worsted yarn by which the long fibers are separated from the short and arranged parallel to each other.

Combing Wool–Wool long enough (at least 2 inches) and strong enough for combing.

Combining Ability–The level of hybrid vigor produced by a breed of animal when crossed. General combining ability may refer to the ability of a breed to produce hybrid vigor when crossed with many other breeds. Specific combining ability may refer to the ability of a breed to produce hybrid vigor when crossed with another specific breed.

Combless Package–A quantity of bees (2 to 5 pounds) with or without a queen, contained in a shipping cage.

Come in(to) Heat–To be in estrus; ready for breeding.

Comfort Zone–The range of temperature within which an animal feels most comfortable and which makes no demand upon the animal's temperaure-regulating mechanism.

Commensal–A nonparasitic organism which lives attached to a host and shares its food.

Commercial–(1) A low-grade market meat. (2) Any commodity produced for the market. (3) A nonpurebred animal.

Commercial Cow-calf Producer–A stockman producing animals from a nonregistered herd.

Commercial Culture–Any bacterial culture which is prepared and sold by a special culture-producing laboratory.

Commercial Herd–A herd of animals that will eventually be slaughtered for meat.

Common–(1) A joint pasture in a village or community to which all members have access for their herds. (2) Legal right of a person in sharing the profit or use of another's land. (3) A low, market grade of meat animals. (4) The American wool grade equivalent to the English 44's; a low grade, only better than braid. (5) A defective grade of lumber useful only in framing, etc.

Common Salt–NaCl, sodium chloride; a chemical used in livestock feeding, in seasoning, and in preservation of food products. It is also an effective nonselective herbicide.

Common Scab of Cattle–Mange; a contagious skin disease caused by the parasite *Psoroptes bovis*, family Psoroptidae; characterized by intense itching on the withers, on the top of the neck, and at the tailhead. Lesions eventually spread over the entire body. Rare in the United States. Also called psoroptic scab, barn itch, cattle mange.

Common Scours–A looseness of bowels in bucket-fed calves from overfeeding, dirty pails, irregular feeding, lack of proper sanitation, or cold, sour, or too-rich milk. It is characterized by listlessness, loss of appetite, bloating, diarrhea, and foul smelling feces. See Bloody Scours, Calf Scours.

Common Wool–A United States wool grade. Next to the coarsest, slightly finer than braid wool.

Common-use Range–Range containing grass, forbs, and browse which allows two or more kinds of stock to graze together during the entire season or separately during part of a season.

Communicable–Readily transferred from one individual to another.

Communis–Gregarious; flocking together, as sheep.

Communitiy Allotment–An allotment upon which several permittees graze their livestock in common.

Community Regulation–See Homeostasis.

Compactness–In animal judging, a closely knit, firmly united animal.

Compatible–Describing two or more chemicals that can be mixed without affecting each other's properties.

Compensatory Gain–Gain in the weight of livestock at an above-normal rate following a period of little or no gain.

Complementability–In crossbreeding animals, the degree to which two or more breeds match so that the strengths of one breed cover the weaknesses of the other.

Complemental Feed–Fodder given to livestock in place of, or as a supplement to, range feed.

Complementary Crop or Livestock–Any crop or livestock yielding a product that contributes to the success of another.

Complete Carcinogen–An agent that can act as both initiator and promoter of cancer.

Complete Protection–The withdrawal of all grazing animals from a given range.

Complete Ration–A single feed (usually a commercially formulated ration) that fulfills all of the nutritional requirements of an animal except for water.

Compound–A chemical term denoting a combination of two or more distinct elements.

Compound Eye–An eye consisting of many individual elements or ommatidia each of which is represented externally by a facet.

Compress–(1) (a) A hydraulic press used to reduce a bale of cotton to about one-third its original size for ease in handling and shipping. (b) The business or building for such a press. (c) To reduce the size of a cotton bale by a hydraulic press. (2) A pad of gauze or other material applied to put pressure on any part of the body to reduce swelling or control hemorrhage.

Conalbumin–Protein found in egg white; has the property of binding iron in an iron-protein complex that turns pink in the presence of oxygen.

Conceive–To become pregnant.

Concentrate–(1) (a) Any feed high in energy (usually grain); sometimes used with reference to other nutrients, such as protein concentrate, etc. (b) Stock feed low in fiber content and high in digestible nutrients. (2) Whole fruit juice, thickened to the consistency of heavy syrup. (3) To increase in strength by removing dilutents.

Concentrate Ratio–The amount of concentrates in comparison to the amount of other feeds fed.

Concentrated Feed–Feeds that have a high caloric density and are high in relatively completely digestible substances such as fats, proteins, starches, and sugars. The definition of concentrates as being those feeds below 18% crude fibre is very arbitrary.

Concentrated Milk–See Evaporated Milk, Plain Condensed Milk, Plain Condensed Skim Milk, Sweetened Condensed Skimmed Milk, Sweetened Condensed Whole Milk.

Conception–The beginning of pregnancy or gestation; first stage in the development of the embryo.

Conception Rate–In cattle breeding, the percentage of first services that conceive.

Concolorous–Of a single uniform color.

Condemned–(1) Describing an animal, carcass, or food which has been declared unfit for human consumption. (2) Referring to real estate property acquired for public purposes under the right of eminent domain.

Condensed Milk–A liquid or semiliquid food made by evaporating a mixture of sweet milk and refined sugar to such point that the finished sweetened condensed milk contains not less than 28 percent of total milk solids and not less than 8.5 percent of milk fat.

Condition–(1) The state of wool regarding the amount of yolk and other foreign matter it contains. (2) The general appearance and/or state of health of an animal, seed, fruit, or flower at a show. (3) To get an animal in good health and appearance by proper feeding and grooming. (4) The degree or amount of fat on a breeding animal.

Confinement Operation–A production system where animals are raised in a relatively small area. Usually, the environment is controlled for the animals' comfort.

Conformation–In judging contests, the type, form, and shape of the live animal, usually with reference to some performance characteristic.

Conformation Score–In the selection of replacement bulls and heifers, a numerical value indicating the overall appearance of the animals. It is figured using numerical values assigned to such physical characteristics as frame size, bone size, degree of muscling, sex, breed and sex character, and structural soundness.

Congenial–Able to cross-fertilize readily; to unite easily.

Congenital–Acquired during prenatal life; certain conditions which exist at birth; often used in the context of birth defects.

Congenital Loco in Poultry–A lethal genetic character of poultry in which the chicks are unable to stand when hatched. The head is held back so that the beak points toward one side, and the chick loses its balance and falls. Mortality rates are high.

Conjugate–Coupled or in pairs.

Conjugation–Side-by-side association (synapis) of homologous chromosomes, as in meiosis.

Conjunctiva–The smooth thin layer of tissue lining the inner surface of the eyelid which continues over the forepart of the eyeball, covering the white of the eye.

Conjunctivitis–Inflammation of the membrane that lines the eyelids and covers the forepart of the eyeball.

Conservative Grazing–(1) The practice of limiting the number of livestock on a grazing range in such proportion as would not exhaust the range in successive seasons. (2) A degree of grazing that causes little or no soil disturbance.

Constipation–The retention of feces in the intestines for a longer than normal period.

Constitution–Used to describe the characteristics of an animal which determine in part its ability to put on flesh, reproduce, and maintain health, vigor, and longevity.

Contact Insecticide–Any substance that kills insects by contact in contrast to a stomach poison, which must be ingested.

Contagion–Spread of disease by direct or indirect contact.

Contagious–Refers to a disease that can be transmitted from one animal to another; infectious.

Contagious Abortion–See Brucellosis.

Contagious Disease–A disease transmitted or spread from animal to animal, person to person, or from plant to plant, by direct or indirect contact with the diseased plant or animal.

Contagious Ecthyma–A highly contagious, viral disease of sheep and goats under one year of age; characterized by pus-filled lesions on the lips and sometimes on the face and ears. Death results from the animals not being able to eat when normal gains should occur. It may cause secondary infections. Also called sore mouth, contagious pustular dermatitis.

Contagious Equine Abortion–A contagious disease among mares caused by *Salmonella abortivoequina*; characterized by abortion at any stage of pregnancy, but usually during the eighth to tenth month.

Contagious Equine Metritis (CEM)–A highly contagious acute veneral disease of horses and other Equidae that notably affects breeding and fertility.

Contaminant–An undesirable substance that is present in foods or feed but is not intentionally added.

Contaminate–To make impure by contact or admixture of harmful bacteria, fungi, or dangerous chemicals, etc.; to render unfit for use.

Contamination–(1) Pollution; the process of being contaminated. (2) Specifically, the addition of bacteria or other foreign substance to milk or other products by means of utensils, containers, exposure to air, etc.

Contemporary Group–A group of cattle that are of the same breed and sex and have been raised in the same management group (same location on the same feed and pasture). Individual animals can then be accurately compared with the others in the group. Contemporary groups should include as many cattle as can be accurately compared.

Contentment–In animal judging, a characteristic of a stable, as contrasted to a nervous, animal.

Continuous Grazing–Grazing for the entire grazing season; synonymous with set stocking.

Contract Feeding–An arrangement for finishing cattle, poultry, or swine for the market. The contribution of each participant depends on a written contract. In the case of cattle, the stock raiser usually furnishes the cattle; the feeder furnishes the feed, equipment, and labor, and usually receives his income on the basis of the increase in weight of the cattle, etc.

Contraindication–Condition of a disease which renders a particular treatment undesirable.

Control–(1) Prevention of losses from plant or animal diseases, insect pests, weeds, etc., by any method. (2) A section of an open water channel where conditions exist that make the water level above it an index of the discharge. (3) A standard entity used for comparative purposes in experimentation. Also called check or check plot.

Controlled Environment–An environment for animals that is kept at the correct temperature and other conditions to maximize animal comfort.

Convalescence–Period of recovery following an illness.

Convulsions–A clinical symptom usually of a deranged function of the brain or spinal cord in an animal. The muscles contract and relax violently, producing aimless body movements.

Cool Out–The reduction of grain in the ration of show animals after the show season; usually using corn and barley with oats and bran to lighten feed.

Cooler–Any chestlike device for reducing the temperature of products, such as milk.

Cooler Shrinkage–A loss in weight of an animal's carcass during storage in a cooler, which usually amounts to 2.5 percent of the carcass weight.

Cooling Basket–A wire basket used for protecting young rabbits during periods of high temperatures.

Coon-footed–Used to describe a horse having long pasterns and shallow heels.

Coop–(1) A small light enclosure, usually made of wire netting, used to confine fowls. (2) A heap, usually of manure.

Copper–Cu; a metallic element found in soils at 1 to 50 parts per million, and in plants up to 100 parts per million. It is necessary for all animal and plant life. High soil phosphorus, zinc, and molybdenum can induce copper deficiency in plants. Also, high copper can reduce plant uptake of phosphorus, iron, zinc, and molybdenum.

Coprogenous–Designating the influence of animal excrement, as the cast of the earthworm in forming soil.

Coprophagous Insect–An insect that consumes dung.

Coprophagy–The eating of its feces by an animal. This is normal in rabbits and horses. See Pica.

Copulation–The sexual union of animals, which results in the deposition of the male gametes (sperm) in close proximity to the female gametes (eggs).

Cord–(1) (a) The spermatic cord. (b) The umbilical cord. (2) A unit of measurement of timber. It contains 128 cubic feet, usually 8 x 4 x 4 feet. A short (face cord is a stack of wood 8 feetlong, 4 feet high, and about 16 inches wide. (3) Twine. (4) To stack in rectangular tiers.

Cordova–Long, coarse wool from Argentina, largely used for carpets.

Corn Chop–A fine or coarse stock feed made by grinding or chopping grains of corn. It contains less than 4 percent foreign material. Also called ground corn, cracked corn.

Corn Cob–The woody core of the corn ears; the highly fibrous axis on which the kernels are borne; used alone or with the grains as ground cattle feed.

Corn Cob Meal–Pulverized corn cobs, used for cattle feed, cleaning furs, burnishing metal, removing oil from tin and metal, sweeping compounds, etc.

Corn Distillers' Dried Grains–A stock feed consisting of the dried residue from the distillation of alcohol from corn or from a grain mixture in which corn predominates.

Corn Fodder–The entire corn plant, cut and fed green, or harvested when mature and dried for future feeding. The term is used also for the stalk remaining after ears have been husked from the plant.

Corn Gluten Meal–A stock feed from the residue of commercial shelled corn, which remains after the extraction of the larger part of the starch and germ, from which the bran has been separated. It may contain corn solubles and corn oil meal. Also called gluten meal.

Corn Meal–Finely ground but unbolted corn, used in corn bread, mush, etc.

Corn Oil Cake–A stock feed consisting of the corn germ, from which oil has been partially pressed.

Corn Screenings–A stock feed consisting of small and broken grains of corn, obtained by screening shelled corn and other material having feed value.

Corn Silage–An excellent stock feed prepared from the entire corn plant. The plant is cut while still green but when the ears have begun to dent. It is then chopped and placed in a silo for fermentation and preservation.

Corn Stover–The dried corn stalk from which the ears have been removed; used as a roughage for livestock.

Corn Stubble– The basal portion of the stems of corn left in the ground after the plants are cut.

Corn-and-Cob Meal– A stock feed consisting of the grain and cob of corn ground together. Also called ground ear corn.

Corn-fed– Designating an animal or fowl fattened on corn prior to marketing; hence, well-finished and of high grade.

Corn-Hog Ratio– Number of bushels of corn that are equal (in value) to 100 lbs. of live hogs; i.e., the price of hogs per hundredweight divided by the price of corn per bushel.

Corned Beef– Cured beef; prepared by packing fresh beef into vats and weighting it down to keep it submerged in the pickling solution of salt, sugar, spices, and water. The treatment requires about twenty-five days, during which the meat is turned to ensure thorough pickling.

Corner– (1) The outer pair of incisor teeth in the upper and lower jaws of a horse. (2) The junction point of boundary liens. (3) To tie up or control all available items of produce for speculation. (4) In lumbering, to cut through the sapwood on all sides of the tree to prevent it from splitting when the tree is felled.

Cornicle– An abortive spur on the leg of a hen that hardens with age. It does not develop into a regular spur, as in the cock.

Cornish– A very heavy, meat-producing breed of chickens originating in Cornwall, England. They have compact bodies, close-fitting plumage, pea comb, yellow skin, and unusually well-fleshed breasts, thighs, and drumsticks.

Cornstalk Disease of Horses– A lethal disease probably caused by the eating of toxic, moldy corn stalks. The affected animal exhibits nervous symptoms such as dullness, excitement, local or general paresis, circling, staggering, and tremors. Jaundice is a common side effect.

Corona Radiata– The granular material adhering to an ovum after ovulation that gives it a sunburst appearance; cumulous oophorus.

Coronet– The part of a horse's hoof where it joins the skin. Also called coronamen, coronary band.

Corpus Albicans– The small bit of scar tissue that remains on the surface of the ovary after the regression of the corpus luteum.

Corpus Hemorrhagicum– The temporary blood clot that forms in the crater formed by the follicle after ovulation and prior to the development of the corpus luteum.

Corpus Luteum– Active tissue that develops on the ovary at the site where an ovum has been shed. If conception does not occur, the tissue gradually disappears. If conception does occur the tissue becomes functional, producing progesterone.

Corral– A small enclosure for handling livestock at close quarters.

Corrected Weight– A means of comparing the growth of animals that are of a different weight. For example to compare lambs in a flock that are weaned at around ninety days, divide the weaning weight of each lamb by its age in days and multiply by 90. This puts all of the lambs on a ninety-day basis.

Corriedale– A popular crossbreed of sheep developed in New Zealand and Australia from the Romney, Lincoln, or Leicester rams and native or Merino ewes. It is a good meat producer and carries a good fleece in grade of three-eighths to one-half blood. The mature rams weigh up to 250 pounds, and mature ewes up to 185 pounds.

Cortex– The outer layer or region of any organ.

Cortisol– A hormone of the adrenal cortex that functions in the metabolism of glucose and the reduction of certain kinds of stress.

Cortisone– A hormone essential in regulating vital functions in the body. It is produced and secreted into the blood by the outer part of the adrenal gland (adrenal cortex) and aids in the mobilization of body proteins and fats and affects carbohydrate metabolism. It also has been synthesized as a chemical compound.

Coryza– In fowls, a cold in the head, which causes an acute catarrhal condition and a mucus discharge. See Infectious Coryza.

Cosset– A lamb raised without the help of its dam.

Costa– (1) A rib (when there is only one), the midvein of a simple leaf or other organ. (2) Less commonly, the rachis of a pinnately compound leaf.

Cote– A shed for sheltering small animals; e.g., a sheep-cote.

Cotswold– A white-faced, long-wool breed of sheep from Gloucestershire, England. Its coarse wool is finer and softer than the Lincoln wool, but lacks the luster of the Leicester wool. Its former usefulness as a good cross-breeder has declined due to the popularity of the Lincoln.

Cottage Cheese– A soft cheese prepared by coagulating raw or pasteurized skimmed milk by lactic culture with or without rennet.

Cottonseed– The seed of cotton after the lint has been removed. It is a source of cottonseed oil, the residue being used as a stock feed.

Cottonseed Cake– The solid residue left after the extraction of oil from cotton seeds. It should contain more than 36 percent protein. It is sold according to its nitrogen or protein content.

Cottonseed Feed– A stock and poultry feed consisting of cottonseed meal and cottonseed hulls. It is sold according to its protein content which is usually between 22 and 36 percent.

Cottonseed Hulls– The outer covering of the cottonseed. It is part of the residue after the extraction of the oil and used extensively for a stock feed. Cottonseed hulls supply no digestible protein but contain 43.7 percent digestible nutrients; they are fed with protein-rich feeds, often cottonseed meal.

Cottonseed Meal– The residue of cottonseed kernels from which oil has been pressed. Usually containing the portions of the fiber, hull, and oil left after processing the seed; it is used as a stock feed and fertilizer.

Cottonseed Meal Poisoning– A livestock poisoning, especially in pigs, caused by excessive use of cottonseed meal high in free gossypol. The animal has difficulty in respiration and/or a "potted" belly due to accumulation of fluid in the abdomen and general weakness. Death results apparently from edema of the heart and lungs or liver. Also called cake poisoning. See Gossypol.

Cotty Wool– Tangled or matted wool on a sheep's back. This condition is caused by insufficient wool grease being produced by the sheep, usually due to breeding, injury, or sickness. This type of defective wool is more common in the medium and coarse wools. The fibers cannot be separated without excessive breakage in manufacturing.

Counterfeit–Referring to deceptive characteristics of good breeding or physique in an animal which it does not actually possess.

Coupling–(1) The loin of an animal. (2) Mating; breeding; copulation. (3) A connecting means for transferring movement from one part of a machine or device to another; it may be hydraulic, mechanical, or electrical.

Cover–(1) A lid placed over something for protection. (2) Woods, underbrush, etc., which may conceal game. (3) Plant life, such as grass, small shrubs, herbs, etc., used to protect soil from erosion. (4) The flesh, hide, and fat on a fattened animal. (5) To buy back future contracts. (6) To copulate with a female, as a bull covers a cow. (7) To incubate, to hatch eggs, as a hen. (8) The proportion of the ground surface under live and dead aerial parts of plants. Also refers to shelter and protection for animals and birds.

Covert–(1) A unit of wildlife cover: e.g., a thicket. (2) One of the small feathers covering the bases of large feathers in fowls.

Covey–A flock of birds: quail, partridge.

Cow–(1) The mature, female bovine. (2) A market term for designating a female bovine that has produced a calf.

Cow, Boss–Individual cows whose temperament causes them to be "piggish" at time of hay feeding or grain feeding at a bunker. Boss cows may walk up and down a line of cows feeding peacefully and butt them away from the feed. A similar action occurs in most species of animals, and in people; known as "bullying" among people and "pecking order" in poultry.

Cow Brute–Cow, steer, bull, calf (southwestern United States).

Cow Calf Operation–A system of raising cattle thats main purpose is the production on calves that are sold at weaning.

Cow Chips–Dried cow manure; used as a fuel in developing countries.

Cow Heifer–A cow that has calved only once.

Cow Horse–A horse or pony trained in rounding up cattle, etc., usually on the western range in the United States.

Cow Index (CI)–A method of evaluating genetic value of dairy cows. In beef cattle selection, a score indication of a cow's calf producing ability. The formula is: (Adjusted Weaning Weight of the Calf divided by the Weight of the Cow at Weaning) multiplied by 100.

Cow Month–The tenure on range or artificial pasture of a cow for one month. The quantity of feed or forage required for the maintenance of a mature cow in good condition for one month. See Sheep Month.

Cow Pat–Cow dung.

Cow Poke–(1) A metal or wooden device used to discourage fence breaking by cattle. It is fastened around the animal's neck and has two hooked arms extending upward and downward. A barb on each of the arms points toward the animal so that when it attempts to put its head through a fence, it receives a sharp prick from the barb. (2) A person who works cattle.

Cow Pox–A contagious eruption appearing on the teats and udders of cows, caused by a filterable virus, similar to the smallpox virus in humans. It spreads through contact with contaminated premises, cattle, or milkers' hands. Cows have been known to contract the disease from people recently vaccinated for smallpox. Humans sometimes contract a poxlike disease from the afflicted cows.

Cow Sense–The ability or intelligence in some horses which show unusual dexterity in working with cattle.

Cow-hocked–Standing with the joints of the hocks bending inward, with the toes pointing outward.

Cowboy–(1) A man who works directly with cattle and horses; specifically, the cattle herders on the large ranches of western United States and Canada. Sometimes called a cowpoke or a cowpuncher. (2) The boy who milks cows (New Zealand).

Cowhide–(1) A hide from a cow, steer, or bull. (2) A coarse, leather whip. (3) To whip an animal with cowhide.

Cowpers Gland–An accessory gland in the male reproductive system that produces a fluid which moves ahead of the seminal fluid, cleansing and neutralizing the urethra.

Cowpuncher–See Cowboy.

Cowy–(1) A bull that lacks masculinity. (2) Milk with an off odor or flavor.

Crab Bit–A bit with prongs extending at the horse's nose. The purpose is to tip the horse's head up and help prevent him from ducking his head, bowing his neck, and pulling hard on the reins.

Crab Meal–A marine poultry feed derived from undecomposed, dried waste of the crab industry. The contents are the shell, viscera, and all of the flesh. The product is more than 25% protein and contains up to 3% common salt.

Cracklings–Crispy pieces of rendered pork fat or pork skins.

Cradle–(1) A handheld implement that was once used to harvest grain. (2) A device made of wood or aluminum worn around the neck of a horse, which prevents him from chewing at sores, blankets, bandages, etc.

Craw–The crop of a chicken or bird.

Crawfish–See Crayfish.

Crayfish–Any of the freshwater crustaceans of the genus *Cambarius*. They are raised for food in some parts of the southern United States. Also known as crawfish or crawdad.

Crazy Chick Disease–A vitamin E deficiency of chicks up to eight weeks of age, which causes nervousness and tremors. Affected chicks sit on the hocks, move backwards by pushing their feet, wheel in circles, fall frequently, and sometimes lie prostrate. Also called nutritional encephalomalacia, epidemic tremors.

Cream–The sweet, fatty liquid or semiliquid separated from cows' milk, containing more than 18 percent milk fat. Also called sweet cream.

Cream Cheese–A soft, uncured cheese prepared from milk or a mixture of cream and milk, used as a spread and constituent in salads. A popular soft cheese in the United States.

Cream Layer–The fatty portion of milk that rises to the top after milk has set for several hours in a bottle or other receptacle.

Cream Separator–A centrifugal machine used for the separation of cream from milk. See Centrifuge.

Creamed Honey–Honey made to crystallize smoothly by seeding with 10 percent crystallized honey and storing at about 57°F.

Creamery–A commercial establishment which buys milk and/or cream and prepares finished dairy products for the market.

Creatinine–A chemical compound containing nitrogen, carbon, hydrogen, and oxygen, which is present in the urine and results from the metabolism of protein.

Creep–(1) Area where young piglets spend most of their time, and which has openings too small to allow the sow to enter. See Creep Feeding. (2) A phosphorus-deficiency disease of animals characterized by anemia, softening of the bones, and a slow gait. Also called bush disease. (3) To spread, as a plant, by means of creeping stems, which root at the nodes. (4) Slow mass movement of soil and soil material down relatively steep slopes primarily under the influence of gravity, but facilitated by saturation with water, strong winds, and by alternate freezing and thawing.

Creep Feeding–A system of feeding young domestic animals by placing a special fence around special feed for the young. The fence excludes mature animals but permits the young to enter.

Creep Grazing–As used in Australia, an area of pasture is closed to adult animals but provision is made for access by young stock during the suckling period.

Creeps–See Rickets.

Cremation–The burning or incineration of dead bodies.

Crest–(1) The tuft of feathers on the heads of certain fowls. (2) A cock's comb. (3) The ridge of an animal's neck. (4) The top of a dam, dike, spillway, or weir. (5) The peak or high water mark of a flood. (6) A toothed or irregular ridge of appendage on the petals or flower cluster of certain plants. (7) The top of the hill.

Cretinism–(1) A severe deficiency of thyroid secretion dating from birth or early life, which results in arrested physical and mental development, dwarfism, sluggishness, thickened tongue. (2) Used to describe the physical characteristics of the deficiency. See Hypothyroidism.

Cribber–A horse that has the nervous habit of biting some stationary object, usually the crib, while holding is neck muscles rigid and noisily sucking air into its stomach. The condition is referred to as wind sucking, cribbing, crib biting.

Crimp–The characteristic form or succession of waves in wool fibers; a desirable crimp is close and distinct.

Crimp-wired Comb Foundation–A nonsagging brood foundation for beehives with steel wires imbedded in each sheet of wax to hold the combs straight. It reduces the number of drone cells.

Crimping–Passing a feed crop through a set of corrugated rollers that are set close together.

Criss Crossing–A method of breeding crossbred animals. Brahman bulls are usually bred to cows with British breeding and British bulls are usually bred to Brahman cows.

Crissum–(1) The area surrounding the cloaca on a bird. (2) Feathers around the cloaca.

Critical Element–Essential element for plants, animals, or people.

Critical Temperature–The environmental temperature below which the heat production in a warm-blooded animal increases to prevent a lowering of body temperature.

Critter–(1) Cattle. (2) Any animal or fowl.

Crop–(1) Any product of the soil. In a narrow sense, the product of a harvest obtained by labor, as distinguished from natural production or wild growth. (2) 10,000 boxes of turpentine, a unit of turpentine orcharding. (3) The young fowls or animals bred on a farm. (4) Craw; a saclike enlargement in the esophagus of many birds to store food. (5) Part of a cow's body just behind the shoulder blades. (6) A whole tanned hide. (7) The honey sac of the honeybee. (8) An identification notch or earmark on an animal. (9) To cut: (a) to clip the ears or tail of an animal, especially a dog; (b) to bite off, as a cow crops grass; (c) to cut off the wattles of a bird; (d) to clip the hair of an animal; to shear. (10) To raise plants usually during one season.

Crop Pasture–A sown crop which is normally harvested but may be used for pasture if necessary, as oats, wheat, soybeans, etc.

Crop-bound–Referring to domestic fowl with a distended crop from which food does not pass out in the regular manner. Also called impacted crop.

Crop-eared–(1) Describing an animal with its ears cut. (2) Designating a small-eared horse.

Cropper–(1) A laborer to whom a farmer assigns a definite tract of land together with all or most of the necessary farming equipment, seed, and fertilizer and a share of the crops grown as remuneration for his labor. He is closely supervised by the farmer, receives free use of a tenant house, and ordinarily is given some opportunity to work for wages when his farming operations do not require his attention. Also called sharecropper. (2) A thrown rider.

Cross–(1) A plant or animal which is the result of mating individuals belonging to different species, races, breeds, varieties, etc. See Crossbred; Hybrid. (2) To mate individuals belonging to different species, varieties, races, breeds, etc. See Backcross, Double Cross, Inbred, Self-Pollination, Single Cross, Top Cross. (3) To plow a field at right angles to the first plowing. Also called cross-cultivation.

Cross Mating–The crossing of two or more strains of animals within the same breed.

Cross-fenced–Describing a farm or ranch enclosed by a fence, with fields or pastures fenced off within the enclosure.

Crossbred–Describing an offspring that resulted from the breeding of two purebred parents of different breeds. See Hybrid.

Crossbred Wool—Wool produced from sheep which results from crossing any of the Merino subbreeds with any of the English long-wool breeds. It is as coarse or coarser than half-blood grade (United States and Australia).

Crossbreeding—The pollinating or breeding of plants or animals that belong to different species, races, breeds, varieties, etc.

Crossing-over—The exchange of parts between homologous chromosomes that occurs during the synapsis of the first division of meiosis.

Crosstie—(1) In the United States, a hewn or sawed piece of timber placed beneath railroad rails for support. Also called sleeper, railroad tie. (2) To tie together three of the feet of an animal as it lies on the ground.

Crotch—(1) A fork formed by the separation of two or more branches of a plant or tree. (2) A small sled made from the fork of a tree used to skid logs. Also called alligator, crazy dray, go-devil, travois, lizard. (3) The area between the hind legs of an animal.

Croup—The rump of a horse.

Crown—(1) The upper part of a tree, which bears branches and leaves. (2) To remove the top of a plant or tree. Sometimes called topping. (3) The place at which the stem and root join in a seed plant: the crown or top of the root. See Collar. (4) The upper surface of a furrow slice. (5) The width of the top of a levee. (6) The neck of a sweep or cultivator point used to fasten the sweep to the cultivator shank. (7) The top of an animal's head. (8) The dent in the cap of a kernel of corn.

Crude Fiber—The part of feeds containing the cellulose, lignin, and other structural carbohydrates as determined by the proximate analysis.

Crude Protein—A measure or estimate of the total protein in a feed determined by multiplying the total nitrogen content by 6.25.

Crumbles—Crushed pellets of feed.

Crumbly—(1) A body or texture defect of butter in which the fat particles lack cohesion. Also called brittle. (2) A defect of ice cream which is dry, open, friable, and tends to fall apart when dipped. (3) Describing raspberries that have not matured properly because of injury to the blossom or immature fruit. (4) A body defect of cheese which tends to fall apart when sliced.

Crupper—A leather strap with a padded semicircular loop. The loop end goes under the tail of a horse and the strap end is affixed at the center of the back band of a harness or the cantle of a saddle to prevent the saddle from slipping over the withers.

Crust—(1) A thin, brittle layer on the bare surface of many dried soils, which sometimes prevents plant emergence. This condition is usually caused by raindrop splash. (2) The outer wall of an animal's hoof. (3) Rough, tanned hide of a sheep or a goat. (4) On desert soils, a hard layer that contains calcium carbonate, gypsum, or other binding material often exposed at the surface.

Crutch—(1) The area between the hind legs of a sheep. Sometimes called crotch. (2) To clip or shear wool from around the crutch of a sheep, removing stained or dingy wool.

Cryptorchid—A male animal in which one or both testicles remained in the body cavity and did not descend into the scrotum during embryonic development. The animal is usually sterile. See Ridgeling.

Crystallized Honey—Honey hardened by formation of dextrose-hydrate crystals. Can be reliquified by gentle heat.

Cube—(1) A sale package of butter that ranges in size from 63 to 80 pounds. (2) Small pellet of compressed stock feed. (3) *Lonchocarpus* spp. (especially *L. nicou*), family Leguminosae; a tropical shrub whose roots are a source of rotenone.

Cubed Steak—A retail cut of meat for broiling, prepared from any of several of the tougher cuts and macerated to tenderize it.

Cubing—Processing a feed by grinding and then forming it into a hard form called a cube. Cubes are larger than pellets.

Cubing Machine—Any device that macerates meat to make it tender.

Cuboni Test—A chemical, urine test to determine pregnancy in a mare.

Cuckoo Bees—Bees of families, particularly the Nomadidae, whose larvae are parasites in other bee's hives.

Cuckoo Lamb—A lamb born after the usual lambing season.

Cud—A small wad of regurgitated feed in a ruminant's mouth, which is rechewed and swallowed.

Cud Chewer—A ruminant.

Cull—(1) Anything worthless or nonconforming which is separated from other similar and better items; the act of removing the inferior items; to cull out. (2) The lowest marketing grade of meat carcasses or dressed poultry. (3) Any animal or fowl eliminated from the herd or flock because of unthriftiness, disease, poor conformation, etc.; a reject. (4) A lumbering term for defective or low-grade timber. (5) Any fruit that fails to meet grading specifications because of defects, maturity, conditions, etc.; e.g., ripe berries that have green tips.

Culling Chute—A chute through which sheep pass in single file so that their fleeces and other qualities can be judged so the poorest animals can be culled out.

Culture—(1) The working of the ground in order to raise crops; cultivation; tillage. (2) Attention and labor given to the growth or propagation of plants or animals, especially with a view to improvement. (3) (a) The growing of microorganisms in a special medium. (b) The microorganisms which are so grown. (4) Bacteria used in making dairy and other products. See Mother Culture. (5) Human-made features of an area. (6) The specific way of life of a given society (community, tribe, nation).

Cultured Buttermilk—The product obtained by souring pasteurized skimmed or partially skimmed milk, which contains not less than 8.5 percent nonfat milk solids, by means of a suitable culture of lactic bacteria. This produces a thicker buttermilk than that which is a by-product of butter making.

Culturing—Artificial propagation of pathogenic or nonpathogenic organisms on nutrient media or living plants.

Cup–(1) The receptacle attached to a tree to collect resin in turpentine orcharding. (2) A notch in a tree trunk made by two downward cuts of an ax for inserting a fatal herbicide. (3) A curve across the face of a piece of board. (4) A mechanical object resembling a drinking cup. (5) In lubrication, a vessel or small funnel for receiving oil and conveying it to a machine part, an oil cup. (6) In grain elevators, a bucket or receptacle with a curved outline.

Cup Waterer–A bowl-like watering device. To receive water, the animal pushes against a valve in the bottom of the cup or bowl.

Cuprocide–copper oxide; used to kill algae.

Curative–(1) A remedy used in the cure of diseases. (2) Relating to the cure of diseases.

Curb Bit–A bridle bit so designed that a slight pull on the reins will exert great pressure on the horse's tongue or jaw.

Curd–(1) When milk becomes sour naturally or by the addition of an acid, it flocculates and settles as soft curd of casein and fat with the surrounding liquid consisting of dissolved substances such as sugars and salt, known as whey. The curd is sold as cottage cheese. (2) The coagulated part of milk that results when milk is clotted by rennet, by natural souring, or by the addition of a starter. (3) The edible white head of cauliflower.

Curdled Mlk–The coagulation or thickening of milk resulting from the development of acids by bacterial action or by the addition of rennet. Sometimes called clabber.

Cure–(1) To preserve a product by drying, smoking, pickling, etc., such as hay, meat, or tobacco. (2) To heal; to restore to health.

Cured Forage–(1) Dry range grasses, harvested or standing, slightly weathered, nutritious, and palatable stock feed. (2) Any forage preserved by drying.

Cured Meat–Meat preserved by smoking, and/or soaking in (salt) brine or other suitable solutions.

Curiosity–Any plant or animal grown for its peculiar properties and not for ornamental or economic reasons; for example, dodder, miniature mules.

Curling Comb–A comb used to untangle and curl an animal's hair coat for a stock show.

Current Cross–A plant or animal offspring developed in the first season of cross-mating.

Curry–(1) To comb and dress the coat of an animal to improve its appearance. (2) An East-Indian dish similar to a stew, but characterized by the pungent flavor of curry powder; a mixture of several spices.

Currycomb–A comb with rows of metallic teeth used in dressing the coats of horses and cattle.

Cushion–(1) The inside face of a ham. (2) The fleshy pads on a horse's lips. (3) The frog of a horse's hoof. (4) The mass of feathers on the rear or back of a bird, especially in the Cochin hen.

Cut–(1) A piece of meat prepared for retail or wholesale trade. (2) A slash wound. (3) The opening made by an ax, saw, etc., on a tree. (4) An excavation in the earth, either human-made or natural. (5) The action of a horse's hooves striking its legs or other hooves in walking or running, interfering with its gait. (6) The yield of certain crops, as wool or lumber. (7) (a) An animal separated from the main herd. (b) To separate an animal from the main herd. (8) A severing of the stem or a part of a plant. (9) The output of a sawmill for a given length of time. (10) (a) A reduction in numbers, amount, size, etc. (b) To reduce in numbers, amount, size, etc. (11) To mow. (12) To castrate; emasculate. (13) To sever the jugular vein of an animal or fowl for slaughter.

Cut Back–Animals removed from a higher to a lower classification in sorting for a specific purpose. Also, animals refused by a buyer because of failure to meet specifications.

Cut Cattle–To separate cattle from a herd.

Cut Comb Honey–Bulk comb honey that is cut into pieces, drained, and wrapped for sale.

Cut Out–(1) The soluble solids concentration, on a percentage basis, in the drained juice of canned fruit. (2) To separate an animal from the herd. (3) To dress an animal. (4) To remove cuts of meat from a carcass. (5) To finish shearing (New Zealand).

Cut-up–A high-flanked animal usually slight in body capacity; also referred to as wasp-waisted.

Cutability–An estimate of the percentage of salable meat (muscle) from a carcass versus the percentage of waste fat. Percentage of retail yield of carcass weight can be estimated by a USDA prediction evaluation that includes hot carcass weight, ribeye area, fat thickness, and estimated percent of kidney, pelvic, and heart fat. Also called yield grade.

Cutaneous–Pertaining to the skin.

Cutaneous Abscess–A pocket of pus and the surrounding diseased tissues on the skin.

Cutaneous Myiasis–An infestation of the skin of animals, particularly sheep, by larvae of species of blowflies, of the family Calliphoridae, such as ***Phormia regina, Phaenicia sericata, P. curpina, Callitroga macellaria***, etc. The condition is seen in various parts of the world, especially in warm climates.

Cuticle–(1) A thin layer of cutin that covers the epidermis of plants above the ground, except where cork has replaced the epidermis. (2) One of the four major parts of an eggshell. (3) The exoskeleton of an insect.

Cutlet–A croquette mixture shaped in flat form; a piece of veal or beef round steak cut in round shape, coated with egg and crumbs and braised.

Cuts–Small groups of animals that have been separated from the main herd.

Cutter–(1) A low grade of beef just above canner. (2) A part of a bean harvester consisting of two broad blades set in a wheeled frame at a 60-degree angle, so as to cut two adjacent rows about 2 inches below the surface. (3) In turpentining, a three-cornered steel tool used to sharpen hacks and pullers. (4) The device that chops forage or other plant material preparatory to placing the item in a silo. (5) A sleigh drawn by one horse.

Cutting–(1) Any part that can be severed from a plant and be capable of regeneration. (2) Mowing. (3) Removing a horse or cow from the main herd. (4) Felling a tree. (5) An area on which trees have been or are to be cut. Also called cutting area, felling area, cutover. (6) The striking of a horse's feet against its joints while running.

Cutting Block–A large block of wood standing so that the grain of the wood is vertical; used in butchering.

Cutting Board–A board or table on which something is placed to be cut, such as bulk honeycomb.

Cutting Chute–A narrow passageway into which animals are driven to remove certain ones from the main herd.

Cutting Fat–Fat removed during the trimming or cutting of animal carcasses, as contrasted with offal fat.

Cutting Horse–A trained horse used to separate animals from the main herd.

CWE–Carcass Weight Equivalent.

Cyanosis–Bluish discoloration of the skin.

Cyclic–(1) In chemistry, atoms linked together to form a ring structure. (2) In animals, coming in estrus (heat) at regular intervals.

Cyprian Bee–A honeybee used in bee breeding in the United States. Native to Cyprus.

Cyst–An abnormally developed, closed sac containing a living organism, a foreign body, or fluid; formed to protect the foreign matter, or to protect the plant or animal in which the sac grows.

Cysticerosis–Infestation of the body with a form of tapeworm called cysticercus, which is sometimes present in raw beef. Beef should be cooked at least to the rare done stage (140°F) to avoid danger.

Cystine–A white crystalline amino acid from proteins, necessary to an animal's diet to develop horn, hair, etc.

Cystitis–Inflammation of the bladder.

Cytology–The scientific study of the structure and function of plant and animal cells, often with particular reference to the chromosomes in the cell.

Cytophagy–The engulfing of cells by other cells.

D Ring–In harness making, a piece of metal shaped like the letter D or O to form a link between two separate bands, straps, or other harness devices.

D-activated Animal Sterol–A livestock feed supplement of plant origin supplying vitamin D_3, obtained by activation of a sterol fraction of animal origin with ultraviolet light or other means.

Dairy–(1) (a) A plant in which milk is processed and where dairy products are manufactured and sold. (b) Pertaining to that which is related to the production, processing, or distribution of milk and its products. (2) A place where milk is kept.

Dairy Breed–Any of the breed of cattle especially developed for milk production such as Holstein, Ayrshire, Jersey, Guernsey, Brown Swiss, Dutch Belted, or Red Dane. See Dairy Type.

Dairy Cow Unified Score Card–A system of classifying dairy cattle on the basis of type or body conformation as it relates to their potential for milk production by awarding points based on observations of twenty-seven factors related to general appearance, dairy character, body capacity, and the mammary system, with 100 a perfect score. Developed originally by the Purebred Dairy Cattle Association in 1943, it does not consider production records.

Dairy Herd Improvement Assocation (DHIA)–A cooperative organization of approximately any twenty-five dairy farmers whose purpose is the testing of dairy cows for milk and fat production and the recording of feed consumed. Each farmer receives a monthly record for each of the cows and a complete, yearly summary of production and feed costs.

Dairy Products–(Weight equivalents) Pounds of milk in: 1 gallon, 8.6; 1 quart, 2.15; 1 pint, 1.075. One pound of butter can be made from 21.1 pounds of whole milk; 1 pound of cheese from 10 pounds; 1 pound of nonfat dry milk from 11 pounds.

Dairy Science–(1) The care, breeding, feeding, and milking of dairy cattle, and the production and sale of milk. (2) The formal study of milk production.

Dairy Shorthorn–A type of the shorthorn breed of cattle that originated in northeastern England. A typical cow weighs from 1,200 to 1,350 pounds and is red, white, or roan. It is the principal dairy cow of England. Also called milking shorthorn. See Shorthorn.

Dairy Type–A cow that indicates an ideal conformation for dairy production. The cow is angular; carries no surplus flesh, but shows evidence of good feeding; has a good development of the udder and milk veins; and shows a marked development of the barrel in proportion to its size. See Dairy Breed.

Daisy Cutter–A horse that seems to skim the surface of the ground at the trot. Such horses are often predisposed to stumbling.

Dallisgrass Poisoning–Cattle poisoning caused by grazing on dallisgrass pastures which have ergot growing on the grass seeds. See Ergot.

Dally–To loop the end of a lariat around the horn of a saddle so that the horse may serve as an anchor in roping animals.

Dam–(1) A quadruped, female parent. (2) An artificial structure which obstructs a stream of water for the purpose of water storage, conservation, water power, flood control, irrigation, recreation, etc.

Damp Wool–Wool that has become damp or wet before or after bagging. The wool may then mildew. This weakens the fibers and seriously affects the spinning properties.

Dance, Bee–A performance given by the worker bee upon returning to the hive after an initial search for pollen or nectar, to indicate to other worker bees the location and distance of the pollen-producing plants.

Dapple–A circular pattern in an animal's coat color in which the outer portion is darker than the center.

Dapple Bay–A term used to describe a horse that has a black mane and tail and light chestnut-colored body, or a light chestnut-colored body covered in part by small rings of darker color.

Dapple Gray–A coat-color pattern of some animals, in which the gray color is overlaid with spots of lighter or darker tones.

Dark Meat–The legs and thighs of cooked fowls.

Dark-chestnut–A term used to describe a brownish-black, mahogany, or liver-colored horse.

Dark-dun–A term used to describe the mouse color of an animal's coat.

Daughter–(1) The female offspring. (2) The primary division or first generation offspring of any plant, regardless of sex.

Daughter Cell–(Doter) A newly formed cell resulting from the division of another cell.

Day-old Chick–Designating a chick which is less than a day old. Chicks are commonly at this age from a hatchery.

DDG–See Distiller Dried Grains.

Deacon–A veal calf that is marketed before it is a week old. Also called bob, bob veal.

Deacon Skin–The hide taken from a veal calf less than a week old.

Dead Mouth–Describing a horse whose mouth is no longer sensitive to direction by rein and bit.

Dead Weight–The weight of an animal after slaughter and when all the offal has been removed (virtually the weight of all the salable meat).

Dead Wool–Wool pulled from dead (not slaughtered) sheep. Wool recovered from sheep that have been dead for some time is sometimes referred to as merrin.

Dead-born–Stillborn.

Deaf Ear–(1) A scab of cereals. (2) The fold in the skin of a fowl just below the ear.

Death Camas–Any perennial, poisonous, bulbous herb, *Zigadenus paniculatrs*, family Liliaceae, occurring mostly in the western United States. Cattle rarely eat these plants, but sheep often do. The characteristics of poisoning are salivation, nausea, vomiting, lowering temperature, staggering, prostration, and finally, a coma in which an animal may remain for days, followed by death. Also called onion poisoning, sage poisoning.

Death Loss–The reduction of the number of animals through death as a result of plant poisoning, accident, or disease; as different from reduction by causes such as straying, theft, or sales.

Debeaker–A device for cutting off the tip of the beak of young chickens and turkeys to reduce pecking, cannibalism, and egg eating.

Debilitation–Loss of strength, or a weakened condition.

Decalcification–(1) Removal of calcium carbonate by leaching. A natural process in soil formation. Technically, it is the replacement of calcium ions by monovalent hydrogen cations. (2) The removal of calcium from bones of animals.

Decay–(1) (a) The decomposition of organic matter by anaerobic bacteria or fungi in which the products are completely oxidized. See Putrefaction. (b) To decompose by aerobic bacteria, fungi, etc., whereby the products are completely oxidized. (2) General disaggregation of rocks, which includes the effects of both chemical and mechanical agents of weathering with stress in the chemical effects. (3) Any chemical or physical process which causes deterioration or disintegration.

Decoction–The residue remaining after a substance has been boiled down.

Decoy Hive–A hive placed to attract stray swarms of honeybees.

Dee Race Bit–A type of English riding bit that is used on thoroughbred race horses.

Deep-litter System–The placing of wood shavings, sawdust, straw, etc., mixed with hydrated lime on the floor of a chicken house to about the depth of 6 inches (15 centimeters). The droppings are not removed, although at intervals the litter is stirred and new material added. Normally the litter is changed only when the birds are sold.

Deer Farming–A commercial operation that raises deer for meat or other purposes. Deer farming occurs in New Zealand and other countries of the world. Deer are efficient producers of lean meat and there are few if any religious taboos against eating venison as there are against pork and beef.

Defecation–Voiding of excrement; movement of the bowels.

Defect–(1) Any blemish, fault, irregularity, imperfection in an animal, fowl, or farm product that reduces its usability or impairs its value. (2) In animals or fowls, a departure from breed or variety specification. Because of a tendency to be inheritable, a serious defect is also a disqualification for registration as a purebred.

Defective Wool–Wool that contains excessive vegetable matter such as burs, seeds, and straw, or which is kempy, cotty, tender, or otherwise faulty. These defects lower the value from a manufacturer's viewpoint.

Deferred Grazing–The keeping of livestock from a pasture until there is enough vegetation to support the animals, or in the western range of United States, until the seeds of the herbage have matured.

Deficiency–(1) An insufficiency in reference to amount, volume, proportion, etc.; a lack; a state of incompleteness. The measure of the deficiency can be useful: e.g., the deficiency of the natural flow of a stream in meeting a given irrigation demand determines the storage necessary, the additional supply necessary, or the limitation of the irrigable area. (2) Absence, deletion, or inactivation of a segment of a chromosome.

Deficiency Disease–A pathological condition in plants or animals which results from a deficiency of a nutrient, mineral, or other necessary element in the food supply; e.g., the kind of goiter in humans or animals which results from a deficiency of iodine. Also called hidden hunger.

Deflea–To remove fleas from an animal, frequently by using an insecticide.

Defluorinated Phosphate–(1) A carrier of phosphorus for use in fertilizers or in livestock feeds. It is produced from rock phosphate by heating with water vapor, or by other treatment, to remove fluorine. (2) A phosphorus feed supplement which is ranked below bone meal in value.

Deformity–Any physical deviation from the normal in animals or plants caused by injury or disease.

Degossypolized Cottonseed Meal–Cottonseed meal in which the gossypol (toxic principle of cottonseed) has been removed or deactivated so as to contain not more than 0.04% free gossypol. It must be so designated at the time of sale and must meet the prescribed quality specifications of cotton seed meal to be used as a livestock feed. See Gossypol.

Degras–The commercial designation given to crude wool grease.

Degrease–To remove the wool oil from wool fiber and to obtain the fat from which lanolin is made. See Lanolin.

Degree of Grazing–A term used to define the closeness of grazing. The degrees are: ungrazed, lightly grazed, moderately grazed, closely grazed, and severely grazed.

Degumming Agents–Used in refining fats to remove mucilaginous matter consisting of gum, resin, proteins, and phosphatides.

Dehorn–(1) To remove the horns from cattle, sheep, and goats or to treat young animals so that horns will not develop. (2) To cut back drastically the larger limbs of a tree.

Dehorning Clippers–A device with long handles attached to sharp blades that are used to remove the horns of a beef animal through a scissorlike action.

Dehydrate–To remove most of the moisture from a substance particularly for the purpose of preservation.

Dehydrated Alfalfa Leaf Meal–A ground feedstuff consisting chiefly of alfalfa leaves. It must be reasonably free from other crop plants and weeds and must not contain more than 18 percenet crude fiber. The freshly cut leaves are artificially dried in such a manner that a temperature of at least 212°F (100°C) is attained for a period of not more than 40 minutes.

Dehydrated Alfalfa Stem Meal–A feedstuff which is the product remaining after the separation of the leafy material from alfalfa hay or meal. It must be reasonably free from other crop plants and weeds. The freshly cut stems are artificially dried in such a manner that a temperature of at least 212°F (100°C) is attained.

Dehydrated Soybean Hay Meal–A feedstuff which is the product obtained from the artificial drying and grinding of the entire soybean plant, including the leaves and beans, but not any stems, straw, or foreign material. It must be reasonably free from other crop plants and weeds and must not contain more than 33 percent crude fiber. It must have been artificially dried when freshly cut.

Delactation–(1) Cessation of giving milk. (2) Cessation of suckling young; weaning.

Delaine Wool–Fine combing wool, usually obtained from the state of Ohio.

Deleterious–Injurious.

Deletion–Absence of a segment of a chromosome involving one or more genes.

Deliming–The soaking of an animal hide in a sulfuric acid bath to remove the lime used in dehairing the hide.

Delousing–The extermination of lice by insecticides.

Demaree–In honeybee keeping, a method of swarm control, by which the queen is separated from most of brood; devised by a man by that name.

Demiluster–Wool that has some luster but not enough to be classified as luster wool. Wool of this type is produced by Romney and similar breeds. Same as semiluster.

Denature–(1) To make a product unfit for human consumption without destroying its value for other purposes: e.g., denatured alcohol. (2) To change the properties of a protein, as to coagulate egg white.

Density–(1) Mass per unit volume. (2) The number of wild animals per unit of area. (3) The degree of closeness with which wool fibers are packed together. (4) In forestry, density of stocking expressed in number of trees, basal area, volume, or other criteria on an acre or hectare basis. See Stocking.

Dental Cup–A small dark-colored depression in the wearing surface of the incisor teeth of horses, having a rim of hard, glistening, white enamel, which disappears as the teeth wear down. Also known as dental star.

Dental Pad–A hard pad in the upper mouth of cattle and other animals that serves in the place of upper teeth.

Denticulate–(1) Diminutive of dentate; with small teeth of the dentate type. (2) A leaf margin that is similar to dentate, but has smaller "teeth."

Dentition–The character, arrangement, and number of teeth in an animal; the formation and growth of teeth.

Denutrition–Lack or withdrawal of nutrition; the failure to transform food elements into nutritional substances.

Deoxyribonucleic Acid (DNA)–A genetic proteinlike nucleic acid on plant and animal genes and chromosomes that controls inheritance. Each DNA molecule consists of two strands in the shape of a double helix. Most inheritance characteristics can be predicted but some cannot, because some genes "jump" (are promiscuous). Such genes can result in resistance to pesticides.

Depauperatum–Stunted.

Depilate–To remove the hair from the skin.

Depilatory–(1) A substance which is applied to the flesh side of pelts to loosen fibers from the skin. (2) An agent for removing hairs from the living body.

Depot Fat–Fat that is accumulated or stored in the body.

Depraved Appetite–A craving in people, animals, and fowls for items not normally eaten by them which may be caused by a diet deficiency. This may include eating clay soil. See Geophagia, Pica.

Depression–(1) A slight following in the flesh of an animal. (2) A hollow in the surface of the land. (3) A severe drop in income and prices for an area or for a nation. (4) The Great Depression denotes the economic reversal that took place in the United States in the 1930s, which was associated with the Stock Market Crash in October 1929.

Dequeen–To deprive a colony of its queen bee.

Dermal–Of the skin; through or by the skin.

Dermal Toxicity–Ability of a chemical to cause injury when absorbed through the skin.

Dermatitis–A nonspecific skin condition affecting both people and animals, characterized by inflammation or infection of the skin. Animals with light-colored skin are most sensitive, although dermatitis may result from external parasites, irritation chemicals or plants, sunlight, or nutritional deficiencies. See Eczema.

Dermatomycosis–Any skin infection caused by a fungus, for example, ringworm.

Dermatophytes–Fungi that infect only the skin.

Dermatosis–Any skin disease.

Dermis–The major layer of the skin, which is located just under the epidermis.

Desiccate–To dry out; to exhaust of water or moisture content.

Determinant Growth–The type of growth that stops in an animal after it reaches a certain age.

Deutectomy–The removal of the yolk sac from newly hatched chicks.

Deutoplasm–The nutritive material or yolk in the cytoplasm or vitellus of an ovum or egg.

Devon–A red, dual-purpose breed of cattle, originating in Devonshire, England. It is one of the oldest English breeds. In the United States, found primarily in New England.

Dewclaw–A vestigial digit on, or just above, the foot of a quadruped, such as on cattle, which does not touch the ground as the animal stands or walks. Also called false hoof.

Dewlap–The pendulous skin fold hanging from the throat of any animal, particularly a member of the ox (cattle) tribe and certain fowl.

Dextrorotatory Honey–Any honey that contains honeydew and whose solutions cause a plane of polarized light to rotate to the right (clockwise).

DHIA–See Dairy Herd Improvement Association.

DHIA Records–Production records of cows tested under the supervision of the tester of a Dairy Herd Improvement Association.

Di-hybrid–An individual who is heterozygous with respect to two pairs of genes.

Diacetyl–The aromatic chemical compound developed in the ripening of cream for butter which is the chief constituent in the aroma of butter.

Diagnosis–The process of identifying a disease by examination and study of its symptoms.

Diagnostic Antigens–Biological agents which are prepared for use in the diagnosis of a specific disease; e.g., tuberculin, which is injected under the skin of cattle to determine if they have tuberculosis. It has little or no effect on healthy subjects, but in an infected animal, it will cause inflammation of the skin around the injection and a possible rise in temperature.

Diagnostician–Any person, especially one trained in diagnosis, who determines the nature and cause of a disease or abnormality in plants or animals and prescribes a treatment.

Diakinesis–A stage of meiosis just before metaphase of the first division: the homologous chromosomes are associated in pairs near the nucleus and have undergone most of the decreases in length.

Diallel Crossing–In animals, a method of progeny testing in which two males, bred at different times to the same females, are compared with the two sets of progeny.

Diamond Skin Disease–See Swine Erysipelas.

Diaphragm–The layer or sheet of muscle and connective tissue that forms the wall between the thoracic and abdominal cavities of mammals and aids in the process of breathing.

Diarrhea–Frequent and profuse fluid defecation commonly caused by an infection of the gastrointestinal tract.

Diastase–Enzyme that aids in converting starch to sugar.

Dickey–A slang term for a donkey or a small bird.

Dicoumarol–An anticoagulant produced by molds on spoiled clovers.

Diecious–(1) Animals that are either male or female; i.e., each individual animal has either male or female reproductive organs but not both. (2) Plant species with male and female organs on separate plants.

Diestrus–The period of the estrous cycle that occurs between metestrus and proestrus.

Diet–The type and amount of food and drink habitually ingested by a person or an animal. See Ration.

Dietary Fiber–Generic name for plant materials that are resistant to the action of normal digestive enzymes.

Diethylstilbestrol (DES)–A synthetic estrogenic hormone that has been used to stimulate faster growth and the deposition of additional fat in steers on feed.

Differentiation–(1) The development of different kinds of organisms in the course of evolution. (2) The development or growth of a cell, organ, or immature organism into a mature organism.

Digest–(1) To convert food within the body into a form that can be assimilated. (2) To soften, dissolve, or alter any substance by heat and chemical action.

Digester Tankage–A feed supplement or fertilizer made from finely ground, dried residue of animal tissues which has been placed in a tank under live steam. It does not include hair, horn, hoof, manure, or the contents of an animal's stomach, except in small, unavoidable traces not in excess of 0.05 percent of any of these. When this product contains more than 4.4 percent phosphorus, the word "bone" must appear as part of the brand name. If more than 5 percent of this product cannot be passed through a 2-millimeter screen, the words "coarsely ground" must be published. Also called meat meal tankage, feeding tankage.

Digester Tankage with Bone–Digester tankage that contains more than 4.4 percent phosphorus. Also called meat and bone meal, digester tankage, feeding tankage with bone.

Digester Tankage with Paunch Contents–Digester tankage to which the processed contents of an animal's stomach have been added.

Digestibility–(1) The readiness with which a substance may be converted into an absorbable form within the body. (2) The rate or amount of a nutrient digested; i.e., the difference between the amount of a nutrient fed and the amount found in the feces, but since some of the nutrient voided in the feces may not have been a part of the nutrient fed, this is considered apparent digestibility. When the amount of a nutrient digested is converted to a percentage, it is called the digestion coefficient.

Digestible–(1) That feed consumed and digested by an animal as opposed to that which is evacuated by the animal. (2) Feed which can be converted into an absorbable form within the body.

Digestible Crude Protein–The total nitrogenous compound protein fed to livestock less the nitrogenous compounds eliminated in the feces.

Digestible Energy (DE)–The proportion of energy in a feed that can be digested and absorbed by an animal.

Digestible Nonnitrogenous Nutrient–The total digestible nutrients less the digestible protein in a feedstuff.

Digestible Nutrient–That portion of a nutrient that may be digested and absorbed into a human or an animal body.

Digestible Protein–The proportion of protein in a feed that can be digested and absorbed by an animal; usually 50 to 80 percent of crude protein.

Digestion–(1) The changes that food undergoes within the digestive tract to prepare it for absorption and use in the body. (2) The reduction of organic matter by biochemical action into more stable organic matter; e.g., the manufacture of sludge from sewage.

Digestion Coefficient–The amount of a given feed that is digested by an animal expressed as a percentage of the gross amount eaten.

Digestion Trial–An experiment with any feed by an animal or fowl to determine the amount of any substance that can be digested.

Digestive Tract–The mouth, esophagus, digestive organs; stomach or stomachs, crop, gizzard, and the small and large intestines, and anus; all of the organs of an animal or fowl through which food passes.

Dilate–To enlarge, expand, to open.

Diluent–Any gaseous, liquid, or solid inert material that serves to dilute or carry an active ingredient, as in an insecticide or fungicide. Also called carrier.

Dilute–(1) To make more liquid by mixing with water, alcohol, etc. (2) To weaken in flavor, brilliancy, force, strength, etc., by mixing with another element.

Diluted Color–A term applied to feather color in poultry. It refers to soft colors such as light tan, cream or buff, light yellow, and light blue.

Diluted Feed–A feed that has a high concentration of roughage or fiber.

Dingy–Wool that is dark or grayish in color and generally heavy in shrinkage.

Dioestrum–That part of female's cycle between periods of estrus (heat).

Dip–(1) Any chemical preparation into which livestock or poultry are submerged briefly to rid them of insets, mites, ticks, etc. See Dipping Vat. (2) Any preservative preparation in which produce is briefly submerged. (3) To remove ice cream from a container with a dipper. (4) To collect oleoresin from a cup in turpentine orchards.

Diphtheria–A highly contagious bacterial disease of mammals caused by the presence of *Corynebacterium diphtheriae*; characterized by the fever, heart weakness, anemia, and prostration; often fatal.

Diploid–(1) Having one genome comprising two sets of chromosomes. Somatic tissues of higher plants and animals are ordinarily diploid in chromosome constitution in contrast with the haploid (monoploid) gametes. (2) An organism or cell with two sets of chromosomes, for example, worker and queen honeybees.

Dipping–(1) A method of preserving seasoned wood by submersion in an open tank of creosote or similar preservative. (2) The submerging of animals or fowls in a liquid bath which contains insecticides, ovicides, repellents, etc.

Dipping Vat–A pit filled with a liquid containing insecticides, ovicides, repellents, etc., through which animals are forced to pass for disinfestation. It is concrete-lined or otherwise waterproofed, rectangular, but so narrow that an animal is unable to turn in it. At the entrance it has a vertical drop of five or more feet while at the exit it has a cleated ramp.

Dipteron–An insect of the order Diptera which usually has two wings. Many, such as the mosquito and tsetse fly, are pests because of their ability to bite and to spread diseases.

Direct Cut Silage–Plants that are cut and chopped for silage in a single operation.

Discharge–(1) The quantity of water, silt, or other mobile substances passing along a conduit per unit of time; rate of flows. (2) An exudate or abnormal material coming from a wound or from any of the body openings: e.g., a bloody discharge from the nose. (3) To remove the electrical energy from a battery.

Discolor–Different colors; off color.

Discriminate Breeder–An animal that will only breed with a certain mate.

Disease–Any deviation from a normal state of health in plants, animals, or people which temporarily impairs vital functions. It may be caused by viruses, pathogenic bacteria, parasites, poor nutrition, congenital or inherent deficiencies, unfavorable environment, or any combination of these.

Disease Control–Any procedure that tends to inhibit the activity or effect of disease-causing organisms, or which modifies conditions favorable to disease.

Disease Resistant–Designating plants, animals, or people not readily susceptible to, or able to withstand, a particular disease.

Disinfect–To destroy or render inert disease-producing microorganisms and to destroy parasites.

Disk–(1) A round, usually sharp-edged, slightly dished, steel plate that cuts the soil as it revolves on a center axis and moves the soil to one side. Several disks are mounted and spaced on a horizontal shaft to make the disk harrow. Also called disk blade. See Disk Harrow. (2) One of a series of metal plates in a centrifugal separator bowl that increases the efficiency of the machine. (3) In botany, (a) an adhesive surface on the tendril ends of creeper plants that enables them to climb along flat surfaces; (b) an enlargement of a flower's receptacle; (c) the center of a composite flower; (d) an organ's surface; (e) the circular valve of a diatom. Also spelled disc.

Dismount–To get down, alight, as from a horse's back.

Disorder–An unwholesome or unnatural physical condition of a plant or animal. See Disease.

Disposition–The temperament or spirit of an animal.

Disqualification–In animal husbandry, a defect of characters of form or breed types that disqualifies an animal from an exhibition or from breed registration.

Dissect–To cut or divide a plant or animal into pieces for examination.

Dissolved Bone–A fertilizer material that consists of ground bone or of bone meal that has been treated with sulphuric acid to make the phosphorus in the bone more readily available to growing plants.

Dissolved Oxygen–The amount of elementary oxygen present in water in a dissolved state. It is commonly reported in parts per million (by weight), or milligrams per liter, or percentage of saturation, of oxygen in the water. Dissolved oxygen is essential for fish and other aquatic life and for aerobic decomposition of organic matter. Dissolved oxygen in surface water bodies should be maintained at a level above the threshold of 3 mg/1 and an optimum of 5 mg/1 for most species of fish.

Distal–Located in a position that is distant from the point of attachment of an organ; for example, the toes are located on the distal part of the leg.

Distemper of Dogs–An acute, widespread, very contagious disease of dogs, especially of young ones, caused by a filterable virus which may cause death. Also called dog disease, canine distemper.

Distemper of Horses–See Strangles.

Distillers' Dried Grains (DDG)–The dried distillers' grains by-product of the grain fermentation process that may be used as a high-protein (28%) animal feed. See Distillers' Grains.

Distillers' Dried Yeast Molasses–The properly dried yeast resulting from the fermentation of molasses and yeast that is separated from the medium, either before or after distillation.

Distillers' Grains–A by-product livestock feed obtained from the manufacture of alcohol and distilled liquors from corn, rye, and sometimes, a mixture of rice and other cereals.

Distributor–(1) Any device that is used to move produce from one place and to scatter it in another; e.g., a fertilizer distributor. (2) An agent or a wholesaler who sells goods in quantity. (3) A device in some bulk milk tanks that spreads the milk over a cooling surface. (4) That part of a motor engine that conducts the secondary current to the spark plugs.

Diuresis–An increased production of urine.

Division Board–Flat board used to separate two colonies or colony of bees into two parts.

Division Board Feeder–A wooden or plastic trough which is placed in a beehive in a frame space to feed the colony honey or sugar syrup.

Division Screen–A wooden frame with two layers of wire screen that serves to separate two colonies of bees within the same hive, one above the other.

Dizygotic Twins–Twins that develop from two separate fertilized ova.

DNA–See Deoxyribonucleic Acid.

Dobbin–An affectionate designation for a gentle horse.

Docile–Refers to an animal that is gentle in nature.

Docile Temperament–Easily managed disposition of an animal.

Dock–(1) To cut short the tail of an animal (most commonly a lamb); usually for sanitary reasons and to facilitate breeding in females. (2) The area around the tail of sheep or other animals. (3) Any plant of the genus *Rumex*; a serious weed pest. (4) A leather case to cover a horse's tail when clipped or cut.

Dockage–(1) Foreign material in harvested grain such as weed seeds, chaff, and dust. (2) The weight deducted from stags and pregnant sows to compensate for unmerchantable parts of an animal.

Doddie–A polled cow.

Doe–An adult female goat, rabbit, or deer.

Dog–(1) Canis *familiaris*, family Canidae; a domesticated, carnivorous animal which is used as a household pet, a watchdog, a herder for sheep, cows, etc. (2) A device, usually consisting of a steel hook and chain, used in skidding logs. (3) A sort of iron hook or bar with one or more sharp fangs that may be fastened into a piece of wood or other heavy article to move it. (4) Any part of a machine acting as a claw or clutch, as an adjustable stop to change the direction of a machine tool. (5) A low-qualitiy beef animal.

Dog Flea–*Ctenocephalides canis*, family Pulicidae; an insect that infests people, rodents, dogs, cats, etc. It is a possible vector for bubonic plague and is an alternate host for the dog tapeworm and the rodent tapeworm, which occasionally parasitizes people.

Dog Follicle Mite–*Demodix canis*, family Demodicidae; a mite infesting dogs which causes red mange.

Dogie–(dogy, dogey, doge) A term used on western ranges in the United States for a motherless calf.

Dogtrot–A slow, gentle trot.

Domestic Fowl–*Gallus domesticus*, family Phasianidae; the chicken.

Domestic Wool–Wools grown in the United States as against foreign wools.

Domesticate–To bring wild animals under the control of humans over a long period of time for the purpose of providing useful products and services; the process involves careful handling, breeding, and care.

Dominance–(1) The tendency for one gene to exert its influence over its partner, after conception occurs and genes exist in pairs. There are varying degrees of dominance, from partial to complete to overdominance. (2) In forestry, the relative basal area of a species to the total basal area of all species in an area. The species having the highest relative basal area is considered the dominant species (syn.: predominant). (3) A term used in various contexts (e.g., in farm programming, input-output budget analysis and decision analysis) to indicate that one alternative is superior to another in the sense of producing higher benefits (output) with equal or lower costs (inputs).

Dominance (Social)–The tendency of one animal in a group to exert its social influence or presence over others in the group. Also referred to as social order or pecking order.

Dominant–(1) In genetics, designating one of any pair of opposite Mendelian characters that dominates the other, when both are present in the germ plasm, and appear in the resultant organism. (2) Designating a plant species that characterizes an area.

Dominant Gene–A gene that prevents its allele from having a phenotypic effect. See Recessive Gene.

Dominique–An American Breed of medium-weight, dual-purpose chickens. The plumage is salt colored, crossed by dark and light bars. The egg shell is brown.

Donkey–(1) A portable engine. (2) An ass.

Dorsal–Refers to the back, or toward the back; the opposite of ventral.

Dorset–(1) A medium-sized breed of sheep originating in southern England. (2) A hard, blue-veined, English cheese made from partly skimmed cow's milk cured with rennet. See Rennet.

Dosage Response Curve–In veterinary medicine, the amount of medicine containing a given chemical as plotted against the killing power of that chemical.

Dose–Proper amount of a medicine to be given at one time.

Dose Syringe–A syringe with a long pipe or nozzle that is used to force medicine down the throats of animals and poultry and to wash out the throats or crops of fowls.

Double–(1) Designating a blossom that has more petals and sepals than normal, usually as a result of plant breeding. Double flowers are often highly prized. (2) Designating a team of horses; e.g., to work horses double.

Double Bit–A set of two bits attached to a horse's bridle: one for guiding the animal, the other for controlling it. The set consists of a curb bit and a straight bit.

Double Cover–Coitus of a mare, cow, or other female animal with a male on successive days to ensure conception. See Double Mating.

Double Fertilization–The usual fertilization in plants, in which the egg nucleus is fertilized by one sperm nucleus and the two fused polar nuclei by another sperm nucleus.

Double Immune–Designating a hog that has been vaccinated with both live virus and antihog cholera serum for protection against hog cholera.

Double Mating–(1) A rarely used breeding practice with Barred Plymouth Rocks to produce birds of exhibition type. (2) The mating of female livestock twice during a given estrous period to the same or different males to ensure conception. See Double Cover.

Double Muscling–An abnormal condition in some beef animals. The term is a misnomer for an undesirable genetically controlled display of gross enlargement of all of the muscles in the animal's body, most noticeably demonstrated by bulging muscles in the round and shoulder. The tail head is set forward and the body is shallow.

Double Reins–Two sets of reins attached to a horse's double bit.

Double-decked Car–A railway car or truck having two main floors, one above the other. It is used for the shipment of sheep, hogs, calves, etc.

Double-rigged Saddle–A saddle that has two cinches.

Double-snaffle Bit–A severe bit for a horse's bridle which consists of two bar bits, each hinged on opposite sides of the middle.

Dourine–A parasitic, contagious disease of the horse and ass caused by *Trypanosoma equiperdum* as a result of infection during coitus. It is characterized in the stallion or jack by a swelling of the prepuce which in time spreads to the scrotum and abdomen, and the testicles become enlarged and sensitive. The penis extends from the sheath, is swollen, and discharges a yellowish fluid. Blisters on the penis and sheath rupture and result in raw ulcers. There is repeated desire to urinate. In the mare or jenny, it is characterized by swollen, external genitals, a gaping vulva, and clitoris being constantly erect. The mammary glands may be swollen. The female is uneasy, switches her tail, voids small quantities of urine, and appears to be in heat. A discharge similar to the male's is evident. Blisters on the inner and outer mucous membranes of the genitals rupture to leave raw ulcers. Both sexes become nervous. The skin lesions appear in the form of urticarial wheals similar to those in hives. The hindquarters may become paralyzed. Also called dollar plague, breeding paralysis, genital glanders, equine syphilis.

Down–(1) A sandhill. See Dune. (2) The soft, furry or feathery covering of young animals and birds that occurs on certain adult birds under the outer feathers and which is used for stuffing pillows, furniture, and comforters, its most important source being the eider duck. (3) The pappus of a plant. (4) Used to describe the individual crop plants that have been blown or forced to the ground by rain, hail, wind, etc.,; e.g., down corn, down wheat.

Down Breeds–One of two English types of sheep that produce a small sheep of the medium wool class. It includes the common breeds: Southdown, Shropshire, Hampshire, Suffolk, Oxford Down, and Dorset Horn.

Down Crossbreed–A sheep that results form crossing a Merino with any of the Down breeds.

Down Wool–Wool of medium fineness produced by such down breeds as Southdown and Shropshire. These wools are lofty and well suited for knitting yarn.

Downer–(1) An animal which, in transit in a truck or railroad car, has fallen or lain down. (2) A diseased animal unable to get up.

Downy–Covered with very short, weak, soft hairs.

Draft–(1) A package deduction, one pound per hundred pounds, allowed a buyer of wool. (2) The horizontal component of pull of an implement parallel to the line of motion. Also called directional pull. (3) Similar to a check but instead of being charged to a signer it is charged to a third person named on the face of the draft. (4) Any feedstuff obtained as a by-product of the distillation of grains.

Draft Animal–Animal used for work stock and, in some countries, is still used for plowing and pulling heavy loads.

Draft Horse–Any horse which, by reason of its size and weight, is used for pulling heavy loads, and which usually weighs over 1,500 pounds and stands up to 17½ hands high. Breeds include: Belgian, Clydesdale, Shire, Suffolk. See Hand.

Drafting Race–A narrow passageway through which sheep can be driven in order to divide them into two or three groups. See Race.

Draining Pen–A pen with a sloping concrete floor adjoining a dipping vat where dipped animals stand while excess fluid drains back into the vat.

Drake–A mature male duck.

Drape–A term applied to a cow or ewe incapable of bearing offspring, especially to such an animal selected for slaughter.

Drawing–(1) The removal of cuttings from a propagation bed. (2) The eviscerating of poultry.

Drawn Comb–Comb having the cells built out (drawn) by honeybees from a sheet of foundation. Cells are about one-half inch deep.

Drench–(drensh) A fluid dose of medication given by introducing it into the mouth, usually with a dose syringe. Frequently used when attempting to remove parasites from the stomach and intestines of people and animals.

Dress–(1) To curry and brush an animal. (2) To remove the feathers and blood from a bird that has been killed. (3) To plane the surface of a board.

Dress Out–(1) To remove the feathers or skin and to cut up and trim the carcass of an animal after slaughter. (2) The percentage of carcass weight to live weight.

Dressage–Designating a horse that has been trained to perform on almost unnoticeable signals by the rider. Also called high school, haute ecole.

Dressed Weight–(1) The weight of a sheep, cow, or hog carcass. (2) The weight of a bird that has had its blood and feathers removed but not its intestines, head, or feet.

Dressing–(1) An application of medicine or bandage made to a wound on a person, animal, fowl, or plant. (2) An application of manure, fertilizer, mulch, etc. See Side Dress, Top Dressing. (3) The preparation of nursery stock for cutting, budding, grafting, etc. (4) The removal of feathers and blood form a bird. (5) The trimming of excess fat and bone form a meat carcass. (6) The external treatment of cheese, including: (a) the assistance given in forming a coat or rind; (b) improvement given to the external appearance of a cheese; (c) the prevention of loss of moisture from a cheese; (d) protection from external microorganisms.

Dressing Comb–A special comb used to prepare an animal's coat for a show. Also called currycomb.

Dressing Defect–A defect of dressed poultry characterized by one or more of the following: pin feathers left on the carcass; incomplete bleeding; a cut, tear, or abrasion in the skin longer than two inches; a broken bone; feed left in the crop; dirty feet; dirty body; dirty vent.

Dressing Loss–The loss in weight between a live animal and its dressed carcass.

Dressing Percent–Carcass weight divided by live weight and multiplied by 100. Usually the cold carcass weight is used. The dressing percentage for cattle averages around 50 to 60 percent; Hogs average around 70 percent

Dressing Weight–The weight of a dressed animal or bird in contrast to its live weight.

Dried Blood–The collected blood of slaughtered animals, dried and ground. It is sold as a fertilizer and contains 8 to 14% nitrogen in organic form (50 to 87.5% crude protein). The better grades are used as an ingredient of feed. See Blood Meal.

Dried Egg Albumen–Preserved egg albumen obtained by the separation of the albumen (white) from the yolk, followed by the bacterial fermentation of the albumen and the drying of the albumen by heat.

Dried Eggs–Preserved eggs obtained by mixing the white with the yolk and removing all but 3 to 5 percent of the moisture by heat.

Dried Manure–See Manure.

Dried Milk–The product resulting from the removal of water from milk. Also called desiccated milk.

Dried Milk By-product–A designation that may be used in the list of ingredients of a mixed feed to indicate the presence of dried skimmed milk, dried buttermilk, or dried whey, or a blend of two or more of these products.

Dried Skimmed Milk–(1) A livestock feed, especially suited for calves, piglets, and chickens, which is produced by evaporating water from clean, sound skimmed milk, and which contains not more than 8% moisture (approximately 34% protein). (2) Also a human food.

Dried Whey–A by-product from the manufacture of cheese or casein that consists of at least 65% lactose (milk sugar) and approx. 12% protein; used as a livestock feed.

Drifting–The return of field bees to colonies other than their own.

Drinking Cup–A metal or porcelain cup just large enough for an animal's muzzle, within which is a valve or float, activated when the animal pushes its muzzle against it.

Drinking Hole–Waterhole.

Drippings–Fat that liquefies during cooking of meat, etc., and runs to the bottom of the pan.

Drive–(1) (a) The moving of livestock under human direction; (b) To herd livestock. (2) In wildlife management, the herding of wild animals past a particular point for counting or shooting. (3) The process of floating logs down a river from a forest area to a mill or to a shipping point. (4) The means for giving motion to a machine part or machine, as a drive chain, a drive wheel.

Drive-ins–Designating those cattle that are herded in to market, as contrasted to those transported there by truck, etc.

Driveway–(1) In the range country, that land which is set aside for the movement of livestock from place to place; e.g., from the home range to the shipping point. Also called stock driveway, stock route. (2) A path or passage, sometimes paved, for the movement of vehicles and/or livestock. (3) The farm road from the building site to the gate at the highway. (4) Vehicle passage into or through a barn.

Drone–The male honeybee hatched from an unfertilized egg. It is larger than a worker bee, does not gather honey, and has no sting.

Drone Comb–Comb with about four cells to the inch and in which drones are reared.

Drone Layer–A queen that lays only unfertilized eggs which always develop into drones. Results from improperly or nonmated queen or an older queen who has run out of sperm.

Drop–(1) A structure in an open or closed conduit that is used for dropping the water to a lower level and dissipating its kinetic energy. (2) A decrease in height or elevation. (3) Any fruit that falls form a tree to the ground because of wind, or other conditions; e.g., diseased, immature, and usually unfertilized fruit; also normal but ripe fruit. (4) A fungal disease of vegetables caused by ***Sclerotinia sclerotiorum***, ***S. minor***, or ***S. intermedia***, family Sclerotiniaceae. (5) To give birth to young, as to drop a calf. (6) To shoot an animal and to cause it to fall.

Drop Band–A group of ewes or nanny goats that is managed separately at the season of bearing young. Also called drop herd.

Droppings–The excrement of animals and birds. See Guano, Manure.

Dropsy–(1) An excessive, intensive, watery enlargement of parts of a plant sometimes due to a superabundance of moisture in the soil. Also called edema. (2) Widespread edema (swelling) throughout the body of an animal.

Drosophila–(Greek; *drosos*, dew; *philein*, to love) A genus of fruit flies common around decaying fruit; used extensively in experimental genetics. One of the most common is ***Drosophila melanogaster***.

Drove–A collection or mass of animals of one species, as a drove of cattle.

Drug Residue–Any amount of drug that is left in an animal's body.

Drum–(1) A cylinder-shaped package or product, as a drum of cheese. (2) A revolving cylinder, as the drum of a seed cleaner. (3) To knock on the sides of a hive to drive the bees upward when transferring them from one hive to another.

Drumstick–The leg of a fowl from the hock joint to the joint next above.

Dry–(1) (a) To cause a pregnant cow to stop giving milk shortly before she drops her calf; (b) Designating a cow who has ceased to give milk shortly after she drops her calf. (2) To preserve a product by dehydration. (3) Designating dressed poultry packed for shipment with dry ice. (4) Designating a time or place of abnormal lack of precipitation, e.g., a dry year. (5) describing a well that yields no water.

Dry Band–A band of sheep without lambs.

Dry Basis–Designating a product, e.g., soil, fertilizer, or feed, which is analyzed for its constituents calculated on the basis of oven-dried material.

Dry Cow–A cow that has ceased to give milk.

Dry Cure–A method of curing pork by rubbing the meat with salt, sugar, and saltpeter, and allowing it to age.

Dry Lot–A bare, fenced-in area used as a place to keep livestock for feeding and fattening.

Dry Matter–The total amount of matter, as in a feed, less the moisture it contains. Dry feeds in storage, such as cereal grains, usually contain about 10 percent water and 90 percent dry matter, wet weight basis.

Dry Off–(1) To reduce the amount of water given a plant, bulb, tuber, or corm, until it becomes dormant and ready to store. (2) To bring the lactation period of a cow to an end.

Dry Period–That time just before parturition when a cow or other lactating animal ceases to give milk.

Dry Pick–To dress a bird without scalding it. It may be bled by cutting the jugular vein, and its medulla oblongata punctured to relax the feather follicles to facilitate plucking.

Dry Rendered–The residues of animal tissues cooked in open, steam-jacketed vessels until the water has evaporated. The fat is removed by draining and pressing the solid residue.

Dry Salt Cure–A method of preserving pork, especially fat back, in which salt is sprinkled heavily over the pieces of meat that are then piled one on top of another. The salted meat is cured from three weeks to a month.

Dry Weight–The weight of a product or material less the weight of the moisture it contains; the weight of the residue of a substance that remains after virtually all the moisture has been removed from it. Also called dry matter.

Dry-packed–Slaughtered poultry that is packed dry and cooled without ice coming in direct contact with the carcasses.

Drying–Reducing the moisture content of a product to the point at which the concentration of the dissolved solids is so high that osmotic pressure will prevent the growth of microorganisms.

Drying-off–The gradual cessation of milk production by the mother after her young have been weaned.

Dual-purpose–Designating a domestic fowl or animal that is bred for two purposes; e.g., a chicken raised for both meat and egg production; a cow grown for milk and meat.

Dual-use Range–A range containing a forage combination of grass, forbs, and browse that allows two or more kinds of stock (such as cattle and sheep) to graze the area to advantage at the same time through the entire season, or separately, during a part of the season.

Dub–To cut the combs and wattles from cockerels when they are twelve to sixteen weeks old, to prevent injury from freezing or from fighting.

Duck–(1) A bird, family Anatidae, especially the domesticated duck raised for meat and eiderdown. (2) The female duck. See Drake. (3) A heavy cotton or linen cloth.

Duckling–A young duck that still has downy plumage.

Ductless Glands–Those glands of the body whose secretions pass directly into the bloodstream or into the lymph, as one of the endocrine glands.

Dumb Rabies–Paralysis that develops in the last stages of rabies. Also called drop jaw. See Rabies.

Dumba–The fur or skin of the Karakul lamb.

Dummy–(1) A sleepy or stupid horse, especially one that has suffered inflammation of the brain. (2) A strongly built frame suggesting the shape of a female animal, often covered with the hide of such a female, which is used to excite a male animal sexually so that it will mount the dummy and accept the use of an artificial vagina so that the semen may be collected for later use in artificial insemination.

Dumpy–(1) Designating an animal or fowl that does not feel very well, but that is not seriously ill. (2) An animal that is short in length and height and is fat.

Dun–An animal's color characterized by a dark dorsal stripe over the withers and shoulders. Recognized shades are mouse dun, buckskin dun, and claybark dun.

Dung–Manure; the feces or excrement of animals and birds. See Guano, Manure.

Dung Locks–On sheep, britch wool locks that are encrusted in hardened dung.

Duodenum–(1) In birds, the part of the small intestine nearest the gizzard. (2) In mammals, the part of the small intestine nearest the stomach.

Duplicate Genes–Factors with the same recessive phenotypic expression.

Durham Cattle–See Polled Shorthorn.

Duroc–A prolific breed of hogs that grow rapidly; red, with drooping ears. Formerly called Duroc-Jersey.

Dust–(1) Fine, dry particles of earth, or other matter, so attenuated that they may be wafted by the wind. (2) Particles less than 0.1 millimeter in diameter that may be bounced off the ground by sand grains moving in saltation. See Saltation. (3) An insecticide, fungicide, etc., which is applied in a dry state, or the application of these to plants, animals, or fowls. (4) (Volcanic) Pyroclastic detritus consisting mostly of particles less than 0.25 millimeter in diameter; i.e., fine volcanic ash. It may be composed of essential, accessory, or accidental material.

Dutch Cattle–Loosely, all black-and-white dairy cattle. See Holstein.

Dwarf–(1) A serious viral disease of other trailing blackberry characterized by a yellowing of the plant and a shortening of the canes. The canes, spindly at first, become stiff, unusually upright and thick. The leaves are dwarfed, crinkled, and mottled with bronze and light green patches. Found in United States Pacific Coast. (2) A plant, especially one that has been intentionally grafted to be a dwarf. (3) Dwarfing may be caused by disease, lack of water, or mineral deficiency. (4) See Dwarf Cattle. (5) Designating any plant disease, one of whose symptoms is a stopping or retarding of growth.

Dwarf Cattle–(Midget) A genetic condition of increasing frequency in most breeds of cattle, especially beef cattle. Calves are of less than normal size at birth, remain small throughout life, and often have abnormal body conformation such as stubby legs, a dished or bulldog face, and pop eyes. Dwarfism is now considered an autosomal recessive character.

Dwindling–In honeybees, the rapid or unusual depletion of hive population, usually in the spring.

Dyad–The univalent chromosome, composed of two chromatids, at meiosis. The pair of cells formed at the end of the first meiotic division.

Dysentery–(1) A term used for many kinds of inflammation of the intestines and frequent stools. The cause may be a chemical irritant, bacteria, protozoa, or parasitic worms. (2) A condition of adult bees resulting from an accumulation of feces. It usually occurs during winter and is caused by unfavorable wintering condition and low-quality food. Its presence is detected by small spots of feces around the entrance and within the hive.

Dysfunction–A lessening in the proper action of part of the body.

Dyspepsia–Impairment of digestion.

Dyspnea–Difficult or labored breathing.

Dystocia–Painful or slow delivery or birth.

Dysuria–Difficult or painful urination.

E. Coli Bacteria–Bacteria that naturally inhabit the human colon.

Ear Canker– An inflamed, swollen, scabby condition of the lower inside ear of rabbits, caused by colonies of rabbit ear mites.

Ear Covert– The very small feathers of a bird that cover the ear.

Ear Implant– A small pellet containing a growth regulator or other medication that is placed beneath the skin of the ear of a beef animal. The ear is used for the implant site because the ear is not used for meat when the animal is slaughtered.

Ear Mange– A mange of cats, dogs, rabbits, and foxes caused by *Otodoectes cynotis* (ear mange mite), family Demodicidae, which infests the skin inside the ear, causing great irritation, impaired hearing or deafness, inability to coordinate movements, and sometimes causing the ear to become packed with dry scabs.

Ear Marking– The process of removing parts of the ears of livestock so as to leave a distinctive pattern; done for the purpose of designating ownership.

Ear Notcher– A punch used to cut notches in the ears of animals for identification.

Ear Tag– A tag fastened in an animal's ear for the purpose of identification.

Eared– (1) Designating the presence of ears or spikes on a plant. (2) Pertaining to an animal restrained by its ears. (3) Designating the presence of earlike tufts of feathers in some birds.

Early Speed– Ability of a horse to obtain his maximum speed in one or two strides.

East Coast Fever– A disease of cattle prevalent in eastern Africa caused by a protozoan transmitted by the biting of the brown dog tick *Rhipicephalus sanguineus*.

Eastern Pulled Wool– Wool from sheep and lambs slaughtered in the east. The wool is pulled from the skins after it has been loosened, usually by a depilatory. Pulled wool should not be confused with dead wool.

Easy– (1) A command to a horse to move slower. (2) Designating an animal that has a low or weak back or pasterns, as easy in the back.

Easy Bit– A horse's bit that restrains the horse lightly. See Curb Bit.

Easy Gait– Any horse's gait that is conveniently slow for the rider.

Easy Keeper– An animal that thrives with a minimum of feed, attention, and care.

Ebrillade– The training of a horse by jerking one rein when he refuses to turn.

EBV– See Estimated Breeding Value.

Eclosion– The process of hatching from an egg.

Economy of Gain– The amount of feed required per unit of gain in the weight of an animal or fowl expressed in cash value, i.e., the pounds of feed per pound of gain converted to dollars.

Ecraseur– A surgical instrument consisting of a fine chain or cord that is looped around the scrotum or a diseased part and gradually tightened, thus severing the enclosed part.

Ectoderm– The outer of the three basic layers of the embryo which gives rise to the skin, hair, and nervous system.

Ectothermic– Deriving heat from without the body, cold-blooded, as lizards and snakes.

Eczema– A nonparasitic, skin disease of people and animals characterized by a red inflammation of the skin, development of papules, vesicles, and pustules, serous discharge, formation of crusts, severe itching, and loss of hair. It may be acute or chronic. It is thought to result from allergy, from hypersensitivity to chemicals, from photosensitization, from poor environment, or from faulting feeding. Also known as mange, paraderatosis, dermatitis, and facial eczema.

Edam– A hard, solid, spherical Dutch cheese prepared from cow's milk containing about 2.5 percent fat. Usually the cheese is highly salted. Before marketing it is dipped in red wax.

Edema– (1) An accumulation, usually abnormal, of serous fluids within the intercellular tissue spaces of the body. If the edema is in the subcutaneous tissue, the affected area will be swollen and will pit with pressure. Also called dropsy. (2) Abnormal swelling of plant parts due to an excessive intake of water.

Edema of the Wattles– A condition in birds, in which the wattles are filled with inflammatory liquid so that they become very much enlarged. The cause is unknown, but *Pasteurella multocida* has been suggested. Also called wattle edema.

Edible– A term applied to food that is fit to eat. It usually refers to food that is suitable for human consumption. The initials E.P. are used to denote the edible portion of a food—e.g., a banana without its skin, a pork chop without the bone, a melon without its seeds and rind.

Effective Progeny Number (EPN)– An indication of the amount of information available for estimation of expected progeny differences in sire evaluation. It is the function of the number of progeny but it is adjusted for their distribution among herds and contemporary groups and for the number of contemporaries by other sires. EPN is less than the actual number because the distribution of progeny is never ideal.

Efferent– Conducting or conveying away from.

Efferent Ducts– Very small ducts located in the testes that connect the rete tests to the epididymis.

Effete– Designating an exhaustion of the ability in animals to produce young or in plants to bear fruit. See Impotence.

Egagropilus– Hair ball in an animal's stomach. See Bezoar, Hair Ball, Stomach Ball.

Egestion– Waste material that is excreted from the digestive tract.

Egg– (1) The reproductive body produced by a female organism: in animals, the ovum; in plants, the germ cell, which after fertilization, develops into the embryo. (2) The oval reproductive body produced by females of birds, reptiles, and certain other animal species, enclosed in a calcerous shell or strong membrane within which the young develop.

Egg Albumen– The white of an egg, a mixture of proteins, which is used in confections, bakery goods, and in manufacturing adhesives.

Egg Candling– The examination of poultry eggs in the shell by means of a bright light, for determining market grade and for the detection of meat spots, blood spots, or air bubbles.

Egg Cart–A rolling device used for transporting stacked trays of eggs to the hatchery.

Egg Cell–A female germ cell.

Egg Cleaner–A device used to clean the surface of egg shells, rendering the eggs more attractive for sale. It consists of a cushion base covered with sandpaper or cloth.

Egg Exchange–A market where eggs are bought and sold in large quantities and where eggs are traded on futures.

Egg Flat–A traylike device used to store eggs.

Egg Grader–A mechanical device used for grading eggs according to weight. As the ungraded eggs are carried along a moving belt, they pass over a weighing device that automatically sorts them into separate bins.

Egg Mash–A mixture of ground cereals or cereal products to which has been added essential protein, vitamin, and mineral supplements in sufficient amounts to provide a balanced ration for laying hens.

Egg Oil–An oil that is obtained from the yolk of a hen's egg.

Egg Pod–A capsule that encloses the egg mass of grasshoppers and that is formed through the cementing of soil particles together by secretions of the ovipositing female.

Egg Powder–Dried eggs.

Egg Production –(1) The tending, feeding, etc., of a flock of hens principally for the eggs they lay. (2) The number of eggs laid by a hen.

Egg Room–A cooled room in which eggs are candled and graded and/or stored.

Egg Tooth–A scale on the tip of the upper mandible of the embryo chick, used as a reinforcement for the beak for breaking open the shell at hatching. Also called pip.

Egg Tube–The oviduct.

Egg Turner–A device used for changing the position of eggs in an incubator.

Egg White–The albumen of an egg.

Egg Yolk–The central, yellow portion of a hen's egg that contains the egg cell.

Ejaculate–(1) Semen which is forced out of the body; specifically that semen taken from a male by use of the artificial vagina or methods other than by coitus for use in artificial insemination. (2) The discharge of semen from the reproductive tract of the male. See Electrical Stimulation.

Elastin–A protein substance that is found in tendons, cartilage, connective tissue, and bone. Elastin is not softened as much as collagen by heat in the presence of water.

Elastration–Bloodless castration in which rubber bands constrict the spermatic cord causing it and the testicles to wither away. It is, however, not always successful.

Elastrator–A tool used in castrating and docking. A tight rubber band is applied to the tail or the scrotum, the circulation is thereby cut off, and the tail or scrotum gradually dries up and falls off. A common method for docking and castrating lambs. Also used in castrating young calves.

Elatus–Tall.

Elbow–The joint in a front leg of an animal that corresponds to the elbow joint of a person's arm.

Electric Exerciser–A device, similar to a merry-go-round, consisting of a steel framework that is turned by a motor. Bulls are chained to the frame at the outer edge so that as the device revolves they are forced to walk in a circle for exercise.

Electric Fence–A fence or enclosing device that consists of a single wire supported by insulators on widely separated posts. The wire is attached to a controller that emits electric current for one-tenth of a second or less, forty-five to fifty-five times per minute. When an animal touches the wire, it receives a sharp, short, but harmless, electrical shock. The fence, when properly installed, is not injurious to human beings.

Electric Prod(der)–A portable, battery-powered, canelike device used to control or drive livestock by giving them a slight electrical shock.

Electrical Stimulation–A method of collecting semen from small animals by means of an electric current. An electrode is inserted about 4 inches into the male's rectum and another electrode is held against moistened skin near the fourth lumbar vertebra. About ten stimuli of 30 volts in 5-second periods result in a satisfactory ejaculation. See Ejaculate.

Electrolyte–(1) A nonmetallic substance that will conduct an electric current by the movement of ions when dissolved in certain solvents or when fused by heat; common salt is an example of an electrolyte. (2) A solution containing salts and energy sources used to feed young animals suffering from scours (diarrhea).

Elephantiasis–The final stage of filariasis caused by a parasitic nematode, ***Wuchereria bancrofti***, carried by a mosquito. Symptoms are excessive thickening and swelling of an animal's legs (usually a horse), which results from various infections or disturbances in circulation. Also present in people.

Eligible to Registry–Designating an animal whose sire and dam are registered in the same breed registry and that meets other rules as to age, color, etc., of that breed.

Emaciation–The wasted condition of an animal's body characterized by slimy degeneration of fatty tissues and serous infiltration of the muscles.

Emasculate–(1) To castrate. (2) To remove the androecium (the stamens and their appendages) from a flower for crossbreeding purposes.

Emasculatome–A pincerlike instrument used for bloodless castrating, which severs or breaks the spermatic cord, causing the testicles to wither away. It may also be used in docking the tails of lambs. See Elastrator.

Emasculator–A tool used for docking and castrating that both cuts and crushes surrounding tissue to prevent excessive bleeding.

Embryo– (1) Any organism in its earliest stages of development. (2) The young, sporophyte that results from the union of male and female cells in a seed plant. Also called seed-germ.

Embryo Sac– (1) The mature female gametophyte in higher plants. (2) A sac that contains the embryo in its very early life in animals. Also called blastodermic vesicle.

Embryo Transfer (Transplant) (ET)– The removal of developing embryos from one female and their transfer to the uterus of another; it usually involves the superovulation of superior females and the transfer of their embryos in an attempt to increase the number of superior offspring.

Embryology– The science that deals with the study of the embryo.

Embryonic– (1) Pertaining to the embryo or its development. (2) Underdeveloped; immature. See Rudimentary.

Embryonic Vesicle– The sac containing the developing embryo.

Emetics– Drugs that produce vomiting.

Emmentaler– Swiss hard cheese with a sweet delicate flavor that is made from whole or semifat milk.

Emphysema– An abnormal presence of air in the tissue of some part of the body, generally the lungs, where a stretching and rupturing of the walls of the smallest air sacs produces abnormal air space.

Emulsion– A mixture in which one liquid is suspended as tiny drops in another liquid, such as oil in water.

Encapsulation– Enclosure in a capsule or sheath.

Encephalitis– Inflammation of the brain that results in various central nervous system disorders, characterized by excitement and irregular movements, depression, paralysis, and death. It may be a symptom of various diseases or it may occur as a primary disease.

Encephalomyelitis– A disease involving an inflammation of the brain or spinal cord that is caused by a pathogenic organism; sleeping sickness.

Encierro– (Spanish) Corral.

Enclosure– A fenced area that confines animals. See Exclosure.

Encrustation– A crust or hard coating on or in a body.

Encyst– To become enclosed in a sac, bladder, or cyst.

Endocarditis– Inflammation of the endocardium or epithelial lining membrane of the heart.

Endocrine Gland– Any gland of the body that secretes a substance or hormone, thereby controlling certain bodily processes: e.g., pituitary, pineal, thyroid, parathyroid, thymus, adrenal, pancreas.

Endometrium– The inside lining of the uterus.

Endoparasite– Any of the various parasites which lives within the body of its host. See Ectoparasite.

Endotoxin– A toxin produced within an organism and liberated only when the organism disintegrates or is destroyed.

Enrich– (1) To add fertilizer or manure to soil. (2) To add a substance or vitamin to food products.

Enrichment– The process by which bodies of water are enriched, especially by nitrogen, phosphorus, and/or carbon, resulting in accelerated growth of undesirable algae and other aquatic vegetation. Enrichment sometimes originates as runoff from domestic animal feedlots, landspreading of sewage sludge, and overfertilized fields.

Ensilage– Any green crop preserved for livestock feed by fermentation in a silo, pit, or stack, usually in chopped form. Ensilage can be made from practically any green crop having the proper moisture content. Also called silage.

Ensile– To place green plant material, such as green crops of grain, grasses, cornstalks, etc., in a silo in such a manner as to bring about proper fermentation for preservation and storage. See Silage.

Enteric– Pertaining to the intestines.

Enteritis– Any inflammatory condition of the lining of the intestines of animals or people. Characteristics of enteritis are frequent evacuations of a liquid or very thin, foul-smelling stool which may or may not contain blood, straining, lethargy, and anorexia. In acute cases there is a rise in body temperature. The condition is seen as a symptom of a number of infectious diseases or it may be caused by specific bacteria or viruses. Other common causes include plant and chemical poisons, parasites, overeating, faulty nutrition, and poor environmental factors.

Enterotoxemia– A disease of calves, sheep, and goats, which results in high mortality and is associated with ***Clostridium perfringens***. It is characterized by a lack of coordination and sudden death. Also called pulpy kidney disease.

Entomologist– A person who specializes in the study of insects.

Entomology– That branch of zoology that deals with insects. See Economic Entomology.

Entozoa– Internal animal parasites; e.g., stomach worms.

Entrails– The visceral organs of the body, particularly the intestines.

Entrance Feeder– A wooden runway that fits into the hive entrance so that bees may obtain syrup from a jar inverted into it.

Entropion– A condition where the eyelids are turned in, usually at birth.

Enucleation– The removal of an organ, tumor, or other body in such a way that it comes out clean and whole as a nut from its shell.

Enurination– Behavior in rabbits where the buck jumps up in some fashion and squirts another rabbit with a stream of urine.

Enzootic Marasmus– Cobalt deficiency.

Enzymatic– Referring to a reaction or process that is catalyzed by an enzyme or group of enzymes.

Enzyme– A large complex protein molecule produced by the body that stimulates or speeds up various chemical reactions without being used up itself; an organic catalyst.

Eolian– (1) Pertaining to action of wind, as soil erosion due to wind, or soil deposits transported by wind. (2) Loosely designating soils that are derived form geologic deposits that are windborne in origin. Also spelled aeolian. See Loess.

EPD– See Expected Progeny Difference.

Epidemic–(1) A widespread invasion or dispersion by an insect or a disease. See Endemic, Epiphytotic. (2) Designating a sudden and widespread attack by a disease or insect infestation. (Most scientists seem to prefer epidemic when referring to all types of disease: people, animal, and plant.)

Epidemiology–(1) A study of the factors determining the frequency and distribution of human diseases. (2) A determination of the causes of localized outbreaks such as infectious hepatitis and of toxic disorders such as nitrate poisoning.

Epidermis–The cellular layer of an organism; the outer skin.

Epididymis–A small, tortuous tube leading from the testicle. A site of sperm storage and maturation.

Epigenetic–As used in reference to cancer, an effect that does not directly involve a change in the sequence of bases in DNA. See DNA.

Epilepsy–A chronic nervous disorder of people and animals, particularly of dogs, which is characterized by a sudden loss of consciousness and convulsions, a stiffened neck, chomping of the jaws, dilated pupils, salivation, distressed breathing, and evacuation of the bladder and bowels. The cause is unknown. Also called fits, falling sickness.

Epinephrine–A hormone, also known as adrenalin, that is produced by the medulla of the adrenal glands in mammals.

Epistasis–(1) The checking of any discharge, secretion, excretion, such as stopping the flow of blood. Also see Epistatic. (2) The type of gene action where genes at one locus affect or control the expression of genes at a different locus.

Epistatic–Designating a condition of genetics in which one factor prevents a factor other than its allelomorph from exhibiting its normal effect on the development of the individual.

Epithelial Layer–Cellular tissue covering all the free body surfaces, cutaneous, mucous, and serous, including the glands and other structures derived therefrom.

Epitheliogenesis Imperfecta–A condition of newborn animals resulting from a lethal recessive genetic factor. There are skin defects on the lower parts of the legs and on the hairless parts of the body. The mucous membranes are defective, and the ears and claws deformed. The animal so affected dies soon after birth.

Epithelium–The dense, cellular tissue that covers all body surfaces and lines all body cavities.

Epizootic–Designating a disease of animals that spreads rapidly and affects many individuals of a kind at the same time, thus corresponding to an epidemic in people. See Endemic, Epidemic, Epiphytotic.

Epizootiology–The study of factors influencing or involved in the occurrence and spread of disease among animals.

Epizooty–See Equine Influenza.

Equine–(1) A horse. (2) Pertaining to, or resembling, a horse or other member of the family Equidae. Horses, mules, and asses are referred to as equines or equine animals.

Equine Encephalomyelitis–A viral disease of horses and people characterized by fever, grinding of the teeth, sleepiness, wobbly gait, difficulty in chewing and swallowing, and frequently death. Also called Borna's disease, horse sleeping sickness.

Equine Infectious Anemia–A widespread, contagious, viral disease of horses and mules characterized by dullness, rapid breathing, weakness of the hind legs, loss of appetite, swollen eyelids, intermittent fever, dropsy, frequent urination, diarrhea, general weakness, loss of the blood's ability to coagulate, emaciation, anemia, and death. Also called pernicious equine anemia, swamp fever, horse malaria, malarial fever, slow fever, mountain fever, creeping fever.

Equine Influenza–The most contagious and most widely spread disease of horses, mules, and asses, caused by a filterable virus with complications attributed to miscellaneous bacteria. It is characterized by fever, extreme weakness, depression, rapid respiration, coughing, watery nasal and eye discharges, a pinkish swelling of the eyelids, edema of the abdomen, legs, and head, and loss of appetite accompanied by excessive thirst. Also called shipping fever, pinkeye, catarrhal fever, epizootic cellulitis, epizooty, stockyard fever.

Equine Piroplasmosis (EP)–A tick-borne disease caused by two blood parasites (*Babesia caballi* and *B. equi*); is clinically indistinguishable from equine infectious anemia (EIA). EP was first reported in the United States in Florida in 1962. It has since been found to be endemic in southern Florida, Puerto Rico, and the United States Virgin Islands. *Dermacentor nitens*, the tropical horse tick, is the principal vector of the disease.

Equine Variola–Horsepox; a contagious, viral disease of horses characterized by fever and pustular eruptions or pox, especially on the pasterns, fetlocks, and mouth. Also called contagious pustular stomatitis.

Equisetum Poisoning–A poisoning of horses and sheep which results from their feed on *Equisetum* spp., especially *E. arvense*, family Equisetaceae, the field horsetail. It is characterized by unthriftiness, loss of weight and muscular control, nervousness, falling, lack of ability to eat, and death. Also known as horsetail poisoning.

Equitation–Horsemanship, horsewomanship.

Erection–The process whereby a body part is made to stand upright; a process where the penis becomes engorged with blood causing it to be firm, turgid, and ready for breeding.

Ergosterol–A cholesterol-like substance found in plants that, when irradiated with ultraviolet light, changes to vitamin D2.

Ergot–(1) A fungal disease of cereals and wild grasses caused by *Claviceps purpurea*, family Clavicepitaceae, which attacks the inflorescence, replacing the grains with black or dark purple, club-shaped, horny structures (sclerotia) that are harvested with the grain and must be removed before the grain is used for flour or feed. Ergot causes a lowered grain yield and ergotism in people and animals. Also called clavus. (2) A drug obtained from such diseased plants used to control bleeding in animals and people. (3) A horny growth behind the fetlock joint of a horse.

Ergotism–A disease of people and lower animals caused from eating grain or grain products contaminated with ergot which is characterized by excessive salivation, redness and blistering of the mouth epithelium, vomiting, colic, diarrhea, and constipation. Also called holy fire, St. Anthony's fire, bread madness, gangrenous ergotism, dry gangrene.

Eructation—The act of belching, or of casting up gas from the stomach.

Erysipelas—A disease caused by a bacillus that usually affects the joints and causes difficult breathing; it primarily affects hogs.

Erythema—Redness or inflammation of the skin caused by congested capillaries.

Erythroblastosis—One manifestation of avian leukosis complex, a viral disease of chickens characterized by paleness or a yellowish discoloration of the comb and wattles, a rapid loss of weight in spite of a good appetite, weakness, prostration, diarrhea, and tightly shut eyes. The infected fowl soon dies. Also called erythroleukosis, erythroblastic leukosis.

Escape—(1) A fowl or animal that has gotten out of its enclosure. (2) Botanically, a cultivated plant that is found growing wild. (3) To become wild after having been in cultivation.

Escape Board—Board with one or more bee escapes on it to permit bees to pass one way.

Escherichia Coli—One of the easily detected species of bacteria in the fecal coliform group. It occurs in large numbers in the gastrointestinal tract and feces of all warm-blooded animals and people. Although not a pathogen, its presence is an indication of the presence of pathogens.

Escutcheon—(1) The scion used in shield budding. (2) That part of a cow which extends upward just above and back of the udder where the hair turns upward in contrast to the normal downward direction of the hair. Also called milk shield, milk mirror.

Esophagus—Gullet; the tube that connects the throat or pharynx with the stomach. It varies greatly in the vertebrates; e.g., in the crop of a bird, it is distended (enlarged) for retention of food.

Essential Amino Acid—Any of the amino acids that cannot be synthesized in the body from other amino acids or substances or that cannot be made in sufficient quantities for the body's use.

Essential Host—A host for one stage in the development of a parasite without which the parasite cannot develop to maturity. See Cedar-Apple Rust.

Estancia—A small farm or small cattle ranch (southwestern United States).

Estimated Breeding Value (EBV)—In beef cattle, an estimate of the value of an animal as a parent, expressed as a ratio with 100 being average. For example, a bull with a yearling weight EBV of 110 would be expected to sire calves with yearling weights 10 percent greater than average.

Estimated Relative Producing Ability (ERPA)—In dairy cattle, a prediction of 305-day, two times per day milking, mature equivalent production compared with the production of other cows in the herd.

Estrapade—The bucking of a horse.

Estray—A wandering domesticated animal of unknown ownership.

Estriol—An estrogenic substance derived form the urine of pregnant animals which, when administered to female animals, causes them to come into heat. Also called theelol.

Estrogen—A hormone or group of hormones produced by the developing ovarian follicle; it stimulates female sex drive and controls the development of feminine characteristics.

Estrone—A product derived form the urine of pregnant animals or from the ovary, which may be administered to female animals to cause them to come into heat. Also called theelin, folliculin, follicular hormone.

Estrous—Pertaining to estrus (heat) in animals.

Estrous Cycle—The reproductive cycle in nonprimates; it is measured from the beginning of estrus or heat period to the beginning of the next.

Estrus—The period of sexual excitement (heat), at which time the female will accept coitus with the male.

Estrus Synchronization—Using synthetic hormones to make a group of females come into heat at the same time. They can then be bred at the same time and all of their calves will be born in a short period, ensuring uniform ages in the calf crop and lower labor and marketing requirements.

ET—See Embryo Transfer.

Ethology—The science of animal behavior related to the environment.

Etiological—Pertaining to the causes of diseases.

Etiology—The science that deals with the origins and causes of disease. Also spelled aetiology.

Eupeptic—Having normal digestion.

Euploid—An organism or cell in which the chromosome number is the exact multiple of the monoploid or haploid number. Terms used for euploid series are: haploid, diploid, triploid, tetraploid, etc.

European Foulbrood—An infectious brood disease of bees considered to be caused by *Bacillus pluton*. Also called black brood. See American Foulbrood.

Evaginated—Turned inside out. Tapeworm larvae often have the head, or scolex, invaginated; when the larvae reach the intestines of the final host, the head is evaginated, or turned inside out, so the suckers can attach to the wall of the intestines.

Evaporated—Designating a product that has had most of the moisture driven off by boiling or other application of heat; e.g., evaporated milk.

Evaporated Buttermilk—The product that results from the removal of a considerable portion of the moisture from clean, sound buttermilk, made from natural cream to which no foreign substance has been added. The product must contain not less than 27 percent total solids, not less than 0.055 percent butterfat for each percent of solids, and not more than 0.14 percent of ash for each percent of solids. Also called concentrated buttermilk, condensed buttermilk.

Evaporated Cultured Skimmed Milk—A product that results from the removal of a considerable portion of moisture from clean, sound skimmed milk which has been made from a suitable culture of lactic bacteria. This product must contain not less than 27 percent of total solids. Also called concentrated cultured skimmed milk, condensed cultured skimmed milk, concentrated sour skimmed milk.

Evaporated Milk–A liquid food product made by evaporating sweet milk to such a point that it contains not less than 7.9 percent of milk fat and not less than 25.9 percent of total milk solids.

Eversion of the Oviduct–The turning inside out and protrusion from the anus of the lower portion of the oviduct of a female bird. The condition renders the bird worthless for egg laying but not for human consumption.

Everted–Turned inside out.

Eviscerate–To remove the entrails, lungs, heart, etc., from a fowl or animal when preparing the carcass for human consumption.

Ewe–A female sheep of any age.

Ewe Index–A factor used in the selection of ewes. It is calculated for ewes giving birth to and raising single lamb: Ewe Index = Adjusted Weight of the Lamb + (3 × fleece Weight). It is calculated for ewes giving multiple birth and raising more than one lamb: Ewe Index = Sum of Adjusted Weight of the Lambs + (3 × Fleece Weight).

Ewe Neck–A neck like that of a sheep, with a dip between the poll and the withers. Also termed a turkey neck and upside-down neck.

Ex–Prefix meaning without or destitute of.

Exclosure–An area of land fenced to prevent all or certain kinds of animals from entering it and used for ecological experiments involving biotic factors such as the grazing pressure of livestock. See Enclosure.

Excluder–A thin grid of wire, wood and wire, sheet plastic, or sheet zinc, with spaces wide enough for worker bees to pass through but not queens or drones. It is used between hive bodies to confine queens to one part of a hive.

Excreta–Matter that is excreted; waste matter discharged from the body; materials cast out of the body. Excrement; waste matters discharged from the bowels; manure.

Exogenous–(1) Produced on the outside of another body. (2) Produced externally, as spores on the tips of hyphae. (3) Growing by outer additions of annual layers, as the wood in dicotyledons. See Endogenous.

Exomphalos–A hernia that has escaped through the umbilicus.

Exoskeleton–Collectively the external plates of the body wall.

Exosmosis–The slow diffusion of the more dense fluid through a membrane to mingle with the less dense fluid.

Exostosis–An outgrowth of a bone: e.g., splints, bone spavins, etc.

Exotic–Foreign, unfamiliar, new, imported; in animal agriculture, it refers to imported breeds, usually of cattle.

Exotic Newcastle Disease Surveillance–Exotic Newcastle disease or velogenic viscerotropic Newcastle disease (VVND) is a contagious and deadly viral disease affecting all species of birds. The disease causes bleeding in the intestines and reproductive glands, along with severe diarrhea. In commercial poultry operations, it kills many of the birds it affects, shortens the lives of others, and reduces egg production. See Newcastle Disease.

Exotoxin–A soluble toxin excreted by specific bacteria and absorbed into the tissues of the host.

Expected Progeny Difference (EPD)–In beef cattle selection, a score used to measure the amount of performance difference in the offspring of a certain sire as compared to the performance level of the herd average. For example, if bull A has a weaning weight EPD of +5 pounds and bull B has a weaning weight EPD of –10, Bull A should produce calves that are 15 pounds heavier at weaning than bull B's calves.

Expectorant–A drug that causes expulsion of mucus from the respiratory tract.

Extender–The liquid in which semen is extended for preservation. It usually consists of egg yolk, sodium citrate, glycerol, water, and antibiotics.

Extract–A solid preparation obtained by evaporating a solution of a drug, the juice of a plant, etc. Vitamin extracts are used to supplement a diet.

Extracted Honey–Liquid honey removed from the comb by means of an extractor or other method of separation. See Extractor.

Extracted Meals–Animal feeds that are fermentation products.

Extractor (honey)–A hand- or power-driven device that removes honey from the comb by centrifugal force.

Exudate–A discharge deposited in or on an organ through pores, injured areas, or natural openings, as a bloody discharge from wounds of animals infested with screw worms, or the gummy discharge from a wound in a tree.

Eye Muscle–The longissimus dorsi muscle of four-footed animals. The major muscle of a rib, loin steak, or chop.

Eyeing–Clipping the wool from around the face of closed-faced sheep to prevent wool blindness.

Eyetooth–Either of the two canine teeth in the upper jaw. Also called dog tooth.

F_1–The first filial generation; the first generation of a given mating.

F_1 **Females**–Female cattle from the first cross of mating of two different purebreeds of cattle.

F_1 **Generation**–The first generation out of a cross of two animals of different breeds.

F_2–The second generation progeny generally produced by crossing two F_1 individuals.

Face–(1) The bare skin on a fowl's head around and below its eyes. (2) The front part of the head of an animal, including the eyes, nose, and mouth. (3) The side of a hill or furrow. (4) The top or bottom layer of produce, especially fruit, which is arranged in a container for display purposes when the container is opened. (5) To arrange a layer of produce in a container for display purposes. (6) In turpentining, the exposed portion of the tree from which the oleoresin exudes. (7) In lumber, the side of a board from which it is graded. See Face Side.

Face Fly–Flies that gather in large numbers on the faces of horses or cattle, especially around the eyes and nose.

Factor–(1) A unit of inheritance occupying a definite locus on one or both members of a definite chromosome pair whose presence is responsible for the development of a certain character or modification of a character of the individual who possesses that genotype; a determiner or gene. (2) An agent, as one who buys and sells a commodity on commission for others. (3) An item in the analysis of a farm business; e.g., labor efficiency. (4) Inherent characteristics of the climatic, nutritional, cultural, or biological environment responsible for the specific performances of plants or animals.

Facultative–Designating an organism that is capable of living under more than one condition: e.g., as a saprophyte and as a parasite; as an aerobic or anaerobic organism.

Facultative Aerobe–A microorganism that lives in the presence of oxygen but may live without it.

Facultative Bacteria–Bacteria that can exist and reproduce under either aerobic or anaerobic conditions.

Facultative Parasite–A parasite that feeds upon an organism until the organism dies and then continues to live on the dead organic material.

Fag–(1) Any tick or fly that attacks sheep. (2) Long coarse grass of the preceding season. (3) See Fagot (2).

Fagot–(1) A pork sausage composed of hog livers, hearts, fresh pork, onions, peppers, and sweet marjoram, molded into a 6-ounce ball encased in hog fat. (2) A small bundle of twigs to be used for fuel.

Fahrenheit Scale–A temperature scale in which the freezing point of water is 32°F and the boiling point is 212°F. Named after Gabriel Daniel Fahrenheit (1686-1736), a German physicist. See Celsius Scale, Kelvin Scale.

Fair Condition–(1) A range-condition class; a range producing only 25 to 50 percent of its potential. The cover consists of early maturing plants of low value for forage or for soil protection. (2) Denoting a medium condition of plants, plant products, or animals.

Falcate–Sickle or scythe-shaped.

Fall Farrow–A pig born in the autumn.

Fall Lamb–A lamb that is usually born in the spring and sold in the fall.

Fall Wool–Wool shorn in the fall (usually in Texas or California) following four to six months' growth. Only a small percentage of the total United States wool production is the product of twice-a-year shearing. Also called fall-shorn wool.

Falling Sickness–See Epilepsy.

Fallopian Tube–One of the two tubes or ducts connected to the uterus of mammals and leading to the ovary; functions in transporting the ovum from the ovary to the uterus.

False Heat–The displaying of estrus by any female animal when she is pregnant or is out of season. It can occur in a healthy animal or may be caused by diseased ovaries.

False Hoof–See Dewclaw.

False Milk Fever–See Acetonemia.

False Molt–The shedding of feathers by a bird due to unnatural causes. See Molt.

False Pregnancy–A condition usually following a sterile mating in which the organs of reproduction go through anatomical and physiological changes similar to those that occur during a real pregnancy. For example, the abdomen may swell as well as the milk glands, which may in fact produce some milk. This is a common problem in the rabbit.

Family Broken–Describing a horse that is gentle and safe for family use.

Family Characteristic–Particular parts of the conformation and/or the temperament that exist among certain families of dairy cattle, etc. Some families are consistently good or bad in straightness of legs, fore udders, depth of body, shape of head, etc. In temperament, there are families that are consistently nervous while others are normally docile. Those identifying marks or characters tend to display relationship.

Fancier–A breeder who shows particular interest in, and the development of, a particular breed or type of animals or plants: e.g., a rose fancier, dog fancier, Hereford fancier.

Fancy–A top-quality grade for many vegetables, fruits, flowers, poultry, and livestock.

Fancy Points–Indications of purity of breeding: e.g., the set of the horns, the carriage of the ear, the color of the hair on certain parts of the body, and the general style of an animal.

Far Side–In horsemanship, the right side of a horse.

Farcy–See Glanders.

Farding Bag–The first stomach, the rumen, of a cow or other ruminant.

Farm Flock–Those animals or fowls, collectively by species, which a farmer raises on his farm, especially a small flock of sheep or chickens.

Farm Manure–The excrement from all or certain of the farm animals, including litter or bedding materials, such as straw. Also called farmyard manure.

Farm Pond–A small reservoir of water on a farm usually formed by constructing a dam across a watercourse or by excavation to collect surface water.

Farmyard Manure–See Farm Manure.

Farrier–(1) A person who shoes horses. (2) An obsolete term for a horse and cattle veterinarian.

Farrow to Finish–A type of farm operation that covers all aspects of breeding, farrowing, and raising pigs to slaughter.

Farrowing Crate–A crate or cage in which a sow is placed at time of farrowing. The crate is so constructed as to prevent the sow from turning around or crushing the newborn pigs as she lies down.

Farrowing House–Any of several types of structures especially designed for a sow and her litter of pigs.

Farrowing Pen–A small pen or enclosure, usually in a farm building, especially designed for a sow that is ready to farrow or has given birth to a litter of pigs. This pen usually includes protective fenders (guard rails) and a brooder for the piglets.

Fascine–(1) A fagot. (2) A long bundle of sticks bound together and used to build a temporary roadway through a marshy soil or to stabilize an unstable slope along an embankment. When used in erosion control, the sticks may be made of willow or other species that sprout. See Gabion.

Fast Breeder–Refers to pigeons, rabbits, and like animals, the parent stock of which are capable of producing a large number of young over a given period such as a year; also referred to as high producers.

Fast Walk–One form of the walk, a horse's gait, which is somewhat more rapid than the walk but slower than the running walk.

Fast-feathering–The maturing of first feathers by certain breeds and strains of chickens at a relatively young age. It is an inherent characteristic that is especially desirable in chickens sold as market broilers.

Fast-reining–Designating a horse that responds readily and quickly to directions given by the rider with the reins.

Fast-stepping Trot–One form of the trot, a horse's gait, characterized by long rapid steps.

Fat–(1) (a) The oily or greasy-substance bearing tissues of an animal. (b) Designating any animal or fowl which abounds in fat. (2) The oily substance of milk; the chief constituent of butter. See Butterfat. (3) The oily or greasy substances found in certain plants; e.g., peanut oil, cottonseed oil. (4) Any food product, e.g., lard or vegetable shortening, which is derived from animal or vegetable fats. (5) Those substances which can be extracted from dry feeds with ether. See Ether Extract. (6) Fattened cattle ready to market. (7) Of, or pertaining to, a prosperous year, as a fat year.

Fat Body–An organ in the insect body with multiple functions in metabolism, food storage, and excretion. Fat body is a misnomer, for protein and glycogen are stored as well as fat.

Fat Content–The amount of butterfat (milk fat) in milk, usually stated in percentages.

Fat Globules–The naturally occurring spheres of butterfat in milk.

Fat Test–A test sometimes applied to determine the fat content of a product such as milk. See Babcock Test.

Fat Type Hog–An antiquated term denoting a hog that yields a high percentage of fat (lard), not necessarily synonymous with the lard-type hog. See Bacon-type Hog, Lard-type Hog, Meat-type Hog.

Fatigue–(1) A weakness of wood or metal that results from a repeated reversal of the load. (2) Loss of energy in people or animals form any of a multitude of causes.

Fatling–A young farm animal that is fattened for slaughter.

Fatten–(1) To add or put on fat, as an animal. (2) To cause an animal or fowl to put on fat.

Fattening–(1) The feeding of animals or fowls so that they put on fat. (2) Designating a feed that when fed to livestock causes them to add fat, as fattening grain.

Fattening Range–A productive range devoted primarily to fattening of livestock for market.

Favor–To protect; to use carefully; as a horse favors a lame leg.

Favus–(1) A fungal disease of poultry caused by *Achorion gallinae*, characterized by yellowish-white, scaly lesions on the unfeathered portion of the head. If it spreads to the neck, the feathers become brittle and break off. (2) A fungal disease of young cats, and sometimes of dogs, caused by *Achorion schoenleini*, characterized by circular, yellowish, or grayish patches that appear on the paws near the claws, the head, and the face. Thick, crusty layers develop. Both of these diseases are transmissible to people.

Fawn–(1) A coat color for an animal that is a soft, grayish-tan. (2) A young deer.

FCM–Fat-corrected milk. A means of evaluating milk-production records of different animals and breeds on a common basis energy-wise. The following formula is used: FCM = $0.4 \times$ milk production – ($15 \times$ pounds of fat produced).

Feather–(1) The epidermal structure partly embedded in follicles of the skin of birds. Feathers vary greatly in color, size, and shape, and generally cover the body of a bird. (2) (plural) Plumage. (3) The long hair of certain breeds of horses that grows below the knees and hocks.

Feather Eating–The pulling out of one another's feathers. It is a vice among poultry and caged birds resulting from irritation by lice and quill mites, from lack of exercise, or from faulty nutrition. If blood is drawn, cannibalism results. Also called feather pulling.

Feather Follicle–The depression in the skin of a bird from which the feather grows.

Feather in Eye–A mark across the eyeball of a horse, not touching the pupil; often caused by an injury.

Feather Picking–The plucking or removing of feathers from a fowl carcass in preparing it for human consumption.

Feather-legged–Designating the breeds of chickens that have feathers on the shanks and toes.

Feathering–(1) A defect of coffee cream characterized by a lack of homogeneity, causing it to rise to the surface of coffee in flocculent masses and form a light, serrated scum. (2) A rough-edged hole that has been bored in wood in which wood fibers remain to project around the perimeter. It usually results from using a dull bit. It is especially bothersome in maple trees used for collection of sap. (3) The scuffing of the tender skin on an early potato in harvesting. (4) The streaks of fat visible on the ribs of a lamb carcass.

Febrile–Pertaining to fever or a rise in body temperature; e.g., febrile period (period of fever), febrile disease (a malady accompanied by fever).

Fecal—Pertaining to excrement, manure, and waste material passed from the bowels.

Fecal Contamination—Pollution with feces, manure, or excrement; e.g., fecal contamination of pastures, food, etc.

Feces—Waste material of the digestive system.

Fecund—Fruitful; fertile; prolific.

Fecundation—Pollination or fertilization.

Fecundity—The ability to reproduce regularly and easily.

Fed Beef—Beef produced from cattle that have been finished on grain rations, as compared with grass-fed or range cattle.

Fed Cattle—Steers or heifers fattened on grain for slaughter.

Fed Lamb—A market lamb that has been fed some grain, in contrast to one that has subsisted entirely on milk.

Feed—(1) Harvested forage, such as hay, silage, fodder, grain, or other processed feed for livestock. See Forage. (2) The quantity of feed in one portion. (3) To furnish with essential nutrients. (4) To deliver, carry, or transport.

Feed Additive Compendium—A publication that lists feed additives in current use and the regulations for their use.

Feed Additives—A material added to livestock feed, usually an antibiotic, that is not a nutrient but enhances the growth efficiency of the animal. Domestic animals gain weight more rapidly because antibiotic feed additives counteract the ill effects of high grain rations, reduce bacterial infections, reduce scouring, stimulate appetite, and stimulate certain enzymes. Other feed additives include hormones to stimulate growth and substances to control bloat, parasites, and feed spoilage.

Feed Analysis—The chemical or material analysis of a commercially prepared feed, printed on a tag and fastened to the bag in which the feed is to be sold.

Feed Bunk—A forage and grain feeding station.

Feed Composition Table—A table showing the nutrients found in feeds.

Feed Conversion Ratio—The rate at which an animal converts feed to meat. If an animal requires four pounds of feed to gain one pound, it is said to have a four to one (4:1) feed conversion ratio.

Feed Crop—Any crop grown as a feed for livestock, as hay, corn, oats, etc.

Feed Efficiency—Term for the number of pounds of feed required for an animal to gain one pound of weight; e.g., 6.5 pounds of feed per pound of gain.

Feed Energy Utilization—The percentage of the energy obtained from a feed that is used for an animal's bodily functions. For example, the feed energy utilization for an average lactating cow is approximately: 30 percent for fecal energy, 20 percent for heat energy, 20 percent for maintenance, 20 percent for milk production, 5 percent for urinary energy, and 5 percent for gaseous energy.

Feed Flavor—A flavor defect of milk or milk products that suggests the taste of cattle feed.

Feed Hopper—A container that allows feed to drop down gradually as more is consumed.

Feed Out—To prepare animals for market by fattening.

Feed Reserve—Feed harvested and stored for future use, standing forage cured on the range, or pasture for future use.

Feed Unit—One pound of corn, or its equivalent in feed value in other feeds, which is fed to cattle under normal farm conditions.

Feeder—(1) A young animal that does not have a high finish but shows evidence of ability to add weight economically. (2) Any device that carries material to or into a machine, as a feeder in a threshing machine or in a cotton gin. (3) A person who fattens livestock for slaughter. (4) A thing that feeds, as a plant. (5) Any of several types of appliances used for feeding sugar syrup to bees. (6) A hopper in which feed is placed for consumption.

Feeder Buyer—(1) One who buys cattle, sheep, or horses, fattens them and offers them for sale. (2) A buyer of feeder livestock.

Feeder Calf—A weaned calf that is under one year of age and is sold to be fed for more growth.

Feeder Lamb—In marketing or stock judging, a lamb that is a feeder.

Feeder Pig—A barrow or gilt carrying enough age and flesh so as to be ready to place in a feedlot for finishing; usually a pig weighing less than 120 poounds.

Feeder Production—The production of cattle for feeders; feeder cattle do not carry enough finish to make the slaughter grades.

Feeders—Animals that have been grown out to a determined size or weight and are ready to be placed in the feedlot finishing to a determined grade.

Feedgrain—Any of several grains, and most commonly used for livestock or poultry feed, such as corn, grain sorghum, oats, and barley.

Feeding—(1) Providing animals with the desirable quality and quantity of feeds; making feed available to animals. (2) The ingestion of feed by an animal or plant.

Feeding Area—An area of a barn, shed, or open lot where cows are fed roughages, water, and sometimes concentrates. This area may or may not include feed storage.

Feeding Fence—A fence so constructed that animals can reach through and eat feed placed along the fence.

Feeding Oat Meal—A livestock feed that is the product obtained in the manufacture of rolled oat groats or rolled oats. It consists of broken rolled oat groats, oat-groat chips, and floury portions of the oat groats, with only such quantity of finely ground oat hulls as is unavoidable in the usual process of commercial milling. It must not contain more than 4 percent of crude fiber.

Feeding Pen—An enclosure in which animals or fowls are fed.

Feeding Ratio—Weight of food consumed divided by increase in weight of an animal, during a given time interval.

Feeding Standard—Established standards that state the amounts of nutrients that should be provided in rations for farm animals of various ages and classes in an attempt more nearly to attain the optimum economy of growth, gain, or production.

Feeding Time–The regular time of the day at which animals and fowls are fed. Regular feeding times result in greater feed efficiency.

Feeding Value–A term referring to the nutritive value of different feeds, i.e., expressing the amount of nutrients furnished by each feed and the degree of their digestibility.

Feedlot ADG/<\d>Gain on Test–A measure of the ability of cattle to gain weight when fed high-energy rations, usually expressed as Average Daily Gain.

Feedstuff–One or a mixture of the substances that form the nutrients; namely, proteins, carbohydrates, fats, vitamins, minerals, and water. A feedstuff is different from a feed in that a feedstuff is not normally fed by itself but is mixed with other feedstuffs to formulate a feed. For example, soybean meal or fish meal.

Fell–(1) The elastic tissue just under the hide of an animal attached to its flesh; facia. (2) To cut down a tree.

Felon Quarter–See Mastitis.

Felt–A cloth made from woven or pressed fibers of wool or wool and hair and made into such articles as hats.

Felting–(1) The property of wool fibers to interlock when rubbed together under conditions of heat, moisture, and pressure. No other fiber can compare with wool in felting properties. (2) The manufacturing of felts from furs of the rabbit and of other animals.

Femoral Artery–The main artery carrying blood to an animal's hind legs.

Femur–The large bone in the pelvic limb between the stifle and the pelvis.

Fence–(1) A hedge or barrier of wood, metal, stone, or plants erected to enclose an area to prevent trespassing or the straying of animals. (2) To enclose an area with a fence. See Drift Fence, Snow Fence.

Fence Line Feed Bunk–A multipurpose structure designed for feeding roughage along a fence in the open. It is primarily a feed bunk with openings through which the animals may feed. The feed may be distributed by dump trucks or feed carriers operated outside of the enclosed area.

Fence Off–To enclose an area with a fence.

Feracious–Fruitful; bearing abundantly.

Feral Species–(1) Nonnative species, or their progeny, which were once domesticated but have since escaped from captivity and are now living as wild animals; such as wild horses, burros, hogs, cats, and dogs. (2) An organism (and its offspring) that has escaped from cultivation or domestication and has reverted to a wild state. Escaped plants are usually referred to as exotics and nonnative animals as feral or exotic species.

Fertile–(1) Productive; producing plants in abundance, as fertile soil. (2) Capable of growing or of development, as a fertile egg. (3) Capable of reproducing viable offspring. (4) Able to produce fruit, as a fertile flower. (5) Plant capable of producing seed.

Fertile Egg–A fertilized, avian egg capable of embryonic development.

Fertile Queen–A queen bee that has mated and is capable of laying fertile eggs.

Fertility–(1) The ability of a plant to mature viable seeds. (2) The ability of an animal or fowl to produce offspring. (3) The quality that enables a soil to provide the proper compounds, in the proper amounts and in the proper balance for the growth of specified plants, when other factors, e.g., light, temperature, and the physical condition of the soil or favorable. See Sterile.

Fertility Test–A test to determine if an animal is fertile.

Fertilization–(1) Union of pollen with the ovule to produce seeds. This is essential in production of edible flower parts such as tomatoes, squash, corn, strawberries, and many other garden plants. (2) Application to the soil of needed plant nutrients, such as nitrogen, phosphorus, and potassium. (3) The union of a sperm and egg.

Fertilize–(1) To supply the necessary mineral and/or organic nutrients to soil or water to aid the growth and development of plants. (2) To fecundate the egg of an animal or plant, or to pollenize the pistil of a flower.

Fescue–Any species of grass of the genus *Festuca*, family Gramineae; grown for pasture, hay, and turf. Native to Eurasia and America.

Fescue, Annual–*Festuca megalura*; native annual bunchgrass adapted to droughty, low-fertility sites. Often called six-weeks fescue. Should be seeded in fall or very early spring. Sets seed and dies in late spring, but reseeds well.

Fescue, Creeping Red–*Festuca rubra*; long-lived, low-growing, competitive (but slow-developing) weakly rhizomatous grass. Performs best on acid soils, actually increasing in productivity with increasing acidity. Well-adapted for roadside seedings as well as a permanent cover crop in orchards. Important persistent grass in erosion control.

Fescue, Hard–*Festuca duriuscula*; a low-growing long-lived competitive bunchgrass adapted to a wide range of climate and soil conditions. Has a dense and voluminous root system. Gives excellent erosion control but is slow in becoming established.

Fescue Lameness–A disease condition of cattle attributed to vitamin A starvation and excessive eating of fully matured, tall fescue grass. It is characterized by lameness and apparent circulatory disturbances of ears, tail, and hind feet. Also called fescue foot.

Feta–A white, pickled, salted, Greek cheese made from ewes', or sometimes goats' milk.

Fetid–Having a disagreeable odor.

Fetlock–The joint of a horse's leg just above the hoof; anatomically, the metatarso-phalangeal articulation. Also the tuft of hair growing there.

Fetter–A shackle for the feet of horses.

Fetus–Later stage of individual development within the uterus. Generally, the new individual is regarded as an embryo during the first half of pregnancy, and as a fetus during the last half.

Fever–(1) A temperature higher than normal in animals, which may be caused by disease organisms, poisonous plants, etc. (2) In some cotton plant diseases, as Texas root rot, affected plants have a so-called fever or higher temperature than normal, just prior to showing severe symptoms, which can be distinctly perceived by feeling the leaves in early morning.

Feverish–(1) Referring to milk or stored cream that has an odor due to poor ventilation of the cow stable. Also called smothered, barny, cowy. (2) Pertaining to animals that have a temperature higher than normal because of disease, etc.

FFA Alumni–A national organization composed of former members and supporters of the National FFA Organization. The purpose is to promote and support agricultural education programs and in particular, programs of the FFA. See FFA; Agricultural Education.

Fibrinogen–A soluble protein present in the blood and body fluids of animals.

Filarial–Pertaining to or caused by filariae, the roundworm parasites occurring outside the alimentary canal.

Filariasis–A disease due to infection with filarial roundworms (nematodes of the superfamily Filarioidea). Usually carried by a mosquito.

Filet–(French) Long strips of boneless meat or fish.

Filial–Refers to the child or offspring; the meaning of the F in F_1 and F_2.

Fill–(1) The soil, sand, gravel, etc., used to fill in a depression in a field, or to build up a terrace or embankment. (2) The substances used in filling tree cavities; e.g., asphalt, concrete, wooden or rubber blocks, etc. Also called filler. (3) The increase in weight and form of livestock that have been watered and fed after arriving at their destination. (4) The shaft of a vehicle. (5) To level a depression in a field or cavities in trees, or to build up an embankment or terrace. (6) To feed and water livestock at the end of the shipment to make up for the loss of weight en route. (7) To enlarge with the enclosed seeds, as the pods of leguminous plants; or to be plump and shriveled when approaching maturity, as the seeds of cereals. (Cereal grains before harvest are often referred to as poorly or well-filled.)

Filler–(1) Material used for packing to prevent breakage. (2) An extra row intercultivated between two regular rows of a crop. (3) An extra, short-lived plant grown between slow-growing, larger plants and removed when the latter approach maturity. (4) Any of the various types of appliances for filling special receptacles, as bottle-filler, silo-filler, etc. (5) Any material, active or inert, added to a mixed fertilizer to increase bulk. (6) The nonessential matter in a manufactured or mixed feed, such as high-fiber materials, oat hulls, screenings, etc.

Filly–A young immature female horse. See Colt.

Filterable Virus–A virus that is capable of passing through the pores of a filter which does not allow passage of the ordinary bacteria. See Virus.

Find–To give birth to young; e.g., a cow finds a calf.

Fine Wool–The finest grade of wool: 64s or finer, according to the Numerical Count Grade. This term is also used in a general way for wool from any of the Merino breeds of sheep.

Fineness–(1) One of the several properties of cotton that determines the grade in which it is classified. Fineness refers to the smallness of the cross section diameter of the fibers (or lint). (2) One of several properties of dry milk. Usually the powders prepared by the spray process are extremely fine, whereas those prepared by the cylinder or drum process are coarser. (3) One of the several properties of high-quality hay that help in the determination of the grade of hay as feed for livestock. (4) The relative smallness of the wool fiber.

Fingerling–A fish from 3 to 6 inches long that is used for stocking ponds or lakes.

Finish–The degree of fatness. This term is often used interchangeably with condition but as finish, the fat should lay smoothly over the body in a proper degree to suit the market.

Finishing–The increased feeding of an animal just prior to butchering, which results in rapid gains and increased carcass quality.

Fire–(1) In several different plant diseases caused by bacteria, fungi, or nutritional deficiency, the final result that gives a burning or scorching appearance. (2) Potash deficiency or leaf scorch is sometimes referred to as firing. (3) The burning off of vegetation on lands of various types. This is apparently beneficial in some cases but is often harmful. (4) To treat a spavin or ringbone on a horse with a strong liniment in an attempt to cure or alleviate lameness.

Fire Ant–*Solenopsis geminata*, family Formicidae; a species of ants that is harmful to plants and whose bite is very painful to people and animals. See Imported Fire Ant.

Firm–(1) An economic unit recognized to be engaged primarily in production. (2) (a) To compact the soil, crushing and pulverizing the lumps to facilitate capillary water movement. (b) Designating well-compacted soil that is not lumpy or powdery. (2) In marketing, designating optimistic conditions. (4) Designating a cheese that feels solid. (5) Designating whites of eggs that are sufficiently viscous to prevent free movement of the yolk. (6) (a) Designating meat that is not soft or soggy. (b) Designating a fruit or vegetable that is not overripe or shriveled.

First–(1) The highest grade of lumber. (2) The second grade, next below extra, for butter. (3) The primary occurrence or the beginning of a series.

First Lock–The portion of a horse's mane immediately behind the foretop and head.

First Meiotic Division–The first of a series of two divisions in the process of producing haploid sex cells or gametes.

Fish Ladder–An inclined trough carrying water down a dam at a velocity against which fish can easily swim upstream to reach their spawning grounds.

Fish Liver and Glandular Meal–A poultry feed obtained by drying and grinding the offal of fish.

Fish Meal–A commercial feed for poultry and other farm animals that consists of the clean, dried, ground tissues of undecomposed whole fish or fish cuttings, with or without the oil extracted.

Fish Oil–Oil, such as obtained from the cod, sardine, halibut, etc., used in preparing feed mixtures containing vitamin A and vitamin D. Also fed to babies as a vitamin supplement.

Fish Pond–A small body of water in which the fish population is managed. On a farm, the pond impounds rainfall runoff, reduces erosion, provides a watering place for livestock, and furnishes a place where fish can be grown.

Fish Pond Fertilizer–Inorganic or organic fertilizer applied to a pond to promote the growth of plankton on which the fish depend directly or indirectly for food.

Fish Scrap–Dried-processed nonedible fish and fish residues from fish canneries. In recent years almost the entire supply has gone into animal feed. The little still sold for fertilizer is bought mostly by organic gardeners. It contains about 9 percent N, 7 percent P_2O_5, small amounts of K_2O, and secondary and micronutrients.

Fishery–(1) All activities connected with propagation, cultivation, and exploitation of fishes in inland and marine waters, as also the management of fish resources. (2) Fishing ground.

Fishiness–A flavor defect of dairy products that suggests a fishy taste.

Fission–(1) A form of reproduction, common among bacteria and protozoa, in which a unit or organism splits into two or more whole units. (2) The splitting of a heavy nucleus such as uranium or plutonium into approximately equal parts, accompanied by the conversion of mass into energy, the release of the energy, and the production of neutrons and gamma rays.

Fission Fungi–Fungi that reproduce only by fission.

Fistula–(1) An unnatural body passage linking a hollow space or abscess to the skin surface or a surface membrane or joining two such abscesses. (2) Surgically established openings between a hollow organ and the skin or between two hollow organs for experimental purposes. Fistulas that communicate with the outside are closed with a mechanical plug.

Fistula of the Teat–A fistula that results frequently from barbed wire cuts on the body of the teat so that the milk flows out of the wound.

Fistulous Withers–An inflamed condition in the region of the withers of a horse, commonly thought to be caused by bruising.

Fit–(1) To notch a tree for felling, and to mark it into log lengths after it is felled. (2) To ring, slit, and peel tanbark. (3) To file and set a saw. (4) To condition livestock for use, sale, or exhibit. (5) To prepare land for sowing, i.e., plowing (or disking), harrowing, and rolling. Land so fitted should have no large, hard clods so that the seeds may be placed at a uniform depth and in contact with moist soil.

Fits–See Epilepsy.

Fitting Cow Ration–A ration fed to cows that are being prepared for exhibition, sale, or calving.

Fitty–Referring to a horse that has fits when overheated.

Five-gaited Saddle Horse–A saddle horse that is trained to use the following gaits: walk, trot, rack, pace, and canter.

Flaccid–Without rigidity; lax and weak.

Flagella–Whiplike appendages of certain single-celled aquatic animals and plants, including some bacteria, the rapid movement of which produces motion.

Flank Cinch–A cinch that is separate from the saddle and pulled tight in front of the hips and under the flanks of a horse. It is used in rodeos to cause the horse to buck.

Flank Streaking–Intramuscular fat visible in the flank of a lamb carcass. It is used in determining the quality grade of the carcass.

Flash Pasteurization–The process of bringing a liquid rapidly to a moderate heat to kill growth of yeasts or molds. Also called flash method, flash process.

Flat–(1) A shallow box containing soil, in which seeds are sown or to which seedlings are transplanted form the seedbed. (2) A level, treeless prairie, especially between hills or mountains. (3) Referring to a defect in the flavor of butter, cheese, or milk, due to errors in processing, as lack of sufficient salt, uncleanliness, etc. The flavor may be insipid or lacking in the usual characteristics of the product. (4) A level landform composed of unconsolidated sediments—usually mud or sand. Flats may be irregularly shaped or elongate and continuous with the shore, whereas bars are generally elongate, parallel to the shore, and separated from the shore by water.

Flat Foot–In a horse, a foot of which the angle is less than 45°, or one in which the sole is not concave, or one with a low, weak heel.

Flat-boned–Reference made to the cannon bone region of a horse's leg, which is constituted of the bone, ligaments, and tendons.

Flatuence–A digestive disturbance in which there is an often painful collection of gas in the stomach or bowels of people or animals.

Flatworm–One of the many organisms that are members of the phylum Platyhelminthes.

Flavor–(1) Odor and taste combined with the feeling of the substance in the mouth. (2) (a) The material added to foods to gain a desired taste. (b) To give taste or flavor to a product by the addition of spices, sugar, etc.

Flavus–(Latin) Yellow.

Flax Straw–The straw left after threshing the flax seed crop; used in the manufacture of various types of paper, including cigarette paper. Flax straw of good quality can be used as a substitute for oat straw roughage in wintering cattle. Also called cattle roughage.

Flaxseed–The seed of flax, known as linseed, which is a source of linseed oil used mainly as a drying agent for paints and varnishes. The residual oil cake is used for livestock feed. See Flax.

Flay–To remove the skin from a carcass.

Flea–Certain active, leaping insects of the family Pulicidae, which feed upon the blood of warmblooded animals. Their legless larvae (maggots) eat organic debris. The most common fleas are found on cats and dogs. Some species attack birds and rats. The human flea, ***Pulex irritans***, is a nuisance in eastern Europe and many other regions.

Flea-bitten Gray–A coat color of horses that is gray with darker hairs throughout.

Flea-bitten White–A coat color of horses that is white with numerous, tiny, dark spots.

Fledge–(Fledgling) To acquire the feathers necessary for flight; for example, a young bird just fledged.

Fleece–(1) The wool from all parts of a single sheep, which consists of the crinkly hair up to 12 inches in length. This waviness enables the wool to be matted together into felt or spun into yarn, twine, or thread. (2) The fluffy mass of cotton that remains after the seeds have been removed by ginning. See Lint Cotton. (3) To shear sheep.

Flehmen–An action by a bull, boar, or ram associated with courtship and sexual activity. The lip curls upward and the animal inhales in the vicinity of urine or the female vulva.

Flesh–(1) The portion of an animal body that consists mainly of muscle. (2) Plumpness or corpulence, especially in such phrases as good flesh, etc. (3) The pulpy or juicy portion of fruits or of storage organs of plants, such as potatoes, etc. (4) To remove adhering fat, flesh, and membrane from the pelt of a butchered animal.

Flesh Fly–Any of a large group of flies that lay their eggs, or place the larvae, in flesh.

Flesh Layer–The innermost layer of a sheepskin, next to the flesh, used for manufacturing chamois leather.

Flesh Side–The side of leather that forms the internal surface of the hide.

Flesh-colored–A coat color of animals in which the hair color is the same as the skin.

Fleshed–(1) Designating muscles or lean meat. (2) Designating a pelt with the flesh or fatty pieces removed from the inner surface.

Fleshy–(1) Designating fruit, leaves, and storage organs of plants with juicy or pulpy tissues. (2) Fat or corpulent. (3) Designating the soft or edible portions of meat.

Flight Coverts–The stiff feathers located at the base forward of the flight feathers (primaries) and covering their base.

Flights–The primary feathers of the wing. The term is sometimes used to denote both primaries and secondaries.

Flitch–(1) A portion of log, sawed on two or more sides, which is intended for sliced or sawed veneer. (2) A pile or bundle of veneer sheets from the same bolt laid together in the sequence of cutting. (3) A side or portion of meat, as a flitch of bacon.

Float–(1) An instrument used for filing animals' teeth. (2) A valve in the cream separator to regulate the flow of milk into the cream-separator bowl; or a similar contrivance for maintaining the desired level of water in a tank. (3) A drag or device for leveling soil. (4) To come to the top of the ground, as with a certain type of plow which automatically adjusts if the bottom strikes an obstruction. (5) To flood irrigate, as in floating a meadow (rare in the United States).

Flock–(1) Several birds or domestic mammals, such as sheep, which are tended as a unit. Also called herd, band. (2) Short fibers sheared from the face of cloth or produced in milling or finishing cloth or obtained by shredding rags to almost a powder. (3) In stock judging, one ram, of any age, and four ewes of varying ages as designated by the show.

Flock Book–The record of breeding and ancestry of sheep, kept privately by the flock owner or officially by the Sheep Breed Association, to register purebred sheep. See Herd Book.

Flock Mating–The indiscriminate breeding of fowls.

Flocking Tendency–The habit of congregating in large flocks, inherent in sheep.

Flow–(1) The quantity of liquid that passes through a pipe, gate, channel, or other conveyance for a given unit of time under given conditions of head or pressure, roughness, etc. Units of measure may be cubic feet per second, acre-feet per day, gallons per minute, etc. (2) The movement of silt, sand, etc., in a channel. (3) Discharge from a pipe, etc. (4) The amount of milk produced per cow herd, etc., at a specified time. (5) The ease or difficulty with which a product can be moved from one place to another.

Flow Rate–The rate at which the cow lets down milk.

Flu–Infectious laryngotracheitis.

Fluff–(1) The downy part of a feather. (2) The soft feathers on the thighs and posteriors of birds.

Fluid Milk–The fluid product of a dairy farm or factory in contrast with the more solid products, such as cream, cheese, butter, and dried milk.

Flukes–Flatworms of the class Trematoda which, at maturity, are internal parasites of vertebrate animals and humans but, in snails, usually have intermediate stages. The mature worms are usually seed-shaped and found in the liver, alimentary canal, and other body cavities, attached usually by two suckers. The eggs of the adults are discharged from the host and hatch under favorable conditions. This resulting intermediate stage enters a snail, passes through several stages of its life cycle and is finally deposited on grasses. Domestic animals and people involved in the food chain from the grasses, are reinfected by the adult flukes. The common liver fluke, ***Fasciola hepatica***, may cause a disease fatal to sheep and cattle.

Fluorides–Gaseous or solid compounds containing fluorine, emitted into the air from a number of industrial processes; fluorides are a major cause of vegetation and, indirectly, livestock toxicity.

Flush–(1) To irrigate a field with just enough water to soften the surface soil crust. (2) To increase the feed allowance to ewes or sows with a protein-rich supplement feed a short time before and during the breeding season. (3) A vigorous or abundant, sudden, new growth. (4) To send out vigorous growth on the twigs of a tree. Thus, in the tropical fruit tree, the mango, most of the twigs flush several times a year. (5) To empty, clean, or wash out any material with a quick, heavy supply of water. (6) To introduce and shortly afterwards withdraw an irrigating solution of mild antiseptic of medicinal value, as to flush the vagina or uterus.

Flushed–An animal that receives extra feed and care prior to breeding.

Flushing–The practice of increasing the feed intake of a female animal just prior to ovulation and breeding. This causes the animal to gain some weight and drop more eggs, often resulting in larger litters.

Fly–Any winged insect, such as a moth, bee, gnat, etc. Specifically a two-winged insect of the family Muscidae. Many flies are blood-sucking pests of people and animals, such as the mosquitoes, horse and deer flies, black flies, punkies or nosee-ums, and some sand flies. Some are vectors of diseases, such as the stable flies,etc. Some flies destroy other insects; some are parasites on plants, as the Hessian fly; others are valuable scavengers. Many flies, as the housefly, pass their larval stage in manure and garbage and upon attaining maturity carry with them the bacteria of filth, thus spreading diseases such as typhoid fever, etc.

Fly Net–A meshed covering made of strips of leather or cords of fabric and placed over an animal, usually a horse, to keep the flies away.

Fly-free Date–The date after which it is safe to plant wheat to avoid serious infestation by the Hessian fly. This date has been determined for each county and is available from the county extension agent.

Flying Stall–A portable stall for a horse in which the animal remains while being loaded or unloaded from a vessel or while in transit.

Foal–(1) The unweaned young of the horse or mule. (2) In stock shows, a horse foaled on or after January 1 of the year shown.

Foal Heat–The estrous period that normally occurs a few days after foaling.

Fodder–Feed for livestock, specifically the dry, cured stalks and leaves of corn and the sorghums. In the case of corn, the ears may be removed from the stalk leaving the stover. See Forage, Roughage.

Foetus–Fetus.

Fold–An enclosure of pen for sheep or cattle.

Fold Unit–A small house, complete with covered-in run, for the controlled grazing of poultry.

Folic Acid–A vitamin found in the leaves of leguminous and other plants, in yeast, liver meal, and wheat. Folic acid is needed in hemoglobin formation and for growth. Also called pteroylglutamic acid.

Follicle–(1) A dry, single-carpel fruit, opening along one side for seed dispersal. (2) A small anatomical cavity; particularly, a small blisterlike development on the surface of the ovary that contains the developing ovum. (3) A small sac, gland, or pit for secretion or excretion. The hairs of an animal grow out of pits called follicles. (4) A one-celled, monocarpellary, dry seed vessel or fruit that splits on the ventral edge. (5) The growth that appears on the surface of the ovary late in the estrous cycle and that contains the developing ovum.

Follicle Mite–*Demodex folliculorum*, family Demodicidae; a mite that infests the hair follicles of people and domestic animals, sometimes causing intense itching. The infested follicles may become infected.

Follicle-stimulating Hormone–A hormone, produced by the pituitary gland, which promotes growth of ovarian follicles in the female and sperm in the male.

Following Cattle–The practice of allowing feeder pigs to run behind feedlot cattle so they may glean unused grains and other nutrients form the cattle manure.

Fomentation–A poultice; the external application of warm moist cloths, or other objects to ease pain.

Fomites–Substances other than food that may harbor or transmit a disease.

Food–Anything which when taken into the body, nourishes the tissues and supplies body heat. Also known as aliment and nutriment.

Food Chamber–Hive body containing honey provided particularly for overwintering bees.

Foot–(1) In people and quadrupeds, the terminal portion of the leg that rests upon the ground. (2) The portion of a cultivator to which the sweep is attached. (3) The organs of locomotion of various invertebrates, as the feet of a caterpillar, the foot of a clam or a snail, etc. (4) The base of a tree, tower, mountain, wall, hill, etc. (5) A unit of linear measure, 12 inches.

Foot Maggot–The larva of some species of the blowfly, family Calliphoridae, which infests the feet of sheep and feeds on wounds. The secondary screwworm, ***Callitroga macellaria***, may also be present.

Foot Mange–A foot-skin disease of animals, especially sheep and cattle, caused by the mite ***Chorioptes bovis***. Also called aphis foot.

Foot Pad–The cushions on the bottom of the feet of such animals as cats and dogs.

Foot Rot–A frequently occurring inflammation in animals' feet, usually followed by pus formation on the soft tissues between the toes. It occurs especially in wet ground in sheep and cattle. One type of the disease is infectious pododermatitis, caused by ***Spherophorus necrophorus***.

Foot-and-Mouth Disease–An acute, highly communicable disease of cloven-footed animals caused by a filterable virus. It is characterized by a short incubation period, high fever, and vesicles in the mouth and on the feet that break to form erosions. Also called aftosa fever, aphthous fever, epizootic aptha.

Forage–(1) That portion of the feed for animals that is secured largely from the leaves and stalks of plants, such as the grasses and legumes used as hays. It may either be for grazing as green or standing dry herbage or be cut and fed green or preserved as dry hay. See Fodder, Roughage. (2) To search for, spread out or seek, for food.

Forage Crops–Those plants or parts of plants that are used for feed before maturing or developing seeds (field corps). The most common forage crops are pasture grasses and legumes.

Forage Feeds–(1) Bulky type feeds composed largely of pasture grasses, hays, silage, etc. (2) A mixture of ground or processed feeds that is composed largely of forages.

Forage Legumes—Any of the legume plants that are grown or used largely as forage for livestock, such as alfalfa, clover, etc.

Forage Mixture—Two or more species grown together for forage production. Usually a mixture of legumes and grasses.

Forage Poisoning—(1) Any poisoning of people or animals from eating food or feed contaminated by the presence of some organism that was not destroyed in the usual processing. Such an organism may produce toxins that are fatal to the animal consuming even small quantities of the forage. the most serious cases are those in which the contaminating organism is *Clostridium botulinum* that can withstand repeated heatings at the temperatures ordinarily used in the home canning of vegetables. (2) Poisoning of animals grazing or eating forage containing a plant chemical such as the glucoside of hydrocyanic acid in Sudangrass or sorghum, or by forages improperly cured, such as sweetclover, where the breakdown of the alkaloid coumarin results in rupture of the small blood capillaries in the animal. See Selenium Poisoning.

Forage Value—(1) The relative importance for grazing purposes of a range plant or plants, as a whole, on a range. (2) The rank of a range plant or plant type for grazing animals under proper management, preferably expressed as proper-use factor. See Grazing Value. (3) The comparative value of the forage portion of an animal's ration compared with the other feeds in the ration.

Forage Volume—(1) That portion of a plant above the ground and within reach of grazing animals. (2) The measure of the forage crops; i.e., the aggregate amount of forage produced on a range area during any one year.

Foramen—A small opening, usually in the bone.

Forced Supersedure—The construction by worker bees, especially in those colonies in which the queen is old and enfeebled, of queen brood cells within which the larvae develop to mature queens. The brood is protected from destruction by the kin queen by the workers who lead off a swarm, leaving the newly emerged queen in the colony.

Forceps—In veterinary medicine, a pliers-like instrument used for grasping, pulling, and compressing.

Foreflank—That section of the body of a hog just behind the lower shoulder or foreleg. It provides the leanest section of bacon.

Forefoot—To rope the front legs of a running animal usually from horseback. The pursued animal is thrown hard when it reaches the end of the rope, allowing the rider to dismount and tie the animal before it fully recovers.

Forehand—The "front" of the horse, including head, neck, shoulders, and forelegs; that portion of the horse in front of the center of gravity.

Forehobble—A strap or rope that is tied around the forelegs of an animal to prevent straying.

Foreign Flavor—A flavor defect of milk or its products that is any flavor not commonly developed in or associated with milk or milk products.

Foreign Matter—Any material, substance, etc., which is unnatural to, or not commonly developed in, a product.

Foreign Odor—Any odor that is not natural to a product.

Foreleg—The lower portion, above the pastern, of either front leg of four-legged animals. It is commonly used in reference to horses.

Forelock—A lock of hair growing above a horse's forehead; the forepart of the mane.

Foremilk—The first 25 to 50 millimeter of milk to be withdrawn from the udder at the beginning of milking in contrast to middle milk and strippings. Foremilk is of poor quality chemically and bacteriologically and should be rejected.

Foremilk Cup—A metal milking cup fitted with a dark shelf. The first milk at each milking is drawn into the cup. If there is udder trouble (mastitis) clots of milk will be seen on the dark shelf.

Forepastern—A term commonly used to describe the portion of the front leg next to the hoof.

Forepunch—A kind of counterpunch used to make a place for the nailheads to be set into a horseshoe.

Forequarters—The front two quarters of an animal.

Forestomachs—The three nonglandular stomachs found in ruminants; specifically, the rumen, reticulum, and omasum.

Forestripping—Removal of a small amount of milk by hand from each teat prior to the milking operation; forestrippings are usually discarded because of high bacterial and low fat content.

Forging—The noisy striking of the foreshoe with the toe of the hind shoe by a horse when walking, trotting, or running. Also called clicking, striking.

Form Board—A device used to assist in the insertion of foundations into frames of beehives.

Formamidine Insecticide—A class of insecticides that is used against eggs and mites.

Fortified—(1) Designating a product to which has been added amounts of a vitamin, as vitamin A or vitamin D. (2) A wine to which additional alcohol has been added.

Foundation Herd—Breeding stock; cows, bulls, and heifers or calves retained for replacement; ewes, rams, and lambs for replacement; and nannies. billies, and kids for replacement.

Founder—An inflammation of the tissue that attaches the hoof to the foot; it may be caused by overfeeding, concussion, or a number of other factors.

Four Footing—Throwing an animal by means of a rope around the feet.

Four-cornered Gait—A horse's gait in which each foot is placed on the ground individually. Also called single-foot, rack.

Four-tooth Sheep—A two-year-old sheep.

Four-way Hybrid—The hybrid that results from mating two single crosses. Also called double cross, See Single Cross.

Fowl—Refers to a bird, usually poultry.

Fowl Cholera—An acute, infectious, septicemic disease caused by the bacterium *Pasteurella multocida*. It is characterized in very acute form by sudden death; in less acute forms by greenish or yellowish diarrhea, listlessness, sleepiness, ruffled feathers, stationary habit, impaired appetite, increased thirst, rapid respiration, accumulation of mucous in upper nasal passages, fever, and purpling of wattles and comb; in chronic forms by emaciation, depression, paleness of wattles, comb, and membranes of the head, and lameness. Also called cholera of fowls, roup of fowls, hemorrhagic septicemia of fowls, avian pasteurellosis, cholera gallinarium, pasteurellosis, chicken septicemia, acute fowl cholera.

Fowl Gapeworm–*Syngamus trachea*; a small roundworm that infests the trachea of birds. Infestation of birds under eight weeks old is characterized by a stretching of the neck with open mouth, sneezing, coughing, shaking of the head, loss of appetite, emaciation, prostration, closed eyes, head drawn back against the body. Also called gapes.

Fowl Leukemia–The avian leukosis complex that attacks the blood cells.

Fowl Paralysis–A type of the avian leukosis complex probably caused by a filterable virus; characterized by lameness or drooping of a wing, prostration, legs held in peculiar positions, limpness of legs and wings, no motor control, withering of the muscles of the affected part and by the feathers in the region of the crop being damp, dark, and permanently discolored. Also called neural lymphomatosis, neuritis of chickens, range paralysis, big liver disease.

Fowl Pest–See Fowl Plague.

Fowl Plague–A highly acute, infectious, viral disease characterized by a sudden onset, depression, droopiness, standing still, head drawn in, eyes closed, irregular gait, difficult respiration, shaking of the head, prostration, and bluish-red, swollen wattles and comb. Also called fowl pest, bird plague, chicken pest, bird pest, Brunswick bird plague.

Fowl Pox–A highly infectious, viral disease of birds, especially chickens. Characterized in the skin type, by small, raised, grayish blisterlike spots on the comb or wattles or other unfeathered parts that later rupture and discharge a sticky fluid. As these sores dry, a dark brownish scab forms. Characterized in the diptheric type, by raised, yellowish patches in the mouth and throat that prevent closure of the mouth, and cause difficult breathing and weight loss. Also called chicken pox, contagious epithelioma, canker, avian diphtheria, sore head, fowl diphtheria.

Fowl Tick–*Argas persicus*, family Argasidae; an ectoparasite of poultry in the south and southwestern parts of United States. It feeds only at night, hiding during the day and is a powerful blood sucker, sometimes killing the bird. Infested birds are characterized by weakness of the legs, droopy wings, paleness of comb and wattles, and cessation of egg laying. Also called adobe tick, chicken tick, dove tick.

Fowl Tuberculosis–See Avian Tuberculosis.

Fowl Typhoid–See Avian Typhoid.

Foxtrot–An uneven, easy-to-ride, four-beat gait in horses, intermediate between the walk and the trot.

Frame–A wooden rectangle that surrounds the comb and hangs within a beehive. It may be referred to as Hoffman, Langstroth, or self-spacing because of differences in size and widened end bars that provide a bee space between the combs. The words frame and comb are often used interchangeably; for example, a comb of brood, a frame of brood.

Frame Score–In cattle a score based on subjective evaluation of height or actual measurement of hip height. This score is related to slaughter weights at which cattle will grade choice or have comparable amounts of fat cover over the loin eye at the twelfth to thirteenth rib.

Free Air Cell–An air cell that moves toward the uppermost point in the egg as the egg is rotated slowly in the candling process. This results in a reduction in grade for the egg. See Candle.

Free Wool–Usually means free from defects, such as vegetable matter.

Free-choice Feeding–A type of feeding routine whereby feed, water, salt, etc., are provided in unlimited quantities and an animal is left to regulate its own intake.

Free-ranging–Allowing animals, especially poultry, to roam freely and eat as they wish without any sort of confinement.

Freemartin–A sterile female calf born as a twin to a normal male calf. It is usually intersexual as a result of a male hormone absorbed from its twin through anastomosed placental vessels.

Freeze Branding–An identification method done by clipping hair from the brand area, wetting skin with alcohol, then applying a branding iron cooled in liquid nitrogen or dry ice and alcohol.

Freidman Test–A test for pregnancy in which a small amount of the urine of the tested animal is injected into the bloodstream of a virgin female rabbit. Pregnancy is indicated by certain changes in the ovaries of the rabbit.

French Combing Wools–Wools intermediate in length between strictly combing and clothing. French combs can handle fine wools from 1¼ to 2½ inches in length. The yarn is softer and loftier than Bradford (worsted) yarn.

Fresh–(1) Designating a cow that has recently dropped a calf. (2) Designating very recently harvested or gathered food products. (3) Designating an egg of good quality. (4) Designating sweet water; i.e., not salty water.

Fresh Ham–Uncured pork from the hindquarters of a pig.

Fresh Manure–Recently excreted animal dung whose direct contact can be harmful to plant tissues because of rapid chemical and fermentive changes that take place.

Freshen–To come into milk, as when a dairy animal gives birth.

Fribs–Short second cuts of wool resulting from faulty shearing; also small-sized dirty or dungy locks.

Friesian–A name sometimes incorrectly applied in the United States to the black-and-white dairy cattle breed the Holstein-Friesian.

Fringed Tapeworm–*Thysanosoma actinioides*; a parasite of domestic sheep and certain related wild ruminants in western North America and Central and South America. The tapeworm lives in the bile ducts of the liver and in the small intestine. It causes thickening, or hypertrophy, of the bile ducts. In the United States, livers infected with this tapeworm are condemned as unfit for human consumption. The life history of this tapeworm is unknown, but is very likely transmitted by some small invertebrate animal.

Frizzle Feather–A term used to denote feathers that are curled and that curve outward and forward, a characteristic of Frizzle chickens.

Frog–(1) That part which holds turning plow bottom parts together; an irregularly shaped piece of metal to which the share, landslide, and moldboard are attached. (2) The triangular, horny pad, located on the posterior-ventral part of the hooves of horses, mules, etc.

Frowsy–A wasty, lifeless-appearing, dry and harsh wool, lacking in character. In direct contrast to lofty.

Fruit–(1) Botanically, the matured ovary of a flower and its contents including any external part that is an integral portion of it. (2) In a popular sense, the fleshy, ripened ovary of a woody plant, tree, shrub, or vine, used as a cooked or raw food; it is not always completely and satisfactorily distinguished from a vegetable. The latter may also include edible leaves, roots, and tubers. A fruit may be considered a dessert and not the principal part of a meal as a vegetable is so considered (such a meaning has accepted usage and has also been confirmed by a court decision). See Vegetable. (3) To bear or produce fruit in any of the senses. (4) A mature ovary, either plant or animal.

Fry–(1) To cook in hot oil or fat. (2) The young stage of fishes, particularly after the yolk sac has been absorbed. (3) Young fish, newly hatched, after yolk has been used up and active feeding commenced.

Fryer–Any young chicken approximately eight to twenty weeks old, of either sex, weighing more than 2½ pounds but not more than 3½ pounds, which is sufficiently soft-meated to be cooked tender by frying.

FSH–See Follicle-stimulating Hormone.

Fuem Board–A general name for any shallow wooden cover used to hold repellents for driving bees from honeycombs.

Full Bloods–A term referring to purebred animals.

Full Feed–A feed or ration being fed to the limit of an animal's appetite.

Full Mouth–A state in sheep or goats when an animal has a full set of permanent teeth. This occurs at approximately the age of four. The animal will continue to have what is known as a full mouth until it loses some teeth or until it loses all of its teeth. See Broken Mouth, Gummer.

Fulling–The operation of shrinking and felting a woolen fabric to make it thicker and denser. The individual yarns cannot be distinguished on a fulled fabric.

Fumagillin–Antibiotic given bees to control nosema disease.

Fumigant–A substance or mixture of substances that produce gas, vapor, fume, or smoke intended to destroy insect and other pests.

Fumigate–To destroy pathogens, insects, etc., by the use of certain poisonous liquids or solids that form vapors. See Fumigant.

Functional Efficiency–In cattle, the production of as much good red meat per unit area as possible. In its broadest sense: fertility, genetic excellence, libido, ability to copulate, estrus, ovulation, fertilization, embryo survival gestation., parturition, and mothering ability of the cow.

Fungal Spray–An herbicide consisting of fungal spores in a liquid solution sprayed on weeds in crops as a means of biological control of the weeds. For reasons not adequately understood, the fungal spores of different fungi are specific for specific hosts, making it possible to control selected plants without damaging other nearby crop plants.

Fungi–Plantlike organisms that have no chlorophyll; they get their nourishment from living or decaying organic matter. Plural of fungus.

Fungistat–An agent or chemical material that prevents the growth and reproduction of, but does not kill, fungi.

Fungous–Pertaining to a fungus, as a fungous disease.

Fungus–A lower order of plant organisms, excluding bacteria, which contains no chlorophyll, has no vascular system, and is not differentiated into roots, stems, or leaves. They are classified in the plant kingdom division Thallophyta, and vary in size from single-celled forms to the huge puffballs of the meadows. Fungi are familiar as molds, rusts, smuts, rots, and mushrooms. For the large part they reproduce prolifically by means of single or multicelled spores of various longevities, which are disseminated by air or water. They are either saprophytic or parasitic and many are considered useful in breaking down dead vegetation and organic matter into humus and as agents of fermentation, as in yeasts; others are also destructive in rotting structural timbers, posts, cloth, leathers, etc. The parasitic forms cause destructive plant diseases; a few are human and animal parasites.

Fur-bearing Sheep–The Karakul; the only breed of sheep in the United States kept primarily for the fur pelts of its young.

Gait–The action of a horse's legs, such as walking or running.

Gaited–Definite rhythmic movement of a horse such as trot, canter, pace, etc.; certain breeds are selected and bred on the basis of their ability to perform the various gaits.

Galactagogue–Any substance that promotes the flow of milk.

Galactose–A white crystalline sugar obtained from lactose (milk sugar) by hydrolysis.

Gallop–A fast, three-beat gait of a horse, in which two diagonal legs are paired, their single beat falling between the successive beats of the other two legs, the hind one of which makes the first beat of the three. A hind foot makes the first beat in the series, the other hind foot and diagonal fore foot make the second beat simultaneously, and the remaining fore foot makes the third beat in the series. Then the body is projected clear of the ground and the hind foot makes the first beat in a new series.

Galloway–A breed of black beef cattle that originated in Scotland.

Gambrel–A wooden or metal rod whose ends are inserted in the hocks of hogs, and which is used to support animals while butchering. Also called gamble. See Beef Pritch.

Gametogenesis–The process in plants or animals, male or female, involving the production of gametes; ovigenesis or spermatogenesis.

Gamma Globulin–A specific protein fraction of the fluid part of the blood in which are included the bodies that protect against certain infections (immune bodies).

Gander–A mature male goose.

Gangrene–Massive death of tissue caused by bacterial infection or interference by injury, infection, freezing, etc., with blood circulation. Dry gangrene, or mummification, may occur without bacterial action; moist gangrene occurs as a result of the action of bacteria.

Gangrenous Mastitis in Ewes–Mastitis of sheep; an inflammation of the udder of ewes, frequently caused by infection with the filth bacteria *Pasteurella mastitidis*, *Staphylococcus aureus*, etc. The condition is characterized by loss of appetite; depression; solitary habit; refusal to suckle her lamb; straddled gait; hot, painful udder; a hardening and caking of the udder; dark, bluish-violet spots on the udder. Also called blue gag, black garget, garget, mammitis.

Gap–(1) A break in a fence or wall that may be used as a gate. (2) A depression forming a break in the continuity of the crest of a mountain ridge.

Garget–(1) The abnormal changes in the udder and its secretion as a result of mastitis. (2) A hog and cattle disease in which the head or throat becomes inflamed.

Gastric–Refers to the stomach.

Gastroenteritis–Inflammation of the stomach and intestines.

Gastrointestinal–Refers to the part of the digestive system made up of the stomach and the intestines.

Gate Cut–Method of dividing a group of cattle by driving through a gate and separating them impartially.

Gaucho–The South American cowboy, in particular, the term is used in the Pampas area of Argentina. The gaucho is considered by many to be the world's finest rough rider.

Gee–A command that directs a horse, mule, or ox to turn to the right; whereas *haw* means turn to the left.

Geese–The plural of goose.

Gel–A colloidal suspension in which the particles have precipitated; e.g., the coagulum in cheese making.

Gelatin–A water-soluble protein prepared from collagen by boiling with water.

Gelbvieh–A large breed of beef cattle that originated in Germany. They range in color from medium red to fawn.

Geld–(1) Designating an animal that is sterile. (2) To render sterile, as in castration.

Gelding–A castrated male horse.

Gene–The simplest unit of inheritance. Physically, each gene is apparently a nucleic acid with a unique structure. It influences certain traits. Sometimes called a trait determiner.

Gene Pool–The genetic base available to animal breeders for stock improvement.

Gene Splicing–The technique of inserting new genetic information in a plasmid. See Plasmid.

Gene Transfer–The process of moving a gene from one organism to another. Biotechnology methods permit the identification, isolation, and transfer of individual genes as a molecule of DNA. These methods make it possible to transfer genes between organisms that would not normally be able to exchange them. See DNA, Gene.

Genera–The plural of genus.

General Use Pesticide–A pesticide that can be purchased and used without obtaining a permit. It is considered safe for general public use. See Restricted Use Pesticide.

Generation–The group of individuals of a given species that have been reproduced at approximately the same time; the group of individuals of the same genealogical rank.

Generation Interval–The period of time between the birth of one generation and the birth of the next.

Generic–A word used to describe a general class of products, such as meats, vegetables, or grains; also refers to an unadvertised brand.

Genesis–(1) Origin, or evolutionary development, as of a soil, plant, or animal. (2) A combining form, to indicate manner or kind of origin, as parthenogenesis, biogenesis, etc.

Genetic Base–The breeding animals available for a producer to use.

Genetic Drift–The gradual change in a plant or animal species because of rearrangement of the genes due to the environment or unknown causes.

Genetic Engineering–Alteration of the genetic components of organisms by human intervention. Also known as biogenetics.

Genetic Index–An estimate of the future Predicted Difference of a young bull. See Predicted Difference.

Genetic Trait Summary (GTS)–The comparative ranking of beef sires derived from evaluating the conformation of the daughters.

Genetics–(1) The science that deals with the laws and processes of inheritance in plants and animals. (2) The study of the ancestry of some special organism or variety of plant or animal. Also called breeding.

Genital–Pertaining to the organs of reproduction.

Genital Eminence–In sexing chicks, a very small, shiny or glistening projection that is the rudimentary male copulatory organ. Also called male process.

Genitalia–The organs of the female reproductive tract including the external genital organs.

Genome–A complete set of chromosomes (hence of genes) inherited as a unit from one parent.

Genotoxicity–The quality of being damaging to genetic material.

Genotype–The genetic constitution (gene makeup), expressed and latent, of an organism. Individuals of the same genotype breed alike. See Phenotype.

Genus–A group of species of plants or animals believed to have descended from a common direct ancestor that are similar enough to constitute a useful unit at this level of taxonomy.

Germ–(1) In reference to animal disease, a small organism, microbe or bacterium that can cause disease. (2) The embryo of a seed.

Germ Cell–A cell capable of reproduction or of sharing in the reproduction of an organism, which may divide to produce new cells in the same organism, as contrasted with the somatic or body cells.

Germ Plasm–Term for the reproductive and hereditary substance of individuals that is passed on from the germ cell in which an individual originates in direct continuity to the germ cells of succeeding generations. By it, new individuals are produced and hereditary characteristics are transmitted.

Germ Spot–The germinal disc on the surface of the yolk of an egg. The blastoderm of a fertilized egg and the point at which embryonic development starts in the making of a chick.

Germicide–Any agent that kills germs.

Gestation Period–The length of time from conception to birth of young in a particular species. Usual gestation periods for farm animals are: mare, 330 to 340 days; cow, 230 to 285 days; ewe, 145 to 150 days; sow, 112 to 115 days; goat, 148 to 152 days; jennet (female ass), 360 to 65 days.

Get–The offspring of an animal, usually the sire. Get of sire: a show classification in which calves of a particular sire are judged against calves of other sires.

Ghee–In India, semifluid butter prepared from the milk of the buffalo, cow, sheep, or goat. After the butter has been extracted from the milk, it is heated to drive off excess water, cooled, and only the more-liquid, oily portion used. It is virtually 100 percent butterfat.

Giblets–The edible small parts of dressed birds: liver, heart, gizzard, etc.

Gid–A disease of sheep and goats, but rarely of cattle, which is caused by the gid bladderworm (*Coenurus cerebralis*) or the larval stage of *Multiceps multiceps*, a tapeworm of dogs. The internal parasites are carried by the bloodstream to various parts of the body, and those that reach the brain complete their full development there as a cyst, reaching the size of a hen's egg. Infestation is characterized by staggering, loss of control of limbs when excited, weaving gait, excitation, walking in circles, prostration, in some cases also by holding the head lowered or elevated, by resting the head on an object, turning backward somersaults, exhaustion, and death. Also called sturdy, turn sickness, staggers.

Gigantism–(1) The production of luxuriant vegetative growth that is usually accompanied by a delay of flowering or fruiting. Also called gone-to-stalk, gone-to-weed. (2) In animals, abnormal overgrowth of a part or all of the body. Also called giantism.

Gigot–A leg of mutton, venison, or veal trimmed ready for the table.

Gill–(1) A unit of liquid measure equal to: 4 fluid ounces, 0.25 liquid pints, 0.118 liters. (2) The breathing mechanism of an aquatic animal such as a fish.

Gilt–A name for a young female pig until it produces its first offspring, when it becomes a sow.

Girth–(1) The circumference of the body of an animal behind the shoulders. (2) A band or strip of heavy leather or webbing that encircles a pack animal's body; used to fasten a saddle or pack on its back. Also called cinch. (3) The circumference of a tree.

Gizzard–The muscular posterior stomach of birds, which has muscular walls and a thick, horny lining; its principal function is the grinding or crushing of coarse feed particles. The presence of grit increases the efficiency of the grinding process.

Gjetost–National cheese of Norway made from the whey of goats' milk. It has a sweet caramel flavor.

Glabrous–Smooth, devoid of hair or surface glands.

Gland–(1) In animals, an organ that secretes substances for the body's use or that excretes waste matter. (2) In plants, any special secreting organ.

Gland Cistern–The internal portion of the mammary gland immediately above each teat into which milk collects as it is secreted by the milk-producing glands of the udder. Each gland cistern is very irregular in shape and capacity.

Glanders–A contagious disease of horses, mules, etc., communicable to people; caused by *Malleomyces mallei (Bacillus mallei)*. Pulmonary glanders is characterized by unthriftiness and difficult breathing. Nasal glanders is characterized by ulcers in the mucous membranes of the nose that discharge a blood-streaked fluid, a swelling of the glands under the jaw, and unthriftiness. Skin glanders is characterized by hard nodules under the skin, which break to form ulcers. There is a thick discharge from the ulcers that become confluent. Also called burr, farcy.

Glass Eye–An animal's eye in which the iris is pearly white.

Glauber Salt–$Na_2SO_4 \cdot 10H_2O$; sodium sulfate decahydrate; used as a laxative.

Globular–Having a round or spherical shape.

Globule–A collection of several molecules of fat that takes on a spherelike appearance, and is insoluble in water.

Glomerata–Dense, compact.

Glossanthrax–A disease of horses and cattle in which the oral cavity, especially the tongue, becomes ulcerated and gangrenous.

Glucagon–A hormone produced by the pancreas that stimulates a rise in blood sugar.

Glutinous–Sticky.

Glyceride–Natural fats and oils formed in plants and animals by the chemical union of glycerin and fatty acids.

Glycerol–One of the components of a fat molecule; a fat molecule is composed of three fatty acids attached chemically to glycerol.

Glycogen–$C_6H_{10}O_5$; a carbohydrate similar to starch, found abundantly in the liver and stored in lesser amounts in other tissues and organs. Also called animal starch.

Glycolytic–Pertaining to the chemical breakdown of sugars to lactic acid.

Go Off Feed–(1) To cease feeding with a normal appetite. (2) To refuse feed in the amounts and kinds previously eaten.

Go Stale–(1) To suffer sperm deterioration, as a bull. (2) To go off flavor, as a product. (3) As a person or animal, not to work at normal standards of production. (4) As an animal, to go off feed.

Goat–Any horned ruminant of the genus *Capra*, family Bovidae, especially the domestic goat, *C. hircus*, which is bred as a source of milk, meat, and wool or hair.

Goat Fever–A photosensitization of sheep and goats resulting from ingestion of *Agave lophantha* var. *poselgeri*, family Amaryllidaceae, or the atamasco lily, and exposure to bright sunlight; characterized by jaundice, liver and kidney lesions, and sometimes by swelling of the face and ears. Also called swell head, big head.

Goat Month–The tenure on range or pasture of a mature goat for one month.

Goatling–Female goat between one and two years of age that has not borne a kid.

Gobby–A lumpy, unattractive condition of the fat covering the body of an animal, such as a sheep or beef animal.

Goiter–The enlargement of the thyroid gland that results from a deficiency of iodine in the diet of people and all farm/ranch animals.

Goitrogen–A food or feed so low in iodine that a steady diet or ration of it may produce goiter in animals or people; e.g., cabbage.

Golden Bay–The rich, yellowish-red of an animal's coat.

Golden Horse–See Palomino.

Gomer Bull–A bull that has been altered so as to render him sterile while leaving the testicles intact. A gomer bull is used to indicate that a cow is in heat by mounting her.

Gonad–The organ in a male or female animal that produces the gametes; an ovary or testis.

Gonadtropin–A hormone that stimulates the gonads.

Gone to Sugar–Designating honey and syrup in which the sugar has crystallized.

Goose–(1) Any large, web-footed bird (intermediate in size between swans and ducks) of the subfamily Anserinae (family Anatidae) including the genus *Anser* and related genera. The domestic goos, *Anser domesticus*, includes a number of breeds that are kept for their fresh and feathers. (2) The female goose as distinguished from the male, or gander.

Goose Feather–Any feather (or feathers) from a goose. Especially prized are the mature, soft, downy feathers from the breast and abdomen, which are used as a filling for pillows, sleeping bags, etc. These soft feathers are often plucked from live geese in the summer prior to molting.

Goose-rumped–An animal having a short, steep croup that narrows at the point of the buttocks.

Goosestep–A peculiar walk or body action, locomotor incoordination (spastic gait) of swine which is caused by a nutritional deficiency of pantothenic acid, B3 complex vitamin.

Gore–To pierce the body with an animal's horns.

Gosling–A very young or recently hatched goose.

Gossypol–A material found in cottonseed that is toxic to swine and certain other simple-stomached animals.

Gout–An uncommon, nutritional disease of mature poultry, characterized by internal deposits of sodium urate in the viscera or joints. Also spelled gowt.

Graafian Follicle–A fluid-filled sac in which an egg develops. Part of the ovary, the graafian follicle, also secretes the female sex hormone estrogen which causes heat in females.

Grade–(1) The slope of a road, channel, or natural ground. (2) The finished slope of a prepared surface of a canal bed, roadbed, top of embankment, or bottom of excavation. (3) Any surface that is prepared for the support of a conduit, paving, ties, rails, etc. (4) Any animal that has one purebred parent and one of unknown or mixed breeding. (5) Designating a herd, flock, brand, etc., of such animals. (6) The classification of a product, animal, etc., by standards of uniformity, size, trueness to type, freedom from blemish or disease, fineness, quality, etc. (7) To smooth the surface of a road. (8) To raise the level of a piece of ground by the addition of earth, gravel, etc.

Grade Animal–An animal with nonpurebred ancestors.

Graded Eggs–Eggs that have been sorted and labeled according to size and quality.

Grading–(1) The classification of products, animals, etc., into grades. (2) The mating of a purebred animal with one of mixed or unknown breeding. (3) The smoothing of the land surface.

Grading Up–The practice of improving a flock whereby purebred sires are mated to grade animals and their offspring. In three generations the offspring will be seven-eighths purebred and in some cases eligible for registration. Upgrading.

Grain-fed–Designating animals, such as cattle, which are being or have been fattened for market largely by the use of grain feeds.

Gram Stain–A staining method devised by a Danish physician, Hans Gram, to aid in the identification of bacteria. Bacteria either resist discoloration with alcohol and retain the initial deep violet stain (gram-positive) or can be decolorized by alcohol and are stained with a contrast stain (gram-negative).

Granualted Honey–Honey in which crystals of a sugar (dextrose) have formed.

Granular–(1) In the form of granules or small particles. (2) Covered with small grains, minutely mealy. (3) A porous soil ped. See Soil Porosity, Soil Structure.

Granular Vaginitis–A disease of cattle characterized by the formation of small granular nodules on the vulvar and vaginal mucous membranes of the cow and prepuce of the bull. The affected mucous membrane may be slightly swollen and sensitive and may bleed easily. Frequent urination and some straining may be seen in the cow. The condition may affect young calves and heifers as well. It may spread by coitus, by the hands of attendants, or perhaps by grooming tools. Also called granular venereal disease, nodular vaginitis, infectious vaginitis, contagious granular vaginitis, bull burn.

Grass Egg–In marketing, an egg with an olive-colored yolk.

Grass Lamb–A lamb that is dropped in the springtime and is raised on pasture in the summer months and butchered in the fall, when pasture is less productive.

Grass Tetany–A magnesium-deficiency disease of cattle characterized by hyperirritability, muscular spasms of the legs, and convulsions. In sheep, it is apparently associated with a calcium and magnesium deficiency. The disease is seen when the animals are turned out to lush spring pastures in some areas. Also called grass staggers.

Grass-fattened–Designating an animal that has been fattened on pasture or range, in contrast to one fattened on grain or other feed concentrate.

Grasser–Cattle marketed directly off grass pastures and not grain-fed.

Gray–(1) A color of an animal's coat that has white hairs mixed with black. (2) A cotton-lint color designation that is the darkest in chroma.

Gray Roan–A coat color for a horse that is roan in combination with gray. See Steel Gray.

Gray Speck–A magnesium-deficiency disease of oats characterized by light green to gray, irregularly shaped flecks on the leaf blades that enlarge, dry out, and turn brown or buff-colored; dwarfing; and reduction of yield. Also called dry leaf spot.

Gray Wool–Fleeces with a few dark fibers, a rather common occurrence in the medium wools produced by Down or black-faced breeds.

Graze–(1) To consume any kind of standing vegetation, as by domestic livestock or wild animals. See Browse. (2) To cause domestic animals to graze.

Graze Off–To cause animals to feed on and almost consume the top growth of herbaceous vegetation.

Graze Out–To allow domestic animals to feed abusively on certain palatable grasses or forbs alone or in combination until the vegetation ceases to exist on a particular pasture or range.

Grazier–A rancher; a person who owns or manages livestock on grazing land.

Grazing–(1) Feed available to animals on ranges and pastures. (2) The process of feeding by livestock on live or standing plants other than browse.

Grazing Bit–A snaffle or easy curb bit that does not prevent a horse from grazing.

Grazing Capacity–In range or pasture management, the ability of a grassed unit to give adequate support to a constant number of livestock for a stated period each year without deteriorating. It is expressed in number of livestock per acre of given kind or kinds, or in number of acres per specified animal. Modifications must be made during years of drought.

Grazing District–In the United States, an administration unit on federal range established by the Secretary of the Interior under the provisions of the Taylor Grazing Act of 1934, as amended; or an administrative unit of state, private, or other range lands, established under state laws.

Grazing Fee–A charge made for livestock grazing on a range on the basis of a certain rate per head for a certain period of time, as distinguished from lease or rental of the land on which animals may be grazed.

Grazing Land–Land used regularly for grazing; not necessarily restricted to land suitable only for grazing: excluding pasture and cropland used as part of farm crop rotation system.

Grazing Pressure–The actual animal-to-forage ratio at a specific time. For example, three animal units per ton of standing forage.

Grazing Unit–(1) The quantity of pasturage used by an average, mature cow or its equivalent in other livestock in a grazing season in a given region. (2) Any division of the range that is used to facilitate range administration or the handling of livestock.

Grazing Value–The worth of a plant or cover for livestock and/or game that is determined by its palatability, nutritional rating, amount of forage produced, longevity, and area of distribution.

Grease–(1) See Fat, Lanolin, Lard. (2) A thick petroleum derivative used for lubrication. (3) Hog fat as distinguished from tallow, which is the fat of cattle and sheep; commercially differentiated by temperature of solidification which is below 40°C for grease and above 40°C for tallow. (4) To lubricate a machine. (5) To apply salve to a wound or irritation.

Grease Heel–In horses, a low-grade infection affecting the hair follicles and skin at the base of the fetlock joint, most frequently the hind legs. It is similar to scratches, but in a more advanced stage.

Grease Mohair–A mohair fleece before it is cleaned. See Mohair.

Grease Wool–Raw wool after it is removed from a sheep and before it is scoured.

Green Algae–Organisms belonging to the class Chlorophyceae and characterized by photosynthetic pigments similar in color to those of the higher green plants. Food manufactured by photosynthesis is stored in algal tissues as starch. See Algae.

Green Broke–A term applied to a horse that has been hitched or ridden only one or two times.

Green Chop–Green forage that is cut with a field chopper and hauled to lots or barns for livestock feed in lieu of pasturing. See Green Chopping.

Green Geese–Geese full fed for fast growth and marketed at ten to thirteen weeks of age when they weigh 10 to 12 pounds (4.5 to 5.4 kilograms); also called junior geese.

Green Hay–(1) Uncured hay. (2) That hay which, on being cured, retains a green color.

Green Hide–Skin that has been cleaned, scraped and dried, but has not yet been permanently tanned.

Green Manure–Crops such as legumes or grasses that are grown to be plowed or spaded into the soil to increase humus content and improve soil structure. See Cover Crop.

Green Pellet–A pellet made from alfalfa meal only, or a complete pellet that contains enough green roughage to color it.

Gregariousness–The tendency within a species population to flock or herd together.

Gristle–Cartilage.

Groin–(1) A structure built from the shore into the water to protect the bank against erosion. (2) The part of the body of a person or animal where the thigh joins the body trunk.

Groom–(1) A person who curries, combs, washes, etc., an animal and cares for it generally. (2) To wash, curry, brush, and generally care for an animal. (3) To trim grass and to make a yard and flower garden neat and trim.

Grooming Chute–A portable chute in which cattle are held while they are being groomed for a show.

Grow–(1) To live and to increase in stature and girth toward maturity. (2) To cultivate plants. (3) To raise animals.

Grow Out–To feed cattle so that the cattle get a certain desired amount of growth without much, if any, fattening.

Growth–(1) The increment in size of a living organism. (2) Plants or plant parts. (3) A tumor, gall, etc. (4) The development of an organism from its earliest stage to maturity. (5) The development or increase of an enterprise or an organization.

Growth Hormone–A hormone that promotes body growth and milk production.

Growthy–A livestock judging term used to describe an animal that is large and well developed for its age.

Grub Hole–A hole or wound in the hide of an animal caused by the larva of the common cattle grub. See Cattle Grub.

Grulla–A coat color of some animals, especially the American quarter horse. It is a slate-blue bordering on a sooty black.

Gruyere–A hard, light, yellow, cooked cheese with holes. It is made in Switzerland, France, Finland, and Argentina.

Guaranteed Analysis–On feed labels or tags, a listing of certain nutrients, usually crude protein, crude fiber, fat, and ash, guaranteeing a minimum or maximum percentage of each in the feed.

Guernsey–A breed of cattle that originated on the island of Guernsey in the English Channel and is highly regarded for its dairy characteristics and qualities. It is widely distributed, with the largest numbers in the United States. Guernseys are fawn or reddish-fawn and white, and medium in size. Guernsey milk is high in color and percentage of butterfat.

Gullet–See Esophagus.

Gummer–A sheep or goat having no teeth. See Broken Mouth, Full Mouth.

Gummy Wool–Grease wool that has an excessive amount of yolk, or scoured wool that still has some yolk in it.

Gummy-legged–A term applied to a horse having legs in which the tendons lack definition, or do not stand out clearly.

Gynandromorph–An individual of which one part of the body exhibits female characteristics and another part male characteristics.

Gyp-rope–The rope used by a trainer to rope or to exercise his horse.

Gyr–A strain of Zebu cattle.

Hackamore–A type of head restraint for horses similar to a halter but provided with a loop or noose that may be placed about the nose of the horse to give additional restraint in handling unbroken horses. Also called jaquima, headstall.

Hackles–(1) The long, narrow, neck plumage of male birds. (2) The erectile hairs on the backs of certain animals.

Hackney–A breed of light harness or carriage horses that originated on the east coast of England, averaging from fifteen to sixteen hands high and having considerable depth through the chest and body. It lifts its forefeet in an exaggerated manner. The preferred coat colors are chestnut and bay.

Hackney Pony–A breed of harness ponies which resulted from crossing a Hackney stallion and a pony mare of Welsh breeding. in general, it is similar to the Hackney except for size, being fourteen hands high or less.

Haemonchosis (Hemonchosis)–The infection of ruminants with the large stomach worm, species of the genus *Haemonchus*, or the disease caused by this parasite. Haemonchosis is usually accompanied by loss of blood and anemia caused by the activities of the worms, and it is one of the most serious helminthic diseases of ruminants.

Hair–The outgrowth of a cell in the epidermis of a plant or animal. In vast numbers it forms the coat of an animal and is frequently used as a fiber, such as wool. See Pubescent, Root Hair.

Hair Ball–Hair an animal has swallowed that has gathered in the stomach in the form of a ball; common in the stomach of cats and in the rumen of ruminants. Also called aegagropila, egagropilus, piliconcretion, trichobezoar. See Bezoar, Stomach Ball.

Hair Slips–An animal hide that has been improperly salted and cured, allowing some decomposition to take place as indicated by slipping patches of hair.

Hairlessness–(1) In genetics, a lethal factor characterized in the Holstein breed by calves being born without hairs except around natural openings and at the end of the tail. (2) An iodine-deficiency disease that causes pigs to be born hairless. See Iodine.

Half Cheek Snaffle Bit–A type of snaffle bit used with racing harness horses. See Snaffle Bit.

Half Sisters–Queen or worker bees produced by a single queen and sired by drones that are not related to each other.

Half-blood Wool–The designation of a grade of wool classified immediately below the fine grade.

Half-bred–Designating a horse that has a Thoroughbred as one parent and a draft horse as the other. A popular, medium-heavyweight horse with considerable action.

Half-brother (Half-sister)–Animals from the same mother but by different sires, or by the same sire from different mothers.

Half-sib–A half-brother or half-sister.

Halfbreed–(1) A cross between two botanical varieties of the same species. (2) A cross between two races.

Halter–A leather, rope, or chain device that is essentially a loop over an animal's nose and another loop behind the ears for leading or restraining.

Ham–A cut of pork that consists of the hindquarters of a swine from the hock to the hip, including the thigh and buttock; specifically, one that has been salt-cured and smoked. See Fresh Ham, Picnic Ham.

Ham Butt–A retail cut of pork that consists of the thigh end of ham.

Ham String–The large tendon above and behind the hock in the hind leg of quadrupeds.

Hamburg(er)–Retail meat which consists of ground lean and fat beef (except that no heart, liver, kidney, etc., is used).

Hame–The wood or metal parts of the harness of a draft animal that fit about the collar to which the traces are attached for pulling.

Hammerhead–A coarse-headed animal.

Hampshire–(1) A medium-wool, black-faced breed of sheep that originated in England. It is a popular mutton breed whose ewes weigh up to 100 pounds. (2) An American breed of black, white-belted swine.

Hamstrung–Designating an injury to the tendon behind the cannon in the hind leg of an animal.

Hand–(1) A laborer who is either permanently employed or migratory, as harvest hand, hired hand. (2) A unit of measurement equal to 4 inches (10 centimeters) that is used to measure the height of horses from the ground to a point at the shoulder. (3) A bunch of tobacco leaves of the same grade that are tired together for easier handling. (4) The near horse in a team used for plowing. (5) A half-whorl-like cluster of bananas attached to the rachis of the spike or bunch. (6) Designating any tool, implement, etc., that is manually operated. (7) Designating any manual labor, as hand chopping.

Hand Breeding (Hand Mating)–A system of animal breeding in which the breeder controls the number of times coitus is performed.

Hand Feeding–A type of feeding routine whereby an animal is fed measured amounts of food, water, salt, etc., at fixed intervals.

Hand Gallop–A restrained or slow gallop of horses. See Canter.

Hand Milker–A person who milks a cow manually.

Hand Strip–(1) To take the last bit of milk from a cow's udder, usually following machine milking. (2) To harvest a seed crop by hand.

Handiness–A characteristic of a horse with good manners who has good coordination, and is not clumsy or awkward in movement or action.

Handpick–(1) To harvest by hand, as contrasted to harvesting by machine. (2) To pluck the feathers of a fowl manually.

Hang–To age meat or game by hanging in a cool unrefrigerated place.

Haploid–An organism or cell with one set of chromosomes; for example, drone bee. See Diploid.

Haploid Number–In genetics, this is half (haploid) the number of chromosomes that are usually present in the nucleus; occurs during reduction division.

Hard Breeder–Designating a female animal that is difficult to breed or has difficulty conceiving.

Hard Feather–A term used in describing a plumage characteristic of game fowl. Hardness is dependent on the narrowness and shortness of the feather, toughness and substance of shaft, substance of the barbs, and the firm closely knitted character of the barbs forming the web and scanty fluff.

Hard Feeder–An animal that stays in a thin condition even though well fed. Sometimes called hard keeper.

Hard Keeper–An animal that is unthrifty and grows or fattens slowly regardless of the quantity and quality of feed. Also called hard feeder.

Hard Milker–A cow that milks slowly due to hardened or constricted sphincter muscles in the end of the teat, or a fleshy udder with limited space for the fast accumulation of milk.

Hard-mouthed–A term used when the membrane of the bars of a horse's mouth where the bit rests have become toughened and the nerves deadened because of the continued pressure of the bit.

Hardiness–The state of being hardy. See Hardy.

Hardware Disease–A condition found in ruminants in which metal objects, such as wire, nails, and screws, are swallowed with feed, and because of their weight, move from the paunch to the second stomach. Lodging there, the metal objects pierce the stomach wall, often causing severe damage, including abscesses, peritonitis, or death. Rod-shaped magnets called rumen magnets are given with a balling gun to an animal to minimize damage to stomach membranes by attracting and holding the metal objects in the paunch (rumen).

Harsh–(1) Designating a fleece that lacks character, as rough hair. (2) Designating vegetation that is rough, hard, or that has some physical characteristic objectionable to livestock.

Harvest–(1) To cut, reap, pick, or gather any crop or product of value, as grain, fruit, or vegetables. (2) The crop or product so harvested.

Hat Racks–Thin cattle; canners.

Hatch– (1) (a) To bring forth young from the egg by natural or artificial incubation. (b) The young produced from one incubation. (2) The access entryway to cellars, attics, haymows, ships, etc.

Hatch Out– To emerge from an egg, as a fully developed chick comes forth from the shell.

Hatchability– (1) In poultry farming, that quality of fertilized eggs when incubated that makes possible normal embryonic development and the emergence of normal young. (2) In incubation practice, the percentage of fertile eggs that hatch.

Hatchery– A place, building, company, etc., where eggs are incubated, usually a commercial establishment where newly hatched young (chicks, poults, ducklings, etc.) are sold.

Hatching Egg– A fertile egg of good form and quality produced by a breeding flock that may be used for hatching.

Haunch– (1) A pivot on the hind feet. It is commonly observed in stock horses used on ranches for culling or cutting. (2) A hindquarter of an animal.

Hay– Any leafy plant material, usually clover, fine-stemmed grasses and sedges, alfalfa, and other legumes, that has been cut and dried principally for livestock feeding. See Fodder, Marsh Hay.

Hay and Dairy Region– A region of the northeastern United States where grasses and legumes are naturally abundant or easily grown, and where the feed supply and barn and market sizes are generally more suitable for dairy cattle and milk production than for other types of farming, and where markets for dairy products are well established.

Hay Bale– A quantity of loose hay compressed usually into a rectangular bale about 3 feet by 18 inches by 14 inches, containing from 40 to 125 pounds depending on the kind of hay, degree of compaction, and moisture content. The compressed hay is held in the bale by baling wire, light metal strips, or heavy twine. More common are the large cylindrical bales weighing up to a ton or more.

Hay Belly– A term applied to animals having a distended barrel due to the excessive feeding of bulky rations, such as hay, straw, or grass.

Hay Fever– Allergic symptoms involving the upper respiratory tract caused by dust and/or pollen grains.

Hay Meadow– A field in which hay is grown.

Haylage– Forage that could have been cut for hay but is stored with a higher moisture content than hay, and with less moisture than silage.

Hays Converter– A breed of beef cattle developed in Canada from Hereford and Brown Swiss.

Hazer– In rodeos, the assistant to the bulldogger who attempts to keep the animal running in a straight line and endeavors to protect the bulldogger from being gored.

Head– (1) Any tightly formed flower cluster, as in members of the family Compositae, or any tightly formed fruit cluster, as the head of wheat or sunflower. (2) A compact, orderly mass of leaves, as a head of lettuce. (3) On a tree, the point or region at which the trunk divides into limbs. (4) The height of water above any point of reference (elevation head). The energy of a given nature possessed by each unit weight of a liquid expressed as the vertical height through which a unit weight would have to fall to release the average energy possessed, used in various compounds, as pressure head, velocity head, lost head, etc. (5) Cows, asses, horses, collectively, as ten head of horses. (6) The part of the body that includes the face, ears, brain, etc. (7) The source of a stream; specifically the highest point upstream at which there is a continuous flow of water, although a channel with an intermittent flow may extend farther. (8) The upstream terminus of a gully. (9) To prune a tree severely. (10) To get in front of a band of sheep, herd of cattle, etc., so as to stop their forward movement (head them off). (11) To place a top on a barrel. (12) That part of an engine that forms the top of the combustion chamber. In many types of modern engines, the exhaust and intake valves are in the head.

Head Shy– Designating a horse on which it is difficult to put a bridle, to lead, or to work around its head.

Headstall– The part of a bridle that encircles a horse's head.

Headstrong– Designating an animal that tends to be stubborn.

Health– The state wherein all body parts of plants, animals, and people are functioning normally.

Heart– (1) The organ of the body that by its rhythmical contractions circulates the blood. It is an edible by-product of slaughter animals and fowls. (2) The center portion of fruits and vegetables.

Heart-girth– A measurement taken around the body just back of the shoulders of an animal, used to indicate fullness of the chest and lung capacity and/or body weight.

Heat– (1) To ferment as a result of wet-stored grains such as wheat, corn, or barley, and forages; sometimes resulting in spontaneous combustion. (2) An animal in heat is ready to breed. See Estrus.

Heat Detection– The process used in determining females that are in estrus. See Gomer Bull.

Heat Mount Detector– A plastic device that is glued to the tailhead of a cow to determine when she comes into heat. Prolonged pressure from a mounting animal's brisket turns the detector a different color.

Heat Period– Estrus; the period during which a female is sexually receptive.

Heat Prostration– Heat stroke; a condition of a person or an animal resulting from excessively hot weather; characterized by lethargy, inability to work, staggering gait, convulsions, and high temperature. Death often occurs.

Heat Spot– A defect of a fertile egg that results from alternating high and low temperatures; characterized by the beginning of the development of the embryo without blood showing.

Heat Synchronization– Causing a group of cows or heifers to exhibit heat together at one time by artificial manipulation of the estrous cycle.

Heat-resistant– (1) Designating a variety or a species that grows under comparatively high temperature conditions; e.g., cotton and rattlesnakes. (2) Any material that is resistant to high temperatures.

Heat-tolerant– Designating the ability of an animal or plant to endure extreme heat conditions.

Heath–(1) Any plant of the genus *Erica*, family Ericaceae. Species are evergreen shrubs and small trees grown in greenhouses and out-of-doors. Also called erica. (2) A natural land feature, an extensive tract of uncultivated land, treeless or nearly so, which is covered by a dense growth of shrubby, ericaceous plants. It may be nearly the same as a high moore. Heaths are generally sandy and the soils strongly acid.

Heaves–An incurable respiratory disease of horses characterized by difficult breathing in which the air cells of the lungs are dilated or ruptured. It is thought to be caused by hard exercise, by consumption of dusty feed, or by heredity. It is found most frequently in old horses and is characterized by forced, double exhaling. Also called broken wind, chronic alveolar emphysema.

Heavy–(1) Designating any material or product that exhibits a comparatively high weight per unit volume. (2) Designating a clay or clayey soil that is difficult to plow. (3) In marketing, designating an abundant supply of a product for sale on one day at one market. (4) The late stages of pregnancy of a cow.

Heavy Breed–A bird (usually referring to a chicken) that has a high meat-to-bone ratio and is therefore suitable for the table. See Broiler.

Heavy Burning–Range firing during the dry, hot season to ensure a fire that will destroy the existing cover, facilitate travel and livestock handling, increase forage for livestock and game, and enhance hunting.

Heavy Grazing–The practice of keeping a large number of animal units on a pasture or range so that the grass or herbage may be closely grazed. Recommended prior to reseeding the pasture or range.

Heavy Wool–Wool that has considerable grease or dirt and will have a high shrinkage in scouring.

Heavyweight Hunter–A classification of a riding horse used for hunting that can carry a rider weighing 190 pounds (86 kilograms) or more.

Heel–(1) The basal end of a plant stem cutting along with a piece of the older stem. (2) See Hock. (3) The end of the branches of a horseshoe. (4) The rear end of the foot.

Heel Fly–The adult form of the cattle grub or ox warble. See Cattle Grub.

Heel of the Round–A retail cut of beef: the heavy fleshing at the rear of the stifle joint that is cut as a roast.

Heifer–The young female of the cattle species; usually applies to the female that has not yet had a calf.

Heiferette–Used to describe a heifer that has calved once, perhaps prematurely, then was "dried up" and fed for slaughter.

Held–(1) The lower end of an ax or other tool handle encased by the metal portion of the tool. (2) Designating the controlling of an animal in a small space or by other controlling devices.

Heliophobous–An organism that grows best in the shade.

Helminths–Worm parasites, such as the flatworms (flukes and tapeworms) and roundworms (hookworms and lungworms).

Helolac–A lake covered by a mat of aquatic plants such as water lilies, water hyacinth, or alligator weed.

Hematoma–Tumor containing effused blood.

Hematopoiesis–Formation or production of blood.

Hematuria–The presence of blood in urine.

Hemi––(Greek) Prefix meaning half.

Hemiparasite–A partial parasite.

Hemizygous–The condition in which only one allele of a pair of genes is present in the cells of an individual plant or animal, the other one being absent.

Hemlock Poisoning–Poisoning of animals resulting from browsing poison hemlock, *Conium maculatum*, family Apiaceae; characterized by cessation of digestion, gas, salivation, dilation of the pupils, rapid pulse and breathing, paralysis, unsteady gait, difficult breathing, and death.

Hemoglobin–The red pigment in the red blood cells of people and animals that carries oxygen from the lungs to other parts of the body. Hemoglobin is a complex chemical compound, made up of iron, carbon, hydrogen, and oxygen, and is essential to life in red-blooded animals.

Hemoglobinuria–The presence of hemoglobin in the urine that is a symptom of some disease: e.g., cattle tick fever and azoturia. Bacillary disease of cattle and sheep is known to occur in western and parts of southern United States. The causative organism is *Clostridium hemolyticum*. See Cattle Tick Fever.

Hemolysis–The destruction of red blood cells and the resulting escape of hemoglobin.

Hemophilia–A hereditary condition of people and animals in which the blood does not coagulate readily and the slightest wound or bruise can cause considerable loss of blood. An animal so affected is often called a bleeder.

Hemorrhage–(1) Any escape or discharge of blood from the blood vessels. (2) To bleed.

Hemorrhagic Enteritis–An inflammatory condition of the intestines in which the small vessels (capillaries) in the lining become engorged and eventually rupture, allowing blood to seep into the intestine. The evacuations are watery and bloody or very dark-colored due to decomposition of the blood present in them.

Hemorrhagic Ova–Immature eggs (ova) that contain blood in the diseased ovary of poultry suffering from pullorum disease or avian typhoid.

Hemorrhagic Septicema–An infectious disease of livestock characterized by numerous, small hemorrhages in the tissues and the presence of bacteria of the pasteurella group and their associated poisons in the blood. Sometimes called shipping fever, stockyard fever, blood poisoning.

Hemostatics–Substances that check internal hemorrhage.

Hen–A female fowl; specifically, the female domestic fowl valued for its egg production.

Hen Battery–A number of individual hen-cage units arranged in a group, usually single or multiple decks, which have provision for watering, feeding, collecting eggs, and disposing of droppings.

Hennery–(1) A poultry farm, particularly one that specializes in the production of market eggs. (2) A building or enclosure where laying hens are kept.

Hepatitis–Inflammation of the liver.

Herbivorous–Designating an animal (herbivore) that feeds, in the native state, on grass and other plants, as cattle, horses, sheep, goats, deer, elk, etc. See Carnivore.

Herd–(1) A group of animals (especially cattle, horses, swine), collectively considered as a unit in farming or grazing practice. See Flock. (2) To tend animals in herds. See Cowboy, Herder, Shepherd.

Herd Book–The recognized, official record of the ancestry of a purebred animal kept by the particular breed association. The first herd book for keeping thoroughbred pedigrees was started in England in 1791. This was followed by Coates Herd Book for Shorthorns started in England in 1822.

Herd Bull Battery–All of the bulls in service in particular herds.

Herd Improvement Registry–A type of registry maintained by certain purebred cattle breeder associations to record the production and yearly records of all producing cows of that breed in a given breeder's herd.

Herd Sire–The male of the species kept for the sole purpose of reproduction. (The term is most commonly applied to cattle and horses but is also used with sheep and swine.) See Bull, Ram, Stallion.

Herd Test–A type of semi-official testing for milk production in which the whole herd of cows of milking age are included. This test was first started by the Ayrshire Breeders Association in 1925 and places the emphasis on the entire herd as a unit rather than on an individual cow.

Herding–The control of animals on the range by guiding their direction and movements to procure grazing and water where and when desired.

Heredity–(1) A study or description of genes passed from one generation to the next through sperm and ova. The heredity of an individual would be the genes received from the sire and dam via the sperm and ovum. (2) Genetic transmission of traits from parents to offspring. (3) The genetic constitution of an individual.

Hereford–A leading breed of beef cattle that originated in England. They are light to dark red with a characteristic white face and underline. Certain strains of the breed are polled (hornless). Also called white face. See Polled Hereford.

Hereford Swine–An American-developed breed of swine that has color markings similar to Hereford cattle.

Heritability–The proportion of the differences among animals, measured or observed, that is transmitted to the offspring. Heritability varies from zero to one. The higher the heritability of a trait, the more accurately does the individual performance predict breeding value and the more rapid should be the response due to selection for that trait.

Heritability Percent Estimates–The percent of a trait that is inherited from an animal's parents and is not controlled by environment.

Hermaphrodite–A bisexual individual that possesses both male and female sex organs. In some species such individuals are capable of reproducing, while in others they are sterile.

Hernia–The protrusion of internal organs through an opening in the body wall of humans or animals, as with a scrotal or umbilical hernia.

Heterogametic–Producing unlike gametes, particularly with regard to the sex chromosome. In species in which the male is of the "X-Y" type, the male is heterogametic, the female homogametic.

Heterogen–A variable group of plants or animals that arise as hybrids, sports, mutations, etc., certain types of which may or may not breed true.

Heterogeneous–Designating elements having unlike qualities.

Heterologous Serum–Serum derived from another species or disease.

Heterosis–The amount of superiority observed or measured in crossbred animals compared with the average of their purebred parents; hybrid vigor.

Heterotrophic–Referring to organisms that for their metabolism are dependent upon organic matter supplied from sources outside of their own bodies. See Autotrophic.

Heteroxenous Parasite–A parasite requiring several or different hosts for its complete development.

Heterozygous–An animal that carries genes for two different characters (impure).

Hexose–Any of various simple sugars that have six carbon atoms per molecule.

Hidden Hunger–Deficiency disease of plants or animals.

Hide–The tanned or untanned skins of animals, especially those of cattle, horses, sheep, and goats.

Hidebound–Designating an animal whose skin is very tightly fastened to its body, often resulting from poor feeding and emaciation.

Hidrosis–Excessive sweating.

High Blower–(1) A horse that has broken wind. (2) A horse that snorts at each exhalation while galloping.

High Lysine Corn–Corn that has a higher than normal content of lysine and tryptophan. This type corn has a better balance of amino acids for monogastric animals.

High School Horse–A horse trained for performing certain relatively highly complicated routines.

High-roller–Designating a horse whose bucking action is higher than usual.

Hilus–The stalk of the ovary that serves as the attachment to the broad ligament.

Hind Cinch–The rear cinch strap that encircles a horse to keep the saddle from tipping up when roping. In normal use, saddles have only one cinch that is fastened just behind the forelegs.

Hindgut–The posterior part of the alimentary canal between the midgut and anus.

Hinny–The offspring of a horse father and a donkey mother.

Hip–(1) The fruit of rose; rose hips. (2) The external angle (ridge) formed by the meeting of two sloping sides of a roof. (3) That region of one of the rear quarters of four-legged animals where the hind leg joins the pelvic region.

Hip Height–A measurement of cattle taken from the ground to the top of the hip. At a given age of the animal, hip height determines the frame score of a bull, steer, or heifer. See Frame Size.

Hippology–The study of the horse.

Hirsute–Covered with coarse hairs.

Hirudin–A substance extracted from the salivary glands of the leech that has the property of preventing coagulation of the blood.

Histamine–An alkaline type of chemical compound composed of carbon, hydrogen, and nitrogen which is probably formed in the animal body, mainly in the intestinal tract, by the action of bacteria on the amino acid, histidine. If it is absorbed into the bloodstream and circulated through the body, it may produce a fall in blood pressure, spastic contraction of the smooth muscles, and other symptoms, some of which may be relieved by antihistamine drugs.

Histology–The science of the microscopic structure of plant and animal cells.

Histopathologic–Designating abnormal changes in body structures, as observed by microscopic examination of sections of abnormal or diseased cells.

Histoplasmosis–A disease of the lymph nodes in the region of the neck caused by a fungus.

Hitch–(1) A catch; anything that holds, as a hook; a knot or noose in a rope, which can be readily undone, intended for a temporary fastening. (2) (a) The connecting of an implement, vehicle, etc., to a source of power, as a tractor, team, etc. (b) The device which is used to make such a connection. (3) A horse or horses used to pull an implement or vehicle. (4) The stride of a horse when one of the hind legs is shorter than the other. (5) To fasten an animal to a post, rail, etc.

Hitch Your Horse to the Ground–To let the reins of a horse drop to the ground for hitching purposes. Horses are trained to stay at a place when the reins are dropped on the ground.

Hive–A home for honeybees that is provided by people; it usually consists of a base, removable supers, and a top.

Hive Body–A single wooden rim or shell that holds a set of frames. When used for the brood nest, it is called a brood chamber; when used above the brood nest for honey storage, it is called a super. It may be of various widths and heights and adapted for comb honey sections.

Hive Cover–The roof or lid of a beehive.

Hive of Bees–A colony of honeybees living in a hive.

Hive Tool–A metal bar used to loosen frames and to separate the parts of a hive.

Hives–In horses, small swellings under or within the skin similar to human hives. They appear suddenly over large portions of the body and can be caused by a change in feed.

Hobble–(1) A thong that couples an animal's forelegs together to restrict its movements. (2) To tie an animal's forelegs so as to prevent straying.

Hock–(1) The region of the tarsal joint in the hind leg of a horse or other quadruped, corresponding to the angle in people. (2) The lower joint section of a ham.

Hock Hobbles–Hobbles that are used to fasten the hocks of the hind legs of a mare to prevent kicking preparatory to coitus.

Hoffman Frame–A self-spacing wooden frame used in Langstroth beehives.

Hog–See Swine.

Hog Bristles–The coarse, stiff hairs on swine used in the manufacture of brushes; now synthetic bristles have largely replaced hog bristles.

Hog Cholera–An acute, contagious, viral disease of swine characterized by sudden onset, fever, high morbidity, and mortality.

Hog Down (Hogging Off)–To pasture hogs on a crop grown for stock feed, thus eliminating the harvesting process. Also called hog off.

Hog Follicle Mite–*Demodex phylloides*, family Demodicidae; the demodex mange mite, a minute, skin parasite of swine that lives in hair follicles and sebaceous glands on the face, base of the tail, and the inner sides of the legs; it produces an inflammation of the skin that appears as small, hard pimples in size up to a marble. Upon breaking, the pimples discharge a yellowish-cheesy pus. Rare in hogs in the United States.

Hog Holder–A device consisting of a metal tube through which a cable is run. A loop on the end of the cable is placed on the snout of a hog to hold the hog for receiving vaccination, etc.

Hog Louse–*Haematopinus suis*, family Haematopinidae; the louse of the European wild boar, also found on domesticated swine in central Europe. *H. adventicius chinensis* is the common sucking louse of domesticated swine in North America.

Hog Ring–A ring that is fastened in the end of a hog's nose to prevent the hog from rooting in the ground.

Hog-Corn Ratio–See Corn-Hog Ratio.

Hog-dressed–Designating a dressed lamb or calf carcass with the head and pelt left on but with the feet viscera removed.

Hogg–A young yearling sheep before shearing.

Hohenheim System of Grazing–An intensive system of grassland management involving the division of the grazing area into a number of paddocks, grazed in rotation to allow a short period for growth between grazings. Nitrogenous fertilizers are usually applied after each grazing.

Hold–(1) To restrain animals to a particular place. (2) To remain on the tree until mature, as a fruit. (3) To retain, as certain soils hold moisture. (4) To maintain condition, as a steer that holds flesh. (5) To store or retain in storage, as to hold eggs. (6) Not to market at harvest time or when an animal is fat, but to wait to sell for a better price.

Hold Up Milk—In a cow, to cause a cessation in milk secretion as a result of undue excitement at milking time. See Let Down Milk.

Holding—An indefinite amount of land that is usually considered large in proportion to the average size of farms or ranches in its vicinity.

Holding Brand—A symbol or number burned into an animal's hide that is a requirement for registration with some cattle breed registry associations. The brand must be recorded in the association office.

Holding Ground—An area where livestock are often held during roundups; also called bunch ground.

Holding Pen—A large pen in which sheep or other animals are held prior to being handled.

Hollow Horn—An imaginary disease arising from the erroneous belief that loss of appetite and listlessness in a cow was due to hollow horns. The remedy was supposed to be (a) boring a hole in each horn just above the horn line, (b) filling the cavity with salt, sugar, and pepper, and (c) plugging the hole with a wooden peg. The belief was that if the cow had hollow horn this remedy would cure her, and if she did not have hollow horn, the remedy would prevent her getting it. See Hollow Tail.

Hollow Tail—An imaginary disease of a cow similar to hollow horn but which affected the tail. The remedy was supposed to be the same as for hollow horn, except that the hole or a slit was made in the tail. Also called wolf in the tail.

Holstein-Friesian—A widely used dairy breed of cattle. The name is American in origin but applies to cattle originally imported largely from the province of Friesland in Holland. They are black and white and the mature cows weigh from 1,100 to 1,800 pounds. They are noted for their high production of milk with a low butterfat content.

Homeostasis—Maintenance of a constant internal environment by a combination of body mechanisms.

Hominy Feed—A livestock feed that is a mixture of corn bran, corn germ, and part of the starchy portion of either white or yellow corn kernels or a mixture thereof as produced in the manufacture of pearl hominy, hominy grits, or table meal. It shall contain not less than 5 percent of crude fat. If prefixed with the words white or yellow, the product must correspond thereto. Also called hominy chop, hominy meal.

Homo—(Latin) (1) Man. (2) Prefix meaning alike or same.

Homogametic—Refers to the particular sex of the species that possesses two of the same kind of sex chromosome such that only one kind of gamete can be produced with respect to the kinds of sex chromosomes it contains; in mammals, the female is the homogametic sex (XX).

Homogeneous—Being of uniform character or nature throughout.

Homogenized Milk—Milk that has been treated in such a way as to break up the particles or globules of fat to a size small enough that they will remain suspended and not rise to the top after standing. This is accomplished by passing the milk through small orifices under high pressure of 1,500 to 4,000 psi (10,341,000 to 27,576,000 pa) to disperse the fat globules so they will not rise as cream. Other substances such as margarine can also be homogenized.

Homolog—One of a pair of structures having similar structure, shape, and function, as with two homologous chromosomes.

Homologous—Organs or parts that exhibit similarity in structure, in position with reference to other parts, and in mode of development, but not necessarily similarity of function, are said to be homologous.

Homologous Chromosomes—Pairs of chromosomes that are the same length, that have their centrioles in the same position, and that pair up during synapsis in meiosis.

Homologous Serum—Serum that is derived from the same species or like disease.

Homothermic—Refers to animals that are able to maintain a fairly constant body temperature; warm-blooded.

Homothermous—Having the same temperature throughout.

Homozygote—An animal whose genotype, for a particular trait or pair of genes, consists of like genes.

Homozygous—Possessing identical genes with respect to any given pair or series of alleles.

Homozygous Recessive—A recessive character that produces two kinds of gametes; one carries the dominant gene, while the other carries the recessive gene.

Honda—A rope, rawhide, or metal loop at one end of a lariat through which the other end of the lariat is pulled to make a noose or large loop.

Honey—(1) An aromatic, viscid, sweet food material derived from the nectar of plants through collection by honeybees; modified by the bees into a denser liquid and finally stored in honeycombs. Of acid reaction, liquid in its original state, it becomes crystalline on standing. Honey consists chiefly of two simple sugars, dextrose and levulose, with occasionally more complex carbohydrates, with levulose usually predominant, and always contains minerals, plant coloring materials, several enzymes, and pollen grains. (2) Legally, the nectar and saccharine exudation of plants, gathered, modified, and stored in the comb by honeybees, which is levorotatory, and contains not more than 25 percent of water, not more than 0.25 percent ash, and not more than 8 percent sucrose.

Honey Bound—A condition under which the queen in a beehive is restricted in her egg-laying by the fact that all or most of the cells in the brood chamber are filled with honey.

Honey Butter—A mixture of creamery butter and 20 to 30 percent table quality, liquid honey used as a spread.

Honey House—Building in which honey is extracted and handled.

Honey Plant—Any plant from which honeybees gather nectar and pollen, especially one which either is in abundance or one which gives a distinctive flavor to honey.

Honey Stomach (Honey Sac)—An enlargement of the posterior end of the esophagus in the bee abdomen. It is the sac in which the bee carries nectar form flower to hive.

Honeybee–*Apis mellifera*, family Apidae; a very important economic insect that produces honey and acts as a pollinator of the flowers of many plants, both wild and cultivated.

Honeycomb–(1) The waxy structure in which honey is stored by bees. (2) A natural arrangement of the soil mass in more or less regular five- or six-sided sections separated by narrow or hairline cracks which is usually found as a surface structure. (3) See Reticulum.

Honeyflow–The incoming of nectar to the beehive, used especially in reference to the periodic changes in quantity related to the blooming periods of dominant flowering plants of the area, such as the clovers.

Hoof–The hard, horny, outer covering of the feet of horses, cattle, sheep, goats, and swine.

Hoof Pick–A tool used to clean the hoof of a horse or other animal.

Hoof-and-Mouth Disease–See Foot-and-Mouth Disease.

Hoofbound–A condition of horses, mules, etc., in which the hoof is dry and contracted, causing pain and occasional lameness.

Hook Bones–The prominent bones on the back of a cow formed by the anterior ends of the ilii (plural of ilium); points of the hip.

Hooks–See Hook Bones.

Hookworm–Any of certain internal, parasitic worms of animals and people that belong to the family Ancylostomatidae. The larvae, which are voracious bloodsuckers of the small intestine, are capable of penetrating the skin to gain entrance to the body, but may also gain entry by ingestion.

Hoose–A verminous bronchitis of cattle, sheep, and goats due to a lungworm infestation (a) of cattle chiefly by *Dictyocaulus viviparus*, (b) of sheep and goats chiefly by *D. filaria*, *Protostrongylus rufescens*, and *Muellerius capillaris*. Symptoms are a husky, spasmodic cough, emaciation, anemia, debility, thick nasal discharge, a breathing through the mouth. Also called husk. See Fowl Gapeworm.

Hopper-feeding–Making available to a rabbit, hog, or other animal a sufficient quantity of feed for several days so the animal may eat as often as it wishes and not be limited to a certain amount.

Horizontal Silo–A silo built with its long dimension parallel to the ground surface rather than perpendicular as in the case of an upright silo. It may be built just below the ground level with openings at both ends to facilitate filling and feeding.

Hormone–A chemical substance formed in some organ of the body, secreted directly into the blood, and carried to another organ or tissue, where it produces a specific effect.

Horn–(1) A natural, bonelike growth or projection on each side of the head of most breeds of cattle, sheep, and goats that is a natural weapon of defense. (2) The pointed end of a blacksmith's anvil used for shaping hot metal. (3) The front, upraised projection of a riding saddle, the snubbing horn. (4) The outer hard covering of the hoofs of horses, cattle, sheep, and swine. (5) A broad term commonly used in describing various shadings of color in the beak of some breeds of fowl.

Horn Fly–*Siphona irritans*, family Muscidae; a bloodsucking fly that infests cattle and goats. It probably causes greater loss in livestock production in the United States than any other bloodsucking fly.

Horn Weight–Ball-shaped weight, usually of lead, which is attached to the tips of the horns of show cattle to assist in the training and shaping of the horn growth.

Horned–Designating an animal that has horns. See Hornless, Polled.

Hornless–Designating a well-defined, polled condition of certain breeds of cattle, such as the Angus and Red Polled. See Polled.

Horny Frog–The semisoft, elastic, V-shaped structure in the sole of a horse's foot. Also called foot pad.

Horse–(1) *Equus caballus*; a quadruped of very ancient domestication used as a beast of burden, a draft animal, and a pleasure animal for riding, and in some areas as a meat animal. Breeds and types vary greatly in size and color from the Shetland pony to the Belgian. It has been important as one of the parents of the mule, and as a source of meat and hides. See Mule. (2) A stallion. (3) Designating an implement that is drawn by a horse.

Horse Bot Fly–*Gasterophilus intestinalis*, family Gasterophilidae; a fly that lays its eggs on the hairs of horses, usually on the front legs but sometimes on the shoulders, belly, or hind legs. When the horse licks himself the larvae cling to his lips or tongue. The larvae then burrow into the mucous membrane of the tongue working their way toward the base of the tongue. They pass to the stomach and remain there from nine to ten months, after which they release their hold on the walls of the stomach and pass to the ground in the excrement. They cause mechanical injury to the tongue, lips, lining of the stomach, and intestines, and they interfere with glandular action. They cause inflamed ulcerated conditions and absorb food that starves the host and may cause complete obstruction from the stomach to the intestines. Badly infested animals are run down, having digestive upsets, and rough coats. Also called common horse bot fly.

Horse Malaria–Equine infectious anemia.

Horse Manure–Dried horse excrement used as a medium for growing mushrooms, for making heat in hot beds, and as a soil amendment.

Horse Meat–In France, Belgium, and other parts of the world, horsemeat is considered a delicacy for human consumption. In the United States, many horses that have outlived their usefulness or that are less valuable for other purposes are processed in modern, sanitary slaughtering plants for pet food.

Horse Sleeping Sickness–See Encephalomyelitis.

Horse Sucking Louse–*Haematopinus asini*, family Haematopinidae; a bloodsucking, insect pest of horses, mules, and asses. Infestation is characterized by scurfy skin, loss of vitality, scratching or rubbing, and raw spots on the skin.

Horse Wrangler–(1) A hostler who keeps the string of extra saddle ponies that accompanies every cattle drive or used during roundup. He keeps them from straying and has them ready when they were needed (southwestern United States). (2) The rider on horseback who accompanies the rodeo rider and keeps the bucking broncho from injuring the contest rider before and after the riding has been done. (3) A person who works with horses during the breaking period.

Horse-biting Louse–*Bovicola equi*, family Trichodectidae; an insect pest of horses, mules, and donkeys that bites the skin, causing irritation.

Horsefly–In a broad sense, any member of the family Tabanidae, but usually restricted to the subfamily Tabaninae.

Horseguard–*Bembix carolina*, family Sphecidae; a black-and-yellow sand wasp found in southern United States where there are horses or cattle. It preys upon horseflies and bottleflies, which it feeds to its larvae.

Horsemanship–The ability to show a high degree of skill in handling horses.

Hoss–(slang) Horse.

Host–Any organism, plant or animal, in or upon which another spends part or all of its existence, and from which it derives nourishment and or protection.

Host Specific–Designating a parasite that can live in or on only one host, to which it is therefore said to be specific.

Hot–(1) Designating a horse with a bad disposition. (2) Designating a fruit, such as a pepper, that has a pungent, strong, lasting flavor. (3) Designating weather. (4) Designating manure that heats upon decomposition. (5) An animal feed that contains a high percentage of concentrate or a feed containing a high level of salt.

Hot Carcass Weight–The weight of a carcass immediately after slaughter before the carcass has had time to age and shrink. See Cold Carcass Weight, Shrinkage, (6).

Hot Iron–The heated iron rod or stamp with a handle used in branding cattle, etc.

Hot Manure–Fresh manure that is going through the process of heating due to fermentation. Horse manure is designated as one of the hot manures. Cow manure is one of the cold manures. See Cold Manure, Cow Manure, Horse Manure.

Hot-blooded–(1) Designating a horse that has some Thoroughbred or Arab blood. (2) Designating a horse that is nervous and at times even vicious. See Cold-blooded.

Hothouse Lamb–A lamb that is dropped in the fall or early winter and is marketed at an age of six to twelve weeks. These are usually sold in periods before the Christmas or Easter holidays to take advantage of higher prices.

Hotis Test–A test for the rapid detection of certain microorganisms in raw milk.

House Bee–A young worker bee, one day to two weeks old, that works only in the hive.

Hover–(1) The sheet metal canopy surrounding a heat source under which incubator chicks are kept warm. (2) To cover chicks, as a hen hovers her chicks.

HQB–High-quality beef.

Humerus–The large bone of the pectoral limb located between the elbow and shoulder joints.

Humilis–Dwarf.

Hump Up–The attitude an animal takes in order to expose less body surface to rain, wind, cold, etc.: to pull the feet together and push the back up.

Hunter–A riding horse of quality conformation and size, especially adapted for the chase and riding to hounds; a Thoroughbred type of breeding, about 16 hands high.

Hurdle–A board made of plywood usually around 3 feet wide and 3 feet long that is used to herd pigs.

Hurdle System–A term sometimes applied to the method of handling sheep by means of a wolf-proof fence.

Husbandry–In its earlier usage, the skill, or art, of tillage, crop production, and rearing of farm animals. Today the word is occasionally used as a synonym for farming. More commonly used in combination, as animal husbandry and poultry husbandry, embracing the art, science, processing, and business of production.

Hutch–A boxlike cage or pen for a small animal; e.g., a rabbit hutch.

Hybrid–(1) An animal produced from the crossing or mating of two animals of different breeds. (2) A plant resulting from a cross between parents that are genetically unlike; more commonly, in descriptive taxonomy, the offspring of two different species.

Hybrid Bees–The offspring resulting from crosses of two or more selected inbred lines (strains) of bees; the offspring of crosses between races of bees.

Hybrid Chick–Chicks that result from crossing two or more inbred lines of the same or different breeds, varieties, or strains.

Hybrid F1–Plants of a first-generation hybrid of two dissimilar parents. Hybrid vigor, insect or disease resistance, and uniformity are qualities of this generation. Seed from hybrid plants should not be saved for future planting. Their vigor and productive qualities are only in the original hybrid seed.

Hybrid Vigor–The increase of size, speed of growth, and vitality of a crossbreed over its parents. See Heterosis.

Hybridization–The production of hybrids by natural crossing or by manipulated crossing.

Hybridize–To create a hybrid.

Hydatid–The larval stage of the dog tapeworm, *Echinococcus granulosus*, which usually develops in the liver and lungs of people, cattle, swine, goats, sheep, and some other mammals. Hydatids are cysts containing the larvae. They may develop to a very large size, displace parts of the organs in which they grow, and cause a disease known as hydatid disease, or echinococcosis.

Hydrocyanic Acid–HCN; one of the most valuable and widely used of the fumigants. Except that it is highly toxic to people, it approaches the ideal in a fumigant. It is found in some plants, particularly sorghums, and may under certain conditions cause poisoning and death to animals. Sometimes called prussic acid.

Hydrogen Peroxide–A chemical substance often used as a bleach to remove color. It is used also in medicine and surgery as an antiseptic agent and as a cleansing agent in mouthwashes, toothpastes and powders. Its antiseptic and cleansing action is due to the fact that it gives off sufficient oxygen to destroy bacteria.

Hydrogenated Lard–Usually, steam-rendered lard to which extra atoms of hydrogen have been added. It is bland, has a high smoke point, is resistant to rancidity, and is whiter than common lard.

Hydrogenated Oils–Oil hardened by treatment with hydrogen in the presence of nickel. Cottonseed, corn, and wheat oils are commonly hardened and used in cooking fats.

Hydrophobia–Literally, fear of water; commonly refers to rabies, which is a misnomer. Rabid animals are not necessarily afraid of water; they often have difficulty drinking water due to paralysis of the tongue. See Rabies.

Hygiene–The science of health; the rules or principles of maintaining health in people and animals; sanitation.

Hyperemia–Excess blood in any part of the body.

Hyperesthesia–Excessive sensitivity of a part of the body to touch or pressure.

Hyperglycemia–An above-normal blood sugar level. See Hypoglycemia.

Hyperimmune–Designating the state or quality of having a high degree of immunity produced by repeated exposure to the same disease agent.

Hyperimmunization–Process of increasing the immunity of an animal by increasing injection of an antigen, subsequent to the establishment of an initial immunity; enhanced immunity.

Hyperparasite–An organism that is parasitic on another parasite.

Hyperphagia–Excessive hunger.

Hyperpituitarism–A condition brought about by excessive production of one or more hormones by the pituitary gland, causing abnormal or excessive growth, as in gigantism.

Hyperplasia–An increase in the number of cells in a tissue or organ.

Hypersaline–Term to characterize waters with salinity greater than 4.0 percent due to land-derived salts.

Hypersensitivity–The violent reaction of an organism to attack by a pathogen, a condition in which the response to a stimulus is unusually prompt or excessive in degree.

Hypersusceptible–Designating a condition of abnormal susceptibility to infection, or to a poison, to which a normal individual is resistant.

Hyperthyroidism–A condition of humans and lower animals in which the thyroid gland secretes excessively; characterized by rapid tissue change and restlessness. See Hypothyroidism.

Hypertrophy–(1) Pertaining to waters of very high nutrient content. (2) Morbid enlargement or overgrowth of an organ or part due to increase in size of its constituent cells.

Hypocalcemia–A below-normal level of calcium in the blood.

Hypodermis–The layer of tissue below the dermis of the skin.

Hypoglycemia–A deficiency of sugar in the blood. See Hyperglycemia.

Hypomagnesemia–A deficiency of magnesium in the blood. See Grass Tetany.

Hypophysectomy–The surgical removal of the pituitary gland.

Hypophysis–Another term for the pituitary gland, a structure located on the underside of the brain and embedded in the sphenoid bone.

Hypothyroidism–A condition of humans and animals in which the thyroid gland secretes inadequately; characterized by cretinism. See Goitrogen, Hyperthyroidism.

Hysterectomy–(1) Technique used in some Specific Pathogen Free laboratories to remove unborn pigs from the sow. The entire uterus is removed with the pigs inside. This operation makes the sow useless, and she is slaughtered immediately after the operation. (2) Surgical removal of all or part of the uterus.

Ice Cream Species–An exceptionally palatable species of forage sought and grazed first by livestock and game animals. Such species are usually overutilized under proper grazing.

Ice Milk–A frozen product that resembles ice cream except that the fat content is lower and the milk solids-not-fat content is usually higher. It contains 3 to 4 percent fat, 10 to 15 percent milk solids-not-fat, about 15 percent sugar, and a stabilizer.

Icterus–(1) Jaundice. (2) In plants, a yellowing of the leaves due to cold, excessive moisture, or other climatic factors.

IDA–International Dairy Arrangement.

Identical Twins–Twins that develop from a single fertilized egg that separates into two parts shortly after fertilization.

Idiopathic–Designating a disease or condition with no apparent cause.

Imago–(Plural, imagoes or imagines) The adult stage of an insect.

Immobilization–The action of rendering an object immovable, as the immobilization of an injured limb in an animal to allow for healing.

Immunity–Having resistance to the action of something, such as a disease. It may be inborn, may result from exposure to a disease, from having had a disease, or from having received an injection of immune serum. Degree of immunity varies in each case. See Immunizing Agent.

Immunize–To render an animal resistant to disease by vaccination or inoculation.

Immunizing Agent–A substance that, when introduced into the body of an animal, will build up antibodies in the blood that will resist or overcome an infection to which most of the same genus or species are susceptible.

Immunoglobulins–Proteinlike materials produced in the body that inactivate or destroy antigens.

Immunology–Science or study of immunity and its factors.

Impaction–The lodgment of undigested food or other objects, such as hair, in the digestive tract. See Bezoar, Hair Ball, Hardware Disease, Stomach Ball.

Impaction of the Bowels–Constipation.

Implantation–The process whereby the mammalian embryo forms an attachment to the uterine wall.

Impotence–Temporary or permanent loss of reproductive power or virility.

Impregnate–To fertilize a female animal or flower.

Imprinting–A kind of behavior common to some newly hatched birds or newborn animals that causes them to adopt the first animal, person, or object they see as their parent.

In Foal–Designating a pregnant mare.

In Hand–Designating an animal under control or immediately available.

In Heat–Designating a female animal at a period when she will accept coitus with a male. Also called estrus. See Estrus.

In Season–(1) That part of the year when a product is normally harvested and, as a result, is cheaper, more plentiful, and more flavorful. (2) See Estrus, In Heat.

In Utero–Within the uterus.

In Vitro–In the test tube, outside the animal body. See In Vivo.

In Vitro Fertilization–Fertilizing an egg with sperm in a test tube or petri dish, then implanting the fertilized egg into the uterus. See Artificial Insemination, Embryo Transfer.

In Vivo–In the living body. See In Vitro.

In-and-In Breeding–The breeding together of closely related plants or animals for a number of successive generations to improve or eliminate certain characteristics. Also called breed in and in, line breeding, inbreeding.

Inappetence–Lack of appetite.

Inbred–(1) An individual with parents who show 50 percent or more of common ancestry in their pedigree. (2) The offspring produced by self-pollination in normally cross-pollinated plants. (3) Of, or pertaining to, a plant or animal produced by breeding between close relatives.

Inbreeding–The mating of very closely related animals such as mother and son, father and daughter, brother and sister. In experienced hands it can be used to selectively maintain certain desirable traits; if used improperly it can produce undesirable traits and downgrade stock.

Incidence–The number of new cases of a disease occurring in a given population during a specific period, divided by the total number of persons at risk of developing the disease during that same period.

Incision–A cut.

Inclusion–A nonliving substance or particle in a cell.

Incompatibility–A condition in either plants or animals in which the viable male gamete will not fertilize the viable female gamete.

Incomplete Dominance–A kind of inheritance where a gene does not completely cover up or modify the expression of its allele; also may be known as codominance or blending inheritance.

Incorporate–To mix pesticides, fertilizers, or plant residues into the soil by plowing or other means.

Incubator–An apparatus or chamber that provides favorable environmental conditions for the development of embryos, the hatching of eggs, or the growth of cultures.

Independent Culling Levels–Selection of culling based on cattle meeting specific levels of performance for each trait included in the breeder's selection program. For example, a breeder could cull all heifers with weaning weights below 400 pounds (or those in the bottom 20 percent of weaning weight) and yearling weights below 650 pounds (or those in the bottom 40 percent).

Index–A system for comparing animals within a herd, or area, based on the average of the group; usually the figure 100 is used for an average index; animals receiving an index of 100 or over are the top end while those indexing less than 100 are the bottom end.

Indian Cattle–See Zebu.

Indian Pony–(1) A small riding pony used by the Indians of southwestern United States, which originated from horses shipped by the Spaniards to Mexico in 1519. Many became wild and their offspring were captured by the Indians. They are the results of natural propagation rather than any selection or planned breeding. (2) Any piebald or pinto pony.

Indigestion–A condition of the digestive system of humans and animals in which the normal digestive process is halted or disturbed. There are many causes, such as infection, spoiled food, overeating, etc.

Indiscriminate Breeder–An animal that will breed with any animal of the same type and the opposite sex.

Individual Claiming Pen–On goat ranches, a pen in which an individual doe that has disowned her kid is kept until she reclaims her offspring.

Individual Drinking Cup–A type of stock-watering equipment in wide use in modern stanchion dairy barns whereby water is supplied by a cup at the stanchion for each individual cow.

Individual Farrowing House–A separate building or quarters equipped so that sows that are to farrow may be kept isolated in warm, clean, and quiet individual pens at farrowing time.

Inedible–A substance that is not fit for food, such as poisonous nuts and plants. Tough skins, seeds, and decayed spots of fruits and vegetables and bones of meat are considered inedible parts because they are not suitable for human consumption.

Inert Ingredient–A substance in a feed, pesticide, etc., that does not act as a feed, pesticide, etc. The substance may serve a purpose but is usually used as a filler, vehicle, etc.

Infect–To cause disease by the introduction of germs, parasites, or fungi. See Infection.

Infection–Invasion of the tissues of the body of a host by disease-producing organisms in such a way that injury results; the presence of multiplying parasites, bacteria, viruses, etc., within the body of a host. See Infestation.

Infection Stage–The period in the course of a disease during which the host responds, symptoms appear, and the disease develops.

Infectious–Designating a communicable disease.

Infectious Abortion–See Brucellosis.

Infectious Anemia in Horses–See Equine Infectious Anemia.

Infectious Bovine Rhinotracheitis (IBR)–A respiratory disease characterized by inflammation, edema, hemorrhage, and necrosis of the mucous membranes of the respiratory passages, and pustular lesions on the genital organs of both male and female animals.

Infectious Bronchitis of Chicks–A respiratory, viral disease characterized by sneezing, coughing, a peculiar breathing sound, nasal exudate, and watery eyes. Mortality rate is high in young chicks. See Bronchitis.

Infectious Coryza–A severe, catarrhal inflammation of the mucous membranes of the upper respiratory tract of poultry caused by *Hemophilus gallinarum*. Symptoms commonly include a nasal discharge, which becomes thick and sticky with an offensive odor; adjacent sinuses become filled with mucus that may change to a dry, cheesy form and cause bulging around the eyes. Affected fowls have difficulty in breathing, shake their heads frequently, are listless, have loss of appetite, and lose strength rapidly. Crowding, dampness, and lack of ventilation are predisposing factors. Also called roup, cold, rhinitis.

Infectious Diarrhea–See Calf Scours.

Infectious Disease–A disease caused by bacteria, protozoa, viruses, or fungi entering the body. It is not necessarily contagious or spread by contact.

Infectious Enteritis in Pigs–A group of enteritis conditions in pigs caused by *Salmonella* bacteria (*Salmonella cholera suis*) and/or a group of other pathogenic enteric organisms such as *Spherophorus necrophorus*, coccidia, and viruses. The conditions are characterized by diarrhea and frequently by the presence of blood in the stool.

Infectious Keratitis–(Bovine) A very common disease of cattle characterized by swollen, red, and congested eyelids, a watery discharge from the eyes which in time contains mucus and pus, cloudy eyeballs, and temporary or permanent blindness. The cause has not been established. Also called pinkeye, contagious ophthalmia in cattle, infectious conjunctivitis.

Infectious Laryngotracheitis–An acute, highly contagious viral disease of fowls characterized by respiratory distress, coughing, watery eyes, sneezing, shaking of the head, sitting with eyes closed, gurgling breathing sound, and a stretching of the neck while inhaling. Also called infectious tracheitis, flu.

Infectious Necrotic Enteritis–See Infectious Enteritis in Pigs.

Infectious Necrotic Hepatitis–An acute disease of sheep caused by *Clostridium novyi* in the presence of liver flukes. The onset is sudden and acute, and apparently healthy animals can be stricken and die overnight. Also called black disease.

Infectious Sinusitis–A viral disease of turkeys characterized by a clear discharge from the nostrils, followed by a foamy condition of the eyes that are closed or partially closed, labored breathing, and loss of weight. Also called swellhead, sinusitis.

Infective–Capable of entering and establishing itself in a host; able to infect a susceptible plant or animal.

Infertile–(1) Designating that which is incapable of reproduction; e.g., a barren female animal, a male animal with nonviable spermatozoa, an unfertilized egg, a flower that will not produce seed. (2) A soil so low or unbalanced in essential nutrients that it will not produce a profitable crop.

Infertile Egg–(1) An ovum that has not united with a spermatozoon to form a zygote, or fertilized egg, hence, incapable of embryonic development. (2) An egg produced by an unmated female; a market egg produced by an unmated hen.

Infest–To assail, attack, overrun, annoy, disturb; as ticks infest a cow, or as an insect infests a plant.

Infestation–(1) Act of infesting, or state of being attacked, molested, vexed, or annoyed by large numbers of insects, etc., as an animal may be subject to an infestation of parasites, such as fleas, ticks, mites, etc. (2) Presence of disease in a population of plants, or of pathogens in a position, as in soil or on seed surfaces, where they have the possibility of producing disease. (Not to be confused with infection, which can be applied only to living, diseased plants and animals.) See Infection.

Infiltration–The flow of a liquid into a substance through pores or other openings, connoting flow into a soil in contradistinction to *percolation*, which connotes flow through a porous substance.

Inflammation–The reaction of a tissue to injury that tends to destroy or limit the spread of the injurious agent and to repair to replace the damaged tissues. The tissue is infiltrated by white blood cells, is red from capillary expansion and hot to the touch.

Influenza in Horses–A highly contagious viral disease characterized by a high temperature, general ill health, coughing, and nasal discharge. Also called shipping fever, pinkeye of horses, horse flu.

Influenza in Swine–An acute, highly contagious, infectious disease caused by the bacterium *Hermophilus influenzae suis* and a filterable virus; characterized by high temperature, thumpy coughing, discharge from the eyes and nose, and listlessness. The death rate is low. Also called hog flu, swine influenza.

Infundibulum–The enlarged, funnel-shaped structure on the end of the fallopian tubes that functions in collecting the ova during ovulation.

Infuse–(1) To steep in liquid so as to extract useful qualities, as tea. (2) To inject or introduce a liquid or medicinal agent, as to infuse the udder with a drug.

Ingest–To eat, or take in food for digestion by way of the mouth.

Ingesta–Contents of the digestive tract. Includes feed, digestive juices, bacteria, etc.

Ingestive Behavior– The mannerisms or habits that an animal uses during the intake of food.

Ingluvies– The crop or craw of a chicken or other bird; also the rumen or first stomach of a ruminating animal, as a cow.

Inguinal Canal– The opening in the abdominal wall through which the testes pass from the body cavity into the scrotum.

Inhalable Particulates– Particulates less than 10 microns that are not filtered from the air, and consequently are inhalable into the upper respiratory system.

Inhalants– Medicinal preparations that are inhaled or drawn into the lungs.

Inheritance– (1) The transmission of genetic factors from parent to offspring. (2) The process or procedure of transferring property, both real and personal, from one generation to the next, either by will or by laws of descent and distribution.

Inherited Characteristic (Trait)– A character, the expression of which is determined by a particular gene or genes.

Inhibit– To suppress, prevent, hinder, restrain.

Inject– To introduce a substance into the body of an animal or plant by mechanical means.

Inner Cover– A cover used under the standard telescoping cover on a beehive.

Inoculation– (1) Introduction into healthy plant or animal tissue of microorganisms to produce a mild form of the disease, followed by immunity. (2) An introduction of nodule-forming bacteria into soil, especially for the purpose of nitrogen fixation. (3) Treatment of seed with bacteria that stimulate development of bacteria nodules on plant roots. Used on legumes such as peas and beans. (4) Bacteria supplied to legumes to "fix" nitrogen from the air. (5) A small amount of bacteria produced from a pure culture that is used to start a new culture.

Insect– An air-breathing animal (phylum Arthropoda) that has a distinct head, thorax, and abdomen. Insects have one pair of antennae on the head, three pairs of legs, and usually two pairs of wings on the thorax. The opening of the reproductive organs is near the posterior end of the body. They may be harmful or useful depending upon their habits. Some infest plants and animals, some are insectivorous, some pollinate plans, and some produce edible products.

Insect Control– The chemical or biological inhibition or killing of insect enemies.

Insect Enemy– (1) Any insect that is destructive or harmful to something desired by humans. (2) An insect, bird, mammal, etc., that preys on other insects.

Insect Growth Regulator– A chemical that interferes with the normal growth pattern of insects causing abnormal development and thus death. In the case of flies, it may be added to chicken feed, passing through the bird with the feces into the manure. Present in the manure, it kills house fly larvae soon after they hatch.

Insect Vector– An insect that carries a virus, bacterium, or the spores or mycelium of a pathogenic fungus and inoculates susceptible plants, animals, and humans.

Insecticide– A substance that kills insects by chemical action, as a stomach poison, contact poison, or fumigant.

Inseminate– To place semen in the vagina of an animal during coitus, or, in the practice of artificial insemination, to introduce semen into the vagina by a method other than coitus.

Inseminating Tube– A rubber, glass, or metal tube, usually with syringe attachment, used in artificial insemination to introduce the semen into the vagina of a female animal. Such instruments vary in character depending on the species being bred.

Inseminator– The technician, in the employ of an artificial breeder's unit, who brings the prepared semen to the farmer's herd and performs the technical service of inseminating the cows.

Insidious Disease– A disease that develops slowly in a stealthy, subtle manner over a long period of time.

Insipid– (1) Designating a flavor defect of cheese characterized by a lack of taste and odor. (2) designating any such flavor.

Insoluble– Not soluble; designating a substance that does not dissolve in another.

Insoluble Ash– That portion of milk ash that is not soluble in water.

Instar– An insect that is between the stages of its molting process.

Instinct– The ability of an animal based upon its genetic makeup to respond to an environmental stimulus; it does not involve a mental decision.

Instrumental Insemination– The act of depositing semen into the oviducts of a queen bee by the use of an instrument.

Insufflation– The blowing of a powder or vapor into a cavity for medication.

Insulin– The hormone from a part of the pancreas that promotes the utilization of sugar in the organism and prevents its accumulation in the blood.

Integrated Pest Management (IPM)– An ecological approach to pest management in which all available necessary techniques are systematically consolidated into a unified program, so that pest populations can be managed in such a manner that economic damage is reduced and adverse side effects are minimized.

Interaction– The process of chemicals being mixed together and having substantially different toxicity than the toxicities of the components. The chemicals may interact to increase or decrease toxicity. See Antagonism, Synergist.

Interbreeding– In livestock breeding, the breeding closely within a family or strain for the purpose of fixing type and desired characters.

Intercellular– Between cells.

Intercrossing– Crossbreeding.

Interdigital– Between the fingers, toes, or claws.

Interfering– The striking of the fetlock or cannon by the opposite foot that is in motion. This condition is predisposed in horses with base-narrow, toe-wide, or splay-footed standing positions.

Intergeneric Cross– A cross between species of different genera (rare).

Intermediate Host–An animal other than the primary host that a parasite uses to support part of its life cycle.

Intermingling Color–A coat color pattern of animals in which the separate colors merge where they meet. It is usually an objectionable color pattern.

Intermittent Attacks–The return of symptoms after a period without symptoms; e.g., recurrent attacks of malaria.

Intermittent Grazing–Grazing a pasture for indefinite periods with periods of rest between grazing.

Intermittent Parasites–Those parasites, such as mosquitoes or bedbugs, that approach the host only when in need of nourishment.

Intersex–Designating an organism that displays primary and secondary sexual characteristics intermediate between male and female.

Interspecific–Referring to events or relationships that occur between individuals of different species.

Interspecific Hybrid–Cross between individuals of different species. Taxonomically identified by listing both species separated by an *x*.

Interstitial–A term referring to the spaces (voids) between particles, as between sand grains.

Intestinal Bloat–Abnormal distention of the intestines with gas.

Intestinal Coccidiosis–See Coccidiosis.

Intestinal Worm–Any internal parasitic worm that inhabits the intestines.

Intestine–The part of the digestive tract between the stomach and anus.

Intoxication–The state of being poisoned or drunk.

Intra–A prefix meaning within, inside; as intrastaminal, inside the (ring of) stamens; intraspecific, within a species.

Intracellular–Within, inside of, a cell.

Intracellular Fungus–A fungus that occurs or grows within a cell.

Intracervical Method–A method of artificial insemination whereby semen is placed directly in the cervix or uterus and not in the vagina.

Intradermal–Within the layers of the skin.

Intradermal Tuberculin Test–A test used to discover tuberculosis in animals, which consists of injecting into the skin of the tail (caudal fold), or elsewhere, a minimum of concentrated tuberculin. If the animal is infected, 72 hours after the injection there will be a swelling at the point of injection.

Intramuscular–Within the muscles.

Intramuscular Injection–Injection of a substance into a muscle.

Intranasal Instillation–The placing of fluid or medicine inside the nose.

Intraocular Vaccination–In poultry, the placement of a vaccine directly into the eye.

Intraperitoneal–Within the cavity of the body that contains the stomach and intestines.

Intraspecific–Referring to events or relationships that occur between individuals of the same species.

Intraspecific Hybrid–Cross between individuals within the same species, but of different genotypes.

Intratracheal–Within the trachea, or windpipe.

Intrauterine–Within the uterus.

Intravenous–In, into, or from within, a vein or veins.

Introduced–Designating a plant, animal, disease, etc., that is not indigenous to an area, but is brought in purposely or accidentally.

Introducing Cage–Small wood and wire cage used to ship queen bees and also sometimes to release them into the colony.

Introgastric–In the stomach.

Introgression, Hybridization–Long-continued interspecific hybridization leading to an infiltration of genes from one specie into another.

Intromittent–Designating the use, in copulation, of the external reproductive parts of many male animals.

Invertase–Enzyme produced by bees that speeds inversion of sucrose to glucose and fructose.

Invertebrate–(1) Any animal with no spinal column. (2) Having no spinal column.

Inverted Nipple–A teat (usually refers to a sow's teat) that appears to be inverted toward the udder. The teat is nonfunctional.

Involution–The return of an organ to normal size after a time of enlargement, as in the case of the uterus after parturition.

Iodine–A trace element in soil, an essential element for the health of humans and animals. A deficiency causes goiter. It is readily absorbed by plants from soil and water, but it is not essential to plants.

Iodine Value–Sometimes referred to as iodine number, the measure of the degree of unsaturation of a fat by the extent of the uptake of iodine (the number of grams of iodine absorbed per 100 grams fat) by the unsaturated double bonds in the fatty acid chain. Hence, butter with an iodine value of 22 to 38 is more highly saturated than cottonseed oil with an iodine value of 104 to 114.

Iodized Salt–That salt prepared by mixing salt with potassium iodide in the proportion of about one part of iodide to 5,000 parts of salt. Iodized salt is of value in the prevention of simple goiter, which is endemic in people and animals in certain parts of the United States.

Ion–An atom or a group of atoms carrying an electrical charge, which may be positive or negative. Ions are usually formed when salts, acids, or bases are dissolved in water. When common salt, sodium chloride, is dissolved in water, positive sodium ions and negative chloride ions are formed. See Anion, Cation.

Ion Exchange–The replacement, in a colloidal system, of one ion by another with a charge of the same sign.

Ionization–The process by which an atom becomes electrically charged, by the removal or addition of one or more of its extranuclear electrons, so that the electrical balance between the electrons and the protons within the atom's nucleus is destroyed. An atom with more

Ionophore–A feed substance or additive that makes the digestive processes in ruminants more efficient.

Iron–Fe; a metallic element essential to people, animals, and plants; very common in some minerals, most rocks, and all soils; an essential constituent of blood hemoglobin where it functions to transport oxygen. Iron is specific for the treatment of anemia in animals. In plants, iron deficiency results in iron chlorosis.

Iron Gray–A coat color term for a horse that (a) in the United States, denotes a white coat with a high percentage of black hairs; (b) in England, a gray coat with reddish hairs throughout.

Irradiated–Designating a food or feed treated with ultraviolet light to increase the vitamin D content.

Irradiated Milk–Market milk treated with ultraviolet light so that the ergosterol is changed to vitamin D.

Irradiated Yeast–Yeast subjected to ultraviolet rays in order to increase its antirachitic potency. It is used as an ingredient in feeds for its vitamin D and B-complex content.

Isinglass–(1) Very pure gelatin prepared from the air bladders of fishes, especially sturgeon; used as a replacement for gelatin in jellies and puddings, and as a clarifying agent. (2) Mica in sheets.

Islet (Island) of Langerhans–The tissue lying in the framework of the pancreas that secretes the hormone insulin.

Isogamy–Morphological similarity of fusing gametes.

Isolate–To place an animal in confinement away from other animals to prevent breeding or spread of disease.

Isotopes–Elements having an identical number of protons in their nuclei, but differing in the number of their neutrons. Isotopes have the same atomic number, differing atomic weights, and almost but not quite the same chemical properties. Different isotopes of the same element have different radioactive behavior.

Isthmus–(1) In biology, a narrow piece of tissue that connects two larger parts; the part of the Fallopian tube between the ampulla and the uterus. (2) In poultry, that part of the oviduct where the two shell membranes are added to the egg.

Italian Bees–A race or variety of honeybee that originated in Italy and has become widely dispersed and crossbred with other races.

Itch Mite–*Sarcoptes scabiei*, family Sarcoptidae; a mite that infests humans and other animals. In humans it causes itch and in animals, mange.

-itis–A suffix denoting inflammation, as dermatitis, inflammation of the skin; tonsillitis, inflammation of the tonsils.

Ivermectin–A broad-spectrum drug that can be injected into livestock for control of both internal and external parasites. It is reputed to kill adult and immature roundworms and lungworms, and external grubs, lice, and mites.

Jack–(1) The male, uncastrated ass. See Jenny. (2) A mechanical, hydraulic device used for lifting heavy objects. (3) See Bone Spavin.

Jackass–See Jack.

Jaundice–A syndrome seen in animals and people characterized by excess bile pigment in the blood and deposition of bile pigments in the mucous membranes and skin resulting in a yellow or yellowish-orange complexion. Also called icterus.

Jejunum–The part of the small intestine between the duodenum and the ileum.

Jenny–A female donkey. Also called jennet.

Jerk–To cut meat, usually beef, into long, thin strips, and dry it in the sun.

Jerked Beef (Jerky)–Beef that has been naturally dried. Also called charqui. See Jerk.

Jersey–A widely used breed of dairy cattle from the Isle of Jersey. They are of extreme dairy type, solid fawn, red, or nearly black, and their milk has the highest butterfat content of any breed. Weight at maturity varies form 800 to 1,000 pounds.

Jersey Buff–A small, well-fleshed, buff-colored variety of turkey. It is a recently developed variety from New Jersey, United States.

Jet Black–A horse color that is a shiny black.

Jib(ber)–A balky horse.

Jig–(1) An uneven gait of a horse, closely associated with prancing and weaving. (2) A device built to aid in construction; used to make duplicate cuts, holes, etc.

Jimmies–A condition of sheep resulting from feeding on the bulbs of cloak fern (*Notholaena sinuata*), family Polypodiaceae. Affected animals arch their backs, walk with a stilted movement of the hind legs, breathe rapidly, and tremble.

Jockey Stick–A stick fastened to the hame of a broken horse and to the bit of a green horse to prevent them from crowding and to keep the green horse in position while being broken.

Johne's Disease–A chronic, infectious disease of cattle, sheep, and goats caused by *Mycobacterium paratuberculosis*. It causes loss of condition, unthriftiness, persistent diarrhea, reduction of milk yield, alternate refusal to eat and ravenous eating, general debility, prostration, and finally death. Also called chronic bacillary diarrhea, pseudotubercular enteritis, paratuberculosis, chronic bacterial dysentery.

Joint–(1) the point or place at which two bones meet: the articulation between the two bones. (2) The pint or meeting of two pipes. (3) A general term for a cut of meat (England). (4) See Node.

Jowl–(1) The cheek of a pig. (2) Meat taken from the cheeks of a hog.

Judas Goat–A trained goat used by a slaughterhouse to lead sheep to the killing pens.

Jug–(1) A small claiming pen where ewes with newborn lambs are kept until the ewe has nursed and claimed her young. (2) A glass bottle holding one-half gallon or a gallon, used to bottle cider, syrup, molasses, etc. (3) A deep vessel of earthenware narrowed to a neck, which is frequently fitted with a finger handle for carrying. (4) A small pen, part of a corral, used to funnel animals down a narrow alley (chute) to facilitate sorting, loading, or catching; used primarily with cattle.

Jughead–A horse with poor characteristics, little sense, and not amenable to training.

Julep–(1) Ancient Arabian name for a cooling drink containing muscilage, opium, etc. (2) An American drink.

Jumbo Hive–A beehive 2½ inches (5.35 centimeters) deeper than standard Langstroth hive.

Jumper–(1) A class of horse, usually a Thoroughbred, able to jump high hurdles and used in shows. (2) Any animal with a tendency to jump over fences and therefore somewhat difficult to restrain.

Junior Calf–In stock judging, a beef or dairy animal that was born after January 1 of the year it is shown.

Junior Champion–In stock judging, the best animal of the younger classes selected from the first prize winners.

Junior Spring Pig–In stock judging, a swine that was farrowed on or after March 15 of the year it is shown.

Junior Yearling–In stock judging: (a) a beef animal calved between January 1 and April 30 of the preceding year; (b) a dairy animal calved between January 1 and June 30 of the preceding year; (c) a swine farrowed between March 1 and August 31 of the preceding year.

Juvenile–(1) Coming to the surface for the first time; fresh, new in origin; applied chiefly to gases and waters. (2) Pertaining to the young of any animal.

Juvenile Hormone–The hormone, secreted by the corpora allata, that maintains the immature form of an insect during early molts.

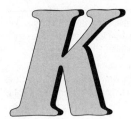

Kak–A kind of saddle bag.

Karakul–A small, light-shearing breed of broadtailed sheep from west-central Asia, bred especially for the lambskins used in the fur pelt industry.

Karatin–Protein found in hair, feathers, horns, and hoofs.

Karyotype–A picture or diagram of the chromosomes of a particular cell as they appear in the metaphase of mitosis arranged in pairs by size and location in the centromeres.

Katabolism–See Catabolism.

Ked–See Sheep Ked.

Keel–(1) The part of a fowl's body that extends backward form the breast. Also called breastbone. (2) The two front, united petals of a pealike flower. (3) In ducks, the pendant fold of skin along the entire underside of the body. (4) In geese, the pendant fold of flesh from the legs forward on the underpart of the body. (5) A central dorsal ridge, like the keel of a boat. (6) The two anterior united petals of a papilionaceous flower.

Keel Disease–See Avian Typhoid.

Keeled–Of, or pertaining to, a ridge (like the bottom of a boat) on an animal or plant part.

Keeping–Designating the storage qualities of a product.

Keet–The young guinea fowl.

Keratin–A complex protein distinguished by high insolubility. It is the substance of which hair, horns, claws, and feathers are composed.

Keratitis–An inflammation of the cornea of the eye.

Ketone Bodies–Chemical substances such as diacetic acid, hydroxybutyric acid, and acetone, produced in the liver of animals. Excessive amounts of such substances in the body result in ketosis. When ketones are eliminated in excessive amounts in the urine, the condition is called ketonuria, acetonemia.

Ketosis–A metabolic disease characterized by loss of appetite, markedly poor condition of skin and flesh, digestive disorders, sometimes nervous symptoms, and foul-smelling milk. It often occurs as a primary metabolic disturbance in cattle, in pregnancy disease of sheep, in diabetes, and certain other conditions. Also called acetonemia, false milk fever, chronic milk fever, acetonuria.

Kicking Hobble–A rope or strap fastened to the rear legs or feet of a horse, ass, or cow to prevent kicking.

Kicking Strap–A strap fastened to the crupper strap of the harness in breaking colts. It extends down each side of the shaft of the carriage and is used to prevent kicking.

Kid–(1) A young goat. (2) To give birth to a kid.

Kid Crop–The number of kids produced by a given number of does, usually expressed in percent of kids weaned of does bred.

Kid House–A small structure designed to give shelter to a newborn kid. The doe is staked so that it cannot abandon its kid.

Kid Mohair–The finest quality of Mohair, taken from young goats.

Kidney–A bean-shaped organ of humans and certain lower animals that excretes urine. The lamb kidney is edible, and kidneys from other animals are converted into edible products.

Kidney Faller–A horse that collapses in the hindquarters when it walks.

Kidney Knob–The fat on a beef carcass that surrounds the area where the kidneys were removed.

Kidney, Pelvic, and Heart Fat–The internal fat of an animal carcass. Specifically, the fat that surrounded the kidneys, pelvic cavity, and heart within the body cavity.

Killing Cattle–Cattle, usually grass-fattened on a ranch or range, of sufficiently good flesh and finish to be slaughtered profitably.

Killing Floor–The place in a slaughterhouse where a stunned animal is killed, bled, skinned, eviscerated, and where the carcass is split, washed, and wrapped.

Kind–(1) All the plants of the same type, accepted as a single vegetable or fruit, as tomato, cabbage, bean, apple, peach, etc. (2) A species, as a cow, sheep, etc.

Kind of Animal–An animal species or species group such as antelope, cattle, deer, elk, goats, horses, etc.

Kindle–(1) To give birth to a litter of rabbits. (2) A young rabbit.

Kine–Cows or cattle.

King Bee–Before the seventeenth century, it was popularly believed that the queen bee was a male who ruled the hive. See Queen Bee.

Kip–A hide taken from a very young calf.

Kippered–Lightly salted and smoked fish.

Kleins' Disease–See Avian Typhoid.

Knee–(1) The joint between the hip and the ankle in humans and quadrupeds (hind leg); in birds, the tarsal joint. (2) The carpal joint (foreleg). (3) The spurlike, root growth of the baldcypress that develops when the tree grows in a swamp.

Knee Banger–A horse that strikes its knees with the opposite front foot while walking or running.

Knee Spavin–A chronic inflammation of the small bones of the carpal joint of horses, characterized by lameness, bony enlargement, a dragging of the toe of the affected leg, and a fusion of the carpal bones.

Knob–(1) A rounded hill or mountain, especially an isolated one. Local in the United States South. (2) The horny protuberance at the juncture of the head and upper bill in African and Chinese geese. (3) A deformity growth on the breastbone, usually at the front, sometimes found in chickens and turkeys; a defect. (4) The rounded protuberant part of the skull in crested fowl.

Knock-kneed–An undesirable conformation or weakness of the front legs of horses in which the legs are not straight due to the knees coming too close together.

Knuckle–The joint of two bones often used in making soup stock.

Koi–Fish of the Carp family that are raised because of their colorful appearance.

KPH–See Kidney, Pelvic, and Heart Fat.

Labium–(1) The posterior mouthpart or lower lip of an insect. (2) A lip or liplike organ. (3) The folds of skin of the vulva.

Labored Breathing–Any abnormal form of breathing in an animal in which the animal appears to be straining or working excessively hard to get air into or out of the lungs. Such breathing may be shallow and rapid, deep and slow, or deep and rapid.

Labrum–The anterior mouthpart or upper lip of an insect.

Laceration–A wound, especially one caused by tearing.

Lactate–To produce and secrete milk.

Lactation–The period of milk secretion. Usually begins at parturition and ends when offspring are weaned, or, in the case of dairy cattle, the animal is dried up.

Lactic Acid–A hydroxypropionic acid; a syrupy liquid generated by the natural souring of milk or the induced fermentation of various food products, such as sugars, starches, etc., by any of various organisms. It is used in cultured milk products, pickles, soft drinks, etc.

Lactic Fermentation–The preserving process that occurs in the making of sauerkraut, dill pickles, fermented string beans, silage, etc.

Lactic Former–Any organism that changes sugars into lactic acid, as *Lactobacilus plantarum*.

Lactiferous–Producing a milky substance, such as the nearly ripe poppy capsule.

Lactogen–The specific hormone that initiates and maintains lactation. Also called gelactin.

Lactogenic Hormone–Any hormone that stimulates lactation, especially prolactin.

Lactoglobulin Fraction–That small portion of milk protein that contains more sulfur and less phosphorus than casein.

Lactometer–An instrument that determines the density of milk. It acts as a specific gravity indicator, from which information on the percentage of milk solids may be calculated.

Lactose–$C_{12}H_{22}O_{11}$; a white crystalline disaccharide made from whey and used in pharmaceuticals, infant foods, bakery products, and confections; also called milk sugar.

Lagoon–(1) Body of shallow water, particularly one possessing a restricted connection with the sea. (2) Water body within an atoll or behind barrier reefs or islands. (3) A body of water more than 1 meter deep established for anaerobic decomposition of organic wastes.

Lagoon, Aerobic–Lagoons larger than one-half acre and deeper than three feet must be stirred mechanically to supply oxygen for aerobic stabilization of wastewater sludge. See Lagoon, Anaerobic.

Lagoon, Anaerobic–A wastewater sludge treatment process larger than one-half acre and deeper than three feet that is not aerated mechanically. See Lagoon, Aerobic.

Lamancha–A breed of dairy goats developed by crossing several pure bred breeds. The breed is noted for their very small, short ears. They may be any color.

Lamarckism–A belief, named after French naturalist Jean Baptiste de Lamarck (1744-1809), that acquired characteristics can be inherited. Now proved false.

Lamb–(1) The young of sheep: specifically, (a) a young ovine that has not yet acquired the front pair of permanent incisor teeth; technically, (b) in stock judging, a sheep born in the same calendar year as shown. (2) To give birth to a lamb.

Lamb Bar–Apparatus for communal feeding of lambs and kids consisting of a container for milk with several artificial teats set into it.

Lamb Crop–The number of lambs produced by a given number of ewes, usually expressed in percent of lambs weaned of ewes bred.

Lamb Flock–In stock judging, one ram lamb and three ewe lambs.

Lamb Hog–A male lamb from weaning time until it is shorn.

Lamb Riblet–A retail cut of lamb for stewing or braising that consists of thin slices of breast along with ribs.

Lamb Shank–A retail cut of lamb for stewing or braising which consists of the shank. Also called trotter.

Lamb's Wool–Short wool taken from lambs not over seven to eight months old, although some wool which is taken from lambs of twelve to fourteen months of age is also graded as lamb's wool. It is soft and has less felting properties than sheep wool.

Lambing–The dropping, or birth, of lambs (in sheep husbandry, a time or regular event in management). See Lambing Time.

Lambing Ground–Range reserved for grazing during the lambing period.

Lambing Loop–A length of smooth or plastic-coated wire used as an aid in difficult lambing.

Lambing Pen–A specially equipped, isolated pen in the sheep barn in which a ewe is placed just before she gives birth to her young.

Lambing Percentage–The number of lambs produced per 100 ewes.

Lambing Range–See Lambing Ground.

Lambing Shed–A shed or barnlike building in which ewes are placed just before they give birth to lambs.

Lambing Time–The season of the year, usually late winter or early spring, when sheep produce their young.

Lamella–(Biology) One of the layers of a cell wall; a thin layer in a shell, like a leaf in a book.

Lameness–Any soreness, tenderness, or unsoundness that is indicated by the unnatural action of feet or legs of animals.

Laminated Wood–A piece of wood built up of layers of wood that have been joined with the grain parallel, either with glue or with mechanical fastenings. The term is most frequently applied where the layers are too thick to be classified as veneer.

Laminitis–Inflammation or congestion of the sensitive tissues, the laminae, which lie immediately below the outer horny wall of a horse's foot. It is characterized in the acute stage by expression of pain, standing still, dilated pupils, bright red membranes of the eyes, and rapid breathing. The pulse at first is very strong and fast but becomes weaker and the temperature is above normal. Among the various causes are: ingestion of excessive amounts of grain (grain founder), drinking cold water while overheated, concussion during hard fast road work or from long truck or train rides, toxemia as a result of pneumonia or metritis, etc. Some cases may be an allergic reaction. Grain founder is usually the most common type. Also called founder. See Founder.

Lampas (Lampers)–A swelling of the mucous membranes in the hard palate immediately behind the upper incisors of the horse. Also called palatitis. See Palatitis.

Land–(1) The total natural and cultural environment within which production must take place. Its attributes include climate, surface configuration, soil, water supply, subsurface conditions, etc., together with its location with respect to centers of commerce and population. Oyster beds and even tracts or bodies of water, as where valuable fishing rights are involved, may be regarded as land. It is often convenient, in fact, to regard land as synonymous with all that nature supplies, external to humans, which is valuable, durable and appropriable, thus including, e.g., waterfalls and other sources of waterpower. (2) In a broad legal sense, any real part of the surface of the earth, including all appurtenances, anything in, on, above, or below the surface. (3) In plowing, a plowed or unplowed space between two furrows. (4) The total width of a strip of land tilled by a farmer, or some designated width, as a perch, $16\frac{1}{2}$ feet. Also called a stitch. (5) Soil. (6) A natural part of the earth's surface characterized by any single factor, or combination, of topography, climate, soil, rocks, vegetation; the natural landscape. (7) Pertaining to agriculture, those areas actually in use or capable of use for the production of farm crops and livestock.

Landrace–A popular, white hog characterized by large drooping ears, a long snout, and good mothering ability.

Lane–A passageway from barnyard to pasture or field areas; a narrow road confined between fences.

Langstroth Frame–A minister from Pennsylvania, L.L. Langstroth, who patented the first beehive incorporating bee space thus providing form removable frames. The modern hive frequently is termed the Langstroth hive and is a simplified version of similar dimensions as patented by Langstroth. The standard frame measures $9\frac{1}{8}$ by $17\frac{5}{8}$ inches. See Bee Space.

Lanolin–The fatty substance removed from sheep wool when it is scoured and cleaned. When refined, it is used extensively in cosmetics and provides a nontoxic carrier for applying plant regulators or other chemicals to the surface of plants.

Lapse Rate–See Adiabatic Lapse Rate.

Lard–The fat rendered from fresh, clean, sound, fatty tissues of hogs in good health at the time of slaughter. It is used as a food product for frying, shortening, etc. In recent years, lard has been largely replaced by vegetable oils.

Lard-type Hog–An antiquated term denoting a fat type of hog that was developed in the United States. Most of the breeds common in the United States were originally classified as this type. It was especially adapted to a high degree of quick fattening, with heavy back and thick sides. See Bacon-type Hog, Meat-type Hog, Fat-type Hog.

Lareat–See Lariat.

Large Egg–A United States grade of eggs having a minimum net weight of 24 ounces per dozen.

Large Roundworm–A common name referring to various species of roundworms belonging to the family Ascaridae, which parasitize humans, domestic livestock, and birds. The scientific names of these large roundworms are: *Parascaris equorum* (equines); *Neoascaris vitulorum* (cattle); *Ascaris lumbricoides* (swine, humans); *Toxocara canis, T. mystax, Toxascaris leonina* (dogs and cats); *Ascaridia galli, A. columbae, A. numidae* (domesticated birds).

Lariat–Rawhide, horsehair, or hemp rope, generally 35 to 45 feet long and arranged in a coil on the saddle. Used to lasso livestock for branding or restraining purposes. See Lasso.

Larva–The immature insect hatching from the egg and up to the pupal stage in orders with complex metamorphosis; the six-legged first instar of mites and ticks.

Larvicide–A chemical used to kill the larval or preadult stages of parasites.

Larynx–The upper portion of the windpipe (trachea).

Lasso–(1) To throw a lariat in such a manner as to catch an animal by the horns, neck, or legs in the loop at the end, generally for restraining purposes. (2) A rope or line with a running loop used for catching animals. Lee Lariat.

Latent–Designating an infection that is present but which is not manifest in the host under consideration. See Dormant.

Lateral–(1) A directional or positional term meaning away from the middle or toward the side. (2) A part of a system that branches out from the main body of the system, such as the tile lateral drain that connects to a main drain in a drainage system. (3) A branch or twig of a tree.

Lateral Canter–A horse's gait characterized by a simultaneous beat of corresponding front and rear feet, as opposed to diagonal front and rear feet.

Lateral Lacunae–The two grooves that border the horny frog of a horse's hoof.

Lateral Ringbone–A bony growth on the side of the pastern bone in a horse's leg. See Ringbone.

Lathyrism–A disease of animals caused by eating roughpea (*Lathyrus* spp.) which is high in selenium.

Lavage–The process of washing out the stomach or intestines. In gastric lavage, a double-way tube is passed down the esophagus into the stomach.

Laxa–Loose.

Laxative–A mild medicine used to relieve constipation.

Lay–To ovulate and deposit an egg or eggs produced within the female generative organs, as a hen lays eggs.

Layer–(1) A mature female fowl that is kept for egg-laying purposes, especially one in current egg production. (2) A course or stratum, as a layer of sand. (3) A plant twig or shoot, tied down and partially covered with earth, so that it can take root while remaining unsevered from the parent stock. (4) To reproduce by layerage.

Laying Ability–The ability of a particular hen to produce eggs, commonly measured by the number of eggs produced in a given time, as in a month or year. It is largely determined by five inherent qualities: (a) earliness of sexual maturity, (b) persistency of production, (c) intensity (or rate) of production, (d) broodiness, and (e) pauses.

Laying Mash–Any of several special feed preparations, conducive to egg production, that is fed to laying flocks of poultry.

Laying Nests–A series of connected cubicles in which hens lay eggs. They are usually made of metal to help keep down problems with parasites.

Laying Worker–A worker bee that lays eggs that produce drones. Laying workers usually develop in queenless colonies.

LC_{50}–See Lethal Concentration.

LD_{50}–See Lethal Dose 50.

Lead Cattle–The cattle at the head of a moving herd.

Lead Line–A bluish line at the margin of the gums of an animal that indicates chronic lead poisoning, usually as a result of ingestion of paint or spray materials. See Lead Poisoning.

Lead Poisoning–Poisoning that results from ingestion of lead, usually of paint containing lead. it is characterized (a) in horses by convulsions, partial paralysis, colic, roaring, thirst, and increased urination; (b) in cattle, by staggering, impaired vision, and unusual postures. See Lead Line.

Leader–(1) The main or dominant stem of a plant. (2) The front animals of a tandem hitch.

Lean–(1) A piece of meat that consists largely of muscles lacking in fat or the proper proportion or distribution of fat. (2) Designating an animal lacking in condition of flesh or finish.

Lean to Fat Ratio–The amount of lean meat in a carcass compared to the amount of fat.

Leather–(1) The cured, tanned skins of animals, especially of the bovines. (2) A pad of leather placed between the shoe and the foot of a horse with a sensitive sole. (3) See Fruit Leather.

Lecithin–One of a group of lipids known as phospholipids. Abundant in brain tissue and egg yolk. Obtained from peanuts, corn, and soybeans for commercial use (as an emulsifier in such products as chocolate).

Lectins–A group of protein substances, natural antibodies; agglutinins for type A red blood cells. Lectin may be obtained from lima beans.

Leg–(1) Any of the limbs of animals that support and move the body, such as foreleg, front leg, hind leg. (2) A cut of meat which is that portion of the leg between the knee and ankle, such as a chicken leg, leg of lamb. (3) To haul or drag a sheep from the pen to the shearing board by a hind leg.

Leg Conformation Grade–A grade assigned to the leg of a lamb or lamb carcass to indicate thickness of muscling. It is used in determining yield grade.

Leg of Lamb–A cut of meat comprising that section of the leg in which the shank of meat is tucked into a pocket made under the vellum on the inside of the leg of a lamb.

Leg of Mutton–A retail cut of meat from a mature sheep that comes from the same portion of the carcass as the leg of lamb.

Leggy–(1) Designating a plant which has unusually long stems. (2) Designating an animal, usually very young, as a colt, whose legs are disproportionally long in relation to its body size.

Leghorn–An egg-type breed of domestic chickens of the Mediterranean class characterized by yellow skin, nonfeathered shanks, nervous disposition, fast feathering, and relative nonbroodiness. Mature males weight about 6 pounds, mature females about 4½ pounds. The eggshell is white. The Single Comb White Leghorn, the outstanding variety of the breed, universally raised, is one of the world's leading and most highly developed breeds for efficient production of market eggs.

Legume Silage–Legume crops, such as alfalfa and Ladino clover, which make satisfactory silage, especially if mixed with grasses and put into the silo in proper condition.

Leishmaniasis–Any disease due to infection with microscopic protozoan parasites of the genus ***Leishmania***.

Leppies–Orphan calves (southwestern United States).

Leprosy–A chronic, transmissible disease due to a specific bacterium, ***Mycobacterium leprae***.

Leptospirosis–A disease of animals and occasionally of humans caused by species of the genus ***Leptospira***. In dogs, it is caused by ***L. canicola*** and is characterized by loss of appetite, depression, fever, thirst, vomiting, loss of weight, nephritis, and death. In cattle, it is caused by ***L. pomona*** or by ***L. icterohemorrhagiae*** and is characterized by loss of appetite, fever, jaundice, abortion, and sometimes death.

Lesion–Injury or diseased condition of tissues or organs.

Let Down Milk–To accelerate greatly the secretion of milk at the regular milking time. The milk glands of the udder are stimulated by massaging the teats and udder or by the sucking of the calf. There is also some influence of stimulating hormones, the most important being prolactin. See Hold Up Milk.

Lethal–Deadly; causing death, as a lethal dose.

Lethal Characters–See Lethal Gene.

Lethal Concentration (LC)–The lethal concentration (written as LC_{10} or LC_{50} or LC_{100} or any percentage) median is the parts per million (ppm) or parts per billion (ppb) of toxicant in water or air that kills 10, or 50, or 100 percent, respectively, of the target species in a 24-hour period. Usually used for fish. See Effective Concentration, Lethal Dose.

Lethal Dose–The lethal dose (written as LD_{10}, LD_{50}, LD_{100}, or any percentage) median is the milligrams of toxicant per kilogram of body weight that kills 10, 50, or 100 percent of the target species. See Effective Concentration, Lethal Concentration.

Lethal Gene–A gene that can cause the death of an individual when it is allowed to express itself.

Letting Down–The adjusting period, with some loss of flesh, which takes place when beef cattle are changed form heavy, drylot feeding to open pastures or range feeding.

Leucocyte–The white cells of the blood that destroy disease germs and help remove foreign matter from the bloodstream or tissue. Found predominantly in pus.

Leucocytozoon Infection–An acute and highly fatal disease of ducklings and poults caused by protozoan parasites: ***Leucocytozoon simondi*** in ducks, ***L. smithi*** in turkeys. Symptoms are loss of appetite, lethargy, thirst, rapid and labored breathing, excitement, loss of equilibrium, convulsions, coma, and foamy discharge from the nose and mouth.

Leukemia–Cancers of the blood-forming organs, characterized by abnormal proliferation and development of leukocytes (white blood cells) and their precursors in the blood, lymph, bone marrow, and lymph glands.

Leukocytes–White blood cells.

Leukosis–Avian leukosis complex.

Levorotatory Honey–A honey that does not contain honeydew. It rotates to the left the beam of polarized light passed through it in a special optical tube. See Dextrorotatory Honey.

Levulose–Noncrystallizing sugar of honey that darkens readily if honey is overheated.

Ley–A temporary pasture. Also spelled lea, a meadow (obsolete).

Leydig Cells–Special cells or tissue located inside the testes that secrete testosterone; also called interstitial cells.

Libido–(Latin) Sexual drive.

Lice–Small, nonflying, biting or sucking insects that are true parasites of humans, animals, and birds. Biting lice belong to the order Mallophaga; the sucking or true lice belong to the order Anoplura.

Lick Tank–A tank containing a liquid feed supplement for cattle. The cattle obtain the feed by licking a wheel or ball that rotates in the liquid. As the wheel or ball turns, the liquid adheres to it and is brought to the surface.

Licks–A United States term given to boggy grounds with salt springs, where cattle go to lick the salt.

Life Cycle–Life history; the changes in the form of life that an organism goes through.

Life History–Habits and changes undergone by an organism from the egg stage to its death as an adult.

Lift–(1) A joint of meat, especially beef, from the thigh. (2) Elevating or pulling power of a pump. (3) To loosen and remove seedlings or transplants from the seedbed or transplant bed prior to transplanting. (4) A hay fork. (5) A fork for lifting heavy loads.

Ligament–Any tough, dense, fibrous band that connects bones or supports viscera.

Ligase–An enzyme that splices segments of DNA together.

Light–(1) The form of radiant energy consisting of wavelengths lying within the limits perceptible by the normal human eye, and, by extension, the shorter and longer wavelengths, the ultraviolet and the infrared light, invisible to the eye but which may be recorded photographically. Light can be absorbed by various substances and transformed into heat. Its excess may produce fading, the destruction of green color in plants, but an insufficiency can also cause lack of chlorophyll production. The coloration of fruits is dependent upon sufficiency of light. The growing parts of plants respond to the stimulus of the direction from which the light comes. See Photosynthesis, Phototropism. (2) To become ignited; to take fire. (3) Designating a deficiency or lack in degree, such as a light rain or a light crop. (4) When applied to food or drink it can have several definitions. It could mean reduced calories, fluffy (full of air), pale, low in sodium, mild in flavor, and/or less alcohol.

Light Bay–A horse color; light tan with a black mane and tail.

Light breed–A smaller and more refined type of a particular kind of livestock, as the Thoroughbred horse, the Leghorn chicken, etc.

Light Chestnut–A horse color; light, yellow-gold.

Light Dun–A horse color; a dun imposed on a sandy bay or sorrel.

Light Feeder–(1) An animal that is being fed for maintenance and normal growth but not for quick finish or fattening. (2) An animal that does not eat as much as most animals of its class.

Light Grazing–Allowing livestock to partially graze an area, such as a pasture or range.

Light Horse–A horse that weighs between 900 and 1,400 pounds at maturity.

Light Meter–A device used to measure light intensity; the measurement is usually in foot candles.

Light Ration–(1) A limited or scanty ration in contrast to a liberal ration. (2) A ration that is loose and bulky in relation to its weight.

Light-handed–Denoting a rider who handles a horse well with a minimum of tugging or jerking on the bridle.

Light-Harness Horse–An American horse produced for speed and performance such as Standard and American Trotter.

Lightning Arrester–A device used across the circuit of a piece of electrical equipment to protect it or its operator from abnormal surges of high voltage, such as from lightning.

Ligule–A strap-shaped organ or body: (a) particularly, a strap-shaped corolla, as in the ray flowers of the Compositae; (b) also, a projection from the top of the sheath in the Gramineae, Palmae, and some other plant families.

Like Produces Like–A rule of thumb in breeding; the offspring will bear a close resemblance to the parents.

Limb–(1) A lateral branch of a tree or shrub. (2) The leg or wing of an animal. (3) To cut off a limb of a tree.

Limberneck–A disease of fowls resulting from ingestion of food contaminated with ***Clostridium botulinum***, characterized by flaccid paralysis of the body. This is the same species of microorganism that causes botulism poisoning, usually from improperly canned meats and vegetables. The toxic chemical is known as botulin or botulismotoxin.

Limburger–A soft cheese within a strong flavor and odor.

Limited Feeding–(1) The feeding of livestock to maintain weight and growth but not to fatten or increase production. (2) Restricting an animal to less than maximum weight increase.

Limosis–Intense hunger.

Limousin–A breed of beef cattle that originated in southern France. They are golden-red and are recognized for their high-yielding carcasses.

Lincoln–A long-wooled breed of sheep from Lincolnshire, England. A large, white-faced mutton breed, it is popular in New Zealand. The mature ewes weigh up to 250 pounds.

Line–(1) The reins of a harness. (2) A rope, cable, string, wire, tube, etc., for tying or hanging objects, or conducting electricity, water, gas, etc., as a power line, gas line, or milk line. (3) A boundary or limit, as a fence line, property line. (4) In marketing, the whole of a herd of sheep. (5) A group of plants or animals that retain their uniform appearance in succeeding generations.

Line Breeding–Mating of selected members of successive generations among themselves in an effort to maintain or fix desirable characteristics.

Line Test–(1) A test in which a series of samples are taken from the milk supply of the whole herd during milking or milk processing; used in the bacteriological control examination of the milk supply. (2) The rapid agglutinin test used to detect abortion infection in herd milk. (3) A test used for the determination of the vitamin D content of milk. (4) A test to determine the degree of calcification of a growing bone; a measure for rickets.

Liniment–A preparation used for bathing or rubbing sprains, bruises, etc., usually containing a counterirritant.

Linkage–The association of characters from one generation to the next due to the fact that the genes controlling the characters are located on the same chromosome linkage group. The genes located on a single chromosome or the characters controlled by such genes.

Linnaean–Conforming the principles of binomial nomenclature of all plants and animals, into genus and species, as advocated by Carl von Linné, a Swedish botanist (1707-1778), who Latinized his name to Carlus Linnaeus.

Linseed Cubes–A livestock feedstuff, consisting of a mixture of linseed oil meal with flaxseed by-products, or both.

Linseed Meal–The product resulting from grinding linseed oil cake produced when flaxseed is pressed to recover linseed oil. Only batches unfit for feed are used as a fertilizer by organic gardeners.

Lip–(1) In a dam, a small wall on the downstream end of the apron to break the flow from the apron. (2) One of the edges of a wound. (3) Either of the two external fleshy folds of the mouth opening. (4) Either one of the inner or outer fleshy folds of the vulva.

Lip and Leg Ulceration–A viral disease of sheep, characterized by lesions on the lips, face, lower portion of the legs, and between the toes. Also called ulcerative dermatosis of sheep.

Lip Strap–The small strap running through the curb chain from one side of the bit shank to the other. Its primary function is to keep the horse from taking the shank or the bit in its teeth.

Lipase–A fat-digesting enzyme.

Lipis–A group of organic substances that are insoluble in water but soluble in such materials as acetone, chloroform, ether, and xylene; e.g., fats, waxes, and steroids including cholesterol.

Lipoid–Concerning fat or fatty tissue.

Lipolysis–The hydrolysis of fats by enzymes, acids, alkalis, or other means, to yield glycerol and fatty acids.

Liquid Manure–The liquid excrement from animals, mainly urine, collected from the gutters of barns into large tanks and hauled to the fields.

Liquid Smoke–A liquid mixture applied to meats in curing, especially to ham and bacon, to replace curing by wood smoke.

Listeriosis–A sporadic, specific bacterial disease of ruminants primarily involving the central nervous system (brain and spinal cord). On occasions it may result in abortions in sheep and cattle. It is caused by *Listeria monocytogenes*, also infective for nonruminant mammals (including people) and chickens. Also called listerelosis, circling disease.

Listless–Lethargic, lifeless, lacking energy. A listless animal would lie around and appear weak and not be aroused by the stimuli that usually arouse it.

Litter–(1) On a forest floor, the uppermost surface layer of debris, leaves, twigs, and other organic matter, undecomposed or slightly altered. In a technical description of a soil profile, it is generally designated by the letter O. (2) Accumulation of leaves, fruits, twigs, branches, and other plant parts on the surface of the soil. See Mulch. (3) A group of young animals born at a single birth, as a litter of pigs, etc. (4) See Bedding.

Litter Floor–The floor of a poultry house that is composed of straw and shavings or ground corncobs, droppings, and other waste materials, and builds up during one laying year.

Little-pig Anemia–See Anemia in Suckling Pigs.

Livability–The inherited stamina, strength, and ability to live and grow; important character of all young animals, such as baby chicks, lambs, pigs, etc.

Live Delivery–A common guarantee by hatcheries, of delivering live chicks by the addition of two or more chicks per order of one hundred.

Live Foal Guarantee–A common guarantee by stallion owners that a live foal will result from the breeding service. In case of failure, the breeding fee is not charged or, if already charged, is refunded.

Live Virus–A virus whose ability to infect has not been altered.

Live Weight–The gross weight of a live animal as compared with the dressed weight after slaughter.

Liver–A glandular organ in people and other animals that secretes bile and performs certain metabolic functions. Liver of various animals and fowls is edible, and its extracts are used medicinally, especially for anemia.

Liver Chestnut–A horse color; a dark shade of chestnut.

Liver Fluke–*Fasciola hepatica*, and *Fascioloides magna*; a macroscopic parasite of sheep, cattle, and swine that lodges in the liver, causing liver rot.

Liverpool Bit–A curb bit used for controlling heavy harness horses. See Curb Bit.

Livestock–Farm animals raised to produce milk, meat, work, and wool; includes beef and dairy cattle, swine, sheep, horses, and goats; may also include poultry.

Livetin–A water-soluble protein found in egg yolk.

Llama–(1) *Auchenia llama*; a ruminant quadruped of South America, allied to the camel, but humpless, smaller (about 3 feet high at the shoulder) and with a long, wooly coat; used in the Andes as a beast of burden. (2) Llama's wool or material made from it.

Loading Chute–An inclined chute that is used for animals to walk from the ground into a truck or trailer.

Loafing Barn–A light type of building usually attached to the main dairy barn or milking parlor where cows may be turned loose after milking for exercise and comfort. Hay and silage are usually supplied in side racks or bunkers.

Lobe–Any segment of an organ, especially if rounded.

Local Effect–An effect that a toxic substance causes at its original contact point with the body, e.g., eye damage.

Lock–(1) A small tuft of cotton, wool, flax, etc., fibers. (2) Small bits of wool that are packed separately for market (Australia). (3) A locule or the cotton in a locule. (4) The single cavity in an ovary.

Locker Beef–Beef that is raised primarily for sale to individuals for home consumption. It is usually bought at the farm.

Lockjaw–Tetanus.

Loco–(Spanish, insane) (1) The name of various poisonous plants in arid regions of the genera *Astragalus, Hosackia, Sophora,* and *Oxytropis*; all of the family Leguminosae. Because they contain excessive selenium they are toxic to horses, cattle, and sheep. (2)

Locoism. Poisoning with loco. Also known as loco disease, and loco poisoning. See Earl Loco, Rocky Mountain Crazyweed.

Locus–The position on or region of a chromosome where a gene is located.

Lofty–Designating wool or a woolen fabric that has springiness when compressed, is bulky in comparison with its weight, is light in condition, and has an even, distinct crimp.

Log–(1) An unhewn, sawed or cut length of a trunk or large limb of a tree. See Saw Log. (2) In the preparation of chip steaks, the molded, frozen piece of meat from which the steaks are cut. (3) To fell trees for lumber.

Loggering–A riding posture in which the rider holds the horn of the saddle rather than sitting free and erect in the saddle.

Loin–(1) The part of the body of animals on each side of the backbone that lies between the floating ribs and hipbone. (2) A wholesale or retail cut of meat that comes from the loin of a carcass. See Short Loin, Tenderloin.

Loin Chop–A retail cut of meat taken form the loin section of the carcass of a lamb, veal, or hog.

Loin Disease of Cattle–Botulism of cattle that results from eating bones or putrid flesh containing toxins produced by the organism *Clostridium botulinum*. (Cattle eat bones or putrid flesh because of a depraved appetite usually resulting form a phosphorus deficiency.) Symptoms of botulism are weakness in the hindquarters and loss of muscle control.

Loin Eye (Rib Eye)–(1) Area of loin eye or rib eye at the twelfth rib; used in carcass evaluation to determine the yield grade of carcass. (2) Cross section of the large muscles that lay on either side of the backbone in the loin area of a carcass.

Lone Star Tick–*Amblyomma americanum*, family Ixodidae; a bloodsucking tick that infests cattle, people, horses, dogs, goats, hogs, and many wild mammals.

Long–(1) Designating a person who has brought more contracts than he has sold. Long hedges are purchases of futures made as a hedge against the sale of the cash commodity (short). (2) Designating a pulse beat that is longer than normal.

Long Feed–(1) Any coarse, unchopped feed for livestock, such as fodder, hay, straw. (2) Feeding for a long period of time as contrasted with a short feed period.

Long Horse–A horse that can travel far and fast (western United States).

Long Wool–Wool from such English breeds as the Lincoln, Leicester, and Cotswold. The wool is large in diameter and up to 12 or 15 inches in length.

Long Yearling–(1) An animal more than a year old. (2) A senior yearling of livestock-show class.

Long-coupled–Too much space between the last rib and the point of the hip of an animal.

Long-nose Cattle Louse–*Linognathus vituli*, family Lingonathidae; a bloodsucking insect that infests cattle.

Long-tailed Sheep–An undocked sheep. common in the United States and Europe.

Long-wool Breeds of Sheep–The principal breeds of long-wool, mutton-type sheep in the United States; Lincoln, Cotswold, Romney Marsh, Black-Faced Highland, and Leicester.

Longeing–Training a horse at the end of a 25- to 30-foot (7.6 to 9 meters) line.

Longevity–Length of life; a long duration of life.

Longhorn Cattle–(1) Cattle of the Spanish type first brought into Mexico in 1521; much of the early cattle of the western range, United States, were derived from this type. (2) A breed of cattle in England probably of Spanish derivation. (3) See Texas Longhorn.

Longwool Crossbreed–A sheep crossbred from two different breeds, one being a long-wool type. A common sheep in western sheep ranches in the United States.

Loose Hay–Hay stored in the hay mow or stack without chopping, baling, or compressing.

Loose Housing–A management system for cattle wherein the adult animals are given unrestricted access to a feeding area, water, a resting area, and an adjoining open lot. In dairies, the lactating animals pass through a milking room at milking time. Other dairy animals may be in separate pens, lots, or buildings.

Loose Rein–A condition in riding or driving in which the reins on the harness of a horse are generally relaxed.

Loose Side–The left side of a beef carcass. See Tight Side.

Lope–In a horse, a slow gallop.

Lordosis–In veterinary medical vernacular, a mating posture assumed by many females in which they lower the forebody toward the ground and slightly elevate the hind end, straightening the arch of the back and tipping the pelvis slightly forward.

Loss of Cud–An old, false idea that a cow sometimes would lose her cud and would have to be supplied with another, such as placing a rag or other material in the mouth. Failure to regurgitate and chew cud is an indication of illness. See Cud.

Lot–(1) A small piece of enclosed land usually adjacent to a barn or shed in which horses, mules, cows, etc., are allowed to exercise. (2) Any particular grouping of animals, plants, seeds, fertilizers, etc., without particular regard to number, as a seed lot. (3) A small tract of land, usually less than an acre, on which a house is constructed.

Louse–A small wingless insect that usually lives as a parasite on humans or domestic animals.

Low Intensity Animal Production–Systems of producing animals that strive to use less capital, energy, and fewer purchased inputs than conventional confinement systems; e.g., a pasture and hutch system for swine production.

Low Wool–Wools of low quarter-blood or lower in quality.

Lugger—A horse that pulls at the bit.

Lumbar—The region of the back between the thorax and the pelvis; refers to the loins.

Lumen—The space in the interior of a tubular structure such as an artery or the intestine.

Lumpy Jaw—See Ray Fungus.

Lumpy Skin Disease—An acute viral disease of cattle, characterized by the eruption of variably sized cutaneous nodules, edema of one or more limbs, and swelling of the superficial lymphatic glands.

Lung Fever—See Hemorrhagic Septicemia.

Lunker—(1) A very big, awkward, heavy-boned horse. (2) Any animal that is considerably larger than the average.

Luster—(1) The property of wood, independent of color, which causes it to reflect light and to exhibit a sheen. (2) The natural gloss or sheen characteristic of the fleeces of long-wool breeds of sheep and Angora goats.

Lutein—A hormone, prepared from the dried corpus luteum of the sow; used to cause ovulation. See Corpus Luteum, Follicle-stimulating Hormone.

Luteinizing Hormone—In animals, a hormone of the anterior pituitary gland that stimulates ovulation and development of the corpus luteum in females and secretion of testosterone by the interstitial cells in males. See Corpus Luteum, Follicle-stimulating Hormone.

Lymph—A nearly colorless fluid made up of the liquid portion of the blood and the white corpuscles but free from the red corpuscles. It is derived from the blood by seeping through the walls of the capillaries (very minute blood vessels) which join the bloodstream in the vicinity of the heart.

Lymphatic—Referring to the system of vessels returning lymph from the tissues to the bloodstream.

Lymphocyte—A kind of white blood cell produced by lymph nodes and certain other tissues, and associated with the production of antibodies.

Lymphoid—Referring to lymph.

Lympholeukosis—See Lymphomatosis.

Lymphoma—A cancer of cells of the immune system (e.g., lymphocytes), where the tumor is confined to lymph glands and related tissues, such as the spleen.

Lymphomatosis—A disease of fowls caused by the formation of tumors of lymphoid tissue in different parts of the body; a form of avian leukosis complex.

Lysin—Antibody that dissolves or disintegrates cells: bacteriolysin dissolves bacteria, hemolysin dissolves red blood cells.

Lysis—(1) The gradual disappearance of the symptoms of a disease. (2) The destruction or dissolving of cells.

Lysozyme—An enzyme present in certain body secretions that can destroy certain kinds of bacteria.

Macro-—Prefix meaning large, long; visibly large.

Macronutrients—Includes primary plant nutrients N, P, and K; and secondary plant nutrients Ca, Mg, and S. See Micronutrient.

Macroorganisms—Plant, animal, or fungal organisms visible to the unaided eye.

Macroparasite—Parasite visible to the naked eye.

Macroscopic—Visible to the human eye without the aid of a microscope.

Mad Itch—(of cattle) See Aujeszky's Disease.

Mad Stone—See Hair Ball.

Madness—See Rabies.

Maggot—A vermiform larva; a larva without legs and without well-developed head capsule; the larva of a fly.

Magnum—The part of the oviduct of a bird located between the infundibulum and the isthmus; the source of the albumin of an egg.

Maiden—(1) A year old, single-stemmed seedling fruit tree used for budding or grafting. See Whip. (2) An unbred female animal.

Maiden Flight—The flight taken by a queen bee during which mating occurs.

Mailing Cage—A shipping container for a queen bee.

Maine Anjou—A breed of large beef cattle originating in France. The animals are horned and vary in color from red with white markings to dark roan.

Maintenance Ration—The amount of feed needed to support an animal when it is doing no work, yielding no product, and gaining no weight.

Malanders—In horses, an eczema on the posterior portion of the foreleg knee. Pustules also appear on the neck.

Malarial Fever—See Equine Infectious Anemia.

Malathion—An organic, phosphorus insecticide. Technical grade *malathion* (95 to 98 percent pure) is a viscous, dark brown liquid with a strong, offensive odor somewhat like that of garlic. Widely used to control pests, it is the least toxic of phosphorus insecticides.

Malde Caderous—An infectious horse disease caused by *Trypanosoma equinum*, family Trypanosomatidae; a protozoan parasite that infests the blood plasma. It is characterized by rapid loss of flesh, red urine, anemia, partial paralysis, and edema. Endemic in South America.

Male–(1) Designating an animal capable of producing spermatozoa, or male sex cells. (2) Designating the stamens of a flower. (3) Designating a staminate plant.

Male Process–Genital eminence.

Malformation–Any unusual, abnormal growth, organ, or part of a plant or animal.

Malignant–Tending to grow worse; exceedingly noxious, dangerous, infectious, or that condition which terminates in death, as a malignant growth.

Malignant Catarrhal Fever–Also known as snotsiekte, an acute generalized disease of cattle and buffaloes characterized by high fever, profuse nasal discharge, severe hyperemia, diffuse necrosis of oral and nasal mucosae, leukopenia, opthalmia, corneal capacity and enlargement of lymph nodes. Four syndromes are recognized: the peracute, intestinal, head and eye, and mild. The natural disease is usually of the head and eye form with low morbidity and high case fatality rates.

Malignant Edema–An acute, fatal wound infection disease caused by *Clostridium septicum*, family Bacillaceae, which affects horses, cattle, sheep, swine, and people. Characterized by swelling and gangrene.

Malignant Pustule–Localized skin lesion of anthrax in people.

Malignant Tumor–A tumor with the potential for invading neighboring tissue and/or metastasizing to distant body sites, or one that has already done so.

Mallein–A diagnostic agent processed as an extract from *Malleomyces mallei* (glanders bacillus); used principally in the diagnosis of glanders in horses. See Glanders.

Malnutrition–An unhealthy condition resulting from either poor feed or lack of feed.

Malocclusion–A deviation of the proper closing or meeting of the upper and lower teeth.

Malt–A grain product, rich in protein and carbohydrates, made by allowing the grain to sprout for a sufficient length of time to produce adequate amounts of enzymes and then dried. The term malt, unqualified, implies barley malt.

Malt Cleanings–A livestock feedstuff obtained from the cleaning of malted barley or from the recleaning of malt that does not meet the minimum protein standard of malt sprouts.

Malt Sprouts–A livestock feedstuff obtained by the removal of the sprouts from malted barley together with the malt hulls, and other parts of malt and foreign material unavoidably present. It should contain not less than 24 percent of protein. The term malt sprouts when applied to a corresponding portion of other malted cereals shall be used in qualified form: e.g., "Rye Malt Sprouts," "Wheat Malt Sprouts," etc.

Maltase–An enzyme that splits maltose into two molecules of glucose.

Maltese–A breed of asses, probably of Arabian origin, first imported into the United States from the Island of Malta at the end of the eighteenth century.

Mammary Gland–Udder or breast. In female mammals, a milk-secreting gland for the nourishment of the young. It is rudimentary in male mammals.

Mammary System–The udder, blood vessels, and teats of an animal, especially of the dairy cow.

Mammary Vein–Any of the numerous, long, tortuous, prominent, large veins or blood vessels found just under the skin along the belly to the udder of a dairy cow.

Mammitis–See Mastitis.

Mammoth Jack–A distinct breed of ass developed in the United States in which the Catalonian ass was an important influence. The Mammoth Jack is larger and heavier boned than other breeds of ass and for this reason is better suited for use in mule breeding. Also called American Jack.

Man-Killer–A vicious horse that will attack its handlers.

Management of Pastures–The handling of animals and plants in such a manner that the stand of desirable forage is not depleted and at the same time provides a sufficient supply of palatable, nutritious forage for the producing or growing animal for as long a time as is possible during the grazing period.

Mandible–The bone of the lower jaw; the lower jaw.

Mane–Long, heavy hair that grows about and on the upper side of the neck of the horse and some other animals.

Mange–A group of contagious skin diseases in livestock caused by certain sarcoptic parasitic mites.

Mange Mite–See Scab Mite.

Manger–A trough or other receptacle of metal, wood, concrete, or stone, in which fodder is placed for cattle to eat.

Manure–(1) Excreta of animals, dung and urine (usually with some bedding), used to fertilize land. (2) In Europe, any material that contains the essential elements of plant nutrients, as chemical fertilizers, excreta of animals, etc. See Compost.

Manure Pit–A storage unit in which accumulations of manure are collected before subsequent handling or treatment, or both, and ultimate disposal. Water may be added in the pit to promote liquefaction so the manure can be spread on fields through a sprinkler irrigation system.

Manure Salts–Crude salts in natural deposits, containing a high percentage of potash (K_2O) in the form of the chloride. Sometimes used as fertilizer.

Manure Spreader–A wagon-type implement for carrying barnyard manure to the field, shredding it and spreading it uniformly on the land. The power for the spreading mechanism and conveyor is supplied from the rear wheels or from a tractor power take-off.

Marbled–Mottled or streaked, like certain kinds of marble. Used to describe the intermuscular fat in meat.

Marbling–The desired distribution of the fat in the muscular tissue of a cut of meat that gives it a spotted appearance. The degree of marbling is used in the grading of beef carcasses.

Marchigiana—A breed of large beef cattle that originated in Italy. They are horned and gray.

Mare—(1) A mature female horse. (2) The insoluble residue left after processing fruits, sugarcane, and sugar beets.

Mare Mule—A female mule. See Mule.

Mariculture—The growing of marine animals such as fish and shrimp under controlled conditions in saltwater rather than in freshwater. See Aquaculture.

Market Barrow—A castrated male hog finished or fattened for meat purposes.

Market Bird—Any fowl produced and fattened for meat purposes.

Market Classes and Grades—Various market classes and grades established by the United States Department of Agriculture to sort livestock according to conformation, finish, quality, use, age, sex, and weight.

Market Grade—A set of descriptive terms, such as prime, choice, etc., used in livestock marketing to designate the comparative value of animals based on differences in type, conformation, degree of finish, etc.

Market Grades of Cows and Bulls—Determined by the age and fatness of the animal. The grades are: commercial, utility, cutter, canner.

Market Grades of Lambs—Prime, choice, utility, canner, cutter, cull.

Market Grades of Slaughter Steers and Heifers—Market grades of beef animals that are to be slaughtered are determined by the age of and the amount of fat on the animal that is indicative of the amount of marbling in the meat. The grades are: prime, choice, select, standard. Cattle are also graded by the percentage of lean retail cuts of meat the animal is expected to yield. See Marbling, Maturity, Yield Grade.

Market Grades of Swine—Various market classes of swine established by the United States Department of Agriculture: United States Number 1, United States Number 2, United States Number 3, and cull.

Marking—(1) Castrating, docking, branding, and ear marking. (2) Selection and indication, usually by blaze or paint spot, of trees that are to be cut or retained in a cutting operation.

Marking Harness—A harness strapped to a ram that marks a ewe's rump when she is bred by the ram.

Marrow—(1) A soft, yellow or red, fatty tissue that fills the cavities of most bones. (2) The pith of plants.

Marrowfat—Any of the several varieties of cultivated, large-seeded peas.

Martin Heifer—See Freemartin.

Martingale—A leather strap attached to the girth of a horse's saddle or harness on one end, with the other divided and passing between the front legs and tied to the noseband of the bridle. It stops the horse from rearing.

Masculine Head—In cattle, the brightness of eye, strong muzzle, horn character, and neck attachment that shows the vigor, alertness, and strength associated with a male.

Masculine Sex Character—In male animals, vigor, body build, alertness, and aggressiveness; desirable characteristics in animals such as bulls and stallions.

Masculine Style—In stock judging, the desirable characteristics of a male animal: i.e., up-headed, straight through the muzzle, wide between the ears, bold and clear in the eyes, wide in body conformation, and vigorous in appearance.

Mash—(1) A mixture of grain and other ingredients with water to prepare wort for brewing operations. (2) A ground feed of cereals and malt, etc., fed in a wet or dry form to livestock and poultry. Also called crowdy. See Wort.

Mash Concentrate—A poultry mash containing 20 to 40 percent protein.

Masked—Designating disease symptoms that are hidden.

Mass Selection—(1) In plant breeding, the selection of a large number of plants for propagation from which the off-type, low-yielding, inferior, and disease-susceptible plants have been eliminated. (2) In animal breeding, the selection for breeding purposes of animals on the basis of their individual performances, type, or conformation.

Massive—(1) Indicating a large-sized and deeply muscled animal; designates an ideal quality in a draft horse. (2) (Soil structure) Large uniform masses of cohesive soil that sometimes have irregular cleavage, as in the C horizons of many fine-textured clay soils. See Soil Structure.

Masticate—To chew; to prepare food for swallowing and digestion.

Mastitis—An infectious or noninfectious inflammation of the udder. All domestic female animals are susceptible, but it is most common among cows, ewes, and goats. It causes serious economic loss. Symptoms include decreased production, varying degrees of abnormal milk, heat, pain, and swelling of the udder, followed by a permanent firmness (fibrosis) of affected parts. Also called garget, felon quarter, weed, bad quarter, mammitis.

Mate—(1) To pair off two animals of opposite sexes for reproduction. Mating may be for a single season or for life. (2) In plants, to be cross pollinated.

Maternal Breeding Value—A prediction of how the daughters of a bull will milk based on weaning weight information. The accuracy figure is the amount of reliability that can be placed on the breeding value.

Maternal Calving Ease of First Calf—Calving ease ratings of daughters of bulls, when they give birth, as reported to breed associations by producers, values above 100 are superior.

Maternal Inheritance—Inheritance from mother to offspring unaffected by inheritance from the father.

Maternal Instinct—The natural instinct of all female mammals to nurse and protect their young.

Maternal Milk—A measure of the amount of calf weight that results form the milk production of a sire's daughters.

Maternal Weaning Weight—Measure of a sire's ability to transmit maternal performance, expressed in weaning weight of his daughters' calves. It is a combination of milk production and growth rate that a sire transmits through his daughters.

Maternity Pen–A special pen in a barn where animals about to bring forth their young may be isolated from the rest of the herd.

Mating Season–The season of the year when animals naturally breed and when conception is normally high. It varies with different species.

Matroclinous–Resembling the female parent. See Patroclinous.

Matron–A mare that has produced a foal.

Matter–(1) Purulent discharge; pus. (2) To discharge matter; to generate pus.

Maturation–(1) Becoming mature or ripe. (2) Changes in cell division, especially in reproductive elements, in which the number of chromosomes in the nucleus of the new cells is half that of the original.

Mature Class–In the show ring, dairy cows five years of age or older that are in milk.

Mature Equivalent Factor–An age conversion formula used to predict the expected mature milk record for a cow, based on a previous year's milk production. It is frequently used in cattle selection and breeding.

Maturity–(1) A state of full or complete growth development or ripeness. (2) The ability of a plant or plant part to withstand the cold. (3) Animals old enough to reproduce. (4) A designation of the age of a beef animal at the time of slaughter. "A" maturity indicates an age range of nine to thirty-six months; "B" maturity indicates thirty to forty-two months; "C" indicates forty-two to seventy months; "D" maturity indicates seventy-two to ninety-six months, and "E" maturity indicates more than ninety-six months. The maturity classification is based on evidence in the carcass such as bone ossification. Maturity combined with marbling gives the basis for carcass quality grade. See Marbling; Quality Grade.

Maverick–(1) Any unbranded animal, particularly a calf. Named after Samuel A. Maverick, a lawyer and a nonconformist Texan (1803-1870) who refused to brand his cattle. (2) A motherless calf (western United States).

Maxilla–The jawbone, particularly the upper one in vertebrates. In insects and arthropods, the two accessory jaws or appendages immediately behind the mandibles.

Mealy Gray–The rusty blue color of a horse's coat.

Measles in Beef–The presence of one or more cyst stages, in the muscles of a beef carcass, of *Taenia saginata*, a parasite of people. Under United States Federal Meat Inspection regulations, carcasses extensively infested with these cysts are condemned.

Measles in Pork–The presence of one or more cyst states, in the muscles of a hog carcass, of *Taenia solium*, a parasite of people. Under Federal Meat Inspection regulations, carcasses extensively infested with these cysts are condemned.

Meat–(1) The edible flesh of an animal. (2) The kernel of a grain or nut. (3) The entire egg except the shell.

Meat Animal–Any animal produced primarily for its meat-producing characteristics, whether for breeding or slaughter.

Meat Bird–A fowl produced primarily for its meat, in contrast to a fowl kept especially as an egg producer. See Broiler.

Meat Curing–Any of several methods of processing meats and meat products so they will keep in usable condition for a satisfactory period of time: salting, pickling, smoking, etc., are the common methods.

Meat Inspector–A graduate veterinarian, engaged by regulatory authorities to inspect all meats as they pass through the slaughterhouse or packing plant, to certify that they are wholesome food products.

Meat Meal–A livestock feedstuff supplement which is the ground, dried, rendered residue from animal tissues exclusive of hoofs, horns, blood, manure, and stomach contents except in such traces as might occur unavoidably in good factory practice. Also called dry-rendered tankage.

Meat-and-Bone Scrap–A livestock feedstuff that is the product obtained by the dry rendering of the meat and bone substance of slaughtered animals in the meat packing industry.

Meat-type Hog–An antiquated term denoting a well-muscled hog that has good length of body and an above average percentage of ham, loin, and shoulder.

Meaty–(1) Designating a well-fleshed animal. (2) Designating a characteristic of good cheese snowing flexibility. It will tear but not break short and is not dry or brittle (3) Any cut of meat that shows a high percentage of good muscle structure. (4) Designating spareribs cut with a heavy layer of the bacon left on.

Mecate–A special kind of riding horse reins often made of horse hair (western United States).

Mechanical Buffer–A machine used in the final operation of plucking feathers from poultry slaughtered for market. Also called buffer.

Mechanical Mainpulation–Collecting semen for artificial breeding purposes by massaging the male's genital organs or by other use of the hand to produce ejaculation.

Mechanical Poultry Picking–The removal of feathers in dressing poultry for the market by the use of machines in contrast to removal by hand labor. See Mechanical Buffer.

Meconium–(1) The first excreta of a newborn animal. (2) The juice of the opium poppy, *Papaver somniferum*.

Media–(1) The middle coat of an artery. (2) The plural of medium. (3) Material for artificial propagation of various microorganisms. (4) Any means through which communication of any type is accomplished. (5) Soil or soil-like material in which plants are grown.

Median Lethal Dose–(LD_{50}) The amount of concentration of a toxic substance that will result in the death of 50 percent of a group of test (target) organisms upon exposure (by ingestion, application, injection or in their surrounding environment) for a specified period of time. (Complementary to median tolerance limit, (TL_{50}.)

Median Plane–A plane that runs from the anterior to posterior ends of an animal that separates it into two equal parts.

Median Tolerance Limit–TL_{50} The concentration of some toxic substance at which just 50 percent of the test (target) animals are able to survive for a specified period of exposures. (Complementary to Median Lethal Dose, LD_{50}.)

Medication–(1) The application of medicines, salves, etc., to an injured or sick animal. (2) The forced introduction of a chemical, usually a water-soluble salt in solution, into the sapstream of a living tree to kill it or make it repellent to insect attack, or into a freshly felled tree to destroy barkbeetle and woodbeetle broods.

Mediterranean Fever–See Undulant Fever.

Medium–(1) Any of a number of natural or artificial substances, pastelike or liquid, in or on which microorganisms, such as bacteria and fungi, can be cultured. (2) A soil or material, such as sand, peat moss, vermiculite, etc., in which plants are raised or cuttings are rooted, especially in the greenhouse. (3) A market grade of roses that has stems 12 to 8 inches in length. (4) One of the six recognized grades of meat, lying between good and common.

Medium Eggs–A United States size grade of eggs, between small and large, which specifies a minimum net weight of 21 ounces per dozen with no individual egg under the rate of 20 ounces per dozen.

Medium Wool Sheep–Any of several breeds of sheep intermediate between the fine wool and long wool breeds, desirable for meat as well as wool; e.g., Cheviot, Columbia, Corriedale, Dorset, Hampshire, Oxford, Panama, Shropshire, Southdown, Suffolk, Tunis.

Medium Wools–Usually refers to one-half blood, three-eighths blood, and one-quarter blood wools, or wools grading 50s to 62s.

Medium-spectrum Antibiotic–An antibiotic that attacks a limited number of gram-positive and gram-negative bacteria.

Medulla–The inner layer or part of an organ.

Mega–(1) A prefix meaning one million. (2) A prefix meaning large or in large numbers.

Megaspore–A spore that has the property of giving rise to a gametophyte (embryo sac) bearing a female gamete. One of the four cells produced by two meiotic divisions of the megaspore-mother-cell (megasporocyte).

Megasporegametophyte–The few-celled haploid generation portion of a seed plant arising from a meiotic division and giving rise through meiosis to the female gametes. Female inflorescence.

Megspore-Mother-Cell–The cell that undergoes two meiotic divisions to produce four megaspores.

Meiosis–Cell division early in the reproductive process, and in the formation of sperm and ova in the testicles and ovaries. Each pair of chromosomes in the cell being divided separates, and one member of each pair goes to each of the two new cells formed.

Melanin–The black or dark brown pigment found in skin and hair cells.

Melanoma–Malignant melanoma is a cancer of the cells that produce the pigment melanin.

Melene–A white, waxy, crystalline hydrocarbon, extracted from certain types of paraffin, and from beeswax by dry distillation.

Melittology–That branch of entomological science concerned with bees.

Membrane–(1) A thin, flexible sheet of vegetable or animal tissue; the thin protoplasmic tissue connecting, covering, or lining a structure, such as a cell of a plant or animal. (2) The layer that surrounds fat globules of milk. Its nature is not definitely known: it may consist mainly of phospholipoids and a protein or proteins not completely identical with casein, albumin, or globulin. (3) A layer of low permeability material: e.g., bentonite (clay) soil, placed in the bottom of a farm pond to reduce seepage losses.

Menadione–Vitamin K3; an antihemorrhagic vitamin that is essential for all animals and people to control bleeding.

Mendel's Law (Medelian Law)–Gregor Johann Mendel (1822-1884), an Austrian monk, naturalist, and plant breeder, first demonstrated dominant and recessive genes in plant breeding, which also applies to some kinds of animal and people inheritance. For example: When a plant with the dominant gene for tallness (TT) is crossed with a plant with a recessive gene for shortness (ss), the first generation, F_1, will be distributed according to the following ratio. One plant, TT, homozygous (pure) true-breeding for tallness. Two plants, Ts, heterozygous for tallness and shortness. One plant, ss, homozygous (pure) true-breeding for shortness. When plants with Ts and Ts genes are crossed, the F_1 progeny will be in the same ratio, i.e., one TT, two Ts, and one ss.

Meningoencephalitis–An inflammation of the brain and its membranes.

Merino–(1) A breed of small sheep that descended from Spanish stock believed to have originated in Africa. Wool from the Merino is of excellent quality, very fine, and about 4 inches in length. Raised chiefly in the United States and Australia for its fleece, it has been crossed with other breeds, especially in New Zealand, to make it a better mutton sheep. (2) A term used in the textile industries to denote wool from the Merino or very high quality wool.

Mesophiles–Parasites and often pathogenic bacteria that grow best at a body temperature of 98.6°F (37°C).

Metabolic–Designating the chemical changes that take place in living plant and animal cells whereby one compound is converted to one or more other compounds.

Metabolizable Energy–The total amount of energy in feed less the losses in feces, combustible gases, and urine. Also called available energy.

Metabolized Milk–Milk produced by cows that have been fed irradiated yeast.

Metacercaria–The encysted larval stage of a trematode (fluke) that develops from another larval stage, the cercaria, and is infective to the final (definitive) host: e.g., the metacercaria of the common liver fluke of sheep and cattle encysts on blades of grass and is swallowed by sheep or cattle while grazing. See Flukes.

Metalimnion–(1) The layer of water between the epilimnion and the hypolimnion in which the temperature exhibits the greatest vertical rate of change, also more frequently termed the thermocline. (2) The zone in which temperature decreases rapidly with depth in a lake when it is thermally stratified in summer. See Epilimnion, Hypolimnion.

Metallic Flavor–An objectionable flavor in milk, butter, cheese, and ice cream, which results from metal contamination, particularly copper and iron.

Metamorphosis–A process by which an organism changes in form and structure in the course of its development, as many insects do. See Complete Metamorphosis.

Metaphase–That stage of cell division in which the chromosomes are arranged in an equatorial plate or plane. It precedes the anaphase stage.

Metaplasia–The abnormal transformation of an adult (mature), full differentiated tissue of one kind into differentiated tissue of another kind.

Metastasis–The spread of a malignancy to distant body sites by cancer cells transported in blood or lymph circulation.

Metestrus–The phase of the estrous cycle of nonprimates following estrus and characterized by the development of the corpus luteum and the preparation of the uterus for pregnancy.

Methane–CH_4; an odorless, colorless, and asphyxiating gas that can explode under certain circumstances; can be produced by manures or solid waste undergoing anaerobic decomposition as in anaerobic lagoons and in silos. There are twelve species of bacteria capable of producing methane gas from manures and other biomes.

Methemoglobin–A rust-colored substance, formed when oxygen unites with an oxidized product of hemoglobin. The oxygen is not easily released form this combination, thus injuring the tissues. Its properties are in contrast to those of oxyhemoglobin, which is bright red and releases its oxygen readily. Methemoglobin results form blood decomposition or from poisoning of blood by nitrates, nitrites, and other substances.

Methionine–A sulfur-containing essential amino acid; indispensable for animals.

Methmoglobinemia–(1) A lack of oxygen in the blood due to oxidation of iron from the ferrous to the ferric state. (2) The presence of methemoglobin in the blood resulting in cyanosis. This can be induced by excessive nitrates, nitrites, and certain drugs, or to a defect on the enzyme NADH.

Methyl Formate–An insect fumigant.

Methyl Red–A chemical indicator; a dye stain used in determining the classification or taxonomy of certain bacteria such as *Coliform bacilli*.

Methylene Blue Reduction Test–A test for the bacteriological quality of milk that consists of placing methylene blue thiocyanate in the milk under controlled conditions. The more rapidly the methylene blue changes color, the greater the number of bacteria in the milk.

Metritis–Uterine inflammation.

Mett–A semidry, 100 percent pork sausage.

Mexican Blindness–An animal disease caused by a filarial parasite, *Onchocerca volvulus*, which is carried by blackflies, buffalo gnats, and turkey gnats. Also known as African river blindness.

Mexican Fruit Fly–*Anastrepha ludens*, family Tephritidae; a serious insect pest of citrus and other fruits in Mexico. At times it has appeared in the Rio Grande Valley of the United States.

Mexican Horse–A breed of horse directly descended from the Andalusian and Arabian breeds brought to Mexico more than 400 years ago by the Spanish conquerors.

Mexican Jack–The small, common burro, seldom used for breeding in the United States. Also called Spanish jack.

Micro—A prefix meaning one millionth (1/1,000,000) or very small.

Microbes–Minute plant or animal life. Some microbes may cause disease, while others are beneficial.

Microbial Insecticide–Microbes that are used to control insects. Of the millions of insects that have been identified, about 350 species have been classified as destructive pests. Like all living organisms, these pests are susceptible to diseases caused by bacteria, fungi, protozoa, and viruses. Research scientists have explored this avenue of control and have developed several strains of microbes that have potential to be effective in the control of insect pests. See Integrated Pest Management.

Microbiologist–A scientist concerned with the study of plant and animal microorganisms.

Microelements–Trace elements; micronutrients.

Micronutrient Elements, Essential–For plants: boron, chlorine, cobalt, copper, iron, manganese, molybdenum, and zinc. For people and animals: chlorine, copper, iron, manganese, molybdenum, zinc, sodium, iodine, selenium, vanadium, chromium, cobalt, nickel, fluorine, silicon, boron, and arsenic.

Microorganism–(1) An organism so small that it cannot be seen clearly without the use of a microscope; a microscopic or submicroscopic organism. (2) Any microscopic animal or plant that may cause a plant disease or have the beneficial effect of decomposing plant and animal residue that becomes humus.

Micropyle–The minute necklike opening in the integuments of an ovule, where the sperms enter.

Microspore–One of the four cells produced by the two meiotic divisions (mitoses) of the microspore-mother-cell (microsporocyte). It gives rise to a gametophyte bearing only male gametes. Also called pollen grain.

Middle Milk–The milk obtained from a cow during the middle of the milking period, as opposed to that obtained in the beginning and at the end. It has a higher fat content and a lower bacteriological count than the other two.

Middlings–(1) The coarse particles separated in the milling of wheat that contain the germ of the wheat grain and fine particles of bran. Higher in protein and digestible nutrients than bran, the middlings are used as a feed chiefly for hogs, calves, and poultry. Also called shorts. (2) The belly portion of a pork carcass from which bacon is cut.

Migratory Beekeeping–Movement of apiaries from one area to another to take advantage of honey flows from different crops and to pollinate the crops.

Mild–(1) Designating cheese that is not sharp or strong in flavor. (2) Designating a gentle animal.

Milk–(1) The natural, whitish or cream-colored liquid discharged by the mammary glands of mammals. Unless otherwise stated, milk usually means cow's milk. (2) The juices or fluids secreted by certain plants, as the milk of the coconut or the milk of immature kernels of corn. (3) To draw milk from the udder of a cow.

Milk Carton–A plastic or wax-coated, disposable, paper container used in retailing milk.

Milk Cistern–A part of a cow's udder; the cavity holding about one-half pint that is drained by the teat ducts.

Milk, Condensed–Milk that has had a large part of its water of constitution removed by evaporation. Sugar has been added as a preservative. See Milk, Evaporated.

Milk Derivative–Any product obtained from milk, such as butter and cheese.

Milk Dry–To draw out the last of the milk from a cow's udder; to strip.

Milk Dryer–Any mechanical device designed to remove most of the water from milk. The water may be evaporated from pans or from steam-heated rollers; or the milk may be sprayed into a stream of hot air. See Drier.

Milk Duct–The cavity in the teat that leads into the milk cistern. Also called teat cistern.

Milk Feed–In poultry husbandry, to feed chickens a fattening mash mixed with milk, six to ten days prior to marketing.

Milk Fermentation–The process by which a change is produced in milk as a result of the activity of one or more species of microorganisms.

Milk Fever–See Parturient Paresis.

Milk Flow–The amount of milk produced by a cow or by the entire herd in a day, week, season, etc.

Milk Goat–A goat bred primarily for milk production, as the Saanen and Toggenbrug. See Alpine, Iamancha, Nubian, Saanen, Toggenburg.

Milk Irradiator–A device that uses an electric arc or mercury vapor lamp to convert ergosterol in milk into vitamin D_2.

Milk Ordinance–A law that sets forth regulations concerning the handling and quality standards of milk sold for public consumption.

Milk Record–A record of the amount of milk produced by a cow during a specified period. Used for evaluation in terms of money, profit, culling, or for registry.

Milk Replacer–A powder that when mixed with water is fed to young animals as the milk portion of their diet.

Milk Sample–A small amount of milk, normally collected at each farm each time the milk is collected; used for making certain tests at the dairy plant to determine butterfat content, bacterial count, etc.

Milk Sanitarian–A professional worker who specializes in the supervision of sanitary milk regulations.

Milk Sickness–(1) A disease of cattle caused by eating white snakeroot. Also called white snakeroot poisoning, trembles, milk sick. (2) A disease in people caused by ingestion of contaminated milk or milk from a cow affected by milk sickness.

Milk Stone–In dairy manufacture, a grayish-white, thin, chalky deposit that sometimes accumulates on heating surfaces, coolers, utensils, and freezers when improper washing methods have been used. It is largely a mixture of the minerals of milk and water.

Milk Substitute–In animal or dairy husbandry, any of a number of gruel feeds or mixtures fed to calves or pigs as a substitute for milk.

Milk Sugar–See Lactose.

Milk Teeth–The temporary teeth of a young animal that are much whiter and smaller than the permanent teeth.

Milk the Crop–To massage the crop of a slaughtered fowl so as to force feed remaining in its crop out of the mouth. It lessens the danger of spoilage of the dressed fowl.

Milk Vein–The subcutaneous, abdominal veins that are a continuation of the mammary veins of the udder of a cow. The largeness and tortuousness of the veins is used in judging dairy cattle.

Milk-borne Disease–Any human disease that results from the intake of pathogens in milk.

Milk-fat Basis–A method of paying for milk at receiving stations on the basis of price per pound of butterfat. See Butterfat.

Milk-fed–Designating animals fed largely on dairy products.

Milk-Feed Price Ratio–The pounds of dairy ration equal in value to one pound of milk, at prices existing at a particular time and place. The United States Department of Agriculture milk-feed price ratio is pounds of dairy concentrates, rather than the entire dairy ration.

Milk-souring Organisms–Groups of organisms, such as lactic streptococci and coliform bacteria, that are responsible for the souring of milk.

Milking–(1) The quantity or quality of milk removed from an animal's udder at one time. (2) The act of removing milk from an animal's mammary gland. (3) Pertaining to milking, as in milking machine.

Milking Machine–A mechanical device replacing hand labor in the milking of cows. The essential parts are teat cups, a vacuum pump, and a milker pail or milk line. Milk is drawn from the udder by application of alternate vacuum and atmospheric pressure.

Milking Parlor–In a dairy, an especially arranged and equipped room where cows are separately fed concentrates and milked by mechanical milking equipment.

Milking Shorthorn–A breed of dairy cattle that originated from the Shorthorn in England. Heavy milk producers, they are popular in England and in the central United States. Predominant colors are red, white, or roan. Individuals weigh from 1,200 to 1,400 pounds.

Milkshed–A designated geographic area of milk production and consumption.

Milky Disinfectant–See Coal-tar Creosote.

Milo Chop–A livestock feedstuff consisting of the entire grain of milo sorghum removed from the head and chopped or ground.

Milo Mill Feed–A livestock feedstuff that is a mixture of milo bran, milo germ, and a part of the starchy portion of the grain, produced in the manufacture of grits from milo grain.

Milt–(1) The roe or spawn of the male fish. (2) To impregnate with milt. (3) A ductless gland near the stomach and intestine of fowls. It contributes to the formation of new blood cells. See Spleen.

Milter–A male fish especially at spawning time.

Mineral–(1) A chemical compound or element of inorganic origin. (2) Designating the inorganic nature of a substance.

Mineral Elements–The inorganic components of soil, plants, or agricultural products.

Mineral Feeder–Any of a number of box or bin devices that automatically supplies a mineral or mineral ration in the feeding of livestock.

Mineral Mixture–Any feed containing salt, limestone, phosphates, minor elements, etc.

Mineral Wool–A substance outwardly resembling wool, presenting a mass of fine interlaced filaments, made by subjecting furnace slag or certain rocks while molten to a strong blast.

Mineralize–(1) To petrify. (2) To impregnate or supply with minerals. (3) To promote the formation of minerals. (4) The microbial breakdown of organic matter to release its minerals.

Mineralized Salt–A salt compound for animal feeding that is composed of a common salt base with the trace mineral elements, such as manganese, copper, iron, iodine, and zinc, added to it.

Miniature Broiler–A fast-growing broiler chicken, produced in six to eight weeks, about 1½ pounds in live weight and about 1 pound in dressed weight. Usually marketed frozen and in individual packages, it is intended to be served whole and generally stuffed with dressing.

Mink Farming–The business of raising minks for their valuable fur. Mink are meat eaters and are often fed meat by-products.

Minnow–A small fish, often used for fish bait.

Minor Elements–See Micronutrient Elements, Essential.

Minorca–The largest breed of chickens of the Mediterranean class. They have long, strong bodies, enamel-white earlobes, and white skin. The eggshell is white. Varieties are Single-comb Black, Rose-comb Black, Single-comb White, Rose-comb White, and Single-comb Buff.

Missouri Mule–A large mule from Missouri, United States, a state famous for mule production.

Missouri River-Bottom Disease–A disease of horses attributed to poisoning from eating arrow crotalaria, a legume found mostly in the southern United States.

Mites–Arachnids; very small spiders belonging to the order Acarina. These parasites constitute a very large group living in all parts of the world and are pests on both plants and animals.

Miticide–Any poisonous substance used to kill mites.

Mitosis–Cell division involving the formation of chromosomes, spindle fibers, and the division of chromosomes by a process of longitudinal splitting. Each of the resulting daughter cells thus has a full set of chromosomes as distinguished from reduction division or meiosis, in which the daughter cells have half the somatic number. See Meiosis.

Mix–(1) A formula; a combination of two or more ingredients, blended together for a specific purposes, such as feedstuff for livestock. (2) To cross or interbreed animals, plant varieties, etc., by chance.

Mixed Breeds–Farm animals that result from crossing different breeds of livestock.

Mixed Calves–A market class for a group of slaughter calves that are not uniform in grade.

Mixed Colored Eggs–Market eggs that are not uniform in color.

Mixed Corn–All corn that does not meet the color rules for yellow corn or white corn.

Mixed Grain Sorghum–A market class that includes all grain sorghums not meeting the separate standards for white, yellow, red, and brown grain sorghums.

Mixed Hay–Hay that consists of two or more forage species, as a mixture of red clover and timothy.

Mixing Pens–In range management, a large pen in which a number of animals are confined. In handling goats, the does and kids are placed together until such time as the kids are able to follow the mothers in grazing.

Mocha Coffee–Arabian coffee.

Mochilla–A large piece of skin or leather covering for a horse saddle. In the last century, mochillas fitted with pockets called continas were used in the western United States by pony express riders.

Moderate Live Virus–A virus that has been changed by passage through an unnatural host, such as passing hog cholera virus through rabbits, so that it no longer possesses the disease-producing characteristics but so that it will stimulate antibody production and immunity when injected into susceptible animals.

Modified Milk–(1) Cow's milk altered to a composition suitable for the special needs of infants. (2) A mineral-vitamin fortified milk for general beverage use. (Sometimes applied to homogenized milk.)

Modifier–(1) Any element that is added to, or taken from a substance that alters its normal appearance or function. (2) A substance that can alter the course of carcinogenesis.

Modifying Gene–A gene that changes the expression of the chief gene, or genes, controlling a character.

Modoc–Wool sheared from range lambs that had been feedlot fed in the central states.

Mohair–(1) The long, lustrous hair from the Angora goat. (2) Cloth made from the hair of Angora goats. See Angora Goat.

Mohair Cincha–A western cinch or saddle girth made of mohair.

Moisture–(1) The total amount of water in any plant or animal product. (2) Any form of water. (3) The total amount of water (exclusive of that in chemical combinations) in the soil, both that which is available and that unavailable for plant growth.

Moisture Pan–A vessel in an incubator that contains the water that provides the humidity necessary for incubation of eggs.

Molar–In mammals, one of the large back teeth with ridged or rounded surface specially adapted for grinding.

Molascuit–A cattle feed made from sugarcane bagasse and molasses.

Molasses Feed–An animal feed, usually a commercial mixture in which molasses is an ingredient; used as a substitute for grain because of its palatability and because it is a cheap source of readily digested carbohydrates.

Molasses Silage–Legume or grass silage to which specified amounts of molasses, such as blackstrap, are added to aid the proper fermentation, increase the carbohydrates, improve palatability, etc.

Mold–(1) A form containing a cavity into which material is poured or pressed to achieve a special shape and design; e.g., molds for Edam and Gouda cheese, butter, ice cream, etc. (2) Any soft, humus soil. (3) Fungi distinguished by the formation of a mycelium (a network of filaments or threads), or by spore masses; usually saprophytes. However, various kinds may do serious damage to fruits, hay, grain, growing crops, and ornamental plants. Also spelled mould. See Compost, Downy Mildew, Mildew, Powdery Mildew.

Mold Poisoning–Harmful effects upon animals from eating certain moldy feeds. The poisoning may be due to the degree of moldiness, or to prolonged feeding when less harmful molds are involved. With horses, the effect may be a nervous disease resulting in blindness and a staggering gait.

Moldiness–In judging and grading cheese, any defect due to the presence, or previous presence, of undesirable molds.

Moldy–Infected by molds (fungi); designating objectionable quality due to appearance or odor, flavor, toxicity, caused by molds.

Molecule–The smallest part of a substance that can exist separately and still retain its chemical properties and characteristic composition; the smallest combination of atoms that will form a given chemical compound.

Mollis–Soft.

Molt–The casting off of old feathers, skin, horns, etc., before a new growth, as with the normal annual renewal of plume of adult chickens, turkeys, etc. Also called the castoff covering (British); moult.

Molting–An interruption of egg laying, either natural or forced.

Molybdenosis–A disease of domestic animals caused by ingesting an excessive amount of high-molybdenum forage. General weakness and diarrhea are the symptoms. So far molybdenosis has not been reported in people. See Teart.

Monday Disease–See Azoturia.

Moniliasis–An infectious mycotic disease of the mouth and especially of the crop of domestic poultry and of the mucous membranes of dogs, calves, colts, and humans. The causative fungus is ***Candida albicans***. Also called thrush crop mycosis.

Monkey Mouth–A condition in animals in which the lower teeth protrude over and beyond the upper teeth.

Mono–Single; only one.

Monoestrous Animal–An animal that has only one estrous (heat) cycle each year.

Monogastric–Refers to an animal that has only one stomach or stomach compartment, such as swine.

Monohybrid–A hybrid whose parents differ in a single character.

Monohybrid Cross–A cross between two individuals that are heterozygous for one pair of genes; an example is Aa × Aa.

Monophagia–(1) Eating only one kind of food. (2) Eating only one meal a day.

Monophagous–Feeding upon only one kind of food, for example one species or one genus of plants.

Monophagous Parasite–A parasite restricted to one species of host.

Monorchid–A male with one testis in the scrotum and one inside the abdominal cavity.

Monotrophic–Designating bees that visit only one kind of flower.

Monoxenous Parasite–A parasite that requires only one host for its complete development.

Moon-blind–Referring to a horse suffering from periodic ophthalmia.

Mop Up Bull–A bull that is put into a pasture with cows that have been artificially inseminated. The purpose is to breed those that are not pregnant.

Morbidity–The condition of being diseased, or the incidence or prevalence of some particular disease. The morbidity rate is equivalent to the incidence rate.

Morgan–An American breed of light, general-purpose horse developed principally as a riding horse. Individuals stand 14.2 to 16 hands high and weigh 800 to 1,200 pounds. Standard colors are bay, brown, black, and chestnut with white markings.

Morphogenesis–The developmental history of organisms or of their parts.

Morphology–(1) A branch of biologic science that deals with the forms, rather than the functions, of plants and animals. (2) Pertaining to pedology or soil science, the study of soil horizons and their arrangement in profiles. (3) Pertaining to land surfaces, the shape or configuration of physical features.

Morrison Standard–A feeding standard formulated by F.B. Morrison of the University of Wisconsin, author of ***Feeds and Feeding***, which is adapted to practical feeding conditions and is used extensively.

Mortality–The number of overall deaths, or deaths from a specific disease, usually expressed as a rate; i.e., the number of deaths from a disease in a given population during a specified period, divided by the average number of people or animals exposed to the disease and at risk of dying from the disease during that time.

Mossy–(1) Designating irregular, dark markings that spoil an otherwise desirable color contrast on the feathers of domestic birds. (2) Covered with moss.

Mossy Horn–(1) An animal horn that has become wrinkled and scaly from age. (2) A term designating a very old animal.

Mossy-coated–In the show ring, designating a beef animal that has hair of a solid color containing darker mottlings.

Most Probable Producing Ability (MPPA)–An estimate of a cow's future productivity for a trait (such as progeny weaning weight ratio) based on her past productivity. For example, a cow's MPPA for weaning ratio is calculated from the cow's average progeny weaning ratio, the number of her progeny with weaning records, and the repeatability of weaning weight.

Motes–(1) In cotton ginning, immature seed and particles of trash. (2) Burs, etc., found in wool. Also called moits.

Mother Culture–In dairy bacteriology, milk or a liquid milk product introduced into a control culture of actively growing bacteria, specific for the production of a definite fermented product.

Mothering Ability–In beef cattle selection, the ability of a cow to wean a large, fast-growing calf. Weaning weight is highly correlated to the dam's ability to produce milk.

Motile–Self-propelling, as spores or sperms, by means of cilia or elaters.

Motility–Active movement in artificial insemination of the sperm in a male's semen.

Mottle–(1) A spot or blotch of a color different from the mass color of a surface, as a mottle caused by a viral disease of a plant. (2) Color difference on a mass of moderately poorly drained soil.

Mottled–(1) Designating bird feathers marked with white tips at the ends or spotted with colors or shades at variance with the ideal. (2) Irregularly marked with spots of different colors, as a mottled soil, butter, etc.

Mottling–A soil mass containing many colors due to poor internal drainage. The colors may be gray, yellow, and/or red in random patterns.

Moufflons–A breed of wild sheep believed to be ancestors of some present-day domestic breeds.

Mould–See Mold.

Moult–See Molt.

Mount–(1) A horse for riding. (2) A riding seat. (3) To copulate, as a male animal. (4) To get into a saddle on the back of a riding horse. (5) A mountain, or a high hill. Used always instead of *mountain* before a proper name.

Mountain Fever–See Equine Infectious Anemia.

Mountain Oyster–The testicles of a lamb, pig, or calf that is used for human consumption.

Mounting Block–Any block of wood, etc., that is designed as a step for easier mounting of a saddle horse.

Mouse Dun–A coat color of some animals, especially horses. It is a dun color imposed on black, seal brown, dark mahogany bay, or dark liver chestnut giving a smoky effect. Sometimes called smoky.

Mouth Speculum–An instrument used by veterinarians for holding an animal's mouth open to facilitate inspection of the mouth, dental work, or the administering of medicines. Also called jaw spreader.

Mouthing–Determining the approximate age of a horse by examining the teeth.

Mouton–A modern fur made by a chemical treatment and processing of sheep pelts carrying designated lengths and qualities of wool on the skin.

Mow-burned–Designating hay that has been damaged or altered by heating and fermentation in the mow. Also called mowburnt.

Mozzarella–Semisoft, mild Italian cheese used in pizza and other cooked foods.

Mucoid–Resembling mucus.

Mucopurulent–A body discharge that contains both mucus and pus.

Mucor–A widely distributed genus of molds of the family Mucoraceae, mainly saprophytic species, abundant in soil, decaying vegetable matter, dung, etc. Some are capable of converting starch into sugar, and thus important industrially. Probably the most frequently met and most troublesome species is the black bread-mold, *Rhizopus nigricans*, which appears wherever starchy substances are found, as on stale bread, and which causes a watery rot of strawberries, and peaches, a soft rot of roots of the sweet potato, and even rot in the tubers of the Irish potato when they have been weakened by cold or bruises or other unfavorable conditions. A few species of *Mucor* are weak parasites in animal tissues. *Choanephora* mold attacks and destroys the blossoms of squash and pumpkin (*Cucurbita*) and of *Hibiscus.*.

Mucosa–See Mucous Membrane.

Mucous Membrane–A form of epithelial tissue that secretes mucus and lives in the body openings and digestive tract.

Mucus–Viscid, watery secretion of the mucous glands, composed of water, mutin, inorganic salts, epithelial cells, leucocytes, and granular matter.

Mud Fever–(1) A variety of erythema that attacks the heels and coronets of horses' feet, causing irritation. It may develop during wet weather from mud caked on the feet. (2) In some sections of the United States, a popular term for avian monocytosis of turkey.

Muffs–The whiskerlike crop of feathers on either side of the face below the eyes of certain fowls.

Mugging–Bulldogging or throwing a calf.

Mule–(1) The hybrid offspring of a jackass and a mare; used as a draft animal. Mules have advantages over horses in that they are more resistant to heat, require less care in feeding, have fewer digestive disturbances, are less nervous, and more surefooted than the horses. (2) The spinning machine used for the manufacture of woolen yarn.

Mule Foot–A condition in which cloven-footed animals such as wine or cattle have no parting of the hoof. The hoof resembles a horse or mule foot. The cause is genetic.

Mule Jack–A jack bred to a mare in contrast to one used for jack stock perpetuation. See Jennet, Jack.

Muley (Mulley)–A cow without horns; generally, any polled or hornless beef or dairy cattle.

Multi–A prefix meaning many, as multiovulate, many-ovuled.

Multiparous–(1) Giving birth to more than one offspring at a time. (2) Having had several offspring resulting from more than one pregnancy.

Multiple Alleles–More than two different genes that can occupy the same locus on homologous chromosomes.

Multiple Cropping–In favorable climates, the growing of two or more crops consecutively on the same field in a single year, such as corn and wheat; soybeans and wheat.

Multiple Farrowing–Arranging the breeding program so that groups of sows farrow at regular intervals throughout the year.

Mummification–(1) In animal reproduction, the drying up and shriveling of the unborn young. (2) In fruit, drying and shriveling caused by the brown rot pathogen.

Mungo–Wool fibers recovered from old and new hard worsteds and woolens of firm structure. The fibers are less than one-half inch in length, and owing to their reduced spinning and felting quality, their use is restricted largely to the cheaper woolen blends. The mungo fiber is usually shorter than the shoddy fiber.

Munsell Color Standards (Munsell Notation)–A color designation system that specifies the relative degrees of the three variables of color: Hue, value, and chroma. The standards may be used in precise comparison of colors of soils, or in standardizing agricultural products: e.g., prime cottonseed cake is 10YR5/5, which means yellow-red with value = 5, and chroma = 5.

Murrain–(1) A disease of Irish potatoes. Murrain is a term applied to the historically famous potato blight disease in Ireland. (2) See Cattle Tick Fever.

Murray Grey–A breed of beef cattle developed in Australia by the systematic crossing of Shorthorn and Angus. They are polled and silver in color with black skin pigmentation.

Murrina–An equine disease caused by *Trypanosoma hippicum*, a protozoan parasite in the blood. The affected animals are emaciated and feverish and may have anemia, conjunctivitis, edema, and paralysis of the hind legs.

Muscle Contracture–A congenital, lethal disease of animals, characterized by stiff joints. The young pigs, lambs, and calves are sometimes affected in the womb and are dead at birth.

Muscovite–A mineral, a member of the mica group, the common white, green, red, or light brown mica of granites, gneisses, and schists. Monoclinic crystal structure. See Mica.

Muscovy–A domesticated, ducklike waterfowl that originated in South America. It is commonly classed as a duck but is of a distinct race and when crossed with other races of ducks its progeny are sterile. The period of incubation for eggs is thirty-five instead of twenty-eight days, as with other ducks. They differ also in that the head and face of both male and female are covered with coruncles. There are two varieties: white and colored.

Muscular Stomach–See Gizzard.

Mushy–Wool that is dry and wasty in manufacturing.

Musk Glands–Glands that secrete an odor as a secondary sexual characteristic, and in the male goat are mainly situated just in front of the horns.

Musquash–The dressed skin or prepared fur of the American muskrat. See Muskrat.

Mustang–(1) In the southwestern United States, a wild descendant of the Spanish horse, generally smaller and inferior to the domestic Spanish horses. (2) A small domestic pony that is largely of Spanish breeding. (3) A range horse that is somewhat wilder than the average. Also called mestang. See Feral.

Mustang Cattle–Wild cattle (southwestern United States).

Mustiness–An offensive, pungent odor caused by molds.

Musty–Designating a defect of milk, cream, butter, or ice cream that has an offensive odor, as that imparted by molds or by a damp, unventilated cellar or basement.

Mutabilis–Variable.

Mutable–Changeable.

Mutagen–A chemical, physical and/or radioactive agent that interacts with DNA to cause a permanent, transmissible change in the genetic material of a cell. See Teratogen.

Mutant, Mutation–A variant, differing genetically and often visibly from its parent or parents and arising rather suddenly or abruptly. Mutation can occur naturally or can be induced by radiation (x rays, gamma rays, or thermal neutrons) or chemically by ethyldemethyl sulfate.

Mutterkorn–See Ergot.

Mutton–The flesh of a grown sheep (at least one year old) as opposed to lamb (less than one year old).

Mutton Buck–A male mutton sheep.

Mutton Goat–A goat sold for its flesh.

Mutton Sheep–Any of several breeds of sheep developed primarily for their meat.

Mutton-withered–An animal that is low in the withers, with heavy shoulder muscling.

Mutualism–Dependency of two organisms upon each other. Among insects, an example is furnished by the cornfield ant and the corn-root aphid.

Mutualistic–Designating a mutually beneficial relationship between organisms; symbiosis, e.g., Rhizobium bacteria and compatible legumes.

Muzzle–(1) The projecting part of an animal's head, comprising the nose, mouth, and jaws, as of a cow, horse, or dog. (2) To cover the muzzle of an animal to prevent eating, biting, etc.

Mycology–The science dealing with fungi.

Mycoplasma–An organism that is between a virus and a bacteria in size. It may possess characteristics of a virus and is not visible under a light microscope.

Mycosis–A disease caused by the growth of fungi in plants or animals.

Mycotic Disease–A disease caused by a fungus.

Mycotic Lymphangitis–A glanderslike horse disease that affects the skin, superficial vessels and lymph nodes. It is attributed to a yeastlike fungus, *Cryptococcus farciminosus*, family Torulopsidaceae. Also called epizootic lymphangitis, lymphangitis epizootica, African glanders, Japanese farcy, Neopolitan farcy.

Mycotoxins–Chemical substances produced by fungi that may result in illness and death of animals and humans when food or feed containing them is eaten.

Myiasis–A disease due to the presence of fly larvae in warm-blooded animals.

Myocarditis–Inflammation of the muscular walls of the heart or myocardium.

Myometrium–The muscular layer of the uterus that brings about the expulsion of the fetus at parturition.

Myricin–The waxy portion of beeswax remaining after the major portion has been dissolved in alcohol.

Myxedema–A disease resulting from insufficient secretion of the thyroid gland. Not common among animals, it is present in some calves of the Dexter breed.

Myxomatosis–A viral disease of the pox group, transmitted by vectors, nonlethal to the South American rabbit, its natural host, but lethal to most European rabbits. Tested by the Commonwealth Scientific and Industrial Research Organization as a method of European rabbit control in Australia and in many European countries in the 1950s. The virus was successful in killing unwanted wild rabbits.

Nag–A horse or pony of nondescript breed.

Nagana–See Trypanosomiasis.

Nanny–A female goat.

Nape–The back of the neck of an animal.

Narrow Nutritive Ration–A feed with high protein content in proportion to its nonnitrogenous fat and carbohydrate nutrients.

Narrow-spectrum Antibiotic–An antibiotic whose activity is restricted to either gram-negative or gram-positive bacteria; e.g., penicillin is active primarily against gram-positive organisms, whereas streptomycin attacks only gram-negative organism.

National Cattlemen's Association–An association of cattle producers that serves as the communications arm of the nation's beef cattle industry. The membership includes cattle breeders, producers, and feeders. This nonprofit organization was officially formed on September 1, 1977, through the consolidation of the American National Cattlemen's Association (founded 1898) and the National Livestock Feeders Association (founded 1946).

National Sire Evaluation–Programs of sire evaluation conducted by breed associations to compare sires on a progeny test basis. Carefully conducted national reference sire evaluation programs give unbiased estimates of expected progeny differences. Sire evaluations based on field data rely on large numbers of progeny per sire to compensate for possible favoritism or bias for sires within herds.

National Wool Act–Legislation that provides price support for shorn wool at an incentive level to encourage production. The law also provides for a payment on sales of unshorn lambs.

Native–(1) Designating a plant that grows naturally in a country or region; one not introduced by people. (2) Designating animals, as cattle, hogs, and horses, which, though originally introduced into a region, have lost some of their original characteristics or have gone wild: a scrub or mongrel. (3) Designating an unbranded beef hide. See Feral, Indigenous.

Native Disease–A disease caused by an indigenous organism.

Natural Cross–Interbreeding or hybridizing that takes place in nature without assistance from people.

Natural Enemy–In nature, any organism that preys or feeds upon another. A natural enemy may be introduced by people for biological control. See Myxomatosis.

Natural Immunity–Immunity to disease, infestation, etc., that results from qualities inherent in plants, animals, or people.

Natural Selection–A natural process by which less-vigorous plants and animals tend to be eliminated from a population in an area without leaving enough descendants to perpetuate their traits.

Natural Selection Theory–A theory of evolution, propounded by Charles Darwin in the nineteenth century, which postulates that the distinctive characteristics of fitness can be inherited. Also called the survival of the fittest theory.

Natural Thickness–A livestock judging term that refers to the amount of muscling on an animal.

Navel–(1) In *Mammalia* spp. the point of connection between the umbilical cord and the fetus. (2) See Navel Orange.

Navel Cord–The umbilical cord.

Navel Hernia—See Umbilical Hernia.

Navel-ill—A disease of newborn animals that results from infection of the umbilical cord shortly after birth. It is characterized by temperature, weakness, listlessness, lameness, hot and painful swelling of the joints, and loss of appetite. The affected animal finally dies.

Navicular Bone—A small oblong bone in the lower part of a horse's leg between the second and third phalanx; the sesamoid of the third phalanx.

Navicular Disease—In horses, an inflammatory disease of the navicular bone and bursa of the front foot. Affected animals go lame and have a short, stubby stride.

NDM—Nonfat dry milk.

Near Side—The left side of a horse; the side from which to mount in riding. See Off Side.

Neat's Foot Oil—An oil obtained by boiling the feet and bones of calves and other cattle; used as a fine lubricant, for dressing leather, and in hoof ointments.

Neck—(1) In humans and animals, the connecting link between the head and the body. (2) To tie cattle neck-to-neck (western United States). (3) (land) An elevated, narrow strip of land between two somewhat parallel streams, or water bodies; a promontory; a peninsula; an isthmus.

Neck Ail—A cobalt-deficiency disease of cattle that results in listlessness and an emaciated appearance.

Neck Piece—Of a meat carcass, that part between the shoulders and head.

Neck Rein—To turn a saddle horse in a desired direction by laying the reins on the neck rather than pulling on the bit.

Neck Rope—(1) A rope tied around the neck of a horse being trained for use in lassoing cattle. (2) A rope used in picketing horses.

Necro—(1) Ulcers of the intestine, necrotic enteritis. (2) Infectious enteritis in pigs. (3) Swollen nose of hogs; bull nose.

Necrobaciloses—Several disease entities in domestic animals, sometimes of uncertain etiology but involving some bacteria, usually the bacillus *Actinomyces*, and such predisposing factors as trauma, excessive moisture, or filth. These include such diseases as foot rot, bull nose of pigs, and necrotic stomatitis.

Necropsy—An examination of the internal organs of a dead body to determine the cause of death. Also called autopsy, post-mortem.

Necros—(Greek) Dead.

Necrosis—Death of plant or animal cells of tissues, usually in localized areas.

Necrosis of the Beak—An inflammation of the beak of poultry chicks caused by particles of finely round, all-mash feed remaining under the tongue and adhering to the edges of the upper and lower mandibles.

Necrotic—Designating a necrosis.

Necrotic Enteritis—Infectious enteritis in pigs.

Nectar—A sweet secretion of flowers of various plants, used by bees to store as honey.

Nectar Carrier—A worker bee actively engaged in carrying nectar from flowers.

Nectar Flow—The period when abundant nectar is available for bees to produce honey for storage in the combs of the hive.

Needle Teeth—The eight sharp teeth present in newborn piglets. Especially in large litters, these teeth can cause injury to other piglets and the sow's udder and should be clipped.

Nematodirus—The internal parasite (nematode) of sheep, a nematode of the genus *Nematodirus*. Also called stomach worms, intestinal worms.

Neomycin—An antibacterial substance produced by the growth of *Streptomyces fradiae*. It is used to treat systemic infections caused by gram-negative microorganisms.

Neonatal—Pertaining to the first four weeks of life.

Neonate—A newborn animal.

Neonicotine—Anabasine, an insecticide with acute toxicity to mammals.

Neontology—Biology; the study of existing life.

Neoplasm—A new growth of tissue with the potential for uncontrolled and progressive growth. A neoplasm may be benign or malignant.

Neopolitan Farcy—See Mycotic Lymphangitis.

Nephritis—Inflammation of the kidney.

Nerve Poison—A poison that is soluble in tissue lipoids in contrast to respiratory poisons, such as cyanide gas, and physical poisons, such as oils and dusts.

Net Livestock Increase (or Decrease)—The clear profit or loss; a figure obtained by subtracting the total of the value of livestock at the beginning of the year and the cost of livestock purchased during the year from the total of the value of livestock at the end of the year and the receipts from livestock sold. If the result is negative, it is a net livestock decrease.

Net-energy Value—The amount of energy that remains after deducting from a feed's total energy value the amount of energy lost in feces, urine, combustible gases, and heat increment. Sometimes called work of digestion.

Neuron—A nerve cell.

Neurotoxicity—The state or condition of being poisonous to the brain and nerves of the body.

Neutered—Designating an animal that has been spayed, caponized, or castrated.

Neutral Breeding Ground—A fenced-off plot, used only for breeding to prevent the spread of venereal diseases.

Neutron—See Atom.

Neutrophil—A phagocytic white blood cell associated with the formation of pus.

New Hampshire—An American breed of intermediate-sized chicken, characterized by chestnut-red plumage, yellow skin, fast feathering, and rapid growth. It is of the same standard weight as the Rhode Island

Red from which it was developed largely as a strain selection. The egg is light brown.

Newberry Castrating Knife–A device that consists of a sharp blade attached to a clamp that is used for castrating calves. The scrotum is slit so the wound will drain properly and heal more quickly.

Newcastle Disease–An acute, rapidly spreading viral disease of poultry, caused by a filterable virus. It occurs all over the world. Also called avian pneumoencephalitis, California flue. See Exotic Newcastle Disease.

Niacin–A vitamin of the B-complex group. Also called nicotinic acid, antipellagra vitamin.

Niche–A term used to describe the status of a plant or animal in its community, that is, its biotic, trophic, and abiotic relationships. All the components of the environment with which an organism or population interacts.

Nick–(Nicking) A term used by livestock producers when the offspring is better than its parents.

Nicked–(1) Descriptive of a horse's tail in which some of the muscles have been severed so that the tail is carried upward. (2) Designating a mating of animals that results in an offspring superior to either of the parents.

Nickel–Ni; a chemical element, a metal which is found in traces in soils. It was once thought to be deleterious rather than beneficial to plant growth, but since 1983 has been suggested as essential for plants. It is required by animals and people.

Nicker–To neigh (as a horse).

Night Blindness–One of several symptoms in livestock, one cause of which is vitamin A deficiency. See Vitamin A.

Night Corral–An enclosure or pen on western United States ranges in which ewes may be placed at night or at lambing time.

Night Hawk–A night wrangler or herder of saddle horses.

Night Horse–A special cow-horse that is surefooted, has good night vision, and a keen sense of direction.

Night Milk–The milk obtained at the evening or night milking.

Night Pasture–A pasture in which domestic animals graze during the night, usually close by the barn or dwelling.

Nightherd–To ride herd at night; to keep cattle or sheep herded at night.

Nigrescent–Blackish.

Nippers–(1) The two central incisor teeth of a horse. (2) Small pincers used for holding, breaking, or cutting.

Nipple–(1) A small, conical elevation. (2) The protuberance of the udder of a mammal that contains the mammary gland. See Teat.

Nipple Pail–A pail with a tube and rubber nipple fastened at the bottom, used in calf feeding, weaning, etc.

Nipple Waterer–An automatic watering system in which the animal pushes a "nipple" in the end of a pipe to get water.

Nit–The egg of a louse.

Nitrate–NO_3; N combined with oxygen. The N form most used by plants. NO_3 is a gas that does not exist alone in fertilizer but is combined, as in ammonium nitrate. All nitrates are water-soluble and, when applied in surplus, move with surface waters to contaminate groundwater.

Nitrate of Potash Poisoning–Poisoning to cattle that results from ingestion of potassium nitrate in oat hay. Also called saltpeter poisoning.

Nitrate Toxicity of Forage–Forage containing more than 6,000 ppm nitrate may be toxic to cattle. Causes of high nitrates in forage include drought, cool temperatures, cloudy weather, acid soils, and heavier-than-recommended nitrogen fertilizer applications. The drying process in making hay does not lower the nitrate level in forage but it is reduced to about half in making silage.

Nitrite–NO_2; a partially oxidized form of nitrogen containing two atoms of oxygen for each atom of nitrogen. Soil and rumen bacteria can change nitrite-nitrogen to nitrate-nitrogen.

Nitrogen–N; a gas that occurs naturally in the air and soil, where it is converted into usable forms for plant use by bacteria and other natural processes. This nutrient is a constituent of proteins and is vital to plant-growing processes. Nitrogen can be added to the soil in any of three fertilizer forms: as urea, ammonia, or nitrates.

Nitrogen-free Extract–The portion of a feed made up primarily of starches and sugars; nitrogen-free extract is determined by subtracting the ether extract, crude fiber, crude protein, ash, and water from the total weight of the feed sample.

Niveus–Snowy-white.

Nodder–A horse that characteristically nods its head when walking, as the Tennessee Walking Horse.

Nodular Worm–A common name applied to nematodes found in cattle, sheep, goats, and swine, whose larvae normally penetrate into the tissues of the intestines and produce nodular lesions there; the adult worms live in the cavity of the large intestines. *Oesophagostomum radiatum* is the nodular worm of cattle; *O. columbianum*, the common nodular worm, and *O. venulosum*, the lesser nodular worm, occur in sheep and goats; four species of nodular worms have been reported in swine in the United States: *O. dentatum, O. longicaudum, O. brevicaudum,* and *O. georgianum*. The symptoms of infestation are called pimply gut, knotty gut.

Nodule–(1) A root tubercle or lump formation on certain leguminous plants produced by the invasion of symbiotic, nitrogen-fixing bacteria. The bacteria furnish the plant with fixed nitrogen compounds and receive nutrient plant juices like carbohydrates. The genus *Rhizobium* and some species of *Azotabacter* and *Clostridium* fix free nitrogen. (2) A small knot, lump, or roundish mass of abnormal tissue. See Nitrogen-fixing Bacteria.

Noil–The short fibers that are removed from the staple wool in combing. Noil is satisfactory for the manufacture of felts and woolens.

Nonadditive Genes–Genes that express themselves in a dominant or epistatic fashion.

Nonessential Amino Acid–Amino acids that can be synthesized by the animal's body.

Nonfat Dry Milk–The product obtained by the removal of fat and water from sweet cows' milk. It contains lactose, milk proteins, and milk minerals in the same proportions as the fresh milk, maximum content of retained moisture being not more than 5 percent by weight and that of fat not over 1½ percent unless otherwise specified. Also called defatted milk solids.

Nonfat Solids–The portion of milk remaining after the water and butterfat have been accounted for; nonfat-dried-milk solids.

Nonmotile–Not capable of locomotion.

Nonnutritive Additive–An additive that has no nutritive or food value; e.g., certain drugs or preservatives.

Nonpalatable–(1) Designating a range plant species not grazed when the range or pasture is properly utilized. (2) Designating feeds not relished by animals.

Nonparasitic–See Abiotic Disease.

Nonpathogenic–Not capable of producing disease.

Nonprotein Nitrogen (NPN)–The nitrates, amides, and amino acids that are the forerunners of protein; toxic to livestock in some of these forms.

Nonreacting Infected Animal–An animal that, though infected by a particular disease, does not react to standard tests for the disease.

Nonreturn–In artificial breeding, the breeding efficiency of bulls expressed as the percentage of cows that conceive on the first service.

Nonruminant–An animal, such as a pig, without a functional rumen. Sometimes called a monogastric.

Nonspecific Dermatitis–A skin disease not due to any particular agent. It may affect various parts of an animal's body.

Nonspecific Immunity–Increase of antibodies or production of immunity, resulting from the injection of some nonspecific antigen.

Nonsweating Species–Species of animals that do not sweat, including cattle, sheep, swine, dogs, and chickens.

Nontoxic–Not poisonous to plant or animal. See Toxic, Toxicity.

Northern Cattle Grub–*Hypoderma bovis*, family Hypodermatidae; an insect that infests cattle in the United States. The adult fly, called a heel, lays eggs on the heels of the animal and the larvae migrate to the back, where they cut small holes through the skin. The flies irritate the animal with consequent decreased production of milk and loss of meat. The grubs damage the hide. Also called bomb fly.

Northern Fowl Mite–*Orithonyssus sylviarum*, family Dermanyssidae; a parasite of fowls, commonly found about the base of the tail and around the vent. It may cause scabs that damage the dressed carcass. Also called northern feather mite.

Nose–(1) The point of a plow share. (2) The part of the face of people and animals that covers the nostrils; snout; muzzle. (3) To round off the end of a log in order to facilitate snaking or skidding.

Nose Bot Fly–*Gasterophilus haemorrhoidalis*, family Gasterophilidae; an insect that lays its eggs in hairs on the lips of the horse. The larvae migrate into the mouth and thence into the intestines. It causes great irritation and nervousness.

Nose Clamp–A device that may be fitted tightly on the nose of an animal and used for its control during shoeing, surgical operations, or various types of training. See Twitch, Nose Twitch.

Nose Fly–Sheep bot fly.

Nose Lead–A removable metal ring that is snapped into a bull's nose. A rope is attached to the ring for leading and controlling the animal. This device is required for showing bulls in most livestock shows.

Nose Ring–(1) A metal ring fastened through the cartilage of the nose of a bull for safe control of the animal. A staff, or metal rod, about 6 feet long may be snapped into the ring for handling the animal. See Nose Lead. (2) A metal ring fastened in the nose of a hog to prevent it from rooting.

Nose Twitch–A looped, light rope or heavy cord attached to a short stick. The loop is placed over the end of the nose of the horse and when twisted acts as a control in breaking, training, or leading. See Nose Clamp.

Nosebag–A feed bag hung with the open end over a horse's nose and attached by straps over the head and back of the ears.

Nosema Disease–An intestinal disorder of bees caused by the parasite *Nosema apis*, family Nosematidae.

Nostril–One of the two outer openings of the nose that serves as a passage for air in breathing.

Notching–(1) Cutting dents on the ears of animals for identification. (2) Removing a V-shaped piece of bark from a branch just above or below a bud.

Notifiable Disease–(reportable) Any disease that must be reported to the government health authorities.

Nubian–A breed of milk goats of Mediterranean origin with long drooping ears, a Roman nose, and commonly colored. The black, red, or tan bucks weigh up to 175 pounds, and the does up to 135 pounds.

Nuclear (Nuc) Box–A small hive used for housing a small colony or nucleus. This type of beehive is used for raising queen bees.

Nuclei Package–In beekeeping, a packaged colony of bees established on combs. See Package Bees.

Nucleus–(1) The central portion of the cell protoplast surrounded by a very thin membrane. It consists of nucleoplasm and includes within itself variously arranged chromatin, nuclear sap, and nutritive substances. It is of crucial significance in metabolism, growth, reproduction and the transmission of the determiners of hereditary characters. (2) A small colony of bees used in queen rearing or in pollination work in greenhouses. (3) A central core around which material collects or is grouped.

Nuclide–Any species of atom that exists for a measurable period of time whose nuclear structure is distinct from that of any other species. Thus each isotope of an element is a separate nuclide.

Number of Contemporaries–The number of animals of similar breed, sex, and age, against which an animal was compared in performance tests. The greater the number of contemporaries, the greater the accuracy of comparisons.

Numdah–A thick, belt blanket placed under a saddle to absorb sweat. Also called namda, nammad.

Nuptial Flight–The flight taken by a queen bee during which mating takes place. She may mate with several drones, but this will be the only mating of her life. She will lay several thousand eggs over a period of two to five years.

Nurse Bees–Three- to ten-day-old adult bees that feed the larvae and perform other tasks in the hive.

Nurse Cow–A milk cow used to supply milk for nursing calves other than her own.

Nursery–(1) Any place where plants, shrubs, and trees are grown either for transplanting or as grafting stocks. (2) A group of young plants or trees in a plantation.

Nursery Deck–A small above-the-floor pen in which the newborn pits are kept to keep them warm and dry.

Nursing Calf–A calf that runs with its mother and is not weaned.

Nursing Mare–A mare with an unweaned foal.

Nursling–A calf that is still suckling the cow.

Nutmeg Liver–Chronic, venous congestion of the liver characterized by a finely mottled appearance; common in old or diseased animals.

Nutrient–(1) A substance that favorably affects the nutritive processes of the body; a food. (2) An element or compound in a soil that is essential for the growth of a plant. (3) In stock feeding, any feed constituent or group of feed constituents of the same general composition that aids in the support of life, as proteins, carbohydrates, fats, minerals, and vitamins.

Nutrient Cycle–The circulation of nutrient elements and compounds in and among the soil, parent rock, streams, plants, animals, and atmosphere.

Nutrient Level–(1) In soils, the amounts and proportions of plant nutrients, such as phosphorus, potassium, and nitrogen in available forms. (2) Specifically, the concentrations of any particular nutrient in the ration of animals.

Nutriment–Nourishment; nutritious substances; food.

Nutrition–The sum of the processes by which an organism utilizes the chemical components of food through metabolism to maintain the structural and biochemical integrity of its cells, thereby ensuring its viability and reproductive potential.

Nutritional Anemia–Anemia in animals that results from a nutritional deficiency, usually iron, copper, or cobalt. See Bush Sickness.

Nutritional Blindness–Blindness of stock that results from a deficiency of vitamin A in the rations.

Nutritional Encephalomalacia–See Crazy Chick Disease.

Nutritional Requirements–Number and quality of complex organic compounds and mineral salts in the diet necessary for optimal development and reproduction of an animal.

Nutritional Roup–See Roup of Fowls.

Nutritive Additive–An additive that has some food value such as a vitamin or mineral.

Nutritive Ratio–In animal feeds, a ratio or proportion between the digestible protein and the digestible nonnitrogenous nutrients found by adding the digestible carbohydrates plus the digestible fat multiplied by 2.25, and dividing the sum by the digestible protein. The ratio is an expression of the energy value of a ration against its body-building power.

Nutritive Value–The relative capacity of a given feed to furnish nutrition for livestock. (Usually prefixed by high, low, etc.)

Nymphomania–Abnormal sexual desire in a female.

O-strain–A strain of the foot-and-mouth disease. See Foot-and-Mouth Disease.

Oak Poisoning–Poisoning or digestive disturbances in cattle resulting from excessive feeding on the ripe acorns of various species of oaks. The symptoms are constipation, black feces, and rose-colored urine. Death may result. Also called acorn poisoning.

Oak-leaf Poisoning–A digestive ailment, observed among cattle on the ranges of the southwestern United States; attributed to excessive ingestion of oak leaves in the spring, before grass makes its appearance. In a few instances the ailment is fatal and otherwise may cause stunted growth.

Oat(s)–(1) Any grass of the genus *Avena*, family Gramineae. Species are grown for forage, as a cover crop, and for their seed used as food and feed. Native to temperate regions. (2) *A. sativa*, the common oat, an annual herb native to Europe and Asia. See Animated Oat, Side Oat, Wild Oat.

Oat Straw–The dry stems and leaves of the mature oat plant left after threshing. It has a somewhat higher feed value than the straw of other small grains and is also used for bedding and industrial purposes.

Oberhasli–A breed of dairy goats originating in Switzerland. Their coat is red in color with black trimmings and the ears are erect. Also known as Oberhasli-brienzer, Graubunden, Chamoise, Brown Alpine, and Rehbraun.

Obligate Aerobe–An organism that lives only in the presence of free oxygen.

Obligate Anaerobe–An organism that lives only in the absence of free oxygen.

Obligate Parasite–An organism that develops and lives only as a parasite, and is confined to a specific host.

Obligate Saprophytes–Microorganisms not related to living cells that secure their nutrients from dead organic tissue or inorganic materials.

Obligate Symbiont–An organism that is dependent on mutual relations with another for its existence.

Observation Hive–A beehive made largely of glass to permit the observation of bees at work.

Obstetrical Chain–A metal chain, 30 to 60 inches long, used by veterinarians to assist animals having difficulty in the delivering of newborn (usually calves or foals). It is used to apply a small amount of traction to the unborn.

Obstetrics–The branch of medicine that deals with the birth of the young and management practices during pregnancy.

Occlusion–(1) The process of healing over or closing of the wound caused by cutting or breaking off of a limb in pruning. (2) The absorption of gases by solids; e.g., the absorption of oxygen by milk powder.

Occult Spavin–A type of spavin that occurs between the bones at the hock joint of a horse. See Spavin.

Ocular Lymphomatosis–One of the aspects of avian leukosis complex. A part of the symptoms includes an absence of pigment in the iris. The affected birds ultimately lose their sight. Also called gray eye, white eye.

Ocular Roup–An advanced stage of coryza of poultry, mainly involving the eyes. It is characterized by a thick, cheesy mass of material that develops in the lower corner of the eye under the eyelid. The disease may result in blindness or death.

Odd Lot–On livestock markets, animals that do not conform uniformly to some particular weight, age, or grade of quality class.

Odontomata–A tumor arising in a tissue that normally produces teeth; sometimes found in horses and cattle.

Oestrus–See Estrous.

Off–(1) In cotton transactions, designating grades below middling, the basic grade. (2) A low-grade or inferior product.

Off Feed–(1) Designating an animal that has digestive disturbances due to excessive or improper feeding. (2) Designating an animal that fails, from any cause, to eat the normal feed.

Off Flavor–Designating milk or any milk product that has a peculiar or unnatural flavor.

Off Grade–Designating an agricultural product that fails to meet requirements of commonly accepted standards or legal or official standards in grading products for sale.

Off Take–The animals removed or harvested from a herd.

Off Type–In plants or animals, any notable deviation from standard or normal.

Off Wool–Any wool of inferior quality; a fleece that is otherwise known as discounts, rejections, or unmerchantable wools.

Offal–(1) The inedible parts of a butchered animal, such as the digestive system, lungs, feet, etc. (2) Anything discarded as useless. (3) In grain milling, the by-products such as wheat bran, shorts, etc.

Offal Fat–In cattle slaughter, the internal fat. Also called intestinal fat, internal fat, killing fat.

Offside–The right side of a horse. See Near Side.

Offspring–The young produced by animals.

Oil–One of several kinds of fatty or greasy liquids that are lighter than water, burn easily, are not soluble in water and are composed principally, if not exclusively, of carbon and hydrogen.

Oil Cake–Stock feed that is a mass of compressed seed from which the oil has been largely extracted, as linseed cake, cottonseed cake.

Oil Gland–The gland at the base of the tail of a chicken or other bird that secretes an oil used by the bird in preening its feathers. The gland is removed in dressing chickens for the market. Also called preen gland, uropygial gland.

Oil-protecting Eggs–The process of protecting eggs by dipping them in a hot liquid for a very short time, partially to coagulate the albumen in the shell membranes and to seal the egg shell. See Oiling Machine.

Oiling Machine–In the poultry industry, a machine used for dipping eggs in oil to form a protective coating and to retard loss of moisture and quality. It is used especially on eggs placed in storage. See Oil-protecting Eggs.

Oily Pork–A swine carcass that is soft or oily because of softening feeds such as peanuts.

Ointment–A salve; an unguent; a medicinal preparation that has a base of some kind of fat or soft unctuous substance, such as the petroleum product, petrolatum.

Old Bird–A stewing chicken, hen or fowl, generally a year or more old.

Old Geese–In marketing, geese over one year old.

Old Hen–In marketing, a female turkey over one year old.

Old Shell–Beef cows in very poor condition; classed as canners and cutters.

Old Skin–Referring to a decrepit horse.

Old Tom–A male turkey over one year old after molt is completed.

Old-crop Lambs–Sheep over one year of age that have lamb teeth.

Old-ewe Disease–See Pregnancy Disease in Ewes.

Oleo Stearin–The compound of fats and oils obtained in the rendering of beef and mutton fats at a temperature below 170°F; it is used in the manufacture of compounds requiring a stearin base.

Oligophagous Parasite–A parasite capable of developing upon a few closely related host species.

Oligotrophia–A condition of inadequate nutrition.

Oligotrophic Waters–Waters with a small supply of nutrients; hence, they support little organic production.

Omasum–The third compartment of the ruminant stomach. Contains a mass of suspended, parallel, rough-surfaced leaves that grind ingesta to a fine consistency. Often called the "manypiles." See Abomasum, Reticulum, Rumen.

Omentum–A fold of the peritoneum connecting the stomach to the adjacent organs. The fat deposited in the folds of the great omentum of cattle, sheep, and swine is in part the source of the suet of commerce.

Omnivorous–Designating animals that feed on both flesh and plants, as people; as applied to insects, voracious, but not necessarily omnivorous. Also called amphivorous. See Carnivorous, Herbivorous.

Omphalitis–An infection, usually fatal, of the navel in baby chicks and turkeys largely due to failure of the navel to close properly and to unsanitary incubation conditions; characterized by listlessness, a puffed and mush condition of the body, and a bluish appearance of the abdomen. Also called mushy-chick disease.

Omphalophlebitis–An inflammation of the umbilical vein and other structures in the umbilicus of young animals caused by filth-borne bacteria. Also called sleepy foal disease, pyosepticemia, navel-ill, joint ill.

On and Off Permit–A grazing permit issued only where movement of livestock is necessitated between national forestland (United States) and adjoining outside range, or where private and forestlands intermingle.

On Feed–(1) Designating livestock kept on farms and ranches that are being fattened for the market. (2) Designating any animal being fed grain for milk or meat production.

On Full Feed–In feeding poultry and livestock, the feeding of the animal all the food necessary for optimum gain.

On Pasture–Designating livestock that are grazing on pasture in contrast to those feeding in the barn or feedlot.

On the Clean Basis–In American wool market reports, referring to quotations given on the cost of the clean fiber that any lot of unscoured wool is estimated to contain.

On the Hoof–Designating a live meat animal.

Onchocerocosis–An ailment of animals characterized by fibrous nodules forming under the skin; caused by worms of the genus *Onchocerca*. See Mexican Blindness.

One Bum Lamp–A blind eye of a horse.

One-ear Bridle–A type of horse bridle that is usually used on working stock horses.

One-season Pasture–Pasture composed of a forage plant, usually an annual, such as Sudangrass, which can furnish grazing only for a single season.

Onion Flavor–A flavor defect in milk and milk products that results from ingestion of the wild onion by dairy cows.

Onion Poisoning–See Death Camas.

Ontogeny–(1) The development of an individual tissue, organ, or organism. (2) The complete developmental history of an organism from the egg, spore, bud, etc., to the adult stage.

Oocyst–The encysted or encapsulated stage of parasites, known as coccidia, passed with droppings of infected animals.

Oocyte–Ovicyte; one of the intermediate cells in the process of ovigenesis.

Oocyte-Egg-Mother-Cell–The cell that undergoes two meiotic divisions, oogenesis, to form the egg cell, as primary oocyte, the stage before completion of the first meiotic division, and secondary oocyte, after completion of the first meiotic division.

Oogenesis–The process by which germ cells are produced by the female.

Oogonium–Ovigonium; the first or primary germ cell from which the female gamete is produced.

Oon Egg–An egg, expelled from the vent of a hen without a shell.

Oophagy–The eating of eggs, said of egg-eating insects.

Open–(1) Designating a female animal that is not pregnant. (2) Designating a body defect of cheese in which the cheese has many mechanical openings. (3) To unfold, as a flower opens. (4) Designating a rural or wilderness area in contrast to the congestion of cities and towns.

Open Breed–A breed in which entry is not restricted to progeny of animals registered in that breed (i.e., opposite of closed breed).

Open Bridle–Bridle without blinds or blinkers covering the eyes.

Open Class–In agricultural shows, an exhibit for which no strict limitations are imposed either for the things exhibited or for the exhibitor.

Open Formula–On feed containers, the statement on the tag about the number of pounds of each ingredient in a ton of mixed feed.

Open Heifer–A nonpregnant heifer. See Bred Heifer.

Open Herding–Allowing a band of sheep or goats to spread freely while grazing.

Open Range–An extensive range area where grazing is unrestricted. Also, ranges that have not been fenced into management units.

Open Round-Up–A round-up of cattle in which corrals or fences are not used (western United States).

Open Side–The left side of a beef carcass. There is space between the kidney knob and abdominal wall on this side. See Closed Side.

Open Wool–Wool that is not dense on the sheep and shows a distinct part down the ridge or middle of the back. Usually found in the coarser wool breeds.

Open-faced–Sheep that naturally have little or no wool covering about the face and eyes. This is a desirable trait as it discourages the problem of wool blindness. See Wool Blindness.

Ophthalmia–Inflammation of the eyeball or conjunctiva.

Opisthotonos–Tetanic spasm in which the head is drawn backward and the back is arched.

Opium Poisoning–The poisoning of animals that results from ingesting the opium poppy or from overdoses of opiates.

Opthalmic–Pertaining to the eye, for example, ophthalmic ointment, one used in the eye.

Opthalmic Tuberculin Test–A test for tuberculosis in which a small quantity of a concentrated tuberculin is placed within the lower eyelid of the animal to be tested. A discharge of mucopurulent material within six hours indicates a positive reaction. The test is commonly used on primates held in zoos; e.g., gorillas, baboons, etc.

Optimum Condition–The ideal environment, with regard to nourishment, light, temperature, etc., for an organism's growth and reproduction.

Optimum Fruitfulness–That favorable condition for growth (especially in fruit trees) in which a plentiful supply of blossom buds is produced. The condition is associated with the carbohydrate-nitrogen relationship.

Optimum Level of Performance–The most profitable or favorable ranges in levels of performance for the economically important traits in a given environment and management system. For example, although many cows produce too little milk, in every management system there is a point beyond which higher levels of milk production may reduce fertility and decrease profit.

Oral Medicine–Medicine taken by mouth, whether in drinking water, feed, or in bolus or capsule form.

Oral Toxicity–The toxicity of a substance when the substance is ingested by mouth.

Orchitis–Inflammation of a testis, which is marked by pain and swelling and a feeling of weight.

Orf–A widespread disease of sheep and goats caused primarily by a virus and secondarily by a bacterial organism, *Fusiformis necrophorus*, family Mycobacteriaceae. In malignant form, ulcers appear inside the mouth and other parts of the body.

Organ–A distinct part of a plant or animal that carries on one or more particular functions; e.g., a leaf, wing of a bird, etc.

Organic–(1) Produced by plants and animals; of plant or animal origin. (2) More inclusively, designating chemical compounds that contain carbon.

Organism–Any living individual whether plant or animal.

Organogen–Any of certain chemical elements without which organisms cannot exist: oxygen, carbon, nitrogen, phosphorus, etc.

Organoleptic–Concerning the sensory impressions, such as temperature, taste, smell, feel, sweet, sour, and salt, associated with eating and drinking.

Orientals–A standard class of certain breeds of chickens native to the Orient or largely of oriental type and breeding, as the Sumatras, Malays, and the Cubalayas.

Orientation Flight–The flight of a young bee. It often begins during the latter part of the nursing period, which may extend slightly beyond thirteen days in case of lack of nurse bees.

Orifice–(1) An opening by which spores, etc., escape; any opening. (2) An opening in a nozzle tip, duster, or granular applicator through which the spray, dust, or granules flow.

Orpington–A breed of general-purpose chickens of the English class. They are large, loose-feathered birds having single combs, white skin, and nonfeathered shanks. Mature cocks weigh about 10 pounds, hens about 8 pounds. The eggshell is brown. Varieties are the Buff, the Black, the White, and the Blue.

Orthotropic–Assuming a vertical position.

Osmophyllic–Organisms that grow in solutions with high osmotic pressure.

Osmosis–The flow of a fluid through a semipermeable membrane separating two solutions, which permits the passage of the solvent but not the dissolved substance. The liquid will flow from a weaker to a stronger solution, thus tending to equalize concentrations.

Osmotic Pressure–The hydrostatic pressure required to stop osmosis or prevent diffusion of molecules of a dilute solution from passing through the walls of a semipermeable membrane into a more-concentrated solution.

Ossification–The process of forming bone. Cartilage is made into bone by the process of ossification. The minerals, calcium, and phosphorus are deposited in the cartilage, changing it into bone.

Ossolets–A soft, warm, sensitive swelling over the front and sometimes sides of the fetlock joint in the horse. It is an inflammation of the bone covering, periosteum, occurring usually only in the front legs. Also called osslets.

Osteocyte–A bone cell, particularly one encased in hard bone.

Osteofibrosis–A loss of calcium salts from the bones that causes them to become fragile. A condition chiefly affecting the horse, it may also appear in pigs, goats, and dogs.

Osteomalacia–A disease in which the bones of animals become softer, supposedly due to a deficiency of phosphorus, calcium, or both, in the diet. Also known as creeping sickness, loin disease of cattle, down in the back, adult rickets.

Osteoporosis–A bone disease caused by calcium deficiency that results in increased porosity and softness of the bone. In the Equidae (horse, etc.), it is a specific disease marked by enlargement, softening, and increased porosity of the bones of the face. In poultry, the condition is commonly referred to as rickets; however, some specialists distinguish between osteoporosis and rickets.

Otolith–Earstone, used by fish for its sense of balance. There is one in each plane of the semicircular canals on each side of the head, making six in all. Fishery biologists use the biggest ones to determine the age of fish.

Outapiary–An apiary located some distance from the beekeeper's home.

Outbreeding–Mating animals distinctly unrelated, usually with diverse type or production traits. See Crossbreeding.

Outcrossing–Mating of individuals that are less closely related than the average of the breed. Commercial breeders and some purebred breeders should be outcrossing by periodically adding new sires that are unrelated to their cow herd. This outcrossing should reduce the possibility of loss of vigor due to inbreeding. See Outbreeding.

Outer Cover–In beekeeping, usually a cover that telescopes over the top of the hive to a depth of an inch or more and is covered with galvanized iron or aluminum sheeting, to protect the hive from the weather.

Outlaw–(1) A horse that refuses or fails to become tractable during the breaking process. (2) A wild, unbroken horse.

Outrider–A cowboy whose duty it is to ride the range and protect his employer's interests (western United States).

Ova–A female egg or gamete.

Ovarian Cyst–A benign cyst, often congenital.

Ovarian Extract–A by-product of packing houses, made form the ovaries of sheep, hogs, and cattle containing female sex hormones, useful in treating ovary disorders or hormone imbalances.

Ovarian Follicle–The small cystlike structures in ovaries which when fully developed contain a mature ovum. Also called Graafian follicle.

Ovary–(1) The portion of the pistil or carpel of a flower that contains one or more ovules. (2) The organ in female animals that produces the egg or ovum.

Overbrowsing–Excessive cropping of shrubs or tree growth, usually by goats, sheep, or game animals.

Overcheck Bridle–A horse bridle used with an overcheck rein.

Overcheck Rein–A checkrein that passes over a horse's head between the ears.

Overfeeding–Consuming excessive amounts of feed. It can cause various digestive disturbances, such as diarrhea, colic, bloat, and founder.

Overgraze–To graze land so heavily as to impair future forage production and to cause range deterioration through consequential damage to plants, soil, or both. Also called overstocking.

Overhaul–(1) The rehandling or repacking of ham during the pickling period to permit a more uniform distribution of pickle. (2) To repair and recondition tools, implements, machines, etc.

Overhead Check–In horses, a rein that extends from the bit over the head of a horse to a hook on the harness backpad to prevent the animal from tossing or lowering its head.

Overo–Denotes a horse that is basically white in color and the spotting is usually roan and extends upwards from the belly. The darker areas are usually small or rather ragged patches; the mane and tail are usually a mixture of color giving a roan effect. Overo horses usually have bald faces, and glass eyes are not uncommon.

Overreach–In a horse, to strike the heel of the forefoot with the front of the hind foot.

Overshot Jaw–A condition where the lower jaw of an animal protrudes beyond the upper one.

Overshot Wheel–A waterwheel operated by the weight of water falling into buckets attached to the periphery of the wheel.

Overstocked–(1) Designating a condition of a stand of trees or of a forest, in which there are more trees than normal or full stocking would require. The overstocking may be to such a degree that growth is slowed down and many trees, including dominants, are suppressed. (2) Designating a pasture, range, or grazing game area that has more animals on it than the vegetation of the area will support. See Overgraze. (3) Designating a locality in which there are too many bees.

Oversummer–To live through the summer.

Ovicide–Any substance that kills parasites or other organisms in the egg stage.

Ovicyte–Same as oocyte, an intermediate cell in the process of ovigenesis.

Oviduct–The tube that leads from the ovary to the uterus or other organs where fertilization or further development of the ovum or egg cell occurs. Also called fallopian tube, tubes.

Ovigenesis–Oogenesis; the process of producing the female gamete.

Ovigonium–The primary germ cell from which the female gamete is produced.

Ovine–(1) A animal of the subfamily Ovidae; sheep, goats, etc. (2) Pertaining to such an animal, commonly to sheep.

Ovine Balano-Posthitis–A veneral disease of sheep causing ulcerations of the penis sheath and vulva, ulcerative vulvitis, thought to be caused by a filterable virus with bacteria as secondary invaders.

Oviposition–The process of laying an egg.

Ovipositor–A tubular structure in female insects used for depositing its eggs.

Ovoid–Referring to a solid body with the shape of a hen's egg, the point of attachment, if any, at the broader end.

Ovotestis–A gonad, part of which is composed of ovarian tissue and testicular tissue.

Ovoviviparous–Refers to animals who produce eggs that are incubated inside the body of the dam and hatch inside the body or shortly after laying. See Oviparous, Viviparous.

Ovulation–The process of releasing eggs or ova from the ovarian follicles.

Ovule–The body that, after fertilization, becomes the seed; the egg-containing unit of the ovary.

Ovum–The female sex cell, produced on the ovary, and carrying a sample half of the genes carried by the female in which it was produced. Plural, ova.

Ox–(Plural: oxen) Any species of the bovine family of ruminants. Specifically, the domesticated and castrated male bovines used for work purposes, as distinguished from steers used for meat, or the uncastrated bulls used for breeding.

Ox Bot–See Cattle Grub.

Ox Tongue–A food product prepared by pickling beef tongues.

Ox Warble Fly–See Cattle Grub, Heel Fly.

Oxalacetate–An intermediate compound formed from sugars and sugar-producing (gluconeogenic) amino acids, essential for the oxidation of carbohydrates, proteins, and fats for energy.

Oxalic Acid Poisoning–A toxic condition of sheep and cattle resulting from the ingestion of large quantities, over a short period of time, of dried plants in which the oxalic acid content is more than 10 percent.

Oxen–See Ox.

Oxford Sheep–A mutton and medium-wool breed of sheep that is a cross of several foundation breeds. Mature ewes weigh 175 to 250

Oxidase–Paint Horse

pounds. First produced in Oxfordshire, England. Also known as Oxford Down.

Oxidase–Oxidizing enzymes. See Enzyme.

Oxidation–Any chemical change that involves the addition of oxygen or its chemical equivalent. Oxidation may affect agricultural products adversely or even cause destruction: e.g., it can be the cause of taints in milk or butter and it can cause spontaneous combustion of hay stored in barns.

Oxidation Pond–A person-made lake or body of water in which wastes are consumed by bacteria. An oxidation pond is the same as a sewage lagoon.

Oxtail–The tail of a beef carcass, one of the edible by-products of packing houses.

Oxygen–The chemical element O; a colorless, odorless gas. The most abundant element in the earth's crust. It accounts for about 47 percent of all elemental material. It is essential in the growth of all crops and for the respiration of most forms of life.

Oxygenation–The absorption by water of elemental oxygen that has: (a) been released into the water by aquatic plants as a waste product of photosynthesis; (b) come from the atmosphere.

Oxyhemoglobin–A substance formed by the union of oxygen with the hemoglobin of the blood, the union being of such a nature that the oxygen is readily given up at times and in places where it is needed.

Oxytocin–The hormone released from the posterior pituitary of the female that causes contractions of the uterus at the time of breeding. These contractions aid the movement of sperm through the cervix and into the uterine horns where fertilization of the egg normally occurs. Oxytocin also aids in parturition and causes milk letdown at milking time.

Oyster–The tenderloin muscle of the poultry carcass. There are two, one on each side, located just in front of the hipbones.

Oyster Shells–Shells of the marine bivalve, genus ***Ostrea***. They are nearly pure calcium carbonate in composition; when finely ground, they make good liming material for soils and a mineral feed for livestock and poultry as a source of calcium.

Ozena–A fetid nasal discharge accompanied by chronic inflammation of the mucous membranes, associated with a disease of bones in the nose.

Pace–(1) A measure of length: the ordinary length of a human step from heel to heel is 2½ feet. The geometric is 5 feet. Distances for land measure are usually stepped according to some definite distance per step, as 3 feet. Land measurements were often paced off, from some known mark as a base, in early land surveying in the United States. (2) Applied to horses, a rapid, two-beat gait in which the lateral fore and hind legs work in pairs. Also called amble, rack, trot.

Pacer–A horse, one of whose gaits is a pace. Also called side-wheeler.

Pack–(1) Fruit, vegetables, meat, etc., which is to be or is packaged, canned, frozen, etc., for the market. (2) The total amount of products that are processed in a season. (3) The manner in which produce is packaged. (4) The load that is carried by a pack animal. (5) To compress or firm soil with a special implement. (6) To damage soil structure by compacting or puddling clay soils, as from the pressure exerted by the wheels of a tractor, by injudicious irrigation, or by excessive rainfall. (7) The sheep given to the shepherd as his share for tending the flock.

Pack Animal–A burro, mule, or horse used for carrying packs and equipment over rough areas and to places inaccessible by other means.

Pack Rat–A bushy-tailed rat, ***Neotoma cinerea***, which collects odds and ends in its burrow. Found in the southwestern United States, it sometimes damages citrus orchards adjacent to canyons with brushy cover.

Package Bees–A quantity of bees (2 to 5 pounds) with or without a queen shipped in a wire-and-wood cage to start or boost colonies.

Packer–(1) One who operates a slaughter and meat-processing business. (2) Pertaining to the business of packing fresh or processed fruits and vegetables or meats. (3) A field tool of the roller type consisting of a set or series of rollers that pack the loose soil after plowing. Also called cultipacker. (4) One who makes a pack, as of vegetables, meat, etc.

Packer Hide–An animal hide prepared for the market as a by-product by meat-packing companies as distinguished from country hides that are not uniformly processed.

Packing House–(1) A slaughtering and meat-distributing organization that buys, slaughters, processes, and distributes livestock products. (2) A separate building or shed adjacent to a range of greenhouses where flowers and greenhouse plants are packed for market. (3) A building equipped and arranged for grading and packing fresh fruit or vegetables.

Packing Sow–A market classification for hogs, the grades based on age, condition, sex, and ratio of lean to fat quality of the animal.

Packsaddle–A carrying unit for camping equipment or other material that is placed on the back of a horse or pack mule, either as separate units or attached to the rider's saddle.

Paddling–A term used to describe a horse that throws the front feet outward as they are picked up. This condition is predisposed in horses with too-narrow or pigeon-toed standing positions.

Paint–A color pattern in horses involving white patches on a dark background.

Paint Horse–A horse, much favored by the American Indians in the western United States, which usually carries odd markings and strong color contrasts. The two common color patterns in paint horses are: piebald, or black-and-white, and skewbald, which is white and any other color than black. Also called calico.

Paint Pony–See Paint Horse.

Paint Roan–A coat color of horses in which the roan color seems to be imposed on other colored areas.

Pair–(1) A male and female of a species. (2) A team of horses or mules frequently spoken of as a well-matched pair regardless of sex.

Palatability–(1) The degree to which a feed is liked or acceptable to an animal. (2) (Forage) Range management usage. The relish that an animal shows for a particular species, plant, or plant part. The characteristics of plants that stimulate a selective grazing or browsing response by animals. Palatability is controlled by the plant factors of chemical composition, proportion of plant parts, growth stage, external form of plant parts, environmental factors such as slope steepness, wind, sun, or shade; and the animal factors of instinct, learning, physiological state, individual behavioral pattern variations, and animal sensory responses.

Palatability Rating–A method of rating the condition of a range, from high to low, depending on the proportion of palatable plants carried by the range. The more such plants the higher the rating.

Palatable–Describing feed that an animal prefers and selects over other feeds.

Palatitis–A painfully swollen condition of the palate of a horse's mouth that may be caused by eating hard food. Also called lampas, lampers.

Pale, Soft, and Exudative (PSE)–Pork that is light or pale in color, is soft, and exudes fluids. The condition is related to porcine stress syndrome (PSS).

Palmetto–Any palm of the genus *Sabal*, family Palmaceae.

Palmitic Acid–One of the fatty acids of butterfat.

Palomino–(1) A western United States horse, doubtfully a distinct breed, characterized by a cream, yellow, or golden color and light-colored mane. Averaging 15 to 16 hands high, the Palominos weigh 1,000 to 1,200 pounds, and are very popular riding horses. Also called caballo de oro, California sorrel, golden horse. (2) A coat color of certain horses which is a golden chestnut, with flax-colored mane and tail. Ideal body color is that of a newly minted gold coin. Purity of color and mane and tail is very important in Palomino selection.

Palpate–To test or examine by feel. Used to determine pregnancy in cattle.

Palpation–A method of pregnancy determination in cattle in which the arm is inserted into the cow's rectum and the reproductive tract is felt for pregnancy indications.

Pancreas–A gland below and behind the stomach that secretes pancreatic juice. The pancreas is commonly known as the sweetbread but should not be confused with the commercial sweetbreads of veal (thymus gland).

Pancreatic Juice–A secretion by the pancreas containing ferments that contribute to digestion of foods.

Pandemic–Designating a disease or organism of worldwide distribution; widely epidemic.

Panzootic–Referring to a widespread epidemic among animals.

Pap–The teat of a cow.

Papilla–Any small nipplelike process.

Papule–A small elevation, usually of the skin, as a pimple.

Parakeratosis–A skin disease found in hogs that results from a zinc deficiency in the ration commonly associated with a high calcium intake. It is characterized by dry, hard, crusted proliferations of the superficial layers of the skin.

Paralysis–(1) Abolition or impairment of function, especially loss of the power of voluntary motion or sensation. It is frequently a symptom or manifestation of various animal diseases. See Fowl Paralysis. (2) A disease afflicting honeybees characterized by trembling, sprawled legs and wings. (3) See Parturient Paresis.

Paraplegia–Paralysis of the posterior limbs of an animal that follows accident or disease resulting in injury to the spinal cord.

Parasite–An organism that lives at least for a time on or in and at the expense of living animals or plants. Some diseases of people and animals are caused by parasites ordinarily classified as protozoan, helminthic, and anthropod species. There are also innumerable species of plant parasites.

Parasiticide–An agent that kills parasites.

Parathormone–The hormone of the parathyroid glands. It helps to maintain the calcium level of an animal's blood by removing calcium from the bones; also called parathyroid hormone.

Parathyroid Dysfunction–Faulty function of the parathyroid gland. The condition may result in a change in the bony structures and in abnormal calcium deposits in various tissues of the body, or in tetany and low blood calcium. See Grass Tetany.

Parathyroid Glands–Small glands about the size of a grain of wheat located near the thyroid that are mainly concerned with calcium metabolism.

Paratuberculosis–Johne's disease, a chronic infectious disease of the digestive tract of cattle, rarely of sheep and goats, caused by the organism *Mycobacterium paratuberculosis*. It is closely related to tuberculosis. Also called chronic bacterial dysentery.

Paratyphoid–See Infectious Enteritis in Pigs, Salmonellosis.

Parental Generation–The P_1 generation; the first generation in a series of crosses; usually involves homozygotes for different alleles.

Parenteral–As applied to drug or vaccine administration, to inject subcutaneously, intramuscularly.

Paresis–Slight or incomplete paralysis that affects the ability of an animal to move.

Paries–A wall of a cavity or hollow body organ in plants and animals.

Parietal–Borne on or belonging to a wall or walls of a cavity.

Paris Green–Acetoarsenite of copper, one of the first arsenical compounds to be widely used in insect sprays as a stomach poison for the destruction of leaf-eating insects. It was almost exclusively used to control the potato beetle. Paris Green is now replaced by insecticides that biodegrade faster.

Parotid–A gland that secretes saliva. This is a paired gland located behind the jaws and somewhat below the ear.

Parrot Mouth–A condition in animals in which the upper teeth protrude over and beyond the lower teeth. See Monkey Mouth.

Parthenogenesis–The development of an individual from an unfertilized egg cell. Known to occur occasionally in turkeys and some lower forms in the animal kingdom but not in mammals. In honeybees, unfertilized eggs produce only drones.

Partial Dominance–A kind of interaction between alleles where one gene is not completely dominant to its allele but where the appearance of the heterozygote is more similar to one of the homozygotes than to the other.

Partido System–In the southwestern United States, a form of operation in which sheep owned by the patron are let out on shares to a partidero, who cares for them and returns part of the increase or income to the owner.

Parts Per Millon–The number of weight or volume units in a million units of a solution or a mixture; a measure of concentration, especially of chemicals in solution: one milligram per liter. Abbreviated ppm.

Parturient Paresis–Partial paralysis that occurs at or near the time of giving birth to young and beginning lactation. The mother mobilizes large amounts of calcium to produce milk to feed the newborn, and blood calcium levels drop below the point necessary for impulse transmission along the nerve tracts. Commonly called milk fever.

Parturition–Giving birth; called calving in cattle, lambing in sheep, farrowing in swine, kidding in goats, and whelping in dogs.

Passive Immunity–A kind of immunity acquired by animals when they are injected with antibodies against some disease.

Pastern–The portion of a horse or other animal's leg that connects the cannon bone and the coffin bone in the hoof. There are two parts to the pastern joint, the long and short pastern. This flexible joint serves as a shock-absorbing mechanism in the action of the leg and foot.

Pasteurization–The process of destroying all or most of the vegetative bacteria in a substance, such as milk or fresh fruit juices, by application of heat of from 140°F to 185°F and then cooling. The word pasteurization is derived from the name of Louis Pasteur (1822-1895), the French scientist who first applied heat for the preservation of wine.

Pasteurized Milk–Milk which has been subjected to a temperature over 142°F for more than 30 minutes or at 160°F for 16 seconds and then promptly cooled to 50°F or lower. There are several systems of milk pasteurization.

Pasturage–All vegetation, grasses, and grasslike plants, forbs, and the fruits and twigs of trees and shrubs upon which grazing animals subsist.

Pasture–(1) A fenced or unfenced tract of land on which farm animals feed by grazing. The pasturage is mainly grass but it may consist of various other herbs, brush, and trees. (2) Nectar and honey plants within flight range of bees of an apiary. (3) To place livestock on a field or area of grass to harvest the crop by grazing.

Pasture Forage–A crop ordinarily grown for pasture but which may be cut for green feed, silage, or cured for hay.

Pasture Improvement–The practice of grazing, clipping, fertilizing, liming, seeding, contour furrowing, or other methods of management that improve the vegetation for grazing purposes.

Pasture Off–To remove a crop by grazing; a common practice on grass seedings and grain fields in late summer or fall.

Pasture Rotation–The practice of moving the herd from one field to another after a few days of intensive pasturing.

Pasture-bred–A cow serviced by a bull in the pasture.

Pastured–Designating an area or crop that has been grazed off by livestock.

Pastureland–(1) Land used primarily for the production of adapted domesticated forage plants to be grazed by livestock. (2) Land producing forage plants, principally introduced species, for animal consumption. Management practices usually include one more more treatments such as reseeding, renovating, mowing, liming, or fertilizing. Native pasture that because of location or soil limitation is treated like rangeland is included as pastureland.

Pasturing–The system of removing plant growth by allowing animals to graze it rather than harvesting by other methods.

Paternal–Pertaining to the male parent. See Maternal.

Pathobiology–The study of disease processes; biology of disease.

Pathogen–In the general sense, anything capable of causing disease, but when referred to by most veterinarians and physicians it signifies a living, microscopic, disease-producing agent such as a bacteria or virus. See Parasite.

Pathology–The science that deals with diseases and the effects that disease have on the structure and function of tissues.

Patothenic Acid–A vitamin of the B-complex group; required by poultry and swine. Also called chick antidermatosis vitamin.

Patroclinous–Resembling the male parent.

Paunch–Another name for the rumen.

Paunching–Removing the entrails from a carcass.

Paunchy–A livestock-judging term used to describe an animal (usually cattle) that has a large paunch.

Pea-size Cake–A livestock feed stuff made from soybean, peanut, or linseed cake that has been cracked and screened to the size of a garden pea.

Peanut Hull–The shell of the peanut. It contains more than one-half fiber and is less nutritious than straw; used chiefly as poultry litter, bedding, and fertilizer. See Peanut Hull Meal.

Peanut Hull Meal–Ground peanut hulls; a very slow-acting fertilizer containing 1.5 to 2.5 percent nitrogen.

Peanut Meal–Finely ground peanut oil cake. See Peanut Oil Meal.

Peanut Oil Cake–A livestock feedstuff obtained as a by-product of the partial extraction of oil by pressure or solvents from peanut kernels.

Peavine Hay–The cured vines of peas used for feeding livestock.

Peavine Silage–Silage usually made form canning refuse consisting of the vines and pods of green peas left after the seeds have been removed for canning. It has a strong odor but is an excellent feed for dairy cows, beef cattle, and sheep.

Pecking Order–The order in which same poultry and wild birds within a flock may peck others without being pecked in return; hens at the top of the order can gain a place at feeders, etc., at will, and others according to rank. The males' social standing and mating order is largely determined by combat.

Pediculus–A genus of sucking lice infesting mammals, including people.

Pedigree–(1) A list of an individual animal's ancestors, usually only those of the five closest generations. (2) A list of the ancestors of a crop plant, as the pedigree of corn.

Pedigree Selection–In plant breeding and improvement, the selection of seed stock form healthy, high-yielding plants. The seeds form each plant are planted in one row so that yield, etc., may be carefully evaluated and comparisons made with other similarly selected plants.

Pedigreed Chicks–Chicks whose parents and female ancestors for at least two generations are known. They are wing-banded for identification.

Pedigreed Stock–Animals that have a pedigree. Pedigreed, purebred, or registered stock are interchangeable terms.

Peewees–Small or stunted lambs.

Peg Pony–A saddle horse trained to change direction rapidly.

Pekin–A heavily meated breed of white ducks, excellent for market and table use. The breed is noted for its ability to make rapid and economical growth, young ducklings commonly attaining a weight of 5 pounds or more in 8 to 10 weeks under commercial feeding. It is used almost exclusively for the production of market ducklings in the United States. Also called White Pekin.

Pelham Bit–A bit with two rings on each side for two controlling reins; used on polo ponies.

Pelham Bridle–A single-bitted, double-reined horse bridle used on pleasure horses.

Pellagra–A nutritional deficiency caused by insufficient niacin, characterized by dermatitis, inflammation of mucous membranes, diarrhea, and psychic disturbances.

Pellet–(1) Mixtures of ground ingredients pressed to form aggregates of convenient size to feed livestock. (2) A mass of indigestible hair and bones regurgitated by carnivorous birds or mammals.

Pelleted Feed–A pill-like or cubical type of animal feed made by forcing the loose, bulky, or dusty feeds into small, uniform pellets by the use of grinding, molding, and compressing machinery. It is used in the manufacture of specialized feeds, such as poultry, calf, rabbit, and dog feeds.

Pelleted Hay–Hay that has been highly compressed by passing through a pelleting machine. It is easy to handle and is free from dust.

Pelt–(1) The natural, whole skin covering, including the hair, wool, or fur of the smaller animals, such as sheep, foxes, etc. A large pelt is more often called a hide. (2) To remove the whole skin or pelt form the body of an animal.

Pelvic Capacity–The dimensions of a female's pelvic area that is an indication of its ability to give birth easily.

Pen–(1) A loose, rectangular stack of fuel wood or pulpwood in layers of two pieces each of varying height and width. (2) In poultry shows, a male and four female birds of the same variety. (3) A small space enclosed by any kind of fence, used for confining pigs, cows, and other animals.

Pen Fattening–Fattening livestock or poultry by keeping them in small pens and giving them full feed.

Pen Lot–A number of animals in an enclosure. In livestock exhibitions, the animals are judged as a lot, or group, in contrast to an exhibition of single individuals.

Pen of Lambs–In the show ring, four lambs of both sexes, owned by the exhibitor.

Pencil Shrink–An assumed percentage deduction taken from the weight of slaughtered animals to allow for uncalculated losses in weight.

Pendulous Udder–A low-hanging, poorly attached udder.

Penetrant–Adjuvant that aids a liquid's ability to enter the pores of a surface. See Adjuvant.

Penicillin–An antibiotic extracted form cultures of certain molds of the genera of *Penicillium* and *Aspergillus* that have been grown on special media. Penicillin is also produced synthetically.

Penis–The male organ of sexual union.

Penta–Greek for five, used in naming chemical compounds.

Pentachlorophenol–A chemical used extensively to treat fenceposts, telephone poles, and bridge planking, against fungal decay. Reports have indicated some hazard to livestock when they lick or chew on posts so treated.

Pepsin–A digestive enzyme secreted by glands in the stomach. Commercially obtained from the lining of the pyloric end of a pig stomach, it is used in medicine as an aid to protein digestion.

Peptide–A compound made up of a series of amino acids; an intermediate in the synthesis or digestion of a protein.

Peptones–Products obtained by the digestive action of enzymes on albuminous matter. Peptones of different kinds are produced form lean muscle tissue or by the action of yogurt bacteria on lactalbumin in milk, and serve as easily assimilated forms of proteins or as cultural media.

Per Os (P.O.)–By the mouth; medicine administered through the mouth.

Peracute–Excessively acute; e.g., when symptoms of a disease occur much earlier than usual and are well marked.

Performance Data–The record of the individual animal for reproduction, production, and possibly carcass merit. Traits included are birth, weaning and yearling weights, calving ease, calving interval, milk production, etc.

Performance Pedigree–A pedigree that includes performance records of ancestors, half and full sibs, and progeny in addition to the usual pedigree information. Also, the performance information is systematically combined to list estimated breeding values on the pedigrees by some breed associations.

Performance Record–The evaluation of an animal's production based on several factors; factors may be number of calves weaned, birth weight, weaning weight, etc.

Performance Testing–The systematic collection of comparative production information for use in decision making to improve efficiency and profitability of beef production. Differences in performance among cattle must be utilized in decision making for performance testing to be beneficial. The most useful performance records for management, selection, and promotion decisions will vary among purebred breeders and for purebred breeders compared with commercial cattle producers.

Pericarditis–In cattle, inflammation of the membranes surrounding the heart that frequently results from hardware disease. See Hardware Disease.

Pericardium–The membrane that encloses the cavity containing the heart.

Period of Lactation–Period of milk production in an animal; the time between the beginning and end of the milk flow.

Periodic Annual Increment–The growth of a stand of trees for any specified period divided by the number of years in the period.

Periople–The thin, outer layer of the hoof of an animal.

Periosteum–The outer membrane or covering of bone.

Periphyton–The association of aquatic organisms attached or clinging to stems and leaves of rooted plants or other surfaces projecting above the bottom of the body of water.

Perishable–Designating any product that is liable to easy or quick destruction by rot, disease, or decomposition, such as fresh fruits, meats, and vegetables.

Peristalsis–(1) The successive wavelike motion of the muscular fibers in the duct walls that forces the egg through the oviduct. (2) The rhythmic contractions and movements by which the alimentary canal propels its contents.

Perithecium–A round or oval, ascus-containing, fungus fruiting body.

Peritoneum–The membrane that lines the abdominal cavity and invests the contained viscera (digestive organs) of an animal.

Peritonitis–Peritoneal inflammation. See Peritoneum.

Permanent Hay–Hay crops, such as alfalfa and perennial grasses, which occupy the land for a long period without intervening crops.

Permanent Parasites–Parasites, such as bloodsucking lice, which spend all life stages on or in the body of the host.

Permanent Pasture–Grazing land in farms occupied by perennial grasses and legumes. It is not a part of a regular rotation of fields and usually remains unplowed for long periods.

Pernicious–Harmful or fatal.

Pernicious Equine Anemia–See Equine Infectious Anemia.

Perosis–A deforming leg weakness in poultry in which the Achilles tendon slips out of its natural groove at the hock joint causing the leg to become permanently bowed or badly twisted. There is usually an enlargement and flattening of the hock joint. It occurs most frequently in rapidly growing chickens due to manganese deficiency and an excess of phosphorus in the diet. Also called slipped tendon, hock disease.

Peroxidase–A heat-resisting enzyme present in milk, especially in abnormal milk.

Peroxide Number–Measure of the oxidative rancidity of fats by the determination of the peroxides present.

Persian Lamb Skin–The skin of the lamb of Karakul sheep. The wool is lustrous and tightly curled, making the skin highly desirable for luxury coats and other apparel.

Persistent Pesticide–A chemical agent used to control pests, which breaks down extremely slowly, remaining toxic to desirable species of wildlife as well as pests, under natural conditions. Some of these include DDT, chlordane, lindane, and dieldrin. Most are now forbidden or restricted in use.

Pessary–Vaginal suppository for administration of drugs.

Pest–Anything, such as an insect, animal, or plant that causes injury, loss or irritation to a crop, stored goods, an animal, or people.

Pest Control–The use of disinfectant, herbicide, pesticide, insecticide, management or cultural practice that controls pests. See Integrated Pest Management.

Pesticide–A substance used to control insect, plant, or animal pests. Pesticides include insecticides, herbicides, fungicides, nematocides, and rodenticides.

Pesticide Residue–Material that remains on a plant after pesticide application.

Pet–Any animal, such as a cat, dog, lamb, bird, etc., that is kept for affection and companionship.

Petechiae–Small, purple or red spots in the skin of animals caused by small hemorrhages in which blood is released from its normal channel in the blood vessels into very minute areas of the surrounding body tissues.

Petite–(1) A grade for canning small peas of $9/32$ inch or less in diameter. (2) Denoting smallness or shortness.

pH–A numerical measure of acidity or hydrogen ion activity of a substance such as food or soil. The neutral point is pH 7.0. All pH values below 7.0 are acid and all above 7.0 are alkaline. The negative logarithm of the hydrogen-ion activity. The degree of acidity (or alkalinity) of a soil as determined by means of a glass, quinhydrone, or other suitable electrode or indicator at a specified moisture content or soil-water ratio, and expressed in terms of the pH scale. See Reaction.

Phagia (Phagy)—A combining form denoting perversion of eating or swallowing; e.g., geophagia (eating earth), aerophagy (swallowing air). See Pica.

Phago—A combining form denoting relationship to eating or consumption by ingestion or engulfing.

Phagocyte—An animal cell capable of ingesting microorganisms or other foreign bodies.

Phagocytic—Designating the ability of certain body cells to assimilate small objects by flowing around them, enclosing them completely, and digesting them, or to render them harmless by eradication. Certain of the white blood cells have this property and use it to destroy germs and worn-out cells and to eliminate foreign particles.

Phagocytosis—The process whereby certain cells such as some white blood cells and amoebae engulf microorganisms and other particles.

Phagomania—An insatiable craving for food.

Phagophobia—A morbid fear of eating.

Pharmaceutical—Any substance used to enhance the health of humans or animals.

Pharynx—The cavity that connects the mouth and nasal cavity to the throat; a passage common to the digestive and respiratory tracts.

Phase—(1) The view that a thing presents to the eye. (2) Any one of the varying aspects or stages through which a disease or process may pass. (3) In colloidal chemistry, the discontinuous portion dispersed in the dispersion medium. (4) In soil taxonomy and soil survey, soil phase terms are surface soil texture, percentage slope, stoniness, saltiness, and erosion. When appropriate these names are added to the soil series name to make a soil mapping unit.

Phase Feeding—Changes in an animal's diet to adjust for age and stage of production, to adjust for season of the year and for temperature change, to account for differences in body weight and nutrient requirements of different strains of animals, or to adjust one or more nutrients as other nutrients are changed for economic or availability reasons.

Phenol—C_6H_5OH; carbolic acid; a colorless, crystalline compound, obtained by the distillation of coal tar; widely used as a disinfectant and as an ingredient in antiseptics.

Phenol Coefficient—A figure representing the relative killing power of a disinfectant, as compared with phenol acting on the same organism for the same length of time.

Phenothiazine—The compound produced by the cyclization of diphenylamine with sulfur. Highly toxic for many insects, it is used for the removal of parasitic worms on livestock.

Phenotype—The observed character of an individual without reference to its genetic nature. Individuals of the same phenotype look alike but may not breed alike. See Genotype.

Pheromone—A substance secreted to the outside of the body by an individual organism that causes a specific reaction by another organism of the same species; e.g., when an earthworm is alarmed it secretes a mucus which is a warning to other earthworms. See Sex Pheromone.

Phildadelphia Dressed—In poultry slaughter, designating fowl bled and plucked but not eviscerated. Also called New York dressed.

Phobia—(Greek) Fear.

Phosphoric Acid Silage—Legume, small grain, and grass silage to which a small amount of commercial phosphoric acid has been added. There is little loss of green color or nutrients, and the available phosphorus in the silage is increased.

Phylum—The highest grouping in the taxonomy of the plant and animal kingdoms, based on assumed common ancestry.

Physical and Mechanical Pest Controls—Direct or indirect (nonchemical) measures to destroy pests outright or to make the environment unsuitable for their entry, dispersal, survival, or reproduction; e.g., steam sterilization to destroy disease organisms, flaming for the control of weeds, cold storage to control pests, metal or other material barriers to prevent pest entry. See Integrated Pest Management.

Physical Poison—Any poisonous material that exerts a physical rather than a biochemical effect, as heavy mineral oils and inert dusts.

Physiologic Races—Pathogens of the same species and variety that are usually structurally indistinguishable but which differ in their physiologic behavior, particularly in their ability to parasitize varieties of a particular host.

Physiological—(1) Referring to or concerning the science of physiology or the branch of biology that deals with life processes and functions. (2) Referring to the functions of the organs of plants and animals.

Physiological Age—The age of an animal that is determined by an examination of the carcass.

Physiological Maturity—The period of advanced age in the cycle of a tree or stand of trees when resistance to adverse influences is so low that death of a tree or net losses in volume of salable wood are likely to occur within a cutting cycle.

Physiology—The science that deals with the function of a plant or animal's body and its organs, systems, tissues, and cells.

Piaffe Step—A high action step to which horses are trained for exhibition.

Pick—(1) The total amount of a crop harvested, or the yield of an individual tree, as the pick of oranges. (2) Small, irregular openings within the body of a cheese. (3) See Pickax. (4) To pull or pluck ripe fruit, as berries, apples, cotton. (5) To pluck the feathers from a fowl in dressing it for the market. (6) To nibble at food.

Picket—To tether or control the grazing range of an animal by a rope.

Picket Rope—Any rope, with a swivel snap on one end of which an animal may be fastened, and with a picket pin on the other which may be driven into the ground to tether the animal while grazing.

Picking Machine—In poultry dressing, a machine consisting of a revolving drum equipped with flexible rubber fingers which picks the feathers from carcasses.

Pickled Pig's Feet–The feet of the hog carcass that, after removal of the toenails, dew claws, and hair are cured in a special pickling solution, which is basically a brine, with certain seasoning ingredients added.

Picnic Ham–The cured meat from the shoulder of a pig carcass.

Pictou Disease–An eastern Canadian cattle poisoning caused by ingestion of the plant ragwort groundsel, *Senecio jacobaea*, either fresh or with hay.

Piebald–Designating a horse with a black-and-white or dark coat; a pinto. Also called pied.

Piebaldism–A skin condition in which the skin is partly white (albinism) and partly brown (vitiligo). A common example is a piebald horse.

Piedmontest–A breed of large beef cattle that originated in Italy. They are grayish-white.

Pig–(1) A young swine weighing less than 120 pounds. (A few markets, as Chicago, United States, set 130 pounds as the maximum weight for an animal of this class.) (2) Any young, unweaned swine.

Pig Eye–In a horse, a small, retracted eye that may cause imperfect vision.

Pig Teeth Nipper–A clipperlike instrument used for clipping the canine teeth (temporary or milk teeth) of suckling pigs.

Pig Typhus–Infectious enteritis in pigs.

Pig-eating Sow–A sow that devours her young at farrowing time presumably as a result of a ration deficiency, an unsuitable environmental situation, or exciting disturbances at farrowing time.

Pig-guard Rail–A rail about 8 to 10 inches above the floor and 8 to 10 inches from the wall of a farrowing pen, beneath which recently farrowed pigs can move and thus be protected from crushing by the sow.

Pigeon–Any bird of the genus *Columba*. In poultry, young pigeons, or squabs, are generally dressed for the market, and old pigeons are sold alive, generally for the Jewish trade. See Squab.

Pigeon Fly–*Pseudolynchia canariensis*, family Hippoboscidae; a blood-sucking fly, somewhat smaller than the house fly, which infests pigeons and related birds. Known in the United States since 1896.

Pigeon Loft–A dovecot, elevated house, or pen atop buildings for raising pigeons.

Pigeon Louse–Either of two species of lice, *Columbicola columbae*, or *Goniocotes bidentatus*, family Philopteridae. Each causes considerable annoyance both to old pigeons and to partially feathered squabs.

Pigeon-toed–Designating an animal or person whose feet turn inward.

Piggin-string–A short rope used for tying down animals, as in calf roping.

Piggy Sow–Any mature female hog that shows advanced pregnancy, as indicated by the swelling of the underline.

Piglet–A young pig of either sex.

Pigment–Any of the natural coloring materials in the cells and tissues of plants and animals. In fruit and vegetables, the green pigment is chlorophyll; orange to red pigments are carotenoids; red to blue colors are anthocyanins; light-yellow pigments are flavoners and flavonols. In meat, the chief pigment producing the pink or red color is myoglobin.

Pigskin–The tanned hides of hogs.

Pigskin Fluff–A food product made from skins removed from fat backs of hog carcasses. The skin are fried in deep fat, or roasted until brittle, and then salted and dried. See Crackling.

Pilchard Oil–The product obtained by extraction of part of the oil from the whole Pacific pilchard fish or from cannery refuse from this species of fish; it is used as a source of vitamins in animal feeds.

Pilgrim–A worn-out, decrepit horse.

Pin–In poultry slaughter, to remove the pin feathers that are embedded in the skin.

Pin Bone–The region on each side of the tail head on the hindquarters of a bovine. These should be wide apart and well-defined for good conformation.

Pin Feather–The young feather embedded in, or just emerging from, the skin of a fowl.

Pin Nipples–Small, underdeveloped nipples on the teats of a pig. They are usually nonfunctional.

Pincers–The incisor teeth of a horse. Also called nippers.

Pine Disease–See Nutritional Anemia.

Pineal Gland–A reddish gland about one-third the size of the pituitary located in a brain cavity behind and just above the pituitary. A by-product of the meat-packing industry, extracts of which are used in medicine to regulate growth.

Pinebarren Deathcamas–*Zigadenus leimanthoides*, family Liliaceae; a bulbous herb that grows in bogs and on wet pine lands and is poisonous to livestock. Native to coastal areas of the southeastern United States. Also called crowpoison.

Piney Woods Cattle–Scrub cattle, or those nondescript in breed (southeastern United States).

Pining–(1) A disease of sheep in certain sections of England and Scotland that results from cobalt deficiency. (2) Designating any unthrifty or unhealthy animal.

Pinion–(1) The outermost part of a bird's wing including the carpus, metacarpus, and phalanges; the part of a wing, corresponding to the forearm, on which the primary flight feathers are borne. (2) To cut off the pinion on one (or both) wings to permanently prevent a bird from flying. (3) A gear that has the teeth formed on the inside of the hub.

Pinkeye–An infectious disease of the eye in cattle, sheep, and goats that is characterized by an inflammation of the cornea and conjunctiva. Also called keratitis, infectious keratitis, contagious ophthalmia in cattle.

Pinning–The sticking of a young lamb's tail to its anus. This will prevent normal bowel action and result in constipation and, if not loosened in time, death.

Pinning Knife–A dull, short-bladed knife that is used in removing pin feathers in the dressing of poultry.

Pinny–In poultry marketing, designating a bird carcass that has an excessive amount of pin feathers, and is therefore of low grade and marketable only at a reduced price.

Pinocytosis–The engulfing or absorption of fluids by cells.

Pins–See Pin Bone.

Pinto–Designating a horse that has a spotted or piebald coat color.

Pinworm–(1) A common name often applied to parasitic nematodes belonging to the family Oxyuridae, that infect people and animals; e.g., *Enterobius vermicularis*, the human pinworm; *Oxyuris equi*, the common horse pinworm; *Skrjabinema ovis*, the sheep and goat pinworm. (2) *Heterakis gallinae*; a cecal, grayish-white worm from ⅕ to ½ inch (0.85 to 1.3 centimeters) long which is found in the intestines of chickens, turkeys, and other fowls. It is involved in the transmission of blackhead disease.

Pinzgauer–A breed of horned beef cattle that originated in Australia. They are chestnut brown with white markings that extend from the underline up over the rump and back.

Pip–(1) A horny, dried condition of the tip of the tongue that develops in poultry in cases of infectious coryza when mouth breathing is long continued. (2) The raised crown or individual rootstock of a plant. (3) The specialized underground bud of lily-of-the-valley.

Piping–(1) A series of shrill sounds made by queen bees. (2) Formation by moving water of subsurface tunnels or pipelike cavities in the soil. See Vertisols.

Pipped Egg–An egg through which the chick has forced its beak in the first step of breaking out of the shell during incubation.

Pipping–The process of breaking the eggshell by a chick before hatching.

Piroplasmosis–Any of a number of tick-borne diseases of domestic animals especially affecting cattle, horses, sheep, and dogs, caused by protozoan parasites of the genus *Babesia*, which invade and destroy red blood cells, giving rise to anemia, elevated body temperature, and sometimes bloody urine. These parasites are transmitted from host to host by ticks of the family Ixodidae. Commonly called tick fever, Texas fever, splenetic fever, southern fever, babesiosis.

Pirouette–A high-school exercise for horses in which the forelegs are held more or less in place while the horse moves his hindquarters around them.

Piscicide–A substance used to kill fish.

Pisciculture–The production of fishing natural or artificial bodies of water under controlled conditions, such as stocking, feeding, and use of chemical fertilizers. See Aquaculture.

Piscis–(Latin) Fish.

Pit Silo–A shallow pit of variable size for storing silage, which is dug in well-drained soil and is frequently walled with lumber or concrete if of a permanent nature. It is very common in the western states of the United States. See Horizontal Silo, Trench Silo.

Pitch–(1) The resin that occurs in the wood of conifers, as the pitch from pines. (2) A heavy, dark, viscous or solid, fusible material obtained by distillation of the tar derived from coal, wood, rosin, and petroleum oils. It consists of many organic compounds, chiefly hydrocarbons, differing according to origin. (3) The jumping action of a horse in its attempt to unseat its rider.

Pithing–A method of animal slaughter in which the spinal cord is severed to cause death.

Pituitary Gland–A small endocrine gland located in the lower part of the brain. Among other functions, it secretes hormones into the bloodstream that influence the growth of the body, stimulate the thyroid gland, the sex organs, and the mammary gland to initiate the secretion of milk after the birth of the young.

Pivot–A leg action in which a horse pivots around his hindquarters while holding one leg in place and side stepping with the other hindfoot.

Pizzle–The penis, especially of a bull.

Placenta–(1) The membranous tissue that envelops the growing fetus in the uterus and establishes communication between the mother and the fetus by the umbilical cord. It is discharged from the uterus at the time of the birth of the young or shortly thereafter. Also called afterbirth, calf bed. (2) In plants, that portion of the ovary on which ovules are borne.

Placental Retention–The undue retention of the placenta and other fetal membranes after the young is born.

Placentome–One of many structures in the pregnant uterus of a ruminant that make up the placenta; composed of a caruncle of the uterus and a cotyledon of the fetal membranes.

Plain–In marketing, designating an animal of poor quality, i.e., one having rough, prominent shoulders, paunchy middle, bare ribs and loin, and small legs.

Plantar Cushion–A pad of fibro-fatty tissue in the foot to the rear of and under the navicular and coffin bones of animals. In horses, it is pronounced and overlies the frog. Also called digital cushion.

Plantation Gait–A natural horse gait, somewhat faster than a walk, which is easy on the horse and rider. Horses with this gait are sometimes called plantation horses, as they are popular on the plantations in the southern United States.

Plantation Horse–A horse that has a plantation gait, as the Tennessee Walking Horse.

Plasma–(1) The liquid portion of blood or lymph. (2) The liquid that may be squeezed from muscle.

Play Flight–Short orientation flight taken by young bees, usually by large numbers at one time and during the warm part of day.

Pleasure Horse–A classification for horses that includes those used for riding, driving, or racing. See Stock Horse, Work Horse.

Pleiotropy—A situation where one gene affects more than one trait.

Plerocercoid—A wormlike larval stage of certain tapeworms.

Plowable Pasture—Land ordinarily kept as pasture but that may be plowed and utilized for other crops.

Pluck—(1) In hog and cattle slaughtering, the organs that lie in the thoracic cavity consisting of the heart, lungs, gullet, and windpipe. (2) In poultry slaughter, to remove the feathers.

Plug—(1) The mass removed by a trier or other special penetrating implement in sampling or testing an agricultural product, as a plug from a cheese, a bale of cotton, or from a watermelon. (2) An old, worn-out horse. (3) A horse with a poor conformation. (4) To repair a leak, as a dam or earth fill. (5) A block of rooted grass that is planted for the purpose of establishing a covering of grass, such as in a law.

Plumage—The feathers of a fowl.

Plump—(1) In poultry slaughter, to create an effect of plumpness by dipping the picked carcass in scalding water. It shrinks the skin and draws up the legs, wings, and neck closer to the body thus imparting a false plumpness. (2) In meat judging, designating a well-muscled carcass or wholesale cut.

Plymouth Rock—A popular breed of dual-purpose chickens of the American class, whose origin is attributed to early crosses of the Dominique and the Black variety of Java. They are characterized by a single comb, nonfeathered shanks, yellow skin, and compact body. Cocks weigh about 9½ pounds, hens about 7½ pounds. the eggshell is brown. The varieties are: Barred, White, Buff, Silver-penciled, Partridge, Columbian, and Blue. The White and the Barred varieties are used extensively for commercial egg and meat production. In earlier times, the Barred Plymouth Rock was commonly called Dominecker by farmers because of its similarity to the earlier Dominique breed.

Pneumoencephalitis—See Newcastle Disease.

Pneumonia—Inflammation of the lungs in people and animals, characterized by fast and labored breathing, elevated body temperature, varying amounts of nasal discharge, and many times by a degree of consolidation of the lungs. Pneumonia is caused by various agents, such as bacteria, viruses, parasites, fungi, hot or cold air, liquids that might be taken into the lungs, and by dust.

Pod—(1) Technically, a dry, many-seeded fruit that splits open, such as a pea pod or bean pod; a legume. (2) To form pods. (3) A flock of animals, birds, etc.

Point Men—Cowboys who ride toward the head of the herd to keep it going in the desired direction (western United States).

Point of Lay—Age at which pullets begin to lay, usually between twenty and twenty-two weeks.

Points—A condition of coat color in animals in which white or a lighter color appears about the muzzle, eyes, feet, and tail.

Poison—Any substance ingested, inhaled, or developed within the body that causes or may cause damage or disturbance of function of plants, animals, or people. See Toxin.

Poison Bait—A poison mixed with wheat bran, molasses, or other attractant used to control cutworms, grasshoppers, and other insects.

Poison Hemlock—*Conium maculatum*, family Umbelliferae; a biennial, rank-growing herb, which is an escape and a weed in various sections of the United States. It is dangerously poisonous to people and animals, being most poisonous to stock in the spring, when the herbage is fresh. Symptoms of poisoning of cattle are loss of appetite, salivation, bloating pain, feeble but rapid pulse, and loss of muscular power. Native to Europe and Asia. Also called deadly hemlock, poison parsley, winter fern.

Poison Ivy—*Toxicodendron radicans*, family Anacardiaceae; a small, erect shrub or climbing vine common throughout the United States in waste places, pastures, woodlands, and along old fencerows. Apparently more dangerous to humans than to animals, contact with the plant among susceptible individuals causes severe blistering and inflammation, especially on the hands, arms, and face. Also called climbing ivy, three-leaved ivy, climath, poisonoak, poison creeper.

Poison Milkweed—*Asclepias galioides*, family Asclepiadaceae; a poisonous, perennial herb found on dry sites on the western ranges in the United States. It is occasionally eaten by animals with fatal results. Sheep appear to be most susceptible to it. The symptoms of poisoning are loss of muscular control, staggering, falling, spasms, bloating, fever, weak but rapid pulse, and respiratory paralysis. The affected animal usually dies. Native to North America.

Poisonous—Containing poison, as a poisonous plant.

Poke—(1) A yoke with an attached sharp spur pointing forward that is placed on the neck of an animal to prevent it from crawling through or jumping a fence. (2) Pokeberry. (3) A paper bag.

Poland China—A large breed of swine that originated in Ohio, United States, early in the nineteenth century, as the result of crossing the native swine with several breeds including Big China, Berkshire, and Irish Crozier swine. The coat is black with white tips on the tail, white feet, and a white dash or spot on the forehead.

Poliomyelitis—Inflammation of the gray matter of the spinal cord; an acute infectious viral disease attended with fever, motor paralysis, and atrophy of groups of muscles.

Poll—(1) The region at the crest or apex of the skull in horses and cattle. (2) To cut back the crown of a tree. (3) To remove the horns of cattle.

Poll Evil—A fistula of the poll between the ears that may follow a severe injury to that part, a common affliction of horses and mules.

Pollard—(1) A tree whose crown has been cut back to invite the production of shoots from the top. Sometimes a tree is so pruned to induce a globelike mass of foliage. (2) A hornless ox, sheep, or goat.

Polled—Designating animals, especially cattle, that normally do not develop horns.

Polled Hereford—A strain of the Hereford breed of beef cattle that was developed in the United States; characterized by the absence of horns.

Polled Shorthorn—A strain of beef cattle within the Shorthorn breed, characterized by the absence of horns. See Shorthorn.

Pollen Basket–An area on a bee's hind leg where pollen is packed and carried with help from a central spine and surrounding hairs.

Pollen Cake–Cake of sugar, water, and pollen or pollen substitute, for bee feed.

Pollen Insert–A device placed in the hive entrance to apply live pollen to outgoing bees for cross pollination, as in apples.

Pollen Substitute–Mixture of water, sugar, and other material, such as soy flour, brewer's yeast, etc., used for bee feed.

Pollen Trap–A grid placed at the entrance to a beehive, which removes pollen from the bees' legs as they enter; the pollen falls into a tray below the grid.

Polo Pony–Any small active horse with the ability to be trained to perform in polo.

Polyestrous–Refers to an animal that has several estrous cycles in a breeding season.

Polygastric–Having many stomach compartments as in the ruminant animal (such as cattle).

Polymorphic–Having two or more forms.

Polymorphonuclear Leucocytes–White blood cells in which the nuclei are constricted into irregular shapes.

Polymyxin–Any of various antibiotics that are derived from cultures of the organism *Bacillus polymyxa*.

Polyneuritis–An inflammation of many nerves at once.

Polyp–A smooth, stalked, or projecting growth from a mucous membrane.

Polyphagia–(1) Voracious appetite. (2) An unnatural craving for many kinds of food. (3) Omnivorousness. See Bulimia, Cynorexia, Sitomania.

Polyphagous Parasite–A parasite capable of parasitizing a considerable number of host species.

Polyphosphates–Salts of polyphosphoric acids such as ammonium polyphosphates and calcium polyphosphates.

Polyphosphoric Acid–Any of a series of phosphoric acids whose molecular structure contains more than one atom of phosphorus such as pryophosphoric acid, tripolyphosphoric acid, and tetrapolyphosphoric acid. See Phosphoric Acids.

Polyploid–An organism with more than two sets of the basic or haploid number of chromosomes; e.g., triploid, tetraploid, pentaploid, hexaploid, heptaploid, octaploid.

Polypnea–Rapid or panting respiration.

Polysaccharide–A large molecular weight carbohydrate made up of many sugar units; e.g., starches, cellulose, and glycogen.

Polyspermy–The entrance of many sperm cells into the ovum at fertilization.

Polyuria–Excessive secretion of urine.

Polyvalent–Designating a stock vaccine made up of many strains of the same organism or different organisms.

Polyvalent Colon Bacteria–A bacterin made from several types of bacteria that inhabit the colon or intestinal tract of animals; sometimes useful in the prevention of calf scours.

Polyvoltine–Designating an animal yielding several broods in a season. Especially applied to certain silkworms producing several broods of cocoons in a year.

Pommel–(1) The knob, ball, or protuberant part serving as a means of grip on anything, as the high, forward part of a saddle. (2) A block of hard wood grooved like a crimping board and employed by curriers to render leather supple and impart a grain to it.

Pony–A horse, under 14.2 hands (about 57 inches) at the withers.

Poor Feeder–Any animal that does not fatten or grow in a satisfactory manner.

Poorly Bled–In poultry slaughter, designating a bird that shows red pin marks on its breast or thighs, or one whose skin is reddened from blood clots.

Poorly Fleshed–(1) In poultry slaughter, designating a bird that has a narrow breast and whose thighs and back are dark-colored from the absence of fat. (2) Designating any meat animal in thin condition.

Pop-eyed–Refers to a horse whose eyes are generally more prominent or bulge out a little more than normal; also to a horse that is "spooky" or attempts to see everything that goes on and is often frightened.

Pop-hole–A hole small enough to allow piglets to creep through to reach their mother.

Popped Knee–In livestock, a general term describing inflammatory conditions affecting the knees, so named because of the sudden swelling that accompanies it.

Porcine–Refers to swine.

Porcine Stress Syndrome (PSS)–A condition in swine characterized by extreme muscling, nervousness, tail tremors, skin blotching, and sudden death.

Pore–(1) In plant and animal membranous tissues, minute openings for absorption and transpiration of matter. (2) In wood anatomy, the cross section of a vessel or a vascular trachea. (3) In soil, the portion of a given volume of soil that is unfilled with solid matter; air spaces, irregular in shape and size.

Pork–The meat of swine.

Pork Roast–A retail cut of pork for roasting made by removing the ribs and backbones from two blade end cuts and placing the inside cut surfaces together.

Pork Sausage–Any ground and seasoned pork product.

Pork Tenderloin–A retail cut of pork from the last rib to the hip joint, which is the major muscle located below the transverse spinal process in the lumbar region of the swine carcass.

Porker–Any young hog.

Porterhouse Steak–A retail cut of beef taken from the posterior end of the beef short loin containing a large section or area of the tenderloin muscle.

Pose—In the show ring, a special stance or position that a horse may be trained to assume. A posed horse stands with his front feet extended and his rear feet back.

Positive Ion—A cation; an ion that carries a positive charge of electricity.

Positron—A subatomic particle equal in mass and weight to the electron and having an equal but opposite charge. Positrons are emitted by some artificially radioactive isotopes.

Post—(1) A short timber used in an upright position for supporting structures. (2) Any timber that supports fencing. It may be round, split, or sawn. (Posts may be of other materials, such as iron and concrete.) (3) The proper position and balance a rider assumes in the saddle in riding a horse while in the trot gait, rising with each second beat of the one-two rhythm of the trot.

Post Cenam—After feeding (in scientific writing often abbreviated p.c.).

Post-legged—An animal having extremely straight hind legs.

Posterior—Hind or rear.

Posterior Paralysis in Pig—A disease characterized by lameness and sometimes complete paralysis of the rear part of the body. Causes are a lack of lime (calcium) in the diet, or an unbalanced ratio of calcium to phosphorus, especially if there is a deficiency in vitamin D. Other mineral and vitamin deficiencies may be involved.

Postmortem—An examination of an animal carcass or human body after death.

Postnatal—Subsequent to birth, relating to an infant immediately after birth.

Postpartem—A period immediately following parturition (giving birth).

Pot-bellied—(1) Designating any animal that has developed an abnormally large abdomen, usually because of improper feeding or nutrition. (2) Designating a type of coal stove used for heating purposes in pioneer stores and homes, so named because of its large, round, lower half.

Potassium Iodide—KI; a white, water-soluble chemical that may be given to farm animals as an iodine supplement.

Potassium Permanganate—$KMNO_4$; a powerful oxidizing compound used as a disinfectant, deodorant, and a reagent in analytical work, especially in the determination of available nitrogen in organic material. Sometimes also used as a fungicide in greenhouses.

Potato Distillers' Dried Residue—A livestock feedstuff that consists of the dried product obtained after the manufacture of alcohol and distilled liquors from potatoes or from mixtures in which potatoes predominate.

Potency—(1) The power of a medicine to produce the desired effects. (2) The ability of an embryo to develop into a viable destiny. (3) The ability of the male of any plant or animal species to fertilize the female germ cells. (4) The degree of toxicity of a chemical.

Poult—The young turkey before its sex can be distinguished. Sometimes applied to the young of other fowls.

Poultice—A hot, wet dressing applied to an injury or swelling for its softening and soothing properties.

Poultry—Any or all domesticated fowls that are raised primarily for their meat, eggs, or feathers, as chickens, turkeys, ducks, and geese.

Poultry Band—A strip of plastic or aluminum marked with numbers, etc., attached to the leg or wing of a bird for identification.

Poultry Bug—*Haematosiphon inodorus*, family Cimicidae; a blood-sucking insect pest of poultry.

Poultry Fattener—(1) Any of several feeds containing ingredients designed especially for fattening poultry for the market. (2) A person engaged in the fattening of poultry for the market. (3) Poultry feed. Any feed supplied to poultry, commonly, specially formulated mashes designed to be fed either alone or in combination with cereal grains.

Poultry House—Any building equipped and used for housing and handling poultry.

Poultry Husbandry—The science and art of the production and distribution of poultry and poultry products, including breeding, incubation, brooding, rearing, housing, feeding, marketing, and poultry farm management.

Poultry Inhalant—A remedy designed to relieve coughs due to colds, and bronchial irritations in poultry. It is usually sprayed in the house while birds are on roosts.

Poultry Knife—A specially designed knife with a very narrow curved blade, used to kill poultry and to sever arteries for bleeding. Also called chicken sticker.

Poultry Lice—Insects of the order Mallophaga that infest fowls.

Poultry Manure—The mixture of feces and urine voided by birds. When freshly produced it contains about 80 percent moisture, 1 percent nitrogen, 0.8 percent phosphoric acid, and 0.5 percent potash. Poultry manure decomposes with a rapid and large loss of moisture and ammonia, especially during warm weather.

Pound—(1) A unit of weight; 16 ounces avoirdupois, 12 ounces troy. The standard British unit of weight equals 7,000 grains, ½, 240 long ton, and 453.59 grams, the weight of 27.692 cubic inches of water at 4°C. (2) An enclosure in which stray animals are legally confined. (3) An enclosure in which groups of animals, as flocks of sheep, may be gathered for shelter, etc. (4) An enclosure used to trap wild animals.

Pound of Gain—In animal feeding, the net gain of weight in pounds derived from a particular number of pounds of feed fed.

Pounding—In the movement of a horse, a heavy contact of the foot with the ground that usually accompanies a high stride.

Pounds of Butterfat—The production of butterfat by a cow for a certain period, such as a day, week, or lactation period, usually quoted in terms of total pounds of butterfat.

Pounds of Milk—The output of milk by a cow, by weight, for a single milking, a day, or a lactation period.

Pounds of Milk Per Acre of Pasture–The feed value of an acre of pasture expressed in terms of pounds of dry grain feed.

Pounds of Retail Cuts Per Day of Age–A measure of cutability and growth combined, it is calculated as follows: cutability times carcass weight divided by age in days.

Powdered Milk–Dry whole milk. See Nonfat Dry Milk.

Powdered Skimmed Milk–See Nonfat Dry Milk.

Prairie–(1) The extensive, nearly treeless and dominantly grass-covered plains of the midwestern United States that lie east of the Rocky Mountains. In a more restricted sense, the tall grasslands with blackish soils; in a more general sense the semiarid shortgrass plains as well. Also called savannah, steppe. (2) In the generally forested eastern part of the United States, any naturally treeless area that is generally dry or naturally well-drained. (3) Wet, treeless, marshy areas. (4) Prehistoric, treeless tracts that resulted from fires.

Prairie Hay–Any hay made from the wild grasses of the prairie.

Prairie Sandreed–*Calamovilfa longifolia*, family Gramineae; a perennial, drought-enduring grass, growing from 2 to 6 feet tall, which is found from Michigan to Colorado, United States, and Alberta, Canada. Important for winter pasture and for hay though grazed but lightly in the summer. It is an important, sand-binding grass on dunes and sand hills. Native to North America. Also called prairie sandgrass.

Prance–To walk with a swagger or nervous action, as a horse.

Prawns–Any of the several genii of crustaceans that are caught or raised for food. Some genii can be raised in either fresh- or salt-water. Grown commercially in the United States.

Precipitin–An antibody developed in the blood serum of an animal that has been injected previously with a foreign protein. Precipitins, present in the blood serum, have the property of bringing about a visible precipitate when brought into contact with a solution of the specific protein that induces their formation.

Preclipping–An agronomic practice, sometimes employed in growing clovers for seed production, in which tops of young plants are clipped before flowering to retard excessive growth and to induce flowering at a time when pollinating insects are numerous.

Preconditioning–Term refers to cattle that have been weaned, castrated, dehorned, vaccinated for several diseases, wormed, treated for grubs, and taught to eat from bunk-type feeders before being shipped to the feedlot.

Precooling–(1) Preliminary cooling of milk immediately after a milking to prevent spoilage. (2) Cooling of fruits immediately after harvesting during periods of hot weather to retard ripening and deterioration. (3) The cooling of meats after slaughter and before cutting.

Predacide–A substance that is used to kill predators.

Predatism–Intermittent parasitism, such as the attacks of mosquitoes and bedbugs upon humans.

Predicted Difference (PD)–The estimated difference of an animal from that of its parents or offspring.

Predisposition–(1) Stress or anything that renders an animal liable to an attack of disease without actually producing it. (2) The effect of one or more environmental factors that makes a plant vulnerable to infection by a pathogen.

Preen–Of a bird, to arrange or dress its feathers with its beak.

Preen Gland–An oil-secreting gland at the root of the tail in most birds whose secretion dresses their feathers. Also called rump gland.

Preference–Selection of palatable plants over others by grazing animals.

Pregnancy–The condition of a female animal having a living fetus in the uterus that occurs after the ovum has been fertilized by the male sperm cell. See Gestation Period.

Pregnancy Disease in Ewes–A usually fatal disease of pregnant ewes near term, associated with a disturbance of carbohydrate metabolism. Commonly occurs in ewes carrying twins or triplets. It is characterized by nervousness, inability to rise, pushing the head against an object, gnashing of the teeth. Also called lambing paralysis, old ewe disease, ketosis, acidosis of pregnant ewes, twin lamb disease.

Pregnancy Mare Serum–A gonadotrophic hormone, the secretion of the uterus, obtained from the blood of pregnant mares. It is used in treating certain types of shy-breeding animals. Its main value lies in its content of follicle-stimulating hormone.

Pregnancy Test–Any of several tests that determine whether or not a female is pregnant. Pregnancy can be determined by these tests in humans, mares, and Rhesus monkeys because of two factors: pregnancy gonadotropins produced by the pregnant female from the 40th to the 150th day of pregnancy, and after that period, by the high amount of estrogens found in the urine. See Friedman Test.

Prehension–The taking of food into the mouth.

Premix–In animal feeding, a previously prepared mixture of small amount, as one containing a vitamin or medicine, which is added to the main feed mixture.

Prenatal–Occurring or existing before birth.

Prenuptial Flight–The flight made by the virgin queen bee supposedly to acquaint herself with landmarks that enable her to return to her own hive after her nuptial flight. See Nuptial Flight.

Preparturient–Occurring before birth.

Prepotency–The ability to transmit characteristics to offspring to an unusual degree.

Prepotent–Designating an animal that transmits its characters to its progeny to a marked or highly uniform degree.

Prepuce–The sheath or foreskin covering the penis or clitoris.

Presentation–An animal giving birth.

Preserve–(1) In wildlife management, a game-shooting area on which game species are propagated or released. (2) A tract of land set aside for preservation of natural conditions, and protected against exploitation or any commercial use. (3) To prepare foods by cooking with some preservative so as to reduce fermentation or decomposition.

Press Cake–Pomace, or the residue left after pressing the juices from fruits, olives, or tomatoes. Dried pomace has some value as animal feed.

Press-cake Meal–A by-product left after the separation of oils, by grinding and distillation, from the pits of peaches, apricots, and cherries. It has some value as a stock feed.

Pressure Necrosis–A sore back on a horse that develops from saddle pressure; an area of necrosis (dead tissue) resulting from excessive pressure.

Prick–To pierce or cut a muscle in the tail of a horse so that the tail will be carried higher; an unlawful act in many states.

Primal Cuts–The most valuable cuts on a carcass. Usually includes leg, loin, and rib.

Primary Feathers–The outermost group of major wingfeathers (usually ten) located on the third joint of a bird's wing, hidden when the wing is folded. Also called flight feathers, flights.

Primary S.P.F. Pig–(Specific Pathogen Free) A pig removed from its mother just before the normal birth date by a surgical process (Cesarean section) and raised in laboratory isolation.

Prime–In the grading of various agricultural products, meat, fruits, beans, cottonseed oil, etc., of first class or high quality; choice.

Prime Cattle–Slaughter cattle that have a high finish, and yield prime meat cuts.

Primordium–A member or organ of a plant in its earliest condition, i.e., the first roots to form on a cutting. Plural, primordia.

Private Herd Number–A number assigned to registered animals by their individual owners. It is required by some breed registry associations in the identification of registered animals.

Proboscis–An elongated nose, such as the snout of a hog, or of some species of insects.

Produce–(1) Commodities produced from or grown in the soil. (2) In animal breeding, a female's offspring.

Produce of Dam–In the show ring: (a) for swine, four breeding animals, the offspring of one sow; (b) for horses, two animals, the offspring of one mare; (c) for dairy cattle, two animals, the offspring of one cow.

Proestrus–The phase of the estrous cycle just before estrus; characterized by the development of the ovarian follicle.

Professional Inspector–A meat inspector who is a graduate of an accredited veterinary college and has passed the required civil service examinations for federal, state, or municipal meat inspection (United States).

Profile–(1) The general outline of an animal's body. (2) See Soil Profile.

Progenitor–An individual animal or plant that is recognized as the source of a certain type or character in its offspring.

Progeny–The offspring of animals or plants.

Progeny Records–The average, comparative performance of the progeny of sires and dams.

Progeny Testing–Determining the breeding value of an animal by studying its progeny.

Progesterone–A hormone produced by the corpus luteum of the ovary that functions in preparing the uterus for pregnancy and maintaining it if it occurs.

Proglottids–The segments or parts of a tapeworm other than the head (scolex) and neck region.

Prognosis–Forecast as to the probable result of an attack of disease, the likelihood of recovery.

Progressive–Of a disease, developing through successive stages, usually in a certain direction, whether improving or deteriorating.

Progressive Robbing–The depredation in which bees from one colony enter another hive to steal honey without antagonizing or destroying the robbed colony.

Prolactin–A hormone of the anterior pituitary gland that functions in stimulating the secretion of milk.

Prolapse–A displacement of a body part from its usual position in relation to other parts, most frequently occurring in tubelike structures such as the anus, where the inner portion slips out and extends beyond the outer portion.

Prolapse of the Uterus–Partial or complete turning inside out of the uterus, usually following parturition.

Prolapse Retainer–A device used to support the prolapsed uterus of a ewe. The uterus is held in place until healing occurs.

Prolific–Having the ability to produce many offspring.

Proper Stocking–Stocking of a range area on the basis of its true grazing capacity in a year of adequate rainfall.

Proper Use–The degree of grazing that an individual plant species, or the total palatable cover of a range area, may endure without damage to the plants or the soil.

Proper Use Factor–As applied to individual range species, the estimated maximum percentage of the total vegetative production of the year within easy reach of the livestock to which a given range species may be grazed without damaging it or associated important palatable plants or the soil; the degree to which each species may be grazed when the range as a whole is properly grazed. The proper use factor for a range type is the weighted average of the proper use factors of the individual plants in the type. See Palatability.

Properties–Characteristics by which a substance may be identified. Physical properties describe its state of matter, color, odor, and density; chemical properties describe its behavior in reaction with other materials; biological properties refer to any life-related characteristics such as biodegradation.

Prophase–The first phase of cell division wherein many of the preparatory steps take place, such as shortening and thickening of the chromosomes, division of the centromeres, disappearance of the nuclear membrane, and formation of the spindle.

Prophylactic–Preventive or protective treatment against disease.

Prophylaxis–Prevention of disease by various measures.

Propolis–A glue or resin collected from trees or other plants by bees; used to close holes and cover surfaces in the hive. Also called bee glue.

Prostaglandins–A large group of chemically related fatty acids that have various physiological effects in an animal's body. Artificial prostaglandins are used to synchronize estrus in cattle.

Prostate–One of the accessory glands of the male reproductive system that encircles the neck of the bladder where it joins the urethra.

Prostration–A symptom of an animal's ailment; pronounced weakness and loss of strength, resulting in the animal lying in prone position unable to rise.

Protective Foods–Foods that furnish additional supplies of certain minerals, vitamins, and proteins for the normal growth of the body and the maintenance of good health.

Protein–Any of a large number of complex, organic compounds of amino acids that has a high molecular weight and is an essential part of all living organisms. Proteins consist largely of carbon, hydrogen, nitrogen, and oxygen; many contain sulfur, and some also contain iron and phosphorus. They constitute a large portion of the protoplasm, and are obtained from foods such as lean meats, and from vegetables such as beans.

Protein Equivalent–An expression used in computing the protein of feedstuffs. In Great Britain, nonprotein, nitrogenous compounds are considered to have one-half the value of true protein. In the United States, protein equivalent is considered to be the sum of the true protein plus one-half of the amount of nonprotein nitrogen.

Protein Palm Nut Oil Meal–A livestock feedstuff that is the ground residue left after the extraction of part of the oil from the fruit of one or more species of palm.

Protein Solvent Extracted Cottonseed Meal–A livestock feedstuff that is the product resulting from grinding solvent extracted cottonseed flakes.

Protein Solvent Extracted Soybean Flakes–A livestock feedstuff that is the product obtained by expelling part of the oil from soybeans by the use of solvents.

Proteolysis–The process by which casein or some insoluble casein derivative is broken down to water-soluble compounds through the acting of organisms.

Proteolytic Bacteria–(1) Bacteria that produce a proteolytic action on the proteins in cheese and cause a strong odor and taste to develop. (2) Bacteria that act on proteins, breaking them down to simpler compounds.

Protoplasm–The gelatinous, colloidal material of plants and animals in which all life activities occur.

Protoplast–A unit of protoplasm in one cell.

Protozoa–A group of one-celled organisms that generally do not contain chlorophyll, including amoebae, paramecia, flagellates, and certain spore-forming organisms; sometimes classified as one-celled animals.

Proven Sire–(Dairy Herd Improvement Association, DHIA). A bull with at least ten daughters that have completed lactation records and are out of dams with completed lactation records.

Provender–All dry feed or fodder for domestic animals.

Provenetriculus–The glandular or true stomach of birds, which is a spindle-shaped organ between the esophagus and gizzard.

Proventriculitis–Inflammation of the glandular (or true) stomach that often occurs in growing chicks reared in confinement and occasionally in adult fowls.

Provitamin–A precursor of a vitamin; a substance from which an animal organism can form a vitamin. Carotene is provitamin A and ergosterol is provitamin D.

Proximal–Opposite of distal; near the point of attachment of reference.

Proximate Analysis–A system of analysis used to determine the total composition of nutrients in feed.

Pruritis–Intense itching.

Prussic Acid Poisoning–Poisoning of livestock from prussic acid (hydrocyanic acid) (HCN) which may result from ingestion of sorghums, such as Johnsongrass, under certain conditions such as drought. Hydrocyanic acid may also be produced from the leaves of several species of wild cherries (***Prunus*** spp.) which are occasionally browsed by livestock.

PSA–Packers and Stockyards Administration.

PSE–See Pale, Soft, and Exudative.

Pseudo–A Greek prefix meaning false or spurious. In most scientific terms it denotes a deceptive resemblance to the substance to whose name it is prefixed, e.g., pseudocarp, false fruit.

Pseudo-albino–A coat color of some animals that is a very light cream to white. In horses there is a silver mane and tail and often glass or blue eyes. It is due to the absence of pigment in hair and skin.

Pseudo-tubercular Enteritis–See Johne's Disease.

Pseudohermaphrodite–Person or animal having internal genital organs of one sex, while its external genital organs and secondary sex characters resemble in whole or in part those of the opposite sex.

Pseudorabies–A highly contagious, herpesviral, lethal disease of swine, cattle, dogs, cats, rats, and most other mammals.

Psittacosis–An acute or chronic viral disease of birds (domestic and wild) transmissible to humans and characterized by systemic reaction and respiratory involvement. Also called parrot fever.

Psomophagia (Psomophagy)–The rapid eating and swallowing (bolting) of food without thorough chewing. Often induced by nervousness and anxiety and usually resulting in obesity.

Psoroptic Scab–See Scab Mite.

PSS–See Porcine Stress Syndrome.

Ptosis–The prolapse of a body organ, specifically the drooping of the upper eyelid.

Ptyalin–An enzyme in saliva that digests starch.

Puberty–The time when sexual maturity is reached. In the female, ova on the ovaries begin to develop. In the male, sperm production is initiated in the testicles.

Pubescent–Strictly, this means covered with soft, short, fine hairs; as commonly used, however, the term means hairy, bearing hairs, in a generalized sense, without reference to the type of hair.

Pubic Bone–See Pubis.

Pubis–One of the paired bones constituting the pelvis that make up the floor of the pelvis.

Pulled Wool–Wool pulled from skins of slaughtered sheep. The wool is pulled from the skins after treatment of the fleshy side of skins with a depilatory. Pulled wool should not be confused with dead wool.

Pullet–An immature female chicken; in poultry shows, a young hen under one year of age.

Pulling Leather–Designating a rider who hangs on to parts of the saddle with the hands to keep from being thrown from a bucking horse.

Pullorum Disease–A widespread, infectious, bacterial disease of poultry caused by the microorganism *Salmonella pullorum*. In baby chicks and poults the disease assumes an acute, septicemic form with a high degree of fatality, most losses occurring in the first two or three weeks of life. Adult fowls are less seriously affected but may harbor the organism and transmit the disease through infected eggs. The disease is largely controlled by the blood-testing of breeding stock to eliminate carriers of the infection, and by incubator and brooder house sanitation. Pullorum disease was formerly inappropriately called bacillary white diarrhea.

Pulmonary Emphysema–An anatomic change in the lungs characterized by a breakdown of the walls of the alveoli, which can become enlarged, lose their resilience, and disintegrate.

Pulpy Kidney Disease–See Enterotoxemia.

Pulse–(1) The edible seed of legumes, such as peas and beans. (2) The expansion and contraction of an artery associated with each heartbeat, which may be felt with the fingers.

Punch–(1) To herd cattle; to take care of cattle. (2) To forcibly drive sheep up hills when they are tired (New Zealand). (3) See Leather Punch.

Punk–(1) A small, scrubby horse. (2) Partly decayed wood.

Punkies–A bloodsucking midge of the genus *Culicoides*, family Tendipedidae, which occurs in some wooded or marsh areas. Also called no-seeum, sand fly.

Pupa–(Plural, pupae) The stage between the larva and the adult in insects with complete metamorphosis, a nonfeeding and usually inactive stage.

Pure Line–A strain of organism that is comparatively pure genetically (homozygous) because of continued inbreeding, etc.

Pure Strain–An animal that is similar to a purebred but the breeding program usually involves a greater degree of inbreeding.

Purebred–Designating an animal belonging to one of the recognized breeds of livestock. Such animals are registered or eligible for registry in the official herdbook of the breed. Purebred, registered, and pedigree stock are often used interchangeably, and Thoroughbred is often improperly used for purebred.

Purebred Breeder–A person who raises purebred animals, that may or may not be pedigreed.

Purebreeding–The practice of breeding animals from within the same breed or line; the production of purebreds.

Purgative–A medicinal agent that actively empties the bowels. Aloes, arecoline, and calomel are examples of laxatives, drastic purgatives, and cholagogue purgatives.

Purpura Hemorrhagica–An acute or subacute toxemic infection of horses marked by a rapid onset, small hemorrhages of the skin and body membranes, and swellings under the skin in many parts of the body.

Purulent–Referring to pus, as a purulent discharge.

Pus–The material produced at the site of an infection consisting of tissue fluids, white blood cells, dead tissue cells, and microorganisms.

Push-in Cage–A cage used to introduce a new queen bee into a colony. A 4-inch square of ordinary screen wire is bent along each edge and the corners clipped to form four sides, making a wire cage that is placed over the queen in an area of emerging brood and pushed into the face of the comb.

Pustule–Small elevation of the skin filled with pus or lymph.

Put Off–To lack the desire to copulate, as a bull.

Putrefaction–Decomposition of animal or vegetable matter, produced by microorganisms in the absence of oxygen.

Putrescible–Organic matter capable of putrefaction.

Putrid–Decomposed, rotten; said of organic materials.

Pyelonephritis–Inflammation of the kidney and the pelvis.

Pyemia–A generalized infection in the bloodstream caused by pyogenic organisms resulting in the occurrence of numerous abscesses in various parts of the body.

Pygostyle–The flesh-covered, triangular bone plate formed by the union of vertebrae at the posterior extremity of a bird's body that supports the main tail feathers. Not found in rumpless fowls.

Pyogenic–Pus-producing.

Pyometra–Condition in which pus is present in a sealed uterus.

Pyridoxine–Vitamin B6, which appears to be necessary for the normal growth of chickens and pigs and for the prevention of a type of nerve disorder. Its distribution in ordinary feeds is fair.

Pyriform–Pear-shaped.

Quadrivalent–In genetics, a group of four associated homologous chromosomes.

Quadruped–An animal that walks on four legs.

Quail–Any of the various species of small upland, gallinaceous game birds belonging to the genus *Coturnix* and allied genera of the family odontophoridae, found in many parts of the world and often erroneously called partridge. Adapted to agricultural lands, they are seldom regarded as a nuisance and are frequently propagated for game purposes. The North American bobwhite quail, *Colinus virginianus*, is common in the central and eastern United States. The California quail, valley quail, and mountain quail are common in the western United States.

Qualitative Traits–Traits having a sharp distinction between phenotypes, and which are usually controlled by only a few genes; e.g., various coat colors and the horned trait in domestic animals.

Quality Grade–Grade given a beef carcass; closely related to marbling, age of the animal, and color of the lean. The most common quality grades are prime, choice, select, and standard.

Quantitative Traits–Traits that do not have a sharp distinction between phenotypes, and usually require some kind of measuring tools to make the distinctions. These traits are normally controlled by many pairs of genes; e.g., growth rate, milk production, and carcass quality. See Genotype, Phenotype.

Quarantine–(1) A regulation under police power for the exclusion or isolation of animal and plant pests or diseases and insects: (a) the isolation of an animal sick with a contagious disease; (b) a place where the sick are detained away form other animals until the danger of spread of a contagious disease has disappeared. In its wider application, the quarantine may be enforced against an individual animal, against all the animals, or all the animals of the same species, in a township, county, or state, and against those in a foreign country. (2) Prohibition to prevent the introduction or spread of any dangerous insects or plant diseases.

Quarter–(1) In slaughtering for meat, one half of the side of beef, as the forequarter or hindquarter. (2) Pertaining to a horseshoe, the branch between the last nail hole and the heel. (3) A unit of weight: (a) one-quarter cwt. (25 pounds) (avoirdupos); (b) eight bushels, formerly one-quarter ton (especially of grain). (4) A quarter section of land, or 160 acres. (5) A section of the bovine udder.

Quarter Boot–A leather piece fitted around a horse's forefoot to prevent self-injury from striking with the hind foot.

Quarter Clip–A clip on the shoe for the hind hoof of a horse. The clips are placed on the quarters of the shoe on the outside or inside to prevent the shoe from shifting laterally on the foot.

Quarter Crack–A vertical split in the wall of the hoof of a horse that results from improper hoof growth or shoeing. Also called sand crack.

Quarters–(1) The parts of the body of a horse or other quadruped above the legs, as the breast and hips. (2) Place used to house workers on plantations or ranches. (3) Farm or ranch housing for domestic animals.

Queen Bee–A fully developed, mated female bee, larger and longer than a worker bee, whose function is to lay eggs. Formerly, it was believed that the queen bee was a male. See Drone, Worker.

Queen Breeder–One who breeds queen bees commercially or experimentally. Queen cells are removed from the cell building colony and introduced into the nuclei, which are placed in queen mating yards.

Queen Cell–A special, elongated, wax cell resembling a peanut shell in which the queen bee is reared. It is usually at the bottom of the comb.

Queen Cup–The beginnings of a queen cell in which the queen may lay a fertile egg to start the rearing of another queen.

Queen Excluder–Device usually made of wood and wire, with an opening of 0.163 inch, to permit worker bees to pass through but excludes queens and drones. Used to restrict the queen to certain parts of the hive.

Queen Substance–Pheromone material secreted from glands in the queen bee and transmitted throughout the colony by workers. It makes the workers aware of the presence of a queen. See Pheromone.

Queen-cage Candy–Candy made of powdered sugar, sugar syrup or honey, kneaded until it forms a stiff dough, used as a food in queen bee shipping cages.

Queenlessness–A condition of a beehive that has lost its queen. This emergency results in swarming.

Queenright Colony–A honeybee colony with a queen.

Quick Coupler–(1) A coupler used with a portable irrigation pipe that, through the use of split gaskets expanding under increased water pressure, effects a water seal between itself and the coupled section of pipe. (2) A device used to connect hydraulic hoses that allows the hoses to be connected quickly.

Quick Freezing–Freezing of products, fruits, vegetables, poultry, meat, dairy products, etc., for preservation under conditions in which the temperature of the product is lowered from 28°F to −15°F within 30 minutes.

Quick Heat–Lack of the usual nervousness, such as riding and other symptoms, that cows display when in heat.

Quill–The hollow, horny part of a feather.

Quittor–A discharging sore on the coronet of a horse's hoof; necrosis of the lateral cartilage of the third phalanx.

Rabbit–Any of certain small mammals of the family Leporidae. Hare and rabbit are often used interchangeably. However the biologist classifies the rabbit as having shorter and smaller legs and ears, and giving birth to naked and helpless young with eyes closed, while the newborn hare has fur and is quite capable of caring for itself.

Rabbit Coop–A wire cage used to contain rabbits for feeding, breeding, and show purposes. Also known as a hutch. See Rabbit Hutch.

Rabbit Feed–A prepared feed which usually consists of ground corn, oats, bran, oil meal, beet pulp, molasses, minerals, and vitamins, usually pressed into pellets.

Rabbit Hutch–An off-the-floor cage or box for raising rabbits.

Rabbit Pox–An acute eruptive disease of laboratory rabbits, caused by a virus related to vacinia virus.

Rabbit Test–(1) A pregnancy test in which virgin female rabbits are used. (2) A test for sweet clover poisoning in which rabbits are fed sweet clover hay and if no injurious effects appear the hay is considered safe for cattle.

Rabbit-proof Fence–A fence of special design which is placed around cultivated fields in New South Wales, Victoria, and Western Australia to keep out rabbits which have overrun the country and have become a very serious pest.

Rabies–An infectious disease caused by a filterable virus which is communicable by means of a bite in which saliva containing the virus enters the wound. It occurs most frequently in dogs, but many other animals and people are quite susceptible. Also called hydrophobia or canine madness.

Race–(1) A group of individual plants which have certain common characteristics because of ancestry. It is generally considered a subdivision of a species; frequently used in plant breeding. (2) Pathogens of the same species and variety which are structurally indistinguishable but which differ in their physiological behavior, especially in their ability to parasitize varieties of a given host. (3) The channel that leads water to or from a waterwheel, the former is the head race and the latter the tail race. (4) A narrow passage or fenced land in a sheep yard for branding, dipping, etc. (5) An elongated white mark on the face of a horse or dog.

Racing Snaffle Bit–A bridle joint with one joint in the center of the mouthpiece which is especially designed for horse racing.

Rack–(1) The gait of a horse in which only one foot touches the ground at any one time, producing a four-beat gait. (2) A frame attached to a truck or wagon for the transportation of hay, tobacco, etc. (3) The rib portion of a sheep carcass. (4) A framework for holding feed for cattle, swine, sheep, etc., with upright partitions so that the animal can insert its head between the partitions and have access to the feed. (5) A frame placed in a stream to prevent the passage of fish. (6) A frame placed at the entrance to a sump pump to remove debris that would clog the pump.

Rack Hay Drying–A method of drying hay in which the freshly cut hay is placed on wooden racks built up off the ground. The method is used most often in small-scale operations in developing countries.

Raddle–The coloring smeared on the chest of a ram to mark ewes when they have mated.

Radical–A group of different elements acting as a single unit in a chemical reaction; normally incapable of separate existence. A radical may be negatively charged, positively charged, or without a charge.

Radioactive Tracers–Small quantities of radioactive isotope mixed with larger amounts of the corresponding stable isotope to be used as labels. Since the stable and radioactive isotope act chemically and biologically in the same manner, as the radioactive one is readily detected.

Radioactivated Milk–Milk treated by cathode rays through special electrical equipment which sterilizes the milk without a sensible rise in the temperature. Also called cold sterilization.

Radius Cruising–The distance between locations at which an individual animal is found at various hours of the day, at various seasons, or at times during various years.

Ragged–Of the fur or hair of animals, shaggy, rough, and hanging in tufts.

Ragged Hips–Irregular or poorly conformed rear quarters of an animal.

Raise–To grow or produce, as to raise corn or cattle.

Raised by Hand–Designating young lambs or other livestock fed from birth out of bottles, specially built pails, etc., in contrast to those nursed from birth by the dam.

Rales–Abnormal lung sounds in cases of pneumonia or lung inflammation.

Ram–A male sheep which has not been castrated, usually used for breeding.

Ram Jacket–A jacket, designed to prevent coitus (breeding), placed on a ram to permit it to run with the flock before the breeding season.

Ram Lamb–A male sheep under one year of age.

Ram Service–The fee charged for the breeding of a ewe.

Rambouillet–A fine wool breed of sheep first developed in France from earlier Spanish Merinos. It is a good wool and mutton breed very popular in the sheep section of the western range in the United States.

Ranch–An expression used mostly in the western United States to describe a tract of land, including land and facilities, used for the production of livestock. Accepted western usage generally refers to the headquarters facilities, pastures, and other land as the ranch, as distinguished from range. Loosely defined, a ranch also may be a small western farm, such as a fruit ranch or a chicken ranch.

Ranch Cattle–Any of the several crossbred varieties of cattle raised as beef animals; e.g., the Santa Gertrudis; the Beefmaster, a three-way cross of Brahman (Zebu), Herefore, and Shorthorn; and the Brangus, an animal which is about 37 percent Brahman (Zebu) and 63 percent Aberdeen Angus. Others include the Braford, a cross between the Brahman (Zebu) and the Hereford; and the Charbray, an animal 12.5 to 25 percent Brahman, the main strain being Charolais. The purpose of this crossbreeding has been to produce beef animals better adapted to tropical climates and more resistant to tick fever.

Rancid–Designating an offensive smell or taste resulting from the chemical transformation or putrefaction of fat, butter, milk, ice cream, and other products.

Random Mating–A mating system with no selection where every male has an equal chance of mating with every female.

Random Sample–A sample taken without bias from an area or from a population in which every part has an equal chance of being taken, in contrast to systematic sampling.

Rang Utilization–(1) For a single plant or species, the degree to which the foliage or herbage has been removed in percentage of the current growth within reach of livestock. (2) For an entire range, the relative amount eaten.

Range Band–A large flock of sheep handled as one unit on the range.

Range Bull–A bull used for breeding purposes on a range.

Range Calving–Permitting cows to drop their calves on the range under approximately natural conditions of shelter and forage.

Range Caterpillar–*Hemileuca oliviae*, family Saturniidae; a range pest on wild grasses in the southwestern United States, which sometimes infests corn and other cultivated plants. Its larvae are covered with coarse, poisonous spines.

Range Cattle–Cattle raised under range conditions.

Range Condition–(1) The state of health or productivity of both soil and forage of a given range in terms of what it could or should be under a normal climate and under the best practicable management. (2) An animal that is in a sufficient state of health or condition to be kept on the range.

Range Count–A census made on a range of the animals using a grazing area as contrasted to feedlot, corral, driveway, or other similar counts.

Range Crane Fly–*Tipula simplex*, family Tipulidae; an insect whose dark, leathery maggots bore into and destroy plant roots. It is sometimes destructive on the ranges of southwestern United States.

Range Ecology–The specialized branch of ecology which deals with vegetational response to environmental factors on rangeland, especially with the effects of grazing.

Range Forage–Forage produced on rangeland. See Cured Forage, Forage, Green Forage.

Range Grasses–Grass vegetation on the range areas. There are three types recognized in western United States; (a) tall grasses, such as big bluestem; (b) medium grasses, such as wheatgrass; (c) short grasses, such as buffalograss and blue grama grass.

Range Improvement–Physical development such as a structure (fencing) or excavation (water holes) to facilitate management of range or livestock. Generally measures to manipulate species composition and density such as revegetation, controlled burning, chemical or mechanical control of undesirable plants to increase the grazing capacity of range or increase its usefulness for watershed, wildlife habitat, or recreation.

Range Indicator–Any plant community portraying the condition of its environment which can be used as an indicator for the condition of a range.

Range Kidding–Permitting does (goats) to drop their kids on the range under approximately natural conditions of shelter and forage. See Kid House.

Range Lambing–Permitting ewes (sheep) to drop their lambs on the range under approximately natural conditions of shelter and forage.

Range Paralysis–See Fowl Paralysis.

Range Readiness–The stage of growth of the important palatable plants on the range and the condition of soil which permits grazing without undue compacting of the soil or endangering the maintenance of the plants.

Range Renovation–Improving rangeland by discing or other mechanical means, chemical treatment, or reseeding.

Range Reseeding–See Range Seeding.

Range Seeding–The process of establishing vegetation by the mechanical dissemination of seed.

Range Sheep–Sheep handled in bands of one to two thousand on a range.

Range Suitability–The adaptability of a range to grazing by livestock and/or game.

Range Type–An area of range which differs from other areas primarily by the difference in plant cover, such as grassland, browse, or conifer. One vegetation group can be distinguished from another group by difference of dominating species.

Range Wool–Wool produced under range conditions in the West and Southwest. With the exception of "Texas" and "California" wools, it is usually classified as "Territory" wool.

Range-raised–Designating livestock raised on the range.

Rangeland–(1) Land on which the natural plant cover is composed principally of native grasses, forbs, or shrubs valuable for forage. (2) Land used for grazing by livestock and big game animals on which the natural potential climax community of plants is dominated by grasses, grasslike plants, forbs, and shrubs.

Rangy–Designating an animal that is long, lean, leggy, and not too muscular in appearance.

Ranny–A poor-quality calf.

Ranting–The restless activity of a boar as it matures sexually and as the breeding season approaches. It is usually accompanied by loss of appetite and weight.

Rasorial–Referring to the animals which scratch the ground to obtain food, as a fowl.

Rat–*Rattus*, family Muridae; a long-tailed rodent which is much larger than a mouse. A serious pest, it is very destructive to stored food and may carry disease.

Rat Bite Fever–A disease caused by the flagellated organism *Spierillum minus* carried by rats and transmitted to people by their bites. Characterized by severe intermittent attacks, the ailment is not contagious. Especially prevalent in Japan.

Rat Poison–(1) A poison for rats obtained from the seed of a shrub, *Chailletia toxicaria*, native to West Africa. (2) *Hamelia erecta*, family Rubiaceae; a tall evergreen shrub, bearing scarlet or orange flowers and small, purple-red fruit, sometimes grown as an ornamental in warm areas. Native to tropical America. Also called scarlet bush.

Rate of Genetic Improvement–Rate of improvement of an animal per unit of time (year). The rate of improvement is dependent on: (a) heritability of traits considered; (b) selection differentials; (c) genetic correlations among traits considered; (d) generation interval in the herd; and (e) the number of traits of which selections are made.

Rate of Growth–(1) The rate at which a tree has laid on wood, measured radially in the trunk or in timber cut from the trunk. (The unit of measure in use is the number of annual growth rings per inch.) (2) The rate at which a young animal increases weight and height.

Ration–The feed allowed an animal during a 24-hour period regardless of whether it is fed at one time or at different times.

Ration of Maintenance–The feed which is necessary to maintain the body of an animal.

Ratios–Performance of an animal compared with its contemporaries, with 100 being average. Ratios greater than 100 are above average, and less than 100 are below average.

Rattail–(1) A slim, hairless tail of a horse. (2) A small, round, tapered file.

Rattle–A wholesale cut of meat made from the arm, shank, brisket, and short plate.

Raw Bone Meal–The dried, ground product suitable for animal feeding obtained by cooking undecomposed bone in water at just enough atmospheric pressure to remove excess fat and meat. Because of incomplete cooking, raw bone meal contains more than 23 percent protein and is lower in calcium and phosphorus than the steamed bone meal. It has limited use as a calcium and phosphorus supplement for livestock.

Raw Milk–Fresh, untreated milk as it comes from the cow.

Raw Wool–Wool prior to the removal of the grease.

Rawhide–Undressed skin of cattle.

Rawhide Bit–A bit made of rawhide used in the breaking of horses to protect the mouth.

Rawhide Hackamore–A bitless bridle or halter made of rawhide, used chiefly in the breaking of horses.

Rawhide Quirt–A riding whip made of rawhide.

Ray Fungus–*Actinomyces bovis*; an organism widely distributed in nature which may enter the tissue of the various organs of the animal causing swelling of either the bone or soft tissue. The infection usually reaches the surface so that fistulae are established and pus-producing organisms gain entrance. It is the cause of lumpy jaw in cattle.

Razorback–(1) A type of hog with long legs and snout, sharp narrow back, and lean body; usually a half-wild mongrel breed (southern United States). (2) A sharp-ridged spur or hill.

Reach–Difference between the average merit of a herd or flock, in one or several traits, and the average merit of those selected to be parents of the next generation.

Reaction–(1) A change in a market trend. (2) The degree of acidity or alkalinity (e.g., of a soil mass) expressed in pH values and in words as follows: extremely acid, below 4.5; very strongly acid, 4.5-5.5; medium acid, 5.6-6.0; slightly acid 6.1-6.5; neutral, 6.6-7.3 (strictly, 7.0); mildly alkaline, 7.4-8.0; strongly alkaline, 8.1-9.0; very strongly alkaline, over 9.1.

Reactor–(1) Especially in testing for a disease, an animal which reacts positively to a foreign substance: e.g., a tuberculous animal would be a reactor to tuberculin. (2) The apparatus in which nuclear fission takes place.

Ready-to-Eat Ham–A ham cooked until the interior reaches 155°F and held at that temperature or above for a further 2 hours.

Reagent–Any substance involved in a chemical action.

Rearing–To care for and support up to maturity, as to raise animals or fish to adults.

Reata–Lariat; a kind of strong Mexican rope made by twisting thongs of hide together, used in roping cattle on western ranges (United States).

Recessive–In genetics, a gene or trait which is masked by a dominant gene.

Reclaimed Wool–Wool that is reclaimed from new or old fabrics.

Reconstitute–To restore to the original form or condition by adding water, as reconstituted milk.

Reconstituted Milk–The product which results from the recombining of milk fat and nonfat dry milk or dried whole milk with water in proportions to yield the constituent percentages occurring in normal milk.

Rectovaginal–Pertaining to the rectum and vagina.

Rectrix–A feather in the tail of a bird.

Rectum–The terminal or lower part of the intestine which ends at the anus.

Recurrent Fever–An infectious viral disease of the horse which is characterized by extreme prostration, swelling in the lower parts of the body, jaundice, a pounding heart, and easy exhaustion. Also called swamp fever, infectious anemia, creeping fever.

Red–A common coat color for several species of animals which may range from dark red, deep rich-red, blood-red, golden-red, light red, or yellow-red. Shades of red are most common in cattle.

Red Danish–A dairy breed of cattle developed in Denmark. In color it is red to brindle. Individuals are heavy milk producers.

Red Dysentery–See Cocciciosis.

Red Mange–A skin disease of dogs caused by ***Demodix canis***, family Demodicidae, a mite that lives deep in the hair follicles and produces bare, inflamed spots about the eyes, ears, and joints. Also called demodectic mange, follicular mange.

Red Meat–Refers generally to the meat of cattle, sheep, hogs, and goats as opposed to that of poultry or fish.

Red Milk–A red color of milk caused by: (a) the bacterium ***Serratia marcescens***, a common soil or dirt contaminant, which grows and produces a red pigment that appears first on the top of the milk and then develops throughout; (b) pink yeasts such as ***Torula rosea*** and ***T. glutinis***, which produce spotty, red pigment in the cream layer; and (c) the presence of blood, as in acute mastitis or udder injury, which tends to settle to the bottom as the milk stands. Red milk caused by yeasts and bacteria is also referred to as red fermentation of milk.

Red Nose–An infectious rhinotracheitis usually found in feeder cattle, characterized by light-colored sometimes bloody nasal discharge, accompanied by drooling, coughing, and a red nose.

Red Polled–A dual-purpose breed of cattle common to central United States which originated in Norfolk and Suffolk counties in England. Animals are red and polled, and mature cows weigh 1,200 to 1,400 pounds with average milk production.

Red Roan–A coat color of horses and cattle, which is a bay, chestnut, or brown with a red background color.

Red Sindhi–A milking strain of Brahman cattle of India used by the United States Department of Agriculture in breeding experiments to develop better dairy cattle for southern United States. See Ranch Cattle.

Red Worms–***Stongylus***, family Strongylidae; nematodes which infest the large intestine of equines. Their red color is due to the presence of hemoglobin in the bodies of the worms, and not to blood sucked from their hosts.

Redia–A larval stage in the development of flukes. Redia of liver flukes of cattle, sheep, and goats occurs in snails. See Liver Flukes.

Reflex Ovulation–Ovulation that is triggered by the sexual act. Only a few animals do this, including the rabbit, and in these animals ovulation generally will not take place without such stimulus.

Registered–An animal whose name, along with the name and number of its sire (father) and dam (mother), has been recorded in the record books of its breed association. The association gives the animal a number, known as a registration number. The association also issues a certificate known as a registration certificate showing that the animal has been registered.

Registration Certificate–A paper issued by a breed association which shows that a particular animal has been registered.

Registry Number–A number assigned to a particular purebred animal that has been registered with a breed association.

Regrassing–Reestablishing grass on areas where it was once prevalent but had been killed out by overgrazing or by some environmental condition, such as drought.

Regression–(1) Destruction of the vegetation, as by fire, grazing, cutting, etc., usually with subsequent deterioration of the site, as by exposure, erosion, or loss of nutrients, to such extent as to give rise to a subsequent simpler vegetative type. It is not a true succession or development from forest to grassland, but a replacement as a consequence of complete destruction of the trees, etc. (2) Measure of the relationship between two variables. The value of one trait can be predicted by knowing the value of the other variable; e.g., easily obtained carcass traits (hot carcass weight, fat thickness, ribeye area, and percent of internal fat) are used to predict percent cutability.

Regurgitate–To return undigested food from the stomach to the mouth, as by ruminants. See Cud.

Rein–To control, check, stop, guide, or back up a horse or horses with the reins.

Reins–The part of a horse's harness fastened to each side of the bit or curb by which the rider or driver directs and controls it.

Reinsemination–To repeat the process of insemination.

Relaxin–An ovarian hormone produced at the time of parturition that is thought to aid in the relaxation of the birth canal.

Remnant Teeth–The first premolar teeth of the horse which usually remain embedded under the gum but which occasionally erupt. Also called wolf teeth.

Remount–A fresh horse used to substitute for another worn out or otherwise incapacitated horse. The term was once used to denote horses raised for government purchase as cavalry horses.

Remuda–A collection or string of broken horses.

Render–To extract, separate, or clarify by melting, as to render lard or fat.

Rendering Wax–Melting old combs and wax cappings and removing refuse to partially refine the beeswax. May be put through a wax press as part of the process.

Rennet–An extract of rennin; a digestive ferment secured from the fourth stomach of suckling calves which is used as a milk coagulant in cheese making. See Rennin.

Rennet Curd–The curd produced in milk by rennet action either as a result of specific bacterial growth or by direct addition of a rennet solution.

Rennet Stomach–The abomasum.

Rennets–The salted or dried stomachs of suckling calves, pigs, or lambs.

Rennin–A coagulant enzyme occurring particularly in the gastric juice of cows, and also in some plants and lower animals. See Rennet.

Repeatability–The tendency of animals to repeat themselves in certain performance traits in successive seasons, pregnancies, or lactations.

Replacement Animal–A young animal that is being raised to take the place of an older animal that is being culled.

Reprocessed Wool–Reprocessed wool comprises scraps and clips of woven and felted fabrics made of previously unused wool. These remnants are "garnetted"; i.e., shredded back into a fibrous state and used in the manufacture of woolens.

Reproduce–(1) Of animals, to bring forth young. (2) Of plants, to bear fruit and seeds.

Reproduction–(1) The production of offspring by organized bodies. (2) The creation of a similar object or situation; duplication; replication. Asexual reproduction; reproduction from vegetative parts. Sexual reproduction; reproduction by the fusion of a female sexual cell and a male sexual cell. Parthogenic reproduction; reproduction by the development of an unfertilized egg.

Reproductive System–The organs of the body, either male or female, concerned with producing offspring.

Repulsion–In genetics, the condition in which an individual heterozygous for two pairs of linked genes receives the dominant member of one pair from one parent and the dominant member of the second pair from the other parent; e.g., AAbb x aaBB.

Requeen–To replace a queen in a hive. Usually to replace an old queen with a young one.

Research–All effort directed toward increased knowledge of natural phenomena and the environment and toward the solution of problems in all fields of science. This includes basic and applied research. Much of the agricultural productivity of the United States is directly the result of applying research.

Reserve–(1) Any tract of land, especially public land, set aside for some special use; e.g., forest reserve, school reserve. Also called reservation. (2) A tree or group of trees left uncut on an area for a period, usually a second rotating. After the stand is reproduced, naturally or artificially, an active stand which is held for future utilization.

Reserve Champion–In the show ring, the best animal in a group of first prize animals after the champion has been selected.

Reservoir Host–An animal in which an infectious agent lives and multiplies and depends upon primarily for survival. The reservoir host is usually not greatly affected by the infectious agent that it harbors.

Residual–(1) Remaining in place after all but the least soluble constituents have been removed. Said of the material eventually resulting from the decomposition of rocks. (2) Standing, as a remnant of a formerly greater mass of rock or area of land, above a surrounding area which has been eroded. Said of some rocks, hills, mountains, mesas, plateaus, and groups of such features. (3) Soil developed in place from underlying bedrock. See Monadnock.

Resistant–Designating a plant or animal capable of withstanding disease, inclement weather, or other adverse environmental conditions.

Respiration–(1) A chemical process that takes place in living cells whereby food (fats, carbohydrates, and proteins) is "burned" (oxidized) to release energy and waste products, mainly carbon dioxide and water. Living things use energy produced through respiration to drive vital life processes such as growth and reproduction. (2) The oxidation of carbohydrates in living organisms and the attendant release of energy and liberation of carbon dioxide and water. (3) In animals, the act of breathing; the drawing of air into the lungs and its exhalation. In small organisms with no special breathing organs, the process takes place over a large part of the body surface.

Responsive Mouth–Designating a horse so trained that it easily obeys commands given by means of the reins.

Rest-rotation Grazing–An intensive system of management whereby grazing is deferred on various parts of the range during succeeding years, allowing the deferred part complete rest for one year. Two or more units are required. Control by fencing is usually necessary on cattle range, but may be obtained by herding on sheep ranges.

Rested Pasture–Pasture ungrazed for an entire growing season.

Resting Area–A sheltered area of loose housing where cows are bedded but not fed. The manure is allowed to accumulate during all or a part of the year. Also called bedded area, loafing area, lounging area.

Resting Pasture–A pasture not grazed by livestock.

Restricted Feeding–A system of feeding poultry whereby feed is provided only during certain periods of the day. See Free-choice Feeding.

Restriction Enzymes–Enzymes used in genetic engineering to remove a gene from a piece of DNA.

Restructured Meats–Meat processed by cutting or shredding it into small flakes and then running the flakes through a special forming machine to produce the desired form. The meat flakes are bound together by extracting meat protein from the flakes, or by using a non-meat binder when running flakes through the forming machine. Tenderness and texture of the meat are influenced by the temperature and flake size. Advantages include cheaper price and quick cooking with dry heat. Forms used include chip steak, steak cutlet, and turkey ham.

Resuscitator–A device that is placed over the mouth and nostrils of a newborn animal to help the animal start breathing.

Retail Cuts–Cuts of meat that are ready for purchase and use by the consumer.

Retail Dairy–A type of dairy plant concerned primarily with the distribution and selling of milk and milk products directly to the consumer.

Retained Afterbirth–See Retained Placenta.

Retained Placenta–The fetal membranes or afterbirth which a mammal mother fails to expel normally within a few hours after her young is born.

Retaining Pen–A pen for holding cattle, sheep, and hogs at the time of dehorning, shearing, dipping, or weighing.

Rete Testis–A network of tubules located inside the testis in the mediastinum connecting the seminiferous tubules to the efferent ducts.

Reticuloendothelial System–A widely spread network of cells in the body concerned with blood cell formation, bile formation, and engulfing or trapping of foreign materials, which includes cells of bone marrow, lymph, spleen, and liver.

Reticulum–The second compartment of the ruminant stomach, where bacterial digestion continues. Has a honeycomb-textured lining, so is often called the honeycomb. See Abomasum, Omasum, Rumen.

Retinol–See Vitamin A.

Retired to Pasture–Designating an animal whose productive life has ended and which is turned out to graze.

Retired to the Stud–Designating a horse which no longer races but is retained for breeding purposes.

Retractor Muscle–Part of the male reproductive system which helps extend the penis from the sheath and draws it back after copulation.

Reused Wool–Also called shoddy; made from old wool which has actually been worn or used, including the rags and miscellaneous old clothing collected by rag dealers. These are cleaned and shredded into fibers again, and then blended to make utility fabrics. The consumer has no way of telling how much the original desirable qualities of wool have been impaired by this previous use.

Reverse Osmosis–(1) An external force is used to reverse normal osmotic flow through a semipermeable membrane, resulting in movement of water from a solution of higher solute concentration to one of lower solute concentration. See Osmosis. (2) A process of desalination of seawater whereby only pure water passes through a fine membrane while the salts cannot pass through.

Reworked Wool–Refers to wool that has been previously used. See Mungo, Reused Wool, Shoddy.

Rheum–See Catarrh.

Rhinitis–Inflammation of the mucous membrane of the nose.

Rhinitis Atrophic–A disease affecting the upper respiratory tract of the pig which causes atrophy of the turbinate bone and frequently distortion of the snout. Affected pigs usually grow at a slower rate than normal.

Rhode Island Red–A breed of domestic chickens of the American class. Single-comb Rhode Island Reds are hardy, dual-purpose birds, bred largely for egg production but with consideration for meat qualities. Individuals are oblong in appearance with relatively long, flat backs, rich red plumage, yellow skin, and yellow shanks tinged with reddish horn. Mature cocks weight about 8½ pounds, mature hens about 6½ pounds. The eggshell is brown. The Rose-Comb variety is less commonly raised.

Rhode Island White–A breed of medium-weight chickens of the American class. Individuals have a rose comb and white plumage. The shape and weight specifications are the same as for the Rhode Island Red. It is not a commonly raised breed.

Riata–See Lariat.

Rib–(1) (a) A cut of meat made from five to eight ribs depending upon the method of cutting the forequarter. (b) Any cut of meat which comes from the rib section, as a rib roast. (2) In judging of livestock, spring of rib indicates conformation and vigor. (3) That which resembles a rib in form or use, as a piece of timber to which boards are fastened. (4) To divide a side of beef into a fore and hind quarter. (5) A prominent vein.

Rib Eye–Main muscle exposed when carcass is separated into front and hindquarters. Area of rib eye, sometimes called loin eye, at 12th rib; used as indication of muscling.

Rib Eye Grid–A clear sheet of plastic that has a grid of dots that represents 0.1 inch per dot. A grader places the grid on the rib eye of a beef carcass and counts the number of dots that are over red meat. The total of the dots counted represents the size of the rib eye in square inches. Rib eye size is used as a factor in determining yield grade.

Ribbed-up–Said of a horse on which the back ribs are well arched and incline well backwards, brining the ends closer to the point of the hip and making the horse shorter in coupling.

Riboflavin–Vitamin B2; lactoflavin; $C_{17}H_{20}N_4O_6$; an essential nutrient for people and animals, riboflavin functions as a coenzyme concerned with oxidative processes. It promotes the growth of rats, prevents the occurrence of a nutritional cataract in rats, and prevents a specific dermatitis in turkeys.

Riboflavin Supplement–A feed material used chiefly for its riboflavin content which shall contain not less than 1,000 mg of riboflavin per pound according to the tentative method of analysis of the Association of Official Agricultural Chemists. The label shall bear a statement of origin.

Ribonucleic Acid (RNA)–The substance in the living cells of all organisms that carries genetic information needed to form protein in the cell.

Rice Hull–A product which consists of the outer covering of the rice. It is low in digestible nutrients and unpalatable, but if well-ground may be used as a low-grade roughage for livestock.

Rice Meal–A livestock feedstuff consisting of ground brown rice.

Rice Straw–The straw remaining after threshing rice. If well cured, it may be fed in the same manner as straw from the other cereals.

Rice Water–A product made by boiling rice grain to a pulp and mixing it with water; especially useful as a drink for horses suffering from diarrhea after exertion.

Rice-Fish Rotation–The alternation of rice and fish crops. Rice fields, which would ordinarily lie fallow and idle, are flooded and made to produce a crop of fish for one or two years. In addition to yielding income from the fish, the land is rested and enriched for the following rice crop.

Rickets–Caused by a deficiency of vitamin D and sunshine, and dysfunction of the parathyroid glands, especially in infancy. The result is subnormal calcium utilization, poor bone development, bowed legs, formation of nodular enlargements on the bones, muscular pain, and sweating of the head.

Rickettsia–A group of microorganisms that are intermediate in size to bacteria and viruses; they survive by reproducing in the cells of larger organisms in much the same way as do viruses.

Rickety–(1) Having rickets. Also called rachitic. (2) Designating an unstable building or old, infirm animal.

Ride–(1) To mount a horse for going from one place to another. (2) To push down a fence, as a cow. (3) To mount a cow, as another cow indicating heat.

Ride Bareback–To ride without a saddle.

Ride Fence–Regularly to inspect range boundary fences for damage or for loss of cattle (western United States).

Ride Herd–To attend cattle on the range.

Ridgeling–Any male animal whose testicles fail to descend normally into the scrotum. See Cryptorchid.

Riding Crop–A short whip used by horseback riders.

Riding Down–Pushing over of small trees, shrubs, and fences by livestock in order to reach and browse in the foliage.

Riding Horse–A horse which is specially bred and trained to be ridden.

Riding Stable–A stable where saddle horses are maintained by the individual owners or are offered for hire by the management.

Rift Valley Fever–An acute arthropod-borne viral disease of sheep, cattle, and goats causing high mortality in young lambs, calves, and kids and abortion in pregnant females.

Rigging Ring–The ring attached to the saddle tree for a cinch.

Rigor Mortis–A physiological process following the death of an animal in which the muscles stiffen and lock into place.

Rind–(1) A hard coating caused by the desiccation of the surface of cheese. (2) Skin of an animal, especially of a hog, as a pork rind. (3) Skin of a fruit or vegetable. (4) Bark of a tree. (5) To remove the skin.

Rinderpest–A specific, acute, lethal, and inoculable viral disease of cattle, characterized by an ulcerative inflammation of the mucous membranes, especially of the alimentary tract, fever, dullness, drooping head, loss of appetite, grinding the teeth, red eyes, and red nasal discharge. Not observed in the United States. Also called cattle plague.

Ring–(1) A cut or girdle around the trunk, branches, or roots of a tree. See Girdle. (2) Annual growth ring of a tree. See Annual Ring. (3) (a) A circular band of metal or wood, as the metal ring in the nose of a bull. (b) To place a ring through the cartilage of the nose of an animal; e.g., to prevent a hog from harmful rooting or to control a bull, etc. (4) (a) A circular, metal or plastic band placed on the leg of a fowl for identification purposes. (b) To place a ring on the leg of a fowl. Also called ringing birds. (5) A ridge which encircles the horns of a cow, the number increasing with age. (6) A circular exhibition place for the showing or sale of livestock or the racing of horses.

Ring Bit–A large ring that passes over the lower jaw of the horse which is used as a bit.

Ring Bone–A bony enlargement, involving the pastern bones just above the hoof, which interferes with the action of the joints and tendons, thus causing lameness to an animal. Usually only seen in horses.

Ring Test–A milk test for brucellosis. See Brucellosis.

Ring-eyed–Designating the condition of an albino horse in which the iris of one or both eyes is devoid of the characteristic coloring matter and in which the dark shining lens appears to be surrounded by a ring of white.

Ring-necked Pheasant–*Phasianus colchicus torquatus*, family Phasianidae; a long-tailed, highly colored, gallinaceous bird with a white ring neck. A highly popular game bird, it is sometimes a nuisance on farms because it pulls young corn plants out of the ground and devours what is left of the kernel. It is also occasionally harmful to vegetable crops, such as tomatoes. Originally from China.

Ringing–(1) Clipping the wool from a breeding ram around the neck, belly, and penis region in order to facilitate proper mating. (2) Putting a ring in the nose of cattle or hogs. (3) Removing a narrow strip of bark from around a branch or tree trunk to encourage fruiting. Only outer bark is removed, and the ring does not extend into the cambium layer.

Ringworm–A skin disease of humans and animals caused by parasitic fungi, usually marked by distinct, circular patches with a scaly appearance.

Ripe–(1) Designating mature seeds which are fit for germination. (2) Designating fruit which has attained full development. (3) In plant propagation, designating wood that will root well. (4) In grafting, designating that wood which is ready for perfect union. (5) Designating the best condition for use, as ripe cheese, ripe wine.

Ripening–(1) Growing to maturity and being fit for food, as ripe fruit or ripe grain. (2) Bringing to a certain condition for use by keeping, as in wine. (3) Preparing milk or batch mixes for making butter or ice cream either by a natural souring or by the addition of starters. (4) Undergoing an aging process, as in meat.

Ripper–An unusually strong, large horse.

Rising Two–Describing a horse, 1½ to 2 years old. Also called coming two.

RNA–See Ribonucleic Acid.

Roan–(1) A coat color of a horse that is chestnut or bay or may be red or strawberry roan, blue roan, or chestnut roan depending upon the intermingling of the background colors. (2) Designating the red-white color phase of Shorthorn cattle.

Roarer–A windbroken animal that makes a loud noise in drawing air into the lungs.

Roaring–A defect in the air passage of a horse which causes him to roar or whistle when respiration is forced.

Roaster–A chicken of either sex that weights between 3½ and 5 pounds and is less than eight months old. See Broiler, Fryer.

Roasting Pig–A pig weighing from 15 to 50 pounds, dressed with the head left on.

Robber Bee–A bee which robs food from another colony.

Robber Fly–A group of flies (family Asilidae) that are predaceous as adults and as larvae. They are generally considered as beneficial except for the species *Sparopogon dispar*, which preys on honey bees.

Robust–Designating a strong vigorous animal or plant.

Rocker Toe Shoe–A special horseshoe which is used on a horse that stumbles.

Rocky Mountain Crazyweed–*Oxytropis saximontana*, family Leguminosae; a poisonous plant found from Montana to Utah, United States, whose foliage is poisonous to cattle. Symptoms of poisoning are dullness, irregularity of gait, lack of appetite, dragging of the feet, a solitary habit, loss of flesh, and shaggy coat. As the animal ceases to eat it dies. Native to North America. See Loco.

Rocky Mountain Spotted Fever–A disease of people and animals characterized by intermittent chills and fever, painful muscles and joints, and red blotches on the skin. Occurring in the Rocky Mountain area, United States, the Rocky Mountain wood tick is the vector.

Rocky Mountain Wood Tick–*Dermacentor andersoni*, family Ixodidae; probably the most important tick vector of disease. This tick transmits Rocky Mountain spotted fever, tularemia, Colorado tick fever, American Q-fever, and encephalomyelitis and, experimentally, anaplasmosis. Most domestic animals, humans, and numerous wild mammals are its hosts. The stages in the life cycle are egg, larva, nymph, and adult. Larvae and nymphs usually live on small wild animals, mainly rodents; adults attack larger animals and people.

Rodent–A classification of mammals, mostly vegetarians, characterized by their single pair of chisel-shaped, upper incisors. Rodents are members of the orders Rodentia (rats, mice, squirrels, etc.) and Lagomorpha (rabbits, etc.).

Rodenticide–Any poison which is lethal to rodents.

Rodeo–(1) The rounding up of cattle. (2) A public performance presenting features of a cattle round-up, as lariat throwing, bronco riding, and bulldogging.

Roe–The eggs or testes of fish. Consisting of two types, the female eggs (hard roe) and the male testes (soft roe), they are widely used for human consumption; e.g., the salted roe of sturgeon (caviar) is highly valued as a delicacy. see Sturgeon.

Rogue–(1) A seedling or plant of inferior or objectionable quality; a variation from type. (2) A horse without inclination to work or cooperate with its handler. (3) To remove and destroy undesirable plants.

Rolcut Secateurs–An instrument for removing the folds of skin from the crotch of ewes, also used in the bloodless emasculation of calves and lambs.

Roll of Honor–A division in the system of Advanced Registration of Ayrshire Cattle in which a cow must meet certain production requirements in 305 days and give birth to a calf within a specified period.

Rolled-toe Shoe–A type of horseshoe which is used to make the foot action of a horse break over easier by improving action at the pivotal point of the shoe.

Rolling–(1) Excessive side motion of shoulders, common in horses with abnormally wide fronts or chests. (2) A part of seed-bed preparation in which the land is rolled to even out the surface. (3) Processing grain through a set of smooth rollers which are close together; sometimes called flaking.

Roman-nosed–Refers to a horse or other animal having a profile that is convex from poll to muzzle.

Room Temperature–In laboratory work, 68° to 70°F (20° to 21°C).

Roost–(1) A resting or lodging place for fowls. (2) A group of roosting fowls. (3) To rest upon a roost or perch.

Rooster–Male chicken. Also called cock, cockerel (when under one year of age).

Rooting–(1) The production of roots by a plant. (2) Digging the earth with the snout, as hogs. (3) A root used for propagation. (4) The production of roots by a cutting.

Rope Bridle–A bridle made of rope instead of leather.

Rope Burn–A skin irritation caused by pressure contact and speedy movement of a rope upon the surface of the skin of a human or animal.

Rope Halter–A halter made of rope instead of leather.

Rope Walking–The leg motion in which the horse swings the striding leg around and in front of the supporting leg in walking or running.

Roped–Designating an animal which has been lassoed or tied.

Roper–A cowboy who is skilled with the lasso.

Roper Curved Cheek Bit–A common type of bit that is used on horses that are used for roping.

Roping Horse–A horse trained for use by a rider wielding a lasso.

Rose Chafer Poisoning–A poultry poisoning which is caused when rose chafers or rose beetles are eaten by chickens under ten weeks old.

Rose-comb–A chicken's comb which is low and solid having the upper surface covered with small, rounded points. There is no indentation in the center and there is a spike termination on the rear. See Single Comb.

Ross Test–A test of urine for the presence of ketone bodies, used in detection of acetonemia in cattle and sheep. See Acetonemia.

Rot–A state of decay caused by bacteria or fungi. See Decay.

Rotary Hog Feeder–A round or many-sided feeder which the hog turns with its snout to make grain feed come down from the hopper to the animal for consumption.

Rotated Pasture–(1) A pasture in the regular crop rotation which is grazed for a few years, usually two or three, and then plowed for other crops. (2) A pasture which is divided into segments by use of fences: the livestock being confined to one segment at a time in a definite rotation pattern.

Rotation Grazing–Grazing forage plants on well-managed pastures in such a manner as to allow for a definite recovery period following each grazing period. This includes alternate use of two or more pastures at regular intervals or the use of temporary fences within pastures to prevent overgrazing.

Rotation-deferred Grazing–Grazing under a system where the key plants in one or more range units are rested at planned intervals throughout the growing season, and no unit is grazed more than half of any growing season or at the same time in successive years.

Rotational Crossbreeding–Systems of crossing two or more breeds where the crossbred females are bred to bulls of the breed contributing the least genes to that female's genotype. Rotation systems maintain relatively high levels of heterosis and produce replacement heifers from within the system. Opportunity to select replacement heifers is greater for rotation systems than for other crossbreeding systems.

Rotted Manure—Animal dung which has undergone decay and is safe to use with growing plants.

Rotten—(1) Designating decomposed or putrid organic matter. (2) Designating ground or soil extremely soft and yielding because of decay, or rocks partially decomposed. (3) Designating sheep attacked by rot.

Rough—(1) To remove the major part of the plumage of a fowl leaving scattered feathers for the finisher. (2) In meat judging, designating uneven contour or uneven distribution of fat on a carcass. (3) Designating an uneven piece of land or a road. (4) Designating food of low quality. (5) Designating a horse not properly trained. (6) A calk for the horse's shoe.

Rough Coat—Coarse, tangled, unkempt animal's coat which may indicate lack of thrift or its remaining in the open during the winter.

Roughage—(1) Any food or feed high in fiber and low in digestible nutrients such as many fruits and vegetables, straw, and low-quality vegetation, hay, haylage, and silage. High-quality grass-legume pasturage and high-protein hay are more properly known as forage. (2) In human nutrition, a coarse food containing considerable indigestible material usually in the form of cellulose, as uncooked celery and lettuce.

Round—The center portion of the hindquarter of a beef animal; the cut of beef taken from the round.

Round Feeder—A round hog feeder with doors around the bottom that the hogs lift to obtain feed.

Round Steak—A retail cut of beef for frying, broiling, or braising which comes from the round of beef.

Round-sausage Casing—A casing made from the small intestines of animals.

Round-up—(1) The deliberate gathering of domestic animals, usually range cattle, for branding, fastening ear tags, injections, pesticide applications, inventory, and potential sales. (2) The brand name of a herbicide used as a foliar spray to kill most plants.

Roundworms—Parasites in humans, animals and plants which may cause disease and great economic loss. They vary in size from a fraction of an inch in length and as thin as a silk thread to over a foot in length and as thick as a lead pencil. In people and animals, they inhabit the intestine, but in completing a complex life cycle may infest the blood stream, lungs, windpipe, liver, kidneys, etc. Symptoms vary but in general are those of unthriftiness. See Nematode.

Roup of Fowls—(1) Infections roup caused by *Hemophilus gallinerum*, characterized by swelling of the sinuses under the eyes and watery discharge from the nostrils. Also called contagious catarrh of fowls. See Infectious Coryza. (2) Nutritional roup, caused by an insufficient amount of vitamin A in the diet, characterized by a whitish exudate in the eye, which may cause serious loss before deficiency is noticed.

Route of Entry—The means by which a hazardous substance enters the body. Common routes are skin contact, eye contact, inhalation, and ingestion.

Rowel—In horseriding, the wheel of a spur with blunt to sharp projecting points.

Royal Jelly—The food supplied by worker bees throughout the developmental period of the larvae destined to become queen bees. Also called bee milk.

Rubber Sleeve—A protective, rubber covering attached to a rubber glove for the hand, arm, and shoulder which is used by veterinarians and artificial inseminators in various phases of their work.

Rubbing Post—Any of several devices consisting mainly of a post set in the ground with a can of oil attached to the top, so arranged that the oil or medicament is allowed to slowly drip or run down the sides of the post so that when hogs or other livestock rub against it a small amount of oil will be deposited on the rubbed or infected area of the skin. Also called hog oilers, livestock oilers.

Rubefaciant—Liniment, plaster, or any substance that produces redness of the skin when applied to it.

Rudimentaries—The teats on a boar.

Rudimentary—Imperfectly developed.

Rudimentary Copulatory Organ—A very small, shiny or glistening eminence in chicks, which is located in the median portion of the fold between the urodaeum and the proteodaeum. By examination of this organ the sex of day-old chicks can be determined.

Rudimentary Teat—An undeveloped teat which may or may not be connected with milk-secreting tissue. It may be attached as a small nipple to the regular teats, an underdeveloped teat behind the regular ones or show up on the male of the species.

Ruffle Fat—The fat that is located between the intestines of an animal.

Rugged—Designating size, strength, and vigor in the body build or conformation of an animal.

Rumen—The largest compartment of the stomach of cattle, sheep, and goats and their relatives; a large amount of bacterial fermentation of feed materials occurs in the rumen; also called the paunch. See Abomasum, Omasum, Reticulum.

Rumen Fistula—A fistula of the rumen or first stomach of a cud-chewing animal.

Rumen Magnet—A smooth oblong magnet that is placed in the rumen to collect small metal objects that are swallowed by the animal during grazing.

Rumenology—A branch of animal science concerned with the study of the rumen.

Rumenotomy—The operation of cutting into the rumen to remove foreign bodies or to observe activity.

Ruminant—Any one of a class of animals including sheep, goats, and cows that have multiple stomachs. They are most efficient feeders because bacterial action in one of the stomachs, the rumen, increases the feed value of low-grade feed.

Rumination–The process of digestion in cattle whereby food is swallowed to the first stomach, the rumen. Later it is regurgitated into the mouth and chewed over again to be swallowed for further processing by the second, third, and fourth stomachs.

Rump–That part of the rear end of an animal which includes the buttocks or fleshy part of the rear quarters.

Rump Roast–A retail cut of meat taken from the rump section of the hind quarter.

Run–(1) A period of time, as a maple syrup run. (2) The amount of sap or sugar produced in a given time. (3) A swiftly moving tributary, rivulet, or mountain stream (eastern United States). (4) The stream outlet of a large spring (Florida, United States). (5) Unrestricted movement, as the colts have the run of the pasture. (6) An area of land or a leasehold (New Zealand). (7) A small, often dry gully or channel carved by water. See Arroyo. (8) A fenced-in pen used for the exercise of animals or poultry. (9) To feed; to graze, as steers run on the open range. (10) To operate, as a plow is set to run at a depth of 6 inches. (11) To cultivate, mow, combine, etc., as to run over a field with a weeder. (12) To maintain animals, as he runs sheep. (13) To work a dog with sheep or cattle. (14) To discharge pus, as a sore runs. (15) To be, as the prices run very high. (16) To move rapidly, as a horse. (17) To turn, as to run a wheel. (18) To grow, as when the vines begin to run. (19) To operate, as to run an engine.

Run Free–To be loose, as animals that are not restrained and are allowed to move about freely in the barn, yard, or pasture.

Run On–To graze or pasture on, as to run on the range.

Run-of-the-Hatchery–Chickens as they come from the hatchery without culling, sorting, or sexing.

Run-out Fleece–A fleece of wool that varies greatly in quality, lacks character, and carries a large percentage of britch and possibly kemp.

Runner–(1) A breed of ducks of very distinctive type, having a long, narrow body and very erect carriage. The breed derives its name from its gait, which is a quick run, quite unlike the waddle of other ducks. The adult drake weighs about 4½ pounds; the adult duck about 4 pounds. The runner is noted as an egg-producing breed and has little or no value for meat production. Its three varieties are the White, the Fawn and White, and the Penciled. Also called Indian Runner. (2) A lateral, aboveground shoot (stolon) of certain plants; e.g., strawberries, which roots and forms young plants at some of the nodes, aiding in propagation. (3) A rope used to increase the mechanical power of a tackle. (4) The upper or rotating stone of a set of millstones. (5) A supporting attachment which slides along the ground, as a sled runner. See Stolon.

Running Out–The condition in which an improved variety of plant or animal is reverting to a former and inferior type or is losing some of its desired qualities.

Running Plate–The aluminum horseshoe worn by race horses.

Running Sheep–The handling and management of a flock of sheep under range conditions.

Running Through–Lactation extended beyond 365 days.

Running Walk–A slow, single-foot or four-beat horse's gait with the break in the impact or rhythm occurring between diagonal fore and hind feet. In the stepping pace, which is also a slow, four-beat gait, the break in the impact occurs between lateral fore and rear feet.

Runt–Any animal smaller than normal that is culled and not used for breeding. Smallness may be due to genetics, injury, or disease. Also called puny.

Rustler–A cattle thief.

Rut–(1) The grooved track left by the wheels of vehicles in soft ground. (2) The season of heightened sexual activity in male mammals that coincides with the season of estrus in the female. See Estrus.

Rutter–A female mammal, such as a cow, which for some abnormality remains constantly in heat. Also called a buller.

Rutting Season–The recurring, usually annual, period when deer, cattle, etc., are in heat. See Estrus Cycle.

Saanen–A breed of milk goats which originated in the Saanen Valley of Switzerland. The animals are white or creamy-white, usually short-haired and hornless. The breed is noted for its persistence in production rather than for high production over short period.

Sacbrood–An infectious disease of the brood of the honey bee caused by a filterable virus, characterized by decomposition of the larval tissue.

Sack–(1) A bag, usually large, oblong, and made of burlap or other stout fabric, used for holding grains, vegetables, etc. (2) To stroke or rub a horse. (A highly nervous horse is frequently sacked by the buster who, from a safe position, soothes the animal by rubbing it with a burlap sack or broom.)

Sacrifice Area–A part of the range that is intentionally overgrazed to obtain efficient overall use of the management area.

Sacrum–That composite bone structure which makes up the vertebral column between the cocygeal and lumbar vertebrae. It is connected directly with, and comprises part of, the pelvis.

Saddle–(1) A seat designed to fit a horse's back and to make riding easy, comfortable, and safe. Saddles are of many types but are usually made of heavy leather over a well-padded frame. They are often raised in the rear to hold the body in place and they may have a firm, raised support or horn in front to attach ropes or gear. They are held to the horse's back by means of one or more wide straps or girths that pass around the body of the horse. (2) The whole upper back section of a meat carcass which in wholesale cuts is commonly divided into fore and hind saddle. In cookery, the rear upper back portion including the two loins. (Sheep, veal calves, goats, and deer.) (3) The rear part of the back of a male chicken extending to the tail, covered by saddle feathers. (4) An apron made of reasonably heavy canvas which is placed on the backs of valuable breeding hens to protect against injury during mating. Saddles are usually held in place by loops which fit under the wings. (5) The uncut stalks of four hills of corn from two rows, tied together, around which corn is shocked. (6) A saddlelike depression in the crest of a ridge. See Saddleback. (7) A specially designed transverse log used in a skid road to guide the moving logs. (8) To put a saddle upon an animal or thing.

Saddle Blanket—A blanket or pad of wool or cotton placed on the back of a horse beneath the saddle to prevent chafing; saddlecloth.

Saddle Breeds of Horses—The principal breeds of saddle horses in the United States are: American Saddle horse, American quarter horse, Appaloosa, Arab(ian), Cleveland Bay, Morgan, Palomino, Tennessee Walking Horse, Thoroughbred.

Saddle Feather—One of a group of long, narrow feathers which grows out of the saddle of a male chicken.

Saddle Fender—A broad piece of leather attached to a stirrup strap to protect the rider's leg from the stirrup straps and perspiration from the side of the horse. Also called sudadero, rosaderos, stirrup fender, leg fender.

Saddle Gait—The foot movement or gait of a saddle horse; a definite manner of walking, running, etc., such as the walk, running walk, trot, canter, rack.

Saddle Gall—A sore or wound on the back of a horse which results from abrasion by a saddle.

Saddle Girth—In horses, a strong band or strap which encircles the body, usually just behind the front legs, and helps fasten a saddle to the back. With western-type saddles it usually consists of two latigo straps on the sides and a webbed band underneath with rings into which the latigo straps fit to tighten and fasten. Also called cinch.

Saddle Horn—A knoblike elevation at the front center of a western saddle, used in roping as an anchor for the lariat.

Saddle Horse—A type of horse especially suitable for riding because of its strong back, style, ease of action, and different gaits. It may include running horses, three- and five-gaited horses, stock horses, hunters, polo ponies, etc.

Saddle Skirt—The leather part of a western-type saddle which extends beyond the sides and to the rear of the saddle seat. On English-type saddles, a broad piece of leather which extends under the saddle, down the side of a horse, and under the stirrup strap.

Saddle-broken—Designating a horse trained to the saddle.

Saddlebag—A bag or pouch of leather or fabric carried at the saddle, especially one of a pair laid across the back of a horse behind the saddle.

Saddlebred—Of, pertaining to, or designating a type of horse which has the breeding and qualities desired in a saddle horse.

Sale Ring—The area which is occupied by the animal or group of animals being offered for sale by the auctioneer at a livestock sale.

Salers—A breed of red beef cattle that originated in France. They were first imported to the United States in 1972.

Saleyard—A yard used for the sale of livestock.

Saline—General term for waters continuing various dissolved salts. The term is restricted to inland waters where the ratios of the salts often vary; the term haline is applied to coastal waters where the salts are roughly in the same proportion as found in undiluted seawater.

Salinity—The quantity of saltness in seawater or freshwater, most commonly expressed in parts of dissolved salt per 1,000 parts of water; e.g., salinity of seawater is 35 parts per thousand. See Alkalinity.

Salinus—Salty.

Saliva—The weakly alkaline fluid consisting of the secretions of the salivary glands; spit.

Salivary Glands—Glands that open into the mouth and secrete a fluid with digestive, irritant, or anticoagulatory properties.

Salivation—A discharge of saliva. In large animals, excessive salivation usually results in the appearance of long strings of a clear, viscid material hanging from the mouth.

Sallow—(1) Goat willow. (2) An unhealthy skin color.

Salmon Liver Oil—A livestock feedstuff supplement; an oil obtained from the livers of salmon. It has properties similar to those of cod liver oil and is used as a vitamin A and D supplement in mixed feeds for livestock. See Cod Liver Oil.

Salmon Oil—A livestock feedstuff supplement which is the product obtained by extraction of part of the oil from the cannery refuse of salmon. It contains important amounts of vitamin A and D.

Salmonella—A large group of bacteria, some of which are associated with food poisoning. Certain salmonellas are sometimes found in raw and dried eggs and in poultry products. They can be destroyed by sufficient heating.

Salmonellosis—Infection by organisms of the genus *Salmonella* causing food poisoning in humans and many diseases in domestic animals.

Salpingitis—(1) Inflammation of a fallopian tube (oviduct). (2) An inflammation of the oviduct of poultry characterized by a discharge from the duct which causes irritation of the vent and a smeary appearance of the feathers below it.

Salt—(1) Sodium chloride (common salt), NaCl; a white crystalline compound occurring abundantly in nature as a solid or in solution. It is a requirement of all livestock and is generally deficient in all natural feed ingredients. It is commonly included in mixed feeds and is otherwise supplied as salt blocks in pastures, barn lots, or stalls. It has wide usage in seasoning, preserving meats and vegetables, tanning hides, etc. In large amounts it is toxic and is frequently used to kill weeds; however, some plants such as Bermudagrass and asparagus tolerate a very high concentration. (2) To add salt as seasoning, preservative, supplement, etc.

Salt Alum Tanning—A simple method of tanning sheep skins whereby the previously salted and fleshed skins are treated overnight with a 2 to 1 mixture of salt and alum and later with neat's-foot oil or glycerine.

Salt Block–A compressed block of common salt which is used to supply salt to livestock. Also called saltera.

Salt Down–To add salt, especially for the preserving or curing of meats, hides, etc. See Salt.

Salt Hay–(1) A mulching material for plants which is cut from coastal marshes. (2) Hay which is made from grasses cut from salt marshes.

Salt Index–An index to compare solubility of chemicals. Most nitrogen and potassium compounds have a high index; phosphate compounds have a low index. When applied too close to seed or foliage, high-index materials may cause plant injury by plasmolysis.

Salt Poisoning–A condition that may result in the death of livestock and poultry caused by ingestion of an unusually large amount of salt. Farm animals that have been deprived of salt over a long period may consume it to excess if unlimited quantities are available.

Salt Pork–Usually pork belly or fat backs that have been cured by the dry salt method.

Salt Sick–A mineral-deficiency disorder which results in loss of appetite, anemia, and even death to cattle and other animals on certain poor, sandy soil areas of Florida and similar soil areas elsewhere. This condition is known to be the result of a critical deficiency in the forage of many of the trace mineral elements, as copper, iron, and cobalt. Also called bush sickness.

Salted–Treated, seasoned, supplied, or preserved with salt.

Salve–Usually a preparation, consisting of a fatty, greasy, or jellylike base with added medications, which has properties of soothing, softening, healing, etc.

Sand Crack–A vertical split which may develop at any part of the wall of a horse's hoof but is commonly seen at the frog in the hind foot and at the inside quarter of the forefoot. Also called quartercrack.

Sand Disease–In horses and cattle, an inflammatory swelling of the stomach and small intestine due to the presence of sand in feed or water. Occurs frequently in flooded areas. See Sand Colic.

Sand-colic–A condition in horses which occurs when they graze on shallow-rooted vegetation and ingest a large quantity of sand or soil. See Sand Disease.

Sandfly–*Culicoides*, family Tendipedidae; a very small, two-winged, blood-sucking midge which is a nuisance in certain areas because of its sharp sting. Also called no-see-ums, punkies.

Sandy Bay–The coat color of a horse which is a light bay.

Sanitary Trap–A part of a vacuum pump installation in a milking machine which is designed for collecting and holding unwanted materials. It can be easily emptied and cleaned.

Sanitation–The developing and practical application of measures designed to maintain or restore healthful conditions, such as the treatment, removal, or destruction of contaminated or infested materials and possible sources of infection or infestation.

Sanitize–To disinfect and make sanitary the utensils used in preparation of food products, as dairy utensils.

Santa Gertrudis–A breed of beef cattle developed at the King Ranch in Texas (United States) from crossbreeding Brahmans (Zebu) and Shorthorns. Cattle of this breed have about $3/8$ Brahman blood and are said to be specially tolerant of subtropical conditions.

Saprogenesis–That period in the life cycle of a pathogen in which it is not directly associated with a living host and in which it may be either dormant or living as a saprophyte. See Saprophyte.

Sapropel–A slimy, fetid sediment on the bottoms of lakes made up of the organic debris from aquatic plants and animals.

Saproplankton–A mass of decaying, aquatic plant and animal material (plankton) which may be found floating on or in stagnant water.

Sarcoma–Cancers of various supporting tissues of the body (e.g., bone cells, blood vessels, fibrous tissue cells, muscle).

Sarcoptic Mange–A persistent and contagious skin disease of humans and animals, including dogs, cattle, sheep, goats, camels, rabbits, and horses. It is caused by the itch mite ***Sarcoptes scabiei***, a small, parasitic mite.

Sarcous–Of, or pertaining to, flesh or muscle.

Sardine Meal–A fish meal resulting as a by-product of the oil extraction from sardines which is used as a protein supplement in animal feeds.

Sardine Oil–A product obtained by the extraction of oil from the whole sardine or from cannery refuse of this species. It is used in animal feeds as a vitamin A and D supplement.

Satiety–Loss of desire to eat or drink (water).

Saturated–Filled to the maximum capacity under existing conditions.

Saturated Fat–A fat whose carbon atoms are associated with the maximum number of hydrogen atoms; no double bonds exist.

Saturation Point–(1) The point beyond which no further additions can be held, utilized, or accommodated; e.g., the water-holding capacity of a particular soil, the flow of goods to a market, or a population of plants. (2) In wildlife, the maximum density of population; the natural condition wherein a species is as abundant as living conditions permit.

Sausage–(1) A ground or chopped meat product of many varieties which is made usually of meat trimmings of pork, beef, veal, etc., often highly seasoned with various blends of spices. It may be a bulk product or enclosed in a tubular casing, such as livestock, intestines which have been specially prepared for this purpose. Sausage may be fresh, smoked, cooked, dry or semidry; e.g., pork sausage, bologna, frankfurters, salami, etc. (2) A long, cylindrical object whose shape suggests a sausage; e.g., the fruit of the sausage-tree.

Sausage Bull–A market classification for mature bulls whose flesh is used mainly in the making of sausage.

Sausage Casing–The small and large intestines of hogs, cattle, and sheep which are soaked, slimed, and scrubbed to be used as a covering for sausage meat.

SBM–Soybean meal.

Scab–(1) The crust formed over a wound or sore. (2) A mangy disease of animals, especially sheep, caused by the itch or scab mite. (3)

Any of the numerous fungal or bacterial diseases of plants characterized by dark or rough, scablike spots, as apple scab, cereal scab, cucumber scab, gladiolus scab, potato scab.

Scab Mite–Any of the certain, almost microscopic, parasitic mites which cause scab or mange of humans and animals. Common scab mites of humans and animals belong to the genera *Sarcoptes*, *Psoroptes*, and *Chorioptes*. These mites puncture the skin with their sharp mouthparts and feed on the raw skin, completely covered over by the scabs which are formed. Affected animals have an intense itchiness, become unthrifty, lose their hair or wool on infested areas, and may even die.

Scabies–A skin disease caused by mites; mange. See Scab Mite.

Scald–(1) A serious burn of plant leaves, fruit, limbs, and trunks which may result from exposure to intense sunlight or extreme heat; e.g., cowpeas, beans, and soybeans may develop tan blotches which eventually cause defoliation; exposed or nearly ripe tomatoes may scald and wither during extreme heat, etc. Winter sun scald of tree trunks is a variation of this condition due to cold temperature following a warm period. A condition resembling scald occurs in stored fruit, but is in fact caused by chemical esters. Also called sun scald. (2) An eroded spot on hillsides where subsoil is exposed. (3) A bare spot in freshwater coastal marshes caused by the killing of plant life by increases in salt content of the soil through encroachment of water from adjacent saltwater marshes. (4) To dip in hot water or to pour on hot liquid, as to scald a killed bird to loosen feathers. (5) To heat a liquid to a temperature just short of the boiling point, as to scald milk.

Scalding Vat–A tank, vat, or barrel used to hold very hot water for the dipping and scalding of slaughtered hogs or poultry to cause hair or feathers to loosen for easy removal.

Scale–(1) Any instrument used to determine weight. (2) Either of the pans of a balance. (3) A measure of dimension, concentration, or intensity. (4) A graduated series of steps or degrees on measuring devices. (5) An instrument or device on which graduated spaces for measurement have been stamped or attached. (6) A graduated list of prices, wages, etc. (7) Any thin flake that peels from a surface, such as from skin. (8) One of the thin, flat, membranous plates which forms the outer, protective covering of certain animals or parts thereof, as the shanks of birds, etc. (9) One of several rudimentary, specialized leaves which protects the buds of plants and deciduous trees in cold climates. (10) A thin, dry, membranous part of a plant or flower. (11) See Scale Insect. (12) The estimated sound content of a log or group of logs scaled by use of a given log rule. Net scale, the scale after deducting for defects. (13) To form into scales. (14) To cover with scales or with an incrustation. (15) To remove in thin layers, as by scraping. (16) To come off in scales. (17) To make according to scale.

Scalping–(1) In flour mills, separating the fuzzy growth from the ends of wheat berries by attrition, with or without the use of suction or blower fans. (2) In milling, partially separating the product from the break rolls into broken wheat and break flour by means of an arrangement of sieves, bolts, or screens of varying degrees of fineness. (3) In horses, a striking of the front of the hind coronet, pastern, or cannon against the front foot when running. (4) Removal of vegetation and a thin layer of soil.

Scaly Leg–A condition in fowls caused by a very small itch mite, ***Knemidokoptes mutans***, family Sarcoptidae, which burrows under the scales on the shanks and feeds on the leg tissue. The severe irritation set up results in the accumulation of grayish, powdery or crusty material under the scales, their loosening and lifting, and the enlarging of the shank, all of which creates a roughened appearance.

Scandinavian Feed-unit System–A feed evaluation system widely used in Scandinavian countries for measuring the relative values of different feeds. In this system, one pound of barley is taken as the standard. The feed-unit value for any other feed is the amount of that feed which is estimated to have the same productive value as a pound of barley.

Scapula–The shoulder blade of an animal; the principal bone of the shoulder girdle.

Scar–(1) An isolated or protruding rock. (2) A steep rocky eminence. (3) A bare place on the side of a mountain or other steep slope. (4) A wound that has healed.

Scarious–Not green but thin, dry, and membranous, often translucent.

Schlachter–A rabbi, who performs the ritualistic killing in the kosher method of slaughtering cattle and other animals for consumption by Orthodox Jews.

Schooled Horse–A horse, usually of a light-weight breed, that is trained or drilled for some special performance. See High School Horse.

Scimitar Boning Knife–A knife of medium length with an upward-curving (scimitar-shaped) blade which is adapted for boning, pelting, and skinning of slaughtered animals.

Scirrhous Cord–In castrated animals, abnormal swelling of the cut end of the spermatic cord caused by infection. It is usually accompanied by discharge of thick, white pus from the castration wound, and may be chronic.

Sclera–The tough, white, supporting covering of the eyeball which encompasses all of the eyeball except the cornea.

Scolex–The anchoring or holdfast organ of a tapeworm which serves for the attachment of the parasite to the intestinal lining. Also called head.

Scorch–(1) The appearance of seemingly burnt or brown irregular patches on foliage, bark, or fruits, as a result of heat, excessive sunburn, lack of water, action of bacteria or fungi, improper use of insecticides, fungicides, or some nutrients. (2) The brownish discoloration or the burning of foods, etc., in the process of cooking, drying, or by other use of heat. (3) See Anthracnose. (4) To burn superficially; to give the appearance of a burn, as to brown with heat.

Scotch Ham–A fresh ham that is skinned, trimmed of most fat, boned, and mildly cured according to a particular formula and practice followed in some parts of Scotland.

Scotch Hobble–A restraining device consisting of a rope passed over a horse's neck in front of the shoulders and tied as a loosely fitting rope collar. The slack end of the rope is passed under a hind pastern and back through the loop, enough slack being taken up to pull the foot from the ground.

Scour–(1) To abrade and flush as by the action of rapidly moving water on stream beds. (2) To cleanse dirt, grease, etc., by rubbing or scrubbing, as to scour wool. (3) To cleanse the bowels of an animal by purging. (4) In plowing, to pass through the ground cleanly without any soil sticking to the moldboard. (5) To rub off the flesh sticking to a hide. (6) See Scours.

Scours–Technically, a bacterial infection in calves and sheep that results in a whitish-yellow, foul-smelling diarrhea. Informally, any diarrhea. See Diarrhea.

Scout Bees–Worker bees searching for nectar or other needs, including suitable location for a swarm to nest.

Scrag End–The nape or back part of the neck, especially in a sheep. Also called scrag.

Scrapie–A disease of sheep and goats caused by a transmissible, filterable, and self-replicating agent which is considered to be a virus with unusual characteristics. The disease causes a progressive degeneration of the central nervous system, which in turn causes the animal to rub, scratch, and become debilitated and uncoordinated.

Scratch Feed–The grain part of a ration for poultry which consists of cracked or whole grain (corn, wheat, oats, etc.) or a mixture of such grains. Scratch feed is fed in hoppers, in the litter or, with range poultry, sometimes on the ground. Also called scratch grain.

Screen–(1) A sieve or grating, such as a frame covered with meshed wire or fabric, or a perforated plate, etc., which is used for separating the finer or coarser parts of soil, gravel, grain products in milling, seeds in cleaning, etc. (2) Any form of grating used to prevent invasion or escape of insects, reptiles, fish, or other kinds of animals. (3) A construction or planting used to conceal an unpleasant sight. (4) To consider a group for the purpose of selecting a relatively small number of individuals. (5) To sift by shaking through a screen. (6) To conceal, as with a screen.

Screenings–The refuse removed from grain harvested by a combine. Screenings include small, broken, or shrunken kernels of grain, small weed seeds, chaff, broken stems, etc. Screenings are sometimes used for feed.

Screw Worm–The larva of the fly *Cochliomyia hominivorax*, family Calliphoridae, occurring in the warmer regions of America which lays its eggs at the edges of sores and wounds of animals. The larvae, or maggots, are able to penetrate the flesh of the host by two hornlike mouthparts and, if left uncontrolled, will go deeply into living tissue causing serious results that are often fatal.

Scrotal Circumference–A measurement of testes size obtained by measuring the distance around the testicles in the scrotum with a tape. Testicle size is related to semen-producing capacity and age at puberty of female progeny.

Scrotal Hernia–A protrusion of loops of intestine or other viscera into the scrotum through an enlarged inguinal ring or through an accidental opening in the inguinal region which results in the scrotum being greatly enlarged.

Scrotum–In most male mammals, the pouch of skin containing the testicles and related structures.

Scrub–(1) Land which under natural conditions supports only shrubs or a dwarfed, stunted growth of trees. The soils are generally infertile, but scrub may appear on various kinds of sites. (2) The stunted trees or shrubs, often in dense stands, on such land. (3) An animal of inferior breeding or condition. (4) Designating an inferior product, animal, etc.

Scud–In tanning, to scrape a dehaired and trimmed hide or pelt to remove any hairs, lime, or other materials remaining in the hair follicles.

Scum–(1) Minute forms of lower plant life and impurities that gather on the surface of stagnant water. (2) Extraneous matter or impurities that rise to the surface of boiling or fermenting liquids. (3) The condensation of solids at the surface of boiled milk upon cooling.

Scurfy–Covered with small scales.

Scurs–A hornlike substance that grows from the skin at the location of the horn pits on polled animals.

Sealed Brood–Immature bees in their late larval and pupal stages within capped cells of the comb.

Seashells and Coral–A source of calcium for feeds. Since seashells are almost pure calcium, they are good sources of calcium for all classes of animals. The shells are finely ground and added to feed.

Season–(1) A portion of the year; e.g., a growing season, when plants grow; the fire season, when danger of fire is greatest; the grazing season, when grazing is possible. (2) To bring to or develop into a condition of greater usefulness, as to dry lumber by natural or artificial means. (3) To flavor food with spices, salt, etc. (4) An animal that is ready for breeding is said to be in season.

Seasonal Breeder–An animal that breeds naturally during a certain season of the year and rarely at other seasons; e.g., sheep, goats, etc., are seasonal breeders in the fall.

Seasonal Distribution–(1) In pasturage, the progressive grazing in a sequence of moves from one part of a range to another as vegetation develops. (2) In meteorology, the amount and kind of precipitation, wind movement, etc., in different seasons.

Seasonal Grazing–The grazing of a range only during a certain period or periods of the year which roughly correspond to one or more of the four seasons.

Seasoning–(1) The process of drying (curing) lumber or other forms of wood for better utilization. Seasoning is natural when the drying is done by air or other natural means; in artificial seasoning, the drying is carried out by means of a kiln, oils, or electrical appliances. (2) Adding salt, pepper, etc., to meat and other food products.

Sebaceous–Referring to anything that secretes or resembles fatty matter; oily, greasy, as the secretions of the sebaceous glands of the skin.

Seborrhea–A group of diseases of the skin involving the sebaceous glands, characterized by an accumulation of dry scurf or by the formation of an excessive oily deposit.

Second Cuts–Short lengths of wool resulting from cutting wool fibers twice under careless shearing. An excessive number of second cuts decreases the average fiber length and hence, depreciates spinning quality.

Second Joint–The thigh or meaty portion of the leg of a fowl.

Second Stomach–See Reticulum.

Secondaries–The long stiff wing feathers growing from the middle wing segments of a fowl.

Secondary Infection–(1) An invasion by a second and different organism after a first organism has become established in a host animal or plant. (2) An infection resulting from inoculum produced by an organism after it is first established in a host plant.

Secondary Insect–Any insect which follows, is associated with a primary form, or feeds upon plant tissue but which is incapable of initiating injury by itself.

Secondary Parasite–A parasite which establishes itself in or upon a host that is a primary parasite.

Secondary Range–A range that is lightly used or unused by livestock under minimal management and will ordinarily not be fully used until the primary range has been overused.

Secondary Screw Worm–The larva of the screw worm fly, *Calitroga hominovorax*, family Calliphoridae, which normally infests unburied carcasses and transfers its attack to living animals only when it has become excessively abundant. It frequently attacks wounds already infested by the true screw-worm and, in southern United States, may be the cause of primary invasion of wounds in animals.

Secretin–A hormone produced in the intestine (duodenum) which controls the secretion of the pancreatic juices or enzymes. See Hormone.

Secretion–The metabolic act or process of synthesis and liberation of substances from cells or glands of animals or plants, as saliva, milk, etc.

Section–(1) A distinct part of a town, country, region, people, etc., as a section of the population of the country. (2) A unit of the township in the United States General Land Office Survey of the public domain. The full section is one square mile, or 640 acres, and the full township consists of 36 sections. (3) In forestry, a logical or natural length of the fire-control line which is handled as a unit during fire suppression. (4) In beekeeping, one of several small frames of basswood placed in a hive in which the bees build surplus honey. A filled frame weighs from 12 to 16 ounces. See Section Comb Honey. (5) Segments making up the interior of a citrus fruit.

Section Comb Honey–Honey in sealed comb produced in thin wooden frames called sections.

Sectioned and Formed Ham–Ham made from pieces of meat trimmed from the hind leg and after curing formed into a loaf, placed in a casing, and cooked and smoked like normal ham.

Secundines–The afterbirth; the placenta.

Sed–(Spanish) Thirst.

Sedative–A drug that calms an animal.

Sedentary–(1) Formed in place without transportation from the underlying rock or by the accumulation of organic material as in peats and mucks; said of some soils. (2) Attached, as an oyster, barnacle, or similar shelled invertebrate.

Seed–(1) The embryo of a plant; also kernels of corn, wheat, etc., which botanically are seedlike fruits as they include the ovary wall. (2) Propagative portions of a plant other than true seeds, as tubers, bulbs, etc. (3) Offspring or progeny. (4) Sperm or semen. (5) To place seed on or in the soil for the production of plants of the same species; to sow seed. (6) Overseed; to sow seed on an area where a crop is established, as clover on wheat, lespedeza on oats. (7) To produce seed. (8) To extract seed, as to seed cherries or grapes.

Seed Stock–(1) Pedigreed or well-bred livestock, which is maintained for breeding purposes. (2) A specially selected strain of plants, or seeds thereof, which are to be used as parents of future generations.

Seed Tick–The newly hatched, six-legged larva of a tick, especially of the cattle tick *Boophilus annulatis*, a one-host tick in which the larva, the nymph, and the adult are all found on cattle. Newly hatched larvae are found on the ground or on grass, weeds, and other objects in fields where infested cattle have pastured.

Seedy–Wool containing an appreciable amount of seeds or other vegetable matter.

Segregation of Genes–The separation of genes on two homologous chromosomes and their distribution to separate gametes during gametogenesis; genes that once existed in pairs in a cell will become separated and distributed to different gametes.

Select–(1) A grade (quality) of a beef carcass having slight marbling at "A" maturity. See Maturity; Marbling. (2) A standard grade of hardwood lumber, being the highest grade with respect to the amount of clear usable material (clear face) that can be cut out of the pieces.

Selection–Choosing certain individuals for breeding purposes in order to propagate or improve some desired quality or characteristic in the offspring.

Selection Difference–The difference between the average for a trait in selected cattle and the average for the group from which they came. Also called reach.

Selection Index–A formula that combines performance records from several traits or different measurements of the same trait into a single value for each animal. Selection indexes weigh the traits for their relative net economic importance and their heritabilities plus the genetic associations among the traits.

Selection Threshold–A trait or standard established by a breeder, below which the breeder will reject potential breeding animals.

Selective Breeding–The breeding of selected plants or animals chosen because of certain desirable qualities or fitness, as contrasted to random or chance breeding.

Selective Grazing—The tendency for livestock and other ruminants to prefer certain plants and to feed on these while grazing little on other species.

Selenium—A nonmetallic element, related to sulfur, rarely present as more than a trace in soil. It is essential for animals, but where it is present in excess it is absorbed by plants and may be toxic to grazing animals. See Blind Staggers.

Selenium Poisoning (Selenosis)—A disease of grazing animals and people caused by ingesting plants grown on irrigated soils containing an excess of selenium. In horses, it is characterized by mange of the mane and tail, overproduction of hoof tissue, and in severe cases it causes death. Recently, arsenic has been found to have some use as an inhibitor of the disease. See Alkali Disease, Blind Staggers.

Self-feeder—(1) In poultry or other livestock farming, any feeding device by means of which the animals can eat at will; e.g., a hopper which supplies feed by gravity to a boxlike trough. (2) Any automatic mechanism for supplying materials to a given point in an operation.

Self-sucker—A cow that has the habit of sucking its own teats.

Semen—A fluid substance produced by the male reproductive system containing spermatozoa suspended in secretions of the accessory glands.

Semen Collector—Any device or receptacle used for the collection of semen from males of domestic animals. It may consist of an artificial cloaca or vagina, a test tube, a small beaker, etc. See Artificial Vagina.

Semen Plasma—The fluid portion of semen.

Semen Tank—A portable tank used to transport or store frozen semen.

Semi-bright—Grease wool that lacks brightness due to the environment under which it is produced, although it is white after scouring.

Seminal Fluid—See Semen.

Seminal Vesicle—A gland attached to the urethra, near the bladder, and which produces fluids to carry and nourish the sperm.

Semipermeable Membrane—A membrane that permits the diffusion of one component of a solution but not the other. In biology, a septum which permits the diffusion of water but not of the solute.

Senepol—A breed of beef cattle that originated by the systematic breeding of Senegalese cattle with Red Poll cattle. They are polled and red.

Senility—Condition of organisms, especially people, that in old age revert to development resembling younger stages.

Senior Yearling—(1) In judging beef cattle, a bovine born on or after May 1 and on or before December 31 of the second preceding year. (2) In judging dairy cattle, a bovine born on or after July 1 and on or before December 31 of the second preceding year. (3) In swine judging, a pig which was farrowed on or after September 1 of the second preceding year and on or before February 28-29 of the preceding year.

Sensitization—Condition of being allergic or sensitive to, for example, a vaccine.

Sensorium—The whole sensory apparatus of the body, principally sight, smell, hearing, taste, and touch (which includes pain). The sensorium gathers all of the external, and some internal, impressions and conveys them to the brain or spinal cord or both for proper evaluation and response by the organism.

Sensus—(Latin) Meaning.

Separator—Mechanical device for separating cream from milk by centrifugal force.

Sepsis—A state of contamination or poisoning by pathogenic bacteria.

Septate—(1) Containing or divided by a septum or septa. (2) Fungus hypha or spores which have cross-walls. (3) Any fruit in which the ovary is divided into sections.

Septic—(1) relating to, or caused by, the presence of disease-producing organisms, or their toxins, in the blood or tissues. (2) Putrefactive; putrefying. (3) A term sometimes used to refer to conditions where dissolved oxygen is absent and decomposition is occurring anaerobically, as in a septic tank.

Septic Sore Throat—A diseased condition of the throat caused by certain strains of streptococcus bacteria. Epidemics of septic sore throat are very largely milk-borne. Milk may be infected by cough-spray of milkers, other persons handling milk, or from the udder of a cow infected from a human source.

Septicemia—(1) Blood poisoning; a disease condition which results from the presence of toxins or poisons of microorganisms in the blood. (2) Usually minor disease of adult bees caused by *Pseudomonas apiseptica*.

Septum—Any dividing membrane or other layer in plants and animals: (a) a partition separating two cavities; (b) a division wall in a compound ovary; (c) a cross-wall in a fungus hypha or spore.

Sericious—Closely covered with soft, silky hair as a sericeous leaf.

Sericulture—The growing of silkworms.

Serological—Designating certain laboratory tests used in detecting diseases which involve the use of the liquid portion of the blood.

Serum—(Plural, sera) (1) The clear portion of any animal liquid which is separated from its cellular elements. It usually applies to the amber-colored liquid called blood serum which separates in the clotting of blood from the clot and corpuscles. (2) The whey of milk. (The term is applied often erroneously to skimmed milk.)

Serum Neutralization—A test employed with certain viral infections to determine the presence or absence of specific antibodies in the blood serum of animals by mixing varying amounts of serum with a constant amount of known virus and determining afterwards by animal inoculation if inactivation of the virus has taken place.

Serve—To copulate with a female; to cover, as a stallion, bull, or other male animal.

Sesame Oil Meal—A livestock feedstuff which is the ground residue obtained from sesame seed after the extraction of part of the oil by pressure.

Sesamoid–A small bony or cartilaginous nodule developing in tendons, as at a joint, such as at the fetlock joint of a horse.

Set–(1) A small propagative part, a bulb, shoot, tuber, etc., which is suitable for setting out or planting, as an onion set. (2) A number of things usually used together, or forming a complete collection, as a set of tools, harness, etc. (3) The initial swelling of the ovary of a flower soon after petal fall. Also called setting, fruit setting. (4) To put in a particular place, position, condition, direction, adjustment, etc. (5) To put eggs under a fowl or in an incubator to hatch them. (6) To put a price or value on something. (7) To put in proper working condition. (8) To adjust the teeth of a saw. (9) To fix a trap or a net to catch animals or fish. (10) To fix in the ground, as a post, tree, or plant. (11) To become stiff, firm, or hard, as cement sets. (12) To adjust into proper apposition the ends of a broken bone, as to set a fracture.

Set Stocking–(1) As used in Australia, synonymous with continuous grazing. (2) Keeping animals in a given area of pasture continuously at a predetermined level of stocking for a defined period.

Setting Hen–A broody hen in the act of incubating eggs.

Settle–(1) To sink gradually to a lower level, as of sediment, etc. (2) To cause a female to become pregnant, as a bull settles the female with which he has had coitus. (3) To occupy land usually for the purpose of farming, ranching, or homesteading.

Settled–(1) Designating suspended particles or impurities in liquids dropped to the bottom. (2) Designating a pregnant cow or mare.

Sex–(Latin, *sexus*) The distinction between male and female plants and animals. Ova (macrogametes) are produced by the female and sperm (microgametes) by the male. The union of these distinctive germ cells results in a new individual.

Sex Alleles–Hereditary characteristics of bees that, in part, determine the sex of the individual bee.

Sex Character–The peculiarity of appearance, other than the presence of sex organs, that distinguishes one sex from another, such as the thick neck and bold, rugged head of a bull as contrasted to the more finely turned neck and head of a cow. Technically, these are secondary sex characteristics.

Sex Chromosomes–A pair of chromosomes in animals that determines the sex of the progeny depending upon which one is distributed; one sex usually has two of the same kind of sex chromosome in its cells while the other sex has two kinds of sex chromosomes; in mammals the female is XX and the male is XY; in birds, the male is ZZ (or XX) and the female is ZW (or XY).

Sex Feathers–(1) Feathers of a chick that designate the sex. The male's feathers are different from the female's. (2) The two top tail-feathers which characterize the mature males of most breeds of duck. These two tail-feathers at their outer ends turn upward and forward making a pronounced curl. This characteristic feathering is useful in sex determination.

Sex Linkage–Association of character with sex due to the fact that the gene for that characteristic is in a sex chromosome.

Sex Pheromone–Sex pheromones are now made synthetically to bait traps and thereby make an estimate of population density of a given species of insect. In this way scientists can determine when control measures should be initiated, known as the economic threshold of insect numbers. Most common has been the use of synthetic female sex hormones to trap male Lepidoptera. A second use of synthetic female sex hormones has been to confuse the male and thereby prevent mating. See Pheromone.

Sex Ratio–The relationship existing between the number of male and female animals within a given herd or band.

Sex-limited Traits–Traits that are expressed only in one sex although the genes for the trait are carried by both sexes.

Sex-linked–Characters developed from genes located on sex chromosomes. The character may be used to determine the sex of an animal; e.g., barring of Barred Plymouth Rock chickens.

Sexed Chicks–Baby chicks separated according to sex at hatching time, making possible the purchase of either pullets or cockerels.

Sexing Baby Chicks–Any method for the determination of sex of baby chicks at hatching time. The method generally used in the United States was introduced by the Japanese who probably acquired it from the Chinese. Briefly, it consists of opening the cloaca of the chick by a slight pressure and in distinguishing differences in the parts thus exposed. With chicks of certain colored breeds, such as the Barred Plymouth Rock, and with certain cross-bred chicks, sex can be determined with reasonable accuracy by differences in color pattern.

Sexual–A method of reproduction which involves the union of sperm and an egg.

Sexual Dimorphism–Differences in form exhibited by males and females of the same species.

Sexual Infantilism–The failure to develop sexually.

Sexual Reproduction–In plants, exchange of genetic material between male and female gametes through fusion of the pollen tube with the ovule releasing the male gamete into the egg cell resulting in a zygote. See Zygote.

Shackling Pen–The pen in a slaughterhouse in which the hind legs of hogs are secured by one end of a shackle, the other end of which is attached to a mechanical hoist.

Shaft–(1) Either of the pair of long bars (thills) sticking out in front of a vehicle between which a horse or other animal is harnessed. (2) The stem of a feather to which the barbs are attached. (3) A connecting rod between a power source and a pulley, gear, or other contrivance to distribute such power.

Shafty Wool–Wool of extra good length, sound, and well grown.

Shampoo Brush–A grooming brush that can be attached to a water hose. It is used in washing cattle for the show. The plastic bristles combined with the running water help remove dirt and grime from the animal's coat.

Shank–(1) That part of the leg joining the knee to the ankle in humans, or the corresponding part in various animals. (2) A market cut of meat from the lower part of the foreleg of a dressed carcass of beef, veal, mutton, or lamb, or a corresponding cut from the hind leg. (3) The tarsus of birds: the part of the leg, below the hock joint, which is covered with scales. (4) That part of a tool, implement, etc., between the acting part and the part by which it is held, as a cultivator-shovel shank, etc. (5) Any connecting part of a plant; a foot stalk, as the part between the stalk and the corn ear.

Shank Feathering–Those feathers which grow on the outer side of the shank of certain birds, as those on the Cochin, Brahma, and Langshan breeds of chicken.

Sharp Freezing–Freezing of products, fruits, vegetables, poultry, meats, dairy products, etc., at low temperatures (–5°F to –20°F).

Shearing–(1) The act or operation of removing wool from a sheep or goat by means of shearing machines. (2) The thinning or heading back of shrubs or trees by removing portions with trimming or pruning shears, as in the trimming of hedges, etc.

Shearing Lamb–In marketing, a subclass of lambs that are in full fleece and sold primarily for their wool.

Shearling (1) A sheep after being shorn for the first time. (2) Pelt of slaughtered sheep carrying one-fourth to one inch growth of wool.

Sheath–(1) The leaf base covering a stem or branch, such as the lower part of leaves of grasses which surround the stem. (2) The enclosing and protecting tubular structure within which the penis of a stallion, bull, and certain other male animals is retracted.

Shed–(1) A simple structure for shelter of livestock, storage of feeds, or both, open on one or more sides, which may be separate from or attached to another structure. (2) To lose or cast off something, as a plant sheds leaves, pollen, seed, etc.; animals shed hair; birds molt or shed feathers; a roof sheds rain, etc.

Shedder–A brushlike device used to remove dead hairs from an animal's coat.

Sheep–Any of a variety of cud-chewing mammals, genus ***Ovis***, related to goats, especially ***Ovis aries***, which includes many breeds domesticated for their heavy wool, edible flesh (mutton), or skin.

Sheep Bell–A bell hung about the neck of a sheep whose tinkling indicates the location of the flock.

Sheep Bot Fly–The maggot or larva of the sheep bot fly, ***Oestrus ovis***, family Oestridae. The eggs are hatched in the body of the female fly and the very small, active maggots are deposited in the nostrils of sheep, goats, and deer. The maggots work their way to the sinus cavities where they lacerate the tissues and feed, ultimately growing to one-half inch or more in length. Affected sheep shake their head, stamp their feet, and hold their noses to the ground. Also called head grub, nose fly, sheep grub.

Sheep Dog–A dog which is trained and used to tend or work sheep, such as an Old English sheepdog or similar dog of shepherd or collie breeding.

Sheep Drench–A liquid, medicinal preparation, usually a vermicide, which is given to sheep by mouth for the removal of stomach worms, tapeworms, hookworms, etc.

Sheep Fescuegrass–*Festuca ovina*, family Gramineae; a perennial grass, some forms of which are native or introduced on western ranges (United States). It is sparingly used for pasture and in some turfgrass mixtures where rough turf is desirable. Native to Europe and Asia; naturalized in North America.

Sheep In-the-wool–Unshorn or unclipped sheep.

Sheep Ked–*Metophagus ovinus*, family Hippoboscidae; a degenerate, wingless, louselike fly that crawls about over the sheep's skin in the wool and feeds by thrusting its sharp mouthparts into the flesh and sucking blood. The irritation causes sheep to rub, bite, and scratch at the wool, thus spoiling the fleece. Also called sheep tick.

Sheep Lice–(1) Lice that commonly infest sheep, primarily the sheep-biting louse, but also ***Hinognathus pedalis***, a blood-sucking body louse. (The latter two are more often pests of other animals, particularly goats.)

Sheep Loco–*Astragalus nothoxys*, family Leguminosae, a poisonous plant found in southern Arizona, United States. Both its green and dried foliage are poisonous, causing dullness, irregularity of gait, lack of appetite, dragging of the feet, a solitary habit, loss of flesh, and shaggy coat in animals. As the animal ceases to eat, it dies. See Loco.

Sheep Manure–The dung of sheep either moist or dried. Dried, it is sold for fertilizer. The approximate analysis is: nitrogen (N) 1.4 percent, phosphoric acid (P_2O_5) 1.0 percent, and potash (K_2O) 3.0 percent.

Sheep Month–The amount of forage or feed which is necessary to maintain a mature sheep or a ewe and its suckling lamb for thirty days. It is commonly figured as one-fifth of a cow month.

Sheep Pox–A contagious sheep disease, similar to smallpox, causing fever and pock marks on the skin. Also called ovinia, sheep pock (British).

Sheep Run–The whole property of a sheep operation, including the residence and other buildings belonging to it, the sock, and freehold and lease-hold country. Originally, the hut, yards, or other buildings where a squatter or his agent stationed himself. See Squatter.

Sheep Scab Mite–*Psoroptes equi* var. ***ovis***, family Psoroptidae; a scab mite that causes severe injury to sheep by puncturing the skin with its sharp mouthparts until the lymph exudes. A scab forms over the larvae, under which they feed, causing added scab formation which eventually lifts the hair out by the roots. Extreme irritation and itching are caused by the work of the mites, and the sheep loses bits of wool on weeds, fences, and other objects as a result of constant rubbing. Psoroptic mange is the most frequent type of mange affecting sheep and is also called scab, scabies. See Scab Mite.

Sheep Sickness–See Braxy.

Sheep Stand (Table)–A platform on which a sheep is placed and held for grooming in preparation for showing.

Sheep Tag–(1) A heavy, dung-covered lock of wool which usually hangs from the crotch of a sheep. (2) A metal tag which is fastened to the ear of a sheep for identification.

Sheep-biting Louse–*Bovicola ovis*, family Trichodectidae; a small louse of sheep and goats which eats off wool fibers and skin scales, tangles and soils the hair, and causes severe irritation to the skin. These lice do not suck blood but when numerous may cause raw sores, scratching, and rubbing. Also called red-headed louse.

Sheeping Down–A method of harvesting and utilizing a crop, as corn, by the use of sheep which are turned into the field to feed. See Hog Down.

Sheepskin–(1) The skin of a sheep, especially one dressed with the fleece on it. Also called sheep pelt. (2) Parchment or leather made from the skin of a sheep, used for clothing, bookbinding, diplomas, etc.

Shelf Life–The length of time fresh or canned fruits remain saleable while on a shelf or in storage.

Shell–(1) A hard outer covering of certain animals, such as turtle, mollusk, or crustacean, which when ground may be used as a poultry feed supplement. (2) The hard outer covering of a fruit or seed, as a nut shell, etc. (3) The hard calcareous exterior covering of an egg. (4) To remove or separate from its shell or pod, as to shell nuts, peas, etc. (5) To drop out of a shell, or out of a cluster, as seeds, grapes, etc. (6) To separate seeds or kernels (of corn, wheat, oats, etc.) from the cob, ear, or husk.

Shell Membranes–Either of the two membranes, the inner and outer shell membranes, surrounding the egg substance and located immediately inside the shell of the egg. At the large end of the egg, the membranes are separated to form the air cell.

Shell Quality–The various characteristics of the shell of an egg, including thickness, porosity, texture, etc., which may vary widely between hens and may be improved through selective breeding.

Shepherd's Crook–The staff used by a shepherd; specifically a staff with a crook at one end which is used in catching and holding a sheep (used by Old World shepherds since ancient times).

Shetland–A breed of small, stocky ponies (native to the Shetland Islands), noted for hardiness, patience, surefootedness, and strength as weight carriers. A desirable height is 39 to 42 inches and the popular colors are black, chestnut, and pinto. The ponies are generally affectionate and sensitive, and because of their small size are popular for children.

Shin–The part of a plow at the forward edge of the moldboard that severs the side of the furrow slice.

Shin Boot–A shaped, leather piece fitted to the skins and cannon bone of horses, especially saddle and running horses, to protect against injury from striking with a hoof when in motion.

Shin Soupbone–A soupbone from the foreshank of a beef carcass.

Ship of the Desert–The camel; either of the two members of the genus *Camelus*, a large, herbivorous mammal which can travel great distances without water. Its large, flat feet enable it to walk over the desert sand without excessive sinking. The one-humped, or Arabian camel (*C. dromedarius*) is the larger of the two; the two-humped, or Bactrian camel (*C. bactrianus*) is better suited to hilly, rocky country and cold climates. Both are used as workhorses throughout the arid areas of North Africa, Arabia, Asia, and northern Australia. Used in India and the Sudan for cavalry. Imported into the United States in the nineteenh century as experimental cavalry in Texas but later abandoned.

Shipping Fever–Any of several infections that may occur in cattle or other animals shipped to stockyards, or from place to place, if they are exposed to pathogenic organisms when their resistance is lowered by the nervous strain and rigors of being shipped.

Shivering in Little Pigs–A blood-sugar deficiency disease in newly born pigs, characterized by shivering, dullness, no desire to nurse, upright hair, and coma. See Hypoglycemia.

Shives–Small particles of vegetable matter other than burrs present in wool.

Shoat–A pig of either sex, usually between 60 and 160 pounds.

Shock–(1) A quantity of a harvested grain crop or corn, etc., set together upright in a field. (2) A small pile of hay. (3) A state of collapse in animals, which results from severe loss of blood flowing surgery, accident, or sometimes under conditions of undue stress. (4) The severe symptoms in plants following acquisition of virus. Later symptoms are absent or milder. (5) A parcel or lot of sixty pieces or units. (6) Twelve sheaves of small grain. (7) To place grain, hay, etc., in shocks.

Shod–Equipped with shoes, as a shod horse.

Shoddy–(1) That leather made by grinding waste leather to a pulp and pressing it into solid sheets with or without the addition of a binding material. (2) Reworked wool that has been recovered from wool cloth trimmings, rags, etc. See Virgin Wool. (3) Anything of an inferior quality.

Shorn–(1) Designating a sheep, lamb, or goat that has had its fleece removed by shearing. (2) Referring to wool that has been sheared.

Shorn Hogget–A lamb after its first shearing and until it gets its first two teeth.

Shorn Wool Clip–In market reports, the total quantity of shorn wool produced in a given year.

Short Ribs–A retail cut of beef for roasting or broiling which consists of short lengths of ribs cut from the plate of a forequarter.

Short Yearling–An animal which is slightly less than a year old.

Short-coupled–A term used to describe a horse having a short distance (usually not more than four fingers width) between the last rib and the point of the hip.

Short-nosed Cattle Louse–*Haematopinus eurysternus*, family Haematopinidae; a bloodsucking cattle louse which pierces the skin, sucks blood, sets up irritation, and causes rubbing. They are found on the head, neck, or along the inner surface of the legs, where they appear as blue patches on the skin.

Short-tailed Sheep–Sheep with naturally short tails, such as the marsh or moorland sheep of northern Europe.

Short-time Feed–A system of fattening cattle for slaughter in which heavy stockers or feeders are fed a heavy grain ration for periods of thirty to ninety days and then sold.

Shortening–Any fat which has the property of making pastry, cakes, etc., friable, easily crumbled, or short.

Shorthorn–A breed of beef cattle which has characteristically short horns, originating in England. These cattle vary in type from beef conformation (Scotch type) to the dual-purpose (English type) which combines to some extent both beef and dairy qualities. The breed is characterized by various colors: red, white, roan, and red and white combined in different patterns. The Polled Shorthorn is a recognized strain.

Shorts–(1) In the florist's trade, roses with short stems (approximately 9 inches in length). (2) Lumber which is shorter than standard lengths. (3) Short pieces or locks of wool that are dropped out while fibers are being sorted. (4) See Wheat Shorts, Wheat White Shorts.

Shote–A young hog of either sex which is weaned but which usually weighs less than 150 pounds. Also spelled shoat.

Shoulder–(1) Of livestock, the upper part of the forelimb and/or the adjacent part of the back; the scapula region. (2) The upper foreleg and/or the adjacent part cut from the carcass of a hog, sheep, or other animal; the upper part of the carcass to which the foreleg is attached. (3) The part of a bird at which the wing is attached. (4) In turpentining, the uppermost corner of a face. (5) A shoulderlike part or projection, as the projection around a tendon. (6) A sudden inward curvature in the outline of something, as the flaring part between side and neck of a milk bottle. (7) The edge of a road adjacent to the traveled way or pavement.

Show Halter–A leather halter that is used on cattle to lead them around the show ring. Show halters look nicer than regular rope halters.

Show Stick–A stick of fiberglass, wood, or aluminum, about 4 to 5 feet long. It has a short hook on the end that is used to place the feet of a beef animal in a show ring. It is also used to straighten the back of the animal and to calm it by scratching the belly.

Showmanship–The art or skill of showing or displaying products, animals, etc., to advantage.

Shrimp Meal–A livestock feedstuff supplement produced from the undecomposed dried wastes of the shrimp industry. It contains the head, hull, or the whole shrimp, either singly or in mixture, and not more than 3 percent salt (NaCl), unless otherwise labeled, and in no case should it contain more than 7 percent salt.

Shrinkage–(1) The process of reducing in dimension, weight, or volume. (2) With livestock, the loss in weight of an animal from the time it is taken off feed to the time it is marketed or slaughtered. (3) The loss in weight of grease wool due to scouring, expressed as a percentage of the original weight. (4) The loss in weight (moisture loss) of grain from harvest time to some later date. Such shrinkage is commonly reflected in the selling price. (5) The contraction of wood caused by drying, usually expressed as a percentage of some specific dimensions (or volume) of the wood when green. (6) Loss in weight of either meat or carcass stored under refrigeration. Also called cooler shrink, freezer shrink.

Shrivel–To contract and wrinkle, as of leaves, grain, fruit, etc.; to dry up.

Shropshire–A widely distributed, English breed of hornless, brown-faced, medium-sized sheep. The breed possesses both mutton and wool qualities but is kept primarily for mutton and lamb production. Individuals are distinguished by having a complete covering of wool from the tip of the nose to the feet. The wool is of medium length and fineness, and grades from (in United States) three-eighths to quarter blood. The average fleece yields between 8 and 10 pounds.

Shroud–In butchery, a sheet of unbleached duck cloth that is immersed in warm water and pinned over the outside of a side of arm beef carcass before it is placed in the cooler so that after chilling the carcass will have a smoother appearance.

Shy–(1) Of, or pertaining to, a horse that is easily frightened or startled; timid; skittish. (2) To act in a startled manner, as a frightened horse may break gait and jump violently to one side, or otherwise show fright until he has a chance to observe the cause of his fear.

Shy Breeder–A male or female of any domesticated livestock that has a low reproductive efficiency. It is associated most often with males or females from highly developed breeds.

Sib–In genetics, a brother or sister.

Sib Testing–A method of livestock selection in which an animal is selected on the basis of the performance of its brothers and sisters.

Sib-Mating–Mating between siblings, two or more individuals having one or both parents in common.

Sibling–One of several sons and/or daughters of the same parents. See Sib.

Sick–(1) Suffering from illness of any kind; not sound or in fit condition. (2) Designating a soil unfit for the profitable production of certain crops as a result of being infected with disease organisms or because of the excess or deficiency of certain elements.

Sickle-hocked–A term applied to an animal when the hind legs set too far forward, giving the impression of a sickle when viewed from the side.

Sickles–The longer, curved tail feathers of a male chicken as distinct from the smaller, similarly curved tail-coverts, the lesser sickles.

Side–(1) In the body of an animal, either of the two lateral surfaces or parts extending from the shoulders to the hips. (2) An entire lateral half of a beef, veal, or pork carcass split lengthwise through the backbone and the corresponding midundersurface. (3) Flesh taken from the lateral surface of an animal carcass, as a slab taken from the side of a hog carcass and used for preparing a side of bacon.

Side Check–A leather strap attached at the side of a horse's bridle to limit side movement of the head. See Checkrein, Overhead Check.

Side Leak–A small, unnatural opening in the wall of a cow's teat; a defect in dairy cows.

Side of Bacon–An entire slab of unsliced bacon.

Side of Beef–An entire lateral half of a beef carcass.

Side Saddle–A saddle for women so constructed that the rider sits with both feet on one (usually the left) side.

Side Step–A stepping sideways; a foot movement expected of a show horse and desirable in a well-trained pleasure horse, as it assists in lining up horses in a show ring. Also called transversal.

Side-bone–Ossification of the lateral cartilages of a horse's foot, noticeable as hard, bony projections immediately above and toward the rear quarter of the hoof head. It is an unsound condition occurring, especially in draft horses, as a result of injury, conformation, excessive use, or perhaps old age. Also called shell bone.

Siding–(1) Boards used to cover the sides of buildings; one board usually being lapped over the upper edge of the board below. Also called clapboard, weatherboard. (2) Removing the hide from the side of an animal, as from a beef or veal carcass.

Sigmoid Flexure–An S-shaped fold in the penis of a bull, ram, or boar that straightens during erection and allows it to extend from the sheath for copulation.

Signal Word–A word on the label of a pesticide container that is required by law to designate the relative toxicity of the chemical "Danger - Poison" denotes a highly toxic compound; "Warning" means that the chemical is moderately toxic; and "Caution" denotes a compound that is slightly toxic.

Silage–A crop that has been preserved in moist, succulent condition by partial fermentation in a tight container (silo) above or below the ground. The chief crops stored in this way are corn, sorghum, and various legumes and grasses. The main use of silage is in cattle feeding.

Silage Corn–Usually a vigorous-growing corn that is planted thicker than grain corn to produce a high yield of forage for ensiling. The kernels should be well dented and the ears should contain about 50 percent moisture at silo-filling time.

Silage Crop–Any of several crops grown and harvested for silage, as corn, sunflowers, sorghum, small grains, legumes, grasses, etc., cut green and stored in a silo.

Silage Harvester–A field implement that harvests standing corn, cuts it into ensilage length, and elevates the chopped fodder into a truck or wagon box ready for transportation to and placement in the silo.

Silk–(1) The long, silky styles with stigmas of the corn plant, each of which is connected to an ovary on the ear and each of which must be pollinated to form a full ear of kernels. The silks are susceptible to pollination before they emerge from the husk and remain so for about two weeks afterwards. (2) The soft, fine, and shiny fiber which is produced by the silkworm.

Silkworm–The larva of the moth *Bombyx mori*, family Bombycidae, which feeds on mulberry leaves and in three or four weeks grows into a large caterpillar about 3 inches long. It then constructs its cocoon by secreting a curious saliva that hardens upon exposure to the air to form soft, delicate threads of remarkable strength and pliability. The cocoons are skillfully unwound by expert operators and the threads are combined to form reels of raw silk. Silkworms that produce a single brood in a season are referred to as monovoltine, those producing more than one brood, polyvoltine. Native to China.

Silkworm Mulberry–*Morus alba* var. *multicaulis*, family Moraceae; a shrubby tree grown for its foliage which is fed to silkworms. Native to China.

Silo–A pit, trench, aboveground horizontal container, or vertical cylindrical structure of relatively air-tight construction into which green crops, such as corn, grass, legumes, or small grain and other feeds are placed and converted into silage for later use as a livestock feedstuff.

Silver-gray–A coat color of a horse which is light grey with a white mane and tail.

Silverleaf Nightshade–*Solanum elaeagnifolium*, family Solanaceae; a perennial, poisonous herb which has silvery-haired stems and leaves and smooth, orange-yellow berries. Poisoning may result if the plant is ingested by livestock. Native to North America. Also called white horse nettle, trompillo. See Nightshade.

Simmental–A breed of large beef cattle that originated in the Simme Valley of Switzerland. They are fawn-colored with white faces.

Singe–To burn lightly, as to subject the carcass of a pig, fowl, etc., to a flame in order to remove bristles or hair, or to sear the ends of stems of flowers (which have a copious flow of sticky sap) to prolong freshness.

Single Comb–Of chickens, a comb consisting of a single, fleshy, serrated (usually five-pointed) formation which extends from the beak backward over the crown of the head. See Rose Comb.

Single Cross–The crossing of one strain, variety, inbred line, or breed of plants or animals with a different strain, variety, inbred line, or breed. The seed, plant, or offspring so produced is designated as a single cross or F_1. See Double Cross, Three-way Cross.

Single-rigged Saddle–A saddle which has one cinch.

Single-service Container–A container for milk, cream, and other food products, made of stout waxed paper. These containers are made in several shapes and with several types of closures and are meant to be used only once.

Sinus–(1) Any of the different cavities in people and some animals, some of which are enclosed by bone, others merely enlargements in a vessel or channel. Among the better known sinuses, because of the frequency with which they become infected in people and birds, are those located in the bones at the sides of the nose or nostrils and above the eyes. (2) A depression between two lobes in a leaf.

Siphon Tubes–Curved tubes about 1.2 to 2 inches in diameter that are used to siphon irrigation water from a ditch, over the bank, and into a furrow.

Sire–(1) The male parent. (2) To father or to beget; to become the sire of, as to sire a fine cow.

Sire Evaluation–An objective program designed and conducted by a breed association to increase the effectiveness of sire selection and to evaluate a bull's true genetic merit. It measures differences in sires using progeny information. It is the best measure of a bull's true transmitting ability.

Sire Index–A mathematical index for measuring a dairy bull's transmitting ability in terms of milk and butterfat. It tends to measure what production may be expected of the daughter when a bull is mated to cows of known productivity.

Sirloin–A loin or part of a loin of beef. In the United States, a steak or roast cut from the rear end of a beef loin, next behind the porterhouse.

Sitomania–Excessive craving for food. (Closely related to cynorexia, bulimia, and polyphagia.)

Sitter–See Setting Hen.

Sitting Breed–A breed of poultry which retains the natural tendency to become broody once or twice a year.

Six-tooth Sheep–A three-year-old sheep.

Skep–A beehive, usually of straw and dome-shaped, that lacks movable frames.

Skewbald–A color pattern in horses involving white spots on any color but black.

Skim–(1) To take something from the top, as to clear scum or matter floating upon the surface of a liquid. (2) To remove cream from milk which has stood for a period of time, by means of a special utensil or apparatus.

Skimmed Milk Powder–See Nonfat Dry Milk.

Skimmed Milk (Skim Milk)–That portion of milk which remains after removal of the fat. See Evaporated Cultured Skimmed Milk, Skimmed Milk Powder.

Skin–(1) The flexible integument which forms the external covering of an animal, especially of vertebrates. (2) Anything that resembles skin in nature or use, as the flexible outer covering or peel of fruits, etc. (3) The pelt of a small animal, such as calf, sheep, goat, fox, mink, etc., which is usually dressed, tanned, or intended for such treatment. (4) To strip the skin from. (5) To rub the skin off. (6) To peel.

Skin Worm–(1) *Cacoecia franciscana*, family Tortricidae; a moth whose larvae burrow underneath the apple skin; found in western United States. (2) *Filaria medinensis*, family Filariida; a thin, very long, parasitic nematode of white color which infests the skins of people and animals in West Africa and the Caribbean area. Also called Guinea worm.

Skirt–(1) The diaphragm muscle, the muscular part of the membrane which separates the abdominal from the thoracic cavity. It is edible meat. (2) To remove inferior grade wool from fleece. (3) See Saddle Skirt.

Skite–To caper about in front of sheep, as a dog (New Zealand). It derives from the Australian slang word meaning to boast.

Slab–(1) The exterior portion of a log which is removed in sawing lumber. (2) An unattractive, overripe or broken, dried fruit (prune, apricot, etc.) which is sorted from a drying tray in commercial fruit drying. (3) A piece of unsliced bacon.

Slat–(1) A thin, narrow strip of wood used in the manufacture of crates, etc. (2) A sheepskin after the wool has been pulled but before other treatment is given it.

Slatted Floors–A manure disposal system in which the floor is composed of wood, metal, plastic, or concrete slats. The slats are close enough together so that the animals can walk across it, but wide enough apart so that droppings can fall between them.

Slaughter–(1) The butchering of cattle, sheep, and other animals for food. See Slaughter Laws. (2) To kill cattle, sheep, and other livestock for food; to butcher. (3) Designating an animal suitable for slaughter.

Slaughter Calves–Calves between three months and one year of age at time of slaughter.

Slaughter Laws–In the United States, state and federal laws have been established dictating the methods to be used in stunning an animal prior to slaughter. Among the various humane methods is the use of carbon dioxide gas as an anaesthesia; a captive bolt which gives the animal an electric shock, stunning it; or a rifle bullet. The knocking hammer is not used in establishments under federal inspection, and tranquilizing drugs are not used, as they leave a residue in the meat. Also called humane slaughter laws.

Sleeper–(1) A wooden cross member supporting railroad rails; a tie. (2) An unbranded calf; a maverick. (3) A horse with equine encephalomyelitis (African sleeping sickness).

Sleepering–Catching a calf before it has left its mother and placing a mark of ownership on it (southwestern United States).

Sleeping Sickness–See Equine Encephalomyelitis.

Slick–Unmarked stock; unbranded calf. Also called a maverick.

Slick Ears–Unbranded range horses or animals having no ear-slits to mark ownership. Also called slicks.

Slightly Onion or Garlic–Designating a mild, but very objectionable flavor frequently found in market milk, especially in the spring and in areas where wild onions or garlic are common.

Slightly Weedy–A flavor defect of market milk or butter resulting from the cows grazing on weedy pastures.

Slime–(1) A bacterial rot of lettuce resulting in a wet, slimy decay of the leaves and heads. It is caused by *Erwinia carotovora*, family Enterobacteriaceae; *Pseudomonas viridilivida*, family Pseudomonadaceae; *P. marginalis*, and other species of bacteria. Slime occurs in the field, during warm, muggy weather; in lettuce not refrigerated in transit, and at markets. Also called bacterial rot of lettuce. (2) In the clarification of milk, a collection or deposit of dirt, leucocytes, cells, and other viscous matter thrown to the walls of the clarifier bowl. (3) In the curing or storage of cheese, a condition frequently produced on the surface by yeasts, bacteria, or other contamination, or faulty handling, causing a slimy surface on the cheese. (4) To remove the slimy or viscous coating from animal intestines, as in the preparation of sausage casings.

Slink–(1) The young of an animal brought forth prematurely or abortively, especially a calf. (2) The flesh or skin of such a calf or other animal.

Slip–(1) A soft-wood or herbaceous cutting from a plant, used for propagation or grafting. (2) An incompletely castrated male. (3) Curdled milk. (4) To take a cutting or cuttings from a plant. (5) To abort.

(6) The downslope movement of a soil mass under wet or saturated conditions; a microlandslide that produces microrelief in soils.

Slipped Tendon–A crippling in young growing birds which results from a nutritional disturbance. See Perosis.

Slobber–The saliva dripping or drooling from the mouth of an animal, usually a condition found following extreme exertion or excitement, or after eating certain types of feed; sometimes a result of infection.

Slop–A homemade, wet feed, mixture made from flour mill by-products, grains, whole or crushed, and other waste or surplus farm products, such as kitchen waste, skim milk, etc. It has been much used with growing swine. See Swill.

Sloping Rump–An undesirable feature of the rump and rear quarters of cattle. The rump should be more nearly level and straight.

Slow Breeding–(1) Delayed conception or settling after insemination. Also called delayed settling. (2) Designating an animal which lacks sexual desire. See Shy Breeder.

Slow Fever–See Equine Infectious Anemia.

Slow Gait–One of the several forward movements or gaits of horses, faster than a walk, but slower than a canter. There are three slow gaits: the running walk, the fox trot, and the slow pace.

Slow Gallop–See Canter.

Slow Poke–The last cow in a herd to come up; a trailer.

Slug–(1) Any of numerous soft, usually grayish or brownish, slimy creatures (gastropods), up to 3 inches or more in length, with unsegmented bodies, related to snails but without a shell. Chiefly terrestrial mollusks of the family Limnacidae, slugs are found in damp places in gardens, fields, and elsewhere under logs, rubbish, and in rotting vegetation. They feed at night by rasping holes in leaves and other plant material and leave slimy trails on surfaces over which they have moved. They may be serious pests in gardens and greenhouses. (2) A wholesale, forequarter cut of lamb or mutton which consists of the chuck, breast, neck, and shanks; all of the forequarter except the rack. Also called rattle.

Slumgum–A dark residue, consisting of brood cocoons and pollen, which is left after wax is rendered by the beekeeper.

Small–A United States grade for eggs having a minimum net weight per dozen of 18 ounces with a minimum weight for individual eggs at the rate of 17 ounces per dozen.

Small Animal Hospital–A veterinary hospital which is specially arranged to handle small animals, such as dogs and cats.

Smear–Material smeared on a surface, usually a small piece of glass (slide), and examined under a microscope. Smears are generally stained with dyes before being examined. Materials suitable for examination from a smear include blood, milk, bacteria from a culture, pus, etc.

Smegma–A dried and hardened, fatty secretion found in the sheaths of stallions and geldings, bulls, etc. When not removed by washing, excessive accumulations may shut off the urinary passageway. Also called bean.

Smoke–To treat meat or meat products, fish, or other foods by prolonged exposure to the smoke of certain nonresinous woods, corn cobs, hardwoods sawdust, etc., for preserving and imparting a desired flavor. Meat products commonly undergo a curing process before being smoked.

Smoke Point–Temperature at which the decomposition products of fat become visible.

Smoking–Process that imparts flavor to meats and fish after pickling. Hardwoods such as oak, elm, and ash are used. Helps preservation by inducing surface dehydration.

Smoky–(1) Of the color of smoke, a brownish or bluish shade of gray; dusky; cloudy. (2) An abnormality in the appearance of a horse's eye, which becomes cloudy, whitish, and pearly-colored. It is indicative of impaired vision. The condition is referred to as smoky eye.

Smooth Mouth–The mouth of a horse whose teeth have lost their natural cups and have become smooth by use and wear, indicating that the horse is ten or more years of age.

Smooth-going–(1) Designating an easy gait or forward movement of a horse used for riding and driving. (2) Designating any practice which proceeds at a brisk, even pace without particular hindrance.

Smooth-mouthed Sheep–An aged sheep, usually a ewe, whose incisor teeth are missing or so badly worn out that it cannot eat normally; a gummer. Also called broken-mouthed.

Smother–(1) To kill by suffocation, as panicky chickens or sheep are smothered when they pile up in a heap. (2) Of plants, to kill by covering thickly with dirt, straw, ice sheet, black sheet plastic, etc. (3) Of plants, to retard in growth or to kill by a dense or more vigorously growing crop. See Smother Crop.

Snaffle–Any type of mouth control used in handling horses, especially in breaking. See Snaffle Bit.

Snaffle Bit–A bridle bit; the mouthpiece of a horse bridle which is composed of two or more linked parts or sometimes made of a single chain which acts as a bit and serves as an effective control.

Snap–(1) A spring fastening or clasp which closes with a clicking sound, used as a fastener on harnesses, gate chains, dog chains, etc. (2) A brief period of cold weather; as a cold snap. (3) To harvest by pulling off with a quick motion, as the ears of corn, the bolls of cotton, etc. (4) In horses, especially in harness or saddle horses, to move briskly with animation; to pick up feet quickly and sharply.

Snap Corn–Ear corn, with all or most of the husks adhering, which is harvested by breaking or snapping from the stalk.

Snip–(1) To remove by clipping, as to snip off a twig, flower, etc. (2) A white marking between the nostrils of a horse.

Snood–The fleshy protuberance at the base of the beak of a turkey. Also known as dew bill.

Snout–(1) The projecting part of an animal's head that contains the nose and jaws, as the snout of a hog. (2) An anterior elongation of the head; a proboscis; a piercing or sucking part of certain insects.

Snub–To control an animal by roping it and tying the rope to an adjacent post or other solid object so as to restrict its movement.

Snub Down–To place the hand on or over the nose of an animal to stop its breathing in order to control it.

Snubbing Post–Any smooth, sturdy post to which animals may be fastened to restrict their movement.

Snubbing Ring–A heavy metal ring attached to the floor, or any other solid object, through which a control rope is passed and by means of which an animal's movement is restricted as in slaughtering or treating for an ailment.

Social Insects–Insects which live in a family society, with parents and offspring sharing a common dwelling place and exhibiting some degree of mutual cooperation; e.g., honeybees, ants, termites.

Sodium Fluoracetate–Compound 1080; a powerful rodenticide developed by the United States Fish and Wildlife Service; a fine, white powder readily soluble in water, used to control rats, mice, and other rodents. It is a restricted pesticide.

Sodium Fluoride–NaF; a white, powdered, chemical compound used for combatting chewing lice on animals and poultry, commonly employed in roach and ant powders, as an anthelmintic used to eliminate ascarids (roundworms) in swine, and as a fungicide.

Soft Hog–Any hog that is fattened on peanuts, soybeans, rice bran, and chufas, producing soft or oil meat carcasses (i.e., soft or oily fat). Also called oily pork.

Soft Pork–A condition of pork in which the meat is soft and of poor consistency. It is caused by feeding hogs liberally on high fat content feeds such as soybeans or peanuts.

Soft-shell Egg–An egg with an incompletely calcified shell which may be due to one of several causes, including nutritional deficiencies of calcium, phosphorus or vitamin D, failure of the shell gland to function, or by violent peristaltic constrictions causing the egg to be forced through the uterus before completion of shell formation.

Soilage–Freshly cut green fodder fed to confined animals.

Soiling–In livestock management, the cutting and bringing of green forage to livestock in place of allowing the animals to eat the green feed where it grows.

Solar Wax Melter–A glass-covered box in which wax honeycombs are melted by sun's rays and wax is recovered in cake form.

Solid Manure–Manure in solid form as contrasted to liquid manure.

Solids-Not-Fat–The solids dissolved or suspended in milk other than butterfat, composed principally of protein (casein), lactose, and minerals.

Solitary–Occurring singly or in pairs, not in colonies.

Solitary Bees–Bees that live alone and whose offspring individually survive the winter, usually in an immature stage in a cell in the ground or a variety of other sites.

Solubility–To be most readily available to plants a nutrient must be at least slightly soluble in the soil solution or be held in an exchangeable form on clay and humus particles. See Available.

Soluble–Capable of changing form or changing into a solution.

Soma–The body, in contrast with the germ or germ plasm.

Somatic–(1) Designating body tissues. (2) Having chromosomes in pairs, one of each pair normally coming from the female parent and one from the male, as contracted with terminal tissue which gives rise to germ cells.

Sorghum, Grain–A cereal grass used mainly for feedgrain or silage. Often grown in corn and wheat areas.

Sorghum Poisoning–A poisoning of livestock which results from ingestion of sorghum containing prussic acid (hydrocyanic acid), a deadly poison which develops in dangerous amounts under certain conditions of plant stress in many of the sorghums.

Sorghum Silage–Silage made from either sweet or grain sorghums ensiled when the seeds are hard and ripe. It is somewhat inferior to corn silage in feed value.

Sorrel–A light shade of chestnut coat color in horses.

Sound–(1) In good condition, not damaged; acceptable. (2) Designating an animal free from blemishes of any kind. (3) Designating wool that has good strength.

Soundness–Freedom from defect or blemishes as: (a) a wool fleece that shows strong wool fiber and no weak spots; (b) an animal that is free from any noticeable defect; (c) an animal that shows good feet and legs.

Sour–(1) Term used to denote a horse that has been overworked or trained to the point that he refuses to perform. (2) A flavor defect most frequently associated with dairy products, such as milk, butter, and ice cream, which is characterized by a sour taste and odor resulting from the reduction of milk sugar to lactic acid by lactic acid-producing bacteria.

Southdown–One of the smallest and oldest breeds of sheep, which had its origin in the Southdown area of England and which is best known for its high-quality mutton carcass. Mature ewes weigh from 135 to 160 pounds.

Sow–(1) A female swine, usually one that shows evidence of having produced pigs or one that is obviously pregnant. (2) To plant seeds by scattering either broadcast or by distribution in a row.

Sow Belly–Salt pork; fat bacon which is not smoked.

Sow Production Breeding Value (SPBV)–A value assigned to a sow based on records on all the litters produced by the sow as well as the estimates of heritability and repeatability of the traits. SPBV is an estimate of the ability of a sow to pass her productivity on to her offspring. See Sow Productivity Index.

Sow Productivity Index–An index used to compare and identify the top producing sows in a herd. The formula is Index = $100 + 6.5 (L - 1) + 1.0 (W - w)$, where L equals the number of pigs born alive for the individual sow, 1 equals the average number of pigs born alive for the herd, W equals the adjusted twenty-one day litter weight for the individual sow, and w equals the average adjusted twenty-one day litter weight for the herd. See Sow Production Breeding Value.

Soybean Oil Meal–A livestock feedstuff consisting of ground soybean oil cake or oil chips. If a name descriptive of the process of manufacture, such as expeller-, hydraulic-, or solvent-extracted be used, the product must correspond thereto. Soybean oil meal has an average protein content of about 45 percent and ranks high as a protein supplement for livestock feeding.

Sp. (Spp.)—The abbreviation, singular (plural) for species; e.g., when one species of a genus of plant or animal is referred to, the name may be written *Canis* sp. meaning one species of dog; or *Canis* spp. meaning more than one species of the genus *Canis*.

Span—(1) A pair of animals usually harnessed together as a team. Also spelled spann. (2) The space between the top of the two walls of a ravine. (3) The length or extent of a bridge between supports. (4) The distance between the thumb and little finger when the hand is fully extended.

Spanish Merino—A breed of sheep developed and bred in Spain which is the ancestor of the present day Merino, a highly valued, fine-wool breed naturalized throughout the world.

Spanish Pony—A small horse common to western United States, which is a descendant of the horses brought in by the early Spanish explorers. It is similar to the other horses of this region called cayuses, mustangs, and Indian ponies.

Spareribs—A retail cut of pork made up of the rib bones and sternum removed from the belly or bacon side.

Spasm—An involuntary, sudden, painful, and violent muscular contraction.

Spasmodic Colic—Acute indigestion in horses.

Spavin—A disease affecting the hock joint of a horse's hind leg, usually a bony growth on the inner, lower part of the hock that causes lameness. See Bog Spavin, Bone Spavin, Occult Spavin.

Spawn—(1) To deposit eggs (of fish). (2) Common term for eggs and sperm. (3) Young fish, usually numerous, in early stages of development.

Spay—To remove the ovaries of a female animal.

Spayed Heifer—A heifer which is unsexed by removal of ovaries. Such heifers make good feeder cattle.

Species—In the naming of plants and animals, Latin is used. Each kind of plant or animal can be identified by genus (plural, genera) and species (both singular and plural); e.g., the generic name (genus) of corn is *Zea* and the species name is *mays*.

Specific—(1) A medicine that cures a particular disease. (2) Pertaining to a species. (3) Produced by a particular microorganism. (4) Restricted by nature to a special animal, thing, etc. (5) Exerting a peculiar influence over any part of the body.

Specific-pathogen Free—Swine that are obtained by laboratory methods designed to eliminate certain diseases of the respiratory and digestive tracts.

Speciosus—Showy, good-looking.

Speckle—A small patch or dot of color.

Speculum—A metal instrument or glass tube inserted into certain organs of the body, as the ear, mouth, nose, rectum, etc. for examination of the interior of the organ by means of a reflected light.

Speculum Method of Insemination—A method of artificial insemination. A metal or glass speculum is inserted into the vagina, and the cervix is located by means of a pen flashlight or head light. When the cervix is located, the loaded insemination tube is then inserted and the semen discharged.

Spent—Of animals, especially horses, completely exhausted; also relates to tired people.

Sperm—The male sex cell, produced by the testicles.

Sperm Cell—Male germ cell.

Sperm Collector—See semen collector.

Spermatheca—Small saclike organ in a queen bee in which sperm are stored.

Spermatic Cord—The cordlike structure by which the testicle is suspended within the scrotum. It contains the vas deferens, the blood vessels, the nerves of the testicle, and a small muscle.

Spermatid—A haploid cell produced from the second division of meiosis in spermatogenesis that has not yet undergone the changes to form a sperm cell.

Spermatocyte—One of the various kinds of cells produced in the development of male gametes in animals. Primary spermatocytes are formed by ordinary mitotic cell divisions from spermatogonia. Meiotic divisions of primary spermatocytes result in the production of secondary spermatocytes which in turn give rise to spermatids by equational cell division.

Spermatogenesis—The development of male gametes or sperm cells.

Spermatogonium—A primary germ cell in the testis that will undergo spermatogenesis to produce spermatozoa. Plural, spermatogonia.

Spermatozoon—A mature male animal germ cell from the testes which impregnates the female ovum to produce another individual of the same species.

Spermiogenesis—That part of the process of spermatogenesis involving the changes that permit spermatids to become spermatozoa.

SPF Pig—See Specific-pathogen Free.

Sphincter—A ring-shaped muscle that closes an orifice; e.g., sphincter muscles at the lower end of a cow's teat. They act to press the teat tissues together and close the opening to the milk duct of the teat.

Spider—(1) Any of various arachnids in the order Araneida. Individuals have eight legs and only two main body divisions. The abdomen has spinerets for spinning silk threads used for constructing webs which serve as nests and as traps for prey. Most spiders are harmless to humans and because of the insects they devour and their aid in pollination, they are exceedingly beneficial. (2) Anything suggestive of a spider in form, as various mechanical parts having radiating members. (3) A tumor or growth sometimes suspended in a cow's teat canal by strands of fibrous tissue which resembles a spider web (colloquial).

Spin—To twist or transform the fibers of wool, cotton, silk, flax, etc., into thread or yarn preparatory to making cloth or fabric.

Spinal—Referring to the spine or vertebral column.

Spindle–(1) The fine threads of achromatic protoplasm arranged in a fusiform mass within the cell during mitosis. (2) In flower-bearing plants, to produce the stalks on which flowers grow.

Spine–(1) A stiff, sharp-pointed outgrowth on a plant or animal. (2) The vertebral column; the backbone.

Spinning Count–Applied to wool, any of the different numbers indicating the fineness of the yarn which can be spun from it.

Spiracles–The openings to an insect's internal breathing tubes, the trachea.

Spiral Stomach Worm–*Dispharnyx nasuta*, family Acuariidae; a short, white roundworm, often curved or even twisted into a spiral, occasionally found in the glandular stomach of chickens, turkeys, and pigeons. If present in considerable numbers they may affect the health of the bird. The sowbug and pillbug may function as intermediate hosts for the spiral roundworm.

Spirochete–Any microorganism that is spiral or wavy in form, highly flexible, and capable of contracting, as the organisms that causes leptospirosis and syphilis.

Spirochetosis–(1) Any infectious disease caused by spirochetes. (2) An acute, highly fatal, septicemic disease of fowls characterized by the presence of the causal organism *Borrelia anserina* in the blood stream during the height of the disease.

Spirometer–An instrument that measures the flow of air in and out of the lungs.

Spit–(1) A small point of land or narrow shoal projecting into a body of water from the shore. (2) A pointed rod on which meat is held while roasting or barbecuing above or in front of open heat. (3) Saliva from animals or humans.

Splayed (Splay-footed)–A common fault found in horses or other animals which means that the front hooves are turned out and the heels turned in.

Spleen–A large, glandlike, ductless organ in the upper part of the body cavity on the left side between the stomach and the diaphragm.

Splenic Fever–See Anthrax.

Splint Bone–The small metacarpal or metatarsal bones found on each side of the cannon bone.

Splints–A hard, bony enlargement on the splint bone which is located on the inside of the fore cannon bone of the leg of a horse. It may appear rarely on the outside of the front legs, but seldom on the hind legs. It is more often a blemish than an unsoundness. See Splint Bone.

Split–(1) A deformed flower. (2) To separate thick hides into layers in the preparation of leather.

Split Litter–A rabbit litter in which some of the rabbits are born, and are followed some time later by the remaining rabbits.

Spoil–(1) To deteriorate by molding or rotting. See Spoilage. (2) Debris or waste material from a coal mine. (3) Dirt or rock that has been removed from its original location, specifically materials that have been dredged from the bottoms of waterways.

Spoilage–(1) Hay or forage that has been improperly cured or stored. (2) Any objectionable change which has occurred in a food, feed, or material. (3) Putrefactive changes which occur in canned goods as a result of underprocessing, causing the growth of vegetative cells, spores, or organisms, or as a result of the growth of organisms entering the can after processing.

Spoilage Organisms–Bacteria, yeasts, and molds that cause food to spoil. They live everywhere: in the air, soil, and water and on food, plants, and animals.

Spool Joint–In grading sheep carcasses, the metacarple bone (the forearm) is struck with a heavy object. If the resulting break occurs at the center of the joint it is referred to as a break joint, and the carcass is classified as a lamb carcass. If the break occurs above the joint and the joint does not separate cleanly it is termed a spool joint. The carcass is then classified as a mutton carcass. See Break Joint.

Sporadic Disease–A disease which occurs in scattered or isolated instances.

Spore Dust–A standardized powder containing spores and used in the biologic control of insect pests. The dust is placed at intervals in the soil, and the spores infect the larvae of certain insects, such as Japanese beetle, causing diseases which destroy them. See Milky Disease.

Sporocyst–In flukes, or trematodes, the intermediate asexual generation or second larval stage and, in some cases, the third larval stage. In protozoa belonging to the group *Coccidia*, a sporocyst is a sac or cyst formed within an oocyst; it contains one or more sporozoites. See Sporozoites.

Sporozoa–A class of protozoa which consists of many parasitic species with complicated life cycles.

Sporozoite–A small, usually elongated, infective stage of sporozoan parasites, such as *coccidia, plasmodia* (malaria), etc.

Sport–A random mutation.

Sporulaton–(1) In bacteria, the formation of spores within the body of the bacterium. The spores represent the inactive resting, or resistant, forms. (2) In coccidia, a kind of reproduction by which the fertilized cell within the oocyst wall splits up into new individuals, called sporozoites. Sporulation of coccidial oocysts usually occurs after the oocyst has been discharged from the body of the host. (3) In plants and animals, the process of spore formation.

Spot Treatment–The application of a pesticide or other material to a restricted or small area of heavier infestation.

Spot Wool–Wool ready for immediate delivery.

Spots–Name given to Spotted Poland China swine.

Spotted Poland China Swine–Swine which originated in the United States, largely in Indiana, shortly before World War I. The crossing of Gloucester Old Spot swine with Poland China swine provided the foundation for the breed. The breed is now known as spots. See Spots.

Spotted Tick Fever–An illness caused by the Rocky Mountain wood tick. See Rocky Mountain Wood Tick.

Sprain–A severe wrench or strain of the parts around a joint which causes pain, swelling, and difficulty in moving.

Spread–(1) A straddle; the difference in price between two delivery months in the same or different markets, or the sale of the one thing against a simultaneous purchase of the other. Straddling between a foreign and the domestic market is often referred to as arbitrage. (2) An extensive tract of land. (3) A ranch, including the buildings and the extent of land grazed by cattle or sheep (western United States). (4) The distribution of a disease. (5) Butter, peanut butter, margarine, or other food mixtures used as a spread on bread to improve its flavor, palatability, and/or nutritive value. (6) To scatter, as to broadcast seed, fertilizer, etc. (7) To disseminate disease.

Spreader–(1) A device used for distributing water uniformly in or from a channel. (2) A spreading agent, or spreader-sticker used to improve the contact between a pesticide and a plant surface. (3) An instrument provided with hooks or flanges used to spread open the incision made in the body wall of a bird. (4) A device that scatters or spreads: e.g., a manure spreader, a fertilizer spreader. (5) An animal, etc., capable of acting as a parasite or disease vector. (6) A species of microorganism which tends to grow profusely over the entire surface of the culture medium.

Sprengel Tube–A U-shaped tube with capillary ends at right angles which is used for determining the specific gravity of milk and fats. The tube is weighed empty, filled with water and then with the fluid, whose specific gravity is given by the ratio of its weight to that of water. See Specific Gravity.

Spring Chicken–A young chicken, usually only a few months old. See Springer.

Spring Lamb–A lamb that is marketed in the spring of the year and prior to July 1; they are usually born in the fall.

Spring-Fall Range–Those grazing areas of the western range which, because of grazing patterns, climatic factors, forage production, quantity of water, etc., are seasonally grazed only in spring and fall. Such grazing may not be necessarily obligatory but it is convenient in a grazing program where summer range at higher elevations is productive and available only during the summer season.

Springer–(1) In the live-poultry industry, a young chicken, larger than a broiler and smaller than a roaster, which commonly weighs from 3¼ to 4¼ pounds; a fryer. (2) In marketing, a pregnant cow or heifer due to calve shortly. (3) A sealed can both ends of which are bulging, either one of which can be easily flattened. (4) A can of spoiled fruits or vegetables.

Springing Cow–A pregnant cow or heifer which shows signs of the approaching birth of its young, as evidenced by a relaxation of the ligaments and muscles on either side of the tail-head and also by a slight elevation of the tail-head.

Spur–(1) A short, stubby shoot, as in some fruit trees where the spurs bear flowers for more than one year or as grapes are pruned to spurs of one to two buds each, etc. (2) A hollow, tubular projection from some part of a flower, very conspicuous in the columbine, usually secreting nectar to induce the visitation of insects. (3) A hornlike protuberance which grows from the inner side of the shank of a fowl. (4) A steel contrivance secured to a rider's heel and used to urge the horse by its pressure. See Spur Rowel.

Spur Rowel–A small wheel with radiating points, attached at the end of a horseman's spur as an added goad.

Sputum–(Latin) Saliva.

Squab–A nestling pigeon raised for its flesh, which may be marketed when fully feathered under the wings and just before it is ready to leave the nest, usually at twenty-five to thirty-five days of age when it may weigh from 12 to 24 ounces.

Squab Broiler–A small broiler that is usually under eight weeks of age, commonly of Leghorn or other lightweight breeds.

Squab Turkey–A young turkey which weighs under five pounds.

Square-gaited–Designating the gait of a horse in which the action is straight on all four feet.

Squeeze Chute–A narrow stall with a hinged side that is used for restraining animals. The animal's head is caught in a head-catch chute, and the sides of the chute are moved against the animal to restrict movements. See Head-catch Chute.

Stable Fly–*Stomoxys calcitrans*, family Muscidae; a small fly, similar to the house fly but with biting mouthparts that enable it to pierce the skin and suck the blood of animals. Due to its irritations animals lose weight, the yield of dairy cattle is reduced, and work animals become unmanageable. It occurs in all parts of the United States and throughout most of the world.

Stag–(1) In animals, a male castrated after reaching sexual maturity which shows pronounced sexual development. (2) A horse which is thick and coarse in the throat latch and crest from late castration. (3) An imperfectly or recently castrated sheep or steer. (4) In marketing, an uncastrated male chicken with flesh slightly darkened and toughened and with comb and spur development showing the bird to be in a state of development between a roasting chicken and a cock. (5) A boar hog usually castrated after having passed breeding usefulness. (6) A wild mature male deer.

Stake–(1) A wooden or metal strip, usually pointed at one end for driving into the ground, used for marking a location in surveying, supporting a vine or plant, tethering an animal, etc. (2) To place stakes as markers. (3) To support plants by tying to stakes. (4) To tether an animal to a stake.

Stall–The space in a barn which is occupied by a single animal, such as a dairy cow or horse, for feeding and handling.

Stall Barn–A barn used for sheltering dairy cattle and/or young stock where the adult animals are confined to stalls by means of stanchions, straps, halters, or chains during part of the year, as in the winter and for milking. Roughages and concentrates may be fed at the individual stalls. None, part, or all of feeds and bedding may be stored in the structure. Also called stanchion barn.

Stall-feeding–A system of management where animals are housed more-or-less continuously (except for exercise), and forage crops are cut and carried to them.

Stallion–A male horse used for breeding purposes; it should be a highly selected animal with very superior breed characteristics.

Stampede–A wild rush of cattle or horses as a result of fright.

Stanchion–A tying or controlling device, usually of wood or metal, used in barns to control animals, usually dairy cows.

Stand–(1) The proper number of uniformly distributed plants per acre. (2) The relative number of plants per area, as a poor, good, or medium stand. In forestry, fully stocked, understocked, pure, mixed, or residual stand. (3) A hive of bees. (4) Density of game per acre, area, etc. (5) A stallion's court. (6) To cease walking or moving; to take or keep a certain position. (7) To rise to the feet. (8) Of a stallion, to be available for breeding purposes.

Standard Cow Day–A feeding index for a cow which is 16 pounds total digestible nutrients per day from grazing.

Standardbred–A breed of horses, trotters and pacers, developed in the United States primarily for speed and endurance as roadsters and racers and named for their ability to trot or pace a mile in standard time. Its origin is attributed to a mixture of the following breeds. Thoroughbred, Norfolk Trotters, Narragansett Pacers, Arabian, and Morgan. In general, the Standardbred in appearance is rather rangy, leggy, deep-chested, narrow, and angular. The physical form is less important than speed in the selection.

Standing Heat–A female that freely accepts the male.

Star–A small white spot in the center of an animal's forehead, especially of a horse. Also called star-faced.

Star and Strip–A marking on the forehead of a horse with a strip to the nasal peak.

Star, Strip, and Snip–A marking on the forehead of a horse with a narrow extension of the nasal peak and opening up again between the nostrils. These may not be connected.

Staring Coat–Hairs standing on end indicative of illness or poor condition.

Starter–(1) A prepared culture of desired organisms used for inoculating milk and cream in the making of various dairy products, such as butter, cheese, etc., chiefly to enhance the flavor. See Mother Culture. (2) The first food provided for young animals.

Stassanization–A heat treatment of milk, common in Europe, in which the product is heated to temperatures somewhat higher and for shorter periods than is commonly employed in the high-temperature, short-time process of heat treatment in the United States.

State Veterinarian–A veterinary official of a State Department of Agriculture responsible for the proper enforcement of all regulatory laws regarding disease control, health and sanitation of livestock, and livestock products of the state.

Steady–(1) A command to a horse to calm him or to make him go more slowly. (2) Regular, constant, uniform; as a steady breeze.

Steak–(1) A thick, usually cross-sectional, slice of meat, its name depending on the part of the carcass from which it is cut; e.g., round steak, sirloin steak. It is usually from the rear quarters of an animal and may be cooked in any of several ways. If not qualified, the term refers to beef steak. (2) A thick, usually cross-sectional, slice of fish; e.g., halibut steak, salmon steak.

Steaming Up–Increased feeding, particularly of concentrates during the latter part of pregnancy.

Steapsin–Pancreatic lipase, a lipolytic enzyme secreted by the pancreas which has the power of hydrolyzing fats to fatty acids and glycerol.

Stearic Acid–One of the fatty acids occurring in combined form in animal and vegetable oils; one of the major fatty acids in butter and most fats. Commercial stearic acid is used in large quantities, in rubber compounding and in the preparation of soaps, greases, and chemicals.

Stearine–The solid material of any fat which is obtained, in the process of refining, by filtration from an animal or vegetable oil after chilling or freezing.

Steel Gray–The coat color of a horse that is an even shade of gray over the body with a black or dark gray mane and tail.

Steer–A male bovine animal castrated before reaching puberty.

Steno—A prefix indicating narrowness.

Steppe–(1) A cow's whole milk cheese with color added, made in Germany, Austria, and Denmark. (2) A treeless grassland which exists under semi-arid climates, such as the east-west zonal belt of plains in the U.S.S.R., lying between the forestland of the north and deserts on the south, or the semi-arid Great Plains of western United States lying east of the Rocky Mountains.

Sterile–(1) In animals, incapable of reproduction; unable to produce normal living young. (2) In soils, unproductive; barren; producing little or nothing, as a sterile soil. (3) In biological products, etc., free from contamination with living bacterial, fungal, or viral organisms; or designating an organism not capable of growing or multiplying.

Sterilization–(1) The destruction of all living organisms. In contrast, disinfection is the destruction of most of the living organism. (2) To make animals infertile.

Sternum–The breastbone of an animal.

Sterol–Any of a group of solid cyclic alcohols, such as cholesterol and phytosterol, with wide distribution among animals and plants. The sterols are neutral and comparatively stable substances which occur partly in the free condition and partly esterified with higher fatty acids. They are of great biological importance since irradiation of some sterols leads to the formation of a form of vitamin D.

Stick–(1) A piece of wood, of indefinite size, that is long as compared with its diameter or cross section. (2) A thick, syrupy by-product obtained by evaporating the tank water produced in the processing of packing-house tankage. Stick is usually mixed with dried, low-grade tankage and used as a fertilizer, or is added to wet rendered tankage to be used as livestock feed. (3) To stab or cut with a knife, as in animal slaughter, to server the jugular veins to cause free bleeding; also with poultry, to pierce the brain with a narrow-bladed knife to cause feather follicles to relax for ease in plucking feathers.

Sticker–(1) A substance added to fungicides and insecticides to improve adherence to plant surfaces. (2) A strip of wood placed between the courses of lumber in a pile and at right angles to the boards to facilitate air circulation. (3) A person whose job is to stick animals in a slaughter house. See Stick.

Sticktight Flea–_Echidnophaga gallinacea_, family Pulicidae; an insect pest of poultry in the southern part of the United States. The fleas attach themselves to the comb, face, earlobes, and wattles of fowls by their mouthparts, engorging themselves with the blood of the host. They are found in clusters and, as their name implies, are removed with difficulty. Also called southern chicken flea.

Stiff-lamb Disease–An ailment of young suckling lambs, generally occurring between one and eight weeks of age, characterized by stiffness, staggering, or inability to walk or to nurse. Because it is associated with a deficiency of vitamin E, supplements of this vitamin are frequently used as a preventive and treatment. Also called white-muscle disease. See White-muscle Disease.

Stifle–The joint next above the hock (tarsus) in the hind leg of animals; comparable to the knee joint of humans. Also called stifle joint.

Stifled–A horse is said to be stifled when the patella (or kneecap) slips out of place and temporarily locks in a location above and to the inside of its normal location. A bull is said to be stifled when the stifle muscle is torn.

Stilbestrol–Diethylstilbestrol, a synthetic estrogenic hormone.

Still-born–Born lifeless; dead at birth.

Stimulus–An activating agent such as heat, moisture, or light.

Stirk–A six-to-twelve-month-old heifer.

Stirrup–An attachment to a riding saddle serving as an aid in mounting and as a footrest for the rider; made of metal, wood, or heavy leather in many different styles.

Stock–(1) Plant or plant part upon which a scion is inserted in propagation. (2) Livestock; domesticated farm animals. (3) Material held for future use or distribution. (4) Material destined to be wrought into finished products, as crate stock, barrel stock, cider stock, etc. (5) A plant or plant part that furnishes cuttings for propagation. Also called stock plant. (6) The main stem or trunk of a plant. (7) A rootstock (rhizome). (8) The original type from which a group of plants or animals has been derived. (9) The base or handle of a whip. (10) The stump of a tree (after the tree is felled). (11) (plural) A small enclosure in which an animal is secured in a standing position during shoeing or an operation. (12) To provide or supply with livestock, as to stock a pasture or range with cattle, sheep, etc. (13) To assemble a supply of materials or commodities. (14) Of, or pertaining to, livestock, as stock barn, stock feed, etc.

Stock Cattle–Usually young steers or cows, light, thin, lacking in maturity and finish. Also called stockers.

Stock Fountain–A watering device, usually consisting of a small receptacle or water cup, which is automatically kept full by a float valve, or filled when a valve is operated by the pressure of the muzzle or snout of the animal.

Stock Horse Type–Horses of the quarter, paint, and Appaloosa breeds, mainly Thoroughbred and Arabian; bred for early speed, agility, disposition, and intelligence.

Stock Prod–See Cattle Prod.

Stock Salt–Common sodium chloride, granulated, in blocks, or in cakes, used for salting livestock.

Stock Solution–A concentrated solution from which a portion is taken and diluted as needed, as a spray, etc.

Stock Tank–Any structure holding water for livestock. Commonly, it is a deep trough, bowl, or tank equipped with a float valve or other means of controlling the water level. Also called basin. In some parts of the country, it also refers to a farm pond.

Stocker–In marketing, a meat-producing animal capable of additional growth and finish and usually considered to be thinner than a feeder.

Stockinet–A knitted, coarse, elastic, netlike type of cotton cloth, similar to stocking material, which is used as a cover for fresh-killed meat to protect it from dirt, from drying out, and from changing color. It is also used as a bandage material.

Stocking–(1) The relative number of livestock per unit area for a specific time. In range management, the relative intensity of animal population, ordinarily expressed as the number of acres of range allowed for each animal for a specific period. (2) In wildlife management, the density of animal population in relation to carrying capacity. (3) In forestry, the density of a stand of trees, such as well-stocked, overstocked, partially stocked. (4) A white leg on an animal.

Stocking Rate–Actual number of animals, expressed in either animal units or animal unit months, on a specified area at a specific time.

Stockyard–(1) A yard for keeping or holding livestock. (2) A series of pens or yards where market animals are collected for sale. It may be only a pen or two along a railroad siding in a small town, or it may be as extensive as the great stockyard systems of Chicago.

Stomach–In humans and certain other vertebrate animals, a large saclike organ into which food passes from the esophagus or gullet for storage while undergoing the early stages of digestion. In true ruminants, e.g., the bovine, the stomach comprises four compartments: the rumen (paunch), reticulum (honeycomb), omasum (manyplies), and abomasum (true stomach). In birds there are commonly two such compartments: the proventriculus (glandular fore-stomach) and the gizzard (muscular posterior stomach). In most invertebrates, there are analogous parts that function as a stomach.

Stomach Ball–An indigestible mass of hair and/or fibrous material which has been retained in the stomach or, particularly in the paunch of cattle, sheep, and goats. See Bezoar, Hair Ball.

Stomach Bloat–Acute indigestion in horses.

Stomach Insecticide–Any insecticide that becomes effective only if taken into the alimentary tract. It is generally applied to control chewing insects.

Stomach-dirt–An accumulation of dirt in the stomach; a commonly occurring condition in horses and cattle grazing on loosely implanted vegetation.

Stone–(1) In land description, a detached rock fragment on the surface of, or embedded in, the soil. (2) In soil survey mapping, a rock which is greater than 10 inches in diameter or more than 15 inches in length. See Boulder. (3) The hard seed of some pulpy fruits, as peach, plum, etc. See Clingstone, Freestone. (4) A unit of weight, 14 pounds (avoirdupois, British). (5) See Testicles. (6) To take the stone or pit out of stone fruits.

Stool–(1) Fecal material; evacuation from the digestive tract. (2) To develop several stems from the crown of a plant or from a stump.

Store–An animal not yet ready for slaughter.

Stot–See Steer.

Stover–(1) The stem and leafy parts of corn fodder after the ears have been removed. (2) Sorghum forage.

Straggler–(1) An animal that wanders or strays from a flock or herd. (2) A plant, branch, etc. that grows irregularly.

Straight Bar Bit–A single, straight, smooth, rounded bar of iron used as a bit on a horse bridle. It is the easiest on the horse of the many bits in use.

Straight-run Chickens–Chicks of both sexes in the order in which they hatch; chicks hatch in about equal numbers of both sexes.

Straight-run Eggs–In marketing, designating eggs which are sold without being graded as to size or quality.

Straightbred–Designating an animal with a straight line of ancestry or pedigree within a recognized breed, such as a registered Thoroughbred horse or a purebred Jersey cow.

Strain–(1) A group of plants of common lineage which, although not taxonomically distinct from others of the species or variety, are distinguishable on the basis of productiveness, vigor, resistance to drought, cold, or diseases; or other ecological or physiological characteristics. (2) A group of individuals within a breed which differ in one or more characters from the other members of the breed; e.g., the Milking Shorthorns or Polled Herefords. (3) An organism or group of organisms which differs in origin or in minor respects from other organisms of the same species or variety. (4) A virus entity whose properties and behavior indicate relationship to a type virus and are sufficiently constant to enable the entity to be recognized whenever isolated. (5) A severe muscular effort on the part of a draft or other animal which may result in muscle, ligament, or other damage. (6) The condition of an overworked part operating above optimum load, such as a belt or motor operating under strain. (7) To filter a liquid and free it of impurities by passing it through some medium or fabric which can retain the solid matter and allow the liquid to pass.

Strain Cross–A cross between members of two different strains.

Strand–(1) Each one of the separate units of fibers, threads, wires, etc., which, when twisted or woven together, may make up a rope, cable, fence wire, etc. (2) Any single filament.

Strangles–An acute, contagious, febrile disease of horses and mules characterized by catarrhal inflammation of the mucous membranes of the nasal passages and pharynx. Also called bastard strangles, distemper of horses, shipping fever of horses.

Strawberry Red Stele–A root-rot disease caused by the fungus ***Phytophthora fragariae*** which is evidenced by early-spring stunting or dwarfing of the plant and wilting and dying of older leaves. In diseased plants, the small fibrous leaves are discolored, or have disappeared, leaving only the long rattail roots. When the central cylinder of the root is exposed by splitting, the core is found to be dark red. Also called brown stele, strawberry red root rot.

Strawberry Roan–The coat color of a horse, predominately red or reddish-bay with white hairs rather uniformly intermixed.

Stray–An animal that has wandered away from the herd, flock, or farm to which it belongs.

Stream–(1) Flowing water in a natural or artificial channel. It may range in volume from a small creek to a major river. (2) A jet of water, as from a nozzle. (3) A continuous flow or succession of anything: air, gas, liquids, light, electricity, persons, animals, and materials.

Strength–(1) The capacity to resist force; solidity or toughness; the quality of bodies by which they endure the application of force without breaking or yielding. (2) Bodily or muscular power; force; vigor. (3) The potency or power of a liquid or other substance; intensity of active properties. (4) The firmness of a market price for a given commodity; a tendency to rise or remain firm in price.

Streptococcus–(Plural, streptococci) Any bacterium of the genus ***Streptococcus***, family Coccaceae. Several species, as ***S. pyogenes, S. scarlatinae***, etc., are responsible for various diseases in humans and animals. Certain others, as ***S. defir*** and ***S. cremoris***, are used in making cheese, etc.

Streptomycin–An antibiotic produced by the soil actinomycete ***Streptomyces griseus***. It is effective against most Gram-negative and acid-fast bacteria and also against some Gram-positive forms. Used mainly to treat tuberculosis.

Stress–(1) Abnormal or adverse conditions and factors to which an animal cannot adapt or adjust satisfactorily, resulting in physiological tension and possible disease; the factors may be physical, chemical, and /or psychological. (2) Plants unable to absorb enough water to replace that lost by transpiration. Results may be wilting, halting of growth, or death of the plant.

Stretch–The amount of body length of an animal; usually referring to the distance between the shoulders and the hips.

Striate–With fine grooves, ridges, or lines of color.

Stricture–An abnormal, localized contraction of any passage or duct of the body; a constriction.

Strictus–Erect.

Stride–The distance from one footprint of a horse to the print of the same foot when it next comes fully to the ground.

String–(1) A cord of small diameter. (2) The several fumigating tents which are used by a single crew of people in fumigating orchards. (3) A group of partly broken horses which are assigned to a cowboy or horse trainer for his personal use or for further breaking or training. (4) A group of horses in a pack train. (5) To unroll wire prior to stretching and fastening to posts, etc.

String-halt–An affliction of a horse causing the hind leg to be raised and suddenly dropped to the ground; known as myoclonus.

Strip–(1) A relatively narrow piece, as a strip of land. (2) Each of the squirts or streams of milk as taken from a cow's teat in milking. (3) To tear off, as the threads from a bolt or nut, the leaves, husks, suckers, from a plant or plant part, or the hide or pelt from a carcass. (4) To take the last of the milk from a cow's udder by hand milking after the machine milker has been removed. (5) A narrow marking extending vertically between the forehead and nostrils of a horse.

Strip Cup–A small metal cup or vessel with a fine wire strainer or inner liner into which the first streams of milk from each teat are milked from the cow for examination to detect any indication of mastitis infection or any other abnormal condition of the milk or udder.

Strip Steak–The steak cut from the loin strip of a beef carcass.

Strip the Crop–To remove the crop content of dressed poultry by a downward stroking and massaging of the crop of a bird that is suspended head downward. The crop material drains from the mouth.

Striped Bass–*Morone saxatilis*, the largest member of the bass family. The fish is deep olive green on the back with a white belly. Seven to eight full length stripes run horizontally on the side. Although they are a salt water fish, striped bass can complete their life cycle in fresh water and are raised to stock lakes as a sports fish.

Strobila–The entire adult tapeworm which has a head (scolex), neck, and a chain of segments. See Tapeworm.

Stroke–(1) The linear distance traveled in one motion by a piston or ram, whether in an engine or pump. (2) A sudden and severe attack, as of paralysis.

Strong–(1) Having a specific quality to a high degree, as strong flavor, etc. (2) Characterized by steady or advancing prices, as a market for a particular commodity.

Strong and Masculine–Indicating a vigorous, healthy male animal with the desired male characteristics.

Strongyle–Any roundworm of the family Strongylidae. They are parasitic in the organs and tissues of people and various animals, often causing severe injuries or death. Some species are specific to certain hosts; e.g., the intestinal srongyles (palisade worms) of horses will not infest cattle, sheep, or swine.

Strontium–A very persistent artificial radionuclide that concentrates in bone tissue. it has a physical half-life in the environment of twenty-eight years. Sources in the human diet include milk and grain products. The more calcium in the soil the less strontium 90 is absorbed by plants.

Structural Soundness–The physical condition of the skeletal structure (especially the feet and legs) of an animal.

Structure of Milk–Normal milk, secreted by the mammary gland which consists of a serum containing fat globules, lactose, salts, albumin, globulin, and solid particles in a colloidal condition in turn consisting of casein and attached calcium phosphate. In addition, as milk comes from the udder it contains some tissue cells, an occasional leucocyte, and perhaps a few harmless bacteria.

Strut–(1) A part of a machine or structure used chiefly to hold things apart. In general, any piece of a frame which resists thrust or pressure in the direction of its own length. See Stay, Tie. (2) In an animal, exaggerated step or gait.

Strychnine–An extremely poisonous vegetable alkaloid which has an intensely bitter taste and is used as poison to kill mice, sparrows, pocket gophers, ground squirrels, etc. It is also used in medicines as a stimulant.

Stub–(1) A short blunt projection, as a blunt projecting stem, branch, or root. (2) A tree with a broken stem less than 20 feet. (3) The quill portion of a short feather appearing on the shanks or toes of otherwise clean-shanked birds.

Stubble Pasture–A field from which a crop of wheat or other grain has been previously harvested and on which animals are placed to consume the crop residues as well as weeds that may follow the grain crop.

Stuck Yolk–The condition of an egg in which the yolk adheres to the inside of the shell. Stuck yolks may occur in eggs that have deteriorated from long holding. Such eggs are classed as inedible.

Stud–(1) A unit of selected animals kept for breeding purposes, usually applied to horses. (2) Abbreviation for stud horse, a stallion; or for stud ram, male sheep.

Stud Book–The official record book of a livestock registry association for horses, ponies, and jacks.

Stud Farm–A livestock farm or establishment sometimes under government supervision, that specializes in breeding and raising herd sires for sale or distribution to smaller breeders or farmers. Also a farm that specializes in stallions for stud purposes.

Stud Flock–A flock of carefully selected breeding ewes kept for raising young rams for future herd sires.

Stud Mating–The breeding of an individual female fowl to a selected male for a desired combination of genetic qualities; used mainly to obtain exhibition individuals.

Stud Ram–A male sheep kept for breeding purposes.

Stunning Ax–A heavy, blunt-nosed ax or sledge used in slaughterhouses for stunning or killing animals.

Stunt–(1) Diseases caused by certain viruses that dwarf a plant and make it unproductive. (2) To check or hinder the growth or development of an animal or plant.

Stupefacient–A drug used to cause birds or other animals to go into a state of stupor so they can be captured and removed.

Sty–A pen where swine are housed and fed. Usually it consists of a low shed and a dirt yard. Also called pigsty.

Style–(1) In the pistil of a flower, the part between the ovary and the stigma; if the style is lacking, the stigma is sessile on the ovary. (2) The manner in which an animal displays itself while at rest or in action. (3) The manifestation of those characteristics which contribute to the general beauty, pleasant appearance, and attractiveness of an animal.

Sub–A prefix meaning either: (a) nearly, somewhat, slightly; e.g., subcordate, nearly cordate; or (b) below, under; e.g., subaxillary, below the axil.

Subacute–A clinical condition intermediate between acute and chronic.

Subbreed–An offshoot or subdivision of a major breed of livestock, such as Polled Jersey.

Subclinical–Designating early or mild stages of a disease before signs or symptoms are noticeable; symptomless or low-grade infections.

Subcutaneous–Situated or occurring beneath the skin. A subcutaneous injection is put just under the skin.

Sublethal–Less than fatal in effect, as a sublethal dosage or application of a toxic substance, etc.

Sublethal Concentration–A concentration in which an organism can survive, but within which adverse physiological changes may be manifested.

Subluxation–A partial dislocation, a sprain of a tendon or ligament which occurs near a joint.

Submaxillary–Situated below the lower jaw.

Subspecies–A major subdivision of a species, ranking between species and variety. It has somewhat varying connotations, depending on the user of the term, and often implies a distinct geographic distribution for the taxon.

Substance of Bone–A judging term used to describe the form or structure of the bone or skeletal framework. Bone and frame are associated with muscling. Substance is generally associated with ruggedness throughout the body.

Succus–Juices or fluids extracted from or secreted by an organism.

Suck–(1) Of an animal, to draw milk from the teat of the udder by application of suction by the mouth; to nurse. (2) To draw a liquid, air, etc., by vacuum-producing action.

Suckle–To obtain or provide milk from mammary glands.

Suckling Pig–A young pig still nursing its mother. When slaughtered at this stage it produces a small carcass for roasting whole.

Sucklings–The young of mammals that are being nursed by the female; unweaned animals.

Suet–Fat from the abdominal cavity of a ruminant animal, especially from cattle or sheep.

Suffolk–(1) A very popular early breed of draft horse originating in the county of Suffolk, England, which always has some shade of chestnut coat color. (2) A large, Down breed of sheep characterized by jet black, polled head and ears, and black legs from the hocks and knees down. It is a mutton-type reed with mature ewes weighing from 160 to 225 pounds. The Suffolk originated in the county of Suffolk, England.

Sugar Cure–Any method of curing meat which includes the use of sugar. The inclusion of sugar in a meat cure counteracts the astringent quality of the salt, makes meat juicier, enhances the flavor, and assists in improving the color of the cured meat.

Sugared Honey–Granulated honey.

Suint–Solid deposits from the perspiration of sheep found in the wool; a source of potassium.

Sulfa Drug–Any of a group of drugs that have a marked bacteriostatic action and sued medicinally largely to enable the host to develop antibodies to combat the invading organisms. Included in the group of sulfa drugs are: sulfapyridine, sulfaguanidine, sulfamerazine, sulfamethazine, sulfaquinoxaline, and others. See Sulfonamides.

Sulfiting–The treatment of foods with sulfur dioxide or certain related compounds. The sulfur combines with enzymes in the food and prevents them from causing the quality to deteriorate.

Sulfonamides–A group of chemicals often used as drugs to suppress the normal growth and reproduction of many disease-producing bacteria, which includes sulfanilamide, sulfapyridine, and sulfathiazole. See Sulfa Drug.

Sulfur–S; an elementary, yellow mineral, insoluble in water, easily fusible and inflammable. Also called brimstone. One of the secondary but important elements in soil fertility and used in relatively large amounts by most plants, it is an important constituent of both protein and protoplasm. The powder form is an effective insecticide for many insects. The dust is used as a fungicide in the control of mildew, etc. When burned it forms sulfur dioxide, a gas which is highly toxic to insects and has long been used as a fumigant as well as a bleaching agent.

Sulfur Bacteria–*Thiobacillur thiooxidons*; bacteria that obtain their metabolic energy by the oxidation of elemental sulfur.

Sulfur Dioxide–SO_2; a compound produced by burning sulfur; it has a suffocating odor and is used as a fumigant for the control of certain insects, for the prevention of molds on dried fruits, and for bleaching wool, straw goods, etc.

Sulfur Oxides–Pungent, colorless gases formed primarily by the combustion of fossil fuels: considered major air pollutants; sulfur oxides may damage the respiratory tract of animals and people as well as be toxic to vegetation.

Sulfuric Acid–H_2SO_4; a dense, heavy, exceedingly corrosive, oily liquid which can decompose animal and vegetable tissue and has a great affinity for water, giving off heat on combining with it. It is used in the manufacture of superphosphate fertilizers by converting the insoluble rock phosphate to a soluble and available form.

Sulking–(1) In commercial mushroom culture, a depression of growth and reduction in yield from an excessive accumulation of carbon dioxide and an unsaturated hydrocarbon gas, given off by the growing mushrooms, which are unable to escape due to insufficient ventilation in mushroom houses. (2) In horses, refusal to obey commands promptly.

Summer Sausage–A dry, uncooked sausage made of coarsely ground meat (usually beef and pork) and seasoning. The average mixture is cured for several days in a cool place below 40°F, then stuffed into casings and smoked. Since this sausage is consumed without cooking, it is necessary to have the inside temperature reach 142°F during the smoking process to comply with federal (United States) regulations. Also called cervelat, saveloy.

Sunburn—(1) Injury to the leaves, fruit, or other parts of plants, or to the skin of animals, as a result of exposure to intense sunlight. See Sunscald. (2) Greening of Irish potatoes, onions, and certain root crops due to exposure to sunlight. (3) A superficial inflammation of the skin of hogs (especially white hogs) on rape pasture, when they are wet from the dew on the plants and exposed to bright sunlight.

Sunflower Silage—Silage from sunflower plants which is usually made when the heads are just showing bloom; prepared chiefly in regions where the season is too short and cool for corn. Sunflower silage is somewhat less palatable and lower in feeding value than corn silage.

Sunflower-seed Oil Meal—A product obtained by grinding the sunflower-seed oil cake remaining after the extraction of oil from sunflower seed. It is used as a supplement in livestock feeds.

Sunstroke—Severe injury or killing of heat-sensitive people and animals by excessive heat during midsummer periods of cloudless skies with temperatures ranging above 100°F or more.

Super—A top or additional compartment added to a beehive in which bees deposit their nectar and honey.

Super-parasitism—(1) Parasitism upon a parasite. (2) An individual attacked by two or more primary parasites, or by one species more than once. See Autoparasitism.

Superfetation—The presence of two fetuses in the uterus that resulted from fertilizations during two different estrous cycles.

Supering—Placing supers of comb or foundation on a hive, either to give more room for brood rearing or for honey storage.

Superovulaton—The stimulation of more than the usual number of ovulations during a single estrous cycle due to the injection of certain hormones.

Supersedure—The act of a young queen bee taking the place of the mother queen while the mother queen is still in the hive.

Supersisters—Queens or worker bees produced by a single queen and sired by identical sperm from a single drone (subfamily).

Superspecies—A group of related species that are geographically isolated; without any implication of natural hybridization among them.

Supine—Prostrate.

Supplement—A feed or feed mixture that is relatively richer in a specific nutrient than the basic feed ingredients in a ration to which it is added. It may be used to supply a single nutrient or may contain a mixture of vitamins, proteins, minerals, and other growth stimulants.

Supplemental Pasture—A pasture to augment range forage, particularly during emergency situations. Supplemental pasture may be provided by annual grasses and/or legumes, or by aftermath of meadows, grain fields, etc.

Suppuration—The formation of pus.

Surcingle—A strap or girth around the body of a beast of burden for holding the saddle or load in place or used for throwing the animal.

Surplus Honey—A term generally used to indicate an excess amount of honey above that amount needed by the bees to survive the winter. This surplus is usually removed by the beekeeper.

Suspended Waterers—Poultry waterers that are suspended from the ceiling. They are raised up as the birds grow taller.

Suspension—Solid particles mixed with but not dissolved in a fluid, which can be filtered off or removed by centrifugal force, as butterfat in milk.

Swallow Bug—*Oeciacus vicarius*, family Cimicidae; an insect that sucks the blood from poultry at night.

Swamp Fever—See Equine Infectious Anemia.

Sward—A closely grazed or mowed area in which the grass and other plant species are close-growing, making an almost complete ground cover. See Meadow, Pasture.

Swarm—A group of worker bees and a queen (usually the old one) that leave the hive to establish a new colony; a word formerly used to describe a hive or colony of bees.

Swarm Box—A box containing combs of pollen and honey used by queen bee breeders to produce starting cells.

Swarm Cell—A specially constructed cell placed along the bottom of the comb to produce young queens for swarming.

Swarm Spore—An asexual reproductive cell which has motion by means of flagella. Also called zospore.

Swarm to Issue—The time of swarming of honeybees which usually is from ten in the morning until two in the afternoon.

Sweat—(1) In animals, moisture exuded through the pores of the skin. (2) Moisture which is given off or collects on the surface of an object.

Sweat Collar—A thick, padded leather collar that is attached to the neck of a beef animal to help shrink a dewlap that is excessive. This is done to make the animal look better in the show ring.

Sweat Scraper—A smooth curved tool used to remove sweat from a horse's coat.

Sweat Shed—A properly ventilated shed or enclosure used to hold sheep so that their natural body heat causes them to sweat. Such a practice immediately preceeding the shearing operation softens the yolk and makes shearing easier.

Sweating Process—The practice of putting sheep skins in a warm, moist room to loosen the wool in preparation for pulling.

Sweed Feed—Refers to feed which is characterized by its sweetness due to the addition of molasses, usually a commercial feed mixture.

Sweeny—A wasting or atrophy of the shoulder muscles of a horse.

Sweet Cherry—*Prunus avium*, family Rosaceae; a deciduous tree of very ancient culture, yielding a globe-shaped fruit, the cherry of commerce. Native to Europe and Asia.

Sweetbread—The thymus and pancreas glands of an animal which are used as food.

Sweetclover Disease of Livestock—A livestock disease which results in a hemorrhage or series of hemorrhages that may be fatal. It is caused by ingestion of sweetclover with molds and mildews upon the natural coumarins in the plant. Eating such plants produces dicoumarol in the animal's body, which antagonizes vitamin K production and thereby prevents the formation of prothrombin needed for blood clotting.

Swell–(1) In a saddle, the part in front of the seat which rises to the horn. Also called fork, front. (2) A tin can containing preserved foods with its ends bulging due to the formation of gas inside.

Swell Head–(1) A turkey disease which affects the sinuses or hollow spaces in the bony structure of the face. See Infectious Sinusitis. (2) A poisoning which occurs among sheep and goats grazing on certain plants, such as littleleaf horsebrush (*Tetradymia glabrata*), *Zephyrantes atamoasco*, etc. Also called big head.

Swill–Liquid food for domestic animals, especially swine, consisting of ground feed mixed with water or milk or liquid garbage form the kitchen. Also called slop.

Swine–Any mammal of the family Suidae. Domesticated species are grown for their edible flesh and fat, for their hides, and for their bristles. Swine are very important in the agricultural economy of the United States. Also called hog, pig.

Swine Brucellosis–A disease of swine caused by the organism *Brucella suis*, and closely related to brucellosis of cattle. It may be characterized by abortions, lameness, posterior paralysis, lowered fertility, or sterility.

Swine Dysentery–An infectious enteritis of swine attributed to *Salmonella*. Also called infectious enteritis in pigs, necro in pigs, pig typhus, salmonellosis suis. See Swine Enteritis.

Swine Enteritis–Inflammation of the intestines which may result from any of several microorganisms, improper feeding or overloading of the digestive tract, or chemical poisons, such as arsenic or mercury. The outstanding symptoms are diarrhea and debility.

Swine Erysipelas–An infectious bacterial disease of pigs characterized by high fever, reddish or purplish spots on the skin and by general debility and lameness. The causative organism is Erysipelothrix rhusiopathiae. Also called diamond skin disease.

Swine Fever–A febrile viral disease of hogs; highly infectious, contagious, and usually fatal. It is characterized by a high fever. More commonly called hog cholera.

Swine Influenza–See Influenza in Swine.

Swine Pellagra–A disease of swine, caused by a protein and niacin (vitamin B-complex) deficiency in feed and characterized by retarded growth, skin lesions, and enteritis. It can be treated by making up the nutritional deficiency.

Swine Plague–A respiratory bacterial disease of swine caused by *Pasteurella multocida*. Also known as hemorrhagic septicemia, pasteurellosis.

Swine Pox–A viral disease of swine characterized by small, red skin lesions, weakness, loss of appetite, chills and fever; transmitted by the hog louse.

Swiss Steak–A cut of beef, usually from the round, one to two inches thick, which is cooked with moist heat.

Switch–The brush of hair on the end of the tail of a dairy cow or other bovine animal.

Swollen Joints–An acute infectious disease, commonly called navel ill, caused by certain bacteria which often gain entrance into the navel soon after birth, becoming septicemic and localizing in the joints.

Symbiosis–The close association of two dissimilar organisms, each known as symbiont. The associations may have five different characterizations as follows: mutualism: beneficial to both species; commensalism: beneficial to one but with no influence on the other; parasitism: beneficial to one and harmful to the other; amensalism: no influence on the other; synnecrosis: detrimental to both species of organisms.

Symbiotic Relationship–A relationship between two different types of organisms that is beneficial to both of them.

Symptom–A perceptible change in any part of the body which indicates disease. A group of symptoms that, considered together, characterize a disease syndrome. A sign, mark, or indication.

Syndrome–A group of signs of symptoms that occur together and characterize a disease.

Syngamy–Union of the gametes in fertilization.

Synthetic–Artificially produced; produced by human effort and design rather than naturally occurring; chemically manufactured.

Synthetic Cream–A creamlike material made by the emulsification of nonmilk fats or hardened oils, such as whale oil, ground nut oil, etc., with dried egg, lecithin, soya, glyceryl, monosterate, and other substances.

Synthetic Food/<\d>Fiber–A food or fiber produced from a nonagricultural raw material; e.g., a nondairy coffee creamer, a synthetic orange juice, an imitation shoe leather, or a human-made fiber.

Synthetic Organic Chemicals–Calcium cyanamid and urea are produced synthetically for use as fertilizers. They contain organic combinations of elements, but behave in the soil like inorganic fertilizers. The nitrogen in cyanamid and urea are defined as "synthetic nonprotein organic nitrogen."

Synthetics–Artificially produced products that may be similar to natural products.

Systemic–(1) Pesticide material absorbed by plants, making them toxic to feeding insects. Also, pertaining to a disease in which an infection spreads throughout the plant. (2) Pertaining to the body as a whole and not confined to one organ or part of the body, as a systemic infection.

Systemic Fungi–Fungi that grow throughout the body tissues of the host.

Systemic Insecticide–An insecticide capable of absorption into plant sap or animal blood and lethal to insects feeding on or within the treated host.

T-bone Steak—A retail cut of beef for broiling taken from the loin section of the carcass. Characterized by its T-shaped bone, it is one of the most desirable cuts.

Tacheture—In animals, a lack of pigmentation in the skin resulting from congenital causes, such as in the piebald horse.

Tachinid Fly—Any fly of the family Tachinidae, whose larvae are beneficial as insect control, as they are parasitic on many noxious insects.

Tachycardia—Excessively rapid heartbeat or pulse rate.

Tack—The riding equipment of a horse, such as the saddle and bridle.

Tack Room—A room in a stable in which saddles, bridles, spurs, boots, harness, etc., are kept.

Tackey Pony—A breed of horses which are excellent riding horses for children. Native to the southeastern United States. Also called marsh pony.

Tackling—Draft horse harness.

Tacky—An animal in poor condition (southern United States).

Taeniacides—Drugs which destroy tapeworms.

Tag—(1) A dung-covered lock of wool. Also called dag, daglock. (2) A plastic or metal piece attached to an animal for identification, or a cardboard or cloth label attached to the container of a product, a feed or fertilizer, giving the content analysis, etc. (3) A lock of cotton fiber which adheres to the boll after picking. (4) To place a tag on a product, animal, etc., for identification.

Tagging—(1) Clipping manured and dirty locks from sheep. (2) The process of attaching identifying tags to animals. See also Brand, Marking.

Tail—(1) The posterior part of the vertebral column of animals. It is usually covered with hair, some of which may be quite long. Also called brush. (2) A fanlike row of rather stiff feathers on the posterior part of a bird. (3) The lowest grade of flour. (4) A wisp of hay not properly tucked into a bale. (5) The bottom layer of produce in a container. (6) The weakest and poorest sheep (Australia and New Zealand). (7) The woody part of a plant which has been propagated by tip layerage. (8) To dock an animal's tail (Australia and New Zealand). (9) To keep sheep together in a flock (Australia and New Zealand). (10) To remove the tail from a carcass. (11) To assist an undernourished cow to its feet by pulling on the tail.

Tail Band—The crupper of harness.

Tail Chewing—The tendency of certain animals to gnaw at their own tails near the anus, sometimes caused by the animal's being infested with worms or afflicted with mange.

Tail Covert—The feathers of a fowl which cover the base of the tail feathers in males and the larger portion of the tail in females. They are curved and pointed in males and oval in females.

Tail End—The poorest quality portion of a group of animals. To tail out is to remove animals from the bottom of the group.

Tail Head—The basal part of an animal's tail.

Tail Riders—Mounted cowboys who, during a drive, stay behind the herd to keep the herd moving (western United States).

Tail Set—(1) A device of leather and metal which, when attached to a horse, causes the horse to hold its tail high. (2) The position of the tail in relationship to the hips of an animal.

Tail Test—See Tuberculin Test.

Tailed—Designating an animal whose tail has been removed. See Docked.

Taint—(1) A contamination or off-flavor of milk or other products. (2) To infect; contaminate.

Take—(1) The uniting of a scion with a stock following grafting or budding. (2) To accept a male in coitus. (3) To result in a mild infection after vaccination.

Tallow—(1) The fat extracted from the fat tissue of cattle and sheep. Used in candle making, soap manufacture, etc. See Suet. (2) Designating various plants which yield flammable waxes, have a tallowy taste, form a fattening feed, or yield greasy substances resembling tallow; e.g., Chinese tallow-tree, tallow-weed, wax myrtle, etc.

Tally—(1) A label or tag attached to a product or animal for identification. (2) The number of products or animals handled or produced. (3) 100 sheep (Australia and New Zealand). (4) The total sheep sheared by each shearer in one day (Australia and New Zealand). (5) To record the number of products handled. (6) To mark for identification.

Talon—(1) The hind part or heel of the foot or hoof of such animals as deer, swine, horses, etc. (2) A claw of a bird, usually of a bird of prey.

Tame—(1) Domesticated. (2) Cultivated. (3) Designating an animal which has been made docile or tractable, as a wild horse is tamed or broken. (4) To domesticate. (5) To make docile.

Tame Hay—Hay produced from sown meadows, as contrasted with the forage from wild areas of native forage plants.

Tamed Iodine—Iodine that is combined with an organic material and is used as a disinfectant, mostly as an antiseptic. The iodine is released slowly from the organic compound, thus it is less irritating than tincture of iodine (iodine in alcohol).

Tamworth—An English breed of swine which has a long neck, body, and legs and whose coat varies from golden-red to dark red. The oldest breed of hogs, it is large and rugged.

Tan—(1) A Japanese unit of land measurement equal to 0.245 acre (0.1 ha). (2) A coat color of some animals, especially dogs. It is a yellowish-brown similar to the color of well-tanned leather, and varies in shade form light to dark. (3) To convert a hide into leather.

Tang Bees—To beat on a pan, etc., to cause a swarm of bees to settle.

Tankage–A fertilizer or animal protein supplement feed consisting of, and restricted to, the rendered, dried, and ground meat and bone by-products of the carcasses of animals which have been slaughtered, or which have died from other causes.

Tanners' Lime–An impure calcium carbonate which is a waste product from the tanning of leather.

Tanning–The conversion of animal hides into leather.

Tansy Ragwort–A yellow flowering plant (*Senecio* spp.) found in the Pacific Northwest that is poisonous to cattle and horses. The whole plant has a strong unpleasant odor when crushed.

Tap Water–Water which comes from a water faucet, as contrasted to well water or rainwater.

Tapadera–A leather hood which covers the front of a stirrup on a saddle.

Tapeworm–A parasitic intestinal worm of a flattened, tapelike form, order Cestoda, composed of separate parts or segments.

Tar–(1) A black, liquid mixture of hydrocarbons and their derivatives obtained by distillation of wood, peat, coal, shale, etc. (2) Tar used as a disinfectant on a sheep accidentally cut by the shearer (Australia and New Zealand).

Target Species–A plant or animal species which a pesticide is intended to kill.

Targhee Sheep–A breed of sheep developed at the United States Sheep Experiment Station at Duboise, Idaho. It is the result of crossbreeding the Rambouillet with Lincoln and certain other longwool breeds, producing a type especially suited to northwestern range conditions. The targhee is a polled, white-faced sheep of intermediate size with mature rams weighing about 200 pounds and ewes about 130.

Tarpan–A miniature horse which roamed wild over Europe for thousands of years and finally became extinct in the latter half of the 19th century. Also called Tartar horse, Prejvalsky horse.

Tarsus–Fifth segment of a bee's leg.

Taste–(1) The flavor of a product as determined by placing the substance in the mouth. (2) A small amount or sample. (3) In judging, to place a morsel or a few drops of a product in the mouth to savor, but often not to swallow.

Tattoo–(1) A letter, number, or other mark, pricked with indelible ink into the skin of an animal for identification. (2) To make such a mark.

Tautonomy–Relations that exist if the same word is used for both the generic and specific name in the name of a species.

Taw–(1) To dress and prepare, as the skins of sheep, lambs, goats, and kids, for gloves, etc., with alum, salt, and other softening and bleaching agents. (2) To prepare hemp by beating.

Taxonomy–(1) The science of classification of organisms and other objects and their arrangement into systematic groups such as species, genus, family, and order. (2) Taxonomy is closely related to classification but it embodies a broader concept. Taxonomy is the science of how to classify and identify. It is the theoretical study of classification including its bases, principles, procedures, and rules. Taxonomy includes classification as well as identification.

TDN–Total digestible nutrients; all nutrients consumed by an animal that are digested and used; generally applied to proteins, carbohydrates, and fats.

Team–(1) Two or more horses, mules, etc., which are harnessed to the same vehicle, plow, etc. (2) Two or more specialists who are jointly investigating a problem. (3) To match two or more draft animals to serve as a team.

Teart of Cattle–Molybdenum poisoning.

Tease–(1) To stimulate an animal to accept coitus. (2) To vex or annoy. (3) Of fibrous materials such as wool or flax, to separate the strands of the fibers, or to prepare the fibers, by combing, for spinning. (4) To form a nap on cloth by stroking the loose fibers in one direction with comblike, natural, or mechanical, teasels.

Teaser–(1) An animal which is used to stimulate sexually one of the opposite sex. (2) Designating a male animal which is used to locate females of the same species in heat.

Teasing–Keeping a ram in sight of, but not in contact with, ewes just prior to breeding. This often stimulates ovulation in the ewes.

Teasing Pole–A pole used to separate a stallion from a mare to determine from the mare's action whether or not she will accept coitus with the stallion.

Teasing Stall–A stall in which a female teaser animal is placed for the purpose of stimulating a male sexually.

Teat–The fleshy protuberance through which milk is drawn from the udder of a mammal. Also called tit. (Teat is used in formal discourse; tit is more common colloquially.) Also called thelium. See Nipple.

Teat Cup–The part of a milking machine that is attached to the cow's teat. The milk is drawn out through suction.

Teat Placement–The placement of teats on a sow's underline; they should be evenly spaced.

Technical–(1) Concerned with a particular science, industrial art, profession, sport, etc. (2) Practicing, using, or pertaining to the technique rather than the theory or underlying principles involved in the execution of a project. (3) Designating the grade of a commodity manufactured in the usual commercial manner. (4) Pertaining to or designating a market where prices are controlled by speculation or manipulation.

Teg–(1) A two-year-old sheep. (2) The fleece of a two-year-old sheep.

Telescoping Cover–A hive cover, used with an inner cover, that extends downward several inches on all four sides of a beehive.

Telophase–The phase of cell division between anaphase and the complete separation of the two daughter cells; includes the formation of the nuclear membrane and the return of the chromosomes to long, threadlike and indistinguishable structure.

Temper–(1) The proper relative condition of moisture in grain preparatory to milling. (2) The relative hardness or softness of the metal in implements and tools. (3) The relative mildness or viciousness of an animal. (4) Milk of lime, etc., added to boiling syrup to clarify it.

Temperament–Disposition of an animal or person.

Temperature–(1) The amount of heat or cold measured in degrees on different scales, as Fahrenheit or Centigrade. At sea level, water freezes at 32°F (0°C) and boils at 212°F or 100°C. (2) The degree of heat in a living body. (3) Abnormal heat in a living body. Also called Fever.

Temperature Zero–The temperature below which certain physiological processes of an organism are carried on at a very slow rate.

Tenderize–To render meat tender by breaking down the fibers during cooking, marination, or pounding.

Tenderized–Designating a cut of meat which has been macerated or made tender by enzyme action, mechanical means, or by the action of certain chemical substances.

Tenderloin–The muscle lying on either side of the backbone of meat animals and considered a very choice cut of meat.

Tendon–The strong tissue terminating a muscle and attached to a bone, for leverage purposes; usually a dense, cordlike structure of various thicknesses.

Tennessee Walking Horse–A saddle-type breed of horse developed in central Tennessee, United States, which stands about 15.2 hands high and weighs from 1,000 to 1,250 pounds. It has a good disposition and a natural running-walk gait. The colors are sorrel, chestnut, black, bay, gray, roan, yellow, and white.

Teratogen–An agent or factor that causes the production of physical defects in the developing embryo of animals and people.

Term–At term; the end of the normal period of gestation of pregnancy when birth is due to occur.

Terminal Sires–Sires used in a crossbreeding system where all progeny, both male and female are marketed.

Terracette–A small hillside step or bench commonly believed to have been caused by livestock. It usually involves slumping on a small scale indicating gradual downhill movement of the entire surface. Also called cattle terrace, sheep track, cat step.

Terramycin–Oxytetrocycline hydrochloride, an antibiotic derived from the organism *Streptomyces rimosus*.

Terrestrial–(1) Referring to earth. (2) Designating a plant which lives in soil as contrasted to one which is epiphytic (growing in air) or one growing in water (aquatic or hydrophytic). (3) Designating a ground bird such as a pheasant, partridge, or chicken, as contrasted to an aerial bird.

Test Adaptation Trial–In pasture and range research, testing the adaptability of forage species to new edaphic and climatic conditions as to complete their life cycle and/or reseed themselves.

Testcross–A kind of genetic cross involving one individual expressing a dominant trait and one expressing the recessive trait, the purpose of the cross being to determine whether the individual expressing the dominant trait is heterozygous or homozygous.

Testicle–Male sex organ which produces sperm after sexual maturity. Plural, testes or testicles.

Testosterone–A hormone produced by the interstitial cells of the testes that functions in stimulating male sex drive, masculine characteristics, development of the male reproductive tract, and spermatogenesis.

Tetanic–Pertaining to severe muscular contractions.

Tetanus–A disease of the nervous system caused by an anaerobic bacillus that results in an inability to control certain muscles, particularly those in the region of the neck and jaw; also called lockjaw.

Tethering–Restraining an animal by tying it to a post, etc., with a long rope, chain, etc., so that restricted grazing is possible.

Tetracycline–A broad-spectrum antibiotic produced by the organism *Streptomyces rimosus* and sold under various brand names, as Tetracyn, Acromycin, Polycycline, etc. A number of other antibiotics are chemically derived from it.

Tetraploid–An organism whose cells contain four haploid (monoploid) sets of chromosomes.

Texas Longhorn–Cattle native to southwestern United States, especially to Texas, which are descendants of cattle first brought into this region by the Spanish explorer Coronado, in 1510. A special characteristic is the very large, long, curving horns. Also called Longhorn.

Texas Steer–Designating a steer hide which has a brand on the side.

Theave–A ewe which has not borne a lamb (British).

Theca–The membranes that form the wall of the ovarian follicle.

Therapeutic–Pertaining to the treatment of disease; curative.

Therapy–The sum total of the treatment given to cure disease in plants, animals, and humans.

Thermal Death Point–The amount of heat required to kill a particular organism.

Thermocline–The layer in a body of water in which the drop in temperature equals or exceeds one degree centigrade for each meter or approximately three feet of water depth.

Thiamin–A member of the vitamin B-complex; vitamin B_2.

Thick Leg–An edematous condition in people and animals. See Elephantiasis.

Thigh–(1) The part of an animal's hind leg between the hock and the trunk. (2) A piece or cut of fowl from the leg between the drumstick and the body.

Thill–Either of the shafts between which a horse or other animal is hitched to a vehicle.

Thin–(1) To reduce the number of plants in a row or area by hoeing, pulling, etc. (2) Designating an animal with little flesh. (3) Designating a pulse which is very feeble.

Thoracic Cavity–Cavity of the chest containing the heart and lungs.

Thorax–The middle body region of an insect to which the wings and legs are attached.

Thoroughbred–A breed of saddle horses developed in the United Kingdom, which stand about 16 hands and weigh about 1,000 pounds. The Thoroughbred is bay, brown, black, chestnut, tan, gray, or roan and is essentially a race horse. The term is sometimes confused with purebred.

Thread-necked Strongyles–Various species of roundworms (nematodes) of the genus *Nemotodirus*, family Trichostrongylidae; parasites in the small intestine of domestic and wild ruminants, as well as certain other mammals. When large numbers are present in the small intestine of a host, they may cause parasitic dysentery.

Threadworms–Small roundworm (nematode) parasites of the genus *Strongyloides*, which live in the small intestine of humans and various domestic and wild animals. Some of the important species are *S. stercoralis* (in humans, dogs); *S. papillasus* (cattle, sheep, goats); *S. ransomi* (swine); *S. westeri* (equines). Infection results from infective larvae penetrating the hosts skin or by being swallowed. Larvae may cause a dermatitis, and adults (female only) in the small intestine may cause parasitic dysentery and anemia.

Three-gaited Horse–A saddle horse which has the following gaits: walk, trot, and canter.

Three-way Cross–The crossing of three different strains or breeds of plants or animals to produce a hybrid. See Double Cross, Single Cross.

Thremmatology–The domestic breeding of plants and animals.

Thriftiness–The capacity to make good use of food or feed.

Throatlatch–The strap on a bridle or halter which passes under the horse's throat.

Throw–(1) In horseback riding, the violent ejection of the rider from the horse's back by action of the horse. (2) To cause an animal, as a horse or cow, to fall to the ground before branding, treating, etc. (3) To wield a lasso.

Throw Up–(1) To hill earth or soil around the crowns of plants. (2) To vomit.

Throwback–(1) In breeding, a reversion to an ancestral type or ancestral characteristic. (2) To revert to an ancestral type or characteristic.

Thrush–(1) A foul-smelling, degenerative condition involving the frog in the sole of a horse's foot. Predisposing causes are generally considered to be unhygienic conditions such as standing in mud and filth. (2) A fungal disease of the digestive tract of poultry caused by *Candida albicans*. About the only clinical signs are depression and emaciation. Also called crop mycosis, intestinal mycosis.

Thumps–An animal ailment resembling hiccoughs in humans which is seen in baby pigs with anemia, swine influenza, and with verminous pneumonia.

Thunk–A wartlike growth in a cow's teat canal.

Thurl–The thigh of a horse.

Thwarter–A disease of sheep, and occasionally of pigs, caused by the tick *Exodes ricinus*, family Ixodidae, characterized by abnormal conformation of the neck, loss of muscular control, paralysis, and often death. Occurs in Scotland and northern England. Also called thwarter-ill, thorter-ill, louping-ill.

Thymus Gland–A ductless, glandular body situated in the thorax which reaches its maximum development early in life. Grayish-red, it usually has two longitudinal lobes. The animal thymus gland, used as food, is called sweetbread.

Thyroid Gland–A ductless gland situated in the neck of humans and animals which secretes thyroxin, a hormone that controls the rate of metabolism and has a profound influence on growth and production.

Thyroid-stimulating Hormone (TSH)–A hormone produced by the anterior pituitary gland that stimulates the thyroid to produce thyroxin.

Thyroiditis–Malfunction of the thyroid of an animal or a person.

Tick–Any of various blood-sucking arachnids which fasten themselves to warm-blooded animals. Some are important vectors of diseases.

Tick Fever–See Cattle Tick Fever, Piroplasmosis.

Tick Paralysis–A disease of sheep, cattle, and people, caused by the ticks *Ixodes* or *Dermacentor*, characterized by a paralysis first affecting the legs, next the chest and neck, and by the ultimate death resulting from the heart becoming involved.

Tick Quarantine–A legal prohibition against shipping livestock from one area to another in an attempt to prevent the spread of ticks.

Tick-infested–Designating an animal or area of land which abounds in ticks.

Tickborne–Transmitted by ticks, as cattle tick fever.

Ticking–Specks or small dots of color on an animal's coat or on the plumage of a fowl.

Tie–(1) A string, rope, metal band, or wire which encircles a bale of cotton, hay, etc., holding the material in a limited space and form and making for ease of handling, storage, and transportation. (2) A rope, twine, etc., for fastening purposes. (3) A stanchion for securing an animal. (4) Depression, usually in the middle of the back of an animal, caused by the skin adhering to the backbone. (5) To fasten, secure, with a rope, string, wire, etc.

Tie Stall–(1) A stanchion to which an animal is fastened. (2) A stall in which an animal is fastened by a halter or chain.

Tie-out Chain–A chain used to fasten an animal to a stake, tree, post, etc.

Tight Side–The right side of a beef carcass. See Loose Side.

Tilapia–A large group of African fresh water fish. Around 70 species of the genus *Tilipia* are distributed around the world. They are cultivated as a food fish.

Tip–(1) The exposed part of the wool fiber in the fleece. (2) The end of a branch, twig, etc. (3) A place where refuse is dumped (England). (4) To remove the ends of branches, stems, twigs, etc.

Tippy Wool–Staples of wool that are encrusted with wool grease and dirt at the weather end.

Tissue–Groups of cells working together to carry out a common function, such as muscle tissue, connective tissue, and epithelial tissue.

Titer—The minimum quantity of a substance required to produce a specific reaction with a given amount of another substance.

Tobiano—Basically, a white horse in which large, smooth, and solid blocks of color originate on the back and rump and extend down. The face is usually marked the same as in other color patterns found in horses.

Toe—(1) Any one of the digits on the foot. (2) The lowest downstream edge of a dam. (3) The bottom of a seed furrow opened by the furrow opener of a grain drill. (4) The tuberous roots of dahlia, used for propagation. (5) The lower edge or edges of a slope. See Jute.

Toe Mark—To punch holes in the membrane between the toes of fowls for identification purposes.

Toed-out—A term used to describe an animal that walks with the feet pointed outward; splay-footed; slew-footed.

Toggenburg—A breed of milk goats, perhaps the most popular and most widely used of any in the United States, varying in color from light brown to chocolate. Its ears are white with a dark spot in the middle, and there are two white streaks on the face. The legs are white from the knees down. Bucks weigh 160 pounds or more and the does 120 pounds or over.

Tom—(1) The male turkey. (2) The male of certain animals, as the tom cat.

Tomentose—Coated with short, matted wooly hair.

Tongue—(1) A part of the mouth, composed chiefly of muscle, on which the taste buds lie. It is one of the organs of taste, and moves the food being chewed around in the mouth and back to the throat for swallowing. (2) A retail cut of beef, veal, mutton, and lamb, consisting of the whole tongue of an animal. (3) The wooden or metal pole which acts as a guide and is fastened to the front end of a vehicle or plow. It separates the two members of a team of draft animals. The eveners for pulling the implement may also be attached to the tongue. (4) To make a slit in the plant's stem in grafting.

Tongue Sensation—The feel or texture that the tongue registers when in contact with water containing various solutes. This is in addition to the sensations of taste and temperature.

Tonic—(1) A medicinal preparation to increase the strength or tone of the body system. (2) Designating a plant exhibiting normal reactions to external influences. (3) Referring to muscular contraction, as tonic convulsions. See Conditioner.

Tooth—(1) In animals, one of the hard, bone like appendages of the jaws used to tear and to masticate food. (2) See Tine. (3) A cog. (4) Any of the piercing projections on the cutting edge of a saw. (5) One of the projecting or piercing points or prongs of a comb.

Tooth Float—A device for filling horses' teeth.

Top—(1) The aboveground parts of certain plants, especially the leaves, as beet tops, turnip tops. See Crown. (2) Scoured, combed, long wool. (3) The highest price paid for a product within a certain period of time. (4) The upper, branchy portion of a felled tree. It is sold as firewood, charcoal source, etc. (5) To cut off the crown and leaves, as of the sugar beet. (6) To remove the upper portion of the crown of a tree. (7) To place the best articles of produce, as eggs, or fruit, on the top layer, in a container so that the whole container appears to include articles of higher quality than is actually the case. (8) To sort out animals that have reached a certain stage of development or finish.

Top Chuck—A retail cut of beef taken from the neck, down to the blade bone, and rolled with sheets of cod fat.

Top Cross—(1) In corn breeding, a cross in which an inbred line is used as the pollen parent and a commercial variety a the seed plant. (2) A cross between purebred males and grade females.

Top End—The highest quality portion of a group of animals. To top out is to select from the best.

Top Grade—A market grade of produce which is either the best or the next best after fancy.

Top Quality—Designating generally, but not specifically, produce of the very best sort.

Topcrossbred—The progeny from the mating of inbred sires with noninbred dams of a different breed.

Topincross—The progeny from the mating of an inbred sire with a noninbred dam of the same breed.

Topknot—A usually decorative crest of feathers on the head of a bird or a tuft of hair on the head of an animal.

Toro—(Spanish) Bull.

Torula—(1) A group of wild yeasts which are of considerable importance in the fermentation of fruit juices. (2) Lactose fermenting yeasts which produce undesirable changes in dairy products.

Total Digestible Nutrients (TDN)—A standard evaluation of the usefulness of a particular feed for livestock which includes all the digestible organic nutrients; protein, fiber, nitrogen-free extract, and fat (the latter being multiplied by 2.25 because its energy value for animals is approximately 2.25 times that of protein or carbohydrates).

Tough—(1) Having the quality of flexibility without brittleness; yielding to force without breaking. (2) Designating market grain which contains moisture in excess of 13.5 to 14.5 percent. (3) Designating plant or animal products that are not brittle, crisp, or tender.

Tower Silo—A cylindrical tower made of wood, concrete, tile, metal, etc., used for storage of silage. It is the most common type of silo in the United States. Also called upright silo.

Toxemia—A generalized blood poisoning which results from absorption of bacterial or other poisons. In this condition only the poisons or toxins are found in the blood stream, and the bacteria or other poison producers are confined to a wound or other affected area.

Toxic—Poisonous; caused by poison.

Toxic Residue—A poisonous residue left on plants, in the soil, or on animals, by a spray or dust.

Toxicant—A substance that injures or kills an organism by physical, chemical, or biological action; e.g., heavy metals, pesticides, and cyanides.

Toxicity—State or degree of being poisonous.

Toxicology–The science which deals with poisons, antidotes, toxins, effects of poisons, and the recognition of poisons.

Toxin–A protein poison produced by some higher plants, certain animals, and pathogenic bacteria. Toxins are differentiated from simple chemical poisons and vegetable alkaloids by their higher molecular weight and antigenicity. See Phytotoxin, Poison.

Toxoid–A toxin which has been chemically altered so that it is no longer toxic but is still capable of uniting with antitoxins and/or stimulating antitoxin formation.

Toy–Designating any small or dwarf variety, as a toy spaniel dog.

Trace–(1) Either of the two leather straps or chains, attached at one end to the hames clamped to the horse collar or breast band, and at the other to tugs on a whippletree. (2) The force to pull the vehicle, etc., as applied to traces. (3) See Trace Element. (4) To follow the course of nutrient elements in plants or animals, as by use of radioisotopes. (5) Designating an artificial flavoring material with little or no true flavoring.

Trachea–The windpipe; in mammals, it extends from the throat to the bronchi.

Tracks–A term used to describe the leg movements of a horse while it is walking.

Tract–(1) An area of land of any size, but bigger than a lot. (2) An anatomical structure as the digestive tract, cerebellospinal tract.

Trail–(1) A pathway made either by repeated passage of people or animals or constructed for easier passage. It is not usable for vehicles. (2) To move cattle or sheep over a long distance to a pasture, a market, or shipping point.

Trail Drive–To drive a herd along the trail to pasture or market.

Trail Herding–Directing and controlling the movement of a group of livestock on restricted overland routes.

Trailing–(1) The driving of livestock from place to place. (2) The voluntary wandering of livestock about a range, usually in search of forage, water, or salt. (3) Designating a plant which puts forth long, recumbent stems.

Trait–Any observable feature or characteristic of an animal.

Trait Ratio–An expression of an animal's performance for a particular trait relative to the herd or contemporary group average. It is usually calculated for most traits by dividing the individual record by the average of animals in the group and multiplying by 100.

Trampling–(1) Treading under feet; the damage to plants or soil brought about by congested movements of livestock, including mechanical injury to tree reproduction and ground cover in woods. (2) Compacting soil in earthen dams and reservoirs by livestock to make the dam or reservoir impervious to water (now replaced by machine compaction).

Transferring–Moving bees and comb from a natural nest in a cavity or container to a movable frame hive.

Transhumance–Seasonal nomadic movement of people and grazing livestock for part of the year combined with some form of permanent lowland farming in the wet season.

Transit Tetany–A condition likely to occur in very warm weather to an animal which is transported over very long distances in confined quarters. Symptoms consist of distress, rapid respiration, high and hard pulse, sweating, and spasms of muscles of the neck and shoulders.

Transition Cell–A cell in a beehive larger than a worker cell, but smaller than a drone cell, where a change from the worker to the drone-size bee occurs.

Transitory Range–Land that is suitable for temporary use for grazing; e.g., on disturbed lands, grass may cover the area for a period of time before being replaced by trees or shrubs.

Translocation–In genetics, change in position of a segment of a chromosome to another part of the same chromosome or to a different chromosome.

Trauma–A wound or injury; a deep emotional shock.

Traumatic–Describing a condition resulting from an injury or wound.

Travis–The partition or half-wall dividing individual stalls in a barn. Also called traverse.

Tread–(1) Any injury to the front legs of a horse caused by its overreaching with a hind foot as it runs. (2) That part of a wheel or tire which rests or bears on the ground. (3) That endless belt on which a crawler tractor moves. Usually called track.

Treat–(1) To care for a sick animal or diseased plant by giving it proper attention and medication. (2) To subject a product to an action or process to improve it in some manner. (3) To subject plants, animals, or soil to various chemicals, practices, etc., in order to learn which are beneficial or harmful.

Tref–Designating a food which does not meet the dietary laws for Orthodox Jews. See Kosher.

Trematode–Parasitic flatworm of the class Trematoda, such as the sheep liver fluke.

Trembles–A disease of cattle and sheep which results from ingestion of white snakeroot, jimmyweed, or burrowweed, all of which contain tremetol, an unsaturated alcohol. It is characterized by a trembling in the muscles about the eyes and legs, depression, inactivity, constipation, nausea, rapid and labored breathing, weakness, prostration, and sometimes death. Humans who drink milk from affected cows may contract this disease. Also called milk sickness.

Tremor–A trembling of the voluntary muscles.

Tremulous Air Cell–An enlarged air cell in poultry eggs which causes the membranes to split apart beyond the edges of the cell. The cell moves slightly as the egg is turned from one side to another. Such an egg is graded C, or the lowest edible grade.

Trench Silo–A trench excavated in a hillside or on firm ground, usually lined with wood or concrete retaining walls. Commonly about 15 to 25 feet wide, 6 to 8 feet deep and as long as the capacity desired. A trench silo must have good drainage and its use is largely limited to arid or semiarid climates.

Tri–A prefix meaning three, as *trilocular*, having three locules.

Tri-purpose Animal–Animal that has been developed and used for draft (work) purposes as well as for meat and milk production. In some countries cows, female water buffalo, and female camels are so used.

Tribe–(1) In botany, a subdivision of a family, roughly equivalent to subfamily. (2) In animal breeding, a group or combination of animals descended through the female line.

Tric–See Bovine Genital Trichomoniasis.

Trick Horse–A horse trained to perform different feats. See High School Horse.

Trihybrid–An individual that is heterozygous for three pairs of genes.

Trim–To remove unwanted or undesirable portions of a product, plant, etc., as to trim fat from a ham or to trim a tree or hedge.

Trimester–Three months, or one-third of the nine months of pregnancy of cattle. The nine months of pregnancy are divided into the first, second, and third trimesters.

Trio–(1) A group consisting of three objects or organisms. (2) A male and two female birds of the same variety which are shown as a unit in exhibitions.

Tripe–A retail cut of beef consisting of the walls of the rumen and the reticulum.

Triple-yolk Egg–An egg which contains three yolks.

Triploidy–A condition where the cells possess three sets of homologous chromosomes rather than two.

Trisomic–Referring to an organism having three chromosomes of one type (chromosome formula 2n/1).

Trocar and Cannula–An instrument used to relieve cattle suffering from indigestion with accumulation of gas.

Trophic–Of or pertaining to nutrition.

Trophic State–Characterization of a body of water in terms of position in a scale ranging from oligotrophy (poor in nutrients) to eutrophy (rich in nutrients).

Trophogenic Region–The superficial layer of a lake in which organic production from mineral substances takes place on the basis of light energy.

Tropholytic Region–The deep layer of the lake where organic dissimilation predominates because of light deficiency.

Tropical Zone–(1) The region of the earth bisected by the equator and extending at low elevations to latitude 23° 27" north (Tropic of Cancer) and south (Tropic of Capricorn). (2) Designating a plant, disease, etc., which flourishes in the tropics or where conditions are made to resemble the tropics in temperature, humidity, length of day, etc.

Trot–A two-beat gait in horses where the diagonal legs move and strike the ground together.

Trots–A diarrheal or abnormally loose condition of the bowels. Sometimes also called skitters.

Trotter–(1) A horse used for sulky racing. (2) Lamb shank.

True–Like the parental type, without change, as a variety which breeds true.

True Stomach–See Abomasum.

True-breeding–Designating varieties which conform to the parental type with respect to certain characteristics, such as color, disease resistance, etc.

Try–To determine if a female is in heat by bringing her and a male in close proximity. They are usually kept separated by a fence, wall, or teasing pole.

Tryer–A grade stallion which is used for the purpose of determining if a mare is in heat. See Try.

Trypanosome–A microscopic, free-swimming, tailed protozoan parasite of the genus *Trypanosoma*, family Trypanosomatidae. Many species are not injurious, but a few cause serious diseases of animals and people, such as *T. brucci* and *T. gambiense*, both of which gain entry to the body by the bite of the tsetse fly.

Trypanosomiasis–The disease affecting mammals caused by the presence in the body of a protozoan parasite of the genus *Trypanosoma*, marked by fever, anemia, and redness of the skin; transmitted by tsetse flies; also contagious to people.

Trypsin–One of the principal proteolytic enzymes of the pancreatic secretion in people and animals.

Tryptophan–One of a group of essential amino acids which constitute fundamental units from which proteins are built and which are needed by people and animals for the building and repair of tissues.

Tsetse Fly–*Glossina mortisans*, family Muscidae; a fly of low elevations in central and southern Africa which is a carrier of the parasite (*Trypanosoma brucei*) that causes the disease nagana in various animals, as horses, cattle, or goats. The parasite is conveyed by the fly's bite. The widespread presence of the fly, during certain seasons, renders some districts completely unsuitable for habitation by such domestic animals. A related species (*G. palpalis*) carries the trypanosome *T. gambiense* that causes sleeping sickness in humans.

Tubbing–Standing a horse's foot in a bucket of hot water with washing soda dissolved in it to soften the hoof.

Tube Dehorner–A device consisting of a metal tube that is sharp on one end and has a round ball-like knob on the other end. When the sharp end of the tube is fitted down over the horn of a calf, the operator gives a quick downward thrust with a twisting motion to remove the horn. The dehorners come in different sizes.

Tubercle–(1) A small tuber. (2) The small nodules ascribed to the action of symbiotic organisms especially on the roots of legumes. (3) A nodule or small eminence in or on soft tissues. (4) A rough, rounded eminence on a bone.

Tubercle Bacilli—Microorganisms that cause tuberculosis.

Tuberculin—A biological agent derived from the growth and further processing of the tubercle bacilli which is used for the detection or diagnosis of tuberculosis in animals and people.

Tuberculosis (TB)—A chronic infectious contagious disease of humans and practically all warm-blooded animals that is caused by *Mycobacterium tuberculosis*.

Tularemia—A bacterial disease occurring mainly in rabbits but also in certain rodents, ungulates, carnivores, birds, livestock, and people; caused by *Francisella tularensis* and transmitted by arthropod vectors (ticks, lice, fleas, biting flies) and by contact of skin with infected material; marked by inflammation of lymph glands, headache, chills, and fever.

Tumor—A swelling; a new growth of cells or tissues governed by factors independent of the laws of growth of the host. It may be either benign or malignant.

Tunica Albugenia—A capsulelike covering of an organ, in particular, the testes.

Tunisian Bee—*Apis nigra*; a ferocious honey bee. Also called Punic bee.

Tunnel—(1) A gallery. (2) To burrow, make a passageway, in plant stems, leaves, roots, etc., as by larvae of certain insects.

Tup—A ram.

Turicata—*Ornithodorus turicata*, family Argasidae; a fever tick, which pesters hogs, cattle, and occasionally humans, causing extreme irritation. Found in Mexico and southwestern United States. Also called relapsing fever tick.

Turken—The female offspring of the turkey cock and domestic hen.

Turkey—*Meleagris gallopavo*; a large, native American bird, now largely domesticated, which is raised for its delicious meat. Roast turkey is traditionally served in the United States at Thanksgiving and Christmastime.

Turkey Crumbles—Large alfalfa pellets that have been crushed and screened. Used to feed turkeys.

Turkey Gnat—*Simulium meridionale*, family Simuliidae; a small, blackish fly that congregates about the eyes, ears, and nostrils of chickens and turkeys, sucking blood from the host. The bite may result in symptoms similar to mastoiditis. It may be a vector of the organism causing onchocerciosis in humans.

Turkey Hen—A mature, female turkey.

Turkey Louse—*Lipeurus caponis*, family Philopteridae; a louse, parasitic on chickens and turkeys, which fastens on to the bird and eats its feathers.

Turkey Poult—A young turkey before its sex can be determined. Also called poult.

Turkey Tom—A mature male turkey.

Turkeymullein—*Eremocarpus setigerus*, family Euphorbiaceae; an annual weedy plant which yields blackish seeds that are nutritionally fattening for the turkeys. Sheep have been killed from grazing on this plant, for the hairs covering the stems and leaves are indigestible. Native to western United States. Also called dove-weed.

Turn—(1) To change the position of an egg in an incubator or in the nest where the hen is incubating it. (2) To restrain an animal, as a fence turns a cow. (3) To plow, so that the lower part of the soil is brought to the surface and the former surface part is covered. (4) To begin to show signs of ripening, as a fruit. (5) To direct an animal to pasture. (6) To sour, as milk. (7) To change, as leaves in the autumn.

Twenty-eight Hour Law—A law that prohibits the transporting of livestock by rail or truck for a longer period than 28 consecutive hours without unloading, feeding, watering, and resting five consecutive hours before resuming transportation.

Twin—(1) Either one of two young born at one birth. (2) To bear two young at a birth when only one is usual.

Twist—The region between the hind legs of an animal where the thigh muscles join. See Twitch.

Twist Him Down—To force a steer or cow to the ground by turning or twisting its head.

Twisted Stomach Worm—*Haemonchus contortus*; a stomach worm which infests sheep, cattle, and goats. Infestation is characterized by a severe anemia, pallid skin, visible mucous membranes, weakness, standing still, a swelling (bottle jaw), and hard dry feces. Also called twisted wire worm, barber pole worm, large stomach worm, common stomach worm.

Twitch—A device used for pinching or squeezing the nose (upper lip) of a horse to take the animal's attention away from some manipulation or minor operation. It is also used as a control measure on some unruly horses. The device may be a loop of rope or chain or a clamp, or may be accomplished with the hand. In some localities, especially where mules are common, the device may be placed on the ear. Also called twist.

Two-grooved Loco—*Astragalus bisulcataus*, family Leguminosae; a poisonous plant found from Manitoba, Canada, to New Mexico, United States.

Typhoid Fever—An acute infectious disease caused by a bacterium, *Salmonella typhi*, characterized by continued fever, inflammation of intestine, intestinal ulcers, a rose spot on the abdomen, and enlarged spleen; food and water-borne but may be transmitted by house flies.

Typhus Fever—A disease of humans caused by a bacterium-like microorganism, *Rickettsia prowazeki*, and transmitted by the body louse *Pediculis humamus humamus*. The disease is characterized by high fever, backache, intense headache, bronchial disturbances, mental confusion, and congested face. Mortality may range from 15 to 75 percent.

Typical–In appraising, that which most frequently occurs or exists in the particular situation under consideration.

Tyrothricin–A combination of the two antibiotics, gramicidin and tyrocidine, derived from the organism *Bacillus brevis*.

U(nited) S(tates) Performance Tested Parent Stock–An official classification under the United States National Poultry Improvement Plan for poultry flocks that qualify through the performances of their progeny in an official random sample test.

U(nited) S(tates) Pullorum-Typhoid Clean–Under the United States National Poultry Improvement Plan, an official classification of poultry flocks in which no pullorum or typhoid reactors were found on the first official blood test, provided that if a reactor or reactors were found on the first test, the flock may be qualified with two consecutive official negative tests at least twenty-one days apart.

U(nited) S(tates) Pullorum-Typhoid Passed–Under the United States National Poultry Improvement Plan, an official classification for poultry flocks in which no pullorum or typhoid reactors were found on the last official test. Birds must be at least five months old when tested and must be tested within twelve months of the time hatching eggs are to be used. The interval between tests must be at least twenty-one days.

U(nited) S(tates) Record of Performance (ROP)–An official designation, under the United States National Poultry Improvement plan, for females that qualify on individual or family trap nest records. Males qualify by being produced from a single-male mating of qualified ROP birds, and by being reasonable representatives of breed or variety and by showing health and vigor and no serious defects.

Ubiquitous–Occurring everywhere, as house flies; house sparrows; weeds.

Udder–The encased mammary gland with teats or nipples, as in a mare, ewe, sow, or cow. Also called bag.

Udder Cannula–A small, plastic, metal, or hard rubber tube designed for the injection of medicines into the udder through the teat canal.

Udder Felon–See Mastitis.

Ulcer–An open sore that discharges pus and is difficult to heal.

Ulcerative Lymphangitis–A contagious disease of a horse's skin caused by *Corynebacterium pseudotuberculosis*, characterized by pustules which, on breaking, discharge a yellow pus that dries, leaving a yellow crust. It involves the limbs primarily and resembles cutaneous glanders. Also called Canadian horse pox, ulcerative cellulitis.

Umbilical Cord–The part of the fetal membranes that connects to the navel of the fetus; it carries the blood vessels that before birth transport fetal blood to and from the placenta.

Umbilicus–The navel.

Unbroken–(1) In marketing, designating an egg free from cracks or breaks in the shell. (2) Designating an untrained horse. (3) Designating soil that has not been plowed. See Virgin Soil.

Uncapping–Cutting a thin layer from a comb surface to remove the wax covering from sealed cells of honey.

Uncapping Knife–A knife, usually heated, for cutting cappings from honeycomb so the honey can be extracted.

Undercut–(1) A notch cut in the trunk of a standing tree below the level of the major cut and on the side to which the tree is to fall. It determines the direction of falling. (2) A saw cut made on the underside of a large branch beyond the point of severance, prior to making the actual primary cut, to prevent splitting or tearing. (3) the harvesting of less timber from a stand than that budged. (4) The tenderloin muscle of beef (British).

Underfeeding–To feed animals less than recommended.

Undergrazing–An intensity of grazing that fails to fully use the forage available for consumption in a given area under a system of conservation range management.

Underground Water–See Aquifer.

Undershot Jaw–A condition in animals where the lower jaw is longer than the upper jaw.

Understocking–Pasturing or grazing a number of livestock less than the carrying capacity of a particular pasture or range.

Undulant Fever–A disease in humans caused by the bacteria, *Brucella* spp., that cause brucellosis in domestic animals. The disease is contracted from handling infected cattle or from drinking unpasteurized, infected milk.

Ungulate–Referring to a hoofed quadruped, such as a cow.

Uni-–A combining form meaning one.

Unicellular–One-celled; refers to an organism the entire body of which consists of a single cell.

Unidentified Growth Factor–A substance occurring in feed that is vital to the normal growth and existence of an animal, but its exact nature is unknown.

Unipera–An animal that regularly produces only one offspring at a parturition.

Unit–(1) A single thing or item of produce. (2) A recognized measure of weight, volume, or distance, such as a bushel, gallon, or mile.

Unite–Combine one colony of bees with another.

Univalent–Designating a chromosome unpaired at meiosis.

Unsealed Brood–Eggs and larvae of the honeybee in open cells before they are sealed by the bees.

Unsex–To remove the ovaries of a female or the testes of a male animal or bird; to castrate; caponize.

Unshod–Designating a horse that has no horseshoes on its hooves.

Unsound–A term designating an animal with a defect that interferes with the complete utilization of the animal's ability to perform a service.

Unthrifty–Refers to an animal that is not progressing well. The animal may be skinny, weak, or sickly or all three.

Upbreeding–The use of a superior breed upon an inferior or mongrel one in order to improve the offspring of the inferior breed.

Upright (Vertical) Silo–A vertical structure used for preserving chopped, green feed silage. Usually it is made of wooden staves, concrete or tile blocks, or metal sheets held together by bands or reinforcements. Diameter and height vary with the capacity desired.

Upset–(1) To enlarge by blows upon the end, as a blacksmith upsets a bar of heated iron by bumping its end upon the anvil. (2) To cause any change in the normal functions of plants or animals.

Upstanding–Refers to an animal that is long legged, rangy, and tall.

Urea–(1) $CO(NH_2)_2$; a nonprotein, organic compound of nitrogen, made synthetically by a combination of ammonia and carbon dioxide, and used in fertilizers and as a livestock feed supplement. (2) The chief compound of nitrogen in the urine of mammals.

Uremia (Uremic Poisoning)–Accumulation of urinary constituents in the blood, and the toxic condition that results.

Urethra–In most mammals, the canal or tube that carries the urine away from the bladder and serves as a duct for the passage of the male's semen.

Urinary Calculi (Urolithiasis)–Stones of phosphates, urates, oxalates, and lime salts that lodge in some part of the urinary tract, as the kidney, ureter, bladder, or urethra. Symptoms of the presence of these stones are obstruction to the flow of urine, passage of a few drops of blood, and convulsive pain when urinating.

Urine–In people and higher animals, a fluid consisting of certain waste products of metabolism. Extracted form the blood by kidneys, it is conveyed to and stored in the bladder, from where it is expelled at intervals through the urethra.

Urodeum–In the cloaca of a bird, that part into which the genital ducts and the ureters empty.

Uropygial Gland–The preen gland of birds.

USDA (U.S.D.A.)–United States Department of Agriculture.

Uterine Capsule–A medication in capsule form placed in the uterus of an animal after parturition. It aids in preventing putrefaction of the retained afterbirth, assists in loosening the afterbirth attachments, and checks various types of uterine infections.

Uterine Insemination–In artificial insemination, the practice of placing the semen in the uterus, as contrasted to placing it in the vagina or in the cervix.

Uterus–The womb; in female mammals, an organ in which the young develops before birth.

V Muscle–The muscles of the chest of a horse that give the appearance of an inverted V.

Vaccination–A process of injecting controlled amounts of microorganisms or microorganism products (vaccine) into an animal in an effort to prevent that animal from contracting a disease caused by that particular organism; the substance injected is called an antigen; it stimulates the production of antibodies that provides some protection to the host from the invading organisms. See Antigen.

Vaccine–A substance that contains live, modified, or dead organisms or their products that is injected into an animal in an attempt to protect the host from a disease caused by that particular organism.

Vacreator–A multiple unit consisting of steam jets and vacuum chambers designed to remove undesirable odor form milk and/or cream.

Vacuolated Cytoplasm–The living substance of cells, exclusive of the nucleus, when filled with or containing bubblelike structures.

Vagina–The canal in female mammals extending from the uterus to the vulva.

Valence–Also called bond or chemical bond; the chemical combining power of an atom. It indicates the number of electrons that can be lost, gained, or shared by an atom in a compound.

Valine–See Amino Acids.

Vane–(1) The thin web part of a bird's feather. (2) A thin plate or strip of metal pivoted on the top of a spire or mast to show the direction of the wind; a weathercock. (3) A blade of a windmill, fan, centrifugal pump, or similar apparatus.

Vaquero–(Spanish) Cowboy.

Variance–A statistical measure of the amount of variation that is observed within or among a group of animals or plants.

Variant–A recognized entity different from normal.

Variation–(1) The angle by which the north end of the compass needle (magnetic north) deviates from true north. (2) One of the laws of organic nature; organisms vary in time, from place to place, and also in one locality with time; they vary also in their appearance (morphology).

Variety–(1) A group of related plants or animals that differs from other similar groups by characters too trivial or inconstant to be recognized as a species; often any category of lower rank than a species. See Cultivar. (2) In domesticated animals, a subdivision of a breed based on some minor character such as color, etc.

Variety Hybrid–A cross between two varieties of the same species.

Vas Deferens–Tube connecting the epididymis of the testicle to the urethra; the sperm-conducting ducts.

Vasectomized–Designating a male whose vas deferens or a portion of it has been surgically removed to make the male sterile or incapable of producing live sperm.

Vasectomy–A surgical method of sterilization in males whereby the vas deferentia are cut to prevent the spermatozoa form being transported from the testes during ejaculation; the ejaculate contains only the seminal plasma. See Gomer Bull.

Vasotocin–An oxytocinlike hormone produced in birds that stimulates oviposition.

Vat–Any large vessel, tub, cistern, etc., used for holding liquids, as a vessel used in large-scale brewing or dipping of domestic animals.

Veal–The meat from calves slaughtered before they are three months old.

Vealer–Veal calf less than three months old.

Vection–The passing of a disease from one plant or animal to another.

Vector–Any agent such as an insect or animal that transmits, carries, or spreads disease from one plant or animal to another.

Vegetarian–(1) One who, because of cultural reasons or personal conviction, abstains from eating meat (in the strictest sense, also milk, butter, and eggs). (2) An herbivorous animal or person.

Veil–A light, metal screen or diaphanous cloth material draped about the head when working with bees to prevent being stung on the face or neck.

Vein–(1) One of the systems of branching tubes, etc., which carry blood back to the heart. (2) One of the fibrovascular bundles that forms the framework of a leaf. (3) In insects, the riblike tubes that strengthen the wings.

Vena Cava–The large veins that carry blood into the right atrium of the heart.

Venereal Disease–A contagious disease that is usually contracted by animals having sexual intercourse.

Vent–(French, wind) (1) A small opening or passage, as an opening for ventilation. (2) The opening of the cloaca; the anus of a bird. (3) A brand mark made to indicate that the animal no longer belongs to the original owner. (4) To cancel an old brand on an animal to show change in ownership.

Vent Shield–A shield used to cover a chicken's vent to prevent picking by other chickens.

Ventral–Pertaining to or relating to the belly or underside; opposite the dorsal or back.

Ventricle–One of the two pumping chambers of the heart.

Vermicide–Any substance that kills internal, parasitic worms; an anthelmintic. See Vermifuge.

Vermiform–Worm-shaped.

Vermifuge–A drug or chemical that expels worms from animals; an anthelmintic. See Vermicide.

Vermin–Any noxious animal; insect, acarid, rodent, etc.

Verminous–Pertaining to or due to worms.

Vernal–Appearing in spring.

Vernis–Of spring.

Vertebra–Any of the bony segments that form the spinal column or backbone.

Vertebrates–Animals with a spinal column or backbone, such as fishes, birds, mammals, and so on.

Vertical Mixing–Vertical circulation of water masses in a lake occurring naturally in temperate climates in spring and autumn.

Vertigo–Staggers; a condition of disturbed equilibrium in which, because of dizziness, an animal is unable to maintain its balance.

Very Heavy Breeds–The chickens of large breeds: Brahma, Langhan, Cochin, Orpington, Cornish, and Jersey Giant.

Vesicle–Blisterlike sac or small bladder containing fluid.

Vesicular Exanthema–In hogs, an acute, highly infectious, viral disease that causes the formation of vesicles or blisters on the snout, mucous membranes of the mouth, the feet between the toes, soles, and dewclaws.

Vesicular Mole–A fleshy mass or tumor formed in the uterus of animals that is cystic and resembles a bunch of grapes.

Vesicular Stomatitis–A viral disease of cattle, swine, and horses that resembles foot-and-mouth disease and vesicular exanthema. It is very common in the southeastern United States.

Vestigial–Imperfectly developed, said of a part or organ that was fully developed and functional in ancestral forms but is now a degenerate relic, usually smaller and less complex than its prototype.

Veterinarian–An authorized practitioner of veterinary science.

Veterinary Science–A branch of knowledge, dealing mainly with domestic animals, that encompasses anatomy, physiology, breeding (including breed improvement), nutrition, animal diseases and treatment (including diseases transmissible to humans), people's use of animals, etc. The first recognized veterinary school was established in 1761 in France; in 1791 England had such a school with others to follow; in North America several schools flourished after 1852, which, however, did not survive when standards were raised after World War I, requiring a high school diploma and four years of college. To that was added two years of preprofessional training, bringing the total up to six years of college study. At present, schools of veterinary science in the United States and Canada are, for the most part, allied within large universities.

Viability–(1) Ability to live (immediately after birth or hatching). (2) The capacity of seeds to germinate. (3) The state of being alive. (4) Pertaining to sperm cells in the semen, capable of living and successfully fertilizing the female gamete.

Viable–(1) Living. (2) Specifically in regard to organisms or agents of disease, able to cause infection in animals. (3) Capable of living. (4) Capable of germinating, as seeds.

Vibriosis–A contagious and infectious disease of cattle and sheep, usually spread venereally, characterized by abortion and infertility. The causative organism is *Vibrio fetus*.

Vice–Any seriously undesirable habit of an animal, such as vent picking and cannibalism in poultry; balking, kicking, shying, etc., in horses.

Vicuna–An expensive fabric made from the delicate wool of a wild South American ruminant of the same name.

Vieth's Ratio–The ratio between lactose, protein, and ash in milk.

Vigor–The desirable state of health of any living thing.

Vilitis–A condition of a horse's hoof in which the soft part of the wall becomes inflamed.

Villi–Microscopic, hairlike extensions or projections of the inner lining of the digestive tract or of the placenta.

Viral–Having the nature of a virus; pertaining to a virus; like a virus.

Viremia–Condition where viral organisms are found in an animal's blood system.

Virgin–(1) Any female that has not had coitus, as a virgin queen bee. (2) In turpentining, the face of cut the first year a tree is bled. (3) Undisturbed, unplowed land, uncut forest.

Virgin Queen–An unmated queen bee.

Virgin Sod–Natural sod, especially that of the prairies of the midwestern and western United States, in contrast to that produced by grasses seeded by humans.

Virgin Soil–Soil in its natural state as distinguished from soil or land that has been plowed or otherwise altered by humans for cultivated crops or other uses.

Virgin Wool–Wool that has never before been processed.

Virile–A male capable of functioning in copulation.

Virology–The study of viruses and viral diseases.

Virucide–A chemical or physical agent that kills or inactivates viruses; a disinfectant.

Virulence–The disease-producing power of an organism.

Virulent–Highly pathogenic; having great disease-producing capacity; deadly; very poisonous or harmful.

Virus–(Plural, viruses) A self-reproducing agent that is considerably smaller than a bacterium and can multiply only within the living cells of a suitable host. Most viruses are too small to be seen even with the aid of the ordinary microscope, but can be photographed with the aid of the electron microscope. Viruses usually are considered to be living agents of microorganisms but some have characteristics of nonliving matter. They are protein-containing bodies of high molecular weight capable of multiplying and acting like living organisms when in living tissue. They are the cause of many animal, human, and plant diseases, such as smallpox, measles, tobacco mosaic, etc. Recovery from some viral diseases confers lasting immunity.

Virus Interaction–The action of a virus in altering the normal development of other viruses or virus strains that is expressed by partial or complete suppression, by synergistic association, by modification of the type of symptoms in the host plant, or by abnormal increase in concentration of one virus.

Virustatic–A substance that prevents the multiplication of a virus.

Viscera–Usually refers to the organs of the abdominal cavity, removed at slaughter, including stomach, intestines, liver, and other accessory organs. Heart and lungs may also be included.

Visceral Lymphomatosis–One type of the avian leukosis complex characterized by the presence of tumorous masses that involve some of the viscera. Diarrhea is frequently present but often the fowl has no outward symptoms.

Viscosity–The resistance of a fluid to flow; thickness of a fluid.

Vitamin–An organic substance that performs specific and necessary functions in relatively small concentrations in an organism. Required for normal growth and maintenance, vitamins are not utilized as building units for the structure of the organism and do not furnish energy, but are essential for the transformation of energy and for the regulation of the metabolism of the organisms.

Vitamin A–Antixerophthalmic factor. It is mainly concerned with the maintenance of healthy epithelial structures.

Vitamin A & D Feeding Oil–A vitamin supplement for feedstuffs, such as fish or fish-liver oil or a blend of two or more of the following: Vitamin A and/or D concentrate, synthetic vitamin D fish oil, fish oil, marine animal oil, or edible vegetable oil. The vitamin potency should be stated in A.O.A.C. (American Association of Agricultural Chemists) chick units of vitamin D and U.S.P. (United States Pharmacopeia) units of vitamin A per pound or units per gram.

Vitamin A Feeding Oil–A livestock feedstuff vitamin supplement that is either fish or fish-liver oil or a blend of two or more of the following; vitamin A concentrate, fish-liver oil, fish oil, marine animal oil, or edible vegetable oil. The vitamin potency should be stated in U.S.P. (United States Pharmacopeia) units of vitamin A per pound or units per gram.

Vitamin B–The original name given to the crude preparations that relieved beriberi and that were later found to contain many factors. Also called water-soluble B. See Vitamin B-complex.

Vitamin B-complex–Factors isolated from yeasts, liver, and other sources of the original water-soluble B, which includes the following vitamins: (a) Biotin, a factor that cures alopecia (baldness) and "spectacle eye" in rats fed a biotin-deficient diet supplemented by raw egg whites. Also called antiegg-white injury factor, vitamin H. (b) Cholin, a factor required for certain metabolic processes. (c) Folic acid, a factor required for growth and blood formation; also called vitamin M, vitamin B10, vitamin B11. (d) Inositol, a factor that prevents alopecia (baldness) and "spectacle eye" in mice and other animals. (3) Niacin, a factor that prevents pellagra. Also called antiblack-tongue factor, nicotinamide, niacinamide, nicotinic acid amide, PP factors. (f) P-aminobenzoic acid, a factor that prevents gray hair in rats and promotes growth in chicks. Also called PABA. (g) Pantothenic acid, a factor that prevents dermatitis in chicks. Also called chick antidermatitis factor, antigray hair factor. (h) Pyridoxine, a factor that prevents a specific dermatitis in young rats and in humans. Also called vitamin B1, vitamin Y. (i) Riboflavin, a component of Warbow's "yellow enzyme" that is essential for cellular oxidation and reduction. Also called vitamin G. (j) Thiamin, an antiberiberi factor. Also called aneurine, antineuritic factor, vitamin B, vitamin F. (k) Vitamin B12, an antipernicious anemia factor. Also called cobalamin, erythrocyte maturation factor.

Vitamin C–Ascorbic acid, a factor essential in the prevention of scurvy. Also called antiscorbic factor, avitamic acid, cevitamic acid.

Vitamin D–The antirachitic vitamin. It is essential in the metabolism of calcium. Several forms of vitamin D are recognized as D_2 found in plant products and D_2 the animal form found in fish oils and in the blood after irradiation. Also the blood-clotting factor. It is essential for the formation of prothrombin. Also called coagulations vitamin (Danish name), antihemorrhagic factor.

Vitamin D Feeding Oil–A vitamin, livestock-feeding supplement that is either fish or fish-liver oil or a blend of two or more of the following: vitamin D concentrate, synthetic vitamin D, fish-liver oil, marine animal oil, or edible vegetable oil. The vitamin potency should be stated in A.O.A.C. (American Association of Agricultural Chemists) chick units of vitamin D per pound or units per gram.

Vitamin D Milk–Milk whose vitamin D content has been increased above that of normal milk by special processes in accordance with public health regulation by feeding cows irradiated yeast, by irradiation of milk or by the addition of vitamin D concentrate to the milk. Also called vitamin-fortified milk.

Vitamin Deficiency–Any disease caused by the lack of one or more vitamins in the diet: e.g., scurvy.

Vitamin E–A factor that is important in the stabilization of vitamin A. It prevents muscular dystrophy in certain animals and is a sterility factor in experimental animals. Also called antisterility factor, the tocopherols.

Vitamin G–See Vitamin B-complex.

Vitamin H–See Vitamin B-complex.

Vitelline Membrane–The thin membrane located inside the zona pellucida of an ovum that contains the cytoplasm or vitellus.

Vitellus–The cytoplasm of a mammalian ovum; it contains relatively large amounts of nutritive material called yolk or deutoplasm.

Vivarium–A glass box resembling an aquarium used to keep or raise animals or plants. See Terrarium.

Vives–See Strangles.

Viviparous–(1) The bringing forth of living offspring from the body, as in mammals, as opposed to the laying and hatching of eggs. (2) Said of seeds that germinate or buds that sprout and form plantlets while still on the parent plant. See Oviparous, Ovoviviparous.

Void–(1) To evacuate feces and/or urine. (2) A general term for pore space or other openings in rocks such as vesicles and solution cavities. (3) Pore spaces in soils. (4) The space between kernels in a bulk of a grain that is usually expressed as percent of total volume.

Volatile Fatty Acids–A group of low-molecular weight acids that form gases rather easily, and are produced by mirobial action in the rumen; e.g., acetic, propionic, and butyric acid.

Vomit–(1) Material ejected from the stomach through the mouth or nostrils. (2) To eject matter from the stomach through the mouth or nostrils; to throw up.

Vulva–The external genital (reproductive) organs of the female: the opening to the vagina. In insects and other invertebrates, the external opening of the oviduct.

Waddy–See Cowboy.

Waif–A stray animal, specifically a lost sheep.

Walk-through–Designating a building or milking room so designed that the traffic moves in one door or gate and out another, rather than in and out the same door.

Walking Horse Bit–A type of English riding bit that is used on walking horses.

Wallow–(1) A mud hole in which hogs lie. (2) To roll, lie, move lazily in a mud hole, as a hog. (3) A land feature resembling a large hog wallow.

War Bridle–Any of several types of halters, hackamores, or bridles with a part of the rope passing through the horse's mouth in such a way that a powerful leverage is placed on the horse's jaw or head. It is used on vicious or unbroken horses.

Warble Fly–Any of the various flies of the family Hypodermatidae, as the ox warble fly, whose larvae live beneath the integument of cattle and certain other mammals and thus lower the quality and value of the meat and hide.

Warm Front–A mass of warm air advancing behind a mass of cool air. Characteristic weather may be several days of heavy overcast and drizzles.

Warm Water Fish–A fish that does not thrive in water colder than 70 degrees.

Warm-blooded Animal–Any animal, such as a bird or mammal, whose body temperature is warmer than its surrounding medium, as contrasted with the cold-blooded animals, such as reptiles.

Warren–An area, place, or enclosure in which rabbits are bred and raised. Also called rabbit warren.

Wart–(1) A tumor on the skin or mucous membrane of an animal composed of fibrous tissue covered over with epithelial cells similar to those of the part of the body on which they are located. For the most part, they are painless and do not interfere with the function of an animal. (2) An unorganized proliferation of plant cells with an appearance like an animal wart. (3) One of the effects of cucumber mosaic.

Wasp–Insects of the family Vespidae. Although sometimes considered a pest because of their painful stings, wasps are truly beneficial insects. They eat large numbers of other insects as adults and some species are parasitic as larvae on other insects.

Wasty–A term used to describe a carcass or live animal that is too fat.

Wasty Wool–Wool that is short, weak, and tangled, which often carries a high percentage of dirt or sand and which is difficult and costly to process.

Water–H_2O; hydrogen oxide; although the liquid may contain associated molecules. The melting and freezing point is 32°F (0°C) and the boiling point 212°F (100°C). The most valuable natural resource and the most limiting factor in crop production.

Water Blister–A collection of watery fluid beneath the epidermis (skin or mucous membrane) which usually results from excessive pressure, heat, or from irritants contacting the skin.

Water Bloom–A sudden increase in the abundance of algae, especially at or near the water surface.

Water Bloom Poisoning–A poisoning of livestock resulting from drinking pond or stream water which is contaminated with certain blue-green algae. This poisoning is rare but very lethal.

Water Bowl–A bowl-shaped livestock waterer with a valve in the bottom. The animal pushes against the valve to receive water.

Water Buffalo–*Bos bubalus*, family Boridae; a large animal found wild in the jungles from India to the Philippines at elevations from sea level to 6,000 feet. The coat color varies from a dark gray to black. The animal has long horns, curving back over the body, which sometimes measure as much as 12 feet along the curve. Widely domesticated, the water buffalo is raised as a beast of burden, for its rich milk, and its hide from which a good-quality leather is manufactured. Also called carabao (Philippines), gamoose, jemoose (Egypt).

Water Carrier–A worker bee which gathers water and transports it to the hive.

Water Conservation–The physical control, protection, management, and use of water resources in such a way as to maintain crop, grazing, and forest lands; vegetal cover; wildlife; and wildlife habitat for maximum sustained benefits to people, agriculture, industry, commerce, and other segments of the national economy. See Soil Conservation Service.

Water Content–(1) The water of the soil or habitat; (physiological) the available water supply; (physical) the total amount of soil water. (2) The percentage of water in a material in relation to oven dry weight.

Water Fallow–The maintenance of a rice field under a cover of water for one or two years during which a crop of fish may be produced.

Water Farm–(1) The cultivation of fish in farm ponds. (2) See Hyroponics.

Water Fountain–(1) Any of several devices which automatically furnish drinking water to animals or birds. (2) A decorative, artificially produced, jet of water usually issuing from a piece of sculpture in or surrounded by a pool.

Waterhole–(1) A natural depression in which water collects or stands, often in the dry bed of a stream; a spring in the desert. (2) A natural spring or pool on the open range where cattle drink (western United States).

Watering–(1) Furnishing water for the consumption of plants or animals. (2) Designating a device for furnishing water to plants or animals.

Wattle–(1) One of the two, thin, leaflike structures suspended from the upper part of the neck of a chicken; they are made of the same kind of tissue as the comb. (2) Fold of skin cut into the dewlap of cattle for identification purposes. (3) A common name for the 800 worldwide species of *Acacia*, family Leguminosae.

Watusi–An African breed of cattle with very large horns. In the United States they are bred for their large horns and for beef.

Wax–(1) Any substance similar to beeswax in composition and use, as carnauba wax, candelilla wax, bayberry wax, japan wax, sugarcane wax, flax wax, cotton wax, esparto wax, cauassu wax, murumuru sect wax, shellac wax, ceresin wax, montan wax, and paraffin. (2) To apply wax to the surface of a fruit, vegetable, foliage, or flower for preservation.

Wax Glands–Glands on the underside of a bee's abdomen from which wax is secreted after the bee has been gorged with food.

Wax Moth–An insect whose larvae feed on and destroy honey bee combs.

Wax-picking–A method of picking poultry. The bird carcasses are coated with hot, melted wax. When cold, the wax is removed, the feathers and pin feathers being carried with it.

Wean–(1) To make a young animal cease to depend on its mother's milk. (2) To accustom partly grown birds to do without artificial heat.

Weaner–(1) Any of many devices used to prevent a young animal from suckling its mother. (2) A weaned lamb, five to twelve months old, before it grows its permanent teeth (Australia). (3) Any lamb from the time it is weaned until it is sheared. (4) A weaned range calf.

Weaning Ring–A ring with spikes that fits on a calf's muzzle to prevent nursing. When the calf approaches the mother, the projections on the ring make the mother move.

Weanling–An animal that has recently been weaned or taken from its mother.

Weather–(1) The atmospheric conditions prevailing at any specified time and place, or those prevailing during any particular period, as shown by meteorological observations and records of air temperature, barometric pressure, wind velocity, humidity, clouds, and precipitation. (2) Bad weather. (3) To be subjected to the influences of atmospheric conditions, as the abrasive action of rainfall, the disintegration due to frost, the deteriorating influences of oxygen and other gases contained in the atmosphere.

Web–(1) The blade of a saw. (2) That portion of an ordinary anvil, between the head and the base, which is of reduced size. (3) The cotton disk used in filtering milk. (4) The fat which surrounds the paunch and intestines of sheep. (5) The disk of a rolling colter. (6) The membrane which unites the fingers and toes, especially of amphibians and water birds.

Wedding Flight–Mating flight of the queen honey bee.

Weed–(1) A plant out of place. Thus, rye growing in a field of wheat is a weed. (2) More popularly, an herbaceous plant which takes possession of fallow fields or finds its way into lawns or planted fields, crowding the vegetation planted there and robbing it of moisture and nutrients. (3) A plant whose usefulness is not recognized or which is undesirable because of odor, spines, prickles, or poisonous characteristics. (4) A tree of inferior value in a forest, or one growing in a street or lawn, where it is not wanted, like the seedlings of the boxelder. (5) Excessive vegetative growth. (6) Any plant that harbors insects, fungi, or viruses that may spread to nearby crop plants. (7) A horse or other domestic animal that is undersized or a misfit. (8) To remove weeds from a lawn, field, or other places where they are not wanted, usually manually but also by machinery or herbicides. (9) Designating a flavor of dairy products, especially milk, which resembles that of certain vegetation (onion, wild garlic, leeks, etc.).

Weedy–(1) Designating an off-flavor of milk or its products which suggests the odor or taste of certain weeds. (2) Designating an abundance of weeds in a field, flower bed, lawn, etc. (3) Designating growth of high vigor of a plant, usually associated with fruiting. (4) Designating a thin or undesirable animal.

Weigh Jars–Containers that hold and measure the milk given by a cow at a single milking. The purpose is to keep production records on each individual cow.

Weight per Day of Age–A measurement of an animal's weight gain; usually from birth to weaning or from birth to one year old.

Weil's Disease–See Leptospirosis.

Well-bred–Designating a plant or animal which has a good pedigree.

Welsh–A breed of horses of the saddle-pony type, originating in Wales, which stands about 12 hands high and weights somewhat less than 500 pounds. The most usual colors are chestnut, bay, gray, and black.

Wen–An encysted skin tumor.

Western Azalea–*Rhododendron occidentale*, family Ericaceae; a deciduous shrub grown for its white or pinkish, yellow-blotched flowers and yellow or scarlet, autumnal foliage. It may be poisonous to sheep and goats. Native to western North America.

Western Equine Encephalitis–A viral disease of horses communicable to people, marked by fever, convulsions, and coma. Transmitted by certain species of mosquitoes.

Western Range–The native livestock grazing areas of the Great Plains, the Rocky Mountains, the Intermountain and the Pacific Coast regions of the United States.

Western Saddle–A heavy, deep-seated saddle with a distinctly raised horn and cantle, as contrasted to the English saddle, which is light and comparatively flat. It is used for working livestock on the ranches in western United States. Also called stock saddle.

Western Snowberry–*Symphoricarpos occidentalis*, family Caprifoliaceae; a deciduous shrub which is a fair browse plant for sheep, goats, and cattle on the western range, and is also grown as an ornamental. Native to North America. Also called buckbrush, wolfberry.

Western Waterhemlock–*Cicuta occidentalis*, family Umbelliferae; an extremely poisonous, perennial herb. Symptoms of poisoning in stock are frothing at the mouth, uneasiness, pain, convulsions, kicking, throwing back of the head, rigid legs, bellowing, and spasmodic contractions of the diaphragm. Death follows ingestion rather rapidly. Native to western North America.

Westsern Chicken Flea–*Ceratophyllus niger*, family Ceratophyllidae; a blood-sucking insect which infests birds in Maine and on the Pacific Coast, United States.

Wet–(1) To dampen; to sprinkle; to furnish water to a plant. (2) Characterized by rain, as wet weather. (3) Moist, covered, soaked, or saturated with water.

Wet Band–A group of ewes or nannies accompanied by nursing young. Also called wet herd.

Wet Sheep–Ewes with suckling lambs. Wet band.

Wet Veal–Slaughter calves under three weeks of age.

Wether–(1) A male sheep castrated before it reaches maturity or develops male characteristics. (2) The fleece from any sheep after the first fleece has been removed (England).

Wettable Powder (Dust)–Any material manufactured in the form of a powder or dust that can be mixed with or dissolved readily in water.

Weymouth Bit–A type of bit for a bridle which has a bar with a U-shaped curve in it and which is so constructed that considerable pressure can be brought to bear on the horse's mouth.

Whang–Tough leather adapted for strings, thongs, and belt laces which is commonly made from calf skin.

Wheat Bran–A livestock feedstuff which is the coarse outer covering of the wheat kernel as separated from cleaned and scoured wheat in the usual process of commercial milling. Also called bran.

Wheat Brown Shorts–A livestock feedstuff which consists mostly of the fine particles of wheat bran, wheat germ, and very little of the fibrous material obtained form the tail of the mill. It is obtained in the commercial milling of wheat and contains less than 7.5 percent of crude fiber. Also called brown shorts, wheat red shorts.

Wheat Germ Oil Cake–A livestock feedstuff which is obtained after oil has been partially expressed from commercial wheat germs. It contains more than 29 percent of protein.

Wheat Gray Shorts–A livestock feedstuff which consists of the fine particles of the outer bran, the inner bran or bee-wing bran, the germ, and the offal or fibrous material obtained from the tail of the mill. It is obtained in the commercial milling of wheat and contains less than 6.0 percent crude fiber. Also called total shorts.

Wheat Poisoning–A condition seen in cattle grazing on lush wheat pasture wherein the animals show a posterior paralysis, tend to drop down, and become semicomatose.

Wheat Red Dog—A livestock feedstuff which is a by-product obtained in commercial flour milling. It consists principally of aleurone with small quantities of wheat flour and fine wheat bran particles and contains less than 4.0 percent crude fiber. Also called red dog flour, wheat red dog flour.

Wheat Shorts—See Wheat Standard Middlings.

Wheat Standard Middlings—A livestock feedstuff which consists mostly of fine particles of wheat bran, wheat germ, and very little of the fibrous material obtained form the tail of the mill. Obtained in the usual commercial process of milling, it contains not more than 9.5 percent crude fiber.

Wheat Straw—The dried, mature wheat stem and leaves used for roughage, bedding in barns, and as poultry litter.

Wheat White Shorts—A livestock feedstuff which consists of a small portion of the fine wheat bran particles, the wheat germ, and a large portion of the fibrous offal obtained from the tail of the mill. Obtained in the usual process of flour milling, it contains no more than 3.5 percent crude fiber.

Whelp—To give birth to a litter of pups; may also refer to the pup.

Whetstone—Natural sandstone shaped into a sharpening stone.

Whey—The watery portion of milk that remains after the curd and cream have been removed; it contains some protein, sugar, and other soluble materials.

Whinny—The gentle, soft cry of a horse.

Whip—(1) Any of several kinds of instruments for lashing animals, usually consisting of a handle and a lash. (2) An unbranched shoot of a woody plant, particularly the first year's growth. (3) A tall, slender tree. (4) To lash an animal. (5) To beat cream, egg white, etc., so that it becomes a froth and holds shape.

Whip Break—To train a horse by use of a whip. (The whip is used sparingly, and never out of anger, so that the horse learns that when it does certain things it is punished and when it does certain other things it is rewarded.)

Whipworm—Any worm of the genus *Trichuri* and related genera, which live internally in the intestines of humans and various domestic animals.

White Crappie—A white perch, *Pomoxis annularis*, that is used to stock lakes as a pan fish. It is smaller and more silvery than the Black Crappie. See Black Crappie.

White Diarrhea—See Pullorum Disease.

White Eggs—In marketing, eggs with white shells, as contrasted to brown-shelled eggs.

White Face—A coat color characteristic of certain breeds of animals, such as Hereford cattle, which have a red body, a white face, and white markings on the brisket, underline, and switch of the tail. See Hereford, Rambouillet.

White Falsehellebore—*Veratrum album*, family Liliaceae; a hardy, perennial herb sometimes grown in a shady border or in a wild garden for its greenish flowers which are white inside. Young livestock and chickens may be fatally poisoned by eating the plant. Symptoms of poisoning are salivation, retching, purging, weakness, general paralysis, spasms, rapid threadlike pulse, lowered temperature, shallow respiration, cold skin, blindness, and death. Native to Europe and northern Asia.

White Heifer Disease—A condition found in white Shorthorn heifers in which an imperforate hymen, poorly developed genitals, or absence of part of the genital organs causes sterility. A genetic factor, it is associated with color character. The condition is occasionally found in other breeds regardless of color.

White Locoweed—See Lambert Crazyweed.

White Lupine—*Lupinus alba*, family Leguminosae; an annual herb grown to some extent for soil improvement. Native to Europe and Asia.

White Meat—The meat from the breast of domesticated chickens and turkeys which is white colored when cooked. See Dark Meat.

White Muscle Disease—A disease resulting from a vitamin E and selenium deficiency.

White Swallowwort—*Cynanchum vincetoxicum*, family Asclepiadaceae; a perennial herb which is a weed in northeastern United States. Cattle, horses, and sheep are reported to have been poisoned from eating it. Native to Europe.

Whole Milk—Milk, especially cows' milk, which has not had any portion of the fat removed.

Whole Stillage—The undried "bottoms" from the beer well comprised of nonfermentable solids, distiller's solubles, and the mashing water.

Wholesale Cuts—The major parts of a carcass that are boxed and sold to wholesale distributors.

Whorl—(1) A swirl, cowlick, in the hair on an animal's coat. (2) Three or more twigs, flowers, leaves, or floral parts, arranged in a circle at one point on a plant.

Wild Bees—Bees living in hollow trees or other abodes not prepared for them by people, as distinguished from the bees raised in a hive.

Wild Calla—*Calla palustris*, family Araceae; an aquatic herb grown along the edge of a pond in mud or shallow water for its oval or heart-shaped leaves and green flowers. It is poisonous for livestock. Symptoms of poisoning are an intense burning feeling in the mucous membranes of the mouth and throat. Native to the north temperate zone. Also called water arum.

Wild Carrot—*Daucus carota*, family Umbelliferae; an annual or biennial herb, which is a serious weed pest but also a beautiful flower plant. Eaten by cows, it imparts an objectionable flavor to the milk. Native to Europe and Asia. Also called bird's nest, lace-flower, devil's plague, bee's-nest plant, Queen Anne's lace.

Wild Hay—Hay made from native or wild, uncultivated grasses and plants.

Wild Horse—An untamed and/or undomesticated horse. In the United States, the wild horses roaming the ranges have descended from the once domesticated horses. Generally, very similar to the common domestic breeds they are distinct from the wild horse of Africa and are captured and used as a source of bucking broncos in rodeos, as riding ponies, and for slaughter.

Wild Pasture—A pasture of native or volunteer grasses and plants. See Range.

Wildling—(1) An escape; a cultivated plant which has become wild and thrives without cultivation. (2) Any wild plant or animal. (3) In forestry, a seedling produced naturally outside of a nursery.

Wilted Silage—Forage that is cut and partially field-cured to reduce moisture content before chopping for silage.

Wiltshire Side—The entire half of a dressed pig, minus the head, shank, shoulder bone, and hip bone. All of the side except the ham and shoulder is sold as bacon.

Wind—(1) The horizontal movement of air on and above the earth's surface. (2) The blast of air from a blower or fan. (3) The breathing of a horse.

Wind Vane Feeder—A type of feeder used in an open pasture to feed cattle a mineral mix or other feed supplement. A large wing-shaped vane is mounted on top of the feeder to cause the feeder to rotate in the direction of the wind so that the opening will be protected from the blowing rain.

Windrow—(1) A long, relatively low, ridge of hay, sheaves of grain, etc., which is made to effect drying, curing, etc. (2) Accumulations of slash left in rows by loggers. (3) To rake hay or to place sheaves in a relatively low, long row for drying and curing.

Wing—(1) A leaflike, dry or membranous expansion or appendage of a plant part, such as along some stems and petioles and of samaras and some capsules. (2) Either of the lateral petals of a pealike flower. An organ of flight for birds, insects, bats, etc. (3) A piece of bird meat consisting or the organ of flight with the feathers removed. (4) The outside corner of the cutting edge of a plow. (5) Fan or blower blade. (6) An extension on a building. (7) The vane of a windmill.

Winged Web Vaccination—In poultry, the process of injecting a vaccine into the skin on the underside of the wing at the elbow.

Winter Cluster—Closely packed colony of bees in winter.

Winter Pause—In poultry, the stoppage in egg production in winter.

Winter Range—The range, usually at low altitudes, used for pasture in the winter months in western United States.

Wintering—The care, feeding, and maintenance of livestock and tender plants through the winter months.

Wired Foundation—In a beehive, foundation comb with strengthening wires embedded in it.

Wired Frames—Frames with wires holding sheets of comb foundation in place within a beehive.

Wiry Wool—Wool that is inelastic and has poor spinning capacity. It is usually straight and is the result of poor breeding.

Wishbone—The forked or V-shaped bone located at the front of the breast of fowls; the clavicle. The tender meat which surrounds it is a popular piece when cooked.

Witch Calf—Designating a cow that is pregnant.

With Calf—Designating a cow that is pregnant.

Withdrawal Period—The length of time a feed additive or drug must not be fed or administered to an animal prior to slaughter.

Withers—The ridge at the apex of the shoulder bones of a horse, cow, sheep, etc.

Wolf Teeth—Needle teeth.

Wolftail—(1) *Lycurus phleoides*, family Gramineae; a slender, perennial grass found from west Texas to Arizona, United States, and south to Mexico at elevations of 4,000 to 8,000 feet; its forage rating is good. (2) An old-time imaginary and fictitious disease of cattle. Also called wolf-in-the-tail.

Womb—Uterus.

Wood Ashes—The residue which remains after burning wood. It has a value as a fertilizer and was used by early settlers in the home manufacture of lye. Sometimes it is used as a source of calcium in livestock feeds, especially for hogs.

Woodland Pasture—(1) Farm woodlands used for grazing. (2) Wooded areas with grass and other grazing plants growing in open spaces among trees.

Wool—The fibers constituting the soft, curly coats of sheep, camel, yak, etc. After cotton, sheep wool is the most important source of natural fiber used in apparel, upholstery, carpets, etc.

Wool Blindness—A condition that develops most often in closed-faced sheep due to irritation of the eyes by wool and particles of chaff contained therein.

Wool Block—An indigestible mass of wool swallowed by a rabbit.

Wool Carding—The process of disentangling and separating the fibers of wool and delivering them in a parallel condition suitable for spinning.

Wool Clip—The annual wool crop.

Wool Combing—The process of dealing with longstapled wools which are to be used in worsted manufacture. The operations are like carding and are effected with a gilling machine.

Wool Fat—The waxy, greasy substance that exudes from the sheep's skin and clings to the wool fibers. After refining, it is sold in the drug trade as lanolin. In the crude form it is often known in commerce as degras. Also called yolk.

Wool Grease—Wool fat.

Wool Pool—A grouping of the wool of many producers into a single total amount so it can be sold on the market as a single unit. Such commodity pools are very helpful in securing a more satisfactory market than the individual producer could command for his small quantity. Wool, cotton, and tobacco are frequently pooled.

Wool Type—Any breed of sheep produced primarily for wool as contrasted to mutton sheep.

Wool Washing—The process necessary to remove the suet (fat), or natural grease, and dirt from the fleece. It is treated first in alkaline liquor which saponifies the grease and then is washed in fresh water.

Wool-ball–Wool ingested by nursing lambs which have been sucking at tags instead of at the teats. It forms a ball in the stomach and may cause serious digestive disturbance. See Hair Ball.

Woolly–Clothed with long, matted hairs.

Woolly Croton–*Croton capitatus*, family Euphorbiaceae; an annual herb which, as hay, is sometimes poisonous to stock. Native to North America.

Woolly Loco–*Astragalus mollissimus*, family Leguminosae; a poisonous plant, found from South Dakota to Arizona, United States. Both its green and dried foliage are poisonous to livestock. Symptoms of poisoning are dullness, irregularity of gait, lack of appetite, dragging of the feet, a solitary habit, loss of flesh, and shaggy coat. As the animal ceases to eat, it dies. Also called purple locoweed, Texas loco, woolly crazyweed.

Woollyleaf Loco–*Astragalus leucophyllus*, family Leguminosae; a perennial herb which is poisonous to livestock. For symptoms of poisoning see Woolly Loco.

Woolsorter's Disease–Anthrax in humans. See Anthrax.

Wooton Loco–*Astragalus wootoni*, family Leguminosae; a poisonous plant, found in southern New Mexico and western Texas, United States, and in northern Mexico. Both its green and dried foliage are poisonous. Symptoms of poisoning are dullness, irregularity of gait, lack of appetite, dragging of the feet, a solitary habit, loss of flesh, and shaggy coat. As the animal ceases to eat, it dies. also called locoweed. See Loco.

Work of Digestion–The loss of energy which occurs in chewing, digesting, and assimilating the food; heat increment.

Worker–The female bee, other than the queen, whose organs of reproduction are only partially developed, and which gathers nectar and pollen, tends to the brood, brings in water, protects the hive, etc. See Drone, Queen Bee.

Worker Cell–One of the cells in a beehive in which a worker is reared.

Worker Comb–Honeycomb with about twenty-five cells per square inch.

Worker Egg–Fertilized bee egg.

Working Dog–(1) A dog trained to drive sheep, cattle, etc., in such a manner that the animals are not harmed. (2) A dog trained to run a treadmill, pull a sled or cart, etc. (3) A seeing-eye dog trained to aid the blind. (4) A hearing dog trained to aid the deaf.

Worm–(1) Any small, soft-bodied, usually limbless animal, such as a larva, grub, maggot, earthworm, silkworm, etc. (2) To rid an animal of internal parasitic worms.

Worm Nest–In cattle, a protuberance or swollen area on the flank or lower portion of the chest which contains filarial worms.

Wormed–Designating an animal which has been given a vermifuge or anthelmintic to kill worms.

Worsted Wool–A smooth compact yarn made from long wool fibres and used for smooth fabrics such as gabardine.

Worsted Yarn–Yarn made from combed wool.

Wound–Any violently caused disruption of the continuity of an internal or external tissue.

Wound Parasite–Any parasite which is not able to enter a host except through wounds or injured tissue.

Wrangler–A herder or handler of wild or range horses on ranches.

Wry Neck–(1) A twisting of the neck and head of fowls into an unnatural position, possibly as a symptom of paralysis and other diseases. (2) An animal with its neck distorted due to its having been carried in a faulty position in the uterus.

Wry Tail–The tail of a fowl or mammal (usually cattle), when it is permanently carried to one side.

Wyandotte–A medium-weight, dual-purpose breed of chicken of the American class. The Wyandotte has a rose comb, yellow skin, nonfeathered shanks, smooth-fitting plumage, and a characteristic curvilinear breed type. Mature cocks weigh about 8½ pounds and mature hens about 6½ pounds. The eggshell is brown. The varieties are the Silver-Laced, Golden-Laced, White, Black, Buff, Partridge, Silver-Penciled, and the Columbian. There are also bantam sorts of most of the above varieties.

Xarque–Jerked (dried) beef, commonly called jerky.

Xerophthalmia–A condition in animals and birds caused by a vitamin A deficiency. It is characterized by a roughened coat; red, sore, puffed eyelids; and an abnormally dry, lusterless condition of the eyeball.

Yak–*Bos grunniens*, family Bovidae; a large, wild or domesticated bovine characterized by short hair on the back and long, wavy hair on the underparts and tail. It is blackish-brown, or sometimes black and white. Domesticated, the yak provides milk, hide, and hair, which is woven into fabrics, and is an important beast of burden in elevated regions of central Asia. Its tail is often used as a fly-flapper (called a chowry) and for decoration. Native to Tibet and central Asia. See Zho.

Yard—(1) A unit of linear measurement equal to 3 feet or 36 inches. (2) The grounds which immediately surround a dwelling. (3) An enclosed area in which stock, fowls, etc., are kept; as a chicken yard. It may be used as a suffix, as in stockyard.

Yardage—The fee charged by a transit company or stockyard for the use of its storage yards.

Yarrow—*Achillea millefolium*, family Compositae; a hardy perennial herb, a weed pest in unworked fields, roadsides, etc., generally in North America. It is grazed moderately by stock, and its forage rating is fair. Native to Europe, Asia, and North America. Also called milfoil, thousand-leaf.

Yean—To give birth to young, especially by goats and sheep.

Yeanling—The newborn or new young of sheep or goats.

Yearling—An animal, usually cattle, approximately one year of age. A short yearling is from 9 to 12 months of age and a long yearling is from 12 to 18 months of age.

Yearling Weight (365 Day Adjusted Weight)—A weight taken as a yearling or long yearling and adjusted within breed and sex for age of calf and age of dam to a standard mature-cow basis of the age group.

Yeld Mare—A mare that is not lactating; a dry mare.

Yellow Corn—(1) Corn which has yellow kernels in contrast to the other common color, white. The yellow corn contains carotenoid pigments which can be converted into vitamin A in the animal body. (2) A marketing class for corn which may not contain more than 5 percent grains of any color other than yellow.

Yellow Dun—A coat color of a horse that is light, lemon-yellow.

Yellow Grease—Yellow fat from a hog carcass, unfit for human consumption, which is used in manufacturing soap, etc.

Yellow Mealworm—*Tenebrio molitor*, family Tenebrionidae; the larva of a beetle which infests stored grain. It may also infest squabs (young pigeons), eating the skin at the neck and vent.

Yellow Teat Disease—See Anaplasmosis.

Yellow Wildindigo—*Baptisia tinctoria*, family Leguminosae; a perennial, tumbleweed-type herb grown for its bright yellow flowers. Stock feeding on it in the wild have been poisoned. Native to North America. Also called clover broom, wild indigo, false indigo.

Yellows—(1) Any plant disease in which yellowing (or chlorosis) and stunting are principal symptoms. Also called xanthoses. (2) Jaundice in people, domestic animals, as cattle, horses, sheep, etc.

Yield—(1) Grade in meat animals, referring to the amount of lean meat produced in a carcass. (2) The quantity of or aggregate products resulting from cultivation or growth, as a yield of 30 bushels of wheat per acre. (3) The percentage of clean wool remaining in a lot after scouring. (4) The ratio of carcass weight to live weight. (5) To produce products as a result of cultivation or growth, as a tree yields fruits.

Yield Grade—A numerical score (with 1 being the highest yielding and 5 being the lowest yielding) given to a beef carcass. The score is based on the estimated carcass weight in boneless, closely trimmed retail cuts from the round, loin, rib, and chuck.

Yogurt—A semisolid, fermented milk product obtained by inoculating milk at body temperature with various yogurt bacteria (*Lactobacillus bulgaricus, Streptococcus lactis, Thermobacterium yoghurti*, etc.) and then incubating it for 8 hours at the same temperature. The bacteria convert the milk sugar and milk proteins to lactic acid, peptones, and amino acids, which produces the characteristic tangy taste of the product. Yogurt contains almost all the known vitamins of the B-complex group, has a low caloric value, and is easily digested, which makes it an ideal food for people of all ages and tastes.

Yoke—(1) A variously shaped wooden frame which is placed on the necks of two oxen to work them together as a team. (2) A clamp which unites two pieces. (3) A team, as a yoke of oxen. (4) A wood or steel bar placed at the end of a tongue connected by rings and leather straps to the hames of the harness of a pair of horses. It holds up the tongue and aids in holding back the wagon going downhill.

Yolk—(1) The yellow part of a fowl's egg that has a germinal disk located on its outer edge from which the embryo develops. The remainder of the yolk, together with the white, provides the nourishment on which the embryo makes its growth. (2) In wool scouring, all the substances, such as wool grease and perspiration, present in the fleece of a sheep when it is sheared.

Yorkshire—A breed of swine originating in England. The hair coat is white, the ears erect, and the face generally dish-shaped.

Zebu—A strain of cattle originating in India. Widely domesticated throughout India, China, and East Africa, they are used as beasts of burden, as meat animals, and to produce milk. The Zebu has a large hump over the shoulders, short horns, drooping ears, and a large pendulous dewlap. It is widely used in crossbreeding programs to breed heat-tolerant beef cattle in hotter areas of Africa, Australia, and the southern United States. The Brahman, an alleged variety of the Zebu, was developed in the United States. However, the names are used interchangeably.

Zero Pasture—Cutting green forage and hauling it to stock in corrals or in dairy barns in lieu of pasturing. Sometimes called zero grazing.

Zho—An animal of the ox family, resulting from a cross between a bull yak and a common cow; used as a domestic animal in northern India, Nepal, and Bhutan. Native to Tibet. Also called zobo, zobu (males), zhomo, jomo (females).

Zinc–Zn; a metallic chemical element, one of the micronutrient elements in soils, essential for both plant and animal growth. Toxic to animals if ingested in too large a quantity. See Micronutrient Elements, Essential.

Zona Pellucida–The relatively thick covering or membrane that forms the outer surface of a mammalian ovum.

Zoned Heating–A system of heating a large building such as a greenhouse or chicken house by means of controls in different zones of the building.

Zooecology–The branch of ecology concerned with the relationships between animals and their environment.

Zoonosis–(1) Any animal disease that can be transmitted from animals to other animals and to people. (2) A disease due to animal parasites.

Zoophagous Parasite–Any parasite that thrives in or on animals.

Zooplankton–(1) The animal constituents of plankton. (2) Tiny animals which drift with the currents. (3) The animal portion of the planktonic organisms. See Phytoplankton, Plankton.

Zootechny–The scientific breeding and/or domestication of animals.

Zygote–A fertilized ovum or egg, it is the diploid cell formed from the union of the sperm with an ovum.

Zymurgy–A branch of applied chemistry dealing with fermentation.

Part 2

Business and Mechanical Technology

A.U.M.–(Abbr.) Animal unit month.

A-frame–A structure resembling the capital letter *A*.

A-Scale Sound Level (A-Weighted Sound Level)–The measurement of sounds approximately the auditory sensitivity of the human ear, used to measure noise levels in decibels; usually written as dBA.

Abney Level–A small hand level suitable for leveling or measuring slope in percent or in degrees.

Abrasion–(1) Wear in farm implements caused by friction, such as abrasion of a plowshare by the soil. (2) A section of skin or mucous membrane from which the surface layers have been rubbed or worn off.

Absentee Owner–A person or corporation who rents land, including housing, to a tenant farmer or who hires a manager.

Absolute Humidity–(1) The mass of water vapor per unit volume of space. (2) The gaseous pressure exerted by water vapor. (3) The number of grams of water in each cubic meter of air.

Absorption Terrace–A ridge which slows the flow of run-off water so it can be absorbed by the soil.

Abstract of Title–A summary document which states the evidences of ownership of real estate and its encumbrances from the time it was taken from the federal or other government down to the present date. In most states, the county office of the Register of Deeds provides the abstract of title.

Abutting–The joining, reaching, or touching of adjoining land. Abutting pieces of land have a common boundary.

Acceleration Clause–A provision in a written mortgage, note, bond, or conditional sales contract requiring that, in the event of default, the whole amount of principal and interest may be declared due and payable at once.

Accidental Sampling–A method of sampling in which the sampled individuals are selected by chance.

Accrual Method–One of the two main methods of accounting used in computing farm earnings in reporting federal income tax. Increases and decreases in inventory items for the year are reflected in income. Farm income is included as income for the year in which it is earned, regardless of when payment is received. Farm expenses are deductible in the taxable year in which incurred, whether paid or not. Complete inventories of livestock, crops, produce, feed, etc., are required. See Cash Method.

Accumulator–A container that stores fluids under pressure as a source of hydraulic power. It may also be used as a shock absorber.

Acoustics–The science of sound.

Acre–A unit of land measure in England and the United States which is equal to 43,560 square feet, or 1/640 of a square mile, or 160 square rods, or 4,840 square yards, or 4,047 square meters. The Scottish acre is 1.26 and the Irish acre is 1.62 times as large. One acre, as used in England and the United States, equals 0.4 hectare, 0.96 feddan (Egypt), 0.31 carreau (Haiti), 1.03 cuerdas (Puerto Rico), 0.03 caballerias (Cuba), 5.00 shih mou (China), 4.08 tan (Japan), 0.84 arpent de Paris (some sections of Canada), 0.37 dessiatine (U.S.S.R).

Acre Foot–The quantity of water necessary to cover one United States or English acre to the depth of one foot: 43,560 cubic feet or 325,850 United States gallons of water.

Acre Inch–The quantity of water necessary to cover one United States or English acre to the depth of one inch: 3,630 cubic feet or 27,154.2 United States gallons of water.

Acreage–An indefinite quantity of land; a collective number of acres.

Acreage Allotment–As established from time to time by Congress, the individual farm's share, based on its previous production, of the national acreage needed to produce sufficient supplies of a particular crop.

Acreage Controls–A provision of many farm programs which attempts to reduce farm output by limiting the acreage that farmers can plant.

Acreage Limitation Program–Established in the 1981 farm bill, this program required participants to limit the acreage of a crop to a specified portion of their base acreage and to divert the remainder to other use.

Acrylite Acrylic–A plastic covering consisting of two layers of material separated by air space to provide insulation.

Activity Budget–A summary of the technical and economic characteristics of a farm's activity.

Activity Gross Income–The value of the output of a farm activity over some accounting period (usually a year), whether that output is sold or not.

Activity Gross Margin–Activity gross income minus the variable expenses attributable to that activity.

Actual Use–The use made of an area by livestock and/or game animals without reference to a recommended utilization; usually expressed in terms of animal units or animal unit months. See Use.

Ad Valorem–Designates an assessment of taxes against property. Literally, according to value.

Adiabatic–Refers to a decrease in temperature, of a rising (expanding, cooling) air mass to an increase in temperature of a descending (contracting, warming) air mass, of 5.4°F/1,000 ft. (9.767°C/km); assuming no external source of heating or cooling.

Adiabatic Lapse Rate–The theoretical rate at which the temperature of the air decreases with altitude or increases as it loses altitude. It is 5.4°F/1,000 ft. in dry air. If a theoretical parcel of air moves from a low altitude to a high one and there is no exchange of heat with its environment, it becomes colder at the adiabatic lapse rate as the pressure on it decreases and allows it to expand.

Adjusted Base Period Price–The average price received by farmers in the most recent ten years, divided by the index of average prices received by farmers for all farm products in the same ten years. Used in parity calculations.

Adjusted Weaning Weight (Adj.205-day Wt.)–Weight of a calf at weaning, adjusted to a standard 205 days of age and adjusted for the age of the dam.

Adjusted Yearling Weight (Adj. 365-day Wt.)–Weight of a calf as a yearling, adjusted to a standard 365 days of age and adjusted for the age of the dam.

Adjustments–(1) Range management:changes in animal numbers, seasons of use, kinds or classes of animals, or management practices as warranted by specific conditions. (2) Ecological: the processes by which an organism becomes better fitted to its environment; functional, never structural. See Adaptation.

Administrative Location of a Farm–An imaginative location of a farm for the purpose of administering farm programs. If all land in a farm is located in one county, the farm shall be administratively located in such county. If the land in a farm is located in more than one county, the farm shall be administratively located in either of such counties as the county committee and the farmer agree. If there is no agreement, the farm shall be located in the county in which the principal dwelling is located or where the major portion of the farm is located, if there is no dwelling.

Adobe–Unburned, sun-dried bricks. Clay and silty deposits found in the desert basins of the southwestern United States and in Mexico, where the material is extensively used for making sun-dried bricks. The composition is a mixture of clay and silt together with minor amounts of other materials such as grass or straw.

Adult Agricultural Education–Organized instruction for persons beyond the age of compulsory school attendance to prepare them for employment or to increase knowledge and skills required in agriculture. An adult class is characterized by flexible scheduling, varied administrative patterns, and content and objectives related specifically to needs of the adults. A common form of such a program is that of young farmer instruction or young farmer education.

Adverse Possession–The right of an occupant of land to acquire title against the real owner, where possession has been actual, continuous, hostile, visible, and distinct for the statutory period.

Advisory Committee (Agriculture)–A group of persons usually from outside the field of education, selected because of their interest, knowledge, and expertise to advise educators regarding vocational programs in agriculture. Committees usually include members from all interested groups: public officials, employers, employees, former students, minority groups, and the public. Such committees may operate at the federal, state, and local levels and often function under names other than that of advisory committee, e.g., advisory council or advisory board.

Adz(e)–A hatchetlike tool for dressing timber. Its bow-shaped blade is set at right angles to the handle.

Aesthetic Value–The value or pleasure that any thing of beauty gives to humans.

Affidavit–A written statement signed and sworn before some person authorized to take an oath.

Age Distribution–The classification of individuals of a certain population according to periods such as pre-reproductive, reproductive, and post-reproductive, or the chronological age classes.

Agent–One who represents another from whom he has derived authority.

Agent Middleman–A food-marketing firm that represents buyers and sellers in the marketplace; agents do not take title to goods for their own account and may not physically handle the food products.

Aggradation–Building up to a uniformity of grade or slope by the addition of material; especially the depositing of sediment in the beds of streams and on the floors of ponds, lakes, etc.

Aggregate–(1) To bring together; to collect or unite into a mass. (2) Composed of a mixture of substances, separable by mechanical means. (3) The mineral material, such as sand, gravel, shells, slag, or broken stone, or combinations thereof, with which cement or bituminous material is mixed to form a mortar or concrete. Fine aggregate may be considered as the material that will pass a 0.25-inch (4.76-millimeter) screen, and coarse aggregate as the material that will not pass such a screen.

Agister–A person who takes charge of animals, properly cares for them, and charges for his or her service.

Agrarian–(1) Pertaining to agriculture. (2) Pertaining to political action or movements for the benefit of farmers. See Rural Population, Urban.

Agrarian Zone–Of or pertaining to the cultivated portion of an area.

Agribusiness–A term referring to the full scale of operations related to the business of agriculture. It connotes the interrelationships of farming, farm services, soil science, agronomy, land grant universities, county extension services, state and federal experiment stations, soil and water conservation services, plant and animal nutrition, plant and animal protection, transportation, finance, and marketing.

Agricola–(Latin) Farmer.

Agricultural Adjustment Act (AAA)–Passed by Congress in 1933 to provide economic relief for farmers by the federal government's making benefit payments to farmers for limiting production (it being assumed that farm prices would then rise), and making rescue loans for several mortgages, as well as marketing loans and providing credit to cooperatives. the basic provisions of the AAA are still in effect.

Agricultural College–An educational institution devoted to study, research, and the dissemination of knowledge in agronomy, horticulture, animal husbandry, agricultural economics, etc. The U.S.D.A. works in conjunction with most agricultural colleges (sometimes called a school or college of agriculture within a university).

Agricultural Commodity–A general term; any product of agriculture.

Agricultural Conservation Program–A program that shares costs with farmers and ranchers to encourage and assist them in carrying out their farms' conserving practices. The program is designed to: (1) restore and improve soil fertility, (2) minimize erosion caused by wind and water, and (3) conserve resources and wildlife.

Agricultural Economics–(1) The branch of economics concerned with farm management and production. (2) A department in a college or university.

Agricultural Education–(1) The general, formal knowledge of agriculture. (2) The course of study (in a college or university or a department of government) to prepare and assist teachers of agriculture in the secondary schools. (3) The term applied to the modern high school course dealing with agriculture. See Vocational Agriculture.

Agricultural Engineering–A course of study in the design, construction, and use of agricultural implements and buildings; soil and water management; rural use of electricity; and processing of agricultural products.

Agricultural Exemption–A clause of the Interstate Commerce Act that exempts from interstate commerce motor carriers that transport raw agricultural commodities.

Agricultural Experiment Station–See Hatch Act.

Agricultural Land–All the land devoted to raising crops and livestock, including farmstead, roadways, drainage and irrigation ditches, ponds, water supply, cropland, and grazing land of every kind. (The term is not strictly synonymous with land in farms, cropland, pasture land, land suitable for crops or land suitable for farming).

Agricultural Management–The decision making involved in planning, directing, and controlling an agricultural business.

Agricultural Marketing Service–The branch of the United States Department of Agriculture (USDA) that provides grading and inspection services for agricultural products.

Agricultural Mechanics Instruction–A combination of subject matter and learning experiences designed to develop knowledge and skills necessary for performing or assisting with the selection, construction, operation, maintenance and repair, or use of agricultural power, agricultural machinery and equipment, structures and utilities, and soil and water management practices.

Agricultural Occupation–An occupation that requires agricultural knowledge and skills. The primary instructional areas which also serve to classify agricultural occupations are: agricultural production, agricultural supplies and services, agricultural mechanics, agricultural products processing and marketing, ornamental horticulture, agricultural resources, and forestry.

Agricultural Production Instruction–The subject matter and planned learning experiences designed to develop knowledge and skills necessary for the production of plants and animals, and to provide practice in making managerial decisions in the science and technology of producing and marketing agricultural products. Sometimes used synonymously with the term production agriculture instruction.

Agricultural Productivity–Refers to operational efficiency in farming; usually measured by the ratio of farm output to farm inputs.

Agricultural Region–A classification of land according to its predominant feature, such as the crop most commonly grown or the type of farming employed.

Agricultural Resource Base–The soil, water, climate, and other natural resources necessary to produce a crop.

Agricultural Resources Instruction–(conservation, utilization and services) The subject matter and planned learning experiences designed to develop knowledge and skills necessary for the conservation and utilization of natural resources (such as air, forests, soil, water, fish, plants, and wildlife) for economic and recreational purposes. Instruction also emphasizes the establishment, management, and operation of lands used for recreational purposes.

Agricultural Stabilization and Conservation Service (ASCS)–A service established by the Secretary of Agriculture on June 5, 1961, under the authority of Reorganization Plan No. 2 of 1593, in accordance with the Reorganization Act of 1949, as amended (5 U.S.C. 901-913). The service carries on the following principal programs from appropriated funds: production adjustment programs, Sugar Act program, agricultural conservation.

Agricultural Structure–Refers to the number, size, ownership, specialization, and other characteristics of farming.

Agricultural Substitute–A product that is manufactured from farm commodities but which is a substitute for a traditional farm food product (e.g., corn-oil margarine or soy-protein steaks).

Agricultural Trade Development and Assistance Act of 1954– Also known as the Food for Peace program, or Public Law 480; the federal program designed to increase United States farm product exports by selling commodities on low-interest loans, exchanging commodities for local currencies, or donating commodities to needy countries.

Agricultural Treadmill–Refers to the situation in which farmers find themselves producing food at a lower price with no increase in profits after having been encouraged to adopt new output-increasing and cost-reducing production technologies.

Agriculturally Related Occupation–An occupation that deals with the processing, marketing, and distribution of agricultural products, or an occupation providing supplies and services to agricultural production.

Agriculture–The broad industry engaged in the production of plants and animals for food and fiber, the provision of agricultural supplies and services, and the processing, marketing, and distribution of agricultural products.

Agriculturist–A person engaged in the production of food and/or fiber; also ancillaries such as teachers of agriculture, farm editors, researchers, etc.

Agroeconomic Zones–Zones which are defined in terms of common features. For different purposes these features will differ but may involve such dimensions as climate, soil resources, land use, ethnic groupings, market access, etc.

AID–Agency for International Development, United States Department of State. An agency that administers agriculturally related as well as other types of development projects in foreign countries. Also known as USAID.

Air Curtain–A method for mechanical containment of oil spills or water by bubbling air through a perforated pipe in front of the advancing oil or water spread. Air curtains are also used to prevent fish from entering a polluted area of water.

Air-cooled Engine–An engine that is cooled without the use of a liquid coolant. Heat is dissipated through the use of fins that are cooled by the air.

Alidade–(1) The part of a surveying instrument consisting of a sighting device, index, and reading or recording devices. (2) A straight-edge ruler carrying a sighting device, such as slot sights or a telescope mounted parallel to the ruler.

Alienated Land–Term used in the United States for lands of one ownership enclosed within boundaries of another ownership. Often refers to land in private ownership within the boundaries of public land.

All-risk Insurance–An insurance policy written by the Federal Crop Insurance Corporation that guarantees the farmer an amount per acre that is equal to his cash expenses.

Allodial Land–Land held in absolute ownership; fee simple, as opposed to feudal land. Land in Louisiana and New York State are by statute allodial.

Allotment–(1) The number of acres, etc., a producer is allowed to grow of a particular crop under a government program. (2) An area designated for the use of a prescribed number of cattle or sheep, or for common use of both.

Allowable Cut–The amount of wood that can be removed from a landowner's property during a given period, without exceeding the net growth during that period on the property.

Allowable Use–The degree of use of a rangeland estimated to be proper until proper use is known. Forty or fifty percent of the annual growth by weight is often used as a rule-of-thumb on ranges in good or excellent condition. Also, the amount of forage planned to be used to accelerate range rehabilitation.

Alodium–Unrestricted ownership of land and now synonymous with *fee simple* in the United Kingdom and United States.

Alternate Agriculture–Any type of agriculture that is not considered to be conventional; e.g., organic farming or the use of insect predators instead of chemicals to control insect pests. See Organic Farming, Sustainable Agriculture.

Alternate Grazing–Changing pastures or ranges so that the forage grows back before it is grazed again. Also called rotational grazing.

Alternating Current (AC)–An electrical current that reverses its direction of flow at regular intervals, usually 60 times in a second (60 cycle). Most farm electrical power that is supplied by the power company is alternating current. See Direct Current.

Altimeter–An aneroid barometer used for determining elevations.

Aluminum Dross–A by-product of refining aluminum metal. It consists of aluminum oxide, aluminum nitride, salt, and various other impurities. The N is slowly available in the soil. Dross is also relatively high in magnesium, copper, manganese, and zinc contents.

American Farm Bureau–A voluntary organization founded in 1919 to protect the economic interests of farmers through legislation and which, in conjunction with the U.S. Extension Service, fosters agricultural education. The bureau is organized on three levels: the national bureau is composed of independent state bureaus, each of which is composed of independent county bureaus.

Ammeter–An instrument on a tractor or other machine or vehicle that measures the electrical flow through a circuit.

Amortization–A specified plan to repay a loan in a specified period of time.

Ampere–A unit of measure for the flow of current in an electrical circuit.

Ampere-Hour–A unit of measure obtained by multiplying the amperes of an electrical current by the time (in hours) during which the current flows; used as a unit of measurement for battery capacity; for example, a battery which provides 5 amperes for 20 hours is said to deliver 100 ampere-hours.

AMS–See Agricultural Marketing Service.

Ancestor Merit–An unbiased estimate of the production of future daughters of a bull based on an accurate evaluation of the bull's sire and maternal grandsire.

Anemometer–An instrument used to measure speed of wind; also called wind gauge. The Dines anemometer measures pressure, which is translated to velocity.

Aneroid Barometer–A device that measures air pressure by its effect on the thin sides of a partially evacuated hollow cylinder.

Angle of Repose–The maximum slope or angle at which a material such as soil or loose rock remains stable. When exceeded, mass movement by slipping as well as by water erosion may be expected.

Angstrom–One hundred millionth of a centimeter; a unit used in measuring the length of light waves.

Animal Month–A month's tenure upon range by one animal.

Animal Unit Conversion Factor–As used in the United States, a numerical figure which allows conversion from one kind or class of animal to another. A conversion factor is satisfactory in respect to the amount of forage required to maintain an animal, but may have no application in determining stocking rates for range use for particular kinds or classes. In the United States, the generally accepted per-month conversion factors are: mature cow with calf, 1.0; mature bull, 1.12; weaned calf, 0.6; yearling over 12 months and under 17 months, 0.7; yearling from 17 to 22 months, 0.75; two-year-old from 22 to 32 months, 0.9; elk, 0.7; white-tail deer, 0.14; mule deer, 0.2; mature ewe with lamb, 0.2; weaned wether, 0.17; doe goat with kid, 0.17; buck, 0.17; weaned wether, 0.14; and grown horse, 1.25.

Animal Unit Months (AUMs)—Amount of grazing required by a 1,000-pound (454 kg) cow or equivalent weight of other domestic animal for one month.

Annual Revaluation Method of Depreciation—A method of determining depreciation where an item is evaluated annually; the present evaluation is subtracted form the preceding evaluation to determine the amount of depreciation.

Anode—A positive electrode.

Antisiphoning Device—A check valve on the filling hose of a sprayer to prevent water form draining back to the source.

Anvil—A block of iron sometimes faced with steel on which metal is shaped by hammering.

AOAC—See Association of Official Analytical Chemists.

Apothecaries Measure—A standard system for measuring fluids. In the United States the gallon contains 8 pints; the pint, 16 fluid ounces; the ounce, 8 drams; the dram, 16 minims. In the United Kingdom the gallon contains 8 pints; the pint, 20 ounces; the ounce, 8 drams; the dram, 60 minims.

Appraised Value—An estimate of the present worth.

Appropriative Right—A doctrine of rights to irrigation water. Rights are based on date of appropriation of water; the right to water is limited by beneficial use on the land as defined by states. The term is used in the western United States. See Riparian Doctrine.

Appurtenance—That which belongs to something else; something which passes as an incident to land, such as right-of-way.

Apron—(1) Canvas or rubberlike material or metal links made into a continuous belt for conveying fruits or vegetables, silage, straw, manure, etc., into bins (especially on grading machines). (2) Folds of flesh on the neck of an animal or fowl or the fatty layer covering its belly. (3) A strip of metal placed to guide the sap into a cup that is suspended from it. (4) A slab of concrete or stone which connects adjoining areas, or over which water is directed to dissipate its energy.

Aquatic Rights—Ownership or permission necessary to navigate or fish rivers, lakes, etc. See Riparian Doctrine.

Arbitrage—Purchase in one market at a low price while selling at a high price in another market.

Arcadian—Simple, rustic; a term applied to a place of contentment, as a family farm or country home.

Arch—(1) A curve. (2) The anterior-posterior curvature of the backs of hogs. (3) A log-supporting device used with a tractor in skidding logs. (4) The curved portion of a dam, bridge, or other structure.

Archimedean Screw—A water lift consisting of a hollow screw or a spiral pipe around an inclined axis. The lower end is submerged in a source of water, and the upper end empties into an irrigation ditch. The water lift is activated by a handle that turns the hollow screw or spiral pipe.

Arcifinious—Having natural boundaries, such as a county bounded by a river.

Are—100 square meters; used as a measure of land in Asia where ownerships are smaller than 1,000 square meters (1 hectare).

Area Reclaim—An area difficult to reclaim after the removal of soil for construction and other uses. Revegetation and erosion control are extremely difficult.

Arithmetic Mean—Obtained when a series of numbers are added and divided by the total numbers; the average.

Arpent—A French measure of land. In Louisiana it is equal to 0.845 acre or a square of 192 feet, as compared with the square of approximately 209 feet for an acre; also called arpen; arpenus.

Arrent—To give license to use, as farmland, at a fee.

ARS—Agricultural Research Service.

ASCS—Agricultural Stabilization and Conservation Service.

Assessed Value—The worth of a property established by a governmental unit as the basis for estimating taxes.

Asset—Any type of property that could be used to pay a debt. See Liability; Liquid Asset.

Assignment—(1) Transfer of one's rights to another. (2) A transfer of notes, mortgages, etc. (3) The gradual payment of a debt, as in the case of a farmer who authorizes a milk dealer to assign a certain portion of his milk check to a bank in payment of a bank loan to buy dairy cows.

Assignment Clause—A clause in an insurance policy that transfers the rights under the policy to a third person.

Association of American Plant Food Control Officials (AAPFCO)—The membership of the association consists of the officers charged by law with the active execution of the laws regulating the sale of commercial fertilizer and fertilizer materials; research workers employed by state, dominion, or federal agencies, who are engaged in the investigation of fertilizers. AAPFCO publishes an annual bulletin giving the official regulations and interpretations.

Association of Official Analytical Chemists (AOAC) (of North America)—A professional scientific society whose primary objective is to obtain, develop, test, and adopt uniform, reliable, state-of-the-art chemical and biological methods for the analysis of food, drugs, cosmetics, pesticides, feeds, fertilizers, hazardous substances, air, water, and any other product or substance affecting the public health and safety, the economic protection of the consumer, or the quality of the environment.

Atmospheric Pressure—The force per unit area exerted by the atmosphere in any part of the atmospheric envelope. Some of the expressions for the normal value of the atmospheric pressure at sea level are: 76.0 cm mercury, 29.92 in. mercury; 1,033.3 cm water; 33.9 ft. water; 1,033.3 g/cm^2; 1,013,250.0 dynes/cm^3; 14.66 lb./in^2; 1.01325 bars (1 bar = 1,000,000 dynes/cm^2); 1,013.25 millibars.

ATO—Agricultural Trade Office

Attorney's Opinion of Title—An instrument written and signed by the attorney who examines the title, stating his/her opinion as to whether a seller may convey good title.

Attrition–(1) Wearing away by friction. (2) The wear and tear that mineral and rock particles in transit in flowing water undergo through mutual rubbing, grinding, knocking, scraping, and bumping with resulting decrease in size.

Auction–A public sale of goods to the highest bidders.

Auctioneer–The agent for the sellers at an auction who sells the goods at the highest price. After the sale is completed he/she becomes the agent for the buyer.

Auger–(1) A conveyor or elevator made on the principle of an Archimedean screw; a spiral or broad-threaded screw in an open tube. (2) A tool for boring holes; especially in agriculture, a broad-threaded screw for boring fencepost holes, etc. See Archimedean Screw.

Avogadro's Law–A statement that equal volumes of different gases at the same temperature and pressure contain the same number of molecules.

Avoirdupois–In English-speaking countries, the established system of weighing materials other than medicines, precious stones, and precious metals. The ounce contains 16 drams; the pound, 16 ounces. But the hundredweight contains 100 pounds in the United States, 112 pounds in the United Kingdom.

Avulsion–The sudden transference of land from the estate of one party to that of another without change of ownership; caused by inundation, or by sudden change in location of the channel of a river or stream.

Away-From-Home Food Market–The market where consumers buy food away from home; includes restaurants, cafeterias, hotels, motels, and other food service operations.

Awl–A small, hand-held tool with a sharp point used to punch holes in leather, heavy cloth, etc.

Axial Flow Combine–A combine that uses a rotating mechanism inside a stationary threshing cage to thresh the grain. The threshing operation is faster and the grain emerges cleaner than in conventional combines. See Combine.

Axle–A rod, usually of steel, upon which a wheel turns.

Back Flow Valve–A valve in an irrigation system that prevents water form flowing backward in the system. This prevents chemicals that are added to the water from contaminating the water supply. See Chemigation.

Backlash–The clearance or "play" between the interworking parts of a machine; e.g., the meshing gears of a transmission or the valves and push rods of an engine.

Backsight–In transit traverse surveying, a sight and reading on a previously occupied instrument station. In level traverse surveying, a sight and reading on a surveying rod held at a point of known elevation to determine the present height of the line of sight on the level.

Balance–(1) The skeletal and muscular makeup of an animal, which gives the animal visual appeal. A well-balanced animal's body parts appear to fit together and blend harmoniously and symmetrically. (2) A device for determining weights. (3) To put in proper proportions, such as balancing a feed ration. (4) A term used to express the ratio between the resources of land, labor, capital, and management that attains the optimum use of each resource in the production of crops and livestock to maximize financial returns and to maintain or improve soil productivity.

Balance of Trade (BOT)–The difference in value between a country's merchandise imports and exports in a specified period. The balance of trade, in the sense of the difference in value between imports and exports, is called favorable when exports exceed imports and unfavorable when the reverse occurs.

Balance Sheet–A list of the assets (current and fixed) and the liabilities (current and other) of an individual or business. The difference between assets and liabilities is the owner's equity, net worth, or financial gain or loss.

Balanced Operation–A range livestock enterprise which provides sufficient feed and forage resources at each season to sustain continuous satisfactory maintenance of its livestock and game throughout the year. Also, an enterprise in which the gross income either equals or exceeds, by a suitable margin for profit, the cost of production.

Bale Accumulator–A trailing attachment for conventional hay balers that collects and automatically unloads about 8 to 12 bales.

Bale Chopper–A tractor-powered implement that chops up bales of hay for use as a feed or as bedding for livestock.

Bale Ejector–An attachment for conventional hay balers that throws bales into a trailing wagon to eliminate hand loading.

Bale Mover–A device for mechanically moving large round bales of hay; may be attached to a tractor 3-point hitch or front-end loader, mounted in a truck bed, or trailed behind a tractor or truck.

Baler–A machine used to compress hay into bales. See Conventional Hay Balers, round Hay Balers.

Balling Hydrometer–A triple scale hydrometer designed to record the specific gravity of a solution containing sugar.

Balloon Frame–A type of barn construction made from a wooden frame without using heavy timbers.

Band Day–Range forage required by a band of sheep of a given size and class during one day.

Bar–(1) The mouth piece of a bridle bit. (2) The landside of a plow. (3) The straight line of a cattle brand. (4) The part of a horse's hoof that is bent inward, extending toward the center of the hoof. (5) Ridges in the roof of a horse's mouth. (6) The space in front of a horse's molars where the bit is placed. (7) A pole used to close a fence gap. (8) 14.50 pounds per square inch of air pressure.

Barb–(1) A breed of horses related to the Arab and native to Barbary; probably introduced into Spain by the Moors and probably smaller and coarser than the Arab; its strain is evident in all known present breeds. (2) A hairlike side branch of a feather. (3) Mucous membrane projections for the openings of submaxillary glands under the tongue of horses and cattle. (4) A pointed projection on a fence wire. (5) In botany, a hooked hair or bristle. (6) The teeth or spines on the awns of grasses, especially barley.

Barbed Wire–Fencing made of twisted wires to form one strand and to which wire barbs are attached at short intervals.

BARC–Beltsville Agricultural Research Center, U.S.D.A.

Bargaining Association–A farm cooperative having as its principal function the influencing of farm prices and other terms of trade.

Bargaining Power–A form of market power denoting the relative strength of buyers and sellers in influencing the terms of exchange in a transaction.

Barn–A farm building used for storage of hay, grain, farm implements, etc., or for housing of domestic animals.

Barn Book–The record book in which breeding and management data on livestock is kept. Commonly included are breeding dates, birth dates, health data, and weight data.

Barn Raising–The construction of a barn. In parts of the United States, this often implies a gathering of friends and neighbors to assist in the construction, at which time much food and drink is served for a pleasurable occasion.

Base Acreage–The average of a crop on the farm used in acreage limitation programs and to calculate farm program acreage; usually the actual planted acreage of the crop on the farm in the previous year, although the previous two-year average could be used; adjusted for disasters, crop rotations, and other factors as necessary to make it fair and equitable.

Base SI Units–The meter, kilogram, and second.

Basic Crops (Commodities)–Crops which, because of acreage, value, climate, etc., are considered most important in an agricultural economy, such as corn, wheat, rice, peanuts, cotton, tobacco, etc.

Basin Irrigation–An efficient system of irrigation in which a field or orchard is divided into basins which are filled with water.

Basis Pricing–A price quotation technique whereby the current cash price of a commodity is described by indicating the basis; for example, "30 cents under" would indicate a $2.70/bushel cash price if the futures price were $3.00.

BATF–Bureau of Alcohol, Tobacco, and Firearms; under the United States Department of Treasury, responsible for the issuance of permits, both experimental and commercial for the production of alcohol. See Methanol.

BCA–Board of Contract Appeals, U.S.D.A.

Beam–(1) The central shaft of a plow, supporting all principal parts, and by which it may be drawn. (2) A heavy piece of timber used to support a building. (3) The balance bar of a scale. (4) In tanning, a sloping board on which hides are dressed.

Beamage–A deduction made when weighing a freshly dressed carcass for the loss of weight by evaporation during cooling.

Bearer–The person holding a check or draft for payment.

Bearing–The direction of a line with reference to the cardinal points of a compass; true bearing is the horizontal angle between a ground line and a geographic meridian. A bearing may be referred to either the south or north point; magnetic bearing is the horizontal angle between a ground line and the magnetic meridian. A magnetic bearing differs from a true bearing by the exact angle of magnetic declination of the locality.

Bearings–That part of a machine on which another part revolves. Bearings are generally designed to take the wear of a machine and can be replaced. Common types are: ball, roller, sleeve, and babbitt.

Bed–(1) A small plot of soil used for growing seedlings, vegetables, flowers or shrubs, and often raised above the level of the surrounding soil. (2) A place for animals to sleep. (3) The hauling platform of a wagon or truck.

Beef Carcass Data Service–A program whereby producers, for a fee, can receive carcass evaluation data on their cattle by using a special "carcass data" eartag for their slaughter animals.

Beef Improvement Federation (BIF)–A federation of organizations, businesses, and individuals interested or involved in performance evaluation of beef cattle. The purposes of BIF are to bring about uniformity of procedures, development of programs, cooperation among interested entities, education of its members and the ultimate consumers of performance evaluation methods, and to build confidence of the beef industry in the principles and potentials of performance testing.

Beef Master–A breed of beef cattle developed in Texas from a systematic crossing of Brahman, Hereford, and Shorthorn cattle. Although there is no set color standard, the predominant colors are red and dun. There are both polled and horned lines.

Beef Middles–The midsection of a cow's or steer's intestine used as sausage casing.

Beef Pritch–A rod used to support a carcass during skinning. See Gambrel.

Beefiness–A cattle-judging term denoting desirable beef characteristics.

Beefy–(1) A term used in judging cattle to designate characteristics of a beef animal as contrasted to a dairy animal; usually refers to thickness of muscling. (2) A term used in poultry judging to designate: (a) that the combs are coarse and overgrown or (b) that the birds are fat and coarse.

Benefit-Cost Ratio–A measure of economic efficiency; computed by dividing total discounted benefits by total discounted costs.

Bilateral Trade Agreement–An agreement between two countries on trading patterns; e.g., the United States and the former Soviet Union grain trading agreements.

Bill of Lading–A contract or receipt signed by a common carrier who agreed to deliver freight to a given person.

Bill of Sale–A formal legal paper for the transfer of goods, real estate, animals, etc. It is a written promise that title will be transferred when the conditions of the sale have been completed.

Bionomics–The study of relations among organisms and the relationships between them and their environment. See Ecology.

Birth Weight–The weight of a calf taken within 24 hours after birth. Heavy birth weights tend to be correlated with calving problems, but the conformation of the calf and the cow are contributing factors.

Bit–(1) The part of a horse bridle that is placed in the mouth; usually made of steel. (2) An earmark for cattle. (3) The blade of an axe or similar tool. (4) A tool used to drill a hole.

Blank–(1) The rough, sawed, or split wood pieces from which a finished product is made, such as handles, chair rounds, patterns, etc. (2) An unstocked area in a tree plantation where few or no trees are growing. (3) An unassembled box or package.

Blanket Mortgage–One mortgage on a number of parcels of real property.

Blast Freezing–Freezing meat with a continuous blast of cold air moving around the meat.

Bleed–(1) The process of removing air from a hydraulic or fuel system of a tractor or other machine. (2) To remove the flood from an animal during the slaughter process.

Blend Price–Price paid a milk producer based on the proportion of milk utilized in each price class, such as fluid milk and manufacturing milk.

Blending–A grain marketing strategy whereby two different qualities of grain are blended in such a way as to raise the total value of both lots.

Block–(1) A piece of wood used as a divider between two bales of hay. (2) A pulley used to increase pulling or hoisting power. (3) In logging, an administrative division of a forest. (4) About 5 pounds of cotton hanks. (5) A portion of hay thrown into a baler. (6) To thin out plants, especially sugar beets, into smaller bunches, which will be again thinned into single plants. (7) To trim fleece to enhance the appearance of a sheep. (8) See Cutting Block.

Blow-by–A leakage or loss of compression past the piston rings between the pistons and cylinders of an engine.

Bob Wire–See Barbed Wire.

Bobbed Wire–See Barbed Wire

Boiling Point–The temperature at which the vapor pressure of a liquid equals the atmospheric pressure. At the boiling point, bubbles of vapor rise continually and break on the surface. The boiling temperature of pure water at sea level (barometric pressure of 30 inches of mercury) is 212°F (100°C). At high altitudes, the boiling point of water is lower because the atmospheric pressure is lower. At 5,000 feet above sea level, for example, the boiling point of water is 203°F (95°C); at 10,000 feet it is 194°F (90°C).

Bolter–(1) A sieve. (2) A circular ripsaw and a mechanically driven carriage used for sawing round bolts into boards. (3) A plant, such as the sugar beet, which produces a seed stalk from the crown of the plant.

Bolting Cloth–A sieve cloth, usually silk.

Bomb Calorimeter–An apparatus used to measure the amount of heat given off by any combustible substance. Used for determining the calorie content of feeds.

Boom–(1) A long beam projecting from a tractor or mast; used for hoisting heavy loads. (2) A pipe or tubing with several nozzles to apply chemicals over a wide area at one time. (3) A barrier made by fastening logs end to end across a river to contain cut timber. (4) A group of floating logs. (5) The spar which projects from a log-loading machine to support logs. (6) An obstruction composed of floating logs. (7) A construction designed to float logs in a certain direction. (8) To rise in volume, as a river booms sufficiently to float logs.

Boot–(1) The hollow metal casting on a planter or a drill through which the seed passes to be planted. (2) Profuse feathering on the shank and toes of fowls. (3) The sheathlike leaf structure on the upper end of grain or grass plants that encloses the inflorescence prior to its emergence. During inflorescence the plant is said to be in the boot stage or in boot. (4) In a grain elevator, the box which contains the lower pulley or sprocket.

BOT–Balance of Trade.

Boundary Tree–An old, tall distinguishable tree standing on a property line, usually blazed or otherwise marked.

Bountyland–Land given as a bounty for military service. According to a United States statute of 1850, lands were granted to persons in military service or their dependents.

Bouyoucos Block–(Named after George John Bouyoucos, Michigan State University) A gypsum or nylon block into which two bare wires are embedded a fixed distance apart. These wires lead to terminals which can be attached to a modified Wheatstone bridge. The gypsum blocks are buried in the soil, usually at a depth of maximum concentration of plant roots. The drier the soil, the greater the resistance to passage of electrical current between the two embedded wires. This resistance reading is calibrated for each soil into: "It is time to irrigate," and "It is time to quit irrigating."

Box–(1) *Buxus sempervirens*; a shrub or small tree; family Buxaceae; one of the most valuable broad-leaved evergreens. Since the days of the Romans it has been planted for hedges and topiary work. It is lethally poisonous to cattle, horses, sheep, and pigs, having emetic and purgative properties. Native to southern Europe. (2) The lowest grade of softwood lumber. (3) A cavity cut into the base of a pine tree to collect sap to make turpentine and rosin. (4) A system for bracing branches of trees. (5) To chip back on the underside of a cut in a tree to prevent the main stem of the tree from splitting when it falls. (6) An accidental mixing of two herds of sheep or cattle (Australia). (7) To corral animals. (8) A device for dividing water in an irrigation system into two or more ditches. (9) A canyon with one entrance and no exits.

Brace–(1) Any device designed to strengthen or support, such as a corner-post brace, wall brace, etc. (2) Specifically, a rod connected to the beam, which reinforces the landside of a plow. (3) A curved tool for holding and rotating bits. (4) A pair, as a brace of pheasants.

Brake Horsepower–Horsepower actually delivered as determined by a dynamometer. See Drawbar Horsepower, Effective Horsepower, Horsepower, Rated Horsepower.

Brand and Brand Name–The American Association of Plant Food Control Officials (AAPFCC) has adopted the following definitions and interpretations: "A brand is a term, design or trademark used in connection with one or several grades of fertilizers....A brand name is a specific designation applied to an individual fertilizer....The grade of a fertilizer should be included with its brand name."

Breach of Contract–Failure, without legal excuse, of one of the parties to a contract to perform according to the contract.

Breadbasket of the Nation–Colloquial name for the Corn and Wheat Belt regions of the north-central United States.

Break In–(1) To discipline; to train, as in a horse. (2) The process of wearing in a desirable fit between the surfaces of two new or reconditioned parts of a machine, as in the pistons and cylinders of an engine.

Breaker–(1) A wave meeting a shore, reef, sandbar, or rock and collapsing. (2) For watering: a widened fan-shaped device screwed on the end of a hose to decrease (break) the velocity of water; used in watering plants. (3) For plowing: an extended moldboard on a turning plow to more completely turn heavy sod upside down. (4) For electrical overloading: an automatic flip switch on an electrical circuit that breaks (shuts off) the current when overloaded. (5) Utility-grade slaughtered cattle in which some degree of marbling is in the meat. The loins and rounds are "broken out" and sold as steaks.

Breast–(1) The front of the body between the neck and the abdomen. (2) The breastbone and flesh surrounding it, especially in fowl, lamb, or veal. (3) The forepart of a moldboard on a plow. (4) The part of a cotton gin consisting of a frame, which contains the huller ribs, the picker rollers, and the roll box. (5) That part of a bulk milk tank which connects the inside lining to the outer wall.

Breather–A vertical section of pipe which is an air vent, preventing pressure buildup. When installed after a steep grade, it is called a relief well.

Breed Trademark–The symbol of a distinguishing characteristic.

Breeding Program Goals–The objective or "direction" of breeder's selection programs. Goals are basic decisions breeders must make to give direction to their breeding program. Goals should vary among breeders because of relative genetic merit of their animals, their resources, and their markets.

Breeding Unit Index–A measure of a breeding herd, including the total number of female animals capable of giving birth, weighted by the production per head in a base period.

Breeding Value–The ability of an animal to transmit the genetic capability to produce meat, milk, eggs, or other economically important products; the value of an animal as a parent. The working definition is twice the difference between a very large number of progeny and the population average when individuals are mated at random within the population and all progeny are managed alike. The difference is doubled because only a sample half (one gene of each pair) is transmitted from a parent to each progeny. Breeding value exists for each trait and is dependent on the population in which the animal is evaluated. For a given trait, an individual can be an above-average producer in one herd and a below-average producer in another herd.

Brick–(1) A rectangular block made of burned clay or concrete usually measuring $4¼ \times 8½ \times 2$ inches ($10.8 \times 21.6 \times 5.08$ centimeters); used extensively in most kinds of durable construction. (2) Any food packaged in the shape of a brick.

Bridge–(1) A part, usually 24 inches or less in width, which extends across a bulk milk tank. Usually in the center, it strengthens the tank and supports the agitator, gauge rods, etc. (2) To form an arch under pressure, as with fertilizer over a distributor in a spreader.

British Thermal Unit (BTU)–The amount of heat required to raise the temperature of one pound of water one degree Fahrenheit under stated conditions of pressure and temperature (equal to 252 calories, 778 foot-pounds, 1,005 joules, and 0.293 watt-hours); the standard unit for measuring quantity of heat energy.

Brix–(1) The percentage of total solids in fruit or sugarcane juice. (2) The percentage of sugar in simple syrup at 20°C (68°F); also known as balling.

Broadax(e)–A wide-bladed ax used to cut logs into beams or to make incisions in trees to obtain turpentine.

Brod–A blacksmith's nail.

Broker–An agent entrusted to buy or sell products or commodities.

Brokerage–The business of bringing buyers and sellers together and arranging contracts for a fee.

BTU–See British Thermal Unit; a measure of heat.

BTU/hr–Quantity of heat needed per hour to maintain a given temperature.

Buckwheater–An incompetent farmer; a novice (United States). So called because buckwheat will grow on the poorest of soils.

Bug–(1) Any insect, specifically of the order Hemiptera. (2) A flaw in the construction and operation of machinery.

Bulk–(1) The major part; signifying volume, amount, or size. (2) Mass or aggregate, e.g., "sold in bulk," meaning not packaged; "bulk density," meaning the density of the entire mass including solid, liquid, and gas, such as the bulk density of a clod or ped of soil.

Bulk Bin–A large storage bin used for storing feed. The feed is put in through the top using an auger from a feed truck or a feed mill. The feed is usually discharged by an auger from the bottom and is moved to the animal's trough or feeder by an auger or conveyer.

Bull Pen–(1) A small wood or metal enclosure for confining bulls. (2) The sale ring at a livestock auction. (3) Living quarters for a group of men. (4) Storage room for machinery.

Bulldozer–A crawler tractor with a blade mounted in front to move earth, snow, gravel, and similar materials, or for clearing land of small trees and shrubs.

Bung–A large plug or cork used as a stopper for a hole in the side of a barrel or cask through which the barrel is filled.

Bunk–(1) A crossbeam of heavy timber that supports logs during transportation. Also called bolster. (2) A car or truck used to transport logs. (3) A sleeping space in a lumber camp or a ranch.

Bunker Silo–A silo for storing silage, consisting of a wide trench constructed in the side of a hill from which surface water has been diverted, such as by a diversion terrace. The vertical walls of the horizontal silo may be form 6 to 8 feet and the length from about 50 to100 feet. When feeding the silage, movable self-feeders can be constructed in either or both ends.

Bunkhouse–A building where rangehands or workers sleep.

Buoyancy–The resultant of upward forces exerted by the water on a submerged or floating body, equal to the weight of the water displaced by this body.

Bureau of Customs–An agency of the United States Treasury Department. The Bureau collects duties and taxes due on imported merchandise and baggage, including countervailing duties to offset foreign subsidies and antidumping duties to offset unfair price competition. Customs agents also attempt to keep harmful insects and diseases out of the United States.

Bureau of Land Management Land–Federal lands administered by the Bureau of Land Management, U.S. Department of the Interior.

Bureau of Reclamation–A federal agency responsible for building dams and canals and providing water to local water districts. The districts then sell water to agricultural producers.

Burlap–A coarse cloth made of jute, flax, etc., used for bags, bales, bundles, and for covering the roots of plants. See Balled and Burlapped.

Burning Index Meter–A device for measuring inflammability and rate of spread of fire. It works by integrating the combined effects of the moisture content of fuel, herbaceous stage, wind velocity, relative humidity, and other factors.

Burr Grinder–A feed grinder with flat, roughened iron plates.

Bushel (U.S. Measure)–(1) A dry measure of 32 quarts; four pecks. (2) A weight of a product assumed to be the equivalent of a bushel: beans 60 pounds; barley, 48 pounds seed corn, 56 pounds; flax-seed, 56 pounds; oats, 32 pounds; peas, 60 pounds; rye, 56 pounds; wheat, 60 pounds. (3) Any basket, tub, or other vessel having the capacity of a bushel. See Imperial Bushel.

Bushwhacker–A hook for cutting brush.

Butane–A gaseous hydrocarbon of the paraffin series, formula C_4H_{10}, used as a heating fuel.

Button–(1) An irregularly shaped berry. (2) A bud. (3) A round seed vessel. (4) Any stunted or immature fruit. (5) A round, firm, cheesy curd of condensed milk; a defect of the body of the milk. (6) A nipple, especially of a hog. (7) A partially dismantled queen bee cell in a beehive, which resembles a small acorn cup. (8) Onion set. (9) A stunted or immature horn growth, as on a calf. (10) A metal clasp used to connect sections of a check row wire. (11) Cartilage on the chine bone of cattle (12) Any shell-like bone construction of the body; also called concha. (13) A leather ring for adjusting a horse's bridle.

Caballeria–(1) A unit of land measurement in Spanish-speaking countries, equal to 33.2 U.S. acres in Cuba; 111.82 in Costa Rica; 111.51 in Guatemala; 111.13 in Honduras; 105.75 in Mexico; 112.41 in Nicaragua; 194.1 in Puerto Rico; and 95.48 in Spain. (2) A place to keep horses.

Cablegation–A method of irrigating a field from an irrigation pipe in which a traveling plug fastened by a cable to a braking mechanism is used to help control the amount of water released by gates or holes in the pipe. It was developed in 1981 at the USDA Snake River Conservation Research Center at Kimberly, Idaho, as a means to automate gravity-flow irrigation systems.

Cadastral Surveys–Surveys which relate to land boundaries and subdivisions, and are made to create or define limitations of titles or ownership.

Cadmium–(Greek; *kadmia*, earth) A bivalent metal similar to tin. Its atomic number is 48 and atomic weight 112.40. A metallic "heavy metal" used in the production of copper, lead, silver, and aluminum alloys, in photography, ceramics, and in insecticides. Cadmium in sewage sludge is of grave concern when sludge is applied to soils used for the production of food and feed crops. Cadmium is a hazardous pollutant to people, domestic animals, and shellfish. It also is an experimental carcinogen.

Calendar Year–A period from January 1 through December 31 of the same year.

Caliper–(1) An instrument used for measuring the diameters of trees. (2) An instrument used for measuring thickness.

Calorie–The heat necessary to raise the temperature of 1 gram of water from 14.5° to 15.5°C is one small calorie; the heat necessary to raise the temperature of 1 kg of water 1°C is one great Calorie (kilocalorie) (Cal), equal to 3.968 BTU. See BTU.

Calorimeter–An instrument for measuring the heat change and the energy in any organic system such as foods or feeds.

Calving Difficulty (Dystocia)–In cattle, abnormal or difficult labor, causing difficulty in delivering the fetus and/or placenta.

Calving Ease Index–A rating for calving ease that combines cleaving ease scores for heifers, second calf and older cows and birth weight as reported to breed associations and published in their annual sire summaries. Some of them label it Ease of Calving Index. The index is used to select replacement heifers.

Calving Season–The time of the year when the calves of a herd are born; occurs at different times in different parts of the country.

Camion–(Spanish) Wagon, truck.

Camp Unit–A subdivision of a sheep allotment on federally owned land.

Campus–(Latin) Field.

Camshaft–A shaft containing lobes on a rotating shaft that operates the valves of an engine.

Cane Grind–A social gathering at a cane mill at crushing time in the southern United States. Similar to sugaring-off parties at maple syrup time in the North.

Cap–(1) A piece of metal placed over the end of a log to make it skid over obstacles. (2) The top member of a trestle or similar support. (3) Milk tooth of an animal. (4) The crown of a kernel of corn. (5) The calyx of the strawberry. (6) To remove the calyx of the berry after it is picked. (7) To place a cover over a stack of hay, bound grain, etc. (8) To seal a cell in the honeycomb with wax, as by a bee.

CAP–Common Agricultural Policy.

Capital Requirements–The amount of fixed and working capital needed to operate a farm or ranch. It includes capital for land, buildings, fences, water supply, drains, machinery, equipment, livestock, current operations, etc.

Capital Resources–The amount of capital available to an individual from different sources.

Capitalization–Determining present value of future net money income.

Capitalization Rate–The ratio of net annual money income to the capitalized money value of property. The rate should probably be near the current rate of interest charged on first mortgages on similar farms in the area. It is influenced by the prevailing money market, the risk both physical and economic, the marketability of the farm in question, and the competition with other forms of investment.

Car Lot–The quantity of produce or number of animals which can be transported in a railroad car.

Carburetor–A device for supplying certain internal combustion engines with a mixture of vaporized fuel and air.

Card–A machine which is used to separate the wool fibers by opening the locks or tufts of wool. The machine contains multiple rolls with teeth. Hand cards are used chiefly in the fitting of show sheep.

Carriage Bolt–A bolt with a rounded head used on a machine, wagon bed, etc. The purpose is to provide a smooth bolt-head surface that will not catch or snag on objects that might come in contract with it.

Carry–(1) To keep an animal on a maintenance ration without obtaining any weight gain or any products from it. (2) To bear, as a pregnant cow carries a calf. (3) To sustain, as a farm carries a debt, or as a range carries stock. (4) To convey during transportation and marketing, as a container carries satisfactorily or a product carries well.

Carrying Capacity–In its true sense, the maximum number of individual animals that can survive the greatest period of stress each year on a given land area. It does not refer to sustained production. In range management, the term has become erroneously synonymous with grazing capacity.

Carryover–The supplies or volume of a farm commodity not yet used at the end of a marketing year. It is the remaining stock carried over into the next year. Marketing years generally start at the beginning of the new harvest for a commodity and extend to the same time in the following year.

Case–(1) A box and sometimes its contents, e.g., a case of eggs. (2) In tobacco leaves, to absorb moisture after curing so that the leaves become pliable enough for handling. (3) In mushroom culture, to cover the prepared bed with its final layer of soil.

Case Hardening–(1) A condition of drying wood in which the surface becomes hard and set. If the interior tensile stresses exceed the strength of the wood, honeycombing results. (2) A condition in the artificial drying of some evergreens when the scales become overdried and hard, usually due to low humidity in the dehydrator. This retards the rate of dehydration of the major portion of the fruit or vegetable. (3) Treatment of iron or steel machine parts with carbon so that the outer surface becomes hard to withstand wear and the center remains soft to withstand shocks.

Cash Basis Operation–Conducting any type of enterprise without using credit; paying in cash only.

Cash Crop–A crop sold directly on the market contrasted to one which his fed to animals or is otherwise used but not sold. Formerly applied mainly to crops which were readily salable, e.g., cotton, wheat, sugarcane, etc.

Cash Farm Expense–Farm expenses which include only cash items or items equivalent to cash: e.g., purchases of animals, expenditures for breeding fees, taxes, farm improvements, machinery, feed, crops, hired labor, interest on farm debts, farm insurance, etc. Principal payments on mortgages and old accounts and expenses on the farm dwelling are not included in the year's farm expenses.

Cash Farm Receipts–Receipts which include sales and trade-in allowances from transactions on the farm; e.g., concerning animals, poultry and dairy products, crops, farm improvements, machinery, work of the farm, etc.

Cash Flow–A payment or receipt in the form of cash (including transactions conducted through a bank).

Cash Flow Budget–A statement of projected farm payments and farm receipts associated with a particular farm plan.

Cash Grain Farm–A farm on which corn, grain sorghum, small grains, soybeans, or field beans and peas account for at least 50 percent of the value of products sold.

Cash Income–Money received by a farmer from his operations. Also called cash farm receipts.

Cash Market–A market where there is an immediate sale and delivery.

Cash Method–A method of accounting where income is credited to an account in the year in which it is received, and the expenses are deducted in the year in which they are paid. See Accrual Method.

Cash Prices–The prices received for commodities as published daily by market news reports.

Cash Value–The amount of cash a policyholder would receive if he surrendered his insurance policy.

Casing–(1) A large pipe sunk into a well to prevent the walls form caving in and within which the pipe for pumping liquids is placed. (2) Cleaned sections of hog, cattle, or sheep intestine used as sausage skins.

Casing Head–In well boring, a driverhead screwed into a casing pipe to take the force of the blows while the pipe is being driven down.

Cask–A barrel, made of tightly fitted wooden staves and hoops, for holding liquids. It is a general term for such items as barrel, pipe, keg, etc. Formerly commonly used as a container for marketed tobacco; and in this connection also, the weight allowed for the container.

Catamaran–(1) A raft for supporting a windlass and a grappling hook for recovering logs sunk in a river. (2) Any raft with at least two logs or floats fastened together.

Cathode–A negative electrode.

Cation–The ion in solution carrying one or more positive charges of electricity depending on its valence. The common soil cations are calcium, magnesium, sodium, potassium, hydrogen, and ammonium.

Cattle Trail–The route on which cattle were driven for long distances in the United States in the nineteenth and early years of the twentieth century.

Cattle Wire–See Barbed Wire.

Cattle Yearlong–The forage or feed required to maintain an animal unit for one year. See Animal Unit, Animal Unit Month.

Caveat Emptor–(Latin) Let the buyer beware.

Caveat Venditor–(Latin) Let the seller beware.

cc–A cubic centimeter; a cube with edges of one centimeter.

Ceiling Price–The highest price allowed for a commodity by the United States federal government.

Cell–(1) A hexagonal unit compartment of a honeycomb. (2) The ultimate functional unit of an organic structure, plant, or animal. It consists of a microscopic mass of protoplasm which includes a nucleus surrounded by a membrane. In most plants it is surrounded by a cell wall. (3) A single element of an electric battery, either primary or secondary, generally consisting of a jar filled with a liquid or a pasty electrolyte, into which the electrodes are inserted or connected. (4) A very small, enclosed compartment.

Cell Grazing Management–A system of pasture rotation whereby pastures are divided into equal-sized segments with respect to carrying capacity and are arranged in a pie-shaped design around a central core area for supplemental feeding, watering, and handling pens. Under this system, livestock are rotated among the pastures using high stocking densities and little time on pasture for any one grazing period.

Celsius Scale–(formerly called centigrade) A temperature scale in which the freezing point of water is 0°C and the boiling point is 100°C. Minus 273.15°C = 0°K: absolute zero. Named after a Swedish astronomer, Anders Celsius (1701-1744). See Fahrenheit Scale, Kelvin Scale.

Census of Agriculture–A count taken by the Census Bureau every five years of the number of farms, land in farms, crop acreage and production, livestock numbers and production, farm spending, farm facilities and equipment, farm tenure, value of farm products sold, farm size, type of farm, and so forth. Data are obtained for states and counties.

Center of Draft (Center of Load)–A point within the plow about which all the forces which act on the plow are balanced.

Center of Pull–The true point of hitch or center of power in machines. On a tractor, this is the point halfway between the wheels, at which the drawbar is attached.

Center Pivot Irrigation System–A large irrigation system that rotates around the terminal end of a large water line. Modern systems are powered by electric motors. The system circles slowly and sprinklers mounted on the system emit irrigation water.

Centigrade Thermometer–A thermometer on whose scale the interval between the freezing and boiling points of water is divided into 100 parts or degrees.

Centrifugal Force–The force which acts upon a body revolving in a circular path, tending to force the body farther from the center of the circle. If the centrifugal force is just sufficient to balance the attraction of the mass around which it revolves, the moving body will continue in a uniform curved path. Should the centrifugal force increase, the body will either take up a larger path farther from the center or else tend to fly off in a straight line.

Centrifugal Pump–A water-lifting device that utilizes the centrifugal force imparted to the water by a rapidly rotating impeller. The water is admitted to the center of the impeller and discharged at its outer periphery.

Centrifuge–A machine, such as a Babcock milk tester, used to separate the fat from the acid-milk mixture.

Ceramic–Pertaining to pottery, including porcelain and terra cotta. Often taken to include all products made by heating natural claylike materials: pottery, chinaware, glass, and bricks.

Certification of Seed–Seed production and marketing under the control of a certifying agency to maintain varietal purity and freedom from seed-borne pests.

Certification Standard–Rules and regulations of a certifying agency concerning the breeding, production, and cleaning of seeds along with the maximum amount of impurities allowed, etc.

Certified Boar–A boar that has sired several animals meeting the requirements for certification as established by a breed association.

Certified Check–A check that has the word *certified* stamped across its face by the bank on which it was drawn. The bank must assume the responsibility for its payment.

Cesium-137–An isotope of the radioactive element cesium having a mass number of 137. One of the important fission products and a constituent of fallout. It has a half-life of thirty-three years.

Cess–A levy or tax, as land cess, usually exacted locally.

Cetane– The measure of the ignition quality of diesel fuel. It is determined by the temperature and pressure at which the fuel will ignite.

cfs– Cubic feet per second. One cubic foot per second equals 4.719×10^4 cubic meters per second.

CFS– Container freight station.

Chain– (1) The legal unit of length for the survey of public lands of the United States; the equivalent of 66 feet or 20.13 meters. The name is derived from Edmund Gunter's chain, which was a series of links connected by rings. There are ten links in a Gunter's chain. Advantage in measuring in chains is that 10 square chains equals 1 acre. (2) Any series of related, interconnected, or similar natural features, e.g., chain of mountains, islands, lakes. (3) A series of metal links forged or welded together and used for fastening, hauling, etc. (4) In chemistry, a series of atoms connected by bonds, forming the skeleton of a number of compounds.

Chain Pump– A machine consisting of an endless chain dipping into water in a shallow well at one end and passing over a revolving wheel at the other. Attached to the chain, at regular intervals, are cups or buckets, which lift the water and drop it into a spout as they pass over the wheel at ground surface. A larger version of this mechanism, using animal power to activate the wheel, is a widely practiced method of land irrigation in Asia (known as a Persian wheel).

Chain Reaction– When a fissionable nucleus is split by a neutron it releases energy and one or more neutrons. These neutrons split other fissionable nuclei, releasing more energy and more neutrons, making the reaction self-sustaining (as long as there are enough fissionable nuclei present).

Chain Saw– A portable, motor-driven saw which consists of an arm on which travels an endless chain with attached sawteeth. Used in felling trees, sawing logs, or in cutting underbrush.

Chain Tie– (1) A section of light chain attached to a manger in a barn that branches into a "Y" to go around both sides of a cow's neck and fasten at the top. It serves to restrain the cow in a manner similar to a stanchion. (2) A light wrapping chain thrown across the top of a load of logs to secure it for transportation.

Chamfer– (1) The flat surface exposed after cutting off the right-angled corner of a block of wood, stone, etc. (2) To bevel.

Chamois– (1) *Rupicapra rupicapra*, a small antelope that resembles a goat. Found in the mountains of Europe and New Zealand. (2) Soft, pliable leather made by rubbing oil into the skins of the chamois, goat, sheep, etc. Also called shammy.

Change Agent– One who advocates or attempts to introduce another method; a teacher, an agricultural extension agent.

Charge– (1) In electrostatics, the amount of electricity, measured in coulombs, present upon any substance which has accumulated electric energy. (2) To restore the electrical ability of a battery by the passage of current through it. (3) The amount of current absorbed by a battery during the operation of charging. It is measured in ampere hours. (4) In a gas engine, the amount of mixture taken into the cylinder during the suction stroke.

Chassis– The essential running gear in automobiles and trucks, consisting of the rectangular framework supported on springs attached to the front and rear axles.

Chattel– Movable personal property, as contrasted with immovable real estate. Certain types of chattel are technically immovable, e.g., a tenant's interest in a growing crop.

Chattel Mortgage– A mortgage on movable property.

Check– (1) In irrigation (a) A basin into which the flow of water is regulated by levees and dykes. (b) An adjustable gate on a canal to regulate the flow of water. (c) Concrete blocks or wooden ties placed on a channel bottom to reduce erosion. (d) A crack which appears in drying soil. (2) In lumbering, a lengthwise separation in the grain, caused by strains during seasoning, which extends across the annual rings. (3) A short crack within the body of a cheese. (4) A narrow crack in a rice grain that may cause it to break during milling. (5) An egg which has a cracked shell, with the inner membrane intact. When the crack is naturally mended, it is called a blind crack. Also called crack, dent. (6) To retard growth.

Chemical Toilets– A type of dry vault toilet in which the sewage is decomposed by adding caustic chemicals, such as quick lime. Such toilets are used by people harvesting agricultural crops.

Chemurgy– Chemical research designed to promote the use of agricultural products for industrial purposes, or to obtain new, valuable products from plants.

Chimney– A vertical opening, a foot or more wide, in a pile of lumber to facilitate circulation of air and drying of the lumber.

Chimney Effect– The movement of air in a greenhouse created when both the top ridge and the side ventilators are open.

Chisel– (1) A machine with penetrating points that are drawn at a depth of 12 in. to loosen subsoil. (2) A hard steel cutting tool with a sharpened blade used for cutting wood or metal.

Chock (Block)– A blocking device used as a wedge to keep a log, truck, etc., from moving.

Choke– (1) An acute condition in livestock brought about by a food mass lodged in the esophagus. (2) Hairy or filamentous undeveloped scales at the base of glove artichoke heads, which are removed before eating. (3) In engine carburetion, to increase the ratio of gasoline to air in the fuel mixture when starting a cold engine. (4) In plants, to kill, dwarf, or stunt through excessive competition for space and nutrients, e.g., the choking of plants by dodder or bindweed. (5) In plowing, the gathering of wet straw, weeds, etc., about the plowshare, which reduces its efficiency.

Chop– (1) Animal feed of coarsely crushed or finely ground cereal grains. (2) A small cut of meat which usually includes a rib, e.g., mutton chop. (3) To cut hay into small portions for easy storage without baling. (4) To hoe a row crop, especially cotton; often the first hoeing is called chopping. (5) To crush grain. (6) Jaw; generally in the plural to denote the jaws forming the mouth. Also called chap. See Green Chop.

Chopper–(1) A cordwood cutter. (2) A machine for cutting forage into small pieces.

Chore–A regular or odd job around a house or farm.

Christmas Tree Farming–The commercial production of evergreen trees to be used as Christmas trees. The most popular species are Scotch pine, Douglas fir, balsam fir, black spruce, and eastern red cedar.

Chuck Wagon–A wagon that carried food and cooking and eating utensils for cattlemen tending livestock in the United States.

Churn–A vessel in which cream is agitated vigorously to obtain butter.

Churning–The beating, shaking, or stirring of whole milk or cream to make butter.

Chute–(1) A narrow passage through which animals are moved for branding, spraying, or loading, or through which grain slides to a lower level. (2) A trough constructed of round timbers in which logs are slid up or down a grade. (3) A stampede of animals (western United States). (4) A high-velocity conduit for carrying water. (5) An inclined drop or fall.

Cigar-leaf States–States of the United States where leaf tobacco for cigars is grown: Connecticut, Florida, Georgia, Wisconsin, Pennsylvania, Ohio, and New York.

Circa–(Latin) About, approximate.

Citizen Suits–The 1970 amendments to the U.S. Clean Air Act permits any citizen to sue a polluter.

City Forestry–The preservation and protection of shade trees in streets and parks.

Claim–A right or supposed right to unsettled land, or the minerals or oil discovered in land within the public domain.

Claw–(1) The sharp nail on the toe of an animal or bird. (2) The slender, extended lower part of the petal, as in the iris, lily, etc. (3) A device on a milking machine to which the stanchion tubes are connected.

Clean Content–The amount of clean, scoured wool remaining after removal of all vegetable and other foreign material, and containing 12 percent by weight of moisture and 1.5 percent by weight of ingredients removable by extractions with alcohol.

Clear Money–Profit.

Clearance–The space allowed for proper operation between two parts of a machine; e.g., the space between a shaft and its bearing.

Cleat–(1) A strip of wood or iron fastened crosswise to something for strength, or to prevent warping. (2) A wooden or metal device having two arms around which turns may be taken with a rope to hold securely and yet be readily released.

Cleavage–(1) In animal or plant reproduction, the splitting of one cell into two identical parts. Each resulting daughter cell matures and may divide again. (2) Tendency of certain minerals or woods to split along particular planes or angles. (3) The weight required to cause splitting in a standard piece of wood three inches long, expressed in pounds per inch of width.

Cleaver–A heavy, wide butcher's knife for cutting bones.

Clevis–A coupling device made of a U-shaped piece of iron whose ends are perforated and through which a pin or pole is placed.

Clinch–Part of a nail bent over to keep it from being pulled out, as in horseshoeing, box manufacture, etc.

Clip–(1) A semicircular metal piece extending from the outer surface of the horseshoe at the toe or side to prevent the shoe from shifting on the hoof. (2) Shears. (3) (a) An inclusive term for shorn wool. (b) The process of shearing wool. (c) The year's production of wool. (4) (a) To shear the hair of an animal close to the skin, as to clip a dog, or a cow's flank and udder. (b) To remove fleece from a sheep or goat. Also called shear. (c) To cut the feathers from a fowl's wing to prevent it from flying. (5) (a) To trim a plant. (b) To cut off the tops of a crop, as clover or alfalfa, at an early stage of growth when the crop is to be harvested for seed. Also called preclipping.

Clod Buster–A tooth type harrow attachment for the rear of a tillage implement; used to break up clods of soil.

Clog–(1) A heavy, wooden block attached to a horse's hind pasterns by a strap to prevent it from kicking when put out to pasture. (2) An impediment, encumbrance, or restraint. (3) A wooden shoe. (4) To stop a machine, as a harvester, by feeding material in too rapidly; to plug up; to choke up.

Clogging–The thickening of lubricating oils due to the absorption of oxygen and the presence of dust. Machinery is said to clog when its lubrication oil becomes thick and dry.

Closing Price–The price of the last contract on an item in a commodity market on a particular day.

Cloud on a Title–A legal defect in the title to property.

Clutch–(1) (a) A nest of eggs in a hatchery. (b) A brood of chickens. Also called cletch. (2) A device to engage or disengage the power from various working parts of machinery.

Co-op–See Cooperative.

Coefficient of Digestibility–The amount of a particular nutrient digested and absorbed by an animal; expressed as a percentage of the amount that was in the animal's feed.

Coefficient of Roughness–A variable factor used to estimate the flow of water in rivers and canals, determined by the roughness or smoothness of channel lining, the mean hydraulic radius, mean velocity of the flow, and slope.

Cohort–(1) A group of people with a defined history of exposure who are studied for a specific length of time to determine cancer incidence or mortality. (2) A group of individuals born within the same time period (usually within five or sometimes ten years of each other). Such groups are called birth cohorts. The diseases among individuals in one birth cohort followed throughout their lifetimes may be different from those in another, implying differences in exposures to environmental factors causing disease.

Cold Chisel–A steel tool strong enough to cut cold metal. Also called cold cutter.

Cold Rating–The cranking load capacity of a battery at low temperatures.

Cold Sterilization–(1) The use of cathode ray or electron beam gun in food processing to kill bacterial or insect life. (2) Chemical sterilization of instruments.

Cold Storage–(1) An insulated storage using mechanical refrigeration to maintain a stable, cold temperature for long-term storage. (2) The treatment given to plants and bulbs to cause certain chemical changes that enable them to respond to forcing treatments.

Collateral–Something of value deposited with a lender as a pledge to secure repayment of a loan.

Collins Dynamometer–A device to measure the tractive pull of tractors or draft horses. It permits the movement of the vehicle to which the tractors or horses are hitched only when the tractive pull equals the weight set on the machine. A pull of a distance of 27½ feet (8.39 meters) is most frequently used to determine horsepower.

Collop–A unit of grazing area which can support a full-grown horse or cow for one year.

Colorimeter–An instrument for chemical analysis of liquids by comparison of the color of the given liquid with standard colors.

Colter–(Middle English, *culter*, knife) A sharp blade or sharp rolling wheel attached to the beam of a turning plow to cut crop residues and clods to reduce trash buildup in front of the plow; also spelled coulter.

Combine–A self-propelled or tractor-drawn machine that cuts, threshes, and cleans the standing crop while moving across the field. It is adapted to harvesting all the small grains, soybeans, grain sorghums, peanuts, rice, beans, etc.

Combustion–The production of heat and light energy through a chemical process—usually oxidation. One of the three basic contributing processes of air pollution, the others being attrition and vaporization.

Combustion Chamber–(1) Primary: The chamber in an incinerator where waste is ignited and burned. Secondary: The chamber of an incinerator where combustible solids, vapors, and gases from the primary chamber are burned and fly ash is settled. (2) That part of an internal combustion engine where the fuel is ignited.

Commensurate Property–Land or controlled livestock water which qualifies a person for grazing preference in other land, either private or public.

Commercial Cow-calf Producer–A stockman producing animals from a nonregistered herd.

Commercial Farming–Farming in which the majority of the farm output is sold, usually also involving appreciable use of purchased inputs.

Commercial Fertilizer–Plant nutrients containing a single essential mineral nutrient or a mixture of essential mineral nutrients or organic materials.

Commercial Herd–A herd of animals that will eventually be slaughtered for meat.

Commercial Species–Tree species suitable for industrial wood products.

Commercial Thinning–Removing trees from a developing young stand so that remaining trees will have more growing space; dead and dying trees will be salvaged; and the operation will, hopefully, make a net profit.

Commercial Timberland–Forest land that is producing or is capable of producing crops of industrial wood and that is not withdrawn from timber utilization by statue or administrative regulation. Note: Areas qualifying as commercial timberland have the capability of producing in excess of 20 cubic feet per acre per year of industrial wood in natural stands. Currently, inaccessible and inoperable areas are included.

Comminution–(1) The reduction of a substance to a fine powder; pulverization; trituration. (2) Mechanical shredding or pulverizing of waste, a process that converts it into a homogeneous and more manageable material. Used in solid-waste management and in the primary stage of waste-water treatment.

Comminutor–(1) An agricultural implement for breaking down the clods in the soil after plowing. (2) A device for the catching and shredding of heavy solid matter in the primary stage of waste treatment.

Commission Merchant–A person entrusted with goods for sale on commission. He may possess and sell the goods in his own name, as distinguished from a broker.

Commodity–A transportable resource product with commercial value; all resource products that are articles of commerce.

Commodity Credit Corporation–An agency of the United States federal government which is authorized to purchase and make loans on farm products, dispose of surplus commodities, and engage in other operations with a view to stabilizing the agricultural market. Organized in 1933 as an affiliate to the Reconstruction Finance Corporation, it was transferred to the U.S. Department of Agriculture in 1939.

Commodity Exchange Authority–An agency of the U.S. Department of Agriculture, established on February 1, 1947, to supervise trading on the seventeen markets. Formally composed of the Commodity Exchange Administration and the Grain Futures Administration.

Commodity Stabilization Service–An agency of the U.S. Department of Agriculture, established on November 2, 1953. It is responsible for: (a) acreage allotments and marketing quotas; (b) soil bank; (c) price support; (d) disposal of government-owned surplus farm products; (3) International Wheat Agreement Acts; (f) storage, shipping, and related service activities; (g) administration of the Sugar Act; and (h) assigned mobilization planning.

Common–(1) A joint pasture in a village or community to which all members have access for their herds. (2) Legal right of a person in sharing the profit or use of another's land. (3) A low, market grade of meat animals. (4) The American wool grade equivalent to the English 44's; a low grade, only better than braid. (5) A defective grade of lumber useful only in framing, etc.

Common-use Range–Range containing grass, forbs, and browse which allows two or more kinds of stock to graze together during the entire season or separately during part of a season.

Community Allotment–An allotment upon which several permittees graze their livestock in common.

Commutator–A contrivance for reversing the direction of the flow of an electric current in a circuit, dynamo, etc.

Compaction–(1) Decrease in volume of sediments, as a result of compressive stress, usually resulting from continued deposition above them, but also from drying and other causes. (2) Reducing the bulk of solid waste by rolling and tamping. (3) Increasing soil bulk density and decreasing porosity caused by the application of mechanical forces to the soil.

Compactor–(1) A vehicle with an enclosed body containing mechanical devices that convey solid waste into the main compartment of the body and compress it. (2) A vehicle equipped with a blade and with rubber tires sheathed in steel or hollow steel cores; both types of wheels are equipped with load concentrations to provide compaction and a crushing effect. A vehicle for working in a sanitary landfill. (3) A machine that reduces the volume of solid wastes by campaction.

Companion Crop–A crop which is grown with another crop, usually applied to a small grain crop sown with a forage crop. Preferred to the term nurse crop. See Nurse Crop.

Comparative Advantage–The placing of emphasis and efforts in the area where the greatest returns will be realized.

Comparative Analysis–Comparison of the performance of a particular farm with some standard such as the average of performance of a group of similar farms.

Comparison Approach–A method of appraising the value of a farm by comparing it in all respects with a previously evaluated farm in the community. The values of the reference farms are established by recent sales or by a group of appraisers.

Compass–(1) An instrument for determining directions, usually by the pointing of a magnetic needle free to turn in a horizontal plane toward magnetic north; sometimes having a clinometer attached for measuring vertical angles. Also, a dip compass, for tracing magnetic iron ore, having a needle hung to move in a vertical plane. (2) An instrument for describing circles, transferring measurements, etc.

Compensatory Payment–A method of supporting agricultural farm prices in which the government pays the farmer directly the difference between the market price and the support price of a particular commodity.

Competing Crops or Livestock–Any two crops or livestock whose production demands the use of the same resources at the same time.

Competitive Crop–(1) A crop planted on a piece of land to force out other forms of plant life. Also called smother crop. (2) Crops that compete with each other for the same time in a farmer's work schedule.

Competitive Exclusion Principle–A generalization that states that two similar species having high competitive capacity rarely, if ever, occupy the same ecological niche.

Complementary Crop or Livestock–Any crop or livestock yielding a product that contributes to the success of another.

Complementary Enterprise–Any enterprises which are mutually contributive to farm income.

Complementary Imports–Agricultural import items not produced in appreciable commercial volume in the United States; e.g., bananas, coffee, rubber, cocoa, tea, spices, and cordage fiber.

Complementary Products–Products that are usually consumed together (e.g., ham and eggs).

Complete Budget–A budget for the entire farming operation or business.

Compound Interest Rate–The rate in interest used in compounding or discounting.

Compounding–Calculation of the future value of a present sum accounting for the rate of compound interest.

Compress–(1) (a) A hydraulic press used to reduce a bale of cotton to about one-third its original size for ease in handling and shipping. (b) The business or building for such a press. (c) To reduce the size of a cotton bale by a hydraulic press. (2) A pad of gauze or other material applied to put pressure on any part of the body to reduce swelling or control hemorrhage.

Compression Ratio–The volume of the combustion chamber of an engine when the piston is at its lowest point compared with the volume of the chamber at the end of the compression stroke.

Condemnation–A legal proceeding to secure land for a public purpose such as a road upon payment of the land's reasonable value. Condemnation proceedings are used when the owner will not voluntarily convey title. Eminent domain proceedings are condemnation proceedings.

Condemned–(1) Describing an animal, carcass, or food which has been declared unfit for human consumption. (2) Referring to real estate property acquired for public purposes under the right of eminent domain.

Conduit–(1) Any channel, open or closed, intended for the conveyance of water; any container for flowing water. (2) A pipe in which wiring is installed.

Cone Guide–A device used on cultivators for precision tillage of crops on bed-shaped land.

Congeal–To change a liquid to a solid by lowering the temperature of the food sufficiently to bring about gelation.

Congelifraction–The mechanical disintegration of minerals and rocks resulting from the pressure exerted by freezing of water contained in their cracks. Freezing water exerts a pressure of 150 tons/square inch (146.48 kg/cm^2).

Connate Water–Water trapped in sediments at the time of deposition. Such water trapped by wells may be highly mineralized and may be a brine or saltwater. See Confined Water, Fossil Water.

Conservation–The control and preservation of natural resources for present and future use. In agriculture, the maintenance and improvement of soil fertility and productivity and the control of erosion.

Conservation District–A public organization created under state enabling law as a special-purpose district to develop and carry out a program of soil, water, and related resource conservation, use, and development within its boundaries, usually a subdivision of state government with a local governing body and always with limited authorities. Often called a soil conservation district or a soil and water conservation district.

Conservation Plan–Includes but is not limited to farm/ranch "conservation plan maps"; water, plant, animal, and other inventory and management information with needed interpretations and evaluations; a record of the decisions made contributing to sound land use and conservation treatment; the alternatives for sound land use(s) and conservation treatment for which conservation decisions have not yet been made (including positive statements about critical problems such as soil erosion, sedimentation, land use, and agricultural pollutants); records of understandings as to cooperative agreements among individuals, groups or government representatives, and resource conservation districts; and other information useful to the decision maker.

Conservation Reserve Program–A program established by the Food Security Act of 1985. It pays farmers to take out of cultivation the most highly erodible cropland.

Consideration–The price or subject matter which induces a contract; may be money, commodity exchange, or a transfer of personal effort.

Consign–To entrust produce or animals to another person for sale on a commission basis.

Consignee–A person entrusted with the sale of produce or animals on commission. See Consignor.

Consignment–(1) The entrusting of a product or animal to a person for sale on commission. (2) Animals or produce consigned for sale.

Consignment Sale–A sale of produce or animals through a commissioned agent.

Consignor–A person who puts produce or animals for sale in charge of a commissioned agent. See Consignee.

Consortium–(1) A group of individual plants of different species, generally belonging to different phyla, which live together in close association. (2) A group of people working on a common problem or project.

Constant Mesh Transmission–A transmission in which the gears are engaged at all times, but shifts are made by sliding collars which lock together two or more gears.

Consumer–Any living thing that is unable to manufacture food from nonliving substances but depends on the energy stored in other living things for its food supply.

Consumer Price Index–General measure of retail prices for goods and services usually bought by urban wage earners and clerical workers. Includes prices of about 400 items, including food, clothing, housing, medical care, and transportation.

Consumer Sovereignty–The proposition that ultimately the consumer should, or does, direct all production and market activities in the country.

Consumption Pattern–The set of products that consumers purchases, as well as the processes by which these products are purchased and prepared for use.

Contemporary Group–A group of cattle that are of the same breed and sex and have been raised in the same management group (same location on the same feed and pasture). Individual animals can then be accurately compared with the others in the group. Contemporary groups should include as many cattle as can be accurately compared.

Contiguous–Touching or in contact; without fusion.

Continuous Grazing–Grazing for the entire grazing season; synonymous with set stocking.

Contour Interval–On a contour map, the difference in elevation, or the vertical distance, between contours.

Contour Line–(1) An imaginary line on the surface of the earth connecting points of the same elevation. (2) A line drawn on a map connecting points of the same elevation.

Contourliner–A leveling device that can be mounted on a tractor to assist the driver in plowing along contour lines (lines of equal elevation).

Contract Farming–An agreement about price and designated products between a farmer and a processor, usually made prior to the growing season. See Vertical Integration.

Contract Feeding–An arrangement for finishing cattle, poultry, or swine for the market. The contribution of each participant depends on a written contract. In the case of cattle, the stock raiser usually furnishes the cattle; the feeder furnishes the feed, equipment, and labor, and usually receives his income on the basis of the increase in weight of the cattle, etc.

Contract Labor–Labor hired for farm operations, whereby the farmer contracts with a family or group leader to care for a certain acreage. Often the laborer's income is increased as the yield per acre goes above a given standard.

Contract Production–Producing crops or livestock under an agreement to deliver specified goods and services in certain quantities and of certain quality at a later time.

Control–(1) Prevention of losses from plant or animal diseases, insect pests, weeds, etc., by any method. (2) A section of an open water channel where conditions exist that make the water level above it an index of the discharge. (3) A standard entity used for comparative purposes in experimentation. Also called check or check plot.

Controlled Atmosphere Storage–A cold storage in which the concentrations of atmospheric gases are adjusted to extend the storage life of fresh produce. Usually oxygen is lowered and carbon dioxide is raised.

Conventional Hay Baler–A hay baler that presses forage into a rectangular bale and ties the bale with twine or wire. See Round Hay Baler.

Conversion Factors–Convenient multipliers that have been calculated for elements and compounds important in fertilizers, soil amendments, and plant nutrition. The calculations are based on atomic weights of each chemical. For example, to determine the conversion factor for changing from a known percentage or mass of ammonia (NH_3) to an unknown percentage or mass of nitrogen (N), divide the atomic weight of NH_3 (17.04) by the atomic weight of N (14.01) = 1.216. Then to convert N to NH_3, multiply N by 1.216.

Converter–A device that converts or changes liquid petroleum (LP) gas from a liquid to a vapor for use by an engine.

Conveyance–Written instrument that evidences transfer of interest in real property from one person to another.

Conveyance Loss–The loss of water from a conduit due to leakage, seepage, and evaporation.

Conveyor–Any mechanical device, as an endless belt, for moving milk bottles, cans, grain, etc., from one place to another.

Cooley Amendment–A term applied to Section 104(e), Public Law 480, because it was introduced by Congressman Harold D. Cooley when he was chairman of the House Committee on Agriculture. It authorizes loans of foreign currencies to the United States and foreign firms operating in foreign countries.

Cooper–A person who makes wooden barrels, kegs, etc.

Cooperative–An enterprise or organization owned by and operated for the benefit of those using its services. In agriculture, such an organization is owned and used by farmers mainly to handle the off-farm part of their business—buying farm supplies, marketing their products, furnishing electric and telephone service, and providing business services—at cost. Essential features are democratic control, limited return on capital, and operation at cost, with distribution of financial benefits to individuals in proportion to their purchases.

Cooperative Extension Service–The Cooperative Extension Service of the Land-Grant colleges and universities was created under federal legislation with the Smith-Lever Act of 1914. Cooperative Extension Service philosophy is to help people identify their own problems and opportunities, and then to provide practical research-oriented information that will help them solve the problems and take advantage of the opportunities. The Cooperative Extension Service is responsible for programs in four major areas: agricultural and natural resources, home economics, community development, and 4-H youth development.

Corn Crib–(1) A building for the storage of ear corn. The roof is watertight, but the sides are usually slatted for ventilation. (2) A circular or rectangular temporary storage, with or without a roof, of woven wire or slats.

Corn Head–An implement mounted on the front of a combine that allows the combine to harvest corn.

Corn Picker–A machine that harvests ears of corn. The stalks are fed into the machine, where the ears are snapped off and loaded into a bin.

Corn-Hog Ratio–Number of bushels of corn that are equal (in value) to 100 lbs. of live hogs; i.e., the price of hogs per hundredweight divided by the price of corn per bushel.

Corner–(1) The outer pair of incisor teeth in the upper and lower jaws of a horse. (2) The junction point of boundary liens. (3) To tie up or control all available items of produce for speculation. (4) In lumbering, to cut through the sapwood on all sides of the tree to prevent it from splitting when the tree is felled.

Corner Tree–A tree, at or nearly adjacent to a land-survey corner, blazed by the original survey party to indicate the corner. Also known as a witness tree.

Corporation Farming–Large-scale farming carried on by hired managers and labor who apply the efficiency procedure developed in business and manufacturing.

Corrected Weight–A means of comparing the growth of animals that are of a different weight. For example to compare lambs in a flock that are weaned at around ninety days, divide the weaning weight of each lamb by its age in days and multiply by 90. This puts all of the lambs on a ninety-day basis.

Corrosion–(1) The process whereby surface or ground waters, by their own solvent action and by the help of solutes they carry, dissolve or chemically alter rock materials with which they come in contact. (2) The action of an agricultural chemical on the metal parts of distributors and containers.

Corrugate (Corrugated)–Wrinkled or in folds.

Corrugated Iron–A furrowed sheet of iron used for roofs or walls of farm buildings, or for the manufacture of large tubes, irrigation pipes, or flumes.

Cost-of-Production–The average amount in dollars per unit used in growing or raising a farm product, including all purchased inputs and sometimes including allowances for management and the use of owned land. May be expressed on a unit, a per-acre, or a per-bushel basis for all farms in an area or in the whole country.

Cost-Price Squeeze–A situation wherein price levels are persistently equal to, or occasionally below, costs of production.

Cost-sharing–A conservation practice in which joint contributions, equal or in some proportion, between the federal government and an agricultural producer are applied to the cost of carrying out a soil, water, woodland or wildlife conservation (in some cases, recreational or pollution control) program.

Cote–A shed for sheltering small animals; e.g., a sheep-cote.

Cotton Belt–The main cotton-growing region in the United States. It comprises the states of Florida, South Carolina, Georgia, Alabama, Arkansas, Texas, Louisiana, Missouri, Tennessee, North Carolina, Mississippi, Oklahoma, New Mexico, Arizona, and California.

Cotton Gin–(1) A machine used to separate the cotton seed from the lint. (2) A plant where the lint cotton is compressed into bales. Also called gin, gin house.

Cotton Linters–Short-staple cotton removed from the seed and husk by a second ginning, or by a linter machine. It is seldom used for textiles; it has other uses, such as in plastics or for mattress padding.

Cotton Picker–A machine that harvests only the mature lint-with-seed (known as seed cotton). A revolving spindle penetrates the cotton plant and winds the seed cotton from the opened boll and carries it into a dropping zone inside the machine. See Cotton Stripper.

Cotton Stripper–A machine designed for pulling (stripping) the entire ripe cotton bolls from the cotton plant. At the cotton gin, the hulls, seed, and lint are separated. See Cotton Picker.

Cotton Wagon–A four-wheel trailer used to haul cotton from the picker to the gin. It is usually pulled with a tractor or truck.

Coulter–See Colter.

County–A political unit in the United States established by a state legislature. A county usually contains sixteen townships. Many counties are irregular in shape and may be smaller or larger than the usual size. The corresponding unit in Louisiana is called a parish.

County Agent–A professional worker—jointly employed by the county, State Cooperative Extension Service, and the U.S. Department of Agriculture—to bring agricultural and homemaking information to local people and to help them solve farm, home, and community problems. Also called extension agent, farm and home advisor, agricultural agent, home demonstration agent, and 4-H or youth agent. See Cooperative Extension Service.

County Drain–A constructed drainage system consisting of an open ditch, a closed conduit, or a combination of both that provides a drainage outlet for agricultural and urban areas. It is administered according to the laws of the individual state.

County Seat–The city, town, or village in which all or most of the governmental offices of the county are located.

Couple–To attach; to hitch: as to couple a trailer to a tractor.

Coupling–(1) The loin of an animal. (2) Mating; breeding; copulation. (3) A connecting means for transferring movement from one part of a machine or device to another; it may be hydraulic, mechanical, or electrical.

Coupling Pole–In a wagon, the shaft that connects the rear axle to the front bolster. Also called reach.

Cover–(1) A lid placed over something for protection. (2) Woods, underbrush, etc., which may conceal game. (3) Plant life, such as grass, small shrubs, herbs, etc., used to protect soil from erosion. (4) The flesh, hide, and fat on a fattened animal. (5) To buy back future contracts. (6) To copulate with a female, as a bull covers a cow. (7) To incubate, to hatch eggs, as a hen. (8) The proportion of the ground surface under live and dead aerial parts of plants. Also refers to shelter and protection for animals and birds.

Cow Calf Operation–A system of raising cattle thats main purpose is the production on calves that are sold at weaning.

Cow Index (CI)–A method of evaluating genetic value of dairy cows. In beef cattle selection, a score indication of a cow's calf producing ability. The formula is: (Adjusted Weaning Weight of the Calf divided by the Weight of the Cow at Weaning) multiplied by 100.

Cow Month–The tenure on range or artificial pasture of a cow for one month. The quantity of feed or forage required for the maintenance of a mature cow in good condition for one month. See Sheep Month.

Cowl–(1) A curved, flaring top fitted to a ventilator, turned by a wind vane in a downwind direction to improve drafts. (2) The hoodcovering of a tractor.

Cradle–(1) A handheld implement that was once used to harvest grain. (2) A device made of wood or aluminum worn around the neck of a horse, which prevents him from chewing at sores, blankets, bandages, etc.

Cranberry Barrel–A standardized container of 5,826-cubic inch capacity, used for marketing cranberries.

Crankcase–The lower housing in which the crankshaft and many other parts of an engine operate.

Crankcase Dilution–A thinning or dilution of the oil in an engine, usually caused by unburned fuel that has gotten by the piston rings and into the crankcase oil.

Crankshaft–The main drive shaft of an engine, which takes reciprocating motion and converts it to rotary motion.

Crawler–(1) A newly hatched insect. (2) A large earthworm. (3) A tractor equipped with tracks, as compared with one equipped with wheels.

Credit–A means of obtaining goods or services now by promising to repay at a later date.

Credit Insurance–Insurance taken out and paid for by the borrower to provide for the payment of the loan in case of the death of the insured borrower. Of the two major types of this insurance, one pays the total amount of the loan and the other pays the unpaid balance.

Credit, Supervised–A technique of providing loans in adequate amounts at low interest combined with intensive supervision provided by a management supervisor to help small farmers and their families upgrade their farming and homemaking.

Creditor–A person or firm extending credit.

Crenate–With shallow, obtuse or rounded teeth; scalloped.

Croker Sack–Gunny sack; toe sack.

Crop Ecology–A science that deals with the study of agricultural crops in relation to their environment, such as soil, climate, plant reaction, etc.

Crop Expense–Denoting the amount spent for seed, seed treatment, fertilizer, lime, marl, spray material, crop insurance, twine, and sometimes custom expenses.

Crop Outlook–An official prediction by the U.S. Department of Agriculture, which attempts to forecast the total amount of the national or state harvest for a particular crop. See Crop Reporter.

Crop Reporter–A person, usually a farmer, who, during the growing season, submits a monthly report to the U.S. Department of Agriculture, estimating the crop outlook in an assigned area, making estimates in percentages of the normal, and often providing other related farm information, such as number of sows farrowing, price of farmland, etc. See Crop Outlook.

Crop Shield–A device attached to a cultivator or sprayer to protect plants from being covered with soil during cultivation or from herbicide during spraying.

Crop Surplus–The portion of a particular crop for national consumption that remains unsold after the normal period for selling, usually at the time of the next harvest.

Crop Year–(1) The span of time from the planting to the harvesting of a crop. (2) The span of time from one harvest to the next.

Crop Yield–The amount of harvest per acre, or other land measure, for a particular crop.

Cropland–Land under cultivation within the past twenty-four months, including cropland harvested, crop failures, cultivated summer fallow, idle cropland used only for pasture, orchards and land in soil-improving crops, but excluding land cultivated in developing improved pasture.

Cropper–(1) A laborer to whom a farmer assigns a definite tract of land together with all or most of the necessary farming equipment, seed, and fertilizer and a share of the crops grown as remuneration for his labor. He is closely supervised by the farmer, receives free use of a tenant house, and ordinarily is given some opportunity to work for wages when his farming operations do not require his attention. Also called sharecropper. (2) A thrown rider.

Cropping Plan–The scheme of growing different crops in succession on the same land, in contrast to a one-crop system or a haphazard change of crops.

Cross-Compliance–A government farm program term meaning that if a farmer wishes to participate in a program for one crop by meeting the qualifications for price supports and loans for that program, the farmer must also meet the program provisions of other major program crops that the farmer grows.

Cross-fenced–Describing a farm or ranch enclosed by a fence, with fields or pastures fenced off within the enclosure.

Cryology–(1) In the United States, the study of refrigeration. (2) In Europe, a synonym for glaciology. (3) The study of ice and snow. (4) The study of sea ice.

CSD–Committee on Surplus Disposal.

CSRS–Cooperative State Research Service.

Cubage–The total number of cubic feet in a building; a term often used in calculating cost of construction. A more common basis for estimating construction costs of a farm or ranch house or barn is square footage of floor space.

Cubic Foot per Second–The standard unit of measurement of water flow in irrigation, which is 1 cubic foot of water flowing past a given point in 1 second.

Cubing Machine–Any device that macerates meat to make it tender.

Cull–(1) Anything worthless or nonconforming which is separated from other similar and better items; the act of removing the inferior items; to cull out. (2) The lowest marketing grade of meat carcasses or dressed poultry. (3) Any animal or fowl eliminated from the herd or flock because of unthriftiness, disease, poor conformation, etc.; a reject. (4) A lumbering term for defective or low-grade timber. (5) Any fruit that fails to meet grading specifications because of defects, maturity, conditions, etc.; e.g., ripe berries that have green tips.

Culling Chute–A chute through which sheep pass in single file so that their fleeces and other qualities can be judged so the poorest animals can be culled out.

Cultipacker–A clod crusher and soil packer with a corrugated roller.

Cultivator–(1) A farm implement used to break the surface of soil in which plants are growing, and remove weeds, consisting of a frame to which several shovels, teeth, disks, or blades are attached. (2) A person who plants, tends, harvests, and improves plants.

Cultural Control–The deliberate manipulation of the environment to make it less favorable for pests by disrupting their reproductive cycles, eliminating their food, or making it more favorable for their natural enemies. See Integrated Pest Management.

Culture–(1) The working of the ground in order to raise crops; cultivation; tillage. (2) Attention and labor given to the growth or propagation of plants or animals, especially with a view to improvement. (3) (a) The growing of microorganisms in a special medium. (b) The microorganisms which are so grown. (4) Bacteria used in making dairy and other products. See Mother Culture. (5) Human-made features of an area. (6) The specific way of life of a given society (community, tribe, nation).

Cup–(1) The receptacle attached to a tree to collect resin in turpentine orcharding. (2) A notch in a tree trunk made by two downward cuts of an ax for inserting a fatal herbicide. (3) A curve across the face of a piece of board. (4) A mechanical object resembling a drinking cup. (5) In lubrication, a vessel or small funnel for receiving oil and conveying it to a machine part, an oil cup. (6) In grain elevators, a bucket or receptacle with a curved outline.

Current–(1) The movement of electricity along a conductor; measured in amperes. See Alternating Current, Direct Current. (2) The flow of a stream of water.

Current Assets–Assets such as cash on hand, bills receivable, and other items that may be converted to cash immediately.

Current Liabilities–A debt that is due or will be due soon.

Current Meter–A device used for measuring the velocity of flow of water in irrigation streams, ditches, and rivers.

Cusec–A cubic foot of flowing water per second.

Custom Mixture–A commercial dry or fluid fertilizer formulated according to specifications, which is sold to a consumer prior to mixing. It is required by state law to be labeled to show the net weight, guaranteed analysis (grade), and the name and address of the distributor. If distributed in bulk a written or printed statement of the information required shall accompany delivery. Custom mixing is often done to apply nutrients required by a soil test.

Custom Work–Specific farm operations performed under contract between the farmer and contractor. The contractor furnishes labor, equipment, and materials to perform the operation. Custom harvesting of grain, spraying and picking of fruit, and sheep shearing are examples.

Cut–(1) A piece of meat prepared for retail or wholesale trade. (2) A slash wound. (3) The opening made by an ax, saw, etc., on a tree. (4) An excavation in the earth, either human-made or natural. (5) The action of a horse's hooves striking its legs or other hooves in walking or running, interfering with its gait. (6) The yield of certain crops, as wool or lumber. (7) (a) An animal separated from the main herd. (b) To separate an animal from the main herd. (8) A severing of the stem or a part of a plant. (9) The output of a sawmill for a given length of time. (10) (a) A reduction in numbers, amount, size, etc. (b) To reduce in numbers, amount, size, etc. (11) To mow. (12) To castrate; emasculate. (13) To sever the jugular vein of an animal or fowl for slaughter.

Cutter–(1) A low grade of beef just above canner. (2) A part of a bean harvester consisting of two broad blades set in a wheeled frame at a 60-degree angle, so as to cut two adjacent rows about 2 inches below the surface. (3) In turpentining, a three-cornered steel tool used to sharpen hacks and pullers. (4) The device that chops forage or other plant material preparatory to placing the item in a silo. (5) A sleigh drawn by one horse.

Cutter Bar–A device attached to the front of a combine, mower, forage harvester, etc. A reciprocating blade works inside finger-shaped guards. The scissorlike action cuts the plants that are to be harvested. Also known as a sickle bar.

Cutting Chute–A narrow passageway into which animals are driven to remove certain ones from the main herd.

Cuttoo–A flap over a wheel that protects the axle from flying mud; a mudguard.

CWE–Carcass Weight Equivalent.

CWT–Abbreviation for hundredweight; 100 pounds (U.S.). A British CWT = 112 pounds.

Cyclone–(1) A device for reducing tomatoes to pulp. Also called pulper. (2) Popularly, but incorrectly, a tornado. (3) A low atmospheric pressure area, frequently several hundred miles in diameter which advances at approximately 25 miles per hours with winds blowing in a counterclockwise direction in the Northern Hemisphere and clockwise in the Southern; associated with warm and cold fronts in the middle latitudes. Also called low pressure area.

Cyclone Seeder–A hand-operated seeder that is strapped over the shoulder. A crank is turned and seeds are broadcast over the ground.

Cylinder–the round chamber in which the piston of an engine or hydraulic system operates.

Cylinder Head–A detachable part of an engine fastened securely to an engine block which contains all or a portion of the combustion chamber. In most modern engines the intake and exhaust valves also operate in the cylinder head.

D Ring–In harness making, a piece of metal shaped like the letter D or O to form a link between two separate bands, straps, or other harness devices.

Dairy Herd Improvement Assocation (DHIA)–A cooperative organization of approximately any twenty-five dairy farmers whose purpose is the testing of dairy cows for milk and fat production and the recording of feed consumed. Each farmer receives a monthly record for each of the cows and a complete, yearly summary of production and feed costs.

Darcy's Law–A volume of water passing through a porous medium such as soil in unit time is proportional to the cross-sectional area and to the difference in hydraulic head and inversely proportional to the thickness of the medium. The proportionality constant is called the hydraulic conductivity.

Dariloid–A sodium alginate, a gel, obtained from the giant Pacific Coast kelp, *Macrocystis pyrifera*, which has remarkable water-absorbing properties. It is used as a constituent in ice cream mixes to stabilize the frozen product against large ice crystal formations.

Datum–(1) Any level surface taken as a surface of reference from which to measure elevations. In surveying, sea level is the basic reference. (2) A figure indicating a fact, as the number, quantity, or weight of an item, used principally in the plural, data.

Dead Center–The extreme top or bottom position of the crankshaft throw of an engine at which the piston is not moving up or down.

Dead Head–(1) A log which is sunken or partially sunken in a stream or body of water. (2) Blighted or dwarfed seed stalks which produce reduced yields of poor seed. (3) A survey crew returning to a known location without surveying.

Dead Man–A log or stone which is partially or completely buried in the ground to serve as an anchor.

Debenture–Bonds issued without security.

Debits–Charges against an account.

Debt Servicing Capacity–Measured as farm net cash flow less cash needed for family living expenses.

Decaliter–Ten liters, equivalent to 610.25 cubic inches or 0.284 bushel (U.S.).

Decare–A unit of metric land measure equal in area to 1,000 sqare meters and 0.247 acre. Also spelled dekare.

Decastere–A unit of measurement which is the equivalent of 10 cubic meters or 13.08 cubic yards.

Decentralization–A market trend that has replaced central market trading by direct sales by farmers to buyers in production areas; also refers to the movement of food-processing plants from cities to farm production areas.

Decibel (dB)–A unit of sound measurement.

Decistere–A unit of measurement containing 0.1 cubic meter or 3.5315 cubic feet.

Declination–The angle, variable with geographic position, between the direction in which the magnetic needle points (magnetic north) and the true meridian (true north).

Decreasing Payment Plan–A plan whereby a fixed amount of principal is paid each year plus interest on the unpaid balance. The payments are larger at the beginning of the payment period than at the end because of the interest that is due.

Deductible Clause–A clause in an insurance policy that specifies that the policyholder will pay a portion of the total claims.

Deed Restriction–A restriction placed on property in the deed.

Default–Nonperformance of a duty arising under a contract.

Defeasance–A provision or condition in a deed or in a separate instrument which, being performed, renders the instrument void.

Deferred Grazing–The keeping of livestock from a pasture until there is enough vegetation to support the animals, or in the western range of United States, until the seeds of the herbage have matured.

Deferred Pricing–A price that is determined sometimes after the product has been transferred from the seller to the buyer.

Deferred-rotation Grazing–The system of range management in which grazing is postponed on various parts of a range during succeeding years, allowing each part successively to rest during the growing season to permit seed production and better vegetative spread.

Deficiency Payment–Direct government payment to producers when the average price received by farmers falls below the target price. For eligible producers the payment is determined by formula: deficiency payment = farm program acreage × farm program payment yield × payment rate.

Deflation–(1) A fall or drop in the general price level. (2) The removal of loose material, such as soil, by the wind, thus leaving a bare surface.

Dehydration–The removal of 95 percent or more of the water from any substance by exposure to high temperature.

Delaney Clause–Legislation passed by the United States Congress in 1958 that forbids the addition to food any additives shown to be carcinogenic in any species of animal or in humans.

Delcivity–A descending slope, as opposed to an acclivity, an ascending slope.

Demand–A schedule of the quantities of products that consumers will buy at alternative prices.

Demand Expansion–A marketing effort that seeks to shift the demand curve for a product of industry to the right so that more can be sold at the same price, or so that a higher price can be obtained for a given quantity of sales.

Demeter–(Greek) Goddess of agriculture. See Ceres.

Demonstration Agent–County Extension Agent; County Agricultural Agent; County Home Demonstration Agent. See Extension Service.

Demonstration Plot–A plot of ground used to demonstrate realistically approved practices, procedures, and techniques in agricultural production. It is not intended for conducting experiments. See Experiment Stations.

Demurrage–A charge made on cars or vessels held beyond a specified time limit by or for consignor or consignee for loading or unloading, for forwarding directions or for any other purpose.

Dendrometer–An instrument for measuring the diameter of a tree, outside bark. If used to measure diameter, continuously, the instrument is known as a dendrograph.

Densimeter–A device which measures the density of a material.

Density–(1) Mass per unit volume. (2) The number of wild animals per unit of area. (3) The degree of closeness with which wool fibers are packed together. (4) In forestry, density of stocking expressed in number of trees, basal area, volume, or other criteria on an acre or hectare basis. See Stocking.

Depreciation–The decrease in value of buildings, other improvements, and machinery which is caused by wear, tear, and obsolescence.

Depression–(1) A slight following in the flesh of an animal. (2) A hollow in the surface of the land. (3) A severe drop in income and prices for an area or for a nation. (4) The Great Depression denotes the economic reversal that took place in the United States in the 1930s, which was associated with the Stock Market Crash in October 1929.

Derived Demand–The relationship of a demand schedule at one market level to a schedule at another market level; for example, the farm demand for hogs is derived (or results from) the consumer demand for pork chops.

Descaling–The removal of milk stone, rust, or other corrosive oxides form metals by means of a weak acid such as acetic acid or 10% hydrochloric acid.

Detergent–(1) A compound of a soaplike nature used in engine oil to remove engine deposits and hold them in suspension in the oil. (2) A chemical (not soap) having the ability to remove oil or grime. Household detergents can be used as surfactants in herbicide sprays. See Surfactants.

Detonation–A premature or too rapid burning or ignition of the fuel mixture in the cylinders of an engine. It causes a knocking sound in the engine.

DHIA–See Dairy Herd Improvement Association.

DHIA Records–Production records of cows tested under the supervision of the tester of a Dairy Herd Improvement Association.

Diameter Tape–A tape that is put around a tree at a 4.5-foot height. The tape is graduated to read the diameter.

Dibble–(1) A small hand tool used to make holes in the soil for planting bulbs and seeds, or for transplanting plants. It usually has a wooden handle and a steel or brass point. Also called dibber. (2) To transplant with a dibble.

Diesel Engine–A type of internal combustion engine that uses a fuel injection system instead of a carburetor and produces combustion temperatures by the heat of compression instead of a spark plug. The fuel is an oil rather than gasoline. The engine was named after its designer Dr. Rudolph Diesel.

Differential–(1) The difference in price paid for grades which are higher or lower than the basic grade, usually fixed by contract. (2) A system of gears in a tractor or other vehicle which permits one wheel of a pair to travel independently of the other when necessary, at the same time causing each wheel to receive its share of the power from the engine.

Differential Leveling–A method of leveling by which the difference in elevation between two points is determined.

Digestion Coefficient–The amount of a given feed that is digested by an animal expressed as a percentage of the gross amount eaten.

Diminishing Balance Method of Depreciation–A fixed amount of the value of an item is depreciated each year until the salvage value is reached.

Diminishing Returns–See Law of Diminishing Returns.

Dipper–(1) A cup that has a long handle. (2) One who collects the oleoresin from turpentine cups, or the tool used to remove oleoresin from the cups.

Direct Buying–The practice whereby food marketing firms purchase directly from farmers or shipping point markets rather than from terminal markets.

Direct Current (DC)–An electrical current that flows steadily in the same direction along a conductor. It is produced by a generator or battery. See Alternating Current.

Direct Drive–The direct engagement between the engine of a tractor or other machine and the driveshaft where the engine crankshaft and the driveshaft turn at the same speed.

Direct Payments–Payments made by the federal government to agricultural producers enrolled in commodity programs. A deficiency payment is the most common form of a direct payment. Deficiency payments can be made in cash or in certificates entitling the producer to receive an equivalent cash value of crops.

Directs–Stock purchased by packers and shipped direct to them for slaughter.

Dirt Farmer–(1) A farmer who does all or part of his own work in the fields and stables, as contrasted to one who hires all labor or rents his land, or to an absentee farm owner who employs a manager to operate his farm. (2) At times, a name used to downgrade a person.

Disc–See Disk.

Discharge–(1) The quantity of water, silt, or other mobile substances passing along a conduit per unit of time; rate of flows. (2) An exudate or abnormal material coming from a wound or from any of the body openings: e.g., a bloody discharge from the nose. (3) To remove the electrical energy from a battery.

Discount–(1) The amount allowed as a deduction for the settlement or payment of a debt before it is due to be paid. (2) The amount of money that is deducted as a fee by banks or brokerage houses from the sale of an issue of securities. (3) The excess money over the par value or face value of a land contract, or other security.

Discount Factor–The value by which a future cash flow must be multiplied to calculate its present value.

Discounted Loan–A loan where the interest is deducted from the principal at the time the loan is made.

Discounting–Calculation of the present value of a future sum accounting for the rate of compound interest.

Disk–(1) A round, usually sharp-edged, slightly dished, steel plate that cuts the soil as it revolves on a center axis and moves the soil to one side. Several disks are mounted and spaced on a horizontal shaft to make the disk harrow. Also called disk blade. See Disk Harrow. (2) One of a series of metal plates in a centrifugal separator bowl that increases the efficiency of the machine. (3) In botany, (a) an adhesive surface on the tendril ends of creeper plants that enables them to climb along flat surfaces; (b) an enlargement of a flower's receptacle; (c) the center of a composite flower; (d) an organ's surface; (e) the circular valve of a diatom. Also spelled disc.

Disk Harrow–A harrow consisting of two or more gangs of disks used: (a) to cut debris, especially vegetable matter on the surface of the oil; (b) to pulverize the top layer of the soil before or after plowing; (c) to prepare plowed ground for planting; (d) to cultivate crops growing in widely spaced rows; (e) to cover seed sown broadcast; (f) for summer fallowing.

Disk Hiller–A cultivator attachment that is composed of a disk that throws a hill of soil around the plants in the row.

Disk Mower–A hay-cutting implement that cuts by means of blades on rotating disks that are mounted on a bar or arm.

Disk Plow–A plow that uses a disk rather than a share and a moldboard to cut and turn the soil. It is preferable to the moldboard plows in very loose soil, hard ground, or land covered with much plant, residue or debris, as it will not scour. Also, it is possible to change the angle of the disk.

Displacement–A measure of the size of an engine. It refers to the volume displaced by one complete stroke or revolution of the pistons. For example a 360 cubic inch displacement (cid) engine.

Displacement Ton–See Ton.

Dispossess–To deprive a person of his/her possession or occupancy of real estate by legal action.

Distillation–The process of separating the components of a mixture by differences in boiling point; a vapor is formed from the liquid by heating the liquid in a vessel and successively collecting and condensing the vapors into liquids.

Distilled Water–Water that has been boiled, converted into steam, and then condensed by cooling. Also known as deionized water.

Distributor–(1) Any device that is used to move produce from one place and to scatter it in another; e.g., a fertilizer distributor. (2) An agent or a wholesaler who sells goods in quantity. (3) A device in some bulk milk tanks that spreads the milk over a cooling surface. (4) That part of a motor engine that conducts the secondary current to the spark plugs.

Ditch–An artificial excavation, as a trench or channel dug to carry irrigation or drainage water.

Ditch Rider–A person appointed by an irrigation company or a governmental unit to examine irrigation ditches, weirs, etc., and to adjust the measuring devices so that the correct amount of water is delivered to each farmer.

Ditching Machine–Used for digging ditches, tile trenches, and making excavations. The machine is powered by an internal combustion engine and has a bucket with a cutting edge attached to a steel frame. It may also have a continuous belt or wheel to which buckets are attached to dig out the soil and deposit it at the side of the trench.

Diversification–The practice of raising many kinds of crops or animals on a farm. See Single-crop Farming.

Diversion Dam–A barrier built to divert part of all of the water from a stream into a different course; e.g., into an irrigation ditch.

Diversion Payments–A per-acre payment available in certain years as an option to producers enrolled in commodity programs who divert land from the production of a program crop in addition to the acreage required by the set-aside provisions of a specific commodity program.

Diversion Terrace–A wide, relatively shallow channel of low gradient with gentle side slopes and ample water capacity, which is constructed across the slope of a field to intercept and change the direction of flow and to reduce the velocity of run-off water and to reduce erosion.

Diviner–One who presumes to discover the location of oil, gas, water, or ore deposits in the earth; a dowser. See Dowsing.

Divining Rod–A rod or switch, often a forked willow or peach twig, which when held tightly in the hands of a dowser (water witch) is supposed to be irresistibly drawn to the earth over an underground supply of water, and hence, is used in locating a place to drill a water well. See Dowsing, Water Witch.

Dockage–(1) Foreign material in harvested grain such as weed seeds, chaff, and dust. (2) The weight deducted from stags and pregnant sows to compensate for unmerchantable parts of an animal.

Dodrans–A full span of a man's hand (from wrist to fingertip), about 9 inches or 23 centimeters.

Dog–(1) Canis *familiaris*, family Canidae; a domesticated, carnivorous animal which is used as a household pet, a watchdog, a herder for sheep, cows, etc. (2) A device, usually consisting of a steel hook and chain, used in skidding logs. (3) A sort of iron hook or bar with one or more sharp fangs that may be fastened into a piece of wood or other heavy article to move it. (4) Any part of a machine acting as a claw or clutch, as an adjustable stop to change the direction of a machine tool. (5) A low-qualitiy beef animal.

Dolly–A platform on rollers on which heavy loads may be moved, usually in a packing plant or warehouse.

Domestic Market–A market within the boundaries of the United States.

Domestic Water Use–Water used for drinking, sanitation, street flushing, fire protection, and lawn and garden irrigation.

Donkey–(1) A portable engine. (2) An ass.

Doodlebug–(1) The larva of several small insects. (2) A homemade tractor constructed from a used automobile.

Double Wheel–A type of wheel used on a planter that consists of two separate wheels having smaller diameters on one side, which are set with the smaller sides touching, forming an acute angle. The wheels serve to cover the seed and compress the soil. Also known as press wheel.

Double-acting Tank Pump–A type of sprayer pump which forces an ejection on both the downward and upward strokes.

Double-bitted Ax(e)–An ax whose head has two cutting edges.

Double-breaking Cart–A cart which usually consists of the hind wheels and gears of a large wagon and a platformlike seat extending out behind the wheels. Its tongue is often made from a pole up to 6 inches in diameter. It is used to train horses and mules to work in teams.

Double-decked Car–A railway car or truck having two main floors, one above the other. It is used for the shipment of sheep, hogs, calves, etc.

Double-disk Harrow–A harrow in which two single-disk harrows are arranged in tandem. The front gangs are set to throw the soil outward, while the rear gangs are set to throw the soil inward to leave the surface level. Also called double-action harrow, tandem disk harrow.

Double-folding Marker–A type of row marker for a planter that consists of two separate rods attached to either side of the planter. On the end of each rod is a disk or spike tooth for marking the soil for the location of the next row. One marker is used as the planter moves across the field, the other on the return trip.

Dousing Rod–Divining rod.

Dower Right–The widow's right in her husband's real estate. At one time, it was a one-third share.

Down Suction–The downward bend of the point of a turning plow share which helps the plow penetrate the soil to the proper depth as it moves forward.

Downspout–A pipe leading downward from a gutter to carry rainwater from a roof.

Dowser–One who uses a divining rod to locate underground water and indicate the place to drive, dig, or drill a well. Also called water witch, water diviner.

Dowsing–(origin of word Unknown) Using a small forked stick known as a divining rod to locate underground water. This technique has never been accepted by scientists.

Draft–(1) A package deduction, one pound per hundred pounds, allowed a buyer of wool. (2) The horizontal component of pull of an implement parallel to the line of motion. Also called directional pull. (3) Similar to a check but instead of being charged to a signer it is charged to a third person named on the face of the draft. (4) Any feedstuff obtained as a by-product of the distillation of grains.

Drag–(1) To pull a heavy object such as a log, across a newly plowed field for the purpose of smoothing the soil surface. (2) To move a fish net through the water.

Drag Chain–An endless belt of chains that is used to move corn, etc., as from a loading platform of the silage cutter.

Drainage District–An organization that operates under legal regulations for financing, constructing, and operating a drainage system.

Drainage Easement–An easement for directing the flow of water. See Easement.

Drawbar–A device at the rear of a tractor to which implements are hitched.

Drawbar Horsepower–A measure of the pulling power of a tractor at the drawbar hitch. See Brake Horsepower, Effective Horsepower, Horsepower, Rated Horsepower.

Dray–(1) A wagon used for heavy hauling. (2) A single sled used in hauling logs in which one end of the log rides on the sled and the other drags on the ground.

Dredge–(1) Oat and barley seed sown together and cultivated to be used for making malt. (2) Any of several different types of machines to deepen channels and clean out ditches, etc. (3) To excavate or deepen and clean stream beds. (4) To coat the surfaces of food with flour, cornmeal, breadcrumbs or other fine substances before cooking.

Drier–Any device used for removing moisture from a product, usually consisting of a heating unit and a large fan for forcing the hot, dry air through the product being dried. Alfalfa is sometimes dried in such a way before being stored. Also called dehydrator.

Drift Fence–A line of fence, open at both ends, used to direct the movement of grazing animals on the ranges of the western United States.

Drifter–(1) A term once used on the open ranges of the western United States for a shepherd, or owner of a band of sheep, who moved about grazing his sheep. (2) A hired hand who moves from place to place with the season or job and has no permanent location.

Drill–(1) A small furrow in which seeds are planted. (2) A row of seeds that have been planted by dibbling in a small furrow. (3) An implement for planting seeds that forms a small furrow, deposits the seed in dibbles, covers the seeds, and packs soil over it. It can also deposit fertilizer, lime, or other soil preparations into the soil, alone or with the seed. (4) A very small, trickling stream or rill.

Drip Gutter–A small gutter inside a greenhouse at the point where the roof and walls meet which carries off water formed by condensation inside the greenhouse.

Drive–(1) (a) The moving of livestock under human direction; (b) To herd livestock. (2) In wildlife management, the herding of wild animals past a particular point for counting or shooting. (3) The process of floating logs down a river from a forest area to a mill or to a shipping point. (4) The means for giving motion to a machine part or machine, as a drive chain, a drive wheel.

Drive Line–The universal joints, drive shaft, and other parts connecting the transmission of a truck, tractor, or other machine with the driving axles. See Universal Joints.

Drive-ins–Designating those cattle that are herded in to market, as contrasted to those transported there by truck, etc.

Driveway–(1) In the range country, that land which is set aside for the movement of livestock from place to place; e.g., from the home range to the shipping point. Also called stock driveway, stock route. (2) A path or passage, sometimes paved, for the movement of vehicles and/or livestock. (3) The farm road from the building site to the gate at the highway. (4) Vehicle passage into or through a barn.

Drosometer–An instrument that measures the quantity of moisture precipitated as dew.

Drum–(1) A cylinder-shaped package or product, as a drum of cheese. (2) A revolving cylinder, as the drum of a seed cleaner. (3) To knock on the sides of a hive to drive the bees upward when transferring them from one hive to another.

Dry Kiln–A structure in which lumber is seasoned artificially or in which pine cones are dried and opened to collect the seed.

Dry Masonry–Stonework which is laid without mortar.

Drying by Sublimation–The removal of water from a product while it is frozen to lock the molecular structure so that changes cannot occur during dehydration and to prevent bacteriological action. Also called freeze-dried.

Dryland Farming–A system of producing crops in semi-arid regions—usually with less than 20 inches (50 centimeters) of annual rainfall—without the use of irrigation. Frequently, in alternate years part of the land will lie fallow to conserve moisture.

Duff Hygrometer–An instrument for measuring the moisture content of litter to determine fire hazard.

Dump–(1) The storage yard of a portable sawmill. (2) A place for refuse. (3) To unload a truck or wagon by raising the bed and allowing the loaded material to fall to the ground. (4) To discard unwanted materials or refuse.

Dumpy Level–A leveling instrument in which the telescope is permanently attached to the leveling base, either rigidly or by a hinge that can be manipulated by means of a micrometer screw.

Duress–Forcing action or inaction against a person's will.

Dwelling–(1) A farm or ranch home or family residence. (2) A barely perceptible pause in the stride of a horse just before it places one of its feet to the ground when approaching a jump.

Dynamite–(Greek; *dynamis*, power) An explosive made from nitroglycerin, ammonium nitrate, or similar material absorbed in a porous material similar to sawdust and used on farms and ranches for blasting out stumps and large rocks and for making drainage ditches.

Dynamometer–An instrument for measuring force over time and distance. Used in horse-pulling and tractor-pulling contests and in tractor-evaluation research.

Easement–(1) The right of access or right-of-way through land which is usually attained by purchase for a specific purpose, such as for the construction and maintenance of a gas line, power line, oil line, or telephone line. (2) A vested or acquired right to use land, other than as a tenant or the owner, for a specific purposes.

Ecological Factor (Limiting Factor)–Any part or condition of the habitat affecting directly or indirectly the life of one more organisms in such a way as to differentiate it from other vegetation; often classified into: (a) climatic, physiographic and edaphic, and biotic factors; (b) direct, indirect, and remote factors.

Economic Entomology–That branch of insect study directed toward preventing human losses and increasing gains through manipulation of insect populations. Examples of economic entomology include the methods of protecting plants, people, and animals from insect-borne diseases by insecticides, drainage, crop rotation, and integrated pest management; and the culture of silkworms, honeybees, and various beneficial parasitic insects.

Economic Injury Level–The point at which the buildup of an insect population starts to cause economic damage to a crop or group of animals.

Economic Maturity–The age and growth rate at which a tree or stand of trees will no longer increase in value fast enough to earn a satisfactory rate of interest. At this time the trees should be marketed. Also called financial maturity.

Economic Poisons–Any of the recommended and approved insecticides, herbicides, rodenticides, etc.

Economic Research Service (ERS)–This USDA service develops and carries out a program of economic research designed to benefit farmers and the general public. The findings of this research are made available to farmers and others through research reports and through economic outlook and situation reports on major commodities, the national economy, and the international economy. The ERS functions through a central office in Washington, D.C., and through a field organization that is involved chiefly in farm and marketing economic research. Much of the research is carried on in cooperation with state experiment stations, state departments of agriculture or marketing, and other state institutions.

Economic Size–Under a specified soil, climate, enterprise, and management, the optimum size of a farm, ranch or other agricultural operation that will yield the highest net return for the operator's labor, capital, and management.

Economic Threshold–The point at which insect control measures should be started before a population of insects becomes dense enough to cause economic loss.

Economy of Gain–The amount of feed required per unit of gain in the weight of an animal or fowl expressed in cash value, i.e., the pounds of feed per pound of gain converted to dollars.

Economy of Scarcity–The theory that limited production and relative scarcity of a commodity will result in greater good for a country's economy than abundance, and consequently lower prices.

Edge Firing–A method of controlled burning in which fires are set around the perimeter of the area to be burned and allowed to spread inward.

Edger–A tool used for cutting the sod along walks, the edges of flower borders, shrub borders, and beds. Also called edge iron.

Effective Field Capacity–The actual work accomplished in acres or hectares per hour by an implement despite loss of time from field end turns, inadequate tractor capacity, deficient tractor or implement preparation, adverse soil conditions, irregular field contours, lack of operator skill, or other factors.

Effective Half-Life–Biological half-life.

Effective Horsepower–A measure of the power of a machine calculated by the formula: drawbar pull (pounds) = speed in feet per minute divided by 33,000. One horsepower = 745.7 watts. See Brake Horsepower, Drawbar Horsepower, Horsepower, Rated Horsepower.

Effective Progeny Number (EPN)–An indication of the amount of information available for estimation of expected progeny differences in sire evaluation. It is the function of the number of progeny but it is adjusted for their distribution among herds and contemporary groups and for the number of contemporaries by other sires. EPN is less than the actual number because the distribution of progeny is never ideal.

Efficiency of Irrigation–The fraction of the water diverted from a river or other source that is consumed by the crop, expressed as percent. Often applied to whole irrigation systems and takes account of conveyance losses. See Consumptive Use.

Egg Cart–A rolling device used for transporting stacked trays of eggs to the hatchery.

Egg Exchange–A market where eggs are bought and sold in large quantities and where eggs are traded on futures.

Egg Money–Income derived from the sale of eggs from a general farm, an enterprise usually supervised by the farmer's wife. It is usually set aside for emergencies or for the purchase of personal luxuries.

Elasticity–The ability of wool to return to its original length after having been stretched. The elasticity varies greatly with the character of the wool. Wool that is sound and has good character has considerable elasticity.

Elasticity of Demand—The tendency of demand for a commodity to be influenced or changed by various factors. *Price elasticity* is the tendency of demand to change as price goes up or down. *Income elasticity* is the tendency of demand to change as the consumer's income goes up or down. Both are related to ability to buy.

Electric Exerciser—A device, similar to a merry-go-round, consisting of a steel framework that is turned by a motor. Bulls are chained to the frame at the outer edge so that as the device revolves they are forced to walk in a circle for exercise.

Electric Fence—A fence or enclosing device that consists of a single wire supported by insulators on widely separated posts. The wire is attached to a controller that emits electric current for one-tenth of a second or less, forty-five to fifty-five times per minute. When an animal touches the wire, it receives a sharp, short, but harmless, electrical shock. The fence, when properly installed, is not injurious to human beings.

Electric Fly Screen—A window screen so charged with electricity that when an insect alights on the screen it is electrocuted.

Electric Prod(der)—A portable, battery-powered, canelike device used to control or drive livestock by giving them a slight electrical shock.

Electric-capacity Moisture Meter—A meter used for determining the moisture content of wood by utilizing the variation in dielectric capacity of wood with changing moisture content.

Electrical-resistance Moisture Meter—A device used for determining moisture content, directly or indirectly, by measuring the electrical resistance of water, or that of the material itself, as soil, in relation to its content of water.

Electrolyte—(1) A nonmetallic substance that will conduct an electric current by the movement of ions when dissolved in certain solvents or when fused by heat; common salt is an example of an electrolyte. (2) A solution containing salts and energy sources used to feed young animals suffering from scours (diarrhea).

Electrostatic Precipitator—A device that collects particulates by placing an electrical charge on them and attracting them onto a collecting electrode. Used for taking pollutants out of the air.

Electrostatic Sprayer—A spraying system that uses electrical forces of attraction to greatly increase the amount of spray that covers the plant. Individual spray droplets are given an intense electrical charge within a specially designed atomizing nozzle and propelled toward the plant. Individual spray droplets are given an intense electrical charge within a specially designed atomizing nozzle and propelled toward the plant. The approaching charged spray cloud induces an opposite electrical charge into the plants. The charges cause the spray droplets to be attracted like a magnet to the plant. Spray coverage is greatly increased.

Elevator—(1) Any of a number of devices, most often consisting of endless belts, chains and buckets, screws, or suction tubes that are used to raise materials from a lower to a higher level. (2) A building designed for the handling and storage of cereal grains, dried beans, and other seed crops. It may also be used for processing and selling products.

Eligible Storage—A term designating cribs or bins which are of such substantial and permanent construction as to afford protection against rodents, other animals, thieves, and weather, and are therefore eligible for use in storing grain for a United States government loan.

Emblement—In law, the profits from sown land.

Emigration—Movement out of an area. See Immigration.

Eminent Domain, Right of—(1) The right or power of government subject to constitutional and statutory limitations to take private property for public use upon making just compensation. (2) The legal right of public agencies to claim private property for public use, if compensation is made to the owner for property so claimed. The use of this power is limited by the "Due Process of the Law" clause in the 14th Amendment of the United States Constitution.

Empirical Yield Table—A table showing the progressive development of a timber stand at periodic intervals. The table covers the greater part of the useful life of the stand. It is prepared based on actual stand conditions.

Encroachment—A fixture, or structure, such as a wall or fence, that invades a portion of a property belonging to another.

Encumbrance—Any lien or liability attached to real property.

End Play—The amount of axial or end-to-end movement in a shaft of a machine that is due to the clearance in the bearings.

Endorsement—An additional signature on a promissory note that indicates that the signer pledges his unsecured property and is willing to pay the note should the borrower fail to do so.

Energy—The capacity to do work. It may take a number of forms, among them mechanical, chemical, and radiant, and can be transformed from one form to another, but cannot be created or destroyed.

Energy Dissipator—A device used to reduce the excess energy of flowing water.

Energy Head—The energy of a unit height of a stream that takes into account the elevation of the hydraulic grade line at any section plus the velocity head of the mean velocity of the water in that section. The energy head may be referred to any datum, or to an inclined plane, such as the bed of a conduit.

Energy of Position—Potential energy, or the energy possessed by a body by virtue of its position as distinguished from the energy of motion or kinetic energy; e.g., water stored in an elevated reservoir represents potential energy, since its release to a lower level may be utilized to do work, as in a hydroelectric power plant.

Engel's Law—A tendency for the share of a family's (or nation's) income spent for food to fall as income rises. This suggests that the income elasticity of food is lower than that of other products.

Enterprise—A project on a farm, such as the production of any crop or livestock.

Entrepreneur–An individual or firm that commits resources to productive activities in pursuit of a profit; a risk taker and profit seeker.

Environmental Impact Statement–A document prepared by a person, an industry, or a political entity on the environmental impact of its proposals for legislation and other major actions significantly affecting the quality of the human environment. Environmental impact statements are used as tools for decision making and are required by the National Environmental Policy Act.

Environmental Protection Agency (EPA)–A federal agency charged by Congress to protect the nation's land, air, and water systems. Under a mandate of national environmental laws focused on air and water quality, solid-waste management and the control of toxic substances, pesticides, noise, and radiation, the EPA strives to formulate and implement actions that lead to a compatible balance between human activities and the ability of natural systems to support and nurture life.

Environmental Quality–An evaluation of environmental quality should include: (a) areas of natural beauty; (b) water, land, and air quality; (c) biological resources and selected ecosystems; (d) geological, archeological, and historical resources; and (e) irretrievable commitments of resources to future use. The sum total of the forces and factors that influence people's satisfactions with their work, leisure, living conditions, and community.

EPA–See Environmental Protection Agency.

EPA Establishment Number–A number assigned to factories that produce pesticides. The numbers are printed on the containers of pesticides to indicate where the pesticide was made.

EPA Registration Number–A number that appears on the label of pesticides to indicate the number under which the pesticide was registered by the manufacturer.

Equitable Rights–Having certain rights because of the ownership of a certain equity.

Equity–The net ownership of a farmer/rancher in his/her business which is the difference between the assets and liabilities of an individual or business as shown on the balance sheet or financial statement. it is comparable to net worth.

Equivalent per Million–An equivalent weight of an ion or salt per one million grams of solution or soil. For solutions, equivalents per million and milliequivalents per liter (meq/1) are numerically identical if the specific gravity of the solution is 1.0.

Equivalent Weight–The weight in grams of an ion or compound that combines with or replaces 1 gram of hydrogen. The atomic weight or formula weight divided by its valence.

Eradicator–(1) Any agent used for the destruction of weeds, insects pests, etc. (2) A device used for scraping the edible flesh from the pineapple shell after the central portion has been removed.

Erect–(1) Designating upright plants, in contrast to prostrate ones. (2) To build or construct buildings, etc.

Erg–Unit of energy measurement. A force of 1 dyne acting through a distance of 1 centimeter.

Ergon–(Greek) Work.

Erodibility Index–An index based on a field's inherent tendency to erode from rain or wind in the absence of a cover crop. It is based on the universal soil loss equation (USLE) and the wind erosion equation (WEE), along with a soil's T value. See T Value, Universal Soil Loss equation, Wind Erosion Equation.

Escheat–The reversion of property to the state in the event the owner thereof dies without leaving a will and has no heirs to whom the property may pass by lawful descent.

Escrow–Money, a deed, or other written instrument delivered to a disinterested third person to be delivered by him or her upon fulfillment of some condition or conditions imposed by written instructions.

Esplees–The yield from farmland; e.g., profit from rent, crops, or remuneration for such.

Establishment–(1) The adjustment of a plant to a new site, consisting of three processes: germination, growth, and reproduction. Also called acesis. (2) A farmstead. (3) An economic unit, generally at a single physical location, where business is conducted or where services or industrial operations are performed.

Estancia–A small farm or small cattle ranch (southwestern United States).

Estate–(1) A relatively large acreage on which are a residence and other buildings, lawns, flower gardens, etc. It connotes wealth and is not usually cultivated for profit. (2) One's entire property. (3) The property left by a person after death.

Estimate–A judgment of the approximate volume or yield of a crop or a timber stand made from incomplete data.

Estimated Breeding Value (EBV)–In beef cattle, an estimate of the value of an animal as a parent, expressed as a ratio with 100 being average. For example, a bull with a yearling weight EBV of 110 would be expected to sire calves with yearling weights 10 percent greater than average.

Estimated Relative Producing Ability (ERPA)–In dairy cattle, a prediction of 305-day, two times per day milking, mature equivalent production compared with the production of other cows in the herd.

Ethnic Group–A segment of a population set aside because of distinct cultural and racial characteristics that is usually of a separate historical origin than the rest of the population.

Eugenics–The application of knowledge of heredity to the improvement of the human race; the study of agencies under social control that may improve or impair the racial qualities of future generations, either physically or mentally.

Evaporative Cooling–Air evaporates water and in the process the air loses heat to the water. Water plus heat equals vapor.

Evaporator–That part of the refrigeration system in which the refrigerant is changed from a liquid to a vapor. The heat required to vaporize the liquid refrigerant is obtained from the product to be cooled.

Even-age Management–The application of a combination of actions designed to create stands in which trees of essentially the same age grow together.

Ex–Prefix meaning without or destitute of.

Excambiator–An exchanger of lands; a broker.

Exempt Carrier–A motor vehicle that hauls raw agricultural commodities and is therefore exempt from Interstate Commerce Commission regulations.

Exempt Stock–In the United States, livestock that are permitted to graze on federal land free of charge. Usually confined to animals actually used for domestic purposes: saddle horses, milk cows, etc.

Exhaust Gas Analyzer–An instrument used to determine the efficiency with which an engine is burning fuel.

Expansion Joint–A device used in connecting long lines of piping, pavement slabs, etc., to permit linear expansion or contraction as the temperature rises or falls.

Experiment–(Latin; *experimentum*, proof from experience) Action to discover or demonstrate general or specific truth.

Experiment Stations–United States Department of Agriculture research facilities associated with state agricultural universities, where new ways of farming, ranching, or rural living are officially tested.

Experimental Design–The plan of an experiment, intended to ensure that the data to be collected will be suitable for statistical analysis.

Exploitive Farming–Crop production carried on for immediate profit and in disregard of long-term soil productivity.

Export-Import Bank of Washington, D.C.–The bank that aids in financing and facilitating exports and imports and the exchange of commodities between the United States and foreign countries or their agencies or nations. The bank supplements rather than competes with private capital. Its loans, generally made for specific purposes, offer reasonable assurance of repayment.

Exports, United States Agricultural–United States shipment of agricultural commodities to foreign countries.

Extension Service–An agency of the United States Department of Agriculture that represents that department in the conduct of cooperative extension work with the various land-grant colleges and universities. See Cooperative Extension Service.

Extensive Farming–Farming in which a relatively small amount of labor is spread over a large tract of land. See Intensive Farming.

F.O.B.–Free on board; an expression indicating that the seller assumes all responsibilities and costs up to the specific point or stage of delivery named, including transportation, packing, insurance, etc. The buyer takes over responsibility and costs at the same point. A wide variety of f.o.b. terms are used, as f.o.b. factory, f.o.b. cars, f.o.b. ship New York, f.o.b. Detroit, etc.

Factor–(1) A unit of inheritance occupying a definite locus on one or both members of a definite chromosome pair whose presence is responsible for the development of a certain character or modification of a character of the individual who possesses that genotype; a determiner or gene. (2) An agent, as one who buys and sells a commodity on commission for others. (3) An item in the analysis of a farm business; e.g., labor efficiency. (4) Inherent characteristics of the climatic, nutritional, cultural, or biological environment responsible for the specific performances of plants or animals.

Factorial Design–An experimental design in which each level of each factor appears with each level of each other factor.

Factory Farm–A term applied to a farm or type of farming that is usually operated on a large scale according to modern business efficiency standards, solely for monetary profit, as contrasted to a so-called family farm, or farming as a way of life.

Fahrenheit Scale–A temperature scale in which the freezing point of water is 32°F and the boiling point is 212°F. Named after Gabriel Daniel Fahrenheit (1686-1736), a German physicist. See Celsius Scale, Kelvin Scale.

Fair–(1) An exhibition of farm products, etc., on a competitive basis, usually with premiums offered for excellence. (2) The next-to-lowest grade of Mexican vanilla beans, hothouse lamb carcasses, and other products.

Fair Condition–(1) A range-condition class; a range producing only 25 to 50 percent of its potential. The cover consists of early maturing plants of low value for forage or for soil protection. (2) Denoting a medium condition of plants, plant products, or animals.

Fair Market Value–In appraising, the summation of the price and terms for which a property should or does exchange within a reasonable time between a seller and an able, willing, and informed buyer with a reasonable cash payment. Also called market value, normal sale value.

Fall–(1) The dropping of a plant part, such as leaves or fruits. (2) The slope of land. (3) The amount of precipitation. (4) (a) The quantity of trees cut, (b) Felling trees. (5) (a) The number of lambs born. (b) Giving birth, as of lambs. (6) One of the three outer segments of an iris flower that is often drooping (usually used in the plural). (7) Autumn. (8) Of, or pertaining to, a plant or fruit that matures in the autumn, as a fall apple.

Fall Short–To fail to attain an expected or desirable yield.

Fallage–(1) The act of cutting trees. (2) Timber that has been felled.

Faller–One who cuts down trees.

Falling Ax–A double-bitted ax designed for felling trees, having a long handle and narrow bits.

Falling Wedge–In lumbering, a wedge that is driven into the cut in a tree behind the saw to direct the fall of the tree away from the saw.

Family–(1) A unit of human organization which varies widely according to the cultural method employed in counting kinship. Generally, it is a local residence unit, such as a household which included male(s) and female(s) of a parent generation and the children or which they are socially responsible. Under other definitions it may include nonlocally resident persons related or assumed to be related by blood through several generations. See Nuclear Family. (2) Any group of related plants or animals that make up a category for classification. It is usually more inclusive than a genus but smaller than an order. (3) See Soil Family. (4) An isolated group of organisms belonging to a single species; often descendants of a single plant.

Family Earnings (Farm)–Compensation for a farm operator's labor and management and the labor of unpaid members of the operator's family. It includes the value of family living from the farm. The sum is computed by deducting an interest charge for the use of farm capital investment from net family farm income.

Family Farm–A farm of such a size and character that it may be operated by the members of the farm family living on it with little if any hired labor and which under good management can return the family a satisfactory living for their effort and investment. (This definition may be considered as establishing the minimum size for a family farm. A family farm may also be described as a two-man farm operated by a father and his son or by the farmer with the aid of one full-time hired man.)

Family Labor–Farm labor that is available within the farm family excluding the farm operator. Its value is determined by the amount of additional labor the operator would have to employ to carry on an equivalent amount of business if family labor were unavailable.

Family Living–The farm value of home-grown food, fuel, and other products used by the farm family, plus the value of the use of the farm dwelling.

Famine–A catastrophic food shortage affecting large numbers of people. Famine is usually the result of crop failure due to drought, flood, insect pests, disease, or war.

Fan–(1) A piece of equipment that will deliver a large volume of air or other gases at low pressures up to one pound per square inch for cooling, heating, etc. (2) Any winnowing machine; specifically, an old term for a kind of basket which, when filled with wheat, was tossed in the air to separate the chaff from the wheat.

Fan-and-Pad Cooling System–A cooling device used in greenhouses. Air is pulled through wet pads by means of fans. As water evaporates, large quantities of heat are absorbed (heat of vaporization).

Fancy–A top-quality grade for many vegetables, fruits, flowers, poultry, and livestock.

Fanning Mill–In seed cleaning, a device equipped with screens to sift out foreign material and a power fan to separate and blow out chaff and other light material.

FAO–See Food and Agriculture Organization.

Farding Deal–The fourth part of an acre of land.

Farm–A place that has annual sales of agricultural products of $1,000 or more.

Farm Accounts–Either of two types of accounts: (a) financial accounts, which are primarily records of inventories and financial transactions plus the production records, usually on a single farm enterprise; (b) cost accounts, which include the following in addition to the financial accounts: feed records; time records of labor, power, and machinery; building records; interest charges; etc., allocated to the different farm enterprises. They may be kept on single enterprises as well as on the entire farm business. Note: The recent popularity of computers on the farm has made possible the keeping of more-detailed accounts.

Farm Appraisal–A definite, written, detailed opinion of the value of a farm that is prepared by a qualified appraiser.

Farm Bloc–A bipartisan group in the United States Congress to secure favorable legislation for the farming industry and farmers as a class.

Farm Budget–(1) A plan for the organization and operation of a farm for a specified period of time, which includes a detailed statement of the anticipated gross income, expenses, and net income. (2) A plan for the future use of land, human labor, power, other resources, and which shows the crops to be grown, the livestock to be kept, and the estimated production, receipts, expenses, and farm income.

Farm Bureau–A free, independent, nongovernmental, voluntary organization of farm and ranch families in the United States and Puerto Rico, united for the purpose of analyzing their problems and formulating action to achieve educational improvement, economic opportunity, and social advancement and dedicated to the preservation of individual opportunity and freedom. It is local, national, and international in its scope and influence, and is nonpartisan, nonsectarian, and nonsecret in character. Farm Bureau policies are determined and implemented in the counties by the members; on the state level by a delegate body comprising representatives chosen by the county Farm Bureau in proportion to their respective membership; at the national level, an annual board of delegates representing the state Farm Bureaus.

Farm Business Record–A record of a year's business on a farm, which includes an inventory of all farm property at the beginning and at the end of the farm year; a record of the number of livestock raised; quantities of livestock products produced; the acreage and production of each crop; and a record of all farm financial transactions, both sales and purchases, made during the year.

Farm Capital–The value of the land, with such permanent improvements as are ordinarily transferred with the title to the land, livestock, machinery and equipment, feed crops, and other supplies which are included in a farm operating unit.

Farm Capital Earnings–Compensation for farm capital that is computed by deducting the value of the operator's labor from net farm income. It is usually expressed as a percentage figure; percent earnings of farm capital; or rate of earnings on the investment.

Farm Cash Surplus–Farm net cash flow adjusted for loans received and interest and principal payments; it represents the amount of cash generated by the farm and available for household use.

Farm Census (United States Census)–Collection of selected information from all the farms every five years. These data are the basis for the "Agricultural Statistics" published by the United States Department of Agriculture.

Farm Conservation Plan–A plan that states a cropping system, livestock system, and land use practices that are designed to reduce the rate of soil erosion and to maintain or improve soil fertility. Usually made with the help of the District Conservationist of the United States Department of Agriculture Soil Conservation Service.

Farm Credit Administration–A credit system established to provide a dependable source of both long- and short-term credit on a sound basis for farmers and stockmen and their cooperative associations. These credit services are supplied in each of the twelve farm credit districts by a federal land bank, a federal intermediate credit bank, and bank for cooperatives. In addition, there is the Central Bank for Cooperatives at Washington, D.C. Farmers obtain individual farm real estate mortgage loans from local National Farm Loan Associations, and loans for production expenses and operating capital from local Production Credit Associations. These associations are largely farm-owned but obtain their financing through the federal land banks and federal intermediate credit banks.

Farm Drainage–The removal, or exclusion, of excess water from farmland by open or closed drains alone or in combination with pumping plants.

Farm Enterprise–A unit of farm business, as a particular crop or a class of livestock.

Farm Equity Capital–Total farm capital less farm borrowings.

Farm Expense–The sum of the annual cash operating expenses, value of unpaid family labor excluding the operator, net decrease in farm inventory including depreciation, and the wages and cost of board furnished to hired laborers.

Farm Flock–Those animals or fowls, collectively by species, which a farmer raises on his farm, especially a small flock of sheep or chickens.

Farm Forestry–The practice of forestry on part of the farmland which is carried on with other farm operations.

Farm Implement–Any mechanical tool used on a farm: e.g., a plow, a tractor.

Farm Income–(1) The net income a farmer receives in operating a farm as a return for the total capital investment and for his/her own labor and management. (2) The collective income of farmers in a county, state, or nation.

Farm Inventory–A list of the amounts and values of all items of farm property as of a given date.

Farm Irrigation Structure–(1) Any structure that is necessary for the proper conveyance and application of irrigation water: e.g., a flume, siphon, suspension, culvert, debris basin, detention dam, etc. These structures require special design, detailed staking and/or on-site assistance. (2) Any small farm irrigation structure for which standard designs are available: e.g., a turnout, check, drop, or measuring device.

Farm Labor–(1) The sum total of all the labor that is used in the operation of a farm. (2) Those people who work on farms as hired hands.

Farm Level–A simple type of surveyor's level, usually consisting of a telescope with crosshairs and a tripod, used in the solution of simple problems of drainage, terracing, irrigation, etc. See Abney Level, Dumpy Level.

Farm Management–(1) The science and art of the organization and operation of farms so as to obtain the maximum amount of continuous net income. It considers the effectiveness of different sizes of operating units and of combinations of productive resources, enterprises, and practices for operating units; programs of adjustment for agricultural areas; and the impact of public policies and programs on economic activities and income on farms. (2) The practical adjustment, organization, and administration of affairs and resources of the farm.

Farm Management Service–An individual or an organization that provides landlords such services as procuring tenants or hired workers and managers for farms, keeping farm records, advising the tenants, workers, or managers, and handling the financial accounts on the farm. Such services are particularly prevalent in the Corn Belt of United States. Also called Commercial Farm Management Services.

Farm Manager–A trained and educated individual who directs the operation of farm enterprises.

Farm Map–An outline map of a farm showing roads, building sites, lanes, and field arrangement, used in farm planning and as a record of the tile lines, lime application, soil tests, etc.

Farm Mechanics–Mechanical activities that are preformed on the farm and in the home, concerning farm power and machinery, farm buildings and conveniences, rural, preparing farm products for sale, storage, or use, and soil and water management.

Farm Net Cash Flow–Farm receipts minus farm payments.

Farm Net Worth–See Farm Equity Capital.

Farm Operator–A person who operates a farm, either by doing or supervising the work and by making the day-to-day operating decisions.

Farm Payments–Cash paid for goods and services purchased for farm use.

Farm Personal Property–The personal, movable property on a farm that is involved in the farm business: e.g., machinery, equipment, livestock, feed, growing crops, crops held for sale, and miscellaneous supplies; also a brooder house or any small building merely resting on the ground that is movable.

Farm Pond–A small reservoir of water on a farm usually formed by constructing a dam across a watercourse or by excavation to collect surface water.

Farm Population–All those persons living on farms, regardless of occupation.

Farm Price–The price a farmer receives for his products at the farm.

Farm Problem–(1) Any problem of farmers collectively on the state or national level: e.g., farm surplus. (2) Any particular problem on a farm: drainage, fertilizer use, marketing, etc.

Farm Program– (1) A plan related to agriculture based upon, or resulting from, the action of the state or national government. (2) A planned program for the operation or management of an individual farm.

Farm Prosperity– A period in which the returns to the farmers are large enough to give them incomes that compare satisfactorily to the incomes of other groups in the United States economy.

Farm Real Estate– The land in a farm including such improvements as the buildings, fences, tile drains, and the permanent features of a water system.

Farm Receipts– The value of cash received from the sale of agricultural output.

Farm Specialization– Any specialization of production activity on the farm, either by commodity, processes, personnel, or geography.

Farm Survey– Data collection from a sample of farms from a given population.

Farm-Retail Price Srpead– The difference between the retail price of a food product and the farm value of an equivalent quantity of food sold by farmers.

Farm-to-Market Road– A hard-surfaced road that serves a farming area and leads to a town or city at which farmers can sell their produce and buy things they need.

Farmer– One who operates a farm, or is engaged in the business of farming; a person who produces agricultural products on a farm. Since there are many different types of farms, a farmer may be designated as a dairy farmer, a cattle farmer, fruit farmer, wheat farmer, sheep farmer, etc. Primarily, a farmer is one who devotes the major portion of his time to farming: i.e., a full-time farmer.

Farmer Cooperative Service– The branch of the United States Department of Agriculture with responsibility for fostering and assisting agricultural cooperatives.

Farmer Owned Lands– Lands owned by a person who operates a farm, either doing the work himself/herself or directly supervising the work.

Farmer-direct Sales– Farm sales of produce made directly to consumers, without the use of traditional middlemen; the farm roadside market is an example.

Farmer-owned Reserve– A federal program designed to encourage farmer storage of wheat and feed grains by providing farmers with low-interest, nonrecourse loans and storage payments.

Farmers Home Administration– An agency of the United States Department of Agriculture that aids farmers financially. The agency also provides emergency loans to farmers in districts suffering from drought, flood, or other disasters. It operates through offices located in all 3,197 counties throughout all the fifty states in the United States. Formerly called Farm Security Administration.

Farmers Institutes– A lyceum type of organization similar to the early teachers institutes which was organized to bring scientific information for better agricultural practices directly to farmers in the United States. They were first suggested by the Massachusetts State Board of Agriculture in 1852. By 1899 over 2,000 institutes were held annually in forty-seven states, with an attendance of more than 500,000. They were the forerunners of the Cooperative Agricultural Extension Service that offers a year-round educational service to all farmers. See Cooperative Extension Service.

Farmers' Market– A building or open space, usually located in a town or city, where farmers may bring their produce for direct sale to the consumer.

Farmers' Share– The farm value of food expressed as a percentage of its retail price.

Farmers' Square– Farmers' market.

Farmers' Union– An abbreviation for the Farmers' Educational and Cooperative Union of America, an organization founded in Texas in 1902. The Farmers' Union is organized in local community unions, county unions, state and territorial unions, and a national union. Among its objectives are those of maintaining and protecting the family-type farm, of expanding local and regional farmers' cooperatives, and of educating farm families in the economic, social, and cultural problems of agriculture. Also called the National Farmers' Union.

Farmhand– A man who is hired to work on a farm by the hour, day, week, month, or year. Also called hired hand.

Farming Operation– (1) The sum total of the activities of a farmer in operating a farm. (2) Any one single activity on the farm; e.g., harvesting, planting, cultivating, fencing, etc.

Farmland– Land that is, or is capable of, being used for raising farm crops without further clearing or draining.

Farmland, Prime– Cropland, pastureland, forestland, or rangeland that has physical and chemical characteristics suitable for producing food, feed, forage, or fiber. Rainfall or irrigation and favorable temperature must be present for economical production with modern farm management. See Farmland, Unique; Soil Survey.

Farmland, Unique– Land other than prime farmland that is used for the production of specific high-value food and fiber crops, such as citrus, cranberries, fruits, vegetables, and olives. See Farmland, Prime; Soil Survey.

Farrier– (1) A person who shoes horses. (2) An obsolete term for a horse and cattle veterinarian.

Farrow– (1) To give birth to a litter of pigs. (2) A litter of pigs.

Farrow to Finish– A type of farm operation that covers all aspects of breeding, farrowing, and raising pigs to slaughter.

Farrowing Crate– A crate or cage in which a sow is placed at time of farrowing. The crate is so constructed as to prevent the sow from turning around or crushing the newborn pigs as she lies down.

Farrowing House– Any of several types of structures especially designed for a sow and her litter of pigs.

FAS– See Foreign Agricultural Service.

Fatigue– (1) A weakness of wood or metal that results from a repeated reversal of the load. (2) Loss of energy in people or animals form any of a multitude of causes.

FCIC–Federal Crop Insurance Corporation.

FCM–Fat-corrected milk. A means of evaluating milk-production records of different animals and breeds on a common basis energy-wise. The following formula is used: FCM = $0.4 \times$ milk production – ($15 \times$ pounds of fat produced).

FDA–See Food and Drug Administration.

Federal Crop Insurance Act–A part of the Agricultural Adjustment Act of 1933 designed to provide crop insurance. The first insurance corporation under the act was empowered to provide insurance for loss of wheat crops.

Federal Crop Insurance Corporation–This is a wholly owned government corporation, within the Department of Agriculture, created to promote the national welfare by improving the economic stability of agriculture through a sound system of crop insurance and providing the means for research and experience helpful in devising and establishing such insurance on a national basis. Crop insurance offered to agricultural producers by the corporation provides protection from losses caused by unavoidable natural hazards such as insect and wildlife damage, plant diseases, fire, drought, flood, wind, and other weather conditions. It does not indemnify producers from losses resulting from negligence or failure to observe good farming practices.

Federal Extension Service–A United States agency created February 26, 1923, to assume leadership of all general educational programs in cooperation with the land-grant colleges and universities of the states and territories and the county governments.

Federal Food Stamp Plan–A plan established in 1939 by the United States Department of Agriculture to dispose of surplus agricultural commodities and to aid low-income families.

Federal Inspection–The inspection by United States Department of Agriculture appointed inspectors of slaughter animals and carcasses, fruits, vegetables, and processed products intended for human consumption, to determine if the animals or products meet established specifications.

Federal Land Bank–Any of several banks, which is a part of the Farm Credit Administration of the United States. The Federal Land Banks through local national farm loan associations make long-term (5 to 40 years), amortized, first-mortgage, farm real estate loans. The Farm Credit Administration divides the United States into twelve districts and each district has a Federal Land Bank. In turn, each district has a number of national farm loan associations. The system is entirely member-owned and does not use government funds in its operations. Money for the loans is obtained from the sale of Federal Land Bank bonds that are secured by the real estate covered by the first mortgages. Members of the Board of Directors of the Farm Credit Administration and of the different districts and associations are usually elected by the member-borrowers.

Federated Cooperative–A group of cooperatives receiving goods and services from a single source (cooperatives) organized for this purpose.

Fee–(1) A charge made for livestock grazing on the basis of a certain rate per head for a certain period of time. (2) A charge made for any services as stud fee, marketing charge, loan fee, etc.

Feed Additive Compendium–A publication that lists feed additives in current use and the regulations for their use.

Feed Analysis–The chemical or material analysis of a commercially prepared feed, printed on a tag and fastened to the bag in which the feed is to be sold.

Feed Bunk–A forage and grain feeding station.

Feed Composition Table–A table showing the nutrients found in feeds.

Feed Conversion Ratio–The rate at which an animal converts feed to meat. If an animal requires four pounds of feed to gain one pound, it is said to have a four to one (4:1) feed conversion ratio.

Feed Efficiency–Term for the number of pounds of feed required for an animal to gain one pound of weight; e.g., 6.5 pounds of feed per pound of gain.

Feed Energy Utilization–The percentage of the energy obtained from a feed that is used for an animal's bodily functions. For example, the feed energy utilization for an average lactating cow is approximately: 30 percent for fecal energy, 20 percent for heat energy, 20 percent for maintenance, 20 percent for milk production, 5 percent for urinary energy, and 5 percent for gaseous energy.

Feed Hopper–A container that allows feed to drop down gradually as more is consumed.

Feed Lot–The enclosed area in which animals are fed for fattening and finishing. Also called feed yard.

Feed Out–To prepare animals for market by fattening.

Feed Reserve–Feed harvested and stored for future use, standing forage cured on the range, or pasture for future use.

Feed Roll(ers)–The pairs of rollers directly in front of a forage harvester cutterhead that regulate the plant material flow to the cutterhead. Adjusting roll speed regulates the length of cut.

Feed Truck–A large truck used to transport feed from the mill to the farm. The feed is unloaded into the storage bin by an auger. (2) A large hand-pushed cart used for distributing feed to animals.

Feed Unit–One pound of corn, or its equivalent in feed value in other feeds, which is fed to cattle under normal farm conditions.

Feeder Buyer–(1) One who buys cattle, sheep, or horses, fattens them and offers them for sale. (2) A buyer of feeder livestock.

Feeders Margin–The difference between the cost per hundredweight of feeder animals and the selling price per hundredweight of the same animals when finished.

Feeding Ratio–Weight of food consumed divided by increase in weight of an animal, during a given time interval.

Feeding Root–A young root, or rootlet, bearing the root hairs that absorb water and mineral nutrients from the soil. Also called Feeder Root.

Feeding Standard–Established standards that state the amounts of nutrients that should be provided in rations for farm animals of various ages and classes in an attempt more nearly to attain the optimum economy of growth, gain, or production.

Feeding Value–A term referring to the nutritive value of different feeds, i.e., expressing the amount of nutrients furnished by each feed and the degree of their digestibility.

Feedlot ADG/Gain on Test–A measure of the ability of cattle to gain weight when fed high-energy rations, usually expressed as Average Daily Gain.

Feeler Gauge–A metal strip or blade used for measuring the clearance between parts of a machine, etc. Such gauges usually come in a set of different blades graduated in thickness by increments of 0.0001 inch.

Felling Head–A tractor-powered implement that grasps a tree and cuts it using a circular bar that has a cutting chain running around the perimeter.

Fellmonger–A dealer in pelts who pulls the wool from the skins, scours the wool, and tans or pickles the skins.

Felly–The outer part or the rim of a wheel to which spokes are attached. Also spelled felloe.

Female Threads–Threads on the inside of a receiving object such as a nut. See Bolt, Male Threads.

Fence–(1) A hedge or barrier of wood, metal, stone, or plants erected to enclose an area to prevent trespassing or the straying of animals. (2) To enclose an area with a fence. See Drift Fence, Snow Fence.

Fence Line Feed Bunk–A multipurpose structure designed for feeding roughage along a fence in the open. It is primarily a feed bunk with openings through which the animals may feed. The feed may be distributed by dump trucks or feed carriers operated outside of the enclosed area.

Fence Pliers–A tool used to cut, bend, and twist wire, remove staples, and do other jobs associated with fence building.

Fence Stretcher–A ratcheting tool used to tighten wire in fence building or repair.

Fencing–(1) Material used in the construction of fences. (2) The use of fences.

Fertilizer Ratio–The relative proportions of primary nutrients in a fertilizer grade divided by the highest common divisor for that grade; e.g., grades 10-6-4 and 20-12-8 have the same ratio of 5-3-2.

Fertilizer Regulatory Service–Usually a service administered by the fifty respective State Departments of Agriculture.

Fertilizer Requirement–The quantity of certain plant nutrient elements needed as determined by a soil test, in addition to the amount supplied by the soil, to increase plant growth to a designated optimum yield. See Yield Goal.

Fertilizer Unit–One percent (20 pounds) of a short ton (2,000 pounds) of fertilizer.

Fertilizer Abstracts–A monthly journal published by the National Fertilizer Development Center, TVA, Muscle Shoals, Alabama, containing summaries of current literature on technology, marketing, use, and related research on fertilizers.

FFA–An organization for students enrolled in high school agricultural education programs. The purpose of the organization is to teach leadership and to provide incentives for learning. Once known as the Future Farmers of America, the organization is now known as the National FFA Organization. See Future Farmers of America.

FFA Alumni–A national organization composed of former members and supporters of the National FFA Organization. The purpose is to promote and support agricultural education programs and in particular, programs of the FFA. See FFA; Agricultural Education.

FGIS–Federal Grain Inspection Service, U.S.D.A.

Fibrillating Current–Excessive current in an electric fence that on contact either stops an animal's breathing by paralyzing the chest muscles or stops the flow of blood.

Field–An area of agricultural land devoted to the production of farm crops, e.g., cereals, vegetables, etc., or an area in which crop-growing has been postponed or abandoned.

Field Capacity–The amount of water held in a saturated soil after the excess or gravitational water has drained away. Also called capillary capacity.

Field Conditioner–A heavy, wheeled, spring-tooth harrow for secondary tillage used to smooth the soil prior to planting.

Field Crate–A container, holding from 10 to 25 kilograms, used to haul fresh fruits or vegetables to the storage or packing house.

Field Cuber–A machine used to produce cubes directly from windrowed hay.

Field Ell–A right-angled pipe joint for main or lateral pipelines to change the direction of the water used in irrigation.

Field Hand–A man employed to work in the fields on a farm. Also called hired man, farm hand.

Field Run–Designating products (potatoes, onions, etc.) harvested in the field and ungraded as to size or quality.

Field Testing–Testing crop varieties, fertility of soils, etc., to determine their effect on yield or quality.

Fifteen-atmosphere Percentage–The moisture percentage, dry-weight basis, of a soil sample that has been wetted and brought to equilibrium in a pressure-membrane apparatus at a pressure of 221 psi (pounds per square inch). This characteristic moisture value for soils approximates the lower limit of water available for plant growth.

File–A steel instrument whose surface is covered with sharp-edged furrows or teeth, used for sharpening, abrading, or smoothing other substances, as metals, wood, tools, etc.

Filterable–Capable of passing through the pores of a filter.

Finance Budget–A budget constructed to show the extent of necessary borrowings and the manner in which interest and principal payments on loans advanced are to be met.

Financial Statement–A statement usually required by a lender from an applicant for credit. It is a list of assets and liabilities of an individual or a business at a specific time. The difference between the assets and liabilities is the net worth at that particular time. The difference between the net worth at the beginning of the year and at the end of the year represents the financial gain or loss for the year.

Fireman–In forestry, a fire guard whose principal function is suppression of fires, and who usually stays at a fixed position awaiting orders to go to a fire. Also called smokechaser.

Firkin–(1) A measure of weight, used for butter and cheese, of 56 pounds avoirdupois. (2) A wooden cask or tub used for holding butter, fish, etc.

Firm–(1) An economic unit recognized to be engaged primarily in production. (2) (a) To compact the soil, crushing and pulverizing the lumps to facilitate capillary water movement. (b) Designating well-compacted soil that is not lumpy or powdery. (2) In marketing, designating optimistic conditions. (4) Designating a cheese that feels solid. (5) Designating whites of eggs that are sufficiently viscous to prevent free movement of the yolk. (6) (a) Designating meat that is not soft or soggy. (b) Designating a fruit or vegetable that is not overripe or shriveled.

Fiscal Year–Any accounting period that consists of twelve successive calendar months, fifty-two weeks, or thirteen four-week periods.

Fission–(1) A form of reproduction, common among bacteria and protozoa, in which a unit or organism splits into two or more whole units. (2) The splitting of a heavy nucleus such as uranium or plutonium into approximately equal parts, accompanied by the conversion of mass into energy, the release of the energy, and the production of neutrons and gamma rays.

Fit–(1) To notch a tree for felling, and to mark it into log lengths after it is felled. (2) To ring, slit, and peel tanbark. (3) To file and set a saw. (4) To condition livestock for use, sale, or exhibit. (5) To prepare land for sowing, i.e., plowing (or disking), harrowing, and rolling. Land so fitted should have no large, hard clods so that the seeds may be placed at a uniform depth and in contact with moist soil.

Fixed Assets–Assets that can not be converted to cash immediately. Lands, barns, and fences are examples of fixed assets.

Fixed Capital–Capital that is invested in land, buildings, fences, tile drains, wells, and other relatively permanent fixtures.

Fixed Cost–That cost which does not change with volume of production.

Fixed Expenses–Expenses that do not change with output. Examples are taxes and depreciation.

Fixed Groundwater–Groundwater held in saturated material with interstices so small that it is attached to the pore walls, and is usually not available as a source of water for pumping. See Aquifer.

Flail–(1) A power implement with swinging knives or blades on a rotating horizontal shaft. The material is usually cut into several pieces as it is struck by succeeding flails and carried from a standing position to the discharge point. (2) A hand tool used for threshing grain. To use a flail to thresh grain.

Flash Point–The specific temperature at which a flammable material will vaporize sufficiently to ignite when touched with a spark or flame.

Flexibility–(1) The quality of bending without breaking. (2) Adjustment to meet changing conditions; essential for farming and ranching.

Flexible Support–A policy of the United States government that favors adjustment of subsidies of price supports for agricultural products as conditions of supply change, in contrast to a policy of fixed or rigid supports of prices.

Flexible Tine Cultivator–A cultivator with sweeps on a vibrant, curved, spring steel shank. It usually works deeper than spike or tine-tooth harrows.

Float–(1) An instrument used for filing animals' teeth. (2) A valve in the cream separator to regulate the flow of milk into the cream-separator bowl; or a similar contrivance for maintaining the desired level of water in a tank. (3) A drag or device for leveling soil. (4) To come to the top of the ground, as with a certain type of plow which automatically adjusts if the bottom strikes an obstruction. (5) To flood irrigate, as in floating a meadow (rare in the United States).

Flock Book–The record of breeding and ancestry of sheep, kept privately by the flock owner or officially by the Sheep Breed Association, to register purebred sheep. See Herd Book.

Flood Classification–Floods are classified by recurrence magnitude, as 100-year flood, 50-year flood, etc.

Flood Control–(1) Any of the various agricultural practices that hold the soil in place and increase its infiltration and water-retention capacity. These practices include the planting of deep-rooted grasses, trees, and shrubs: no-till crop farming; contouring; terracing; and keeping heavy machinery off the soil when it is too wet to support such loads. (2) Any of the various engineering practices used to reduce flood damage resulting from overflowing rivers and streams. It includes deepening the existing channels, digging diversion canals, building levees and dams to impound the water in numerous small ponds or retention basins in the upper parts of the watershed, or by large detention basins farther downstream.

Flood Control Act–An act of United States Congress passed in 1944 to control floods in the Missouri River basin.

Flood Relief–(1) Assistance given by the International Red Cross, local governments, etc., to persons who have suffered loss in floods. (2) Flood control.

Flood Stage–The stage in which the level of the water rises above the tops of stream banks or dikes. Frequently, an arbitrary level established, beyond which the stream or river is considered to be at flood stage.

Florist–A person who sells plants and cut flowers, often growing these plants and flowers himself/herself. Sales of accessories, such as fungicides, insecticides, seeds, fertilizers, pots, pot labels, etc., are often a part of the florist's service.

Flotation– (1) The ability of tractor or implement tires to stay on the top of the soil surface. It is usually related to soil condition, tractor or implement weight, and contact area between tires and soil surface. (2) Separation of materials by mixing with a liquid of a different specific gravity; one of which will float, the other sink. Overmature peas may be separated from green peas in this manner; frozen oranges may be separated from sound fruit in running water. See Floats, Flotation Concentrates.

Flue– (1) A vertical space, usually 6 inches or less in width, between two adjacent tiers of stock in a lumber pile. (2) A square or round ventilating passageway. (3) The passage in a chimney for conducting smoke, flame, or hot air. (4) The special passageways for heated air in a barn for curing tobacco.

Fluid– A liquid or a gas composed of particles that freely change their relative position without separating. See Fluid Fertilizer.

Fluid Drilling– A technique for planting germinated seeds without injuring the tiny shoots or roots by mixing previously germinated seed with a protective gel and pumping it into the soil behind the planter shoe through tubing, usually plastic, from a holding tank on the planter. Originally developed at Great Britain's National Vegetable Research Station, the technique was first used commercially in the United States to fluid-drill tomato seedlings.

Fluid Drive– A means of power transmission by the use of a pair of vaned rotating elements held in position close to each other without touching. Rotation is transmitted to the driven part by the driving part through the resistance of a body of oil between the parts.

Fluid Fertilizers– Liquid, slurry, or suspension fertilizers.

Fluidity– The property of flowing possessed by gases and liquids. Fluidity and viscosity are opposite (inverse) characteristics. See Viscosity.

Flux– (1) State of change. (2) Substance that reduces the melting point of a mixture. (3) Passage across a physical boundary such as CO_2 from atmosphere to hydrosphere, or across a chemical boundary as CO_2 from atmosphere to organic matter.

FMD– Foot-and-mouth disease.

FmHA– Farmers Home Administration, USDA.

FNS– Food and Nutrition Service.

Foil– (1) A thin metal 0.006 inch (0.1524 millimeter) thick, such as aluminum foil. Materials with greater thickness are known as a sheet.

Food Additive– Any substance that is added to food, either directly or indirectly, to improve nutrition, taste, or shelf life.

Food and Agriculture Organization (FAO)– An agency of the United Nations, formally organized in 1945, for supporting worldwide studies of agricultural production and food supplies. It is also empowered to lend technical assistance to nations in the solution of problems in agricultural production and human nutrition.

Food and Drug Administration (FDA)– A division of the United States Department of Health and Human Services, charged with the enforcement of federal laws and inspection services pertaining to adulteration and misbranding of foods, feeds, and drugs in intrastate commerce.

Food Constituent– Any essential nutrient (protein, carbohydrate, fat, etc.) contained in a food.

Food Extender– A cheaper product that can be added to a traditional food to expand its volume (e.g., "Hamburger Helper").

Food Poisoning– Any sudden, painful, intestinal disorder in people and animals caused by some harmful bacteria or fungal organism taken into the body with the food. The organism multiplies in the alimentary canal and produces poisonous substances that may bring about death. ***Clostridium botulinum***, one of the bacteria, will cause the disease known as botulism. Also known as food intoxication.

Food Preservation– The treatment of foods in various ways to keep them wholesome and free from decomposition; e.g., refrigeration, canning, smoking, adding of certain antiseptic chemicals, drying, exclusion of air, and irridiation.

Food Technology– The science of food qualities, values, preparation, processing, and marketing.

Foot– (1) In people and quadrupeds, the terminal portion of the leg that rests upon the ground. (2) The portion of a cultivator to which the sweep is attached. (3) The organs of locomotion of various invertebrates, as the feet of a caterpillar, the foot of a clam or a snail, etc. (4) The base of a tree, tower, mountain, wall, hill, etc. (5) A unit of linear measure, 12 inches.

Foot Pound– A measure of the amount of energy or work needed to lift one pound the distance of one foot.

Foot-candle– Standard measure of light. The light of one candle falling on a surface 1 foot away from the candle.

Footing– (1) The rock or soil on which a foundation rests. (2) The condition of the ground surface as it relates to the pulling capacity of a draft animal without slipping.

Forage Blower– A fan type conveyer used for placing forage into storage structures.

Forage Box– A self unloading wagon box for handling silage, haylage, green chop, ear corn, etc. Most forage boxes unload either through a front conveyer or a rear door.

Forage Harvester– A machine that chops hay or forage into short lengths for easy storage or handling. With different header attachments they can cut standing crops, pick up windrows, snap ear corn, or gather stover.

Forage Value– (1) The relative importance for grazing purposes of a range plant or plants, as a whole, on a range. (2) The rank of a range plant or plant type for grazing animals under proper management, preferably expressed as proper-use factor. See Grazing Value. (3) The comparative value of the forage portion of an animal's ration compared with the other feeds in the ration.

Forage Volume– (1) That portion of a plant above the ground and within reach of grazing animals. (2) The measure of the forage crops; i.e., the aggregate amount of forage produced on a range area during any one year.

Forage Wagon–A four-wheel wagon that receives forage from a forage harvester and transports the forage. The forage is unloaded using a conveyor and/or auger.

Forbes Scale–*Aspidiotus forbesi*, family Diaspididae; a scale insect that occurs on fruit trees and bushes in the eastern United States. It is the chief apple scale in some regions.

Force–Action that attempts to change or changes the motion of a body, usually expressed in pounds per square inch or pascals. One pound per square inch = 6,894.757 pascals.

Forecast–(1) A statement of the anticipated weather conditions in a given region, usually for a period of 12, 24, or 36 hours; 5 days or 30 days. (2) The predicted future price of a product.

Foreclosure–A procedure whereby a lender takes steps toward attaining ownership of property given as security for a loan. The usual steps are: default on the mortgage or contract by the buyer; court proceedings; and sale of the land by the sheriff in case of a real estate mortgage after expiration of the legal redemption period.

Foreign Agricultural Market Development–Includes all activities—in the United States as well as in foreign countries—to influence the flow of United States farm products to foreign consumers. Market development encompasses measures by government agencies, farmers, processors, exporters, and others to improve acceptability of United States farm commodities in the foreign markets. It includes such services as information and grading. It includes government and international agencies to lower trade barriers. It involves market promotion activities carried on cooperatively by government and trade groups, such as participation in trade fairs, advertising, distribution of samples, sponsored visits of foreign buyers to the United States, and the like. Market development, in brief, covers activities undertaken all the way along the "marketing chain."

Foreign Agricultural Service–An agency of the United States Department of Agriculture which is responsible for developing foreign markets for United States farm products.

Foreign Currencies–An expression used frequently in connection with operations under Title I, public law (P.L.) 480, which authorized sales of surplus United States farm products for foreign currencies. Title I also spells out uses to markets for United States farm products; to obtain military equipment, materials, facilities, and services; to finance the purchase of goods or services for other friendly countries; to promote economic development and trade of developing countries; to pay United States obligations abroad; to finance educational exchange activities; and others.

Foreign Market–A market outside the boundaries of the United States.

Forest Economics–The money aspects of forest management that deals with the needs, values, and policies of forest lands in relation to public or individual gains. See Forest Management.

Forest Experiment Station–A United States Forest Service organization that consists of research stations in each major forest region. It conducts research on the protection, management, use, and survey of forest resources. Created within the McNary-McSweeney Act, 1928, and administered under the Secretary of the United States Department of Agriculture.

Forest Industry Lands–Lands owned by companies or individuals operating wood-using plants.

Forest Land–(1) Land at least 10 percent occupied by forest trees of any size, or formerly having had such tree cover and not currently developed for nonforest use. (2) Lands that are at least 10 percent stocked by trees capable of producing timber or other wood products or that exert an influence on the climate or water regime.

Forest Management–The application of business methods and technical principles to the operator of a forest property. It involves the computation of income from forest lands; the establishment of cutting cycles; the conservation of cover, land and water; and the formulation and conduct of long-range plans of operations.

Forester–A person professionally trained in the science of managing trees and woodlands.

Fork–(1) Any of various tools or implements consisting of a handle, shank, and two or more tines that is used for digging soil, grasping or lifting hay, manure, etc. (2) The juncture of two roads, rivers, or tree branches, etc.

Fork Lift–A vehicle with two forks or runners attached to the front that is used to lift and transport heavy loads.

Formula–Any general equation; a rule, or principle expressed in algebraic symbols; e.g., the statement of the quantity and kind of stock materials that are used in making a fertilizer, feed, or other mixture.

Formulation–A term used synonymously with product. It contains the pesticide or fertilizer in a form that can be: (a) dissolved or suspended in a carrier and distributed in solution or suspension by sprayers, (b) distributed dry by dusters or spreaders, or (c) easily vaporized for application.

Forty–Forty acres of land, as a forty, one-sixteenth of a section of land.

Forward Price Contract–A contract farmers may enter into, prior to the harvesting of a crop, which fixes the price in advance.

Forward Pricing–A method of price support designed to announce a guaranteed minimum price far enough in advance so that production can be adjusted. As contrasted with parity prices, forward prices would be set close to the expected free market price under reasonably full employment conditions.

Fossil Fuel–A deposit of organic material containing stored solar energy that can be used as fuel. The most important are coal, natural gas, and oil; oil shale and tar sand have future potential as fossil fuels.

Foulbrood–A general name for infectious diseases of immature bees that cause them to die and their remains to smell bad. The term most often refers to American Foulbrood. See American Foulbrood, European Foulbrood.

Found–Room, board, and laundry, as well as necessary equipment for a hired farm/ranch worker that may be offered in addition to, or as part of, his/her salary.

Four-Cycle (Stroke) Engine–An internal combustion engine that takes four movements of the piston to complete the power cycle. Fuel is brought into the cylinder, ignited, power is produced, and exhaust fumes are removed in the completion of the piston moving up and down two times. See Two-cycle Engine.

Four-Wheel-Drive Tractor–A tractor that will engage power to all four wheels instead of the conventional two; in other words, the engine turns all four wheels. Four-wheel-drive tractors are usually more powerful and have better traction than conventional tractors.

Fractional Distillation–A process of separating alcohol and water (or other mixtures).

Fraud–A misstatement of a material fact made with intent to deceive or made with reckless disregard of the truth, and that actually does deceive.

Free Market–A marketplace with minimum direct involvement of government in market decisions.

Free-ranging–Allowing animals, especially poultry, to roam freely and eat as they wish without any sort of confinement.

Free-wheeling Clutch–A device on a tractor or other machine that engages the driving member to deliver motion to a driven member in one direction but not in the other direction. Also known as an overrunning clutch.

Freehold–An estate in land held for life.

Freeze–(1) A condition of the weather in which the air temperature at plant level falls to 32°F (0°C) or lower, with the result that tender plants are frozen. See Frost. (2) To reach the temperature of freezing, 32°F (0°C). (3) Of a plant, to die or be impaired as a result of cold. (4) To preserve food products by rapidly reducing the temperature to about 0°F and maintaining the temperature well below 32°F.

Freeze Drying–A method of drying in which the material is frozen and a high vacuum applied. The cooling effect of the evaporation keeps the material frozen while the water distills off as a vapor. Freeze-dried material is very porous and occupies the same volume as the original. The process is applied to foods with advantage as they reconstitute rapidly, with a minimum loss of flavor and texture.

Freezer Burn–(1) An undesirable and unattractive condition frequently found with dressed poultry that is kept or stored in quick or deep freezers; caused by an improper control of temperatures and humidity within the freezer. (2) Small, white dehydrated areas that occur on improperly wrapped frozen foods.

Freight Ton–See Ton.

Froe–See Frow.

Front-end Loader–An implement mounted to the front end of a tractor. It is raised and lowered hydraulically and is used to load materials such as feed, fertilizer, manure, hay, soil, etc., onto a truck or trailer.

Frontage–That portion of real estate which lies immediately adjacent to a stream, lake, street, or highway.

Frost Alarm–An alarm that rings by an electric mechanism when the temperature falls to a predetermined level near freezing. It is used as a warning of impending frost.

Frow–A cleaving tool with a wedge-shaped blade used by coopers in splitting staves for casts, etc., and in making shingles and clapboards. The handle is at right angles to the back of the blade and is held in the left hand, while the mallet that drives the wedge into the log is held in the right hand. Also spelled froe.

FS–Forest Service, USDA.

FSW–Farm Sale Weight.

Fuel Cell–A device for converting chemical energy into electrical energy.

Fuel Loading–In forestry, the amount of fuel (burnable materials) expressed as weight of fuel per unit area; generally expressed in tons per acre.

Fuel Management–The practice of planning and executing the treatment or control of living and dead vegetative material, primarily for wildfire hazard reduction.

Fuel-moisture Indicator Stick–A scientifically selected wooden stick of known moisture-absorbing properties used to determine relative flammability of forest fuels. Of known dry weight, the stick is exposed to the weather on the forest floor, weighted periodically to determine absorption of moisture, and the gain or loss in moisture to predict forest fire hazard.

Functional Efficiency–In cattle, the production of as much good red meat per unit area as possible. In its broadest sense: fertility, genetic excellence, libido, ability to copulate, estrus, ovulation, fertilization, embryo survival gestation., parturition, and mothering ability of the cow.

Furlong–A distance equal to 1/8 mile; 40 rods; 220 yards; 660 feet; or 201.17 meters.

Furnish–To supply a farmer, especially a sharecropper, with food for his family, seeds, fertilizer, feed, etc., to enable him to raise a crop. The various items are charged to his account, and after the crop is harvested, the account with interest must be paid before any money received from the crop is paid to the farmer. It has generally been proven an unsatisfactory system and has largely fallen into disuses. See Sharecropper.

Furrow Slice–The soil that is cut, raised, and inverted by the moldboard plow.

Furrow Wheel–A tractor or implement wheel that runs in a furrow from a previous implement pass. This helps the operator guide the implement in the proper direction and distance from the previous pass.

Fuse–A replaceable safety device that prevents the overload of an electrical circuit. Fuses are used in the wiring circuits of a building or in the electrical system of a machine.

Fusel Oil–A clear, colorless, poisonous liquid mixture of alcohols obtained as a by-product of grain fermentation; generally amyl, isoamyl, propyl, isopropyl, butyl, isobutyl alcohols, and acetic and lactic acids.

Fusion–(1) The combination of certain light nuclei, such as deuterium and tritium, forming a heavier nucleus and releasing energy.

(2) Act of melting or rendering liquid by heat. (3) State of being melted or dissolved by heat. (4) Union or blending of things as if melted together.

Future Farmers of America (FFA)–The national organization of students enrolled in high school agricultural education programs. The national FFA organization is composed of state FFA associations; state associations are composed of local FFA chapters. The FFA chapters are located in public school offering instruction in agricultural education. The FFA activities are an integral part of other instructional programs under provisions of the national vocational education acts. The primary purpose of this youth organization is to develop leadership, cooperation, and citizenship. Collegiate FFA chapters are located in colleges and universities for students in agricultural education programs. In 1988, delegates to the National Convention voted to change the name to the National FFA Organization. See FFA.

Futures Contract–An agreement between two people, one who sells and agrees to deliver and one who buys and agrees to receive a certain kind, quality, and quantity of product to be delivered during a specified delivery month at a specified price.

Futures Market–A market at which contracts for future delivery of a commodity are bought and sold.

Futures Price–The current price of a futures contract.

Futures Trading–A means of buying or selling agricultural commodities that are to be produced or are in the process of being produced.

Gallon–(1) A liquid measure: in the United States, 8.3359 pounds of water at 40°F (4°C), 231 cubic inches (3.7853 liters). (2) A dry measure, in some areas, for the sale of berries, cherries, etc., of 268.8 cubic inches.

Gallonage–The number of gallons of a mixed or finished spray that is used to cover an acre, a tree, or other unit.

Gallons per Minute–An index or unit of flow that is used to indicate the rate of flow of water, especially in pump drainage and in irrigation, as 1,000 gallons per minute. One gallon per minute = 0.227 cubic meters per hour = 0.00227 hectarecentimeters per hour.

Galton's Law–The theory of inheritance expounded by Sir Francis Galton (1822-1911). According to this genetic theory, the individual's inheritance is determined as follows: one-fourth by its sire and one-fourth by its dam; one-sixteenth by each of the four grandparents; one-sixteenth by each of the eight great grandparents; and on and on, with each ancestor contributing just one-fourth as much to the total inheritance as did the one a generation nearer to the individual. Galton's law is correct in the sense that the relationship between ancestor and descendant is halved with each additional generation that intervenes between them. It is not correct in the sense that the individual's heredity is completely determined by the heredity of its ancestors. Rather, in a random-bred population, the individual is one-fourth determined by each parent and one-half determined by chance in Mendelian segregation. Determination by more remote ancestors is included in the determination by the parent. Galton's law is often used as a stamina index by thoroughbred breeders.

Galvanize–(Named after Luigi Galvani, 1737-1798, an Italian physiologist) (1) The coating of iron with zinc (by the electroplating process) to preserve it. (2) To stimulate muscular action by electricity. (3) To rouse to action; to startle.

Gang Bolt–The shaft on which the blades of a disk harrow are mounted.

Gang Disk Plow–A disk plow that has two or more sets of disks.

Gang Plow–A turning plow with two or more bottoms attached to the frame.

Gauge Wheel–A wheel mounted on a tillage implement or planter to control the working depth and to improve stability.

Gear–(1) A round cylinder, ring, or cone-shaped machine part having teeth on one or more surfaces that mate with and engage the teeth on another part. (2) Any or all of the tools, materials, etc., that go together to care for a horse or other animal.

Gear Ratio–In a machine, the ratio of the number of teeth on the larger gear to the number of teeth on the smaller gear.

Gear Sack–A type of duffle bag used by cowboys to carry their personal belongings when traveling on horseback. It is usually tied behind the saddle.

General Land Office–Formerly, a division of the United States Department of the Interior, constituted by Congress in 1812, which had charge of public lands. In 1946 it was combined with the United States Grazing Service to form the Bureau of Land Management.

Generator–A device that converts mechanical energy into electrical energy.

Genetic Index–An estimate of the future Predicted Difference of a young bull. See Predicted Difference.

Genetic Trait Summary (GTS)–The comparative ranking of beef sires derived from evaluating the conformation of the daughters.

Geological Survey–A United States government service established in 1879 for the purpose of: making topographical maps; mapping earth structures; studying groundwater; and appraising energy and mineral resources.

Gill–(1) A unit of liquid measure equal to: 4 fluid ounces, 0.25 liquid pints, 0.118 liters. (2) The breathing mechanism of an aquatic animal such as a fish.

Gin–(1) The entire machine processes, considered as a unit, which remove cotton lint from the cottonseed and bales lint cotton. (2) The grounds, buildings, management, etc., of a gin. Also called cotton gin. (3) The alcoholic beverage obtained by distilling pure spirits over various flavoring materials, principally juniper berries, but also coriander seed, angelica root, and other plant products.

Glean–To gather the useful bits left in the field after harvest.

Gleaners–(Ancient Order of) A rural, fraternal organization founded in Caro, Michigan, in 1894. One of its principal purposes was the promotion of cooperatives for the benefit of farmers.

Glebe–An archaic and poetic term for land assigned to a clergyman for his use.

Glut on the Market–Designating an oversupply of a product on the market usually causing a depression in its price.

GNP–Gross National Product.

Go-devil–(1) Any of several types of sled cultivators or of various other similar implements used on farms. (2) A rough sled used for hauling a log out of a woods; one end of the log is placed on the sled and the other allowed to drag on the ground. (3) A stoneboat. (4) Any old car made over into a tractorlike machine. (5) Any homemade device that is particularly handy for a given job.

Goatherd–A person who tends a flock of goats. See Shepherd.

Governing Body, Conservation District–The appointed or elected supervisors (directors or cmmissioners) of a (soil or soil and water) conservation district established according to state law. Name of district, number and method of naming members of governing body, and tenure vary with state laws; usual number is five.

Government Farm Credit–Farm credit provided by the federal government, United States, particularly through an agency such as the Farmer's Home Administration or the Commodity Credit Corporation. It differs from credit provided by a government-sponsored farm credit agency, such as the Farm Credit Administration, in that units of the agency are member-owned and obtain their loan funds from the sale of bonds and debentures on the open money market.

Governor–A device on the engine of a tractor or other machine that controls and regulates the speed of the engine.

GPA–Gallons per acre.

GPM–Gallons per minute.

Grade–(1) The slope of a road, channel, or natural ground. (2) The finished slope of a prepared surface of a canal bed, roadbed, top of embankment, or bottom of excavation. (3) Any surface that is prepared for the support of a conduit, paving, ties, rails, etc. (4) Any animal that has one purebred parent and one of unknown or mixed breeding. (5) Designating a herd, flock, brand, etc., of such animals. (6) The classification of a product, animal, etc., by standards of uniformity, size, trueness to type, freedom from blemish or disease, fineness, quality, etc. (7) To smooth the surface of a road. (8) To raise the level of a piece of ground by the addition of earth, gravel, etc.

Graded Terrace–A terrace with a constant lengthwise slope sufficient to cause runoff to flow at a nonerosive velocity.

Grader Blade–An implement mounted, usually on the rear of a tractor, that is used for leveling and grading soil, scraping manure, etc.

Grading–(1) The classification of products, animals, etc., into grades. (2) The mating of a purebred animal with one of mixed or unknown breeding. (3) The smoothing of the land surface.

Grading Cotton–A system of classifying cotton according to grade, composed of three factors: color, foreign matter, and ginning preparation.

Grain Alcohol–See Ethanol.

Grain Drier–A device used to lower the moisture content of grain. It can utilize circulating air, or heat generated by LP gas, electricity, or the sun.

Grain Drill–A tractor-drawn implement that plants grain seed in a series of closely spaced rows. A small furrow is opened, the seed deposited, and the furrow closed. See Air Drill, Press Drill, Stubble-Mulch Press Drill.

Grain Probe–A sampling device commonly used by elevator operators when purchasing grain to obtain a sample of a load, bin, or gab in order to determine grain quality and dockage content. It consists of a long, hollow tube that is inserted into grain to obtain a representative sample of the load. See Grain Trier.

Grain Trier–An instrument or device usually made of brass and consisting of two tubes from 6 inches to 8 feet in length. These tubes are slotted on the sides, and one is closely fitted by freely moving within the other. When closed this instrument may be inserted in a bag or bin of grain and then opened to admit a representative sample from each layer of the grain. After filling, the trier is closed and withdrawn and the sample discharged upon an examination cloth for inspection or grading. Also called grain sampler.

Grain Wagon–A truck-or tractor-drawn four-wheel wagon that transports grain from a harvester to a storage bin or to market.

Gram–(1) A unit of weight used in the metric system; one gram being equivalent to 0.035 ounces avoirdupois, or to 0.032 apothecary weight. (2) Mung bean. (3) An Indian gram. See Gram Chickpea.

Grange, The National–An organization dedicated primarily to the improvement of rural life, founded in the United States in 1867, as The Order of the Patrons of Husbandry; membership included both men and women. The Grange has exerted a strong influence on agricultural legislation and rural education and has promoted cooperative enterprises designed to protect the farmer from economic exploitation.

Grant–(as used in Public Law 480 Programs) The transfer, by the United States to foreign governments, of foreign currencies acquired by the United States through sales of United States farm products. *Grant* also is used in connection with the transfer, by the United States to foreign governments, of United States agricultural commodities as distinguished from *donation* of farm products to United States voluntary agencies carrying on foreign relief operations. Grant and donation, however, are often used interchangeably.

Grantor–A person who conveys real estate by deed; the seller.

Grazing Capacity–In range or pasture management, the ability of a grassed unit to give adequate support to a constant number of livestock for a stated period each year without deteriorating. It is expressed in number of livestock per acre of given kind or kinds, or in number of acres per specified animal. Modifications must be made during years of drought.

Grazing District–In the United States, an administration unit on federal range established by the Secretary of the Interior under the provisions of the Taylor Grazing Act of 1934, as amended; or an administrative unit of state, private, or other range lands, established under state laws.

Grazing Fee–A charge made for livestock grazing on a range on the basis of a certain rate per head for a certain period of time, as distinguished from lease or rental of the land on which animals may be grazed.

Grazing Land–Land used regularly for grazing; not necessarily restricted to land suitable only for grazing: excluding pasture and cropland used as part of farm crop rotation system.

Grazing Pressure–The actual animal-to-forage ratio at a specific time. For example, three animal units per ton of standing forage.

Grazing Unit–(1) The quantity of pasturage used by an average, mature cow or its equivalent in other livestock in a grazing season in a given region. (2) Any division of the range that is used to facilitate range administration or the handling of livestock.

Grazing Value–The worth of a plant or cover for livestock and/or game that is determined by its palatability, nutritional rating, amount of forage produced, longevity, and area of distribution.

Grease–(1) See Fat, Lanolin, Lard. (2) A thick petroleum derivative used for lubrication. (3) Hog fat as distinguished from tallow, which is the fat of cattle and sheep; commercially differentiated by temperature of solidification which is below 40°C for grease and above 40°C for tallow. (4) To lubricate a machine. (5) To apply salve to a wound or irritation.

Grid–A small, rigid, rectangular frame for delimiting a sampling area.

Grindstone–A large circular stone made from sandstone and formerly used quite extensively for the sharpening of many different tools and instruments. Now largely replaced by artificial vitrified abrasives.

Grist Mill–A grain-grinding mill equipped with large, heavy grinding stones or burrs usually run by waterwheel power. Grains such as wheat, rye, and corn are ground into meal and flour. Such mills were common in early colonial days in the United States.

Gross Energy (GE)–The total amount of heat or energy in an organic substance, such as feed, measured by complete combustion in a bomb calorimeter.

Gross Income–The receipts and net increases from the farm/ranch business for the year, which include the cash receipts from the sale of dairy products, eggs, forest products, machine work off farm, labor off farm, and other receipts of a minor nature and net increases from dairy cattle, beef cattle, swine, sheep, poultry, crops, agriculture payments, and orchards, etc. (Net decreases in any of the foregoing items are subtracted from the total gross income.)

Gross Weight–(1) The weight of an article together with the weight of its container and the material caused for packing. (2) As applied to a carload: the weight of a railcar or truck together with the weight of its entire contents.

Ground–(1) The surface of the earth. (2) The soil; land. (3) (a) An electrical connection to the earth. (b) To make such an electrical connection. (4) In the electrical circuit of a tractor or other machine, the break in the circuit that occurs when any part of the circuit unintentionally touches the metal body of the tractor.

Grout–(1) A watery mixture of mortar that will flow between stones to make a more stable mass. (2) A coarse grain meal of porridge.

Grub Wagon–A horse- or mule-drawn wagon that carried food supplies for cowboys on roundups (western United States, but now rare). Also called chuck wagon.

Grubbing–The process of removing roots, stumps, and low-growing vegetation.

Grubbing Hoe (Grub Hoe)–A tool used for digging in hard ground; a mattock.

Guaranteed Analysis–On feed labels or tags, a listing of certain nutrients, usually crude protein, crude fiber, fat, and ash, guaranteeing a minimum or maximum percentage of each in the feed.

Gunny Sack–A coarse fabric bag manufactured from jute fiber that is commonly used as a container for feeds, seeds, potatoes, and various other products. Gunny fibres come from one of two tropical jute plants: *Corchorus olitorius* or *C. capsularis*. Also called toe sack, burlap sack.

Gunter's Chain–A steel-linked chain 66 feet (20.13 meters) long used in the original land survey of the United States. Modern Gunter's chains are steel tapes. There are exactly 80 Gunter's chains in a mile.

Gutter–(1) A channel in the floor behind cows in stanchion barns used to catch the manure and to simplify cleaning. (2) A ditch, dug in the path of a forest fire into which burning materials fall, designed to arrest the fire. (3) In turpentine orcharding, the groove or channel along which the oleoresin travels form the tree to the cup. (4) A metal channel (eave trough) mounted to intercept water flowing from a roof.

Guy–(1) To anchor a post, plant, limb, tower, etc., by the installation of cables or wires from the member to a firmly established object. (2) Designating a cable or wire so used.

Gypsum Block–See Tensionmeter.

Ha-Ha Fence–(1) A sunken wall that encloses a lawn or pasture; one that does not obscure the view. (2) A boundary trench; a ditch or moat.

Hacienda– (1) The principal dwelling or headquarters of a ranch. (2) The ranch itself. (Southwestern Unitd States).

Half-section– Three hundred and twenty acres of land.

Hame– The wood or metal parts of the harness of a draft animal that fit about the collar to which the traces are attached for pulling.

Hammer Mill– A feed-grinding device or mill in which hammerlike projections are mounted on the surface of a cylinder that revolves at a high speed within a heavy perforated metal enclosure, and shatters the feed material by beating it to pieces.

Hammer Mill Magnet– A magnet that is placed on the apron of a hammer mill to attract bits of iron, wire, etc., to prevent them from entering the mill and damaging the hammer.

Hamper– A basket in which vegetables are shipped.

Hand– (1) A laborer who is either permanently employed or migratory, as harvest hand, hired hand. (2) A unit of measurement equal to 4 inches (10 centimeters) that is used to measure the height of horses from the ground to a point at the shoulder. (3) A bunch of tobacco leaves of the same grade that are tired together for easier handling. (4) The near horse in a team used for plowing. (5) A half-whorl-like cluster of bananas attached to the rachis of the spike or bunch. (6) Designating any tool, implement, etc., that is manually operated. (7) Designating any manual labor, as hand chopping.

Hand Gun– A spray gun held in the hands during operation.

Hand Labor– Any labor or production effort that is produced entirely by the hands or use of the human body. (The term often denotes any unskilled a farm laborer working with his hands or with simple tools, such as a hoe.)

Hand Level– A leveling instrument without tripod or stand that is held in the hand while making a horizontal line of sight to establish relative elevations. See Abney Level.

Handler– A person or company that buys, sells, sorts, packages, or stores a product.

Hank– (1) A measure of length for yarn (hank of worsted yarn contains 560 yards (512 meters). (2) A bundle of animal casings.

Hardening– The process of tempering steel-edge tools. The tool, heated to a cherry red, is plunged into brine or clear water, the sudden quenching rendering it hard enough to scratch glass. Subsequent reheating to a point determined by experience is necessary to draw the temper to the requisite hardness.

Hardness of Wood– A property of wood measured by the load applied at a standard rate required to imbed a 0.444-inch (1.13-centimeter) steel ball halfway.

Hardness Scale– The empirical scale by which the hardness of a mineral is determined as compared with a standard. The Mohs scale is as follows: 1. talc; 2. gypsum; 3. calcite; 4. fluorite; 5. apatite; 6. orthoclase; 7. quartz; 8. topaz; 9. corundum; 10. diamond.

Hardware Cloth– Galvanized, square-meshed, wire screening made in several widths with openings of various sizes up to about 1 inch (2.5 centimeter) square; used in the construction of poultry and small animal cages, incubator and dehydrator trays, various types of screens, etc.

Harness– The straps, bands, collars, hames, lines, and attachments that are necessary to equip a draft animal, such as a horse, properly to pull or move a load. The harness, of many types, is usually largely made of leather. Also called gear.

Harrow– A farm implement used to level the ground and crush clods, to stir the soil, and to prevent and destroy weeds. Six principal kinds are the disk, spike-tooth, spring-tooth, rotary, cross harrow, and soil surgeon.

Harvest Hand– A person, usually a temporary or unskilled worker, who assists in harvesting.

Harvest Interval– The period of time required by law between the application of a pesticide and harvest of the crop.

Harvestsed Acres– Acres actually harvested for a particular crop, usually somewhat smaller at the national level than planted acres because of abandonment brought on by weather damage or other disasters or market prices too low to cover harvesting costs.

Hatch– (1) (a) To bring forth young from the egg by natural or artificial incubation. (b) The young produced from one incubation. (2) The access entryway to cellars, attics, haymows, ships, etc.

Hatch Act– An act of the United States Congress in 1887 that allocated $15,000 annually to each state and territory of the development and operation of an agricultural experiment station. This act was amended in 1955 to include subsequent acts of Congress that appropriated funds for the operation of these experiment stations. The consolidation or amended Hatch Act includes the original Hatch Act, the Adams Act of 1906, the Purnell Act of 1925, the Bankhead-Jones Act of 1935, and Title I, section 9 of the Amendment of 1946 to the Bankhead-Jones Act. Currently valid.

Hatchet– A small ax used for splitting kindling or chopping small trees.

Hay Bale– A quantity of loose hay compressed usually into a rectangular bale about 3 feet by 18 inches by 14 inches, containing from 40 to 125 pounds depending on the kind of hay, degree of compaction, and moisture content. The compressed hay is held in the bale by baling wire, light metal strips, or heavy twine. More common are the large cylindrical bales weighing up to a ton or more.

Hay Baler– A tractor-powered implement that picks up hay from a windrow, presses it into bales, ties and ejects the bales. See Round Baler.

Hay Conditioner– (1) A mechanical device consisting of two closely spaced, parallel, smooth surface rollers that crush the freshly cut stems of hay to facilitate drying and curing. (2) A mechanical device consisting of two closely spaced, parallel rollers with corrugations resembling gear teeth paralleling the axle that kinks the stems of hay to break them open. Both methods results in a more even and rapid drying of the hay and less loss of leaves than conventional swath curing methods. (3) Any chemical that is supposed to be beneficial in preventing storage losses and/or increasing palatability.

Hay Rake–An implement that is used to rake mowed hay into windrows so it can be picked up by a baler.

Hayseed–(1) Bits of straw, chaff, and seeds that cling to the garments in haymaking. (2) A degrading term for a rustic farmer or countryman (obsolete).

Hazardous Occupation–An occupation in which the daily labor is, or may be, hazardous to life and limb, including some in agriculture, such as the pesticide applicator. Certain persons, primarily youth, are prohibited by federal and state labor laws form accepting employment in such an occupation.

Hazel Hoe–In forestry, a fire-trenching or digging tool that resembles a grub hoe but has a shorter, broader, and lighter blade, a round or oval eye, and usually a straight, picklike head.

Head–(1) Any tightly formed flower cluster, as in members of the family Compositae, or any tightly formed fruit cluster, as the head of wheat or sunflower. (2) A compact, orderly mass of leaves, as a head of lettuce. (3) On a tree, the point or region at which the trunk divides into limbs. (4) The height of water above any point of reference (elevation head). The energy of a given nature possessed by each unit weight of a liquid expressed as the vertical height through which a unit weight would have to fall to release the average energy possessed, used in various compounds, as pressure head, velocity head, lost head, etc. (5) Cows, asses, horses, collectively, as ten head of horses. (6) The part of the body that includes the face, ears, brain, etc. (7) The source of a stream; specifically the highest point upstream at which there is a continuous flow of water, although a channel with an intermittent flow may extend farther. (8) The upstream terminus of a gully. (9) To prune a tree severely. (10) To get in front of a band of sheep, herd of cattle, etc., so as to stop their forward movement (head them off). (11) To place a top on a barrel. (12) That part of an engine that forms the top of the combustion chamber. In many types of modern engines, the exhaust and intake valves are in the head.

Head Flume–A flume, chute, trough, or lined channel used at the head of a gully or at the lower end of a terrace outlet to reduce soil erosion.

Head Gasket–A sealing device between the cylinder head and block of an engine that prevents the leaking of oil, coolant, and compression.

Head Loss–Energy loss due to friction, eddies, changes in velocity, or direction of flow of water or other liquids through a pipe.

Head Pressure–Air pressure in a water pipe.

Head-catch Chute–A device at the end of a narrow alley that catches and holds the heads of cattle while they are treated. See Cutting Chute, Squeeze Chute.

Headgate–Water control structure; the gate at the entrance to a conduit.

Headland–(1) A type of farm road established at frequent intervals in sugarcane or sugar beet fields to permit the best possible movement of heavy loads with as little travel over the rows as possible. (2) That portion at the end of the field that is reserved for travel and for turning the tillage implement around after completing a furrow or row.

Headspace–The space between the top of food in a container and the container lid or closure.

Heat Gradient–Change in temperature of the earth with depth; approximately 30°C per kilometer in the upper part of the earth's crust.

Heat Increment–The amount of heat or energy produced by an animal in eating, digesting, absorbing, and metabolizing food; also called the work of digestion.

Heat of Condensation–The same as the heat of vaporization, except that the heat is given up as the vapor condenses to a liquid at its boiling point.

Heat of Vaporization–The heat input required to change a liquid at its boiling point (water at 212°F or 100°C) to a vapor at the same temperature.

Heating Value–The amount of heat obtainable from a fuel and expressed, for example, in Btu/lb. or joules/kg.

Heavy–(1) Designating any material or product that exhibits a comparatively high weight per unit volume. (2) Designating a clay or clayey soil that is difficult to plow. (3) In marketing, designating an abundant supply of a product for sale on one day at one market. (4) The late stages of pregnancy of a cow.

Hectare–A metric unit of land measurement equal to 100 ares, 100,000 square meters, or 2,471 acres.

Hedge–(1) A fence or barrier formed by bushes, shrubs, or small trees growing close together in a line, sometimes with interwoven branches used as a screen. (2) A managed belt of shrubs or small trees usually placed across fields or along field or property boundaries for wildlife rather than for wind control. (3) To buy or sell futures to protect from loss due to a rise or fall in prices.

Hedgers–People who buy and sell futures contracts in an attempt to reduce the hazard of loss.

Height of Instrument–The elevation of the line of sight of a surveying instrument, which is obtained by adding the backsight to the elevation of a known point.

Held–(1) The lower end of an ax or other tool handle encased by the metal portion of the tool. (2) Designating the controlling of an animal in a small space or by other controlling devices.

Hemi-–(Greek) Prefix meaning half.

Herd Book–The recognized, official record of the ancestry of a purebred animal kept by the particular breed association. The first herd book for keeping thoroughbred pedigrees was started in England in 1791. This was followed by Coates Herd Book for Shorthorns started in England in 1822.

Herd Improvement Registry–A type of registry maintained by certain purebred cattle breeder associations to record the production and yearly records of all producing cows of that breed in a given breeder's herd.

Herder–One who tends livestock on a range. Usually applied to the man herding a band of sheep or goats.

Herdsman–One who cares for and manages a herd of livestock, usually cattle.

Heritability–The proportion of the differences among animals, measured or observed, that is transmitted to the offspring. Heritability varies from zero to one. The higher the heritability of a trait, the more accurately does the individual performance predict breeding value and the more rapid should be the response due to selection for that trait.

Heritability Percent Estimates–The percent of a trait that is inherited from an animal's parents and is not controlled by environment.

Hermetic–Food containers that do not permit gas or microorganisms to enter the container or to escape from it. A properly sealed tin can is an hermetic container.

Hertz–An international term for sound frequency in cycles per second. One hertz equals one cycle per second.

Hex Bolt–A bolt that has a six-sided head designed to accept a wrench.

High Dump Wagon–A wagon used to haul forage and other products that are unloaded by hydraulically raising and tipping the box to one side to empty the entire load at one time.

High-density Bale–(1) A bale of cotton similar to a standard-density bale except the average density is 36 pounds per cubic feet. (2) A highly compacted bale of hay with a density of 12 to 15 pounds per cubic feet.

High-pressure Sprayer–A sprayer utilizing a high-pressure pump to force the spray through nozzles for both atomization and delivery to the plant.

Highest Profit Combination–In the production of a single product with varying and fixed resources, the combination of the factors of production in which (a) additional units of the variable factors exactly pay for themselves; i.e., when marginal factor cost equals the marginal value product and (b) the revenue derived from the added units is declining.

Hillbilly–(1) A person who lives in rural, hill, or mountainous area (jocular and derogatory). (2) A cottontail rabbit.

Hip–(1) The fruit of rose; rose hips. (2) The external angle (ridge) formed by the meeting of two sloping sides of a roof. (3) That region of one of the rear quarters of four-legged animals where the hind leg joins the pelvic region.

Hip Roof–A roof so constructed that a rain falling on it drains off all four sides. If the building is square, the four hip rafters meet at the center.

Hitch–(1) A catch; anything that holds, as a hook; a knot or noose in a rope, which can be readily undone, intended for a temporary fastening. (2) (a) The connecting of an implement, vehicle, etc., to a source of power, as a tractor, team, etc. (b) The device which is used to make such a connection. (3) A horse or horses used to pull an implement or vehicle. (4) The stride of a horse when one of the hind legs is shorter than the other. (5) To fasten an animal to a post, rail, etc.

Hive Loader–A mechanically operated boom and cradle for manipulating beehives and placing them on a truck.

HNIS–Human Nutrition and Information Service, USDA.

Hoe Culture–A primitive type of agriculture in which the hoe was the principal implement or tool.

Hoedad–A heavy bladed hoe with a short, stout handle used to open the ground to plant tree seedlings.

Hoedown–An old-fashioned, riotous country dance. Also called breakdown, barn dance.

Hog Bristles–The coarse, stiff hairs on swine used in the manufacture of brushes; now synthetic bristles have largely replaced hog bristles.

Hog Holder–A device consisting of a metal tube through which a cable is run. A loop on the end of the cable is placed on the snout of a hog to hold the hog for receiving vaccination, etc.

Hog-Corn Ratio–See Corn-Hog Ratio.

Hogshead–(1) A liquid measure that contains 63 United States gallons or 52½ imperial gallons. (2) A cask that contains from 100 to 140 gallons, as used for molasses, etc.

Hohenheim System of Grazing–An intensive system of grassland management involving the division of the grazing area into a number of paddocks, grazed in rotation to allow a short period for growth between grazings. Nitrogenous fertilizers are usually applied after each grazing.

Hold–(1) To restrain animals to a particular place. (2) To remain on the tree until mature, as a fruit. (3) To retain, as certain soils hold moisture. (4) To maintain condition, as a steer that holds flesh. (5) To store or retain in storage, as to hold eggs. (6) Not to market at harvest time or when an animal is fat, but to wait to sell for a better price.

Home Range–The area around an animal's established home that is traversed in its normal activities. See Territory.

Home-grown–(1) Designating crops, vegetables, or animals that are raised on one's property, as contrasted to purchased products. (2) Designating produce raised locally in contrast to that shipped in from a distance.

Homestead–(1) A home place; especially a farm/ranch dwelling and the surrounding ground. (2) In the United States, a unit of public land, usually not more than 160 acres, acquired by an individual under the Homestead Act of Congress, 1862, and subsequent acts. See Homesteading.

Homestead Rights–The widow's or widower's rights to the family dwelling.

Homesteading–In the United States, the process of securing a unit of virgin, undeveloped federal land under the regulations of the Federal Homestead Act of 1862. See Homestead.

Homogeneity of Products–A characteristic of a set of products denoting that, in the eyes of traders, are prefect substitutes for each other.

Hone–(1) The whetstone, carborundum stone, metal, or other device used to sharpen knives, scissors, or other cutting tools. (2) To sharpen a knife.

Honey Extractor–See Extractor.

Honey House—Building in which honey is extracted and handled.

Honeysuckle—*Lonicera*; bush honeysuckles are large shrubs growing to 10 to 15 feet (3 to 5 meters) in height. The branches are spreading. Amur honeysuckle is a large spreading shrub with fragrant conspicuous white flowers. It bears fruit late in summer and holds fruit well into the winter. Tartarian is a large shrub that bears pink flowers in the spring and a berry crop in summer, but does not hold fruit into the winter.

Hoof Pick—A tool used to clean the hoof of a horse or other animal.

Hook-up—See Hitch.

Horizontal Bench Terrace—A type of bench terrace that has no measurable slope from the back to the front of the bench.

Horizontal Integration—Combining similar marketing functions and decisions at the same market level into a single firm; e.g., one food processor buying another food processing company.

Horizontal Silo—A silo built with its long dimension parallel to the ground surface rather than perpendicular as in the case of an upright silo. It may be built just below the ground level with openings at both ends to facilitate filling and feeding.

Horn—(1) A natural, bonelike growth or projection on each side of the head of most breeds of cattle, sheep, and goats that is a natural weapon of defense. (2) The pointed end of a blacksmith's anvil used for shaping hot metal. (3) The front, upraised projection of a riding saddle, the snubbing horn. (4) The outer hard covering of the hoofs of horses, cattle, sheep, and swine. (5) A broad term commonly used in describing various shadings of color in the beak of some breeds of fowl.

Horse Wrangler—(1) A hostler who keeps the string of extra saddle ponies that accompanies every cattle drive or used during roundup. He keeps them from straying and has them ready when they were needed (southwestern United States). (2) The rider on horseback who accompanies the rodeo rider and keeps the bucking broncho from injuring the contest rider before and after the riding has been done. (3) A person who works with horses during the breaking period.

Horsepower—The rate at which work is accomplished when a resistance (weight) of 33,000 pounds is moved 1 foot in 1 second. One horsepower working for one hour = 745.7 watt-hours, = 1,980,000 foot pounds. Engines and motors are rated in horsepower, based on tests made on a machine known as the dynamometer. See Brake Horsepower, Drawbar Horsepower, Effective Horsepower, Rated Horsepower.

Hoseboy—A mechanical device used for watering cut flower crops in a greenhouse or storage area.

Hostler—One who takes care of horses.

Hot Iron—The heated iron rod or stamp with a handle used in branding cattle, etc.

Hot Walker—A person who leads horses in order to cool them slowly after a workout or race.

Hover—(1) The sheet metal canopy surrounding a heat source under which incubator chicks are kept warm. (2) To cover chicks, as a hen hovers her chicks.

Humidistat—A sensing device used in a greenhouse to maintain humidity at a certain level.

Hundredweight—A unit of weight: a short hundredweight equals 100 pounds (45.4 kilograms) in the United States and a long hundredweight equals 112 pounds. (50.8 kilograms avoirdupois in England.)

Hurdle—A board made of plywood usually around 3 feet wide and 3 feet long that is used to herd pigs.

Hurdle Plots—(1) A small fenced-in plot of land, from 4×4 to 20×20 feet, on which the herbage is clipped at predetermined intervals during the growing season to determine forage composition and yield. (2) Small plots grazed by means of hurdles.

Husbandman—A person engaged in the production of animals or crops by tillage or other use of land; a farmer.

Husbandry—In its earlier usage, the skill, or art, of tillage, crop production, and rearing of farm animals. Today the word is occasionally used as a synonym for farming. More commonly used in combination, as animal husbandry and poultry husbandry, embracing the art, science, processing, and business of production.

Husking Bee—An early American gathering of neighbors and friends in which the men husked harvested corn for a particular farmer. The women, children, and sweethearts accompanied the men and prepared sumptuous meals. It was a delightful social gathering.

Hydrator—Vegetable crisper; a drawerlike section in refrigerators that protects fresh fruits and vegetables form drying out during refrigerator storage.

Hydraulic—Designating fluids (usually water) that are moving or at rest under forces of gravity or pressure.

Hydraulic Elements—Factors determining the rate of flow of water or other fluid in a pipe or a channel are: area, depth, slope, velocity, energy, roughness, viscosity, temperature, pressure, and fluid characteristics.

Hydraulic Gradient—(1) The slope of the hydraulic grade line. (2) The slope of the free surface of water flowing in an open channel.

Hydraulic Pressure—The pressure exerted through the medium of a liquid. In tractors and machinery, hydraulic pressure is used to raise and lower implements, assist in steering, etc.

Hydraulic Ram—A mechanical device that utilizes momentum of water flowing down an incline to raise a part of that water to a higher elevation.

Hydraulic Sprayer—A machine that applies pesticides by using water at a high pressure and high volume.

Hydraulics—(1) The mechanics of fluids; hydromechanics. Commonly used to designate that body of hydromechanical principles given practical application by the engineer in dealing with water. (2) That branch of service or of engineering that deals with water or other fluid in motion.

Hydrogen—H; an element; a colorless gas; one of the essential elements for plant and animal growth.

Hydrogen-ion Concentration–A measure of acidity or alkalinity, expressed in terms of the pH of the solution. A pH, of 7 is neutral, from 1 to less than 7 is acid, and from 7 to 14 is alkaline. See pH, Reaction.

Hydrogenation–The addition of hydrogen to any unsaturated compound. Oils are changed to solid fats by hydrogenation.

Hydrogenesis–A process of natural condensation of moisture in the air spaces in the surface soil or rock.

Hydrogenic Soil (Hydromorphic)–Soil that was developed under the dominant influence of water.

Hydrograph–A graph that shows the fluctuation of the flow of water with respect to time.

Hydrography–The measuring, recording, and analyzing of the flow of water, e.g., measuring and mapping water courses, shorelines, and navigable waters.

Hydrology–The science that deals with the waters of the earth, their occurrence, distribution, and circulation through the unending hydrologic cycle of precipitation, runoff, infiltration and storage, eventual evaporation, transpiration, and reprecipitation. A simple distinction between hydrology and hydraulics is that in hydraulics the source of the water is immaterial, attention being confined to its motion.

Hydrolysis–Chemical reaction in which a compound reacts with water to produce a weak acid, a weak base, or both.

Hydrolyze–The splitting of a compound into smaller units by the addition of water.

Hydrometer–A long-stemmed glass tube with a weighted bottom; it floats at different levels depending on the relative weight (specific gravity) of the liquid; the specific gravity or other information is read where the calibrated stem emerges from the liquid.

Hydrostatic Pressure–The pressure exerted by water at any given point in the body of water at rest. The hydrostatic pressure of groundwater is generally due to the weight of water at higher levels in the zone of saturation.

Hygric (Hydric)–A soil environment with a high moisture supply. See Mesic, Xeric.

Hygrometer–An instrument used for determining the relative humidity of the air. See Humidity, Relative.

Hygroscopic–Capable of expanding in the presence or contracting in the absence of moisture.

Hygroscopic Coefficient–The moisture, in percentage of dry weight, that a dry soil will absorb in saturated air at a given temperature.

Hygroscopicity–The tendency of salts to adsorb water whenever the vapor pressure of moisture in the air exceeds that of a saturated solution of the salt.

Hygrothermograph–A device that continuously records both temperature and relative humidity.

Hypochlorite–A sodium hypochlorite solution used for sterilizing milking equipment by release of free chlorine.

Hypsometer–Any instrument used to measure heights of trees, based either on geometric or trigonometric principles.

ICA–International Coffee Agreement.

IDA–International Dairy Arrangement.

Idle Farmland–Includes former croplands, orchards, improved pastures and farm sites not tended within the past two years, and presently less than 10 percent stocked with trees.

Idle Land–(1) Land that has been managed or exploited for some particular use but is now in a state of disuse; abandoned land. (2) Farmland that is capable of producing but is not in use; by extension, any land that has potentiality, but is not being put to any productive use. (3) Land being fallowed.

Imbricate–Overlapping, as shingles on a roof.

Immersion Heater–A type of electric water heater used for warming small quantities of water in which the heating element is placed directly in the water to be heated. These are used in livestock watering troughs to prevent freezing.

Immigration–Movement of people, animals, or plants into an area. See Emigration.

Imperial Bushel–A dry measure used in Great Britain, slightly larger than the standard American bushel: 1 imperial bushel is equal to 1.0321 American bushels. See Bushel.

Implement–Any tool that aids a person to make work and effort more productive and effective.

Improvement Project–(Agriculture) A project planned and conducted by a student enrolled in agricultural education as a part of the student's supervised agricultural experience program that results in an increase in the value or income of a business, or in the improvement and increased convenience of the home or its surroundings.

In-and-In Breeding–The breeding together of closely related plants or animals for a number of successive generations to improve or eliminate certain characteristics. Also called breed in and in, line breeding, inbreeding.

Incentive Payment–A form of compensatory payment in which the support price is set at a level high enough to encourage the increase in production of a particular commodity to a desired level.

Income Elasticity–The responsiveness of food consumption (measured in quantity or expenditures) to changes in consumers' income.

Income Support Payment–Funds paid to farmers when farm prices are below support levels; arrived at by subtracting from the target price, or the total support level, the higher of (a) the loan rate or (b) the national average price of a commodity during the first five months of the marketing year calendar year price for cotton). Generally, the federal government pays this difference to a farmer who qualifies (by meeting all farm program conditions) for that portion of the farmer's production specified in the farm program. The payments are sometimes called deficiency payments.

Increment–The increase in volume, weight, etc., of an animal on pasture, of a tree growing in a stand, etc.

Index–A system for comparing animals within a herd, or area, based on the average of the group; usually the figure 100 is used for an average index; animals receiving an index of 100 or over are the top end while those indexing less than 100 are the bottom end.

Individual Claiming Pen–On goat ranches, a pen in which an individual doe that has disowned her kid is kept until she reclaims her offspring.

Individual Drinking Cup–A type of stock-watering equipment in wide use in modern stanchion dairy barns whereby water is supplied by a cup at the stanchion for each individual cow.

Individual Farrowing House–A separate building or quarters equipped so that sows that are to farrow may be kept isolated in warm, clean, and quiet individual pens at farrowing time.

Inert Gas–Also called noble or rare gas; one that does not react with other substances under ordinary conditions.

Inertia–(1) Sluggishness. (2) A property of matter by which it remains at rest or in uniform motion in a straight line.

Infiltration–The flow of a liquid into a substance through pores or other openings, connoting flow into a soil in contradistinction to ***percolation***, which connotes flow through a porous substance.

Infiltrometer–A device used in measuring the rate of movement into the soil of water applied uniformly over an area at a given rate or in a given volume.

Initial Water Deficiency–The amount by which the actual water content of a given soil zone (usually the root zone) is less than field capacity at the beginning of a season or specified time period.

Injection Pump–A device on a diesel and sometimes a gasoline engine that meters and delivers the correct amount of fuel to the injectors at the proper time. See Injector.

Injection Well–A well into which surface water is pumped to increase subsurface water volume.

Injector–(1) A device on a diesel engine that receives a measured amount of fuel from the injection pump and injects a charge of fuel into the combustion chamber at the correct time. (2) A device that meters the correct amount of fertilizer, pesticide, etc., into an irrigation system. See Injection Pump.

Injunction–A legal writ or command issued by a court and directed to a particular person or corporation, requiring that the person or corporation stop certain actions.

Inlet–(Hydraulics) (1) A surface connection to a closed drain. (2) A structure at the diversion end of a conduit. (3) The upstream end of any structure through which water may flow.

Input Shaft–A shaft that is driven by an engine or other power source that provides power to the transmission.

Inputs–Items purchased to carry out a farm's operation. Such items include fertilizers, pesticides, seed, fuel, and animal feed.

Insectary–A place used for raising insects for scientific study.

Inseminating Tube–A rubber, glass, or metal tube, usually with syringe attachment, used in artificial insemination to introduce the semen into the vagina of a female animal. Such instruments vary in character depending on the species being bred.

Inseminator–The technician, in the employ of an artificial breeder's unit, who brings the prepared semen to the farmer's herd and performs the technical service of inseminating the cows.

Installment–(1) A partial payment of a debt often at regular intervals such as monthly or yearly installments. (2) A partial delivery of a given lot of produce, etc.

Insulator–(1) A substance that offers very high resistance to the passage of electricity; a nonconductor. (2) A device made of some nonconducting material, such as glass, used for fastening or supporting a conductor such as electric fence wire. (3) A substance used to prevent heat or sound transfer.

Intake–(1) The headworks of a conduit, the place of diversion. (2) Entry of water into soil. See Infiltration, Percolation. (3) The part of an internal combustion engine that admits air.

Integrated Pest Management (IPM)–An ecological approach to pest management in which all available necessary techniques are systematically consolidated into a unified program, so that pest populations can be managed in such a manner that economic damage is reduced and adverse side effects are minimized.

Integration–The combination (under the management of one firm or farmer) of two or more of the processes in the production and marketing of a particular product—generally the processes are capable of being operated as separate businesses. Diversification, on the other hand, is the production of two or more farm products by one firm or farmer.

Intensity–The force, energy, and concentration with which a farm business is operated, usually based on the input of labor per acre. The higher the labor requirement per acre the greater the intensity.

Intensive Farming–Farming in which a comparatively large amount of labor and working capital is used per tillable acre of farm land. Dairy, poultry, vegetable and fruit crops are examples of intensive enterprises.

Interest–The cost paid for borrowing money.

Internal-combustion Engine–An engine in which both the heat energy and the ensuing mechanical energy are produced inside the engine proper.

International Code of Botanical Nomenclature–A system, adopted by European botanists and some American botanists, for applying the Latin names to plants. Established by a congress in Vienna in 1905, it was revised in 1931 when it was universally accepted, replacing, in the United States, the American Code of Botanical Nomenclature. It provided that most plants shall retain the name applied by Linnaeus in his Species Plantarum (1753). it accepted the idea of ***nomen conservandum*** and provided that after January 1, 1931, publication of all new species shall be accompanied by a Latin name. Also called Vienna Code, Vienna Rules.

International Rule–One of the official log rules used by the United States Forest Service. It allows a 0.5 inch taper for each 4 foot of log length and 1/16 inch in shrinkage for each 1 inch board. The assumed saw kerf varies from 1/8 inch to 1/4 inch.

International Trade Barriers–Regulations used by governments to restrict imports from other countries; e.g., tariffs, embargoes, import quotas, and unnecessary sanitary restrictions.

Interregional Competition–The competition of one agricultural region with another for the market of the same or similar agricultural product or products; e.g., Maine, Michigan, and Idaho potatoes competing in the same market. Production costs, yields, quality, prices, and transportation are important factors in determining the comparative advantage of one region over another.

Intestate–Legal designation of a person who has died without leaving a valid will.

Inventory–A list of property that shows the amounts and values of all physical assets involved; e.g., in a farm business, a list of amounts and values of land, buildings, equipment, etc. In more strict accounting terms, farm inventory is restricted to livestock, growing feed crops, and supplies. Buildings, machinery, etc., are included in the depreciation schedule.

Investment Appraisal–An evaluation of the profitability of some investment. Commonly involves net present value or internal rate of return calculations.

Investment Capital–Value of inputs (purchased or owned) that are allocated to an enterprise with the expectation of a return at a later point in time.

Investments–An outlay of money for income and profit.

IOOA–International Olive Oil Agreement.

Irrigating Head–(1) The measure of stored-up water ready to be used in irrigation. (2) The depth of a body of water in covering land, as by flooding. The proper wetting of the whole ground. (3) The indication on a measuring device of the rate of flow of irrigation water.

Irrigation District–A cooperative, self-governing, public corporation set up as a subdivision of the state, with definite geographic boundaries, organized to obtain and distribute water for irrigation of lands within a district. A district may have taxing power under authority of the state legislature with the consent of a designated fraction of the landowners or citizens (United States).

Irrigation Efficiency–The ratio or percentage of the water consumed by crops in an irrigated farm or project to the water diverted from a river or other source into the farm or project canals.

Irrigation Requirement–The quantity of water, exclusive or precipitation, required for crop production, including unavoidable wastes.

ISA–International Sugar Agreement.

Isobar–A line on a chart or diagram that connects points having the same barometric pressure, most commonly drawn on weather maps to show relative pressure in terms of inches of mercury or millibars of mercury.

Isocline Equation–The equation specifying the least-cost combination of a set of input factors for production of any specified quantity of output.

Isodyne–A line drawn on a map of a cultivated field that connects points having equal dynamometer readings of the force required to pull a cultivating implement.,

Isogonic–Lines drawn on a map designating equal magnetic declination.

Isohyet–A line drawn on a map designating equal magnetic declination.

Isolation Strip–An area surrounding a sample plot treated the same as the plot.

Isomers–Two or more substances having the same chemical composition but different properties.

Isoneph–Lines drawn on a map designating equal cloudiness.

Isotherm–A line on a chart or diagram drawn through places or points having equal temperature.

ITC–International Trade Commission

IU (International Unit)–A unit of measurement of a biological (e.g., a vitamin, hormone, antibiotic, antitoxin) as defined and adopted by the International Conference for Unification of Formulas. The potency is based on bioassay that produces a particular biological effect agreed on internationally.

IWA–International Wheat Agreement.

Jack–(1) The male, uncastrated ass. See Jenny. (2) A mechanical, hydraulic device used for lifting heavy objects. (3) See Bone Spavin.

Jacob's Staff–A single straight rod, pointed and iron-shod at the bottom, and having a brass ball-and-socket at the top; used instead of a tripod for supporting a compass.

Jet Nozzle–A flask-shaped nozzle, containing a removable inner core with spiral flanges, that imparts a swirling motion to the liquid and breaks it up into coarse droplets before it leaves the orifice. Used in sprayers.

Jet Pump–(1) Shallow well type: a combined centrifugal and jet pump with the jet located in the pump above ground. A single suction pipe connects the well to the pump. (2) Deep well type consists of a simple centrifugal pump above ground and a jet (or injector) in the well. When the impeller is operated at normal speed it draws water up the suction pipe. Some of this water is forced down the pressure pipe to a nozzle, or injector, where it passes through a restricted opening at high velocity, which creates a vacuum that in turn causes atmospheric pressure to force water into the chamber from the well through a foot valve. The incoming water is mixed with this water coming through the jet and is carried up the suction pipe under pressure.

Jig–(1) An uneven gait of a horse, closely associated with prancing and weaving. (2) A device built to aid in construction; used to make duplicate cuts, holes, etc.

Job Lot–A form of contract that has a smaller unit of trading than the regular contract.

Jobber–A middleman who sells to retailers in relatively small lots.

Jockey Stick–A stick fastened to the hame of a broken horse and to the bit of a green horse to prevent them from crowding and to keep the green horse in position while being broken.

Joint–(1) the point or place at which two bones meet: the articulation between the two bones. (2) The pint or meeting of two pipes. (3) A general term for a cut of meat (England). (4) See Node.

Joint Tenancy–(With rights of survivorship) A legal arrangement for ownership of real property by two or more persons in which, upon the death of one of the persons, the title to the property goes to the surviving party or parties of the joint tenancy rather than to the estate of the deceased. A joint tenancy may be dissolved by either of the parties during his or her lifetime.

Joist–(1) A piece of dimension lumber 2 to 5 inches thick and 4 or more inches wide, commonly used to support the floor of a building. (2) A piece of lumber 8 feet or more long, 1½ to 4½ inches thick, and 6 to 12 inches wide.

Jolly Balance–A sensitive spring balance used primarily for measuring specific gravity (density) by weighing a specimen in air and again when immersed in water.

Joule–A unit of work equal to 0.7375 foot pounds or 0.2390 calories.

Judge–(1) An experienced person selected to make an official placement of entries in exhibits at a show. (2) To consider the qualities, breed, and conformation of animals, plants, and products at a show or exhibit.

Judging Team–A selected group of persons, usually youths, who represent an organization, such as a 4-H Club, FFA, or a college of agriculture. They compete with similar groups at shows or exhibits in judging and placing classes of crops, soils, or livestock.

Judgment–The official and authentic decision of a court of justice concerning the respective rights and claims of the parties to an action or suit.

Judicial Ditch–An intercounty drainage ditch maintained under the jurisdiction of the circuit court (United States).

Jug–(1) A small claiming pen where ewes with newborn lambs are kept until the ewe has nursed and claimed her young. (2) A glass bottle holding one-half gallon or a gallon, used to bottle cider, syrup, molasses, etc. (3) A deep vessel of earthenware narrowed to a neck, which is frequently fitted with a finger handle for carrying. (4) A small pen, part of a corral, used to funnel animals down a narrow alley (chute) to facilitate sorting, loading, or catching; used primarily with cattle.

K–The chemical symbol for potassium.

Keel–(1) The part of a fowl's body that extends backward form the breast. Also called breastbone. (2) The two front, united petals of a pealike flower. (3) In ducks, the pendant fold of skin along the entire underside of the body. (4) In geese, the pendant fold of flesh from the legs forward on the underpart of the body. (5) A central dorsal ridge, like the keel of a boat. (6) The two anterior united petals of a papilionaceous flower.

Keep–(1) Board and room furnished a hired hand. (2) Salary. (3) To maintain animals for use or profit, as to keep cows, bees, etc. (4) To sell or to have in a store for sale, as to keep tobacco. (5) To store, as apples keep well. (6) To withhold from sale. (7) To maintain business records, as to keep books.

Keg–A small cask or small barrel, usually of 10-gallon (38-liter) capacity or less. A keg of nails weights 100 pounds (45.4 kilograms).

Kelvin Scale–The absolute scale of temperature. Below 0° Kelvin, all thermal motion of atoms and molecules ceases. One degree on the Kelvin scale is equal to one degree on the Celsius scale. The conversions are: $-273.15°C = .459.67°F$. Named after William Thompson (Lord Kelvin) (1824-1907), an English physicist. See Celsius Scale, Fahrenheit Scale.

Kerf–(1) The width of cut of a saw; also the channel cut. (2) That which is cut. (3) The act or process of cutting. (4) The cut of a cloth-cutting machine.

Key Area–(1) An area upon which the success of the ranching operations is largely dependent. (2) Critical areas of rangeland, which represent range that is most likely to be overgrazed; used as criteria or indices of the proper use of the range. (3) A guide to the general management of the entire area of which it is a part.

Kibbutz–A collective farm in Israel.

Kid House–A small structure designed to give shelter to a newborn kid. The doe is staked so that it cannot abandon its kid.

Killer–(1) A person who slaughters animals. (2) A meat animal. (3) A pesticide. (4) A predator.

Kiln–An oven, furnace, or large heated room for the curing of lumber, tile, bricks, etc.

Kilo—A combining form used in naming units of measurement to indicate a quantity one thousand (10^3) times the unit designated by the root with which it is combined; e.g. one kilogram = 1,000 grams.

Kilometer—A length of 1,000 meters, equal to 3,280.839 feet, or 0.621 of a mile. The chief unit for long distances in the metric system.

Kinetic Energy—Energy possessed by a mass because of its motion; for example, water falling over a dam has the energy to put turbines into motion.

Kinetic Theory—The theory that a material body is not one continuous, uninterrupted mass but is made up of minute, invisible particles (molecules) that are in constant motion, oscillating, bumping into each other, and bouncing back. Increasing the temperature causes a raise in the speed of the molecules, thus increasing the kinetic energy of the molecules.

King Cotton—A phrase testifying to the importance of cotton in the United States.

Kinkaid Act—An act of the United States Congress, passed in 1904, which increased the size of homesteads to 640 acres. It was restricted in application to the grazing lands of western Nebraska.

Knapsack Seeder—A device for broadcast seeding that consists of a canvas sack fastened to a seeding mechanism. A crank, turned by hand, revolves a wheel having radial ribs that throw seeds to the front and sides. It is sometimes used for spreading fertilizer and lime.

Knapsack Sprayer—A kidney-shaped tank carried on the back of the operator. A pump is used to build up pressure in the tank, and an extension tube connects a flexible hose from the tank to a nozzle for spraying.

Knee Cushion—A pad fastened to a person's knee to enable him/her to work in some comfort while kneeling. Also called knee pad.

Knife Guards—Fingerlike projections on the cutterbar of a mowing machine, combine, etc., that protect the blade from solid objects and guide the plants into the blade. See Cutterbar.

Knotter—A device in a hay baler that automatically holds and ties twine or wire around a bale when the bale reaches the proper size.

Kosher—Designating any food produced, killed, or prepared according to Jewish dietary laws.

Label—A tag of wood, metal, paper, plastic, or other material, fastened to a plant, animal, or product for identification, ownership, composition, etc.

Labor Budget—A budget comparing labor requirements with labor available, usually constructed on a seasonal basis.

Labor Chart—A form of labor budget constructed as a figure with a calendar of working days recorded on the horizontal axis and with number of workers recorded on the vertical axis; the chart shows the number of workers assigned to each task and the duration of that task.

Labor Expense—The hired labor costs, plus the estimated value of unpaid family labor, and a charge for the operator's labor.

Labor Income—Farm income after all expenses, depreciation, unpaid family labor other than the operator, and an interest charge on the total capital investment have been deducted.

Labor Profile—The seasonal pattern of labor requirements for a given farm activity.

Lactometer—An instrument that determines the density of milk. It acts as a specific gravity indicator, from which information on the percentage of milk solids may be calculated.

Lacuna—(1) A cavity, hole, gap. (2) A gap in essential information.

Lambing Loop—A length of smooth or plastic-coated wire used as an aid in difficult lambing.

Lambing Pen—A specially equipped, isolated pen in the sheep barn in which a ewe is placed just before she gives birth to her young.

Lambing Percentage—The number of lambs produced per 100 ewes.

Lambing Shed—A shed or barnlike building in which ewes are placed just before they give birth to lambs.

Land—(1) The total natural and cultural environment within which production must take place. Its attributes include climate, surface configuration, soil, water supply, subsurface conditions, etc., together with its location with respect to centers of commerce and population. Oyster beds and even tracts or bodies of water, as where valuable fishing rights are involved, may be regarded as land. It is often convenient, in fact, to regard land as synonymous with all that nature supplies, external to humans, which is valuable, durable and appropriable, thus including, e.g., waterfalls and other sources of waterpower. (2) In a broad legal sense, any real part of the surface of the earth, including all appurtenances, anything in, on, above, or below the surface. (3) In plowing, a plowed or unplowed space between two furrows. (4) The total width of a strip of land tilled by a farmer, or some designated width, as a perch, 16½ feet. Also called a stitch. (5) Soil. (6) A natural part of the earth's surface characterized by any single factor, or combination, of topography, climate, soil, rocks, vegetation; the natural landscape. (7) Pertaining to agriculture, those areas actually in use or capable of use for the production of farm crops and livestock.

Land Area—The area of dry land and land temporarily or partially covered by water such as marshes, swamps, and river floodplains (omitting tidal flats below mean high tide); streams, sloughs, estuaries, and canals less than one-eighth of a statute mile in width; and lakes, reservoirs, and ponds less than 40 acres in area.

Land Bank—Federal Land Bank.

Land Capability—A measure of the suitability of land for use without damage. In the United States, it usually expresses the effect of physical land conditions, including climate, on the total suitability for agricultural use without damage. Arable soils are grouped according to their limitations for sustained production of the common cultivated crops without soil deterioration. Nonarable soils are grouped according to their limitations for the production of permanent vegetation and their risks of soil damage if mismanaged.

Land Capability Class–One of eight classes of land distinguished according to potentiality for agricultural use. Class I consists of lands that are nearly level and can be cultivated continuously with little erosion. Class II consists of lands that are nearly level soils that require only simple practices such as contour tillage to control erosion. Class III consists of lands that require such practices as terraces and contour tillage to keep them productive for row cropping. Class IV consists of lands that require contour plowing, terracing, and the planting of sod-like crops every two to three years to control erosion. Class V consists of lands that cannot be planted to cultivated crops without extensive practices. Such soils should be maintained in sod crops. Class VI consists of lands that are too steep to be used for any crop except sod crops. However, even sod crops are difficult to establish and maintain. Class VII consists of lands recommended for use as watershed or woodland. Class VIII consists of lands recommended for recreational uses only.

Land Capability Map–(1) A map that shows land capability units, classes, and subclasses. (2) A soil conservation survey map that is colored to show land-capability classes.

Land Contract–An agreement between buyer and seller that states the terms of the purchase and sale of an item of real estate. The initial down payment on the property involved is usually small, often from 5 to 20 percent of the purchase price. Subsequent payments are scheduled for regular stated dates. The seller often retains the deed and title of the property until the purchase price is entirely paid; or the deed and title may be transferred at some earlier period agreed to in the contract. It is an important method used by young farmers in attaining farm ownership. Sometimes called installment land contract.

Land Grants–(1) Tracts of land given by the United States government to states, colleges, railroads, canal companies, and others for the purposes of aid to education and promotion of economic development of the unoccupied public domain. (2) Grants of land, to individuals or groups, that were made by French, Spanish, and Mexican governments in North America before the territory became a part of the United States. Such grants may still have a legal status. (3) Tracts of land awarded to individuals in the British Empire and in the United States by the governments as recognition of their services.

Land Patent–A deed or record title issued by a government or state for the conveyance of some portion of the public domain.

Land Plane–A large, tractor-drawn machine designed for planing or smoothing land for more efficient use of irrigation water or for easier tillage of land not irrigated. It consists of a long steel frame mounted on wheels, near the center of which is attached a large, long, adjustable, combination steel blade and scraper to remove soil from high points and to convey it to depressions. See Land Forming.

Land Policy (United States)–The guiding attitude or political philosophy in relation to land use that is based upon analysis of facts and pressures arising from problems in land use. It is a consistently followed, fundamental course of action, though open to revision.

Land Shark–A tricky or unscrupulous person who deals in land.

Land Tenure–The holding of land and the rights that go with such holding which includes everything from fee simple title, embracing all possible rights within the general limits imposed by the government, down to the most restricted forms of tenancy (holding and/or operating land under the ownership of another).

Land Type–A geographic division of land based upon some one or a combination of natural factors as soil, relief, vegetation, and climate. The term is often loosely used to indicate use of the land such as agricultural, grazing, mining, forest, urban, etc.

Land-Grant Universities–State colleges and universities started from federal government grants of land to each state to encourage further practical education in agriculture, homemaking, and the mechanical arts. The mission of these universities is to conduct programs in teaching, service and research. implemented by the Morrill Acts of 1862 and 1890. See Morrill Act.

Land-Poor–(1) Designating a person who owns a great deal of unproductive land. (2) Designating one who owns ample land but makes inefficient use of it or is obliged to borrow money to pay taxes, interest, etc.

Landed–Designating a person, or group, that owns a large estate of rural land.

Landing–A place at which logs are assembled for transportation in loads or rafts.

Landlord–One who owns land or property and leases it to a tenant.

Landscape Architect–A person trained in the art and science of arranging land and objects upon it for human use and enjoyment.

Landscape Design–The profession concerned with the planning and planting of outdoor space to secure the most desirable relationship between land forms, architecture, and plants to best meet human needs for function and beauty.

Lane–A passageway from barnyard to pasture or field areas; a narrow road confined between fences.

Lareat–See Lariat.

Lariat–Rawhide, horsehair, or hemp rope, generally 35 to 45 feet long and arranged in a coil on the saddle. Used to lasso livestock for branding or restraining purposes. See Lasso.

Lasher–A piece of rope used for binding or making one thing fast to another.

Lasso–(1) To throw a lariat in such a manner as to catch an animal by the horns, neck, or legs in the loop at the end, generally for restraining purposes. (2) A rope or line with a running loop used for catching animals. Lee Lariat.

Latex– (1) A usually white or yellowish fluid produced by the cells of some plants, as *Asclepias* and *Euphorbia*; often referred to as milky sap. (2) Designates a paint that is water-soluble in the fluid state.

Latitude– (1) Distance north or south on the earth's surface form the equator (latitude zero), measured in degrees of the meridian (North Pole is 90° north latitude and the South Pole is 90° south latitude). (2) In plane surveying, the perpendicular distance in a horizontal plane of a point from an east-west axis of reference.

Law of Adverse Possession– The acquiring of property by having it in possession for a time prescribed by law.

Law of Demand– An economic principle that states that, everything else being equal, consumers can be expected to buy more of a product as its price falls and less as its price rises.

Law of Diminishing Returns– An economic law that states that as an increasing amount of a variable factor of production, e.g., humanpower, is applied to a fixed factor of production, e.g., land, the result is a decrease in average returns per variable unit applied.

Law of One Price– A marketing principle that holds that, under perfectly competitive market conditions, all prices within a market will be uniform after the costs of adding place, time, and form utility are taken into consideration.

Law of Original Horizontality– A general law of geology. Water-laid sediments are deposited in strata that are horizontal and parallel to the earth's surface.

Law of Supply– An economic principle suggesting that, everything else being equal, producers will offer to sell more of a product at a higher price than at a lower price.

Law of the Minimum– Liebig's Law of the Minimum: "When a process is conditioned as to its rapidity by a number of separate factors, the rate of the process is limited by the pace of the slowest factor." Mitscherlich's restatement: "The increase of crop produced by unit increment of the lacking factor is proportional to decrement from the maximum." Our restatement: "Plant growth is restricted by any one essential factor available in least relative supply."

Law of Universal Gravitation– Newton's law of gravitation. Every particle of mass attracts every other particle with a force inversely proportional to the square of the distance between them.

LCL– Less than a carload lot.

Lease– (1) A legal written agreement entered into between the owner and the tenant for the use of a given property or acreage of land for a given period of time. (2) To rent by written contract.

Leased Land– Land leased by its owner, sometimes from a state or federal government, for farming or grazing purposes.

Least Cost Ration– The ration for an animal that is the most economical yet still provides all the essential nutrients needed by the animal.

Ledger– (1) A book in which farm/ranch accounts are kept. All credits and debits, etc., are registered under appropriate headings. (2) A horizontal board that forms the top rail of a simple fence, etc.

Left Bank– The left bank of a flowing body of water is on your left as you face downstream. Because of the coriolis effect, in the Northern Hemisphere the left bank is usually less steep and has more sediment accumulation than the right bank; and vice versa in the Southern Hemisphere. See Coriolis Effect, Right Bank.

Legal Control– Control of pests through the enactment of legislation that enforces control measures or imposes regulations, such as quarantines, to prevent the introduction or spread of pests or disease.

Legal Description– The exact, geographical survey description of real estate used in abstracts of title and on the legal real estate tax rolls.

Legend– Explanation of the symbols and patterns shown on a map.

Length-of-Run– In irrigation, the distance water must travel in furrows or over the surface of a field from one head ditch to another or to the end of a field.

Less-developed Countries– Countries that have not yet achieved sustained economic growth. They also are referred to as underdeveloped, developing, emerging, or third world. Most less-developed countries are in Asia, Africa, and Latin America.

Lessee– One who leases property from the owner or another lessee. See Lessor.

Lessor– One who leases property to someone else. See Lessee.

Levee– (1) An earth dam or dike built along a river or the sea to serve as flood protection. (2) A continuous ridge built around a field to retain flood irrigation water. (3) An embankment built along the banks of a river to reduce the hazard of flooding.

Levee District– A geographical area, subject to flooding by a stream and its tributaries, in which the landowners have incorporated to build and maintain levees. It is a quasigovernmental unit that assesses fees by taxation. Common along the lower Mississippi River.

Level– (1) An instrument used to ascertain the horizontal lines at different altitudes as a guide in terracing, ditching, foundation laying, etc. (2) The amount of a component in a mixture, as the nitrate level. (3) The amount, usually expressed in parts per million of an element, of fertility in a soil solution. (4) To make a field approximately horizontal. (5) Designating a piece of land that is nearly on a horizontal plane.

Level Transit– An instrument used in running grade lines or elevations for laying tile and drainage ditches, designing terraces, etc. See Abney Level, Hand Level.

Leveler– A drag or any other device used for smoothing land for irrigation or drainage. See Land Plane.

Liability– Money, goods, or services that a person owes to another.

Lick Tank– A tank containing a liquid feed supplement for cattle. The cattle obtain the feed by licking a wheel or ball that rotates in the liquid. As the wheel or ball turns, the liquid adheres to it and is brought to the surface.

Liebig's Law of the Minimum– See Law of the Minimum.

Lien– A legal claim on a property or asset for the security of a debt, such as a lien on a crop to secure payment of money advanced to grow it.

Lift– (1) A joint of meat, especially beef, from the thigh. (2) Elevating or pulling power of a pump. (3) To loosen and remove seedlings or transplants from the seedbed or transplant bed prior to transplanting. (4) A hay fork. (5) A fork for lifting heavy loads.

Light–(1) The form of radiant energy consisting of wavelengths lying within the limits perceptible by the normal human eye, and, by extension, the shorter and longer wavelengths, the ultraviolet and the infrared light, invisible to the eye but which may be recorded photographically. Light can be absorbed by various substances and transformed into heat. Its excess may produce fading, the destruction of green color in plants, but an insufficiency can also cause lack of chlorophyll production. The coloration of fruits is dependent upon sufficiency of light. The growing parts of plants respond to the stimulus of the direction from which the light comes. See Photosynthesis, Phototropism. (2) To become ignited; to take fire. (3) Designating a deficiency or lack in degree, such as a light rain or a light crop. (4) When applied to food or drink it can have several definitions. It could mean reduced calories, fluffy (full of air), pale, low in sodium, mild in flavor, and/or less alcohol.

Light Oil–Any lubricating oil with a viscosity number of ten or less, as graded by the Society of Automotive Engineers.

Lightning Rod–A metallic rod set upright on the roof of a building, a tall chimney, etc., and grounded by a wire conductor. It serves to conduct the lightning to the earth, thus minimizing damage to the structure. Also called lightning conductor.

Limit–The top or bottom price that a commission house will allow its buyers to pay for a product at a certain place for a certain period of time.

Limited Feeding–(1) The feeding of livestock to maintain weight and growth but not to fatten or increase production. (2) Restricting an animal to less than maximum weight increase.

Limited Partnership–A partnership in which one or more of the partners, but not all of them, has a limited liability to the partnership creditors.

Limiting Factor–Any influence or material that tends to slow down growth and productivity in an ecosystem; either too much or too little of these critical factors will limit production. See Law of the Minimum.

Line–(1) The reins of a harness. (2) A rope, cable, string, wire, tube, etc., for tying or hanging objects, or conducting electricity, water, gas, etc., as a power line, gas line, or milk line. (3) A boundary or limit, as a fence line, property line. (4) In marketing, the whole of a herd of sheep. (5) A group of plants or animals that retain their uniform appearance in succeeding generations.

Line Camp–A livestock camp away from headquarters; used by riders for temporary shelter (southwestern United States).

Line Fence–A boundary fence, usually of better quality and more permanent than the cross fences.

Line Intercept Method–A method of sampling vegetation by recording the plants intercepted by a measured line placed close to the ground, or by vertical projection of plants to the line. See Line Transect, Transect.

Line of Credit–The amount of credit granted an individual or business by a lending institution that may or may not all be used at any one time by the borrower.

Line Post–Any of several fenceposts used in a line fence.

Line Rider–A man who rides horseback along the outskirts or line fences of a cattle range to see that all is in order (western United States).

Line Transect–A method of recording plant species in a range study in which a tape or chain is stretched on the ground and the plants touched by the tape are recorded. See Line Intercept Method.

Line-plot Survey–A system of sampling vegetation in which plots of uniform size are located at regular intervals along a line.

Linear Evaluation–A method of evaluating the degree of a trait in an animal. It involves assigning a numerical value to a trait so deviations from the average are descriptive by degree rather than subjective. This system is used extensively in the dairy industry.

Linear Programming–A computer-based procedure used for solving allocation problems such as farm planning and formulation of livestock rations.

Lineolate–Marked with fine lines.

Link–A unit of linear measure one one-hundredth of a gunter's chain; equivalent to 7.92 inches (20.116 centimeters).

Lint Index–The weight of lint cotton produced by 100 seeds.

Liquefaction–The change in the phase of a substance to the liquid state; in the case of fermentation, the conversion of water-insoluble carbohydrate to water-soluble carbohydrate.

Liquefied Petroleum (LP) Gas–A fuel that has been made usable as a fuel for internal combustion engines by compressing volatile petroleum gas into liquid form. It must be kept under pressure when stored.

Liquid Asset–Cash or any asset that can be readily converted to cash.

Liquid Limit–The moisture content at which the soil passes from a plastic to a liquid state.

Liquid Smoke–A liquid mixture applied to meats in curing, especially to ham and bacon, to replace curing by wood smoke.

Liquidation–The sale of a previously bought contract, otherwise known as long liquidation. It may also be the repurchase of a previously sold contract, generally referred to as short covering.

Liquidation Value–(1) In appraising, the price at which a property can be sold within a short period for cash or its practical equivalent and in which the seller forces the sale and the buyer dictates the terms. (2) The value of a forest estimated on the assumption that all merchantable trees will be cut.

LISA–See Low Input Sustainable Agriculture.

Lister–A double moldboard plow that throws the soil on both sides; used in semiarid regions for planting crops in the bottom of the furrow. Also called lister plow, middlebreaker, middlebuster.

Listing Agreement–Complete information concerning a tract of land and authority to offer a property for sale.

Liter–A standard measure for liquids denoting a capacity of 61.025 cubic inches or 1.9567 United States liquid quarts.

Litre–See Liter.

Litter–(1) On a forest floor, the uppermost surface layer of debris, leaves, twigs, and other organic matter, undecomposed or slightly altered. In a technical description of a soil profile, it is generally designated by the letter O. (2) Accumulation of leaves, fruits, twigs, branches, and other plant parts on the surface of the soil. See Mulch. (3) A group of young animals born at a single birth, as a litter of pigs, etc. (4) See Bedding.

Litter Floor–The floor of a poultry house that is composed of straw and shavings or ground corncobs, droppings, and other waste materials, and builds up during one laying year.

Live Delivery–A common guarantee by hatcheries, of delivering live chicks by the addition of two or more chicks per order of one hundred.

Livestock Commission House–An organization of livestock buyers that acts as an agent for the sale of livestock of many different producers at a given market for a fee or commission. Most livestock marketing in the United States is handled by such houses.

Livestock Enterprise–A farm activity that consists of one phase of the livestock business, such as the dairy enterprise, laying flock enterprise, etc.

Livestock Gross Income–The value of livestock production in the form of animals and produce, adjusted for inventory changes and net of the value of any livestock purchased or obtained as gifts.

Livestock Market–A market that specializes in transactions of livestock for processing and distribution, as Chicago and Kansas City livestock markets or the numerous farmers' markets in the United States.

Livestock Mortality Policy–An insurance policy protecting the owner of an animal against loss by death of the animal.

Livestock Production–The weight, number of animals, etc., that a particular range, pasture, or management system produces. May also refer to the business of producing livestock.

Livestock Ranch–An area of land usually larger than a farm and largely in grass where livestock, such as beef cattle or sheep, are produced.

Livestock Registry Association–A purebred breeding association with which animals of known and recorded ancestry may be registered. See Breed Association.

Livestock Shipping Association–An association of several producers who pool their livestock for the purpose of shipping and marketing; usually a cooperative.

Livestock Show–A show at which livestock are exhibited according to breed and class and are judged according to standards of perfection. Also called cattle show, swine show, etc.

Load–(1) The work sustained by a machine. (2) The resistance offered to a motor by the machinery it drives, apart from the friction of its own parts. (3) In the export trade, a unit of lumber measurement, used particularly in India; it comprises 40 cubic feet of round timber or 50 cubic feet of squared timber, equaling approximately a long ton. (4) The weight carried by a beam, girder, truss span, or structure of any sort, or any part of such structure, including its own weight. (5) A burden; a weight; as a heavy load. (6) The work done by a prime mover. (7) To place a load on a wagon, truck, etc. (8) In erosion and corrosion the material transported may be called the load. The load is transported by two methods; a portion is moved along the bottom. (9) The sediment moved by a stream, whether in suspension or at the bottom, is its load. (10) The quantity of material actually transported by a current.

Loading Chute–An inclined chute that is used for animals to walk from the ground into a truck or trailer.

Loafing Barn–A light type of building usually attached to the main dairy barn or milking parlor where cows may be turned loose after milking for exercise and comfort. Hay and silage are usually supplied in side racks or bunkers.

Loan Rate–The price per unit (bushel, bale, pound) at which the government will provide loans to farmers to enable them to hold their crops for later sale.

Loan Value–In appraising, the value of real estate on which lenders base the amount of first mortgage credit they will advance.

Local–Limited to one region, area, or place, rather than widely spread; e.g., local pain, local anesthetic, local infection, local markets, local customs.

Lock Washer–A split ring of steel that is used under a nut or bolt to prevent the nut or bolt from vibrating loose.

Locker–A unit in a cold-storage plant that usually holds from 100 to 250 pounds of quick-frozen products, such as meats, fruits, and vegetables. It is usually rented by the user, but often is owned by a cooperative.

Log Chain–A long steel-link chain with hooks on each end that is used for securing logs. Around the farm, it is often used as a general purpose chain.

Log Mark–A letter or sign stamped or chopped on logs indicating ownership.

Log Rule–A table showing the estimated amount of lumber that can be sawed from logs of given length and diameter. See Doyle Rule, Scribner Rule.

Logging Residues–The unused portions of poletimber and sawtimber trees cut or killed by logging.

Long–(1) Designating a person who has brought more contracts than he has sold. Long hedges are purchases of futures made as a hedge against the sale of the cash commodity (short). (2) Designating a pulse beat that is longer than normal.

Long Hedge–A hedge that initially is established by buying in the futures market. See Hedge.

Long Ton–See Ton.

Long-term Cash Flow Budget–A cash flow budget constructed for a planning horizon of ten years or so with intermediate cash balances normally calculated at annual intervals. See Short-term Cash Flow Budget.

Long-term Credit–Credit usually advanced for five to thirty-three years for the purchase of real estate. See Short-term Loan.

Long-term Liabilities–Liabilities that will be due within the next seven to thirty years.

Long-term Loans–Loans made for a period of seven to thirty years.

Lookout–(1) A person employed to detect and report forest fires. (2) A tower or elevated post used primarily in the detection of forest fires.

Loose Housing–A management system for cattle wherein the adult animals are given unrestricted access to a feeding area, water, a resting area, and an adjoining open lot. In dairies, the lactating animals pass through a milking room at milking time. Other dairy animals may be in separate pens, lots, or buildings.

Lopping Shears–A long-handled set of cutters used for removing branches from trees and other plants.

Lost Corner–In land surveying, a corner whose position cannot be definitely determined from traces of the monument or by reliable testimony relating to it; it can be restored only by surveying from an established survey point.

Louver–A slatted opening designed to permit the passage of air but to exclude rain, direct sunlight, and vision. Used in homes, barns, and animal housing.

Low Input Sustainable Agriculture (LISA)–A production concept where expenses, inputs, etc., are kept to a minimum in order to allow the operation to continue indefinitely with the least negative impact on the environment. For instance, planting nitrogen-fixing plant species, crop rotation as a means of pest control, and special diversity of plants could be used to limit inputs.

Low Intensity Animal Production–Systems of producing animals that strive to use less capital, energy, and fewer purchased inputs than conventional confinement systems; e.g., a pasture and hutch system for swine production.

LTL–Less than a truckload lot.

Luxury Consumption–A phenomenon brought about by the fertilization of plants with excessive quantities of such compounds as ammonia, ammonium nitrate, or potassium chloride, which causes the plant to absorb ammonium nitrate and potassium ions in excess of those amounts required for normal growth and reproduction. This may not be harmful to the plant, but may be a waste of fertilizer.

Lyophilization–The evaporation of water from a frozen product with the aid of high vacuum. The process is used for the preservation of stocks of bacteria and other microorganisms.

Lysimeter–A device used to measure the quantity or rate of water movement through or from a block of soil or other material, such as solid waste, or used to collect percolated water for quality analysis.

Machete–A long, heavy knife designed for cutting small brush and for hand-harvesting of sugarcane.

Machine–A device that uses mechanical energy to get work done.

Machinery Expense–The total operating cost, including depreciation and repairs, for farm machinery, which includes trucks and the farm share of the family car.

Macro–Prefix meaning large, long; visibly large.

Macroeconomics–Economic studies or statistics of groups of commodities or subjects such as total consumption, total employment, or total income. See Microeconomics.

Magnetic Compass–An instrument having a freely pivoted magnetic needle that aligns with the earth's magnetic field such that one end of the needle points to the magnetic north.

Main–The principal pipeline in a water supply or drainage system.

Main Crop–The principal or most important crop grown on a farm or in an area.

Maintaining Forage Reserves–The reservation of native forage supplies for such emergencies as drought, fire, or other unforeseen circumstances; obtained by the exclusion of livestock or by light use of grazing areas.

Maintenance–(1) The servicing of facilities of equipment of farms that is required for upkeep, replacement of worn or broken parts on machinery, lubrication, etc. (2) The upkeep of gardens and grounds.

Male Threads–Threads on the outside of a projecting object such as a bolt. See Bolt, Female Threads.

Malthusian Theory–A theory or doctrine expounded by Thomas R. Malthus (1766-1834), an English clergyman, according to which human population tends to increase at a greater rate than food supply. The corollary is that unless the birth rate is controlled, poverty and famine are inevitable.

Man-Day–A unit of measurement of labor input or requirement, usually assumed to represent the work accomplished by an adult male worker in eight hours.

Management–The human factor within a production process (or firm) that delimits problems, accumulates information relevant to their solution, analyzes that information, reaches decisions, acts on those decisions, and bears responsibility for the consequences of those actions. See Farm Management.

Management of Pastures–The handling of animals and plants in such a manner that the stand of desirable forage is not depleted and at the same time provides a sufficient supply of palatable, nutritious forage for the producing or growing animal for as long a time as is possible during the grazing period.

Manger–A trough or other receptacle of metal, wood, concrete, or stone, in which fodder is placed for cattle to eat.

Manifest–To show, to appear.

Manning's Formula–(Hydraulics) A formula used to predict the velocity of water flow in an open channel or pipeline: $V = 1.486/n \, (r^{2/3} S^{1/2})$. Where is the mean velocity of flow in feet per second; r is the hydraulic radius; S is the slope of the energy gradient or for assumed uniform flow the slope of the channel in feet per foot; and n is the roughness coefficient or retardance factor of the channel lining.

Manometer–A device for measuring a vacuum.

Manual Watering–Watering plants by hand with a hose, bucket, or some other manual means.

Manure Bucket–A hydraulic attachment for a tractor, used to lift and load manure. Also called manure scoop.

Map Scale–The ratio of the distance between two points shown on a map and the actual distance between the points on the earth's surface. Scale is commonly expressed as a representative ratio as 1:1,000.

Marginal Land–Agricultural land definitely not first class, and near the margin between profitable and unprofitable use.

Market–(1) A place, usually public, where goods and commodities are offered for sale or exchange. (2) The demand for a product; or a country, region, etc., in which there is such a demand. (3) To buy and sell produce.

Market Basket of Farm Foods–Average quantities of United States farm foods purchased annually per household in a given period, usually a base period. Retail cost of these foods used as a basis for computing an index of retail prices for domestically produced farm foods. Excluded are fishery products, imported foods, and meals eaten away from home.

Market Bird–Any fowl produced and fattened for meat purposes.

Market Classes and Grades–Various market classes and grades established by the United States Department of Agriculture to sort livestock according to conformation, finish, quality, use, age, sex, and weight.

Market Development–Marketing activities and efforts designed to enhance the value of food products to consumers and in the process expand sales and profits.

Market Fluctuation–Variation in prices that are received or paid for products placed on the market.

Market Garden–A farm that produces vegetables to be sold on a roadside market or in a nearby city.

Market Grade–A set of descriptive terms, such as prime, choice, etc., used in livestock marketing to designate the comparative value of animals based on differences in type, conformation, degree of finish, etc.

Market Grades of Cows and Bulls–Determined by the age and fatness of the animal. The grades are: commercial, utility, cutter, canner.

Market Grades of Lambs–Prime, choice, utility, canner, cutter, cull.

Market Grades of Slaughter Steers and Heifers–Market grades of beef animals that are to be slaughtered are determined by the age of and the amount of fat on the animal that is indicative of the amount of marbling in the meat. The grades are: prime, choice, select, standard. Cattle are also graded by the percentage of lean retail cuts of meat the animal is expected to yield. See Marbling, Maturity, Yield Grade.

Market Grades of Swine–Various market classes of swine established by the United States Department of Agriculture: United States Number 1, United States Number 2, United States Number 3, and cull.

Market Information–Any form of information relevant to a market decision.

Market News–Descriptive information on current market conditions, including prices, stocks, demand, and so on.

Market Order–An order to buy or sell that is to be executed at the best possible price as soon as received in the trading ring or pit.

Market Pack–A fiber or plastic container used to grow bedding plants; the container holds from six to twelve plants.

Market Performance–The economic results that market participants (farmers, consumers, middlemen) and society expect from the food marketing system.

Market Power–The ability to influence markets, market behavior, or market results.

Market Price–The price for a product, animal, etc., established by buyers and sellers in competition in the marketplace.

Market Risk–The possibility of loss—through product deterioration in quantity or quality or value change—while a product is being produced, stored, or marketed.

Market Segmentation–The marketing technique of developing separate products and marketing programs to appeal to different consumer classes.

Market Standard–The official standard governing weights, grades, and qualities of products offered for sale on a market, such as No. 2 corn, grade A eggs, etc., set by the United States Department of Agriculture.

Marketable Surplus–The production of an individual or a society that exceeds that needed or desired for personal consumption; this surplus is then available for sale to other individuals or countries.

Marketing Agreement Act–An act of the United States Congress authorizing the Secretary of Agriculture to establish orderly marketing conditions for agricultural commodities to help achieve parity. It gives the Secretary power to enter into agreements with processors and handlers of any product affecting interstate commerce.

Marketing Agreements–A voluntary contract between an agency of the United States Department of Agriculture and a handler of an agricultural product.

Marketing Bill–The total dollar expenditures going to food marketing firms to pay for all marketing activities.

Marketing Channels–Alternative routes of product flows from producers to consumers.

Marketing Concept–A management philosophy that holds that all company planning begins with an analysis of consumer wants, and that all company decisions should be based upon the profitable satisfaction of consumer wants.

Marketing Margin–The portion of the consumers' food dollar paid to food marketing firms for their services and value-adding activities; the "price" of all food marketing activities.

Marketing Mix–The unique way in which a firm or industry combines its price, promotion, product, and distribution channel strategies to appeal to consumers.

Marketing Myopia–A term referring to the tendency of some firms to define their business too narrowly in terms of a specific product; for example, a dairy firm may define its business as dairy products or, more broadly, as fluid beverages.

Marketing Orders and Agreements–(Federal) A means (authorized by, and based on, enabling legislation) to permit agricultural producers collectively to influence the supply, demand and/or price for a particular crop or commodity in order to improve the orderly marketing of the crop or commodity. Once approved by a required number of producers—usually two-thirds—of the regulated commodity, the marketing order is binding on all handlers of the commodity in the area of regulation. A marketing agreement may contain more diversified provisions, but it is enforceable with respect to those producers or handlers who voluntarily enter into the agreement with the Secretary of Agriculture.

Marketing Process–The sequence of events and actions that coordinate the flow of food and the value-adding activities in the food marketing system.

Marketing Quota–That quantity of a crop that will provide adequate and normal market supplies. This quantity is translated into terms of acreage needed to grow that amount and allotted among individual farms, based on their previous production of that commodity. When marketing quotas are in effect (only after approval by two-thirds or more of the eligible producers voting in a referendum), growers who produce in excess of their farm acreage allotments are subject to marketing penalties on the "excess" production and are ineligible for government price support loans. For certain tobaccos, a poundage limitation is applicable as well as acreage allotments, when approved by grower referendum.

Marketing Spread–The difference between the retail price of a product and the farm value of the ingredients in the product. This farm-retail spread includes the charges made by marketing firms for assembling, sorting, processing, transporting, and distributing the products.

Marketing Strategy–A plan to achieve a market goal, for example, one grocery store may use the strategy of low prices to attract consumers; another might employ the strategy of high quality.

Marketing Year–The year beginning at harvestime during which a crop moves to market.

Marking Harness–A harness strapped to a ram that marks a ewe's rump when she is bred by the ram.

Mastica–A plastic material used in the place of putty for glazing greenhouse glass and hotbed sash.

Maternal Breeding Value–A prediction of how the daughters of a bull will milk based on weaning weight information. The accuracy figure is the amount of reliability that can be placed on the breeding value.

Maternal Calving Ease of First Calf–Calving ease ratings of daughters of bulls, when they give birth, as reported to breed associations by producers, values above 100 are superior.

Maternal Weaning Weight–Measure of a sire's ability to transmit maternal performance, expressed in weaning weight of his daughters' calves. It is a combination of milk production and growth rate that a sire transmits through his daughters.

Mattock–A hand tool that consists of a short handle with a two-bladed steel head. One blade, which is perpendicular to the handle, is flat and is used for grubbing. The second blade may be pointed like a pick or it may be flat and ground to a sharp edge like an ax for cutting.

Mature Class–In the show ring, dairy cows five years of age or older that are in milk.

Mature Equivalent Factor–An age conversion formula used to predict the expected mature milk record for a cow, based on a previous year's milk production. It is frequently used in cattle selection and breeding.

Maverick–(1) Any unbranded animal, particularly a calf. Named after Samuel A. Maverick, a lawyer and a nonconformist Texan (1803-1870) who refused to brand his cattle. (2) A motherless calf (western United States).

Maverickers–Men who steal and put their brands on unbranded cattle (western United States).

MBF–Thousand board feet.

McLean County Sanitation System–A system developed in McLean County, Illinois, by veterinarians of the United States Bureau of Animal Industry for the prevention and control of filth-borne diseases of swine. It has particular reference to internal parasites.

Mean–A value determined by the addition of several values and dividing by the number of values added; the average. See Median, Mode.

Mean Annual Increment–In forestry, the total growth (board feet or cubic feet) of a stand of trees divided by the total age in years.

Mean High Water–The average height of the high water over a specified number of years.

Mean Low Water–The average height of the low water over a specified number of years.

Meander Line–The waterline or shore of a body of water, as a large lake, established at the time of an official land survey, as that of the United States General Land Office. Legally, it is important in determining the ownership and use of shoreland.

Mechanic's Lien–Claim for purpose of securing priority payment for work and/or materials furnished in erecting or repairing a building.

Mechanical Aeration–Mechanical energy used to inject air into water, causing the waste stream to absorb more oxygen. See Lagoon, Aerated.

Mechanical Buffer–A machine used in the final operation of plucking feathers from poultry slaughtered for market. Also called buffer.

Mechanical Control–(1) Control of pests by mechanical means such as window screens, earth barriers, and so on. (2) Human-built structures used to control erosion, such as terraces, dams, retards, baffles, etc., in contrast to vegetative control.

Mechanical Energy–(1) The ratio between the indicated horsepower of an engine and the brake horsepower of an engine. See Brake Horsepower. (2) Energy in a form that can do work directly.

Mechanical Poultry Picking–The removal of feathers in dressing poultry for the market by the use of machines in contrast to removal by hand labor. See Mechanical Buffer.

Mechanical Refrigeration–Cold condition created by a machine compressing certain low boiling point gases which replaces cold cellars, the use of ice, and other older methods of preserving foods and products at low temperature.

Medial–A directional term that means toward the middle.

Median–The central value of a series of numbers, when ranked numerically, such that an equal number of entries lie on either side of it. See Mean, Mode.

Medium-term Cash Flow Budget–A cash flow budget extending over three or four years with the intermediate cash balances calculated at quarterly or half-yearly intervals.

Mega–(1) A prefix meaning one million. (2) A prefix meaning large or in large numbers.

Memorandum of Understanding–A written document showing the intent of two or more parties to cooperate in carrying out an undertaking that will result in mutual benefit to the parties concerned. It specifies precisely what each party is to do.

Mensuration–The science dealing with the measurement of volume, growth, and development of individual trees and stands, and the determination of various products obtainable from them.

Mercator Map Projection–A map projection of the so-called cylindrical type. The equator is represented by a strain line true to scale; the geographic meridians are represented by parallel straight lines perpendicular to the line representing the equator; they are spaced according to their distance apart at the equator. The geographic parallels are represented by a second system of straight lines perpendicular to the family of lines representing the meridians, and therefore parallel to the equator. The greatest distortion is at the north and south poles.

Merchant Middleman–A food marketing firm that provides a variety of marketing functions, including taking title to products.

Merchant Wholesaler–A wholesaling middleman who physically handles and takes title to products.

Merchantable Height–The length of the tree stem from the top of the stump to the top of the last merchantable section. Usually expressed in feet or number of logs of a specified length.

Merchantable Timber–A tree or stand of trees that may be converted into salable products.

Merchantable Volume–The amount of wood in a single tree or forest stand that is considered salable.

Meridian Line–A true north and south line; a line from which range lines are established in the General Land Office survey (United States).

Meridian, Principal–(United States public-land surveys) A line extending north and south along the astronomical meridian passing through the initial point, along which township, section, and quarter-section corners are established. The principal meridian is the line from which is initiated the survey of the township boundaries along the parallels.

Mesh–One of the openings or spaces in a screen. The value of the mesh is usually given as the number of openings per linear inch. This gives no recognition to the diameter of the wire, so that the mesh number does not always have a definite relation to the size of the openings.

Metabolizable Energy–The total amount of energy in feed less the losses in feces, combustible gases, and urine. Also called available energy.

Metal–(1) Any of a class of substances that typically are fusible and opaque, are good conductors of electricity, and show a unique metallic luster, as gold, copper, bronze, and aluminium. Most metals are also malleable and comparatively heavy, and all except mercury are solid at ordinary temperatures. Metals constitute over three-fourths of the recognized elements. They form oxides and hydroxides that are basic, and they may exist in solution as positive ions (cations). (2) Ore from which a metal is derived. (3) Molten glass. (4) Railway rails.

Metayer–A tenant farmer who works a piece of land for a share in the crop. See Sharecropper.

Meter–A unit of length equivalent to 39.37 inches, 3.28 feet, and 1.09 years. Also spelled metre.

Metes and Bounds–In the United States, a means of defining the legal boundaries of a parcel of land by giving the bearings and distances from points of reference. Often a tree, a stone, or some natural feature is used in the description; the line from one point to another is not necessarily straight but may be curved. Land descriptions were commonly made by metes and bounds prior to the establishment of the rectangular system of land surveying by the General Land Office of the United States.

Metric–(1) Pertaining to measures based on the meter. (2) Having the meter as a basis.

Metric Ton–See Ton.

Metrology–The science of weights and measures.

Micro–A prefix meaning one millionth (1/1,000,000) or very small.

Microeconomics–Economic studies or statistics of individual commodities or subjects such as the demand for wheat or employment on farms. See Macroeconomics.

Microgram–One millionth of a gram.

Micrometer–A device for measuring width, length, or thickness with precision, usually in conjunction with a microscope or telescope.

Micron–A unit of measurement, approximately 1/25,000 inches (0.001 millimeters) used for measuring spores, bacteria, fat globules, soil particles, and other microscopic objects.

Micronaire–An instrument used to measure fineness of cotton, and other fibers.

Microwave–A very short electromagnetic wave of high-frequency energy produced by the oscillation of an electric charge. Microwave energy is converted into heat when it is absorbed by the food. Microwaves are about 5 inches long, in contrast to radio waves, which average about 0.3 mile in length. A short wave has a greater frequency, or vibrations per second, than a longer wave has. Electronic ovens have a frequency of 2,450 megacycles (million cycles) per second. Ordinary AC electricity, which vibrates 60 times per second, has a frequency of 60 cycles per second.

Mid-pivot Mower Conditioner–A mower conditioner that swivels in the middle for greater maneuverability. See Mower Conditioner.

Middle Atlantic Truck Crop Belt–One of the principal agricultural regions of the United States: the coastal belt, extending from South Carolina to Long Island, New York, in which the production of truck or garden crops is a major industry.

Middlebuster–In tillage of soil, a double-moldboard plow designed to make a bed, especially for cotton and corn. Also called middle breaker.

Middleman–A merchant, broker, jobber, or dealer who performs a marketing service between the producer and the retailer.

Midget Cattle–See Dwarf Cattle.

Migrant–Designating a plant, animal, or person that changes, or has changed, its natural location.

Migratory Worker–A farm laborer who moves from one locality to another according to variations in time of harvesting crops, or one who lives in a locality only for the time required to carry out operations in the production and harvesting of one or more crops. Usually, the family of the laborer is also employed on the farm.

Milacre–A sample plot, 1/1,000th of an acre (usually 1/10 chain square), used in plant reproduction or vegetation surveys.

Milk Dryer–Any mechanical device designed to remove most of the water from milk. The water may be evaporated from pans or from steam-heated rollers; or the milk may be sprayed into a stream of hot air. See Drier.

Milk Irradiator–A device that uses an electric arc or mercury vapor lamp to convert ergosterol in milk into vitamin D_2.

Milk Ordinance–A law that sets forth regulations concerning the handling and quality standards of milk sold for public consumption.

Milk Record–A record of the amount of milk produced by a cow during a specified period. Used for evaluation in terms of money, profit, culling, or for registry.

Milk Sanitarian–A professional worker who specializes in the supervision of sanitary milk regulations.

Milk-fat Basis–A method of paying for milk at receiving stations on the basis of price per pound of butterfat. See Butterfat.

Milk-Feed Price Ratio–The pounds of dairy ration equal in value to one pound of milk, at prices existing at a particular time and place. The United States Department of Agriculture milk-feed price ratio is pounds of dairy concentrates, rather than the entire dairy ration.

Milker–(1) A milk-giving cow. (2) (a) One who milks cows. (b) A machine that milks cows.

Milking Machine–A mechanical device replacing hand labor in the milking of cows. The essential parts are teat cups, a vacuum pump, and a milker pail or milk line. Milk is drawn from the udder by application of alternate vacuum and atmospheric pressure.

Milking Parlor–In a dairy, an especially arranged and equipped room where cows are separately fed concentrates and milked by mechanical milking equipment.

Milkshed–A designated geographic area of milk production and consumption.

Mill–(1) A building with machinery used in grinding, pressing, processing, etc., e.g., grain mill, cane mill, cider mill, saw mill, etc. (2) To process a product. (3) To move around in a circle, as a herd of cattle.

Mill Work–Remanufactured lumber products, such as sash, doors, and molding.

Miller–(1) The operator of a mill, especially of a flour mill. (2) Any moth with wings that appear dusty or powdery.

Milli—A prefix meaning 1/1,000.

Milliequivalent–One thousandth of an equivalent weight.

Milliequivalent Per Liter–A milliequivalent of an ion or a compound in one liter of solution.

Millimeter–One thousandth of a meter.

Millimho–A measure of electrical conductivity in proportion to salinity, used in expressing the comparative salinity of soils and water.

Millrace–A sluiceway through which water runs to drive the wheel of a water mill. Common in old-fashioned grist mills. See Millstone.

Millstone–A large circular stone grooved from the outer rim to a hole in the center. It is used for grinding cereals into flour and meal and it was in fairly common use in water grist mills in the United States up to the end of the nineteenh century. The stones were quarried from a very hard, coarse sandstone formation or a cellular quartz conglomerate. Still common in most developing countries.

Miner's Inch–A unit of irrigation water flow for small users established by statute in several western states and in British Columbia.

Mineral Rights–Rights pertaining to ownership of subsurface land minerals and to access in their exploitation. An owner of land may sell only surface rights and retain ownership of mineral rights or sell only mineral rights.

Minimum Returns Analysis–A procedure for assessing risky production alternatives by examining their worst possible net returns and selecting that alternative whose worst return or whose average return for its worst possibilities is highest among the alternatives being considered.

Miscible–Designating two or more substances that, when mixed together, form a uniformly homogenous solution.

Mist–(1) Liquid particles in a fine state of division but of perceptible size that make up spray material. (2) Transparent or translucent suspended water particles near the surface of the earth. See Fog.

Mixing Nozzle–A nozzle in which the extruded liquid is mixed with a stream of air before it finally leaves the orifice.

Mode–The value of the random variable that occurs most frequently. See Mean, Median.

Model–A simplified representation of reality built to reflect those features of a farm, enterprise, process, etc., that are of most importance in the context of a particular study.

Module–(1) A unit used as a standard or ratio of measurement. (2) In the design of gears in the United States, it is the pitch diameter in inches divided by the number of gear teeth.

Mohs' Scale of Hardness–The hardness scale consisting of 10 minerals from talc (hardness 1), the softest, through gypsum, calcite, fluorite, apatite, orthoclase, quartz, topaz, corundum, to the hardest, diamond (hardness 10).

Moisture Meter–A device that is buried in the soil at plant root depth and used to determine how much moisture is in the soil. See Tensiometer.

Moisture Tension–The force at which water is held by soil, usually expressed as the equivalent of a unit column of water in centimeters: 1,000 centimeters being equal to one atmospheric tension. Moisture tension increases with dryness and indicates the degree of work required to remove soil moisture for use by plants.

Moisture-proofing–Any treatment that reduces the absorption or adsorption of moisture by wood. It is usually accomplished by the impregnation of wood so that some of its hygroscopic moisture is replaced by a moisture-repellent substance.

Moisture-vapor-resistant–Packing materials that protect foods from moisture loss during freezer storage; e.g., freezer wraps (paper, plastic, or foil), plastic bags, waxed freezer cartons.

Mold–(1) A form containing a cavity into which material is poured or pressed to achieve a special shape and design; e.g., molds for Edam and Gouda cheese, butter, ice cream, etc. (2) Any soft, humus soil. (3) Fungi distinguished by the formation of a mycelium (a network of filaments or threads), or by spore masses; usually saprophytes. However, various kinds may do serious damage to fruits, hay, grain, growing crops, and ornamental plants. Also spelled mould. See Compost, Downy Mildew, Mildew, Powdery Mildew.

Moldboard–The part of a turning plow just back of the share that receives the furrow slice from the share and turns it partially or completely over. Also called breastboard, earthboard.

Molecular Ratio–The relative proportion between two or more chemical substances in terms of the number of gram molecules of each. Also called mole ratio.

Money Crop–In a farm enterprise, the crop that is a major source of income, as cotton, wheat, potatoes. Other corps may be grown only as subsidiaries for producing feed for livestock or for maintaining or increasing the yield of the money crop. Also called cash crop.

Money Field Price–(1) (of an input) The purchase price of a unit of an input factor plus other direct expenses (such as transportation costs) per unit of input incurred in using the input factor. (2) (of an output) The market price of a unit of product minus harvest, storage, transportation and marketing costs, and quality discounts.

Mono-–Single; only one.

Monoculture–Cultivation of a single crop, as wheat or cotton, to the exclusion of other possible uses of the land.

Monolithic Concrete Silo–A concrete silo made as a poured casting, in contrast to one made of concrete staves.

Monomer–A simple molecule that is capable of combining with a number of like or unlike molecules to form a polymer.

Moonarian–(1) A farmer or gardener who plants and cultivates crops according to phases of the moon. (2) A nonscientific person.

Morrill Act–An Act of the United States Congress approved July 2, 1862, which provided for the establishment of colleges of agriculture and mechanical arts in the various states. The bill was introduced by Representative (later Senator) Morrill of Vermont in 1857. The main feature of this and later acts was the donation of public lands for the support of the land-grant educational institutions in each state. A subsequent act (the second Morrill Act), passed in 1890. allocated an annual, federal grant of $25,000 to each land-grant college and university. See Land-grant University.

Morrison Standard–A feeding standard formulated by F.B. Morrison of the University of Wisconsin, author of ***Feeds and Feeding***, which is adapted to practical feeding conditions and is used extensively.

Mortality–The number of overall deaths, or deaths from a specific disease, usually expressed as a rate; i.e., the number of deaths from a disease in a given population during a specified period, divided by the average number of people or animals exposed to the disease and at risk of dying from the disease during that time.

Mortgage–A lien on land, buildings, personal property, etc., which is given by a borrower to the lender as security for a loan.

Mortgage Insurance–An insurance plan sponsored by the United States federal government for insuring the payment of real estate mortgage loans. First used by the Federal Housing Administration, it was later adapted to farm-ownership loans available through the Farmers Home Administration. In this program the farm purchaser must make a 10 percent down payment; the mortgage insurance is on the 90 percent balance of the reasonable value of the farm including necessary repairs and improvements.

Mortgagee–One to whom a mortgage is given as security for money he has loaned to another.

Mortgagor–One who, having all or part of title to property, pledges that property as security for a debt.

Most Probable Producing Ability (MPPA)–An estimate of a cow's future productivity for a trait (such as progeny weaning weight ratio) based on her past productivity. For example, a cow's MPPA for weaning ratio is calculated from the cow's average progeny weaning ratio, the number of her progeny with weaning records, and the repeatability of weaning weight.

Motor–A device for changing electrical energy into mechanical energy. This term should not be used when referring to the engine of a machine.

Mottling–A soil mass containing many colors due to poor internal drainage. The colors may be gray, yellow, and/or red in random patterns.

Mould–See Mold.

Mounted Plow–A plow that is mounted on the tractor as opposed to a trailing plow. See Trailing Plow.

Movable Dam–A water barrier that may be opened in whole or in part. The movable part may consist of gates, stop logs, flash boards, wickets, or any other device whereby the area for water flow through or over the dam may be controlled.

Mow–(1) The place in a barn where hay is stored. (2) A pile of hay. (3) To cut hay or the grass of a lawn.

Mowdrying–The use of a blower and duct system in the hay mow to provide forced ventilation to reduce the moisture content of the hay. Hay having approximately 40 percent moisture may be put in the mow and the curing completed with the aid of the blower system. Normal outside air or artificially heated air may be used.

Mower–(1) A machine with a mowing sickle cutting bar that is designed to cut forage for hay, weeds, etc. (2) A machine with a cutting reel or rotating blade for cutting the grass of a lawn or weeds or tall grass in a pasture.

Mower Conditioner–A combination mower and conditioner that has conditioning rolls that runs the length of the cutter bar. The hay is cut and conditioned in one pass.

Mud Cap–A mass of mud placed on an explosive charge on top of an object to be shattered, as a large boulder. The mud tends to confine the explosive force.

Mud Fence–A fence made of clay, adobe soil, etc. To reduce raindrop erosion, the top of the mud fence is usually protected with brush, boards, or rocks. Because of the hazard of raindrop erosion, mud fences are more common in semiarid and arid regions.

Mule Skinner–(1) A mule-team driver, especially in the western United States. Also called muleteer. (2) A whip used on a mule team.

Multi–A prefix meaning many, as multiovulate, many-ovuled.

Multiple Cropping–In favorable climates, the growing of two or more crops consecutively on the same field in a single year, such as corn and wheat; soybeans and wheat.

Multiple Farrowing–Arranging the breeding program so that groups of sows farrow at regular intervals throughout the year.

Multiple Use–Use of the land for two or more purposes such as grazing, wildlife, recreation, and watershed protection.

Multiple-price System–A proposal for variable prices, presumably set by the federal government (United States), for agricultural products to overcome inequities in production and to deal with surpluses: e.g., one price for domestic consumption and another price when the same product is to be exported. Also called two-price system.

Multivariable Production Function–A production function involving several variable inputs.

Munsell Color Standards (Munsell Notation)–A color designation system that specifies the relative degrees of the three variables of color: Hue, value, and chroma. The standards may be used in precise comparison of colors of soils, or in standardizing agricultural products: e.g., prime cottonseed cake is 10YR5/5, which means yellow-red with value = 5, and chroma = 5.

NAL–National Agricultural Library.

Narrow Base Terrace–A terrace for controlling soil erosion. It is similar to a broad-base terrace in all respects except the width of ridge and channel: the base of a narrow terrace is usually 4 to 8 feet wide. it is subject to frequent failures and has not been widely accepted in the United States. See Broad Base Terrace, Nichol's Terrace.

NASS–National Agricultural Statistical Service.

National Cattlemen's Association–An association of cattle producers that serves as the communications arm of the nation's beef cattle industry. The membership includes cattle breeders, producers, and feeders. This nonprofit organization was officially formed on September 1, 1977, through the consolidation of the American National Cattlemen's Association (founded 1898) and the National Livestock Feeders Association (founded 1946).

National Cooperative Soil Survey–See Soil Survey.

National Cotton Council of America–A council organized in 1938 to promote cotton from field to fabric. Seven cotton segments have combined to form this council: producers, ginners, warehouse managers, crushers, cooperatives, merchants, and manufacturers.

National Council of Farmer Cooperatives–In the United States, a national federation dedicated to promote the interests of farmer cooperatives through its influence on various governmental and other agencies. The council provides an avenue through which cooperatives are advised of current economic, technological, legal, and other developments, and also provides a forum through which better understanding among cooperatives can be attained.

National Farm Loan Associations–In the United States, local borrower-owned cooperative agencies that initiate and service farm mortgage loans made through the Federal Land Banks.

National Farm Program Acreage–The number of harvested acres of feed grains, wheat, and cotton needed nationally to meet domestic and export use and to accomplish any desired increase or decrease in carryover levels. Program acreage for an individual farm is based on the producer's share (historic farm production) of the national farm program acreage.

National Forest System–Units of federally owned forest, range, and related lands throughout the United States and its territories dedicated to the long-term benefit for present and future generations. The National Forest System includes all national forestlands acquired through purchase, exchange, donation, or other means, the National grasslands, and land utilization projects administered under Title III of the Bankhead-Jones Farm Tenant Act and other lands, waters, or interests therein which are administered by the Forest Service or are designated for administration through the United States Department of Agriculture Forest Service as a part of the system.

National Grange–See Grange, The National.

National Grassland–Land, mainly grass and shrub cover, administered by the United States Department of Agriculture Forest Service as part of the National Forest System for promotion of grassland agriculture, watersheds, grazing, wildlife, and recreation.

National Program Acreage–The acreage that the Secretary of Agriculture estimates will produce the desired quantity of a crop if average farm program payment yields are realized.

National Sire Evaluation–Programs of sire evaluation conducted by breed associations to compare sires on a progeny test basis. Carefully conducted national reference sire evaluation programs give unbiased estimates of expected progeny differences. Sire evaluations based on field data rely on large numbers of progeny per sire to compensate for possible favoritism or bias for sires within herds.

National Wool Act–Legislation that provides price support for shorn wool at an incentive level to encourage production. The law also provides for a payment on sales of unshorn lambs.

Natural Area–In the United States, an area permanently preserved in unmodified condition as representative of the virgin growth of a major forest or range type, primarily for the purposes of science, research, and education. Timber cutting and grazing are prohibited.

Natural Boundary–Any feature not made by people, such as a river, or a mountain ridge, which separates states, countries, or tracts of land.

Naval Stores–Products, such as turpentine and rosin, obtained from the distillation of crude pine resin.

Nave–The hub of a wheel to which spokes are attached.

Neck Finisher–A mechanical buffer or picking machine for finishing the dressing of poultry for the market. It cleans blood and feed from the head and removes feathers and pin feathers from the neck.

Neck Rope–(1) A rope tied around the neck of a horse being trained for use in lassoing cattle. (2) A rope used in picketing horses.

Neckyoke–A bar, usually of wood, that connects two draft animals abreast by the neck and supports the forward end of the tongue of a wagon, plow, or harrow.

Nervous Market–A market in which traders expect a sharp break momentarily and are easily influenced by rumors.

Nester–A homesteader or squatter who legally or illegally occupies land (western United States).

Net Cash Income–The cash income after the cash expenses are deducted from the gross cash income.

Net Duty of Water–The amount of irrigation water delivered to the land to produce a crop. It is measured at the point of delivery to the field.

Net Farm Earnings–The net cash income plus an increase (or minus a decrease) in total farm investment other than bare lands, plus the value of farm products furnished to the farm family, and less the value of unpaid family labor (excluding operator's labor).

Net Farm Income–The farm income left after expenses are deducted from gross income.

Net Income–The total amount earned after all of the operating costs and overhead have been subtracted.

Net Increment–In forestry, the addition to tree growth that represents an increase in usable timber.

Net Livestock Increase (or Decrease)–The clear profit or loss; a figure obtained by subtracting the total of the value of livestock at the beginning of the year and the cost of livestock purchased during the year from the total of the value of livestock at the end of the year and the receipts from livestock sold. If the result is negative, it is a net livestock decrease.

Net Present Value (NPV)–The net total of the discounted values of the payments and receipts associated with a given project or farm plan.

Net Sales–Gross sales less deductions for freight, handling, discounts, etc.

Net Weight–(1) The weight of an article, or a mass of anything, after deducting from the gross weight the weight of the container or covering. (2) The weight of a slaughtered animal after removal of hide, viscera, etc.

Net Worth–The net ownership of a farmer or individual in his/her business; the difference between the assets and liabilities; the owner's equity in the business.

Net Yield–The measured yield per hectare or acre in the field, minus harvest losses and storage losses where appropriate.

Net-energy Value–The amount of energy that remains after deducting from a feed's total energy value the amount of energy lost in feces, urine, combustible gases, and heat increment. Sometimes called work of digestion.

Neutron—See Atom.

Neutron Probe—A field device used for measuring soil water percentage. As neutrons emitted from the probe collide with hydrogen in water they are deflected and slowed. The slowed neutrons are deflected to the counter. The more water in the soil, the more the neutrons measured by the meter.

New Parity Prices—Parity prices based upon the most recent ten-year period for an individual commodity and farm wages paid to labor.

New York Cotton Exchange—A world-famous central exchange organized in 1870 in New York City, for trading, buying, selling, and speculating in cotton.

Newton—A unit of force required to accelerate a mass of one kilogram one meter per second per second.

Nichol's Terrace—A once common small terrace for disposal of surplus water. It has a comparatively deep, narrow channel and a low, flat ridge with a slope that merges quite closely with the downhill side. Modern terraces are usually broad-based to allow machinery to operate over them. See Broad-base Terrace, Narrow-base Terrace.

Nickel—Ni; a chemical element, a metal which is found in traces in soils. It was once thought to be deleterious rather than beneficial to plant growth, but since 1983 has been suggested as essential for plants. It is required by animals and people.

Night Corral—An enclosure or pen on western United States ranges in which ewes may be placed at night or at lambing time.

Nippers—(1) The two central incisor teeth of a horse. (2) Small pincers used for holding, breaking, or cutting.

Nipple Pail—A pail with a tube and rubber nipple fastened at the bottom, used in calf feeding, weaning, etc.

Nipple Waterer—An automatic watering system in which the animal pushes a "nipple" in the end of a pipe to get water.

Nitre—(1) A precipitate of malic acid formed in making maple syrup. (2) Nitrate of potash (saltpeter), KNO_3, used as a fertilizer and in the manufacture of explosives. (3) Sodium nitrate, or chile saltpeter.

Nitric Acid—HNO_3; a strong mineral acid which, combined with metals or alkalies, forms nitrates. It is now made synthetically on a large scale by passing ammonia (NH_3) and air through a platinum gauze catalyst, whereby the ammonia oxidizes. Nitric acid is used in the production of nitrate fertilizer compounds, including nitric phosphates. Some nitric acid is used as an oxidant for carbonaceous material that causes the black color in liquid fertilizers made with commercial phosphoric acid.

Nomen—(Latin) Name.

Noncommercial Forest Land—Forestland withdrawn from its commercial use for timber because (a) it is utilized for such purposes as parks, game refuges, military reservations, or reservoir protection or (b) its poor growing conditions or inaccessibility render its commercial use unprofitable.

Nonconforming Uses—Uses of land permitted by zoning ordinances in areas in which such uses are not otherwise permissible.

Nonmoney Farm Income—A statistical allowance used in farm income compilations to credit farmers with income for the value of farm products used on the farm (instead of being sold for cash) and the rental value of farm dwellings. It assumes farmers otherwise live rent-free on their farm business premises.

Nonrecourse Loan—Participants in federal commodity programs may obtain loans from the Commodity Credit Corporation (CCC) by pledging planted or stored crops as collateral. These loans enable producers to pay for planting costs or to store crops for later sale. The producer can settle the loan by paying it back with interest or by turning the stored crop over to the CCC when the loan period ends. Loans are generally paid off when market prices rise above loan rates. Crops are frequently forfeited to the CCC at the end of the loan period when prices are below loan rates.

Nonrenewable—Natural resources that once used up are gone forever.

Nonslip Loop—A loop of rope so knotted that it will not slip or tighten when pulled; used on a foal being trained to lead.

Normal Yield Table—In forestry, an accepted standard yield table with which to compare actual yields. The statements of a normal yield table are derived as an average from the best-producing, fully stocked areas for particular species and sites.

Norms—Customs, ethics, moral codes, laws, and institutions. (Most sociologists regard norms as an ideal standard for behavior, but some would include statements about average or typical behavior patterns.)

Norris-Doxey Act—An act of United States Congress, approved in 1937, which authorizes the Secretary of Agriculture to cooperate with land-grant colleges and universities and state forestry agencies for the development of farm forestry in the states and territories of the United States. An important provision rules that these bodies produce or procure and distribute tree and shrub planting stock to farm owners on condition that the cooperator make available the land for planting without charge.

Northwestern Box—A standard type of container for packaging apples that has the capacity of one bushel.

Nose—(1) The point of a plow share. (2) The part of the face of people and animals that covers the nostrils; snout; muzzle. (3) To round off the end of a log in order to facilitate snaking or skidding.

Nose Clamp—A device that may be fitted tightly on the nose of an animal and used for its control during shoeing, surgical operations, or various types of training. See Twitch, Nose Twitch.

Nose Lead—A removable metal ring that is snapped into a bull's nose. A rope is attached to the ring for leading and controlling the animal. This device is required for showing bulls in most livestock shows.

Nose Ring–(1) A metal ring fastened through the cartilage of the nose of a bull for safe control of the animal. A staff, or metal rod, about 6 feet long may be snapped into the ring for handling the animal. See Nose Lead. (2) A metal ring fastened in the nose of a hog to prevent it from rooting.

Nose Twitch–A looped, light rope or heavy cord attached to a short stick. The loop is placed over the end of the nose of the horse and when twisted acts as a control in breaking, training, or leading. See Nose Clamp.

Nostrum–A quack remedy or worthless patent medicine.

Notch–(1) The opening in a dam or spillway for the passage of water. (2) A gap, pass, or defile between mountains (chiefly in northeastern United States). (3) An undercut; in logging, a cut in the trunk of a tree to govern direction of fall. (4) A gap cut into the ear of an animal for identification.

Notched Colter–A rotating, circular attachment for the forward part of the plow that has a cutter edge with a serrated perimeter for cutting the soil surface and trash in front of the plowshare. Also called cutaway disk, cut-away colter. See Colter.

Notifiable Disease–(reportable) Any disease that must be reported to the government health authorities.

Novale–Land newly plowed and brought under cultivation.

Novalis–In civil law, land that has been fallow for a year after the first plowing.

Nozzle–The pouring end of a spout; the orifice through which liquids are discharged in a stream or spray.

Nuclear Family–The fundamental unit of human organization, specifically a male, female, and their unmarried children, all of whom live together.

Number of Contemporaries–The number of animals of similar breed, sex, and age, against which an animal was compared in performance tests. The greater the number of contemporaries, the greater the accuracy of comparisons.

Numdah–A thick, belt blanket placed under a saddle to absorb sweat. Also called namda, nammad.

Nursery Deck–A small above-the-floor pen in which the newborn pits are kept to keep them warm and dry.

Nursery Knife–A sharp knife that is used to prune, graft, or take cuttings from plants.

Nut Sweeper–A machine that harvests nuts by sweeping them up off the ground.

Nutritional Labeling–Labels that provide consumers with information about the nutritional values of products.

Nutritive Ratio–In animal feeds, a ratio or proportion between the digestible protein and the digestible nonnitrogenous nutrients found by adding the digestible carbohydrates plus the digestible fat multiplied by 2.25, and dividing the sum by the digestible protein. The ratio is an expression of the energy value of a ration against its body-building power.

Obsolescence–In appraisal, the impairment of desirability and usefulness of a structure, machine, or equipment consequent upon new developments in art, design, process, or any other circumstances.

Obstetrical Chain–A metal chain, 30 to 60 inches long, used by veterinarians to assist animals having difficulty in the delivering of newborn (usually calves or foals). It is used to apply a small amount of traction to the unborn.

Obtuse–(1) An angle that is greater than 90 degrees, but less than 180 degrees. (2) A leaf apex that is similar in shape to an obtuse angle.

Occupational Safety and Health Act (OSHA)–Public Law 91-596, December 29, 1970. The purpose of this law is "to assure so far as possible every working man and woman in the nation safe and healthful working conditions and to preserve our human resources." The act provides references and standards of compliance for the various segments of business and industry, including agriculture, in the entire nation. The principal provisions of the act stipulate standards, enforcement, penalties for noncompliance, appeals, and research.

Octane–A measurement of the antiknock (detonation) quality of a gasoline or LP gas fuel.

Ocular Estimate–Estimate by sight only. (1) In forestry, the determination of the approximate volume and quantity of standing timber without the use of measuring instruments. (2) On a range, the qualitative procedures for determining the degree of cropping of forage plants. Observations of a general reconnaissance nature are made visually: (a) by examining small random plots at the end of each grazing season.

Odd Area–In farm planning, a small area, such as bare knob, fence corner, sink hole, blowout, borrow pit, or an irregularly shaped area that is unsuitable for cultivated crops and is best used for wildlife crops.

Odd Lot–On livestock markets, animals that do not conform uniformly to some particular weight, age, or grade of quality class.

Off Grade–Designating an agricultural product that fails to meet requirements of commonly accepted standards or legal or official standards in grading products for sale.

Off Season–In farming, the season of the year when the production of a particular crop is difficult or impossible.

Off-set Tractor–A tractor that has the operator seat and controls mounted to one side of the tractor instead of in the center of the tractor.

Offset-disk Harrow–A harrow adapted for use in orchards and vineyards, which can be set to run to the side of the tractor and thus cultivate under branches too low for the tractor to pass under.

Ohm– A unit of electrical resistance equal to the resistance of a current in which an electromotive force of one volt maintains a current of one ampere. See Ampere, Voltage, Wattage.

Ohmmeter– A device used to measure the electrical resistance of a circuit.

Oil– One of several kinds of fatty or greasy liquids that are lighter than water, burn easily, are not soluble in water and are composed principally, if not exclusively, of carbon and hydrogen.

Oil Mill– A mill or factory in which oils are extracted from vegetable seeds, such as cottonseed and soybean, by hydraulic, expeller, or solvent method of extraction.

Oiling Machine– In the poultry industry, a machine used for dipping eggs in oil to form a protective coating and to retard loss of moisture and quality. It is used especially on eggs placed in storage. See Oil-protecting Eggs.

Oilstone– A fine granite stone used with oil for sharpening tools.

Oligopoly– A market situation with relatively few sellers who are mutually interdependent in their marketing activities; some food-processing industries are oligopolistic.

Oligopsony– A market situation where there are a few large buyers of a product.

Ombrometer– A rain gauge. See Pluviometer.

On and Off Permit– A grazing permit issued only where movement of livestock is necessitated between national forestland (United States) and adjoining outside range, or where private and forestlands intermingle.

On Contract– Designating a crop, as vegetables for processing, grown under an agreement whereby the producer sells to a processor usually at prices previously agreed upon.

On the Block– (1) Up for auction. (2) Designating carcasses ready for cutting for sale. See On the Hoof.

On the Clean Basis– In American wool market reports, referring to quotations given on the cost of the clean fiber that any lot of unscoured wool is estimated to contain.

On the Hoof– Designating a live meat animal.

One-crop Farming– A system of farming in which the producer grows a single crop as a source of income; e.g., wheat, corn, or cotton.

Open Flume– An uncovered passageway for irrigation water.

Open Listing– A listing under which the principal (owner) reserves the right to list the property he/she is attempting to sell with other brokers.

Open Range– An extensive range area where grazing is unrestricted. Also, ranges that have not been fenced into management units.

Open-code Dating– Food labels providing consumers with information on when food was processed and packaged, when it should be sold or withdrawn from the market, or when the product is no longer acceptable for sale.

Open-ditch Drainage– Drainage of excess water from land by open ditches as opposed to tile drainage.

Open-kettle Canning– A procedure whereby food is cooked in an ordinary kettle, then packed into hot jars and sealed. Jars of food receive no additional heat processing. This is a dangerous practice, as spoilage organisms may enter the jar during the transfer of food from kettle to jar.

Opening– (1) A treeless or very sparsely timbered area in a forested region. See Oak Opening. (2) In cranberry culture, the first swelling of a terminal bud. (3) The unfolding of a flower or boll of cotton. (4) Designating the first price offered for a commodity when a market day begins.

Operating Capital– The amount of money used to run a business.

Operator– One who does the appropriate work and management necessary to conduct a business, as farming.

Optimum Condition– The ideal environment, with regard to nourishment, light, temperature, etc., for an organism's growth and reproduction.

Optimum Level of Performance– The most profitable or favorable ranges in levels of performance for the economically important traits in a given environment and management system. For example, although many cows produce too little milk, in every management system there is a point beyond which higher levels of milk production may reduce fertility and decrease profit.

Optimum Temperature– That certain temperature at which a particular plant or animal grows satisfactorily, other conditions being favorable for growth.

Optimum Water Content– The amount of water in a soil needed by a plant for its optimum growth, varying from 40 to 60 percent the moisture-holding capacity (field capacity).

Option– Right to purchase property within a definite time at a specified price. No obligation to purchase, but seller is obligated to sell if the option holder exercises his/her right to purchase.

Optional Prepayment Plan– A policy that permits the lender to pay all or part of a loan at any time.

Orchard Heater– Any of several types of heaters used in the orchards to reduce frost injury.

Order of Terms– The order in which constituents of a mixed fertilizer appear on the printed formula: nitrogen, phosphorus, potassium.

Organic Gardening (Farming)– A system of farming or home gardening that utilizes organic wastes and composts to the exclusion of chemical fertilizers. Advocates of the system teach that chemical fertilizers are injurious to health and that organic composts give higher yields, better quality and better taste of produce, less plant damage by insects and disease, reduction of weed menace, and stronger seeds that germinate better and produce successively stronger plants. None of these claims have so far been confirmed by reproducible (scientific) proof. Much publicity of the concept has resulted in a number of followers but it is significant that the leading followers of the concept do not depend entirely on agriculture for their livelihood. The organic gardening concept persists because of the basic truth that organic matter in soils is beneficial to agriculture. However, the sole scientific foundation for economic production of abundant and healthful foods consist of the liberal use of organic matter plus chemical fertilizers applied according to a soil test.

Organized Market–A market where trading is supervised and a large volume of a product is traded.

Orifice–(1) An opening by which spores, etc., escape; any opening. (2) An opening in a nozzle tip, duster, or granular applicator through which the spray, dust, or granules flow.

Orometer–A sensitive aneroid barometer that is calibrated in feet and/or meters of elevation above sea level. Used for obtaining approximate elevations at various points of observation during reconnaissance mapping.

Oscillator–In nursery practice, a mechanical device, worked by hydraulic pressure, for automatically turning overhead sprinkler lines through an arc of varying degrees up to 180 degrees. It produces an even spread of water.

OSHA–See Occupational Safety and Health Act.

Other Forestland–Forestland incapable of producing 20 cubic feet per acre of industrial wood under natural conditions because of adverse site conditions such as infertile soils, dry climate, poor drainage, high elevation, steepness, or rockiness.

Out of Season–Designating products, as fresh fruits and vegetables, available at a time at which they are not normally produced locally.

Outer Cover–In beekeeping, usually a cover that telescopes over the top of the hive to a depth of an inch or more and is covered with galvanized iron or aluminum sheeting, to protect the hive from the weather.

Outfit–(1) A group of cowhands and their equipment (western United States). (2) To supply the necessary equipment for an expedition, a practice, a business, etc.

Outhouse–(1) Any small building on a farm, usually other than the barn, in which tools or small animals, etc., are kept. Also called outbuilding. (2) A small building over a pit dug in the ground, used as an outdoor waterless toilet. Also known as backhouse or privy.

Outlook–A forecast of future market events, such as supplies, prices, and so on.

Output–A marketable product of a farming operation, such as cash crops, livestock, etc.

Output Shaft–The shaft or gear that delivers power from the transmission of a tractor or other machine. See Input Shaft.

Outrider–A cowboy whose duty it is to ride the range and protect his employer's interests (western United States).

Outside Man–(1) A man who attends roundups on ranges away from those of the home ranch (western United States). (2) A farm laborer who works in the fields as contrasted to one who works in the barn, etc.

Outside Quotation–The highest price for a commodity on a particular market for a particular day.

Oven-dried–Designating wood or other material dried by exposure to 212°F (100°C), or slightly higher temperatures, until ceasing to lose weight, usually 24 to 48 hours.

Overall–A suit of clothes, usually one piece, made of heavy cotton; often worn by farmers over their regular clothes for added warmth or to protect their clothes when engaging in particularly messy work.

Overflow(ed) Land–(1) Land that is subject to overflow or flooding; generally floodplains of rivers. (2) Loosely, swampland. (3) Legally, land that is covered by nonnavigable waters, but not including land covered by the normal flow of tides.

Overhaul–(1) The rehandling or repacking of ham during the pickling period to permit a more uniform distribution of pickle. (2) To repair and recondition tools, implements, machines, etc.

Overhead–The general expenses of operating a farm, such as taxes, insurances, etc., which cannot be charged directly to any particular crop or livestock. In cost accounts, overhead expenses are prorated to the productive enterprises, such as crops and livestock, etc., in proportion to one of the major inputs or to the total debits of each enterprise.

Overlay–Any transparency, containing supplemental information, superimposed over a map or a data sheet to show this information more clearly.

Overpopulation–A population density exceeding the capacity of the environmental resources of an ecosystem to supply the requirements of the individual organisms occupying it; usually accompanied by a high mortality rate because of inadequate nutrition, insufficient shelter, and increased predation, disease, or parasitism.

Overseer–(1) A manager or supervisor of a ranch, farm, etc., who has supervisory duties over the laborers. (2) In New Zealand, an undermanager.

Owner-samples DHIA Plan–A system of records for dairy cows in which the owner weighs the milk and takes the samples that are later picked up by the tester who runs the tests and compiles the records. This system gives the owner the same information, is more economical, and the tester can handle many more herds each month, but is less official than the standard DHIA plan. See Dairy Herd Improvement Association.

Ownership–The property owned by one owner, including all parcels of land in the United States.

Oxygen–The chemical element O; a colorless, odorless gas. The most abundant element in the earth's crust. It accounts for about 47 percent of all elemental material. It is essential in the growth of all crops and for the respiration of most forms of life.

P/E Ratio–Precipitation/evaporation ratio. This is an aridity index.

Pace–(1) A measure of length: the ordinary length of a human step from heel to heel is 2½ feet. The geometric is 5 feet. Distances for land measure are usually stepped according to some definite distance per step, as 3 feet. Land measurements were often paced off, from some known mark as a base, in early land surveying in the United States. (2) Applied to horses, a rapid, two-beat gait in which the lateral fore and hind legs work in pairs. Also called amble, rack, trot.

Packer–(1) One who operates a slaughter and meat-processing business. (2) Pertaining to the business of packing fresh or processed fruits and vegetables or meats. (3) A field tool of the roller type consisting of a set or series of rollers that pack the loose soil after plowing. Also called cultipacker. (4) One who makes a pack, as of vegetables, meat, etc.

Packsaddle–A carrying unit for camping equipment or other material that is placed on the back of a horse or pack mule, either as separate units or attached to the rider's saddle.

Palatability Rating–A method of rating the condition of a range, from high to low, depending on the proportion of palatable plants carried by the range. The more such plants the higher the rating.

Palea (Palet)–A small, chaffy bract, especially: (a) one of the chaffy scales on the surface of the receptacle of the flowerhead in many Compositae, or (b) the inner of the two bracts or glumes enclosing a flower in the Gramineae, the outer being the lemma.

Paling Fence–A light fence made of narrow slats or poles. The palings may be sharpened and driven into the ground a few inches apart or supported at top and bottom by horizontal strips nailed to each pale. It is used for gardens, backyards, and small enclosures.

Pallet–A platform of any size supported by two or three runners or stringers used as a base for stacking and transporting a box or several packages as a unit.

Palm–Any tropical or subtropical tree or shrub of the family Palmaceae. Species are grown for their wood, edible fruits, resins, fiber, and as a source of oils and drugs.

Panel Plot–A small area protected from grazing during a part of the growing or grazing season by a movable fence. Also called hurdle plot, panel.

Pannage–(1) Pasturing swine in a forest to feed on acorns, etc. (2) The right or privilege to pasture swine in this manner. (3) The payment made to the landowner for the privilege of pasturing swine in his forest.

Paper Grain–Grain purchased on contract on an Exchange or Board of Trade for speculation or hedging. No grain is actually handled; only the paper contracts actually change hands.

Paper Profit–The profit that would be realized if open contracts were liquidated at a certain time at a certain price.

Paraffin–A tasteless, odorless, waxlike substance obtained mostly as a residue in the distillation of petroleum; used in sealing the tops of jars of jellies, preserved fruits, as a coating for paper milk bottles, cheeses, and for various other similar purposes.

Paraplow–A chisel plow that leaves most crop residues on the soil surface while tilling about 15 inches (35 centimeters) deep.

Parasitic Advertising–Advertising by one group that takes sales away from another group; e.g., beef advertisements may reduce pork sales; orange advertisements may reduce apple sales.

Parchment–(1) The sheetlike fiber found in the pods of unimproved beans and pea varieties. (2) A paper used in food wrapping; e.g., in the packaging of butter.

Pare–To peel or trim off outside covering, as with fruits and vegetables.

Parity–A legislative formula designed to maintain a just balance between the prices a farmer receives for his products and the payment he makes for the essentials of production such as seed, fertilizer, and machinery.

Parity Price–Price per bushel (or pound or bale) that would be necessary for a bushel today to buy the same quantity of goods that a bushel would have bought in the base period at the prices then prevailing.

Parshall Flume–A Venturi-type device commonly used in irrigation for measuring the discharge of small streams and ditches. See Venturi Tube.

Part-time Farming–Farming in which the operator is employed much of his/her time in occupations other than farming and who derives a substantial part of his/her income from other occupations. As defined by the United States census, farming in which the operator spends 100 or more days off the farm, or in which the nonfarm income received by him/her and members of his/her family is greater than the value of the farm products sold.

Particle Density–The average density of the soil particles. Particle density is usually expressed in grams per cubic centimeter and is sometimes referred to as real density, grain density, or specific gravity.

Particle-size Analysis–Determination of the amounts of different particle sizes in a soil sample, usually by sedimentation, sieving, micrometry, or a combination of these methods.

Particle-size Distribution–The amount of the various soil separates, sands, silt, and clay in a soil sample, expressed as dry weight percentages.

Particle-size Histogram–A graphic method of presenting the particle-size distribution of sediments as a series of vertical bars whose heights are proportional to the frequency in each class. The term itself is standard statistical usage for such diagrams.

Partido System–In the southwestern United States, a form of operation in which sheep owned by the patron are let out on shares to a partidero, who cares for them and returns part of the increase or income to the owner.

Partnership–An association of two or more persons as co-owners of a profit-making business. Chief criteria of a partnership are participation in management, sharing profits, sharing losses, ownership of assets together, and having a firm name, a single joint bank account, and a single set of farm records. No one factor is controlling. All are applied to the particular arrangement to determine whether in the sum it is a partnership. In most states of the United States, the law states that the partnership name must be filed with the county clerk in the court house. Most father and son farming arrangements are legal partnerships regardless of whether they are operating under a written or an oral agreement.

Parts Per Millon– The number of weight or volume units in a million units of a solution or a mixture; a measure of concentration, especially of chemicals in solution: one milligram per liter. Abbreviated ppm.

Party Wall– Wall erected on a line between adjoining properties for the use of both properties.

Pascal– A unit of pressure equal to one newton per square meter. See Newton.

Pasture Improvement– The practice of grazing, clipping, fertilizing, liming, seeding, contour furrowing, or other methods of management that improve the vegetation for grazing purposes.

Patent– (1) A grant made by public authority to an inventor or discoverer entitling him/her to certain exclusive rights. in the United States patents may be granted by the United States Patent Office to originators of strains or some variation from existing varieties of plants. Many flowers, especially roses and some fruits, have been patented. (2) Open or unoccluded, as a patent duct or orifice.

Payee– The person to whom a payment is made.

Payment Limitation– A limitation set by law on the amount of money any one person may receive in farm program payments each year under the feed grain, wheat, cotton, rice, and other approved commodities payment plans.

Payment-in-Kind Certificates (PIK)– Used by Commodity Credit Corporation (CCC) in both export and domestic commodity programs, the PIK certificates are issued, or made available, to producers, buyers, and exporter. The certificates, expressed as a dollar value, may be redeemed either for specified commodities and products from CCC stocks, or in face value cash equivalent.

Peak– (1) In turpentining, the upper point of the V-shaped streak at the top of the incision. (2) In hydrology, the maximum rate of flow recorded at a gauging station during a flood.

Peak Year– The year of greatest production in the life of a fruit tree, a bed of berries, asparagus, etc.

Peanut Harvester– A machine that picks up peanuts vines, separates the peanuts, and conveys them to a bin.

Peck– A unit of measure: eight quarts or one-fourth of a bushel by volume of such materials as grains, fruits, or vegetables.

Peel– (1) The outer covering or skin of fruits or vegetables such as the apple, orange, etc. Also called peeling, rind. (2) To remove the skin from a fruit or vegetable, as to peel a banana.

Pelham Bit– A bit with two rings on each side for two controlling reins; used on polo ponies.

Pelham Bridle– A single-bitted, double-reined horse bridle used on pleasure horses.

Pen– (1) A loose, rectangular stack of fuel wood or pulpwood in layers of two pieces each of varying height and width. (2) In poultry shows, a male and four female birds of the same variety. (3) A small space enclosed by any kind of fence, used for confining pigs, cows, and other animals.

Pen Fattening– Fattening livestock or poultry by keeping them in small pens and giving them full feed.

Pen Lot– A number of animals in an enclosure. In livestock exhibitions, the animals are judged as a lot, or group, in contrast to an exhibition of single individuals.

Pencil Shrink– An assumed percentage deduction taken from the weight of slaughtered animals to allow for uncalculated losses in weight.

Penetrometer– A device that measures the force required to push a probe rod into the soil. It can be used to measure the density or degree of compaction in a soil.

Penny– (nail designation) Of obscure British origin, whether based on weight or price. (a) Based on weight; e.g., 1,000 fourpenny (4d) nails weigh 4 pounds (1.8 kilograms). (b) Based on price; e.g., 100 tenpenny (10d) nails cost 10 cents (British).

Penta–– Greek for five, used in naming chemical compounds.

Per Capita– Per person.

Per Capita Food Consumption– The average quantity of food eaten per person within a time period, usually a year; calculated by dividing the total food available for consumption by the population.

Per Person Indemnities– The maximum amount agreed to be paid by an insurance company per person in an accident.

Percentage Lease– Lease in which all or part of rental is a specified percentage of gross income from total sales made upon the premises.

Percentage Share– The percent of cost shared between the federal government and farmer or rancher in carrying out an approved soil and water conservation practice.

Percentile Taper– The relative taper of a tree in terms of diameter at regular intervals along the stem which is expressed in percent of diameter at breast height.

Perch– (1) A unit of land measure equal in length to a rod (16.5 feet; 5.029 meters). (2) Also used to designate a square rod. (3) A pole or limb on which chickens and wild birds roost.

Perforated Polyethelene Tubing– Plastic tubing 18 to 24 inches wide, with regularly spaced holes on each side that is used to distribute air in a greenhouse.

Performance Data– The record of the individual animal for reproduction, production, and possibly carcass merit. Traits included are birth, weaning and yearling weights, calving ease, calving interval, milk production, etc.

Performance Pedigree– A pedigree that includes performance records of ancestors, half and full sibs, and progeny in addition to the usual pedigree information. Also, the performance information is systematically combined to list estimated breeding values on the pedigrees by some breed associations.

Performance Record–The evaluation of an animal's production based on several factors; factors may be number of calves weaned, birth weight, weaning weight, etc.

Performance Testing–The systematic collection of comparative production information for use in decision making to improve efficiency and profitability of beef production. Differences in performance among cattle must be utilized in decision making for performance testing to be beneficial. The most useful performance records for management, selection, and promotion decisions will vary among purebred breeders and for purebred breeders compared with commercial cattle producers.

Permanent Pasture–Grazing land in farms occupied by perennial grasses and legumes. It is not a part of a regular rotation of fields and usually remains unplowed for long periods.

Permanent Wilting Percentage–The soil moisture content at which plants remain permanently wilted unless water is added to the soil. Soil water potential at wilting can vary from −5 to −20 bars. Because of the shape of the water potential-water content drying curve, large changes in water potential at higher tensions accompany minor decreases in water content, so permanent water for plant growth is approximately 15 bars.

Permeability–The capacity of soil or rock for transmitting a fluid. Degree of permeability depends upon the size and shape of the pores, the size and shape of their interconnections, and the extent of the latter. It is measured by the rate at which a fluid of standard viscosity can move a given distance through a given interval of time. The unit of permeability is the darcy. See Darcy's Law.

Permissible Hydraulic Velocity–The highest velocity at which water may be carried safely in a channel or other conduit. The highest velocity that can exist through a substantial length of a conduit and not cause scouring of the channel. Safe or noneroding velocity.

Permissible Velocity–The highest velocity at which water may be carried safely in a canal or other conduit without scouring the channel sides.

Permittee–A person who is legally allowed to graze a certain number of livestock on a particular area of public range.

Person-day–The equivalent of one person working for eight hours.

Person-Hour–The equivalent of one person working for one hour.

Personal Property–The rights and interests a person has in all things subject to ownership except a freehold interest in land.

pF Value–The logarithm of the height, in centimeters, of a water column necessary to produce a force equal to the energy with which moisture is held by a soil. The "p" indicates a common logarithm, and the "F" suggests force or energy.

pH–A numerical measure of acidity or hydrogen ion activity of a substance such as food or soil. The neutral point is pH 7.0. All pH values below 7.0 are acid and all above 7.0 are alkaline. The negative logarithm of the hydrogen-ion activity. The degree of acidity (or alkalinity) of a soil as determined by means of a glass, quinhydrone, or other suitable electrode or indicator at a specified moisture content or soil-water ratio, and expressed in terms of the pH scale. See Reaction.

Photometer–An instrument for measuring luminous (light) intensity or brightness by comparison of two unequal lights from different sources.

Phytometer–(1) A plant used to measure the physiological activities of the habitat. (2) A device used for measuring water transpiration of a plant.

Pick–(1) The total amount of a crop harvested, or the yield of an individual tree, as the pick of oranges. (2) Small, irregular openings within the body of a cheese. (3) See Pickax. (4) To pull or pluck ripe fruit, as berries, apples, cotton. (5) To pluck the feathers from a fowl in dressing it for the market. (6) To nibble at food.

Pick-Up–(1) A light automobile truck usually with a load capacity of 0.5 to 1.0 ton that is adaptable to many different uses on farms. (2) A rotating drum mechanism with flexible teeth for lifting and conveying a windrow from the ground to a baler, forage harvester, combine, etc.

Pick-Up Baler–A power-driven, automatic, self-tying hay baler with a windrow pick-up and with or without cross conveyor attachment similar to that used on combines. It consists of a pick-up rotating cylinder, an elevator mechanism, and a cross conveyor to carry the hay to the self-feeder in the baler.

Pickax–A hand tool with a short handle and a double blade. One part of the blade is narrow, curved, and sharpened on the face like an ax, the other is pointed. It is used in grubbing roots or working in stony or hard soil.

Picker–(1) One who harvests a crop by removing it from the plant; e.g., an apple picker. (2) A mechanical device that removes the fruit from the plant. (3) A type of machine potato planter that is provided with metal forks (pickers) attached to a vertical revolving disk. The picker moves through the hopper and spears the seed potato which is then punched off by a special ejector into the seed spout.

Picket–To tether or control the grazing range of an animal by a rope.

Picket Fence–A light, wooden fence made of narrow pickets or slats that are usually woven into a fence by the use of wires at the top and bottom. It is often used for small enclosures, such as gardens, chicken yards, etc.

Picket Pin–A stake to which a horse or mule is tied in picketing (western United States).

Picket Rope–Any rope, with a swivel snap on one end of which an animal may be fastened, and with a picket pin on the other which may be driven into the ground to tether the animal while grazing.

Picking Machine–In poultry dressing, a machine consisting of a revolving drum equipped with flexible rubber fingers which picks the feathers from carcasses.

Picking Season–The season of the year when a crop is harvested, especially such crops as cucumbers, string beans, cotton, etc., where the picking period extends over a number of days or weeks. The picking season for cotton may extend from August to January in some sections of the southern United States.

Pie Chart–A figure in the form of a circle that is divided into segments such that the size of each segment (angle) is proportional to the magnitude or frequency of that class.

Pig Teeth Nipper–A clipperlike instrument used for clipping the canine teeth (temporary or milk teeth) of suckling pigs.

Pig-guard Rail–A rail about 8 to 10 inches above the floor and 8 to 10 inches from the wall of a farrowing pen, beneath which recently farrowed pigs can move and thus be protected from crushing by the sow.

Piling–Round timbers or steel cylinders driven into the ground to support structures such as buildings or piers. The steel cylinders, after being driven into place, are filled with concrete. These supports may extend below the water table in soils or through unstable soils into stable soils below, as through muck into mineral soil.

Pinch Point–The place where two parts of a mechanism move toward one another, creating a potentially hazardous situation.

Pinhooker–A speculator in tobacco prices.

Pinion–(1) The outermost part of a bird's wing including the carpus, metacarpus, and phalanges; the part of a wing, corresponding to the forearm, on which the primary flight feathers are borne. (2) To cut off the pinion on one (or both) wings to permanently prevent a bird from flying. (3) A gear that has the teeth formed on the inside of the hub.

Pinning Knife–A dull, short-bladed knife that is used in removing pin feathers in the dressing of poultry.

Piston–A cylindrical engine part that fits into a cylindrical opening. The piston is closed on one end and connected to a rod that is connected to the crankshaft on the other end. The ignition of fuel in the combustion chamber forces the piston down, thereby causing the connecting rod to move the crankshaft.

Piston Ring–An expanding ring placed in the groove of a piston to seal off the passage of fluid or gas past the piston. See Piston.

Pit–(1) The endocarp of a drupe; the seed-stone of a fruit, as the pit of a peach or prune. (2) An excavation in soil in which vegetables, such as potatoes, carrots, and parsnips are placed and covered over for storage (seldom practiced now in the United States). (3) A place on the floor of an exchange in which traders stand when dealing in wheat, cotton, and other commodities. (4) In botany, a small hollow or depression in a cell wall. Various types are recognized in wood anatomy as blind, bordered, primordial, simple.

Pit Silo–A shallow pit of variable size for storing silage, which is dug in well-drained soil and is frequently walled with lumber or concrete if of a permanent nature. It is very common in the western states of the United States. See Horizontal Silo, Trench Silo.

Pitch Hay–To handle loose hay with a pitchfork.

Pitcher Pump–A hand-driven water pump with a cylinder that is part of the pump assembly. It has a short handle and a short suction stroke and is suitable for lifting water only a short vertical distance (about 30 feet), as from a cistern or shallow well.

Pitchfork–A farm implement consisting of a long handle usually with three or four long, curved, sharp-pointed tines used in handling loose hay or straw.

Pitman–A device that converts rotary motion to reciprocating motion to drive a reciprocating cutterbar. See Cutterbar.

Pitman-Robertson Act–An act of the United States Congress, passed in 1937, for aiding the states in the selection and improvement of land and water areas for wildlife restoration. Farmlands are included for improvement of wildlife habitats under the provisions of the Act.

Pitter–A machine for removing the seeds or stones from drupes, such as peaches or cherries.

Pitting–(1) The development of little cavities in metal, especially in aluminum cans, caused by the development of lactic acid by bacteria in milk residues. Pitting can also be caused from corrosion by brine. (2) In fruit processing, removal of seeds or stones, as from preaches and cherries.

Plane Survey–A survey in which the curvature of the earth is disregarded, as in ordinary field and topographic surveying.

Planetary Gears–A system of gearing in which a pinion is surrounded by an internal ring gear and plant gears are in mesh between the ring gear and pinion around which all revolves in much the same way the planets revolve around the sun; hence the name. See Pinion, Ring Gear.

Planimeter–An instrument for measuring the area of any plane figure by passing a tracer around the boundary.

Plant Analysis–Analytical procedures to determine the concentration of nutrients in plants.

Plant Material Centers–In the United States there are twenty-five plant material centers managed by the United States Department of Agriculture Soil Conservation Service. They receive superior grasses, legumes, forbs, shrubs, and trees. These plant materials are reproduced for distribution.

Plant Patent–A patent granted by the United States Patent Office to originators of varieties, strains, or some variation from existing varieties of asexually reproduced plants.

Plant Variety Protection Act–This act, passed in 1970, offers legal protection to developers of new varieties of plants that reproduce sexually; that is, through seed. Developers of plants that reproduce asexually have received protection by the United States Patent Office since 1930. The law states that protection will be extended to a "novel variety" if it has these three qualifications: (a) distinctiveness: the variety must differ from all known prior varieties by one or more identifiable morphological, physiological, or other characteristics; (b) uniformity: if any variations exist in the safety, they must be describable, predictable, and commercially acceptable; (c) stability: when sexually produced, the variety must remain unchanged in its essential and distinctive characteristics to a degree expected of similarly developed varieties.

Plant-exploration Service–A very important activity of the Agricultural Research Service, United States Department of Agriculture. Plant explorers are sent all over the world to collect plants, seeds, or plant materials of desired species for breeding and development of new crops for agricultural and other uses.

Plantation– (1) A large-scale agricultural unit, especially one devoted to the production of cotton or sugarcane. In pre-Civil War times in the United States, a large manorial estate on which cotton, tobacco, and other crops were produced with slave labor. (2) An artificially reforested area established by planting or by direct seeding.

Planter– (1) A mechanical device used for the rapid, efficient, and uniform planting of seeds. Planters are of many different kinds, from the simplest hand planter to the large highly mechanized, multirow, power-driven machines. (2) A farmer, especially a cotton farmer. (In its earliest use, it generally implied wealth, and denoted the owner of a large cotton or tobacco plantation.) (3) A person who plants. (4) A container, usually rectangular, containing soil or vermiculite, which is used for growing foliage plants indoors. It may be made of plastic, pottery, wood, metal, or other material and may be movable or a permanent part of the building.

Plastic Mulch– Thin polyethylene film, which may be clear or black, that is used as a mulch, especially for vegetables. Benefits include moisture retention, increased soil temperature, and, with black plastic, complete weed control.

Plasticity Index– The numerical difference between the liquid limit and the plastic limit; the range of moisture content within which the soil remains plastic.

Plat– A diagram drawn to scale showing all essential data pertaining to the boundaries and subdivisions of a tract of land, as determined by a plane survey.

Plat Book– A record of recorded subdivisions of land.

Plot– In agricultural research, a small parcel of land, usually rectangular and of a definite size, used in comparing yields of crop varieties, testing different applications of fertilizers, comparing methods of tillage, etc.

Plow– (1) The whole implement, of various types, used to cut, break, or turn a soil layer in preparation for planting, seeding, or other agricultural practices. More specifically, the removable metal point, share and moldboard or disk, attached to a plow stock or frame. (2) Any such implement not used primarily for agricultural purposes, such as a snow plow. (3) To make a furrow or to turn over a layer of soil. (4) To cultivate; e.g., plow corn. Also spelled plough by the British.

Plow Bolt– A flat-headed bolt designed to fasten a sweep or plowshare to the frame of the plow. The head of the bolt fits flush with the surface of the sweep or plowshare to prevent soil from catching on the bolt head.

Plow Draft– The amount of force required to pull a plow bottom through a soil under specified conditions. The draft may be measured by a dynamometer.

Plow Sole– See Plow Pan.

Plow Tip– The tip end of a plowshare. It is subject to the most severe wear and is replaceable by welding on a new tip in place of the worn tip on the old share. Also called plow point.

Plowshare– The cutting edge of a moldboard plow.

Plug– (1) The mass removed by a trier or other special penetrating implement in sampling or testing an agricultural product, as a plug from a cheese, a bale of cotton, or from a watermelon. (2) An old, worn-out horse. (3) A horse with a poor conformation. (4) To repair a leak, as a dam or earth fill. (5) A block of rooted grass that is planted for the purpose of establishing a covering of grass, such as in a law.

Plumb– Vertical. See Plumb Bob.

Plumb Bob– A pointed weight hanging by a string from the center of a transit level. Its purpose is to center the transit exactly over the benchmark.

Pluviometer– A rain gauge. See Ombrometer.

Point– In the marketplace, 1/10 of 1 cent; 1/1,000 of a dollar.

Point of Lay– Age at which pullets begin to lay, usually between twenty and twenty-two weeks.

Poke– (1) A yoke with an attached sharp spur pointing forward that is placed on the neck of an animal to prevent it from crawling through or jumping a fence. (2) Pokeberry. (3) A paper bag.

Polarity– A term applied to the positive (+) and the negative (−) ends of a magnet or electrical mechanism such as a battery or coil.

Pole– (1) A young tree 4 inches or more in diameter at breast height. The maximum size of poles is usually, though not invariably, between 8 and 12 inches. See Sapling. (2) A round timber used to support telephone and power lines. (3) A linear measurement of land, by statute, 16½ feet. Also called perch, rod.

Pole Ax– An axhead with a cutting edge on one end and squared off on the other with a point or claw bending downward, or projecting from the back of the head, which is fixed to a pole or handle. This type differs form the doublebit ax, which has two cutting edges.

Pole Barn– A form of barn construction in which natural poles are used in the framing instead of the regular sawed lumber or dimensions material generally used in such framing.

Pole Fence– A fence similar to a zigzag rail fence but made of poles instead of split rails.

Pole Pruner– A tree-pruning implement; a pole saw or clipper used for removing small limbs beyond the reach of a hand saw or hand shears.

Pole Saw– See Pole Pruner.

Police Powers– The powers of government to regulate property for promoting the public's safety, health, morals, and general welfare.

Polymer– A substance made of molecules comprised of long chains or cross-linked simple molecules.

Polystrene Foam– A very lightweight synthetic, plastic material made into flakes or beads and used in artificial plant growth media.

Polyvinyl Chloride (PVC)– A common plastic material that releases hydrochloric acid when burned.

Pommel– (1) The knob, ball, or protuberant part serving as a means of grip on anything, as the high, forward part of a saddle. (2) A block of hard wood grooved like a crimping board and employed by curriers to render leather supple and impart a grain to it.

Pool Auction–A cooperative method of marketing in which individually owned products are pooled and sold to the highest bidder.

Portland Cement–Obtained by burning an intimate mixture of pulverized materials containing lime, silica, and alumina in varying proportions but within certain narrow limits, and by pulverizing the clinker that results.

Positive Ion–A cation; an ion that carries a positive charge of electricity.

Post–(1) A short timber used in an upright position for supporting structures. (2) Any timber that supports fencing. It may be round, split, or sawn. (Posts may be of other materials, such as iron and concrete.) (3) The proper position and balance a rider assumes in the saddle in riding a horse while in the trot gait, rising with each second beat of the one-two rhythm of the trot.

Post Driver–(1) A hydraulically operated instrument used to drive fenceposts into the ground. (2) A metal cylinder with a weighted end and handles on the side. The cylinder is slipped over the top of a post and used to drive the post into the ground.

Post Hole Digger–(1) A hand-operated implement for digging circular post holes that consists of two semicylindrical cutting blades, hinged together but each mounted on separate handles. The soil is cut or loosened by plunging the blades downward and removed by moving the handles outward and lifting. (2) A tractor-powered or small gas engine-powered auger that bores a hole in the ground for putting in fenceposts.

Pot-bellied–(1) Designating any animal that has developed an abnormally large abdomen, usually because of improper feeding or nutrition. (2) Designating a type of coal stove used for heating purposes in pioneer stores and homes, so named because of its large, round, lower half.

Potential Energy–Energy inherent in a mass because of its position with reference to other masses; e.g., a rock at the edge of a precipice has potential energy. Water behind a dam also has potential energy.

Potometer–An instrument to measure the rate at which a growing plant absorbs moisture.

Poultry House–Any building equipped and used for housing and handling poultry.

Poultry Husbandry–The science and art of the production and distribution of poultry and poultry products, including breeding, incubation, brooding, rearing, housing, feeding, marketing, and poultry farm management.

Pound–(1) A unit of weight; 16 ounces avoirdupois, 12 ounces troy. The standard British unit of weight equals 7,000 grains, ½, 240 long ton, and 453.59 grams, the weight of 27.692 cubic inches of water at 4°C. (2) An enclosure in which stray animals are legally confined. (3) An enclosure in which groups of animals, as flocks of sheep, may be gathered for shelter, etc. (4) An enclosure used to trap wild animals.

Pound of Gain–In animal feeding, the net gain of weight in pounds derived from a particular number of pounds of feed fed.

Pounds of Grain Per Acre of Pasture–The feed value of an acre of pasture expressed in terms of pounds of dry grain feed.

Pounds of Milk Per Acre of Pasture–The feed value of an acre of pasture expressed in terms of pounds of dry grain feed.

Pounds of Retail Cuts Per Day of Age–A measure of cutability and growth combined, it is calculated as follows: cutability times carcass weight divided by age in days.

Pounds per Square Inch Gauge–Expressed as a quantity measured from above atmospheric pressure.

Pour Point–The lowest temperature at which a fluid, such as oil, will flow under specific conditions.

Power Loader–A fork, or fork and scoop combination, mounted at the front or rear of a tractor and operated by hydraulic power to load manure, bales of cotton, baled hay, pallet boxes, soil, gravel, etc.

Power of Attorney–An authorization by a person to another person to act for him/her on his/her legal behalf.

Power Shift Transmission–A tractor transmission in which the gears are selected manually but are power actuated; no master clutch is involved.

Power Subsoiler–A tractor-drawn plow that has a long, narrow shank with a wedge-shaped point for deep penetration to break up stiff clay subsoils and hardpans. See Hardpan, Plowpan.

Power Take-off–An attachment, usually consisting of a shaft and two or more universal joints, which is used to transmit power from a tractor to an attached unit, such as a combine, hay baler, mower, etc.

Power Trencher–A machine used on agricultural land in digging narrow trenches in which drain tiles are laid. Usually the power is supplied by an internal combustion engine.

PPM–See Parts Per Million.

Prairie Breaker–A long, low, moldboard plow with a gradual twist that completely inverted the furrow slice with a minimum of breakup and complete coverage of vegetation. It was used especially to cut and turn the tough virgin sod of the prairies of the Midwest.

Precipitate–(1) To cause a substance in solution to settle out in solid particles. (2) Occurring with undue rapidity.

Precipitation–(1) The amount of water, hail, sleet, snow, or other moisture received from clouds. Snow is also reported in its equivalent of liquid water. Precipitation is classified by the conditions that produce the rising column of unsaturated air which is antecedent to precipitation. Convection precipitation is the result of uneven heating of the ground, which causes the air to rise and expand, vapor to condense, and precipitation to occur. This is the major type of precipitation during the summer, producing high-intensity, short-duration storms. Orographic precipitation is caused by topographic barriers that force in the moisture-laden air to rise and cool. Cyclonic precipitation is related to large low-pressure systems that require five or six days to cross the United States from the northwest or Gulf of Mexico. These systems are the major source of winter precipitation. (2) The phenomenon of a solution or suspension that is flocculated. (3) The electrostatic or other means of removal of polluting particulates from the air. See Acid Rain.

Precipitation-Effectiveness (P-E) Index–The sum of twelve months of precipitation divided by the same period of evaporation.

Precipitators–Any of a number of devices using mechanical, electrical, or chemical means to collect particulates for measurement, analysis, or control. See Particulates.

Precursor–A compound from which another is made or synthesized.

Predicted Difference (PD)–The estimated difference of an animal from that of its parents or offspring.

Preemption Act–A law passed by the United States Congress in 1841 providing for the sale of public lands at $1.25 per acre under certain conditions. It was employed in conjunction with the later Homestead Act in many western states to increase the acreage available to an individual and to secure land with water facilities by ranch owners. It recognized the vested interest of squatters on the public domain and their right to preempt or buy the land they occupied.

Preference Permit–A license valid up to ten years issued by the United States Forest Service to permit domestic livestock grazing in a National Forest. The permit is renewable annually.

Preignition–Engine ignition that occurs earlier than intended; e.g., a piece of hot carbon in the combustion chamber can cause the fuel mixture to ignite before the spark plug ignites. The result is an engine knock and in severe cases, engine damage.

Premium Price–A price for a product, above the average for the market, paid for prize livestock, the first bale of cotton, superior produce, etc.

Prepayment Clause in a Mortgage–Statement of the terms upon which the mortgagor may pay the entire or stated amount on the mortgage principal at some time prior to the due date.

Prescription–(1) Legal title to land obtained by long possession. See Squatter's Right. (2) A written direction for the preparation and administration of a remedy.

Present Market Value–In farm appraisal, the price for which a property can be sold under present conditions with a substantial down payment, and a reasonable amount of effort, by a willing but not forced seller to a typical, desirous, but not anxious purchaser.

Present Value–In farm appraisals when applied to farm buildings, the replacement cost plus major improvements less depreciation and obsolescence.

Press–(1) A device for extracting juice from fruits or oil from seeds, etc. (2) Any device that compresses a loose or bulky product for packing or storage, such as a press box in a cotton gin.

Press Grain Drill–A grain drill that has press wheel gangs mounted on the rear of the drill to firm the soil over the planted seeds. See Air Drill, Grain Drill, Stubble Mulch Drill.

Press Wheels–Wheel attachments to field seed drills that follow the drills or tubes and press the loose soil over the seeds for better covering and compaction, usually resulting in increased germination.

Pressing–(1) (a) The extraction of the juice from fruit, as in cider making or in the production of orange juice. (b) The solid material left after such a process. (2) The process of bringing the small loose pieces of curds together in a shaped solid body in cheese making by means of a press.

Pressure Regulator–A device on a spray machine that adjusts and maintains the pressure desired and prevents excessive pressure from being built up, especially when the nozzles are closed.

Price Cycle–Any regularly recurring movement of prices that extends over a period of more than a year.

Price Elasticity of Demand–The relationship between the change in the price of a commodity and the accompanying change in the quantity that can be sold at that price.

Price Indexes–An indicator of the average price change for a group of commodities that compares prices for the same commodities in some other period, commonly called the base period. Monthly price indexes computed by the United States Department of Agriculture are the Index of Prices Received by Farmers and the Index of Prices Paid by Farmers for Commodities and Services, Interest, Taxes, and Farm Rates, referred to as the Parity Index when expressed in the base selected.

Price Level–Weighted average of prices that arise from all business transactions during a specified period of time. The United States Bureau of Labor Statistics index of wholesale prices and consumer price index are commonly used as indicators of the general price level in the United States.

Price Support Level–The price for a unit of a farm commodity (bushel, bale, pound) which the government will support through price support payments. Price support levels are determined by law and are set by the Secretary of Agriculture.

Price-spread–The difference between the price paid to the producer and that paid by the consumer for a product.

Prices-received Index–A measure computed on the basis of prices farmers received usually at the farm or in small local markets.

Prills–Spherical pellets or granules formed when molten material or a solution melt is sprayed through cool air. See Granulation.

Prime Agricultural Lands–The most productive lands for raising the common food and fiber crops; whereas unique agricultural lands are those most productive for the less common but high-value-per-acre crops such as rice, cranberries, citrus, etc. See Unique Agricultural Lands.

Primitive Area–(1) In the United States Forest Service, wilderness area. (2) In the United States National Park Servcie, an area of indeterminate size in which no commercial development nor the construction of any roads for motorized transportation is permitted.

Primocane–A biennial shoot or cane, particularly of a bramble (*Rubus* spp.), during its first year of growth and before flowering.

Principal Meridian–A north and south line accurately located and used as a basis from which to construct township and section lines as used in the United States Public Land Survey.

Principle of Comparative Advantage–The economic principle implying that various crops and livestock should be produced in those areas where the soils, climate, and human and market resources are best suited for their production.

Principle of Diminishing Physical and Economic Returns–The economic principle that variable resources should be added to fixed resources as long as the added return expected from the last unit of variable resource used is just sufficient to cover the added cost of that unit.

Principle of Independent Assortment–Mendel's genetic theory that differentiation characters (e.g., dominant and recessive characters), recombine at random. He proved by experiment that a certain mathematical relation exists among all the resulting combinations. See Mendel's Law.

Principle of Marginality The economic principle that choices about the use of resources should be made such that the marginal gain from the slightest possible change in resource use is equal to the marginal loss implied by the change.

Priority–(Water rights) The rights to special consideration and prior claim, over other individuals, to the use of a quantity of water.

Prism Wedge–A small hand-held prism used in forestry to estimate the basal area of a stand of trees. See Basal Area.

Private Herd Number–A number assigned to registered animals by their individual owners. It is required by some breed registry associations in the identification of registered animals.

Probability of Response–A soil test level given to reflect a "probability of response" for the applied nutrient. Expressing soil test results as a probability of response would assign a level according to the ability of each crop to produce a predicted yield following a recommended fertilizer application.

Productivity Rating–The productivity of the various soils on a farm based on the expected yield of the major crop or crops with known management. The rating is a percentage based on standards developed by research and statistical records of crop yields in the region. The best soils have a productivity rating of 100. In some instances, tables have been established showing the relation of crop yield to land values, with fixed management. See Soil Survey.

Profile Leveling–A method of leveling that is used to secure the elevation of a series of points located along a line. Profile leveling is employed in laying out a terrace.

Profit Margin–The amount of money made when expenses are subtracted from income.

Profits–The financial returns from a business, enterprise, or transaction above all costs, including both actual receipts and actual or estimated appreciation on the capital involved.

Progeny Records–The average, comparative performance of the progeny of sires and dams.

Progeny Testing–Determining the breeding value of an animal by studying its progeny.

Programmed Harvest–Timber scheduled for harvest for a specified year.

Promissory Note–The primary document in most credit transactions, the signed promise of the borrower to repay a loan. It states the date and amount of the loan, the interest rate, and terms of repayment.

Proof–A measure of ethanol alcohol content; 1 percent equals 2 proof.

Proper Use Factor–As applied to individual range species, the estimated maximum percentage of the total vegetative production of the year within easy reach of the livestock to which a given range species may be grazed without damaging it or associated important palatable plants or the soil; the degree to which each species may be grazed when the range as a whole is properly grazed. The proper use factor for a range type is the weighted average of the proper use factors of the individual plants in the type. See Palatability.

Properties–Characteristics by which a substance may be identified. Physical properties describe its state of matter, color, odor, and density; chemical properties describe its behavior in reaction with other materials; biological properties refer to any life-related characteristics such as biodegradation.

Proprietary–A medicine or food made by a person or company having the sole right to manufacture and sell it.

Protection Forestry–The practice of forestry with the primary objectives of: (a) conserving water supplies; (b) maintaining desirable streamflow; (c) increasing groundwater storage; (d) reducing erosion and reducing sedimentation; (e) providing high-quality water and reducing pollution; (f) ameliorating adverse climatic conditions, especially wind.

Protective Gear–Clothes and other equipment used to guard workers against poisoning when working with toxic materials.

Proton–See Atom.

Proven Sire–(Dairy Herd Improvement Association, DHIA). A bull with at least ten daughters that have completed lactation records and are out of dams with completed lactation records.

Proximal–Opposite of distal; near the point of attachment of reference.

Pruning Saw–A curved bladed saw usually mounted on a long pole that is used for removing limbs from a tree.

Pruning Snips–A hand-held pair of scissorlike cutters that is used for cutting small branches from plants.

PSA–Packers and Stockyards Administration.

PSI–Pounds per square inch.

Psychroenergetics–Science dealing with the effect of ambient temperature and humidity upon conversion of feed into bodily heat or energy. It is important in the study of housing for livestock and its effect on production.

Psychrometer–An instrument used to measure relative humidity of the atmosphere.

Public Domain–All lands over which a national government exercises proprietary rights, including national parks, grazing lands, forests, and military and other reservations.

Public Food Program–Publicly supported programs that attempt to increase the demand for food, improve the level of the diet, or otherwise influence food consumption; e.g., the food stamp program, the school lunch program.

Public Lands–The general public domain; lands belonging to a national government that are subject to sale or disposal and that are not reserved for any special governmental or public purpose.

Public Law 480–A law passed by the Congress in 1954, often referred to as "P.L. 480" or the "Food for Peace" program. Primary purposes are to expand foreign markets for United States agricultural products and use United States agricultural abundance to combat hunger and encourage economic development in the developing countries. The program makes United States agricultural commodities available at low interest, long-term credit under Title I of the act, and as donations for famine or other emergency relief under Title II. Under Title I, the recipient country agrees to undertake agricultural development projects to improve its own food production or distribution.

Public Liability–The responsibility of the owner for the safety of a person on his/her property, with or without the owner's consent.

Pulaski Tool–A combination chopping and cutting tool widely used in fire line construction for the protection of forests. It is a light, single-edged ax with a straight handle, having a narrow, adz-like trenching blade attached to its head.

Pulling–A hand procedure for stapling cotton or determining length of fiber in which a tuft for a sample is pulled into two parts, of which one part is manipulated by fingers and thumbs properly to ready it for measurement of the staple length.

Pulling Contest–An exhibit or public contest between two or more matched teams of horses or tractors to determine their pulling ability.

Pump–(1) A device that converts mechanical power into hydraulic power. (2) A device that lifts or pushes a liquid. The types of pump designs are: gear, vane, and piston units.

Purlin–A structural unit that is used to support the sash bars of a greenhouse or other building. It is built lengthwise along the greenhouse.

Pyrolysis–The decomposition of a substance by heat.

Pyrometer–An instrument for measuring temperatures, especially those beyond the range of mercurial thermometers, as by means of the change of electric resistance, the production of thermoelectric current, the expansion of gases, the specific heat of solids, or the intensity of the heat or light radiated.

Quadrat–A small plot or sample area frequently one square meter (one milacre) in size. In ecological field studies, a unit area in which vegetation changes are recorded.

Quadrate–Nearly square in form.

Quantum–The smallest indivisible quantity of radiant energy; a photon.

Quarter–(1) In slaughtering for meat, one half of the side of beef, as the forequarter or hindquarter. (2) Pertaining to a horseshoe, the branch between the last nail hole and the heel. (3) A unit of weight: (a) one-quarter cwt. (25 pounds) (avoirdupos); (b) eight bushels, formerly one-quarter ton (especially of grain). (4) A quarter section of land, or 160 acres. (5) A section of the bovine udder.

Quarter Clip–A clip on the shoe for the hind hoof of a horse. The clips are placed on the quarters of the shoe on the outside or inside to prevent the shoe from shifting laterally on the foot.

Quarter Section–According to the survey of public lands by the general Land Office (United States), one of four parts of a section of land, generally 160 acres. The square section of land was divided into four equal parts by a north-south and east-west line, and in a legal land description the quarters were designated by the points of the compass as NE, NW, SE, and SW.

Quarters–(1) The parts of the body of a horse or other quadruped above the legs, as the breast and hips. (2) Place used to house workers on plantations or ranches. (3) Farm or ranch housing for domestic animals.

Quick Tests–Certain standard chemical tests devised for the very rapid determination of the amounts of nutrient elements in a soil. When properly performed and interpreted such tests are useful in making recommendations for fertilizers and lime.

Quit-claim Deed–A document by which the seller conveys all his/her rights to the real estate in question but does not give any warranties as to the title. The use of the quit-claim deed is usually reserved for clearing title and settling the rights of persons who have claim to the property being transferred.

Quonset–A type of curved building or structure that takes its name from the Quonset hut of World War II. The Quonset shape is used in the construction of greenhouses, hog houses, shops, and storage buildings.

Quotation–The current market price of a commodity.

Race–(1) A group of individual plants which have certain common characteristics because of ancestry. It is generally considered a subdivision of a species; frequently used in plant breeding. (2) Pathogens of the same species and variety which are structurally indistinguishable but which differ in their physiological behavior, especially in their ability to parasitize varieties of a given host. (3) The channel that leads water to or from a waterwheel, the former is the head race and the latter the tail race. (4) A narrow passage or fenced land in a sheep yard for branding, dipping, etc. (5) An elongated white mark on the face of a horse or dog.

Rack–(1) The gait of a horse in which only one foot touches the ground at any one time, producing a four-beat gait. (2) A frame attached to a truck or wagon for the transportation of hay, tobacco, etc. (3) The rib portion of a sheep carcass. (4) A framework for holding feed for cattle, swine, sheep, etc., with upright partitions so that the animal can insert its head between the partitions and have access to the feed. (5) A frame placed in a stream to prevent the passage of fish. (6) A frame placed at the entrance to a sump pump to remove debris that would clog the pump.

Rad–Acronym for radiation absorbed dose; an amount of radiation sufficient to cause 1 gram of animal tissue to absorb 100 ergs of energy. The biological effect depends on the kind of radiation to which the tissue is exposed.

Radiant Energy–The energy of electromagnetic waves such as radiowaves, x rays, gamma rays, and visible light. Radiant energy is transmitted in a straight line from the sun to the earth and can be absorbed or reflected.

Radiate–(1) Standing on and spreading from a common center. (2) Having ray flowers, as in the family Compositae.

Radiation (Radioactivity)–The emission of very fast atomic particles or rays by nuclei. Some elements are naturally radioactive while others become radioactive after bombardment with neutrons or other particles. The three major forms of radiation are alpha, beta, and gamma.

Radiational Cooling–The loss of heat from plant leaves by means of radiation through the greenhouse glass. The loss is most intense on clear cloudless nights.

Radioacative Tracers–Small quantities of radioactive isotope mixed with larger amounts of the corresponding stable isotope to be used as labels. Since the stable and radioactive isotope act chemically and biologically in the same manner, as the radioactive one is readily detected.

Radioactive Element–An element capable of changing spontaneously into another element by the emission of charged particles from the nuclei of its atoms. For some elements, e.g., uranium, all known isotopes are radioactive; for others, e.g., potassium, only one of the several isotopes is radioactive. Radioactive isotopes of most elements can be prepared artificially, but only a few elements are naturally radioactive.

Radioactivity–The nuclear energy released when the nucleus of an atom disintegrates.

Radioecology–The study of the effects of radiation on species of plants and animals in natural communities.

Radiosonde–An instrument carried aloft by a balloon which transmits information by radio on atmospheric temperature, pressure, and humidity from upper altitudes.

Rafter–A beam which gives slope to a roof and supports the roofing.

Rain Gauge–(United States Weather Service) A container with a funnel-shaped top 8 inches in diameter which intercepts rainfall and drains into a tube with exactly one tenth the cross section of the collector. The amount of rain is measured by removing the top and inserting a measuring stick in the tube. A reading of 15 inches (38 centimeters) would, therefore, be a rainfall of 1.5 inches (3.8 centimeters).

Rain-tight–Referring to a construction which, when exposed to rain, does not allow water to enter.

Raise–To grow or produce, as to raise corn or cattle.

Raising–Erecting the assembled framing or making the structure ready for putting on the roofing and siding. In former times it was regarded as a social gathering.

Rake–(1) A farm implement which consists of several bars or head pieces or wheels equipped with tines which are caused to rotate and move swath hay or other material into a windrow for ease in picking up with a loader or baler. (2) A hand tool consisting of teeth inserted in a headpiece attached to a long handle; used for collecting hay and leaves and smoothing of the soil.

Ram Service–The fee charged for the breeding of a ewe.

Rammed Earth Wall–An earthen wall made by packing moist clay between heavy plank forms. Some type of protective cover of the earth wall is constructed to decrease raindrop splash erosion.

Ramp–An inclined plane serving as a passageway between two different levels; commonly used in loading livestock and machinery on trucks or trailers.

Ranch–An expression used mostly in the western United States to describe a tract of land, including land and facilities, used for the production of livestock. Accepted western usage generally refers to the headquarters facilities, pastures, and other land as the ranch, as distinguished from range. Loosely defined, a ranch also may be a small western farm, such as a fruit ranch or a chicken ranch.

Ranch Hand–A person hired to work on a ranch tending animals, making hay, mending fences, etc.

Ranch Headquarters–The usual place of residence of the rancher and his place of business.

Ranch House–The main residential building on a ranch.

Ranch Unit–An operating unit of grazing land in the western United States consisting of individually owned land and permits to use public land issued by the United States Forest Service under the Taylor Grazing Act.

Rancher–One who owns, occupies, operates, or works on a ranch.

Rancho–(Spanish) Ranch.

Random Sample–A sample taken without bias from an area or from a population in which every part has an equal chance of being taken, in contrast to systematic sampling.

Range Allotment–A specific range area to which certain livestock of a permittee or group of permittees are assigned. Also called grazing allotment.

Range Appraisal–A definite, written, detailed opinion by a qualified appraiser of the value of range land. Among the considerations involved are its grazing capacity, accessibility, facilities for handling livestock, availability of feed sources, and income-producing ability.

Range Condition–(1) The state of health or productivity of both soil and forage of a given range in terms of what it could or should be under a normal climate and under the best practicable management. (2) An animal that is in a sufficient state of health or condition to be kept on the range.

Range Conservation Plan–A statement of objectives in range conservation and management on a given ranch or range together with recommended course of action to achieve them, supported by inventory data and analysis.

Range Economics–The scientific study of the relationship of the use of grazing lands by livestock to the material and social welfare of individuals, communities, and the nation.

Range Inventory–An itemized list of resources of a management area such as range sites, range condition classes, range condition trends, range use, estimated proper stocking rates, physical developments, and natural conditions such as water, barriers, etc.

Range Management–The art and science of planning and directing range use to obtain sustained maximum forage for animal production consistent with uses of the land for other important purposes.

Range Rights–In the United States Old West, assumed rights to the range for grazing livestock by the ranchers in opposition to the farmer and small-herd owner.

Range Site–An area of land having a combination of edaphic, climatic, topographic, and natural biotic factors that is significantly different from adjacent areas. Various sites are significantly different in their potential forage production and/or different in management requirements for proper land use.

Range Suitability–The adaptability of a range to grazing by livestock and/or game.

Range Survey–(1) A determination by inspection of the carrying capacity of extensive areas of natural vegetation, based on the density and palatability of the forage. (2) an inventory which assembles important facts needed for perfecting a sound management plan taking into account the number of animals grazed, grazing capacity, period grazed, salt, needed water development, areas in need of revegetation, better livestock distribution, and the location of special problem areas.

Range Technician–A person who assists farmers and ranchers in making the best use of their ranges by collecting and compiling all information pertaining to proper range management, analyzing it, and collaborating with the operator in the formulation of range use plans which are technically and economically sound.

Range Trend–The direction of change in range condition, either deterioration or improvement.

Rangeland–(1) Land on which the natural plant cover is composed principally of native grasses, forbs, or shrubs valuable for forage. (2) Land used for grazing by livestock and big game animals on which the natural potential climax community of plants is dominated by grasses, grasslike plants, forbs, and shrubs.

Ranger–An administrative officer in charge of a unit of forest land, usually a subdivision of a public forest or park. Various classifications are recognized, as a forest ranger, district ranger, county ranger, etc.

Rasp–A coarse filelike tool used to wear away excess when trimming an animal's hoof.

Rate–The amount of active ingredient applied to a unit area, such as the amount of a pesticide that is applied per acre.

Rate Earned on Investment–An accounting figure determined by dividing the net income from the capital investment by the total investment in the farm business at the beginning of the year and multiplying the result by 100. This total represents the return for the capital and management expressed in percent.

Rate of Genetic Improvement–Rate of improvement of an animal per unit of time (year). The rate of improvement is dependent on: (a) heritability of traits considered; (b) selection differentials; (c) genetic correlations among traits considered; (d) generation interval in the herd; and (e) the number of traits of which selections are made.

Rated Horsepower–The value used by the manufacturer of an engine to rate the power of the engine, allowing for safe loads, etc. See Brake Horsepower, Drawbar Horsepower, Effective Horsepower, Horsepower.

Ratios–Performance of an animal compared with its contemporaries, with 100 being average. Ratios greater than 100 are above average, and less than 100 are below average.

Rattail–(1) A slim, hairless tail of a horse. (2) A small, round, tapered file.

Rawhide–Undressed skin of cattle.

Rawhide Bit–A bit made of rawhide used in the breaking of horses to protect the mouth.

Rawhide Hackamore–A bitless bridle or halter made of rawhide, used chiefly in the breaking of horses.

Rawhide Quirt–A riding whip made of rawhide.

REA–Rural Electrification Administration.

Reach–Difference between the average merit of a herd or flock, in one or several traits, and the average merit of those selected to be parents of the next generation.

Reaction–(1) A change in a market trend. (2) The degree of acidity or alkalinity (e.g., of a soil mass) expressed in pH values and in words as follows: extremely acid, below 4.5; very strongly acid, 4.5-5.5; medium acid, 5.6-6.0; slightly acid 6.1-6.5; neutral, 6.6-7.3 (strictly, 7.0); mildly alkaline, 7.4-8.0; strongly alkaline, 8.1-9.0; very strongly alkaline, over 9.1.

Reactor–(1) Especially in testing for a disease, an animal which reacts positively to a foreign substance: e.g., a tuberculous animal would be a reactor to tuberculin. (2) The apparatus in which nuclear fission takes place.

Real Property Mortgage–A specific listing of real estate set aside to guarantee payment of a debt.

Real Property (Real Estate)–Real property, or real estate as it is often called, consists of: land; anything affixed to it as to be regarded as a permanent part of the land; that which is appurtenant to the land; and that which is immovable by law.

Reata–Lariat; a kind of strong Mexican rope made by twisting thongs of hide together, used in roping cattle on western ranges (United States).

Receipts–The number of livestock, cattle, hogs, and sheep which arrive at a public market for a specified time, as by days or months.

Receiver–Court-appointed custodian who holds property for the court, pending final disposition of the matter before the court.

Reciprocal Trade–The lowering of trade barriers by one country in consideration of similar treatment granted by other countries. The objective is to expand trade in the interest of all trading partners.

Reciprocating Engine–An engine in which the up-and-down motion of a piston is transformed into the circular motion of the crankshaft.

Reclamation Service–A bureau in the United States Department of Interior concerned mainly with the arid lands of western United States: classification, irrigation, improvement, administration, and with power development.

Recommendation Domain–A group of farmers within an agroeconomic zone whose farms are sufficiently similar and who follow sufficiently similar practices that a given recommendation is applicable to the entire group.

Recommended Reduction–The percentage difference between the national base acreage and the national program acreage for a crop.

Reconnaissance–(1) A cruise of forest property to obtain general information of the forest conditions. (2) An extensive range survey which is carried out to estimate average density and composition of range vegetation within a type or subtype without use of systematically established plots. (3) A type of survey in which land features are examined at wide intervals and are not delineated in detail. It is intended to furnish information primarily for extensive or overall planning. See Soil Map.

Rectangular System of Land Division–A system of land division which is based upon permanent meridianal and latitudinal lines yielding rectangular-shaped land areas as nearly as possible: 6 miles square divided into 36 secondary units called sections.

Red Grain Sorghum–A market class of grain sorghums as provided by the Official Grain Standard of the United States which includes all varieties of red grain sorghums and may include not more than 10 percent of grain sorghums of other colors.

Reducer–A coupling which links together pipes of different sizes. It is commonly used in sprinkler irrigation and plumbing.

Reductase–An enzyme which has the power of reducing or decolorizing methylene blue.

Reduction–(1) The process of removing oxygen from a compound; e.g., hematite is reduced to metallic iron. (2) The addition of electrons to an atom or ion.

Reeve–To pass the end of a rope through a ring, a hole in a block, etc.

Reference Sire–A bull designated to be used as a benchmark in progeny testing other bulls (young sires). Progeny by reference sires in several herds enable comparisons to be made between bulls not producing progeny in the same herd(s).

Refractometer–An instrument for measuring the percent of soluble solids in a solution, designed to read directly in percent water; used for measuring the percent water in honey and nectar.

Refrigeration–Artificial cooling, either by the application of ice or by utilizing the principle of the latent heat of evaporation.

Register Ton–See Ton.

Registered Seed–The progeny of foundation or registered seed, approved and certified by the certifying agency, that is so handled as to maintain satisfactory genetic identity and purity.

Registration Certificate–A paper issued by a breed association which shows that a particular animal has been registered.

Regression–(1) Destruction of the vegetation, as by fire, grazing, cutting, etc., usually with subsequent deterioration of the site, as by exposure, erosion, or loss of nutrients, to such extent as to give rise to a subsequent simpler vegetative type. It is not a true succession or development from forest to grassland, but a replacement as a consequence of complete destruction of the trees, etc. (2) Measure of the relationship between two variables. The value of one trait can be predicted by knowing the value of the other variable; e.g., easily obtained carcass traits (hot carcass weight, fat thickness, ribeye area, and percent of internal fat) are used to predict percent cutability.

Regulation Cut–The determination, periodically or in advance of cutting, of the volume of timber to be felled under the objectives of a given management plan.

Regulator–(1) Any chemical, other than a nutrient, which will increase or decrease the growth of a plant or a plant part. (2) A mechanical device which regulates an operation or condition, as a temperature regulator.

Reins–The part of a horse's harness fastened to each side of the bit or curb by which the rider or driver directs and controls it.

Reject–Any product, such as lumber, milk, wool, etc., that does not measure up to recognized grade specifications.

Relative Risk–An estimate obtained by dividing the incidence of disease in the exposed group by the incidence in the corresponding unexposed or control group.

Relay–An electrical switch that opens and closes a circuit automatically.

Release–To relinquish an interest in or a claim to a piece of property.

Remote Hydraulic Cylinder–A hydraulic cylinder that is attached to a tractor-drawn or operated implement. The cylinder is "remote" from the tractor and is used to assist in the operation of the implement.

Remote Sensing–(1) The measurement of some property or phenomenon by a recording device that is not in physical or intimate contact with the object under study. The technique employs such devices as the camera, lasers, infrared and ultraviolet detectors, microwave and radio frequency receivers, and radar systems. (2) The practice of data collection in the wavelengths from ultraviolet to radio regions. This restricted sense is the practical outgrowth from airborne photography. Remote sensing is sometimes used for rapid reconnaissance of crop conditions and for soil mapping.

Rent–(1) A return paid or received for the use of property. (2) To lease property for money or for a share of the produce.

Rental–The amount for which a property will rent.

Renter–One who rents farm or pasture land from an owner. See Rent.

Repeatability–The tendency of animals to repeat themselves in certain performance traits in successive seasons, pregnancies, or lactations.

Replacement Value–The estimated cost at current prices of replacing an animal, a building, a tree, or some other object.

Research–All effort directed toward increased knowledge of natural phenomena and the environment and toward the solution of problems in all fields of science. This includes basic and applied research. Much of the agricultural productivity of the United States is directly the result of applying research.

Reserve–(1) Any tract of land, especially public land, set aside for some special use; e.g., forest reserve, school reserve. Also called reservation. (2) A tree or group of trees left uncut on an area for a period, usually a second rotating. After the stand is reproduced, naturally or artificially, an active stand which is held for future utilization.

Reservoir–(1) Any storage place. (2) A place where a quantity of water, usually large, is collected and stored for use when required, as to supply a city, for irrigation, etc. Such a body of water is often held from flow by a dam.

Residue Management–Of crops. Use of that portion of the plant or crop left in the field after harvest for protection or improvement of the soil. See Conservation Tillage.

Retail Dairy–A type of dairy plant concerned primarily with the distribution and selling of milk and milk products directly to the consumer.

Return on Farm Capital–The net farm income less the value of the operator's labor which represents the return on the farm capital.

Return to Total Capital–Net farm income less the value of family labor used on the farm, usually expressed as a percentage of total farm capital.

Riata–See Lariat.

Ribband–In barn construction, a horizontal strip of wood notched into the studs in a balloon frame, which forms a support for joists above the first floor. Also called ribbon.

Rick–(1) Hay or grain carefully placed in a rectangular pile and ridged at the top which may be thatched for protection from the weather. (2) A pile of cordwood for use as firewood. It is 8 feet long, 4 feet high, and has a variable width. (3) A pile of cordwood, stave bolts, etc., split from short logs.

Rickety–(1) Having rickets. Also called rachitic. (2) Designating an unstable building or old, infirm animal.

Ridge–(1) A divide between two hollows; specifically, the strip of land thrown up by a plow or left between two furrows, or anything which resembles the elevation left by the plow. (2) A crest, as the summit of an abrupt ascent; a long hill. (3) The highest part of the roof of a building where the upper ends of the rafters meet. (4) The earth levee that is thrown up in terrace construction to confine run-off water.

Riding Crop–A short whip used by horseback riders.

Riding Stable–A stable where saddle horses are maintained by the individual owners or are offered for hire by the management.

Rig–(1) An apparatus or outfit of tools for any particular purpose, more especially for well boring. (2) The complete assembly of beam, shanks, and shovels of the corn cultivator. (3) The equipment used for the application of spray material for the control of diseases, insects, and weeds. (4) A carriage with its horses. (5) A ridge made in plowing. (6) The line down through the middle of the back on coarse wools where the wool separates, some falling down one side, the rest falling down the other side.

Rigging–That part of the saddle construction which has to do with the securing of the cinch around the horse.

Rigging Ring–The ring attached to the saddle tree for a cinch.

Right Bank–The right bank of any stream, brook, or river is on your right as you face downstream. Because of the coriolis effect, in the Northern Hemisphere, the right bank is usually more eroded than the left bank; and vice versa in the Southern Hemisphere. See Coriolis effect, Left Bank.

Right-handed Plow–A plow that throws the furrow slice to the right.

Right-of-Way–(1) A privilege which one person or persons may have for passage over the land of another. (2) A strip of land used for highways, railroads, pipe lines, etc. It may be an easement, or may have been purchased outright.

Right-to-Farm Laws–Laws passed by many states designed to help protect producers from lawsuits initiated by city dwellers who move to the country.

Right-to-Know Law–A federal law that requires producers to inform workers of any hazardous or toxic materials that the workers may come in contact with while performing their duties.

Rim–(1) A metal strip on the outside of a wheel. (2) The edge of a tank, gutter, natural depression, etc.

Ring–(1) A cut or girdle around the trunk, branches, or roots of a tree. See Girdle. (2) Annual growth ring of a tree. See Annual Ring. (3) (a) A circular band of metal or wood, as the metal ring in the nose of a bull. (b) To place a ring through the cartilage of the nose of an animal; e.g., to prevent a hog from harmful rooting or to control a bull, etc. (4) (a) A circular, metal or plastic band placed on the leg of a fowl for identification purposes. (b) To place a ring on the leg of a fowl. Also called ringing birds. (5) A ridge which encircles the horns of a cow, the number increasing with age. (6) A circular exhibition place for the showing or sale of livestock or the racing of horses.

Ring Bit–A large ring that passes over the lower jaw of the horse which is used as a bit.

Ring Gear–A gear that surrounds or rings the sun and planet gears in a planetary gear system. Also the term for the spiral bevel gear in a differential. See Differential, Planetary Gear.

Riparian Doctrine–Under common law (United States), the owner of land along a stream is entitled to have the streamflow to his land undiminished in quantity and unimpaired in quality by upstream riparian owners, except that such owners are entitled to use water for domestic purposes. In some states, it is held that the riparian right includes the right to make use of the water for irrigation purposes and other uses, and that such right is a property right entitled to protection. See Aquatic Rights.

Riparian Forest–Tree growth adjacent to streams or other watercourses whose roots are in or close to the zone of saturation due to the proximity of surface or underground water.

Riparian Habitat–That portion of a watershed or shoreline influenced by surface or subsurface waters, including stream or lake margins, marshes, drainage courses, springs, and seeps.

Riparian Owner–One who owns land adjacent to a stream or other body of water.

Riparian Rights–The rights accruing to a landowner on the bank of a natural watercourse, lake, or ocean. These rights vary with state laws. Riparian rights cease at the water's edge and do not interfere with use of the water area by others offshore.

Ripper Hipper–A tractor-drawn implement used to pulverize the soil and create high beds for planting. Used in planting cotton.

Riprap–Broken rock, cobbles, or boulders placed on earth surfaces, such as the face of a dam or the bank of a stream, for protection against the action of water (waves); also applied to brush or pole mattresses, or brush and stone, or other similar materials used for soil erosion control.

Rise of Salts–Soluble salts, previously distributed in the soil, which is brought to the surface with the rise of capillary water and left as a surface incrustation when the moisture evaporates. This accumulation may cause crop injury. See Waterlogged.

Risk Attitude–Extent to which a person seeks to avoid or is willing to face risk.

Risk Rating–An estimate of the credit and responsibility of an individual or business concern.

Roadside Market–A food-marketing system in which farmers sell directly to consumers.

Rock Salt–Crude, common salt (sodium chloride).

Rocker Toe Shoe–A special horseshoe which is used on a horse that stumbles.

Rod–(1) A linear measure of land equal to 16½ feet. In square measure, a plot of land 16½ feet on each side; a square rod. Also called perch, pole. (2) An old, English measure of land which varied from 6 to 8 yards.

Rod Weeder–A secondary tillage implement used primarily for fallow-land weed control with minimum soil surface disturbance and moisture loss. The device employs round or square weeding rods powered from gauge wheels, rotating under the soil surface.

Rodeo–(1) The rounding up of cattle. (2) A public performance presenting features of a cattle round-up, as lariat throwing, bronco riding, and bulldogging.

Rolcut Secateurs–An instrument for removing the folds of skin from the crotch of ewes, also used in the bloodless emasculation of calves and lambs.

Roll Bar–A heavy bar across the top of a tractor or other machine that is in place to protect the operator in case the machine overturns.

Rolled-toe Shoe–A type of horseshoe which is used to make the foot action of a horse break over easier by improving action at the pivotal point of the shoe.

Roller Drier–A type of drier for milk in which the milk is dried in a thin film on the surface of a horizontal steel cylinder heated by steam. The dried milk is removed by a scraper knife.

Roller Foundation Mill–A roller mill used for the manufacture of honeycomb foundation used in the production of honey.

Roller Grader–For the grading of fruit, two rollers which revolve from each other and are closer together at the upper end than at the lower. As the fruit passes along the rollers the small fruit drops through first and the large last. This type of grader is used for whole, unpeeled fruit.

Roller Harrow–A preplanting implement with one row of heavy rollers, two rows of spring teeth, then another row of rollers. It is used to smooth the soil for planting.

Roller Mulcher–A tractor-drawn implement that conditions a plowed field prior to planting. The plowed ground is packed to help retain moisture in the seed zone.

Rolling–(1) Excessive side motion of shoulders, common in horses with abnormally wide fronts or chests. (2) A part of seed-bed preparation in which the land is rolled to even out the surface. (3) Processing grain through a set of smooth rollers which are close together; sometimes called flaking.

Rolling Colter–A circular, flat, steel disk, sharpened on the edge and suspended on a shank and yoke from the beam of a plow, so constructed that it can be adjusted up and down for depth and sideways for width of cut. It cuts the furrow slice loose from the furrow wall, especially when plowing under heavy plant growth with a moldboard plow.

Room Temperature–In laboratory work, 68° to 70°F (20° to 21°C).

Root Spade–A spade with a long, narrow blade that is used for digging up plants that are to be balled and burlapped. See Balled and Burlapped.

Rope–(1) A thick, strong cord made by twisting or intertwining metal strands or fiber, as iron, steel, and manila hemp. (2) Form of bacteria that can survive baking of bread; under conditions of warmth and moisture, may convert mass of the bread into starchy patches.

Rope Bridle–A bridle made of rope instead of leather.

Rope Halter–A halter made of rope instead of leather.

Rope Surcingle–(1) A belt, band, or girth made of rope, passed under the horse and over a load on its back to hold the load fast. (2) A rope girth for a rider to hang on to when riding a bucking horse.

Rope Wick Applicator–See Wick Applicator.

Roper Curved Cheek Bit–A common type of bit that is used on horses that are used for roping.

Rotary Hoe–An implement consisting of rigid, curved teeth, mounted on "wheels" that roll over the ground, penetrate almost straight down, and lift soil as they emerge.

Rotary Hog Feeder–A round or many-sided feeder which the hog turns with its snout to make grain feed come down from the hopper to the animal for consumption.

Rotary Mower–A mower that cuts by means of rotating blades. Sometimes called a bush hog.

Rotary Plow (Tiller)–Cutting knives (tines) or an auger mounted on a horizontal or vertical power-driven shaft whose revolving tines loosen the soil and destroy weeds. Also called rotary tiller.

Rotary Sprinkler–A revolving head spray sprinkler used on sprinkler irrigation systems.

Rotation Irrigation–A system of irrigation through which the irrigator receives his allotted quantity of water not at a continuous rate but as a large flow at stated intervals: e.g., a number of irrigators receiving water from the same lateral may agree among themselves to rotate the water, each taking the entire flow in turn for a limited period.

Rotation-deferred Grazing–Grazing under a system where the key plants in one or more range units are rested at planned intervals throughout the growing season, and no unit is grazed more than half of any growing season or at the same time in successive years.

Rotten–(1) Designating decomposed or putrid organic matter. (2) Designating ground or soil extremely soft and yielding because of decay, or rocks partially decomposed. (3) Designating sheep attacked by rot.

Roughing Machine–A mechanical feather picker, usually consisting of a rapidly revolving drum equipped with flexible rubber fingers against which the previously lightly scalded fowl is held to remove feathers.

Round Bale Feeder–A feeder that allows cattle to eat hay from a round bale, but prevents them from trampling and destroying the bale.

Round Barn–A circular or multisided barn with a litter alley around the wall and cows arranged in a single row around the periphery facing toward the center. A silo might be constructed in the center.

Round Feeder–A round hog feeder with doors around the bottom that the hogs lift to obtain feed.

Round Hay Baler–A machine that rolls up a windrow of hay to form bales weighing 1,000 to 3,000 pounds.

Round-up–(1) The deliberate gathering of domestic animals, usually range cattle, for branding, fastening ear tags, injections, pesticide applications, inventory, and potential sales. (2) The brand name of a herbicide used as a foliar spray to kill most plants.

Roustabout–A person who does all kinds of work.

Row Marker–An extension device attached to a row planter or fertilizer or lime spreader that makes a continuous mark to guide the tractor driver on the next round to avoid skips and overlaps. On the extended end the device may have a spike, a small shovel, or a disk. Sometimes a white foamy substance is exuded from the end to serve as a continuous marker.

Rowel–In horseriding, the wheel of a spur with blunt to sharp projecting points.

RPM–Revolutions per minute.

RTC–Ready-to-cook; refers to a broiler or turkey marketed as eviscerated, with blood and feathers removed.

Rubber Sleeve–A protective, rubber covering attached to a rubber glove for the hand, arm, and shoulder which is used by veterinarians and artificial inseminators in various phases of their work.

Rubbing Post–Any of several devices consisting mainly of a post set in the ground with a can of oil attached to the top, so arranged that the oil or medicament is allowed to slowly drip or run down the sides of the post so that when hogs or other livestock rub against it a small amount of oil will be deposited on the rubbed or infected area of the skin. Also called hog oilers, livestock oilers.

Rumen Magnet–A smooth oblong magnet that is placed in the rumen to collect small metal objects that are swallowed by the animal during grazing.

Run–(1) A period of time, as a maple syrup run. (2) The amount of sap or sugar produced in a given time. (3) A swiftly moving tributary, rivulet, or mountain stream (eastern United States). (4) The stream outlet of a large spring (Florida, United States). (5) Unrestricted movement, as the colts have the run of the pasture. (6) An area of land or a leasehold (New Zealand). (7) A small, often dry gully or channel carved by water. See Arroyo. (8) A fenced-in pen used for the exercise of animals or poultry. (9) To feed; to graze, as steers run on the open range. (10) To operate, as a plow is set to run at a depth of 6 inches.

(11) To cultivate, mow, combine, etc., as to run over a field with a weeder. (12) To maintain animals, as he runs sheep. (13) To work a dog with sheep or cattle. (14) To discharge pus, as a sore runs. (15) To be, as the prices run very high. (16) To move rapidly, as a horse. (17) To turn, as to run a wheel. (18) To grow, as when the vines begin to run. (19) To operate, as to run an engine.

Run-down—Designating anything that has been allowed to deteriorate, such as a fence, a set of farm buildings, or a field that has been poorly farmed and has lost productivity.

Run-off Modulus—The depth of water in inches over the underdrained area which must be removed by tile drains in 24 hours. It is a measure of the maximum rate at which the water will move through the soils to the laterals.

Runner—(1) A breed of ducks of very distinctive type, having a long, narrow body and very erect carriage. The breed derives its name from its gait, which is a quick run, quite unlike the waddle of other ducks. The adult drake weighs about 4½ pounds; the adult duck about 4 pounds. The runner is noted as an egg-producing breed and has little or no value for meat production. Its three varieties are the White, the Fawn and White, and the Penciled. Also called Indian Runner. (2) A lateral, aboveground shoot (stolon) of certain plants; e.g., strawberries, which roots and forms young plants at some of the nodes, aiding in propagation. (3) A rope used to increase the mechanical power of a tackle. (4) The upper or rotating stone of a set of millstones. (5) A supporting attachment which slides along the ground, as a sled runner. See Stolon.

Running Plate—The aluminum horseshoe worn by race horses.

Rural Electrification Administration—In the United States, an agency which organizes loans for central station electric service to people who do not have it, for rural electric and telephone system loans, and loans to finance the wiring of farmsteads and installation of electric appliances and plumbing. Loans are not made directly to consumers but to R.E.A. borrowers for relending to members. Created by Executive Order, May 11, 1935, and later reorganized and revised to become a part of the Department of Agriculture July 1, 1939.

Rural Land—Land which is occupied by farmers or used for agricultural purposes as distinguished from urban land, park or recreational land, and wilderness.

Rural Population—Those people who do not live in cities. (A city is arbitrarily defined in the United States Census statistics as a community of 2,500 persons or more.)

Rural Sociology—The science of rural human relationships.

Rural-Farm—One of three residence categories used by the United States Bureau of the Census to encompass all persons living on farms without regard to occupation. See Rural-Nonfarm, Urban.

Rural-Farm Population—Persons who live on farms (as defined by United States Census) located in rural territory without regard to occupation.

Rural-Nonfarm—One of the three residence categories used by the United States Bureau of the Census which includes persons living outside urban areas (with certain exceptions) but not residing on farms. See Rural-Farm, Urban.

Rurban—The transition zone between the corporate limits of a city, or thickly populated suburbs, and the open country. Occupied land which is neither definitely urban nor rural.

Rut—(1) The grooved track left by the wheels of vehicles in soft ground. (2) The season of heightened sexual activity in male mammals that coincides with the season of estrus in the female. See Estrus.

S-tine Cultivator—A cultivator that has an S-shaped foot to which the sweep is attached. The S shape gives the cultivator a spring action. See Cultivator, Sweep.

Saccharimeter—Any device used to determine the concentration of sugar in a solution, particularly some form of polarimeter. See Syrup Hydrometer.

Sack—(1) A bag, usually large, oblong, and made of burlap or other stout fabric, used for holding grains, vegetables, etc. (2) To stroke or rub a horse. (A highly nervous horse is frequently sacked by the buster who, from a safe position, soothes the animal by rubbing it with a burlap sack or broom.)

Sacked—Designating grain, produce, etc., which has been placed in sacks.

Saddle—(1) A seat designed to fit a horse's back and to make riding easy, comfortable, and safe. Saddles are of many types but are usually made of heavy leather over a well-padded frame. They are often raised in the rear to hold the body in place and they may have a firm, raised support or horn in front to attach ropes or gear. They are held to the horse's back by means of one or more wide straps or girths that pass around the body of the horse. (2) The whole upper back section of a meat carcass which in wholesale cuts is commonly divided into fore and hind saddle. In cookery, the rear upper back portion including the two loins. (Sheep, veal calves, goats, and deer.) (3) The rear part of the back of a male chicken extending to the tail, covered by saddle feathers. (4) An apron made of reasonably heavy canvas which is placed on the backs of valuable breeding hens to protect against injury during mating. Saddles are usually held in place by loops which fit under the wings. (5) The uncut stalks of four hills of corn from two rows, tied together, around which corn is shocked. (6) A saddlelike depression in the crest of a ridge. See Saddleback. (7) A specially designed transverse log used in a skid road to guide the moving logs. (8) To put a saddle upon an animal or thing.

Saddle Blanket–A blanket or pad of wool or cotton placed on the back of a horse beneath the saddle to prevent chafing; saddlecloth.

Saddle Fender–A broad piece of leather attached to a stirrup strap to protect the rider's leg from the stirrup straps and perspiration from the side of the horse. Also called sudadero, rosaderos, stirrup fender, leg fender.

Saddle Girth–In horses, a strong band or strap which encircles the body, usually just behind the front legs, and helps fasten a saddle to the back. With western-type saddles it usually consists of two latigo straps on the sides and a webbed band underneath with rings into which the latigo straps fit to tighten and fasten. Also called cinch.

Saddle Horn–A knoblike elevation at the front center of a western saddle, used in roping as an anchor for the lariat.

Saddle Mule–A mule suitable for riding.

Saddle Skirt–The leather part of a western-type saddle which extends beyond the sides and to the rear of the saddle seat. On English-type saddles, a broad piece of leather which extends under the saddle, down the side of a horse, and under the stirrup strap.

Saddle Soap–A special soap used to clean and preserve saddles, boots, and leather goods.

Saddlebag–A bag or pouch of leather or fabric carried at the saddle, especially one of a pair laid across the back of a horse behind the saddle.

Saddler–(1) One who makes, repairs, or deals in saddles or saddlery. (2) A saddle horse.

Safety Belt–A belt used to fasten a climber securely to a tree, pole, tower, or seat so as to free his hands for work.

Safety Release Hitch–An attachment designed to release a pulled implement, such as a plow, mower, etc., from a tractor when the implement strikes an obstruction.

Safety Trips–Mechanical devices on plows and other tillage implements that release backward if the plow hits a solid obstruction such as a rock or stump.

Sales Contract–An agreement entered into by two or more parties to sell and/or buy property.

Sales Pavilion–A building, large tent, or other covered structure in which livestock sales are held, especially auction sales.

Saliferous–Producing or containing a large proportion of salt.

Saline–General term for waters continuing various dissolved salts. The term is restricted to inland waters where the ratios of the salts often vary; the term haline is applied to coastal waters where the salts are roughly in the same proportion as found in undiluted seawater.

Salinity–The quantity of saltness in seawater or freshwater, most commonly expressed in parts of dissolved salt per 1,000 parts of water; e.g., salinity of seawater is 35 parts per thousand. See Alkalinity.

Salinometer–Hydrometer used to measure concentration of salt solutions.

Salometer–A device, usually a hydrometer, with special graduations in degrees or percentages, used to determine the strength or salinity of a particular salt solution, such as a meat pickle or a pickle for cucumbers, etc. It is graduated so that 4 degrees on the salometer correspond to each 1 percent of sodium chloride in a solution at a temperature of 60°F. When used for calcium chloride brine, the salometer shows a reading nearly $1/10$ larger for equal percentages of strength. Also called salimeter, salinometer.

Salt–(1) Sodium chloride (common salt), NaCl; a white crystalline compound occurring abundantly in nature as a solid or in solution. It is a requirement of all livestock and is generally deficient in all natural feed ingredients. It is commonly included in mixed feeds and is otherwise supplied as salt blocks in pastures, barn lots, or stalls. It has wide usage in seasoning, preserving meats and vegetables, tanning hides, etc. In large amounts it is toxic and is frequently used to kill weeds; however, some plants such as Bermudagrass and asparagus tolerate a very high concentration. (2) To add salt as seasoning, preservative, supplement, etc.

Salt Index–An index to compare solubility of chemicals. Most nitrogen and potassium compounds have a high index; phosphate compounds have a low index. When applied too close to seed or foliage, high-index materials may cause plant injury by plasmolysis.

Sample–(1) Anything taken or shown as representative of the whole, as samples of cotton from a bale, seed, or fertilizer from a bag, etc. (2) In cotton buying, two parts taken from either side of the bale, which weigh about 6 ounces and have a face 4 to 6 inches wide. (3) To take a portion of some commodity and to test it, or offer it for test, as a representative of the whole.

Sample Plot–An accurately measured and outlined area used for sampling, or an area of land used for experimentation.

Sampler–(1) Any of the several different devices used for taking a sample from a bag, a car of grain, a can of milk, a tub of butter, a cheese, etc. (2) One who takes samples for inspection or testing. See Sampling Tube.

Sampling Error–The probable maximum error of an estimated total or average that arises from taking a sample rather than making a complete inventory or measurement.

Sampling Fertilizers–The State Department of Agriculture in each of the fifty states maintains a fertilizer inspection service that samples and analyzes fertilizers to determine if the fertilizer grade on the bag or in the bulk is properly designated to conform to the respective state laws. In the interest of uniform sampling procedures among the states, the Association of American Plant Food Control Officials has sponsored joint regional conferences of fertilizer inspectors and fertilizer industry personnel for the purpose of exchanging information on the many problems connected with sampling. An Inspector's Manual has been developed and demonstrations of sampling procedures have been held.

Sampling Tube–(1) Any of the several tubelike devices, commonly made of brass or stainless steel, of open or closable types used for extracting representative samples of grains, feeds, seeds, fertilizers, liquids, etc. (2) Any of the several kinds of tubes used for collecting samples of soils.

Sand Filter–A method of water purification by filtration in which the impure water is allowed to trickle downward through particles of sand which sift out the suspended impurities.

Sash–(1) The frame which holds the glass in a window. (2) A glazed frame used to cover a hotbed or cold frame.

Saturated–Filled to the maximum capacity under existing conditions.

Saturation Point–(1) The point beyond which no further additions can be held, utilized, or accommodated; e.g., the water-holding capacity of a particular soil, the flow of goods to a market, or a population of plants. (2) In wildlife, the maximum density of population; the natural condition wherein a species is as abundant as living conditions permit.

Saw–(1) A cutting device of various types, sizes, and shapes which consists essentially of a serrated metal blade; used for cutting wood, metal, bone, etc. (2) To cut with a saw.

Saw Kerf–The width of a cut made by a saw.

Saw Log–A log meeting minimum standards of diameter, length, and defect, including logs at least 8 feet long, sound and straight, and with a minimum diameter inside bark for soft woods of 6 inches (8 inches for hardwoods) or other combinations of size and defect specified by regional standards.

Saw Timber–Trees that will yield logs suitable in size and quality for lumber.

Sawbuck–An X-shaped frame or saw horse used for supporting poles and small logs to be sawed.

Sawdust–Dust and small particles of wood produced by the cutting action of saws. A by-product of the lumber and wood industries. It is used as a bedding material for livestock or as soil mulch. Mixed with soil, it has no fertilizer value until decomposed and may depress plant growth in its fresh state unless nitrogen fertilizer is added to it.

Sawmill–A plant at which logs are sawed into salable products.

Sawtimber Stands–Stands at least 10 percent occupied with growing-stock trees, with half or more of total stocking in sawtimber or poletimber trees, and with sawtimber stocking at least equal to poletimber stocking.

Scale–(1) Any instrument used to determine weight. (2) Either of the pans of a balance. (3) A measure of dimension, concentration, or intensity. (4) A graduated series of steps or degrees on measuring devices. (5) An instrument or device on which graduated spaces for measurement have been stamped or attached. (6) A graduated list of prices, wages, etc. (7) Any thin flake that peels from a surface, such as from skin. (8) One of the thin, flat, membranous plates which forms the outer, protective covering of certain animals or parts thereof, as the shanks of birds, etc. (9) One of several rudimentary, specialized leaves which protects the buds of plants and deciduous trees in cold climates. (10) A thin, dry, membranous part of a plant or flower. (11) See Scale Insect. (12) The estimated sound content of a log or group of logs scaled by use of a given log rule. Net scale, the scale after deducting for defects. (13) To form into scales. (14) To cover with scales or with an incrustation. (15) To remove in thin layers, as by scraping. (16) To come off in scales. (17) To make according to scale.

Scale Caliper–A log scale stick fitted with calipers graduated for the volumes of logs of different lengths and middle diameters.

Scandinavian Feed-unit System–A feed evaluation system widely used in Scandinavian countries for measuring the relative values of different feeds. In this system, one pound of barley is taken as the standard. The feed-unit value for any other feed is the amount of that feed which is estimated to have the same productive value as a pound of barley.

Scantling–A piece of yard lumber 2 inches thick and less than 8 inches wide.

Scarf–(1) The tapered or notched ends of pieces of wood that fit together to form a scarf joint. (2) The beveled cut on a log or stump which results from undercutting a tree in felling. (3) The incision on a stock on which the scion is inserted in propagation by budding.

Schlachter–A rabbi, who performs the ritualistic killing in the kosher method of slaughtering cattle and other animals for consumption by Orthodox Jews.

School Lands–In the United States, public lands of a state set apart by the state or by the federal government for the establishment and maintenance of public schools. See Land-Grant College.

School Lunch Program–In the United States, a federal program which began as a means of supplementing the lunches of needy children. It became a permanent law after the passage by Congress of the National School Lunch Act of 1946. Under this act the following types of government assistance are authorized: (a) cash reimbursement for a part of the school's expenditure for purchases of food from local sources; (b) foods of special nutritive value purchased by the United States Department of Agriculture and distributed to participating schools in accordance with need; (c) part of the food acquired by the United States Department of Agriculture under its surplus removal operations, distributed to all schools serving lunches on a nonprofit basis.

School Milk Programs–In the United States, the special school milk programs which were established under the Agricultural Act of 1954, to move more milk directly into consumption channels by assisting states and local communities to increase the serving of milk in schools. This is accomplished by reimbursing schools for a portion of the cost of additional milk served. A base is established for reach participating school representing former consumption of milk by children in the school. Schools are reimbursed up to a maximum of four cents for each half-pint served in addition to the base. Schools not serving lunches under the federal school lunch program are reimbursed up to a maximum of three cents per half-pint served to children.

School Section–In the United States, a section of land, 640 acres, given by the federal government to a state for the support of public schools. In the General Land Office Survey this was Section 16 of a township.

Scimitar Boning Knife–A knife of medium length with an upward-curving (scimitar-shaped) blade which is adapted for boning, pelting, and skinning of slaughtered animals.

Scoop– (1) A large ladle; a bucket or vessel with a handle, used for dipping, gathering, skimming, etc. (2) A shovel-like implement used for dipping, gathering, digging out, or shoveling, as a grain scoop, cranberry scoop, curd scoop, etc. (3) A small implement (usually curved) used for cutting, gouging, or removing an amount of material, as an ice cream scoop. (4) A mechanical implement with a large bucket used for the removal or conveying of earth, liquids, etc. (5) The amount of material which is obtained with a scoop. (6) To remove, to take out from, to gather, or to empty with a scoop.

Scoop Shovel– An implement consisting of a broad, flat, moderately deep ladle fitted to a D handle used for handling grain, coal, sand, gravel, etc.

Score– (1) A mark, scratch, line, incision, etc., especially one made for keeping tally. (2) A numerical valuation or rating as to the merit of an animal, a product, etc. (3) The record of points made by contestants in a contest or test. (4) To keep account by notches, lines, or figures on a tally or score card. (5) To keep score in a contest. (6) To mark with lines, notches, grooves, gashes, furrows, etc., as to score timber for hewing or squaring, hides in skinning, meat in cookery, etc.

Score Card– A numerical standard of excellence with definite values assigned to the various sections or aspects of an animal or product; used as a basis in judging.

Scotch Comb– A T-shaped comb used to groom the hair or switch of an animal for show.

Scotch Hobble– A restraining device consisting of a rope passed over a horse's neck in front of the shoulders and tied as a loosely fitting rope collar. The slack end of the rope is passed under a hind pastern and back through the loop, enough slack being taken up to pull the foot from the ground.

Scouting– The inspection of a field for pests (insects, weeds, pathogens). Scouting is a basic component of integrated pest management programs. See Integrated Pest Management.

Scrape– (1) A tillage implement; a sweep. (2) A spreading wing attachment for a cultivator. (3) In turpentine operations, oleoresin remaining on a tree from which volatile oils have evaporated. (4) To collect scrape from trees in turpentine operations.

Scraper– (1) A tool or device with which something is cleaned, leveled, etc., as a shaped piece of wood or metal used to remove sweat from a horse, an attachment which removes soil from a disk or roller while it is in operation, or a metal piece to clean butcher blocks, etc. (2) A large steel pan or scoop used in excavation or to even the surface, which is drawn, dragged, or pushed over a surface by tractors.

Screen– (1) A sieve or grating, such as a frame covered with meshed wire or fabric, or a perforated plate, etc., which is used for separating the finer or coarser parts of soil, gravel, grain products in milling, seeds in cleaning, etc. (2) Any form of grating used to prevent invasion or escape of insects, reptiles, fish, or other kinds of animals. (3) A construction or planting used to conceal an unpleasant sight. (4) To consider a group for the purpose of selecting a relatively small number of individuals. (5) To sift by shaking through a screen. (6) To conceal, as with a screen.

Screw Pump– A pump which consists of an inclined or vertical shaft with many propeller-like blades attached at the lower end of the shaft which rotates in a cylindrical casing. It is particularly adapted to low-lift conditions, such as exist in coastal drainage districts and organic soil areas. Also called propeller pump.

Scribe– A specially designed wood chisel for scoring the surface of a witness tree or post on which is marked the section corner or square corner. Used in all Untied States Government Land Office Surveys in forested areas.

Scribner Rule– In forestry, a diagram rule for determining the board measure of logs, one of the oldest in existence, which assumes 1-inch boards and ¼-inch kerf, makes a liberal allowance for slabs and disregards taper. (It is the official rule in many parts of the United States). See Doyle Rule, Doyle-Scribner Rule.

SCS– Soil Conservation Service.

Scuffle Hoe– A garden hand tool which consists of a straight, flat blade attached to a long handle. It is pushed back and forth so that the blade cuts just beneath the surface of the soil to break up the crust and destroy weeds.

Scythe– A long, sharp, curved blade attached to a long bent handle, used for mowing or cutting grass and weeds by hand.

Sealed Bid Sale– A sale method sometimes used with livestock, etc., in which sealed bids are submitted to the seller with the privilege of purchase going to the highest bidder if the terms of sale are met. Also known as silent auction.

Seasonal Grazing– The grazing of a range only during a certain period or periods of the year which roughly correspond to one or more of the four seasons.

Seasonal Labor– In farming, the labor required at a particular season; as at fruit picking, grain harvesting, sheep shearing, etc.

Second/Foot– A measuring unit for the volume of the flow of water expressed in cubic feet per second.

Section– (1) A distinct part of a town, country, region, people, etc., as a section of the population of the country. (2) A unit of the township in the United States General Land Office Survey of the public domain. The full section is one square mile, or 640 acres, and the full township consists of 36 sections. (3) In forestry, a logical or natural length of the fire-control line which is handled as a unit during fire suppression. (4) In beekeeping, one of several small frames of basswood placed in a hive in which the bees build surplus honey. A filled frame weighs from 12 to 16 ounces. See Section Comb Honey. (5) Segments making up the interior of a citrus fruit.

Sectioned 32 Program– In the United States, this section amended in 1935 the Agricultural Adjustment Act by allocating 30 percent of the collections from tariffs on all imports to the Secretary of Agriculture for his use to encourage the export of agricultural commodities and to increase domestic consumption by diverting surplus commodities to low-income groups. The three primary subsidy programs developed under Section 32 have been Food Stamp Program, School Lunch Program, and the Relief and School Milk Distribution Program.

Selection Difference–The difference between the average for a trait in selected cattle and the average for the group from which they came. Also called reach.

Selection Index–A formula that combines performance records from several traits or different measurements of the same trait into a single value for each animal. Selection indexes weigh the traits for their relative net economic importance and their heritabilities plus the genetic associations among the traits.

Selection Threshold–A trait or standard established by a breeder, below which the breeder will reject potential breeding animals.

Self-feeder–(1) In poultry or other livestock farming, any feeding device by means of which the animals can eat at will; e.g., a hopper which supplies feed by gravity to a boxlike trough. (2) Any automatic mechanism for supplying materials to a given point in an operation.

Self-sufficing Farming–A type of farming in which the production is mainly for the use of the farm family. See Subsistence Farmer.

Semen Collector–Any device or receptacle used for the collection of semen from males of domestic animals. It may consist of an artificial cloaca or vagina, a test tube, a small beaker, etc. See Artificial Vagina.

Semen Tank–A portable tank used to transport or store frozen semen.

Sending Unit–A device, usually located in an engine or other system, that transmits information to a gauge on an instrument panel.

Sensus–(Latin) Meaning.

Separator–Mechanical device for separating cream from milk by centrifugal force.

Sequoia–Either of the two tall trees, giant sequoia and redwood, genus *Sequoia*, family Pinaceae, characterized by oval, drooping cones and woody scales.

Seral–A biotic community that is in a developmental, transitory stage.

Serape–(1) A Mexican saddle cloth. (2) A small, often brightly colored blanket, worn as an outer garment (Mexico and southwestern United States).

Set–(1) A small propagative part, a bulb, shoot, tuber, etc., which is suitable for setting out or planting, as an onion set. (2) A number of things usually used together, or forming a complete collection, as a set of tools, harness, etc. (3) The initial swelling of the ovary of a flower soon after petal fall. Also called setting, fruit setting. (4) To put in a particular place, position, condition, direction, adjustment, etc. (5) To put eggs under a fowl or in an incubator to hatch them. (6) To put a price or value on something. (7) To put in proper working condition. (8) To adjust the teeth of a saw. (9) To fix a trap or a net to catch animals or fish. (10) To fix in the ground, as a post, tree, or plant. (11) To become stiff, firm, or hard, as cement sets. (12) To adjust into proper apposition the ends of a broken bone, as to set a fracture.

Set Aside–The percentage of a commodity program acreage base that must be idled in a given year. The purpose is to help reduce commodity supplies and limit the cost of farm programs.

Set Screw–A machine screw used to fasten one part to another, as to fasten gears, pulleys, etc., to a shaft.

Set Stocking–(1) As used in Australia, synonymous with continuous grazing. (2) Keeping animals in a given area of pasture continuously at a predetermined level of stocking for a defined period.

Settle–(1) To sink gradually to a lower level, as of sediment, etc. (2) To cause a female to become pregnant, as a bull settles the female with which he has had coitus. (3) To occupy land usually for the purpose of farming, ranching, or homesteading.

Settleable Solids–Suspended solids that will settle in quiescent water or sewage in a reasonable period. Such period is commonly, though arbitrarily, taken as two hours.

Settler–a person who establishes residence, especially one who settles in a new country.

Severance Tax–A tax due at the time of harvest of timber or other natural resources.

Sewer–A pipe or conduit, generally closed, but normally not flowing full, which is used for carrying off sewage and other liquid wastes.

Shackle–(1) A device, such as a ring, chain, or band, used to restrict free movement of an animal. Particularly useful in slaughterhouses. (2) A ring, clevis, chain, hobbles, etc., used for attaching or coupling, so as to leave some degree of movement.

Shade Cloth–A fabric woven from saran fibers to provide shade levels ranging from less than 20 percent to more than 90 percent.

Shade Device–Plastic netting, latex paint, whitewash, cheesecloth, or other material used to reduce light and heat entering a plant-growing structure.

Shade House–A structure that is covered with a shade cloth, laths, etc., for the purpose of blocking out part of the sun's rays in order to protect young growing plants.

Shadoof–A pivoted pole with a bucket attached at one end and a counterweight at the other, which facilitates the raising of water, used for irrigation purposes in the Near East.

Shaft–(1) Either of the pair of long bars (thills) sticking out in front of a vehicle between which a horse or other animal is harnessed. (2) The stem of a feather to which the barbs are attached. (3) A connecting rod between a power source and a pulley, gear, or other contrivance to distribute such power.

Shake–(1) A lengthwise separation of wood (usually caused by wind) that usually occurs between and parallel to the growth layers. (2) A thin section split from a bolt of wood and used for roofing or siding; a split shingle.

Shank–(1) That part of the leg joining the knee to the ankle in humans, or the corresponding part in various animals. (2) A market cut of meat from the lower part of the foreleg of a dressed carcass of beef, veal, mutton, or lamb, or a corresponding cut from the hind leg. (3) The tarsus of birds: the part of the leg, below the hock joint, which is covered with scales. (4) That part of a tool, implement, etc., between the acting part and the part by which it is held, as a cultivator-shovel shank, etc. (5) Any connecting part of a plant; a foot stalk, as the part between the stalk and the corn ear.

Shard–(1) A piece of broken flowerpot which is placed over the hole in the bottom of another flowerpot to aid in providing drainage. Also called crock. (2) To provide or use a shard.

Share–(1) That portion of the crop or livestock received by either the tenant or the landlord on a farm or from farmland. (2) That part which provides the front cutting edge for a plow. The principal parts of the share are the point, the wing, and the cutting edge or throat.

Sharecropper–A tenant who shares crops, livestock, or livestock products with the landowner, who, in turn, often extends credit to and closely supervises the tenant. The sharecropper generally supplies only labor.

Sharp Practice–Hard bargaining; a deal or transaction in which advantage is taken or sought through unscrupulous methods or trickery; e.g., operating upon, or tampering with, an animal for the purpose of concealing faults.

Shear Bolt (Pin)–A bolt or pin that connects two parts of a machine. It is designed to break or shear rather than allow damage to the machine.

Shear Line–The point at which the ground near a ditch or river begins to cave in when subjected to a heavy weight such as a tractor.

Sheathing–(1) An undercovering of boards or waterproof material fitted to the framework of an outside wall or roof of a frame building. (2) Materials which are used for sheathing, such as lumber other than siding, sheets of waterproof material, etc.

Shed–(1) A simple structure for shelter of livestock, storage of feeds, or both, open on one or more sides, which may be separate from or attached to another structure. (2) To lose or cast off something, as a plant sheds leaves, pollen, seed, etc.; animals shed hair; birds molt or shed feathers; a roof sheds rain, etc.

Shedder–A brushlike device used to remove dead hairs from an animal's coat.

Sheep Bell–A bell hung about the neck of a sheep whose tinkling indicates the location of the flock.

Sheep Month–The amount of forage or feed which is necessary to maintain a mature sheep or a ewe and its suckling lamb for thirty days. It is commonly figured as one-fifth of a cow month.

Sheep Stand (Table)–A platform on which a sheep is placed and held for grooming in preparation for showing.

Sheep Tag–(1) A heavy, dung-covered lock of wool which usually hangs from the crotch of a sheep. (2) A metal tag which is fastened to the ear of a sheep for identification.

Shell–(1) A hard outer covering of certain animals, such as turtle, mollusk, or crustacean, which when ground may be used as a poultry feed supplement. (2) The hard outer covering of a fruit or seed, as a nut shell, etc. (3) The hard calcareous exterior covering of an egg. (4) To remove or separate from its shell or pod, as to shell nuts, peas, etc. (5) To drop out of a shell, or out of a cluster, as seeds, grapes, etc. (6) To separate seeds or kernels (of corn, wheat, oats, etc.) from the cob, ear, or husk.

Shelling Percentage–The weight of corn kernels (shelled corn), expressed as a percentage of the weight of the ear corn prior to shelling.

Shelterbelt–A long series of trees planted in semi-arid regions to moderate winds and reduce erosion.

Shelterbelt Project–In the mid-1930s and early 1940s, a United States government plan to plant trees on about 1,000,000 acres of land in a 100-mile-wide belt extending from the Canadian boundary to the Texas Panhandle.

Shepherd–A flockmaster or herdsman who feeds and guards a flock of sheep or goats.

Shepherd's Crook–The staff used by a shepherd; specifically a staff with a crook at one end which is used in catching and holding a sheep (used by Old World shepherds since ancient times).

Shim–Thin sheets, usually of metal, used as spacers between two parts of a machine.

Shingle–(1) (a) A thin, oblong piece of wood sawed or split, with one end thinner than the other, which is lapped lengthwise in covering roofs and outer walls of buildings. (b) A piece of any kind of material, as asbestos, used in the same way. (2) Loose, rather sharp shale or similar rock, common to certain rocky areas in New Zealand. (3) A mass of loose, coarse, rock fragments of cobble or small boulder size, rounded or flattish, found on the seashore and on the beaches of large inland lakes. (4) A thin slab of slate stone, cut or chipped into a section used as a shingle. (5) To cover with shingles.

Shock–(1) A quantity of a harvested grain crop or corn, etc., set together upright in a field. (2) A small pile of hay. (3) A state of collapse in animals, which results from severe loss of blood flowing surgery, accident, or sometimes under conditions of undue stress. (4) The severe symptoms in plants following acquisition of virus. Later symptoms are absent or milder. (5) A parcel or lot of sixty pieces or units. (6) Twelve sheaves of small grain. (7) To place grain, hay, etc., in shocks.

Shoddy–(1) That leather made by grinding waste leather to a pulp and pressing it into solid sheets with or without the addition of a binding material. (2) Reworked wool that has been recovered from wool cloth trimmings, rags, etc. See Virgin Wool. (3) Anything of an inferior quality.

Shoe–(1) A plate or rim of metal, usually iron, nailed to the underside of a horse's or other such animal's hoof as a protection from injury or to assist in obtaining footholds. (2) A part or parts of certain farm implements, particularly parts that travel in or through the soil, such as the metal plate on the bottom of a subsoil plow, furrow openers on certain seeders, etc. (3) The shoelike runner which supports the outer end of the cutter bar of a mower. (4) Small pieces of the woody stalk of the hemp plant which are produced when the retted stalks are broken to remove fiber or lint. (5) To fit and attach a shoe, as on a horse's hoof.

Shomar–The Jewish official who supervises the production and handling of special milk, such as Kosher; kedassia for Orthodox Jews.

Shoot—(1) A stem of a plant, including the leaves. (2) A young stem arising from or near the crown of a plant. (3) A new leafy growth developed from a bud. See Sprout. (4) An area of land inhabited by game (grouse, etc.), and kept especially for the shooting of such game as a sport; the shooting rights to such an area. (5) To develop new stems. (6) To detonate a charge of dynamite or other explosive. (7) See Chute.

Short—Designating one who has sold a futures contract that does not liquidate a previously bought contract for the same delivery month. Short hedges are sales of futures made as hedges against holdings of the spot commodity, or products thereof. Short interest is the sum total of all contracts or the owners of such contracts.

Short Circuit—An electrical problem resulting when one part of a circuit comes in contact with another part of the same circuit, diverting the flow of current from its desired path.

Short Ton—See Ton.

Short-handed—Designating insufficient help, an inadequate labor force.

Short-hold Products—Products of a perishable nature which will permit only short-term storage without loss.

Short-term Cash Flow Budget—A cash flow budget normally constructed over a twelve-month planning horizon with the intermediate cash balance calculated at monthly or bimonthly intervals.

Short-term Loans—Loans made for less than two years.

Short-time Feed—A system of fattening cattle for slaughter in which heavy stockers or feeders are fed a heavy grain ration for periods of thirty to ninety days and then sold.

Shorts—(1) In the florist's trade, roses with short stems (approximately 9 inches in length). (2) Lumber which is shorter than standard lengths. (3) Short pieces or locks of wool that are dropped out while fibers are being sorted. (4) See Wheat Shorts, Wheat White Shorts.

Shoulder—(1) Of livestock, the upper part of the forelimb and/or the adjacent part of the back; the scapula region. (2) The upper foreleg and/or the adjacent part cut from the carcass of a hog, sheep, or other animal; the upper part of the carcass to which the foreleg is attached. (3) The part of a bird at which the wing is attached. (4) In turpentining, the uppermost corner of a face. (5) A shoulderlike part or projection, as the projection around a tendon. (6) A sudden inward curvature in the outline of something, as the flaring part between side and neck of a milk bottle. (7) The edge of a road adjacent to the traveled way or pavement.

Shovel—(1) A scoop or spadelike tool, consisting of a broad metal blade attached to a handle, used for lifting grain, soil, litter, etc. (2) A strap-shaped, V-pointed attachment which is used for various plowstocks, cultivators, grain drills, and soil scarifiers.

Show—(1) A gathering together of livestock, other farm products, flowers, etc., for public exhibition, judging, etc. (2) To display livestock, farm products, etc., for rating by judges or for inspection by prospective buyers.

Show Halter—A leather halter that is used on cattle to lead them around the show ring. Show halters look nicer than regular rope halters.

Show Stick—A stick of fiberglass, wood, or aluminum, about 4 to 5 feet long. It has a short hook on the end that is used to place the feet of a beef animal in a show ring. It is also used to straighten the back of the animal and to calm it by scratching the belly.

Showmanship—The art or skill of showing or displaying products, animals, etc., to advantage.

Shrinkage—(1) The process of reducing in dimension, weight, or volume. (2) With livestock, the loss in weight of an animal from the time it is taken off feed to the time it is marketed or slaughtered. (3) The loss in weight of grease wool due to scouring, expressed as a percentage of the original weight. (4) The loss in weight (moisture loss) of grain from harvest time to some later date. Such shrinkage is commonly reflected in the selling price. (5) The contraction of wood caused by drying, usually expressed as a percentage of some specific dimensions (or volume) of the wood when green. (6) Loss in weight of either meat or carcass stored under refrigeration. Also called cooler shrink, freezer shrink.

Shutter—A wooden cover of the same size as a cold-frame sash which is used as protection from cold for plants.

Sib Testing—A method of livestock selection in which an animal is selected on the basis of the performance of its brothers and sisters.

Sib-Mating—Mating between siblings, two or more individuals having one or both parents in common.

Sickle—(1) A sharp, curved metal blade fitted with a short handle; used for cutting weeds, grass, etc. It is one of the earliest hand implements used for harvesting small grain. Also called reap hook. (2) The cutting section of a hay or grain-harvesting machine.

Sickle Bar—See Cutter Bar.

Side Camp—In forestry, a small camp set up to accommodate a crew working away from a main camp or headquarters.

Side Check—A leather strap attached at the side of a horse's bridle to limit side movement of the head. See Checkrein, Overhead Check.

Side Draft—A sidewise pull produced when the center of a load or resistance is out of the true line of hitch or draft.

Side Saddle—A saddle for women so constructed that the rider sits with both feet on one (usually the left) side.

Sideboard—One of the boards which form the long side of a wagon or truck box; an additional board to raise the height of a wagon-box side, as those used in corn husking to prevent overthrow. Sometimes called bang board.

Siding—(1) Boards used to cover the sides of buildings; one board usually being lapped over the upper edge of the board below. Also called clapboard, weatherboard. (2) Removing the hide from the side of an animal, as from a beef or veal carcass.

Sieve—A frame of metal or fabric with meshes, through which the finer particles or substances are passed to separate them from the coarser forms. Sieves of special types are used in preparing many food products and also in the screening of seed, sands, gravels, and fertilizers.

Sieved–Designating products, usually food products (fruits, vegetables, etc.) that have been reduced to a pulp and then passed through a fine-screened finisher to remove small pieces of fiber and to give a fine-grained, smooth puree. Such products are used chiefly as baby foods. Also called strained.

Sifter–Any device used for sifting, such as a mechanical sifter (sieve) used to separate small from large particles in the grain-milling process.

Signature Loans–A loan requiring only the signature of the borrower as collateral.

Silage Harvester–A field implement that harvests standing corn, cuts it into ensilage length, and elevates the chopped fodder into a truck or wagon box ready for transportation to and placement in the silo.

Silent Auction–See Sealed Bid Sale.

Silo–A pit, trench, aboveground horizontal container, or vertical cylindrical structure of relatively air-tight construction into which green crops, such as corn, grass, legumes, or small grain and other feeds are placed and converted into silage for later use as a livestock feedstuff.

Silviculture–(1) The science and art of growing and tending forest crops, based on a knowledge of silvics. See Silvics. (2) More particularly the theory and practice of controlling the establishment, composition, constitution, and growth of forests.

Single-action Disk Harrow–A harrow with two opposed disk gangs that work side by side.

Single-crop Farming–Farming which depends largely upon one crop as a source of farm income. Also called one-crop farming. See Diversification.

Single-entry System–A bookkeeping system requiring only one entry, either an expense or income, for each transaction.

Single-rigged Saddle–A saddle which has one cinch.

Single-service Container–A container for milk, cream, and other food products, made of stout waxed paper. These containers are made in several shapes and with several types of closures and are meant to be used only once.

Siphon Tubes–Curved tubes about 1.2 to 2 inches in diameter that are used to siphon irrigation water from a ditch, over the bank, and into a furrow.

Sire Evaluation–An objective program designed and conducted by a breed association to increase the effectiveness of sire selection and to evaluate a bull's true genetic merit. It measures differences in sires using progeny information. It is the best measure of a bull's true transmitting ability.

Sire Index–A mathematical index for measuring a dairy bull's transmitting ability in terms of milk and butterfat. It tends to measure what production may be expected of the daughter when a bull is mated to cows of known productivity.

Sisal–*Agave sisalana*, family Amaryllidaceae; a tropical plant with a short stem and long, succulent leaves which yield a strong, durable fiber used for making rope, twine, etc. Probably native to America. Also called Bahama hemp, sisal hemp.

Sitology–(1) All knowledge of food, diet, and nutrition. (2) Food and nutrition science.

Skid–A plank, roller, frame, etc., on which something heavy may rest or on which it may be slid or pushed along. (2) In forestry, logs or poles which are used in pairs to form a skidway. (3) To pull logs from the felling site to a loading point or mill.

Skid Steer–A relatively small, self-powered machine used for loading manure, fertilizer, hay, soil, etc. The machine is guided by locking the wheels on one side and "skidding" around.

Skidder–(1) A person who skids logs. (2) A yarding machine which skids logs by a cable.

Skidway–A prepared place for pulling logs prior to loading which consists of two parallel supports at right angles to the road generally raised at the end nearest the road.

Skip-row Planting–A custom followed by many cotton producers and to a lesser extent by producers of other commodities. It involves planting in uniform spaces one or more rows to a commodity, then skipping one or more rows. Some of the most common patterns followed are: plant one, skip one; plant two, skip two, skip four; and plant four, skip four.

Slab–(1) The exterior portion of a log which is removed in sawing lumber. (2) An unattractive, overripe or broken, dried fruit (prune, apricot, etc.) which is sorted from a drying tray in commercial fruit drying. (3) A piece of unsliced bacon.

Slack-cooperage–A barrel or keg suitable only for nonliquid products. See Tight-cooperage.

Slack-filled–Designating a can in which the level of the contents is approximately 10 percent or more below the top. In fruit or vegetable products canning, a slacked-filled container prevents damage to the contents from drastic changes in volume due to changes in temperature.

Slat–(1) A thin, narrow strip of wood used in the manufacture of crates, etc. (2) A sheepskin after the wool has been pulled but before other treatment is given it.

Slatted Floors–A manure disposal system in which the floor is composed of wood, metal, plastic, or concrete slats. The slats are close enough together so that the animals can walk across it, but wide enough apart so that droppings can fall between them.

Slaughter Calves–Calves between three months and one year of age at time of slaughter.

Slaughter Laws–In the United States, state and federal laws have been established dictating the methods to be used in stunning an animal prior to slaughter. Among the various humane methods is the use of carbon dioxide gas as an anaesthesia; a captive bolt which gives the animal an electric shock, stunning it; or a rifle bullet. The knocking hammer is not used in establishments under federal inspection, and tranquilizing drugs are not used, as they leave a residue in the meat. Also called humane slaughter laws.

Sleeper–(1) A wooden cross-member supporting railroad rails; a tie. (2) An unbranded calf; a maverick. (3) A horse with equine encephalomyelitis (African sleeping sickness).

Slide–(1) The slipping of a mass of earth or rock down the slope of a hill or mountain; a landslide. (2) A smooth surface for sliding; a chute.

Sliding Gear Transmission–A transmission in which the gears are moved on their shafts to change gear ratios.

Sliding Poise–A movable weight mounted on the bar of a scale beam.

Sling Blade–A tool used for cutting weeds. It consists of a sharp, double-edged blade mounted on a handle. It is used by swinging back and forth. Also known as a swing blade.

Slippage–The reduced traction of a tractor's drive wheels caused by soil conditions. Excessive slippage wastes time and fuel and increases tire wear, but some slippage is desirable to cushion the tractor engine and drive train from sudden overloads.

Slope–The inclination of the land surface from the horizontal. Percentage of slope is the vertical distance divided by horizontal distance, then multiplied by 100. Thus, a slope of 20 percent is a drop of 20 feet in 100 feet of horizontal distance.

Slope Aspect–The direction a slope faces.

Slope Characteristics–Slopes may be characterized as concave (decrease in steepness in lower portion), uniform, or convex (increase in steepness at base). Erosion is strongly affected by slope.

Slope Designations–Through custom, road engineers designate slopes as ratios of level (horizontal) distance to vertical rise; such as 3:1 slope, meaning 3 feet on the level (horizontal) to 1 foot rise. Slopes are also designated in degrees (0° equals level; 90° equals vertical) and in percent (e.g., 5 feet rise or fall for each 100 feet horizontal equals 5 percent slope).

Slope Drains–Permanent or temporary devices that are used to carry water down cut or embankment slopes. May be pipe, half sections, paved, or have special plastic lining.

Slope Orientation–See Slope Aspect.

Slough–A narrow stretch of backwaters, frequently the abandoned channel of a river, in an alluvial plain, a narrow, marshy depression. See Oxbow.

Slow-moving Vehicle–Any vehicle such as a tractor that moves slower than twenty miles an hour on a public road must carry a danger sign. This consists of a yellow-orange triangle with a dark red border and with the outer corners clipped.

Sluice–(1) A conduit which carries water at high velocity. (2) An opening in a structure for passing debris. (3) A water gate. (4) A channel which carries or drains off surplus water. (5) To cause water to flow at high velocities for wastage, for excavation, ejecting debris, etc. (6) To move or float logs over or through a dam from the water above into the stream below.

Small–A United States grade for eggs having a minimum net weight per dozen of 18 ounces with a minimum weight for individual eggs at the rate of 17 ounces per dozen.

Small Animal Hospital–A veterinary hospital which is specially arranged to handle small animals, such as dogs and cats.

Smith-Hughs Act–A United States federal law passed by Congress in 1917 which gives federal aid for vocational education in the secondary schools. It established a national system of vocational education of broad scope in agriculture, home economics, trades, and industries.

Smith-Lever Act–An act of United States Congress which provides funds for extension work in agriculture cooperatively between the agricultural universities of the several states and the United States Department of Agriculture. The act was approved May 8, 1914.

Smoker–A steel container with an attached bellows in which burning materials furnish smoke to repel and subdue honeybees.

Smoothing Harrow–A tooth-type harrow attachment for the rear of disk harrows and plows. The purpose is to break up clods and smooth the soil prior to planting.

Smudge Fire–A fire made from damp straw, smoldering manure, or any other fuel which produces a very heavy smoke that hangs low to the ground, erroneously believed to prevent frost injury to crops. See Smudge Pot.

Smudge Pot–A type of open pot or vessel in which an oil fire is maintained to heat the air in orchards and vegetable plots in hope of reducing frost damage.

Snaffle Bit–A bridle bit; the mouthpiece of a horse bridle which is composed of two or more linked parts or sometimes made of a single chain which acts as a bit and serves as an effective control.

Snag–(1) A sharp point of wood which results from a break. (2) A broken tree trunk more than 20 feet tall.

Snap–(1) A spring fastening or clasp which closes with a clicking sound, used as a fastener on harnesses, gate chains, dog chains, etc. (2) A brief period of cold weather; as a cold snap. (3) To harvest by pulling off with a quick motion, as the ears of corn, the bolls of cotton, etc. (4) In horses, especially in harness or saddle horses, to move briskly with animation; to pick up feet quickly and sharply.

Snipe–To round off the end of a log so that it will drag easily in snaking or skidding.

Snippers–A small hand tool used for cutting, as for cutting wire.

Snorting Pole–See Teasing Pole.

Snubbing Post–Any smooth, sturdy post to which animals may be fastened to restrict their movement.

Snubbing Ring–A heavy metal ring attached to the floor, or any other solid object, through which a control rope is passed and by means of which an animal's movement is restricted as in slaughtering or treating for an ailment.

Soap–A cleansing agent made usually by treating fats or fatty acids with sodium or potassium salts. Soap is sometimes used as a component of insect sprays for its insecticidal effect, as an emulsifier, wetter, or spreader, and for enhancing the toxicity of nicotine sprays for aphids.

Soft Soap–A smeary, semiliquid soap that is usually made by treating an oil high in oleic acid with potash as the alkali.

Soil Auger–A tool used for boring into the soil and withdrawing a sample for field or laboratory observation. Soil augers are of two general types: those with worm-type bits and those having a hollow cylinder with a cutting edge in place of the worm bit.

Soil Complex–A mapping unit used in detailed soil surveys where two or more defined taxonomic units are so intimately associated geographically that they cannot be separated by boundaries on the scale used. See Soil Association.

Soil Conservation–The efficient use and stability of each area of soil that is needed for use at its optimum level of developed productivity according to the specific patterns of soil and water resources of individual farms, ranches, forests, and other land-managment units. The term includes the positive concept of improvement of soils for use as well as their protection and preservation.

Soil Conservation District–A legal subdivision of state government, with a locally elected governing body, responsible for developing and carrying out a program of soil and water conservation within a geographic boundary usually coinciding with county lines or watersheds. The many districts have varying names—soil conservation, soil and water conservation, natural resources, resource conservation, or conservation districts. They help individual landowners, local groups, and other find help in natural resource management from United States Department of Agriculture and many other agencies at all levels.

Soil Conservation Service–A bureau of the United States Department of Agriculture established by the Soil Conservation Act of 1935. Its basic purpose is to aid in bringing about physical adjustments in land use and treatment that will conserve natural resources, establish a permanent and balanced agriculture, and reduce the hazards of floods and sedimentation. Its program is carried out principally under memoranda of understanding with wool conservation districts. It has major responsibilities in watershed management, and it supervises the agricultural phase of certain specified water utilization programs in the western states. See Soil Erosion Service.

Soil Erosion Service–In the United States, an emergency agency first authorized in 1933 which was placed in the Department of Interior, and subsequently in 1935 transferred to the Department of Agriculture, as the Soil Conservation Service. See Soil Conservation Service.

Soil Incorporation–The mechanical mixing of a pesticide or fertilizer into the soil.

Soil Loss Equation, Universal–See Universal Soil Loss Equation.

Soil Map–(1) A map which shows the location and extent of the different soils in an area. Soil maps are prepared and published by various agencies of state and local governments, in cooperation with the United States Department of Agriculture Soil Conservation Service or the United States Department of Agriculture Forest Service. (2) A map showing the distribution of soil mapping units in relation to the prominent physical and cultural features of the earth's surface. The following kinds of soil maps are recognized in the United States: detailed, detailed reconnaissance, generalized, and reconnaissance.

Soil Map, Detailed–United States Department of Agriculture Soil Conservation Service usage. Referring to a soil map at a publication scale commonly of 1:190,080 on which soil phases and soil series are the main units delineated on an aerial map base. The smallest unit size shown on such maps is about 1 acre. Survey traverses are usually made at one-quarter mile, or more frequent, intervals. The unit boundaries on detailed soil maps should have been seen throughout their course and their placement on the map should be accurate to at least 100 feet. The maximum amount of unlike soil inclusions in mapped units is 15 percent.

Soil Map, Detailed Reconnaissance–A reconnaissance soil map on which some areas or features are shown in greater detail than usual, or than others.

Soil Map, Generalized–United States Department of Agriculture Soil Conservation Service usage. Referring to a soil map with a publication scale commonly of 1 inch = 1 mile on which soil associations and miscellaneous land types are the delineated units. The smallest unit size shown is 3½ acres. The maximum amount of unlike soil inclusions in such mapping units is 15 percent.

Soil Map, Reconnaissance–United States Department of Agriculture Soil Conservation Service usage. Referring to a soil map of highly variable publication scale (1 inch = 1 mile to 1 inch = 8 miles). The most detailed units commonly shown are miscellaneous land types and soil associations or one or more phases of soil families. The smallest unit that can be shown on such a soil map at a scale of 1 inch = 1 mile is about 3½ acres and at 1 inch = 4 miles about 280 acres. Map unit boundaries are plotted where they cross field survey traverses. Traverses are at intervals varying from about one-half mile to several miles. Between these points of field observation most boundaries are sketched from the appearance of ground patterns on aerial photographs and the general appearance of the landscape.

Soil Moisture Capacity–The amount of water expressed in percentage of dry weight that a soil can retain against the pull of gravity. See Field Capacity.

Soil Moisture Tension–The force with which the soil holds onto the soil moisture present. As soil moisture declines, soil moisture tension increases.

Soil Probe–A T-shaped metal tube that is forced into the ground for the purpose of obtaining a sample of the soil.

Soil Science–The science which deals with the scientific study of soil as a natural body and an economic resource. See Pedology.

Soil Slope–An incline of a land surface. See Slope.

Soil Solution–(1) The liquid part of soil that surrounds soil particles and contains the elements in solution that are absorbed by plant roots. It is not constant in amount or composition. (2) Technically, the aqueous solution which exists in equilibrium with the soil particles at a particular moisture tension.

Soil Survey–(1) In the United States, this refers to the National Cooperative Soil Survey conducted by the Soil Conservation Service, United States Department of Agriculture, in cooperation with state agricultural universities and other public and private agencies and organizations. (2) Systematic examination of soils in the field and in laboratories; publishing of descriptions and classifications; mapping of kinds of soils; and interpretation of soils according to their adaptability to various crops such as field crops, fruits, vegetables, and trees.

Soil Survey Report–A written statement accompanying a soil map which describes the geography of the area surveyed, the characteristics and capabilities of the soil-mapping units. It also discusses the principal factors responsible for soil development.

Sole–(1) The bottom of the furrow on which the plow bottom slides. (2) The bottom of a plow. Also called slide. (3) The bottom of an animal's or person's foot.

Sole Ownership–The owning of property, particularly real estate, in one person's name in fee simple.

Solubility–To be most readily available to plants a nutrient must be at least slightly soluble in the soil solution or be held in an exchangeable form on clay and humus particles. See Available.

Soluble–Capable of changing form or changing into a solution.

Solution–A mixture of two or more substances which is chemically and physically homogeneous.

Solvent–(1) A liquid that is used to dissolve a substance. (2) Designating an individual or business which is in good financial condition; i.e., able to pay all debts and other liabilities.

Spade–A long, flat, sharp-bladed, earth-excavating, hand tool made in several shapes, which is pushed into the soil by foot pressure to turn over, excavate, and loosen soil in small areas. Equivalent of tillage.

Spading Fork–A short-handled spade in which the spade blade is made of six, pointed, flat tines used for turning sandy or loose soil and/or for removing tubers from the hill or row, etc. Also called fork, garden fork.

Spall–(1) Relatively thin, commonly curved and sharp-edged pieces of rock produced by exfoliation. (2) To break off in layers parallel to a surface. (3) A piece of rock chipped off with a hammer.

Span–(1) A pair of animals usually harnessed together as a team. Also spelled spann. (2) The space between the top of the two walls of a ravine. (3) The length or extent of a bridge between supports. (4) The distance between the thumb and little finger when the hand is fully extended.

Spar–A tall-topped tree or guyed tall pole, used for fastening the elevated cables for skyline and high-lead logging of a forest.

Spark Plug–A device that fits into the cylinder head of an internal combustion engine that provides ignition for the gases within the engine cylinder. The spark is generated by a high voltage electrical current.

Spatulate–Oblong, with the basal and attenuated like a chemist's spatula.

Spaulding Rule–A diagram rule for estimating the lumber content of logs which disregards taper and allows for an $11/32$ inch saw kerf, with size of slab varying with size of log. It is the statute rule for California and is widely used elsewhere on the Pacific Coast (United States). Also called Columbia River rule.

Specialty Crops–Crops with a limited number of producers and limited demand; or crops with high per acre production costs and value.

Specific–(1) A medicine that cures a particular disease. (2) Pertaining to a species. (3) Produced by a particular microorganism. (4) Restricted by nature to a special animal, thing, etc. (5) Exerting a peculiar influence over any part of the body.

Specific Gravity–The ratio of the mass of a solid or liquid to the mass of an equal volume of distilled water at $4°C$.

Specific Heat–Heat or thermal capacity of a substance in relation to that of water.

Specific-pathogen Free–Swine that are obtained by laboratory methods designed to eliminate certain diseases of the respiratory and digestive tracts.

Spectrophotometer–An optical instrument for comparing the intensities of the corresponding colors of two spectra. Used for determining soil phosphorus.

Speculators–People who buy and/or sell and attempt to make a profit on the fluctuation of prices.

Speleology–The scientific study or exploration of caverns and related features.

Spike–(1) A very large nail or anything shaped like a nail, as the pointed iron teeth of a spike-toothed harrow, etc. (2) A flower cluster or type of inflorescence in which the individual flowers are sessile (stalkless or nearly so) on a central axis, as in the common plantain. (3) The spikelike panicle of wheat, barley, rye, certain grasses, and other plants. (4) A thick, upright stem which carries one or more flowers, as the stems of gladiolus, foxglove, and delphinium.

Spike-tooth Harrow–A secondary tillage implement for seedbed preparation. It consists of rigid square or diamond-shaped teeth extending down from several rows of frame bars.

Spile–(German, *speil*) (1) A spout for conducting water through an irrigation levee to a row. (2) A spout for conducting sugar maple sap into a container. (3) A spout for draining liquid from a barrel.

Spin–To twist or transform the fibers of wool, cotton, silk, flax, etc., into thread or yarn preparatory to making cloth or fabric.

Spirometer–An instrument that measures the flow of air in and out of the lungs.

Spite Walls and Spite Fences–Structures erected to interfere with the vision and air of an adjacent property owner. An owner may sometimes legally stop the erection of these structures if he can prove malicious state of mind on the part of the builder.

Splice–To join the ends of two ropes or wires by the interweaving of strands.

Split Shingle–A rough, flat, roofing shingle, split from a shingle bolt, usually by a broadax or a froe; a shake.

Split-rail–(1) A long bar of timber split from a log or large limb and used largely in the construction of rail fences. (2) Any wooden bar that is split from a log, limb, bolt, or the like.

Spoke–One of the set of bars or rods radiating from the hub of a wheel and supporting the rim.

Sponge–The porous, elastic mass of interlacing fibers forming the internal skeleton of certain marine animals of tropical seas, and having great power of absorbing water. Gathered, treated, and sold on a commercial basis, pieces of the above substance are used in various ways, as for washing, cleaning, etc.

Spontaneous Combustion–An internal heating in a product or substance which reaches an ignition point without an external source of heat. In stored hay, temperatures within the hay of approximately 160°F are considered to be reaching the danger point of internal ignition.

Spot Basis–In marketing in the grain trade, the difference between the present cash price and the futures price of a particular option in a specified market.

Spot Checking–The inspection of an occasional bale of cotton or other commodity to see if the quality of the product is up to the intended grade or specification.

Spot Market–The market for goods available for immediately delivery.

Spot Price–The current price at which the spot or cash commodity is selling. In grain trading, it is called the cash price.

Spout–(1) A pipe, tube, or similar conduit, through which a liquid, grain, seed, or other loose material flows and is discharged. (2) A trough or chute by which grain, coal, sand, or the like is discharged from a conveyor or bin to a truck, etc. (3) A pipe or projection (as of a sprinkling can) by which a liquid is discharged or poured. (4) A spigot. (5) A spile which is used in the collection of sap. (6) To shoot out, as a liquid, etc., from a spout; to flow or shoot out with force.

Spray–(1) A small, growing or detached shoot, twig, or branch of a plant with leaves, flowers, berries, etc.; a horizontal arrangement of flowers or greens. (2) A solution or suspension of material, as insecticides, fungicides, etc., which is broken up into fine, liquid particles driven through the air by mechanical means, coating surfaces of plants or other objects that it contacts. (3) A jet of fine liquid particles forced under pressure from a spray nozzle of a spraying device. (4) A device for shooting out a jet or spray. See Sprayer. (5) To scatter a liquid in fine particles. (6) To sprinkle or treat with a spray, as to direct a spray of fine particles of a liquid upon plants or other objects.

Spray Boom–On a spray rig, an assembly of several spray nozzles on a horizontal pipe used to extend the area of coverage of a single operation, as for spraying several rows of potatoes, etc. A similar arrangement on a vertical pipe is used for spraying trees.

Sprayer–A device or mechanism by means of which a solution or suspension of a material in liquid is broken up into droplets of a predetermined size range and distributed by hydraulics, by air blast, by gravity from an airplane, or by a combination of the above. It is used to sprinkle or treat plants, trees, etc., with a spray.

Spread–(1) A straddle; the difference in price between two delivery months in the same or different markets, or the sale of the one thing against a simultaneous purchase of the other. Straddling between a foreign and the domestic market is often referred to as arbitrage. (2) An extensive tract of land. (3) A ranch, including the buildings and the extent of land grazed by cattle or sheep (western United States). (4) The distribution of a disease. (5) Butter, peanut butter, margarine, or other food mixtures used as a spread on bread to improve its flavor, palatability, and/or nutritive value. (6) To scatter, as to broadcast seed, fertilizer, etc. (7) To disseminate disease.

Spreader–(1) A device used for distributing water uniformly in or from a channel. (2) A spreading agent, or spreader-sticker used to improve the contact between a pesticide and a plant surface. (3) An instrument provided with hooks or flanges used to spread open the incision made in the body wall of a bird. (4) A device that scatters or spreads: e.g., a manure spreader, a fertilizer spreader. (5) An animal, etc., capable of acting as a parasite or disease vector. (6) A species of microorganism which tends to grow profusely over the entire surface of the culture medium.

Sprengel Tube–A U-shaped tube with capillary ends at right angles which is used for determining the specific gravity of milk and fats. The tube is weighed empty, filled with water and then with the fluid, whose specific gravity is given by the ratio of its weight to that of water. See Specific Gravity.

Spring–(1) A natural flow of water from the ground. The source may be either shallow or deep-seated, and the spring may be at the base of a slope where a pervious stratum overlies a relatively impervious one, or it may be an artesian spring issuing at the surface through rock fracture or along fault lines, or a subterranean stream issuing through a solution opening as in a limestone terrain. (2) The season between winter and summer. (3) An elastic device or part which recovers its shape after being stretched, compressed, bent, etc., as the curved steel blades of a spring-tooth harrow. (4) The act of springing or leaping. (5) Referring to the spring season of the year in which a crop is harvested, or a flower blooms, as an early-spring crop, or a spring flower. (6) To indicate proximity of calving by a swelling of the udder, as in a cow.

Spring-tooth Harrow–A secondary tillage implement used for deep seedbed preparation and for destroying persistent weeds. It has vibrant spring-steel shanks that hold a variety of soil-working tools.

Spring-wheat Region–The region in which spring wheat is commonly grown: North Dakota, South Dakota, Montana, and areas in bordering states to the south and east, United States.

Sprinkler–(1) A device for spraying water on plants or lawn; a sprinkling unit attached to a water pipe or hose. (2) A sprinkling can.

Sprinkler Irrigation–Irrigation by means of above-ground applicators which project water outward through the air making it reach the soil in droplet form approaching rainfall. Applicators commonly used are rotary or fixed sprinklers, oscillating or perforated pipe.

Sprocket–A wheel or gear with projections that interact with a drive chain to turn part of a machine. Some planters, combines, etc. use sprockets and drive chains for transmitting power.

Spud–(1) A narrow, sharp spadelike tool usually fitted with a long handle used for undercutting and lifting out deep-rooted weeds, for removing bark from trees, for tree planting, etc. (2) An Irish potato. (3) In well drilling, to raise and lower the drilling tool.

Spur–(1) A short, stubby shoot, as in some fruit trees where the spurs bear flowers for more than one year or as grapes are pruned to spurs of one to two buds each, etc. (2) A hollow, tubular projection from some part of a flower, very conspicuous in the columbine, usually secreting nectar to induce the visitation of insects. (3) A hornlike protuberance which grows from the inner side of the shank of a fowl. (4) A steel contrivance secured to a rider's heel and used to urge the horse by its pressure. See Spur Rowel.

Spur Rowel–A small wheel with radiating points, attached at the end of a horseman's spur as an added goad.

Square–(1) Of flooring or roofing, an area of 100 square feet. (2) As applied to shingles or other roofing, that number which will cover 100 square feet of surface. (3) An L-shaped instrument used in carpentry to lay out or test square work, etc. (4) An unopened flower bud of cotton with its subtending involucre bracts.

Squatter–One who settles on land, especially on public land, without right or title (United States).

Squatter's Right–In the United States, the right to occupancy of land created by virtue of long and undisturbed use in the absence of legal title or arrangement.

Squeeze Chute–A narrow stall with a hinged side that is used for restraining animals. The animal's head is caught in a head-catch chute, and the sides of the chute are moved against the animal to restrict movements. See Head-catch Chute.

Stable–(1) (a) A building used for the lodging and feeding of horses, usually one fitted with stalls, mangers, feed boxes, and a place to keep harness or riding gear. (b) A building equipped with stanchions, racks, etc., in which cattle, goats, etc., are kept: e.g., a cow stable. (2) To put an animal into a stable. (3) Referring to things, chemical compounds, or conditions, not readily changed.

Stable Manure–The excreta of livestock intermixed with straw or other bedding material which commonly includes stalks and uneaten, coarse stems of forage, corn cobs, etc.; an important organic fertilizer. See Manure.

Stable Price–(1) A price usually referred to as being firm. The stable price might be in terms of a fixed number of dollars to be exchanged for a given commodity or unit of a commodity. (2) A price which maintains a specified rate of exchange between two commodities or between one commodity and all other commodities.

Staddle–(1) Anything which serves for a support; a prop. (2) A small tree, especially a forest tree.

Stadia–A method of measuring distance with a surveying instrument. The distance is determined by reading the length in feet on the level rod that is intercepted between the stadia hairs on the instrument.

Staff Cage–A graduated scale mounted on a plank, pier, wall, or other like object from which the water surface elevation may be read.

Stain–(1) A discoloration produced by foreign matter; a soiling, as of eggs by contact with damp litter, etc. (2) A dye or mixture of dyes which renders minute or transparent structures visible or which differentiates tissue elements by coloring with selective stains for microscopic examination.

Stake–(1) A wooden or metal strip, usually pointed at one end for driving into the ground, used for marking a location in surveying, supporting a vine or plant, tethering an animal, etc. (2) To place stakes as markers. (3) To support plants by tying to stakes. (4) To tether an animal to a stake.

Stall–The space in a barn which is occupied by a single animal, such as a dairy cow or horse, for feeding and handling.

Stall Barn–A barn used for sheltering dairy cattle and/or young stock where the adult animals are confined to stalls by means of stanchions, straps, halters, or chains during part of the year, as in the winter and for milking. Roughages and concentrates may be fed at the individual stalls. None, part, or all of feeds and bedding may be stored in the structure. Also called stanchion barn.

Stanchion–A tying or controlling device, usually of wood or metal, used in barns to control animals, usually dairy cows.

Standard Barrel–A standardized container of 7,056 cubic inches capacity used for marketing produce. For cranberries, the standard barrel contains 5,826 cubic inches.

Standard Cow Day–A feeding index for a cow which is 16 pounds total digestible nutrients per day from grazing.

Standard Length–In softwood lumber, lengths of boards in multiples of even feet; in hardwood lumber, lengths from 3 to 20 feet, including both odd and even number of feet.

State Department of Agriculture–In the United States, a department of the state government established to administer regulatory laws relating to agricultural and consumer interests, primarily with matters within the state to which federal legislation does not apply. These state laws include feed, seed, fertilizer, and lime regulations.

State Soil Conservation Committee, Commission, or Board–The state agency established by state soil conservation district enabling legislation to assist with the administration of the provisions of the state soil conservation district's law. The official title may vary from the above as new or amended state laws are made.

State Trading–Government control over the importation and exportation of a farm product or products; e.g., the governments of several countries control imports of tobacco: among them France, Austria, and Japan. Canada's Wheat Board control of wheat, oats, and barley is a state trading enterprise. Communist countries, which control major commercial transactions, are also state traders. Also called state or government monopoly.

State Veterinarian–A veterinary official of a State Department of Agriculture responsible for the proper enforcement of all regulatory laws regarding disease control, health and sanitation of livestock, and livestock products of the state.

Static Head–(1) The vertical distance that the top of a column of fluid is above a reference level. (2) The total head without deduction for velocity head or losses; e.g., the difference in elevation of headwater and tailwater of a power plant.

Statistical Reporting Service–This United States Department of Agriculture service gives coordinated leadership to the statistical reporting research and service programs of the department. It provides a channel for the orderly flow of statistical intelligence about the agricultural economy of this country. The primary responsibilities of this service are the nationwide crop and livestock statistical estimates, coordination and improvement in the department's statistical requirements, and special surveys of market potentials for agricultural products.

Status–(1) In anthropology, the rights and duties of an individual in reference to another individual where they stand in a reciprocal relation. (2) In sociology, a hierarchy of positions in a society and a summation of a person's position in socioeconomic terms.

Statute–A rule or measurement established by law.

Statute Inch–In northern California, United States, a water flow of 1.5 cubic feet per minute, or 1/40 cubic feet per second.

Stave–Any long, narrow strip, usually of wood, set edge to edge, used in making fences or enclosures, such as silos, tubs, and buckets.

Stave Silo–A large, round bin or receptacle, made of narrow strips of wood or concrete staves, placed edge to edge and held together with steel bands, in which chopped green forage is stored. See Silo.

Stay–(1) A guy rope, cable, etc., that supports a vertical pole, mast, etc. (2) A prop or tie piece that supports or holds parts together, or contributes to stiffness, as a timber supporting the frame of a structure, the posts of a fence, etc.

Steady Market–A market in which there is no distinct tendency for a rise or fall of prices.

Steam Sterilization–The use of steam to sterilize soil, pots, tools, and other materials used to produce plants in a greenhouse. A temperature of 160°F (57°C) is held for thirty minutes.

Steam-pressure Canner–A large, heavy metal pan having a tight-fitting cover which is fitted with safety valve, steam vent or petcock, and a gauge, either weighted or dial; used for processing low-acid foods under pressure at high temperatures in order to ensure their safety.

Steel–(1) A very hard steel rod very finely ground (usually fluted), and fitted with a handle, used for sharpening kitchen and butcher knives. (2) To sharpen with a steel.

Steno–A prefix indicating narrowness.

Stere–A stack of cordwood $1 \times 1 \times 1$ meter (1 cubic meter 35.3 cubic feet)

Stick–(1) A piece of wood, of indefinite size, that is long as compared with its diameter or cross section. (2) A thick, syrupy by-product obtained by evaporating the tank water produced in the processing of packing-house tankage. Stick is usually mixed with dried, low-grade tankage and used as a fertilizer, or is added to wet rendered tankage to be used as livestock feed. (3) To stab or cut with a knife, as in animal slaughter, to server the jugular veins to cause free bleeding; also with poultry, to pierce the brain with a narrow-bladed knife to cause feather follicles to relax for ease in plucking feathers.

Sticker–(1) A substance added to fungicides and insecticides to improve adherence to plant surfaces. (2) A strip of wood placed between the courses of lumber in a pile and at right angles to the boards to facilitate air circulation. (3) A person whose job is to stick animals in a slaughter house. See Stick.

Stile–A set of steps, ladder, etc., which allows passage of a person over or through a fence or wall while at the same time forming a barrier for livestock.

Still–A vessel in which a liquid or semisolid substance is submitted to distillation; i.e., its more volatile constituents are evaporated and led to another vessel or apparatus in which they are condensed. Such devices are used by mint growers in the production of peppermint and spearmint oil. (Sometimes the term is used to include the whole apparatus.) See Distillation.

Stimulus–An activating agent such as heat, moisture, or light.

Stirrup–An attachment to a riding saddle serving as an aid in mounting and as a footrest for the rider; made of metal, wood, or heavy leather in many different styles.

Stock–(1) Plant or plant part upon which a scion is inserted in propagation. (2) Livestock; domesticated farm animals. (3) Material held for future use or distribution. (4) Material destined to be wrought into finished products, as crate stock, barrel stock, cider stock, etc. (5) A plant or plant part that furnishes cuttings for propagation. Also called stock plant. (6) The main stem or trunk of a plant. (7) A rootstock (rhizome). (8) The original type from which a group of plants or animals has been derived. (9) The base or handle of a whip. (10) The stump of a tree (after the tree is felled). (11) (plural) A small enclosure in which an animal is secured in a standing position during shoeing or an operation. (12) To provide or supply with livestock, as to stock a pasture or range with cattle, sheep, etc. (13) To assemble a supply of materials or commodities. (14) Of, or pertaining to, livestock, as stock barn, stock feed, etc.

Stock Car–A slatted or open freight car for the handling and shipping of livestock. Some are double-decked for handling hogs and sheep.

Stock Cattle–Usually young steers or cows, light, thin, lacking in maturity and finish. Also called stockers.

Stock Country–Any region of extensive grasslands suitable for livestock grazing, as the western and southwestern range areas of the United States.

Stock Fountain–A watering device, usually consisting of a small receptacle or water cup, which is automatically kept full by a float valve, or filled when a valve is operated by the pressure of the muzzle or snout of the animal.

Stock Pile–A quantity of some material kept for future sale or use.

Stock Prod–See Cattle Prod.

Stock Tank–Any structure holding water for livestock. Commonly, it is a deep trough, bowl, or tank equipped with a float valve or other means of controlling the water level. Also called basin. In some parts of the country, it also refers to a farm pond.

Stocker–In marketing, a meat-producing animal capable of additional growth and finish and usually considered to be thinner than a feeder.

Stocking–(1) The relative number of livestock per unit area for a specific time. In range management, the relative intensity of animal population, ordinarily expressed as the number of acres of range allowed for each animal for a specific period. (2) In wildlife management, the density of animal population in relation to carrying capacity. (3) In forestry, the density of a stand of trees, such as well-stocked, overstocked, partially stocked. (4) A white leg on an animal.

Stocking Rate–Actual number of animals, expressed in either animal units or animal unit months, on a specified area at a specific time.

Stockyard–(1) A yard for keeping or holding livestock. (2) A series of pens or yards where market animals are collected for sale. It may be only a pen or two along a railroad siding in a small town, or it may be as extensive as the great stockyard systems of Chicago.

Stone–(1) In land description, a detached rock fragment on the surface of, or embedded in, the soil. (2) In soil survey mapping, a rock which is greater than 10 inches in diameter or more than 15 inches in length. See Boulder. (3) The hard seed of some pulpy fruits, as peach, plum, etc. See Clingstone, Freestone. (4) A unit of weight, 14 pounds (avoirdupois, British). (5) See Testicles. (6) To take the stone or pit out of stone fruits.

Storage–(1) Placing or holding in reserve in usable form and condition, as the storage of nutrients by plants in roots, for future use; the storage of honey by bees, during periods of abundance. (2) The holding of products for future sale or use, as in a granary, root cellar, cold storage warehouse, etc.

Storage Payments–Annual payments per bushel or by weight made to individuals and corporations for the storage of commodities in the Farmer Held Reserve or placed under loan to the Commodity Credit Corporation.

Straight Bar Bit–A single, straight, smooth, rounded bar of iron used as a bit on a horse bridle. It is the easiest on the horse of the many bits in use.

Straight Interest–The same interest rate as the stated rate; often called simple interest.

Straight-line Method of Depreciation–A method of figuring depreciation by dividing the original cost of an asset, less salvage value, by the number of years the asset is expected to last.

Straightbred–Designating an animal with a straight line of ancestry or pedigree within a recognized breed, such as a registered Thoroughbred horse or a purebred Jersey cow.

Strain–(1) A group of plants of common lineage which, although not taxonomically distinct from others of the species or variety, are distinguishable on the basis of productiveness, vigor, resistance to drought, cold, or diseases; or other ecological or physiological characteristics. (2) A group of individuals within a breed which differ in one or more characters from the other members of the breed; e.g., the Milking Shorthorns or Polled Herefords. (3) An organism or group of organisms which differs in origin or in minor respects from other organisms of the same species or variety. (4) A virus entity whose properties and behavior indicate relationship to a type virus and are sufficiently constant to enable the entity to be recognized whenever isolated. (5) A severe muscular effort on the part of a draft or other animal which may result in muscle, ligament, or other damage. (6) The condition of an overworked part operating above optimum load, such as a belt or motor operating under strain. (7) To filter a liquid and free it of impurities by passing it through some medium or fabric which can retain the solid matter and allow the liquid to pass.

Strake–An iron band fastening wheel fellies to each other; it differs from the tire in being made of separate pieces.

Straw Boss–A subordinate foreman in a logging camp or other work crew.

Stream–(1) Flowing water in a natural or artificial channel. It may range in volume from a small creek to a major river. (2) A jet of water, as from a nozzle. (3) A continuous flow or succession of anything: air, gas, liquids, light, electricity, persons, animals, and materials.

Strength–(1) The capacity to resist force; solidity or toughness; the quality of bodies by which they endure the application of force without breaking or yielding. (2) Bodily or muscular power; force; vigor. (3) The potency or power of a liquid or other substance; intensity of active properties. (4) The firmness of a market price for a given commodity; a tendency to rise or remain firm in price.

Stretcher–A hand-operated mechanical device of several types used in stretching fence, such as barbed or woven wire, to a desired tautness. It consists principally of a stretcher bar, a ratchet device, stretcher arms and a stretcher clamp.

String–(1) A cord of small diameter. (2) The several fumigating tents which are used by a single crew of people in fumigating orchards. (3) A group of partly broken horses which are assigned to a cowboy or horse trainer for his personal use or for further breaking or training. (4) A group of horses in a pack train. (5) To unroll wire prior to stretching and fastening to posts, etc.

Strip–(1) A relatively narrow piece, as a strip of land. (2) Each of the squirts or streams of milk as taken from a cow's teat in milking. (3) To tear off, as the threads from a bolt or nut, the leaves, husks, suckers, from a plant or plant part, or the hide or pelt from a carcass. (4) To take the last of the milk from a cow's udder by hand milking after the machine milker has been removed. (5) A narrow marking extending vertically between the forehead and nostrils of a horse.

Stripcropping–The practice of growing crops in a systematic arrangement of strips, or bands. Commonly cultivated crops and sod crops are alternated in strips to protect the soil and vegetation against running water or wind. The alternate strips are laid out approximately on the contour on erodible soils or at approximate right angles to the prevailing direction of the wind where soil blowing is a hazard.

Stripper–(1) A device or machine used for stripping cotton bolls and lint from the plant at harvest time. (2) A hand-operated implement or a power-driven machine used for gathering grass seed by combing or stripping the seed from the plant stems. (3) A machine for clearing new ground. See Cotton Picker, Cotton Stripper.

Stroke–(1) The linear distance traveled in one motion by a piston or ram, whether in an engine or pump. (2) A sudden and severe attack, as of paralysis.

Strong–(1) Having a specific quality to a high degree, as strong flavor, etc. (2) Characterized by steady or advancing prices, as a market for a particular commodity.

Strut–(1) A part of a machine or structure used chiefly to hold things apart. In general, any piece of a frame which resists thrust or pressure in the direction of its own length. See Stay, Tie. (2) In an animal, exaggerated step or gait.

Stub–(1) A short blunt projection, as a blunt projecting stem, branch, or root. (2) A tree with a broken stem less than 20 feet. (3) The quill portion of a short feather appearing on the shanks or toes of otherwise clean-shanked birds.

Stub Bars–Short frame extensions for such multiple-framebar implements as field cultivators and chisel plows.

Stubble Mulch Grain Drill–A grain drill used to plant seed in a field that has stubble left from the last crop. The concept is to use minimum tillage methods of production. Also called minimum tillage drill. See Air Drill, Grain Drill, Press Drill.

Stud Book–The official record book of a livestock registry association for horses, ponies, and jacks.

Stud Farm–A livestock farm or establishment sometimes under government supervision, that specializes in breeding and raising herd sires for sale or distribution to smaller breeders or farmers. Also a farm that specializes in stallions for stud purposes.

Stump Fence–A fence built of pulled tree stumps laid on edge so that the pronglike horizontal roots interlock. (It was a common form of fencing for fields following the lumbering of the pine forests in parts of Michigan and some other northern states in the United States.)

Stump Grinder–A power-driven device that chips or grinds up a stump so that the stump is below the surface of the ground.

Stump Puller–An appliance of several types, usually involving the use of a cable and winch; used for extracting stumps from the soil.

Stunning Ax–A heavy, blunt-nosed ax or sledge used in slaughterhouses for stunning or killing animals.

Sty–A pen where swine are housed and fed. Usually it consists of a low shed and a dirt yard. Also called pigsty.

Sub–A prefix meaning either: (a) nearly, somewhat, slightly; e.g., subcordate, nearly cordate; or (b) below, under; e.g., subaxillary, below the axil.

Subangular–A roundness grade in which definite effects of wear are shown, the fragments retaining their original form and the faces virtually untouched, but the edges and corners rounded off so some extent.

Sublimation–The characteristic of a solid to become a gas without becoming a liquid, or a solid becoming a gas without becoming a liquid; e.g., the evaporation of snow and the formation of frost.

Subsidence, Land–A downward movement of the ground surface caused by solution and collapse of underlying soluble deposits, rearrangement of particles upon removal of underground mineral deposits, reduction of fluid pressures within an aquifer or petroleum reservoir, or decomposition of organic matter.

Subsidiary Crop–A crop planted with a main crop, such as clover in small grain.

Subsidy–A government grant of money to aid or encourage a private enterprise that serves to benefit the public. In agriculture, the subsidy may be granted to the total group or to one component group, such as the dairy farmer or the wheat grower. It may be effected through tariffs on an imported agricultural commodity, through price support programs, direct payments, or the providing of services and information.

Subsistence Farming–Farming in which the majority of the output is used by the farm family.

Subsoiling–Tilling a soil below normal plow depth, ordinarily to shatter a hardpan or claypan.

Suburban–Designating areas adjacent to cities or towns whose population is largely urban in outlook and habits, usually deriving its livelihood from the nearby cities or towns rather than from agriculture or local industry.

Succession–(1) The progressive development of vegetation toward its highest ecological expression, the climax. The replacement of one plant community by another. (2) Transfer of operation and of ownership of a farm from one generation to the next, as from father to son.

Suck–(1) Of an animal, to draw milk from the teat of the udder by application of suction by the mouth; to nurse. (2) To draw a liquid, air, etc., by vacuum-producing action.

Sulfur Dioxide–SO_2; a compound produced by burning sulfur; it has a suffocating odor and is used as a fumigant for the control of certain insects, for the prevention of molds on dried fruits, and for bleaching wool, straw goods, etc.

Sulfuric Acid–H_2SO_4; a dense, heavy, exceedingly corrosive, oily liquid which can decompose animal and vegetable tissue and has a great affinity for water, giving off heat on combining with it. It is used in the manufacture of superphosphate fertilizers by converting the insoluble rock phosphate to a soluble and available form.

Sump–A tank or pit receiving drainage water which is then pumped to discharge it. Water in a sump is at too low an elevation for drainage by gravity flow.

Sunken Fence–A ditch with a retaining wall used to divide lands without defacing the landscape.

Super–A top or additional compartment added to a beehive in which bees deposit their nectar and honey.

Supervised Agricultural Experience Program (SAEP)–A series of related learning experiences, which are an integral part of the instructional program of a student enrolled in vocational agriculture, designed to develop knowledge and skills in agriculturally related fields. These supervised learning experiences may be provided by utilizing facilities of the home, farm, school, or an agricultural business. Programs may include any of the following types of experiences: school farm or school laboratory activities, supervised farming program, placement for farm experience, or on-the-job agricultural training.

Surcingle–A strap or girth around the body of a beast of burden for holding the saddle or load in place or used for throwing the animal.

Surface Tension–Tendency of a liquid to contract until its area is the smallest possible for a given volume of liquid. This makes it appear to be covered by an elastic membrane.

Survey–The use of geometry and trigonometry to determine and lay out land lines and boundaries. See Transit.

Suspended Waterers–Poultry waterers that are suspended from the ceiling. They are raised up as the birds grow taller.

Sustainable Agriculture–An agricultural system in which all the inputs, outputs, and planning is aimed toward a profitable enterprise that can be continued indefinitely with little or no harmful impact on the environment.

Sustained Yield Management–Controlled exploitation of a forest unit in such a way that the annual or periodic yield of timber or other products can be maintained in perpetuity.

Suture–(1) A line along which two things or parts are united. (2) In botany, a line of union or a line of dehiscence. (3) In horticulture, a longitudinal protuberance along one side of drupaceous fruits, as in the peach. (4) To unite or join by sewing.

Swamper–A worker who clears away obstructions in logging.

Swarm Box–A box containing combs of pollen and honey used by queen bee breeders to produce starting cells.

Swath–A single strip cut by an implement or machine in cutting or harvesting hay or grain.

Sweat Collar–A thick, padded leather collar that is attached to the neck of a beef animal to help shrink a dewlap that is excessive. This is done to make the animal look better in the show ring.

Sweat Scraper–A smooth curved tool used to remove sweat from a horse's coat.

Sweat Shed–A properly ventilated shed or enclosure used to hold sheep so that their natural body heat causes them to sweat. Such a practice immediately preceeding the shearing operation softens the yolk and makes shearing easier.

Sweep–(1) A soil-tillage device resembling a cultivator shovel which has attached wings; used for cultivating soil growing cotton, etc. (2) A gradual bend in a log, pole, or piling; a defect.

Sweet Water–(1) Water from wells, springs, and surface water which is fresh, in contrast to that which is salty or alkali. (2) In cooling systems, refrigerated freshwater in contrast to brine.

Swell–(1) In a saddle, the part in front of the seat which rises to the horn. Also called fork, front. (2) A tin can containing preserved foods with its ends bulging due to the formation of gas inside.

Swing Men–(1) In trail herding of cattle, the riders who ride behind the front rider and on the sides of the herd to keep the animals from spreading out and straying (western United States). (2) Dairy plant personnel whose function is to relieve regular employees when on vacation, etc.

Symbiotic Bacteria–Bacteria in nodules or tubercles growing on roots of legumes which have the power to fix free nitrogen from the air into forms that can be utilized by plants.

Synchromesh Transmission–A transmission gearing which aids the meshing of two gears by causing the speed of both gears to coincide.

Synergist–A chemical substance that when used with an insecticide, drug, etc. will result in greater total effect than the sum of their individual effects.

Syrup Hydrometer–A floating instrument with a special scale for measuring the specific gravity of concentrated sugar solutions. Used to determine the strength of syrups, as in the making of maple syrup.

Systematics–A study of similarities and differences in organisms and their relationships to each other. It includes taxonomy and classification.

Systems Analysis–An integrated, step-by-step approach for helping a decision maker to choose a course of action by investigating the full problem, searching out alternatives, comparing them in light of their consequences, and using an appropriate framework to bring objective judgment and intuition to bear on the solution of the problem.

Sythe–A single-edge blade mounted on a crooked handle, sued to cut weeds by swinging with both hands. See Sickle.

T Factor–A measure of the amount of erosion in tons per acre per year that a soil can tolerate without losing productivity. For most cropland soils, T values fall in the range of 3 to 5 tons per acre per year.

Table Use–Fit for human consumption and desirable enough to be served as food.

Tachometer–A device on a tractor or other machine that measures the number of revolutions per minute.

Tack–The riding equipment of a horse, such as the saddle and bridle.

Tack Room–A room in a stable in which saddles, bridles, spurs, boots, harness, etc., are kept.

Tag–(1) A dung-covered lock of wool. Also called dag, daglock. (2) A plastic or metal piece attached to an animal for identification, or a cardboard or cloth label attached to the container of a product, a feed or fertilizer, giving the content analysis, etc. (3) A lock of cotton fiber which adheres to the boll after picking. (4) To place a tag on a product, animal, etc., for identification.

Tail Band–The crupper of harness.

Tailboard–The gate at the rear end of a wagon or truck which can be let down or removed to facilitate loading or unloading. Also called end board, end gate, shoveling board, tailgate.

Tally–(1) A label or tag attached to a product or animal for identification. (2) The number of products or animals handled or produced. (3) 100 sheep (Australia and New Zealand). (4) The total sheep sheared by each shearer in one day (Australia and New Zealand). (5) To record the number of products handled. (6) To mark for identification.

Tamp–(1) Any tool used for firming soil. (2) To firm soil which has been dug, as around a plant after transplanting or around a fence post.

Tan–(1) A Japanese unit of land measurement equal to 0.245 acre (0.1 ha). (2) A coat color of some animals, especially dogs. It is a yellowish-brown similar to the color of well-tanned leather, and varies in shade form light to dark. (3) To convert a hide into leather.

Tandem–One behind the other, as the hitching of horses or tractors one behind the other, a tandem team of animals.

Tandem Disk Harrow–An X-shaped disk harrow with two opposed front gangs and two opposed rear gangs.

Tandem Hitch–Two or more implements hitched, one behind another.

Tank–A pond or pool, natural or artificial, which serves as a supply of drinking water for livestock (western United States). (2) A metal or wooden receptacle used to hold liquids, as the tank of a sprayer, a water tank, a gasoline tank.

Tanker–(1) A specialized truck which carries a very large, specially built tank for the transporting of bulk milk. (2) A specialized truck with a large tank of water, a power pump, and base, used to fight forest fires.

Tap–(1) In lumbering, a cut made from the inside of a log. (2) A wooden basket which is used for packing figs. (3) A faucet on a pipe or container containing liquid. (4) To remove a taproot from a plant. (5) To make a cut in a tree for obtaining sap for use in making maple sugar or turpentine. (6) To cut threads in metal that will receive a bolt or screw.

Tapadera–A leather hood which covers the front of a stirrup on a saddle.

Tar Paper–A heavy paper, coated or impregnated with a bituminous derivative, used as a temporary covering or an undercovering for siding and roofing, the top half of drainage-tile joints, or to cover a frame, as a tar-paper shack.

Tare–(1) Weed seeds. (2) Deduction made in the weight of packaged products to allow for the weight of the container; gross weight minus the net weight. (3) Deduction made in the weight of sugar beets to allow for earth, crown tops, etc., which might adhere to the beets. (4) Any impurity in seed crops, such as dirt, chaff, weed seeds, broken seeds, etc.

Target Price–A commodity-specific price per unit of production (bushel or pound) for certain program commodities that is set by Congress and administered by the USDA. Target prices are usually above market prices. They are used to determine deficiency payments.

Tariff–An excise tax levied by a government on imported products.

Tarp–See Tarpaulin.

Tarpaulin–A large, waterproof, canvas or plastic sheet, used to protect produce, etc., from inclement weather.

Tax–A compulsory charge which is levied by a governmental, national, state, or local unit against income or wealth for the common good.

TDN–Total digestible nutrients; all nutrients consumed by an animal that are digested and used; generally applied to proteins, carbohydrates, and fats.

Team–(1) Two or more horses, mules, etc., which are harnessed to the same vehicle, plow, etc. (2) Two or more specialists who are jointly investigating a problem. (3) To match two or more draft animals to serve as a team.

Tease–(1) To stimulate an animal to accept coitus. (2) To vex or annoy. (3) Of fibrous materials such as wool or flax, to separate the strands of the fibers, or to prepare the fibers, by combing, for spinning. (4) To form a nap on cloth by stroking the loose fibers in one direction with comblike, natural, or mechanical, teasels.

Technical–(1) Concerned with a particular science, industrial art, profession, sport, etc. (2) Practicing, using, or pertaining to the technique rather than the theory or underlying principles involved in the execution of a project. (3) Designating the grade of a commodity manufactured in the usual commercial manner. (4) Pertaining to or designating a market where prices are controlled by speculation or manipulation.

Tedder–A tractor-powered implement consisting of a series of forks mounted on rotating wheels. It is used to stir cut hay so the hay will cure faster and more evenly.

Teepees–Pyramidal structures such as tripods or quadripods of lath, lumber, poles, pipe, or bamboo spread at the base and lashed together at the top to provide support for climbing vines.

Temper–(1) The proper relative condition of moisture in grain preparatory to milling. (2) The relative hardness or softness of the metal in implements and tools. (3) The relative mildness or viciousness of an animal. (4) Milk of lime, etc., added to boiling syrup to clarify it.

Tempering–(1) Accustoming planting stock gradually to a material change in temperature. (2) Treatment of tools and implements to impart a degree of hardness or softness to the metal.

Temporary License or Permit–A document authorizing grazing a given number of animals on public lands during an emergency, or for a certain period.

Tenancy–The status of being a tenant. Farm tenancy as a study deals with (a) the conditions and circumstances under which land is rented and operated and (b) the rights and responsibilities of farm landlords and tenants.

Tenancy from Year to Year–A situation in which a landlord and a farm tenant agree that the tenant rents and operates the farm on a one-year basis, but with the understanding that the lease will be automatically extended for another year unless either party serves written notice on or before an agreed date, upon the other party, informing him/her of the termination of the lease at the end of the year.

Tenancy in Common–A legal arrangement in which each party owns an undivided share in property.

Tenancy-by-the-Entirety–A joint-tenancy between husband and wife which differs from joint-tenancy (which can be between any relatives or nonrelatives) in that neither husband nor wife can break the arrangement without the other's consent.

Tenant–(1) A person usually called a cash tenant, fixed-rent tenant, who rents land from another for a fixed amount of cash or a fixed amount of a commodity to be paid as rent; or (2) a person, other than a sharecropper, usually called a share tenant, who rents land from another person and pays as rent a share of the crops or proceeds.

Tenant Farmer–A farmer who rents or leases land from another person, or one who works it on shares.

Tend–To care for, as to tend a garden, to tend a flock of sheep.

Tender–In agricultural commodity trade, a notice of intent to buy or sell. A tender may be quite simple. Also, it may be complex, spelling out in detail quantities that will be purchased, quality desired, type of packaging required, method of delivery, etc. Sometimes referred to as an offer.

Tender Annual–Any annual plant which cannot withstand a frost and whose seeds will not live outdoors over winter in the regions where subfreezing temperatures prevail in winter.

Tennessee Valley Authority–A corporation created by an act of Congress, May 18, 1933, to take custody of the Wilson Dam and appurtenant plant at Muscle Shoals, Alabama, and to operate them in the interest of national defense and for the development of new types of fertilizers for use in agricultural programs, and for the development of the Tennessee River and its tributaries in the interest of navigation, the control of floods, and the generation and disposition of hydroelectric power.

Tensiometer–Any of several types of devices which measure moisture, tension, or condition of water in soil. Used to determine when to irrigate.

Tension Spring–A spring used to keep tension on a fence wire.

Tenure–In rural sociology and agricultural economics, the nature of property rights under which land is held and utilized. (In the United States, almost all property rights in land are lodged in the landowner through the system of ownership in fee simple. The state retains the right of eminent domain.)

Term License or Permit–A document authorizing grazing on government lands for a stated number of years contrasted with annual or temporary license.

Terminal–(1) A station for delivery or receipt of produce. (2) Designating growing or being located at the end of a branch or stem.

Terminal Sires–Sires used in a crossbreeding system where all progeny, both male and female are marketed.

Termite–A pale, soft-bodied social insect, mostly of tropical and subtropical regions, of the order Isoptera which feeds on moist or west cellulose. Three main group of termites are recognized: (a) dry-wood termites that live above the soil and attack live wood; (b) damp-wood termites that live above the soil but work in wet wood, mostly decayed; (c) subterranean termites that live in damp soil but are capable of building tunnels of soil to reach nearby dead wood. Principal families Mastotermitidae and Termitidae.

Termite Shield–A shield, usually of sheet metal, which is placed in or on a foundation wall or other mass of masonry or around pipes to prevent the passage of termites.

Terrace Outlet Structure–A structure usually installed in or at the end of a terrace outlet to reduce erosion.

Test Weight per Bushel–The actual weight per measured bushel of grain. Any given grain may vary in weight per bushel among the varieties, depending on harvesting conditions.

Therm–A megacalorie, equivalent to 1,000 Kcal or 1,000,000 calories.

Thermister–A thermally sensitive electrical resistor used for measuring soil temperature.

Thermostat–(1) An electrical control device. It can turn on or off a heater, fan, or heating cable at a set temperature. (2) A device used to control and regulate the cooling system of an engine.

Thixotropy–The property exhibited by some gels of becoming fluid when shaken. The change is reversible. Some fine clays exhibit thixotropy. See Vertisols.

Thole–A wood or metal pin or peg used to fasten the handle to a scythe; an oar to the oar lock, etc.

Thong–A strap of leather.

Thorium–A heavy element. When bombarded with neutrons, thorium changes into uranium, becomes fissionable, and thus is a source of atomic energy.

Thornthwaite Index of Precipitation–A measure of the amount of precipitation corrected for temperature. As the temperature increases, evaporation and transpiration increase. This results in a decrease in the effectiveness of the precipitation. The monthly index of effective precipitation is figured as follows: $(P/T-10)\ 10/9$, where P equals mean monthly precipitation in inches and T equals mean monthly temperature in degrees F.

Threshold–The minimum concentration of a substance or condition necessary to produce a measurable physiological or psychological result.

Threshold Limit Value (TLV)—The maximum air concentration of a chemical, expressed as milligrams per cubic meter, in which a worker may perform his/her duties 8 hours per day, 40 hours per week, with no adverse health effects.

Threshold Velocity—The minimum velocity at which wind will begin moving particles of sand or other soil material.

Tie—(1) A string, rope, metal band, or wire which encircles a bale of cotton, hay, etc., holding the material in a limited space and form and making for ease of handling, storage, and transportation. (2) A rope, twine, etc., for fastening purposes. (3) A stanchion for securing an animal. (4) Depression, usually in the middle of the back of an animal, caused by the skin adhering to the backbone. (5) To fasten, secure, with a rope, string, wire, etc.

Tie Stall—(1) A stanchion to which an animal is fastened. (2) A stall in which an animal is fastened by a halter or chain.

Tie-out Chain—A chain used to fasten an animal to a stake, tree, post, etc.

Tier—(1) In grafting and budding, one who ties the bud or scion to the stem or stock. (2) One of the horizontal layers in the vertical stacking of boards in a well-defined pile for seasoning. (3) A layer of lumber, boxes, crates, bags, etc., in storage.

Tight-cooperage—Wooden barrels so constructed that they will hold liquid products.

Tile—(1) A short length of pipe made of fired shale, clay, plastic, or concrete, used to carry away excess water, to distribute sewage from the septic tank and to the drain field. Also called drainage tile. (2) To drain an area by means of tile.

Tile Bat—A broken piece of tile.

Tile Spade—A spade used by tilers which has a long, narrow blade and a short handle.

Tillage Equipment—Field tools and machinery which are designed to lift, invert, stir, or pack soil, reduce the size of clod, and uproot weeds; i.e., plows, harrows, disks, and cultivators.

Tiller—(1) An erect shoot arising from the crown of a grass. (2) One who tills. (3) An implement for tilling.

Tiller Keel—A flat circular disk at the rear of a disk tiller to keep the tiller running straight; also called coulter wheel.

Timber Marking—The selection and indication, usually by blaze or paint spot, of trees which are to be cut or retained in a cutting operation.

Timber Products—Roundwood products and plant by-products. Timber products output includes roundwood products cut from growing stock on commercial forest land; from other sources, such as cull trees, salvable dead trees, limbs, and saplings; from trees on noncommercial and nonforest lands; and from plant by-products.

Timber Resource System—A United States Forest Service system whose role is to grow and make available wood for the nation on a continuing basis. The system includes those activities necessary to (a) protect, improve, grow, and harvest timber from forest land and (b) protect, process, and utilize wood and wood-related products. In addition to wood, the system produces other goods and services, either by design or incidentally. The six "systems" established by the United States Forest Service to have a systematic, orderly way to view and evaluate its many diverse but interrelated activities are: land and water, timber resource, outdoor recreation and wilderness, rangeland grazing, wildlife and fish habitat, and human and community development.

Tine—One of the long, slender projections of a rake, pitchfork, etc. Also called tooth, prong, spike.

Titer—The minimum quantity of a substance required to produce a specific reaction with a given amount of another substance.

Title—The legal interest of an owner or owners in property, usually real estate. It may vary from fee simple, the highest type of ownership, to tenancy in common, joint tenancy, or contingent interest. The title of ownership is transferred from one party to another by means of a deed. The history and present status of the title is shown by its abstract.

Title Insurance—Insurance which is designed to pay the holder for loss sustained for reason of defects in a title, up to and including the policy limits. Available only after a certified survey.

TLV—See Threshold Limit Value.

TNTC—Too numerous to count.

Tobacco Barn—A tall building in which tobacco is hung for curing.

Tobacco Harvester—A machine with lateral platforms on which people ride and harvest tobacco as they approach the plants in the row.

Tobacco Spear—A spearlike device fitted over the end of a lath or stick to aid in piercing the butt ends of tobacco plants and in spacing them on the lath before hanging in the barn for curing.

Tobacco Stick—A lath on which tobacco leaves or plant are hung for curing.

Tobacco Warehouse—A large building where tobacco is stored and is later auctioned off.

Toe Sack—Gunny sack. See Jute.

Toll—(1) The part of the flour or meal a miller keeps as payment for grinding the grain. (2) The fee paid for use of a bridge, ferry, road, etc.

Ton—(1) A unit of weight. A long ton equals 2,240 pounds or 20 long hundredweights (used in England); a short ton weighs 2,000 pounds or 20 short hundredweights (used in the United States); a metric ton equals 1.1 short tons. See Hundredweight. (2) A unit of volume. A register ton equals 100 cubic feet of space occupied in the interior of a ship. A displacement ton is calculated based on the volume of the weight of a long ton of sea water. It is used in determining the displacement of a ship and is equal to 35 cubic feet. A freight ton is equal to 40 cubic feet of cargo.

Tonnage—The total weight in tons of any particular produce.

Tool—(1) Any manually operated implement or instrument, as an ax, hoe, saw, lathe, etc. (2) Any of the machinery on the farm including both hand and mechanical tools.

Tool Bar—A heavy frame on a tractor to which tools can be attached for various tillage operations.

Tooth–(1) In animals, one of the hard, bone like appendages of the jaws used to tear and to masticate food. (2) See Tine. (3) A cog. (4) Any of the piercing projections on the cutting edge of a saw. (5) One of the projecting or piercing points or prongs of a comb.

Tooth Float–A device for filling horses' teeth.

Top–(1) The aboveground parts of certain plants, especially the leaves, as beet tops, turnip tops. See Crown. (2) Scoured, combed, long wool. (3) The highest price paid for a product within a certain period of time. (4) The upper, branchy portion of a felled tree. It is sold as firewood, charcoal source, etc. (5) To cut off the crown and leaves, as of the sugar beet. (6) To remove the upper portion of the crown of a tree. (7) To place the best articles of produce, as eggs, or fruit, on the top layer, in a container so that the whole container appears to include articles of higher quality than is actually the case. (8) To sort out animals that have reached a certain stage of development or finish.

Top Grade–A market grade of produce which is either the best or the next best after fancy.

Top Quality–Designating generally, but not specifically, produce of the very best sort.

Topographic Factors–Physiographic and edaphic factors which may cause changes in vegetation.

Topographic Map–A scale representation, by means of conventional signs, of a part of the earth's surface, showing the culture, relief, and elevations above a datum, hydrography, and, frequently, the vegetation.

Torque–The amount of twisting or turning power, usually measured in foot pounds.

Torque Converter–A device that transmits power from a tractor engine to the transmission through hydraulic action.

Torque Wrench–A wrench that has a built-in indicator to measure the applied turning force or torque. It is designed to tighten bolts to the proper specifications.

Torrens System–A system of registration of title to land in which the government guarantees the title. It is in effect in some states in the United States, in Australia, South Africa, to some extent in Canada, and in central Europe.

Torsion Balance–The torsion balance measures small forces, such as gravitational or electrical, by determining the amount of torsion or twisting they cause in a slender wire or filament.

Total Digestible Nutrients (TDN)–A standard evaluation of the usefulness of a particular feed for livestock which includes all the digestible organic nutrients; protein, fiber, nitrogen-free extract, and fat (the latter being multiplied by 2.25 because its energy value for animals is approximately 2.25 times that of protein or carbohydrates).

Total Supply–The national, indicated production of a product, plus carryover from the previous year or years, plus the estimated imports.

Tote Road–Any road used to haul supplies to a logging camp.

Tough–(1) Having the quality of flexibility without brittleness; yielding to force without breaking. (2) Designating market grain which contains moisture in excess of 13.5 to 14.5 percent. (3) Designating plant or animal products that are not brittle, crisp, or tender.

Tow–(1) The short fibers of flax and hemp after they have been combed out of the stalk. (2) The parts of flax, jute, and hemp when made ready for spinning. (3) A strong rope.

Tower Silo–A cylindrical tower made of wood, concrete, tile, metal, etc., used for storage of silage. It is the most common type of silo in the United States. Also called upright silo.

Township–A unit of land consisting of 36 sections or 36 square miles.

Township Survey–The unit of survey of the public lands of the United States and of Canada. Normally a quadrangle approximately 6 miles on a side with boundaries conforming to meridians and parallels. It is further subdivided into 36 sections, each approximately one mile square. A political township may be one or more survey townships.

Toxic Substances Control Act (TSCA)–A law passed in 1977 that gave the Environmental Protection Agency authority to obtain information from industry on the production, use, health effects, and other matters concerning chemical mixtures and substances.

Trace–(1) Either of the two leather straps or chains, attached at one end to the hames clamped to the horse collar or breast band, and at the other to tugs on a whippletree. (2) The force to pull the vehicle, etc., as applied to traces. (3) See Trace Element. (4) To follow the course of nutrient elements in plants or animals, as by use of radioisotopes. (5) Designating an artificial flavoring material with little or no true flavoring.

Tractor–(1) A vehicle with an automotive or diesel-type engine used to supply power to other machines in one of five ways: (a) pulling at the drawbar; (b) belt power from the belt pulley; (c) rotary power from the power take-off shaft; (d) hydraulic power of the operation of hydraulic cylinders or other positioning mechanisms; (e) electric power when a generator is mounted on the tractor. (2) Designating an implement pulled or powered by a tractor.

Trade Name–A registered trade mark of the company manufacturing the product.

Trafficability–The capability of a terrain to bear traffic. It refers to the extent to which the terrain will permit continued movement of any and/or all types of traffic. Soil type has much to do with the trafficability of the terrain; e.g., vertisols have zero trafficability when saturated, and oxisols can bear some traffic when wet.

Trail–(1) A pathway made either by repeated passage of people or animals or constructed for easier passage. It is not usable for vehicles. (2) To move cattle or sheep over a long distance to a pasture, a market, or shipping point.

Trail Herding–Directing and controlling the movement of a group of livestock on restricted overland routes.

Trailer–A vehicle with one, two, four, or more wheels, designed to be pulled by another vehicle. For agricultural purposes, it is used to carry heavy loads.

Trailerette–A one-wheeled trailer.

Trailing Plow–A plow that is pulled by a tractor rather than the plow being mounted directly on the tractor. See Mounted Plow.

Trailjackal–Any man who by dishonest means lived off the trail herds or who preyed on them (southwestern United States).

Trait Ratio–An expression of an animal's performance for a particular trait relative to the herd or contemporary group average. It is usually calculated for most traits by dividing the individual record by the average of animals in the group and multiplying by 100.

Transient–Having only a brief existence; shortlived.

Transit–A surveying instrument with the telescope mounted so that it can be reversed in direction without dismounting; called also a transit theodolite.

Trap–(1) A two-wheeled, one-horse cart. (2) Any of various devices used for catching living insects, animals, birds, etc. (3) A device which consists of a U-shaped pipe or chamber so that liquid flowing through it always forms a seal against a flow-back of gas. (4) A chamber to collect sediment flowing in a pipe.

Trash Bars–Flat bars which prevent the buildup of soil and trash between the disks of tillage implements.

Trash Farming–Stubble mulch farming. See Conservation Tillage, No Till.

Traverse–A plane land survey consisting of compass bearings (or angles) and distances and error of closure back at the starting point.

Travis–The partition or half-wall dividing individual stalls in a barn. Also called traverse.

Tray–(1) A wide, flat-bottomed, topless, shallow container used for picking, carrying, handling, drying, or storing produce. (2) A short piece of heavy wrapping paper on which seedless grapes are sun dried.

Tread–(1) Any injury to the front legs of a horse caused by its overreaching with a hind foot as it runs. (2) That part of a wheel or tire which rests or bears on the ground. (3) That endless belt on which a crawler tractor moves. Usually called track.

Tree Baler–A device used to put a netting around a Christmas tree for shipping.

Tree Caliper–(1) A caliperlike device used to measure diameters of tree trunks and logs. (2) The diameter of a tree trunk; in forestry measured at 4.5 feet above the ground; in horticulture, measured at one-foot height.

Tree Planter–A tractor-drawn implement that is used for planting tree seedlings. An operator rides on the planter and places a seedling in the furrow opened up by the implement.

Tree Planting Bar–A heavy wedge shaped bar of steel attached to a handle that is used for opening the ground to plant tree seedlings.

Tree Processor–A machine that trims and delimbs trees that have been harvested. The trees are fed through the machine and emerge with the limbs removed.

Tree Shaker–(1) A machine that grasps a nut tree and shakes the tree causing the nuts to fall to the ground where they can be picked up by a nut sweeper. See Nut Sweeper. (2) A machine that shakes harvested Christmas trees to remove dead needles, debris, etc.

Tree Spade–A large hydraulically operated machine that scoops into the ground around a tree and removes the tree from the ground for transplanting.

Tref–Designating a food which does not meet the dietary laws for Orthodox Jews. See Kosher.

Trellis–A latticework frame or wires supported by posts on which vines or flowers are trained. It may be constructed as a bower, arbor, etc.

Trench–(1) A long, relatively narrow excavation; a ditch. (2) To dig a ditch; to make a long, relatively narrow excavation.

Trench Silo–A trench excavated in a hillside or on firm ground, usually lined with wood or concrete retaining walls. Commonly about 15 to 25 feet wide, 6 to 8 feet deep and as long as the capacity desired. A trench silo must have good drainage and its use is largely limited to arid or semiarid climates.

Trends–A term usually applied to changes in production and prices of agricultural products.

Trespass–To enter unlawfully on a property.

Tret–An allowance formerly made to a purchaser of a product to allow for waste due to transportation, miscounting, etc., which consisted of giving 4 pounds of the product free for every purchase of 100 pounds after deducting the tare.

Tri—A prefix meaning three, as *trilocular*, having three locules.

Trial Run–A preliminary test to estimate the performance of a machine, procedure, or of any device involving motion.

Triangulation–The laying out and accurate measurement of a network of triangles, especially on the surface of the earth, as in surveying.

Trickling Filter–A 5 foot diameter pile of cinders or small rocks over which water containing animal or human effluent or food processing plant wastes are sprayed to achieve aerobic microbial decomposition. A widely used practice.

Tripod–A three-legged stand upon which a surveying instrument is placed.

Triturated–Ground or rubbed into fine particles.

Trough–(1) A long, relatively narrow, open-topped vessel used to hold water or feed for livestock. (2) A shallow conduit, usually of V or U section, open at the top.

Trowel–(1) A short-handled tool having a relatively narrow, concave and pointed blade. It is used for transplanting, planting bulbs, etc. (2) A flat hand tool used in concrete work and handling mortar in bricklaying.

Troy Weight–A system of weights based on 1 pound equals 12 ounces; 1 ounce equals 480 grains (20 pennyweights).

Truck–(1) A strong vehicle, especially one with four wheels, a wagon or an automotive vehicle used for transporting machinery, freight, and other heavy articles. (2) A low vehicle similar to a wheelbarrow but having two small, stout wheels in place of one, and a forward lip, used for moving barrels, boxes, etc., by hand; a barrow-truck. (3) A swiveling carriage, having two, four, or six wheels, placed under a locomotive or car. (4) Vegetables raised for sale. (5) To transport produce by means of an automotive truck.

Truck Crop–A vegetable crop usually raised on a relatively large acreage and under intensive methods of farming. See Market Garden.

Trustee–A person who holds property in trust; a person or agent to whom legal title has been transferred and who has the responsibility of managing the property for the benefit of a second party. The second party is entitled to all the rents or benefits and is the real owner.

Truth-in-Fabric Laws–Those laws which require the manufacturer of all fabrics to label each product to show the percentage of each fabric used.

Tube Dehorner–A device consisting of a metal tube that is sharp on one end and has a round ball-like knob on the other end. When the sharp end of the tube is fitted down over the horn of a calf, the operator gives a quick downward thrust with a twisting motion to remove the horn. The dehorners come in different sizes.

Tuber Indexing–Growing a seed piece of a potato under conditions suitable for virus determination. According to the reaction, the remainder of the tubers produced by that plant are either condemned as unfit to plant or certified as disease-free. Also called hill indexing.

Tule Lands–Marshy land which is occupied chiefly by bulrushes of the genus *Scirpus*, especially such land in the lower part of the Sacramento Valley in California, United States. Such land, when reclaimed by drainage, has been highly productive agricultural land. Also called tulare.

Turbine–An engine that forces a stream of gas or liquid through jets at high pressure against the curved blades of a wheel, thus forcing the blades to turn.

Turbocharger–A blower or pump that forces air into engine cylinders at higher than atmospheric pressure. The increased pressure forces more air into the cylinder enabling more fuel to be burned and more power produced.

Turf Farm–A farm that produces turf grass for transplanting in lawns, golf courses, etc. The turf is harvested and sold in blocks or plugs.

Twenty-eight Hour Law–A law that prohibits the transporting of livestock by rail or truck for a longer period than 28 consecutive hours without unloading, feeding, watering, and resting five consecutive hours before resuming transportation.

Twitch–A device used for pinching or squeezing the nose (upper lip) of a horse to take the animal's attention away from some manipulation or minor operation. It is also used as a control measure on some unruly horses. The device may be a loop of rope or chain or a clamp, or may be accomplished with the hand. In some localities, especially where mules are common, the device may be placed on the ear. Also called twist.

Two Cycle Engine–An internal combustion engine that takes two movements of the piston to complete the power cycle. Fuel is brought into the cylinder, ignited, power is produced, and exhaust fumes are removed in the completion of the piston moving up and down one time.

Two-way Plow–A plow with two sets of plow bottoms, or one set of disks used to throw soil in opposite directions. Bottoms or disks are alternated at each end of the field. It is primarily used in irrigated land where back furrows and dead furrows would impede the desired water flow.

Type of Farming–The form of organization of a farm business, primarily the kinds and amounts of the different crop and livestock enterprises managed therein.

U(nited) S(tates) Approved–An official designation for breeding flocks, hatching eggs and chicks qualifying under the rules and standards of the U(nited) S(tates) National Poultry Improvement Plan. For United States Approved classification, all males and females in breeding flocks are selected by an authorized state agency. See United States Certified for Eggs.

U(nited) S(tates) Certified for Eggs–An official designation for breeding flocks, hatching eggs and chicks qualifying under the rules and standards of the United States National Poultry Improvement Plan.

U(nited) S(tates) National Poultry Improvement Plan (NPIP)–A plan established in 1935 with approval of the United States Secretary of Agriculture and an appropriation by Congress. Objectives of the plan are: (a) to improve production and meat qualities of poultry; (b) reduce losses from hatchery-disseminated diseases; (c) establish uniform terminology for grades; and (d) control advertising of poultry produces. The United States Department of Agriculture cooperates with state authorities in the administration of health and production regulations for the poultry industry. Participation in the plan is optional with the states, and with the hatcheries and their flocks within each state.

U(nited) S(tates) Performance Tested Parent Stock–An official classification under the United States National Poultry Improvement Plan for poultry flocks that qualify through the performances of their progeny in an official random sample test.

U(nited) S(tates) Pullorum-Typhoid Clean–Under the United States National Poultry Improvement Plan, an official classification of poultry flocks in which no pullorum or typhoid reactors were found on the first official blood test, provided that if a reactor or reactors were found on the first test, the flock may be qualified with two consecutive official negative tests at least twenty-one days apart.

U(nited) S(tates) Pullorum-Typhoid Passed–Under the United States National Poultry Improvement Plan, an official classification for poultry flocks in which no pullorum or typhoid reactors were found on the last official test. Birds must be at least five months old when tested and must be tested within twelve months of the time hatching eggs are to be used. The interval between tests must be at least twenty-one days.

U(nited) S(tates) Record of Performance (ROP)–An official designation, under the United States National Poultry Improvement plan, for females that qualify on individual or family trap nest records. Males qualify by being produced from a single-male mating of qualified ROP birds, and by being reasonable representatives of breed or variety and by showing health and vigor and no serious defects.

Ultra-low Volume (ULV)–The spraying by air or ground of a pesticide undiluted and in a very concentrated liquid form. The usual dosage is from 2 to 16 fluid ounces per acre but is always less than one-half gallon per acre.

Ultrasonics–The use of high-frequency sound waves to measure fat thickness and loin eye area of animals. It is also used to test for pregnancy in animals.

Ultraviolet Light–That portion of the spectrum composed of light waves just shorter than violet light. It is used in irradiation, disinfection, and sterilization.

Uncapping Knife–A knife, usually heated, for cutting cappings from honeycomb so the honey can be extracted.

Uni-–A combining form meaning one.

Unified Soil Classification System–(Engineering) A classification system based on the identification of soils according to their particle size, gradation, plasticity index, and liquid limit. Indicated for each soil-mapping unit in modern soil survey reports.

Uniformitarianism–The concept that the present is a key to the past, and that past geologic events are to be explained by those same physical principles that govern the present.

Unilateral–An action taken by a country acting singly; e.g., the imposition of an import quota by one country without consulting other countries affected by the quota.

Unique Agricultural Lands–Land that is particularly suited for high production of a crop; e.g., the orange-growing soils of Florida and the cranberry bogs in Massachusetts. See Prime Agricultural Lands.

Unit–(1) A single thing or item of produce. (2) A recognized measure of weight, volume, or distance, such as a bushel, gallon, or mile.

United Nations (UN)–An international organization aimed at maintaining peace and security among nations. It functions in the economic, social, cultural, and humanitarian areas. It was organized in October 1945. Headquarters: New York City. The Food and Agricultural Organization (FAO) is a branch of the United Nations.

Universal Joint–A joint that connects the shaft of a power unit to an implement, or transmits power between moving parts of a machine. The universal joint allows lateral movement of the shaft.

Universal Soil Loss Equation–An equation used for the design of water erosion control systems: A = RKLSPC wherein A = average annual soil loss in tons per acre per year; R = rainfall factor; K = soil erodibility factor; L = length of slope; S = percent of slope; P = conservation practice factor; and C = cropping and management factor. (T = soil loss tolerance value that has been assigned each soil series expressed in tons/acre/year.) See T Factor.

Unpaid Family Labor–Labor used in conducting the farm business that is furnished by the members of the farm family, other than the operator, for which no direct wage is paid. Its value is determined by the amount of additional labor the operator would have had to hire at current wages to carry on the same size business had the family labor not been available.

Unsweetened Pack–Fruit packed for freezing without any sweetening added. It may be packed dry or covered with water.

Upkeep–The cost of maintenance.

Upset–(1) To enlarge by blows upon the end, as a blacksmith upsets a bar of heated iron by bumping its end upon the anvil. (2) To cause any change in the normal functions of plants or animals.

Urban–(1) Belonging to or residing in a town or city rather than the countryside. (2) In the United States, designating any of the following residence categories, used by the Bureau of the Census: (a) places having 2,500 persons or more, incorporated as cities, boroughs, and villages; (b) incorporated towns of 2,500 persons or more, except in New England, New York, and Wisconsin, where towns are simply minor civil divisions; (c) the densely populated areas, incorporated and unincorporated, around cities with a population of 50,000 or more; and (d) unincorporated places with a population of 2,500 or more outside any urban fringe. See Rural-Farm, Rural-Nonfarm.

Urbanized Area–An area identified by the United States Bureau of the Census as having a population over 50,000 or by the Office of Management and Budget as a standard metropolitan statistical area. Small urban areas are those areas that have a population of 5,000 to 50,000.

USDA (U.S.D.A.)–United States Department of Agriculture.

Use–The proportion of the current season's growth available for grazing that is consumed or destroyed by grazing animals; usually expressed as a percentage of biomass, but may be estimated as overuse, proper-use, or underuse; it may be applied to a single species or to the entire rangeland.

Usuary–Charging more than the rate of interest allowed by law.

Vacreator–A multiple unit consisting of steam jets and vacuum chambers designed to remove undesirable odor form milk and/or cream.

Vacuum Cooling–A cooling system for fresh leafy vegetables, such as lettuce. The product is put into a vacuum chamber, and the atmospheric pressure is lowered. As water evaporates, the heat of vaporization quickly removes heat from the product.

Valley Train–Deposits of comminuted rock material that were carried in prehistoric times by streams originating from the melting ice of glaciers. The deposits are confined to valleys and thus are distinguished from outwash plains. (A physical feature of parts of the glacial region of the United States.) The valley train often constitutes a distinctive land type of agricultural significance.

Valorization–The establishment and maintenance of an arbitrary price for a product by government action through subsidies, loans, government purchase, etc.

Valuable Consideration–Any service, goods, etc., given in payment or partial payment. (The term is often used in a deed, as "one dollar and other valuable considerations," to prevent disclosure of the full price.)

Value of Farm Operator's Labor–An allowance used primarily in computing the rate of return on the farm capital investments for the services of the farm operator for his/her labor at the rate at which he/she would have to pay another person to perform these functions. It is exclusive of the family living from the farm. It is used in computing the rate earned on the capital invested in the farm business and the total and per tillable acre charge for farm labor.

Valve–A device that controls the flow of liquids or gases.

Van–A large, covered truck or wagon.

Vane–(1) The thin web part of a bird's feather. (2) A thin plate or strip of metal pivoted on the top of a spire or mast to show the direction of the wind; a weathercock. (3) A blade of a windmill, fan, centrifugal pump, or similar apparatus.

Vapor Lock–An engine problem that occurs when the fuel boils in the fuel system, forming bubbles that retard or stop the flow of fuel to the carburetor.

Vapor Pressure–The pressure at any given temperature of a vapor in equilibrium with its liquid or solid form.

Vaporize–To change from a liquid or a solid to a vapor, as in heating water to steam.

Vara–A Spanish-American measure of length, equal to about 33 inches, which was used in description of Mexican land grants (southwestern United States).

Variable Costs–The portion of total cash production costs used for inputs needed to produce a specific yield of a specific crop. Variable costs typically include fertilizers, seed, pesticides, hired labor, fuel, repairs, and animal feed.

Variable Expenses–Those expenses that may be different from year to year such as fuel costs, fertilizer costs, etc.

Variance–A statistical measure of the amount of variation that is observed within or among a group of animals or plants.

Variant–A recognized entity different from normal.

Variation–(1) The angle by which the north end of the compass needle (magnetic north) deviates from true north. (2) One of the laws of organic nature; organisms vary in time, from place to place, and also in one locality with time; they vary also in their appearance (morphology).

Vat–Any large vessel, tub, cistern, etc., used for holding liquids, as a vessel used in large-scale brewing or dipping of domestic animals.

Vehicle–A car, carriage, wagon, truck, trailer, or other conveyance.

Veil–A light, metal screen or diaphanous cloth material draped about the head when working with bees to prevent being stung on the face or neck.

Velocity–The speed of flow in feet per minute or miles per hour.

Vendee–A person who buys something, especially a land buyer.

Vendor–A person who sells something; a seller. Also spelled vender.

Vent–(French, wind) (1) A small opening or passage, as an opening for ventilation. (2) The opening of the cloaca; the anus of a bird. (3) A brand mark made to indicate that the animal no longer belongs to the original owner. (4) To cancel an old brand on an animal to show change in ownership.

Vent Shield–A shield used to cover a chicken's vent to prevent picking by other chickens.

Ventilation–The movement of air that brings about an exchange of air in the material or space being ventilated. Louvers or other divides are used to keep rainfall out of the openings. The airflow may result from wind pressure, convection, or fan forces.

Venturi Tube–A closed conduit that is gradually contracted to a throat causing a reduction of pressure head by which the velocity through the throat may be determined. The contraction is generally followed, but not necessarily, by gradual enlargement to original size. Piezometers connected to the pipe above the contracting section and at the throat indicate the drop in the pressure head, which is an index of flow.

Verge–(1) In a city or town, the strip of grass lying between the street and the sidewalk. (2) The edge of a flowerbed.

Vernier–An auxiliary scale used in conjunction with the main scale of a measuring device to obtain one more significant figure of a particular measurement.

Vertical Integration–The control by a single firm of two or more stages in the chain of production, processing, and distribution. This chain extends from the supply of production resources to the point at which the commodity reaches the consumer. Recent developments in agriculture include extensive integration in the broiler industry and with some integration in laying flocks, poultry, hog fattening, and cattle feeding.

Vertical Interval–In terrace farming, the vertical distance in feet from the center of one terrace line, ridge, or channel to the corresponding point on an adjacent terrace. The interval of the first terrace is the vertical distance from the top of the hill to the staked channel of the terrace.

Veterinary Science–A branch of knowledge, dealing mainly with domestic animals, that encompasses anatomy, physiology, breeding (including breed improvement), nutrition, animal diseases and treatment (including diseases transmissible to humans), people's use of animals, etc. The first recognized veterinary school was established in

1761 in France; in 1791 England had such a school with others to follow; in North America several schools flourished after 1852, which, however, did not survive when standards were raised after World War I, requiring a high school diploma and four years of college. To that was added two years of preprofessional training, bringing the total up to six years of college study. At present, schools of veterinary science in the United States and Canada are, for the most part, allied within large universities.

Via–(Latin) Way; road; by way of.

Viscosity–The resistance of a fluid to flow; thickness of a fluid.

Visible Supply–In the United States grain marketing, a statistical compilation of the amounts of the different kinds of grain in elevators east of the Rocky Mountains that receive and ship grain without converting it into manufactured products. It indicates the amount of grain that is potentially in merchandising channels.

Vitrified Tile–Clay or shale tile that has been so burned as to produce a hard, glassy product.

Vocational Advisory Committee–(Agriculture) A local group, whose members are selected from the local community, appointed to advise the school on matters pertaining to teaching agricultural occupations.

Vocational Agriculture–Generally refers to the curriculum or program in agricultural education designed to offer students at the secondary level the opportunity to explore and prepare for agricultural occupations. The name of the program has now been changed to Agricultural Education. Also, postsecondary and adult programs are recognized as legal components of Agricultural Education. See Agricultural Education.

Void–(1) To evacuate feces and/or urine. (2) A general term for pore space or other openings in rocks such as vesicles and solution cavities. (3) Pore spaces in soils. (4) The space between kernels in a bulk of a grain that is usually expressed as percent of total volume.

Volatile–A compound is volatile when it evaporates or vaporizes (changes from a liquid to a gas) at ordinary temperatures on exposure to air.

Volatile Solids–That portion of the total solids driven off as volatile (combustible) gases at a specified temperautre and time (usually 600°C for at least 1 hour).

Volatility–The tendency for a fluid to evaporate rapidly; e.g., gasoline is more volatile than diesel fuel because gasoline evaporates at a lower temperature.

Voltage–A measurement (in volts) of the pressure of electricity flowing through a conductor. See Amperes, Wattage.

Voltmeter–A device used to measure the amount of pressure or force of electricity.

Waiver–A relinquishment of a claim; e.g., a landlord may wavier his/her lien on a tenant's crop in favor of a lender who wants to make the tenant a loan.

Walk-through–Designating a building or milking room so designed that the traffic moves in one door or gate and out another, rather than in and out the same door.

Walking a Farm–The procedure followed in the preliminary stages of appraising the value of a farm in which the appraiser walks over all parts of the farm to observe and record the physical features including soil, ponds, drainage, crops, pastures, and woodlots, which affect the value of the farm.

Walking Horse Bit–A type of English riding bit that is used on walking horses.

WAOB–World Agricultural Outlook Board.

War Bridle–Any of several types of halters, hackamores, or bridles with a part of the rope passing through the horse's mouth in such a way that a powerful leverage is placed on the horse's jaw or head. It is used on vicious or unbroken horses.

Warrant–In land allotments, a certificate authorizing ownership of public tracts of land to the holder. Also called land warrants. In some presettlement land surveys in Pennsylvania, United States, the land was subdivided into blocks of 1,000 acres called warrants.

Warranty Deed–A deed that commits the grantor to guarantee the buyer a clean title and obligates him to protect the buyer against adverse claims to property.

Warren Hoe–A triangular hoe used for weeding plants that are planted closely together.

Waste Gate–A water gate in a dam or reservoir for the discharge of excess water.

Water Application Efficiency–The percentage of irrigation water applied that can be accounted for as moisture increase in the soil occupied by the rooting system of the crop.

Water Application Rate–The rate in inches per hour that irrigation water is applied to fields.

Water Control–(Soil and water conservation) The physical control of water by such measures as conservation practices on the land, channel improvement, and installation of structures for water retardation and sediment detention (does not refer to legal control or water rights).

Water Deficit–The amount of water lacking for given purposes.

Water Depletion Allowance–A provision of the tax law that allows for a tax deduction based on the depletion of certain aquifers used for agricultural irrigation.

Water Disposal System–The complete system for removing excess water from land with minimum erosion. For sloping land, it may include a terrace system, terrace outlet channels, dams, and grassed waterways. For level land, it may include only surface drains or both surface and subsurface drains. For homes and factories it may consist of a septic tank and drain field. For barnlots it may consist of a lagoon.

Water Elevator–(1) Any mechanism used for raising water, especially to a considerable height above ground, as a windmill operating a pump. (2) A vertical or inclined conveyor whose buckets dip up water from a source of supply and deliver it into a conduit or elevated receptacle, thus fulfilling the function of a pump.

Water Finder–See Dowser.

Water Gang–A constructed channel used for drainage or irrigation.

Water Gap–(1) A gorge cut through a ridge by a stream. (2) The gap below a fence where it crosses a small stream.

Water Gauge–(1) A staff gauge used to indicate the water level in an open channel above a specified datum. (2) A pressure gauge used to indicate the pressure in a closed water system. (3) Any device used to indicate the water level in an open channel.

Water, Nonconsumptive Use–(1) Those uses of resources that do not reduce the supply, such as many types of recreation. (2) A use of an area or resource which does not alter the area or resource, and which the fact of one person partaking of this use does not reduce the quality for another user. (3) Some consumptive uses of water are irrigation and domestic and industrial use, while nonconsumptive uses would include direct power generation as well as boating and swimming.

Water Quality Criteria (Water Quality Standard)–The levels of pollutants that affect the suitability of water for a given use. Water use classification includes: public water supply, recreation, propagation of fish and other aquatic life, irrigation, and industrial use.

Water Repellent–Designating a cloth or product which, due to an application of certain substances, sheds most of the water that falls on it.

Water Requirement–The quantity of water, regardless of its source, required by a crop in a given period of time for its normal growth under field conditions. It includes surface evaporation and other unavoidable wastes. Usually it is expressed as depth (volume per unit area) for a given time; e.g., acre inches/hour.

Water Resource Region–The twenty-one major hydrologic regions into which the United States is delineated.

Water Right–A legal right to use the water of a natural stream, furnished by a canal for general or for specific purposes. It may entitle a person to use the canal to full capacity, to a measured extent, or for a definite period of time, and to change the place of diversion, storage, or use of water as long as it does not infringe upon the rights of other people. In some states in the United States, the rights to water may be sold and transferred separately from the land. See Riparian Doctrine.

Water Spreading–(1) The artificial application of water to lands for the purpose of storing it in the ground for subsequent withdrawal by pumps for crops. (2) Irrigation by surplus waters out of cropping season. (3) The diversion of run-off water from gullies or watercourses and its distribution on adjacent, gently sloping, grazing lands needing additional water. The volume of water flowing down the channels is reduced and the moisture absorbed by the spreading area increases the growth of vegetation.

Water Witch–One who claims to be able to locate underground water with a divining instrument. This instrument may be a tree fork, a Y-shaped wire, or almost anything. See Dowsing.

Water Year–A special grouping of the periods of the year to facilitate water supply studies. The United States Geological Survey uses October 1 to September 30 as the water year. See Climate Year.

Water Yield–(1) The total outflow of a drainage basin through either surface channels and/or subsurface aquifers. (2) The surplus of precipitation over infiltration and canopy interception. Also known as run-off water.

Waterhole–(1) A natural depression in which water collects or stands, often in the dry bed of a stream; a spring in the desert. (2) A natural spring or pool on the open range where cattle drink (western United States).

Waterproof–To apply a covering or coating which will shed water and reduce water absorption to a minimum.

Waterway–A natural or artificially constructed course for the concentrated flow of water.

Wattage–A measurement of electrical power (in watts), calculated as amperage times voltage. Absolute equals 0.001 kilowatt equals 0.00134 horse power. See Amperage, Voltage.

Wax–(1) Any substance similar to beeswax in composition and use, as carnauba wax, candelilla wax, bayberry wax, japan wax, sugarcane wax, flax wax, cotton wax, esparto wax, cauassu wax, murumuru sect wax, shellac wax, ceresin wax, montan wax, and paraffin. (2) To apply wax to the surface of a fruit, vegetable, foliage, or flower for preservation.

Weaning Ring–A ring with spikes that fits on a calf's muzzle to prevent nursing. When the calf approaches the mother, the projections on the ring make the mother move.

Weather Vane–Any of several devices used to show the direction from which the wind is blowing. It is often a decoration for a barn or house roof.

Weatherproof–Constructed or protected so that exposure to the weather will not interfere with successful operation, storage, preservation, etc.

Web–(1) The blade of a saw. (2) That portion of an ordinary anvil, between the head and the base, which is of reduced size. (3) The cotton disk used in filtering milk. (4) The fat which surrounds the paunch and intestines of sheep. (5) The disk of a rolling colter. (6) The membrane which unites the fingers and toes, especially of amphibians and water birds.

Wedge– A piece of hard metal or wood, tapering from the butt to the front end, which can be driven or forced into a narrow opening. It is used to split logs, wood, etc., and to hold a tree off the saw in felling. A wedge may also be used to tighten a loose connection or to reinforce a weak joint in wood or metal.

Weed Control– Any device or system used for the destruction, checking of growth and spreading of weeds by cultivation, by the use of herbicides, cultural practices, etc.

Weed Hook– A device used on moldboard plows to hold tall weeds and bulky trash against the furrow slice for improved covering of the weeds by the soil.

Weeder– (1) A light spring-tooth harrow which has long, flexible teeth. (2) A hand tool consisting of a short handle and several curved prongs, used to remove weeds from flower beds. (3) Any machine used to go over an entire field, regardless of rows, in the early stages of development of both crops and weeds. Specifically a light, spring-tined machine which cultivates lightly over the entire field, regardless of rows, when weeds are small and easily dislodged without undue damage to crop plants. (4) A rotary hoe.

Weighbridge– The platform of a scale, flush with the road or street, on which vehicles can drive to be weighed; or a smaller platform used for weighing sacks of grain or cans of cream.

Weight per Day of Age– A measurement of an animal's weight gain; usually from birth to weaning or from birth to one year old.

Weights– (Farm/ranch) Gross weight is the weight, e.g., of a truck full of grain. Tare weight is the weight of the empty truck. Net weight is the weight of the grain only (gross minus tare equals net weight).

Well Casing– The wooden, stone, brick, metal, concrete, or plastic pipe lining of a well.

Well Curbing– The platform which is a part of the raised, enclosing frame for a well.

Well Point– A perforated pipe covered with a sand screen which is sunk into sand to permit the pumping of groundwater and the exclusion of sand.

Western Saddle– A heavy, deep-seated saddle with a distinctly raised horn and cantle, as contrasted to the English saddle, which is light and comparatively flat. It is used for working livestock on the ranches in western United States. Also called stock saddle.

Wetlands– Those areas inundated by surface or groundwater often enough to support the kind of vegetation or aquatic life that requires saturated (or seasonally saturated) soil conditions for growth and reproduction. Wetlands generally include swamps, marshes, bogs, and similar areas such as sloughs, potholes, wet meadows, river overflows, mud flats, and natural ponds.

Weymouth Bit– A type of bit for a bridle which has a bar with a U-shaped curve in it and which is so constructed that considerable pressure can be brought to bear on the horse's mouth.

Whang– Tough leather adapted for strings, thongs, and belt laces which is commonly made from calf skin.

Wheat Belt– A region in which wheat is the principal crop. In the United States, it extends from northern Texas and western Oklahoma to North Dakota and the Canadian border. The southern part of the United States Wheat Belt is the winter wheat belt and the northern part is the spring wheat belt.

Wheel Track Planting– Planting a crop, as corn, on plowed fields, in the tractor wheel tracks. The weight of the wheels crush and firm the soil. The rough surface left reduces the amount of potential soil erosion.

Whip– (1) Any of several kinds of instruments for lashing animals, usually consisting of a handle and a lash. (2) An unbranched shoot of a woody plant, particularly the first year's growth. (3) A tall, slender tree. (4) To lash an animal. (5) To beat cream, egg white, etc., so that it becomes a froth and holds shape.

Wholesale Price Index– Measure of average changes in prices of commodities sold in primary United States Markets. Wholesale refers to sales in large quantities by producers, not to prices received by wholesalers, jobbers, or distributors. In agriculture, it is the average price received by farmers for their farm commodities at the first point of sale when the commodity leaves the farm.

Wick– (1) A thick, heavy cloth, fiberglass, etc., placed in the bottom of a flower pot with one end in water, which furnishes a constant supply of water to the plant. (2) A loosely woven cloth which, by capillary attraction, draws up a proper amount of oil in a lamp.

Wick Applicator– See Weed Wiper.

Wicket– (1) A small gate set within a larger one; a half door. (2) A small sluice or opening for letting the water out of a canal lock. (3) A valve or throttle in the chute of a waterwheel for regulating the flow of water.

Will– A legal document, signed by a person and two unrelated witnesses, containing instructions for the disposition of his/her property after his/her death.

Wilting Coefficient– The moisture, in percentage of dry weight, which remains in the soil within the root-feeding zone as plants reach a condition of permanent wilting. Also know as wilting percentage, permanent wilting point.

Winch– (1) A hand crank for turning a machine which revolves. (2) A drumlike device, turned manually or mechanically by a crank, on which a cable is wound; used for hauling or hoisting. See Windlass. (3) A device mounted on the front of a truck that is used to pull the vehicle when it becomes immobilized by mud, etc.

Wind– (1) The horizontal movement of air on and above the earth's surface. (2) The blast of air from a blower or fan. (3) The breathing of a horse.

Wind Erodibility Equation– An equation expressing the erodibility potential of a soil due to wind. The equation is E equals f(I,K,C,L,V) where E equals total erosion in tons per acre per year; f indicates that erosion is a function of the various terms; I equals soil erodibility based on texture and aggregation; K equals surface roughness; C equals climate factors (windspeed and soil moisture); L equals effect of field size (length); and V equals equivalent quantity of vegetative cover.

Wind Indicator– A device that indicates the direction and velocity of the surface wind. See Vane.

Wind Vane Feeder– A type of feeder used in an open pasture to feed cattle a mineral mix or other feed supplement. A large wing-shaped vane is mounted on top of the feeder to cause the feeder to rotate in the direction of the wind so that the opening will be protected from the blowing rain.

Windfall– (1) A tree uprooted or broken off by wind. (2) An area on which the trees have been blown down by wind. (3) A fruit that falls from a tree before the crop is harvested, usually associated with wind. Also called drop. (4) Unexpected income or tax break.

Windlass– A hand crank that is used to lift or pull loads. It is used to raise and lower curtains in poultry houses.

Windmill– A device which consists of propellerlike blades, sometimes called sails, attached to a gear head and connected to a central shaft and a rear vane to guide the wheel (fan) into the wind. The action of the wind causes the blades to turn the shaft. The energy of the wind is used most frequently for pumping water and generating electricity. (The best wind velocity for driving windmills is from 15 to 20 miles per hour.)

Windmill Pump– A well pump of the reciprocating type actuated through a crank driven by a windmill. A reservoir or supply tank is generally erected adjacent to the windmill derrick, a float in this tank throwing the pump out of gear when full, and engaging it once more as the water level decreases.

Windmill Sails– The propellerlike blades of a windmill which are usually approximately rectangular with a length of five times their breadth, the total amount of surface being about one-fourth the area of the circle described by their motion.

Windrow– (1) A long, relatively low, ridge of hay, sheaves of grain, etc., which is made to effect drying, curing, etc. (2) Accumulations of slash left in rows by loggers. (3) To rake hay or to place sheaves in a relatively low, long row for drying and curing.

Windrower– A machine for cutting forage or grain in which material is carried to the center and fed into a narrow conditioner or dropped onto the ground in a windrow.

Wing– (1) A leaflike, dry or membranous expansion or appendage of a plant part, such as along some stems and petioles and of samaras and some capsules. (2) Either of the lateral petals of a pealike flower. An organ of flight for birds, insects, bats, etc. (3) A piece of bird meat consisting or the organ of flight with the feathers removed. (4) The outside corner of the cutting edge of a plow. (5) Fan or blower blade. (6) An extension on a building. (7) The vane of a windmill.

Winnow– To separate grain, etc., from the husks by means of an air current or by wind.

Winter Storage– (1) Storage of produce during the winter. (2) Storage of nutrients within a plant for winter maintenance and/or for the following growing season.

Wire– (1) A slender thread of metal, usually flexible, which may be used for such purposes as to carry electric current, tie bales, construct fences, make temporary repairs on machinery, etc. (2) See Barbed Wire.

Wire Stretcher– Any of several devices used to draw fence wire taut before the fence is fastened to posts.

Wired Foundation– In a beehive, foundation comb with strengthening wires embedded in it.

Wired Frames– Frames with wires holding sheets of comb foundation in place within a beehive.

Witch– See Water Witch.

Withdrawal Period– The length of time a feed additive or drug must not be fed or administered to an animal prior to slaughter.

Witness Corner– A marker such as a rock, mound of earth, or a stake set on a property line leading to a corner, used where it would be impracticable to maintain a monument at the corner itself.

Witness Tree– A living tree marked by a land surveyor to indicate the location of a corner of a tract of land. For the public domain, surveyed by the United States General Land Office, the section and quarter-section corners were so marked. Also called bearing tree.

Wood Alcohol– See Methanol.

Wood Ashes– The residue which remains after burning wood. It has a value as a fertilizer and was used by early settlers in the home manufacture of lye. Sometimes it is used as a source of calcium in livestock feeds, especially for hogs.

Wood Charcoal– A carbon fuel prepared by heating or charring wood out of contact with air, either in retorts or in stacks in which a retarded combustion is effected. The carbon thus obtained is nearly pure; it is light and extremely porous possessing the power of absorbing gases. It is used as a fuel and is fed to animals, especially hogs, because it absorbs gases in passing through the intestines. It is also fed to poultry as an additive to mash or scratch feeds.

Wood Distillation– The decomposition of wood in a closed retort by heat so that wood alcohol, acetone, acetic acid, water, and tar are driven off, and only charcoal remains in the retort.

Wood Shavings– Very thin strips of wood resulting from planing boards. Sometimes used as bedding for animals, for a plant mulch, and for kindling.

Wood Technology– The study of wood in all its aspects; the science of wood, including its anatomy, chemistry, physical properties, treatment, and uses.

Wood Turpentine– The essential oil obtained from pine stumps or other resinous wood by destructive or steam distillation.

Wood-wool– A wood product resembling wool, made from waste pulp or fine hairlike shavings. Used primarily as insulating and packing material.

Woodland Management– The management of existing woodlands and plantations that have passed the establishment stage including all measures designed to improve the quality and quantity of woodland growing stock and to maintain litter and herbaceous ground cover for soil and water conservation. It includes all such measures as planting, improvement-cutting, thinning, pruning, slash disposal, fire protection, and grazing control.

Wool Carding–The process of disentangling and separating the fibers of wool and delivering them in a parallel condition suitable for spinning.

Wool Pool–A grouping of the wool of many producers into a single total amount so it can be sold on the market as a single unit. Such commodity pools are very helpful in securing a more satisfactory market than the individual producer could command for his small quantity. Wool, cotton, and tobacco are frequently pooled.

Work–(1) The performance of any productive operation, act, etc. (2) The productive operation, act, matter, etc., upon which a person expends energy. (3) (plural) The inner machinery, as the works of a combine. (4) The capacity for performing productive operations, etc., as the horse has a lot of work left in him. (5) To perform a productive operation, act, etc. (6) To expend energy in any productive act, etc. (7) To operate in a machine, as to work the mower. (8) To engage in a vocation or profession, as he works in Extension. (9) To proceed or move slowly, as water works down into the soil. (10) To respond to exertion of energy, as sandy land works well. (11) To cause to labor, as to work horses. (12) To cultivate the soil. (13) To bud or to graft. (14) To ferment, as cider works. (15) To drive livestock on horseback or by means of a dog.

Work of Digestion–The loss of energy which occurs in chewing, digesting, and assimilating the food; heat increment.

Working Assets–Assets that are more liquid than fixed assets and include such items as equipment, breeding stock, and vehicles.

Working Capital–Capital that is used to operate a business or from which a person is receiving some returns.

Working Dog–(1) A dog trained to drive sheep, cattle, etc., in such a manner that the animals are not harmed. (2) A dog trained to run a treadmill, pull a sled or cart, etc. (3) A seeing-eye dog trained to aid the blind. (4) A hearing dog trained to aid the deaf.

World Food Program (WFP)–A multinational food aid program administered by the Food and Agriculture Organization of the United Nations. Headquarters: Rome, Italy.

Wrangler–A herder or handler of wild or range horses on ranches.

Writ of Execution–A court order which authorizes and directs the proper officer of the court (usually the sheriff) to carry into effect the judgment or decree of the court.

Wye Level–A leveling instrument having the telescope with attached spirit level supported in wyes (Ys), in which it may be rotated about its longitudinal axis (collimation axis), and from which it may be lifted and reversed, end for end.

Xylometer–A calibrated tank into which wood or any other substance can be submerged to measure its volume by the amount of water it displaces.

Yard–(1) A unit of linear measurement equal to 3 feet or 36 inches. (2) The grounds which immediately surround a dwelling. (3) An enclosed area in which stock, fowls, etc., are kept; as a chicken yard. It may be used as a suffix, as in stockyard.

Yardage–The fee charged by a transit company or stockyard for the use of its storage yards.

Yearling Weight (365 Day Adjusted Weight)–A weight taken as a yearling or long yearling and adjusted within breed and sex for age of calf and age of dam to a standard mature-cow basis of the age group.

Yield, Economic Maximum–The most that can be produced on full efficient application of technology presently known by all farmers. Assumes there are no limitations on management, materials, equipment, capital, and experience.

Yield Gap–The difference between actual farm yields and either potential farm yields or experiment station yields.

Yield Goal–In soil testing, the estimated and desired crop yield level that is set for the purpose of recommending lime an d fertilizer to reach that particular level of yield.

Yield Grade–A numerical score (with 1 being the highest yielding and 5 being the lowest yielding) given to a beef carcass. The score is based on the estimated carcass weight in boneless, closely trimmed retail cuts from the round, loin, rib, and chuck.

Yoke–(1) A variously shaped wooden frame which is placed on the necks of two oxen to work them together as a team. (2) A clamp which unites two pieces. (3) A team, as a yoke of oxen. (4) A wood or steel bar placed at the end of a tongue connected by rings and leather straps to the hames of the harness of a pair of horses. It holds up the tongue and aids in holding back the wagon going downhill.

Young Farmer Association–A formally organized association of young adults engaged in agricultural occupations and enrolled in "young farmer" educational programs. Such programs ordinarily include both group and individualized instruction. The primary purposes of the organization include providing educational activities, leadership training, community service, cooperative activities, recreation, and social activities.

Zoned Heating–A system of heating a large building such as a greenhouse or chicken house by means of controls in different zones of the building.

Zoning–Township, city, village, county, or state laws regulating land uses and development; used to implement and enforce plans to protect public health and welfare and to attain the "best use" of available land. Can be used to protect shoreland areas from overdevelopment, for preserving wildlife habitat, and reserving lands for agricultural purposes. Agricultural zoning usually pertains to crop and livestock production, while rural zoning pertains to both agricultural and nonagricultural use in unincorporated or relatively sparsely populated areas.

Part 3

Forestry and Natural Resources Management

A Horizon—The leached upper member of a soil profile; the eluvial layer.

A.U.M.—(Abbr.) Animal unit month.

Abatement—(1) Decrease in the action of, such as pollution abatement, odor abatement, and noise abatement; e.g., the storm has abated. (2) Alleviation; mitigation; suppression; termination. (3) In law, wrongful entry on land; a decrease in the legacies of a will when assets are insufficient; a voiding; a failure. (4) In medicine, a decrease in the severity of pain or symptoms.

Abiotic—The nonliving elements (factors) of the environment; i.e., soil, climate, physiography.

Ablation—(1) Wearing away of the surface of rocks or of glaciers by the kinetic energy of running or dropping water. (2) All processes whereby ice or snow wastes away.

Aborescent—Approaching the size and habit of a tree.

Aborist—A specialist in the field of tree culture.

Absorber—A kind of scrubber that physically attracts solid or liquid pollutants and thereby permits their removal.

Absorption Terrace—A ridge which slows the flow of run-off water so it can be absorbed by the soil.

Abutment—A retaining wall that holds back unstable areas and prevents land slippage.

Abutting—The joining, reaching, or touching of adjoining land. Abutting pieces of land have a common boundary.

Accelerated Erosion—Excessive wearing away of soil or rock brought about by changes in the natural cover or ground conditions, including changes due to human activity.

Acclivity—A land surface that rises from a point of reference. See Declivity.

Accretion—(1) The gradual addition of new land to old by the deposition of sediment carried by the water of a stream. (2) The process by which inorganic bodies grow larger, by the addition of fresh particles to the outside. (3) The process of illuviation of soils is usually one of the addition of minerals by accretion. (4) Increase of height, diameter, quality, and value of a tree or woods; increment; growth. (5) Increase of soil water. (6) Process of recovering land from the sea by diking and draining. See Relict.

Accretion Cutting—The thinning of a forest to allow the proper growth of the remaining trees.

Accumulating Pesticide—A chemical that can build up in animals or in the environment.

Accumulator Plant—A plant that absorbs certain elements and accumulates them in its tissues to a much higher degree than most plants; e.g., selenium accumulators.

Acid—(1) A substance containing hydrogen that dissociates to form hydrogen ions when dissolved in water (or which reacts with water to form hydronium ions). (2) A substance capable of donating protons to other substances. (3) A term applied to igneous rocks having a higher percentage of silica (66%) than orthoclase. (4) Term applied to any substance with a pH less than 7.0. See Acetic Acid.

Acid Mine Drainage—Drainage from certain mines of water containing minerals and having a low pH. The low pH is commonly caused by oxidation of iron sulfide to sulfuric acid. MIne water usually contains a high concentration of iron.

Acid Organic Material—An organic material that leaves an acid residue in the growing medium, e.g., sphagnum peat.

Acid Rain (Precipitation)—Rain or other precipitation that contains a higher than normal amount of acid. The condition is caused by raindrops absorbing substances from air pollution. Acid precipitation (acid rain) is generally considered to be harmful to the environment.

Acid Soil—A soil giving an acid reaction (precisely, below pH 7.0; practically, below pH 6.6) throughout most or all of the portion occupied by roots. A soil having a preponderance of hydrogen ions over hydroxyl ions in the soil solution. In common parlance, a "sour" soil, as opposed to a "sweet" soil.

Acid-fast—Property of not being readily decolorized by acids.

Acidulation—The process of treating a material with an acid to make it more soluble. The most common acidulation process is the treatment of phosphate rock with an acid such as sulfuric acid (H_2SO_4), nitric acid (HNO_3), or phosphoric acid (H_3PO_4).

Acre—A unit of land measure in England and the United States which is equal to 43,560 square feet, or 1/640 of a square mile, or 160 square rods, or 4,840 square yards, or 4,047 square meters. The Scottish acre is 1.26 and the Irish acre is 1.62 times as large. One acre, as used in England and the United States, equals 0.4 hectare, 0.96 feddan (Egypt), 0.31 carreau (Haiti), 1.03 cuerdas (Puerto Rico), 0.03 caballerias (Cuba), 5.00 shih mou (China), 4.08 tan (Japan), 0.84 arpent de Paris (some sections of Canada), 0.37 dessiatine (U.S.S.R).

Acre Per Animal Unit Month—The estimated number of acres necessary to provide forage for one animal unit for one month under proper use.

Acreage—An indefinite quantity of land; a collective number of acres.

ACS—Agricultural Cooperative Extension Service.

Actinomycetes—A large group of moldlike microorganisms which give off an odor characteristic of rich earth and are the significant organisms involved in the stabilization of solid wastes by composting. They are common in the soil. Selected strains are used for the production of certain antibiotics.

Action—The effects of environmental factors such as heat or light on organisms.

Action Threshold–The level of competition or plant cover that triggers control action. For example, the point at which one plant or type of plant outgrows another and the larger plant crowds out the smaller.

Actionable Fire–A conflagration (fire) started or allowed to spread in violation of law or regulation.

Actium–A plant-animal community on a rocky seashore.

Activated Charcoal–Charcoal heated with steam or carbon dioxide to 800° to 900°C (1,472° to 1,652°F), producing a porous structure. In this state, it is used to absorb gases and colloidal solids.

Activated Sludge–A process of waste treatment used to biologically degrade organic matter in a dilute water suspension. Diffusion of air at a high rate through the liquid promotes the growth of bacterial and other organisms, which, acting on the organic matter in the presence of dissolved oxygen, produce a sludge floccule.

Activator–A chemical added to a pesticide to increase its activity.

Active Agent (Active Ingredient)–That part of an insecticide, herbicide, or fungicide formula that has toxic properties to target species.

Actophilous–Refers to organisms that grow well on rocky seashores.

Actual Analysis–The composition of a material based on a laboratory chemical analysis rather than a generalized guaranteed analysis.

Acute Toxicity–The potential of a substance to cause injury or illness when given in a single dose or in multiple doses over a period of 24 hours or less. In aquatic studies, exposure to a given concentration for 96 hours or less.

Adaptability–The capability of an organism to make changes that make it more fit to its environmental conditions.

Adaptation–(1) A measure of physiological fitness of the organism to one or several conditions of its environment. (2) The structures or activities of an organism, or of one or more of its parts, which tend to fit it better for life in its environment or for particular functions. See Adjustments.

Adder–(1) The harmless North American milk snake, ***Lampropeltis triangulum***, so called because it is sometimes found in dairy buildings, where it goes in search of mice rather than milk. (2) The harmless North American puffing adder, ***Heterodan contortix***; also called blowing adder, flat-headed adder, hognose snake, spreading adder, sand viper. (3) The common venomous European snake ***Vipera berus***. (4) Any species of the venomous European snake family Viperidae. (5) The Australian venomous death adder, ***Acanthophis antarctius***. (6) The Indian banded adder or cobralike krait, ***Bunganis coeruleus***. (7) The African venomous puff adder, ***Bitus arietans*** or ***Clotbo arietans***. (8) The venomous Near Eastern horned adder, ***Cerastes cornutes***. (9) The venomous Indian asp or serpent ***Naje Haje***. (10) The sea adder, ***Spinachia spinachia***, or pipe-fish of the genus ***Nerophis***, family Syngnathidae.

Adherence–The act of sticking to a surface.

Adhesion–Molecular attraction which holds the surfaces of two unlike substances in contact, such as water and soil particles. See Cohesion.

Adjustments–(1) Range management:changes in animal numbers, seasons of use, kinds or classes of animals, or management practices as warranted by specific conditions. (2) Ecological: the processes by which an organism becomes better fitted to its environment; functional, never structural. See Adaptation.

Adjuvant–(1) Any solid or liquid added to a substance, such as a pesticide or a fertilizer, to increase its effectiveness, e.g., solvents, diluents, carriers, emulsifiers, stickers, spreaders, or sometimes a pesticide to another pesticide or a fertilizer to another fertilizer. (2) A carrier for a biological that releases the biological into the bloodstream over an extended period, thus serving the function of a series of booster shots; consequently, the adjuvant helps lengthen the period of immunity provided by the biological.

Adobe Soil–Any clayey soil of the arid and semiarid regions of the western United States Soils included under the term may vary widely in different localities. Also called dobe, dobie, dobby, doby, adaubi, adabe, dogie.

Adsorption–The increased concentration of molecules or ions at a surface, including exchangeable cations and anions on soil particles. Adsorption is an advanced way of treating wastes in which activated carbon removes organic matter from waste water. It is also used in flavor control in the milk industry.

Adulterant–An impurity not allowed by law in a food, plant, animal, fertilizer, or pesticide formulation.

Adulterate–To lower the quality of a product by mixing in another substance; as to adulterate milk by adding water to it. The federal government and most states legislate against the adulteration of food, drugs, fertilizers, and pesticides.

Advective Frost–Frost produced by cold air that has moved downslope; as in a frost pocket.

Adventive Species–Organisms which have invaded from a distance and become more-or-less naturalized.

Adverse Possession–The right of an occupant of land to acquire title against the real owner, where possession has been actual, continuous, hostile, visible, and distinct for the statutory period.

Adverse Weather–Weather not suitable for farm/ranch operations, such as making hay.

Aeolian–Refers to soil materials which are subject to wind movement or have been moved by the wind action. See Loess.

Aeolus–(Latin) God of the wind.

Aerate–(1) To force a thin layer of cooled air over milk to remove odors. (2) To cause air to pass throughout or around a substance.

Aeration–The process of being supplied or impregnated with air. Aeration is used in waste water treatment to foster biological and chemical purification.

Aeration, Soil–The process by which air in the soil is replenished by air from the atmosphere. In a well-aerated soil, the air in the soil is similar in composition to the atmosphere above the soil. Poorly aerated soils usually contain a much higher percentage of carbon dioxide and a correspondingly lower percentage of oxygen. The rate of aeration depends largely on the size, volume, and continuity of pores in the soil.

Aerial Dusting–Spreading insecticides, fungicides, or herbicides in the form of powder from an airplane. Commonly called crop dusting.

Aerial Fertilization–The broadcast distribution of fertilizers from aircraft.

Aerify–To cultivate (turfgrass) by loosening soil, removing cores, and leaving holes or cavities in the turf.

Aerobe–Bacteria or other organisms which live only in free oxygen.

Aerobic–Pertaining to organisms that grow only in the presence of oxygen, as bacteria in a properly prepared compost.

Aerobiology–The study of microorganisms carried in the air.

Aerogens–Gas-producing bacteria.

Aerology–The science of atmosphere at all elevations but sometimes limited to the upper atmosphere.

Aeroplankton–Small plants, spores, pollen grains, insects, bacteria, and small animals that are suspended in the atmosphere. See Plankton.

Aerosol–Finely atomized spray or smoke with particles ranging in size between 0.1 and 50 microns. The particles are produced by blasts of heated air, or exhaust gases, or rapid volatilization of a liquefied gas containing a nonvolatile chemical solution. Aerosols include insecticides, antibiotics, germicides, and deodorants.

Aesthetic Insults–A degradation of environmental beauty through mistreatment.

Aesthetic Value–The value or pleasure that any thing of beauty gives to humans.

Aesthetics–Evaluation of elements of the environment in relation to human perceptual qualities of sight, sound, smell, touch, tastes, and freedom of movement.

Aestival–Pertaining to summer.

Aestival Pond–A pond which contains some water throughout the open season but freezes to the bottom in winter.

Affinity–The relationship between organisms that indicates a common origin; used occasionally to denote certain similarities of plant and animal communities.

Affluent–(1) A stream that flows into another stream or lake; a tributary stream. (2) A plant that flowers abundantly.

Afforestation–The establishment of trees where they never existed before, such as on a prairie.

Aftermath–The regrowth of range or artificial pasture forage after grazing or mowing. More commonly used to refer to grazing of forage or cropped areas after harvest where there may not always be regrowth but just crop residues.

Age Stand–The average age of the trees that compose a stand. In practice, applied to even-aged stands by obtaining the average age of representative dominant trees.

Age Tree–The number of years elapsed since the germination of the seed, or the budding of the sprout or root sucker of a tree.

Agent–One who represents another from whom he has derived authority.

Agent Orange–An herbicidal mixture of 2, 4-D and 2, 4, 5-T made by Dow Chemical Company and used by the United States military in Vietnam in the 1960s to defoliate vegetation serving as hiding places for the enemy. More than 11 million gallons (41.6 million liters) were used. An estimated 220 pounds (100 kg) of dioxin, a very potent toxin, was a contaminant in the herbicide. Although now more restricted, this combination of 2, 4-D and 2, 4, 5-T is used extensively by farmers and ranchers to kill broad-leaved plants, including poison ivy. See Dioxin.

Ager–(Latin) Field.

Agglomeration–A processing step in the granulation of fertilizers. The assembling of small particles into larger, stable granules.

Aggregate–(1) To bring together; to collect or unite into a mass. (2) Composed of a mixture of substances, separable by mechanical means. (3) The mineral material, such as sand, gravel, shells, slag, or broken stone, or combinations thereof, with which cement or bituminous material is mixed to form a mortar or concrete. Fine aggregate may be considered as the material that will pass a 0.25-inch (4.76-millimeter) screen, and coarse aggregate as the material that will not pass such a screen.

Aggregate Soil–Many fine particles held in a single mass or cluster. Natural soil aggregates, such as granules, blocks, or prisms, are called peds. Clods are aggregates produced by tillage or logging.

Aggregated Subarea–Subdelineations of water resource regions; also based upon hydrologic boundaries.

Aggregation–(1) The process resulting in the grouping of organisms or soil particles either through active movement or as a result of offspring clusters about the parents.

Aging Cycle of a Lake–The life cycle of a lake, whereby it gradually fills in with sediment and plant debris (over decades or millennia, depending on the size of the lake) until it becomes a swamp and gradually upland soil.

Agitation–The process of stirring or mixing in a sprayer.

Agitator–Revolving paddles which keep a liquid, powder, or gas in motion to maintain a proper mixture within a tank. Used in dairying, spraying, fertilizer drill, etc.

Agonic Line–A line passing through points on the earth's surface at which the direction of the magnetic needle is truly north and south; a line of no magnetic declination. A line drawn on a map joining points of zero magnetic declination (magnetic variation).

Agrarian–(1) Pertaining to agriculture. (2) Pertaining to political action or movements for the benefit of farmers. See Rural Population, Urban.

Agrarian Zone–Of or pertaining to the cultivated portion of an area.

Agrestal; (Agrestial)–Uncultivated; growing wild.

Agribusiness–A term referring to the full scale of operations related to the business of agriculture. It connotes the interrelationships of farming, farm services, soil science, agronomy, land grant universities, county extension services, state and federal experiment stations, soil and water conservation services, plant and animal nutrition, plant and animal protection, transportation, finance, and marketing.

Agrichemicals–A term used to designate chemical materials used in agriculture, such as herbicides, insecticides, fungicides, and fertilizers.

Agricola–(Latin) Farmer.

Agricultural Chemistry–The branch of the science of chemistry which is concerned with the composition and transformation of the plants and animals on which the economy of the farmer rests.

Agricultural College–An educational institution devoted to study, research, and the dissemination of knowledge in agronomy, horticulture, animal husbandry, agricultural economics, etc. The U.S.D.A. works in conjunction with most agricultural colleges (sometimes called a school or college of agriculture within a university).

Agricultural Conservation Program–A program that shares costs with farmers and ranchers to encourage and assist them in carrying out their farms' conserving practices. The program is designed to: (1) restore and improve soil fertility, (2) minimize erosion caused by wind and water, and (3) conserve resources and wildlife.

Agricultural Land–All the land devoted to raising crops and livestock, including farmstead, roadways, drainage and irrigation ditches, ponds, water supply, cropland, and grazing land of every kind. (The term is not strictly synonymous with land in farms, cropland, pasture land, land suitable for crops or land suitable for farming).

Agricultural Pollution–The liquid and solid wastes from all types of farming, including runoff from pesticides, fertilizers, and feedlots; erosion and dust from plowing; animal manure and carcasses; and crop residues and debris. Old cars and trucks are a part of aesthetic agricultural pollution.

Agricultural Region–A classification of land according to its predominant feature, such as the crop most commonly grown or the type of farming employed.

Agricultural Resource Base–The soil, water, climate, and other natural resources necessary to produce a crop.

Agricultural Resources Instruction–(conservation, utilization and services) The subject matter and planned learning experiences designed to develop knowledge and skills necessary for the conservation and utilization of natural resources (such as air, forests, soil, water, fish, plants, and wildlife) for economic and recreational purposes. Instruction also emphasizes the establishment, management, and operation of lands used for recreational purposes.

Agricultural Slag–(1) A term applied to a fused silicate whose calcium and magnesium contents are capable of neutralizing soil acidity and which is sufficiently fine to react readily in the soil. (2) A low-grade agricultural lime. See Lime.

Agricultural Stabilization and Conservation Service (ASCS)–A service established by the Secretary of Agriculture on June 5, 1961, under the authority of Reorganization Plan No. 2 of 1593, in accordance with the Reorganization Act of 1949, as amended (5 U.S.C. 901-913). The service carries on the following principal programs from appropriated funds: production adjustment programs, Sugar Act program, agricultural conservation.

Agricultural Sulfur–A coarsely ground mineral which increases the acidity of soil or corrects sulfur deficiency. Also called brimstone, flowers of sulfur. See Sulfur.

Agriculture–The broad industry engaged in the production of plants and animals for food and fiber, the provision of agricultural supplies and services, and the processing, marketing, and distribution of agricultural products.

Agriculturist–A person engaged in the production of food and/or fiber; also ancillaries such as teachers of agriculture, farm editors, researchers, etc.

Agroeconomic Zones–Zones which are defined in terms of common features. For different purposes these features will differ but may involve such dimensions as climate, soil resources, land use, ethnic groupings, market access, etc.

Agroecosystem–The relatively artificial ecosystem in an agricultural field, orchard, or pasture.

Agroforestry–The practice of raising trees, forage, and livestock on the same ground, at the same time. Common associations are cattle and trees or sheep and trees.

Agrology–That branch of agricultural science dealing with the origin, analysis, and classification of soils in relation to crop production. (A word used widely in Canada but not in the United States.) See Agronomy.

Agronomist–A specialist in soil and crop sciences.

Agstone–A term widely used to denote agricultural limestone, in contrast to limestone for making cement or used as building stones. See Lime.

Air–So-called pure air is a mixture of gases containing about 78 percent nitrogen; 21 percent oxygen; less than 1 percent of carbon dioxide, argon, and other inert gases; and varying amounts of water vapor.

Air Cleaner–A device that is used for cleaning, filtering and removing dirt, dust, etc., from the air before it enters and is used by an engine.

Air Curtain–A method for mechanical containment of oil spills or water by bubbling air through a perforated pipe in front of the advancing oil or water spread. Air curtains are also used to prevent fish from entering a polluted area of water.

Air Drainage–The flow of cold air down a slope and warm air up a slope that is caused by gravity rather than the wind.

Air Plowing–Aeration by the pumping of air into the lower, oxygen-depleted layer of a water body to encourage the mixing of deep and surface waters and/or the oxidation of bottom sediments.

Air Pollution–Human-made contamination of the atmosphere, beyond that which is natural.

Air Pollution Episode–The occurrence of abnormally high concentrations of air pollutants usually due to low winds and temperature inversion, resulting in illness and sometimes death of humans and animals. See Temperature Inversion.

Air Porosity–The proportion of the bulk volume of soil that is filled with air at any given time or under a given condition, such as a specified moisture condition.

Air Quality Standard–As the federal government uses the term, the prescribed level of a pollutant in the outside air that cannot legally be exceeded during a specified time in a specified geographical area.

Air Well–In desert areas, a depression filled with rocks that condense moisture from the air. Water collects as a result of the condensation.

Air-slaked Lime–Burned limestone that has been allowed to weather. See Lime.

Airshed–A term, now little used, denoting a geographical area the whole of which, because of topography, meteorology, and climate, shares the same air.

Alba, Albidus, Albus–Terms meaning white.

Albedo–(Latin; albedo, whiteness.) The percentage of the sunlight reflected from a surface in relation to the total amount of light falling on that surface. Typical values are: snow, 55 to 80 percent; grass, 25 percent; forest, 10 to 20 percent.

Alburnum–The part of a tree trunk between the bark and the wood, consisting of the cambium and the phloem cells. The alburnum of some species of trees; e.g., the Ponderosa Pine (*Pinus ponderosa*) of the Northwest was eaten by American Indians during times of famine. See Sapwood.

Alcohol–The family name of a group of organic chemical compounds composed of carbon, hydrogen, and oxygen; a series of molecules that vary in chain length and are composed of a hydrocarbon plus a hydroxyl group, CH_2-$(CH_2)N$-OH; includes methanol, ethanol, isopropyl alcohol, and others.

Alcohol Slime Flux–A white, frothy seepage from the bark or sapwood near the base of a tree trunk. Rich in starches, sugars, and proteins. Over long periods, it is fatal to the tree. See Brown Slime Flux, Slime Flux.

Alcoholic Fermentation–The transformation of simple hexose sugars, especially glucose, into alcohol and carbon dioxide. Useful as a method of preservation.

Aldehydes–Any of a class of highly reactive organic chemical compounds obtained by oxidation of primary alcohols, characterized by the common group CHO, and used in the manufacture of resins, dyes, and organic acids.

Alder–Any tree or shrub of the genus *Alnus*, of the family Betulaceae, some species are called tag alders. The lumber from alders is used in cabinet making.

Alder Buckthorn–*Rhamnus frangula;* a shrub cultivated for its lustrous green leaves, which turn bright yellow in autumn.

Alder Flea Beetle–*Altica ambiens*, family Chrysomelidae; it feeds on the leaves of alder, willow, and poplar trees.

Alder Witches' Broom–A European fungal disease of alders. *Taphrina epiphylla*, family Taphrinaceae; it causes profuse branching.

Aldrin–A white crystalline, chlorinated hydrocarbon insecticide.

Alevin–(1) Newly hatched salmon or related fish (usually) with a yolk sac attached, before it emerges from the spawning gravel to begin swimming freely. (2) Newly hatched, incompletely developed fishes (usually salmonids) still in nest or inactive on the bottom, living off stored yolk. (3) The stage from hatching to end of dependence on yolk sac as primary source of nutrition (usually salmonids).

Alfisols–A soil order with gray to brown surface horizons, medium to high supply of bases, and B horizons of alluvial clay accumulation. These soils form mostly under forest or savanna vegetation in climates with slight to pronounced seasonal moisture deficit.

Algae–(singular, alga) Comparatively simple plants containing photosynthetic pigments. A majority are aquatic, and many are microscopic. They grow in sunlit waters. They are food for fish and small aquatic animals and, like all green plants, put oxygen in the water when carrying on photosynthesis. Although some forms of algae are necessary and desirable, excessive concentrations tend to discolor water and cause objectionable tastes and odors, severely limiting all uses of the water.

Algal "Bloom"–A term for a rapid increase in numbers of algae in a body of water, often caused by a sudden increase in available nitrogen, phosphorus, temperature, and/or sunshine. A density of 500 or more individual algae per milliliter of water is often considered a "bloom." Excess algae are not desirable because they physically reduce the use of bodies of water for swimming and boating; at night they deplete oxygen for fish and other aquatic organisms, and upon decomposition have a foul odor of hydrogen sulfides, H_2S.

Algicide–A chemical highly toxic to algae and satisfactory for application to water, commonly copper sulfate. See Copper Sulfate.

Alkali–(1) A chemical compound of oxygen and hydrogen with one element, such as sodium, potassium, calcium, magnesium, or the ammonium radical, capable of neutralizing acids. (2) A general term denoting salts of sodium, calcium, potassium, and magnesium which injure plant growth. (3) A general term for caustic soda, sodium hydroxide, caustic potash, sodium carbonate, etc., all cleansing agents in food processing plants.

Alkali Soil–A soil containing alkali salts, usually sodium carbonate (with a pH value of 8.5 and higher). The term frequently includes both alkali and saline soil as here defined. White alkali is applied to saline soil in some localities, and black alkali to alkali soils. See Alkaline Soil.

Alkaline–A chemical term referring to basic reaction where the pH reading is above 7, as distinguished from acidic reaction where the pH reading is below 7.

Alkaline Cleaners–Group of general cleaners used to clean milking equipment; includes such substances as caustic soda (lye), soda ash, baking soda, and metasilicate of soda.

Alkaline Soil–A soil that gives an alkaline reaction, precisely, a reading above pH 7.0, but in practice, readings as high as pH 7.3 may be considered neutral. In humid regions alkalinity most commonly comes from calcium and magnesium carbonates; in drier regions nitrates and salts of sodium and potassium as well as calcium and magnesium may be the source. See Acid Soil, Alkali Soil, pH.

Alkaloids–Substances found in plants, many having powerful pharmacologic action, and characterized by content of nitrogen and the property of combining with acids to form salts. Alkaloids may be beneficial (atrophine) or toxic (from *Senecio*spp).

All-aged–Applied to a stand of trees in which trees of all ages up to and including those of the felling age are found.

Allegheny Mound Ant–*Formica exsectoides;* a mound-building ant, family Formicidae; common in the northeastern United States. It kills trees, shrubs, etc., within a radius of 20 to 30 feet., that threaten to shade its mound.

Allelopathy–The harmful influence of one living plant on another living plant by the secretion of a toxic substance. Examples are juglone secreted by black walnut roots and leaves and an unidentified toxin secreted by certain species of sedges that depresses rice plants.

Allergy–An exaggerated susceptibility to a substance harmless to most members of the same species.

Allethrin–A synthetic insecticide similar to pyrenthrins.

Allochoric–Refers to a species occurring in two or more similar regional communities.

Allowable Cut–The amount of wood that can be removed from a landowner's property during a given period, without exceeding the net growth during that period on the property.

Alluvial–Pertaining to material that is transported and deposited by running water.

Alluvial Community–A boundary zone between water and land which is subject to inundation.

Alluvial Cone–A delta-form feature composed of soil and rock detritus, deposited as storm wash at the mouths of streams. Similar to alluvial fan, except that the angle of slope is greater.

Alluvial Fan–Soil and rock fragments moved by swiftly flowing water and deposited at the mouth of a stream in the form of fans or deltas. Alluvial fans are most common in arid regions at the bases of mountain slopes. Several fans may coalesce to form terraces and piedmont plains. The agricultural value of the soils in alluvial fans varies.

Alluvial Soil–Soil developed from transported and relatively recent water-deposited material (alluvium), characterized by little or no modification of the original material by soil-forming processes.

Alluvio Maris–Soil formed by the washing-up of the sea; sand or any other material adjoining land near the sea.

Alluvion–Land added by accretion, as that built up by deposition from sea, lake, or river, or that left by recession of water or by the natural shifting of the channel of a river. It always belongs to the owner of the land to which it is an accretion. The word is often erroneously used as the equivalent of alluvium.

Alluvium–Matter transported and deposited by streams; usually composed of mixed inorganic (mineral) and organic particles. The word most commonly refers to the recent deposits on the flood plains; more inclusively it embraces deposits on second bottoms or river valley terraces, with the detritus spread out as fans. See Alluvial Fan.

Alpha-mesosaprobic Zone–Area of active decomposition, partly aerobic, partly anaerobic, in a stream heavily polluted with organic wastes.

Alpine–(1) Resembling the conditions in the Alps. Implies high elevation, particularly above tree line, and cold climate. (2) A breed of dairy goats originating in Europe. They have erect ears and may range from pure black to pure white with a variety of color patterns. Colors may also include fawn, brown, gray, and red. Some have horns and some are polled.

Alum–See Aluminum Sulfate.

Aluminum–A widely distributed element, commonly found as a silicate in various clays and rocks. While aluminum may be essential to the growth of some plants, the amount required, if any, is very small. The supply in all soils is abundant. Some acid soils contain sufficient aluminum in soluble or exchangeable form to kill certain plants. See Aluminum Sulfate.

Aluminum Dross–A by-product of refining aluminum metal. It consists of aluminum oxide, aluminum nitride, salt, and various other impurities. The N is slowly available in the soil. Dross is also relatively high in magnesium, copper, manganese, and zinc contents.

Aluminum Phosphate–($AlPO_4$) Occurs in quantity in the colloidal or soft phosphates found in Florida. It also occurs in small quantities in all grades of phosphate rock. It is insoluble in water and its phosphorus content is but slightly available in the soil unless treated with an acid. Much aluminum phosphate in phosphate rock is objectionable because of the tendency of the superphosphate made from such rock to become moist and sticky.

Aluminum Sulfate–A mineral soluble in water and usually made by treating bauxite with sulfuric acid. It is sometimes applied to soils, especially in the West, to make them less alkaline. In the East, it is used to produce an acid condition for such plants as rhododendrons, azaleas, camellias, and blueberries. Also used in tanning leather, sizing paper, and purifying water.

Alunite–A hydrated sulfate of aluminum and potassium mineral. There are large deposits in Utah and other western states. The potassium is insoluble in water, but is rendered soluble by roasting. Roasted material marketed in the West averages about 6.5 percent K_2O.

Alvar–In Sweden, a vegetation type consisting of dwarf shrubs resembling steppe. See Steppe.

Amber–A fossil resin from coniferous trees.

Ambient–The prevalent surrounding environment, usually expressed as functions of temperature, pressure, and humidity.

Ambient Air–The air which surrounds an object. See Atmosphere.

Ammonia–NH_3; a chemical compound composed of 82.25 percent nitrogen and 17.75 percent hydrogen. At ordinary temperatures, it is a colorless, pungent gas about one-half as heavy as air. Liquid ammonia is used as a fertilizer. Sometimes added to livestock feed to increase nutritional value.

Ammonia Liquor–NH_4OH, aqua ammonia; ammonia and water combine readily to form ammonium hydroxide (NH_4OH). Aqua ammonia made from anhydrous, ammonia usually contains from 27 to 30 percent NH_3 (22.2 to 24.7% N). The solution is strongly alkaline and unites with strong acids to form ammonium fertilizer salts; for example, with phosphoric acid (H_3PO_4) to form ammonium phosphates; with sulfuric acid (H_2SO_4) to form ammonium sulfate; and with nitric acid (HNO_3) to form ammonium nitrate.

Ammonia Oxidation–When ammonia (NH_3) and air are mixed and passed through a platinum alloy gauze, the ammonia burns, combines with the oxygen of the air, and forms nitric oxide (NO). This oxide is converted to nitrogen dioxide (NO_2), which is absorbed in water to form nitric acid (HNO_3). This process is used commercially not only to manufacture ammonium nitrate fertilizer, but also in the chamber process for the manufacture of sulfuric acid.

Ammonia Volatilization–The loss of gaseous ammonia to the atmosphere. This occurs during the application of urea fertilizer, anhydrous fertilizer, or any ammonia salt under these conditions: (1) when applied on the soil surface, (2) when applied on alkaline soil, or (3) when applied under high humidity and high temperatures.

Ammoniated Superphosphate–When superphosphate of any grade is brought into contact with free ammonia (NH_3), the ammonia immediately reacts with any free acid and acidic phosphates present. Gypsum, when present, also enters the reactions, and thus the fertilizers, dicalcium phosphate, monoammonium phosphate, and ammonium sulfate are formed.

Ammonification–The formation of ammonia in soil by microorganisms.

Ammonium Chloride–(NH_4Cl) The ammonium salt of hydrochloric acid. Similar in fertilizer properties to ammonium sulfate, but more expensive and may acidify the soil more. Also called sal-ammoniac, muriate of ammonia.

Ammonium Nitrate–An ammonium salt with a nitrate radical, excellent as a source of nitrogen fertilizer. Made in large quantities from the atmospheric nitrogen and contains 30 to 35 percent nitrogen. Dangerous to work with in the pure state because of its explosive nature.

Ammonium Nitrate Limestone–A fertilizer mixture of ammonium nitrate and finely pulverized calcic or dolomitic limestone in a ratio of 3:2 prepared in pellet or prilled form suitable for mixing or direct application. It is not explosive and contains 20.57% N.

Ammonium Phosphate Nitrate–A fertilizer mixture of ammonium phosphate and ammonium nitrate. It is produced by ammoniating the solution separated from phosphate rock that has been acidulated with an excess of nitric acid. The average composition of the fertilizer is 27% N and 15% P_2O_5.

Ammonium Phosphate Sulfate–A double salt of ammonium phosphate and ammonium sulfate or a mixture of these two salts. When first marketed in 1923 this material was called Ammophos-B. It is guaranteed to contain 16 percent N and 20 percent available P_2O_5. It contains 45 percent ammonium sulfate. The 13-39-0 grade contains about 20 percent ammonium sulfate.

Ammonium Phosphates–Both monoammonium phosphate (MAP) and diammonium phosphate (DAP) are valuable fertilizers. See also Ammonium Polyphosphate (APP), Monoammonium Phosphate ($NH_4H_2PO_4$) (MAP). Pure monoammonium phosphate contains 12.17 percent nitrogen and 61.71 percent phosphoric oxide. Some fertilizer materials containing 12 percent N and 61 percent available P_2O_5 are on the market for the manufacture of liquid fertilizers and starter solutions. The crude salt, originally sold under the trade name Ammophos-A (11-48-0), is made by neutralizing crude phosphoric acid with ammonia and evaporating the solution.

Ammonium Polyphosphate (APP)–Any ammonium salt of a polyphosphoric acid such as triammonium pyrophosphate [$(NH_4)_3HP_2O_7$] and pentamonium tripolyphosphate [$(NH_4)_5P_3O_{10}$]. Fertilizer grades of ammonium polyphosphate generally are phosphate mixtures containing a substantial proportion of one or more ammonium polyphosphates and can be made by ammoniation of superphosphoric acid or by thermal dehydration of ammonium orthophosphate. Aqueous solutions of ammonium polyphosphates are used widely in the production of liquid fertilizers. The usual grades of these solutions are 10-34-0 and 11-37-0. A 12-44-0 is also available. Granular 15-62-0, made by ammoniating superphosphoric acid, is suitable for direct application, bulk blending, or as an intermediate in producing liquid fertilizers.

Ammonium Polysulfide (APS)–A combination fertilizer and soil acidifier containing 20% N and 36% S.

Ammonium Sulfate Nitrate–Formerly imported under the trade name saltpeter, this double salt, less hygroscopic than ammonium nitrate alone, is produced in many modern granulated fertilizers when hot solutions or slurries are mixed, or formed in processing, and dried. It contains 26% N.

Ammonium Sulfate (Sulfate of Ammonia)–[$(NH_4)2SO_4$]; a white or grayish crystalline salt made by neutralizing 30 to 50 percent sulfuric acid with ammonia in a saturator. Ammonium sulfate is ammonia in a saturator. Ammonium sulfate is an ingredient of most mixed fertilizers. It is also widely used as a separate fertilizer in the western part of the country and on rice. It is seldom used on acid soils, because of its high physiological acidity (See Acidity). Where sulfur content in the soil is low it is valuable for its sulfur content also.

Amorphous–Without form; applied to rocks and minerals having no definite crystalline structure; a plant of indeterminate shape.

Amphibian–A cold-blooded animal with legs, feet, and lungs that breathes by means of gills in the early stages and by means of lungs in the later stages of life.

Amphoteric–Capable of reacting as either an acid or a base, as casein.

Amplitude–(Ecological) The range of an environmental factor or complex of conditions in which an organism can exist or that it can tolerate.

Anabatic Wind–Local breezes caused by heating of the land surface by the sun creating warmer air currents, which rise, moving up valleys and up hills. Upon cooling, the winds move down valleys and hills. Anabatic winds are very important in creating or reducing "frost pockets," which often kill fruit buds in the spring.

Anadromous–Pertaining to fishes that leave seas and ascend freshwater streams to spawn; e.g., salmon.

Anaerobe–Organisms (usually bacteria) which live and multiply without free oxygen.

Anaerobic–(1) Living or active in the absence of free oxygen. (2) Pertaining to or induced by organisms that can live in the absence of free oxygen.

Anaerobic Bacteria–Bacteria not requiring the presence of free or dissolved oxygen for metabolism. Strict anaerobes are hindered or completely blocked by the presence of dissolved oxygen and sometimes by the presence of highly oxidized substances, such as sodium nitrates, nitrites, and, perhaps, sulfates. End-product gases include methane and hydrogen sulfide.

Anaerobic Decomposition–Reduction of the net energy level and change in chemical composition of organic matter caused by microorganisms in an anaerobic environment.

Analysis–The percentage composition of fertilizers, feeds, etc., as determined by chemical analysis, expressed in terms specified by law.

Anastomosing–(1) Of a stream; branching, interlacing, intercommunicating, thereby producing a netlike or braided appearance. (2) Netted; intervened; said of leaves marked by cross veins forming a network; sometimes the vein branches meet only at the margin.

Anchored Dune–Sand dune stabilized by growth of vegetation.

Andisols–The eleventh soil order in the United States system of soil taxonomy approved in 1989. It consists of a black mineral soil developed from volcanic ejecta including 5% glass.

Anemochore–(anemochorous) An ecological term which denotes a plant, such as milkweed, with seeds distributed by wind.

Anemometer–An instrument used to measure speed of wind; also called wind gauge. The Dines anemometer measures pressure, which is translated to velocity.

Anemoscope–Any device used to indicate and/or record wind direction.

Aneroid Barometer–A device that measures air pressure by its effect on the thin sides of a partially evacuated hollow cylinder.

Angle of Repose–The maximum slope or angle at which a material such as soil or loose rock remains stable. When exceeded, mass movement by slipping as well as by water erosion may be expected.

Angleworm–See Earthworm.

Angleworm Cast–The aboveground accumulation of soil caused by the underground activity of the earthworm.

Anhydride–An oxide that will react with water to form the corresponding acid or base; e.g., P_2O_5 is the anhydride of H_3PO_4; CaO is the anhydride of $Ca(OH)_2$.

Anhydrite–($CaSO_4$) Anhydrous calcium sulfate; used for the same purposes as land plaster or gypsum, i.e., as a soil amendment; also used in India and the United Kingdom for making sulfuric acid.

Anhydrous–A compound that does not contain water either absorbed on its surface or as water of crystallization; e.g., anhydrous ammonia.

Anhydrous Ammonia–A gas used as a chemical fertilizer containing approximately 82 percent nitrogen; normally sold as liquid under high pressure. Main advantages are low cost and high concentration, thus less handling. Changes to a gas when not under pressure, so it must be injected into soil and sealed.

Animal Sign–The evidence of an animal's presence in an area.

Animal Unit–Measurement based on the amount of feed eaten and manure produced by an average mature horse, cow, or the equivalent; as two heifers over one year, four calves under one year, seven ewes or bucks, two and one-half brood sows or boars, five hogs raised to 200 pounds each, or 100 hens.

Animal Unit Months (AUMs)–Amount of grazing required by a 1,000-pound (454 kg) cow or equivalent weight of other domestic animal for one month.

Anion–The ion solution carrying one or more negative electrical charges depending on its valence or combining power with positively charge cations; e.g., NO_3, H_2PO_4, SO_{42}. Anions and cations are always present in the liquid phase of fertilizers; an ion carrying a negative charge of electricity. See Ion.

Annual Layer–Sedimentary layer deposited or presumed to have been deposited during the course of a year; e.g., glacial varve.

Annual Precipitation–The water, usually expressed in inches of depth, deposited on the earth in one year by rain, snow, dew, hail, fog, mist, sleet, or other precipitation.

Annual ring–(1) The growth layer of a perennial plant which represents the growth of one year, as viewed on the cross-section of a stem, branch, or root. (2) The layer of xylem (wood) formed by one year's growth of cambium.

Annual Snowfall–The snow falling in one year, usually expressed in inches in the United States. For conversion to its approximate water equivalent, the depth in inches is divided by 10. (In some other countries, for example, Australia, the ratio used is 12:1. However, the ratio can vary from 5:1 to 50:1 depending on many factors).

Annulation–The ring growths visible in the cross-section of a log or around the body of a worm, etc.

Annulus–The annual mark or zone on the scales, vertebrae, or other hard portion of a fish, which is formed once each year.

Anoxia–A deficiency of oxygen.

Ant–Any insect of the family Formicidae; ants modify natural soils; certain species damage crops, and some are lawn pests.

Antagonism–The loss of activity of a chemical when exposed to another chemical.

Antarctic Zone–Lies between the Antarctic Circle (66°33' S) and the South Pole (90°0' S). Very little agriculture occurs in this zone. See Arctic Zone, Temperate Zone, Tropical Zone.

Anthoecology–The study of flowers in relation to their environment.

Anthography–Flower description.

Anthraxylon–The vitreous-appearing components of coal, which in thin section are shown to be derived from the woody tissues of plants, such as stems, limbs, branches, twigs, roots, including both wood and cortex, changed and broken up in fragments of greatly varying sizes through biological decomposition and weathering during the peat stage, and later flattened and transformed into coal through the coalification process, but still present as definite units.

Anthropic–Of or pertaining to people; similar in meaning to anthropogenic.

Anthropic Soils–Soils that have been under cultivation for a long period of time.

Anthropogenic–Changes in soils caused by action of people, such as plowing.

Anti-caking Agent–See Conditioners.

Antibacterial–Any substance that has the ability, even in dilute solutions, to destroy or inhibit the growth or reproduction of bacteria and other microorganisms; used especially in the treatment of infectious diseases of people, animals, and plants.

Antibiosis–An association between two organisms in which one harms the other.

Anticarcinogen–A substance that inhibits or eliminates the activity of a carcinogen (cancer-producing substance).

Anticline–Folded rocks that are convex upwards, with older rocks in their core. See Syncline.

Anticyclone–A large, relatively high-pressure, relatively cool air mass with winds which move downward and outward in a clockwise direction in the Northern Hemisphere and counterclockwise in the Southern Hemisphere. Fair weather is usually associated with anticyclones. See Cyclone.

Antidote Statement–A required statement on the containers of chemicals explaining methods that may be used to counteract the effects of the chemical.

Antidune–A transient form of ripple on a stream bed analogous to that of a sand dune. An antidune progressively moves upstream.

Antifoaming Agents–Substances that reduce foaming caused by the presence of dissolved proteins or other substances.

APA–Available phosphoric anhydride (P_2O_5, frequently called available phosphoric acid.

Apatite–$(CaF)Ca_4(PO_4)_3$; phosphate of lime with fluorine; present as a mineral in very small amounts, but it is the primary source of phosphate deposits that are mined for use as fertilizers.

APHIS–Animal and Plant Health Inspection Service.

Appulsion–The tearing away by storm waters and wind of river banks, which are then carried downstream.

Aqua Ammonia–A solution of ammonia dissolved in water with a usual nitrogen concentration of 20 to 21 percent; used as a fertilizer for direct injection into the soil.

Aqua Humus–A fertilizer concentration of water-dispersible (about 60 percent) humic and fulvic acid derivatives. Aqua Humus 60 percent Concentrate is used as a base in preparing various liquid and dry blends of nitrogen, phosphorus, and potassium. Aqua Humus has a grade of 12-9-6 and is a specialty fertilizer which contains 36 percent water-dispersible humic and fulvic acid derivatives.

Aquaculture–Underwater agriculture, commonly called fish farming, that includes ordinarily the raising of water animals such as fish and shrimp, but which also includes the growing of water vegetation such as kelp. Mariculture is a broad category within the field of aquaculture, but is limited to the raising of marine animals in saltwater rather than in fresh water. See Pisciculture.

Aquada–A sink hole in limestone areas that holds water during the rainy season.

Aquagium–A canal, ditch, or watercourse which runs through marshy ground.

Aquatic–Living in water.

Aqueous–(1) Pertaining to water. (2) Pertaining to sediment deposited by water. (3) Ammonia in water; also called ammonium hydroxide (NH_4OH) solution. A low-pressure solution, about 20 to 21 percent N, used as a fertilizer.

Aquiclude–A geologic formation that will not transmit water fast enough to furnish an appreciable supply for a well or spring. See Aquifer.

Aquifer–A geologic formation or structure that transmits water in sufficient quantity to supply the needs for a water development, such as a well. The term ***water-bearing*** is sometimes used synonymously with aquifer when a stratum furnishes water for a specific use. Aquifers are usually saturated sands, gravel, fractured rock, or cavernous and vesicular rock.

Aquifuge–A rock which contains no interconnected openings or interstices and therefore neither absorbs nor transmits water.

Arable Land–Land so located that production of cultivated crops is economical and practical.

Arbor–(Latin, tree) (1) A bower; small structure of latticework to support vines and provide a shady retreat. (2) A tree, as distinguished from a shrub. (3) A wooden platform used for sun-drying fibers. (4) The central or supporting beam or rod, as the arbor bolt on a disk gang harrow.

Arbor Day–In the United States, a ceremonial occasion for planting trees and shrubs; usually in early May, but varying by state.

Arboraceous (Arboreous)–From Latin arbor; treelike, or pertaining to trees.

Arboreal–Living in, on, or among trees.

Arboretum–A garden where trees and shrubs are grown for study and exhibition.

Arboriculture–Cultivation of woody plants, particularly those used for decoration and shade.

Arcifinious–Having natural boundaries, such as a county bounded by a river.

Arctic Alpine–Term used for plants of arctic and alpine distribution but found only south of the Arctic zone.

Arctic Zone—The region within the Arctic Circle (66°30' N) and the North Pole (90° N). Little agriculture occurs in the Arctic zone. See Antarctic zone, Temperate Zone; Tropical zone.

Area Reclaim—An area difficult to reclaim after the removal of soil for construction and other uses. Revegetation and erosion control are extremely difficult.

Area Source of Pollution—In air and water pollution, a large and diffuse source of emission of pollution, such as a feedlot, in contrast to a point source.

Arenaceous—Applied to rocks that have been derived from sand or that contain sand.

Arenarius—Of sandy places.

Areole—(1) A small area, especially the open space between anastomosing veins. (2) A spine-bearing sunken or raised spot on the stem in cacti.

Argentine Ant—*Iridomyrmex humilis*, family Formicidae; a household and orchard insect. It also attacks shade trees, bees, their broods, and hives. Found in the southeastern United States and California.

Argillaceous—Applied to all rocks or substances composed of clay minerals, or having a notable proportion of clay in their composition, as shale, slate, etc. Argillaceous rocks are readily distinguished by the peculiar, "earthy" odor when dampened.

Argillic Horizon—The soil horizon where clay accumulates, usually the B horizon. See Soil Horizon.

Arid Climate—A dry desert, or semidesert climate with only sparse vegetation. Precipitation varies, with an upper limit for cool regions of 10 inches or less and for tropical regions of as much as 15 or 20 inches. (The Thornthwaite precipitation effectiveness [P-E] index ranges between 0 and 16.) See Precipitation-effectiveness.

Aridisols—A soil order with pedogenic horizons, low in organic matter, that are never moist as long as three consecutive months. They have an ochric epipedon that is normally soft when dry or that has distinct structure. In addition, they have one or more of the following diagnostic horizons: argillic, natric, cambric, calcic, petrocalcic, gypsic or salic, or duripan. See Soil Orders, Soil Taxonomy.

Aridity—The state of a region in respect to its dryness or lack of moisture. The amount of rainfall is not a sure index, for the aridity of a region depends in part on temperature, winds, and relative humidity.

Army Ants—Ants which group together to find and capture prey.

Arroyo—An incised channel; usually dry, occurring in alluvial slopes at the bases of mountains; found in arid regions of the southwestern United States; also spelled arroya, aroya.

Arsenic Poisoning—A condition which results from ingestion or absorption through the skin of an arsenical; acute cases are characterized by trembling, weakness, prostration, and diarrhea.

Arsenicals—Pesticides containing arsenic.

Artesian—Refers to ground water under sufficient hydrostatic head to rise above the aquifer containing it.

Artesian Well—Generally refers to a flowing well that does not have to be pumped. Actually, an artesian well is any well in which the water rises above the top of the aquifer but not necessarily above the surface of the ground. The terms ***artesian aquifer*** and ***confined aquifer*** mean the same.

Artificial Revegetation—The establishment of vegetation by mechanical methods.

Artificial Soil—A mixture of materials designed for container growing of vegetable, flower, and woody plants. Artificial soils are lightweight, take water easily, drain rapidly, and do not shrink away from the sides of the container; less likely than garden soils to harbor diseases.

Artificial Substrate—A device placed in the water for a period extending to a few weeks that provides living spaces for a multiplicity of drifting and natural-born organisms that would not otherwise be at the particular spot because of limiting physical habitat. Examples include tiles, bricks, wooden shingles, concrete blocks, and brush piles.

Ascomycetes—A group of fungi producing a sacklike ascus in which ascospores are borne. This group contains some of the most destructive fungi, but few cause wood decay; e.g., chestnut blight, nectria canker, larch canker, needle cast, and blights of conifers.

Aseptic—Being free from infectious microorganisms.

Atmosphere—The gaseous envelope surrounding the earth. It consists of 78 percent nitrogen, 21 percent oxygen, 0.9 percent argon, 0.03 percent carbon dioxide.

Atmospheric Drought—Transpiration from a plant in excess of the plant's ability to obtain moisture from the soil; e.g., a plant may wilt under hot dry winds, although there is moisture in the soil sufficient under normal conditions to supply the needs of the plant.

Atom—The smallest unit of an element to retain the chemical characteristics of that element. It consists of negatively charged particles called electrons orbiting around the nucleus and, within the small mass of the nucleus, other particles—protons (which are positively charged and balance the extranuclear electrons) and neutrons (which have no charge). It is the number and arrangement of the atom's electrons and protons that make one element differ from another; the number of neutrons distinguishes one isotope from another of the same element.

Atomic Energy—Energy released in nuclear reactions. Of particular interest is the energy released when a neutron splits an atom's nucleus into smaller pieces (fission) or when two nuclei are joined together at hundreds of millions of degrees of heat (fusion). Atomic energy is really a popular misnomer; it is more correctly called nuclear energy.

Atomic Weight—The average relative weight of an atom of an element as compared with another element (usually oxygen) that is taken as a standard. Isotopes of the same element have different atomic weights because the number of neutrons differs from isotope to isotope; a quoted atomic weight is the average weight unless a specific isotope is named.

Atomization—The process of breaking a liquid into a fine spray.

Atomize—To reduce a liquid into very fine particles in a sprayer.

Atropine–A poisonous, crystalline alkaloid used in medicine; a specific antidote for poisoning by organic phosphate insecticides.

Attapulgite Clay–A fuller's earth. The main constituent is a hydrous magnesium aluminum silicate having high adsorptive and swelling properties; used in fertilizer production, including conditioning of fertilizer products, and as a suspending agent in suspension fertilizers.

Atypical–Disagreeing with the form, state, or situation usually found under similar circumstances; not typical.

Aureus (Australia)–Golden.

Austral–Southern.

Austrlian Cockroach–*Periplaneta australasiae*, family Blattidae; an insect found throughout the world, particularly in southern United States.

Autarky–People and animals eating only what is grown locally.

Autocidal Control–A technique that is used primarily for insect control. It involves the rearing and release of insects that are sterile or altered genetically in order to suppress members of their own species that are causing pest problems. A type of autocidal control is the sterile male method whereby large numbers of male insects are artificially sterilized by irradiation or chemical sterilants. This method is most popularly exemplified in the eradication of the screwworm fly in the Southwest and parts of Mexico.

Autoclave–Vessel in which high temperatures can be reached by using high pressure; e.g., the domestic pressure cooker.

Autoparasitism–A parasite growing on a parasite; also called secondary parasitism or superparasitism.

Autotrophic Nutrition–Also known as lithotrophic nutrition; the ability of an organism to manufacture its own food from inorganic sources, using CO_2 as the sole carbon source. Autotrophs are of two general types: photoautotrophs (photolithotrophs) whose energy is derived from sunlight, and chemoautotrophs or chemolithotrophs, which obtain energy for growth and reproduction from oxidation of inorganic materials. Important among the photoautotrophs are all higher plants, algae, and a few genera of bacteria. Principal chemoautotrophs are all higher plants, algae, and a few genera of bacteria. Principal chemoautotrophs of vital concern in food and agriculture (with their substrate) are: Nitrobacter (nitrite), Nitrosomonas (ammonium), and Theobacillus (sulfur).

Autumn–(1) Fall; the season between summer and winter. (2) (Phenological) A cooling period usually following maxiumum development and activity of the majority of plants and animals, during which maturity, dormancy, and/or death occur. Does not necessarily coincide with the calendar autumn.

Available Soil Water–The amount of water retained in a soil between field capacity and the permanent wilting percentage.

Avalanche–A large mass of snow or ice, sometimes accompanied by other material, moving rapidly down a mountain slope. Avalanches are usually classified by the type of snow involved as climax, combination, damp snow, delayed action, direct action, dry snow, hangfire, and windslab avalanche.

Avalanche Cone–Material such as rocks, snow, ice, and trees deposited at the base of the path of an avalanche.

Avicide–A substance used to kill birds.

Aviculture–The caring for and raising of birds.

Azotobacter–A genus of free-living nonsymbiotic, aerobic, motile, and oval bacteria occurring in soils. They are capable of fixing atmospheric N_2 and belong to the family Azotobacteraceae.

B Horizon–See Horizon, Soil.

Bacillus–A genus of single-celled, rod-shaped organisms, family Bacillaceae. Although most species are harmless, they do cause chemical changes in animal and vegetable matter.

Bacillus thuringiensis–A bacterial insecticide effective against the army worm, cabbage worm, and the gypsy moth; not injurious to people, animals, or plants.

Back Cut–In the process of felling a tree, the final cut, made on the opposite side of the tree from the face cut (undercut).

Back Furrow–A ridge of soil made when a turning plow throws soil on top of a previously plowed ridge made while moving in the opposite direction. This is the usual method of starting to plow a field with a one-way turning plow.

Back Slope–See Cut Slope

Backburn–A fire set deliberately to burn against the wind as a means of controlling advancing grass or forest fire. See Backfire.

Backfire–Controlled fire set ahead of a forest fire to create a firebreak by reducing fuel in the path of the main fire. See Backburn.

Background Level of Air Pollution–The amounts of pollutants present in the ambient air due to natural causes.

Bacteria–Single-celled microorganisms; some cause human, animal, or plant diseases; others are beneficial.

Bacteria, Facultative–Bacteria capable of adapting to an aerobic or anaerobic environment.

Bacterial Neutralization of Pesticides–The use of genetically altered bacteria to neutralize and render safe such chemicals as parathion, diazinon, and malathion. The bacteria break the pesticides into simpler, less toxic forms.

Bactericide–Anything which destroys bacteria; a germicide.

Badger–A nocturnal carnivorous mammal; *Taxidea taxus*, family Mustelidae; has a black and gray coat, with a white streak running from the muzzle to a point behind the ears, short legs, a heavy body, and a round back; native to North America.

Badlands–A region nearly devoid of vegetation where erosion has cut the land into an intricate maze of narrow ravines and sharp crests and pinnacles.

Baffle–The planting and growing of plants to make a wall or partition.

Baffles–Pieces of material, usually wood or plastic, which are placed in an air vent to prevent direct draft without impeding the air flow.

Bag Limit–The number of game animals allowed each hunter under the law.

Bait–A food lure used to capture or destroy birds, beasts, fish, or insects.

Bait Insecticide–An appetizing mixture of poison and food which, when eaten by an insect, results in its death.

Bait Shyness–The tendency for rodents, birds, or other pests to avoid a poisoned bait.

Bald–(1) A small treeless area in a forested hilly or mountainous region, especially a crest, as bald knob. (2) Beardless, as bald wheat.

Bald Eagle–*Haliaeetus luecocephalus*; a large eagle native to North America. The mature birds have a white tail and white head that gives the appearance of being bald. Once considered destructive because of the lambs and chickens they ate, they are now protected and considered beneficial because of the large number of rodents they eat.

Baldcypress–*Taxodium distichum*; a deciduous swamp tree, family Pinaceae; very durable in water or moist conditions; grown as an ornamental and used in outdoor construction.

Balk–(1) A ridge of untilled land missed in plowing. (2) The act of an animal stopping and refusing to move.

Balling–(1) The clustering of bees tightly around a queen bee, usually in an attempt to kill her. (2) A measurement of sugar percentage in simple syrup at fixed temperature (20° C); also called brix.

Banco–A bank or deposit of sand or silt exposed due to shifts in location of river channels; a term used in the southwestern United States.

Band Day–Range forage required by a band of sheep of a given size and class during one day.

Banking–The mounding of earth or soil to preserve moisture or coolness or to cover the crowns of cultivated plants.

Barium Carbonate–A solid material used as a rat poison and as a water purifier; a mineral known as witherite in its natural state.

Barium Fluosilicate–$BaSiF_6$; a compound used as an insecticide and a rodent poison.

Barking Squirrel–A prairie dog.

Barn Owl–*Tyto alba*, family Tytonidae; an owl that frequents barns and other such buildings and feeds on rodents, etc.

Barn Swallow–A migratory bird; *Hirundo rustica*, family Hirundidae; native to North America; beneficial to farms and ranches because it builds its nest in barns and eats many flying insects.

Barometer–An instrument used for measuring atmospheric pressure.

Barometric Elevation–In surveying, an elevation above mean sea level established by the use of instruments to measure the difference in the air pressure between the point in question and some reference base of known value whose elevation is based on a more precise type of data.

Barren–(1) Denoting an area unable to support vegetation. (2) Incapable of producing offspring, seed or fruit.

Barrens–An area relatively free of vegetation in comparison with adjacent areas, because of adverse soil or climatic conditions, wind, or other adverse environmental factors; e.g., sand barrens or rock barrens.

Basal Area–(1) The area in square feet of the cross-section at breast height of a single tree or of all trees in a stand, usually expressed as square feet per acre. (2) The area or proportion of soil surface occupied by the stems and root crowns of range plants; sometimes called basal cover.

Basalt–(1) An extrusive rock composed primarily of calcic plagioclase and pyroxene, with or without olivine. (2) Any fine-grained, dark-colored igneous rock. Soils developed from basalt are usually dark-colored and productive.

Base–(1) In chemistry, a substance which reacts with an acid to form a salt; a substance which gives off hydroxyl ions when dissolved in water, as contrasted with an acid, which gives off hydrogen ions. (2) In botany, the part of a leaf or branch attached to a stem or trunk. (3) The portion of a machine on which the operating parts are mounted.

Base Exchange–(1) In soil science, denoting a physical-chemical process in which cations adsorbing by soils can be replaced by other cations in chemically equivalent quantities. (2) A term used in medical science usually in relation to the acid-base exchanges of the blood necessary to keep the blood and tissues at a physiologic constant.

Base Exchange Capacity–(1) In soil science, the cations that a soil can absorb; expressed in milli-equivalent per 100 grams of soil. (2) In medical science, the maximum capacity of the blood to exchange base (alkali) for acid in maintaining a physiologic balance.

Base Flow–That portion of the water flowing in a stream that is due to ground water seeping into the ground.

Base Level–The point at which erosion by running water is halted, e.g., sea level.

Base Line–East and west survey lines located at intervals of 24 miles, which intersect at right angles the north and south principal meridian.

Basic–On the basic side of neutral pH (above 7.0).

Basic Slag–A finely ground by-product of steel mills containing 12 to 25 percent P_2O_5, 40 to 50 percent CaO, and 5 to 15 percent SiO_2. It has some value as fertilizer because of the total phosphorus content.

Basin–(1) A hollow for holding water, as around a tree, made by forming a ridge or levee of earth on all sides. (2) The depression in a pome fruit such as apple and pear, in which calyx lobes are located; also called eye. (3) An extensive depressed area into which the adjacent land drains with no surface outlet; use of the term is almost wholly confined to the arid West. (4) The drainage or catchment area of a stream or lake.

Basin Irrigation—An efficient system of irrigation in which a field or orchard is divided into basins which are filled with water.

Basswood—*Tilia americana*; a deciduous tree commonly grown as a shade or street tree or as a bee plant. The wood is used for making veneer core for plywood, slack-cooperage, excelsior, boxes, apiary supplies, piano keys, etc. Also known as American linden, lime, whitewood, linwood, American basswood.

Bat—Any of the winged, flying mammals of the order Chiroptera. The animals are nocturnal and are very beneficial to agriculture because of the large number of insects they devour.

Bat Guano (Bat Manure)—Bat excreta used for fertilizer.

Bauxite—The principal ore of aluminum.

Bay—(1) A horse with a body color ranging from tan to red to reddish brown. The mane and tail are black, and the lower legs are usually black. (2) The laurel or its berries. (3) A shallow swamp supporting dense tree and shrub vegetation, and containing peat or muck soil. See Pocosin. (4) A coastal inlet. (5) A compartment in a barn for storage or for a special use, as a horse barn.

Bayou—(1) A secondary creek or water course which is often sluggish with muddy water. (2) A channel of a creek or river closed or partially closed at both ends. (3) An abandoned channel of meandering rivers, especially the lower course of the Mississippi River. (4) A narrow navigable channel connecting two rivers or lakes.

Bearing—The direction of a line with reference to the cardinal points of a compass; true bearing is the horizontal angle between a ground line and a geographic meridian. A bearing may be referred to either the south or north point; magnetic bearing is the horizontal angle between a ground line and the magnetic meridian. A magnetic bearing differs from a true bearing by the exact angle of magnetic declination of the locality.

Beaufort Wind Scale—A system of estimating wind velocities, originally based (1806) by its inventor, Admiral Sir Francis Beaufort of the British Navy, on the effects of various wind speeds on the amount of canvas that a full-rigged frigate of the early nineteenth century could carry; since modified and widely used in international meteorology.

Beaver—A large semi-aquatic rodent of the genus *Castor*. They have a rich fur that was once in demand for making hats. They can be a pest because they dam up streams and cause flooding.

Bedding—(1) Plowing or otherwise elevating surface soil into a series of crowned parallel beds or lands with shallow surface drains separating them. (2) A method of surface drainage consisting of narrow plowed lands in which dead furrows run parallel to the prevailing slope; the land between adjacent dead furrows is called a bed. (3) Straw, leaves, sawdust, sand, peat moss, etc., to make a bed for an animal. It is used where an animal may lie to absorb its urine. (4) Flowers that are appropriate for growing in flower beds for a massed decoration effect.

Bedding Planes—Fine stratifications, less than 0.2 inches (5 mm) thick, in unconsolidated alluvial, eolian, lacustrine, or marine sediments.

Bedload—The sediment that moves by sliding, rolling, or bounding on or near the streambed; sediment moved mainly by gravitational forces, or both, but at velocities less than the surrounding flow.

Bedrock—Unweathered hard rock that lies directly beneath the soil layers or beneath superficial geological deposits, such as glacial drift.

Bee—(1) Any insect of the families Prospidae, Colletidae, Megachillidae, Xylocopidae, Ceratinidae, Nomadidae, Andrenidae, Anthophoridae, Bomidae, or Apidae. They are generally beneficial, being insect parasites, predators, pollenizers, or producers of honey and beeswax. (2) See Honeybee, Drone, Worker, Queen Bee, Bumblebee. (3) A social gathering of neighbors, especially in rural areas, to husk corn, to raise a barn, or to quilt in the home, etc.

Bee Gum—Usually a hollow log used as a hive. The term is also used to refer to any beehive.

Bee Martin—Any birds of the genus *Tyrannus* which catch and kill bees in large numbers; also called kingbird.

Bee Pasture—Vegetation, attractive to bees, which is within flying distance of a hive; also called bee pasturage.

Bee Tree—(1) A hollow tree which is occupied by a colony of bees. (2) A tree whose flowers are a source of nectar or pollen.

Bee Wolf—A solitary wasp, *Philanthus triangulum*, which captures honeybees and stores its nest with them as food for its larvae. Where plentiful, it can cause beekeepers considerable trouble. Found in Europe; related species have been observed in the United States preying on native bees but not on honeybees.

Beech—Any tree of the genus *Fagus*, family Fagaceae; lumber from the tree is valued for its density and hardness.

Beehive—A domicile prepared for a colony of honeybees.

Belt—(1) A geographical term denoting an area of similar soil and climate particularly suited to specific crops or animals, such as the wheat belt. (2) A band of hair or skin of another color, often white, around an animal's body, as in Dutch belted cattle. (3) In machinery, a broad flexible band passed around two wheels to transmit motion from one to the other.

Bench Terrace—A steplike embankment of earth with a flat top and a steep or vertical downhill face constructed along land contours to control runoff and erosion.

Benchlands—(1) Terraces or shelflike land features representing former water levels or shorelines of lakes, rivers, or seas; usually composed of alluvium or unconsolidated coarse sediments. (2) A foothill below a mountain.

Benchmark—A relatively permanent object, natural or artificial, bearing a marked point whose elevation above or below an adapted datum (such as sea level) is known. The usual designation is B.M. or P.B.M. (permanent benchmark). A temporary or supplemental benchmark (T.B.M.) is less permanent, and the elevation may be less precise.

Benthic Region—The bottom of a body of water.

Bentonite–A porous rock of clay minerals derived from weathered volcanic ash or tuff. It is used: (a) as a dust diluent and carrier for insecticides; (b) in clarifying or refining wines, fruit juices, etc.; (c) as a wetting agent; and (d) as a seepage retardant in the bottom of a water storage reservoir.

Benzene–C_6H_6; a volatile and inflammable hydrocarbon which is a derivative of coal tar and is used: (a) in degreasing bones, etc., in fertilizer manufacture, and (b) for destroying larvae. It is particularly effective against the screwworm fly.

Berm–(1) A horizontal mound of earth erected across a slope to divert water and reduce erosion, sometimes made more effective with a ditch constructed along the upper side. In this meaning, a berm is similar to a terrace. (2) The shoulder of a highway. (3) The ledge on a beach. (4) On the shore of an ocean, the horizontal areas parallel to the edge of the ocean built by the different levels of high tide. (5) The edge of a drainage ditch. (6) The area from the upper edge of a drainage ditch to the edge of the spoil bank.

BGR–"Big Gam Repellent," a putrefied-egg product originally developed by Weyerhaeuser Company scientists; now sold under various trade names as a repellent to animal browse.

Biltmore Stick–A rule graduated so that the height and diameter of a standing tree may be calculated; used to estimate board feet in a tree.

Binomial–By international agreement, all plants and animals have two Latin names: genus and species.

Bioaccumulation–The process by which plants and animals accumulate substances, especially pollutants, that may not be injurious to that organism but may injure other organisms that eat them. For example, nitrates may accumulate in corn and oats and be injurious to animals feeding on them but not to the plants themselves. In a like manner, fish may accumulate DDT or PCB in their fat, which may be toxic to an animal eating the fish but not to the fish themselves.

Bioassay–The quantitative or qualitative determination of herbicides by the use of sensitive indicator plants or other organisms.

Biochemical Oxygen Demand (BOD)–The amount of oxygen required by the biological population of a water sample to oxidize the organic matter in that water. It is usually determined over a five-day period under standardized laboratory conditions and hence may not represent actual field conditions. Also known as Biological Oxygen Demand.

Bioconcentration–The increase in the concentration of a chemical within organisms as it moves up the food chain.

Biodegradable–Substances capable of being degraded into their constituent elements. This term is used especially in reference to toxins, such as pesticides, being degraded to nontoxins.

Bioecology–A collective term that includes plant and animal ecology as one discipline. See Ecology.

Biogenesis–(1) Formation by the action of organisms. (2) The doctrine that all life has been derived from previously living organisms.

Biogeography–The study of geographical distribution of plants and animals and the reasons for their distribution.

Biological–Products derived from a living process or living matter, such as sera, vaccines, bacterins, and antitoxins, etc. Also called biologics.

Biological Community–All of the living things in a given area or environment.

Biological Control–A method of pest control by the use of predatory insects, fungi, or viruses; as contrasted to control by chemical pesticides. See Biophage.

Biological Environmental Indicator–(1) A particular type of plant or animal whose presence in a certain location or situation is a fairly certain sign or symptom that particular environmental conditions are also present. For example, the presence of *Escherichia coli* bacteria in water is used to indicate probable pollution by human fecal material. (2) Index species—a plant or animal species so highly adapted to a particular kind of environment that its mere presence is sufficient indication that specific conditions are also present. Elm, ash, and basswood, for example, indicate a productive soil.

Biological Erosion–Erosion by water or wind as a result of soil being exposed by the burrowing of rodents, destruction of vegetation by insects, etc.

Biological Growth Potential–The average net growth attainable in a fully stocked natural forest stand.

Biological Lag–The time period necessary for a changed production decision to influence market supplies, owing to the biological nature of agricultural products.

Biological Magnification–The concentration of certain substances up a food chain. A very important mechanism in concentrating pesticides and heavy metals in organisms such as fish.

Biological Mineralization–The conversion of an element occurring in organic compounds to the inorganic form through biological decomposition.

Biological Potential–The amount of living matter potentially producible by the unit being discussed without being fertilized or irrigated.

Biologist–A person who studies living organisms as a career.

Biology–The field of study dealing with living organisms. It may be divided into the study of plants (botany) and of animals (zoology).

Biomass–The amount of matter of biological origin in a given area; e.g., the living and decaying matter in the soil as opposed to the inorganic mineral components such as sand, silt, and clay.

Biome–A term derived form the Greek, *bios*, meaning relation to life; used in ecology to include major life in the area, such as tundra biome, tropical rainforest biome, and grassland biome.

Biometry–The application of statistical methods to the study of biological problems.

Bionomics–The study of relations among organisms and the relationships between them and their environment. See Ecology.

Biophages–Organisms that obtain nourishment from other organisms, e.g., predators, parasites, and pathogens. See Biological Control.

Biopsy–The microscopic or chemical analysis of tissue removed from a living body, usually to discover the cause of illness.

Biosphere–(1) The zone on the earth's surface where all life exists. (2) All living organisms of the earth.

Biostress–Difficulties that plants and animals have in obtaining the necessities of life: food, water, and living space.

Biota–All species of plants and animals occurring in a specified area.

Biotechnology–Technology concerning the application of biological and engineering techniques to microorganisms, plants, and animals, sometimes used in the narrower sense of genetic engineering.

Biotic–Pertaining to life; biological.

Biotic Community–All of the plant and animal populations occupying a given area, usually named after the dominant plants or animals in the area.

Biotic Factors–Factors of a biological nature such as availability of food and water, competition among species, or predator-prey relationships.

Biotic Influence–Biological influences on plant and animal life as contrasted to climatic influences.

Biotic Potential–(1) The maximum reproduction power or ability. The inherent ability of an organism to reproduce and survive in greater numbers. (2) The ability of an organism to reproduce in an optimum, unrestricted, and noncompetitive environment.

Biotype–Groups of plants or animals primarily distinguishable on the basis of interaction with relatively genetically stable varieties or clones of host plants; a strain of plant or animal species.

Bipyridyliums–A group of synthetic organic pesticides that includes the herbicide paraquat.

Birch–Any tree of the genus *Betula*, family Betulaceae; species are valuable as timber having a hard, close-grain wood; also used for ornamentals.

Bird–(1) Any member of the class Avis of the phylum Vertebrata, the body of which is covered with feathers. (2) Domesticated birds: chickens, turkeys, ducks, geese, guineas, etc.

Bird Guano–See Guano.

Bird Peck–A small hole or distortion in the grain of wood resulting from sapsuckers attacking a tree; may be associated with discoloration, such as mineral stain.

Bird's-eye Maple–The lumber of the sugar maple having natural grain markings resembling eyes.

Bittern–A concentrated solution of salts which remains after the common table salt has been refined out. Sometimes used as an herbicide.

Bitternut Hickory–*Carya cordiformis*; a deciduous tree, family Juglandaceae; grown for shade, and for its wood, which is used in making tool handles, ladders, furniture, sporting goods, implements, woodenware, and for fuel and smoking meat.

Biuret–A compound, toxic to some crops, formed by thermal decomposition of urea; to be avoided in the manufacture of fertilizer urea. See Urea.

Black Alder–See European Alder.

Black Alkali–An obsolete term meaning a soil high in exchangeable sodium. The modern term is *sodic soil*.

Black Ash–*Fraxinus nigra*; a deciduous tree, family Oleaceae, which grows in a swampy, moist soil; used in cabinet making, and basket making; found in eastern North America; also called brown ash, hoop ash, basket ash, swamp ash.

Black Blizzards–Clouds of dust consisting of organic matter and the finer particles of soil from bare fields and grazing land denuded of vegetation; prevalent in the southwestern United States.

Black Carpenter Ant–*Camponotus herculeanus pennsylvanicus*, family Formicidae; the largest American common ant; attacks tree stumps, tree trunks, telephone poles, windowsills, and old timbers; found in the eastern United States.

Black Cherry–*Prunus serotina*, a deciduous tree, family Rosaceae; an important timber tree used in making furniture; the leaves are poisonous to livestock; also called wild black cherry, black choke.

Black Crappie–A sunfish, *Pomoxis nigro-maculatus*, that is used for stocking lakes in the eastern United States. Also known as Calico Bass. See White Crappie.

Black Fallow–A field left bare (no crop) to accumulate enough soil moisture from precipitation to grow a crop.

Black Frost–A blackening of vegetation which occurs when the temperature drops low enough below the freezing point to destroy vegetation, including staple crops; there is no actual deposit of frost crystals.

Black Mangrove–*Avicennia marina*; an evergreen tree, family Verbenaceae, whose flower is a good source of honey in Florida.

Black Maple–*Acer nigrum*; a hardy deciduous tree, family Aceraceae; one of the American sugar maples. Grown in highway plantings and in city yards. Also used in hardwood distillation and for railroad ties, veneer (especially good for birdseye), shoe lasts, bowling pins, fuel, furniture, flooring, and boxes. Also called black sugar maple.

Black Oak–*Quercus velutina*; a hardy, columnar, deciduous tree, family Fagaceae; the lumber is used for railroad ties, cooperage, fence posts, mine timbers, pilings, veneers, firewood, flooring, mill products, and furniture. Native to North America. Also called tanbark oak, quercitron, yellowbark oak.

Black Snake–A harmless, common snake, which is black on the dorsal side and light on the ventral; may exceed five feet in length; in North America, this name is applied to *Coluber constrictor*; in Jamaica to *Ocyophis ater*; in Australia, to *Pseudechis porphyriaceus*, *Hoplocephalus superbus*, and *Notechis scutalus*.

Black Sugar Maple–See Black Maple.

Black Swallowtail–*Papilio palyxenes asterius*, family Papilionideae; a butterfly whose larvae feed on celery, carrots, dill, parsley, carroway, parsnips, and related plants. Found east of the Rocky Mountains. Also called celeryworm, parsleyworm.

Black Tupelo–*Nyssa sylvatica*; a deciduous tree, family Nyssaceae; grown for its fine, red autumnal foliage. Used in making veneer, pulp, cabinet work, boxes, etc. Native to North America. Also called black gum (forestry name), sour gum, pepperidge, tupelo.

Black Turpentine Beetle–*Dendroctonus terebrans*; an insect pest of pine trees. The larvae feed on the inner bark of the trees, causing the trees to die.

Black Walnut–See Eastern Black Walnut.

Blackland–Generally, areas having dark-brown or black surface soils. Specifically, the prairies of central Texas, the soils of which are black, waxy, clayey. See Vertisols.

Blade–(1) The expanded portion of a leaf. (2) A cutting tool, as the blade of an axe or the blade of a bulldozer. (3) A retail cut of meat from the forequarter or shoulder. (4) The hind part of a fowl's single comb.

Blank–(1) The rough, sawed, or split wood pieces from which a finished product is made, such as handles, chair rounds, patterns, etc. (2) An unstocked area in a tree plantation where few or no trees are growing. (3) An unassembled box or package.

Blanket Algae–A mass of filamentous algae floating as a visible mat on the surface of the water.

Blast-furnace Slag–The dross from the smelting of iron ore, which contains calcium silicate, a low-grade source of lime for agricultural use in the southeastern United States. It is also a carrier of low-grade total phosphorus.

Bleaching Clay–Any clay in its natural state or after chemical activation that has the capacity for adsorbing coloring matter; generally montmorillonitic clays are used.

Blind Area–In forestry, an area which cannot be seen from a lookout station.

Blind Furrow–A double furrow caused by plowing two adjacent furrows and throwing the soil in opposite directions; also called a dead furrow.

Blind Inlet–Inlet to a drain in which entrance of water is by percolation rather than by open flow channels.

Blind Knot–An overgrown knot that forms a pronounced swelling on the trunk of a tree; also called blind cork. See Knot.

Blind Valley–(1) A feature in karst (soluble limestone) areas where a stream flows into a cave at the closed end of a valley. (2) A type of valley in which a spring emerges from an underground channel to form a surface stream whose valley is enclosed at the head by steep and possibly precipitous walls.

Blizzard–A severe storm of high wind, low temperatures, and heavy snow.

Blocky Structure (Soil)–The arrangement of particles which indicates that soil units exceed 1 cc.

Bloodtwig Dogwood–*Cornus sanguinea*; a showy shrub, family Cornaceae; grown for its white flowers, black fruit, and blood-red twigs. Native to Eurasia. Also called dog tree, red dogwood.

Bloom–A word with several contrasting meanings: (1) Plants in the state of flowering. (2) Plants reproducing and growing. (3) Fruits with a white covering. (4) Minerals that absorb one wavelength of light and reflect another. (5) A very fine, often waxy, powdery coating on the surface of certain leaves, stems, fruits, or other organs; usually whitish, grayish, or bluish, and easily rubbed off, as the bloom on a plum. See Algal Bloom.

Blow-out–(1) An excavation by the wind of loose soil, usually sand. (2) A hole in soil above a tile line caused by excessive pressure in the drain. (3) To walk or exercise a horse either to loosen its muscles for further exercise, or to prevent chilling and stiffening after a hard workout.

Blue Dove–The wild pigeon, *Columba livia*, derived from the domestic pigeon. It is bluish-gray, and has a purple breast and white rump. Found on rocky coasts.

Blue Gill–A common warm water sunfish, *Lepomis Machrochris*, grown in the south as a game fish.

Blue Racer–*Coluber constrictor flaviventris*; a subspecies of the harmless black snake. Found in the southern United States.

Blue Spruce–See Colorado Spruce.

Blue-green Algae–The group Myxophyceae; characterized by simplicity of structure and reproduction, with cells in a slimy matrix and containing no starch, nucleus, or plastids and with a blue pigment present in addition to the green chlorophyll. It is capable of fixing atmospheric nitrogen.

Blue-stain–Blue coloration of light-colored wood caused by species of the fungi genera *Ceratocystis*, *Aureobasidium*, and *Lasiodiplodia*.

Bluebeech–See American Hornbeam.

Bluff–(1) Any high headland or bank presenting a precipitous front. (2) In the United States, the name given to the high vertical banks of certain rivers. (3) A high steep bank or cliff.

Board–(1) Lumber less than 2 inches thick and more than 8 inches wide. (2) Lumber 1½ inches thick, 6 or more inches wide, and 8 or more feet long (British). (3) Lumber of all widths 1 inch thick; widths of less than 6 inches are sometimes called strips. (4) Plank floor on which sheep are sheared (New Zealand).

Board Foot–A lumber measurement; a board 1 foot long, 1 foot wide, 1 inch thick, based on the original cut before planing and surfacing; also called super foot (Australia).

Board Mill–A sawmill that cuts 1- and 2-inch lumber, as compared with a timber mill, which cuts thicker lumber.

Bob–(1) Veal from calves slaughtered when less than four to six weeks old; also called slunk. (2) The docked tail of a horse. (3) In transporting lumber, a single pair of runners on which the forward ends of the logs are placed. (4) To transport logs on a bob.

Bob Cat–*Lynx rufus*, a common wildcat found in most areas of North America. They are generally considered beneficial to agriculture because of the number of rodents they eat.

BOD–See Biochemical Oxygen demand.

Bodewash–The dried dung of buffalo and cattle, which was used for fuel in the early days on the treeless frontier parts of the Great Plains of the United States. See Buffalo Chips.

Bog–A wetland usually developing in a depression. Often a lake with poor drainage. Generally characterized by extensive peat deposits, acidic water, floating sedge or sphagnum mats, and heath shrubs and often by the presence of coniferous trees such as black spruce, black ash, and white cedar.

Bog Iron–A spongy variety of hydrated oxide of iron (limonite). Found in layers and lumps on level, sandy, acid soils which have been covered with a swamp or bog.

Boiling Point–The temperature at which the vapor pressure of a liquid equals the atmospheric pressure. At the boiling point, bubbles of vapor rise continually and break on the surface. The boiling temperature of pure water at sea level (barometric pressure of 30 inches of mercury) is 212°F (100°C). At high altitudes, the boiling point of water is lower because the atmospheric pressure is lower. At 5,000 feet above sea level, for example, the boiling point of water is 203°F (95°C); at 10,000 feet it is 194°F (90°C).

Bois d'Arc (Osage Orange)–*Malcura poomifera*; a tree used extensively in the United States Great Plains for a shelterbelt and for fence posts.

Bolander Waterhemlock–*Cicuta bolanderi*; a perennial California herb, family Umbelliferae; it is poisonous to stock, causing vomiting, diarrhea, dilated pupils, difficult respiration, frothing at the mouth, weak but rapid pulse, convulsions.

Bole–The trunk of a tree. It may extend to the top of the tree as in some conifers, or it may be lost in the ramification of the crown, as in deciduous species.

Bolling–A tree whose branches have been removed.

Bolson–A drainage basin surrounded by high land and flanked by alluvial fans in the arid regions of the southwestern United States and Mexico.

Bolt–(1) A section sawed or split from a short log from which blocks, shingles, staves, etc., are made. (2) A short log used in making pulpwood or veneer. (3) A strong, metal pin with a head at one end and threads at the other. (4) To sift coarse elements from fine, as bran from flour. (5) To flower or to produce seed stalks, often prematurely. (6) To eat rapidly. (7) To run away.

Bora–(Latin, boreas-north wind) A cold, dry, violent downslope wind, named after the north winds that move rapidly down the Alps in Yugoslavia to the Adriatic Sea. The name is now used for a similar wind any place in the world.

Borax–Sodium tetraborate, $Na_2B_4O_7 \times 10\ H_2O$. It is sometimes applied directly to soil, or added to commercial fertilizer in very small quantities (toxic to plants in high concentrations) to supply boron. It is also used as an herbicide and an insecticide, especially to kill the maggots of flies in manure and refuse piles. Borax is 11 percent boron. Also called borac, tincal.

Border Dyke–Ridges of earth constructed to hold irrigation water within certain limits in a field.

Border Irrigation–A system of irrigating land which has a slope of 2 to 4 feet per 1,000 feet. Parallel levees are built, and the area between levees is flooded.

Border Strip–A zone or strip surrounding a field research plot usually given the same treatment as the plot but not included in sampling or study.

Boreal–Of high latitudes, including arctic-alpine; more or less coincident with the needle-leaf forest formations.

Boreal Forest–The forest consisting chiefly of conifers extending across northern North America from Newfoundland to Alaska.

Borealis–Northern.

Borer–An insect that bores into and feeds on the woody sections or bark of trees.

Boron–An element essential for plant growth. The original source of boron is tourmaline, but available forms are from soil organic matter.

Borrow–Earth material acquired from an off-site location for use in grading on a site.

Bosque–A name given to a wooded thicket, usually of tamarisk, willow, and cottonwood, on the flood plains of the southwestern United States.

Bottled Gas–Popular name for liquefied petroleum gas kept under pressure in a metal container; usually butane (C_4H_{10}), propane (C_3H_8), or any mixture of the two. Formerly used for home lighting systems, it is used now for cooking, heating, and for internal combustion engine fuels, particularly pumping engines and farm tractors. Butane is also widely used in many synthetic-producing industries.

Bottom–(1) (Often plural) Low-lying land adjacent to a river, usually rich in alluvial deposits, e.g., Mississippi River bottoms. (2) A section of a plow consisting of the moldboard, share, frame, and landside. (2) Stamina in a horse.

Bottomland–Lowland along a river (sometimes flooded).

Boulder–In soil and geologic literature, a boulder is a coarse mineral fragment larger than 10 inches (25 centimeters) in diameter.

Boulder Clay–Massive, compact deposits of gritty clay of glacial origin, containing imbedded gravel and boulders. It is the parent material for a number of highly productive soils.

Bound Water–Water adsorbed by colloids and therefore not easily freezable; also called unfree water, hygroscopic water.

Boundary Tree–An old, tall distinguishable tree standing on a property line, usually blazed or otherwise marked.

Box–(1) *Buxus sempervirens*; a shrub or small tree; family Buxaceae; one of the most valuable broad-leaved evergreens. Since the days of the Romans it has been planted for hedges and topiary work. It is lethally poisonous to cattle, horses, sheep, and pigs, having emetic and purgative properties. Native to southern Europe. (2) The lowest grade of softwood lumber. (3) A cavity cut into the base of a pine tree to collect sap to make turpentine and rosin. (4) A system for bracing branches of trees. (5) To chip back on the underside of a cut in a tree to prevent the main stem of the tree from splitting when it falls. (6) An accidental mixing of two herds of sheep or cattle (Australia). (7) To corral animals. (8) A device for dividing water in an irrigation system into two or more ditches. (9) A canyon with one entrance and no exits.

Box Elder–*Acer negundo*; a deciduous, fast-growing shade tree, family Aceraceae; its branches are brittle; the wood is used for paper pulp and woodenware; also called ashleaf maple.

Box Scraper–An implement mounted on the rear of a tractor that is used in grading and leveling soil.

Box the Heart–To cut boards from the sides of a log so that the center or heartwood is left as one piece of timber.

Brackish–(1) Term applied to waters whose saline content is intermediate between that of fresh streams and seawater; neither fresh nor salty, but in between. (2) Salty, generally less so than seawater, variously defined as less than 15 to 30 parts per thousand salinity.

Brae–A steep bank; a slope; hillside.

Brake–(1) An irrigation check. (2) A place overgrown with bracken, brush, wood, cane, etc., e.g., a cane brake, a thicket.

Bramble–(1) Any plant of the genus ***Rubus***, family Rosaceae, as the blackberry, raspberry, and dewberry, called bramble fruits. (2) Any prickly, clinging shrub.

Branch–(1) A lateral stem arising from the main stem of a plant, bough, or limb. (2) Part of a horseshoe from the first nailhole to the end of the heel. (3) A tributary stream. (4) An ancillary pipe attached to a main pipe.

Branch Wilt of Walnut–A fungal disease of Persian walnut (***Juglans regia***) caused by ***Hendersonula toruloidea***, family Sphaeropsidaceae; characterized by sudden wilting of the leaves in summer. It appears on small twigs and branches, but eventually involves the larger limbs as well.

Brand Tillage–A system of farming in which new land is periodically cleared for plowing, while abandoned fields are allowed to regain their natural cover. Common in the tropics. See Shifting Cultivation.

Brasada–Brush country or thicket vegetation (southwestern United States).

Brashy Soil–Soil composed almost entirely of coarse, angular rock fragments, with fine soil matter filling the interstices.

Brasswood–A common name for osage orange. See Bois d'Arc, Osage Orange.

Breadfruit Tree–*Artocarpus communis*; a tropical tree, family Moraceae that has heavy foliage of thick oval leaves, small flowers, and bears an edible fruit, called breadfruit. Native to Polynesia; grown rarely in the extreme southern tip of Florida.

Break Ground–To plow land for the first time.

Breaking–Plowing a native or planted sodlike crop.

Breaks–(1) Bluff land or steeply sloping escarpments broken by stream dissection, as in the eastern edge of the High Plains, south of the Arkansas River to the Texas Panhandle. This area is known as Breaks of the Plain. (2) The broken land at the border of an upland that is dissected by ravines. (3) An area in rolling land eroded by small ravines and gullies; also used to indicate any sudden change in topography, as from a plain to hilly country.

Breakwater–A fixed or floating structure that protects a shore area, harbor, anchorage, or basin by intercepting waves.

Breast High Diameter–In forestry, the point at which the diameter of a tree is measured. Custom has fixed this point at 4½ feet above ground level. Also called diameter at breast height.

Breather–A vertical section of pipe which is an air vent, preventing pressure buildup. When installed after a steep grade, it is called a relief well.

Brewer Oak–*Quercus oestediana*, family Fagaceae; an oak used for forage on the western ranges of the United States.

Brimstone–See Sulfur.

Broad-base Terrace–An erosion control system for steep slopes. It consists of a horizontal ridge of earth 10 to 30 inches high, 15 to 30 feet in base width with gently sloping slides, a rounded crown, and a broad, shallow water channel along the upper side. It may be level or have a slight fall toward one or both ends. Crops can be planted on the terrace. See Narrow-Based Terrace, Nichols Terrace.

Broad-leaved Evergreen–Evergreen plants which are not coniferous, e.g., the box, rhododendron, eucalyptus, wattles, etc.

Broad-leaved Plants–Botanically, those classified as dicotyledons. Morphologically, those that have broad, usually compound leaves.

Broad-spectrum Pesticide–A pesticide that kills a wide variety of insects. It may kill many beneficial insects, fish, birds, and mammals as well as target pests.

Broadcast–(1) To scatter seed or fertilizers uniformly over the soil surface rather than placing it in rows. (2) To sow seed in all directions by scattering. (3) To scatter manure, lime, etc. (3) To plow, throwing the soil in one direction, so that the field is left level.

Broadcast Burning–Allowing a prescribed fire to burn over a designated area within well-defined boundaries for the reduction of fuel hazard, as a silvicultural treatment, or both.

Broadcast Seeding–Scattering seed on the surface of the soil; contrast with drill seeding, which places the seed in rows in the soil.

Broads–In Canada, lumber 12 or more inches wide.

Broken–(1) Describing land plowed for the first time. (2) Of plants or flowers, having irregular color distribution.

Broken Flat–Designating land which has been broadcast plowed. See Broadcast.

Broken Land–Terrain with numerous sharply cut valleys.

Broom Pine–Long-leaf pine.

Broom, Scotch–*Cytisus scoparius*; a nitrogen-fixing, small, leguminous, tap-rooted shrub adapted to well-drained, low-fertility soils in the mild climate west of the Cascades. It is used for secondary stabilization on coastal sand and for roadside or erosion control plantings. It has a prolific crown of yellow flowers in early summer.

Broomsedge–A grass, ***Andropogon virginicus***, family Gramineae; a widespread weed on poor land in the southeastern United States. It is an inferior livestock forage, but a covering of it protects poor soils from erosion. Also called broom grass, yellowsedge, bluestem.

Brow–The edge of the top of a hill or mountain; the point at which a gentle slope changes to an abrupt one; the top of a bluff or cliff.

Brown Rat–***Rattus norvegicus***, family Muridae; the common rat; also called Norway rat, house rat, wharf rat, sewer rat.

Browse–Leaves, small twigs, and shoots of shrubs, seedling, and sapling trees, and vines available for forage for livestock and wildlife.

Brush Control–Control of woody plants.

Brush Matting–(1) A matting of branches placed on badly eroded land to conserve moisture and reduce erosion while trees or other vegetative cover are being established. (2) A matting of mesh wire and brush used to retard streambank erosion.

Brush Monkey–A lumber term for someone who cuts down small trees so that logs may be removed.

Brushland–An area of little commercial value except perhaps for grazing goats and sheep.

Buck Moth–***Hemileuca maia***, family Saturniidae; a moth whose caterpillars infest oaks and willows in the autumn and feed voraciously on the leaves.

Buckbrush Ceanothus–***Ceanothus cuneatus***, a shrub, family Rhamnaceae; a forage for livestock on the western ranges of the United States. Also called wedgeleaf ceanothus.

Bucking–Cutting a felled tree into specified log lengths.

Buckshot Soil–(1) Soils containing hard, round iron and manganese concretions of buckshot size. Occurs in the southern United States. (2) Wet clay soil, which upon drying disintegrates into aggregates about the size of buckshot. Descriptive of the soils of the Yazoo bottomland of the Mississippi River in the southern United States. Most common soil is Sharkey clay.

Buffalo–The American buffalo, which is better known as the American bison. The bison has been crossed with cattle to produce cattalo or beefalo.

Buffalo Chips–Dried dung of buffalo and cattle used in early times in treeless areas as fuel. Cattle dung is used as fuel today in many developing countries. See Cow Chips.

Buffalo Cholera–A disease of buffalo, which may spread as an epidemic among domestic animals.

Buffalo Gnat–A small insect of the genus ***Simulium***, found in the lower Mississippi Valley. Many bites may kill a domestic animal.

Buffalo Wallows–Shallow clayey depressions on the Great Plains of the United States. Popularly thought to have been bathing holes for buffalo.

Buffer–(1) A substance in soil, such as organic matter, clay, carbonates, or phosphates, which resists changes of soil pH. (2) Animals which serve as food for predators, thus reducing danger to game. (3) A tool used to cut clinches from horseshoe nails before removing the shoe; also called clinch cutter.

Buffer Capacity of Soils–The ability of a soil to resist a change in its hydrogen-ion concentration; or to resist the tendency to become more acid upon the addition of an acid (or an acid-forming material), or more alkaline upon addition of a base (or a base-forming material).

Buffer Solution–A solution to which large amounts of acid or base may be added with only a very small resultant change in the hydrogen-ion concentration.

Buffer Strip–Rows of vegetation planted along contours to reduce erosion.

Buffer Zone–Public land surrounding and supplementing private range lands.

Buffering Agent–Substance that adjusts the pH level of a spray mixture on a plant or soil surface. Often these materials can make a pesticide work better and last longer.

Bug Kill–Trees or timber stands killed by insects.

Bulk Density–The relative weight of a soil aggregate (air space plus solid soil particles) compared with the weight of the same volume of water.

Bulkhead–A structure or partition placed on a bank or bluff to retain or prevent sliding of the land and to protect the inland area against damage from wave action. See Sea Wall.

Bull Frogs–***Rana catesbeiana***, large frogs that are grown commercially for food. The legs are considered to be a delicacy.

Bull Pine–See Ponderosa Pine.

Bull Snake–A harmless snake of the genus ***Pituophis*** which eats rodents. Also called gopher snake, pine snake.

Bulldozer–A crawler tractor with a blade mounted in front to move earth, snow, gravel, and similar materials, or for clearing land of small trees and shrubs.

Bumblebee–An insect of the genus ***Bombus***; valued for flower pollination and honey, which it often stores in underground nests. Its colonies are small.

Bunch–(1) A cluster of plants or fruits, as a bunch of grapes. (2) In pigs, an improperly healed wound following castration. (3) To skid logs together for hauling. (4) To pile harvested crops in the field.

Bunding–A centuries-old soil erosion control method in India; building a low ridge of soil along the contour of a field's lower edge.

Bur Oak–***Quercus macrocarpa***, family Fagaceae; a very hardy massive oak, with its acorns nearly covered by big, fringed, scaly cups. Its wood is pest-free and very durable, and is commonly used for railroad ties, agricultural implements, furniture, etc. Also called mossy-cup oak.

Buried Soil–The soil of an original land surface which is buried to considerable depths by subsequent geologic deposition. Such soils have been observed in sections of thick alluvial deposits; soils of interglacial periods, e.g., found in glacial deposits; and soils in regions of volcanic action.

Burl– (1) A knotlike growth on the trunks of trees or plants; often produced with adventitious buds. (2) A distorted grain in lumber surrounding the pits of undeveloped buds. Considered valuable, if of sufficient size.

Burn– (1) An area in which the trees and/or grass have been destroyed by fire. (2) A branding iron. (3) In plants, to give a scorched appearance by excessive use of pesticides. (4) To blister the skin of an animal by friction with a rope or hobble, etc. (5) To become withered due to lack of moisture, as the foliage of a tree. (6) To become discolored because of insufficient drying, as in lumber.

Burned Lime– Limestone that has been heated until it forms a powder.

Burning Index Meter– A device for measuring inflammability and rate of spread of fire. It works by integrating the combined effects of the moisture content of fuel, herbaceous stage, wind velocity, relative humidity, and other factors.

Burning, Prescribed– See Prescribed Burning.

Bush– (1) A grove of sugar maple trees. See Sugar Bush. (2) A forest wilderness or vast scrubland (Canada, Alaska, New Zealand, and Australia; formerly used in the northern United States also). (3) A shrub. (4) A method of training plants and fruit trees to assume a desired shape by means of stakes, ties, wires, and other supports. (5) The brush of an animal; the tail. (6) To force a seller to accept a lower price for a horse than was bid in the auction ring.

Bush-fallow– An Appalachian region term for growing in rotation a crop on cleared land before allowing wild vegetation to cover it. In ten years or so the trees are cleared and a cultivated crop is again grown. See Shifting Cultivation.

Butt– (1) The thick base or lower end of a tree or plant. (2) A stump, especially a walnut stump. (3) In butchering, the upper half of a ham or shoulder. (4) In Australia, a package of wool containing 196 pounds greasy or 112 pounds scoured, with a tare weight of more than 11 pounds. (5) To strike with the head or horns, as do cattle, goats, or rams.

Butt Cut– (1) In tree harvesting, the first log to be cut above the stump. (2) Bark removed from the butt of a tree before it is felled; specifically for tanbark; also called butt log.

Butte– A high, flat-topped hill with steeply sloping sides, similar to but smaller than a mesa. Common in the arid and semi-arid regions of the western United States.

Buttress– Swelling (ridges) of a tree at its base near the soil, common among trees of the humid tropics and subtropics, especially on wet soils.

By-product– A product of significantly less value than the major product. In beef cattle, the major product is meat; by-products include the hide and other items.

By-product Lime– Any material containing calcium and/or magnesium resulting from the manufacture of another product, e.g., iron smelting, sugar factory lime.

C Horizon– Soils are composed of one or more horizons. The A horizon is the surface, the B horizon is the subsoil, and the C horizon is the underlying parent material above the R horizon, bedrock. See Soil Horizon.

Caballeria– (1) A unit of land measurement in Spanish-speaking countries, equal to 33.2 U.S. acres in Cuba; 111.82 in Costa Rica; 111.51 in Guatemala; 111.13 in Honduras; 105.75 in Mexico; 112.41 in Nicaragua; 194.1 in Puerto Rico; and 95.48 in Spain. (2) A place to keep horses.

Cablegation– A method of irrigating a field from an irrigation pipe in which a traveling plug fastened by a cable to a braking mechanism is used to help control the amount of water released by gates or holes in the pipe. It was developed in 1981 at the USDA Snake River Conservation Research Center at Kimberly, Idaho, as a means to automate gravity-flow irrigation systems.

Cactus– Any desert plant of the family Cactaceae. Mostly spiny, some are regarded as ornamentals and others as food for range livestock. Native to America, except the genus ***Rhipsalis (Hatiora)***, which is sometimes found in Madagascar.

Cadmium– (Greek; *kadmia*, earth) A bivalent metal similar to tin. Its atomic number is 48 and atomic weight 112.40. A metallic "heavy metal" used in the production of copper, lead, silver, and aluminum alloys, in photography, ceramics, and in insecticides. Cadmium in sewage sludge is of grave concern when sludge is applied to soils used for the production of food and feed crops. Cadmium is a hazardous pollutant to people, domestic animals, and shellfish. It also is an experimental carcinogen.

Cairn– A pile of rocks that has been made to mark a survey point or a foot trail.

Caking– Caking is a process by which a fertilizer loses its desirable free-flowing property. The principal reason for caking is increase of moisture content. When damp fertilizers are dried, the salts that were in solution in the film of moisture on the surfaces of the particles are crystallized. These crystals adhere tightly and tend to knit the particles together. Plastic substances cake under pressure. Chemical reactions between ingredients of a mixture may also cause severe caking. Caking can be prevented or lessened by granulation, maintenance of low moisture content in the fertilizer at all times, or by coating the particles with conditioning agents such as oils, waxes, and clays.

Calabash Tree– *Crescentia cujete*, family Bignoniaceae; grown for its large, hard-rinded fruit whose shells or husks are used for dippers, utensils, etc. Native to tropical America.

Calabazilla–*Cucurbita foetidissima*; an annual, trailing, herbaceous vine, family Cucurbitaceae, grown in warm areas for its fruits, which when dried are used as ornamental gourds. Its roots were once used as a cleanser and its leaves medicinally. Native to North America. Also called Arizona gourd, buffalo gourd, Missouri gourd, wild pumpkin, calibassa, calabazza, calabaza.

Calcareous–Formed of calcium carbonate or magnesium carbonate or both by biological deposition or inorganic precipitation in sufficient quantities to effervesce carbon dioxide visibly when treated with cold 0.1 normal hydrochloric acid. Calcareous sands are usually formed of a mixture of fragments of mollusk shell, echinoderm spines and skeletal material, coral, foraminifera, and algal platelets.

Calcic–(1) Containing calcium, as calcic lime, calcic plagioclase, calcic pyroxene. Also said of igneous rocks containing such minerals. (2) Refers to igneous rocks having an alkali-lime index of more than 61.

Calcic Horizon–A soil layer high in calcium carbonate formed by precipitation from high lime layers above dissolving and reprecipitating.

Calcicole–A plant growing in soil rich in calcium. Synonym, calciphile. See Calcifuge.

Calciferol–Another name for vitamin D_2 obtained from irradiated ergosterol.

Calciferous–Producing or containing carbonate of lime.

Calcification–(1) A soil-forming process, under low rainfall conditions, whereby a calcareous layer is formed at or below the soil surface. (2) Replacement of the original hard parts of an animal or plant by calcium carbonate.

Calcifuge–A plant that grows best in acid soil. See Acidophilous, Calciphile, Calciphobe.

Calcify–To deposit or secrete lime salts, which harden, as when gristle (cartilage) becomes bone.

Calcination–The heating of a substance to effect a physical and/or chemical change. A common example in agriculture is the heating of limestone ($CaCO_3$, to produce calcium oxide (quicklime) (CaO) and carbon dioxide (CO_2). Quicklime is applied to soils when a rapid increase in pH is needed.

Calcined Brucite–A mineral high in magnesium that has been burned to a powder; e.g., serpentine, dolomite.

Calcined Clay–A clay that has been heated to reduce its shrink-sell properties when wet-dry. Calcined clay exhibits the physical properties of a fine sand.

Calcined Phosphate–A general term for several materials produced by heating phosphate rock, with or without reagents. Simply heating to 1,520°F (827°C) in a rotary kiln destroys all organic matter, converts the calcium carbonate to the oxide, and drives out some of the fluorine. This calcined phosphate is used in the manufacture of high-grade phosphoric acid. When heated with silica and moisture to 2,700°F (1,482°C) the fluorine is almost entirely expelled, and the phosphate rock is converted to tricalcium phosphate, also known as fused tricalcium phosphate. It contains about 24 percent P_2O_5 available by the 2% citric acid method. A similar product is sold to the livestock feed trade as defluorinated phosphate.

Calciphobe–An acidophilous plant, e.g., blueberries. See Calcifuge, Calciphile.

Calcite–A mineral, $CaCO_3$, calcium carbonate, in crystalline form. Sometimes used as a diluent for insecticidal dusts.

Calcitic Dolomite–A carbonate rock in which the percentage of calcite is between 10 and 50, and the percentage of dolomite between 50 and 90.

Calcium–Ca; a chemical element present in variable amounts in all soils. It is essential for plant and animal growth and is the principal mineral element in bones.

Calcium Ammonium Nitrate–A trade name for an ammonium nitrate-limestone mixture (20.5% N), made in Europe.

Calcium Ammonium Nitrate Solution–An aqueous solution of calcium nitrate and ammonium nitrate containing 20.5% N. It is used extensively in the western coastal regions of the United States.

Calcium Carbonate–Calcium limestone and oyster shells are composed largely of calcium carbonate ($CaCO_3$). In such forms it is used extensively to neutralize soil acids. It is also a principal component of dolomite. Marble (proper) differs from common limestone in being harder and more compact as a result of metamorphism. Chalk is a soft limestone deposited in saltwater. Pure calcium carbonate contains 56 percent calcium oxide (CaO) and 44 percent carbon dioxide (CO_2).

Calcium Chloride–$CaCl_2$; a chemical used as a bleaching agent and for controlling dust on roads.

Calcium Cyanide–$Ca(CN)_2$; a very effective fumigant for controlling insects and rodents.

Calcium Hydroxide–$Ca(OH)_2$; a calcium compound, useful in raising the pH of soils. It is of greater neutralizing power than ground limestone. Also called slaked lime, hydrated lime.

Calcium Lactate–A calcium salt of lactic acid, used to induce thickening and more rapid clotting of the blood.

Calcium Nitrate–$Ca(NO_3)_2$; the calcium salt of nitric acid sometimes used as a fertilizer.

Calcium Nitrate-Urea–$Ca(NO_3)_2 \times CO(NH_2)_2$; as crystallized from solution this material consists of calcium nitrate with four molecules of urea of crystallization, replacing the water of crystallization. It contains about 33% N and 9% Ca.

Calcium Oxide–CaO; a salt produced by heating limestone, marble, etc. In the dehydrated form, it is used as a dust diluent, calcium carrier, and for raising the pH of soil. Also known under many other names.

Calcium Phosphate–The phosphatic part of good superphosphate consists largely of monocalcium phosphate ($CaH_4P_2O_8$), which is water-soluble. Precipitated bone is largely dicalcium phosphate ($CaHPO_4$), which is citrate-soluble. Bone meal contains calcium phosphate-carbonate ($3Ca_3[PO_4]_2 \times CaCO_3$). Apatite is calcium fluorphosphate or chlorophosphate, and phosphate rock usually contains a complex calcium fluorphosphate. The term ***bone phosphate of lime*** (BPL), first applied as a name when animal bones were the principal source of phosphorus fertilizer, is commonly used to express the content of tricalcium phosphate ($Ca_3(PO_4)_2$) in phosphate rock.

Calcium Polyphosphate–Any calcium salt of a polyphosphoric acid such as calcium dihydrogen pyrophosphate and dicalcium pyrophosphate. Fertilizer grades of calcium polyphosphate generally refer to phosphate mixtures containing a substantial proportion of one or more calcium polyphosphates and can be made by thermal dehydration of triple superphosphate.

Calcium Sulfate–$CaSO_4$. See Gypsum.

Caldera–A large basinlike depression with steep sides in the top of a volcanic mountain.

Calendar Year–A period from January 1 through December 31 of the same year.

Calibration–In field agriculture, the setting of a distributor or spreader to meter the predetermined amount of pesticide, lime, or fertilizer.

Caliche–(1) In Chile and Peru, impure native nitrate of soda. (2) A desert soil formed by the near-surface crystallization of calcite and/or other soluble minerals by upward-moving solutions. (3) In Chile, whitish clay in the selvage of veins. (4) In Mexico, feldspar, a white clay. (5) A compact transition limestone. (6) In Colombia, a mineral vein recently discovered. (7) In placer mining, a bank composed of clay, sand, and gravel. (8) In the southwestern United States, gravel, sand, or desert debris cemented by porous calcium carbonate; also the calcium carbonate itself.

California Buckthorn–*Rhamnus californica*, a shrub, family Rhamnaceae, which is planted as a bee tree in California. Found in the western United States.

California Incensecedar–*Libocedrus decurrens*, family Pinaceae; an evergreen tree grown as an ornamental and for its valuable timber. Also called white cedar.

California Privet–*Ligustrum ovalifolium*, family Oleaceae; a hedge plant with small clusters of flowers, planted extensively in Asia and the United States. Native to Japan.

Cambric Horizon–A soil horizon that has developed the color and/or structure of a subsoil horizon but does not qualify as an argillic or spodic horizon.

Campestrian–Of plains or open country; specifically of the Great Plains of North America.

Camphor Tree–*Cinnamomum camphora*, family Lauraceae; an aromatic evergreen tree with alternate leaves, whose wood contains the camphor gum of commerce, obtained by steaming and sublimating the bark. Native to China.

Campos–Grasslands or savanna located south of the equatorial forests in Brazil.

Canada Hemlock–*Tsuga canadensis*, family Pinaceae; a hardy evergreen tree, grown as an ornamental and for its valuable lumber used in pulpwood and general construction. Native to North America. Also called hemlock, eastern hemlock, hemlock spruce, tanbark tree.

Canada Wildrye–*Elymus canadensis*, family Gramineae; a perennial forage grass, widely distributed throughout the United States and Canada.

Canada Yew–*Taxus canadensis*, family Taxaceae; an evergreen, low-growing shrub, useful as a ground cover under evergreen trees. Highly poisonous. Also called ground hemlock, American yew.

Canal–(1) An artificial watercourse cut through a land area for navigation, irrigation, or drainage. (2) A long, narrow arm of the sea extending far inland. (3) On the Atlantic Coast, a sluggish coastal stream. (4) A long, fairly straight, natural channel with steeply sloping sides, generally a mile or more in width. (5) A cave passage partly filled with water.

Cannon–(1) That portion of an animal's leg, ankle, or pastern from the knee to the fetlock, in the front legs, or the hock to the fetlock, ankle, or pastern in the rear legs. (2) A horse's bit. (3) To lift and swing a log crosswise to a load of logs.

Canopy–The uppermost vegetation layer consisting of crowns of trees or shrubs in a forest or woodland.

Canopy Trees–Trees with crowns in the uppermost layer of forest or woodland.

Cant Hook–A hand-operated wooden lever with a variable iron hook on one end; used for handling logs.

Canvas Hose Irrigation–A method of irrigation which originated in Michigan. A porous canvas hose, in which water is flowing under pressure, is placed on the ground alongside each row. The water seeps out of the canvas hose along its length and is absorbed by the soil. Synthetic fibers often replace canvas.

Canyon–(1) A steep-walled chasm, gorge, or ravine; a channel cut by running water in the surface of the earth, the sides of which are composed of cliffs or series of cliffs rising from its bed. Sometimes spelled canon. (2) (Oceanography) A deep submarine depression of a valley with relatively steep sides.

Cap–(1) A piece of metal placed over the end of a log to make it skid over obstacles. (2) The top member of a trestle or similar support. (3) Milk tooth of an animal. (4) The crown of a kernel of corn. (5) The calyx of the strawberry. (6) To remove the calyx of the berry after it is picked. (7) To place a cover over a stack of hay, bound grain, etc. (8) To seal a cell in the honeycomb with wax, as by a bee.

Capacity, Carrying–In wildlife or livestock management, the optimum density of animals that a given environment or range is capable of sustaining permanently.

Capillarity–The attractive force between two unlike molecules, illustrated by the rising of water in capillary tubes of small diameters or the drawing-up of water in small interstices, as those between the grains of soil or a rock.

Capillary Capacity–The amount of moisture which is held in the soil by capillary attraction after excess or free water has drained away.

Capillary Fringe–A belt of capillary interstices in the soil belt that overlie the zone of saturation, or water table. Some or all of these are filled with water, held by capillary action against the force of gravity. The fringe water may be used by deep rooting or phreatic plants.

Capillary Mat–A fiber mat designed to distribute water to potted plants in a greenhouse; the plants take up the water through capillary action.

Capillary Water–(1) Underground water that is held by the soil above the water table by capillary attraction. (2) The water held in the "capillary" or small pores of a soil, usually with tension greater than 60 centimeters (23.63 inches) of water. Much of this water is considered to be readily available to plants.

Caragana–*Caragana arborescens*; also known as Siberian pea shrub. A very hardy, deciduous, leguminous shrub or small tree to 30 feet (10 meters). It has pinnate leaves with up to eighteen small leaflets and large pealike yellow flowers in early spring. Planted in eastern Oregon as a windbreak shrub. Widely adapted, drought- and cold-tolerant. Has excellent value for windbreak, fair value of wildlife food and cover, and fair to excellent ornamental value.

Carbamate–A synthetic organic pesticide containing carbon, hydrogen, nitrogen, and sulfur.

Carbon–An essential chemical element component in plants and animals. It is present in soils, in humus, plant residues, charcoal, and particles of coal, carbonaceous shale, etc.

Carbon Bisulfide (Carbon Disulfide)–CS_2; a highly flammable colorless liquid, boiling at 115°F (46°C). Used for degreasing seed meals, such as castor and cottonseed meals and as a seed fumigant to kill insects.

Carbon Cycle–The sequence of transformation undergone by carbon utilized by organisms. Absorbed from plants and other sources by an organism, it is later liberated upon the organism's death and decomposition, and returned to its original form of carbon dioxide gas, to be again taken up by plants.

Carbon Dioxide–CO_2; a colorless, odorless gas constituting 0.03 percent of unpolluted air. It is absorbed by green plants through the leaf stomata and is used as the source of carbon for manufacturing sugars, starches, proteins, and fats. The burning of fossil fuels has put so much CO_2 in congested cities that the concentration may reach more than three times the background (normal) level. This is causing a "greenhouse effect," resulting in a slow warming of the earth's surface.

Carbon Dioxide Evolution–The liberation of gaseous carbon dioxide from soil by biological processes.

Carbon Disulfide–See Carbon Bisulfide.

Carbon Monoxide–CO; a colorless, odorless, very toxic gas with a faint metallic odor and taste, produced by any process that involves the incomplete combustion of carbon-containing substances. One of the major air pollutants, it is emitted primarily through the exhaust of gasoline-powered vehicles. It is also produced during the thermal degradation and microbial decomposition of organic solid wastes when the oxygen supply is limited. It also concentrates in silage in silos.

Carbon: Nitrogen Ratio–The value obtained by dividing the percentage of organic carbon by the percentage of total nitrogen in a soil or in an organic material. A narrow ratio results in the rapid decomposition of organic materials.

Carbon Tetrachloride–CCl_4, a chemical that is an excellent solvent for grease. It is not inflammable and can be used as a fire extinguisher. Frequently used as a fumigant, though not in large-scale operations; toxic when inhaled.

Carbon-14 Dating–The use of radioactive carbon, which has an atomic mass of 14 and an approximate half-life of 5,000 years, for determining approximately the age of organic materials in soils, buried materials such as wood, and other organic materials.

Carbonaceous–(1) Pertaining to, or composed largely of, carbon. (2) The carbonaceous sediments include original organic tissues and subsequently produced derivatives of which the composition is chemically organic.

Carbonate of Lime–Calcium carbonate.

Carbonation–(1) Process of introducing carbon dioxide into water. (2) A process of chemical weathering by which minerals are replaced by carbonates.

Carcinogen–A chemical, physical, or biological agent that increases the incidence of cancer.

Carcinoma–Malignant cancerous growth.

Cardinal Points–The four principal points of the compass: North, East, South, and West.

Caribou–The reindeer of North America.

Carnivore–A member of a large order of mammals which customarily eat flesh; e.g., dogs, cats, bears, and seals.

Carnivorous Plants–Literally, meat-eating plants; usually equivalent to insectivorous. As many as 450 species of plants have been reported to trap insects. The most common are Venus flytrap, pitcher plant, and sundew. About 150 species of fungi are known to trap nematodes.

Carolina Hemlock–*tsuga caroliniana*, a beautiful evergreen tree, family Pinaceae; commonly grown for ornament, especially in the southern United States. Native to North America. Also called spruce pine.

Carolina Horsenettle–*Solanum carolinense*, a perennial weed herb, family Solanaceae; may be poisonous to livestock feeding on its herbage or berries. Commonly found in the southern United States. Also called sand brier, bull nettle, apple-of-Sodom.

Carolina Laurelcherry–*Prunus caroliniana*, family Rosaceae; a small, evergreen tree, which has small cream-white flowers; planted for shade. Native to the southeastern United States. Also called evergreen cherry, American cherry laurel, wildorange, mock orange, cherry laurel.

Carp–Any of a large number of fresh water species of fish of the genus *Cyprinus*. They are large fish with soft fins and large scales. Some species are used in fish ponds to control weeds.

Carriage–(1) A frame on which a log is held. (2) Manner or bearing of an animal; its physical control and behavior.

Carrier–(1) A genetic term that refers to an animal that expresses the dominant trait but is heterozygous for the recessive gene. (2) Animal or person in apparently good health who harbors pathogenic microorganisms. (3) The liquid or solid material added to a chemical compound to facilitate uniformity of application in the field.

Carrion–Putrefying carcass.

Carrying Capacity–In its true sense, the maximum number of individual animals that can survive the greatest period of stress each year on a given land area. It does not refer to sustained production. In range management, the term has become erroneously synonymous with grazing capacity.

Cartography–(1) The science and art of expressing graphically, by means of maps and charts, the visible physical features of the earth's surface, both natural and human-made. (2) The science and art of map construction.

Casing–(1) A large pipe sunk into a well to prevent the walls form caving in and within which the pipe for pumping liquids is placed. (2) Cleaned sections of hog, cattle, or sheep intestine used as sausage skins.

Casing Head–In well boring, a driverhead screwed into a casing pipe to take the force of the blows while the pipe is being driven down.

Cat–(1) *Felis libyca domestica*; a lithe, carnivorous, domesticated animal used as a pet. Probably native to Egypt. (2) A general term for an animal of the family Felidae, e.g., the lion, leopard, tiger, etc.

Catalpa–Any tree of the genus *Catalpa*, family Bignoniaceae; attractive flowering trees grown as street trees; grown to produce caterpillars that feed on the leaves. The caterpillars are used for fish bait. Native to America and Asia.

Catamaran–(1) A raft for supporting a windlass and a grappling hook for recovering logs sunk in a river. (2) Any raft with at least two logs or floats fastened together.

Catamount–Any wild animal of the family Felidae; usually a panther, puma, or cougar.

Cataract–(1) An opaque or cloudy condition of the lens of the eye, or of the capsule for the lens, which obstructs vision. (2) A waterfall. See Cascade.

Catch Basin–A pond designed to minimize the admission of grit and detritus into a tile drainage system or sewer.

Catena–(1) A sequence of soils of about the same age, derived from similar parent material, and occurring under similar climatic conditions but having different characteristics due to variation in relief and drainage.

Caterpillar–The wormlike larva of a butterfly, moth, or other insect. All larvae have several pairs of legs, short antennae, and strong cutting jaws. Some are covered with hairs. They feed on the succulent parts of the plant and are known as canker worm, army worm, cutworm, etc.

Catface–(1) A healing or healed wound on the trunk of a tree; a blemish. (2) Fruit abnormalities of tomato, attributed to various unknown growth disturbances, partially climatic. (3) A scar or irregular growth on an apple or peach that results from insect stings.

Cation Exchange–The interchange among cations in soil solution and cations on the surface of clay, humus, and/or plant roots. For example, in acid soils three calcium ions can exchange with six hydrogen ions or two aluminum ions. In sodic soil reclamation, one calcium ion replaces two sodium ions.

Cation Exchange Capacity (CEC)–(Formerly called "base exchange capacity") A measure of the total amount of exchangeable cations that can be held by a given mass of soil, expressed in milliequivalents (meq) per 100 grams of soil at neutrality (pH 7) or at some other stated pH value. The exchange capacity of soils usually ranges between 2 and 50 meq/100 grams of soil. Cations are interchanged primarily on the surfaces of roots, clay, and humus.

Catsteps–Very small, irregular terraces, on steep hillsides, especially in pastures, formed by cattle trails and/or slippage of saturated fine-textured soil.

Cattail–(1) Any marshy reed of the genus *Typha*, family Typhaceae, notable for its long, cylindrical, brown spike containing innumerable, tiny, petalless flowers. Generally 5 or 6 feet high, the plant has an astringent rootstock and long, thin, stiff leaves. It is the principal plant of bogs and marshes, and its young shoots are sometimes eaten in England and the U.S.S.R. The floss from the spike is useful for insulating purposes and filling mattresses. The spike itself is sometimes used as a cut flower for decoration. The leaves are used in making mats, baskets, and chair seats. (2) *Typha latifolia*, the common cattail. Also called bulrush, cat's-tail reed, cattail flag, cooper's flag, cooper's fly, ditch-down, reed mace, cattail rush.

Cattle Chips–Dried cow manure; formerly used as fuel in the United States and still used in developing countries. Also called cow chips.

Cattle Country–(1) The grazing area of other western plains and mountains in the United States where cattle are one of the principal sources of income. (2) A country whose principal product is cattle, e.g., Argentina.

Causeway–A raised road across a wet place or shallow water.

Caustic Lime–See Burnt Lime, Calcium Oxide.

Cave–(1) A natural cavity, recess, chamber, or series of chambers and galleries beneath the surface of the earth, within a mountain, a ledge of rocks, etc.; sometimes a similar cavity artificially excavated. (2) Any hollow cavity. (3) A cellar or underground room. (4) The ash pit in a glass furnace. (5) The partial or complete falling in of a mine; called also cave-in. (6) Underground opening, generally produced by solution of limestone, large enough to be entered by a human. See Cavern, Karst, Thermokarst.

Cavern–A subterranean hollow; an underground cavity; a cave. Often used, as distinguished from cave, with the implication of largeness or indefinite extent. See Cave, Karst, Thermokarst.

Cavity–(1) A hollow space within a body, e.g., the buccal cavity. (2) Any hole in the trunk or branches of a tree.

Cedar–(1) Any tree of the genus *Cedrus*, family Pinaceae, especially the deodar cedar, Atlas cedar, Cyprian cedar, cedar-of-Lebanon. (2) Commonly but not accurately, any tree of the genus *Juniperus*, family

Pinaceae, as the southern red cedar, eastern red cedar, Rocky Mountain juniper. (3) Commonly but incorrectly, certain trees of the genus *Cedrela*, family Meliaceae, as cigarbox cedrela. (4) Commonly but incorrectly, any tree of the genus ***Libocedrus***, family Pinaceae, as California torryea. (6) Commonly but incorrectly, any tree of the genus ***Chamaecyparis***, family Pinaceae, as white cedar, false cypress.

Cedar Apples–Hard, brown, spherelike excrescences of *Juniperus* spp. caused by the rust fungus Gymnosporangium *juniperi-virginianae*, family Pucciniaceae, up to 2 inches in diameter, scattered over the tree, which sometimes bend the branches with their weight. They represent the telial stage in the development of the fungus, which may also spend part of its development on the apple or other members of the family Rosaceae. Also called cedar balls, cedar galls, cedar flowers. See Apple Rust, Cedar-Apple Rust.

Cedar Galls–Small swellings on branches, twigs, and leaves of junipers and related trees and shrubs. See Cedar Hawthorn Rust.

Cedar Hawthorn Rust–A fungal disease of hawthorn and cedar caused by ***Gymnosporangium globosum***, family Pucciniaceae, which alternately infests junipers and hawthorns, rarely apples. It produces small, irregularly shaped galls on the junipers, which have wedge-shaped, gelatinous, orange spore masses. Its symptoms are similar to apple rust on the apple.

Cedar, Incense–*Libocedrus decurrens*. Trees, generally to 90 ft. (30 m), native to southwest Oregon and western Cascades. Valuable for wood products, windbreaks, and ornamental plantings. Slow-growing, seedling mortality can be high if improperly managed.

Cedar Rust–See Apple Rust, Cedar Hawthorn Rust.

Cedar Tree Borer–*Semanotus ligneus*, family Cerambycidae; a beetle whose larvae infests arborvitae, redwood, Douglas-fir, and Monterey pine, making winding galleries in the inner bark and sapwood.

Cedar-of-Lebanon–*Cedrus libani*, family Pinaceae; the famous evergreen tree of the Bible. It is one of the traditional cedars, growing up to 100 feet high and sometimes bearing branches 40 to 50 feet long. Native to Turkey; grown in the southern United States.

Cellar–See Cold Cellar, Storm Cellar.

Celsius Scale–(formerly called centigrade) A temperature scale in which the freezing point of water is 0°C and the boiling point is 100°C. Minus 273.15°C = 0°K: absolute zero. Named after a Swedish astronomer, Anders Celsius (1701-1744). See Fahrenheit Scale, Kelvin Scale.

Cement Rock–An argillaceous limestone used in the manufacture of natural hydraulic cement. Contains lime, silica, and alumina in varying proportions, and usually magnesia.

Cemented–Indurated; having a hard, brittle consistency because the particles are held together by cementing substances such as colloidal organic matter (humus), calcium carbonate, or the oxides of silicon, iron, and aluminum. The hardness and brittleness persist even when wet.

Cenote–A type of sink developed in limestone areas by the collapse of caverns, which cuts off natural channels of circulation and allows water to fill the depression. See Karst.

Center Fire–A method of controlled burning in which initial fires are set in the center of an area to be broadcast-burned to create a strong draft. Later, additional fires are set near the outer control lines of the area to act as backfires to check outward spread.

Center Pivot Irrigation System–A large irrigation system that rotates around the terminal end of a large water line. Modern systems are powered by electric motors. The system circles slowly and sprinklers mounted on the system emit irrigation water.

Ceramic–Pertaining to pottery, including porcelain and terra cotta. Often taken to include all products made by heating natural claylike materials: pottery, chinaware, glass, and bricks.

Ceratostomella–A fungus genus, family Ceratostomataceae. *Ceratostomella pilifera* is the cause of blue stain in wood.

Cerrero–Describing a wild, untamed horse (western United States).

Cesspool–A hole in the ground usually lined or filled with loose rock into which household effluent is piped. The liquid seeps into the surrounding soil, and the solids decompose anaerobically.

Chain–(1) The legal unit of length for the survey of public lands of the United States; the equivalent of 66 feet or 20.13 meters. The name is derived from Edmund Gunter's chain, which was a series of links connected by rings. There are ten links in a Gunter's chain. Advantage in measuring in chains is that 10 square chains equals 1 acre. (2) Any series of related, interconnected, or similar natural features, e.g., chain of mountains, islands, lakes. (3) A series of metal links forged or welded together and used for fastening, hauling, etc. (4) In chemistry, a series of atoms connected by bonds, forming the skeleton of a number of compounds.

Chain Saw–A portable, motor-driven saw which consists of an arm on which travels an endless chain with attached sawteeth. Used in felling trees, sawing logs, or in cutting underbrush.

Chalk–(1) Soft, white, fine-grained lime which is highly fossiliferous. As the parent rock, it lends distinctive character to soils. Locally impure chalks may be called marl. (2) To spread chalk over land.

Chamaecyparis–A genus of pine trees, family Pinaceae, which yields valuable timber. Native to North America, Japan.

Chamaedaphne–A genus of evergreen shrubs grown in rock gardens, family Ericaceae; found in bogs of northern Europe, northern Asia, North America.

Chamiso–(1) *Adenostoma fasiculatum*, family Rosaceae; an evergreen shrub grown in warm, humid regions for its needlelike foliage and clusters of white flowers. The most characteristic shrub of the California chaparral. (2) *Atriplex canescens*, family Chenopodiaceae; a semidesert shrub found in New Mexico in the United States. Also called greasewood chamise.

Champaign–An expanse of flat, open country.

Chance–In logging, a place in a forest suitable for one operation, e.g., logging chance, cutting chance, etc.

Chance Seedling–A fruit or nut tree which grows from seed scattered by wind, birds, etc.; sometimes bears fruit. Many varieties of fruits originated as chance seedlings.

Channel–(1) A canal or small river. (2) The deeper parat of a river or harbor usually caused by the main current. (3) A passage of water joining two bodies of water. (4) Any narrow passage through which liquid or air flows.

Channelization–The straightening and deepening of streams to permit water to move faster, to reduce flooding, or to drain marshy acreage for farming. However, channelization may reduce the organic waste assimilation capacity of the stream and may disturb fish breeding and destroy the stream's natural beauty.

Channery–Small, thin, flattish fragments of limestone, standstone, or schist found in soils.

Chapparal–A dense growth of shrubby vegetation or stunted and dwarf trees in arid and semi-arid regions of the southwestern United States. Characteristic shrubs of the coast ranges of California are chamiza, ceanothus, and manzanita. In other places dense thickets of mesquite (*Prosopis* spp.) and thickets of oaks may be called chaparral.

Char–The solid, carbonaceous residue that results from incomplete combustion (pyrolysis) of organic material. It can be burned for its energy content or, if free from large amounts of impurities, processed further for production of activated carbon for use as a filtering medium. Char produced from coal is generally called coke, while that produced from wood or bone is called charcoal.

Char-pitting–A method of removing stumps by burning. A fire is built on the stump and kept covered until the wood is consumed.

Charco–(1) A structure similar to an artificial pond, but differs in that it is made of two parts, a desilting basin and a holding basin. The water runs into the desilting basin and then flows to the holding basin through a pipe. (2) A waterhole, or a standing pool of water, which remains following rains in adobe regions of the southwestern United States.

Charcoal–A black, porous, carbonaceous substance, obtained by the partial burning of an animal or vegetable substance in the absence of air. A common fuel, it is also a mild antiseptic and deodorant.

Charring–Burning partly; scorching.

Chasm–(1) A deep breach in the earth's surface; an abyss; a gorge; a deep canyon. (2) A deep recess extending below the floor of a cave.

Chasmophyte–A plant growing in a rock crevice.

Chat–Tailings from lead and zinc mines consisting essentially of dolomite or dolomitic limestone. It is widely used in Missouri, Tennessee, and Illinois as a liming material. It contains about 0.5 percent manganese and 0.016 percent zinc in addition to 21 percent calcium and 11.3 percent magnesium.

Chausee–(1) A levee used for retaining water. (2) A term for a levee also used as a road in Louisiana.

Check–(1) In irrigation (a) A basin into which the flow of water is regulated by levees and dykes. (b) An adjustable gate on a canal to regulate the flow of water. (c) Concrete blocks or wooden ties placed on a channel bottom to reduce erosion. (d) A crack which appears in drying soil. (2) In lumbering, a lengthwise separation in the grain, caused by strains during seasoning, which extends across the annual rings. (3) A short crack within the body of a cheese. (4) A narrow crack in a rice grain that may cause it to break during milling. (5) An egg which has a cracked shell, with the inner membrane intact. When the crack is naturally mended, it is called a blind crack. Also called crack, dent. (6) To retard growth.

Check Irrigation–Application of a comparatively large stream of water in level plots surrounded by levees.

Chelates–Certain organic chemicals, known as chelating agents, form ring compounds in which a polyvalent metal is held between two or more atoms. Such rings are chelates. Among the best chelating agents known are ethylenediaminotetraacetic acid (EDTA), hydroxyethylenediaminetriacetic acid (HEDTA), and diethylenetriaminepentaacetic acid (DTPA). In the absence of chelates in the soil, iron, copper, manganese, and zinc are all converted to insoluble and unavailable hydroxides or other basic salts even in acid soils. Chelates keep these micronutrients available in soils or solutions of pH up to 8 or 9. Manure is an example of a chelate.

Chemical–(1) A substance obtained by or used in a chemical process. (2) In dairying, an uncommon flavor defect which occurs when cheese has been made from contaminated milk.

Chemical Additives–Substances added to foods to improve their flavor, color, texture, or keeping quality.

Chemical Brown Stain–A brown discoloration of chemical origin that sometimes develops on wood during air seasoning or kiln drying; probably due to the oxidation of extractives in the wood.

Chemical Fertilizer–See Commercial Fertilizer

Chemical Gardening–See Hydroponics.

Chemical Oxygen Demand (COD)–The quantity of oxygen required to oxidize organic matter in a sample of waste under specific conditions of oxidizing agent, temperature, and time. Potassium dichromate dissolved in 50% sulfuric acid is the oxidizing agent. Silver sulfate is added as a catalyst, and mercuric sulfite is added to remove interference of chlorides. Excess dichromate is titrated with standard ferrous ammonium sulfate, using an orthophenanthroline-ferrous complex as the indicator.

Chemical Toilets–A type of dry vault toilet in which the sewage is decomposed by adding caustic chemicals, such as quick lime. Such toilets are used by people harvesting agricultural crops.

Chemical Weed Control–The application of herbicides as preemergence and postemergence sprays for the control of weeds in crops.

Chemical Wood–Wood cut or prepared for the manufacture, distillation, or extraction of chemicals, charcoals, gases, or other products.

Chemigation–The application of agricultural crop chemicals such as herbicides or fertilizers through an irrigation system.

Chemistry–(Greek, chemeia) Science of compounds, elements, and atoms.

Chemolysis–Decomposition of organic matter brought about by chemical agents.

Chemosphere–The atmospheric zone about 15 to 60 miles (25 to 100 kilometers) above the earth's surface containing a concentration of ozone.

Chemuck–A food consisting mainly of ground acorns. Formerly used by Indians in the northwestern United States.

Cheniers–Low, narrow ridges in the coastal marshes of Louisiana in the United States. Usually 5 or 6 feet above the surrounding level, they were home sites and farm lands for the early settlers.

Cherokee Outlet–A strip of land in northwestern Oklahoma, United States, ceded to the Cherokee tribe of Indians in 1828 by the federal government to provide them with an outlet to buffalo hunting grounds in the west.

Cherry–Species of the genus *Prunus* grown for its edible fruit or for ornament. Cultivated cherries are native to Europe and are classified in three groups: sour cherries, sweet cherries, and dukes, which are hybrids between the first two. One wild variety, native to eastern North America, the black cherry (*P. serotina*) is a very large tree, the wood of which is used for fence posts, rails, cross-ties, furniture, and fuel. All leaves are toxic to livestock.

Cherry, Bitter–*Prunus emarginata*; a deciduous shrub growing 15 to 20 feet (5 to 7 meters) tall. Is often straggling or depressed in habitat, growing on dry rocky slopes, but not limited to those areas. It bears a prolific crop of small bright or dark red berries during the autumn.

Chert–A very dense, cryptocrystalline, flintlike form of silica, which breaks with a splintery fracture. It is very resistant to decomposition, remaining as inert angular fragments in the residual mass of weathering. Since chert is harder than steel, plows and disks are dulled in cherty soil. See Flint.

Chestnut–(1) See American Chestnut. (2) A horse color of brown with red-yellow hues which lacks brilliance. (3) The sweet, edible nut produced by *Castanea* spp., family Fagaceae. (4) The horny projection found on the lower inner part of the forearm and hock joint of a horse.

Chestnut Blight–A fungal disease caused by *Endothia parasitica*, family Diaporthaceae, which attacks the bark and cambium of the twigs, branches, and main trunk of chestnut trees. The cankers formed girdle the stems resulting eventually in the death of the tree. The blight has almost completely destroyed the American chestnut, *Castanea dentata*. However, the East Asian varieties have shown some degree of resistance to it. Also called endothia canker, chestnut canker, chestnut-bark disease.

Chestnut Oak–*Quercus montana*, family Fagaceae; a tall deciduous tree found on dry hillsides in the eastern United States. It becomes crooked if not trained upright. The wood is used for railroad ties, railroad cars, fences, pilings, timbers, veneer, flooring, vehicles, planing-mill products, furniture, cooperage boxes, crates, agricultural implements, caskets, coffins, handles, etc. Native to America. Also called tanbark oak, rock chestnut oak.

Chigger–*Eutrombicula alfreddugesi*, family Trombiculidae; a larva infesting humans, domestic animals, some birds, snakes, turtles, and rodents. Its bite results in inflamed spots accompanied by intense itching. Also called red bug, jigger.

Chigoe–*Tunga penetrans*, family Tungidae; a small, reddish-brown flea which burrows into the skin, especially between the toes and under the toenail of humans and animals, causing intense pain, itching, and sometimes chronic sores. Sometimes wrongly called chigger.

Chigoo–*Sacropsylla penetrans*, type of the family Sacropsyllidae. The female of this flea genus attacks humans and animals by getting under the skin of feet and hands causing severe sores. Also called jigger.

Chimney–A vertical opening, a foot or more wide, in a pile of lumber to facilitate circulation of air and drying of the lumber.

China Wood Oil Tree–See Tung Oil Tree.

Chinaberry–*Melia azedarach*, family Meliaceae; a semi-evergreen tree grown for shade in the United States. Its seeds are sometimes used for rosaries; the drupes are poisonous, producing complete paralysis, irregular respiration, and suffocation. Also called china tree, bead tree, pride of India.

Chinook Wind–A northwestern United States term for a hot, dry wind which, with its moisture precipitated, descends from the Rocky Mountain crests and spreads over the lower land eastward, evaporating the snow rapidly. It is beneficial in winter, but may desiccate crops during the growing season.

Chip–(1) A small piece of wood or stone. (2) To cut wood or stone into small pieces. (3) To rewind the face of a scar on a pine tree to renew the flow of oleoresin in the turpentining industry. (4) To hoe or cultivate (Australia).

Chips–(1) Small pieces of wood or bark which have fallen into the cup or box of a turpentined tree or which have been removed with the scrape from the scarified face above the box. (2) Pieces of dried dung used for fuel.

Chiseling–Tillage with an implement having one or more soil-penetrating points that loosen the subsoil and bring clods to the surface. A form of emergency tillage to control soil blowing.

Chitin–A nitrogenous polysaccharide occurring in the cuticle of arthropods and certain other invertebrates. Probably occurs naturally only in chemical combination with protein.

Chlordane–A chlorinated hydrocarbon insecticide. Once used as a termite control chemical, it has now been banned for most applications because it remains effective in the soil for at least twenty years.

Chlorides (Muriates)–Salts of hydrochloric acid, which may be formed by the action of hydrochloric acid on an alkali or metal.

Chlorinate–(1) To treat with chlorine, as a water system. (2) Chlorinated hydrocarbons: a class of generally long-lasting, broad-spectrum pesticides of which the best known is DDT, first used for insect control during World War II. Other similar compounds include aldrin, dieldrin, heptachlor, chlordane, lindane, endrin, mirex, benzene hexachloride (BHC), and toxaphene. Many chlorinated hydrocarbons are now banned from use in the United States because they last too long.

Chlorinated Lime–Calcium hypochlorite; a powerful disinfectant and deodorant. Also called bleaching powder, bleach, chloride of lime, calcium chloride.

Chlorination–Disinfecting a substance, retarding decomposition, or oxidizing organic matter with chlorine or bleaching powder.

Chlorinator–A device for adding chlorine gas to sewage to kill infectious organisms.

Chlorine–The chemical element Cl, a greenish-yellow, poisonous gas. Chlorine compounds are used in disinfectants and deodorants.

Cholinesterase Testing–A pesticide-monitoring program that monitors a person's blood at various intervals during periods that he/she is using pesticides to determine the degree to which the person has been exposed to the pesticide.

Chronosequence–A sequence of soils whose properties are functionally related to time as a soil formation factor.

Chrysalis–The pupa, or the resting stage (intermediate between larva and adult), of butterflies.

Cide–A suffix which indicates a killer; e.g., amoebacide, that which kills amoeba.

Cienaga–(1) An area where the water table is at or near the surface of the ground. The term is usually applied to areas ranging in size from several hundred square feet to several hundred or more acres. Sometimes springs or small streams originate in the cienaga and flow from it for short distances. (2) An elevated or hillside marsh containing springs. Local in the southwestern United States.

Ciguatoxin–A toxin, sometimes fatal to people, occurring in such fish as sea bass, red snapper, barracuda, and grouper.

Cinchona–A large genus of trees and shrubs, family Rubiaceae, characterized by opposite leaves and small panicled flowers. The bark from several of its species yields the commercial quinine. Native to South America; grown in Southeast Asia and the West Indies.

Cinder Soil–Soil composed of the ejects of volcanic eruptions, generally fragmental matter. It forms considerable aggregate area in parts of the western United States.

Circa–(Latin) About, approximate.

Circumneutral–Term applied to water with a pH of 6.5 to 7.4.

Cirque–A deep, steep-walled recess in a mountain, caused by glacial erosion.

Cistern–An underground tank for storing rainwater for use by a farm or ranch home. Water from home or barn roofs is caught by gutter and directed into the cistern. It is a soft water, meaning it is low in calcium and magnesium.

Citrus Belt–The citrus-producing regions of the world lying at low elevations generally within 35° north and south of the equator.

City Forestry–The preservation and protection of shade trees in streets and parks.

Clam Shells–Finely ground clam shells are used as a liming material along most sea costs. The mean composition is 38 percent calcium and 45 percent carbon dioxide (CO_2).

Clarification–In waste-water treatment, the removal of turbidity and suspended solids by settling, often aided by centrifugal action and chemically induced coagulation.

Clarify–To remove undesirable, solid substances from a liquid, such as milk or fruit juice, by ordinary or by centrifugal filtration.

Class–A division of the plant or animal kingdom lower than a phylum and higher than an order; e.g., the class Insecta.

Classification–The forming, sorting, apportioning, grouping, or dividing of objects into classes to form an ordered arrangement of items having a defined range of characteristics. Classification systems may be taxonomic, mathematical, or other types, depending on the purpose to be served.

Classify–(1) To systematically categorize plants or animals according to a set scheme. (2) To sort individuals together into groups having common characteristics or attributes.

Clay–(1) A size term denoting particles, regardless of mineral composition, with diameter less than 2 microns (agriculture) or 4 microns (geology). (2) A group of hydrous alumino-silicate minerals related to the micas (clay minerals). (3) A sediment of soft, plastic consistency composed primarily of fine-grained minerals. (4) In engineering, any surficial material that is unconsolidated.

Clay Eaters–Animals (and sometimes people) that eat clay because of diseases or inadequate nutrition.

Clay Film–A thin coating of clay on the surface of a soil aggregate or lining soil pores or root channels. Syn.: clay coat, clay skin. Common in some B horizons of soil.

Claypan–A dense, compact layer in the subsoil having a much higher clay content than the overlying material, from which it is separated by a sharply defined boundary; formed by downward movement of clay or by synthesis of clay in place during soil formation. Claypans are usually hard when dry, and plastic and sticky when wet. Also, they usually impede the movement of water and air and the growth of plant roots. They can occur naturally or be induced by heavy traffic. See Fragipan.

Clean Culture–Intensive cultivation of a field so as to remove all weeds, etc. Also called clean cultivation, clean tillage.

Clean Water Zone–That area of water, in a polluted stream, in which self-purification has been completed.

Clean-burning–A backfire set close to the fire edge to control a forest fire.

Clear–(1) In milling, the residual wheat flour left after the separation of patent flour. (2) In land reclamation, to remove trees, brush, stones, or other obstacles to tillage.

Clear Length–The limb-free portion of the stem of a tree; i.e., from the ground to the lowest branch or branch stub.

Clearcutting–An area on which the entire timber stand has been cut. Removal of the entire stand in one cut. Tree reproduction is obtained with or without planting or artificial seeding.

Clematis, Western–*Clematis ligusticifolia*; a vigorous, climbing, deciduous native vine with pinnate leaves and feathery or densely villous white seeds. It is valuable wildlife habitat and ground cover in riparian areas and along fencelines.

Cliff–A high, steep face of rock; a precipice.

Climate–The sum total of all atmospheric or meteorological influences, principally temperature, moisture, wind, pressure, and evaporation, which combine to characterize a region and give it individuality by influencing the nature of its land forms, soils, vegetation, and land use.

Climate Year–A year which is selected for presentation of data on streamflow, precipitation, etc. The climate year of the U.S. Geological Survey, called a water year, extends from October 1 to September 30 following. Also called climatic year.

Climatic Classification–Classification of the climates of the different regions of the earth's surface, based on one or more of the climatic elements, such as temperature, rainfall, humidity, wind, temperature and rainfall, nearness to land and sea, and many others. Classifications may also be based on the distribution of vegetation, on physiological effects, or may be on any basis suitable for the particular purpose or investigation.

Climatology–The science of climates.

Climax Vegetation–(1) The group of plant species which is the culminating stage in plant succession for a given set of environmental conditions. (2) A relatively stable type of vegetation in equilibrium with is environment and with good self-perpetuating reproduction of the dominant plant species.

Climbing Bog–Regions characterized by a short summer and a considerable amount of rainfall, in which sphagnum moss frequently extends upward from the original level of the swamp, carrying the marsh conditions to higher land.

Climbing Dune–An active sand dune capable of moving over obstructions.

Climograph–An expression of the climatic conditions of a particular area by a graph of which one coordinate is the mean monthly temperature and the other the mean monthly precipitation.

Clinical–Concerning the investigation of decease on a living subject by observation, as contrasted to a controlled experiment.

Clinical Evidence–Any symptom of disease that can be determined by direct observation, such as fever, lack of appetite, swellings, paralysis, etc.

Clinker–(1) Burnt-looking, vitrified, or slaggy material thrown out by a volcano. (2) Rough, jagged lava, generally basic, typically occurring at the surface of lava flows. (3) Slaggy or vitreous masses of coal ash.

Clinometer–A handheld instrument used to measure angles and slopes. It is used in forestry to determine the height of a tree.

Clinosequence–A sequence of soils whose properties are functionally related to the amount of slope on which they are formed.

Clod–A compact, coherent mass of soil ranging in size from 0.2 to 0.4 inches (5 to 10 millimeters) to as much as 8 to 10 inches (200 to 250 millimeters); produced artificially, usually by plowing or digging, especially when these operations are performed on clay soils that are either too wet or too dry for normal tillage operations. See Ped.

Closed Drain–Conduits laid to grade beneath the surface to remove excess water from wet land. The joints are sealed to make a closed line on steep slopes or in areas where tree roots can penetrate and clog the line.

Clothes Moth–Several species of moths of the order Lepidoptera, family Tineidae, especially ***Tineola bisselliella***, the webbing clothes moth, and ***Tinea pellionella***, the casemaking clothes moth. The female lays the eggs in clothes and rugs, etc., which hatch in about five days. The larvae, depending on the food supply, temperature, and humidity, may require six weeks to four years to reach maturity and, in the meantime feed on hair fiber, wool, silk, felt, feathers, etc., in clothing, carpets, or upholstery.

Cloud Seeding–The production of rain by use of artificial nuclei. Favorable clouds are seeded with dry ice, silver iodide, or other substances to produce nucleation and thereby precipitation. The success of such procedures remains controversial.

Cloudburst–An intense rainstorm which usually lasts for a short time.

Clouding–An undesirable translucent condition of clear liquids.

Coal–A readily combustible mineral containing more than 50 percent by weight and more than 70 percent by volume of carbonaceous material including inherent moisture, formed from compaction and induration of variously altered plant remains similar to those in peat. Differences in the kinds of plant materials (type), in degree of metamorphism (rank), and in the range of impurity (grade) are characteristic of coal and are used in their classification.

Coal Gas–The fuel gas produced from a high-volatile bituminous coal. The average composition of coal gas by volume is: 50 percent hydrogen, 30 percent methane, 8 percent carbon monoxide, 4 percent other hydrocarbons, and 8 percent carbon dioxide, nitrogen, and oxygen.

Coal-tar Creosote–A dark green or brown distillate oil frequently containing naphthalene and anthracene, which is used as a wood preservative and disinfectant. It is poisonous and a skin irritant.

Coarse Texture–(1) In soils, the sand, loamy sand, and sandy loam (except the very fine sandy loam) textural classes. (2) (a) A body of defect of sherbet and ice cream caused by insufficient sugar or stabilizer, bad refrigeration, or failure to put them in the hardening room immediately after removal from a freezer. (b) A body and textural defect of ice cream, characterized by a lack of smoothness. (3) A textural condition of a fruit, vegetable, or other plant or plant produce.

Coast–The strip of land, of indefinite width (up to several miles), that extends from the shoreline inland to the first major change in terrain features.

Cobble–A rock fragment between 25 and 100 inches (64 and 256 millimeters) in diameter, thus larger than a pebble and smaller than a boulder, rounded or otherwise abraded in the course of aqueous (water), eolian (wind), or glacial (ice) transport.

Coccus–A spherical bacterium, as of the family Coccaceae.

Cockroach– An insect pest of the family Blattidae. Species are found in homes, stores, and almost anywhere. They can transmit many diseases.

Coconut Palm– *Cocos nucifera*, family Palmaceae; a widely distributed, pinnate-leaved tropical tree, grown for its drupe fruit, the coconut of commerce. Native to Asia and Pacific islands.

Codominant– (1) A tree which grows close enough to other trees as to receive very little sunlight from the sides. (2) Any plant or animal species that shares dominance with other species in an area.

Coefficient of Roughness– A variable factor used to estimate the flow of water in rivers and canals, determined by the roughness or smoothness of channel lining, the mean hydraulic radius, mean velocity of the flow, and slope.

COH– Abbreviation for coefficient of haze, a unit of measurement of visibility interference.

Cohabitation– Two or more organisms (plants or animals) living together.

Cohort– (1) A group of people with a defined history of exposure who are studied for a specific length of time to determine cancer incidence or mortality. (2) A group of individuals born within the same time period (usually within five or sometimes ten years of each other). Such groups are called birth cohorts. The diseases among individuals in one birth cohort followed throughout their lifetimes may be different from those in another, implying differences in exposures to environmental factors causing disease.

Cold Front– The movement of a relatively cold air mass advancing behind a relatively warm air mass. Characteristic weather includes short but violent thunderstorms, heavy rain squalls, and occasionally hail and tornados. Weather behind the passage of the cold front will usually be cooler and dryer than before, with clear skies.

Cold Pocket– A low area into which cold air settles from adjoining slopes. See Frost Pocket.

Coldframe– An enclosed, unheated but covered frame useful for growing and protecting young plants in early spring. The top is covered with glass or plastic and located so it is heated by sunlight.

Colemanite– A natural calcium borate ore that contains 32 to 50% B_2O_3. It is insoluble in water, but readily available to plants. See Borax, Boron.

Colloid– (1) An insoluble substance consisting of particles small enough to remain suspended indefinitely in a medium. (2) A mineral particle less than 0.002 millimeters in diameter. (3) A substance which does not diffuse readily through animal or vegetable membranes. Its presence does not affect the freezing point or vapor tension of the solution.

Colloidal Materials– Gas, liquid, or solid particles suspended in a gas, liquid, or solid that are intermediate in size between true solutions and suspensions.

Colloidal Peat– The unctuous, gelatinous mass of organic residue formed as a bottom deposit in lakes. It may lie at the base of a peat deposit, but does not conform in texture and appearance to the material commonly designated as commercial peat. See Peat.

Colloidal Phosphate– A finely divided raw (unprocessed) mineral phosphate or phosphatic clay. Occurs in large quantity mixed with Florida hard rock, from which it is removed by washing, carried by the wash water to settling basins or ponds and remains after the water has evaporated. Although sometimes applied directly to the soil, its phosphoric oxide is only slightly available, except in strongly acid soils. It contains from 50 to 58% tricalcium phosphate and from 9 to 12% iron and aluminum oxides.

Collop– A unit of grazing area which can support a full-grown horse or cow for one year.

Colluvium– A general term applied to loose and incoherent deposits, usually at the foot of a slope or cliff and brought there chiefly by gravity. Talus and cliff debris are included in such deposits. See Scree, Talus.

Colmatage– (Latin; *culmen*, ridge; Indo-European; *colen*, hill) (1) The practice of collecting soil sediments from low-lying places and spreading them on upland soils of low fertility to enhance plant growth. A common practice in all developing countries such as India and China. (2) The gradual filling in of lowland by siltation.

Colonization– A natural phenomenon whereby a plant or animal species invades an area previously unoccupied by that species and becomes established.

Colony– (1) An aggregate of worker bees, drones, and a queen bee, living together in a hive as a unit. (2) A cluster of bacteria or fungi grown on a culture medium, usually originating from a single bacterium, spore, or inoculation transfer. (3) A group of people bound by communal or religious ideals that occupies tracts of land, often wild, for permanent settlement, especially in the early development of the United States and Canada.

Color Profile– A soil profile described by the aid of the Munsell color chart according to the colors of its separate horizons. See Munsell Color Standards.

Colorado Spruce– *Picea pungens*, family Pinaceae; a hardy evergreen tree grown as an ornamental. The most widely grown variety is the blue spruce.

Columnar– Describing a plant or a soil structure whose outline resembles a column; i.e., one formed by relatively perpendicular sides.

Combined Sewer– A sewer system that carries both sewage and storm water runoff.

Commensal– A nonparasitic organism which lives attached to a host and shares its food.

Commercial Species– Tree species suitable for industrial wood products.

Commercial Thinning– Removing trees from a developing young stand so that remaining trees will have more growing space; dead and dying trees will be salvaged; and the operation will, hopefully, make a net profit.

Commercial Timberland– Forest land that is producing or is capable of producing crops of industrial wood and that is not withdrawn from timber utilization by statue or administrative regulation. Note: Areas qualifying as commercial timberland have the capability of producing in excess of 20 cubic feet per acre per year of industrial wood in natural stands. Currently, inaccessible and inoperable areas are included.

Comminution–(1) The reduction of a substance to a fine powder; pulverization; trituration. (2) Mechanical shredding or pulverizing of waste, a process that converts it into a homogeneous and more manageable material. Used in solid-waste management and in the primary stage of waste-water treatment.

Common–(1) A joint pasture in a village or community to which all members have access for their herds. (2) Legal right of a person in sharing the profit or use of another's land. (3) A low, market grade of meat animals. (4) The American wool grade equivalent to the English 44's; a low grade, only better than braid. (5) A defective grade of lumber useful only in framing, etc.

Communalism–The biotic relations among individual species of plant and animal life, both beneficial and harmful, at a given place.

Communicable–Readily transferred from one individual to another.

Community–A group of plants growing together or all of the plants and animals of an area.

Community Air–The outside air shared by an entire community.

Commutatus–Changing.

Compaction–(1) Decrease in volume of sediments, as a result of compressive stress, usually resulting from continued deposition above them, but also from drying and other causes. (2) Reducing the bulk of solid waste by rolling and tamping. (3) Increasing soil bulk density and decreasing porosity caused by the application of mechanical forces to the soil.

Compass–(1) An instrument for determining directions, usually by the pointing of a magnetic needle free to turn in a horizontal plane toward magnetic north; sometimes having a clinometer attached for measuring vertical angles. Also, a dip compass, for tracing magnetic iron ore, having a needle hung to move in a vertical plane. (2) An instrument for describing circles, transferring measurements, etc.

Compatibility Agent–An adjuvant that helps unlike chemicals mixed together for even application. See Adjuvant.

Compatible–Describing two or more chemicals that can be mixed without affecting each other's properties.

Compatible Pesticides–Compounds or formulations that can be mixed and applied together without undesirably altering their separate effects. Sometimes such a mixture makes one or both pesticides more effective.

Compensation–The condition when one or a group of environmental factors, usually limiting to the life of an organism, is counteracted by an excess of other factors. For example, cold winters in Alaska are compensated by long hours of daylight.

Competition–The struggle for existence that results when two or more organisms have similar requirements in excess of the supply.

Competitor–An organism competing with one or more other organisms.

Complete Carcinogen–An agent that can act as both initiator and promoter of cancer.

Complete Fertilizer–A fertilizer that contains some of each of the three essential nutrients, N, P, and K. Ratios of each can vary depending upon the formulation.

Completely Saturated–(1) Designating clay or humus containing all of the basic ions plus hydrogen ions that it is capable of holding. (2) Any substance, such as soil, which can hold no more water.

Complex Slope–Irregular or variable slope in a field. Planning or constructing terraces, diversions, and other water-control measures is difficult.

Complex Soil–A map unit of two or more kinds of soil occurring in such an intricate pattern that they cannot be shown separately on a soil map at the selected scale of mapping and publication.

Compositae–A very large family of plants, herbs, and trees, which includes flowers, such as marigold, dahlia, chrysanthemum; vegetables, such as chicory and lettuce; and weeds, as burdock, etc.

Composition–(1) The relative production of various plant species in a given area. (2) The components used in any mixture of feed, fertilizers, hay, seeds, etc.

Compost–(1) Organic residues or their mixture, such as peat, manure, or discarded plant material and soil, placed in a pit or enclosure, moistened, and allowed to become decomposed. Sometimes lime and chemical fertilizers are also added. Used as a fertilizer. (2) To cause vegetable matter to become decomposed as a fertilizer.

Compound–A chemical term denoting a combination of two or more distinct elements.

Compound Fertilizer–A mixed fertilizer or formulation containing at least two of the primary plant nutrients (N_1 P_2O_5, and K_2O) formed by intimately mixing two or more fertilizer materials by dry or fluid bulk blends. Compound fertilizers are made in registered grades approved by the laws of the respective states.

Compression Wood–Abnormal wood formed on the lower side of branches and inclined boles of conifer trees.

Concentrate Sprayer–A sprayer designed to deliver pesticides to a crop at normal amounts per acre of active ingredients but in much lower volumes of water.

Concrete Frost Structure–A type of frost in which the soil becomes virtually solid. See Granular Frost Structure.

Concretion–(1) An inorganic body formed in a human or animal body cavity, as a kidney stone. (2) Grains, pellets, or nodules of various sizes, shapes, and colors consisting of concentrated compounds or cemented soil grains. The composition of most concretions is unlike that of the surrounding soil. Calcium carbonate and iron oxide are common compounds in soil concretions.

Condensation–The process of turning a vapor into a liquid; e.g., dew is formed by condensation.

Conditioners–Conditioning materials, anticaking agents. Finely divided, dry, bulky, inert powders such as diatomaceous earth, siliceous dusts, and clays in common use as coating agents to decrease the caking tendency of fertilizers. Oils, organic amines, and plastic coatings are used in a few specialty fertilizers. Conditioning agents will prevent caking only when the moisture content of the product is sufficiently low to inhibit the formation of crystalline bridges between particles.

Conduit–(1) Any channel, open or closed, intended for the conveyance of water; any container for flowing water. (2) A pipe in which wiring is installed.

Cone of Depression–The downward slope of the water table surface around a well as a result of pumping or around a drainage line as a result of drainage.

Cone Penetrator–A rod-shaped instrument with a cone-shaped end, used for penetrating the soil to measure soil strength.

Conifer–(1) A tree or shrub that bears cones, as pines and firs. (2) Any plant belonging to the family Coniferae, a large group of trees and shrubs, which includes the families Taxaceae, Pinaceae, Cycadaceae, and Ginkgoaceae. Conifers bear cones or strobili, and because many of the species produce new leaves before all the old leaves are shed they are looked upon as synonymous with evergreens. One principal exception is the bald cypress, a deciduous conifer.

Confined Water–Underground water trapped (as petroleum is) in geologic structures. It is usually highly mineralized and once drawn out is not renewable since such water is not a part of the general hydrologic circulatory system. Also known as fossil water. See Fossil Water.

Confluence–The point where two streams meet.

Confluent–Merging or blending together.

Congelifraction–The mechanical disintegration of minerals and rocks resulting from the pressure exerted by freezing of water contained in their cracks. Freezing water exerts a pressure of 150 tons/square inch (146.48 kg/cm^2).

Conservation Plan–Includes but is not limited to farm/ranch "conservation plan maps"; water, plant, animal, and other inventory and management information with needed interpretations and evaluations; a record of the decisions made contributing to sound land use and conservation treatment; the alternatives for sound land use(s) and conservation treatment for which conservation decisions have not yet been made (including positive statements about critical problems such as soil erosion, sedimentation, land use, and agricultural pollutants); records of understandings as to cooperative agreements among individuals, groups or government representatives, and resource conservation districts; and other information useful to the decision maker.

Conservation Pool–The unusable, undrainable space below the outlet in a reservoir. It is often allocated to silt storage, fish, waterfowl, recreation, etc.

Conservation Practice–A land treatment measure to protect or conserve soil, water, woodland, or wildlife resources, or the installation of a structure or other measure for this purpose.

Conservation Reserve Program–A program established by the Food Security Act of 1985. It pays farmers to take out of cultivation the most highly erodible cropland.

Conservation Tillage–Any of several farming methods that provide for seed germination, plant growth, and weed control yet maintain effective ground cover throughout the year and disturb the soil as little as possible. The aim is to reduce soil loss and energy use while maintaining crop yields and quality. Kinds of conservation tillage are: chisel-plow, no-till (zero-till), plow-plant, ridge-plant, strip-tillage, sweep-tillage, till-plant, and wheel-track-plant.

Conservative Grazing–(1) The practice of limiting the number of livestock on a grazing range in such proportion as would not exhaust the range in successive seasons. (2) A degree of grazing that causes little or no soil disturbance.

Consistence (Consistency, Soil Strength)–The feel of the soil and the ease with which a lump can be crushed by the fingers. Terms commonly used to describe consistence are: *loose*—noncoherent when dry or moist; does not hold together in a mass. *Friable*—when moist, crushes easily under gentle pressure between thumb and forefinger and can be pressed together into a lump. *Firm*—when moist, crushes under moderate pressure between thumb and forefinger, but resistance is distinctly noticeable. *Plastic*—when wet, readily deformed by moderate pressure but can be pressed into a lump; will form a "wire" when rolled between thumb and forefinger. *Sticky*—when wet, adheres to other material and tends to stretch somewhat and pull apart rather than to pull free from other material. *Hard*—when dry, moderately resistant to pressure; can be broken with difficulty between thumb and forefinger. *Soft*—when dry, breaks into powder or individual grains under very slight pressure. *Cemented*—hard; little affected by moistening.

Consorting–A species of trees that, although it may be found in pure stands, is most often found as a major segment in a mixed stand, that is, one of the more abundant species.

Constituent–The components or elements of a substance; e.g., the constituents of soil.

Consumptive Use–The water used by plants in transpiration and growth, plus water vapor lost from adjacent soil or snow or from intercepted precipitation in any specified time. Usually expressed as equivalent depth of free water per unit of time, such as acre-inches per week.

Contaminant–An undesirable substance that is present in foods or feed but is not intentionally added.

Contaminate–To make impure by contact or admixture of harmful bacteria, fungi, or dangerous chemicals, etc.; to render unfit for use.

Contamination–(1) Pollution; the process of being contaminated. (2) Specifically, the addition of bacteria or other foreign substance to milk or other products by means of utensils, containers, exposure to air, etc.

Continental Climate–Climate typical of great land masses. It is characterized by a great range of temperature occurring in parts of a continent that are not affected materially by nearness to the sea.

Continental Drift–The concept that the continents can drift on the surface of the earth because of the weakness of the suboceanic crust, much as ice can drift through water. As proposed by the German meteorologist Alfred Wegener in 1912, the theory that the continents were once joined into a landmass which broke apart into several landmasses, which then drifted, their shapes changing somewhat, and eventually arriving at their present positions.

Contour–(Latin; *contra*, against; *tornare*, turn) A line on a map joining points on the land surface of equal elevation. A series of such lines depicts the topography of an area.

Contour Border Irrigation–A method of irrigating gently sloping fields. The whole area is divided into strips by ridges along the contours and cross ridges. The ridges confine the water to a particular strip until it is completely full, before letting it flow to the next lower strip. Also called contour check irrigation.

Contour Ditch–A ditch laid out approximately on the contour.

Contour Farming–Field operations, such as plowing, planting, cultivating, and harvesting, on the contour, or at right angles to the natural slope to reduce soil erosion, protect soil fertility, and use water more efficiently.

Contour Furrow–(1) A narrow furrow plowed along a contour or at a uniform grade. (2) A level furrow made in a field or pasture to reduce water runoff and soil loss, and to increase water infiltration in soil.

Contour Interval–On a contour map, the difference in elevation, or the vertical distance, between contours.

Contour Line–(1) An imaginary line on the surface of the earth connecting points of the same elevation. (2) A line drawn on a map connecting points of the same elevation.

Contour Map–A map showing relative elevations by a series of lines, each line connecting points of equal elevation. Each map has a uniform contour interval. Topography can be interpreted from a contour map.

Contour Planting–The planting or drilling of crops in rows along contour lines in contrast to plantings which run parallel to field boundaries or up- and downhill on sloping land.

Contour Plowing–Plowing a field on a contour.

Contour Row–(1) A row, all points of which have the same elevation within a given tolerance. (2) A level row that runs at right angles to the line of slope regardless of the irregularities of the landscape.

Contour Strip Cropping–The production of crops in long, variable-width strips, which are placed approximately on the contour, and crosswise to the line of slope.

Contour Tillage–A system of farming in which the various operations are performed on the contour. See Conservation Tillage.

Contracted Weir–A device for measuring the flow of water. The sides produce a contraction in the cross-sectional width of the water channel.

Control–(1) Prevention of losses from plant or animal diseases, insect pests, weeds, etc., by any method. (2) A section of an open water channel where conditions exist that make the water level above it an index of the discharge. (3) A standard entity used for comparative purposes in experimentation. Also called check or check plot.

Control Flume–An open conduit or artificial channel arranged for measuring the flow of water. See Parshall Flume.

Control of a Fire–The surrounding of a fire with control lines and backfiring of any unburned surfaces adjacent to the inner edge of the control lines.

Control Section–The part of the soil on which classification is based. The thickness varies among different kinds of soil, but for many it is 10 to 30 centimeters.

Controlled Burning–Setting fire to land cover under conditions that presumably will accomplish specific silvicultural, wildlife, grazing, or fire hazard reduction purposes. See Prescribed Burning.

Controlled Release Fertilizers–Fertilizers in which one or more of the nutrients have limited solubility in the soil solution, so that they become available to the growing plant over a controlled period. The ideal in such a fertilizer would be the release of nutrients at a rate exactly equal to the needs of the plant. Manures and sewage sludges are decomposed by microbes and the result is a controlled release of plant nutrients approximately in harmony with the growing plant.

Convection–The transfer of heat through a liquid or gas by the actual movement of the molecules.

Convection Downdraft–Downward movements of cold air in a building caused by warm air rising by convection and being cooled by contact with uninsulated surfaces.

Conversion–(1) Sawing or cutting timber to any shape. (2) The change from one reproduction method to another. (3) The transformation of a forest from one dominant type to another.

Conversion Factors–Convenient multipliers that have been calculated for elements and compounds important in fertilizers, soil amendments, and plant nutrition. The calculations are based on atomic weights of each chemical. For example, to determine the conversion factor for changing from a known percentage or mass of ammonia (NH_3) to an unknown percentage or mass of nitrogen (N), divide the atomic weight of NH_3 (17.04) by the atomic weight of N (14.01) = 1.216. Then to convert N to NH_3, multiply N by 1.216.

Conveyance Loss–The loss of water from a conduit due to leakage, seepage, and evaporation.

Cooperage–The barrels, casks, and tubs made by staves and hoops usually from wood; made by a cooper. Tight-cooperage is made from such nonporous woods as white oak to hold liquids such as whiskey. Slack-cooperage is made form porous but tough wood such as ash and elm for holding nonliquids such as rice and cranberries.

Copper–Cu; a metallic element found in soils at 1 to 50 parts per million, and in plants up to 100 parts per million. It is necessary for all animal and plant life. High soil phosphorus, zinc, and molybdenum can induce copper deficiency in plants. Also, high copper can reduce plant uptake of phosphorus, iron, zinc, and molybdenum.

Copper Carbonate–A compound used as a seed disinfectant.

Copper Oxide–A chemical used in the seed treatment sprays of dusts for sugar beets and vegetables. It controls damping off, rose leaf diseases, celery blight, tobacco and hops downy mildew. Also called cuprocide.

Copper Poisoning–A livestock poisoning usually in sheep, from consuming too much salt to which copper sulfate has been added for control of stomach worms. Symptoms are loss of appetite, jaundice, and often bloody urine.

Copper Sulfate–A common chemical used extensively to kill algae in waste waters or natural surface waters. Syn.: blue vitriol, bluestone, blue copperas, and cupric sulfate.

Copperas–See Ferrous Sulfate.

Coppice Forest–A forest of second-growth sprouts. Also called sprout forest.

Coprogenous–Designating the influence of animal excrement, as the cast of the earthworm in forming soil.

Coprolites–(1) Round lumps of fossilized excrements of prehistoric animals, rich in phosphate of lime. (2) Phosphate nodules which are not fossil excrement.

Coprophagous Insect–An insect that consumes dung.

Cord–(1) (a) The spermatic cord. (b) The umbilical cord. (2) A unit of measurement of timber. It contains 128 cubic feet, usually 8 x 4 x 4 feet. A short (face cord is a stack of wood 8 feetlong, 4 feet high, and about 16 inches wide. (3) Twine. (4) To stack in rectangular tiers.

Corduroy Road–A passage made by laying saplings or small poles close together across a road. Used in swampy or wet areas.

Core–(1) Central part of the earth beginning at a depth of about 2,900 kilometers, probably consisting of iron-nickel alloy; divisible into an outer core that may be liquid and an inner core about 1,300 kilometers in radius that may be solid. (2) Sample of rock obtained in core drilling. (3) The central part of something, especially the filling of a hollow object. (4) The heart of a nuclear reactor, where energy is released.

Coriolis Effect–The defective force of the earth's rotation. In the Northern Hemisphere, moving bodies are deflected to the right, and to the left in the Southern Hemisphere. (Named after the French mathematician, G. G. de Coriolis, 1792-1843.)

Cork Oak–*Quercus suber*, family Fagaceae; a fairly small tree, whose thick and corky bark is the commercial source of cork. Native to southern Europe.

Corn and Winter Wheat Belt–That area of the Midwestern United States where soft red winter wheat is grown in rotation with corn. Includes Ohio, Indiana, Iowa, and Illinois.

Corn Belt–Central part of the United States, where corn is extensively cultivated. Because of increasing diversification of crops, the area is becoming less distinctive, and the term is losing some of its value as a geographic referrent.

Corn Dance–A ceremonial dance held by American Indians at the planting and harvesting of corn.

Corner–(1) The outer pair of incisor teeth in the upper and lower jaws of a horse. (2) The junction point of boundary liens. (3) To tie up or control all available items of produce for speculation. (4) In lumbering, to cut through the sapwood on all sides of the tree to prevent it from splitting when the tree is felled.

Corrasion–The wearing away of rock material by running water, glaciers, winds, waves, or mass movements.

Correction Strip–In strip cropping, a strip of irregular width placed between uniformly wide contour strips of cultivated crops to reduce soil erosion.

Corrosion–(1) The process whereby surface or ground waters, by their own solvent action and by the help of solutes they carry, dissolve or chemically alter rock materials with which they come in contact. (2) The action of an agricultural chemical on the metal parts of distributors and containers.

Corrugation System–A method of irrigation used primarily for small grain and hay crops. Small irrigation furrows are placed 18 to 36 inches apart to wet the soil between them.

Cottontail–The common American rabbit, *Syulvilagus floridamus*, family Leporidae; a pest which often injures crops, shrubs, bushes, etc., by feeding on them. It is also a carrier of tularemia.

Cottony Maple Scale–*Pulvinaria innumerabilis*, family Coccidae; a destructive insect that infests the soft maple in the United States and Canada. Cottony masses of scale appear along the undersides of twigs and branches of the infested trees in May and June. The entire foliage of the tree turns yellow, and the heavily infested branches die. It also attacks the linden, Norway maple, willow, apple, pear, poplar, grape, hackberry, sycamore, honey locust, beech, elm, plum, peach, gooseberry, Virginia creeper, currant, and sumac.

Coulee–(1) A small creek or run; a sharply cut narrow valley carrying only storm water. Also called arroyo (southwestern United States). (2) A steep-walled, dry, glacial stream valley; a dry canyon as in the scablands of the Columbia Plateau region of the northwestern United States. (3) A short, blocky, steep-sided lava flow, generally of glassy rhyolite or obsidian, issuing from the flank of a volcanic dome or from the summit crater of a volcano.

County Drain–A constructed drainage system consisting of an open ditch, a closed conduit, or a combination of both that provides a drainage outlet for agricultural and urban areas. It is administered according to the laws of the individual state.

Cove–(1) A small, flat-bottomed valley with a steep head (eastern United States). (2) A narrow, limestone valley, associated with mountainous ridges of less valuable land in the Appalachian Valley of the eastern United States. (3) A narrow strip of prairie land that extends into a forest. (4) A small inlet, or creek, protruding into a shore.

Cover–(1) A lid placed over something for protection. (2) Woods, underbrush, etc., which may conceal game. (3) Plant life, such as grass, small shrubs, herbs, etc., used to protect soil from erosion. (4) The flesh, hide, and fat on a fattened animal. (5) To buy back future contracts. (6) To copulate with a female, as a bull covers a cow. (7) To incubate, to hatch eggs, as a hen. (8) The proportion of the ground surface under live and dead aerial parts of plants. Also refers to shelter and protection for animals and birds.

Cover Crop–A crop growing close to the ground for the chief purpose of protecting the soil from erosion and also for the improvement of its productivity, between periods of regular production of the main crops, or between trees and vines in orchards and vineyards.

Cover Material–Soil that is used to cover compacted solid waste in a sanitary landfill.

Cover Type–The present vegetation occupying an area; based on plant dominance (devised by foresters and wildlife managers).

Covert–(1) A unit of wildlife cover: e.g., a thicket. (2) One of the small feathers covering the bases of large feathers in fowls.

Covert Escape–Vegetation which serves as a refuge for animals.

Cowl–(1) A curved, flaring top fitted to a ventilator, turned by a wind vane in a downwind direction to improve drafts. (2) The hoodcovering of a tractor.

Coyote–*Canis latrans*; a species of wolf that is a predator of sheep, goats, and chickens. However, it also keeps the rabbit population in check.

Cracking–(1) Development of fissures as wet clay soil dries and bakes in the sun. (2) Development of fissures in the skin of fruits, which usually occurs during rainy periods at or just before ripening. Also called fruit cracking.

Cradle Knoll–A mound or ridge of earth on a forest floor, often no more than a foot or two high; probably caused by the overthrow of large trees and consequent accumulation of the soil in the root mat through rainfall, frost action, and decay of the wood (Lake States, United States).

Cranberry Bog–A low-lying, wet area of peat, muck, or gray, sandy, acid soil capable of being flooded and drained periodically; used for the production of cranberries.

Crappie–See Black Crappie.

Crater, Volcanic–A steep-walled depression at the top of a volcanic cone or on the flanks of a volcano, directly above a pipe or vent that feeds the volcano, and out of which volcanic materials are ejected. In its simplest form, usually a flat-bottomed or pointed, inverted cone more or less circular in plan. The diameter of the floor is seldom over 300 meters; the depth may be as much as several hundred meters (yards). Primarily the result of explosions or collapse at the top of a volcanic conduit.

Crawfish–See Crayfish.

Crawler–(1) A newly hatched insect. (2) A large earthworm. (3) A tractor equipped with tracks, as compared with one equipped with wheels.

Crayfish–Any of the freshwater crustaceans of the genus *Cambarius*. They are raised for food in some parts of the southern United States. Also known as crawfish or crawdad.

Crayfish Land (Crawfish Land)–Flat, wet land, underlaid by gray, clayey soil, which is characterized by holes and tiny mounds made by crayfish, the freshwater crustacean. (Used loosely as a term for poor agricultural land.)

Creek–A small stream or river. The term is relative, since in one section of the United States a stream of water called a creek may be called a river in another.

Creep–(1) Area where young piglets spend most of their time, and which has openings too small to allow the sow to enter. See Creep Feeding. (2) A phosphorus-deficiency disease of animals characterized by anemia, softening of the bones, and a slow gait. Also called bush disease. (3) To spread, as a plant, by means of creeping stems, which root at the nodes. (4) Slow mass movement of soil and soil material down relatively steep slopes primarily under the influence of gravity, but facilitated by saturation with water, strong winds, and by alternate freezing and thawing.

Crest–(1) The tuft of feathers on the heads of certain fowls. (2) A cock's comb. (3) The ridge of an animal's neck. (4) The top of a dam, dike, spillway, or weir. (5) The peak or high water mark of a flood. (6) A toothed or irregular ridge of appendage on the petals or flower cluster of certain plants. (7) The top of the hill.

Cretaceous–The most recent geological period of the Mesozoic era, beginning about 135 million years ago and lasting for about 50 million years.

Crevasse–(1) A wide crack in the ice formed under the influence of various strains. (2) A nearly vertical fissure (crack) in a glacier. (3) A break in a levee or stream embankment.

Crevice–(1) A shallow fissure in the bedrock under a gold placer, in which small but highly concentrated deposits of gold are found. (2) The fissure containing a vein. As employed in the Colorado statute relative to a discovery shaft, a crevice is a mineral-bearing vein. (3) Narrow, deep opening in a cavern floor. A narrow, high passageway. (4) An enlarged joint whether mineralized or not. (5) Corruption of crevasse.

Crickets–A large group of leaping insects, of the family Gryllidae, some of which are nocturnal and known for their chirping. The most common is the black field cricket.

Critical Slope–See Angle of Repose.

Crop Adaptation–The harmonious use of the hereditary and environmental factors of plants in particular areas and seasons so that the plants will grow to maturity. See Crop Ecology.

Crop Ecology–A science that deals with the study of agricultural crops in relation to their environment, such as soil, climate, plant reaction, etc.

Crop Residue–That part of a plant left in the field after harvest, leaves, stubble, roots, straw, etc.

Crop Residue Management–The system of retaining crop residue on land between harvest and replanting to reduce erosion, ensuring future crop production.

Crop Rotation–Growing annual plants in a different location in a systematic sequence. This helps control insects and diseases, improves the soil structure and productivity, and decreases erosion.

Crop Tree–An individual tree that is best adapted for a crop, usually as a saw log because of species, form, and condition.

Crop Yield Index–Crop yield per acre for a farm compared with crop yield per acre for the county or other comparable area, usually for a particular year.

Cross Section–(1) A cut made at right angles to ground level to permit examination of earth layers, underlying strata, etc. (2) In surveying, a grid system linking points of equal elevation and thus determining land contours.

Crosstie–(1) In the United States, a hewn or sawed piece of timber placed beneath railroad rails for support. Also called sleeper, railroad tie. (2) To tie together three of the feet of an animal as it lies on the ground.

Crotch–(1) A fork formed by the separation of two or more branches of a plant or tree. (2) A small sled made from the fork of a tree used to skid logs. Also called alligator, crazy dray, go-devil, travois, lizard. (3) The area between the hind legs of an animal.

Crow–Any of the large birds of the genus *Corvus*. They are glossy black in color and can be considered a pest in newly planted fields where they eat seeds and uproot seedlings.

Crow Drip–(1) That part of rain or melted snow intercepted by vegetation that falls to the ground in large drops. (2) The area of ground normally fertilized.

Crown–(1) The upper part of a tree, which bears branches and leaves. (2) To remove the top of a plant or tree. Sometimes called topping. (3) The place at which the stem and root join in a seed plant: the crown or top of the root. See Collar. (4) The upper surface of a furrow slice. (5) The width of the top of a levee. (6) The neck of a sweep or cultivator point used to fasten the sweep to the cultivator shank. (7) The top of an animal's head. (8) The dent in the cap of a kernel of corn.

Crown Fire–A fire that runs through the tops of living trees or brush.

Cruise–A forest survey made to estimate quantity and quality of timber.

Crumb–A natural structure in soil; small, spheroidal, very porous, easily crushed aggregates.

Crust–(1) A thin, brittle layer on the bare surface of many dried soils, which sometimes prevents plant emergence. This condition is usually caused by raindrop splash. (2) The outer wall of an animal's hoof. (3) Rough, tanned hide of a sheep or a goat. (4) On desert soils, a hard layer that contains calcium carbonate, gypsum, or other binding material often exposed at the surface.

Crustaceae–Aquatic animals with a rigid outer covering, jointed appendages, and gills. Included are the water fleas such as *Daphnia* and the Copepoda such as *Cyclops*.

Crustaceous–Hard and brittle.

Cryology–(1) In the United States, the study of refrigeration. (2) In Europe, a synonym for glaciology. (3) The study of ice and snow. (4) The study of sea ice.

Cryopedology–The study of frozen ground, including soils of arctic and antarctic regions.

Cryosphere–All of the earth's surface that is permanently frozen. See Permafrost.

Cryoturbation–Frost action, including frost heaving.

Crystal System–One of the six subdivisions of the thirty-two possible crystal classes. Each system is characterized by the relative lengths and inclinations of the crystallographic axes. The six systems are isometric, tetragonal, hexagonal, orthorhombic, monoclinic, and triclinic.

Crystallography–The science of the interatomic arrangement of solid matter, its causes, its nature, and its consequences.

Cuban Pine–Slash pine.

Cubic Foot per Second–The standard unit of measurement of water flow in irrigation, which is 1 cubic foot of water flowing past a given point in 1 second.

Cucaracha–(Spanish) Cockroach.

Cuesta–Used in southwestern United States for a plain with a gentle slope on one side and a steep slope on the other.

Cul-de-Sac–(1) A cavern passage that connects with another passage at only one point and ends abruptly in a rock wall or is blocked by cave fill or debris. (2) A dead end. (3) A turnaround at a dead end.

Culitgen–Any horticultural variety not known in the wild state.

Cultivation–(1) The planting, tending, harvesting, and improving of plants. (2) Tillage of the soil to promote crop growth after the plant has germinated and appeared above ground. (3) Loosening of the soil and removal of weeds from among desirable plants.

Cultural Control–The deliberate manipulation of the environment to make it less favorable for pests by disrupting their reproductive cycles, eliminating their food, or making it more favorable for their natural enemies. See Integrated Pest Management.

Culture–(1) The working of the ground in order to raise crops; cultivation; tillage. (2) Attention and labor given to the growth or propagation of plants or animals, especially with a view to improvement. (3) (a) The growing of microorganisms in a special medium. (b) The microorganisms which are so grown. (4) Bacteria used in making dairy and other products. See Mother Culture. (5) Human-made features of an area. (6) The specific way of life of a given society (community, tribe, nation).

Culvert–A conduit to convey water underneath a road, through an embankment, etc.

Cumulose Soil–Sedentary soil formed by the accumulation of decayed organic remains; organic soils. See Muck Soil, Peat.

Cuprocide–copper oxide; used to kill algae.

Curiosity–Any plant or animal grown for its peculiar properties and not for ornamental or economic reasons; for example, dodder, miniature mules.

Current–(1) The movement of electricity along a conductor; measured in amperes. See Alternating Current, Direct Current. (2) The flow of a stream of water.

Current Annual Increment–The growth of a stand of trees for a specific year.

Cusec–A cubic foot of flowing water per second.

Cut–(1) A piece of meat prepared for retail or wholesale trade. (2) A slash wound. (3) The opening made by an ax, saw, etc., on a tree. (4) An excavation in the earth, either human-made or natural. (5) The action of a horse's hooves striking its legs or other hooves in walking or running, interfering with its gait. (6) The yield of certain crops, as wool or lumber. (7) (a) An animal separated from the main herd. (b) To separate an animal from the main herd. (8) A severing of the stem or a part of a plant. (9) The output of a sawmill for a given length of time. (10) (a) A reduction in numbers, amount, size, etc. (b) To reduce in numbers, amount, size, etc. (11) To mow. (12) To castrate; emasculate. (13) To sever the jugular vein of an animal or fowl for slaughter.

Cut and Fill–A process of earth moving by excavating part of an area and using the excavated material for adjacent embankments or fill areas.

Cut Slope–(1) The slope of an excavated bank on roadsides or drainage ditches, expressed as a ratio, degree, or percent. (2) On roadways built on hillsides, the upper slope, as contrasted to the lower slope known as the fill slope.

Cut-off–(1) An artificial channel for a stream constructed to straighten the channel or to reduce the possibility of flooding. (2) A natural channel made when a stream cuts through the neck of an oxbow meander. (3) A wall, collar, etc., constructed to reduce the seepage of water along otherwise smooth surfaces or through porous strata.

Cut-off System–The location of tile on a slope or hillside to intercept water from seeps or springs which would otherwise flow onto adjacent bottom land.

Cut-off Trench–An excavation in the base of a dam or other structure which is filled with relatively impervious materials such as fine clay to reduce percolation loss of water.

Cut-off Wall–A wall or diaphragm, made of concrete, steel, wood or clay, through the center of a fill to reduce seepage of water.

Cutan–A concentration of clay or humus along natural parting surfaces within the soil; clay skins.

Cutback Furrow Irrigation–A system of surge irrigation consisting of alternating the release of irrigation water from one furrow to the adjoining furrow. This allows more uniform infiltration throughout the length of the furrow.

Cutback Saddle–A long flat saddle that rests low on the horse's back and is designed to place the rider's weight toward the rear. The name is derived from the U-shaped cutaway slot for the withers. It is used primarily for riding in shows.

Cutbank–The concave wall of a meandering stream that is maintained as a steep or even overhanging cliff by the eroding water at its base.

Cutover Land–Forest land that has been logged over, leaving only stumps, unmarketable trees, slashings, and other evidence of recent lumbering operations. The term may be applied several years after lumbering, even though considerable new growth has appeared and evidence of original lumbering is no longer apparent.

Cutting Cycle–The planned lapse of time between major felling of trees in a forest.

Cyanobacteria–Blue-green algae that are capable of fixing atmospheric nitrogen; similar in function to legume bacteria that fix nitrogen from the air.

Cyanogenetic Plants–Plants that produce hydrocyanic (prussic) acid when under stress such as drought. A number of such plants, including the sorghums, Sudangrass and Johnsongrass, may, under these conditions, cause hydrocyanic or prussic acid poisoning in the animals which eat them.

Cycling–The flow of energy and material from one living thing to another and from one environment to another.

Cyclone–(1) A device for reducing tomatoes to pulp. Also called pulper. (2) Popularly, but incorrectly, a tornado. (3) A low atmospheric pressure area, frequently several hundred miles in diameter which advances at approximately 25 miles per hours with winds blowing in a counterclockwise direction in the Northern Hemisphere and clockwise in the Southern; associated with warm and cold fronts in the middle latitudes. Also called low pressure area.

Cynips–The small gall wasps of the family Cynipidae, which form galls on oak trees.

Cypress–Any tree of the genus *Cupressus*, family Pinaceae; grown as an ornamental in warm regions. See Baldcypress.

Cyprian Bee–A honeybee used in bee breeding in the United States. Native to Cyprus.

Cypriere–Swamp (Louisiana).

D Plus Rule–A rule of thumb used in thinning trees to estimate the desired spacing. A given number is added to the diameter breast height (dbh) of the crop tree. A D+4 rule would mean that a 16-inch dbh tree would need 16+4 or 20 feet of growing space.

Dale–(1) A valley. In common usage there is no distinction between dale, valley, vale, and dell, except that dale is not applied to a large river valley. (2) A piece or share of common land.

Dam–(1) A quadruped, female parent. (2) An artificial structure which obstructs a stream of water for the purpose of water storage, conservation, water power, flood control, irrigation, recreation, etc.

Damselfly Nymph–(Odonata) The immature damselfly. This aquatic insect nymph has an enormous grasping lower jaw and three flat leaflike gill plates that project from the posterior of the abdomen. Nymphs live most of their lives searching for food among submerged plants in still water; a few cling to plants near the current's edge; and a very few cling to rocks in flowing water. The carnivorous adults capture lesser insects on the wing.

Dark of the Moon–That period of the lunar month during which the moon does not shine. Superstition has it that certain crops should or should not be planted during this period.

Datum–(1) Any level surface taken as a surface of reference from which to measure elevations. In surveying, sea level is the basic reference. (2) A figure indicating a fact, as the number, quantity, or weight of an item, used principally in the plural, data.

DBH–See Diameter Breast Height.

DDT–The first of the modern, long-lasting, chlorinated hydrocarbon insecticides. It has a half-life of fifteen years, and its residues can become concentrated in the fatty tissues of certain organisms, especially fish. Because of its persistence in the environment and its ability to accumulate and magnify in the food chain, EPA has banned the registration and interstate sale of DDT for nearly all uses in the United States effective December 31, 1972.

Dead Head–(1) A log which is sunken or partially sunken in a stream or body of water. (2) Blighted or dwarfed seed stalks which produce reduced yields of poor seed. (3) A survey crew returning to a known location without surveying.

Dead Lake–An aging lake, usually overgrowing with water plants.

Dead Reckoning–Determining a position by knowing the speed, direction, and elapsed time since being at a known position.

Dead Water–(1) Absolutely still water. (2) The eddy water just behind the stern of a boat.

Dealkalization–The removal of alkali from the soil, usually by amendments and by leaching. Technically, it is the replacement of monovalent metallic ions, such as sodium, by alkaline earth cations, e.g., calcium, or by hydrogen ions.

Debouchure–(1) The mount of a river or channel; the point from which a spring bursts. (2) Point at which tubular passages connect with larger passages or chambers. (3) Point of issuance of an underground stream.

Debris–(1) The coarse material resulting from the decay and disintegration of rocks occurring either in the place where it was produced, or the place of deposition after transportation by wind, streams, glaciers, or gravity. (2) Plant residues left on a field after harvest. (3) A mixture of straw particles, dirt, shattered pupal cases, etc., found in the wool of sheep infested with sheep keds.

Debris Cone–A fan-shaped deposit of soil, sand, gravel, and boulders which is built up at the point where a mountain stream flows into a valley, or where its velocity is otherwise reduced sufficiently to cause such deposits.

Debris Dam–A barrier built across a stream channel principally to retain rock, sand, gravel, silt, or other material, such as trash or leaves. See Catch Basin.

Debris Guard–A screen or grate at the intake of a channel, drainage, or pump structure for the purpose of stopping debris.

Decalcification–(1) Removal of calcium carbonate by leaching. A natural process in soil formation. Technically, it is the replacement of calcium ions by monovalent hydrogen cations. (2) The removal of calcium from bones of animals.

Decay–(1) (a) The decomposition of organic matter by anaerobic bacteria or fungi in which the products are completely oxidized. See Putrefaction. (b) To decompose by aerobic bacteria, fungi, etc., whereby the products are completely oxidized. (2) General disaggregation of rocks, which includes the effects of both chemical and mechanical agents of weathering with stress in the chemical effects. (3) Any chemical or physical process which causes deterioration or disintegration.

Decomposition–(1) The breaking down of complex chemical compounds to form simpler ones. (2) In soils, the breaking down of minerals and rocks by chemical weathering.

Decreaser Plant Species–Plant species of the original vegetation that decreases in density or cover with continued overuse. Often termed decreases.

Deep Percolation Loss–Water that permeates downward through the soil beyond the reach of plant roots, as a result of excessive irrigation or rainfall and a permeable soil.

Deep Soil–In soil judging, a soil that is more than 36 inches (91 centimeters) deep.

Deep Well–Any well whose water level during pumping is below the practical suction lift of 25 feet (under sea-level conditions).

Deer Farming–A commercial operation that raises deer for meat or other purposes. Deer farming occurs in New Zealand and other countries of the world. Deer are efficient producers of lean meat and there are few if any religious taboos against eating venison as there are against pork and beef.

Deeryard–An area where deer gather in the winter for feed and protection from the weather.

Deferred-rotation Grazing–The system of range management in which grazing is postponed on various parts of a range during succeeding years, allowing each part successively to rest during the growing season to permit seed production and better vegetative spread.

Deflation–(1) A fall or drop in the general price level. (2) The removal of loose material, such as soil, by the wind, thus leaving a bare surface.

Deflocculate–To separate or break down soil aggregates of clay into their individual particles; e.g., the dispersion of the particles of a granulated colloid to form a clay which tends to run or puddle.

Deflocculating Agent–A chemical additive that produces deflocculation (dispersion); e.g., sodium carbonate added to a water suspension of clay.

Defoliant–Any substance which, when applied to a plant, causes the foliage to drop off.

Defoliation–Removal of the leaves of plants.

Deforestation–Removal of all of the trees from a forest area.

Deglacial–Pertaining to the postglacial ice-melting period.

Degradable Wastes–Substances that are changed in form and/or reduced in quantity by the biological, chemical, and physical phenomena characteristic of natural waters. Biodegradable is a a term specifically referring to decomposition by biological processes. See Biodegradable.

Degree of Aggregation–Soil aggregation is the binding together (cohering) of groups of soil particles into clusters or masses such as a clod or ped. The degree of aggregation in soils refers to the extent of water-stable aggregates which will not disintegrate easily and are of special importance to soil structure for desirable plant growth.

Degree of Availability–A soil test level given to represent a degree of nutrient availability.

Degree of Grazing–A term used to define the closeness of grazing. The degrees are: ungrazed, lightly grazed, moderately grazed, closely grazed, and severely grazed.

Delcivity–A descending slope, as opposed to an acclivity, an ascending slope.

Deleterious–Injurious.

Deliquescent–(1) To ramify into fine divisions, as the veins of a leaf or the trunk or branches of a tree. Having a large number of branches or branching so that the stem is lost in the branches, as maple or oak stems. (2) Of, or pertaining to, the ability of a substance to absorb moisture from the air, thus being reduced to a liquid state.

Dell–A small, wooded valley.

Delta–(1) An alluvial deposit at the mouth of a river which is usually fanlike or triangular, such as the Mississippi River Delta below Cairo, Illinois. The soils are generally high in fertility. (2) Deposits which were spread out at the mouths of streams during the Glacial Period.

Deltoid–(1) Shaped like the Greek letter delta. (2) Refers to a leaf shape that is triangular.

Deltoides–Triangular, like the leaf of a cottonwood.

Demersal–Refers to fish and other animals that live on or adjacent to the sea bottom and feed on benthonic organisms.

Demersed–Situated or growing under water, especially aquatic plants.

Den–A compartment in which the acid-rock slurry obtained in the manufacture of superphosphate is allowed to degas, solidify, and cure.

Den Tree–A hollow tree which is used as a home by a mammal.

Dendritic–(1) Marked by a branching habit resembling that of a shrub or tree; usually said of river systems, various plants, and of the leaf veins of many higher plants. (2) Marked by the pattern presented by certain apple scab lesions; having a treelike marking. (3) A drainage pattern characterized by irregular branching in all directions with the tributaries joining the main stream at all angles.

Dendrochronology–Study and matching of tree rings with the object of dating events in the recent past.

Dendrolatry–(Greek; *dendron*, tree) The worship of trees. See Druids.

Dendrology–The study of trees, including the identification of trees by external features.

Dendrometer–An instrument for measuring the diameter of a tree, outside bark. If used to measure diameter, continuously, the instrument is known as a dendrograph.

Denitrification–The process by which nitrates or nitrites in the soil or organic deposits are transformed by anaerobic bacteria into ammonia or free nitrogen and results in the escape of nitrogen gas into the air.

Density–(1) Mass per unit volume. (2) The number of wild animals per unit of area. (3) The degree of closeness with which wool fibers are packed together. (4) In forestry, density of stocking expressed in number of trees, basal area, volume, or other criteria on an acre or hectare basis. See Stocking.

Denudation–(1) A laying bare; the process of washing away of the covering of soil, subsoil, and substrate. (2) The process which, if continued far enough, would reduce all surface inequalities of the globe to a uniform base level.

Deodorant–Any substance which is used to destroy, mask, or combat offensive odors.

Depergelation–The act or process of thawing permanently frozen ground. See Permafrost.

Depletion–The utilization of a natural renewable resource at a rate greater than the rate of replenishment.

Deposition–(1) The addition of sediment, as by flowing water. (2) The process of accumulation of rock material or other debris when dropped due to slackening of transporting agencies, as water, waves, winds, glaciers, or gravitational mass movement. Alluvial fans, offshore bars, dunes, glacial moraines, mudflow debris are such depositional features. (3) In pesticide application, the dust or spray which remains on the plant surface.

Depth to Rock–Bedrock at a depth that adversely affects the specified use of the soil.

Desalination–(1) Salt removal from sea or brackish water. (2) Salt removed from salty soil.

Desert–A region so devoid of vegetation as to be incapable of supporting any considerable population. Four kinds of deserts may be distinguished: (1) the polar ice and snow deserts, marked by perpetual snow cover and intense cold; (2) the middle latitude deserts, in the basinlike interiors of the continents, such as the Gobi, characterized by scant rainfall and high summer temperatures; (3) the trade wind deserts, notably the Sahara, the distinguishing features of which are negligible precipitation and a large daily temperature range; and (4) coastal deserts, where there is a cold ocean current on the western coast of a large land mass such as occurs in Peru and western California.

Desert Crust–A hardpan of calcium carbonate, calcium sulfate (gypsum), or other binding material which is exposed in desert regions by wind or water erosion.

Desert Garden–A flower garden in which plants adapted to desert areas are planted, e.g., a cactus garden.

Desert Varnish–A glossy covering of dark-colored compounds, probably composed largely of iron and manganese, found on exposed pebbles, stones, and large rock surfaces in hot deserts.

Desertification–(1) The decline in productivity of arid and semi-arid soils caused by either natural or person-made stresses. (2) Person-induced soil degradation resulting in increased aridity of the microclimate; e.g., overgrazing.

Desilting Area–An area covered by vegetation such as grasses or bushes used solely for the deposition of silt and other debris from flowing water, located above a reservoir, pond, or field, which needs protection from accumulation of sediment.

Destructive Distillation–The decomposition of organic compounds by heat in a closed container and the collection of resulting volatile liquids and oils, as in the production of wood tar or coal tar.

Desulfurization–The process of sulfur removal, as from flue gases.

Detention Dam–A dam constructed for the purpose of temporary storage of streamflow or surface runoff and for releasing the stored water at controlled rates.

Detergent–(1) A compound of a soaplike nature used in engine oil to remove engine deposits and hold them in suspension in the oil. (2) A chemical (not soap) having the ability to remove oil or grime. Household detergents can be used as surfactants in herbicide sprays. See Surfactants.

Detoxified–Having the toxic, or poisonous, quality of a substance removed.

Detritus–(1) Material produced by the disintegration and weathering of rocks that has been moved from its site of origin. (2) A deposit of such material. (3) Any fine particulate debris, usually of organic origin, but sometimes defined as organic and inorganic debris.

Devil's Darning Needle–See Dragonfly Nymph.

Dew–The condensed, atmospheric moisture which has collected, usually at night, on plants, soil, rocks, and other surfaces which have cooled below the condensation point of the moisture in the atmosphere. In humid areas, the amount of water contributed by dew may be as much as 5 inches (13 centimeters) a year. The amount is also important for plant growth in semi-arid or arid regions. See Dewpoint.

Dew Point–The temperature at which a given percent of vapor moisture in the air condenses into droplets of water.

Dew Pond–A human-made pond used for watering livestock, so called because the only apparent source of water seems to be dew. However, it must be artificially filled first in order to function (England).

Dewfall–The condensation of atmospheric moisture as dew.

Diameter Breast Height (DBH)–The diameter of a tree outside of the bark at roughly breast height. Normally measured 4.5 feet off the ground on the uphill side of the tree.

Diameter Limit Cutting–A system of selection harvest based on cutting all trees in the stand over a specified diameter. This eliminates marking individual trees.

Diameter Tape–A tape that is put around a tree at a 4.5-foot height. The tape is graduated to read the diameter.

Diammonium Phosphate (DAP)–Diammonium phosphate has an economic advantage over monoammonium phosphate because the same amount of acid fixes twice as much ammonia. The DAP pure salt is not stable and loses ammonia when damp. The Tennessee Valley Authority has developed processes that have been commercially developed on a larger scale. The pure salt contains 21.21% nitrogen and 53.76% phosphoric oxide. The fertilizer grade material contains from 18 to 21% N, and 46 to 54% P_2O_5. The 18-46-0 is a popular grade.

Diapersant–A chemical product that acts to lower the surface tension between a solid and a liquid. A diapersant added to a pesticide or foliar fertilizer spray makes the substance stick to plant leaves, where it is more effective.

Diatomaceous Earth–A siliceous sedimentary rock consisting of opaline residues of diatoms, a one-celled aquatic plant related to algae. Used as a filtering agent and in paints, rubber, and plastics.

Diatoms–Microscopic algae that belong to the family Chlorophyceae. They are found in both fresh and salt water and are distinguished by their siliceous cell walls, which sink to the bottom of the water after the living organism dies, forming diatomaceous earth. Diatoms are used as a scouring powder, as a filler in insecticides, and in the manufacture of dynamite and pottery glaze.

Dicalcium Phosphate–$(Ca_2[HPO_4]_2)$ A manufactured product consisting chiefly of a dicalcium salt of phosphoric acid, used as a supplement to stock feed and as a constituent of commercial fertilizers.

Differential Weathering–When rocks are not uniform in character because the minerals vary in hardness of solubility, an uneven surface may be developed; in desserts by the action of the wind and in moist regions by solution.

Diffraction–The bending of light rays passing through microscopic droplets of water vapor, which causes halos around the sun or moon. Also causes the breaking up of light into bands of color, resulting in rainbows following a shower.

Diffuse-porous Wood–Wood in which the pores are of fairly uniform or of only gradually changing size and distribution throughout the growth ring. See Ring-porous Wood.

Diffused Runoff–Rainfall which runs off simultaneously over the whole area of a slope causing surface erosion, in contrast to runoff concentrated in a single channel, which causes gully erosion.

Dig–(1) To turn up the earth. (2) To excavate; e.g., to dig a ditch. (3) To remove from the earth; e.g., to dig potatoes; to dig strawberry plants for transplanting.

Digger Pine–*Pinus sabiniana*, family Pinaceae; a coniferous tree that produces an edible nut. However, it is often a weed on the California foothills when it invades grazing lands. Also called grayleaf pine.

Digitate–Similar to, or resembling, the fingers on a hand, as the leaflets of an American chestnut tree. See American Chestnut.

Dike–(1) A ridge of earth that is thrown up to impound water, as in irrigation, or to divert water, as in soil erosion control. (2) An embankment or levee that is constructed to prevent inundation of low land, as on the seacoast or in the flood plain of a river valley. Also spelled dyke. See Levee.

Dilation–The expansion of ice from the freezing of water in fissures. When water freezes to ice it expands about 11 percent in volume.

Dingle–A small narrow valley.

Dingo–*Canis dingo*, family Canidae; a wild dog, found in Australia, which is a predator of cattle, sheep, and fowl. It is usually yellowish-red and has erect, triangular ears, a bushy tail with a white tip, and white, splayed feet.

Dioxin–A highly toxic substance that was a contaminant of agent orange, a mixture of 2, 4-D and 2, 4, 5-T used as a defoliant in Vietnam in the 1960s. See Agent Orange.

Dip Slope–A slope of a land surface that conforms approximately to the angle and direction of the underlying rocks.

Directed Application–The application of a pesticide or other material to a specific kind of plant or parts of a plant.

Dirt–(1) Soil out of place. (2) In excavation engineering, any unconsolidated, geologic deposit.

Disc Pan–A compacted soil layer produced by the bottom cutting edges of a disc plow.

Discharge–(1) The quantity of water, silt, or other mobile substances passing along a conduit per unit of time; rate of flows. (2) An exudate or abnormal material coming from a wound or from any of the body openings: e.g., a bloody discharge from the nose. (3) To remove the electrical energy from a battery.

Disease–Any deviation from a normal state of health in plants, animals, or people which temporarily impairs vital functions. It may be caused by viruses, pathogenic bacteria, parasites, poor nutrition, congenital or inherent deficiencies, unfavorable environment, or any combination of these.

Disintegration–A term often applied to the natural mechanical breaking down of a rock on weathering.

Disjunctive Symbiosis–(1) The living together of two distinct, unconnected organisms without apparent contribution of benefit of one upon the other, as the living of ants in hollow thorns of some species of acacia. (2) That symbiosis involving only temporary association of the symbionts, as in the cross-pollination of flowers by insects.

Dispersal–(1) The actual transfer or movement of organisms from one place to another. It does not constitute migration, but is a necessary antecedent to it. (2) The history of the movement of a group of organisms. See Migrant.

Disperse–To scatter or distribute over an area or to separate a substance into smaller parts.

Dispersed Soil–Dispersed soils usually have low permeability. They tend to shrink, crack, and become hard on drying and to slake and become plastic on wetting.

Dissected Land–Surfaces that have been cut into valleys and ridges by stream erosion. Sharpness of sculpturing and close spacing of ravines and valleys are usually implied. Dissection may be deep or shallow. In some plateau regions the interstream areas or ridges may be fat-topped, as in the Ozark Plateau.

Dissolved Oxygen–The amount of elementary oxygen present in water in a dissolved state. It is commonly reported in parts per million (by weight), or milligrams per liter, or percentage of saturation, of oxygen in the water. Dissolved oxygen is essential for fish and other aquatic life and for aerobic decomposition of organic matter. Dissolved oxygen in surface water bodies should be maintained at a level above the threshold of 3 mg/l and an optimum of 5 mg/l for most species of fish.

Distributary–(1) An outflowing branch of a river, such as occurs characteristically on a delta. (2) A river branch flowing away from the main stream and not rejoining it. See Tributary.

Distributing Stand–Any device used in irrigation practices from which water is distributed into furrows and basins. Usually it is a part of a riser outlet from an underground pipe system.

Ditch–An artificial excavation, as a trench or channel dug to carry irrigation or drainage water.

Diveresion–(1) A ridge of earth, generally a terrace, built to protect downslope areas by diverting runoff from its natural course. (2) A ditch constructed across the slope for the purpose of intercepting surface runoff; changing the accustomed course of all or part of a stream.

Diversion Terrace–A wide, relatively shallow channel of low gradient with gentle side slopes and ample water capacity, which is constructed across the slope of a field to intercept and change the direction of flow and to reduce the velocity of run-off water and to reduce erosion.

Divert–(1) To turn a river or stream aside from its natural channel into a new one. (2) To take water from a stream and direct it into another channel by gravity or pumping.

Divide–(1) The land area which separates two watersheds or drainage systems. (2) To separate a colony of bees so as to produce two or more colonies. (3) To separate the crown or underground parts of a plant into two or more portions so as to obtain more plants.

Divining Rod–A rod or switch, often a forked willow or peach twig, which when held tightly in the hands of a dowser (water witch) is supposed to be irresistibly drawn to the earth over an underground supply of water, and hence, is used in locating a place to drill a water well. See Dowsing, Water Witch.

Division Box–A structure for dividing and diverting all or part of a flow of irrigation water into other channels. See Weir.

Dobe House–A house made of sun-dried mud cakes or crude bricks. Also, one whose walls are made of mud, sometimes mixed with straw and cast in a frame or form to dry and harden in the sun. It is a very ancient form of building structure of the Middle East and was also used in arid regions of the southwestern United States by early settlers. See Adobe Soil.

Doe–An adult female goat, rabbit, or deer.

Dokuchaiev Formula–The formula given for soil formation by the Russian pedologist, V. V. Dokuchaiev, in 1899, as follows: $P=f(K,O,G,V,)$ when P is soil, f is function of; K, climate; O, organisms; G, "subsoil" or parent material; V, age or time factor.

Doldrums–(1) An area astride the equator characterized by constant low pressure, warm and rising air masses, and gentle surface winds. (2) A depressed state of mind.

Dolomite–A common mineral, $CaMg(CO_3)_2$ usually with ferrous iron and manganese replacing some magnesium. In dolomite the calcium to magnesium ratio must be lower than 1:5. See Dolomitic Limestone.

Dolomitic Limestone–A limestone in which the calcium to magnesium ratio can be from less than 1:5 to greater than 1:60. Any less magnesium is known as calcic limestone. See Dolomite.

Domestic Water Use–Water used for drinking, sanitation, street flushing, fire protection, and lawn and garden irrigation.

Dominance (Social)–The tendency of one animal in a group to exert its social influence or presence over others in the group. Also referred to as social order or pecking order.

Dominant Trees–Trees with crowns extending above the general level of the crown cover and receiving full light from above and partly from the sides.

Dornick–A piece of rock or a stone that is not too large to be thrown by a person. Also, a rock of iron ore.

Dote–An early decay of wood characterized by patches or streaks of lighter or darker color than normal. Also called doze.

Double Cropping–The growing of two crops at different times in one year on the same field.

Double Hacking–The girdling of a tree to kill it by making parallel, overlapping, downward, ax cuts and the removal of the chips between the cuts.

Double Superphosphate–A commercial fertilizer that contains two and one-half to three times as much available phosphoric acid as superphosphate. The phosphate rock is treated with phosphoric acid, and in the manufacturing process the gypsum is removed. Also called triple superphosphate, treble superphosphate.

Douglas Fir–Any tree of the genus *Pseudotsuga*, family Pinaceae, but specifically *Pseudotsuga menziesii*. It sometimes reaches a height of 300 feet and is a timber tree of greatest commercial value for lumber in northwestern United Sstates. It is also grown as a shade and shelterbelt tree and certain horticultural varieties are grown as ornamentals east of the Rocky Mountains, United States.

Douglas Waterhemlock–*Cicuta douglasi*, family Umbelliferae; a poisonous, perennial herb. In grazing animals, the symptoms of poisoning are stomach pain, diarrhea, dilated pupils, labored breathing, frothing at the mouth, a weak but rapid pulse, and violent convulsions. Death results from respiratory failure. Roots and root stocks are the most poisonous parts of the plant; the leaves and fruits may be eaten in hay without danger. Found from Alaska to California, United States.

Douse (Douce, Dowse)–(1) To beat out or to extinguish a fire. (2) To search for deposits of ore or water with the dowsing or divining rod. See Dowsing.

Dove–See Mourning Dove.

Down–(1) A sandhill. See Dune. (2) The soft, furry or feathery covering of young animals and birds that occurs on certain adult birds under the outer feathers and which is used for stuffing pillows, furniture, and comforters, its most important source being the eider duck. (3) The pappus of a plant. (4) Used to describe the individual crop plants that have been blown or forced to the ground by rain, hail, wind, etc.,; e.g., down corn, down wheat.

Downpour–A heavy rainstorm.

Downs–(1) A term usually applied to ridges of sand thrown up by the sea, or the wind along the seacoast. It is also a general name for any undulating tract of upland too coarse-textured for cultivation and covered with short grass. (2) A hill; especially a bank or hillock of sand thrown up by the wind in or near the shore; a flattish-topped hill. A tract of open upland, often undulating and covered with a fine turf that serves chiefly for the grazing of sheep.

Downspout–A pipe leading downward from a gutter to carry rainwater from a roof.

Downstream–Toward the mouth of a river or stream from the point of reference.

Doyle Rule–In the lumber trade, a simple method of estimating the amount of lumber that can be obtained from a small log. The length in feet multiplied by the diameter of the log from which 4 inches has been subtracted, and the result divided by 4 and squared.

Doyle-Scribner Rule–A combining log scale wherein the Doyle rule is used for logs up to 28 inches (71 centimeters) in diameter and the Scribner rules for larger logs. See Doyle Rule, Scribner Rule.

Dragonfly Nymph–The immature dragonfly. This aquatic insect nymph has gills on the inner walls of its rectal respiratory chamber. It has an enormous grasping lower jaw that it can extend forward to a distance several times the length of its head. Although many of these nymphs climb among aquatic plants, most sprawl in the mud where they lie in ambush to await their prey. The carnivorous adults capture lesser insects on the wing.

Drain–(1) An artificial or natural channel that receives and carries off excess or unwanted water from an area of land. It does not carry sewage and is therefore distinguished from a sewer. (2) The loss of growing trees in a forest due to cutting, fire, or natural causes.

Drain Tiles–Pipes of burned clay or concrete, in short lengths, which are laid with open joints to collect and remove drainage water. Plastic pipe with drain holes has replaced other forms of drain tiles in recent years.

Drain Well–A well into which drainage water is directed. The excavation is deep enough to reach pervious substrate, such as sand and gravel, or a porous or cavernous limestone. In many states this type of drainage is prohibited because of pollution hazard.

Drain-tile Breathers—Small-diameter pipes extending from inside a tile line to the soil surface to prevent the development of a vacuum and thus to assure uniform flow of water in the drain. The breathers should be located at the junction where the grade of the tile line changes. Also called relief pipe.

Drainage—(1) The removal of surplus ground or surface water by artificial means. (2) The manner in which the waters of an area are removed. (3) The area from which waters are drained; a drainage basin. In soil science, the natural drainage under which the soil developed.

Drainage Area (Drainage Basin)—(1) The area (square meters, acres, etc.) of a drainage basin. (2) A catchment area; drainage basin.

Drainage System—(1) An artificial network of furrows, ditches, tile drains, or combinations of these that provide drainage for farmland. (2) The natural means by which a river and its tributaries carry water from the land.

Drainage Water—(1) The water that the soil is unable to retain against the force of gravity. Also called gravitational water, free water, excess water. (2) Water flowing in a drain derived from ground, surface, or storm water.

Draw—(1) A small natural head tributary, drainage hollow, without a permanent stream in its bottom. (In local usage, it is not always sharply differentiated from ravine, hollow, gulch, or coulee.) (2) A slip or shoot of a sweet potato. (3) To shape and build.

Draw Well—A well in which a bucket and rope are used to lift the water to the ground surface.

Drawdown—(1) The lowering of the water table or piezometric surface caused by pumping (or artesian flow). (2) The distance that the water is lowered in a well or at any point in the aquifer as a result of pumping.

Dredge—(1) Oat and barley seed sown together and cultivated to be used for making malt. (2) Any of several different types of machines to deepen channels and clean out ditches, etc. (3) To excavate or deepen and clean stream beds. (4) To coat the surfaces of food with flour, cornmeal, breadcrumbs or other fine substances before cooking.

Dredging—The deepening of streams, swamps, or coastal waters by scraping and removing solids from the bottom.

Dressed Size Lumber—The dimensions of lumber after planing, usually 3/8 inch less than the normal or rough size.

Dried Kelp—A livestock feedstuff supplement that contains over 30% mineral matter and from 0.15 to 0.20% iodine, consisting of dried and comminuted brown seaweeds, of the genus *Macrocystis* or related genera. Also called kelp meal.

Drift—Movement by wind of fine particles of a dust or spray to an area not intended to be treated.

Drift Control Agent—Adjuvant used in liquid spray mixtures to create a stickier spray solution that reduces drift. See Adjuvant.

Drift Fence—A line of fence, open at both ends, used to direct the movement of grazing animals on the ranges of the western United States.

Drill—(1) A small furrow in which seeds are planted. (2) A row of seeds that have been planted by dibbling in a small furrow. (3) An implement for planting seeds that forms a small furrow, deposits the seed in dibbles, covers the seeds, and packs soil over it. It can also deposit fertilizer, lime, or other soil preparations into the soil, alone or with the seed. (4) A very small, trickling stream or rill.

Drillability—A property of fertilizers that affects their rate of flow and uniformity of distribution from distributors. Fertilizers with good drillability flow rapidly and evenly when dispensed, and those of poor drillability flow slowly and irregularly. The angle of repose of a fertilizer is a useful measure of drillability. An angle below 40 degrees indicates excellent drillability. An angle above 50 degrees indicates poor drillability. See Angle of Repose.

Drilled Well—A well that is sunk by percussion tools, by rotating bits, or by jets of water; the most common type of water well.

Drilling Mud—A suspension, generally aqueous, used in rotary drilling for oil and gas that is pumped down through the drill pipe to seal off porous zones and to counterbalance the pressure of oil and gas; consists of various substances in a finely divided state among which bentonite clay and ground barite mineral are common. Oil may be used as a base instead of water.

Drip—Moisture from fog or clouds that condenses on leaves or twigs or plants and drips onto the ground.

Drip Irrigation—Watering plants so that only soil in the plant's immediate vicinity is moistened. Water is supplied from a thin plastic tube at a low flow rate. The technique sometimes is called trickle irrigation.

Drip of a Tree—The places at which much water runs off the tree during rains; the periphery of the tree. The usual area fertilized.

Driven Well—A well that is sunk by forcing a metal pipe into the ground until it extends below the surface of the water table.

Driving Rain—Rain accompanied by a high wind.

Drizzle—Precipitation from stratiform clouds in the form of numerous, minute droplets, which appears to float in the air. It is heavier than a mist but less intense than a shower.

Drop—(1) A structure in an open or closed conduit that is used for dropping the water to a lower level and dissipating its kinetic energy. (2) A decrease in height or elevation. (3) Any fruit that falls form a tree to the ground because of wind, or other conditions; e.g., diseased, immature, and usually unfertilized fruit; also normal but ripe fruit. (4) A fungal disease of vegetables caused by *Sclerotinia sclerotiorum, S. minor*, or *S. intermedia*, family Sclerotiniaceae. (5) To give birth to young, as to drop a calf. (6) To shoot an animal and to cause it to fall.

Drop Off—(1) To fall to the ground, as a leaf or blossom. (2) To decrease in number, quality, demand, etc.

Drosometer—An instrument that measures the quantity of moisture precipitated as dew.

Drought–(1) A period of insufficient rainfall for normal plant growth which begins when soil moisture is so diminished that vegetation roots cannot absorb enough water to replace that lost by transpiration. A drought can be arbitrarily defined on the basis of amount of precipitation during a certain period, although such a measure has no uniform significance because of variable amounts of moisture in soils when the measurement period begins. The intensity or damage form drought varies with the plant since some plants are much more tolerant of dry conditions than others. (2) Any dry spell. (3) A period when precipitation is not adequate to meet the needs of the human population. (4) In an arid area, a period when there is an especial shortage of water for irrigation. Also spelled drouth. (5) Drought is an imprecise term. On a coarse-textured soil with a deep water table, plants may need water every week but on a deep, fine-textured soil, only every two weeks. High temperatures and high winds also determine water needs.

Droughty–Designating a soil that has low water-holding capacity; e.g., deep sands or gravel.

Drown–(1) To deprive a plant or animal of life by immersion in water so that the supply of oxygen is cut off. (2) To submerge land with water, whether by a rise in the level of a water table, a lake, ocean, or river, or by a sinking of the land as the lowering of a coastal region drowns the lower courses of the rivers and connects their valleys into estuaries.

Drowned Topography–(Geology) Depression whereby the lower coursers of most of the rivers are submerged beneath the sea.

Drumlin–An elongated, oval hill or ridge that is composed of glacial drift, normally compact and unstratified, usually with its longer axis conforming to the direction of the movement of the ice responsible for its deposition. Drumlins are present in parts of the glaciated region of the United States especially in parts of New York, Michigan, and Wisconsin, where they are not only striking relief features, but also have agricultural significance such as being excellent sites for orchards.

Drunken Forest–A forest in which the trees have leaning positions due to displacement caused by landslides, high winds, or from freezing and thawing of soil when supersaturated.

Dry–(1) (a) To cause a pregnant cow to stop giving milk shortly before she drops her calf; (b) Designating a cow who has ceased to give milk shortly after she drops her calf. (2) To preserve a product by dehydration. (3) Designating dressed poultry packed for shipment with dry ice. (4) Designating a time or place of abnormal lack of precipitation, e.g., a dry year. (5) describing a well that yields no water.

Dry Basis–Designating a product, e.g., soil, fertilizer, or feed, which is analyzed for its constituents calculated on the basis of oven-dried material.

Dry Farming–This term applies mostly to nonirrigated semi-arid areas where short-season, drought-resistant crops are grown on fertile soil to make the most efficient use of scarce soil water. Level terraces with closed ends are sometimes used to hold all precipitation in each field for use by crops. Also known as rainfed agriculture.

Dry Fog–A haze that is due to the presence of dust or smoke in the air; not water vapor.

Dry Ice–(1) Solid carbon dioxide that is purified, liquefied, expanded to form snow, and finally pressed into blocks. It is used as a refrigerant. (2) Bare glacial ice with no standing water. (3) Ice with temperature several degrees below the freezing point.

Dry Land–(1) Farmland in arid or semi-arid regions. See Dry Farming. (2) Land that normally has a dry surface, as contrasted to marsh or swampland that is usually saturated.

Dry Lightning–Lightning in a thunderstorm accompanied by insufficient rain to moisten fuels and prevent fire from a stroke of lightning.

Dry Meadow–A meadow, dominated by grass, which becomes moderately dry by midsummer.

Dry Pergelisol (Dry Permafrost)–Soil material having the requisite mean temperature to be permanently frozen but lacking water. See Pergelisol.

Dry Rot–(1) Any of many types of decay but especially of that which, when in an advanced stage, permits timber to be easily crushed to a dry powder. The term is actually a misnomer for any decay, since all fungi require considerable moisture for growth. (2) A metabolic defect that causes ill health in animals.

Dry Spell–In humid regions, a period longer than usual with little or no rainfall that is less serious in consequence than a drought.

Dry Valley–A valley in which the stream that once ran through it has dried up or sunk into the alluvium of the streambed. See Karst.

Dry Weather–A period during any part of the year that has relatively little or no precipitation.

Dry Weight–The weight of a product or material less the weight of the moisture it contains; the weight of the residue of a substance that remains after virtually all the moisture has been removed from it. Also called dry matter.

Dry Well–(1) A well that contains little or no water. (2) A covered pit with open-jointed linings through which drainage from roofs or basement floors may seep or leach into the surrounding porous soil.

Duff–The organic layers on a forest floor, including freshly fallen leaves and twigs and other organic residues in all stages of decomposition.

Duff Hygrometer–An instrument for measuring the moisture content of litter to determine fire hazard.

Dug Well–A well that is sunk by men digging with picks and shovels to the water table or just below it. Such as well is usually 4 feet in diameter and walled with bricks or stones.

Dug-out–(1) A crude pioneer dwelling in the western United States which was often little more than an excavation in the side of a hill with a door and a hole for a chimney. (2) An excavated pond.

Dump–(1) The storage yard of a portable sawmill. (2) A place for refuse. (3) To unload a truck or wagon by raising the bed and allowing the loaded material to fall to the ground. (4) To discard unwanted materials or refuse.

Dune–A mound or ridge of loose sand piled up by the wind that is common where loose, bare sand is abundant, as along seashores,

shores of the Great Lakes, in river valleys, and in desert areas. Under drought conditions dune may form from the topsoil of cultivated land lying unprotected by vegetation.

Dungworm–A worm found in cowdung, used as bait for fishing. Also called fishworm.

Duricrust–The case-hardened crust of soil; usually formed in arid or semi-arid climates, but the plinthites of lateritization have also been called duricrust, as have some silcretes, both of which form in areas of higher rainfall.

Duripan–A horizon in a soil characterized by cementation by silica or other minerals. See Fragipan.

Dust–(1) Fine, dry particles of earth, or other matter, so attenuated that they may be wafted by the wind. (2) Particles less than 0.1 millimeter in diameter that may be bounced off the ground by sand grains moving in saltation. See Saltation. (3) An insecticide, fungicide, etc., which is applied in a dry state, or the application of these to plants, animals, or fowls. (4) (Volcanic) Pyroclastic detritus consisting mostly of particles less than 0.25 millimeter in diameter; i.e., fine volcanic ash. It may be composed of essential, accessory, or accidental material.

Dust Bowl–A part of the Great Plains region, United States, which is subject to severe droughts, during which dry soil is picked up by strong winds to form great clouds of dust. Includes parts of southeastern Colorado and adjacent parts of Kansas, Oklahoma, and northwestern Texas.

Dust Devil–A violent, dust-laden whirlwind.

Dust Mulch–A shallow layer of loose dry surface soil, which may be natural or created by tillage. Theoretically such a covering lessens losses of soil moisture resulting from direct evaporation, but in practice, it is of questionable value as a soil conservation measure.

Dust Soils–Soils, observed most commonly in arid regions, which are normally pulverulent. They are penetrated by water with extreme slowness and thus constitute a problem in management under irrigation.

Dust Storm–(1) A strong wind carrying large clouds of dust (or silt), common in desert and plains regions; its influence may be felt thousands of miles from the source. China, the United States, Egypt, the Sahara, the Gobi, and numerous other parts of the world are subject to dust storms. (2) For the development of a dust storm, there are three essentials: (a) an ample supply of fine dust or dry silt, (b) relatively strong winds to stir it up, and (c) a steep lapse rate of temperature in the dust-carrying air.

Dustproof–Describing anything constructed or protected so that an accumulation of dust will not interfere with its successful operation.

Dutch Ashes–Ashes that result from the burning of peat.

Dutch Elm Disease–A fungal disease of elm trees. See Smaller European Elm Bark Beetle.

Dy–Limnic peat, a dark lake bottom sediment rich in colloidal humic substances formed by plant decay in saprobic condition.

Dystrophic Lakes–Shallow lakes with low nutrient content; in contrast to eutrophic lakes with high nutrient content. Both dystrophic and eutrophic lakes have high biochemical oxygen demand (BOD).

E. Coli Bacteria–Bacteria that naturally inhabit the human colon.

E Horizon–A mineral horizon, mainly a residual concentration of sand and silt high in content of resistant minerals as a result of the loss of silicate clay, iron, aluminium, or a combination of these.

Early Spring–(1) In the Northern Hemisphere, March to May; in the Southern Hemisphere, October to December, depending on the latitude and elevation. (2) A growing season that starts earlier than is usual for the latitude, elevation, or soil.

Early Wood–The less dense, larger-celled, first-formed part of a growth layer along the trunk and branches of a tree (spring wood).

Earth–(1) Soil, as black earth. (2) Ground. (3) Soil-like material, as ocher. (4) Certain minerals, the rare earths, thorium, zirconium, etc. (5) Land, or land areas. (6) The underground hole of a burrowing animal. (7) To place earth over the roots or around a plant.

Earth Dam–A barrier built across a water course composed of clay, silt, sand, and gravel, or a combination of soil and rock, as contrasted to a dam made of concrete.

Earth Flow–The moderately rapid flowage of large masses of soil and rock materials that is common on the edge of clay terraces, in certain glaciated valleys and in some mountainous regions of shale bedrock; as in parts of West Virginia, Pennsylvania, and Ohio, United States. Also called flow slides, mud-rock flow.

Earth Tide–The rising and falling of the surface of the solid earth in response to the same forces that produce the tides of the sea. Semidaily earth tides fluctuate between 3 and 6 inches (7 and 15 centimeters).

Earthquake–A sudden motion or trembling in the earth caused by the abrupt release of slowly accumulated strain (by faulting or by volcanic activity). See Richter Scale.

Earthworm–Any of the several species of the genus ***Lumbricus***, family Lumbricidae, which are commonly, and often abundantly, represented in the fauna of surface soils, especially in the humid regions. They are generally regarded as beneficial because they improve the structure and fertility of soil; however, they may do minor damage by eating plants.

Earthworm Farm–A farm that raises earthworms for fish bait or other purposes.

Earwig–See European Earwig.

Easter–A high wind- or rainstorm coming from an easterly direction.

Eastern Black Walnut–*Juglans nigra*, family Juglandaceae; a hardy, deciduous tree grown for its fruit, the black walnut of commerce, which is used in confections. Its lumber is one of the finest of the domestic cabinet woods, used also for veneer, interior paneling, furniture. Native to North America. See Walnut.

Eastern Larch–*Larix laricina*, family Pinaceae; a hardy deciduous tree grown for its valuable timber and as an ornamental. Native to North America. Also called American larch, hackmatack, epinette, and, among foresters, tamarack.

Eastern Mistletoe–Also known as Christmas American mistletoe, ***Phoradendron serotinum***, family Loranthaceae. A plant that is a parasite on trees, even though the plant contains chlorophyll.

Eastern Redbud–*Cercis canadensis*, family Leguminosae; a deciduous tree grown for its early spring, rosy-pink flowers. Native to North America. Also called American Judas tree, American Judas tree, American redbud, Judas tree. See Redbud/.

Eastern Redcedar–*Juniperus virginiana*, family Pinaceae; a very hardy evergreen tree grown as an ornamental and as the source of cedar wood linings, millwork, pencil slats, woodenware, and water buckets. Native to North America. Also called red cedar, Tennessee red cedar.

Eastern White Pine–*Pinus strobus*, family Pinaceae; an evergreen, coniferous tree which is one of the outstanding timber pines and ornamentals. It sometimes grows to 200 feet at maturity and is often referred to as the "Queen of the Forest." Its lumber is soft and is put to a wide variety of uses. Native to North America. Also called white pine, northern white pine, American white pine, apple pine, deal pine.

Eatern Red Oak–Quercus borealis var. ***maxima***, family Fagaceae; a tall, relatively quick-growing tree grown as an ornamental shade tree and for its wood, which is used in furniture and interior finishing. Native to North America.

Eccentric–Used to describe an irregular or unsymmetrical tree trunk.

Ecdysone–A hormone that is essential in the molting process of insects. See Molting.

Ecoclimate–The sum total of the meteorological factors within a habitat.

Ecological Competition–The competition of two or more organisms for a limited supply of an environmental resource.

Ecological Dominance–The condition in a plant community or in vegetational strata in which one or more species, by means of their abundance, coverage, size, or vigor have considerable influence or major control upon the existing conditions of associated species.

Ecological Factor (Limiting Factor)–Any part or condition of the habitat affecting directly or indirectly the life of one more organisms in such a way as to differentiate it from other vegetation; often classified into: (a) climatic, physiographic and edaphic, and biotic factors; (b) direct, indirect, and remote factors.

Ecology–The totality or pattern of the interrelationship of organisms and their environment, and the science that is concerned with that interrelationship. the subject is often categorized into plant, animal, marine, terrestrial, freshwater, desert, tundra, and tropical ecology.

Economic Entomology–That branch of insect study directed toward preventing human losses and increasing gains through manipulation of insect populations. Examples of economic entomology include the methods of protecting plants, people, and animals from insect-borne diseases by insecticides, drainage, crop rotation, and integrated pest management; and the culture of silkworms, honeybees, and various beneficial parasitic insects.

Economic Injury Level–The point at which the buildup of an insect population starts to cause economic damage to a crop or group of animals.

Economic Maturity–The age and growth rate at which a tree or stand of trees will no longer increase in value fast enough to earn a satisfactory rate of interest. At this time the trees should be marketed. Also called financial maturity.

Ecosphere–The layer of earth and troposphere inhabited by or suitable for the existence of living organisms.

Ecosystem–The entire system of life and its environmental and geographical factors that influence all life, including the plants, animals, and the environmental factors.

Ecotone–A transitional area between two types of vegetation, as between forest and prairie.

Ecotype–A variation in plants brought about by a particular type of environment, such as smaller leaves in full sunlight.

Ectodynamorphic Soil–A soil whose dominant character has been determined by external forces, such as climate and vegetation. See Endodynamorphic Soil.

Ectomycorrhizal–Subsisting on the surface, that is, beneficial ectomycorrhizal fungi that live on the surface of tree roots, attached to the epidermal cell layer, and penetrating intercellular space of cortical cells. See Endomycorrhizal.

Ectoparasite–A parasite that lives on the outside of the body of its host; e.g., a tick.

Edaphology–That science which is concerned with the study of the relationships between plants and soils.

Eddy-chamber–A chamber in a nozzle in which liquid is atomized.

Edge–The cover afforded by the vegetative margin between different types of wildlife habitats, as between fields and woodlands.

Edge Firing–A method of controlled burning in which fires are set around the perimeter of the area to be burned and allowed to spread inward.

EDTA–Ethylenediamine tetraacetic acid; a chelate (sequestering agent) used to correct iron, copper, and zinc deficiencies in plants. It can be applied to the soil or to the plant foliage.

Effective Concentration (EC)–The effective concentration (written as EC 10 or 50 or 100) or any concentration (median) is the parts per million (ppm) or parts per billion (ppb) of a toxicant that produces a designated effect on 10 percent or 50 percent or 100 percent of the target species. See Lethal Concentration (LC), Lethal Dose (LD).

Effective Field Capacity–The actual work accomplished in acres or hectares per hour by an implement despite loss of time from field end turns, inadequate tractor capacity, deficient tractor or implement preparation, adverse soil conditions, irregular field contours, lack of operator skill, or other factors.

Effective Half-Life–Biological half-life.

Effective Precipitation–That portion of total precipitation that becomes available for plant growth. It does not include precipitation lost to deep percolation below the root zone, to surface runoff, or to plant interception.

Efferent–Conducting or conveying away from.

Effervescence–The vigorous escape of gas bubbles as a result of chemical action, such as the carbon dioxide escape from a carbonate rock when hydrochloric acid is added.

Efficiency of Irrigation–The fraction of the water diverted from a river or other source that is consumed by the crop, expressed as percent. Often applied to whole irrigation systems and takes account of conveyance losses. See Consumptive Use.

Efficiency of Water Application–The fraction of the water delivered to the farm that is stored in the soil in the plant root zone for use by the crop, expressed as percent.

Efflorescence–(1) The period or state of full bloom. (2) The crusts or coatings of any soluble salts, such as the sulfates and chlorides of calcium, magnesium, and sodium that appear on the bare surface of soil, on clods, in cracks of natural soils, or on heavily fertilized soils as a result of the evaporation of moisture in the soil.

Effluent–(1) Flowing out, as lava through fissures in the side of a volcano, as a river from a lake. (2) Anything that flows forth. (3) A discharge of pollutants into the environment, partially or completely treated or in its natural state. Generally used in regard to discharges into waters.

Elder–Any plant of the genus ***Sambucus***, family Caprifoliaceae. Species are occasionally cultivated for their flowers, and the fruit of wild and cultivated plants is used in making wine, pies, and jelly. Also called elderberry, elder blow.

Electron–See Atom.

Electrostatic Sprayer–A spraying system that uses electrical forces of attraction to greatly increase the amount of spray that covers the plant. Individual spray droplets are given an intense electrical charge within a specially designed atomizing nozzle and propelled toward the plant. Individual spray droplets are given an intense electrical charge within a specially designed atomizing nozzle and propelled toward the plant. The approaching charged spray cloud induces an opposite electrical charge into the plants. The charges cause the spray droplets to be attracted like a magnet to the plant. Spray coverage is greatly increased.

Element–(1) A substance that cannot be decomposed into other substances. (2) A substance all of whose atoms have the same atomic number. The first definition was accepted until the discovery of radioactivity (1896), and is still useful in a qualitative sense. It is no longer strictly correct, because (a) natural radioactive decay involves the decomposition of one element into others; (b) one element may be converted into another by bombardment with high-speed particles; (c) an element can be separated into its isotopes. The second definition is accurate, but it has little relevance to ordinary chemical reactions or to geologic processes.

Elevation–(1) Height above sea level or height above any selected base point. (2) A land feature, such as a hill, which is of greater height than the surrounding land area.

Elm Blight–See Dutch Elm Disease.

Elm Casebearer–*Coleophora limosipennella*, family Coleophoridae; an insect that infests the elms of the northeastern United States by mining (eating) between the principal veins of the leaves.

Elm Lacebug–*Corythucha ulmi*, family Tingidae; an insect that infests the American elm, causing the leaves to turn yellowish in the early summer.

Elm Leaf Beetle–*Galerucella xanthomelaena*, family Chrysomelidae; an insect pest of American, Scotch, English, and Camperdown elms throughout the United States, which skeletonizes the leaves on which it feeds, sometimes defoliating the tree.

Elm Sawfly–*Cimbex americana*, family Cimbicidae; an insect that lays eggs in the leaf tissue, forming blisters on the underside of leaves. The larvae feed on the foliage, defoliating twigs and branches. They infest elm, willow, alder, basswood, birch, poplar, and maple. Common throughout the northern United States and west to Colorado. Also called giant American sawfly.

Eluviation–The movement of soil material from one place to another in solution or in suspension by natural soil-forming processes. Soil horizons that have lost material through eluviation are referred to as eluvial, and those that have received materials as illuvial. Designated as the E horizon (formerly A2).

Embankment–A raised bank or any structure for confining a river, etc. The act of embanking.

Embouchure–The mouth of a river or stream.

Emergency Spillway–A point of discharge used in a dam to carry runoff water that exceeds a given design flood resulting from rainfall.

Emergency Tillage–The cultivation of the soil by listing, duckfooting, chiseling, or pitting in order to roughen the soil or to bring clods to the surface for the reduction of wind erosion.

Emergent Vegetation–Various aquatic plants usually rooted in shallow water and having most of their vegetative growth above water, such as cattails and bulrushes.

Emission Standard–The maximum amount of a pollutant that is permitted to be discharged from a single polluting source, e.g., the number of pounds of fly ash per cubic foot of air per day that may be emitted from a coal-fired boiler.

Emitters–The openings through which water flows from an irrigation system.

Empirical Yield Table–A table showing the progressive development of a timber stand at periodic intervals. The table covers the greater part of the useful life of the stand. It is prepared based on actual stand conditions.

Empty-cell Process–Any process for impregnating wood with preservatives or chemicals, in which air is imprisoned in the wood under the pressure of the entering preservative, which expands when the pressure is released, driving out part of the injected preservative.

Emulsifiable Concentrate–A concentrated pesticide that contains ingredients that will allow the pesticide to become suspended in water.

Emulsifying Agents–Substances such as gums, soaps, agar, lecithin, glycerol monostearate, alginates, and Irish moss that aid to uniform dispersion of oil in water. Examples of emulsions formed: margarine, salad dressing, and ice cream.

Emulsion–A mixture in which one liquid is suspended as tiny drops in another liquid, such as oil in water.

Encapsulated Formulation–A pesticide that is enclosed in a material that causes the pesticide to be released at a desired rate.

Endangered Species–Any species of animal or plant that is in danger of extinction throughout all or a significant portion of its range.

Endemic–Native or confined naturally to a particular and unusually restricted area or region; biologically the site of origin of a plant or animal. An endemic taxon remains in the site of origin after other taxa have spread beyond that site.

Endodynamorphic Soil–A soil whose character has been determined largely by or inherited from the parent material or underlying rock with less influence from the external environment. See Ectodynamorphic Soil.

Endogen–The internal growth of new wood in a plant stem.

Endomychorrhizal–Subsisting within the interior of the plant cell, that is, endomycorrhizal fungi penetrate the interior of tree cortical root cells. See Mycorrhizae; Ectomycorrhizal.

Endothal–A chemical compound used as an herbicide, a defoliant, a drying agent, and as a preemergent herbicide.

Endotoxin–A toxin produced within an organism and liberated only when the organism disintegrates or is destroyed.

Endrin–The endo-endo isomer of the insecticide dieldrin.

Energy Head–The energy of a unit height of a stream that takes into account the elevation of the hydraulic grade line at any section plus the velocity head of the mean velocity of the water in that section. The energy head may be referred to any datum, or to an inclined plane, such as the bed of a conduit.

Energy Resources–The natural supply of energy available for use. Several major sources include: earth's internal heat (geothermal energy), fossil fuels (principally coal, natural gas, and oil), hydropower (rivers, ocean currents, tides, and waves), nuclear energy, solar energy, and wind.

Engelmann Spruce–*Picea engelmanni*, family Pinaceae; an evergreen, coniferous tree, one of the hardiest spruces, which grows up to a height of 150 feet It is grown as an ornamental for its bluish-green to steel-blue needles and for its lumber, which has a wide variety of uses including pulpwood. Native to the Rocky Mountains of western North America.

English Oak–*Quercus robur*, family Fagaceae; a hardy, deciduous tree having many varieties. It is valuable for its timber and as a shade tree. Its foliage and young buds when browsed in large quantities are poisonous to cattle, sheep, and goats. Symptoms of poisoning are constipation, emaciation, edema, blood in the feces, and subnormal temperature. Native to Europe.

English Sparrow–*Passer domesticus*; a small bird that feeds on seeds, buds, insects, etc., and is often a pest. Native to Europe. Also called house sparrow.

English Walnut–Persian walnut, *Juglans regia*, family Juglandaceae. Grown for its high yielding flavorful nut, the tree is an important agricultural crop in such states as California.

Enrichment–The process by which bodies of water are enriched, especially by nitrogen, phosphorus, and/or carbon, resulting in accelerated growth of undesirable algae and other aquatic vegetation. Enrichment sometimes originates as runoff from domestic animal feedlots, landspreading of sewage sludge, and overfertilized fields.

Entisols–A soil order that has no diagnostic pedogenic horizons. They may be found in virtually any climate on very recent geomorphic surfaces, either on steep slopes that are undergoing active erosion or on fans and floodplains where the recently eroded materials are deposited. They may also be on older geomorphic surfaces if the soils have been recently disturbed to such depths that the horizons have been destroyed or if the parent materials are resistant to alteration, as in quartz sands. See Soil Orders.

Environment–The sum total of all the external conditions that may act upon an organism or community to influence its development or existence. For example, the surrounding air, light, moisture, temperature, wind, soil, and organisms are all parts of the environment.

Environmental Design–A design (e.g., of a highway) that includes consideration of the impact of the facility on the community or region based on aesthetic, ecological, cultural, sociological, economic, historical, conservation, and other factors.

Environmental Disease–Disease caused by improper environmental conditions, e.g., lack of oxygen, temperature variations, etc.

Environmental Impact Statement–A document prepared by a person, an industry, or a political entity on the environmental impact of its proposals for legislation and other major actions significantly affecting the quality of the human environment. Environmental impact statements are used as tools for decision making and are required by the National Environmental Policy Act.

Environmental Protection Agency (EPA)–A federal agency charged by Congress to protect the nation's land, air, and water systems. Under a mandate of national environmental laws focused on air and water quality, solid-waste management and the control of toxic substances, pesticides, noise, and radiation, the EPA strives to formulate and implement actions that lead to a compatible balance between human activities and the ability of natural systems to support and nurture life.

Environmental Quality–An evaluation of environmental quality should include: (a) areas of natural beauty; (b) water, land, and air quality; (c) biological resources and selected ecosystems; (d) geological, archeological, and historical resources; and (e) irretrievable commitments of resources to future use. The sum total of the forces and factors that influence people's satisfactions with their work, leisure, living conditions, and community.

Environs–Environment, neighborhood.

Enzootic–Referring to the prevalence of an animal disease in a particular district or region.

Eolation–All the direct, geologic activities of the wind that either tear down or build up; the process by which wind modifies land surfaces, both directly by transportation of dust and sand, by the abrasion of sandblasts, and indirectly by wave action on shores.

EPA–See Environmental Protection Agency.

EPA Establishment Number–A number assigned to factories that produce pesticides. The numbers are printed on the containers of pesticides to indicate where the pesticide was made.

EPA Registration Number–A number that appears on the label of pesticides to indicate the number under which the pesticide was registered by the manufacturer.

Ephemeral–Designating a very short-lived plant or animal.

Ephemeral Stream–A stream or portion of a stream that flows only in direct response to precipitation.

Epigent Weathering–The rock weathering of the surface crust of the earth, including such end products as soils.

Epilimnion–That region of a body of water that extends from the surface to the thermocline and does not have a permanent temperature stratification. See Hypolimnion, Metalin, Thermocline.

Epsom Salts–$MgSO_4 \cdot 7H_2O$, hydrated magnesium sulphate which is used as a laxative or purgative. Originally obtained from the mineral waters at Epsom, England, but now principally from the mineral kieserite, a potassium/magnesium salt.

Equilibrium Moisture Content–(1) The equilibrium established between the vapor pressures of air and a product, such as grain, hay, popcorn, etc., at identical temperatures. (2) The moisture content at which wood neither gains nor loses moisture when surrounded by air at a given relative humidity and temperature.

Equivalent per Million–An equivalent weight of an ion or salt per one million grams of solution or soil. For solutions, equivalents per million and milliequivalents per liter (meq/1) are numerically identical if the specific gravity of the solution is 1.0.

Eradicant–A fungicide in which a chemical has been added to rid a host of a pathogen.

Eradicate–To destroy or abolish, as a disease, insect, or weed pest.

Eremacausis–The natural burning out (oxidation) of organic matter, which results in a very low content of humus in soils of hot, arid regions.

Erode–(1) To wear away or carry away the surface of the earth, as water erodes a rock formation. See Erodible, Erosion. (2) To wear away, eat into the cells or layer of cells.

Eroded Area–A place that is worn away. In an animal's body this may be mechanical, as the wearing away of teeth or outer skin, or by the action of chemicals or germs that destroy living cells.

Erodibility Index–An index based on a field's inherent tendency to erode from rain or wind in the absence of a cover crop. It is based on the universal soil loss equation (USLE) and the wind erosion equation (WEE), along with a soil's T value. See T Value, Universal Soil Loss equation, Wind Erosion Equation.

Erodible–Susceptible to erosion, as erodible soil. See Erosive.

Erosion–The group of processes whereby earthy or rock material is worn away, loosened or dissolved and removed from any part of the earth's surface. It includes the processes of weathering, solution, corrasion, and transportation. Erosion is often classified by the eroding agent (wind, water, wave, or raindrop erosion) and/or by the appearance of the erosion (sheet, rill, or gully erosion) and/or by the location of the erosional activity (surface or shoreline) or by the material being eroded (soil erosion or beach erosion). Relations between erosion terms: *raindrop erosion* always takes the form of *sheet erosion*, though sheet erosion can also be caused by wind action or the movement of thin sheets of water over the ground surface. *Sheet, gully* and *rill erosion* are all forms of soil erosion. Sheet and rill erosion are the two forms that *surface* erosion may take. *Beach erosion* is always *shoreline erosion*—though, because not all shorelines are beaches, shoreline erosion is not always beach erosion. The term *accelerated erosion* is used in comparing erosion caused by human activities with that occurring at natural rates, called geologic erosion.

Erosion Pavement–A residue of pebbles and stones on a land surface formed by the removal of the finer surface particles by wind or water; common in deserts.

Erosion-resistant Crop–Those crops which, because of dense foliage, extensive root system, and/or heavy litter, provide effective protection against soil loss by erosion.

Erosive–Refers to wind or water having sufficient velocity to cause erosion. Not to be confused with erodible as a quality of soil. See Erodible.

Escarpment–A steep descent or almost vertical steep slope.

Esker–A minor land feature of the glaciated region of the United States; i.e., a low, narrow ridge with curved or rounded slopes composed of gravel and sand showing evidence of deposition in glacial water. Also called osar, hogsback. Usually a good source of building sand and gravel.

Essential Element–Any element necessary for growth and reproduction of a plant or animal. The sixteen essential elements for seed-bearing plants are: carbon, hydrogen, oxygen, phosphorus, potassium, nitrogen, sulfur, calcium, iron, magnesium, boron, manganese, copper, zinc, molybdenum, and chlorine. The twenty essential elements for humans and animals are: carbon, hydrogen, oxygen, phosphorus, potassium, nitrogen, sulfur, calcium, iron, magnesium, manganese, copper, zinc, sodium, iodine, selenium, vanadium, chromium, chlorine, and cobalt.

Established– (1) Designating a plant or plant community growing naturally in healthy adjustment to the environment. (2) Designating a pest or disease that is thriving. (3) Designating a plantation of trees that requires little or no cultivation, weeding, etc., and in which the young trees are almost certain to survive. (4) Designating a seeding of meadow or pasture plants that is thriving.

Establishment– (1) The adjustment of a plant to a new site, consisting of three processes: germination, growth, and reproduction. Also called acesis. (2) A farmstead. (3) An economic unit, generally at a single physical location, where business is conducted or where services or industrial operations are performed.

Esthetic Woods– A woodland established and maintained for its beauty and esthetic values, as contrasted with forests or woods maintained for commercial purposes or for windbreaks. See Arboretum, Pinetum.

Esthetics– See Aesthetics.

Estimate– A judgment of the approximate volume or yield of a crop or a timber stand made from incomplete data.

Estival– Pertaining to summer.

Estuary– All or part of the mouth of a navigable river or stream or other body of water having unimpaired natural connection with the open sea within which the seawater is measurably diluted with freshwater derived from land runoff.

Ethanol– C_2H_5OH; the alcohol product of fermentation that is used in alcohol beverages and for industrial purposes; chemical formula blended with gasoline to make gasohol; also known as ethyl alcohol or grain alcohol.

Ether Extract– The fatty substances in food and feed that are soluble in ether; one of the procedures in the proximate analysis of feeds.

Ethyl Acetate– $CH_3COOC_2H_5$, an inflammable fumigant that leaves a characteristic odor on treated grain.

Ethyl Alcohol– Also known as ethanol or grain alcohol. See Ethanol.

Ethyl Formate– A fumigant.

Ethyl Mercury Phosphate– A fungicide.

Ethylene– C_2H_4; a colorless, inflammable, unsaturated hydrocarbon gas used to hasten ripening of harvested fruits, especially bananas, pears, and green tomatoes, and to hull Persian walnuts and other nuts. It is used as a coloring agent for harvested citrus fruits, especially oranges, and as a defoliation agent for fall-dug nursery stock, especially rose bushes. It is also used for blowtorch welding and cutting.

Ethylene Dibromide– A heavy liquid used as a soil fumigant, especially for the control of nematodes (wireworms) in the irrigated land of the western United States. Emulsified, it is effective against Japanese beetle grubs in grasslands and against the European chafer in soils. It is also used to fumigate honeycombs for control of wax moth.

Ethylene Dichloride– A soil fumigant and nematocide.

Ethylene Oxide– C_2H_4O; a chemical used as a fumigant.

Ethylmercuric Chloride– A fungicide for seed and bulb treatment.

Ethylmercuric Phosphate– A fungicide for seed and bulb treatment.

Ethylmercuric-P-Toluence Sulfonilimide– A fungicide for seed and bulb treatment.

Eucalyptus– Any tree of the genus *Eucalyptus*, family Myrtaceae, of Australia. Some species are gigantic, some highly aromatic, and some are evergreen. They are grown as ornamentals and as bee trees in warm regions. Also called eucalypt, gum tree, stringy bark.

European Alder– *Alnus glutinosa*, family Corylaceae; a deciduous tree easily grown as an ornamental in swampy locations. Tolerates strongly acid soils. Native to Eurasia and North Africa. Also called black alder, dog tree.

European Beachgrass– See Beachgrass, European.

European Earwig– *Forficula auricularia*, family Forficulidae, an insect whose nymphs feed on the shoots of plants and eat holes in the leaves, while the adults feed on blossoms and ripening fruit. The adults often enter houses. Found in the northern United States.

Euryvalent– A species that extends over many habitats.

Eutrophic Water– Waters that have a high concentration of plant nutrients such as nitrogen and phosphorus.

Eutrophication– A condition in stagnant pools and lakes characterized by an abundant accumulation of nutrients that support a dense growth of plant and animal life, the decay of which depletes the shallow waters of oxygen, especially during warm weather. Growth of water plants receiving a sufficient supply of nutrients in a natural process and is desirable for the production of fish, but excessive growth of algae and other plants leads to a deterioration in water quality. The latter condition is one of the many public concerns about environmental pollution. Nitrogen and phosphorus in fertilizers are receiving much attention as possible pollutants of surface and ground waters that reach lakes and streams and cause eutrophication. Also of special concern is nitrate nitrogen in water for domestic use which, in excessive concentration, can induce methemoglobinemia in infants and numerous disorders in livestock. The United States Public Health Standards limit nitrate nitrogen to 10 ppm (45 ppm of NO_3) in water for domestic use.

Eutropic Peat– Productive peat or muck soil nearly neutral or alkaline in reaction that is relatively high in bases and essential mineral plant nutrients.

Evaluation Plantation– A tree plantation established to evaluate one or more genetic characteristics, whether individual trees or their parents.

Evaporation– The changing of a liquid into a gas. Industrially it may mean merely drying or driving off the moisture of a product, and not complete vaporization. It may be loosely used as the equivalent of dehydration. See Transpiration.

Evaporation Pan– (United States Weather Service) For determining the amount of evaporation, a circular tank 4 feet (122 centimeters) in diameter and 10 inches (25.4 centimeters) in depth is set in an open field on 2-inches (5-centimeters) supports with the top 12 inches (30.5 centimeters) from the ground. The water level in the tank is maintained between 1.5 inches and 2.5 inches (3.8 and 6.35 centimeters) from the top of the pan. At regular intervals the water level is measured and recorded as the rate of surface evaporation from an open-water surface. In irrigated areas, the frequency of applying irrigation water can be determined by the rate of water evaporation in the pan.

Evapotranspiration–That part of the root zone moisture that is consumed by evaporation and transpiration combined, including all water consumed by plants plus the water evaporated form bare land and water surface. See Evaporation, Transpiration.

Even-age Management–The application of a combination of actions designed to create stands in which trees of essentially the same age grow together.

Everglade–A tract of swampy land covered mostly with tall grass; a swamp or inundated tract of lowland. Local in the South.

Everglades–A 4,000 to 5,000 square mile tract of low-lying marshland in Florida, United States, that extends from Lake Okeechobee to the southern tip of Florida. The area, less than 20 feet above sea level, is covered with water the greater part of the year, and its characteristic vegetation is sawgrass (*Cladium effusum*). The soil is mainly peat, but extensive marl prairies occur. Only a small percentage of the area is used for agriculture, while a large tract has been set aside as a national park. See Marl.

Evergreen–(1) A plant that retains its leaves or needles longer than one growing season so that leaves are present throughout the whole year. See Deciduous. (2) The leafy branches of evergreen plants that are used for decoration. Also called greens.

Evolution–Gradual change in succeeding generations of plants and animals brought about by variations in reproductive habits, mutations, recombinations, crossbreeding, selection, migrations, and genetic drift. See Genetic Drift.

Excavation–Any activity by which earth, sand, gravel, rock, or any other similar material is dug into, cut, quarried, uncovered, removed, displaced, relocated, or bulldozed and includes the conditions resulting therefrom.

Excess Alkali–Excess exchangeable sodium. The resulting poor physical and chemical properties restrict the growth of plants.

Excess Water–(1) Legally, surplus water of a stream, that which is not adjudicated, or that which is in excess of the needs of those who have prior rights to its use. (2) Any water, particularly rainfall, over and above that needed for plant growth that results in ponded water or runoff. See Drainage.

Excessively Drained–Water is removed from the soil very rapidly. Excessively drained soils are commonly very coarse, rocky, or shallow. Some are steep. All are free of the mottling related to wetness. See Soil Drainage Classes.

Exchangeable Cation–A cation that is absorbed on the exchange complex and which is capable of exchange with other cations.

Exchangeable Sodium Percentages–The degree of saturation of the soil exchange complex with sodium. It may be calculated by the formula: ESP = exchangeable sodium (meq/100g soil) ÷ cation − exchange capacity (meq/100 g soil) × 100.

Excrescence–An unusual or wartlike growth, as on the stem of a tree.

Excurrent–Stem of a tree that extends from base to tip without dividing, as spruce or hemlock stems.

Exfoliation–The process of chipping off of rock particles in thin sheets from rock surfaces as a result of mostly mechanical weathering. Changes in heat result in changes in rock volume and thus of a surface layer of rock scaling off. Ice wedging in cracks and under rock scales assists the mechanical chipping. Chemical and biochemical changes in the minerals in the rock also accelerate exfoliation. Lichens assist in the biochemical processes by generating carbonic acid to solubilize minerals. See Lichens.

Exogenous–(1) Produced on the outside of another body. (2) Produced externally, as spores on the tips of hyphae. (3) Growing by outer additions of annual layers, as the wood in dicotyledons. See Endogenous.

Exotic Species–An organism that is not native to the region in which it occurs.

Exploitive Farming–Crop production carried on for immediate profit and in disregard of long-term soil productivity.

Exposure–The direction a slope faces, such as northern exposure. Also known as aspect.

Face Cord–A pile of wood 8 feet long and 4 feet high, regardless of the length of the pieces; usually 12 to 24 inches long. See Cord.

Face Side–The side of a board from which the grade is determined. Softwoods are most frequently graded from the better side; hardwoods, the poorer side.

Face-of-Tree–In log grading and sawing, one-fourth of the circumference of a log for its entire length.

Facilitation, Social–The effect of the presence of one organism upon the behavior of another.

Factor–(1) A unit of inheritance occupying a definite locus on one or both members of a definite chromosome pair whose presence is responsible for the development of a certain character or modification of a character of the individual who possesses that genotype; a determiner or gene. (2) An agent, as one who buys and sells a commodity on commission for others. (3) An item in the analysis of a farm business; e.g., labor efficiency. (4) Inherent characteristics of the climatic, nutritional, cultural, or biological environment responsible for the specific performances of plants or animals.

Fahrenheit Scale–A temperature scale in which the freezing point of water is 32°F and the boiling point is 212°F. Named after Gabriel

Daniel Fahrenheit (1686-1736), a German physicist. See Celsius Scale, Kelvin Scale.

Fall–(1) The dropping of a plant part, such as leaves or fruits. (2) The slope of land. (3) The amount of precipitation. (4) (a) The quantity of trees cut, (b) Felling trees. (5) (a) The number of lambs born. (b) Giving birth, as of lambs. (6) One of the three outer segments of an iris flower that is often drooping (usually used in the plural). (7) Autumn. (8) Of, or pertaining to, a plant or fruit that matures in the autumn, as a fall apple.

Fall Line–A line connecting the points where rivers leave the uplands for the lowlands, marked by an increased slope and waterfalls.

Fall Overturn–A physical phenomenon that may take place in a deep body of water during the early autumn. The sequence of events leading to fall overturn include: (a) cooling of surface waters, (b) density change in surface waters producing convection currents from top to bottom, (c) circulation of the total water volume by wind action, and (d) vertical temperature equality. The overturn results in a uniformity of the physical and chemical properties of the water.

Fallage–(1) The act of cutting trees. (2) Timber that has been felled.

Fallout–Dust particles that contain radioactive fission products resulting from a nuclear explosion. The wind can carry fallout particles many miles.

Family, Soil–In soil classification one of the categories between the soil subgroups and the soil series. See Soil Family.

Fancy Price–(1) An unusually high price. (2) Prices paid for top-grade products.

Fanega–(1) A measure of land, used in Spanish settlements in America, which consisted of 8.81 acres. (2) A dry measure, of Spanish origin, once used in southern United States, equivalent to about two and one-half United States bushels.

Fang–(1) To form a forked root, especially of root crops, such as sugar beets. (2) Any tooth that is long and pointed.

Farm Forestry–The practice of forestry on part of the farmland which is carried on with other farm operations.

Farm Pond–A small reservoir of water on a farm usually formed by constructing a dam across a watercourse or by excavation to collect surface water.

Farm Survey–Data collection from a sample of farms from a given population.

Fascine–(1) A fagot. (2) A long bundle of sticks bound together and used to build a temporary roadway through a marshy soil or to stabilize an unstable slope along an embankment. When used in erosion control, the sticks may be made of willow or other species that sprout. See Gabion.

Fast Intake–The rapid movement of water into the soil. See Infiltration.

Fastigiate Tree–Any tree whose limbs are in an upright position; e.g., the Lombardy poplar, ***Populus nigra***, var. ***betulifolia***, family Salicaceae.

Fat Pine–(1) A term used in many parts of southern United States, to denote branches and knots of pine that are rich in resinous material, and not to denote any particular species of pine. (2) See Longleaf Pine.

Fattening Range–A productive range devoted primarily to fattening of livestock for market.

Fauna–The total wild animal life of an area in the broadest sense, including wild mammals, birds, fish, reptiles, insects, and smaller animal life. See Flora.

Fawn–(1) A coat color for an animal that is a soft, grayish-tan. (2) A young deer.

Fe–The chemical symbol for iron.

Fee Simple Title–Complete ownership of land.

Feld–(German) Field.

Feldspars–A group of silicate minerals present in soils, igneous rocks, granites, etc. The feldspars are the primary source of potassium, sodium, and kaolinitic clay of soils. Its two principal groups are orthoclase, potash feldspars; and plagioclase, soda-lime feldspars.

Feline Enteritis (Distemper)–A highly contagious disease of cats caused by a filterable virus. It is associated with a high fever, loss of appetite, rapid loss of weight, depression, diarrhea, vomiting, soreness of the abdomen, and sometimes a nasal and eye discharge. Also called infectious gastroenteritis, feline distemper, croupous enteritis, epizootic enteritis, malignant panleucopenia, infectious feline agranulocytosis, feline typhus, feline influenza. See Enteritis.

Fell–(1) The elastic tissue just under the hide of an animal attached to its flesh; facia. (2) To cut down a tree.

Felling Head–A tractor-powered implement that grasps a tree and cuts it using a circular bar that has a cutting chain running around the perimeter.

Fen–Wet, marshy land around freshwater lakes or in estuaries. It is rich in humus, muck, and/or peat.

Fence–(1) A hedge or barrier of wood, metal, stone, or plants erected to enclose an area to prevent trespassing or the straying of animals. (2) To enclose an area with a fence. See Drift Fence, Snow Fence.

Fericulture–The propagation of game and fur-bearing animals as a separate commercial enterprise or as an incidental or supplemental one to farming.

Ferment–(1) To undergo catalytic decomposition of complex compounds into simpler ones, as in the souring of milk, the formation of vinegar from cider, etc. (2) Any agent which can cause fermentation. See Enzyme, Fermentation.

Fermentation–(1) The processing of food by means of yeasts, molds, or bacteria; the catalytic decomposition of complex compounds into simple ones, as in the souring of milk, the production of vinegar from cider, etc. The most important products of fermentation are alcohol, acetic acid, lactic acid, butyl acid, citric acid, and acetone. (2) An enzymatic curing of stored tobacco leaves which improves the quality of the product. (3) Digestion in the rumen by microorganisms.

Fermentation Ethanol–Ethyl alcohol produced from the enzymatic transformation of organic substances. See Ethanol.

Ferric Oxide–Fe_2O_3; oxide of iron; iron is widely distributed in nature. It is found in phosphate rock where its action is similar to alumina (Al_2O_3). More than 3 to 4 percent of oxide of iron will make "sticky" superphosphate. See Iron.

Ferric Phosphate–A chemical compound containing 25 percent iron and 13.9 percent phosphorus used in mineral feeds for livestock. Also known as iron phosphate.

Ferric Sulfate–$Fe_2(SO_4)_3 \cdot 7H_2O$; a yellow-brown salt and a quickly available source of iron. Fertilizer-grade material contains about 2.2 percent Fe (30.3 percent Fe_2O_3). See Micronutrient Fertilizers.

Ferruginous Soil–Soil containing a large amount of iron minerals, especially limonite and hematite, characterized by a vivid red color or various shades of yellow and brown. See Ironstone.

Ferrum–(Latin) Iron.

Fertile–(1) Productive; producing plants in abundance, as fertile soil. (2) Capable of growing or of development, as a fertile egg. (3) Capable of reproducing viable offspring. (4) Able to produce fruit, as a fertile flower. (5) Plant capable of producing seed.

Fertilized Pond–A pond to which fertilizer is regularly applied to promote the growth of algae and other aquatic plankton for fish food.

FFA Alumni–A national organization composed of former members and supporters of the National FFA Organization. The purpose is to promote and support agricultural education programs and in particular, programs of the FFA. See FFA; Agricultural Education.

Fidelity–An ecological concept that expresses the limitation of species to specific habitats.

Field Capacity–The amount of water held in a saturated soil after the excess or gravitational water has drained away. Also called capillary capacity.

Field Mice–Various kinds of mice that live among grasses and weeds, mostly of the genus *Microtus*, family Cricetidae.

Field Percolation Test–A technique of determining the rate of percolation in a soil. The test consists of digging holes about 2 feet deep, filling them with water, and measuring the rate of water movement downward.

Field Windbreak–A barrier of trees and shrubs grown to check the force of wind for protection of orchards, feedlots, and fields, and to check drifting snow and retain it to increase soil moisture. See Windbreak, Shelterbelt.

Figure Grain of Wood–The pattern produced on a wood surface by irregular coloration, growth layers, rays, knots, and such deviations from regular grain as bird's-eye, interlocked, and wavy grain.

Fill–(1) The soil, sand, gravel, etc., used to fill in a depression in a field, or to build up a terrace or embankment. (2) The substances used in filling tree cavities; e.g., asphalt, concrete, wooden or rubber blocks, etc. Also called filler. (3) The increase in weight and form of livestock that have been watered and fed after arriving at their destination. (4) The shaft of a vehicle. (5) To level a depression in a field or cavities in trees, or to build up an embankment or terrace. (6) To feed and water livestock at the end of the shipment to make up for the loss of weight en route. (7) To enlarge with the enclosed seeds, as the pods of leguminous plants; or to be plump and shriveled when approaching maturity, as the seeds of cereals. (Cereal grains before harvest are often referred to as poorly or well-filled.)

Fill Dirt–Soil used to change the grade or elevation of an area.

Filter Blanket–A layer of sand and/or gravel around a drain line designed to reduce the movement of fine-grained soils into the drain.

Filter Crop–Any close-growing crop planted across a slope to retard runoff and cause deposition of the sediment load. See Filter Strip.

Filter Strip–A strip of permanent vegetation of sufficient width and vegetative density above farm ponds, division terraces, and other structures which retards the flow of run-off water, causing the flowing water to deposit soil, thereby reducing the rate of silting of the reservoir below.

Fine Clayey–A soil family with a clay content between 35 and 59 percent in the subsoil. See Soil Family.

Fine Granular–Designating a soil aggregate that is less than 5 millimeters in diameter. See Soil Texture.

Fine Particulate Matter–A general term referring to particulate matter less than 10 microns in diameter.

Fine Sand–In the mechanical analysis of soils, soil separates (mineral particles) ranging in diameter from 0.25 to 0.10 millimeters (United States Department of Agriculture scheme). See Soil Texture.

Fine Sandy Loam–A class of soil which contains 30 percent or more of fine sand or less than 30 percent of very fine sand (or) between 15 and 30 percent very coarse, coarse, and medium sands. The clay content is 20 percent or less and the silt content less than 50 percent. See Soil Texture.

Fine-textured Soil–Sandy clay, silty clay, and clay. The obsolete term is heavy textured. See Soil Texture.

Fineness–(1) One of the several properties of cotton that determines the grade in which it is classified. Fineness refers to the smallness of the cross section diameter of the fibers (or lint). (2) One of several properties of dry milk. Usually the powders prepared by the spray process are extremely fine, whereas those prepared by the cylinder or drum process are coarser. (3) One of the several properties of high-quality hay that help in the determination of the grade of hay as feed for livestock. (4) The relative smallness of the wool fiber.

Fines–Fine particulates; aerosols.

Fingerling–A fish from 3 to 6 inches long that is used for stocking ponds or lakes.

Fire–(1) In several different plant diseases caused by bacteria, fungi, or nutritional deficiency, the final result that gives a burning or scorching appearance. (2) Potash deficiency or leaf scorch is sometimes referred to as firing. (3) The burning off of vegetation on lands of various types. This is apparently beneficial in some cases but is often harmful. (4) To treat a spavin or ringbone on a horse with a strong liniment in an attempt to cure or alleviate lameness.

Fire Ant–*Solenopsis geminata*, family Formicidae; a species of ants that is harmful to plants and whose bite is very painful to people and animals. See Imported Fire Ant.

Fire Danger–In forestry, the result of both constant and variable factors that determine whether fires will start, spread, and damage; determining as well, the difficulty of control. Constant factors are those that are relatively unchanging, e.g., normal risk of ignition, topography, all fuels, and exposure to prevailing wind. Variable factors change from day to day, season to season, and year to year; e.g., all weather elements, moisture content of fuel, and variable risks of ignition by lightning and people.

Fire Farming–Clearing patches of land by destroying wild cover by fire, cropping for a few years, and then allowing the area to remain fallow for natural revegetation and restoration. Practiced mainly in humid, tropical regions. Also called milpa and canuca in Latin America, langland in Indochina, caingin in the Philippines, and shifting cultivation in most other countries. See Shifting Cultivation.

Fire Flap–A flat flexible piece of material attached to a long handle that is used for putting out a fire. Also called a fire swatter.

Fire Line–In forestry, the narrow portion of a control line from which inflammable materials have been removed by scraping or digging down to mineral soil.

Fire Plow–A heavy duty, usually specialized machine, either of the share or disk type, which is designed solely for abusive work in the woods and is used to construct firebreaks and fire lines.

Fire Presuppression–Those forest fire control activities concerned with the organization, training, instruction, and management of the fire-control organization, and with the inspection and maintenance of fire control improvements, equipment, and supplies to ensure effective fire suppression.

Fire Prevention–Those fire control activities concerned with the attempt to reduce the number of forest fires through education, hazard reduction, and law enforcement.

Fire Retardant Chemical–Any chemical that when injected into or applied on wood will reduce inflammability.

Fire Wound–The fresh or healed injury of the cambium of a tree, caused by fire.

Firebreak–An existing barrier, or one constructed before a fire occurs, from which all or most of the inflammable materials have been removed. It is designed to stop or check creeping or running fire but not to stop crown fires, to serve as a line from which to work, and to facilitate the movement of men and equipment in fire suppression.

Firebreak, Living–A firebreak on which is maintained nonflammable, green vegetation such as grass (kept nonflammable by fertilization and close grazing), forest trees (whose lower branches are kept pruned by natural or artificial means, such as *Eucalyptus gmelina* in Australia), and the semitropical iceplant *Mesembryanthemum* spp. from South Africa and the Mediterranean region but widely naturalized in southern California, especially the species *M. deule*.

Firewood–Logs cut short enough and split into convenient sizes for use as fuel in fireplaces or wood-burning stoves.

First–(1) The highest grade of lumber. (2) The second grade, next below extra, for butter. (3) The primary occurrence or the beginning of a series.

First Bottom–In a river valley, the first alluvial plain level above the channel of the stream that is most subject to inundation during flood stages of the river. See Floodplain.

First Law of Thermodynamics–The first law of thermodynamics introduces the concept of internal energy of a system and expresses the fact that the change of energy of the system is equal to the amount of energy received from the external world. The energy received from the external world is equal to the heat taken in by the system and the work done on the system.

First-order Stream–A headwater stream with no tributaries. Two first-order streams combine to form a second-order stream; two second-order streams combine to form a third-order stream, etc.

Fish Ladder–An inclined trough carrying water down a dam at a velocity against which fish can easily swim upstream to reach their spawning grounds.

Fish Manure–An organic fertilizer made from fish scrap.

Fish Pomace–A fertilizer, the fish refuse after the oil has been extracted.

Fish Pond–A small body of water in which the fish population is managed. On a farm, the pond impounds rainfall runoff, reduces erosion, provides a watering place for livestock, and furnishes a place where fish can be grown.

Fish Pond Fertilizer–Inorganic or organic fertilizer applied to a pond to promote the growth of plankton on which the fish depend directly or indirectly for food.

Fishery–(1) All activities connected with propagation, cultivation, and exploitation of fishes in inland and marine waters, as also the management of fish resources. (2) Fishing ground.

Fixation of Nitrogen–Fixation of atmospheric nitrogen in the soil by bacteria, both symbiotic and nonsymbiotic. Symbiotic bacteria grow in nodules on the roots of legumes; e.g., alfalfa and clovers; the nitrogen fixed by them may become available to the host plant, may be released directly to the soil, or through decomposition of the legume may become available to other plants.

Fixed Ammonium–Ammonium, held in a soil, which is neither water soluble nor readily exchangeable.

Fixed Groundwater–Groundwater held in saturated material with interstices so small that it is attached to the pore walls, and is usually not available as a source of water for pumping. See Aquifer.

Fixed Phosphorus–Phosphorus in soil that is unavailable to plants; soluble phosphorus of fertilizers which, as a result of reactions taking place in the soil with calcium and iron, has become relatively insoluble and unavailable as a plant nutrient.

Fixed Potassium–Potassium held by a soil, which is neither water soluble nor readily exchangeable or available to plants.

Fjeld–(Norwegian) An Arctic, stony dessert.

Flagella–Whiplike appendages of certain single-celled aquatic animals and plants, including some bacteria, the rapid movement of which produces motion.

Flaggy–Designating a soil characterized by coarse rock fragments that are flat, thin, and angular, with dimensions (arbitrarily assigned) of 6 to 15 inches.

Flat–(1) A shallow box containing soil, in which seeds are sown or to which seedlings are transplanted form the seedbed. (2) A level, treeless prairie, especially between hills or mountains. (3) Referring to a defect in the flavor of butter, cheese, or milk, due to errors in processing, as lack of sufficient salt, uncleanliness, etc. The flavor may be insipid or lacking in the usual characteristics of the product. (4) A level landform composed of unconsolidated sediments—usually mud or sand. Flats may be irregularly shaped or elongate and continuous with the shore, whereas bars are generally elongate, parallel to the shore, and separated from the shore by water.

Flatwoods–Any of various types of flat, wooded land usually of inferior quality for agriculture. The land may be wet, or intermittently wet and dry, and it is generally either a clay or a sand soil. Flatwoods may be designated by the kind of native vegetation as post-oak flatwoods, saw-palmetto flatwoods, etc. (Florida and other parts of the southern United States).

Flint–A dense, amorphous, or cryptocrystalline rock composed essentially of silica (SiO_2) and not different in composition from chert. It remains as a residue in the weathering of chalks and limestones and hence is locally abundant in soils, thus characterizing land, as flint hills, flint ridge, flinty soil. See Chert.

Flitch–(1) A portion of log, sawed on two or more sides, which is intended for sliced or sawed veneer. (2) A pile or bundle of veneer sheets from the same bolt laid together in the sequence of cutting. (3) A side or portion of meat, as a flitch of bacon.

Floating Gardens–A technique consisting of planting vegetation on soil on rafts floating in water; known also as chinampa cultivation in parts of Mexico.

Floating Plant–A nonanchored plant that floats freely in the water or on the surface; e.g., water hyacinth (***Eichhornia crassipes***) or common duckweed (***Lemna minor***).

Floating-leaved Plant–A rooted, herbaceous hydrophyte with some leaves floating on the water surface; e.g., white water lily (***Nymphaea odorata***), floating-leaved pondweed (***Potamogeton natans***). Plants such as yellow water lily (***Nuphar luteum***), which sometimes have leaves raised above the surface, are considered floating-leaved plants or emergents, depending on their growth habit at a particular site.

Floats–Finely ground rock, rich in phosphate, used as a fertilizer. It is separated from impurities by flotation.

Flocculation–(1) A process whereby particles in solution or suspension contract and adhere to make larger sized clusters. (Syn.: coagulation, agglomeration) (2) A process of converting a stable colloidal suspension into an unstable system of larger particles that settle. This can be accomplished by adding an appropriate electrolyte. Flocculation is very important in the stabilization of clay in soils. Clays saturated with hydrogen or calcium are both flocculated and stable. Colloids in sewage effluents are often flocculated and thereby stabilized by ferric sulfate, aluminum sulfate, or calcium oxide.

Flood–(1) The overflow by water from natural or human causes into fields, farms, cities, etc. (2) To run water upon a field to the depth of a few inches to supply sufficient moisture for the growth of the crop, as rice, cranberries, etc., to control weeds, or to protect the crop from injury by frost.

Flood Control–(1) Any of the various agricultural practices that hold the soil in place and increase its infiltration and water-retention capacity. These practices include the planting of deep-rooted grasses, trees, and shrubs: no-till crop farming; contouring; terracing; and keeping heavy machinery off the soil when it is too wet to support such loads. (2) Any of the various engineering practices used to reduce flood damage resulting from overflowing rivers and streams. It includes deepening the existing channels, digging diversion canals, building levees and dams to impound the water in numerous small ponds or retention basins in the upper parts of the watershed, or by large detention basins farther downstream.

Flood Irrigation–A system of irrigation consisting of adding water at the highest point in a field and allowing the water to cover the soil.

Flood Peak–The highest value of the stage or discharge attained by a flood, thus, peak stage or peak discharge.

Flood Stage–The stage in which the level of the water rises above the tops of stream banks or dikes. Frequently, an arbitrary level established, beyond which the stream or river is considered to be at flood stage.

Flooding–The temporary covering of soil with water from overflowing streams, runoff from adjacent slopes, and tides. Frequency, duration, and probable dates of occurrence are estimated. Frequency is expressed as none, rare, occasional, and frequent. None means that flooding is not probable; rare that it is unlikely but possible under unusual weather conditions; occasional that it occurs on an average of once or less in two years; and frequent that it occurs on an average of more than once in two years. Duration is expressed as very brief if less than two days, brief if two to seven days, and long if more than seven days. Probable dates are expressed in months; November-May, for example, means that flooding can occur during the period of November through May. Water standing for short periods after rainfall or commonly covering swamps and marshes is not considered flooding.

Floodplain–(1) That portion of a stream valley, adjacent to the channel, which is built of sediments during the present and geologic times and which is covered with water when the stream overflows its banks at flood stages. A river has only one floodplain (currently subject to flooding) and may have one or more terraces (historically subject to flooding). (2) The nearly level land situated on either side of a channel that is subject to overflow flooding. (3) The extent of a floodplain obviously fluctuates with the size of overbank stream flows. Thus, no simple, absolute floodplain commonly exists. As a consequence, floodplains are delineated in terms of some specified flood size (e.g., the fifty-year floodplain—the area that would be flooded by the largest stream flow that will, on the average, occur once within a fifty-year period). Such expected flood-return frequencies are estimated from historic records of stream flows. The largest, absolute floodplain that is ever likely to occur is sometimes referred to as the flood basin.

Floodway–The space between a river bank and a levee or other natural elevation on either side of a stream. It acts as an overflow channel.

Floor, Forest–All dead vegetable matter on the soil surface including litter and unincorporated humus.

Flora–(1) The aggregate of plants that grow without cultivation in a given area within a stated period of time. (2) The bacteria in or on an animal, a plant, or the soil, usually referred to as bacterial flora. See Fauna.

Flotation Concentrates–The high-grade phosphate fertilizer that has been separated by the flotation process from the clay, sand, etc., present in crushed phosphate rock.

Flotation Sulfur–An extremely fine powdered sulfur, more active chemically than "flowers of sulfur"; produced by the oxidation of hydrogen sulfide (H_2S) recovered in aqueous alkali from coal gas fumes. See Flowers of Sulfur.

Floury–Designating a fine-textured soil, consisting primarily of silt size which when dry is incoherent, smooth, and dustlike. See Soil Texture.

Flow–(1) The quantity of liquid that passes through a pipe, gate, channel, or other conveyance for a given unit of time under given conditions of head or pressure, roughness, etc. Units of measure may be cubic feet per second, acre-feet per day, gallons per minute, etc. (2) The movement of silt, sand, etc., in a channel. (3) Discharge from a pipe, etc. (4) The amount of milk produced per cow herd, etc., at a specified time. (5) The ease or difficulty with which a product can be moved from one place to another.

Flow Line–(1) The position of the water surface of a running stream for a normal or specified rate of discharge. (2) The bed of a stream or culvert.

Flowers of Sulfur–A very light, floury form of elemental sulfur produced by sublimation and not by crushing or grinding. It is used as a fungicide, especially against the powdery mildews and also in the control of certain insects or mites. Also called sublimed sulfur. See Flotation Sulfur.

Flowstone–Deposits of calcium carbonate that have accumulated on the walls in many places where water trickles from limestone rock.

Fluffy–(1) Describing a soil in which the aggregates of the surface are loose, light in weight, and fine; a desirable structure for seed germination and plant growth. (2) Describing a body defect of ice cream characterized by large air cells and open texture.

Flume–An inclined, open conduit or troughlike structure for the conduction of water and floating objects, as logs, sugarcane, etc. An aqueduct is a special-purpose flume.

Fluor-Apatite–$Ca_{10}F_2(PO_4)_6$; a mineral that is a source of phosphorus, but because of its fluorine content is not recommended as a stock feed supplement without removing the fluorine. See Apatite.

Fluorides–Gaseous or solid compounds containing fluorine, emitted into the air from a number of industrial processes; fluorides are a major cause of vegetation and, indirectly, livestock toxicity.

Fluorine–A pale greenish, gaseous, acid-forming element, having the symbol F, small quantities of which are detrimental to plants and to animals since it is a cumulative poison. Occasionally tested as a fumigant of plants against insects, it is dangerous to handle without extreme care. Many of the compounds of fluorine are used for insecticides or rodenticides. In sufficiently dilute solutions in drinking water, some fluorine compounds reduce the occurrence of dental cavities, but care must be taken not to exceed certain limits or the teeth may be harmed by being pitted and brown. Minute amounts of fluorine are present in normal animal tissues, bones, teeth, and skin, but it has never been known to be essential to plant life. It is widely distributed in rocks, soils, and water, and is present in the rock phosphates that are mixed in feed rations. Soluble fluorides are toxic to germinating seeds, and chronic fluorosis may occur in people and animals. See Fluorosis.

Fluorine Recovery–The process of removing silicon tetrafluoride (SiF_4) and hydrofluoric acid (HF) from the off-gases of the phosphate plant to form fluosilicic acid (H_2SiF_6) and fluosilicate salts, which are useful intermediates in producing other valuable fluorine compounds.

Fluorosis–A toxic condition due to an excess of fluorine, usually in the drinking water, which manifests itself mainly in mottling of teeth of people and animals.

Flush–(1) To irrigate a field with just enough water to soften the surface soil crust. (2) To increase the feed allowance to ewes or sows with a protein-rich supplement feed a short time before and during the breeding season. (3) A vigorous or abundant, sudden, new growth. (4) To send out vigorous growth on the twigs of a tree. Thus, in the tropical fruit tree, the mango, most of the twigs flush several times a year. (5) To empty, clean, or wash out any material with a quick, heavy supply of water. (6) To introduce and shortly afterwards withdraw an irrigating solution of mild antiseptic of medicinal value, as to flush the vagina or uterus.

Flush Cut–To prune a branch so that the cut is approximately even and parallel with the branch or trunk from which it was removed. The latest recommendation is to prune a branch on the outside of the swelling at the bole of the tree.

Fluvial–Of, or pertaining to, rivers; growing or living in streams or ponds; produced by river action, as a fluvial plain.

Flux–(1) State of change. (2) Substance that reduces the melting point of a mixture. (3) Passage across a physical boundary such as CO_2 from atmosphere to hydrosphere, or across a chemical boundary as CO_2 from atmosphere to organic matter.

Fly–Any winged insect, such as a moth, bee, gnat, etc. Specifically a two-winged insect of the family Muscidae. Many flies are blood-sucking pests of people and animals, such as the mosquitoes, horse and deer flies, black flies, punkies or nosee-ums, and some sand flies.

Some are vectors of diseases, such as the stable flies, etc. Some flies destroy other insects; some are parasites on plants, as the Hessian fly; others are valuable scavengers. Many flies, as the housefly, pass their larval stage in manure and garbage and upon attaining maturity carry with them the bacteria of filth, thus spreading diseases such as typhoid fever, etc.

Fly Ash–The particulate impurities resulting from the burning of coal and other material, which are exhausted into the air from stacks.

Fly-free Date–The date after which it is safe to plant wheat to avoid serious infestation by the Hessian fly. This date has been determined for each county and is available from the county extension agent.

Foam Suppressant–Adjuvant for reducing both surface foam and trapped air; used in fluid fertilizers and liquid pesticides.

Foaming Agent–Surface-active substance that forms a fast-draining foam. This can provide maximum contact of spray with plant surface, insulate the surface, reduce the rate of evaporation, or mark sprayed surfaces to guide implements across a field.

Foehn Wind–A wind descending down the Alps that becomes warmer and drier in its descent. A similar westerly wind in the United States Rocky Mountains is called chinook. See Chinook.

Fog–(1) Almost microscopically small water drops suspended in air that reduce horizontal visibility to less than one mile. (2) Heat-generated and gas-propelled aerosols, in which the particles of the liquid are 50 microns or less in diameter. See Aerosol. (3) The rather luxuriant regrowth of grass following the first mowing (northern England, Scotland). (4) Rotting and decaying grass in water.

Fog Belt–A narrow belt approximately a mile or two wide, immediately adjacent to certain coasts, in which fogs abound due to the closeness of the ocean, the direction of the prevailing winds, and the high atmospheric humidity.

Fog Drip–The moisture from fog which has collected on the leaves of trees, shrubs, etc., and which drips to the ground. In some regions the amount may be equivalent to a light rain, and thus is a considerable factor in supplying moisture to plants. Also called wet fog.

Food–Anything which when taken into the body, nourishes the tissues and supplies body heat. Also known as aliment and nutriment.

Food Chain–(1) A series of spatially associated species, each of which lives (at least in part) as a predator, parasite, or absorber of the next lower down in the series. This concept is useful for analyzing the feeding relations among associated animals. (2) The transfer of food energy from the initial source in plants through a series of organisms by repeated eating and being eaten. Food chains are interlocking patterns known as the food web.

Food Web–The dependence of organisms upon others in a series for food. The chain begins with plants or scavenging organisms and ends with the largest carnivores.

Foodway–A society's behavior pattern of producing, selling, buying, sharing, and consuming food.

Foot–(1) In people and quadrupeds, the terminal portion of the leg that rests upon the ground. (2) The portion of a cultivator to which the sweep is attached. (3) The organs of locomotion of various invertebrates, as the feet of a caterpillar, the foot of a clam or a snail, etc. (4) The base of a tree, tower, mountain, wall, hill, etc. (5) A unit of linear measure, 12 inches.

Foot Pad–The cushions on the bottom of the feet of such animals as cats and dogs.

Foothills–The hills or dissected, sloping land at the base of a mountain or mountain range. Such land differs in native vegetation, potentialities for agriculture, grazing, and other uses from the higher slopes and crests of the mountains.

Footing–(1) The rock or soil on which a foundation rests. (2) The condition of the ground surface as it relates to the pulling capacity of a draft animal without slipping.

Forced Burning–In forestry, a technique in broadcast or spot burning: the selection of a time when the lower layers of duff and flashy fuels are damp but which can be made to burn by setting a large number of fires in heavier concentrations of debris over the entire area. Sometimes slash disposal is accomplished by progressive burning.

Forebay–The impounded waters just above a dam.

Foredune–A low dune, often occupied by a sandbinding grass, bordering the sandy shore of a sea or lake.

Forelle–(French) Trout.

Forest–A plant association predominantly of trees and other woody vegetation that occupies an extensive area of land. See Forest Land, Woodlot.

Forest Cover–All trees and associated plants that cover the ground in a forest.

Forest Entomology–The branch of zoology that deals with insects in their relation to forests and forest products.

Forest Harvesting–United States Forest Service usage of group, patch, strip, and stand cutting. These are essentially size designations of the same logging approach—commonly known as clearcutting. All require the removal of all merchantable trees except a few that may sometimes be carried over where needed for nesting sites and animal food caches. (a) Group cuttings are on areas which are smaller than the size normally recognized in type and condition class mapping. These should not exceed 10 acres, nor usually be less than one-fifth acre. (b) Patch cuttings are on areas that are of a size generally mapped for type and condition and for control, but which do not include the entire stand of which they are a part. (c) Strip cuttings are on areas running through a stand and usually of a width equal to one to two times the general stand height. (d) Stand cuttings are on areas large enough to be practical for management.

Forest Industry Lands–Lands owned by companies or individuals operating wood-using plants.

Forest Influences–The effects of forests upon water, soil, climate, and health.

Forest Land–(1) Land at least 10 percent occupied by forest trees of any size, or formerly having had such tree cover and not currently developed for nonforest use. (2) Lands that are at least 10 percent stocked by trees capable of producing timber or other wood products or that exert an influence on the climate or water regime.

Forest Litter–Leaves, dead branches, bits of bark, rotting wood, etc., found on the forest floor.

Forest Management–The application of business methods and technical principles to the operator of a forest property. It involves the computation of income from forest lands; the establishment of cutting cycles; the conservation of cover, land and water; and the formulation and conduct of long-range plans of operations.

Forest Mensuration–A branch of forestry concerned with tree measurements, log scaling, and sawn timber volume determination.

Forest Pathology–The science that pertains to diseases of forest trees or stands and to the deterioration of forest products by organisms.

Forest Tent Caterpillar–*Malacosoma disstria*, family Lasiocampidae; a caterpillar that feeds voraciously on leaves of many shade and forest trees and also on field and truck crops.

Forestation–(1) Establishing a forest on an area, regardless of whether a forest at one time occupied the tract. (2) A word which means both afforestation (not previously forested) and reforestation (previously forested).

Forestry–The sciences, arts, and business practices of crating, conserving, and managing natural resources on lands designated as forests.

Formaldehyde–A gas, readily soluble in water, which is generally used in aqueous solutions as an insecticide, a fungicide, a soil fumigant, a tool disinfectant, a storage house disinfectant, in disease treatment, and as a disinfectant for seeds, tubers, and bulbs.

Formalin–A 40 percent aqueous solution of the gas formaldehyde.

Formation–(1) Any igneous, sedimentary, or metamorphic rock that is represented as a unit in geological mapping. (2) A major division of the climax vegetation of a continent.

Fossil–The remains or traces of animals or plants that have been preserved by natural causes in the earth's crust exclusive of organisms that have been buried since the beginning of historic time.

Fossil Fuel–A deposit of organic material containing stored solar energy that can be used as fuel. The most important are coal, natural gas, and oil; oil shale and tar sand have future potential as fossil fuels.

Fossil Soil–A soil developed upon an old land surface and later covered by younger formations.

Fossil Water–Water stored in the soil during geologic times. It is not renewed by precipitation.

Fossilize–To convert into a fossil; to petrify.

Fox–*Vulpes vulpes, V. fulva, Urocyon cinereoargenteus*, etc.; any of several doglike animals of the family Canidae, which under certain conditions are farm predators, especially on fowls.

Fragipan–A loamy, brittle subsurface soil horizon low in porosity, low in organic matter, and low or moderate in clay but high in silt and/or very fine sand. A fragipan appears cemented and restricts roots. When dry, it is hard or very hard and has a higher bulk density than the horizon or horizons above. When moist, it tends to rupture suddenly under pressure rather than to deform slowly. See Induration, Claypan.

Frass–Excrement of larvae and/or refuse left by boring insects.

Free Flight–A form of wind erosion where the soil particles are small enough to be carried long distances and hence form the spectacular dust storms.

Free Water–That water the soil is unable to retain against the force of gravity. Also called gravitational water, excess water, drainage water, water of percolation.

Free-ranging–Allowing animals, especially poultry, to roam freely and eat as they wish without any sort of confinement.

Freeze–(1) A condition of the weather in which the air temperature at plant level falls to $32°F$ ($0°C$) or lower, with the result that tender plants are frozen. See Frost. (2) To reach the temperature of freezing, $32°F$ ($0°C$). (3) Of a plant, to die or be impaired as a result of cold. (4) To preserve food products by rapidly reducing the temperature to about $0°F$ and maintaining the temperature well below $32°F$.

French Alpine–A breed of domestic milk goats that originated in the European Alps. No distinct color has been established for the breed.

Fresh–(1) Designating a cow that has recently dropped a calf. (2) Designating very recently harvested or gathered food products. (3) Designating an egg of good quality. (4) Designating sweet water; i.e., not salty water.

Fresh Manure–Recently excreted animal dung whose direct contact can be harmful to plant tissues because of rapid chemical and fermentive changes that take place.

Fresh Water–Water with less than 0.2 percent salinity (2,000 ppm). This may not be the same as potable (drinkable) water.

Freshet–(1) A pool or stream of fresh water. (2) An overflow of a river caused by rain or melted snow.

Friable–(soil) Generally refers to a soil that crumbles when handled. A loam soil with physical properties that provide good aeration and drainage, easily tilled. Friable condition is improved or maintained by annual applications of organic matter.

Frill–A series of overlapping ax cuts, made through the bark and into the sapwood, that encircles a tree trunk to kill the tree. Also called ring, girdle. See Girdle.

Frill Treatment–The placement of a herbicide into a series of overlapping ax cuts made through the bark in a ring around the trunk of a tree.

Fringewater–Mobile water, suspended in the capillary fringe above the water table and completely filling the smaller interstices, which moves with the fluctuations of the water table.

Frit–Fritted micronutrients; also called frits. Frits are special glasses that provide controlled release with as high as 50 percent of micronutrient elements incorporated. The micronutrients are available to plants under conditions that render water-soluble forms unavailable. Potassium, phosphorus, calcium, and magnesium are also fritted and can be included in these glasses.

Front– The boundary between a warm and a cold mass of air.

Frontage– That portion of real estate which lies immediately adjacent to a stream, lake, street, or highway.

Frontier– In the settlement of the United States, the margin between occupied land and the wilderness.

Frost– (1) A condition of the weather in which the air temperature at plant level falls below 32°F (0°C) due to a local cooling of the air during a calm, cloudless night, causing the ground and other objects to be coated by frozen dew. (2) A freeze. (3) Of, or pertaining to, a frozen condition of soil, as in the expression, "Frost is still in the ground." See Frost Line.

Frost Action– The weathering process caused by repeated cycles of freezing and thawing.

Frost Control– Various measures taken to prevent frost damage to more valuable crops, such as citrus, or certain other orchard or truck crops. Preventive measures are: the use of any of several types of heaters; irrigating or sprinkling water on the plants; using propeller blades to mix the upper warm air with the lower cold air; use of coverings of paper, cloth, or plastics.

Frost Crack– (1) An opening in the soil produced by the development of an ice wedge. (2) A longitudinal crack or splits in the trunk or large limbs of a tree, extending from the bark into the sapwood, which results from a sudden extreme drop in temperature during winter. Also called trunk splitting.

Frost Creep– Soil creep resulting from frost action.

Frost Heaving– The pushing up of the soil surface by the growth of ice crystals. In north temperate and arctic regions, this is a serious source of winter killing of newly planted trees and shrubs and new seedings of crop plants. Also a common cause of road and sidewalk breakup.

Frost Line– The maximum depth to which soil is frozen in the winter; this may be an inch or two, or may be several feet. The soil thaws completely in temperate regions but may remain permanently frozen a short distance below the surface in the Arctic and Antarctic regions. See Permafrost.

Frost Pocket– A depression or hollow in either a hilly terrain or a pitted plain into which cold night air descends, creating a frost hazard for sensitive crops. See Cold Pocket.

Frost Splitting– Breaking of rock by water freezing in its cracks.

Frost-free– (1) Designating that period or season of the year between the last killing frost of spring and the first killing frost of autumn. See Killing Frost. (2) Designating a region in which frosts or temperatures below 32°F (0°C) are not experienced.

Fuel Loading– In forestry, the amount of fuel (burnable materials) expressed as weight of fuel per unit area; generally expressed in tons per acre.

Fuel Management– The practice of planning and executing the treatment or control of living and dead vegetative material, primarily for wildfire hazard reduction.

Fuelbreak– A wide strip of range or forest where the fuel is reduced or altered to act as a fire-control measure.

Full Coverage Spray– The application of a spray over the top of a crop that covers all parts of the plants.

Full Moon– The instant when the moon is directly opposite the sun, in elevation 180°, and therefore fully illuminated, appearing as a bright circular disc. Some people erroneously believe that certain plants should, and others should not, be planted in the full of the moon. See Moonarian.

Full-cell Process– Any process for impregnating wood with preservatives or chemicals in which a vacuum is created to remove air from the wood before admitting the preservative.

Fuller's Earth– A massive or shalelike clay deposit. The clay mineral is mainly montmorillonite. It is employed in degreasing wool and in clarifying mineral and vegetable oils.

Fungal Spray– An herbicide consisting of fungal spores in a liquid solution sprayed on weeds in crops as a means of biological control of the weeds. For reasons not adequately understood, the fungal spores of different fungi are specific for specific hosts, making it possible to control selected plants without damaging other nearby crop plants.

Fungi– Plantlike organisms that have no chlorophyll; they get their nourishment from living or decaying organic matter. Plural of fungus.

Fungistat– An agent or chemical material that prevents the growth and reproduction of, but does not kill, fungi.

Fungous– Pertaining to a fungus, as a fungous disease.

Fungus– A lower order of plant organisms, excluding bacteria, which contains no chlorophyll, has no vascular system, and is not differentiated into roots, stems, or leaves. They are classified in the plant kingdom division Thallophyta, and vary in size from single-celled forms to the huge puffballs of the meadows. Fungi are familiar as molds, rusts, smuts, rots, and mushrooms. For the large part they reproduce prolifically by means of single or multicelled spores of various longevities, which are disseminated by air or water. They are either saprophytic or parasitic and many are considered useful in breaking down dead vegetation and organic matter into humus and as agents of fermentation, as in yeasts; others are also destructive in rotting structural timbers, posts, cloth, leathers, etc. The parasitic forms cause destructive plant diseases; a few are human and animal parasites.

Furnace Acid– See Phosphoric Acid.

Furrow– (1) The opening left in the soil after the furrow slice has been turned by the turning plow. (2) To make a furrow with a plow.

Furrow Drain– To drain surface water from a field by means of open furrows made with a turning plow.

Furrow Irrigation– A method of irrigating in which water is run in small ditches, furrows, or corrugations, usually spaced close enough together for lateral penetration between them.

Gabion–A wire basket filled with stones placed on a stream or shorebank to control erosion. Also known as pannier. See Fascine.

Gale–A wind that has a velocity of about 40 miles (65 kilometers) per hour or more at the height of 32 feet (9.8 meters) above the ground. On the Beaufort scale a gale is a wind force of 8. See Beaufort Wind Scale.

Gallates–Salts and esters of gallic acid, found in many plants. Propyl, octyl, and dodecyl gallates are legally permitted antioxidants.

Gallery–A passage or burrow, excavated by an insect under bark or in wood for feeding or egg-laying purposes.

Game–Wild birds, fish, or animals hunted for sport or for use as food.

Game Fish–(1) Any fish caught for enjoyment or recreation. (2) Fish species traditionally regarded as game, especially salmonids.

Game Management–The manipulation of game populations and their food supply, habitat, competitors, etc., so that a given supply is available for harvest.

Game Refuge–An area closed to hunting in order that its excess game populations may flow out and restock surrounding areas.

Gamete–A "sex cell," capable of uniting with another gamete to produce a cell (fertilized egg, or zygote) that in turn is capable of developing into a new individual.

Gap–(1) A break in a fence or wall that may be used as a gate. (2) A depression forming a break in the continuity of the crest of a mountain ridge.

Garden State–A nickname that has been applied to New Jersey because of its highly specialized gardening and truck crop industries.

Gasohol–Registered trade names for a blend of 90 percent unleaded gasoline with 10 percent fermentation ethanol. Also spelled gasahol. See Ethanol, Methanol.

Gastropoda–A class of the phylum Mollusca; commonly known as gastropods, snails.

Gene Pool–The genetic base available to animal breeders for stock improvement.

General Use Pesticide–A pesticide that can be purchased and used without obtaining a permit. It is considered safe for general public use. See Restricted Use Pesticide.

Generic–A word used to describe a general class of products, such as meats, vegetables, or grains; also refers to an unadvertised brand.

Genesis–(1) Origin, or evolutionary development, as of a soil, plant, or animal. (2) A combining form, to indicate manner or kind of origin, as parthenogenesis, biogenesis, etc.

Genetic Drift–The gradual change in a plant or animal species because of rearrangement of the genes due to the environment or unknown causes.

Genetic Engineering–Alteration of the genetic components of organisms by human intervention. Also known as biogenetics.

Genotoxicity–The quality of being damaging to genetic material.

Gently Sloping–In soil judging, a slope of 1 to 3 feet rise or fall in 100 feet (1 to 3 percent).

Genus–A group of species of plants or animals believed to have descended from a common direct ancestor that are similar enough to constitute a useful unit at this level of taxonomy.

Geologic or Natural Erosion–Natural erosion caused by geological processes acting over long geologic periods and resulting in the wearing away of mountains and the building up of floodplains or coastal plains.

Geological Survey–A United States government service established in 1879 for the purpose of: making topographical maps; mapping earth structures; studying groundwater; and appraising energy and mineral resources.

Geology–The study of the earth, earthy materials, their history, processes, and products. This includes water, air, minerals, and rocks.

Geophagous–Pertaining to organisms feeding on soil. See Pica.

Geoponics–The tillage of the earth. See Hydroponics.

Geosere–The development of vegetation of the earth from the first appearance of communities of marine algae to the highest form of seed-bearing plants.

Geotropism–A growth of shoots and roots of plants in response to the stimulus of gravity that is positive when the growth curvature is toward the center of the earth (as plant roots) or negative when the direction of growth is opposite to the pull of gravity (as plant shoots).

Geotype–A geotypic population including races and subspecies occurring in a habitat that is partly isolated by topographic barriers; subspecies. (Syn.: geocotype)

Germ–(1) In reference to animal disease, a small organism, microbe or bacterium that can cause disease. (2) The embryo of a seed.

Germicide–Any agent that kills germs.

Giant Ragweed–*Ambrosia trifida*; a common weed in grain crops. While the seed is an important food for birds, it contaminates grain.

Giant Sequoia–*Sequoia gigantea* (Syn.: ***Sequoiadendron giganteum***), family Pinaceae; perhaps the world's largest coniferous tree. Its growth is limited to small areas. Native to California, United States. Also called big tree. See Redwood.

Gigantism–(1) The production of luxuriant vegetative growth that is usually accompanied by a delay of flowering or fruiting. Also called gone-to-stalk, gone-to-weed. (2) In animals, abnormal overgrowth of a part or all of the body. Also called giantism.

Gilgai–A succession of microbasins and microknolls on flat clayey soils; similar to hog-wallow land. Characteristic of vertisols. See Vertisols.

Ginkgo–*Ginkgo biloba*, family Ginkgoaceae; a deciduous tree with fan-shaped leaves, the female of which produces a fleshy, unpleasant-smelling fruit. The male is grown as a street tree. Native to China. Also called maidenhair tree.

Giraffe–A mechanism used for elevating a worker to do pruning in an orchard. Also called orchard giraffe, steel squirrel.

Girdle–A wound encircling a living plant that involves the absence of tissues as deep or deeper than the cambium layer. The encircling cuts are deep enough to check the flow of elaborated foods or to cause the death of the plant above the point of the girdle. A simple cut by people or animals, or even the removal of a narrow strip of bark during dormancy or very early in the season will check the rate of growth above that point; if done after the first flush of growth, it will result in an accumulation of elaborated foods above the girdle. Wide girdling may cause death of the plant above that point.

Girdling Root–A large root that grows so close to the main trunk or another large root of a plant that it chokes or constricts the other roots.

Girth–(1) The circumference of the body of an animal behind the shoulders. (2) A band or strip of heavy leather or webbing that encircles a pack animal's body; used to fasten a saddle or pack on its back. Also called cinch. (3) The circumference of a tree.

Glacial Drift–Coarse and fine mineral deposits left as a result of glaciation.

Glacial Outwash–Sandy material carried by a stream of meltwater from a glacier and deposited in the form of plains, deltas, and valley trains.

Glacial Soil–Soil developed from materials transported and deposited by glacial action; e.g., till plains, moraines, drumlins, kames, and eskers.

Glacial Till–The unconsolidated, heterogeneous mass of clay, sand, pebbles, and boulders deposited by receding ice sheets.

Glaciation–The geological action of glaciers; the overspreading of a large land surface with glacial ice and the consequent production of glacial phenomena; e.g., glacial drift, grooves and striae on rocks, moraines, drainage modifications, etc.

Glade–(1) A grassy tract in a forest due to shallow soil. (2) An open wet tract, generally covered with grasses and sedges.

Glass Wool–An insulating material made of fine threads of glass.

Glauconite–A green, natural sedimentary mineral; hydrated aluminum silicate of iron, sodium, magnesium, and potassium, which has been used to a small extent as a very slow-acting fertilizer chiefly for its potassium and magnesium. Found abundantly in greensand beds. Also called green earth. See Greensand.

Glaze–A smooth coating of ice deposited by rain or drizzle on terrestrial objects that have a temperature below 32°F (0°C).

Glei–(gley) A soil horizon in which the material is bluish-gray, sticky, compact clay with massive structure. It has developed under the influence of continuously excessive moisture over hundreds of years.

Glen–A narrow valley in mountainous or hilly terrain which usually contains a stream.

Glossina–A genus of the tsetse flies, family Muscidae. The principal species is *Glossina mortisans* that transmits trypanosomiasis (African sleeping sickness) to people and animals. Very common at low elevations in humid Africa.

Glycophyte–A plant that does not grow well in a salty soil solution when the osmotic pressure of the solution rises above two atmospheres.

Gnarled–Pertaining to trees or shrubs with a twisted, bent, or distorted appearance.

Gnat–Any small dipterous, biting insect.

Go-down–(1) A cut or break in the bank of a river or creek through which animals may pass to get water or to cross the stream (western United States). (2) A storeroom (British).

Golden Eagle–*Aquila chrysaetos*, a large eagle found in many areas of the northern hemisphere. Once considered destructive because of the lambs and chickens they ate, they are now protected and considered beneficial because of the large number of rodents they eat.

Goose Nest Land–That part of Kentucky, United States, characterized by limestone and a high frequency of solution sinkholes. See Karst.

Gopher, Pocket–(1) A rodent, *Thomomys* spp., family Geomyidae. It feeds on the roots, stems, and bark of plants and damages cultivated crops. Sometimes called ground squirrel. (2) The land turtle, *Gopherus polyphemus*. Gopher is a common name applied to this reptile, especially in Florida, United States. It makes deep burrows in dry sand soils, sometimes damaging the trees in citrus groves.

Gorge–(1) A deep valley with nearly vertical rock sides. (2) To overeat.

Grade–(1) The slope of a road, channel, or natural ground. (2) The finished slope of a prepared surface of a canal bed, roadbed, top of embankment, or bottom of excavation. (3) Any surface that is prepared for the support of a conduit, paving, ties, rails, etc. (4) Any animal that has one purebred parent and one of unknown or mixed breeding. (5) Designating a herd, flock, brand, etc., of such animals. (6) The classification of a product, animal, etc., by standards of uniformity, size, trueness to type, freedom from blemish or disease, fineness, quality, etc. (7) To smooth the surface of a road. (8) To raise the level of a piece of ground by the addition of earth, gravel, etc.

Graded Terrace–A terrace with a constant lengthwise slope sufficient to cause runoff to flow at a nonerosive velocity.

Gradient–(1) The rate of change of elevation, velocity, pressure, temperature, or other characteristics per unit length. (2) Slope.

Grain–(1) The seed of the cereal crops. See Cereal, Small Grain. (2) Commercially, or as listed on boards of trade, buckwheat, soybeans, and flaxseed, in addition to the cereals. (3) In buttermaking, the butter particles after the emulsion has broken. (4) The arrangement, direction, quality, etc., of the fibers in wood. See Annual Ring.

Granite Chips–Small particles of granite used as a growth medium in gravel culture.

Granite Dust–Fine rock granite particles used by some organic gardeners as a source of potassium.

Granite Soil–Soil that overlies granite rocks or is derived from the residue of weathering of such rocks. It has no single distinctive universal character but is variable, depending upon local environmental factors.

Granular–(1) In the form of granules or small particles. (2) Covered with small grains, minutely mealy. (3) A porous soil ped. See Soil Porosity, Soil Structure.

Granular Frost Structure–Granules of ice that are scattered through porous and easily spaded soil. It is common in frozen soils high in organic matter. See Concrete Frost Structure.

Grassed Waterway–A natural or constructed waterway, usually broad and shallow, covered with erosion-resistant grasses, used to conduct surface water from cropland.

Grassland–(1) Land that is in grass but not under cultivation. (As commonly used, it is not restricted to true grasses, but may include legumes and other nongrasses.) (2) Land that produced grass just previous to being planted to a particular crop. (3) Any of the great natural regions of the world in which the vegetation is mainly grass or other herbaceous plants. Climatic conditions are intermediate between those of forest regions and of deserts. Prairie, steppe, savanna, and pampas have been recognized as subdivisions of grasslands. The Great Plains constitute the major grassland region of the United States.

Gravel–(1) Accumulation of rounded water-worn pebbles. The word *gravel* is generally applied when the size of the pebbles does not much exceed that of an ordinary hen's egg. The finer varieties are called sand, while the coarser varieties are called shingles. (2) An accumulation of rounded rock or mineral pieces larger than 2 millimeters in diameter. Divided into granule, pebble, cobble, and boulder gravel. The individual grains are usually rounded. (3) Accumulation of uncemented pebbles. Pebble gravel. May or may not include interstitial sand ranging from 50 to 70 percent of total mass.

Gravel Envelope–A layer of gravel or coarsely ground limestone surrounding a newly laid tile drain to stabilize it and to provide large pore spaces for underground water to move quickly into the tile drainage system.

Gravitational Water–Water that drains away in large pores in the soil under the force of gravity when under-drainage is free. Also known as free water and percolation water.

Gravitropism–Also known as geotropism; the growth of a plant in response to the force of gravity; e.g., most plant roots grow downward and most plant tops grow upward. If such a growing plant in a pot were inverted, roots would still grow downward toward the rim of the pot and tops would grow upward toward the bottom of the pot.

Gravity Water–(1) Water that moves through soil under the influence of gravity. (2) A gravity supply of water as distinguished from a pumped supply.

Grazing Pressure–The actual animal-to-forage ratio at a specific time. For example, three animal units per ton of standing forage.

Great Basin Region–A vast geographic region of the United States, lying between the Rocky Mountains and the Sierra Nevada Mountains, which is mostly arid or semiarid; a region of mountains and wide, intermountain, alluvial plains and valleys, which sustains mostly bunchgrasses and sagebrush. It is largely sheep and cattle grazing land, and in part nearly barren desert shrub and alkali and salt flat wastes, but a small part is cultivated either under irrigation or dryland farming.

Great Divide–In the United States, the hypothetical line drawn along the highest points of the Rocky Mountains. Precipitation falling to the east of it flows into the Gulf of Mexico, while that falling to the west flows into the Gulf of California or into the Pacific Ocean.

Great Group–A category in soil taxonomy between suborder and subgroup. The total categories are: order, suborder, great group, subgroup, family, and series.

Great Plains–A vast, almost treeless region of plains (prairie and steppe) that occupies the central western United States. It may be somewhat arbitrarily delimited on the east by the 98th meridian and thence westward with increasing elevation to the foothills of the Rocky Mountains. North to south, it extends from the Canadian border of Montana and North Dakota to Texas. The climate is semiarid to subhumid, and the region is subject to periods of disastrous drought. In use, it is grazing land; in part dry farming mostly wheat and sorghums; and a very small (in relative acreage) part is under irrigation.

Grecian Ibex–A species of wild goat believed to be ancestors of some of today's domestic breeds.

Green–(1) A small piece of land covered with grass, as a golf green, or a village green. (2) Unmatured; immature, as a fruit, seed, honey. (3) Uncured, as green lumber, green cheese. (4) Untrained, as a green colt. (5) Designating a plant whose foliage is a shade of green, as light green foliage. (6) Designating a flavor defect of milk characterized by a weak or acid taste and by a repulsive odor.

Green Algae–Organisms belonging to the class Chlorophyceae and characterized by photosynthetic pigments similar in color to those of the higher green plants. Food manufactured by photosynthesis is stored in algal tissues as starch. See Algae.

Green Ash–*Fraxinus pennsylvanica* var. *lanceolata*, family Oleaceae; a hardy, deciduous tree grown in shelterbelts in the Great Plains, United States, as a shade tree and for its lumber used in handles, furniture, sporting goods, boxes, implements, boats, woodenware, cooperage, etc. Native to North America.

Green Lumber–(1) Lumber with the moisture content greater than that of air-dried lumber. (2) Unseasoned lumber, boards from logs processed through mill before drying.

Greenbelt–(1) A strip of land of varying size, used for esthetics, recreation or farming, which encircles a community and protects it from the intrusion of objectionable real property uses from adjacent communities and/or provides the community with space of orderly expansion. (2) A similar strip around a commercial enterprise into which it can expand without the invasion of residential property.

Greenhouse Effect–The thermal effect resulting when comparatively short wavelengths of solar radiation penetrate the atmosphere rather freely, only to be largely absorbed near and at the earth's surface, whereas the resulting long-wavelength terrestrial radiation thus formed passes upward with great difficulty. The effect is because the absorption bands of water vapor, ozone, and carbon dioxide are more prominent in the wavelengths occupied by terrestrial radiation than the short wavelengths of solar radiation. Hence the lower atmosphere is almost perfectly transparent to incoming radiation, but partially opaque to outgoing long-wave radiation. The effect is increased when the atmospheric content of carbon dioxide, particulates, water vapor, or ozone is increased. The net result is a warmer land surface.

Grit–(1) Sand, especially coarse sand. (2) Coarse-grained sandstone. (3) Sandstone with angular grains. (4) Sandstone with grains of varying size producing a rough surface. (5) Sandstone suitable for grindstones. Stones fed to chickens to assist them in grinding feed in their craw.

Grocer's Itch–An infestation of a person's skin by flour mites, *Sarcoptes scabiei*, family Sarcoptidae, and cheese mites, which are similar to chigger bites.

Groin–(1) A structure built from the shore into the water to protect the bank against erosion. (2) The part of the body of a person or animal where the thigh joins the body trunk.

Ground Cover–(1) Any of many different plants, usually perennials, that grow well on sites on which grass does not thrive, as on banks, terraces, in the shade, etc. (2) All herbaceous plants and low-growing shrubs in a forest. (3) Any vegetation that grows close to the ground, producing protection for the soil. See Cover Crop.

Ground Fire–A fire that not only consumes most of the organic materials on the soil surface but also burns into organic materials of the underlying soil itself; e.g., a peat fire.

Ground Limestone–A commercial soil amendment; the product obtained by grinding either high calcic or dolomitic limestone so that all materials will pass state fineness and purity laws. See Lime.

Ground Moraine–The earthy material deposited from a glacier on the ground surface over which the glacier has moved. It is bordered by lateral and/or end moraines. See Till Plain.

Ground Shell Marl–A commercial soil amendment; the product obtained by grinding natural deposits of shell marl so that it conforms with state fineness laws. See Lime.

Ground Squirrel–Any of several different burrowing rodents.

Groundhog–See Woodchuck.

Grounds–(1) The land surrounding a house, building, etc., in lawns, flowers, parks, which is usually carefully tended. (2) Sediment, as coffee grounds.

Groundwater–Water within the earth that supplies wells and springs. Specifically, water in the zone of saturation, where all openings in soils and rocks are filled with water, the upper surface of which forms the water table. See Phreatic Water.

Groundwater Flow–The movement of groundwater under hydraulic gradient; the water that comes to the surface in springs or causes streams and rivers to flow after periods of surface runoff have ended. See Groundwater.

Groundwater Level–See Water Table.

Groundwater Reservoir–An aquifer.

Groundwater Surface–The level below which the rock and subsoil voids are full of water, is known as the groundwater level, groundwater surface, or water table.

Grouse–Any of the numerous birds from the family Tetraonidae. Members include the prairie chicken and the ruffled grouse. They are raised and released as a game bird.

Grove–(1) A stand of trees, smaller than a forest, which grows naturally and has little undergrowth. (2) A small woods. (3) An orchard, or a planting of trees, as an orange grove, pecan grove.

Growing Conditions–The environmental conditions of soil, precipitation, sunlight, temperature, cultivation, etc., which affect the growth of a plant or animal.

Growth Layer (Annual Ring)–A layer of xylem and phloem produced in woody stems usually during each growing season. See Spring Wood, Summer Wood.

Growth Rate–With reference to wood, the rate at which wood has been added to the tree at any particular point, usually expressed in the number of annual rings per inch. May also be stated as annual leader growth.

Growth Regulators–Synthetic or natural organic compounds such as indoleacetic acid, gibberellin, and napthalene acetic acid, that promote, inhibit, or modify plant growth processes. Commonly used in rooting cuttings.

Growth Ring–See Annual Ring.

Grub Prairie–Forest land that has become grassland due to annual burning.

Grub Wagon–A horse- or mule-drawn wagon that carried food supplies for cowboys on roundups (western United States, but now rare). Also called chuck wagon.

Guano–The dry excrement of marine birds, which is a fertilizer rich in nitrogen and phosphorus. Guano was first imported form Peru, where it was collected form arid islands off the coast. The term now has no specific meaning and has been used in some parts of the world for any kind of fertilizer, organic or inorganic.

Guayule–*Parthenium argentatum*, family Compositae; small, drought-tolerant shrub of northern Mexico and southwestern Texas that is a potential source of natural rubber.

Guinea Pig–A rodent, *Cavia porcella*, native to Peru; raised in the United States for laboratory animals and for pets.

Gulch–A deep, narrow, sharply cut trench or gorgelike feature that may or may not contain a stream in its bed. Deeper than a gully or arroyo, it has less magnitude than a canyon or gorge, although local usage of the term is not consistent in this respect (western United States).

Gulf Coast Tick–*Amblyomma maculatum*, family Ixodidae; an insect pest that sucks blood from people and animals. It has been ranked second to cattle tick in cattle destruction.

Gully–(1) A gash cut into a slope of unconsolidated sediments by concentration of rainfall runoff into a channel. Gullies usually have a transverse V-shaped profile and generally are relatively shallow, although locally the cutting may extend to a depth of more than 100 feet and have sides and flat floors. One of the most destructive forms of erosion. (2) A furrow, channel, or miniature valley that is cut by running water during and immediately after rains or during the melting of snow. At other times it usually does not contain water. Special forms may be described as: dendritic arborescent, resembling the branching of a tree or shrub; linear, resembling a line, rather long, narrow, and of uniform width. The distinction between gully and rill is primarily one of depth. A gully is a channel of sufficient depth that it would not be obliterated by normal tillage operation, whereas a rill would be smoothed by ordinary tillage. (3) To make gullies; to form small drainage channels by the action of running water.

Gully Control Plantings–The planting of forage, legume, or woody plant seeds, seedlings, cuttings, or transplants in gullies to establish or reestablish a vegetative cover adequately for the purpose of gully control.

Gully Erosion–Severe erosion of the soil in which large gullies are washed out. Gully erosion is caused by running water. See Gully.

Gumbo–(1) A name current in western and southern states for those soils that yield a sticky mud when wet; in southwest Missouri a puttylike clay associated with lead and zinc deposits; in Texas, a clay encountered in drilling for oil and sulfur. (2) The stratified portion of the lower till of the Mississippi Valley. (3) A vertisol when wet. See Vertisols. (4) A dish usually containing okra.

Gummosis–The formation of clear or amber-colored exudates that set into solid masses upon the surface of affected plant parts and that may be caused by mechanical injury, pathogenic fungi, insects, or excessive nitrogen in the soil.

Gust–A sudden blast of wind.

Gutter–(1) A channel in the floor behind cows in stanchion barns used to catch the manure and to simplify cleaning. (2) A ditch, dug in the path of a forest fire into which burning materials fall, designed to arrest the fire. (3) In turpentine orcharding, the groove or channel along which the oleoresin travels form the tree to the cup. (4) A metal channel (eave trough) mounted to intercept water flowing from a roof.

Guzzler–A surfaced area to provide water runoff and storage for upland game birds.

Gyprock–(1) A rock composed chiefly of gypsum ($CaSO_4$). (2) A driller's term for a rock of any kind in which he has trouble making holes. See Gypsum.

Gypsum (Land Plaster)–A mineral calcium sulfate, combined with water of hydration. Pure gypsum has the formula $CaSO_4 \cdot 2H_2O$. Gypsum occurs in large deposits of soft crystalline rock and as a sand. Normal superphosphate is approximately half gypsum and/or anhydrite. In making concentrated superphosphate, most of the gypsum is removed. Gypsum is widely used for growing peanuts and sometimes for other legumes. In arid regions, large tonnages are used to treat sodic soils. The calcium sulfate reacts with sodium or potassium carbonate to form corresponding sulfates and calcium carbonate. In irrigation agriculture it is used to increase permeability of soils. Following gypsum applications, infiltration can be increased from 50 to 100 percent. Sometimes used as a source of calcium. See Liming Materials.

Gypsy Moth–*Lymantria dispar*, family Lymantriidae; an important insect whose larvae defoliate apple, pear, basswood, birch, hawthorn, oak, poplar, willow, cherry, elm, hickory, hornbeam, larch, maple, and sassafras.

Habit of Growth–(1) The general form or arrangement of stems, roots, and branches, or of the entire plant which is possessed in common by a species in a given habitat. (2) The erectness of the stems of young plants of the small grains as a means of distinguishing between winter and spring varieties.

Habitant–(1) Any plant or animal that is a permanent resident of a place or section. (2) A settler or a descendant of settlers of French descent in Canada or Louisiana, United States, who belongs to the farming class.

Habitat–(1) The place where a plant or animal normally lives and grows under natural conditions, such as a forest habitat, marsh habitat, grassland habitat, etc. (2) Environment. (3) In parasitism, the host.

Habituation–The process of becoming accustomed to a particular environment. See Vernalization.

Hail–Precipitation in the form of hard pellets of ice, called hailstones, which often falls from cumulo-nimbus clouds and accompanies thunderstorms. Hailstones are formed when raindrops are swept up by strong air currents into regions where the temperature is below the freezing point. In falling, the hailstone grows by condensation from the warm, moist air which it encounters and may increase greatly in size and cause severe damage to greenhouses, buildings, livestock, and standing crops. See Sleet.

Half Lands–Tracts of land reserved by the United States government for half-breed Indians.

Half-life–A means of classifying the rate of decay of radioisotopes according to the time it takes them to lose half their strength (intensity). Half-lives range from fractions of seconds to billions of years. Cobalt-60, for example, has a half-life of 5.3 years. A radioactive material loses half its strength when its age is equal to its half-life.

Halide–A compound characterized by a halogen such as chlorine, iodine, or bromine as the anion. Rock salt is a familiar example, called halite, NaCl.

Haline–Term used to indicate dominance of ocean salt.

Halite Rock Salt–A mineral, NaCl. A common mineral of evaporates.

Halogens–A group of elements, including fluorine, chlorine, bromine, and iodine, which are powerful oxidizing agents and have disinfectant properties. An alcoholic solution of iodine (tincture of iodine) is an example of an iodine-containing disinfectant; chlorinated lime is an example of a chlorine-containing disinfectant.

Halophytic Vegetation–Salt-loving or salt-tolerant vegetation that usually has fleshy leaves or thorns and resembles desert vegetation.

Hammock–(1) A vegetational land feature, recognized in peninsular Florida and westward along the Gulf Coast to Mississippi, United States, whose tree vegetation is, or was originally, in contrast to that of the prevailing pine land. (2) A small isolated tree-covered area in extensive marshes. Hammocks may be designated specifically according to tree growth, as oak hammocks, palm hammocks, etc. (3) In some parts of Mississippi, alluvial second bottoms.

Hand–(1) A laborer who is either permanently employed or migratory, as harvest hand, hired hand. (2) A unit of measurement equal to 4 inches (10 centimeters) that is used to measure the height of horses from the ground to a point at the shoulder. (3) A bunch of tobacco leaves of the same grade that are tired together for easier handling. (4) The near horse in a team used for plowing. (5) A half-whorl-like cluster of bananas attached to the rachis of the spike or bunch. (6) Designating any tool, implement, etc., that is manually operated. (7) Designating any manual labor, as hand chopping.

Hand Planting–A reforestation method of planting seedlings by hand.

Haploid–An organism or cell with one set of chromosomes; for example, drone bee. See Diploid.

Haploid Number–In genetics, this is half (haploid) the number of chromosomes that are usually present in the nucleus; occurs during reduction division.

Harbor–To serve as a habitat for, as to harbor insects.

Hard–Water containing soluble salts of calcium and magnesium and sometimes iron. Hardness caused by bicarbonate salts of these metals is known as temporary hardness, because boiling expels the carbon dioxide and converts the bicarbonate to the insoluble carbonate, forming incrustation on the walls and bottom of the container. Hardness from chlorides and sulfates of calcium and magnesium is not affected by boiling.

Hard Land–Land that consists of compact clayey soil in contrast to more loamy or sandy soil (Great Plains, United States).

Hard on the Land–Designating a crop that draws heavily on or exhausts the fertility of the soil or leaves it in a bad physical condition; often said of sorghums.

Hard Scabble Land–Land that has thin, stony soil or is so unproductive that it barely yields a poor living for the occupant.

Hard Water–Water that contains certain minerals, usually calcium and magnesium sulfates, chlorides, or carbonates, in solution in amounts that cause a curd or precipitate instead of a lather when soap is added. Generally defined as containing 60 ppm of calcium carbonate. Very hard water may cause precipitates in some pesticidal sprays.

Hard Winter–A winter characterized by severely cold temperatures and/or heavy precipitation in the form of snow.

Hardness of Wood–A property of wood measured by the load applied at a standard rate required to imbed a 0.444-inch (1.13-centimeter) steel ball halfway.

Hardness Scale–The empirical scale by which the hardness of a mineral is determined as compared with a standard. The Mohs scale is as follows: 1. talc; 2. gypsum; 3. calcite; 4. fluorite; 5. apatite; 6. orthoclase; 7. quartz; 8. topaz; 9. corundum; 10. diamond.

Hardpan–A dense, compacted layer of soil under the surface that may interfere with the downward penetration of both roots and water. Shallow hardpans restrict the roots, especially of plants with taproots, causing them to grow laterally and often resulting in a less hardy and less productive plant. Plowing at the same depth each year may create an artificial hardpan called a plowsole. See Fragipan.

Hardwood–(1) (a) A term used by lumbermen designating any tree that has broad leaves, in contrast to the conifers. (b) The wood or lumber of such a tree. (2) Designating several kinds of timber that resemble teak, used in buildings, etc. (Australia).

Hardwood Ashes–The ashes resulting from the burning of hardwood that contain 2 to 8 percent or more of potash (K_2O), principally as a carbonate; often used as fertilizer. See Wood Ashes.

Hardwood Land–Land that has, or originally supported, a native growth of deciduous forest trees, as distinguished from pine land.

Hare–*Lepus* spp.; an animal related to, but larger than, the rabbit. Its young have fur and are able to care for themselves from birth. Some important species are the varying hare, ***Lepus americanus***, also called the snowshoe hare; European hare, ***L. europas***. and jack rabbit of the western plains. The hare may be a pest of growing plants and trees and shrubs in the winter.

Harrier–Any of the hawks of the genus *Circus*. They are slender-bodied hawks that feed on small rodents and insects; they are considered to be beneficial.

Harvest–(1) To cut, reap, pick, or gather any crop or product of value, as grain, fruit, or vegetables. (2) The crop or product so harvested.

Harvest Moon–The full moon occurring nearest to the autumn equinox on September 21, which furnishes light to harvesters working after sunset.

Haul Road–A temporary road, generally unimproved, used to transport material to and from highway construction, borrow pits, timber harvest areas, and waste areas.

Haw–(1) The third eyelid of animals. Also called hook. (2) See Hawthorn. (3) A command to a horse to turn left.

Hay and Dairy Region–A region of the northeastern United States where grasses and legumes are naturally abundant or easily grown, and where the feed supply and barn and market sizes are generally more suitable for dairy cattle and milk production than for other types of farming, and where markets for dairy products are well established.

Haydite–Shale and clay fused at high temperatures, then ground or crushed, and used in making lightweight cement blocks and bricks.

Hazardous Air Pollutant–By law a pollutant so toxic no ambient air quality standard applies; including mercury, asbestos, and beryllium.

Head–(1) Any tightly formed flower cluster, as in members of the family Compositae, or any tightly formed fruit cluster, as the head of wheat or sunflower. (2) A compact, orderly mass of leaves, as a head of lettuce. (3) On a tree, the point or region at which the trunk divides into limbs. (4) The height of water above any point of reference (elevation head). The energy of a given nature possessed by each unit weight of a liquid expressed as the vertical height through which a unit weight would have to fall to release the average energy possessed, used in various compounds, as pressure head, velocity head, lost head, etc. (5) Cows, asses, horses, collectively, as ten head of horses. (6) The part of the body that includes the face, ears, brain, etc. (7) The source of a stream; specifically the highest point upstream at which there is a continuous flow of water, although a channel with an intermittent flow may extend farther. (8) The upstream terminus of a gully. (9) To prune a tree severely. (10) To get in front of a band of sheep, herd of cattle, etc., so as to stop their forward movement (head them off). (11) To place a top on a barrel. (12) That part of an engine that forms the top of the combustion chamber. In many types of modern engines, the exhaust and intake valves are in the head.

Head Flume–A flume, chute, trough, or lined channel used at the head of a gully or at the lower end of a terrace outlet to reduce soil erosion.

Headfire–A fire that is extended by the wind.

Headgate–Water control structure; the gate at the entrance to a conduit.

Headward Erosion (Head Erosion)–Headwater erosion; a gully is lengthened at its upper end and is cut back by the water that flows in at its head, the direction being determined by the greatest column of water that enters it and the relative erodibility of the soil.

Headwater–(1) The small tributary streams that join to form a river; the water source of a river. (2) The water upstream from a structure, such as a dam.

Health–The state wherein all body parts of plants, animals, and people are functioning normally.

Heart Rot–A tree decay characteristically confined to the heartwood. It usually originates in the living tree.

Heartwood–(1) The inner core of a woody stem that is wholly composed of nonliving cells and is usually differentiated from the outer, enveloping layer (sapwood) by its darker color. Also called duramen. (2) In wood anatomy, the various tannins, gums, resins, and other materials that are deposited in the cell cavities and walls when transformation from sapwood to heartwood takes place. It serves to differentiate sapwood and heartwood. See Sapwood.

Heat Island Effect–The phenomenon of air circulation peculiar to cities, in which warm air builds up in the center, rises, spreads out over the town, and as it cools, sinks at the edges of the city; while cooler air from the outskirts flows in toward the center of the city to repeat the flow pattern. In this way a self-contained circulation system is put in motion that can be broken only by relatively strong winds.

Heat-resistant–(1) Designating a variety or a species that grows under comparatively high temperature conditions; e.g., cotton and rattlesnakes. (2) Any material that is resistant to high temperatures.

Heath–(1) Any plant of the genus *Erica*, family Ericaceae. Species are evergreen shrubs and small trees grown in greenhouses and out-of-doors. Also called erica. (2) A natural land feature, an extensive tract of uncultivated land, treeless or nearly so, which is covered by a dense growth of shrubby, ericaceous plants. It may be nearly the same as a high moore. Heaths are generally sandy and the soils strongly acid.

Heaving–An upward movement of soil caused by freezing and thawing of free water in the soil, thus involving expansion and contraction. Damage, and sometimes destruction of plants, may result form the lifting action, and fence posts may also be pushed upward. Also called frost heaving, frost lifting. See Ice Damage.

Heavy Burning–Range firing during the dry, hot season to ensure a fire that will destroy the existing cover, facilitate travel and livestock handling, increase forage for livestock and game, and enhance hunting.

Heavy Grazing–The practice of keeping a large number of animal units on a pasture or range so that the grass or herbage may be closely grazed. Recommended prior to reseeding the pasture or range.

Heavy Metals–Metallic elements with high molecular weight, usually toxic in low concentrations, including mercury, cadmium, chromium, arsenic, and lead.

Heavy Soil–Generally, a clayey soil in contrast with a sandy one. The reference is to resistance and need for greater power in pulling a plow rather than to actual weight or specific gravity that is less for dry clay than for dry sand.

Hedgerow–A single row of shrubs or trees that provides a screen or wildlife food and cover, improves the landscape, or serves as a fence or a windbreak.

Heliophilous–An organism that is adapted to life in full sunlight.

Helios–(Greek) Sun.

Helispot–A temporary landing place for a helicopter, usually related to forest fire fighting.

Hell Fence–A fence made of stumps laid upside down or on edge.

Hellgrammites–Dobson-fly larvae (family Corydalidae). Full-grown larvae are 2 to 3 inches in length; they have a dark-brown rough-looking skin, large jaws, and posterior hooks. The aquatic larval stage lasts 2 to 3 years. They are secretive and predaceous, living under rocks and debris in flowering water. These larvae are considered one of the finest live baits by fishermen. Pupation occurs on shore, under rocks and debris near the stream edge. The terrestrial adults are short-lived.

Helolac–A lake covered by a mat of aquatic plants such as water lilies, water hyacinth, or alligator weed.

Hematite–Fe_2O_3; anhydrous sesquioxide of iron, a mineral that occurs as deposits of iron ore and is widely distributed in rocks and soils where, in an amorphous condition, it causes a reddish color in the soil.

Hematophagous–Feeding or subsisting on blood, as female mosquitoes.

Hemi––(Greek) Prefix meaning half.

Hemic Soil Material–Organic soil material intermediate in degree of decomposition between the less decomposed fibric and the more decomposed sapric material. See Histosols.

Hemlock Poisoning–Poisoning of animals resulting from browsing poison hemlock, *Conium maculatum*, family Apiaceae; characterized by cessation of digestion, gas, salivation, dilation of the pupils, rapid pulse and breathing, paralysis, unsteady gait, difficult breathing, and death.

Hemlock, Western–*Tsuga heterophylla*; a native conifer that thrives in mild, humid climates of the Pacific Northwest. Important timber species but useful for windbreaks and landscape plantings. Young seedlings are shade-tolerant.

Heptachlor–A chlorinated hydrocarbon insecticide that is closely related to chlordane.

Herbicide Modifier–A chemical substance (safener, extender, etc.) used with herbicides to change their properties.

Herbicides–Chemicals used to kill plants. They are used in contact with the seed, stem, or leaf of a plant. Herbicides are further divided into nonselective (kill all plants) and a selective (kill only certain species).

Herbivorous–Designating an animal (herbivore) that feeds, in the native state, on grass and other plants, as cattle, horses, sheep, goats, deer, elk, etc. See Carnivore.

Herbland–Any lands on which herbaceous species dominate the vegetation.

Herbosa–Vegetation dominated by nonwoody plants.

Heterogeneous–Designating elements having unlike qualities.

Hexachlorophene–A chemical compound that is an exceptionally persistent insecticide.

Hibernation–Winter dormancy. The dormant state in which certain animals pass the winter. It is characterized by narcosis (insensibility), very low metabolic activity, and low body temperature. See Estivation.

Hickory Land–Land on which hickory (*Carya* spp.) was a dominant or characteristic growth and was valued accordingly as agricultural land by early settlers of eastern United States.

Hiemal–Pertaining to winter.

High–A mass of air whose barometric pressure is higher than the pressure of surrounding air. Characteristic weather within a high-pressure system is usually clear.

High Grading–The removal from the stand of only the best trees, often resulting in a poor-quality residual stand.

High Magnesic Products–Any material in which more than 10 percent of the total calcium and magnesium oxide content consists of magnesium oxide. See Lime.

High-lime Soil–A soil that contains lime in sufficient amount to cause effervescence when the soil is tested with cold 10 percent hydrochloric acid. Loosely used in humid regions for any soil that gives an alkaline reaction (greater than pH 7.0), presumably because of its lime content.

High-moor Peat–A group of peat soils that are raw or undecomposed and usually strongly acidic, as opposed to low-moor peat which includes the decomposed and less acidic or alkaline peats. Also called hochmoor peat. See Histosols.

Highly Erodible Land–Land that has an erodibility index of greater than 8. See Erodibility Index.

Hill–(1) A natural relief feature, a rounded elevation a few feet to a few hundred feet above the adjacent lowland, which is distinguished from other features such as mesa, butte, and peak by its form and from mountain by its lesser magnitude. (2) An artificially raised elevation; e.g., a row in a field or garden that is slightly raised above the surrounding soil or a small mound of soil in which seed may be planted. (3) A plant or a cluster of plants, as a hill or potatoes, a hill of beans. (4) To make soil into a hill or to form soil up around a plant.

Hill Country–Hilly or mountainous regions.

Hillock–A small hill.

Histosols–(Peats and mucks) A soil order in the United States system of soil taxonomy comprising organic soils. Of the four suborders, Fibrists, Folists, Hemists, and Saprists, only Fibrists have the physical characteristics essential for use as a filler and conditioner of fertilizers. Fibrists have a density of <0.1 g/cm^3 (<6.2 $lb./ft.^3$). See Peat.

Hoar Frost–Frost accompanied by a visible deposit of frozen moisture on trees as well as on ground vegetation.

Hoedad–A heavy bladed hoe with a short, stout handle used to open the ground to plant tree seedlings.

Hog-wallow Land–Flatland characterized by natural microbasins and mounds; frequently no more than 2 or 3 feet in relief. (A vernacular term, mostly southern United States, without a very precise meaning; used especially in Texas to describe parts of black clay land prairies.) See Gilgai, Vertisols.

Hogback–(1) A low, narrow, but often rounded, crested ridge, such as an esker in the glaciated region of the United States. (2) In western United States, any hill or mountain ridge with a sharp, rocky crest. (3) Of, or pertaining to, a horse with an arched or convex back. Also called swayback.

Hole–(1) In the western United States, a depression that results from a natural widening or expansion of a deep valley. (2) A small bay, inlet, or creek. (3) See Karst, Sinkhole.

Hollow–(1) A relatively small, open depression between hills, or a small reentrant tributary valley in hill country that may or may not contain a stream. (2) Any of variously shaped, shallow concavities in a plain surface.

Holocoenotic–The simultaneous action and interaction of many factors in the environment.

Homo–(Latin) (1) Man. (2) Prefix meaning alike or same.

Homogeneity–Refers to the regularity in the distribution and abundance of the species in a community or area.

Homogeneous–Being of uniform character or nature throughout.

Homothermic–Refers to animals that are able to maintain a fairly constant body temperature; warm-blooded.

Honeycomb Frost Structure–Frozen soil that has a granular structure or is easily broken into fragments, associated with the shallow freezing and granular soils high in organic matter.

Horizon, Soil–A layer of soil, approximately parallel to the surface, having distinct characteristics produced by soil-forming processes. The major horizons of mineral soil are as follows: (1) O horizon: an organic layer, fresh and decaying plant residue, at the surface of a mineral soil. (2) A horizon: the mineral horizon, formed or forming at or near the surface, in which an accumulation of humified organic matter is mixed with the mineral material. Also, a plowed surface horizon most of which was originally part of a B horizon. (3) E horizon: a mineral horizon, mainly a residual concentration of sand and silt high in content of resistant minerals as a result of the loss of silicate clay, iron, aluminum, or a combination of these. (4) B horizon: the mineral is in part a layer of change from the overlying A or E to the underlying C horizon. The B horizon also has distinctive characteristics caused (a) by accumulation of clay, sesquioxides, humus, or a combination of these; (b) by prismatic or blocky structure; (c) by redder or browner colors than those in the A horizon; or (d) by a combination of these. The combined A, E, and B horizons are generally called the solum, or true soil. If a soil lacks a B horizon, the A or E horizon alone is the solum. (5) C horizon: the mineral horizon or layer, excluding indurated bedrock, that is little affected by soil-forming processes and does not have the properties typical of the A or B horizon. The material of a C horizon may be either like or unlike that from which the solum is presumed to have formed. If the material is known to differ from that in the solum, the Roman numeral II precedes the letter C. (6) R layer: consolidated rock beneath the soil. The rock commonly underlies a C horizon, but can be directly below an A, E, or B horizon.

Hornet–A large wasp of the family Vespidae. The insects build large nests from wood pulp. Although their sting is quite painful, they are considered to be beneficial because of the large number of insects they eat.

House Fly–*Musca domestica*, family Muscidae; a worldwide insect pest of houses, people, and animals. It is the vector or suspected vector of more than twenty human and domestic animal diseases and intestinal worms.

Humectant–Substance that absorbs moisture; used to maintain strength of materials such as baking powder.

Humic Acid–(1) Any of certain substances related to humus that is formed by heating carbohydrates with water under pressure. These substances are weak acids that form salts. Part of the natural humus is acid in nature and is generally present in the soil in the form of a salt that, when dissolved by an alkali and the solution neutralized by an acid, precipitates humic acid. (2) Acid formed by the decomposition of organic matter.

Humid Climate–(1) Generally, a climate that has sufficient precipitation to support a forest vegetation. The lower limit of precipitation in cool regions may be as little as 20 inches (51 centimeters), whereas in hot regions it may be as much as 60 inches (152 centimeters). The Thornthwaite precipitation effectiveness index ranges between 64 and 128. (2) A climate in which the average relative humidity, as measured by the hygrometer, is very high. See Relative Humidity.

Humidify–(1) To add water vapor to the atmosphere. (2) To add water vapor or moisture to any material.

Humidity, Absolute–The weight of water vapor in a given volume of air.

Humidity, Relative–The ratio of the actual amount of water vapor present in the atmosphere to the quantity that would be there if it were saturated.

Humin–The organic matter in the soil that is not dissolved upon extraction of soil with a dilute alkali.

Hummock–(1) A minor feature of a land surface that is a low, rounded elevation of less magnitude than a hill or knoll; a common feature produced by wind-drifting of sand. (2) See Hammock. (3) The small mound built by grasses and sedges in a swamp or bog. See Tussock.

Humus–(1) Organic matter in the soil that has reached an advanced stage of decomposition and has become colloidal in nature. It is usually characterized by a dark color, a considerable nitrogen content, and chemical properties such as a high cation-exchange capacity. (2) Any organic matter in the surface layer of soil. (3) Commercially, peat and muck, regardless of degree of decomposition. (4) Leaf mold and duff of a forest floor.

Hurdle System–A term sometimes applied to the method of handling sheep by means of a wolf-proof fence.

Hurricane–A tropical cyclone; a cyclonic whirl that has a calm central core or eye. Winds reach velocities ranging from 75 to 200 miles per hour. The central pressure may drop well below 28.50 inches of mercury. It is accompanied by lightning and excessive rainfall. Called typhoon in the Pacific Ocean.

Hydrate–A compound formed by the union of one or more molecules of water with a salt or a mineral.

Hydrated–Chemically combined with water.

Hydrated Lime–$Ca(OH)_2$; a dry product that consists chiefly of the hydroxide of calcium and oxide-hydroxide of magnesium, sometimes used as a soil amendment. Also called slaked lime; calcium hydroxide. See Lime.

Hydraulic Elements–Factors determining the rate of flow of water or other fluid in a pipe or a channel are: area, depth, slope, velocity, energy, roughness, viscosity, temperature, pressure, and fluid characteristics.

Hydraulic Gradient–(1) The slope of the hydraulic grade line. (2) The slope of the free surface of water flowing in an open channel.

Hydric–Tending to be wet. Hydrophytic plants grow in water or in very wet soils. See Mesic, Xeric.

Hydric Soil–Soil that is wet long enough to periodically produce anaerobic conditions, thereby influencing the growth of plants.

Hydrocarbon–Any of a vast family of compounds containing carbon and hydrogen in various combinations; found especially in fossil fuels. Some of the hydrocarbon compounds are major air pollutants; they may be carcinogenic or active participants in the photochemical process.

Hydrochloric Acid–HCl; A colorless, incombustible, pungent gas, commonly known in the form of its aqueous solution. It is used in treating phosphate rock in the manufacture of superphosphate fertilizer. See Sulfuric Acid.

Hydroclimate–The climate of water bodies, lakes, etc. See Marine Climates.

Hydrocyanic Acid–HCN; one of the most valuable and widely used of the fumigants. Except that it is highly toxic to people, it approaches the ideal in a fumigant. It is found in some plants, particularly sorghums, and may under certain conditions cause poisoning and death to animals. Sometimes called prussic acid.

Hydrofluoric Acid–HF; produced in the off-gases from phosphate rock acidulation.

Hydrogen–H; an element; a colorless gas; one of the essential elements for plant and animal growth.

Hydrogen Peroxide–A chemical substance often used as a bleach to remove color. It is used also in medicine and surgery as an antiseptic agent and as a cleansing agent in mouthwashes, toothpastes and powders. Its antiseptic and cleansing action is due to the fact that it gives off sufficient oxygen to destroy bacteria.

Hydrogen Sulfide–H_2S; a poisonous gas with the odor of rotten eggs that is produced from the reduction of sulfates in, and the putrefaction of, a sulfur-containing organic material.

Hydrogenic Soil (Hydromorphic)–Soil that was developed under the dominant influence of water.

Hydrologic Cycle–The complete cycle through which water passes, commencing as atmospheric water vapor, passing into liquid and solid forms (ice and snow) as precipitation, into the ground surface, and finally again returning in the form of atmospheric water vapor by means of evaporation and transpiration. Syn.: water cycle.

Hydrologic Soil Groups–Refers to soils grouped according to their runoff-producing characteristics. The chief consideration is the inherent capacity of soil bare of vegetation to permit infiltration. The slope and the kind of plant cover are not considered, but are separate factors in predicting runoff. Soils are assigned to four groups. In group A are soils having a high infiltration rate when thoroughly wet and having a low runoff potential. They are mainly deep, well-drained, and sandy or gravelly. In group D, at the other extreme, are soils having a very slow infiltration rate and thus a high runoff potential. They have a claypan or clay layer at or near the surface, have a permanent high water table, or are shallow over nearly impervious bedrock or other material. A soil is assigned to two hydrologic groups if part of the acreage is artificially drained and part is undrained.

Hydromorphic Soil (Hydrogenic)–A soil that has been formed under a strong influence from water, as one developed in a bog or swamp.

Hydrophyte–(1) Any plant growing in water or on a substrate that is at least periodically deficient in oxygen as a result of excessive water content. (2) Plants typically found in wet habitats.

Hydroponics–Growing plants in nutrient solutions rather than soil. Also called soilless gardening. See Geoponics.

Hydroscopic–Designating the ability or tendency to absorb or condense moisture from the air.

Hydroscopic Water–(1) Water remaining in the soil after gravitational and capillary water have been removed. (2) Water held as thin films on soil grains, especially colloids.

Hydrosphere–(1) The water portion of the earth, as distinguished from the solid part, which is called the lithosphere, and the gaseous part, which is the atmosphere. (2) In a more inclusive sense, the water vapor in the atmosphere, the sea, the rivers, and the groundwaters. (3) The liquid and solid water that rests on the lithosphere, including the solid, liquid, and gaseous materials that are suspended or dissolved in the water.

Hydroxide of Lime–See Hydrated Lime.

Hymenium–The spore-bearing layer of a fungous fruiting body.

Hymenoptera–An order to which all bees belong, as well as ants, wasps, and certain parasitic insects.

Hypersaline–Term to characterize waters with salinity greater than 4.0 percent due to land-derived salts.

Hypertrophy–(1) Pertaining to waters of very high nutrient content. (2) Morbid enlargement or overgrowth of an organ or part due to increase in size of its constituent cells.

Hypolimnion–The region of a body of water that extends from the thermocline to the bottom of the lake and is removed from surface influence. See Epilimnion, Metalimnion, Thermocline.

Hypsometer–Any instrument used to measure heights of trees, based either on geometric or trigonometric principles.

Iatroagriculture–A word coined by the authors to mean the environmentally enhancing management activities practiced by the best farmers and ranchers throughout the world.

Ice–(1) Water that has been transformed to a solid state by low temperature; frozen water. the density of ice is 0.9166. Since the density of water is 1.0 at 39.2°F (4°C), freezing water expands 10.91 percent in turning to ice. Being less dense, ice will always float. (2) Any frozen food product made from water or sweet skimmed milk or whole milk and sugar, with or without eggs, fruit juices, or other natural flavoring.

Ice Lenses–Wedges of ice that occur in a tundra soil. See Tundra.

Ice Storm–A heavy deposit of glaze or frozen rain that causes much breakage in the tops of trees and damage to other plants.

Idle Farmland–Includes former croplands, orchards, improved pastures and farm sites not tended within the past two years, and presently less than 10 percent stocked with trees.

Idle Land–(1) Land that has been managed or exploited for some particular use but is now in a state of disuse; abandoned land. (2) Farmland that is capable of producing but is not in use; by extension, any land that has potentiality, but is not being put to any productive use. (3) Land being fallowed.

Igneous Rock–A primary rock formed by the solidification of molten material.

Ignis–(Latin) Fire.

Illite–A group of micalike clay minerals that may be present in the clay constituent of soils, and that has chemical properties different from other clay groups, such as kaolinite and montmorillonite. The mineral is a hydrous silicate of aluminum, containing potassium, iron, and magnesium, but is variable because of replacement of aluminum by iron and magnesium.

Illuvial Horizon–A soil horizon that has received material from the A horizon through the process of eluviation. Usually called B horizon. See Horizon.

Imbibition–The process by which solids, especially organic substances in the gel state, take up liquids and in so doing, swell. The imbibing substance may develop great pressures during the swelling; e.g., absorption of water by seeds, wood, gelatin; absorption of ether by rubber. It is important in the physiology of the plant as one of the means of absorbing water by roots and in the uptake of water by seeds and the splitting of seed coats in the process of germination.

Imperial Valley–A geographic, agricultural region in southern California, United States, noted for its prolific production of truck crops, melons, fruits, etc. It was once a desert area, but with modern irrigation practices the valley was rendered productive. The basin is filled with alluvium; the greater part of the land surface is flat and below sea level.

Impermeability–(1) In groundwater hydrology, the property of rock or earth which does not permit water to move through it readily. (2) Being impervious, not easily penetrated, as by oil or water, either because of low porosity, very small individual pores, or pores that are disconnected. (3) Having the property of not permitting the passage of liquids or gases.

Impervious Soil–A subsurface soil layer that is resistant to the free, downward movement of water or to penetration by plant roots. See Sodic Soil.

Imported Fire Ant–*Solenopsis saevissima* var. *richteri*, family Formicidae; an ant imported from South America; widespread in southern United States and rapidly moving northward. Its sting is harmful to humans, and it is a serious menace because of its destructive feeding on crop plants, its attacks on young animals, and the nuisance of ant hills in fields. See Fire Ant.

Impoundment–A body of water, such as a pond, confined by a dam, dike, floodgate, or other barrier.

In Situ–In its natural position or place. Said specifically of a rock, soil, or fossil when it is in the same location in which it was originally formed or deposited.

Inceptisols–One of the eleven soil orders in the United States system of soil taxonomy. Soils that are usually parent materials but not of illuviation (materials moving in). Generally, the direction of soil development is not yet evident from the marks left by the various soil-forming processes or the marks are too weak to classify in another soil order. See Soil Orders.

Incineration–The controlled process by which solid, liquid, or gaseous combustible wastes are burned to destroy them.

Incinerator–An apparatus used to burn waste substances and in which all the combustion factors of temperature, retention time, turbulence, and combustion air can be controlled.

Incipient Erosion–The early stages of erosion, especially with reference to gullying. See Erosion.

Inclement Weather–Harsh, disagreeable weather.

Incorporate–To mix pesticides, fertilizers, or plant residues into the soil by plowing or other means.

Increment Borer–An instrument used for boring toward the center of a tree to extract a small core that can be used for studying annual growth as well as depth of penetration of a preservative.

Index Species–An organism that is so exactly adapted to its habitat that it is unable to survive elsewhere.

Indian Lands–Tribal lands held in fee by the federal government but administered for Indian tribal groups and Indian trust allotments.

Indian Summer–A brief period in late autumn, usually after a killing frost, when the weather is unseasonably warm and the atmosphere is hazy.

Indicator Plant–A native plant that indicates, in general, and often in a specific manner, the nature of soil conditions with regard to moisture and salinity. Dominant species are the most important indicators.

Indigenous–Produced, growing, or living naturally in a particular region or environment.

Induced Crusts–A surface compaction of the soil caused by trampling by people or livestock or by the use of heavy wheeled traffic on a wet soil.

Induced Pans–A dense layer of soil at about 6 to 12 inches deep that is caused by heavy traffic on a wet soil.

Induration–The process of hardening of soil or sediments through cementation, pressure, heat, and/or other causes.

Industrial Wood–All commercial roundwood products except fuel wood.

Inert–Relatively inactive.

Inert Gas–Also called noble or rare gas; one that does not react with other substances under ordinary conditions.

Inert Ingredient–A substance in a feed, pesticide, etc., that does not act as a feed, pesticide, etc. The substance may serve a purpose but is usually used as a filler, vehicle, etc.

Inertial Separators–Air pollution control equipment that uses the principle of inertia to remove particulate matter from a stream of air or gas.

Infertile–(1) Designating that which is incapable of reproduction; e.g., a barren female animal, a male animal with nonviable spermatozoa, an unfertilized egg, a flower that will not produce seed. (2) A soil so low or unbalanced in essential nutrients that it will not produce a profitable crop.

Infest–To assail, attack, overrun, annoy, disturb; as ticks infest a cow, or as an insect infests a plant.

Infestation–(1) Act of infesting, or state of being attacked, molested, vexed, or annoyed by large numbers of insects, etc., as an animal may be subject to an infestation of parasites, such as fleas, ticks, mites, etc. (2) Presence of disease in a population of plants, or of pathogens in a position, as in soil or on seed surfaces, where they have the possibility of producing disease. (Not to be confused with infection, which can be applied only to living, diseased plants and animals.) See Infection.

Infiltration Rate–The maximum rate at which a soil, in a given condition at a given time, can absorb water, commonly expressed in inches of depth per hour. According to the National Cooperative Soil Survey administered by the USDA Soil Conservation Service, the movement of water into the soil (infiltration rate) is classified as follows: (1) Very low; soils with infiltration rates of less than 0.1 inches (0.25 centimeters) per hour; soils in this group are very high in percentage of clay. They may also be soils that have been abused by excessive tillage. (2) Low; soils with infiltration rates of 0.1 to 0.5 inches (0.25 to 2.5 centimeters) per hour; soils in this group are loams and silts. (4) Soils with infiltration rates of greater than 1.0 inches (2.5 centimeters) per hour; these are deep sands, well-aggregated silt loams, tropical sesquioxide clay soils with high porosity, and well-managed vertisols.

Infiltrometer–A device used in measuring the rate of movement into the soil of water applied uniformly over an area at a given rate or in a given volume.

Influent–(1) A tributary stream or river. (2) A stream is influent with respect to groundwater if it contributes water to the zone of saturation. (3) A liquid flowing into a containing space. See Effluent.

Inherited Soil Characteristic–Any soil characteristic that is due directly to the nature of the parent material, as contrasted with those partly or wholly due to the process of soil formation; e.g., the red color of a soil is said to be inherited if it is due entirely to the fact that the parent material is red.

Inhibit–To suppress, prevent, hinder, restrain.

Inhibiting Vegetation–Vegetative cover sufficiently dense to control the establishment of undesirable vegetation. See Cover Crop.

Inhibitor–Any agent, chemical, etc., that hinders, stops, suppresses, as a rust inhibitor.

Initial Water Deficiency–The amount by which the actual water content of a given soil zone (usually the root zone) is less than field capacity at the beginning of a season or specified time period.

Injection Well–A well into which surface water is pumped to increase subsurface water volume.

Inland Saltgrass–*Distichlis stricta*, family Gramineae; a widely distributed grass of the western and midwestern states, United States, whose general forage value is fair. Also called desert saltgrass.

Inland Waters–Lakes, reservoirs, and ponds over 2 acres in size, and all waterways.

Inlet–(Hydraulics) (1) A surface connection to a closed drain. (2) A structure at the diversion end of a conduit. (3) The upstream end of any structure through which water may flow.

Inoculation–(1) Introduction into healthy plant or animal tissue of microorganisms to produce a mild form of the disease, followed by immunity. (2) An introduction of nodule-forming bacteria into soil, especially for the purpose of nitrogen fixation. (3) Treatment of seed with bacteria that stimulate development of bacteria nodules on plant roots. Used on legumes such as peas and beans. (4) Bacteria supplied to legumes to "fix" nitrogen from the air. (5) A small amount of bacteria produced from a pure culture that is used to start a new culture.

Inorganic–(1) Not made up of or derived from plant or animal materials. (2) Mineral content of the soil. In reference to fertilizers, those produced chemically. Not arising from natural growth.

Inorganic Matter–Compounds that do not contain carbon and hydrogen.

Inorganic Soil–A mineral soil; a soil in which the solid matter is dominantly rock minerals in contrast to organic soils, such as peats and mucks.

Inquiline–An animal that lives in the nest or home of another species.

Insect–An air-breathing animal (phylum Arthropoda) that has a distinct head, thorax, and abdomen. Insects have one pair of antennae on the head, three pairs of legs, and usually two pairs of wings on the thorax. The opening of the reproductive organs is near the posterior end of the body. They may be harmful or useful depending upon their habits. Some infest plants and animals, some are insectivorous, some pollinate plans, and some produce edible products.

Insect Growth Regulator–A chemical that interferes with the normal growth pattern of insects causing abnormal development and thus death. In the case of flies, it may be added to chicken feed, passing through the bird with the feces into the manure. Present in the manure, it kills house fly larvae soon after they hatch.

Insect Vector–An insect that carries a virus, bacterium, or the spores or mycelium of a pathogenic fungus and inoculates susceptible plants, animals, and humans.

Insecticide–A substance that kills insects by chemical action, as a stomach poison, contact poison, or fumigant.

Insectigation–The application of insecticides through an irrigation system.

Insectivorous–Designating any animal that feeds on insects. Also applied to those plants that capture insects and absorb nutrients from them. The pitcher plant and the roundleaf sundew are sometimes so designated.

Insolation–(1) Solar radiation received by the earth. (2) The rate of delivery of solar radiation per unit of horizontal surface.

Insoluble–Not soluble; designating a substance that does not dissolve in another.

Intake–(1) The headworks of a conduit, the place of diversion. (2) Entry of water into soil. See Infiltration, Percolation. (3) The part of an internal combustion engine that admits air.

Integrated Pest Management (IPM)–An ecological approach to pest management in which all available necessary techniques are systematically consolidated into a unified program, so that pest populations can be managed in such a manner that economic damage is reduced and adverse side effects are minimized.

Interaction–The process of chemicals being mixed together and having substantially different toxicity than the toxicities of the components. The chemicals may interact to increase or decrease toxicity. See Antagonism, Synergist.

Interception–The process by which precipitation is caught and held by foliage, twigs, and branches of trees, shrubs, and other vegetation. Often used for "interception loss" or the amount of water evaporated from the precipitation intercepted.

Interception Channel (Ditch)–A channel excavated at the top of earth cuts, at the foot of slopes, or at other critical places to intercept surface flow; a catch drain; a diversion channel.

Interflow–That portion of precipitation that infiltrates into the soil and moves laterally under its surface until intercepted by a stream channel or until it resurfaces downslope from its point of infiltration.

Interfluve–The area between two adjacent streams with the same direction of flow.

Interior Live Oak–*Quercus wislizeni*, family Fagaceae; a broad-leaved, evergreen tree browsed on the western range, United States. Its browse rating is good. Native to California, United States, and Mexico.

Intermittent Grazing–Grazing a pasture for indefinite periods with periods of rest between grazing.

Intermittent Stream–A stream or portion of a stream that flows only in direct response to precipitation. It receives little or no water from springs and no long-continued supply from melting snow or other sources. It is dry for a large part of the year, ordinarily for more than three months.

Intermountain Region–The land area lying between the Rocky Mountains and the Sierra Nevada Mountains, United States. Its climate varies from hot and arid to cool and humid. It has regions of irrigated farmlands, orchards, dryland farms, ranges, and desert wastes.

Internal Drainage–The relative degree of downward movement of water in a soil. Also called permeability.

Interphase–The period in the life of a cell between mitotic divisions.

Interplant–(1) To set out young trees among other existing young trees, planted or natural. It is applicable also to planting land partly occupied by brushwood. (2) To plant seeds of one species between those of another species, as to plant soybeans with corn. (3) To plant one crop in rows and another between the rows; e.g., peanuts grown between rows of cotton or corn.

Intervales–The valley lands that occur in a hilly country; includes the terraces of river valleys.

Intolerance–The inability of a woody plant to grow well in a particular environment; commonly used to denote trees that cannot endure heavy shade.

Intra–A prefix meaning within, inside; as intrastaminal, inside the (ring of) stamens; intraspecific, within a species.

Intractability–The characteristic of a soil that causes it to be difficult to work.

Introduced–Designating a plant, animal, disease, etc., that is not indigenous to an area, but is brought in purposely or accidentally.

Inundated–Flooded or covered completely; filled to capacity.

Invading Plant–A species of plant that comes in on lands after the more stable, climax plants have been diminished in the stand by drought, fire, or overgrazing.

Invasion–The action or process by which plants or animals successfully establish themselves in a new area or environment in competition with the native or previous occupants.

Inversion–The phenomenon of a layer of cool air trapped by a layer of warmer air above it so that the bottom layer cannot rise. A special problem in polluted areas because the contaminating substances cannot be dispersed by dilution.

Invert Emulsion–A thick liquid mixture resulting from dispersing water into oil rather than oil in water.

Invertebrate–(1) Any animal with no spinal column. (2) Having no spinal column.

Ion–An atom or a group of atoms carrying an electrical charge, which may be positive or negative. Ions are usually formed when salts, acids, or bases are dissolved in water. When common salt, sodium chloride, is dissolved in water, positive sodium ions and negative chloride ions are formed. See Anion, Cation.

Ion Exchange–The replacement, in a colloidal system, of one ion by another with a charge of the same sign.

Ion Exchanger–A water-softening system, such as the zeolite type, which functions through the exchange of ions, especially of sodium for calcium and magnesium; in water, the sodium salts of certain compounds readily release sodium and take in its place calcium and magnesium. See Hard Water, Soft Water.

Ionization–The process by which an atom becomes electrically charged, by the removal or addition of one or more of its extranuclear electrons, so that the electrical balance between the electrons and the protons within the atom's nucleus is destroyed. An atom with more than its normal complement of electrons has a negative charge; with less, it has a positive charge.

Iron–Fe; a metallic element essential to people, animals, and plants; very common in some minerals, most rocks, and all soils; an essential constituent of blood hemoglobin where it functions to transport oxygen. Iron is specific for the treatment of anemia in animals. In plants, iron deficiency results in iron chlorosis.

Iron Pan–A type of hardpan, in which a considerable amount of iron oxide is present. See Ironstone.

Iron Phosphate–$FePO_4$; occurs in small quantities in practically all phosphate rock and in rather large quantities in some of the lower grades of rock. It is insoluble in water, and its phosphoric acid content is practically unavailable unless treated with a mineral acid such as sulfuric acid. When present in large amounts, it impairs the physical and chemical condition and nutritional value of superphosphate. See Iron.

Iron Sulfate–See Ferrous Sulfate, Micronutrient Fertilizers.

Ironstone–(1) Any rock containing a substantial proportion of an iron compound from which the metal may be smelted commercially. (2) An iron-rich sedimentary rock, either deposited directly as a ferruginous sediment or resulting from chemical replacement.

Irradiated–Designating a food or feed treated with ultraviolet light to increase the vitamin D content.

Irrigable Area–The area under an irrigation system capable of being irrigated, as determined principally upon the quality of soil and elevation of land.

Irrigate–To furnish water to the soil for plant growth in place of, or in addition to, natural precipitation, by surface flooding or sprinkling and subirrigation methods, using surface water or that from underground sources.

Irrigated–Designating land or crops that have received water from other sources as well as from natural precipitation.

Irrigating Head–(1) The measure of stored-up water ready to be used in irrigation. (2) The depth of a body of water in covering land, as by flooding. The proper wetting of the whole ground. (3) The indication on a measuring device of the rate of flow of irrigation water.

Irrigation–The artificial application of water to soil for the purpose of increasing plant production.

Irrigation Efficiency–The ratio or percentage of the water consumed by crops in an irrigated farm or project to the water diverted from a river or other source into the farm or project canals.

Irrigation Methods–The manner in which water is artificially applied to an area. The principal methods and the manner of applying the water are as follows: Border strip: the water is applied at the upper end of a strip with earth borders to confine the water to the strip. Check basin: the water is applied rapidly to relatively level plots surrounded by levees. The basin is a small check. Corrugation: the water is applied to small, closely spaced furrows, frequently in grain crops, pastures, and forage crops, to direct the flow. Drip (trickle): the water is applied as a drip from "emitters" along a plastic tubing line that usually lies on the soil surface. Flooding: the water from field ditches is allowed to flood over the land. Furrow: the water is applied between crop rows in small ditches. Sprinkler: the water is sprayed over the soil surface through nozzles from a pressure system. Subirrigation: the water is applied in open ditches or tile lines until the water table is raised sufficiently to wet the plant root zone. Wild flooding: the water is released at high points in the field, and distribution is largely uncontrolled.

Irrigation Requirement–The quantity of water, exclusive or precipitation, required for crop production, including unavoidable wastes.

Irrigation, Winter–Off-season irrigation during fall, winter, or early spring to store water in the soil for later use by growing plants. Winter irrigation is practiced in areas where water is available in the winter but less available prior to time of planting.

Isinglass–(1) Very pure gelatin prepared from the air bladders of fishes, especially sturgeon; used as a replacement for gelatin in jellies and puddings, and as a clarifying agent. (2) Mica in sheets.

Isoneph–Lines drawn on a map designating equal cloudiness.

Isotherm–A line on a chart or diagram drawn through places or points having equal temperature.

Isotopes–Elements having an identical number of protons in their nuclei, but differing in the number of their neutrons. Isotopes have the same atomic number, differing atomic weights, and almost but not quite the same chemical properties. Different isotopes of the same element have different radioactive behavior.

Jack Pine–*Pinus banksiana*, family Pinaceae; a hardy but small evergreen that grows farther north in Canada than any other native pine.

Jack Rabbit–Any of the long-eared, large hares of other genus *Lepus*, family Leporidae, found in the western United States. Species have long hind legs and often feed on crops and damage citrus trees.

January Thaw–A period of warm weather that occurs irregularly during the month of January, especially in the eastern United States.

Japanese Chestnut–*Castanea crenata*, family Fagaceae; a deciduous tree, grown for its edible nuts and for use in breeding blight-free chestnut trees. Native to Japan.

Jeffrey Pine–*Pinus jeffreyi*, family Pinaceae; an evergreen tree grown for lumber and as an ornamental. Native to North America.

Jetty–A dike built of piles, rock, or other material, extending into a stream or the sea at the mouths of rivers to induce scouring or to build banks and levees.

Jigger–See Chigger.

Jungle–(1) Any dense, nearly impenetrable growth of vegetation, especially in the tropics. (2) A Hindi name for any woodland.

Juvenile Hormone–The hormone, secreted by the corpora allata, that maintains the immature form of an insect during early molts.

K–The chemical symbol for potassium.

K_2O–Potassium oxide. In present-day fertilizer labeling, the amount of K in fertilizer is expressed as pounds of the oxide per 100 pounds total. To find actual K, multiply K_2O equivalent by 0.83.

Kalmia–See Lambskill Kalmia.

Kame–A short, irregular ridge, hill, or hillock of stratified glacial drift. Most kames are interspersed with depressions, sometimes known as kettles, having no surface drainage. See Kettle.

Kangaroo–An herbivorous marsupial, family Macropodidae, with a small head, upright ears, strong hind legs, and a thick, round tail. In large numbers, kangaroos may damage grasslands. Found in Australia and surrounding islands.

Kaolin–A white or yellowish aluminum silicate clay with low shrink-swell potential. It is used for making high-quality porcelain ceramics as an absorbent in the alimentary tract of animals. Also known as argilla, bolus alba, and China clay. Named for a hill in southeastern China, Kao-ling.

Karroo–An open vegetation type in South Africa consisting of succulent and sclerophyllous shrubs; found where the annual precipitation amounts to 70 to 350 millimeters and falls mostly in the summer.

Karst–A type of topography that is formed over limestone, dolomite, or gypsum by dissolving. It is characterized by closed depressions or sinkholes, caves, and underground drainage. See Thermokarst.

Katabatic Winder–A wind that flows down slopes that are cooled by radiation, the direction of flow being controlled orographically. Such winds are the result of downward convection of cooled air. Syn.: mountain wind, canyon wind, gravity wind. See Anabatic Wind.

Katarobic Zone–That area of a stream that is free of organic pollution and its products.

Keel–(1) The part of a fowl's body that extends backward form the breast. Also called breastbone. (2) The two front, united petals of a pealike flower. (3) In ducks, the pendant fold of skin along the entire underside of the body. (4) In geese, the pendant fold of flesh from the legs forward on the underpart of the body. (5) A central dorsal ridge, like the keel of a boat. (6) The two anterior united petals of a papilionaceous flower.

Keeled–Of, or pertaining to, a ridge (like the bottom of a boat) on an animal or plant part.

Kelp–Any of the seaweeds of the families Laminariaceae and Fucaceae; at one time used as a source of potassium for fertilizers by organic gardeners. Dried kelp contains 1.6 to 3.3 percent nitrogen, 1.0 to 2.0 percent phosphoric acid, and 15 to 20 percent potash. Kelp has been used in animal feeds as a source of iodine.

Kernite–An important ore of boron.

Kerosene–A petroleum derivative with a higher flashing point than gasoline; used as a fuel and as a solvent.

Kettle–(1) A depression of variable size and depth, sometimes containing lakes and swamps. It is a feature of moraines in the glaciated regions of the United States, especially in Wisconsin and Michigan. (2) A pot used for heating water.

Key Area–(1) An area upon which the success of the ranching operations is largely dependent. (2) Critical areas of rangeland, which represent range that is most likely to be overgrazed; used as criteria or indices of the proper use of the range. (3) A guide to the general management of the entire area of which it is a part.

Kieserite–Magnesium sulphate, a source of Epsom salts. Sometimes used alone as a soil amendment and in fertilizer mixtures as a source of magnesium.

Kill–(1) Purposeful killing of insects, plants, or wild or domestic animals. (2) The total number of wild animals killed in a period.

Killer–(1) A person who slaughters animals. (2) A meat animal. (3) A pesticide. (4) A predator.

Killing Frost–A temperature condition sufficiently low to kill most staple crops.

Kiln Burn–A browning of the surfaces of lumber that may occur during kiln drying.

Kiln-dried Lumber–Lumber dried by artificial heat to a moisture content less than that obtained through natural, air seasoning.

Kind–(1) All the plants of the same type, accepted as a single vegetable or fruit, as tomato, cabbage, bean, apple, peach, etc. (2) A species, as a cow, sheep, etc.

Kind of Animal–An animal species or species group such as antelope, cattle, deer, elk, goats, horses, etc.

Kindle–(1) To give birth to a litter of rabbits. (2) A young rabbit.

Kippered–Lightly salted and smoked fish.

Kjeldahl–Refers to a procedure for determining the amount of total nitrogen in organic materials; named for Kjeldahl, the Danish chemist who developed it.

Knee–(1) The joint between the hip and the ankle in humans and quadrupeds (hind leg); in birds, the tarsal joint. (2) The carpal joint (foreleg). (3) The spurlike, root growth of the baldcypress that develops when the tree grows in a swamp.

Knob–(1) A rounded hill or mountain, especially an isolated one. Local in the United States South. (2) The horny protuberance at the juncture of the head and upper bill in African and Chinese geese. (3) A deformity growth on the breastbone, usually at the front, sometimes found in chickens and turkeys; a defect. (4) The rounded protuberant part of the skull in crested fowl.

Knoll–(1) A submerged elevation of rounded shape rising from the ocean floor, but less prominent than a seamount. (2) A small rounded hill.

Knot–(1) The interlacing of string or rope pulled into a tie to fasten something. (2) A swelling or protuberance on the stem of a cut flower through which water moves very slowly to the flower. See Node. (3) The knob attached to a check wire. (4) That portion of a tree branch that has become incorporated in the body of the tree. (5) The unit of speed used in navigation. It is equal to 1 nautical mile (6076.118 feet, 1852 meters) per hour. (6) The meeting point of two or more mountain chains. (7) A dark-colored, hard, somewhat round blemish in a board where a limb grew out of the tree.

Knothole–A hole in a piece of lumber or a tree caused by a knot slipping out.

Krilium–A synthetic organic chemical related to the slime in earthworms. Both krilium and earthworm slime stabilize soil aggregates.

Krummholz–The twisted and distorted woody vegetation characteristic of mountain timberlines.

Kwashiorkor–Syndrome in children produced by severe protein deficiency and characterized by edema (swollen tissues), retarded growth, dermatoses, and necrosis (dead cells) and fibrosis (fibrous tissue formation) of the liver. Other symptoms are a peevish mental apathy, atrophy of the pancreas, and gastrointestinal disorders. Common in the tropics. Related to marasmus.

Labile–Applied to particles in a rock or other substance that decomposes easily.

Lacuna–(1) A cavity, hole, gap. (2) A gap in essential information.

Lacustrine–(1) Pertaining to a lake; e.g. lucastrine sands deposited on the bottom of a lake, or a lacustrine terrace formed along the margin of a lake. (2) Growing in lakes; e.g., lucustrine fauna. (3) Said of a region characterized by lakes; e.g., a lacustrine desert containing the remnants of numerous lakes that are now dry.

Lacustrine Deposit–Sediments deposited in lake waters.

Ladder Fuels–Burnable materials, living and dead, which tend to be continuous between ground fuels and tree crowns. Thus they form a "ladder" that can allow fire to climb into the tree crowns.

Lagg–The depressed margin of a raised peat bog.

Lagoon–(1) Body of shallow water, particularly one possessing a restricted connection with the sea. (2) Water body within an atoll or behind barrier reefs or islands. (3) A body of water more than 1 meter deep established for anaerobic decomposition of organic wastes.

Lagoon, Aerobic–Lagoons larger than one-half acre and deeper than three feet must be stirred mechanically to supply oxygen for aerobic stabilization of wastewater sludge. See Lagoon, Anaerobic.

Lagoon, Anaerobic–A wastewater sludge treatment process larger than one-half acre and deeper than three feet that is not aerated mechanically. See Lagoon, Aerobic.

Laguna–(Spanish) (1) Shallow ephemeral lakes fed by streams whose sources are in the neighboring mountains, and that flow only during rainstorms or melting snow. (2) A lake or pond. (3) A pseudokarst feature; large shallow sinks with clay bottoms; developed in impure soluble limestone in the south-central United States. See Karst.

Lake–(1) Any standing body of inland water. (2) A pool of other fluid substance such as oil or asphalt.

Lake Deposits–Soils that formed in the bottom of lakes as the result of running water from melted glaciers.

Lake Shore Disease–See Cobalt Deficiency.

Lake States–States of the United States that border on at least one of the Great Lakes: Minnesota, Wisconsin, Illinois, Indiana, Michigan, Ohio, Pennsylvania, and New York.

Laminated Soil–Designating soil arranged in very thin plates or layers, less than 1 millimeter thick, which lie horizontally or parallel to the soil surface, usually fragile and of medium to soft consistency. See Varve.

Lamination–(1) The layering or bedding less than 1 centimeter in thickness in a sedimentary rock. (2) The more-or-less distinct alternation of material in soils and rocks that differ one from the other in grain size or composition.

Land–(1) The total natural and cultural environment within which production must take place. Its attributes include climate, surface configuration, soil, water supply, subsurface conditions, etc., together with its location with respect to centers of commerce and population. Oyster beds and even tracts or bodies of water, as where valuable fishing rights are involved, may be regarded as land. It is often convenient, in fact, to regard land as synonymous with all that nature supplies, external to humans, which is valuable, durable and appropriable, thus including, e.g., waterfalls and other sources of waterpower. (2) In a broad legal sense, any real part of the surface of the earth, including all appurtenances, anything in, on, above, or below the surface. (3) In plowing, a plowed or unplowed space between two furrows. (4) The total width of a strip of land tilled by a farmer, or some designated width, as a perch, 16½ feet. Also called a stitch. (5) Soil. (6) A natural part of the earth's surface characterized by any single factor, or combination, of topography, climate, soil, rocks, vegetation; the natural landscape. (7) Pertaining to agriculture, those areas actually in use or capable of use for the production of farm crops and livestock.

Land Area–The area of dry land and land temporarily or partially covered by water such as marshes, swamps, and river floodplains (omitting tidal flats below mean high tide); streams, sloughs, estuaries, and canals less than one-eighth of a statute mile in width; and lakes, reservoirs, and ponds less than 40 acres in area.

Land Capability–A measure of the suitability of land for use without damage. In the United States, it usually expresses the effect of physical land conditions, including climate, on the total suitability for agricultural use without damage. Arable soils are grouped according to their limitations for sustained production of the common cultivated crops without soil deterioration. Nonarable soils are grouped according to their limitations for the production of permanent vegetation and their risks of soil damage if mismanaged.

Land Capability Class–One of eight classes of land distinguished according to potentiality for agricultural use. Class I consists of lands that are nearly level and can be cultivated continuously with little erosion. Class II consists of lands that are nearly level soils that require only simple practices such as contour tillage to control erosion. Class III consists of lands that require such practices as terraces and contour tillage to keep them productive for row cropping. Class IV consists of lands that require contour plowing, terracing, and the planting of sod-like crops every two to three years to control erosion. Class V consists of lands that cannot be planted to cultivated crops without extensive practices. Such soils should be maintained in sod crops. Class VI consists of lands that are too steep to be used for any crop except sod crops. However, even sod crops are difficult to establish and maintain. Class VII consists of lands recommended for use as watershed or woodland. Class VIII consists of lands recommended for recreational uses only.

Land Capability Map–(1) A map that shows land capability units, classes, and subclasses. (2) A soil conservation survey map that is colored to show land-capability classes.

Land Degradation–The result of one or more processes that lessen the current and potential capability of soil to produce (quantitatively or qualitatively) goods or services; e.g., soil erosion.

Land Disturbance–Any activity involving the clearing, grading, or filling of land, and any other activity that causes land to be exposed to the danger of erosion.

Land Drainage–(1) The removal of surface water or excess groundwater by artificial means, such as ditching or tilling. (2) Natural drainage, or removal of water by creeks, rivers, in natural water channels.

Land Forming–A general term that means reshaping, smoothing, and/or reforming the soil surface to facilitate irrigation, drainage, or erosion and sediment control. Land forming for drainage is usually known as land grading. The implement used is usually known as a land plane. See Land Plane.

Land Plane–A large, tractor-drawn machine designed for planing or smoothing land for more efficient use of irrigation water or for easier tillage of land not irrigated. It consists of a long steel frame mounted on wheels, near the center of which is attached a large, long, adjustable, combination steel blade and scraper to remove soil from high points and to convey it to depressions. See Land Forming.

Land Reclamation–Making soil capable of more intensive use such as draining wet soils, irrigating dry soils, and removing excessive salt from saline, saline-sodic, and/or sodic soils.

Land Roller–A heavy implement that is used for smoothing ground and forcing rocks deeper into the soil.

Land Stones–Loose stones that occur naturally on a land surface or in a field, as contrasted with those that are quarried, cut, or broken by humans. Also called fieldstones.

Land Subsidence–The lowering of the land surface due to the removal of groundwater, oil, or the decomposition of cultivated peat and muck soils. See Histosols, Karst, Thermokarst.

Land Type–A geographic division of land based upon some one or a combination of natural factors as soil, relief, vegetation, and climate. The term is often loosely used to indicate use of the land such as agricultural, grazing, mining, forest, urban, etc.

Landform–One of the multitudinous features that taken together make up the surface of the earth. It includes all broad features, such as plain, plateau, and mountain, and also all the minor features, such as hill, valley, slope, canyon, arroyo, and alluvial fan.

Landlocked–Designating a piece of land belonging to one person that is surrounded by land belonging to other persons so that it cannot be approached except over their land.

Landslide–(1) Downslope movement of soil/rock, usually because of supersaturation, associated sometimes with building roads on a sidehill and timber harvesting. (2) That part of the plow bottom that slides along the face of the furrow wall. It helps to counteract the side pressure exerted by the furrow slice on the moldboard, and to steady the plow while being operated.

Langbeinite–A mineral, $K_2Mg_2(SO_4)_3$. Isometric crystal structure. Occurs in potassium salt deposits; mined as a source of potassium sulfate (K_2SO_4) fertilizer.

Large Mouth Bass–A fresh water game fish *Cropterus salmoides*, that inhabits warm still waters of the United States. It is grown commercially to stock lakes for fishing.

LASA–Acronym for large aperture seismic array; a geophone array in Montana set up to detect nuclear explosions and to distinguish them from earthquakes.

Late–(1) Designating a variety of plant that flowers or fruits at a later time in the year than others of its species. (2) Designating a season that comes at a later time in the year than normal, as a late spring.

Late Wood–The denser, smaller-celled, later-formed part of a growth layer along the trunk and branches of a tree; summerwood.

Latent Effect–An effect that occurs a considerable time after exposure to a toxic substance.

Lateral–(1) A directional or positional term meaning away from the middle or toward the side. (2) A part of a system that branches out from the main body of the system, such as the tile lateral drain that connects to a main drain in a drainage system. (3) A branch or twig of a tree.

Laterite–(Soil) The zonal group of soils having very thin organic and inorganic mineral layers over leached soil that rests upon highly weathered material, rich in aluminia or iron oxide, or both, and poor in silica. The soils are usually red. Laterite soils are developed under the tropical forest or in a subtropical climate. This group of soils is now called Oxisols. See Soil Orders.

Lava–Fluid rock that issues from a volcano or a fissure in the earth's surface; also the same material solidified by cooling.

Lay-of-Land–The topography of a tract of land, as level, sloping, rolling, hilly, etc.

LC_{50}–See Lethal Concentration.

LD_{50}–See Lethal Dose 50.

Leachate–Liquid that has percolated through solid waste or other medium and may contain extracted, dissolved, or suspended materials from the medium.

Leached Soil–A natural soil from which most of the more soluble constituents have been removed throughout the entire profile or removed from one part of the profile and accumulated below the rhizosphere (root zone).

Leached Wood Ashes–Wood ashes with part of the plant nutrients removed by exposure to rain or snow. See Wood Ashes.

Leaching–(1) The removal of soluble constituents from soils or other materials by percolating water. (2) The removal of soluble salts from soils by abundant irrigation combined with drainage.

Leaching Requirement–The amount of water that must pass through the root zone in order to leach out excess soil salinity.

Lead Line–A bluish line at the margin of the gums of an animal that indicates chronic lead poisoning, usually as a result of ingestion of paint or spray materials. See Lead Poisoning.

Lead Poisoning–Poisoning that results from ingestion of lead, usually of paint containing lead. it is characterized (a) in horses by convulsions, partial paralysis, colic, roaring, thirst, and increased urination; (b) in cattle, by staggering, impaired vision, and unusual postures. See Lead Line.

Leaded Gasoline–Gasoline containing tetraethyl lead to raise the octane value.

Leaf–(1) A flattened outgrowth from a plant stem, varying in size and shape, usually green, which is concerned primarily with the manufacture of carbohydrates by photosynthesis. (2) To put forth leaves as in the spring.

Leaf Blight–(1) A symptom of any of several plant diseases generally characterized by a killing or partial killing of the leaves. A blight of twigs usually accompanies leaf blight. (2) A symptom of black rot in which many small purple specks appear on the leaves. As the spots enlarge, the margin remains purple but the center turns brown or yellowish-brown. Also called frog-eye leaf spot. (3) A symptom of fire blight in which the leaves are completely or partially killed. (4) A symptom of currant anthracnose in which small, dark brown, circular spots appear, most obviously in the upper leaf surfaces. If numerous, the spots coalesce and result in larger, dead, leaf surfaces. There may be a chlorosis between the dead areas. (5) A disease of sweet potato caused by *Phyllosticta batatas*, family Sphaerioidaceae; characterized by small, brownish spots with dark margins. Although widespread, leaf blight is not a serious disease.

Leaf Canopy–Crown cover of a plant.

Leaf-cast Disease of Conifers–The premature shedding of the leaves (needles) mostly of pine, larch, fir, spruce, etc., usually due to the effect of parasitic fungi growing within the leaves. These are most often species of the genera *Hypoderma, Hypodermella, Lophodermium*, family Phacidiaceae.

Ledge–(1) A narrow shelf of rock much longer than its width, formed along the face of a cliff by differential wave action. (2) A rocky cliff. (3) A natural outcrop of a mineral deposit.

Lee (Leeward)–(1) The sheltered side turned away from the wind. (2) Chiefly nautical: the quarter or region toward which the wind blows. (3) Opposite from windward. See Windward.

Leeches–*Hirudinea*; segmented worms, flat from top to bottom, with terminal suckers that are used for attachment and locomotion. Various species may be parasites, predators, or scavengers; most are aquatic.

Left Bank–The left bank of a flowing body of water is on your left as you face downstream. Because of the coriolis effect, in the Northern Hemisphere the left bank is usually less steep and has more sediment accumulation than the right bank; and vice versa in the Southern Hemisphere. See Coriolis Effect, Right Bank.

Legal Control–Control of pests through the enactment of legislation that enforces control measures or imposes regulations, such as quarantines, to prevent the introduction or spread of pests or disease.

Leghemoglobinemia–A disease especially of babies caused by drinking water high in nitrates. Water with more than 10 parts per million of nitrate nitrogen should not be used. Boiling increases nitrate concentration of water.

Length-of-Run–In irrigation, the distance water must travel in furrows or over the surface of a field from one head ditch to another or to the end of a field.

Lethal Concentration (LC)–The lethal concentration (written as LC_{10} or LC_{50} or LC_{100} or any percentage) median is the parts per million (ppm) or parts per billion (ppb) of toxicant in water or air that kills 10, or 50, or 100 percent, respectively, of the target species in a 24-hour period. Usually used for fish. See Effective Concentration, Lethal Dose.

Lethal Dose–The lethal dose (written as LD_{10}, LD_{50}, LD_{100}, or any percentage) median is the milligrams of toxicant per kilogram of body weight that kills 10, 50, or 100 percent of the target species. See Effective Concentration, Lethal Concentration.

Leucite–$KAL(SiO2)2$; potassium aluminum silicate, a feldspathoid mineral that is a minor source of potassium for fertilizer. In the United

States, its principal occurrence is in the rock Wyomingite found in Wyoming.

Ley– A temporary pasture. Also spelled lea, a meadow (obsolete).

Lichen Line– The edge of lichens on undisturbed rock surface; the clean portion of the rock indicating the extent to which the soil has been eroded from around its base.

Lichens– Fungi, usually of the class Ascomycetes (rarely of the class Basidiomycetes) that grow in mutual symbiosis with algae, forming a joint structure in which the algal cells are embedded in the network of fungous hyphae.

Licks– A United States term given to boggy grounds with salt springs, where cattle go to lick the salt.

Lie Bare– To remain unplanted, as improved or cultivated land left fallow or subject to the ill effects of wind and water erosion for a period of time without a crop cover.

Life Cycle– Life history; the changes in the form of life that an organism goes through.

Life History– Habits and changes undergone by an organism from the egg stage to its death as an adult.

Light– (1) The form of radiant energy consisting of wavelengths lying within the limits perceptible by the normal human eye, and, by extension, the shorter and longer wavelengths, the ultraviolet and the infrared light, invisible to the eye but which may be recorded photographically. Light can be absorbed by various substances and transformed into heat. Its excess may produce fading, the destruction of green color in plants, but an insufficiency can also cause lack of chlorophyll production. The coloration of fruits is dependent upon sufficiency of light. The growing parts of plants respond to the stimulus of the direction from which the light comes. See Photosynthesis, Phototropism. (2) To become ignited; to take fire. (3) Designating a deficiency or lack in degree, such as a light rain or a light crop. (4) When applied to food or drink it can have several definitions. It could mean reduced calories, fluffy (full of air), pale, low in sodium, mild in flavor, and/or less alcohol.

Light Burning– (1) In forestry, a controlled burning to check disease, remove fire hazard, or to favor a certain type of growth. (2) The method of reducing the volume of inflammable material in forests. Fires are deliberately set at a time when inflammability is low and the likelihood of damage is at a minimum; designed to lessen the danger of damage when fires run more freely.

Light Freeze– See Light Frost.

Light Frost– A frost that does not kill but may cause some injury to the more tender and susceptible plants in exposed places. See Black Frost, Killing Frost.

Light Grazing– Allowing livestock to partially graze an area, such as a pasture or range.

Light Soil– Soil that is easy to cultivate, retains little moisture, and has a sandy (coarse) texture.

Lightly Buffered– Designating a soil such as a sand that has a small cation exchange capacity.

Lightning Injury– The injury to a plant, most often a tree, in which some bark is stripped off, limbs are broken, etc., caused by lightning. Frequently herbaceous plants or crops in a circular area are killed by a single streak of lightning.

Lightning Ring– A band of abnormal wood in a growth ring of a tree that results from mild lightning shock to the cambium.

Lightweight Fertilizer– Specialty fertilizers containing highly soluble plant nutrient compounds absorbed on lightweight materials such as peat, expanded vermiculite, or perlite.

Lightwood– (1) Conifer wood, especially pine and Douglas fir, abnormally inflammable because of its high resin content. (2) The highly resinous heartwood, stumps, knots, and dead roots of pines, which are used mainly for kindling. Also called lighterwood. (3) Blackwood acacia.

Lignite– A brownish-black coal in which the alteration of vegetal material has proceeded further than in peat but not so far as in subbituminous coal.

Lime– (1) Strictly, calcium oxide, CaO, but, as commonly used in agricultural terminology, calcium carbonate, $CaCO_3$, and calcium hydroxide, $Ca(OH)_2$, are included. Agricultural lime refers to any of these compounds, with or without magnesium carbonate, which are used as amendments chiefly for acid soils. The most common agricultural limes are calcic lime ($CaCO_3$) and dolomitic lime ($CaMgCO_3$). (2) *Citrus aurantifolia*, family Rutaceae; a small, spiny, evergreen tree grown for its edible, very acid fruit, the lime of commerce. Native to Asia. Also called acid lime. (3) To apply lime to excrement to reduce odor and hasten decomposition.

Lime (Calcium) Requirement– The amount of agricultural limestone required per acre to a soil depth of 6 inches (15 centimeters) or on about 2 million pounds (910,000 kilograms) of soil to raise the pH of the soil to a desired value under field conditions.

Lime Chlorosis– A chlorosis that is associated with high lime soils because soil iron is made less available. See Chlorosis.

Lime, Hydrated– See Hydrated Lime.

Lime Nitrate– See Calcium Nitrate.

Lime Pan– In soils, a subsurface layer cemented by calcium carbonate. See Caliche.

Lime, Waste– An industrial by-product or waste that contains calcium and/or magnesium in forms that will neutralize soil acidity. It is usually known by the name of the industry from which it comes as gas lime or beet-sugar waste lime. It usually consists of calcium hydroxide and/or calcium carbonate with various impurities. Lactic acid and soda waste limes are high in gypsum. Oxygen waste lime is largely CaO. Gas-house lime contains cyanides and sulfites, which are harmful to plants but decompose in the soil. The latter kind should, therefore, be applied well in advance of planting.

Limestone– (1) A bedded sedimentary deposit consisting chiefly of calcium carbonate ($CaCO_3$) that yields lime (calcium oxide) (CaO) when burned. Limestone is the most important and widely distributed of the carbonate rocks and is the consolidated equivalent of limy mud, marl, chalk, calcareous sand, or shell fragments. (2) A general term for the class of rocks that contain at least 80 percent of the carbonates of calcium or calcium and magnesium.

Liming—The application of lime to land, primarily to reduce soil acidity, reduce toxic aluminium, and supply calcium for plant growth. Dolomitic limestone supplies both calcium and magnesium. May also improve soil structure, organic matter content, and nitrogen content of the soil by encouraging the growth of legumes and soil microorganisms. Liming an acid soil to a pH value of about 6.5 is desirable for maintaining a high degree of availability of most of the nutrient elements required by plants.

Liming Materials—Agricultural limestone used in the United States totals about 21 million tons annually in recent years. The American Association of Plant Food Control Officials (AAPFCO) official interpretations and definitions of liming materials follow: Agricultural liming materials are those, the calcium and magnesium content of which are in forms that are capable of reducing soil acidity. Air-slaked lime is a product composed of varying proportions of the oxide, hydroxide, and carbonate of calcium, or of calcium and magnesium, and derived from exposure to the atmosphere of quick lime. High calcic liming materials are products 90 percent or more of whose total calcium and magnesium expressed as oxides, is calcium oxide, and contains 35 percent or more of calcium oxide equivalent. High magnesic liming materials are those containing 10 percent or more of magnesium oxide equivalent. Hydrated lime is a dry product consisting chiefly of calcium and magnesium hydroxides. Quicklime, burned lime, caustic lime, lump lime, and unslaked lime are calcined materials comprised chiefly of calcium oxide in natural association with lesser amounts of magnesium oxide, and which are capable of slaking with water. Ground limestone (corse ground limestone) is calcitic or dolomitic limestone ground sufficiently fine for effective use as a liming material. Pulverized limestone (fine ground limestone) is the product obtained by grinding either calcitic or dolomitic limestone so that all the materials will pass an 850-micron (number 20) sieve and at least 75 percent will pass a 150-micron (number 100) sieve. Ground shells is the product obtained by grinding the shells of mollusks so that not less than 50 percent shall pass a 150-micron (number 100) mesh sieve. The product shall also carry the name of the mollusk from which said product is made. Marl is a granular or loosely consolidated earthy material comprised largely of shell fragments and calcium carbonate precipitated in freshwater ponds. Ground shell marl is the product obtained by grinding natural deposits of shell marl so that at least 75 percent shall pass a 150-micron (number 100) mesh sieve. Waste lime (by-product lime) is any industrial waste or by-product containing calcium or calcium and magnesium in forms that will neutralize acids. It may be designated by prefixing the name of the industry or process by which it is produced, i.e., gas-house lime, tanners' lime, acetylene lime-waste, lime-kiln ashes, or blast-furnace slag. Agricultural slag is a fused silicate whose calcium and magnesium content is capable of neutralizing soil acidity and that is sufficiently fine to react readily in soil. Labels of artificial mixture of two or more liming materials or of gypsum with liming materials shall include a list of ingredients used. See also Cation Exchange.

Limiting Factor—Any influence or material that tends to slow down growth and productivity in an ecosystem; either too much or too little of these critical factors will limit production. See Law of the Minimum.

Limnetic Zone—(1) The open water of a lake. (2) Ponds and inland seas.

Limnic—Formed in fresh waters.

Limnology—The scientific study of fresh waters, especially that of ponds and lakes. In its broadest sense it deals with all physical, chemical, meteorological, and biological conditions pertaining to such a body of water.

Limonite—$2Fe_2O_3 \times 3H_2O$; a field term for a group of brown, amorphous, naturally occurring hydrous ferric oxides. May consist of the minerals goethite, hematite, and lepidocrocite. Also called bog iron ore.

Limy—Designating a soil having a high content of lime, usually in the form of calcium carbonate.

Lindane—The pure gamma isomer of benzene hexachloride; used as an insecticide.

Linden—Any tree of the genus *Tilia*, family Tiliaceae. Species are grown for shade, lumber, and as ornamentals. Also called basswood.

Line—(1) The reins of a harness. (2) A rope, cable, string, wire, tube, etc., for tying or hanging objects, or conducting electricity, water, gas, etc., as a power line, gas line, or milk line. (3) A boundary or limit, as a fence line, property line. (4) In marketing, the whole of a herd of sheep. (5) A group of plants or animals that retain their uniform appearance in succeeding generations.

Lining—A protective covering over all or part of the perimeter of a reservoir or a conduit to reduce seepage losses, withstand pressure, resist erosion, and reduce friction, or otherwise improve conditions of flow.

Lip—(1) In a dam, a small wall on the downstream end of the apron to break the flow from the apron. (2) One of the edges of a wound. (3) Either of the two external fleshy folds of the mouth opening. (4) Either one of the inner or outer fleshy folds of the vulva.

Liquid Creosote—An oil obtained from coal-tar creosote by the extraction of part of the compounds that crystallize at normal atmospheric temperatures. See Creosote.

Liquid Fertilizers—Water-soluble essential plant nutrients supplied in liquid form and applied by pumping, air pressure, or gravitational flow from a spray applicator. See Fluid Fertilizer.

Liquid Manure—The liquid excrement from animals, mainly urine, collected from the gutters of barns into large tanks and hauled to the fields.

Liquid Nitrogen—Nitrogen in its liquid state. It is extremely cold and is used for storing frozen semen and for cooling irons used in freeze branding. See Freeze Branding.

Liquid Sludge–Sludge that contains sufficient water (ordinarily above 80 percent) to permit it to flow by gravity.

Lithification–The complex processes that convert a newly deposited sediment into an indurated (cemented) rock. It may occur soon or a long time after deposition. The "cement" may be silica, iron, aluminum, lime, or gypsum.

Lithochromic Soil–A soil that owes its color to parent material, or rock from which the detrital material is derived, rather than to color developed in soil-forming processes.

Lithology–(1) The physical character of a rock, generally as determined megascopically (with the eyes only) or with the aid of a low-power magnifier. (2) The microscopic study and description of rocks.

Lithophyte–A plant that grows on rocks in little or no soil, deriving its nourishment chiefly from the atmosphere, as some Orchidaceae. See Lichens.

Lithosequence–A sequence of soils whose properties are functionally related to differences in the parent rock.

Lithosols–A soil group characterized by an incomplete solum of no clearly expressed soil morphology and consisting of freshly and imperfectly weathered rock or rock fragments.

Lithosphere–The outer shell of the earth that is about 62 miles (100 kilometers) thick and is composed predominantly of rocks whether coherent or incoherent, including the disintegrated rock materials, soils and subsoils, together with everything inside this rocky crust. Other materials (chiefly water and gases that are found in the air) are intermingled, but rocks and soil predominate.

Litmus–An acid-alkali indicator derived from various species of lichens, principally *Lecanora subfusea* var. *variolosa*; the main constituents are azolitmin and erythrolitmin. The commercial indicator has a pH range of 4.5 to 8.3. The acid color is red and the alkaline color blue. Usually it is used as litmus paper or as an indicator in milk to show acid or alkaline reduction by microorganisms. Slips of litmus paper were formerly used to determine whether soils were acid or alkaline; now glass electrode potentiometers are standard for determining soil pH.

Litter–(1) On a forest floor, the uppermost surface layer of debris, leaves, twigs, and other organic matter, undecomposed or slightly altered. In a technical description of a soil profile, it is generally designated by the letter O. (2) Accumulation of leaves, fruits, twigs, branches, and other plant parts on the surface of the soil. See Mulch. (3) A group of young animals born at a single birth, as a litter of pigs, etc. (4) See Bedding.

Little Black Ant–*Monomorium minimum*, family Formicidae; a North American ant that is a general feeder on fruits, sweets, honeydew, vegetables, and meats.

Little Fire Ant–*Wasmannia auropunctata*, family Formicidae; a small ant pest living in citrus orchards on honeydew whose vicious sting can make hand harvest almost impossible.

Littoral Vegetation–Vegetation growing along the banks of a large lake or the ocean. Vegetation along a small lake, river, swamp, or spring is known as riparian, riverine, riverain, or phreatophytes.

Littoral Zone–(1) Strictly, zone bounded by high and low tide levels. (2) Loosely, zone related to the shore, extending to some arbitrary shallow depth of water.

Live Fence–A hedge made by planting closely spaced thorny or other plants, such as Osage orange, and maintaining them in a living condition.

Live Lime–Burnt lime; CaO.

Live Oak–*Quercus virginiana*, family Fagaceae; a broad-leaved, evergreen, tender tree grown for shade and ornament and for its wood used in boats and articles requiring strength and toughness. It also furnishes some browse in the southwestern United States. Also called American live oak, encina.

Living Mulch–An understory of vegetation that helps reduce soil erosion and adds organic matter to the soil, but does not compete heavily with the crop for water and nutrients.

Llano–(Spanish) (1) An extensive plain with or without vegetation. (2) The term is an exact equivalent of the English word ***plain***, and by Spanish-speaking persons is so used. Generally the term is applied to the vast treeless plains of Texas, Mexico, and South America.

Llano Estacado–The level grassland comprising the High Plains division of the Great Plains in western Texas, United States.

Load–(1) The work sustained by a machine. (2) The resistance offered to a motor by the machinery it drives, apart from the friction of its own parts. (3) In the export trade, a unit of lumber measurement, used particularly in India; it comprises 40 cubic feet of round timber or 50 cubic feet of squared timber, equaling approximately a long ton. (4) The weight carried by a beam, girder, truss span, or structure of any sort, or any part of such structure, including its own weight. (5) A burden; a weight; as a heavy load. (6) The work done by a prime mover. (7) To place a load on a wagon, truck, etc. (8) In erosion and corrosion the material transported may be called the load. The load is transported by two methods; a portion is moved along the bottom. (9) The sediment moved by a stream, whether in suspension or at the bottom, is its load. (10) The quantity of material actually transported by a current.

Loam–Soil that consists of less than 52 percent sand, 28 to 50 percent silt, and 7 to 27 percent clay, resulting in a soil texture ideal for gardening.

Loamy–Intermediate in texture and properties between fine-textured and coarse-textured soils. Includes all textural classes with the word loam as a part of the class name, such as clay loam. See Loam, Soil Texture.

Loblolly Pine–*Pinus taeda*, family Pinaceae; an evergreen tree, valuable for timber and pulpwood. Native to southeastern North America.

Local Effect–An effect that a toxic substance causes at its original contact point with the body, e.g., eye damage.

Lodgepole Pine–*Pinus contoria* var. *latifolia*, family Pinaceae; an evergreen tree valued for its lumber used in ties, mine timbers, poles, fuel, rough construction, and planing mill products. Native to North America.

Loess–A massive deposit of tan or buff-colored silt whose particles are typically angular and uniform in size; it is usually calcareous, often contains concretions of calcium carbonate, and sometimes shows lamination or bedding. Where deeply eroded, loess areas have a distinctive topography characterized by vertical walls, sometimes many feet high. Some deposits in the United States are believed to have been formed from finely ground rock material produced during the Ice Age and transported by water and winds to surrounding highlands.

Log–(1) An unhewn, sawed or cut length of a trunk or large limb of a tree. See Saw Log. (2) In the preparation of chip steaks, the molded, frozen piece of meat from which the steaks are cut. (3) To fell trees for lumber.

Log Drive–A large collection of logs floating down a river.

Log Rule–A table showing the estimated amount of lumber that can be sawed from logs of given length and diameter. See Doyle Rule, Scribner Rule.

Logged–Designating an area from which trees have been cut.

Loggerhead–A large snapping turtle found in the southeastern United States. Loggerheads can be a serious pest in farm fish ponds. Also called alligator snapping turtle.

Logging–The harvesting of trees for lumber.

Logging Residues–The unused portions of poletimber and sawtimber trees cut or killed by logging.

Long Butt–A section of the lower end of a butt log cut off to remove serious basal defect, such as deep fire scar and advanced butt rot.

Long-lasting Fertilizer–Any fertilizer that is water-insoluble, e.g., animal manures, oil meals, and composts.

Longitude–(1) Distance east or west on the earth's surface, measured by the angle which the meridian through a place makes with some standard meridian, e.g., that of Greenwich. (2) A coordinate distance, linear or angular, from a north-south reference line.

Longleaf Pine–*Pinus palustris*, family Pinaceae; a coniferous evergreen tree; an abundant native forest species in the Atlantic and Gulf Coastal plains of the United States, where it is an important source of lumber and naval stores. Also called fat pine, southern pine, fatwood, broom pine, Georgia pine. See Naval Stores.

Loose Rock Dam–A dam built of rock without the use of mortar. The rock is usually confined by woven wire. See Gabion.

Lost Corner–In land surveying, a corner whose position cannot be definitely determined from traces of the monument or by reliable testimony relating to it; it can be restored only by surveying from an established survey point.

Lot–(1) A small piece of enclosed land usually adjacent to a barn or shed in which horses, mules, cows, etc., are allowed to exercise. (2) Any particular grouping of animals, plants, seeds, fertilizers, etc., without particular regard to number, as a seed lot. (3) A small tract of land, usually less than an acre, on which a house is constructed.

Lotic–Living in rapidly moving waters.

Low–An area where the barometric pressure is lower than the surrounding areas. Usually caused by the interaction of a warm and cold air mass.

Low Flow Augmentation–Increasing of an existing flow. The total flow of a stream can seldom be increased, but its ability to assimilate waste can generally be improved by storage of floodflows and their subsequent release when natural flows are low and water-quality conditions are poor.

Luster–(1) The property of wood, independent of color, which causes it to reflect light and to exhibit a sheen. (2) The natural gloss or sheen characteristic of the fleeces of long-wool breeds of sheep and Angora goats.

Lysimeter–A device used to measure the quantity or rate of water movement through or from a block of soil or other material, such as solid waste, or used to collect percolated water for quality analysis.

Macro–Prefix meaning large, long; visibly large.

Macronutrients–Includes primary plant nutrients N, P, and K; and secondary plant nutrients Ca, Mg, and S. See Micronutrient.

Macroorganisms–Plant, animal, or fungal organisms visible to the unaided eye.

Macrophytic Algae–Algal plants large enough either as individuals or communities to be readily visible without the aid of optical magnification.

Macroscopic–Visible to the human eye without the aid of a microscope.

Made Land–An area consisting of artificial fills of earth material, or dumps of waste, refuse, debris of any kind, smoothed over and stabilized; e.g., reclaimed land from strip mining.

Magma–Naturally occurring molten (melted) rock material, generated within the earth, from which igneous rocks have been derived through solidification. See Lava.

Magnesia–Magnesium oxide (MgO), obtained by heating or burning the hydroxide or the carbonate. A commercial product, containing a high percentage of MgO, is used as a liming material for soils. Magnesium oxide is also used in the treatment of grass tetany. See Grass Tetany.

Magnesian Limestone–A limestone that contains magnesium carbonate. See Dolomitic Limestone.

Magnesite–A natural mineral (magnesium carbonate: $MgCO_3$). See Magnesia.

Magnesium–Mg; a white metal, essential for people and animals and for plant growth because it is a constituent of chlorophyll.

Magnesium Ammonium Phosphate–$MgNH_4PO_4 \times 6H_2O$. This material is sometimes formed in mixed fertilizers when a magnesium compound (dolomite or magnesium sulfate), superphosphate, and an ammoniacal solution are mixed together. It is also sold commercially as MgAmP. The pure salt contains about 44 percent water in combined form, magnesium (Mg) 9.90 percent, nitrogen (N) 5.70 percent, and phosphoric oxide (P_2O_5) 28.93 percent.

Magnesium Deficiency–(1) A physiological disease of plants grown on soil deficient in available magnesium, generally characterized by chlorosis and necrosis and a reduction of growth. In tomatoes, the veins remain green. In cabbages, the lower leaves pucker and turn white at the margin and center. Also called sand drown. (2) The cause of grass tetany in cattle. See Grass Tetany.

Magnesium Oxide–MgO. See Magnesia.

Magnesium Sulfate–$MgSO_4$; a soluble salt obtained from kieserite that is added to mixed fertilizers to supply the magnesium essential to plant growth.

Magnetic Compass–An instrument having a freely pivoted magnetic needle that aligns with the earth's magnetic field such that one end of the needle points to the magnetic north.

Magnetic Declination–The angle between the direction of the magnetic and geographic meridians. In nautical and aeronautical navigation the term magnetic variation is preferred.

Magnetic North–Direction indicated by the magnetic needle of a compass.

Magnitude–(of an earthquake) A quantity characteristic of the total energy released by an earthquake, as contrasted to intensity, which describes its effects at a particular place. The Richter Scale is related to the logarithm of an observed displacement on a calibrated instrument and its distance from the epicenter. Each step of one magnitude represents a ten-fold increase in observed amplitude. See Richter Scale.

Main Canal–In irrigation and other systems, the principal conduit feeding water from the supply source to the lateral conduits.

Maintaining Forage Reserves–The reservation of native forage supplies for such emergencies as drought, fire, or other unforeseen circumstances; obtained by the exclusion of livestock or by light use of grazing areas.

Malachite Green–Copper carbonate.

Malaria–An infectious febrile disease of people caused by Protozoa of the genus *Plasmodium* that invade the red blood corpuscles and are transmitted by mosquitoes of the genus *Anopheles*.

Malathion–An organic, phosphorus insecticide. Technical grade *malathion* (95 to 98 percent pure) is a viscous, dark brown liquid with a strong, offensive odor somewhat like that of garlic. Widely used to control pests, it is the least toxic of phosphorus insecticides.

Mammal–A group of animals that secrete milk, and nurse their young, grow hair on their bodies, and possess a diaphragm between the thoracic and abdominal cavities.

Manganese–Mn; a metallic element, found in soils (determined as manganous oxide) from a mere trace to as much as 15 percent, but when present it is often in forms unavailable to plants. It is regarded as essential to normal plant growth and is often applied, usually as manganous sulfate, to soils deficient in this element. It is also essential to growth, reproduction, and lactation in animals.

Manganese Agstone–A considerable deposit of limestone occurring in Arkansas that contains 10 to 15 percent Mn soluble in dilute organic acids. It is in the form of manganese carbonate.

Manganese Chlorosis–A discoloring disease of plants caused by lack of manganese.

Manganese Deficiency–A nutrition deficiency that results in dwarfing of plants, chlorosis of the upper leaves often becoming spotted. Crops are frequently unmarketable. In animals, it causes perosis (slipped tendon) in poultry and impaired growth in swine.

Manganese Oxide–There are several oxides of manganese. Two of the important ones are manganous oxide (MnO) and manganese dioxide (MnO_2). The latter occurs as the ore pyrolusite and is used in producing manganese sulfate and manganous oxide.

Mangrove–Woody tropical plants that grow along saline shores. The black mangrove, *Avicennia nitida*, is a valuable source of wild honey. The red mangrove, *Rhizophora mangle*, is used for tanbark. See Tanbark.

Mantis–Any of the large insects, family Mantidae, predaceous on other insects, capturing their prey with their unusually well-developed front legs. Species are considered beneficial in gardens. See Praying Mantis.

Manure–(1) Excreta of animals, dung and urine (usually with some bedding), used to fertilize land. (2) In Europe, any material that contains the essential elements of plant nutrients, as chemical fertilizers, excreta of animals, etc. See Compost.

Manure Pit–A storage unit in which accumulations of manure are collected before subsequent handling or treatment, or both, and ultimate disposal. Water may be added in the pit to promote liquefaction so the manure can be spread on fields through a sprinkler irrigation system.

Map–A chart of any portion of the earth's surface that shows its location, shape, and extent of physical, political, and other features. See Soil Map.

Maple–Any tree or shrub of the genus *Acer*, family Aceraceae. Species are grown as ornamentals, for shade, lumber, and for the sap that yields maple syrup and maple sugar. Some types are: bigleaf maple, box elder, Japanese maple, Norway maple, planetree maple, red maple, silver maple, striped maple, sugar maple, vine maple, and flame maple.

Maple Syrup–A sweet, concentrated food product made in the spring by boiling the natural sap of the sugar maple tree (*Acer saccarum*) to the proper concentration. Further concentration will cause crystallization of the syrup into maple sugar.

Marginal Land–Agricultural land definitely not first class, and near the margin between profitable and unprofitable use.

Mariculture–The growing of marine animals such as fish and shrimp under controlled conditions in saltwater rather than in freshwater. See Aquaculture.

Marijuana–A preparation of the dried leaves and flowers of male or female plants of wild hemp, ***Cannabis sativa***, family Cannabaceae. Used in cigarettes for its euphoric properties. Illegal to grow in the United States. See Hemp.

Marine Climates–Climatic types determined by oceans. The climate of oceanic islands and of windward shores of continents, contrasted with interior or continental climates, which usually have a greater range in daily and annual temperatures, i.e., colder winters and warmer summers.

Marking–(1) Castrating, docking, branding, and ear marking. (2) Selection and indication, usually by blaze or paint spot, of trees that are to be cut or retained in a cutting operation.

Marl–(1) A natural, earthy deposit, consisting chiefly of calcium carbonate mixed with clay or other impurities in varying proportions. (2) Incoherent sands containing shells. (3) Soft, impure limestone. (4) Soft, mudlike deposits originating mostly in lakes and found in underlying bogs widely distributed in the glaciated region of the United States. Sometimes called bog lime. Marls are used as an amendment for soils deficient in lime. See Lime.

Marsh–An area of soft, wet, or periodically submerged land, generally treeless and usually characterized by grasses and other low vegetation.

Marsh Gas–Impure methane (CH_4), the chief constituent of natural gas. Results from the partial decay of plants in swamps.

Mass Wasting (Mass Movement)–(1) The slow downslope movement of rock debris. (2) A general term for a variety of processes by which large masses of earth material are moved by gravity either slowly or quickly form one place to another.

Massive–(1) Indicating a large-sized and deeply muscled animal; designates an ideal quality in a draft horse. (2) (Soil structure) Large uniform masses of cohesive soil that sometimes have irregular cleavage, as in the C horizons of many fine-textured clay soils. See Soil Structure.

Mat–A layer of algae, generally of the filamentous type. The layer may be either floating on the water or covering a rock or other substrate.

Mattress–A blanket of brush or poles interwoven or otherwise lashed together and placed to cover an area subject to river bank erosion. It is weighted with rock, concrete blocks, or otherwise held in place.

Mature Soil–(1) A soil that has well-developed characteristics produced by the natural processes of soil formation and that is in equilibrium with its environment. (2) A soil with an A, B, and C horizon. See Soil Profile.

Maximum Temperature–In plant physiology, the temperature beyond which growth ceases. It varies with plant species and the duration of high temperature.

Mayfly Naiads–Family Ephemeridae; the immature mayfly. The terrestrial adults lack functional mouth parts and live only a few hours.

McLean County Sanitation System–A system developed in McLean County, Illinois, by veterinarians of the United States Bureau of Animal Industry for the prevention and control of filth-borne diseases of swine. It has particular reference to internal parasites.

Meadow–(1) A field cut for hay. (2) An opening in a forest where grass is the dominant vegetation. (3) In arid and semiarid regions, a productive grassy area usually with a high water table. (4) At high elevations, an alpine meadow.

Meadow Foxtail–A less desirable grass, adapted to wet soils that are subject to frequent and/or prolonged flooding. Produces less forage than other grasses. Tolerant of frost and prolonged snow cover in high-altitude areas. Meadow foxtail is very undesirable in seed-producing areas, as its seed is extremely difficult to separate from some of the more important seed crops.

Meadow Mouse–A rodent of the genus ***Microtus*** that is larger than the common mouse but smaller than a rat. In fields it forms paths in the grass, which connect holes and tunnels. It also damages many trees by girdling them.

Meadow Nematode–Species of the genus ***Pratylenchus***, family Tylenchoidae, a widely distributed, wormlike root nematode found in the cortical tissue of the roots causing sloughing off and the formation of fine roots near the surface of the ground. It infests many trees, such as apple, cherry, walnut, and almond, and causes brown root rot of tobacco, alfalfa, cotton, and other plants. See Nematode.

Meadow Soil–A mineral soil that developed under a grass-sedge or herbaceous vegetation in a wet habitat, such as a mountain meadow. Also called Wiesenboden.

Meadowfoam–*Limnanthes alba*; a low-growing winter annual native to California, southern Oregon, and Vancouver Island, British Columbia. The seeds of the plant contain oil that has a potential use as lubricants, polymers, and waxes.

Mean Annual Increment–In forestry, the total growth (board feet or cubic feet) of a stand of trees divided by the total age in years.

Mean Annual Precipitation–The mean measured depth of all forms of moisture received from the atmosphere in one year. Snow depth is converted to equivalent water depth, usually 10 inches (25 centimeters) of snow equals 1 inch (2.5 centimeters) of water.

Mean Annual Temperature–The mean of the yearly maximum and minimum means of temperature for a period of years in a specific locality.

Mean Depth–(Hydraulics) Average depth; cross-sectional area of a stream or channel divided by its surface width.

Mean Sample Tree–Any tree that represents the average of a group of trees in any given class or stand.

Mean Sea Level–The mean level of the sea at a given station; i.e., the mean elevation of the two daily high tides and the two daily low tides caused by the attraction of the moon and sun.

Mean Velocity–The velocity at a given section of a stream, obtained by dividing the discharge of the stream by a cross-sectional area at that location.

Meander–(1) One of a series of somewhat regular and looplike bends in the course of a stream. (2) A land survey traverse along the bank of a permanent body of water.

Meander Line–The waterline or shore of a body of water, as a large lake, established at the time of an official land survey, as that of the United States General Land Office. Legally, it is important in determining the ownership and use of shoreland.

Measurable Precipitation–Any precipitation equivalent to over 0.005 inches of rainfall. Less than 0.005 inches is considered a trace.

Meat Inspector–A graduate veterinarian, engaged by regulatory authorities to inspect all meats as they pass through the slaughterhouse or packing plant, to certify that they are wholesome food products.

Mechanical Aeration–Mechanical energy used to inject air into water, causing the waste stream to absorb more oxygen. See Lagoon, Aerated.

Mechanical Practices–Soil and water conservation practices that primarily change the surface of the land or that store, convey, regulate, or dispose of run-off water without excessive erosion.

Mechanical Soil Analysis–See Particle Size Distribution.

Media–(1) The middle coat of an artery. (2) The plural of medium. (3) Material for artificial propagation of various microorganisms. (4) Any means through which communication of any type is accomplished. (5) Soil or soil-like material in which plants are grown.

Median Lethal Dose–(LD_{50}) The amount of concentration of a toxic substance that will result in the death of 50 percent of a group of test (target) organisms upon exposure (by ingestion, application, injection or in their surrounding environment) for a specified period of time. (Complementary to median tolerance limit, (TL_{50}.)

Median Tolerance Limit–TL_{50} The concentration of some toxic substance at which just 50 percent of the test (target) animals are able to survive for a specified period of exposures. (Complementary to Median Lethal Dose, LD_{50}.)

Mediterranean Climate–Any warm-temperature climate that is relatively dry in the summer and relatively moist in the winter.

Medium–(1) Any of a number of natural or artificial substances, pastelike or liquid, in or on which microorganisms, such as bacteria and fungi, can be cultured. (2) A soil or material, such as sand, peat moss, vermiculite, etc., in which plants are raised or cuttings are rooted, especially in the greenhouse. (3) A market grade of roses that has stems 12 to 8 inches in length. (4) One of the six recognized grades of meat, lying between good and common.

Medium Sand–In soil analysis, a separate that includes particles ranging in diameter from 0.5 to 0.25 millimeters. See Soil Textures.

Medium-textured Soil–Very fine sandy loam, loam, silt loam, or silt.

Mega–(1) A prefix meaning one million. (2) A prefix meaning large or in large numbers.

Megathermic Tree–Any tree that requires a warm climate and long growing season, such as palm, laurel, magnolia, live oak, tupelo gum, redwood, baldcypress, slash pine, longleaf pine, and other southern pines.

Melanic Epipedon–A black surface soil horizon high in organic matter from grass residues. See Soil Taxonomy.

Melanoma–Malignant melanoma is a cancer of the cells that produce the pigment melanin.

Mellow Soil–(1) A nonscientific term describing a soil, especially one containing a high percentage of humus, that is soft and loose. (2) A soil that pulverizes easily into a good seedbed.

Membrane–(1) A thin, flexible sheet of vegetable or animal tissue; the thin protoplasmic tissue connecting, covering, or lining a structure, such as a cell of a plant or animal. (2) The layer that surrounds fat globules of milk. Its nature is not definitely known: it may consist mainly of phospholipoids and a protein or proteins not completely identical with casein, albumin, or globulin. (3) A layer of low permeability material: e.g., bentonite (clay) soil, placed in the bottom of a farm pond to reduce seepage losses.

Mendel's Law (Medelian Law)–Gregor Johann Mendel (1822-1884), an Austrian monk, naturalist, and plant breeder, first demonstrated dominant and recessive genes in plant breeding, which also applies to some kinds of animal and people inheritance. For example: When a plant with the dominant gene for tallness (TT) is crossed with a plant with a recessive gene for shortness (ss), the first generation, F_1, will be distributed according to the following ratio. One plant, TT, homozygous (pure) true-breeding for tallness. Two plants, Ts, heterozygous for tallness and shortness. One plant, ss, homozygous (pure) true-breeding for shortness. When plants with Ts and Ts genes are crossed, the F_1 progeny will be in the same ratio, i.e., one TT, two Ts, and one ss.

Merchantable Height–The length of the tree stem from the top of the stump to the top of the last merchantable section. Usually expressed in feet or number of logs of a specified length.

Merchantable Timber–A tree or stand of trees that may be converted into salable products.

Merchantable Volume–The amount of wood in a single tree or forest stand that is considered salable.

Mere–A sheet of shallow standing water, a pool, sometimes a marsh.

Meridian Line–A true north and south line; a line from which range lines are established in the General Land Office survey (United States).

Meridian, Principal–(United States public-land surveys) A line extending north and south along the astronomical meridian passing through the initial point, along which township, section, and quarter-section corners are established. The principal meridian is the line from which is initiated the survey of the township boundaries along the parallels.

Mesa–(Spanish, table) A tableland; a flat-topped mountain or other elevation bounded on at least one side by a steep cliff. Local in the Southwest.

Mesic–Moderately moist where dry periods are not of long duration. Mesophytes are plants that grow in soils that contain moderate amounts of moisture. See Hydric, Xeric.

Mesophyte–A plant whose normal habitat is neither wet nor dry but intermediate in soil moisture.

Mesotherm–(1) Any plant that grows under intermediate conditions of temperature along with abundant moisture. (2) Any plant that grows under conditions of alternating high and low temperatures.

Mesotrophic–Descriptive of a type of lake intermediate in nutrient concentration between oligotrophic (poor) and eutrophic (rich). See Eutrophic, Oligotrophic.

Mesquite–Any plant of the genus *Prosopis*, family Leguminosae. Species are usually thorny shrubs, the foliage and seed pods of which have browse value in the desert, but the invasion of the plant on open grassland lowers the grazing capacity of the land. One of the more useful mesquites is *P. juliflora*, also called algarroba, an important bee and forage plant. Also called mesquit, musquit, muzquit.

Mesquite American Mistletoe–*Phoradendron californicum*; family Loranthaceae; a parasitic shrub that grows on trees. Its berries are used for decoration chiefly at Christmastime. Native to North America. See Mistletoe.

Metal–(1) Any of a class of substances that typically are fusible and opaque, are good conductors of electricity, and show a unique metallic luster, as gold, copper, bronze, and aluminium. Most metals are also malleable and comparatively heavy, and all except mercury are solid at ordinary temperatures. Metals constitute over three-fourths of the recognized elements. They form oxides and hydroxides that are basic, and they may exist in solution as positive ions (cations). (2) Ore from which a metal is derived. (3) Molten glass. (4) Railway rails.

Metaldehyde–A crystalline polymer of acetaldehyde, used at two to three percent as an attractant in poison baits for snails and slugs.

Metalimnion–(1) The layer of water between the epilimnion and the hypolimnion in which the temperature exhibits the greatest vertical rate of change, also more frequently termed the thermocline. (2) The zone in which temperature decreases rapidly with depth in a lake when it is thermally stratified in summer. See Epilimnion, Hypolimnion.

Metamorphic Rock–Any rock that has undergone profound changes by the combined action of pressure, heat, and water, as schist, slate, and marble.

Metamorphosis–A process by which an organism changes in form and structure in the course of its development, as many insects do. See Complete Metamorphosis.

Metaphosphoric Acid–HPO_3. See Phosphoric Acid.

Meteorology–(1) The study of the physical processes that occur in the atmosphere and of the related processes of the lithosphere and hydrosphere. (Long usage has caused the term to be used almost solely in connection with phenomena of the weather). (2) The science of the atmosphere.

Methane–CH_4; an odorless, colorless, and asphyxiating gas that can explode under certain circumstances; can be produced by manures or solid waste undergoing anaerobic decomposition as in anaerobic lagoons and in silos. There are twelve species of bacteria capable of producing methane gas from manures and other biomes.

Methanesulphonyl Fluoride–A highly effective greenhouse fumigant.

Methanol–A light volatile, flammable, poisonous, liquid alcohol, formed in the destructive distillation of wood or made synthetically and used especially as a fuel, a solvent, an antifreeze, or a denaturant for ethyl alcohol, and in the synthesis of other chemicals; methanol can be used as fuel for motor vehicles; also known as methyl alcohol or wood alcohol. See Gasohol.

Methoxychlor–A chlorinated hydrocarbon insecticide, closely related to DDT and TDE.

Methyl Alcohol–Also known as methanol or wood alcohol. See Methanol.

Methyl Bromide–CH_3Br; an insect fumigant. It is also used to sterilize the soil.

Methyl Formate–An insect fumigant.

Metrox–See Copper Oxide.

Mexican Pinyon Pine–*Pinus cembroides*, family Pinaceae; a low tree that bears an edible seed. Native to North America. Also called pinyon, Mexican stone pine, nut pine, pinon. See Nut Pine.

Mica–A common mineral in soils; muscovite, biotite, and a number of others. They are not easily soluble minerals of the mica group residual in soils derived form the weathering of granites, gneisses, and schistose rocks. When present in considerable quantity, mica imparts looseness and a greasy or soapy feel to a soil. See Vermiculite.

Micro–A prefix meaning one millionth (1/1,000,000) or very small.

Microbes–Minute plant or animal life. Some microbes may cause disease, while others are beneficial.

Microbial Insecticide–Microbes that are used to control insects. Of the millions of insects that have been identified, about 350 species have been classified as destructive pests. Like all living organisms, these pests are susceptible to diseases caused by bacteria, fungi, protozoa, and viruses. Research scientists have explored this avenue of control and have developed several strains of microbes that have potential to be effective in the control of insect pests. See Integrated Pest Management.

Microbiologist–A scientist concerned with the study of plant and animal microorganisms.

Microclimate–Climate of a small area or locality as compared to a county or state. For example, the climate adjacent to the north side of a home, or influence of a lake on a portion of a county.

Microelements–Trace elements; micronutrients.

Microflora–(1) Microscopic plant, such as bacteria. (2) In plant geography, the vegetation of a small area in contrast to that of a region. (3) Microbial life characteristic of a region, such as the bacteria and protozoa populating the rumen.

Micronized Sulfur–Sulfur in an extremely fine state of division with a particle size mostly under 10 microns. Used as dusts and sprays for control of plant diseases.

Micronutrient Elements, Essential–For plants: boron, chlorine, cobalt, copper, iron, manganese, molybdenum, and zinc. For people and animals: chlorine, copper, iron, manganese, molybdenum, zinc, sodium, iodine, selenium, vanadium, chromium, cobalt, nickel, fluorine, silicon, boron, and arsenic.

Microorganism–(1) An organism so small that it cannot be seen clearly without the use of a microscope; a microscopic or submicroscopic organism. (2) Any microscopic animal or plant that may cause a plant disease or have the beneficial effect of decomposing plant and animal residue that becomes humus.

Microthermic Tree–A tree that tolerates a cold climate and short growing season: e.g., aspen, birch, alder, willow, mountain ash, jack pine, Scotch pine, lodgepole pine, spruce, larch, and alpine fir.

Middens–(1) The castings of organisms that live and work in the soil, especially of earthworms. (2) A pile of refuse; a manure pile.

Middles–The spaces between the rows of trees or a crop which may be cultivated, planted to an interrow crop, or left in sod.

Midge–A loose name for any small, mosquitolike insect; specifically applied to those of family Chironomidae.

Migrant–Designating a plant, animal, or person that changes, or has changed, its natural location.

Milky Disease–A disease of insects caused by ***Bacillus popillae*** that is used in the biological control of the Japanese beetle.

Mill Pond–A small body of water retained by a dam above a mill that is driven by water power. See Millrace.

Milleped(e)s–Arthropods of the division Diplopoda, similar to centipedes but having a greater number of legs. They are known to be abundant in some soils and may be a factor in the rapid breakdown of leaves. They seldom injure plants but occasionally enter houses. Also called thousand legged worms, galleyworms.

Miller–(1) The operator of a mill, especially of a flour mill. (2) Any moth with wings that appear dusty or powdery.

Milpa–A cultivated field, specifically a cornfield (presumably originated from the Indians of the southwestern United States).

Milt–(1) The roe or spawn of the male fish. (2) To impregnate with milt. (3) A ductless gland near the stomach and intestine of fowls. It contributes to the formation of new blood cells. See Spleen.

Milter–A male fish especially at spawning time.

Mine–(1) To eat out a gallery or passageway, as a larva mines a fruit. (2) An open pit or shaft excavation for obtaining mineral ores.

Miner's Inch–A unit of irrigation water flow for small users established by statute in several western states and in British Columbia.

Mineral–(1) A chemical compound or element of inorganic origin. (2) Designating the inorganic nature of a substance.

Mineral Cycle–The movement of a mineral ion in nature through a complete cycle-uptake, retention, restitution, and leaching within the soil-plant system.

Mineral Elements–The inorganic components of soil, plants, or agricultural products.

Mineral Ion–Electrically charged atoms or groups of atoms formed by disassociation of a molecule, and the form by which most mineral elements enter plant roots. Ions move from outside into the plant root cells as a result of ionic diffusion. See Cation-exchange Capacity, Essential Elements.

Mineral Land–A class of land valued chiefly for its subsurface or surface mineral resources; a use classification.

Mineral Oil–Oil obtained from petroleum and other hydrocarbon minerals, as contrasted with vegetable oils.

Mineral Rights–Rights pertaining to ownership of subsurface land minerals and to access in their exploitation. An owner of land may sell only surface rights and retain ownership of mineral rights or sell only mineral rights.

Mineralize–(1) To petrify. (2) To impregnate or supply with minerals. (3) To promote the formation of minerals. (4) The microbial breakdown of organic matter to release its minerals.

Mineralogy–The science of the study of minerals.

Minimum Temperature–(1) A degree of warmth at which plant growth first begins. See Cool-season Crops, Warm-season Crops. (2) The lowest temperature recorded during a given period, such as the minimum daily temperature.

Minimum Tillage–A soil management system in which the residue of the previous crop remains on the soil surface and the next crop is planted with little or no tillage.

Mink–A semiaquatic, weasel-like mammal of the genus ***Mustela***, especially ***M. vison***, which has thick, soft, usually dark brown fur with lustrous outside hairs.

Minnow–A small fish, often used for fish bait.

Minor Elements–See Micronutrient Elements, Essential.

Mire–A small muddy marsh or bog; wet, spongy earth; soft, deep mud.

Mississippi Delta–The fertile alluvial bottomlands that lie along the lower course of the Mississippi River northward from the delta proper to southeastern Missouri in the United States.

Mist Fertilizer–A low-nutrient solution applied to newly propagated vegetative cuttings in a greenhouse.

Mistletoe–(1) Any of a number of segmented, flowering plants of the family Loranthaceae. Species are parasitic on trees and other woody plants. The mistletoes are divided into: (a) a dwarf group including members of the genus ***Arceuthobium***, all without apparent leaves, occurring only on conifers. Also called lesser mistletoes; (b) a leafy group including, in temperate North and South America, members of the genus ***Pharadendron***, mostly with thick, flat, green leaves but sometimes without apparent leaves. This group usually occurs on hardwoods, but some species grow on conifers. See Christmas American Mistletoe. (2) *Viscum album*, the common mistletoe of Europe. The plant bears small yellow flowers and occurs on oak, apple, etc. Native to Europe and Asia. See Eastern Mistletoe.

Miticide–Any poisonous substance used to kill mites.

Mitosis–Cell division involving the formation of chromosomes, spindle fibers, and the division of chromosomes by a process of longitudinal splitting. Each of the resulting daughter cells thus has a full set of chromosomes as distinguished from reduction division or meiosis, in which the daughter cells have half the somatic number. See Meiosis.

Mixed Fertilizers–Two or more fertilizer materials mixed or granulated together into individual pellets. The term includes dry mixed powders, granulates (bulk blends), granulated mixtures, and clear liquid mixed fertilizers, suspensions, and slurries. They were at one time called complete fertilizers, but are now referred to as multinutrient mixtures.

Mixed Prairie–A phytogeographic region west of the tall-grass prairie in which midgrasses form the upper story and short grasses the lower story of the dominant vegetation.

Mixing Depth–The expanse in which warm air rises from the earth and mixes with the air above it until it meets air of the same temperature.

Mockernut Hickory–*Carya tomentosa*, family Juglandaceae; a deciduous tree grown for shade and for its valuable wood used in tool handles, ladders, furniture, sporting goods, implements, woodenware, and for fuel and smoking meat. Also called big bud hickory, bird's-eye hickory, black hickory, curly hickory, whiteheart hickory.

Moderate Live Virus–A virus that has been changed by passage through an unnatural host, such as passing hog cholera virus through rabbits, so that it no longer possesses the disease-producing characteristics but so that it will stimulate antibody production and immunity when injected into susceptible animals.

Modifier–(1) Any element that is added to, or taken from a substance that alters its normal appearance or function. (2) A substance that can alter the course of carcinogenesis.

Mohave Desert–A hot, extremely arid region of southern California, United States, characterized by complete absence of rain during summer and by only scant winter rain. The natural vegetation is mostly desert shrub types. The area includes highly productive soils under irrigation, and crops such as alfalfa, citrus, cotton, and dates are grown.

Moisture–(1) The total amount of water in any plant or animal product. (2) Any form of water. (3) The total amount of water (exclusive of that in chemical combinations) in the soil, both that which is available and that unavailable for plant growth.

Moisture Penetration–(1) The depth to which moisture penetrates in soil following an irrigation or rain before the rate of downward movement becomes negligible. (2) Downward movement of water into a soil (a soil may be open, porous, and permit free penetration or it may be compact and tight and thus prevent free downward movement).

Moisture Tension–The force at which water is held by soil, usually expressed as the equivalent of a unit column of water in centimeters: 1,000 centimeters being equal to one atmospheric tension. Moisture tension increases with dryness and indicates the degree of work required to remove soil moisture for use by plants.

Moisture-retentive Soil–A soil that retains a high percentage of moisture or water. In mineral soils, retentiveness bears a direct relation to clay or colloid content; e.g., a clayey soil is more retentive than a sandy soil.

Mole–(1) One of several species of burrowing, insectivorous mammals, characterized by soft, dark fur, long nose, very small eyes, and large, strongly developed front feet. The moles belong mainly to the family Talpidae. A common species in the eastern United States is *Scalopus aquaticus*. They can damage lawns and gardens with their underground burrows but otherwise are not serious pests. (2) A natural levee or depositional ridge on the bank of a large river. (3) A plow used in establishing mole drainage. See Mole Drainage.

Mole Drainage–A method of draining land by making open channels in the soil at a depth of 2 to 3 feet. These are constructed with a mole plow consisting of a two-wheel frame and a projectile-shaped ball varying in diameter from 2 to 8 inches and chain linked behind a sharp-pointed, sharp-edged shank. The plow is pulled along the course of the desired drain with the ball at depths and grades to direct the water toward an outlet drain.

Molecule–The smallest part of a substance that can exist separately and still retain its chemical properties and characteristic composition; the smallest combination of atoms that will form a given chemical compound.

Mollic Epipedon–A thick, dark surface soil horizon that has developed under grass.

Mollis–Soft.

Mollisols–A soil order of black, organic-rich surface horizons and a high supply of bases. They have mollic epipedons and base saturation greater than 50 percent (NH_4OAc) in any cambic or argillic horizon. They lack the characteristics of vertisols and must not have oxic or spodic horizons. See Soil Orders.

Mollusk–Any animal of the phylum Mollusca, as a slug, snail, etc.; characterized by a soft, unsegmented body. Some species are protected wholly or in part by a calcareous shell.

Molybdenum–Mo; a gray metallic element, essential in very small amounts to the growth of plants, but usually present in sufficient amounts. Deficiencies have been discovered in a few highly acid soils. When these soils are limed, ample molybdenum is usually released from slowly soluble forms in the soil. It is one of the micronutrients and is applied to soils usually as sodium molybdate, known in the trade as Moly. The normal amount of Mo in the soil is 1 to 3 ppm. More than 10 ppm in the soil may be injurious to some plants and animals. Recent tests have indicated a positive soybean response to molybdenum in most of the soybean-growing areas of the Midwestern and some of the southern states.

Monadnock– A residual rock, resistant hill, or mountain standing above a plain.

Mono-- Single; only one.

Monolith– An undisturbed (not broken up) sample of a soil profile.

Monsoon– A wind that blows fairly steadily along the Pacific Asiatic Coast. During the winter it blows from the northeast and is dry; during the summer it blows from the southwest and brings rain. See Rainy Seasons.

Montane– Of or pertaining to mountains.

Monte– (1) A tract of dense mesquite brush. (2) The chaparral of the foothills of southern California, United States.

Montmorillonite– A group of clay minerals whose formulas may be derived by ion substitution. They are characterized by swelling when wet. Mineral members of the group are montmorillonite, nontronite, saponite, hectorite, sauconite. Also referred to as the smectite group.

Moor– A bleak, treeless wasteland. In Scotland and England, moors are generally characterized by peat soil and heather (***Calluna*** spp.), and in wetter spots by sedges and grasses.

Mopping Up– In forestry, making a fire safe, after it is controlled, by extinguishing or removing burning material along or near the control line and by felling snags.

Mor Humus– A type of forest humus layer of unincorporated organic material, usually matted or compacted or both, distinctly delimited from the mineral soil, unless the latter has been blackened by washing in organic matter. Syn.: raw humus.

Moraine– The geological deposition formed on the margins of glaciers or beneath moving ice sheets. In the glaciated region of the United States, moraines vary greatly according to kind: terminal, ground, etc., in rock composition relief and minor topographic features, but they are significant locally in soil and land classification.

Morass– Any murky land, bog, marsh, or swamp.

Morphogenesis– The developmental history of organisms or of their parts.

Morphology– (1) A branch of biologic science that deals with the forms, rather than the functions, of plants and animals. (2) Pertaining to pedology or soil science, the study of soil horizons and their arrangement in profiles. (3) Pertaining to land surfaces, the shape or configuration of physical features.

Morphometry– The physical shape and form of a water body.

Mose– A marsh.

Mosquito– An insect belonging to the family Culicidae. The proboscis of the female has a needle-sharp tube in it with which it pierces the skin of people or animals to suck blood. Mosquitoes are pernicious pests and transmit disease to both animals and humans.

Moss– A flowerless plant intermediate in character between algae and ferns. Mosses of the Tundra have a grazing value, some have value as ornamental plants, and some have a commercial value for packing and rooting plants.

Moss Peat– Peat composed mainly of the identifiable remains of mosses, especially those of the genera ***Sphagnum, Polytrichum***, and ***Hypnum***. Moss peats are natural soils and are used commercially for mulches, pot and greenhouse soils, poultry litter, barn bedding, paper making, fuel, etc. Also called highmoor peat, peat moss, sphagnum peat. See Histosols.

Mossy– (1) Designating irregular, dark markings that spoil an otherwise desirable color contrast on the feathers of domestic birds. (2) Covered with moss.

Moth– An insect belonging to the order Lepidoptera. The adults have wings covered with scales and differ from butterflies in being night fliers and in having heavy, hairy bodies. In the caterpillar stage, many moths are destructive; they may be injurious and do tremendous damage to cloth, crops, trees, and stored grain.

Moth Ball– See Naphthalene.

Mother Ditch– A main canal that carries irrigation water.

Mother Tree– In plant breeding, the original tree from which varieties are developed.

Motte– An isolated, small clump of trees or shrubs; a minor vegetational feature of the natural landscape in Texas and other states (United States). Also called mott.

Mound Prairie– In some parts of the United States, a presettlement land feature comprising a vast expanse of treeless grassland marked by enclosures, called mounts, believed to have been built by Indians.

Mount– (1) A horse for riding. (2) A riding seat. (3) To copulate, as a male animal. (4) To get into a saddle on the back of a riding horse. (5) A mountain, or a high hill. Used always instead of ***mountain*** before a proper name.

Mountain– Any portion of the earth's crust rising considerably above the surrounding surface. The term is usually applied to heights of more than 2,000 feet (656 meters), all beneath that amount being regarded as hills, and when low, as hillocks.

Mountain Meadow– A glade or a moist or wet treeless tract that supports sedges, grasses, and other herbaceous vegetation, usually above the timberline at high altitudes.

Mountain Pine Beetle– ***Dendroctonus monticolae***, family Scolytidae; a bark beetle that is especially damaging to lodgepole pine and western white pine in the northwestern United States.

Mourning Dove– A dove, ***Zenaidura macroura carolinenesis***, native to the United States. They are so named because of their mournful call. They are hunted as a game bird, particularly in the south.

Mouse– Any of the numerous, smaller rodents of the family Muridae and related families, the most common of which is ***Mus musculus***, the house mouse, which is typically 6 inches long including the tail.

Movable Dam– A water barrier that may be opened in whole or in part. The movable part may consist of gates, stop logs, flash boards, wickets, or any other device whereby the area for water flow through or over the dam may be controlled.

Muck–(1) Highly decomposed peat often used for raising crops, or sold in bags as "potting soil" or "organic soil." (2) Feces and urine from domestic animals in a wet state. (3) Something of questionable, dirty, trashy, or antisocial value. (4) In surface mining, the useless overburden. See Histosols.

Muck Ditch–In flood control engineering, a trench excavated under the center of a levee and refilled with impervious material to form a cut-off wall and to prevent the seepage of water along the foundation plane of the levee. Also called cut-off trench.

Mucor–A widely distributed genus of molds of the family Mucoraceae, mainly saprophytic species, abundant in soil, decaying vegetable matter, dung, etc. Some are capable of converting starch into sugar, and thus important industrially. Probably the most frequently met and most troublesome species is the black bread-mold, *Rhizopus nigricans*, which appears wherever starchy substances are found, as on stale bread, and which causes a watery rot of strawberries, and peaches, a soft rot of roots of the sweet potato, and even rot in the tubers of the Irish potato when they have been weakened by cold or bruises or other unfavorable conditions. A few species of *Mucor* are weak parasites in animal tissues. *Choanephora* mold attacks and destroys the blossoms of squash and pumpkin (*Cucurbita*) and of *Hibiscus*..

Mud–(1) Fine-textured soil so saturated with water as to be semifluid or in a sticky, plastic condition. (2) Freshly deposited alluvium of flood waters. (3) Flows of ash from the heavy rains occurring after volcanic eruptions.

Mud Cracks–Irregular or polygonal cracks, caused by shrinkage from drying, on the bare surface of clay soil or on flats of freshly deposited alluvium.

Mud Flat–Mud, or very fine sediments, built up under water at the mouth of an estuary along a seacoast. It is usually bordered inland by saltwater marsh and exposed only at low tide. Such land when reclaimed has a high agricultural potential. See Tidal Flat.

Mud-Rock Flow–A great mass of soupy earth and boulders carried down a steep slope by heavy rains or spewed in torrents form the mouth of a canyon. It is a phenomenon of mountainous regions. Also called slumgullion.

Mudflow–The movement of water-saturated soil and other unconsolidated deposits downslope under the force of gravity. See Mud-Rock Flow.

Mulberry–Any of the several plants of the genus *Morus*, family Moraceae. Species are grown for their edible fruit and for foliage used as food for silkworms.

Mulch–(1) Soil, straw, wood chips, peat, or any other loose material placed on the ground to conserve soil moisture or prevent undesirable plant growth or soil erosion. (2) Material, such as straw, placed over plants or plant parts to protect them from cold or heat, as the mulch for strawberries. (3) To modify soil, as by cultivation, so that it serves as a mulch, or to supply mulching materials to the soil.

Mulched-basin Irrigation–In soil irrigation, the building of permanent basins with substantial levees and covering each basin with a mulch of organic material, such as barnyard manure, bean straw, or alfalfa hay.

Multi-–A prefix meaning many, as multiovulate, many-ovuled.

Multiple Annual Ring–In tree growth, an annual ring that consists of two or more false annual rings.

Muriate of Potash–KCl; potassium chloride containing 50 to 62 percent potash (K_2O). It may be used as a fertilizer, singly or in combinations with other fertilizers.

Muskeg–Areas with acid, organic soils and very poor drainage. The swamp or bog usually contains ericaceous plants such as leatherleaf and bog rosemary. Most muskeg bogs quiver when walked on.

Muskrat–*Ondatra zibethicus*, an aquatic rodent. Raising and trapping muskrats for their fur provides a supplemental source of income for some farmers in parts of the United States. The propagation of this animal is one of the multiple uses of certain kinds of marsh lands. Also called marsh rabbit.

Musquash–The dressed skin or prepared fur of the American muskrat. See Muskrat.

Mutabilis–Variable.

Mutable–Changeable.

Mutagen–A chemical, physical and/or radioactive agent that interacts with DNA to cause a permanent, transmissible change in the genetic material of a cell. See Teratogen.

Mutant, Mutation–A variant, differing genetically and often visibly from its parent or parents and arising rather suddenly or abruptly. Mutation can occur naturally or can be induced by radiation (x rays, gamma rays, or thermal neutrons) or chemically by ethyldemethyl sulfate.

Mutual Drainage–A soil drainage system agreed to and of benefit to all persons involved.

Mutualism–Dependency of two organisms upon each other. Among insects, an example is furnished by the cornfield ant and the corn-root aphid.

Mutualistic–Designating a mutually beneficial relationship between organisms; symbiosis, e.g., Rhizobium bacteria and compatible legumes.

Mycelial Fan–A thin layer of fungous growth composed of mycelial strands spread in a fan-shaped pattern, occurring either in a layer over the surface of decaying timbers or as small patches between bark layers or between the bark and wood of trees.

Mycology–The science dealing with fungi.

Myrmecophilous–(1) Designating insects, other than ants, that live in ant hills. (2) Designating plants that live symbiotically with ants or are fertilized by them.

Myxomatosis–A viral disease of the pox group, transmitted by vectors, nonlethal to the South American rabbit, its natural host, but lethal to most European rabbits. Tested by the Commonwealth Scientific and Industrial Research Organization as a method of European rabbit control in Australia and in many European countries in the 1950s. The virus was successful in killing unwanted wild rabbits.

Nanism–Dwarf growth.

Nannoplankton–Unattached aquatic organisms that are so small that very high magnification with the microscope is required to make them clearly visible. The magnification commonly used for them is 430 to 1,200 x.

Naphthalene–A chemical derived form coal tar and widely used as a soil fumigant, as a repellent, and in the form of moth balls to repel clothes moths.

Naphthol–$C_{10}H_7OH$; either of the two coal tar derivatives from naphthalene, alpha-naphthol and betanaphthol. Alpha-naphthol is a pleasant-smelling solid with a melting point of 122°F (50°C). It is the standard material for impregnating tree bands used to control the codling moth. The beta-naphthol is an antiseptic.

Nappe–(1) A sheet of water flowing over a weir or dam. (2) A thin sheet of water flowing over the soil surface. (3) A thin lava flow.

Narrow Base Terrace–A terrace for controlling soil erosion. It is similar to a broad-base terrace in all respects except the width of ridge and channel: the base of a narrow terrace is usually 4 to 8 feet wide. it is subject to frequent failures and has not been widely accepted in the United States. See Broad Base Terrace, Nichol's Terrace.

National Forest System–Units of federally owned forest, range, and related lands throughout the United States and its territories dedicated to the long-term benefit for present and future generations. The National Forest System includes all national forestlands acquired through purchase, exchange, donation, or other means, the National grasslands, and land utilization projects administered under Title III of the Bankhead-Jones Farm Tenant Act and other lands, waters, or interests therein which are administered by the Forest Service or are designated for administration through the United States Department of Agriculture Forest Service as a part of the system.

National Grassland–Land, mainly grass and shrub cover, administered by the United States Department of Agriculture Forest Service as part of the National Forest System for promotion of grassland agriculture, watersheds, grazing, wildlife, and recreation.

National Park–A large area of land owned by the federal government and managed in the public interest, which is set aside for its scenic beauty, human interest, and value for recreation.

Native–(1) Designating a plant that grows naturally in a country or region; one not introduced by people. (2) Designating animals, as cattle, hogs, and horses, which, though originally introduced into a region, have lost some of their original characteristics or have gone wild: a scrub or mongrel. (3) Designating an unbranded beef hide. See Feral, Indigenous.

Native Disease–A disease caused by an indigenous organism.

Native Forage–Indigenous vegetation suitable as feed for livestock and game.

Native Species–A species that is a part of an area's original fauna or flora.

Natric Horizon–A clayey soil horizon formed in an arid environment and containing more than 15 percent exchangeable sodium content.

Natural Area–In the United States, an area permanently preserved in unmodified condition as representative of the virgin growth of a major forest or range type, primarily for the purposes of science, research, and education. Timber cutting and grazing are prohibited.

Natural Boundary–Any feature not made by people, such as a river, or a mountain ridge, which separates states, countries, or tracts of land.

Natural Control–Nature's method of maintaining a biotic balance, as in the reduction of harmful insects through the action of heat, cold, rain, drought, parasites, predators, and disease. See Integrated Pest Management.

Natural Enemy–In nature, any organism that preys or feeds upon another. A natural enemy may be introduced by people for biological control. See Myxomatosis.

Natural Habitat–The place a plant or animal naturally occupies; natural environment.

Natural Immunity–Immunity to disease, infestation, etc., that results from qualities inherent in plants, animals, or people.

Natural Land Type–A unit of land area capable of being mapped upon a combination or integration of natural characteristics, such as soils, vegetation, relief, and climate, as contrasted with types based upon use, money value, or modifications by people. See Land Type.

Natural Landscape–That aspect of the land surface resulting from the effect of relief, water, and vegetational features, produced by nature, which is essentially unmodified by people's activities. See Landscape.

Natural Levee–A low ridge, generally of sands, on the rim of the bank of a large stream, formed by the initial deposition of sediment in the natural flooding of the bottomland.

Natural Pan–A soil formed in nature that has a clay pan horizon, a fragipan horizon, or a horizon cemented by iron, aluminum, silica, calcium carbonate, gypsum, or colloidal humus. See Induced Pan.

Natural Reproduction–Renewal of plants by self-sown seeds, sprouts, rhizomes, etc.

Natural Reseeding–In range management, the restoration of depleted grazing land to permit natural revegetation by reseeding of desirable forage species.

Natural Resources—The elements of supply inherent to an area that can be used to satisfy needs of people, including air, soil, water, native vegetation, minerals, wildlife, etc.

Natural Selection—A natural process by which less-vigorous plants and animals tend to be eliminated from a population in an area without leaving enough descendants to perpetuate their traits.

Natural Selection Theory—A theory of evolution, propounded by Charles Darwin in the nineteenth century, which postulates that the distinctive characteristics of fitness can be inherited. Also called the survival of the fittest theory.

Natural System (Drainage)—The system of land drainage in which the tile lines are laid in the natural drainage depressions. Also called random system.

Natural Thinning—The death of trees in a stand as a result of competition.

Natural Water—Water (H_2O) as it occurs in nature with extremely variable proportions of solid, liquid, and gaseous materials in solution and suspension. In common usage, the term *water* refers to natural water rather than to chemically pure water, deionized water, or distilled water.

Natural Watercourse—A natural stream that flows in a well-defined bed or channel. It is distinguished from an artificial watercourse, such as a dug drainage ditch or a canal.

Naturalization—The process of successful adaptation by plants and animals introduced to an area where they are not indigenous.

Naturalized Disease—A disease whose causal organism is of foreign origin but which has become established in a new habitat and has reached a state of balance with native organisms.

Naval Stores—Products, such as turpentine and rosin, obtained from the distillation of crude pine resin.

Navigable Stream—A stream that maintains an average width of 30 feet from its mouth upstream.

Neap Tide (Low Tide)—A tide having about 10 to 30 percent less range than the average, occurring about the time of quarter moons.

Neat's Foot Oil—An oil obtained by boiling the feet and bones of calves and other cattle; used as a fine lubricant, for dressing leather, and in hoof ointments.

Nebraska Potash—Salts obtained from the brines of lakes in the semiarid regions of western Nebraska that contain from 20 to 30 percent of K_2O in the form of carbonate and sulfate of potassium. Not used extensively as a fertilizer.

Neck—(1) In humans and animals, the connecting link between the head and the body. (2) To tie cattle neck-to-neck (western United States). (3) (land) An elevated, narrow strip of land between two somewhat parallel streams, or water bodies; a promontory; a peninsula; an isthmus.

Necros—(Greek) Dead.

Needle Cast—A fungal disease of spruces and other conifers caused by *Lophodermium filiforme*, family Phacidiaceae; characterized by the lower branches of red spruce and black spruce being completely defoliated.

Needle-leaved Deciduous—Woody gymnosperms (trees or shrubs) with needle-shaped or scalelike leaves that are shed during the cold or dry season; e.g., baldcypress (***Taxodium distichum***).

Needle-leaved Evergreen—Woody gymnosperms with green, needle-shaped, or scalelike leaves that are retained by plants throughout the year; e.g., black spruce (***Picea mariana***).

Negligible Residue—The amount of a pesticide that is allowed on a food or feed crop at harvest. The trace amount that is on the crop is the result of indirect contact with the chemical.

Nematocide—Any substance used to kill parasitic worms called nematodes, abundant in many soils. Many nematodes attack and destroy plant roots. See Nematode.

Nematode—(1) Microscopic, wormlike, transparent organisms that can attack plant roots or stems to cause stunted or unhealthy growth. (2) Any of the round, threadlike, unsegmented animal worms of the phylum Nematoda, ranging in size from microscopic to 1 meter long. Nematodes may be saprophytes or parasites of plants and animals. They are responsible for important animal and plant diseases resulting in much economic loss. The animal parasites are designated as roundworm of horses and swine, and hookworm of people. Also called eelworms, roundworms, and nemas. See Root Knot.

Nematology—A branch of zoology concerned with the study of nematodes usually including their relation to plants and animals.

Nemoral—Of, or pertaining to, woods or forest, as nemoral culture.

Neoeluvium—A soil developed from the eluviation and illuviation of recent deposits, such as alluvium.

Neonicotine—Anabasine, an insecticide with acute toxicity to mammals.

Neontology—Biology; the study of existing life.

Neritic—Of or pertaining to the shallow waters adjoining the seacoast; e.g., neritic fish.

Nerve Poison—A poison that is soluble in tissue lipoids in contrast to respiratory poisons, such as cyanide gas, and physical poisons, such as oils and dusts.

Net Increment—In forestry, the addition to tree growth that represents an increase in usable timber.

Neutral Soil—A soil that is not definitely acid or alkaline; specifically, a soil with a pH between 6.6 and 7.3. See pH.

Neutron—See Atom.

Neutron Probe—A field device used for measuring soil water percentage. As neutrons emitted from the probe collide with hydrogen in water they are deflected and slowed. The slowed neutrons are deflected to the counter. The more water in the soil, the more the neutrons measured by the meter.

New Ground—Land recently cleared of shrubs, trees, grasses, forbs, stones, stumps, etc., and put under cultivation for the first time.

New Wood—In horticulture and forestry, the current year's growth.

Niche–A term used to describe the status of a plant or animal in its community, that is, its biotic, trophic, and abiotic relationships. All the components of the environment with which an organism or population interacts.

Nichol's Terrace–A once common small terrace for disposal of surplus water. It has a comparatively deep, narrow channel and a low, flat ridge with a slope that merges quite closely with the downhill side. Modern terraces are usually broad-based to allow machinery to operate over them. See Broad-base Terrace, Narrow-base Terrace.

Nicotine–A very poisonous alkaloid with an odor of pyridine from the leaves of the tobacco plant and also produced synthetically. Formerly widely used as an external parasiticide and as an insecticide.

Night Soil–Human feces and urine, an unsanitary potent manure for cropland. It is rarely used in the United States but remains in use in some other parts of the world, especially the Orient.

Nigrescent–Blackish.

Nimbus–(Latin) Rain cloud.

Nit–The egg of a louse.

Niter–Saltpeter; a mineral; KNO_3.

Nitrate–NO_3; N combined with oxygen. The N form most used by plants. NO_3 is a gas that does not exist alone in fertilizer but is combined, as in ammonium nitrate. All nitrates are water-soluble and, when applied in surplus, move with surface waters to contaminate groundwater.

Nitrate of Ammonia–See Ammonium Nitrate.

Nitrate of Lime–See Calcium Nitrate.

Nitrate of Potash–KNO_3; the potassium salt of nitric acid that occurs in nature only in small quantities. The pure compound contains about 46.6 percent potash (K_2O). Used as a fertilizer or as a component in fertilizer formulations, commercial grades contain about 13 percent nitrogen and 44 percent potash.

Nitrate of Soda–

Nitre–(1) A precipitate of malic acid formed in making maple syrup. (2) Nitrate of potash (saltpeter), KNO_3, used as a fertilizer and in the manufacture of explosives. (3) Sodium nitrate, or chile saltpeter.

Nitric Acid–HNO_3; a strong mineral acid which, combined with metals or alkalies, forms nitrates. It is now made synthetically on a large scale by passing ammonia (NH_3) and air through a platinum gauze catalyst, whereby the ammonia oxidizes. Nitric acid is used in the production of nitrate fertilizer compounds, including nitric phosphates. Some nitric acid is used as an oxidant for carbonaceous material that causes the black color in liquid fertilizers made with commercial phosphoric acid.

Nitric Phosphate–Fertilizers made by processes involving treatment of phosphate rock with nitric acid or a mixture of nitric and/or sulfuric or phosphoric acid, or all three acids, usually followed by ammoniation.

Nitrification–The process of formation of nitrates (NO_3) in the soil, from other compounds of nitrogen, by various microorganisms. Many plants, especially during their early growth, can utilize nitrogen in the form of ammonium, but all plants can metabolize nitrate nitrogen. Nitrification is very rapid in warm, moist, neutral soils. It practically ceases when the soil temperature falls below 4°C (40°F).

Nitrite–NO_2; a partially oxidized form of nitrogen containing two atoms of oxygen for each atom of nitrogen. Soil and rumen bacteria can change nitrite-nitrogen to nitrate-nitrogen.

Nitro-Chalk–See Ammonium Nitrate-Limestone.

Nitrobacter–A genus of microorganisms occurring in productive soil that oxidize nitrites to nitrates. This genus includes two species: *Nitrobacter agilis* and *N. winogradskyii*.

Nitrobacteria–Minute, rodlike bacilli that convert nitrogen into soluble nitrates in soils.

Nitrogen–N; a gas that occurs naturally in the air and soil, where it is converted into usable forms for plant use by bacteria and other natural processes. This nutrient is a constituent of proteins and is vital to plant-growing processes. Nitrogen can be added to the soil in any of three fertilizer forms: as urea, ammonia, or nitrates.

Nitrogen Carrier–Any of a great number of inorganic or organic compounds that contain nitrogen and can be classed as a fertilizer.

Nitrogen Cycle–The circulation of nitrogen in nature. Dead organic matter is converted to ammonium compounds during decay in the soil. These are converted to nitrates, through nitrification as a result of bacterial action, which can be utilized again by plants. *Nitrosomonas* oxidize ammonium ions to nitrites and *Nitrobacter* oxidize nitrites to nitrates.

Nitrogen Excesses–The presence of nitrogen, as nitrates or ammonium salts in soils, in excessive amounts, which results in injury to plants, as burning or stunting. The excess may occur naturally or may result from excessive applications of nitrogenous fertilizers. In tree fruits, it may cause excessive vigor resulting in reduced fruiting.

Nitrogen Oxides–Gases formed in great part from atmospheric nitrogen and oxygen when combustion takes place under conditions of high temperature and high pressure; e.g., in internal combustion engines; considered major air pollutants. Written as NOx when the oxides have not been identified, or as NO-nitric oxide and NO_2-nitrogen dioxide.

Nitrogen Solutions–Solutions of nitrogenous fertilizer chemicals in water. Solutions are used in manufacturing liquid fertilizers and are applied to the soil either with special applicators or in irrigation water. A typical nitrogen solution is characterized by a code number, e.g., 414 (19-66-6) indicating that it contains 41.4 percent total nitrogen, 19 percent free ammonia, 66 percent ammonium nitrate, 6 percent urea, and by difference, 9 percent water.

Nitrogen-fixing Bacteria–Species of the genus *Rhizobium*, family Rhizobiaceae, that live symbiotically in the root nodules of leguminous plants, upon which they are dependent. They are capable of extracting nitrogen from the air and converting it to a form utilizable by the plant. Species of the genus *Azotobacter* and *Clostridium* are free-living and act independently (nonsymbiotically). Some plants that are not legumes also have a species of symbiotic bacteria. See Nodule.

Nitrogenous Materials–Materials that contain nitrogen, whether in organic or inorganic form. Some are readily available, others, even though they contain fairly high percentages of total nitrogen, such as hoof, hair, and plastic and leather scraps, are not readily available.

Nitrosamine–Any of a group of N-nitroso derivatives of secondary amines. Some nitrosamines are carcinogenic.

Nitrosobacterium–A microorganism that oxidizes nitrites (NO_2) to nitrates (NO_3).

Nitrosospira–A genera of two species of microorganisms that oxidize ammonia (NH_3), to nitrate (NO_2) very slowly: ***Nitrosospira antarctica*** **and *N. briensis*.**

Niveus–Snowy-white.

No-tillage–A system of growing crops in which the seeds are planted in the ground with little or no disturbing of the soil. Weed and pest control is accomplished through the use of chemicals. This method of farming is used to control erosion. See Conversation Tillage.

Noble Fir–*Abies procera*, family Pinaceae; a huge evergreen tree grown in the United States West Coast as an ornamental and for its lumber, used in general construction, boxes, pulp, and planing-mill products. Native to North America.

Nocturnal–Of the night; e.g., a nocturnal plant is one whose flowers open at night; a nocturnal animal is one which is active at night. See Diurnal.

Noise Pollution–(1) Sound that lacks agreeable quality, is noticeably loud, harsh, or discordant. (2) The addition of energy in the form of sound to the environment beyond what would naturally occur. The degree of noise pollution is measured in terms of intensity, duration, frequency of occurrence, and wavelength or pitch. (3) Sound that unreasonably interferes with the enjoyment of life or property.

Nomadism–A primitive unsettled pastoral type of existence, especially in steppe and desert regions; the life led by people who wander in search of grazing for their livestock and food for themselves.

Noncommercial Forest Land–Forestland withdrawn from its commercial use for timber because (a) it is utilized for such purposes as parks, game refuges, military reservations, or reservoir protection or (b) its poor growing conditions or inaccessibility render its commercial use unprofitable.

Nondegradable Wastes–Substances that are not changed in form and/or reduced in quantity by the biological, chemical, and physical phenomena characteristic of natural waters. Although nondegradable wastes may be diluted by receiving water, they are not reduced in total quantity. (Sometimes known as nonbiodegradable wastes.)

Nonionic Surfactants–Substances that have neither a positive nor a negative electrical charge and are compatible with most pesticides.

Nonpoint Pollution Sources–Those sources of pollution that are diffuse in both origin and in time and points of discharge, and depend heavily on weather conditions such as rainstorms or snowmelt. Pollutants can originate from natural source areas as well as areas affected by people's activities.

Nonporous Wood–Wood devoid of pores or vessels, as the conifers and a few broadleaf species.

Nonrenewable–Natural resources that once used up are gone forever.

Nonsaline-Sodic Soil–A soil that contains sufficient exchangeable sodium to interfere with the growth of most crop plants and does not contain appreciable quantities of soluble salts. The exchangeable-sodium-percentage is greater than 14, and the electrical conductivity of the saturation extract is less than 4 millimhos per centimeter (at 25°C). The pH reading of the saturated soil paste is usually greater than 8.5.

Nonselective Herbicide–A herbicide that will kill any plant with which it comes in contact. See Selective Herbicide.

Nonsymbiotic Nitrogen Fixation–Fixation of nitrogen in soils by free-living microorganisms in contrast to that fixed through the agency of symbiotic organisms. See Nitrogen-fixing Bacteria.

Nontoxic–Not poisonous to plant or animal. See Toxic, Toxicity.

Normal Diameter–In logging, the diameter of a tree at breast height above the root swell.

Normal Erosion–The erosion that takes place on the land surface in its natural environment undisturbed by human activity. It includes (a) rock erosion, or erosion of rocks on which there is little or no developed soil, as in stream channels and rocky mountains, and (b) normal soil erosion, or the erosion of the soil under its natural condition or native vegetative cover undisturbed by human activity.

Normal Fire Season–A forest fire season in which weather, rated fire danger, and number and distribution of fires are average.

Normal Kill–In wildlife management, the average number of animals or birds that can be killed yearly on a particular area without diminishing future productivity.

Normal Moisture Capacity–The amount of water that is normally held by a soil after the free or gravitational water has drained. See Field Capacity.

Normal Soil–A soil having a profile in equilibrium with the forces of the environment, such as native vegetation and climate. Usually developed on gently undulating upland, with good drainage, from parent material (not of extreme texture or chemical composition) in place long enough for the soil-forming process to exert their full effect.

Normal Stand–In forestry, a stand of trees fully stocked, in proper growing condition and having normal increment conforming to a yield table.

Norther–A strong wind from the north that usually brings cold, sleet, rain, or dust, especially in the Plains of the southwestern United States.

Northern Leaf Blight–A fungal disease of corn, Sudangrass, and sorghums caused by ***Helminthosporium turcicum***, family Dematiaceae; characterized by linear or irregular, elliptical, watersoaked lesions on the leaves. The lesions turn brown, and then black. Sometimes the entire leaf dies.

Northern Rat Flea–*Nosopsyllus fasciatus*, family Dolichopsyllidae; a vector of diseases among people and animals, especially bubonic plague.

Northern Red Oak–*Quercus rubra* var. ***borealis***, family Fagaceae; a northern variety of red oak. Lumber from the northern red oak is valuable in the making of furniture.

Northern White Cedar–*Thuja occidentalis*, family Pinaceae; an evergreen tree valuable for timber, posts, and as an ornamental. Native to North America.

Norway Pine–See Red Pine.

Norway Rat–*Rattus norvegicus*, family Muridae; a native rodent of northern Europe, now common in many civilized countries and harmful to people in various ways. Also called house rat, brown rat, wharf rat, sewer rat.

Norway Spruce–*Picea abies*, family Pinaceae; a hardy evergreen tree, probably the most widely grown ornamental evergreen in America; it has a large number of varieties differing in height, form, and foliage. Dwarf varieties are 1 to 2 feet high while other varieties reach a height of 150 feet.

Notch–(1) The opening in a dam or spillway for the passage of water. (2) A gap, pass, or defile between mountains (chiefly in northeastern United States). (3) An undercut; in logging, a cut in the trunk of a tree to govern direction of fall. (4) A gap cut into the ear of an animal for identification.

Notifiable Disease–(reportable) Any disease that must be reported to the government health authorities.

Novale–Land newly plowed and brought under cultivation.

Novia–An ancient Mideastern system of irrigation that uses the force of the stream or river current to turn a paddlewheel water lift.

NPK–Symbols for three primary nutrients needed by plants. N is for total nitrogen, P for available phosphorus (reported as P_2O_5), K for water-soluble potassium (reported as K_2O). Percentage of these nutrients in a fertilizer package is always listed in that order and is known as the fertilizer grade. See Fertilizer Grade.

Nucleus–(1) The central portion of the cell protoplast surrounded by a very thin membrane. It consists of nucleoplasm and includes within itself variously arranged chromatin, nuclear sap, and nutritive substances. It is of crucial significance in metabolism, growth, reproduction and the transmission of the determiners of hereditary characters. (2) A small colony of bees used in queen rearing or in pollination work in greenhouses. (3) A central core around which material collects or is grouped.

Nuclide–Any species of atom that exists for a measurable period of time whose nuclear structure is distinct from that of any other species. Thus each isotope of an element is a separate nuclide.

Nudation–The removal of all of the vegetation of an area.

Nuisance Birds–Starlings, sparrows, crows, redwinged blackbirds, pigeons, etc., that damage the farm, orchard, and garden crops, or become objectionable because of their roosts, or the grain they consume.

Nullah–(Hindi) Watercourse, wet or dry.

Nurse Crop–A companion crop grown to protect some other crop sown with it, as small grain is sometimes seeded with clover. The nurse crop can sometimes become competitive and harmful also.

Nut–(1) An indehiscent, one-celled and one-seeded, hard and bony fruit, as the acorn of ***Quercus***. (2) As frequently and loosely used, a drupe with relatively thin fleshy exocarp and a large stone (pyrene), or the pyrene itself, as the walnut (***Juglans***) and hickory nut (***Carya***).

Nut Pine–(1) Any of various pines that bear edible nuts. (2) ***Pinus cembroides*** var. ***edulis***, family Pinaceae; a hardy variation of the Mexican pinyon pine. This term may include var. ***monophylla***.

Nut Structure–The aggregation of fine particles of soil into small subangular, blocky, or roughly rounded, masses of definite size or shape. See Soil Structure.

Nutlet–A small or diminutive nut, similar to an achene but with a harder and thicker wall.

Nutria–*Myocaster coypus*; an aquatic rodent valued for its pelt that is produced to a limited extent in fur framing in parts of the United States. The animal exists in the wild state, as an escape, in the marshes of Louisiana and Texas, and is trapped annually in considerable numbers. Native to South America.

Nutrient–(1) A substance that favorably affects the nutritive processes of the body; a food. (2) An element or compound in a soil that is essential for the growth of a plant. (3) In stock feeding, any feed constituent or group of feed constituents of the same general composition that aids in the support of life, as proteins, carbohydrates, fats, minerals, and vitamins.

Nutrient Cycle–The circulation of nutrient elements and compounds in and among the soil, parent rock, streams, plants, animals, and atmosphere.

Nutrient Level–(1) In soils, the amounts and proportions of plant nutrients, such as phosphorus, potassium, and nitrogen in available forms. (2) Specifically, the concentrations of any particular nutrient in the ration of animals.

Nutrient Solution–A solution in which plants may be grown, made by dissolving salts containing the essential elements for growth in water, which is used in hydroponics and in experimental work in laboratories. See Hydroponcis.

Nutriment–Nourishment; nutritious substances; food.

Nutrition–The sum of the processes by which an organism utilizes the chemical components of food through metabolism to maintain the structural and biochemical integrity of its cells, thereby ensuring its viability and reproductive potential.

Nymph–A stage in the development of some insects and related forms, such as ticks, immediately preceding the adult stage.

O Horizon–(Soil) Organic horizons above mineral soil. Horizons: (1) formed or forming above the mineral part; (2) dominated by fresh or partly decomposed organic material; and (3) containing more than 30 percent organic matter if the mineral fraction is more than 50 percent clay or more than 20 percent organic matter, if the mineral fraction has no clay. Proportional percentages are between these extremes. See Organic Soil Horizon.

Oak–Any deciduous, evergreen, or partly evergreen tree or shrub of the genus *Quercus*, family Fagaceae. Species are grown as ornamentals, as shade and lawn trees, for bark, lumber, acorns, and browse. See Oak-leaf Poisoning.

Oak Barrens–In the eastern United States, land that supports only a scrubby or sparse growth of oaks, especially scrub oak, *Quercus ilicifolia*, and blackjack oak, *Q. marilandica*; land of low agricultural value.

Oak Opening–A prairie area in an oak forest, or an area only very sparsely occupied by oaks.

Oak Skeletonizer–*Bucculatrix ainsliella*, family Lyonetiidae; an insect whose larvae eat oak leaves, leaving a skeletonized surface.

Oak Wilt–A very serious fungal disease of various species of oak in the North Central states of the United States, caused by *Endoconidiosphora fagacearum*, known in some publications by the name of its imperfect stage, *Chalara quercina*, family Ophiostomataceae. It is related to the fungus causing the Dutch elm disease, and like this disease is carried by various insects that feed on the pads of fungus mycelium that are formed under the bark and project outward through cracks. It is also spread by the natural grafting of the roots of diseased and healthy trees.

Oak-leaf Poisoning–A digestive ailment, observed among cattle on the ranges of the southwestern United States; attributed to excessive ingestion of oak leaves in the spring, before grass makes its appearance. In a few instances the ailment is fatal and otherwise may cause stunted growth.

Obligate Aerobe–An organism that lives only in the presence of free oxygen.

Obligate Anaerobe–An organism that lives only in the absence of free oxygen.

Obligate Hydrophytes–Species that are found only in wetlands, e.g., cattail (*Typha latifolia*), as opposed to ubiquitous species that grow either in wetland or on upland, e.g., red maple (*Acer rubrum*).

Obligate Parasite–An organism that develops and lives only as a parasite, and is confined to a specific host.

Obligate Saprophytes–Microorganisms not related to living cells that secure their nutrients from dead organic tissue or inorganic materials.

Obligate Symbiont–An organism that is dependent on mutual relations with another for its existence.

Obnoxious Weed–(1) In dairying, any weed which, ingested by a cow, imparts an obnoxious odor or flavor to milk, cream, or butter. (2) Any serious weed pest. See Noxious Weed.

Oceanic Climate–The characteristic type of climate of land areas near oceans, which have a moderating influence on the vegetation.

Ochric Epipedon–A surface soil horizon that is too light colored or too thin to be a mollic epipedon. See Soil Taxonomy.

Octillo–*Fouquieria splendens*, family Fouquieriaceae; a perennial, tall, woody, cactuslike plant producing leaves for short periods whenever moisture conditions are favorable, and grown for its showy, scarlet flowers. Characteristic of certain soil conditions in the desert regions of the southwestern United States and adjacent Mexico. Native to North America. Also called coachwhip, vine cactus, candlewood, Jacob's staff, Spanish bayonet.

Ocular Estimate–Estimate by sight only. (1) In forestry, the determination of the approximate volume and quantity of standing timber without the use of measuring instruments. (2) On a range, the qualitative procedures for determining the degree of cropping of forage plants. Observations of a general reconnaissance nature are made visually: (a) by examining small random plots at the end of each grazing season.

Odor–The property of a substance, receptive to the human nose, which permits pleasant or unpleasant sensations of fragrance or smells that can be recognized.

Odor Threshold–The lowest concentration of an airborne odor that a human can detect.

Odorous House Ant–*Tapinoma sessile*, family Formicidae; a household ant that is a pest over much of the United States.

Oecology–See Ecology.

Ogalla Aquifer–A large aquifer extending from Nebraska to the panhandle of Texas. It contains mostly fossil water. See Aquifer; Fossil Water.

Oil Nut–See Butternut.

Oil Palm–*Elaeis* spp. A tree grown in many African and Asian countries for the oil that it produces. Two kinds of oil are produced; a very stable oil from the fleshy part of the fruit and palm kernel oil from inside the hard kernel. Palm oil is used as a frying fat, in margarines, and in soups and gravies. Palm kernel oil is not edible but is used in paints and varnishes.

Old Growth–A forest that has never been changed by management or harvesting. This term is misapplied by many to describe any forest that appears to be old. Individual trees in this type of forest are usually over 200 years old, and there are large standing and fallen dead trees throughout the stand.

Old Soils–In pedologic terminology, soils on flat land surfaces that have fully developed profiles but are stagnant, i.e., undergoing changes with extreme slowness. (2) Soils buried by erosion debris or volcanic detritus.

Old Wood–A shoot or branch of a woody plant that is at least one year old.

Old-field Stand–A stand of trees grown up on land once used for crops or pasture.

Old-house Borer–*Hylotrupes bajulus*, family Cerambycidae; an insect that infests well-seasoned wood in buildings.

Oleoresin of Pyrethrum–A product used in insecticides, obtained by treating finely ground pyrethrum with ethylene dichloride and then evaporating.

Oligohaline–Term to characterize water with salinity of 0.5 percent to 5.0 percent, due to ocean-derived salts.

Oligophagous Parasite–A parasite capable of developing upon a few closely related host species.

Oligosaprobic Zone–The area of a stream that contains the mineralized products of self-purification from organic pollution but with none of the organic pollutants remaining.

Oligotrophic–(1) Pertaining to plants that grow in areas of poor soil with respect to the nutrients. (2) Swamps or water bodies poor in humus or plant nutrients.

Oligotrophic Lakes–Deep lakes with low nutrient content and low biochemical oxygen demand (BOD). See BOD.

Oligotrophic Peat–Peat soil that is strongly acid in reaction and low or wanting in bases, such as calcium, and other mineral plant nutrients.

Oligotrophic Waters–Waters with a small supply of nutrients; hence, they support little organic production.

Omnivorous–Designating animals that feed on both flesh and plants, as people; as applied to insects, voracious, but not necessarily omnivorous. Also called amphivorous. See Carnivorous, Herbivorous.

On the Contour–See Contour Row.

One-third-atmosphere Percentage (Soil)–The moisture percentage, dry-weight basis of a soil sample that has been air-dried, screened, wetted, and brought to hydraulic equilibrium with a permeable membrane at a soil-moisture tension of 344 centimeters of water. This retentivity value closely approximates the moisture equivalent value of many soils at field capacity.

Oophagy–The eating of eggs, said of egg-eating insects.

Oospore–A resting spore produced by sexual reproduction in the downy mildews and related fungi.

Open–(1) Designating a female animal that is not pregnant. (2) Designating a body defect of cheese in which the cheese has many mechanical openings. (3) To unfold, as a flower opens. (4) Designating a rural or wilderness area in contrast to the congestion of cities and towns.

Open Flume–An uncovered passageway for irrigation water.

Open Formula Fertilizer–A fertilizer that has both the analysis and the formula or materials used in it printed on a tag or on the bag container.

Open Range–An extensive range area where grazing is unrestricted. Also, ranges that have not been fenced into management units.

Open Weather–Pleasant weather with mild temperatures and an absence of fog or severe storms., especially during the winter.

Open Woodland–A parkland type of vegetation in which trees form a closed canopy. There is very little undergrowth.

Open-ditch Drainage–Drainage of excess water from land by open ditches as opposed to tile drainage.

Opening–(1) A treeless or very sparsely timbered area in a forested region. See Oak Opening. (2) In cranberry culture, the first swelling of a terminal bud. (3) The unfolding of a flower or boll of cotton. (4) Designating the first price offered for a commodity when a market day begins.

Optimum Condition–The ideal environment, with regard to nourishment, light, temperature, etc., for an organism's growth and reproduction.

Optimum Temperature–That certain temperature at which a particular plant or animal grows satisfactorily, other conditions being favorable for growth.

Optimum Water Content–The amount of water in a soil needed by a plant for its optimum growth, varying from 40 to 60 percent the moisture-holding capacity (field capacity).

Orchard Cover Crop–Grasses, legumes, or any herbaceous plant, grown in an orchard to form a cover to reduce soil erosion, and to improve the soil through the addition of organic matter. It may also serve to use up available nitrogen so the trees will harden for winter.

Order–(1) In botanical classification of plants, a category in between class and family. (2) In soils, one of the eleven highest categories.

Organ–A distinct part of a plant or animal that carries on one or more particular functions; e.g., a leaf, wing of a bird, etc.

Organic–(1) Produced by plants and animals; of plant or animal origin. (2) More inclusively, designating chemical compounds that contain carbon.

Organic Acids–Acids containing only carbon, hydrogen, and oxygen. Among the best known are citric acid (in citrus fruits) and acetic acid (in vinegar).

Organic Fertilizers–Although not specifically defined by the leaders of organic gardening, organic fertilizers usually include only natural, organic, proteinaceous materials of plant and animal origin and exclude synthetic, organic, nonproteinaceous materials such as urea [$CO(NH_2)_2$].

Organic Gardening (Farming)–A system of farming or home gardening that utilizes organic wastes and composts to the exclusion of chemical fertilizers. Advocates of the system teach that chemical fertilizers are injurious to health and that organic composts give higher yields, better quality and better taste of produce, less plant damage by insects and disease, reduction of weed menace, and stronger seeds that germinate better and produce successively stronger plants. None of these claims have so far been confirmed by reproducible (scientific) proof. Much publicity of the concept has resulted in a number of followers but it is significant that the leading followers of the concept do not depend entirely on agriculture for their livelihood. The organic gardening concept persists because of the basic truth that organic matter in soils is beneficial to agriculture. However, the sole scientific foundation for economic production of abundant and healthful foods consist of the liberal use of organic matter plus chemical fertilizers applied according to a soil test.

Organic Matter–(1) Matter found in, or produced by, living animals and plants, which contains carbon, hydrogen, oxygen, and often nitrogen and sulfur. (2) Any matter, defined as organic incorporated in or on the surface of soil. It may include undecomposed plant matter as well as that which is highly humified. Sometimes used interchangeably with humus.

Organic Phosphate Insecticide–A chemical compound used as an insecticide, such as tetraethyl pyrophosphate, ethyl metaphosphate, and tetraethyl dithio-pyrophosphate.

Organic Soil Horizon–A surface horizon of a soil formed from organic litter derived from plants and animals, and designated as the O horizons. These horizons do not include horizons formed by illuviation of organic material into mineral material, nor do they include horizons high in organic matter formed by a decomposing root mat below the surface of a mineral material, usually designated as O_1 and O_2.

Organic Soils–Histosols; soils having organic soil materials that extend from the surface to: (a) a depth within 10 centimeters or less of a lithic or paralithic contact, provided the thickness of the organic soil materials is more than twice that of the mineral soil above the contact; (b) any depth if the organic material rests on fragmental material and the interstices are filled with organic materials; or (c) have organic materials that have an upper boundary within 40 centimeters of the surface; (i) having a bulk density of 0.1 g/cc; (ii) the organic soil is saturated with water six months of the year. See Soil Taxonomy.

Organism–Any living individual whether plant or animal.

Organogen–Any of certain chemical elements without which organisms cannot exist: oxygen, carbon, nitrogen, phosphorus, etc.

Organoleptic–Concerning the sensory impressions, such as temperature, taste, smell, feel, sweet, sour, and salt, associated with eating and drinking.

Organophosphates–A group of nonpersistent pesticides containing phosphorus, including malathion and parathion. Malathion has low human and animal toxicity; parathion has high toxicity to humans and animals.

Oriental Rat Flea–*Xenopsylla cheopis*, family Pulicidae; a vector in the spread of plague from rats to people. It may also affect both domestic and wild animals.

Orophytes–Plants that grow in mountains only.

Osmunda Peat–A potting material made from the roots of fern of the genus *Osmunda*.

Other Forestland–Forestland incapable of producing 20 cubic feet per acre of industrial wood under natural conditions because of adverse site conditions such as infertile soils, dry climate, poor drainage, high elevation, steepness, or rockiness.

Otolith–Earstone, used by fish for its sense of balance. There is one in each plane of the semicircular canals on each side of the head, making six in all. Fishery biologists use the biggest ones to determine the age of fish.

Outcropping–A geological stratum that is exposed on the surface of the earth; as an outcropping of rock.

Outfall–The point where water is discharged from a conduit; the mouth or outlet of a drain or sewer.

Outlet Channel–In soil conservation engineering, a waterway provided to receive and carry away the runoff discharge from terrace channels and diversions. Also called terrace outlet channel.

Outwash Plain–A physical land feature of glaciofluvial origin. The plains are smooth, or where pitted are usually low in relief. The deposits are mainly sandy or coarse textured. The plains constitute a large total area of land and a large number of separate soil and land types in the glaciated region of the United States.

Ovalhead Sedge–*Carex festivella*, family Cyperaceae; a grasslike, herbaceous plant, found at elevations of 6,500 to 12,000 feet from New Mexico to California (United States) and north to Alberta (Canada). Its forage value is good and its fibrous roots afford protection against soil erosion.

Overbrowsing–Excessive cropping of shrubs or tree growth, usually by goats, sheep, or game animals.

Overburden–Material of any nature, consolidated or unconsolidated, that overlies a deposit of useful materials, ores, or coal, especially those deposits that are mined from the surface by open pits.

Overcutting–The cutting of a quantity of timber in excess of the mean annual growth.

Overfall–An abrupt change in stream channel elevation; the part of a dam or weir over which the water flows.

Overflow–(1) In cotton gins with a belt distributor system, the surplus cotton carried on belts to the feeders, which drops to the floor when the feeders are full. (2) The floodwater of swollen streams that spreads over bottomland.

Overflow(ed) Land–(1) Land that is subject to overflow or flooding; generally floodplains of rivers. (2) Loosely, swampland. (3) Legally, land that is covered by nonnavigable waters, but not including land covered by the normal flow of tides.

Overgraze–To graze land so heavily as to impair future forage production and to cause range deterioration through consequential damage to plants, soil, or both. Also called overstocking.

Overgrown–Designating abundant growth on a field, usually of some undesirable vegetation as weeds or brush.

Overhead Sprinkling System–A system of irrigating a small acreage by sprinkling from permanently installed overhead pipes.

Overirrigation–Application of water in excess of the need of a crop and of the water-holding capacity of the soil; it damages the plants in addition to wasting the water.

Overlay–Any transparency, containing supplemental information, superimposed over a map or a data sheet to show this information more clearly.

Overlength–In milling logs, an allowance for trim; an extra length left on a log so that it may be trimmed to an exact length.

Overseeding–(1) Seeding in a crop already established, such as seeding a forage crop in the spring on a field of winter wheat established the previous fall. (2) Using an amount of seed in excess of that necessary to assure a good stand.

Overstocked–(1) Designating a condition of a stand of trees or of a forest, in which there are more trees than normal or full stocking would require. The overstocking may be to such a degree that growth is slowed down and many trees, including dominants, are suppressed. (2) Designating a pasture, range, or grazing game area that has more animals on it than the vegetation of the area will support. See Overgraze. (3) Designating a locality in which there are too many bees.

Overstopped Trees (Suppressed)–Trees with crowns entirely below the general level of the crown cover receiving no direct light either from above or from the sides.

Overstory–That portion of trees in a forest stand that forms the upper crown cover. Also called overwood.

Oversummer–To live through the summer.

Overtopping–In river flood control, the passage of water over the top of a levee during a flood.

Overturn–(1) The mixing of the waters of a lake in the temperate zone when the summer thermal stratification ends. Most northern lakes also experience a spring overturn. (2) The period of mixing of previously stratified water masses that occurs in the spring and autumn when water temperatures in the lake are uniform.

Overuse–The utilization of key forage species on a range to such an extent that they fail to reproduce and maintain themselves over a period of time.

Overwinter–To hibernate; to live in a dormant condition through the winter months.

Oviposition–The process of laying an egg.

Ovipositor–A tubular structure in female insects used for depositing its eggs.

Ovoviviparous–Refers to animals who produce eggs that are incubated inside the body of the dam and hatch inside the body or shortly after laying. See Oviparous, Viviparous.

Oxbow–(1) A U-shaped wooden collar, part of the harness of an ox, placed on the neck of the animal with the open ends passing through the yoke. See Yoke. (2) A bend in the channel of a river resembling an oxbow. (A large river may cut across the narrow neck of a nearly circular meander and thus leave an abandoned channel, as a lake, swamp, or bayou.) Also called oxbow meander.

Oxic Horizon–A highly weathered subsurface soil horizon (endopedon) high in iron and aluminum formed in humid, tropical, old land surfaces. See Oxisols, Soil Order.

Oxidase–Oxidizing enzymes. See Enzyme.

Oxidation Pond–A person-made lake or body of water in which wastes are consumed by bacteria. An oxidation pond is the same as a sewage lagoon.

Oxisols–One of the eleven soil orders, they are soils with residual accumulations of inactive clays, free oxides, kaolin, and quartz. They occur mostly in tropical climates on old land surfaces.

Oxygen–The chemical element O; a colorless, odorless gas. The most abundant element in the earth's crust. It accounts for about 47 percent of all elemental material. It is essential in the growth of all crops and for the respiration of most forms of life.

Oxygenation–The absorption by water of elemental oxygen that has: (a) been released into the water by aquatic plants as a waste product of photosynthesis; (b) come from the atmosphere.

Oyster–The tenderloin muscle of the poultry carcass. There are two, one on each side, located just in front of the hipbones.

Oyster Shells–Shells of the marine bivalve, genus *Ostrea*. They are nearly pure calcium carbonate in composition; when finely ground, they make good liming material for soils and a mineral feed for livestock and poultry as a source of calcium.

Ozarks–A high plateau region that embraces a large part of southern Missouri and northern Arkansas as well as a small part of Oklahoma and Kansas (United States). The terrain is generally hilly and deeply dissected by streams. The region contains fertile bottomlands and productive soils on slopes, but the greater part is inferior agriculturally because of steep slopes and cherty soils. The productive soils of most bottomlands have recently been covered with reservoirs. The most productive soils remaining lie on relatively level ridge tops. The area is a prime retirement and recreational "paradise." It has four seasons, adequate and evenly distributed rainfall, and many reservoirs for water-related sports. The greater part of the land is primarily suited for forestry and grazing of beef and dairy cattle.

Ozone–O_3; a pungent, colorless, toxic gas. As a product of the photochemical process, it is a major air pollutant.

P–Chemical symbol for phosphorus.

P_2O_5–Phosphorus pentoxide; in current fertilizer labeling, the amount of P in fertilizer is expressed as percent of P_2O_5. To find actual P, multiply P_2O_5 equivalent by 0.44.

P/E Ratio–Precipitation/evaporation ratio. This is an aridity index.

Pacific Coast Tick–*Dermacentor occidentalis* family Ixodidae; a tick that is one of the vectors of Rocky Mountain spotted fever.

Pacific Hemlock–*Tsuga heterophylla*, family Pinaceae; used for its lumber and as pulp, for general construction and planing-mill products. Native to the Pacific Coast from Alaska to California. Also called western hemlock (forestry name), West Coast hemlock, hemlock spruce.

Paddy Soil–The soil of a rice paddy: it is wet and subject partially to anaerobic conditions.

Paha–Low, rounded, glacial ridges of silt and clay with a loess covering in the area of Iowan glacial drift in northeastern Iowa.

Painted Hickory Borer–*Megacyllene caryae*, family Cerambycidae; an insect pest that infests hickory, walnut, butternut, Osage orange, and hackberry.

Paleo-–A combining form meaning old, ancient, used to denote: (a) remote in the past, (b) early, primitive, archaic. Before vowels usually pale-, palae-.

Paleopedology–That branch of pedology that deals with buried soils, fossil soils, or those formed in past eras.

Palmetto Flatwoods–A kind of land, constituting extensive areas in Florida (United States), in which the ground cover or vegetation is characteristically sawplametto, although various other plants are present and the trees are mainly pines. The land is periodically very wet and very dry. The soils are leached sands commonly underlaid by organic hardpans.

Palouse–A region of about 8,000,000 acres in eastern Washington adjacent to part of Idaho and the north-central part of Oregon. It has a semiarid and subhumid climate, and undulating to hilly and dunelike topography, productive silty soil, mostly deep, but locally severely eroded and shallow over basalt rock. it was originally grassland characterized by bunch wheatgrass but is now largely in farms with wheat as the principal crop.

Palustrine–(1) Designating a swampy or boggy soil. (2) Designating a plant that grows in a swamp.

Pampas–Vast, rolling plains of grasses without trees, especially in Argentina.

Pan–(1) A compact, indurated, subsurface layer in a soil profile. See Hardpan. (2) A depression in a saltwater marsh. (3) In forestry, a wide, steel sheet used to support the end of a log in skidding. (4) A flower pot about one-half the depth of the standard flower pot; used for growing shallow-rooted plants and bulbs. Also called bulb pan, seed pan. See Fragipan.

Pancake Land–Flat, prairie land with low-lime clay soils that has a compact baked surface (eastern Texas, United States).

Pandemic–Designating a disease or organism of worldwide distribution; widely epidemic.

Panphytotic–Designating worldwide distribution of a plant disease.

Panplane–A nearly flat surface brought about by the lateral erosion of streams, which erodes the divides and causes a coalescing of all of the flood plains of a region.

Paper Birch–*Betula papyrifera*, family Betulaceae; a deciduous, average-sized tree grown as an ornamental in the United States for its white, paperlike bark. Native to North America. Also called canoe birch, white birch.

Paper Mulch–Heavy paper rolled out in strips to cover the soil as a mulch between rows of plants, or squares of paper with slits through which individual plants grow. The paper may be as effective as usual mulches of straw, manure, leaves, etc., in conserving moisture and preventing weed growth.

Paper Pulp–The product used in the manufacture of paper that is made by mechanical and chemical treatment of wood or other cellulose material.

Paradichlorobenzene–A white crystalline substance used in beehives to fumigate combs and to repel wax moths. It is also used as a soil fumigant.

Paraeluvium–The residual product that results from the eluviation of the weathering products of sedimentary rocks. See Eluviation.

Parasite–An organism that lives at least for a time on or in and at the expense of living animals or plants. Some diseases of people and animals are caused by parasites ordinarily classified as protozoan, helminthic, and anthropod species. There are also innumerable species of plant parasites.

Parasitic Seed Plant–A plant that obtains part or all of its sustenance from other plants; e.g., dodder and mistletoe.

Parasiticide–An agent that kills parasites.

Parent Element–The radioactive element from which a daughter element is produced by radioactive decay; e.g., radium is the parent element of radon.

Parent Material–(1) The horizon of weathered rock or partly weathered soil material from which the soil is formed. Horizon C of a soil profile. (2) The unconsolidated material from which a soil develops. See Soil Horizon.

Parent Rock–(1) The original rock from which sediments were derived to form later rocks. (2) The rock from which parent materials of soils are formed; the R horizon. See Soil Horizon.

Parental Generation–The P_1 generation; the first generation in a series of crosses; usually involves homozygotes for different alleles.

Paris Green–Acetoarsenite of copper, one of the first arsenical compounds to be widely used in insect sprays as a stomach poison for the destruction of leaf-eating insects. It was almost exclusively used to control the potato beetle. Paris Green is now replaced by insecticides that biodegrade faster.

Park–(1) A natural opening of grassland or a parklike area in a forested region. (2) In the Rocky Mountain region, an enclosed valley that may be partly grassland. (3) The land enclosed in the oxbow meander of a river. (4) Land set apart to be used for public recreation.

Particle Density–The average density of the soil particles. Particle density is usually expressed in grams per cubic centimeter and is sometimes referred to as real density, grain density, or specific gravity.

Particle-size Analysis–Determination of the amounts of different particle sizes in a soil sample, usually by sedimentation, sieving, micrometry, or a combination of these methods.

Particle-size Distribution–The amount of the various soil separates, sands, silt, and clay in a soil sample, expressed as dry weight percentages.

Particulates–(1) Of or relating to particles or occurring as minute particles. (2) Finely divided solid or liquid particles in the air or in an emission. Particulates include dust, smoke, fumes, mist, spray, and fog.

Party Wall–Wall erected on a line between adjoining properties for the use of both properties.

Pass–(1) A gap, defile, or other relatively low break in a mountain range through which a road or trail may pass; an opening in a ridge forming a passageway. (2) A navigable channel, especially at a river's mouth. (3) A narrow connecting channel between two bodies of water; an inlet. (4) An opening through a barrier reef, atoll, or sand bar.

Pastoral–(1) Rural; rustic; bucolic. (2) In farming, referring to a kind of livelihood dependent almost entirely upon pasturing or herding animals, as sheep, goats, cattle, reindeer, etc.

Pastureland–(1) Land used primarily for the production of adapted domesticated forage plants to be grazed by livestock. (2) Land producing forage plants, principally introduced species, for animal consumption. Management practices usually include one more more treatments such as reseeding, renovating, mowing, liming, or fertilizing. Native pasture that because of location or soil limitation is treated like rangeland is included as pastureland.

Patana–A grassy slope with a moderate supply of moisture, resembling a savanna.

Patch–(1) A small piece of land of indefinite size used for cultivated plants, as garden patch, melon patch. (2) A relatively small island like tract named after the vegetation growing on it, as a brier patch, patch of woods, etc.

Pathobiology–The study of disease processes; biology of disease.

Pathogen–In the general sense, anything capable of causing disease, but when referred to by most veterinarians and physicians it signifies a living, microscopic, disease-producing agent such as a bacteria or virus. See Parasite.

Pathology–The science that deals with diseases and the effects that disease have on the structure and function of tissues.

Pavement Ant–*Tetramorium caespitum*, family Formicidae; a yellowish ant, a pest in lawns, gardens, and greenhouses.

PCBs–See Polychlorinated Biphenyls.

Pea-sick Soil–A soil that is infested with fungi that cause the root rot of peas.

Peak–(1) In turpentining, the upper point of the V-shaped streak at the top of the incision. (2) In hydrology, the maximum rate of flow recorded at a gauging station during a flood.

Peat–(1) A dark-brown or black residuum produced by the partial decomposition and disintegration of mosses, sedges, trees, and other plants that grow in wet places. (2) Fibrous, partly decayed fragments of vascular plants that retain enough structure so that the peat can be identified as originating from certain plants (e.g., sphagnum peat or sedge peat). See Histosols, Muck.

Peat Bog–A bog in which the soil or cumulose deposit is peat, a common relief and vegetational feature in the northern glaciated region of the United States from Minnesota eastward. The bog may be treeless or it may support a growth of trees, such as black spruce.

Peat Lover–A plant that grows in peat; a plant that is tolerant or adapted by nature to highly acid and other conditions found in natural peat soils.

Peat Moss–Moss plants that grow on heath bogs, such as species of ***Sphagnum*** and ***Polytrichum***.

Peavey–In lumbering, a stout lever, 5 to 7 feet long, which is fitted with a socket, spike, and a curved steel hook that works on a bolt hinge. It is used in handling logs. Also spelled peevy.

Pecan Weevil–*Curculio caryae*, family Curculionidae; a weevil that infests newly formed pecans causing them to shrivel and drop. Also called hickory nut weevil.

Peccary, Collared–*Pecari angulatus*; a piglike mammal ranging in the wild from Texas to Paraguay, South America.

Pecky Wood–Wood having pockets of decay caused by fungi or pits or holes caused by insects and birds. It may be used for rustic interior paneling.

Ped–A unit of soil structure such as an aggregate, crumb, prism, block, or granule, formed by natural processes (in contrast with a clod, which is formed artificially by compression of a wet clay soil).

Pedistalled–Plants and stones perched on small columns (pedistals) of soil, the surrounding soil having been lost by raindrop splash erosion.

Pedogenesis–Soil formation.

Pedologist–One versed in pedology; a soil scientist.

Pedology–The science that deals with the origin, fundamental nature, and classification of soils.

Pedon–The smallest volume that can be called a soil. It has three dimensions, and it extends downward to the depth of plant roots or to the lower limit of the genetic soil horizons. Its lateral cross section is roughly hexagonal and ranges form 1 to 10 square meters, depending on the variability in the horizons.

Pedosphere–The outer soil layer of the earth in and on which organic life exists.

Pelage–The fur coat or covering of an animal, as in rabbits. See Pelt.

Pelagic–A term designating a fish or other animal that lives in the sea and far from land.

Pelophyte–A plant that grows in clayey soil.

Pelt–(1) The natural, whole skin covering, including the hair, wool, or fur of the smaller animals, such as sheep, foxes, etc. A large pelt is more often called a hide. (2) To remove the whole skin or pelt form the body of an animal.

Pemmican–A concentrated food used by American Indians and explorers that consists of powdered, dried strips of meat mixed with dried fruits and vegetables.

Peneplain–(1) A land surface worn down by erosion to a nearly flat or broadly undulating plain; the penultimate stage of old age of the land produced by the forces of erosion. (2) By extension, such a surface uplifted to form a plateau and subjected to renewed degradation and discussion.

Penetrant–Adjuvant that aids a liquid's ability to enter the pores of a surface. See Adjuvant.

Penetrometer–A device that measures the force required to push a probe rod into the soil. It can be used to measure the density or degree of compaction in a soil.

Penstock–(1) A sluice for regulating the flow of water. (2) A conduit for conducting water.

Penta—Greek for five, used in naming chemical compounds.

Pentachlorophenol–A chemical used extensively to treat fenceposts, telephone poles, and bridge planking, against fungal decay. Reports have indicated some hazard to livestock when they lick or chew on posts so treated.

Percentile Taper–The relative taper of a tree in terms of diameter at regular intervals along the stem which is expressed in percent of diameter at breast height.

Perched Water Table–(1) A body of water that has been retarded in its downward course by an impermeable or nearly impermeable bed to such an extent that it forms an upper zone of saturation overlying but separated from a lower zone. (2) An area at the bottom of a pot or bench filled with saturated growing medium.

Percolation–The downward movement of water through the soil in response to the pull of gravity.

Perennial Stream–A stream that flows throughout the year.

Pergelisol–See Permafrost, Thermokarst.

Periodic Annual Increment–The growth of a stand of trees for any specified period divided by the number of years in the period.

Peripheral–Pertaining to the surface of the body of a plant, especially a tree.

Periphyton–The association of aquatic organisms attached or clinging to stems and leaves of rooted plants or other surfaces projecting above the bottom of the body of water.

Perishable–Designating any product that is liable to easy or quick destruction by rot, disease, or decomposition, such as fresh fruits, meats, and vegetables.

Perlite–A volcanic glass having numerous concentric cracks that give rise to perlitic structure. A high proportion of all perlites are rhyolitic in composition. Used in greenhouses as a synthetic soil mix. See Vermiculite.

Permafrost–Permanently frozen ground. Permafrost areas are divided into more northern areas in which permafrost is continuous, and those more southern areas in which patches of permafrost alternate with unfrozen ground. See Thermokarst.

Permafrost Table–An irregular surface that represents the upper limit of permafrost.

Permanent Dam–A dam made of concrete or masonry.

Permanent Parasites–Parasites, such as bloodsucking lice, which spend all life stages on or in the body of the host.

Permanent Soil Sterilization–Sterilization of soil by chemical treatment lasting for more than a year to suppress all undesirable vegetation on roadsides and ditch banks.

Permanent Water–A watering place that supplies water at all times throughout the year or grazing season.

Permanent Wilting Percentage–The soil moisture content at which plants remain permanently wilted unless water is added to the soil. Soil water potential at wilting can vary from –5 to –20 bars. Because of the shape of the water potential-water content drying curve, large changes in water potential at higher tensions accompany minor decreases in water content, so permanent water for plant growth is approximately 15 bars.

Permatodes–Roundworms that live in soil or water.

Permeability–The capacity of soil or rock for transmitting a fluid. Degree of permeability depends upon the size and shape of the pores, the size and shape of their interconnections, and the extent of the latter. It is measured by the rate at which a fluid of standard viscosity can move a given distance through a given interval of time. The unit of permeability is the darcy. See Darcy's Law.

Permeable–Refers to the ability of soil to be penetrated by air and water.

Perpelic–Referring to plants that grow on clay soils.

Persian Walnut–*Juglans regia*, family Juglandaceae; a deciduous tree grown for its edible nut, the English walnut of commerce. Native to Europe and Asia. Also called Circassian walnut, English walnut, European walnut.

Persimmon–*Diospyros virginiana*, family Ebenaceae; a hardy deciduous tree that bears an edible fruit and is also a bee plant. The wood has been used in the manufacture of wooden golf clubs. Also called American persimmon.

Persistent–(1) Of, or pertaining to, the presence of the calyx lobes on a ripened pome fruit, especially pear. (2) Designating fruits, flowers, leaves, etc., that remain on a plant after frost. (3) Designating a disease that is difficult to control.

Persistent Emergent–Emergent water plants that normally remain standing at least until the beginning of the next growing season; e.g., cattails (*Typha* spp.) or bulrushes (*Scirpus* spp.).

Persistent Herbicide– A herbicide that will harm susceptible crops planted in normal rotation after the harvesting of the treated crop.

Persistent Pesticide– A chemical agent used to control pests, which breaks down extremely slowly, remaining toxic to desirable species of wildlife as well as pests, under natural conditions. Some of these include DDT, chlordane, lindane, and dieldrin. Most are now forbidden or restricted in use.

Peruvian Guano– The dry excrement of marine birds obtained from islands off the coast of Peru. See Guano.

Pervious Soil– An open soil; one with large pore spaces that permits a ready flow of water through it. It is generally a sandy or gravelly soil.

Pest– Anything, such as an insect, animal, or plant that causes injury, loss or irritation to a crop, stored goods, an animal, or people.

Pest Control– The use of disinfectant, herbicide, pesticide, insecticide, management or cultural practice that controls pests. See Integrated Pest Management.

Pesticide– A substance used to control insect, plant, or animal pests. Pesticides include insecticides, herbicides, fungicides, nematocides, and rodenticides.

Pesticide Residue– Material that remains on a plant after pesticide application.

Pet– Any animal, such as a cat, dog, lamb, bird, etc., that is kept for affection and companionship.

Petra– (Greek) Rock.

Petrify– To become stone. Organic substances, such as shells, bones, and wood, embedded in sediments, become converted into stone by the gradual replacement of their tissues, particle by particle, with corresponding amounts of infiltrated mineral matter such as lime or silica.

Petrography– A branch of geology directed toward the description and systematic classification of rocks, especially igneous and metamorphic rocks by means of microscopic examination of thin sections.

Petrology– That branch of geology dealing with the origin, occurrence, structure, and history of rocks, mainly igneous and metamorphic rocks.

Petromyzontidae– The major family of the group of vertebrates comprising the lampreys and borers, eel-like, scaleless, aquatic creatures found in both fresh- and saltwater. The freshwater species, such as the brook lampreys, are harmless, nonparasitic and dwarfed, and furnish, along with their larvae, food for freshwater game fish. The sea lamprey has become established in the Great Lakes in the United States, and is a destructive parasite of salt- and freshwater fish.

pF Value– The logarithm of the height, in centimeters, of a water column necessary to produce a force equal to the energy with which moisture is held by a soil. The "p" indicates a common logarithm, and the "F" suggests force or energy.

pH– A numerical measure of acidity or hydrogen ion activity of a substance such as food or soil. The neutral point is pH 7.0. All pH values below 7.0 are acid and all above 7.0 are alkaline. The negative logarithm of the hydrogen-ion activity. The degree of acidity (or alkalinity) of a soil as determined by means of a glass, quinhydrone, or other suitable electrode or indicator at a specified moisture content or soil-water ratio, and expressed in terms of the pH scale. See Reaction.

Phase– (1) The view that a thing presents to the eye. (2) Any one of the varying aspects or stages through which a disease or process may pass. (3) In colloidal chemistry, the discontinuous portion dispersed in the dispersion medium. (4) In soil taxonomy and soil survey, soil phase terms are surface soil texture, percentage slope, stoniness, saltiness, and erosion. When appropriate these names are added to the soil series name to make a soil mapping unit.

Pheasant– Any brightly colored, gallinaceous bird of the genus *Phasianus*. Some species, especially the ring-necked pheasant are farm game birds, and are also raised in limited numbers for the market as a luxury meat fowl. Locally, they may be regarded as a minor pest because of the damage they do to some truck and garden crops.

Phenol– C_6H_5OH; carbolic acid; a colorless, crystalline compound, obtained by the distillation of coal tar; widely used as a disinfectant and as an ingredient in antiseptics.

Phenol Coefficient– A figure representing the relative killing power of a disinfectant, as compared with phenol acting on the same organism for the same length of time.

Phenology– The science that deals with the time of appearance of characteristic periodic phenomena in the life cycles of natural organisms, especially in relation to climate and other environmental factors.

Phenotype– The observed character of an individual without reference to its genetic nature. Individuals of the same phenotype look alike but may not breed alike. See Genotype.

Phobia– (Greek) Fear.

Phosphate– A term commonly used to indicate a fertilizer that supplies phosphorus. A major element in fertilizers.

Phosphate Rock– Any rock or mineral that contains phosphorus in sufficient quantity to make it useful directly, or indirectly through manufacture, as fertilizers. Most phosphate rock is tricalcium phosphate, $Ca_3(PO_4)_2$, with various impurities. See Apatite.

Phosphates– Salts of phosphoric acid, such as those made by combining phosphoric acid with sodium, potassium, and calcium. The phosphate most commonly used in fertilizers is made by treating phosphate rock with sulfuric acid.

Phosphatide– A complex organic compound that contains choline phosphoric acid or amino-ethyl phosphoric acid. It occurs in milk, is easily destroyed by heat and oxidation, and is an important factor in the auto-oxidation of milk fat and the development of taints (off taste) in milk.

Phosphoric Acid– H_3PO_4; orthophosphoric acid. To be made available to plants, it is converted into soluble acid salts by processing with sulfuric acid. The amounts of available phosphorus in fertilizers are usually expressed in terms of percentage of weight of available P_2O_5.

Phosphorite– A variety of apatite which is fibrous and concretionary. See Apatite.

Phosphorized Limestone—A phosphatic limestone used as a mineral supplement in stock feeds. Fluorine is usually present in such phosphatic rocks.

Phosphorus—P; a chemical element found in soils in various mineral forms, but only small amounts are readily available to plants at any one time. It stimulates early growth and root development, and hastens grain maturity. In the water environment it enhances the growth of aquatic plants.

Phosphorus Deficiency—In soils, insufficient amounts of phosphorus in available form for optimal growth of plants, or in insufficient amounts to produce an economic or desired yield of a crop.

Photic Zone—Surface waters that are penetrated by sunlight.

Photobiotic—Living only in the presence of light, such as green plants. See Photosynthesis.

Photochemical Process—The chemical changes brought about by the radian energy of the sun acting upon various polluting substances. The products are known as photochemical smog.

Photodecomposition—The breakdown of a substance, especially a chemical compound, into simpler compounds by the action of radiant energy.

Photolysis—(1) Chemical changes brought about by the absorption of light. (2) In botany, the grouping of the chloroplasts in relation to the amount of light the plant receives.

Photoperiod—Length of the light period in a day. See Day-neutral Plant, Long-day Plant, Short-day Plant.

Photoperiodic Adaptation—The adjustment of plants in their native or artificial habitat to a definite length of daily exposure to light. See Photoperiodism.

Photoperiodicity—The response of plants or animals to the length of daily exposure to light.

Photoperiodism—The reaction of plants to periods of daily exposure to light which is generally expressed in formation of blossoms, tubers, fleshy roots, runners, etc. See Day-neutral Plant, Long-day Plant, Short-day Plant.

Photosensitization—A condition in which external parts of the body, usually the skin, become sensitive to light. This condition may occur in cattle after eating certain substances that have the ability to sensitize light-colored areas of the skin to sunlight. It can result in swellings and fluid-containing vesicles on unpigmented skins. The exudate from the vesicles forms a crust on the skin which may permanently harm the outer part of the affected skin.

Photosynthesis—Process by which green plants, using chlorophyll and the energy of sunlight, produce carbohydrates from water and carbon dioxide, and release oxygen.

Photosynthetic Rate—The time required for a plant to manufacture a given quantity of foods such as proteins, carbohydrates, and fats.

Photothermal Induction—The inducing of a plant process, particularly reproduction, by a combination of photoperiods and temperature. Also called thermophotoperiodic induction.

Phototropism—The response of a plant to the stimulus of sunlight in which the plant or its parts seem to turn to face the light. The parts of the plant receiving the direct rays grow more slowly and the plant appears to turn.

Phreatic Water—A term that originally was applied only to water that occurs in the upper part of the zone of saturation under watertable conditions; now it means all water in the zone of saturation, thus making it an exact synonym of groundwater. See Groundwater.

Phreatophyte—Any plant that sends its roots deep into the soil to the level of groundwater or to the capillary fringe above the water table. In semiarid and arid regions, these plants may be objectionable because they rob more shallow-rooted, less hardy but more desirable plants of water. However, alfalfa, a deep-rooted plant, under some conditions may be considered a phreatophyte.

Phycomycetes—Algae and fungi bearing nonsepate, branching filaments not organized into compact bodies of definite form. Containing several damping-off fungi.

Phylliform Soil Structure—A thin, leaflike layer of soil that is less distinct and thinner than platy. Where this condition is present in the C horizon, as in soils developed from thin-bedded sediments, the term *laminated* is used.

Phylum—The highest grouping in the taxonomy of the plant and animal kingdoms, based on assumed common ancestry.

Physical and Mechanical Pest Controls—Direct or indirect (nonchemical) measures to destroy pests outright or to make the environment unsuitable for their entry, dispersal, survival, or reproduction; e.g., steam sterilization to destroy disease organisms, flaming for the control of weeds, cold storage to control pests, metal or other material barriers to prevent pest entry. See Integrated Pest Management.

Physical Poison—Any poisonous material that exerts a physical rather than a biochemical effect, as heavy mineral oils and inert dusts.

Physical Properties—(Of soil) Different characteristics such as color, texture, consistency, depth, etc.

Physical Weathering—In soil formation, the natural disintegration of rocks and minerals by alternate heating and cooling, expansion by water freezing in crevices, corrosion by moving ice and water, splitting of rocks by roots of trees, and other physical factors, as contrasted with chemical weathering.

Physiogenic Disease—Any disease caused by unfavorable environmental factors.

Physiognomy—(1) The general, outward appearance of a plant community as determined by the life of the dominant species. (2) The physical appearance of a landscape created by a combination of land forms or relief features. (3) The external, facial features of a human or an animal.

Physiographic—Physical geography or a general description of nature or natural phenomena.

Physiographic Location—The location of a stand of trees with respect to slope aspect, slope position, and slope inclination, and the slope conformation relative to the overall terrain.

Physiologic Races–Pathogens of the same species and variety that are usually structurally indistinguishable but which differ in their physiologic behavior, particularly in their ability to parasitize varieties of a particular host.

Physiological–(1) Referring to or concerning the science of physiology or the branch of biology that deals with life processes and functions. (2) Referring to the functions of the organs of plants and animals.

Physiological Drought–Inability of a plant to obtain water from soil although the water may be present in it, as when a soil is frozen, or by reason of weak osmotic force of plant roots.

Physiological Maturity–The period of advanced age in the cycle of a tree or stand of trees when resistance to adverse influences is so low that death of a tree or net losses in volume of salable wood are likely to occur within a cutting cycle.

Physiology–The science that deals with the function of a plant or animal's body and its organs, systems, tissues, and cells.

Physiooogical Fuel Values–Units, expressed in calories, used to measure food energy in human nutrition.

Phytobezoar–Undigestible plant fibers ingested by an animal that have gathered in the rumen or stomach in the form of balls. See Bezoar, Hair Ball, Stomach Ball.

Phytoedaphon–The microflora of the soil.

Phytogenesis–History of plant evolution; the origin of vegetation.

Phytogenic Soil–See Phytomorphic Soil.

Phytohormone–See Growth Regulator.

Phytome–The vegetable substance composing a plant.

Phytometer–(1) A plant used to measure the physiological activities of the habitat. (2) A device used for measuring water transpiration of a plant.

Phytomorphic Soil–In pedology, a group of soils that includes those developed under the predominant influence of vegetation.

Phytopathogenic–Capable of causing plant disease.

Phytopathology–The science of plant disease.

Phytophagous–Feeding upon plants.

Phytophthora–A genus of fungi of the family Pythiaceae that includes a number of plant-parasitic species many of which can also live as saprophytes. At the exterior of the infected host plants the fungus sends out a downy or felty mass of hyphae, which is the downy mildew phase of the *Phytophthora*. Within the host plant the advancing hyphae of the fungus cause the death and decay of the invaded tissues, the rot phase of the infection. If infections occur on young stem tissues of leaves, lesions of varying sizes are produced and the blight stage results. The stem may be rotted off, and if this occurs near or just below the soil surface, the condition is called damping-off. If the conidia infect a fruit this may result in fruit rot. Storage regions underground, such as tubers, bulbs, corns, roots, may be infected by asexual spores carried into the depths of the soil by rain, as in the tuber-rot phase of late blight of potatoes. Fourteen species of genus *Phytophthora* are known to cause disease of about fifty economically important plants.

Phytoplankton–Plant microorganisms, such as certain algae, living unattached in the water. See Plankton, Zooplankton.

Phytotomy–Plant anatomy; dissection of vegetation.

Phytotoxin–A poisonous substance of plant origin; e.g., abrin from rosary pea, *Abrus precatorius*; ricin from castor bean, *Ricinus communis*; crotin from croton, *Codiaeum variegatum*; and robin from black locust, *Robinia pseudoacacia*. See Poison, Toxin.

Pica–An unnatural craving for unusual or excessive amounts of food usually associated with nervousness, hysteria, pregnancy, malnourishment, or anxiety created by hearing or observing dramatic events or highly competitive sports. See Geophagia, Phagomania.

Piedmont–Meaning literally at the foot of the mountain.

Piedmont Plateau Region–A physiographic division of the eastern United States; the eastward-sloping, dissected, upland plain lying between the Appalachian Mountains and the Atlantic Coastal Plain. One of the older agricultural regions of the eastern United States, it is notable in its central and southern sections for tobacco, cotton, and fruit production, red soils, and for severe soil erosion.

Pigeon–Any bird of the genus *Columba*. In poultry, young pigeons, or squabs, are generally dressed for the market, and old pigeons are sold alive, generally for the Jewish trade. See Squab.

Pigeon Horntail–*Tremex columba*, family Siricidae; a wasplike insect whose larvae bear into the trunks of diseased and dying elm, sugar maple, and other trees. Also called pigeon tremen.

Pigeon Loft–A dovecot, elevated house, or pen atop buildings for raising pigeons.

Pigeon Louse–Either of two species of lice, *Columbicola columbae*, or *Goniocotes bidentatus*, family Philopteridae. Each causes considerable annoyance both to old pigeons and to partially feathered squabs.

Pigment–Any of the natural coloring materials in the cells and tissues of plants and animals. In fruit and vegetables, the green pigment is chlorophyll; orange to red pigments are carotenoids; red to blue colors are anthocyanins; light-yellow pigments are flavoners and flavonols. In meat, the chief pigment producing the pink or red color is myoglobin.

Pigmy Forest–A small tree and shrub type of rangeland that borders other shrublands and grasslands in the western states. Also known as the pinyon-juniper vegetation type.

Pilchard Oil–The product obtained by extraction of part of the oil from the whole Pacific pilchard fish or from cannery refuse from this species of fish; it is used as a source of vitamins in animal feeds.

Pillbug–*Armadillidium vulgare*; not an insect but a creature closely related to the crayfish that lives in damp places, as under rotting boards. Mainly a greenhouse pest, it may eat roots of seedlings and do other damage. See Sowbug.

Pin Cherry–*Prunus pennsylvanica*, family Rosaceae; a deciduous tree or shrub that produces small, sour, bright red cherries on long slender stalks. Found in the United States, from the Atlantic to the Rocky Mountains and from Canada to Virginia and Tennessee, it is especially abundant on burned-over land. The foliage is sometimes lethal when browsed by livestock. Native to North America. Also called fire cherry.

Pin Oak–*Quercus palustris*, family Fagaceae; a hardy deciduous tree found from Massachusetts west to Michigan and Illinois and south to Arkansas. Easily transplanted, it is grown as a shade tree on city streets and in landscaping for its autumnal foliage. Its lumber is used for ties, fuel, flooring, planing-mill products, furniture, boats, etc. Native to North America.

Pine–Any cone-bearing, evergreen tree with needle-shaped leaves, of the genus *Pinus*, family Pinaceae. Species number about 50 in the United States and constitute valuable timber trees, a source of turpentine and rosin, ornamentals, and Christmas trees.

Pine Barrens–Land, embracing various soils, but generally sandy, and low in fertility, on which pines were the dominant growth.

Pine Engraver–*Ips pini*, family Scolytidae; a beetle that infests all species of pine, burrowing small circular holes into branches and trunks of trees. Native to the northern United States.

Pine Littleleaf–A fungal disease of shortleaf pines in the southern United States caused by *Phytophthora cinnamomi*, family Mycosphaerellaceae; characterized by short, yellow needles at the ends of branches and death of the fine roots of the tree.

Pine Meadows–A vegetational land feature of the southern Mississippi along the Gulf Coast (United States), characterized by flat topography, wetness, a scrubby growth of longleaf pine, cypress, and a ground cover of sedgy vegetation.

Pine Straw–The fallen dead needles of pine trees. The loose, open mat at the base of trees may represent the fall of several years. It gradually becomes broken and decomposed at the bottom.

Pinery–A forest area that supports a heavy growth of pines. (A localism, especially in Wisconsin and Michigan, United States.) See Pinetum.

Pinetum–A small groove or planting of pine trees kept for its aesthetic and educational value rather than as a forest.

Piney Woods–Native forest that consists predominantly of open stands of pine, especially longleaf, in contrast to deciduous and swamp forests. (Colloquial especially in the Gulf Coast region of the United States).

Pink Locust–Pink-flowering bristly locust; *Robinia hispida*, family Leguminosae; an excellent erosion-control shrub because of its profuse root system.

Pinon–Mexican pinyon pine that bears edible seeds.

Pioneer Plants–Herbaceous annual and seedling perennial plants that colonize bare areas as a first stage in secondary succession.

Pioneer Settlers–In the early settlement of the United States, those settlers who were the first to enter new country or virgin wilderness to establish farm homes.

Piping–(1) A series of shrill sounds made by queen bees. (2) Formation by moving water of subsurface tunnels or pipelike cavities in the soil. See Vertisols.

Piscary–The right of fishing. Common of piscary is the right of fishing in waters belonging to another person.

Piscicide–A substance used to kill fish.

Pisciculture–The production of fishing natural or artificial bodies of water under controlled conditions, such as stocking, feeding, and use of chemical fertilizers. See Aquaculture.

Piscis–(Latin) Fish.

Pit–(1) The endocarp of a drupe; the seed-stone of a fruit, as the pit of a peach or prune. (2) An excavation in soil in which vegetables, such as potatoes, carrots, and parsnips are placed and covered over for storage (seldom practiced now in the United States. (3) A place on the floor of an exchange in which traders stand when dealing in wheat, cotton, and other commodities. (4) In botany, a small hollow or depression in a cell wall. Various types are recognized in wood anatomy as blind, bordered, primordial, simple.

Pit Recharge–The draining of excess water into a large pit; the pit must permit the water to enter the water table.

Pitahaya–*Lemaireocereus thurberi*, family Cactaceae; a tall, treelike cactus of the arid region of Arizona, California (United States), and Mexico, which yields an edible fruit. Native to North America. Also called sweet pitahaya, aya pitaya.

Pitch–(1) The resin that occurs in the wood of conifers, as the pitch from pines. (2) A heavy, dark, viscous or solid, fusible material obtained by distillation of the tar derived from coal, wood, rosin, and petroleum oils. It consists of many organic compounds, chiefly hydrocarbons, differing according to origin. (3) The jumping action of a horse in its attempt to unseat its rider.

Pitch Pine–(1) *Pinus rigida*, family Pinaceae; a hardy evergreen, rugged but beautiful tree sometimes grown as an ornamental. Native to North America. (2) Any pine tree with an abundance of pitch in the wood.

Pitch Pocket–A well-defined, lens-shaped opening between or within annual growth layers of conifer wood which usually contains solid or liquid pitch.

Pitch Streak–A well-defined accumulation of pitch (resin) that is in the form of a regular streak in the wood of certain conifers.

Plains–Flat, or nearly level, extensive areas of land. Some major physiographic divisions named plains may have broken, hilly topography, and may exhibit strong relief locally. They may be classified into various kinds, as alluvial, coastal, grassland, pine, desert, high, low, etc.

Plane Survey–A survey in which the curvature of the earth is disregarded, as in ordinary field and topographic surveying.

Plankton–The small (usually less than 2 millimeters long) floating or drifting life forms in water bodies, plankton includes both plants (phytoplankton) and animals (zooplankton) that are carried passively in the water currents. Those that can swim do so to change or adjust their depth in the water, not to move from place to place. Plankton is one of the three main divisions of aquatic life, the others being nekton (the animals that swim actively and may move long distances for feeding or breeding) and the benthos (organisms that crawl about on the bottom). See Phytoplankton, Zooplankton.

Plant Analysis–Analytical procedures to determine the concentration of nutrients in plants.

Plant Association–A unit of natural vegetation essentially uniform in habitat, aspect, general appearance, ecological structure, and floristic composition; a plant community. An area may be dominated by some one of several species with minor species associated, such as a sugar maple-beech association or a bunchgrass association.

Plant Community–An assemblage of plants living together under the same environmental conditions.

Plant Cover–(1) The vegetal mantle covering the soil. (2) All plants found on a particular range regardless of their availability, palatability, toxicity, or other characteristics.

Plant Disease–A condition that affects physiologic activities of a plant so as to check normal development, lead to abnormal formations or to premature death of all or a part of a plant.

Plant Doctor–A plant pathologist; a person skilled in the treatment of plant diseases.

Plant Ecology–The science that deals with plants in relation to their environment. See Ecology.

Plant Immunity–Resistance of a plant to a pathogen so that a disease does not cause appreciable damage.

Plant Indicator–A plant that indicates soil characteristics; e.g., black walnut (*Juglans nigra*): high-lime, fine-textured, well-drained soil; longleaf pine (*Pinus palustris*): acid, coarse-textured, well-drained; basswood (*Tilia americana*): moderately acid, medium-textured, well-drained; blueberries (*Vaccinium* spp.): acid, coarse-textured, well-drained; cypress (*Taxodium* spp.): mildly acid, fine-textured, water-saturated. Trees that indicate areas frequently burned over are: jack pine, *Pinus banksiana*; lodgepole pine, *Pinus contorta*; pin cherry, *Prunus pennsylvanica*; trembling aspen, *Populus tremuloides*.

Plant Kingdom–One of the two great categories of living things, the other being the animal kingdom.

Plant Life–Plant organisms of any kind from lowest to highest forms, as distinguished from animal organisms.

Plant Material Centers–In the United States there are twenty-five plant material centers managed by the United States Department of Agriculture Soil Conservation Service. They receive superior grasses, legumes, forbs, shrubs, and trees. These plant materials are reproduced for distribution.

Plant Patent–A patent granted by the United States Patent Office to originators of varieties, strains, or some variation from existing varieties of asexually reproduced plants.

Plant Pathology–That branch of science that deals with the diseases of plants.

Plant Physiology–The science that deals with the response of a plant to its environment including moisture, temperature, light, and nutrients.

Plant Pigments–The colors that are in plants. Some of the pigments, chlorophylls (green), carotenes (yellow/orange), and xanthophylls (red/brown), are essential in the photosynthetic process. Other plant pigments, anthocynins (red in acid solutions and blue in alkaline solutions), anthoxanthins (yellow), and betacyanins (red), attract insects that cross pollinate flowers and are esthetically attractive. Fall colors in temperate regions are the result of these pigments.

Plant Propagation–The practice of producing new plants that may be asexual means, such as division, graftage, cuttage, layerage, etc., or by the use of seeds.

Plant Quarantine–The isolation of or restriction placed on the movement of plant materials to prevent either the plant or plant-carried insects and diseases from contaminating an agricultural area free from them. The Plant Inspection Branch of the Agricultural Research Service in the United States Department of Agriculture processes all plant materials from foreign sources.

Plant Regeneration–The development of volunteer vegetation from seed or by other natural reproductive processes from plants existing nearby.

Plant Residue–(1) Plant material, such as leaves, roots, straw, corn stover, grass, and weeds left after harvesting; any plant material remaining after any harvesting or other process. (2) Wood materials from manufacturing plants not utilized for some product; e.g., slabs, edgings, trimmings, miscuts, sawdust, shavings, veneer cores and clippings, and pulp screening.

Plant Retrogression–The process of vegetational deterioration whereby the same area becomes successively occupied by different plant communities of lower ecological order. See Plant Succession.

Plant Succession–The process of vegetational development whereby an area becomes successively occupied by different plant communities of higher ecological order. See Plant Retrogression.

Plant Variety Protection Act–This act, passed in 1970, offers legal protection to developers of new varieties of plants that reproduce sexually; that is, through seed. Developers of plants that reproduce asexually have received protection by the United States Patent Office since 1930. The law states that protection will be extended to a "novel variety" if it has these three qualifications: (a) distinctiveness: the variety must differ from all known prior varieties by one or more identifiable morphological, physiological, or other characteristics; (b) uniformity: if any variations exist in the safety, they must be describable, predictable, and commercially acceptable; (c) stability: when sexually produced, the variety must remain unchanged in its essential and distinctive characteristics to a degree expected of similarly developed varieties.

Plant-exploration Service–A very important activity of the Agricultural Research Service, United States Department of Agriculture. Plant explorers are sent all over the world to collect plants, seeds, or plant materials of desired species for breeding and development of new crops for agricultural and other uses.

Plant-tissue Test–A test, usually rapid and not precisely quantitative, using various chemical reagents that are made upon tissue and sap of growing plants to determine the adequacy of essential plant nutrients.

Plantation–(1) A large-scale agricultural unit, especially one devoted to the production of cotton or sugarcane. In pre-Civil War times in the United States, a large manorial estate on which cotton, tobacco, and other crops were produced with slave labor. (2) An artificially reforested area established by planting or by direct seeding.

Planting Easement–An easement for reshaping roadside areas, power lines, etc., and establishing, maintaining, and controlling plant growth thereon.

Plasmid–A circular piece of DNA found outside the chromosome in bacteria. Plasmids are the principal tool for inserting new genetic information into microorganisms or plants.

Plastic Limit–The moisture content at which a soil changes from a semisolid to a plastic state. A very useful test to determine trafficability.

Plasticity Index–The numerical difference between the liquid limit and the plastic limit; the range of moisture content within which the soil remains plastic.

Plat–A diagram drawn to scale showing all essential data pertaining to the boundaries and subdivisions of a tract of land, as determined by a plane survey.

Plat Book–A record of recorded subdivisions of land.

Plateau–An elevated area of comparatively flat land with at least one side having a steep descent to lower land.

Platelike–Describing a soil structure that resembles a stack of plates, usually resulting from compaction caused by heavy machinery.

Platy–Designating laminated soil aggregates predominantly developed along the horizontal axes. See Structure, Soil.

Playa–A flat basin or sump area on the floor of a desert valley in the western United States. The sediments of the playa left by flooding are generally clayey, highly charged with salts, and such areas are nearly devoid of vegetation.

Pleiotropy–A situation where one gene affects more than one trait.

Plinthite–The sesquioxide-rich, humus-poor, highly weathered mixture of clay with quartz and other diluents that commonly appears as red mottles, usually in platy, polygonal, or reticulate patterns. Plinthite changes irreversibly to an ironstone hardpan or to irregular aggregates on exposure to repeated wetting and drying, especially if it is exposed also to heat from the sun. In a moist soil, plinthite can be cut with a spade, whereas ironstone cannot be cut but can be broken or shattered with a spade. Plinthite is one form of the material that has been called laterite. See Laterite.

Plot–In agricultural research, a small parcel of land, usually rectangular and of a definite size, used in comparing yields of crop varieties, testing different applications of fertilizers, comparing methods of tillage, etc.

Plow–(1) The whole implement, of various types, used to cut, break, or turn a soil layer in preparation for planting, seeding, or other agricultural practices. More specifically, the removable metal point, share and moldboard or disk, attached to a plow stock or frame. (2) Any such implement not used primarily for agricultural purposes, such as a snow plow. (3) To make a furrow or to turn over a layer of soil. (4) To cultivate; e.g., plow corn. Also spelled plough by the British.

Plow Down–To bury material lying on the surface of a field by plowing; e.g., to plow in a surface application of fertilizer, or to plow under a cover crop.

Plow Pan–A compacted layer formed in the soil immediately below plow depth. It is attributed to the sliding action and weight pressure of the plow bottom. Also called plow sole.

Plow Planting–The plowing and planting of land in a single operation. See Conservation Tillage.

Plow Under–To bury and incorporate in the soil a cover of living or dead plants by the use of the plow. See Plow Down.

Plow Up–(1) To cut a furrow slice and partly or completely to invert it. (2) To lift out of the ground unwanted plants more with the purpose of destroying than preserving. (3) To bring the surface vegetation that was turned under during a previous plowing.

Plowable–Designating land suitable, in its present condition, for plowing.

Plowable Pasture–Land ordinarily kept as pasture but that may be plowed and utilized for other crops.

Plug–(1) The mass removed by a trier or other special penetrating implement in sampling or testing an agricultural product, as a plug from a cheese, a bale of cotton, or from a watermelon. (2) An old, worn-out horse. (3) A horse with a poor conformation. (4) To repair a leak, as a dam or earth fill. (5) A block of rooted grass that is planted for the purpose of establishing a covering of grass, such as in a law.

Plutonic Rock–An igneous rock formed at great depth by magmatic crystallization or chemical alteration, such as granite.

Pluvial–Due to the action of rain; pertaining to deposits by rainwater or ephemeral streams.

Plywood–A sheet of wood, commonly 4 x 8 feet. Made by gluing together three or more thin layers of wood in such a way that the wood grain of one layer is at right angles to the one it is glued against; this increases the strength.

Pneumatophore–Stilt roots or root "knees" developed by some woody plants growing in water to facilitate respiration; e.g., baldcypress (***Taxodium distichum***), tupelo (***Nyssa sylvatica***), and mangroves (principal genera: ***Rhizophora, Avicennia, Conocarpus***, and ***Laguncularia***).

Poach–To catch and carry off fish or animals illegally, or by unsportsmanlike methods.

Pocket–(1) A relatively small basin-depression in a land surface. (2) A short cove-type valley or hollow in a mountainous terrain. Often used in connection with water drainage and air drainage of land. See Kettle.

Pocket Gopher–A small, stout-bodied, shortlegged rodent of the family Geomyidae, which builds nests underground in fields and orchards. Its burrows may constitute a nuisance as it has been known to girdle citrus trees below the ground surface, commit minor damage to corn, and destroy roots and bulbs in gardens. Commonly found in the United States and Central America. Also called pouched rat. See Salamander.

Pocosin–A minor, natural land feature of the flat Atlantic coastal plain of the eastern United States. It is a swamp occupying shallow basins and is thus properly differentiated from creek swamps. (The term is largely restricted to the Carolinas.) Also spelled pocoson.

Pod–(1) Technically, a dry, many-seeded fruit that splits open, such as a pea pod or bean pod; a legume. (2) To form pods. (3) A flock of animals, birds, etc.

Poikilothermic–Refers to animals such as reptiles, amphibians, fish, and insects that are not able to maintain a constant body temperature; the body temperature normally varies with that of the environment.

Point of Runoff–A spray application that is just heavy enough to produce runoff from the leaves.

Point Source–Pollution that occurs from a single source.

Poison–Any substance ingested, inhaled, or developed within the body that causes or may cause damage or disturbance of function of plants, animals, or people. See Toxin.

Poison Bait–A poison mixed with wheat bran, molasses, or other attractant used to control cutworms, grasshoppers, and other insects.

Poison Hemlock–*Conium maculatum*, family Umbelliferae; a biennial, rank-growing herb, which is an escape and a weed in various sections of the United States. It is dangerously poisonous to people and animals, being most poisonous to stock in the spring, when the herbage is fresh. Symptoms of poisoning of cattle are loss of appetite, salivation, bloating pain, feeble but rapid pulse, and loss of muscular power. Native to Europe and Asia. Also called deadly hemlock, poison parsley, winter fern.

Poison Ivy–*Toxicodendron radicans*,family Anacardiaceae; a small, erect shrub or climbing vine common throughout the United States in waste places, pastures, woodlands, and along old fencerows. Apparently more dangerous to humans than to animals, contact with the plant among susceptible individuals causes severe blistering and inflammation, especially on the hands, arms, and face. Also called climbing ivy, three-leaved ivy, climath, poisonoak, poison creeper.

Poisonoak–See Pacific Poisonoak, Poison Ivy.

Poisonous–Containing poison, as a poisonous plant.

Polders–Flat tracts in the Netherlands below the level of the sea or the nearest river, such as a lake that has been drained and brought under cultivation. The surplus water is pumped out. They are protected form inundation by embankments called dikes. Half of the people in the Netherlands live below sea level. Similar to the fens of England.

Pollard–(1) A tree whose crown has been cut back to invite the production of shoots from the top. Sometimes a tree is so pruned to induce a globelike mass of foliage. (2) A hornless ox, sheep, or goat.

Pollutant–A substance, medium, or agent that causes physical impurity.

Pollution–The presence of substances in a body of water, soil, or air to impair the usefulness or render it offensive to the senses of sight, taste, or smell. Contamination may accompany pollution. In general, a public-health hazard is created, but, in some instances, only economy or esthetics are involved, as when waste salt brines contaminate surface waters or when foul odors pollute the air. Pollution is also defined as a resource out of place.

Pollution, Nonpoint Source–Cultural pollution of air or water from any physical, chemical, biological, or radiological source other than a point source. See Point Source, Pollution.

Pollution, Point Source–A discharge of pollution into air or water from a discrete source such as a pipe, ditch, well, confined animal feeding lot, floating craft, or motor vehicle. See Pollution, Nonpoint Source.

Pollution, Thermal–Discharge of heated water into surface waters at temperatures harmful to aquatic life.

Polychlor–Any of the group of the chlorinated class of insecticides, such as DDT.

Polychlorinated Biphenyls (PCBs)–A group of aromatic organic compounds each with two 6-carbon unsaturated rings, with chlorine atoms substituted on each ring and more than two such chlorine atoms per molecule of PCB. These compounds, used in the manufacture of plastics, are very stable, resist chemical and microbiological degradation, and are very toxic to people and animals.

Polyculture–The simultaneous planting of two or more crops in the same area at the same density of each as they would be planted separately in monoculture.

Polyelectrolytes–Synthetic chemicals used to speed the removal of solids from sewage. The chemicals cause the solids to flocculate or clump together more rapidly than traditional chemicals such as alum or lime.

Polygenetic Soil–Soil produced under conditions that have changed importantly with time.

Polygons, Mud Crack–Mud cracks (sun cracks, shrinkage cracks) form as sediments dry out. The cracks form polygons, which vary in number of sides and dimensions of angles between the sides. Cracks are rarely straight, and polygons may be bounded by three to eight cracks.

Polyhalite–A mineral, commonly in pink, red, or gray masses in potassium salt deposits.

Polyhedral Disease–A disease caused by a microorganism that attacks the gypsy moth and cabbage looper larvae and results in their decimation. However, people have not yet been able to spread the infection artificially for biologic control.

Polyphagia–(1) Voracious appetite. (2) An unnatural craving for many kinds of food. (3) Omnivorousness. See Bulimia, Cynorexia, Sitomania.

Polyphagous Parasite–A parasite capable of parasitizing a considerable number of host species.

Polyphosphates–Salts of polyphosphoric acids such as ammonium polyphosphates and calcium polyphosphates.

Polyphosphoric Acid–Any of a series of phosphoric acids whose molecular structure contains more than one atom of phosphorus such as pryophosphoric acid, tripolyphosphoric acid, and tetrapolyphosphoric acid. See Phosphoric Acids.

Polyploid–An organism with more than two sets of the basic or haploid number of chromosomes; e.g., triploid, tetraploid, pentaploid, hexaploid, heptaploid, octaploid.

Polysaprobic Zone–That area of a grossly polluted stream that contains the complex organic waste matter that is decomposing primarily by anaerobic processes.

Pond–A small, sometimes stagnant, body of water. It may be a natural feature, or human-made, as a farm pond made by impounding runoff water by a dam or by excavating. (In some parts of the United States, lake and pond are used for natural water bodies without any consistent distinction.)

Pond Baldcypress–*Taxodium ascendens*, family Pinaceae; a deciduous, coniferous tree grown for its valuable lumber used in shingles, buckets, window boxes, coffins, outdoor seats, etc., and sometimes as an ornamental. Native to the southern United States. See Baldcypress.

Pond Management–The adoption of suitable measures for the protection of ponds from siltation and for the production of fish and other wildlife.

Ponderosa Pine–*Pinus ponderosa*, family Pinaceae, an evergreen tree, large or small depending upon the site, and valued chiefly for its timber. Native to British Columbia, Canada, and the western United States. Also called western yellow pine, Arizona pine, bull pine, Arizona longleaf pine.

Poor Drainage–A natural condition of standing water on the surface of a saturated soil. (The term usually denotes that an excess of water is present for most desired agricultural uses of the land; for farm ponds the same drainage may be necessary.)

Poor Land–Any land that is inferior in quality for some particular use, as for agriculture. (Commonly the term is loosely used: land may be poor either because of some unfavorable relief, or because of some physical or chemical condition of the soil. Often land and soil are used interchangeably.)

Poor Man's Manure–Snow, which in some localities may be beneficial on plowed land because usually all of the water from snow goes into the soil.

Poor Outlets–Surface or subsurface drainage outlets difficult or expensive to install.

Poor Soil–A soil unsuitable for some particular plant growth because of deficiency in fertility or some unfavorable physical condition. A poor soil, however, may be valueless for one plant but quite suitable for the growth of another; e.g., a sandy, acid soil is desirable for blueberries but not suitable for alfalfa.

Poor Tilth–Poor physical condition of soil such as hard or cloddy.

Poorly Graded–Refers to soil material consisting mainly of particles of nearly the same size. Because there is little difference in the size of the particles, density can be increased only slightly by compaction.

Poplar–Any tree of the genus ***Populus***, family Salicaceae. Species are mostly quick-growing, soft-wooded trees grown as ornamentals, for pulpwood, and other wood uses. Hybrid poplars are fast-growing and make good shade trees.

Population–A group of plants, animals, or people of the same species living in a defined area.

Population Density–The number of units, persons, families, dwellings, etc., per given area. Rural density is usually expressed on a square mile basis.

Population Equivalent–A means of expressing the amount of a pollutant by equating it to the amount of pollutant from an equivalent number of persons.

Population Pressure–In wildlife, the force of increasing numbers that results in dispersion and endangers perpetuation of a species in a particular area.

Porosity–Refers to the extent of voids or openings in the soil that exist between soil particles and soil peds or clods. These pores hold water and air for absorption by plant roots. About half of soil volume which is in a good physical condition for plant growth is pore space.

Positive Ion–A cation; an ion that carries a positive charge of electricity.

Positron–A subatomic particle equal in mass and weight to the electron and having an equal but opposite charge. Positrons are emitted by some artificially radioactive isotopes.

Post Oak–*Quercus stellata*, family Fagaceae; a deciduous tree, whose wood is suitable for posts, ties, tight-cooperage, veneer, fuel, flooring, planing-mill products, etc. Native to central and southern United States. Also called brash oak.

Postemergent Herbicide–Applying a herbicide after the weeds begin to grow.

Potable–Referring to water that is drinkable as a result of being free of pathogens, toxic materials, unpleasant tastes, objectionable odors, color, and other undesirable physical, chemical, and biological characteristics.

Potamology–The science of streams and rivers; a branch of hydrology.

Potassium–K; the chemical element, an alkali metal, which occurs widely in minerals; e.g., in the orthoclase feldspars of granites and in salt deposits as the chlorides and sulfates. Regarded as an essential plant nutrient, potassium is present naturally in some form in all soils but in extremely variable amounts, and is likely to be in largest amounts in clay soils and in least amounts in highly silicious soils and in peats.

Potassium Chloride–KCl; the salt most commonly used as a source of potassium in fertilizers. Its commercial form is known as muriate of potash.

Potassium Fixation–The process of converting exchangeable and water-soluble potassium to slowly available potassium. Wetting and drying of clay soils tends to make available potassium less available.

Potassium Magnesium Sulfate–See Sulfate of Potash-Magnesia.

Potassium Metaphosphate–KPO_3; a chemical containing 40 percent potash (K_2O), which is used as a fertilizer. Commercial grades contain about 37 percent potash and 55 percent phosphoric acid.

Potassium Nitrate–KNO_3; saltpeter, nitrate of potash; the potassium salt of nitric acid. It is manufactured by the direct reaction of potassium chloride with concentrated nitric acid to produce potassium nitrate and chlorine.

Potassium Permanganate–$KMNO_4$; a powerful oxidizing compound used as a disinfectant, deodorant, and a reagent in analytical work, especially in the determination of available nitrogen in organic material. Sometimes also used as a fungicide in greenhouses.

Potassium Phosphate–Any of the various phosphates of potassium, especially potassium metaphosphate, used as fertilizers.

Potassium Phosphate Solutions–They have been used successfully as liquid fertilizers in California. They contain an average 10.6 percent P_2O_5 and 14.5 percent K_2O. The pH in a 1 to 400 dilution is 10, indicating that the salt in solution is dipotassium phosphate (K_2HPO_4).

Potassium Sulfate–K_2SO_4; one of the carriers of potash used in the manufacture of mixed fertilizers. The salt occurs in nature, but the commercial product is manufactured by treating potassium chloride with sulfate of magnesia or sulfuric acid. The product contains about 60 percent potassium (K_2O).

Potency–(1) The power of a medicine to produce the desired effects. (2) The ability of an embryo to develop into a viable destiny. (3) The ability of the male of any plant or animal species to fertilize the female germ cells. (4) The degree of toxicity of a chemical.

Potential Energy–Energy inherent in a mass because of its position with reference to other masses; e.g., a rock at the edge of a precipice has potential energy. Water behind a dam also has potential energy.

Potential Soil Acidity–Those hydrogen ions on the negatively charged soil particles that are not disassociated, and therefore not active in the soil solution; the amount of exchangeable hydrogen ions in a soil that can be released by cation exchange. Also called reserve soil acidity.

Pothole–(1) A hole worn into solid rock by strong currents whereby sand, gravel, and stones are spun around by the force of the current. (2) In Death Valley, a circular opening 2 to 4 feet in diameter filled with brine and lined with salty crystals. (3) A rounded, steep-sided depression resulting from downward surface solution. (4) A rounded cavity in the roof of a mine caused by a fall of rock, coal, or ore. (5) A hole in the ground from which clay for pottery has been taken. (6) A depression between sand dunes that contains water. (7) Vertical erosion in vertisols and in thermafrost. See Thermafrost, Vertisols.

Poudrette–Dried night soil mixed with charcoal powder, gypsum, lime or peat; used as a fertilizer. Used in European countries from early times, it has never been used to any extent in the United States.

Pound–(1) A unit of weight; 16 ounces avoirdupois, 12 ounces troy. The standard British unit of weight equals 7,000 grains, ½, 240 long ton, and 453.59 grams, the weight of 27.692 cubic inches of water at 4°C. (2) An enclosure in which stray animals are legally confined. (3) An enclosure in which groups of animals, as flocks of sheep, may be gathered for shelter, etc. (4) An enclosure used to trap wild animals.

Powder-post Beetle–Any of the beetles of the families Ptinidae, Anobiidae, Bostrichidae, or Lyctidae, which breed in old, dry wood and reduce it to powder. The greatest damage is done by species of *Lyctus* which confine their infestation to the seasoned sapwood of a number of hardwoods, especially oak, ash, and hickory.

Power Subsoiler–A tractor-drawn plow that has a long, narrow shank with a wedge-shaped point for deep penetration to break up stiff clay subsoils and hardpans. See Hardpan, Plowpan.

Prairie–(1) The extensive, nearly treeless and dominantly grass-covered plains of the midwestern United States that lie east of the Rocky Mountains. In a more restricted sense, the tall grasslands with blackish soils; in a more general sense the semiarid shortgrass plains as well. Also called savannah, steppe. (2) In the generally forested eastern part of the United States, any naturally treeless area that is generally dry or naturally well-drained. (3) Wet, treeless, marshy areas. (4) Prehistoric, treeless tracts that resulted from fires.

Prairie Dog–Any animal of the genus *Cynomys*. Species are small, stubby, burrowing rodents found in abundance on the prairies and treeless plains of the midwestern and western United States by early settlers and explorers. They are considered a nuisance by western cattlemen largely because their burrows constitute a danger to horses and their riders. They eat green crops and some insects, such as the Mormon cricket. They live in colonies or towns and through their extensive burrowing they have been a considerable factor in modifying the soil. Native to North America. Also called barking squirrel.

Prairie Sandreed–*Calamovilfa longifolia*, family Gramineae; a perennial, drought-enduring grass, growing from 2 to 6 feet tall, which is found from Michigan to Colorado, United States, and Alberta, Canada. Important for winter pasture and for hay though grazed but lightly in the summer. It is an important, sand-binding grass on dunes and sand hills. Native to North America. Also called prairie sandgrass.

Prairie Schooner–A covered wagon which was used by early settlers in their westward movement and occupation of the prairies and plains of the midwestern United States.

Praying Mantis–Predaceous insects of the genus *Mantis*, family Mantidae, which prey upon other insects and are generally beneficial to humans. They wait for their prey with front legs raised like hands in prayer and seize it when it comes within reach.

Precipitated Phosphate–A by-product from the manufacture of monocalcium phosphate that is sufficiently soluble in the soil solution to be of value as a fertilizer.

Precipitation–(1) The amount of water, hail, sleet, snow, or other moisture received from clouds. Snow is also reported in its equivalent of liquid water. Precipitation is classified by the conditions that produce the rising column of unsaturated air which is antecedent to precipitation. Convection precipitation is the result of uneven heating of the ground, which causes the air to rise and expand, vapor to condense, and precipitation to occur. This is the major type of precipitation during the summer, producing high-intensity, short-duration storms. Orographic precipitation is caused by topographic barriers that force in the moisture-laden air to rise and cool. Cyclonic precipitation is related to large low-pressure systems that require five or six days to cross the United States from the northwest or Gulf of Mexico. These systems are the major source of winter precipitation. (2) The phenomenon of a solution or suspension that is flocculated. (3) The electrostatic or other means of removal of polluting particulates from the air. See Acid Rain.

Preclimax–The stage of succession of an area that has never supported climax vegetation but is in the process of succession toward climax vegetation for a particular climate. Thus, an extremely shallow soil or soil material in the true prairie association may support preclimax blue gramagrass rather than climax bluestem, because soil development has not yet reached a stage at which it will support the latter. See Climax Vegetation.

Precommercial Thinning–Any type of thinning (cutting or otherwise killing trees in a stand) that takes place before the size or condition of the trees makes them of sufficient value to cover the costs of the activity.

Precooling–(1) Preliminary cooling of milk immediately after a milking to prevent spoilage. (2) Cooling of fruits immediately after harvesting during periods of hot weather to retard ripening and deterioration. (3) The cooling of meats after slaughter and before cutting.

Predacide–A substance that is used to kill predators.

Predatism–Intermittent parasitism, such as the attacks of mosquitoes and bedbugs upon humans.

Predator–An animal that attacks and feeds on other animals, usually smaller and weaker than itself.

Predisposition–(1) Stress or anything that renders an animal liable to an attack of disease without actually producing it. (2) The effect of one or more environmental factors that makes a plant vulnerable to infection by a pathogen.

Preemergence Application–Applying a herbicide to the soil to kill weed seeds before they germinate, or after a crop is planted but before it germinates and seedlings emerge above the soil's surface.

Preference Permit–A license valid up to ten years issued by the United States Forest Service to permit domestic livestock grazing in a National Forest. The permit is renewable annually.

Preformed Terrace–A terrace or bench constructed or sloped before a planting, as in setting out an orchard.

Preirrigation–In rice culture, the application of irrigation water prior to the usual irrigating after emergence of the plants.

Preplant Application–A treatment with a pesticide on the soil before planting or transplanting.

Preplant Soil Incorporation–A treatment with a pesticide by tilling the chemical in the soil before planting or transplanting.

Prescribed Burning–Setting fire scientifically under "safe" conditions of surface fuel, weather, and soil moisture to: (a) reduce uncontrolled fire hazards, (b) control "weed" species, (c) increase grazing, and/or (d) encourage designated forest tree species.

Preservative–(1) Any material, as salt or sugar, that delays or prevents spoilage and decay of food products, etc. (2) A substance, such as creosote, which when suitably applied to wood makes it resistant to infestation by fungi, insects, and marine borers.

Preserve–(1) In wildlife management, a game-shooting area on which game species are propagated or released. (2) A tract of land set aside for preservation of natural conditions, and protected against exploitation or any commercial use. (3) To prepare foods by cooking with some preservative so as to reduce fermentation or decomposition.

Pressure Pan–(Traffic sole, plow pan, tillage pan, traffic pan, plow sole, compacted layer) An induced subsurface soil horizon or layer having a higher bulk density and lower total porosity than the soil material directly above and below, but similar in particle size analysis and chemical properties. The pan is usually found just below the maximum depth of normal plowing and frequently restricts root development and water movement. See Plow Pan.

Prevailing Wind–The wind that comes most frequently in an area from a particular point of the compass, either seasonally or annually.

Preying Mantis–A misnomer for praying mantis occasionaly used because of the preying habit of the insect. See Praying Mantis.

Pricklypear–Any of various cacti of the genus *Opuntia*, family Cactaceae. Species are prostrate to treelike forms, which contain both spines and cushions or barblike hairs. Some species spread very rapidly and in some places have greatly depreciated the value of the land. With the spines and barbs burned off, the cacti have been eaten by cattle during periods of drought and scarcity of any other feed. The pearlike fruit when preserved is edible for humans. Also called devil's fig.

Primary Succession–Plant migration on newly formed soils or upon soils that have never borne vegetation, such as a sand bar in a river.

Primary Treatment (Sewage)–Removes the material that floats or will settle in sewage. It is accomplished by using screens to catch the floating objects and tanks for the heavy matter to settle in.

Primate–A member of a group of mammals including humans, monkeys, and apes.

Prime Agricultural Lands–The most productive lands for raising the common food and fiber crops; whereas unique agricultural lands are those most productive for the less common but high-value-per-acre crops such as rice, cranberries, citrus, etc. See Unique Agricultural Lands.

Primeval Forest–The forest as it existed before the advent of humans.

Primitive Area–(1) In the United States Forest Service, wilderness area. (2) In the United States National Park Servcie, an area of indeterminate size in which no commercial development nor the construction of any roads for motorized transportation is permitted.

Principal Meridian–A north and south line accurately located and used as a basis from which to construct township and section lines as used in the United States Public Land Survey.

Prism Wedge–A small hand-held prism used in forestry to estimate the basal area of a stand of trees. See Basal Area.

Prismatic Soil Structure–Blocky soil structure in which the vertical axis of a block is longer than the horizontal axis; the upper ends of the block (column) are not rounded.

Proboscis–An elongated nose, such as the snout of a hog, or of some species of insects.

Product–The herbicide as it is sold commercially. It contains not only the active ingredients but also various solvents, surfactants, carriers, and other adjuvants that are designated as inert ingredients.

Productivity Rating–The productivity of the various soils on a farm based on the expected yield of the major crop or crops with known management. The rating is a percentage based on standards developed by research and statistical records of crop yields in the region. The best soils have a productivity rating of 100. In some instances, tables have been established showing the relation of crop yield to land values, with fixed management. See Soil Survey.

Profile Leveling–A method of leveling that is used to secure the elevation of a series of points located along a line. Profile leveling is employed in laying out a terrace.

Profundal Zone–The deep- and bottom-water area beyond the depth of effective light penetration. All of the lake floor beneath the hypolimnion. See Hypolimnion.

Progenitor–An individual animal or plant that is recognized as the source of a certain type or character in its offspring.

Prognosis–Forecast as to the probable result of an attack of disease, the likelihood of recovery.

Programmed Harvest–Timber scheduled for harvest for a specified year.

Progressive–Of a disease, developing through successive stages, usually in a certain direction, whether improving or deteriorating.

Propagation–(1) Increasing the number of plants by planting seed or by vegetative means from cuttings, division, grafting, or layering. (2) Construction of drainage ditches by use of dynamite. charges are closely spaced so that detonation of one charge causes the explosion of the next charge and so on until the complete series of charges is exploded. A water-filled soil is most desirable for this method.

Proper Stocking–Stocking of a range area on the basis of its true grazing capacity in a year of adequate rainfall.

Properties–Characteristics by which a substance may be identified. Physical properties describe its state of matter, color, odor, and density; chemical properties describe its behavior in reaction with other materials; biological properties refer to any life-related characteristics such as biodegradation.

Prophylactic–Preventive or protective treatment against disease.

Prophylaxis–Prevention of disease by various measures.

Proportional Weir–(Irrigation) A weir so shaped that the flow rate of water varies directly with the water head.

Protectants–Chemicals containing heavy metals, sulfur, or organic compounds used as sprays, dusts, or dips on seeds, stems, leaves, or wounds of living plants to reduce entrance of fungi or bacteria, and on fabric or wood products to reduce decay.

Protection Forestry–The practice of forestry with the primary objectives of: (a) conserving water supplies; (b) maintaining desirable streamflow; (c) increasing groundwater storage; (d) reducing erosion and reducing sedimentation; (e) providing high-quality water and reducing pollution; (f) ameliorating adverse climatic conditions, especially wind.

Protection Strip–In white pine blister rust control, a belt or zone outside the pine stand that is included in the control unit and is freed of *Ribes* spp., the alternate host of blister rust.

Proteolysis–The process by which casein or some insoluble casein derivative is broken down to water-soluble compounds through the acting of organisms.

Protista–A taxonomic kingdom comprising bacteria, algae, slime molds, fungi, and protozoa. It includes all of the single-celled organisms, some of which are plantlike, some animal-like, and others different from either plants or animals.

Protopam Choloride–An antidote used for certain organo-phosphate poisons.

Protozoa–A group of one-celled organisms that generally do not contain chlorophyll, including amoebae, paramecia, flagellates, and certain spore-forming organisms; sometimes classified as one-celled animals.

Protozoacide–(1) Any chemical agent that kills protozoan parasites. (2) Any agent that is employed for the prevention, suppression, or cure of infection by protozoan parasites.

Provenance–The ultimate natural origin of a tree or a group of trees. Trees having a common center of origin.

Provenetriculus–The glandular or true stomach of birds, which is a spindle-shaped organ between the esophagus and gizzard.

Provenience–Origin; particularly the geographic origin of seed and inoculum.

Pseudo––A Greek prefix meaning false or spurious. In most scientific terms it denotes a deceptive resemblance to the substance to whose name it is prefixed, e.g., pseudocarp, false fruit.

Psittacosis–An acute or chronic viral disease of birds (domestic and wild) transmissible to humans and characterized by systemic reaction and respiratory involvement. Also called parrot fever.

Public Lands–The general public domain; lands belonging to a national government that are subject to sale or disposal and that are not reserved for any special governmental or public purpose.

Puddle–(1) Immersing bare roots of trees and shrubs in a mixture of clay soil and water during transplanting to prevent the roots from drying out. (2) Destroying the desirable soil structure if the soil is worked or cultivated when too wet.

Puffballs–Certain globose fungi, of the family Lycoperdaceae, whose ripe fruit emits clouds of spores when disturbed. The spore fruits of most species of the family are edible when young.

Pulp–(1) In paper making, the product obtained by digesting wood in a slightly alkaline or neutral sodium sulfite cooking liquor. (2) The juicy or fleshy tissue of a fruit.

Pulverize–(Soil) To reduce clods, or peds, to a fine granular state in preparation for planting seeds.

Pulverized Limestone–The product obtained by grinding either calcareous or dolomitic limestone so that all material will pass a 20-mesh sieve and half of the material, a 100-mesh sieve. It is applied to land to reduce soil acidity and to supply calcium and magnesium as a nutrient. Also called finely ground limestone. See Lime.

Punk–(1) A small, scrubby horse. (2) Partly decayed wood.

Punky–Designating a soft, weak, often spongy, wood condition caused by decay.

Pupa–(Plural, pupae) The stage between the larva and the adult in insects with complete metamorphosis, a nonfeeding and usually inactive stage.

Puparium–The hardened larval skin or protective case that encloses the living pupa.

Pupate–The change from an active, immature insect into the inactive pupal stage.

Pure Culture–(1) A crop of one species, variety, etc., grown in contrast to a customary mixture; e.g., timothy grown from a pure seeding rather than mixed with clover. (2) A bacteriological culture that contains only a single species, or one that contains those desired for a particular purpose, as a yeast culture for making vinegar.

Pure Stand–A stand in which at least 80 percent of the trees in the main crown canopy are of a single species.

Putrefaction–Decomposition of animal or vegetable matter, produced by microorganisms in the absence of oxygen.

Putrescible–Organic matter capable of putrefaction.

Pycnidium–In certain fungi, the flasklike fruiting body containing conidia.

Pyr–(Greek) Fire.

Pyrethrins–The active ingredients (pyrethrins I and II) of pyrethrum, which are derived from plants belonging to the genus *Chrysanthemum*, family Compositae. One of the oldest and most widely used insecticides, they control insects affecting humans, animals, crops, and households.

Pyrethrum Powder–An insecticide made from the ground flower heads of various species of chrysanthemums whose use as an insecticide is reported to have originated in the Trans-Caucasus region of Asia about 1800.

Pyrheliograph–A device for measuring the amount and intensity of the incoming radiation from the sun.

Pyrie–(1) Designating the fire factor in an environment. (2) Designating modifications of soil and vegetation caused by fire.

Pyriform–Pear-shaped.

Pyrite–Ironpyrite; FeS_2; the most common form of several metal pyrites, which, on roasting, yield sulfur oxides for the production of sulfuric acid.

Pyroligneous Acid–An acid liquor obtained by the destructive distillation of wood, especially of certain species of hardwoods. It is a complex mixture containing 80 to 90 percent water and many organic compounds including acetic acid and methanol. Also called pyracetic acid, wood vinegar.

Pyrology–The scientific art of protecting forests from fire.

Pyrophyte–A tree with a thick, fire-resisting bark, or one not subject to fatal damage from ordinary forest fires, such as longleaf pine.

Quadrat–A small plot or sample area frequently one square meter (one milacre) in size. In ecological field studies, a unit area in which vegetation changes are recorded.

Quadrivalent–In genetics, a group of four associated homologous chromosomes.

Quagmire–Soft, miry ground, as a bog, marsh, swamp, or morass.

Quail–Any of the various species of small upland, gallinaceous game birds belonging to the genus *Coturnix* and allied genera of the family odontophoridae, found in many parts of the world and often erroneously called partridge. Adapted to agricultural lands, they are seldom regarded as a nuisance and are frequently propagated for game purposes. The North American bobwhite quail, *Colinus virginianus*, is common in the central and eastern United States. The California quail, valley quail, and mountain quail are common in the western United States.

Quaking Aspen–*Populus tremuloides*, family Salicaceae; a small to fairly tall, short-lived, deciduous tree. It is the most widely distributed tree in North America. It occurs from New Jersey to Alaska. It grows in dense stands following logging or burning of forestlands. Its wood is now valued chiefly as a source of pulp for paper making, and it furnishes browse for sheep, goats, and deer, and food for beaver. Native to North America.

Qualitative Traits–Traits having a sharp distinction between phenotypes, and which are usually controlled by only a few genes; e.g., various coat colors and the horned trait in domestic animals.

Quantitative Traits–Traits that do not have a sharp distinction between phenotypes, and usually require some kind of measuring tools to make the distinctions. These traits are normally controlled by many pairs of genes; e.g., growth rate, milk production, and carcass quality. See Genotype, Phenotype.

Quarantine–(1) A regulation under police power for the exclusion or isolation of animal and plant pests or diseases and insects: (a) the isolation of an animal sick with a contagious disease; (b) a place where the sick are detained away form other animals until the danger of spread of a contagious disease has disappeared. In its wider application, the quarantine may be enforced against an individual animal, against all the animals, or all the animals of the same species, in a township, county, or state, and against those in a foreign country (2) Prohibition to prevent the introduction or spread of any dangerous insects or plant diseases.

Quarter–(1) In slaughtering for meat, one half of the side of beef, as the forequarter or hindquarter. (2) Pertaining to a horseshoe, the branch between the last nail hole and the heel. (3) A unit of weight: (a) one-quarter cwt. (25 pounds) (avoirdupos); (b) eight bushels, formerly one-quarter ton (especially of grain). (4) A quarter section of land, or 160 acres. (5) A section of the bovine udder.

Quarter Section–According to the survey of public lands by the general Land Office (United States), one of four parts of a section of land, generally 160 acres. The square section of land was divided into four equal parts by a north-south and east-west line, and in a legal land description the quarters were designated by the points of the compass as NE, NW, SE, and SW.

Quartz–SiO_2; a mineral that is a crystalline form of silica. It is the most common mineral component of sand soils, especially in humid regions. The earth contains about 47 percent oxygen and 28 percent silica.

Quick Lime–See Lime.

Quick Tests–Certain standard chemical tests devised for the very rapid determination of the amounts of nutrient elements in a soil. When properly performed and interpreted such tests are useful in making recommendations for fertilizers and lime.

Quick-Growing–Designating a plant in a particular climatic region that matures in a comparatively short period; e.g., buckwheat and aspen are quick-growing among crops and trees in a cool climate.

Quicksand–Any loose fine sand that is mobile or semifluid when supersaturated with water. It is encountered in excavations, especially in localities where soils are derived from lacustrine sediments or river alluvium. Also called running sand.

Quinine–An alkaloid obtained from the bark of the tropical tree *Cinchona officinalis*. It is a white crystalline powder that is odorless with a very bitter taste. Its principal use is in medicine where it is specific for all forms of malaria. Also used for its analgesic, oxytocic, antipyretic, and sclerosing properties. Quinine can be made synthetically.

R Horizon–Underlying consolidated bedrock, such as granite, sandstone, or limestone. See Soil Horizon.

R Layer–In soil profile description, the rock on which the soil rests. See Soil Horizon.

Rabbit–Any of certain small mammals of the family Leporidae. Hare and rabbit are often used interchangeably. However the biologist classifies the rabbit as having shorter and smaller legs and ears, and giving birth to naked and helpless young with eyes closed, while the newborn hare has fur and is quite capable of caring for itself.

Rabbit Control–See Myxomatosis.

Rabbit Pox–An acute eruptive disease of laboratory rabbits, caused by a virus related to vacinia virus.

Rabbit-proof Fence–A fence of special design which is placed around cultivated fields in New South Wales, Victoria, and Western Australia to keep out rabbits which have overrun the country and have become a very serious pest.

Rabies–An infectious disease caused by a filterable virus which is communicable by means of a bite in which saliva containing the virus enters the wound. It occurs most frequently in dogs, but many other animals and people are quite susceptible. Also called hydrophobia or canine madness.

Raccoon–*Procyon lotor*, family Procyonidae; a nocturnal mammal, which is gray, with black and white facial markings. It eats and does some damage to corn especially when the kernels are well-developed but soft.

Race–(1) A group of individual plants which have certain common characteristics because of ancestry. It is generally considered a subdivision of a species; frequently used in plant breeding. (2) Pathogens of the same species and variety which are structurally indistinguishable but which differ in their physiological behavior, especially in their ability to parasitize varieties of a given host. (3) The channel that leads water to or from a waterwheel, the former is the head race and the latter the tail race. (4) A narrow passage or fenced land in a sheep yard for branding, dipping, etc. (5) An elongated white mark on the face of a horse or dog.

Rack–(1) The gait of a horse in which only one foot touches the ground at any one time, producing a four-beat gait. (2) A frame attached to a truck or wagon for the transportation of hay, tobacco, etc. (3) The rib portion of a sheep carcass. (4) A framework for holding feed for cattle, swine, sheep, etc., with upright partitions so that the animal can insert its head between the partitions and have access to the feed. (5) A frame placed in a stream to prevent the passage of fish. (6) A frame placed at the entrance to a sump pump to remove debris that would clog the pump.

Radical–A group of different elements acting as a single unit in a chemical reaction; normally incapable of separate existence. A radical may be negatively charged, positively charged, or without a charge.

Radioactive Tracers–Small quantities of radioactive isotope mixed with larger amounts of the corresponding stable isotope to be used as labels. Since the stable and radioactive isotope act chemically and biologically in the same manner, as the radioactive one is readily detected.

Radioactive Element–An element capable of changing spontaneously into another element by the emission of charged particles from the nuclei of its atoms. For some elements, e.g., uranium, all known isotopes are radioactive; for others, e.g., potassium, only one of the several isotopes is radioactive. Radioactive isotopes of most elements can be prepared artificially, but only a few elements are naturally radioactive.

Radioactive Soil–A soil which possesses a minute degree of radioactivity. Radioactive phosphorus and other plant nutrients are sometimes added to the soil in research work to discover more facts about their functions in plant growth.

Radioactivity–The nuclear energy released when the nucleus of an atom disintegrates.

Radiocarbon Dating–The determination of the age of organic remains by measuring its radioactivity caused by the presence of C14 which has a half-life of 5,568 years and maintains a constant rate of disintegration after the death of an organism.

Radius Cruising–The distance between locations at which an individual animal is found at various hours of the day, at various seasons, or at times during various years.

Ragged–Of the fur or hair of animals, shaggy, rough, and hanging in tufts.

Rain–Precipitation in the form of water condensed from the atmosphere which falls in drops.

Rain Belt–A region of heavy or above-average rainfall.

Rain Forest–A low-altitude, closed, evergreen (usually broad-leaf) forest, in equatorial regions where annual rainfall exceeds 80 inches with no or very short dry seasons and where the temperature is high and constant and does not approach freezing at any time.

Rain Shadow–Refers to an area in which little or no rain falls because it is located to the leeward or dry side of mountains which on the opposite side are exposed to moisture-laden winds.

Rainbow Trout–A highly valued game fish, *Salmo gairdnerii*, that inhabits cool, clear streams and lakes. The fish are greenish in the upper body, white on the belly and have a pink or red stripe down the side. They are grown commercially for stocking lakes and streams.

Raindrop Erosion–Soil splash resulting from the impact of raindrops on bare supersaturated soil. The result is surface sealing and sheet erosion.

Rainfall–The amount of precipitation, expressed in the form of water but including snow, sleet, and hail which falls in a given time. In the United States, it is usually measured in inches of water.

Rainfall Distribution–The distribution of annual rainfall over the different seasons or months.

Rainfall Intensity–The rate at which rain is falling at any given instant, usually expressed in inches per hour.

Rainfall Interception–The amount of rainfall retained by leaves, branches, trunks, and surface vegetation, both living and dead, which never reaches the soil.

Rainfall Penetration–The depth below the surface of the soil reached by a given quantity of rainfall.

Rainmaker–(1) A rain simulator used for applying water in drops upon a research plot surface. (2) An individual who claims to be able to increase the rainfall of an area.

Rainstorm–A storm accompanied by rain.

Rainwater–Water that falls from the atmosphere in the form of rain, at times being caught and directed into cisterns for further use.

Rainy Season–In the tropics, the annual season of heavy or more-than-normal rainfall as contrasted to the dry season of little or no rainfall.

Raise–To grow or produce, as to raise corn or cattle.

Rammed Earth Wall–An earthen wall made by packing moist clay between heavy plank forms. Some type of protective cover of the earth wall is constructed to decrease raindrop splash erosion.

Ranch Cattle–Any of the several crossbred varieties of cattle raised as beef animals; e.g., the Santa Gertrudis; the Beefmaster, a three-way cross of Brahman (Zebu), Herefore, and Shorthorn; and the Brangus, an animal which is about 37 percent Brahman (Zebu) and 63 percent Aberdeen Angus. Others include the Braford, a cross between the Brahman (Zebu) and the Hereford; and the Charbray, an animal 12.5 to 25 percent Brahman, the main strain being Charolais. The purpose of this crossbreeding has been to produce beef animals better adapted to tropical climates and more resistant to tick fever.

Random Sample–A sample taken without bias from an area or from a population in which every part has an equal chance of being taken, in contrast to systematic sampling.

Random Searching–The hypothesis that postulates that an organism obtains food, suitable niches, and mates by entirely unorganized search, in contrast to systematic searching.

Rang Utilization–(1) For a single plant or species, the degree to which the foliage or herbage has been removed in percentage of the current growth within reach of livestock. (2) For an entire range, the relative amount eaten.

Range–(1) Uncultivated land, including forest land, which produces forage suitable for livestock grazing. (2) Specifically, a unit of grazing land used by an integral herd of livestock. (3) Ecologically, the geographic area of natural occurrence of certain plants or animals. (4) In the United States land survey, a tier of townships, according to number east or west of a principal meridian line. See Township. (5) The difference between extremes.

Range Allotment–A specific range area to which certain livestock of a permittee or group of permittees are assigned. Also called grazing allotment.

Range Appraisal–A definite, written, detailed opinion by a qualified appraiser of the value of range land. Among the considerations involved are its grazing capacity, accessibility, facilities for handling livestock, availability of feed sources, and income-producing ability.

Range Caterpillar–*Hemileuca oliviae*, family Saturniidae; a range pest on wild grasses in the southwestern United States, which sometimes infests corn and other cultivated plants. Its larvae are covered with coarse, poisonous spines.

Range Condition–(1) The state of health or productivity of both soil and forage of a given range in terms of what it could or should be under a normal climate and under the best practicable management. (2) An animal that is in a sufficient state of health or condition to be kept on the range.

Range Count–A census made on a range of the animals using a grazing area as contrasted to feedlot, corral, driveway, or other similar counts.

Range Crane Fly–*Tipula simplex*, family Tipulidae; an insect whose dark, leathery maggots bore into and destroy plant roots. It is sometimes destructive on the ranges of southwestern United States.

Range Ecology–The specialized branch of ecology which deals with vegetational response to environmental factors on rangeland, especially with the effects of grazing.

Range Fire–On rangeland, any fire which is not being used as a tool in range management.

Range Indicator–Any plant community portraying the condition of its environment which can be used as an indicator for the condition of a range.

Range Paralysis–See Fowl Paralysis.

Range Plant–Any herbaceous or shrubby plant which grows on a range.

Range Plant Cover–The ground vegetation composed of all herbaceous and shrubby plants which are within easy reach of livestock on a range area regardless of whether or not they constitute forage.

Range Readiness–The stage of growth of the important palatable plants on the range and the condition of soil which permits grazing without undue compacting of the soil or endangering the maintenance of the plants.

Range Renovation–Improving rangeland by discing or other mechanical means, chemical treatment, or reseeding.

Range Site–An area of land having a combination of edaphic, climatic, topographic, and natural biotic factors that is significantly different from adjacent areas. Various sites are significantly different in their potential forage production and/or different in management requirements for proper land use.

Range Suitability–The adaptability of a range to grazing by livestock and/or game.

Range Survey–(1) A determination by inspection of the carrying capacity of extensive areas of natural vegetation, based on the density and palatability of the forage. (2) an inventory which assembles important facts needed for perfecting a sound management plan taking into account the number of animals grazed, grazing capacity, period grazed, salt, needed water development, areas in need of revegetation, better livestock distribution, and the location of special problem areas.

Range Trend–The direction of change in range condition, either deterioration or improvement.

Range Type–An area of range which differs from other areas primarily by the difference in plant cover, such as grassland, browse, or conifer. One vegetation group can be distinguished from another group by difference of dominating species.

Range Woodland–Woodlands also used as range.

Rangeland–(1) Land on which the natural plant cover is composed principally of native grasses, forbs, or shrubs valuable for forage. (2) Land used for grazing by livestock and big game animals on which the natural potential climax community of plants is dominated by grasses, grasslike plants, forbs, and shrubs.

Raptors–Birds of prey, such as eagles, falcons, hawks, owls.

Rare Species–A rare species is one that, although not presently threatened with extinction, exists in such small numbers throughout its range that it may be endangered if its environment worsens.

Rat–*Rattus*, family Muridae; a long-tailed rodent which is much larger than a mouse. A serious pest, it is very destructive to stored food and may carry disease.

Rat Bite Fever–A disease caused by the flagellated organism *Spierillum minus* carried by rats and transmitted to people by their bites. Characterized by severe intermittent attacks, the ailment is not contagious. Especially prevalent in Japan.

Rat Mange–A mange caused by an infestation of *Notoedres muris*, a very small mite just visible to the naked eye.

Rat Poison–(1) A poison for rats obtained from the seed of a shrub, *Chailletia toxicaria*, native to West Africa. (2) *Hamelia erecta*, family Rubiaceae; a tall evergreen shrub, bearing scarlet or orange flowers and small, purple-red fruit, sometimes grown as an ornamental in warm areas. Native to tropical America. Also called scarlet bush.

Rat Typhus–A disease caused by the parasitic organism *Rickettsia prowazekii*; symptoms are weakness, headache, chills, and fever. It is transmitted to humans by the oriental rat flea *Xenopsylla cheopis*, family Pulicidae. Also called murine typhus, flea typhus.

Rate of Spread–In forestry, the quantitative increase in area per unit of time, used to describe forest fires and serious new forest disease.

Ratio Kill–The proportion or percentage of wildlife population which can be killed yearly without diminishing subsequent yield.

Ravine–A sharply cut, natural depression, V-shaped in transverse profile, usually without a permanent stream in its bottom. It has less width and size than a valley, but is larger than a gully.

Raw Land–Virgin land; an uncleared tract which is not ready for cultivation.

Raw Water–Water which is available as a supply for use but which has not yet been treated or purified.

Raw-rock Phosphate–Phosphate rock, finely ground but otherwise untreated, which is sometimes used as a slow-acting fertilizer on acid soils.

Ray–(1) In wood anatomy, a ribbon-shaped strand of tissue formed by the cambium and extending in a radial direction across the grain which serves to store food and transport it horizontally in the tree. (2) In botany, one of the small flowers radiating out from the margin of a dense inflorescence.

Ray Fungus–*Actinomyces bovis*; an organism widely distributed in nature which may enter the tissue of the various organs of the animal causing swelling of either the bone or soft tissue. The infection usually reaches the surface so that fistulae are established and pus-producing organisms gain entrance. It is the cause of lumpy jaw in cattle.

Razorback–(1) A type of hog with long legs and snout, sharp narrow back, and lean body; usually a half-wild mongrel breed (southern United States). (2) A sharp-ridged spur or hill.

Reaction–(1) A change in a market trend. (2) The degree of acidity or alkalinity (e.g., of a soil mass) expressed in pH values and in words as follows: extremely acid, below 4.5; very strongly acid, 4.5-5.5; medium acid, 5.6-6.0; slightly acid 6.1-6.5; neutral, 6.6-7.3 (strictly, 7.0); mildly alkaline, 7.4-8.0; strongly alkaline, 8.1-9.0; very strongly alkaline, over 9.1.

Reaeration–The absorption of oxygen in water from the atmosphere. This phenomenon enables self-purification of streams by providing the necessary oxygen to bacteria.

Reagent–Any substance involved in a chemical action.

Receiving Waters–Rivers, lakes, oceans, or other water courses that receive treated or untreated waste waters.

Recent Soil–A soil which has a profile without definite horizons of eluviation or illuviation; e.g., recently deposited alluvium. See Eluviation, Illuviation.

Receptive Hypha–A fungous hypha in a pycnium of rusts, with a female sexual function, ready to be fertilized.

Recessive–In genetics, a gene or trait which is masked by a dominant gene.

Recharge–The processes by which water is added or the amount added to the water table.

Reclamation–(1) Mining usage. The filling in of open pits, grading of the mined area, reduction of high walls, replacement of topsoil, planting, revegetation, and such other work as is necessary to restore an area of land disturbed by surface mining operations. (2) The process of reconverting mined land to other forms of productive uses. (3) Removing excess soluble salts from soils.

Reconnaissance–(1) A cruise of forest property to obtain general information of the forest conditions. (2) An extensive range survey which is carried out to estimate average density and composition of range vegetation within a type or subtype without use of systematically established plots. (3) A type of survey in which land features are examined at wide intervals and are not delineated in detail. It is intended to furnish information primarily for extensive or overall planning. See Soil Map.

Recovery Zone–The area of a stream in which active, primarily aerobic, decomposition of the pollutants is occurring.

Recreation, Outdoor–(1) Leisure-time activity such as swimming, picnicking, boating, hunting, and fishing. (2) Use of leisure time for personal satisfaction and enjoyment. It may be undertaken individually or with others. It may be planned or spontaneous. It may be passive or active, may or may not require skills and training, and may or may not require a designated area. (3) Outdoor recreation involves the protection, preservation, development, public use, and enjoyment of scenery, water, primitive or natural landscape including roadless areas), wildlife, natural phenomena (e.g., petrified wood), and archeological and historical sites.

Recycling–The process by which waste materials are transformed into new products in such a manner that the original products may lose their identity.

Red Algae–A family of algae (Rhodophyceae) most members of which are marine. They contain a red or violet pigment in addition to chlorophyll.

Red and Black Harlequin Bug–See Harlequin Bug.

Red Ash–*Fraxinus pennsylvanica*, family Oleaceae; a hardy deciduous tree, grown especially in the prairie states of the United States for shade, elsewhere for its lumber. Its wood is used in handles, furniture, oars, baseball bats, cabinets, boat buildings, implements, woodenware, novelties, cooperage, planing-mill products. Also called river ash.

Red Carpenter Ant–*Camponotus herculeanus* and *C. pennsylvanicus*, family Formicidae; an ant which lives in large colonies in the north temperate zone under stones, stumps, logs, and often far under ground.

Red Fir–(1) *Abies magnifica*, family Pinaceae; an evergreen tree grown for its valuable lumber used for general construction, boxes, crates, pulpwood, and planing-mill products. Native to North America. Also called California red fir. (2) See Douglas-fir.

Red Harvester Ant–*Pogonomyrmex barbatus*, family Formicidae; an ant about ¼ inch long, with a black head and legs, and red abdomen. It lives in large colonies and makes bare circles 2 to 12 feet in diameter, with a deep nest. It is injurious to grasslands in southwestern United States because it feeds on grass seed. Also called agricultural ant.

Red Haw–Downy hawthorn, *Crataeus* spp. See Hawthorn.

Red Hickory–*Carya ovalis*, family Juglandaceae; a deciduous tree grown for its valuable wood, used in tool handles, ladders, furniture, sporting goods, implements, woodenware, and for fuel and smoking meat. It is also grown as a shade tree. Native to North America. Also called mockernut, oval pignut hickory.

Red Maple–*Acer rubrum*, family Aceraceae; a large tree, grown as an ornamental and as a shade tree for its brilliant scarlet and yellow, autumnal foliage. Native to North America.

Red Mulberry–*Morus rubra*, family Moraceae; a tree which produces red or purplish-red, edible fruit about one inch long. Native to North America. Also called American mulberry. See Mulberry.

Red Oak–Northern red oak, *Quercus rubra*; southern red oak, *Q. falcata*; a tree whose lumber is valued for its strength and durability. It is also valued as lumber for constructing furniture.

Red Pine–*Pinus resinosa*, family Pinaceae; a quick-growing, hardy, evergreen tree, grown as an ornamental and for its valuable lumber. Native to North America. Also called Norway pine.

Red Spruce–*Picea rubens*, family Pinaceae; an evergreen, coniferous tree which is a chief source of paper pulp. Native to North America.

Red Tailed Hawk–*Buteo jamaicensis*, a hawk widely distributed across the eastern United States that is characterized by a short red colored tail. They eat large numbers of rodents and are generally considered to be beneficial.

Red Worms–*Stongylus*, family Strongylidae; nematodes which infest the large intestine of equines. Their red color is due to the presence of hemoglobin in the bodies of the worms, and not to blood sucked from their hosts.

Redbud–*Cercis canadensis*; a wild shrub that blooms in early spring. See Eastern Redbud.

Redbug–See Chigger.

Redcedar–Eastern, *Juniperus virginiana*; western, *Thuja plicata*; a tree that produces lumber valued for its fragrance and decay-resisting qualities.

Redd–(1) A gravel bed in a river in which salmon lay their eggs. (2) The spawning nest which is excavated in the gravel or stones of the stream bed, filled with the eggs, and then partially refilled with coarse stones. Applied especially to the nests of the family Salmonidae.

Redox Potential–The expression for the relative state of oxidation.

Reduction–(1) The process of removing oxygen from a compound; e.g., hematite is reduced to metallic iron. (2) The addition of electrons to an atom or ion.

Redwood–*Sequoia sempervirens*, family Pinaceae; an evergreen tree which grows to 340 feet and is valued as an important source of lumber. Native to California and Oregon, United States. See Giant Sequoia.

Reed–(1) Any tall, slender plant, usually having coarse and jointed stems, including certain grasses and grasslike plants. (2) *Phragmites communis*, family Gramineae; the common reed, a widespread, perennial marsh grass with erect culms 10 to 15 feet tall, and stout, creeping rhizomes; used for making mats, thatching, cordage, and carrying nets. Native to north temperate zone and South America.

Reed Peat–A class of peat developed from reeds.

Reedgrass–Any grass of the genus *Calamagrostis*, family Gramineae. Species are tall, perennial grasses with creeping rhizomes, valuable as range grass and for wild hay.

Reentry Period–The length of time that is required by law between the application of certain hazardous pesticides and the entrance of people into the area without protective clothing.

Reforestation–The natural or artificial restocking with forest trees of an area previously under forest.

Refuge–(1) A tract of land set aside for wildlife protection, particularly for game animals and wildfowl. In practice, the protection afforded in a refuge is more limited than in a sanctuary. (2) In nature, a place affording protection against enemies; a shelter; a covert.

Regenerative Agriculture–A system of farming and ranching whereby resources can be used economically in perpetuity. See Iatro-agriculture.

Regolith–Soil and mantle rock; saprolith. The layer or mantle of loose, incoherent rock material, of whatever origin, that forms the surface of the land and rests on bedrock. It comprises rock waste of all sorts, including volcanic ash and glacial drift.

Regression–(1) Destruction of the vegetation, as by fire, grazing, cutting, etc., usually with subsequent deterioration of the site, as by exposure, erosion, or loss of nutrients, to such extent as to give rise to a subsequent simpler vegetative type. It is not a true succession or development from forest to grassland, but a replacement as a consequence of complete destruction of the trees, etc. (2) Measure of the relationship between two variables. The value of one trait can be predicted by knowing the value of the other variable; e.g., easily obtained carcass traits (hot carcass weight, fat thickness, ribeye area, and percent of internal fat) are used to predict percent cutability.

Regulation Cut–The determination, periodically or in advance of cutting, of the volume of timber to be felled under the objectives of a given management plan.

Reindeer–Any of several species of the genus *Rangifer*, a type of large deer with branching horns used in Arctic regions for food.

Rejuvenate–The use of lime and fertilizer to enhance soil productivity.

Relative Age–In soil development, the appearance of the profile features; e.g., a soil along a stream may receive floodwater sediment every year or two and therefore remain relatively young in profile development.

Relative Humidity–The ratio of the mass of moisture actually present in any volume of air at a given temperature to the maximum amount possible at that temperature and pressure, usually expressed in percentage.

Relief–(1) The variation in elevation of the ground surface. It is indicated on maps by hachures, shading, or more accurately, by contour lines. (2) The physical or geomorphological features of a land surface. (3) The local difference in elevation of land features expressed in feet or in such terms as strong, mild, low, rugged.

Relief Well–A small-diameter riser extending from inside a tile drainage system to the surface air, especially at the beginning of a steeper gradient to permit uniform flow when the tile line is running full of water. The steeper gradient will cause faster water flow; this will produce a partial vacuum and restricted flow unless more atmospheric air is supplied by a relief well. Also known as relief pipe or breather.

Renewable Natural Resources–Resources such as forests, rangeland, soil, and water that can be restored and improved to produce the food, fiber, and other things humans need on a sustained basis.

Renewable Resources–Renewable energy; resources that can be replaced after use through natural means; e.g., solar energy, wind energy, energy from growing plants.

Renovate–To renew, as by pruning old wood to induce fresh growth; to fertilize and spray old lawns to produce new growth; to repair and again make usable as to renovate a machine or building.

Repellent–(1) Any substance obnoxious to insects, which prevents them from injuring their hosts or laying eggs for hatching future generations. (2) Any substance disliked by a nuisance animal, used to discourage its visits.

Reproduction Period–The time required or normally decided upon for the renewal of a stand of trees by natural or artificial regeneration.

Repulsion–In genetics, the condition in which an individual heterozygous for two pairs of linked genes receives the dominant member of one pair from one parent and the dominant member of the second pair from the other parent; e.g., AAbb x aaBB.

Research–All effort directed toward increased knowledge of natural phenomena and the environment and toward the solution of problems in all fields of science. This includes basic and applied research. Much of the agricultural productivity of the United States is directly the result of applying research.

Reserve–(1) Any tract of land, especially public land, set aside for some special use; e.g., forest reserve, school reserve. Also called reservation. (2) A tree or group of trees left uncut on an area for a period, usually a second rotating. After the stand is reproduced, naturally or artificially, an active stand which is held for future utilization.

Reserve Food–Plant food stored in the various parts of any plant, especially that stored in the roots and crowns of biennial and perennial grasses and legumes.

Reserve Soil Acidity–Potential soil acidity. See Reaction.

Resident Species–Species common to an area without distinction as to being native or introduced.

Residual–(1) Remaining in place after all but the least soluble constituents have been removed. Said of the material eventually resulting from the decomposition of rocks. (2) Standing, as a remnant of a formerly greater mass of rock or area of land, above a surrounding area which has been eroded. Said of some rocks, hills, mountains, mesas, plateaus, and groups of such features. (3) Soil developed in place from underlying bedrock. See Monadnock.

Residual Effect–The effect of a particular product or substance that remains after its initial use. Some insecticides and fertilizers remain effective for some time after they are applied.

Residual Stand–Trees, often of saw log size, left in a stand after thinning to grow until the next harvest. Also called reserve stand.

Residual Value–Of fertilizers, limes, and manures. The value of the fertilizer, lime, or manure to succeeding crops after it has been in the soil for one cropping season.

Residue–A deposit of an pesticide which persists on an exterior surface or which is absorbed by plant or animal tissue following treatment.

Residue Management–Of crops. Use of that portion of the plant or crop left in the field after harvest for protection or improvement of the soil. See Conservation Tillage.

Residues–(1) Coarse residues: plant residues suitable for chipping, such as slabs, edgings, and ends. (2) Fine residues: plant residues not suitable for chipping such as sawdust, shavings, and veneer clippings. (3) Plant residues: wood materials from primary manufacturing plants that are not used for any product. (4) Logging residues: the unused portions of sawtimber and poletimber trees cut or killed by logging. (5) Urban residues: wood materials from urban areas, such as newspapers, lumber, and plywood from building demolition, and used packaging and shipping wood materials.

Resinosis–An abnormal exudation of resin or pitch from conifers or the abnormal impregnation of their tissues by resin. Also called resin-flux.

Resins–A class of flammable, amorphous, vegetable substances secreted by certain plants and trees, and characterizing the wood of many coniferous species. They are oxidation or polymerization products of the terpenes, and consist of mixtures of aromatic acids and esters. Produced form oleoresins.

Resources–The available means for production. Land, labor, and capital are the basic means of production on farms.

Respirable Particulate Matter–That portion of the total particulate matter that has an especially long residence time in the atmosphere and penetrates deeply into the lungs. These characteristics are due to size (smaller than 2 to 3 microns).

Respiration–(1) A chemical process that takes place in living cells whereby food (fats, carbohydrates, and proteins) is "burned" (oxidized) to release energy and waste products, mainly carbon dioxide and water. Living things use energy produced through respiration to drive vital life processes such as growth and reproduction. (2) The oxidation of carbohydrates in living organisms and the attendant release of energy and liberation of carbon dioxide and water. (3) In animals, the act of breathing; the drawing of air into the lungs and its exhalation. In small organisms with no special breathing organs, the process takes place over a large part of the body surface.

Respirator–A device worn over the nose or mouth to protect the respiratory tract during the spraying of pesticides, working in dusty conditions, etc.

Rest Period–A period of quiescence or inactivity in plants even though moisture and temperature conditions are favorable for growth. See Dormancy.

Rest-rotation Grazing–An intensive system of management whereby grazing is deferred on various parts of the range during succeeding years, allowing the deferred part complete rest for one year. Two or more units are required. Control by fencing is usually necessary on cattle range, but may be obtained by herding on sheep ranges.

Rested Pasture–Pasture ungrazed for an entire growing season.

Resting Land–See Fallow.

Resting Pasture–A pasture not grazed by livestock.

Restock–To stock again; to add new or additional livestock, game, or fish.

Restorer Line–Parent plants used to restore adequate levels of fertility in hybrid crops such as hybrid sorghum or hybrid cotton.

Restricted Use Pesticide–A pesticide that can be applied only by a certified applicator. These pesticides are toxic to humans and/or pose a potential threat to the environment.

Restriction Enzymes–Enzymes used in genetic engineering to remove a gene from a piece of DNA.

Retention Reservoir–A reservoir used strictly for flood control. One or more openings at the base of the dam, open at all times, are of such size that the discharge is never more than the capacity of the stream below.

Reticulate Mottling–In oxisols, a network of coarse streaks of different colors characteristic of parent materials.

Revegetation–Reestablishment of vegetation which may take place naturally or be induced by humans through seeding or transplanting.

Reverse Osmosis–(1) An external force is used to reverse normal osmotic flow through a semipermeable membrane, resulting in movement of water from a solution of higher solute concentration to one of lower solute concentration. See Osmosis. (2) A process of desalination of seawater whereby only pure water passes through a fine membrane while the salts cannot pass through.

Reverted Phosphoric Oxide–The phosphoric oxide in phosphate fertilizers may be divided into three parts: (a) water-soluble, (b) insoluble in water but soluble in neutral ammonium citrate, and (c) insoluble in water or neutral ammonium citrate. The addition of lime, limestone, cyanamide, ammonia, or similar basic materials may cause some reversion of phosphoric oxide to less-soluble forms. Reverted phosphoric oxide is that which has been changed from the water-soluble to the citrate-soluble form or from the citrate-soluble to that phosphoric oxide that is neither water-soluble nor citrate-soluble.

Revetment–A facing of stone or other material, either permanent or temporary, placed along the edge of a stream to stabilize the bank and protect it from the erosive action of the stream.

Rhizobia–Bacteria that live symbiotically in roots of legumes and fix nitrogen from the air.

Rhododendron–Certain north temperate zone, evergreen or deciduous shrubs of the genus ***Rhododendron***, family Ericaceae. Species are grown as ornamentals. They require acid soils.

Rice-Fish Rotation–The alternation of rice and fish crops. Rice fields, which would ordinarily lie fallow and idle, are flooded and made to produce a crop of fish for one or two years. In addition to yielding income from the fish, the land is rested and enriched for the following rice crop.

Ridge Land–Any broad or narrow, crested, elongated land feature which rises above or separates lowlands plains or valley basins, sometimes constituting a drainage divide.

Ridging–(1) Pulling the soil into a low ridge at the base of plants. Potatoes are commonly ridged or hilled to keep the tubers covered and prevent greening caused by exposure to sunlight. (2) Making small embankments or borders in fields to control irrigation water, conserve runoff from rainfall, or to assist in drainage.

Riding Down–Pushing over of small trees, shrubs, and fences by livestock in order to reach and browse in the foliage.

Riffles–Shallow rapids in an open stream where the water surface is broken into waves by obstructions wholly or partly submerged.

Riga Pine–Scotch pine.

Rill–(1) A very small trickling stream of water; a very small brook. (2) A minute stream that flows away from a beach as a wave subsides. (3) A small channel made by circulating water in the wall, floor, or ceiling of a cave. An erosion channel a few centimeters deep.

Rill Erosion–An erosion process in which numerous small channels of only several centimeters (inches) in depth are formed; occurs mainly on recently cultivated soils.

Rime–A white, icy coating on grass and leaves, formed by hoar frost, white frost, etc.

Ring–(1) A cut or girdle around the trunk, branches, or roots of a tree. See Girdle. (2) Annual growth ring of a tree. See Annual Ring. (3) (a) A circular band of metal or wood, as the metal ring in the nose of a bull. (b) To place a ring through the cartilage of the nose of an animal; e.g., to prevent a hog from harmful rooting or to control a bull, etc. (4) (a) A circular, metal or plastic band placed on the leg of a fowl for identification purposes. (b) To place a ring on the leg of a fowl. Also called ringing birds. (5) A ridge which encircles the horns of a cow, the number increasing with age. (6) A circular exhibition place for the showing or sale of livestock or the racing of horses.

Ring-necked Pheasant–***Phasianus colchicus torquatus***, family Phasianidae; a long-tailed, highly colored, gallinaceous bird with a white ring neck. A highly popular game bird, it is sometimes a nuisance on farms because it pulls young corn plants out of the ground and devours what is left of the kernel. It is also occasionally harmful to vegetable crops, such as tomatoes. Originally from China.

Ring-porous Wood–Any of a group of hardwoods in which the pores are comparatively large at the beginning of each annual ring and decrease in size more-or-less abruptly toward the outer portion of the ring, thus forming a distinct inner zone of pores known as the spring wood and an outer zone with smaller pores known as the summer wood.

Rio–(Spanish) River.

Ripa–(1) A stream or river bank. (2) Land which borders the water line of a river or sea.

Riparian–(1) Designating land which borders a stream or body of water. (2) Pertaining to the banks of a body of water; a riparian owner is one who owns to the edge of a body of water; a riparian right is the right to control and use water by virtue of the ownership to the bank or banks of a body of water.

Riparian Doctrine–Under common law (United States), the owner of land along a stream is entitled to have the streamflow to his land undiminished in quantity and unimpaired in quality by upstream riparian owners, except that such owners are entitled to use water for domestic purposes. In some states, it is held that the riparian right includes the right to make use of the water for irrigation purposes and other uses, and that such right is a property right entitled to protection. See Aquatic Rights.

Riparian Forest–Tree growth adjacent to streams or other watercourses whose roots are in or close to the zone of saturation due to the proximity of surface or underground water.

Riparian Habitat–That portion of a watershed or shoreline influenced by surface or subsurface waters, including stream or lake margins, marshes, drainage courses, springs, and seeps.

Riparian Owner–One who owns land adjacent to a stream or other body of water.

Riparian Rights–The rights accruing to a landowner on the bank of a natural watercourse, lake, or ocean. These rights vary with state laws. Riparian rights cease at the water's edge and do not interfere with use of the water area by others offshore.

Riparian Vegetation–Vegetation growing along the banks of a small lake, river, swamp, or spring; also known as phreatophytes, and riverine and riverain vegetation. Vegetation growing along a large lake or ocean are termed littoral.

Ripe–(1) Designating mature seeds which are fit for gemination. (2) Designating fruit which has attained full development. (3) In plant propagation, designating wood that will root well. (4) In grafting, designating that wood which is ready for perfect union. (5) Designating the best condition for use, as ripe cheese, ripe wine.

Ripe Snow–Snow which during the process of melting and settling has attained its maximum power of water suspension without loss of water. Snow is overripe when it has exceeded its ability to hold water and is losing it. The difference in density between ripe and overripe snow does not exceed 10 percent.

Riprap–Broken rock, cobbles, or boulders placed on earth surfaces, such as the face of a dam or the bank of a stream, for protection against the action of water (waves); also applied to brush or pole mattresses, or brush and stone, or other similar materials used for soil erosion control.

Rise of Salts–Soluble salts, previously distributed in the soil, which is brought to the surface with the rise of capillary water and left as a surface incrustation when the moisture evaporates. This accumulation may cause crop injury. See Waterlogged.

River–A stream of water bearing the waste of the land from higher to lower ground, and as a rule to the sea. A trunk stream and all the branches that joint it constitute a river system. Stream is a general term, with little relation to size. Rill, rivulet, brook, and creek apply to streams of small or moderate size. River is generally applied to a trunk stream or to the larger branches of a river system.

River Ash–See Red Ash.

River Bank–(1) The nearly vertical or steeply sloping ground, generally alluvium, which forms the side of the river channel. Generally, it is not as high as a bluff, although there is little distinction between the two. Banks are subject to continuous, slow erosion by river flow, or to rapid undercutting during flood periods. (2) The brim of a river or land bordering its edge.

River Basin–The area of land drained by a river and its tributaries. See Watershed.

River Bottomland–Low land which consists of alluvial deposits along a river. Generally, but not everywhere, it is composed of soil more fertile and potentially more productive than adjacent upland.

Riverain–Of or pertaining to a river.

Riverine–Pertaining to, or formed by, a stream or river systems, such as riverine erosion or riverine vegetation along the banks of a river.

Riverwash–Alluvial deposition in the channel of rivers.

Rock–(1) Any consolidated and relatively hard naturally formed mass of mineral matter; stone. (2) A peak or cliff of rock, usually bare, and considered as one mass, as the Rock of Gibraltar. (3) To the engineer, the term rock signifies firm and coherent or consolidated substances that cannot normally be excavated by manual methods alone.

Rock Cedar–*Juniperus mexicana*, family Cupressaceae; a tall evergreen tree grown as an ornamental and for its aromatic, brownish lumber used for making boxes, pencils, etc. Native to southwestern United States and Mexico.

Rock Disintegration–The natural process by which hard rock is disrupted and comminuted. The factors involved are: expansion and contraction due to change in temperature, the chemical and mechanical work of plant roots, movements of stone fragments down slopes through force of gravity, a partial chemical solution, abrasion by moving water, ice, and lichens. See Lichens.

Rock, Igneous–This class of rock types is often subdivided, for convenience, into plutonic rocks (those which were formed by the solidification of molten materials below the ground surface, e.g., granite) and volcanic rocks (those formed by solidification of molten materials which have been extruded onto the ground surface).

Rock, Metamorphic–Rocks which have been formed in the solid state under the conditions of high pressure, high temperature, and the introduction of new chemical substances that, in general occur at great depths within the earth, e.g., slate, marble, jade, and schist.

Rock Outcrop–A surface exposure of bedrock which may be a feature of stony land but which is distinguished from boulders or detached fragments of rock.

Rock Phosphate–The natural mineral deposit essentially tricalcium phosphate ($Ca_3(PO_4)_2$), which is the principal source of the phosphorus of commercial fertilizers. The raw rock finely ground is also applied to acid soils as a fertilizer. The principal mining (United States) at present is in Florida and Tennessee, but large deposits occur in Wyoming, Idaho, Utah, and Montana, and smaller deposits occur in South Carolina, Kentucky, and Arkansas.

Rock, Sedimentary–Rocks which have been formed from deposits of sediment, whether from fragments of other rock transported from their sources and deposited by water, e.g., sandstone or shale, or by precipitation from solution or fixation by organisms, e.g., rock salt, gypsum, and limestone.

Rock Weathering–The natural processes which cause the physical and chemical disintegration and decomposition of rock; the residual mass of weathering in the parent material of mineral soils.

Rocky Mountain Crazyweed–*Oxytropis saximontana*, family Leguminosae; a poisonous plant found from Montana to Utah, United States, whose foliage is poisonous to cattle. Symptoms of poisoning are dullness, irregularity of gait, lack of appetite, dragging of the feet, a solitary habit, loss of flesh, and shaggy coat. As the animal ceases to eat it dies. Native to North America. See Loco.

Rocky Mountain Spotted Fever–A disease of people and animals characterized by intermittent chills and fever, painful muscles and joints, and red blotches on the skin. Occurring in the Rocky Mountain area, United States, the Rocky Mountain wood tick is the vector.

Rocky Mountain Wood Tick–*Dermacentor andersoni*, family Ixodidae; probably the most important tick vector of disease. This tick transmits Rocky Mountain spotted fever, tularemia, Colorado tick fever, American Q-fever, and encephalomyelitis and, experimentally, anaplasmosis. Most domestic animals, humans, and numerous wild mammals are its hosts. The stages in the life cycle are egg, larva, nymph, and adult. Larvae and nymphs usually live on small wild animals, mainly rodents; adults attack larger animals and people.

Rodent–A classification of mammals, mostly vegetarians, characterized by their single pair of chisel-shaped, upper incisors. Rodents are members of the orders Rodentia (rats, mice, squirrels, etc.) and Lagomorpha (rabbits, etc.).

Rodenticide–Any poison which is lethal to rodents.

Roe–The eggs or testes of fish. Consisting of two types, the female eggs (hard roe) and the male testes (soft roe), they are widely used for human consumption; e.g., the salted roe of sturgeon (caviar) is highly valued as a delicacy. see Sturgeon.

Roll-over Terrace–A terrace which has little height and a broad base, hence gently sloping. Under certain conditions it is effective in conserving water and reducing erosion, and it has the advantage of offering no obstacle to the movement of farm equipment over fields. See Broad-base Terrace.

Rolling–(1) Excessive side motion of shoulders, common in horses with abnormally wide fronts or chests. (2) A part of seed-bed preparation in which the land is rolled to even out the surface. (3) Processing grain through a set of smooth rollers which are close together; sometimes called flaking.

Rolling Land–A land surface which has rounded undulations, and relatively low relief.

Roof Rat–*Rattus rattus alexandrinus*, one of the three types of rats found in the United States, about 15 inches from tip of nose to end of tail. Its fur on the back and sides is gray to gray-brown and the undersurface of the body is almost white. It breeds four times a year with an average of six young in a litter. Also called Alexandrine rat. See Rat.

Rookery–A place where large numbers of wild birds or animals congregate and breed.

Rosin–Solidified amber-colored sap (resin) mostly from longleaf pine (*Pinus palustris*), occurring near the Gulf of Mexico in southern United States. Chemically it is mostly abietic acid anhydride ($C_{44}H_{62}O_4$). It is used as a stiffening agent in plasters and ointments, on violins, and on the hand of baseball pitchers. Rosin was formerly known as colophony.

Ross–To remove bark from, or to smooth a log.

Rot–A state of decay caused by bacteria or fungi. See Decay.

Rotated Pasture–(1) A pasture in the regular crop rotation which is grazed for a few years, usually two or three, and then plowed for other crops. (2) A pasture which is divided into segments by use of fences: the livestock being confined to one segment at a time in a definite rotation pattern.

Rotation–(1) In cropping, the growing of two or more crops on the same piece of land in different years in sequence and according to a definite plan. One of the most widely recommended crop rotations consists of (a) a tilled crop, often corn; (b) a small grain; (c) a legume or grass crop or a mixture of legumes and grasses often for more than one year. (2) In forestry, the period of years required to establish and grow timber crops to a specified condition of maturity.

Rotation Burning–The burning of brush and meadows every three years to provide more pasture grasses.

Rotenone–The common name of a botanical insecticide; the main toxic constituent in the roots of certain leguminous plants, such as *Derris elliptica* and *Lonchocarpus utilis* and *L. urucu*; moderately toxic to mammals, acute oral LD_{50} for rats 132 mg/kg; nontoxic to plants.

Rotifer–One of the Rotifera, a class of generally microscopic, many-celled animals abundant in stagnant and fresh water and in the soil.

Rotten–(1) Designating decomposed or putrid organic matter. (2) Designating ground or soil extremely soft and yielding because of decay, or rocks partially decomposed. (3) Designating sheep attacked by rot.

Rotten Cull Trees—Live trees of commercial species that do not contain a saw log now or prospectively, primarily because of rot (e.g., when rot accounts for more than 50 percent of the total cull volume.)

Rotten Wood—Wood in a state of decay produced by an attack of bacteria or fungi.

Rough Broken Land—Land naturally so deeply and minutely dissected by stream cutting that it is nonarable. Technically, it is differentiated from badlands and gullied land.

Rough Ground—An uneven, uncultivated land.

Rough Land—Nonarable land on steep, broken slopes. It may be either stony or highly dissected by erosion.

Rough Lumber—Lumber as it comes from the saw and before it is planed.

Rough Trees—(1) Live trees of commercial species that do not contain at least one 12-foot saw log, or two noncontiguous saw logs, each 8 feet or longer, now or prospectively, primarily because of roughness, poor form, splits, and cracks, and with less than one-third of the gross tree volume in sound material. (2) All live trees of noncommercial species.

Roundwood Products—Logs, bolts, and other round sections cut from trees for industrial or consumer uses. Included are saw logs, veneer logs and bolts, cooperage logs and bolts, pulpwood, fuelwood, piling, poles and posts, hewn ties, mine timbers, and various other round, split, or hewn products.

Rubber Tree—*Hevea brasiliensis*; a large tree that grows in tropical areas near the equator. It provides the principal source of the natural rubber of commerce.

Rubble—(1) Water-worn stones on a beach. (2) Rough stones in or from a quarry. (3) Any debris resulting from violent action such as a blast or an earthquake.

Rubble Dam—An unmortared dam, similar to a loose rock dam, which is made of broken stones, bricks, etc., and used in gully control in soil conservation.

Ruderal—Designating weeds which spring up on uncultivated or abandoned land.

Run—(1) A period of time, as a maple syrup run. (2) The amount of sap or sugar produced in a given time. (3) A swiftly moving tributary, rivulet, or mountain stream (eastern United States). (4) The stream outlet of a large spring (Florida, United States). (5) Unrestricted movement, as the colts have the run of the pasture. (6) An area of land or a leasehold (New Zealand). (7) A small, often dry gully or channel carved by water. See Arroyo. (8) A fenced-in pen used for the exercise of animals or poultry. (9) To feed; to graze, as steers run on the open range. (10) To operate, as a plow is set to run at a depth of 6 inches. (11) To cultivate, mow, combine, etc., as to run over a field with a weeder. (12) To maintain animals, as he runs sheep. (13) To work a dog with sheep or cattle. (14) To discharge pus, as a sore runs. (15) To be, as the prices run very high. (16) To move rapidly, as a horse. (17) To turn, as to run a wheel. (18) To grow, as when the vines begin to run. (19) To operate, as to run an engine.

Run On—To graze or pasture on, as to run on the range.

Run-off Modulus—The depth of water in inches over the underdrained area which must be removed by tile drains in 24 hours. It is a measure of the maximum rate at which the water will move through the soils to the laterals.

Runner—(1) A breed of ducks of very distinctive type, having a long, narrow body and very erect carriage. The breed derives its name from its gait, which is a quick run, quite unlike the waddle of other ducks. The adult drake weighs about 4½ pounds; the adult duck about 4 pounds. The runner is noted as an egg-producing breed and has little or no value for meat production. Its three varieties are the White, the Fawn and White, and the Penciled. Also called Indian Runner. (2) A lateral, aboveground shoot (stolon) of certain plants; e.g., strawberries, which roots and forms young plants at some of the nodes, aiding in propagation. (3) A rope used to increase the mechanical power of a tackle. (4) The upper or rotating stone of a set of millstones. (5) A supporting attachment which slides along the ground, as a sled runner. See Stolon.

Running Fire—In forestry, a fire which spreads rapidly with a well-defined head but little crowning.

Runoff—(1) The total stream discharge of water, including both surface and subsurface flow, usually expressed in acre feet. (2) The rate at which water is discharged from a drainage area, usually expressed in cubic feet per second per square mile of drainage area.

Runup—The rush of water up a beach or structure, associated with the breaking of a wave. The amount of runup is measured according to the vertical height above still water level that the rush of water reaches.

Russian Olive—*Elaeagnus angustifolia*, family Elaeagnaceae; a hardy, wind-resistant, spiny shrub or small tree grown for its ornamental fruit and silvery leaves in hedges and windbreaks. Native to Europe and Asia. Also called oleaster.

Rut—(1) The grooved track left by the wheels of vehicles in soft ground. (2) The season of heightened sexual activity in male mammals that coincides with the season of estrus in the female. See Estrus.

Rutter—A female mammal, such as a cow, which for some abnormality remains constantly in heat. Also called a buller.

Rutting Season—The recurring, usually annual, period when deer, cattle, etc., are in heat. See Estrus Cycle.

Sacrifice Area—A part of the range that is intentionally overgrazed to obtain efficient overall use of the management area.

Saddle–(1) A seat designed to fit a horse's back and to make riding easy, comfortable, and safe. Saddles are of many types but are usually made of heavy leather over a well-padded frame. They are often raised in the rear to hold the body in place and they may have a firm, raised support or horn in front to attach ropes or gear. They are held to the horse's back by means of one or more wide straps or girths that pass around the body of the horse. (2) The whole upper back section of a meat carcass which in wholesale cuts is commonly divided into fore and hind saddle. In cookery, the rear upper back portion including the two loins. (Sheep, veal calves, goats, and deer.) (3) The rear part of the back of a male chicken extending to the tail, covered by saddle feathers. (4) An apron made of reasonably heavy canvas which is placed on the backs of valuable breeding hens to protect against injury during mating. Saddles are usually held in place by loops which fit under the wings. (5) The uncut stalks of four hills of corn from two rows, tied together, around which corn is shocked. (6) A saddlelike depression in the crest of a ridge. See Saddleback. (7) A specially designed transverse log used in a skid road to guide the moving logs. (8) To put a saddle upon an animal or thing.

Saddleback–(1) A curved depression which connects two higher elevations in a ridge. (2) Resembling or suggesting a saddle in shape or marking; saddlebacked.

Saddleback Caterpillar–*Sibine stimulea*, family Limacodidae; a caterpillar with a green patch in the middle of the back, widely distributed through the Atlantic states of the United States, which feed on oak and cherry but sometimes on other plants.

Safener–Chemical coatings applied to seeds before planting that cause the emerging seedling to detoxify soil-applied herbicides after germination. Such coated seeds allow farmers to use stronger herbicides with less risk of crop injury. Sorghum was the first seed for which a commercial safener became available.

Sag–A depression in a nearly flat or undulating land surface; a microrelief feature.

Sage Tick–Spotted fever tick.

Sago Palm–Any of the palms the pith of whose trunks is a source of sago; especially ***Metroxylon rumphii***, family Palmaceae, which flourishes in marshy areas and is native to Indonesia. Some of the other sago palms are gomuti palm (***Arenga pinnata***), kittoolpalm (***Caryota urens***), and cabbage palm (***Corypha umbraculifera***), all native to Indonesia.

Saguaro–*Cereus giganteus*, family Cactaceae; a huge cactus which produces an edible fruit used in sweetmeats. Found in southeastern California, southern Arizona, and Mexico. Also called giant cactus, sahuaro, suwarro.

Salamander–(1) ***Geomys tuza***; a very common rodent in the dry sandy pine land of the Gulf Coast (United States). Because of its burrowing, it is considered a factor modifying these sand soils. (2) A portable stove which is used in construction work to provide heat, in winter, to keep newly poured concrete from freezing while curing. (3) Any of the small amphibians of the order Caudata. They resemble lizards but are scaleless.

Salicetum–An arboretum of willow trees.

Saline–General term for waters continuing various dissolved salts. The term is restricted to inland waters where the ratios of the salts often vary; the term haline is applied to coastal waters where the salts are roughly in the same proportion as found in undiluted seawater.

Saline Seep–In semi-arid regions, a down-slope location where salt has accumulated from upper slopes.

Saline Soil–A soil containing soluble salts in such quantities that they interfere with the growth of most crop plants. The electrical conductivity of the saturation extract is greater than 4 mmhos per centimeter (at 25°C), and the exchangeable-sodium percentage is less than 15. The pH readings of the saturated soil is usually less than 8.5.

Saline-sodic Soil–A soil containing sufficient exchangeable sodium to interfere with the growth of most crop plants and containing appreciable quantities of soluble salts. The exchangeable-sodium percentage is greater than 15, and the electrical conductivity of the saturation extract is greater than 4 mmhos per centimeter at (77°F or 25°C). The pH reading of the saturated soil is usually less than 8.5. See Saline Soil, Sodic Soil.

Salinity–The quantity of saltness in seawater or freshwater, most commonly expressed in parts of dissolved salt per 1,000 parts of water; e.g., salinity of seawater is 35 parts per thousand. See Alkalinity.

Salinization–The process of accumulation of soluble salts in soil.

Salinus–Salty.

Salmonella–A large group of bacteria, some of which are associated with food poisoning. Certain salmonellas are sometimes found in raw and dried eggs and in poultry products. They can be destroyed by sufficient heating.

Salmonellosis–Infection by organisms of the genus ***Salmonella*** causing food poisoning in humans and many diseases in domestic animals.

Salt–(1) Sodium chloride (common salt), NaCl; a white crystalline compound occurring abundantly in nature as a solid or in solution. It is a requirement of all livestock and is generally deficient in all natural feed ingredients. It is commonly included in mixed feeds and is otherwise supplied as salt blocks in pastures, barn lots, or stalls. It has wide usage in seasoning, preserving meats and vegetables, tanning hides, etc. In large amounts it is toxic and is frequently used to kill weeds; however, some plants such as Bermudagrass and asparagus tolerate a very high concentration. (2) To add salt as seasoning, preservative, supplement, etc.

Salt Bottom–A flat, saline, relatively low piece of land (western United States).

Salt Grounds–The several locations on a range where salt is provided for the livestock usually to influence their movement and to bring about a more uniform utilization of the range.

Salt Hay–(1) A mulching material for plants which is cut from coastal marshes. (2) Hay which is made from grasses cut from salt marshes.

Salt Index–An index to compare solubility of chemicals. Most nitrogen and potassium compounds have a high index; phosphate compounds have a low index. When applied too close to seed or foliage, high-index materials may cause plant injury by plasmolysis.

Salt Lick–A place where salt occurs naturally on the surface of the ground and to which wild animals go to obtain salt needed in their diet.

Salt Marsh–A flat, grass- and shrub-covered piece of land, occasionally inundated by saltwater.

Salt Pan–A depression in a salt marsh, usually bare of vegetation.

Salt-marsh Caterpillar–*Estigmene acrea*, family Arctiidae; the larvae of a moth which is generally distributed and at times may be a severe pest infesting various types of vegetation.

Salt-marsh Peat–The peat occurring in salt marshes mainly along the seacoasts which typically consists of coarse, fibrous, only partially decomposed plants and large amounts of admixed mineral sediments. See Histosols, Peat.

Salt-susceptible–Plants that will not grow on soils containing appreciable quantities of soluble salts.

Saltation–(1) The leaping movements of eroding soil grains arising largely in combinations of friction with inertia. (2) The process by which a particle, picked up by the stream current, is flung upward after which, being too heavy to remain in suspension, it drops to the stream floor again at a spot downstream.

Saltpeter–(1) Potassium nitrate (KNO_3), a white crystalline salt of potassium and nitrogen whose principal agricultural use is as a fertilizer and as a preservative in curing and preserving of meats. Also spelled saltpetre. (2) An incrustation of certain salts occasionally formed on tobacco leaves during processing.

Saltwater Encroachment–The phenomenon occurring when a body of saltwater, because of its greater density, invades a body of freshwater. It can occur either in surface or groundwater bodies.

Salvage–Recycling of solid wastes, such as aluminium cans.

Sampling Fertilizers–The State Department of Agriculture in each of the fifty states maintains a fertilizer inspection service that samples and analyzes fertilizers to determine if the fertilizer grade on the bag or in the bulk is properly designated to conform to the respective state laws. In the interest of uniform sampling procedures among the states, the Association of American Plant Food Control Officials has sponsored joint regional conferences of fertilizer inspectors and fertilizer industry personnel for the purpose of exchanging information on the many problems connected with sampling. An Inspector's Manual has been developed and demonstrations of sampling procedures have been held.

Sanctuary–A reservation where animals and birds are sheltered and protected from hunting or molestation. See Refuge.

Sand–(1) Any of the small, loose, granular fragments which are the remains of disintegrated rocks. It may contain a great variety of minerals and rocks but the most common mineral is quartz. (2) In soil science, a group of textural classes in which the particles are finer than gravel but coarser than silt, ranging in size from 2.00 to 0.5 millimeters in diameter. It is the textural class of any soil that contains 85 percent or more of sand and not more than 10 percent of clay. (3) An area or deposit of this material, as a sand shore, a sand bank, etc. (4) In maple sugar processing, a granular deposit, mainly of silica and calcium malate, formed as the sap is condensed. (5) To cover with sand.

Sand Bag–A bag of burlap or other strong fabric which, when filled with sand or earth, is used in flood control to increase the height of or to reinforce a levee, etc.

Sand Bar–A ridge or narrow shoal of sand formed in a river or along a shore by the action of currents or tides.

Sand Binder–Any plant which can hold sand in its place, and reduce its shifting, by the abundance of its rootstocks. Grasses of the genera *Ammophila* and *Calamovilfa* are commonly used for such a purpose.

Sand Dune–Loose sand drifted by wind and piled up into hills and hummocks. It is commonly found on seashores, the shores of large inland lakes, and in deserts.

Sandfly–*Culicoides*, family Tendipedidae; a very small, two-winged, blood-sucking midge which is a nuisance in certain areas because of its sharp sting. Also called no-see-ums, punkies.

Sandstone–A sedimentary rock composed of sand particles, mainly quartz, whose particles are bound together by natural cementing materials as silica, iron, and/or lime carbonate. Soils resulting from the weathering of such rocks are generally sands or sandy loams in texture. Sandstones of certain textures are used in the manufacture of sharpening stones and grinding wheels.

Sandstorm–A windstorm which carries such large amounts of sand that visibility is drastically reduced.

Sandy Clay–Soil which contains more than 45 percent sand, more than 35 percent clay, and less than 20 percent silt.

Sandy Clay Loam–A soil which contains from 20 to 35 percent clay, less than 28 percent silt, and 45 percent or more of sand.

Sandy Loam–Gritty, loamy soil which contains less than 20 percent clay, less than 50 percent silt, and between 43 and 52 percent sand.

Sandy Soil–In common usage, the sand and sandy loam textural classes; any soil which contains more than 70 percent sand and less than 15 percent clay.

Sanitary Landfill–A site where solid waste materials are disposed of on land, supposedly in a manner which prevents their escape into, or pollution of, the surrounding environment. The waste is spread in layers, then compacted to the smallest practical volume and covered with compacted soil at the end of each working day.

Sanitation–The developing and practical application of measures designed to maintain or restore healthful conditions, such as the treatment, removal, or destruction of contaminated or infested materials and possible sources of infection or infestation.

Sanitize–To disinfect and make sanitary the utensils used in preparation of food products, as dairy utensils.

Sapling-Seedling Stands–Stands at least 10 percent stocked with growing stock trees of which more than half are saplings and/or seedlings.

Saplings–Live trees of commercial species 1.0 to 5.0 inches in diameter at breast height and of good form and vigor.

Sapodilla–*Achras zapota (Lucuma mammosa)*, family Sapotaceae; a large evergreen tree which bears an edible fruit, and whose latex yields chicle, the chief natural ingredient of chewing gum. Native to tropical areas of North America. Also called sapote, marmalade plum, naseberry.

Saponification–The chemical conversion of fats into soap.

Saponin–Any of several glucosides or glycerides present in many plants; e.g., soapwort, which has soaplike qualities. Used in detergents, to reduce the surface tension of water in emulsifying oils, and to produce foam in fire extinguishers and beverages.

Saprobic–Pertaining to an environment rich in organic matter and relatively low in oxygen.

Saprogen–An organism capable of producing decay in nonliving organic matter, such as dead wood, etc.

Saprogenesis–That period in the life cycle of a pathogen in which it is not directly associated with a living host and in which it may be either dormant or living as a saprophyte. See Saprophyte.

Saprolite–A soft, earthy, clay-rich, thoroughly decomposed rock formed in place by chemical weathering of igneous or metamorphic rocks. Forms in humid, tropical, or subtropical climates; commonly red or brown.

Sapropel–A slimy, fetid sediment on the bottoms of lakes made up of the organic debris from aquatic plants and animals.

Saprophyte–Plants, including certain bacteria and molds, capable of obtaining nutrients and energy from dead organic matter.

Saproplankton–A mass of decaying, aquatic plant and animal material (plankton) which may be found floating on or in stagnant water.

Sapsucker–(1) Any of several small woodpeckers, ***Sphyrapicus***, especially ***S. varius***, the yellow-bellied sapsucker which makes a horizontal series of holes around trees, such as apple, birch, pine, etc. It returns from time to time to feed on the sap and insects from those holes. (2) Any insect that sucks sap from the roots, stem, or leaves of plants, as the corn-root aphid, cinch bug, etc.

Sapwood–That portion of the wood of a tree more recently formed which contains living cells intermingled with nonliving, woody tissue. It is located between the bark and heartwood of trees. Also called Alburnum, splintwood. See Heartwood.

Saturated–Filled to the maximum capacity under existing conditions.

Saturated Air–Air containing as much water vapor as it can hold without condensation. In saturated air, the partial pressure of the water vapor is equal to the vapor pressure of water at the existing temperature. see Relative Humidity.

Saturation Point–(1) The point beyond which no further additions can be held, utilized, or accommodated; e.g., the water-holding capacity of a particular soil, the flow of goods to a market, or a population of plants. (2) In wildlife, the maximum density of population; the natural condition wherein a species is as abundant as living conditions permit.

Savannah–Also spelled savana. (1) An area of grassland usually flat and devoid of trees or containing only scattered trees, especially marshy or wet expanses in the southeastern United States. (2) Dry grassland which contains isolated or scattered trees and shrubs.

Savannah, Derived–A mixed tree/grass vegetation growing on soils that formerly supported only "high forest" because of such anthropogenic disturbances as tree cutting, fire, cultivation, and overgrazing.

Saw Log–A log meeting minimum standards of diameter, length, and defect, including logs at least 8 feet long, sound and straight, and with a minimum diameter inside bark for soft woods of 6 inches (8 inches for hardwoods) or other combinations of size and defect specified by regional standards.

Saw Timber–Trees that will yield logs suitable in size and quality for lumber.

Sawdust–Dust and small particles of wood produced by the cutting action of saws. A by-product of the lumber and wood industries. It is used as a bedding material for livestock or as soil mulch. Mixed with soil, it has no fertilizer value until decomposed and may depress plant growth in its fresh state unless nitrogen fertilizer is added to it.

Sawtimber Stands–Stands at least 10 percent occupied with growing-stock trees, with half or more of total stocking in sawtimber or poletimber trees, and with sawtimber stocking at least equal to poletimber stocking.

Scabland–Land which is characterized by bare rock surface or a very thin covering of soil, especially the dry lava plateau country of southeastern Oregon (United States).

Scale–(1) Any instrument used to determine weight. (2) Either of the pans of a balance. (3) A measure of dimension, concentration, or intensity. (4) A graduated series of steps or degrees on measuring devices. (5) An instrument or device on which graduated spaces for measurement have been stamped or attached. (6) A graduated list of prices, wages, etc. (7) Any thin flake that peels from a surface, such as from skin. (8) One of the thin, flat, membranous plates which forms the outer, protective covering of certain animals or parts thereof, as the shanks of birds, etc. (9) One of several rudimentary, specialized leaves which protects the buds of plants and deciduous trees in cold climates. (10) A thin, dry, membranous part of a plant or flower. (11) See Scale Insect. (12) The estimated sound content of a log or group of logs scaled by use of a given log rule. Net scale, the scale after deducting for defects. (13) To form into scales. (14) To cover with scales or with an incrustation. (15) To remove in thin layers, as by scraping. (16) To come off in scales. (17) To make according to scale.

Scalping–(1) In flour mills, separating the fuzzy growth from the ends of wheat berries by attrition, with or without the use of suction or blower fans. (2) In milling, partially separating the product from the break rolls into broken wheat and break flour by means of an arrangement of sieves, bolts, or screens of varying degrees of fineness. (3) In horses, a striking of the front of the hind coronet, pastern, or cannon against the front foot when running. (4) Removal of vegetation and a thin layer of soil.

Scar–(1) An isolated or protruding rock. (2) A steep rocky eminence. (3) A bare place on the side of a mountain or other steep slope. (4) A wound that has healed.

Scarlet Oak–*Quercus coccinea*, family Fagaceae; a rather cylindrical tree grown as a shade tree and for its autumnal scarlet foliage. Native to North America.

Scarp–An escapment, cliff, or steep slope of some extent along the margin of a plateau, mesa, terrace, or bench.

Scat–Animal excrement; especially applied to insect and wild animal droppings.

Scatology–The study of feces as a means of identifying the presence of wild animals.

Scavenger–A mammal, bird, or other organism that feeds on carrion or dead organic matter.

Scenic Rivers–Rivers or sections of rivers free of impoundments, with shorelines or watersheds still largely primitive and shorelines largely undeveloped, but accessible in places by roads.

Schist–A medium or coarse-grained metamorphic rock with subparallel orientation of the micaceous minerals which dominate its composition.

Schizont–A stage in the life cycle of certain protozoan parasites in which a cell, without fertilization, gives rise to a number of new individuals, called merozoites.

Schwedler Maple–*Acer platonoides*; a horticultural variety of Norway maple which has dark red foliage throughout the early growing period. It is grown as a shade tree and as an ornamental.

Sciophyte–A plant which grows well on, or is adapted to, shady sites.

Scoliid Wasp–Any wasp of the genus *Scolia*, all species of which lay their eggs on the larvae of various beetles. These wasps are important in the control of certain destructive beetles, such as the Asiatic or Oriental beetle. See Integrated Pest Management.

Scoria–(1) The slaglike clinker left after the smelting of metals. (2) A cellular or cinderlike rock of volcanic origin.

Scorpion–Any of the various arachnids of tropical and subtropical regions. They have a segmented body up to 4 or 5 inches long, with poisonous sting at the tip of the tail. Living on insects, they are probably more beneficial than harmful.

Scotch Pine–*Pinus sylvestris*, family Pinaceae; a hardy evergreen tree grown as an ornamental, a specimen, in windbreaks, for Christmas trees, and for its lumber. Pyramidal when young, it becomes irregular with age. It grows rapidly on acid, poor, sandy soil, and is resistant to cold, drought, and dry winds. Native to Europe and Asia. Also called Danzig pine, Riza pine, Scots pine.

Scots Pine–See Scotch Pine.

Scour–(1) To abrade and flush as by the action of rapidly moving water on stream beds. (2) To cleanse dirt, grease, etc., by rubbing or scrubbing, as to scour wool. (3) To cleanse the bowels of an animal by purging. (4) In plowing, to pass through the ground cleanly without any soil sticking to the moldboard. (5) To rub off the flesh sticking to a hide. (6) See Scours.

Scouting–The inspection of a field for pests (insects, weeds, pathogens). Scouting is a basic component of integrated pest management programs. See Integrated Pest Management.

Scrape–(1) A tillage implement; a sweep. (2) A spreading wing attachment for a cultivator. (3) In turpentine operations, oleoresin remaining on a tree from which volatile oils have evaporated. (4) To collect scrape from trees in turpentine operations.

Scree–A heap of rock waste at the base of a cliff or a sheet of coarse debris on a mountain slope. By most writers scree is considered to be a synonym of talus, but it is a more inclusive term. Whereas talus is an accumulation of material at the base of a cliff, scree also includes loose material lying on slopes without cliffs. See Colluvium, Talus.

Screen–(1) A sieve or grating, such as a frame covered with meshed wire or fabric, or a perforated plate, etc., which is used for separating the finer or coarser parts of soil, gravel, grain products in milling, seeds in cleaning, etc. (2) Any form of grating used to prevent invasion or escape of insects, reptiles, fish, or other kinds of animals. (3) A construction or planting used to conceal an unpleasant sight. (4) To consider a group for the purpose of selecting a relatively small number of individuals. (5) To sift by shaking through a screen. (6) To conceal, as with a screen.

Scribner Rule–In forestry, a diagram rule for determining the board measure of logs, one of the oldest in existence, which assumes 1-inch boards and ¼-inch kerf, makes a liberal allowance for slabs and disregards taper. (It is the official rule in many parts of the United States). See Doyle Rule, Doyle-Scribner Rule.

Scrub–(1) Land which under natural conditions supports only shrubs or a dwarfed, stunted growth of trees. The soils are generally infertile, but scrub may appear on various kinds of sites. (2) The stunted trees or shrubs, often in dense stands, on such land. (3) An animal of inferior breeding or condition. (4) Designating an inferior product, animal, etc.

Scrub Forest–A forest of small trees occurring on ridges or hillsides with little shade and with little accumulation of litter.

Scrub Pine–(1) *Callitris calcarata*, family Cupressaceae; a black or red pine which grows on stony hill slopes and ridges. Twigs of this tree have been used as a vermifuge for horses. Native to Australia. (2) Any of several dwarf pines of the United States which often thrive on stony or arid land.

Scrubber–A device that uses a liquid spray to remove aerosol and gaseous pollutants from an air stream. The gases are removed either by absorption or chemical reaction. Solid and liquid particulates are removed through contact with the spray.

Scum–(1) Minute forms of lower plant life and impurities that gather on the surface of stagnant water. (2) Extraneous matter or impurities that rise to the surface of boiling or fermenting liquids. (3) The condensation of solids at the surface of boiled milk upon cooling.

Sea Lamprey–*Petromyzon marinus*, family Petromyzontidae; an eel-like, external parasite of fish that reaches a maximum size of 3 feet. It attaches itself to the fish with its sucker mouth and drains the blood. Although by nature a marine creature, the sea lamprey has become established in the Great Lakes (United States), and has done considerable damage to the game and food fish of this area. Attempts have been made to control the increase of these parasites by trapping and destroying the adults as they enter freshwater streams to spawn.

Sealing–The action of high-intensity rainfall which causes disintegration of unprotected soil granules, a consequent separation of colloids which fill spaces between larger particles, thus producing a seal or relatively impervious surface crust.

Seaside Bent–A strain of creeping bentgrass with long stolons and narrow, stiff, oppressed blades, usually densely matted, which is used for lawns and putting greens of golf courses, etc. Native to the marshy areas of the Atlantic and Pacific coasts of United States and Canada.

Season–(1) A portion of the year; e.g., a growing season, when plants grow; the fire season, when danger of fire is greatest; the grazing season, when grazing is possible. (2) To bring to or develop into a condition of greater usefulness, as to dry lumber by natural or artificial means. (3) To flavor food with spices, salt, etc. (4) An animal that is ready for breeding is said to be in season.

Seasonal Distribution–(1) In pasturage, the progressive grazing in a sequence of moves from one part of a range to another as vegetation develops. (2) In meteorology, the amount and kind of precipitation, wind movement, etc., in different seasons.

Seasoning–(1) The process of drying (curing) lumber or other forms of wood for better utilization. Seasoning is natural when the drying is done by air or other natural means; in artificial seasoning, the drying is carried out by means of a kiln, oils, or electrical appliances. (2) Adding salt, pepper, etc., to meat and other food products.

Seawall–A structure separating land and water areas, primarily designed to prevent erosion and other damage due to wave action. See also Bulkhead.

Seawater–Water in the seas and oceans that has a salt content of approximately 3.5 percent.

Seaweed–See Kelp.

Second Bottom–The first terrace level of a stream valley which lies above the floodplain. It is rarely or never flooded.

Second Growth–(1) A forest which originates naturally after removal of a previous stand. (2) The smaller trees which remain after a cutting or the residual trees available for another logging on the same area.

Secondary Growth–(1) Growth which may arise from a secondary bud if the first shoots are destroyed, as with shrubs or trees. (2) A shoot or spur that grows from a bud or growing point the same year that bud or growing point is produced, as with shrubs and trees. (3) Growth from the cambium.

Secondary Tillage–Soil tillage that follows primary tillage to prepare the soil for planting; usually not as deep as primary tillage.

Secondary Treatment–The second step in most waste treatment systems in which aerobic bacteria consume the organic parts of the wastes before they are discharged into the environment.

Sedentary–(1) Formed in place without transportation from the underlying rock or by the accumulation of organic material as in peats and mucks; said of some soils. (2) Attached, as an oyster, barnacle, or similar shelled invertebrate.

Sedge Peat–A brown to black peat, mostly the partially decomposed remains of sedge vegetation. It is generally less acid than most peat and of lower water-absorbing capacity than moss peat. See Histosols.

Sediment–(1) Matter that separates itself from a liquid and settles to the bottom. (2) Earthy or organic matter deposited by water.

Sediment Detention Basin–A reservoir which retains flows sufficiently to cause deposition of transported sediment.

Sedimentary Peat–(1) Fine-textured organic material deposited in lakes. Finer in texture than the ordinary peat, it is of minor significance as soil amendment. (2) Admixed or alternate layers of peat and mineral sediments formed in coastal marshes and alluvial valleys.

Sedimentary Rock–A rock composed of particles transported or deposited by water. The chief groups of sedimentary rocks are conglomerates (from gravel), sandstones (from sands), shales (from clays), and limestones (from calcium carbonate deposits).

Sedimentation–That portion of the metamorphic cycle from the separation of the particles from the parent rock, no matter what its origin or constitution, to and including their consolidation into another rock.

Sedimentation Rate–The speed, measured in minutes, at which particles, usually red blood cells, settle in a tube. The sedimentation rate of red blood cells in an index of the variations in the cell count, blood viscosity, and other chemical changes in the blood.

Seed Tree–(1) A tree that produces seed. (2) In forestry, a tree reserved, in a cutting operation, to supply seed for reforestation.

Seep–(1) A spot where water oozes out slowly from the soil and gathers in a pool or produces merely a wet place, usually on a hillside or at a hill base. (2) To soak through pores. (3) To lose liquid by drainage. See Saline Seep.

Selection Cutting–Removal of mature timber, usually the oldest or largest trees, either as single scattered trees or in small groups at relatively short intervals, commonly five to twenty years, repeated indefinitely, by means of which the continuous establishment of natural reproduction is maintained.

Selective Grazing–The tendency for livestock and other ruminants to prefer certain plants and to feed on these while grazing little on other species.

Selective Herbicide–Any of certain chemical toxicants (usually applied as a spray) that will kill weeds in a growing crop or turf and produce little or no injury on the crop plant or turf. See Nonselective Herbicide.

Self-cleaning Velocity–The threshold velocity in sewers, terraces, drainage ditches, and irrigation canals to keep solids in suspension.

Self-mulching Soil–(1) A soil that cracks deeply and becomes so granular at the surface when very dry that the granular mulch washes into the cracks when rains begin, the whole soil swelling enough as it becomes moist to force material upward between the former cracks. (2) A soil in which the surface layer becomes so well-aggregated that it does not crust and seal under the impact of rain. See Vertisols.

Self-pruning–Natural pruning; the natural death and fall of branches from live plants resulting from light and food deficiencies, decay, insect attack, snow, wind, and ice.

Self-purification–The process by which a stream is purified some time after receiving a waste discharge. This occurs in the decomposition of organic matter by oxygen-using bacteria.

Self-seeding–(1) The act of disseminating its seed by a plant or plant species. (2) Designating a plant which scatters its seed and reestablishes its kind.

Semi-arid–A term applied to regions or climates where moisture is normally greater than under arid conditions but still definitely limiting to the growth of most crops. Dryland farming methods or irrigation generally is required for crop production. The upper limit of average annual precipitation in the cool semi-arid regions is as low as 15 inches (38 centimeters), whereas in tropical regions it is as high as 45 or 50 inches (114 to 127 centimeters).

Seminarium–(Latin) Tree nursery.

Semipermeable Membrane–A membrane that permits the diffusion of one component of a solution but not the other. In biology, a septum which permits the diffusion of water but not of the solute.

Senescence–The process of growing old. Sometimes used to refer to lakes nearing extinction.

Senility–Condition of organisms, especially people, that in old age revert to development resembling younger stages.

Sensus–(Latin) Meaning.

Sentry Maple–A horticultural variety of the sugar maple that has upward-turning branches, parallel sides, and narrow form. It is suitable for planting on narrow streets.

Sepsis–A state of contamination or poisoning by pathogenic bacteria.

Septage–(1) Sludge produced in a septic tank. (2) That which is pumped from a septic tank in cleaning. (3) Effluent in a cesspool.

Septic–(1) relating to, or caused by, the presence of disease-producing organisms, or their toxins, in the blood or tissues. (2) Putrefactive; putrefying. (3) A term sometimes used to refer to conditions where dissolved oxygen is absent and decomposition is occurring anaerobically, as in a septic tank.

Septic Tank–A large container that holds solids from human effluents until they are decomposed by anaerobic bacteria, while permitting surplus liquids to flow through the container to a drain field where they are absorbed by the soil.

Sequestration (Chelation)–A chemical complexing of certain metallic cations with inorganic compounds. Very useful in preparing fertilizer formulations with available micronutrients such as iron, magnesium, zinc, copper, and manganese.

Sere–(1) The series of stages that follow one another in ecological plant succession, comprising development from pioneer stage to climax under natural influences. (2) The series of plant communities that follow one another in an ecologic succession.

Series, Soil–A group of soils, formed from a particular kind of parent material, having horizons that, except for the texture of the A (surface) horizon, are similar in all profile characteristics and in arrangement in the soil profile. Among these characteristics are color, texture, structure, reaction, consistence, and mineralogical and chemical composition.

Serotiny–Tree cones that require heat to open and release seeds.

Serpentine–A secondary rock mineral, a hydrous magnesium silicate. Soils residual from the weathering of the rock have the reputation of being infertile and are locally known as barrens.

Serrate–(1) Pertaining to the rocky summit of a mountain having a sawtooth profile; a small sierra-shaped ridge. Local in the Southwest. (2) Said of a leaf margin when saw-toothed, with the teeth pointing forward.

Sessile–(1) Of leaves, flowers, fruits, etc., attached directly by the base without a stem or stalk, such as leaves of grasses, sedges, and certain other plants. (2) An organism that is attached to an object or is fixed in place.

Seston–The living and nonliving bodies of plants or animals that float or swim in the water.

Set–(1) A small propagative part, a bulb, shoot, tuber, etc., which is suitable for setting out or planting, as an onion set. (2) A number of things usually used together, or forming a complete collection, as a set of tools, harness, etc. (3) The initial swelling of the ovary of a flower soon after petal fall. Also called setting, fruit setting. (4) To put in a particular place, position, condition, direction, adjustment, etc. (5) To put eggs under a fowl or in an incubator to hatch them. (6) To put a price or value on something. (7) To put in proper working condition. (8) To adjust the teeth of a saw. (9) To fix a trap or a net to catch animals or fish. (10) To fix in the ground, as a post, tree, or plant. (11) To become stiff, firm, or hard, as cement sets. (12) To adjust into proper apposition the ends of a broken bone, as to set a fracture.

Settle–(1) To sink gradually to a lower level, as of sediment, etc. (2) To cause a female to become pregnant, as a bull settles the female with which he has had coitus. (3) To occupy land usually for the purpose of farming, ranching, or homesteading.

Settleable Solids–Suspended solids that will settle in quiescent water or sewage in a reasonable period. Such period is commonly, though arbitrarily, taken as two hours.

Settled–(1) Designating suspended particles or impurities in liquids dropped to the bottom. (2) Designating a pregnant cow or mare.

Settling Basin–An enlargement in a stream or conduit to permit the settling of debris carried in suspension.

Sewage–The spent water supply after it has received the various household, industrial, and other wastes of a community.

Sewage Sludge–An organic product, remaining after the treatment of sewage in sewage-disposal plants, sometimes used as a fertilizer. Almost half of the sewage sludge produced in the United States is spread on land. The other half is burned, buried, or dumped in the ocean. See Activated Sewage.

Sewage Treatment–Any artificial process to which sewage is subjected in order to remove or reduce its objectional constituents.

Sewer–A pipe or conduit, generally closed, but normally not flowing full, which is used for carrying off sewage and other liquid wastes.

Sewer Gas–A mixture of gasses originating from anaerobic decomposition of sewage. It contains carbon dioxide (toxic in high concentration), methane (very toxic and flammable), hydrogen sulfide (toxic to people at threshold concentrations of 0.07 percent). Similar to gas generated from animal manures in anaerobic lagoons on farms and ranches. In confined places this gas will kill livestock and people.

Sewerage–Sewage plus all of the essential pipes, pumps, tanks, and treating plant necessary to process sewage to make it safe to release into the environment.

Sewered Population–The population served by waste water collecting sewers.

Sex Pheromone–Sex pheromones are now made synthetically to bait traps and thereby make an estimate of population density of a given species of insect. In this way scientists can determine when control measures should be initiated, known as the economic threshold of insect numbers. Most common has been the use of synthetic female sex hormones to trap male Lepidoptera. A second use of synthetic female sex hormones has been to confuse the male and thereby prevent mating. See Pheromone.

Shade Plant–(1) A plant that grows in shade or can tolerate shade; as moss, woodland flora, etc. (2) A plant grown to provide shade for crops that require it.

Shade Tree–Any tree grown to provide shade; specifically, one of luxuriant foliage, such as a street tree or a park tree.

Shagbark Hickory–*Carya ovata*, family Juglandaceae; a deciduous tree so named because its shaggy, gray bark peels off in long strips. It is valued for timber, fire wood, and its edible nuts. Native to North America.

Shale–(1) A sedimentary rock which is composed of clay and silt finely laminated in structure. There are many variations in color and many gradations into hard, thinly-bedded sandstones and limestones and into slates. (2) The sliver-like lining of the carpel or bur which before the opening of the cotton boll serves to separate the locks of cotton. The presence of fine particles of shale in ginned cotton is highly objectionable as they are difficult to remove from the lint.

Shallow Cultivation–Scraping the surface, scarifying, or cultivating to shallow depths, normally less than 5 inches. Fewer of the crop roots are injured.

Shallow Well–A well from which a supply of water is obtained at little depth, commonly 20 feet or less below the surface.

Shamrock–A plant having trifoliate leaves, traditionally believed to have been used by St. Patrick as illustrative of the doctrine of the trinity, and hence adopted as Ireland's national emblem. The name shamrock is now commonly applied to the emblematic plant worn by the Irish on St. Patrick's day.

Sheet Erosion–Removal by rainwater and snowmelt runoff of a uniform layer of soil or comminuted material from a portion of the land surface, as contrasted to gullying and stream erosion caused by surface water flowing in channels. See Erosion.

Shellbark Hickory–*Carya laciniosa*, family Juglandaceae; a deciduous tree grown especially for use in making sporting goods, implements, woodenware, for fuel and for smoking meats. It may be used as a shade tree, and its nuts are edible and sweet. Native to North America. Also called bigleaf shagbark hickory, big hickory, thick shellbark hickory.

Shellfish–Crustaceans, which include crab, lobster, and shrimp; mollusks, which include abalone, clams, oysters, and scallops.

Shelterwood–A forest management method of securing natural reproduction under the partial shelter of seed trees which are gradually removed by successive cutting to admit the amount of light and heat that the seedlings will require.

Shenandoah Valley–A part of the Great Valley of the Appalachian Region of the United States; a great limestone belt of lowland plains and mountains lying directly west of the Blue Ridge Mountains in western Virginia, which is traversed in part by the Shenandoah River. It is notable for fertile soils, farm products, livestock, and apple orchards.

Shifting Cultivation–A farming system, common in the tropics, in which land is cleared, the debris burned, and crops grown for a relatively short period until yields decline. The land is then abandoned. The original land is cleared and cropped again after an uncontrolled fallow period of three to twenty years, usually when soil fertility has been naturally restored by woody vegetation.

Shingle–(1) (a) A thin, oblong piece of wood sawed or split, with one end thinner than the other, which is lapped lengthwise in covering roofs and outer walls of buildings. (b) A piece of any kind of material, as asbestos, used in the same way. (2) Loose, rather sharp shale or similar rock, common to certain rocky areas in New Zealand. (3) A mass of loose, coarse, rock fragments of cobble or small boulder size, rounded or flattish, found on the seashore and on the beaches of large inland lakes. (4) A thin slab of slate stone, cut or chipped into a section used as a shingle. (5) To cover with shingles.

Shinnery–A dense thicket which, because the growth is relatively new or because growth conditions are unfavorable, is scarcely more than tall shrubbery. Although shinnery is usually made up of mixed woody species, it is generally thought of as oak shinnery because of the usual predominance of dwarf oak.

Shipmast Locust–*Robinia psuedoacocia*; a tall, straight-growing clone of the black locust that is supposed to be resistant to the black locust borer. At one time the trees were used as masts for sailing ships.

Shore–A narrow strip of land in immediate contact with the water, including the zone between high and low water. See also Backshore, Foreshore.

Shore Erosion–Loss of soil from the land area adjacent to oceans or large lakes which results from the action of waves and wind. Bank erosion may occur along streams as a result of undercutting, etc. Also called beach erosion.

Shortleaf Pine–*Pinus echinata*, family Pinaceae; an evergreen tree used for pulp, poles, ties, slack-cooperage, veneer, structural timber, planing-mill products, etc. Native to North America.

Shot-hole Borer–*Scolytus rugulosus*, family Scolytidae; a small beetle which makes small holes in the bark of healthy fruit trees and in the branches and trunks of weakened trees. On stone fruits, the shot holes are usually covered with gum. The beetles and larvae also make galleries in the sapwood of trees which they infest. If the beetles are abundant, the foliage yellows and wilts, and the tree may die.

Shoulder–(1) Of livestock, the upper part of the forelimb and/or the adjacent part of the back; the scapula region. (2) The upper foreleg and/or the adjacent part cut from the carcass of a hog, sheep, or other animal; the upper part of the carcass to which the foreleg is attached. (3) The part of a bird at which the wing is attached. (4) In turpentining, the uppermost corner of a face. (5) A shoulderlike part or projection, as the projection around a tendon. (6) A sudden inward curvature in the outline of something, as the flaring part between side and neck of a milk bottle. (7) The edge of a road adjacent to the traveled way or pavement.

Shower–A local rain, usually of short duration.

Shrink-Swell–The shrinking of soil when dry and the swelling when wet. Shrinking and swelling can damage roads, dams, building foundations, and other structures. It can also damage plant roots. See Vertisols.

Shrub Live Oak–*Quercus turbinella*, family Fagaceae; an evergreen shrub or small tree growing to 14 feet in height at elevations up to 8,000 feet in Texas, Arizona, California, Colorado, New Mexico, and northern Mexico. It is an abundant species of the chaparral and is browsed to some extent by livestock. Native to North America.

Siccideserta–Dry desert.

Sick–(1) Suffering from illness of any kind; not sound or in fit condition. (2) Designating a soil unfit for the profitable production of certain crops as a result of being infected with disease organisms or because of the excess or deficiency of certain elements.

Side Camp–In forestry, a small camp set up to accommodate a crew working away from a main camp or headquarters.

Side Slope–The slope of the sides of a hill, canal, dam, or embankment. Usually expressed as the ratio of the horizontal distance to the vertical such as two to one, four to one, etc.

Sieve Analysis–Determination of the percentage distribution of particle size by passing a measured sample of soil through standard sieves of various sizes.

Signal Word–A word on the label of a pesticide container that is required by law to designate the relative toxicity of the chemical "Danger - Poison" denotes a highly toxic compound; "Warning" means that the chemical is moderately toxic; and "Caution" denotes a compound that is slightly toxic.

Silica–SiO_2; silica dioxide; present in quartz, flint, etc.,; one of the most abundant minerals in the earth's crust.

Silicon–Si; a nonmetallic element which always occurs in a combined form in nature, where it is exceeded in abundance only by oxygen with which it usually combines to form silica. Found in all soils, it is a constituent of plants but is not regarded as an essential plant nutrient.

Silkoak Grevillea–*Grevillea robusta*, family Proteaceae; an ornamental tree grown for its showy orange flowers and handsome foliage with a fernlike appearance. It is a popular street tree in frost-free areas, and is grown by florists as a pot plant. Its timber is sometimes used in furniture making. Native to Australia. Planted extensively as shade for tea plants in India.

Silo–A pit, trench, aboveground horizontal container, or vertical cylindrical structure of relatively air-tight construction into which green crops, such as corn, grass, legumes, or small grain and other feeds are placed and converted into silage for later use as a livestock feedstuff.

Silo Gas–Dangerous gas, primarily carbon dioxide and nitrogen dioxide, that can build up to lethal levels inside silos and are almost impossible to detect.

Silt–(1) Small, mineral, soil particles, ranging in diameter from 0.05 to 0.002 millimeters in the United States Department of Agriculture system or 0.02 to 0.002 millimeters in the International System. (2) A textural class of soils that contains 80 percent or more of silt and less than 12 percent clay. (3) Water-borne sediment with diameters of individual grains approaching that of silt. (4) To fill or obstruct a channel, etc., with water-borne sediment; to become filled or choked up with silt or sediment, as a channel, riverbed, etc. See Soil Separate.

Silt Loam–A textural class of soil that contains 50 percent or more of silt and 12 to 27 percent of clay, or 50 to 80 percent silt and less than 12 percent clay.

Silty Clay–A textural class of soil that contains 40 percent or more of clay and 40 percent or more of silt.

Silty Clay Loam–A textural class of soil that contains 27 to 40 percent of clay and less than 20 percent sand.

Silva–(Latin) Wood, forest.

Silver Thaw–A weather phenomenon in which great quantities of ice collect on trees and other vegetation, often causing much breakage.

Silverfish–*Lepisma saccharina*, family Lepismatidae; a small, silvery or greenish-gray, wingless insect which prefers damp places, such as basement rooms, and feeds on papers, cards, wallpaper, or any starchy material. It may deface bookbindings, etc.

Silvicide–Any chemical which is capable of killing trees, usually by contact, as by spraying or injection.

Silvics–(1) The natural science which deals with the laws underlying the growth and development of single trees and of the forest as a biological unit. (2) The pure science upon which silviculture is based. See Silviculture.

Silvicultural Rotation–The rotation through which stands a timber should be grown to maintain maximum vigor of growth and reproduction.

Silviculture–(1) The science and art of growing and tending forest crops, based on a knowledge of silvics. See Silvics. (2) More particularly the theory and practice of controlling the establishment, composition, constitution, and growth of forests.

Single-grained Structure–A soil in which there is no aggregation of particles, such as sand. See Soil Structure.

Sinker–A log whose specific gravity is greater than that of water and will not float.

Sinkhole–A basin depression, often deep and circular, present in many parts of the United States, which is underlaid by limestone or gypsum. The sink results from the collapse of the roofs of subterranean solution caverns. Diagnostic of karst and permafrost topography.

Site–(1) Location for a specific use. (2) A location for a forest planting or for the natural development of a forest. (3) An area, considered as to its ecological factors with reference to capacity to produce forests or other vegetation; the combination of biotic, climatic, and soil conditions of an area. (4) The situation of a fruit planting regarding elevation, topography, soil, and nearness to large bodies of water.

Site Classes–A classification of forest land in terms of potential growth in cubic feet per year of fully stocked natural stands.

Site Index–A designation of the quality of a forest site based on the height of the dominant stand at an arbitrarily chosen age; e.g., if the average height attained by dominant and codominant trees in a fully stocked stand at the age of fifty years is 75 feet, the site index is 75 feet.

Site Preparation–A general term for removing unwanted vegetation, slash, roots, etc. from a site before reforestation. It is accomplished by using mechanical, chemical, biological means, or fire.

Sitka Spruce–*Picea sitchensis*, family Pinaceae; an evergreen tree grown as an ornamental and as a timber tree. Native to North America.

Skidding Tongs–A hinged, clawlike device used to grasp a log for skidding. A cable or chain is fastened to the tongs and to a skidder or tractor. As the cable or chain is pulled the tongs grip the logs tighter.

Skidway–A prepared place for pulling logs prior to loading which consists of two parallel supports at right angles to the road generally raised at the end nearest the road.

Skim–(1) To take something from the top, as to clear scum or matter floating upon the surface of a liquid. (2) To remove cream from milk which has stood for a period of time, by means of a special utensil or apparatus.

Skin–(1) The flexible integument which forms the external covering of an animal, especially of vertebrates. (2) Anything that resembles skin in nature or use, as the flexible outer covering or peel of fruits, etc. (3) The pelt of a small animal, such as calf, sheep, goat, fox, mink, etc., which is usually dressed, tanned, or intended for such treatment. (4) To strip the skin from. (5) To rub the skin off. (6) To peel.

Skunk Spruce–See White Spruce.

Sky Pond–A pond constructed near the ridge or top of a hill or mountain. Since there is little watershed, it depends largely on direct rainfall or snowfall for filling it, and is sometimes supplemented by runoff from farm buildings.

Skyline–In forestry, the cable suspended between the head spar tree and the trial tree on which the trolley travels in cableway logging.

Slab–(1) The exterior portion of a log which is removed in sawing lumber. (2) An unattractive, overripe or broken, dried fruit (prune, apricot, etc.) which is sorted from a drying tray in commercial fruit drying. (3) A piece of unsliced bacon.

Slack Water–(1) The interval between high and low water when the tide is in a state of rest. (2) The apparently motionless water in the upper reaches of a dam. (3) In a log jam, the resulting sluggish flow of water.

Slag–The dross from iron ore furnaces and a by-product of the steel industry which can be converted into forms for agricultural use, as fertilizer and soil amendments. Some prepared material from slag, containing little or no phosphorus, but calcium in the form of silicates, may have value primarily as a liming material and a soil conditioner. Other slags, as basic slag, have some value as fertilizer because of the total phosphorus which they contain.

Slaked Lime–(1) CaO; quicklime or burnt lime, treated with water to produce hydrated lime, which in its pure form has the composition $Ca(OH)_2$. It is marketed in a dry, powdered form and is used as a soil amendment in addition to other uses. Also called calcium hydroxide, hydrated lime. (2) Burnt lime which has been exposed to the open air and thereby lost its causticity.

Slash–The branches, bark, tops, chunks, cull logs, uprooted stumps, and broken or uprooted trees which remain on the ground after logging.

Slash Disposal–Treatment or handling of slash to reduce the fire hazard or for other purposes which includes broadcast burning, strip burning, piling and burning, lopping, scattering, pulling tops away, etc.

Slash Pine–*Pinus elliottii*, family Pinaceae; an evergreen tree grown for its lumber used in general construction and as a source of turpentine and rosin. Native to North America. Also called Cuban pine.

Slashing–Forest area which has been logged off and upon which the slash remains.

Sleet–(1) Fine ice particles, or partly frozen rain that freezes as it falls. (2) The coating of thin ice that sometimes forms on trees, buildings, etc., when rain or sleet falls at a low temperature. (3) A storm or shower of sleet.

Sliced Veneer–Very thin sections of wood sliced off by moving a log or bolt of wood against a knife.

Slick Spots–Clayey soils that are black because of humus highly dispersed by sodium and, when plowed with a turning plow, turn over in "rubbery," shiny, slices. Such spots seldom produce a crop because of too little soil oxygen, too little available water, and toxic sodium.

Slickensides—Polished and grooved surfaces produced by one mass sliding past another. In soils, slickensides may occur at the bases of slip surfaces on the steeper slopes; on faces of blocks, prisms, and columns; and in swelling clayey soils, where there is marked change in moisture content. See Vertisols.

Slide—(1) The slipping of a mass of earth or rock down the slope of a hill or mountain; a landslide. (2) A smooth surface for sliding; a chute.

Slime—(1) A bacterial rot of lettuce resulting in a wet, slimy decay of the leaves and heads. It is caused by ***Erwinia carotovora***, family Enterobacteriaceae; ***Pseudomonas viridilivida***, family Pseudomonadaceae; ***P. marginalis***, and other species of bacteria. Slime occurs in the field, during warm, muggy weather; in lettuce not refrigerated in transit, and at markets. Also called bacterial rot of lettuce. (2) In the clarification of milk, a collection or deposit of dirt, leucocytes, cells, and other viscous matter thrown to the walls of the clarifier bowl. (3) In the curing or storage of cheese, a condition frequently produced on the surface by yeasts, bacteria, or other contamination, or faulty handling, causing a slimy surface on the cheese. (4) To remove the slimy or viscous coating from animal intestines, as in the preparation of sausage casings.

Slime Flux—A persistent, fermenting exudate of seepage from trunk or branch wounds in deciduous trees. In some cases pathogenic organisms are associated with the flux.

Slip—(1) A soft-wood or herbaceous cutting from a plant, used for propagation or grafting. (2) An incompletely castrated male. (3) Curdled milk. (4) To take a cutting or cuttings from a plant. (5) To abort. (6) The downslope movement of a soil mass under wet or saturated conditions; a microlandslide that produces microrelief in soils.

Slope Drains—Permanent or temporary devices that are used to carry water down cut or embankment slopes. May be pipe, half sections, paved, or have special plastic lining.

Slow-release Fertilizers—See Controlled-release Fertilizers.

Sludge—(1) The residual solids which remain after sewage treatment; the sewage solids deposited by sedimentation in sewage treatment; often used as an organic fertilizer. See Activated Sewage. (2) Borings from a well. (3) A composition of oxidized petroleum products along with an emulsion formed by the mixture of water and oil. This forms a pasty substance that clogs oil lines and passages and interferes with engine lubrication.

Sludge Digestion—The microbiological decomposition of organic matter in sludge, either aerobically or anaerobically or both, resulting in liquefaction, mineralization, reduction in offensive odors, and fewer pathogens.

Slug—(1) Any of numerous soft, usually grayish or brownish, slimy creatures (gastropods), up to 3 inches or more in length, with unsegmented bodies, related to snails but without a shell. Chiefly terrestrial mollusks of the family Limnacidae, slugs are found in damp places in gardens, fields, and elsewhere under logs, rubbish, and in rotting vegetation. They feed at night by rasping holes in leaves and other plant material and leave slimy trails on surfaces over which they have moved. They may be serious pests in gardens and greenhouses. (2) A wholesale, forequarter cut of lamb or mutton which consists of the chuck, breast, neck, and shanks; all of the forequarter except the rack. Also called rattle.

Sluice—(1) A conduit which carries water at high velocity. (2) An opening in a structure for passing debris. (3) A water gate. (4) A channel which carries or drains off surplus water. (5) To cause water to flow at high velocities for wastage, for excavation, ejecting debris, etc. (6) To move or float logs over or through a dam from the water above into the stream below.

Slump—A landslide.

Slurry—(1) A thin mixture of water and any of several fine insoluble materials, as clay, derris or cube powder, and so on. (2) The thick liquid formed by the mixing of dung and urine. (3) A thick suspension of chemicals used in coating seed with a fungicide. (4) A supersaturated, fluid fertilizer.

Small Mouth Bass—A game fish, ***Micropterus dolomieu***, that inhabits cool, clear lakes and rivers.

Smear—Material smeared on a surface, usually a small piece of glass (slide), and examined under a microscope. Smears are generally stained with dyes before being examined. Materials suitable for examination from a smear include blood, milk, bacteria from a culture, pus, etc.

Smectite—A clay mineral group similar to the montmorillonite group. See Montmorillonite.

Smelter Injury—Toxic effects on plant life which are caused by flue gasses or dusts discharged by smelters. It is similar to that caused by smoke in highly industrial areas.

Smoke Chaser—A member of a forest-fire protective unit that goes to reported fires to extinguish them.

Smoky—(1) Of the color of smoke, a brownish or bluish shade of gray; dusky; cloudy. (2) An abnormality in the appearance of a horse's eye, which becomes cloudy, whitish, and pearly-colored. It is indicative of impaired vision. The condition is referred to as smoky eye.

Smoldering Fire—A forest fire still smoking, not spreading appreciably, and burning with little or no flame.

Smothering Disease of Coniferous Seedlings—A disease of coniferous seedlings caused by the smothering fungus ***Thelephora laciniata***, family Thelephoraceae.

Smudge—A smoky fire or the dense smoke used to drive insects away.

Smudge Fire—A fire made from damp straw, smoldering manure, or any other fuel which produces a very heavy smoke that hangs low to the ground, erroneously believed to prevent frost injury to crops. See Smudge Pot.

Snag—(1) A sharp point of wood which results from a break. (2) A broken tree trunk more than 20 feet tall.

Snail—An organism that typically possesses a coiled shell and crawls on a single muscular foot. Air-breathing snails, called pulmonates, do not have gills but typically obtain oxygen through a "lung" or pulmonary cavity. At variable intervals most pulmonate snails come to the surface of the water for a fresh supply of air. Gill-breathing snails possess an internal gill through which dissolved oxygen is removed from the surrounding water.

Snake–(1) Any of the limbless vertebrates of the order Serpentes. Although a few species are poisonous, most snakes are beneficial to humans by the number of rodents they consume. (2) To drag a log or other object across the ground using a rope, chain, or cable.

Snow–Small ice crystals usually aggregated in the form of flakes which form during freezing temperature by the congealing of atmospheric vapor.

Snow Fence–A slate-and-wire fence, used in winter to intercept drifting snow, which protects roadways, railways, and other areas from snow drifts. See Snowbreak.

Snow Line–On mountain slopes, a natural line above which snow lies during certain seasons of the year. In rangelands, it marks the upper limits for effective grazing. See Timber Line.

Snow Pack–(1) Accumulations of snow in mountains, of vital importance in supplying snowmelt water to be caught behind dams for irrigation in arid and semi-arid regions. (2) Accumulations of snow at ski resorts.

Snowbreak–(1) An artificial or natural barrier such as a shelterbelt or windbreak to intercept blowing snow and thereby reduce excessive accumulations on roadways or farmsteads. Almost equivalent to a snow fence or a windbreak. (2) The breaking of tree limbs by the accumulation of snow. See Snow Fence.

Snoweater–See Chinook Wind, Foehn Wind.

Snowstorm–A weather condition in which falling snow flakes or particles are blown about by the wind. It usually implies a storm less severe than a blizzard.

Social–Living in more-or-less organized communities of individuals, such as termites or bees.

Social Insects–Insects which live in a family society, with parents and offspring sharing a common dwelling place and exhibiting some degree of mutual cooperation; e.g., honeybees, ants, termites.

Sod–(1) A tight, closely knit growth of grass or other plant species which may or may not be closely cut at frequent intervals. (2) A surface layer of soil containing grass or other plant species with their matted root systems.

Sod Crop–(1) Any crop or vegetative cover which quickly forms a heavy, close-knit, top growth over the surface of the soil and a root system which binds the soil particles together, thus forming a sod, as white clover, bluegrass, etc. (2) Any forage grasses or legumes.

Sod Culture–A soil management practice of maintaining permanent sod in orchards as opposed to annual tillage.

Sod Grasses–Stoloniferous or rhizomatous grasses that form a sod or turf.

Sod Species–Any grass with rhizomes or stolons, such as Bermudagrass.

Sod Strip–(1) A band or narrow strip of sod used for checking erosion in waterways. (2) A narrow band of grass or other close-rooted crop placed across the channel of a gully to spread and retard the flow of water. (3) In strip cropping, a strip that is in sod.

Sodbound–Designating a condition in which the roots are so restricted and form such a dense mat that growth of the plant is impeded.

Sodding–Removing sod from one area and placing it on a bare soil area in another location.

Sodic Soil–A soil with an exchangeable sodium percentage of 15 percent or greater, a saline content below 4 mmhos, and a pH between 9.5 and 10.0. (Obsolete name: black alkali.)

Sodium–Na; pure metallic sodium is a soft waxy material at ordinary temperatures; it burns in air to form sodium oxide (NaO) which readily takes up water to form sodium hydroxide (NaOH). Combined with chlorine, it forms sodium chloride (NaCl) (common salt), and with nitric acid, it forms sodium nitrate ($NaNO_3$). Sodium can substitute for or replace part of the potassium needed by some plants. It increases the availability of soil phosphorus and reduces fixation of the phosphorus applied in superphosphate. Sodium appears to have some value of its own for such crops as oats and table beets. This, however, may be due to the fact that it can release potassium, calcium, and magnesium by cationic exchange and thus temporarily increase these available cations in the soil.

Sodium Alginate–A sodium salt of the seaweed, algae. It has excellent water-binding properties and is used by some manufacturers to prevent water crystallization or iciness in ice cream.

Sodium Borate–See Borax.

Sodium Fluoracetate–Compound 1080; a powerful rodenticide developed by the United States Fish and Wildlife Service; a fine, white powder readily soluble in water, used to control rats, mice, and other rodents. It is a restricted pesticide.

Sodium Fluosilicate–A fluorine compound sometimes substituted for sodium fluoride for the control of lice, roaches, and certain other insects. Also called silicofluoride.

Sodium Hypochlorite–NaOCl; a chlorine compound which is commonly sold in a powdered form or in solutions under various trade names for use in the disinfection of processing equipment and as a household disinfectant and bleach.

Sodium Nitrate–$NaNO_3$; nitrate of soda; Chile saltpeter; a salt recovered from natural deposits in Chile, or produced synthetically by reacting nitric acid with sodium carbonate. Pure sodium nitrate contains 16.48 percent N and 27.05 percent sodium (Na).

Sodium Selenate–A highly poisonous, water-soluble salt sometimes applied in solution to greenhouse soil to control red spider mites and aphids. The compound is taken up from the soil by the plants, and the plant juices become toxic to the insects feeding on them.

Soft Phosphate with Colloidal Clay–A soft, clayey, low-grade, phosphatic rock, not suitable for use in the manufacture of superphosphate. It is the fine-grained waste which results from the washing of hard rock and pebble phosphate, collected in ponds, subsequently dried and sold as a low-grade fertilizer or used as a filler for mixed fertilizers. It has some value on acid sandy soils. Also called colloidal phosphate, waste-pond phosphate, phosphatic clay.

Soft Water—Water relatively free from various hardening compounds, particularly water free from a significant amount of carbonate of calcium.

Softwood—(1) Any wood of light texture which is easily cut. (2) In forestry, any tree having such wood, particularly a coniferous tree. (3) Immature, succulent wood, such as is used for a softwood cutting. See Hardwood, Softwood Cutting.

Soil—The mineral and organic surface of the earth capable of supporting upland plants. It has been (and is being) formed by the active factors of climate and biosphere exerting their influence on passive parent material and topography over neutral time.

Soil Acidifier—A material or mixture used, especially in semi-arid western United States to reduce soil alkalinity. Sulfuric acid, phosphoric acid, liquid sulfur dioxide, aluminium sulfate, sulfur, and calcium polysuflide are common soil acidifiers.

Soil Acidity—Excess of hydrogen (H) ions over hydroxyl (OH) ion concentration in the soil, consisting of: (a) active, which is estimated by pH below 7.0, and (b) inactive or reserve, less successfully estimated by base exchange determinations, as opposed to soil alkalinity. See pH, Reaction.

Soil Aggregate—A single mass consisting of many soil particles held together. See Clod, Ped.

Soil Alkalinity—Excess hydroxyl (OH) ions over hydrogen (H) ion concentration in the soil, as opposed to soil acidity.

Soil Amendment—Any material, such as lime, gypsum, sawdust, compost, animal manures, crop residues, or sewage sludge, that is worked into the soil or applied on the surface to enhance plant growth. Amendments may contain important fertilizer elements, but the term commonly refers to added materials other than those used primarily as fertilizers.

Soil Analysis—Chemical, mechanical, and mineralogical determinations of the separate components of a soil which may be complete or partial depending upon the objectives.

Soil Association—A group of soils with or without common characteristics, geographically associated in an individual pattern. It may constitute a natural land type or a geographical unit in soil mapping. A soil association is described in terms of the taxonomic units included, their relative proportions, and their pattern of association. Sometimes called natural land type. See Land Type, Soil Complex.

Soil Auger—A tool used for boring into the soil and withdrawing a sample for field or laboratory observation. Soil augers are of two general types: those with worm-type bits and those having a hollow cylinder with a cutting edge in place of the worm bit.

Soil Bacteria—Two large groups of bacteria, based on their energy requirements: (a) the heterotrophic bacteria, which obtain their energy and carbon from complex organic substances; (b) the autotrophic bacteria, which can obtain their energy from the oxidation of inorganic elements or compounds, their carbon from carbon dioxide, and their nitrogen and other minerals from inorganic compounds. See Bacteria.

Soil Buffer Compounds—Clay, organic matter, and such compounds as carbonates and phosphates which enable the soil to resist appreciable change in pH value.

Soil Bulk Density—Ratio of dry weight of a soil mass to its volume.

Soil Chemistry—That branch of chemistry which deals with investigations into the chemical nature of soil components and their interrelationships.

Soil Class—See Land Capability Class, Soil Textural Class.

Soil Classification—The systematic grouping of soils into categories from higher to lower. It may be done on a genetic basis, on the basis of the intrinsic properties of different soils, or a combination of the two. The smaller taxonomic units of a scientific system may be regrouped for various purposes, such as agriculture, highway construction, and forestry. See Soil Taxonomy.

Soil Climate—The moisture and temperature conditions existing within the soil.

Soil Colloid—Inorganic and organic matter of very small size having a relatively large surface area per unit of mass. Under certain conditions, soil colloids form a stable suspension in water and are distinguishable from true solutions in that all particles have not dispersed to the molecular state. One micron, or 0.001 millimeter has been arbitrarily set as the upper limit for size of a soil colloid.

Soil Complex—A mapping unit used in detailed soil surveys where two or more defined taxonomic units are so intimately associated geographically that they cannot be separated by boundaries on the scale used. See Soil Association.

Soil Conservation—The efficient use and stability of each area of soil that is needed for use at its optimum level of developed productivity according to the specific patterns of soil and water resources of individual farms, ranches, forests, and other land-managment units. The term includes the positive concept of improvement of soils for use as well as their protection and preservation.

Soil Consistence—The relative neutral attraction of the particles in the whole soil mass or their resistance to separation or deformation, as evidenced in cohesion and plasticity. Soil consistence is described by such general terms as loose or open; slightly, moderately, or very compact; mellow; friable; crumbly; plastic; sticky; soft; firm; hard; and cemented.

Soil Creep—The very slow downward movement of soil, distinguished from landslides, slippage, earth flows, avalanches, and wash from rainfall runoff.

Soil Depth—(1) Technically the thickness of the solum (A and B horizons), as distinguished from the C and R horizons of the profile. (2) The thickness of the humus layer or plowed soil. See Soil Profile.

Soil Deterioration—Loss of potential productive capacity of a soil which results from destructive processes accelerated by the activities of people; e.g., through soil erosion, waterlogging, and excessive accumulation of salts.

Soil Development–The work of the natural evolutionary process of soil formation which leads to maturity or complete differentiation into soil horizons. See Soil Profile.

Soil Drainage Classes–Classes of soils grouped according to their ability to hold water. The seven classes are: excessively drained, somewhat excessively drained, well drained, moderately well drained, somewhat poorly drained, poorly drained, and very poorly drained.

Soil Erosion–Removal of soil material from a land surface by wind or water, including normal soil erosion and accelerated soil erosion. (Sometimes used loosely in reference to accelerated erosion only.) See Accelerated Erosion, Erosion, Normal Erosion.

Soil Erosion Service–In the United States, an emergency agency first authorized in 1933 which was placed in the Department of Interior, and subsequently in 1935 transferred to the Department of Agriculture, as the Soil Conservation Service. See Soil Conservation Service.

Soil Family–A group of soils having similar amounts of sand, silt, and clay; similar kinds of minerals and temperatures; and similar other properties important to plant growth. There are about 2,000 soil families recognized.

Soil Fertility–The capability of a soil to supply all essential plant nutrients in balanced and adequate amounts and with proper timeliness. See Soil Productivity.

Soil Formation Factors–The independent variables that define the soil system. Five main groups of soil formation factors are generally recognized by soil scientists: parent material, climate, organisms, topography, and time.

Soil Genesis–The mode of origin of the soil, particularly the processes which are responsible for the development of the solum from the parent material.

Soil Heaving–See Heaving.

Soil Horizon–A natural layer of a vertical section of the soil profile. See Soil Profile.

Soil Injection–Placing pesticide or fertilizer below the surface of the soil.

Soil Loss Equation, Universal–See Universal Soil Loss Equation.

Soil Loss Tolerance–In conservation farming, the maximum average annual soil loss in tons per acre per year that should be permitted on a given soil. The rate of soil formation should equal or exceed the rate of soil erosion loss. See T Factor.

Soil Management–The combination of all tillage operations, cropping practices, fertilizer, lime, and other treatments applied to the soil for the production of plants.

Soil Map–(1) A map which shows the location and extent of the different soils in an area. Soil maps are prepared and published by various agencies of state and local governments, in cooperation with the United States Department of Agriculture Soil Conservation Service or the United States Department of Agriculture Forest Service. (2) A map showing the distribution of soil mapping units in relation to the prominent physical and cultural features of the earth's surface. The following kinds of soil maps are recognized in the United States: detailed, detailed reconnaissance, generalized, and reconnaissance.

Soil Map, Detailed–United States Department of Agriculture Soil Conservation Service usage. Referring to a soil map at a publication scale commonly of 1:190,080 on which soil phases and soil series are the main units delineated on an aerial map base. The smallest unit size shown on such maps is about 1 acre. Survey traverses are usually made at one-quarter mile, or more frequent, intervals. The unit boundaries on detailed soil maps should have been seen throughout their course and their placement on the map should be accurate to at least 100 feet. The maximum amount of unlike soil inclusions in mapped units is 15 percent.

Soil Map, Detailed Reconnaissance–A reconnaissance soil map on which some areas or features are shown in greater detail than usual, or than others.

Soil Map, Generalized–United States Department of Agriculture Soil Conservation Service usage. Referring to a soil map with a publication scale commonly of 1 inch = 1 mile on which soil associations and miscellaneous land types are the delineated units. The smallest unit size shown is 3½ acres. The maximum amount of unlike soil inclusions in such mapping units is 15 percent.

Soil Map, Reconnaissance–United States Department of Agriculture Soil Conservation Service usage. Referring to a soil map of highly variable publication scale (1 inch = 1 mile to 1 inch = 8 miles). The most detailed units commonly shown are miscellaneous land types and soil associations or one or more phases of soil families. The smallest unit that can be shown on such a soil map at a scale of 1 inch = 1 mile is about 3½ acres and at 1 inch = 4 miles about 280 acres. Map unit boundaries are plotted where they cross field survey traverses. Traverses are at intervals varying from about one-half mile to several miles. Between these points of field observation most boundaries are sketched from the appearance of ground patterns on aerial photographs and the general appearance of the landscape.

Soil Microbiology–The branch of science which deals with the study of microorganisms in the soil, such as algae, fungi, bacteria, and protozoa.

Soil Mixture–A mixture which consists of soil with varying proportions of one or more of the following: manure, peat moss, leaf mold, and sand used for growing seedlings, pot plants, plants in beds in greenhouses, and as a medium for the rooting of cuttings. Also called potting soil.

Soil Moisture–The water in soil normally expressed as a percentage of the weight of water to the oven-dry weight of soil.

Soil Moisture Capacity–The amount of water expressed in percentage of dry weight that a soil can retain against the pull of gravity. See Field Capacity.

Soil Moisture Tension–The force with which the soil holds onto the soil moisture present. As soil moisture declines, soil moisture tension increases.

Soil Morphology–The kinds and arrangements of horizons in a soil profile.

Soil Mottling–A variation of color in a soil horizon, as mottles, spots, splotches, or veins of one color in a background mass of another; usually gray splotches with a brown to red matrix.

Soil Mulch–(1) The natural cover of dead plant residue on a soil. (2) The corresponding loose matter on the surface of a grassland soil, in contrast to an artificial mulch. See Mulch.

Soil Orders–The classifications of soils. According to the United States Department of Agriculture system of soil taxonomy, the eleven soil orders are: entisols, vertisols, inceptisols, aridisols, mollisols, spodosols, alfisols, ultisols, oxisols, histosols, and andisols.

Soil Organic Matter–The organic fraction of the soil that includes plant and animal residues at various stages of decomposition, cells and tissues of soil organisms, and substances synthesized by the soil population. Commonly determined as the amount of organic carbon contained in a soil sample passed through a 2-millimeter sieve.

Soil Particle–A soil separate in the mechanical analysis of a soil.

Soil Pasteurization–The heat or chemical treatment of soil to destroy all harmful organisms but not necessarily all soil organisms.

Soil Phase–In soil taxonomy and soil mapping, a subdivision of soil series such as surface soil texture, slope, stoniness, saltiness, and erosion. Phase may also be used for a soil order, suborder, great group, subgroup, or family.

Soil Physics–That branch of science which is concerned with the application of the laws, knowledge, and techniques of physics to the study of soils.

Soil Porosity–The degree to which the soil is permeated with pores or cavities. It is expressed in percent of the volume of the soil unoccupied by soil particles.

Soil Probe–A T-shaped metal tube that is forced into the ground for the purpose of obtaining a sample of the soil.

Soil Productivity–The ability of a soil to produce crops under a particular condition of management. See Soil Fertility.

Soil Profile–A vertical section of a soil. The section, or face of an exposure made by a cut, may exhibit with depth a succession of separate layers although these may not be separated by sharp lines of demarcation. See Horizon, Soil.

Soil Reaction–The degree of acidity or alkalinity of the soil mass expressed in pH values. See pH.

Soil Salinity–The amount of soluble salts in a soil.

Soil Science–The science which deals with the scientific study of soil as a natural body and an economic resource. See Pedology.

Soil Separate–Any of a group of soil particles which has definite size limits. The names and average diameters of separates recognized in the United States are: very coarse sand, 2.0 to 1.0 millimeters; coarse sand, 1.0 to 0.05 millimeters; medium sand, 0.05 to 0.25 millimeters; fine sand 0.25 to 0.10 millimeters; very fine sand 0.10 to 0.05 millimeters; silt, 0.05 to 0.002 millimeters; and clay, less than 0.002 millimeters. The class of separates recognized by the International Society of Soil Science are: I, 2.0 to 0.2 millimeters; II, 0.2 to 0.02 millimeters; III, 0.02 to 0.002 millimeters; IV, less than 0.002 millimeters.

Soil Series–A group of soils having horizons similar in characteristics and arrangement in the soil profile, except for the texture of the surface. They are given proper names from place names within the areas where they occur. Thus, Norfolk, Miami, and Houston are names of well-known soil series.

Soil Slaking–The phenomenon expressed by crumbling, flaking, or complete disintegration of a mass of dry soil when immersed in water.

Soil Slip–A downward movement of a mass of soil on hillsides under conditions of complete saturation. The distance covered and the mass of the movement in a soil slip are less than in a landslide or avalanche.

Soil Slope–An incline of a land surface. See Slope.

Soil Solution–(1) The liquid part of soil that surrounds soil particles and contains the elements in solution that are absorbed by plant roots. It is not constant in amount or composition. (2) Technically, the aqueous solution which exists in equilibrium with the soil particles at a particular moisture tension.

Soil Strength–A characteristic of how well soil particles stick together or resist fragmentation. It is used to predict a soil's adaptation to cultivation or to support the foundation of a building.

Soil Structure–See Structure, Soil.

Soil Survey–(1) In the United States, this refers to the National Cooperative Soil Survey conducted by the Soil Conservation Service, United States Department of Agriculture, in cooperation with state agricultural universities and other public and private agencies and organizations. (2) Systematic examination of soils in the field and in laboratories; publishing of descriptions and classifications; mapping of kinds of soils; and interpretation of soils according to their adaptability to various crops such as field crops, fruits, vegetables, and trees.

Soil Survey Report–A written statement accompanying a soil map which describes the geography of the area surveyed, the characteristics and capabilities of the soil-mapping units. It also discusses the principal factors responsible for soil development.

Soil Taxonomy–Classification according to natural relationships. Comparable to plant taxonomy and animal taxonomy. The latest system of soil taxonomy was implemented in the United States in 1965. See Soil Classification, Soil Orders.

Soil Temperature–The temperature in the soil mass which varies according to depth and local climate. It is an important factor in the growth of plants.

Soil Testing–Any of the various laboratory and field examinations of soil samples to determine the amounts, kinds, and availability of plant nutrients. Tests may also be made in relation to plant growth, agriculture, highway and building construction, such as acidity, alkalinity, moisture, porosity, penetrability, etc.

Soil Textural Class–The relative proportion in a soil of the various size groups of individual soil grains. The coarseness or fineness of the soil depends on the predominance of one or the other of these groups, which are silt, clay, and sands. See Soil Separate.

Soil Tilth–Structure or physical condition favorable for germination and development of crop plants. It is usually produced by loosening the soil with plows, harrows, or other agricultural implements, and by other agronomic practices.

Soil Water–(1) Any or all of the various forms of water which may be contained in the soil, as free, capillary, combined. (2) The water below the ground surface in the zone of aeration which may be discharged as evaporation, plant transpiration, and seepage.

Soils, Poorly Drained–Water is not removed because of a compact subsoil or the soil is too low for the water to drain out. The soil is usually a gray color within 6 inches of the surface.

Soils, Well Drained–Water moves readily through the subsoil. They are not discolored but show uniform, bright colors throughout the soil profile.

Sol–(Spanish) Sun.

Soldier–In termites, sterile males or females with large heads and mandibles; they function to protect the colony.

Solid Manure–Manure in solid form as contrasted to liquid manure.

Solid Waste–Useless, unwanted, or discarded material with insufficient liquid content to be free-flowing. Agricultural: the solid waste that results from the rearing and slaughtering of animals and the processing of animal products and orchard and field crops.

Solubility–To be most readily available to plants a nutrient must be at least slightly soluble in the soil solution or be held in an exchangeable form on clay and humus particles. See Available.

Soluble–Capable of changing form or changing into a solution.

Solum–The upper part of the soil profile above the parent material, in which the processes of soil formation have taken/are taking place. In mature soils this includes the A and B horizons which may be, and usually are, greatly unlike the parent material beneath. Living roots and life processes are largely confined to the solum. See A Horizon, B Horizon, Parent Material.

Solution Potholes–Includes all the holes that are formed primarily by solution action. Such holes are more numerous in soluble rocks, notably limestones. See Karst, Permafrost, Thermokarst.

Songbird–A singing bird, generally beneficial or relatively harmless on farms in contrast to nuisance birds, such as the crow, and game birds, such as quail and pheasants.

Soot–Very finely divided carbon particles clustered together in long chains. Sometimes used as a low-grade fertilizer.

Soporific–See Stupefacient.

Sorption–A term including both adsorption and absorption. Sorption is basic to many processes including the attraction and release of cations from clay and humus, to remove pollutants from the environment, and to clean up oil spills.

Sounding–Testing the depth or quality of bottom water.

Sour Soil–An acid soil.

Source–(1) The point of origin of a stream, river, etc. (2) A place or an object from which something may be obtained; e.g., a source of supply; grapes are a source of wine. (3) Special areas around the earth over which cold or warm masses of air originate.

Southern House Mosquito–*Culex quinquefasciatus*, family Culicidae; a blood-sucking insect that in some parts of the world is an important vector of a filarial parasite which lives in the blood and lymph of birds and horses.

Southern Magnolia–*Magnolia grandiflora*, family Magnoliaceae; a magnificent evergreen tree, grown in relatively frost-free areas for its large, fragrant, cup-shaped, white flowers. Its wood is used for furniture, fixtures, Venetian blinds, interior finish, sash, doors, and millwork. Native to North America. Also called evergreen magnolia, bull bay. The state tree of Mississippi.

Southern Pine Beetle–*Dendroctonus frontalis*, family Scolytidae; an insect that infests pine and makes winding egg galleries in the inner bark. It kills healthy trees, as well as attacking weakened or felled specimens. It is a serious pest of pine from Pennsylvania to Florida and west to Texas and Oklahoma, United States.

Sp. (Spp.)–The abbreviation, singular (plural) for species; e.g., when one species of a genus of plant or animal is referred to, the name may be written ***Canis*** sp. meaning one species of dog; or ***Canis*** spp. meaning more than one species of the genus ***Canis***.

Spaulding Rule–A diagram rule for estimating the lumber content of logs which disregards taper and allows for an $11/32$ inch saw kerf, with size of slab varying with size of log. It is the statute rule for California and is widely used elsewhere on the Pacific Coast (United States). Also called Columbia River rule.

Spawn–(1) To deposit eggs (of fish). (2) Common term for eggs and sperm. (3) Young fish, usually numerous, in early stages of development.

Specialty Fertilizers–Fertilizers recommended or used principally for growing roses, lawns, home gardens, organic gardens, house plants, or for any purpose other than growing farm crops.

Species–In the naming of plants and animals, Latin is used. Each kind of plant or animal can be identified by genus (plural, genera) and species (both singular and plural); e.g., the generic name (genus) of corn is ***Zea*** and the species name is ***mays***.

Specific–(1) A medicine that cures a particular disease. (2) Pertaining to a species. (3) Produced by a particular microorganism. (4) Restricted by nature to a special animal, thing, etc. (5) Exerting a peculiar influence over any part of the body.

Speckled Alder–*Alnus incana*, family Betulaceae; a deciduous shrub or small tree of several varieties, grown as ornamentals. Native to North America, Europe, and Asia. Also called hoary alder, European alder. See Alder.

Speleology–The scientific study or exploration of caverns and related features.

Spent Bone Black–Bone black or bone charcoal left after use in the refining of sugar. It is valued by fertilizer manufacturers chiefly because of its phosphorus content.

Spider–(1) Any of various arachnids in the order Araneida. Individuals have eight legs and only two main body divisions. The abdomen has spinerets for spinning silk threads used for constructing webs which serve as nests and as traps for prey. Most spiders are harmless to humans and because of the insects they devour and their aid in pollination, they are exceedingly beneficial. (2) Anything suggestive of a spider in form, as various mechanical parts having radiating members. (3) A tumor or growth sometimes suspended in a cow's teat canal by strands of fibrous tissue which resembles a spider web (colloquial).

Spillway–A grassed waterway, channel, or structure to convey water from the stream above the reservoir into the stream below the dam.

Spiracles–The openings to an insect's internal breathing tubes, the trachea.

Spit–(1) A small point of land or narrow shoal projecting into a body of water from the shore. (2) A pointed rod on which meat is held while roasting or barbecuing above or in front of open heat. (3) Saliva from animals or humans.

Splash Erosion (Raindrop Erosion)–The spattering of small soil particles caused by the impact of raindrops on bare wet soils. The loosened and spattered particles may or may not be subsequently removed by surface runoff.

Split Application–Breaking up the application of fertilizer into two or more applications throughout the growing season. Split applications are intended to supply nutrients more evenly and at times when the crop can most effectively use them.

Split Crotch–Of a tree, a split or separation at the juncture of two large limbs.

Split Ditch–A lateral ditch used in the sugarcane system of surface land drainage in Louisiana, United States. Such ditches are usually parallel and spaced 100 to 250 feet apart with rows of sugarcane parallel to them.

Spodic Horizon–A subsurface soil horizon high in iron, aluminum, and humus but low in clay. See Spodosols.

Spodosols–Soils with illuvial accumulations of amorphous materials in subsurface horizons. The amorphous material is organic matter and compounds of aluminum and usually iron. These soils are formed in acid parent materials and are mainly coarse-textured sands in humid and mostly cool or temperate climates. Also spodosols develop in humid, sandy Florida.

Spoil–(1) To deteriorate by molding or rotting. See Spoilage. (2) Debris or waste material from a coal mine. (3) Dirt or rock that has been removed from its original location, specifically materials that have been dredged from the bottoms of waterways.

Spoil Bank–A small to large ridge of soil excavated from, and piled alongside, a drainage ditch. In good drainage practice, this spoil is leveled, i.e., scattered over a considerable area alongside the ditch.

Spoilage Organisms–Bacteria, yeasts, and molds that cause food to spoil. They live everywhere: in the air, soil, and water and on food, plants, and animals.

Spore–The one to many-celled unit of a fern, fungus, bacterium, or protozoan that has entered the resting state and is capable of growth and reproduction when conditions become favorable. Spores in lower forms correspond to seeds in higher plants.

Spore Dust–A standardized powder containing spores and used in the biologic control of insect pests. The dust is placed at intervals in the soil, and the spores infect the larvae of certain insects, such as Japanese beetle, causing diseases which destroy them. See Milky Disease.

Sport–A random mutation.

Sporulaton–(1) In bacteria, the formation of spores within the body of the bacterium. The spores represent the inactive resting, or resistant, forms. (2) In coccidia, a kind of reproduction by which the fertilized cell within the oocyst wall splits up into new individuals, called sporozoites. Sporulation of coccidial oocysts usually occurs after the oocyst has been discharged from the body of the host. (3) In plants and animals, the process of spore formation.

Spot Burning–A method of slash disposal in which only the heaviest accumulations of slash are fired, and the fire is not allowed to spread over the entire cut-over area.

Spot Fire–A fire set in advance of, or away from, the main fire by flying sparks or embers.

Spot Treatment–The application of a pesticide or other material to a restricted or small area of heavier infestation.

Spotted Black Bass–A game fish, *Micropterus pseudoplites*, of the central United States that is grown for stocking lakes.

Spotted Waterhemlock–*Cicuta maculata*, family Umbelliferae; a poisonous, fleshy-rooted, perennial herb, which occurs chiefly in wet meadows and pastures, and along ditches and streams, from eastern North America westward to the Great Plains. The plant, especially the roots, contains a violent poison which causes serious or fatal results if eaten by humans or livestock. Native to North America. Also called beaver poison, cowbane, musquash root, water hemlock.

Spout–(1) A pipe, tube, or similar conduit, through which a liquid, grain, seed, or other loose material flows and is discharged. (2) A trough or chute by which grain, coal, sand, or the like is discharged from a conveyor or bin to a truck, etc. (3) A pipe or projection (as of a sprinkling can) by which a liquid is discharged or poured. (4) A spigot. (5) A spile which is used in the collection of sap. (6) To shoot out, as a liquid, etc., from a spout; to flow or shoot out with force.

Spray Cultivation–The destruction of weeds in growing crops by the use of chemical weed killers (herbicides). See Herbicide.

Spray Drift–The movement of air-borne spray particles from the target plants or target area of application.

Spray Injury–Injury resulting from the application of a chemical compound, as a spray to the foliage or other parts of a plant. The injurious effects may include damage to leaves, blossoms, fruit, twigs, or to the entire plant, sometimes resulting in general necrosis and death.

Spray Irrigation–Irrigation by means of spray from pipes or pipe projections above the soil surface. Also called sprinkler irrigation. See Sprinkler Irrigation.

Spreader–(1) A device used for distributing water uniformly in or from a channel. (2) A spreading agent, or spreader-sticker used to improve the contact between a pesticide and a plant surface. (3) An instrument provided with hooks or flanges used to spread open the incision made in the body wall of a bird. (4) A device that scatters or spreads: e.g., a manure spreader, a fertilizer spreader. (5) An animal, etc., capable of acting as a parasite or disease vector. (6) A species of microorganism which tends to grow profusely over the entire surface of the culture medium.

Spreader-Sticker–A surfactant closely related to wetting agents that facilitates spreading and increases sticking of an herbicide on vegetation. See Surfactant.

Spreading of Water–Spreading excessive run-off water over a large land area to replenish the underground water supply or to provide enough surface water in dry areas to allow grasses to grow at least in a limited area.

Spring–(1) A natural flow of water from the ground. The source may be either shallow or deep-seated, and the spring may be at the base of a slope where a pervious stratum overlies a relatively impervious one, or it may be an artesian spring issuing at the surface through rock fracture or along fault lines, or a subterranean stream issuing through a solution opening as in a limestone terrain. (2) The season between winter and summer. (3) An elastic device or part which recovers its shape after being stretched, compressed, bent, etc., as the curved steel blades of a spring-tooth harrow. (4) The act of springing or leaping. (5) Referring to the spring season of the year in which a crop is harvested, or a flower blooms, as an early-spring crop, or a spring flower. (6) To indicate proximity of calving by a swelling of the udder, as in a cow.

Spring Crop–Any crop that is planted or starts rapid growth in the early spring and is ready for harvest in the spring season, as turnips, lettuce, and radishes.

Spring Migrant–An animal, bird, insect, or plant that migrates or changes habitat in the spring of the year.

Spring Overturn–The period of mixing of previously stratified water masses which occurs in the spring when water temperatures in the lake are uniform.

Spring Tide–The highest high and lowest low tides during the lunar month.

Spring Wood–The less dense, large-celled part of a growth layer formed in woody plants by rapid growth during the spring. See Summer Wood.

Spring-Fall Range–Those grazing areas of the western range which, because of grazing patterns, climatic factors, forage production, quantity of water, etc., are seasonally grazed only in spring and fall. Such grazing may not be necessarily obligatory but it is convenient in a grazing program where summer range at higher elevations is productive and available only during the summer season.

Springtail–Any one of a large number of minute, wingless insects of the order Collembola which has chewing or piercing mouthparts and which jumps long distances by means of a forked appendage, the furcula, folded forward under the abdomen when at rest. Found in damp places, some species infest seeds, seedlings, mushrooms, and other plants or plant parts close to the ground.

Springtime–The season of the year following winter and preceding summer. For the north temperate zone it begins with the vernal equinox and ends with the summer solstice, and is generally recognized to be from March 21 to June 21.

Sprinkle–(1) A light rain. (2) To scatter water, sand, seeds, etc., in drops or particles. (3) To irrigate by a hose and nozzle.

Sprinkler Irrigation–Irrigation by means of above-ground applicators which project water outward through the air making it reach the soil in droplet form approaching rainfall. Applicators commonly used are rotary or fixed sprinklers, oscillating or perforated pipe.

Spruce–Any tree of the genus *Picea*, family Pinaceae. Species are handsome evergreen trees with dense conical crowns; widely used for timber and pulpwood and valuable for horticultural planting; native to the cooler regions of the Northern Hemisphere. Principal species of spruce are: black spruce, Colorado spruce, Englemann spruce, Norway spruce, red spruce, silver spruce, Sitka spruce, and white spruce.

Spruce Aphid–*Aphis abictina*, family Aphidae, a dull-green insect that is very destructive to Sitka spruce in Oregon and Washington, United States.

Spruce Budworm–*Choristoneura fumiferana*, family Tortricidae; a moth whose larvae infest confiners feeding on the terminal shoots which they have webbed together to form shelters. The tops of infested trees first appear as though scorched by fire, and if heavily infested the trees may die. The budworm is one of the most destructive insects in the forests of northern United States and southern Canada.

Spruce Sawfly–*Neodiprion abietis*, family Diprionidae; a sawfly whose larvae defoliate spruce and pine trees, causing great destruction. In the 1930s, this fly reached epidemic proportions in Canadian forests, before being controlled by spruce sawfly parasites from Europe. Found first in Canada.

Spruce Spider Mite–*Oligonychus sununguis*, family Tetranychidae; a black mite which sucks sap from foliage and spins webbing between needles of evergreens. These mites may kill young spruces the first season and cause older trees to die progressively.

Spur Dike–A dike of rock or other material built from the bank into the channel for bank protection or for channel improvement. Also called jetty.

Spur Terrace–In erosion control, a short terrace which is used to collect and hold or divert runoff.

Squall–Sudden and violent successive gusts of wind often accompanied by rain, sleet, or snow.

Squaw Winter–A short spell of snow and cold weather in the early autumn.

Stability–The atmospheric condition existing when the temperature of the air increases rather than falls with altitude. There is little or no vertical air movement.

Stabilization Pond–An enclosure for sewage designed to promote the intensive growth of algae. These organisms provide oxygen that stimulates transformation of the wastes into inoffensive end products.

Stabilized Grade–A construction slope at which neither erosion nor silting occurs. See Angle of Repose.

Stable Manure–The excreta of livestock intermixed with straw or other bedding material which commonly includes stalks and uneaten, coarse stems of forage, corn cobs, etc.; an important organic fertilizer. See Manure.

Stag–(1) In animals, a male castrated after reaching sexual maturity which shows pronounced sexual development. (2) A horse which is thick and coarse in the throat latch and crest from late castration. (3) An imperfectly or recently castrated sheep or steer. (4) In marketing, an uncastrated male chicken with flesh slightly darkened and toughened and with comb and spur development showing the bird to be in a state of development between a roasting chicken and a cock. (5) A boar hog usually castrated after having passed breeding usefulness. (6) A wild mature male deer.

Stag-headed–A tree dead at the top as a result of injury, disease, deficient moisture, or deficient nutrients.

Stage–The elevation of a water surface above any chosen datum plane, often above an established low-water plane; gauge height.

Stagnant–Water standing still, as in a pool.

Stalacite Frost Structure–Frozen soil which contains tiny icicles or columns of ice separating soil sheets; heaved soil. It is associated with partial thawing and refreezing. See Heaving.

Stand–(1) The proper number of uniformly distributed plants per acre. (2) The relative number of plants per area, as a poor, good, or medium stand. In forestry, fully stocked, understocked, pure, mixed, or residual stand. (3) A hive of bees. (4) Density of game per acre, area, etc. (5) A stallion's court. (6) To cease walking or moving; to take or keep a certain position. (7) To rise to the feet. (8) Of a stallion, to be available for breeding purposes.

Stand Improvement–Measures such as thinning, pruning, release cutting, girdling, weeding, or poisoning of unwanted trees aimed at improving growing conditions for the remaining trees.

Standard Atmosphere–The International Standard Atmosphere which is used as the basis of graduation of altimeters assumes at mean sea level a temperature of 15°C, a pressure of 760 millimeters mercury (Hg) (1,013.2 millibars) and a lapse rate of 6.5°C/kilometers from sea level up to 11 kilometers in elevation.

Standard Length–In softwood lumber, lengths of boards in multiples of even feet; in hardwood lumber, lengths from 3 to 20 feet, including both odd and even number of feet.

Standing Water–Any water that collects or stands for a period of time on the surface of soil following a heavy rainfall or fast-melting snow.

Starling–*Sturnus vulgaris*, a bird that is native to Europe and naturalized to the United States. The birds live in large colonies and can be a serious pest.

Static Head–(1) The vertical distance that the top of a column of fluid is above a reference level. (2) The total head without deduction for velocity head or losses; e.g., the difference in elevation of headwater and tailwater of a power plant.

Station–The geographical locality of occurrence of an organism, a community or stand. See Habitat.

Stationary Front–The front or boundary between a cold and warm air mass which remains stationary over the earth for a period of hours or days.

Steno––A prefix indicating narrowness.

Stenovalent–A species which is restricted to one habitat.

Steppe–(1) A cow's whole milk cheese with color added, made in Germany, Austria, and Denmark. (2) A treeless grassland which exists under semi-arid climates, such as the east-west zonal belt of plains in the U.S.S.R., lying between the forestland of the north and deserts on the south, or the semi-arid Great Plains of western United States lying east of the Rocky Mountains.

Sterile–(1) In animals, incapable of reproduction; unable to produce normal living young. (2) In soils, unproductive; barren; producing little or nothing, as a sterile soil. (3) In biological products, etc., free from contamination with living bacterial, fungal, or viral organisms; or designating an organism not capable of growing or multiplying.

Sterilization–(1) The destruction of all living organisms. In contrast, disinfection is the destruction of most of the living organism. (2) To make animals infertile.

Stilling Basin–An open-top structure or excavation at the foot of a waterfall, drop structure, or rapids, used to dissipate the energy of the descending current.

Stilling Well–A protected, miniature well, beside and connected with a body of water, used for accurately determining the changes in water level. It is usually used in connection with the measurement of discharge water in a canal, stream, river, or flume.

Stimulus–An activating agent such as heat, moisture, or light.

Sting–(1) The resulting wound or welt caused by the sharp-pointed organ of certain insects, scorpions, etc. (2) The smart or irritation produced by touching certain plants; e.g., the nettle. (3) The puncture or wound in the skin or rind of fruit, etc., made by the ovipositor or mouthparts of an insect.

Stock–(1) Plant or plant part upon which a scion is inserted in propagation. (2) Livestock; domesticated farm animals. (3) Material held for future use or distribution. (4) Material destined to be wrought into finished products, as crate stock, barrel stock, cider stock, etc. (5) A plant or plant part that furnishes cuttings for propagation. Also called stock plant. (6) The main stem or trunk of a plant. (7) A rootstock (rhizome). (8) The original type from which a group of plants or animals has been derived. (9) The base or handle of a whip. (10) The stump of a tree (after the tree is felled). (11) (plural) A small enclosure in which an animal is secured in a standing position during shoeing or an operation. (12) To provide or supply with livestock, as to stock a pasture or range with cattle, sheep, etc. (13) To assemble a supply of materials or commodities. (14) Of, or pertaining to, livestock, as stock barn, stock feed, etc.

Stock Country–Any region of extensive grasslands suitable for livestock grazing, as the western and southwestern range areas of the United States.

Stocking–(1) The relative number of livestock per unit area for a specific time. In range management, the relative intensity of animal population, ordinarily expressed as the number of acres of range allowed for each animal for a specific period. (2) In wildlife management, the density of animal population in relation to carrying capacity. (3) In forestry, the density of a stand of trees, such as well-stocked, overstocked, partially stocked. (4) A white leg on an animal.

Stone–(1) In land description, a detached rock fragment on the surface of, or embedded in, the soil. (2) In soil survey mapping, a rock which is greater than 10 inches in diameter or more than 15 inches in length. See Boulder. (3) The hard seed of some pulpy fruits, as peach, plum, etc. See Clingstone, Freestone. (4) A unit of weight, 14 pounds (avoirdupois, British). (5) See Testicles. (6) To take the stone or pit out of stone fruits.

Stony Land–Land which contains enough stone, either detached fragments in or on the soil or rock outcrop, to interfere with tillage.

Storm–A violent disturbance of the lower atmosphere, frequently accompanied by wind, rain, snow, sleet, or hail and usually by lightning and thunder during summer.

Storm Cellar–An underground structure designed to protect people from tornadoes and severe storms.

Storm Sewer–A sewer which carries off only surface run-off water and is therefore different in function from a sanitary sewer.

Storm Track–The path across the earth of a low-pressure system. Storm tracks across the United States usually come from the Northwest, dip Southeastward into the Plains states, then change direction and travel northeasterly across the East Coast and New England toward the North Atlantic ocean.

Strain–(1) A group of plants of common lineage which, although not taxonomically distinct from others of the species or variety, are distinguishable on the basis of productiveness, vigor, resistance to drought, cold, or diseases; or other ecological or physiological characteristics. (2) A group of individuals within a breed which differ in one or more characters from the other members of the breed; e.g., the Milking Shorthorns or Polled Herefords. (3) An organism or group of organisms which differs in origin or in minor respects from other organisms of the same species or variety. (4) A virus entity whose properties and behavior indicate relationship to a type virus and are sufficiently constant to enable the entity to be recognized whenever isolated. (5) A severe muscular effort on the part of a draft or other animal which may result in muscle, ligament, or other damage. (6) The condition of an overworked part operating above optimum load, such as a belt or motor operating under strain. (7) To filter a liquid and free it of impurities by passing it through some medium or fabric which can retain the solid matter and allow the liquid to pass.

Strata–(Singular, stratum) (1) In geology, layers or beds of rock which are nearly lithologic units. (2) In ecology, different vertical layers of vegetation, as ground cover, herbaceous growth, shrubs, and trees.

Stratification–(1) The rest period which some seeds must have before they will germinate; generally the seeds must be exposed to a chilling temperature during this period before they will germinate. (2) A method of cross-breeding sheep of different types. (3) Arrangement in layers. The term refers to geologic material. Layers in soils that result from the processes of soil formation are called horizons; those inherited from the parent material are called strata. See Soil Profile.

Stream–(1) Flowing water in a natural or artificial channel. It may range in volume from a small creek to a major river. (2) A jet of water, as from a nozzle. (3) A continuous flow or succession of anything: air, gas, liquids, light, electricity, persons, animals, and materials.

Stream Bank–Either of the usual nonflooded boundaries of a stream channel. See River Bank.

Stream Duration–(a) Perennial: a water source which has not gone dry within historic times and which flows continuously throughout the year. (b) Intermittent: source containing water only at certain times during the year when it receives water from springs or runoff. The term may be arbitrarily restricted to sources containing water for periods of at least one month per year. (c) Ephemeral: sources which contain water only in direct response to precipitation and may be arbitrarily restricted to sources containing water for periods of less than one month per year. See Gully.

Stream Erosion–The erosion caused by a river or stream cutting into its natural banks, usually as a result of excess volume or shift in stream current.

Stream Gauging–The quantitative determination of stream flow using gauges, current meters, weirs, or other measuring instruments at selected locations.

Stream Load–Quantity of solid and dissolved material carried by a stream. See Sediment.

Stream Runoff–The total outflow of a drainage basin through surface channels; or subsurface channels and surface channels if there are exposed subsurface aquifers.

Streambank Wheatgrass–*Agropyron riparium*, family Gramineae; a forage grass found in dry or moist meadows and hills from North Dakota to Alberta, Canada, and Washington, south to Oregon and Colorado in the United States that is sometimes used in range reseeding. Native to North America.

Strength–(1) The capacity to resist force; solidity or toughness; the quality of bodies by which they endure the application of force without breaking or yielding. (2) Bodily or muscular power; force; vigor. (3) The potency or power of a liquid or other substance; intensity of active properties. (4) The firmness of a market price for a given commodity; a tendency to rise or remain firm in price.

Stress–(1) Abnormal or adverse conditions and factors to which an animal cannot adapt or adjust satisfactorily, resulting in physiological tension and possible disease; the factors may be physical, chemical, and/or psychological. (2) Plants unable to absorb enough water to replace that lost by transpiration. Results may be wilting, halting of growth, or death of the plant.

String Out Tile–To lay a number of tiles end to end in the field, prior to placement in the ground; as in a drainage line.

Strip Burning–Controlled use of fire to destroy surface material; e.g., an area to be broadcast burned may be successively burned in strips or the method may involve only the burning of inflammable material along roads, firelines, railroads, or highways. See Slash Disposal.

Strip Grazing–See Rotational Grazing.

Strip Rotation–A crop rotation for contour strips within a field, using erosion-resistant crops alternating with cultivated crops.

Strip Sodding–The laying of sod in strips separated by unsodded space.

Stripcropping–The practice of growing crops in a systematic arrangement of strips, or bands. Commonly cultivated crops and sod crops are alternated in strips to protect the soil and vegetation against running water or wind. The alternate strips are laid out approximately on the contour on erodible soils or at approximate right angles to the prevailing direction of the wind where soil blowing is a hazard.

Striped Bass–*Morone saxatilis*, the largest member of the bass family. The fish is deep olive green on the back with a white belly. Seven to eight full length stripes run horizontally on the side. Although they are a salt water fish, striped bass can complete their life cycle in fresh water and are raised to stock lakes as a sports fish.

Strong–(1) Having a specific quality to a high degree, as strong flavor, etc. (2) Characterized by steady or advancing prices, as a market for a particular commodity.

Strong Land–Land composed of clay or loam soils. Such soils generally are durable under cultivation and their fertility is less quickly depleted by cropping than that of sandy soils.

Structure, Soil–The arrangement of primary soil particles into compound particles or aggregates that are separated from adjoining aggregates. The principal forms of soil structure are: platy (laminated), prismatic (vertical axis of aggregates longer than horizontal), columnar (prisms with rounded tops), blocky (angular or subangular), and granular. Structureless soils are either single-grained (each grain by itself, as in dune sand) or massive (the particles adhering without any regular cleavage, as in many silty and clayey hardpans).

Strychnine–An extremely poisonous vegetable alkaloid which has an intensely bitter taste and is used as poison to kill mice, sparrows, pocket gophers, ground squirrels, etc. It is also used in medicines as a stimulant.

Stubble–(1) The ungrazed bases of grasses and forbs left on the range after grazing. (2) Standing crop residues.

Stubble Crop–(1) A crop developed from the stubble of the previous season. (2) A crop sown on grain stubble after harvest for plowing under, as green manure the following spring.

Stubble Harvesting–The harvesting of a second crop of rice from the root system of plantings which have been previously harvested. See Ratoon.

Stubble Mulch–A protective cover provided by leaving plant residues of any previous crop as a mulch on the soil surface when preparing for and planting the succeeding crop. See Conservation tillage, No-till.

Stubble Mulch Grain Drill–A grain drill used to plant seed in a field that has stubble left from the last crop. The concept is to use minimum tillage methods of production. Also called minimum tillage drill. See Air Drill, Grain Drill, Press Drill.

Stump–The basal part of a tree or plant left after the bole and top have been cut.

Stump Sucker–A shoot arising from a stump and developing into a tree. See Coppice.

Stump Wood–The resinous stump and roots of longleaf or slash pine used for extraction of turpentine and rosin.

Stumpage–The value of standing timber based on cubic volume offered for sale, usually at an agreed-upon log scale.

Stupefacient–A drug used to cause birds or other animals to go into a state of stupor so they can be captured and removed.

Sturgeon–Any of the large, long-lived, noncommunal, fresh- or saltwater fishes of the family Acipenseridae. The adults are caught as they enter shallow, coastal waters, rivers, or streams to spawn. The flesh is edible, the air bladder used in the manufacture of synthetic isinglass, and the roe made into caviar. The latter is a fairly important industry in the United States, Germany, Norway, the U.S.S.R., and Sweden. The common Atlantic sturgeon is *Acipenser sturio*. Native to the north temperate zone.

Sub–A prefix meaning either: (a) nearly, somewhat, slightly; e.g., subcordate, nearly cordate; or (b) below, under; e.g., subaxillary, below the axil.

Subalpine–The mountain zone just below the timberline.

Subarctic–(1) The region immediately south of the Arctic Circle and those which have a similar climate. (2) That region in which arctic and nonarctic surface waters join. (3) The regions in which mean temperature is not higher than 10°C for more than four months of the year and the mean temperature of the coldest month not more than 0°C.

Subgrade–The final level to which an excavation or fill is made before applying a surfacing material.

Subhumid Climate–A climate intermediate between semi-arid and humid which has sufficient precipitation to support a moderate to dense growth of tall and short grasses but in most instances insufficient to support a dense deciduous forest. Some subhumid areas, where the rainfall comes mostly during the growing season, have scattered deciduous trees with grass vegetation in between. The annual rainfall in subhumid climates may be as low as 20 inches in cool regions and as high as 60 inches in hot areas. The precipitation-effectiveness indexes are 32 to 48 for the dry subhumid and 48 to 64 for the moist subhumid.

Subirrigation–Water supplied to the soil from ditches or through underground tile lines, perforated pipe lines, or by natural subsoil moisture, in sufficient amounts to maintain a water table sufficiently close to the soil surface to supply adequate water quantities for crop needs. A low-permeability layer in the soil profile is essential to the maintenance of the watertable. Plants and planted seeds in containers, such as pots, are frequently subirrigated by setting them in a water container.

Sublethal–Less than fatal in effect, as a sublethal dosage or application of a toxic substance, etc.

Sublethal Concentration–A concentration in which an organism can survive, but within which adverse physiological changes may be manifested.

Sublimation–The characteristic of a solid to become a gas without becoming a liquid, or a solid becoming a gas without becoming a liquid; e.g., the evaporation of snow and the formation of frost.

Submarginal Land–Land incapable of sustaining a certain use or ownership status economically. See Marginal Land.

Suborder–A category next below an order and above a family in a classification of animals and plants.

Subsidence–(1) A sinking of a large part of the earth's crust. (2) Movement in which surface material is displaced vertically downward with little or no horizontal component. (3) A lowering of the surface by decomposition of peats and muck soil under cultivation.

Subsidence, Land–A downward movement of the ground surface caused by solution and collapse of underlying soluble deposits, rearrangement of particles upon removal of underground mineral deposits, reduction of fluid pressures within an aquifer or petroleum reservoir, or decomposition of organic matter.

Subsoil–(1) That part of the profile which lies below the usual plow depth without any specific limitation in depth or kind of material; a B horizon, the first change with depth in texture or structure; e.g., a clay layer underlying a sandy surface. (2) To plow deeply into the subsoil. See Soil Profile.

Substrate–(1) The material on which a plant lives, as soil or rock. (2) The material upon which an enzyme or fermenting agent acts; the material in or upon which a fungus grows or to which it is attached; the matrix.

Substratum–(1) The base or substance upon which an organism is growing or attached. (2) A vague term for the C horizon and/or R horizon of a soil. See Soil Profile.

Subsurface Flow–Water which enters the ground, but which emerges as seepage into streams, ditches, rivers, or springs. See Surface Runoff.

Subterranean–Existing under the surface of the earth.

Subterranean Stream–A body of flowing water that passes through large openings such as a cave, cavern, or a group of large communicating interstices; common in karst regions. See Karst, Thermokarst.

Subtropical–Refers to the region between the tropics and the temperate zone, having distinct seasons of summer and winter and with greater heat than the temperate zone.

Succession–(1) The progressive development of vegetation toward its highest ecological expression, the climax. The replacement of one plant community by another. (2) Transfer of operation and of ownership of a farm from one generation to the next, as from father to son.

Succession Planting–The growing of two or more crops per season in the same space; e.g., planting soybeans after wheat is harvested from the land.

Sugar Factory Lime–Lime that has been used in sugar refining. On a dry basis it will usually test about 80 percent calcium carbonate ($CaCO_3$). It has use for liming soils and is usually sold by the cubic yard. Two yards, air dry, have a lime equivalent of about one ton of high-grade ground limestone.

Sugar Maple–*Acer saccharum*, family Aceraceae; a large deciduous tree, chiefly of northern United States and Canada from Quebec to Pennsylvania, westward to Minnesota and Kansas, whose sweet sap may be collected in early spring and used to make syrup and sugar. The tree is sometimes planted for shade and for roadside plantings. Its wood is used in furniture, flooring, etc. Native to North America.

Sugar Palm–Any palm tree that can yield sugar, especially palms of the genus *Arenga*, family Palmaceae.

Sugar Pine–*Pinus lambertiana*, family Pinaceae; a tall-growing evergreen tree whose easily-worked wood is used for millwork, door, sash, interior and exterior trim, foundry patterns, etc. Native to North America.

Sugar Weather–Weather (toward the beginning of spring) in which the temperature during the day rises above 32°F and during the night drops below the freezing point causing the sap to start flowing in sugar maple trees and indicating that the time for making maple sugar is at hand.

Sugar-tree Land–A type of agricultural land with a native growth of sugar maple (*Acer saccharum*). It was recognized by early settlers in United States, especially in Indiana and Ohio, as first class land for farming.

Sugarloaf–A conical hill or mountain comparatively bare of timber.

Sul-Po-Mag–See Sulfate of Potash-Magnesia.

Sulfate of Potash-Magnesia–A double sulfate of potash and magnesia; a salt containing not less than 25 percent potash (K_2O, nor less than 25 percent sulfate of magnesia, and not more than 2.5 percent chlorine. Used as a low-grade fertilizer. Also called double manure salt.

Sulfite Yeast–The yeast grown on the waste liquor from the sulfite wood-pulping process. Produced under sanitary conditions and suitable for animal feeding, it contains not less than 40 percent protein.

Sulfofication–The oxidation of sulfur, free from organic or inorganic compounds, largely through the action of microorganisms. In the soil, the oxides are quickly changed into sulfites and sulfates.

Sulfur–S; an elementary, yellow mineral, insoluble in water, easily fusible and inflammable. Also called brimstone. One of the secondary but important elements in soil fertility and used in relatively large amounts by most plants, it is an important constituent of both protein and protoplasm. The powder form is an effective insecticide for many insects. The dust is used as a fungicide in the control of mildew, etc. When burned it forms sulfur dioxide, a gas which is highly toxic to insects and has long been used as a fumigant as well as a bleaching agent.

Sulfur Bacteria–*Thiobacillur thiooxidons*; bacteria that obtain their metabolic energy by the oxidation of elemental sulfur.

Sulfur Oxides–Pungent, colorless gases formed primarily by the combustion of fossil fuels: considered major air pollutants; sulfur oxides may damage the respiratory tract of animals and people as well as be toxic to vegetation.

Sulfuric Acid–H_2SO_4; a dense, heavy, exceedingly corrosive, oily liquid which can decompose animal and vegetable tissue and has a great affinity for water, giving off heat on combining with it. It is used in the manufacture of superphosphate fertilizers by converting the insoluble rock phosphate to a soluble and available form.

Summer Fallow–The plowing of a field prior to or during the summer and cultivating enough to control weeds and to accumulate soil moisture in preparation for a later crop.

Summer Wood–The portion of the annual growth ring of a tree formed during the latter part of the yearly growth period. It is usually more dense and stronger mechanically than spring wood. See Ring Porous Wood, Spring Wood.

Sunfish–A bright colored fresh water fish of the family Centrarchidae.

Sunstroke–Severe injury or killing of heat-sensitive people and animals by excessive heat during midsummer periods of cloudless skies with temperatures ranging above 100°F or more.

Super Slurper–Starch-based absorbents that absorb 300 to more than 1,000 times their own dry weight of water and are used to coat seeds. Originally developed by the United States Department of Agriculture, such absorbents have potential for helping seeds germinate faster.

Super-parasitism–(1) Parasitism upon a parasite. (2) An individual attacked by two or more primary parasites, or by one species more than once. See Autoparasitism.

Superphosphate–A fertilizer product which results from treating rock phosphate with sulfuric acid. The grade shows the available phosphoric acid; e.g., 20 percent superphosphate contains 20 percent phosphoric acid.

Superphosphoric Acid–A phosphoric acid mixture containing a substantial proportion of one or more polyphosphoric acids. It can be made by thermal dehydration of orthophosphoric acid or by absorption of P_2O_5 in water. The most common in fertilizer manufacture is for production of ammonium polyphosphate solution. See Ammonium Polyphosphate. Superphosphoric acid made by dehydration of wet-process phosphoric acid usually contains 68 to 72 percent P_2O_5.

Superspecies–A group of related species that are geographically isolated; without any implication of natural hybridization among them.

Supine–Prostrate.

Supplemental Irrigation–Irrigation carried on in areas where annual rainfall is normally high enough for good yields but is insufficient for some crops during part of a growing season.

Surface Fire–A fire which runs over the forest floor and burns only the surface litter, the loose debris, and the smaller vegetation. It does not burn tree crowns.

Surface Irrigation–Irrigation distribution of water over the soil surface by flooding or in furrows for storage in the soil for plant use.

Surface Runoff–The portion of runoff which flows to stream channels over the surface of the ground. It is quite distinct from the runoff absorbed by the soil, lost by evaporation, or retained as surface storage. The usual sources of surface runoff are rainfall occurring with intensities exceeding the rate of infiltration and snow upon the ground melting at rates exceeding the rate of infiltration. Surface runoff is the principal source of flood-flow occurring in streams at times of rainstorm or melting snow. See Subsurface Flow.

Surface Sealing–Orientation and packing of dispersed soil particles in the surface layer of soil that makes it almost impermeable to water.

Surface Soil–The soil ordinarily moved in tillage, or its equivalent in uncultivated soil, ranging in depth from 4 to 10 inches (10 to 25 centimeters). Frequently designated as the plow layer, or the A horizon. See Soil Horizon, Soil Profile.

Surface Storage–(1) The storage of water in surface reservoirs, such as constructed reservoirs, ponds, swamps, and lakes to reduce flood flow, increase low-water flow, and retain runoff for irrigation. (2) The retention of water in minute depressions on the soil during a period of runoff. Also called depression storage.

Surface Tension–Tendency of a liquid to contract until its area is the smallest possible for a given volume of liquid. This makes it appear to be covered by an elastic membrane.

Surface Tillage–Cultivating or working the soil to a shallow depth or barely below the surface.

Surface Water–Water on the surface, as lakes and rivers, in contrast to that underground. Surface water properly includes snow and ice.

Surface-active Agent–See Surfactant.

Surfactant–A material that improves the emulsifying, dispersing, spreading, wetting, and other surface-modifying properties of pesticide formulations.

Surficial–Superficial; occurring on the earth's surface; especially of unconsolidated residual, eolian, alluvial, or glacial deposits on bedrock. Soils are a surficial media for plant growth.

Surge Irrigation–Water applied in a series of regulated pulses or surges into the furrow rather than in the conventional manner of a continuous stream. Surging is done by switching the flow from one furrow to another and back again at regular intervals. Developed at Utah State University, advantages include irrigating more acres in less time, getting more-uniform application without excessive runoff, and making it easier to apply a small amount of water.

Surinam Cockroach–*Pycnoscelus surinamensis*, family Blattidae; a large insect which eats seedlings, injures flowers, and girdles plants in the greenhouse.

Suspension Fertilizer–Liquid fertilizer in which the solids are held in suspension (prevented from settling) by the use of a suspending agent, usually a swelling type clay. It must be fluid enough to be mixed, pumped, and applied to the soil without major alterations in the application equipment, and yet remain homogeneous during application. See Fluid Fertilizers, Liquid Fertilizers.

Sustained Yield Management–Controlled exploitation of a forest unit in such a way that the annual or periodic yield of timber or other products can be maintained in perpetuity.

Suture–(1) A line along which two things or parts are united. (2) In botany, a line of union or a line of dehiscence. (3) In horticulture, a longitudinal protuberance along one side of drupaceous fruits, as in the peach. (4) To unite or join by sewing.

Swale–A small, shallow, wet sag in land surface; a natural drainage way which does not necessarily contain a watercourse or permanent stream.

Swamp–(1) A natural area which has standing water on the surface for all or a considerable period of the year and which has a dense cover of native vegetation. See Bog, Marsh. (2) To clear the ground of underbrush, fallen trees, and other obstructions to facilitate subsequent logging operations, such as skidding.

Sweet Gum–*Liquidamber styraciflua*; a large, fast-growing tree that is native to the southern United States. The trees are grown for shade trees. The lumber is used to build furniture frames.

Sweet Water–(1) Water from wells, springs, and surface water which is fresh, in contrast to that which is salty or alkali. (2) In cooling systems, refrigerated freshwater in contrast to brine.

Swimmer's Itch–A rash produced on bathers by a parasitic flatworm in the cercarial stage of its life cycle. The organism is killed by the human body as soon as it penetrates the skin; however, the rash may persist for a period of about two weeks.

Sylvinite–A potassium salt, a mixture of the minerals sylvite and halite. It is a minor source of potassium chloride used in commercial fertilizers. Also called hardsalt.

Symbiosis–The close association of two dissimilar organisms, each known as symbiont. The associations may have five different characterizations as follows: mutualism: beneficial to both species; commensalism: beneficial to one but with no influence on the other; parasitism: beneficial to one and harmful to the other; amensalism: no influence on the other; synnecrosis: detrimental to both species of organisms.

Symbiotic Relationship–A relationship between two different types of organisms that is beneficial to both of them.

Syncline–A geologic structure formed by strata from opposite sides dipping downward toward a common line.

Syndrome–A group of signs of symptoms that occur together and characterize a disease.

Synecology–The study of plant communities, their taxonomy, life history, and relationship to other biotic communities.

Syneresis–Shrinkage and cracking of a wet gel such as wet clay upon drying.

Syngameon–The sum total of species or semi-species linked by frequent or occasional hybridization in nature; hybridizing group of species.

Synoptic–(Meteorology) Atmospheric conditions existing at a given time over an extended region, e.g., a synoptic weather map, which is drawn from observations taken simultaneously at a network of stations over a large area, thus giving a general view of current weather conditions.

Synthetic–Artificially produced; produced by human effort and design rather than naturally occurring; chemically manufactured.

Synthetic Detergent–A nonsoap chemical compound which has cleansing properties.

Synthetic Food/Fiber–A food or fiber produced from a nonagricultural raw material; e.g., a nondairy coffee creamer, a synthetic orange juice, an imitation shoe leather, or a human-made fiber.

Synthetic Manure–(1) Organic material, such as leaves, grass, and straw, to which a mineral fertilizer and lime have been added to hasten decomposition. After decomposition it is spread on, or worked into, the soil in the manner of barnyard manure. Also called artificial manure, compost. (2) Synthetic chemical fertilizers, such as anhydrous ammonia and urea.

Synthetic Materials–Materials that are manufactured chemically (by synthesis) from their elements or other chemicals as contrasted to those found ready-made in nature. Synthetic ammonia (NH_3) is now made on a large scale from its elements, nitrogen (from the air) and hydrogen (mostly from natural gas, CH_4); and synthetic urea from ammonia and carbon dioxide.

Synthetic Organic Chemicals–Calcium cyanamid and urea are produced synthetically for use as fertilizers. They contain organic combinations of elements, but behave in the soil like inorganic fertilizers. The nitrogen in cyanamid and urea are defined as "synthetic nonprotein organic nitrogen."

Synthetics–Artificially produced products that may be similar to natural products.

Systemic–(1) Pesticide material absorbed by plants, making them toxic to feeding insects. Also, pertaining to a disease in which an infection spreads throughout the plant. (2) Pertaining to the body as a whole and not confined to one organ or part of the body, as a systemic infection.

Systemic Fungi–Fungi that grow throughout the body tissues of the host.

Systemic Insecticide–An insecticide capable of absorption into plant sap or animal blood and lethal to insects feeding on or within the treated host.

Systemic Pesticide–A pesticide that is capable of translocation within the plant.

T Factor– A measure of the amount of erosion in tons per acre per year that a soil can tolerate without losing productivity. For most cropland soils, T values fall in the range of 3 to 5 tons per acre per year.

Tableland– A broad, elevated area of land having steep slopes or cliffs. See Mesa.

Tachinid Fly– Any fly of the family Tachinidae, whose larvae are beneficial as insect control, as they are parasitic on many noxious insects.

Taiga– (1) Subarctic coniferous forests of northern United States, southern Canada, northern Europe, and northern Asia, interspersed with muskeg bogs. (2) An area of land lying between the north temperate zone and the Arctic zone.

Tail– (1) The posterior part of the vertebral column of animals. It is usually covered with hair, some of which may be quite long. Also called brush. (2) A fanlike row of rather stiff feathers on the posterior part of a bird. (3) The lowest grade of flour. (4) A wisp of hay not properly tucked into a bale. (5) The bottom layer of produce in a container. (6) The weakest and poorest sheep (Australia and New Zealand). (7) The woody part of a plant which has been propagated by tip layerage. (8) To dock an animal's tail (Australia and New Zealand). (9) To keep sheep together in a flock (Australia and New Zealand). (10) To remove the tail from a carcass. (11) To assist an undernourished cow to its feet by pulling on the tail.

Tail Tree– In forestry power skidding, a tree at the end of a run to which the tackle is fastened.

Tail Water– The water immediately downstream from a structure such as at the end of a row in furrow irrigation.

Tailing Pile– The residue left from mining.

Tailrace– A channel which conducts water away from a waterwheel; an afterbay.

Talc– A finely ground powder, from the mineral talc, a hydrated magnesium metasilicate, which is used as a carrier for poison dusts and in pharmaceutical products as a filler.

Talik– (Russian) (1) A layer of unfrozen ground between the seasonal frozen ground (active layer) and the permafrost. (2) An unfrozen layer within the permafrost. See Permafrost, Thermokarst.

Tall Grasses– An association of tall-stemmed native grasses found in prairie areas, such as the big bluestem, ***Andropogon***. See Midgrass, Shortgrass.

Tall Timber– Wild, uninhabited country wholly or partially covered with forest (colloquial).

Tall-grass Prairie– A region in central North America, lying east of the 100th meridian in the Mississippi Valley, and some 250,000,000 acres in area. Its native vegetation was the tall grasses, but much of it is now farmland.

Talon– (1) The hind part or heel of the foot or hoof of such animals as deer, swine, horses, etc. (2) A claw of a bird, usually of a bird of prey.

Talus– The heap or fragments of rock and soil material which collects at the foot of a cliff or very steep slope chiefly as a result of rock weathering and gravitational force. See Colluvium, Scree.

Tame– (1) Domesticated. (2) Cultivated. (3) Designating an animal which has been made docile or tractable, as a wild horse is tamed or broken. (4) To domesticate. (5) To make docile.

Tame Hay– Hay produced from sown meadows, as contrasted with the forage from wild areas of native forage plants.

Tamed Iodine– Iodine that is combined with an organic material and is used as a disinfectant, mostly as an antiseptic. The iodine is released slowly from the organic compound, thus it is less irritating than tincture of iodine (iodine in alcohol).

Tanbark– The bark of hemlock, chestnut, oak, mimosa (wattle), or tanoak, etc., which is used as a source of tannin. The residue is often used as a ground covering in circus lots, race tracks, livestock pavilions. Also called tanner's bark.

Tannic Acid– Tannin; gallic acid, which is obtained from various trees and wood. It is soluble in water and is capable of combining with the proteins in animal skins to convert the skins into leather.

Tannin– See Tannic Acid.

Tap– (1) In lumbering, a cut made from the inside of a log. (2) A wooden basket which is used for packing figs. (3) A faucet on a pipe or container containing liquid. (4) To remove a taproot from a plant. (5) To make a cut in a tree for obtaining sap for use in making maple sugar or turpentine. (6) To cut threads in metal that will receive a bolt or screw.

Tap Water– Water which comes from a water faucet, as contrasted to well water or rainwater.

Taper– The gradual reduction in diameter of the trunk of a tree from the base to the top. Also called rise.

Tar– (1) A black, liquid mixture of hydrocarbons and their derivatives obtained by distillation of wood, peat, coal, shale, etc. (2) Tar used as a disinfectant on a sheep accidentally cut by the shearer (Australia and New Zealand).

Target Species– A plant or animal species which a pesticide is intended to kill.

Tarn– A small mountain lake.

Taste– (1) The flavor of a product as determined by placing the substance in the mouth. (2) A small amount or sample. (3) In judging, to place a morsel or a few drops of a product in the mouth to savor, but often not to swallow.

Taungya System–(1) Raising a forest crop in conjunction with an agricultural crop, usually in a modified shifting cultivation system. (2) Intercropping with woody and herbaceous plants.

Tautonomy–Relations that exist if the same word is used for both the generic and specific name in the name of a species.

Taxadjuncts–Soils that cannot be classified in a series recognized in the classification system. Such soils are named for a series they strongly resemble and are designated as taxadjuncts to that series because they differ in ways too small to be of consequence in interpreting their use or management.

Taxon–(Plural, taxa) A general term used for a taxonomic group of any rank without being specific, and in the plural (taxa) to refer to more than one such group collectively even if of different ranks.

Taxonomy–(1) The science of classification of organisms and other objects and their arrangement into systematic groups such as species, genus, family, and order. (2) Taxonomy is closely related to classification but it embodies a broader concept. Taxonomy is the science of how to classify and identify. It is the theoretical study of classification including its bases, principles, procedures, and rules. Taxonomy includes classification as well as identification.

Teak–*Tectona grandis*, family Verbanaceae; a tall tree grown for its hard, durable, yellowish-brown lumber used in shipbuilding, furniture manufacture, etc. Native to the East Indies.

Teart–(1) A soil or plant that contains a large amount of molybdenum. (2) Molybdenosis caused by a domestic animal ingesting an excess of molybdenum, usually from eating molybdenum-rich forage grown on soils high in molybdenum. The symptoms are diarrhea and general debility (weakness).

Technical–(1) Concerned with a particular science, industrial art, profession, sport, etc. (2) Practicing, using, or pertaining to the technique rather than the theory or underlying principles involved in the execution of a project. (3) Designating the grade of a commodity manufactured in the usual commercial manner. (4) Pertaining to or designating a market where prices are controlled by speculation or manipulation.

Tectonics–The study of the broader structural features of the earth and their causes.

Telophase–The phase of cell division between anaphase and the complete separation of the two daughter cells; includes the formation of the nuclear membrane and the return of the chromosomes to long, threadlike and indistinguishable structure.

Temperate Rain Forest–A woodland usually with a dominant species; found in a temperate zone in areas having little frost.

Temperature–(1) The amount of heat or cold measured in degrees on different scales, as Fahrenheit or Centigrade. At sea level, water freezes at $32°F$ ($0°C$) and boils at $212°F$ or $100°C$. (2) The degree of heat in a living body. (3) Abnormal heat in a living body. Also called Fever.

Temperature Inversion–An unusual temperature condition in the atmosphere in which the temperature rises with increased elevation. A common condition when a radiation type frost forms.

Temperature Zero–The temperature below which certain physiological processes of an organism are carried on at a very slow rate.

Tensiometer–Any of several types of devices which measure moisture, tension, or condition of water in soil. Used to determine when to irrigate.

Tent Caterpiller–See Eastern Tent Caterpiller.

Teratogen–An agent or factor that causes the production of physical defects in the developing embryo of animals and people.

Terminal Moraine–A moraine formed across the course of a glacier at its farthest advance, at or near a relatively stationary edge, or at places marking the termination of important glacial advances.

Termite–A pale, soft-bodied social insect, mostly of tropical and sub-tropical regions, of the order Isoptera which feeds on moist or west cellulose. Three main group of termites are recognized: (a) dry-wood termites that live above the soil and attack live wood; (b) damp-wood termites that live above the soil but work in wet wood, mostly decayed; (c) subterranean termites that live in damp soil but are capable of building tunnels of soil to reach nearby dead wood. Principal families Mastotermitidae and Termitidae.

Termite Runway–An earth-covered tunnel built by subterranean termites to connect their nests with their food supply.

Terra–(Latin) Soil, earth.

Terra Cotta–(Italian, baked earth) Kiln-burnt reddish-brown clay fashioned into vases, statuettes, and other moldings.

Terra Culta–(Latin) Cultivated land.

Terra Rossa–(Italian, red earth) Residual red clay developed on limestone bedrock in the Mediterranean region.

Terrace–(1) A long, low embankment or ridge of earth constructed across a slope with a flat or graded channel to control runoff and minimize soil erosion by scouring. Two general types are bench terrace and ridge terrace. Variations of these types are the horizontal bench terrace, sloping bench terrace, broad-base terrace, narrow-base terrace, Nichol's terrace. (2) A raised, level area of earth supported on one side by a wall, bank, etc., as a terraced yard. Often there is one above another on a hillside or slope. (3) A natural land feature which is an elevated, relatively long, flat upland with a short scarp or embankment-like front. The plain may be narrow and the terrace steep or benchlike, or it may be wide and properly designated a terrace plain. Terraces represent former sea, lake, and river levels, and are locally significant in soil and land classification and use. (4) To construct a terrace, especially one across a slope to control erosion and runoff.

Terrace Channel–The depression along the upper side of the terrace ridge in which the water flows to the outlet, or is retained when the channel is not graded so that the water does not have an outlet other than by evaporation and through the soil by infiltration and lateral movement.

Terrace Crown–The highest part of the terrace ridge; the top of the terrace.

Terrace Grade–The slope of the terrace channel in inches or feet per 100 feet length. Terrace grades may be variable or uniform.

Terrace Height–The vertical difference in elevation between the bottom of the terrace channel and the crown of the terrace ridge at adjacent points.

Terrace Interval–Horizontal: the horizontal distance in feet from the center of one terrace line, ridge, or channel to the corresponding point on an adjacent terrace. Vertical: the vertical distance in feet from the center of one terrace line, ridge, or channel to the corresponding point on the adjacent terrace. (The interval of the first terrace is the vertical distance from the top of the hill to the terrace line of the terrace.)

Terrace Outlet–The water channel at the end of the terrace into which the flow from one or more terraces is discharged and conveyed from the field usually through a grassed outlet.

Terrace Outlet Structure–A structure usually installed in or at the end of a terrace outlet to reduce erosion.

Terracette–A small hillside step or bench commonly believed to have been caused by livestock. It usually involves slumping on a small scale indicating gradual downhill movement of the entire surface. Also called cattle terrace, sheep track, cat step.

Terracing–A practice of constructing a ditch or channel, with a ridge below, across the slope at various vertical intervals to intercept runoff. The channel and ridge are usually constructed in a way that will permit contour operations with farm equipment.

Terrain–(1) Area of ground considered as to its extent and natural features in relation to its use for a particular operation. (2) The tract of ground immediately under observation.

Terrarium–A tightly fitted, glass-enclosed, indoor garden, resembling an aquarium, in which plants are grown for decoration, plant propagation, nature study, etc. Also called Wardian case, bottle garden, crystal garden, glass garden.

Terrestrial–(1) Referring to earth. (2) Designating a plant which lives in soil as contrasted to one which is epiphytic (growing in air) or one growing in water (aquatic or hydrophytic). (3) Designating a ground bird such as a pheasant, partridge, or chicken, as contrasted to an aerial bird.

Terrigenous–Derived from or originating on the land (usually referring to sediments) as opposed to material or sediments produced in the ocean (marine) or as a result of biologic activity (biogenous).

Terriherbosa–Prairie and steppe vegetation.

Territory–The area occupied by an individual or group of organisms.

Tertiary Treatment–Treatment of sewage beyond the secondary stage to accomplish a very high degree of nutrient and/or biochemical oxygen demand (BOD) reduction.

Test Adaptation Trial–In pasture and range research, testing the adaptability of forage species to new edaphic and climatic conditions as to complete their life cycle and/or reseed themselves.

Test Pit–A hole dug in the ground to determine the character of the subsoil or substratum prior to ditching, laying footings, etc.

Tetrachlorocethane–A chemical compound used as a soil fumigant, nematicide, and herbicide.

Texas Bluegrass–*Poa arachnifera*, family Gramineae; a perennial, vigorous, forage grass, found in the United States from North Carolina to Texas. It is sometimes cultivated for winter pasture. Native to North America.

Texas Leaf-cutting Ant–*Atta texans*, family Formicidae; a serous insect pest of garden and field crops which cuts leaves off the plants and invades houses to steal seed and farinaceous (starchy) foods. It is found in eastern and southern Texas and western Louisiana, United States.

Texas Panhandle–The high plains region of northwest Texas, United States, important agriculturally both as range country and for farm crops.

Textural Family–Soil groupings according to particle size. Clayey: contains more than 35 percent clay by weight and less than 35 percent rock fragments by volume; fine-silty; less than 15 percent fine sand (0.25 to 0.1 millimeter diameter) or coarser by weight, including fragments up to 7.5 centimeters diameter, 18 to 34 percent clay; fine-loamy: by weight 15 percent or more fine sand (0.25 to 0.1 millimeter) or coarser by weight, including fragments up to 7.5 centimeter diameter, 18 to 34 percent clay; coarse-loamy: same as fine loamy except has less than 18 percent clay; sandy: sand or loamy sand but not loamy very fine sand or very fine sand, rock fragments less than 35 percent.

Texture–(1) The physical properties of a product; that is, the structure and arrangement of the parts which make up the whole. (2) The relative proportions of sand, silt, and clay particles in a mass of soil. The basic textural classes, in order of increasing proportion of fine particles: sand, loamy sand, sandy loam, loam, silt, silt loam, sandy clay loam, clay loam, silty clay loam, sandy clay, silty clay, and clay. The sand, loamy sand, and sandy loam classes may be further divided by specifying coarse, medium, fine, or very fine.

Thallium–A rodenticide, also used to control ants.

Thallophyta–Division of nonvascular plants, those without differentiated roots, stems, or leaves; includes algae and fungi.

Thaw–(1) A period of weather following ice and snow which is warm enough to cause melting. (2) To melt frozen products. (3) To melt ice, snow, and frost, in soil.

Thaw Depression–A hollow formed by the melting of ice in perennially frozen ground. See Permafrost, Thermokarst.

Theobromine–A toxic alkaloid occurring in chocolate, cocoa, tea, and cola nuts that is closely related to caffeine. In practical terms, theobromine is a stimulant and in excess may cause death in dogs and humans.

Therapeutic–Pertaining to the treatment of disease; curative.

Therapy–The sum total of the treatment given to cure disease in plants, animals, and humans.

Thermal Belt–A well-defined area on the sides of valleys, as on the eastern slope of the San Joaquin Valley in California, United States, in which descending cool air is warmed by compression as it nears the lower elevation, while at the lowest elevation cold air accumulates. These belts are of great agricultural importance, for in them the danger of frost is minimized.

Thermal Death Point–The amount of heat required to kill a particular organism.

Thermal Pollution–Degradation of water quality by the introduction of a heated effluent. Primarily a result of the discharge of cooling waters from industrial processes, particularly from electrical power generation.

Thermal Turbulence–Air movement and mixing caused by convection.

Thermocline–The layer in a body of water in which the drop in temperature equals or exceeds one degree centigrade for each meter or approximately three feet of water depth.

Thermokarst–In permafrost areas, karstlike topography resulting from removal of vegetation and organic soil layer and subsequent melting of ground ice and the settling of the soil. Sinkholes, solution caverns, and underground streams develop in much the same way as in limestone karst topography; e.g., in the Ozark Plateau. See Karst, Permafrost.

Thermoperiodicity–The response of plants to changes in day and night temperatures.

Thermophyllic–Plants capable of growing at high temperatures.

Thicket–A dense growth of shrubs or small trees.

Thief Ant–*Solenopsis molesta*, family Formicidae; a common house ant pest which shows a preference for protein foods. It may also damage germinating grain seeds.

Thin–(1) To reduce the number of plants in a row or area by hoeing, pulling, etc. (2) Designating an animal with little flesh. (3) Designating a pulse which is very feeble.

Thin Land–Soil which is unproductive because of a lack of depth or fertility.

Thin Stillage–The water-soluble fraction of a fermented mash plus the mashing water.

Thinning–(1) The removal of plants in a stand to reduce crowding. (2) Removing certain flowers or clusters of flowers or individual fruits after fruit has set and natural dropping has occurred. (3) Removing live branches in a tree crown. (4) In woodland management, the cutting made in immature stands after the sapling stage to increase the rate of growth of the remaining trees.

Thixotropy–The property exhibited by some gels of becoming fluid when shaken. The change is reversible. Some fine clays exhibit thixotropy. See Vertisols.

Thorax–The middle body region of an insect to which the wings and legs are attached.

Thornthwaite Index of Precipitation–A measure of the amount of precipitation corrected for temperature. As the temperature increases, evaporation and transpiration increase. This results in a decrease in the effectiveness of the precipitation. The monthly index of effective precipitation is figured as follows: $(P/T-10)\ 10/9$, where P equals mean monthly precipitation in inches and T equals mean monthly temperature in degrees F.

Threatened Species–Any species of animal or plant which is likely to become an endangered species within the foreseeable future throughout all or a portion of its range.

Three-field System–A system of fallowing in which land tracts are divided into three fields, with one field fallowed each year. The practice dates back to the Middle Ages in Europe. See Fallow.

Three-leaved Ivy–See Poison Ivy.

Throughfall–In a forest, all of the precipitation that reaches the soil surface directly plus that which is intercepted by the leaves and drips to the ground. Some authorities include stemflow (down the tree trunk into the soil) in throughfall and some do not.

Throwback–(1) In breeding, a reversion to an ancestral type or ancestral characteristic. (2) To revert to an ancestral type or characteristic.

Thunderstorm–A rainstorm, usually intense but of short duration, accompanied by lightning and thunder.

Tick–Any of various blood-sucking arachnids which fasten themselves to warm-blooded animals. Some are important vectors of diseases.

Tick Quarantine–A legal prohibition against shipping livestock from one area to another in an attempt to prevent the spread of ticks.

Tick-infested–Designating an animal or area of land which abounds in ticks.

Tickborne–Transmitted by ticks, as cattle tick fever.

Tidal Day–The time of the rotation of the earth with respect to the moon, or the interval between two successive upper transits of the moon over the meridian of a place, about 24.84 solar hours (24 hours and 50 minutes) in length or 1.035 times as great as the mean solar day.

Tidal Marsh (Tidal Flat)–Low, flat marshland traversed by interlacing channels and tidal sloughs, inundated by high tides. It supports a cover of saltmarsh vegetation, usually low-growing plants, with areas of rushes and reeds.

Tide–The periodic rising and falling of the ocean caused by the gravitational pull of the moon and the sun acting on the rotating earth. The average time interval between tides is 12.42 hours from high tide to high tide. A tidal day is twice this time interval, i.e., 24.84 hours.

Tideland–(1) The area adjacent to the sea or ocean which is covered during floodtide. (2) The shallow sea bottom (in some instances several miles from shore).

Tier–(1) In grafting and budding, one who ties the bud or scion to the stem or stock. (2) One of the horizontal layers in the vertical stacking of boards in a well-defined pile for seasoning. (3) A layer of lumber, boxes, crates, bags, etc., in storage.

Tierra– (Spanish) Soil, earth.

Tight Soil– A compact clay soil difficult to plow and till and permitting a very slow infiltration of water.

Tiled– Designating an area of land which has been underdrained with tile, as contrasted to open ditches.

Till– (1) A deposit or mixture of earth, sand, gravel, and boulders which has been transported by, and deposited under, glaciers. Till is generally unstratified. Also called boulder clay. (2) To cultivate, plow, fit, and sow land. See Tilth.

Till Plain– A level or undulating land surface covered by glacial till. See Till.

Tillable Acres– That part of the farm land that can be used for cultivated crops without additional drainage, clearing, or irrigation.

Tillage– The mechanical manipulation of soil for any purpose; but in agriculture it is usually restricted to the modifying of soil conditions for crop production.

Tillage Pan– A compacted layer of soil, usually just below the depth of usual tillage. Tillage pans are very serious because they reduce the permeability of the soil to air and water. Crop yields are low on these soils.

Tilth– The artificial condition of soil structure brought about by tillage (cultivation).

Timber– (1) Forest stands. (2) Sawed lumber 5×5 inches or more. (3) Sawed lumber 4½ × 6 inches or more (United Kingdom). (4) Wood in form for heavy construction.

Timber Line– The highest extension of woody or forest growth on mountains above which further tree growth is prevented because of unfavorable environmental influences. The elevation at which the timber line occurs varies with latitude and aspect. See Snow Line.

Timber Marking– The selection and indication, usually by blaze or paint spot, of trees which are to be cut or retained in a cutting operation.

Timber Poisonvetch– *Astragalus convallarius*, family Leguminosae; a poisonous plant found on mountain slopes from British Columbia, Canada, to Arizona, United States. Characteristics of poisoning in animals are dullness, irregularity in gait, solitary habit, loss of nervous sensibility, loss of flesh, shaggy coat, loss of appetite, and may result in death.

Timber Products– Roundwood products and plant by-products. Timber products output includes roundwood products cut from growing stock on commercial forest land; from other sources, such as cull trees, salvable dead trees, limbs, and saplings; from trees on noncommercial and nonforest lands; and from plant by-products.

Titer– The minimum quantity of a substance required to produce a specific reaction with a given amount of another substance.

Toad– (1) Any of several tailless amphibians of the genus ***Bufo*** or similar genera, of the family Bufonidae. Species are generally terrestrial and insectivorous. (2) An animal that is short and squatty.

Toadstool– The fruiting structure of a fungus of class Basidiomycetes, order Agaricales. The vegetative part of the fungus consists of fine, branching, usually white threads of cells which grow in the soil, in wood (often causing decay in it), in manure, or in other vegetable organic substances. The fruit, produced in the air, is usually fleshy, more rarely tough or even woody consisting of a cap (pileus), the underside of which has radiating plates (gills) on which the spores are produced. More often the cap may sit on a central stalk (stipe) but this may be lateral or wanting in those toadstools that are attached by one edge. Popularly, but not scientifically, the name toadstool is applied to poisonous species, and mushroom to edible ones.

Tobacco Sick Soil– Soil infested with the root rot fungus which attacks the tobacco plant. The sickness is commonly ascribed to depletion of available plant nutrients in the soil by continued cultivation of tobacco.

Toe Slope– The outermost inclined surface at the base of a hill; part of a foot slope.

Tolerance– (1) The ability of a tree to grow in the shade of other trees and in competition with them. (2) The acquired ability of an animal or insect to take poison without ill effects. (3) The ability of a species to develop and survive under unfavorable environmental conditions. (4) A percentage of off-grade or off-sized produce which is allowed in a particular grade, or containers in a lot, that fail to meet grade specifications, especially of fruits and vegetables. (5) The amount of toxic residue permitted by federal law to be present on or in food and forage products that are to be marketed for consumption. (6) The accuracy limits specified in the construction of a machine part as plus or minus 0.005 inch.

Tone– The condition of normal balance between an organism and its environment.

Tonkabean Wood– *Alyxia buxifolia*, family Apocynaceae; a small, straggling, evergreen shrub sometimes grown for its fragrant, tonkabeanlike odor. Native to Tasmania. Also called scentwood, boxleaf alyxia.

Toonea– (1) *Toona ciliata*, family Meliaceae; a tall, nearly evergreen tree useful for its flowers, which yield a dye, and its soft, reddish wood valuable for making furniture. Native to the Himalaya mountains. Also called Indian mahogany. (2) Panama gumtree.

Toothed Euphorbia– *Euphorbia dentata*, family Euphorbiaceae, a plant which secretes a poisonous, milky juice. The effects of poisoning are swelling of the eyes and mouth, abdominal pains, collapse, excessive scours, and death. Native to North America. Also called toothed spurge.

Top– (1) The aboveground parts of certain plants, especially the leaves, as beet tops, turnip tops. See Crown. (2) Scoured, combed, long wool. (3) The highest price paid for a product within a certain period of time. (4) The upper, branchy portion of a felled tree. It is sold as firewood, charcoal source, etc. (5) To cut off the crown and leaves, as of the sugar beet. (6) To remove the upper portion of the crown of a tree. (7) To place the best articles of produce, as eggs, or fruit, on the top layer, in a container so that the whole container appears to include articles of higher quality than is actually the case. (8) To sort out animals that have reached a certain stage of development or finish.

Top Diameter– The diameter at the uppermost end of a log or salable piece of timber.

Top Kill– (1) A portion of the top of a tree which is dead, while the remainder of the tree is alive. See Staghead. (2) To kill the top growth of a plant with chemicals.

Topographic Factors– Physiographic and edaphic factors which may cause changes in vegetation.

Topographic Map– A scale representation, by means of conventional signs, of a part of the earth's surface, showing the culture, relief, and elevations above a datum, hydrography, and, frequently, the vegetation.

Topography– The shape of the ground surface as determined by such major features as hills, mountains, or large plains. ***Steep*** and ***flat*** (topography) commonly indicate slope of the land.

Toposequence– A sequence or chain of soils whose properties vary with topographic position. Theoretically, the slope is the variable, and other factors of soil formation remain the same.

Topsoil– Surface soils and subsurface soils which presumably are fertile soils, rich in organic matter or humus debris. Topsoil is found in the uppermost soil layer called the A horizon. See Soil Horizon, Soil Profile.

Tornado– A violent whirlwind with a destructive diameter usually ranging from 50 to 200 yards. The average forward speed is 25 to 40 miles per hour. It is usually associated with cold fronts, squall lines, and severe thunderstorms. In the United States, the paths are most frequently from southwest to northeast, and from south to north along the Atlantic coast. Not to be confused with cyclone. Also called twister. See Cyclone.

Torrens System– A system of registration of title to land in which the government guarantees the title. It is in effect in some states in the United States, in Australia, South Africa, to some extent in Canada, and in central Europe.

Torrent– A swift stream of water.

Torrential– Refers to the fast and violent movement of materials such as water in a stream or heavy rainfall.

Toxaphene– A chlorinated hydrocarbon insecticide.

Toxic– Poisonous; caused by poison.

Toxic Residue– A poisonous residue left on plants, in the soil, or on animals, by a spray or dust.

Toxicant– A substance that injures or kills an organism by physical, chemical, or biological action; e.g., heavy metals, pesticides, and cyanides.

Toxicarol– A rotenoid obtained from the roots of ***Tephrosia toxicaria*** (fishdeath tephrosia); a poisonous compound.

Toxicity– State or degree of being poisonous.

Toxicology– The science which deals with poisons, antidotes, toxins, effects of poisons, and the recognition of poisons.

Toxin– A protein poison produced by some higher plants, certain animals, and pathogenic bacteria. Toxins are differentiated from simple chemical poisons and vegetable alkaloids by their higher molecular weight and antigenicity. See Phytotoxin, Poison.

Toxiphobia– (Greek) Fear of being poisoned.

Toxoid– A toxin which has been chemically altered so that it is no longer toxic but is still capable of uniting with antitoxins and/or stimulating antitoxin formation.

Trace Element– Any of certain chemical elements necessary in minute quantities for optimum growth and development of plants and animals.

Trace of Precipitation– Precipitation in amounts too small to be measured in a gauge. Less than 0.005 inch for rain and less than 0.05 inch for snow.

Tracer– A radioisotope mixed with a stable substance, by means of which a material can be traced as it undergoes physical and chemical changes. In agricultural research, radioactive isotopes of phosphorus in a chemical fertilizer can be traced through the plant as it is absorbed by the roots.

Tract– (1) An area of land of any size, but bigger than a lot. (2) An anatomical structure as the digestive tract, cerebellospinal tract.

Tractive Current– The current in water that transports sediment along the bottom, as in a river, contrasted with a turbidity current or current not in contact with the bottom.

Traffic Pan– A compacted soil layer resulting from wheel traffic compressing the soil when wet.

Trafficability– The capability of a terrain to bear traffic. It refers to the extent to which the terrain will permit continued movement of any and/or all types of traffic. Soil type has much to do with the trafficability of the terrain; e.g., vertisols have zero trafficability when saturated, and oxisols can bear some traffic when wet.

Trailing– (1) The driving of livestock from place to place. (2) The voluntary wandering of livestock about a range, usually in search of forage, water, or salt. (3) Designating a plant which puts forth long, recumbent stems.

Trailing Sumac– See Poison Ivy.

Tramp– (1) To compact soil usually around a transplant by treading on it or by pounding it with a tool. See Tamp. (2) To compact silage in the silo by walking on the material.

Trampling– (1) Treading under feet; the damage to plants or soil brought about by congested movements of livestock, including mechanical injury to tree reproduction and ground cover in woods. (2) Compacting soil in earthen dams and reservoirs by livestock to make the dam or reservoir impervious to water (now replaced by machine compaction).

Transhumance– Seasonal nomadic movement of people and grazing livestock for part of the year combined with some form of permanent lowland farming in the wet season.

Transient– Having only a brief existence; shortlived.

Transitory Range– Land that is suitable for temporary use for grazing; e.g., on disturbed lands, grass may cover the area for a period of time before being replaced by trees or shrubs.

Translocate—The transfer of the products of metabolism, etc., from one part of a plant to another.

Translocated Herbicide—A herbicide which, when applied to one portion of a plant, travels to another part of the plant.

Transpiration Ratio—The ratio of the weight of water transpired by a plant to the dry weight of the plant.

Transported Soil—A soil formed by the consequent or subsequent weathering of materials transported and deposited by some agency, such as water, wind, ice, or gravity.

Trap—(1) A two-wheeled, one-horse cart. (2) Any of various devices used for catching living insects, animals, birds, etc. (3) A device which consists of a U-shaped pipe or chamber so that liquid flowing through it always forms a seal against a flow-back of gas. (4) A chamber to collect sediment flowing in a pipe.

Trap Nest—A fowl nest so arranged that the hen may enter it, but not leave until released by an attendant. It is used for checking egg production.

Trash Farming—Stubble mulch farming. See Conservation Tillage, No Till.

Traumatic—Describing a condition resulting from an injury or wound.

Traverse—A plane land survey consisting of compass bearings (or angles) and distances and error of closure back at the starting point.

Travertine—Calcium carbonate, $CaCO_3$; of light color and usually concretionary and compact, deposited from solution in ground and surface waters. Extremely porous or cellular varieties are known as calcareous tufa, calcareous sinter, or spring deposit. Compact, banded varieties, capable of taking a polish, are called onyx marble. Travertine forms the stalactites and stalagmites of limestone caves, and the filling of veins in hot springs.

Treated Area—A place where a pesticide has been applied.

Treatment Plant—Waste treatment facilities placed between the sewers and a water body in order to purify wastes and sewage by physical, chemical, and biological processes.

Treble Superphosphate—Double superphosphate.

Tree—Any woody, perennial plant which normally has one well-defined stem and a definitely formed crown. It is usually considered to have a minimum mature height of 15 feet. See Shrub.

Tree Caliper—(1) A caliperlike device used to measure diameters of tree trunks and logs. (2) The diameter of a tree trunk; in forestry measured at 4.5 feet above the ground; in horticulture, measured at one-foot height.

Tree Dressing—Paint or paste used to cover and protect wounds of a tree caused by limb breakage or pruning.

Tree Farm—A tract of land on which trees, especially forest species, are grown as a managed crop.

Tree Injection—The introduction of a chemical into the sap stream of a tree (a) to kill it, (b) to prevent disease, (c) to prevent chlorosis, (d) to color the wood prior to sawing, or (e) to destroy insects.

Tree Line—See Timber Line.

Tree Paint—Any one of several types of wound dressings for trees, such as orange shellac, asphalt paints, creosote paints, grafting waxes, house paint, Bordeaux paste, commercial tree paints, etc. Ideally, it should disinfect, prevent fungus entrance, prevent checking, permit callus growth, be toxic to insects, be easily applied, allow excess moisture to evaporate, and should not crack on drying, or injure the tissues.

Tree Planter—A tractor-drawn implement that is used for planting tree seedlings. An operator rides on the planter and places a seedling in the furrow opened up by the implement.

Tree Planting—To grow trees, shrubs, vines, or cuttings to prevent excessive runoff and soil loss. It includes plantings for gully control, post lots, sand dune fixation, bank protection, reinforcement of existing woodlands, windbreaks, etc. Another purpose of tree planting is to obtain wood products or trees to sell for ornament, as Christmas trees, etc.

Tree Processor—A machine that trims and delimbs trees that have been harvested. The trees are fed through the machine and emerge with the limbs removed.

Tree Protector—Any of the several kinds of material for wrapping loosely around the trunk of a tree to protect it from sunburn or injury by rodents.

Tree Size Classes—A classification of growing stock trees according to diameter at breast height outside bark, including sawtimber trees, poletimber trees, saplings, and seedlings.

Tree Slash—The debris remaining after logs have been removed during the harvesting of a forest.

Tree Spade—A large hydraulically operated machine that scoops into the ground around a tree and removes the tree from the ground for transplanting.

Tree Species—(1) Commercial species. Tree species currently or prospectively suitable for industrial wood products; excludes so-called weed species such as blackjack oak and blue beech. (2) Hardwoods or dicotyledonous trees, usually broadleaved and deciduous. (3) Softwoods or coniferous trees, usually evergreen, having needles or scale-like leaves.

Tree Surgery—The after-care of trees which includes pruning, repair of injury, spraying, and fertilizing when necessary.

Trench—(1) A long, relatively narrow excavation; a ditch. (2) To dig a ditch; to make a long, relatively narrow excavation.

Trench Fever—A disease of humans caused by a bacterium-like microorganism, ***Rickettsia quintana***, nonfatal and characterized by sudden onset of fever, headache, dizziness, and pains in muscles and bones; transmitted by the body louse through its feces.

Tri-—A prefix meaning three, as ***trilocular***, having three locules.

Tribasic Copper Sulfate—A copper salt used as a fungicide.

Tribasic Phosphate of Soda—Trisodium phosphate, a sodium salt which in a 2.5 percent solution has an appreciable disinfecting value.

Tributary—A branch stream which joins a main stream or river. See Distributary.

Tricalcium Phosphate–A salt of phosphoric acid, present in phosphate rock and bones. It is the source of almost all phosphorus-containing substances. Reaction of tricalcium phosphate and sulfuric acid yield the widely used fertilizer marketed as superphosphate. Also called bone phosphate of lime, BPL, calcium phosphate, phosphate of lime.

Trickle Irrigation–The application of small quantities of water directly to the root zone through various types of delivery systems on a daily basis. Also known as drip irrigation.

Trickling Filter–A 5 foot diameter pile of cinders or small rocks over which water containing animal or human effluent or food processing plant wastes are sprayed to achieve aerobic microbial decomposition. A widely used practice.

Trio–(1) A group consisting of three objects or organisms. (2) A male and two female birds of the same variety which are shown as a unit in exhibitions.

Trockentorf–A relatively undecomposed, peatlike deposit occurring on the surface of well-drained soils under forest cover. It is composed of the remains of leaves and fragments of wood.

Trona–A whitish mineral that occurs in thick saline beds and is mined as a major source of sodium of commerce.

Trophic–Of or pertaining to nutrition.

Trophic State–Characterization of a body of water in terms of position in a scale ranging from oligotrophy (poor in nutrients) to eutrophy (rich in nutrients).

Trophogenic Region–The superficial layer of a lake in which organic production from mineral substances takes place on the basis of light energy.

Tropholytic Region–The deep layer of the lake where organic dissimilation predominates because of light deficiency.

Tropical Zone–(1) The region of the earth bisected by the equator and extending at low elevations to latitude 23° 27" north (Tropic of Cancer) and south (Tropic of Capricorn). (2) Designating a plant, disease, etc., which flourishes in the tropics or where conditions are made to resemble the tropics in temperature, humidity, length of day, etc.

Tropism–A growth reaction of a plant to various external or internal stimuli such as phototropism, the increased growth toward or away from light; geotropism, growth in response to gravity; chemotropism, plant response to chemicals; hydrotropism, plant response to water.

Troposphere–The innermost part of the air encircling the earth; it extends about 5 miles (8 km) high at the poles and about 10 miles (16 km) high at the equator.

True Mountain Mahogany–*Cercocarpus montanus*, family Rosaceae; a tender evergreen shrub grown as an ornamental for its 3 to five inch long, feathery-tailed fruit. It is found on dry ridges at altitudes of 4,000 to 10,000 feet from northern Montana to New Mexico, United States, and furnishes a large amount of high-quality browse.

Truncated Soil Profile–A soil profile from which one or more of the upper horizons normally present have been removed by accelerated erosion or by other means. The profile may have lost part or all of the A and sometimes the B horizon, leaving as soil only the poor, undeveloped parent material or C horizon. A comparison of eroded soil profiles of the same area, soil series, and slope conditions, indicates the degree of truncation.

Trunk–The main unbranched body, stalk, or stem of a vine or tree.

Trypanosome–A microscopic, free-swimming, tailed protozoan parasite of the genus *Trypanosoma*, family Trypanosomatidae. Many species are not injurious, but a few cause serious diseases of animals and people, such as *T. brucci* and *T. gambiense*, both of which gain entry to the body by the bite of the tsetse fly.

Tsetse Fly–*Glossina mortisans*, family Muscidae; a fly of low elevations in central and southern Africa which is a carrier of the parasite (*Trypanosoma brucei*) that causes the disease nagana in various animals, as horses, cattle, or goats. The parasite is conveyed by the fly's bite. The widespread presence of the fly, during certain seasons, renders some districts completely unsuitable for habitation by such domestic animals. A related species (*G. palpalis*) carries the trypanosome *T. gambiense* that causes sleeping sickness in humans.

Tube Well–A small well for obtaining water from shallow strata which consists of a pointed pipe driven into the ground without boring. See Driven Well.

Tubificidae–A family name for aquatic segmented worms that exhibit marked population increases in aquatic environments containing organic decomposable wastes.

Tuff–Fine volcanic detritus in various states of stratification and consolidation. *Tufa* applies to similar rocks, but more especially to a kind of porous rock formed as a deposit from springs or streams; usually applied to calcareous deposits as traverteine and calcareous tufa. See Detritus.

Tularemia–A bacterial disease occurring mainly in rabbits but also in certain rodents, ungulates, carnivores, birds, livestock, and people; caused by *Francisella tularensis* and transmitted by arthropod vectors (ticks, lice, fleas, biting flies) and by contact of skin with infected material; marked by inflammation of lymph glands, headache, chills, and fever.

Tule Lands–Marshy land which is occupied chiefly by bulrushes of the genus *Scirpus*, especially such land in the lower part of the Sacramento Valley in California, United States. Such land, when reclaimed by drainage, has been highly productive agricultural land. Also called tulare.

Tulip Tree–*Liriodendron tulipifera*, family Magnoliaceae; a deciduous, magnificent, broadly pyramidal tree, up to 150 feet tall, grown for its size, shape, and for its lovely, tuliplike flowers. Its wood has a number of commercial uses including veneer, cabinet making, boxes, etc. Native to North America. Also called yellow poplar (forestry name), white wood, tulip poplar.

Tumble Windmillgrass–*Chloris verticillata*, family Gramineae; a perennial weedy grass, the inflorescence of which breaks away at maturity and rolls as a tumbleweed; however it is sometimes grazed by livestock early in the season. Also called windmill grass.

Tumbler–The pupae of the mosquito.

Tumbleweed–Any weed that breaks away from the ground, usually at or near the surface, at the end of the growing season and because of its form is often blown considerable distances, scattering seeds en route.

Tumulus–(1) Nest mound of wild bees. (2) Any artificial mound.

Tundra–One of the level or undulating nearly treeless plains characteristic of arctic regions, having a black muck soil and a permanently frozen subsoil. See Permafrost.

Tung Oil Tree–*Aleurites fordi*, family Euphorbiaceae; a small tropical tree grown for its nutlike fruit, the seeds of which yield the tung oil used in dyes, varnishes, waterproofing agents, etc. Native to central Asia. Also called China wood oil tree, tung.

Tunisian Bee–*Apis nigra*; a ferocious honey bee. Also called Punic bee.

Tunnel–(1) A gallery. (2) To burrow, make a passageway, in plant stems, leaves, roots, etc., as by larvae of certain insects.

Turbidity–Cloudiness of water caused by the presence of colloidal matter or other finely divided suspended matter.

Turf–(1) A close-growing, well-knit, usually fine-leaved growth of a grass, mixture of grasses, or other plant species, which is best maintained by mowing, fertilizing, and watering so as to present a pleasing appearance. It is useful for lawns, golf courses, horse-racing tracks, athletic fields, etc. (2) A slab of peat used for fuel (British). (3) The surface layer of a peat soil. (4) To produce a turf by the seeding, sodding, or other vegetative propagation of grass or other plant species.

Turkeymullein–*Eremocarpus setigerus*, family Euphorbiaceae; an annual weedy plant which yields blackish seeds that are nutritionally fattening for the turkeys. Sheep have been killed from grazing on this plant, for the hairs covering the stems and leaves are indigestible. Native to western United States. Also called dove-weed.

Turn–(1) To change the position of an egg in an incubator or in the nest where the hen is incubating it. (2) To restrain an animal, as a fence turns a cow. (3) To plow, so that the lower part of the soil is brought to the surface and the former surface part is covered. (4) To begin to show signs of ripening, as a fruit. (5) To direct an animal to pasture. (6) To sour, as milk. (7) To change, as leaves in the autumn.

Turning Under–The covering or burying of the surface soil, trash, or a green manure crop, by plowing with a turning plow.

Turpentine–(1) Oleoresin. (2) The essential oil derived from the distillation of oleoresin obtained from longleaf and slash pines in the southeastern United States. (3) To tap a tree to obtain oleoresin.

Turpentine Orchard–A stand of coniferous trees from which oleoresin is taken.

Turtle Backing–Preparation of a wide bed, in the form of a low rounded ridge, bordered on each side by furrows; a tillage method used in the planting of sugarcane on flat, wet lands. Known also as drainage-by-beds.

Tussock–A dense, heavy tuft or matted growth of grass or sedge which forms a small hillock.

Twister–See Tornado.

Two,Four,Five-T (2,4,5-T)–A herbicide used effectively to kill woody vegetation. Its use is restricted.

Two,Four-D (2,4-D)–A once widely used herbicide that controls most broad-leaf plants. Its use is now restricted or prohibited because of the damage that can occur to plants that are not intended to be killed.

Two-grooved Loco–*Astragalus bisulcataus*, family Leguminosae; a poisonous plant found from Manitoba, Canada, to New Mexico, United States.

Typhoid Fever–An acute infectious disease caused by a bacterium, *Salmonella typhi*, characterized by continued fever, inflammation of intestine, intestinal ulcers, a rose spot on the abdomen, and enlarged spleen; food and water-borne but may be transmitted by house flies.

Typhus Fever–A disease of humans caused by a bacterium-like microorganism, *Rickettsia prowazeki*, and transmitted by the body louse *Pediculis humamus humamus*. The disease is characterized by high fever, backache, intense headache, bronchial disturbances, mental confusion, and congested face. Mortality may range from 15 to 75 percent.

Typical–In appraising, that which most frequently occurs or exists in the particular situation under consideration.

U-shaped Valley–A valley carved by glacial erosion and having a characteristic parabolic cross section.

Ubiquitous–Occurring everywhere, as house flies; house sparrows; weeds.

Ultisols–Soils that are low in supply of bases and have subsurface horizons of illuvial clay accumulation. They are usually moist, but during the warm season some are dry part of the time. The balance between liberation of bases by weathering and removal by leaching is normally such that a permanent agriculture is impossible without fertilizers, lime, or shifting cultivation. See Soil Order.

Ultra-low Volume (ULV)–The spraying by air or ground of a pesticide undiluted and in a very concentrated liquid form. The usual dosage is from 2 to 16 fluid ounces per acre but is always less than one-half gallon per acre.

Umbra–(Latin) Shade.

Umbric Epipedon–A thick, dark surface soil horizon that is too acid and/or has too wide carbon to nitrogen ratio to be a mollic epipedon. See Mollic Epipedon.

Unbroken–(1) In marketing, designating an egg free from cracks or breaks in the shell. (2) Designating an untrained horse. (3) Designating soil that has not been plowed. See Virgin Soil.

Unconfined Groundwater–Groundwater under atmospheric pressure wherein the watertable rises or falls as the volume of stored water changes.

Uncontrolled Burning–See Prescribed Burning, Wildfire.

Underbrush–In humid regions, woody plants, usually short, growing in a forest, underneath tree species. In arid and semiarid regions, woody plants, grasses, and forbs.

Undercut–(1) A notch cut in the trunk of a standing tree below the level of the major cut and on the side to which the tree is to fall. It determines the direction of falling. (2) A saw cut made on the underside of a large branch beyond the point of severance, prior to making the actual primary cut, to prevent splitting or tearing. (3) the harvesting of less timber from a stand than that budged. (4) The tenderloin muscle of beef (British).

Undercutting–Cutting away at the base or underpart of a steep slope or cliff, as by a stream, glacier, wind, or wave erosion, thereby steepening the slope or producing an overhanging cliff.

Underdrain–Drain tiles or drain tubes placed in trenches and covered with soil to such a depth that the surface soil can be cultivated and the plant root profile adequately drained.

Underflow–(1) The water flowing beneath the dry beds of rivers, especially in arid regions. (2) The rate of flow or discharge of subsurface water, as in a spring.

Undergrazing–An intensity of grazing that fails to fully use the forage available for consumption in a given area under a system of conservation range management.

Underground Runoff–Water that flows toward stream channels after infiltration into the ground. See Subsurface Flow, Underflow.

Undergrowth–See Underbrush.

Understocked Stand–A stand of trees in which the growing space is not effectively occupied by crop trees.

Understocking–Pasturing or grazing a number of livestock less than the carrying capacity of a particular pasture or range.

Understory–(1) Grasses and small shrubs growing beneath a tree canopy. (2) Plants growing in the shade of other plants.

Undulating–Designating a relief, landscape aspect, or lay-of-land, characterized by successive rolls or rounded hills.

Uni-–A combining form meaning one.

Unified Soil Classification System–(Engineering) A classification system based on the identification of soils according to their particle size, gradation, plasticity index, and liquid limit. Indicated for each soil-mapping unit in modern soil survey reports.

Uniformitarianism–The concept that the present is a key to the past, and that past geologic events are to be explained by those same physical principles that govern the present.

Unique Agricultural Lands–Land that is particularly suited for high production of a crop; e.g., the orange-growing soils of Florida and the cranberry bogs in Massachusetts. See Prime Agricultural Lands.

Unit–(1) A single thing or item of produce. (2) A recognized measure of weight, volume, or distance, such as a bushel, gallon, or mile.

Universal Soil Loss Equation–An equation used for the design of water erosion control systems: $A = RKLSPC$ wherein A = average annual soil loss in tons per acre per year; R = rainfall factor; K = soil erodibility factor; L = length of slope; S = percent of slope; P = conservation practice factor; and C = cropping and management factor. (T = soil loss tolerance value that has been assigned each soil series expressed in tons/acre/year.) See T Factor.

Unleached Wood Ashes–Wood ashes that have had no part of their plant nutrients removed and that contain 4 percent or more of water soluble potash. (K_2O).

Unslaked Lime (CaO)–See Burnt Lime, Calcium Oxide.

Up-country–The country that lies inland from the seacoast.

Upland–(1) Highland, an elevated plain in association with or in contrast to a valley plain or lowland. (2) Designating crops or crop varieties grown on upland in contrast to lowland areas. See Upland Cotton.

Urban–(1) Belonging to or residing in a town or city rather than the countryside. (2) In the United States, designating any of the following residence categories, used by the Bureau of the Census: (a) places having 2,500 persons or more, incorporated as cities, boroughs, and villages; (b) incorporated towns of 2,500 persons or more, except in New England, New York, and Wisconsin, where towns are simply minor civil divisions; (c) the densely populated areas, incorporated and unincorporated, around cities with a population of 50,000 or more; and (d) unincorporated places with a population of 2,500 or more outside any urban fringe. See Rural-Farm, Rural-Nonfarm.

Urbanized Area–An area identified by the United States Bureau of the Census as having a population over 50,000 or by the Office of Management and Budget as a standard metropolitan statistical area. Small urban areas are those areas that have a population of 5,000 to 50,000.

Urea–(1) $CO(NH_2)_2$; a nonprotein, organic compound of nitrogen, made synthetically by a combination of ammonia and carbon dioxide, and used in fertilizers and as a livestock feed supplement. (2) The chief compound of nitrogen in the urine of mammals.

Urea Phosphate–$CO(NH_2)_2H_3PO_4$; a crystalline compound formed in processing mixtures containing urea and phosphoric acid or other phosphatic materials.

Urea-Ammonia Liquors–Any of several solutions of urea in crude aqueous ammonia, used in ammoniating superphosphate fertilizer.

Urea-Ammonium Phosphate–Intimate mixtures containing ammonium orthophosphates and urea. The reaction product of commercial phosphoric acid and ammonia is mixed with molten urea. Heat of reaction dries the granular product. Grades produced include 28-28-0, 36-18-0, and a variety of other grades with total plant nutrient contents up to 60 percent. Potassium can be included in these products. Modified treatment produces urea-ammonium polyphosphate for use as a solid or in liquid fertilizers.

Urea-Ammonium Sulfate–Granulated homogeneous mixtures of urea and ammonium sulfate containing 30 to 40 percent N and 4 to 13 percent S. The products have greater granule strength than urea alone.

Uric Acid–A white, crystalline compound, derived from guano. It contains about 33 percent available nitrogen.

USDA (U.S.D.A.)–United States Department of Agriculture.

Use–The proportion of the current season's growth available for grazing that is consumed or destroyed by grazing animals; usually expressed as a percentage of biomass, but may be estimated as overuse, proper-use, or underuse; it may be applied to a single species or to the entire rangeland.

Use Hazard–In land use, any serious inconvenience or danger incident to occupancy, as poisonous plants, landslides, floods, avalanches, insects, or dust.

Ustoll–A suborder name of a group of soils in semiarid and subhumid climates with complex profiles suggesting several climatic cycles of humidity and aridity.

Uvala–A natural land feature in a karst region. It is characterized by depression larger than a sinkhole, often formed by the coalescence of several sink holes. See Karst, Thermokarst.

Vadose–Water held in soil, or other surficial geological formations, above the level of permanent groundwater.

Vagabond Bee–Any worker bee that lacks an individual foraging area.

Valence–Also called bond or chemical bond; the chemical combining power of an atom. It indicates the number of electrons that can be lost, gained, or shared by an atom in a compound.

Valley–(1) An elongated, erosional depression usually occupied by a stream that has a downward slope conforming to the direction of flow of the occupying stream, and includes both bottomland and slopes. Some features designated as valleys may be primarily structural, due to rock folding and faulting rather than erosional; some may be dry or without streams; some in arid regions may be basins that do not have the usual form of valleys in humid regions. (2) The drainage area of some rives, as the Mississippi Valley, or a combination of flowland, and mountainous ridges as the Great Appalachian Valley of eastern United States.

Valley Train–Deposits of comminuted rock material that were carried in prehistoric times by streams originating from the melting ice of glaciers. The deposits are confined to valleys and thus are distinguished from outwash plains. (A physical feature of parts of the glacial region of the United States.) The valley train often constitutes a distinctive land type of agricultural significance.

Vapor Drift–The movement of pesticidal vapors from the target area of application.

Var.–The abbreviation for the word variety; as in *Sarcoptes scabiei* var. *bovis*. See Cultivar.

Variable Grade–A terrace channel having a variable slope that decreases as the distance from the outlet increases; e.g., a terrace 1,200 feet long may be built in four 300-feet sections each having different grades, the outlet end having 0.40 feet fall per 100 feet; the next section, 0.30; the third, 0.20, and the fourth (upper) section, 0.10 foot per 100 feet.

Variable Oak Leaf Caterpillar–*Heterocampa manteo*, family Notodontidae; a caterpillar that voraciously feeds on the foliage of the oak, basswood (linden), walnut, birch, elm, hawthorn, and persimmon.

Variance–A statistical measure of the amount of variation that is observed within or among a group of animals or plants.

Variant–A recognized entity different from normal.

Variation–(1) The angle by which the north end of the compass needle (magnetic north) deviates from true north. (2) One of the laws of organic nature; organisms vary in time, from place to place, and also in one locality with time; they vary also in their appearance (morphology).

Varve–(1) An annual layer of dust, sand, silt, or clay deposited, regardless of origin. (2) The layers of uniform-sized soil materials deposited by glacial streams, whereby a coarse sediment is deposited by fast-moving water in spring and summer and fine sediment deposited by slow-moving waters in fall and winter when glacial meltwaters were less.

Vection–The passing of a disease from one plant or animal to another.

Vector–Any agent such as an insect or animal that transmits, carries, or spreads disease from one plant or animal to another.

Vega–(1) A fertile, grass-covered tract or extensive plain (southwestern United States). (2) Irrigated land from which a single crop per year is produced.

Vegetarian–(1) One who, because of cultural reasons or personal conviction, abstains from eating meat (in the strictest sense, also milk, butter, and eggs). (2) An herbivorous animal or person.

Vegetated Channel–See Grassed Waterway.

Vegetation–Any group or association of plants; the sum of vegetable life; plants in general.

Veld (Veldt)–A term used in South Africa for grasslands at the higher elevations and for shrub or savannah at the lower elevations.

Venison–The edible flesh of deer.

Venom–Poisonous matter secreted by such creatures as snakes, bees, and scorpions.

Ventral–Pertaining to or relating to the belly or underside; opposite the dorsal or back.

Verdant–Designating fields or forests that are green with growing vegetation.

Vermiform–Worm-shaped.

Vermifuge–A drug or chemical that expels worms from animals; an anthelmintic. See Vermicide.

Vermin–Any noxious animal; insect, acarid, rodent, etc.

Vernacular Names–The names given to plants and animals in an area. They are often called common names.

Vernal–Appearing in spring.

Vernal Pond–Pond that exists for a limited period in the spring.

Vernis–Of spring.

Vertebra–Any of the bony segments that form the spinal column or backbone.

Vertebrates–Animals with a spinal column or backbone, such as fishes, birds, mammals, and so on.

Vertical Interval–In terrace farming, the vertical distance in feet from the center of one terrace line, ridge, or channel to the corresponding point on an adjacent terrace. The interval of the first terrace is the vertical distance from the top of the hill to the staked channel of the terrace.

Vertical Mixing–Vertical circulation of water masses in a lake occurring naturally in temperate climates in spring and autumn.

Vertisols–Fine clayey soils with high shrink-swell potential that have wide, deep cracks when dry. Most of these soils have distinct wet and dry periods throughout the year. See Soil Orders.

Very Coarse Sand–See Soil Separates.

Very Fine Sand–See Soil Separates.

Viability–(1) Ability to live (immediately after birth or hatching). (2) The capacity of seeds to germinate. (3) The state of being alive. (4) Pertaining to sperm cells in the semen, capable of living and successfully fertilizing the female gamete.

Viable–(1) Living. (2) Specifically in regard to organisms or agents of disease, able to cause infection in animals. (3) Capable of living. (4) Capable of germinating, as seeds.

Vigor–The desirable state of health of any living thing.

Viral–Having the nature of a virus; pertaining to a virus; like a virus.

Virgin–(1) Any female that has not had coitus, as a virgin queen bee. (2) In turpentining, the face of cut the first year a tree is bled. (3) Undisturbed, unplowed land, uncut forest.

Virgin Forest–A mature or overmature forest that has grown entirely uninfluenced by human activity.

Virology–The study of viruses and viral diseases.

Virucide–A chemical or physical agent that kills or inactivates viruses; a disinfectant.

Virulence–The disease-producing power of an organism.

Virulent–Highly pathogenic; having great disease-producing capacity; deadly; very poisonous or harmful.

Virus–(Plural, viruses) A self-reproducing agent that is considerably smaller than a bacterium and can multiply only within the living cells of a suitable host. Most viruses are too small to be seen even with the aid of the ordinary microscope, but can be photographed with the aid of the electron microscope. Viruses usually are considered to be living agents of microorganisms but some have characteristics of nonliving matter. They are protein-containing bodies of high molecular weight capable of multiplying and acting like living organisms when in living tissue. They are the cause of many animal, human, and plant diseases, such as smallpox, measles, tobacco mosaic, etc. Recovery from some viral diseases confers lasting immunity.

Virus Interaction–The action of a virus in altering the normal development of other viruses or virus strains that is expressed by partial or complete suppression, by synergistic association, by modification of the type of symptoms in the host plant, or by abnormal increase in concentration of one virus.

Virustatic–A substance that prevents the multiplication of a virus.

Vitreous–Resembling glass in hardness or brittleness.

Vitriol–Sulfuric acid or some of its compounds. Blue vitriol is hydrous copper sulfate; green vitriol, copperas; red vitriol is either a sulfate of cobalt or a ferric sulfate; white vitriol is hydrated zinc sulfate.

Vivarium–A glass box resembling an aquarium used to keep or raise animals or plants. See Terrarium.

Vivianite–A hydrous ferrous phosphate, a mineral, and a number of oxidation forms of it, which is found in soils especially those in swamps and bogs. Also called blue ochre, blue iron earth.

Vly (Vlei, Vley)–(1) A small swamp, usually open and continuing a pond. (2) A valley where water collects. Local in the United States Middle Atlantic States.

Volcanic Ash–The finely comminuted ejecta of volcanic eruptions which, in thick deposits, are the parent material of soil. The volcanic ash carried by winds is widely distributed in soils.

Volcanic Soil–A soil that is derived from or consists of volcanic rocks, especially ash and other fragmentary ejecta.

Vole–*Microtus*; a small rodent resembling a mouse that has a blunt nose, short tail and ears, and a stout body. It often does much damage to crops. Also called meadow mouse, field mouse.

Volunteer–Any plant that grows from self-sown seed.

Wadi–A desert watercourse which contains water only immediately after a heavy rain (Sahara and Arabian countries).

Wainable–Designating that which may be plowed or manured or is tillable.

Wall Tree–An espaliered tree; a tree, shrub, or vine pruned and trained to grow flat against a wall.

Walnut–Any tree of the genus *Juglans*, family Juglandaceae. Species are tall deciduous, valuable nut and timber trees, including butternut, black walnut, Persian walnut.

Walnut Caterpillar–*Datana integerrina*, family Notodonidae; a moth whose larvae feed in colonies on walnut, hickory, and pecans in the United States from Maine to Florida and west to Kansas.

Wandering Dune–An unstabilized sand dune; one that is being shifted about by wind action.

Waney –(1) Designating lumber which has bark on a square-edged piece. (2) An imperfect board.

Warm Front–A mass of warm air advancing behind a mass of cool air. Characteristic weather may be several days of heavy overcast and drizzles.

Warm-blooded Animal–Any animal, such as a bird or mammal, whose body temperature is warmer than its surrounding medium, as contrasted with the cold-blooded animals, such as reptiles.

Warp–(1) Yarn which runs the long way in a woven fabric. (2) In lumber, any variation of a board from a plane surface often caused by a too rapid loss of moisture in the curing process. (3) To vary from a plane surface, as twisted or buckled board. (4) Sediment deposited by water which acts as a fertilizer.

Wash Land–Any land which is periodically overflowed.

Washout–(1) The erosion of a portion of a levee, railroad bed, road, etc., by water. (2) Designating a channel made by erosion in the deposit of a certain sediment and filled by fresh material. (3) The flushing out of a pipe, etc.

Wasp–Insects of the family Vespidae. Although sometimes considered a pest because of their painful stings, wasps are truly beneficial insects. They eat large numbers of other insects as adults and some species are parasitic as larvae on other insects.

Waste Bank–The ridge of excavated earth parallel to an open drainage ditch. Also called spoil bank.

Waste Gate–A water gate in a dam or reservoir for the discharge of excess water.

Waste Lime–Any industrial waste or by-product containing calcium or calcium and magnesium in forms that will neutralize soil acids. It may be designated by a prefix suggesting the name of the industry or process by which it is produced, as gas-house lime, tanner's lime, etc. Also called by-product lime. See Lime.

Waste Water–The spent or used water from a home or an industry containing dissolved and suspended matter capable of polluting the environment.

Wasteland–Land essentially incapable at present of producing anything of value (not used to describe an idle farm or forest land).

Watch Tower–In forestry, a tower usually constructed on a hilltop or prominence for a fire warden or ranger to observe the surrounding area and keep a lookout for fires.

Water–H_2O; hydrogen oxide; although the liquid may contain associated molecules. The melting and freezing point is 32°F (0°C) and the boiling point 212°F (100°C). The most valuable natural resource and the most limiting factor in crop production.

Water Application Efficiency–The percentage of irrigation water applied that can be accounted for as moisture increase in the soil occupied by the rooting system of the crop.

Water Application Rate–The rate in inches per hour that irrigation water is applied to fields.

Water Bloom–A sudden increase in the abundance of algae, especially at or near the water surface.

Water Bloom Poisoning–A poisoning of livestock resulting from drinking pond or stream water which is contaminated with certain blue-green algae. This poisoning is rare but very lethal.

Water Buffalo–*Bos bubalus*, family Boridae; a large animal found wild in the jungles from India to the Philippines at elevations from sea level to 6,000 feet. The coat color varies from a dark gray to black. The animal has long horns, curving back over the body, which sometimes measure as much as 12 feet along the curve. Widely domesticated, the water buffalo is raised as a beast of burden, for its rich milk, and its hide from which a good-quality leather is manufactured. Also called carabao (Philippines), gamoose, jemoose (Egypt).

Water Conditioner–Any of several devices or materials which change the composition of water to a desirable state, as a water softener, and acidifier, etc.

Water Conservation–The physical control, protection, management, and use of water resources in such a way as to maintain crop, grazing, and forest lands; vegetal cover; wildlife; and wildlife habitat for maximum sustained benefits to people, agriculture, industry, commerce, and other segments of the national economy. See Soil Conservation Service.

Water Content–(1) The water of the soil or habitat; (physiological) the available water supply; (physical) the total amount of soil water. (2) The percentage of water in a material in relation to oven dry weight.

Water Cushion–A pool of water maintained to break the impact of water outflowing from a dam, chute, or other spillway structure.

Water Deficit–The amount of water lacking for given purposes.

Water Disposal System–The complete system for removing excess water from land with minimum erosion. For sloping land, it may include a terrace system, terrace outlet channels, dams, and grassed waterways. For level land, it may include only surface drains or both surface and subsurface drains. For homes and factories it may consist of a septic tank and drain field. For barnlots it may consist of a lagoon.

Water Elevator–(1) Any mechanism used for raising water, especially to a considerable height above ground, as a windmill operating a pump. (2) A vertical or inclined conveyor whose buckets dip up water from a source of supply and deliver it into a conduit or elevated receptacle, thus fulfilling the function of a pump.

Water Erosion–Erosion by water. See Gully Erosion, Rill Erosion, Sheet Erosion, Splash Erosion.

Water Extract–Whatever can be removed or dissolved out of a substance with water. A substance like sugar is completely soluble in water, whereas when yeast is shaken up with water only a small portion of it goes into solution.

Water Gang–A constructed channel used for drainage or irrigation.

Water Gap–(1) A gorge cut through a ridge by a stream. (2) The gap below a fence where it crosses a small stream.

Water Harvesting–Application of managerial techniques that increase run-off water yield from the land. This includes the use of plastic sheets to divert water to a place where it is needed most.

Water Head–(1) The source of a stream. (2) A pond created to furnish irrigation water for a small plot, or to furnish limited power for a mill. (3) The pressure created by the depth of water in such a pond.

Water Hickory–*Carya aquatica*, family Juglandaceae; a deciduous tree grown for its valuable wood used in tool handles, sporting goods, ladders, furniture, implements, woodenware, and for fuel and smoking meat. Native to North America. Also called bitter pecan, swamp hickory.

Water Hyacinth–*Eichhornia crassipes*; a large floating aquatic plant that causes clogging of water areas in the southern United States.

Water Level–(1) The surface of water which is used in measurement of surface or stream water depth. (2) The free water level in the soil. (3) See Water Table.

Water Mark–(1) Any mark, such as stain from suspended sediment, which indicates the level to which surface water has risen. (2) A water-soaked spot on citrus fruit resulting from a moderate freezing temperature.

Water, Nonconsumptive Use–(1) Those uses of resources that do not reduce the supply, such as many types of recreation. (2) A use of an area or resource which does not alter the area or resource, and which the fact of one person partaking of this use does not reduce the quality for another user. (3) Some consumptive uses of water are irrigation and domestic and industrial use, while nonconsumptive uses would include direct power generation as well as boating and swimming.

Water Plant–See Aquatic.

Water Pollution–The addition of sewage, industrial wastes, or other objectionable material to water in sufficient quantities to result in measurable degradation of water quality for specified uses.

Water, Potable–Water that is fit for human consumption. Drinking water standards are established and enforced by states and by the United States Environmental Protection Agency. Furthermore, bottled water sold for drinking is regulated by the United States Food and Drug Administration. Water scientists claim that water from the tap at most municipalities is healthful.

Water Quality Criteria (Water Quality Standard)–The levels of pollutants that affect the suitability of water for a given use. Water use classification includes: public water supply, recreation, propagation of fish and other aquatic life, irrigation, and industrial use.

Water Requirement–The quantity of water, regardless of its source, required by a crop in a given period of time for its normal growth under field conditions. It includes surface evaporation and other unavoidable wastes. Usually it is expressed as depth (volume per unit area) for a given time; e.g., acre inches/hour.

Water Resource Region–The twenty-one major hydrologic regions into which the United States is delineated.

Water Retained–Field capacity of a soil. Any surplus water is lost as surface runoff or deep percolation.

Water Right–A legal right to use the water of a natural stream, furnished by a canal for general or for specific purposes. It may entitle a person to use the canal to full capacity, to a measured extent, or for a definite period of time, and to change the place of diversion, storage, or use of water as long as it does not infringe upon the rights of other people. In some states in the United States, the rights to water may be sold and transferred separately from the land. See Riparian Doctrine.

Water Shoot–See Water Sprout.

Water Snake–(1) Any of the snakes of the genus *Natrix*, especially *N. sipedon* of North America, and other allied genera, inhabiting fresh water and feeding on aquatic animals such as fish. Severe in appearance and temperament, their sting, though painful, is not generally poisonous. A poisonous species, the water moccasin, *Agkistrodon piscivorous*, family Crotalidae, with a hollow depression between the eyes and nostril, is found in southern United States. (2) A second group of water snakes, family Homalopsinae, characterized by a tapered tail, back teeth, and valvelike nostrils on the snout, are found in Australia and Polynesia.

Water Spreader–A terrace, dike, or other structure intended to distribute surface water runoff and increase the area of infiltration into the soil for plant use.

Water Spreading–(1) The artificial application of water to lands for the purpose of storing it in the ground for subsequent withdrawal by pumps for crops. (2) Irrigation by surplus waters out of cropping season. (3) The diversion of run-off water from gullies or watercourses and its distribution on adjacent, gently sloping, grazing lands needing additional water. The volume of water flowing down the channels is reduced and the moisture absorbed by the spreading area increases the growth of vegetation.

Water Table, Apparent–A thick zone of free water in the soil. An apparent water table is indicated by the level at which water stands in an uncased borehole after adequate time is allowed for percolation into the surrounding soil.

Water Table, Artesian–A water table under hydrostatic head, generally beneath an impermeable layer. When this layer is penetrated, the water level rises in an uncased borehole.

Water Table, Perched–A water table standing above an unsaturated zone. In places an upper, or perched, water table is separated from a lower one by a dry zone.

Water Table, True–The upper limit of the soil or underlying rock that is wholly saturated with water.

Water Year–A special grouping of the periods of the year to facilitate water supply studies. The United States Geological Survey uses October 1 to September 30 as the water year. See Climate Year.

Water Yield–(1) The total outflow of a drainage basin through either surface channels and/or subsurface aquifers. (2) The surplus of precipitation over infiltration and canopy interception. Also known as run-off water.

Water-ground–(1) Designating corn which has been soaked in warm water and passed through crushing mills so regulated that the hulls, endosperms, and embryos are mechanically separated without damage. (2) Designating corn meal that is ground at a water grist mill in contrast to that ground at a conventional mill.

Water-holding Capacity–The ability of a soil and crop system to hold water in the root zone.

Watercourse–(1) A channel, usually natural, through which a stream of water flows and which may be dry during unusual drought periods. (2) Legally, a natural stream flowing in a well-defined channel. Water flowing underground in a known and well-defined channel also constitutes a watercourse and is governed by law applicable to surface streams. See Karst.

Watercress–*Nasturtium officinale*, family Cruciferae; a hardy, perennial, aquatic herb grown as a salad plant for its edible leaves. Native to Europe.

Waterfront Zone–A strip of land of varying width along a lake, pond, stream, ocean, or other water body, on which the scenic values are safeguarded from impairment by excluding or restricting the utilization of natural resources, commercial development, and occupancy.

Waterhemlock–Any plant of the genus *Cicuta*, family Umbelliferae. Some species are among the most poisonous of plants.

Waterhole–(1) A natural depression in which water collects or stands, often in the dry bed of a stream; a spring in the desert. (2) A natural spring or pool on the open range where cattle drink (western United States).

Waterlogged–(1) Water saturation of any object. (2) Water saturation of soil to the extent upland plants will not grow normally. The cause of waterlogging may be a naturally poorly drained soil, excess irrigation, too much precipitation, flooding by a stream or river, or seepage from a dam or irrigation canal.

Watershed–(Catchment) (1) The total land area, regardless of size, above a given point on a waterway that contributes run-off water to the flow at that point. A major subdivision of a drainage basin. On the basis of this concept, the United States is generally divided into 18 major drainage areas, 160 principal river drainage basins, containing some 12,700 smaller watersheds. (2) The area contained within a drainage divide above a specified point on a stream. See Catchment.

Waterway–A natural or artificially constructed course for the concentrated flow of water.

Weak Soil Structure–Soil aggregates (clods or peds) that disrupt easily when it rains.

Weather–(1) The atmospheric conditions prevailing at any specified time and place, or those prevailing during any particular period, as shown by meteorological observations and records of air temperature, barometric pressure, wind velocity, humidity, clouds, and precipitation. (2) Bad weather. (3) To be subjected to the influences of atmospheric conditions, as the abrasive action of rainfall, the disintegration due to frost, the deteriorating influences of oxygen and other gases contained in the atmosphere.

Weather Hazard–A chance that bad weather will occur at an inappropriate time.

Weathered–(1) Designating the physical and chemical disintegration or decomposition of rocks under natural conditions. (2) Designating a product which has been exposed to a variety of weather conditions and has been changed as a result.

Weathering–(1) The action of all natural forces on a product left exposed to the weather. (2) The deterioration of the unprotected surface layer of wood which has been exposed to the weather. (3) Atmospheric action on rock surfaces producing decomposition, disintegration, or alteration of rocks at or close to the earth's surface. Weathering implies no removal of material other than that in solution in drainage waters.

Web-footed–Designating any of certain water fowls whose toes are joined together by a membrane, as a duck.

Weed–(1) A plant out of place. Thus, rye growing in a field of wheat is a weed. (2) More popularly, an herbaceous plant which takes possession of fallow fields or finds its way into lawns or planted fields, crowding the vegetation planted there and robbing it of moisture and nutrients. (3) A plant whose usefulness is not recognized or which is undesirable because of odor, spines, prickles, or poisonous characteristics. (4) A tree of inferior value in a forest, or one growing in a street or lawn, where it is not wanted, like the seedlings of the boxelder. (5) Excessive vegetative growth. (6) Any plant that harbors insects, fungi, or viruses that may spread to nearby crop plants. (7) A horse or other domestic animal that is undersized or a misfit. (8) To remove weeds from a lawn, field, or other places where they are not wanted, usually manually but also by machinery or herbicides. (9) Designating a flavor of dairy products, especially milk, which resembles that of certain vegetation (onion, wild garlic, leeks, etc.).

Weed Killer–See Herbicide.

Weephole–A hole through an abutment or retaining wall to provide for the passage of seepage water.

Weeping–Designating a tree with pendant (drooping) branches; e.g., weeping willow, weeping cherry.

Weir–(1) Stationary fish trap which acts as a barrier to fish movements and leads fish into pots. (2) Dam or barrier to raise level of water for different purposes.

Weir Head–The depth of water flowing over the weir crest as measured at a point at least 2.5 and preferably 4 times the depth of water flowing over the weir, upstream from the weir. The zero point of the measuring scale should be placed at the exact level of the weir crest.

Weir Notch–The opening in a weir for the passage of water.

Well, Bored–A water well established by hand tools, powered augers, or auger buckets in rock-free soil. This is in contrast to a driven well which is pounded into rocky soil. See Well, Driven.

Well, Driven–A water well established by pounding a well point into the earth below the permanent water table. This is in contrast to a bored well established in rock-free soil by hand tools, powered augers, or auger buckets. See Well, Bored.

Well Water–Water which is drawn or pumped from a well, in contrast to ditch water, creek water, rain water, spring water, etc.

Well-drained–Designating a soil (in humid regions) that normally contains only capillary moisture, in contrast to a swampy or waterlogged condition which also contains some gravitational water.

Well-graded–Refers to a soil or soil material consisting of particles well-distributed over a wide range in size or diameter. Such a soil normally can be easily increased in density and bearing properties by compaction.

West–(1) That portion of the United States lying west of the 100th meridian. Sometimes arbitrarily defined as that part west of the 100th meridian, that part west of the Mississippi River, or the Rocky Mountains region and westward. The geographic location of where the West begins has been variable according to the time in history and according to context. (2) Western Australia.

West Indian Locust–*Hymenaea courbaril*, family Leguminosae; a large, hardwood tree, bearing white flowers and woody pods. The brown wood is used for construction, the pods yield an edible pulp, and a resin obtained from the tree has medicinal and wood-finishing value. Native to tropical America. Also called algarroba, courbaril.

West Indies Mahogany–*Swietenia mahogoni*, family Meliaceae; an evergreen tree grown for its hard, red wood, used especially in furniture. Native to tropical America.

Western Bracken–*Pteridium aquilinum*, family Polypodiaceae; a fern which is poisonous to livestock. After about two months of feeding on it cattle become sick with very high temperature, difficult breathing, salivation, nosebleed and hemorrhage. Native to western North America. Also called brake fern, hog brake, bracken, bracken fern. See Bracken, Eastern Bracken.

Western Equine Encephalitis–A viral disease of horses communicable to people, marked by fever, convulsions, and coma. Transmitted by certain species of mosquitoes.

Western Fir–See Douglas-fir.

Western Larch–*Larix occidentalis*, family Pinaceae; a deciduous tree, important as a source of wood used as ties, poles, in building construction, planing-mill products, boxes, crates, etc. Native to North America. Also called western tamarack.

Western Pine Beetle–*Dendroctonus brevicomis*, family Scolytidae; a most important insect pest of ponderosa pine and coulter pine along the United States West Coast, causing great losses each year. The adults construct egg galleries between the bark and sapwood.

Western Pitch Pine–Ponderosa pine.

Western Range–The native livestock grazing areas of the Great Plains, the Rocky Mountains, the Intermountain and the Pacific Coast regions of the United States.

Western Spruce–Sitka spruce; *Picea sitchensis*, family Pinaceae.

Western Tamarack–See Western Larch.

Western Waterhemlock–*Cicuta occidentalis*, family Umbelliferae; an extremely poisonous, perennial herb. Symptoms of poisoning in stock are frothing at the mouth, uneasiness, pain, convulsions, kicking, throwing back of the head, rigid legs, bellowing, and spasmodic contractions of the diaphragm. Death follows ingestion rather rapidly. Native to western North America.

Western Wheatgrass–*Agropyron smithii*, family Gramineae; a tall, coarse, smooth-stemmed grass bearing whitish flowers; widely cultivated as a forage crop. Native to western United States. Also called bluestem, bluestem wheatgrass.

Western White Fir–White fir.

Western White Pine–*Pinus monticola*, family Pinaceae; an evergreen tree important for its wood used for matches, boxes, construction, millwork, fixtures, etc. Native to western Canada and United States.

Western Yarrow–*Achillea lanulosa*, family Compositae; a perennial herb of fair forage value. It is grazed somewhat on the western ranges, United States.

Western Yellow Pine–See Ponderosa Pine.

Wet–(1) To dampen; to sprinkle; to furnish water to a plant. (2) Characterized by rain, as wet weather. (3) Moist, covered, soaked, or saturated with water.

Wet Climate–The climate in which precipitation effectiveness is such that rainforest vegetation prevails.

Wet Down–(1) To water a plant so that the soil surrounding its roots is soaked. (2) To bring soil to field capacity by rain, as it wet down six inches.

Wet Meadow–A meadow where the surface remains wet or moist throughout the summer, usually characterized by vegetation of sedges and rushes.

Wet Pocket–A low area in a field, woods, etc., in which water collects and remains for considerable periods of time.

Wet Season–(1) The season or time of year during which most of the precipitation occurs. See Dry Season. (2) A season in which the precipitation is greatly in excess of normal.

Wet-Dry Climate–Any climate which has alternating wet and dry seasons, such as a wet summer and a dry winter, or the reverse. Sometimes the climatic condition in the trade-wind belt, where daily showers are interspersed with dry sunny weather, is also considered to be wet-dry.

Wetlands–Those areas inundated by surface or groundwater often enough to support the kind of vegetation or aquatic life that requires saturated (or seasonally saturated) soil conditions for growth and reproduction. Wetlands generally include swamps, marshes, bogs, and similar areas such as sloughs, potholes, wet meadows, river overflows, mud flats, and natural ponds.

Wettable Powder (Dust)–Any material manufactured in the form of a powder or dust that can be mixed with or dissolved readily in water.

Wettable Sulfur–That sulfur so treated as to be readily miscible in water. It is widely used in spray mixture, especially for its fungicidal effect.

Wetting Agent–Material included in pesticide solutions that reduces surface tension and helps to completely cover the surface or foliage area of the plant being sprayed. See Spreader.

Weymouth Pine–See Eastern White Pine.

Whale Oil–Lubricating oil obtained principally from the fat or blubber of the Greenland whale, which is largely used for oiling wool during combing, in batching flax and other fibers, in currying, and in dressing chamois leathers. The oil of commerce may be obtained from the blubber of any whale or dolphin. Also called train oil.

Whelp–To give birth to a litter of pups; may also refer to the pup.

Whirlwind–A windstorm, a rapidly rotating inward and upward spiral motion of the air, covering a very small area. It is often seen on hot, summer days, especially on open or prairie land. Also called devil duster.

White Ant–See Termite.

White Ash–*Fraxinus americana*, family Oleaceae, a large, deciduous tree grown for its lumber used for handles, furniture, vehicle parts, sporting goods (bats, oars, paddles, etc.), and other articles. It is one of the better street trees in places where the soil is favorable. Native to eastern North America.

White Crappie–A white perch, *Pomoxis annularis*, that is used to stock lakes as a pan fish. It is smaller and more silvery than the Black Crappie. See Black Crappie.

White Ironbark Eucalyptus–*Eucalyptus leucoxylon*, family Myrtaceae; a tall tree, with smooth, pale to dark gray bark, which is of value in California, United States, as a lumber tree, bee tree, and an ornamental. Native to Australia. Also called fat cake, white ironbark.

White Locoweed–See Lambert Crazyweed.

White Lupine–*Lupinus alba*, family Leguminosae; an annual herb grown to some extent for soil improvement. Native to Europe and Asia.

White Oak–*Quercus alba*, family Fagaceae; a roundheaded, important deciduous, timber tree, which is one of the slowest growing and longest lived trees in the eastern United States. It is a magnificent tree, but it is not grown ordinarily as an ornamental due to difficulty in transplanting. The commercial uses of its wood are extensive.

White Pine–See Eastern White Pine.

White Snakeroot–*Eupatorium rugosum (E. urticaefolium)*, family Compositae; a perennial herb grown for its branched clusters of white flowers which last from midsummer to frost. A poisonous plant, it is found wild from eastern North America to Minnesota and Texas.

Symptoms of poisoning in livestock are trembling of the legs, depression, inactivity, constipation, nausea, rapid breathing, and collapse. The poison may be transmitted to humans by milk from a cow which has eaten the plant. Also called richweed, white sanicle, white snakeweed, Indian sanicle.

White Spruce–*Picea glauca*, family Pinaceae; a hardy, evergreen tree, which is grown as an ornamental. Its wood is used in pulp, cooperage, boxes, general building construction, planing-mill products, furniture, ladders, etc. Native to North America. Also called catpine, skunk spruce.

White-marked Tussock Moth–*Hemerocampa leucostigma*, family Lymantriidae; a moth whose larva feeds on the foliage of deciduous trees: elm, linden, maple, horsechestnut, poplar, sycamore, buckeye, willow, apple, pear, quince, plum, etc., skeletonizing the leaves and scarring the fruit.

White-pine Aphid–*Cinara strobi*, family Aphidae; an aphid which infests the eastern white pine east of the Mississippi River, sometimes killing small trees. A sooty mold develops in the honeydew secreted by the aphids.

White-pine Blister Rust–A destructive, fungal disease of eastern and western white pines, caused by the rust *Cronartium ribicola*, family Melampsoraceae. Hyphal threads of the fungus penetrate the sapwood and inner bark, causing white blisters to appear on the bark. The uredial and telial stages of this rust occur on currants and gooseberries. Occurs in eastern and Pacific United States and Europe. Also called blister rust of pine.

White-pine Cone Beetle–*Conophthorus corniperda*, family Scolytidae; a beetle whose larva is destructive to cones of the eastern white pine, feeding on the scales, seeds, and other parts of the cone. Occurs in Canada and northern United States.

White-pine Sawfly–*Neodiprion pinetum*, family Diprionidae; the larva of an insect that infests the eastern white pine and sometimes other pines in northeastern United States.

White-pine Weevil–*Pissodes strobi*, family Curculionidae; an insect that infests eastern white pine and other pines and spruces in eastern United States. The leader on an infested plant becomes brown in contrast to the rest of the tree as a result of girdling by the weevils.

Whitecedar (Atlantic)–*Chamaecyparis thyoides*; an evergreen, swamp tree whose wood is used for poles, shingles, woodenware, planing-mill products, water tanks, boats, boxes, fences, etc. Native to eastern United States. Also called white cedar, Atlantic white cedar, southern white cedar.

Whitecedar (Northern)–*Thuja occidentalis*; an evergreen swamp tree native to Michigan, southern New York, central Vermont, and New Hampshire. Used for fence posts and poles. The foliage is good winter browse for deer. Also known as aborvitae.

Wild–(1) Uncultivated; not cultivated by humans; growing without the care of people. (2) Not domesticated, as a wild animal.

Wild Areas–Areas of forest lands similar to wilderness area which are subject to the same restrictions and regulations but are smaller.

Wild Calla–*Calla palustris*, family Araceae; an aquatic herb grown along the edge of a pond in mud or shallow water for its oval or heart-shaped leaves and green flowers. It is poisonous for livestock. Symptoms of poisoning are an intense burning feeling in the mucous membranes of the mouth and throat. Native to the north temperate zone. Also called water arum.

Wild Carrot–*Daucus carota*, family Umbelliferae; an annual or biennial herb, which is a serious weed pest but also a beautiful flower plant. Eaten by cows, it imparts an objectionable flavor to the milk. Native to Europe and Asia. Also called bird's nest, lace-flower, devil's plague, bee's-nest plant, Queen Anne's lace.

Wild Flower–Any plant, with attractive or interesting blossoms, leaves, habits, etc., which flourishes untended and uncultivated in its natural habitat.

Wild Fowl–(1) Wild birds collectively. (2) Wild, aquatic birds; sometimes only the game species, such as ducks and geese.

Wild Hay–Hay made from native or wild, uncultivated grasses and plants.

Wild Horse–An untamed and/or undomesticated horse. In the United States, the wild horses roaming the ranges have descended from the once domesticated horses. Generally, very similar to the common domestic breeds they are distinct from the wild horse of Africa and are captured and used as a source of bucking broncos in rodeos, as riding ponies, and for slaughter.

Wild Leek–*Allium tricoccum*, family Liliaceae; a bulbous, perennial herb, which is a weed pest. Cows feeding on it give milk which has an onion off-flavor. Also called ramps.

Wild Pasture–A pasture of native or volunteer grasses and plants. See Range.

Wilderness Area–An area of land providing isolation from the sights and sounds associated with modern living; having the general appearance of being unaltered from its natural state by commercial, industrial, or agricultural activities and characterized by primitive conditions of transportation and the absence of permanent habitations.

Wildfire–(1) A bacterial leaf-spot disease of tobacco and soybeans caused by ***Pseudomonas tabaci*,** family Pseudomonadaceae, characterized on tobacco by yellow lesions with white spots of dead tissue in the center. In soybeans, it is characterized by light brown to dark necrotic spots each surrounded by a yellow, sometimes indistinct, halo. Severe infestation can cause severe shedding. (2) The promiscuous, indiscriminate, or accidental burning of vegetation. It involves no responsibility for damage to property resulting from escape of the flames.

Wildflooding–Irrigation water released in a high point in a field without controlled distribution.

Wildlife–The complete fauna, exclusive of domesticated animals. In common usage, the mammals and birds of rural and wilderness areas.

Wildlife Conservation–(1) Preservation of rare or vanishing species. (2) Reasonable use and management of environmental factors so as to allow maximum breeding populations and cropping of game, furbearers, etc.

Wildling–(1) An escape; a cultivated plant which has become wild and thrives without cultivation. (2) Any wild plant or animal. (3) In forestry, a seedling produced naturally outside of a nursery.

Wilds–Country which has not been cleared and cultivated: woods, uninhabited prairies, remote areas, etc.

Willamette Valley–The drainage basin of the Willamette River and its tributaries in western Oregon, United States. It is mainly a flat alluvial plain partly covered by hills of basalt rock. Part of the land is excessively wet, but the arable soils are very productive, yielding many crops.

Willow Oak–*Quercus phellos*, family Fagaceae; a deciduous tree grown for shade and ornament in warm areas. Native to North America.

Wind–(1) The horizontal movement of air on and above the earth's surface. (2) The blast of air from a blower or fan. (3) The breathing of a horse.

Wind Direction–The pint of the compass from which wind comes, not the direction it is moving.

Wind Erodibility Equation–An equation expressing the erodibility potential of a soil due to wind. The equation is E equals $f(I,K,C,L,V)$ where E equals total erosion in tons per acre per year; f indicates that erosion is a function of the various terms; I equals soil erodibility based on texture and aggregation; K equals surface roughness; C equals climate factors (windspeed and soil moisture); L equals effect of field size (length); and V equals equivalent quantity of vegetative cover.

Wind Erosion–The removal of the soil by action of the wind. This is a serious problem in arid areas.

Wind Gap–A natural land feature: a narrow gorge, or notch in a ridge, which may have been cut by a stream, but in which the stream is no longer present.

Wind Rose–A circular diagram made for a specific location where data are available that show the relative seasonal frequency percentage of the directions from which the wind blows, including the periods of calm winds.

Wind Shadow–That portion of a scarp or slope which is protected form the direct action of the wind blowing over it.

Wind Strip Cropping–The production of crops in long, relatively narrow strips which are placed crosswise of the direction of the prevailing winds without regard to the contour of the land.

Wind Tide–A change in elevation of any surface of water due to locally unequal atmospheric pressures and transport by wind.

Wind-borne–Designating diseases, seeds, spores, etc., distributed by air movements. Also called air-borne.

Wind-firm–Designating a tree which withstands a heavy wind.

Wind-pollinated–Designating those plants which have the pollen carried by the wind instead of by insects, as the pines, grasses, etc.

Windbreak–(1) Any object which serves as an obstacle to free movement of surface winds. In forestry, trees serve such a purpose. Tree windbreaks are classified according to their general arrangement as rows and hedgerows, belts and shelterbelts of three or more rows, and groves, or in the most extensive case, forests. They may be of natural or artificial origin. (2) A vegetative barrier used to reduce or check the force of the wind. See Shelterbelt.

Windfall–(1) A tree uprooted or broken off by wind. (2) An area on which the trees have been blown down by wind. (3) A fruit that falls from a tree before the crop is harvested, usually associated with wind. Also called drop. (4) Unexpected income or tax break.

Windward–(1) Designating the point from which the wind blows. (2) Situated toward the point from which the wind blows. See Leeward.

Winter–(1) The cold season of the year in the temperate and arctic zones. (2) To spend the winter and retain viability or vigor throughout the winter period.

Winter Cover Crop–Any of several plants, such as winter rye, wheat, or oats, planted in the early autumn. It makes sufficient growth before winter comes to aid in protecting the soil from wind and water erosion. In the spring it is usually plowed under and used as green manure.

Winter Irrigation–The application of water to soil in late fall or early spring to store for subsequent use by plants.

Winter Range–The range, usually at low altitudes, used for pasture in the winter months in western United States.

Winter Sun Scald–Of trees, bark injury on the sun-exposed side of tree trunks resulting from periodic freezing of the tissues. It is characterized by loose, brown, dead bark, sometimes splitting and cracking. Eventually the bark peels away to produce an open wound or dead area on the trunk.

Witchweed–A pest under control action by the United States Department of Agriculture. Distribution in United States: North Carolina, South Carolina. Principal hosts: corn, sorghum, sugarcane, and other plants of the grass family.

Withdrawal Use–Water that is taken from a source, used, and then returned to a source for reuse.

Wold–Rolling upland, or hill country (England).

Wolf–Certain large, doglike, carnivorous animals of the genus *Canis*, especially *C. occidentalis*, the gray or timber wolf. At one time in the United States it was a serious predator of domesticated animals, particularly sheep.

Wolf Plant–A plant that, though of a species generally considered palatable, is not grazed by livestock and becomes rank in growth with much accumulation of stem and leaf material. It competes with more-palatable plants.

Wolf Tree–A tree which occupies more space than its silvicultural value warrants, thus curtailing the space of better species.

Wood Preservative–Any of several chemical compounds used to reduce decay in wood by sealing the exposed surfaces or by penetrating the wood; e.g., creosote.

Wood Rat–Any rodent of the genus *Neotoma*. Species are soft-furred rats sometimes destructive to fruit trees. Found mostly in California, United States, and southern North America.

Wood Shavings–Very thin strips of wood resulting from planing boards. Sometimes used as bedding for animals, for a plant mulch, and for kindling.

Woodchuck–A marmot, *Marmota monax*; an animal which occasionally becomes a nuisance on farms because of its burrows in pastures, orchards, and fields. Also called groundhog.

Wooded–Concerning, or designating an area full of trees.

Woodgate Rust–A rust of the Scotch pine caused by the aecial stage of a form of *Cronartium quercuum*, which produces large, round galls on the woody twigs and young branches of the plant. Occurs in New Hampshire, New York, and Michigan, United States.

Woodland Management–The management of existing woodlands and plantations that have passed the establishment stage including all measures designed to improve the quality and quantity of woodland growing stock and to maintain litter and herbaceous ground cover for soil and water conservation. It includes all such measures as planting, improvement-cutting, thinning, pruning, slash disposal, fire protection, and grazing control.

Woodland Pasture–(1) Farm woodlands used for grazing. (2) Wooded areas with grass and other grazing plants growing in open spaces among trees.

Woodlot–A tract of woodland, a part of a farm and usually only a few acres in size, which may have a single or multiple use, as a source of fire wood and logs for lumber, as a woodland pasture, and as a means of providing food and protective cover for wildlife.

Woolly–Clothed with long, matted hairs.

Woollybear–A very hairy caterpillar belonging to the family Arctiidae, the tiger moths. In folklore, this caterpillar is used to predict the severity of the coming weather.

Woolsorter's Disease–Anthrax in humans. See Anthrax.

Wooton Loco–*Astragalus wootoni*, family Leguminosae; a poisonous plant, found in southern New Mexico and western Texas, United States, and in northern Mexico. Both its green and dried foliage are poisonous. Symptoms of poisoning are dullness, irregularity of gait, lack of appetite, dragging of the feet, a solitary habit, loss of flesh, and shaggy coat. As the animal ceases to eat, it dies. also called locoweed. See Loco.

Worm–(1) Any small, soft-bodied, usually limbless animal, such as a larva, grub, maggot, earthworm, silkworm, etc. (2) To rid an animal of internal parasitic worms.

Wriggler–The larva of a mosquito.

Xeric–A soil and climatic environment with a low moisture content or supply. As applied to plants, xeric means those plants that are tolerant of very dry soil or arid conditions. Other words in the same series are hydric and mesic. See Hydric, Mesic.

Xerophyte–Plant which habitually grows where the evaporation stress is high and the water supply is low, as in desert regions, where the soil dries to the wilting coefficient at frequent intervals.

Xerophytic Vegetation–Vegetation which is characteristic of the desert regions, such as thorny brush, cacti, shrubs, and small flowering annual and perennial plants.

Yak–*Bos grunniens*, family Bovidae; a large, wild or domesticated bovine characterized by short hair on the back and long, wavy hair on the underparts and tail. It is blackish-brown, or sometimes black and white. Domesticated, the yak provides milk, hide, and hair, which is woven into fabrics, and is an important beast of burden in elevated regions of central Asia. Its tail is often used as a fly-flapper (called a chowry) and for decoration. Native to Tibet and central Asia. See Zho.

Yeast–(1) A yellowish substance composed of microscopic, unicellular fungi of family Saccharomycetaceae, which induces fermentation in juices, worts, doughs, etc. The fungi may reproduce asexually by the division and separation of the cells into two equal parts, as in the fission yeasts, or more frequently, in the budding yeast, by the budding out of small outgrowths which enlarge until they attain full size and then separate form the parent cell. Most yeasts produce enzymes which bring about the fermentation of carbohydrates to produce alcohols and carbon dioxide. The most common of such yeasts is *Saccharomyces cerevisiae* whose fementary process produces the carbon dioxide which leavens bread dough, as well as the alcohol used for beverages (beer, wines, etc.). This is the so-called bread-yeast or brewers'-yeast. In yeast inoculation of a substance, the surface or top yeast ferments it at a faster rate than the sedimentary yeasts at the bottom. (2) The asporogenous yeasts, family Torulopsidaceae, in which the production of internal spores (ascospores) is lacking. Some of them (especially in the genus *Candida*) are parasites of people or animals and may cause various diseases, such as thrush. In recent years, especially in Germany, some yeasts have been cultivated on sugar-containing derivatives of sawdust and have become valuable sources of animal feed. Also called pseudo-yeast.

Yellow Birch–*Betula alelghaniensis*, family Betulaceae; a hardy deciduous tree which is important for its timber used in veneer, hardwood distillation, railroad ties, slack-cooperage, furniture, planing-mill products, woodenware, scientific instruments, etc. Native to the Lake States and the northeastern Untied States. Also known as birch, gray birch, silver birch, and swamp birch.

Yellow Buckeye–*Aesculus octandra*, family Hippocastanaceae; a hardy, deciduous tree grown for shade and in lawns. Its leaves are poisonous to stock and its nutlike seeds have poisoned children. Symptoms of poisoning are inflammation of the mucous membranes, vomiting, depression, weakness, and paralysis. Native to North America. Also called sweet buckeye.

Yellow Cypress–See Baldcypress.

Yellows of Walnut–A zinc-deficiency disease of Persian walnuts characterized by a yellowish-green mottle between the lateral veins of the leaves; a narrow, small leaf; bunching of the leaves; a tendency of the leaves to stand erect; and death of twigs and small branches.

Yew–Any one of the several evergreen, hardy shrubs or trees of the genus *Taxus*, family Taxaceae. Species are valuable, slow-growing ornamentals and are among the most popular of all evergreens. The wood, bark, leaves, and seed are very poisonous.

Young Soil–Recently deposited or exposed soil material on which the soil profile development has begun but is in the initial stages. See Soil Horizon, Soil Profile.

Zanja–An irrigation ditch (southwestern United States).

Zanjero–An irrigator (southwestern United States).

Zero Tillage–See Conservation Tillage.

Zero Water Level–(1) Water level used as a standard for comparison. (2) Mean sea level.

Zinc–Zn; a metallic chemical element, one of the micronutrient elements in soils, essential for both plant and animal growth. Toxic to animals if ingested in too large a quantity. See Micronutrient Elements, Essential.

Zinc Nitrate–$ZnNO_3$; a highly soluble compound used in solution form as a source of zinc in liquid fertilizers.

Zinc Oxide–ZnO; a compound used as a fungicide, a seed fumigant, a source of zinc in livestock rations, and as a fertilizer.

Zinc Oxysulfate–A soluble form of zinc, also known as basic zinc sulfate. It can be added to fertilizers or applied as a spray or dust to correct zinc deficiencies. It contains 52 percent zinc. See Micronutrient Fertilizer.

Zone of Saturation–The zone of groundwater, the upper surface of which is the water table. Water from this zone feeds springs, streams, and wells. Generally, it is beneath the depth of penetration of plant roots.

Zone of Weathering–The surface crust of the earth in which weathering, or subaerial decay of rocks, takes place. Lying above the water table, it includes soil and the parent materials of soil.

Zooecology–The branch of ecology concerned with the relationships between animals and their environment.

Zoophagous Parasite–Any parasite that thrives in or on animals.

Zooplankton–(1) The animal constituents of plankton. (2) Tiny animals which drift with the currents. (3) The animal portion of the planktonic organisms. See Phytoplankton, Plankton.

Zymogenic Flora–Lower forms of plant life that reproduce in abundance in the soil following the application of large amounts of rapidly biodegradable organic matter such as animal manures, compost, and sewage effluent.

Zymurgy–A branch of applied chemistry dealing with fermentation.

Part 4

Plant Science

A Horizon—The leached upper member of a soil profile; the eluvial layer.

A.U.M.—(Abbr.) Animal unit month.

A-Harrow—A spike-tooth harrow, shaped like the capital letter *A*, which is dragged from a hitch at the apex to smooth plowed ground.

Abacterial—Free from bacteria.

Abaxial—Away from or facing away from the axis; dorsal; said of the surface or a part of a lateral organ. For example, the lower surface of a leaf is abaxial.

Aberration—In genetics, an irregularity in chromosome distribution during cell division that results in deviation from normal.

Abiotic Disease—A disease caused by an inanimate agent; a nonparasitic disease. Also called physiogenic disease; e.g., a mineral deficiency in plants or animals.

Abnormal—Deviating from that which is typical.

Aborescent—Approaching the size and habit of a tree.

Aborist—A specialist in the field of tree culture.

Aborvitae—Any plant of the genus *Thuja*, family Pinaceae; the species are valuable evergreen trees used for ornament and in the timber industry. White cedar or Thuja oil is used as a vermifuge. See Northern White Cedar.

Abrasion—(1) Wear in farm implements caused by friction, such as abrasion of a plowshare by the soil. (2) A section of skin or mucous membrane from which the surface layers have been rubbed or worn off.

Abscess—Localized collection of pus in a cavity formed by the disintegration of tissue.

Abscission—The droppng off of a leaf, fruit or flower; shedding.

Abscission Layer—A narrow, transverse band of cells at the base of leaf, flower, stalk, fruit stalk, or a portion of stem, in which the cell walls dissolve, leaving the vascular bundles for support. These eventually break, and the leaf, flower, fruit, or even a portion of the stem falls. Also called absciss layer, separation layer.

Absorb—(1) To assimilate or incorporate one substance into another, as a blotter absorbs ink. (2) In botany, to assimilate water and nutrients.

Absorption—(1) The processes by which water enters the earth. (2) The passage of digested food from the alimentary tract into the circulatory system. (3) The intake of water, gases, nutrients or other substances by plants.

Absorption Rate—The rate at which a chemical enters the body.

Abundance—The number of plants or animals per unit area.

Acacia—Any shrub or tree of the genus *Acacia*, family Leguminosae; although basically tropical and short-lived, many species are grown in southern regions of the United States as ornamentals for their showy, dense clusters of small, usually yellow flowers.

Acariasis—The condition of plants or animals infested with mites or ticks.

Acaricide—A substance, solution, or paste which kills mites or ticks.

Acarid—Any mite or tick of the family Acaridae or order Acarina.

Acarine—Any mite or tick.

Acarology—The branch of science that deals with mites and ticks.

Acaulis—Stemless.

Acceptable Daily Intake (ADI)—The daily dosage of a drug or a chemical residue that appears to present no appreciable risk to health during the entire lifetime of a human being or animal.

Accessory Fruit—A fruit or assemblage of fruits in which the conspicuous, fleshy parts are not derived from the pistil; as in the strawberry (*Fragaria*), where the soft, red, edible flesh is the enlarged receptacle and the true fruits (seeds) are embedded in its surface.

Acclimate—To become conditioned to a new climate or different growing environment, usually applied to plants. (See Hardening-off.) May also be applied to animals adjusting to a new environment.

Accrescent—Refers to plants that continue to grow after flowering.

Accretion—(1) The gradual addition of new land to old by the deposition of sediment carried by the water of a stream. (2) The process by which inorganic bodies grow larger, by the addition of fresh particles to the outside. (3) The process of illuviation of soils is usually one of the addition of minerals by accretion. (4) Increase of height, diameter, quality, and value of a tree or woods; increment; growth. (5) Increase of soil water. (6) Process of recovering land from the sea by diking and draining. See Relict.

Accumulator Plant—A plant that absorbs certain elements and accumulates them in its tissues to a much higher degree than most plants; e.g., selenium accumulators.

Acetic Acid—An organic acid, CH_3COOH, which is the chief active component of vinegar; also important as a product in lactic acid fermentation and therefore, an important constituent of flavor in many milk products.

Acetic Bacteria—The bacteria that produce acetic acid (vinegar acid) from alcohol.

Acetic Fermentation—The process by means of which acetic acid is formed from ethyl alcohol in weak solution through the action of bacterial ferments; e.g., the formation of vinegar from hard cider or from wine.

Acetobacter—A bacterium, family Nitobacteriaceae, which grows in beer and wine.

Acetone—CH_3COCH_3, a ketone; a clear, rapidly evaporating liquid which: (1) in nature is obtained from fermentation of sugar and starch and is present in diabetic urine, breath, blood. Its presence in the breath of a lactating animal (cattle, sheep, milk goats) indicates her failure to oxidize the fatty material of her feed, making her deficient in carbohydrates; (2) is commercially obtained from distillation of wood and has wide application as a solvent.

Acetylene Reduction–A technique of measuring the change from acetylene (C_2H_2) to ethylene (C_2H_4) as a measure of the amount of atmospheric nitrogen (N_2) fixation.

Achene–Any small, dry fruit having but one seed whose pericarp does not burst when the fruit is ripe.

Aciculifruticeta–Needle-leaved plant.

Aciculilignosa–Narrow-leaf sclerophyll forest and bush.

Aciculisilvae–Needle-leaved forests.

Acid–(1) A substance containing hydrogen that dissociates to form hydrogen ions when dissolved in water (or which reacts with water to form hydronium ions). (2) A substance capable of donating protons to other substances. (3) A term applied to igneous rocks having a higher percentage of silica (66%) than orthoclase. (4) Term applied to any substance with a pH less than 7.0. See Acetic Acid.

Acid Food–Food with a pH of 4.6 or below. An acid food can be safely processed for canning in a boiling-water bath for specified times. Includes most fruits, some tomatoes, and pickled vegetables. See Low-Acid Food.

Acid Hydrolysis–Decomposition or alteration of a chemical substance by acid.

Acid Organic Material–An organic material that leaves an acid residue in the growing medium, e.g., sphagnum peat.

Acid Rain (Precipitation)–Rain or other precipitation that contains a higher than normal amount of acid. The condition is caused by raindrops absorbing substances from air pollution. Acid precipitation (acid rain) is generally considered to be harmful to the environment.

Acid Soil–A soil giving an acid reaction (precisely, below pH 7.0; practically, below pH 6.6) throughout most or all of the portion occupied by roots. A soil having a preponderance of hydrogen ions over hydroxyl ions in the soil solution. In common parlance, a "sour" soil, as opposed to a "sweet" soil.

Acid-fast–Property of not being readily decolorized by acids.

Acid-forming Fertilizer–Fertilizer that increases the residual acidity of the soil, e.g., ammonium sulfate.

Acid-tolerant Crops–Crops, such as oats and corn, which will grown in moderately high acid soils.

Acidification of Soils–The process by which soils become acid. In humid regions, soils become acid because of accelerated erosion, the use of fertilizers, lime removed in harvested crops, and acid precipitation. In arid regions, soils too alkaline can be acidified by applying sulfur, lime-sulfur solution, iron sulfate, or aluminum sulfate.

Acidified Silage–Silage preserved by the addition of acid, such as commercial phosphoric acid, sulfuric acid, or hydrochloric acid, or a combination of the latter two.

Acidity–The measure of how many hydrogen ions a solution contains.

Acidophilus–Refers to organisms that grow well or exclusively in an acidic soil or medium.

Acidulation–The process of treating a material with an acid to make it more soluble. The most common acidulation process is the treatment of phosphate rock with an acid such as sulfuric acid (H_2SO_4), nitric acid (HNO_3), or phosphoric acid (H_3PO_4).

Aciniform–(1) Having the shape of a bunch of grapes. (2) Containing small grapelike kernels.

Acorn–The fruit or seed of oaks, any species or the genus *Quercus*; eaten by domesticated grazing animals, cattle, hogs, and goats, and by several kinds of wildlife, both birds and mammals, and to a limited extent by humans.

Acorn Fruit–A symptom of a virus disease, *Citrivir pertinaciae*, citrus trees, characterized by acorn-shaped fruit; as much as one-half the production of a tree may be so deformed. the fruit has very thin skin toward the stylar (away from the stem) end. Also called pink nose, blue nose.

Acorn Squash–Hubbard group. A squash that is characterized by a small, deeply fluted diamond or acorn-shaped shell.

Acquired Character–A change in character of a plant or animal, morphological or physiological, due to the environment, which is not passed on to the next generation; not a genetic change.

Acre–A unit of land measure in England and the United States which is equal to 43,560 square feet, or 1/640 of a square mile, or 160 square rods, or 4,840 square yards, or 4,047 square meters. The Scottish acre is 1.26 and the Irish acre is 1.62 times as large. One acre, as used in England and the United States, equals 0.4 hectare, 0.96 feddan (Egypt), 0.31 carreau (Haiti), 1.03 cuerdas (Puerto Rico), 0.03 caballerias (Cuba), 5.00 shih mou (China), 4.08 tan (Japan), 0.84 arpent de Paris (some sections of Canada), 0.37 dessiatine (U.S.S.R).

Acre Inch–The quantity of water necessary to cover one United States or English acre to the depth of one inch: 3,630 cubic feet or 27,154.2 United States gallons of water.

Acreage–An indefinite quantity of land; a collective number of acres.

Acreage Allotment–As established from time to time by Congress, the individual farm's share, based on its previous production, of the national acreage needed to produce sufficient supplies of a particular crop.

Acrophytia–Plant communities in alpine regions.

Acrospire–The young spiral shoot of germinating seeds.

Acrostalagmus Fungi–A fungus belonging to the family Dematiaceae. (1) *Acrostalagmus albus* Pr., a fungus which attacks brown scale on citrus. (2) *Acrostalagmus aphidum*, a fungus which attacks the citrus-inhabiting aphid, cotton aphid, green peach aphid, and the black citrus aphid.

ACS–Agricultural Cooperative Extension Service.

Actinomorphic–Regular, radially symmetrical; capable of being divided vertically and in more than one plane into two essentially equal halves, as the flower of Tulipa or Rosa.

Actinomycetes—A large group of moldlike microorganisms which give off an odor characteristic of rich earth and are the significant organisms involved in the stabilization of solid wastes by composting. They are common in the soil. Selected strains are used for the production of certain antibiotics.

Actinomycosis of Potato—See Corky Scab of Potato.

Action Threshold—The level of competition or plant cover that triggers control action. For example, the point at which one plant or type of plant outgrows another and the larger plant crowds out the smaller.

Actium—A plant-animal community on a rocky seashore.

Activator—A chemical added to a pesticide to increase its activity.

Active Agent (Active Ingredient)—That part of an insecticide, herbicide, or fungicide formula that has toxic properties to target species.

Actophilous—Refers to organisms that grow well on rocky seashores.

Actual Analysis—The composition of a material based on a laboratory chemical analysis rather than a generalized guaranteed analysis.

Acuminate—Gradually tapering to a pointed apex.

Acute—(1) Usually refers to a disease that runs a short, severe course. (2) Having the shape of an acute angle.

Acute Toxicity—The potential of a substance to cause injury or illness when given in a single dose or in multiple doses over a period of 24 hours or less. In aquatic studies, exposure to a given concentration for 96 hours or less.

Adaptability—The capability of an organism to make changes that make it more fit to its environmental conditions.

Adaptation—(1) A measure of physiological fitness of the organism to one or several conditions of its environment. (2) The structures or activities of an organism, or of one or more of its parts, which tend to fit it better for life in its environment or for particular functions. See Adjustments.

Adapted Variety—Varieties of plants that have been developed or adapted to grow in a particular area or climate.

Additivies—Materials added to food to help manufacture and preserve it and to improve nutritive value, palatability, and eye appeal. May be classified as emulsifiers, flavors, thickeners, curing agents, humectants, colors, nutrients, or as mold, yeast, or bacterial inhibitors. Amounts used in food are regulated by law.

Adherence—The act of sticking to a surface.

Adhesion—Molecular attraction which holds the surfaces of two unlike substances in contact, such as water and soil particles. See Cohesion.

Adhesions—The adherence of separate tissues or organs of the body to each other as the result of an inflammation.

Adhesive Agent—A material that acts as a cementing agent between two solids. In spraying, it causes the dispersed material to cling to leaves and fruit.

Adjustments—(1) Range management:changes in animal numbers, seasons of use, kinds or classes of animals, or management practices as warranted by specific conditions. (2) Ecological: the processes by which an organism becomes better fitted to its environment; functional, never structural. See Adaptation.

Adjuvant—(1) Any solid or liquid added to a substance, such as a pesticide or a fertilizer, to increase its effectiveness, e.g., solvents, diluents, carriers, emulsifiers, stickers, spreaders, or sometimes a pesticide to another pesticide or a fertilizer to another fertilizer. (2) A carrier for a biological that releases the biological into the bloodstream over an extended period, thus serving the function of a series of booster shots; consequently, the adjuvant helps lengthen the period of immunity provided by the biological.

Adnation—The vertical fusion of flower parts in different whorls.

Adobe Soil—Any clayey soil of the arid and semiarid regions of the western United States Soils included under the term may vary widely in different localities. Also called dobe, dobie, dobby, doby, adaubi, adabe, dogie.

Adsorption—The increased concentration of molecules or ions at a surface, including exchangeable cations and anions on soil particles. Adsorption is an advanced way of treating wastes in which activated carbon removes organic matter from waste water. It is also used in flavor control in the milk industry.

Adulterant—An impurity not allowed by law in a food, plant, animal, fertilizer, or pesticide formulation.

Adulterate—To lower the quality of a product by mixing in another substance; as to adulterate milk by adding water to it. The federal government and most states legislate against the adulteration of food, drugs, fertilizers, and pesticides.

Advective Frost—Frost produced by cold air that has moved downslope; as in a frost pocket.

Adventitious—Out of place; not in the normal, expected place. Arising from the secondary meristem. See Aerial Root.

Adventitious Buds—Buds that develop in an abnormal position and are usually caused by a wound to the plant.

Adventitious Root—A root growing from stem or branch, not from primary root tissue.

Adventive Species—Organisms which have invaded from a distance and become more-or-less naturalized.

Adverse Weather—Weather not suitable for farm/ranch operations, such as making hay.

Aeolian—Refers to soil materials which are subject to wind movement or have been moved by the wind action. See Loess.

Aerate—(1) To force a thin layer of cooled air over milk to remove odors. (2) To cause air to pass throughout or around a substance.

Aeration, Soil—The process by which air in the soil is replenished by air from the atmosphere. In a well-aerated soil, the air in the soil is similar in composition to the atmosphere above the soil. Poorly aerated soils usually contain a much higher percentage of carbon dioxide and a correspondingly lower percentage of oxygen. The rate of aeration depends largely on the size, volume, and continuity of pores in the soil.

Aerial Bulblet–A bulb capable of producing a plant, found in the leaf axis or in the place of flowers on certain plants, as in the tiger lily and onion. Also called bulblet, bood bud, bulbil, bulbel.

Aerial Dusting–Spreading insecticides, fungicides, or herbicides in the form of powder from an airplane. Commonly called crop dusting.

Aerial Fertilization–The broadcast distribution of fertilizers from aircraft.

Aerial Propagation–The production of roots on a cutting, under moss or the like, without a covering of sand or soil. See Air Layering.

Aerial Root–(1) The root of an epiphyte or air plant, which anchors the plant and, in some cases, functions in photosynthesis. (2) A root growing from the stem that is exposed to the air. Also called adventitious root.

Aerial Seeding–Broadcast seeding from aircraft, especially used in seeding rice.

Aerify–To cultivate (turfgrass) by loosening soil, removing cores, and leaving holes or cavities in the turf.

Aerobe–Bacteria or other organisms which live only in free oxygen.

Aerobic–Pertaining to organisms that grow only in the presence of oxygen, as bacteria in a properly prepared compost.

Aerobiology–The study of microorganisms carried in the air.

Aerogens–Gas-producing bacteria.

Aerology–The science of atmosphere at all elevations but sometimes limited to the upper atmosphere.

Aeroplankton–Small plants, spores, pollen grains, insects, bacteria, and small animals that are suspended in the atmosphere. See Plankton.

Aerosol–Finely atomized spray or smoke with particles ranging in size between 0.1 and 50 microns. The particles are produced by blasts of heated air, or exhaust gases, or rapid volatilization of a liquefied gas containing a nonvolatile chemical solution. Aerosols include insecticides, antibiotics, germicides, and deodorants.

Aestilignosa–Broad-leafed, summer-green forests and shrubs; deciduous.

Aestisilvae–Summer-green forests.

Aestival–Pertaining to summer.

Aestivation–(1) The manner in which parts of a plant are folded up before expansion. (2) The passing of summer in an inactive or dormant state.

Affluent–(1) A stream that flows into another stream or lake; a tributary stream. (2) A plant that flowers abundantly.

Afforestation–The establishment of trees where they never existed before, such as on a prairie.

Aflatoxin–The substance produced by some strains of the fungus *Aspergillus Flavus*, the most potent natural carcinogen yet discovered; a persistent contaminant of corn, other grains, and peanuts.

African Millet–See Pearl Millet.

African Oil Palm–*Elaeis guineensis*; a palm of the family Palmaceae, important for the African oil trade, but an ornament in the southern United States.

African Violet–*Saintpaulia ionantha*; an herbaceous plant of the family Gesneriaceae, native to Africa. Grown outside the tropics as a house plant or in greenhouses. The flowers are usually deep violet-blue; in some varieties they are streaked with white or are entirely white. Not related to the true violet.

After-ripening–The biochemical and physical changes occurring in tubers, seeds, fruits, and bulbs after harvest; necessary for germination.

Aftermath–The regrowth of range or artificial pasture forage after grazing or mowing. More commonly used to refer to grazing of forage or cropped areas after harvest where there may not always be regrowth but just crop residues.

Agave–See Lechuguilla Agave, Sacahuista.

Age Stand–The average age of the trees that compose a stand. In practice, applied to even-aged stands by obtaining the average age of representative dominant trees.

Age Tree–The number of years elapsed since the germination of the seed, or the budding of the sprout or root sucker of a tree.

Agent–One who represents another from whom he has derived authority.

Agent Orange–An herbicidal mixture of 2, 4-D and 2, 4, 5-T made by Dow Chemical Company and used by the United States military in Vietnam in the 1960s to defoliate vegetation serving as hiding places for the enemy. More than 11 million gallons (41.6 million liters) were used. An estimated 220 pounds (100 kg) of dioxin, a very potent toxin, was a contaminant in the herbicide. Although now more restricted, this combination of 2, 4-D and 2, 4, 5-T is used extensively by farmers and ranchers to kill broad-leaved plants, including poison ivy. See Dioxin.

Ager–(Latin) Field.

Agglomeration–A processing step in the granulation of fertilizers. The assembling of small particles into larger, stable granules.

Agglutination–(1) A joining, as by adhesion. (2) The gathering together of the fat globules in milk or of bacteria suspended in immune sera. (3) The sticking together of bacteria, other organisms, or cells suspended in a fluid by antibodies called agglutinins.

Aggregate Fruit–A "fruit" composed of several separate ripened ovaries of a single flower, as in blackberry (*(Rubus)*.

Aggregate Soil–Many fine particles held in a single mass or cluster. Natural soil aggregates, such as granules, blocks, or prisms, are called peds. Clods are aggregates produced by tillage or logging.

Agitation–The process of stirring or mixing in a sprayer.

Agitator–Revolving paddles which keep a liquid, powder, or gas in motion to maintain a proper mixture within a tank. Used in dairying, spraying, fertilizer drill, etc.

Agostadero–A pasture used only in the summer (southwestern United States).

Agrarian–(1) Pertaining to agriculture. (2) Pertaining to political action or movements for the benefit of farmers. See Rural Population, Urban.

Agrarian Zone–Of or pertaining to the cultivated portion of an area.

Agrestal; (Agrestial)–Uncultivated; growing wild.

Agribusiness–A term referring to the full scale of operations related to the business of agriculture. It connotes the interrelationships of farming, farm services, soil science, agronomy, land grant universities, county extension services, state and federal experiment stations, soil and water conservation services, plant and animal nutrition, plant and animal protection, transportation, finance, and marketing.

Agrichemicals–A term used to designate chemical materials used in agriculture, such as herbicides, insecticides, fungicides, and fertilizers.

Agricola–(Latin) Farmer.

Agricultural Chemistry–The branch of the science of chemistry which is concerned with the composition and transformation of the plants and animals on which the economy of the farmer rests.

Agricultural College–An educational institution devoted to study, research, and the dissemination of knowledge in agronomy, horticulture, animal husbandry, agricultural economics, etc. The U.S.D.A. works in conjunction with most agricultural colleges (sometimes called a school or college of agriculture within a university).

Agricultural Gypsum–See Gypsum.

Agricultural Land–All the land devoted to raising crops and livestock, including farmstead, roadways, drainage and irrigation ditches, ponds, water supply, cropland, and grazing land of every kind. (The term is not strictly synonymous with land in farms, cropland, pasture land, land suitable for crops or land suitable for farming).

Agricultural Lime–See Calcium Carbonate, Lime.

Agricultural Pollution–The liquid and solid wastes from all types of farming, including runoff from pesticides, fertilizers, and feedlots; erosion and dust from plowing; animal manure and carcasses; and crop residues and debris. Old cars and trucks are a part of aesthetic agricultural pollution.

Agricultural Slag–(1) A term applied to a fused silicate whose calcium and magnesium contents are capable of neutralizing soil acidity and which is sufficiently fine to react readily in the soil. (2) A low-grade agricultural lime. See Lime.

Agricultural Sulfur–A coarsely ground mineral which increases the acidity of soil or corrects sulfur deficiency. Also called brimstone, flowers of sulfur. See Sulfur.

Agriculture–The broad industry engaged in the production of plants and animals for food and fiber, the provision of agricultural supplies and services, and the processing, marketing, and distribution of agricultural products.

Agriculturist–A person engaged in the production of food and/or fiber; also ancillaries such as teachers of agriculture, farm editors, researchers, etc.

Agrobiology–The science of plant nutrition and growth, crop production, and soil management.

Agrology–That branch of agricultural science dealing with the origin, analysis, and classification of soils in relation to crop production. (A word used widely in Canada but not in the United States.) See Agronomy.

Agronomist–A specialist in soil and crop sciences.

Agronomy–The specialization of agriculture concerned with the theory and practice of field-crop production and soil management. The scientific management of land. See Agrology.

Agropyron Mosaic–A viral disease of wheatgrass (*Agropyron* spp.) characterized by leaf mottling and striping, necrosis, and stunting.

Agrostology–The branch of systematic botany dealing with grasses and grasslike plants.

Agstone–A term widely used to denote agricultural limestone, in contrast to limestone for making cement or used as building stones. See Lime.

Air Drainage–The flow of cold air down a slope and warm air up a slope that is caused by gravity rather than the wind.

Air Drill–A planter for small grains that uses an air stream created by a high-speed fan to carry seed from fluted seed-metering feed cups under the seed hopper through hoses to drop into the furrow behind double-disk or hoe-type openers.

Air Layering–A method of propagation by rooting branches of woody plants. An incision is made in the stem to be rooted, and sphagnum moss is wrapped around the wounded stem. The wounded area is moistened and covered tightly with an oil paper, cheesecloth, wire screen, or plastic film, and left until sufficient roots have established themselves in the sphagnum moss. Then the branch is cut away from the main stem and transplanted. Also called Chinese layerage.

Air Plowing–Aeration by the pumping of air into the lower, oxygen-depleted layer of a water body to encourage the mixing of deep and surface waters and/or the oxidation of bottom sediments.

Air Porosity–The proportion of the bulk volume of soil that is filled with air at any given time or under a given condition, such as a specified moisture condition.

Air-Dry–To dry or dehumidify forage, hay, wood, etc., by means of natural air movement.

Air-slaked Lime–Burned limestone that has been allowed to weather. See Lime.

Airpotato Yam–*Dioscorea bulbifera;* an East Indian, tender, climbing, herbaceous vine, of the family Dioscoreaceae, grown for ornament and for its 8- to 12-inch long edible tubers, which are formed in the axils of the leaves.

Airshed–A term, now little used, denoting a geographical area the whole of which, because of topography, meteorology, and climate, shares the same air.

Ajowan—Fruit of the Eurasian plant *Carum copticum*, family Ammiaceae; i.e., valued as a seasoning and for medicinal properties. Antiseptic ajowan oil is made from it; also called Javanese seed, ajava.

Akala—*Rubus macraci*, family Rosaceae; a shrub grown in the Hawaiian Islands for its large, red, edible fruits.

Alang Grass—*Imperata cylindrica*; a grass used in paper making and thatching roofs. Grown in southern Europe.

Alaska Spruce Beetle—*Dendroctonus borealis*, family Scolytidae; a beetle which attacks the white and Englemann spruce in Alaska and northwestern Canada. It bores into the inner bark, and its larvae bore into the wood of the tree.

Alate—Winged; having wings.

Alba, Albidus, Albus—Terms meaning white.

Albedo—(Latin; albedo, whiteness.) The percentage of the sunlight reflected from a surface in relation to the total amount of light falling on that surface. Typical values are: snow, 55 to 80 percent; grass, 25 percent; forest, 10 to 20 percent.

Albino—(1) A human or animal with pink eyes and white skin and hair, caused by a congenital deficiency of pigment in the skin. (2) In botany, plants with a chlorophyll deficiency.

Albino Cherry—A viral disease of sweet cherry found in the Rogue River Valley, Oregon. The disease spreads rapidly and kills trees. The foliage turns bronze-green; the fruit turns white and fails to mature.

Albizzia—(1) A genus of trees belonging to the family Mimosaceae. These mimosa trees have twice-pinnated leaves, and bear clusters of pink and white flowers. Native to southern Europe.

Alburnum—The part of a tree trunk between the bark and the wood, consisting of the cambium and the phloem cells. The alburnum of some species of trees; e.g., the Ponderosa Pine (*Pinus ponderosa*) of the Northwest was eaten by American Indians during times of famine. See Sapwood.

Alcohol—The family name of a group of organic chemical compounds composed of carbon, hydrogen, and oxygen; a series of molecules that vary in chain length and are composed of a hydrocarbon plus a hydroxyl group, $CH_2\text{-}(CH_2)N\text{-}OH$; includes methanol, ethanol, isopropyl alcohol, and others.

Alcohol Slime Flux—A white, frothy seepage from the bark or sapwood near the base of a tree trunk. Rich in starches, sugars, and proteins. Over long periods, it is fatal to the tree. See Brown Slime Flux, Slime Flux.

Alcoholic Fermentation—The transformation of simple hexose sugars, especially glucose, into alcohol and carbon dioxide. Useful as a method of preservation.

Aldehydes—Any of a class of highly reactive organic chemical compounds obtained by oxidation of primary alcohols, characterized by the common group CHO, and used in the manufacture of resins, dyes, and organic acids.

Alder—Any tree or shrub of the genus *Alnus*, of the family Betulaceae, some species are called tag alders. The lumber from alders is used in cabinet making.

Alder Buckthorn—*Rhamnus frangula*; a shrub cultivated for its lustrous green leaves, which turn bright yellow in autumn.

Alder Flea Beetle—*Altica ambiens*, family Chrysomelidae; it feeds on the leaves of alder, willow, and poplar trees.

Alder Witches' Broom—A European fungal disease of alders. *Taphrina epiphylla*, family Taphrinaceae; it causes profuse branching.

Aldrin—A white crystalline, chlorinated hydrocarbon insecticide.

Aleppo Grass—See Johnsongrass.

Aletophyte—A weed growing in a mesic habitat.

Aleukia—(1) Absence of leukocytes in the blood. (2) Alimentary toxic aleukia. Mycotoxicosis in people and some animals caused by ingesting one or more of the following genera of fungi that develop on damp grain crops: *Actinomyces, Alternaria, Cladosporium, Fusarium, Penicillium, Piptocephalis, Rhizopus, Thammidium, Tricoderma, Trichothecium,* and *Verticillium*.

Aleurone—Protein grains found in the endosperm of ripe seeds.

Aleurone Layer—In wheat and corn (maize), the outer differentiated layer of cells of the endosperm.

Alfalfa—*Medicago sativa*; a perennial, leguminous herb of the family Leguminosae; a native of central Asia. Used as forage, green manure, and hay. Also called lucerne, purple medic, purple alfalfa.

Alfalfa Caterpillar—Colias philodice eurytheme, family Pieridae; an insect pest in the southwestern United States, which feeds principally on leaves.

Alfalfa Dwarf—A viral disease of alfalfa characterized by dwarfing of the plant.

Alfalfa Gall Midge—*Asphordylia websteri*, family Itonididae; an insect pest of alfalfa in Arizona and New Mexico which attacks the floral head, causing the seed pods to become swollen.

Alfalfa Group—The group of *Rhizobium* bacteria used to inoculate seed of alfalfa and the clovers.

Alfalfa Leaf Meal—A stock feed consisting chiefly of ground, leafy materials separated from alfalfa hay or meal. It is reasonably free from other crop plants and weeds, and does not contain more than 18 percent of crude fiber.

Alfalfa Meal—A stock feed of ground alfalfa hay. It is reasonably free from other crop plants and weeds, and does not contain more than 33 percent of crude fiber.

Alfalfa Root Rot—*Rhizoctonia crocorum*; the imperfect stage of the more rarely found *Helicobasidium purpureum*, family Auriculariaceae; a fungal, killing disease. The affected plants become yellow, wilt, and the tops and roots die.

Alfalfa Tea—Made from alfalfa leaves; credited with medicinal value and possible a slight tonic effect. Used in China.

Alfalfa Weevil–*Hypera postica*, family Curculionidae; a common insect pest in western and many eastern states of the United States. It destroys first cuttings and delays the subsequent cuttings.

Alfalfa Wilt–*Corynebacterium insidiosum*, family Bacteriaceae; a disease of alfalfa in which diseased plants are stunted and have excessively weak stems; the leaves curl up, turn yellow, and die; the taproot has a wet, yellow zone under the bark. Causes serious damage in the United States. Other organisms, such as *Fusarium* spp, and *Cylindrocarpon* spp; also cause wilt.

Alfalfa Yellow Leaf Blotch–*Pseudopeziza jonesi* (=*Pyrenopeziza medicaginis*), family Mollisiaceae; a worldwide, fungal disease which causes defoliation.

Alfalfa Yellows–A boron deficiency disease characterized by pale leaflets or bronzed leaves with green midribs, or, in some cases, a stopping of terminal growth. The margins of the leaves may dry up and shrivel. Also called alfalfa mosaic.

Alfileria–*Erodium cicutarium*; an annual herb, of the family Geraniaceae; native to southern Europe; an important forage plant in California and the southwest United States. Also called storksbill, pin grass, filaree, alfilaria, alfillerilla, red-stem filaree.

Alfisols–A soil order with gray to brown surface horizons, medium to high supply of bases, and B horizons of alluvial clay accumulation. These soils form mostly under forest or savanna vegetation in climates with slight to pronounced seasonal moisture deficit.

Algae–(singular, alga) Comparatively simple plants containing photosynthetic pigments. A majority are aquatic, and many are microscopic. They grow in sunlit waters. They are food for fish and small aquatic animals and, like all green plants, put oxygen in the water when carrying on photosynthesis. Although some forms of algae are necessary and desirable, excessive concentrations tend to discolor water and cause objectionable tastes and odors, severely limiting all uses of the water.

Algodon–(Spanish) Cotton.

Algology–The study of algae.

Alkali–(1) A chemical compound of oxygen and hydrogen with one element, such as sodium, potassium, calcium, magnesium, or the ammonium radical, capable of neutralizing acids. (2) A general term denoting salts of sodium, calcium, potassium, and magnesium which injure plant growth. (3) A general term for caustic soda, sodium hydroxide, caustic potash, sodium carbonate, etc., all cleansing agents in food processing plants.

Alkali Chlorosis–A yellowing or blanching of plant leaves caused by excessive alkali.

Alkali Disease–Selenium poisoning of animals caused by grazing on plants containing excessive amounts of selenium. Characterized by emaciation, loss of hair (mane and tail among horses), deformed hoofs, and blind staggers.

Alkali Muhly–*Muhlenbergia asperifolia*, family Gramineaae; a native grass of the northern and western United States; sometimes an undesirable weed pest in grazing lands. Frequently referred to as *Sporobolus asperifolius*. Also called fine-top salt grass, scratchgrass.

Alkali Sacaton–*Sporobolus airoides*, a grass native to the United States, of the family Gramineae; used in reseeding grazing lands. Also called alkali dropseed, tussock grass.

Alkali Sida–*Sida hederacea*, of the family Malvaceae; a weed pest.

Alkali Soil–A soil containing alkali salts, usually sodium carbonate (with a pH value of 8.5 and higher). The term frequently includes both alkali and saline soil as here defined. White alkali is applied to saline soil in some localities, and black alkali to alkali soils. See Alkaline Soil.

Alkaline–A chemical term referring to basic reaction where the pH reading is above 7, as distinguished from acidic reaction where the pH reading is below 7.

Alkaline Soil–A soil that gives an alkaline reaction, precisely, a reading above pH 7.0, but in practice, readings as high as pH 7.3 may be considered neutral. In humid regions alkalinity most commonly comes from calcium and magnesium carbonates; in drier regions nitrates and salts of sodium and potassium as well as calcium and magnesium may be the source. See Acid Soil, Alkali Soil, pH.

Alkaloids–Substances found in plants, many having powerful pharmacologic action, and characterized by content of nitrogen and the property of combining with acids to form salts. Alkaloids may be beneficial (atrophine) or toxic (from *Senecio*spp).

All-aged–Applied to a stand of trees in which trees of all ages up to and including those of the felling age are found.

Allegheny Blackberry–Rubus allegheniensis; an erect blackberry, of the family Rosaceae; common in the northern United States; the source of many cultivated varieties of blackberries.

Allele–(1) The alternative forms of genes having the same place in homologous chromosomes which influence the development of alternative traits or characters. (2) A pair of Mendelian genes at the same locus as a pair of homologous chromosomes. Also called allelomorph.

Allelopathy–The harmful influence of one living plant on another living plant by the secretion of a toxic substance. Examples are juglone secreted by black walnut roots and leaves and an unidentified toxin secreted by certain species of sedges that depresses rice plants.

Allergy–An exaggerated susceptibility to a substance harmless to most members of the same species.

Allethrin–A synthetic insecticide similar to pyrenthrins.

Alliaceous–Having the smell or taste of onion or garlic.

Allicin–An antibiotic derived from garlic.

Allite–Soils characterized by a silica-alumina ratio less than 2:1 in the hydrochloric acid extract. A soil from which silica has been removed during formation, and in whose clay fractions aluminum is in the form of AL_2O_3, as in bauxite and laterite.

Allogamy–Cross-fertilization.

Allogenic–Succession in which a plant community is replaced by another as a result of a change in the environment due to nonbiotic factors, e.g., improved drainage or irrigation.

Allohexaploid–Having six genomes or basic sets of chromosomes, with one or more sets derived from a species different from that of the other sets. In common wheat (*Triticum aestivum*), for instance, two each of three different chromosome sets are present, derived from three different parent species.

Allopolyploid–A polyploid containing genetically different chromosome sets, e.g., from two or more species.

Allosomes–Chromosomes distinguished by peculiarities of behavior or sometimes by a difference in size or shape.

Allotment–(1) The number of acres, etc., a producer is allowed to grow of a particular crop under a government program. (2) An area designated for the use of a prescribed number of cattle or sheep, or for common use of both.

Allspice–*Pimenta officinalis*; a tropical American tree, of the family Myrtaceae; so named because its spice resembles the favor of a mixture of cinnamon, nutmeg, and cloves. Grown in the United States only in Florida, and there as an ornament. Also called pimento allspice.

Alluvial–Pertaining to material that is transported and deposited by running water.

Alluvial Community–A boundary zone between water and land which is subject to inundation.

Alluvial Cone–A delta-form feature composed of soil and rock detritus, deposited as storm wash at the mouths of streams. Similar to alluvial fan, except that the angle of slope is greater.

Alluvial Fan–Soil and rock fragments moved by swiftly flowing water and deposited at the mouth of a stream in the form of fans or deltas. Alluvial fans are most common in arid regions at the bases of mountain slopes. Several fans may coalesce to form terraces and piedmont plains. The agricultural value of the soils in alluvial fans varies.

Alluvial Soil–Soil developed from transported and relatively recent water-deposited material (alluvium), characterized by little or no modification of the original material by soil-forming processes.

Alluvio Maris–Soil formed by the washing-up of the sea; sand or any other material adjoining land near the sea.

Alluvion–Land added by accretion, as that built up by deposition from sea, lake, or river, or that left by recession of water or by the natural shifting of the channel of a river. It always belongs to the owner of the land to which it is an accretion. The word is often erroneously used as the equivalent of alluvium.

Alluvium–Matter transported and deposited by streams; usually composed of mixed inorganic (mineral) and organic particles. The word most commonly refers to the recent deposits on the flood plains; more inclusively it embraces deposits on second bottoms or river valley terraces, with the detritus spread out as fans. See Alluvial Fan.

Almond–*Prunus amygdalus (= Amygdalus communis)*; an Asiatic, deciduous, small-nut tree, of the family Rosaceae; cultivated for its popular edible kernel nut, the almond of commerce.

Almond Cherry–*Prunus gladulosa;* an Asiatic, deciduous shrub, of the family Rosaceae; grown for its clusters of white or pinkish-white flowers. Also called flowering almond.

Aloe Vera–Family Liliaceae; a succulent herb grown in a pot or tub as a specimen plant. The dehydrated sap of the leaves of ***Aloe vera*** and several other species of the genus ***Aloe*** is called bitter aloe. It has a medicinal value as a tonic, a purgative, and an agent promoting menstrual discharge. Native to the Mediterranean region. Also called Barbados aloe.

Alpestrine–Denoting plants growing on mountains below the timberline.

Alpine Fescue–*Festuca ovina* var. *brachyphylla,* family Gramineae; a variety of sheep fescue, found from the Arctic to California, in the Rocky Mountains to New Mexico, and in the higher mountains of New England and New York.

Alpine Fir–*Abies lasiocarpa;* an ornamental, pyramid-shaped evergreen tree, of the family Pinaceae, of western America; also called Arizona fir, mountain balsam, Rocky Mountain fir.

Alpine Hemlock–*Tsuga mertensiana,* family Pinaceae; a tall evergreen tree valued for its timber. Found from Alaska to Montana, Idaho, and California.

Alpine Meadow Grassland–Summer pasturage (meadows or parks) near or above timberline.

Alpine Timothy–*Phleum alpinum*, family Gramineae; a smaller grass than cultivated timothy, common in alpine meadows from Alaska to northern New England and northern Michigan.

Alsike Clover–A short-lived perennial clover, ***Trifolium hybridum,*** used in much the same manner as red clover. Suitable for soils that are too wet, cold, or acid for red clover. See Red Clover.

Alter–(1) To neuter by castrating or spaying. Also called cut, geld, emasculate. (2) To prune. (3) To remove the comb from a beehive.

Alternate–Arranged singly at different heights and on different sides of the axis or stem, as alternate leaves or branches.

Alternate Host–A plant or animal upon which a disease organism exists for only a part of its life cycle; e.g., cedar and apple for rust.

Alternation of Generations–Reproduction in which common characteristics are found only in every second generation, e.g., one generation reproduces sexually and the next asexually.

Alternifolius–Alternate-leaved.

Altherbosa–Tall herbage.

Alum–See Aluminum Sulfate.

Aluminum–A widely distributed element, commonly found as a silicate in various clays and rocks. While aluminum may be essential to the growth of some plants, the amount required, if any, is very small. The supply in all soils is abundant. Some acid soils contain sufficient aluminum in soluble or exchangeable form to kill certain plants. See Aluminum Sulfate.

Aluminum Dross–A by-product of refining aluminum metal. It consists of aluminum oxide, aluminum nitride, salt, and various other impurities. The N is slowly available in the soil. Dross is also relatively high in magnesium, copper, manganese, and zinc contents.

Aluminum Phosphate–($AlPO_4$) Occurs in quantity in the colloidal or soft phosphates found in Florida. It also occurs in small quantities in all grades of phosphate rock. It is insoluble in water and its phosphorus content is but slightly available in the soil unless treated with an acid. Much aluminum phosphate in phosphate rock is objectionable because of the tendency of the superphosphate made from such rock to become moist and sticky.

Aluminum Sulfate–A mineral soluble in water and usually made by treating bauxite with sulfuric acid. It is sometimes applied to soils, especially in the West, to make them less alkaline. In the East, it is used to produce an acid condition for such plants as rhododendrons, azaleas, camellias, and blueberries. Also used in tanning leather, sizing paper, and purifying water.

Alunite–A hydrated sulfate of aluminum and potassium mineral. There are large deposits in Utah and other western states. The potassium is insoluble in water, but is rendered soluble by roasting. Roasted material marketed in the West averages about 6.5 percent K_2O.

Alyce Clover–*Alysicarpus vaginalis*; Asiatic, tropical, annual herb, of the family Leguminosae; grown for forage, hay, and green manure.

Ambient Air–The air which surrounds an object. See Atmosphere.

Ambosexous–Bisexual.

Amendment–Any material other than fertilizers such as lime, manure, or sewage sludge that is worked into the soil to make it more productive.

Ament–A catkin or dry, scaly spike, usually unisexual, such as the inflorescence of willows, birches, etc., and the staminate inflorescence in hickories.

American Beech–*Fagus grandifolia*, a hardy, deciduous tree, of the family Fagaceae; cultivated ornamentally and used commercially for hardwood distillation, crossties, pulp, slack-cooperage, veneer, etc.

American Bittersweet–*Celastrus scandens;* a woody vine of the family Celastraceae produces orange berries that remain on the vine throughout the winter; poisonous to livestock. Used as a winter decoration indoors. Found from eastern Canada to South Dakota and from North Carolina to New Mexico. Also called false bittersweet, waxswork, stafftree, climbing bittersweet.

American Chestnut–*Castanca dentata;* a tree of the family Fagaceae, now nearly exterminated by chestnut blight; once found from Maine to Alabama, west to Michigan. It bears a sweet, edible nut, and its wood is used in making tannin, a semichemical pulp, poles, fence posts, crossties, slack-cooperage, and furniture. Also called chestnut.

American Elder–*Sambucus canadensis;* a shrub of the family Caprifoliaceae grown for its flowers and fruit, which are used in wine and jelly. The leaves, opening buds, and young shoots are lethal to cattle or sheep. Also called elderberry, sweet elder.

American Elm–*Ulmus americana;* a beautiful, long-lived deciduous tree of the family Ulmaceae; a popular street tree and used in shelter belts. The trees are subject to Dutch Elm disease. Used in slack-cooperage, crates, veneer, furniture, and wooden ware. Also called white elm, water elm.

American Holly–*Ilex opaca*; an evergreen tree of the family Aquifoliaceae. A popular Christmas decoration; used in inlay wood, engravings, scrollwork, carvings, scales, rules, piano keys, and other wooden ware. Also called white holly.

American Hophornbeam–*Ostrya virginiana;* a North American deciduous tree of the family Betulaceae; grown for its fruiting clusters. Also called ironwood.

American Hornbeam–*Carpinus caroliniana;* a native, small, deciduous tree of the family Betulaceae; cultivated for ornament. Slow-growing but hardy in shady and protected places. Used in handles, vehicle parts, and fuel wood. Also called bluebeech, ironwood, water beech.

American Mistletoe–Any epiphytic plant of the genus **Phoradendron**, especially the American Christmas mistletoe.

Amino Acids–Organic substances from which organisms build proteins, or the end product of protein decomposition.

Amitosis–Cell division without formation and splitting of chromosomes. See Mitosis.

Ammonia–NH_3; a chemical compound composed of 82.25 percent nitrogen and 17.75 percent hydrogen. At ordinary temperatures, it is a colorless, pungent gas about one-half as heavy as air. Liquid ammonia is used as a fertilizer. Sometimes added to livestock feed to increase nutritional value.

Ammonia Liquor–NH_4OH, aqua ammonia; ammonia and water combine readily to form ammonium hydroxide (NH_4OH). Aqua ammonia made from anhydrous, ammonia usually contains from 27 to 30 percent NH_3 (22.2 to 24.7% N). The solution is strongly alkaline and unites with strong acids to form ammonium fertilizer salts; for example, with phosphoric acid (H_3PO_4) to form ammonium phosphates; with sulfuric acid (H_2SO_4) to form ammonium sulfate; and with nitric acid (HNO_3) to form ammonium nitrate.

Ammonia Oxidation–When ammonia (NH_3) and air are mixed and passed through a platinum alloy gauze, the ammonia burns, combines with the oxygen of the air, and forms nitric oxide (NO). This oxide is converted to nitrogen dioxide (NO_2), which is absorbed in water to form nitric acid (HNO_3). This process is used commercially not only to manufacture ammonium nitrate fertilizer, but also in the chamber process for the manufacture of sulfuric acid.

Ammonia Volatilization–The loss of gaseous ammonia to the atmosphere. This occurs during the application of urea fertilizer, anhydrous fertilizer, or any ammonia salt under these conditions: (1) when applied on the soil surface, (2) when applied on alkaline soil, or (3) when applied under high humidity and high temperatures.

Ammoniated Superphosphate–When superphosphate of any grade is brought into contact with free ammonia (NH_3), the ammonia immediately reacts with any free acid and acidic phosphates present. Gypsum, when present, also enters the reactions, and thus the fertilizers, dicalcium phosphate, monoammonium phosphate, and ammonium sulfate are formed.

Ammonification–The formation of ammonia in soil by microorganisms.

Ammonium Chloride–(NH_4Cl) The ammonium salt of hydrochloric acid. Similar in fertilizer properties to ammonium sulfate, but more expensive and may acidify the soil more. Also called sal-ammoniac, muriate of ammonia.

Ammonium Nitrate–An ammonium salt with a nitrate radical, excellent as a source of nitrogen fertilizer. Made in large quantities from the atmospheric nitrogen and contains 30 to 35 percent nitrogen. Dangerous to work with in the pure state because of its explosive nature.

Ammonium Nitrate Limestone–A fertilizer mixture of ammonium nitrate and finely pulverized calcic or dolomitic limestone in a ratio of 3:2 prepared in pellet or prilled form suitable for mixing or direct application. It is not explosive and contains 20.57% N.

Ammonium Phosphate Nitrate–A fertilizer mixture of ammonium phosphate and ammonium nitrate. It is produced by ammoniating the solution separated from phosphate rock that has been acidulated with an excess of nitric acid. The average composition of the fertilizer is 27% N and 15% P_2O_5.

Ammonium Phosphate Sulfate–A double salt of ammonium phosphate and ammonium sulfate or a mixture of these two salts. When first marketed in 1923 this material was called Ammophos-B. It is guaranteed to contain 16 percent N and 20 percent available P_2O_5. It contains 45 percent ammonium sulfate. The 13-39-0 grade contains about 20 percent ammonium sulfate.

Ammonium Phosphates–Both monoammonium phosphate (MAP) and diammonium phosphate (DAP) are valuable fertilizers. See also Ammonium Polyphosphate (APP), Monoammonium Phosphate ($NH_4H_2PO_4$) (MAP). Pure monoammonium phosphate contains 12.17 percent nitrogen and 61.71 percent phosphoric oxide. Some fertilizer materials containing 12 percent N and 61 percent available P_2O_5 are on the market for the manufacture of liquid fertilizers and starter solutions. The crude salt, originally sold under the trade name Ammophos-A (11-48-0), is made by neutralizing crude phosphoric acid with ammonia and evaporating the solution.

Ammonium Polyphosphate (APP)–Any ammonium salt of a polyphosphoric acid such as triammonium pyrophosphate [$(NH_4)_3HP_2O_7$] and pentamonium tripolyphosphate [$(NH_4)_5P_3O_{10}$]. Fertilizer grades of ammonium polyphosphate generally are phosphate mixtures containing a substantial proportion of one or more ammonium polyphosphates and can be made by ammoniation of superphosphoric acid or by thermal dehydration of ammonium orthophosphate. Aqueous solutions of ammonium polyphosphates are used widely in the production of liquid fertilizers. The usual grades of these solutions are 10-34-0 and 11-37-0. A 12-44-0 is also available. Granular 15-62-0, made by ammoniating superphosphoric acid, is suitable for direct application, bulk blending, or as an intermediate in producing liquid fertilizers.

Ammonium Polysulfide (APS)–A combination fertilizer and soil acidifier containing 20% N and 36% S.

Ammonium Sulfate Nitrate–Formerly imported under the trade name saltpeter, this double salt, less hygroscopic than ammonium nitrate alone, is produced in many modern granulated fertilizers when hot solutions or slurries are mixed, or formed in processing, and dried. It contains 26% N.

Ammonium Sulfate (Sulfate of Ammonia)–[$(NH_4)_2SO_4$]; a white or grayish crystalline salt made by neutralizing 30 to 50 percent sulfuric acid with ammonia in a saturator. Ammonium sulfate is ammonia in a saturator. Ammonium sulfate is an ingredient of most mixed fertilizers. It is also widely used as a separate fertilizer in the western part of the country and on rice. It is seldom used on acid soils, because of its high physiological acidity (See Acidity). Where sulfur content in the soil is low it is valuable for its sulfur content also.

Ammophilous–Denoting plants or insects in sand.

Amoeba–A genus of unicellular protozoan organisms of microscopic size, existing in nature in large numbers; many live as parasites; some species are pathogenic to humans.

Amorphophyte–Any plant which bears irregular flowers.

Ampherotoky–Production of both male and female from a parthenogenic ovum, seed, or spore.

Amphicarpogenous–Fruit, such as the peanut, which grows below ground before maturity.

Amphicarpus–Bearing two varieties of fruit which may mature at the same time.

Amphigean–Native of both the Old and New Worlds.

Amphiphyte–A plant growing in the border zone of wet land and water, with amphibious characteristics.

Amphispermous–A plant, such as grass, in which the pericarp tightly encloses the seed.

Amphivorous–Omnivorous.

Amphoteric–Capable of reacting as either an acid or a base, as casein.

Anabatic Wind–Local breezes caused by heating of the land surface by the sun creating warmer air currents, which rise, moving up valleys and up hills. Upon cooling, the winds move down valleys and hills. Anabatic winds are very important in creating or reducing "frost pockets," which often kill fruit buds in the spring.

Anabolism–Constructive process by which simple substances are converted by living cells into more-complex compounds. See Catabolism.

Anacardiaceae–The cashew or sumac family, which includes approximately sixty tree and shrub genera of which many are of economic value; e.g., the mango.

Anaerobe–Organisms (usually bacteria) which live and multiply without free oxygen.

Anaerobic–(1) Living or active in the absence of free oxygen. (2) Pertaining to or induced by organisms that can live in the absence of free oxygen.

Anaerobic Bacteria–Bacteria not requiring the presence of free or dissolved oxygen for metabolism. Strict anaerobes are hindered or completely blocked by the presence of dissolved oxygen and sometimes by the presence of highly oxidized substances, such as sodium nitrates, nitrites, and, perhaps, sulfates. End-product gases include methane and hydrogen sulfide.

Anaerobic Decomposition–Reduction of the net energy level and change in chemical composition of organic matter caused by microorganisms in an anaerobic environment.

Anagreeta–A term used in the United States in Revolutionary or pre-Revolutionary times for corn gathered before maturity and dried in an oven or hot sun.

Analog Food–A food product that resembles, and is a substitute for, a traditional farm food product (e.g., a steak made out of soybean meal, with a plastic bone).

Analogous–Refers to an organ of one organism that corresponds in function to an organ of another animal or plant but which is not homologous. See Homologous.

Analysis–The percentage composition of fertilizers, feeds, etc., as determined by chemical analysis, expressed in terms specified by law.

Anastomosing–(1) Of a stream; branching, interlacing, intercommunicating, thereby producing a netlike or braided appearance. (2) Netted; intervened; said of leaves marked by cross veins forming a network; sometimes the vein branches meet only at the margin.

Anatomy–The branch of biology that deals with the structure of organisms.

Ancestor–An individual from whom an animal or person is descended.

Anchored Dune–Sand dune stabilized by growth of vegetation.

Andisols–The eleventh soil order in the United States system of soil taxonomy approved in 1989. It consists of a black mineral soil developed from volcanic ejecta including 5% glass.

Androecium–The mail element, the stamens, whether one or many, as a unit of the flower.

Androgametophore–A plant having male sexual organs.

Androgenesis–Development of offspring with the paternal chromosomes only.

Androgynous–In botany, a plant having stamens and pistils.

Androperianth–The fusion of perianth and stamen.

Anemochore–(anemochorous) An ecological term which denotes a plant, such as milkweed, with seeds distributed by wind.

Anemophilous–Pertaining to plants pollinated by wind.

Anemoscope–Any device used to indicate and/or record wind direction.

Aneroid Barometer–A device that measures air pressure by its effect on the thin sides of a partially evacuated hollow cylinder.

Aneuploid–An organism or cell which has a chromosome number other than an exact multiple of the monoploid or basic number; i.e., Hyperploids=higher; hypoploid=lower.

Angelica–Any plant of the genus *Angelica*, family Umbelliferae; usually aromatic herbs; livestock forage throughout the western United States; the garden angelica, *A.archangelica*, was formerly blanched and eaten like celery; the tender stalks were cut for candying.

Angiosperm–The subdivision of Spermatophytes (seed plants) in which seeds are produced within the ovary; includes monocotyledons and dicotyledons. See Gymnosperms.

Angleworm–See Earthworm.

Angleworm Cast–The aboveground accumulation of soil caused by the underground activity of the earthworm.

Angola Grass–See Paragrass.

Angular Leaf Spot–An infectious disease of cucumbers caused by the bacterium *Pseudomonas lachrymans*, in which the leaves look watersoaked and later turn brown.

Angular Winged Katydid–*Microcentrum retinerve*, family Tettigoniidae; a long-horned grasshopper that feeds on many plants.

Angustifolius–Narrow-leaved.

Anhydride–An oxide that will react with water to form the corresponding acid or base; e.g., P_2O_5 is the anhydride of H_3PO_4; CaO is the anhydride of $Ca(OH)_2$.

Anhydrite–($CaSO_4$) Anhydrous calcium sulfate; used for the same purposes as land plaster or gypsum, i.e., as a soil amendment; also used in India and the United Kingdom for making sulfuric acid.

Anhydrous–A compound that does not contain water either absorbed on its surface or as water of crystallization; e.g., anhydrous ammonia.

Anhydrous Ammonia–A gas used as a chemical fertilizer containing approximately 82 percent nitrogen; normally sold as liquid under high pressure. Main advantages are low cost and high concentration, thus less handling. Changes to a gas when not under pressure, so it must be injected into soil and sealed.

Anion–The ion solution carrying one or more negative electrical charges depending on its valence or combining power with positively charge cations; e.g., NO_3, H_2PO_4, SO_{42}. Anions and cations are always present in the liquid phase of fertilizers; an ion carrying a negative charge of electricity. See Ion.

Anise–*Pimpinella anisum*; an annual herb, family Umbelliferae; grown for its aromatic seed, which causes the expulsion of stomach and intestinal gas. Native to eastern Mediterranean lands.

Ankee–Barnyard grass.

Annual Bearing–Producing fruit each year.

Annual Bluegrass–*Poa annua*, family Gramineae; native to Eurasia; an annual grass which may become a weed pest in turf; also called annual meadow grass, early meadow grass, low spear grass, annual poa, winter grass, goose grass.

Annual Layer–Sedimentary layer deposited or presumed to have been deposited during the course of a year; e.g., glacial varve.

Annual Precipitation–The water, usually expressed in inches of depth, deposited on the earth in one year by rain, snow, dew, hail, fog, mist, sleet, or other precipitation.

Annual ring–(1) The growth layer of a perennial plant which represents the growth of one year, as viewed on the cross-section of a stem, branch, or root. (2) The layer of xylem (wood) formed by one year's growth of cambium.

Annual Ryegrass–A fast-growing, competitive, winter annual grass used extensively as a cover crop, for erosion control, and for short-term forage production; quite tolerant of wet soils; very useful as an interim crop between permanent pasture seedings. Tetraploid annual ryegrasses have good early spring production, are quite palatable to livestock, and tolerate a wide variety of soil conditions.

Annual Snowfall–The snow falling in one year, usually expressed in inches in the United States. For conversion to its approximate water equivalent, the depth in inches is divided by 10. (In some other countries, for example, Australia, the ratio used is 12:1. However, the ratio can vary from 5:1 to 50:1 depending on many factors).

Annual Vernalgrass–*Anthoranthum aristatum*; an a aromatic grass, family Gramineae; sometimes a weed pest.

Annual White Sweetclover–See Hubam Sweetclover.

Annual Wildrice–Zizania aquatica; an annual marsh grass, family Gramineae; cultivated for its edible grain and for an ornament; widely planted by sportsmen as feed for waterfowl.

Annual Yellow Sweetclover–*Meliotus indica*; an herb, family Leguminosae; useful for pasture, hay, silage, soiling, soil improvement, and as a bee plant. Uncontrolled, it can become a weed pest. Also called sourclover.

Annuals–Plants living one year or less. During this time the plant grows, flowers, produces seeds, and dies; e.g., beans, peas, sweet corn, squash.

Annulation–The ring growths visible in the cross-section of a log or around the body of a worm, etc.

Anoxia–A deficiency of oxygen.

Ant–Any insect of the family Formicidae; ants modify natural soils; certain species damage crops, and some are lawn pests.

Antagonism–The loss of activity of a chemical when exposed to another chemical.

Antennae–Slender jointed feelers, which bear certain sense organs, on head of insects.

Anther–The saclike part of the stamen on seed-producing plants which develops and contains the pollen.

Antheridium–The male organ in cryptogams, producing male gametes (sperm), and corresponding to the anther in flowering plants.

Anthesis–Full blossom; the opening of a flower bud; by extension, the duration of life of a flower, from the opening of the bud to the setting of fruit.

Anthocyanin–Pigments in sap responsible for scarlet to purple or blue coloration in plants. Also called anthocyan. See Anthoxanthin.

Anthoecology–The study of flowers in relation to their environment.

Anthography–Flower description.

Anthophorous–A plant that bears flowers.

Anthopilous–Flower-eating or living amongst flowers.

Anthos–(Greek) Flower.

Anthotaxy–Arrangement of flower blooms.

Anthoxanthin–Yellow crystalline pigments in sap responsible for pale yellow to ivory white coloration in plants. See Anthocyanin.

Anthracnose–A fungal disease of plants characterized by spots that appear on the leaves and/or fruit.

Anthropic Soils–Soils that have been under cultivation for a long period of time.

Anthropogenic–Changes in soils caused by action of people, such as plowing.

Anthurium–*Anthurium scherzerianum*; a tropical perennial herb, family Araceae; grown intensively in greenhouses for its bright scarlet spathe and its coiled yellow spadix. Many varieties of different sizes and colors are cultivated. Native to Central America; also called flamingo flower.

Anti-caking Agent–See Conditioners.

Antibacterial–Any substance that has the ability, even in dilute solutions, to destroy or inhibit the growth or reproduction of bacteria and other microorganisms; used especially in the treatment of infectious diseases of people, animals, and plants.

Antibiosis–An association between two organisms in which one harms the other.

Antibiotic–Germ-killing substance produced by a bacterium or mold.

Anticarcinogen–A substance that inhibits or eliminates the activity of a carcinogen (cancer-producing substance).

Anticyclone–A large, relatively high-pressure, relatively cool air mass with winds which move downward and outward in a clockwise direction in the Northern Hemisphere and counterclockwise in the Southern Hemisphere. Fair weather is usually associated with anticyclones. See Cyclone.

Antidote Statement–A required statement on the containers of chemicals explaining methods that may be used to counteract the effects of the chemical.

Antifoaming Agents–Substances that reduce foaming caused by the presence of dissolved proteins or other substances.

Antisepsis–Prevention of infection by the exclusion, inhibition, or destruction of the causative organisms.

Antiseptic–An agent that destroys or severely inhibits microorganisms that cause disease, decomposition, or fermentation.

APA–Available phosphoric anhydride (P_2O_5, frequently called available phosphoric acid.

Apache Plume–*Fallugia paradoxa*; an evergreen shrub, family Rosaceae; good as browse for goats. Found in the southwestern United States and Mexico.

Apatite–$(CaF)Ca_4(PO_4)_3$; phosphate of lime with fluorine; present as a mineral in very small amounts, but it is the primary source of phosphate deposits that are mined for use as fertilizers.

Apetalous–Without petals. See Naked Flower.

Apex–The tip of leaf, twig, or other plant part.

Aphid–A small insect, family Aphidae, which sucks plant juices, often doing great damage. Many aphids secrete honeydew, of which ants are fond; common agents for conveying viral diseases of plants; also called plant louse.

APHIS–Animal and Plant Health Inspection Service.

Aphosphorosis–Phosphorus deficiency.

Aphyllous–Leafless.

Apiarist–A beekeeper.

Apiary–Bee colonies, hives, and other honeybee equipment assembled in one location; also called bee yard.

Apical–Relating to the apex or tip.

Apiculture–The science and art of studying and using honeybees for human benefit.

Apidae–The bee family, including the subfamilies of the common bee (Apinae) and the stingless bee (Meliponinae).

Apiology–The scientific study of bees, especially honeybees.

Apis–The genus to which the honeybee belongs.

Apis cerana–Scientific name of the Eastern honeybee, the honey producer of South Asia; also called *Apis indica*.

Apis dorsata–Scientific name for the large honeybee of Asia, which builds open-air nests of single comb suspended from tree branches, rocky ledges, etc.

Apis florea–Scientific name for the small honeybee of Asia.

Apis mellifera–Scientific name of the Western honeybee.

Apogamous–Developed without fertilization; parthenogenetic.

Apomixis–Reproduction of many types in which a nonsexual process has replaced the sexual, and no fusion of male and female gemetes is involved. In many plants, seeds regularly develop from unfertilized egg cells or from cells other than egg cells; the resultant plants, termed apomicts, are genetically identical with the mother plant.

Apoplast–The total, nonliving continuum in a plant, including cell walls, intercellular spaces, and the xylem and phloem vessels that form a system and phloem through which water and solutes flow.

Apple–(1) The tree or fruit of the genus *Malus* (or *Pyrus*), family Rosaceae, especially *M. pumila*, variously designated as *P. malus*, *M. malus*, *M. sylvestris*, and *M. domestica*; specifically the fruit of the tree which is cultivated throughout the temperate zones. (2) A name applied to other plants, the fruit of which resembles the common apple, such as custard apple, mayapple, pondapple, roseapple.

Apple Butter–A sauce made from stewing apples in cider, or by allowing apple juice to thicken by evaporation.

Apple Cider–The juice pressed from apples; unless pasteurized it ferments rapidly. Unfermented juice is called sweet cider, and fermented juice is called hard cider.

Apple Cider Vinegar–Vinegar produced from long-fermented apple cider; also called cider vinegar.

Apple Cider Yeast–*Saccharomyces malei*, family Saccharomycetaceae; a yeast which occurs naturally in fruits; used in making sparkling hard cider.

Apple Maggot–The larvae of a fly, *Rhagoletic pomonella*, family Tephritidae, which bore through the flesh of apples, making winding galleries. It also infests wild crab apples, blueberries, European plums, and cherries. Also called railroad worm.

Apple of Love–Another term for the tomato.

Apple Redbug–*Lygidea mendax*, family Miridae; an important sucking insect that feeds on the sap of the leaves and the juice of apples, leaving the fruit misshapen and pitted with russeted spots; found east of the Mississippi River, especially in New England, New York, and southeastern Canada.

Apple Rust–A very serious fungal disease caused by *Gymnosporangium juniperi-virginianae*, family Pucciniaceae; characterized by rust spots on the leaves, rust lesions on the twigs, lesions on the fruit; results in defoliation, dwarfing, and reduction in the quality of the fruit. Found east of the Rocky Mountains. The fungus lives alternately on *Juniperus* spp. and on apples; also called cedar rust, cedar-rust disease of apples, cedar apple rust.

Apple Scab–A widespread parasitic fungal disease caused by *Venturia inaequalis*, family Pleosporaceae; it attacks species of *Malus*; characterized by lesions of the leaves, blossoms, and fruit; dwarfed or misshapen fruit, and premature dropping of fruit.

Apple Scald–A serious, physiological disorder caused by alcoholic esters emanating from the apples; characterized by browning of the skin after harvest, in storage, or after storage; usually associated with immaturity at harvest; more prevalent on green-skinned varieties.

Apricot–*Prunus armeniaca*; deciduous popular Asiatic tree, family Rosaceae; bears a drupe fruit, the apricot of commerce; resembles the peach.

Apterous–Wingless.

Aqua Ammonia–A solution of ammonia dissolved in water with a usual nitrogen concentration of 20 to 21 percent; used as a fertilizer for direct injection into the soil.

Aqua Humus–A fertilizer concentration of water-dispersible (about 60 percent) humic and fulvic acid derivatives. Aqua Humus 60 percent Concentrate is used as a base in preparing various liquid and dry blends of nitrogen, phosphorus, and potassium. Aqua Humus has a grade of 12-9-6 and is a specialty fertilizer which contains 36 percent water-dispersible humic and fulvic acid derivatives.

Aqueous–(1) Pertaining to water. (2) Pertaining to sediment deposited by water. (3) Ammonia in water; also called ammonium hydroxide (NH_4OH) solution. A low-pressure solution, about 20 to 21 percent N, used as a fertilizer.

Aquiclude–A geologic formation that will not transmit water fast enough to furnish an appreciable supply for a well or spring. See Aquifer.

Aquifer–A geologic formation or structure that transmits water in sufficient quantity to supply the needs for a water development, such as a well. The term *water-bearing* is sometimes used synonymously with aquifer when a stratum furnishes water for a specific use. Aquifers are usually saturated sands, gravel, fractured rock, or cavernous and vesicular rock.

Aquifuge–A rock which contains no interconnected openings or interstices and therefore neither absorbs nor transmits water.

Arabian Coffee–*Coffee arabica*; coffee made from the fruit of an evergreen shrub, family Rubiaceae, native to Yemen; grown in Florida as a curiosity. Also called mocha coffee.

Arable Land–Land so located that production of cultivated crops is economical and practical.

Arachnid–A group of arthropods having four pairs of legs and one or two body segments; mites, ticks, and spiders are examples.

Arachnoid–Cobwebby; of slender entangled hairs.

Aranose–Like a spider web.

Arbor–(Latin, tree) (1) A bower; small structure of latticework to support vines and provide a shady retreat. (2) A tree, as distinguished from a shrub. (3) A wooden platform used for sun-drying fibers. (4) The central or supporting beam or rod, as the arbor bolt on a disk gang harrow.

Arbor Day–In the United States, a ceremonial occasion for planting trees and shrubs; usually in early May, but varying by state.

Arboraceous (Arboreous)–From Latin arbor; treelike, or pertaining to trees.

Arborator–A tree tender.

Arboreal–Living in, on, or among trees.

Arboretum–A garden where trees and shrubs are grown for study and exhibition.

Arboriculture–Cultivation of woody plants, particularly those used for decoration and shade.

Archetype–In biology, the antecedent of a group of plants or animals from which certain typical characteristics have been inherited; a progenitor.

Arctic Alpine–Term used for plants of arctic and alpine distribution but found only south of the Arctic zone.

Arctic Timothy–*Alopecurus alpinus*, family Gramineae; a grass grown by Eskimos for stock feed.

Arctic Zone–The region within the Arctic Circle (66°30' N) and the North Pole (90° N). Little agriculture occurs in the Arctic zone. See Antarctic zone, Temperate Zone; Tropical zone.

Arena–(1) Sand. (2) Sphere of action, such as a livestock arena.

Arenarius–Of sandy places.

Argentine Ant–*Iridomyrmes humilis*, family Formicidae; a household and orchard insect. It also attacks shade trees, bees, their broods, and hives. Found in the southeastern United States and California.

Argillaceous–Applied to all rocks or substances composed of clay minerals, or having a notable proportion of clay in their composition, as shale, slate, etc. Argillaceous rocks are readily distinguished by the peculiar, "earthy" odor when dampened.

Argillic Horizon–The soil horizon where clay accumulates, usually the B horizon. See Soil Horizon.

Arid Climate–A dry desert, or semidesert climate with only sparse vegetation. Precipitation varies, with an upper limit for cool regions of 10 inches or less and for tropical regions of as much as 15 or 20 inches. (The Thornthwaite precipitation effectiveness [P-E] index ranges between 0 and 16.) See Precipitation-effectiveness.

Aridisols–A soil order with pedogenic horizons, low in organic matter, that are never moist as long as three consecutive months. They have an ochric epipedon that is normally soft when dry or that has distinct structure. In addition, they have one or more of the following diagnostic horizons: argillic, natric, cambric, calcic, petrocalcic, gypsic or salic, or duripan. See Soil Orders, Soil Taxonomy.

Aridity–The state of a region in respect to its dryness or lack of moisture. The amount of rainfall is not a sure index, for the aridity of a region depends in part on temperature, winds, and relative humidity.

Arizona Cypress–*Cupressus arizonica*; an evergreen tree, family Cupressaceae; grown as a windbreak and an ornamental in the southern and southwestern United States.

Arizona Fescue–*Festuca arizonica*, family Gramineae; an excellent forage grass growing at elevations of 6,000 to 10,000 feet. in southwestern United States.

Arizona Mistletoe–*Phoradendron macrophyllum*, family Loranthaceae; a parasite on alder, ash, poplar, hackberry, sycamore, walnut, and willow; also called bigleaf mistletoe.

Army Ants–Ants which group together to find and capture prey.

Army Cutworm–The larva of the moth *Chorizagrotis auxiliaris*, family Noctuidae, which infests the roots of many kinds of green vegetation. Found in the western United States.

Aroostock–A county in Maine characterized by rolling topography and by a mellow, silty loam soil; world-famous for growing Irish potatoes.

Arrow Crotalaria–*Crotalaria sagittalis*; a poisonous annual herb, family Leguminosae; eating it green or in hay may kill a horse. Found in the southern and eastern United States. Also called rattlebox; rattleweed; wild pea.

Arrowroot—Starchy substance obtained from the root of the arrowroot plant; almost pure starch.

Arsenic Poisoning—A condition which results from ingestion or absorption through the skin of an arsenical; acute cases are characterized by trembling, weakness, prostration, and diarrhea.

Arsenicals—Pesticides containing arsenic.

Artemisia—See Sagebrush.

Arthropod—A phylum or division of the animal kingdom; includes insects, spiders, and Crustacea; characterized by a coating which serves as an external skeleton and by legs with distinct movable segments or joints.

Artichoke—*Cynara scolymus*; a perennial herb, family Compositae; the edible portions are the bracts of the unripe flower head; native to southern Europe; also called bur artichoke, globe artichoke. See Jerusalem Artichoke.

Articular—Pertaining to the joints of plants and animals.

Articulation—(1) The point of attachment of a leaf stem to the main stalk. (2) Any thickened portion of the plant stem which resembles such a point.

Artificial Manure—See Compost.

Artificial Pasture—Grazing lands under relatively intensive management, usually supporting introduced forage species and receiving periodic culture treatment, such as tillage, fertilization, mowing, and irrigation. See Pasture.

Artificial Revegetation—The establishment of vegetation by mechanical methods.

Artificial Soil—A mixture of materials designed for container growing of vegetable, flower, and woody plants. Artificial soils are lightweight, take water easily, drain rapidly, and do not shrink away from the sides of the container; less likely than garden soils to harbor diseases.

Artificial Substrate—A device placed in the water for a period extending to a few weeks that provides living spaces for a multiplicity of drifting and natural-born organisms that would not otherwise be at the particular spot because of limiting physical habitat. Examples include tiles, bricks, wooden shingles, concrete blocks, and brush piles.

Arundinaceous—Reedlike.

Arviculture—The study of field crops.

Arzun—(1) The Italian millet *Setaria italica*, family Gramineae; cultivated as forage and cereal. (2) The American millet is *Panicum miliaceum*, family Gramineae.

Ascending—Growing obliquely or indirectly upward from point of attachment.

Ascomycetes—A group of fungi producing a sacklike ascus in which ascospores are borne. This group contains some of the most destructive fungi, but few cause wood decay; e.g., chestnut blight, nectria canker, larch canker, needle cast, and blights of conifers.

Ascorbic Acid (Vitamin C)—A chemical compound, $C_6H_8O_6$, which occurs in fruits and vegetables and prevents scurvy in mammals.

Aseptic—Being free from infectious microorganisms.

Aseptic Canning—A process that involves rapid heating of food to destroy food spoilage organisms, then transferring the cooked food into sterile cans by procedures that prevent the reentry of microorganisms into the cooked food during the filling and sealing operations.

Asexual Reproduction—Reproduction without fertilization, which includes various forms of vegetative reproduction, layering, stump sprouts, root sprouts; also includes grafting and reproduction by cuttings.

Ash—(1) Any tree of the genus *Fraxinus*, family Oleaceae; grown mainly for timber, sometimes for ornament. (2) Residue after combustion. (3) Specifically, minerals left after combustion has freed food from carbon. (4) In coal, the inorganic residue after burning.

Asparagus—*Asparagus officinalis*, family Liliaceae; a perennial herb whose tender new shoots are edible; native to Europe; also called garden asparagus.

Aspect—(1) Physical appearance of a plant type. (2) Seasonal appearance. (3) Direction in which a slope faces; also called exposure.

Aspen, Quaking—*Populus tremuloides*; the most widely distributed tree species in North America; moisture-loving and growing to 75 feet (25 meters) high; valuable browse species; easily started from cuttings.

Asper—Rough.

Asperous—Rough or harsh to the touch.

Asphyxiation—(1) Suffocation or death from the lack of oxygen. (2) A systemic condition of plants, etc., brought on by oxygen deficiency; characterized by seeds not germinating, blighting of plants, and breakdown of fruits, tubers, etc.

Assassin Bug—Any of the insects of the family Reduviidae which prey on other insects. Some species also attack mammals.

Association—(1) In botany, a grouping of plants having common growth requirements; the association is identified according to the dominant species, such as coniferous association, etc. (2) A grouping of soils to form a land type. (3) Persons united for a common end.

Association of American Plant Food Control Officials (AAPFCO)—The membership of the association consists of the officers charged by law with the active execution of the laws regulating the sale of commercial fertilizer and fertilizer materials; research workers employed by state, dominion, or federal agencies, who are engaged in the investigation of fertilizers. AAPFCO publishes an annual bulletin giving the official regulations and interpretations.

Assurgent—The curving upward of a plant stem.

Astringent—A drug, such as tannic acid, alum, and zinc oxide or zinc sulphate, that causes contraction of tissues.

Asymmetrical—Without proper proportion of parts; unsymmetrical.

Atavism—Reversion; reappearance of a characteristic or disease after a lapse of one or more generations.

Athel Tamarix—*Tamarix aphylla*; an Asiatic tree, family Tamaricaceae, whose blossoms provide nectar for bees; grown in beach and dessert gardens in the southwestern United States. Also called evergreen tamarix.

Atheroma–A fatty degeneration of the inner walls of arteries.

Atherosclerosis–A disease involving the fatty degeneration of the inner walls of the arteries; a form of arteriosclerosis.

Atlas–The first cervical vertebra; it connects to the occipital bone of the cranium.

Atmometer–An instrument for measuring evaporation; the Livingston porous cup style.

Atmospheric Drought–Transpiration from a plant in excess of the plant's ability to obtain moisture from the soil; e.g., a plant may wilt under hot dry winds, although there is moisture in the soil sufficient under normal conditions to supply the needs of the plant.

Atom–The smallest unit of an element to retain the chemical characteristics of that element. It consists of negatively charged particles called electrons orbiting around the nucleus and, within the small mass of the nucleus, other particles—protons (which are positively charged and balance the extranuclear electrons) and neutrons (which have no charge). It is the number and arrangement of the atom's electrons and protons that make one element differ from another; the number of neutrons distinguishes one isotope from another of the same element.

Atomic Energy–Energy released in nuclear reactions. Of particular interest is the energy released when a neutron splits an atom's nucleus into smaller pieces (fission) or when two nuclei are joined together at hundreds of millions of degrees of heat (fusion). Atomic energy is really a popular misnomer; it is more correctly called nuclear energy.

Atomic Weight–The average relative weight of an atom of an element as compared with another element (usually oxygen) that is taken as a standard. Isotopes of the same element have different atomic weights because the number of neutrons differs from isotope to isotope; a quoted atomic weight is the average weight unless a specific isotope is named.

Atomization–The process of breaking a liquid into a fine spray.

Atomize–To reduce a liquid into very fine particles in a sprayer.

Atropine–A poisonous, crystalline alkaloid used in medicine; a specific antidote for poisoning by organic phosphate insecticides.

Atrovirens–Dark green.

Attapulgite Clay–A fuller's earth. The main constituent is a hydrous magnesium aluminum silicate having high adsorptive and swelling properties; used in fertilizer production, including conditioning of fertilizer products, and as a suspending agent in suspension fertilizers.

Attenuate–To dilute, thin down, enfeeble or reduce. Specifically, to reduce the virulence of an organism; to destroy the power of disease-producing bacteria by means of chemical action, or of viruses by passing them through unnatural hosts, etc. This principle is used to develop certain bacterins and vaccines.

Attractant–Specifically, any substance that attracts or draws insects or other animals. A substance identified as gyptol, obtained from the terminal segments of the female gypsy moths, is an attractant for male gypsy moths.

Atypical–Disagreeing with the form, state, or situation usually found under similar circumstances; not typical.

Aubergine–The edible, blackish-purple fruit of the eggplant.

Aujeszky's Disease–A lethal viral disease of cattle; less so of dogs, cats, and pigs; characterized by intense itching, paralysis and convulsions. Death occurs within 36 to 48 hours after symptoms become evident; also called pseudorabies, infectious bulbar paralysis, mad itch.

Auriculate–A leaf base that is deeply lobed. The lobes are rounded.

Australian Saltbush–*Atriplex semibaccata;* an alkaline- and saline-tolerant herb, family Chenopodiaceae; it is of some feed value to livestock. Native to Australia but has spread by escape along the coast of California; also called creeping saltbush.

Austrian Pine–*Pinus nigra*; an evergreen tree, family Pinaceae; one of the most widely cultivated of all ornamental pines; used in shelterbelt planting.

Austrian Winter Pea–An annual legume often sown with small grains when an interim crop is being used between permanent pasture seedings. Good drainage is needed. Austrian peas are somewhat more winter-hardy than other field pea varieties but may still suffer substantial winter kill in prolonged freezing weather.

Autocidal Control–A technique that is used primarily for insect control. It involves the rearing and release of insects that are sterile or altered genetically in order to suppress members of their own species that are causing pest problems. A type of autocidal control is the sterile male method whereby large numbers of male insects are artificially sterilized by irradiation or chemical sterilants. This method is most popularly exemplified in the eradication of the screwworm fly in the Southwest and parts of Mexico.

Autoclave–Vessel in which high temperatures can be reached by using high pressure; e.g., the domestic pressure cooker.

Autoecious Rust–A rust having all stages of its life cycle on a single species of a plant.

Autogamous–Self-fertilizing.

Autolysis–Self-digestion. The natural softening process of fruits or vegetables after picking, or meat after slaughtering.

Autoparasitism–A parasite growing on a parasite; also called secondary parasitism or superparasitism.

Autopollination–The automatic transfer of pollen from anthers to stigma within a flower as it opens.

Autosomes–All the chromosomes except the sex chromosomes.

Autotrophic Nutrition–Also known as lithotrophic nutrition; the ability of an organism to manufacture its own food from inorganic sources, using CO_2 as the sole carbon source. Autotrophs are of two general types: photoautotrophs (photolithotrophs) whose energy is derived from sunlight, and chemoautotrophs or chemolithotrophs, which obtain energy for growth and reproduction from oxidation of inorganic materials. Important among the photoautotrophs are all higher plants, algae, and a few genera of bacteria. Principal chemoautotrophs are all higher plants, algae, and a few genera of bacteria. Principal chemoautrotrophs of vital concern in food and agriculture (with their substrate) are: Nitrobacter (nitrite), Nitrosomonas (ammonium), and Theobacillus (sulfur).

Autumn Olive–*Elaeagnus umbellata*; a deciduous, open growing shrub reaching 10 to 15 feet (3 to 5 meters) in height. It grows on low-fertility soil because it fixes atmospheric nitrogen, and is also adapted to acidic soils. It bears a prolific crop of red berries, which are used by wildlife; not adapted to poorly drained, shallow, or fine-textured soils; "Cardinal" is an improved variety.

Auxins–Plant hormones; substances that promote the growth of plants.

Available–Available plant nutrients is intended to mean that which is in a form capable of being assimilated by growing plants or of being converted into such a form in the soil during the current growing season; all essential elements in the soil solution and all of those adsorbed on clay and humus particles are available to plants.

Available Soil Water–The amount of water retained in a soil between field capacity and the permanent wilting percentage.

Avicide–A substance used to kill birds.

Avidin–A protein material that can combine with the B vitamin biotin, causing the vitamin to be unavailable to the body. Cooking renders avidin inactive.

Avirulent–Without the ability to produce disease.

Avocado–*Persea americana*; a green to greenish black tropical fruit that has a yellow flesh; used in salads.

Avoidance–A response whereby organisms prolong their dormancy, thereby achieving lesser vulnerability to environmental stresses.

Awn–A hairlike appendage found on glumes and lemmas in barley, oats, and many other grasslike plants; also called beard.

Axil–Upper angle between a leaf or other plant part and the stem to which it is attached.

Axillary–Developing in the axil of a leaf, as a bud or flower.

Axis–(1) The main stem of a plant. (2) The second cervical vertebra.

Azolla–A tropical fern that fixes atmospheric nitrogen; of the genus *Azolla*.

Azotobacter–A genus of free-living nonsymbiotic, aerobic, motile, and oval bacteria occurring in soils. They are capable of fixing atmospheric N_2 and belong to the family Azotobacteraceae.

Aztec Tobacco–*Nicotiana rustica*, family Solanaceae; a wild tobacco of high nicotine content; sometimes cultivated in flower gardens and used in the manufacture of insecticides; tobacco first grown by the Indians in the eastern United States; also called emetic weed, Indian tobacco, wild tobacco, eyebright.

B Horizon–See Horizon, Soil.

B. T.–See Bacillus thuringiensis.

Baccate–Berrylike, pulpy or fleshy.

Baccharis–A species of plant that invades the semi-arid regions of the United States and competes with forage grasses.

Bacciferous–Producing berries.

Bacillus–A genus of single-celled, rod-shaped organisms, family Bacillaceae. Although most species are harmless, they do cause chemical changes in animal and vegetable matter.

Bacillus Larvae–A bacterial organism causing a disease called American Foulbrood in bees. See American Foulbrood.

Bacillus thuringiensis–A bacterial insecticide effective against the army worm, cabbage worm, and the gypsy moth; not injurious to people, animals, or plants.

Back Flow Valve–A valve in an irrigation system that prevents water form flowing backward in the system. This prevents chemicals that are added to the water from contaminating the water supply. See Chemigation.

Back Furrow–A ridge of soil made when a turning plow throws soil on top of a previously plowed ridge made while moving in the opposite direction. This is the usual method of starting to plow a field with a one-way turning plow.

Backburn–A fire set deliberately to burn against the wind as a means of controlling advancing grass or forest fire. See Backfire.

Backfire–Controlled fire set ahead of a forest fire to create a firebreak by reducing fuel in the path of the main fire. See Backburn.

Bacteria–Single-celled microorganisms; some cause human, animal, or plant diseases; others are beneficial.

Bacteria, Facultative–Bacteria capable of adapting to an aerobic or anaerobic environment.

Bacterial Gummosis–Exudation of gumlike liquid resulting from bacterial infection, particularly bacterial canker of stone fruits.

Bacterial Neutralization of Pesticides–The use of genetically altered bacteria to neutralize and render safe such chemicals as parathion, diazinon, and malathion. The bacteria break the pesticides into simpler, less toxic forms.

Bacterial Wilt of Alfalfa–One of the most destructive diseases of alfalfa, caused by *Corynebacterium insidiosum*; characterized by yellowing and dwarfing of the stems and leaves; sometimes by a rot of the crown, a yellow ring just under the epidermis, and a yellowing of the roots. The disease is most serious in wet areas.

Bacterial Wilt of Corn–A disease which is particularly troublesome on sweet corn. It is caused by ***Bacterium stewartii***, which infects the whole plant; characterized by long, wavy streaks on the leaves, which emit a yellow exudate; also called Stewart's disease.

Bacterial Wilts–Plant diseases in which the causative bacteria produce slime that plugs the water-conducting tissue of the invaded plant.

Bactericide–Anything which destroys bacteria; a germicide.

Bacteriophage–A viruslike, bacteria destroying agent that can propagate itself only in the presence of young, active, susceptible bacteria.

Bacteriostat–A compound which inhibits growth and reproduction of or kills certain bacteria.

Baffle–The planting and growing of plants to make a wall or partition.

Bag Worm–***Thyridopteryx ephemeraeformis***, family Psychidae; the larvae of this moth spin a silken bag or sack about themselves for protection. They infest nearly all species of trees. Found in the eastern United States.

Bagasse–Residue from sugarcane stalks after the juice has been extracted by crushing. Used in the manufacturing of wall boards, insulating materials, and in mixtures with molasses for livestock feeds; also denotes sugar beet, sorghum, and sisal residue.

Bagging–Cotton, jute, or other material used in covering cotton bales; any cloth used for making bags.

Bahia Grass–***Paspalum notatum***, family Gramineae; a rapidly growing palatable range grass whose seed is low in germination. It is grown as a forage in the south and is also used to control erosion.

Bait Insecticide–An appetizing mixture of poison and food which, when eaten by an insect, results in its death.

Bait Shyness–The tendency for rodents, birds, or other pests to avoid a poisoned bait.

Balanced Fertilizer–A soil additive containing the proper proportions of each essential mineral element to develop a plant or crop.

Bald–(1) A small treeless area in a forested hilly or mountainous region, especially a crest, as bald knob. (2) Beardless, as bald wheat.

Baldcypress–***Taxodium distichum***; a deciduous swamp tree, family Pinaceae; very durable in water or moist conditions; grown as an ornamental and used in outdoor construction.

Baldwin–A good quality, large, red, winter apple grown in New York and New England.

Bale–A large, tightly pressed and bound bundle of produce, as a bale of cotton, straw, hay, wool, etc.

Balk–(1) A ridge of untilled land missed in plowing. (2) The act of an animal stopping and refusing to move.

Ball–(1) The compact earth adhering to the roots of a plant in transplanting. (2) A large pill. (3) A mass of bees formed tightly around a queen bee.

Ball Planting–A method of transplanting in which plants are taken from the ground with enough compact soil on the roots to provide nourishment and help their growth in a new location.

Balled and Burlapped–The covering with sacking of the compact mass of earth left on the roots of a plant in transplanting.

Balling–(1) The clustering of bees tightly around a queen bee, usually in an attempt to kill her. (2) A measurement of sugar percentage in simple syrup at fixed temperature (20° C); also called brix.

Balsam Fir–***Abies balsamea***; an evergreen tree used as Christmas trees, and in the making of cooperage, boxes, etc. Native to the northern United States.

Bamboo–A woody stemmed perennial of the genus ***Bambusa***, family Gramineae; apparently the most primitive grass tribe known. More than 20 genera and 200 species are recognized, some growing 60 feet or more in height. The hollow stems may die down to the ground or remain living for many years. They are often thick enough to be used in building construction, etc.; in the tropics and subtropics, the succulent young shoots of some species are used for food.

Banana–A herbaceous treelike perennial, ***Musa paradisiaca*** var. ***sapientum***, bearing clusters of pulpy fruit ranging from 4 to 9 inches in length and usually 1 inch in diameter; native to southeastern Asia, but there are also many varieties in other tropical zones cultivated for their high food value and agreeable flavor.

Band Application–(1) A herbicide application made in a narrow band, usually over or alongside a row. (2) A method of applying fertilizer in bands near plant rows, where the fertilizer will be more efficiently used rather than applying it in an application to the entire soil surface.

Band Seeding–(1) Planting seed in close proximity to a row of fertilizer that has been drilled in at a shallow depth and covered over. (2) The surface sowing of legumes and grasses on drilled ground by using a special attachment to the grain drill.

Banded Cucumber Beetle–***Diabrotica balteata***, family Chrysomelidae; a pest of beans, legumes, and cucurbits.

Banding–(1) The marking of animals or birds by fastening bands to their legs or ears or by painting their fathers for identification. (2) The placing of a collar or band around the stem of a plant or the trunk of a tree to poison, repel, or trap insects. (3) A method of castration in which a tight rubber band is placed around the scrotum; circulation is stopped and the testicles atrophy. (4) The placement of fertilizer or herbicide in a band along a row.

Banking–The mounding of earth or soil to preserve moisture or coolness or to cover the crowns of cultivated plants.

Banner Crop–A crop which is high in yield, quality, or both.

Baobab Tree–***Adansonia digitata***, family Bombacae; a huge tropical tree which bears a large edible fruit called monkey's bread; fibers are used in making cloth, rope, and paper; native to Africa and long cultivated in India; also called cream-of-tartar tree.

Barb–(1) A breed of horses related to the Arab and native to Barbary; probably introduced into Spain by the Moors and probably smaller and coarser than the Arab; its strain is evident in all known present breeds. (2) A hairlike side branch of a feather. (3) Mucous membrane projections for the openings of submaxillary glands under the tongue of horses and cattle. (4) A pointed projection on a fence wire. (5) In botany, a hooked hair or bristle. (6) The teeth or spines on the awns of grasses, especially barley.

Barberry—Any shrub of the genus *Berberis*, family Berberidaceae; often grown for its ornamental foliage and for its berries, which are of great value to birds.

Bare Fallow—Land unsown for a season but kept in cultivation or sprayed with herbicides to keep down weeds; also called bare summer fallow, naked fallow.

Bare Root—A method of transplanting in which plants are taken from the ground with little soil left on the roots.

Bare Root Grafting—A method of joining a scion to a plant by grating a plant onto bare root stock before growth has started.

Barium Carbonate—A solid material used as a rat poison and as a water purifier; a mineral known as witherite in its natural state.

Barium Fluosilicate—$BaSiF_6$; a compound used as an insecticide and a rodent poison.

Bark—(1) The exterior of a woody stem containing phloem tubes and usually some bast fibers. Cork cells may develop in the outer layer. (2) To bark is to remove the external layer form a log or tree. (3) To cure or dye with bark extract; to tan. (4) To enclose or surround with bark.

Bark Grafting—A method of joining plants in which the scion is inserted between the bark and the xylem of the stock.

Barley—One of the cereals; *Hordeum* spp., family Gramineae; cultivated since prehistoric times for human and animal consumption, for malting, or as nurse or smother crop; *Hordeum vulgare* is the commonly cultivated barley. See Six-Row Barley.

Barley Corn—A grain of barley.

Barley Malt—Fermented barley used in the manufacturing of beverages, foods, and medicines.

Barley Yellow Dwarf—A viral disease of cereals, marked by leaves rapidly turning light green and yellow, beginning at the tips; transmitted by certain species of aphids.

Barn Curing—The artificial drying of hay with natural or heated air, usually to reduce exposure time in the field and possible weather damage.

Barn Swallow—A migratory bird; *Hirundo rustica*, family Hirundidae; native to North America; beneficial to farms and ranches because it builds its nest in barns and eats many flying insects.

Barnyard Grass—*Echinochloa crus-galli*, family Gramineae; an annual grass which is often a weed in cultivated grassland; can be cut for hay if special care can be given to the curing, as it is a host for ergot; also called ankee, barn grass; cockspur grass, water grass, barnyard millet. See Ergot.

Barrel—(1) The trunk of a domesticated animal. (2) A measure of corn on the ear: one barrel equals one bushel of shelled corn. (3) In the earlier period of flour milling in the United States, the wooden barrel, net weight 196 pounds of flour, was a common container for shipment. (4) A cylindrical container. (5) A liquid measure of 42 United States gallons.

Barrens—An area relatively free of vegetation in comparison with adjacent areas, because of adverse soil or climatic conditions, wind, or other adverse environmental factors; e.g., sand barrens or rock barrens.

Bartlett—A widely grown variety of pear; genus *Pyrus*, family Rosaceae; introduced to the United States by Enoch Bartlett of Dorchester, Massachusetts, from England where it is known as bon chretien.

Basal—Leaves growing or arising at the base of a plant or stem.

Basal Application—The application of a herbicide to the lower portion of the stems of woody plants.

Basal Area—(1) The area in square feet of the cross-section at breast height of a single tree or of all trees in a stand, usually expressed as square feet per acre. (2) The area or proportion of soil surface occupied by the stems and root crowns of range plants; sometimes called basal cover.

Basal Meristem—Latent or actively growing plant cells at the base of plants.

Basalt—(1) An extrusive rock composed primarily of calcic plagioclase and pyroxene, with or without olivine. (2) Any fine-grained, dark-colored igneous rock. Soils developed from basalt are usually dark-colored and productive.

Base—(1) In chemistry, a substance which reacts with an acid to form a salt; a substance which gives off hydroxyl ions when dissolved in water, as contrasted with an acid, which gives off hydrogen ions. (2) In botany, the part of a leaf or branch attached to a stem or trunk. (3) The portion of a machine on which the operating parts are mounted.

Base Exchange—(1) In soil science, denoting a physical-chemical process in which cations adsorbing by soils can be replaced by other cations in chemically equivalent quantities. (2) A term used in medical science usually in relation to the acid-base exchanges of the blood necessary to keep the blood and tissues at a physiologic constant.

Base Exchange Capacity—(1) In soil science, the cations that a soil can absorb; expressed in milli-equivalent per 100 grams of soil. (2) In medical science, the maximum capacity of the blood to exchange base (alkali) for acid in maintaining a physiologic balance.

Base Flow—That portion of the water flowing in a stream that is due to ground water seeping into the ground.

Base Goods—Various grades of fertilizer mixtures that can be mixed with other materials to produce the grade of fertilizer requested by the customer.

Basic—On the basic side of neutral pH (above 7.0).

Basic Slag—A finely ground by-product of steel mills containing 12 to 25 percent P_2O_5, 40 to 50 percent CaO, and 5 to 15 percent SiO_2. It has some value as fertilizer because of the total phosphorus content.

Basidiomycetes—A group of fungi containing the wood and root rots and the rusts that form a clublike structure on which spores are borne after nuclear fusion and meiosis. Some fungi in the group are responsible for the formation of beneficial root mycorrhizae. See Mycorrhizae.

Basin—(1) A hollow for holding water, as around a tree, made by forming a ridge or levee of earth on all sides. (2) The depression in a pome fruit such as apple and pear, in which calyx lobes are located; also called eye. (3) An extensive depressed area into which the adjacent land drains with no surface outlet; use of the term is almost wholly confined to the arid West. (4) The drainage or catchment area of a stream or lake.

Basipetal—Toward the base of a plant organ, generally downward in shoots and upward in roots.

Basswood—*Tilia americana*; a deciduous tree commonly grown as a shade or street tree or as a bee plant. The wood is used for making veneer core for plywood, slack-cooperage, excelsior, boxes, apiary supplies, piano keys, etc. Also known as American linden, lime, whitewood, linwood, American basswood.

Bastard—In botany, a hybrid; an offspring between species that do not ordinarily mate.

Bastard Clover—See Alsike Clover.

Bat—Any of the winged, flying mammals of the order Chiroptera. The animals are nocturnal and are very beneficial to agriculture because of the large number of insects they devour.

Bat Guano (Bat Manure)—Bat excreta used for fertilizer.

Batting—Cotton, flax, or wool loosely matted and not spun.

Bauxite—The principal ore of aluminum.

Bay—(1) A horse with a body color ranging from tan to red to reddish brown. The mane and tail are black, and the lower legs are usually black. (2) The laurel or its berries. (3) A shallow swamp supporting dense tree and shrub vegetation, and containing peat or muck soil. See Pocosin. (4) A coastal inlet. (5) A compartment in a barn for storage or for a special use, as a horse barn.

Bay Leaves—The leaves of the bay-rum tree (*Pimenta acris*), which are used for flavoring and as a source for bay rum.

Beachgrass, European—*Ammophila arenaria*; a tall, stout, perennial grass adapted to coastal sand. It has long rhizomes. The panicle is thicker in the middle and tapers somewhat to the tip. The leaves have a long membranous ligule at the collar. Propagation is by transplanting of culms. Seed propagation is not feasible because very fewer viable seeds are produced and because seed on coastal dunes is often removed by wind before establishment.

Beachpea, Purple—*Lathyrus japonicus*; slow-developing, long-lived native legume that spreads by rhizomes. Used to seed into beachgrass after initial coastal dune stabilization. Commercial seed is unavailable and must be collected from native stands. The seeds are often severely affected by pea moth larvae.

Beak—(1) The upper and lower mandibles or nibs of a fowl. (2) The awn on the outer chaff of wheat. (3) Any part of a plant resembling a bird's beak.

Bean—(1) A vegetable grown for its highly nutritive value to people and animals, and eaten when green, dried, or ground into meal. The kidney-shaped (or in the broad bean the flattened) seeds of leguminous annual or perennial herbs, which are more often climbing or trailing. (2) Any plant bearing such seeds. When a bean sprouts, the bean seeds emerge through the soil; whereas, when a pea sprouts, the pea seed stays underground. (3) Protuberance on the upper part of the beak of certain water fowl.

Bean Group—The group of beans including navy, garden, and kidney beans that are capable of being inoculated by the same rhizobial bacteria.

Bean Pole—A stick usually 6 to 10 feet tall used to support climbing beans.

Bean Rust—A fungal disease caused by *Uromyces phaseoli* var. *typica*, family Pucciniaceae; characterized by reddish-brown pustules on the leaves; most prevalent in humid areas.

Bean Weevil—*Acanthoseelides obtectus*; a stout beetle that feeds on beans until the pods are formed and then lays its eggs in the seed; the larvae burrow into and devour the dried beans. It attacks the growing kidney bean, lima bean, and cowpea, and beans, peas, and lentils in storage.

Bearberry—Arctostaphylos uva-ursi; an evergreen, trailing shrub, family Ericaceae; grown to control erosion on steep exposed slopes because its branches quickly take root; also grown for its red berries in winter; found in the United States and Eurasia.

Beard—(1) Hairy appendages on the face of a man or animal, such as the whiskers of a goat. (2) A group of feathers hanging from the throat of certain breeds of chickens. (3) A tuft of bristly hair projecting from the upper part of a turkey's breast. (4) A long awn or bristlelike hair, as in the inflorescence of some grasses. (5) A tuft, line, or zone of pubescence, as on the falls of the bearded irises. (6) The part of a horse's lower jaw which carries the curb of the bridle.

Beardless Bluebunch Wheatgrass—*Argopyron inerme*; a good perennial forage plant; also grown to control soil erosion; found in the western United States. Sometimes called beardless wheatgrass.

Bearing Tree—See Witness Tree.

Beaufort Wind Scale—A system of estimating wind velocities, originally based (1806) by its inventor, Admiral Sir Francis Beaufort of the British Navy, on the effects of various wind speeds on the amount of canvas that a full-rigged frigate of the early nineteenth century could carry; since modified and widely used in international meteorology.

Bed—(1) A small plot of soil used for growing seedlings, vegetables, flowers or shrubs, and often raised above the level of the surrounding soil. (2) A place for animals to sleep. (3) The hauling platform of a wagon or truck.

Bed Shaper–An implement that makes wide beds on which up to four rows of crops are planted. Furrows between beds are used for irrigation and to guide cultivators.

Bedder–See Lister.

Bedding–(1) Plowing or otherwise elevating surface soil into a series of crowned parallel beds or lands with shallow surface drains separating them. (2) A method of surface drainage consisting of narrow plowed lands in which dead furrows run parallel to the prevailing slope; the land between adjacent dead furrows is called a bed. (3) Straw, leaves, sawdust, sand, peat moss, etc., to make a bed for an animal. It is used where an animal may lie to absorb its urine. (4) Flowers that are appropriate for growing in flower beds for a massed decoration effect.

Bedding Planes–Fine stratifications, less than 0.2 inches (5 mm) thick, in unconsolidated alluvial, eolian, lacustrine, or marine sediments.

Bedding Plants–Flower and vegetable plants planted in beds.

Bedrock–Unweathered hard rock that lies directly beneath the soil layers or beneath superficial geological deposits, such as glacial drift.

Bee–(1) Any insect of the families Prospidae, Colletidae, Megachillidae, Xylocopidae, Ceratinidae, Nomadidae, Andrenidae, Anthophoridae, Bomidae, or Apidae. They are generally beneficial, being insect parasites, predators, pollenizers, or producers of honey and beeswax. (2) See Honeybee, Drone, Worker, Queen Bee, Bumblebee. (3) A social gathering of neighbors, especially in rural areas, to husk corn, to raise a barn, or to quilt in the home, etc.

Bee Gum–Usually a hollow log used as a hive. The term is also used to refer to any beehive.

Bee Martin–Any birds of the genus *Tyrannus* which catch and kill bees in large numbers; also called kingbird.

Bee Pasture–Vegetation, attractive to bees, which is within flying distance of a hive; also called bee pasturage.

Bee Tree–(1) A hollow tree which is occupied by a colony of bees. (2) A tree whose flowers are a source of nectar or pollen.

Beech–Any tree of the genus *Fagus*, family Fagaceae; lumber from the tree is valued for its density and hardness.

Beehive–A domicile prepared for a colony of honeybees.

Beet–(1) Any plant of the genus *Beta*, family Chenopodiaceae. (2) *Beta vulagaris*; the garden beet, grown for its root and young edible leaves.

Beet Webworm–*Loxostege sticticalis*, family Pyraustidae; a caterpillar which feeds on cabbage, beets, sugar beets, beans, peas, potatoes, spinach, cucurbits, and other vegetables, as well as field crops. Found in the western United States.

Beetle–Any insect of the order Coleoptera. The upper pair of its four wings form hard cases to protect the true wings when folded.

Bell Glass–A glass covering which is sometimes used in greenhouses for covering special plants and cuttings.

Belladonna–*Atropa belladonna*; a perennial herb, family Solanaceae; its berries are poisonous, and its sap yields alkaloid atropine from which the drug is made. Also called death herb, deadly nightshade, doftberry. Native to Eurasia; grown in the United States, where it is sometimes found escaped, particularly in the eastern United States.

Belt–(1) A geographical term denoting an area of similar soil and climate particularly suited to specific crops or animals, such as the wheat belt. (2) A band of hair or skin of another color, often white, around an animal's body, as in Dutch belted cattle. (3) In machinery, a broad flexible band passed around two wheels to transmit motion from one to the other.

Belted–(1) Any fruit which has a retarded growth area extending completely around its middle; often due to frost injury when the fruit is very young; band is usually shrunken; also called banded fruit. (2) Any animal having a band of different color around its body.

Bench Terrace–A steplike embankment of earth with a flat top and a steep or vertical downhill face constructed along land contours to control runoff and erosion.

Benchlands–(1) Terraces or shelflike land features representing former water levels or shorelines of lakes, rivers, or seas; usually composed of alluvium or unconsolidated coarse sediments. (2) A foothill below a mountain.

Benders–Cotton staple of intermediate length.

Beneficial Insect–Insects that are of economic or other value; often refers to insects that prey on other insects; also refers to insects that pollinate or produce honey, such as the honeybee.

Benign–In reference to disease or disease processes, mild or nonmalignant, such as a benign tumor or growth, in contrast to malignant or cancerous tumor or growth.

Bentgrass, Colonial–*Agrostis tenuis*; long-lived, creeping turfgrass, adapted to a wide variety of soil conditions; has both stolons and rhizomes; not suited for improved pastures because of poor production, but it does provide good erosion control on road cuts and fills.

Bentgrass, Creeping–*Agrostis palustris*; very similar to colonial bentgrass, creeping bentgrass spreads aggressively by stolons, is especially moisture-tolerant, and provides excellent stabilization at the waterline of drainage channels and irrigation ditches, where it is compatible with big trefoil and white clover.

Bentgrass, Redtop–Agrostis alba; widely adapted to wet, acid-to-neutral, low-fertility soils; can withstand short summer droughts; provides a low-growing dense cover; develops more rapidly than other bentgrasses, but is shorter lived.

Bentonite–A porous rock of clay minerals derived from weathered volcanic ash or tuff. It is used: (a) as a dust diluent and carrier for insecticides; (b) in clarifying or refining wines, fruit juices, etc.; (c) as a wetting agent; and (d) as a seepage retardant in the bottom of a water storage reservoir.

Benzene–C_6H_6; a volatile and inflammable hydrocarbon which is a derivative of coal tar and is used: (a) in degreasing bones, etc., in fertilizer manufacture, and (b) for destroying larvae. It is particularly effective against the screwworm fly.

Bermuda Onion–Any species of mild onion introduced into the United States from Bermuda. Now commonly grown in the southwestern United States.

Bermudagrass–*Cynodon dactylon*, family Gramineae; a perennial grass providing excellent forage, but becoming a weed in cultivated fields; also excellent for making lawns; commonly found in the southern United States. Also called scutch grass, bahama grass, devil grass, doob, couchgrass. Hybrids are now available.

Berry–(1) Any small pulpy fruit, as the raspberry, strawberry, etc.; usually edible. (2) Any fruit which has either a fleshy or pulpy pericarp, such as the banana, cranberry, tomato, etc. (3) The dry seed of certain plants, as the coffee berry, which is really a seed contained in the fleshy true berry.

Berry Picker–A machine that straddles berry rows and harvests the berries through the use of long thin flails that shake the vines, causing the berries to fall onto a conveyer.

Berseem Clover–See Egyptian Clover.

Bessey Cherry–*Prunus besseyi*, family Rosaceae; a shrub bearing a small, sweet, juicy, edible, fruit, which is purplish-red. Also grown as an ornamental. Found in the central United States.

BGR–"Big Gam Repellent," a putrefied-egg product originally developed by Weyerhaeuser Company scientists; now sold under various trade names as a repellent to animal browse.

Bicolor–Having two colors.

Biennial–A plant that lives for two years and then dies.

Big Bluegrass–*Poa ampla*, family Gramineae; this grass is used in reseeding rangeland. Found in the Yukon Territory of Alaska and southward to New Mexico and California. See Bluegrass.

Big Bluestem–*Andropogon furcatus*, family Gramineae; a tall grass, producing good forage. It grows in the central and southwestern United States in dry places, on prairies, and in open woods.

Big Trefoil–A good legume for soils that are wet the year around or subject to prolonged flooding or ponding in the winter. It can be used for pasture or hay. It is more grazing-tolerant than birdsfoot trefoil because of its rhizomes; not winter-hardy in cold areas; requires a special inoculum (not the same as birdsfoot inoculum).

Bigleaf Maple–*Acer macrophyllum*; a deciduous tree, family Aceraceae; grown as a shade and park tree. Its large leaves turn bright orange in the fall. Native to North America; also known as Oregon maple.

Bilateral–Having two sides.

Bilateral Symmetry–Similarity of form, one side with the other.

Biltmore Stick–A rule graduated so that the height and diameter of a standing tree may be calculated; used to estimate board feet in a tree.

Binder Tobacco Leaf–Tobacco wrapped around the filler leaves and segments of a cigar to hold the cigar together.

Bindweed–Twining or creeping species of ***Convolvulus***, family Convolvulaceae; also called glorybind, wild morning glory.

Binomial–By international agreement, all plants and animals have two Latin names: genus and species.

Bioaccumulation–The process by which plants and animals accumulate substances, especially pollutants, that may not be injurious to that organism but may injure other organisms that eat them. For example, nitrates may accumulate in corn and oats and be injurious to animals feeding on them but not to the plants themselves. In a like manner, fish may accumulate DDT or PCB in their fat, which may be toxic to an animal eating the fish but not to the fish themselves.

Bioassay–The quantitative or qualitative determination of herbicides by the use of sensitive indicator plants or other organisms.

Biodegradable–Substances capable of being degraded into their constituent elements. This term is used especially in reference to toxins, such as pesticides, being degraded to nontoxins.

Biogenesis–(1) Formation by the action of organisms. (2) The doctrine that all life has been derived from previously living organisms.

Biological–Products derived from a living process or living matter, such as sera, vaccines, bacterins, and antitoxins, etc. Also called biologics.

Biological Control–A method of pest control by the use of predatory insects, fungi, or viruses; as contrasted to control by chemical pesticides. See Biophage.

Biological Erosion–Erosion by water or wind as a result of soil being exposed by the burrowing of rodents, destruction of vegetation by insects, etc.

Biological Growth Potential–The average net growth attainable in a fully stocked natural forest stand.

Biological Lag–The time period necessary for a changed production decision to influence market supplies, owing to the biological nature of agricultural products.

Biological Magnification–The concentration of certain substances up a food chain. A very important mechanism in concentrating pesticides and heavy metals in organisms such as fish.

Biological Mineralization–The conversion of an element occurring in organic compounds to the inorganic form through biological decomposition.

Biologist–A person who studies living organisms as a career.

Biology–The field of study dealing with living organisms. It may be divided into the study of plants (botany) and of animals (zoology).

Biomass–The amount of matter of biological origin in a given area; e.g., the living and decaying matter in the soil as opposed to the inorganic mineral components such as sand, silt, and clay.

Biome–A term derived form the Greek, ***bios***, meaning relation to life; used in ecology to include major life in the area, such as tundra biome, tropical rainforest biome, and grassland biome.

Biometry–The application of statistical methods to the study of biological problems.

Bionomics—The study of relations among organisms and the relationships between them and their environment. See Ecology.

Biophages—Organisms that obtain nourishment from other organisms, e.g., predators, parasites, and pathogens. See Biological Control.

Biopsy—The microscopic or chemical analysis of tissue removed from a living body, usually to discover the cause of illness.

Biostress—Difficulties that plants and animals have in obtaining the necessities of life: food, water, and living space.

Biotechnology—Technology concerning the application of biological and engineering techniques to microorganisms, plants, and animals, sometimes used in the narrower sense of genetic engineering.

Biotic—Pertaining to life; biological.

Biotic Influence—Biological influences on plant and animal life as contrasted to climatic influences.

Biotic Potential—(1) The maximum reproduction power or ability. The inherent ability of an organism to reproduce and survive in greater numbers. (2) The ability of an organism to reproduce in an optimum, unrestricted, and noncompetitive environment.

Biotype—Groups of plants or animals primarily distinguishable on the basis of interaction with relatively genetically stable varieties or clones of host plants; a strain of plant or animal species.

Bipinnate—Twice pinnate, the primary pinnae or leaflets being again divided into secondary leaflets; often written 2-pinnate.

Bipyridyliums—A group of synthetic organic pesticides that includes the herbicide paraquat.

Birch—Any tree of the genus *Betula*, family Betulaceae; species are valuable as timber having a hard, close-grain wood; also used for ornamentals.

Bird—(1) Any member of the class Avis of the phylum Vertebrata, the body of which is covered with feathers. (2) Domesticated birds: chickens, turkeys, ducks, geese, guineas, etc.

Bird Guano—See Guano.

Bird Peck—A small hole or distortion in the grain of wood resulting from sapsuckers attacking a tree; may be associated with discoloration, such as mineral stain.

Bird's-eye Maple—The lumber of the sugar maple having natural grain markings resembling eyes.

Birdsfoot Trefoil—A long-lived, deep-rooted legume suitable for hay or pasture in areas of wet soils or low soil pH. Very winter-hardy and tolerant of dry summer conditions. Useful on irrigated or dry land. Nonbloating legume suitable for sheep or cattle but not a good horse pasture, because of the presence of tannins. Not tolerant of early spring grazing or continuous grazing. It establishes slowly and with some difficulty, but is vigorous once established. Requires a special inoculum.

Bisexual—An animal or plant that produces both eggs and sperm, or a flower that bears both stamens and pistils.

Bitterbrush—*Purshia tridentata*; a valuable forage plant, family Rosaceae; found in the high mountain areas of the western United States.

Bittern—A concentrated solution of salts which remains after the common table salt has been refined out. Sometimes used as an herbicide.

Bitternut Hickory—*Carya cordiformis*; a deciduous tree, family Juglandaceae; grown for shade, and for its wood, which is used in making tool handles, ladders, furniture, sporting goods, implements, woodenware, and for fuel and smoking meat.

Bitterweed Actinea—*Actineae odorata*; an annual herb, family Compositae; poisonous to sheep, which may eat it during a shortage of forage. Poisoning is characterized by loss of appetite, cessation of chewing of the cud, abdominal pain, bloating, frothing at the mouth, and a green nasal discharge; found on overgrazed ranges in the western United States; also called bitter rubberweed, Colorado rubberweed, bitterweed.

Biuret—A compound, toxic to some crops, formed by thermal decomposition of urea; to be avoided in the manufacture of fertilizer urea. See Urea.

Black Alder—See European Alder.

Black Alkali—An obsolete term meaning a soil high in exchangeable sodium. The modern term is *sodic soil*.

Black Ash—*Fraxinus nigra*; a deciduous tree, family Oleaceae, which grows in a swampy, moist soil; used in cabinet making, and basket making; found in eastern North America; also called brown ash, hoop ash, basket ash, swamp ash.

Black Blizzards—Clouds of dust consisting of organic matter and the finer particles of soil from bare fields and grazing land denuded of vegetation; prevalent in the southwestern United States.

Black Carpenter Ant—*Camponotus herculeanus pennsylvanicus*, family Formicidae; the largest American common ant; attacks tree stumps, tree trunks, telephone poles, windowsills, and old timbers; found in the eastern United States.

Black Cherry—*Prunus serotina*, a deciduous tree, family Rosaceae; an important timber tree used in making furniture; the leaves are poisonous to livestock; also called wild black cherry, black choke.

Black Cloth—A means of providing artificial short days in a greenhouse so that short-day plants will bloom.

Black Cutworm—*Agrotis ypsilon*, family Phalaenidae; a surface cutworm which attacks truck crops; found throughout the United States.

Black Fallow—A field left bare (no crop) to accumulate enough soil moisture from precipitation to grow a crop.

Black Frost—A blackening of vegetation which occurs when the temperature drops low enough below the freezing point to destroy vegetation, including staple crops; there is no actual deposit of frost crystals.

Black Gramagrass—*Bouteloua eriopoda*, family Gramineae; an excellent forage grass; found in the southern desert ranges of the United States.

Black Mangrove—*Avicennia marina*; an evergreen tree, family Verbenaceae, whose flower is a good source of honey in Florida.

Black Maple–*Acer nigrum*; a hardy deciduous tree, family Aceraceae; one of the American sugar maples. Grown in highway plantings and in city yards. Also used in hardwood distillation and for railroad ties, veneer (especially good for birdseye), shoe lasts, bowling pins, fuel, furniture, flooring, and boxes. Also called black sugar maple.

Black Medic–*Medicaago lupulina*; an annual prostrate herb, family Leguminosae; regarded as a weed in lawns, but has value as a pasture plant; also called hop medic, nonesuch, blackseed, hop clover, yellow trefoil.

Black Mustard–*Brassica nigra*; an annual herb, family Crucifereae; largely grown for its seed used in the manufacture of mustard. Escaped, it is a weed pest. Native to Europe.

Black Nightshade–*Solanum nigrum*; an annual herb, family Solanaceae; a common weed pest. Poisonous to livestock and to bees. Poisoning of livestock is characterized by salivation, vomiting, diarrhea, and bloating. Also called poison berry, deadly nightshade, petty morel, and stubbleberry.

Black Oak–*Quercus velutina*; a hardy, columnar, deciduous tree, family Fagaceae; the lumber is used for railroad ties, cooperage, fence posts, mine timbers, pilings, veneers, firewood, flooring, mill products, and furniture. Native to North America. Also called tanbark oak, quercitron, yellowbark oak.

Black Pepper–*Piper nigrum*; a woody vine, family piperaceae; grown for its fruits, which yield white and black pepper. Native to the East Indies.

Black Rot–(1) A fungal disease of apples, pears, quinces, and other hosts caused by ***Physalospora malorum***, family Pleosporaceae; characterized by cankers on the twigs and limbs, by a leafspot, and by rotting of the fruit as it approaches maturity or when in storage. Also called frog eye, ring rot, blossom end rot, brown rot. The symptoms on twigs and limbs are also known as dieback, twig blight, apple canker, black rot canker, New York apple tree canker. (2) A fungal disease of citrus fruit, particularly oranges, caused by ***Alternaria citri***, family Dematiaceae; the infected fruits turn deep orange. They appear sound but are decayed near the stylar end. The inside is brown to black. (3) A general marketing term describing bad eggs.

Black Rot of Crucifers–A bacterial disease caused by ***Xanthomonas campestris***. Characterized, especially in cabbage, by yellowing of the leaves and blackening of the veins at the leaf margins and developing toward the petiole and the main stem. Once established, the disease rapidly affects the entire plant, causing the leaves to wilt, turn yellow, dry up, and become thin and parchmentlike. Also called bacteriosis, bacterial rot, brown rot.

Black Sage (Brush)–*Salvia mellifera*; a perennial herb, family Labiateae; important as a bee plant and sometimes as forage. Found on the Pacific Coast of the United States.

Black Shank of Tobacco–A disease caused by the parasitic fungus ***Phytophthora parasitica*** var. ***nicotianae***, family Pythiaceae; characterized by blackened and dead roots, by decay at the base of the stalk, brown blotches on the leaves, and by damping-off of the seedlings.

Black Spot–(1) Bull's-eye rot; a disease of apple caused by ***Neofabraea malicorticis***, family Mollisiaceae. (2) A fungal disease of citrus caused by ***Phoma citricarpa***, family Sphaeropsidaceae; characterized by reddish-brown spots on the fruit. Reddish-brown halos form, and the brown centers become depressed. The spots then turn dark, sometimes black. (3) A disease of potatoes. The tissue beneath the skin breaks down and turns black or blue-black. The blackening frequently does not show up until the potatoes have been removed from storage. Heavy fertilization with nitrogen and excessive soil moisture seem to increase blackening following storage. (4) A fungal disease of roses caused by ***Diplocarpon rosae***, family Mollisiaceae; characterized by black spots having rays and fibrils, which sometimes cover the leaves. In some cases, entire leaves turn yellow and defoliation follows. Similar spots also appear on canes. Inconspicuous spots on the petioles, stipules, flower receptacles, sepals, and petals may cause distortion of the flower. (5) A minor bacterial disease of sugar beets caused by ***Pseudomonas aptata***, which blights plants that are grown for seed. Also called black streak. (6) A low-grade infection resulting from mechanical injury to the teat of a cow. See Apple Scab, Peach Scab.

Black Spruce–*Picea mariana*; a hardy, slow-growing, often stunted, evergreen, coniferous tree, family Pinaceae; grown in moist soils, it is a most common species in strongly acid peat bogs. Used for pulpwood, boxes, planing-mill products, furniture, woodenware, etc. Native to North America. Also called double spruce, eastern spruce, bog spruce.

Black Sugar Maple–See Black Maple.

Black Tupelo–*Nyssa sylvatica*; a deciduous tree, family Nyssaceae; grown for its fine, red autumnal foliage. Used in making veneer, pulp, cabinet work, boxes, etc. Native to North America. Also called black gum (forestry name), sour gum, pepperidge, tupelo.

Black Turpentine Beetle–*Dendroctonus terebrans*; an insect pest of pine trees. The larvae feed on the inner bark of the trees, causing the trees to die.

Black Walnut–See Eastern Black Walnut.

Blackberry–*Rubus* spp.; vigorous growing, rambling, thorny, deciduous vines. The edible berry crop is well utilized by many species of animals, and the rambling growth forms a soil-protecting cover. Easily started from cuttings.

Blackberry Leaf and Can Spot–A fungal disease caused by ***Septoria rubi***. Characterized by light-colored, small spots with red or purple borders on the leaves. The spots also appear on the canes and result in premature defoliation of the fruiting cane. Found mostly on dewberry in Oregon and Idaho and in the coastal regions of California.

Blackeyed Peas–A variety of the cowpea, ***Vigna sinensis***; grown mainly for livestock forage and for green manure. The seeds are eaten as a vegetable.

Blackland–Generally, areas having dark-brown or black surface soils. Specifically, the prairies of central Texas, the soils of which are black, waxy, clayey. See Vertisols.

Blackstrap–The liquid by-product of crystallization of sugar from sugarcane or sugarbeet juice; used as a feed for livestock having about 75 percent of the feed value of corn. Also called blackstrap molasses.

Blade–(1) The expanded portion of a leaf. (2) A cutting tool, as the blade of an axe or the blade of a bulldozer. (3) A retail cut of meat from the forequarter or shoulder. (4) The hind part of a fowl's single comb.

Blanch–(1) To bleach or whiten a vegetable as it is growing by wrapping the stalk and leaves with paper or by mounding soil around the portion to be whitened, as celery is blanched. Also called etiolate. (2) To heat vegetables and fruit in water, live steam, or dry heat to inactivate enzymes preparatory to processing.

Bland–Mild-flavored, not stimulating to the taste; smooth, soft-textured.

Blast-furnace Slag–The dross from the smelting of iron ore, which contains calcium silicate, a low-grade source of lime for agricultural use in the southeastern United States. It is also a carrier of low-grade total phosphorus.

Blastula–A mass of cells with a cavity that occurs from the dividing of a fertilized egg. From this stage the cells begin to differentiate.

Bleaching Clay–Any clay in its natural state or after chemical activation that has the capacity for adsorbing coloring matter; generally montmorillonitic clays are used.

Blight–Disease symptoms, parasite, fungi, bacteria, viruses, or unfavorable atmospheric conditions which cause the withering and death of a plant.

Blind Area–In forestry, an area which cannot be seen from a lookout station.

Blind Furrow–A double furrow caused by plowing two adjacent furrows and throwing the soil in opposite directions; also called a dead furrow.

Blind Knot–An overgrown knot that forms a pronounced swelling on the trunk of a tree; also called blind cork. See Knot.

Blizzard–A severe storm of high wind, low temperatures, and heavy snow.

Block–(1) A piece of wood used as a divider between two bales of hay. (2) A pulley used to increase pulling or hoisting power. (3) In logging, an administrative division of a forest. (4) About 5 pounds of cotton hanks. (5) A portion of hay thrown into a baler. (6) To thin out plants, especially sugar beets, into smaller bunches, which will be again thinned into single plants. (7) To trim fleece to enhance the appearance of a sheep. (8) See Cutting Block.

Blocky Structure (Soil)–The arrangement of particles which indicates that soil units exceed 1 cc.

Bloodtwig Dogwood–*Cornus sanguinea*; a showy shrub, family Cornaceae; grown for its white flowers, black fruit, and blood-red twigs. Native to Eurasia. Also called dog tree, red dogwood.

Bloom–A word with several contrasting meanings: (1) Plants in the state of flowering. (2) Plants reproducing and growing. (3) Fruits with a white covering. (4) Minerals that absorb one wavelength of light and reflect another. (5) A very fine, often waxy, powdery coating on the surface of certain leaves, stems, fruits, or other organs; usually whitish, grayish, or bluish, and easily rubbed off, as the bloom on a plum. See Algal Bloom.

Blossom–(1) The flower or bloom of a seed plant; to flower. (2) A horse with a sorrel or bay coat mixed with white.

Blowings–Cotton waste from gins.

Blowout Grass–*Redfieldia flexuosa*, family Gramineae; used in reseeding ranges and establishing pastures and to control blow-outs. It grows naturally on deep, loose sands in the Great Plains of the southwestern United States.

Blue Gramagrass–*Bouteloua gracilis*; a native American grass, family Gramineae, used for reseeding in the western ranges of the United States.

Blue Loco–*Astragalus diphysus*; a common poisonous herb, family Leguminosae; causes death in cattle and horses. Symptoms are characterized by a shaggy coat and by loss of muscular control in walking, nervous sensibility, and loss of weight and appetite. It grows extensively in Arizona and New Mexico during January and February; also called blue locoweed.

Blue Mold of Tobacco–A disease in plant beds caused by *Peronospora tabicina*, family Peronosporaceae; characterized by cupped leaves with a whitish or violet-colored moldy growth over the lower surfaces. It kills young plants and the leaf tissues of older plants. Found in tobacco areas throughout the United States except in Wisconsin.

Blue Ribbon–A prize or award for the best livestock or other agricultural commodity in a class; made of inscribed blue cloth.

Blue Spruce–See Colorado Spruce.

Blue Wildrye–*Elymus glaucus*; a fairly good forage grass, family Gramineae; with strong seed habits. Found up to 9,000 feet in altitude from Ontario and Michigan to Alaska and south as far as South Dakota, New Mexico, and California.

Blue-stain–Blue coloration of light-colored wood caused by species of the fungi genera *Ceratocystis*, *Aureobasidium*, and *Lasiodiplodia*.

Bluebeech–See American Hornbeam.

Blueberry–About 50 species of *Vaccinium*, family Ericaceae, of which the most important are the highbush blueberry, V. *corymbosum*, the large-fruited blueberry of the northeastern and eastern north-central United States; the rabbiteye blueberry, V. *virgatum*, of the southeastern United States; and the low-bush blueberry, V. *angustifolium*, of Canada and the northeastern and northcentral United States (often incorrectly called huckleberry).

Bluegrass–(1) Any grass of the genus *Poa*, family Gramineae. Valued for hay and lawns and as the most palatable range and pasture grasses. (2) Grasses of the genus *Andropogon* in Australia. (3) *Agropyron* *scabrum* in New Zealand. Alpine, ***Poa alpina***; annual, ***Poa annua***; big. ***Poa ampla***; Canada, ***Poa compressa***; Kentucky, ***Poa pratensis***; mutton, ***Poa fendleriana***; roughstalk, ***Poa trivialis***; Texas, ***Poa arachnifera***.

Bluegrass, Kentucky–*Poa pratensis*; a major lawn and sod grass, adapted to cool climates and moist growing conditions. Although characteristically low-yielding, it may persist and outyield other forage species at high altitudes. It is an excellent erosion control grass where adapted.

Bluejoint Reedgrass–*Calamagrostis canadensis*, family Gramineae; a perennial grass; fair forage. Found throughout the United States up to 12,000 feet and usually in marshy or wet soil. Also called bluejoint, bluetop.

Board–(1) Lumber less than 2 inches thick and more than 8 inches wide. (2) Lumber 1½ inches thick, 6 or more inches wide, and 8 or more feet long (British). (3) Lumber of all widths 1 inch thick; widths of less than 6 inches are sometimes called strips. (4) Plank floor on which sheep are sheared (New Zealand).

Board Foot–A lumber measurement; a board 1 foot long, 1 foot wide, 1 inch thick, based on the original cut before planing and surfacing; also called super foot (Australia).

Board Mill–A sawmill that cuts 1- and 2-inch lumber, as compared with a timber mill, which cuts thicker lumber.

Bob–(1) Veal from calves slaughtered when less than four to six weeks old; also called slunk. (2) The docked tail of a horse. (3) In transporting lumber, a single pair of runners on which the forward ends of the logs are placed. (4) To transport logs on a bob.

Bobtail Barley–*Hordeum jubatum* var. *caespitosum*, family Gramineae; a subspecies of foxtail barley with short awns; of fair forage value. Found from North Dakota to Alaska, south to Kansas and Arizona.

Body–(1) The physique of an animal. (2) The consistency of a substance, such as honey, butter, cheese, etc. (3) The fullness of flavor, especially of a beverage.

Boer Lovegrass–*Eragrostis chloromelas*, family Gramineae; a grass used for reseeding for forage on the western ranges of the United States. Native to South Africa.

Bog Iron–A spongy variety of hydrated oxide of iron (limonite). Found in layers and lumps on level, sandy, acid soils which have been covered with a swamp or bog.

Bogie–Tobacco leaves twisted into small rolls.

Boil Down–To concentrate a liquid, such as a fruit juice or sap of a sugar maple tree into a thick syrup or sugar.

Boiling Point–The temperature at which the vapor pressure of a liquid equals the atmospheric pressure. At the boiling point, bubbles of vapor rise continually and break on the surface. The boiling temperature of pure water at sea level (barometric pressure of 30 inches of mercury) is 212°F (100°C). At high altitudes, the boiling point of water is lower because the atmospheric pressure is lower. At 5,000 feet above sea level, for example, the boiling point of water is 203°F (95°C); at 10,000 feet it is 194°F (90°C).

Boiling-water-bath Canner–A large kettle with lid, rack, and cover; must be deep enough to allow jars to be covered with 1 to 2 inches of water and still have additional height for water to boil actively. Suitable for processing acid foods.

Bois d'Arc (Osage Orange)–*Malcura poomifera*; a tree used extensively in the United States Great Plains for a shelterbelt and for fence posts.

Bolander Waterhemlock–*Cicuta bolanderi*; a perennial California herb, family Umbelliferae; it is poisonous to stock, causing vomiting, diarrhea, dilated pupils, difficult respiration, frothing at the mouth, weak but rapid pulse, convulsions.

Bole–The trunk of a tree. It may extend to the top of the tree as in some conifers, or it may be lost in the ramification of the crown, as in deciduous species.

Boll–(1) The leathery capsule that contains the seed and lint of the cotton plant. (2) The seed pod of flax.

Boll Period–That phase in the growth when cotton bolls have formed but not opened.

Boll Weevil–*Anthonomus grandis*, family Curculionidae; a beetle that causes cotton losses of more than $100 million annually. The larvae feed only on cotton, but the adults may feed on okra, hollyhock, and hibiscus. Native to Mexico and Central America.

Bollies–Unopened bolls of cotton that contain usable lint.

Bolling–A tree whose branches have been removed.

Bollworm–*Heliothis zea*, family Phalaenidae; an insect whose larvae feed on cotton bolls, corn, tobacco, tomatoes, beans, vetch, alfalfa, and other garden plants and flowers. Also called corn earworm; tomato fruitworm.

Bolly–A boll of cotton which remains unopened after frost.

Bolly Cotton–The lint extracted from the bollies.

Bolster–(1) The padded undersection of a saddle. (2) In growing plants, the excrescence at the leafstalk where it joins the axis. (3) A cross member of a vehicle on which the bed, rack, or load rests.

Bolt–(1) A section sawed or split from a short log from which blocks, shingles, staves, etc., are made. (2) A short log used in making pulpwood or veneer. (3) A strong, metal pin with a head at one end and threads at the other. (4) To sift coarse elements from fine, as bran from flour. (5) To flower or to produce seed stalks, often prematurely. (6) To eat rapidly. (7) To run away.

Bolter–(1) A sieve. (2) A circular ripsaw and a mechanically driven carriage used for sawing round bolts into boards. (3) A plant, such as the sugar beet, which produces a seed stalk from the crown of the plant.

Bolting–Production of flowers and seeds by such plants as spinach, lettuce, and radishes, generally occurring when days are long and temperatures warm.

Bolting Cloth–A sieve cloth, usually silk.

Bone Ash–The highly calcic and phosphatic residue of bones burned in air; of limited use as a fertilizer. Also used in making assay cups and in cleaning jewelry.

Bone Black–The residue from heating bones in a closed retort similar to that of making coke. Bone black is used in refining sugar, oil, etc. After use, it is sold as spent bone black for fertilizer manufacture. Also called bone charcoal, bone char, animal charcoal, animal black.

Bone Meal–(1) The product of drying and grinding animal bones not previously steamed under pressure. The composition is nitrogen 3.3 to 4.1 percent, bone phosphate 43 to 50 percent or phosphorus 10 to 12 percent, and calcium 21 to 24 percent. Used as a fertilizer and stock feed. (2) (Steamed) A product of grinding animal bones previously steamed under pressure. The composition is nitrogen 1.65 to 2.5 percent and calcium 24 to 33 percent. Used as a fertilizer and stock feed.

Bone Phosphate–The calcium phosphate obtained from bones; also in commerce, applied to calcium phosphate obtained from phosphatic rocks, e.g., those of Florida.

Bone Products–The skeleton of vertebrates (20 to 30 percent total P_2O_5); earliest source of fertilizer phosphorous, now used largely in animal feeds and, to some extent, in specialty fertilizers such as for growing roses.

Boot–(1) The hollow metal casting on a planter or a drill through which the seed passes to be planted. (2) Profuse feathering on the shank and toes of fowls. (3) The sheathlike leaf structure on the upper end of grain or grass plants that encloses the inflorescence prior to its emergence. During inflorescence the plant is said to be in the boot stage or in boot. (4) In a grain elevator, the box which contains the lower pulley or sprocket.

Boot Leaf–The leaf arising from the protective sheath enclosing the young inflorescence of grain.

Bora–(Latin, boreas-north wind) A cold, dry, violent downslope wind, named after the north winds that move rapidly down the Alps in Yugoslavia to the Adriatic Sea. The name is now used for a similar wind any place in the world.

Borage–(French, bourrache) *Borago officinalis*; an herb with rough spiny leaves and blue flowers, excellent for flavoring lettuce salads.

Borax–Sodium tetraborate, $Na_2B_4O_7 \times 10\ H_2O$. It is sometimes applied directly to soil, or added to commercial fertilizer in very small quantities (toxic to plants in high concentrations) to supply boron. It is also used as an herbicide and an insecticide, especially to kill the maggots of flies in manure and refuse piles. Borax is 11 percent boron. Also called borac, tincal.

Border Dyke–Ridges of earth constructed to hold irrigation water within certain limits in a field.

Border Irrigation–A system of irrigating land which has a slope of 2 to 4 feet per 1,000 feet. Parallel levees are built, and the area between levees is flooded.

Border Strip–A zone or strip surrounding a field research plot usually given the same treatment as the plot but not included in sampling or study.

Boreal Forest–The forest consisting chiefly of conifers extending across northern North America from Newfoundland to Alaska.

Borer–An insect that bores into and feeds on the woody sections or bark of trees.

Boron–An element essential for plant growth. The original source of boron is tourmaline, but available forms are from soil organic matter.

Boron Deficiency–A soil deficiency. In vegetables, the terminal buds and phloem tissue are killed, and the development of lateral buds is arrested. In citrus trees, the surface and main veins of the leaves cork or split, and the leaves curl downward; long, narrow splits occur on the trunk bark of small citrus trees. In deciduous trees, an internal bark necrosis occurs, which can often be detected on the surface.

Boron Injury–Slowed or arrested germination, killing or stunting of plants, fading of color, premature ripening, or small yields, caused by an excess of boron in the soil.

Boscage–A thicket of shrubs.

Bosque–A name given to a wooded thicket, usually of tamarisk, willow, and cottonwood, on the flood plains of the southwestern United States.

Botanical Garden–A garden of plants that have scientific interest.

Botanical Name–The scientific name of plants, which includes the genus and species.

Botanical Pesticide–A pesticide made from plants; also called a plant-derived pesticide.

Botany–The science of plants; anatomy, cytology, ecology, morphology, mycology, paleobotany, pathology, physiology, phytogeography, taxonomy, etc.

Botryose–Used to describe a plant whose flowers grow in clusters.

Botrytis–A genus of parasitic fungi, family Moniliaceae, whose various species cause spots on flowers, leaves, and stems of plants. In wet weather, the spots spread and are covered by a gray, moldy, sporal growth of the fungus.

Bottom–(1) (Often plural) Low-lying land adjacent to a river, usually rich in alluvial deposits, e.g., Mississippi River bottoms. (2) A section of a plow consisting of the moldboard, share, frame, and landside. (2) Stamina in a horse.

Bottom Heat–The use of heating cables or mats to achieve the hoptimum growing media temperature for producing plants.

Bottomland–Lowland along a river (sometimes flooded).

Botulism–A food poisoning caused by the bacterium *Clostridium botulinum* in preserved foods or feed for animals. The poison can be fatal to humans or animals. Called limberneck in poultry. Also called food poisoning, allantiasis.

Boulder Clay–Massive, compact deposits of gritty clay of glacial origin, containing imbedded gravel and boulders. It is the parent material for a number of highly productive soils.

Bound Water–Water adsorbed by colloids and therefore not easily freezable; also called unfree water, hygroscopic water.

Boundary Tree–An old, tall distinguishable tree standing on a property line, usually blazed or otherwise marked.

Bourbon Cotton–*Gossypium purpurascens*; a perennial shrub or small tree, family Malvaceae; grown in the tropics as a source of commercial cotton. Also called Puerto Rico cotton, Siam cotton.

Box–(1) *Buxus sempervirens*; a shrub or small tree; family Buxaceae; one of the most valuable broad-leaved evergreens. Since the days of the Romans it has been planted for hedges and topiary work. It is lethally poisonous to cattle, horses, sheep, and pigs, having emetic and purgative properties. Native to southern Europe. (2) The lowest grade of softwood lumber. (3) A cavity cut into the base of a pine tree to collect sap to make turpentine and rosin. (4) A system for bracing branches of trees. (5) To chip back on the underside of a cut in a tree to prevent the main stem of the tree from splitting when it falls. (6) An accidental mixing of two herds of sheep or cattle (Australia). (7) To corral animals. (8) A device for dividing water in an irrigation system into two or more ditches. (9) A canyon with one entrance and no exits.

Box Elder–*Acer negundo*; a deciduous, fast-growing shade tree, family Aceraceae; its branches are brittle; the wood is used for paper pulp and woodenware; also called ashleaved maple.

Box Scraper–An implement mounted on the rear of a tractor that is used in grading and leveling soil.

Box the Heart–To cut boards from the sides of a log so that the center or heartwood is left as one piece of timber.

Boysenberry–(Boysen) *Rubus* spp., family Rosaceae; a large, sweet, deep-red, high-quality berry, the hybrid of blackberry, loganberry, and raspberry. Commercially important on the West Coast of the United States.

Brace Roots–The supporting roots of corn, etc., which spring from the first two or three nodes above the ground; also called aerial roots, prop roots, adventitious roots.

Brachiate–Being branched in pairs that sprout from opposite sides of a stem, as on all species of ash and maple trees.

Brachyism–A dwarfing of plants by shortening of the internodes.

Bracken–*Pteridium aquilinum*; a widely distributed fern, family Polypodiaceae; sometimes grown in gardens, but, if escaped, is a weed pest. Also called female fern, fern brake.

Bract–A small leaf at the base of the flower.

Brake–(1) An irrigation check. (2) A place overgrown with bracken, brush, wood, cane, etc., e.g., a cane brake, a thicket.

Bramble–(1) Any plant of the genus *Rubus*, family Rosaceae, as the blackberry, raspberry, and dewberry, called bramble fruits. (2) Any prickly, clinging shrub.

Bran–The broken seed coat of the cereal grains, most commonly of wheat, separated by fine sieves or finer cloth after grinding of the grain into flour. The bran is a desirable "bulk" feed for livestock or fiber food for people.

Branch–(1) A lateral stem arising from the main stem of a plant, bough, or limb. (2) Part of a horseshoe from the first nailhole to the end of the heel. (3) A tributary stream. (4) An ancillary pipe attached to a main pipe.

Branch Wilt of Walnut–A fungal disease of Persian walnut (*Juglans regia*) caused by *Hendersonula toruloidea*, family Sphaeropsidaceae; characterized by sudden wilting of the leaves in summer. It appears on small twigs and branches, but eventually involves the larger limbs as well.

Brand Tillage–A system of farming in which new land is periodically cleared for plowing, while abandoned fields are allowed to regain their natural cover. Common in the tropics. See Shifting Cultivation.

Brashy Soil–Soil composed almost entirely of coarse, angular rock fragments, with fine soil matter filling the interstices.

Brassica–A genus of plants, family Crucifereae, which includes the common cabbage, turnip, black mustard, etc. Native to Asia, Africa, and Europe.

Brasswood–A common name for osage orange. See Bois d'Arc, Osage Orange.

Bread Grain–A cereal, such as wheat or rye, which yields flour for leavened bread. Other cereals are called coarse grains or feed grains.

Breadbasket of the Nation–Colloquial name for the Corn and Wheat Belt regions of the north-central United States.

Breadfruit Tree–*Artocarpus communis*; a tropical tree, family Moraceae that has heavy foliage of thick oval leaves, small flowers, and bears an edible fruit, called breadfruit. Native to Polynesia; grown rarely in the extreme southern tip of Florida.

Breadroot Scurfpea–*Psoralea esculenta*; a perennial herb, family Leguminosae, which has large, tuberous roots. Native to western North America, and once extensively eaten by the American Indians. Also called pomme de prairie, pomme blanche, prairie potato, meadow potato.

Breadstuffs–Various grains used for making bread.

Break Ground–To plow land for the first time.

Breaker–(1) A wave meeting a shore, reef, sandbar, or rock and collapsing. (2) For watering: a widened fan-shaped device screwed on the end of a hose to decrease (break) the velocity of water; used in watering plants. (3) For plowing: an extended moldboard on a turning plow to more completely turn heavy sod upside down. (4) For electrical overloading: an automatic flip switch on an electrical circuit that breaks (shuts off) the current when overloaded. (5) Utility-grade slaughtered cattle in which some degree of marbling is in the meat. The loins and rounds are "broken out" and sold as steaks.

Breaking–Plowing a native or planted sodlike crop.

Breaks–(1) Bluff land or steeply sloping escarpments broken by stream dissection, as in the eastern edge of the High Plains, south of the Arkansas River to the Texas Panhandle. This area is known as Breaks of the Plain. (2) The broken land at the border of an upland that is dissected by ravines. (3) An area in rolling land eroded by small ravines and gullies; also used to indicate any sudden change in topography, as from a plain to hilly country.

Breast High Diameter–In forestry, the point at which the diameter of a tree is measured. Custom has fixed this point at 4½ feet above ground level. Also called diameter at breast height.

Breathing Pore–See Lenticel, Stoma.

Breed–(1) Animals having a common origin and distinguishing characteristics. (2) To improve, through control, characteristics in plants and animals.

Breeder–(1) The owner of the dam when she was bred. The definition holds true in registering all classes of livestock. (2) A specialist in breeding. (3) An animal or plant used for breeding. (4) Also known as a converter; a nuclear reactor that converts nonfissionable atoms of one element into fissionable atoms of the same or a different element. An advanced breeder is a converter that produces more fissionable fuel than it consumes.

Breeder Seed–Seed directly controlled by the originating or sponsoring plant-breeding institution, person or designee thereof; the source for the production of the foundation, registered, and certified classes of seed.

Brewer Oak–*Quercus oestediana*, family Fagaceae; an oak used for forage on the western ranges of the United States.

Brewers' Dried grains–The dried residue left after wort has been extracted form barley malt; used in mixed feeds and equals corn gluten in content of digestible protein; it is unpalatable; also called dried brewers' grains, beer grains.

Brewers' Dried Yeast–A source of B complex vitamins from yeast filtered from fermented beer. It has roughly 50 percent protein and 1.5 percent phosphorus.

Bridge Graft(ing)–A method of preserving trees which have suffered from winter injury, rodents, or disease. A split with loose ends is made in the bark above and below the wound. Then a scion longer than the affected spot is cut and inserted in the slit. The ends are finally covered with wax.

Brier (Briar)–(1) Any plant or bush bearing thorns or prickles on a woody stem, as *Rosa, Rubus, Similax*, etc. (2) A thorn, twig, or branch of a brier. (3) *Erica arborea*; a shrub whose roots are used in making pipe bowls. Native to southern Europe.

Bright Tobacco–Cured light-yellow tobacco.

Brimstone–See Sulfur.

Brindle–(1) An animal which is tawny, gray, or brown, having irregular, dark streaks or spots. (2) Tobacco mosaic.

Brine–A salt solution of water used for food preservation, refrigeration, etc.

Brine Poisoning–Excessive consumption of salt by animals, resulting in severe vomiting, thirst, salivation, diarrhea, flatulence, dilation of the pupils, and hypersensitivity to touch. Convulsions and blindness occur in severe attacks. Also called salt poisoning.

Bristle–Stiff, sharp hair or hairlike parts which grow on animals and plants.

Broad-base Terrace–An erosion control system for steep slopes. It consists of a horizontal ridge of earth 10 to 30 inches high, 15 to 30 feet in base width with gently sloping slides, a rounded crown, and a broad, shallow water channel along the upper side. It may be level or have a slight fall toward one or both ends. Crops can be planted on the terrace. See Narrow-Based Terrace, Nichols Terrace.

Broad-leaved Evergreen–Evergreen plants which are not coniferous, e.g., the box, rhododendron, eucalyptus, wattles, etc.

Broad-leaved Plants–Botanically, those classified as dicotyledons. Morphologically, those that have broad, usually compound leaves.

Broad-spectrum Pesticide–A pesticide that kills a wide variety of insects. It may kill many beneficial insects, fish, birds, and mammals as well as target pests.

Broadbean–*Vicia fabia*, family Leguminosae; one of the most ancient cultivated food plants. It has a strong, upright stalk with small leaflets, and fruits in a pod, ½ to 1 foot long, which contains edible seeds. Native to North Africa and southwestern Asia; seldom grown in the United States. Also called horsebean, Windsor bean.

Broadcast–(1) To scatter seed or fertilizers uniformly over the soil surface rather than placing it in rows. (2) To sow seed in all directions by scattering. (3) To scatter manure, lime, etc. (3) To plow, throwing the soil in one direction, so that the field is left level.

Broadcast Burning–Allowing a prescribed fire to burn over a designated area within well-defined boundaries for the reduction of fuel hazard, as a silvicultural treatment, or both.

Broadcast Seeding–Scattering seed on the surface of the soil; contrast with drill seeding, which places the seed in rows in the soil.

Broadleaf–(1) Cigar tobaccos with broad leaves. (2) Tobacco leaf used for wrapping cigars. (3) A tree or a plant, usually a weed, with wide leaves.

Broads–In Canada, lumber 12 or more inches wide.

Broccoli–*Brassica oleracea italica*; an herb, family Crucifereae; the thickened flower branches are popular as a food; also called asparagus broccoli, sprouting broccoli, branching broccoli, Italian broccoli, calabrese.

Broken–(1) Describing land plowed for the first time. (2) Of plants or flowers, having irregular color distribution.

Broken Flat–Designating land which has been broadcast plowed. See Broadcast.

Broken Knees–A term used to describe horses whose knees have scars on them, indicating that the horse has fallen. Often scars are an indication that the horse is awkward and inclined to stumble.

Brome Mosaic–A viral disease of smooth bromegrass, caused by *Marmor graminis*; characterized by local lesions, yellow stripes on the foliage, and death of new growth. Occurs in Kansas and Nebraska.

Bromegrass, Field–*Bromus arvensis*; a winter annual bunchgrass which develops a fibrous mass of roots useful in cover cropping to control erosion and improve soil tilth. Seedlings are winter-hardy to -30°F (-34°C) and produce tall rank growth, tillering profusely in the spring. Should be fall planted.

Bromegrass, Smooth–*Bromus inermis*; highly variable, cool-season, cross-pollinated, palatable, long-lived, sod-forming grass. Long grown for hay and pasture, it is also very useful in erosion control seedings. Adapted to fertile, well-drained soils. Often planted in mixture with alfalfa.

Broom Pine–Long-leaf pine

Broom, Scotch–*Cytisus scoparius*; a nitrogen-fixing, small, leguminous, tap-rooted shrub adapted to well-drained, low-fertility soils in the mild climate west of the Cascades. It is used for secondary stabilization on coastal sand and for roadside or erosion control plantings. It has a prolific crown of yellow flowers in early summer.

Broomcorn–*Sorghum vulgare* var. *technicum*; a stiff grass, family Gramineae; used in making brooms and brushes. The mature seed is fed to poultry.

Broomrape–A root-parasitic herb of the genus *Orobanche*, family Orobanchaceae; subsists on herbaceous and woody plants. *Orobanche ramosa*, native to Eurasia, injures tobacco, tomato, muskmelon, etc.

Broomsedge–A grass, *Andropogon virginicus*, family Gramineae; a widespread weed on poor land in the southeastern United States. It is an inferior livestock forage, but a covering of it protects poor soils from erosion. Also called broom grass, yellowsedge, bluestem.

Brown Grain Sorghum–A market classification which includes all varieties of brown grain sorghum and very few other varieties.

Brown Hay–Hay discolored by fermentation because of insufficient drying, which decreases its nutrient value.

Brown Heart–A common disease of beets and turnips caused by a boron deficiency.

Brown Lacewing–An insect, family Chrysopidae, that preys on citrus mites.

Brown Leaf Rust–A fungal disease of rye caused by *Puccinia dispersa*, family Pucciniaceae; a widespread, injurious rust of rye. Long brown or reddish-brown, granular pustules burst through any part of a plant, but mostly on the stem and leaf sheath.

Brown Leaf Spot of Tobacco–A fungal disease occurring in periods of high temperature and humidity; caused by *Alternaria tenuis*, family Dematiaceae; large, circular brown spots appear on the mature leaves after harvesting.

Brown Lint–Brown fibers which occasionally appear on any variety of cotton.

Brown Patch–A fungal disease of grasses occurring during periods of high temperature and humidity, caused by *Pellicularia filamentosa*, family Thelephoraceae; spots appear over the entire turf area; in time the turf turns brown and dies. See Dollarspot.

Brown Rat–*Rattus norvegicus*, family Muridae; the common rat; also called Norway rat, house rat, wharf rat, sewer rat.

Brown Rice–Rice with the hull removed but not polished. It retains the bran layers normally removed by milling.

Brown Slime Flux–A foul-smellling seepage from cuts in shade trees, usually birch, elm, and maple of declining vigor. The clear sap, which contains nutrients, is fed upon by fungi, bacteria, and insects, and then turns brown and slimy.

Brown Spot of Rice–A fungal disease caused by **Helminthosporium oryzae**, family Dematiaceae, which kills the seedling plants, discolors leaves with reddish-brown spots, and rots the stalk at the base of the head. It specks the seeds, reducing the yield. Serious in the United States. Also called rotten neck of rice.

Brown Spot of Smooth Bromegrass–A common disease caused by **Helminthosporium bromi**, family Dematiaceae; characterized by small, dark brown, oblong spots on the leaves in spring. The spots coalesce into large, yellow areas, the leaves later turn brown from tip to base and die. The disease attacks plants in the United States in wet June weather.

Brown Spot of Soybeans–A fungal disease caused by *Septoria glycines*, family Sphaeropsidaceae; it causes defoliation, especially in wet seasons, with eventual spotting of stems and pods. Widespread in Europe, Asia, and North America.

Brown Spot of Sugarcane–A widespread, minor disease, caused by *Cercospora longipes*, family Dematiaceae; narrow spots appear on the leaves, enlarging into ovals with straw-colored centers. The leaf blades die.

Brown Stem Rot–A fungal disease of soybeans, caused by *Cephalosporium gregatum*, family Moniliceae; characterized by a sudden blighting of leaves which resembles frost damage in fall. Leaf tissues become brown, and the entire leaf soon withers. A badly infested field has a brownish appearance.

Browning Root Rot of Cereals–A fungal disease of cereals caused by *Pythium arrhenomanes*, family Pythiaceae; characterized by rotting of root; common in central North America in marshy areas. The same organism attacks sugarcane in Louisiana and Hawaii.

Browntail Moth–*Nygmia phaeorrhoea*, family Lymantriidae; this moth is destructive to foliage of many trees. Its hair carries a poison which irritates the skin of humans.

Browse–Leaves, small twigs, and shoots of shrubs, seedling, and sapling trees, and vines available for forage for livestock and wildlife.

Brush Control–Control of woody plants.

Brush Matting–(1) A matting of branches placed on badly eroded land to conserve moisture and reduce erosion while trees or other vegetative cover are being established. (2) A matting of mesh wire and brush used to retard streambank erosion.

Brush Monkey–A lumber term for someone who cuts down small trees so that logs may be removed.

Brushland–An area of little commercial value except perhaps for grazing goats and sheep.

Brussels Sprouts–*Brassica oleracea* var. *gemmifera*; an annual or biennial herb, family Cruciferae; grown for its axillary buds, which develop into small, edible heads. An important vegetable plant related to the cabbage.

Bryology–That branch of botany dealing with mosses.

Bucare Coralbean–*Erythrina poeppigiana*, a common coral tree, family Leguminosae; grown for its striking cinnabar-red flowers. Also used in the tropics to shade young coffee and cacao plants. Probably native to Peru; grown in Florida and California.

Buck–(1) A male goat, sheep, rabbit, deer, or antelope. (2) Wool from rams. See Bucks. (3) To saw felled trees into logs. (4) The common beech or its nuts. (5) Of a horse, to quickly leap with back arched and head held low.

Buck Moth–*Hemileuca maia*, family Saturniidae; a moth whose caterpillars infest oaks and willows in the autumn and feed voraciously on the leaves.

Buckbrush Ceanothus–*Ceanothus cuneatus*, a shrub, family Rhamnaceae; a forage for livestock on the western ranges of the United States. Also called wedgeleaf ceanothus.

Buckhorn Plantain–*Plantago lanceolata*, family Plantaginaceae; a common, serious weed found in lawns and fields. Eaten with some relish by sheep. Native to Europe, but widespread in the United States. Also called buckhorn, English plantain, narrow-leaved plantain, ribwort, ripple-grass, rattail, ribgrass.

Bucking–Cutting a felled tree into specified log lengths.

Buckra–Denoting whiteness of certain vegetables.

Buckshot Soil–(1) Soils containing hard, round iron and manganese concretions of buckshot size. Occurs in the southern United States. (2) Wet clay soil, which upon drying disintegrates into aggregates about the size of buckshot. Descriptive of the soils of the Yazoo bottomland of the Mississippi River in the southern United States. Most common soil is Sharkey clay.

Buckskin–(1) A disease of citrus, usually grapefruit and oranges, caused by the mite *Phyllocoptes oleivorous*; characterized by a crusted, leathery condition of the rind. (2) A term for a grayish-brown horse. (3) The skin of a buck deer. (4) A disease of sweet cherry, the symptoms of which vary according to the strain of virus and the stock. In general, the fruit loses its luster and drops or fails to ripen. (5) A horse that is light brown and has a black mane and tail.

Buckthorn–(1) *Rhamnus cathartica*, family Rhamnaceae; a poisonous shrub once used for hedges but now escaped in fence rows and pastures in the eastern United States. Its bark, leaves, and berries contain a purgative, but being bitter, are seldom eaten by livestock. Native to Eurasia. Also called Hart's thorn, waythorn, Rhineberry. (2) *Ceanothus sordiatus*; a thorny shrub, also called jimbrush. It is the alternate host for the fungus causing the crown rust of oats.

Buckwheat–*Fagopyrum sagittatum*, family Polygonaceae; a fleshy, annual herb whose grain is sometimes erroneously considered a cereal. It is used in making griddle cakes and as a stock and poultry feed. It is harmful to unpigmented stock, which developed photosensitization after eating it. Native to Asia; a minor grain crop in the United States. Also called fat hen, brank.

Bud–(1) A protuberance containing miniature leaves or flowers, located terminally or laterally on a stem. (2) An undeveloped shoot or stem. (3) To graft by inserting a bud in a slot in the stem of a different plant. It is economical, as only a single bud, instead of a scion, is used. Also called budding.

Bud Blasting–The death of the developing flower bud, usually because of water stress.

Bud Cutting–A method of quick propagation of plants. Parts of a stem are cut and split lengthwise so that each cutting has one bud. These are then buried under a shallow covering of soil to germinate. Also called single eye cutting, leaf bud cutting.

Bud Graft–The union of two plants by inserting the bud of one into the stem of the other.

Bud Mutation–A genetic change in the tissues of a shoot of a grafted bud. See Bud Variation, Sport.

Bud Variation–Abnormal variations in stems, leaves, or fruit which can be transmitted by asexual reproduction. See Bud Mutation, Sport.

Budburst–In woody plants, the time in the spring when flower or leaf buds begin their annual growth.

Budcap–A piece of paper or other suitable material attached to a young seedling, covering the terminal bud to prevent animal browse.

Budding–A form of grafting whereby a bud from a parent plant is placed in a normal position on the cambium of the stock.

Budworm–Larvae of the spruce budworm (*Choristoneura fumiferana*) and the black-headed budworm (*Acleris variana*), both serious forest pests in Canada and the United States.

Buffalobur Nightshade–*Solanum rostratum*, family Solanaceae; an annual herb commonly found in waste places with light, sandy soil. Its berries or herbage may poison animals. Native to North America. Also called buffalobur, beaked nightshade, prickly nightshade, Texas thistle, Colorado bur, sand bur, yellowbloom sticker weed.

Buffalograss–*Buchloe dactyloides*, family Gramineae; a short, valuable, native forage grass of the Great Plains of the West, occasionally used as a lawn grass in that area. Also called early mesquite.

Buffer–(1) A substance in soil, such as organic matter, clay, carbonates, or phosphates, which resists changes of soil pH. (2) Animals which serve as food for predators, thus reducing danger to game. (3) A tool used to cut clinches from horseshoe nails before removing the shoe; also called clinch cutter.

Buffer Capacity of Soils–The ability of a soil to resist a change in its hydrogen-ion concentration; or to resist the tendency to become more acid upon the addition of an acid (or an acid-forming material), or more alkaline upon addition of a base (or a base-forming material).

Buffer Solution–A solution to which large amounts of acid or base may be added with only a very small resultant change in the hydrogen-ion concentration.

Buffer Strip–Rows of vegetation planted along contours to reduce erosion.

Buffer Zone–Public land surrounding and supplementing private range lands.

Buffering Action–Resistance to change in acidity or alkalinity. Also called buffering.

Buffering Agent–Substance that adjusts the pH level of a spray mixture on a plant or soil surface. Often these materials can make a pesticide work better and last longer.

Bug–(1) Any insect, specifically of the order Hemiptera. (2) A flaw in the construction and operation of machinery.

Bug Kill–Trees or timber stands killed by insects.

Bulb–(1) The subterranean bud of some plants, which has a short stem bearing overlapping, membranelike leaf bases, as in onions and tulips. It stores food for reproduction and represents the inactive stage of the plant. (2) Any plant or flower shaped like a bulb. (3) The upper part of a horse's heel.

Bulb Cellar–A room used to give bulb crops a cold temperature treatment; also called a rooting room.

Bulb Planter–A device consisting of a tube connected to a handle that is used for making an opening in the ground to plant a bulb.

Bulb Waterhemlock–*Cicuta bulbifera*; a perennial herb, family Umbelliferae, which causes stock poisoning, the symptoms of which are nausea, difficult respiration, dilated pupils, frothing, weak and rapid pulse, and convulsions. Native to North America.

Bulbel–A small bulb produced above the ground, among the flowers or in the axil of a leaf. It is capable of reproducing the plant. Also called bulbil, brood bud.

Bulblet (Bulbil)–A small bulb or bulblike structure, usually borne in leaf axils, or among or in the place of flowers, or in other unusual places, as in pineapple.

Bulbous Bluegrass–*Poa bulbosa*, family Gramineae; a perennial grass with a bulblike growth; occasionally found in pastures. Native to Europe. Also called winter bluegrass.

Bulgur–Wheat which has been parboiled, dried, and partially de-branned for later use in either cracked or whole-grain form.

Bulk Density–The relative weight of a soil aggregate (air space plus solid soil particles) compared with the weight of the same volume of water.

Bulk Fertilizer–Commercial fertilizer delivered to the purchaser, either in the solid or liquid state, in a nonpackaged form to which a label cannot be attached.

Bull Pine–See Ponderosa Pine.

Bull Thistle–*Cirsium lanceolatum*; a weed found in pastures and waste places. Its prickly leaves are an annoyance and, when numerous, interfere with grazing.

Bullace–(1) A European plum, *Prunus domestica institia*. (2) Any wild grape or muscadine (southeastern United States).

Bulrush, Alkali–*Scirpus paludosus*; a coarse, perennial marsh plant growing 2 to 4 feet (61 to 122 centimeters) tall with underground, tuber-bearing rhizomes. It has triangular stems and bears a cluster of small, conelike seedheads at the apex of the top leaves. It is adapted to saturated soils and areas of shallow water. It is choice food for some waterfowl.

Bumblebee–An insect of the genus *Bombus*; valued for flower pollination and honey, which it often stores in underground nests. Its colonies are small.

Bunch–(1) A cluster of plants or fruits, as a bunch of grapes. (2) In pigs, an improperly healed wound following castration. (3) To skid logs together for hauling. (4) To pile harvested crops in the field.

Bunch Beans–Green beans that grow on a bush rather than on a vine.

Bunch Peanut–A type of peanut whose plant is bushy and erect; largely grown for forage. See Runner Peanut.

Bunchgrass–A grass with bunch or tussock habit of growth due to the absence of rhizomes and stolons.

Bunt of Wheat–A fungal disease caused by infection at germination by parasites. *Tilletia caries* and *T. foetida*. The stricken plant and its head remain undersized. The glumes tend to be loose; the head becomes bluish-green and never regains the normal color of ripened wheat. individual grains in the head are filled with an oily black powder, mainly smut spores and, if broken, give an offensive odor, hence the name stinking smut. In bearded varieties, awns are often shed. Also called high smut, low smut, pepper brand, smut ball.

Bur–(1) The rough, sticky, or prickly envelope of any fruit. (2) A mature hull on the cotton boll. (3) The cone on the hop plant at flowering. (4) A weed with burrs. (5) To remove burrs, as in wool cleansing.

Bur Artichoke–The flower head of artichoke; used as a vegetable along with its oval bracts.

Bur Oak–*Quercus macrocarpa*, family Fagaceae; a very hardy massive oak, with its acorns nearly covered by big, fringed, scaly cups. Its wood is pest-free and very durable, and is commonly used for railroad ties, agricultural implements, furniture, etc. Also called mossy-cup oak.

Bur Sage–Any shrub of the genus *Franseria*, family Compositae; the characteristic vegetation of the southwestern deserts in the United States; often a cause of hay fever.

Burdekin Plum–*Pleiogynium solandri*, family Anacardiacae; an evergreen tree with clusters of greenish flowers. It bears an edible, plumlike, red fruit. Native to Australia; grown in Florida and California.

Burdock—Plants of the genus *Arctium*, family Compositae, all of which are weeds.

Burgeon—(1) To bring forth; to sprout buds. (2) To grow forth, as an appendage or a disease in an animal. (3) A young bud.

Buried Soil—The soil of an original land surface which is buried to considerable depths by subsequent geologic deposition. Such soils have been observed in sections of thick alluvial deposits; soils of interglacial periods, e.g., found in glacial deposits; and soils in regions of volcanic action.

Burl—(1) A knotlike growth on the trunks of trees or plants; often produced with adventitious buds. (2) A distorted grain in lumber surrounding the pits of undeveloped buds. Considered valuable, if of sufficient size.

Burley—A strain of tobacco, very light-colored when cured. Grown in Kentucky and Tennessee.

Burn—(1) An area in which the trees and/or grass have been destroyed by fire. (2) A branding iron. (3) In plants, to give a scorched appearance by excessive use of pesticides. (4) To blister the skin of an animal by friction with a rope or hobble, etc. (5) To become withered due to lack of moisture, as the foliage of a tree. (6) To become discolored because of insufficient drying, as in lumber.

Burn of Potato—A disease caused by heat, intense light, and a lack of moisture. The tips of potato leaflets become wilted, turn yellow, and later brown. The browning kills the tip tissues and spreads further in the vascular system of the leaf, destroying its various portions.

Burned Lime—Limestone that has been heated until it forms a powder.

Burning, Prescribed—See Prescribed Burning.

Burrel—(1) A variety of pear with soft, tasty flesh. (2) Blue-colored wild sheep found in the Himalayas.

Burroweed—*Suaeda fruticosa*, family Chenopodiaceae; a common weed which has lowered the grazing capacity of large areas in the western United States.

Burrowing Nematode—*Radopholus similis*, family Tylenchidae; a worm which attacks citrus, banana, sweet potato, sugarcane, edible canna, etc.

Bush—(1) A grove of sugar maple trees. See Sugar Bush. (2) A forest wilderness or vast scrubland (Canada, Alaska, New Zealand, and Australia; formerly used in the northern United States also). (3) A shrub. (4) A method of training plants and fruit trees to assume a desired shape by means of stakes, ties, wires, and other supports. (5) The brush of an animal; the tail. (6) To force a seller to accept a lower price for a horse than was bid in the auction ring.

Bush Bean—Beans cultivated for food which grow in bush form and do not send out climbing runners.

Bush-fallow—An Appalachian region term for growing in rotation a crop on cleared land before allowing wild vegetation to cover it. In ten years or so the trees are cleared and a cultivated crop is again grown. See Shifting Cultivation.

Buster—(1) One who trains animals for saddle or work. (2) Middle-buster (a lister) that moves soil to the right and left simultaneously.

Butt—(1) The thick base or lower end of a tree or plant. (2) A stump, especially a walnut stump. (3) In butchering, the upper half of a ham or shoulder. (4) In Australia, a package of wool containing 196 pounds greasy or 112 poundsscoured, with a tare weight of more than 11 pounds. (5) To strike with the head or horns, as do cattle, goats, or rams.

Butt Cut—(1) In tree harvesting, the first log to be cut above the stump. (2) Bark removed from the butt of a tree before it is felled; specifically for tanbark; also called butt log.

Butter Bean—Any of the several types of lima bean found in the southern United States.

Butternut—(1) *Juglans cinerea*, family Juglandaceae; a tall, deciduous nut tree. The husk of the fruit was once used to dye cloth yellow. The nut meat is oily, rich, and spicy. The wood is used for furniture, cabinet work, interior trim, and novelties. Found from New Brunswick to Georgia and west to the Dakotas and south to Arkansas. Also called white walnut. (2) In the early settlement period of the eastern United States, a backwoodsman who wore clothing dyed yellow with butternut bark.

Button—(1) An irregularly shaped berry. (2) A bud. (3) A round seed vessel. (4) Any stunted or immature fruit. (5) A round, firm, cheesy curd of condensed milk; a defect of the body of the milk. (6) A nipple, especially of a hog. (7) A partially dismantled queen bee cell in a beehive, which resembles a small acorn cup. (8) Onion set. (9) A stunted or immature horn growth, as on a calf. (10) A metal clasp used to connect sections of a check row wire. (11) Cartilage on the chine bone of cattle (12) Any shell-like bone construction of the body; also called concha. (13) A leather ring for adjusting a horse's bridle.

Button Clover—*Medicago orbicularis*, family Leguminosae; a forage plant introduced into the United States from Europe. Also called button medic.

Button Weed—*Malva rotundifolia*; a common weed found in fields and around buildings.

Buttoning—The formation of tiny heads of cole crops on small plants, caused by the late transplanting of overly mature seedlings.

Buttress—Swelling (ridges) of a tree at its base near the soil, common among trees of the humid tropics and subtropics, especially on wet soils.

By-product—A product of significantly less value than the major product. In beef cattle, the major product is meat; by-products include the hide and other items.

By-product Lime—Any material containing calcium and/or magnesium resulting from the manufacture of another product, e.g., iron smelting, sugar factory lime.

C Horizon–Soils are composed of one or more horizons. The A horizon is the surface, the B horizon is the subsoil, and the C horizon is the underlying parent material above the R horizon, bedrock. See Soil Horizon.

Cabbage–(1) *Brassica oleracea* var. *capitata*, family Cruciferae; a widely cultivated plant with a mass of enlarged leaves packed together around a short stem, forming a large budlike structure, the cabbage head; an extremely valuable vegetable. Developed from an ancient coastal weed of Europe, *B. oleracea*. See Broccoli, Brussels Sprouts, Cauliflower, Collard, Kale, Kohlrabi. (2) A cheese whose holes are so large and numerous that the cheese between is often paper thin. Its cross-section has a cabbagelike appearance.

Cabbage Looper–*Trichoplusia ni*, family Phalaenidae; a larva which infests the leaves of crucifers, lettuce, spinach, beet, pea, celery, parsley, potato, tomato, carnation, nasturtium, and mignonette.

Cabbage Maggot–*Hylemya brassicae*, family Muscidae; a larva which infests the roots of cabbage, cauliflower, broccoli, Brussels sprouts, radish, turnip, beets, cress, and celery. The adult, a two-winged fly, lays the eggs near the plant stem. After hatching, the larvae promptly eat into the roots, honeycombing them with tunnels, and thus rotting them. Also called cabbage root maggot.

Cablegation–A method of irrigating a field from an irrigation pipe in which a traveling plug fastened by a cable to a braking mechanism is used to help control the amount of water released by gates or holes in the pipe. It was developed in 1981 at the USDA Snake River Conservation Research Center at Kimberly, Idaho, as a means to automate gravity-flow irrigation systems.

Cacao–*Theobroma cacao*, family Sterculiaceae; a tree from which the cocoa of commerce is harvested. Its yellow flowers grow on the trunk and larger branches of the tree and produce seed-containing oval pods several inches long. The fat extracted from these seeds is the cocoa-butter and the remainder of the seed, dried and ground, is the cocoa. The seeds are also used to make chocolate. Native to America; cultivated extensively in high rainfall areas of Asia and Africa.

Cactus–Any desert plant of the family Cactaceae. Mostly spiny, some are regarded as ornamentals and others as food for range livestock. Native to America, except the genus *Rhipsalis (Hatiora)*, which is sometimes found in Madagascar.

Cadmium–(Greek; *kadmia*, earth) A bivalent metal similar to tin. Its atomic number is 48 and atomic weight 112.40. A metallic "heavy metal" used in the production of copper, lead, silver, and aluminum alloys, in photography, ceramics, and in insecticides. Cadmium in sewage sludge is of grave concern when sludge is applied to soils used for the production of food and feed crops. Cadmium is a hazardous pollutant to people, domestic animals, and shellfish. It also is an experimental carcinogen.

Cake–A residue from the pressed kernels of cottonseed, linseed, or soybean, etc., from which the oil has been removed. It is pressed into hard cakelike masses and is used as a feed.

Caking–Caking is a process by which a fertilizer loses its desirable free-flowing property. The principal reason for caking is increase of moisture content. When damp fertilizers are dried, the salts that were in solution in the film of moisture on the surfaces of the particles are crystallized. These crystals adhere tightly and tend to knit the particles together. Plastic substances cake under pressure. Chemical reactions between ingredients of a mixture may also cause severe caking. Caking can be prevented or lessened by granulation, maintenance of low moisture content in the fertilizer at all times, or by coating the particles with conditioning agents such as oils, waxes, and clays.

Calabash Tree–*Crescentia cujete*, family Bignoniaceae; grown for its large, hard-rinded fruit whose shells or husks are used for dippers, utensils, etc. Native to tropical America.

Calabazilla–*Cucurbita foetidissima*; an annual, trailing, herbaceous vine, family Cucurbitaceae, grown in warm areas for its fruits, which when dried are used as ornamental gourds. Its roots were once used as a cleanser and its leaves medicinally. Native to North America. Also called Arizona gourd, buffalo gourd, Missouri gourd, wild pumpkin, calibassa, calabazza, calabaza.

Caladium–(1) Any tuberous-rooted herb of the genus *Caladium*, family Araceae. Species are grown for their brightly colored leaves. (2) Caladium bicolor, a tropical American herb grown for its varied colored, large leaves.

Calcareous–Formed of calcium carbonate or magnesium carbonate or both by biological deposition or inorganic precipitation in sufficient quantities to effervesce carbon dioxide visibly when treated with cold 0.1 normal hydrochloric acid. Calcareous sands are usually formed of a mixture of fragments of mollusk shell, echinoderm spines and skeletal material, coral, foraminifera, and algal platelets.

Calcic–(1) Containing calcium, as calcic lime, calcic plagioclase, calcic pyroxene. Also said of igneous rocks containing such minerals. (2) Refers to igneous rocks having an alkali-lime index of more than 61.

Calcic Horizon–A soil layer high in calcium carbonate formed by precipitation from high lime layers above dissolving and reprecipitating.

Calcicole–A plant growing in soil rich in calcium. Synonym, calciphile. See Calcifuge.

Calciferol–Another name for vitamin D_2 obtained from irradiated ergosterol.

Calciferous–Producing or containing carbonate of lime.

Calcification–(1) A soil-forming process, under low rainfall conditions, whereby a calcareous layer is formed at or below the soil surface. (2) Replacement of the original hard parts of an animal or plant by calcium carbonate.

Calcifuge–A plant that grows best in acid soil. See Acidophilous, Calciphile, Calciphobe.

Calcify–To deposit or secrete lime salts, which harden, as when gristle (cartilage) becomes bone.

Calcination–The heating of a substance to effect a physical and/or chemical change. A common example in agriculture is the heating of limestone ($CaCo_3$, to produce calcium oxide (quicklime) (CaO) and carbon dioxide (CO_2). Quicklime is applied to soils when a rapid increase in pH is needed.

Calcined Brucite–A mineral high in magnesium that has been burned to a powder; e.g., serpentine, dolomite.

Calcined Clay–A clay that has been heated to reduce its shrink-sell properties when wet-dry. Calcined clay exhibits the physical properties of a fine sand.

Calcined Phosphate–A general term for several materials produced by heating phosphate rock, with or without reagents. Simply heating to 1,520°F (827°C) in a rotary kiln destroys all organic matter, converts the calcium carbonate to the oxide, and drives out some of the fluorine. This calcined phosphate is used in the manufacture of high-grade phosphoric acid. When heated with silica and moisture to 2,700°F (1,482°C) the fluorine is almost entirely expelled, and the phosphate rock is converted to tricalcium phosphate, also known as fused tricalcium phosphate. It contains about 24 percent P_2O_5 available by the 2% citric acid method. A similar product is sold to the livestock feed trade as defluorinated phosphate.

Calciphile–A plant that grows best in calcareous soil, e.g., black walnut. See Calcifuge, Calciphobe.

Calciphobe–An acidophilous plant, e.g., blueberries. See Calcifuge, Calciphile.

Calcite–A mineral, $CaCO_3$, calcium carbonate, in crystalline form. Sometimes used as a diluent for insecticidal dusts.

Calcitic Dolomite–A carbonate rock in which the percentage of calcite is between 10 and 50, and the percentage of dolomite between 50 and 90.

Calcium–Ca; a chemical element present in variable amounts in all soils. It is essential for plant and animal growth and is the principal mineral element in bones.

Calcium Ammonium Nitrate–A trade name for an ammonium nitrate-limestone mixture (20.5% N), made in Europe.

Calcium Ammonium Nitrate Solution–An aqueous solution of calcium nitrate and ammonium nitrate containing 20.5% N. It is used extensively in the western coastal regions of the United States.

Calcium Carbonate–Calcium limestone and oyster shells are composed largely of calcium carbonate ($CaCO_3$). In such forms it is used extensively to neutralize soil acids. It is also a principal component of dolomite. Marble (proper) differs from common limestone in being harder and more compact as a result of metamorphism. Chalk is a soft limestone deposited in saltwater. Pure calcium carbonate contains 56 percent calcium oxide (CaO) and 44 percent carbon dioxide (CO_2).

Calcium Chloride–$CaCl_2$; a chemical used as a bleaching agent and for controlling dust on roads.

Calcium Cyanide–$Ca(CN)_2$; a very effective fumigant for controlling insects and rodents.

Calcium Deficiency–A physiological condition of plants or animals which results from insufficient calcium in the body. In plants, new fibrous roots die within four weeks, then the terminal bud, and ultimately the plant. A deficiency of vitamin D in birds and mammals makes it impossible for them to utilize the calcium in their diets. In such cases, hens may lay normal eggs at the expense of calcium needed for bone development, resulting in crooked breastbones. In animals, it produces rickets, abnormal and fragile bones, and low blood calcium.

Calcium Hydroxide–$Ca(OH)_2$; a calcium compound, useful in raising the pH of soils. It is of greater neutralizing power than ground limestone. Also called slaked lime, hydrated lime.

Calcium Lactate–A calcium salt of lactic acid, used to induce thickening and more rapid clotting of the blood.

Calcium Nitrate–$Ca(NO_3)_2$; the calcium salt of nitric acid sometimes used as a fertilizer.

Calcium Nitrate-Urea–$Ca(NO_3)_2 \times CO(NH_2)_2$; as crystallized from solution this material consists of calcium nitrate with four molecules of urea of crystallization, replacing the water of crystallization. It contains about 33% N and 9% Ca.

Calcium Oxide–CaO; a salt produced by heating limestone, marble, etc. In the dehydrated form, it is used as a dust diluent, calcium carrier, and for raising the pH of soil. Also known under many other names.

Calcium Phosphate–The phosphatic part of good superphosphate consists largely of monocalcium phosphate ($CaH_4P_2O_8$), which is water-soluble. Precipitated bone is largely dicalcium phosphate ($CaHPO_4$), which is citrate-soluble. Bone meal contains calcium phosphate-carbonate ($3Ca_3[PO_4]_2 \times CaCO_3$). Apatite is calcium fluorphosphate or chlorophosphate, and phosphate rock usually contains a complex calcium fluorphosphate. The term *bone phosphate of lime* (BPL), first applied as a name when animal bones were the principal source of phosphorus fertilizer, is commonly used to express the content of tricalcium phosphate ($Ca_3(PO_4)_2$) in phosphate rock.

Calcium Polyphosphate–Any calcium salt of a polyphosphoric acid such as calcium dihydrogen pyrophosphate and dicalcium pyrophosphate. Fertilizer grades of calcium polyphosphate generally refer to phosphate mixtures containing a substantial proportion of one or more calcium polyphosphates and can be made by thermal dehydration of triple superphosphate.

Calcium Sulfate–$CaSO_4$. See Gypsum.

Calculi (singular, Calculus)–Stones of calcium carbonate or calcium oxalate which may form inside the bodies of people or animals due to stagnant fluids and other secretions.

Caldera–A large basinlike depression with steep sides in the top of a volcanic mountain.

Calendar Year– A period from January 1 through December 31 of the same year.

Calibration– In field agriculture, the setting of a distributor or spreader to meter the predetermined amount of pesticide, lime, or fertilizer.

Caliche– (1) In Chile and Peru, impure native nitrate of soda. (2) A desert soil formed by the near-surface crystallization of calcite and/or other soluble minerals by upward-moving solutions. (3) In Chile, whitish clay in the selvage of veins. (4) In Mexico, feldspar, a white clay. (5) A compact transition limestone. (6) In Colombia, a mineral vein recently discovered. (7) In placer mining, a bank composed of clay, sand, and gravel. (8) In the southwestern United States, gravel, sand, or desert debris cemented by porous calcium carbonate; also the calcium carbonate itself.

Calico– (1) (a) Any cotton cloth from India. (b) Coarse printed cotton cloth from the United States. (c) Plain white cotton cloth from the United Kingdom. (2) A spotted horse. (3) A viral disease of the potato, characterized by irregular blotches of various shades on the leaflets. It is one of the potato mosaics. (4) Tobacco mosaic.

California Buckthorn–*Rhamnus californica*, a shrub, family Rhamnaceae, which is planted as a bee tree in California. Found in the western United States.

California Bur Clover–*Medicago hispida*, family Leguminosae. A Eurasian forage which is good grazing throughout most of the year in the southwestern United States. It may also be used as a cover crop and for green manure. Also called bur clover.

California Danthonia–*Danthonia californica*, a perennial grass, family Gramineae; used for reseeding western ranges of the United States. Also called California oatgrass.

California Harvester Ant–*Pogonomyrmex californicus*; an ant which collects seeds. Often bites humans and animals severely. Found in the southwestern United States.

California Incensecedar–*Libocedrus decurrens*, family Pinaceae; an evergreen tree grown as an ornamental and for its valuable timber. Also called white cedar.

California Privet–*Ligustrum ovalifolium*, family Oleaceae; a hedge plant with small clusters of flowers, planted extensively in Asia and the United States. Native to Japan.

Callus– (1) In plants, protective covering that forms over a wounded surface. (2) Undifferentiated or unorganized tissue that grows from a plant cell or a piece of leaf when it is placed on media containing certain hormones; an early event in the regeneration of a whole plant from engineered cells. (3) In animals, an area of skin which becomes horny and thickens. (4) A hard exudate which forms at the end of a broken bone and eventually becomes part of it.

Calyx– The outer, usually green, leaflike parts of a flower.

Cambium– The actively growing cells between the bark and the wood in a tree or shrub. They give rise to secondary xylem and phloem of dicotyledonous stems.

Cambric Horizon– A soil horizon that has developed the color and/or structure of a subsoil horizon but does not qualify as an argillic or spodic horizon.

Camphor Tree–*Cinnamomum camphora*, family Lauraceae; an aromatic evergreen tree with alternate leaves, whose wood contains the camphor gum of commerce, obtained by steaming and sublimating the bark. Native to China.

Canada Blueberry–*Vaccinium canadense*, family Ericaceae; a species of wild blueberry with white flowers and dark blue fruits. Native to North America.

Canada Bluegrass–*Poa compressa*, family Gramineae; a perennial European grass, found throughout temperate and colder zones. Used widely in permanent pastures and lawns. Also called English bluegrass.

Canada Garlic–*Allium canadense*, family Liliaceae; an annual, bulbous weed. Cows feeding on it produce onion-flavored milk. Also called wild onion.

Canada Goldenrod–*Solidago canadensis*; a weed that produces extensive rhizomes and is a troublesome pest in pastures.

Canada Hemlock–*Tsuga canadensis*, family Pinaceae; a hardy evergreen tree, grown as an ornamental and for its valuable lumber used in pulpwood and general construction. Native to North America. Also called hemlock, eastern hemlock, hemlock spruce, tanbark tree.

Canada Potato– Jerusalem artichoke, *Helianthus tuberosa*, family Compositae; a wild sunflower which bears tubers on its roots.

Canada Rice– Annual wildrice.

Canada Wildrye–*Elymus canadensis*, family Gramineae; a perennial forage grass, widely distributed throughout the United States and Canada.

Canada Yew–*Taxus canadensis*, family Taxaceae; an evergreen, low-growing shrub, useful as a ground cover under evergreen trees. Highly poisonous. Also called ground hemlock, American yew.

Canal– (1) An artificial watercourse cut through a land area for navigation, irrigation, or drainage. (2) A long, narrow arm of the sea extending far inland. (3) On the Atlantic Coast, a sluggish coastal stream. (4) A long, fairly straight, natural channel with steeply sloping sides, generally a mile or more in width. (5) A cave passage partly filled with water.

Canarygrass– (1) *Phalaris canariensis*, family Gramineae; an annual grass grown for its seed, the canary seed of commerce, the most commonly used bird seed. It has escaped as a weed in many states of the United States. (2) See Canarygrass, Reed.

Canarygrass, Reed–*Phalaris arundinacea*; coarse, vigorous, productive, long-lived sodgrass; frost-tolerant, suited to wet soils (but also somewhat drought-tolerant). Once established it can withstand continuous inundation for seventy days in cool weather. Recommended for pasture on wetlands or in filter field for waste disposal water.

Cancer– (1) A malignant tumor of humans and animals. (2) In plants, a growth similar to that caused by crown gall.

Candy–A pastelike mixture fed to queen bees in transit. Made by kneading confectioners' sugar into honey or invert sugar. In Europe, candy for feeding bees during the winter is made by evaporating sugar syrup and recrystallizing it in blocks.

Cane–(1) Woody stems of any small fruits, as of the grape or raspberry. (2) Stems of reeds and large grasses, such as the bamboos and sorghums. Also sometimes applied to the stems of rosebushes and some small palms. (3) Any hollow or pithy stem which is flexible and slender. (4) Sugarcane. See Bramble.

Cane Brake–A thicket or dense growth of native bamboos, species of *Arundinaria*, found in the southern United States, especially Mississippi, Louisiana, Alabama. They are a good forage.

Cane Fruit–Any small fruit, such as the blackberry, raspberry, dewberry, which bears its fruit on canes. See Bramble.

Cane Hay–Hay produced from sweet sorghum.

Cane Pruning–Systematic cutting back of grapevines in order to limit the fruit yield, improve the quality, and keep the plant shapely and manageable.

Cane Renewal–Replacement of 2-year-old grape canes by new canes that will produce the next season's crop. Practiced in all systems of grape pruning.

Cane Sugar–The carbohydrate sucrose or saccharose in the juices of many plants, especially sugarcane; extracted and refined to form the sugar of commerce, $C_{12}H_{22}O_{11}$.

Canescent–A leaf surface that is covered with dense, grayish-white hairs.

Canker–(1) A diseased lesion of the bark and underlying tissue in woody plants. (2) An ulcer in the mouth. (3) A serious disease of the horse's hoof. The frog frequently discharges a stinking fluid, and ultimately the sole and the frog rot.

Canners' Alkali–Sodium carbonate-sodium hydroxide mixture; used to loosen fruit skin to make peeling easier, as in the peach.

Canning–Preserving food in airtight rigid containers. Microorganisms are destroyed by heat-processing containers of food at the temperature and time specified for each food. It is essential to follow reliable canning instructions exactly, to ensure a safe canned product that is free from botulism-causing bacteria and spoilage organisms.

Canning Factory Silage–Husks, cobs, and undesirable kernels of sweet corn, pea vines, pear wastes, etc., which are unfit for canning, but usable as a stock feed.

Canning-ripe–Fruit and vegetables of full size and good flavor; suitable for canning but not ripe enough for table use.

Canola–A cultivar of rape, *Brassica napus*, family Cruciferae. See Rape.

Canopy–The uppermost vegetation layer consisting of crowns of trees or shrubs in a forest or woodland.

Canopy Trees–Trees with crowns in the uppermost layer of forest or woodland.

Cant Hook–A hand-operated wooden lever with a variable iron hook on one end; used for handling logs.

Cantaloupe–(1) *Cucumis melo* var. *cantalupensis*; an annual herb with trailing vines, family Cucurbitaceae; grown in Europe for its edible fruit with a hard rind. (2) For American cantaloupe, see Muskmelon.

Canvas Hose Irrigation–A method of irrigation which originated in Michigan. A porous canvas hose, in which water is flowing under pressure, is placed on the ground alongside each row. The water seeps out of the canvas hose along its length and is absorbed by the soil. Synthetic fibers often replace canvas.

Caper Euphorbia–*Euphorbia lathyrus*, family Euphorbiaceae; a poisonous, annual herb; contact with the milky sap causes a dermatitis with blisters and inflammation. Animals will not eat the growing plant, but sometimes eat hay containing the dry plants, which causes swellings on the head and excessive scours and may eventually prove fatal. Native to Europe; occasionally grown in gardens. Its pods are sometimes used for pickles and seeds as a purgative. Also called English capers, Capuchin capers, caper spurge, molewood, mole plant, false caper.

Capers–The unopened flower buds of a plant grown in Greece, northern Africa, and southern Europe. The buds are pickled in salt and vinegar and used as a seasoning and a condiment.

Capillarity–The attractive force between two unlike molecules, illustrated by the rising of water in capillary tubes of small diameters or the drawing-up of water in small interstices, as those between the grains of soil or a rock.

Capillary Capacity–The amount of moisture which is held in the soil by capillary attraction after excess or free water has drained away.

Capillary Fringe–A belt of capillary interstices in the soil belt that overlie the zone of saturation, or water table. Some or all of these are filled with water, held by capillary action against the force of gravity. The fringe water may be used by deep rooting or phreatic plants.

Capillary Mat–A fiber mat designed to distribute water to potted plants in a greenhouse; the plants take up the water through capillary action.

Capillary Water–(1) Underground water that is held by the soil above the water table by capillary attraction. (2) The water held in the "capillary" or small pores of a soil, usually with tension greater than 60 centimeters (23.63 inches) of water. Much of this water is considered to be readily available to plants.

Capitate–(1) Headlike, formed like a head. (2) In heads, aggregated into a very dense or compact cluster.

Capitulum–A head; an inflorescence composed of a dense cluster of usually sessile flowers, as in clover (*Trifolium*) and the Compositae.

Capsicum–An important genus of tropical plants, family Solanaceae, which includes the common red pepper, long pepper, etc. The plants bear small, whitish flowers, and produce berrylike fruits which are used as vegetables and for pickling and flavoring. Native to America.

Capsule–(1) A fruit (dry at maturity) of more than one carpel but not necessarily of more than one locule, which dehisces at maturity, permitting the escape of the enclosed seeds. Typical capsules are found in the iris, lily, poppy, cotton. (2) Gelatine covering for medicines.

Car Lot–The quantity of produce or number of animals which can be transported in a railroad car.

Caragana–*Caragana arborescens*; also known as Siberian pea shrub. A very hardy, deciduous, leguminous shrub or small tree to 30 feet (10 meters). It has pinnate leaves with up to eighteen small leaflets and large pealike yellow flowers in early spring. Planted in eastern Oregon as a windbreak shrub. Widely adapted, drought- and cold-tolerant. Has excellent value for windbreak, fair value of wildlife food and cover, and fair to excellent ornamental value.

Caraway–(1) *Carum carvi*, family Umbelliferae; an annual or biennial herb whose aromatic seeds are used as a condiment. (2) Any cheese spiced with caraway seeds.

Carbamate–A synthetic organic pesticide containing carbon, hydrogen, nitrogen, and sulfur.

Carbohydrate–Any of certain organic chemical compounds of carbon, hydrogen, and oxygen, which include sugars and starches. Formed in plants by photosynthesis, carbohydrates make up a large part of animal feed.

Carbon–An essential chemical element component in plants and animals. It is present in soils, in humus, plant residues, charcoal, and particles of coal, carbonaceous shale, etc.

Carbon Bisulfide (Carbon Disulfide)–CS_2; a highly flammable colorless liquid, boiling at 115°F (46°C). Used for degreasing seed meals, such as castor and cottonseed meals and as a seed fumigant to kill insects.

Carbon Dioxide–CO_2; a colorless, odorless gas constituting 0.03 percent of unpolluted air. It is absorbed by green plants through the leaf stomata and is used as the source of carbon for manufacturing sugars, starches, proteins, and fats. The burning of fossil fuels has put so much CO_2 in congested cities that the concentration may reach more than three times the background (normal) level. This is causing a "greenhouse effect," resulting in a slow warming of the earth's surface.

Carbon Dioxide Evolution–The liberation of gaseous carbon dioxide from soil by biological processes.

Carbon Disulfide–See Carbon Bisulfide.

Carbon: Nitrogen Ratio–The value obtained by dividing the percentage of organic carbon by the percentage of total nitrogen in a soil or in an organic material. A narrow ratio results in the rapid decomposition of organic materials.

Carbon Tetrachloride–CCl_4, a chemical that is an excellent solvent for grease. It is not inflammable and can be used as a fire extinguisher. Frequently used as a fumigant, though not in large-scale operations; toxic when inhaled.

Carbon-14 Dating–The use of radioactive carbon, which has an atomic mass of 14 and an approximate half-life of 5,000 years, for determining approximately the age of organic materials in soils, buried materials such as wood, and other organic materials.

Carbonaceous–(1) Pertaining to, or composed largely of, carbon. (2) The carbonaceous sediments include original organic tissues and subsequently produced derivatives of which the composition is chemically organic.

Carbonate Cider–Aerated apple cider.

Carbonate of Lime–Calcium carbonate.

Carbonation–(1) Process of introducing carbon dioxide into water. (2) A process of chemical weathering by which minerals are replaced by carbonates.

Carcinogen–A chemical, physical, or biological agent that increases the incidence of cancer.

Carcinoma–Malignant cancerous growth.

Cardamom–*Elettaria cardamomum*, family Zingiberaceae; an herb grown in southern Florida. Its seed is used in medicine and as a condiment. Native to East Indies.

Cardoon–*Cynara cardunuclus*; a vegetable of the thistle family. The tender stalks and roots are used in soups and as a vegetable.

Careless Weed–See Redroot Amaranth.

Carex–A large genus of grassy-looking plants with hard stems, family Cyperaceae, commonly called sedges, a few of which are occasionally cultivated in gardens. Fiber from some of its species issued in matting. As a weed the plants are difficult to control.

Carnauba Wax–A hard wax derived from the leaves of a palm tree, *Copernicia cerifera*, family Palmaceae, of Brazil. Used in shoe polish, varnish, floor wax, etc.

Carniolan Bee–A dark honeybee race originating in southeastern Europe.

Carnivorous Plants–Literally, meat-eating plants; usually equivalent to insectivorous. As many as 450 species of plants have been reported to trap insects. The most common are Venus flytrap, pitcher plant, and sundew. About 150 species of fungi are known to trap nematodes.

Carob–*Geratonia siliqua*; a Mediterranean, evergreen tree, family Leguminosae; grown for its edible fruit, which is used as fodder for farm animals. Thought to be the "locusts" eaten by John the Baptist, and the "husks" eaten by the prodigal son. Also called St. John's bread, algaroba, bean tree.

Carolina Buckthorn–*Rhamnus caroliniana*, family Rhamnaceae; a tender shrub grown for its fruit, which turns black when ripe. Native to North America. Also called yellow bush, Indian cherry.

Carolina Grasshopper–*Dissosteira carolina*, family Arcrididae; a common grasshopper, 2 inches long.

Carolina Hemlock–*tsuga caroliniana*, a beautiful evergreen tree, family Pinaceae; commonly grown for ornament, especially in the southern United States. Native to North America. Also called spruce pine.

Carolina Horsenette–*Solanum carolinense*, a perennial weed herb, family Solanaceae; may be poisonous to livestock feeding on its

Carolina Laurelcherry–*Prunus caroliniana*, family Rosaceae; a small, evergreen tree, which has small cream-white flowers; planted for shade. Native to the southeastern United States. Also called evergreen cherry, American cherry laurel, wildorange, mock orange, cherry laurel.

Carolina Mantis–*Stagmomantis carolina*, family Mantidae; the praying mantis, so called because the front legs resemble hands folded in prayer. It preys on other insects but is not injurious to humans or other animals. Also called rearhorse.

Carotenase–An enzyme in the human body capable of converting carotene into vitamin A. See Carotene.

Carotene–An orange or red pigment found in such foods as carrots, sweet potatoes, milk fat, and egg yolk and also in green leaves. Carotene may be converted in the animal body into vitamin A; hence called the precursor of vitamin A.

Carpel–One of the units composing a pistil or ovary. A simple pistil has one carpel, a compound pistil has two or more united carpels.

Carpellate, Carpelled–Possessing or composed of carpels; when written 2-carpellate or 3-carpellate, meaning composed of two or three carpels, respectively.

Carpenter Ant–Ant species that bore into wood or a tree to build their nests. The American species are *Camponotus abdominalis* and *C. herculeanus*.

Carpenterworms–Larvae of carpenter moths, *Prionoxystus robiniae* and *P. macmurtrei*, family Cossidae, which burrow into trees and weaken them. Their presence is marked by sawdust or dark-colored sap coming out of small holes and discoloring the trunk. They attack the black and red oak, elm, locust, poplar, willow, maple, ash, chestnut, etc., and are distributed throughout central North America.

Carpet Weed–*Mollugo verticillata*; a common weed of gardens and cultivated fields. They form mats on the surface of the soil.

Carpetgrass–*Axonopus affinus*, family Gramineae; a stoloniferous, perennial grass sometimes used as a turf grass. Its forage rating is good. Native to the United States, found from Florida to Louisiana. Also called true carpetgrass.

Carpology–The study of structural anatomy of fruits and seeds.

Carpoptosis–Abnormal fruit drop.

Carriage–(1) A frame on which a log is held. (2) Manner or bearing of an animal; its physical control and behavior.

Carrier–(1) A genetic term that refers to an animal that expresses the dominant trait but is heterozygous for the recessive gene. (2) Animal or person in apparently good health who harbors pathogenic microorganisms. (3) The liquid or solid material added to a chemical compound to facilitate uniformity of application in the field.

Carrot–*Daucus carota* var. *sativa*, family Umbilliferae; an annual herb grown for its edible root used in salads and stews and as a vegetable; a very important root crop. Native to Eurasia.

Carrot Beetle–*Ligyrus gibbosus*, family Scarabaeidae; a black insect that feeds on the roots of carrot, parsnip, celery, beet, potato, cabbage, corn, cotton, sunflower, dahlia, amaranthus, and other plants. It is widespread throughout the United States except in the northern states.

Carrot Yellows–A serious viral disease of carrots; characterized by a yellowing of the young leaves, side shoots, and a bronzing or reddening and twisting of older, outer leaves. Later the crown turns black and the plant dies. This is the same virus as that of aster yellows and is transmitted by leaf hoppers, especially *Macrosteles fascifrons*, family Cicadellidae.

Caryopsis–Small, one-seeded, dry fruit with a thin pericarp surrounding and adhering to the seed, as in barley, wheat, etc.

Casaba Melon–Winter melon.

Cascade–A series of small, closely spaced waterfalls or very steep rapids.

Case–(1) A box and sometimes its contents, e.g., a case of eggs. (2) In tobacco leaves, to absorb moisture after curing so that the leaves become pliable enough for handling. (3) In mushroom culture, to cover the prepared bed with its final layer of soil.

Case Hardening–(1) A condition of drying wood in which the surface becomes hard and set. If the interior tensile stresses exceed the strength of the wood, honeycombing results. (2) A condition in the artificial drying of some evergreens when the scales become overdried and hard, usually due to low humidity in the dehydrator. This retards the rate of dehydration of the major portion of the fruit or vegetable. (3) Treatment of iron or steel machine parts with carbon so that the outer surface becomes hard to withstand wear and the center remains soft to withstand shocks.

Cash Crop–A crop sold directly on the market contrasted to one which his fed to animals or is otherwise used but not sold. Formerly applied mainly to crops which were readily salable, e.g., cotton, wheat, sugarcane, etc.

Cash Grain Farm–A farm on which corn, grain sorghum, small grains, soybeans, or field beans and peas account for at least 50 percent of the value of products sold.

Cashew–*Anacardium occidentale*, family Anacardiaceae; a tropical evergreen tree with small, sweet-smelling, pale pink flowers. It is chiefly known for its kidney-shaped fruit, the cashew nut of commerce, which is edible after roasting. Its bark has medicinal properties. Native to tropical America.

Cassava–*Manihot esculenta*, a woody shrub, family Euphorbiaceae; grown for its roots in the warmest parts of the United States as a stock feed. It is of importance as a source of tapioca starch and as a food plant in tropical America. The rootstock contains a poison which must be dissolved or heated out before it is edible. Also called tapioca plant, manioc, mandioca, bitter cassava, sweet potato tree.

Cassia–A large genus of trees and plants of the senna family (Caesalpinaceae) which is grown in warm regions. The leaves of many species are used in medicinal tea. The pods of *Cassia fistula* contain a pulp used as a laxative. Also called pudding tree, canafistula, drumstick tree. Seeds of *C. occidentalis* called coffee senna and Mogdad are marketed as an adulterant of coffee. *C. alata* is used in treating ringworm.

Cast–(1) The number of young produced. (2) A caterpillar which has had its entire larval body filled and displaced with a fungus which replaces any normal host structures, as by the fungus *Cordyceps*. (3) Device for immobilization of fractured bones. (4) A sheep which is unable to rise (New Zealand). (5) Wool which has fallen off a sheep before shearing (New Zealand). (6) Angleworm excreta. (7) To throw off a horseshoe, as by a horse when the nails become worn or loosened, etc. (9) To force to the ground, as to cast an animal to the ground to control it. (10) To broadcast seed or fertilizer.

Castings–(1) Excrement of earthworms, an important factor in the enrichment of soils. (2) Fecal pellets of the animal kingdom.

Castor Bean–*Ricinus communis*, family Euphorbiaceae; a tall, annual herb, with alternate leaves, the only species of genus *Ricinus*; grown for its large, striking foliage and seeds, which yield oil. The plant itself is poisonous, causing nausea, gastric pain, diarrhea, thirst, and the dullness of vision. Its ingestion in large amounts is lethal. Native to tropical Asia and Africa.

Castor Oil–The vegetable oil from the seeds of castor bean cultivated in the East generally. it is used as a purgative, a lubricant, and for dressing leather.

Castor Pomace–Residue of castor bean left after the oil has been extracted; an organic fertilizer containing up to 7 percent nitrogen, 1.5 percent phosphoric acid, and 1.5 percent potash.

Castrate–(1) To remove the testicles or to destroy their use; to geld. See also Spay. (2) To remove the stamens from a flower.

Cat Grape–*Vitis palmata*, family Vitaceae; a variety of grape from the central and southern United States. Also called Missouri grape.

Catalase–Enzyme in plants and animals; splits hydrogen peroxide into water and gaseous oxygen.

Catalpa–Any tree of the genus *Catalpa*, family Bignoniaceae; attractive flowering trees grown as street trees; grown to produce caterpillars that feed on the leaves. The caterpillars are used for fish bait. Native to America and Asia.

Catalpa Sphinx–*Ceratomia catalpae*, family Sphingidae; an insect which is a very serious pest of the catalpa leaves; distributed over the United States from New York to Colorado.

Catalyst–An agent that promotes interaction between or among other chemical substances without itself being changed.

Catch–(1) A sheep whose fleece is cut after the day's work has stopped (New Zealand). (2) A successful establishment from a seeding, e.g., a catch of grass or grain. (3) To germinate, sprout, and become established as a crop after sowing. (4) To conceive after breeding, as a mare.

Catch Crop–(1) A quick-growing crop, planted and harvested between two regular crops in consecutive seasons, or between two patches of regular crops in the same season. (2) Such a crop planted after another has failed, or when the season is too late for the usual crops.

Catclaw Acacia–*Acacia greggi*, family Leguminosae; a forage plant of poor pasture rating that grows in semi-arid areas of the western United States. It is not very palatable, but is utilized in dry years when other feed is scarce. Also called devil's claw, cat's claw.

Catena–(1) A sequence of soils of about the same age, derived from similar parent material, and occurring under similar climatic conditions but having different characteristics due to variation in relief and drainage.

Caterpillar–The wormlike larva of a butterfly, moth, or other insect. All larvae have several pairs of legs, short antennae, and strong cutting jaws. Some are covered with hairs. They feed on the succulent parts of the plant and are known as canker worm, army worm, cutworm, etc.

Catface–(1) A healing or healed wound on the trunk of a tree; a blemish. (2) Fruit abnormalities of tomato, attributed to various unknown growth disturbances, partially climatic. (3) A scar or irregular growth on an apple or peach that results from insect stings.

Cation–The ion in solution carrying one or more positive charges of electricity depending on its valence. The common soil cations are calcium, magnesium, sodium, potassium, hydrogen, and ammonium.

Cation Exchange–The interchange among cations in soil solution and cations on the surface of clay, humus, and/or plant roots. For example, in acid soils three calcium ions can exchange with six hydrogen ions or two aluminum ions. In sodic soil reclamation, one calcium ion replaces two sodium ions.

Cation Exchange Capacity (CEC)–(Formerly called "base exchange capacity") A measure of the total amount of exchangeable cations that can be held by a given mass of soil, expressed in milliequivalents (meq) per 100 grams of soil at neutrality (pH 7) or at some other stated pH value. The exchange capacity of soils usually ranges between 2 and 50 meq/100 grams of soil. Cations are interchanged primarily on the surfaces of roots, clay, and humus.

Catkin–An inflorescence; a spike or raceme having a very flexible rachis, which often makes it appear to be hanging. It has a scaly appearance in plants that lack petals on the flowers. However, this scaliness is not noticeable when the petals are formed.

Catnip–*Nepeta cataria*, family Labiatae; a perennial herb grown for medicinal purposes. Its fresh or dried leaves are very attractive to cats, and the oil is used as a lure in trapping wild animals of the feline family. Also called cat-mint.

Cattail–(1) Any marshy reed of the genus *Typha*, family Typhaceae, notable for its long, cylindrical, brown spike containing innumerable, tiny, petalless flowers. Generally 5 or 6 feet high, the plant has an astringent rootstock and long, thin, stiff leaves. It is the principal plant of bogs and marshes, and its young shoots are sometimes eaten in England and the U.S.S.R. The floss from the spike is useful for insulating purposes and filling mattresses. The spike itself is sometimes used as a cut flower for decoration. The leaves are used in making mats, baskets, and chair seats. (2) *Typha latifolia*, the common cattail. Also called bulrush, cat's-tail reed, cattail flag, cooper's flag, cooper's fly, ditch-down, reed mace, cattail rush.

Cattle Saltbush–*Atriplex polycarpa*, family Chenopodiaceae; a forage plant found in the dried areas of desert shrub vegetation in North America. Also called desert saltbush.

Caudex–(1) The stout, persistent, usually underground stem base of a perennial herb, from which the annual stems arise. (2) The trunk of a palm.

Caulescent–Refers to a plant which has a readily perceived stem above ground.

Cauliflorous–Refers to a woody plant that produces an inflorescence directly from the trunk or one of the chief branches; e.g., the fig tree.

Cauliflower–*Brassica oleracea* var. *botrytis*, family Cruciferae; a stemless annual or biennial herb, grown for its white or green, enlarged, edible inflorescence. It is an important vegetable crop. Native to Europe. See Broccoli, Cabbage.

Caustic Lime–See Burnt Lime, Calcium Oxide.

Cavity–(1) A hollow space within a body, e.g., the buccal cavity. (2) Any hole in the trunk or branches of a tree.

Cayenne Pepper–A condiment prepared by grinding the seeds of several plants of the genus *Capsicum*, e.g., redpepper, chili, etc.

Ceanothus–A large genus of beautiful flowering plants, family Rhamnaceae, a few species of which are grown for their clusters of small blue or white flowers. Native to North America. Also called wild lilac.

Cedar–(1) Any tree of the genus *Cedrus*, family Pinaceae, especially the deodar cedar, Atlas cedar, Cyprian cedar, cedar-of-Lebanon. (2) Commonly but not accurately, any tree of the genus *Juniperus*, family Pinaceae, as the southern red cedar, eastern red cedar, Rocky Mountain juniper. (3) Commonly but incorrectly, certain trees of the genus *Cedrela*, family Meliaceae, as cigarbox cedrela. (4) Commonly but incorrectly, any tree of the genus *Libocedrus*, family Pinaceae, as California torreya. (6) Commonly but incorrectly, any tree of the genus *Chamaecyparis*, family Pinaceae, as white cedar, false cypress.

Cedar Apples–Hard, brown, spherelike excrescences of *Juniperus* spp. caused by the rust fungus Gymnosporangium *juniperi-virginianae*, family Pucciniaceae, up to 2 inches in diameter, scattered over the tree, which sometimes bend the branches with their weight. They represent the telial stage in the development of the fungus, which may also spend part of its development on the apple or other members of the family Rosaceae. Also called cedar balls, cedar galls, cedar flowers. See Apple Rust, Cedar-Apple Rust.

Cedar Galls–Small swellings on branches, twigs, and leaves of junipers and related trees and shrubs. See Cedar Hawthorn Rust.

Cedar Hawthorn Rust–A fungal disease of hawthorn and cedar caused by *Gymnosporangium globosum*, family Pucciniaceae, which alternately infests junipers and hawthorns, rarely apples. It produces small, irregularly shaped galls on the junipers, which have wedge-shaped, gelatinous, orange spore masses. Its symptoms are similar to apple rust on the apple.

Cedar, Incense–*Libocedrus decurrens*. Trees, generally to 90 ft. (30 m), native to southwest Oregon and western Cascades. Valuable for wood products, windbreaks, and ornamental plantings. Slow-growing, seedling mortality can be high if improperly managed.

Cedar Rust–See Apple Rust, Cedar Hawthorn Rust.

Cedar Tree Borer–*Semanotus ligneus*, family Cerambycidae; a beetle whose larvae infests arborvitae, redwood, Douglas-fir, and Monterey pine, making winding galleries in the inner bark and sapwood.

Cedar-Apple Rust–A fungal disease that lives on alternate hosts, *Juniperus* (cedar) and *Malus* (apple).

Cedar-of-Lebanon–*Cedrus libani*, family Pinaceae; the famous evergreen tree of the Bible. It is one of the traditional cedars, growing up to 100 feet high and sometimes bearing branches 40 to 50 feet long. Native to Turkey; grown in the southern United States.

Celariac–*Apium graveolens* var. *rapaceum*; a Eurasian, biennial herb, family Umbelliferae, grown for its edible, thickened, celery-flavored root. It is a minor vegetable. Also called turnip-rooted celery, knob-rooted celery.

Celastrus–A woody vine genus, family Celastraceae, to which belong the sudforific American bittersweet and East Indian *Celastrus paniculatus*, which is the source of the oil oleum nigrum.

Celery–*Apium graveolens*, famly, Umbelliferae; A salad plant that grows best in a cool climate.

Celery Salt–A mixture of 25 percent ground celery seed, 73 percent common salt, and 2 percent calcium phosphate; used as a condiment.

Celery Seed–A condiment made of the dried fruit of celery, *Apium graveolens*. The seeds are also used in preparing oil of celery.

Cell–(1) A hexagonal unit compartment of a honeycomb. (2) The ultimate functional unit of an organic structure, plant, or animal. It consists of a microscopic mass of protoplasm which includes a nucleus surrounded by a membrane. In most plants it is surrounded by a cell wall. (3) A single element of an electric battery, either primary or secondary, generally consisting of a jar filled with a liquid or a pasty electrolyte, into which the electrodes are inserted or connected. (4) A very small, enclosed compartment.

Cell Wall–The membranous covering of a cell secreted by the cytoplasm in growing plants. It consists largely of cellulose, but may contain chitin in some fungi and silica in some algae.

Cellar–See Cold Cellar, Storm Cellar.

Cellar Gardening–(1) The forcing of plants, e.g., rhubarb, asparagus, French endive, etc., to grow in a cool room. (2) The indoor growing of mushrooms, chard, parsley, etc., during the winter.

Cellulase–An enzyme capable of splitting cellulose. See Cellulose.

Cellulose–An inert, complex carbohydrate which makes up the bulk of the cell walls of plants. Derived from stalks, straws, etc., it is an important farming by-product, and is used for making paper, fabrics, and building materials.

Cement Rock–An argillaceous limestone used in the manufacture of natural hydraulic cement. Contains lime, silica, and alumina in varying proportions, and usually magnesia.

Cemented–Indurated; having a hard, brittle consistency because the particles are held together by cementing substances such as colloidal organic matter (humus), calcium carbonate, or the oxides of silicon, iron, and aluminum. The hardness and brittleness persist even when wet.

Centeotl–Aztec goddess of growing vegetation. See Ceres.

Center Fire–A method of controlled burning in which initial fires are set in the center of an area to be broadcast-burned to create a strong draft. Later, additional fires are set near the outer control lines of the area to act as backfires to check outward spread.

Center Pivot Irrigation System–A large irrigation system that rotates around the terminal end of a large water line. Modern systems are powered by electric motors. The system circles slowly and sprinklers mounted on the system emit irrigation water.

Centipede–Any myriapods of the class Chilopoda, close relatives of the true insects. They are beneficial because they prey on harmful insects.

Centipede Grass–*Eremochloa ophiuroides*; a creeping, perennial grass, family Gramineae, introduced from China. It has proved useful in stabilizing waterways in soil erosion work in the southern United States and as a lawn grass because it quickly forms a dense turf.

Centriole–A small organelle located near the nuclear membrane of cells that divides during mitosis and forms the centers toward which the chromosomes move upon division of the cell.

Centromere–A small structure located on a chromosome that appears to form an attachment to the spindle fibers during cell division.

Centrosome–A minute protoplasmic body sometimes held to be the dynamic center of mitotic activity.

Centrum–The central portion; here used specifically for the large central air space in hollow stems such as those of the genus *Equisetum* (horsetail).

Cepaceous–Smelling or tasting like onion or garlic.

Cephus–A genus of sawflies, family Cephidae, whose larvae burrow into plant stems. *Cephus pygmaeus* infests and destroys wheat; *C. cinctus* is harmful to some grasses and grains.

Ceratostomella–A fungus genus, family Ceratostomataceae. *Ceratostomella pilifera* is the cause of blue stain in wood.

Cercariae–The tailed, immature stage of a parasitic flatworm.

Cercis–A genus of redbud shrubs, family Leguminosae, grown for their showy red-purple or pink flowers, which blossom in early spring.

Cercospora–An imperfect fungus genus, family Dematiaceae, of which over 1,000 species have been given names, their distinguishing characters being largely based on the species of the plants infected (host plans) and the type of lesions produced. The species produce elongated, slender spores or tufted conidiophores, which emerge from the stomata of the infected spots. These spores are borne by air currents or are splashed by raindrops and carried to other leaves, where germination of the spore occurs. A few named species of *Cercospora* grown under controlled conditions on special culture media have produced the perfect stage of reproduction.

Cereal–(1) Any grass grown for its edible grain. (2) A prepared foodstuff from grain. The major cereals are wheat, corn, barley, rye, sorghum, and oats.

Cereal Scab–A fungal disease of the heads of wheat, barley, rye, and ears of corn, caused by *Gibberella zeae*, family Nectriaceae, which makes the grain unwholesome. The fungus may also bring about stem and root rots of these plants, so that the infected stems may become sufficiently weakened to break over in a strong wind.

Ceres–Roman goddess of corn and growing vegetation, adapted from the Greek goddess Demeter in the fifth century B.C. She was yearly honored with sacrifices and other rituals just after the seeding time and before harvesting. The word cereal is derived from her name.

Cerrero–Describing a wild, untamed horse (western United States).

Certification of Seed–Seed production and marketing under the control of a certifying agency to maintain varietal purity and freedom from seed-borne pests.

Certification Standard–Rules and regulations of a certifying agency concerning the breeding, production, and cleaning of seeds along with the maximum amount of impurities allowed, etc.

Certified Seed–The class of crop seed that is the progeny of registered seed. This class is the most widely available class. The seed carry a blue tag and meet high standards of genetic identity and purity. See Registered Seed; Breeder Seed; Foundation Seed; Certification of Seed.

Cespitose–Turflike; having dense growth as in turf.

Chaff–(1) The outer covering of grain, the glumes, husks, separated from the seed by threshing. (2) Finely cut hay or straw used as feed for cattle. (3) The scales associated with the seeds in the heads of some composite plants.

Chaffy Corn–Immature corn cut and dried to 17 percent or less of moisture content.

Chai–(Russian) Tea.

Chain Saw–A portable, motor-driven saw which consists of an arm on which travels an endless chain with attached sawteeth. Used in felling trees, sawing logs, or in cutting underbrush.

Chalaza–(1) Either of two cordlike, opalescent, albuminous strands in the white of a fowl's egg, which are attached to the yolk and prolonged toward the ends of the egg. The chalazas aid in keeping the yolk in proper position. (2) The place in an ovule or seed at which the integuments diverge from the nucellus.

Chalcid Fly–Any small insect of the superfamily Chalcidoideae, which in its early stage exists in or on other larvae or pupae, thus destroying many undesirable pests.

Chalk–(1) Soft, white, fine-grained lime which is highly fossiliferous. As the parent rock, it lends distinctive character to soils. Locally impure chalks may be called marl. (2) To spread chalk over land.

Chamaecyparis–A genus of pine trees, family Pinaceae, which yields valuable timber. Native to North America, Japan.

Chamaedaphne–A genus of evergreen shrubs grown in rock gardens, family Ericaceae; found in bogs of northern Europe, northern Asia, North America.

Chamiso–(1) *Adenostoma fasiculatum*, family Rosaceae; an evergreen shrub grown in warm, humid regions for its needlelike foliage and clusters of white flowers. The most characteristic shrub of the California chaparral. (2) *Atriplex canescens*, family Chenopodiaceae; a semidesert shrub found in New Mexico in the United States. Also called greasewood chamise.

Chamiza–Fourwing saltbush.

Champignon–(1) A mushroom, especially *Marasmius oreades*, the fairy ring mushroom. (2) Inflammation of a horse's spermatic cord.

Chance Seedling–A fruit or nut tree which grows from seed scattered by wind, birds, etc.; sometimes bears fruit. Many varieties of fruits originated as chance seedlings.

Changa–*Scapteriscus vicinus*, family Gryllidae; a mole cricket that attacks garden crops, peanuts, and tobacco by uprooting seedlings, cutting the roots off, and eating into the underground parts. Native to South America and the West Indies. Also called Puerto Rican mole cricket, mole cricket.

Channery–Small, thin, flattish fragments of limestone, standstone, or schist found in soils.

Chanterelle–*Cantharellus cibarius*; a common, deep yellow mushroom.

Chapparal–A dense growth of shrubby vegetation or stunted and dwarf trees in arid and semi-arid regions of the southwestern United States. Characteristic shrubs of the coast ranges of California are chamiza, ceanothus, and manzanita. In other places dense thickets of mesquite (*Prosopis* spp.) and thickets of oaks may be called chaparral.

Char-pitting–A method of removing stumps by burning. A fire is built on the stump and kept covered until the wood is consumed.

Charcoal Rot–(1) A fungal disease caused by *Macrophomina phaseoli*, family Sphaeropsidaceae, which attacks over thirty species of plants. Characterized by a rotting of the roots and of the stalk just above the roots, which stunts the plant and reduces the yield. (2) A decay of limes (*Citrus aurantifolia*) caused by *Diplodia natalensis*, family Sphaeropsidaceae, characterized by a brownish spot on the mature fruit. The fruit dries and becomes filled with a black, hard mass of mycelium. (3) Jet-black rot of the root system of sweet potatoes caused by the imperfect fungus *Sclerotium bataticola*.

Chard–Leaf beet; Swiss chard. A common garden plant grown for greens.

Charlock–*Brassica kaber*, family Cruciferae; a weed, bearing yellow flowers, found in grainfields and among other cereal crops. Feeds containing large amounts of its seed may result in chronic enteritis, hemorrhagic diarrhea, colic, abortion, nephritis, apathy, etc., in animals. Native to Eurasia. Also called wild mustard, corn mustard.

Charred Peat–Peat, dried at such a temperature as to cause partial decomposition.

Charring–Burning partly; scorching.

Chasmophyte–A plant growing in a rock crevice.

Chat–Tailings from lead and zinc mines consisting essentially of dolomite or dolomitic limestone. It is widely used in Missouri, Tennessee, and Illinois as a liming material. It contains about 0.5 percent manganese and 0.016 percent zinc in addition to 21 percent calcium and 11.3 percent magnesium.

Cheatgrass–*Bromus* spp., family Gramineae; an annual grass weed, palatable in the early stages of growth. Native to Europe; a troublesome weed in the forage fields of the midwestern United States. Also called annual downy chess, downy brome, wild oats.

Check–(1) In irrigation (a) A basin into which the flow of water is regulated by levees and dykes. (b) An adjustable gate on a canal to regulate the flow of water. (c) Concrete blocks or wooden ties placed on a channel bottom to reduce erosion. (d) A crack which appears in drying soil. (2) In lumbering, a lengthwise separation in the grain, caused by strains during seasoning, which extends across the annual rings. (3) A short crack within the body of a cheese. (4) A narrow crack in a rice grain that may cause it to break during milling. (5) An egg which has a cracked shell, with the inner membrane intact. When the crack is naturally mended, it is called a blind crack. Also called crack, dent. (6) To retard growth.

Check Irrigation–Application of a comparatively large stream of water in level plots surrounded by levees.

Check Plot–In field demonstrations or research plots, a small area of soil either not treated or given a standard treatment and used for comparing results of one or more treatments assumed to be superior.

Checkerberry–*Gaultheria procumbens*, family Ericaceae; an evergreen spreading shrub bearing white or pinkish, bell-shaped flowers and purplish berries, called checkerberries. its fragrant foliage yields a medicinal oil, commonly known as wintergreen oil or gaultheria oil. Native to North America. Also called wintergreen, teaberry, ground holly, spiceberry, aromatic wintergreen, partridge berry, boxberry, creeping wintergreen.

Checkrow–A row with seeds planted at regular intervals by a checkrow planter. Series of checkrows are established crosswise in two directions at right angles to each other to form squares to facilitate cultivation.

Chelates–Certain organic chemicals, known as chelating agents, form ring compounds in which a polyvalent metal is held between two or more atoms. Such rings are chelates. Among the best chelating agents known are ethylenediaminotetraacetic acid (EDTA), hydroxyethylenediaminetriacetic acid (HEDTA), and diethylenetriaminepentaacetic acid (DTPA). In the absence of chelates in the soil, iron, copper, manganese, and zinc are all converted to insoluble and unavailable hydroxides or other basic salts even in acid soils. Chelates keep these micronutrients available in soils or solutions of pH up to 8 or 9. Manure is an example of a chelate.

Chemical–(1) A substance obtained by or used in a chemical process. (2) In dairying, an uncommon flavor defect which occurs when cheese has been made from contaminated milk.

Chemical Additives–Substances added to foods to improve their flavor, color, texture, or keeping quality.

Chemical Bond–See Valence.

Chemical Brown Stain–A brown discoloration of chemical origin that sometimes develops on wood during air seasoning or kiln drying; probably due to the oxidation of extractives in the wood.

Chemical Energy–The energy contained in the chemical bond between atoms; it can be released into the environment by a chemical reaction, e.g., combustion.

Chemical Fertilizer–See Commercial Fertilizer

Chemical Gardening–See Hydroponics.

Chemical Stimulation–In turpentining, the application of chemicals, usually acids, to the chipped face of a pine tree to increase the rate and duration of the oleoresin flow from the tree.

Chemical Weed Control–The application of herbicides as preemergence and postemergence sprays for the control of weeds in crops.

Chemical Wood–Wood cut or prepared for the manufacture, distillation, or extraction of chemicals, charcoals, gases, or other products.

Chemigation–The application of agricultural crop chemicals such as herbicides or fertilizers through an irrigation system.

Chemistry–(Greek, chemeia) Science of compounds, elements, and atoms.

Chemolysis–Decomposition of organic matter brought about by chemical agents.

Cherokee Tickclover–*Desmodium tortuosum*, family Leguminosae; a tropical plant grown extensively in Florida and adjacent states of the United States as a cover crop and for green manure. The plant is resistant to strains of the rootknot nematode. Native to the United States. Also called Florida beggarweed.

Cherry–Species of the genus ***Prunus*** grown for its edible fruit or for ornament. Cultivated cherries are native to Europe and are classified in three groups: sour cherries, sweet cherries, and dukes, which are hybrids between the first two. One wild variety, native to eastern North America, the black cherry (***P. serotina***) is a very large tree, the wood of which is used for fence posts, rails, cross-ties, furniture, and fuel. All leaves are toxic to livestock.

Cherry, Bitter–*Prunus emarginata*; a deciduous shrub growing 15 to 20 feet (5 to 7 meters) tall. Is often straggling or depressed in habitat, growing on dry rocky slopes, but not limited to those areas. It bears a prolific crop of small bright or dark red berries during the autumn.

Cherry Pit–The stone of the cherry, in which the seed is enclosed.

Cherry Tomato–*Lycopersicon esculentum* var. ***cerasiforme***; an annual herb, family Solanaceae, grown for its very small, edible, tomatolike fruits used in salads. Native to South America. Also called plum tomato.

Chert–A very dense, cryptocrystalline, flintlike form of silica, which breaks with a splintery fracture. It is very resistant to decomposition, remaining as inert angular fragments in the residual mass of weathering. Since chert is harder than steel, plows and disks are dulled in cherty soil. See Flint.

Chess Brome–*Bromus secalinus*, family Gramineae; a weed pest found in grain fields; especially troublesome in wheat fields when its seed becomes mixed with the threshed wheat, thus lowering the grade. Grown for hay in some regions of the northwest Pacific. (The name has also been applied to darnel ryegrass, ***Lolium temulentum***.)

Chester White–A large breed of white swine which originated in Chester County, Pennsylvania, from crossing Yorkshire, Lincolnshire, and Cheshire hogs.

Chestnut–(1) See American Chestnut. (2) A horse color of brown with red-yellow hues which lacks brilliance. (3) The sweet, edible nut produced by ***Castanea*** spp., family Fagaceae. (4) The horny projection found on the lower inner part of the forearm and hock joint of a horse.

Chestnut Blight–A fungal disease caused by ***Endothia parasitica***, family Diaporthaceae, which attacks the bark and cambium of the twigs, branches, and main trunk of chestnut trees. The cankers formed girdle the stems resulting eventually in the death of the tree. The blight has almost completely destroyed the American chestnut, ***Castanea dentata***. However, the East Asian varieties have shown some degree of resistance to it. Also called endothia canker, chestnut canker, chestnut-bark disease.

Chestnut Oak–*Quercus montana*, family Fagaceae; a tall deciduous tree found on dry hillsides in the eastern United States. It becomes crooked if not trained upright. The wood is used for railroad ties, railroad cars, fences, pilings, timbers, veneer, flooring, vehicles, planing-mill products, furniture, cooperage boxes, crates, agricultural implements, caskets, coffins, handles, etc. Native to America. Also called tanbark oak, rock chestnut oak.

Chewing Insects–Insects, such as worms, grasshoppers, and Japanese beetles, with chewing mouthparts.

Chianophile–A plant that requires or can endure snowcover during the winter and spring.

Chick-pea–See Gram Chick-pea.

Chickasaw Plum–*Prunus angustifolia*, family Rosaceae; a small tree or shrub from which many cultivated varieties of plum in the southern United States have been derived. Native to North America.

Chickweed–*Stellaria media*; a weed pest found throughout the United States in cultivated fields, gardens, lawns, waste ground and around ornamental shrubs. Native to Eurasia. Also called satin flower, tongue grass, starwort, starweed, winterweed, birdweed, chickenwort.

Chicle–The sap of *Manilkara zapota*, collected from a tree in Central America and used to make chewing gum.

Chicory–(1) *Cichorium intybus*, family Compositae; a perennial, blue-flowered herb cultivated for its roots, which are used as an additive to coffee, and for its leaves, which are used as a potherb and salad green. It is an escaped weed along roadsides in the United States. Native to Europe. Also called succory, blue dandelion, coffee weed, blue sailor. (2) A French variety of chicory with blanched leaves, used raw as a salad, is called French endive or witloof.

Chile Dodder–*Cuscuta suaveolens*, family Convolvulaceae; a parasitic plant which attacks both clover and alfalfa. Native to South America; common in red clover and alfalfa seed imported to the United States from there.

Chilean Deervetch–*Lotus subpinnatus*, family Leguminosae; a valuable species of herbaceous forage plants found in California, United States. Also called fineleaf lotus.

Chilean Tarweed–*Madia sativa*, family Compositae; an annual herb grown for its heavy-scented, brownish-yellow flowers. Native to America.

Chilies–Hot red and green peppers in fresh and powdered form; used for cayenne pepper, chili powder, and for sauces, pickles, and meat dishes.

Chill Requirement–Number of hours that deciduous fruits require below 45°F (7.2°C) before normal growth will resume in spring. Without adequate chilling, blossoming and foliage development are delayed; e.g., Elberta peaches require 900 hours below 45°F and Floridabelle peaches only 150 hours.

Chilling–(1) The development of a sweetish taste in stored potatoes kept at temperatures approaching the freezing point. It is not a serious injury, since the tubers become normal upon exposure to higher temperatures. (2) Exposure of a plant to low temperature to promote development.

Chilling Injury–Damage to certain horticultural products, such as banana, papaya, cucumber, and sweet potato, which results from exposure to cold, but above-freezing temperatures.

Chilling Room–An insulated, refrigerated compartment especially constructed for the storage or transport of perishable food in which the temperature is maintained at a few degrees above the freezing point. Meat and delicate fruits keep fresh and sweet in it for about a month; quality is not impaired as in a cold storage with temperatures below the freezing point.

Chimera–A plant or part of it, composed of tissues of two or more genetically different types. If tissue of one genetic type is external to and surrounds another genetic type, it is a periclinal chimera. If the tissue extends from the surface into and is deeply seated in another genetic type, it is a sectorial chimera. When the tissues are greatly intermingled, it is hyperchimera.

Chimerism–The quality of being a chimera; i.e., in genetics, the presence in an individual plant or animal of cells from a different species, a different genotype, or with different antigens, caused by radiation, grafting, or mutation; e.g., a fruit, half orange and half lemon; an apple with sweet and sour flesh; peaches with half fuzzy and half smooth skin.

Chimney–A vertical opening, a foot or more wide, in a pile of lumber to facilitate circulation of air and drying of the lumber.

Chimney Effect–The movement of air in a greenhouse created when both the top ridge and the side ventilators are open.

China Wood Oil Tree–See Tung Oil Tree.

Chinaberry–*Melia azedarach*, family Meliaceae; a semi-evergreen tree grown for shade in the United States. Its seeds are sometimes used for rosaries; the drupes are poisonous, producing complete paralysis, irregular respiration, and suffocation. Also called china tree, bead tree, pride of India.

Chinch Bug–*Blissus leucopterus*, family Lygaeidae; a black-and-white insect, destructive to small grains, corn, and sorghum; found all over North America below central Canada, particularly in the central United States.

Chinese–A breed of medium-sized, white or brown, domestic geese from China, having long arched necks and large knobs at the base of the beak. The adult gander weighs as much as 12 pounds, the goose as much as 10 pounds.

Chinese Cabbage–A leafy green *Brassica chinensis* grown as a vegetable. It is not a true cabbage, but is more closely related to mustard. See Petsai.

Chinese Insect Wax–A yellowish-white wax deposited on trees by a scale insect, *Ericerus pela*, family Coccidae; common in China and India. It is used for candles, polishing wax, sizing paper, etc., in place of beeswax. Also called insect wax, vegetable spermaceti.

Chinese Lespedeza–*Lespedeza cuneata*, family Leuminosae; a coarse, perennial herb which, if cut young, makes fair hay for cattle, but because of its tannin content, becomes extremely unpalatable later on. It is grown in the southern United States for pasture, hay, and soil improvement. Native to eastern Asia. Also called sericea lespedeza.

Chinese Mantis–*Tenodera aridifolia sinensis*, family Mantidae; a large, important insectivorous insect which devours immense numbers of insects. It was introduced in the United States from Asia. Also called praying mantis.

Chinese Sumac–Tree of heaven; *Ailanthus altissima*, family Simaroubaceae; a rapid-growing tree that grows to 60 feet The tree withstands urban smog.

Chinese Tallowtree–*Sapium sebiferum*, family Euphorbiaceae; a tree whose wood is used for engraving work and the waxy covering of whose seeds yields a tallowlike substance used in soap and candle making. Native to China and Japan; grown in United States.

Chingma Abutilon–*Abutilon theophrasti*, family Malvaceae; an annual herb, grown for its yellow flowers and beautiful foliage. In India, its coarse fiber is used for making ropes and bags. Escaped, it can be a pernicious weed. Native to India; naturalized in North America. Also called American jute, Indian mallow, buttonweed, velvetleaf.

Chinkapin–(1) Certain trees of the genus *Castanea*, family Fagaceae, and their edible nuts. (2) See Giant Evergreen Chinkapin.

Chinkapin Oak–Either of two plants of the family Fagaceae: (a) the tall tree *Quercus muhlenbergii*, also called yellow chestnut oak; (b) the shrubby oak *Quercus prinoides*, whose sweet acorns are edible. Native to North America.

Chinook Wind–A northwestern United States term for a hot, dry wind which, with its moisture precipitated, descends from the Rocky Mountain crests and spreads over the lower land eastward, evaporating the snow rapidly. It is beneficial in winter, but may desiccate crops during the growing season.

Chip–(1) A small piece of wood or stone. (2) To cut wood or stone into small pieces. (3) To rewind the face of a scar on a pine tree to renew the flow of oleoresin in the turpentining industry. (4) To hoe or cultivate (Australia).

Chip Budding–A scion, beveled at both ends, which consists of one bud, bark, and a small amount of wood. It is inserted into a mortise in the stock and is usually held in place by tying.

Chips–(1) Small pieces of wood or bark which have fallen into the cup or box of a turpentined tree or which have been removed with the scrape from the scarified face above the box. (2) Pieces of dried dung used for fuel.

Chiseling–Tillage with an implement having one or more soil-penetrating points that loosen the subsoil and bring clods to the surface. A form of emergency tillage to control soil blowing.

Chit–(1) Second- or third-grade rice. (2) A young animal. (3) A plant shoot.

Chitin–A nitrogenous polysaccharide occurring in the cuticle of arthropods and certain other invertebrates. Probably occurs naturally only in chemical combination with protein.

Chive–*Allium schoenoprasum*, a perennial, bulbous herb, family Liliaceae, grown for its onion-flavored leaves. Used in salads and as a flavoring. Native to Eurasia. Also called cive.

Chloranthus–Describing a plant with green flowers.

Chlorellin–An antibiotic derived from a freshwater alga of the genus *Chlorella*.

Chlorinated Lime–Calcium hypochlorite; a powerful disinfectant and deodorant. Also called bleaching powder, bleach, chloride of lime, calcium chloride.

Chlorinator–A device for adding chlorine gas to sewage to kill infectious organisms.

Chlorophyll–A substance present in all green plants; it evidences itself as the green coloring in leaves. Chlorophyll transforms light energy from the sun into chemical energy for the manufacture of plant food from carbon dioxide, water, and essential soil minerals. This process is called photosynthesis. Four different chlorophylls have been identified as a, b, c, and d. All four contain carbon, hydrogen, oxygen, nitrogen, and magnesium.

Chloropicrin–Insecticidal fumigant for treatment of stored grain and cereal products and for soil treatment to control insects, nematodes, weeds; very toxic to plants when injected into soil.

Chloroplast–Minute objects within plant cells which contain the green pigment, chlorophyll.

Chlorosis–Yellowing or whitening of normally green leaves caused by lack of nutrients, air pollutants, or diseases.

Chocolate Tree–See Cacao, Cocoa.

Choke–(1) An acute condition in livestock brought about by a food mass lodged in the esophagus. (2) Hairy or filamentous undeveloped scales at the base of glove artichoke heads, which are removed before eating. (3) In engine carburetion, to increase the ratio of gasoline to air in the fuel mixture when starting a cold engine. (4) In plants, to kill, dwarf, or stunt through excessive competition for space and nutrients, e.g., the choking of plants by dodder or bindweed. (5) In plowing, the gathering of wet straw, weeds, etc., about the plowshare, which reduces its efficiency.

Chokecherry–*Prunus virginiana*, family Rosaceae; a treelike, deciduous shrub of little decorative value, but dangerously poisonous to humans and animals. The poisoned animal becomes uneasy, staggers, collapses, has convulsions, breathes with difficulty; its eyes roll, and finally, within an hour, it bloats and dies. The reddish-black fruit has a high content of astringent substance, largely tannin. The plant is a host of the "X" disease of stone fruits. Native to North America.

Cholinesterase Testing–A pesticide-monitoring program that monitors a person's blood at various intervals during periods that he/she is using pesticides to determine the degree to which the person has been exposed to the pesticide.

Chomophyte–Any plant which grows in rocks.

Chop–(1) Animal feed of coarsely crushed or finely ground cereal grains. (2) A small cut of meat which usually includes a rib, e.g., mutton chop. (3) To cut hay into small portions for easy storage without baling. (4) To hoe a row crop, especially cotton; often the first hoeing is called chopping. (5) To crush grain. (6) Jaw; generally in the plural to denote the jaws forming the mouth. Also called chap. See Green Chop.

Chorion–(1) The outermost membrane that encloses the unborn fetus in mammals. (2) The tender, fleshy substance of the original nucleus of a plant seed.

Chou–(Chinese) Cabbage.

Chou-fleur–(French) Cauliflower.

Chowchow–A preserve which is a mixture of green tomato, peppers, cauliflower, onion, and pickles.

Christmas American Mistletoe–*Phoradendron flavescens*, family Loranthaceae; a woody plant parasitic on various deciduous trees of the eastern United States, which causes the limbs to become malformed. Its branches, preferably bearing berries, are a Christmas decoration. Also called eastern mistletoe.

Christmas Rose–*Helleborus niger*, family Ranunculaceae; an evergreen perennial herb, grown for its late-fall or winter, white or purple berries with green flowers. The plant is poisonous to stock, and contact with its bruised parts may also result in severe dermatitis in human skin. Also called black hellebore, felon grass.

Christmas Tree Farming–The commercial production of evergreen trees to be used as Christmas trees. The most popular species are Scotch pine, Douglas fir, balsam fir, black spruce, and eastern red cedar.

Christmasberry–*Heteromeles arbutifolia*, family Rosaceae; an evergreen shrub widely planted in the western United States for its dense, white flower clusters and bright red berries. Poisonous to stock because its leaves contain prussic acid. Poisoned animals become uneasy, stagger, have difficulty in breathing, roll their eyes, and extend their tongues. An hour after ingestion, the animal becomes quiet, bloats, and dies. Native to California. Also called toyon, redberry, California holly.

Chroma–(Greek) Color. See Mundell Color Standard.

Chromatid–One strand of a doubled chromosome seen in the prophase and metaphase of mitosis.

Chromoseres–The smallest particles identifiable by characteristic size and position in the chromosome thread. They are minute subdivisions of chromatin arranged in a linear, beadlike manner on the chromosome.

Chromosome–A microscopic, dark-staining body, visible in the nucleus of the cell at the time of nuclear division, which carries the genes, arranged in linear order. Its number in any species is usually constant, and it serves as the bridge of inheritance, i.e., the sole connecting link between two succeeding generations.

Chronic Water Deficiency–A plant condition caused by water-deficient soil. The plant is stunted, the roots are undersized, and the leaves become discolored. In cereals, the grain is shriveled. In fruit crops, the fruit is spotted, deformed, and shriveled. Excess light and heat injury may accompany the deficiency.

Chronosequence–A sequence of soils whose properties are functionally related to time as a soil formation factor.

Chrysalis–The pupa, or the resting stage (intermediate between larva and adult), of butterflies.

Chrysanthemum–Any plant of the genus *Chrysanthemum*, family Compositae, whose annual or perennial species originated in Eurasia, but which are now found in most parts of the world. Some species are grown as ornamentals, some as medicinal plants, while others are pernicious weeds. They flower in all colors except true purple and blue.

Chufa–A sedgelike plant with creeping rootstalks that produces small tubers or nuts. In some places it is considered a pernicious weed and is called yellow nutgrass. Chufa is sometimes grown as a forage. The tubers or nuts are fed to hogs, and the tops can be used for hay.

Churchmouse Threeawn Grass–*Aristida dichotoma*, family Gramineae; a perennial, tufted grass of dry, sandy soils. Its sharp awns are injurious to grazing animals. Native to the United States. Also called poverty grass.

Chytrid Diseases–Plant diseases caused by one of the *Chytridiales*, an order of fungi; e.g., clubroot of cabbage, powdery scab of potato, Physoderma brown spot, potato wart, cranberry gall, crown wart, beet root tumor.

Cicada–Large wide-bodied insects of the family Cicadidae. They have a broad, blunt head and transparent wings.

Cicoria–(Italian) Dandelion.

Cide–A suffix which indicates a killer; e.g., amoebacide, that which kills amoeba.

Cider–See Apple Cider.

Cider Vinegar–The product of alcoholic and subsequent acetous fermentation of apple juice.

Cider Yeast–*Saccharomyces malei*, family Saccharomycetaceae; a yeast which causes rather slow fermentation; useful in the manufacture of sparkling hard cider.

Cigar Casebearer–*Coleophora occidentis*, family Coleophoridae; a small insect whose larvae weave tough, silken cases about themselves, which they drag about while feeding on leaves, buds, and fruit of apple, pear, quince, plum, and cherry.

Cigar Filler–Tobacco leaves placed in the core of a cigar. The leaf fragments are as long as the cigar in superior brands, or short or shredded in low-priced cigars.

Cigar Wrapper–A tobacco leaf (usually imported or shade-grown domestic) wrapped spirally around the filler.

Cigar-leaf States–States of the United States where leaf tobacco for cigars is grown: Connecticut, Florida, Georgia, Wisconsin, Pennsylvania, Ohio, and New York.

Cigarette Beetle–*Lasioderma serricorne*, family Anobiidae; a major pest of stored tobacco. It cuts holes in the plant leaves and builds nests in them; also destructive to upholstered furniture, seeds, and dried spices, etc.

Cilantro–Fresh green coriander; also called Chinese parsley.

Cilia–(1) Minute hairlike formations on the cells of many animals. Constantly vibrating, they serve as means of locomotion in aquatic unicellular animals and help remove mucus and fluid residue in higher forms. (2) Tender, sensitive hair forming a fringe along the edges of leaves.

Ciliate–Bearing cilia, fringed with hairs, bearing hairs on the margin.

Cinchona–A large genus of trees and shrubs, family Rubiaceae, characterized by opposite leaves and small panicled flowers. The bark from several of its species yields the commercial quinine. Native to South America; grown in Southeast Asia and the West Indies.

Cinder Soil–Soil composed of the ejects of volcanic eruptions, generally fragmental matter. It forms considerable aggregate area in parts of the western United States.

Cinnamon–*Camphora*, family Lauraceae; a flavoring made from the fragrant dried bark of cultivated varieties of cinnamon. The bark is stripped from the trees, piled up, and permitted to undergo fermentation to soften the outer skin for removal. It is then dried in the sun.

Cinnamon Fern–*Osmunda cinnamomea*, family Osmundaceae; a corse, easily grown fern, found in wet, low woods. Its fibrous root is the source of most orchid peats used in potting. Native to Asia and America.

Cion–See Scion.

Cipolla–(Italian) Onion.

Circa–(Latin) About, approximate.

Circinate–Rolled or coiled from the top downward, with the apex nearest the center of the coil, as an unexpanded fern frond.

Circumneutral–Term applied to water with a pH of 6.5 to 7.4.

Citrioli–(Italian) Cucumber.

Citron–*Citrus medica*, family Rutaceae; a large, thorny shrub grown in very warm areas for its highly aromatic fruit, the citron of commerce. The thick fruit rind is candied and used as a preserve. It is one of the oldest citrus fruits; native to eastern Asia. Also called cedrat.

Citronne–(French) Anything that has the taste or flavor of lemon.

Citrus–An important genus of trees, family Rutaceae, extensively cultivated in tropical and subtropical areas for its fruits, such as the lime, orange, bergamot, lemon, citron, and grapefruit. Native to Asia.

Citrus Belt–The citrus-producing regions of the world lying at low elevations generally within 35° north and south of the equator.

Citrus Thrips–*Scirtothrips citri*, family Thripidae; an American pest of deciduous fruits and other plants in California and Arizona. It infests the buds, new growth, and the fruit.

Citrus White Fly–*Dialeurodes citri*, family Aleurodidae, a pest which sucks the sap on the underside of leaves of citrus and other plants, especially the camellia, causing stunting of the trees and undersized fruit of poor color. Found in the Gulf States of the Untied States, including Florida, where it attacks the chinaberry.

City Forestry–The preservation and protection of shade trees in streets and parks.

Cive–See Chive.

Cladophyll–A leaflike structure which may bear flowers, fruits, and temporary leaves.

Clam Shells–Finely ground clam shells are used as a liming material along most sea costs. The mean composition is 38 percent calcium and 45 percent carbon dioxide (CO_2).

Clambering–Vinelike, climbing, often without the aid of tendrils or twining stems.

Clambering Monkshood (Wild Monkshood)–*Aconitum uncinatum*; a poisonous, perennial herb, family Ranunculaceae, found along streams in shady places from Pennsylvania to Georgia in the United States. Symptoms of poisoning are weakness, irregular and difficult breathing, weak pulse, bloat, attempted swallowing, belching.

Clammy Locust–*Robinia visoca*, family Leguminosae; a deciduous tree grown for its ornamental foliage and pink and yellow flowers. The leaves and bark, covered with sticky glands, are poisonous (sometimes fatally) to horses, cattle, sheep, and chickens.

Clarify–To remove undesirable, solid substances from a liquid, such as milk or fruit juice, by ordinary or by centrifugal filtration.

Clary Sage–*Salvia sclarea*, family Labiatae; a hardy, biennial medicinal herb with bluish-white flowers and grayish-haired leaves. The leaves are used for flavor preserves and omelets and to perfume sachets. Tea and a cordial called clary water are made from the flowers, and an oil is extracted from the plant for use in perfumes. Native to southern Europe and western Asia.

Clasping–Partially or completely surrounding the stem, as the bases of some leaves.

Class–A division of the plant or animal kingdom lower than a phylum and higher than an order; e.g., the class Insecta.

Class of Seedling Stock–The age of nursery stock denoted by two or more figures, as 2-0 or 2-1. The first figure shows the number of years in the seedbed, and the succeeding figure the number of years in the transplant bed.

Classification–The forming, sorting, apportioning, grouping, or dividing of objects into classes to form an ordered arrangement of items having a defined range of characteristics. Classification systems may be taxonomic, mathematical, or other types, depending on the purpose to be served.

Classify–(1) To systematically categorize plants or animals according to a set scheme. (2) To sort individuals together into groups having common characteristics or attributes.

Clavate–Club-shaped, with a long body thickened toward the top, like a baseball bat.

Claw–(1) The sharp nail on the toe of an animal or bird. (2) The slender, extended lower part of the petal, as in the iris, lily, etc. (3) A device on a milking machine to which the stanchion tubes are connected.

Clay–(1) A size term denoting particles, regardless of mineral composition, with diameter less than 2 microns (agriculture) or 4 microns (geology). (2) A group of hydrous alumino-silicate minerals related to the micas (clay minerals). (3) A sediment of soft, plastic consistency composed primarily of fine-grained minerals. (4) In engineering, any surficial material that is unconsolidated.

Clay Film–A thin coating of clay on the surface of a soil aggregate or lining soil pores or root channels. Syn.: clay coat, clay skin. Common in some B horizons of soil.

Claypan–A dense, compact layer in the subsoil having a much higher clay content than the overlying material, from which it is separated by a sharply defined boundary; formed by downward movement of clay or by synthesis of clay in place during soil formation. Claypans are usually hard when dry, and plastic and sticky when wet. Also, they usually impede the movement of water and air and the growth of plant roots. They can occur naturally or be induced by heavy traffic. See Fragipan.

Clean Content–The amount of clean, scoured wool remaining after removal of all vegetable and other foreign material, and containing 12 percent by weight of moisture and 1.5 percent by weight of ingredients removable by extractions with alcohol.

Clean Cotton–Seed cotton, free from extraneous hulls, bolls, or trash.

Clean Culture–Intensive cultivation of a field so as to remove all weeds, etc. Also called clean cultivation, clean tillage.

Clean-burning–A backfire set close to the fire edge to control a forest fire.

Clear–(1) In milling, the residual wheat flour left after the separation of patent flour. (2) In land reclamation, to remove trees, brush, stones, or other obstacles to tillage.

Clear Length–The limb-free portion of the stem of a tree; i.e., from the ground to the lowest branch or branch stub.

Clearcutting–An area on which the entire timber stand has been cut. Removal of the entire stand in one cut. Tree reproduction is obtained with or without planting or artificial seeding.

Cleavage–(1) In animal or plant reproduction, the splitting of one cell into two identical parts. Each resulting daughter cell matures and may divide again. (2) Tendency of certain minerals or woods to split along particular planes or angles. (3) The weight required to cause splitting in a standard piece of wood three inches long, expressed in pounds per inch of width.

Cleft–Divided to or nearly to the middle into lobes, as a palmately or pinnately cleft ear.

Cleft Graft–A simple method of grafting in which large trees are used for stock. The branch is sawed squarely across and split lengthwise, and two scions are inserted into the cleft next to the cambium layer. A thin coat of wax is spread over the cut surface and down both sides of the graft.

Clematis, Western–*Clematis ligusticifolia*; a vigorous, climbing, deciduous native vine with pinnate leaves and feathery or densely villous white seeds. It is valuable wildlife habitat and ground cover in riparian areas and along fencelines.

Click Beetle–The adult of a wireworm.

Climacteric Peak–The maximum point of the respiration rate of mature fruit, especially apples. The term is usually applied to harvested fruit, but it may apply to fruit still on the tree.

Climax Vegetation–(1) The group of plant species which is the culminating stage in plant succession for a given set of environmental conditions. (2) A relatively stable type of vegetation in equilibrium with is environment and with good self-perpetuating reproduction of the dominant plant species.

Climber–Any plant having long, relatively slender branches which, in order to support its branches, may attach itself to other plants, fences, walls, etc., by means of tendrils, disks, entwining, etc.

Climbing Fig–*Ficus pumila*, family Moraceae; a tender, woody vine grown in the northern United States in greenhouses and in warm areas out of doors as an ornamental and a basket plant. Native to Asia and Australia.

Clingstone–(1) Any fruit, especially varieties of the peach, whose flesh adheres strongly to the seed, or stone, even at maturity. (2) The stone of such a fruit. See Freestone.

Clinical–Concerning the investigation of decease on a living subject by observation, as contrasted to a controlled experiment.

Clinical Evidence–Any symptom of disease that can be determined by direct observation, such as fever, lack of appetite, swellings, paralysis, etc.

Clinosequence–A sequence of soils whose properties are functionally related to the amount of slope on which they are formed.

Clip–(1) A semicircular metal piece extending from the outer surface of the horseshoe at the toe or side to prevent the shoe from shifting on the hoof. (2) Shears. (3) (a) An inclusive term for shorn wool. (b) The process of shearing wool. (c) The year's production of wool. (4) (a) To shear the hair of an animal close to the skin, as to clip a dog, or a cow's flank and udder. (b) To remove fleece from a sheep or goat. Also called shear. (c) To cut the feathers from a fowl's wing to prevent it from flying. (5) (a) To trim a plant. (b) To cut off the tops of a crop, as clover or alfalfa, at an early stage of growth when the crop is to be harvested for seed. Also called preclipping.

Clitocybe Root Rot–A fungal disease of pear, peach, apple, and other woody plants caused by *Clitocybe tabescens*, family Agaricaceae. Found on citrus in Florida and other trees throughout the southern United States. Its characteristics are similar to mushroom root rot. See Mushroom Root Rot.

Cloche–(1) A French term for a bell glass. (2) A miniature greenhouse used outdoors to protect plants against low temperatures. They are often placed close together along a row of plants to be protected.

Clod–A compact, coherent mass of soil ranging in size from 0.2 to 0.4 inches (5 to 10 millimeters) to as much as 8 to 10 inches (200 to 250 millimeters); produced artificially, usually by plowing or digging, especially when these operations are performed on clay soils that are either too wet or too dry for normal tillage operations. See Ped.

Clonal Variety–One clone or several closely similar clones that are propagated by cuttings, tubers, corms, bulbs, rhizomes, grafts, or seed produced by obligate apomixis; e.g., the Elberta peach and the Higgins buffelgrass.

Clone–A group of plants originating from a single plant, propagated vegetatively rather than by seeds, as cuttings from a rose.

Close Cropping–The grazing by animals of grasses and other forage dangerously close to the crown of the plant, which can ruin grazing lands. Also called close grazing.

Close-growing–Pertaining to plants that have many stems to the square foot.

Cloth House–A structure resembling a greenhouse, but covered with light cloth instead of glass. It is used for growing certain flower crops, such as asters and chrysanthemums to maturity and provides reduced light, reduced temperature, protection from wind, slightly higher humidity, and protection from insects to the plants. Also used for growing tobacco to maturity when a certain quality is desired.

Clove–(1) A small division of a separable bulb, as a clove of garlic. (2) A weight unit of 8 pounds of wool, cheese, etc. (England). (3) The dried flower bud of *Syzgium aromaticum (Eugenia aromatica, E. caryophyllata)*, family Myrtaceae, a tropical tree native to the East Indies, but now extensively cultivated throughout the tropical regions. It is used in flavoring, perfumes, and medicine.

Clover–(1) A general term for low herbs of the genus *Trifolium*, family Leguminosae, with compound, usually trifoliolate, leaves. Clovers are excellent forage, serve as good bee plants, and fix nitrogen in soil. (2) A name applied with a qualifying word to plants of the genera *Medicago* (bur clovers), *Lespedeza* (bush clovers), *Petalostemum* (prairie clovers), *Melilotus* (sweet clovers), and *Alysicarpus* (alyceclovers), etc.

Clover, Alsike–*Trifolium hybridum*; short-lived, perennial legume suited for hay or pasture under irrigation or on dry land where the effective precipitation is 18 inches (45 centimeters) or more; adapted for use on poorly drained, acid soils, especially in cool areas; also tolerant of moderately alkaline conditions.

Clover Dodder–*Cuscuta epithymum*, family Convolvulaceae; a yellow, threadlike plant, parasitic on various species of clover, alfalfa, and some other legumes. It is an introduced species occurring throughout North America. Also called epithyme, fairies' hair, thyme dodder, loveyine.

Clover Leafhopper–*Aceratagallia sanguinolenta*, family Cicadellidae; an insect that infests many plants, especially the clovers, alfalfa, grains, and grasses. Heavily infested plants show a lack of thriftiness; the leaves become whitened and mottled and may turn yellow, red, or brown. it is especially destructive to clover during the seedling stage and after cutting.

Clover Mite–*Bryobia praetiosa*, family Tetranychidae; a mite that feeds on clover, grass, and weeds. In winter it may invade houses and become a nuisance, although it does not bite humans, transmit disease, or damage household furnishings.

Clover, Red–(*Trifolium pratense*); short-lived, perennial legume suited primarily for hay and silage under irrigation, or on dry land where the effective precipitation is 25 inches (62 centimeters) or more. Requires well-drained soil. Produces best under medium acid to neutral soil conditions. It is compatible with white clover in pasture mixtures.

Clover Root Borer–*Hylastinus obscurus*, family Scolytidae; a small beetle that infests the roots of clover, sweetclover, alfalfa, peas, and vetch. Infested plants turn brown, wilt, and die. Found in the northern United States and eastern Canada.

Clover, Rose–*Trifolium hirtum*; a densely hairy, low-growing winter-annual legume. Is adapted to thin dry soils and mild climates. In most areas it is less productive than other annual legumes.

Clover Rusts–Fungi of the genus *Uromyces*, family Pucciniaceae, which attack clovers. The species concerned are *Uromyces elegans* on Carolina clover in the southern United States *U. trifolii* var. *fallens* on red clover, and *U. trifolii* var. *trifolii-repentis* and *U. nerviphilus* on white clover. These rust species are autoceious, i.e., all known stags of the rust occur on the same host. The telial stage (and often the uredial stage) of the rust form rounded, brown pustules on the lower side of the leaf and on petioles and stems, sometimes leading to defoliation.

Clover Seed Chalcid–*Bruchophagus gibbus*, family Eurytomidae; an important insect pest of alfalfa and clover seed. It lays its eggs in the seed before it is hardened, and the larvae consume the entire inside of the seed. It does not affect the production of hay. Also called alfalfa seed chalcid.

Clover, Strawberry–*Trifolium fragiferum*; a spreading, pasture-type, perennial legume suited to eastern Oregon for use under irrigation or semiwet soils, and strongly to very strongly sodic conditions. Less productive than white clover where the latter can be grown.

Clover, Subterranean–*Trifolium subterraneum*; a winter-annual legume ideally suited for foothill and nonirrigated pastures. Also good for fall-seeded erosion-control plantings. Will volunteer freely for many years if managed properly. Requires well-drained soils.

Clover Taint–An off-flavor of milk produced from cows that have grazed on clover prior to milking.

Clover, White–*Trifolium repens*; long-lived, stoloniferous perennial legume suited primarily for pasture, but can be used for hay and silage. Can be grown under irrigation or on dry land where the effective precipitation is 18 in. (45 cm) or more. Requires medium to high fertility.

Club Wheat–*Triticum compactum*, family Gramineae; an important annual cereal plant with flat leaves and club-shaped spikes. Grown for its grains, which are a principal source of common bread. Probably native to western Asia. Also called cluster wheat.

Clubroot of Cabbage–A fungal disease of crucifers caused by *Plasmodiophora brassicae*, family Plasmodiophoraceae. Because it attacks the underground parts of the plant, the disease may run its course before host symptoms appear. Root tissues are infested with soft-rot organisms, filling the root with gall and deforming it. This results in a slow reduction in growth, wilting, and premature death of the plant. Also called clump foot, clubbing, finger and toe disease, anbury, club, dactylorhiza, club foot.

Clump–(1) A cluster of plants. (2) A mass of roots or rootstocks, as a clump of iris.

Cluster–(1) The form in which bees cling together in the hive after swarming or during winter. Also called clustering. (2) A large number of plants grouped, arranged, or growing in close proximity to one another. (3) An inflorescence.

Coarse Texture–(1) In soils, the sand, loamy sand, and sandy loam (except the very fine sandy loam) textural classes. (2) (a) A body of defect of sherbet and ice cream caused by insufficient sugar or stabilizer, bad refrigeration, or failure to put them in the hardening room immediately after removal from a freezer. (b) A body and textural defect of ice cream, characterized by a lack of smoothness. (3) A textural condition of a fruit, vegetable, or other plant or plant produce.

Coated Seeds—Seed that have been coated with clay or other substances. Seeds are treated in this manner to make them easier to handle and plant. For instance, some seeds are so tiny that it is difficult to plant the individual seeds in the proper spacing.

Cob—(1) The chaffy axis upon which the kernels of corn grow. Also called corn cob. (2) A stocky, short-legged horse used for pulling light carts. (3) Male swan.

Cocaine Tree—Any shrub or small tree of the genus *Erythroxylum (Erythroxylon)*, family Erythroxylaceae. Native to South America.

Coccus—A spherical bacterium, as of the family Coccaceae.

Cockle—See Corn Cockle.

Cocklebur—Any plant of the genus *Xanthium*, family Compositae. Species are troublesome, poisonous weeds having seeds with spiny hooked barbs, and large leaves. Animals, after eating the plant, become depressed, nauseated, stagger, and have rapid and weak pulse with a drop in temperature. Also called clotbur.

Cockspur Hawthorn—*Crataegus crus-galli*, family Rosaceae; a small thorny tree whose wood is useful for fenceposts and fuel. Native to North America. Also called cockspur thorn.

Cocoa—A food product obtained from the beans of cacao. The cacao fruit pod resembles a small melon with well-defined segments. The seeds from the pods are fermented, which destroys the spongy pulp about the bean and eliminates a bitter taste, and are then dried at temperatures less than 125°F. The dried beans are screened, roasted, crushed, and separated from shells by winnowing. The crushed product is ground, made into a thick paste, and colored to set into solid slabs (bitter chocolate). To make cocoa, the slabs are warmed and pulverized by hydraulic presses to remove cocoa butter (fat). Also called chocolate. See Cacao, Chocolate Tree.

Cocoa Shell—The residual hull of the cacao beans, after the cocoa has been extracted; used for animal feed and bedding, and to make cocoa tea.

Cocoa Shell Meal—A fertilizer obtained by grinding cacao bean husks. It contains 2.5% nitrogen, 1.0% phosphoric acid, 2.5% potash.

Cocoa Tankage—The cocoa residue resulting from chemical processing of ground cocoa cake, containing approximately 4% N, 1.15% available P_2O_5, 2% K_2O, and 0.06% Mg. It may contain 20 percent lime (CaO).

Coconut—The coconut palm fruit. Its fibrous skin yields coir (a coarse fiber for making mats), and the meat inside, after drying, yields commercial copra (ground or shredded coconut). See Copra.

Coconut Oil—The oil extracted from the dried fruit of the coconut (copra); used in making soap and coconut butter. Also called copra oil.

Coconut Oil Cake—A stock feed, a by-product of dried coconut meal after coconut oil has been extracted. It averages 21.3% protein; it equals corn gluten in feed value. Also called copra oil cake.

Coconut Palm—*Cocos nucifera*, family Palmaceae; a widely distributed, pinnate-leaved tropical tree, grown for its drupe fruit, the coconut of commerce. Native to Asia and Pacific islands.

Cocoon—The silky outer covering which the larvae of certain insects spin around themselves before transforming to the pupae stage.

Codlin—A hard cooking apple. Also called codling or codding.

Codling Moth—*Carpocapsa pomonella*, family Olethreutidae; a destructive insect whose larva chews its way into apples. It also attacks the pear, quince, wild haw, crabapple, and English walnut.

Codominance—A kind of gene action where one allele does not exhibit complete dominance over the other.

Codominant—(1) A tree which grows close enough to other trees as to receive very little sunlight from the sides. (2) Any plant or animal species that shares dominance with other species in an area.

Codominant Genes—Genes that are neither dominant nor recessive.

Coenzyme—A partner needed by some enzymes to accomplish a biochemical change.

Coffee—Any evergreen shrub or tree of the genus *Coffea*, family Rubiaceae, which has opposite leaves, cream-colored flowers, and a berry-like fruit. The seeds inside the berries are freed of the pulp, dried in the sun, and used green or are roasted and ground to make coffee. Native to Asia and Africa.

Coffee Chaff—Consists of the finely ground dried pericarps from coffee beans. It has been used as a fertilizer conditioner in nongranular goods.

Coffee Substitute—A food product that contains no coffee; generally prepared from a roasted and ground mixture of bran, wheat, and molasses.

Coffeeweed—*Daubentonia puncea*; a large perennial weed pest of the southeastern United States. The seeds are poisonous.

Cohesion—(1) The force that attracts or holds like molecules (liquids or solids) together, owing to the attraction between like molecules. (2) The lateral fusion of flower parts within a single whorl. See Adhesion.

Coincidence—In genetics, a term used in estimating the distance between two genes. It is the ratio between the actual percentage of double crossing over and the percentage expected on the assumption that each crossing over is an independent event.

Coir—The fiber from the husks of the coconut; used for cordage and coarse matting.

Colchicine—A poisonous alkaloid extracted from the common autumn crocus; used in medicine and in plant breeding to serve as an intracellular poison to block anaphase separation during cellular division, thereby producing polyploidy and doubling the number of chromosomes. It often increases plant size.

Cold Cellar—An underground room for the storage of vegetables, fruits, etc. The temperature should remain about 38 to 40°F.

Cold Front—The movement of a relatively cold air mass advancing behind a relatively warm air mass. Characteristic weather includes short but violent thunderstorms, heavy rain squalls, and occasionally hail and tornados. Weather behind the passage of the cold front will usually be cooler and dryer than before, with clear skies.

Cold Manure–Farmyard manure which does not ferment and heat unduly while in storage, such as cattle manure. See Hot Manure.

Cold Pack–Raw, unheated food packed into canning containers and covered with boiling syrup, juice, or water.

Cold Pit–A deep pit with a cover which is used for the winter storage of bulbs and hardy plants.

Cold Pocket–A low area into which cold air settles from adjoining slopes. See Frost Pocket.

Cold Storage–(1) An insulated storage using mechanical refrigeration to maintain a stable, cold temperature for long-term storage. (2) The treatment given to plants and bulbs to cause certain chemical changes that enable them to respond to forcing treatments.

Cold Test–(1) A test for viability in which sample seeds are placed on wet soil, sand, cloth, or blotter at a constant temperature of 40°F for three days or longer. The sample is then placed in a warm place and the percentage of germination recorded. (2) Synonymous with cold shrinkage. After wool scouring, the hot wool is allowed to stand in a cool room until it has taken on a normal amount of moisture from the air. The wool is then weighed, and the shrinkage figured to give the cold shrinkage, or cold test.

Cole (Cabbage Family)–Any plant of the genus *Brassica*, family Cruciferrae. See Broccoli, Brussels Sprouts, Cabbage, Cauliflower, Collard, Kohlrabi.

Colemanite–A natural calcium borate ore that contains 32 to 50% B_2O_3. It is insoluble in water, but readily available to plants. See Borax, Boron.

Collar–(1) The padded and reinforced part of a harness which girdles the neck. Straps for pulling are attached to it by means of hames. (2) A closely bound retail cut of rolled meat. (3) The part of a tree or other higher plant at the line of union of the roots and the trunk.

Collards–*Brassica oleracea*; a cool-season, leafy green vegetable grown primarily in the South.

Collected–(1) A collected horse has full control over its limbs at all gaits and is ready and able to respond to the signals or aids of its rider. (2) Plants dug from the wild and offered for sale.

Collection–(1) The balance of a horse and rider which makes for instinctive coordinated movement. (2) The accumulation of sperm for use in artificial insemination. (3) In flower shows, an exhibit of several varieties of plants, flowers, fruits, or vegetables. A collection differs from a group in that the number of varieties is an important factor. Artistic arrangement is not necessarily mandatory.

Colloid–(1) An insoluble substance consisting of particles small enough to remain suspended indefinitely in a medium. (2) A mineral particle less than 0.002 millimeters in diameter. (3) A substance which does not diffuse readily through animal or vegetable membranes. Its presence does not affect the freezing point or vapor tension of the solution.

Colloidal Materials–Gas, liquid, or solid particles suspended in a gas, liquid, or solid that are intermediate in size between true solutions and suspensions.

Colloidal Peat–The unctuous, gelatinous mass of organic residue formed as a bottom deposit in lakes. It may lie at the base of a peat deposit, but does not conform in texture and appearance to the material commonly designated as commercial peat. See Peat.

Colloidal Phosphate–A finely divided raw (unprocessed) mineral phosphate or phosphatic clay. Occurs in large quantity mixed with Florida hard rock, from which it is removed by washing, carried by the wash water to settling basins or ponds and remains after the water has evaporated. Although sometimes applied directly to the soil, its phosphoric oxide is only slightly available, except in strongly acid soils. It contains from 50 to 58% tricalcium phosphate and from 9 to 12% iron and aluminum oxides.

Colluvium–A general term applied to loose and incoherent deposits, usually at the foot of a slope or cliff and brought there chiefly by gravity. Talus and cliff debris are included in such deposits. See Scree, Talus.

Colmatage–(Latin; *culmen*, ridge; Indo-European; *colen*, hill) (1) The practice of collecting soil sediments from low-lying places and spreading them on upland soils of low fertility to enhance plant growth. A common practice in all developing countries such as India and China. (2) The gradual filling in of lowland by siltation.

Colonum–Cultivated.

Colony–(1) An aggregate of worker bees, drones, and a queen bee, living together in a hive as a unit. (2) A cluster of bacteria or fungi grown on a culture medium, usually originating from a single bacterium, spore, or inoculation transfer. (3) A group of people bound by communal or religious ideals that occupies tracts of land, often wild, for permanent settlement, especially in the early development of the United States and Canada.

Color Profile–A soil profile described by the aid of the Munsell color chart according to the colors of its separate horizons. See Munsell Color Standards.

Colorado Loco Vetch–See Lambert Crazyweed.

Colorado Potato Beetle–*Leptinotarsa decemlineata*, family Chrysomelidae; a striped beetle whose adults and larvae feed on the leaves and retard growth of the potato. It also infests the tomato, tobacco, pepper, ground cherry, henbane, horsenettle, belladonna, thistle, and mullein. Found east of the Rocky Mountains, United States.

Colorado Spruce–*Picea pungens*, family Pinaceae; a hardy evergreen tree grown as an ornamental. The most widely grown variety is the blue spruce.

Columnar–Describing a plant or a soil structure whose outline resembles a column; i.e., one formed by relatively perpendicular sides.

Colza Oil (Rapeseed Oil)–A pale yellow oil, specific gravity 0.912 to 0.920 at 60°F, pressed from the seeds of the winter rape. It is used in making soaps and blackening, as a cooking medium, an illuminant, and a lubricant. For the latter purpose, air is blown through it while it is heating; the blown oil may be used alone or mixed with mineral oil.

The residue of crushed seeds is made into oil cake. See Canola, Rapeseed Meal.

Coma—(1) A tuft of soft hairs on a seed, as in milkweeds (*Asclepias*). (2) A tuft of leaves or bracts at the apex of an inflorescence, as in pineapple (*Ananas*). (3) A leafy crown or head, as in many palms. (4) Insensibility of an animal or person caused by poison, injury, or disease.

Comate Seed—A seed covered with long silky hairs or bristles.

Comfrey—(1) Any plant of the genus ***Symphytum***, family Boraginaceae. The species are hardy, perennial herbs bearing purple, blue, or yellow flowers; native to Eurasia. (2) ***Symphytum officinale***, the common comfrey, grown for its yellowish, rose, or white, flower clusters. The plant is eaten by stock, the root is used in medicines, and the young leaves and blanched stalks and roots have been used also as human "survival" food.

Commercial Culture—Any bacterial culture which is prepared and sold by a special culture-producing laboratory.

Commercial Fertilizer—Plant nutrients containing a single essential mineral nutrient or a mixture of essential mineral nutrients or organic materials.

Commercial Species—Tree species suitable for industrial wood products.

Commercial Thinning—Removing trees from a developing young stand so that remaining trees will have more growing space; dead and dying trees will be salvaged; and the operation will, hopefully, make a net profit.

Commercial Timberland—Forest land that is producing or is capable of producing crops of industrial wood and that is not withdrawn from timber utilization by statue or administrative regulation. Note: Areas qualifying as commercial timberland have the capability of producing in excess of 20 cubic feet per acre per year of industrial wood in natural stands. Currently, inaccessible and inoperable areas are included.

Commercial Wheat Germ Meal—A by-product of flour milling that consists of the wheat germ and some bran and middlings. It contains 31.1 percent protein and 9.7 percent fat, and is used as a supplement in poultry feeds and calf meals. Also called wheat germ oil meal.

Commerical Variety—See Cultivar.

Common Scab of Irish Potato—See Corky Scab of Potato.

Community—A group of plants growing together or all of the plants and animals of an area.

Community Regulation—See Homeostasis.

Compaction—(1) Decrease in volume of sediments, as a result of compressive stress, usually resulting from continued deposition above them, but also from drying and other causes. (2) Reducing the bulk of solid waste by rolling and tamping. (3) Increasing soil bulk density and decreasing porosity caused by the application of mechanical forces to the soil.

Companinon Crop—A crop which is grown with another crop, usually applied to a small grain crop sown with a forage crop. Preferred to the term nurse crop. See Nurse Crop.

Compatibility—In fruit plants, a condition in which the male gamete will fertilize the female gamete.

Compatibility Agent—An adjuvant that helps unlike chemicals mixed together for even application. See Adjuvant.

Compatible—Describing two or more chemicals that can be mixed without affecting each other's properties.

Compatible Pesticides—Compounds or formulations that can be mixed and applied together without undesirably altering their separate effects. Sometimes such a mixture makes one or both pesticides more effective.

Compensation Point—The light intensity or temperature at which the rates of photosynthesis and respiration in plants are equal.

Competitive Crop—(1) A crop planted on a piece of land to force out other forms of plant life. Also called smother crop. (2) Crops that compete with each other for the same time in a farmer's work schedule.

Complementary Crop or Livestock—Any crop or livestock yielding a product that contributes to the success of another.

Complementary Genes—Genes that are similar in their individual effect but which together interact to produce a new character; e.g., in the sweet pea, either of two dominant factors, C and P, individually produce white flowers, whereas, when present together, the same two factors produce purple-flowered plants.

Complete Carcinogen—An agent that can act as both initiator and promoter of cancer.

Complete Fertilizer—A fertilizer that contains some of each of the three essential nutrients, N, P, and K. Ratios of each can vary depending upon the formulation.

Complete Flower—A flower containing sepals, petals, stamens, and at least one pistil.

Complete Metamorphosis—Metamorphosis in which the insect develops by four distinct stages; namely egg, larva, pupa, and adult.

Completely Saturated—(1) Designating clay or humus containing all of the basic ions plus hydrogen ions that it is capable of holding. (2) Any substance, such as soil, which can hold no more water.

Complex Slope—Irregular or variable slope in a field. Planning or constructing terraces, diversions, and other water-control measures is difficult.

Complex Soil—A map unit of two or more kinds of soil occurring in such an intricate pattern that they cannot be shown separately on a soil map at the selected scale of mapping and publication.

Compositae—A very large family of plants, herbs, and trees, which includes flowers, such as marigold, dahlia, chrysanthemum; vegetables, such as chicory and lettuce; and weeds, as burdock, etc.

Composition—(1) The relative production of various plant species in a given area. (2) The components used in any mixture of feed, fertilizers, hay, seeds, etc.

Compost–(1) Organic residues or their mixture, such as peat, manure, or discarded plant material and soil, placed in a pit or enclosure, moistened, and allowed to become decomposed. Sometimes lime and chemical fertilizers are also added. Used as a fertilizer. (2) To cause vegetable matter to become decomposed as a fertilizer.

Compound–A chemical term denoting a combination of two or more distinct elements.

Compound Fertilizer–A mixed fertilizer or formulation containing at least two of the primary plant nutrients (N_1 P_2O_5, and K_2O) formed by intimately mixing two or more fertilizer materials by dry or fluid bulk blends. Compound fertilizers are made in registered grades approved by the laws of the respective states.

Compound Leaf–A leaf composed, usually, of two or more leaflets. A leaf is digitately or palmately compound when three or more leaflets arise from a common point at the end of the petiole; pinnately compound when three or more leaflets arise from a common point at the end of the petiole; pinnately compound when one or more pairs of leaflets are arranged along the sides of the axis, with (odd-pinnate) or without (even-pinnate) a terminal leaflet; alternately compound when the leaflets or the divisions of the leaf occur in threes.

Compound Ovary–The fusion of the flower carpels.

Compressed Bale–(1) A bale of cotton which has been reduced in volume to a density of 22½ pounds per cubic foot. (2) A bale of hay reduced in size by extra pressure.

Compression Wood–Abnormal wood formed on the lower side of branches and inclined boles of conifer trees.

Concentrate–(1) (a) Any feed high in energy (usually grain); sometimes used with reference to other nutrients, such as protein concentrate, etc. (b) Stock feed low in fiber content and high in digestible nutrients. (2) Whole fruit juice, thickened to the consistency of heavy syrup. (3) To increase in strength by removing dilutents.

Concentrate Sprayer–A sprayer designed to deliver pesticides to a crop at normal amounts per acre of active ingredients but in much lower volumes of water.

Conch–The bracket-shaped fruiting body of certain fungi on the bark, branches, or the crown of a tree. Also called punk, conk.

Concolorous–Of a single uniform color.

Concrete Frost Structure–A type of frost in which the soil becomes virtually solid. See Granular Frost Structure.

Concretion–(1) An inorganic body formed in a human or animal body cavity, as a kidney stone. (2) Grains, pellets, or nodules of various sizes, shapes, and colors consisting of concentrated compounds or cemented soil grains. The composition of most concretions is unlike that of the surrounding soil. Calcium carbonate and iron oxide are common compounds in soil concretions.

Condemned–(1) Describing an animal, carcass, or food which has been declared unfit for human consumption. (2) Referring to real estate property acquired for public purposes under the right of eminent domain.

Condiment Plant–A plant which yields spices or other seasoning material.

Condition–(1) The state of wool regarding the amount of yolk and other foreign matter it contains. (2) The general appearance and/or state of health of an animal, seed, fruit, or flower at a show. (3) To get an animal in good health and appearance by proper feeding and grooming. (4) The degree or amount of fat on a breeding animal.

Conditioners–Conditioning materials, anticaking agents. Finely divided, dry, bulky, inert powders such as diatomaceous earth, siliceous dusts, and clays in common use as coating agents to decrease the caking tendency of fertilizers. Oils, organic amines, and plastic coatings are used in a few specialty fertilizers. Conditioning agents will prevent caking only when the moisture content of the product is sufficiently low to inhibit the formation of crystalline bridges between particles.

Conducting Tissue–The plant tissue that is primarily concerned with the movement of water and food. Xylem and phloem are usually considered as conducting tissues.

Cone Penetrator–A rod-shaped instrument with a cone-shaped end, used for penetrating the soil to measure soil strength.

Confectioner's Sugar–A very finely pulverized sugar combined with cornstarch, used for cake icings, etc.

Confier–(1) A tree or shrub that bears cones, as pines and firs. (2) Any plant belonging to the family Coniferae, a large group of trees and shrubs, which includes the families Taxaceae, Pinaceae, Cycadaceae, and Ginkgoaceae. Conifers bear cones or strobili, and because many of the species produce new leaves before all the old leaves are shed they are looked upon as synonymous with evergreens. One principal exception is the bald cypress, a deciduous conifer.

Confused Flour Beetle–*Tribolium confusum*, family Tenebrionidae; a very common insect of stored products such as grains, flour, beans, peas, baking powder, ginger, dried plant roots, dried fruits, bran, nuts, chocolate snuff, cayenne pepper, etc. Also called bran bug.

Congeal–To change a liquid to a solid by lowering the temperature of the food sufficiently to bring about gelation.

Congelifraction–The mechanical disintegration of minerals and rocks resulting from the pressure exerted by freezing of water contained in their cracks. Freezing water exerts a pressure of 150 tons/square inch (146.48 kg/cm^2).

Congenial–Able to cross-fertilize readily; to unite easily.

Congou–(Chinese) Black tea.

Conilignosa–Vegetation which is dominated by trees and shrubs with needlelike foliage.

Conjugate–Coupled or in pairs.

Conjugation–Side-by-side association (synapis) of homologous chromosomes, as in meiosis.

Conk–A hard, spore-bearing structure of a wood-destroying fungus, which projects beyond the bark of a tree.

Connate—United; used especially of like structures, such as leaves, joined from the start.

Conservation Practice—A land treatment measure to protect or conserve soil, water, woodland, or wildlife resources, or the installation of a structure or other measure for this purpose.

Conservation Tillage—Any of several farming methods that provide for seed germination, plant growth, and weed control yet maintain effective ground cover throughout the year and disturb the soil as little as possible. The aim is to reduce soil loss and energy use while maintaining crop yields and quality. Kinds of conservation tillage are: chisel-plow, no-till (zero-till), plow-plant, ridge-plant, strip-tillage, sweep-tillage, till-plant, and wheel-track-plant.

Conservatory—A glass house constructed for the accommodation of plants for display as contrasted to a structure for propagation.

Consistence (Consistency, Soil Strength)—The feel of the soil and the ease with which a lump can be crushed by the fingers. Terms commonly used to describe consistence are: *loose*—noncoherent when dry or moist; does not hold together in a mass. *Friable*—when moist, crushes easily under gentle pressure between thumb and forefinger and can be pressed together into a lump. *Firm*—when moist, crushes under moderate pressure between thumb and forefinger, but resistance is distinctly noticeable. *Plastic*—when wet, readily deformed by moderate pressure but can be pressed into a lump; will form a "wire" when rolled between thumb and forefinger. *Sticky*—when wet, adheres to other material and tends to stretch somewhat and pull apart rather than to pull free from other material. *Hard*—when dry, moderately resistant to pressure; can be broken with difficulty between thumb and forefinger. *Soft*—when dry, breaks into powder or individual grains under very slight pressure. *Cemented*—hard; little affected by moistening.

Consorting—A species of trees that, although it may be found in pure stands, is most often found as a major segment in a mixed stand, that is, one of the more abundant species.

Consortium—(1) A group of individual plants of different species, generally belonging to different phyla, which live together in close association. (2) A group of people working on a common problem or project.

Constant Level Method—A method of watering plants in a greenhouse by a channel having a V-shaped bottom (or a flat bottom covered by half tile). This is covered with pea gravel. A layer of sand, topped by a soil layer, is placed over the gravel. The water table is held constant, usually in the sand layer, by use of a float valve.

Constituent—The components or elements of a substance; e.g., the constituents of soil.

Consumptive Use—The water used by plants in transpiration and growth, plus water vapor lost from adjacent soil or snow or from intercepted precipitation in any specified time. Usually expressed as equivalent depth of free water per unit of time, such as acre-inches per week.

Contact Herbicide—An herbicide that kills a plant primarily by contact with plant tissue rather than by internal absorption.

Contact Insecticide—Any substance that kills insects by contact in contrast to a stomach poison, which must be ingested.

Contact Pesticide—A pesticide that kills primarily by contact with insect or plant tissue rather than as a result of ingestion or translocation by the pest.

Container Planting—A system of raising and planting forest tree seedlings in such containers as baskets, blankets, tubes, "bullets," pots, or porous blocks.

Container Stock—Plants grown and marketed in the same container.

Contaminant—An undesirable substance that is present in foods or feed but is not intentionally added.

Contaminate—To make impure by contact or admixture of harmful bacteria, fungi, or dangerous chemicals, etc.; to render unfit for use.

Contamination—(1) Pollution; the process of being contaminated. (2) Specifically, the addition of bacteria or other foreign substance to milk or other products by means of utensils, containers, exposure to air, etc.

Continuous Layerage—A method of propagation of woody plants by rooting the branches which are still attached to the parent plant. The whole shoot is bent down flat in a shallow trench, running radially from the plant, and is covered with soil. However, the end remains exposed. This method is used for the shrubs and vines whose buds will grow readily even though covered with soil. The snowball European viburnum is often propagated this way.

Contour Border Irrigation—A method of irrigating gently sloping fields. The whole area is divided into strips by ridges along the contours and cross ridges. The ridges confine the water to a particular strip until it is completely full, before letting it flow to the next lower strip. Also called contour check irrigation.

Contour Ditch—A ditch laid out approximately on the contour.

Contour Farming—Field operations, such as plowing, planting, cultivating, and harvesting, on the contour, or at right angles to the natural slope to reduce soil erosion, protect soil fertility, and use water more efficiently.

Contour Furrow—(1) A narrow furrow plowed along a contour or at a uniform grade. (2) A level furrow made in a field or pasture to reduce water runoff and soil loss, and to increase water infiltration in soil.

Contour Planting—The planting or drilling of crops in rows along contour lines in contrast to plantings which run parallel to field boundaries or up- and downhill on sloping land.

Contour Plowing—Plowing a field on a contour.

Contour Row—(1) A row, all points of which have the same elevation within a given tolerance. (2) A level row that runs at right angles to the line of slope regardless of the irregularities of the landscape.

Contour Strip Cropping—The production of crops in long, variable-width strips, which are placed approximately on the contour, and crosswise to the line of slope.

Contour Tillage–A system of farming in which the various operations are performed on the contour. See Conservation Tillage.

Contracted Weir–A device for measuring the flow of water. The sides produce a contraction in the cross-sectional width of the water channel.

Contraindication–Condition of a disease which renders a particular treatment undesirable.

Control–(1) Prevention of losses from plant or animal diseases, insect pests, weeds, etc., by any method. (2) A section of an open water channel where conditions exist that make the water level above it an index of the discharge. (3) A standard entity used for comparative purposes in experimentation. Also called check or check plot.

Control of a Fire–The surrounding of a fire with control lines and backfiring of any unburned surfaces adjacent to the inner edge of the control lines.

Control Section–The part of the soil on which classification is based. The thickness varies among different kinds of soil, but for many it is 10 to 30 centimeters.

Controlled Atmosphere Storage–A cold storage in which the concentrations of atmospheric gases are adjusted to extend the storage life of fresh produce. Usually oxygen is lowered and carbon dioxide is raised.

Controlled Burning–Setting fire to land cover under conditions that presumably will accomplish specific silvicultural, wildlife, grazing, or fire hazard reduction purposes. See Prescribed Burning.

Controlled Release Fertilizers–Fertilizers in which one or more of the nutrients have limited solubility in the soil solution, so that they become available to the growing plant over a controlled period. The ideal in such a fertilizer would be the release of nutrients at a rate exactly equal to the needs of the plant. Manures and sewage sludges are decomposed by microbes and the result is a controlled release of plant nutrients approximately in harmony with the growing plant.

Convergent Lady Beetle–*Hippodamia convergens*, family Coccinellidae; a small, beautifully colored beetle which feeds on certain insects and their eggs. It is helpful in the destruction of various plant insects and scales.

Conversion–(1) Sawing or cutting timber to any shape. (2) The change from one reproduction method to another. (3) The transformation of a forest from one dominant type to another.

Conversion Factors–Convenient multipliers that have been calculated for elements and compounds important in fertilizers, soil amendments, and plant nutrition. The calculations are based on atomic weights of each chemical. For example, to determine the conversion factor for changing from a known percentage or mass of ammonia (NH_3) to an unknown percentage or mass of nitrogen (N), divide the atomic weight of NH_3 (17.04) by the atomic weight of N (14.01) = 1.216. Then to convert N to NH_3, multiply N by 1.216.

Converted Rice–Rice steamed to cause the vitamins from the hulls to penetrate the endosperm. This changes the characteristics and appearance of milled rice, but when cooked by boiling, the kernels appear as white as raw rice and retain their shape better.

Cool House–A house, in a range of greenhouses, which is maintained at a day temperature of about 55° to 60°F (13° to16°C) and a night temperature of about 50° to 55° F (10° to 13°C). Chrysanthemum, carnation, sweet pea, snapdragon, and other plants do well in these temperatures. Also called cool greenhouse.

Cool-season Crops–Vegetables that thrive best in the cool season, such as cabbage, English peas, lettuce, or spinach.

Cooler–Any chestlike device for reducing the temperature of products, such as milk.

Cooling Pad–A system of cooling greenhouses and livestock facilities. Water is filtered through a pad where evaporation occurs and helps cool the building.

Cooperage–The barrels, casks, and tubs made by staves and hoops usually from wood; made by a cooper. Tight-cooperage is made from such nonporous woods as white oak to hold liquids such as whiskey. Slack-cooperage is made form porous but tough wood such as ash and elm for holding nonliquids such as rice and cranberries.

Copper–Cu; a metallic element found in soils at 1 to 50 parts per million, and in plants up to 100 parts per million. It is necessary for all animal and plant life. High soil phosphorus, zinc, and molybdenum can induce copper deficiency in plants. Also, high copper can reduce plant uptake of phosphorus, iron, zinc, and molybdenum.

Copper Carbonate–A compound used as a seed disinfectant.

Copper Oxide–A chemical used in the seed treatment sprays of dusts for sugar beets and vegetables. It controls damping off, rose leaf diseases, celery blight, tobacco and hops downy mildew. Also called cuprocide.

Copper Poisoning–A livestock poisoning usually in sheep, from consuming too much salt to which copper sulfate has been added for control of stomach worms. Symptoms are loss of appetite, jaundice, and often bloody urine.

Copper Sulfate–A common chemical used extensively to kill algae in waste waters or natural surface waters. Syn.: blue vitriol, bluestone, blue copperas, and cupric sulfate.

Copperas–See Ferrous Sulfate.

Coppice Forest–A forest of second-growth sprouts. Also called sprout forest.

Copra–The dried coconut kernel from which coconut oil and shredded coconut is obtained. See Coconut, Coconut Oil.

Copra Oil–Coconut oil.

Coprogenous–Designating the influence of animal excrement, as the cast of the earthworm in forming soil.

Cord–(1) (a) The spermatic cord. (b) The umbilical cord. (2) A unit of measurement of timber. It contains 128 cubic feet, usually 8 x 4 x 4 feet. A short (face cord is a stack of wood 8 feetlong, 4 feet high, and about 16 inches wide. (3) Twine. (4) To stack in rectangular tiers.

Cordate–Refers to a leaf that is shaped like a heart.

Cordgrass–*Spartina alterniflora*; grows in salt marshes.

Cordiform–Shaped like a heart.

Coriander–*Coriandrum sativum*, family Umbelliferae; an annual herb grown for its seeds, which are used as a condiment and in liqueurs. Its fruit is used in medicines and for seasoning. The oil obtained from the seeds is used in toilet waters.

Cork–(1) The thick external suberized tissue of some woody plants, especially the cork oak, produced by the activity of the phellogen, used commercially for floats, stoppers, etc. (2) Plant tissue similar to oak bark in composition or structure. (3) A spot disease of apples, which causes internal tissue to become dry and punky, but does not show any external malformations such as depressions or ridges. The large dry rot areas may become hollow, reducing the size of the apple. Also called punky disease, York spot, hollow apple, crinkle, confluent bitter pit.

Cork Oak–*Quercus suber*, family Fagaceae; a fairly small tree, whose thick and corky bark is the commercial source of cork. Native to southern Europe.

Corky Scab of Potato–A bacterial disease caused by *Actinomyces scabies*. Scablike areas appear on the surface of the tuber. Also called brown scab, Oospora scab, American scab, deep scab, actinomycosis of potato.

Corm–Enlarged fleshy base of a stem, bulblike but solid, in which food accumulates. Propagated by division of the cloves; e.g., Dasheen (taro), garlic, and shallots.

Cormel (Cormlet)–A small, leafless corm produced by an old corm.

Corn–(1) *Zea mays*, family Gramineae; an annual herb; a cereal grain which is one of the most important agricultural crops in the United States; grown for human and animal consumption. It is also a source of vegetable oil. Also called Indian corn, maize. (2) The seed or the plant of any of the cereal crops, as wheat, rye, etc. (United Kingdom). (3) A growth on the foot of a fowl afflicted with bumblefoot. (4) An afflicted part of a horse's forehoof. An injury to the soft tissue of the sole causes blood to diffuse into the horny sole between the bar and the outside wall. (5) To preserve beef by seasoning with salt, sugar, baking soda, saltpeter.

Corn and Winter Wheat Belt–That area of the Midwestern United States where soft red winter wheat is grown in rotation with corn. Includes Ohio, Indiana, Iowa, and Illinois.

Corn Belt–Central part of the United States, where corn is extensively cultivated. Because of increasing diversification of crops, the area is becoming less distinctive, and the term is losing some of its value as a geographic referrent.

Corn Cob–The woody core of the corn ears; the highly fibrous axis on which the kernels are borne; used alone or with the grains as ground cattle feed.

Corn Cockle–*Agrostemma githago*; a common weed found in grain fields. The poisonous seeds are ripe at harvest time and are carried with the grain seed. Flour made from wheat containing a quantity of cockle is poisonous and dangerous for human food.

Corn Crib–(1) A building for the storage of ear corn. The roof is watertight, but the sides are usually slatted for ventilation. (2) A circular or rectangular temporary storage, with or without a roof, of woven wire or slats.

Corn Distillers' Dried Grains–A stock feed consisting of the dried residue from the distillation of alcohol from corn or from a grain mixture in which corn predominates.

Corn Earworm–*Heliothis zea*, order Lepidoptera, family Noctuidae; an insect pest that infests corn, cotton, tobacco, tomatoes, sorghum, vetch, alfalfa, etc. Also called tomato fruitworm, bollworm.

Corn Fodder–The entire corn plant, cut and fed green, or harvested when mature and dried for future feeding. The term is used also for the stalk remaining after ears have been husked from the plant.

Corn Gluten Meal–A stock feed from the residue of commercial shelled corn, which remains after the extraction of the larger part of the starch and germ, from which the bran has been separated. It may contain corn solubles and corn oil meal. Also called gluten meal.

Corn Grits–See Hominy Grits.

Corn Leaf Aphid–*Rhopalosiphum maidis*, family Aphidae; an aphid which sucks juices from the leaves, stems, and flowering parts of corn, barley, sugarcane, millet, broomcorn, sorghum, Sudangrass and other grasses, and coats them with honeydew. It is especially destructive to sorghum. Common throughout the Corn Belt, but more abundant in the southern United States.

Corn Meal–Finely ground but unbolted corn, used in corn bread, mush, etc.

Corn Oil–A yellow, edible oil obtained from the germ of corn. Used as a salad and cooking oil, and in soap manufacture.

Corn Oil Cake–A stock feed consisting of the corn germ, from which oil has been partially pressed.

Corn Picker–A machine that harvests ears of corn. The stalks are fed into the machine, where the ears are snapped off and loaded into a bin.

Corn Planthopper–*Peregrinus maidis*, family Delphacidae; a leafhopper-like insect which infests the buds and leaves of late-planted corn. It is found in the southern United States.

Corn Screenings–A stock feed consisting of small and broken grains of corn, obtained by screening shelled corn and other material having feed value.

Corn Shuck–See Cornhusk.

Corn Silage–An excellent stock feed prepared from the entire corn plant. The plant is cut while still green but when the ears have begun to dent. It is then chopped and placed in a silo for fermentation and preservation.

Corn Silk–The pistillate floral part of the corn plant, protruding from the husk, which catches the pollen and serves as the avenue for the fertilization of the egg cell, which develops into the kernel. Each silk is the much elongated style and stigma of the flower.

Corn Smut–A fungal disease caused by *Ustilago maydis*, family ustilaginaceae, which accounts for up to 2 percent of the crop losses over wide areas in the United States. It produces galls on the parts of the plant, usually pea-sized on leaves, but up to about 6 inches in diameter on other parts. The galls appear greenish-white, but on enlargement, they rupture and expose a powdery, black spore mass. The young galls while still firm are often cooked and eaten. Also called boil smut.

Corn Stover–The dried corn stalk from which the ears have been removed; used as a roughage for livestock.

Corn Stubble–The basal portion of the stems of corn left in the ground after the plants are cut.

Corn Syrup–Commercial glucose; obtained chiefly by acid hydrolysis of cornstarch. It is an uncrystallizable syrup, used as a sweetening agent.

Corn Tassel–The terminal inflorescence of the corn plant, borne at the top of the stalk, which bears the male pollen grains.

Corn-Hog Ratio–Number of bushels of corn that are equal (in value) to 100 lbs. of live hogs; i.e., the price of hogs per hundredweight divided by the price of corn per bushel.

Cornborer–See European Cornborer.

Corner–(1) The outer pair of incisor teeth in the upper and lower jaws of a horse. (2) The junction point of boundary liens. (3) To tie up or control all available items of produce for speculation. (4) In lumbering, to cut through the sapwood on all sides of the tree to prevent it from splitting when the tree is felled.

Cornfield–A grain field. In the United States, corn is the same as maize. In England wheat, barley, or rye are known as corn.

Cornhusk–The bract or outer, leafy covering which encloses the corn ear. Also called corn shuck.

Cornlet–A small or immature ear of corn.

Cornstarch–Starch extracted from the corn kernel; used for puddings, thickening of foods, etc.

Corolla–In flowering plants, the inner circle or second whorl of floral envelope; if the parts are separate, they are petals, and the corolla is said to be choripetalous or polypetalous; if they are to any degree united, the corolla is said to be gamopetalous or sympetalous, and the parts are evident only as teeth or lobes, or may be undifferentiated.

Coronal Root–A root arising just at or slightly above the soil surface from the nodes at the crown of a plant of the grass family.

Corrasion–The wearing away of rock material by running water, glaciers, winds, waves, or mass movements.

Correction Strip–In strip cropping, a strip of irregular width placed between uniformly wide contour strips of cultivated crops to reduce soil erosion.

Corrugation System–A method of irrigation used primarily for small grain and hay crops. Small irrigation furrows are placed 18 to 36 inches apart to wet the soil between them.

Cortex–The outer layer or region of any organ.

Cos Lettuce–*Lactuca sativa* var. ***longifolia***, family Compositae; a horticultural variety of lettuce, with long erect leaves forming a loose head, which is grown as a salad plant; used in regions too warm to grow other types of lettuce. Also called romaine lettuce.

Cotton–A soft, white fibrous substance composed of elliptical cells surrounding the seeds of various freely branching tropical plants. The two principal cottons grown in the United States are of the mallow family (Malvaceae): upland cotton (***Gossypium hirsutum***) and American Egyptian cotton (***Gossypium barbadense***). Upland cotton is grown throughout the Cotton Belt and varies in staple length from about $1^{3}/_{16}$ inches to about $1^{3}/_{8}$ inches. American Egyptian cotton is grown in the western states (Arizona, California, New Mexico, and Texas) and varies in staple length from $1^{1}/_{8}$ to $1^{3}/_{4}$ inches.

Cotton Anthracnose–A serious disease caused by the fungus ***Glomerella gossypii***, which infests all the aboveground parts of the plant. It causes lesions and cankers on the leaves, stem, and bolls, and is sometimes lethal. Also called boll rot.

Cotton Aphid–*Aphis gossypii*, family Aphidae; an aphid that sucks the juice of cotton, cucurbits, okra, and citrus plants. Also called melon aphid.

Cotton Bale–A bound package of compressed cotton lint. It averages 500 pounds in gross weight.

Cotton Belt–The main cotton-growing region in the United States. It comprises the states of Florida, South Carolina, Georgia, Alabama, Arkansas, Texas, Louisiana, Missouri, Tennessee, North Carolina, Mississippi, Oklahoma, New Mexico, Arizona, and California.

Cotton Bollworm–See Corn Earworm.

Cotton Burs–Mature cotton boles remaining after the lint and seeds have been removed.

Cotton Fleahopper–*Psallus seriatus*, family Miridae; an insect that damages cotton plants, especially along the coast of Texas, by puncturing the young squares, which turn brown or black and fall from the plant. It also causes abnormal growth and excessive branching.

Cotton Fuzz–See Linters.

Cotton Gin–(1) A machine used to separate the cotton seed from the lint. (2) A plant where the lint cotton is compressed into bales. Also called gin, gin house.

Cotton Lint–See Lint.

Cotton Linters–Short-staple cotton removed from the seed and husk by a second ginning, or by a linter machine. It is seldom used for textiles; it has other uses, such as in plastics or for mattress padding.

Cotton Picker–A machine that harvests only the mature lint-with-seed (known as seed cotton). A revolving spindle penetrates the cotton plant and winds the seed cotton from the opened boll and carries it into a dropping zone inside the machine. See Cotton Stripper.

Cotton Root Rot–An important fungal disease caused by ***Phymatotrichum omnivorum***, family Moniliceae, which attacks over 2,000 species of plants, especially cotton. The leaves become yellow, and later, wilt and turn brown. In the early stage, white or tan threads of the fungus mycelium appear on the roots and break down the outer cells. As the disease progresses, it destroys the cortical tissues and penetrates into the central cylinder giving it a tan or buff color. In the advanced stage the cortical tissues are completely disintegrated and the central cylinder becomes red or brown as contrasted with the whitish color of the healthy tissue. Common in the southwestern United States and Mexico. Also called Texas root rot.

Cotton Rust–A potash deficiency in cotton, which causes bronzing and marginal browning of the leaves. Later, the leaves become reddish-brown and die; the bolls are dwarfed, do not open, and contain inferior fiber.

Cotton Square–The unopened cotton flower bud with its surrounding bracts.

Cotton Stripper–A machine designed for pulling (stripping) the entire ripe cotton bolls from the cotton plant. At the cotton gin, the hulls, seed, and lint are separated. See Cotton Picker.

Cotton Wagon–A four-wheel trailer used to haul cotton from the picker to the gin. It is usually pulled with a tractor or truck.

Cotton Wilt–A fungal disease of cotton caused by *Fusarium vasinfectum*, which yellows and blights the foliage and kills the plant. Also called black root, black heart, frenching. See Fusarium.

Cottonseed–The seed of cotton after the lint has been removed. It is a source of cottonseed oil, the residue being used as a stock feed.

Cottonseed Cake–The solid residue left after the extraction of oil from cotton seeds. It should contain more than 36 percent protein. It is sold according to its nitrogen or protein content.

Cottonseed Hulls–The outer covering of the cottonseed. It is part of the residue after the extraction of the oil and used extensively for a stock feed. Cottonseed hulls supply no digestible protein but contain 43.7 percent digestible nutrients; they are fed with protein-rich feeds, often cottonseed meal.

Cottonseed Meal–The residue of cottonseed kernels from which oil has been pressed. Usually containing the portions of the fiber, hull, and oil left after processing the seed; it is used as a stock feed and fertilizer.

Cottonseed Oil–The oil extracted from cottonseed, graded and used for human and animal consumption. Within the grades of crude, prime, and choice, there are, e.g., crude cottonseed oil, prime summer white cottonseed oil, and choice summer yellow cottonseed oil.

Cottony Maple Scale–*Pulvinaria innumerabilis*, family Coccidae; a destructive insect that infests the soft maple in the United States and Canada. Cottony masses of scale appear along the undersides of twigs and branches of the infested trees in May and June. The entire foliage of the tree turns yellow, and the heavily infested branches die. It also attacks the linden, Norway maple, willow, apple, pear, poplar, grape, hackberry, sycamore, honey locust, beech, elm, plum, peach, gooseberry, Virginia creeper, currant, and sumac.

Cotyledon–(1) The first leaf to be developed by the embryo in seed plants. Also called seed leaf. (2) The areas of attachment of the fetal placenta to the maternal placenta (carunde) in certain types of ruminants.

Coumarin–A bitter organic chemical present in most sweetclovers, associated with a hemorrhagic (bleeding) disease of livestock that eat spoiled sweetclover hay. It is also found in tonkabeans and is used in scenting tobacco.

Cover–(1) A lid placed over something for protection. (2) Woods, underbrush, etc., which may conceal game. (3) Plant life, such as grass, small shrubs, herbs, etc., used to protect soil from erosion. (4) The flesh, hide, and fat on a fattened animal. (5) To buy back future contracts. (6) To copulate with a female, as a bull covers a cow. (7) To incubate, to hatch eggs, as a hen. (8) The proportion of the ground surface under live and dead aerial parts of plants. Also refers to shelter and protection for animals and birds.

Cover Crop–A crop growing close to the ground for the chief purpose of protecting the soil from erosion and also for the improvement of its productivity, between periods of regular production of the main crops, or between trees and vines in orchards and vineyards.

Cover Type–The present vegetation occupying an area; based on plant dominance (devised by foresters and wildlife managers).

Cowpea–(1) Any bean of the genus *Vigna*, family Leguminosae. (2) *Vigna sinensis*, a hot-weather legume grown for food, forage, and green manure. Two of the several edible groups are the blackeyed pea group and the cream crowder pea group. The latter are crowded in the pod resulting in blunted ends and are about one-third the size of the former. Also called cornfield pea, blackeyed pea, blackeyed cowpea, cowbean.

Cowpea Weevil–*Callosobruchus maculatus*, family Bruchidae; a weevil that is very destructive to cowpeas. The female deposits her eggs in the cowpeas in the field, and the larvae eat out the insides of the seed. It also attacks other beans and peas in storage.

Coyote Tobacco–*Nicotiana attenuata*, family Solanaceae; a poisonous plant widespread in the western United States in dry, sandy soils and along stream beds. Animals eating it suffer from vomiting, retching, salivation, bloating, diarrhea, colic, weakness, staggering, falling, lameness, spasms, paleness, stupor, and violent palpitation of the heart. Also called wild tobacco.

Crabapple–*Malus*; small trees or shrubs native to North America, Europe, and Asia. Several species and horticultural varieties are a valuable for wildlife food and landscape plantings.

Crabgrass–Any of the several species of the genus *Digitaria*, family Gramineae. Species are serious weed pests of lawns and fields.

Cracked Stem of Celery–A boron-deficiency disease characterized by brown mottling of the leaves, brown stripes on the stalks, breaking and curling of the epidermis, and browning and dying back of the roots.

Cracking–(1) Development of fissures as wet clay soil dries and bakes in the sun. (2) Development of fissures in the skin of fruits, which usually occurs during rainy periods at or just before ripening. Also called fruit cracking.

Cranberry–*Vaccinium macrocarpum*, family Ericaceae; a bog plant grown for its edible fruit; used in jellies, jams, fruit juice drinks, and salads. Native to North America.

Cranberry Barrel–A standardized container of 5,826-cubic inch capacity, used for marketing cranberries.

Cranberry Bog–A low-lying, wet area of peat, muck, or gray, sandy, acid soil capable of being flooded and drained periodically; used for the production of cranberries.

Crawler–(1) A newly hatched insect. (2) A large earthworm. (3) A tractor equipped with tracks, as compared with one equipped with wheels.

Crayfish Land (Crawfish Land)–Flat, wet land, underlaid by gray, clayey soil, which is characterized by holes and tiny mounds made by crayfish, the freshwater crustacean. (Used loosely as a term for poor agricultural land.)

Crazy Weed–See Lambert Crazy Weed.

Creep–(1) Area where young piglets spend most of their time, and which has openings too small to allow the sow to enter. See Creep Feeding. (2) A phosphorus-deficiency disease of animals characterized by anemia, softening of the bones, and a slow gait. Also called bush disease. (3) To spread, as a plant, by means of creeping stems, which root at the nodes. (4) Slow mass movement of soil and soil material down relatively steep slopes primarily under the influence of gravity, but facilitated by saturation with water, strong winds, and by alternate freezing and thawing.

Creep Grazing–As used in Australia, an area of pasture is closed to adult animals but provision is made for access by young stock during the suckling period.

Creeping Bentgrass–*Agrostis palustris*, family Gramineae; a perennial grass used on putting greens of golf courses, and mixed with other grasses for lawns and turf. Native to North America.

Crenate–With shallow, obtuse or rounded teeth; scalloped.

Crenshaw Melon–A horticultural variety of winter melon. See Muskmelon.

Crenulate–A leaf margin that is similar to crenate, but has more "teeth." See Crenate.

Creosote–An antiseptic liquid; a wood preservative; obtained by the distillation of wood tar or coal tar. It is also a mild insecticide.

Cresols–Oily liquids or solids derived from coal tar or wood, commonly used as disinfectants and insecticides. Classified chemically as metacresols, paracresols, and orthocresols. Obtained commerically from the coal tar and petroleum industry.

Crest–(1) The tuft of feathers on the heads of certain fowls. (2) A cock's comb. (3) The ridge of an animal's neck. (4) The top of a dam, dike, spillway, or weir. (5) The peak or high water mark of a flood. (6) A toothed or irregular ridge of appendage on the petals or flower cluster of certain plants. (7) The top of the hill.

Crested Wheatgrass–*Agropyron cristatum*, family Gramineae; a hardy, perennial bunchgrass that produces an abundance of leaves. It grows well in the higher elevations of the western United States; its palatability and high feed value make it one of the most valuable of the forage grasses. Native to Russia.

Crickets–A large group of leaping insects, of the family Gryllidae, some of which are nocturnal and known for their chirping. The most common is the black field cricket.

Crimson Clover–*Trifolium incarnatum*, family Leguminosae; an annual herb grown chiefly in the southeastern United States and along the Atlantic seaboard as a cover crop and green manure crop. Also called annual clover, scarlet clover, incarnate.

Crinkled Garden Lettuce–*Lactuca sativa* var. *crispa*, family Compositae; an annual herb grown popularly as a salad plant. Also called garden lettuce, leaf lettuce.

Critical Element–Essential element for plants, animals, or people.

Crookneck Group–A group of squash or pumpkins, ***Cucurbita pepo***, family Cucurbitaceae, which has edible, golden-yellow or white, warted fruit with long curving necks; a popular vegetable.

Crop–(1) Any product of the soil. In a narrow sense, the product of a harvest obtained by labor, as distinguished from natural production or wild growth. (2) 10,000 boxes of turpentine, a unit of turpentine orcharding. (3) The young fowls or animals bred on a farm. (4) Craw; a saclike enlargement in the esophagus of many birds to store food. (5) Part of a cow's body just behind the shoulder blades. (6) A whole tanned hide. (7) The honey sac of the honeybee. (8) An identification notch or earmark on an animal. (9) To cut: (a) to clip the ears or tail of an animal, especially a dog; (b) to bite off, as a cow crops grass; (c) to cut off the wattles of a bird; (d) to clip the hair of an animal; to shear. (10) To raise plants usually during one season.

Crop Adaptation–The harmonious use of the hereditary and environmental factors of plants in particular areas and seasons so that the plants will grow to maturity. See Crop Ecology.

Crop Ecology–A science that deals with the study of agricultural crops in relation to their environment, such as soil, climate, plant reaction, etc.

Crop Expense–Denoting the amount spent for seed, seed treatment, fertilizer, lime, marl, spray material, crop insurance, twine, and sometimes custom expenses.

Crop Pasture–A sown crop which is normally harvested but may be used for pasture if necessary, as oats, wheat, soybeans, etc.

Crop Plant–A plant cultivated for its fruit, fiber, root, flower, etc.

Crop Residue–That part of a plant left in the field after harvest, leaves, stubble, roots, straw, etc.

Crop Residue Management–The system of retaining crop residue on land between harvest and replanting to reduce erosion, ensuring future crop production.

Crop Rotation–Growing annual plants in a different location in a systematic sequence. This helps control insects and diseases, improves the soil structure and productivity, and decreases erosion.

Crop Shield–A device attached to a cultivator or sprayer to protect plants from being covered with soil during cultivation or from herbicide during spraying.

Crop Surplus–The portion of a particular crop for national consumption that remains unsold after the normal period for selling, usually at the time of the next harvest.

Crop Tolerance–The degree or the ability of a crop to be treated with a chemical and not be injured.

Crop Tree–An individual tree that is best adapted for a crop, usually as a saw log because of species, form, and condition.

Crop Year–(1) The span of time from the planting to the harvesting of a crop. (2) The span of time from one harvest to the next.

Crop Yield–The amount of harvest per acre, or other land measure, for a particular crop.

Crop Yield Index–Crop yield per acre for a farm compared with crop yield per acre for the county or other comparable area, usually for a particular year.

Cropland–Land under cultivation within the past twenty-four months, including cropland harvested, crop failures, cultivated summer fallow, idle cropland used only for pasture, orchards and land in soil-improving crops, but excluding land cultivated in developing improved pasture.

Cross–(1) A plant or animal which is the result of mating individuals belonging to different species, races, breeds, varieties, etc. See Crossbred; Hybrid. (2) To mate individuals belonging to different species, varieties, races, breeds, etc. See Backcross, Double Cross, Inbred, Self-Pollination, Single Cross, Top Cross. (3) To plow a field at right angles to the first plowing. Also called cross-cultivation.

Cross Pollination–Transfer of pollen between plants that are not of identical genetic material.

Cross Section–(1) A cut made at right angles to ground level to permit examination of earth layers, underlying strata, etc. (2) In surveying, a grid system linking points of equal elevation and thus determining land contours.

Crossbred–Describing an offspring that resulted from the breeding of two purebred parents of different breeds. See Hybrid.

Crossbreeding–The pollinating or breeding of plants or animals that belong to different species, races, breeds, varieties, etc.

Crossing-over–The exchange of parts between homologous chromosomes that occurs during the synapsis of the first division of meiosis.

Crosstie–(1) In the United States, a hewn or sawed piece of timber placed beneath railroad rails for support. Also called sleeper, railroad tie. (2) To tie together three of the feet of an animal as it lies on the ground.

Crotalaria–Any plant of the genus *Crotalaria*, family Leguminosae; some species of which are used for improving soils in the southern United States. Native to the southeastern United States and the tropics.

Crotch–(1) A fork formed by the separation of two or more branches of a plant or tree. (2) A small sled made from the fork of a tree used to skid logs. Also called alligator, crazy dray, go-devil, travois, lizard. (3) The area between the hind legs of an animal.

Crow–Any of the large birds of the genus *Corvus*. They are glossy black in color and can be considered a pest in newly planted fields where they eat seeds and uproot seedlings.

Crow Drip–(1) That part of rain or melted snow intercepted by vegetation that falls to the ground in large drops. (2) The area of ground normally fertilized.

Crown–(1) The upper part of a tree, which bears branches and leaves. (2) To remove the top of a plant or tree. Sometimes called topping. (3) The place at which the stem and root join in a seed plant: the crown or top of the root. See Collar. (4) The upper surface of a furrow slice. (5) The width of the top of a levee. (6) The neck of a sweep or cultivator point used to fasten the sweep to the cultivator shank. (7) The top of an animal's head. (8) The dent in the cap of a kernel of corn.

Crown Fire–A fire that runs through the tops of living trees or brush.

Crown Gall–A destructive plant disease caused by a bacterium *Agrobacterium tumefaciens*. Tumors or galls are formed on the plant.

Crownvetch–*Coronilla varia*, family Leguminosae; a sprawling, perennial herb grown for its dense clusters of pink and white flowers. Planted extensively to stabilize banks and roadside slopes. It is not a true vetch. Native to Europe. Also called axseed.

Crowpoison–*Amianthium muscaetoxicum*, family Liliaceae (sometimes called *Zigadenus muscaetoxicus*); an herbaceous, perennial plant from a coated bulb with a raceme of white to greenish flowers. At one time the poisonous bulbs were used in the manufacture of a fly poison. Native to North America, it grows in low sandy grounds, bogs, etc., in the Atlantic Coastal states from Florida to New York and westward to Missouri and Oklahoma. Also called stagger grass, fly poison.

Crucifer–(1) Any plant of the family Cruciferae (Brassicaceae), such as cabbage, mustard, cress, radish, candytuft, sweetalyssum, stocks, cauliflower, horseradish, broccoli, Brussels sprouts, turnip, kohlrabi, the mustard family, etc.

Cruise–A forest survey made to estimate quantity and quality of timber.

Crumb–A natural structure in soil; small, spheroidal, very porous, easily crushed aggregates.

Crumbly–(1) A body or texture defect of butter in which the fat particles lack cohesion. Also called brittle .(2) A defect of ice cream which is dry, open, friable, and tends to fall apart when dipped. (3) Describing raspberries that have not matured properly because of injury to the blossom or immature fruit. (4) A body defect of cheese which tends to fall apart when sliced.

Crust–(1) A thin, brittle layer on the bare surface of many dried soils, which sometimes prevents plant emergence. This condition is usually caused by raindrop splash. (2) The outer wall of an animal's hoof. (3) Rough, tanned hide of a sheep or a goat. (4) On desert soils, a hard layer that contains calcium carbonate, gypsum, or other binding material often exposed at the surface.

Crustaceous–Hard and brittle.

Cryptogam–A plant reproducing by spores instead of by seeds, as ferns, mosses, algae, fungi.

Cryptostegia–A genus of the family Asclepiadaceae (milkweed), whose woody vines are a small source of rubber. Native to tropical Africa.

Cuban Pine–Slash pine.

Cube–(1) A sale package of butter that ranges in size from 63 to 80 pounds. (2) Small pellet of compressed stock feed. (3) *Lonchocarpus* spp. (especially *L. nicou*), family Leguminosae; a tropical shrub whose roots are a source of rotenone.

Cucumber–*Cucumis sativus*, family Cucurbitaceae; a sprawling, tendril-bearing vine which produces the commercial cucumber used in salads, pickles, and relishes. Also called earth apple.

Cucumber Beetle–A striped beetle of the family Chrysomelidae that causes damage to vegetable crops. The larvae feed on the fruit of the crop.

Cucurbit–A plant belonging to the gourd family, Cucurbitaceae, such as pumpkin, squash, and cucumber.

Cucurbit Anthracnose–A serious disease of muskmelon, watermelon, cucumber, and gourd caused by the fungus *Colletotrichum lagenarium*, family Melanconiaceae, characterized by lesions and cankers on the leaves, stems, and fruit.

Culitgen–Any horticultural variety not known in the wild state.

Cull–(1) Anything worthless or nonconforming which is separated from other similar and better items; the act of removing the inferior items; to cull out. (2) The lowest marketing grade of meat carcasses or dressed poultry. (3) Any animal or fowl eliminated from the herd or flock because of unthriftiness, disease, poor conformation, etc.; a reject. (4) A lumbering term for defective or low-grade timber. (5) Any fruit that fails to meet grading specifications because of defects, maturity, conditions, etc.; e.g., ripe berries that have green tips.

Culm–The stem of the Gramineae (grasses and bamboos), usually hollow except at the swollen nodes; applied often also to the Cyperaciae (sedges), though they are usually solid throughout.

Cultivar–A cultivated variety. A group of cultivated plants that are distinguished by any significant character and that retain their distinguishing features when reproduced sexually or asexually.

Cultivation–(1) The planting, tending, harvesting, and improving of plants. (2) Tillage of the soil to promote crop growth after the plant has germinated and appeared above ground. (3) Loosening of the soil and removal of weeds from among desirable plants.

Culture–(1) The working of the ground in order to raise crops; cultivation; tillage. (2) Attention and labor given to the growth or propagation of plants or animals, especially with a view to improvement. (3) (a) The growing of microorganisms in a special medium. (b) The microorganisms which are so grown. (4) Bacteria used in making dairy and other products. See Mother Culture. (5) Human-made features of an area. (6) The specific way of life of a given society (community, tribe, nation).

Culturing–Artificial propagation of pathogenic or nonpathogenic organisms on nutrient media or living plants.

Cumin–*Cuminum cyminum*, family Umbelliferae; an annual herb grown for its aromatic fruit, whose seeds are used to flavor cheese, bread, kraut, and as an ingredient of curry powder. Oil distilled from the seeds is used to flavor liqueurs. Native to the Mediterranean.

Cumulose Soil–Sedentary soil formed by the accumulation of decayed organic remains; organic soils. See Muck Soil, Peat.

Cuneate–Wedge-shaped.

Cup–(1) The receptacle attached to a tree to collect resin in turpentine orcharding. (2) A notch in a tree trunk made by two downward cuts of an ax for inserting a fatal herbicide. (3) A curve across the face of a piece of board. (4) A mechanical object resembling a drinking cup. (5) In lubrication, a vessel or small funnel for receiving oil and conveying it to a machine part, an oil cup. (6) In grain elevators, a bucket or receptacle with a curved outline.

Cupping–(1) A turning upward of the edges of a leaf. (2) In turpentine orcharding, attaching a cup to a tree to collect the resin.

Curative–(1) A remedy used in the cure of diseases. (2) Relating to the cure of diseases.

Cure–(1) To preserve a product by drying, smoking, pickling, etc., such as hay, meat, or tobacco. (2) To heal; to restore to health.

Cured Forage–(1) Dry range grasses, harvested or standing, slightly weathered, nutritious, and palatable stock feed. (2) Any forage preserved by drying.

Curing of Fertilizers–Mixing and storing fertilizer in piles to facilitate certain chemical reactions before it is ground, screened, and bagged.

Curiosity–Any plant or animal grown for its peculiar properties and not for ornamental or economic reasons; for example, dodder, miniature mules.

Curled Mustard–*Brassica juncea crispifolia*, family Cruciferae; an edible herb with crinkled foliage. Native to Asia.

Curled Toe Paralysis–A vitamin deficiency of riboflavin in chicks that causes twisted or flexed toes.

Curly Top–A viral disease of sugar beets, beans, tomatoes, and other plants, transmitted by the beet leafhopper.

Curlymesquite–*Hilaria belangeri*, family Gramineae; a leafy, wiry, perennial grass of excellent forage value which is drought-resistant and withstands close grazing. Found from Texas and Arizona, to Central America, between altitudes of 1,500 and 5,500 feet. Also called mesquitegrass.

Currant–(1) Any of certain plants of the genus *Ribes*, family Saxifragaceae; popular as ornamentals and for the fruit they produce. (2) Dried seedless raisins, varieties of *Vitis vinifera*, family Vitaceae; used mainly in breads and pastries. Largely produced in Corinth, Greece.

Currant Tomato–*Lycopersicon pimpinellifolium*, family Solanaceae; a weak, perennial plant grown for its small, red, tomato fruit, used in salads. Native to Peru.

Current Annual Increment–The growth of a stand of trees for a specific year.

Current Cross–A plant or animal offspring developed in the first season of cross-mating.

Curry–(1) To comb and dress the coat of an animal to improve its appearance. (2) An East-Indian dish similar to a stew, but characterized by the pungent flavor of curry powder; a mixture of several spices.

Cushaw–*Cucurbita moschata*, family Cucurbitaceae; a crookneck squash. Also called Canada crookneck squash, winter croookneck squash.

Cuspidate–A leaf apex that is concave and elongated with a sharp pointed tip.

Custom Mixture–A commercial dry or fluid fertilizer formulated according to specifications, which is sold to a consumer prior to mixing. It is required by state law to be labeled to show the net weight, guaranteed analysis (grade), and the name and address of the distributor. If distributed in bulk a written or printed statement of the information required shall accompany delivery. Custom mixing is often done to apply nutrients required by a soil test.

Cut–(1) A piece of meat prepared for retail or wholesale trade. (2) A slash wound. (3) The opening made by an ax, saw, etc., on a tree. (4) An excavation in the earth, either human-made or natural. (5) The action of a horse's hooves striking its legs or other hooves in walking or running, interfering with its gait. (6) The yield of certain crops, as wool or lumber. (7) (a) An animal separated from the main herd. (b) To separate an animal from the main herd. (8) A severing of the stem or a part of a plant. (9) The output of a sawmill for a given length of time. (10) (a) A reduction in numbers, amount, size, etc. (b) To reduce in numbers, amount, size, etc. (11) To mow. (12) To castrate; emasculate. (13) To sever the jugular vein of an animal or fowl for slaughter.

Cut Flower–A flower that has been harvested by cutting it off with the stem attached.

Cut Oak Groats–A livestock feed of cut, cracked, or ground oat groats. Also called cracked oat groats.

Cut Out–(1) The soluble solids concentration, on a percentage basis, in the drained juice of canned fruit. (2) To separate an animal from the herd. (3) To dress an animal. (4) To remove cuts of meat from a carcass. (5) To finish shearing (New Zealand).

Cut-leaf Nightshade–*Solanum triflorum*, family Solanaceae; a poisonous plant found from Ontario to British Columbia, Canada, and south to Kansas and Arizona, United States.

Cutan–A concentration of clay or humus along natural parting surfaces within the soil; clay skins.

Cutback Furrow Irrigation–A system of surge irrigation consisting of alternating the release of irrigation water from one furrow to the adjoining furrow. This allows more uniform infiltration throughout the length of the furrow.

Cuticle–(1) A thin layer of cutin that covers the epidermis of plants above the ground, except where cork has replaced the epidermis. (2) One of the four major parts of an eggshell. (3) The exoskeleton of an insect.

Cutin–A noncellular, waxy material secreted in or on the walls of epidermal cells and, less frequently, on subepidermal cells of plants.

Cutover Land–Forest land that has been logged over, leaving only stumps, unmarketable trees, slashings, and other evidence of recent lumbering operations. The term may be applied several years after lumbering, even though considerable new growth has appeared and evidence of original lumbering is no longer apparent.

Cutter–(1) A low grade of beef just above canner. (2) A part of a bean harvester consisting of two broad blades set in a wheeled frame at a 60-degree angle, so as to cut two adjacent rows about 2 inches below the surface. (3) In turpentining, a three-cornered steel tool used to sharpen hacks and pullers. (4) The device that chops forage or other plant material preparatory to placing the item in a silo. (5) A sleigh drawn by one horse.

Cutting–(1) Any part that can be severed from a plant and be capable of regeneration. (2) Mowing. (3) Removing a horse or cow from the main herd. (4) Felling a tree. (5) An area on which trees have been or are to be cut. Also called cutting area, felling area, cutover. (6) The striking of a horse's feet against its joints while running.

Cutting Bench–A bench, usually in a greenhouse, containing a medium, such as sand, peat moss, or vermiculite, in which cuttings are rooted. Also called cutting table.

Cutting Cycle–The planned lapse of time between major felling of trees in a forest.

Cutting Wood–That part of a plant which is to be used as a cutting. See Cutting.

Cutting-graftage–A method of plant propagation in which a scion is grafted on a cutting, which in turn is allowed to form roots.

Cutworm–The larval stage of moths of the family Noctuidae, which are soft-skinned caterpillars up to 1½ inches long. Some climb up plants and devour the leaves, some chew their way into fruits, and some feed at or below ground level. They infest most garden vegetables, flowers, field crops, and fruit trees.

Cyanobacteria–Blue-green algae that are capable of fixing atmospheric nitrogen; similar in function to legume bacteria that fix nitrogen from the air.

Cyanogenetic Plants–Plants that produce hydrocyanic (prussic) acid when under stress such as drought. A number of such plants, including the sorghums, Sudangrass and Johnsongrass, may, under these conditions, cause hydrocyanic or prussic acid poisoning in the animals which eat them.

Cyclic–(1) In chemistry, atoms linked together to form a ring structure. (2) In animals, coming in estrus (heat) at regular intervals.

Cyclone Seeder–A hand-operated seeder that is strapped over the shoulder. A crank is turned and seeds are broadcast over the ground.

Cyme–A broad, more or less flat-topped determinate inflorescence in which the central flowers bloom first.

Cynips–The small gall wasps of the family Cynipidae, which form galls on oak trees.

Cypress–Any tree of the genus *Cupressus*, family Pinaceae; grown as an ornamental in warm regions. See Baldcypress.

Cystolith–In a plant cell, a hardened deposit of calcium carbonate.

Cytase–An enzyme that dissolves cellulose, found especially in grass seeds.

Cytology–The scientific study of the structure and function of plant and animal cells, often with particular reference to the chromosomes in the cell.

Cytophagy–The engulfing of cells by other cells.

Cytoplasm–The living substance within a plant or animal cell excluding the nucleus. See Protoplasm.

D Plus Rule–A rule of thumb used in thinning trees to estimate the desired spacing. A given number is added to the diameter breast height (dbh) of the crop tree. A D+4 rule would mean that a 16-inch dbh tree would need 16+4 or 20 feet of growing space.

Dallisgrass–*Paspalum dilatatum*, family Gramineae; a deep-rooted, perennial bunchgrass of excellent forage rating, grown in the southern and western United States. Native to South America.

Damping-off–Any of several fungal diseases that attack and cause rotting of seedlings and cuttings. The fungi concerned are various species of *Phythium* or *Phytophthora*, family Thelephoraceae. See Seedling Blight.

Damselfly Nymph–(Odonata) The immature damselfly. This aquatic insect nymph has an enormous grasping lower jaw and three flat leaflike gill plates that project from the posterior of the abdomen. Nymphs live most of their lives searching for food among submerged plants in still water; a few cling to plants near the current's edge; and a very few cling to rocks in flowing water. The carnivorous adults capture lesser insects on the wing.

Dandeliion–*Taraxacum officinale*, family Compositae; a common yard and field weed pest. A perennial herb, it is sometimes used as a potherb (edible greens), especially improved forms, which may be of commercial value. Native to Europe. Also called dindle.

Dark Air-cured–A tobacco classification denoting its method of curing in which the leaves have been air cured and are dark in color.

Dark Fire-cured–A tobacco classification denoting its method of curing in which heated air is used, darkening the leaves.

Dark Northern Spring Wheat–A subclass of hard red spring wheat which contains more than 75 percent dark, hard, vitreous kernels.

Dark Tobacco–Tobacco which is partially air cured and partially heat cured to give special aroma to the leaf. See Dark Air-cured, Dark Fire-cured.

Dark Winter Wheat–A subclass of hard red winter wheat in which there is more than 75 percent dark, hard, vitreous kernels. It is used for making bread.

Dasheen–*Colocasia esculenta*, family Araceae; a tropical herb that produces a starchy, edible tuber. It is widely grown in the Pacific islands and occasionally in the southern United States. Native to tropical Asia. Also called taro, eddo, tanier.

Date Sugar–Sugar prepared from the fruit of the date palm.

Daughter–(1) The female offspring. (2) The primary division or first generation offspring of any plant, regardless of sex.

Daughter Cell–(Doter) A newly formed cell resulting from the division of another cell.

Day Length–The length of daylight in hours during 24 hours which is an important factor in the growing of plants. See Day-neutral Plant, Long-day plant, Short-day Plant.

Day-neutral Plant–A plant in which the flowering period or some other process is not influenced by length of daily exposure to light. Also called neutral-day plant. See Photoperiodism.

DBH–See Diameter Breast Height.

DDG–See Distiller Dried Grains.

DDT–The first of the modern, long-lasting, chlorinated hydrocarbon insecticides. It has a half-life of fifteen years, and its residues can become concentrated in the fatty tissues of certain organisms, especially fish. Because of its persistence in the environment and its ability to accumulate and magnify in the food chain, EPA has banned the registration and interstate sale of DDT for nearly all uses in the United States effective December 31, 1972.

Dead Furrow–A double furrow or trench made by a plow throwing two furrow slices in opposite directions. Also known as blind furrow.

Dead Ripe–Describing fruit, grain, etc., which is absolutely ripe.

Deaf Ear–(1) A scab of cereals. (2) The fold in the skin of a fowl just below the ear.

Dealkalization–The removal of alkali from the soil, usually by amendments and by leaching. Technically, it is the replacement of monovalent metallic ions, such as sodium, by alkaline earth cations, e.g., calcium, or by hydrogen ions.

Death Camas–Any perennial, poisonous, bulbous herb, *Zigadenus paniculatrs*, family Liliaceae, occurring mostly in the western United States. Cattle rarely eat these plants, but sheep often do. The characteristics of poisoning are salivation, nausea, vomiting, lowering temperature, staggering, prostration, and finally, a coma in which an animal may remain for days, followed by death. Also called onion poisoning, sage poisoning.

Deathcup Amanita–*Amanita phalloides*; a poisonous mushroom that is very often lethal to people.

Deblossom–To remove blossoms from a plant to increase its vigor vegetatively.

Debris–(1) The coarse material resulting from the decay and disintegration of rocks occurring either in the place where it was produced, or the place of deposition after transportation by wind, streams, glaciers, or gravity. (2) Plant residues left on a field after harvest. (3) A mixture of straw particles, dirt, shattered pupal cases, etc., found in the wool of sheep infested with sheep keds.

Decalcification–(1) Removal of calcium carbonate by leaching. A natural process in soil formation. Technically, it is the replacement of calcium ions by monovalent hydrogen cations. (2) The removal of calcium from bones of animals.

Decapitation–A method of pruning trees for rejuvenation in which most of the branches are severely pruned. Also called deheading.

Decay–(1) (a) The decomposition of organic matter by anaerobic bacteria or fungi in which the products are completely oxidized. See Putrefaction. (b) To decompose by aerobic bacteria, fungi, etc., whereby the products are completely oxidized. (2) General disaggregation of rocks, which includes the effects of both chemical and mechanical agents of weathering with stress in the chemical effects. (3) Any chemical or physical process which causes deterioration or disintegration.

Deciduous–(1) Pertaining to a woody plant whose leaves fall at the end of the growing season. (2) Pertaining to the falling of leaves, fruits, and petals of flowers. (3) Pertaining to the falling of the calyx lobes of a pome fruit, especially pear, before the fruit is ripe. (4) Pertaining to teeth, horns, hair, etc., which are shed at maturity of an animal, or at certain seasons, as in a molting animal.

Declination–The angle, variable with geographic position, between the direction in which the magnetic needle points (magnetic north) and the true meridian (true north).

Decoction–The residue remaining after a substance has been boiled down.

Decomposition–(1) The breaking down of complex chemical compounds to form simpler ones. (2) In soils, the breaking down of minerals and rocks by chemical weathering.

Decore–To remove the core of a fruit.

Decorticate–To remove bark, husk, peel, etc.

Decreaser Plant Species–Plant species of the original vegetation that decreases in density or cover with continued overuse. Often termed decreases.

Decumbent–Denoting a plant whose stems lie along the ground with only the extremities rising vertically.

Decurrent–Extending down as the leaves of some plants extend down the stem of the plant.

Decussate–Leaf pairs growing at a right angle to leaf pairs growing directly above and below.

Deep Percolation Loss–Water that permeates downward through the soil beyond the reach of plant roots, as a result of excessive irrigation or rainfall and a permeable soil.

Deep Plunging–The placing of cut flowers and greens in deep containers full of water so that the distance the water must move by capillary action and plant processes is materially lowered. This reduces the exposed leaf surface, thereby maintaining turgidity and freshness.

Deep Soil–In soil judging, a soil that is more than 36 inches (91 centimeters) deep.

Deep-rooted–Designating a plant, such as alfalfa, which roots deeply into the soil.

Deering Velvetbean–*Stizolobium deeringianum*, family Leguminosae; a tropical vine grown for its clusters of purple flowers, as a cover crop, as a green manure crop, and for stock feed. Native to Asia. Also called Florida velvetbean. See Velvetbean.

Defect–(1) Any blemish, fault, irregularity, imperfection in an animal, fowl, or farm product that reduces its usability or impairs its value. (2) In animals or fowls, a departure from breed or variety specification. Because of a tendency to be inheritable, a serious defect is also a disqualification for registration as a purebred.

Deferred-rotation Grazing–The system of range management in which grazing is postponed on various parts of a range during succeeding years, allowing each part successively to rest during the growing season to permit seed production and better vegetative spread.

Deficiency–(1) An insufficiency in reference to amount, volume, proportion, etc.; a lack; a state of incompleteness. The measure of the deficiency can be useful: e.g., the deficiency of the natural flow of a stream in meeting a given irrigation demand determines the storage necessary, the additional supply necessary, or the limitation of the irrigable area. (2) Absence, deletion, or inactivation of a segment of a chromosome.

Deficiency Disease–A pathological condition in plants or animals which results from a deficiency of a nutrient, mineral, or other necessary element in the food supply; e.g., the kind of goiter in humans or animals which results from a deficiency of iodine. Also called hidden hunger.

Deflocculate–To separate or break down soil aggregates of clay into their individual particles; e.g., the dispersion of the particles of a granulated colloid to form a clay which tends to run or puddle.

Deflocculating Agent–A chemical additive that produces deflocculation (dispersion); e.g., sodium carbonate added to a water suspension of clay.

Deflorate–To remove the flowers or blossoms from a plant.

Defoliant–Any substance which, when applied to a plant, causes the foliage to drop off.

Defoliation–Removal of the leaves of plants.

Deforestation–Removal of all of the trees from a forest area.

Deformity–Any physical deviation from the normal in animals or plants caused by injury or disease.

Degerm–To separate germ cells from cereal grains such as corn and wheat.

Deglutinate–To remove gluten from a cereal.

Degossypolized Cottonseed Meal–Cottonseed meal in which the gossypol (toxic principle of cottonseed) has been removed or deactivated so as to contain not more than 0.04% free gossypol. It must be so designated at the time of sale and must meet the prescribed quality specifications of cotton seed meal to be used as a livestock feed. See Gossypol.

Degree Days–A unit that represents one degree of declination from a given point in the mean outdoor temperature of one day.

Degree of Aggregation–Soil aggregation is the binding together (cohering) of groups of soil particles into clusters or masses such as a clod or ped. The degree of aggregation in soils refers to the extent of water-stable aggregates which will not disintegrate easily and are of special importance to soil structure for desirable plant growth.

Degree of Availability–A soil test level given to represent a degree of nutrient availability.

Dehiscence–(1) The bursting of certain dry fruits at maturity. (2) The opening and discharge of pollen from the anther. See Indehiscent.

Dehorn–(1) To remove the horns from cattle, sheep, and goats or to treat young animals so that horns will not develop. (2) To cut back drastically the larger limbs of a tree.

Dehull (Dehusk)–To separate hulls from grains, seeds, etc.

Dehydrate–To remove most of the moisture from a substance particularly for the purpose of preservation.

Dehydrated Alfalfa Leaf Meal–A ground feedstuff consisting chiefly of alfalfa leaves. It must be reasonably free from other crop plants and weeds and must not contain more than 18 percenet crude fiber. The freshly cut leaves are artificially dried in such a manner that a temperature of at least 212°F (100°C) is attained for a period of not more than 40 minutes.

Dehydrated Alfalfa Stem Meal–A feedstuff which is the product remaining after the separation of the leafy material from alfalfa hay or meal. It must be reasonably free from other crop plants and weeds. The freshly cut stems are artificially dried in such a manner that a temperature of at least 212°F (100°C) is attained.

Dehydrated Foods–Products from which most of the water has been removed in order to improve their stability during storage.

Dehydrated Potatoes–A food made from chopped or thinly sliced potatoes that have been subjected to heat for the removal of moisture; used as human food and a stockfeed.

Dehydrated Soybean Hay Meal–A feedstuff which is the product obtained from the artificial drying and grinding of the entire soybean plant, including the leaves and beans, but not any stems, straw, or foreign material. It must be reasonably free from other crop plants and weeds and must not contain more than 33 percent crude fiber. It must have been artificially dried when freshly cut.

Delayed Dormant Spray–An orchard spray applied during the period from swollen bud to late green tip of bud development; often called swollen bud stage in stone fruits.

Deleterious–Injurious.

Deletion–Absence of a segment of a chromosome involving one or more genes.

Delignification–The removal of lignin from wood or other cellulose material.

Delint–To remove the linters (short lint) from cottonseed. See Linters.

Deliquescent–(1) To ramify into fine divisions, as the veins of a leaf or the trunk or branches of a tree. Having a large number of branches or branching so that the stem is lost in the branches, as maple or oak stems. (2) Of, or pertaining to, the ability of a substance to absorb moisture from the air, thus being reduced to a liquid state.

Deltoid–(1) Shaped like the Greek letter delta. (2) Refers to a leaf shape that is triangular.

Deltoides–Triangular, like the leaf of a cottonwood.

Denature–(1) To make a product unfit for human consumption without destroying its value for other purposes: e.g., denatured alcohol. (2) To change the properties of a protein, as to coagulate egg white.

Dendritic–(1) Marked by a branching habit resembling that of a shrub or tree; usually said of river systems, various plants, and of the leaf veins of many higher plants. (2) Marked by the pattern presented by certain apple scab lesions; having a treelike marking. (3) A drainage pattern characterized by irregular branching in all directions with the tributaries joining the main stream at all angles.

Dendrium–An orchard community.

Dendrochronology–Study and matching of tree rings with the object of dating events in the recent past.

Dendrolatry–(Greek; *dendron*, tree) The worship of trees. See Druids.

Dendrology–The study of trees, including the identification of trees by external features.

Dendrometer–An instrument for measuring the diameter of a tree, outside bark. If used to measure diameter, continuously, the instrument is known as a dendrograph.

Denitrification–The process by which nitrates or nitrites in the soil or organic deposits are transformed by anaerobic bacteria into ammonia or free nitrogen and results in the escape of nitrogen gas into the air.

Density–(1) Mass per unit volume. (2) The number of wild animals per unit of area. (3) The degree of closeness with which wool fibers are packed together. (4) In forestry, density of stocking expressed in number of trees, basal area, volume, or other criteria on an acre or hectare basis. See Stocking.

Dent Corn–*Zea mays indentata*, family Gramineae; corn that has a dent in the top of the grain. This is the most widely cultivated of all corn.

Dentate–(1) Of, or pertaining to, a leaf with regular serrations on the margins. (2) With sharp, spreading, rather coarse teeth that are perpendicular to the margin.

Denticulate–(1) Diminutive of dentate; with small teeth of the dentate type. (2) A leaf margin that is similar to dentate, but has smaller "teeth."

Denudation–(1) A laying bare; the process of washing away of the covering of soil, subsoil, and substrate. (2) The process which, if continued far enough, would reduce all surface inequalities of the globe to a uniform base level.

Denutrition–Lack or withdrawal of nutrition; the failure to transform food elements into nutritional substances.

Deoxyribonucleic Acid (DNA) –A genetic proteinlike nucleic acid on plant and animal genes and chromosomes that controls inheritance. Each DNA molecule consists of two strands in the shape of a double helix. Most inheritance characteristics can be predicted but some cannot, because some genes "jump" (are promiscuous). Such genes can result in resistance to pesticides.

Depauperatum–Stunted.

Dependent Plant–A plant that does not make its own food but lives on other living things or on dead things. See Parasitic Seed Plant, Symbiosis.

Deposition–(1) The addition of sediment, as by flowing water. (2) The process of accumulation of rock material or other debris when dropped due to slackening of transporting agencies, as water, waves, winds, glaciers, or gravitational mass movement. Alluvial fans, offshore bars, dunes, glacial moraines, mudflow debris are such depositional features. (3) In pesticide application, the dust or spray which remains on the plant surface.

Depression–(1) A slight following in the flesh of an animal. (2) A hollow in the surface of the land. (3) A severe drop in income and prices for an area or for a nation. (4) The Great Depression denotes the economic reversal that took place in the United States in the 1930s, which was associated with the Stock Market Crash in October 1929.

Depulp–To remove the fleshy substance from around the seeds of berries or fruits. Also called macerate.

Derived–Designating an invading plant, or one coming into a plant community from another plant community.

Derris–A powder or liquid derived from the roots and stems of the plant ***Derris elliptica***, family Leguminosae, which contains rotenone, an insecticide not harmful to people.

Desalination–(1) Salt removal from sea or brackish water. (2) Salt removed from salty soil.

Desert Garden–A flower garden in which plants adapted to desert areas are planted, e.g., a cactus garden.

Desertification–(1) The decline in productivity of arid and semi-arid soils caused by either natural or person-made stresses. (2) Person-induced soil degradation resulting in increased aridity of the microclimate; e.g., overgrazing.

Desiccate–To dry out; to exhaust of water or moisture content.

Desilting Area–An area covered by vegetation such as grasses or bushes used solely for the deposition of silt and other debris from flowing water, located above a reservoir, pond, or field, which needs protection from accumulation of sediment.

Detassel–To remove the tassels from corn, a practice prevalent in hybrid corn seed production.

Determinant Tomato–Stem growth stops when the terminal bud becomes a flower bud. Tomato plants of this type are also known as self-topping or self-pruning. See Indeterminant Tomato.

Detritus–(1) Material produced by the disintegration and weathering of rocks that has been moved from its site of origin. (2) A deposit of such material. (3) Any fine particulate debris, usually of organic origin, but sometimes defined as organic and inorganic debris.

Deuteromycetes–A large miscellaneous, artificial group of fungi in which sexual reproduction does not occur or has not been found. Contains most of the wilts and some damping-off fungi.

Devil's Darning Needle–See Dragonfly Nymph.

Devilgrass–Bermudagrass, ***Cynodon dactylon***. So named because of the difficulty in controlling it as a weed. See Bermudagrass.

Dew Leaf–Any leaf so cupped that it retains dew.

Dew Retting–The exposing of jute, hemp, and flax to the sun, dew, rain, or pond to free the fibers. Also called dew rot, dew ret.

Dewberry–Certain trailing prostrate plants of the genus ***Rubus***, family Rosaceae, cultivated for their blackberry-like fruits. The lucretia dewberry (***R. flagellaris*** var ***roribaccus***) is raised in the northeastern United States and varieties of the southern dewberry (***R. trivialis***) are raised in the southeastern United States. The grapeleaf California dewberry (***R. ursinus*** var. ***vitifolius***) is believed to be one parent of the loganberry. Native to North America.

Dewing–To detach the wings from seeds such as in pines and maples.

Dextrin–A gummy, soluble carbohydrate; an intermediate product in the hydrolysis of starch to sugar.

Dextrose–Also known as glucose; one of the principal sugars of honey.

Di-hybrid–An individual who is heterozygous with respect to two pairs of genes.

Diagnosis–The process of identifying a disease by examination and study of its symptoms.

Diagnostician–Any person, especially one trained in diagnosis, who determines the nature and cause of a disease or abnormality in plants or animals and prescribes a treatment.

Diakinesis–A stage of meiosis just before metaphase of the first division: the homologous chromosomes are associated in pairs near the nucleus and have undergone most of the decreases in length.

Dialdehyde Starch–A chemical derivative of starch derived from cereal grains, used to improve wet strength of paper products and tanning leather and other uses.

Diameter Breast Height (DBH)–The diameter of a tree outside of the bark at roughly breast height. Normally measured 4.5 feet off the ground on the uphill side of the tree.

Diameter Limit Cutting–A system of selection harvest based on cutting all trees in the stand over a specified diameter. This eliminates marking individual trees.

Diameter Tape–A tape that is put around a tree at a 4.5-foot height. The tape is graduated to read the diameter.

Diammonium Phosphate (DAP)–Diammonium phosphate has an economic advantage over monoammonium phosphate because the same amount of acid fixes twice as much ammonia. The DAP pure salt is not stable and loses ammonia when damp. The Tennessee Valley Authority has developed processes that have been commercially developed on a larger scale. The pure salt contains 21.21% nitrogen and 53.76% phosphoric oxide. The fertilizer grade material contains from 18 to 21% N, and 46 to 54% P_2O_5. The 18-46-0 is a popular grade.

Diapersant–A chemical product that acts to lower the surface tension between a solid and a liquid. A diapersant added to a pesticide or foliar fertilizer spray makes the substance stick to plant leaves, where it is more effective.

Diaspore–Any spore, seed, fruit, bud, or other portion of plant that constitutes its active dispersal phase and is capable of reproducing a new plant.

Diatomaceous Earth–A siliceous sedimentary rock consisting of opaline residues of diatoms, a one-celled aquatic plant related to algae. Used as a filtering agent and in paints, rubber, and plastics.

Diatoms–Microscopic algae that belong to the family Chlorophyceae. They are found in both fresh and salt water and are distinguished by their siliceous cell walls, which sink to the bottom of the water after the living organism dies, forming diatomaceous earth. Diatoms are used as a scouring powder, as a filler in insecticides, and in the manufacture of dynamite and pottery glaze.

Dibble–(1) A small hand tool used to make holes in the soil for planting bulbs and seeds, or for transplanting plants. It usually has a wooden handle and a steel or brass point. Also called dibber. (2) To transplant with a dibble.

Dicalcium Phosphate–($Ca_2[HPO_4]_2$) A manufactured product consisting chiefly of a dicalcium salt of phosphoric acid, used as a supplement to stock feed and as a constituent of commercial fertilizers.

Dicaryon Cells–In some groups of fungi, especially in the *Ascomycetes* and *Basidiomycetes*, a cell that contains two haploid nuclei (of separate sexual origin), the result being a cell with the two sets of genes in separate nuclei, instead of one diploid nucleus as in the majority of plants undergoing sexual reproduction.

Dice–(1) To cut into small cubes, as to dice carrots. (2) Used in the past tense to denote that which has been cut into small cubes; e.g., diced beets.

Dichotomous–Forking regularly and repeatedly, the two branches of each fork usually essentially equal.

Dicot–(Dicotyledon) Plant whose seeds have two cotyledons or seed leaves, such as beans.

Dicoumarol–An anticoagulant produced by molds on spoiled clovers.

Dieback–The progressive dying, from the tip downward, of terminal twigs or branches of woody or herbaceous plants, which may be caused by disease, winter injury, etc. See Exanthema.

Diecious–(1) Animals that are either male or female; i.e., each individual animal has either male or female reproductive organs but not both. (2) Plant species with male and female organs on separate plants.

Dietary Factors–Substances that are present in or characteristics that are associated with the diet; for example, the amount of total fat, dietary fiber, the ratio of saturate versus unsaturated fat, and the method of cooking.

Differentiation–(1) The development of different kinds of organisms in the course of evolution. (2) The development or growth of a cell, organ, or immature organism into a mature organism.

Differnetial Grasshopper–*Melanoplus differentialis*, family Locustidae; a large, destructive grasshopper that infests all kinds of crops, but is especially destructive to corn.

Diffraction–The bending of light rays passing through microscopic droplets of water vapor, which causes halos around the sun or moon. Also causes the breaking up of light into bands of color, resulting in rainbows following a shower.

Diffuse-porous Wood–Wood in which the pores are of fairly uniform or of only gradually changing size and distribution throughout the growth ring. See Ring-porous Wood.

Dig–(1) To turn up the earth. (2) To excavate; e.g., to dig a ditch. (3) To remove from the earth; e.g., to dig potatoes; to dig strawberry plants for transplanting.

Digger Pine–*Pinus sabiniana*, family Pinaceae; a coniferous tree that produces an edible nut. However, it is often a weed on the California foothills when it invades grazing lands. Also called grayleaf pine.

Digitalis–A valuable drug, *Digitalis* spp., family Scrophulariaceae, having diuretic properties; made from the dried leaves of the foxglove and used in the treatment of heart diseases.

Digitate–Similar to, or resembling, the fingers on a hand, as the leaflets of an American chestnut tree. See American Chestnut.

Dill–*Anethum graveolens*, family Umbelliferae; an annual or biennial herb whose seed is used for seasoning and medicinal purposes. Native to Europe.

Dill Pickle–A cucumber pickle prepared by fermentation in a diluted brine solution to which dill seeds and spices have been added. It is marketed in this solution rather than in vinegar.

Diluent–Any gaseous, liquid, or solid inert material that serves to dilute or carry an active ingredient, as in an insecticide or fungicide. Also called carrier.

Dilute–(1) To make more liquid by mixing with water, alcohol, etc. (2) To weaken in flavor, brilliancy, force, strength, etc., by mixing with another element.

Dimorphic–Occurring in two different forms, as the leaves of those ferns in which the fertile fronds or segments have a form different from that of the sterile ones, or as the juvenile and adult forms of foliage in *Hedera* and some *Eucalyptus* and *Juniperus*.

Dioecious–A plant that produces flowers that have either stamens or pistils, but not both on the same plant.

Dioxin–A highly toxic substance that was a contaminant of agent orange, a mixture of 2, 4-D and 2, 4, 5-T used as a defoliant in Vietnam in the 1960s. See Agent Orange.

Dip–(1) Any chemical preparation into which livestock or poultry are submerged briefly to rid them of insets, mites, ticks, etc. See Dipping Vat. (2) Any preservative preparation in which produce is briefly submerged. (3) To remove ice cream from a container with a dipper. (4) To collect oleoresin from a cup in turpentine orchards.

Diplodia Disease of Citrus–A fungal disease caused by *Diplodia natalensis*, family Sphaerioidaceae; characterized by the death of the bark on branches of all sizes, an oozing of gum, and by a stem-end rot of the fruit.

Diplodia Disease of Corn–A fungal disease caused by *Diplodia zeae*, family Sphaerioidaceae; characterized by a seedling blight and by a dry rot of the ears. The dry rot of the ears reduces the yield, the market value, and the quality of the corn for seed purposes. Also called mold, moldy corn, mildew rot, ear rot, dry rot, Diplodia dry rot, Diplodia ear rot.

Diplodia Tip Blight–A fungal disease of pine, fir, and spruce caused by *Diplodia pinea*, family Sphaerioidaceae; characterized by stunting of new growth and browning of the needles. Lower branches are heavily affected.

Diploid–(1) Having one genome comprising two sets of chromosomes. Somatic tissues of higher plants and animals are ordinarily diploid in chromosome constitution in contrast with the haploid (monoploid) gametes. (2) An organism or cell with two sets of chromosomes, for example, worker and queen honeybees.

Dipoldia Boll Rot–A fungal disease of cotton caused by *Diplodia* spp., family Sphaerioidaceae; characterized by the diseased bolls turning black and becoming covered with pycnidia so as to appear smutty.

Dipping–(1) A method of preserving seasoned wood by submersion in an open tank of creosote or similar preservative. (2) The submerging of animals or fowls in a liquid bath which contains insecticides, ovicides, repellents, etc.

Direct Cut Silage–Plants that are cut and chopped for silage in a single operation.

Direct Seeding–Planting seeds as opposed to setting out started plants.

Directed Application–The application of a pesticide or other material to a specific kind of plant or parts of a plant.

Dirty–In marketing, an egg which has dirt adhering to the shell, or a shell with stained, soiled spots covering more than one-eighth of the shell surface but with no adhering dirt.

Disaccharides–The class of compound sugars that yield two monosaccharide units upon hydrolysis; e.g., sucrose, mannose, and lactose.

Disbud–(1) To remove the small horns of an animal. (2) To remove the flower buds and/or shoot buds from a plant.

Disc Pan–A compacted soil layer produced by the bottom cutting edges of a disc plow.

Discolor–Different colors; off color.

Disease–Any deviation from a normal state of health in plants, animals, or people which temporarily impairs vital functions. It may be caused by viruses, pathogenic bacteria, parasites, poor nutrition, congenital or inherent deficiencies, unfavorable environment, or any combination of these.

Disease Control–Any procedure that tends to inhibit the activity or effect of disease-causing organisms, or which modifies conditions favorable to disease.

Disease Resistant–Designating plants, animals, or people not readily susceptible to, or able to withstand, a particular disease.

Disinfect–To destroy or render inert disease-producing microorganisms and to destroy parasites.

Disk–(1) A round, usually sharp-edged, slightly dished, steel plate that cuts the soil as it revolves on a center axis and moves the soil to one side. Several disks are mounted and spaced on a horizontal shaft to make the disk harrow. Also called disk blade. See Disk Harrow. (2) One of a series of metal plates in a centrifugal separator bowl that increases the efficiency of the machine. (3) In botany, (a) an adhesive surface on the tendril ends of creeper plants that enables them to climb along flat surfaces; (b) an enlargement of a flower's receptacle; (c) the center of a composite flower; (d) an organ's surface; (e) the circular valve of a diatom. Also spelled disc.

Disorder–An unwholesome or unnatural physical condition of a plant or animal. See Disease.

Dispersed Soil–Dispersed soils usually have low permeability. They tend to shrink, crack, and become hard on drying and to slake and become plastic on wetting.

Dissect–To cut or divide a plant or animal into pieces for examination.

Dissemination–(1) The spreading or transporting of inoculum from a diseased individual to a healthy one. (2) The broadcast sowing of seed. (3) The natural scattering or distribution of weed seeds.

Disseminule–A detachable part of a plant that is capable of dispersal and of giving rise to a new independent plant.

Dissolved Bone–A fertilizer material that consists of ground bone or of bone meal that has been treated with sulphuric acid to make the phosphorus in the bone more readily available to growing plants.

Distal—Located in a position that is distant from the point of attachment of an organ; for example, the toes are located on the distal part of the leg.

Distributing Stand—Any device used in irrigation practices from which water is distributed into furrows and basins. Usually it is a part of a riser outlet from an underground pipe system.

Ditch—An artificial excavation, as a trench or channel dug to carry irrigation or drainage water.

Diurnal—(1) Pertaining to a day; the action performed within 24 hours. (2) Designating a plant whose blossoms open during the day and close at night.

Divide—(1) The land area which separates two watersheds or drainage systems. (2) To separate a colony of bees so as to produce two or more colonies. (3) To separate the crown or underground parts of a plant into two or more portions so as to obtain more plants.

Division—Propagation of plants by cutting them into sections as is done with plant crowns, rhizomes, stem tubers, and tuberous roots. Each section must have at least one head or stem; e.g., rhubarb.

Divot—A small, irregular piece of turf loosened by the swing of a golf club.

DNA—See Deoxyribonucleic Acid.

Dockage—(1) Foreign material in harvested grain such as weed seeds, chaff, and dust. (2) The weight deducted from stags and pregnant sows to compensate for unmerchantable parts of an animal.

Dodder—Any plant of the genus *Cuscuta*, family Convolvulaceae. These are parasitic vines whose seeds germinate in the ground and, as soon as they establish themselves on a host plant, draw all their food from the host. They have a reputation of being poisonous to livestock and to bees. Also called strangleweed, love vine, goldthread, pull down, clover dodder, devil's-hair, hell bind, devil's-guts, hairweed.

Dog Fennel—*Anthemis cotula*; a common weed in pastures, grain, alfalfa, and hay. They cause damage by crowding out desirable plants.

Dokuchaiev Formula—The formula given for soil formation by the Russian pedologist, V. V. Dokuchaiev, in 1899, as follows: $P=f(K,O,G,V,)$ when P is soil, f is function of; K, climate; O, organisms; G, "subsoil" or parent material; V, age or time factor.

Dollarspot—(1) A fungal disease, particularly of the bent grasses, caused by *Sclerotinia homeocarpa*, family Sclerotiniaceae, which is characterized by dark, water-soaked spots on the turf. These small spots then turn brown and later become straw-colored. (2) Used to describe the piebald coat of a horse.

Dolomite—A common mineral, $CaMg(CO_3)_2$ usually with ferrous iron and manganese replacing some magnesium. In dolomite the calcium to magnesium ratio must be lower than 1:5. See Dolomitic Limestone.

Dolomitic Limestone—A limestone in which the calcium to magnesium ratio can be from less than 1:5 to greater than 1:60. Any less magnesium is known as calcic limestone. See Dolomite.

Domesticate—To bring wild animals under the control of humans over a long period of time for the purpose of providing useful products and services; the process involves careful handling, breeding, and care.

Dominance—(1) The tendency for one gene to exert its influence over its partner, after conception occurs and genes exist in pairs. There are varying degrees of dominance, from partial to complete to overdominance. (2) In forestry, the relative basal area of a species to the total basal area of all species in an area. The species having the highest relative basal area is considered the dominant species (syn.: predominant). (3) A term used in various contexts (e.g., in farm programming, input-output budget analysis and decision analysis) to indicate that one alternative is superior to another in the sense of producing higher benefits (output) with equal or lower costs (inputs).

Dominant—(1) In genetics, designating one of any pair of opposite Mendelian characters that dominates the other, when both are present in the germ plasm, and appear in the resultant organism. (2) Designating a plant species that characterizes an area.

Dominant Gene—A gene that prevents its allele from having a phenotypic effect. See Recessive Gene.

Dominant Trees—Trees with crowns extending above the general level of the crown cover and receiving full light from above and partly from the sides.

Doob (Doub)—See Bermudagrass.

Doodlebug—(1) The larva of several small insects. (2) A homemade tractor constructed from a used automobile.

Dormancy—(1) (seed) A physiologic state where the seed embryo is incapable of growth; a state when the metabolic processes are slowed. Especially applies to respiration (syn.: afterripening). (2) (vegetative) A physical state when the metabolic processes are slowed; when deciduous plants, the period without leaves.

Dormant—Designating a period in a plant, animal, or insect's life when physiological activities are slowed to a minimum. Also called resting, sleeping, inactive, quiescent.

Dormant Bud—(1) A bud that is ordinarily inactive but that may begin to develop as a result of certain conditions; e.g., when the stem beyond it is removed. (2) A bud that is inactive during the winter due to unfavorable conditions for growth.

Dormant Grafting—Grafting performed on a plant during the period the plant is physiologically inactive.

Dormant Pruning—Pruning carried on during the period of a plant's physiological inactivity.

Dormant Season—That portion of the year when frosts occur. In areas of no frost, dormancy may be induced by dry weather.

Dormant Spray—An insecticide, fungicide, etc., applied to a plant during its period of physiological inactivity.

Dormant-budded—Designating a plant that has been bud-grafted when it is physiologically inactive.

Dorsal—Refers to the back, or toward the back; the opposite of ventral.

Dote–An early decay of wood characterized by patches or streaks of lighter or darker color than normal. Also called doze.

Dothichiza Canker–A fungal disease of the Lombardy poplar, balsam, California poplar, and eastern cottonwood caused by *Dothichiza populea*, family Sphaerioidaceae; characterized by elongated dark, sunken lesions on the trunk, limbs, and twigs. The bark and cambium are destroyed, and when the lesions girdle the trunk or limbs, the distal portion dies.

Dothiorella Rot–A citrus fruit rot caused by the fungus *Botryosphaeria ribis*, family Dothioraceae; usually characterized by a rot beginning at the style end of the fruit, resulting in a leathery, drab-brown, pliable decay. An amber juice is evident and as the rot continues, the color becomes dark-olive to black. It occurs on fruit in transit or in storage.

Dothiorella Wilt–A fungal disease of elms caused by *Dothiorella ulmi*, family Sphaerioidaceae; characterized by a drooping and wilting of the leaves. the leaves become yellowed, then brown, roll upward to the midrib and drop off. There is an extensive formation of suckers below the point of infection. Also called Verticillium wilt, Cephalosporium wilt.

Double–(1) Designating a blossom that has more petals and sepals than normal, usually as a result of plant breeding. Double flowers are often highly prized. (2) Designating a team of horses; e.g., to work horses double.

Double Crenate–A leaf margin that is slashed irregularly, fairly deeply and sharply (between toothed and lobed).

Double Cropping–The growing of two crops at different times in one year on the same field.

Double Cross–The resultant seed secured from mating two single cross hybrids such as corn. It is planted extensively in hybrid corn production.

Double Fertilization–The usual fertilization in plants, in which the egg nucleus is fertilized by one sperm nucleus and the two fused polar nuclei by another sperm nucleus.

Double Hacking–The girdling of a tree to kill it by making parallel, overlapping, downward, ax cuts and the removal of the chips between the cuts.

Double Planting–See Interplant.

Double Row–A planting system for small vegetables and fruits in which the seeds or plants are set closely together in two parallel lines with a wider spacing between the pairs of rows.

Double Superphosphate–A commercial fertilizer that contains two and one-half to three times as much available phosphoric acid as superphosphate. The phosphate rock is treated with phosphoric acid, and in the manufacturing process the gypsum is removed. Also called triple superphosphate, treble superphosphate.

Double-disk Harrow–A harrow in which two single-disk harrows are arranged in tandem. The front gangs are set to throw the soil outward, while the rear gangs are set to throw the soil inward to leave the surface level. Also called double-action harrow, tandem disk harrow.

Dough Stage–That stage in the development of a cereal grain in which the inside of the kernel has a consistency like bread dough.

Douglas Fir–Any tree of the genus *Pseudotsuga*, family Pinaceae, but specifically *Pseudotsuga menziesii*. It sometimes reaches a height of 300 feet and is a timber tree of greatest commercial value for lumber in northwestern United Sstates. It is also grown as a shade and shelterbelt tree and certain horticultural varieties are grown as ornamentals east of the Rocky Mountains, United States.

Douglas Waterhemlock–*Cicuta douglasi*, family Umbelliferae; a poisonous, perennial herb. In grazing animals, the symptoms of poisoning are stomach pain, diarrhea, dilated pupils, labored breathing, frothing at the mouth, a weak but rapid pulse, and violent convulsions. Death results from respiratory failure. Roots and root stocks are the most poisonous parts of the plant; the leaves and fruits may be eaten in hay without danger. Found from Alaska to California, United States.

Down–(1) A sandhill. See Dune. (2) The soft, furry or feathery covering of young animals and birds that occurs on certain adult birds under the outer feathers and which is used for stuffing pillows, furniture, and comforters, its most important source being the eider duck. (3) The pappus of a plant. (4) Used to describe the individual crop plants that have been blown or forced to the ground by rain, hail, wind, etc.,; e.g., down corn, down wheat.

Down Grain–Ripening grain that has been blown to the ground by a storm or that has fallen because of overripening, weakness of the straw/stalk, or heaviness of the ripened heads of grain.

Downy–Covered with very short, weak, soft hairs.

Downy Andromeda–*Andromeda glaucophylla*, family Ericaceae; a poisonous shrub which is sometimes eaten by sheep when other browse is unavailable. Symptoms of poisoning are salivation, flow of tears, a flow from the nostrils, convulsions, and paralysis. Found in acid bogs and uplands of northeastern U.S. and southeastern Canada.

Downy Mildew–Various plant diseases caused by fungi of the families Pythiaceae and Peronosporacea. These fungi infect the leaves, herbaceous stems, flowers, and fruits of the host plants which are mostly herbaceous, although the herbaceous parts of some woody plants may be affected (rose, grape, hackberry). The chief genera which are called downy mildews are *Peronospora*, *Plasmopara*, *Rhysotheca*, *Bremia*, *Bremiella*, *Pseudoperonospora*, *Basidiophora*, and *Sclerospora*, all belonging to the family Peronosporaceae; and *Phtophthora*, family Pythiaceae.

Doyle Rule–In the lumber trade, a simple method of estimating the amount of lumber that can be obtained from a small log. The length in feet multiplied by the diameter of the log from which 4 inches has been subtracted, and the result divided by 4 and squared.

Doyle-Scribner Rule–A combining log scale wherein the Doyle rule is used for logs up to 28 inches (71 centimeters) in diameter and the Scribner rules for larger logs. See Doyle Rule, Scribner Rule.

Dragonfly Nymph–The immature dragonfly. This aquatic insect nymph has gills on the inner walls of its rectal respiratory chamber. It has an enormous grasping lower jaw that it can extend forward to a distance several times the length of its head. Although many of these nymphs climb among aquatic plants, most sprawl in the mud where they lie in ambush to await their prey. The carnivorous adults capture lesser insects on the wing.

Drain–(1) An artificial or natural channel that receives and carries off excess or unwanted water from an area of land. It does not carry sewage and is therefore distinguished from a sewer. (2) The loss of growing trees in a forest due to cutting, fire, or natural causes.

Drainage–(1) The removal of surplus ground or surface water by artificial means. (2) The manner in which the waters of an area are removed. (3) The area from which waters are drained; a drainage basin. In soil science, the natural drainage under which the soil developed.

Drainage System–(1) An artificial network of furrows, ditches, tile drains, or combinations of these that provide drainage for farmland. (2) The natural means by which a river and its tributaries carry water from the land.

Drained Weight–The weight of canned fruit or vegetables after removing the liquid.

Draw–(1) A small natural head tributary, drainage hollow, without a permanent stream in its bottom. (In local usage, it is not always sharply differentiated from ravine, hollow, gulch, or coulee.) (2) A slip or shoot of a sweet potato. (3) To shape and build.

Drawing–(1) The removal of cuttings from a propagation bed. (2) The eviscerating of poultry.

Dredge–(1) Oat and barley seed sown together and cultivated to be used for making malt. (2) Any of several different types of machines to deepen channels and clean out ditches, etc. (3) To excavate or deepen and clean stream beds. (4) To coat the surfaces of food with flour, cornmeal, breadcrumbs or other fine substances before cooking.

Dress–(1) To curry and brush an animal. (2) To remove the feathers and blood from a bird that has been killed. (3) To plane the surface of a board.

Dressed Size Lumber–The dimensions of lumber after planing, usually ⅜ inch less than the normal or rough size.

Dressing–(1) An application of medicine or bandage made to a wound on a person, animal, fowl, or plant. (2) An application of manure, fertilizer, mulch, etc. See Side Dress, Top Dressing. (3) The preparation of nursery stock for cutting, budding, grafting, etc. (4) The removal of feathers and blood form a bird. (5) The trimming of excess fat and bone form a meat carcass. (6) The external treatment of cheese, including: (a) the assistance given in forming a coat or rind; (b) improvement given to the external appearance of a cheese; (c) the prevention of loss of moisture from a cheese; (d) protection from external microorganisms.

Dried–Designating a product which, for purposes of preservation, has had most of the moisture removed, as dried apples, raisins, etc.

Dried Citrus Pulp–A livestock feed which is the dried and ground peel, residue of the inside portions, and occasional cull fruits of the citrus family with or without the extraction of part of the oil of the peel.

Dried Fruit–Any preserved fruits, such as apples, apricots, figs, peaches, pears, prunes, raisins, dates, etc., which have had enough moisture removed by semidrying or by artificial dehydration so that they no longer support the growth of decay microorganisms.

Dried Kelp–A livestock feedstuff supplement that contains over 30% mineral matter and from 0.15 to 0.20% iodine, consisting of dried and comminuted brown seaweeds, of the genus *Macrocystis* or related genera. Also called kelp meal.

Dried Sweet Potatoes–A livestock feedstuff, rich in nitrogen-free extract (carbohydrate) and low in fiber, which is made from shredded and dehydrated sweet potatoes. See Sweet Potato Meal.

Dried Tomato Pomace–A livestock feed consisting of a dried mixture of tomato skins, pulp, and crushed seeds, produced by extracting the juice from tomatoes.

Drier–Any device used for removing moisture from a product, usually consisting of a heating unit and a large fan for forcing the hot, dry air through the product being dried. Alfalfa is sometimes dried in such a way before being stored. Also called dehydrator.

Drillability–A property of fertilizers that affects their rate of flow and uniformity of distribution from distributors. Fertilizers with good drillability flow rapidly and evenly when dispensed, and those of poor drillability flow slowly and irregularly. The angle of repose of a fertilizer is a useful measure of drillability. An angle below 40 degrees indicates excellent drillability. An angle above 50 degrees indicates poor drillabillity. See Angle of Repose.

Drip Gutter–A small gutter inside a greenhouse at the point where the roof and walls meet which carries off water formed by condensation inside the greenhouse.

Drip Irrigation–Watering plants so that only soil in the plant's immediate vicinity is moistened. Water is supplied from a thin plastic tube at a low flow rate. The technique sometimes is called trickle irrigation.

Drip of a Tree–The places at which much water runs off the tree during rains; the periphery of the tree. The usual area fertilized.

Droop–To hang downward from exhaustion, lack of water, or nourishment, etc., as a plant droops in the hot sun.

Drop–(1) A structure in an open or closed conduit that is used for dropping the water to a lower level and dissipating its kinetic energy. (2) A decrease in height or elevation. (3) Any fruit that falls form a tree to the ground because of wind, or other conditions; e.g., diseased, immature, and usually unfertilized fruit; also normal but ripe fruit. (4) A fungal disease of vegetables caused by *Sclerotinia sclerotiorum*, *S. minor*, or *S. intermedia*, family Sclerotiniaceae. (5) To give birth to young, as to drop a calf. (6) To shoot an animal and to cause it to fall.

Drop Off–(1) To fall to the ground, as a leaf or blossom. (2) To decrease in number, quality, demand, etc.

Droppings–The excrement of animals and birds. See Guano, Manure.

Dropseed—Any plant of the genus ***Sporobolus***, family Gramineae. Species are perennial, rarely annual grasses. See Alkali Sacaton, Sand Dropseed.

Dropsy—(1) An excessive, intensive, watery enlargement of parts of a plant sometimes due to a superabundance of moisture in the soil. Also called edema. (2) Widespread edema (swelling) throughout the body of an animal.

Drosophila—(Greek; *drosos*, dew; *philein*, to love) A genus of fruit flies common around decaying fruit; used extensively in experimental genetics. One of the most common is ***Drosophila melanogaster***.

Drought—(1) A period of insufficient rainfall for normal plant growth which begins when soil moisture is so diminished that vegetation roots cannot absorb enough water to replace that lost by transpiration. A drought can be arbitrarily defined on the basis of amount of precipitation during a certain period, although such a measure has no uniform significance because of variable amounts of moisture in soils when the measurement period begins. The intensity or damage form drought varies with the plant since some plants are much more tolerant of dry conditions than others. (2) Any dry spell. (3) A period when precipitation is not adequate to meet the needs of the human population. (4) In an arid area, a period when there is an especial shortage of water for irrigation. Also spelled drouth. (5) Drought is an imprecise term. On a coarse-textured soil with a deep water table, plants may need water every week but on a deep, fine-textured soil, only every two weeks. High temperatures and high winds also determine water needs.

Drought-enduring Plants—Species of plants that require little moisture to grow and reproduce; e.g., cacti. Also called xerophytes.

Droughty—Designating a soil that has low water-holding capacity; e.g., deep sands or gravel.

Drown—(1) To deprive a plant or animal of life by immersion in water so that the supply of oxygen is cut off. (2) To submerge land with water, whether by a rise in the level of a water table, a lake, ocean, or river, or by a sinking of the land as the lowering of a coastal region drowns the lower courses of the rivers and connects their valleys into estuaries.

Drunken Forest—A forest in which the trees have leaning positions due to displacement caused by landslides, high winds, or from freezing and thawing of soil when supersaturated.

Drupe—A single-seeded, fleshy fruit that does not split open, e.g., cherry, peach, plum, or olive. Also called stone fruit.

Drupelet—A small drupe, often one of many in a cluster or group, e.g., the pulpy grain of the fruit of the dewberry, raspberry, blackberry, etc.

Dry Basis—Designating a product, e.g., soil, fertilizer, or feed, which is analyzed for its constituents calculated on the basis of oven-dried material.

Dry Farming—This term applies mostly to nonirrigated semi-arid areas where short-season, drought-resistant crops are grown on fertile soil to make the most efficient use of scarce soil water. Level terraces with closed ends are sometimes used to hold all precipitation in each field for use by crops. Also known as rainfed agriculture.

Dry Matter—The total amount of matter, as in a feed, less the moisture it contains. Dry feeds in storage, such as cereal grains, usually contain about 10 percent water and 90 percent dry matter, wet weight basis.

Dry Off—(1) To reduce the amount of water given a plant, bulb, tuber, or corm, until it becomes dormant and ready to store. (2) To bring the lactation period of a cow to an end.

Dry Rot—(1) Any of many types of decay but especially of that which, when in an advanced stage, permits timber to be easily crushed to a dry powder. The term is actually a misnomer for any decay, since all fungi require considerable moisture for growth. (2) A metabolic defect that causes ill health in animals.

Dry Weight—The weight of a product or material less the weight of the moisture it contains; the weight of the residue of a substance that remains after virtually all the moisture has been removed from it. Also called dry matter.

Drying—Reducing the moisture content of a product to the point at which the concentration of the dissolved solids is so high that osmotic pressure will prevent the growth of microorganisms.

Dryland Farming—A system of producing crops in semi-arid regions—usually with less than 20 inches (50 centimeters) of annual rainfall—without the use of irrigation. Frequently, in alternate years part of the land will lie fallow to conserve moisture.

Dual-use Range—A range containing a forage combination of grass, forbs, and browse that allows two or more kinds of stock (such as cattle and sheep) to graze the area to advantage at the same time through the entire season, or separately, during a part of the season.

Duff—The organic layers on a forest floor, including freshly fallen leaves and twigs and other organic residues in all stages of decomposition.

Duff Hygrometer—An instrument for measuring the moisture content of litter to determine fire hazard.

Dumb Cane—See Seguin Tuftroot.

Dune Grass—Grass that is adapted to sand dunes, e.g., beach grass, ***Ammobila arenaria***.

Duplicate Genes—Factors with the same recessive phenotypic expression.

Duricrust—The case-hardened crust of soil; usually formed in arid or semi-arid climates, but the plinthites of laterization have also been called duricrust, as have some silcretes, both of which form in areas of higher rainfall.

Duripan—A horizon in a soil characterized by cementation by silica or other minerals. See Fragipan.

Durra—(1) ***Sorghum vulgare*** var. ***durra***, family Gramineae; a variety of sorghum that is widely grown throughout the world for its grain and for forage. (2) ***Triticum durum***, family Gramineae; an annual spring-planted grain grown as a livestock feed, and for the high-quality flour obtained from it and used in the manufacture of macaroni, etc. Also called macaroni wheat, hard amber durum, amber durum.

Durum Wheat–A type of wheat grown through the upper Great Plains. It contains a high percent of gluten but it is considerably lower quality than the hard red winter and spring wheats. Durum is used for making pastas like macaroni and spaghetti.

Dust–(1) Fine, dry particles of earth, or other matter, so attenuated that they may be wafted by the wind. (2) Particles less than 0.1 millimeter in diameter that may be bounced off the ground by sand grains moving in saltation. See Saltation. (3) An insecticide, fungicide, etc., which is applied in a dry state, or the application of these to plants, animals, or fowls. (4) (Volcanic) Pyroclastic detritus consisting mostly of particles less than 0.25 millimeter in diameter; i.e., fine volcanic ash. It may be composed of essential, accessory, or accidental material.

Dust Mulch–A shallow layer of loose dry surface soil, which may be natural or created by tillage. Theoretically such a covering lessens losses of soil moisture resulting from direct evaporation, but in practice, it is of questionable value as a soil conservation measure.

Dust Soils–Soils, observed most commonly in arid regions, which are normally pulverulent. They are penetrated by water with extreme slowness and thus constitute a problem in management under irrigation.

Dutch Ashes–Ashes that result from the burning of peat.

Dutch Elm Disease–A fungal disease of elm trees. See Smaller European Elm Bark Beetle.

Dwarf–(1) A serious viral disease of other trailing blackberry characterized by a yellowing of the plant and a shortening of the canes. The canes, spindly at first, become stiff, unusually upright and thick. The leaves are dwarfed, crinkled, and mottled with bronze and light green patches. Found in United States Pacific Coast. (2) A plant, especially one that has been intentionally grafted to be a dwarf. (3) Dwarfing may be caused by disease, lack of water, or mineral deficiency. (4) See Dwarf Cattle. (5) Designating any plant disease, one of whose symptoms is a stopping or retarding of growth.

Dy–Limnic peat, a dark lake bottom sediment rich in colloidal humic substances formed by plant decay in saprobic condition.

Dyad–The univalent chromosome, composed of two chromatids, at meiosis. The pair of cells formed at the end of the first meiotic division.

E Horizon–A mineral horizon, mainly a residual concentration of sand and silt high in content of resistant minerals as a result of the loss of silicate clay, iron, aluminium, or a combination of these.

Ear–(1) The grain-bearing spike of a cereal plant, e.g., corn, rye, wheat, etc. (2) To produce or form fruiting spikes.

Ear Corn–Unshelled corn.

Ear Leaf–A cotyledon or seed leaf.

Eared–(1) Designating the presence of ears or spikes on a plant. (2) Pertaining to an animal restrained by its ears. (3) Designating the presence of earlike tufts of feathers in some birds.

Earl Loco–*Astragalus earlei*, family Leguminosae; a perennial, poisonous herb. Symptoms of poisoning are dullness, irregularity of gait, lack of appetite, dragging of the feet, a solitary habit, loss of flesh, and a shaggy coat. The animal ceases to eat and may die. Found in western Texas, United States. Also called locoweed.

Early–(1) Designating a variety that regularly produces mature fruit, leaves, etc., earlier than normal. (2) Designating a period for planting or harvesting that comes before the normal or usual time.

Early Soil–A soil that warms up early in the spring. It is usually well drained with a sand or sandy loam texture.

Early Spring–(1) In the Northern Hemisphere, March to May; in the Southern Hemisphere, October to December, depending on the latitude and elevation. (2) A growing season that starts earlier than is usual for the latitude, elevation, or soil.

Early Wood–The less dense, larger-celled, first-formed part of a growth layer along the trunk and branches of a tree (spring wood).

Earth–(1) Soil, as black earth. (2) Ground. (3) Soil-like material, as ocher. (4) Certain minerals, the rare earths, thorium, zirconium, etc. (5) Land, or land areas. (6) The underground hole of a burrowing animal. (7) To place earth over the roots or around a plant.

Earth Up–To cultivate in such a manner that soil is thrown toward plants, as in the hilling of potatoes.

Earthing–Banking up the soil around plants, as in blanching celery.

Earthworm–Any of the several species of the genus *Lumbricus*, family Lumbricidae, which are commonly, and often abundantly, represented in the fauna of surface soils, especially in the humid regions. They are generally regarded as beneficial because they improve the structure and fertility of soil; however, they may do minor damage by eating plants.

Earthworm Farm–A farm that raises earthworms for fish bait or other purposes.

East Indian Millet–See Pearl Millet.

Easter Lily–(1) *Lilium longiflorum*, family Liliaceae; a bulbous herb grown in flower gardens and forced by florists, for its fragrant, trumpet-shaped, pure-white flowers. Native to Japan. Also called white-trumpet lily. (2) Any lily placed on the market principally during Easter week, although it may be sold at other times. The most popular is the Easter lily. Others include the speciosum lily, the Madonna lily, and the atomasco lily.

Eastern Arborvitae–*Thuja occidentalis*, family Pinaceae; an evergreen tree from which innumerable horticultural ornamental varieties have been derived. Found in eastern North America. Also called American arborvitae, white cedar, Queen Victoria arborvitae.

Eastern Black Walnut–*Juglans nigra*, family Juglandaceae; a hardy, deciduous tree grown for its fruit, the black walnut of commerce, which is used in confections. Its lumber is one of the finest of the domestic cabinet woods, used also for veneer, interior paneling, furniture. Native to North America. See Walnut.

Eastern Bracken–*Pteridium aquilinum*, family Polypodiaceae; a fern poisonous to livestock at certain stages of its development although it is sometimes used as fodder or bedding. Cattle become sick with very high temperature, having difficulty in breathing, salivate, and have nosebleed and hemorrhage. Found chiefly in northeastern United Sstates and Canada. Also called brake fern, eagle fern, bracken. See Bracken, Western Bracken.

Eastern Field Wireworm–*Limonius agonus*, family Elateridae; the larva of a click beetle that infests the seed, underground stems, and roots of tobacco, corn, potato, beets, carrots, radishes, and onions. Commonly found in the eastern United States.

Eastern Larch–*Larix laricina*, family Pinaceae; a hardy deciduous tree grown for its valuable timber and as an ornamental. Native to North America. Also called American larch, hackmatack, epinette, and, among foresters, tamarack.

Eastern Mistletoe–Also known as Christmas American mistletoe, *Phoradendron serotinum*, family Loranthaceae. A plant that is a parasite on trees, even though the plant contains chlorophyll.

Eastern Pear and Apple Rust–A leaf rust caused by *Gymnosporangium globosum*, family Pucciniaceae, which is common to apple, pear, and hawthorn, with alternate stages in its life cycle on pomaceous species (aecial stage) and on junipers (telial stage). It is common in North America where eastern red cedars are adjacent to susceptible pomaceous species or varieties. Also called cedar quince rust, cedar-apple rust.

Eastern Redbud–*Cercis canadensis*, family Leguminosae; a deciduous tree grown for its early spring, rosy-pink flowers. Native to North America. Also called American Judas tree, American Judas tree, American redbud, Judas tree. See Redbud/.

Eastern Redcedar–*Juniperus virginiana*, family Pinaceae; a very hardy evergreen tree grown as an ornamental and as the source of cedar wood linings, millwork, pencil slats, woodenware, and water buckets. Native to North America. Also called red cedar, Tennessee red cedar.

Eastern Snow Mold–A fungal disease of fine turf caused by *Fusarium nivale*, family Tuberculariaceae. It appears soon after snow melts and is distinguished by whitish-gray dead areas of irregular circular outline. Also called snow mold.

Eastern Subterannean Termite–*Reticulitermes flavipes*, family Rhinotermitidae; an insect that infests the woodwork of buildings, sometimes weakening joists and floors.

Eastern Tent Caterpillar–*Malacosoma americanum*, family Lasiocampidae; an insect whose larvae gather in the forks of the limbs of various trees, spinning a large weblike nest from which they emerge during the day to feed on the foliage. They are serious pests found in North America from the Rocky Mountains eastward. Also called apple-tree tent caterpillar.

Eastern Wahoo–*Euonymus atropurpureus*, family Celastraceae; a shrub or small tree grown for its foliage and scarlet fruit. It is poisonous to sheep and goats, causing nausea, prostration, and cold sweat. Found in the eastern United States and as far west as Texas. Also called burningbush, wahoo, eastern burningbush.

Eastern White Pine–*Pinus strobus*, family Pinaceae; an evergreen, coniferous tree which is one of the outstanding timber pines and ornamentals. It sometimes grows to 200 feet at maturity and is often referred to as the "Queen of the Forest." Its lumber is soft and is put to a wide variety of uses. Native to North America. Also called white pine, northern white pine, American white pine, apple pine, deal pine.

Eatage–(1) An edible growth of grass for livestock, especially the second mowing. (2) The privilege of using grassland for pasture.

Eatern Red Oak–Quercus borealis var. *maxima*, family Fagaceae; a tall, relatively quick-growing tree grown as an ornamental shade tree and for its wood, which is used in furniture and interior finishing. Native to North America.

Eating–Designating food in regard to its quality, as in eating apple is one that has sufficient quality for eating out-of-hand.

Eating Ripe–Designating fruit that is somewhat riper than that suitable for processing; a fruit that has attained a near optimum in degree of ripeness.

Ecad–A plant or species of plant that has been changed morphologically from the normal by environment.

Eccentric–Used to describe an irregular or unsymmetrical tree trunk.

Ecdysone–A hormone that is essential in the molting process of insects. See Molting.

Ecesis–The adjustment of a plant to a new habitat, involving germination, growth, and reproduction.

Economic Entomology–That branch of insect study directed toward preventing human losses and increasing gains through manipulation of insect populations. Examples of economic entomology include the methods of protecting plants, people, and animals from insect-borne diseases by insecticides, drainage, crop rotation, and integrated pest management; and the culture of silkworms, honeybees, and various beneficial parasitic insects.

Economic Injury Level–The point at which the buildup of an insect population starts to cause economic damage to a crop or group of animals.

Economic Maturity–The age and growth rate at which a tree or stand of trees will no longer increase in value fast enough to earn a satisfactory rate of interest. At this time the trees should be marketed. Also called financial maturity.

Economic Poisons–Any of the recommended and approved insecticides, herbicides, rodenticides, etc.

Ecotone–A transitional area between two types of vegetation, as between forest and prairie.

Ecotype–A variation in plants brought about by a particular type of environment, such as smaller leaves in full sunlight.

Ectodynamorphic Soil–A soil whose dominant character has been determined by external forces, such as climate and vegetation. See Endodynamorphic Soil.

Ectomycorrhizal–Subsisting on the surface, that is, beneficial ectomycorrhizal fungi that live on the surface of tree roots, attached to the epidermal cell layer, and penetrating intercellular space of cortical cells. See Endomycorrhizal.

Ectoparasite–A parasite that lives on the outside of the body of its host; e.g., a tick.

Edaphology–That science which is concerned with the study of the relationships between plants and soils.

Eddish–Crop aftermath.

Edema–(1) An accumulation, usually abnormal, of serous fluids within the intercellular tissue spaces of the body. If the edema is in the subcutaneous tissue, the affected area will be swollen and will pit with pressure. Also called dropsy. (2) Abnormal swelling of plant parts due to an excessive intake of water.

Edge–The cover afforded by the vegetative margin between different types of wildlife habitats, as between fields and woodlands.

Edge Firing–A method of controlled burning in which fires are set around the perimeter of the area to be burned and allowed to spread inward.

Edger–A tool used for cutting the sod along walks, the edges of flower borders, shrub borders, and beds. Also called edge iron.

Edging–(1) A border of flowers, stones, metal, or wooden strips, etc., outlining the boundary of a flower bed, lawn, drive, etc. See Edging Plant. (2) Waste strips of wood that result from squaring boards.

Edging Plant–Any low-growing plant that is used along the edges of a flower bed. See Edging.

Edible–A term applied to food that is fit to eat. It usually refers to food that is suitable for human consumption. The initials E.P. are used to denote the edible portion of a food—e.g., a banana without its skin, a pork chop without the bone, a melon without its seeds and rind.

Edible Coating–Edible liquid coatings applied to meat that dry or harden quickly and help lengthen shelf life.

Edible Oil–Any vegetable oil that is fit for humans to eat.

Edible-podded Pea–A type of garden pea whose seed and pod are edible. There are several varieties, the most common being *Pisum sativum*, family Leguminosae. Also called sugar pea, sugar pod.

EDTA–Ethylenediamine tetraacetic acid; a chelate (sequestering agent) used to correct iron, copper, and zinc deficiencies in plants. It can be applied to the soil or to the plant foliage.

Eelworm Diseases–Plant diseases caused by many species of root knot nematodes. See Root Knot.

Effective Field Capacity–The actual work accomplished in acres or hectares per hour by an implement despite loss of time from field end turns, inadequate tractor capacity, deficient tractor or implement preparation, adverse soil conditions, irregular field contours, lack of operator skill, or other factors.

Effective Precipitation–That portion of total precipitation that becomes available for plant growth. It does not include precipitation lost to deep percolation below the root zone, to surface runoff, or to plant interception.

Efferent–Conducting or conveying away from.

Effete–Designating an exhaustion of the ability in animals to produce young or in plants to bear fruit. See Impotence.

Efficiency of Irrigation–The fraction of the water diverted from a river or other source that is consumed by the crop, expressed as percent. Often applied to whole irrigation systems and takes account of conveyance losses. See Consumptive Use.

Efficiency of Water Application–The fraction of the water delivered to the farm that is stored in the soil in the plant root zone for use by the crop, expressed as percent.

Efflorescence–(1) The period or state of full bloom. (2) The crusts or coatings of any soluble salts, such as the sulfates and chlorides of calcium, magnesium, and sodium that appear on the bare surface of soil, on clods, in cracks of natural soils, or on heavily fertilized soils as a result of the evaporation of moisture in the soil.

Egg–(1) The reproductive body produced by a female organism: in animals, the ovum; in plants, the germ cell, which after fertilization, develops into the embryo. (2) The oval reproductive body produced by females of birds, reptiles, and certain other animal species, enclosed in a calcerous shell or strong membrane within which the young develop.

Egg Pod–A capsule that encloses the egg mass of grasshoppers and that is formed through the cementing of soil particles together by secretions of the ovipositing female.

Eggplant–*Solanum melongena*, family Solanaceae; a tropical, perennial herb or small shrub grown as an annual for its edible, blackish-purple fruit. Native to the East Indies. Also called garden eggplant, egg apple, eggfruit, aubergine, brinjal, mad-apple, Jew's-apple.

Eggplant Blight–A fungal disease caused by ***Phomopsis vexans***, family Sphaerioidaceae; characterized by a damping off or seedling blight, brown leaf spots, and circular, depressed, discolored fruit lesions of rotted tissue, which may cause the entire fruit to rot.

Egyptian Bee–*Apis fasciata*; a honeybee that is nervous and ill-tempered, but prolific and useful in crossbreeding. It is a good honey producer but not suited to comb-honey production. Native to Egypt.

Egyptian Clover–*Trifolium alexandrinum*, family Leguminosae; an annual herb grown in arid and semi-arid regions for soil improvement and for forage. It is tolerant of saline soils. Native to Egypt and Syria. Also called berseem, berseem clover, winter lucerne.

Egyptian Corn–A variety of the durra grain sorghums.

Egyptian Cotton–An agricultural variety of cotton noted for its long staple. Native to the Western Hemisphere.

Egyptian Millet–See Johnsongrass, Pearl Millet.

Egyptian Pea–Gram chickpea, ***Cicer arietinum***, family Leguminosae. Also known as garbanzo bean.

Einkorn–***Triticum monococcum***, family Gramineae; perhaps the most primitive of the wheats, which usually has only one kernel to the spikelet.

Elaborated Sap–A watery solution containing dissolved foods such as sugars, organic acids, etc., produced by the green parts of a plant and carried mainly in the phloem tissues to those regions where these foods are to be used or stored.

Elatus–Tall.

Elder–Any plant of the genus ***Sambucus***, family Caprifoliaceae. Species are occasionally cultivated for their flowers, and the fruit of wild and cultivated plants is used in making wine, pies, and jelly. Also called elderberry, elder blow.

Electrostatic Sprayer–A spraying system that uses electrical forces of attraction to greatly increase the amount of spray that covers the plant. Individual spray droplets are given an intense electrical charge within a specially designed atomizing nozzle and propelled toward the plant. Individual spray droplets are given an intense electrical charge within a specially designed atomizing nozzle and propelled toward the plant. The approaching charged spray cloud induces an opposite electrical charge into the plants. The charges cause the spray droplets to be attracted like a magnet to the plant. Spray coverage is greatly increased.

Elephantgrass–(1) ***Typha elephantina***, family Typhaceae; a marsh plant whose leaves are used in basketmaking and whose pollen is used in making bread in India. Also called elephant's grass. Native to southern Eurasia. (2) See Napiergrass.

Elevator–(1) Any of a number of devices, most often consisting of endless belts, chains and buckets, screws, or suction tubes that are used to raise materials from a lower to a higher level. (2) A building designed for the handling and storage of cereal grains, dried beans, and other seed crops. It may also be used for processing and selling products.

Elliptical–Refers to a leaf shape like an ellipse (circular cone). It is broader in the middle and tapers to the apex and petiole.

Elm Blight–See Dutch Elm Disease.

Elm Casebearer–***Coleophora limosipennella***, family Coleophoridae; an insect that infests the elms of the northeastern United States by mining (eating) between the principal veins of the leaves.

Elm Lacebug–***Corythucha ulmi***, family Tingidae; an insect that infests the American elm, causing the leaves to turn yellowish in the early summer.

Elm Leaf Beetle–***Galerucella xanthomelaena***, family Chrysomelidae; an insect pest of American, Scotch, English, and Camperdown elms throughout the United States, which skeletonizes the leaves on which it feeds, sometimes defoliating the tree.

Elm Sawfly–***Cimbex americana***, family Cimbicidae; an insect that lays eggs in the leaf tissue, forming blisters on the underside of leaves. The larvae feed on the foliage, defoliating twigs and branches. They infest elm, willow, alder, basswood, birch, poplar, and maple. Common throughout the northern United States and west to Colorado. Also called giant American sawfly.

Eluviation–The movement of soil material from one place to another in solution or in suspension by natural soil-forming processes. Soil horizons that have lost material through eluviation are referred to as eluvial, and those that have received materials as illuvial. Designated as the E horizon (formerly A2).

Emarginate–A leaf apex that has a broad, deep indentation. See Retuse.

Embryo–(1) Any organism in its earliest stages of development. (2) The young, sporophyte that results from the union of male and female cells in a seed plant. Also called seed-germ.

Embryo Rescue–The use of tissue culture techniques to propagate an embryo that otherwise would not develop into an individual. In plants, embryo rescue is used primarily with excised immature embryos, usually to allow crosses to be made between sexually incompatible species. It may be used also with unfertilized egg cells to produce haploid plants.

Embryo Sac–(1) The mature female gametophyte in higher plants. (2) A sac that contains the embryo in its very early life in animals. Also called blastodermic vesicle.

Embryology–The science that deals with the study of the embryo.

Embryonic–(1) Pertaining to the embryo or its development. (2) Underdeveloped; immature. See Rudimentary.

Emergence–(1) The appearance of the first leaves of the crop plant above the ground. (2) The escaping of an insect from its cocoon, pupal case, etc.

Emergency Tillage–The cultivation of the soil by listing, duckfooting, chiseling, or pitting in order to roughen the soil or to bring clods to the surface for the reduction of wind erosion.

Emergent Vegetation–Various aquatic plants usually rooted in shallow water and having most of their vegetative growth above water, such as cattails and bulrushes.

Emmer–***Triticum dicoccum***, family Gramineae; an annual cereal herb, a wheat, which is grown as a source of bread flour in southern Europe, but is grown only as a stock feed in the United States. Native to western Asia and the Mediterranean area. Also called amelcorn.

Empirical Yield Table–A table showing the progressive development of a timber stand at periodic intervals. The table covers the greater part of the useful life of the stand. It is prepared based on actual stand conditions.

Empty-cell Process–Any process for impregnating wood with preservatives or chemicals, in which air is imprisoned in the wood under the pressure of the entering preservative, which expands when the pressure is released, driving out part of the injected preservative.

Emulsifiable Concentrate–A concentrated pesticide that contains ingredients that will allow the pesticide to become suspended in water.

Emulsifying Agents–Substances such as gums, soaps, agar, lecithin, glycerol monostearate, alginates, and Irish moss that aid to uniform dispersion of oil in water. Examples of emulsions formed: margarine, salad dressing, and ice cream.

Emulsion–A mixture in which one liquid is suspended as tiny drops in another liquid, such as oil in water.

Enation–An outgrowth from the surface of an organ of a plant usually caused by the parasitic invasion of fungi, bacteria, or viruses.

Encapsulated Formulation–A pesticide that is enclosed in a material that causes the pesticide to be released at a desired rate.

Encapsulation–Enclosure in a capsule or sheath.

Endive–*Cichorium endivia*, family Compositae; an annual or biennial herb grown for its edible leaves used in salads. Native to Asia.

Endocarp–The inner layer of a multiple-layered pericarp, e.g., the shell of a cherry stone.

Endodynamorphic Soil–A soil whose character has been determined largely by or inherited from the parent material or underlying rock with less influence from the external environment. See Ectodynamorphic Soil.

Endogen–The internal growth of new wood in a plant stem.

Endogenous–Growing throughout the substance of the stem, instead of by superficial layers. See Exogenous.

Endomychorrhizal–Subsisting within the interior of the plant cell, that is, endomycorrhizal fungi penetrate the interior of tree cortical root cells. See Mycorrhizae; Ectomycorrhizal.

Endosperm–The nutritive portion in some seeds that originates in the embryo sac, but that is outside the embryo. It usually follows the fertilization of the two primary endosperm nuclei of the embryo sac by one of the two male sperms. In a diploid organism, the endosperm is triploid. It serves as food for the growing plant during seed germination, often being entirely consumed in the process. The cereals have seeds that are referred to as endosperm type seeds wherein some of this nutritive material remains within the seed. Also called nurse embryo.

Endothal–A chemical compound used as an herbicide, a defoliant, a drying agent, and as a preemergent herbicide.

Endotoxin–A toxin produced within an organism and liberated only when the organism disintegrates or is destroyed.

Endoxerosis–An internal necrosis of the lemon characterized by a loss of yellow at the stylar end of the fruit, a depressed nipple, and pinkish to rust-brown masses in the albedo of the rind. It is thought to be related to drying within the tree and fruit, brought about by inadequate irrigation. Also called internal decline, blossom-end decline, blossom-end decay, dry tip, yellow tips, pink tip, tip deterioration.

Endrin–The endo-endo isomer of the insecticide dieldrin.

Enfleurage–The extraction of flower scents from picked flowers with cold fat (lard and tallow). The alcoholic washings of the perfumed fat (floral extracts) may be concentrated to any strength desired.

Engelmann Spruce–*Picea engelmanni*, family Pinaceae; an evergreen, coniferous tree, one of the hardiest spruces, which grows up to a height of 150 feet It is grown as an ornamental for its bluish-green to steel-blue needles and for its lumber, which has a wide variety of uses including pulpwood. Native to the Rocky Mountains of western North America.

Engelmann Spruce Weevil–*Pissodes engelmanni*, family Curculionidae; an insect that attacks the Engelmann spruce in western North America, killing or seriously injuring terminal shoots, causing the infested trees to have worthless, crooked trunks.

English Grain Aphid–*Macrosiphum granarium*, family Aphidae; a sucking insect pest of small grains and cultivated grasses which clusters in large numbers on the bracts of cereals, especially wheat. Infested plants have shriveled seed and, if infested early in the growing season, may die.

English Ivy–*Hedera helix*, family Araliaceae; a climbing, evergreen vine, a variable species popularly used out-of-doors to cover masonry walls, in greenhouses as a pot plant and for cut greens, and in homes as a pot plant. Also called ivy, evergreen ivy.

English Oak–*Quercus robur*, family Fagaceae; a hardy, deciduous tree having many varieties. It is valuable for its timber and as a shade tree. Its foliage and young buds when browsed in large quantities are poisonous to cattle, sheep, and goats. Symptoms of poisoning are constipation, emaciation, edema, blood in the feces, and subnormal temperature. Native to Europe.

English Pea–Garden pea, *Pisum satireum*, family Leguminosae.

English Walnut–Persian walnut, *Juglans regia*, family Juglandaceae. Grown for its high yielding flavorful nut, the tree is an important agricultural crop in such states as California.

Engraver Beetles–*Ips* spp.; a group of insects that attack southern pine trees. Larvae cut or "engrave" channels in the inner bark of the trees causing the death of the trees.

Enology–The art and science of winemaking.

Enphytotic–Designating a plant disease peculiar to a locality: e.g., an enphytotic fungus disease.

Enrich–(1) To add fertilizer or manure to soil. (2) To add a substance or vitamin to food products.

Enriched Cereals–Processed cereals or cereal products to which have been added certain minerals, proteins, and vitamins which may have been removed in the ordinary processing.

Enriched Rice–Rice to which vitamins and nutrients have been added to compensate for the many nutrients lost in milling.

Enrichment–The process by which bodies of water are enriched, especially by nitrogen, phosphorus, and/or carbon, resulting in accelerated growth of undesirable algae and other aquatic vegetation. Enrichment sometimes originates as runoff from domestic animal feedlots, landspreading of sewage sludge, and overfertilized fields.

Ensilage–Any green crop preserved for livestock feed by fermentation in a silo, pit, or stack, usually in chopped form. Ensilage can be made from practically any green crop having the proper moisture content. Also called silage.

Ensile—To place green plant material, such as green crops of grain, grasses, cornstalks, etc., in a silo in such a manner as to bring about proper fermentation for preservation and storage. See Silage.

Entire—A leaf margin that is smooth.

Entire Leaf—Any leaf that is in no way divided and whose margin is not toothed or serrated.

Entisols—A soil order that has no diagnostic pedogenic horizons. They may be found in virtually any climate on very recent geomorphic surfaces, either on steep slopes that are undergoing active erosion or on fans and floodplains where the recently eroded materials are deposited. They may also be on older geomorphic surfaces if the soils have been recently disturbed to such depths that the horizons have been destroyed or if the parent materials are resistant to alteration, as in quartz sands. See Soil Orders.

Entomologist—A person who specializes in the study of insects.

Entomology—That branch of zoology that deals with insects. See Economic Entomology.

Entomophagous Insect—Any insect that eats other insects.

Entomophagous Parasite—An insect or fungus that parasitizes insects.

Entomophilous—Designating flowers that are fertilized by insects, which carry the pollen from plant to plant.

Entomophthora Fungus—A fungus of the genus *Entomophthora*, as *Entomophthora fumosa*, which infects and effectively controls the citrus mealybug in Florida, United States, especially during summer rains.

Entyloma—A fungus that causes leaf smut of rice.

Envelope—Any pod, leaf, cover, etc., of a plant part.

Enzootic Marasmus—Cobalt deficiency.

Enzymatic—Referring to a reaction or process that is catalyzed by an enzyme or group of enzymes.

Enzyme—A large complex protein molecule produced by the body that stimulates or speeds up various chemical reactions without being used up itself; an organic catalyst.

Eolation—All the direct, geologic activities of the wind that either tear down or build up; the process by which wind modifies land surfaces, both directly by transportation of dust and sand, by the abrasion of sandblasts, and indirectly by wave action on shores.

Eolian—(1) Pertaining to action of wind, as soil erosion due to wind, or soil deposits transported by wind. (2) Loosely designating soils that are derived form geologic deposits that are windborne in origin. Also spelled aeolian. See Loess.

Ephedra—A genus of plants of the family Ephedraceae; a source of ephedrine. Tea made from the plants are supposed to have medicinal value. Species browsed on the range are rated as good. Also called Mormon tea, Brigham tea, canatillo, teamster's tea.

Ephedrine—$C_{10}H_{15}NO$; an alkaloid used in treating hay fever and asthma; made from the plant genus *Ephedra*.

Ephemer—An introduced plant that is unable to persist and thus soon disappears.

Ephemeral—Designating a very short-lived plant or animal.

Ephemeral Fever—A slight fever lasting a day or two.

Epicarp—The outermost layer, or exocarp, of a fruit.

Epicormic Branch—A shoot arising from a dormant bud on a bole. See Sprout.

Epicotyl—The part of the axis of an embryo above the region of attachment of the cotyledons.

Epidemic—(1) A widespread invasion or dispersion by an insect or a disease. See Endemic, Epiphytotic. (2) Designating a sudden and widespread attack by a disease or insect infestation. (Most scientists seem to prefer epidemic when referring to all types of disease: people, animal, and plant.)

Epidermis—The cellular layer of an organism; the outer skin.

Epigynous—Classification of a flower in which the perianth and stamens are attached above the ovary.

Epinasty—Stronger growth on the upper surface than on the under surface of a plant member, which causes the member, such as a leaf, to curl downward.

Epipetalous—The fusion of stamens and petals.

Epiphyte—A plant growing on another plant or object but not nourished therefrom; such as Spanish moss on a limb of a tree or on a telephone wire, and lichens on tree bark or a rock. When young, the banyan tree (*Ficus benghalensis*) is classed as an epiphyte.

Epiphytotic—(1) That which is common or prevalent among plants. (2) Designating a sudden and widespread attack of a plant disease. See Endemic, Epidemic, Epizootic.

Epistatic—Designating a condition of genetics in which one factor prevents a factor other than its allelomorph from exhibiting its normal effect on the development of the individual.

Epithelial Layer—Cellular tissue covering all the free body surfaces, cutaneous, mucous, and serous, including the glands and other structures derived therefrom.

Equilibrium Moisture Content—(1) The equilibrium established between the vapor pressures of air and a product, such as grain, hay, popcorn, etc., at identical temperatures. (2) The moisture content at which wood neither gains nor loses moisture when surrounded by air at a given relative humidity and temperature.

Equinoctial Plant—A plant whose flowers open and close at a particular hour of the day, e.g., wild morning glory, *Convolvulus septium*, family Convolvulaceae.

Equisetum Poisoning—A poisoning of horses and sheep which results from their feed on *Equisetum* spp., especially *E. arvense*, family Equisetaceae, the field horsetail. It is characterized by unthriftiness, loss of weight and muscular control, nervousness, falling, lack of ability to eat, and death. Also known as horsetail poisoning.

Equitant–Leaves with bases that overlap.

Equivalent per Million–An equivalent weight of an ion or salt per one million grams of solution or soil. For solutions, equivalents per million and milliequivalents per liter (meq/1) are numerically identical if the specific gravity of the solution is 1.0.

Eradicant–A fungicide in which a chemical has been added to rid a host of a pathogen.

Eradicate–To destroy or abolish, as a disease, insect, or weed pest.

Eradicator–(1) Any agent used for the destruction of weeds, insects pests, etc. (2) A device used for scraping the edible flesh from the pineapple shell after the central portion has been removed.

Erect–(1) Designating upright plants, in contrast to prostrate ones. (2) To build or construct buildings, etc.

Erectum–Upright.

Eremacausis–The natural burning out (oxidation) of organic matter, which results in a very low content of humus in soils of hot, arid regions.

Eremium–A desert plant community.

Ergosterol–A cholesterol-like substance found in plants that, when irradiated with ultraviolet light, changes to vitamin D2.

Ergot–(1) A fungal disease of cereals and wild grasses caused by *Claviceps purpurea*, family Clavicepitaceae, which attacks the inflorescence, replacing the grains with black or dark purple, club-shaped, horny structures (sclerotia) that are harvested with the grain and must be removed before the grain is used for flour or feed. Ergot causes a lowered grain yield and ergotism in people and animals. Also called clavus. (2) A drug obtained from such diseased plants used to control bleeding in animals and people. (3) A horny growth behind the fetlock joint of a horse.

Ergotism–A disease of people and lower animals caused from eating grain or grain products contaminated with ergot which is characterized by excessive salivation, redness and blistering of the mouth epithelium, vomiting, colic, diarrhea, and constipation. Also called holy fire, St. Anthony's fire, bread madness, gangrenous ergotism, dry gangrene.

Erica–Heath; heather, family Ericacea. There are about 1,900 species worldwide. See Blueberry, Cranberry.

Erinose–A plant disease, especially of the grape, walnut, and mountain maple, caused by mites of the genus *Eriophyes*, family Eriophyidae, which is characterized by an abnormal development of hairs or trichomes from the surface of the leaves, suggesting a fetlike patch.

Erode–(1) To wear away or carry away the surface of the earth, as water erodes a rock formation. See Erodible, Erosion. (2) To wear away, eat into the cells or layer of cells.

Eroded Area–A place that is worn away. In an animal's body this may be mechanical, as the wearing away of teeth or outer skin, or by the action of chemicals or germs that destroy living cells.

Erodibility Index–An index based on a field's inherent tendency to erode from rain or wind in the absence of a cover crop. It is based on the universal soil loss equation (USLE) and the wind erosion equation (WEE), along with a soil's T value. See T Value, Universal Soil Loss equation, Wind Erosion Equation.

Erodible–Susceptible to erosion, as erodible soil. See Erosive.

Erose–Designating a leaf or flower which is irregularly notched on its outer margin, as if it were partially eaten away.

Erosion–The group of processes whereby earthy or rock material is worn away, loosened or dissolved and removed from any part of the earth's surface. It includes the processes of weathering, solution, corrasion, and transportation. Erosion is often classified by the eroding agent (wind, water, wave, or raindrop erosion) and/or by the appearance of the erosion (sheet, rill, or gully erosion) and/or by the location of the erosional activity (surface or shoreline) or by the material being eroded (soil erosion or beach erosion). Relations between erosion terms: *raindrop erosion* always takes the form of *sheet erosion*, though sheet erosion can also be caused by wind action or the movement of thin sheets of water over the ground surface. *Sheet, gully* and *rill erosion* are all forms of soil erosion. Sheet and rill erosion are the two forms that *surface* erosion may take. *Beach erosion* is always *shoreline erosion*—though, because not all shorelines are beaches, shoreline erosion is not always beach erosion. The term *accelerated erosion* is used in comparing erosion caused by human activities with that occurring at natural rates, called geologic erosion.

Erosion Pavement–A residue of pebbles and stones on a land surface formed by the removal of the finer surface particles by wind or water; common in deserts.

Erosion-resistant Crop–Those crops which, because of dense foliage, extensive root system, and/or heavy litter, provide effective protection against soil loss by erosion.

Erosive–Refers to wind or water having sufficient velocity to cause erosion. Not to be confused with erodible as a quality of soil. See Erodible.

Escape–(1) A fowl or animal that has gotten out of its enclosure. (2) Botanically, a cultivated plant that is found growing wild. (3) To become wild after having been in cultivation.

Escarole–*Cichorium endivie*, family Compositae; a variety of endive having broad, thick, entire-edged leaves used as a salad plant, as a potherb, and in soups and stews.

Eschallot–Shallot, *Allium cepa*, family Liliaceae; a winter onion.

Esculent–(1) Edibility, especially where it modifies plants or plant parts. (2) A plant suitable for human food.

Escutcheon–(1) The scion used in shield budding. (2) That part of a cow which extends upward just above and back of the udder where the hair turns upward in contrast to the normal downward direction of the hair. Also called milk shield, milk mirror.

Espalier–(1) A method of training a fruit tree in which the tree is usually planted against a wall and the main branches trained in a plane parallel to the wall in a geometric design. Plums, apples, pears, cherries, and peaches are suited to this type of training. (2) A tree so trained. (3) The trellis upon which an espalier tree is trained.

Essence–A substance that contains to a high degree the essential oils or other stored products of a plant, etc., as essence of peppermint.

Essential Element–Any element necessary for growth and reproduction of a plant or animal. The sixteen essential elements for seed-bearing plants are: carbon, hydrogen, oxygen, phosphorus, potassium, nitrogen, sulfur, calcium, iron, magnesium, boron, manganese, copper, zinc, molybdenum, and chlorine. The twenty essential elements for humans and animals are: carbon, hydrogen, oxygen, phosphorus, potassium, nitrogen, sulfur, calcium, iron, magnesium, manganese, copper, zinc, sodium, iodine, selenium, vanadium, chromium, chlorine, and cobalt.

Essential Host–A host for one stage in the development of a parasite without which the parasite cannot develop to maturity. See Cedar-Apple Rust.

Essential Oil–Any volatile, aromatic oil derived from plants that is characteristic of the plant in odor and sometimes in other properties. It is used in essences, perfumes, attractants, repellents, etc. It is distinguished from the so-called fatty oils in plant tissue by the fact that it evaporates or volatilizes in contact with air and gives off an aromatic odor or flavor such as peppermint, camphor, etc. Also called ethereal oil, volatile. oil.

Established–(1) Designating a plant or plant community growing naturally in healthy adjustment to the environment. (2) Designating a pest or disease that is thriving. (3) Designating a plantation of trees that requires little or no cultivation, weeding, etc., and in which the young trees are almost certain to survive. (4) Designating a seeding of meadow or pasture plants that is thriving.

Establishment–(1) The adjustment of a plant to a new site, consisting of three processes: germination, growth, and reproduction. Also called acesis. (2) A farmstead. (3) An economic unit, generally at a single physical location, where business is conducted or where services or industrial operations are performed.

Ester–An organic salt formed by the interaction of an alcohol (base) and an organic acid.

Esthetic Woods–A woodland established and maintained for its beauty and esthetic values, as contrasted with forests or woods maintained for commercial purposes or for windbreaks. See Arboretum, Pinetum.

Esthetics–See Aesthetics.

Estimate–A judgment of the approximate volume or yield of a crop or a timber stand made from incomplete data.

Estival–Pertaining to summer.

Estivation–(1) Dormancy in the summer months. (2) The arrangement of the parts of a flower in the bud.

Ethanol–C_2H_5OH; the alcohol product of fermentation that is used in alcohol beverages and for industrial purposes; chemical formula blended with gasoline to make gasohol; also known as ethyl alcohol or grain alcohol.

Ether Extract–The fatty substances in food and feed that are soluble in ether; one of the procedures in the proximate analysis of feeds.

Etherization–The subjecting of a plant to one part of ether to 400 or 500 parts of air for the purpose of shortening the dormant period of seed and hastening germination.

Ethyl Acetate–$CH_3COOC_2H_5$, an inflammable fumigant that leaves a characteristic odor on treated grain.

Ethyl Alcohol–Also known as ethanol or grain alcohol. See Ethanol.

Ethyl Formate–A fumigant.

Ethyl Mercury Phosphate–A fungicide.

Ethylene–C_2H_4; a colorless, inflammable, unsaturated hydrocarbon gas used to hasten ripening of harvested fruits, especially bananas, pears, and green tomatoes, and to hull Persian walnuts and other nuts. It is used as a coloring agent for harvested citrus fruits, especially oranges, and as a defoliation agent for fall-dug nursery stock, especially rose bushes. It is also used for blowtorch welding and cutting.

Ethylene Dibromide–A heavy liquid used as a soil fumigant, especially for the control of nematodes (wireworms) in the irrigated land of the western United States. Emulsified, it is effective against Japanese beetle grubs in grasslands and against the European chafer in soils. It is also used to fumigate honeycombs for control of wax moth.

Ethylene Dichloride–A soil fumigant and nematocide.

Ethylene Oxide–C_2H_4O; a chemical used as a fumigant.

Ethylmercuric Chloride–A fungicide for seed and bulb treatment.

Ethylmercuric Phosphate–A fungicide for seed and bulb treatment.

Ethylmercuric-P-Toluence Sulfonilimide–A fungicide for seed and bulb treatment.

Etiolate–To prevent a plant, such as cauliflower, or plant part from developing chlorophyll by shielding it from light; to blanch.

Etiolated Seedlings–Stretchy, leggy plant seedlings resulting from exposure to a very low light.

Etiological–Pertaining to the causes of diseases.

Etiology–The science that deals with the origins and causes of disease. Also spelled aetiology.

Eucalyptus–Any tree of the genus *Eucalyptus*, family Myrtaceae, of Australia. Some species are gigantic, some highly aromatic, and some are evergreen. They are grown as ornamentals and as bee trees in warm regions. Also called eucalypt, gum tree, stringy bark.

Eucharis–Any plant of the genus *Eucharis*, family Amaryllidaceae. Species are bulbous herbs useful in greenhouses as pot plants. Species are evergreen trees and shrubs used as ornamentals in Florida and California, United States, and in other places with a similar climate. In the tropics, some species are grown for their fruit and for the preparation of some spices.

Euploid–An organism or cell in which the chromosome number is the exact multiple of the monoploid or haploid number. Terms used for euploid series are: haploid, diploid, triploid, tetraploid, etc.

European Alder–*Alnus glatinosa*, family Corylaceae; a deciduous tree easily grown as an ornamental in swampy locations. Tolerates strongly acid soils. Native to Eurasia and North Africa. Also called black alder, dog tree.

European Beachgrass–See Beachgrass, European.

European Cornborer–*Pyrausta numilalis*, family Pyraustidae; an insect pest of corn whose larvae bore into the corn stalk and ear shank. Infested plants show broken tassels, bent stalks, and "sawdust" outside small holes.

European Filbert–*Corylus avellana*, family Betulaceae; a hardy shrub grown in the United States principally for ornament, and in Europe for its edible nuts. It is the source of several varieties. Native to Europe. Also called European hazel.

European Glorybind–*Convolvulus arnensis*, family Convolvulaceae; a slender, herbaceous vine, an extremely competitive weed pest over the northern half of United States, which twines itself around other plants. Native to Europe. Also called field bindweed, small-flowered morning-glory, small bindweed, hedge-bells, lap-love, creeping Charlie, creeping Jenny, wild morning-glory, English bindweed, lesser bindweed, bindweed. See Morning Glory.

European Grape–*Vitis vinifera*, family Vitaceae; a woody vine that is the source of all the best wine grapes, commonly called vinifera grapes. Grown in Europe, Asia, and principally in California in the United States, it is of vast economic importance. Native to Eurasia.

European Pine Shoot Moth–*Rhyacionia buoliana*, family Olethreutidae; a moth that lays its eggs on twigs, needle sheaths, or bark of red, mugho, Scotch, and perhaps other pines. Its larvae bore into the needle base and winter in the buds, destroying them. It is found in United States from Massachusetts to Virginia and west to Illinois.

European Red Mite–*Metatetranychus ulmi*, family Tetranychidae; a serious fruit pest in the United States that infests many deciduous trees and shrubs, especially plums, prunes, and apples.

European Spruce Sawfly–*Diprionhercyniae*, family Diprionidae; an insect that infests white, red, black, and Norway spruce in New England, New York, and New Jersey in the United States. The larvae feed on the foliage.

Eutropic–Pertaining to the turning of a plant toward the sun's rays.

Eutropic Peat–Productive peat or muck soil nearly neutral or alkaline in reaction that is relatively high in bases and essential mineral plant nutrients.

Evaluation Plantation–A tree plantation established to evaluate one or more genetic characteristics, whether individual trees or their parents.

Evaporated–Designating a product that has had most of the moisture driven off by boiling or other application of heat; e.g., evaporated milk.

Evaporation–The changing of a liquid into a gas. Industrially it may mean merely drying or driving off the moisture of a product, and not complete vaporization. It may be loosely used as the equivalent of dehydration. See Transpiration.

Evaporation Pan–(United States Weather Service) For determining the amount of evaporation, a circular tank 4 feet (122 centimeters) in diameter and 10 inches (25.4 centimeters) in depth is set in an open field on 2-inches (5-centimeters) supports with the top 12 inches (30.5 centimeters) from the ground. The water level in the tank is maintained between 1.5 inches and 2.5 inches (3.8 and 6.35 centimeters) from the top of the pan. At regular intervals the water level is measured and recorded as the rate of surface evaporation from an open-water surface. In irrigated areas, the frequency of applying irrigation water can be determined by the rate of water evaporation in the pan.

Evapotranspiration–That part of the root zone moisture that is consumed by evaporation and transpiration combined, including all water consumed by plants plus the water evaporated form bare land and water surface. See Evaporation, Transpiration.

Even Grain–Designating wood whose grain shows little contrast between early-season and late-season (spring and summer) growth rings.

Even-age Management–The application of a combination of actions designed to create stands in which trees of essentially the same age grow together.

Everbearing–Designating any small fruit plant that produces fruit throughout most of the season.

Evergreen–(1) A plant that retains its leaves or needles longer than one growing season so that leaves are present throughout the whole year. See Deciduous. (2) The leafy branches of evergreen plants that are used for decoration. Also called greens.

Everted–Turned inside out.

Ex–Prefix meaning without or destitute of.

Exanthema–Dieback of citrus trees characterized by gummy tissues and brown-stained leaves; caused by malnutrition.

Excess Alkali–Excess exchangeable sodium. The resulting poor physical and chemical properties restrict the growth of plants.

Excess Water–(1) Legally, surplus water of a stream, that which is not adjudicated, or that which is in excess of the needs of those who have prior rights to its use. (2) Any water, particularly rainfall, over and above that needed for plant growth that results in ponded water or runoff. See Drainage.

Excessively Drained–Water is removed from the soil very rapidly. Excessively drained soils are commonly very coarse, rocky, or shallow. Some are steep. All are free of the mottling related to wetness. See Soil Drainage Classes.

Exchangeable Cation–A cation that is absorbed on the exchange complex and which is capable of exchange with other cations.

Exchangeable Sodium Percentages–The degree of saturation of the soil exchange complex with sodium. It may be calculated by the formula: $ESP = $ exchangeable sodium (meq/100g soil) ÷ cation − exchange capacity (meq/100 g soil) $\times 100$.

Excrescence–An unusual or wartlike growth, as on the stem of a tree.

Excurrent–Stem of a tree that extends from base to tip without dividing, as spruce or hemlock stems.

Exfoliate—To peel off in shreds, thin layers, or plates, as the bark in sycamore and eastern red cedar.

Exocarp—The outer skinlike region of the fruit pericarp.

Exogenous—(1) Produced on the outside of another body. (2) Produced externally, as spores on the tips of hyphae. (3) Growing by outer additions of annual layers, as the wood in dicotyledons. See Endogenous.

Exoskeleton—Collectively the external plates of the body wall.

Exotic Plant—A plant that is out of the ordinary. Many exotic plants are used for house plants.

Exotoxin—A soluble toxin excreted by specific bacteria and absorbed into the tissues of the host.

Exposure—The direction a slope faces, such as northern exposure. Also known as aspect.

Expression—The action of squeezing out fixed oils by pressing the seeds of certain plants such as sesame.

Extract—A solid preparation obtained by evaporating a solution of a drug, the juice of a plant, etc. Vitamin extracts are used to supplement a diet.

Exudate—A discharge deposited in or on an organ through pores, injured areas, or natural openings, as a bloody discharge from wounds of animals infested with screw worms, or the gummy discharge from a wound in a tree.

Exuviae—The cast-off skin of immature insects.

F_1—The first filial generation; the first generation of a given mating.

F_2—The second generation progeny generally produced by crossing two F_1 individuals.

Fabiform—Having the shape of a bean.

Face—(1) The bare skin on a fowl's head around and below its eyes. (2) The front part of the head of an animal, including the eyes, nose, and mouth. (3) The side of a hill or furrow. (4) The top or bottom layer of produce, especially fruit, which is arranged in a container for display purposes when the container is opened. (5) To arrange a layer of produce in a container for display purposes. (6) In turpentining, the exposed portion of the tree from which the oleoresin exudes. (7) In lumber, the side of a board from which it is graded. See Face Side.

Face Cord—A pile of wood 8 feet long and 4 feet high, regardless of the length of the pieces; usually 12 to 24 inches long. See Cord.

Face Side—The side of a board from which the grade is determined. Softwoods are most frequently graded from the better side; hardwoods, the poorer side.

Face-of-Tree—In log grading and sawing, one-fourth of the circumference of a log for its entire length.

Factor—(1) A unit of inheritance occupying a definite locus on one or both members of a definite chromosome pair whose presence is responsible for the development of a certain character or modification of a character of the individual who possesses that genotype; a determiner or gene. (2) An agent, as one who buys and sells a commodity on commission for others. (3) An item in the analysis of a farm business; e.g., labor efficiency. (4) Inherent characteristics of the climatic, nutritional, cultural, or biological environment responsible for the specific performances of plants or animals.

Facultative—Designating an organism that is capable of living under more than one condition: e.g., as a saprophyte and as a parasite; as an aerobic or anaerobic organism.

Facultative Aerobe—A microorganism that lives in the presence of oxygen but may live without it.

Facultative Bacteria—Bacteria that can exist and reproduce under either aerobic or anaerobic conditions.

Facultative Parasite—A parasite that feeds upon an organism until the organism dies and then continues to live on the dead organic material.

Fag—(1) Any tick or fly that attacks sheep. (2) Long coarse grass of the preceding season. (3) See Fagot (2).

Fahrenheit Scale—A temperature scale in which the freezing point of water is 32°F and the boiling point is 212°F. Named after Gabriel Daniel Fahrenheit (1686-1736), a German physicist. See Celsius Scale, Kelvin Scale.

Fair—(1) An exhibition of farm products, etc., on a competitive basis, usually with premiums offered for excellence. (2) The next-to-lowest grade of Mexican vanilla beans, hothouse lamb carcasses, and other products.

Fair Condition—(1) A range-condition class; a range producing only 25 to 50 percent of its potential. The cover consists of early maturing plants of low value for forage or for soil protection. (2) Denoting a medium condition of plants, plant products, or animals.

Fairy Ring—(1) A darker green, more luxuriant growth that occurs as a ring or circle in a turf area or meadow, due to the presence of any fungus of the class Basidiomycetes. At certain seasons of the year mushrooms appear on the periphery. It was once believed to be a dancing place for fairies. Also called fairy circle; fairy green. (2) A fungal disease of carnations caused by *Heterosporium echinulatum*, family Dematiaceae, which is characterized by round, bleached spots on the foliage. Black spores appear in concentric rings within the spots.

Fairy-ring Mushroom—(1) Any mushroom producing or causing fairy rings. (2) The fungus *Marasmius oreades*.

Falcate—Sickle or scythe-shaped.

Fall–(1) The dropping of a plant part, such as leaves or fruits. (2) The slope of land. (3) The amount of precipitation. (4) (a) The quantity of trees cut, (b) Felling trees. (5) (a) The number of lambs born. (b) Giving birth, as of lambs. (6) One of the three outer segments of an iris flower that is often drooping (usually used in the plural). (7) Autumn. (8) Of, or pertaining to, a plant or fruit that matures in the autumn, as a fall apple.

Fall Armyworm–*Laphygma frugiperda*, family Noctuidae; the larva of a moth that travels in great numbers, consuming alfalfa, grass, corn, and other plants. It appears in the northern United States in autumn. Also called budworm.

Fall Webworm–*Hyphantria cunea*. family Arctiidae; an insect that infests a wide variety of plants, feeding in May and June, but more destructively from July through September. the insects build unsightly, weblike nests and defoliate a whole branch.

Fallage–(1) The act of cutting trees. (2) Timber that has been felled.

Fallow–Cropland left idle in order to restore productivity through accumulation of moisture. Summer fallow is common in regions of limited rainfall where cereal grains are grown. The soil is tilled or herbicides are used for at least one growing season for weed control and decomposition of plant residue.

False-packed Cotton–Cotton in a bale: (a) containing entirely foreign substances; (b) containing damaged cotton in the interior with or without any indication of such damage upon the exterior; (c) composed of good cotton upon the exterior and decidedly inferior cotton in the interior in such a manner as not to be detected by customary examination; or (d) containing pickings or linters worked into the bale.

Family Characteristic–Particular parts of the conformation and/or the temperament that exist among certain families of dairy cattle, etc. Some families are consistently good or bad in straightness of legs, fore udders, depth of body, shape of head, etc. In temperament, there are families that are consistently nervous while others are normally docile. Those identifying marks or characters tend to display relationship.

Family, Soil–In soil classification one of the categories between the soil subgroups and the soil series. See Soil Family.

Fan Training–A method of training grapevines and fruit trees, especially dwarfs, that have practically no trunk. Branches from near the ground are trained obliquely upright so that the plant grows in one plane, fanlike in shape, against a wall or building.

Fancier–A breeder who shows particular interest in, and the development of, a particular breed or type of animals or plants: e.g., a rose fancier, dog fancier, Hereford fancier.

Fancy–A top-quality grade for many vegetables, fruits, flowers, poultry, and livestock.

Fang–(1) To form a forked root, especially of root crops, such as sugar beets. (2) Any tooth that is long and pointed.

Fanning Mill–In seed cleaning, a device equipped with screens to sift out foreign material and a power fan to separate and blow out chaff and other light material.

Farina–(1) A food product that is the middlings of hard wheat, except durum. (2) Potato starch. (3) Originally, a flour or meal from any grain, seed, or starchy root. (4) A coarse meal made from manioc (cassava). (5) French origin for flour.

Farinaceous–Starchy.

Farinha–(Portuguese) Coarse flour from manioc (cassava).

Farm Forestry–The practice of forestry on part of the farmland which is carried on with other farm operations.

Fasciation–The condition in plants in which stems and branches have grown together to form a flattened, abnormal, and malformed structure.

Fascicle–A small bundle; the bundle of piercing stylets of insects with piercing-sucking mouthparts.

Fascine–(1) A fagot. (2) A long bundle of sticks bound together and used to build a temporary roadway through a marshy soil or to stabilize an unstable slope along an embankment. When used in erosion control,the sticks may be made of willow or other species that sprout. See Gabion.

Fast Intake–The rapid movement of water into the soil. See Infiltration.

Fastigiate Tree–Any tree whose limbs are in an upright position; e.g., the Lombardy poplar, ***Populus nigra***, var. ***betulifolia***, family Salicaceae.

Fat–(1) (a) The oily or greasy-substance bearing tissues of an animal. (b) Designating any animal or fowl which abounds in fat. (2) The oily substance of milk; the chief constituent of butter. See Butterfat. (3) The oily or greasy substances found in certain plants; e.g., peanut oil, cottonseed oil. (4) Any food product, e.g., lard or vegetable shortening, which is derived from animal or vegetable fats. (5) Those substances which can be extracted from dry feeds with ether. See Ether Extract. (6) Fattened cattle ready to market. (7) Of, or pertaining to, a prosperous year, as a fat year.

Fat Body–An organ in the insect body with multiple functions in metabolism, food storage, and excretion. Fat body is a misnomer, for protein and glycogen are stored as well as fat.

Fat Pine–(1) A term used in many parts of southern United States, to denote branches and knots of pine that are rich in resinous material, and not to denote any particular species of pine. (2) See Longleaf Pine.

Feathering–(1) A defect of coffee cream characterized by a lack of homogeneity, causing it to rise to the surface of coffee in flocculent masses and form a light, serrated scum. (2) A rough-edged hole that has been bored in wood in which wood fibers remain to project around the perimeter. It usually results from using a dull bit. It is especially bothersome in maple trees used for collection of sap. (3) The scuffing of the tender skin on an early potato in harvesting. (4) The streaks of fat visible on the ribs of a lamb carcass.

Fecund–Fruitful; fertile; prolific.

Fecundation–Pollination or fertilization.

Fecundity–The ability to reproduce regularly and easily.

Feed Crop—Any crop grown as a feed for livestock, as hay, corn, oats, etc.

Feeding Gun—See Feeding Needle.

Feeding Needle—A long nozzle attached to a tank by a hose used to place soluble fertilizer in the ground around trees and shrubs.

Feeding Oat Meal—A livestock feed that is the product obtained in the manufacture of rolled oat groats or rolled oats. It consists of broken rolled oat groats, oat-groat chips, and floury portions of the oat groats, with only such quantity of finely ground oat hulls as is unavoidable in the usual process of commercial milling. It must not contain more than 4 percent of crude fiber.

Feeding Oil—Any vegetable oil used as a feed supplement or additive, as cottonseed oil.

Feedstock—The base raw material that is the source of sugar for fermentation.

Fell—(1) The elastic tissue just under the hide of an animal attached to its flesh; facia. (2) To cut down a tree.

Felling Head—A tractor-powered implement that grasps a tree and cuts it using a circular bar that has a cutting chain running around the perimeter.

Fenestration—The arranging of plants to provide space between them. The spacing may be horizontal, round, or vertical depending upon the type of plants used.

Feracious—Fruitful; bearing abundantly.

Ferment—(1) To undergo catalytic decomposition of complex compounds into simpler ones, as in the souring of milk, the formation of vinegar from cider, etc. (2) Any agent which can cause fermentation. See Enzyme, Fermentation.

Fermentation—(1) The processing of food by means of yeasts, molds, or bacteria; the catalytic decomposition of complex compounds into simple ones, as in the souring of milk, the production of vinegar from cider, etc. The most important products of fermentation are alcohol, acetic acid, lactic acid, butyl acid, citric acid, and acetone. (2) An enzymatic curing of stored tobacco leaves which improves the quality of the product. (3) Digestion in the rumen by microorganisms.

Fermentation Ethanol—Ethyl alcohol produced from the enzymatic transformation of organic substances. See Ethanol.

Fern Brake—(1) Bracken. (2) A dense, wild growth of fern plants. Also called fern scrub, bracken fern. See Bracken.

Ferric Oxide—Fe_2O_3; oxide of iron; iron is widely distributed in nature. It is found in phosphate rock where its action is similar to alumina (Al_2O_3). More than 3 to 4 percent of oxide of iron will make "sticky" superphosphate. See Iron.

Ferric Phosphate—A chemical compound containing 25 percent iron and 13.9 percent phosphorus used in mineral feeds for livestock. Also known as iron phosphate.

Ferric Sulfate—$Fe_2(SO_4)_3 \cdot 7H_2O$; a yellow-brown salt and a quickly available source of iron. Fertilizer-grade material contains about 2.2 percent Fe (30.3 percent Fe_2O_3). See Micronutrient Fertilizers.

Ferruginous Soil—Soil containing a large amount of iron minerals, especially limonite and hematite, characterized by a vivid red color or various shades of yellow and brown. See Ironstone.

Fertigation—The simultaneous application of soluble fertilizers in irrigation waters.

Fertile—(1) Productive; producing plants in abundance, as fertile soil. (2) Capable of growing or of development, as a fertile egg. (3) Capable of reproducing viable offspring. (4) Able to produce fruit, as a fertile flower. (5) Plant capable of producing seed.

Fertility—(1) The ability of a plant to mature viable seeds. (2) The ability of an animal or fowl to produce offspring. (3) The quality that enables a soil to provide the proper compounds, in the proper amounts and in the proper balance for the growth of specified plants, when other factors, e.g., light, temperature, and the physical condition of the soil or favorable. See Sterile.

Fertilization—(1) Union of pollen with the ovule to produce seeds. This is essential in production of edible flower parts such as tomatoes, squash, corn, strawberries, and many other garden plants. (2) Application to the soil of needed plant nutrients, such as nitrogen, phosphorus, and potassium. (3) The union of a sperm and egg.

Fertilize—(1) To supply the necessary mineral and/or organic nutrients to soil or water to aid the growth and development of plants. (2) To fecundate the egg of an animal or plant, or to pollenize the pistil of a flower.

Fertilized Pond—A pond to which fertilizer is regularly applied to promote the growth of algae and other aquatic plankton for fish food.

Fertilizer—(1) Any organic or inorganic material added to soil or water to provide plant nutrients and to increase the growth, yield, quantity, or nutritive value of the plants grown therein. (2) Bees and other organisms that bring about the fertilization of a flower.

Fertilizer Analysis—The exact percentage composition as determined in a chemical laboratory, expressed in terms of total nitrogen (N), available phosphoric acid (P_2O_5), and water-soluble potassium (K_2O). (Not to be confused with fertilizer grade). See Fertilizer Grade.

Fertilizer Banding—The application of fertilizer in a band instead of applying it broadcast over all of the soil.

Fertilizer Formula—The quantity and grade of crude stock materials used in making a fertilizer formulation.

Fertilizer Grade—The guaranteed minimum analysis, in percent, of the major plant nutrient elements contained in a fertilizer material or in a mixed fertilizer. A 20-10-5 fertilizer refers to the percentage of N-P_2O_5-K_2O, respectively. When a grade is less than the analysis by a specified legal amount, the company is fined. See Fertilizer Analysis.

Fertilizer Material—Any substance used to supply plant nutrients (primary, secondary, or micronutrient) either for direct application, or for bulk mixing or processing with other materials.

Fertilizer Ratio—The relative proportions of primary nutrients in a fertilizer grade divided by the highest common divisor for that grade; e.g., grades 10-6-4 and 20-12-8 have the same ratio of 5-3-2.

Fertilizer Requirement–The quantity of certain plant nutrient elements needed as determined by a soil test, in addition to the amount supplied by the soil, to increase plant growth to a designated optimum yield. See Yield Goal.

Fertilizer Spike–A slow-acting fertilizer molded in the shape of a spike that can be driven in the ground to supply nutrients to plant roots over a prolonged period of time.

Fertilizer Unit–One percent (20 pounds) of a short ton (2,000 pounds) of fertilizer.

Fescue Lameness–A disease condition of cattle attributed to vitamin A starvation and excessive eating of fully matured, tall fescue grass. It is characterized by lameness and apparent circulatory disturbances of ears, tail, and hind feet. Also called fescue foot.

Fescue, Sheep–*Festuca ovina*; a long-lived short-growing bunchgrass with short leaf blades. It is more drought resistant than other small-leaved fescues. Production of tops is low, but ground cover is excellent, and root production is outstanding.

Fescue, Tall–*Festuca arundinacea*; a long-lived, high-producing bunchgrass suited for use under a wide range of soil and climatic conditions. It is tolerant of strongly acid to strongly alkaline conditions. Suited to irrigation, subirrigation, or moderately wet conditions as well as dryland areas where the effective precipitation is over 18 inches (45 centimeters).

Feterita–A group of grain sorghums.

Fetid–Having a disagreeable odor.

FFA Alumni–A national organization composed of former members and supporters of the National FFA Organization. The purpose is to promote and support agricultural education programs and in particular, programs of the FFA. See FFA; Agricultural Education.

Fiber–(1) (a) In plants, the elongated, elastic cells, often tapered at each end, which serve various functions, such as support of the stem, as in the case of cotton fibers on the seed; protection, as in fiber-tracheids. (b) Similar cells or structures in animals. (2) (a) Commercially, true fiber, a single cell, or aggregation of such cells, as in flax, or even fibrovascular bundles used in the production of textiles. (b) Similar synthetic structures, as in rayon, nylon, etc. (3) In the analysis of animal feeds, the insoluble residue left after the sugars, acids, proteins, and ash are accounted for. It consists mainly of cellulose and lignified cell walls.

Fiber Crop–Crop grown for its fiber, as cotton and flax.

Fiber-tracheid–(1) In plants, a fiberlike tracheid, commonly thick-walled, with a small lumen, pointed ends, and small bordered pits having lenticular or slitlike apertures. (2) The late-wood tracheids of certain gymnosperms as well as the fibrous tracheids of dicotyledons. A fiber-tracheid combines the functions of fibers for support and tracheids for transport of water. See Fiber, Traceid.

Fibrin–A whitish, insoluble protein formed from fibrinogen by the action of a ferment, thrombin.

Fibrous–Like fiber. The fine roots of grass.

Fibrous Root System–A root system that is composed of profusely branched roots with many lateral rootlets.

Fibrovascular Bundles–Strands of water-conducting cells (tracheary tissue) and food-conducting cells (sieve cells) with some associated protective and supporting wood or bast fibers, which are united into branching (elongated bundles extending from the roots) through the stem and into the leaves and other parts of the plant. They serve to conduct food from the leaves or from storage organs to all living parts of the plant and to carry from the roots the water absorbed from the soil. In the leaves they are called veins. In some plants where they are easily separable form the leaf or stem tissues (by the process called retting) the whole fibrovascular bundles are popularly called fibers, as in sisal or hemp.

Field–An area of agricultural land devoted to the production of farm crops, e.g., cereals, vegetables, etc., or an area in which crop-growing has been postponed or abandoned.

Field Bean–(1) A term applied in the United States to varieties of the kidney bean; ***Phaseolus vulgaris***, family Leguminosae, e.g., navy, kidney, marrow, pea beans. They are grown for human consumption of their dry seeds. (2) A term in southern and central Europe applied to varieties of the broadbean. ***Vicia faba***, family Leguminosae, e.g., the Windsor or flat bean. See Bean.

Field Bee–A worker bee that is about sixteen or more days old and that works in the field to collect nectar, pollen, water, and propolis. See Honeybee.

Field Capacity–The amount of water held in a saturated soil after the excess or gravitational water has drained away. Also called capillary capacity.

Field Corn–Any variety of corn that is grown extensively in large fields primarily for livestock feed, as contrasted with the horticultural varieties, such as sweet corn or popcorn. Most field corn is of the dent variety.

Field Crate–A container, holding from 10 to 25 kilograms, used to haul fresh fruits or vegetables to the storage or packing house.

Field Crops–Feed plants grown primarily for their seeds. For example, corn, wheat, oats, soybeans, etc. See Forage Crops.

Field Curing–Permitting hay or forage to dry naturally in a swath or windrow prior to chopping or baling.

Field Dodder–*Cuscuta pentagona*, family Cuscutaceae; a native species of dodder that is found on wild and cultivated plants, especially alfalfa, throughout North America, but most commonly east of the Mississippi River. (This common name is also applied to ***Cuscuta campestris***.) See Dodder.

Field Garlic–*Allium vineale*, family Liliaceae; a pernicious, perennial, bulbous herb that is a pest of fields and pastures. Native to southern Europe. Also called wild garlic, wild onion, crow onion, crow garlic.

Field Horsetail–*Equisetum arvense*, family Equisetaceae; a perennial plant that possesses poisonous characteristics. Horses feeding on it fresh or in hay will show signs of poisoning; e.g., unthriftiness, loss of weight and muscular control, staggering, falling, inability to rise, lack of ability to eat, extreme nervousness, death. native to the United Statess and Eurasia. Also called cornfield horsetail, horsetail fern, meadow pine, bottle brush, snake pipes, mare's tail.

Field Pea–*Pisum sativum* var *arvense (Pisum arvense)*, family Leguminosae; a tall, annual herb, grown for fodder, as a vegetable, and for its seeds, used dried for split-pea soup. Native to Europe.

Field Percolation Test–A technique of determining the rate of percolation in a soil. The test consists of digging holes about 2 feet deep, filling them with water, and measuring the rate of water movement downward.

Field Run–Designating products (potatoes, onions, etc.) harvested in the field and ungraded as to size or quality.

Field Windbreak–A barrier of trees and shrubs grown to check the force of wind for protection of orchards, feedlots, and fields, and to check drifting snow and retain it to increase soil moisture. See Windbreak, Shelterbelt.

Fiery Hunter–*Calosoma calidum*, family Carabidae; a black, ground beetle, which feeds voraciously on cutworms, larvae of potato beetles, and other succulent larvae.

Figure Grain of Wood–The pattern produced on a wood surface by irregular coloration, growth layers, rays, knots, and such deviations from regular grain as bird's-eye, interlocked, and wavy grain.

Filament–(1) The part of the stamen of a flower that is below the anther and supports it. (2) A series of cells, especially of some algae, bacteria, and fungi, attached to each other forming a slender object or a long cylindrical cell.

Filbert–See European Filbert, Hazelnut.

Filial–Refers to the child or offspring; the meaning of the F in F_1 and F_2.

Fill–(1) The soil, sand, gravel, etc., used to fill in a depression in a field, or to build up a terrace or embankment. (2) The substances used in filling tree cavities; e.g., asphalt, concrete, wooden or rubber blocks, etc. Also called filler. (3) The increase in weight and form of livestock that have been watered and fed after arriving at their destination. (4) The shaft of a vehicle. (5) To level a depression in a field or cavities in trees, or to build up an embankment or terrace. (6) To feed and water livestock at the end of the shipment to make up for the loss of weight en route. (7) To enlarge with the enclosed seeds, as the pods of leguminous plants; or to be plump and shriveled when approaching maturity, as the seeds of cereals. (Cereal grains before harvest are often referred to as poorly or well-filled.)

Filler–(1) Material used for packing to prevent breakage. (2) An extra row intercultivated between two regular rows of a crop. (3) An extra, short-lived plant grown between slow-growing, larger plants and removed when the latter approach maturity. (4) Any of the various types of appliances for filling special receptacles, as bottle-filler, silo-filler, etc. (5) Any material, active or inert, added to a mixed fertilizer to increase bulk. (6) The nonessential matter in a manufactured or mixed feed, such as high-fiber materials, oat hulls, screenings, etc.

Filter Blanket–A layer of sand and/or gravel around a drain line designed to reduce the movement of fine-grained soils into the drain.

Filter Crop–Any close-growing crop planted across a slope to retard runoff and cause deposition of the sediment load. See Filter Strip.

Filter Strip–A strip of permanent vegetation of sufficient width and vegetative density above farm ponds, division terraces, and other structures which retards the flow of run-off water, causing the flowing water to deposit soil, thereby reducing the rate of silting of the reservoir below.

Filterable Virus–A virus that is capable of passing through the pores of a filter which does not allow passage of the ordinary bacteria. See Virus.

Fimbriate–A leaf surface with a fringelike appearance occurring all along the margin.

Fine Clayey–A soil family with a clay content between 35 and 59 percent in the subsoil. See Soil Family.

Fine Granular–Designating a soil aggregate that is less than 5 millimeters in diameter. See Soil Texture.

Fine Particulate Matter–A general term referring to particulate matter less than 10 microns in diameter.

Fine Sand–In the mechanical analysis of soils, soil separates (mineral particles) ranging in diameter from 0.25 to 0.10 millimeters (United States Department of Agriculture scheme). See Soil Texture.

Fine Sandy Loam–A class of soil which contains 30 percent or more of fine sand or less than 30 percent of very fine sand (or) between 15 and 30 percent very coarse, coarse, and medium sands. The clay content is 20 percent or less and the silt content less than 50 percent. See Soil Texture.

Fine-textured Soil–Sandy clay, silty clay, and clay. The obsolete term is heavy textured. See Soil Texture.

Fineleaf Actinea–*Actinea linearifolia*, family Compositae; an annual herb, sometimes used for fattening cattle in the southwestern United States. Also called tallow weed.

Fineness–(1) One of the several properties of cotton that determines the grade in which it is classified. Fineness refers to the smallness of the cross section diameter of the fibers (or lint). (2) One of several properties of dry milk. Usually the powders prepared by the spray process are extremely fine, whereas those prepared by the cylinder or drum process are coarser. (3) One of the several properties of high-quality hay that help in the determination of the grade of hay as feed for livestock. (4) The relative smallness of the wool fiber.

Fines–Fine particulates; aerosols.

Fire–(1) In several different plant diseases caused by bacteria, fungi, or nutritional deficiency, the final result that gives a burning or scorching appearance. (2) Potash deficiency or leaf scorch is sometimes referred to as firing. (3) The burning off of vegetation on lands of various types. This is apparently beneficial in some cases but is often harmful. (4) To treat a spavin or ringbone on a horse with a strong liniment in an attempt to cure or alleviate lameness.

Fire Ant–*Solenopsis geminata*, family Formicidae; a species of ants that is harmful to plants and whose bite is very painful to people and animals. See Imported Fire Ant.

Fire Blight–A serious, bacterial disease of susceptible varieties of apple, pear, quince, and many other rosaceous hosts caused by *Erwina amylovorus* (Syn.: *Bacillus amylovorus*), family Bacteriaceae, characterized by the rapid invasion and death of blossoms or succulent leaves and shoots. The bacteria enter and invade the cambium, killing twigs, branches, or entire limbs of trees by invasion or girdling, until bacterial multiplication is checked by maturity of the tissues. The dead blossoms and leaves persist giving the tree a fire-scorched appearance. Native to North America and has spread to Japan, Italy, and New Zealand. Also called apple blight, blossom blight, twig blight, blight canker, body blight.

Fire Danger–In forestry, the result of both constant and variable factors that determine whether fires will start, spread, and damage; determining as well, the difficulty of control. Constant factors are those that are relatively unchanging, e.g., normal risk of ignition, topography, all fuels, and exposure to prevailing wind. Variable factors change from day to day, season to season, and year to year; e.g., all weather elements, moisture content of fuel, and variable risks of ignition by lightning and people.

Fire Farming–Clearing patches of land by destroying wild cover by fire, cropping for a few years, and then allowing the area to remain fallow for natural revegetation and restoration. Practiced mainly in humid, tropical regions. Also called milpa and canuca in Latin America, langland in Indochina, caingin in the Philippines, and shifting cultivation in most other countries. See Shifting Cultivation.

Fire Line–In forestry, the narrow portion of a control line from which inflammable materials have been removed by scraping or digging down to mineral soil.

Fire Plow–A heavy duty, usually specialized machine, either of the share or disk type, which is designed solely for abusive work in the woods and is used to construct firebreaks and fire lines.

Fire Presuppression–Those forest fire control activities concerned with the organization, training, instruction, and management of the fire-control organization, and with the inspection and maintenance of fire control improvements, equipment, and supplies to ensure effective fire suppression.

Fire Prevention–Those fire control activities concerned with the attempt to reduce the number of forest fires through education, hazard reduction, and law enforcement.

Fire Retardant Chemical–Any chemical that when injected into or applied on wood will reduce inflammability.

Fire Wound–The fresh or healed injury of the cambium of a tree, caused by fire.

Fire-cured–Designating tobacco that has been hung in the smoke and heat of fires built on the curing-barn floor; a process that dries out the tobacco and halts further enzymatic decomposition.

Firebreak–An existing barrier, or one constructed before a fire occurs, from which all or most of the inflammable materials have been removed. It is designed to stop or check creeping or running fire but not to stop crown fires, to serve as a line from which to work, and to facilitate the movement of men and equipment in fire suppression.

Firebreak, Living–A firebreak on which is maintained nonflammable, green vegetation such as grass (kept nonflammable by fertilization and close grazing), forest trees (whose lower branches are kept pruned by natural or artificial means, such as *Eucalyptus gmelina* in Australia), and the semitropical iceplant *Mesembryanthemum* spp. from South Africa and the Mediterranean region but widely naturalized in southern California, especially the species *M. deule*.

Fireweed–*Erechtites hieracifolia*; a common weed of waste places, especially in burned-over land and abandoned fields.

Firewood–Logs cut short enough and split into convenient sizes for use as fuel in fireplaces or wood-burning stoves.

Firing–(1) A physiological condition of tobacco in which the leaves are scorched due to the plants being grown in soil containing an excess of boron compounds. (2) A change of color of the lower leaves of green corn usually attributed to a lack of plant nutrients or moisture deficiency.

Firm–(1) An economic unit recognized to be engaged primarily in production. (2) (a) To compact the soil, crushing and pulverizing the lumps to facilitate capillary water movement. (b) Designating well-compacted soil that is not lumpy or powdery. (2) In marketing, designating optimistic conditions. (4) Designating a cheese that feels solid. (5) Designating whites of eggs that are sufficiently viscous to prevent free movement of the yolk. (6) (a) Designating meat that is not soft or soggy. (b) Designating a fruit or vegetable that is not overripe or shriveled.

First–(1) The highest grade of lumber. (2) The second grade, next below extra, for butter. (3) The primary occurrence or the beginning of a series.

First Meiotic Division–The first of a series of two divisions in the process of producing haploid sex cells or gametes.

Fish Manure–An organic fertilizer made from fish scrap.

Fish Pomace–A fertilizer, the fish refuse after the oil has been extracted.

Fission–(1) A form of reproduction, common among bacteria and protozoa, in which a unit or organism splits into two or more whole units. (2) The splitting of a heavy nucleus such as uranium or plutonium into approximately equal parts, accompanied by the conversion of mass into energy, the release of the energy, and the production of neutrons and gamma rays.

Fission Fungi–Fungi that reproduce only by fission.

Fit–(1) To notch a tree for felling, and to mark it into log lengths after it is felled. (2) To ring, slit, and peel tanbark. (3) To file and set a saw. (4) To condition livestock for use, sale, or exhibit. (5) To prepare land for sowing, i.e., plowing (or disking), harrowing, and rolling. Land so fitted should have no large, hard clods so that the seeds may be placed at a uniform depth and in contact with moist soil.

Five Finger–*Pontentilla* spp.; a trailing, yellow-flowered weed usually found in infertile dry pastures.

Fixation of Nitrogen–Fixation of atmospheric nitrogen in the soil by bacteria, both symbiotic and nonsymbiotic. Symbiotic bacteria grow in nodules on the roots of legumes; e.g., alfalfa and clovers; the nitrogen fixed by them may become available to the host plant, may be released directly to the soil, or through decomposition of the legume may become available to other plants.

Fixed Ammonium–Ammonium, held in a soil, which is neither water soluble nor readily exchangeable.

Fixed Oils–Fatty oils, usually bland, that leave a stain and do not evaporate (e.g., sesame oil); may be obtained by expression or solvent extraction, but not through distillation.

Fixed Phosphorus–Phosphorus in soil that is unavailable to plants; soluble phosphorus of fertilizers which, as a result of reactions taking place in the soil with calcium and iron, has become relatively insoluble and unavailable as a plant nutrient.

Fixed Potassium–Potassium held by a soil, which is neither water soluble nor readily exchangeable or available to plants.

Flagella–Whiplike appendages of certain single-celled aquatic animals and plants, including some bacteria, the rapid movement of which produces motion.

Flame Cultivation–The use of flares of burning liquid petroleum gas to kill young weeds between crop rows.

Flat–(1) A shallow box containing soil, in which seeds are sown or to which seedlings are transplanted form the seedbed. (2) A level, treeless prairie, especially between hills or mountains. (3) Referring to a defect in the flavor of butter, cheese, or milk, due to errors in processing, as lack of sufficient salt, uncleanliness, etc. The flavor may be insipid or lacking in the usual characteristics of the product. (4) A level landform composed of unconsolidated sediments—usually mud or sand. Flats may be irregularly shaped or elongate and continuous with the shore, whereas bars are generally elongate, parallel to the shore, and separated from the shore by water.

Flat Pea–*Lathyrus sylvestris*, family Leguminosae; a perennial, herbaceous, climbing or straggling herb with flowers that resemble the sweetpea (*Lathyrus odoratus*), but lack its fragrance. Sometimes used as a forage plant or as green manure. Native to Europe. Also called Wagner flat pea.

Flatwoods–Any of various types of flat, wooded land usually of inferior quality for agriculture. The land may be wet, or intermittently wet and dry, and it is generally either a clay or a sand soil. Flatwoods may be designated by the kind of native vegetation as post-oak flatwoods, saw-palmetto flatwoods, etc. (Florida and other parts of the southern United States).

Flavedo–Colored outer peel layer of citrus fruits. Also called the epicarp. Contains oil sacs and pigments (chlorophyll, carotene, and xanthophyll).

Flavescence–See Chlorosis.

Flavor–(1) Odor and taste combined with the feeling of the substance in the mouth. (2) (a) The material added to foods to gain a desired taste. (b) To give taste or flavor to a product by the addition of spices, sugar, etc.

Flavus–(Latin) Yellow.

Flax–(1) (***Linum usitatissimum***, family Linaceae; an annual, herbaceous, crop plant, grown for its fiber from which linen is made; for its seed from which is extracted a valuable oil (linseed oil) used extensively in the paint and other industries; to a very limited extent, as a human food; and for the oil-cake (seeds with the oil expressed) which is a valuable livestock feed. Native to Eurasia. Also called Baltic hemp. (2) Wild species of the genus ***Linum***, common in the western United States; sometimes grazed by sheep and cattle. Some species (New Mexican flax and stiff-stem flax) contain a cyanogenetic glucoside known as linamarin, and are poisonous to animals eating them.

Flax Straw–The straw left after threshing the flax seed crop; used in the manufacture of various types of paper, including cigarette paper. Flax straw of good quality can be used as a substitute for oat straw roughage in wintering cattle. Also called cattle roughage.

Flax Wilt–A disease caused by the fungus ***Fusarium lini***, family Tuberculariaceae, which multiplies in soil where flax is grown for several years. The soil is called flax-sick. The control is to plant wilt-resistant varieties of flax and to rotate crops.

Flaxseed–The seed of flax, known as linseed, which is a source of linseed oil used mainly as a drying agent for paints and varnishes. The residual oil cake is used for livestock feed. See Flax.

Fleabane–*Erigeron annuus*; a common weed of hay fields that reduces the quality of the hay.

Fleece–(1) The wool from all parts of a single sheep, which consists of the crinkly hair up to 12 inches in length. This waviness enables the wool to be matted together into felt or spun into yarn, twine, or thread. (2) The fluffy mass of cotton that remains after the seeds have been removed by ginning. See Lint Cotton. (3) To shear sheep.

Flesh–(1) The portion of an animal body that consists mainly of muscle. (2) Plumpness or corpulence, especially in such phrases as good flesh, etc. (3) The pulpy or juicy portion of fruits or of storage organs of plants, such as potatoes, etc. (4) To remove adhering fat, flesh, and membrane from the pelt of a butchered animal.

Fleshy–(1) Designating fruit, leaves, and storage organs of plants with juicy or pulpy tissues. (2) Fat or corpulent. (3) Designating the soft or edible portions of meat.

Fleshy Fruits–Classification of fruits that includes the berry, grape, muskmelon (pepo), citrus (hesperidium), peach (drupe), and apple (pome). They have a pericarp that is soft and fleshy at maturity.

Fleshy Root–An enlarged root such as the sweet potato.

Flint Corn–*Zea mays* var. *indurata*, family Gramineae; a type of Indian corn having a soft, starchy endosperm, completely surrounded by a very hard, horny, vitreous, outer layer. The kernels are somewhat flat and rounded rather than pointed. Some varieties are of early maturity, others medium, and the tropical flints are late-maturing. See Dent Corn.

Flinty–Consisting of the hard, translucent endosperm of corn or wheat kernels in contrast to the softer, more mealy portions.

Flitch–(1) A portion of log, sawed on two or more sides, which is intended for sliced or sawed veneer. (2) A pile or bundle of veneer sheets from the same bolt laid together in the sequence of cutting. (3) A side or portion of meat, as a flitch of bacon.

Floating Gardens–A technique consisting of planting vegetation on soil on rafts floating in water; known also as chinampa cultivation in parts of Mexico.

Floating Plant–A nonanchored plant that floats freely in the water or on the surface; e.g., water hyacinth (***Eichhornia crassipes***) or common duckweed (***Lemna minor***).

Floating-leaved Plant–A rooted, herbaceous hydrophyte with some leaves floating on the water surface; e.g., white water lily (***Nymphaea odorata***), floating-leaved pondweed (***Potamogeton natans***). Plants such as yellow water lily (***Nuphar luteum***), which sometimes have leaves raised above the surface, are considered floating-leaved plants or emergents, depending on their growth habit at a particular site.

Flock–(1) Several birds or domestic mammals, such as sheep, which are tended as a unit. Also called herd, band. (2) Short fibers sheared from the face of cloth or produced in milling or finishing cloth or obtained by shredding rags to almost a powder. (3) In stock judging, one ram, of any age, and four ewes of varying ages as designated by the show.

Flood–(1) The overflow by water from natural or human causes into fields, farms, cities, etc. (2) To run water upon a field to the depth of a few inches to supply sufficient moisture for the growth of the crop, as rice, cranberries, etc., to control weeds, or to protect the crop from injury by frost.

Flood Irrigation–A system of irrigation consisting of adding water at the highest point in a field and allowing the water to cover the soil.

Flood Stage–The stage in which the level of the water rises above the tops of stream banks or dikes. Frequently, an arbitrary level established, beyond which the stream or river is considered to be at flood stage.

Floodplain–(1) That portion of a stream valley, adjacent to the channel, which is built of sediments during the present and geologic times and which is covered with water when the stream overflows its banks at flood stages. A river has only one floodplain (currently subject to flooding) and may have one or more terraces (historically subject to flooding). (2) The nearly level land situated on either side of a channel that is subject to overflow flooding. (3) The extent of a floodplain obviously fluctuates with the size of overbank stream flows. Thus, no simple, absolute floodplain commonly exists. As a consequence, floodplains are delineated in terms of some specified flood size (e.g., the fifty-year floodplain—the area that would be flooded by the largest stream flow that will, on the average, occur once within a fifty-year period). Such expected flood-return frequencies are estimated from historic records of stream flows. The largest, absolute floodplain that is ever likely to occur is sometimes referred to as the flood basin.

Floor, Forest–All dead vegetable matter on the soil surface including litter and unincorporated humus.

Flora–(1) The aggregate of plants that grow without cultivation in a given area within a stated period of time. (2) The bacteria in or on an animal, a plant, or the soil, usually referred to as bacterial flora. See Fauna.

Floral–(1) Referring to the structure and character of a flower. (2) Referring to the flora of a region. (3) Composed of, or characterized by, flowers.

Floral Nectary–The secretory glands of a flower that assist in pollination by producing the nectar that attracts the insects. The glands may be united into a complete ring near the base of the corolla or calyx, or may occur in special parts of the flower, as in the floral spurs of columbine, some orchids, etc.

Floret–An individual flower on a spike, such as a gladiolus floret.

Floribunda–Designating a plant that blooms abundantly, as the Japanese flowering crabapple and the floribunda rose.

Floricane–In the bramble fruits, the second year's growth that bears the flowers and fruits and then dies. Sometimes called second-year cane.

Floriculture–The cultivation of plants for their flowers.

Florida Gummosis–A disease of orange and grapefruit trees in Florida, United States, possibly caused by a virus; characterized by longitudinal cracks in the bark of the trunk and large limbs. A gum exudes from the cracks, runs down the bark, and hardens in a mass.

Florida Harvester Ant–***Pogonomyrmex badius***, family Formicidae; a troublesome, stinging pest that infests groves and gardens in the southeastern United States, feeding on seeds.

Florida Poisontree–***Metopium toxiferum (Rhus metopium)***, family Anacardiaceae; a shrub or small tree, which, on contact, causes severe dermatitis to susceptible persons. Also called poison wood, doctor gum, and coral sumac.

Florida Red Scale–***Chrysomphalus aonidum***, family Diaspididae; a scale insect that is harmful to citrus twigs, leaves, and fruits, to many other cultivated plants in the Gulf States, United States, and to greenhouse plants in cooler regions.

Florida Shade-grown Wrapper Leaf–A type of tobacco grown for its large leaves used for wrappers in cigar making.

Florist–A person who sells plants and cut flowers, often growing these plants and flowers himself/herself. Sales of accessories, such as fungicides, insecticides, seeds, fertilizers, pots, pot labels, etc., are often a part of the florist's service.

Floss–The lint of cotton and other plants, resembling fine wool, which consists of the longer hairs from the surface of the seed intermingled with short fuzz.

Flotation Sulfur–An extremely fine powdered sulfur, more active chemically than "flowers of sulfur"; produced by the oxidation of hydrogen sulfide (H_2S) recovered in aqueous alkali from coal gas fumes. See Flowers of Sulfur.

Flour–The fine-ground product obtained in the commercial milling of wheat, which consists essentially of the starch and gluten of the endosperm. It is used for baking bread, pastry, cakes, etc.

Flour Beetles–Beetles of the genus *Tribolium (T. confusum* and *T. castaneum)*, family Tenebrionidae, which infest stored grains and other food products.

Flour Corn–A type of corn in which nearly all the starch is soft, with only the shell being the more vitreous material. It was widely grown by the Indians in the United States as it could be easily made into corn meal in their primitive mills. The blue and variegated types are called squaw corn.

Flour Mite–Any of the several species of mites of the genus *Acarus (Tyroglyphus)*, family Acaridae, that infest flour and other cereal products. These mites may also infest sugar, dried fruits, cheese, etc. Also called cheese mite. See Grocer's Itch.

Flour Moths–Any of several types of small moths including the Angoumois grain moth, *Sitotroga cerealella*, family Gelechiidae. They lay their eggs on starchy grains, breakfast food, flour, and ripe grains of corn on the cob, etc. Endemic in Europe and America.

Flour Strength–The quality in flour that enables the baker to produce a loaf of bread of a particular volume and texture by the use of proper ingredients, together with proper mixing, fermentation, and baking. High-strength flour is used in making ordinary fermented bread. Low-strength flour is preferred in making crackers and pastry, as it lacks toughness and yields a flaky, brittle, or crumbly product.

Floury–Designating a fine-textured soil, consisting primarily of silt size which when dry is incoherent, smooth, and dustlike. See Soil Texture.

Flower–(1) The reproductive structure of a seed-bearing plant, consisting of the male and/or female organs that are surrounded by one or two series of outer coverings (calyx and corolla). The calyx, the outermost series, consists of sepals, generally greenish, which cover the unopened bud. The inner series, corolla, consists of petals which give the flower its characteristic color. These two enclose the essential reproductive system. The female organ, the pistil, consists of one or more carpels, and contains ovules, each of which produces the pollen grains (or microspores). On coming into contact with the pistil, the pollen grains send out a tube that grows down to an ovule and carries the enclosed male cell whose nucleus unites with the egg and fertilizes it. The fertilized egg enlarges and undergoes repeated divisions to form the embryo. Flowers may have both pistil and stamens in the same flower or in separate flowers. A perfect flower contains both corolla and calyx, although some flowers have only colored sepals. See Dioecious, Monoecious.

Flower Bud–A bud which, when mature and open, reveals a flower. On woody plants, the bud may be initiated in the preceding summer or fall and then, surrounded by tough bud scales, lie dormant until the next spring. In some bulbous plants, as some lilies, the bud is formed inside the subterranean bulb in the fall, but does not emerge form the ground until spring. In most of the other plants, the flower buds do not show until shortly before blossom time.

Flower Spot of Azalea–An infectious fungal disease characterized by small, rapidly enlarging spots that eventually ruin the whole flower. The perfect stage of this fungus develops the fallen, overwintered petals of the previous year's infected flowers and produces its spores (ascospores) in cup-shaped structures (apothecia) at the time the new blossoms open. The infected petals carry the imperfect stage of the fungus, producing a large number of asexual spores (conidia) which rapidly spread the disease. It is injurious to azaleas and rhododendrons in the southern United States.

Flower Thrips–*Frankliniella tritici*, family Thripidae; small insects that feed on portions of the flower in many cultivated plants, such as wheat, strawberry, cotton, and citrus, etc.

Flowering Bulb–(1) A subterranean bud; a short, enlarged stem, variously shaped, which emits roots below and bears a number of overlapped, fleshy or membranelike scales. Sometimes called tunicated bulb. Usually corms are miscalled bulbs, as are some tubers. See Bulb. (2) Any of several species of flowering plants, mainly of the families Liliaceae and Amaryllidaceae, which produce true bulbs; grown as ornamentals. In general practice, any plant producing a bulbous root, whether a true bulb, a corm, or a tuber, grown for its flowers, is called a flowering bulb.

Flowering Dogwood–*Cornus florida*, family Cornaceae; a small tree that sometimes attains a height of 40 feet, grown for its large white flowers in the early spring. (Occasionally a pink variety occurs wild.) Its scarlet, drupe fruit in the fall is eaten by bob-white, ruffed grouse, and wild turkey. fall foliage is a beautiful dark red. Native to the eastern United States. Also called American dogwood, boxwood, American cornelian tree.

Flowers of Sulfur–A very light, floury form of elemental sulfur produced by sublimation and not by crushing or grinding. It is used as a fungicide, especially against the powdery mildews and also in the control of certain insects or mites. Also called sublimed sulfur. See Flotation Sulfur.

Flue–(1) A vertical space, usually 6 inches or less in width, between two adjacent tiers of stock in a lumber pile. (2) A square or round ventilating passageway. (3) The passage in a chimney for conducting smoke, flame, or hot air. (4) The special passageways for heated air in a barn for curing tobacco.

Fluffy–(1) Describing a soil in which the aggregates of the surface are loose, light in weight, and fine; a desirable structure for seed germination and plant growth. (2) Describing a body defect of ice cream characterized by large air cells and open texture.

Fluid Drilling–A technique for planting germinated seeds without injuring the tiny shoots or roots by mixing previously germinated seed with a protective gel and pumping it into the soil behind the planter shoe through tubing, usually plastic, from a holding tank on the planter. Originally developed at Great Britain's National Vegetable Research Station, the technique was first used commercially in the United States to fluid-drill tomato seedlings.

Fluid Fertilizers–Liquid, slurry, or suspension fertilizers.

Flume–An inclined, open conduit or troughlike structure for the conduction of water and floating objects, as logs, sugarcane, etc. An aqueduct is a special-purpose flume.

Fluorescent Bacteria–Species of the genus *Pseudomonas*, family Bacteriaceae, which produce a fluorescent pigment.

Fluorides–Gaseous or solid compounds containing fluorine, emitted into the air from a number of industrial processes; fluorides are a major cause of vegetation and, indirectly, livestock toxicity.

Fluorine–A pale greenish, gaseous, acid-forming element, having the symbol F, small quantities of which are detrimental to plants and to animals since it is a cumulative poison. Occasionally tested as a fumigant of plants against insects, it is dangerous to handle without extreme care. Many of the compounds of fluorine are used for insecticides or rodenticides. In sufficiently dilute solutions in drinking water, some fluorine compounds reduce the occurrence of dental cavities, but care must be taken not to exceed certain limits or the teeth may be harmed by being pitted and brown. Minute amounts of fluorine are present in normal animal tissues, bones, teeth, and skin, but it has never been known to be essential to plant life. It is widely distributed in rocks, soils, and water, and is present in the rock phosphates that are mixed in feed rations. Soluble fluorides are toxic to germinating seeds, and chronic fluorosis may occur in people and animals. See Fluorosis.

Flush–(1) To irrigate a field with just enough water to soften the surface soil crust. (2) To increase the feed allowance to ewes or sows with a protein-rich supplement feed a short time before and during the breeding season. (3) A vigorous or abundant, sudden, new growth. (4) To send out vigorous growth on the twigs of a tree. Thus, in the tropical fruit tree, the mango, most of the twigs flush several times a year. (5) To empty, clean, or wash out any material with a quick, heavy supply of water. (6) To introduce and shortly afterwards withdraw an irrigating solution of mild antiseptic of medicinal value, as to flush the vagina or uterus.

Flush Cut–To prune a branch so that the cut is approximately even and parallel with the branch or trunk from which it was removed. The latest recommendation is to prune a branch on the outside of the swelling at the bole of the tree.

Fluted–Regularly marked by alternating ridges and grooves.

Fly–Any winged insect, such as a moth, bee, gnat, etc. Specifically a two-winged insect of the family Muscidae. Many flies are blood-sucking pests of people and animals, such as the mosquitoes, horse and deer flies, black flies, punkies or nosee-ums, and some sand flies. Some are vectors of diseases, such as the stable flies,etc. Some flies destroy other insects; some are parasites on plants, as the Hessian fly; others are valuable scavengers. Many flies, as the housefly, pass their larval stage in manure and garbage and upon attaining maturity carry with them the bacteria of filth, thus spreading diseases such as typhoid fever, etc.

Fly Net–A meshed covering made of strips of leather or cords of fabric and placed over an animal, usually a horse, to keep the flies away.

Fly Speck–(1) A common disease of the apple fruit, caused by the imperfect stage of the fungus *Leptothyrium pomi*, family Leptostromataceae, characterized by minute fruiting bodies of fungus on the fruit, which resemble the spots made by flies on defecation. (2) The small, dark specks that are dried fecal excrement of flies.

Fly-free Date–The date after which it is safe to plant wheat to avoid serious infestation by the Hessian fly. This date has been determined for each county and is available from the county extension agent.

Flyings–A trade term for the cured, lowermost leaves of the tobacco plant.

Foam Suppressant–Adjuvant for reducing both surface foam and trapped air; used in fluid fertilizers and liquid pesticides.

Foaming Agent–Surface-active substance that forms a fast-draining foam. This can provide maximum contact of spray with plant surface, insulate the surface, reduce the rate of evaporation, or mark sprayed surfaces to guide implements across a field.

Fodder–Feed for livestock, specifically the dry, cured stalks and leaves of corn and the sorghums. In the case of corn, the ears may be removed from the stalk leaving the stover. See Forage, Roughage.

Fog Drip–The moisture from fog which has collected on the leaves of trees, shrubs, etc., and which drips to the ground. In some regions the amount may be equivalent to a light rain, and thus is a considerable factor in supplying moisture to plants. Also called wet fog.

Foliaceous–Leafy or leaflike.

Foliage–The leaves, collectively, of a plant; leafage.

Foliage Burn–A browning and dying of leaves, caused by deficiency of water, injury from harmful chemicals from spray materials, or the poisonous secretions from the mouth parts of sucking insects such as the insects causing tip burn.

Foliage Plant–A plant grown for the color and/or shape of its foliage; e.g., coleus.

Foliar–Refers to leaves.

Foliar Diagnosis (Foliar Analysis)–Estimation of the plant-nutrient status of a plant or the plant-nutrient requirements of the soil for producing a crop through chemical analyses or color manifestations of plant leaves or by both methods together.

Foliar Fertilization–Application of soluble fertilizer in the form of spray on the foliage of plants. See Foliar Spray.

Foliar Nematode–*Aphelenchoides olesistus*, family Anguillulinidae, and related species, known as nematodes or eelworms. These almost microscopic worms gain entry to the leaves of many greenhouse plants through the stomata (breathing ores), or they enter the roots from infested soil and work their way up through the stem into the leaves. Necrotic (dead) areas are produced in the affected part that sometimes involves the whole leaf.

Foliar Spray–(1) Any spray applied to plants when leaves are present, as contrasted to dormant spray. (2) A spray that supplies fertilizer to plants by spraying liquid fertilizer directly onto the leaves where it is absorbed and utilized by the plant. See Foliar Fertilization.

Foliation–The process of producing foliage by a plant; the coming into leaf.

Folic Acid–A vitamin found in the leaves of leguminous and other plants, in yeast, liver meal, and wheat. Folic acid is needed in hemoglobin formation and for growth. Also called pteroylglutamic acid.

Foliocellosis–A mottling of leaves of citrus in which the areas between the lateral nerves are paler; probably caused by zinc deficiency.

Folium–(Latin) Leaf.

Follicle–(1) A dry, single-carpel fruit, opening along one side for seed dispersal. (2) A small anatomical cavity; particularly, a small blisterlike development on the surface of the ovary that contains the developing ovum. (3) A small sac, gland, or pit for secretion or excretion. The hairs of an animal grow out of pits called follicles. (4) A one-celled, monocarpellary, dry seed vessel or fruit that splits on the ventral edge. (5) The growth that appears on the surface of the ovary late in the estrous cycle and that contains the developing ovum.

Food–Anything which when taken into the body, nourishes the tissues and supplies body heat. Also known as aliment and nutriment.

Foot Rot of Citrus–A disease in which the basal portion of the trunk and the roots of trees are infected with species of *Phytophthora*, especially *P. citrophthora*, family Pythiaceae. The same fungus may attack the fruit and cause brown rot gummosis.

Foot Rot of Plants–A fungus-induced disease of the roots or basal portions of the stems of many different plants. Usually various tissues of the roots and the cambium at the base of the stem are destroyed, resulting in the ultimate death of the whole plant. If not woody, the plant may topple over as the roots and stem bases die and lose their turgor, a phenomenon called damping off. Several types of fungi are involved in the disease; e.g., various species of the genera *Phytophthora* and *Pythium*, family Pythiaceae, in woody plants, and possibly the fungus *Cercosporella herpotrichoides*, family Moniliaceae.

Foots–Settlings or sediments, as in oil or molasses.

Forage–(1) That portion of the feed for animals that is secured largely from the leaves and stalks of plants, such as the grasses and legumes used as hays. It may either be for grazing as green or standing dry herbage or be cut and fed green or preserved as dry hay. See Fodder, Roughage. (2) To search for, spread out or seek, for food.

Forage Crops–Those plants or parts of plants that are used for feed before maturing or developing seeds (field corps). The most common forage crops are pasture grasses and legumes.

Forage Crops Blind Seed Disease–A fungal disease of perennial ryegrass cuaed by *Phialea temulenta*, family Helotiaceae, whose small, stalked apothecia grow out of the diseased seeds lying on or shallowly buried in the soil and discharge their ascospores into the air, thus infecting the flower of the ryegrass and sterilizing the developed seeds.

Forage Feeds–(1) Bulky type feeds composed largely of pasture grasses, hays, silage, etc. (2) A mixture of ground or processed feeds that is composed largely of forages.

Forage Legumes–Any of the legume plants that are grown or used largely as forage for livestock, such as alfalfa, clover, etc.

Forage Mixture–Two or more species grown together for forage production. Usually a mixture of legumes and grasses.

Forage Poisoning–(1) Any poisoning of people or animals from eating food or feed contaminated by the presence of some organism that was not destroyed in the usual processing. Such an organism may produce toxins that are fatal to the animal consuming even small quantities of the forage. the most serious cases are those in which the contaminating organism is ***Clostridium botulinum*** that can withstand repeated heatings at the temperatures ordinarily used in the home canning of vegetables. (2) Poisoning of animals grazing or eating forage containing a plant chemical such as the glucoside of hydrocyanic acid in Sudangrass or sorghum, or by forages improperly cured, such as sweetclover, where the breakdown of the alkaloid coumarin results in rupture of the small blood capillaries in the animal. See Selenium Poisoning.

Forb–A palatable, broad-leaved, flowering herbaceous plant whose stem does not become woody.

Forced Burning–In forestry, a technique in broadcast or spot burning: the selection of a time when the lower layers of duff and flashy fuels are damp but which can be made to burn by setting a large number of fires in heavier concentrations of debris over the entire area. Sometimes slash disposal is accomplished by progressive burning.

Forcing–Bringing a plant to a specific stage in its development or to maturity earlier in the season than normal, by growing the plant in a greenhouse, under artificial light, etc.

Foreign Matter–Any material, substance, etc., which is unnatural to, or not commonly developed in, a product.

Foreign Odor–Any odor that is not natural to a product.

Forest–A plant association predominantly of trees and other woody vegetation that occupies an extensive area of land. See Forest Land, Woodlot.

Forest Cover–All trees and associated plants that cover the ground in a forest.

Forest Entomology–The branch of zoology that deals with insects in their relation to forests and forest products.

Forest Harvesting–United States Forest Service usage of group, patch, strip, and stand cutting. These are essentially size designations of the same logging approach—commonly known as clearcutting. All require the removal of all merchantable trees except a few that may sometimes be carried over where needed for nesting sites and animal food caches. (a) Group cuttings are on areas which are smaller than the size normally recognized in type and condition class mapping. These should not exceed 10 acres, nor usually be less than one-fifth acre. (b) Patch cuttings are on areas that are of a size generally mapped for type and condition and for control, but which do not include the entire stand of which they are a part. (c) Strip cuttings are on areas running through a stand and usually of a width equal to one to two times the general stand height. (d) Stand cuttings are on areas large enough to be practical for management.

Forest Industry Lands–Lands owned by companies or individuals operating wood-using plants.

Forest Influences–The effects of forests upon water, soil, climate, and health.

Forest Land–(1) Land at least 10 percent occupied by forest trees of any size, or formerly having had such tree cover and not currently developed for nonforest use. (2) Lands that are at least 10 percent stocked by trees capable of producing timber or other wood products or that exert an influence on the climate or water regime.

Forest Litter–Leaves, dead branches, bits of bark, rotting wood, etc., found on the forest floor.

Forest Management–The application of business methods and technical principles to the operator of a forest property. It involves the computation of income from forest lands; the establishment of cutting cycles; the conservation of cover, land and water; and the formulation and conduct of long-range plans of operations.

Forest Mensuration–A branch of forestry concerned with tree measurements, log scaling, and sawn timber volume determination.

Forest Pathology–The science that pertains to diseases of forest trees or stands and to the deterioration of forest products by organisms.

Forest Tent Caterpillar–*Malacosoma disstria*, family Lasiocampidae; a caterpillar that feeds voraciously on leaves of many shade and forest trees and also on field and truck crops.

Forestation–(1) Establishing a forest on an area, regardless of whether a forest at one time occupied the tract. (2) A word which means both afforestation (not previously forested) and reforestation (previously forested).

Forestry–The sciences, arts, and business practices of crating, conserving, and managing natural resources on lands designated as forests.

Formaldehyde–A gas, readily soluble in water, which is generally used in aqueous solutions as an insecticide, a fungicide, a soil fumigant, a tool disinfectant, a storage house disinfectant, in disease treatment, and as a disinfectant for seeds, tubers, and bulbs.

Formamidine Insecticide–A class of insecticides that is used against eggs and mites.

Formation–(1) Any igneous, sedimentary, or metamorphic rock that is represented as a unit in geological mapping. (2) A major division of the climax vegetation of a continent.

Fortified–(1) Designating a product to which has been added amounts of a vitamin, as vitamin A or vitamin D. (2) A wine to which additional alcohol has been added.

Fossil Soil–A soil developed upon an old land surface and later covered by younger formations.

Fossil Water–Water stored in the soil during geologic times. It is not renewed by precipitation.

Foundation Planting–Any plants used decoratively to conceal the juncture of a building with the ground.

Foundation Seed–Seed that is the progeny of breeder or foundation seed produced under the control of the originator or sponsoring plant-breeding institution, or person, or designee thereof. As applied to certified seed, foundation seed is a class of certified seed that is produced under procedures established by the certifying agency for the purpose of maintaining genetic purity and identity.

Four-way Hybrid–The hybrid that results from mating two single crosses. Also called double cross, See Single Cross.

Fourwing Poisonvetch–*Astragalus tetrapterus*, family Leguminosae; a plant in which the foliage, green or dried, is poisonous. Symptoms of poisoning are dullness, irregularity of gait, lack of appetite, dragging of the feet, a solitary habit, loss of flesh, and a shaggy coat. As the animal ceases to eat, it dies. Found in Nevada and southern Utah, United States. Also erroneously called locoweed.

Fourwing Saltbush–*Atriplex canescens*, family Chaneopodiaceae; an alkali-tolerant, evergreen shrub extensively browsed by cattle, sheep, and goats on arid lands of western ranges, United States. Also called cenzio, chamiza.

Fox Grape–*Vitis labrusca*, family Vitaceae; a hardy, woody vine that produces an edible, thick-skinned, sweet, musky berry. It is the origin of many American grape varieties. Also called northern fox grape.

Foxglove–(1) Any herb of the genus **Digitalis**, family Scrophulariaceae. (2) **Digitalis purpurea**, a biennial herb grown for its purple, white, or yellow flowers, particularly in Michigan. Also grown commercially as a source of the drug digitalis. Native to Europe and central Asia. Also called fairy finger, fairy cap, finger flower, finger root, fairy bell, fairies' petticoat, common foxglove, fairy glove, dog's finger, deadman's bells. See Foxglove Poisoning. (3) A common name for several other plants.

Foxglove Poisoning–A poisoning of animals feeding on foxglove, characterized by gastrointestinal irritation, dullness, loss of appetite, nausea, contracted pupils, and slow but steady pulse.

Foxtail Barley–*Hordeum jubatum*, family Gramineae; a biennial or perennial grass that is a weed pest in meadows, pastures, roadsides, and waste places in the western United States. It is harmful to animals feeding on it when the awns get under the tongue or gums or into the eyes. Also called squirreltail grass, wild barley, skunk barley.

Foxtail, Creeping Meadow–*Alopecurus arundinaceus*; a long-lived cool-season, sod-forming grass; possesses strong rhizomes, forming a dense sod. It is tolerant of strongly sodic soil conditions, and responds to high fertility.

Foxtail, Meadow–*Alopecurus pratensis*; a grass that is a long-lived, weak sod former, especially adapted to wet soils and to land subject to flooding in winter or early spring. It is suited for pasture and hay and is tolerant of frost and prolonged snow cover. Also tolerant of strongly sodic soil conditions. Responds to high fertility. The meadow foxtails are difficult to seed. They require a carrier for the seed if sown through a standard drill.

Foxtail Millet–*Setaria italica*, family Gramineae; an annual grass grown in the United States for its grain to be fed to hogs, poultry, and birds, etc., but grown in the Orient for human food.

Fragipan–A loamy, brittle subsurface soil horizon low in porosity, low in organic matter, and low or moderate in clay but high in silt and/or very fine sand. A fragipan appears cemented and restricts roots. When dry, it is hard or very hard and has a higher bulk density than the horizon or horizons above. When moist, it tends to rupture suddenly under pressure rather than to deform slowly. See Induration, Claypan.

Fragrance–Sweetness or agreeableness of smell; a pleasant odor.

Frass–Excrement of larvae and/or refuse left by boring insects.

Free Water–That water the soil is unable to retain against the force of gravity. Also called gravitational water, excess water, drainage water, water of percolation.

Freestone–(1) Any of several varieties of peaches and nectarines characterized by the easy separation of the fleshy part of the fruit from the stone, i.e., the mesocarp separates readily from the endocarp. (2) Designating the stone of such a fruit.

Freezer Burn–(1) An undesirable and unattractive condition frequently found with dressed poultry that is kept or stored in quick or deep freezers; caused by an improper control of temperatures and humidity within the freezer. (2) Small, white dehydrated areas that occur on improperly wrapped frozen foods.

Freezing Injury–(1) Any damage to living plant or animal tissue caused by low temperature. (2) Damage to fruit and vegetable tissues that are exposed to temperatures at or below their freezing points, usually occurring after harvest.

French Endive–A variety of chicory grown for a winter salad whose roots are dug from the ground in the autumn and stored as a root crop. As they are needed, they are placed in pots and forced. Short, crisp, flavorful leaves develop that are served in salads. Also called witloof.

Fresh–(1) Designating a cow that has recently dropped a calf. (2) Designating very recently harvested or gathered food products. (3) Designating an egg of good quality. (4) Designating sweet water; i.e., not salty water.

Friable–(soil) Generally refers to a soil that crumbles when handled. A loam soil with physical properties that provide good aeration and drainage, easily tilled. Friable condition is improved or maintained by annual applications of organic matter.

Frijole–Any cultivated bean of the genus *Phaseolus*, but especially a black-seeded variety of the kidney bean that is grown for its edible, dried seed (Mexico and the southwestern United States).

Frill–A series of overlapping ax cuts, made through the bark and into the sapwood, that encircles a tree trunk to kill the tree. Also called ring, girdle. See Girdle.

Frill Treatment–The placement of a herbicide into a series of overlapping ax cuts made through the bark in a ring around the trunk of a tree.

Fringed Sagebrush–*Artemisia frigida*, family Compositae; a perennial herb grown for its silvery-white, aromatic foliage. It is found wild over a large part of the western United States, where it is browsed by sheep, goats, and cattle. Its forage rating is good for goats and sheep; fair for cattle. Also called mountain sage, wormwood sage, wild sage, mountain fringe, fringed wormwood.

Fringewater–Mobile water, suspended in the capillary fringe above the water table and completely filling the smaller interstices, which moves with the fluctuations of the water table.

Frit–Fritted micronutrients; also called frits. Frits are special glasses that provide controlled release with as high as 50 percent of micronutrient elements incorporated. The micronutrients are available to plants under conditions that render water-soluble forms unavailable. Potassium, phosphorus, calcium, and magnesium are also fritted and can be included in these glasses.

Frog–(1) That part which holds turning plow bottom parts together; an irregularly shaped piece of metal to which the share, landslide, and moldboard are attached. (2) The triangular, horny pad, located on the posterior-ventral part of the hooves of horses, mules, etc.

Frogeye–(1) More-or-less rounded leaf spots of various plants in which the necrotic, central portion is darker than the outer portion; caused by any of several different fungi. Frogeye on apple leaves is due to infection by *Physalospora obtussa*, family Pleosporaceae, which, if it attacks the fruit, causes black rot. On tobacco leaves frogeye spots are due to *Cercospora nicotianae*, family Dematiaceae. In periods of variable moisture and temperature, the fungus produces concentric or zonate lesion resulting in frogeye or target board.

Frond–(1) The leaves of true ferns (order Filicales); less often, the large leaves of palm trees. (2) Sometimes, the rounded, flattened stems of duckweed.

Frost–(1) A condition of the weather in which the air temperature at plant level falls below 32°F (0°C) due to a local cooling of the air during a calm, cloudless night, causing the ground and other objects to be coated by frozen dew. (2) A freeze. (3) Of, or pertaining to, a frozen condition of soil, as in the expression, "Frost is still in the ground." See Frost Line.

Frost Action–The weathering process caused by repeated cycles of freezing and thawing.

Frost Control–Various measures taken to prevent frost damage to more valuable crops, such as citrus, or certain other orchard or truck crops. Preventive measures are: the use of any of several types of heaters; irrigating or sprinkling water on the plants; using propeller blades to mix the upper warm air with the lower cold air; use of coverings of paper, cloth, or plastics.

Frost Crack–(1) An opening in the soil produced by the development of an ice wedge. (2) A longitudinal crack or splits in the trunk or large limbs of a tree, extending from the bark into the sapwood, which results from a sudden extreme drop in temperature during winter. Also called trunk splitting.

Frost Creep—Soil creep resulting from frost action.

Frost Heaving—The pushing up of the soil surface by the growth of ice crystals. In north temperate and arctic regions, this is a serious source of winter killing of newly planted trees and shrubs and new seedings of crop plants. Also a common cause of road and sidewalk breakup.

Frost Line—The maximum depth to which soil is frozen in the winter; this may be an inch or two, or may be several feet. The soil thaws completely in temperate regions but may remain permanently frozen a short distance below the surface in the Arctic and Antarctic regions. See Permafrost.

Frost Pocket—A depression or hollow in either a hilly terrain or a pitted plain into which cold night air descends, creating a frost hazard for sensitive crops. See Cold Pocket.

Frost Splitting—Breaking of rock by water freezing in its cracks.

Frost-free—(1) Designating that period or season of the year between the last killing frost of spring and the first killing frost of autumn. See Killing Frost. (2) Designating a region in which frosts or temperatures below 32°F (0°C) are not experienced.

Frost-proof—Designating a variety, plant species, or a product that is not harmed, injured, or impaired by freezing conditions.

Fructification—Production of spores by fungi, algae, and other plants on the organ producing the spores or seeds. See Fruiting Body.

Fructose—Fruit sugar ($C_8H_{12}O_6$) found in all sweet fruits, corn, honey, and in Jerusalem artichoke tubers.

Fruit—(1) Botanically, the matured ovary of a flower and its contents including any external part that is an integral portion of it. (2) In a popular sense, the fleshy, ripened ovary of a woody plant, tree, shrub, or vine, used as a cooked or raw food; it is not always completely and satisfactorily distinguished from a vegetable. The latter may also include edible leaves, roots, and tubers. A fruit may be considered a dessert and not the principal part of a meal as a vegetable is so considered (such a meaning has accepted usage and has also been confirmed by a court decision). See Vegetable. (3) To bear or produce fruit in any of the senses. (4) A mature ovary, either plant or animal.

Fruit Bud—See Blossom Bud.

Fruit Leather—A food that is prepared by drying fruit, pressing the dried fruit and rolling it out in a thin sheet. It is then stored in jars or bags for later use.

Fruit Spur—A short, stout branch on which the flowers and fruit of a fruit tree are borne.

Fruiting Body—A complex fungus structure containing or bearing spores. Important types are apothecia, perithecia, conidiophores, coremia, sporangia, pycnia, aecia, pycnidia, acervuli, and sporodochia. At maturity the spores are disseminated by the fruiting body.

Frutescent—Shrublike in appearance or habit.

Fruticetum—A growing collection of shrubs similar to an arboretum. See Arboretum.

Fruticose—Designating a woody plant that has the characteristics of a shrub.

Fuel Loading—In forestry, the amount of fuel (burnable materials) expressed as weight of fuel per unit area; generally expressed in tons per acre.

Fuel Management—The practice of planning and executing the treatment or control of living and dead vegetative material, primarily for wildfire hazard reduction.

Fuelbreak—A wide strip of range or forest where the fuel is reduced or altered to act as a fire-control measure.

Fugacious—Soon perishing, as the flower of a plant.

Full Coverage Spray—The application of a spray over the top of a crop that covers all parts of the plants.

Full-cell Process—Any process for impregnating wood with preservatives or chemicals in which a vacuum is created to remove air from the wood before admitting the preservative.

Full-Slip—In harvesting melons, the easy separation of the fruit from the vine in which a cleavage layer develops around the peduncle at the base of the fruit so that as the fruit is removed from the vine; the fruit slips off easily and leaves a large, clean scar.

Fuller's Earth—A massive or shalelike clay deposit. The clay mineral is mainly montmorillonite. It is employed in degreasing wool and in clarifying mineral and vegetable oils.

Fumigant—A substance or mixture of substances that produce gas, vapor, fume, or smoke intended to destroy insect and other pests.

Fumigate—To destroy pathogens, insects, etc., by the use of certain poisonous liquids or solids that form vapors. See Fumigant.

Fungal Spray—An herbicide consisting of fungal spores in a liquid solution sprayed on weeds in crops as a means of biological control of the weeds. For reasons not adequately understood, the fungal spores of different fungi are specific for specific hosts, making it possible to control selected plants without damaging other nearby crop plants.

Fungi—Plantlike organisms that have no chlorophyll; they get their nourishment from living or decaying organic matter. Plural of fungus.

Fungicide—A pesticide chemical used to control plant diseases caused by fungi such as molds and mildew. See Pesticide.

Fungistat—An agent or chemical material that prevents the growth and reproduction of, but does not kill, fungi.

Fungous—Pertaining to a fungus, as a fungous disease.

Fungus—A lower order of plant organisms, excluding bacteria, which contains no chlorophyll, has no vascular system, and is not differentiated into roots, stems, or leaves. They are classified in the plant kingdom division Thallophyta, and vary in size from single-celled forms to the huge puffballs of the meadows. Fungi are familiar as molds, rusts, smuts, rots, and mushrooms. For the large part they reproduce prolifically by means of single or multicelled spores of various longevities, which are disseminated by air or water. They are either saprophytic or parasitic and many are considered useful in breaking down dead vegetation and organic matter into humus and as agents of fermentation, as in yeasts; others are also destructive in rotting structural timbers, posts, cloth, leathers, etc. The parasitic forms cause destructive plant diseases; a few are human and animal parasites.

Furnace Acid–See Phosphoric Acid.

Furrow–(1) The opening left in the soil after the furrow slice has been turned by the turning plow. (2) To make a furrow with a plow.

Furrow Drain–To drain surface water from a field by means of open furrows made with a turning plow.

Furrow Irrigation–A method of irrigating in which water is run in small ditches, furrows, or corrugations, usually spaced close enough together for lateral penetration between them.

Furrow Wheel–A tractor or implement wheel that runs in a furrow from a previous implement pass. This helps the operator guide the implement in the proper direction and distance from the previous pass.

Fusarium–A genus of fungi of the family Tuberculariaceae, of which many species have been studied and named. Most of these are saprophytes, but many are important plant parasites. The hyphae of some parasitic species of Fusarium may invade the water-conducting tissues of plants resulting in wilting.

Gall–An abnormal growth of plant tissues induced by the presence and stimulus of an animal or another plant.

Gallates–Salts and esters of gallic acid, found in many plants. Propyl, octyl, and dodecyl gallates are legally permitted antioxidants.

Gallery–A passage or burrow, excavated by an insect under bark or in wood for feeding or egg-laying purposes.

Gamete–A "sex cell," capable of uniting with another gamete to produce a cell (fertilized egg, or zygote) that in turn is capable of developing into a new individual.

Gametogenesis–The process in plants or animals, male or female, involving the production of gametes; ovigenesis or spermatogenesis.

Gametophyte–That generation or stage in the life history of a plant that produces gametes; in ferns, it is a minute thalluslike body bearing archegonia and antheridia, which produce female and male gametes respectively; in angiosperms, it is the pollen tube, which develops from a pollen grain and produces male gametes, and the embryo sac, which develops within the ovule and produces a female gamete. See Sporophyte.

Garbanzo–Gram chickpea, *Clicer arietinum*, family Leguminosae.

Garden–(1) A plot of ground, usually less than an acre, devoted to the growing of vegetables, flowers, herbs, fruits, etc., which is often adjacent to the home. See Botanic Garden. (2) The plants in a garden collectively. (3) To cultivate plants in a garden.

Garden Pea–*Pisum sativum*, family Leguminosae; an annual herb grown for its edible seed; it is a very important and popular vegetable. Native to Eurasia. Also called English pea. See Field Pea; Sugarpod Garden Pea.

Garlic–*Allium sativum*, family Liliaceae; a perennial herb grown for its strong, pungent, distinctive-flavored bulb used in seasoning. Native to Eurasia.

Garnish–A food used to enhance or decorate the dish, such as sprigs of watercress or parsley; radishes, carrots, and vegetables; lemon rind, croutons, nutmeats, or chocolate curls.

Gelatin–A water-soluble protein prepared from collagen by boiling with water.

Gelatinization–The rupture of starch granules by temperature that forms a gel of soluble starch and dextrins.

Gelatinous Matrix–Jellylike intercellular substance of a tissue; a semisolid material surrounding the cell wall of some algae.

Geld–(1) Designating an animal that is sterile. (2) To render sterile, as in castration.

Gene–The simplest unit of inheritance. Physically, each gene is apparently a nucleic acid with a unique structure. It influences certain traits. Sometimes called a trait determiner.

Gene Pool–The genetic base available to animal breeders for stock improvement.

Gene Splicing–The technique of inserting new genetic information in a plasmid. See Plasmid.

Gene Transfer–The process of moving a gene from one organism to another. Biotechnology methods permit the identification, isolation, and transfer of individual genes as a molecule of DNA. These methods make it possible to transfer genes between organisms that would not normally be able to exchange them. See DNA, Gene.

Genera–The plural of genus.

General Use Pesticide–A pesticide that can be purchased and used without obtaining a permit. It is considered safe for general public use. See Restricted Use Pesticide.

Generation–The group of individuals of a given species that have been reproduced at approximately the same time; the group of individuals of the same genealogical rank.

Generation Interval–The period of time between the birth of one generation and the birth of the next.

Generic–A word used to describe a general class of products, such as meats, vegetables, or grains; also refers to an unadvertised brand.

Genesis–(1) Origin, or evolutionary development, as of a soil, plant, or animal. (2) A combining form, to indicate manner or kind of origin, as parthenogenesis, biogenesis, etc.

Genetic Drift–The gradual change in a plant or animal species because of rearrangement of the genes due to the environment or unknown causes.

Genetic Engineering–Alteration of the genetic components of organisms by human intervention. Also known as biogenetics.

Genetics–(1) The science that deals with the laws and processes of inheritance in plants and animals. (2) The study of the ancestry of some special organism or variety of plant or animal. Also called breeding.

Genome–A complete set of chromosomes (hence of genes) inherited as a unit from one parent.

Genotoxicity–The quality of being damaging to genetic material.

Genotype–The genetic constitution (gene makeup), expressed and latent, of an organism. Individuals of the same genotype breed alike. See Phenotype.

Gently Sloping–In soil judging, a slope of 1 to 3 feet rise or fall in 100 feet (1 to 3 percent).

Genus–A group of species of plants or animals believed to have descended from a common direct ancestor that are similar enough to constitute a useful unit at this level of taxonomy.

Geoponics–The tillage of the earth. See Hydroponics.

Geotropism–A growth of shoots and roots of plants in response to the stimulus of gravity that is positive when the growth curvature is toward the center of the earth (as plant roots) or negative when the direction of growth is opposite to the pull of gravity (as plant shoots).

Germ–(1) In reference to animal disease, a small organism, microbe or bacterium that can cause disease. (2) The embryo of a seed.

Germ Cell–A cell capable of reproduction or of sharing in the reproduction of an organism, which may divide to produce new cells in the same organism, as contrasted with the somatic or body cells.

Germ Plasm–Term for the reproductive and hereditary substance of individuals that is passed on from the germ cell in which an individual originates in direct continuity to the germ cells of succeeding generations. By it, new individuals are produced and hereditary characteristics are transmitted.

Germ Tube–(1) The hypha produced by a germinated fungus spore. (2) The tube produced by a germinating pollen grain by means of which the sperm nuclei are brought into contact with the egg and endosperm nuclei.

German Knotweed–*Scleranthus annuss*; a weed closely related to the chickweed that is a pest among winter-growing plants.

Germicide–Any agent that kills germs.

Germination–Sprouting of a seed, and beginning of plant growth.

Germination Test–Any of several different types of tests conducted to determine the percentage of viable seed in a particular lot. State Departments of Agriculture set the standard for germination percentages for specified seeds.

Germinative Energy–The relative strength of sprouts and roots at the time of germination. One lot of seeds may exhibit strong, healthy sprouts, whereas another of the same variety may have weak ones.

Germinator–Any of several devices used to test seed for germination prior to planting.

Gherkin–Any small cucumber used for pickling.

Giant Cane–*Arundinaria gigantea*, family Gramineae; a tall, bamboolike grass with stems growing 12 to 25 feet high and leaves nearly 12 inches long with rough and cutting edges. It is the chief plant of the canebreaks form Virginia southward in the United States. Native to North America.

Giant Horsetail–*Equisetum telmateia*, family Equisetaceae; a poisonous, perennial herb found in wet meadows in the Pacific Northwest. Native to North America. Also called ivory horsetail. See Quisetum Poisoning.

Giant Ragweed–*Ambrosia trifida*; a common weed in grain crops. While the seed is an important food for birds, it contaminates grain.

Giant Sequoia–*Sequoia gigantea* (Syn.: *Sequoiadendron giganteum*), family Pinaceae; perhaps the world's largest coniferous tree. Its growth is limited to small areas. Native to California, United States. Also called big tree. See Redwood.

Giant Wildrye–*Elymus condensatus*, family Gramineae; a perennial grass growing to a height of 12 feet, which is grazed closely by cattle and horses but only to a limited extent by sheep. Its forage value is fair. Found from Manitoba to British Columbia, Canada, and south to New Mexico, United States.

Gibberellins–A group of substances originally isolated from *Gibberella fujikuroi*, an organism that causes elongation of rice shoots, and later from *Fusarium moniliforme*. Three chemically different substances have been isolated. Certain plants treated with these compounds increase in height much greater than normally; some treated biennial plants produce seed in one season.

Gigantism–(1) The production of luxuriant vegetative growth that is usually accompanied by a delay of flowering or fruiting. Also called gone-to-stalk, gone-to-weed. (2) In animals, abnormal overgrowth of a part or all of the body. Also called giantism.

Gin–(1) The entire machine processes, considered as a unit, which remove cotton lint from the cottonseed and bales lint cotton. (2) The grounds, buildings, management, etc., of a gin. Also called cotton gin. (3) The alcoholic beverage obtained by distilling pure spirits over various flavoring materials, principally juniper berries, but also coriander seed, angelica root, and other plant products.

Gin Bale–A bale of compressed lint cotton as it comes from a gin, weighing about 500 pounds.

Ginger–(1) *Zingiber officinale*, family Zingiberaceae; a perennial herb grown in India, East Indies, and China, as a source of the ginger of commerce. In the United States it is grown only in very warm sections as an outside ornamental for its yellowish-green flowers and as a pot plant. Native to southeastern Asia. (2) The dried or peeled rhizome of ginger that is grown in Jamaica, China, Malaysia, West Coast of Africa, and South America. The roots are dried after boiling in water for about 10 minutes to prevent them from sprouting. The flavor of ginger is due to a volatile oil, gingerol.

Ginkgo–*Ginkgo biloba*, family Ginkgoaceae; a deciduous tree with fan-shaped leaves, the female of which produces a fleshy, unpleasant-smelling fruit. The male is grown as a street tree. Native to China. Also called maidenhair tree.

Girdle–A wound encircling a living plant that involves the absence of tissues as deep or deeper than the cambium layer. The encircling cuts are deep enough to check the flow of elaborated foods or to cause the death of the plant above the point of the girdle. A simple cut by people or animals, or even the removal of a narrow strip of bark during dormancy or very early in the season will check the rate of growth above that point; if done after the first flush of growth, it will result in an accumulation of elaborated foods above the girdle. Wide girdling may cause death of the plant above that point.

Girdling Root–A large root that grows so close to the main trunk or another large root of a plant that it chokes or constricts the other roots.

Girth–(1) The circumference of the body of an animal behind the shoulders. (2) A band or strip of heavy leather or webbing that encircles a pack animal's body; used to fasten a saddle or pack on its back. Also called cinch. (3) The circumference of a tree.

Glacial Drift–Coarse and fine mineral deposits left as a result of glaciation.

Glacial Outwash–Sandy material carried by a stream of meltwater from a glacier and deposited in the form of plains, deltas, and valley trains.

Glacial Soil–Soil developed from materials transported and deposited by glacial action; e.g., till plains, moraines, drumlins, kames, and eskers.

Glacial Till–The unconsolidated, heterogeneous mass of clay, sand, pebbles, and boulders deposited by receding ice sheets.

Glade–(1) A grassy tract in a forest due to shallow soil. (2) An open wet tract, generally covered with grasses and sedges.

Gland–(1) In animals, an organ that secretes substances for the body's use or that excretes waste matter. (2) In plants, any special secreting organ.

Glandular–A leaf surface with small glands that secrete oil, tar, or resin.

Glasshouse–See Greenhouse.

Glauconite–A green, natural sedimentary mineral; hydrated aluminum silicate of iron, sodium, magnesium, and potassium, which has been used to a small extent as a very slow-acting fertilizer chiefly for its potassium and magnesium. Found abundantly in greensand beds. Also called green earth. See Greensand.

Glaucous–Covered with a bloom (a fine whitish, grayish, or pale bluish powder), which is often waxy in nature and easily rubbed off.

Glei–(gley) A soil horizon in which the material is bluish-gray, sticky, compact clay with massive structure. It has developed under the influence of continuously excessive moisture over hundreds of years.

Globose–Spherical, globular.

Globular–Having a round or spherical shape.

Globule–A collection of several molecules of fat that takes on a spherelike appearance, and is insoluble in water.

Glomerata–Dense, compact.

Glucose–A common monosaccharide sugar that serves as the building block for many complex carbohydrates; blood sugar.

Glume–A small shaftlike bract, especially in the Gramineae and related plants; in particular, an empty glume. An empty glume is one of two sterile bracts at the base of a grass spikelet, which are usually referred to merely as the glumes. A fertile or flowering glume is another term for a lemma. A sterile flowering glume is a lemma whose flower is staminate or obsolete.

Gluten–The protein-rich product derived from cereal grains. Gluten is spoken of loosely as the protein constituent obtained in separating starch from corn in the wet-milling process. For wheat, the term is more specifically applied to the viscous and semi-elastic substance that gives adhesiveness and the rising quality to bread dough.

Glutinous–Sticky.

Glycolytic–Pertaining to the chemical breakdown of sugars to lactic acid.

Glycophyte–A plant that does not grow well in a salty soil solution when the osmotic pressure of the solution rises above two atmospheres.

Glycosides–Physiologically active substances found in plants; when treated with acids or enzymes, they yield glucose or other sugars as well as nonsugar components.

Gnarled–Pertaining to trees or shrubs with a twisted, bent, or distorted appearance.

Golden Bee–A honeybee that was developed in the United States, probably from Cyprian or Syrian and Italian races.

Golden Groundsel–*Senecio aureus*, family Compositae; a hardy, perennial herb grown in flower borders for its clusters of yellow flowers. It is found wild in wet meadows of the northwestern United States, and it may be poisonous to livestock ingesting large amounts. Also called golden ragwort.

Golden Honey Plant–*Actinomeris squarrosa* (*A. alternifolia*) family Compositae; a hardy, perennial herb sometimes grown in the wild garden for its yellow flowers. It is a very good bee plant. Native to North America.

Golden-eye Lacewing–*Chrysopa oculata*, family Chrysopidae; an insect whose larvae are predators to aphids, mealybugs, cottony cushion scale, and sometimes to thrips and mites.

Golden-nematode Disease of Irish Potato–A pernicious disease prevalent in Europe for many years; caused by the golden nematode, *Heterodera rostochiensis*, family Heteroderidae. Now established in a limited area in the United States, it is characterized by delay in sprouting, stunting, unthriftiness, and the production of small tubers. Also called potato sickness.

Goldenseal–*Hydrastis canadensis*, family Ranunculaceae; a perennial herb grown in the wild garden for its showy leaves and red fruit and for its rootstock used in tonics. It is poisonous to livestock. Native to North America. Also called yellow puccoon.

Goober–See Peanut.

Good Tilth–Physical condition of a soil favorable to seed germination and plant growth.

Goose Grass–*Eleusine indica*; a common annual grass that crowds out desired plants if left uncontrolled.

Gossypol–A material found in cottonseed that is toxic to swine and certain other simple-stomached animals.

Gourd–Any cucurbitous fruit of various shapes and sizes, of the genera *Lagenaria* and *Cucurbita*, family Cucurbitaceae, that is hard-shelled and used as an ornamental or for making domestic utensils. (2) In England, a generic name for species of genus *Cucurbita*.

Gourdseed Corn–A late-maturing variety of corn (*Zea mays*) grown by early southern United States settlers. The gourdseed kernels are white, soft, deeply dented, large, and pointed, resembling gourd seeds. The dent varieties of today may be the result of intentional or accidental crossing with the old gourdseed corns. Also called Virginia gourdseed corn.

Grabbling–(Potatoes) Harvesting the larger tubers from a potato hill without killing the plant; the soil is replaced, permitting the remaining tubers to mature. It is an old practice used by home gardeners to obtain early or new potatoes.

Gracilis–Slender.

Grade–(1) The slope of a road, channel, or natural ground. (2) The finished slope of a prepared surface of a canal bed, roadbed, top of embankment, or bottom of excavation. (3) Any surface that is prepared for the support of a conduit, paving, ties, rails, etc. (4) Any animal that has one purebred parent and one of unknown or mixed breeding. (5) Designating a herd, flock, brand, etc., of such animals. (6) The classification of a product, animal, etc., by standards of uniformity, size, trueness to type, freedom from blemish or disease, fineness, quality, etc. (7) To smooth the surface of a road. (8) To raise the level of a piece of ground by the addition of earth, gravel, etc.

Graded Terrace–A terrace with a constant lengthwise slope sufficient to cause runoff to flow at a nonerosive velocity.

Grading–(1) The classification of products, animals, etc., into grades. (2) The mating of a purebred animal with one of mixed or unknown breeding. (3) The smoothing of the land surface.

Grading Cotton–A system of classifying cotton according to grade, composed of three factors: color, foreign matter, and ginning preparation.

Graft–(1) A scion or bud taken from one plant and inserted into a cut on a root, stem, or branch of another or the same plant on which it continues to grow. (2) To make a graft. (3) The growth that results from this process.

Graft Hybrid–A stem arising from an adventitious bud formed at a graft union that contains the genetically different tissues of both the stock and scion. See Chimera.

Grafting–(1) The inserting of a piece of one plant into another or the same plant with the intention that it shall grow there. See Graft. (2) A process of removing a worker bee larva from its cell and placing it in an artificial queen cup, for the purpose of having it reared into a queen.

Grain–(1) The seed of the cereal crops. See Cereal, Small Grain. (2) Commercially, or as listed on boards of trade, buckwheat, soybeans, and flaxseed, in addition to the cereals. (3) In buttermaking, the butter particles after the emulsion has broken. (4) The arrangement, direction, quality, etc., of the fibers in wood. See Annual Ring.

Grain Corn–Shelled corn, as contrasted to silage corn or corn on the cob.

Grain Drier–A device used to lower the moisture content of grain. It can utilize circulating air, or heat generated by LP gas, electricity, or the sun.

Grain Drill–A tractor-drawn implement that plants grain seed in a series of closely spaced rows. A small furrow is opened, the seed deposited, and the furrow closed. See Air Drill, Press Drill, Stubble-Mulch Press Drill.

Grain Dust–Small particles of broken grain and dust from grain generated when grain is stored. Such material is highly explosive.

Grain Probe–A sampling device commonly used by elevator operators when purchasing grain to obtain a sample of a load, bin, or gab in order to determine grain quality and dockage content. It consists of a long, hollow tube that is inserted into grain to obtain a representative sample of the load. See Grain Trier.

Grain Sorghum–*Sorghum bicolor*. Any variety of the species of sorghum that is grown principally for the grain it produces as contrasted with the sorghum grown for forage or syrup.

Grain Trier–An instrument or device usually made of brass and consisting of two tubes from 6 inches to 8 feet in length. These tubes are slotted on the sides, and one is closely fitted by freely moving within the other. When closed this instrument may be inserted in a bag or bin of grain and then opened to admit a representative sample from each layer of the grain. After filling, the trier is closed and withdrawn and the sample discharged upon an examination cloth for inspection or grading. Also called grain sampler.

Grain Wagon–A truck-or tractor-drawn four-wheel wagon that transports grain from a harvester to a storage bin or to market.

Gram Chickpea–*Cicer arietinum*, family Leguminosae; an annual herb that has long been cultivated in Europe for its edible seeds used for food and livestock feed. Grown in California, United States, the herbage is said to cause livestock poisoning. Native to Asia. Also called chestnut bean, chickpea, garbanzo, Egyptian pea, fasels, coffee pea.

Gram Stain–A staining method devised by a Danish physician, Hans Gram, to aid in the identification of bacteria. Bacteria either resist discolorization with alcohol and retain the initial deep violet stain (gram-positive) or can be decolorized by alcohol and are stained with a contrast stain (gram-negative).

Grama–Any grass of the genus *Bouteloua*, family Gramineae. A large number of species are native on the Great Plains and elsewhere in the western United States and are rated as good to excellent forage on ranges and pasture lands.

Graminoid–Grasslike in appearance, with leaves mostly very narrow or linear in outline.

Granary–(1) A storehouse, especially for threshed or husked grain. (2) Any storeroom where food or supplies are kept or stored. (3) A large food-producing area: e.g., the central west of the United States is sometimes called the food granary of the world.

Granary Weevil–*Sitophilus granarius*, family Curculionidae; an insect that infests stored grain and flour.

Granite Chips–Small particles of granite used as a growth medium in gravel culture.

Granite Dust–Fine rock granite particles used by some organic gardeners as a source of potassium.

Granite Soil–Soil that overlies granite rocks or is derived from the residue of weathering of such rocks. It has no single distinctive universal character but is variable, depending upon local environmental factors.

Granular–(1) In the form of granules or small particles. (2) Covered with small grains, minutely mealy. (3) A porous soil ped. See Soil Porosity, Soil Structure.

Granular Frost Structure–Granules of ice that are scattered through porous and easily spaded soil. It is common in frozen soils high in organic matter. See Concrete Frost Structure.

Granulation–(1) The formation of dextrose hydrate crystals in honey and crystals in other sugar syrups. (2) The cementation of particles into masses, as grains, aggregates, or clumps. (3) The formation of excess or scar tissue in early wound healing and repair.

Granules–A formulation in which the pesticide is attached to particles of some inert carrier, such as clay or ground corn cobs. Size of most presently used granules is 15-40 mesh.

Granuloblastosis–One type of avian leukosis complex affecting fowl over six months of age; characterized by a yellowish discoloration of the comb and unfeathered parts, weakness, emaciation, sleepiness, and diarrhea.

Granville Wilt–A bacterial disease of tobacco and other solanaceous plants caused by *Xanthomonas solanacearum*; characterized by sudden wilting of the leaves, which turn yellow and brown, and by death of the plant. Also called southern wilt, brown rot, bacterial wilt of tobacco.

Grape–See European Grape.

Grape Flea Beetle–*Altica chalybea*, family Chrysomelidae; an insect pest of grapes that feeds on the buds as they unfold in the spring, resulting in ragged, tattered foliage. It also infests plum, apple, quince, beech, elm, and Virginia creeper.

Grape Leaf Skeletonizer–*Harrisina americana*, family Zygaenidae; a caterpillar that feeds on upper leaf surfaces of wild and cultivated grapes.

Grape Mealybug–*Pseudococcus maritimus*, family Pseudococcidae; an insect pest of buckcoleus, columbine, lemon, orange, passionflower, pear, potato, Japanese flowering quince, strawberry, Persian walnut, and Japanese jew, which sucks plant juices and secretes honeydew in which disease may become established. Also called Baker mealybug.

Grass–(1) In its collective usage, grass includes 169 genera, 1,398 different species, and an uncounted number of varieties, family Gramineae, of annual and perennial herbs (rarely woody, such as bamboo) with hollow or solid stems. (2) Ranked, usually parallel-veined leaves that consists of two parts; a flat bladelike portion and sheath that enclose the stem. At the junction between the two parts of the leaf, and on the inside, is an appendage called the ligule which is either hyaline or hairy and membranaceous. The flowers are prefect (rarely unisexual) with no distinct perianth, borne in spikelets, and aggregated into spikes and panicles at the end of the stems. The flowers may have one to six stamens, but usually three with two-celled anthers; one pistil, one-ovulated ovary, and the fruit is a caryopsis with starchy endosperm. The grasses include the cereals, sugarcane, sorghum, millets, and bamboo, which produce much of our human food and, in addition, furnish the bulk of forage for domestic animals. This group of plants is important not only as food for people and animals, but also for turf, soil-holding, and ornament. (3) Unscientifically, any low-growing, herbaceous, grasslike plant, as a rich green pasture, the forage in a meadow, the herbage of the range, a well-kept lawn, or golf course.

Grass Culm–The vertical stem of a grass.

Grass Hay–Any hay totally or primarily from a grass crop.

Grass Killer–See Herbicide.

Grass Pea–*Lathyrus sativus*, family Fabaceae; an annual herb used as forage in Europe and Asia. Its seeds, used for human food in India and elsewhere for stock feed, are poisonous if consumed over a period of time. A symptom of poisoning is paralysis of the legs. Also called vetchling, grass vetch, chickling, chickling vetch.

Grass Sorghum–A class of sorghums that includes Sudangrass, Tunisgrass, and Johnsongrass. Sudangrass is the most important and is used for hay, pasture, and sometimes silage. See Broomcorn, Grain Sorghum, Sorgo.

Grass Thrips–*Anaphothrips obscurus*, family Thripidae; an insect pest of grains and grasses that sometimes destroys the softer parts of the leaves, flowers, and kernels.

Grass-seed Nematode Disease–A disease affecting grass seed caused by the nematode *Anguina agrostis*; characterized by the seeds being transformed into purple or black galls that are poisonous to livestock.

Grassed Waterway–A natural or constructed waterway, usually broad and shallow, covered with erosion-resistant grasses, used to conduct surface water from cropland.

Grasshopper–Any insect of the order Orthoptera, family Acididae (Locustidae). Species have long legs for jumping and may be voracious feeders on many kinds of foliage.

Grassland–(1) Land that is in grass but not under cultivation. (As commonly used, it is not restricted to true grasses, but may include legumes and other nongrasses.) (2) Land that produced grass just previous to being planted to a particular crop. (3) Any of the great natural regions of the world in which the vegetation is mainly grass or other herbaceous plants. Climatic conditions are intermediate between those of forest regions and of deserts. Prairie, steppe, savanna, and pampas have been recognized as subdivisions of grasslands. The Great Plains constitute the major grassland region of the United States.

Grasslike Plant–A plant of the Cyperaceae (sedges) or Juncaceae (rushes) families that vegetatively resembles a true grass of the Gramincac family.

Gravel–(1) Accumulation of rounded water-worn pebbles. The word *gravel* is generally applied when the size of the pebbles does not much exceed that of an ordinary hen's egg. The finer varieties are called sand, while the coarser varieties are called shingles. (2) An accumulation of rounded rock or mineral pieces larger than 2 millimeters in diameter. Divided into granule, pebble, cobble, and boulder gravel. The individual grains are usually rounded. (3) Accumulation of uncemented pebbles. Pebble gravel. May or may not include interstitial sand ranging from 50 to 70 percent of total mass.

Gravel Culture–Growing plants in artificial nutrient solutions with gravel as the medium for supporting the roots.

Gravitational Water–Water that drains away in large pores in the soil under the force of gravity when under-drainage is free. Also known as free water and percolation water.

Gravitropism–Also known as geotropism; the growth of a plant in response to the force of gravity; e.g., most plant roots grow downward and most plant tops grow upward. If such a growing plant in a pot were inverted, roots would still grow downward toward the rim of the pot and tops would grow upward toward the bottom of the pot.

Gravity Water–(1) Water that moves through soil under the influence of gravity. (2) A gravity supply of water as distinguished from a pumped supply.

Gray–(1) A color of an animal's coat that has white hairs mixed with black. (2) A cotton-lint color designation that is the darkest in chroma.

Gray Oats–A market classification of oats based on the grayish color of the lemma and palea of the matured grain. Other classifications are: white oats, red oats, black oats, mixed oats.

Greasewood–Any low-growing bush that burns easily (western United States), such as white sage, black greasewood, chamiza.

Great Group–A category in soil taxonomy between suborder and subgroup. The total categories are: order, suborder, great group, subgroup, family, and series.

Great Northern–A white variety of the kidney bean that is grown for its edible, dried seeds. It is distinguished from the navy or pea bean by its larger size.

Green–(1) A small piece of land covered with grass, as a golf green, or a village green. (2) Unmatured; immature, as a fruit, seed, honey. (3) Uncured, as green lumber, green cheese. (4) Untrained, as a green colt. (5) Designating a plant whose foliage is a shade of green, as light green foliage. (6) Designating a flavor defect of milk characterized by a weak or acid taste and by a repulsive odor.

Green Algae–Organisms belonging to the class Chlorophyceae and characterized by photosynthetic pigments similar in color to those of the higher green plants. Food manufactured by photosynthesis is stored in algal tissues as starch. See Algae.

Green Ash–*Fraxinus pennsylvanica* var. *lanceolata*, family Oleaceae; a hardy, deciduous tree grown in shelterbelts in the Great Plains, United States, as a shade tree and for its lumber used in handles, furniture, sporting goods, boxes, implements, boats, woodenware, coopcragc, etc. Native to North America.

Green Asparagus–Asparagus that has not been blanched.

Green Chop–Green forage that is cut with a field chopper and hauled to lots or barns for livestock feed in lieu of pasturing. See Green Chopping.

Green Chopping–The practice of chopping and hauling green forage to dairy and beef cattle during the pasture period. The most common practice is to chop and haul forage once or twice daily. Green-chopped forage may be fed in self-feeding wagons filled directly by the field chopper or in bunks. Also called zero grazing.

Green Fescue–*Festuca viridula*, family Gramineae; a perennial grass found from British Columbia to Alberta, Canada, and south to California and Idaho, United States, at 6,500 to 10,000 feet elevation whose forage rating is good. Native to North America. Also called greenleaf fescue, mountain bunchgrass.

Green Hay–(1) Uncured hay. (2) That hay which, on being cured, retains a green color.

Green June Beetle–*Cotinus nitida*, family Scarabaeidae; an insect pest, the adults of which feed on the foliage and fruit of a number of trees and plants; the larvae damage lawns, vegetable roots, and tobacco seedlings.

Green Lacewing–*Chrysopa californica*, family Chrysopidae; an important insect predator on the citrophilus mealybug.

Green Lumber–(1) Lumber with the moisture content greater than that of air-dried lumber. (2) Unseasoned lumber, boards from logs processed through mill before drying.

Green Manure–Crops such as legumes or grasses that are grown to be plowed or spaded into the soil to increase humus content and improve soil structure. See Cover Crop.

Green Needlegrass–*Stipa viridula*, family Gramineae, a perennial bunchgrass widely distributed on dry slopes in North America from New York to Alberta, Canada, south to New Mexico. Its forage rating is good.

Green Olives–Olives that are harvested while green, and then pickled. They are first soaked in a lye solution, washed, and placed in barrels with a brine solution for fermentation.

Green Onion–An immature garden onion, *Allium cepa*, family Amaryllicaceae, which is harvested before the bulb has begun to take on a spherical shape; a very popular salad vegetable.

Green Pellet–A pellet made from alfalfa meal only, or a complete pellet that contains enough green roughage to color it.

Green Pepper–The sweetbell, *Capsicum annuum*, var. *annuum*, family Solanaceae, picked after it has attained full growth, but before it turns red; used in salads, cooking, etc.

Green Stink Bug–*Acrosternum hilare*, family Pentatomidae; an insect that infests peach, nectarine, apple, cabbage, catalpa, corn, dogwood, elderberry, basswood, maple, okra, orange, pea, tomato, turnip, and cotton..

Green Up–To become green again following the winter period or in one season, as a plant after a period of drought followed by rain.

Green Vegetable–Any vegetable whose edible portions are green, such as spinach, garden peas, mustard, lettuce, and broccoli.

Greenbelt–(1) A strip of land of varying size, used for esthetics, recreation or farming, which encircles a community and protects it from the intrusion of objectionable real property uses from adjacent communities and/or provides the community with space of orderly expansion. (2) A similar strip around a commercial enterprise into which it can expand without the invasion of residential property.

Greenbrier–*Smilax rotundifolia*, family Liliaceae; a prickly, green-stemmed, wiry, climbing vine that is often a nuisance to eradicate once it has become established. Native to North America. Also called catbrier, horse brier, roundleaf, bullbrier, devil's-hopvine.

Greenbug–*Toxoptera graminum*, family Aphidae; an aphid that infests small grains, rice, corn, and sorghum and is a vector of sugarcane mosaic disease.

Greenhouse–Any of several different types of heated, glass- or plastic covered structures used for the growing of plants.

Greening–(1) The turning green of the Irish potato tuber due to exposure to sunlight. This green potato is bitter and toxic. Also called sunburn. (2) Any of several varieties of green-colored apples.

Greenouse Whitefly–*Trialeurodea vaporariorum*, family Aleyrodidae; a sucking insect pest of greenhouse plants. Infested plants lack vigor, wilt, turn yellow, and die.

Greens–(1) The green leaves of various plants, both wild and domesticated, which are boiled for human food, usually with fat meats, such as bacon or hog jowls. Wild plants, such as lambsquarter, docks, and young shoots of pokeweed and the wild blackberry were mixed together and boiled, especially in spring, by early settlers of the United States. The leaves of cultivated corps such as turnips and spinach are also considered as greens. Also called potherb. (2) Any foliage that is added to a floral arrangement, such as bouquets, funeral pieces, etc.

Greensand–(Glauconite) Essentially a hydrated silicate of iron, magnesium, and potassium. The potassium is, however, insoluble in water and only slightly available as a fertilizer material without special treatment. Sometimes used as a "nonchemical" fertilizer by "organic" gardeners. See Glauconite.

Grex–A collective term for cultivars of the same hybrid origin, in some cases further divisible into groups.

Grimes–A golden-yellow, fine-flavored, variety of eating apple. Also called Grimes Golden.

Grist–(1) A quantity of grain brought to a mill for grinding. (2) The product from such grinding.

Grit–(1) Sand, especially coarse sand. (2) Coarse-grained sandstone. (3) Sandstone with angular grains. (4) Sandstone with grains of varying size producing a rough surface. (5) Sandstone suitable for grindstones. Stones fed to chickens to assist them in grinding feed in their craw.

Grits–(1) A corn food product that consists of the hulled, ground grain of even-sized particles larger than corn meal. Also called hominy grits. See Grist, Groats.

Groats–(1) The hulled seeds of cereals and buckwheat that are ground into particles of even size but larger than flour particles. (2) Unground oat kernels.

Groom–(1) A person who curries, combs, washes, etc., an animal and cares for it generally. (2) To wash, curry, brush, and generally care for an animal. (3) To trim grass and to make a yard and flower garden neat and trim.

Ground–(1) The surface of the earth. (2) The soil; land. (3) (a) An electrical connection to the earth. (b) To make such an electrical connection. (4) In the electrical circuit of a tractor or other machine, the break in the circuit that occurs when any part of the circuit unintentionally touches the metal body of the tractor.

Ground Bed–A plot of soil at or near ground level in greenhouses, in contrast to soil in raised benches.

Ground Cover–(1) Any of many different plants, usually perennials, that grow well on sites on which grass does not thrive, as on banks, terraces, in the shade, etc. (2) All herbaceous plants and low-growing shrubs in a forest. (3) Any vegetation that grows close to the ground, producing protection for the soil. See Cover Crop.

Ground Limestone–A commercial soil amendment; the product obtained by grinding either high calcic or dolomitic limestone so that all materials will pass state fineness and purity laws. See Lime.

Ground Moraine–The earthy material deposited from a glacier on the ground surface over which the glacier has moved. It is bordered by lateral and/or end moraines. See Till Plain.

Ground Pea–Peanut, *Arachis hypogaea*, family Fabaceae.

Ground Rot–A rot of melons, originating at the point of contact with the ground, caused by the fungus *Sclerotium rolfsii*; characterized by a heavy growth of white mold and the formation of many brown bodies suggesting buckshot.

Ground Shell Marl–A commercial soil amendment; the product obtained by grinding natural deposits of shell marl so that it conforms with state fineness laws. See Lime.

Groundnuts–The word for peanuts in the United Kingdom and some Commonwealth countries.

Grounds–(1) The land surrounding a house, building, etc., in lawns, flowers, parks, which is usually carefully tended. (2) Sediment, as coffee grounds.

Groundwater–Water within the earth that supplies wells and springs. Specifically, water in the zone of saturation, where all openings in soils and rocks are filled with water, the upper surface of which forms the water table. See Phreatic Water.

Groundwater Flow–The movement of groundwater under hydraulic gradient; the water that comes to the surface in springs or causes streams and rivers to flow after periods of surface runoff have ended. See Groundwater.

Groundwater Level–See Water Table.

Groundwater Reservoir–An aquifer.

Groundwater Surface–The level below which the rock and subsoil voids are full of water, is known as the groundwater level, groundwater surface, or water table.

Group–As used semitechnically in the nomenclature of cultivated plants, an assemblage of similar cultivars within a species of interspecific hybrid, as the Cepa, Aggregatum, and Proliferum groups in onion (*Allium cepa*).

Grove–(1) A stand of trees, smaller than a forest, which grows naturally and has little undergrowth. (2) A small woods. (3) An orchard, or a planting of trees, as an orange grove, pecan grove.

Grow–(1) To live and to increase in stature and girth toward maturity. (2) To cultivate plants. (3) To raise animals.

Growing Conditions–The environmental conditions of soil, precipitation, sunlight, temperature, cultivation, etc., which affect the growth of a plant or animal.

Growing Medium–Soil or soil substitute prepared by combining such materials as peat, vermiculite, sand, or weathered sawdust. Used for growing potted plants or germinating seed.

Growing Point–One of the sections of a plant body in which growth is localized especially the tips of stems and roots.

Growing Season–The number of days from the average date of the last freeze in the spring to the average date of the first freeze in the fall; the period in which most plants grow. The length of the period varies according to the hardiness of the plant and geographic location.

Growing Tray–A tray that is made of small cells in which growing medium is placed and seeds are planted. When the seedlings are the proper size they are transplanted.

Growth–(1) The increment in size of a living organism. (2) Plants or plant parts. (3) A tumor, gall, etc. (4) The development of an organism from its earliest stage to maturity. (5) The development or increase of an enterprise or an organization.

Growth Layer (Annual Ring)–A layer of xylem and phloem produced in woody stems usually during each growing season. See Spring Wood, Summer Wood.

Growth Rate–With reference to wood, the rate at which wood has been added to the tree at any particular point, usually expressed in the number of annual rings per inch. May also be stated as annual leader growth.

Growth Regulators–Synthetic or natural organic compounds such as indoleacetic acid, gibberellin, and napthalene acetic acid, that promote, inhibit, or modify plant growth processes. Commonly used in rooting cuttings.

Growth Ring–See Annual Ring.

Grub–(1) Roots and crown that remain in the soil after a tree or shrub has been cut off or killed. (2) The larva of any of several beetles that are common in soil. (3) To dig out a plant by the roots, as to grub out persimmon sprouts. (4) A slang term for food.

Grub Prairie–Forest land that has become grassland due to annual burning.

Grubworm–The larva of a June beetle. See June Beetle.

Gruel–A food or feed prepared by mixing a ground feed with hot or cold water.

Guano–The dry excrement of marine birds, which is a fertilizer rich in nitrogen and phosphorus. Guano was first imported form Peru, where it was collected form arid islands off the coast. The term now has no specific meaning and has been used in some parts of the world for any kind of fertilizer, organic or inorganic.

Guar–*Cyanopsis proraloides (C. tetragonoloba)*, family Leguminosae; an annual herb cultivated for its edible beans. Grown experimentally or only to a very limited extent in the United States, the seeds yield a starch that has some potentiality for use in the adhesive industry. Native to the East Indies.

Guard Cell–One of two epidermal cells in a plant leaf or needle that encloses a stoma. The stoma when open permit CO_2 and O_2 to enter the leaf interior and H_2O to evaporate.

Guayule–*Parthenium argentatum*, family Compositae; small, drought-tolerant shrub of northern Mexico and southwestern Texas that is a potential source of natural rubber.

Guinea–A domesticated bird, about the size of a small adult chicken, which descended from a wild, North African species, *Numida meleagris*. The most popular variety is the Pearl, which has purplish-gray plumage regularly dotted with white. Other varieties are white and lavendar. They are not raised extensively, but some farmers keep a few because they utter sharp cries if hawks, crows, or other natural enemies appear near the premises. Guinea meat is dark and has a gamey flavor.

Guineagrass–*Panicum maximum*, family Gramineae; a drought-resistant, coarse, perennial bunchgrass; an important forage grass in tropical America grown only to a limited extent in parts of Florida and Texas. Native to Africa; introduced at an early period into the West Indies.

Gully–(1) A gash cut into a slope of unconsolidated sediments by concentration of rainfall runoff into a channel. Gullies usually have a transverse V-shaped profile and generally are relatively shallow, although locally the cutting may extend to a depth of more than 100 feet and have sides and flat floors. One of the most destructive forms of erosion. (2) A furrow, channel, or miniature valley that is cut by running water during and immediately after rains or during the melting of snow. At other times it usually does not contain water. Special forms may be described as: dendritic arborescent, resembling the branching of a tree or shrub; linear, resembling a line, rather long, narrow, and of uniform width. The distinction between gully and rill is primarily one of depth. A gully is a channel of sufficient depth that it would not be obliterated by normal tillage operation, whereas a rill would be smoothed by ordinary tillage. (3) To make gullies; to form small drainage channels by the action of running water.

Gully Control Plantings–The planting of forage, legume, or woody plant seeds, seedlings, cuttings, or transplants in gullies to establish or reestablish a vegetative cover adequately for the purpose of gully control.

Gully Erosion–Severe erosion of the soil in which large gullies are washed out. Gully erosion is caused by running water. See Gully.

Gum–Viscous exudates from plants, best known of which are gum arabic (*Acacia arabica*), Sudan gum arabic (*A. senegal*), gum tragacanth (*Astragalus gummifera*), Indian tragacanth (*Sterculia urens*), and carob seed gum (*Ceratonia siliqua*). Best known of the gum resins are asafetida (*Ferula* spp.), myrrh (*Commiphora* spp.) and frankincense (*Boswellia carteri*). In the United States, oleoresin, the exudate of longleaf (*Pinus palustris*) and slash pines (*P. elliottii*) is also known as *gum*. Gum turpentine is obtained from distilling this oleoresin.

Gumbo–(1) A name current in western and southern states for those soils that yield a sticky mud when wet; in southwest Missouri a puttylike clay associated with lead and zinc deposits; in Texas, a clay encountered in drilling for oil and sulfur. (2) The stratified portion of the lower till of the Mississippi Valley. (3) A vertisol when wet. See Vertisols. (4) A dish usually containing okra.

Gummosis–The formation of clear or amber-colored exudates that set into solid masses upon the surface of affected plant parts and that may be caused by mechanical injury, pathogenic fungi, insects, or excessive nitrogen in the soil.

Gur–Cane juice, concentrated nearly to dryness by boiling over an open fire, without centrifuging and with no other purification than by skimming. This ancient process is still used for producing a large share of the sugar consumed in India and some other countries. The crude product is high in glucose and correspondingly low in sucrose.

Guttation–The forcing out of droplets of water or cell sap upon leaves of plants, a normal physiological process.

Gymnosperm–(1) The family of conifers (needle-leaved trees and shrubs), in contrast to angiosperms, the family of broad-leaved plants. (2) Naked seed not enclosed in an ovary, or a plant having such seeds. (3) Any of the seed plant species. See Angiosperm.

Gynandromorph–An individual of which one part of the body exhibits female characteristics and another part male characteristics.

Gynoecium–The female parts of a flower collectively, including the carpels; the pistil or pistils of a flower collectively. Also called gynecium.

Gynomonoecious–Type of sex expression where plants contain perfect as well as imperfect pistillate flowers on the same plant.

Gypsophilous–Refers to plants growing characteristically on soils rich in gypsum.

Gypsum (Land Plaster)–A mineral calcium sulfate, combined with water of hydration. Pure gypsum has the formula $CaSO_4 \cdot 2H_2O$. Gypsum occurs in large deposits of soft crystalline rock and as a sand. Normal superphosphate is approximately half gypsum and/or anhydrite. In making concentrated superphosphate, most of the gypsum is removed. Gypsum is widely used for growing peanuts and sometimes for other legumes. In arid regions, large tonnages are used to treat sodic soils. The calcium sulfate reacts with sodium or potassium carbonate to form corresponding sulfates and calcium carbonate. In irrigation agriculture it is used to increase permeability of soils. Following gypsum applications, infiltration can be increased from 50 to 100 percent. Sometimes used as a source of calcium. See Liming Materials.

Gypsy Moth–*Lymantria dispar*, family Lymantriidae; an important insect whose larvae defoliate apple, pear, basswood, birch, hawthorn, oak, poplar, willow, cherry, elm, hickory, hornbeam, larch, maple, and sassafras.

Habit–The general appearance of a plant.

Habit of Growth–(1) The general form or arrangement of stems, roots, and branches, or of the entire plant which is possessed in common by a species in a given habitat. (2) The erectness of the stems of young plants of the small grains as a means of distinguishing between winter and spring varieties.

Hair–The outgrowth of a cell in the epidermis of a plant or animal. In vast numbers it forms the coat of an animal and is frequently used as a fiber, such as wool. See Pubescent, Root Hair.

Hairy Gramagrass–*Bouteloua hirsuta*, family Gramineae; a perennial bunchgrass that has a good forage rating on the western range. Found from Wisconsin to South Dakota, United States, and south to Mexico.

Hairy Vetch–*Vicia villosa*, family Leguminosae; a hardy annual or biennial, climbing herb grown for forage, as a cover crop, and for green manure. Native to Eurasia. Also called winter vetch, sand vetch, Russian vetch.

Half Hardy–Designating a plant or plant part that is able to withstand a moderately low temperature but which is injured by severe freezing.

Half-shrub–A perennial plant having stems that are woody at the base.

Half-slip–In harvesting melons, a stage of ripeness in which, as the fruit is pulled from the vine, only a portion of the stem separates easily form the base of the fruit leaving a smaller scar on the fruit than when harvested at full-slip.

Haline–Term used to indicate dominance of ocean salt.

Halite Rock Salt–A mineral, NaCl. A common mineral of evaporates.

Halls Panicum–*Panicum balli*, family Gramineae; a grass that produces abundant forage on bottomlands and in irrigated fields and whose forage rating is fair. Native to North America; found in Texas and Arizona, United States.

Halo–(1) A bright ring or system of rings sometimes seen surrounding the sun or moon. The reflection from the ice crystals of cirrus clouds, it is an indicator of a probable change in the weather. Also called sun dog, moon dog. (2) A narrow, light-colored band of diseased tissue that sometimes surrounds a lesion on a plant.

Halo Blight of Beans–A bacterial disease caused by *Pseudomonas phaseolicola*, family Pseudomonadaceae, characterized in cool weather by a halolike zone of greenish-yellow tissue around a water-soaked spot on leaves.

Halo Blight of Oats–A widespread, bacterial disease of oats, wheatgrass, and bromegrass caused by *Pseudomas coronafaciens*, family Pseudomonadaceae, characterized by egg-shaped, small, water-soaked spots on the leaf sheaths and floral bracts. The spots change from green to light brown, a light yellow halo forms around the spot, and the spots coalesce to form irregular halo areas.

Halogens–A group of elements, including fluorine, chlorine, bromine, and iodine, which are powerful oxidizing agents and have disinfectant properties. An alcoholic solution of iodine (tincture of iodine) is an example of an iodine-containing disinfectant; chlorinated lime is an example of a chlorine-containing disinfectant.

Halophyte–A plant tolerant of salt (sodium chloride) in the soil.

Halophytic Vegetation–Salt-loving or salt-tolerant vegetation that usually has fleshy leaves or thorns and resembles desert vegetation.

Hammock–(1) A vegetational land feature, recognized in peninsular Florida and westward along the Gulf Coast to Mississippi, United States, whose tree vegetation is, or was originally, in contrast to that of the prevailing pine land. (2) A small isolated tree-covered area in extensive marshes. Hammocks may be designated specifically according to tree growth, as oak hammocks, palm hammocks, etc. (3) In some parts of Mississippi, alluvial second bottoms.

Hamper–A basket in which vegetables are shipped.

Hand Planting–A reforestation method of planting seedlings by hand.

Hand Pollination–The pollination of flowers by hand, especially for the production of hybrid seed.

Hand Strip–(1) To take the last bit of milk from a cow's udder, usually following machine milking. (2) To harvest a seed crop by hand.

Handpick–(1) To harvest by hand, as contrasted to harvesting by machine. (2) To pluck the feathers of a fowl manually.

Hanging Basket–A basket that contains a growing ornamental plant that is usually suspended in the air.

Haploid Number–In genetics, this is half (haploid) the number of chromosomes that are usually present in the nucleus; occurs during reduction division.

Harbor–To serve as a habitat for, as to harbor insects.

Hard Cider–Fermented juice, as of apple, pear, etc.

Hard Dough Stage–An intermediate stage between the milk and hard or ripe stage of ripening grain, especially corn or wheat.

Hard Land–Land that consists of compact clayey soil in contrast to more loamy or sandy soil (Great Plains, United States).

Hard on the Land–Designating a crop that draws heavily on or exhausts the fertility of the soil or leaves it in a bad physical condition; often said of sorghums.

Hard Pinch–A pinch that removes enough terminal growth of a plant so that the part removed can be used as a cutting for propagating a new plant.

Hard Red Spring Wheat–In marketing, a wheat producing hard, flinty kernels rich in protein and possessing the quality for making good bread.

Hard Red Winter Wheat–A red-berried wheat sown in the fall that matures the following year producing hard, flinty kernels and yielding a flour of superior baking qualities.

Hard Scabble Land–Land that has thin, stony soil or is so unproductive that it barely yields a poor living for the occupant.

Hard Winter Wheat–A market grade of hard red winter wheat that contains more than 25 percent and less than 75 percent of dark, hard, and vitreous kernels.

Hardening-off–Adapting plants to outdoor conditions by withholding water, lowering the temperature, or gradually eliminating the protection of a coldframe, hotbed, or greenhouse. This conditions plants for survival when transplanted outdoors.

Hardiness–The state of being hardy. See Hardy.

Hardness of Wood–A property of wood measured by the load applied at a standard rate required to imbed a 0.444-inch (1.13-centimeter) steel ball halfway.

Hardness Scale–The empirical scale by which the hardness of a mineral is determined as compared with a standard. The Mohs scale is as follows: 1. talc; 2. gypsum; 3. calcite; 4. fluorite; 5. apatite; 6. orthoclase; 7. quartz; 8. topaz; 9. corundum; 10. diamond.

Hardpan–A dense, compacted layer of soil under the surface that may interfere with the downward penetration of both roots and water. Shallow hardpans restrict the roots, especially of plants with taproots, causing them to grow laterally and often resulting in a less hardy and less productive plant. Plowing at the same depth each year may create an artificial hardpan called a plowsole. See Fragipan.

Hardwood–(1) (a) A term used by lumbermen designating any tree that has broad leaves, in contrast to the conifers. (b) The wood or lumber of such a tree. (2) Designating several kinds of timber that resemble teak, used in buildings, etc. (Australia).

Hardwood Ashes–The ashes resulting from the burning of hardwood that contain 2 to 8 percent or more of potash (K_2O), principally as a carbonate; often used as fertilizer. See Wood Ashes.

Hardwood Cutting–A mature shoot of last season's growth that is removed from the plant after the leaves have fallen to be used in propagating new plants. See Softwood Cutting.

Hardwood Land–Land that has, or originally supported, a native growth of deciduous forest trees, as distinguished from pine land.

Hardy–Designating a plant that withstands various environmental factors, especially the low temperatures of a given region. See Half Hardy, Tender.

Hardy Annual–(1) Any annual plant that may be sown where it is to grow, as contrasted to a plant that must be transplanted. (2) An annual flowering plant whose seeds will live outdoors over winter.

Hardy Plants–Plants adapted to winter temperatures or other climatic conditions of an area. Half hardy indicates some plants may be able to survive local conditions with a certain amount of protection.

Hardy Tree Fruits–Trees, such as apple, pear, cherry, peach, plum, and apricot, which are temperate-zone plants as contrasted to tropical and subtropical fruit plants such as coconut and orange.

Haricot–(French) Bean.

Harlequin Bug–*Murgantia histrionica*, family Pentatomidae; the most important insect enemy of cabbage and related crops in the southern half of the United States. It also infests many other plants occurring as far north as Michigan. Also called harlequin cabbage bug, fire bug, calico back.

Harness–The straps, bands, collars, hames, lines, and attachments that are necessary to equip a draft animal, such as a horse, properly to pull or move a load. The harness, of many types, is usually largely made of leather. Also called gear.

Harness Horse–A lighter and faster horse than the draft horse that is especially suited to carriage and light driving use. There are two recognized types: the carriage or heavy harness horse and the roadster or light harness horse.

Harsh–(1) Designating a fleece that lacks character, as rough hair. (2) Designating vegetation that is rough, hard, or that has some physical characteristic objectionable to livestock.

Harvest–(1) To cut, reap, pick, or gather any crop or product of value, as grain, fruit, or vegetables. (2) The crop or product so harvested.

Harvest Interval–The period of time required by law between the application of a pesticide and harvest of the crop.

Harvest Season–The time of the year when the matured crop is gathered, especially in the autumn months.

Harvestsed Acres–Acres actually harvested for a particular crop, usually somewhat smaller at the national level than planted acres because of abandonment brought on by weather damage or other disasters or market prices too low to cover harvesting costs.

Hashish–Wild hemp, *Cannabis sativa*, family Moraceae. Also called marijuana, marihuana.

Hassock–A separate tuft of coarse grass that grows somewhat dispersed in a bog or meadow and is frequently elevated above the soil surface by an accumulation of dead leaves and stems. Also called tussock.

Hastate–A leaf base with pointed lobes that are at right angles.

Haulm–(1) The culm of a grass. (2) The stem of an herb. (3) The stalk of beans and peas that is cured for fodder or hay. Also called halm, haum.

Haw–(1) The third eyelid of animals. Also called hook. (2) See Hawthorn. (3) A command to a horse to turn left.

Hawthorn–A shrub or tree of the genus *Crataegus*, family Rosaceae. Species are grown as ornamentals for their (most often) white flowers. The state flower of Missouri. Also called haw, thorn, thornapple.

Hawthorn Rust–A fungal disease of apples, pears, and hawthorn caused by *Gymnosporangium globosum*, family Pucciniaceae; characterized by small lesions on the leaves, rarely on the fruit. Species of the genus *Juniperus* are the alternate hosts for this fungus. See Apple Rust, Cedar-Apple Rust.

Hay–Any leafy plant material, usually clover, fine-stemmed grasses and sedges, alfalfa, and other legumes, that has been cut and dried principally for livestock feeding. See Fodder, Marsh Hay.

Hay and Dairy Region–A region of the northeastern United States where grasses and legumes are naturally abundant or easily grown, and where the feed supply and barn and market sizes are generally more suitable for dairy cattle and milk production than for other types of farming, and where markets for dairy products are well established.

Hay Meadow–A field in which hay is grown.

Haycock–A small, conical pile of hay in the field used for curing and protection of the hay prior to baling and transporting it to a barn or haystack. Now rare.

Hayden Poisonvetch–*Astragalus haydenianus*, family Leguminosae (Fabaceae); a plant whose green and dried foliage are poisonous. Symptoms of poisoning are dullness, irregularity of gait, lack of appetite, dragging of feet, a solitary habit, loss of flesh, and a shaggy coat. Found in southern Wyoming, United States. Also called milk vetch.

Haydite–Shale and clay fused at high temperatures, then ground or crushed, and used in making lightweight cement blocks and bricks.

Haylage–Forage that could have been cut for hay but is stored with a higher moisture content than hay, and with less moisture than silage.

Hayseed–(1) Bits of straw, chaff, and seeds that cling to the garments in haymaking. (2) A degrading term for a rustic farmer or countryman (obsolete).

Hazlenut–(Filbert) *Corylus avellana* and *C. maxima*. Edible nuts produced in Oregon and in the Mediterranean area.

Head–(1) Any tightly formed flower cluster, as in members of the family Compositae, or any tightly formed fruit cluster, as the head of wheat or sunflower. (2) A compact, orderly mass of leaves, as a head of lettuce. (3) On a tree, the point or region at which the trunk divides into limbs. (4) The height of water above any point of reference (elevation head). The energy of a given nature possessed by each unit weight of a liquid expressed as the vertical height through which a unit weight would have to fall to release the average energy possessed, used in various compounds, as pressure head, velocity head, lost head, etc. (5) Cows, asses, horses, collectively, as ten head of horses. (6) The part of the body that includes the face, ears, brain, etc. (7) The source of a stream; specifically the highest point upstream at which there is a continuous flow of water, although a channel with an intermittent flow may extend farther. (8) The upstream terminus of a gully. (9) To prune a tree severely. (10) To get in front of a band of sheep, herd of cattle, etc., so as to stop their forward movement (head them off). (11) To place a top on a barrel. (12) That part of an engine that forms the top of the combustion chamber. In many types of modern engines, the exhaust and intake valves are in the head.

Head Flume–A flume, chute, trough, or lined channel used at the head of a gully or at the lower end of a terrace outlet to reduce soil erosion.

Head Lettuce–*Lactuca sativa* var. *capitata*, family Compositae; a commercially important annual herb grown for its edible, compact head of leaves. It is popular as a salad plant. Native to Eurasia. Also called lettuce.

Head Smut–A fungal disease of corn and sorghums, caused by *Sphacelotheca reiliana*, characterized by large smut balls replacing the tassel and ear in corn and the panicle in sorghums.

Head Smut of Grasses–A smutting of the heads of cereal grains caused by several fungi, most important of which are *Ustilago* spp., *Sorosporium* spp., family Ustilaginaceae.

Head Smut of Millet–A fungal disease caused by *Sphacelotheca destruens*, family Ustilaginaceae. The disease is evident as the panicles emerge, the entire inflorescence becoming enclosed by a grayish-white false membrane that ruptures as the plant matures, exposing the dark brown spore masses.

Heading Back–A type of pruning cut where the terminal portion of the shoot is removed but the basal portion is not.

Headward Erosion (Head Erosion)–Headwater erosion; a gully is lengthened at its upper end and is cut back by the water that flows in at its head, the direction being determined by the greatest column of water that enters it and the relative erodibility of the soil.

Headwater–(1) The small tributary streams that join to form a river; the water source of a river. (2) The water upstream from a structure, such as a dam.

Health–The state wherein all body parts of plants, animals, and people are functioning normally.

Hear Corn Grow–On hot, humid, summer nights, to hear popping noises from growing corn, indicating optimum growing conditions.

Heart–(1) The organ of the body that by its rhythmical contractions circulates the blood. It is an edible by-product of slaughter animals and fowls. (2) The center portion of fruits and vegetables.

Heart Rot–A tree decay characteristically confined to the heartwood. It usually originates in the living tree.

Heartwood–(1) The inner core of a woody stem that is wholly composed of nonliving cells and is usually differentiated from the outer, enveloping layer (sapwood) by its darker color. Also called duramen. (2) In wood anatomy, the various tannins, gums, resins, and other materials that are deposited in the cell cavities and walls when transformation from sapwood to heartwood takes place. It serves to differentiate sapwood and heartwood. See Sapwood.

Heat–(1) To ferment as a result of wet-stored grains such as wheat, corn, or barley, and forages; sometimes resulting in spontaneous combustion. (2) An animal in heat is ready to breed. See Estrus.

Heat-resistant–(1) Designating a variety or a species that grows under comparatively high temperature conditions; e.g., cotton and rattlesnakes. (2) Any material that is resistant to high temperatures.

Heath–(1) Any plant of the genus *Erica*, family Ericaceae. Species are evergreen shrubs and small trees grown in greenhouses and out-of-doors. Also called erica. (2) A natural land feature, an extensive tract of uncultivated land, treeless or nearly so, which is covered by a dense growth of shrubby, ericaceous plants. It may be nearly the same as a high moore. Heaths are generally sandy and the soils strongly acid.

Heaving–An upward movement of soil caused by freezing and thawing of free water in the soil, thus involving expansion and contraction. Damage, and sometimes destruction of plants, may result form the lifting action, and fence posts may also be pushed upward. Also called frost heaving, frost lifting. See Ice Damage.

Heaving Out–The forcing of a plant crown above the soil line by heaving.

Heavy–(1) Designating any material or product that exhibits a comparatively high weight per unit volume. (2) Designating a clay or clayey soil that is difficult to plow. (3) In marketing, designating an abundant supply of a product for sale on one day at one market. (4) The late stages of pregnancy of a cow.

Heavy Burning–Range firing during the dry, hot season to ensure a fire that will destroy the existing cover, facilitate travel and livestock handling, increase forage for livestock and game, and enhance hunting.

Heavy Soil–Generally, a clayey soil in contrast with a sandy one. The reference is to resistance and need for greater power in pulling a plow rather than to actual weight or specific gravity that is less for dry clay than for dry sand.

Hedge–(1) A fence or barrier formed by bushes, shrubs, or small trees growing close together in a line, sometimes with interwoven branches used as a screen. (2) A managed belt of shrubs or small trees usually placed across fields or along field or property boundaries for wildlife rather than for wind control. (3) To buy or sell futures to protect from loss due to a rise or fall in prices.

Hedgerow–A single row of shrubs or trees that provides a screen or wildlife food and cover, improves the landscape, or serves as a fence or a windbreak.

Hedging–Repeated browsing of trees and shrubs that results in low branching and thick growth, as with a clipped hedge.

Heel–(1) The basal end of a plant stem cutting along with a piece of the older stem. (2) See Hock. (3) The end of the branches of a horseshoe. (4) The rear end of the foot.

Heel-In–To temporarily store young plants by burying the roots in moist soil or other suitable material until they are planted in a permanent location.

Heliophilous–An organism that is adapted to life in full sunlight.

Heliophobous–An organism that grows best in the shade.

Heliotropism–See Phototropism.

Helminthic–Of, or pertaining to, helminths, or worm parasites.

Helminthosporium Blight of Rice–A worldwide, fungal disease caused by *Helminthosporium oryzae*, family Dematiaceae; characterized by yellow or brown spots on the leaves that elongate and turn reddish-brown or red. There is a gray spot in the center of the lesion.

Helminthosporium Leaf Blight of Grasses–A fungal disease of sorghums, Sudangrass, corn, and Johnsongrass caused by *Helminthosporium turcicum*, family Dematiaceae.

Helolac–A lake covered by a mat of aquatic plants such as water lilies, water hyacinth, or alligator weed.

Hematite–Fe_2O_3; anhydrous sesquioxide of iron, a mineral that occurs as deposits of iron ore and is widely distributed in rocks and soils where, in an amorphous condition, it causes a reddish color in the soil.

Hemi-–(Greek) Prefix meaning half.

Hemic Soil Material–Organic soil material intermediate in degree of decomposition between the less decomposed fibric and the more decomposed sapric material. See Histosols.

Hemicellulose–A complex carbohydrate that occurs widely in plants as a structural part of their cell walls. It differs chemically from cellulose by being subject to hydrolysis with dilute mineral acids.

Hemicryptophytes–Perennial bunchgrasses and similar plants whose shots die back to the ground surface at the beginning of the unfavorable growth season, leaving the reproductive buds protected by the upper soil layer and the remains of the dead and weathered plant.

Hemiepiphyte–A plant growing on another plant but not nourished by it, which later sends roots into the soil., e.g., strangular fig (banyan tree, *Ficus benghalensis*) of the tropics. Note: When young, the banyan tree is classed as an epiphyte (not attached to the soil).

Hemiparasite–A partial parasite.

Hemispherical Scale–*Saissetia hemisphaerica*, family Coccidae; a sucking insect pest of oleander, bay, vinca, croton, cyclamen, fern, palm, figs, citrus, and other plants.

Hemizygous–The condition in which only one allele of a pair of genes is present in the cells of an individual plant or animal, the other one being absent.

Hemlock Poisoning–Poisoning of animals resulting from browsing poison hemlock, *Conium maculatum*, family Apiaceae; characterized by cessation of digestion, gas, salivation, dilation of the pupils, rapid pulse and breathing, paralysis, unsteady gait, difficult breathing, and death.

Hemlock, Western–*Tsuga heterophylla*; a native conifer that thrives in mild, humid climates of the Pacific Northwest. Important timber species but useful for windbreaks and landscape plantings. Young seedlings are shade-tolerant.

Hemp–*Cannabis sativa*, family Moraceae; an annual herb that yields hemp fiber (not Manila hemp), hashish, and bird seed. It is the oldest cultivated fiber plant which, escaped, is a weed pest. It also presents serious problems because of its use by drug addicts; the stems and leaves when smoked have a narcotic effect. It is an important crop in Russia and India. Native to Asia. Also called Indian hemp, marijuana, red-root, gallows-grass, hashish, locoweed. See Marijuana.

Hemp Sesbania–*Sesbania exaltata*, family Leguminosae; an annual, tropical herb having beautiful yellow flowers that is grown in warm areas as a cover crop and for green manure. Escaped, it can become a weed pest. Found from Missouri to Louisiana, United States. Also called tall indigo, Colorado river hemp.

Henbit–*Lamium amplexicaule*; a weed related to mint that grows in early spring.

Heptachlor–A chlorinated hydrocarbon insecticide that is closely related to chlordane.

Heptane–C_7H_{16}; a colorless liquid obtained from the oleoresin of the Jeffrey pine, the Digger pine, and from petroleum; used as a constituent of test fuel, a solvent, insecticide, and drug.

Herb–(1) A plant with leaves, stems, roots (and often with flowers), not markedly woody, which dies down to the root crown at the end of its growing season. The roots may or may not survive for another season. (2) A plant that is usually grown in a garden for culinary seasoning or medicinal use.

Herb Garden–A garden in which plants are grown for use in flavoring, for fragrance, for historic, legendary, or other purposes.

Herbaceous–(1) Not woody, dying back to the ground each year, such as rhubarb and asparagus (applied to a plant or stems). (2) Leaflike in color and texture (applied to plant parts). See Forb.

Herbaceous Perennials–Plants with soft, succulent stems whose tops are killed back by frost in many temperate and colder climates, but whose roots and crowns remain alive and send out top growth when favorable growing conditions return.

Herbaceous-Cuttings–Cuttings made from succulent herbaceous plants, such as chrysanthemums, coleus, and geraniums.

Herbage–(1) Herbaceous plants, collectively, specially with reference to their value as animal feed. (2) Nonwoody plants in contrast to trees and shrubs.

Herbal–A book or manuscript that describes plants, often with illustrations, written mostly by the early herbalists.

Herbarium–(Plural, herbaria) (1) A collection of dried or otherwise preserved plant specimens used for reference and for the verification of names of plants. It is usually an adjunct to a college or university where botany is studied and taught. Such a collection of dried, pressed plants formerly was called a *hortus siccus*. (2) The building in which such a collection is displayed.

Herbicide Modifier–A chemical substance (safener, extender, etc.) used with herbicides to change their properties.

Herbicides–Chemicals used to kill plants. They are used in contact with the seed, stem, or leaf of a plant. Herbicides are further divided into nonselective (kill all plants) and a selective (kill only certain species).

Herbivorous–Designating an animal (herbivore) that feeds, in the native state, on grass and other plants, as cattle, horses, sheep, goats, deer, elk, etc. See Carnivore.

Herbland–Any lands on which herbaceous species dominate the vegetation.

Herbosa–Vegetation dominated by nonwoody plants.

Heredity–(1) A study or description of genes passed from one generation to the next through sperm and ova. The heredity of an individual would be the genes received from the sire and dam via the sperm and ovum. (2) Genetic transmission of traits from parents to offspring. (3) The genetic constitution of an individual.

Hesperidia–Fruits with a leathery rind.

Hessian Fly–*Phytophaga destructor*, family Cecidomyiidae; an important insect pest of wheat, barley, and rye, but especially of wheat. Its larva sucks plant juices and causes the infested plants to lodge. A few wheats have been bred to resist Hessian fly damage. See Fly-free Date.

Heterocyst–A specialized vegetative cell in certain filamentous blue-green algae; larger, clearer, and thicker-walled than the regular vegetative cells.

Heteroecious Rust–A rust having different stages of its life cycle on two unlike types of host plants, as stem rust on cereals and barberry; cedar-apple rust on cedar and apple.

Heterogametic–Producing unlike gametes, particularly with regard to the sex chromosome. In species in which the male is of the "X-Y" type, the male is heterogametic, the female homogametic.

Heterogen–A variable group of plants or animals that arise as hybrids, sports, mutations, etc., certain types of which may or may not breed true.

Heterogeneous–Designating elements having unlike qualities.

Heterosis–The amount of superiority observed or measured in cross-bred animals compared with the average of their purebred parents; hybrid vigor.

Heterotrophic–Referring to organisms that for their metabolism are dependent upon organic matter supplied from sources outside of their own bodies. See Autotrophic.

Hexachlorophene–A chemical compound that is an exceptionally persistent insecticide.

Hexamitiasis–An acute, infectious disease of turkey poults, caused by the protozoan organism *Hexamita meleagridis*, characterized by listlessness, droopiness, stilted gait, lowered temperature, watery and foamy diarrhea, and death. Birds that survive the disease are stunted and emaciated.

Hexose–Any of various simple sugars that have six carbon atoms per molecule.

Hibernation–Winter dormancy. The dormant state in which certain animals pass the winter. It is characterized by narcosis (insensibility), very low metabolic activity, and low body temperature. See Estivation.

Hibiscus–See Kenaf Hibiscus.

Hickory Land–Land on which hickory (*Carya* spp.) was a dominant or characteristic growth and was valued accordingly as agricultural land by early settlers of eastern United States.

Hickory Tussock Moth–*Orygia caryae*, family Aarctiidae; a moth which, in the caterpillar stage, infests deciduous trees and shrubs, especially walnut, butternut, apple, pear, and hickory.

Hidden Hunger–Deficiency disease of plants or animals.

High Grading–The removal from the stand of only the best trees, often resulting in a poor-quality residual stand.

High Lysine Corn–Corn that has a higher than normal content of lysine and tryptophan. This type corn has a better balance of amino acids for monogastric animals.

High Magnesic Products–Any material in which more than 10 percent of the total calcium and magnesium oxide content consists of magnesium oxide. See Lime.

High-density Bale–(1) A bale of cotton similar to a standard-density bale except the average density is 36 pounds per cubic feet. (2) A highly compacted bale of hay with a density of 12 to 15 pounds per cubic feet.

High-lime Soil–A soil that contains lime in sufficient amount to cause effervescence when the soil is tested with cold 10 percent hydrochloric acid. Loosely used in humid regions for any soil that gives an alkaline reaction (greater than pH 7.0), presumably because of its lime content.

High-moor Peat–A group of peat soils that are raw or undecomposed and usually strongly acidic, as opposed to low-moor peat which includes the decomposed and less acidic or alkaline peats. Also called hochmoor peat. See Histosols.

Highbush Blueberry–*Vaccinium corymbosum*, family Ericaceae; a bushy shrub found in swamps and low woods. It produces a large, edible berry in the wild state and is the source of horticultural varieties in commercial production of blueberries. Found in the northeastern United States. Also called high blueberry, swamp blueberry, tall blueberry, tall bilberry. See Blueberry.

Highly Erodible Land–Land that has an erodibility index of greater than 8. See Erodibility Index.

Hill–(1) A natural relief feature, a rounded elevation a few feet to a few hundred feet above the adjacent lowland, which is distinguished from other features such as mesa, butte, and peak by its form and from mountain by its lesser magnitude. (2) An artificially raised elevation; e.g., a row in a field or garden that is slightly raised above the surrounding soil or a small mound of soil in which seed may be planted. (3) A plant or a cluster of plants, as a hill or potatoes, a hill of beans. (4) To make soil into a hill or to form soil up around a plant.

Hilum–In the seed, the scar or mark indicating the point of attachment.

Hip–(1) The fruit of rose; rose hips. (2) The external angle (ridge) formed by the meeting of two sloping sides of a roof. (3) That region of one of the rear quarters of four-legged animals where the hind leg joins the pelvic region.

Hirsute–Covered with coarse hairs.

Histology–The science of the microscopic structure of plant and animal cells.

Histosols–(Peats and mucks) A soil order in the United States system of soil taxonomy comprising organic soils. Of the four suborders, Fibrists, Folists, Hemists, and Saprists, only Fibrists have the physical characteristics essential for use as a filler and conditioner of fertilizers. Fibrists have a density of <0.1 g/cm^3 (<6.2 lb./ft.^{3}). See Peat.

Hoar Frost–Frost accompanied by a visible deposit of frozen moisture on trees as well as on ground vegetation.

Hoarhound–*Marrubium vulgare*; a weed that was once used for medicinal purposes, e.g., as a cough medicine.

Hog-wallow Land–Flatland characterized by natural microbasins and mounds; frequently no more than 2 or 3 feet in relief. (A vernacular term, mostly southern United States, without a very precise meaning; used especially in Texas to describe parts of black clay land prairies.) See Gilgai, Vertisols.

Holiday Disease–See Azoturia.

Hollow Heart–A condition of cauliflower caused by a boron deficiency.

Hominy–Corn soaked in an alkali; sometimes ground and served as grits. See Hominy Feed, Hominy Grits.

Hominy Grits–A stock feed and human food that consists of the fine or medium-sized, hard, flinty portions of sound corn containing little or none of the bran or germ. Also called corn grits.

Homo–(Latin) (1) Man. (2) Prefix meaning alike or same.

Homogametic–Refers to the particular sex of the species that possesses two of the same kind of sex chromosome such that only one kind of gamete can be produced with respect to the kinds of sex chromosomes it contains; in mammals, the female is the homogametic sex (XX).

Homogamy–(1) The simultaneous maturation of male and female elements of the same flower. (2) In flowering plants, bearing one kind of flower throughout, as members of the family Cichoriaceae.

Homogeneous–Being of uniform character or nature throughout.

Homolog–One of a pair of structures having similar structure, shape, and function, as with two homologous chromosomes.

Homologous–Organs or parts that exhibit similarity in structure, in position with reference to other parts, and in mode of development, but not necessarily similarity of function, are said to be homologous.

Homologous Chromosomes–Pairs of chromosomes that are the same length, that have their centrioles in the same position, and that pair up during synapsis in meiosis.

Homothallic–Designating a fungus whose mycelium produces hyphae functionally male or female.

Homozygous–Possessing identical genes with respect to any given pair or series of alleles.

Homozygous Recessive–A recessive character that produces two kinds of gametes; one carries the dominant gene, while the other carries the recessive gene.

Honey–(1) An aromatic, viscid, sweet food material derived from the nectar of plants through collection by honeybees; modified by the bees into a denser liquid and finally stored in honeycombs. Of acid reaction, liquid in its original state, it becomes crystalline on standing. Honey consists chiefly of two simple sugars, dextrose and levulose, with occasionally more complex carbohydrates, with levulose usually predominant, and always contains minerals, plant coloring materials, several enzymes, and pollen grains. (2) Legally, the nectar and saccharine exudation of plants, gathered, modified, and stored in the comb by honeybees, which is levorotatory, and contains not more than 25 percent of water, not more than 0.25 percent ash, and not more than 8 percent sucrose.

Honey Plant–Any plant from which honeybees gather nectar and pollen, especially one which either is in abundance or one which gives a distinctive flavor to honey.

Honeycomb Frost Structure–Frozen soil that has a granular structure or is easily broken into fragments, associated with the shallow freezing and granular soils high in organic matter.

Honeydew–(1) A sweetish, sticky exudate that drops from the leaves and petioles of certain plants. It is attributed to infestation by sucking insects, excessive plant turgidity, or to a superabundance of sap. It often occurs during hot weather. (2) The excretion of certain homopterans, such as aphids, leafhoppers, and scale insects.

Honeydew Melon–A very famous variety of the muskmelon, ***Cucumis melo***, family Cucurbitaceae. See Muskmelon.

Honeyflow–The incoming of nectar to the beehive, used especially in reference to the periodic changes in quantity related to the blooming periods of dominant flowering plants of the area, such as the clovers.

Hop–(1) Any vine of the genus ***Humulus***, family Moraceae. (2) ***H. lupulus***, a perennial, tall vine grown for its ripened and dried pistillate cones, the hops of commerce, used in beer manufacture and in medicines. Native to Eurasia.

Hop Clover–Certain members of the genus ***Trifolium***, family Leguminosae, whose flower heads resemble those of the hop.

Hopperburn–A disease of potato, alfalfa, and other plants resulting from the feeding of the potato leafhopper, a toxicogenic insect.

Horizon, Soil–A layer of soil, approximately parallel to the surface, having distinct characteristics produced by soil-forming processes. The major horizons of mineral soil are as follows: (1) O horizon: an organic layer, fresh and decaying plant residue, at the surface of a mineral soil. (2) A horizon: the mineral horizon, formed or forming at or near the surface, in which an accumulation of humified organic matter is mixed with the mineral material. Also, a plowed surface horizon most of which was originally part of a B horizon. (3) E horizon: a mineral horizon, mainly a residual concentration of sand and silt high in content of resistant minerals as a result of the loss of silicate clay, iron, aluminum, or a combination of these. (4) B horizon: the mineral is in part a layer of change from the overlying A or E to the underlying C horizon. The B horizon also has distinctive characteristics caused (a) by accumulation of clay, sesquioxides, humus, or a combination of these; (b) by prismatic or blocky structure; (c) by redder or browner colors than those in the A horizon; or (d) by a combination of these. The combined A, E, and B horizons are generally called the solum, or true soil. If a soil lacks a B horizon, the A or E horizon alone is the solum. (5) C horizon: the mineral horizon or layer, excluding indurated bedrock, that is little affected by soil-forming processes and does not have the properties typical of the A or B horizon. The material of a C horizon may be either like or unlike that from which the solum is presumed to have formed. If the material is known to differ from that in the solum, the Roman numeral II precedes the letter C. (6) R layer: consolidated rock beneath the soil. The rock commonly underlies a C horizon, but can be directly below an A, E, or B horizon.

Horizontal Resistance–A plant's ability to uniformly resist all strains of a pathogen.

Hormone–A chemical substance formed in some organ of the body, secreted directly into the blood, and carried to another organ or tissue, where it produces a specific effect.

Hornworm–See Tobacco Hornworm, Tomato Hornworm.

Horse Manure–Dried horse excrement used as a medium for growing mushrooms, for making heat in hot beds, and as a soil amendment.

Horse Nettle–*Solanum carolinense*; a perennial weed that has sharp spines and a ripe fruit resembling small tomatoes about the size of a marble.

Horseradish–*Armoracia lapathifolia (A. rusticana)*, family Cruciferae; a perennial herb grown for its edible, pungent root used as a condiment and in flavoring; native to Eurasia.

Horsetail–See Equisetum Poisoning.

Horseweed–*Erigeron canadensis*; a common weed of waste places and abandoned fields.

Horticulture–The science of agriculture that relates to the cultivation of gardens or orchards, including the growing of vegetables, fruits, flowers, and ornamental shrubs and trees.

Hortus–(Latin) Garden.

Host–Any organism, plant or animal, in or upon which another spends part or all of its existence, and from which it derives nourishment and or protection.

Host Plant–Plant on which an insect or a disease-causing organism lives.

Host Plant Resistance–The development of plant resistance to pests by selective breeding. Components of host resistance may include nonpreference, antibiosis (abnormal biological effects on the pest), and tolerance.

Host Specific–Designating a parasite that can live in or on only one host, to which it is therefore said to be specific.

Hot–(1) Designating a horse with a bad disposition. (2) Designating a fruit, such as a pepper, that has a pungent, strong, lasting flavor. (3) Designating weather. (4) Designating manure that heats upon decomposition. (5) An animal feed that contains a high percentage of concentrate or a feed containing a high level of salt.

Hot Caps–Waxpaper cones, paper sacks, cardboard boxes, or plastic jugs with bottoms removed placed over individual plants in spring for frost and wind protection.

Hot Manure–Fresh manure that is going through the process of heating due to fermentation. Horse manure is designated as one of the hot manures. Cow manure is one of the cold manures. See Cold Manure, Cow Manure, Horse Manure.

Hot Pack–Food heated in syrup, water, steam, or juice, and packed hot into canning jars.

Hot Pepper–Any of certain varieties of redpepper (***Capsicum*** spp.) which has a distinctive, pungent, and persistent flavor.

Hotbed–Same type of structure as a coldframe but heated, as with an electric cable or by using horse manure, which heats upon decomposition.

Hothouse–See Greenhouse.

House Plant–An ornamental plant that is grown for its beauty and kept in the house.

Howard Scale–*Aspidiotus howardi*, family Diaspididae; a scale insect that infests almond, apple, ash, peach, plum, prune, and other deciduous fruits and vegetables, but especially pear. It is found in Colorado and New Mexico, United States.

Hubam Sweetclover–*Melilotus alba* var. ***annua***, family Leguminosae; an annual herb developed by Professor Hughes at Ames, Iowa, United States, from plants found growing wild in Alabama, which is sometimes sued for pasture, hay, silage, soiling or soil improvement, and as a bee plant. Also called annual white sweetclover. See Sweetclover; White Sweetclover.

Hubbard–Group of squash, ***Cucurbita*** spp., family Cucurbitaceae; a group of varieties of the winter squash that are generally ovoid and pointed at both ends. Their surfaces vary from smooth to warted, and their colors range from deep green to orange and gray.

Huckleberry–(1) Any of several wild species of the genus ***Gaylussacia***, family Ericaceae. Species are seldom cultivated but are sometimes harvested from the wild for their edible fruits used at home or occasionally sold on markets. (2) Incorrectly called the lowbush blueberry.

Hull–(1) The outer shell or covering of any fruit. Also called husk. (2) The enlarged calyx of the fruit of the strawberry. (3) An excessively thin animal. (4) The outer shell or covering of any device or product. (5) To remove the hulls from seeds, or calyxes (caps) from strawberries.

Hulled Seeds–Seed from which the pods, glumes, or other outer coverings have been removed.

Humectant–Substance that absorbs moisture; used to maintain strength of materials such as baking powder.

Humic Acid–(1) Any of certain substances related to humus that is formed by heating carbohydrates with water under pressure. These substances are weak acids that form salts. Part of the natural humus is acid in nature and is generally present in the soil in the form of a salt that, when dissolved by an alkali and the solution neutralized by an acid, precipitates humic acid. (2) Acid formed by the decomposition of organic matter.

Humilis–Dwarf.

Humin–The organic matter in the soil that is not dissolved upon extraction of soil with a dilute alkali.

Humus–(1) Organic matter in the soil that has reached an advanced stage of decomposition and has become colloidal in nature. It is usually characterized by a dark color, a considerable nitrogen content, and chemical properties such as a high cation-exchange capacity. (2) Any organic matter in the surface layer of soil. (3) Commercially, peat and muck, regardless of degree of decomposition. (4) Leaf mold and duff of a forest floor.

Husk–(1) The dry, harsh, or rough, outer envelope of seeds or fruit, as a corn husk. (2) To remove the husks form anything, especially from mature corn.

Hybrid–(1) An animal produced from the crossing or mating of two animals of different breeds. (2) A plant resulting from a cross between parents that are genetically unlike; more commonly, in descriptive taxonomy, the offspring of two different species.

Hybrid F1–Plants of a first-generation hybrid of two dissimilar parents. Hybrid vigor, insect or disease resistance, and uniformity are qualities of this generation. Seed from hybrid plants should not be saved for future planting. Their vigor and productive qualities are only in the original hybrid seed.

Hybrid Seeds–Seeds produced from crossing two varieties.

Hybrid Vigor–The increase of size, speed of growth, and vitality of a crossbreed over its parents. See Heterosis.

Hybridization–The production of hybrids by natural crossing or by manipulated crossing.

Hybridize–To create a hybrid.

Hydrarch–The type of plant succession that originates in lakes or ponds and progresses toward a mesic plant community.

Hydrate–A compound formed by the union of one or more molecules of water with a salt or a mineral.

Hydrated–Chemically combined with water.

Hydrated Lime–$Ca(OH)_2$; a dry product that consists chiefly of the hydroxide of calcium and oxide-hydroxide of magnesium, sometimes used as a soil amendment. Also called slaked lime; calcium hydroxide. See Lime.

Hydrator–Vegetable crisper; a drawerlike section in refrigerators that protects fresh fruits and vegetables form drying out during refrigerator storage.

Hydric–Tending to be wet. Hydrophytic plants grow in water or in very wet soils. See Mesic, Xeric.

Hydric Soil–Soil that is wet long enough to periodically produce anaerobic conditions, thereby influencing the growth of plants.

Hydrochloric Acid–HCl; A colorless, incombustible, pungent gas, commonly known in the form of its aqueous solution. It is used in treating phosphate rock in the manufacture of superphosphate fertilizer. See Sulfuric Acid.

Hydrocyanic Acid–HCN; one of the most valuable and widely used of the fumigants. Except that it is highly toxic to people, it approaches the ideal in a fumigant. It is found in some plants, particularly sorghums, and may under certain conditions cause poisoning and death to animals. Sometimes called prussic acid.

Hydrofluoric Acid–HF; produced in the off-gases from phosphate rock acidulation.

Hydrogen Peroxide–A chemical substance often used as a bleach to remove color. It is used also in medicine and surgery as an antiseptic agent and as a cleansing agent in mouthwashes, toothpastes and powders. Its antiseptic and cleansing action is due to the fact that it gives off sufficient oxygen to destroy bacteria.

Hydrogen Sulfide–H_2S; a poisonous gas with the odor of rotten eggs that is produced from the reduction of sulfates in, and the putrefaction of, a sulfur-containing organic material.

Hydrogenated Oils–Oil hardened by treatment with hydrogen in the presence of nickel. Cottonseed, corn, and wheat oils are commonly hardened and used in cooking fats.

Hydrogenic Soil (Hydromorphic)–Soil that was developed under the dominant influence of water.

Hydrologic Cycle–The complete cycle through which water passes, commencing as atmospheric water vapor, passing into liquid and solid forms (ice and snow) as precipitation, into the ground surface, and finally again returning in the form of atmospheric water vapor by means of evaporation and transpiration. Syn.: water cycle.

Hydrologic Soil Groups–Refers to soils grouped according to their runoff-producing characteristics. The chief consideration is the inherent capacity of soil bare of vegetation to permit infiltration. The slope and the kind of plant cover are not considered, but are separate factors in predicting runoff. Soils are assigned to four groups. In group A are soils having a high infiltration rate when thoroughly wet and having a low runoff potential. They are mainly deep, well-drained, and sandy or gravelly. In group D, at the other extreme, are soils having a very slow infiltration rate and thus a high runoff potential. They have a claypan or clay layer at or near the surface, have a permanent high water table, or are shallow over nearly impervious bedrock or other material. A soil is assigned to two hydrologic groups if part of the acreage is artificially drained and part is undrained.

Hydromorphic Soil (Hydrogenic)–A soil that has been formed under a strong influence from water, as one developed in a bog or swamp.

Hydrophyte–(1) Any plant growing in water or on a substrate that is at least periodically deficient in oxygen as a result of excessive water content. (2) Plants typically found in wet habitats.

Hydroponics–Growing plants in nutrient solutions rather than soil. Also called soilless gardening. See Geoponics.

Hydroxide of Lime–See Hydrated Lime.

Hygiene–The science of health; the rules or principles of maintaining health in people and animals; sanitation.

Hymenium–The spore-bearing layer of a fungous fruiting body.

Hypersaline–Term to characterize waters with salinity greater than 4.0 percent due to land-derived salts.

Hypertrophy–(1) Pertaining to waters of very high nutrient content. (2) Morbid enlargement or overgrowth of an organ or part due to increase in size of its constituent cells.

Hypha–The simple or branched threadlike filaments that compose the weblike mycelium of fungi.

Hypobaric Storage–A cold-storage principle in which atmospheric pressure is reduced. The subatmospheric pressure lowers respiration by reducing both oxygen and ethylene concentration in plant tissues.

Hypocotyl–The short stem of an embryo seed plant, the portion of the axis of the embryo seedling between the attachment of the cotyledons and the radicle.

Hypogeous–Growing or remaining underground.

Hypogynous–Classification of a flower in which the sepals, petals, and stamens are attached to the receptacle below the ovary.

Hypsometer–Any instrument used to measure heights of trees, based either on geometric or trigonometric principles.

Ice Cream Species–An exceptionally palatable species of forage sought and grazed first by livestock and game animals. Such species are usually overutilized under proper grazing.

Ice Damage–(1) The breakage of tops, branches, or stems of plants by the action of ice storms. (2) Smothering of perennial plants, such as alfalfa, by ice sheet formation over the crowns of the plants. The heaving of plants out of the ground. See Heaving.

Ice Scald–A physiological condition of wrapped peaches that have been stored at too low a temperature.

Ice Smothering–A suffocation of a plant due to a covering of ice remaining on the plant and the surrounding soil for a prolonged period.

Icterus–(1) Jaundice. (2) In plants, a yellowing of the leaves due to cold, excessive moisture, or other climatic factors.

Idaho Fescue–*Festuca idahoensis*, family Gramineae; a perennial bunchgrass whose forage rating is excellent. Found from Alberta, Canada, south to Arizona and west to California, United States, at elevations between 5,000 and 10,000 feet.

Idiopathic–Designating a disease or condition with no apparent cause.

Idle Land–(1) Land that has been managed or exploited for some particular use but is now in a state of disuse; abandoned land. (2) Farmland that is capable of producing but is not in use; by extension, any land that has potentiality, but is not being put to any productive use. (3) Land being fallowed.

Illite–A group of micalike clay minerals that may be present in the clay constituent of soils, and that has chemical properties different from other clay groups, such as kaolinite and montmorillonite. The mineral is a hydrous silicate of aluminum, containing potassium, iron, and magnesium, but is variable because of replacement of aluminum by iron and magnesium.

Illuvial Horizon–A soil horizon that has received material from the A horizon through the process of eluviation. Usually called B horizon. See Horizon.

Imago–(Plural, imagoes or imagines) The adult stage of an insect.

Imbibition–The process by which solids, especially organic substances in the gel state, take up liquids and in so doing, swell. The imbibing substance may develop great pressures during the swelling; e.g., absorption of water by seeds, wood, gelatin; absorption of ether by rubber. It is important in the physiology of the plant as one of the means of absorbing water by roots and in the uptake of water by seeds and the splitting of seed coats in the process of germination.

Imbricate–Overlapping, as shingles on a roof.

Imbricated Snout Beetle–*Epicaerus imbricatus*, family Curculionidae; an insect pest that infests plants and is generally especially destructive to strawberries.

Immunity–Having resistance to the action of something, such as a disease. It may be inborn, may result from exposure to a disease, from having had a disease, or from having received an injection of immune serum. Degree of immunity varies in each case. See Immunizing Agent.

Imperfect Flowers–Flowers lacking either stamens or pistils.

Imperfect Fungi–In the classification of fungi, that large group (*Fungi imperfecti*), in which the sexual stage may be lacking.

Impermeability–(1) In groundwater hydrology, the property of rock or earth which does not permit water to move through it readily. (2) Being impervious, not easily penetrated, as by oil or water, either because of low porosity, very small individual pores, or pores that are disconnected. (3) Having the property of not permitting the passage of liquids or gases.

Impervious Soil–A subsurface soil layer that is resistant to the free, downward movement of water or to penetration by plant roots. See Sodic Soil.

Imported Fire Ant–*Solenopsis saevissima* var. *richteri*, family Formicidae; an ant imported from South America; widespread in southern United States and rapidly moving northward. Its sting is harmful to humans, and it is a serious menace because of its destructive feeding on crop plants, its attacks on young animals, and the nuisance of ant hills in fields. See Fire Ant.

Impotence–Temporary or permanent loss of reproductive power or virility.

Impotent Pollen–Pollen that is nonviable and incapable of fertilization.

Impregnate–To fertilize a female animal or flower.

In Season–(1) That part of the year when a product is normally harvested and, as a result, is cheaper, more plentiful, and more flavorful. (2) See Estrus, In Heat.

In Situ–In its natural position or place. Said specifically of a rock, soil, or fossil when it is in the same location in which it was originally formed or deposited.

In Vitro–In the test tube, outside the animal body. See In Vivo.

In Vivo–In the living body. See In Vitro.

Inarching–A method of plant propagation in which a plant, still attached to its own roots, is grafted to another such plant. The cambium regions of both plants are exposed and fastened together so that they will unite. After the union is accomplished, one plant may be severed from its roots. Also called approaching.

Inceptisols–One of the eleven soil orders in the United States system of soil taxonomy. Soils that are usually parent materials but not of illuviation (materials moving in). Generally, the direction of soil development is not yet evident from the marks left by the various soil-forming processes or the marks are too weak to classify in another soil order. See Soil Orders.

Inchworm–Any larva of a moth of the family Geometridae. Also called measuring worm, cankerworm.

Incipient Erosion–The early stages of erosion, especially with reference to gullying. See Erosion.

Incision–A cut.

Inclusion–A nonliving substance or particle in a cell.

Incompatibility–A condition in either plants or animals in which the viable male gamete will not fertilize the viable female gamete.

Incomplete Dominance–A kind of inheritance where a gene does not completely cover up or modify the expression of its allele; also may be known as codominance or blending inheritance.

Incomplete Fertilizer–A commercial fertilizer that contains only one or two of the three major elements essential to plant growth: nitrogen, phosphorus, and potassium.

Incomplete Flower–A flower that lacks one or more of the four organs: sepals, petals, stamens, or pistils.

Incorporate–To mix pesticides, fertilizers, or plant residues into the soil by plowing or other means.

Increaser Plant Species–Plant species of the original vegetation that increase in relative amount, at least for a time, under overuse. Commonly termed increasers. See Decreaser Plant Species.

Increment Borer–An instrument used for boring toward the center of a tree to extract a small core that can be used for studying annual growth as well as depth of penetration of a preservative.

Incross–The progeny from the cross of individuals of inbred lines of the same breed.

Incubation–The process of development of a fertile poultry egg within the shell. This is accomplished through the use of a mechanical incubator that controls the egg's environment with regard to temperature and humidity.

Incubation Period–(1) The period between the setting and hatching out of eggs. The following are average incubation periods for some common birds: chickens, 21 days; most ducks, 28 days; Muscovy ducks, 33 to 35 days; geese, 30 to 34 days; turkey, 28 days; guinea fowl, 26 to 28 days; pheasant, 23 to 25 days; pea fowl, 28 to 30 days; pigeons, 17 to 19 days. (2) The elapsed time between exposure to infection and the appearance of disease symptoms.

Indehiscent–(1) Designating certain dry fruits that do not split open when mature. (2) Not opening, or not opening by valves or along regular lines. See Dehiscence.

Indeterminant Tomato–Terminal bud is always vegetative, thus the stem grows indefinitely. Indeterminant plants can be trained on a trellis, a stake, or in wire cages. See Determinate Tomato.

Indeterminate Growth–A type of growth (in plants) that continues as long as the plant lives.

India Rubber–See Latex.

Indian Corn–See Maize.

Indian Millet–See Pearl Millet.

Indian Physic–*Gillenia stipulata*, family Rosaceae; a perennial herb grown in the wild garden for its loose clusters of white flowers and its emetic root. Native to North America. Also called American ipecac, Hemp dogbane.

Indian Ricegrass–*Oryzopsis hymenoides*, family Gramineae; a perennial bunchgrass found from Manitoba to British Columbia in Canada, and south to Texas, United States. Its general forage rating is good.

Indian Tobacco (Lobelia)–*Lobelia inflata*, family Lobeliaceae; an herb that is poisonous to livestock. Symptoms of poisoning are nausea, exhaustion, prostration, dilated pupils, stupor, coma, convulsions, and death. Found from Canada to Georgia and Nebraska to Arkansas, United States.

Indiancurrant (Coralberry)–*Symphoricarpos orbiculatus*, family Caprifoliaceae; a hardy, deciduous shrub grown for its white flowers, reddish-purple fruit, and its foliage, which turns crimson in autumn. Native to North America. Also called Indian currant.

Indianfig–*Opuntia ficus-indica*, family Cactaceae; a bushy or treelike cactus grown in arid, warm areas for its edible, red, juicy fruit. Native to Mexico.

Indicator Plant–A native plant that indicates, in general, and often in a specific manner, the nature of soil conditions with regard to moisture and salinity. Dominant species are the most important indicators.

Indifferent Plant–See Day-neutral Plant.

Induced Crusts–A surface compaction of the soil caused by trampling by people or livestock or by the use of heavy wheeled traffic on a wet soil.

Induced Pans–A dense layer of soil at about 6 to 12 inches deep that is caused by heavy traffic on a wet soil.

Induction Period–The number of times or periods of a given daylength to which a plant must be exposed to cause flowering. See Photoperiodism.

Induration–The process of hardening of soil or sediments through cementation, pressure, heat, and/or other causes.

Industrial Wood–All commercial roundwood products except fuel wood.

Inedible–A substance that is not fit for food, such as poisonous nuts and plants. Tough skins, seeds, and decayed spots of fruits and vegetables and bones of meat are considered inedible parts because they are not suitable for human consumption.

Inequalateral–A leaf base that is slanted with unequal sides.

Inert Ingredient–A substance in a feed, pesticide, etc., that does not act as a feed, pesticide, etc. The substance may serve a purpose but is usually used as a filler, vehicle, etc.

Infect–To cause disease by the introduction of germs, parasites, or fungi. See Infection.

Infection–Invasion of the tissues of the body of a host by disease-producing organisms in such a way that injury results; the presence of multiplying parasites, bacteria, viruses, etc., within the body of a host. See Infestation.

Infection Stage–The period in the course of a disease during which the host responds, symptoms appear, and the disease develops.

Infectious–Designating a communicable disease.

Infectious Chlorosis–A graft-transmissible disease of green plants that causes irregular variegations, yellowing, or blanching of the leaves. The symptoms vary considerably among different species of plants; however, affected leaves may be stunted, and sometimes the plant dies. The disease is known to occur in the following plant families: Malvaccac, Cclastraceae, Oleaceae, Leguminosae, Cornaceae, Rutaceae, and Rosaceae.

Infectious Disease–A disease caused by bacteria, protozoa, viruses, or fungi entering the body. It is not necessarily contagious or spread by contact.

Infectious Variegation–A form of psorosis virus disease of citrus characterized by crinkled, chlorotically variegated leaves (and misshapen fruit in lemons).

Infective–Capable of entering and establishing itself in a host; able to infect a susceptible plant or animal.

Infertile–(1) Designating that which is incapable of reproduction; e.g., a barren female animal, a male animal with nonviable spermatozoa, an unfertilized egg, a flower that will not produce seed. (2) A soil so low or unbalanced in essential nutrients that it will not produce a profitable crop.

Infest–To assail, attack, overrun, annoy, disturb; as ticks infest a cow, or as an insect infests a plant.

Infestation–(1) Act of infesting, or state of being attacked, molested, vexed, or annoyed by large numbers of insects, etc., as an animal may be subject to an infestation of parasites, such as fleas, ticks, mites, etc. (2) Presence of disease in a population of plants, or of pathogens in a position, as in soil or on seed surfaces, where they have the possibility of producing disease. (Not to be confused with infection, which can be applied only to living, diseased plants and animals.) See Infection.

Infiltration–The flow of a liquid into a substance through pores or other openings, connoting flow into a soil in contradistinction to ***percolation***, which connotes flow through a porous substance.

Infiltration Rate–The maximum rate at which a soil, in a given condition at a given time, can absorb water, commonly expressed in inches of depth per hour. According to the National Cooperative Soil Survey administered by the USDA Soil Conservation Service, the movement of water into the soil (infiltration rate) is classified as follows: (1) Very low; soils with infiltration rates of less than 0.1 inches (0.25 centimeters) per hour; soils in this group are very high in percentage of clay. They may also be soils that have been abused by excessive tillage. (2) Low; soils with infiltration rates of 0.1 to 0.5 inches (0.25 to 2.5 centimeters) per hour; soils in this group are loams and silts. (4) Soils with infiltration rates of greater than 1.0 inches (2.5 centimeters) per hour; these are deep sands, well-aggregated silt loams, tropical sesquioxide clay soils with high porosity, and well-managed vertisols.

Inflorescence–(1) The mode of arrangement of the flowers on a plant. (2) The flowering part of a plant. (3) The coming into flower of a plant.

Inheritance–(1) The transmission of genetic factors from parent to offspring. (2) The process or procedure of transferring property, both real and personal, from one generation to the next, either by will or by laws of descent and distribution.

Inherited Characteristic (Trait)–A character, the expression of which is determined by a particular gene or genes.

Inherited Soil Characteristic–Any soil characteristic that is due directly to the nature of the parent material, as contrasted with those partly or wholly due to the process of soil formation; e.g., the red color of a soil is said to be inherited if it is due entirely to the fact that the parent material is red.

Inhibit–To suppress, prevent, hinder, restrain.

Inhibiting Vegetation–Vegetative cover sufficiently dense to control the establishment of undesirable vegetation. See Cover Crop.

Initial Water Deficiency–The amount by which the actual water content of a given soil zone (usually the root zone) is less than field capacity at the beginning of a season or specified time period.

Injection Well–A well into which surface water is pumped to increase subsurface water volume.

Inland Saltgrass–*Distichlis stricta*, family Gramineae; a widely distributed grass of the western and midwestern states, United States, whose general forage value is fair. Also called desert saltgrass.

Inlaying–A graft that consists of removing an area of the stock and preparing a scion that will fit it exactly. See Grafting.

Inlet–(Hydraulics) (1) A surface connection to a closed drain. (2) A structure at the diversion end of a conduit. (3) The upstream end of any structure through which water may flow.

Inoculation–(1) Introduction into healthy plant or animal tissue of microorganisms to produce a mild form of the disease, followed by immunity. (2) An introduction of nodule-forming bacteria into soil, especially for the purpose of nitrogen fixation. (3) Treatment of seed with bacteria that stimulate development of bacteria nodules on plant roots. Used on legumes such as peas and beans. (4) Bacteria supplied to legumes to "fix" nitrogen from the air. (5) A small amount of bacteria produced from a pure culture that is used to start a new culture.

Inorganic–(1) Not made up of or derived from plant or animal materials. (2) Mineral content of the soil. In reference to fertilizers, those produced chemically. Not arising from natural growth.

Inorganic Matter–Compounds that do not contain carbon and hydrogen.

Inorganic Soil–A mineral soil; a soil in which the solid matter is dominantly rock minerals in contrast to organic soils, such as peats and mucks.

Insect–An air-breathing animal (phylum Arthropoda) that has a distinct head, thorax, and abdomen. Insects have one pair of antennae on the head, three pairs of legs, and usually two pairs of wings on the thorax. The opening of the reproductive organs is near the posterior end of the body. They may be harmful or useful depending upon their habits. Some infest plants and animals, some are insectivorous, some pollinate plans, and some produce edible products.

Insect Control–The chemical or biological inhibition or killing of insect enemies.

Insect Enemy–(1) Any insect that is destructive or harmful to something desired by humans. (2) An insect, bird, mammal, etc., that preys on other insects.

Insect Growth Regulator–A chemical that interferes with the normal growth pattern of insects causing abnormal development and thus death. In the case of flies, it may be added to chicken feed, passing through the bird with the feces into the manure. Present in the manure, it kills house fly larvae soon after they hatch.

Insect Vector–An insect that carries a virus, bacterium, or the spores or mycelium of a pathogenic fungus and inoculates susceptible plants, animals, and humans.

Insecticide–A substance that kills insects by chemical action, as a stomach poison, contact poison, or fumigant.

Insectigation–The application of insecticides through an irrigation system.

Insidious Disease–A disease that develops slowly in a stealthy, subtle manner over a long period of time.

Insipid–(1) Designating a flavor defect of cheese characterized by a lack of taste and odor. (2) designating any such flavor.

Insoluble–Not soluble; designating a substance that does not dissolve in another.

Inspissosis–A fruit rot of citrus caused by either *Nematospora coryli* or *N. gosspii*, family Saccharomycetaceae; characterized by a thickening and drying of the internal tissues of the fruit with no external symptoms. Also called dry rot.

Instar–An insect that is between the stages of its molting process.

Integrated Pest Management (IPM)–An ecological approach to pest management in which all available necessary techniques are systematically consolidated into a unified program, so that pest populations can be managed in such a manner that economic damage is reduced and adverse side effects are minimized.

Integument–Any covering layer, as the skin of animals, the body wall of an insect, or the shell or outer layer of the ovules of plants.

Interaction–The process of chemicals being mixed together and having substantially different toxicity than the toxicities of the components.

The chemicals may interact to increase or decrease toxicity. See Antagonism, Synergist.

Intercellular–Between cells.

Interception–The process by which precipitation is caught and held by foliage, twigs, and branches of trees, shrubs, and other vegetation. Often used for "interception loss" or the amount of water evaporated from the precipitation intercepted.

Interception Channel (Ditch)–A channel excavated at the top of earth cuts, at the foot of slopes, or at other critical places to intercept surface flow; a catch drain; a diversion channel.

Intercrop–To grow two or more crops simultaneously, as by alternate rows in the same field. See Catch Crop, Companion Crop.

Interfertility–The ability of a plant to produce viable seeds by cross pollination. Also called cross fertility.

Interfruitfulness–The ability of a variety of a fruit plant to mature fruit following pollination by another specific variety. Also called cross fruitfulness.

Intergeneric Cross–A cross between species of different genera (rare).

Interior Live Oak–*Quercus wislizeni*, family Fagaceae; a broad-leaved, evergreen tree browsed on the western range, United States. Its browse rating is good. Native to California, United States, and Mexico.

Intermediate Host–An animal other than the primary host that a parasite uses to support part of its life cycle.

Intermingling Color–A coat color pattern of animals in which the separate colors merge where they meet. It is usually an objectionable color pattern.

Intermittent Parasites–Those parasites, such as mosquitoes or bedbugs, that approach the host only when in need of nourishment.

Internal Bark Necrosis–A nontransmissible condition of pome fruit trees characterized by elevations scattered over the surface, brown necrotic areas in outer bark, stunted growth, and roughened appearance on the limbs. Studies indicate it may be due to a mineral imbalance.

Internal Browning of Cauliflower–A boron-deficiency disease characterized by poor growth, yellowing of the terminal growth, and death of the terminal buds. The terminal growth becomes brittle, breaks easily, and shows vascular discoloration. There are light brown spots of dead tissue within the fruit.

Internal Cork of Apples–A boron-deficiency disease characterized by poor growth, yellowing of the terminal growth, and death of the terminal buds. The terminal growth becomes brittle, breaks easily, and shows vascular discoloration. There are light brown spots of dead tissue within the fruit.

Internal Drainage–The relative degree of downward movement of water in a soil. Also called permeability.

International Code of Botanical Nomenclature–A system, adopted by European botanists and some American botanists, for applying the Latin names to plants. Established by a congress in Vienna in 1905, it was revised in 1931 when it was universally accepted, replacing, in the United States, the American Code of Botanical Nomenclature. It provided that most plants shall retain the name applied by Linnaeus in his Species Plantarum (1753). it accepted the idea of ***nomen conservandum*** and provided that after January 1, 1931, publication of all new species shall be accompanied by a Latin name. Also called Vienna Code, Vienna Rules.

Internode–The portion of a stem or other structure between two nodes.

Interphase–The period in the life of a cell between mitotic divisions.

Interplant–(1) To set out young trees among other existing young trees, planted or natural. It is applicable also to planting land partly occupied by brushwood. (2) To plant seeds of one species between those of another species, as to plant soybeans with corn. (3) To plant one crop in rows and another between the rows; e.g., peanuts grown between rows of cotton or corn.

Interrow Tillage–The tillage between the rows of growing plants to condition the soil and reduce weed growth.

Intersex–Designating an organism that displays primary and secondary sexual characteristics intermediate between male and female.

Interspecific–Referring to events or relationships that occur between individuals of different species.

Interspecific Hybrid–Cross between individuals of different species. Taxonomically identified by listing both species separated by an *x*.

Intersterility–The inability of a variety of fruit to produce viable seeds if only the pollen of another variety is used. Also called cross sterility.

Interstitial–A term referring to the spaces (voids) between particles, as between sand grains.

Interunfruitfulness–The inability of a variety of plant to mature fruit if only the pollen of another variety is used. Also called cross unfruitfulness.

Interveinal–Between the veins, especially of a leaf.

Intolerance–The inability of a woody plant to grow well in a particular environment; commonly used to denote trees that cannot endure heavy shade.

Intra-–A prefix meaning within, inside; as intrastaminal, inside the (ring of) stamens; intraspecific, within a species.

Intracellular–Within, inside of, a cell.

Intractability–The characteristic of a soil that causes it to be difficult to work.

Intraspecific–Referring to events or relationships that occur between individuals of the same species.

Intraspecific Hybrid–Cross between individuals within the same species, but of different genotypes.

Introduced–Designating a plant, animal, disease, etc., that is not indigenous to an area, but is brought in purposely or accidentally.

Introgression, Hybridization—Long-continued interspecific hybridization leading to an infiltration of genes from one specie into another.

Inulin—A polymeric carbohydrate comprised of fructose monomers found in the roots of many plants, particularly Jerusalem artichokes.

Invading Plant—A species of plant that comes in on lands after the more stable, climax plants have been diminished in the stand by drought, fire, or overgrazing.

Inverted T—In budding, a method of making an incision on the stock in which a vertical cut about 1½ inch long is made at a point where the bud is to be inserted. At the base of this cut a horizontal cut is made so that the two cuts resemble an inverted T.

Ion—An atom or a group of atoms carrying an electrical charge, which may be positive or negative. Ions are usually formed when salts, acids, or bases are dissolved in water. When common salt, sodium chloride, is dissolved in water, positive sodium ions and negative chloride ions are formed. See Anion, Cation.

Ion Exchange—The replacement, in a colloidal system, of one ion by another with a charge of the same sign.

Ion Exchanger—A water-softening system, such as the zeolite type, which functions through the exchange of ions, especially of sodium for calcium and magnesium; in water, the sodium salts of certain compounds readily release sodium and take in its place calcium and magnesium. See Hard Water, Soft Water.

Ionization—The process by which an atom becomes electrically charged, by the removal or addition of one or more of its extranuclear electrons, so that the electrical balance between the electrons and the protons within the atom's nucleus is destroyed. An atom with more than its normal complement of electrons has a negative charge; with less, it has a positive charge.

Irish Cobbler—An early-maturing variety of potato widely cultivated in the United States. See Potato.

Irish Potato—See Potato.

Iron—Fe; a metallic element essential to people, animals, and plants; very common in some minerals, most rocks, and all soils; an essential constituent of blood hemoglobin where it functions to transport oxygen. Iron is specific for the treatment of anemia in animals. In plants, iron deficiency results in iron chlorosis.

Iron Chlorosis—A yellowing of the foliage of plants due to an iron deficiency.

Iron Pan—A type of hardpan, in which a considerable amount of iron oxide is present. See Ironstone.

Iron Phosphate—$FePO_4$; occurs in small quantities in practically all phosphate rock and in rather large quantities in some of the lower grades of rock. It is insoluble in water, and its phosphoric acid content is practically unavailable unless treated with a mineral acid such as sulfuric acid. When present in large amounts, it impairs the physical and chemical condition and nutritional value of superphosphate. See Iron.

Iron Sulfate—See Ferrous Sulfate, Micronutrient Fertilizers.

Ironstone—(1) Any rock containing a substantial proportion of an iron compound from which the metal may be smelted commercially. (2) An iron-rich sedimentary rock, either deposited directly as a ferruginous sediment or resulting from chemical replacement.

Irradiated—Designating a food or feed treated with ultraviolet light to increase the vitamin D content.

Irradiated Yeast—Yeast subjected to ultraviolet rays in order to increase its antirachitic potency. It is used as an ingredient in feeds for its vitamin D and B-complex content.

Irrigable Area—The area under an irrigation system capable of being irrigated, as determined principally upon the quality of soil and elevation of land.

Irrigate—To furnish water to the soil for plant growth in place of, or in addition to, natural precipitation, by surface flooding or sprinkling and subirrigation methods, using surface water or that from underground sources.

Irrigated—Designating land or crops that have received water from other sources as well as from natural precipitation.

Irrigating Head—(1) The measure of stored-up water ready to be used in irrigation. (2) The depth of a body of water in covering land, as by flooding. The proper wetting of the whole ground. (3) The indication on a measuring device of the rate of flow of irrigation water.

Irrigation—The artificial application of water to soil for the purpose of increasing plant production.

Irrigation Efficiency—The ratio or percentage of the water consumed by crops in an irrigated farm or project to the water diverted from a river or other source into the farm or project canals.

Irrigation Methods—The manner in which water is artificially applied to an area. The principal methods and the manner of applying the water are as follows: Border strip: the water is applied at the upper end of a strip with earth borders to confine the water to the strip. Check basin: the water is applied rapidly to relatively level plots surrounded by levees. The basin is a small check. Corrugation: the water is applied to small, closely spaced furrows, frequently in grain crops, pastures, and forage crops, to direct the flow. Drip (trickle): the water is applied as a drip from "emitters" along a plastic tubing line that usually lies on the soil surface. Flooding: the water from field ditches is allowed to flood over the land. Furrow: the water is applied between crop rows in small ditches. Sprinkler: the water is sprayed over the soil surface through nozzles from a pressure system. Subirrigation: the water is applied in open ditches or tile lines until the water table is raised sufficiently to wet the plant root zone. Wild flooding: the water is released at high points in the field, and distribution is largely uncontrolled.

Irrigation Requirement—The quantity of water, exclusive or precipitation, required for crop production, including unavoidable wastes.

Irrigation, Winter—Off-season irrigation during fall, winter, or early spring to store water in the soil for later use by growing plants. Winter irrigation is practiced in areas where water is available in the winter but less available prior to time of planting.

Isoenzyme–Analysis of plant proteins by substituting known enzymes to a plant substrate to determine whether a reaction will occur. To determine presence of various enzymes within plant tissue.

Isogamy–Morphological similarity of fusing gametes.

Isotopes–Elements having an identical number of protons in their nuclei, but differing in the number of their neutrons. Isotopes have the same atomic number, differing atomic weights, and almost but not quite the same chemical properties. Different isotopes of the same element have different radioactive behavior.

Italian Broccoli–See Broccoli.

Italian Kale–A variety of turnip with no tuberous thickening, grown as a potherb.

Italian Millet–See Foxtail Millet.

Italian Ryegrass–*Lolium multiflorum*, family Gramineae; a short-lived perennial grass used as a pasture grass and in lawn mixtures. Native to Europe. Also called lovitto ryegrass, Australian ryegrass.

Ivy–(1) Any plant of the genus *Hedera*, family Araliaceae. Species are evergreen climbing, woody vines.

Jack Pine–*Pinus banksiana*, family Pinaceae; a hardy but small evergreen that grows farther north in Canada than any other native pine.

Jack Spavin–See Bone Spavin.

Jackfruit–*Artocarpus heterophyllus (A. integer)*, family Moraceae; a tropical tree grown for its large, edible fruit. Native to Asia. Also called jack.

Jackstock–A male ass used as a stud on a mulebreeding farm. See Mule.

January Thaw–A period of warm weather that occurs irregularly during the month of January, especially in the eastern United States.

Japan Clover–See Lespedeza.

Japan Wood Oil Tree–*Aleurites cordata*, family Euphorbiaceae; a small tropical tree grown for a drying oil obtained from its seed. Native to Asia. See Tung Oil Tree.

Japanese Barberry–*Berberis thunbergii*, family Berberidaceae; a hardy shrub widely grown in the shrub border for its yellow, red-tinged flowers, bright red fruit and brilliant scarlet, fall foliage. It is resistant to wheat rust. Native to Japan.

Japanese Cane–A variety of sugarcane used for fodder.

Japanese Chestnut–*Castanea crenata*, family Fagaceae; a deciduous tree, grown for its edible nuts and for use in breeding blight-free chestnut trees. Native to Japan.

Japanese Field Mint–*Mentha arvensis* var. *piperascens*, family Labiatae; a perennial herb grown as a source of peppermint oil. Native to Eurasia and North America.

Japanese Rose–*Rosa multiflora*, family Rosaceae. See Multiflora Rose.

Japanse Beetle–*Popillia japonica*, family Scarabaeidae; a beetle that feeds on plants in the eastern United States. The adults attack the leaves and the fruit; the larvae feed on grass, vegetables, and nursery stock.

Jeffrey Pine–*Pinus jeffreyi*, family Pinaceae; an evergreen tree grown for lumber and as an ornamental. Native to North America.

Jerseytea Ceanothus–*Ceanothus americanus*, family Rhamnaceae; a hardy, low deciduous shrub grown for its clusters of white flowers. Its leaves were used as a substitute for tea during the American Revolution. Native to North America. Also called New Jersey tea, Indian tea, Walpole tea, redroot shrub. See Ceanothus.

Jerusalem Artichoke–*Helianthus tuberosus*, family Compositae; a perennial, strong-growing herb, with edible tubers containing levulose, which were used for food by American Indians. Grown in some localities for hog feed; it can become a weed. Native to North America. Also called earth apple, girasole, Canada potato, Canadian bur, sunchoke. See Artichoke.

Jimmyweed–*Aplopappus heterophyllus*, family Compositae; a perennial herb, poisonous to livestock. Symptoms of poisoning are trembling, depression, inactivity, constipation, nausea, labored breathing, and weakness. Native to North America.

Jimsonweed–*Datura stramonium*, family Solanaceae; an annual herb, which is poisonous and is a weed pest. Livestock feeding on it have nausea, vertigo, extreme thirst, dilated pupils, blindness, loss of voluntary movement, convulsions, and finally die. Native to north America. Also called Jamestown weed, thornapple, dewtry, devil's-apple.

Johnny Appleseed–The nickname of John Chapman, an eccentric in the United States, who in the early part of the last century distributed apple and fennel seeds in his wanderings through the Ohio Valley.

Johnsongrass–*Sorghum halepense*, family Gramineae; a perennial grass growing up to 6 feet in height that produces heavy yields on fertile soils. It is an important livestock feed in the southern United States, but a serious weed pest in cultivated fields. Native to the Mediterranean area. Also called Alabama guinea grass, Aleppo grass, Arabian millet.

Jonathan–One of the older and better-known varieties of apple; ripens in late autumn; the fruit is bright red.

June Beetle (June Bug)–A common leaf-eating beetle. The larva is one of the most destructive of soil pests. Adults feed at night on the leaves of several trees. Larvae feed entirely underground on the roots of plants.

June Bloom Fruit–An orange fruit produced as a result of blossoming shortly after the beginning of the summer, rainy season in Florida, United States. It ripens after the main crop is harvested.

June Budding–Shield budding, usually in June, in warm areas in which it is possible to force the bud to grow in the same season it is set. The top of the stock (plant) is bent or partially broken about ten days after budding to force the growth of the bud.

June Drop–A shedding of tree fruits during the early summer; believed often to be caused by, or associated with, embryo abortion.

Jungherr's Disease–Trichomoniasis of the upper digestive tract. See Trichomoniasis.

Juniper–Any of the several species of evergreen trees belonging to the genus *Juniperus*. Most species give off a very distinctive odor.

Juniper American Mistletoe–*Phoradendron juniperinum*, family Loranthaceae; a parasite plant that infests junipers in central North America. See Mistletoe.

Juniper Webworm–*Dichomeris marginella*, family Gelechiidae; a larva that webs twigs and needles of junipers together and feeds within the web.

Jute–Tropical herbaceous plants, *Corchorus capsularis* and *C. olitorius*, whose fibers are used in making burlap bags.

Juvenile Hormone–The hormone, secreted by the corpora allata, that maintains the immature form of an insect during early molts.

K–The chemical symbol for potassium.

K_2O–Potassium oxide. In present-day fertilizer labeling, the amount of K in fertilizer is expressed as pounds of the oxide per 100 pounds total. To find actual K, multiply K_2O equivalent by 0.83.

Kafir–*Sorghum vulgare* var. *caffrorum*, family Gramineae; a tall, annual grass, a sorghum grown principally in the semiarid sections of the United States for its grain, forage, or silage. Also called kaffir corn.

Kale–*Brassica oleracea* var. *acephala*, family Cruciferae; an annual or biennial herb grown as a potherb.

Kapok–*Ceiba pentandra*, family Bombacaceae; a tropical tree grown for its cottonlike fiber, the kapok of commerce. Native to America. Also called silk-cotton tree.

Karst–A type of topography that is formed over limestone, dolomite, or gypsum by dissolving. It is characterized by closed depressions or sinkholes, caves, and underground drainage. See Thermokarst.

Karyotype–A picture or diagram of the chromosomes of a particular cell as they appear in the metaphase of mitosis arranged in pairs by size and location in the centromeres.

Katydid–A long-horned grasshopper of the family Tettigoniidae.

Keel–(1) The part of a fowl's body that extends backward form the breast. Also called breastbone. (2) The two front, united petals of a pealike flower. (3) In ducks, the pendant fold of skin along the entire underside of the body. (4) In geese, the pendant fold of flesh from the legs forward on the underpart of the body. (5) A central dorsal ridge, like the keel of a boat. (6) The two anterior united petals of a papilionaceous flower.

Keeled–Of, or pertaining to, a ridge (like the bottom of a boat) on an animal or plant part.

Keeping–Designating the storage qualities of a product.

Kelp–Any of the seaweeds of the families Laminariaceae and Fucaceae; at one time used as a source of potassium for fertilizers by organic gardeners. Dried kelp contains 1.6 to 3.3 percent nitrogen, 1.0 to 2.0 percent phosphoric acid, and 15 to 20 percent potash. Kelp has been used in animal feeds as a source of iodine.

Kelp Meal–See Dried Kelp.

Kemp–A chalky-white, brittle, weak wool fiber that, when found mixed with normal fibers of a fleece, constitutes a very serous defect. Kemp will not take dyes.

Kenaf Hibiscus–*Hibiscus cannabinus*, family Malvaceae; a woody plant grown as a source of fiber for cordage and coarse cloth. Native to India and Indonesia. In the United States, grown as an ornamental flowering shrub. Also called Deccan hemp and Indian hemp.

Kentucky Bluegrass–*Poa pratensis*, family Gramineae; a perennial grass widely grown in the north-central states, United States, as a pasture and turf grass. Native to Eurasia.

Kentucky Wonder–A very well-known type of green bean.

Kernel–(1) The portion of a seed contained within the seed coat. (2) The whole grain of corn, wheat, etc.

Kernite–An important ore of boron.

Kettle–(1) A depression of variable size and depth, sometimes containing lakes and swamps. It is a feature of moraines in the glaciated regions of the United States, especially in Wisconsin and Michigan. (2) A pot used for heating water.

Key Fruit–A one-seeded fruit that does not split open and has a membranous wing attached, as in the maple; a samara.

Key Species–Plant species that endure moderately close grazing and are abundant.

Kibbled Grain–Grain that has been chopped into pieces rather than ground.

Kid-glove Orange–Tangerine.

Kieserite–Magnesium sulphate, a source of Epsom salts. Sometimes used alone as a soil amendment and in fertilizer mixtures as a source of magnesium.

Killing Frost– A temperature condition sufficiently low to kill most staple crops.

Kiln Burn– A browning of the surfaces of lumber that may occur during kiln drying.

Kiln-dried Lumber– Lumber dried by artificial heat to a moisture content less than that obtained through natural, air seasoning.

Kind– (1) All the plants of the same type, accepted as a single vegetable or fruit, as tomato, cabbage, bean, apple, peach, etc. (2) A species, as a cow, sheep, etc.

King Cotton– A phrase testifying to the importance of cotton in the United States.

Kitchen Garden– A small garden in which vegetables and fruits are grown for home use.

Kiwi– (1) A commercially grown edible tropical fruit, *Actinidia chinensis*, family Actinidiaceae. The fruit is about 1 inch (2.5 centimeters) in diameter, dark brown on the outside and greenish-yellow in the inside. Kiwi plants have been developed for temperate regions. (2) A bird native to New Zealand, *Apteryx* spp., family Apterygidae. It is about the size of a chicken.

Kjeldahl– Refers to a procedure for determining the amount of total nitrogen in organic materials; named for Kjeldahl, the Danish chemist who developed it.

Klendusity– The ability of a susceptible plant to escape infection under conditions in which other susceptible plants contract the disease.

Kleptoparasite– A parasite that steals food from another parasite.

Knapsack Seeder– A device for broadcast seeding that consists of a canvas sack fastened to a seeding mechanism. A crank, turned by hand, revolves a wheel having radial ribs that throw seeds to the front and sides. It is sometimes used for spreading fertilizer and lime.

Knee– (1) The joint between the hip and the ankle in humans and quadrupeds (hind leg); in birds, the tarsal joint. (2) The carpal joint (foreleg). (3) The spurlike, root growth of the baldcypress that develops when the tree grows in a swamp.

Knockdown– The killing ability and speed of a toxic material, especially an insecticide or an herbicide.

Knot– (1) The interlacing of string or rope pulled into a tie to fasten something. (2) A swelling or protuberance on the stem of a cut flower through which water moves very slowly to the flower. See Node. (3) The knob attached to a check wire. (4) That portion of a tree branch that has become incorporated in the body of the tree. (5) The unit of speed used in navigation. It is equal to 1 nautical mile (6076.118 feet, 1852 meters) per hour. (6) The meeting point of two or more mountain chains. (7) A dark-colored, hard, somewhat round blemish in a board where a limb grew out of the tree.

Knothole– A hole in a piece of lumber or a tree caused by a knot slipping out.

Kochia– *Kochia scoparia*, family Chenopodiaceae; an annual, herbaceous plant frequently cultivated for its bright autumn color; locally a weed. It has been used as a forage crop during drought periods in the northern and southern Great Plains. Also known as summer cypress and belreedere.

Kohlrabi– *Brassica oleracea* var. *gongylodes*, family Cruciferae; a biennial herb, related to the cabbage, grown for it edible, enlarged fleshy stem. Also called turnip cabbage, turnip-rooted cabbage, stem cabbage.

Korean Lespedeza– *Lespedeza stipulacea*, family Leguminosae; an annual herb widely grown in warm areas as a pasture, hay, and green manure crop. See Lespedeza.

Krona Pepper– A mild red pepper. See Pepper.

Kudzu– *Pueraria lobata* and *P. phaseoloides*, family Leguminosae; a very rapidly growing vine, once valued for soil conservation, that has now become a serious pest in the southern part of the United States.

Kumquat– Any tree of the genus *Fortunella*, family Rutaceae. Species are grown for their edible fruit. They resemble a small orange.

Labellum– The specialized petal (lip) of an orchid that may be tubular (as in vanilla), pouchlike, or otherwise shaped differently from the other petals.

Labiate– (1) Any plant of the mint family. Species have divided calyx and corolla. (2) Designating the form of a flower; e.g., the snapdragon, in which the corolla or calyx is divided into two parts of unequal size, one overhanging the other.

Labium– (1) The posterior mouthpart or lower lip of an insect. (2) A lip or liplike organ. (3) The folds of skin of the vulva.

Labrum– The anterior mouthpart or upper lip of an insect.

Lac– A resin secreted by the tropical Southeast Asian insect *Laccifer lacca*. The resin is the principal ingredient in shellac, but is used also in varnishes and waxes. The insect feeds primarily on the twigs of these tree genera: *Acacia* (many species), *Butea* (Bengal kino), *Schleichera* (lac-tree), and *Zizyphus* (jujube).

Lacewing– Any insect of the family Chrysopidae. They feed on aphids and other insects and are considered a helpful insect. See Golden-eye Lacewing.

Lacustrine– (1) Pertaining to a lake; e.g. lucustrine sands deposited on the bottom of a lake, or a lacustrine terrace formed along the margin of a lake. (2) Growing in lakes; e.g., lucustrine fauna. (3) Said of a region characterized by lakes; e.g., a lacustrine desert containing the remnants of numerous lakes that are now dry.

Lacustrine Deposit–Sediments deposited in lake waters.

Ladder Fuels–Burnable materials, living and dead, which tend to be continuous between ground fuels and tree crowns. Thus they form a "ladder" that can allow fire to climb into the tree crowns.

Ladino Clover–A large, rapidly growing, perennial type of common white clover, which spreads by creeping stems that root at the node; grown primarily as a pasture crop but also for hay, silage, and seed. Introduced to the United States from Italy. See White Clover.

Lady Beetle–Any of the insects of the family Coccinellidae; beneficial because they prey upon aphids, scale insects, and mealybugs. Sometimes they are collected and introduced into particular locations because of their value in biologic control. Also called ladybug, ladybird beetle.

Lagg–The depressed margin of a raised peat bog.

Laid By–Designating the last cultivation or working of a row crop for the season.

Lamarckism–A belief, named after French naturalist Jean Baptiste de Lamarck (1744-1809), that acquired characteristics can be inherited. Now proved false.

Lambert Crazyweed–*Oxytropis lambertii*, family Leguminosae; a plant, the foliage of which is poisonous both green and dried. Symptoms of poisoning are dullness, irregularity of gait, lack of appetite, dragging of the feet, a solitary habit, loss of flesh, and a shaggy coat. As the animal ceases to eat, it dies. Found in the Great Plains area of the United States, and south to Mexico. Also called crazyweed, white locoweed, rattleweed, Colorado loco vetch.

Lambkill Kalmia–*Kalmia angustifolia*, family Ericaceae; an evergreen shrub of thin, open habit, usually growing 2 to 3 feet high. It is considered to be injurious to sheep and other grazing animals and may occur as a troublesome wild growth in cranberry bogs. Also called sheep laurel, wicky, narrow-leaved laurel, calfkill, dwarf laurel.

Lambsquarters Goosefoot–*Chenopodium album*, family Chenopodiaceae; an annual herb that is sometimes a weed pest and sometimes used as a potherb. Native to Europe. Also called lamb's quarters, fat hen, chou grass, poulette grass.

Lamella–(Biology) One of the layers of a cell wall; a thin layer in a shell, like a leaf in a book.

Laminated Soil–Designating soil arranged in very thin plates or layers, less than 1 millimeter thick, which lie horizontally or parallel to the soil surface, usually fragile and of medium to soft consistency. See Varve.

Lanceolate–Refers to a leaf that is narrow with the widest point at the base. It is shaped like a lance. See Obolanceolate.

Land–(1) The total natural and cultural environment within which production must take place. Its attributes include climate, surface configuration, soil, water supply, subsurface conditions, etc., together with its location with respect to centers of commerce and population. Oyster beds and even tracts or bodies of water, as where valuable fishing rights are involved, may be regarded as land. It is often convenient, in fact, to regard land as synonymous with all that nature supplies, external to humans, which is valuable, durable and appropriable, thus including, e.g., waterfalls and other sources of waterpower. (2) In a broad legal sense, any real part of the surface of the earth, including all appurtenances, anything in, on, above, or below the surface. (3) In plowing, a plowed or unplowed space between two furrows. (4) The total width of a strip of land tilled by a farmer, or some designated width, as a perch, 16½ feet. Also called a stitch. (5) Soil. (6) A natural part of the earth's surface characterized by any single factor, or combination, of topography, climate, soil, rocks, vegetation; the natural landscape. (7) Pertaining to agriculture, those areas actually in use or capable of use for the production of farm crops and livestock.

Land Capability–A measure of the suitability of land for use without damage. In the United States, it usually expresses the effect of physical land conditions, including climate, on the total suitability for agricultural use without damage. Arable soils are grouped according to their limitations for sustained production of the common cultivated crops without soil deterioration. Nonarable soils are grouped according to their limitations for the production of permanent vegetation and their risks of soil damage if mismanaged.

Land Capability Class–One of eight classes of land distinguished according to potentiality for agricultural use. Class I consists of lands that are nearly level and can be cultivated continuously with little erosion. Class II consists of lands that are nearly level soils that require only simple practices such as contour tillage to control erosion. Class III consists of lands that require such practices as terraces and contour tillage to keep them productive for row cropping. Class IV consists of lands that require contour plowing, terracing, and the planting of sod-like crops every two to three years to control erosion. Class V consists of lands that cannot be planted to cultivated crops without extensive practices. Such soils should be maintained in sod crops. Class VI consists of lands that are too steep to be used for any crop except sod crops. However, even sod crops are difficult to establish and maintain. Class VII consists of lands recommended for use as watershed or woodland. Class VIII consists of lands recommended for recreational uses only.

Land Capability Map–(1) A map that shows land capability units, classes, and subclasses. (2) A soil conservation survey map that is colored to show land-capability classes.

Land Degradation–The result of one or more processes that lessen the current and potential capability of soil to produce (quantitatively or qualitatively) goods or services; e.g., soil erosion.

Land Drainage–(1) The removal of surface water or excess groundwater by artificial means, such as ditching or tilling. (2) Natural drainage, or removal of water by creeks, rivers, in natural water channels.

Land Forming–A general term that means reshaping, smoothing, and/or reforming the soil surface to facilitate irrigation, drainage, or erosion and sediment control. Land forming for drainage is usually known as land grading. The implement used is usually known as a land plane. See Land Plane.

Land Plane–A large, tractor-drawn machine designed for planing or smoothing land for more efficient use of irrigation water or for easier tillage of land not irrigated. It consists of a long steel frame mounted on wheels, near the center of which is attached a large, long, adjustable, combination steel blade and scraper to remove soil from high points and to convey it to depressions. See Land Forming.

Land Roller–A heavy implement that is used for smoothing ground and forcing rocks deeper into the soil.

Land Stones–Loose stones that occur naturally on a land surface or in a field, as contrasted with those that are quarried, cut, or broken by humans. Also called fieldstones.

Land Subsidence–The lowering of the land surface due to the removal of groundwater, oil, or the decomposition of cultivated peat and muck soils. See Histosols, Karst, Thermokarst.

Land Type–A geographic division of land based upon some one or a combination of natural factors as soil, relief, vegetation, and climate. The term is often loosely used to indicate use of the land such as agricultural, grazing, mining, forest, urban, etc.

Landscape–(1) In soil geography, the sum total of the characteristics that distinguish a certain area of the earth's surface from other areas, as soil series, vegetation, rock formations, hills, valleys, streams, cultivated fields, roads, and buildings. These characteristics are the result not only of natural forces but also of human occupancy and use of the land, and together give the area its pattern. (2) In a broad sense, the complex pattern of an extensive area, such as the rural landscape, or the mountain landscape. (3) To beautify terrain as with plantings of trees, shrubs, and flowering herbs; with ornamental features, such as terraces, rock gardens, bog gardens, pools, walks, drives, etc.

Landscape Architect–A person trained in the art and science of arranging land and objects upon it for human use and enjoyment.

Landscape Design–The profession concerned with the planning and planting of outdoor space to secure the most desirable relationship between land forms, architecture, and plants to best meet human needs for function and beauty.

Landscape Types, Visual–Landscape is defined as the sum total of the characteristics that distinguish a certain area on the earth's surface from other areas. These characteristics are a result of natural forces and human occupancy and use of the land. The seven recognized visual landscape types are: canopied, cultural, detailed, enclosing, feature-dominated, focal, and panoramic.

Landslide–(1) Downslope movement of soil/rock, usually because of supersaturation, associated sometimes with building roads on a sidehill and timber harvesting. (2) That part of the plow bottom that slides along the face of the furrow wall. It helps to counteract the side pressure exerted by the furrow slice on the moldboard, and to steady the plow while being operated.

Langbeinite–A mineral, $K_2Mg_2(SO_4)_3$. Isometric crystal structure. Occurs in potassium salt deposits; mined as a source of potassium sulfate (K_2SO_4) fertilizer.

Larkspur–Any annual or perennial herb of the genus *Delphinium*, family Ranunculaceae. Species are very popular flower garden plants; poisonous to livestock. See Menzies Larkspur, Sierra Larkspur.

Larva–The immature insect hatching from the egg and up to the pupal stage in orders with complex metamorphosis; the six-legged first instar of mites and ticks.

Larvicide–A chemical used to kill the larval or preadult stages of parasites.

Late–(1) Designating a variety of plant that flowers or fruits at a later time in the year than others of its species. (2) Designating a season that comes at a later time in the year than normal, as a late spring.

Late Wood–The denser, smaller-celled, later-formed part of a growth layer along the trunk and branches of a tree; summerwood.

Latent–Designating an infection that is present but which is not manifest in the host under consideration. See Dormant.

Latent Bud–A bud that remains dormant or semidormant in relation to those actively producing shoots.

Lateral–(1) A directional or positional term meaning away from the middle or toward the side. (2) A part of a system that branches out from the main body of the system, such as the tile lateral drain that connects to a main drain in a drainage system. (3) A branch or twig of a tree.

Lateral Bud–A bud attached to the side of a branch or spur. See Terminal Bud.

Lateral Root–A root that is confined to a horizontal growth pattern in the soil following the surface soil horizons or if developed at a lower level continues a lateral growth. See Tap Roots.

Laterite–(Soil) The zonal group of soils having very thin organic and inorganic mineral layers over leached soil that rests upon highly weathered material, rich in aluminia or iron oxide, or both, and poor in silica. The soils are usually red. Laterite soils are developed under the tropical forest or in a subtropical climate. This group of soils is now called Oxisols. See Soil Orders.

Latex–(1) A usually white or yellowish fluid produced by the cells of some plants, as *Asclepias* and *Euphorbia*; often referred to as milky sap. (2) Designates a paint that is water-soluble in the fluid state.

Latexosis–The production of latex by a plant.

Lath House–A structure for the propagating and raising of tropical plants or plants needing shade, which consists of a frame covered with slats or laths, with about one inch separations between pieces. This allows about one-half the amount of light received out-of-doors, and provides shelter from the wind.

Latifolia–(Latin) Broad leaf.

Laurel, English–*Prunus laurocerasus*; common laurelcherry is a vigorous, coarse-textured, broadleaved evergreen shrub commonly used for hedging. Useful for ornamental windbreaks. Tolerates a wide range of soils, but especially thrives on moderately wet soils.

Lava–Fluid rock that issues from a volcano or a fissure in the earth's surface; also the same material solidified by cooling.

Lawn–Any open-ground area about a house or building planted to fine grasses and maintained in good turf especially for its esthetic value.

Lawn and Garden Fertilizers–Specialty fertilizers, such as tablets and liquids for potted house plants, "lightweight" fertilizers for lawns and gardens, fertilizer-pesticide mixtures, and special formulations and mixtures for organic gardeners. Controlled-release fertilizers are featured in such products.

Lawn Grass–Any of the grasses, alone or in combination (including certain legumes), planted in lawns.

Laxa–Loose.

Lay-by–To cease cultivating a crop, as corn, because of stage of growth, lack of weeds, etc.

Lay-of-Land–The topography of a tract of land, as level, sloping, rolling, hilly, etc.

Layer–(1) A mature female fowl that is kept for egg-laying purposes, especially one in current egg production. (2) A course or stratum, as a layer of sand. (3) A plant twig or shoot, tied down and partially covered with earth, so that it can take root while remaining unsevered from the parent stock. (4) To reproduce by layerage.

Layerage (Layering)–The method of propagating woody plants by covering portions of their stems or branches with moist soil or sphagnum moss so that they take root while still attached to the parent plant.

LC$_{50}$–See Lethal Concentration.

LD$_{50}$–See Lethal Dose 50.

Leachate–Liquid that has percolated through solid waste or other medium and may contain extracted, dissolved, or suspended materials from the medium.

Leached Soil–A natural soil from which most of the more soluble constituents have been removed throughout the entire profile or removed from one part of the profile and accumulated below the rhizosphere (root zone).

Leached Wood Ashes–Wood ashes with part of the plant nutrients removed by exposure to rain or snow. See Wood Ashes.

Leaching–(1) The removal of soluble constituents from soils or other materials by percolating water. (2) The removal of soluble salts from soils by abundant irrigation combined with drainage.

Leaching Requirement–The amount of water that must pass through the root zone in order to leach out excess soil salinity.

Leader–(1) The main or dominant stem of a plant. (2) The front animals of a tandem hitch.

Leaf–(1) A flattened outgrowth from a plant stem, varying in size and shape, usually green, which is concerned primarily with the manufacture of carbohydrates by photosynthesis. (2) To put forth leaves as in the spring.

Leaf Beet–*Beta cicla*, family Chenopodiaceae; a biennial herb popularly grown as an annual in the vegetable garden as a potherb. Native to Europe and Asia. Also called chard. See Swiss Chard.

Leaf Canopy–Crown cover of a plant.

Leaf Crumpler–*Acrobasis indigenella*, family Phycitidae; a caterpillar that infests various fruit trees, particularly the apple, feeding on the buds, shoots, and leaves.

Leaf Cutting–A method of plant propagation in which a leaf is taken from a plant and placed in or on a rooting medium. Different treatment is preferable for each species; leaf cutting is especially adapted to gloxinia, begonia, and members of the family Geseriaceae.

Leaf Dimorphism–Difference between early leaves produced from previous year's buds, and late leaves, leaves produced by current season buds; late-season leaves are usually larger. Present in sweet gum, yellow poplar, aspen, cottonwood, birch.

Leaf Gap–An area in the vascular region of a stem where parenchyma instead of vascular tissue differentiates; located immediately above a leaf trace where a vascular bundle connects the vascular system of a leaf with that of the stem. The leaf bud develops into a stem with leaves.

Leaf Lettuce–Crinkled garden lettuce. See Lettuce.

Leaf Miner–An insect that lives in and feeds upon the leaf cells between the upper and lower surfaces of a leaf.

Leaf Mold–Partially decayed leaves useful for improving soil structure and fertility. See Compost.

Leaf Roll–(1) A viral disease of potatoes characterized by dwarfing, rolling of the leaves, uprightness, rigidity, and chlorosis. (2) A European viral disease of grapes. (3) See Rhizoctonia Disease of Potatoes. (4) A symptom of improper root functioning, or drought, in some plants, such as corn.

Leaf Rust of Cereals–A disease of cereals, caused in barley by *Puccinia anomala*, in corn by *P. sorghi*, in rye by *P. rubigo-vera*, and in wheat by *P. rubigo-vera tritici*, fungi of family Pucciniaceae; characterized by the appearance of orange-brown pustules on the leafblades and sheaths of the plants.

Leaf Sheath–In grass and grasslike plants, the lower part of the leaf that encircles the stem.

Leaf Skeletonizer–Any caterpillar that feeds by eating out the tissue of the leaves, leaving the epidermis and the veins.

Leaf Smut of Grasses–A widespread disease of noncereal grasses caused by *Ustilago striiformis*, family Ustilaginaceae; known to occur on over 75 grass species including 24 genera, among the most important of which are *Agropyron, Agrostis, Bromus, Calamaguostis, Elymus, Festuca, Phleum,* and *Poa*. The fungous fruits are mainly in the leaves, leaf sheaths, stalks, and sometimes the inflorescence. Life history studies of the fungus show great variability depending largely upon the host differences.

Leaf Spot of Peanuts–A widespread, fungal disease caused by *Mycosphaerella berkleyi*, family Mycosphaerellaceae; characterized by a brown to black leaf spot without a distinct halo. Lesions also appear on petioles, stems, and pods.

Leaf Spot of Pear–A widespread, fungal disease caused by *Mycosphaerella sentina*, family Mycosphaerellaceae; characterized by small, light-colored spots on the leaves enlarging to ¼ inch and becoming grayish-white with well-defined margins. If the lesions are numerous and coalesce, defoliation results.

Leaf Spot of Strawberry–A widespread, fungal disease caused by *Mycosphaerella fragariae*, family Mycosphaerellaceae; characterized by indefinite deep purple spots on the leaflets, petioles, fruit, and fruit stems. The spot enlarges and turns white or light brown with a purple halo. If the spots are numerous, the leaflet may be killed. Also called common leaf spot, leaf blight, rust, spot disease.

Leaf Spot of Tomato–A general, fungal disease caused by *Deptoria lycopersici*, family Sphaerioidaceae; characterized by water-soaked leaf spots that turn brown; the leaf yellows, rolls, and drops. The disease attacks the lower leaves first.

Leaf Stalk (Leaf Stem)–Petiole.

Leaf Tier–Any caterpillar that fastens leaves together with silk and feeds within this shelter.

Leaf Tip Burn–A necrosis or scorching of tips and edges of leaves; causes may include dry air, high soluble salt levels, and pesticide injury.

Leaf Tobacco–Harvested tobacco before the leaves have been cut or shredded; the chief form in which the crop is transported prior to manufacturing. The diverse classes and grades depend on variety, curing, country of origin, etc.

Leaf-cast Disease of Conifers–The premature shedding of the leaves (needles) mostly of pine, larch, fir, spruce, etc., usually due to the effect of parasitic fungi growing within the leaves. These are most often species of the genera *Hypoderma, Hypodermella, Lophodermium*, family Phacidiaceae.

Leaf-Scorch–A symptom of potassium deficiency of woody plants characterized by dead spots on the leaves, which sometimes appear to have been caused by drought.

Leafhopper–Any of the many small insects of the family Cidadellae that suck juices from plants. They are so named because they jump from leaf to leaf on plants.

Leaflet–(1) A small, immature leaf. (2) A separate division of a compound leaf.

Leafy Euphorbia (Leafy Spurge)–*Euphorbia esula*, family Euphorbiaceae; a perennial herb that is a weed pest found in waste spaces across the northern United States. It is eaten by sheep but avoided by cattle. Contact with its juice may cause dermatitis in humans and death to fish. Native to North America.

League–(1) A unit of linear measure used mostly in Texas. A land league equals 3 statute miles or 15,840 feet (4.8280 kilometers). A nautical league equals 3 geographical miles or 18,240.78 feet (5.5597 kilometers). (2) An area embraced in a square 5,000 varas on each side. It contains 4428.40 acres, 6.919 square miles (17.92 square kilometers).

Lechuguilla Agave–*Agave lophantha* var. *poselgeri*, family Amaryllidaceae; a succulent, desert plant that is a source of fiber, also grown as a pot plant for its bluish-green leaves, pale-banded along the midrib. It is poisonous to sheep and goats, causing goat fever. See Sacahuista.

Lecithin–One of a group of lipids known as phospholipids. Abundant in brain tissue and egg yolk. Obtained from peanuts, corn, and soybeans for commercial use (as an emulsifier in such products as chocolate).

Lectins–A group of protein substances, natural antibodies; agglutinins for type A red blood cells. Lectin may be obtained from lima beans.

Leek–*Allium porrum*, family Liliaceae; a perennial herb grown as a mild, onion-flavored vegetable. Native to Europe and Asia.

Leggy–(1) Designating a plant which has unusually long stems. (2) Designating an animal, usually very young, as a colt, whose legs are disproportionally long in relation to its body size.

Leghemoglobin–A red pigment occurring in the root nodules of legumes.

Legume–A family of plants, including many valuable food and forage species, such as peas, beans, soybeans, peanuts, clovers, alfalfas, sweetclovers, lespedezas, vetches, and kudzu. With aid of symbiotic bacteria, they can convert nitrogen from the air to build up nitrogen in the soil. Many of the nonwoody species are used as a cover crop and are plowed under for improvement of the soil.

Legume Inoculation–The application of a dust or liquid containing species of the bacterium *Rhizobium* to the seeds of legumes prior to planting so that nodules will form on the roots to fix atmospheric nitrogen.

Legume Silage–Legume crops, such as alfalfa and Ladino clover, which make satisfactory silage, especially if mixed with grasses and put into the silo in proper condition.

Leguminous–Having the nature of or bearing a legume.

Lehmann Lovegrass–*Eragrostis lehmanniana*; a grass of the family Gramineae, used for forage, especially where mean annual precipitation is 20 inches (51 centimeters) or less.

Lemma–The lower of the two bracts enclosing the flower in the grasses, formerly called the flowering glume.

Lemon–*Citrus limonia*, family Rutaceae; a tropical evergreen tree grown for its very valuable, edible fruit, the lemon of commerce. Native to Southeast Asia.

Lemongrass–*Cymbopogon citratus*, family Gramineae; a perennial, tropical grass with lemon-scented foliage grown in southern Florida and southern California, United States, and India; origin unknown. It yields lemongrass oil used in perfume.

Length of Staple–The normal measurement, without regard to quality or value, of a typical portion of lint cotton staple fibers under a relative humidity of 65 percent and a temperature of 70°F.

Length-of-Run–In irrigation, the distance water must travel in furrows or over the surface of a field from one head ditch to another or to the end of a field.

Lenticel–An opening through the bark or outer covering of fruits, etc., that permits exchange of gases from the inner tissues with the surrounding air. It originates as a loose mass of rounded, corky cells formed beneath a stoma and appears as the light-colored, lens-shaped structure visible in the outer bark of the cherry, the birch, and other woody plants, and on potatoes, carrots, etc.

Lentil–*Lens culinaris*, family Leguminosae; an annual, herbaceous legume grown for its edible, pealike seeds. Native to Europe.

Leprosis–A disease of citrus, especially of sweet oranges and less so of grapefruit and sour oranges, thought to be caused by a virus; characterized by spots on the leaves, twigs, and fruit. On the leaves, the spots are chestnut to auburn, raised, and become glazed, hard, and brittle. On the fruits, the spots are chestnut-brown with a lemon-yellow halo. Also called nailhead rust, scaly bark.

Lesion–Injury or diseased condition of tissues or organs.

Lespedeza–(1) Any annual or perennial herb or shrub of the genus *Lespedeza*, family Leguminosae. Species are grown for hay, pasture, soil improvement, green manure, and ornament. (2) *L. striata*, an annual herb that is important as a hay, pasture, soil improvement, and green manure crop. Native to China and Japan. Also called common lespedeza, Japan clover.

Lespedeza Meal–A livestock feedstuff obtained from the grinding of lespedeza hay without the addition of any lespedeza stems, lespedeza straw, foreign material, or the removal of leaves. It must be reasonably free from other crop plants and weeds and must not contain more than 28 percent of crude fiber.

Lespedeza Stem Meal–A livestock feedstuff, the ground product remaining after the separation of the leafy material from lespedeza hay or meal. It must be reasonably free from other crop plants and weeds.

Lesser Grain Borer–*Rhyzopertha dominica*, family Bostrichidae; an insect pest (both in the larval and adult stages) of grains, seeds, drugs, dry roots, and cork; especially destructive to wheat.

Lesser Wax Moth–*Achroia grisella*, family Galeriidae, an insect pest of beehives that tunnels in and feeds on the combs.

Lethal Concentration (LC)–The lethal concentration (written as LC_{10} or LC_{50} or LC_{100} or any percentage) median is the parts per million (ppm) or parts per billion (ppb) of toxicant in water or air that kills 10, or 50, or 100 percent, respectively, of the target species in a 24-hour period. Usually used for fish. See Effective Concentration, Lethal Dose.

Lethal Dose–The lethal dose (written as LD_{10}, LD_{50}, LD_{100}, or any percentage) median is the milligrams of toxicant per kilogram of body weight that kills 10, 50, or 100 percent of the target species. See Effective Concentration, Lethal Concentration.

Lettuce–(1) Any herb of the genus *Lactuca*, family Compositae. Most species are weedy plants, but one, *L. sativa*, an annual herb, is widely grown for its leaves, used in salads. Native to Eurasia.

Leucite–$KAL(SiO2)2$; potassium aluminum silicate, a feldspathoid mineral that is a minor source of potassium for fertilizer. In the United States, its principal occurrence is in the rock Wyomingite found in Wyoming.

Levant Cotton–*Gossypium herbaceum*, family Malvaceae; an annual herb widely grown in Asia and the Near East as a fiber plant for its lint, the cotton of commerce. Native to Eurasia. Also called Arabian cotton, Asian cotton. See Cotton.

Liberian Coffee–*Coffee liberica*, family Rubiaceae; a tropical shrub sometimes grown for its white, fragrant flowers and red fruit. Native to Africa.

Lichen Line–The edge of lichens on undisturbed rock surface; the clean portion of the rock indicating the extent to which the soil has been eroded from around its base.

Lichens–Fungi, usually of the class Ascomycetes (rarely of the class Basidiomycetes) that grow in mutual symbiosis with algae, forming a joint structure in which the algal cells are embedded in the network of fungous hyphae.

Licorice–*Glycyrrhiza glabra*, family Leguminosae; a perennial herb grown as a source of the flavoring licorice, used in cough remedies, tobacco, candy, etc. Native to the Mediterranean region.

Lie Bare–To remain unplanted, as improved or cultivated land left fallow or subject to the ill effects of wind and water erosion for a period of time without a crop cover.

Lie Dormant–Not to germinate, as seeds of certain plants that remain in the soil for several years before germinating; characteristic of many weed seeds.

Life Cycle–Life history; the changes in the form of life that an organism goes through.

Life History–Habits and changes undergone by an organism from the egg stage to its death as an adult.

Ligase–An enzyme that splices segments of DNA together.

Light–(1) The form of radiant energy consisting of wavelengths lying within the limits perceptible by the normal human eye, and, by extension, the shorter and longer wavelengths, the ultraviolet and the infrared light, invisible to the eye but which may be recorded photographically. Light can be absorbed by various substances and transformed into heat. Its excess may produce fading, the destruction of green color in plants, but an insufficiency can also cause lack of chlorophyll production. The coloration of fruits is dependent upon sufficiency of light. The growing parts of plants respond to the stimulus of the direction from which the light comes. See Photosynthesis, Phototropism. (2) To become ignited; to take fire. (3) Designating a deficiency or lack in degree, such as a light rain or a light crop. (4) When applied to food or drink it can have several definitions. It could mean reduced calories, fluffy (full of air), pale, low in sodium, mild in flavor, and/or less alcohol.

Light Burning–(1) In forestry, a controlled burning to check disease, remove fire hazard, or to favor a certain type of growth. (2) The method of reducing the volume of inflammable material in forests. Fires are deliberately set at a time when inflammability is low and the likelihood of damage is at a minimum; designed to lessen the danger of damage when fires run more freely.

Light Compensation Point–The lowest light intensity at which the amount of food produced by photosynthesis in a plant equals the energy used up in respiration.

Light Freeze–See Light Frost.

Light Frost–A frost that does not kill but may cause some injury to the more tender and susceptible plants in exposed places. See Black Frost, Killing Frost.

Light Period–See Photoperiodism.

Light Saturation Point–The intensity of light beyond which a plant cannot use a brighter light for photosynthesis.

Light Soil–Soil that is easy to cultivate, retains little moisture, and has a sandy (coarse) texture.

Light-seeded–(1) Designating those seeds that are relatively light in weight so that care must be taken in sowing or harvesting. (2) Designating a plant that produces few seeds.

Lightly Buffered–Designating a soil such as a sand that has a small cation exchange capacity.

Lightning Injury–The injury to a plant, most often a tree, in which some bark is stripped off, limbs are broken, etc., caused by lightning. Frequently herbaceous plants or crops in a circular area are killed by a single streak of lightning.

Lightning Ring–A band of abnormal wood in a growth ring of a tree that results from mild lightning shock to the cambium.

Lightweight Fertilizer–Specialty fertilizers containing highly soluble plant nutrient compounds absorbed on lightweight materials such as peat, expanded vermiculite, or perlite.

Lightwood–(1) Conifer wood, especially pine and Douglas fir, abnormally inflammable because of its high resin content. (2) The highly resinous heartwood, stumps, knots, and dead roots of pines, which are used mainly for kindling. Also called lighterwood. (3) Blackwood acacia.

Ligneous–Like wood; woody.

Lignification–The process in which plant cells become woody by conversion of certain constituents of the cell wall into lignin, generally considered to include the hardening, strengthening, and cementing of the cell walls in the formation of wood.

Lignin–A complex, highly indigestible material associated with cellulose and the other fibrous parts of a plant.

Lignocellulose–The substance of which woody tissue is mainly composed.

Ligule–A strap-shaped organ or body: (a) particularly, a strap-shaped corolla, as in the ray flowers of the Compositae; (b) also, a projection from the top of the sheath in the Gramineae, Palmae, and some other plant families.

Like Produces Like–A rule of thumb in breeding; the offspring will bear a close resemblance to the parents.

Lilac, Common–*Syringa vulgaris*; an upright, deciduous shrub growing to 10 feet (3 meters). It is a hardy, drought-tolerant shrub or small tree with showy fragrant flowers. Is well suited for use as windbreaks. It suckers freely and is relatively free of insects and diseases.

Lima Bean–(1) *Phaseolus limensis*, family Leguminosae; a tropical, twining perennial herb grown as an annual for its edible seeds, the Lima beans of commerce. Native to America.

Limb–(1) A lateral branch of a tree or shrub. (2) The leg or wing of an animal. (3) To cut off a limb of a tree.

Lime–(1) Strictly, calcium oxide, CaO, but, as commonly used in agricultural terminology, calcium carbonate, $CaCO_3$, and calcium hydroxide, $Ca(OH)_2$, are included. Agricultural lime refers to any of these compounds, with or without magnesium carbonate, which are used as amendments chiefly for acid soils. The most common agricultural limes are calcic lime ($CaCO_3$) and dolomitic lime ($CaMgCO_3$). (2) *Citrus aurantifolia*, family Rutaceae; a small, spiny, evergreen tree grown for its edible, very acid fruit, the lime of commerce. Native to Asia. Also called acid lime. (3) To apply lime to excrement to reduce odor and hasten decomposition.

Lime (Calcium) Requirement–The amount of agricultural limestone required per acre to a soil depth of 6 inches (15 centimeters) or on about 2 million pounds (910,000 kilograms) of soil to raise the pH of the soil to a desired value under field conditions.

Lime Chlorosis–A chlorosis that is associated with high lime soils because soil iron is made less available. See Chlorosis.

Lime, Hydrated–See Hydrated Lime.

Lime Nitrate–See Calcium Nitrate.

Lime Pan–In soils, a subsurface layer cemented by calcium carbonate. See Caliche.

Lime, Waste–An industrial by-product or waste that contains calcium and/or magnesium in forms that will neutralize soil acidity. It is usually known by the name of the industry from which it comes as gas lime or beet-sugar waste lime. It usually consists of calcium hydroxide and/or calcium carbonate with various impurities. Lactic acid and soda waste limes are high in gypsum. Oxygen waste lime is largely CaO. Gas-house lime contains cyanides and sulfites, which are harmful to plants but decompose in the soil. The latter kind should, therefore, be applied well in advance of planting.

Lime-hater–Any plant that grows well only in an acid soil.

Limequat–A hybrid tree resulting from crossing the lime (*Citrus aurantifolia*) with the kumquat (*Fortunella* spp.); grown for its edible fruits but more importantly for breeding purposes.

Limestone–(1) A bedded sedimentary deposit consisting chiefly of calcium carbonate ($CaCO_3$) that yields lime (calcium oxide) (CaO) when burned. Limestone is the most important and widely distributed of the carbonate rocks and is the consolidated equivalent of limy mud, marl, chalk, calcareous sand, or shell fragments. (2) A general term for the class of rocks that contain at least 80 percent of the carbonates of calcium or calcium and magnesium.

Liming–The application of lime to land, primarily to reduce soil acidity, reduce toxic aluminium, and supply calcium for plant growth. Dolomitic limestone supplies both calcium and magnesium. May also improve soil structure, organic matter content, and nitrogen content of the soil by encouraging the growth of legumes and soil microorganisms. Liming an acid soil to a pH value of about 6.5 is desirable for maintaining a high degree of availability of most of the nutrient elements required by plants.

Liming Materials–Agricultural limestone used in the United States totals about 21 million tons annually in recent years. The American Association of Plant Food Control Officials (AAPFCO) official interpretations and definitions of liming materials follow: Agricultural liming materials are those, the calcium and magnesium content of which are in forms that are capable of reducing soil acidity. Air-slaked lime is a product composed of varying proportions of the oxide, hydroxide, and carbonate of calcium, or of calcium and magnesium, and derived from exposure to the atmosphere of quick lime. High calcic liming materials are products 90 percent or more of whose total calcium and magnesium expressed as oxides, is calcium oxide, and contains 35 percent or more of calcium oxide equivalent. High magnesic liming materials are those containing 10 percent or more of magnesium oxide equivalent. Hydrated lime is a dry product consisting chiefly of calcium and magnesium hydroxides. Quicklime, burned lime, caustic lime, lump lime, and unslaked lime are calcined materials comprised chiefly of calcium oxide in natural association with lesser amounts of magnesium oxide, and which are capable of slaking with water. Ground limestone (corse ground limestone) is calcitic or dolomitic limestone ground sufficiently fine for effective use as a liming material. Pulverized limestone (fine ground limestone) is the product obtained by grinding either calcitic or dolomitic limestone so that all the materials will pass an 850-micron (number 20) sieve and at least 75 percent will pass a 150-micron (number 100) sieve. Ground shells is the product obtained by grinding the shells of mollusks so that not less than 50 percent shall pass a 150-micron (number 100) mesh sieve. The product shall also carry the name of the mollusk from which said product is made. Marl is a granular or loosely consolidated earthy material comprised largely of shell fragments and calcium carbonate precipitated in freshwater ponds. Ground shell marl is the product obtained by grinding natural deposits of shell marl so that at least 75 percent shall pass a 150-micron (number 100) mesh sieve. Waste lime (by-product lime) is any industrial waste or by-product containing calcium or calcium and magnesium in forms that will neutralize acids. It may be designated by prefixing the name of the industry or process by which it is produced, i.e., gas-house lime, tanners' lime, acetylene lime-waste, lime-kiln ashes, or blast-furnace slag. Agricultural slag is a fused silicate whose calcium and magnesium content is capable of neutralizing soil acidity and that is sufficiently fine to react readily in soil. Labels of artificial mixture of two or more liming materials or of gypsum with liming materials shall include a list of ingredients used. See also Cation Exchange.

Limonite–$2Fe_2O_3 \times 3H_2O$; a field term for a group of brown, amorphous, naturally occurring hydrous ferric oxides. May consist of the minerals goethite, hematite, and lepidocrocite. Also called bog iron ore.

Limy–Designating a soil having a high content of lime, usually in the form of calcium carbonate.

Lindane–The pure gamma isomer of benzene hexachloride; used as an insecticide.

Linden–Any tree of the genus *Tilia*, family Tiliaceae. Species are grown for shade, lumber, and as ornamentals. Also called basswood.

Line Fleece–A fleece of wool midway between two grades in quality and length, which can be thrown into either grade.

Linear–Refers to a leaf with a length approximately eight times longer than the width of the leaf. The sides are parallel.

Linecross–A cross between two inbred lines.

Linen–A type of cloth made form the fibers of the flax plant. See Flax.

Lineolate–Marked with fine lines.

Liners–(1) Plants growing in a line in a field and produced from seeds or cuttings. (2) Small plants produced by greenhouse operators and sold to other growers.

Linkage–The association of characters from one generation to the next due to the fact that the genes controlling the characters are located on the same chromosome linkage group. The genes located on a single chromosome or the characters controlled by such genes.

Linnaean–Conforming the principles of binomial nomenclature of all plants and animals, into genus and species, as advocated by Carl von Linné, a Swedish botanist (1707-1778), who Latinized his name to Carlus Linnaeus.

Linseed–The mature seed of *Linum usitatissimum*, a cool-season herbaceous crop used as a livestock protein feed. Also called flaxseed, linum.

Linseed Cubes–A livestock feedstuff, consisting of a mixture of linseed oil meal with flaxseed by-products, or both.

Linseed Meal–The product resulting from grinding linseed oil cake produced when flaxseed is pressed to recover linseed oil. Only batches unfit for feed are used as a fertilizer by organic gardeners.

Linseed Oil–The drying oil expressed from the seed of flax, used in paints, printer's ink, etc. See Flax, Flaxseed.

Linseed Oil Cake–A protein supplement for livestock feeds obtained after the partial removal of oil from ground flaxseed. It should not contain more than 0.5 percent acid-insoluble ash.

Linsey-Woolsey–A cloth made of a mixture of linen and wool.

Lint–Cotton fiber; the epidermal hairs of the cotton seed.

Lint Cotton–The cotton fibers removed from the seed as contrasted to seed cotton, which contains both fibers and seeds.

Lint Index–The weight of lint cotton produced by 100 seeds.

Linters–Short cotton fibers that remain on the cotton seed after ginning. These are removed at cotton oil mills and are an important source of cellulose; used in manufacturing felts, upholstery, artificial fibers, etc.

Lipis– A group of organic substances that are insoluble in water but soluble in such materials as acetone, chloroform, ether, and xylene; e.g., fats, waxes, and steroids including cholesterol.

Liquid Creosote– An oil obtained from coal-tar creosote by the extraction of part of the compounds that crystallize at normal atmospheric temperatures. See Creosote.

Liquid Fertilizers– Water-soluble essential plant nutrients supplied in liquid form and applied by pumping, air pressure, or gravitational flow from a spray applicator. See Fluid Fertilizer.

Liquid Manure– The liquid excrement from animals, mainly urine, collected from the gutters of barns into large tanks and hauled to the fields.

Liquid Nitrogen– Nitrogen in its liquid state. It is extremely cold and is used for storing frozen semen and for cooling irons used in freeze branding. See Freeze Branding.

Lithification– The complex processes that convert a newly deposited sediment into an indurated (cemented) rock. It may occur soon or a long time after deposition. The "cement" may be silica, iron, aluminum, lime, or gypsum.

Lithochromic Soil– A soil that owes its color to parent material, or rock from which the detrital material is derived, rather than to color developed in soil-forming processes.

Lithophyte– A plant that grows on rocks in little or no soil, deriving its nourishment chiefly from the atmosphere, as some Orchidaceae. See Lichens.

Lithos– (Greek) Stone.

Lithosequence– A sequence of soils whose properties are functionally related to differences in the parent rock.

Lithosols– A soil group characterized by an incomplete solum of no clearly expressed soil morphology and consisting of freshly and imperfectly weathered rock or rock fragments.

Litmus– An acid-alkali indicator derived from various species of lichens, principally *Lecanora subfusea* var. *variolosa*; the main constituents are azolitmin and erythrolitmin. The commercial indicator has a pH range of 4.5 to 8.3. The acid color is red and the alkaline color blue. Usually it is used as litmus paper or as an indicator in milk to show acid or alkaline reduction by microorganisms. Slips of litmus paper were formerly used to determine whether soils were acid or alkaline; now glass electrode potentiometers are standard for determining soil pH.

Litter– (1) On a forest floor, the uppermost surface layer of debris, leaves, twigs, and other organic matter, undecomposed or slightly altered. In a technical description of a soil profile, it is generally designated by the letter O. (2) Accumulation of leaves, fruits, twigs, branches, and other plant parts on the surface of the soil. See Mulch. (3) A group of young animals born at a single birth, as a litter of pigs, etc. (4) See Bedding.

Little Barley–*Hordeum pusillum*, family Gramineae; an annual grass that is a plant pest on the western ranges, United States. The dried awns penetrate the tissues of the mouth parts or eyes of grazing animals causing wounds and infections. Native to North America.

Little Black Ant–*Monomorium minimum*, family Formicidae; a North American ant that is a general feeder on fruits, sweets, honeydew, vegetables, and meats.

Little Bluestem–*Andropogon scoparius*, family Gramineae; a perennial grass that grows throughout most of the United States. Locally valuable for grazing and wild hay, its general forage rating is fair. Native to North America.

Little Fire Ant–*Wasmannia auropunctata*, family Formicidae; a small ant pest living in citrus orchards on honeydew whose vicious sting can make hand harvest almost impossible.

Little-leaf Disease– Any of several diseases of trees and vines that result in small leaves. The causes are various, such as zinc deficiency, nematodes, root rots, and viruses.

Littleseed Alfalfa Dodder–*Cuscuta planiflora*, family Convolvulaceae; a species of dodder that infests alfalfa. See Dodder.

Littoral Vegetation– Vegetation growing along the banks of a large lake or the ocean. Vegetation along a small lake, river, swamp, or spring is known as riparian, riverine, riverain, or phreatophytes.

Live Fence– A hedge made by planting closely spaced thorny or other plants, such as Osage orange, and maintaining them in a living condition.

Live Lime– Burnt lime; CaO.

Live Oak–*Quercus virginiana*, family Fagaceae; a broad-leaved, evergreen, tender tree grown for shade and ornament and for its wood used in boats and articles requiring strength and toughness. It also furnishes some browse in the southwestern United States. Also called American live oak, encina.

Live Seed– Viable seed.

Live Virus– A virus whose ability to infect has not been altered.

Living Mulch– An understory of vegetation that helps reduce soil erosion and adds organic matter to the soil, but does not compete heavily with the crop for water and nutrients.

Loam– Soil that consists of less than 52 percent sand, 28 to 50 percent silt, and 7 to 27 percent clay, resulting in a soil texture ideal for gardening.

Loamy– Intermediate in texture and properties between fine-textured and coarse-textured soils. Includes all textural classes with the word loam as a part of the class name, such as clay loam. See Loam, Soil Texture.

Lobelia– See Indian Tobacco.

Loblolly Pine–*Pinus taeda*, family Pinaceae; an evergreen tree, valuable for timber and pulpwood. Native to southeastern North America.

Local Effect– An effect that a toxic substance causes at its original contact point with the body, e.g., eye damage.

Lock–(1) A small tuft of cotton, wool, flax, etc., fibers. (2) Small bits of wool that are packed separately for market (Australia). (3) A locule or the cotton in a locule. (4) The single cavity in an ovary.

Loco–(Spanish, insane) (1) The name of various poisonous plants in arid regions of the genera *Astragalus, Hosackia, Sophora,* and *Oxytropis*; all of the family Leguminosae. Because they contain excessive selenium they are toxic to horses, cattle, and sheep. (2) Locoism. Poisoning with loco. Also known as loco disease, and loco poisoning. See Earl Loco, Rocky Mountain Crazyweed.

Locoed–Designating an animal that has the symptoms of loco disease.

Locule–(1) A small cavity, as of a pollen sac. (2) A single cavity of a compound ovary. Also called lock.

Locus–The position on or region of a chromosome where a gene is located.

Locust–(1) Any of several species of migratory grasshopper, usually of the genus *Melanoplus*, family Acrididae. This pest can have devastating effects on crops in many places in the world. (2) Any shrub or tree of the genus *Robinia*, family Leguminosae. Species are grown as ornamentals and for their hard, durable wood.

Locust Borer–*Megacyllene robiniae*, family Cerambycidae; a beetle larva that tunnels into the sapwood and heartwood of the black locust. Infested trees show swollen areas on the trunk, cracked bark, and exposed tunnels.

Lodge–Pertaining to field crops, to break, bend over, or lie flat on the ground, sometimes forming a tangle. Lodging may be caused by high nitrogen level in the soil and lush growth, wind and heavy rainfall, and plant diseases.

Lodgepole Pine–*Pinus contoria* var. *latifolia*, family Pinaceae; an evergreen tree valued for its lumber used in ties, mine timbers, poles, fuel, rough construction, and planing mill products. Native to North America.

Loess–A massive deposit of tan or buff-colored silt whose particles are typically angular and uniform in size; it is usually calcareous, often contains concretions of calcium carbonate, and sometimes shows lamination or bedding. Where deeply eroded, loess areas have a distinctive topography characterized by vertical walls, sometimes many feet high. Some deposits in the United States are believed to have been formed from finely ground rock material produced during the Ice Age and transported by water and winds to surrounding highlands.

Log–(1) An unhewn, sawed or cut length of a trunk or large limb of a tree. See Saw Log. (2) In the preparation of chip steaks, the molded, frozen piece of meat from which the steaks are cut. (3) To fell trees for lumber.

Log Drive–A large collection of logs floating down a river.

Log Rule–A table showing the estimated amount of lumber that can be sawed from logs of given length and diameter. See Doyle Rule, Scribner Rule.

Loganberry–*Rubus ursinus* var. *loganobaccus*, family Rosaceae; a cane that produces large sweet berries and that is grown commercially.

Logged–Designating an area from which trees have been cut.

Logging–The harvesting of trees for lumber.

Logging Residues–The unused portions of poletimber and sawtimber trees cut or killed by logging.

Long Butt–A section of the lower end of a butt log cut off to remove serious basal defect, such as deep fire scar and advanced butt rot.

Long Redpepper–*Capsicum frutescens* var. *longum*, family Lolanaceae; a tropical, woody plant grown for its edible, long, red or yellow fruit with a sharp, pungent flavor, used in flavoring, condiments, etc. It is the source of cayenne pepper. See Pepper.

Long-day Plant–A plant in which the flowering period or some other process is accelerated by a relatively long, daily exposure to light. See Photoperiodism.

Long-grain Type–Varieties of rice with grains longer than normal, such as Texas, Patra, Bluebonnet, Prelude, Nira, etc.

Long-lasting Fertilizer–Any fertilizer that is water-insoluble, e.g., animal manures, oil meals, and composts.

Long-Staple–Designating cotton fibers 1⅛ inches or more in length.

Long-tailed Mealybug–*Pseudococcus adonidum*, family Pseudococcidae; a sucking insect, a pest of greenhouse plants and citrus.

Longleaf Pine–*Pinus palustris*, family Pinaceae; a coniferous evergreen tree; an abundant native forest species in the Atlantic and Gulf Coastal plains of the United States, where it is an important source of lumber and naval stores. Also called fat pine, southern pine, fatwood, broom pine, Georgia pine. See Naval Stores.

Longtom–*Paspalum lividum*, family Gramineae; a marsh grass sometimes grown in the southern United States as a tame pasture grass. Native to North America.

Loopers–Certain caterpillars of the family Geometridae. Some species have a looping motion when crawling. Also called measuring worm, span worm, inch worm.

Loose Hay–Hay stored in the hay mow or stack without chopping, baling, or compressing.

Lop–(1) To chop branches, tops, or small trees after felling so that the slash will be close to the ground to rot faster. (2) To cut the limbs from a tree or shrub. (3) To fall or spread over, as a plant lops over the space between rows.

Lopping–(1) A method of forcing budded trees; the top of the stock is cut almost through above the bud and left partially attached for some time. the cut is made on the bud side, and the top is bent over to rest on the ground. (2) Pruning or cutting away parts of trees.

Lot–(1) A small piece of enclosed land usually adjacent to a barn or shed in which horses, mules, cows, etc., are allowed to exercise. (2) Any particular grouping of animals, plants, seeds, fertilizers, etc., without particular regard to number, as a seed lot. (3) A small tract of land, usually less than an acre, on which a house is constructed.

Love Vine–A pernicious weed of the genera *Cuscuta*. All species are true seed-bearing, leafless annual herbs with yellowish or reddish filamentous stems. The vine entwines around other plants and develops suckers that attach to the stems of the other plants and become entirely parasitic. Also known as dodder and strangleweed.

Lovegrass–See Lehmann Lovegrass.

Low-acid Food–Food with a pH above 4.6. For canning, a low-acid food requires processing at high temperatures under pressure to destroy microorganisms and ensure a safe canned product. Includes all vegetables and some tomatoes. See Acidic Food.

Lupine–Any plant of the genus ***Lupinus***, family Leguminosae. Species are grown in the flower garden, as cover crops, for green manuring, and to a small extent, as a feed crop. Also called blue pea. The Blue Lupine (bluebonnet) is the state flower of Texas. See Silvery Lupine.

Luster–(1) The property of wood, independent of color, which causes it to reflect light and to exhibit a sheen. (2) The natural gloss or sheen characteristic of the fleeces of long-wool breeds of sheep and Angora goats.

Luxuriant–Designating a lush or abundant growth of plant life, such as a heavy-yielding field of grass or hay.

Lycopene–Red pigment in tomato; sometimes used as a coloring material.

Lye–Any strong liquid alkali, e.g., hydroxide of soda, hydroxide of potash, or a mixture of the tow, used for cleaning, peeling of fruits, in the canning industry, etc.

Lye Peeling–The use of lye to remove the skin of fruits for canning or for the removal of the hard outer covering of the corn kernel in the preparation of hominy.

Lysimeter–A device used to measure the quantity or rate of water movement through or from a block of soil or other material, such as solid waste, or used to collect percolated water for quality analysis.

Lysis–(1) The gradual disappearance of the symptoms of a disease. (2) The destruction or dissolving of cells.

Macadamia Nut–the edible seeds of the trees of ***Macadamia integrifolia*** and ***M. tetraphylla***, family Proteaceae. Native to Australia; naturalized in Hawaii, where its cultivation is estimated to be five times that in Australia. Also called Australian nut, bopple nut, bopple-popple, bush nut, Queensland nut.

Macaroni Wheat–See Durum Wheat.

Macro--Prefix meaning large, long; visibly large.

Macronutrients–Includes primary plant nutrients N, P, and K; and secondary plant nutrients Ca, Mg, and S. See Micronutrient.

Macroorganisms–Plant, animal, or fungal organisms visible to the unaided eye.

Macroparasite–Parasite visible to the naked eye.

Macrophyte–Large plants, macroscopic in size.

Macroscopic–Visible to the human eye without the aid of a microscope.

Magma–Naturally occurring molten (melted) rock material, generated within the earth, from which igneous rocks have been derived through solidification. See Lava.

Magnesia–Magnesium oxide (MgO), obtained by heating or burning the hydroxide or the carbonate. A commercial product, containing a high percentage of MgO, is used as a liming material for soils. Magnesium oxide is also used in the treatment of grass tetany. See Grass Tetany.

Magnesian Limestone–A limestone that contains magnesium carbonate. See Dolomitic Limestone.

Magnesite–A natural mineral (magnesium carbonate: $MgCO_3$). See Magnesia.

Magnesium–Mg; a white metal, essential for people and animals and for plant growth because it is a constituent of chlorophyll.

Magnesium Ammonium Phosphate–$MgNH_4PO_4 \times 6H_2O$. This material is sometimes formed in mixed fertilizers when a magnesium compound (dolomite or magnesium sulfate), superphosphate, and an ammoniacal solution are mixed together. It is also sold commercially as MgAmP. The pure salt contains about 44 percent water in combined form, magnesium (Mg) 9.90 percent, nitrogen (N) 5.70 percent, and phosphoric oxide (P_2O_5) 28.93 percent.

Magnesium Deficiency–(1) A physiological disease of plants grown on soil deficient in available magnesium, generally characterized by chlorosis and necrosis and a reduction of growth. In tomatoes, the veins remain green. In cabbages, the lower leaves pucker and turn white at the margin and center. Also called sand drown. (2) The cause of grass tetany in cattle. See Grass Tetany.

Magnesium Oxide–MgO. See Magnesia.

Magnesium Sulfate–$MgSO_4$; a soluble salt obtained from kieserite that is added to mixed fertilizers to supply the magnesium essential to plant growth.

Maiden–(1) A year old, single-stemmed seedling fruit tree used for budding or grafting. See Whip. (2) An unbred female animal.

Maidencane–*Panicum hemitomon*, family Gramineae; a perennial marsh grass sometimes a weed pest, but also one of the better forage grasses of the freshwater marshes of the Gulf Coast, United States. Native to North America.

Main Crop–The principal or most important crop grown on a farm or in an area.

Maintaining Forage Reserves–The reservation of native forage supplies for such emergencies as drought, fire, or other unforeseen circumstances; obtained by the exclusion of livestock or by light use of grazing areas.

Maize—*Zea mays*, family Gramineae; an annual cereal grass of several varieties and strains, widely grown for its edible seed. Cultivated North and South America for several thousand years, it was introduced into Europe and Africa around the fifteenth century and its cultivation spread rapidly throughout the Old World. Also called corn (United States and Canada), Indian corn.

Maize Billbug—*Calendra maidis*, family Curculionidae; an insect pest of corn that feeds inside the stalk. Found in the southern United States.

Mal Secco—A serious disease of citrus trees in Italy, Greece, Turkey, and Palestine caused by the fungus ***Deuterophoma tracheiphila***, family Sphaerioidaceae, characterized by a sudden wilting and drying of leaves and twigs.

Malachite Green—Copper carbonate.

Malathion—An organic, phosphorus insecticide. Technical grade *malathion* (95 to 98 percent pure) is a viscous, dark brown liquid with a strong, offensive odor somewhat like that of garlic. Widely used to control pests, it is the least toxic of phosphorus insecticides.

Male—(1) Designating an animal capable of producing spermatozoa, or male sex cells. (2) Designating the stamens of a flower. (3) Designating a staminate plant.

Malformation—Any unusual, abnormal growth, organ, or part of a plant or animal.

Malic Acid—Organic acid widely found in many fruits; abundant in apples.

Malling (Root) Stock—Any of sixteen clonal understocks for apples designated as Malling I, Malling II, etc., standardized at East Malling Research Station, East Malling, Kent, England. They are used to produce trees of different sizes, such as dwarf, semidwarf, and standard.

Malt—A grain product, rich in protein and carbohydrates, made by allowing the grain to sprout for a sufficient length of time to produce adequate amounts of enzymes and then dried. The term malt, unqualified, implies barley malt.

Malt Cleanings—A livestock feedstuff obtained from the cleaning of malted barley or from the recleaning of malt that does not meet the minimum protein standard of malt sprouts.

Malt Sprouts—A livestock feedstuff obtained by the removal of the sprouts from malted barley together with the malt hulls, and other parts of malt and foreign material unavoidably present. It should contain not less than 24 percent of protein. The term malt sprouts when applied to a corresponding portion of other malted cereals shall be used in qualified form: e.g., "Rye Malt Sprouts," "Wheat Malt Sprouts," etc.

Malta Fever—See Undulant Fever.

Maltase—An enzyme that splits maltose into two molecules of glucose.

Malting—The germinating of moistened barley grains especially for brewing or distilling.

Mammoth Clover—*Trifolium pratense*, family Fabaceae; a variety of giant red clover that is a very important forage crop.

Mammoth Wildrye—*Elymus giganteus*, family Gramineae; a grass, developed from a Siberian grass by the Soil Conservation Service (United States), and used to stabilize sand in the Pacific Northwest. Also called Volga wildrye.

Manganese—Mn; a metallic element, found in soils (determined as manganous oxide) from a mere trace to as much as 15 percent, but when present it is often in forms unavailable to plants. It is regarded as essential to normal plant growth and is often applied, usually as manganous sulfate, to soils deficient in this element. It is also essential to growth, reproduction, and lactation in animals.

Manganese Agstone—A considerable deposit of limestone occurring in Arkansas that contains 10 to 15 percent Mn soluble in dilute organic acids. It is in the form of manganese carbonate.

Manganese Chlorosis—A discoloring disease of plants caused by lack of manganese.

Manganese Deficiency—A nutrition deficiency that results in dwarfing of plants, chlorosis of the upper leaves often becoming spotted. Crops are frequently unmarketable. In animals, it causes perosis (slipped tendon) in poultry and impaired growth in swine.

Manganese Oxide—There are several oxides of manganese. Two of the important ones are manganous oxide (MnO) and manganese dioxide (MnO_2). The latter occurs as the ore pyrolusite and is used in producing manganese sulfate and manganous oxide.

Mango—*Mangifera indica*, family Anacardiaceae; a fruit tree, grown throughout the tropics for its edible, large, usually green, red, or yellowish-orange, aromatic, very juicy fruit. In the United States its culture is limited to California and southern Florida. Native to India and southeast Asia.

Mangrove—Woody tropical plants that grow along saline shores. The black mangrove, ***Avicennia nitida***, is a valuable source of wild honey. The red mangrove, ***Rhizophora mangle***, is used for tanbark. See Tanbark.

Manson's Eye Worm—*Oxyspirura mansoni*, a slender, colorless roundworm about ¾ inch long. It is a parasite found under the transparent membrane over the eyeball of the fowl. infestation causes puffiness around the eye, continuous winking, and sometimes blindness.

Mantis—Any of the large insects, family Mantidae, predaceous on other insects, capturing their prey with their unusually well-developed front legs. Species are considered beneficial in gardens. See Praying Mantis.

Manual Watering—Watering plants by hand with a hose, bucket, or some other manual means.

Manufacturing Tobacco—One of the general class of tobacco. It includes the Virginia sun-cured, white burley, flue-cured, and dark fire-cured.

Manure—(1) Excreta of animals, dung and urine (usually with some bedding), used to fertilize land. (2) In Europe, any material that contains the essential elements of plant nutrients, as chemical fertilizers, excreta of animals, etc. See Compost.

Manure Pit–A storage unit in which accumulations of manure are collected before subsequent handling or treatment, or both, and ultimate disposal. Water may be added in the pit to promote liquefaction so the manure can be spread on fields through a sprinkler irrigation system.

Manure Salts–Crude salts in natural deposits, containing a high percentage of potash (K_2O) in the form of the chloride. Sometimes used as fertilizer.

Manure Spreader–A wagon-type implement for carrying barnyard manure to the field, shredding it and spreading it uniformly on the land. The power for the spreading mechanism and conveyor is supplied from the rear wheels or from a tractor power take-off.

Maple–Any tree or shrub of the genus *Acer*, family Aceraceae. Species are grown as ornamentals, for shade, lumber, and for the sap that yields maple syrup and maple sugar. Some types are: bigleaf maple, box elder, Japanese maple, Norway maple, planetree maple, red maple, silver maple, striped maple, sugar maple, vine maple, and flame maple.

Maple Cream–The sap of the sugar maple, processed to a consistency between syrup and crystallized sugar; used as a spread. Also called maple butter.

Maple Sugar–A crystallized, sweet food product, largely saccharose in nature, made by evaporating the sweet sap of certain maple trees. A food delicacy now, at one time it was an essential human food in the humid northern parts of the United States.

Maple Syrup–A sweet, concentrated food product made in the spring by boiling the natural sap of the sugar maple tree (*Acer saccarum*) to the proper concentration. Further concentration will cause crystallization of the syrup into maple sugar.

Marcescent–Designating leaves that wither on a plant without falling off.

Marcot–A branch that, for purposes of propagation, is air-layered by having a rooting medium bound to it.

Margarine–A food containing 80 percent fat from refined oils of cottonseed, soybean, corn, peanut, palm, sunflower, or safflower fortified with vitamin A. The type of oil used must be indicated on the package. Also known as oleomargarine.

Margin–The edge, border or borderline, as margin of a leaf.

Marijuana–A preparation of the dried leaves and flowers of male or female plants of wild hemp, ***Cannabis sativa***, family Cannabaceae. Used in cigarettes for its euphoric properties. Illegal to grow in the United States. See Hemp.

Marinate–To soak or steep in a marinade of vinegar, lemon juice or wine, oil, and seasonings to give added flavor, to tenderize, or to preserve.

Market Garden–A farm that produces vegetables to be sold on a roadside market or in a nearby city.

Market Pack–A fiber or plastic container used to grow bedding plants; the container holds from six to twelve plants.

Marking–(1) Castrating, docking, branding, and ear marking. (2) Selection and indication, usually by blaze or paint spot, of trees that are to be cut or retained in a cutting operation.

Marl–(1) A natural, earthy deposit, consisting chiefly of calcium carbonate mixed with clay or other impurities in varying proportions. (2) Incoherent sands containing shells. (3) Soft, impure limestone. (4) Soft, mudlike deposits originating mostly in lakes and found in underlying bogs widely distributed in the glaciated region of the United States. Sometimes called bog lime. Marls are used as an amendment for soils deficient in lime. See Lime.

Marl Frenching–Certain disease symptoms of citrus trees characterized by yellowing of foliage; attributed to deficiencies of manganese and zinc. Also called marl chlorosis.

Marsh Grass–Any water-tolerant plant that grows on a marsh, including sedges and rushes as well as true grasses. Also called watergrass.

Marsh Hay–Any of the swamp or marsh grasses and sedges, usually coarse in nature and low in feeding value, that are sometimes cut for hay. Also used for packing crockery, glassware, etc., and for making grass ropes. These were the only sources of hay for early settlers in the forested regions of the United States.

Marsh Muhly–*Muhlenbergia racemosa*, family Gramineae; a perennial grass found in moist places from Newfoundland to Arizona, United States. Its forage rating is good. Native to North America.

Mash–(1) A mixture of grain and other ingredients with water to prepare wort for brewing operations. (2) A ground feed of cereals and malt, etc., fed in a wet or dry form to livestock and poultry. Also called crowdy. See Wort.

Mash Concentrate–A poultry mash containing 20 to 40 percent protein.

Mashing–The converting of starch in cereals, potatoes, etc., into sugar by the action of amylase enzymes of malt, and the accompanying breakdown of proteins by the protease enzymes of the malt, in the manufacture of alcohol or alcoholic beverages.

Masid–A leaf margin that is cut or cleft close to the midrib.

Masked–Designating disease symptoms that are hidden.

Mass Selection–(1) In plant breeding, the selection of a large number of plants for propagation from which the off-type, low-yielding, inferior, and disease-susceptible plants have been eliminated. (2) In animal breeding, the selection for breeding purposes of animals on the basis of their individual performances, type, or conformation.

Massive–(1) Indicating a large-sized and deeply muscled animal; designates an ideal quality in a draft horse. (2) (Soil structure) Large uniform masses of cohesive soil that sometimes have irregular cleavage, as in the C horizons of many fine-textured clay soils. See Soil Structure.

Mast–Nuts and seeds accumulated on the forest floor and often serving as food for livestock and wildlife, especially nuts of oak (acorns) and beach, and seeds of certain pines such as longleaf and pinyon.

Mat Sandbur–*Cenchrus pauciflorus*, family Gramineae; an annual grass considered a weed. It produces burs that easily drop off or cling to clothing of people and to the hair of passing animals. Native to North America. Also called sandbur, loose bur, sandspear, bur grass.

Mataxenia–Direct effect of pollen on parts of the seed and fruit that lie outside the embryo and endosperm.

Matroclinous–Resembling the female parent. See Patroclinous.

Maturation–(1) Becoming mature or ripe. (2) Changes in cell division, especially in reproductive elements, in which the number of chromosomes in the nucleus of the new cells is half that of the original.

Mature Soil–(1) A soil that has well-developed characteristics produced by the natural processes of soil formation and that is in equilibrium with its environment. (2) A soil with an A, B, and C horizon. See Soil Profile.

Maximum Temperature–In plant physiology, the temperature beyond which growth ceases. It varies with plant species and the duration of high temperature.

May Beetle–Species of the genera *Phyllophaga* or *Lachnosterna*, family Scarabaeidae; the adults of the white grub that are brown or brownish black, appear May to July, fly at night, and are attracted by lights. They are widely distributed. Also called June bugs, daw bugs.

May Troubles–The difficulties encountered in the spring by milk manufacturers, due to chemical and bacteriological changes in milk when the cows go from dry feeds of winter to green pasture.

Mayfly Naiads–Family Ephemeridae; the immature mayfly. The terrestrial adults lack functional mouth parts and live only a few hours.

Mayweed–See Dog Fennel.

McIntosh–A hardy, aromatic, late variety of bright red, striped apple, with thin skin and tender, white flesh. Grown in Canada and northern states of the United States.

Meadow–(1) A field cut for hay. (2) An opening in a forest where grass is the dominant vegetation. (3) In arid and semiarid regions, a productive grassy area usually with a high water table. (4) At high elevations, an alpine meadow.

Meadow Barley–*Hordeum brachyantherum*, family Gramineae; a perennial bunchgrass with scant foliage and short awns, found in moist areas from Alaska to New Mexico. It is grazed closely by livestock until time of heading, but its forage value is only fair.

Meadow Fescue–*Festuca elatior*, family Gramineae; an upright, tufted, perennial grass grown for pasture and hay. Also called alta fescue, tall fescue, English bluegrass.

Meadow Foxtail–A less desirable grass, adapted to wet soils that are subject to frequent and/or prolonged flooding. Produces less forage than other grasses. Tolerant of frost and prolonged snow cover in high-altitude areas. Meadow foxtail is very undesirable in seed-producing areas, as its seed is extremely difficult to separate from some of the more important seed crops.

Meadow Nematode–Species of the genus *Pratylenchus*, family Tylenchoidae, a widely distributed, wormlike root nematode found in the cortical tissue of the roots causing sloughing off and the formation of fine roots near the surface of the ground. It infests many trees, such as apple, cherry, walnut, and almond, and causes brown root rot of tobacco, alfalfa, cotton, and other plants. See Nematode.

Meadow Soil–A mineral soil that developed under a grass-sedge or herbaceous vegetation in a wet habitat, such as a mountain meadow. Also called Wiesenboden.

Meadow Spittlebug–*Philaenus leucophthalmus*, family Cercopidae; an insect pest that infests clover, alfalfa, and other forage crops, strawberries, and flowers in gardens and greenhouses.

Meadow Stink–The name applied to the odor of a lateritic (ironstone) soil immediately after a rain. See Laterite.

Meadow Strip–A sloping strip of grassed land which, in addition to yielding a hay crop, acts as a broad, shallow water channel during periods of runoff. Often used as a terrace outlet channel; usually much wider than a grassed waterway.

Meadowfoam–*Limnanthes alba*; a low-growing winter annual native to California, southern Oregon, and Vancouver Island, British Columbia. The seeds of the plant contain oil that has a potential use as lubricants, polymers, and waxes.

Meal–(1) A coarsely ground, often unbolted grain, such as cornmeal, oatmeal. (2) Any dried and ground feed material, as fish meal.

Mealy Potato–An Irish potato, high in starch content, that produces a light, flaky, uniform product when cooked.

Mealybug–Species of the genus *Pseudococcus*, family Pseudococcidae; widely distributed, small, elliptical, waxy-covered insects that infest various fruit trees and greenhouse plants.

Mean Annual Increment–In forestry, the total growth (board feet or cubic feet) of a stand of trees divided by the total age in years.

Mean High Water–The average height of the high water over a specified number of years.

Mean Low Water–The average height of the low water over a specified number of years.

Mean Sample Tree–Any tree that represents the average of a group of trees in any given class or stand.

Means Grass–See Johnsongrass.

Measles of Apples–A condition in apples characterized by small rough bumps in the bark of apple trees that cause the appearance of scabs. It may be caused by bacteria, fungus, or insects.

Measurable Precipitation–Any precipitation equivalent to over 0.005 inches of rainfall. Less than 0.005 inches is considered a trace.

Mechanical Practices–Soil and water conservation practices that primarily change the surface of the land or that store, convey, regulate, or dispose of run-off water without excessive erosion.

Mechanical Soil Analysis–See Particle Size Distribution.

Meconium–(1) The first excreta of a newborn animal. (2) The juice of the opium poppy, *Papaver somniferum*.

Media–(1) The middle coat of an artery. (2) The plural of medium. (3) Material for artificial propagation of various microorganisms. (4) Any means through which communication of any type is accomplished. (5) Soil or soil-like material in which plants are grown.

Median Lethal Dose–(LD_{50}) The amount of concentration of a toxic substance that will result in the death of 50 percent of a group of test (target) organisms upon exposure (by ingestion, application, injection or in their surrounding environment) for a specified period of time. (Complementary to median tolerance limit, TL_{50}.)

Median Tolerance Limit–TL_{50} The concentration of some toxic substance at which just 50 percent of the test (target) animals are able to survive for a specified period of exposures. (Complementary to Median Lethal Dose, LD_{50}.)

Medication–(1) The application of medicines, salves, etc., to an injured or sick animal. (2) The forced introduction of a chemical, usually a water-soluble salt in solution, into the sapstream of a living tree to kill it or make it repellent to insect attack, or into a freshly felled tree to destroy barkbeetle and woodbeetle broods.

Medics–Certain plants of the genus *Medicago*, family Leguminosae. Some have a high forage value; some are regarded as weeds.

Mediterranean Fruit Fly–*Ceratitis capitata*, family Tephritidae; a fly that infests a number of fruits, especially citrus fruits. Common in the Mediterranean region, it has invaded the United States, notably the citrus groves of Florida and California.

Medium–(1) Any of a number of natural or artificial substances, pastelike or liquid, in or on which microorganisms, such as bacteria and fungi, can be cultured. (2) A soil or material, such as sand, peat moss, vermiculite, etc., in which plants are raised or cuttings are rooted, especially in the greenhouse. (3) A market grade of roses that has stems 12 to 8 inches in length. (4) One of the six recognized grades of meat, lying between good and common.

Medium Sand–In soil analysis, a separate that includes particles ranging in diameter from 0.5 to 0.25 millimeters. See Soil Textures.

Medium-textured Soil–Very fine sandy loam, loam, silt loam, or silt.

Mega–(1) A prefix meaning one million. (2) A prefix meaning large or in large numbers.

Megaspore–A spore that has the property of giving rise to a gametophyte (embryo sac) bearing a female gamete. One of the four cells produced by two meiotic divisions of the megaspore-mother-cell (megasporocyte).

Megasporegametophyte–The few-celled haploid generation portion of a seed plant arising from a meiotic division and giving rise through meitosis to the female gametes. Female inflorescence.

Megathermic Tree–Any tree that requires a warm climate and long growing season, such as palm, laurel, magnolia, live oak, tupelo gum, redwood, baldcypress, slash pine, longleaf pine, and other southern pines.

Megspore-Mother-Cell–The cell that undergoes two meiotic divisions to produce four megaspores.

Meiosis–Cell division early in the reproductive process, and in the formation of sperm and ova in the testicles and ovaries. Each pair of chromosomes in the cell being divided separates, and one member of each pair goes to each of the two new cells formed.

Melanic Epipedon–A black surface soil horizon high in organic matter from grass residues. See Soil Taxonomy.

Melassic Acid–An acid generated in molasses and glucose by alkali action.

Melaxuma–Any of several diseases causing dark to black bark cankers, as the fungal disease of the walnut caused by *Dothiorella gregaria*.

Mellow–Designating soft, tender, juicy, ripe fruit.

Mellow Soil–(1) A nonscientific term describing a soil, especially one containing a high percentage of humus, that is soft and loose. (2) A soil that pulverizes easily into a good seedbed.

Melon–The usually round, fleshy fruit of *Cucumis melo* and *Citrullus vulgaris*, family Cucurbitaceae, and their variations. See Cantaloupe, Muskmelon, Watermelon.

Membrane–(1) A thin, flexible sheet of vegetable or animal tissue; the thin protoplasmic tissue connecting, covering, or lining a structure, such as a cell of a plant or animal. (2) The layer that surrounds fat globules of milk. Its nature is not definitely known: it may consist mainly of phospholipoids and a protein or proteins not completely identical with casein, albumin, or globulin. (3) A layer of low permeability material: e.g., bentonite (clay) soil, placed in the bottom of a farm pond to reduce seepage losses.

Mendel's Law (Medelian Law)–Gregor Johann Mendel (1822-1884), an Austrian monk, naturalist, and plant breeder, first demonstrated dominant and recessive genes in plant breeding, which also applies to some kinds of animal and people inheritance. For example: When a plant with the dominant gene for tallness (TT) is crossed with a plant with a recessive gene for shortness (ss), the first generation, F_1, will be distributed according to the following ratio. One plant, TT, homozygous (pure) true-breeding for tallness. Two plants, Ts, heterozygous for tallness and shortness. One plant, ss, homozygous (pure) true-breeding for shortness. When plants with Ts and Ts genes are crossed, the F_1 progeny will be in the same ratio, i.e., one TT, two Ts, and one ss.

Menzies Larkspur–*Delphinium menziesi*, family Ranunculaceae; a perennial, poisonous herb, which, when eaten by cattle, causes uneasiness, staggering, constipation, nausea, bloat, and rigid limbs. Native to North America. See Larkspur.

Merchantable Height–The length of the tree stem from the top of the stump to the top of the last merchantable section. Usually expressed in feet or number of logs of a specified length.

Merchantable Timber–A tree or stand of trees that may be converted into salable products.

Merchantable Volume–The amount of wood in a single tree or forest stand that is considered salable.

Mericarp–One-half of the dry, dehiscent (splitting-open) fruit (schizocarp) found in most members of the parsley family.

Meristem–Plant tissue capable of cell division and therefore responsible for growth.

Mescalbean Sophora–*Sophora secudiflora*, family Leguminosae; an evergreen shrub or small tree that causes poisoning of cattle, sheep, and goats when eaten in large quantities. It is sometimes planted for its fragrant, violet-blue flowers. Native to the southwestern ranges of the United States. Also called mescal bean.

Mescalbutton Peyote–*Lophophora williamsi*, family Cactaceae; a small, globe-shaped cactus that yields the mescal-button, a bulb having narcotic properties. Native to North America. Also called hikuli, peyote.

Mesocarp–The layer between the endocarp and the epicarp of a multiple-layered pericarp. In stone-fruits, it is thick, fleshy, and edible.

Mesophyll–The parenchyma tissue between the upper and lower epidermis of a leaf; the cells usually contain chloroplasts.

Mesophyte–A plant whose normal habitat is neither wet nor dry but intermediate in soil moisture.

Mesotherm–(1) Any plant that grows under intermediate conditions of temperature along with abundant moisture. (2) Any plant that grows under conditions of alternating high and low temperatures.

Mesquite–Any plant of the genus *Prosopis*, family Leguminosae. Species are usually thorny shrubs, the foliage and seed pods of which have browse value in the desert, but the invasion of the plant on open grassland lowers the grazing capacity of the land. One of the more useful mesquites is *P. juliflora*, also called algarroba, an important bee and forage plant. Also called mesquit, musquit, muzquit.

Mesquite American Mistletoe–*Phoradendron californicum*; family Loranthaceae; a parasitic shrub that grows on trees. Its berries are used for decoration chiefly at Christmastime. Native to North America. See Mistletoe.

Mesquite Pod–The beanlike pod or seed of mesquite (*Prosopis* spp.) rich in sugar and a valuable feed for stock.

Metabolic–Designating the chemical changes that take place in living plant and animal cells whereby one compound is converted to one or more other compounds.

Metaldehyde–A crystalline polymer of acetaldehyde, used at two to three percent as an attractant in poison baits for snails and slugs.

Metamorphic Rock–Any rock that has undergone profound changes by the combined action of pressure, heat, and water, as schist, slate, and marble.

Metamorphosis–A process by which an organism changes in form and structure in the course of its development, as many insects do. See Complete Metamorphosis.

Metaphase–That stage of cell division in which the chromosomes are arranged in an equatorial plate or plane. It precedes the anaphase stage.

Metaphosphoric Acid–HPO_3. See Phosphoric Acid.

Metcalfe Bean–*Phaseolus metcalfei*, family Leguminosae; a perennial herb with a long trailing stem, grown mostly for forage. Native to North America.

Methane–CH_4; an odorless, colorless, and asphyxiating gas that can explode under certain circumstances; can be produced by manures or solid waste undergoing anaerobic decomposition as in anaerobic lagoons and in silos. There are twelve species of bacteria capable of producing methane gas from manures and other biomes.

Methanesulphonyl Fluoride–A highly effective greenhouse fumigant.

Methanol–A light volatile, flammable, poisonous, liquid alcohol, formed in the destructive distillation of wood or made synthetically and used especially as a fuel, a solvent, an antifreeze, or a denaturant for ethyl alcohol, and in the synthesis of other chemicals; methanol can be used as fuel for motor vehicles; also known as methyl alcohol or wood alcohol. See Gasohol.

Methoxychlor–A chlorinated hydrocarbon insecticide, closely related to DDT and TDE.

Methyl Alcohol–Also known as methanol or wood alcohol. See Methanol.

Methyl Bromide–CH_3Br; an insect fumigant. It is also used to sterilize the soil.

Methyl Formate–An insect fumigant.

Methyl Red–A chemical indicator; a dye stain used in determining the classification or taxonomy of certain bacteria such as **Coliform bacilli**.

Metrox–See Copper Oxide.

Mexican Bean Beetle–*Epilachna varivestis*, family Coccinellidae; a brownish, spotted beetle whose larvae feed on kidney beans, soybeans, and cowpeas. Also called bean ladybird.

Mexican Black Bean–See Frijole.

Mexican Jumping Bean–The seeds of certain shrubs of the genus *Sapium* or *Sebastiania*, family Euphorbiaceae, which have been infested by a small moth, genus *Carpocapsa*, family Olethreutidae. The moth lays its eggs in the seeds, and the action of the larvae spinning cocoons within cause the seeds to jump about.

Mexican Milkweed–*Asclepias mexicana*, family Asclepiadaceae; a perennial herb, poisonous to sheep and poultry. Native to North America, found from Washington to Mexico. Also called narrow-leaved milkweed, Mexican whorled milkweed.

Mexican Pinyon Pine–*Pinus cembroides*, family Pinaceae; a low tree that bears an edible seed. Native to North America. Also called pinyon, Mexican stone pine, nut pine, pinon. See Nut Pine.

Mexican Teosinte–*Euchlaena mexicana*, family Gramineae; a grass closely related to corn. Native to Central America.

Mexican Vanilla–*Vanilla planifolia*, family Orchidaceae; a tropical tall-climbing, fleshy-stemmed vine grown in frost-free areas for its aromatic seed pods, the source of the vanilla of commerce used in flavoring. Native to Mexico.

Mexican Weed–*Caperonia palustris*, a troublesome weed in rice fields in the southern United States.

Mica–A common mineral in soils; muscovite, biotite, and a number of others. They are not easily soluble minerals of the mica group residual in soils derived form the weathering of granites, gneisses, and schistose rocks. When present in considerable quantity, mica imparts looseness and a greasy or soapy feel to a soil. See Vermiculite.

Micro–A prefix meaning one millionth (1/1,000,000) or very small.

Microbes–Minute plant or animal life. Some microbes may cause disease, while others are beneficial.

Microbial Insecticide–Microbes that are used to control insects. Of the millions of insects that have been identified, about 350 species have been classified as destructive pests. Like all living organisms, these pests are susceptible to diseases caused by bacteria, fungi, protozoa, and viruses. Research scientists have explored this avenue of control and have developed several strains of microbes that have potential to be effective in the control of insect pests. See Integrated Pest Management.

Microbiologist–A scientist concerned with the study of plant and animal microorganisms.

Microelements–Trace elements; micronutrients.

Microflora–(1) Microscopic plant, such as bacteria. (2) In plant geography, the vegetation of a small area in contrast to that of a region. (3) Microbial life characteristic of a region, such as the bacteria and protozoa populating the rumen.

Micronized Sulfur–Sulfur in an extremely fine state of division with a particle size mostly under 10 microns. Used as dusts and sprays for control of plant diseases.

Micronutrient Elements, Essential–For plants: boron, chlorine, cobalt, copper, iron, manganese, molybdenum, and zinc. For people and animals: chlorine, copper, iron, manganese, molybdenum, zinc, sodium, iodine, selenium, vanadium, chromium, cobalt, nickel, fluorine, silicon, boron, and arsenic.

Microorganism–(1) An organism so small that it cannot be seen clearly without the use of a microscope; a microscopic or submicroscopic organism. (2) Any microscopic animal or plant that may cause a plant disease or have the beneficial effect of decomposing plant and animal residue that becomes humus.

Micropropagation–A propagation technique that uses a single cell of the meristematic tissue of a plant to produce a new plant. The procedure is also known as cloning and tissue culture.

Micropyle–The minute necklike opening in the integuments of an ovule, where the sperms enter.

Microspore–One of the four cells produced by the two meiotic divisions (mitoses) of the microspore-mother-cell (microsporocyte). It gives rise to a gametophyte bearing only male gametes. Also called pollen grain.

Microsporegametophyte–The few-celled haploid generation portion of a seed plant arising from a meiotic division and giving rise through meiosis to the male gametes. Male inflorescence.

Microthermic Tree–A tree that tolerates a cold climate and short growing season: e.g., aspen, birch, alder, willow, mountain ash, jack pine, Scotch pine, lodgepole pine, spruce, larch, and alpine fir.

Middens–(1) The castings of organisms that live and work in the soil, especially of earthworms. (2) A pile of refuse; a manure pile.

Middles–The spaces between the rows of trees or a crop which may be cultivated, planted to an interrow crop, or left in sod.

Middlings–(1) The coarse particles separated in the milling of wheat that contain the germ of the wheat grain and fine particles of bran. Higher in protein and digestible nutrients than bran, the middlings are used as a feed chiefly for hogs, calves, and poultry. Also called shorts. (2) The belly portion of a pork carcass from which bacon is cut.

Midgrass–(range) An area on the western plains of the United States, where the prevailing growth height of native grasses is less than in the tall grass prairie or that is characterized by species of the genus ***Andropogon***, such as small and big bluestems. The midgrasses include the wheatgrasses and Junegrass.

Midrib–The main vein of a leaf; located halfway between the two edges. It is a continuation of the leaf stalk.

Mildew–Plant disease caused by several fungi, recognized by the white cottony coating on plants.

Milk, Evaporated–Milk that has had part of its water of constitution evaporated by heating. It does not have sugar added. See Milk, Condensed.

Milk Stage–The period in the development of a grain when the contents of the seed are still fluid and of the color of milk. See Dough Stage.

Milk Vetch–See Hayden Poisonvetch.

Milkweed–(1) Any plant of the genus ***Asclepias***, family Asclepiadaceae. The plants are weedy, perennial herbs characterized by milky juice. Some species are poisonous to stock, and some yield a good quality of light-colored honey. The white silky floss of the mature seed pod has been harvested experimentally as a substitute for kapok and other uses. (2) ***A. syriaca***, a perennial, weed pest of waste ground and untilled fields. Found in the central and western United States and Canada. Also called silkweed, cottonweed. See Showy Milkweed.

Milky Disease–A disease of insects caused by ***Bacillus popillae*** that is used in the biological control of the Japanese beetle.

Milled Grain–Grain processed in a mill; e.g., wheat that has been made into flour.

Milled Rice–Rice grains that have been run through hullers for removal of the bran coat.

Milleped(e)s–Arthropods of the division Diplopoda, similar to centipedes but having a greater number of legs. They are known to be abundant in some soils and may be a factor in the rapid breakdown of leaves. They seldom injure plants but occasionally enter houses. Also called thousand legged worms, galleyworms.

Millet–Any of certain annual grasses and cereals with little seeds of the family Gramineae. Species are grown as field crops for grain and forage. Native to Europe and Asia. The principal kinds grown in America are: prosso, foxtail millet, and barnyardgrass.

Millet, Foxtail–*Setaria italica*; a coarse, erect annual with large, cylindrical, spikelike panicles. The panicles are lobed, yellow to purplish and awned. It is adapted to fertile, well-drained soils and to deep, fine-textured, slowly-permeable soils. It has excellent seedling vigor and is early maturing. It is a preferred winter food for pheasants.

Millet, Japanese–*Echinochloa crusgallis* var. *frumentacia*; a coarse, annual grass similar to barnyardgrass, except that it is awnless and later maturing. Seed maturity is approximately 110 days after planting, depending on variety and climate. It is a valuable wildlife food.

Millet, Proso–*Panicum miliaceum*; a coarse, erect, branching, warm-season annual grass, growing from 1.5 to 6 feet (½ to 2 millimeters) tall. It has a large drooping panicle. Seed is reasonably resistant to shattering. The forage is not palatable to livestock. Some seeds retain viability over the winter. It is best adapted to deep, moist, fine sand, or loam soils. It should be planted from May 1 to July 1 for use by ducks, doves, pheasants, and other birds.

Millet Woodrush–*Luzula parviflora*, family Juncaceae; a grasslike plant with soft, flat leaves that grows on very moist or wet sites. Its forage rating is fair.

Milling–The processing of grain, or other products, by hulling, chopping, crushing, etc.; e.g., the conversion of wheat into flour in rolling mills.

Milling Quality–(1) The percentage of whole kernels and total milled rice that is obtained from a given quantity of rough rice. (2) Any wheat that will yield a high percentage of white, sound flour.

Milo–*Sorghum vulgare* var. *subglabrescens*, family Gramineae; a drought-resistant, annual grass, one of the older varieties of grain sorghums, which is grown mainly in the southern Great Plains Region of the United States mostly for livestock feed. Native to tropical Africa. See Grain Sorghum, Sorghum.

Milo Chop–A livestock feedstuff consisting of the entire grain of milo sorghum removed from the head and chopped or ground.

Milo Disease–A destructive root rot affecting milo, probably caused by the soil-dwelling phycomycete *Pythium arrhenomanes* and other fungi in conjunction.

Milo Grain–In stock feeding, the grain or seed of the milo sorghum, in contrast to the whole head cluster.

Milo Mill Feed–A livestock feedstuff that is a mixture of milo bran, milo germ, and a part of the starchy portion of the grain, produced in the manufacture of grits from milo grain.

Miner's Lettuce–*Claytonia perfoliata*, family Portulacaceae; an annual herb grown for its small white flowers as a potherb or salad plant for its edible leaves. Native to North America. Also called winter purslane, Indian lettuce.

Mineral–(1) A chemical compound or element of inorganic origin. (2) Designating the inorganic nature of a substance.

Mineral Elements–The inorganic components of soil, plants, or agricultural products.

Mineral Ion–Electrically charged atoms or groups of atoms formed by disassociation of a molecule, and the form by which most mineral elements enter plant roots. Ions move from outside into the plant root cells as a result of ionic diffusion. See Cation-exchange Capacity, Essential Elements.

Mineralize–(1) To petrify. (2) To impregnate or supply with minerals. (3) To promote the formation of minerals. (4) The microbial breakdown of organic matter to release its minerals.

Minimum Temperature–(1) A degree of warmth at which plant growth first begins. See Cool-season Crops, Warm-season Crops. (2) The lowest temperature recorded during a given period, such as the minimum daily temperature.

Minimum Tillage–A soil management system in which the residue of the previous crop remains on the soil surface and the next crop is planted with little or no tillage.

Minor Elements–See Micronutrient Elements, Essential.

Mint–Any herb of the genus *Mentha*, family Labiatae. Species are grown for their aromatic plant parts and as a field crop; source of the essential oil used in flavoring and for medicinal purposes; e.g., apple mint, bergamot mint, field mint, peppermint, pennyroyal mint, spearmint.

Mire–A small muddy marsh or bog; wet, spongy earth; soft, deep mud.

Mist Fertilizer–A low-nutrient solution applied to newly propagated vegetative cuttings in a greenhouse.

Miticide–Any poisonous substance used to kill mites.

Mitosis–Cell division involving the formation of chromosomes, spindle fibers, and the division of chromosomes by a process of longitudinal splitting. Each of the resulting daughter cells thus has a full set of chromosomes as distinguished from reduction division or meiosis, in which the daughter cells have half the somatic number. See Meiosis.

Mix–(1) A formula; a combination of two or more ingredients, blended together for a specific purposes, such as feedstuff for livestock. (2) To cross or interbreed animals, plant varieties, etc., by chance.

Mixed Corn–All corn that does not meet the color rules for yellow corn or white corn.

Mixed Fertilizers–Two or more fertilizer materials mixed or granulated together into individual pellets. The term includes dry mixed powders, granulates (bulk blends), granulated mixtures, and clear liquid mixed fertilizers, suspensions, and slurries. They were at one time called complete fertilizers, but are now referred to as multinutrient mixtures.

Mixed Grain Sorghum–A market class that includes all grain sorghums not meeting the separate standards for white, yellow, red, and brown grain sorghums.

Mixed Hay–Hay that consists of two or more forage species, as a mixture of red clover and timothy.

Mockernut Hickory–*Carya tomentosa*, family Juglandaceae; a deciduous tree grown for shade and for its valuable wood used in tool handles, ladders, furniture, sporting goods, implements, woodenware, and for fuel and smoking meat. Also called big bud hickory, bird's-eye hickory, black hickory, curly hickory, whiteheart hickory.

Moderate Live Virus–A virus that has been changed by passage through an unnatural host, such as passing hog cholera virus through rabbits, so that it no longer possesses the disease-producing characteristics but so that it will stimulate antibody production and immunity when injected into susceptible animals.

Modified Stem–Any type of plant stem, such as a tuber, rhizome, runner, or stolon, that differs from the upright stem.

Modifier–(1) Any element that is added to, or taken from a substance that alters its normal appearance or function. (2) A substance that can alter the course of carcinogenesis.

Modifying Gene–A gene that changes the expression of the chief gene, or genes, controlling a character.

Modiola–*Modiola procumbens*, family Malvaceae; a plant grown in California, United States, for pasturage on sodic soils. Escaped it can become a weed. Native to Chile.

Mogote–A thick patch of shrubbery (southwestern United States).

Moisture–(1) The total amount of water in any plant or animal product. (2) Any form of water. (3) The total amount of water (exclusive of that in chemical combinations) in the soil, both that which is available and that unavailable for plant growth.

Moisture Content–For grains, the percentage of moisture expressed as wet weight. For soils, the percentage of moisture expressed as oven-dry weight.

Moisture Penetration–(1) The depth to which moisture penetrates in soil following an irrigation or rain before the rate of downward movement becomes negligible. (2) Downward movement of water into a soil (a soil may be open, porous, and permit free penetration or it may be compact and tight and thus prevent free downward movement).

Moisture Tension–The force at which water is held by soil, usually expressed as the equivalent of a unit column of water in centimeters: 1,000 centimeters being equal to one atmospheric tension. Moisture tension increases with dryness and indicates the degree of work required to remove soil moisture for use by plants.

Moisture-retentive Soil–A soil that retains a high percentage of moisture or water. In mineral soils, retentiveness bears a direct relation to clay or colloid content; e.g., a clayey soil is more retentive than a sandy soil.

Molascuit–A cattle feed made from sugarcane bagasse and molasses.

Molasses–(1) The dark brown syrup obtained as a by-product in the manufacture of sugar from sugarcane. (2) Loosely, a syrup obtained by evaporation of the sweet juices of any plant, as sorghum, corn, sugar beets, sugar maple, etc.

Molasses Feed–An animal feed, usually a commercial mixture in which molasses is an ingredient; used as a substitute for grain because of its palatability and because it is a cheap source of readily digested carbohydrates.

Molasses Silage–Legume or grass silage to which specified amounts of molasses, such as blackstrap, are added to aid the proper fermentation, increase the carbohydrates, improve palatability, etc.

Molassesgrass–*Melinis minuliflora*, family Gramineae; a tall, perennial grass grown as a forage crop in the United States. Native to Africa.

Mold–(1) A form containing a cavity into which material is poured or pressed to achieve a special shape and design; e.g., molds for Edam and Gouda cheese, butter, ice cream, etc. (2) Any soft, humus soil. (3) Fungi distinguished by the formation of a mycelium (a network of filaments or threads), or by spore masses; usually saprophytes. However, various kinds may do serious damage to fruits, hay, grain, growing crops, and ornamental plants. Also spelled mould. See Compost, Downy Mildew, Mildew, Powdery Mildew.

Mold Poisoning–Harmful effects upon animals from eating certain moldy feeds. The poisoning may be due to the degree of moldiness, or to prolonged feeding when less harmful molds are involved. With horses, the effect may be a nervous disease resulting in blindness and a staggering gait.

Moldiness–In judging and grading cheese, any defect due to the presence, or previous presence, of undesirable molds.

Moldy–Infected by molds (fungi); designating objectionable quality due to appearance or odor, flavor, toxicity, caused by molds.

Mole–(1) One of several species of burrowing, insectivorous mammals, characterized by soft, dark fur, long nose, very small eyes, and large, strongly developed front feet. The moles belong mainly to the family Talpidae. A common species in the eastern United States is ***Scalopus aquaticus***. They can damage lawns and gardens with their underground burrows but otherwise are not serious pests. (2) A natural levee or depositional ridge on the bank of a large river. (3) A plow used in establishing mole drainage. See Mole Drainage.

Mole Cricket–Any of the several insects of the family Gryllidae that eat plant roots, other insects, and earthworms.

Mole Drainage–A method of draining land by making open channels in the soil at a depth of 2 to 3 feet. These are constructed with a mole plow consisting of a two-wheel frame and a projectile-shaped ball varying in diameter from 2 to 8 inches and chain linked behind a sharp-pointed, sharp-edged shank. The plow is pulled along the course of the desired drain with the ball at depths and grades to direct the water toward an outlet drain.

Molecule–The smallest part of a substance that can exist separately and still retain its chemical properties and characteristic composition; the smallest combination of atoms that will form a given chemical compound.

Mollic Epipedon–A thick, dark surface soil horizon that has developed under grass.

Mollis–Soft.

Mollisols–A soil order of black, organic-rich surface horizons and a high supply of bases. They have mollic epipedons and base saturation greater than 50 percent (NH_4OAc) in any cambic or argillic horizon. They lack the characteristics of vertisols and must not have oxic or spodic horizons. See Soil Orders.

Molybdenum–Mo; a gray metallic element, essential in very small amounts to the growth of plants, but usually present in sufficient amounts. Deficiencies have been discovered in a few highly acid soils. When these soils are limed, ample molybdenum is usually released from slowly soluble forms in the soil. It is one of the micronutrients and is applied to soils usually as sodium molybdate, known in the trade as Moly. The normal amount of Mo in the soil is 1 to 3 ppm. More than 10 ppm in the soil may be injurious to some plants and animals. Recent tests have indicated a positive soybean response to molybdenum in most of the soybean-growing areas of the Midwestern and some of the southern states.

Mongrel Seed–Seed used for planting that is a mixture of seeds of several strains or varieties, as opposed to pure seed or that of a single variety.

Mono–Single; only one.

Monocotyledon (Monocot)–Plant having a single cotyledon or seed leaf such as corn. See Dicotyledon.

Monoecious–Plants that have male and female sex organs in different flowers on the same plant, such as cucumbers and squash.

Monogerm–A seed with only one germ or capable of only one germination.

Monohybrid–A hybrid whose parents differ in a single character.

Monohybrid Cross–A cross between two individuals that are heterozygous for one pair of genes; an example is Aa × Aa.

Monolith–An undisturbed (not broken up) sample of a soil profile.

Monophagous–Feeding upon only one kind of food, for example one species or one genus of plants.

Monophagous Parasite–A parasite restricted to one species of host.

Monophyllous–Designating that plant part which is made up of a single leaf: e.g., certain calyces. Sometimes used in reference to plants bearing only one leaflet.

Monopodial–Designating a plant in which the growth of the main stem is continued indefinitely by the terminal bud.

Monopterous–Designating a single-winged seed as in ash and maple trees. See Samara.

Monosaccharide–The simplest form of sugar; a single sugar unit.

Monoxenous Parasite–A parasite that requires only one host for its complete development.

Montmorillonite–A group of clay minerals whose formulas may be derived by ion substitution. They are characterized by swelling when wet. Mineral members of the group are montmorillonite, nontronite, saponite, hectorite, sauconite. Also referred to as the smectite group.

Moonseed–*Menispermum canadense*, family Menispermaceae; a woody vine bearing grapelike fruit and attractive foliage, grown as an ornamental in the United States. It is a poisonous plant: the rootstock contains bitter alkaloids, and cases of poisoning, attributed to wild grapes, have occurred from eating moonseed fruit by mistake. When eaten, the pits with their sharp ridges may also cause mechanical injury to the intestines. Native to North America. Also called mooncreeper.

Mopping Up–In forestry, making a fire safe, after it is controlled, by extinguishing or removing burning material along or near the control line and by felling snags.

Mor Humus–A type of forest humus layer of unincorporated organic material, usually matted or compacted or both, distinctly delimited from the mineral soil, unless the latter has been blackened by washing in organic matter. Syn.: raw humus.

Morbidity–The condition of being diseased, or the incidence or prevalence of some particular disease. The morbidity rate is equivalent to the incidence rate.

Morel–An edible mushroom: the fruit body of *Morchella* spp., especially *M. esculenta*, family Helvellaceae.

Morello–A pomological group of varieties in the sour cherry species. The fruit and juice of these varieties are dark. Some common varieties are English Morello and Wragg.

Mormon Cricket–*Anabrus simplex*, family Tetigoniidae; an insect found in Utah and other western states (United States), which can be very destructive to nearly all field, orchard, and garden crops.

Morning Glory–(1) Any of several plants of the genus *Ipomoea*, family Convolvulaceae. Species are mostly annual and perennial, twining vines grown as ornamentals on trellises, fences, etc. The sweet potato, however, is grown for its edible tuber. (2) *I. purpurea*, an annual, twining, tender plant grown for its purple, blue, or pink flowers with pale tubes. Native to tropical America. Also called common morning glory, bindweed. As weeds, the plants are difficult to control.

Morning Glory Sphinx–The larva of certain moths of the genus *Herse*, that feeds on plants of *Ipomea* spp. *H. convolvuli* attacks the morning glories and *H. cingulata* attacks sweet potatoes (southern United States).

Morphogenesis–The developmental history of organisms or of their parts.

Morphology–(1) A branch of biologic science that deals with the forms, rather than the functions, of plants and animals. (2) Pertaining to pedology or soil science, the study of soil horizons and their arrangement in profiles. (3) Pertaining to land surfaces, the shape or configuration of physical features.

Mortality–The number of overall deaths, or deaths from a specific disease, usually expressed as a rate; i.e., the number of deaths from a disease in a given population during a specified period, divided by the average number of people or animals exposed to the disease and at risk of dying from the disease during that time.

Mosaic–Viral disease that damages or kills plants, often giving the foliage a mottled appearance. Some mosaics can be spread by sucking insects, and some by handling the plants or by use of tools around the plants.

Moss–A flowerless plant intermediate in character between algae and ferns. Mosses of the Tundra have a grazing value, some have value as ornamental plants, and some have a commercial value for packing and rooting plants.

Moss Peat–Peat composed mainly of the identifiable remains of mosses, especially those of the genera *Sphagnum*, *Polytrichum*, and *Hypnum*. Moss peats are natural soils and are used commercially for mulches, pot and greenhouse soils, poultry litter, barn bedding, paper making, fuel, etc. Also called highmoor peat, peat moss, sphagnum peat. See Histosols.

Mossy–(1) Designating irregular, dark markings that spoil an otherwise desirable color contrast on the feathers of domestic birds. (2) Covered with moss.

Motes–(1) In cotton ginning, immature seed and particles of trash. (2) Burs, etc., found in wool. Also called moits.

Moth–An insect belonging to the order Lepidoptera. The adults have wings covered with scales and differ from butterflies in being night fliers and in having heavy, hairy bodies. In the caterpillar stage, many moths are destructive; they may be injurious and do tremendous damage to cloth, crops, trees, and stored grain.

Moth Bean–*Phaseolus aconitifolius*, family Leguminosae; a trailing, annual herb with yellow flowers, cultivated for its edible seeds and for forage. Probably native to India. Also called mat bean.

Mother–In a chick hatchery, a shelter for newly hatched chicks.

Mother Cane–A cane or branch, especially of grape, used in propagation by continuous or modified-continuous layerage.

Mother Plant–(1) In sugar-beet seed production, first-year root which is dug, stored over winter, and transplanted to the seed-growing field the following spring. (2) Any plant from which parts are removed for vegetative propagation.

Mother Root–In the propagation of the sweet potato, the root that is bedded for propagation of the new crop.

Mother Seed Potato–The Irish potato tuber or slice containing eyes, which is planted for the propagation of the new crop.

Mother Tree–In plant breeding, the original tree from which varieties are developed.

Mother-of-Vinegar–The thick film of bacteria *Acetobacter* spp. that may form on the surface of vinegar during its production.

Motte–An isolated, small clump of trees or shrubs; a minor vegetational feature of the natural landscape in Texas and other states (United States). Also called mott.

Mottle–(1) A spot or blotch of a color different from the mass color of a surface, as a mottle caused by a viral disease of a plant. (2) Color difference on a mass of moderately poorly drained soil.

Mottled–(1) Designating bird feathers marked with white tips at the ends or spotted with colors or shades at variance with the ideal. (2) Irregularly marked with spots of different colors, as a mottled soil, butter, etc.

Mottling–A soil mass containing many colors due to poor internal drainage. The colors may be gray, yellow, and/or red in random patterns.

Mould–See Mold.

Mound–(1) In tillage, to heap up or ridge the soil about a plant or row of plants. (2) To cover portions of branches in mound layering.

Mound Layering–The rooting of branches of woody plants or shrubs by leaving the branches upright and mounding soil about the basal portions of the stems.

Mound Planting–A method of planting on wet ground, the seeds or young trees being planted on mounds, ridges, or hills; hill planting; ridge planting.

Mound Storage–Storage of vegetables on the ground surface under a covering of straw and earth. Now obsolete.

Mountain Brome–*Bromus carinatus*, family Gramineae; a bunchgrass of excellent forage value. Native to the western United States. Also called California brome.

Mountain Pine Beetle–*Dendroctonus monticolae*, family Scolytidae; a bark beetle that is especially damaging to lodgepole pine and western white pine in the northwestern United States.

Mouse Barley–*Hordeum murinum*, family Gramineae; an annual grass common in the foothills and valleys of California, United States, that has a fair forage rating. Native to Europe.

Mow-burned–Designating hay that has been damaged or altered by heating and fermentation in the mow. Also called mowburnt.

Mucilage–A sticky or gelatinous substance manufactured by certain plants, as the marshmallow.

Mucilage Cell–In wood anatomy, a specialized parenchyma cell that secretes mucilage.

Muck–(1) Highly decomposed peat often used for raising crops, or sold in bags as "potting soil" or "organic soil." (2) Feces and urine from domestic animals in a wet state. (3) Something of questionable, dirty, trashy, or antisocial value. (4) In surface mining, the useless overburden. See Histosols.

Muck Ditch–In flood control engineering, a trench excavated under the center of a levee and refilled with impervious material to form a cut-off wall and to prevent the seepage of water along the foundation plane of the levee. Also called cut-off trench.

Mucor–A widely distributed genus of molds of the family Mucoraceae, mainly saprophytic species, abundant in soil, decaying vegetable matter, dung, etc. Some are capable of converting starch into sugar, and thus important industrially. Probably the most frequently met and most troublesome species is the black bread-mold, ***Rhizopus nigricans***, which appears wherever starchy substances are found, as on stale bread, and which causes a watery rot of strawberries, and peaches, a soft rot of roots of the sweet potato, and even rot in the tubers of the Irish potato when they have been weakened by cold or bruises or other unfavorable conditions. A few species of ***Mucor*** are weak parasites in animal tissues. ***Choanephora*** mold attacks and destroys the blossoms of squash and pumpkin (***Cucurbita***) and of ***Hibiscus***.

Mucronate–A leaf apex that has a short, abrupt point.

Mud–(1) Fine-textured soil so saturated with water as to be semifluid or in a sticky, plastic condition. (2) Freshly deposited alluvium of flood waters. (3) Flows of ash from the heavy rains occurring after volcanic eruptions.

Mud Cracks–Irregular or polygonal cracks, caused by shrinkage from drying, on the bare surface of clay soil or on flats of freshly deposited alluvium.

Mud-in–(1) To seed grain, such as oats and corn, in excessively wet soil. (2) To transplant by dipping the bare roots of the plant in mud.

Mudding–Dipping the roots of a transplant in a tub of fluid mud made from fine-textured soil. Also called puddling.

Mulberry–Any of the several plants of the genus ***Morus***, family Moraceae. Species are grown for their edible fruit and for foliage used as food for silkworms.

Mulch–(1) Soil, straw, wood chips, peat, or any other loose material placed on the ground to conserve soil moisture or prevent undesirable plant growth or soil erosion. (2) Material, such as straw, placed over plants or plant parts to protect them from cold or heat, as the mulch for strawberries. (3) To modify soil, as by cultivation, so that it serves as a mulch, or to supply mulching materials to the soil.

Mulch Fork–A hand-held implement consisting of a row of long tines attached to a handle. It is used for handling mulches in the garden.

Mulched-basin Irrigation–In soil irrigation, the building of permanent basins with substantial levees and covering each basin with a mulch of organic material, such as barnyard manure, bean straw, or alfalfa hay.

Mule Jack–A jack bred to a mare in contrast to one used for jack stock perpetuation. See Jennet, Jack.

Multi––A prefix meaning many, as multiovulate, many-ovuled.

Multiflora Rose–***Rosa polyantha***, ***R. multiflora***, family Rosaceae; a vigorous plant with white flowers that was once used to control erosion. It has now escaped and is a weed pest.

Multiflorum–Many-flowered.

Multiline–Crop seed composed of a mixture of several breeding lines of the same variety, each containing a different resistance gene to a specific host. A single breeding line is vulnerable to crop failure when a pathogen mutates and regains virulence over the single resistance gene. In contrast, the mixing of different resistance traits in a multiline greatly reduces the probability of crop damage.

Multiple Alleles–More than two different genes that can occupy the same locus on homologous chromosomes.

Multiple Annual Ring–In tree growth, an annual ring that consists of two or more false annual rings.

Multiple Fruit–A "fruit" formed by the connation of the individual fruits of several flowers in a cluster, as the pineapple (***Ananas***) or mulberry (***Morus***).

Multiplication Plot–A field used for growing a crop to produce seed in large quantities from a comparatively small lot of selected seed.

Multiplier Onion–***Allium cepa*** var. ***solaninum***, family Liliaceae; a bulbous herb that produces separable bulbs used in propagation; grown for its edible bulb and leaves. Also called potato onion.

Mum–See Chrysanthemum.

Mummification–(1) In animal reproduction, the drying up and shriveling of the unborn young. (2) In fruit, drying and shriveling caused by the brown rot pathogen.

Mung Bean–***Phaseolus aureus***, family Leguminosae; an annual herb whose small, green or yellow seeds are used for food, or to make beansprouts. The plants are used for forage or green manure. Native to Asia. Also called back gram, golden gram, green gram, gram, mungo bean.

Munsell Color Standards (Munsell Notation)–A color designation system that specifies the relative degrees of the three variables of color: Hue, value, and chroma. The standards may be used in precise comparison of colors of soils, or in standardizing agricultural products: e.g., prime cottonseed cake is 10YR5/5, which means yellow-red with value = 5, and chroma = 5.

Muriate of Potash–KCl; potassium chloride containing 50 to 62 percent potash (K_2O). It may be used as a fertilizer, singly or in combinations with other fertilizers.

Murrain–(1) A disease of Irish potatoes. Murrain is a term applied to the historically famous potato blight disease in Ireland. (2) See Cattle Tick Fever.

Mushmelon–See Muskmelon.

Mushroom–(1) A general term for the fruiting bodies of fungi of the class Basidiomycetes. See Toadstool. (2) More specifically, the fleshy, edible fruiting bodies of fungi such as the morels or cup fungi of class ascomycetes, order Pezizales; truffles of order Tuberals, of the same class; puffballs, of class Basidiomycetes, order Lycoperdales; and the sulfur mushroom of the order Polyporales, of the same class. (3) In the popular sense, ***Agaricus campestris***, the extensively cultivated common mushroom, and other wildgrowing, edible species of genus ***Agaricus***. See Morel, Puff-ball.

Mushroom Root Rot–An important, sometimes fatal, disease of fruit and forest trees caused by the fungus ***Armillaria mellea (Armillariella mellea)***, family Agaricaceae. It is characterized by a rot in the roots, by mycelium between bark and wood in the rotted portion, and by fungus strands growing out into the soil. There is stunting of growth and yellowing and wilting of foliage. Also called oak fungus disease, oak root fungus disease, honey-mushroom root rot, shoestring root rot, fungus root rot, crown rot, toadstool disease, Armillaria root rot.

Mushroom Soil, Spent–Well-rotted horse manure that has been used to grow mushrooms, but is no longer satisfactory for this purpose; is used as a fertilizer. See Manures.

Mushroom Truffle Disease–A disease, caused by a weed mold, that affects the production of mushroom spawn. It is controlled by growing the mushroom crop at temperatures below 65°F.

Muskmelon–***Cucumis melo***, family Cucurbitaceae; an annual, trailing herb, of which there are several commercially valuable varieties bearing a round, fleshy, edible fruit. There are many cultivated forms of this species and their classification is difficult. The botanical variety ***C. melo*** var. ***reticulatus*** includes the musk-scented types to which muskmelon is more restrictedly applied. They are also called netted melon or nutmeg melon, and in the trade are commonly called cantaloupe. The true cantaloupe, however, belongs to ***C. melo*** var. ***cantalupensis***. The fruits of ***C. melo*** var. ***reticulatus*** usually have a netted skin, shallow to moderate ribbing, and flesh from light green to reddish-orange. The botanical variety ***C. melo*** var. ***inodorus*** is the wintermelon. The fruit is relatively nonodorous and has a smooth or ridged skin, ripens late, and keeps well. The commercial production of the wintermelon is largely confined to California and Arizona where honeydew, Casaba, Crenshaw, and Persian are representative horticultural varieties. Native to Asia.

Must–The fresh juice of grapes before fermentation.

Mustang Grape–***Vitis candicans***, family Vitaceae; a wild grape that grows very profusely in southern United States. The fruit has a pungent taste and is gathered for making jelly. Also called bird grape.

Mustard–Common name for some of the herbs of the genus ***Brassica***, family Cruciferae. Species are grown for use as potherbs and for their seed used in the manufacture of the condiment mustard.

Mustiness–An offensive, pungent odor caused by molds.

Mutabilis–Variable.

Mutable–Changeable.

Mutagen–A chemical, physical and/or radioactive agent that interacts with DNA to cause a permanent, transmissible change in the genetic material of a cell. See Teratogen.

Mutant, Mutation–A variant, differing genetically and often visibly from its parent or parents and arising rather suddenly or abruptly. Mutation can occur naturally or can be induced by radiation (x rays, gamma rays, or thermal neutrons) or chemically by ethyldemethyl sulfate.

Mutterkorn–See Ergot.

Mutton Bluegrass–***Poa fendleriana***, family Gramineae; a perennial, bunch bluegrass, 1 to 2 feet high, growing at high elevations in the western United States. It is a range plant of excellent forage value. Native to North America.

Mutton Cane–***Panicum dichotomum***, family Gramineae; a slender grass browsed by sheep (eastern United States).

Mutton Corn–Ears of green corn, as roasting ears (southern United States).

Mutual Drainage–A soil drainage system agreed to and of benefit to all persons involved.

Mutualism–Dependency of two organisms upon each other. Among insects, an example is furnished by the cornfield ant and the corn-root aphid.

Mutualistic–Designating a mutually beneficial relationship between organisms; symbiosis, e.g., Rhizobium bacteria and compatible legumes.

Mycelial Fan–A thin layer of fungous growth composed of mycelial strands spread in a fan-shaped pattern, occurring either in a layer over the surface of decaying timbers or as small patches between bark layers or between the bark and wood of trees.

Mycelium–A collective term for the vegetative hyphae of a fungus.

Mycology–The science dealing with fungi.

Mycoplasma–An organism that is between a virus and a bacteria in size. It may possess characteristics of a virus and is not visible under a light microscope.

Mycorrhizae–A symbiotic relationship between a specific fungus and roots of a particular plant species, especially trees and shrubs. More common in soils of low fertility, mycorrhizae help plant roots to grow by aiding absorption of minerals such as phosphorus.

Mycosis–A disease caused by the growth of fungi in plants or animals.

Mycosphaerella Leaf Spot–Any leaf spot disease caused by species of ***Mycosphaerella***, family Mycosphaerellaceae, on cotton, citrus, and some other plants.

Mycotic Disease–A disease caused by a fungus.

Myristicin–A toxic substance found in the fruit of certain plants of the nutmeg family. Together with other constituents it is responsible for the hallucinogenic action of nutmeg. In large doses these substances produce excitement. See Nutmeg.

Myrrh–An aromatic gum resin produced by several trees or shrubs of the genus ***Commiphora***, family Burseraceae, especially ***C. abyssinica*** and ***C. myrrha***, native to northeast Africa and Arabia. Myrrh is used in the manufacture of perfumes, dentifrices, and as an ingredient in medicinal tonics.

Naked Flower–A flower without petals (apetalous) or sepals.

Naked Oats–*Avena nude*, family Gramineae; a cereal that possesses a lemma and a palea readily separated form the kernel. Also called hulless oats.

Naked Seed–See Achene.

Nanism–Dwarf growth.

Nap–Matted masses of fibers that contribute to the rough appearance of ginned cotton.

Naphthol–$C_{10}H_7OH$; either of the two coal tar derivatives from naphthalene, alpha-naphthol and betanaphthol. Alpha-naphthol is a pleasant-smelling solid with a melting point of 122°F (50°C). It is the standard material for impregnating tree bands used to control the codling moth. The beta-naphthol is an antiseptic.

Napiergrass–*Pennisetum purpureum*, family Gramineae; a perennial grass that yields an enormous amount of green fodder and is cut several times in a season. In the United States it is grown in Florida and Georgia. Native to Africa. Also called Merker grass, elephant grass, Carter grass.

Nappiness–The degree of naps or neps found in cotton. Naps are large clumps or matted masses of fibers, while neps are small tangled knots of fibers. Also known as neppiness.

Nappy–Coarse and sticky lint cotton.

Narrow Base Terrace–A terrace for controlling soil erosion. It is similar to a broad-base terrace in all respects except the width of ridge and channel: the base of a narrow terrace is usually 4 to 8 feet wide. it is subject to frequent failures and has not been widely accepted in the United States. See Broad Base Terrace, Nichol's Terrace.

Nascent–In the act of being formed; nascent tissue. See Meristem.

Nasturtium–(1) Any of the several herbs of the genus *Tropaeolum*, family Tropaeolaceae. Species are grown for their flowers, for salad plants, and for seeds used in sauces as a condiment. (2) *T. majus*, an annual herb with red-spotted, yellow-orange flowers, grown as a covering for trellises and as an ornamental. Its young flower buds and fruits are used as seasoning. Native to South America.

National Cotton Council of America–A council organized in 1938 to promote cotton from field to fabric. Seven cotton segments have combined to form this council: producers, ginners, warehouse managers, crushers, cooperatives, merchants, and manufacturers.

National Forest System–Units of federally owned forest, range, and related lands throughout the United States and its territories dedicated to the long-term benefit for present and future generations. The National Forest System includes all national forestlands acquired through purchase, exchange, donation, or other means, the National grasslands, and land utilization projects administered under Title III of the Bankhead-Jones Farm Tenant Act and other lands, waters, or interests therein which are administered by the Forest Service or are designated for administration through the United States Department of Agriculture Forest Service as a part of the system.

National Grassland–Land, mainly grass and shrub cover, administered by the United States Department of Agriculture Forest Service as part of the National Forest System for promotion of grassland agriculture, watersheds, grazing, wildlife, and recreation.

Native–(1) Designating a plant that grows naturally in a country or region; one not introduced by people. (2) Designating animals, as cattle, hogs, and horses, which, though originally introduced into a region, have lost some of their original characteristics or have gone wild: a scrub or mongrel. (3) Designating an unbranded beef hide. See Feral, Indigenous.

Natric Horizon–A clayey soil horizon formed in an arid environment and containing more than 15 percent exchangeable sodium content.

Natural Control–Nature's method of maintaining a biotic balance, as in the reduction of harmful insects through the action of heat, cold, rain, drought, parasites, predators, and disease. See Integrated Pest Management.

Natural Cross–Interbreeding or hybridizing that takes place in nature without assistance from people.

Natural Enemy–In nature, any organism that preys or feeds upon another. A natural enemy may be introduced by people for biological control. See Myxomatosis.

Natural Land Type–A unit of land area capable of being mapped upon a combination or integration of natural characteristics, such as soils, vegetation, relief, and climate, as contrasted with types based upon use, money value, or modifications by people. See Land Type.

Natural Pan–A soil formed in nature that has a clay pan horizon, a fragipan horizon, or a horizon cemented by iron, aluminum, silica, calcium carbonate, gypsum, or colloidal humus. See Induced Pan.

Natural Reproduction–Renewal of plants by self-sown seeds, sprouts, rhizomes, etc.

Natural Reseeding–In range management, the restoration of depleted grazing land to permit natural revegetation by reseeding of desirable forage species.

Natural Selection–A natural process by which less-vigorous plants and animals tend to be eliminated from a population in an area without leaving enough descendants to perpetuate their traits.

Natural Selection Theory–A theory of evolution, propounded by Charles Darwin in the nineteenth century, which postulates that the distinctive characteristics of fitness can be inherited. Also called the survival of the fittest theory.

Natural Thinning–The death of trees in a stand as a result of competition.

Natural Water–Water (H$_2$O) as it occurs in nature with extremely variable proportions of solid, liquid, and gaseous materials in solution and suspension. In common usage, the term *water* refers to natural water rather than to chemically pure water, deionized water, or distilled water.

Naval Stores–Products, such as turpentine and rosin, obtained from the distillation of crude pine resin.

Navel–(1) In *Mammalia* spp. the point of connection between the umbilical cord and the fetus. (2) See Navel Orange.

Navel Orange–A variety, usually seedless, of the sweet orange with a small navel-like formation at the apex of the fruit. It originated in Brazil as a budsport and was introduced in the United States in 1870. Widely grown in California.

Navy Bean–White varieties of the kidney bean, grown for their edible seed, generally known as pea bean.

Near Wilt of Pea–A fungal disease of the field pea caused by *Fusarium oxysporum f. pisi* race two, family Sphaerioidaceae, which remains indefinitely in the soil. The symptoms are similar to wilt, except that there is less stunting of the plant and the vascular discoloration is brick-red and occurs farther up the stream.

Nebraska Potash–Salts obtained from the brines of lakes in the semi-arid regions of western Nebraska that contain from 20 to 30 percent of K2O in the form of carbonate and sulfate of potassium. Not used extensively as a fertilizer.

Necrobiosis–A viral infection of plants; one of three types of phloem alteration characterized by gradual swelling of the cell walls in either healthy or diseased plants.

Necros–(Greek) Dead.

Necrosis–Death of plant or animal cells of tissues, usually in localized areas.

Necrotic–Designating a necrosis.

Necrotic Ring Spot–A viral disease of sour cherry caused by one or more viruses or strains of the same virus. The disease is widespread on *Prunus* spp. The symptoms are most severe after initial infection, consisting of necrotic rings or spots, and dieback of spurs and shoots. After this, the virus becomes latent without symptoms. The effect on fruit production is controversial.

Nectar–A sweet secretion of flowers of various plants, used by bees to store as honey.

Nectar Flow–The period when abundant nectar is available for bees to produce honey for storage in the combs of the hive.

Nectarine–*Prunus persica* var. *nectarina*, family Rosaceae; a deciduous tree grown for its edible fruit, the nectarine of commerce. The fruit is quite similar to the peach except that the skin is fuzzless. Native to China.

Needle Cast–A fungal disease of spruces and other conifers caused by *Lophodermium filiforme*, family Phacidiaceae; characterized by the lower branches of red spruce and black spruce being completely defoliated.

Needle-leaved Deciduous–Woody gymnosperms (trees or shrubs) with needle-shaped or scalelike leaves that are shed during the cold or dry season; e.g., baldcypress (*Taxodium distichum*).

Needle-leaved Evergreen–Woody gymnosperms with green, needle-shaped, or scalelike leaves that are retained by plants throughout the year; e.g., black spruce (*Picea mariana*).

Needlegrass–Certain species of the genus *Stipa*, family Gramineae; perennial bunchgrasses with numerous, mostly basal, narrow or rolled leaves, open or narrow panicles, narrow, long-awned spikelets, and hardened lemmas ending in a long, bent, twisted, and sometimes feathery awn. Two species, green needlegrass and needle-and-thread are of good grazing value on western ranges (United States). Porcupine grass produces long-tailed, sharp-pointed seeds that may penetrate the skin of animals and cause their death.

Neem–*Azadirachta indica*, family Meliaceae; a large tree that grows in India and several other tropical countries. Its leaves are used as an insecticide. Neem oil is made from the seeds and the fruits and is effective as an anthelmintic. See Anthelmintic.

Negative Geotropism–A natural reaction in which a plant grows upward away from gravity.

Negligible Residue–The amount of a pesticide that is allowed on a food or feed crop at harvest. The trace amount that is on the crop is the result of indirect contact with the chemical.

Nematocide–Any substance used to kill parasitic worms called nematodes, abundant in many soils. Many nematodes attack and destroy plant roots. See Nematode.

Nematode–(1) Microscopic, wormlike, transparent organisms that can attack plant roots or stems to cause stunted or unhealthy growth. (2) Any of the round, threadlike, unsegmented animal worms of the phylum Nematoda, ranging in size from microscopic to 1 meter long. Nematodes may be saprophytes or parasites of plants and animals. They are responsible for important animal and plant diseases resulting in much economic loss. The animal parasites are designated as roundworm of horses and swine, and hookworm of people. Also called eelworms, roundworms, and nemas. See Root Knot.

Nematode Disease of Wheat–A disease caused by the nematode *Agnuina tritici*, family Tylenchidae. It invades the inflorescence and transforms the kernels into galls resembling smut balls. Also called bunted, earcockle.

Nematology–A branch of zoology concerned with the study of nematodes usually including their relation to plants and animals.

Nemoral–Of, or pertaining to, woods or forest, as nemoral culture.

Neoeluvium–A soil developed from the eluviation and illuviation of recent deposits, such as alluvium.

Neontology–Biology; the study of existing life.

Net Increment–In forestry, the addition to tree growth that represents an increase in usable timber.

Net Necrosis–A disease of the Irish potato caused by a virus, or sometimes a frost, in which cut tubers show a network of brown strands, especially near the surface and stem end; associated with leaf roll. See Leaf Roll.

Net Russeting of Apples–A discoloration of the skin of apples caused by the powdery mildew, ***Podoshaera leucotricha***, family Erysiphaceae. The russeting may be spread all over the fruit or may be in the form of small patches, rings, or bands.

Net Veined–A leaf type with veins that form a net pattern in the leaf as in the leaves of an oak tree.

Neuter–Of a plant, asexual; having neither stamen nor pistil.

Neutral Soil–A soil that is not definitely acid or alkaline; specifically, a soil with a pH between 6.6 and 7.3. See pH.

Neutral-day Plant–See Day-neutral Plant.

Neutron–See Atom.

Nevada Bluegrass–***Poa nevadensis***, family Gramineae; a tufted, grayish-green, perennial grass that occurs on western ranges of the United States at elevations of 3,000 to 11,000 feet. It has good forage value.

Nevada Ephedra–***Ephedra nevadensis***, family Gnetaceae; a woody plant with very small leaves, growing in semiarid regions of Utah, California, and other states, United States, which is good browse for cattle and sheep in late fall and winter. Native to the United States. Also called Nevada jointfir.

New Ground–Land recently cleared of shrubs, trees, grasses, forbs, stones, stumps, etc., and put under cultivation for the first time.

New Mexico Flax–***Linum neomexicanum***, family Linaceae; an annual herb whose ingestion is apparently poisonous to livestock. Native to the southwestern United States. Also called yellow pine-flax.

New Mexico Locust–***Robinia neomexicana***, family Leguminosae; a small to large spiny shrub, grown as a browse plant for goats and horses. The pods after falling to the ground are eaten by sheep and cattle. Native to the United States.

New Potato–(1) The tuber of the Irish potato taken from a potato hill before full maturity. (2) The potato of the newly harvested crop in contrast to stored potatoes.

New Wood–In horticulture and forestry, the current year's growth.

New York Weevil–***Ithycerus noveboracensis***, family Curculionidae; a chewing insect that infests the twigs and small branches of apples, especially of the young trees.

New Zealand Spinach–***Tetragonia expansa***, family Aizoaceae; an herbaceous annual vine grown as a potherb for its edible leaves. It is a hot-weather substitute for spinach. Native to East Asia, Australia, and South America. Also called New Zealand ice plant.

Niacin–A vitamin of the B-complex group. Also called nicotinic acid, antipellagra vitamin.

Niche–A term used to describe the status of a plant or animal in its community, that is, its biotic, trophic, and abiotic relationships. All the components of the environment with which an organism or population interacts.

Nichol's Terrace–A once common small terrace for disposal of surplus water. It has a comparatively deep, narrow channel and a low, flat ridge with a slope that merges quite closely with the downhill side. Modern terraces are usually broad-based to allow machinery to operate over them. See Broad-base Terrace, Narrow-base Terrace.

Nickel–Ni; a chemical element, a metal which is found in traces in soils. It was once thought to be deleterious rather than beneficial to plant growth, but since 1983 has been suggested as essential for plants. It is required by animals and people.

Nicotiana–A genus of solanaceous annual plants native to tropical America from which tobacco is derived. Named after Jean Nicot de Villemain (1530-1600) who introduced the habit of chewing tobacco to Catherine de Medici. See Nicotine.

Nicotine–A very poisonous alkaloid with an odor of pyridine from the leaves of the tobacco plant and also produced synthetically. Formerly widely used as an external parasiticide and as an insecticide.

Night Soil–Human feces and urine, an unsanitary potent manure for cropland. It is rarely used in the United States but remains in use in some other parts of the world, especially the Orient.

Nightshade–Any plant of the 1,700 species of the genus ***Solanum***, family Solanaceaae. Some species are flowers, some are shrubs, some vegetables (Irish potatoes), and some are poisonous plants. See Silverleaf Nightshade.

Nigrescent–Blackish.

Nigrospora Cob Rot–A fungal diseases of corn caused by ***Nigrospora oryzae***, family Dematiaceae; widely distributed throughout the United States Corn Belt. It is characterized by a shredding and gray discoloration of the cob. The pith of the cob may completely disintegrate. Also known as nigrospora dry rot.

Nip–(1) A knot in cotton fiber. (2) A sudden check in the plant or animal growth due to severe cold or frost.

Niter–Saltpeter; a mineral; KNO_3.

Nitrate–NO_3; N combined with oxygen. The N form most used by plants. NO_3 is a gas that does not exist alone in fertilizer but is combined, as in ammonium nitrate. All nitrates are water-soluble and, when applied in surplus, move with surface waters to contaminate groundwater.

Nitrate of Ammonia–See Ammonium Nitrate.

Nitrate of Lime–See Calcium Nitrate.

Nitrate of Potash–KNO_3; the potassium salt of nitric acid that occurs in nature only in small quantities. The pure compound contains about 46.6 percent potash (K_2O). Used as a fertilizer or as a component in fertilizer formulations, commercial grades contain about 13 percent nitrogen and 44 percent potash.

Nitrate of Soda–

Nitre–(1) A precipitate of malic acid formed in making maple syrup. (2) Nitrate of potash (saltpeter), KNO_3, used as a fertilizer and in the manufacture of explosives. (3) Sodium nitrate, or chile saltpeter.

Nitric Acid–HNO_3; a strong mineral acid which, combined with metals or alkalies, forms nitrates. It is now made synthetically on a large scale by passing ammonia (NH_3) and air through a platinum gauze catalyst, whereby the ammonia oxidizes. Nitric acid is used in the production of nitrate fertilizer compounds, including nitric phosphates. Some nitric acid is used as an oxidant for carbonaceous material that causes the black color in liquid fertilizers made with commercial phosphoric acid.

Nitric Phosphate–Fertilizers made by processes involving treatment of phosphate rock with nitric acid or a mixture of nitric and/or sulfuric or phosphoric acid, or all three acids, usually followed by ammoniation.

Nitrification–The process of formation of nitrates (NO_3) in the soil, from other compounds of nitrogen, by various microorganisms. Many plants, especially during their early growth, can utilize nitrogen in the form of ammonium, but all plants can metabolize nitrate nitrogen. Nitrification is very rapid in warm, moist, neutral soils. It practically ceases when the soil temperature falls below 4°C (40°F).

Nitro-Chalk–See Ammonium Nitrate-Limestone.

Nitrobacter–A genus of microorganisms occurring in productive soil that oxidize nitrites to nitrates. This genus includes two species: *Nitrobacter agilis* and *N. winogradskyii*.

Nitrobacteria–Minute, rodlike bacilli that convert nitrogen into soluble nitrates in soils.

Nitrogen–N; a gas that occurs naturally in the air and soil, where it is converted into usable forms for plant use by bacteria and other natural processes. This nutrient is a constituent of proteins and is vital to plant-growing processes. Nitrogen can be added to the soil in any of three fertilizer forms: as urea, ammonia, or nitrates.

Nitrogen Carrier–Any of a great number of inorganic or organic compounds that contain nitrogen and can be classed as a fertilizer.

Nitrogen Cycle–The circulation of nitrogen in nature. Dead organic matter is converted to ammonium compounds during decay in the soil. These are converted to nitrates, through nitrification as a result of bacterial action, which can be utilized again by plants. *Nitrosomonas* oxidize ammonium ions to nitrites and *Nitrobacter* oxidize nitrites to nitrates.

Nitrogen Excesses–The presence of nitrogen, as nitrates or ammonium salts in soils, in excessive amounts, which results in injury to plants, as burning or stunting. The excess may occur naturally or may result from excessive applications of nitrogenous fertilizers. In tree fruits, it may cause excessive vigor resulting in reduced fruiting.

Nitrogen Oxides–Gases formed in great part from atmospheric nitrogen and oxygen when combustion takes place under conditions of high temperature and high pressure; e.g., in internal combustion engines; considered major air pollutants. Written as NOx when the oxides have not been identified, or as NO-nitric oxide and NO_2-nitrogen dioxide.

Nitrogen Solutions–Solutions of nitrogenous fertilizer chemicals in water. Solutions are used in manufacturing liquid fertilizers and are applied to the soil either with special applicators or in irrigation water. A typical nitrogen solution is characterized by a code number, e.g., 414 (19-66-6) indicating that it contains 41.4 percent total nitrogen, 19 percent free ammonia, 66 percent ammonium nitrate, 6 percent urea, and by difference, 9 percent water.

Nitrogen-fixing Bacteria–Species of the genus *Rhizobium*, family Rhizobiaceae, that live symbiotically in the root nodules of leguminous plants, upon which they are dependent. They are capable of extracting nitrogen from the air and converting it to a form utilizable by the plant. Species of the genus *Azotobacter* and *Clostridium* are free-living and act independently (nonsymbiotically). Some plants that are not legumes also have a species of symbiotic bacteria. See Nodule.

Nitrogenous Materials–Materials that contain nitrogen, whether in organic or inorganic form. Some are readily available, others, even though they contain fairly high percentages of total nitrogen, such as hoof, hair, and plastic and leather scraps, are not readily available.

Nitrosamine–Any of a group of N-nitroso derivatives of secondary amines. Some nitrosamines are carcinogenic.

Nitrosobacterium–A microorganism that oxidizes nitrites (NO_3) to nitrates (NO_3).

Nitrosospira–A genera of two species of microorganisms that oxidize ammonia (NH_3), to nitrate (NO_2) very slowly: *Nitrosospira antarctica* and *N. briensis*.

Niveus–Snowy-white.

No-tillage–A system of growing crops in which the seeds are planted in the ground with little or no disturbing of the soil. Weed and pest control is accomplished through the use of chemicals. This method of farming is used to control erosion. See Conversation Tillage.

Noble Fir–*Abies procera*, family Pinaceae; a huge evergreen tree grown in the United States West Coast as an ornamental and for its lumber, used in general construction, boxes, pulp, and planing-mill products. Native to North America.

Node–The place upon a stem that normally bears a leaf or whorl of leaves; the solid constriction in the culm of a grass; a knoblike enlargement.

Nodulated–Designating plants that have nodules on their roots in which nitrogen-fixing bacteria live.

Nodule–(1) A root tubercle or lump formation on certain leguminous plants produced by the invasion of symbiotic, nitrogen-fixing bacteria. The bacteria furnish the plant with fixed nitrogen compounds and receive nutrient plant juices like carbohydrates. The genus *Rhizobium* and some species of *Azotabacter* and *Clostridium* fix free nitrogen. (2) A small knot, lump, or roundish mass of abnormal tissue. See Nitrogen-fixing Bacteria.

Nonadditive Genes–Genes that express themselves in a dominant or epistatic fashion.

Noncommercial Forest Land–Forestland withdrawn from its commercial use for timber because (a) it is utilized for such purposes as parks, game refuges, military reservations, or reservoir protection or

(b) its poor growing conditions or inaccessibility render its commercial use unprofitable.

Nondegradable Wastes—Substances that are not changed in form and/or reduced in quantity by the biological, chemical, and physical phenomena characteristic of natural waters. Although nondegradable wastes may be diluted by receiving water, they are not reduced in total quantity. (Sometimes known as nonbiodegradable wastes.)

Noninfectious Chlorosis of Strawberries—A physiological condition caused by a deficiency of nitrogen, magnesium, manganese, boron, or oxygen in a water-logged or alkaline soil; characterized by yellowing or loss of normal green color.

Nonionic Surfactants—Substances that have neither a positive nor a negative electrical charge and are compatible with most pesticides.

Nonpalatable—(1) Designating a range plant species not grazed when the range or pasture is properly utilized. (2) Designating feeds not relished by animals.

Nonparasitic—See Abiotic Disease.

Nonpathogenic—Not capable of producing disease.

Nonporous Wood—Wood devoid of pores or vessels, as the conifers and a few broadleaf species.

Nonprotein Nitrogen (NPN)—The nitrates, amides, and amino acids that are the forerunners of protein; toxic to livestock in some of these forms.

Nonsaline-Sodic Soil—A soil that contains sufficient exchangeable sodium to interfere with the growth of most crop plants and does not contain appreciable quantities of soluble salts. The exchangeable-sodium-percentage is greater than 14, and the electrical conductivity of the saturation extract is less than 4 millimhos per centimeter (at 25°C). The pH reading of the saturated soil paste is usually greater than 8.5.

Nonselective Herbicide—A herbicide that will kill any plant with which it comes in contact. See Selective Herbicide.

Nonsymbiotic Nitrogen Fixation—Fixation of nitrogen in soils by free-living microorganisms in contrast to that fixed through the agency of symbiotic organisms. See Nitrogen-fixing Bacteria.

Nontoxic—Not poisonous to plant or animal. See Toxic, Toxicity.

Nootka Falsecypress—*Chamaecyparis nootkatensis*, family Pinaceae; an evergreen tree grown as an ornamental and as a valuable timber tree. Its wood is used for poles, interior finish, furniture, caskets, hulls, oars, greenhouses, acid tanks, sounding boards, etc. Native to North America. Also called Alaska cedar, Sitka falsecypress.

Normal Diameter—In logging, the diameter of a tree at breast height above the root swell.

Normal Erosion—The erosion that takes place on the land surface in its natural environment undisturbed by human activity. It includes (a) rock erosion, or erosion of rocks on which there is little or no developed soil, as in stream channels and rocky mountains, and (b) normal soil erosion, or the erosion of the soil under its natural condition or native vegetative cover undisturbed by human activity.

Normal Fire Season—A forest fire season in which weather, rated fire danger, and number and distribution of fires are average.

Normal Moisture Capacity—The amount of water that is normally held by a soil after the free or gravitational water has drained. See Field Capacity.

Normal Soil—A soil having a profile in equilibrium with the forces of the environment, such as native vegetation and climate. Usually developed on gently undulating upland, with good drainage, from parent material (not of extreme texture or chemical composition) in place long enough for the soil-forming process to exert their full effect.

Normal Stand—In forestry, a stand of trees fully stocked, in proper growing condition and having normal increment conforming to a yield table.

Normal Yield—A term designating the average historic yield established for a particular farm or area. Can also describe average yield. Normal production would be the normal acreage planted to a commodity multiplied by the normal yield.

Normal Yield Table—In forestry, an accepted standard yield table with which to compare actual yields. The statements of a normal yield table are derived as an average from the best-producing, fully stocked areas for particular species and sites.

Northern Anthracnose—A fungal disease of red clover in the cooler areas of the United States, caused by *Kabatiella caulivora*. It produces brown lesions on the stems and petioles and a cracking of the stem tissues.

Northern Corn Leaf Blight—A fungal disease caused by *Helminthosporium turticum*, family Dematiaceae; characterized by lesions beginning on the lower leaves and appearing progressively upward. Heavily infected fields may appear dry and fired.

Northern Corn Rootworm—The larva of a beetle, *Diabrotica longicornis*, family Chrysomelidae, which burrows through the roots of corn, leaving small, brown tunnels. It is a very serious corn pest in the upper Mississippi Valley of the United States.

Northern Hornworm—See Tomato Hornworm.

Northern Leaf Blight—A fungal disease of corn, Sudangrass, and sorghums caused by *Helminthosporium turticum*, family Dematiaceae; characterized by linear or irregular, elliptical, watersoaked lesions on the leaves. The lesions turn brown, and then black. Sometimes the entire leaf dies.

Northern Red Oak—*Quercus rubra* var. *borealis*, family Fagaceae; a northern variety of red oak. Lumber from the northern red oak is valuable in the making of furniture.

Northern Spy—One of the older, very well known and popular varieties of apple, a juicy winter variety that originated in New York, United States.

Northern White Cedar—*Thuja occidentalis*, family Pinaceae; an evergreen tree valuable for timber, posts, and as an ornamental. Native to North America.

Norway Pine—See Red Pine.

Norway Spruce–*Picea abies*, family Pinaceae; a hardy evergreen tree, probably the most widely grown ornamental evergreen in America; it has a large number of varieties differing in height, form, and foliage. Dwarf varieties are 1 to 2 feet high while other varieties reach a height of 150 feet.

Notch–(1) The opening in a dam or spillway for the passage of water. (2) A gap, pass, or defile between mountains (chiefly in northeastern United States). (3) An undercut; in logging, a cut in the trunk of a tree to govern direction of fall. (4) A gap cut into the ear of an animal for identification.

Notching–(1) Cutting dents on the ears of animals for identification. (2) Removing a V-shaped piece of bark from a branch just above or below a bud.

Notifiable Disease–(reportable) Any disease that must be reported to the government health authorities.

Novia–An ancient Mideastern system of irrigation that uses the force of the stream or river current to turn a paddlewheel water lift.

Noxious Seed–Weed seed that are present in a quantity of crop seed. The presence of noxious seed in crop seed lowers the value of the crop seed.

Noxious Weed–A weed that crowds out desirable crops, robs them of plant nutrients and moisture, and causes extra labor in cultivation. States designate specific weeds as noxious.

NPK–Symbols for three primary nutrients needed by plants. N is for total nitrogen, P for available phosphorus (reported as P_2O_5), K for water-soluble potassium (reported as K_2O). Percentage of these nutrients in a fertilizer package is always listed in that order and is known as the fertilizer grade. See Fertilizer Grade.

Nubbin–(1) A viral disease of the cucumber, characterized by yellow-green systemic mottling of the leaves with small, distorted plants, and mottled and misshapen fruit, called crook cucumber. The disease is also called mottled-leaf mosaic, wart disease, white pickle. (2) An underdeveloped ear of corn that is not fully covered with kernels. (3) A small ear of corn. (4) A strawberry irregular in shape and smaller than normal.

Nucleus–(1) The central portion of the cell protoplast surrounded by a very thin membrane. It consists of nucleoplasm and includes within itself variously arranged chromatin, nuclear sap, and nutritive substances. It is of crucial significance in metabolism, growth, reproduction and the transmission of the determiners of hereditary characters. (2) A small colony of bees used in queen rearing or in pollination work in greenhouses. (3) A central core around which material collects or is grouped.

Nuclide–Any species of atom that exists for a measurable period of time whose nuclear structure is distinct from that of any other species. Thus each isotope of an element is a separate nuclide.

Nudation–The removal of all of the vegetation of an area.

Nuisance Birds–Starlings, sparrows, crows, redwinged blackbirds, pigeons, etc., that damage the farm, orchard, and garden crops, or become objectionable because of their roosts, or the grain they consume.

Nurse Crop–A companion crop grown to protect some other crop sown with it, as small grain is sometimes seeded with clover. The nurse crop can sometimes become competitive and harmful also.

Nursery–(1) Any place where plants, shrubs, and trees are grown either for transplanting or as grafting stocks. (2) A group of young plants or trees in a plantation.

Nursery Knife–A sharp knife that is used to prune, graft, or take cuttings from plants.

Nut–(1) An indehiscent, one-celled and one-seeded, hard and bony fruit, as the acorn of *Quercus*. (2) As frequently and loosely used, a drupe with relatively thin fleshy exocarp and a large stone (pyrene), or the pyrene itself, as the walnut (*Juglans*) and hickory nut (*Carya*).

Nut Pine–(1) Any of various pines that bear edible nuts. (2) *Pinus cembroides* var. *edulis*, family Pinaceae; a hardy variation of the Mexican pinyon pine. This term may include var. *monophylla*.

Nut Structure–The aggregation of fine particles of soil into small subangular, blocky, or roughly rounded, masses of definite size or shape. See Soil Structure.

Nutgrass Flatsedge–*Cyperus rotundus*, family Cyperaceae; a sedge that is widely distributed as an agricultural weed. The tubers were used as food by North American Indians. Native to tropical and subtropical regions.

Nutlet–A small or diminutive nut, similar to an achene but with a harder and thicker wall.

Nutmeg–*Myristica fragrans*, family Myristicaceae; a tree grown for its seed, the nutmeg of commerce used in food flavoring. Native to Indonesia. See Myristicin.

Nutmeg Pelargonium–*Pelargonium odoratissimum*, family Geraniaceae; a tender, perennial herb grown for its nutmeg-scented, white flowers. Native to South Africa. Also called apple geranium.

Nutrient–(1) A substance that favorably affects the nutritive processes of the body; a food. (2) An element or compound in a soil that is essential for the growth of a plant. (3) In stock feeding, any feed constituent or group of feed constituents of the same general composition that aids in the support of life, as proteins, carbohydrates, fats, minerals, and vitamins.

Nutrient Cycle–The circulation of nutrient elements and compounds in and among the soil, parent rock, streams, plants, animals, and atmosphere.

Nutrient Level–(1) In soils, the amounts and proportions of plant nutrients, such as phosphorus, potassium, and nitrogen in available forms. (2) Specifically, the concentrations of any particular nutrient in the ration of animals.

Nutrient Solution–A solution in which plants may be grown, made by dissolving salts containing the essential elements for growth in water, which is used in hydroponics and in experimental work in laboratories. See Hydroponcis.

Nutrient Spray–Liquid fertilizer sprayed directly on the foliage and absorbed therefrom. Also known as foliar spray. See Foliar Fertigation, Foliar Spray.

Nutriment–Nourishment; nutritious substances; food.

Nutrition–The sum of the processes by which an organism utilizes the chemical components of food through metabolism to maintain the structural and biochemical integrity of its cells, thereby ensuring its viability and reproductive potential.

Nutsedge–*Cyperus rotundus*. A serious perennial weed pest that propagates by deep rootstocks and by small tubers on the roots. Also known as nutgrass.

Nuttall Blister Beetle–*Lytta nuttallii*, family Meloidae; an insect that feeds on leguminous crops; the larvae feed on grasshopper eggs.

Nuttall Saltbush–*Atriplex nuttalli* (*A. gardneri*), family Chenopodiaceae; a useful browse plant found on alkaline or saline soils on western ranges (United States). Native to North America. Also called Gardner saltbush. See Saltbush.

Nuxvomica Poisonnut–*Strychnos nuxvomica*, family Loganiaceae; a tropical tree whose fruit yields the poison strychnine. Native to India. Also called dog button.

Nymph–A stage in the development of some insects and related forms, such as ticks, immediately preceding the adult stage.

O Horizon–(Soil) Organic horizons above mineral soil. Horizons: (1) formed or forming above the mineral part; (2) dominated by fresh or partly decomposed organic material; and (3) containing more than 30 percent organic matter if the mineral fraction is more than 50 percent clay or more than 20 percent organic matter, if the mineral fraction has no clay. Proportional percentages are between these extremes. See Organic Soil Horizon.

Oak–Any deciduous, evergreen, or partly evergreen tree or shrub of the genus *Quercus*, family Fagaceae. Species are grown as ornamentals, as shade and lawn trees, for bark, lumber, acorns, and browse. See Oak-leaf Poisoning.

Oak Gall–Swelling of various shapes that commonly occurs on the leaves and stems of oaks; usually caused by insects.

Oak Leaf Blotch–Disease of the leaves, leaf sheaths, and culms of oats; caused by the fungus *Helminthosporium avenae*, family Dematiaceae, which begins by producing spots on the leaves, spreads to the sheaths, and finally blackens the stems, even filling the hollow of the stem with a white mycelium. In severe cases the stems break at the third or fourth internodal region.

Oak Skeletonizer–*Bucculatrix ainsliella*, family Lyonetiidae; an insect whose larvae eat oak leaves, leaving a skeletonized surface.

Oak Wilt–A very serious fungal disease of various species of oak in the North Central states of the United States, caused by *Endoconidiosphora fagacearum*, known in some publications by the name of its imperfect stage, *Chalara quercina*, family Ophiostomataceae. It is related to the fungus causing the Dutch elm disease, and like this disease is carried by various insects that feed on the pads of fungus mycelium that are formed under the bark and project outward through cracks. It is also spread by the natural grafting of the roots of diseased and healthy trees.

Oak-leaf Poisoning–A digestive ailment, observed among cattle on the ranges of the southwestern United States; attributed to excessive ingestion of oak leaves in the spring, before grass makes its appearance. In a few instances the ailment is fatal and otherwise may cause stunted growth.

Oat(s)–(1) Any grass of the genus *Avena*, family Gramineae. Species are grown for forage, as a cover crop, and for their seed used as food and feed. Native to temperate regions. (2) *A. sativa*, the common oat, an annual herb native to Europe and Asia. See Animated Oat, Side Oat, Wild Oat.

Oat Straw–The dry stems and leaves of the mature oat plant left after threshing. It has a somewhat higher feed value than the straw of other small grains and is also used for bedding and industrial purposes.

Oatmeal–Hulled oats boiled and crushed or rolled; used as a breakfast food and in feed mixtures. Also called porridge (British).

Obligate Aerobe–An organism that lives only in the presence of free oxygen.

Obligate Anaerobe–An organism that lives only in the absence of free oxygen.

Obligate Parasite–An organism that develops and lives only as a parasite, and is confined to a specific host.

Obligate Saprophytes–Microorganisms not related to living cells that secure their nutrients from dead organic tissue or inorganic materials.

Obligate Symbiont–An organism that is dependent on mutual relations with another for its existence.

Obnoxious Weed–(1) In dairying, any weed which, ingested by a cow, imparts an obnoxious odor or flavor to milk, cream, or butter. (2) Any serious weed pest. See Noxious Weed.

Obolanceolate–Refers to a leaf that is narrow with the widest point near the apex. The petiole is attached at the narrow end. See Lanceolate.

Obovate–Refers to a leaf shape that resembles the ovate but the stem is at the narrow end. See Ovate.

Obtuse–(1) An angle that is greater than 90 degrees, but less than 180 degrees. (2) A leaf apex that is similar in shape to an obtuse angle.

Ochric Epipedon–A surface soil horizon that is too light colored or too thin to be a mollic epipedon. See Soil Taxonomy.

Octillo–*Fouquieria splendens*, family Fouquieriaceae; a perennial, tall, woody, cactuslike plant producing leaves for short periods whenever moisture conditions are favorable, and grown for its showy, scarlet flowers. Characteristic of certain soil conditions in the desert regions of the southwestern United States and adjacent Mexico. Native to North America. Also called coachwhip, vine cactus, candlewood, Jacob's staff, Spanish bayonet.

Ocular Estimate–Estimate by sight only. (1) In forestry, the determination of the approximate volume and quantity of standing timber without the use of measuring instruments. (2) On a range, the qualitative procedures for determining the degree of cropping of forage plants. Observations of a general reconnaissance nature are made visually: (a) by examining small random plots at the end of each grazing season.

Off–(1) In cotton transactions, designating grades below middling, the basic grade. (2) A low-grade or inferior product.

Off Grade–Designating an agricultural product that fails to meet requirements of commonly accepted standards or legal or official standards in grading products for sale.

Off Type–In plants or animals, any notable deviation from standard or normal.

Off Year–The year during which an alternate or biennial-bearing fruit tree produced little or no fruit.

Offshoot–A lateral shoot or branch that rises from one of the main stems of a plant.

Ogalla Aquifer–A large aquifer extending from Nebraska to the panhandle of Texas. It contains mostly fossil water. See Aquifer; Fossil Water.

Oil–One of several kinds of fatty or greasy liquids that are lighter than water, burn easily, are not soluble in water and are composed principally, if not exclusively, of carbon and hydrogen.

Oil Cake–Stock feed that is a mass of compressed seed from which the oil has been largely extracted, as linseed cake, cottonseed cake.

Oil Cell–In wood anatomy, a specialized and usually much enlarged wood parenchyma or ray cell that secretes an essential oil.

Oil Meal–A kind of feed obtained from soybeans, cottonseed, flaxseed, and certain other seeds after the oil has been removed, and that contains a relatively large amount of protein.

Oil Nut–See Butternut.

Oil of Lemon–An essential oil used in flavorings and in perfumes, extracted from the rind of the lemon.

Oil Palm–*Elaeis* spp. A tree grown in many African and Asian countries for the oil that it produces. Two kinds of oil are produced; a very stable oil from the fleshy part of the fruit and palm kernel oil from inside the hard kernel. Palm oil is used as a frying fat, in margarines, and in soups and gravies. Palm kernel oil is not edible but is used in paints and varnishes.

Oil Seed Crops–Primarily soybeans, peanuts, cottonseed, sunflower seeds, and flaxseed used for the production of oils for cooking, protein meals, and nonfood uses. Lesser oil crops are safflower, castor beans, rapeseed, and sesame.

Oil Sprays–Compounds of mineral or vegetable oils used to control scale and other insects on trees and shrubs.

Oil-bearing Crops–Any crop grown primarily for its oil content, such as soybeans, peanuts, cottonseed, and linseed (flaxseed).

Okra–*Abelmoschus (Hibiscus) esculentus*, family Malvaceae; a tropical annual, grown as a garden vegetable all over the warmer parts of the world. The mucilaginous, immature pods of the plant are cooked and eaten by themselves or in soups. Native to Africa and Asia. Also known as gumbo.

Old Growth–A forest that has never been changed by management or harvesting. This term is misapplied by many to describe any forest that appears to be old. Individual trees in this type of forest are usually over 200 years old, and there are large standing and fallen dead trees throughout the stand.

Old Soils–In pedologic terminology, soils on flat land surfaces that have fully developed profiles but are stagnant, i.e., undergoing changes with extreme slowness. (2) Soils buried by erosion debris or volcanic detritus.

Old Wood–A shoot or branch of a woody plant that is at least one year old.

Old-field Stand–A stand of trees grown up on land once used for crops or pasture.

Old-house Borer–*Hylotrupes bajulus*, family Cerambycidae; an insect that infests well-seasoned wood in buildings.

Oleic Acid–A shoot or branch of a woody plant that is at least one year old.

Oleomargarine–A product of certain animal or vegetable fats, sometimes fish and coloring matter, emulsified, churned, or mixed with cream, milk, water, or other liquid and made in the semblance of butter. Some of the vegetable fats used are cottonseed, coconut, corn, soybean oil, palm oil, and sunflower seed oil. Also called margarine.

Oleoresin–A natural combination of resinous substances and essential oils that occurs in, or exudes from, plants. Upon distillation, the oleoresins obtained by chipping the living trees of longleaf pine (*Pinus palustris*) and slash pine (*P. elliottii*

Oleoresin of Pyrethrum–A product used in insecticides, obtained by treating finely ground pyrethrum with ethylene dichloride and then evaporating.

Olericulture–The production, storage, processing, and marketing of vegetables.

Oligohaline–Term to characterize water with salinity of 0.5 percent to 5.0 percent, due to ocean-derived salts.

Oligophagous Parasite–A parasite capable of developing upon a few closely related host species.

Oligosaprobic Zone–The area of a stream that contains the mineralized products of self-purification from organic pollution but with none of the organic pollutants remaining.

Oligotrophia–A condition of inadequate nutrition.

Oligotrophic–(1) Pertaining to plants that grow in areas of poor soil with respect to the nutrients. (2) Swamps or water bodies poor in humus or plant nutrients.

Oligotrophic Peat–Peat soil that is strongly acid in reaction and low or wanting in bases, such as calcium, and other mineral plant nutrients.

Olive–(1) Any shrub or tree of the genus *Olea*, family Oleaceae. (2) *O. europaea*, an evergreen tree grown for centuries for its edible fruit, the olive of commerce, and as a source of olive oil. Native to Europe and Asia.

Olive Oil–The edible oil derived from the fruit of the olive by crushing, pressing, and filtering. Chemically, olive oil consists of olein, palmitin, some sterin, and a small amount of linolein.

Olive Scale–*Parlatoria oleae*, family Diaspididae, a scale insect pest that infests and damages not only the olive, but a great number of other plant species in California, United States.

Omnivorous–Designating animals that feed on both flesh and plants, as people; as applied to insects, voracious, but not necessarily omnivorous. Also called amphivorous. See Carnivorous, Herbivorous.

Omnivorous Looper–*Sabulodes caberata*, family Geometridae; an insect whose larvae are very destructive to avocado trees in California, United States, often defoliating the entire tree. They also infest many other cultivated trees including the orange.

On Contract–Designating a crop, as vegetables for processing, grown under an agreement whereby the producer sells to a processor usually at prices previously agreed upon.

On the Cob–(1) Designating unshelled corn. (2) Designating green, or immature corn, cooked and eaten with the kernels attached to the cob.

On the Contour–See Contour Row.

On the Dry Side–In greenhouse culture, designating a reduction in the amount of water ordinarily given to plants to keep the soil comparatively dry when growth is inactive.

On Year–Referring to the year during which an alternate, or biennially fruit-bearing tree produced fruit. See Off Year.

One-season Pasture–Pasture composed of a forage plant, usually an annual, such as Sudangrass, which can furnish grazing only for a single season.

One-third-atmosphere Percentage (Soil)–The moisture percentage, dry-weight basis of a soil sample that has been air-dried, screened, wetted, and brought to hydraulic equilibrium with a permeable membrane at a soil-moisture tension of 344 centimeters of water. This retentivity value closely approximates the moisture equivalent value of many soils at field capacity.

Onion–(1) Any of several herbs of the genus ***Allium***, family Liliaceae. Species include some of the very important vegetables. (2) *A. cepa*; a perennial herb grown as an annual salad plant and as a potherb for its strong-tasting bulb and leaves. It is a very ancient vegetable. Native to western Asia. Also called garden onion. (3) See Chive, Garlic, Leek, Shallot.

Onion Set–A small, immature onion bulblet that is grown from seed, harvested from a mature onion plant.

Oocyte–Ovicyte; one of the intermediate cells in the process of ovigenesis.

Oocyte-Egg-Mother-Cell–The cell that undergoes two meiotic divisions, oogenesis, to form the egg cell, as primary oocyte, the stage before completion of the first meiotic division, and secondary oocyte, after completion of the first meiotic division.

Oogenesis–The process by which germ cells are produced by the female.

Oogonium–Ovigonium; the first or primary germ cell from which the female gamete is produced.

Oophagy–The eating of eggs, said of egg-eating insects.

Oospore–A resting spore produced by sexual reproduction in the downy mildews and related fungi.

Open–(1) Designating a female animal that is not pregnant. (2) Designating a body defect of cheese in which the cheese has many mechanical openings. (3) To unfold, as a flower opens. (4) Designating a rural or wilderness area in contrast to the congestion of cities and towns.

Open Formula Fertilizer–A fertilizer that has both the analysis and the formula or materials used in it printed on a tag or on the bag container.

Open Pedigree–Among corn breeders, those pedigrees of inbred lines that are published or available to the general public.

Open Woodland–A parkland type of vegetation in which trees form a closed canopy. There is very little undergrowth.

Open-center Tree–An orchard tree pruned so as to have a vase-shaped top in contrast to one having a central leader.

Open-ditch Drainage–Drainage of excess water from land by open ditches as opposed to tile drainage.

Open-pollinated–Designating normal or natural pollination. Applied to corn, the term has two meanings: (a) all corn grown before there was any production of hybrid corn; (b) experimental fields where there is no control of the distribution of pollen from or to any of the plants within the field. See Pollination.

Opening–(1) A treeless or very sparsely timbered area in a forested region. See Oak Opening. (2) In cranberry culture, the first swelling of a terminal bud. (3) The unfolding of a flower or boll of cotton. (4) Designating the first price offered for a commodity when a market day begins.

Opening Date–The date on which an established grazing season begins.

Opium Poppy–*Papaver somniferum*, family Papaveraceae; an annual herb grown as a source of opium, a drug, and for its seeds used on the crust of breads and rolls. Native to Asia and the Near East.

Opposite Leaves–Leaves that are arranged directly opposite of each other on the stem. See Alternate.

Optimum Condition–The ideal environment, with regard to nourishment, light, temperature, etc., for an organism's growth and reproduction.

Optimum Fruitfulness–That favorable condition for growth (especially in fruit trees) in which a plentiful supply of blossom buds is produced. The condition is associated with the carbohydrate-nitrogen relationship.

Optimum Temperature–That certain temperature at which a particular plant or animal grows satisfactorily, other conditions being favorable for growth.

Optimum Water Content–The amount of water in a soil needed by a plant for its optimum growth, varying from 40 to 60 percent the moisture-holding capacity (field capacity).

Orange–Certain tropical or subtropical trees of the genus ***Citrus*,** family Rutaceae. Species are grown for their edible fruits, the oranges of commerce.

Orange Milkweed–*Asclepias tuberosa*, family Asclepiadaceae; a perennial herb that causes poisoning in animals eating its leaves. Native to Mexico, Florida, and Arizona in the United States. Also called butterfly weed, pleurisy-root, butterfly milkweed. See Milkweed.

Orange Rust–A fungal disease of blackberries and black raspberries caused by ***Gymnoconia interstitialis*,** family Pucciniaceae; characterized by a bright orange, powdery mass of spores on the lower leaf surfaces early in summer, stunting, production of little or no fruit or spines, no tillering, and premature defoliation.

Orange Tristerza–A viral disease, similar to orange-tree quick decline, which is very destructive in orange groves. It probably originated in South Africa but has spread to other countries including the United States.

Orange Worms–The larvae of four species of small moths, ***Argyrotaenia citrana, Pyroderces rileyi, Holcocera iceryaella*,** and ***Platynota sultana*,** which infest the orange in certain areas of southern California, United States, feeding on the fruit.

Orange-tree Quick Decline–A viral disease that affects sweet orange plants grafted on sour orange rootstocks. Young trees may collapse and the leaves and fruit suddenly dry up. Older trees decline more slowly.

Orangery–An artificially heated house, the predecessor of the greenhouse, used as early as the fourteenth or fifteenth century in Europe primarily for growing oranges and citrons in climates too cold for outdoor growth.

Orchard–(1) An acreage of fruit trees. (2) A grove of nut trees, as pecans. (3) An olive grove. (4) A grove of sugar maple trees. (5) In the southern United States, a tract of pine trees that is being bled for turpentine.

Orchard Cover Crop–Grasses, legumes, or any herbaceous plant, grown in an orchard to form a cover to reduce soil erosion, and to improve the soil through the addition of organic matter. It may also serve to use up available nitrogen so the trees will harden for winter.

Orchard Heater–Any of several types of heaters used in the orchards to reduce frost injury.

Orchard Mite–A spider mite, ***Tetranychus pacificus*,** family Tetranychidae; a serious pest along the Pacific Coast of the United States. It infests fruit trees and other plants, heavily webbing the foliage. There may also be extensive defoliation. Also called Pacific mite.

Orchard-run–Fruit as it is picked from an orchard and before it is graded, sorted, or processed.

Orchardgrass–*Dactylis glomerata*; a long-lived, high-producing bunchgrass adapted to well-drained soils. Can be grown under irrigation or on dry land where the effective precipitation is 16 inches (40 centimeters) or more. Is shade-tolerant. Suited for pasture, hay, silage, and erosion control.

Orchardist–A manager of orchards.

Order–(1) In botanical classification of plants, a category in between class and family. (2) In soils, one of the eleven highest categories.

Oregon Wireworm–*Melanotus oregonensis*, family Elateridae; the larva of a click beetle, common to the West Coast area of the United States, a destructive pest of grains, grasses, root crops, vegetables, and seeds.

Organ–A distinct part of a plant or animal that carries on one or more particular functions; e.g., a leaf, wing of a bird, etc.

Organic–(1) Produced by plants and animals; of plant or animal origin. (2) More inclusively, designating chemical compounds that contain carbon.

Organic Acids–Acids containing only carbon, hydrogen, and oxygen. Among the best known are citric acid (in citrus fruits) and acetic acid (in vinegar).

Organic Fertilizers–Although not specifically defined by the leaders of organic gardening, organic fertilizers usually include only natural, organic, proteinaceous materials of plant and animal origin and exclude synthetic, organic, nonproteinaceous materials such as urea $[CO(NH_2)_2]$.

Organic Gardening (Farming)–A system of farming or home gardening that utilizes organic wastes and composts to the exclusion of chemical fertilizers. Advocates of the system teach that chemical fertilizers are injurious to health and that organic composts give higher yields, better quality and better taste of produce, less plant damage by insects and disease, reduction of weed menace, and stronger seeds that germinate better and produce successively stronger plants. None of these claims have so far been confirmed by reproducible (scientific) proof. Much publicity of the concept has resulted in a number of followers but it is significant that the leading followers of the concept do not depend entirely on agriculture for their livelihood. The organic gardening concept persists because of the basic truth that organic matter in soils is beneficial to agriculture. However, the sole scientific foundation for economic production of abundant and healthful foods consist of the liberal use of organic matter plus chemical fertilizers applied according to a soil test.

Organic Matter–(1) Matter found in, or produced by, living animals and plants, which contains carbon, hydrogen, oxygen, and often nitrogen and sulfur. (2) Any matter, defined as organic incorporated in or on the surface of soil. It may include undecomposed plant matter as well as that which is highly humified. Sometimes used interchangeably with humus.

Organic Phosphate Insecticide–A chemical compound used as an insecticide, such as tetraethyl pyrophosphate, ethyl metaphosphate, and tetraethyl dithio-pyrophosphate.

Organic Soil Horizon–A surface horizon of a soil formed from organic litter derived from plants and animals, and designated as the O horizons. These horizons do not include horizons formed by illuviation of organic material into mineral material, nor do they include horizons high in organic matter formed by a decomposing root mat below the surface of a mineral material, usually designated as O_1 and O_2.

Organic Soils–Histosols; soils having organic soil materials that extend from the surface to: (a) a depth within 10 centimeters or less of a lithic or paralithic contact, provided the thickness of the organic soil materials is more than twice that of the mineral soil above the contact; (b) any depth if the organic material rests on fragmental material and the interstices are filled with organic materials; or (c) have organic materials that have an upper boundary within 40 centimeters of the surface; (i) having a bulk density of 0.1 g/cc; (ii) the organic soil is saturated with water six months of the year. See Soil Taxonomy.

Organism–Any living individual whether plant or animal.

Organogen–Any of certain chemical elements without which organisms cannot exist: oxygen, carbon, nitrogen, phosphorus, etc.

Organophosphates–A group of nonpersistent pesticides containing phosphorus, including malathion and parathion. Malathion has low human and animal toxicity; parathion has high toxicity to humans and animals.

Oriental Fruit Fly–*Dacus dorsalis*, family Tephritidae; an insect pest, especially serious in areas of mild winters, which infests all citrus fruits as well as the banana, avocado, pineapple, and a host of other plants.

Oriental Fruit Moth–*Grapholitha molesta*, family Olethreutidae; a most destructive insect pest whose larvae first bore into tender twigs of fruit trees and later generations attack the fruit making it wormy.

Oriental Leaf Disease–A fungal disease caused by *Hemileia vastatrix*, family Pucciniaceae, which seriously affects the coffee plant in Asia and Africa.

Oriental Moth–*Cindocampa flavescens*, family Limacodidae; a Japanese insect largely confined to Massachusetts in the United States, whose larva feeds preferably on the Norway maple, sycamore maple, buckthorn, black birch, cherry, apple, pear, and plum, but may also infest other trees.

Orifice–(1) An opening by which spores, etc., escape; any opening. (2) An opening in a nozzle tip, duster, or granular applicator through which the spray, dust, or granules flow.

Origanum–*Origanum vulgare*, family Labiatae; a hardy, perennial herb grown for its aromatic foliage used as a flavoring, in sachets, and with tobacco. Native to Europe. Also called marjoram, pot marjoram, wild marjoram, winter sweet.

Ornamental Plants–Plants produced for their beauty.

Ortet–The original parent plant from which a clone has been derived.

Orthotropic–Assuming a vertical position.

Osage Orange–*Maclura pomifera*, family Moraceae, a thorny tree grown as a hedge plant, for fenceposts, and for its roots, a source of dye. Native to North America. Also called Bois d'arc, bodark, brasswood, mockorange.

Osmophyllic–Organisms that grow in solutions with high osmotic pressure.

Osmosis–The flow of a fluid through a semipermeable membrane separating two solutions, which permits the passage of the solvent but not the dissolved substance. The liquid will flow from a weaker to a stronger solution, thus tending to equalize concentrations.

Osmotic Pressure–The hydrostatic pressure required to stop osmosis or prevent diffusion of molecules of a dilute solution from passing through the walls of a semipermeable membrane into a more-concentrated solution.

Osmunda Peat–A potting material made from the roots of fern of the genus *Osmunda*.

Other Forestland–Forestland incapable of producing 20 cubic feet per acre of industrial wood under natural conditions because of adverse site conditions such as infertile soils, dry climate, poor drainage, high elevation, steepness, or rockiness.

Out of Season–Designating products, as fresh fruits and vegetables, available at a time at which they are not normally produced locally.

Outgrasses–Any of the grasses of the genus *Arrhenatherum*, family Gramineae.

Outwash Plain–A physical land feature of glaciofluvial origin. The plains are smooth, or where pitted are usually low in relief. The deposits are mainly sandy or coarse textured. The plains constitute a large total area of land and a large number of separate soil and land types in the glaciated region of the United States.

Ovalhead Sedge–*Carex festivella*, family Cyperaceae; a grasslike, herbaceous plant, found at elevations of 6,500 to 12,000 feet from New Mexico to California (United States) and north to Alberta (Canada). Its forage value is good and its fibrous roots afford protection against soil erosion.

Ovary–(1) The portion of the pistil or carpel of a flower that contains one or more ovules. (2) The organ in female animals that produces the egg or ovum.

Ovate–Refers to a leaf shape that is wide and broad at the base.

Overcutting–The cutting of a quantity of timber in excess of the mean annual growth.

Overflow–(1) In cotton gins with a belt distributor system, the surplus cotton carried on belts to the feeders, which drops to the floor when the feeders are full. (2) The floodwater of swollen streams that spreads over bottomland.

Overgrown–Designating abundant growth on a field, usually of some undesirable vegetation as weeds or brush.

Overhead Sprinkling System–A system of irrigating a small acreage by sprinkling from permanently installed overhead pipes.

Overirrigation–Application of water in excess of the need of a crop and of the water-holding capacity of the soil; it damages the plants in addition to wasting the water.

Overlength–In milling logs, an allowance for trim; an extra length left on a log so that it may be trimmed to an exact length.

Overmaturity–The period in the life cycle of trees, fruits, and vegetables when growth or value is declining. See Maturity, Overripe.

Overripe–(1) Designating fruit which is mature, soft, or decayed to the degree that it is inedible or unmarketable. (2) Designating a degree of maturity, such as fruit, that is too ripe for canning or shipping but is not too ripe for table use.

Overseeding–(1) Seeding in a crop already established, such as seeding a forage crop in the spring on a field of winter wheat established the previous fall. (2) Using an amount of seed in excess of that necessary to assure a good stand.

Overstocked–(1) Designating a condition of a stand of trees or of a forest, in which there are more trees than normal or full stocking would require. The overstocking may be to such a degree that growth is slowed down and many trees, including dominants, are suppressed. (2) Designating a pasture, range, or grazing game area that has more animals on it than the vegetation of the area will support. See Overgraze. (3) Designating a locality in which there are too many bees.

Overstopped Trees (Suppressed)–Trees with crowns entirely below the general level of the crown cover receiving no direct light either from above or from the sides.

Overstory–That portion of trees in a forest stand that forms the upper crown cover. Also called overwood.

Oversummer–To live through the summer.

Ovicide–Any substance that kills parasites or other organisms in the egg stage.

Ovine Ecthyma–Soremouth; an acute, highly contagious disease of sheep caused by a filterable virus. Lambs are most frequently affected. The virus produces vesicles or blisters, which later turn into wartlike crusts, on the inner and outer part of the lips and interfere with normal feeding. Immunity may be conferred through vaccination.

Oviparous–Refers to animals who produce eggs that are incubated inside the body of the dam and hatch inside the body or shortly after laying. See Oviparous, Viviparous.

Ovipositor–A tubular structure in female insects used for depositing its eggs.

Ovulation–The process of releasing eggs or ova from the ovarian follicles.

Ovule–The body that, after fertilization, becomes the seed; the egg-containing unit of the ovary.

Ovum–The female sex cell, produced on the ovary, and carrying a sample half of the genes carried by the female in which it was produced. Plural, ova.

Oxalacetate–An intermediate compound formed from sugars and sugar-producing (gluconeogenic) amino acids, essential for the oxidation of carbohydrates, proteins, and fats for energy.

Oxalic Acid–An organic acid that occurs in many plants. It is probably the acid secreted by crustaceous lichens having the power to dissolve minerals of hard rock surfaces; e.g., granites. It is produced synthetically by the fusion of sawdust with caustic soda or potash.

Oxic Horizon–A highly weathered subsurface soil horizon (endopedon) high in iron and aluminum formed in humid, tropical, old land surfaces. See Oxisols, Soil Order.

Oxidase–Oxidizing enzymes. See Enzyme.

Oxisols–One of the eleven soil orders, they are soils with residual accumulations of inactive clays, free oxides, kaolin, and quartz. They occur mostly in tropical climates on old land surfaces.

Oxygen–The chemical element O; a colorless, odorless gas. The most abundant element in the earth's crust. It accounts for about 47 percent of all elemental material. It is essential in the growth of all crops and for the respiration of most forms of life.

Oyster Shells–Shells of the marine bivalve, genus *Ostrea*. They are nearly pure calcium carbonate in composition; when finely ground, they make good liming material for soils and a mineral feed for livestock and poultry as a source of calcium.

Oystershell Scale–*Lepidosaphes ulmi*, family Diaspididae; a scale isnnect, about one-tenth inch long, which covers twigs and branches. The pests overwinter in the egg stage under the scales. The young crawling stage appears in late May. This scale is a pest of apples and a number of shade and ornamental plants.

P–Chemical symbol for phosphorus.

P_2O_5–Phosphorus pentoxide; in current fertilizer labeling, the amount of P in fertilizer is expressed as percent of P_2O_5. To find actual P, multiply P_2O_5 equivalent by 0.44.

P/E Ratio—Precipitation/evaporation ratio. This is an aridity index.

Pacific Bunchgrass Association—One of the major grassland associations of the United States whose grasses have a characteristic bunch-growth. Regionally, it embraces Idaho, Montana, Nevada, Washington, Oregon, and California.

Pacific Dogwood—*Cornus nutalli*, family Cornaceae; a deciduous, tender tree grown for its white or pinkish flowers. Native to the Pacific coasts of Canada and the United States.

Pacific Flathead Borer—*Chrysobothris mali*, family Buprestidae; a beetle that bores into the bark and wood of nearly all forest, shade, and fruit trees in the western United States.

Pacific Hemlock—*Tsuga heterophylla*, family Pinaceae; used for its lumber and as pulp, for general construction and planing-mill products. Native to the Pacific Coast from Alaska to California. Also called western hemlock (forestry name), West Coast hemlock, hemlock spruce.

Pacific Poisonoak—*Toxicodendron diversilobum*, family Anacardiaceae; a woody shrub or vine that on contact with susceptible humans causes a severe dermatitis. Native to the western United States and Canada. See Poison Ivy.

Pack—(1) Fruit, vegetables, meat, etc., which is to be or is packaged, canned, frozen, etc., for the market. (2) The total amount of products that are processed in a season. (3) The manner in which produce is packaged. (4) The load that is carried by a pack animal. (5) To compress or firm soil with a special implement. (6) To damage soil structure by compacting or puddling clay soils, as from the pressure exerted by the wheels of a tractor, by injudicious irrigation, or by excessive rainfall. (7) The sheep given to the shepherd as his share for tending the flock.

Paddy Rice—Rice as harvested from the fields before it is milled.

Paddy Soil—The soil of a rice paddy: it is wet and subject partially to anaerobic conditions.

Pahala Blight—A manganese-deficiency disease of sugarcane.

Painted Hickory Borer—*Megacyllene caryae*, family Cerambycidae; an insect pest that infests hickory, walnut, butternut, Osage orange, and hackberry.

Pakchoi—*Brassica chinensis*, family Cruciferae; an annual or biennial herb grown as a potherb and as a salad plant. Native to Asia. Also called Chinese cabbage.

Palatability—(1) The degree to which a feed is liked or acceptable to an animal. (2) (Forage) Range management usage. The relish that an animal shows for a particular species, plant, or plant part. The characteristics of plants that stimulate a selective grazing or browsing response by animals. Palatability is controlled by the plant factors of chemical composition, proportion of plant parts, growth stage, external form of plant parts, environmental factors such as slope steepness, wind, sun, or shade; and the animal factors of instinct, learning, physiological state, individual behavioral pattern variations, and animal sensory responses.

Palay Rubbervine—*Cryptostegia grandiflora*, family Asclepiadaceae; a tropical, woody vine grown in warm areas or in the greenhouse for its habit and its purplish flowers. It is also a source of small quantities of rubber. Native to Africa. Also called India-rubber vine.

Pale Agoseris—*Agoseris glauca*, family Compositae; a perennial, leafy herb widespread throughout the northern United States. Species of this genus contain poisonous alkaloids that may cause death to livestock, especially sheep. Native to North America.

Pale-striped Flea Beetle—*Systena blanda*, family Chrysomelidae; the larva of a flea beetle that infests corn seed so that the seed often fails to sprout, or produces a pale, weak plant. The adult infests watermelon, pumpkin, pea, bean, and eggplant.

Palisade Layers—Tissues just below the epidermis of the leaf of a plant; most photosynthesis takes place in the palisade layers.

Palisade Mesophyll—A leaf tissue composed of slightly elongated cells containing chloroplasts; located just beneath the upper leaf epidermis, and above the spongy mesophyll in broad-leaved plants and in some conifers.

Palm—Any tropical or subtropical tree or shrub of the family Palmaceae. Species are grown for their wood, edible fruits, resins, fiber, and as a source of oils and drugs.

Palm Kernel Oil Meal—A livestock feedstuff that is the ground residue from the extraction of part of the oil, by pressure or solvents, from the kernel of the fruit of the African oil palm. See Oil Palm.

Palm Oil—See Oil Palm.

Palmate—With three or more nerves, lobes, or leaflets radiating fanwise from a common basal point of attachment.

Palmerworm—*Dichomeris ligulella*, family Gelechiidae; a caterpillar, widespread from Maine to Texas (United States) that skeletonizes the leaves of trees and sometimes eats into the fruit of apple, cherry, hazel, oak, pear, and plum.

Palmetto Flatwoods—A kind of land, constituting extensive areas in Florida (United States), in which the ground cover or vegetation is characteristically sawplametto, although various other plants are present and the trees are mainly pines. The land is periodically very wet and very dry. The soils are leached sands commonly underlaid by organic hardpans.

Palouse—A region of about 8,000,000 acres in eastern Washington adjacent to part of Idaho and the north-central part of Oregon. It has a semiarid and subhumid climate, and undulating to hilly and dunelike topography, productive silty soil, mostly deep, but locally severely eroded and shallow over basalt rock. it was originally grassland characterized by bunch wheatgrass but is now largely in farms with wheat as the principal crop.

Palustrine—(1) Designating a swampy or boggy soil. (2) Designating a plant that grows in a swamp.

Pampas—Vast, rolling plains of grasses without trees, especially in Argentina.

Pan– (1) A compact, indurated, subsurface layer in a soil profile. See Hardpan. (2) A depression in a saltwater marsh. (3) In forestry, a wide, steel sheet used to support the end of a log in skidding. (4) A flower pot about one-half the depth of the standard flower pot; used for growing shallow-rooted plants and bulbs. Also called bulb pan, seed pan. See Fragipan.

Pancake Land– Flat, prairie land with low-lime clay soils that has a compact baked surface (eastern Texas, United States).

Pandemic– Designating a disease or organism of worldwide distribution; widely epidemic.

Panel Plot– A small area protected from grazing during a part of the growing or grazing season by a movable fence. Also called hurdle plot, panel.

Panerogam– A true seed-bearing plant.

Pangolagrass– *Digitaria decumbens*, family Gramineae; a perennial, extensively stoloniferous or creeping, tender grass grown in Florida and southern California, United States, as a pasture grass. It is tolerant of highly acid to neutral soils but not of saline soils, and is adapted to sandy soils when fertilized. Native to South Africa.

Panicle– A type of inflorescence that has a main stem and subdivided branches. It may be compact and spikelike, as in timothy, or open and spreading, as in the oat.

Pannage– (1) Pasturing swine in a forest to feed on acorns, etc. (2) The right or privilege to pasture swine in this manner. (3) The payment made to the landowner for the privilege of pasturing swine in his forest.

Panphytotic– Designating worldwide distribution of a plant disease.

Panplane– A nearly flat surface brought about by the lateral erosion of streams, which erodes the divides and causes a coalescing of all of the flood plains of a region.

Papain– An enzyme present in the green papaya fruit used as a digestant and meat tenderizer.

Papaya– *Carica papaya*, family Caricaceae; a tropical, unbranched, rapidly growing tree. The mature fruit is green to yellow and resembles a smooth-skinned muskmelon.

Paper Birch– *Betula papyrifera*, family Betulaceae; a deciduous, average-sized tree grown as an ornamental in the United States for its white, paperlike bark. Native to North America. Also called canoe birch, white birch.

Paper Doll– A method of testing seed for germination in which seeds are placed on wet paper towels. The towels are then rolled and incubated.

Paper Mulch– Heavy paper rolled out in strips to cover the soil as a mulch between rows of plants, or squares of paper with slits through which individual plants grow. The paper may be as effective as usual mulches of straw, manure, leaves, etc., in conserving moisture and preventing weed growth.

Paper Pulp– The product used in the manufacture of paper that is made by mechanical and chemical treatment of wood or other cellulose material.

Papershell– Designating any variety of pecan that produces a large nut with a thin shell.

Papilionaceous– Literally, butterflylike; applied to a type of corolla characteristic of the subfamily Fabaceae of the Leguminosae, having five petals, with the uppermost one outside and usually largest, the two laterals (the wings) paired and usually clawed, and the two lower-most united along their lower margin and forming a sheath (the keel) enclosing the stamens and pistil.

Papilla– Any small nipplelike process.

Paprika– The powdered ripe fruit of sweet red peppers which is used as coloring matter in ketchups and relishes, as a flavoring, and a garnish.

Para Rubbertree– *Hevea brasiliensis*, family Euphorbiaceae; a tree grown as the chief source of the world's supply of natural crude rubber. Native to the Brazilian Amazon jungle but naturalized in Malaysia.

Parachute Fruit– A seed or achene with a spreading tuft of hairs, bristles, or scales at one end, which enables it to be carried by wind and become widely distributed: e.g., a weed seed, such as that of the dandelion.

Paradichlorobenzene– A white crystalline substance used in beehives to fumigate combs and to repel wax moths. It is also used as a soil fumigant.

Paraeluvium– The residual product that results from the eluviation of the weathering products of sedimentary rocks. See Eluviation.

Paragrass– *Panicum purpurascens*, family Gramineae; a tropical, heavy, perennial grass grown in the southern United States for forage. It is adapted to wet sites. Native to Africa; introduced into the United States via Brazil.

Parallel Veined– A leaf type that has veins running parallel to each other as in a blade of grass.

Parasite– An organism that lives at least for a time on or in and at the expense of living animals or plants. Some diseases of people and animals are caused by parasites ordinarily classified as protozoan, helminthic, and anthropod species. There are also innumerable species of plant parasites.

Parasitic Seed Plant– A plant that obtains part or all of its sustenance from other plants; e.g., dodder and mistletoe.

Parasiticide– An agent that kills parasites.

Parch Blight– A condition in conifers where the foliage turns brown in late winter or early spring as a result of desiccation due to low-temperature injury, also known as winterkill, physiological drought.

Parchment– (1) The sheetlike fiber found in the pods of unimproved beans and pea varieties. (2) A paper used in food wrapping; e.g., in the packaging of butter.

Pare– To peel or trim off outside covering, as with fruits and vegetables.

Parent Element–The radioactive element from which a daughter element is produced by radioactive decay; e.g., radium is the parent element of radon.

Parent Material–(1) The horizon of weathered rock or partly weathered soil material from which the soil is formed. Horizon C of a soil profile. (2) The unconsolidated material from which a soil develops. See Soil Horizon.

Parent Plant–A plant from which cuttings, bulbs, etc., is taken to produce new plants.

Parent Rock–(1) The original rock from which sediments were derived to form later rocks. (2) The rock from which parent materials of soils are formed; the R horizon. See Soil Horizon.

Parent Seed–(1) In hybrid corn production, the single-cross seed used by a grower for producing hybrid seed corn. (2) Any seedstock from which an increase of seed is secured.

Parental Generation–The P_1 generation; the first generation in a series of crosses; usually involves homozygotes for different alleles.

Paris Green–Acetoarsenite of copper, one of the first arsenical compounds to be widely used in insect sprays as a stomach poison for the destruction of leaf-eating insects. It was almost exclusively used to control the potato beetle. Paris Green is now replaced by insecticides that biodegrade faster.

Parry Larkspur–*Delphinium parryi*, family Ranunculaceae; a very poisonous herb found in the coastal regions and foothills of California (United States). Symptoms of poisoning are loss of appetite, uneasiness, staggering, prostration with rigidly extended limbs, constipation, bloating, and nausea. See Larkspur.

Parthenogenesis–The development of an individual from an unfertilized egg cell. Known to occur occasionally in turkeys and some lower forms in the animal kingdom but not in mammals. In honeybees, unfertilized eggs produce only drones.

Parthenogenic–Fruit produced without fertilization of the ovule(s). Usually seedless. See Fertilization.

Partial Dominance–A kind of interaction between alleles where one gene is not completely dominant to its allele but where the appearance of the heterozygote is more similar to one of the homozygotes than to the other.

Particle Density–The average density of the soil particles. Particle density is usually expressed in grams per cubic centimeter and is sometimes referred to as real density, grain density, or specific gravity.

Particle-size Analysis–Determination of the amounts of different particle sizes in a soil sample, usually by sedimentation, sieving, micrometry, or a combination of these methods.

Particle-size Distribution–The amount of the various soil separates, sands, silt, and clay in a soil sample, expressed as dry weight percentages.

Parts Per Millon–The number of weight or volume units in a million units of a solution or a mixture; a measure of concentration, especially of chemicals in solution: one milligram per liter. Abbreviated ppm.

Parvifolius–Small-leaved.

Paspalum–Any grass of the genus ***Paspalum***, family Gramineae. Species are most abundant in the tropics. A number are native in the southeastern United States and have forage value. See Smooth Paspalum.

Passion Fruit–*Passiflori* spp., family Passifloriaceae; a tropical vine with large purple flowers and edible fruit.

Pasturage–All vegetation, grasses, and grasslike plants, forbs, and the fruits and twigs of trees and shrubs upon which grazing animals subsist.

Pasture–(1) A fenced or unfenced tract of land on which farm animals feed by grazing. The pasturage is mainly grass but it may consist of various other herbs, brush, and trees. (2) Nectar and honey plants within flight range of bees of an apiary. (3) To place livestock on a field or area of grass to harvest the crop by grazing.

Pasture Forage–A crop ordinarily grown for pasture but which may be cut for green feed, silage, or cured for hay.

Pasture Improvement–The practice of grazing, clipping, fertilizing, liming, seeding, contour furrowing, or other methods of management that improve the vegetation for grazing purposes.

Pastured–Designating an area or crop that has been grazed off by livestock.

Pastureland–(1) Land used primarily for the production of adapted domesticated forage plants to be grazed by livestock. (2) Land producing forage plants, principally introduced species, for animal consumption. Management practices usually include one more more treatments such as reseeding, renovating, mowing, liming, or fertilizing. Native pasture that because of location or soil limitation is treated like rangeland is included as pastureland.

Pasturing–The system of removing plant growth by allowing animals to graze it rather than harvesting by other methods.

Patana–A grassy slope with a moderate supply of moisture, resembling a savanna.

Patch–(1) A small piece of land of indefinite size used for cultivated plants, as garden patch, melon patch. (2) A relatively small island like tract named after the vegetation growing on it, as a brier patch, patch of woods, etc.

Patch Budding–Budding by cutting out a rectangular piece of bark from the stock and replacing it with a similar piece of bark containing the bud from another plant.

Pathobiology–The study of disease processes; biology of disease.

Pathogen–In the general sense, anything capable of causing disease, but when referred to by most veterinarians and physicians it signifies a living, microscopic, disease-producing agent such as a bacteria or virus. See Parasite.

Pathology–The science that deals with diseases and the effects that disease have on the structure and function of tissues.

Pavement Ant–*Tetramorium caespitum*, family Formicidae; a yellowish ant, a pest in lawns, gardens, and greenhouses.

PCBs–See Polychlorinated Biphenyls.

Pea–Any annual or perennial herb of the genus *Pisum*, family Leguminosae. Species are grown for food and feed. When a pea sprouts, the pea seed stays underground. In contrast, when a bean sprouts, the bean seed emerges through the soil.

Pea Aphid–*Macrosiphum pisi*, family Aphidae; a large green aphid present in home gardens but which does its most serious damage in fields of canning peas.

Pea Bean–See Navy Bean.

Pea Moth–*Laspeyresia nigricana*, family Olethreutidae; a moth that emerges about the time the peas come into blossom and places its eggs on almost any part of the plant. The small, active larvae locate the pods and bore into them, feeding on the developing seeds and spinning a very light web in the process. The insect attacks all varieties of garden peas, sweet peas, and vetch and is a serious pest in the northern pea-canning areas of the United States.

Pea Powdery Mildew–A mildew of pea caused by the fungus *Erysiphe polygone*, family Erysiphaceae; characterized by a white to grayish growth on the surface of the plant parts.

Pea Shrub–Siberian pea shrub; *Caragana arborescens*; a shrub used in shelter belts and windbreak plantings.

Pea Weevil–*Bruchus pisorum*, family Bruchidae, a beetle whose larvae bore into the pea seeds, feed, and later pupate there. It is one of the most serious insect pests of field peas.

Pea-sick Soil–A soil that is infested with fungi that cause the root rot of peas.

Peach–*Prunus persica*, family Rosaceae; a deciduous tree grown for its edible, sweet fruit, the peach of commerce, eaten fresh, dehydrated, sun-dried, frozen, or canned. It forms one of the most widely distributed and valuable orchard crops in the United States. Double-flowered and purple-leaved varieties are grown as ornamentals. Native to China. See Nectarine.

Peach Bacterial Spot–A bacterial disease caused by *Xanthomonas pruni*, family Pseudomonadaceae, characterized by severe leaf drop. The fruit becomes unfit for market.

Peach Leaf Curl–A curl of peach leaves caused by *Taphrina deformans*, family Taphrinaceae. Affected leaves are noticeably reddened or paler than normal and become much curled, puckered, or distorted. Later the leaves acquire a grayish appearance, gradually turn brown, and finally wither and falls from the tree. Young twigs and fruit may also be involved. the disease is of worldwide distribution. Also called curl, leaf blister.

Peach Mosaic–A viral disease of peach prevalent in the southwestern United States. It is characterized by leaf mottling, retarded foliation, rosetting, and the breaking of color in the blossoms. The virus has a wide range of hosts in stone fruits, and there are a number of variable strains. The disease causes large losses, and quarantines have been set up for it.

Peach Rosette–A viral disease of peach which is indigenous to the southern United States. Infected trees turn yellow, followed by the formation of new shoots bearing many small leaves. Peach trees die shortly after symptoms appear. Some plums and apricots are symptomless carriers of the virus.

Peach Scab–A fungal disease of peach and apricot caused by *Cladosporium carpophilum*, family Dematiaceae, which on the fruit causes superficial, dark, sooty freckles. When numerous, they may emerge into dark blotches beneath which the skin may dry and crack. Small lesions may be formed on twigs and on the underside of leaves. The disease is widely distributed and may occur on other stone fruits also. Also called freckles, black spot.

Peach Tree Borer–*Sanninoidea exitiosa* Say, family Aegeriidae; a moth whose larva is a very serious pest of the peach. It bores into the base of the tree from a foot above to 2 or 3 inches below the surface of the soil causing masses of gum to exude. These borers produce dead areas in the bark and may eventually kill the tree.

Peak–(1) In turpentining, the upper point of the V-shaped streak at the top of the incision. (2) In hydrology, the maximum rate of flow recorded at a gauging station during a flood.

Peak Year–The year of greatest production in the life of a fruit tree, a bed of berries, asparagus, etc.

Peanut–*Arachis hypogaea*, family Leguminosae; an annual herb whose pods ripen underground and contain usually two large, oily seeds, the peanuts of commerce, that are edible when roasted. The seeds are valuable as food, and serve as a source of peanut oil. The plant is widely cultivated as a cash crop in warm regions, where it is harvested for nuts and forage. Also called American earthnut, earth almond, earth pea, goober, ground nut, pinda pindar.

Peanut Butter–A food made of roasted peanuts ground until soft and smooth. It is a wholesome and highly nutritious food, containing 2,825 calories per pound, and about 29 percent protein, 46 percent fat, and 17 percent carbohydrate.

Peanut Hull–The shell of the peanut. It contains more than one-half fiber and is less nutritious than straw; used chiefly as poultry litter, bedding, and fertilizer. See Peanut Hull Meal.

Peanut Hull Meal–Ground peanut hulls; a very slow-acting fertilizer containing 1.5 to 2.5 percent nitrogen.

Peanut Meal–Finely ground peanut oil cake. See Peanut Oil Meal.

Peanut Oil–The oil expressed from peanuts; used primarily as a food. Also called earthnut oil.

Peanut Oil Cake–A livestock feedstuff obtained as a by-product of the partial extraction of oil by pressure or solvents from peanut kernels.

Peanut Oil Meal–A livestock feedstuff which is ground peanut oil cake.

Peanut Seed Rot–A rot of peanut seed that occurs usually within the first week after planting, caused mainly by species of *Fusarium, Rhizopus, Mucor, Diplodia, Penicillium,*, and *Aspergillus* fungi. Their entrance is facilitated by broken seedcoats and by other mechanical injuries produced by the sheller. The disease is most destructive under conditions that retard germination: cool, damp weather; abnormally deep planting; and water-logged soils.

Peanut Southern Blight–A blight of peanuts caused by the fungus *Sclerotium rolfsii*, which may appear on plants at any time during the growing season but is more likely to cause damage in late summer or early fall. It causes wilting and eventual death of the plant above the point of infection.

Pear–(1) Any tree or shrub of the genus *Pyrus*, family Rosaceae. Species are grown for their edible fruits and as ornamentals. (2) *P. communis*; a deciduous tree grown for its edible fruit, the pear of commerce. Native to Europe and Asia.

Pear Blight–See Fire Blight.

Pear Decline–A disease of pear trees of north-central Washington (United States), in which many of the finer feeding roots are infected with various fungi, the most abundant of which is *Pythium debaryanum*, family Pythiaceae. It occurs especially in roots that are in waterlogged soil.

Pear Gray Mold–A serious disease of pear fruits in storage caused by the fungus *Botrytis cinerea*, family Sclerotiniaceae; characterized by the appearance of a slate-gray mold on the surface of the skin and rotting of the fruit. In advanced stages the fruit often has a sour odor.

Pear Hawthorn–One of several hundred species of hawthorn, *Crataegus calpodendron*, family Rosaceae; a thorny, deciduous shrub or small tree grown for its small, white flowers, and small, pear-shaped, orange-red fruit. Native to North America. Also called pear haw, black thorn.

Pear Leaf Blight–A fungal disease of pears and quinces caused by *Fabraea maculata*, family Mollisiaceae; characterized by several purple dots extending to purple or dark brown lesions on the leaves, fruit, and shoots. Also called leaf-spot, fruit spot, leaf scald.

Pear Psylla–*Psylla pyricola*, family Pysllidae; an insect that infests pears feeding on sap sucked from leaves, fruit buds, and the young fruit. Heavy infestation lowers the vitality of the tree, and fruits do not develop properly.

Pear Rust–A fungal disease of pear caused by *Gymnosporangium sabinae*, family Pucciniaceae, of which the juniper tree is the alternate host. It can be controlled by the removal of junipers from within 500 feet of the pear trees.

Pear Slug–*Caliroa cerasi*, family Tenthredinidae; an insect pest that infests cherry, plum, pear, hawthorn, Juneberry, mountain ash, and quince, skeletonizing the foliage.

Pear Thrip–*Taeniothrips inconsequens*, family Thripidae; a very small insect that infests the pear, cherry, apple, and other fruit trees. It attacks the buds of fruit trees early in the spring causing them to shrivel and turn brown. Mature fruit is very seriously damaged by scabbing, russeting, and deforming.

Pear Tomato–*Lycopersicon esculentum* var. *pyriforme*; a variety of the tomato whose fruit is small, yellow, and pear-shaped.

Pearl Millet–*Pennisetum glaucum*, family Gramineae; a tall annual grass grown in the tropics as grain for food, and in warm areas, generally, for forage. Also called cattail millet, Indian millet, African millet, African cane, Egyptian millet, Egyptian wheat, East India millet, ragmillet.

Pearl Popcorn–One of the types of popcorn whose kernels are rounded in contrast to the shoe-peg or rice-shaped kernels of other types. See Popcorn.

Pearl-barley–Barley ground down to small rounded grains.

Peat–(1) A dark-brown or black residuum produced by the partial decomposition and disintegration of mosses, sedges, trees, and other plants that grow in wet places. (2) Fibrous, partly decayed fragments of vascular plants that retain enough structure so that the peat can be identified as originating from certain plants (e.g., sphagnum peat or sedge peat). See Histosols, Muck.

Peat Bog–A bog in which the soil or cumulose deposit is peat, a common relief and vegetational feature in the northern glaciated region of the United States from Minnesota eastward. The bog may be treeless or it may support a growth of trees, such as black spruce.

Peat Lover–A plant that grows in peat; a plant that is tolerant or adapted by nature to highly acid and other conditions found in natural peat soils.

Peat Moss–Moss plants that grow on heath bogs, such as species of *Sphagnum* and *Polytrichum*.

Peat Pot–A container made of compressed peat usually used for growing seedlings in preparation for transplanting without removal from the pot.

Peavey–In lumbering, a stout lever, 5 to 7 feet long, which is fitted with a socket, spike, and a curved steel hook that works on a bolt hinge. It is used in handling logs. Also spelled peevy.

Peavine Hay–The cured vines of peas used for feeding livestock.

Peavine Silage–Silage usually made form canning refuse consisting of the vines and pods of green peas left after the seeds have been removed for canning. It has a strong odor but is an excellent feed for dairy cows, beef cattle, and sheep.

Pebrine–A destructive, contagious disease of silkworms caused by a microsporidian protozoan, *Nosema bombycis*, family Nosematidae; characterized by black spots and stunted growth.

Pecan–*Carya illinoiensis*, family Juglandaceae; a tender deciduous tree grown for its edible nut, the pecan of commerce, and for its lumber used in tool handles, ladders, furniture, sporting goods, woodenware, novelties, etc. Native to North America. Also called Mississippi nut. Illinois nut. See Paper Shell.

Pecan Carpenterworm–*Cossula magnifica*, family Cossidae; the larva of a moth that bores into the trunks and branches of the pecan, oak, and hickory in the southern United States. Also called hickory cossid.

Pecan Nut Casebearer–*Acrobasis caryae*, family Phycitidae; a moth whose larva does serious damage to the pecan crop by feeding on buds and nuts.

Pecan Rosette–A nutritional disease of pecans caused by a deficiency of zinc in the soil and characterized by a yellow mottling of the leaves in the tops of the trees, a dwarfing of the leaves, and death of the twigs and branches. Zinc sulfate or zinc oxide applied to the soil can cure this deficiency.

Pecan Scab–A fungal disease of pecans caused by ***Cladosporium effusum***, family Dematiaceae, characterized by the blackening and shriveling of the nuts.

Pecan Weevil–*Curculio caryae*, family Curculionidae; a weevil that infests newly formed pecans causing them to shrivel and drop. Also called hickory nut weevil.

Pecky Wood–Wood having pockets of decay caused by fungi or pits or holes caused by insects and birds. It may be used for rustic interior paneling.

Pectin–Any of the fruit juice substances that form a colloidal suspension with water and are derived from pectose (protopectin) in ripening processes or other forms of hydrolysis. Pectin is derived from citrus fruits and apple wastes and is used in jelly making to firm the body of the product.

Ped–A unit of soil structure such as an aggregate, crumb, prism, block, or granule, formed by natural processes (in contrast with a clod, which is formed artificially by compression of a wet clay soil).

Peddler–(1) An itinerant seller or merchant without a fixed place of business who carries his goods with him and sells form place to place. (2) The larva of the tortoise beetle, ***Cassida bivittata***, family Chrysomelidae, which looks like a moving bit of dirt or excrement.

Pedicel–(1) A small stalk in plants, especially one of the several stalks attached to the peduncle, each supporting a flower in a branched inflorescence. (2) Any of the various small stalklike structures in animals. Also called peduncle.

Pedigree–(1) A list of an individual animal's ancestors, usually only those of the five closest generations. (2) A list of the ancestors of a crop plant, as the pedigree of corn.

Pedistalled–Plants and stones perched on small columns (pedistals) of soil, the surrounding soil having been lost by raindrop splash erosion.

Pedogenesis–Soil formation.

Pedologist–One versed in pedology; a soil scientist.

Pedology–The science that deals with the origin, fundamental nature, and classification of soils.

Pedon–The smallest volume that can be called a soil. It has three dimensions, and it extends downward to the depth of plant roots or to the lower limit of the genetic soil horizons. Its lateral cross section is roughly hexagonal and ranges form 1 to 10 square meters, depending on the variability in the horizons.

Pedosphere–The outer soil layer of the earth in and on which organic life exists.

Peduncle–The stalk of a flower cluster, or of a solitary flower when the inflorescence consists of only one flower. Also called pedicel.

Peel–(1) The outer covering or skin of fruits or vegetables such as the apple, orange, etc. Also called peeling, rind. (2) To remove the skin from a fruit or vegetable, as to peel a banana.

Peg–The plant mechanism of the peanut by which the fertilized peduncle forces the ovary under the ground where it develops into the pod.

Pelleted Hay–Hay that has been highly compressed by passing through a pelleting machine. It is easy to handle and is free from dust.

Pelleted Seed–Seed coated with an inert material such as fertilizer, fungicide, etc., and shaped into a uniform unit like a pill or pellet. Pelleting enables better preservation of the seed, precision planting, and a great saving in seed.

Pelletized Fertilizer–Fertilizer compressed into small grains or pellets of uniform size and shape. These pellets can be more easily handled and more exactly placed by fertilizer machinery than the loose material, and there is less loss by deterioration in storage.

Pelophyte–A plant that grows in clayey soil.

Peltate–Refers to a leaf shape similar to a shield with the petiole attached in the center of the blade.

Peneplain–(1) A land surface worn down by erosion to a nearly flat or broadly undulating plain; the penultimate stage of old age of the land produced by the forces of erosion. (2) By extension, such a surface uplifted to form a plateau and subjected to renewed degradation and discussion.

Penetrant–Adjuvant that aids a liquid's ability to enter the pores of a surface. See Adjuvant.

Penetrometer–A device that measures the force required to push a probe rod into the soil. It can be used to measure the density or degree of compaction in a soil.

Penta–Greek for five, used in naming chemical compounds.

Pentachlorophenol–A chemical used extensively to treat fenceposts, telephone poles, and bridge planking, against fungal decay. Reports have indicated some hazard to livestock when they lick or chew on posts so treated.

Pepper–Any herb, shrub, woody vine, or tree of the genus ***Piper***, family Piperaceae.

Pepper Grass–*Lepidium virginicum*; a winter annual that is troublesome in grain and other overwinter crops.

Pepper Maggot–*Zonosemata electa*, family Tephritidae; an insect that feeds on the core of red peppers and spoils the pod.

Pepper Wilt–A bacterial disease caused by ***Pseudomonas solanacearum***, family Pseudomonadaceae, which attacks red peppers and other plants in the warmer sections of the United States; characterized by a drooping of the leaves soon followed by wilting and death of the plant. The inner tissues of the stem have a dark, water-soaked appearance and when pressed there is a gray, slimy exudate.

Peppercorn–(1) The dried fruit of the black pepper ***Piper nigrum***, used whole or ground as seasoning. (2) In English law, a minimal rent for land.

Peppermint–The dried leaves and tops of the herbaceous plant ***Mentha piperita***, used as an oil, spirit, or water extract to flavor candies and drugs.

Peracute–Excessively acute; e.g., when symptoms of a disease occur much earlier than usual and are well marked.

Percentile Taper–The relative taper of a tree in terms of diameter at regular intervals along the stem which is expressed in percent of diameter at breast height.

Perched Water Table–(1) A body of water that has been retarded in its downward course by an impermeable or nearly impermeable bed to such an extent that it forms an upper zone of saturation overlying but separated from a lower zone. (2) An area at the bottom of a pot or bench filled with saturated growing medium.

Percolation–The downward movement of water through the soil in response to the pull of gravity.

Perennial–A plant that lives for more than two years. There are two classes of perennials; herbaceous and woody. An herbaceous perennial is one having succulent aboveground parts that are usually killed by winter cold, the roots living more than two years. A woody perennial is one having aboveground parts that are woody and in the temperate zones normally withstands winter cold. Both roots and aboveground parts live more than two years.

Perennial Crop–A crop that continues to produce without annual or biennial reseeding or replanting, such as alfalfa, many grasses, asparagus, rhubarb, artichoke, Jerusalem artichoke, and certain other varieties of tree or shrub crops, etc.

Perennial Peavine–***Lathyrus latifolius***, family Leguminosae; a perennial, climbing herb grown for its habit and its flowers of various colors. Native to Europe. Also called everlasting pea, perennial pea.

Perennial Ryegrass–See Ryegrass, Perennial.

Perfect Flower–A flower with both stamens and a pistil or pistils. See Imperfect Flower.

Perfect Stage–Sexual reproductive stage of a fungus.

Pergelisol–See Permafrost, Thermokarst.

Perianth–The floral envelopes of a flowering plant: the calyx, corolla, or both.

Pericarp–The outer layer of the ovary wall around the seeds. It sometimes consists of three separate structural layers: the endocarp, the mesocarp, and the epicarp.

Periderm–(1) The outer skin of potatoes and other vegetables. (2) The corky layers of plant stems.

Periodic Annual Increment–The growth of a stand of trees for any specified period divided by the number of years in the period.

Periodical Cicada–***Magicicada septendecim***, family Cicadidae; a cicada found in the northeastern part of the United States, that has a life cycle of seventeen years, with a number of broods overlapping. Injury to plants is caused by the female cicada depositing her eggs in roughened punctures, 1 to 4 inches long, in the twigs and branches of apple and many other trees. Native to North America. Also called seventeen-year locust.

Peripheral–Pertaining to the surface of the body of a plant, especially a tree.

Periphyton–The association of aquatic organisms attached or clinging to stems and leaves of rooted plants or other surfaces projecting above the bottom of the body of water.

Perique–A kind of tobacco, black, strongly flavored, cured in its own juice and mixed in small amounts with some smoking tobaccos. In the United States, it is grown mainly in Louisiana.

Perishable–Designating any product that is liable to easy or quick destruction by rot, disease, or decomposition, such as fresh fruits, meats, and vegetables.

Perlite–A volcanic glass having numerous concentric cracks that give rise to perlitic structure. A high proportion of all perlites are rhyolitic in composition. Used in greenhouses as a synthetic soil mix. See Vermiculite.

Permafrost–Permanently frozen ground. Permafrost areas are divided into more northern areas in which permafrost is continuous, and those more southern areas in which patches of permafrost alternate with unfrozen ground. See Thermokarst.

Permafrost Table–An irregular surface that represents the upper limit of permafrost.

Permanent Cover–Any vegetation that grows continuously on the land.

Permanent Grass–Any of the grasses that in nature reseed or renew themselves form year to year, as buffalograss, bluestem, and other prairie grasses.

Permanent Hay–Hay crops, such as alfalfa and perennial grasses, which occupy the land for a long period without intervening crops.

Permanent Meadow–A meadow planned and handled so as to produce a regular crop of grass for several years without plowing and reseeding.

Permanent Parasites–Parasites, such as bloodsucking lice, which spend all life stages on or in the body of the host.

Permanent Pasture–Grazing land in farms occupied by perennial grasses and legumes. It is not a part of a regular rotation of fields and usually remains unplowed for long periods.

Permanent Roots–In plants, the top or lateral roots that make up the bulk of the root system during the life of the plant in contrast to the primary roots that are temporary and function only in the early stage.

Permanent Soil Sterilization–Sterilization of soil by chemical treatment lasting for more than a year to suppress all undesirable vegetation on roadsides and ditch banks.

Permanent Water–A watering place that supplies water at all times throughout the year or grazing season.

Permanent Wilting Percentage–The soil moisture content at which plants remain permanently wilted unless water is added to the soil. Soil water potential at wilting can vary from -5 to -20 bars. Because of the shape of the water potential-water content drying curve, large changes in water potential at higher tensions accompany minor decreases in water content, so permanent water for plant growth is approximately 15 bars.

Permatodes–Roundworms that live in soil or water.

Permeability–The capacity of soil or rock for transmitting a fluid. Degree of permeability depends upon the size and shape of the pores, the size and shape of their interconnections, and the extent of the latter. It is measured by the rate at which a fluid of standard viscosity can move a given distance through a given interval of time. The unit of permeability is the darcy. See Darcy's Law.

Permeable–Refers to the ability of soil to be penetrated by air and water.

Perpelic–Referring to plants that grow on clay soils.

Persian Clover–*Trifolium resupinatum*, family Leguminosae; an annual herb grown as a winter and spring pasture and hay crop. Native to southern Europe and Turkey. also called annual clover.

Persian Insect Powder–A pyrethrum powder. Also called Dalmatian powder.

Persian Walnut–*Juglans regia*, family Juglandaceae; a deciduous tree grown for its edible nut, the English walnut of commerce. Native to Europe and Asia. Also called Circassian walnut, English walnut, European walnut.

Persimmon–*Diospyros virginiana*, family Ebenaceae; a hardy deciduous tree that bears an edible fruit and is also a bee plant. The wood has been used in the manufacture of wooden golf clubs. Also called American persimmon.

Persistent–(1) Of, or pertaining to, the presence of the calyx lobes on a ripened pome fruit, especially pear. (2) Designating fruits, flowers, leaves, etc., that remain on a plant after frost. (3) Designating a disease that is difficult to control.

Persistent Emergent–Emergent water plants that normally remain standing at least until the beginning of the next growing season; e.g., cattails (*Typha* spp.) or bulrushes (*Scirpus* spp.).

Persistent Herbicide–A herbicide that will harm susceptible crops planted in normal rotation after the harvesting of the treated crop.

Persistent Pesticide–A chemical agent used to control pests, which breaks down extremely slowly, remaining toxic to desirable species of wildlife as well as pests, under natural conditions. Some of these include DDT, chlordane, lindane, and dieldrin. Most are now forbidden or restricted in use.

Peruvian Guano–The dry excrement of marine birds obtained from islands off the coast of Peru. See Guano.

Pervious Soil–An open soil; one with large pore spaces that permits a ready flow of water through it. It is generally a sandy or gravelly soil.

Pest–Anything, such as an insect, animal, or plant that causes injury, loss or irritation to a crop, stored goods, an animal, or people.

Pest Control–The use of disinfectant, herbicide, pesticide, insecticide, management or cultural practice that controls pests. See Integrated Pest Management.

Pest-resistant–Designating a plant that naturally developed through plant breeding and is resistant to infestation by specific pests.

Pesticide–A substance used to control insect, plant, or animal pests. Pesticides include insecticides, herbicides, fungicides, nematocides, and rodenticides.

Pesticide Residue–Material that remains on a plant after pesticide application.

Petal–A division of a flower inside the calyx; a unit of the corolla, consisting of petioles, which usually surrounds the pistil and stamens.

Petal Fall Spray–A spray applied for the control of apple scab and codling moth immediately after the fall of the flower petals. Also called calyx spray.

Petalody–The transformation of a calyx bracts or other floral structures into petals; a morphologic change that may result from overnutrition of the plant.

Peteca–A pitting or sinking of the rind of lemons in small discrete areas that may be caused by suboxidation. It is made worse by cold weather, heavy oil sprays, or waxes.

Petiolate–Having petioles (stems of leaves), not sessile.

Petiole–The stem of any leaf.

Petite–(1) A grade for canning small peas of $9/32$ inch or less in diameter. (2) Denoting smallness or shortness.

Petsai–*Brassica pekinensis*, family Cruciferae; a biennial herb grown as a potherb and salad plant for its loose, celerylike head. Native to China. Also called celery cabbage, Chinese cabbage.

Petty Euphorbia–*Euphorbia peplus*, family Euphorbiaceae; an annual herb that is a weed pest of cultivated spaces. Contact with its milky sap may result in a severe dermatitis. Native to Europe. Also called petty spurge.

Peyote–See Mescalbutton Peyote.

pH–A numerical measure of acidity or hydrogen ion activity of a substance such as food or soil. The neutral point is pH 7.0. All pH values below 7.0 are acid and all above 7.0 are alkaline. The negative logarithm of the hydrogen-ion activity. The degree of acidity (or alkalinity) of a soil as determined by means of a glass, quinhydrone, or other suitable electrode or indicator at a specified moisture content or soil-water ratio, and expressed in terms of the pH scale. See Reaction.

Phanerogam–A plant with flowers in which stamens and pistils are developed. See Cryptogam.

Phase–(1) The view that a thing presents to the eye. (2) Any one of the varying aspects or stages through which a disease or process may pass. (3) In colloidal chemistry, the discontinuous portion dispersed in the dispersion medium. (4) In soil taxonomy and soil survey, soil phase terms are surface soil texture, percentage slope, stoniness, saltiness, and erosion. When appropriate these names are added to the soil series name to make a soil mapping unit.

Phenol–C_6H_5OH; carbolic acid; a colorless, crystalline compound, obtained by the distillation of coal tar; widely used as a disinfectant and as an ingredient in antiseptics.

Phenol Coefficient–A figure representing the relative killing power of a disinfectant, as compared with phenol acting on the same organism for the same length of time.

Phenospermy–The production of seeds that cannot reproduce.

Phenotype–The observed character of an individual without reference to its genetic nature. Individuals of the same phenotype look alike but may not breed alike. See Genotype.

Phenylmercuric Acetate–A fungi-killing, chemical compound used in cottonseed treatment, for the mildew-proofing of textiles and as a spray for apple, strawberry, etc. Also used as an herbicide for control of crabgrass.

Pheromone–A substance secreted to the outside of the body by an individual organism that causes a specific reaction by another organism of the same species; e.g., when an earthworm is alarmed it secretes a mucus which is a warning to other earthworms. See Sex Pheromone.

Philippine Downy Mildew–A downy mildew of sugarcane caused by Sclerospora philippinensis, family Peronosporaceae.

Philodendron–*Philodendron scandens*, a plant with heart-shaped leaves grown for house plants.

Phlobatannin–The iron greening tannins that are widely distributed in plants and fruits. Hemlock tannin is a phlobatannin.

Phloem–Inner bark; the principle tissue concerned with the translocation of elaborated food produced in the leaves, or other areas, downward in the branches, stem, and roots.

Phloem Necrosis–(1) A viral disease highly destructive to elm trees, characterized by the phloem tissue being killed, the leaves drooping, curling upward at the margins, turning yelllow, becoming stiff and brittle and falling off. (2) An internal disease of the potato characterized by a network of brown strands extending throughout the flesh of the tuber.

Phloem Parenchyma Necrosis–A necrosis of the Irish potato characterized by spotlike rather than blotchy necrotic areas in the tubers and showing no foliage symptoms. Also called pseudo-net necrosis.

Phoma Root Rot–A fungal disease of celery and other truck crops, such as carrots and parsnips, caused by *Phoma apicola*, family Sphaerioidaceae.

Phony Peach–A viral disease of peaches that causes the infected trees to leaf out earlier and ripen fruit earlier than normal. The fruits are reduced in size and number but are more highly colored.

Phosphate–A term commonly used to indicate a fertilizer that supplies phosphorus. A major element in fertilizers.

Phosphate Rock–Any rock or mineral that contains phosphorus in sufficient quantity to make it useful directly, or indirectly through manufacture, as fertilizers. Most phosphate rock is tricalcium phosphate, $Ca_3(PO_4)_2$, with various impurities. See Apatite.

Phosphates–Salts of phosphoric acid, such as those made by combining phosphoric acid with sodium, potassium, and calcium. The phosphate most commonly used in fertilizers is made by treating phosphate rock with sulfuric acid.

Phosphatide–A complex organic compound that contains choline phosphoric acid or amino-ethyl phosphoric acid. It occurs in milk, is easily destroyed by heat and oxidation, and is an important factor in the auto-oxidation of milk fat and the development of taints (off taste) in milk.

Phosphoric Acid–H_3PO_4; orthophosphoric acid. To be made available to plants, it is converted into soluble acid salts by processing with sulfuric acid. The amounts of available phosphorus in fertilizers are usually expressed in terms of percentage of weight of available P_2O_5.

Phosphoric Acid Silage–Legume, small grain, and grass silage to which a small amount of commercial phosphoric acid has been added. There is little loss of green color or nutrients, and the available phosphorus in the silage is increased.

Phosphorite–A variety of apatite which is fibrous and concretionary. See Apatite.

Phosphorized Limestone–A phosphatic limestone used as a mineral supplement in stock feeds. Fluorine is usually present in such phosphatic rocks.

Phosphorus–P; a chemical element found in soils in various mineral forms, but only small amounts are readily available to plants at any one time. It stimulates early growth and root development, and hastens grain maturity. In the water environment it enhances the growth of aquatic plants.

Phosphorus Deficiency–In soils, insufficient amounts of phosphorus in available form for optimal growth of plants, or in insufficient amounts to produce an economic or desired yield of a crop.

Photobiotic–Living only in the presence of light, such as green plants. See Photosynthesis.

Photochemical Process–The chemical changes brought about by the radian energy of the sun acting upon various polluting substances. The products are known as photochemical smog.

Photodecomposition–The breakdown of a substance, especially a chemical compound, into simpler compounds by the action of radiant energy.

Photolysis–(1) Chemical changes brought about by the absorption of light. (2) In botany, the grouping of the chloroplasts in relation to the amount of light the plant receives.

Photoperiod–Length of the light period in a day. See Day-neutral Plant, Long-day Plant, Short-day Plant.

Photoperiodic Adaptation–The adjustment of plants in their native or artificial habitat to a definite length of daily exposure to light. See Photoperiodism.

Photoperiodicity–The response of plants or animals to the length of daily exposure to light.

Photoperiodism–The reaction of plants to periods of daily exposure to light which is generally expressed in formation of blossoms, tubers, fleshy roots, runners, etc. See Day-neutral Plant, Long-day Plant, Short-day Plant.

Photosensitization–A condition in which external parts of the body, usually the skin, become sensitive to light. This condition may occur in cattle after eating certain substances that have the ability to sensitize light-colored areas of the skin to sunlight. It can result in swellings and fluid-containing vesicles on unpigmented skins. The exudate from the vesicles forms a crust on the skin which may permanently harm the outer part of the affected skin.

Photosynthesis–Process by which green plants, using chlorophyll and the energy of sunlight, produce carbohydrates from water and carbon dioxide, and release oxygen.

Photosynthetic Rate–The time required for a plant to manufacture a given quantity of foods such as proteins, carbohydrates, and fats.

Photothermal Induction–The inducing of a plant process, particularly reproduction, by a combination of photoperiods and temperature. Also called thermophotoperiodic induction.

Phototropism–The response of a plant to the stimulus of sunlight in which the plant or its parts seem to turn to face the light. The parts of the plant receiving the direct rays grow more slowly and the plant appears to turn.

Phreatophyte–Any plant that sends its roots deep into the soil to the level of groundwater or to the capillary fringe above the water table. In semiarid and arid regions, these plants may be objectionable because they rob more shallow-rooted, less hardy but more desirable plants of water. However, alfalfa, a deep-rooted plant, under some conditions may be considered a phreatophyte.

Phycomycetes–Algae and fungi bearing nonsepate, branching filaments not organized into compact bodies of definite form. Containing several damping-off fungi.

Phylliform Soil Structure–A thin, leaflike layer of soil that is less distinct and thinner than platy. Where this condition is present in the C horizon, as in soils developed from thin-bedded sediments, the term *laminated* is used.

Phyllody–The transformation of floral organs into leaflike structures.

Phylogeny–The evolutionary history of a species, genus, or type of plant.

Phylum–The highest grouping in the taxonomy of the plant and animal kingdoms, based on assumed common ancestry.

Phymatotrichum Root Rot of Cotton–A fungal disease of cotton caused by ***Phymatotrichum omnivorum***, family Moniliaceae; characterized by the leaves turning yellow or bronze-colored, wilting, and becoming dry. The bark and cambium of the roots turn brown and root rot appears.

Physalospora Ear Rot–A fungal ear rot of corn caused by ***Physalospora zeicola***, family Mycosphaerellaceae; characterized by stalk lesions and the slate-gray color of the rotted ears. Endemic in the Gulf States of the United States.

Physical and Mechanical Pest Controls–Direct or indirect (non-chemical) measures to destroy pests outright or to make the environment unsuitable for their entry, dispersal, survival, or reproduction; e.g., steam sterilization to destroy disease organisms, flaming for the control of weeds, cold storage to control pests, metal or other material barriers to prevent pest entry. See Integrated Pest Management.

Physical Poison–Any poisonous material that exerts a physical rather than a biochemical effect, as heavy mineral oils and inert dusts.

Physical Properties–(Of soil) Different characteristics such as color, texture, consistency, depth, etc.

Physical Weathering–In soil formation, the natural disintegration of rocks and minerals by alternate heating and cooling, expansion by water freezing in crevices, corrosion by moving ice and water, splitting of rocks by roots of trees, and other physical factors, as contrasted with chemical weathering.

Physiogenic Disease–Any disease caused by unfavorable environmental factors.

Physiognomy–(1) The general, outward appearance of a plant community as determined by the life of the dominant species. (2) The physical appearance of a landscape created by a combination of land forms or relief features. (3) The external, facial features of a human or an animal.

Physiographic Location–The location of a stand of trees with respect to slope aspect, slope position, and slope inclination, and the slope conformation relative to the overall terrain.

Physiologic Races–Pathogens of the same species and variety that are usually structurally indistinguishable but which differ in their physiologic behavior, particularly in their ability to parasitize varieties of a particular host.

Physiological–(1) Referring to or concerning the science of physiology or the branch of biology that deals with life processes and functions. (2) Referring to the functions of the organs of plants and animals.

Physiological Drought–Inability of a plant to obtain water from soil although the water may be present in it, as when a soil is frozen, or by reason of weak osmotic force of plant roots.

Physiological Maturity–The period of advanced age in the cycle of a tree or stand of trees when resistance to adverse influences is so low that death of a tree or net losses in volume of salable wood are likely to occur within a cutting cycle.

Physiology–The science that deals with the function of a plant or animal's body and its organs, systems, tissues, and cells.

Phytobezoar–Undigestible plant fibers ingested by an animal that have gathered in the rumen or stomach in the form of balls. See Bezoar, Hair Ball, Stomach Ball.

Phytoedaphon–The microflora of the soil.

Phytogenesis–History of plant evolution; the origin of vegetation.

Phytogenic Soil–See Phytomorphic Soil.

Phytohormone–See Growth Regulator.

Phytome—The vegetable substance composing a plant.

Phytometer—(1) A plant used to measure the physiological activities of the habitat. (2) A device used for measuring water transpiration of a plant.

Phytomorphic Soil—In pedology, a group of soils that includes those developed under the predominant influence of vegetation.

Phytopathogenic—Capable of causing plant disease.

Phytopathology—The science of plant disease.

Phytophagous—Feeding upon plants.

Phytophthora—A genus of fungi of the family Pythiaceae that includes a number of plant-parasitic species many of which can also live as saprophytes. At the exterior of the infected host plants the fungus sends out a downy or felty mass of hyphae, which is the downy mildew phase of the ***Phytophthora***. Within the host plant the advancing hyphae of the fungus cause the death and decay of the invaded tissues, the rot phase of the infection. If infections occur on young stem tissues of leaves, lesions of varying sizes are produced and the blight stage results. The stem may be rotted off, and if this occurs near or just below the soil surface, the condition is called damping-off. If the conidia infect a fruit this may result in fruit rot. Storage regions underground, such as tubers, bulbs, corns, roots, may be infected by asexual spores carried into the depths of the soil by rain, as in the tuber-rot phase of late blight of potatoes. Fourteen species of genus ***Phytophthora*** are known to cause disease of about fifty economically important plants.

Phytoplankton—Plant microorganisms, such as certain algae, living unattached in the water. See Plankton, Zooplankton.

Phytotomy—Plant anatomy; dissection of vegetation.

Phytotoxin—A poisonous substance of plant origin; e.g., abrin from rosary pea, ***Abrus precatorius***; ricin from castor bean, ***Ricinus communis***; crotin from croton, ***Codiaeum variegatum***; and robin from black locust, ***Robinia pseudoacacia***. See Poison, Toxin.

Pick—(1) The total amount of a crop harvested, or the yield of an individual tree, as the pick of oranges. (2) Small, irregular openings within the body of a cheese. (3) See Pickax. (4) To pull or pluck ripe fruit, as berries, apples, cotton. (5) To pluck the feathers from a fowl in dressing it for the market. (6) To nibble at food.

Picked Cotton—Cotton lint plucked from the boll, as contrasted with that obtained from bolls snapped off the stalk.

Picking Season—The season of the year when a crop is harvested, especially such crops as cucumbers, string beans, cotton, etc., where the picking period extends over a number of days or weeks. The picking season for cotton may extend from August to January in some sections of the southern United States.

Pickle—(1) A solution of salt in water used for curing or preserving many different products, such as meat, vegetables, etc. There are many different types of pickling solutions with many different seasonings or condiments added for the particular product being pickled or preserved. (2) The cucumber, any vegetable, fruit, or meat that has been processed through a brine solution. (3) To place various products, such as cucumbers, fruits, or meat in a brine solution for preservation; one of the oldest methods of food preservation.

Pickle Brine—A salt solution, generally about 10 percent salt, in which cucumbers, fruits, vegetables, etc., undergo a preliminary fermentation.

Pie Fruit—The trimmed, overripe, and otherwise low-quality fruit canned especially for use in pies. It is wholesome but not choice in quality for table use. Also known as water grade.

Pie Plant—See Rhubarb.

Piedmont—Meaning literally at the foot of the mountain.

Piedmont Plateau Region—A physiographic division of the eastern United States; the eastward-sloping, dissected, upland plain lying between the Appalachian Mountains and the Atlantic Coastal Plain. One of the older agricultural regions of the eastern United States, it is notable in its central and southern sections for tobacco, cotton, and fruit production, red soils, and for severe soil erosion.

Pigeon Horntail—***Tremex columba***, family Siricidae; a wasplike insect whose larvae bear into the trunks of diseased and dying elm, sugar maple, and other trees. Also called pigeon tremen.

Pigeon Plum—(1) Any tree of the genus ***Coccolobin***, especially *C. laurifolia*, family Polygonaceae, bearing edible, berrylike fruit. Native to the tropical and subtropical regions of North America. (2) ***Chrysobalanus ellipticus***, family Rosaceae; a small tropical tree grown for its edible, drupaceous fruit. Native to Africa.

Pigment—Any of the natural coloring materials in the cells and tissues of plants and animals. In fruit and vegetables, the green pigment is chlorophyll; orange to red pigments are carotenoids; red to blue colors are anthocyanins; light-yellow pigments are flavoners and flavonols. In meat, the chief pigment producing the pink or red color is myoglobin.

Pigmy Forest—A small tree and shrub type of rangeland that borders other shrublands and grasslands in the western states. Also known as the pinyon-juniper vegetation type.

Pignolias—Pine nuts; seeds of pine cones from the stone pine.

Pigweed—See Red Root.

Pillbug—***Armadillidium vulgare***; not an insect but a creature closely related to the crayfish that lives in damp places, as under rotting boards. Mainly a greenhouse pest, it may eat roots of seedlings and do other damage. See Sowbug.

Pimiento Pepper—A type of sweetbell red pepper whose varieties are thick-walled and more-or-less pointed at the apex. It is raised primarily for canning. Also called pimento.

Pin Cherry—***Prunus pennsylvanica***, family Rosaceae; a deciduous tree or shrub that produces small, sour, bright red cherries on long slender stalks. Found in the United States, from the Atlantic to the Rocky Mountains and from Canada to Virginia and Tennessee, it is especially abundant on burned-over land. The foliage is sometimes lethal when browsed by livestock. Native to North America. Also called fire cherry.

Pin Oak–*Quercus palustris*, family Fagaceae; a hardy deciduous tree found from Massachusetts west to Michigan and Illinois and south to Arkansas. Easily transplanted, it is grown as a shade tree on city streets and in landscaping for its autumnal foliage. Its lumber is used for ties, fuel, flooring, planing-mill products, furniture, boats, etc. Native to North America.

Pinching–Removal of the terminal shoots off plants by pinching off, as with the fingers, to produce stocky, bushy plants, to produce long-stemmed flowers for exhibition or market, and to prevent flower production in early stages of growth of a plant.

Pine–Any cone-bearing, evergreen tree with needle-shaped leaves, of the genus ***Pinus***, family Pinaceae. Species number about 50 in the United States and constitute valuable timber trees, a source of turpentine and rosin, ornamentals, and Christmas trees.

Pine Barrens–Land, embracing various soils, but generally sandy, and low in fertility, on which pines were the dominant growth.

Pine Beetle–Any of the tree-killing bark beetles of the genus ***Dendroctonus***, family Scolytidae, which infests pines.

Pine Blister Rust–A very serious rust disease of pines, especially the white pine, caused by the fungus ***Cronartium ribicola***, family Melampsoraceae. Cultivated and wild species of the genus ***Ribes***, currants and gooseberries are alternate host plants. Pines are infected only from spore forms produced on ***Ribes*** spp., which in turn are infected only from spores produced on ***Pinus*** spp.

Pine Bluegrass–*Poa scrabrella*, family Gramineae; a highly palatable, bunchgrass of western ranges (United States), which grows at low to medium altitudes from Montana and southern Washington to California.

Pine Disease–See Nutritional Anemia.

Pine Dropseed–*Blepharoneuron tricholepis*, family Gramineae; a range grass of good forage rating found from Colorado and Utah to Arizona and western Texas. Native to western North America. Also called Beardless pinegrass.

Pine Engraver–*Ips pini*, family Scolytidae; a beetle that infests all species of pine, burrowing small circular holes into branches and trunks of trees. Native to the northern United States.

Pine Littleleaf–A fungal disease of shortleaf pines in the southern United States caused by ***Phytophthora cinnamomi***, family Mycosphaerellaceae; characterized by short, yellow needles at the ends of branches and death of the fine roots of the tree.

Pine Meadows–A vegetational land feature of the southern Mississippi along the Gulf Coast (United States), characterized by flat topography, wetness, a scrubby growth of longleaf pine, cypress, and a ground cover of sedgy vegetation.

Pine Straw–The fallen dead needles of pine trees. The loose, open mat at the base of trees may represent the fall of several years. It gradually becomes broken and decomposed at the bottom.

Pineapple–*Ananas comosus*, family Bromeliaceae; a plant grown in frost-free areas for its edible fruit, the pineapple of commerce. It is widely consumed fresh and canned as a dessert fruit. Grown to a very limited extent in Florida. Most of that sold in the United States is produced in Hawaii and the West Indies. Native to tropical America.

Pineapple Mealybug–*Pseudococcus brevipes*, family Pseudococcidae; a tropical insect that sometimes infests pineapple, banana, and sugarcane, causing a condition known as pineapple wilt.

Pinebarren Deathcamas–*Zigadenus leimanthoides*, family Liliaceae; a bulbous herb that grows in bogs and on wet pine lands and is poisonous to livestock. Native to coastal areas of the southeastern United States. Also called crowpoison.

Pinegrass–*Calamagrostis rubescens*, family Gramineae; a tall, perennial, range bunchgrass that grows in the ponderosa and lodgepole pine belts of the western United States. It is rated as only fair in grazing value. Native to western North America.

Pineland Threeawn–*Aristida stricta*, family Gramineae; a perennial, erect grass characteristic of the open, sandy pine land of southern Georgia and northern Florida, United States. It is extensively grazed by cattle, but only the young growth is palatable. Also called wiregrass.

Pinery–A forest area that supports a heavy growth of pines. (A localism, especially in Wisconsin and Michigan, United States.) See Pinetum.

Pinetum–A small groove or planting of pine trees kept for its aesthetic and educational value rather than as a forest.

Piney Woods–Native forest that consists predominantly of open stands of pine, especially longleaf, in contrast to deciduous and swamp forests. (Colloquial especially in the Gulf Coast region of the United States).

Pingue–*Actinea richardsoni*, family Compositae; an herb that contains a toxic saponin. It also contains a low percentage of rubber. Native to California (United States).

Pink Boll Rot–See Cotton Anthracnose.

Pink Boll Worm–*Pectinophora gossypiella*, family Gelechiidae; one of the most injurious insect pests of cotton. Eggs deposited on various parts of the plant hatch into small, pink caterpillars, which penetrate into the boll or square and later enter the seed.

Pink Color Salt Method–A method of testing seed corn for viability in which kernels of corn are soaked in lukewarm water for about 12 hours, then split the long way and perpendicularly to the germ side. These split kernels are then placed in a weak, colorless solution of a salt, tetraxolium chloride. The action of this salt on living germs in the soaked, split kernels turns the exposed parts pink in 2 hours or less.

Pink Fruit–A viral disease of sour cherry which is similar to and possibly a strain of western X-disease. It has caused serious losses in western Washington, United States.

Pink Locust–Pink-flowering bristly locust; ***Robinia hispida***, family Leguminosae; an excellent erosion-control shrub because of its profuse root system.

Pink-eyed Potato–An Irish potato variety that has pink eyes.

Pink-fleshed Grapefruit–Any variety or strain of the grapefruit (*Citrus paradisi*), such as the Foster and Thompson, which has pink flesh. Extensively grown in Texas and Florida, United States.

Pinnate–Constructed somewhat like a feather, with the parts (e.g., veins, lobes, branches) arranged along both sides of an axis, as in pinnate venation. A pinnate leaf is compound, with the leaflets arranged on both sides of the rachis.

Pinnate Tansymustard–*Descuraninia pinnata*, family Cruciferae; an annual herb which grows on the arid and semiarid range lands of western United States. Eaten in the blossoming stage by stock, it is a cause of poisoning referred to as paralyzed tongue. Native to North America.

Pinocytosis–The engulfing or absorption of fluids by cells.

Pinon–Mexican pinyon pine that bears edible seeds.

Pinto Bean–A variety of the kidney bean characterized by mottled seeds. It is a medium-sized bean grown mainly in Colorado, Idaho, Nebraska, and New Mexico, United States.

Pioneer Plants–Herbaceous annual and seedling perennial plants that colonize bare areas as a first stage in secondary succession.

Pip–(1) A horny, dried condition of the tip of the tongue that develops in poultry in cases of infectious coryza when mouth breathing is long continued. (2) The raised crown or individual rootstock of a plant. (3) The specialized underground bud of lily-of-the-valley.

Piperine–A colorless, crystalline alkaloid found in black and white pepper.

Pippin–A famous apple of which there are a number of varieties, such as the Newtown, Albermarle, Missouri.

Pistachios–Kernels of the nut of the turpentine tree, *Pistacia vera*, used for flavoring and garnishing galantines, sweets, etc.

Pistil–The female element of a flower; composed of stigma, style, and ovary.

Pistillate–Designating a flower that has a pistil or pistils but lacks stamens; an imperfect flower. See Staminate.

Pistillody–The change of stamens into carpels.

Pit–(1) The endocarp of a drupe; the seed-stone of a fruit, as the pit of a peach or prune. (2) An excavation in soil in which vegetables, such as potatoes, carrots, and parsnips are placed and covered over for storage (seldom practiced now in the United States). (3) A place on the floor of an exchange in which traders stand when dealing in wheat, cotton, and other commodities. (4) In botany, a small hollow or depression in a cell wall. Various types are recognized in wood anatomy as blind, bordered, primordial, simple.

Pitahaya–*Lemaireocereus thurberi*, family Cactaceae; a tall, tree-like cactus of the arid region of Arizona, California (United States), and Mexico, which yields an edible fruit. Native to North America. Also called sweet pitahaya, aya pitaya.

Pitch–(1) The resin that occurs in the wood of conifers, as the pitch from pines. (2) A heavy, dark, viscous or solid, fusible material obtained by distillation of the tar derived from coal, wood, rosin, and petroleum oils. It consists of many organic compounds, chiefly hydrocarbons, differing according to origin. (3) The jumping action of a horse in its attempt to unseat its rider.

Pitch Pine–(1) *Pinus rigida*, family Pinaceae; a hardy evergreen, rugged but beautiful tree sometimes grown as an ornamental. Native to North America. (2) Any pine tree with an abundance of pitch in the wood.

Pitch Pocket–A well-defined, lens-shaped opening between or within annual growth layers of conifer wood which usually contains solid or liquid pitch.

Pitch Streak–A well-defined accumulation of pitch (resin) that is in the form of a regular streak in the wood of certain conifers.

Pith–The spongy center of exogenous plants. In the corn plant the pith fills the center cavity of the stalk and serves as a storehouse for moisture and food.

Pitting–(1) The development of little cavities in metal, especially in aluminum cans, caused by the development of lactic acid by bacteria in milk residues. Pitting can also be caused from corrosion by brine. (2) In fruit processing, removal of seeds or stones, as from preaches and cherries.

Placenta–(1) The membranous tissue that envelops the growing fetus in the uterus and establishes communication between the mother and the fetus by the umbilical cord. It is discharged from the uterus at the time of the birth of the young or shortly thereafter. Also called afterbirth, calf bed. (2) In plants, that portion of the ovary on which ovules are borne.

Plain-leaved Parsley–A variety of curly garden parsley that has a plain leaf in contrast to the more popular and common curled-leaf variety.

Plains–Flat, or nearly level, extensive areas of land. Some major physiographic divisions named plains may have broken, hilly topography, and may exhibit strong relief locally. They may be classified into various kinds, as alluvial, coastal, grassland, pine, desert, high, low, etc.

Plane Survey–A survey in which the curvature of the earth is disregarded, as in ordinary field and topographic surveying.

Plankton–The small (usually less than 2 millimeters long) floating or drifting life forms in water bodies, plankton includes both plants (phytoplankton) and animals (zooplankton) that are carried passively in the water currents. Those that can swim do so to change or adjust their depth in the water, not to move from place to place. Plankton is one of the three main divisions of aquatic life, the others being nekton (the animals that swim actively and may move long distances for feeding or breeding) and the benthos (organisms that crawl about on the bottom). See Phytoplankton, Zooplankton.

Plant–(1) An organism distinguished from the animals in that it takes nutrients entirely in liquid solution, rather than in solid form. (2) To set in the soil young plants, as seedling trees in forest plantings, cabbage plants in gardening. (3) To sow seed. (4) To set out roots, etc.

Plant Analysis–Analytical procedures to determine the concentration of nutrients in plants.

Plant Association–A unit of natural vegetation essentially uniform in habitat, aspect, general appearance, ecological structure, and floristic composition; a plant community. An area may be dominated by some one of several species with minor species associated, such as a sugar maple-beech association or a bunchgrass association.

Plant Band–Short strip of heavy paper that may be folded or rolled into a potlike unit to replace claypots for growing young plants, which are later transplanted direct to the field without the removal of the plant band. See Peat Pot.

Plant Bed–A specially prepared bed in which seeds are planted, the young plants being later transplanted into field areas; e.g., tobacco and cabbage plant beds.

Plant Breeding–The improvement of plants by selection, inbreeding, hybridization, and other scientific procedures.

Plant Certification–Certification by a public or private agency that a given plant conforms to certain standards as to varietal characteristics, freedom from disease, etc.

Plant Community–An assemblage of plants living together under the same environmental conditions.

Plant Cover–(1) The vegetal mantle covering the soil. (2) All plants found on a particular range regardless of their availability, palatability, toxicity, or other characteristics.

Plant Disease–A condition that affects physiologic activities of a plant so as to check normal development, lead to abnormal formations or to premature death of all or a part of a plant.

Plant Doctor–A plant pathologist; a person skilled in the treatment of plant diseases.

Plant Ecology–The science that deals with plants in relation to their environment. See Ecology.

Plant Food–Organic compounds manufactured within the growing plant, such as sugars, and used by the plant of growth and reproduction. Because of its sales appeal, the term *plant food* is used incorrectly in the fertilizer trade as a synonym for plant nutrient, essential element, or fertilizer materials such as ammonium salts, nitrate salts, superphosphates, potassium salts, or any fertilizer offered for sale.

Plant Hoppers–Any of a number of leaping insects with piercing-sucking mouths that belong to the order Homoptera.

Plant Hybrid–A plant which results from cross pollination of two genetically different plants. See Hybrid.

Plant Immunity–Resistance of a plant to a pathogen so that a disease does not cause appreciable damage.

Plant Indicator–A plant that indicates soil characteristics; e.g., black walnut (*Juglans nigra*): high-lime, fine-textured, well-drained soil; longleaf pine (*Pinus palustris*): acid, coarse-textured, well-drained; basswood (*Tilia americana*): moderately acid, medium-textured, well-drained; blueberries (*Vaccinium* spp.): acid, coarse-textured, well-drained; cypress (*Taxodium* spp.): mildly acid, fine-textured, water-saturated. Trees that indicate areas frequently burned over are: jack pine, *Pinus banksiana*; lodgepole pine, *Pinus contorta*; pin cherry, *Prunus pennsylvanica*; trembling aspen, *Populus tremuloides*.

Plant Kingdom–One of the two great categories of living things, the other being the animal kingdom.

Plant Life–Plant organisms of any kind from lowest to highest forms, as distinguished from animal organisms.

Plant Material Centers–In the United States there are twenty-five plant material centers managed by the United States Department of Agriculture Soil Conservation Service. They receive superior grasses, legumes, forbs, shrubs, and trees. These plant materials are reproduced for distribution.

Plant Nutrient–Any element taken in by a plant from the air and soil. These are essential to growth and are used by the plant in the elaboration of its foods and tissues.

Plant Nutrition–The processes by which plants use essential elements from air, water, soil, and fertilizer, and chlorophyll, and energy from the sun to grow and reproduce. Most crop plants are composed of the following sixteen essential elements in roughly the proportions indicated: oxygen 45 percent; carbon 44 percent; hydrogen 6 percent; nitrogen 2 percent; phosphorus 0.5 percent; potassium 1 percent; calcium 0.6 percent; sulfur 0.4 percent; magnesium 0.3 percent; boron 0.005 percent; chlorine 0.015 percent; copper 0.001 percent; iron 0.02 percent; manganese 0.05 percent; molybdenum 0.0001 percent; zinc 0.010 percent.

Plant Patent–A patent granted by the United States Patent Office to originators of varieties, strains, or some variation from existing varieties of asexually reproduced plants.

Plant Pathology–That branch of science that deals with the diseases of plants.

Plant Physiology–The science that deals with the response of a plant to its environment including moisture, temperature, light, and nutrients.

Plant Pigments–The colors that are in plants. Some of the pigments, chlorophylls (green), carotenes (yellow/orange), and xanthophylls (red/brown), are essential in the photosynthetic process. Other plant pigments, anthocynins (red in acid solutions and blue in alkaline solutions), anthoxanthins (yellow), and betacyanins (red), attract insects that cross pollinate flowers and are esthetically attractive. Fall colors in temperate regions are the result of these pigments.

Plant Propagation–The practice of producing new plants that may be asexual means, such as division, graftage, cuttage, layerage, etc., or by the use of seeds.

Plant Quarantine–The isolation of or restriction placed on the movement of plant materials to prevent either the plant or plant-carried insects and diseases from contaminating an agricultural area free from them. The Plant Inspection Branch of the Agricultural Research Service in the United States Department of Agriculture processes all plant materials from foreign sources.

Plant Regeneration–The development of volunteer vegetation from seed or by other natural reproductive processes from plants existing nearby.

Plant Regulator (Growth Regulator)–A chemical that increases, decreases, or changes the normal growth or reproduction of a plant.

Plant Residue–(1) Plant material, such as leaves, roots, straw, corn stover, grass, and weeds left after harvesting; any plant material remaining after any harvesting or other process. (2) Wood materials from manufacturing plants not utilized for some product; e.g., slabs, edgings, trimmings, miscuts, sawdust, shavings, veneer cores and clippings, and pulp screening.

Plant Retrogression–The process of vegetational deterioration whereby the same area becomes successively occupied by different plant communities of lower ecological order. See Plant Succession.

Plant Succession–The process of vegetational development whereby an area becomes successively occupied by different plant communities of higher ecological order. See Plant Retrogression.

Plant Variety Protection Act–This act, passed in 1970, offers legal protection to developers of new varieties of plants that reproduce sexually; that is, through seed. Developers of plants that reproduce asexually have received protection by the United States Patent Office since 1930. The law states that protection will be extended to a "novel variety" if it has these three qualifications: (a) distinctiveness: the variety must differ from all known prior varieties by one or more identifiable morphological, physiological, or other characteristics; (b) uniformity: if any variations exist in the safety, they must be describable, predictable, and commercially acceptable; (c) stability: when sexually produced, the variety must remain unchanged in its essential and distinctive characteristics to a degree expected of similarly developed varieties.

Plant-exploration Service–A very important activity of the Agricultural Research Service, United States Department of Agriculture. Plant explorers are sent all over the world to collect plants, seeds, or plant materials of desired species for breeding and development of new crops for agricultural and other uses.

Plant-tissue Test–A test, usually rapid and not precisely quantitative, using various chemical reagents that are made upon tissue and sap of growing plants to determine the adequacy of essential plant nutrients.

Plantation–(1) A large-scale agricultural unit, especially one devoted to the production of cotton or sugarcane. In pre-Civil War times in the United States, a large manorial estate on which cotton, tobacco, and other crops were produced with slave labor. (2) An artificially reforested area established by planting or by direct seeding.

Planter–(1) A mechanical device used for the rapid, efficient, and uniform planting of seeds. Planters are of many different kinds, from the simplest hand planter to the large highly mechanized, multirow, power-driven machines. (2) A farmer, especially a cotton farmer. (In its earliest use, it generally implied wealth, and denoted the owner of a large cotton or tobacco plantation.) (3) A person who plants. (4) A container, usually rectangular, containing soil or vermiculite, which is used for growing foliage plants indoors. It may be made of plastic, pottery, wood, metal, or other material and may be movable or a permanent part of the building.

Planting–(1) An area seeded to a crop. (2) The sum of plants in a seeded area; e.g., a strawberry planting. See Plant.

Planting Bar–A hand tool used in making a slithole in the soil in which tree seedlings are planted.

Planting Easement–An easement for reshaping roadside areas, power lines, etc., and establishing, maintaining, and controlling plant growth thereon.

Planting Season–The period of the year when planting and/or transplanting is considered advisable from the standpoint of successful establishment.

Planting Stock Age–The age of the planting stock from seed with distinction between periods spent in seedling and transplant beds, as 2 to 0 for two year seedlings and 2 to 2 for stock two years in the seedbed and two years in the transplant bed.

Plasmid–A circular piece of DNA found outside the chromosome in bacteria. Plasmids are the principal tool for inserting new genetic information into microorganisms or plants.

Plasmolysis–A separation of the protoplasm from the cell wall by osmotic action when the cell is in contact with the liquids of stronger salt content. Wilting and in extreme cases death of a plant results.

Plastic Limit–The moisture content at which a soil changes from a semisolid to a plastic state. A very useful test to determine trafficability.

Plastic Mulch–Thin polyethylene film, which may be clear or black, that is used as a mulch, especially for vegetables. Benefits include moisture retention, increased soil temperature, and, with black plastic, complete weed control.

Plasticity Index–The numerical difference between the liquid limit and the plastic limit; the range of moisture content within which the soil remains plastic.

Plastid–A body in a plant cell that contains photosynthetic pigments.

Plate Budding–A method employed in the propagation of plants in which a rectangular incision of the bark is made by two vertical cuts and one cut across the top, and a rectangle of bark containing the bud to be fitted is cut to fit the face exposed on the stock. The flap of bark on the stock is then brought up over the bud and tied. See Patch Budding, Shield Grafting.

Platelike–Describing a soil structure that resembles a stack of plates, usually resulting from compaction caused by heavy machinery.

Platy–Designating laminated soil aggregates predominantly developed along the horizontal axes. See Structure, Soil.

Pleach–A method of pruning trees to produce a wall-like hedge.

Pleiotropy–A situation where one gene affects more than one trait.

Plinthite–The sesquioxide-rich, humus-poor, highly weathered mixture of clay with quartz and other diluents that commonly appears as red mottles, usually in platy, polygonal, or reticulate patterns. Plinthite changes irreversibly to an ironstone hardpan or to irregular aggregates on exposure to repeated wetting and drying, especially if it is exposed also to heat from the sun. In a moist soil, plinthite can be cut with a spade, whereas ironstone cannot be cut but can be broken or shattered with a spade. Plinthite is one form of the material that has been called laterite. See Laterite.

Plot–In agricultural research, a small parcel of land, usually rectangular and of a definite size, used in comparing yields of crop varieties, testing different applications of fertilizers, comparing methods of tillage, etc.

Plow–(1) The whole implement, of various types, used to cut, break, or turn a soil layer in preparation for planting, seeding, or other agricultural practices. More specifically, the removable metal point, share and moldboard or disk, attached to a plow stock or frame. (2) Any such implement not used primarily for agricultural purposes, such as a snow plow. (3) To make a furrow or to turn over a layer of soil. (4) To cultivate; e.g., plow corn. Also spelled plough by the British.

Plow Down–To bury material lying on the surface of a field by plowing; e.g., to plow in a surface application of fertilizer, or to plow under a cover crop.

Plow Pan–A compacted layer formed in the soil immediately below plow depth. It is attributed to the sliding action and weight pressure of the plow bottom. Also called plow sole.

Plow Planting–The plowing and planting of land in a single operation. See Conservation Tillage.

Plow Under–To bury and incorporate in the soil a cover of living or dead plants by the use of the plow. See Plow Down.

Plow Up–(1) To cut a furrow slice and partly or completely to invert it. (2) To lift out of the ground unwanted plants more with the purpose of destroying than preserving. (3) To bring the surface vegetation that was turned under during a previous plowing.

Plowable–Designating land suitable, in its present condition, for plowing.

Plowable Pasture–Land ordinarily kept as pasture but that may be plowed and utilized for other crops.

Plug–(1) The mass removed by a trier or other special penetrating implement in sampling or testing an agricultural product, as a plug from a cheese, a bale of cotton, or from a watermelon. (2) An old, worn-out horse. (3) A horse with a poor conformation. (4) To repair a leak, as a dam or earth fill. (5) A block of rooted grass that is planted for the purpose of establishing a covering of grass, such as in a law.

Plum–The tree and fruit of certain species of the genus *Prunus*, family Rosaceae. Varieties of domesticated species are widely grown for their edible fruits, fresh and cured (as prunes); some species are grown for ornament; and some native wild species, such as the beach plum, *Prunus maritima*, which grows in dune areas of the East Coast of United States, especially Cape Cod, where the fruit is very popular for use in preserves. Four main groups are under cultivation: Domesticas or European plums, Damsons, Japanese, and varieties of native American species.

Plum Blotch–A fungal disease of plum in which tiny, gray or brown, angular spots appear on the leaves. The fruit is blotched with brown or gray spots. Caused by *Phyllosticta congesta*.

Plum Curculio–*Conotrachelus nenuphar*, family Curculionidae; a snout beetle that infests plums, apples, peaches, cherries, and apricots. The beetles first feed on leaves and petals and later feed upon, puncture, and lay eggs in the fruits, making them grubby, misshapen, and unmarketable. Found east of the Rocky Mountains, United States. Also called plum weevil.

Plump Grain–A quality designation for small grains; wheat, barley, and oats, which denotes fullness of grain, better quality, and greater weight per bushel, as contrasted with a shriveled grain.

Plumule–In a germinating seed plant, the primary bud that develops into the primary stem. In a dicotyledon, such as beans, it is located between the cotyledons.

Plunge–To bury a plant-containing pot in soil. It is a common practice in floriculture, as by this operation the plant can have some of the advantages of growing free in soil and can be lifted without repotting.

Pluvial–Due to the action of rain; pertaining to deposits by rainwater or ephemeral streams.

Pneumatophore–Stilt roots or root "knees" developed by some woody plants growing in water to facilitate respiration; e.g., baldcypress (*Taxodium distichum*), tupelo (*Nyssa sylvatica*), and mangroves (principal genera: *Rhizophora, Avicennia, Conocarpus*, and *Laguncularia*).

Pod–(1) Technically, a dry, many-seeded fruit that splits open, such as a pea pod or bean pod; a legume. (2) To form pods. (3) A flock of animals, birds, etc.

Pod Corn–*Zea mays* var. *tunicata*, family Gramineae; a curious variety of corn in which each kernel is enclosed in a pod or husk whose ear is also enclosed in husks. It is sometimes grown as a curiosity, and some mutants, which have striped and purple leaves, are cultivated as ornamental plants.

Point of Runoff–A spray application that is just heavy enough to produce runoff from the leaves.

Poison–Any substance ingested, inhaled, or developed within the body that causes or may cause damage or disturbance of function of plants, animals, or people. See Toxin.

Poison Bait–A poison mixed with wheat bran, molasses, or other attractant used to control cutworms, grasshoppers, and other insects.

Poison Creeper–See Poison Ivy.

Poison Dogwood–See Poison Sumac.

Poison Hemlock–*Conium maculatum*, family Umbelliferae; a biennial, rank-growing herb, which is an escape and a weed in various sections of the United States. It is dangerously poisonous to people and animals, being most poisonous to stock in the spring, when the herbage is fresh. Symptoms of poisoning of cattle are loss of appetite, salivation, bloating pain, feeble but rapid pulse, and loss of muscular power. Native to Europe and Asia. Also called deadly hemlock, poison parsley, winter fern.

Poison Ivy–*Toxicodendron radicans*, family Anacardiaceae; a small, erect shrub or climbing vine common throughout the United States in waste places, pastures, woodlands, and along old fencerows. Apparently more dangerous to humans than to animals, contact with the plant among susceptible individuals causes severe blistering and inflammation, especially on the hands, arms, and face. Also called climbing ivy, three-leaved ivy, climath, poisonoak, poison creeper.

Poison Milkweed–*Asclepias galioides*, family Asclepiadaceae; a poisonous, perennial herb found on dry sites on the western ranges in the United States. It is occasionally eaten by animals with fatal results. Sheep appear to be most susceptible to it. The symptoms of poisoning are loss of muscular control, staggering, falling, spasms, bloating, fever, weak but rapid pulse, and respiratory paralysis. The affected animal usually dies. Native to North America.

Poison Sumac–*Toxicodendron vernix*, family Anacardiaceae; a shrub or small tree bearing greenish-white flowers, contact with which causes a severe blistering and inflammation to the skin of susceptible persons. Native to North America. Also called poison ash, poison dogwood.

Poisonoak–See Pacific Poisonoak, Poison Ivy.

Poisonous–Containing poison, as a poisonous plant.

Poisonvetch–See Timber Poisonetch.

Pokeberry–(1) Any plant of the genus *Phytolacca*, family Phytolaccaceae. Also called pokeweed. (2) *P. americana*, a tall, stout, perennial weed that has purplish or greenish flowers in racemes and black or dark wine-colored fruit. Its young asparagus-like shoots are sometimes eaten as greens; the roots and seeds are poisonous. Native to North America. Also called foxglove, pokeweed, common pokeberry, poke, scoke, garget, pigeon berry.

Poll–(1) The region at the crest or apex of the skull in horses and cattle. (2) To cut back the crown of a tree. (3) To remove the horns of cattle.

Pollard–(1) A tree whose crown has been cut back to invite the production of shoots from the top. Sometimes a tree is so pruned to induce a globelike mass of foliage. (2) A hornless ox, sheep, or goat.

Pollen–The male element that carries the spores in the fertilization of the egg nucleus in the ovule of a flower. The pollen is borne by the anthers and is usually a yellowish, dustlike mass of separate grains.

Pollen Sterile–A variety, strain, or species of a plant whose pistils are fertile but the stamens sterile, so that inbreeding does not occur. The necessary pollen is secured from the desired male parent that produces viable pollen.

Pollen Trap–A grid placed at the entrance to a beehive, which removes pollen from the bees' legs as they enter; the pollen falls into a tray below the grid.

Pollination–The transfer of the pollen from the anther to the stigma of a flower, the first step in producing a fruit or seed. See Cross Pollination.

Pollination, Self–Transfer of pollen from the male part of one flower to the female part of the same flower, or to another flower on the same plant.

Pollinizer–(1) An insect that carries pollen from flower to flower and thereby brings about the fertilization necessary for the production of the fruit or seed of the plant. (2) A plant (especially a tree) in a fruit planting which provides pollen for self-unfruitful varieties. Also known as pollinator.

Poly–-A prefix meaning many, as polycarpic, fruiting many times (as opposed to monocarpic).

Polyandrous–A plant with a large, indefinite number of stamens.

Polyanthes, Polyanthus–Many-flowered.

Polybrid–A plant hybrid that originated from a cross between two particular species, varieties, or genera.

Polychlor–Any of the group of the chlorinated class of insecticides, such as DDT.

Polyculture–The simultaneous planting of two or more crops in the same area at the same density of each as they would be planted separately in monoculture.

Polygamous–Having both perfect and imperfect flowers borne on the same individual plant.

Polygenetic Soil–Soil produced under conditions that have changed importantly with time.

Polygons, Mud Crack–Mud cracks (sun cracks, shrinkage cracks) form as sediments dry out. The cracks form polygons, which vary in number of sides and dimensions of angles between the sides. Cracks are rarely straight, and polygons may be bounded by three to eight cracks.

Polyhedral Disease–A disease caused by a microorganism that attacks the gypsy moth and cabbage looper larvae and results in their decimation. However, people have not yet been able to spread the infection artificially for biologic control.

Polymorphic–Having two or more forms.

Polynutrient Fertilizers–Fertilizers containing two or more essential elements. Also called multinutrient fertilizers.

Polyphagia–(1) Voracious appetite. (2) An unnatural craving for many kinds of food. (3) Omnivorousness. See Bulimia, Cynorexia, Sitomania.

Polyphagous Parasite–A parasite capable of parasitizing a considerable number of host species.

Polyphosphates–Salts of polyphosphoric acids such as ammonium polyphosphates and calcium polyphosphates.

Polyphosphoric Acid–Any of a series of phosphoric acids whose molecular structure contains more than one atom of phosphorus such as pryophosphoric acid, tripolyphosphoric acid, and tetrapolyphosphoric acid. See Phosphoric Acids.

Polyploid–An organism with more than two sets of the basic or haploid number of chromosomes; e.g., triploid, tetraploid, pentaploid, hexaploid, heptaploid, octaploid.

Polysaccharide–A large molecular weight carbohydrate made up of many sugar units; e.g., starches, cellulose, and glycogen.

Polystrene Foam–A very lightweight synthetic, plastic material made into flakes or beads and used in artificial plant growth media.

Pomace–(1) The crushed apple pulp left after the extraction of juice in cider making. (2) Residue of fruit and plant parts after the expression of juices.

Pomace Fly–*Drosophila* spp., family Drosophilidae; an insect that infests bruised or decaying fruit. Also called sour gnat.

Pomato–A horticultural curiosity; a chimera that originated by grafting a tomato scion on a potato stock. The fruit is fragrant, juicy, and tomatolike. Also called potomato, topato.

Pome–A fleshy fruit having several seeds instead of a stone, as apple, pear, and quince.

Pomegranate–*Punica granatum*, family Punicaceae; a shrub or small tree grown in warm areas for its edible, red, orange, or yellow fruit about the size of an orange with astringent rind and juicy, slightly acid flesh. Native to Asia, it is occasionally grown outdoors in Florida, Georgia, and other parts of the United States for its fruit or ornamental value. Some purely decorative forms are grown in greenhouses. The juice of the fruit is used in making grenadine syrup.

Pomelo–Grapefruit.

Pomiculture–Fruit culture.

Pomodoro–(Italian) Tomato.

Pomology–The art and science of growing and handling fruits, especially tree fruits.

Pomona–The Roman goddess of tree fruits.

Pompelmous–Grapefruit.

Pond Baldcypress–*Taxodium ascendens*, family Pinaceae; a deciduous, coniferous tree grown for its valuable lumber used in shingles, buckets, window boxes, coffins, outdoor seats, etc., and sometimes as an ornamental. Native to the southern United States. See Baldcypress.

Ponderosa Pine–*Pinus ponderosa*, family Pinaceae, an evergreen tree, large or small depending upon the site, and valued chiefly for its timber. Native to British Columbia, Canada, and the western United States. Also called western yellow pine, Arizona pine, bull pine, Arizona longleaf pine.

Pone–An oval-shaped cake of corn bread.

Pony Grass–*Calamagrostis neglecta*, family Gramineae; a perennial grass that is valuable forage in the Rocky Mountains, United States. Native to the north temperate zone. Also called slimstem reedgrass.

Poophyte–A meadow plant that grows under mesic or intermediate moisture conditions.

Poor Drainage–A natural condition of standing water on the surface of a saturated soil. (The term usually denotes that an excess of water is present for most desired agricultural uses of the land; for farm ponds the same drainage may be necessary.)

Poor Land–Any land that is inferior in quality for some particular use, as for agriculture. (Commonly the term is loosely used: land may be poor either because of some unfavorable relief, or because of some physical or chemical condition of the soil. Often land and soil are used interchangeably.)

Poor Man's Alfalfa–Lespedeza, so named because of its capacity to make a fairly satisfactory growth on soil unsuitable for alfalfa. See Lespedeza.

Poor Man's Manure–Snow, which in some localities may be beneficial on plowed land because usually all of the water from snow goes into the soil.

Poor Soil–A soil unsuitable for some particular plant growth because of deficiency in fertility or some unfavorable physical condition. A poor soil, however, may be valueless for one plant but quite suitable for the growth of another; e.g., a sandy, acid soil is desirable for blueberries but not suitable for alfalfa.

Poor Stand–A seeding of any crop that germinates poorly and produces an uneven stand of young plants or seedlings.

Poor Tilth–Poor physical condition of soil such as hard or cloddy.

Poorly Graded–Refers to soil material consisting mainly of particles of nearly the same size. Because there is little difference in the size of the particles, density can be increased only slightly by compaction.

"Pop-Up" Fertilizer–See Starter Fertilizer.

Popcorn–A type of corn whose dried kernels explode into large, fluffy masses when heated. Most popcorns have small, hard, flinty kernels and small cobs.

Poplar–Any tree of the genus *Populus*, family Salicaceae. Species are mostly quick-growing, soft-wooded trees grown as ornamentals, for pulpwood, and other wood uses. Hybrid poplars are fast-growing and make good shade trees.

Poplar Borer–*Saperda calcarata*, family Cerambycidae; an insect, the larva of which causes black and swollen scars on the limbs and trunks and sawdust at the base of poplars. Damage by this insect frequently makes it impossible to grow certain species of poplars in some localities.

Porcupine Grass–*Stipa spartea*, family Gramineae; a tall perennial grass, whose sharp awns cause severe injury to mouths and tongues of grazing animals. Native to the United States.

Pore–(1) In plant and animal membranous tissues, minute openings for absorption and transpiration of matter. (2) In wood anatomy, the cross section of a vessel or a vascular trachea. (3) In soil, the portion of a given volume of soil that is unfilled with solid matter; air spaces, irregular in shape and size.

Porosity–Refers to the extent of voids or openings in the soil that exist between soil particles and soil peds or clods. These pores hold water and air for absorption by plant roots. About half of soil volume which is in a good physical condition for plant growth is pore space.

Positive Ion–A cation; an ion that carries a positive charge of electricity.

Positron–A subatomic particle equal in mass and weight to the electron and having an equal but opposite charge. Positrons are emitted by some artificially radioactive isotopes.

Post Oak–*Quercus stellata*, family Fagaceae; a deciduous tree, whose wood is suitable for posts, ties, tight-cooperage, veneer, fuel, flooring, planing-mill products, etc. Native to central and southern United States. Also called brash oak.

Postemergent Herbicide–Applying a herbicide after the weeds begin to grow.

Pot–(1) To place plants in various kinds of containers for growing indoors. (2) To grow seedlings in paper pots or other containers for transplanting without disturbing roots. (3) See Flower Pot.

Pot Binding–The growing of a plant in a pot until its root system has become densely matted; a method practiced by florists to promote flowering.

Pot Layerage–Air layering in which two halves of flowerpots are placed around the branch to be rooted, tied together, and filled with sphagnum moss or soil.

Pot-bound–Plants whose roots completely fill a container and surround the soil ball in which they are growing, restricting normal top growth of the plant.

Potagerie–A kitchen garden in which various kinds of vegetables are grown.

Potassium–K; the chemical element, an alkali metal, which occurs widely in minerals; e.g., in the orthoclase feldspars of granites and in salt deposits as the chlorides and sulfates. Regarded as an essential plant nutrient, potassium is present naturally in some form in all soils but in extremely variable amounts, and is likely to be in largest amounts in clay soils and in least amounts in highly silicious soils and in peats.

Potassium Chloride–KCl; the salt most commonly used as a source of potassium in fertilizers. Its commercial form is known as muriate of potash.

Potassium Fixation–The process of converting exchangeable and water-soluble potassium to slowly available potassium. Wetting and drying of clay soils tends to make available potassium less available.

Potassium Magnesium Sulfate–See Sulfate of Potash-Magnesia.

Potassium Metaphosphate–KPO_3; a chemical containing 40 percent potash (K_2O), which is used as a fertilizer. Commercial grades contain about 37 percent potash and 55 percent phosphoric acid.

Potassium Nitrate–KNO_3; saltpeter, nitrate of potash; the potassium salt of nitric acid. It is manufactured by the direct reaction of potassium chloride with concentrated nitric acid to produce potassium nitrate and chlorine.

Potassium Permanganate–$KMNO_4$; a powerful oxidizing compound used as a disinfectant, deodorant, and a reagent in analytical work, especially in the determination of available nitrogen in organic material. Sometimes also used as a fungicide in greenhouses.

Potassium Phosphate–Any of the various phosphates of potassium, especially potassium metaphosphate, used as fertilizers.

Potassium Phosphate Solutions–They have been used successfully as liquid fertilizers in California. They contain an average 10.6 percent P_2O_5 and 14.5 percent K_2O. The pH in a 1 to 400 dilution is 10, indicating that the salt in solution is dipotassium phosphate (K_2HPO_4).

Potassium Sulfate–K_2SO_4; one of the carriers of potash used in the manufacture of mixed fertilizers. The salt occurs in nature, but the commercial product is manufactured by treating potassium chloride with sulfate of magnesia or sulfuric acid. The product contains about 60 percent potassium (K_2O).

Potato–*Solanum tuberosum*, family Solanaceae; a perennial herb grown as an annual for its edible tuber, the potato of commerce. The potato is indigenous to South America and is believed to have been brought to England in 1586. It first became an important food crop in Ireland and was reintroduced into America from that country, hence the name Irish potato. It is now extensively grown throughout the world and is one of the important agricultural crops. Also called earth apple, spud, white potato.

Potato Aphid–*Macrosiphum solanifolii*, family Aphidae; a green or pinkish-winged or wingless aphid that infests the potato, tomato, eggplant, pepper, and many other plants. It may occur in epidemic numbers and do tremendous damage as a sucking insect and as a vector of viral diseases. Also called pink and green aphid.

Potato Beetle–See Colorado Potato Beetle.

Potato Distillers' Dried Residue–A livestock feedstuff that consists of the dried product obtained after the manufacture of alcohol and distilled liquors from potatoes or from mixtures in which potatoes predominate.

Potato Flea Beetle–*Epitrix cucumeris*, family Chrysomelidae; an insect whose larvae feed on the roots and tubers of the potato, causing scurfy or pimply potatoes. It infests numerous other plants also and is one of the most destructive garden pests.

Potato Flour–A flour made from cooked and dried potatoes or from sliced raw potatoes, sulfured, dehydrated, ground, and bolted. Used in breads, it compares favorably with wheat flour in food value, but is lower in protein.

Potato Late Blight–A serious disease of the potato caused by the fungus *Phytophthora infestans*, family Pythiaceae, which is characterized by a moldlike growth affecting the plant tissue and killing the tops. The fungus may also infect the tuber causing a rapid breakdown.

Potato Latent Mosaic–A potato mosaic attributed to virus X. The virus is latent in a great many varieties of potatoes and may cause little or no damage. A very severe necrosis may be caused depending upon the particular strain of the virus.

Potato Leaf Roll–One of the several degenerative or viral diseases of the potato characterized by a pronounced rolling of the leaves and reduction in size of plants and tubers. Apparently it is caused by more than one race of viruses. Aphids are vectors. Also called potato phloem necrosis virus, potato virus, Solanum virus, Corium solani.

Potato Leafhopper–*Empoasca fabae*, family Cicadellidae; one of the most injurious insect pests of potatoes in the eastern half of the United States. It also infests alfalfa, other legumes, and deciduous nursery stock.

Potato Mosaic–Any mosaic disease of potato caused by any of several distinct viruses or virus complexes. Three important mosaics, latent mosaic, virus A, and vein banding mosaic are transmitted mainly by aphids.

Potato Rugose Mosaic–A composite infection of potatoes caused by both virus X and the vein banding mosaic virus. It may be more severe than either one of the above two virus infections. It is characterized by rugosity or wrinkled surface of the leaf, by necrotic spots and streaks, and by the dropping of the leaves in the later stages.

Potato Scab–See Corky Scab of Potato.

Potato Seed–Seed from the flower of the potato, as distinguished from a seed potato that is the tuber or part of a tuber normally used for planting and reproduction.

Potato Sickness–Golden-nematode disease of potato.

Potato Stalk Borer–*Trichobaris trinotata*, family Curculionidae; a snout beetle whose grubs hollow out the stems of potato causing wilting and death. Occasionally it destroys entire fields of potatoes.

Potato Tuber Necrosis–A disease of potatoes that results from infection by the same virus that causes alfalfa mosaic. The necrosis starts at the stem end of the tuber and may extend into all tissues.

Potato Tuberworm–*Gnorimoschema operculella*, family Gelechiidae; the larva of a moth that burrows into the potato tuber, often riddling it with slender, dirty-looking, silk-lined burrows. The larvae also infest tobacco leaves.

Potato Vein Banding Mosaic–A mosaic, caused by virus Y, which is distinguished by rugosity or wrinkled surfaces of the leaves in early stages, and necrotic spots and streaks in later stages. The disease is responsible for serious reductions in yields of potatoes. Also called potato virus Y, Solanum virus 2, Marmor upsilon.

Potato Verticillium Wilt–A fungal disease of potatoes caused by ***Verticillium albo-atrum***, family Moniliaceae; characterized by wilting of the tops and a vascular discoloration of the stems, tubers, and roots. It may affect one or all of the stems in a hill, often causing death of the plant.

Potency–(1) The power of a medicine to produce the desired effects. (2) The ability of an embryo to develop into a viable destiny. (3) The ability of the male of any plant or animal species to fertilize the female germ cells. (4) The degree of toxicity of a chemical.

Potential Soil Acidity–Those hydrogen ions on the negatively charged soil particles that are not disassociated, and therefore not active in the soil solution, the amount of exchangeable hydrogen ions in a soil that can be released by cation exchange. Also called reserve soil acidity.

Potherb–Greens; any plant yielding foliage that is edible when cooked, such as spinach, kale, chard.

Potomato–See Pomato.

Potting–The practice of transplanting very young plants from seed frames to individual flower pots. These are selected for size and provided with soil mixture and drainage according to the needs of the particular plant.

Potting Mixture–Combination of soil and other ingredients such as peat, sand, perlite, or vermiculite designed for starting seed or growing plants in containers.

Poudrette–Dried night soil mixed with charcoal powder, gypsum, lime or peat; used as a fertilizer. Used in European countries from early times, it has never been used to any extent in the United States.

Poultry Manure–The mixture of feces and urine voided by birds. When freshly produced it contains about 80 percent moisture, 1 percent nitrogen, 0.8 percent phosphoric acid, and 0.5 percent potash. Poultry manure decomposes with a rapid and large loss of moisture and ammonia, especially during warm weather.

Poultry Netting–Wire netting with hexagonal meshes, made with 1-, 1½, or 2-inch mesh openings, used for fencing, constructing wire partitions, cages, trellises, etc.

Poultry Range–Pasture that provides green feed and exercise for poultry.

Pounds of Grain Per Acre of Pasture–The feed value of an acre of pasture expressed in terms of pounds of dry grain feed.

Powder-post Beetle–Any of the beetles of the families Ptinidae, Anobiidae, Bostrichidae, or Lyctidae, which breed in old, dry wood and reduce it to powder. The greatest damage is done by species of ***Lyctus*** which confine their infestation to the seasoned sapwood of a number of hardwoods, especially oak, ash, and hickory.

Powdered Sugar–White sugar partially reduced to powder.

Powdery Mildew–A common disease of many plants, caused by one of the several species of the fungus family Erysiphaceae, which are obligate parasites that grow on the surface of the leaf, stem, or fruit, obtaining their sustenance by means of rootlike projections that penetrate the plant cells. Infections are local rather than systemic. Little moisture is necessary for infection, and the disease often prevails under relatively dry conditions. Sulfur sprays or dusts and resistant varieties of plants have frequently been effective where control is desired. Sometimes also called ovidium disease.

Powdery Scab of Potato–A fungal disease of potatoes caused by ***Spongospora subterranea***, family Plasmodiophoraceae; characterized by small to medium-sized, round pimples on the tubers that turn brown and burst open at maturity, allowing the powdery mass of spores to escape. The tubers and roots may be infected. Also called corky scab, corky end of potato.

Power Subsoiler–A tractor-drawn plow that has a long, narrow shank with a wedge-shaped point for deep penetration to break up stiff clay subsoils and hardpans. See Hardpan, Plowpan.

Prairie–(1) The extensive, nearly treeless and dominantly grass-covered plains of the midwestern United States that lie east of the Rocky Mountains. In a more restricted sense, the tall grasslands with blackish soils; in a more general sense the semiarid shortgrass plains as well. Also called savannah, steppe. (2) In the generally forested eastern part of the United States, any naturally treeless area that is generally dry or naturally well-drained. (3) Wet, treeless, marshy areas. (4) Prehistoric, treeless tracts that resulted from fires.

Prairie Hay–Any hay made from the wild grasses of the prairie.

Prairie June Grass–*Koeleria cristata*, family Gramineae; a tufted, perennial grass, found from Ontario to British Columbia in Canada and south to Pennsylvania, Oklahoma, and California, United States, whose forage rating is good. Native to North America.

Prairie Sandreed–*Calamovilfa longifolia*, family Gramineae; a perennial, drought-enduring grass, growing from 2 to 6 feet tall, which is found from Michigan to Colorado, United States, and Alberta, Canada. Important for winter pasture and for hay though grazed but lightly in the summer. It is an important, sand-binding grass on dunes and sand hills. Native to North America. Also called prairie sandgrass.

Praying Mantis–Predaceous insects of the genus *Mantis*, family Mantidae, which prey upon other insects and are generally beneficial to humans. They wait for their prey with front legs raised like hands in prayer and seize it when it comes within reach.

Prebloom–Designating a plant in bud. Clover or alfalfa cut in the prebloom stage of growth makes high-quality hay.

Prechilling–In nursery practice, a seed pretreatment method to overcome dormancy. Seed is placed on moist blotting paper in standard germinators, and the whole apparatus is held at near freezing temperatures for a predetermined time.

Precipitated Phosphate–A by-product from the manufacture of monocalcium phosphate that is sufficiently soluble in the soil solution to be of value as a fertilizer.

Precipitation–(1) The amount of water, hail, sleet, snow, or other moisture received from clouds. Snow is also reported in its equivalent of liquid water. Precipitation is classified by the conditions that produce the rising column of unsaturated air which is antecedent to precipitation. Convection precipitation is the result of uneven heating of the ground, which causes the air to rise and expand, vapor to condense, and precipitation to occur. This is the major type of precipitation during the summer, producing high-intensity, short-duration storms. Orographic precipitation is caused by topographic barriers that force in the moisture-laden air to rise and cool. Cyclonic precipitation is related to large low-pressure systems that require five or six days to cross the United States from the northwest or Gulf of Mexico. These systems are the major source of winter precipitation. (2) The phenomenon of a solution or suspension that is flocculated. (3) The electrostatic or other means of removal of polluting particulates from the air. See Acid Rain.

Preclimax–The stage of succession of an area that has never supported climax vegetation but is in the process of succession toward climax vegetation for a particular climate. Thus, an extremely shallow soil or soil material in the true prairie association may support preclimax blue gramagrass rather than climax bluestem, because soil development has not yet reached a stage at which it will support the latter. See Climax Vegetation.

Precommercial Thinning–Any type of thinning (cutting or otherwise killing trees in a stand) that takes place before the size or condition of the trees makes them of sufficient value to cover the costs of the activity.

Precook–In food preservation, to heat thoroughly before canning.

Precooling–(1) Preliminary cooling of milk immediately after a milking to prevent spoilage. (2) Cooling of fruits immediately after harvesting during periods of hot weather to retard ripening and deterioration. (3) The cooling of meats after slaughter and before cutting.

Predisposition–(1) Stress or anything that renders an animal liable to an attack of disease without actually producing it. (2) The effect of one or more environmental factors that makes a plant vulnerable to infection by a pathogen.

Preemergence Application–Applying a herbicide to the soil to kill weed seeds before they germinate, or after a crop is planted but before it germinates and seedlings emerge above the soil's surface.

Preference–Selection of palatable plants over others by grazing animals.

Preference Permit–A license valid up to ten years issued by the United States Forest Service to permit domestic livestock grazing in a National Forest. The permit is renewable annually.

Preformed Terrace–A terrace or bench constructed or sloped before a planting, as in setting out an orchard.

Pregermination–A special seed treatment applied prior to testing or sowing to increase the rapidity or completeness of germination. it usually involves seed exposure for a month or more to temperatures between 30° and 40°F while in contact with moist peat or sand.

Preirrigation–In rice culture, the application of irrigation water prior to the usual irrigating after emergence of the plants.

Prepink–Designating the stage of growth of an apple blossom bud when the leaves have spread away from the blossoms but before the stems of the flower cluster have separated. Very little if any pink is showing in the blossoms.

Prepink Spray–A spray applied to apples in the prepink state. See Prepink.

Preplant Application–A treatment with a pesticide on the soil before planting or transplanting.

Preplant Soil Incorporation–A treatment with a pesticide by tilling the chemical in the soil before planting or transplanting.

Prescribed Burning–Setting fire scientifically under "safe" conditions of surface fuel, weather, and soil moisture to: (a) reduce uncontrolled fire hazards, (b) control "weed" species, (c) increase grazing, and/or (d) encourage designated forest tree species.

Preservative–(1) Any material, as salt or sugar, that delays or prevents spoilage and decay of food products, etc. (2) A substance, such as creosote, which when suitably applied to wood makes it resistant to infestation by fungi, insects, and marine borers.

Preserve–(1) In wildlife management, a game-shooting area on which game species are propagated or released. (2) A tract of land set aside for preservation of natural conditions, and protected against exploitation or any commercial use. (3) To prepare foods by cooking with some preservative so as to reduce fermentation or decomposition.

Press Cake–Pomace, or the residue left after pressing the juices from fruits, olives, or tomatoes. Dried pomace has some value as animal feed.

Press Grain Drill–A grain drill that has press wheel gangs mounted on the rear of the drill to firm the soil over the planted seeds. See Air Drill, Grain Drill, Stubble Mulch Drill.

Press Wheels–Wheel attachments to field seed drills that follow the drills or tubes and press the loose soil over the seeds for better covering and compaction, usually resulting in increased germination.

Press-cake Meal–A by-product left after the separation of oils, by grinding and distillation, from the pits of peaches, apricots, and cherries. It has some value as a stock feed.

Pressing–(1) (a) The extraction of the juice from fruit, as in cider making or in the production of orange juice. (b) The solid material left after such a process. (2) The process of bringing the small loose pieces of curds together in a shaped solid body in cheese making by means of a press.

Pressure Pan–(Traffic sole, plow pan, tillage pan, traffic pan, plow sole, compacted layer) An induced subsurface soil horizon or layer having a higher bulk density and lower total porosity than the soil material directly above and below, but similar in particle size analysis and chemical properties. The pan is usually found just below the maximum depth of normal plowing and frequently restricts root development and water movement. See Plow Pan.

Preying Mantis–A misnomer for praying mantis occasionaly used because of the preying habit of the insect. See Praying Mantis.

Pricklypear–Any of various cacti of the genus *Opuntia*, family Cactaceae. Species are prostrate to treelike forms, which contain both spines and cushions or barblike hairs. Some species spread very rapidly and in some places have greatly depreciated the value of the land. With the spines and barbs burned off, the cacti have been eaten by cattle during periods of drought and scarcity of any other feed. The pearlike fruit when preserved is edible for humans. Also called devil's fig.

Primary Bud–The largest and strongest bud in the compound bud on the shoot or cane of a grape vine.

Primary Plant Nutrients–Nitrogen, phosphorus, and potassium. See Plant Nutrient.

Primary Root–The main descending axis of a plant; the pole of the embryo opposite the shoot.

Primary Succession–Plant migration on newly formed soils or upon soils that have never borne vegetation, such as a sand bar in a river.

Primeval Forest–The forest as it existed before the advent of humans.

Priming Method–A method of harvesting tobacco by picking the individual leaves from the stem, as they ripen.

Primordium–A member or organ of a plant in its earliest condition, i.e., the first roots to form on a cutting. Plural, primordia.

Prismatic Soil Structure–Blocky soil structure in which the vertical axis of a block is longer than the horizontal axis; the upper ends of the block (column) are not rounded.

Privet Lippia–*Lippia ligustrina*, family Verbenaceae; an aromatic shrub, a troublesome invader in the grasslands of the southwestern United States. Native to America. Also called whitebush.

Proboscis–An elongated nose, such as the snout of a hog, or of some species of insects.

Procumbent–Designating a prostrate plant that differs from a creeper in that the parts that lie on the soil do not take root.

Produce–(1) Commodities produced from or grown in the soil. (2) In animal breeding, a female's offspring.

Producers–Primarily green plants, the basic link in any food web. By means of photosynthesis, plants manufacture the food on which all other living things ultimately depend.

Product–The herbicide as it is sold commercially. It contains not only the active ingredients but also various solvents, surfactants, carriers, and other adjuvants that are designated as inert ingredients.

Productivity Rating–The productivity of the various soils on a farm based on the expected yield of the major crop or crops with known management. The rating is a percentage based on standards developed by research and statistical records of crop yields in the region. The best soils have a productivity rating of 100. In some instances, tables have been established showing the relation of crop yield to land values, with fixed management. See Soil Survey.

Profile–(1) The general outline of an animal's body. (2) See Soil Profile.

Progenitor–An individual animal or plant that is recognized as the source of a certain type or character in its offspring.

Prognosis–Forecast as to the probable result of an attack of disease, the likelihood of recovery.

Programmed Harvest–Timber scheduled for harvest for a specified year.

Progressive–Of a disease, developing through successive stages, usually in a certain direction, whether improving or deteriorating.

Proliferous–(1) Pertaining to a new branch that arises from an older one, or one head or cluster of flowers from another. (2) Reproducing fully by offsets, buds, or other vegetative means, as in most species of cactus genus *Echinopsis*. (3) Developing shoots from flowers or fruit as in some pricklypears.

Prong Budding–A kind of budding or plant grafting similar to shield budding except that a short spur or prong is used in place of a simple bud.

Prop Roots–A leafless stem that produces aerial roots, as in the genus *Selaginells*.

Propagation–(1) Increasing the number of plants by planting seed or by vegetative means from cuttings, division, grafting, or layering. (2) Construction of drainage ditches by use of dynamite. charges are closely spaced so that detonation of one charge causes the explosion of the next charge and so on until the complete series of charges is exploded. A water-filled soil is most desirable for this method.

Propagule–A plant part such as a shoot, bud, root, or tuber used to propagate a new individual vegetatively.

Proper Stocking–Stocking of a range area on the basis of its true grazing capacity in a year of adequate rainfall.

Proper Use–The degree of grazing that an individual plant species, or the total palatable cover of a range area, may endure without damage to the plants or the soil.

Proper Use Factor–As applied to individual range species, the estimated maximum percentage of the total vegetative production of the year within easy reach of the livestock to which a given range species may be grazed without damaging it or associated important palatable plants or the soil; the degree to which each species may be grazed when the range as a whole is properly grazed. The proper use factor for a range type is the weighted average of the proper use factors of the individual plants in the type. See Palatability.

Properties–Characteristics by which a substance may be identified. Physical properties describe its state of matter, color, odor, and density; chemical properties describe its behavior in reaction with other materials; biological properties refer to any life-related characteristics such as biodegradation.

Prophase–The first phase of cell division wherein many of the preparatory steps take place, such as shortening and thickening of the chromosomes, division of the centromeres, disappearance of the nuclear membrane, and formation of the spindle.

Prophylactic–Preventive or protective treatment against disease.

Prophylaxis–Prevention of disease by various measures.

Proportional Weir–(Irrigation) A weir so shaped that the flow rate of water varies directly with the water head.

Proso–*Panicum miliaceum*, family Gramineae; an annual herb grown for its seed used for birdseed, poultry and stock feed. Native to Asia. Also called broom corn millet, broom millet, hog millet.

Prostrate–A general term for lying flat on the ground as in some plants such as sweet potatoes.

Protandry–Pollen dehiscence (shedding) before the stigma is receptive.

Protectants–Chemicals containing heavy metals, sulfur, or organic compounds used as sprays, dusts, or dips on seeds, stems, leaves, or wounds of living plants to reduce entrance of fungi or bacteria, and on fabric or wood products to reduce decay.

Protection Forestry–The practice of forestry with the primary objectives of: (a) conserving water supplies; (b) maintaining desirable streamflow; (c) increasing groundwater storage; (d) reducing erosion and reducing sedimentation; (e) providing high-quality water and reducing pollution; (f) ameliorating adverse climatic conditions, especially wind.

Protection Strip–In white pine blister rust control, a belt or zone outside the pine stand that is included in the control unit and is freed of *Ribes* spp., the alternate host of blister rust.

Protein–Any of a large number of complex, organic compounds of amino acids that has a high molecular weight and is an essential part of all living organisms. Proteins consist largely of carbon, hydrogen, nitrogen, and oxygen; many contain sulfur, and some also contain iron and phosphorus. They constitute a large portion of the protoplasm, and are obtained from foods such as lean meats, and from vegetables such as beans.

Protein Equivalent–An expression used in computing the protein of feedstuffs. In Great Britain, nonprotein, nitrogenous compounds are considered to have one-half the value of true protein. In the United States, protein equivalent is considered to be the sum of the true protein plus one-half of the amount of nonprotein nitrogen.

Protein Palm Nut Oil Meal–A livestock feedstuff that is the ground residue left after the extraction of part of the oil from the fruit of one or more species of palm.

Protein Solvent Extracted Cottonseed Meal–A livestock feedstuff that is the product resulting from grinding solvent extracted cottonseed flakes.

Protein Solvent Extracted Soybean Flakes–A livestock feedstuff that is the product obtained by expelling part of the oil from soybeans by the use of solvents.

Protein Supplement–Animal feed that contains approximately 20 percent or more protein.

Proteolysis–The process by which casein or some insoluble casein derivative is broken down to water-soluble compounds through the acting of organisms.

Protogyny–Receptivity of the stigma of flowers before pollen dehiscence (shedding).

Protopectin–The parent substance from which pectin is derived. Also called pectose.

Protoplasm–The gelatinous, colloidal material of plants and animals in which all life activities occur.

Protoplast–A unit of protoplasm in one cell.

Protozoa–A group of one-celled organisms that generally do not contain chlorophyll, including amoebae, paramecia, flagellates, and certain spore-forming organisms; sometimes classified as one-celled animals.

Provenance–The ultimate natural origin of a tree or a group of trees. Trees having a common center of origin.

Provenience–Origin; particularly the geographic origin of seed and inoculum.

Provitamin–A precursor of a vitamin; a substance from which an animal organism can form a vitamin. Carotene is provitamin A and ergosterol is provitamin D.

Proximal–Opposite of distal; near the point of attachment of reference.

Pruinose–Having a dusty or chalky appearance, as a leaf surface.

Prune–(1) To trim or cut back a plant or plant part. (2) A dried plum.

Pruning–Removing branches or twigs to control the size or shape of a plant, to control fruiting, to remove dead or broken branches, or to strengthen or improve the appearance of a plant.

Pruning Saw–A curved bladed saw usually mounted on a long pole that is used for removing limbs from a tree.

Prussic Acid Poisoning–Poisoning of livestock from prussic acid (hydrocyanic acid) (HCN) which may result from ingestion of sorghums, such as Johnsongrass, under certain conditions such as drought. Hydrocyanic acid may also be produced from the leaves of several species of wild cherries (***Prunus*** spp.) which are occasionally browsed by livestock.

Pseudo–A Greek prefix meaning false or spurious. In most scientific terms it denotes a deceptive resemblance to the substance to whose name it is prefixed, e.g., pseudocarp, false fruit.

Pseudo Yeast–Any yeast that does not form spores.

Pseudogamy–Parthenogenetic development of the female gamete that requires the stimulation of pollination but not complete syngamy. See Syngamy.

Psyllid Yellows–A disease of potatoes prevalent in the western United States, characterized by marked stunting of the plant, yellowing, and upward rolling of the leaves. The disease may be transmitted to other solanaceous plants, such as tomato and eggplant.

Pubescent–Strictly, this means covered with soft, short, fine hairs; as commonly used, however, the term means hairy, bearing hairs, in a generalized sense, without reference to the type of hair.

Public Lands–The general public domain; lands belonging to a national government that are subject to sale or disposal and that are not reserved for any special governmental or public purpose.

Puddle–(1) Immersing bare roots of trees and shrubs in a mixture of clay soil and water during transplanting to prevent the roots from drying out. (2) Destroying the desirable soil structure if the soil is worked or cultivated when too wet.

Puffballs–Certain globose fungi, of the family Lycoperdaceae, whose ripe fruit emits clouds of spores when disturbed. The spore fruits of most species of the family are edible when young.

Pulp–(1) In paper making, the product obtained by digesting wood in a slightly alkaline or neutral sodium sulfite cooking liquor. (2) The juicy or fleshy tissue of a fruit.

Pulpwood–Wood cut or prepared primarily for manufacture into wood pulp for subsequent manufacture into paper, fiber board, etc., depending largely on the species cut and the pulping process.

Pulse–(1) The edible seed of legumes, such as peas and beans. (2) The expansion and contraction of an artery associated with each heartbeat, which may be felt with the fingers.

Pulverize–(Soil) To reduce clods, or peds, to a fine granular state in preparation for planting seeds.

Pulverized Limestone–The product obtained by grinding either calcareous or dolomitic limestone so that all material will pass a 20-mesh sieve and half of the material, a 100-mesh sieve. It is applied to land to reduce soil acidity and to supply calcium and magnesium as a nutrient. Also called finely ground limestone. See Lime.

Pummelo–***Citrus grandis***, family Rutaceae; a tree widely grown in subtropical regions for its large pale yellow, edible fruit similar to but larger and coarser than grapefruit.

Pumpelly Brome–***Bromus pumpellianus***, family Gramineae; a stout perennial grass, native at elevations of 5,000 to 9,000 feet in the Black Hills of South Dakota, in Colorado, Idaho, and other states (United States), where it furnishes good grazing for all kinds of livestock.

Pumpkin–***Cucurbita pepo***, family Cucurbitaceae; an annual, prostrate, herbaceous vine of unknown origin grown for its edible fruit, the pumpkin of commerce.

Puncture Vine–***Tribulus terrestris***; a serious annual weed pest in the deep south. The vines may grow to 6 feet in length and have sharp spines on the burs.

Punk–(1) A small, scrubby horse. (2) Partly decayed wood.

Punky–Designating a soft, weak, often spongy, wood condition caused by decay.

Pupa–(Plural, pupae) The stage between the larva and the adult in insects with complete metamorphosis, a nonfeeding and usually inactive stage.

Puparium–The hardened larval skin or protective case that encloses the living pupa.

Pupate–The change from an active, immature insect into the inactive pupal stage.

Pure Culture–(1) A crop of one species, variety, etc., grown in contrast to a customary mixture; e.g., timothy grown from a pure seeding rather than mixed with clover. (2) A bacteriological culture that contains only a single species, or one that contains those desired for a particular purpose, as a yeast culture for making vinegar.

Pure Line–A strain of organism that is comparatively pure genetically (homozygous) because of continued inbreeding, etc.

Pure Live Seed–The product of the percentage of germination plus the hard seed and the percentage of pure seed, divided by 100.

Pure Stand–A stand in which at least 80 percent of the trees in the main crown canopy are of a single species.

Puree–A smooth, pulpy food product from which the rough fiber has been removed by sieving or other means.

Purple Nutsedge–***Cyperus rotundus***; a common, hard-to-control, perennial weed. The plant is spread by rhizomes. It is dark green with a reddish-purple seed head.

Purple Top–***Triodia flava***; a serious pasture weed pest that crowds out more-desirable plants.

Purple Vetch–*Vicia atropurpurea*, family Leguminosae; a viny, nonhardy, winter annual herb grown in the milder parts of California, United States, for seed, as a green manure crop, and for hay. Native to southern Europe.

Purpurea–Purple.

Purslane–A common low-growing weed pest that is characterized by red stems.

Purslane Sesuvium–*Sesuvium portulacastrum*, family Aizoaceae; a small, alkali-tolerant, prostrate, blood-red or purple, maritime herb with succulent stems, which is commonly found in the damp, coastal sands in warm areas of the United States, especially from Florida to Texas.

Pussy Willow–*Salix discolor*, family Salicaceae; a deciduous shrub or small tree grown as an ornamental for the female catkins that are hairy and resemble a cat's paw. Native to North America.

Putamen–The hard endocarp or pit of a stone fruit, as in cherries, peaches, and plums.

Putrefaction–Decomposition of animal or vegetable matter, produced by microorganisms in the absence of oxygen.

Putrescible–Organic matter capable of putrefaction.

Pycnidium–In certain fungi, the flasklike fruiting body containing conidia.

Pyramidal–Designating a tree or shrub whose silhouette is generally triangular; like a pyramid or cone.

Pyrethrins–The active ingredients (pyrethrins I and II) of pyrethrum, which are derived from plants belonging to the genus *Chrysanthemum*, family Compositae. One of the oldest and most widely used insecticides, they control insects affecting humans, animals, crops, and households.

Pyrethrum Powder–An insecticide made from the ground flower heads of various species of chrysanthemums whose use as an insecticide is reported to have originated in the Trans-Caucasus region of Asia about 1800.

Pyrie–(1) Designating the fire factor in an environment. (2) Designating modifications of soil and vegetation caused by fire.

Pyriform–Pear-shaped.

Pyroligneous Acid–An acid liquor obtained by the destructive distillation of wood, especially of certain species of hardwoods. It is a complex mixture containing 80 to 90 percent water and many organic compounds including acetic acid and methanol. Also called pyracetic acid, wood vinegar.

Pyrology–The scientific art of protecting forests from fire.

Pyrophyllite–Hydrous aluminum silicate, a mineral used as a diluent or carrier for insecticide and fungicide dusts.

Pyrophyte–A tree with a thick, fire-resisting bark, or one not subject to fatal damage from ordinary forest fires, such as longleaf pine.

Pythium Root Rot–One of the fungal root rots caused by species of the genus *Pythium*, family Pythiaceae. Usually more severe in moist soils, it infests wheat, oats, barley, corn, sorghums, etc.

Pythium Seedling Blight–A seedling blight and root rot of corn, tobacco, alfalfa, etc., caused by the fungus *Pythium arrhenomanes*, family Pythiaceae.

Quack Grass–*Agropyron repens*, family Gramineae; a perennial grass, widespread in the northeastern and central states, United States. It has become established on farm fields and is generally regarded as a pernicious weed. Spreading mainly by rhizomes (rootstocks or underground stems), it is difficult to eradicate. However, it has some value as a forage and hay plant and as a cover plant to control soil erosion. Native to Eurasia. Also called couch grass, couch wheat, dog grass, false wheat (grass), quitch grass, quake grass.

Quadrivalent–In genetics, a group of four associated homologous chromosomes.

Quagmire–Soft, miry ground, as a bog, marsh, swamp, or morass.

Quaking Aspen–*Populus tremuloides*, family Salicaceae; a small to fairly tall, short-lived, deciduous tree. It is the most widely distributed tree in North America. It occurs from New Jersey to Alaska. It grows in dense stands following logging or burning of forestlands. Its wood is now valued chiefly as a source of pulp for paper making, and it furnishes browse for sheep, goats, and deer, and food for beaver. Native to North America.

Qualitative Traits–Traits having a sharp distinction between phenotypes, and which are usually controlled by only a few genes; e.g., various coat colors and the horned trait in domestic animals.

Quantitative Traits–Traits that do not have a sharp distinction between phenotypes, and usually require some kind of measuring tools to make the distinctions. These traits are normally controlled by many pairs of genes; e.g., growth rate, milk production, and carcass quality. See Genotype, Phenotype.

Quarantine–(1) A regulation under police power for the exclusion or isolation of animal and plant pests or diseases and insects: (a) the isolation of an animal sick with a contagious disease; (b) a place where the sick are detained away form other animals until the danger of spread of a contagious disease has disappeared. In its wider application, the quarantine may be enforced against an individual animal, against all the animals, or all the animals of the same species, in a township, county, or state, and against those in a foreign country. (2) Prohibition to prevent the introduction or spread of any dangerous insects or plant diseases.

Quartz–SiO_2; a mineral that is a crystalline form of silica. It is the most common mineral component of sand soils, especially in humid regions. The earth contains about 47 percent oxygen and 28 percent silica.

Quick Cure–A pickling solution containing sodium nitrite or a combination of sodium nitrite and sodium nitrate.

Quick Decline–A viral disease of sweet orange or sour orange stock in California, characterized by the ashen color of foliage and curling of the leaves lengthwise and upward. See Tristeza.

Quick Freezing–Freezing of products, fruits, vegetables, poultry, meat, dairy products, etc., for preservation under conditions in which the temperature of the product is lowered from 28°F to –15°F within 30 minutes.

Quick Lime–See Lime.

Quick Tests–Certain standard chemical tests devised for the very rapid determination of the amounts of nutrient elements in a soil. When properly performed and interpreted such tests are useful in making recommendations for fertilizers and lime.

Quick-Growing–Designating a plant in a particular climatic region that matures in a comparatively short period; e.g., buckwheat and aspen are quick-growing among crops and trees in a cool climate.

Quicksand–Any loose fine sand that is mobile or semifluid when supersaturated with water. It is encountered in excavations, especially in localities where soils are derived from lacustrine sediments or river alluvium. Also called running sand.

Quiescence–See Dormancy.

Quince–*Cydonia oblonga (Pyrus cydonia)*, family Rosaceae; a deciduous tree grown for its edible fruit, the quince of commerce, used in making preserves and jellies, and as an ornamental for its white flowers. Native to Asia. Also called apple of Hesperides.

R Horizon–Underlying consolidated bedrock, such as granite, sandstone, or limestone. See Soil Horizon.

R Layer–In soil profile description, the rock on which the soil rests. See Soil Horizon.

Rabbiteye Blueberry–*Vaccinium virgatum*, family Ericaceae; a bushy shrub which is more heat- and drought-resistant and requires a shorter cold rest period than the high bush blueberry. It is grown in the southeastern United States for its edible fruit and as a source of improved blueberry strains. Native to North America.

Rabbitfoot Clover–*Trifolium arvense*; a common weed clover of fields and waste places. This clover should not be fed to livestock when the fuzzy heads are on the plant because the fuzz may cause intestinal problems.

Rabi–In India, the important group of crops, including cereals (especially wheat), fodders, mustard, etc., sown in October and November and reaped in March and April.

Race–(1) A group of individual plants which have certain common characteristics because of ancestry. It is generally considered a subdivision of a species; frequently used in plant breeding. (2) Pathogens of the same species and variety which are structurally indistinguishable but which differ in their physiological behavior, especially in their ability to parasitize varieties of a given host. (3) The channel that leads water to or from a waterwheel, the former is the head race and the latter the tail race. (4) A narrow passage or fenced land in a sheep yard for branding, dipping, etc. (5) An elongated white mark on the face of a horse or dog.

Raceme–A flower cluster in which the pedicels or supports of the flowers are of equal length and attached along a common axis, the sequence of blooming being upward.

Rachis–(1) The elongated axis to which the pedicels of an inflorescence are attached. (2) The extension of the petiole to which the leaflets of a compound leaf are attached.

Rack Hay Drying–A method of drying hay in which the freshly cut hay is placed on wooden racks built up off the ground. The method is used most often in small-scale operations in developing countries.

Radiate–(1) Standing on and spreading from a common center. (2) Having ray flowers, as in the family Compositae.

Radical–A group of different elements acting as a single unit in a chemical reaction; normally incapable of separate existence. A radical may be negatively charged, positively charged, or without a charge.

Radicle–The embryonic roots of seed plants.

Radioactive Element–An element capable of changing spontaneously into another element by the emission of charged particles from the nuclei of its atoms. For some elements, e.g., uranium, all known isotopes are radioactive; for others, e.g., potassium, only one of the several isotopes is radioactive. Radioactive isotopes of most elements can be prepared artificially, but only a few elements are naturally radioactive.

Radioactive Soil–A soil which possesses a minute degree of radioactivity. Radioactive phosphorus and other plant nutrients are sometimes added to the soil in research work to discover more facts about their functions in plant growth.

Radish–A garden vegetable eaten raw or used as a garnish; *Raphanus sativus*.

Rag–A parchmentlike membrane found in certain fruits; e.g., in dates around the pit and in oranges around the sections.

Rag Doll–A method of germinating corn kernels or seeds of other crops by placing them on a strip of cloth which is then rolled, soaked in water, and stored at the proper temperature for germination. After a few days the cloth is unrolled and the percentage germination is determined.

Rag Sumpweed–*Iva zanthifolia*, family Compositae; an annual weed with straight erect branches and drooping heads of greenish-white flowers. Ingested by cattle its leaves can cause dermatitis and a bitter flavor in the milk. The pollen may cause hay fever in humans. Native to North America. Also called marsh elder.

Ragmillet–See Pearl Millet.

Ragweed–*Ambrosia artemisifolia*, family Compositae; an annual weed with a bitter taste, which bears greenish-yellow flowers and produces an abundance of pollen during the late summer and early fall which causes hay fever. Native to North America. Also called Roman wormwood, hog weed, bitter weed, sarriette.

Rain–Precipitation in the form of water condensed from the atmosphere which falls in drops.

Rain Belt–A region of heavy or above-average rainfall.

Rain Forest–A low-altitude, closed, evergreen (usually broad-leaf) forest, in equatorial regions where annual rainfall exceeds 80 inches with no or very short dry seasons and where the temperature is high and constant and does not approach freezing at any time.

Rain Shadow–Refers to an area in which little or no rain falls because it is located to the leeward or dry side of mountains which on the opposite side are exposed to moisture-laden winds.

Raindrop Erosion–Soil splash resulting from the impact of raindrops on bare supersaturated soil. The result is surface sealing and sheet erosion.

Rainfall Interception–The amount of rainfall retained by leaves, branches, trunks, and surface vegetation, both living and dead, which never reaches the soil.

Rainfall Penetration–The depth below the surface of the soil reached by a given quantity of rainfall.

Rainfed Agriculture–See Dry Farming.

Rainwater–Water that falls from the atmosphere in the form of rain, at times being caught and directed into cisterns for further use.

Rainy Season–In the tropics, the annual season of heavy or more-than-normal rainfall as contrasted to the dry season of little or no rainfall.

Raisin–The dried fruit of certain varieties of grapes.

Ramet–An individual member of a clone, descended through vegetative propagation from an ortet. See Ortet.

Ramie–*Boehmeria nivea*, family Urticaceae; a plant growing 6 to 8 feet and about one-half inch in diameter and bearing between the bark and the woody core a white, lustrous, strong fiber, which can be spun and woven. Native to China and Japan, it is an important crop there. Also grown to a very limited extent in southern Florida.

Ramsons–*Allium ursinum*, family Liliaceae; a bulbous herb with a pungent taste and odor. Its growth is especially objectionable in pastures for milk cows because the milk produced has an off flavor. It is most common in the south central United States in the spring. Also called buckram, ramps. See Wild Leek.

Rancid–Designating an offensive smell or taste resulting from the chemical transformation or putrefaction of fat, butter, milk, ice cream, and other products.

Random Sample–A sample taken without bias from an area or from a population in which every part has an equal chance of being taken, in contrast to systematic sampling.

Randomized Block–An experimental design for testing crop varieties or strains by dividing them into groups and assigning treatments at random to the units within each group. The allotment is done so that all units in the same group or block are as nearly identical as possible.

Rang Utilization–(1) For a single plant or species, the degree to which the foliage or herbage has been removed in percentage of the current growth within reach of livestock. (2) For an entire range, the relative amount eaten.

Range–(1) Uncultivated land, including forest land, which produces forage suitable for livestock grazing. (2) Specifically, a unit of grazing land used by an integral herd of livestock. (3) Ecologically, the geographic area of natural occurrence of certain plants or animals. (4) In the United States land survey, a tier of townships, according to number east or west of a principal meridian line. See Township. (5) The difference between extremes.

Range Caterpillar–*Hemileuca oliviae*, family Saturniidae; a range pest on wild grasses in the southwestern United States, which sometimes infests corn and other cultivated plants. Its larvae are covered with coarse, poisonous spines.

Range Condition–(1) The state of health or productivity of both soil and forage of a given range in terms of what it could or should be under a normal climate and under the best practicable management. (2) An animal that is in a sufficient state of health or condition to be kept on the range.

Range Count–A census made on a range of the animals using a grazing area as contrasted to feedlot, corral, driveway, or other similar counts.

Range Crane Fly–*Tipula simplex*, family Tipulidae; an insect whose dark, leathery maggots bore into and destroy plant roots. It is sometimes destructive on the ranges of southwestern United States.

Range Ecology–The specialized branch of ecology which deals with vegetational response to environmental factors on rangeland, especially with the effects of grazing.

Range Forage–Forage produced on rangeland. See Cured Forage, Forage, Green Forage.

Range Grasses–Grass vegetation on the range areas. There are three types recognized in western United States; (a) tall grasses, such as big bluestem; (b) medium grasses, such as wheatgrass; (c) short grasses, such as buffalograss and blue grama grass.

Range Improvement–Physical development such as a structure (fencing) or excavation (water holes) to facilitate management of range or livestock. Generally measures to manipulate species composition and density such as revegetation, controlled burning, chemical or mechanical control of undesirable plants to increase the grazing capacity of range or increase its usefulness for watershed, wildlife habitat, or recreation.

Range Indicator–Any plant community portraying the condition of its environment which can be used as an indicator for the condition of a range.

Range Plant–Any herbaceous or shrubby plant which grows on a range.

Range Plant Cover–The ground vegetation composed of all herbaceous and shrubby plants which are within easy reach of livestock on a range area regardless of whether or not they constitute forage.

Range Readiness–The stage of growth of the important palatable plants on the range and the condition of soil which permits grazing without undue compacting of the soil or endangering the maintenance of the plants.

Range Renovation–Improving rangeland by discing or other mechanical means, chemical treatment, or reseeding.

Range Reseeding–See Range Seeding.

Range Seeding–The process of establishing vegetation by the mechanical dissemination of seed.

Range Suitability–The adaptability of a range to grazing by livestock and/or game.

Range Type–An area of range which differs from other areas primarily by the difference in plant cover, such as grassland, browse, or conifer. One vegetation group can be distinguished from another group by difference of dominating species.

Range Woodland–Woodlands also used as range.

Rangeland–(1) Land on which the natural plant cover is composed principally of native grasses, forbs, or shrubs valuable for forage. (2) Land used for grazing by livestock and big game animals on which the natural potential climax community of plants is dominated by grasses, grasslike plants, forbs, and shrubs.

Rank–(1) Designating vegetative growth that is luxuriant, vigorous, and often coarse. (2) A vertical row; leaves that are two-ranked are in two vertical rows, and may be alternate or opposite. (3) In nomenclature, the position of a taxon in the taxonomic hierarchy.

Rape–*Brassica napus*, family Cruciferae; an annual herb grown for its seed and as a temporary pasture. The leaves are eaten by people as a green vegetable. See Canola, Colza Oil, Winter Rape.

Rape-seed Meal–A livestock feedstuff obtained from grinding the cake produced when winter rape seed is pressed to recover the oil. It contains approximately 5.7 percent nitrogen, 1.7 percent phosphoric acid and 1.2 percent potash. See Colza Oil.

Raspberry–Any of the various species of ***Rubus***, family Rosaceae, consisting of brambly canes which yield red, black, purple, and yellow berrylike fruit used fresh or frozen for canning, jams, and preserves. See Red Raspberry.

Raspberry Late Rust–A fungal disease of raspberry leaves caused by ***Pucciniastrum americanum***, family Melampsoraceae; characterized by the infected leaves being covered with light yellow to orange masses of fungus spores turning brown to black due to the development of the teliospores. It occurs only on red raspberries.

Raspberry Verticillium Wilt–A soil-borne fungal disease caused by ***Verticillium alboatrum***, family Moniliaceae; characterized by wilting, yellowing, and defoliation. Black streaks are found in the wood.

Rat–*Rattus*, family Muridae; a long-tailed rodent which is much larger than a mouse. A serious pest, it is very destructive to stored food and may carry disease.

Rat Poison–(1) A poison for rats obtained from the seed of a shrub, ***Chailletia toxicaria***, native to West Africa. (2) ***Hamelia erecta***, family Rubiaceae; a tall evergreen shrub, bearing scarlet or orange flowers and small, purple-red fruit, sometimes grown as an ornamental in warm areas. Native to tropical America. Also called scarlet bush.

Rate of Growth–(1) The rate at which a tree has laid on wood, measured radially in the trunk or in timber cut from the trunk. (The unit of measure in use is the number of annual growth rings per inch.) (2) The rate at which a young animal increases weight and height.

Rate of Seeding–The amount of seed sown or planted per unit area, usually an acre.

Rate of Spread–In forestry, the quantitative increase in area per unit of time, used to describe forest fires and serious new forest disease.

Ratio of a Fertilizer–The percentage in a fertilizer reduced to a smaller expression; e.g., a 4-10-20 fertilizer has a ratio of 1-2.5-4.

Ratoon–(1) A basal sucker used for propagation, as in sugarcane, pineapple, and banana. (2) Second and subsequent crops grown from the root systems of previous plantings of sugarcane and of rice in the tropics.

Rattail Radish–*Raphanus sativus* var. ***caudatus***, family Cruciferae; an annual or biennial herb, grown for its edible, long, curved, often twisted pod, eaten either raw or pickled. Native to Europe and Asia.

Raw Cane Sugar–The crude, yellowish or brown unrefined cane sugar.

Raw Cotton–Unprocessed cotton.

Raw-rock Phosphate–Phosphate rock, finely ground but otherwise untreated, which is sometimes used as a slow-acting fertilizer on acid soils.

Ray–(1) In wood anatomy, a ribbon-shaped strand of tissue formed by the cambium and extending in a radial direction across the grain which serves to store food and transport it horizontally in the tree. (2) In botany, one of the small flowers radiating out from the margin of a dense inflorescence.

Ray Floret–(1) The marginal flowers of an inflorescence distinct from the disk. These are the florets in a flower, such as a sunflower, which have well-developed petals. (2) Flowers with florets that have only pistils.

Reaction–(1) A change in a market trend. (2) The degree of acidity or alkalinity (e.g., of a soil mass) expressed in pH values and in words as follows: extremely acid, below 4.5; very strongly acid, 4.5-5.5; medium acid, 5.6-6.0; slightly acid 6.1-6.5; neutral, 6.6-7.3 (strictly, 7.0); mildly alkaline, 7.4-8.0; strongly alkaline, 8.1-9.0; very strongly alkaline, over 9.1.

Reagent–Any substance involved in a chemical action.

Reap–To cut or harvest, as a crop of grain.

Recent Soil–A soil which has a profile without definite horizons of eluviation or illuviation; e.g., recently deposited alluvium. See Eluviation, Illuviation.

Receptacle–(1) The swollen part of a plant stem which forms the base of a flower (floral receptacle). Also called thalamus, torus. (2) The stem of an inflorescence (receptacle of inflorescence).

Recessive–In genetics, a gene or trait which is masked by a dominant gene.

Reconnaissance–(1) A cruise of forest property to obtain general information of the forest conditions. (2) An extensive range survey which is carried out to estimate average density and composition of range vegetation within a type or subtype without use of systematically established plots. (3) A type of survey in which land features are examined at wide intervals and are not delineated in detail. It is intended to furnish information primarily for extensive or overall planning. See Soil Map.

Recurrent Selection–In corn breeding, a method to determine which plants of a composite make the best hybrids in combination with the particular single cross used as a tester.

Red and Black Harlequin Bug–See Harlequin Bug.

Red Ash–*Fraxinus pennsylvanica*, family Oleaceae; a hardy deciduous tree, grown especially in the prairie states of the United States for shade, elsewhere for its lumber. Its wood is used in handles, furniture, oars, baseball bats, cabinets, boat buildings, implements, woodenware, novelties, cooperage, planing-mill products. Also called river ash.

Red Cabbage Group–One of the eight groups of cabbage distinguished by its purplish-red leaves.

Red Clover–*Trifolium pratense*, family Leguminosae; a forage herb. Native to Europe, it is grown extensively in the northeastern quarter of the United States and locally in the northern Rocky Mountain and Pacific Coast states. A perennial, it behaves under cultivated conditions as a biennial. There are two types (a) the widely grown double cut, *T. pratense*, medium red or June clover, the first crop being used for hay and the second for either hay or seed, depending upon seasonal conditions; (b) the single cut, *T. pratense* var. *perenne*, called mammoth red, bull, sapling or peavine clover, producing only one crop a year which may be harvested for either hay or seed. The stems are coarser and the quality of forage usually lower than the double cut. The red clover is also called broad-leaved clover.

Red Clover Powdery Mildew–A fungal disease of red clover caused by *Erysiphe polygoni*, family Erysiphaceae, characterized by patches of fine, white, cobwebby mycelium developing on the upper surface of the leaves until the entire field appears white.

Red Clover Southern Anthracnose–A fungal disease of red clover caused by *Colletotrichum trifolii*, family Melanconiaceae, characterized by irregular, dark brown spots on the leaves spreading to the stems and petioles as small water-soaked spots, lengthening to form depressed, black lesions and browning of the entire stem. It is a major disease in the southern red clover section of the United States.

Red Clover Vein-Mosaic–A viral disease of red clover causing vein chlorosis, mild to severe mottling, and chlorotic patches between the veins. It is more prevalent during the cooler growing season. Aphids are very important carriers. Damage ranges from little to severe.

Red Cluster Pepper–*Capsicum frutescens* var. *fasciculatum*, family Solanaceae; a variety of bush red pepper, used as a condiment.

Red Core–See Strawberry Red Stele.

Red Durum Wheat–A class of spring wheat grown primarily for livestock feed and not used for the manufacture of macaroni as are other durum wheats.

Red Fescuegrass–*Festuca rubra*, family Gramineae; a stoloniferous, perennial grass with weak underground stems, which is an excellent lawn grass especially on sandy, acid, low-fertility soils and for shady areas. It is sometimes used in pasture mixtures on poorer soils. Native to Europe, Asia, and North America.

Red Fir–(1) *Abies magnifica*, family Pinaceae; an evergreen tree grown for its valuable lumber used for general construction, boxes, crates, pulpwood, and planing-mill products. Native to North America. Also called California red fir. (2) See Douglas-fir.

Red Fungus of Citrus White Fly–*Aschersonia aleyroidis*, family Zythiaceae; a fungus that under favorable conditions attacks and destroys the citrus white fly and related species in the tropical and subtropical regions. Its rarely found perfect stage is *Hypocrella libera*, family Hypocreaceae. This fungus has been cultured in pure culture on sweet potato strips and the spores sprayed on trees infested with the white fly.

Red Grain Sorghum–A market class of grain sorghums as provided by the Official Grain Standard of the United States which includes all varieties of red grain sorghums and may include not more than 10 percent of grain sorghums of other colors.

Red Harvester Ant–*Pogonomyrmex barbatus*, family Formicidae; an ant about ¼ inch long, with a black head and legs, and red abdomen. It lives in large colonies and makes bare circles 2 to 12 feet in diameter, with a deep nest. It is injurious to grasslands in southwestern United States because it feeds on grass seed. Also called agricultural ant.

Red Haw–Downy hawthorn, *Crataeus* spp. See Hawthorn.

Red Hickory–*Carya ovalis*, family Juglandaceae; a deciduous tree grown for its valuable wood, used in tool handles, ladders, furniture, sporting goods, implements, woodenware, and for fuel and smoking meat. It is also grown as a shade tree. Native to North America. Also called mockernut, oval pignut hickory.

Red Maple–*Acer rubrum*, family Aceraceae; a large tree, grown as an ornamental and as a shade tree for its brilliant scarlet and yellow, autumnal foliage. Native to North America.

Red Mite–One of the spider mites of the family Tetranychidae, which is a plant feeder and which may do serious damage to orchard trees, field crops, and greenhouse plants.

Red Mulberry–*Morus rubra*, family Moraceae; a tree which produces red or purplish-red, edible fruit about one inch long. Native to North America. Also called American mulberry. See Mulberry.

Red Oak–Northern red oak, *Quercus rubra*; southern red oak, *Q. falcata*; a tree whose lumber is valued for its strength and durability. It is also valued as lumber for constructing furniture.

Red Oats–A market class of oats designated by the color of the lemma and palea of the matured grain in the Handbook of Official Grain Standards of the United States.

Red Onion–See Onion.

Red Pepper–Any plant of the genus *Capsicum*, family Solanaceae. The edible, usually red fruits are very pungent except in the so-called sweet peppers. In their tropical homes they are woody shrubs, but in temperate-zone cultivation they are treated as annuals, being killed by heavy frosts. The dried fruits of the pungent kinds are ground and sold as red pepper, Cayenne pepper, capsicum, etc. Native to Central America.

Red Pine–*Pinus resinosa*, family Pinaceae; a quick-growing, hardy, evergreen tree, grown as an ornamental and for its valuable lumber. Native to North America. Also called Norway pine.

Red Raspberry–(1) *Rubus idaeus*, family Rosaceae; a shrubby plant, grown for its edible red fruit. It is the parent of many cultivated varieties. Also called redcap. (2) Any variety of raspberry which produces a red fruit. See Raspberry.

Red Raspberry Mosaic–A viral disease, prevalent in mild or latent form in red raspberries, which produces more severe mottling and splotches of yellow and green on the leaves of black raspberries. Crops suffer loss of quality and yield. Also called green mottle mosaic.

Red Rice–*Oryza rufipogon*, family Gramineae; an annual cereal herb, origin unknown, which has a red color in the grain; a very serious weed pest in rice fields as it has many of the same characteristics as white rice and is very difficult to eradicate. Its presence lowers the yield, grade, and value of the threshed grain of rice.

Red Rot of Cole Crops–A physiological disease which apparently results from a boron deficiency in cole crops, as cauliflower and cabbage. First symptoms appear as a water-soaked area in the stem changing to a rusty brown, causing thickening and downward curl of older leaves.

Red Spider Mite–*Tetranychus telaris*, family Tetranychidae; a mite which makes mealy cobwebs on the undersides of the leaves of a large number of plants, sometimes covering an entire new shoot. Also called two-spotted spider mite.

Red Spring Wheat–A subclass of hard red spring wheat established by the United States Grain Standards Act, United States Department of Agriculture, which contains 25 percent or less of dark, hard, and vitreous kernels.

Red Spruce–*Picea rubens*, family Pinaceae; an evergreen, coniferous tree which is a chief source of paper pulp. Native to North America.

Red Tart Cherry–See Sour Cherry.

Red Three-Awn–*Aristida longiseta*, family Gramineae; a densely cushioned bunchgrass with sharp, pointed leaves, which has some forage rating depending upon season pastured. Native to southwestern United States.

Red Turnip Beetle–*Entomoscelis americana*, family Chrysomelidae, an insect which feeds on turnips, cabbage, radishes, and other garden crops.

Red Water Disease–See Cattle Tick Fever.

Red Winter Wheat–A subclass of the soft red winter wheat class, established by the United States Grain Standards Act, which produces the flour for the making of cake and biscuits.

Red Worms–*Stongylus*, family Strongylidae; nematodes which infest the large intestine of equines. Their red color is due to the presence of hemoglobin in the bodies of the worms, and not to blood sucked from their hosts.

Red-backed Cutworm–*Euxoa ochrogaster*, family Noctuidae; an insect whose larvae feed on garden vegetables and other plants, cutting them off near the soil surface. Widely distributed in Canada and the northern states of the United States.

Red-headed Pine Sawfly–*Neodiprion lecontei*, family Diprionidae; an insect whose young larvae destroy the needles of numerous species of pines. Found in eastern United States, it is often very destructive.

Red-stemmed Peavine–*Astragalus emoryanus*, family Leguminiosae; a vetchlike herb with alternate compound leaves and small, pealike flowers in clusters which sometimes causes poisoning of livestock. Native to Texas, United States.

Red-winged Pine Beetle–*Dendroctonus frontalis*, family Scolytidae; an insect pest of pine. The adult beetles lay eggs in tunnels between the bark and wood, and the very small larvae or grubs feed on the inner bark tissue. When present in very large numbers, they may eventually girdle the tree.

Redbud–*Cercis canadensis*; a wild shrub that blooms in early spring. See Eastern Redbud.

Redcedar–Eastern, *Juniperus virginiana*; western, *Thuja plicata*; a tree that produces lumber valued for its fragrance and decay-resisting qualities.

Redheart of Lettuce–A physiological disease of lettuce characterized by chestnut-brown discoloration and breakdown of inner leaves, with outer leaves developing brown pits on the mid-rib and veins. It is caused by a lack of sufficient oxygen due to poor aeration or by prolonged exposure to low temperatures during shipment and storage.

Redolent–Fragrant, as flowers.

Redroot Amaranth–*Amaranthus retroflexus*, family Amaranthaceae; an annual herb which is a weed pest of gardens, field crops, and waste ground. Cases of livestock being poisoned from feeding on the plant have been reported. Also called careless weed, Chinaman's redberry, rough pigweed, green amaranth.

Redtop–*Agrostis alba*, family Gramineae; a stoloniferous, creeping, perennial grass which produces a coarse, loose turf, and is tolerant of wet acid soils. It is used in mixtures for pasture, lawns, and golf courses. Native to Europe; widespread in the northeastern quarter of the United States. Also called Burden's grass, fiorin, herd's grass, meadow redtop.

Redwood–*Sequoia sempervirens*, family Pinaceae; an evergreen tree which grows to 340 feet and is valued as an important source of lumber. Native to California and Oregon, United States. See Giant Sequoia.

Reed–(1) Any tall, slender plant, usually having coarse and jointed stems, including certain grasses and grasslike plants. (2) *Phragmites communis*, family Gramineae; the common reed, a widespread, perennial marsh grass with erect culms 10 to 15 feet tall, and stout, creeping rhizomes; used for making mats, thatching, cordage, and carrying nets. Native to north temperate zone and South America.

Reed Canarygrass–*Phalaris arundinacea*; used in swampy areas that are under water for long periods. Uniform grazing and haying will be difficult, but reed canarygrass can be very productive if grazed under careful management. Can become a serious weed in drainage and irrigation ditches. Gets quite coarse if allowed to overmature.

Reed Mace–See Cattail.

Reed Meadow Grass–*Glyceria grandis*, family Gramineae; a tall, coarse, perennial pasture grass growing in wet areas. Native to North America.

Reedgrass–Any grass of the genus *Calamagrostis*, family Gramineae. Species are tall, perennial grasses with creeping rhizomes, valuable as range grass and for wild hay.

Reentry Period–The length of time that is required by law between the application of certain hazardous pesticides and the entrance of people into the area without protective clothing.

Reference Sire–A bull designated to be used as a benchmark in progeny testing other bulls (young sires). Progeny by reference sires in several herds enable comparisons to be made between bulls not producing progeny in the same herd(s).

Refined Foods–Food products from which some of the minerals, vitamins, and/or other elements have been removed during processing and manufacturing, as white flour, degerminated corn products, and polished rice.

Refined Sugar–Sugar freed from impurities.

Reforestation–The natural or artificial restocking with forest trees of an area previously under forest.

Refuge Cover–Vegetation of sufficient density that can afford an unusual degree of protection to wildlife from hunters or predators.

Registered Seed–The progeny of foundation or registered seed, approved and certified by the certifying agency, that is so handled as to maintain satisfactory genetic identity and purity.

Regolith–Soil and mantle rock; saprolith. The layer or mantle of loose, incoherent rock material, of whatever origin, that forms the surface of the land and rests on bedrock. It comprises rock waste of all sorts, including volcanic ash and glacial drift.

Regrassing–Reestablishing grass on areas where it was once prevalent but had been killed out by overgrazing or by some environmental condition, such as drought.

Regression–(1) Destruction of the vegetation, as by fire, grazing, cutting, etc., usually with subsequent deterioration of the site, as by exposure, erosion, or loss of nutrients, to such extent as to give rise to a subsequent simpler vegetative type. It is not a true succession or development from forest to grassland, but a replacement as a consequence of complete destruction of the trees, etc. (2) Measure of the relationship between two variables. The value of one trait can be predicted by knowing the value of the other variable; e.g., easily obtained carcass traits (hot carcass weight, fat thickness, ribeye area, and percent of internal fat) are used to predict percent cutability.

Regrowth–Herbage that grows after grazing or after the plants have gone through a period of dormancy. See Aftermath.

Regulation Cut–The determination, periodically or in advance of cutting, of the volume of timber to be felled under the objectives of a given management plan.

Rejuvenate–The use of lime and fertilizer to enhance soil productivity.

Relative Age–In soil development, the appearance of the profile features; e.g., a soil along a stream may receive floodwater sediment every year or two and therefore remain relatively young in profile development.

Relative Humidity–The ratio of the mass of moisture actually present in any volume of air at a given temperature to the maximum amount possible at that temperature and pressure, usually expressed in percentage.

Relict–A plant or a community of plants that has survived some important change in natural conditions, such as climate, etc.

Remontant–(1) Having a second blooming period the same growing season, as with certain roses, delphiniums, and some other plants. It can be induced by special care, as feeding and pruning. (2) Describing certain hybrid perpetual roses.

Rendering Plant–Establishment for the rendering of dead, condemned, or diseased animals unfit for food.

Reniform–Refers to a leaf that is shaped like a kidney or bend with a round apex.

Renovate–To renew, as by pruning old wood to induce fresh growth; to fertilize and spray old lawns to produce new growth; to repair and again make usable as to renovate a machine or building.

Repens–Reptans, creeping.

Replication–The systematic laying out of several test rows or plats to test soil fertility and other environmental factors.

Repot–To place a growing plant in a larger pot to prevent roots from becoming potbound. See Potbound.

Reproduce–(1) Of animals, to bring forth young. (2) Of plants, to bear fruit and seeds.

Reproduction–(1) The production of offspring by organized bodies. (2) The creation of a similar object or situation; duplication; replication. Asexual reproduction; reproduction from vegetative parts. Sexual reproduction; reproduction by the fusion of a female sexual cell and a male sexual cell. Parthogenic reproduction; reproduction by the development of an unfertilized egg.

Reproduction Period–The time required or normally decided upon for the renewal of a stand of trees by natural or artificial regeneration.

Repulsion–In genetics, the condition in which an individual heterozygous for two pairs of linked genes receives the dominant member of one pair from one parent and the dominant member of the second pair from the other parent; e.g., AAbb x aaBB.

Research–All effort directed toward increased knowledge of natural phenomena and the environment and toward the solution of problems in all fields of science. This includes basic and applied research. Much of the agricultural productivity of the United States is directly the result of applying research.

Reseed–To seed a second time on an area where the first seeding has failed due to unfavorable seasonal conditions, overgrazing, or other environmental conditions.

Reserve–(1) Any tract of land, especially public land, set aside for some special use; e.g., forest reserve, school reserve. Also called reservation. (2) A tree or group of trees left uncut on an area for a period, usually a second rotating. After the stand is reproduced, naturally or artificially, an active stand which is held for future utilization.

Reserve Food–Plant food stored in the various parts of any plant, especially that stored in the roots and crowns of biennial and perennial grasses and legumes.

Reserve Soil Acidity–Potential soil acidity. See Reaction.

Residual–(1) Remaining in place after all but the least soluble constituents have been removed. Said of the material eventually resulting from the decomposition of rocks. (2) Standing, as a remnant of a formerly greater mass of rock or area of land, above a surrounding area which has been eroded. Said of some rocks, hills, mountains, mesas, plateaus, and groups of such features. (3) Soil developed in place from underlying bedrock. See Monadnock.

Residual Effect–The effect of a particular product or substance that remains after its initial use. Some insecticides and fertilizers remain effective for some time after they are applied.

Residual Stand–Trees, often of saw log size, left in a stand after thinning to grow until the next harvest. Also called reserve stand.

Residual Value–Of fertilizers, limes, and manures. The value of the fertilizer, lime, or manure to succeeding crops after it has been in the soil for one cropping season.

Residue–A deposit of an pesticide which persists on an exterior surface or which is absorbed by plant or animal tissue following treatment.

Residue Management–Of crops. Use of that portion of the plant or crop left in the field after harvest for protection or improvement of the soil. See Conservation Tillage.

Residues–(1) Coarse residues: plant residues suitable for chipping, such as slabs, edgings, and ends. (2) Fine residues: plant residues not suitable for chipping such as sawdust, shavings, and veneer clippings. (3) Plant residues: wood materials from primary manufacturing plants that are not used for any product. (4) Logging residues: the unused portions of sawtimber and poletimber trees cut or killed by logging. (5) Urban residues: wood materials from urban areas, such as newspapers, lumber, and plywood from building demolition, and used packaging and shipping wood materials.

Resinosis–An abnormal exudation of resin or pitch from conifers or the abnormal impregnation of their tissues by resin. Also called resinflux.

Resins–A class of flammable, amorphous, vegetable substances secreted by certain plants and trees, and characterizing the wood of many coniferous species. They are oxidation or polymerization products of the terpenes, and consist of mixtures of aromatic acids and esters. Produced form oleoresins.

Resistant–Designating a plant or animal capable of withstanding disease, inclement weather, or other adverse environmental conditions.

Respiration–(1) A chemical process that takes place in living cells whereby food (fats, carbohydrates, and proteins) is "burned" (oxidized) to release energy and waste products, mainly carbon dioxide and water. Living things use energy produced through respiration to drive vital life processes such as growth and reproduction. (2) The oxidation of carbohydrates in living organisms and the attendant release of energy and liberation of carbon dioxide and water. (3) In animals, the act of breathing; the drawing of air into the lungs and its exhalation. In small organisms with no special breathing organs, the process takes place over a large part of the body surface.

Respirator–A device worn over the nose or mouth to protect the respiratory tract during the spraying of pesticides, working in dusty conditions, etc.

Respond to–To show increased growth or favorable reaction, due to an application of fertilizer or any other treatment.

Rest Period–A period of quiescence or inactivity in plants even though moisture and temperature conditions are favorable for growth. See Dormancy.

Rest-rotation Grazing–An intensive system of management whereby grazing is deferred on various parts of the range during succeeding years, allowing the deferred part complete rest for one year. Two or more units are required. Control by fencing is usually necessary on cattle range, but may be obtained by herding on sheep ranges.

Rested Pasture–Pasture ungrazed for an entire growing season.

Resting Land–See Fallow.

Resting Pasture–A pasture not grazed by livestock.

Resting Spore–A spore, often thick-walled, that can remain dormant for some time before germinating and becoming capable of initiating growth or infection.

Restorer Line–Parent plants used to restore adequate levels of fertility in hybrid crops such as hybrid sorghum or hybrid cotton.

Restricted Use Pesticide–A pesticide that can be applied only by a certified applicator. These pesticides are toxic to humans and/or pose a potential threat to the environment.

Restriction Enzymes–Enzymes used in genetic engineering to remove a gene from a piece of DNA.

Ret–To prepare flax, hemp, and some other fiber-producing plants in order to separate the fibers from the straw. In dew retting the flax is spread out in the field and wetted by dew, the moisture stimulating the action of bacteria and molds. Water retting is accomplished by soaking the flax in water at summer temperature for several days.

Reticulate Mottling–In oxisols, a network of coarse streaks of different colors characteristic of parent materials.

Retuse–A leaf apex that has a shallow indentation. See Emarginate.

Revegetation–Reestablishment of vegetation which may take place naturally or be induced by humans through seeding or transplanting.

Reverse Osmosis–(1) An external force is used to reverse normal osmotic flow through a semipermeable membrane, resulting in movement of water from a solution of higher solute concentration to one of lower solute concentration. See Osmosis. (2) A process of desalination of seawater whereby only pure water passes through a fine membrane while the salts cannot pass through.

Reversion–The interaction of a soluble plant nutrient with the soil which causes a precipitation of the nutrient in a less-soluble form, usually restricted to the conversion of monocalcium phosphate to the less-soluble dicalcium phosphate.

Reverted Phosphoric Oxide–The phosphoric oxide in phosphate fertilizers may be divided into three parts: (a) water-soluble, (b) insoluble in water but soluble in neutral ammonium citrate, and (c) insoluble in water or neutral ammonium citrate. The addition of lime, limestone, cyanamide, ammonia, or similar basic materials may cause some reversion of phosphoric oxide to less-soluble forms. Reverted phosphoric oxide is that which has been changed from the water-soluble to the citrate-soluble form or from the citrate-soluble to that phosphoric oxide that is neither water-soluble nor citrate-soluble.

Rhizobia–Bacteria that live symbiotically in roots of legumes and fix nitrogen from the air.

Rhizoctonia Disease–A disease with various symptoms occurring in many different plants, caused by the fungus usually called *Rhizoctonia solani*, which is the imperfect (asexual) stage classified as belonging to class Fungi Imperfecti, order Mycelia Sterilia. This may infect the underground growing tips of roots or germinating seeds or stems near the surface of the ground causing damping-off. Various aboveground parts of the plant may be infected, such as stems, leaves, and fruits. Under certain conditions the fungus may cover parts of the stems and leaves with a grayish-white powdery growth on which may develop the perfect (sexual) stage, classified variously as *Corticium vagum*, which then bears the name *Pellicularia filamentosa*, and *Thanatephorus cumeris*, family Tehelephoraceae.

Rhizoctonia Disease of Potato–A widespread fungal disease of the Irish potato caused by *Corticium vagum*, family Thelephoraceae, characterized by the production of stem lesions below the ground level, the appearance on the tubers of black specks that will not wash off; various minor injuries and secondary or accompanying effects of rather uncertain relationship. Also called black scab, black scurf, black speck, brown stem, damping-off, dartrose, dirt that will not wash off, little potatoes, russet, stem canker. See Rhizoctonia Disease.

Rhizoid–A rootlike structure which is capable of attaching itself to a surface.

Rhizomes–Elongate underground stems or branches of a plant which send off shoots above and roots below and are often tuber-shaped. These contain deposits of reserve food material and are used for vegetative propagation of plants.

Rhizophora–The true mangroves, tropical and semitropical trees and shrubs of the family Rhizophoraceae. *Rhizophora mangle*, the common or American mangrove is native in marshes under sea influence, tidal shores, and shoals from southern Florida (United States) through Central and South America. In such regions the mangrove acts as a protection for shores and as an aid in building organic soil, and affords suitable background for the development of property.

Rhizophore–A leafless stem that produces roots, as in *Selanginella*.

Rhizosphere–The root zone, the area where microorganisms are most active in increasing the availability of nutrients for plants.

Rhodesgrass–*Chloris gayana*, family Gramineae; an erect, perennial grass, propagated by seed and by running branches that root at the joints, grown for forage and as an ornamental in warm areas. Native to South Africa.

Rhododendron–Certain north temperate zone, evergreen or deciduous shrubs of the genus *Rhododendron*, family Ericaceae. Species are grown as ornamentals. They require acid soils.

Rhomboid–A leaf that is shaped like a diamond. The petiole is attached at one of the sharper angles.

Rhubarb–An herb of the genus *Rheum*, family Polygonaceae. Species are grown for their edible leaf petioles used in sauces, pies, etc., and as ornamentals for their striking foliage.

Ribbon Cane–A variety of sugarcane, marked by a longitudinal reddish stripe on the stalk which is grown for syrup production in some southern states of the United States.

Ribes Eradication–In blister rust control, the removal of currant and gooseberry plants from stands of the five-needled pines for protection against white-pine blister rust.

Ribonucleic Acid (RNA)–The substance in the living cells of all organisms that carries genetic information needed to form protein in the cell.

Rice–*Oryza sativa*, family Gramineae; an annual marsh grass, one of the oldest cultivated crops grown for its seeds, the rice of commerce, one of the most important starchy foods of the world. There are two types, lowland and upland, the first being irrigated. Rice is also classified as long, short, and medium grain depending upon the length of grain. Native to Indonesia. It has been grown in China for at least 4,000 years and is now grown throughout the world, especially in eastern and southeastern Asia. Louisiana, Texas, California, and Arkansas are the principal producing states in the United States.

Rice Bran–A livestock feedstuff consisting of the pericarp or bran layer of the rice and only such quantity of hull fragments as is unavoidable in the regular milling of rice.

Rice Hull–A product which consists of the outer covering of the rice. It is low in digestible nutrients and unpalatable, but if well-ground may be used as a low-grade roughage for livestock.

Rice Meal–A livestock feedstuff consisting of ground brown rice.

Rice Polish(ings)–A livestock feedstuff consisting of the by-product obtained in the milling operation of brushing the grain to polish the kernels.

Rice Straw–The straw remaining after threshing rice. If well cured, it may be fed in the same manner as straw from the other cereals.

Rice Weevil–*Sitophilus oryza*, family Curculionidae; a destructive grain insect, inhabiting farmer's bins, grain elevators, and ships where grain lies undisturbed for some time.

Rice-Fish Rotation–The alternation of rice and fish crops. Rice fields, which would ordinarily lie fallow and idle, are flooded and made to produce a crop of fish for one or two years. In addition to yielding income from the fish, the land is rested and enriched for the following rice crop.

Ricegrass, Indian–*Oryzopsis hymenoides*; a cool-season, drought-resistant, perennial native bunchgrass adapted to sandy soils and desert ranges. Seed formation occurs throughout the season due to its indeterminate flowering habit. The seeds enhance forage value because of a high protein and fat content.

Rick–(1) Hay or grain carefully placed in a rectangular pile and ridged at the top which may be thatched for protection from the weather. (2) A pile of cordwood for use as firewood. It is 8 feet long, 4 feet high, and has a variable width. (3) A pile of cordwood, stave bolts, etc., split from short logs.

Ridge Land–Any broad or narrow, crested, elongated land feature which rises above or separates lowlands plains or valley basins, sometimes constituting a drainage divide.

Ridge Planting–The practice of plowing two furrows together and planting on the ridge so formed. A practice on flat land which would otherwise be excessively wet for cropping.

Ridging–(1) Pulling the soil into a low ridge at the base of plants. Potatoes are commonly ridged or hilled to keep the tubers covered and prevent greening caused by exposure to sunlight. (2) Making small embankments or borders in fields to control irrigation water, conserve runoff from rainfall, or to assist in drainage.

Riga Pine–Scotch pine.

Rill Erosion–An erosion process in which numerous small channels of only several centimeters (inches) in depth are formed; occurs mainly on recently cultivated soils.

Rim Fire–A potash deficiency disease of plants, characterized by necrosis of the leaf along its margins.

Rime–A white, icy coating on grass and leaves, formed by hoar frost, white frost, etc.

Rind–(1) A hard coating caused by the desiccation of the surface of cheese. (2) Skin of an animal, especially of a hog, as a pork rind. (3) Skin of a fruit or vegetable. (4) Bark of a tree. (5) To remove the skin.

Ring–(1) A cut or girdle around the trunk, branches, or roots of a tree. See Girdle. (2) Annual growth ring of a tree. See Annual Ring. (3) (a) A circular band of metal or wood, as the metal ring in the nose of a bull. (b) To place a ring through the cartilage of the nose of an animal; e.g., to prevent a hog from harmful rooting or to control a bull, etc. (4) (a) A circular, metal or plastic band placed on the leg of a fowl for identification purposes. (b) To place a ring on the leg of a fowl. Also called ringing birds. (5) A ridge which encircles the horns of a cow, the number increasing with age. (6) A circular exhibition place for the showing or sale of livestock or the racing of horses.

Ring Rot–A very destructive, highly infectious, bacterial disease of the potato tuber caused by *Corynebacterium sepedonicum*, family Corynebacteriaceae. A black ring discoloration in the outer flesh often results in only a shell around the soft decayed center.

Ring Spot–(1) A fungal disease of cauliflower and other crucifers caused by *Mycosphaerella brassicaecola*. (2) A disease of agaves, including sisal, caused by the fungus *Colletotrichum agaves*, forming pale circular lesions.

Ring Spot of Stone Fruits–A widespread, viral disease of stone fruits reported from Europe and North America, and frequently latent. The disease is characterized by chlorotic rings and necrotic spots on leaves and occasionally by die-back of shoots. Severe symptoms are usually limited to a time shortly following acquisition of the virus, after which they are rarely evident. There are numerous strains of the virus or similar viruses resulting in a gradient of symptoms from mild to severe. The virus may be mechanically transmitted to cucurbits and tobacco.

Ring-porous Wood–Any of a group of hardwoods in which the pores are comparatively large at the beginning of each annual ring and decrease in size more-or-less abruptly toward the outer portion of the ring, thus forming a distinct inner zone of pores known as the spring wood and an outer zone with smaller pores known as the summer wood.

Ringing–(1) Clipping the wool from a breeding ram around the neck, belly, and penis region in order to facilitate proper mating. (2) Putting a ring in the nose of cattle or hogs. (3) Removing a narrow strip of bark from around a branch or tree trunk to encourage fruiting. Only outer bark is removed, and the ring does not extend into the cambium layer.

Riparian Forest–Tree growth adjacent to streams or other watercourses whose roots are in or close to the zone of saturation due to the proximity of surface or underground water.

Ripe–(1) Designating mature seeds which are fit for gemination. (2) Designating fruit which has attained full development. (3) In plant propagation, designating wood that will root well. (4) In grafting, designating that wood which is ready for perfect union. (5) Designating the best condition for use, as ripe cheese, ripe wine.

Ripe Snow–Snow which during the process of melting and settling has attained its maximum power of water suspension without loss of water. Snow is overripe when it has exceeded its ability to hold water and is losing it. The difference in density between ripe and overripe snow does not exceed 10 percent.

Ripening–(1) Growing to maturity and being fit for food, as ripe fruit or ripe grain. (2) Bringing to a certain condition for use by keeping, as in wine. (3) Preparing milk or batch mixes for making butter or ice cream either by a natural souring or by the addition of starters. (4) Undergoing an aging process, as in meat.

Rise of Salts–Soluble salts, previously distributed in the soil, which is brought to the surface with the rise of capillary water and left as a surface incrustation when the moisture evaporates. This accumulation may cause crop injury. See Waterlogged.

Risky Treadsoftly–*Cnidoscolus texanus*, family Euphorbiaceae; an herbaceous plant with long perennial roots and stout, branched stems, whose bristly hair contains a strong irritant which upon contact produces a dermatitis. Native to North America. Also called bull nettle, spurge nettle, stinging nettle.

River Ash–See Red Ash.

River Bottomland–Low land which consists of alluvial deposits along a river. Generally, but not everywhere, it is composed of soil more fertile and potentially more productive than adjacent upland.

Riverwash–Alluvial deposition in the channel of rivers.

RNA–See Ribonucleic Acid.

Roasting Ear–An ear of field corn having its kernels well filled, plump, soft, or in the milky stage when it is ready to be eaten boiled or roasted; sometimes applied to sweet corn in the same stage of development.

Robust–Designating a strong vigorous animal or plant.

Rock Cedar–*Juniperus mexicana*, family Cupressaceae; a tall evergreen tree grown as an ornamental and for its aromatic, brownish lumber used for making boxes, pencils, etc. Native to southwestern United States and Mexico.

Rock Disintegration–The natural process by which hard rock is disrupted and comminuted. The factors involved are: expansion and contraction due to change in temperature, the chemical and mechanical work of plant roots, movements of stone fragments down slopes through force of gravity, a partial chemical solution, abrasion by moving water, ice, and lichens. See Lichens.

Rock Garden–A garden which contains alpine and rock plants displayed in an appropriate rock setting: a rockery.

Rock Phosphate–The natural mineral deposit essentially tricalcium phosphate $(Ca_3(PO_4)_2)$, which is the principal source of the phosphorus of commercial fertilizers. The raw rock finely ground is also applied to acid soils as a fertilizer. The principal mining (United States) at present is in Florida and Tennessee, but large deposits occur in Wyoming, Idaho, Utah, and Montana, and smaller deposits occur in South Carolina, Kentucky, and Arkansas.

Rock Plants–Plants which will grow well in rock gardens.

Rocky Mountain Crazyweed–*Oxytropis saximontana*, family Leguminosae; a poisonous plant found from Montana to Utah, United States, whose foliage is poisonous to cattle. Symptoms of poisoning are dullness, irregularity of gait, lack of appetite, dragging of the feet, a solitary habit, loss of flesh, and shaggy coat. As the animal ceases to eat it dies. Native to North America. See Loco.

Rod Weeder–A secondary tillage implement used primarily for fallow-land weed control with minimum soil surface disturbance and moisture loss. The device employs round or square weeding rods powered from gauge wheels, rotating under the soil surface.

Rodent–A classification of mammals, mostly vegetarians, characterized by their single pair of chisel-shaped, upper incisors. Rodents are members of the orders Rodentia (rats, mice, squirrels, etc.) and Lagomorpha (rabbits, etc.).

Rodenticide–Any poison which is lethal to rodents.

Rogue–(1) A seedling or plant of inferior or objectionable quality; a variation from type. (2) A horse without inclination to work or cooperate with its handler. (3) To remove and destroy undesirable plants.

Roll-over Terrace–A terrace which has little height and a broad base, hence gently sloping. Under certain conditions it is effective in conserving water and reducing erosion, and it has the advantage of offering no obstacle to the movement of farm equipment over fields. See Broad-base Terrace.

Rolled Oats–Oat groats which have been pressed between heated rollers.

Rolling–(1) Excessive side motion of shoulders, common in horses with abnormally wide fronts or chests. (2) A part of seed-bed preparation in which the land is rolled to even out the surface. (3) Processing grain through a set of smooth rollers which are close together; sometimes called flaking.

Rolling Land–A land surface which has rounded undulations, and relatively low relief.

Rome Beauty–A famous variety of apple whose fruit is large, red-striped, and of medium quality. The trees are small, very productive, and grow well southwest of New York, United States.

Roof Garden–A garden site, often elaborate in design, on the roof, terrace, or set back of a high building.

Room Temperature–In laboratory work, 68° to 70°F (20° to 21°C).

Root–The lower portion of a plant bearing neither leaves nor reproductive organs which mostly develops underground and anchors the plant in the soil. It bears the root hairs, which absorb water and mineral nutrients. Some roots are aerial or partly aerial. In some aquatic plants, the roots do not always anchor the plants as the plants float.

Root Absorption–Taking in of water and nutrients from the soil by the roots of a plant.

Root Antagonism–Injury to the roots of a species which is grown near certain other species or when one is grown after the other on the same plot, as root antagonism exists between tomato and cucumber roots.

Root Aphids–Certain sucking insects of the family Aphidae, which feed on the roots of many plants, as corn, cotton, grasses, and ornamentals, often causing serious damage. Ants collect the aphid eggs in the fall and store them over winter. In spring the young are carried to roots where, through a complex life cycle, many generations are produced. The aphids excrete honeydew which the ants use as food.

Root Cap–The extreme tip of the root consisting of a group of cells that slough off and are replaced as the tip moves through the soil. It protects the growing region of the root.

Root Climber–Any plant, as the common ivy, fig, etc., which uses its unusually located (adventitious) roots for climbing.

Root Collar (Root Crown)–The transition zone between stem and root usually recognizable in trees and seedlings by the presence of a slight swelling.

Root Crops–Any of a number of field and garden crops whose underground roots are used as food for people and animals, as turnips, beets, carrots, and sweet potatoes.

Root Cutting–The portion of the root of a plant used for propagation.

Root Grafting–A system of grafting in which the scion is grafted onto the root of a plant.

Root Hair–A hairlike growth on an epidermal cell of the root. It absorbs water and mineral nutrients for the plant.

Root Knot–A disease of plants caused by the penetration of nematodes, genus *Meloidogyne*, family Heteroderidae, which stimulates cell development, resulting in enlargement of the root system, and the formation of galls. Many crops, including potatoes, tomatoes, celery, spinach, cabbage, tobacco, cucumbers, sugar beets, and cotton, are hosts. Infested plants are paler, wilt more easily, and die early. Also called eelworm disease. See Nematode.

Root Lesion Nematode–Genus *Pratylenchus*; nematodes which infest at least 100 plant species including strawberry, coffee, cotton, tobacco, beets, corn, various small grain crops, and many legumes. They attack the roots of the plants causing wilting, yellowing, poor yield, and often death.

Root Pressure–The force with which water is pushed up through the intact stem of a plant or exuded from holes made in the stem or from fresh stumps of branches or stems. Also called pressure, endodermal pressure, bleeding pressure, exudation pressure.

Root Pruning–Any cutting back of the roots: (a) during potting of plants to contain roots in a smaller volume of soil; (b) before planting trees to reduce roots to the size of the hole already dug; (c) to reduce top growth; (d) to remove a root and promote branching for later removal of the tree with a ball of soil attached by undercutting the roots with a tree digger or power knife.

Root Reserves–The storage of plant food in the form of proteins, carbohydrates, etc., in the root.

Root Spade–A spade with a long, narrow blade that is used for digging up plants that are to be balled and burlapped. See Balled and Burlapped.

Root Spread–The extent of growth of the root system of a tree or other plant. Some trees have a root spread twice as great as the spread of the branches.

Root Sucker–A sprout which rises from a root.

Root Tubercle–A small, rounded swelling on a root.

Root Zone (Rhizosphere)–The part of the soil that can be penetrated by plant roots.

Root-bound–When plants have grown in a container too long the roots become a mass of fibers and no longer support desired top growth. See Potbound.

Root-hardy–Designating a plant whose roots can survive the winter.

Root-nodule Bacteria–See Rhizobia.

Root-tipping–Used for the tips of arching or trailing stems which root on touching the ground.

Rootbed–The soil depth modified by tillage or amendments in which plant roots are or will be growing.

Rooting–(1) The production of roots by a plant. (2) Digging the earth with the snout, as hogs. (3) A root used for propagation. (4) The production of roots by a cutting.

Rooting Media–Materials such as peat, sand, or vermiculite in which cuttings are placed during the development of roots.

Rootlet–A small root.

Rootstock–Root system upon which named varieties of fruit have been grafted; e.g., apple varieties are grafted onto dwarfing or size-controlling rootstocks.

Rope–(1) A thick, strong cord made by twisting or intertwining metal strands or fiber, as iron, steel, and manila hemp. (2) Form of bacteria that can survive baking of bread; under conditions of warmth and moisture, may convert mass of the bread into starchy patches.

Rope Wick Applicator–See Wick Applicator.

Rosary–A rose garden.

Rose Chafer–*Macrodactylus subspinosus*, family Scarabaeidae; a beetle that feeds on the leaves of the rose and many other plants including grape, apple, peach, cherry, corn, and beans.

Rose Hip–In rose plants, the much enlarged, fleshy, berrylike stem end upon which the flower is borne. The exact nature of hips varies with the particular variety of rose.

Rose Leaf Beetle–*Nodonota puncticollis*, family Chrysomelidae; an insect that infests buds, blossoms, and foliage of roses and fruit trees as well as clover and other meadow plants in the spring.

Rose Leafhopper–*Edwardsiana rosae*,, family Cicadellidae; an insect that infests plants of the rose family, especially the apple and rose.

Rosemary–*Rosmarinus officinalis*, family Labiatae; a perennial shrub bearing evergreen, strongly scented leaves and light blue flowers, grown as an ornamental and for use in seasonings and scents. Native to the Mediterranean area.

Rosette–An arrangement of leaves radiating from a crown or center and usually at or close to the earth, as in the dandelion.

Rosette Disease of Blackberries and Dewberries–A fungal disease of blackberries and dewberries caused by *Cercosporella rubi*, family Moniliaceae, characterized by an abnormal production of multiple, short, bunchy shoots from the overwintering canes.

Rosette Disease of Peanut–A viral disease of peanuts transmitted by *Aphis medicaginis, A. leguminosae*, and *A. craccivorae*, characterized by mottling and chlorosis of the leaves and stems. Internode elongation suppression results in a rosette or witches' broom growth habit.

Rosette Mosaic of Peach and Plum–A soilborne, viral disease of peaches and plums characterized by delayed foliation, chlorotic mottling of the leaves, and a shortening of the internodes which give a rosette appearance to the shoots.

Rosette of Peach–A viral disease of peaches prevalent in southeastern United States. The leaf buds develop compact tufts or rosettes which may contain several hundred small leaves on axes not more than two or three inches long. Partial infection soon spreads over the entire tree and death results.

Rosette of Pecan and Apple Tree–A zinc-deficiency disease of apple and pecan trees characterized by narrow and small, bunchy and erect leaves.

Rosin–Solidified amber-colored sap (resin) mostly from longleaf pine (*Pinus palustris*), occurring near the Gulf of Mexico in southern United States. Chemically it is mostly abietic acid anhydride ($C_{44}H_{62}O_4$). It is used as a stiffening agent in plasters and ointments, on violins, and on the hand of baseball pitchers. Rosin was formerly known as colophony.

Ross–To remove bark from, or to smooth a log.

Rot–A state of decay caused by bacteria or fungi. See Decay.

Rot in the Ground–Not to germinate, as seeds, due to one or more fungi in wet, cool weather.

Rot of Pineapple–A fungal disease of pineapple caused by *Ceratostomella paradoxa*, family Synchytriaceae, characterized by a destructive rot beginning at the stem end of the fruit. Also called blackheart of pineapple.

Rotated Pasture–(1) A pasture in the regular crop rotation which is grazed for a few years, usually two or three, and then plowed for other crops. (2) A pasture which is divided into segments by use of fences: the livestock being confined to one segment at a time in a definite rotation pattern.

Rotation–(1) In cropping, the growing of two or more crops on the same piece of land in different years in sequence and according to a definite plan. One of the most widely recommended crop rotations consists of (a) a tilled crop, often corn; (b) a small grain; (c) a legume or grass crop or a mixture of legumes and grasses often for more than one year. (2) In forestry, the period of years required to establish and grow timber crops to a specified condition of maturity.

Rotation Burning–The burning of brush and meadows every three years to provide more pasture grasses.

Rotation Grazing–Grazing forage plants on well-managed pastures in such a manner as to allow for a definite recovery period following each grazing period. This includes alternate use of two or more pastures at regular intervals or the use of temporary fences within pastures to prevent overgrazing.

Rotation Irrigation–A system of irrigation through which the irrigator receives his allotted quantity of water not at a continuous rate but as a large flow at stated intervals: e.g., a number of irrigators receiving water from the same lateral may agree among themselves to rotate the water, each taking the entire flow in turn for a limited period.

Rotenone–The common name of a botanical insecticide; the main toxic constituent in the roots of certain leguminous plants, such as *Derris elliptica* and *Lonchocarpus utilis* and *L. urucu*; moderately toxic to mammals, acute oral LD_{50} for rats 132 mg/kg; nontoxic to plants.

Rothrock Gramagrass–*Bouteloua rothrocki*, family Gramineae; a short-lived, perennial grass which grows in small bunches and resembles blue gramagrass. Occurring in the southwestern United States on mesas and dry slopes, its forage value is fair to good. Native to north America.

Rott Off–See Damping-off.

Rotted Manure–Animal dung which has undergone decay and is safe to use with growing plants.

Rotten Cull Trees–Live trees of commercial species that do not contain a saw log now or prospectively, primarily because of rot (e.g., when rot accounts for more than 50 percent of the total cull volume.)

Rotten Wood–Wood in a state of decay produced by an attack of bacteria or fungi.

Rough Broken Land–Land naturally so deeply and minutely dissected by stream cutting that it is nonarable. Technically, it is differentiated from badlands and gullied land.

Rough Land–Nonarable land on steep, broken slopes. It may be either stony or highly dissected by erosion.

Rough Lemon–A horticultural variety of lemon introduced by the Spanish into Florida, United States, where it is often used as a stock for other citrus fruits. Also called Florida rough lemon, French lemon mazoe.

Rough Lumber–Lumber as it comes from the saw and before it is planed.

Rough Rice–The hull-covered rice as it leaves the thresher. Also known as paddy rice.

Rough Trees–(1) Live trees of commercial species that do not contain at least one 12-foot saw log, or two noncontiguous saw logs, each 8 feet or longer, now or prospectively, primarily because of roughness, poor form, splits, and cracks, and with less than one-third of the gross tree volume in sound material. (2) All live trees of noncommercial species.

Rough-bearded Barley–Any barley with tiny barbs projecting from the beards. The barbs are very irritating to both humans and animals.

Roughstalk Bluegrass–*Poa trivialis*, family Gramineae; a perennial grass grown in pastures and as a shady-lawn grass. Native to Europe. Also called rough bluegrass, rough meadow grass.

Round-up–(1) The deliberate gathering of domestic animals, usually range cattle, for branding, fastening ear tags, injections, pesticide applications, inventory, and potential sales. (2) The brand name of a herbicide used as a foliar spray to kill most plants.

Roundhead Borer–Any of the large group of very small to very large, oblong beetles, called longicorns, having extremely long antennae and belonging to the family Cerambycidae. The larvae are white, long, slender, usually legless grubs, with enlarged thoracid segments, with a horny plate on the top surface of the first segment. Most of these are wood borers or cambium miners in injured, felled, dying, or dead trees. Some, however, infest and develop in living trees. Also called longhorn beetles, sawyers.

Roundhead Lespedeza–*Lespedeza capitata*, family Leguminosae; a weedy herb which grows even in the poorest soils and is used for pasturage and soil improvement. Native to North America. Also called bush clover, dusty clover.

Roundheaded Apple Tree Borer–*Saperdacandida*, family Cerambycidae; an insect larva which attacks the sapwood and heartwood of the apple, serviceberry, wild crab, hawthorn, mountain ash, quince, pear, and plum. Feeding on the cambium and heartwood, it weakens the tree and often kills it.

Roundwood Products–Logs, bolts, and other round sections cut from trees for industrial or consumer uses. Included are saw logs, veneer logs and bolts, cooperage logs and bolts, pulpwood, fuelwood, piling, poles and posts, hewn ties, mine timbers, and various other round, split, or hewn products.

Row Crop–A crop, as corn, beans, sugar beets, cotton, sorghum, etc., which is usually grown and cultivated in rows.

Row Mulching–Mulch placed on the row so that the plants emerge through the mulch.

Rowan–In New England, the regrowth of a cut hay crop. Also known as aftermath.

Rubber Rabbitbrush–*Chrysothamnus nauseousus*, family Carduaceae; a widely distributed shrub whose presence is an indication of range depletion. Native to western North America.

Rubber Tree–*Hevea brasiliensis*; a large tree that grows in tropical areas near the equator. It provides the principal source of the natural rubber of commerce.

Rubidium–A rare element known to appear in traces in soils and in plants. It probably enters into the same reactions in soil and plants as does potassium. In very small quantities it appears to stimulate plant growth.

Ruderal–Designating weeds which spring up on uncultivated or abandoned land.

Rudimentary–Imperfectly developed.

Rudimentary Ear–The female flower of the corn plant before fertilization.

Rugose–In plants, designating a condition of wrinkling.

Run–(1) A period of time, as a maple syrup run. (2) The amount of sap or sugar produced in a given time. (3) A swiftly moving tributary, rivulet, or mountain stream (eastern United States). (4) The stream outlet of a large spring (Florida, United States). (5) Unrestricted movement, as the colts have the run of the pasture. (6) An area of land or a leasehold (New Zealand). (7) A small, often dry gully or channel carved by water. See Arroyo. (8) A fenced-in pen used for the exercise of animals or poultry. (9) To feed; to graze, as steers run on the open range. (10) To operate, as a plow is set to run at a depth of 6 inches. (11) To cultivate, mow, combine, etc., as to run over a field with a weeder. (12) To maintain animals, as he runs sheep. (13) To work a dog with sheep or cattle. (14) To discharge pus, as a sore runs. (15) To be, as the prices run very high. (16) To move rapidly, as a horse. (17) To turn, as to run a wheel. (18) To grow, as when the vines begin to run. (19) To operate, as to run an engine.

Run On–To graze or pasture on, as to run on the range.

Run-off Modulus–The depth of water in inches over the underdrained area which must be removed by tile drains in 24 hours. It is a measure of the maximum rate at which the water will move through the soils to the laterals.

Runnel–A tree that has been pruned so that all of its main branches are cut back to the trunk.

Runner–(1) A breed of ducks of very distinctive type, having a long, narrow body and very erect carriage. The breed derives its name from its gait, which is a quick run, quite unlike the waddle of other ducks. The adult drake weighs about 4½ pounds; the adult duck about 4 pounds. The runner is noted as an egg-producing breed and has little or no value for meat production. Its three varieties are the White, the Fawn and White, and the Penciled. Also called Indian Runner. (2) A lateral, aboveground shoot (stolon) of certain plants; e.g., strawberries, which roots and forms young plants at some of the nodes, aiding in propagation. (3) A rope used to increase the mechanical power of a tackle. (4) The upper or rotating stone of a set of millstones. (5) A supporting attachment which slides along the ground, as a sled runner. See Stolon.

Runner Peanut–Any variety of peanut whose pods are scattered along procumbent rooting stems, as contrasted to the bunch type in which the pods are clustered at the base of the plant. See Bunch Peanut.

Runner Plant–A plant which is capable of being propagated by runners; a plant produced from a runner.

Running Fire–In forestry, a fire which spreads rapidly with a well-defined head but little crowning.

Running Out–The condition in which an improved variety of plant or animal is reverting to a former and inferior type or is losing some of its desired qualities.

Runoff–(1) The total stream discharge of water, including both surface and subsurface flow, usually expressed in acre feet. (2) The rate at which water is discharged from a drainage area, usually expressed in cubic feet per second per square mile of drainage area.

Russet–(1) Brownish, roughened areas on the skins of fruit, tubers of potatoes, etc., resulting from abnormal production of cork tissue which may be caused by diseases, insects, other injury, or by a natural varietal character. (2) Any of the varieties of apples having natural, brownish-colored skins.

Russet Scab–See Rhizoctonia Disease of Potato.

Russian Olive–*Elaeagnus angustifolia*, family Elaeagnaceae; a hardy, wind-resistant, spiny shrub or small tree grown for its ornamental fruit and silvery leaves in hedges and windbreaks. Native to Europe and Asia. Also called oleaster.

Russian Thistle–*Salsola kali*, family Chenopodiaceae; an annual weed. When the plant dies it breaks away from the ground blowing about as a tumbleweed. Native to Europe and Asia. Also called Russian tumbleweed.

Russian Vetch–See Hairy Vetch.

Rust of Alfalfa–A fungal disease caused by *Uromyces striatus medicaginis*, family Pucciniaceae, characterized by reddish-brown spots on the leaves and stems. See Rusts.

Rust of Peanut–A fungal disease caused by *Puccinia arachidis*, family Pucciniaceae, which appears sporadically. See Rusts.

Rust of Sorghum–A fungal disease caused by *Puccinia purpurea*, family Pucciniaceae, which occurs in limited amounts on both surfaces of the leaves of all the sorghums. See Rusts.

Rust of Sugarcane–A fungal disease caused by *Puccinia kuehnii*, family Pucciniaceae, characterized by elongated, orange-brown spots on both surfaces of the leaves. See Rusts.

Rust-resistant–Designating the quality possessed by certain varieties of plants to resist the attack of rust.

Rusts–Fungi which are obligate parasites on the higher classes of plants (ferns and seed plants), i.e., they are not capable of growth and development apart from their living hosts. The fungus is internal, consisting of fine, colorless, branching threads (hyphae) creeping between the cells of the host, from where they draw nourishment by means of small suckers, haustroia, which penetrate the walls of the host cells.

Rusty Tussock Moth–*Orgyia antiqua*, family Lymantriidae; a moth whose larvae feed on many kinds of fruit trees as well as on beech, birch, oak, willow, etc.

Rutabaga–*Brassica napobrassica*, family Cruciferae; a biennial herb grown for its yellowish, edible root used for human and animal consumption. Native to Europe and Asia. Also called macomber, Swede, Swedish turnip.

Rutin–A yellow plant pigment isolated from buckwheat and flue-cured, bright tobacco. It is used by the medical profession in treating cases of fragility of blood capillaries.

Rydberg Clover–*Trifolium rydbergii*, family Leguminosae; a wild clover of the Rocky Mountain region from Idaho to Arizona between elevations of 4,500 and 13,000 feet whose forage value is excellent. Native to the United States.

Rye–*Secale cereale*, family Gramineae; a hardy winter annual cereal grown as a cover crop, for forage, a bread grain, feed for livestock, and for use in the manufacture of whiskey. It is one of the principal bread grains of northern Europe. Native to Europe and Asia.

Rye Bran–A livestock feedstuff consisting of the coarse outer covering of the rye kernel separated from the cleaned and scoured rye.

Rye Distillers' Dried Grains–A livestock feedstuff consisting of the dried residue obtained in the manufacture of alcohol and distilled liquors from rye, or from a grain mixture in which rye predominates.

Rye Ergot–A fungal disease of rye caused by *Claviceps purpurea*, family Hypocreaceae, characterized at first by the appearance of the honeydew stage when the exudate accumulates as droplets on the surface of the floral structure and later by the blue-black sclerotial body replacing the kernels in the infected flowers. Some of the chemical compounds of the sclerotia have medicinal value. See Ergot.

Rye Red Dog–A by-product obtained in the process of milling rye flour, consisting principally of aleurone with small quantities of flour and fine bran particles and containing not more than 3.5 percent of crude fiber. See Aleurone.

Ryegrass–Any grass of the genus *Lolium*, family Gramineae. Species are grown in lawns and pastures. Also called ray grass. See Ryegrass, Perennial.

Ryegrass, Annual–*Lolium multiflorum*; a vigorous, winter-active annual grass adapted west of the Cascades to a wide variety of soil conditions. Can be grown under irrigation or on dry land where the effective precipitation is comparable to 12 inches (30 centimeters) or more. Makes a good winter cover crop or temporary seeding on disturbed areas. May be seeded with red clover for hay in short rotations. Establishes rapidly, is strongly competitive, and retards establishment of perennial grasses and legumes if it is seeded too heavily in a mixture.

Ryegrass, Hybrid–*Lolium hybridum*; though very similar in performance to perennial ryegrass, hybrid ryegrass has demonstrated seedling vigor equal to that of annual ryegrass. It is more compatible than annual or perennial ryegrass with long-lived grasses and legumes in erosion-control seedings. However, seeding rates above 3 pounds per acre in a mixture usually retard development of slower growing plants.

Ryegrass, Perennial–*Lolium perenne*; a relatively short-lived, rapidly developing vigorous perennial bunchgrass adapted to a wide variety of soil conditions. Can be grown under irrigation or on dry land where the effective precipitation is 15 inches (38 centimeters) or more. Well adapted to short rotations with clover. Retards establishment of other perennials if it is seeded too heavily in a mixture. Has good recovery after grazing in the spring but tends to go dormant in summer.

Sacahuista–*Nolina microcarpa*, family Liliaceae; a perennial plant with short woody stems found on certain Texas (United States) ranges. Sheep and goats gazing upon this plant have developed a disease similar to that produced by ingestion of lechuguilla agave. The leaves can be used in cordage and basketry.

Sacaton–*Sporobolus wrightii*, family Gramineae; a stout, perennial bunchgrass which grows from 2 to 8 feet tall on alluvial fans and in arroyos from souther California to northern Texas (United States). It is grazed by all stock, especially when it is green, and has a fair forage rating.

Saccharification–Conversion of polysaccharides in wood or other cellulosic material into sugars by hydrolysis with acids.

Saccharin–A white crystalline compound ($C_7H_5NO_3S$) several hundred times sweeter than sucrose ($C_{12}H_{22}O_{11}$). Used as a noncaloric sweetening agent.

Saccharomyces–A class of single-cell yeasts which selectively consume simple sugars.

Saccharose–Sucrose.

Sacked–Designating grain, produce, etc., which has been placed in sacks.

Sacred Cotton–*Gossypium hirsutum* var. *religiosum*, family Malvaceae; a variety of upland cotton that bears rust-colored lint.

Saddle–(1) A seat designed to fit a horse's back and to make riding easy, comfortable, and safe. Saddles are of many types but are usually made of heavy leather over a well-padded frame. They are often raised in the rear to hold the body in place and they may have a firm, raised support or horn in front to attach ropes or gear. They are held to the horse's back by means of one or more wide straps or girths that pass around the body of the horse. (2) The whole upper back section of a meat carcass which in wholesale cuts is commonly divided into fore and hind saddle. In cookery, the rear upper back portion including the two loins. (Sheep, veal calves, goats, and deer.) (3) The rear part of the back of a male chicken extending to the tail, covered by saddle feathers. (4) An apron made of reasonably heavy canvas which is placed on the backs of valuable breeding hens to protect against injury during mating. Saddles are usually held in place by loops which fit under the wings. (5) The uncut stalks of four hills of corn from two rows, tied together, around which corn is shocked. (6) A saddlelike depression in the crest of a ridge. See Saddleback. (7) A specially designed transverse log used in a skid road to guide the moving logs. (8) To put a saddle upon an animal or thing.

Saddle Graft–A graft in which a wedge-shaped stock fits into a deep cleft cut in the scion. It is used more on herbaceous than on woody plants.

Saddleback Caterpillar–*Sibine stimulea*, family Limacodidae; a caterpillar with a green patch in the middle of the back, widely distributed through the Atlantic states of the United States, which feed on oak and cherry but sometimes on other plants.

Safener–Chemical coatings applied to seeds before planting that cause the emerging seedling to detoxify soil-applied herbicides after germination. Such coated seeds allow farmers to use stronger herbicides with less risk of crop injury. Sorghum was the first seed for which a commercial safener became available.

Safflower–*Carthamus tinctorius*, family Compositae; an herb, resembling a thistle, cultivated for its orange-colored flower heads which yield a drug and red dye, and for its seed, a source of a rapid-drying oil used in food and medicine and by the paint industry. Native to Europe and Asia. Also called false saffron.

Safflower Meal–The by-product obtained after oil has been pressed from safflower seeds.

Saffron–A deep orange-colored product consisting of the dried stigmas of the saffron crocus, used mainly to color and flavor foods.

Sage–(1) *Salvia officinalis*, family Labiatae; a perennial herb grown for its grayish-green, dried leaves commonly used to flavor meats, in dressings, etc. Native to the Mediterranean area. Also called garden sage. (2) See Sagebrush.

Sagebrush–Any plant of the genus *Artemisia*, family Compositae. Species are herblike, bushy plants relatively important as browse for goats and sheep and to some extent for cattle especially in late fall, winter, and early spring, when there is a scarcity of other forage. Growing wild in western and southwestern United States, it often becomes a pest by invading better grazing lands that are overgrazed. Native to North America.

Sagittate–A leaf base that is similar in shape to an arrowhead and is triangular with the basal lobes pointing downward.

Sago—A food made from the starchy pith of the sago palm or some other plants, as cycads, jaggery, etc. In granulated or powdered form, the latter known as the sago flour, it is used as an item of diet or in making cakes, soups, and puddings. Also used as a stiffening agent in the textile industry.

Sago Palm—Any of the palms the pith of whose trunks is a source of sago; especially ***Metroxylon rumphii***, family Palmaceae, which flourishes in marshy areas and is native to Indonesia. Some of the other sago palms are gomuti palm (***Arenga pinnata***), kittoolpalm (***Caryota urens***), and cabbage palm (***Corypha umbraculifera***), all native to Indonesia.

Sagrain—One of several important hybrids between grain sorghums and sweet sorghums; grown for seed and forage.

Saguaro—***Cereus giganteus***, family Cactaceae; a huge cactus which produces an edible fruit used in sweetmeats. Found in southeastern California, southern Arizona, and Mexico. Also called giant cactus, sahuaro, suwarro.

Sainfoin—***Onobrychis viciaefolia***, family Leguminosae; a perennial herb occasionally cultivated for fodder. Native to Europe and northern Asia. Also called common sainfoin, holy clover, esparsette.

Sake—An alcoholic beverage produced in Japan by a process which involves steaming, cooling, and inoculating rice with spores of the mold ***Aspergillus oryzae***, the mass being allowed to incubate and more steamed rice mixed with water being added at different progressive stages of the chemical action. The percentage of alcohol in sake is 18 to 24.

Salad Plant—Any food plant grown primarily for its edible leaves, stalk, or other vegetative parts used in salads. Commonly included are such plants as lettuce, parsley, endive, cress, chicory, salad chervil, watercress, and celery.

Salanine—A bitter-tasting alkaloid which is toxic to humans. Salanine is present when white (Irish) potatoes are exposed to sunlight and the skin turns green. Green-skinned white potatoes should not be eaten.

Salicetum—An arboretum of willow trees.

Salicylic Acid—$C_6H_4(OH)CO_2H$; a colorless acid found in many plants and fruits, it may be obtained especially from oil of wintergreen. Manufactured commercially by processing dry sodium phenate (C_6H_2ONa); it is used as a preservative and as a fungicide.

Saline—General term for waters continuing various dissolved salts. The term is restricted to inland waters where the ratios of the salts often vary; the term haline is applied to coastal waters where the salts are roughly in the same proportion as found in undiluted seawater.

Saline Seep—In semi-arid regions, a down-slope location where salt has accumulated from upper slopes.

Saline Soil—A soil containing soluble salts in such quantities that they interfere with the growth of most crop plants. The electrical conductivity of the saturation extract is greater than 4 mmhos per centimeter (at 25°C), and the exchangeable-sodium percentage is less than 15. The pH readings of the saturated soil is usually less than 8.5.

Saline-sodic Soil—A soil containing sufficient exchangeable sodium to interfere with the growth of most crop plants and containing appreciable quantities of soluble salts. The exchangeable-sodium percentage is greater than 15, and the electrical conductivity of the saturation extract is greater than 4 mmhos per centimeter at (77°F or 25°C). The pH reading of the saturated soil is usually less than 8.5. See Saline Soil, Sodic Soil.

Salinity—The quantity of saltness in seawater or freshwater, most commonly expressed in parts of dissolved salt per 1,000 parts of water; e.g., salinity of seawater is 35 parts per thousand. See Alkalinity.

Salinization—The process of accumulation of soluble salts in soil.

Salinus—Salty.

Sallow—(1) Goat willow. (2) An unhealthy skin color.

Salsify (Salsifis)—***Tragopogon porrifolius***; an edible plant; sometimes called oyster-plant.

Salt—(1) Sodium chloride (common salt), NaCl; a white crystalline compound occurring abundantly in nature as a solid or in solution. It is a requirement of all livestock and is generally deficient in all natural feed ingredients. It is commonly included in mixed feeds and is otherwise supplied as salt blocks in pastures, barn lots, or stalls. It has wide usage in seasoning, preserving meats and vegetables, tanning hides, etc. In large amounts it is toxic and is frequently used to kill weeds; however, some plants such as Bermudagrass and asparagus tolerate a very high concentration. (2) To add salt as seasoning, preservative, supplement, etc.

Salt Hay—(1) A mulching material for plants which is cut from coastal marshes. (2) Hay which is made from grasses cut from salt marshes.

Salt Index—An index to compare solubility of chemicals. Most nitrogen and potassium compounds have a high index; phosphate compounds have a low index. When applied too close to seed or foliage, high-index materials may cause plant injury by plasmolysis.

Salt Pan—A depression in a salt marsh, usually bare of vegetation.

Salt-susceptible—Plants that will not grow on soils containing appreciable quantities of soluble salts.

Saltation—(1) The leaping movements of eroding soil grains arising largely in combinations of friction with inertia. (2) The process by which a particle, picked up by the stream current, is flung upward after which, being too heavy to remain in suspension, it drops to the stream floor again at a spot downstream.

Saltbush—An herb or shrub of the genus ***Atriplex***, family Chenopodiaceae. Species occupy large areas of the arid and alkaline soils of western and southwestern United States and Australia. They grow at low and intermediate elevations under desert conditions. The plants have a salty taste which adds to their palatability for stock. Some species, especially shadscale saltbush, have a fairly high rating as browse plants.

Salted—Treated, seasoned, supplied, or preserved with salt.

Saltgrass—Designating a grass of the genus ***Distichlis***, family Gramineae. Species are low, leafy, perennial grasses with strong running rootstocks which occupy alkaline and saline lands and which are high in salt content. All saltgrasses are rather low in nutritive value and rate only fair for general forage purposes.

Saltpeter–(1) Potassium nitrate (KNO_3), a white crystalline salt of potassium and nitrogen whose principal agricultural use is as a fertilizer and as a preservative in curing and preserving of meats. Also spelled saltpetre. (2) An incrustation of certain salts occasionally formed on tobacco leaves during processing.

Samara–A single-seeded, winged fruit, whose wing helps in seed dissemination, like that of the maple, ash, and elm.

Sampling Fertilizers–The State Department of Agriculture in each of the fifty states maintains a fertilizer inspection service that samples and analyzes fertilizers to determine if the fertilizer grade on the bag or in the bulk is properly designated to conform to the respective state laws. In the interest of uniform sampling procedures among the states, the Association of American Plant Food Control Officials has sponsored joint regional conferences of fertilizer inspectors and fertilizer industry personnel for the purpose of exchanging information on the many problems connected with sampling. An Inspector's Manual has been developed and demonstrations of sampling procedures have been held.

San Jose Scale–*Quadraspidiotus perniciosus*, family Diaspididae; an important scale insect of woody plants which may infest most of the cultivated fruits and a large number of ornamental trees and shrubs. It is often lethal if left unchecked. Trees lightly infested show small, round, grayish specks, about the size of a pin head, with the bark or fruit often reddened around each scale. On more heavily infested trees, the bark is covered by a gray layer of overlapping scales.

Sand–(1) Any of the small, loose, granular fragments which are the remains of disintegrated rocks. It may contain a great variety of minerals and rocks but the most common mineral is quartz. (2) In soil science, a group of textural classes in which the particles are finer than gravel but coarser than silt, ranging in size from 2.00 to 0.5 millimeters in diameter. It is the textural class of any soil that contains 85 percent or more of sand and not more than 10 percent of clay. (3) An area or deposit of this material, as a sand shore, a sand bank, etc. (4) In maple sugar processing, a granular deposit, mainly of silica and calcium malate, formed as the sap is condensed. (5) To cover with sand.

Sand Binder–Any plant which can hold sand in its place, and reduce its shifting, by the abundance of its rootstocks. Grasses of the genera *Ammophila* and *Calamovilfa* are commonly used for such a purpose.

Sand Corn–*Zigadenus paniculatus*, family Liliaceae; an herbaceous plant, occasionally cultivated for its greenish-white, racemose, flower clusters. The leaves are poisonous to cattle. Native to western United States. Also called foothill deathcamas.

Sand Dropseed–*Sporobolus cryptandrus*, family Gramineae; a perennial bunchgrass that grows in small clumps and is recognized by the tufts of long, white hairs at the top of the sheath. It is adapted to sandy or gravelly soils at altitudes up to 5,000 feet. Regarded as a fair forage grass, it is used sometimes in the west-central states of the United States to reduce erosion along highways. Found in western United States.

Sand Drown–A chlorophyll-deficiency condition in tobacco and other plants. In tobacco, symptoms in midseason are a progressive loss of normal green color on lower leaves which results in a lower market grade of affected leaves. It appears on sandy soils, especially in the coastal plain of South Carolina, United States.

Sand Lovegrass–*Eragrostis trichodes*, family Gramineae; a grass frequently used in the reseeding of sandy upland pastures in the Great Plains region of the southwestern United States. It provides useful forage and is used largely as a summer pasture grass.

Sand Pear–*Pyrus pyrifolia*, family Rosaceae; a deciduous tree which is resistant to fire blight and has edible fruits with many grit cells. Its small-fruited forms were called late pears (*P. serotina*); while large-fruited forms were the *P. serotina culta* to which the name sand pear was applied. Horticultural varieties of the large-fruited forms are raised in Japan and formerly in America. Crossed with the common pear, *P. communis*, it has yielded the Kieffer, Leconte, and Garber varieties of hybrids. Native to China. Also called Chinese pear, Japanese pear.

Sand Sagebrush–*Artemisia filifolia*, family Compositae; a shrub that is a weed pest on grazing land in the southern Great Plains area of the United States.

Sand Vetch–Hairy vetch.

Sand Wireworm–*Horistonotus uhlerii*, family Elateridae; an insect which infests corn, cotton, and potato plants below the ground surface. Also called cotton wireworm, corn wireworm.

Sandberg Bluegrass–*Poa secunda*, family Gramineae; a perennial bunchgrass which occurs in lower elevations and exposed sites in western areas of the United States and Canada. It endures grazing better than most bluegrasses and is rated good in general forage value.

Sandbur–Any of the various weeds, especially the annuals of the genus *Cenchrus*, which bear prickly fruit and grow in sandy or waste places. Also called bur-grass, sand-spur.

Sandiness–(1) A condition of a product resembling or suggesting sand; grittiness. (2) The defect in foods experienced in tasting, when a food, such as ice cream or condensed milk, contains tiny crystals of sugar or ice.

Sandreed–Any of a small group of rigid, tall, perennial grasses with creeping rhizomes, belonging to the genus *Calamovilfa* or *Ammophila*. The more important species are (a) the drought-enduring prairie sandweed, *Calamovilfa longifolia*, family Gramineae; a plant with tapering, long, slender leaves which grows from 2 to 6 feet tall. It is grazed lightly in the summer; in the western United States, it is important as winter pasture and for hay; (b) American beachgrass, *Ammophila breviligulata*, family Gramineae, found on the Atlantic shore and the Great Lakes area of the United States. They are important sand-binding grasses on dunes and sandhills.

Sandy Clay–Soil which contains more than 45 percent sand, more than 35 percent clay, and less than 20 percent silt.

Sandy Clay Loam—A soil which contains from 20 to 35 percent clay, less than 28 percent silt, and 45 percent or more of sand.

Sandy Loam—Gritty, loamy soil which contains less than 20 percent clay, less than 50 percent silt, and between 43 and 52 percent sand.

Sandy Soil—In common usage, the sand and sandy loam textural classes; any soil which contains more than 70 percent sand and less than 15 percent clay.

Sanitation—The developing and practical application of measures designed to maintain or restore healthful conditions, such as the treatment, removal, or destruction of contaminated or infested materials and possible sources of infection or infestation.

Sanitize—To disinfect and make sanitary the utensils used in preparation of food products, as dairy utensils.

Sap—(1) The juices or fluid constituents of a plant, especially a woody plant. (2) The moisture and its soluble constituents in unseasoned wood.

Sapling-Seedling Stands—Stands at least 10 percent stocked with growing stock trees of which more than half are saplings and/or seedlings.

Saplings—Live trees of commercial species 1.0 to 5.0 inches in diameter at breast height and of good form and vigor.

Sapodilla—*Achras zapota (Lucuma mammosa)*, family Sapotaceae; a large evergreen tree which bears an edible fruit, and whose latex yields chicle, the chief natural ingredient of chewing gum. Native to tropical areas of North America. Also called sapote, marmalade plum, naseberry.

Saponification—The chemical conversion of fats into soap.

Saponin—Any of several glucosides or glycerides present in many plants; e.g., soapwort, which has soaplike qualities. Used in detergents, to reduce the surface tension of water in emulsifying oils, and to produce foam in fire extinguishers and beverages.

Saprobic—Pertaining to an environment rich in organic matter and relatively low in oxygen.

Saprogenesis—That period in the life cycle of a pathogen in which it is not directly associated with a living host and in which it may be either dormant or living as a saprophyte. See Saprophyte.

Sapsucker—(1) Any of several small woodpeckers, *Sphyrapicus*, especially *S. varius*, the yellow-bellied sapsucker which makes a horizontal series of holes around trees, such as apple, birch, pine, etc. It returns from time to time to feed on the sap and insects from those holes. (2) Any insect that sucks sap from the roots, stem, or leaves of plants, as the corn-root aphid, cinch bug, etc.

Sapwood—That portion of the wood of a tree more recently formed which contains living cells intermingled with nonliving, woody tissue. It is located between the bark and heartwood of trees. Also called Alburnum, splintwood. See Heartwood.

Sapwood Rot—The decay of sapwood caused by various fungi, as white spongy rot of the sapwood of horse chestnut caused by the toadstool *Collybia velutipes*. The same fungus produces a similar rot in basswood and American elm.

Sarsaparilla—*Aralia hispida*; the flavoring extract which is obtained form the roots of various tropical species of *Similax*, family Liliaceae, especially *S. aristolochiifolia* and *S. officinalis*.

Sassafras—(1) Any of the several aromatic plants of the genus *Sassafras*, family Lauraceae. Also called cinnamonwood, saxafras. (2) *S. allbidum*, a deciduous tree grown for its gorgeous, scarlet fall foliage. Its dried root bark has been used in flavoring and as a medicine. It is now known to cause cancer. The wood is used for fence posts, general millwork, rails, wooden pails, small boats. Escaped, it can be an invading pest. Native to North America.

Saturated—Filled to the maximum capacity under existing conditions.

Saturated Fat—A fat whose carbon atoms are associated with the maximum number of hydrogen atoms; no double bonds exist.

Saturation Point—(1) The point beyond which no further additions can be held, utilized, or accommodated; e.g., the water-holding capacity of a particular soil, the flow of goods to a market, or a population of plants. (2) In wildlife, the maximum density of population; the natural condition wherein a species is as abundant as living conditions permit.

Sauerkraut—A food obtained by fermentation of prepared and shredded cabbage in the presence of not less than 2 percent nor more than 3 percent salt. It contains, upon completion of fermentation, not less than 1.5 percent acid, expressed as lactic acid.

Savory—(1) Any aromatic annual or perennial herb of the genus *Satureia*, family Labiatae. (2) The dried leaves and flowering tops of summer savory used as a flavoring.

Savoy Cabbage—A hardy variety of cabbage having wrinkled leaves.

Saw Log—A log meeting minimum standards of diameter, length, and defect, including logs at least 8 feet long, sound and straight, and with a minimum diameter inside bark for soft woods of 6 inches (8 inches for hardwoods) or other combinations of size and defect specified by regional standards.

Saw Timber—Trees that will yield logs suitable in size and quality for lumber.

Sawdust—Dust and small particles of wood produced by the cutting action of saws. A by-product of the lumber and wood industries. It is used as a bedding material for livestock or as soil mulch. Mixed with soil, it has no fertilizer value until decomposed and may depress plant growth in its fresh state unless nitrogen fertilizer is added to it.

Sawfly—Any of a large group of insects belonging to families Tenthredinidae, Pamphiliidae, Cimbicidae, Diprionidae, and certain other related families. The adult females have sawlike organs for puncturing foliage and twigs of plants to deposit their eggs.

Sawpalmetto—*Serenoa repens (S. serrulata)*, family Palmaceae; a palm usually almost stemless, and having a prickly leaf stalk. In flat wood areas, especially in Florida, United States, it is commonly an undergrowth with pines. Native to North America. Also called scrub palmetto.

Sawtimber Stands–Stands at least 10 percent occupied with growing-stock trees, with half or more of total stocking in sawtimber or poletimber trees, and with sawtimber stocking at least equal to poletimber stocking.

SBM–Soybean meal.

Scabrous–A leaf surface covered with minute projections that is rough to the touch.

Scaffold Branches–The lateral branches produced when young fruit trees are pruned to obtain the desired size, shape, and strength of framework of the top.

Scald–(1) A serious burn of plant leaves, fruit, limbs, and trunks which may result from exposure to intense sunlight or extreme heat; e.g., cowpeas, beans, and soybeans may develop tan blotches which eventually cause defoliation; exposed or nearly ripe tomatoes may scald and wither during extreme heat, etc. Winter sun scald of tree trunks is a variation of this condition due to cold temperature following a warm period. A condition resembling scald occurs in stored fruit, but is in fact caused by chemical esters. Also called sun scald. (2) An eroded spot on hillsides where subsoil is exposed. (3) A bare spot in freshwater coastal marshes caused by the killing of plant life by increases in salt content of the soil through encroachment of water from adjacent saltwater marshes. (4) To dip in hot water or to pour on hot liquid, as to scald a killed bird to loosen feathers. (5) To heat a liquid to a temperature just short of the boiling point, as to scald milk.

Scale–(1) Any instrument used to determine weight. (2) Either of the pans of a balance. (3) A measure of dimension, concentration, or intensity. (4) A graduated series of steps or degrees on measuring devices. (5) An instrument or device on which graduated spaces for measurement have been stamped or attached. (6) A graduated list of prices, wages, etc. (7) Any thin flake that peels from a surface, such as from skin. (8) One of the thin, flat, membranous plates which forms the outer, protective covering of certain animals or parts thereof, as the shanks of birds, etc. (9) One of several rudimentary, specialized leaves which protects the buds of plants and deciduous trees in cold climates. (10) A thin, dry, membranous part of a plant or flower. (11) See Scale Insect. (12) The estimated sound content of a log or group of logs scaled by use of a given log rule. Net scale, the scale after deducting for defects. (13) To form into scales. (14) To cover with scales or with an incrustation. (15) To remove in thin layers, as by scraping. (16) To come off in scales. (17) To make according to scale.

Scale Insect–Any of the several members of a widely diverse group of usually small, sucking insects, so called because of the appearance of the more widely known members, such as the San Jose scale. The males sometimes have wings; the females are scalelike, gall-like, or grublike in form, and clothed in wax. The forms most commonly seen (adult female scales) generally remain fixed on the host. The families Coccidae, Diaspididae, and Pseudococcidae are typical of scale insects and their allies. Scale insects infest most cultivated fruits, a large number of ornamental trees and shrubs, and greenhouse plants, and feed by inserting their slender mouthparts into the plant tissue and sucking the juices. Heavy scale infestations often kill branches or entire trees or shrubs. See San Jose Scale.

Scalecide–A spray material especially formulated for the control of scale insects, which is generally applied as a dormant spray in early spring before buds open.

Scalping–(1) In flour mills, separating the fuzzy growth from the ends of wheat berries by attrition, with or without the use of suction or blower fans. (2) In milling, partially separating the product from the break rolls into broken wheat and break flour by means of an arrangement of sieves, bolts, or screens of varying degrees of fineness. (3) In horses, a striking of the front of the hind coronet, pastern, or cannon against the front foot when running. (4) Removal of vegetation and a thin layer of soil.

Scandent–Climbing, as a climbing vine.

Scar–(1) An isolated or protruding rock. (2) A steep rocky eminence. (3) A bare place on the side of a mountain or other steep slope. (4) A wound that has healed.

Scarcrow–An object, usually a crude human representation, which is set up in fields to frighten crows and other birds away from crops.

Scarf–(1) The tapered or notched ends of pieces of wood that fit together to form a scarf joint. (2) The beveled cut on a log or stump which results from undercutting a tree in felling. (3) The incision on a stock on which the scion is inserted in propagation by budding.

Scarify–(1) to abrade, scratch, or modify the surface for increasing water absorption, as to scarify an impervious seed coat. (2) To scratch, loosen to shallow depth, or crush a hard uneven surface to provide a more tillable soil. (3) To make superficial incisions or punctures in the skin.

Scarious–Not green but thin, dry, and membranous, often translucent.

Scarlet Clover–See Crimson Clover.

Scarlet Oak–*Quercus coccinea*, family Fagaceae; a rather cylindrical tree grown as a shade tree and for its autumnal scarlet foliage. Native to North America.

Schizocarp–A dry fruit that splits at maturity into one or more closed carpels, each containing one or more seeds.

Schwedler Maple–*Acer platonoides*; a horticultural variety of Norway maple which has dark red foliage throughout the early growing period. It is grown as a shade tree and as an ornamental.

Scion–An unrooted portion of a plant (scionwood) having one or more buds, used for grafting or budding on to rootstock (rooted portion of plant). Also spelled cion. See Rootstock.

Sciophyte–A plant which grows well on, or is adapted to, shady sites.

Sclerophyll–Plants with stiff, leathery, evergreen leaves heavily cutinized either broad-leaved as *Ficus* and *Olea* or narrow-leaved as *Pinus*. See Chaparral.

Sclerotia–Resting bodies of fungi consisting of hardened masses of mycelium from which fruiting bodies may develop.

Sclerotinia Drop–A fungal disease of lettuce caused by ***Sclerotinia sclerotiorum*** and ***S. minor***, family Sclerotiniaceae, characterized by a soft, watery rot of the stem and leaf base, and sudden collapse of the plant. The entire plant is a soft, watery mass and finally dies.

Sclerotinia Rot–Any rot caused by ***Sclerotinia***, family Sclerotiniaceae, including certain rots of fruits, garden vegetables, etc.

Scoliid Wasp–Any wasp of the genus ***Scolia***, all species of which lay their eggs on the larvae of various beetles. These wasps are important in the control of certain destructive beetles, such as the Asiatic or Oriental beetle. See Integrated Pest Management.

Scorch–(1) The appearance of seemingly burnt or brown irregular patches on foliage, bark, or fruits, as a result of heat, excessive sunburn, lack of water, action of bacteria or fungi, improper use of insecticides, fungicides, or some nutrients. (2) The brownish discoloration or the burning of foods, etc., in the process of cooking, drying, or by other use of heat. (3) See Anthracnose. (4) To burn superficially; to give the appearance of a burn, as to brown with heat.

Scorch Disease–See Anthracnose.

Scoria–(1) The slaglike clinker left after the smelting of metals. (2) A cellular or cinderlike rock of volcanic origin.

Scotch Pine–***Pinus sylvestris***, family Pinaceae; a hardy evergreen tree grown as an ornamental, a specimen, in windbreaks, for Christmas trees, and for its lumber. Pyramidal when young, it becomes irregular with age. It grows rapidly on acid, poor, sandy soil, and is resistant to cold, drought, and dry winds. Native to Europe and Asia. Also called Danzig pine, Riza pine, Scots pine.

Scots Pine–See Scotch Pine.

Scour–(1) To abrade and flush as by the action of rapidly moving water on stream beds. (2) To cleanse dirt, grease, etc., by rubbing or scrubbing, as to scour wool. (3) To cleanse the bowels of an animal by purging. (4) In plowing, to pass through the ground cleanly without any soil sticking to the moldboard. (5) To rub off the flesh sticking to a hide. (6) See Scours.

Scouringrush–***Equisetum hyemale***, family Equisetaceae; an evergreen, perennial, rushlike herb, found wild on moist, springy grasslands, and along roadsides. The epidermis of its stems is very harsh because of high silicon content and was formerly used for scouring purposes. It is poisonous to animals, especially when fed in hay. Native to the north temperate zone.

Scouting–The inspection of a field for pests (insects, weeds, pathogens). Scouting is a basic component of integrated pest management programs. See Integrated Pest Management.

Scrap Grade–A grade of tobacco that is essentially a by-product from handling leaf tobacco including floor sweepings and other waste tobacco material except stems.

Scrape–(1) A tillage implement; a sweep. (2) A spreading wing attachment for a cultivator. (3) In turpentine operations, oleoresin remaining on a tree from which volatile oils have evaporated. (4) To collect scrape from trees in turpentine operations.

Screen–(1) A sieve or grating, such as a frame covered with meshed wire or fabric, or a perforated plate, etc., which is used for separating the finer or coarser parts of soil, gravel, grain products in milling, seeds in cleaning, etc. (2) Any form of grating used to prevent invasion or escape of insects, reptiles, fish, or other kinds of animals. (3) A construction or planting used to conceal an unpleasant sight. (4) To consider a group for the purpose of selecting a relatively small number of individuals. (5) To sift by shaking through a screen. (6) To conceal, as with a screen.

Screenings–The refuse removed from grain harvested by a combine. Screenings include small, broken, or shrunken kernels of grain, small weed seeds, chaff, broken stems, etc. Screenings are sometimes used for feed.

Scribner Rule–In forestry, a diagram rule for determining the board measure of logs, one of the oldest in existence, which assumes 1-inch boards and ¼-inch kerf, makes a liberal allowance for slabs and disregards taper. (It is the official rule in many parts of the United States). See Doyle Rule, Doyle-Scribner Rule.

Scrub–(1) Land which under natural conditions supports only shrubs or a dwarfed, stunted growth of trees. The soils are generally infertile, but scrub may appear on various kinds of sites. (2) The stunted trees or shrubs, often in dense stands, on such land. (3) An animal of inferior breeding or condition. (4) Designating an inferior product, animal, etc.

Scrub Forest–A forest of small trees occurring on ridges or hillsides with little shade and with little accumulation of litter.

Scrub Pine–(1) ***Callitris calcarata***, family Cupressaceae; a black or red pine which grows on stony hill slopes and ridges. Twigs of this tree have been used as a vermifuge for horses. Native to Australia. (2) Any of several dwarf pines of the United States which often thrive on stony or arid land.

Scum–(1) Minute forms of lower plant life and impurities that gather on the surface of stagnant water. (2) Extraneous matter or impurities that rise to the surface of boiling or fermenting liquids. (3) The condensation of solids at the surface of boiled milk upon cooling.

Scuppernong–A horticultural variety of the native bullace grape cultivated since early colonial times in southeastern United States. This variety has a strong-growing vine, nearly round, triangular-toothed leaves, nonforked tendrils, and short, dense flower clusters. It bears an edible, amber-green fruit and is grown on arbors for its edible fruit and shade.

Scurf–(1) Any scaly matter or incrustation on a surface. (2) A roughening or browning of a smooth plant surface which is more pronounced than russeting; e.g., apple scab. (3) A disease of sweet potato, which results in brown discoloration on the surface of underground parts, caused by the fungus ***Monilochaetes infuscans***, family Dematiaceae.

Scurfpea–See Slimflower, Scurfpea.

Scutch Grass–See Bermudagrass.

Scutellum–The cotyledon of the grasses, sometimes referred to as the seed leaf.

Sea Island Cotton–*Gossypium barbadense*, family Malvaceae; a woody plant grown as an annual for its long staple lint cotton used in fine cotton fabrics. Native to tropical America. Also called American cotton, Andes cotton.

Sealing–The action of high-intensity rainfall which causes disintegration of unprotected soil granules, a consequent separation of colloids which fill spaces between larger particles, thus producing a seal or relatively impervious surface crust.

Seaside Bent–A strain of creeping bentgrass with long stolons and narrow, stiff, oppressed blades, usually densely matted, which is used for lawns and putting greens of golf courses, etc. Native to the marshy areas of the Atlantic and Pacific coasts of United States and Canada.

Season–(1) A portion of the year; e.g., a growing season, when plants grow; the fire season, when danger of fire is greatest; the grazing season, when grazing is possible. (2) To bring to or develop into a condition of greater usefulness, as to dry lumber by natural or artificial means. (3) To flavor food with spices, salt, etc. (4) An animal that is ready for breeding is said to be in season.

Seasonal Distribution–(1) In pasturage, the progressive grazing in a sequence of moves from one part of a range to another as vegetation develops. (2) In meteorology, the amount and kind of precipitation, wind movement, etc., in different seasons.

Seasonal Grazing–The grazing of a range only during a certain period or periods of the year which roughly correspond to one or more of the four seasons.

Seasoning–(1) The process of drying (curing) lumber or other forms of wood for better utilization. Seasoning is natural when the drying is done by air or other natural means; in artificial seasoning, the drying is carried out by means of a kiln, oils, or electrical appliances. (2) Adding salt, pepper, etc., to meat and other food products.

Second Bottom–The first terrace level of a stream valley which lies above the floodplain. It is rarely or never flooded.

Second Growth–(1) A forest which originates naturally after removal of a previous stand. (2) The smaller trees which remain after a cutting or the residual trees available for another logging on the same area.

Second-year Cane–In bramble fruits (*Rubus*), the canes which in their second year make fruiting shoots, bloom, bear fruit, and die. Sometimes called floricane. See Primocane.

Secondary Growth–(1) Growth which may arise from a secondary bud if the first shoots are destroyed, as with shrubs or trees. (2) A shoot or spur that grows from a bud or growing point the same year that bud or growing point is produced, as with shrubs and trees. (3) Growth from the cambium.

Secondary Infection–(1) An invasion by a second and different organism after a first organism has become established in a host animal or plant. (2) An infection resulting from inoculum produced by an organism after it is first established in a host plant.

Secondary Insect–Any insect which follows, is associated with a primary form, or feeds upon plant tissue but which is incapable of initiating injury by itself.

Secondary Parasite–A parasite which establishes itself in or upon a host that is a primary parasite.

Secondary Plant Nutrients–Calcium, magnesium, and sulfur are the secondary nutrient elements, so called because they are essential to plant growth in lesser quantity than the primary nutrients nitrogen, phosphorus, and potassium, and in greater quantity than the micronutrients.

Secondary Range–A range that is lightly used or unused by livestock under minimal management and will ordinarily not be fully used until the primary range has been overused.

Secondary Root–a lateral branch of a primary or main root.

Secondary Tillage–Soil tillage that follows primary tillage to prepare the soil for planting; usually not as deep as primary tillage.

Secretion–The metabolic act or process of synthesis and liberation of substances from cells or glands of animals or plants, as saliva, milk, etc.

Sedentary–(1) Formed in place without transportation from the underlying rock or by the accumulation of organic material as in peats and mucks; said of some soils. (2) Attached, as an oyster, barnacle, or similar shelled invertebrate.

Sedge–Any member of the botanical family Cyperaceae which is related to the grasses but is generally inferior for forage. Sometimes erroneously applied to grasses, as broomsedge.

Sedge Peat–A brown to black peat, mostly the partially decomposed remains of sedge vegetation. It is generally less acid than most peat and of lower water-absorbing capacity than moss peat. See Histosols.

Seed–(1) The embryo of a plant; also kernels of corn, wheat, etc., which botanically are seedlike fruits as they include the ovary wall. (2) Propagative portions of a plant other than true seeds, as tubers, bulbs, etc. (3) Offspring or progeny. (4) Sperm or semen. (5) To place seed on or in the soil for the production of plants of the same species; to sow seed. (6) Overseed; to sow seed on an area where a crop is established, as clover on wheat, lespedeza on oats. (7) To produce seed. (8) To extract seed, as to seed cherries or grapes.

Seed Analyst–A person who examines samples of seed to determine the percentage of pure seed, the kind and percentage of inert material, and the seed viability.

Seed Association–An organized group of seed growers which produces seed of known ancestry, guaranteed to be pure, viable, and reasonably free from seed-borne diseases. Such seed is commonly offered for sale as registered or certified seed.

Seed Classes–The different divisions given to seeds grown under conditions so as to group them into a certain classification. The recognized classes of seed in the United States are: Breeder, Foundation, Registered, and Certified.

Seed Coat–(1) The hard outer layer of a seed; the protective covering, or integument. Also called testa. (2) The bran of milled cereals.

Seed Cotton–Unginned cotton which consists of the seed with the attached lint as picked from the boll.

Seed Dressing–Most seeds sold commercially have been dusted (dressed) with a fungicide or insecticide. A few have been dressed with an animal/bird repellent. Occasionally some seeds, such as legumes, have been treated with a micronutrient fertilizer such as molybdenum and with *Rhizobium* bacteria.

Seed Farm–A farm on which the production of seed is the most important part of the farm business.

Seed Flat–A receptacle, usually a shallow, flat, plastic box filled with soil in which seeds are planted for later transplanting.

Seed Fly–*Agromyza lantanae*, family Agromyzidae; a fly brought to the Hawaiian Islands from Mexico for the biological control of a thorny pest plant, *Lantana camara*. This seed fly infests only the lantana, and its larvae eat the berries so that birds do not spread the seed.

Seed Inoculation–The treating of seeds of alfalfa, sweet clover, and certain other legumes with cultures of nitrogen-fixing bacteria. These beneficial organisms live on the roots of legumes and store up nitrogen gathered from the air in nodules on the roots of the host plant to the benefit of the host and subsequent crops. Also see Rhizobium.

Seed Label–A label or tag, provided by law for each lot or parcel of seed exposed for sale, which gives the purchaser pertinent information as to the kind of seed and the approximate percentage by weight of pure seed present. Most seed laws require the seed label to show the percentage by weight of inert matter, of seeds other than agricultural seeds, and to establish definite limits on the percentage of noxious weed seeds that may be present.

Seed Leaf–A cotyledon, or a leaflike part of a seed, which appears as the original leaf of an embryo.

Seed Leaves–See Cotyledon.

Seed Orchard–A plantation of selected clones or tree progenies. The orchard is isolated or managed to avoid or reduce pollination from outside sources. The object is to produce frequent, abundant and easily harvested crops of seed.

Seed Pellet–A mixture of clay, fertilizer, rodent and insect repellents, and a single seed which is pressed together to form a small hard unit like a shot for precision planting.

Seed Piece–Any plant segment including a node which is used for planting, as with sugarcane, potatoes, etc.

Seed Pod–(1) Technically, any dry seed vessel or fruit that splits open, as a pea pod. (2) In common usage, any dry seed vessel or fruit that contains seed.

Seed Potato–An Irish potato used for propagation. In planting, it is used either as a whole tuber or, more commonly, halved or quartered into seed pieces of 1½ to 2 ounces.

Seed Production–(1) The growing of a crop primarily for its seed. (2) The developing of seed by fruiting plants. With annual plants, such as corn, seed is produced and the life cycle is completed in one season. Biennial plants require two seasons to attain full development and bear seed, after which they die; e.g., common sweet clover and sugar beet. Perennial plants may live for several years and may produce seed each year after attaining maturity.

Seed Protectant–A seed preparation used to treat disease, such as seedling blight and damping-off.

Seed Purity–The percentage of the desired species of seeds in relation to the total quantity, including other species, weed seeds, and foreign matter.

Seed Rot of Peanut–A seed rot caused by any of numerous fungi (*Fusarium, Rhizopus, Penicillium, Aspergillus*), which cause seed to rot when poor-quality or mechanically damaged seed is retarded in germination by cold, wet soils or deep planting.

Seed Stock–(1) Pedigreed or well-bred livestock, which is maintained for breeding purposes. (2) A specially selected strain of plants, or seeds thereof, which are to be used as parents of future generations.

Seed Treatment–The treating of seeds with chemical dusts, chemical solutions, heat, etc., for the control of seed-borne infections, such as smuts of cereals, or scab of potato, and for the protection of seedlings from damping-off fungi, etc.

Seed Tree–(1) A tree that produces seed. (2) In forestry, a tree reserved, in a cutting operation, to supply seed for reforestation.

Seed-borne Disease–A plant disease which may be transmitted by seed, as the fungal diseases of corn caused by *Diplodia* and *Fusarium* and some viral diseases, as the common bean mosaic.

Seed-corn Beetle–*Agonoderus lecontei*, family Carabidae; an insect which eats out the contents of seed corn and is sometimes the cause of corn failing to sprout.

Seed-corn Maggot–Family Muscidae; the maggot of the cornfly, which burrows into seeds of peas, beans, and corn, infests shoots of melon, cucumber, potato, young plants of cabbage, radish, onion, and several other garden vegetables. It is especially injurious in cold, wet seasons on land rich in organic matter.

Seed-testing Laboratory–A laboratory, usually operated by a federal or state agricultural department, in which samples of seed are tested for purity, viability, and for conformity to seed laws and regulations.

Seedbed–The upper portion of the soil prepared by natural or artificial means to receive seed and to promote germination and growth.

Seeded–(1) Planted or supplied with seed; sown. (2) Supplied or planted with spores, enzymes, cultures, etc. (3) Having the seed or seeds extracted; as seeded cherries.

Seeding–(1) The act of sowing seed. (2) The kind or amount of seed sown or to be sown, as a light seeding of oats for a nurse crop, or the stand of a seeding of alfalfa, etc. (3) In a broad sense, all methods of planting regardless of whether the crop is reproduced by seed or by vegetative parts. (4) The act of extracting seed or seeds, as the seeding of raisins. (5) The addition of crystalline nuclei to induce mass crystallization, as the seeding of sweetened condensed milk.

Seeding Time–The time of the year at which seed of any given sort is sown. Spring is commonly regarded as seeding time in the temperate zones although seeds are sown at various times of the year, as forage crops, or the fall sowing of wheat, rye, etc.

Seedling–(1) The early growth stage of a plant grown from seed as it emerges above the ground surface. (2) In forestry, a tree which has not been transplanted. (3) A tree which has not reached a height of 3 feet. (4) In horticulture, a fruit tree grown from seed which has not been budded or grafted.

Seedling Blight–A disease which attacks and kills germinating seeds and seedlings before or after emerging from the soil. Seedling blight may be caused by species of ***Fusarium, Diplodia, Gibberella***, and other fungi, and may at times result in serious damage to cereal grains, cotton, flax, garden and greenhouse plants, and other seedlings. See Damping-off.

Seedling Rootstock–A rootstock grown from seed which has been grafted or budded.

Seepwillow Baccharis–*Baccharis glutinosa*, family Compositae; a deep-rooted shrub which grows in arid and semi-arid regions of southwestern United States. Because of its thicket-developing habit, it is supposed to be an indicator of water. Also called water wally, botamote.

Segment–Any one of the parts into which an object can be naturally separated or divided; a section of an orange, a flower, or an insect.

Sego–*Calochortus nuttallii*, family Liliaceae; a perennial herb bearing an edible bulb; also grown for its white, purple-lined flowers. The sego lily is Utah's state flower. Native to western United States.

Segregation of Genes–The separation of genes on two homologous chromosomes and their distribution to separate gametes during gametogenesis; genes that once existed in pairs in a cell will become separated and distributed to different gametes.

Seguin Tuftroot–*Dieffenbachia seguine*, family Araceae; a foliage plant grown in the greenhouse for its yellow-blotched foliage. Ingestion of the leaves results in temporary speechlessness. Native to West Indies. Also called dumb cane.

Selection Cutting–Removal of mature timber, usually the oldest or largest trees, either as single scattered trees or in small groups at relatively short intervals, commonly five to twenty years, repeated indefinitely, by means of which the continuous establishment of natural reproduction is maintained.

Selective Breeding–The breeding of selected plants or animals chosen because of certain desirable qualities or fitness, as contrasted to random or chance breeding.

Selective Grazing–The tendency for livestock and other ruminants to prefer certain plants and to feed on these while grazing little on other species.

Selective Herbicide–Any of certain chemical toxicants (usually applied as a spray) that will kill weeds in a growing crop or turf and produce little or no injury on the crop plant or turf. See Nonselective Herbicide.

Selenium–A nonmetallic element, related to sulfur, rarely present as more than a trace in soil. It is essential for animals, but where it is present in excess it is absorbed by plants and may be toxic to grazing animals. See Blind Staggers.

Self Incompatibility–Inability of a plant to set seed without cross pollination.

Self-blanching–Any of certain varieties of celery that loses its chlorophyll as it matures.

Self-cleaning Velocity–The threshold velocity in sewers, terraces, drainage ditches, and irrigation canals to keep solids in suspension.

Self-fertility–The ability of a variety of fruit to produce viable seeds without the aid of pollen from another variety.

Self-mulching Soil–(1) A soil that cracks deeply and becomes so granular at the surface when very dry that the granular mulch washes into the cracks when rains begin, the whole soil swelling enough as it becomes moist to force material upward between the former cracks. (2) A soil in which the surface layer becomes so well-aggregated that it does not crust and seal under the impact of rain. See Vertisols.

Self-pollination–The transfer of pollen from the anther to the stigma of the same flower or to flowers of the same plant or other plants of identical genetic material such as apple varieties, clones of wild blueberries, etc.

Self-pruning–Natural pruning; the natural death and fall of branches from live plants resulting from light and food deficiencies, decay, insect attack, snow, wind, and ice.

Self-seeding–(1) The act of disseminating its seed by a plant or plant species. (2) Designating a plant which scatters its seed and reestablishes its kind.

Self-sterility–The inability of a plant variety to produce viable seeds if only the pollen of that plant or variety is available for pollination.

Selfed–See Selfing, Self-pollination.

Selfing–In plant breeding, the pollination of a stigma with pollen from the anthers of the same flower, plant, or variety. Also called inbreeding. See Self-pollination.

Semi-annual–(1) Of plants, those which bloom or grow for a half year. (2) Recurring every six months.

Semi-arid–A term applied to regions or climates where moisture is normally greater than under arid conditions but still definitely limiting to the growth of most crops. Dryland farming methods or irrigation generally is required for crop production. The upper limit of average annual precipitation in the cool semi-arid regions is as low as 15 inches (38 centimeters), whereas in tropical regions it is as high as 45 or 50 inches (114 to 127 centimeters).

Semihardy–Referring to cold resistance: designating a plant that will not endure as much cold temperature as the more hardy species do. It is a relative term designating plants between tender and hardy in a given climate.

Seminal Root–One of the several small roots which grow laterally form a germinating seed of corn and other cereals.

Seminarium–(Latin) Tree nursery.

Semipermeable Membrane–A membrane that permits the diffusion of one component of a solution but not the other. In biology, a septum which permits the diffusion of water but not of the solute.

Semolina–A food made from that part of durum wheat which is resistant to millstone action. The purified, coarse middlings are used in making macaroni and comparable edible pastes. Also called semola bran.

Sensitivefern–*Onoclea sensibilis*, family Polypodiaceae; a hardy perennial, rather coarse fern which grows to a height of 2 to 4 feet. It is widely distributed and is common in low, wet meadows, swamps, and marshes. Instances of poisoning to horses have been reported in New England (United States), where sensitivefern is frequently associated with hay containing brake fern and equisetum. Also called meadow brake, polypody brake.

Sensus–(Latin) Meaning.

Sentry Maple–A horticultural variety of the sugar maple that has upward-turning branches, parallel sides, and narrow form. It is suitable for planting on narrow streets.

Sepal–One of the separate units of a calyx, usually green and foliaceous.

Separation Layer–Abscission layer.

Septate–(1) Containing or divided by a septum or septa. (2) Fungus hypha or spores which have cross-walls. (3) Any fruit in which the ovary is divided into sections.

Septic–(1) relating to, or caused by, the presence of disease-producing organisms, or their toxins, in the blood or tissues. (2) Putrefactive; putrefying. (3) A term sometimes used to refer to conditions where dissolved oxygen is absent and decomposition is occurring anaerobically, as in a septic tank.

Septoria Leaf Spot–A fungal disease of the tomato and other plants caused by *Septoria lycopersici*, family Sphaerioidaceae; characterized by numerous, water-soaked spots that become circular with gray centers and later show dark specks.

Septum–Any dividing membrane or other layer in plants and animals: (a) a partition separating two cavities; (b) a division wall in a compound ovary; (c) a cross-wall in a fungus hypha or spore.

Sequestration (Chelation)–A chemical complexing of certain metallic cations with inorganic compounds. Very useful in preparing fertilizer formulations with available micronutrients such as iron, magnesium, zinc, copper, and manganese.

Sequoia–Either of the two tall trees, giant sequoia and redwood, genus *Sequoia*, family Pinaceae, characterized by oval, drooping cones and woody scales.

Sere–(1) The series of stages that follow one another in ecological plant succession, comprising development from pioneer stage to climax under natural influences. (2) The series of plant communities that follow one another in an ecologic succession.

Sericea Lespedeza–Chinese lespedeza, *Lespedeza sericea*.

Sericious–Closely covered with soft, silky hair as a sericeous leaf.

Series, Soil–A group of soils, formed from a particular kind of parent material, having horizons that, except for the texture of the A (surface) horizon, are similar in all profile characteristics and in arrangement in the soil profile. Among these characteristics are color, texture, structure, reaction, consistence, and mineralogical and chemical composition.

Serotiny–Tree cones that require heat to open and release seeds.

Serpentine Layering–Plant propagation in which several portions of a branch are covered with soil leaving the intervening portions above the soil until rooting takes place. Usually used with vines. Also called compound layering.

Serrate–(1) Pertaining to the rocky summit of a mountain having a sawtooth profile; a small sierra-shaped ridge. Local in the Southwest. (2) Said of a leaf margin when saw-toothed, with the teeth pointing forward.

Serrulate–A leaf margin that has small saw-like teeth (minutely serrate). See Serrate.

Sesame–Sesame refers to the oil-bearing seeds and to the plant *Sesamum indicum*. The oil is used for cooking.

Sesame Oil–A fixed or nonvolatile oil expressed from sesame seed whose chief uses are for food, and as the source of the product sesamin. Also called benne oil.

Sesame Oil Meal–A livestock feedstuff which is the ground residue obtained from sesame seed after the extraction of part of the oil by pressure.

Sesamin–A white crystalline compound which is present in sesame oil. Sesamin concentrates, obtained from sesame oil in the process of refining for use as an edible oil, are use din pyrethrum insecticides. Sesamin itself is not insecticidal, but it greatly increases the effectiveness of the pyrethrums.

Sessile–(1) Of leaves, flowers, fruits, etc., attached directly by the base without a stem or stalk, such as leaves of grasses, sedges, and certain other plants. (2) An organism that is attached to an object or is fixed in place.

Set–(1) A small propagative part, a bulb, shoot, tuber, etc., which is suitable for setting out or planting, as an onion set. (2) A number of things usually used together, or forming a complete collection, as a set of tools, harness, etc. (3) The initial swelling of the ovary of a flower soon after petal fall. Also called setting, fruit setting. (4) To put in a particular place, position, condition, direction, adjustment, etc. (5) To put eggs under a fowl or in an incubator to hatch them. (6) To put a price or value on something. (7) To put in proper working condition. (8) To adjust the teeth of a saw. (9) To fix a trap or a net to catch animals or fish. (10) To fix in the ground, as a post, tree, or plant. (11) To become stiff, firm, or hard, as cement sets. (12) To adjust into proper apposition the ends of a broken bone, as to set a fracture.

Seta–A stiff hair, bristle, or bristle-like body or appendage, as the stalk supporting the capsule of mosses; a bristle occurring on or in some fungus fruiting structures; or the hairs (setae) of a caterpillar, etc.

Sets–Small onion bulbs used for early planting.

Settle–(1) To sink gradually to a lower level, as of sediment, etc. (2) To cause a female to become pregnant, as a bull settles the female with which he has had coitus. (3) To occupy land usually for the purpose of farming, ranching, or homesteading.

Seventeen-year Locust—See Periodical Cicada.

Sewage Sludge—An organic product, remaining after the treatment of sewage in sewage-disposal plants, sometimes used as a fertilizer. Almost half of the sewage sludge produced in the United States is spread on land. The other half is burned, buried, or dumped in the ocean. See Activated Sewage.

Sex—(Latin, *sexus*) The distinction between male and female plants and animals. Ova (macrogametes) are produced by the female and sperm (microgametes) by the male. The union of these distinctive germ cells results in a new individual.

Sex Linkage—Association of character with sex due to the fact that the gene for that characteristic is in a sex chromosome.

Sex Pheromone—Sex pheromones are now made synthetically to bait traps and thereby make an estimate of population density of a given species of insect. In this way scientists can determine when control measures should be initiated, known as the economic threshold of insect numbers. Most common has been the use of synthetic female sex hormones to trap male Lepidoptera. A second use of synthetic female sex hormones has been to confuse the male and thereby prevent mating. See Pheromone.

Sex-limited Traits—Traits that are expressed only in one sex although the genes for the trait are carried by both sexes.

Sex-linked—Characters developed from genes located on sex chromosomes. The character may be used to determine the sex of an animal; e.g., barring of Barred Plymouth Rock chickens.

Sexual—A method of reproduction which involves the union of sperm and an egg.

Sexual Dimorphism—Differences in form exhibited by males and females of the same species.

Sexual Propagation—Producing plants from seeds. See Asexual Reproduction.

Sexual Reproduction—In plants, exchange of genetic material between male and female gametes through fusion of the pollen tube with the ovule releasing the male gamete into the egg cell resulting in a zygote. See Zygote.

Sexupara—That form of female organism that produces new individuals without fertilization. See Parthenogenesis.

Shade—(1) A spot or area protected from direct sunshine. (2) To protect seedlings and other plants from intense sunshine by screens, spaced strips on frames, etc.

Shade Cloth—A fabric woven from saran fibers to provide shade levels ranging from less than 20 percent to more than 90 percent.

Shade Device—Plastic netting, latex paint, whitewash, cheesecloth, or other material used to reduce light and heat entering a plant-growing structure.

Shade House—A structure that is covered with a shade cloth, laths, etc., for the purpose of blocking out part of the sun's rays in order to protect young growing plants.

Shade Plant—(1) A plant that grows in shade or can tolerate shade; as moss, woodland flora, etc. (2) A plant grown to provide shade for crops that require it.

Shade Tree—Any tree grown to provide shade; specifically, one of luxuriant foliage, such as a street tree or a park tree.

Shade-grown Wrapper Leaf—A leaf of the Cuban tobacco group used in cigar making which is produced on plants shaded by a screen of open-mesh cotton fabric during growth.

Shade-tolerant—Designating a plant that grows well in open sunshine and can also show satisfactory growth in moderate shade.

Shadow Jumper—A nervous, skittish horse.

Shadscale Saltbrush—*Atriplex confertifolia*, family Saxifragaceae; a scurfy, gray shrub, which ranks well as a browse plant. Native to the western United States. See Saltbush.

Shagbark Hickory—*Carya ovata*, family Juglandaceae; a deciduous tree so named because its shaggy, gray bark peels off in long strips. It is valued for timber, fire wood, and its edible nuts. Native to North America.

Shale—(1) A sedimentary rock which is composed of clay and silt finely laminated in structure. There are many variations in color and many gradations into hard, thinly-bedded sandstones and limestones and into slates. (2) The sliver-like lining of the carpel or bur which before the opening of the cotton boll serves to separate the locks of cotton. The presence of fine particles of shale in ginned cotton is highly objectionable as they are difficult to remove from the lint.

Shallot—*Allium ascalonicum*, family Liliaceae; a perennial, bulbous, onionlike herb grown for its edible, mild-flavored, white stalk and green leaves. Native to Europe and Asia.

Shallow Cultivation—Scraping the surface, scarifying, or cultivating to shallow depths, normally less than 5 inches. Fewer of the crop roots are injured.

Shallow Well—A well from which a supply of water is obtained at little depth, commonly 20 feet or less below the surface.

Shampoo Brush—A grooming brush that can be attached to a water hose. It is used in washing cattle for the show. The plastic bristles combined with the running water help remove dirt and grime from the animal's coat.

Shamrock—A plant having trifoliate leaves, traditionally believed to have been used by St. Patrick as illustrative of the doctrine of the trinity, and hence adopted as Ireland's national emblem. The name shamrock is now commonly applied to the emblematic plant worn by the Irish on St. Patrick's day.

Shank—(1) That part of the leg joining the knee to the ankle in humans, or the corresponding part in various animals. (2) A market cut of meat from the lower part of the foreleg of a dressed carcass of beef, veal, mutton, or lamb, or a corresponding cut from the hind leg.

(3) The tarsus of birds: the part of the leg, below the hock joint, which is covered with scales. (4) That part of a tool, implement, etc., between the acting part and the part by which it is held, as a cultivator-shovel shank, etc. (5) Any connecting part of a plant; a foot stalk, as the part between the stalk and the corn ear.

Sharp Freezing–Freezing of products, fruits, vegetables, poultry, meats, dairy products, etc., at low temperatures (−5°F to −20°F).

Shatter–To scatter or disperse as leaves, grains, seeds, fruits, etc.; e.g., leaves fall from dry hay, seeds drop out of husks or pods from over-ripeness, etc.

Shearing–(1) The act or operation of removing wool from a sheep or goat by means of shearing machines. (2) The thinning or heading back of shrubs or trees by removing portions with trimming or pruning shears, as in the trimming of hedges, etc.

Sheath–(1) The leaf base covering a stem or branch, such as the lower part of leaves of grasses which surround the stem. (2) The enclosing and protecting tubular structure within which the penis of a stallion, bull, and certain other male animals is retracted.

Sheath and Culm Blight of Rice–A fungal disease of rice caused by *Rhizoctonia oryzae*, characterized by lesions with reddish-brown margins and light yellow or greenish-yellow centers on the leaves and culms.

Shed–(1) A simple structure for shelter of livestock, storage of feeds, or both, open on one or more sides, which may be separate from or attached to another structure. (2) To lose or cast off something, as a plant sheds leaves, pollen, seed, etc.; animals shed hair; birds molt or shed feathers; a roof sheds rain, etc.

Sheep Fescuegrass–*Festuca ovina*, family Gramineae; a perennial grass, some forms of which are native or introduced on western ranges (United States). It is sparingly used for pasture and in some turfgrass mixtures where rough turf is desirable. Native to Europe and Asia; naturalized in North America.

Sheep Loco–*Astragalus nothoxys*, family Leguminosae, a poisonous plant found in southern Arizona, United States. Both its green and dried foliage are poisonous, causing dullness, irregularity of gait, lack of appetite, dragging of the feet, a solitary habit, loss of flesh, and shaggy coat in animals. As the animal ceases to eat, it dies. See Loco.

Sheep Manure–The dung of sheep either moist or dried. Dried, it is sold for fertilizer. The approximate analysis is: nitrogen (N) 1.4 percent, phosphoric acid (P_2O_5) 1.0 percent, and potash (K_2O) 3.0 percent.

Sheep Sorrel–*Rumex acetosella*; a common weed found on thin soil. It tolerates acid soil better than most crops and pasture grasses.

Sheeping Down–A method of harvesting and utilizing a crop, as corn, by the use of sheep which are turned into the field to feed. See Hog Down.

Sheet Erosion–Removal by rainwater and snowmelt runoff of a uniform layer of soil or comminuted material from a portion of the land surface, as contrasted to gullying and stream erosion caused by surface water flowing in channels. See Erosion.

Shelf Life–The length of time fresh or canned fruits remain saleable while on a shelf or in storage.

Shell–(1) A hard outer covering of certain animals, such as turtle, mollusk, or crustacean, which when ground may be used as a poultry feed supplement. (2) The hard outer covering of a fruit or seed, as a nut shell, etc. (3) The hard calcareous exterior covering of an egg. (4) To remove or separate from its shell or pod, as to shell nuts, peas, etc. (5) To drop out of a shell, or out of a cluster, as seeds, grapes, etc. (6) To separate seeds or kernels (of corn, wheat, oats, etc.) from the cob, ear, or husk.

Shellbark Hickory–*Carya laciniosa*, family Juglandaceae; a deciduous tree grown especially for use in making sporting goods, implements, woodenware, for fuel and for smoking meats. It may be used as a shade tree, and its nuts are edible and sweet. Native to North America. Also called bigleaf shagbark hickory, big hickory, thick shellbark hickory.

Shelling Percentage–The weight of corn kernels (shelled corn), expressed as a percentage of the weight of the ear corn prior to shelling.

Shelterbelt–A long series of trees planted in semi-arid regions to moderate winds and reduce erosion.

Shelterwood–A forest management method of securing natural reproduction under the partial shelter of seed trees which are gradually removed by successive cutting to admit the amount of light and heat that the seedlings will require.

Shepherd's Purse–*Capsella bursa-pastoris*; an annual winter weed that crowds winter-growing crops such as wheat.

Shield Budding–A bud graft made by inserting a bud of a specified variety, with an attached segment of bark and wood, cut in the shape of a shield, into an opening in the bark of a stem or branch of a different stock. The opening in the bark of the stem is a T-shaped cut. On some fruit trees the T is reversed for a better graft. Also called T-budding, bud, budding, budgraft.

Shield Graft(ing)–A side graft in which a T-shaped incision is made in the bark of the stock as for budding. Into this incision a wedge-shaped scion is inserted and tied in place. The stock is eventually cut off above a scion. Also called sprig graft(ing). See Bud.

Shifting Cultivation–A farming system, common in the tropics, in which land is cleared, the debris burned, and crops grown for a relatively short period until yields decline. The land is then abandoned. The original land is cleared and cropped again after an uncontrolled fallow period of three to twenty years, usually when soil fertility has been naturally restored by woody vegetation.

Shinnery–A dense thicket which, because the growth is relatively new or because growth conditions are unfavorable, is scarcely more than tall shrubbery. Although shinnery is usually made up of mixed woody species, it is generally thought of as oak shinnery because of the usual predominance of dwarf oak.

Shipmast Locust–*Robinia psuedoacocia*; a tall, straight-growing clone of the black locust that is supposed to be resistant to the black locust borer. At one time the trees were used as masts for sailing ships.

Shock–(1) A quantity of a harvested grain crop or corn, etc., set together upright in a field. (2) A small pile of hay. (3) A state of collapse in animals, which results from severe loss of blood flowing surgery, accident, or sometimes under conditions of undue stress. (4) The severe symptoms in plants following acquisition of virus. Later symptoms are absent or milder. (5) A parcel or lot of sixty pieces or units. (6) Twelve sheaves of small grain. (7) To place grain, hay, etc., in shocks.

Shoddy–(1) That leather made by grinding waste leather to a pulp and pressing it into solid sheets with or without the addition of a binding material. (2) Reworked wool that has been recovered from wool cloth trimmings, rags, etc. See Virgin Wool. (3) Anything of an inferior quality.

Shoe Boil–A soft, flabby swelling at the point of the elbow of a horse. It is usually caused by contact with the shoe when the horse is lying down. Also called capped elbow.

Shoot–(1) A stem of a plant, including the leaves. (2) A young stem arising from or near the crown of a plant. (3) A new leafy growth developed from a bud. See Sprout. (4) An area of land inhabited by game (grouse, etc.), and kept especially for the shooting of such game as a sport; the shooting rights to such an area. (5) To develop new stems. (6) To detonate a charge of dynamite or other explosive. (7) See Chute.

Shore Podgrass–*Triglochin maritima*, family Juncaceae; a perennial, poisonous marsh herb. Symptoms of poisoning in cattle are abnormal breathing, trembling, spasms, and convulsions causing death. Native to North America. Also called goose grass, arrowgrass, sour-grass.

Short Grass–Grass or grasses of short stature, such as buffalograss and blue gramagrass which are characteristic of and associated with certain range areas of the Great Plains of western United States. These are commonly the dominant grasses of certain low-rainfall regions west of the tall-grass prairies, and the area is known as short-grass country; plains.

Short Shoots–Lateral shoots that do not produce branches, but produce buds and leaves in successive years. Short shoots bear buds that either are completely inhibited or do not develop, or which develop into flowers also known as spur shoots.

Short-day Plant–A plant in which the flowering period or some other process is accelerated by a relatively short daily exposure to light. See Day-neutral Plant, Long-day Plant, Photoperiodism.

Short-season Vegetables–Vegetables ready for harvest after one to two months following planting.

Short-staple–Lint cotton whose fiber length is ⅞ to 1 inch.

Shortening–Any fat which has the property of making pastry, cakes, etc., friable, easily crumbled, or short.

Shortleaf Pine–*Pinus echinata*, family Pinaceae; an evergreen tree used for pulp, poles, ties, slack-cooperage, veneer, structural timber, planing-mill products, etc. Native to North America.

Shorts–(1) In the florist's trade, roses with short stems (approximately 9 inches in length). (2) Lumber which is shorter than standard lengths. (3) Short pieces or locks of wool that are dropped out while fibers are being sorted. (4) See Wheat Shorts, Wheat White Shorts.

Shot-hole Borer–*Scolytus rugulosus*, family Scolytidae; a small beetle which makes small holes in the bark of healthy fruit trees and in the branches and trunks of weakened trees. On stone fruits, the shot holes are usually covered with gum. The beetles and larvae also make galleries in the sapwood of trees which they infest. If the beetles are abundant, the foliage yellows and wilts, and the tree may die.

Shoulder–(1) Of livestock, the upper part of the forelimb and/or the adjacent part of the back; the scapula region. (2) The upper foreleg and/or the adjacent part cut from the carcass of a hog, sheep, or other animal; the upper part of the carcass to which the foreleg is attached. (3) The part of a bird at which the wing is attached. (4) In turpentining, the uppermost corner of a face. (5) A shoulderlike part or projection, as the projection around a tendon. (6) A sudden inward curvature in the outline of something, as the flaring part between side and neck of a milk bottle. (7) The edge of a road adjacent to the traveled way or pavement.

Showy Chloris–*Chloris virgata*, family Gramineae; a grass, widespread in the United States, which is grazed on by livestock. Its forage rating is fair. In certain areas it is a weed. Native to tropical America. Also called feather fingergrass.

Showy Crotalaria–*Crotalaria spectabilis*, family Leguminosae; a tropical annual herb grown as a cover crop and as a soil-improving plant. Conditions of poisoning are reported with quail, chickens, swine, and cattle that have eaten the seeds. The green or dried plants are equally poisonous and, when eaten, frequently result in death. Native to tropical America.

Showy Milkweed–*Asclepias speciosa*, family Asclepiadaceae; a coarse herb which grows up to 6 feet high, mostly in moist or somewhat shady situations, from Minnesota to British Columbia and south to Texas, Arizona, and California (United States). Although not generally eaten, instances of poisoning have been reported with sheep and cattle grazing on the leaves and stems. See Milkweed.

Shrink-Swell–The shrinking of soil when dry and the swelling when wet. Shrinking and swelling can damage roads, dams, building foundations, and other structures. It can also damage plant roots. See Vertisols.

Shrinkage–(1) The process of reducing in dimension, weight, or volume. (2) With livestock, the loss in weight of an animal from the time it is taken off feed to the time it is marketed or slaughtered. (3) The loss in weight of grease wool due to scouring, expressed as a percentage of the original weight. (4) The loss in weight (moisture loss) of grain from harvest time to some later date. Such shrinkage is commonly reflected in the selling price. (5) The contraction of wood caused by drying, usually expressed as a percentage of some specific dimensions (or volume) of the wood when green. (6) Loss in weight of either meat or carcass stored under refrigeration. Also called cooler shrink, freezer shrink.

Shrivel–To contract and wrinkle, as of leaves, grain, fruit, etc.; to dry up.

Shrub–A plant that has persistent, woody stems and a relatively low growth habit, and that generally produces several basal shoots instead of a single bole. It differs from a tree by its low stature and nonarborescent form.

Shrub Live Oak–*Quercus turbinella*, family Fagaceae; an evergreen shrub or small tree growing to 14 feet in height at elevations up to 8,000 feet in Texas, Arizona, California, Colorado, New Mexico, and northern Mexico. It is an abundant species of the chaparral and is browsed to some extent by livestock. Native to North America.

Shuck–(1) The outer covering of certain nuts, as of hickory, chestnut, etc., or the outer covering of an ear of corn; a husk or pod. (2) The calyx of the flower of stone fruits (as of the peach) which for an interval encloses the fruit but gradually dries, splits, and is ultimately pushed off by the expanding fruit. (3) To remove the shucks from corn, nuts, etc.

Sick–(1) Suffering from illness of any kind; not sound or in fit condition. (2) Designating a soil unfit for the profitable production of certain crops as a result of being infected with disease organisms or because of the excess or deficiency of certain elements.

Sickle Alfalfa–*Medicago sativa* var. *falcata*, family Leguminosae; a perennial, hardy herb which in its pure form has yellow flowers and sickle-shaped seed pods. It has been used extensively in hybridization for creating improved varieties more resistant to cold. Native to Europe. Also called Siberian alfalfa. See Alfalfa.

Side Camp–In forestry, a small camp set up to accommodate a crew working away from a main camp or headquarters.

Side (Graft(ing)–A method of grafting used outdoors and in the greenhouse. A rather long downward diagonal cut is made in the stock and a scion with a lower end shaped into a thin wedge is inserted within its bark, the cut side being kept close to the wood. Tying and waxing may or may not be done. See Shield Graft(ing).

Side Oat–*Avena sativa* var. *orientalis*, family Gramineae; an oat very much like the common oat (*A. sativa*) except that the branches of its panicles tend to develop on one side. Not grown extensively because its yields are lower than those of the common oat.

Side Spur–A short, woody branch, commonly one which bears a blossom, bud or fruit, as in the apple tree.

Side Whip Grafting–A modification of whip graft(ing) in which the scion is placed on the side of a stock rather than on the end of it. Also called bottle grafting.

Sidedressing–Applying fertilizer on soil surface close enough to a plant so cultivating or watering would carry the fertilizer to the plant's roots.

Sideoats Gramagrass–*Bouteloua curtipendula*, family Gramineae; a loosely bunched, perennial, widely distributed grass having a good forage rating on ranges and sometimes cut for hay. Native to western North America.

Sierra Larkspur–*Delphinium scopulorum* var. *glaucum*, family Ranunculaceae; a variety of the tall mountain larkspur, a poisonous plant, which is found in the higher mountains of western Canada and Alberta, Washington to Nevada and California (United States). Also called mountain larkspur. See Larkspur.

Sieve Analysis–Determination of the percentage distribution of particle size by passing a measured sample of soil through standard sieves of various sizes.

Sieved–Designating products, usually food products (fruits, vegetables, etc.) that have been reduced to a pulp and then passed through a fine-screened finisher to remove small pieces of fiber and to give a fine-grained, smooth puree. Such products are used chiefly as baby foods. Also called strained.

Signal Word–A word on the label of a pesticide container that is required by law to designate the relative toxicity of the chemical "Danger - Poison" denotes a highly toxic compound; "Warning" means that the chemical is moderately toxic; and "Caution" denotes a compound that is slightly toxic.

Silage–A crop that has been preserved in moist, succulent condition by partial fermentation in a tight container (silo) above or below the ground. The chief crops stored in this way are corn, sorghum, and various legumes and grasses. The main use of silage is in cattle feeding.

Silage Corn–Usually a vigorous-growing corn that is planted thicker than grain corn to produce a high yield of forage for ensiling. The kernels should be well dented and the ears should contain about 50 percent moisture at silo-filling time.

Silage Crop–Any of several crops grown and harvested for silage, as corn, sunflowers, sorghum, small grains, legumes, grasses, etc., cut green and stored in a silo.

Silage Harvester–A field implement that harvests standing corn, cuts it into ensilage length, and elevates the chopped fodder into a truck or wagon box ready for transportation to and placement in the silo.

Silica–SiO_2; silica dioxide; present in quartz, flint, etc.,; one of the most abundant minerals in the earth's crust.

Siliceous–Containing minute particles of silica, as the stems of *Equisetum*.

Silicon–Si; a nonmetallic element which always occurs in a combined form in nature, where it is exceeded in abundance only by oxygen with which it usually combines to form silica. Found in all soils, it is a constituent of plants but is not regarded as an essential plant nutrient.

Silk–(1) The long, silky styles with stigmas of the corn plant, each of which is connected to an ovary on the ear and each of which must be pollinated to form a full ear of kernels. The silks are susceptible to pollination before they emerge from the husk and remain so for about two weeks afterwards. (2) The soft, fine, and shiny fiber which is produced by the silkworm.

Silkoak Grevillea–*Grevillea robusta*, family Proteaceae; an ornamental tree grown for its showy orange flowers and handsome foliage with a fernlike appearance. It is a popular street tree in frost-free areas, and is grown by florists as a pot plant. Its timber is sometimes used in furniture making. Native to Australia. Planted extensively as shade for tea plants in India.

Silkworm Mulberry–*Morus alba* var. *multicaulis*, family Moraceae; a shrubby tree grown for its foliage which is fed to silkworms. Native to China.

Silky Sophora–*Sophora sericea*, family Leguminosae; a small perennial herb with white flowers and thick, leathery, hairy pods with few or several seeds. The seeds contain a toxic alkaloid and if eaten in large quantities may cause poisoning of livestock. Native to North America.

Silt–(1) Small, mineral, soil particles, ranging in diameter from 0.05 to 0.002 millimeters in the United States Department of Agriculture system or 0.02 to 0.002 millimeters in the International System. (2) A textural class of soils that contains 80 percent or more of silt and less than 12 percent clay. (3) Water-borne sediment with diameters of individual grains approaching that of silt. (4) To fill or obstruct a channel, etc., with water-borne sediment; to become filled or choked up with silt or sediment, as a channel, riverbed, etc. See Soil Separate.

Silt Loam–A textural class of soil that contains 50 percent or more of silt and 12 to 27 percent of clay, or 50 to 80 percent silt and less than 12 percent clay.

Silty Clay–A textural class of soil that contains 40 percent or more of clay and 40 percent or more of silt.

Silty Clay Loam–A textural class of soil that contains 27 to 40 percent of clay and less than 20 percent sand.

Silva–(Latin) Wood, forest.

Silver Beardgrass–*Andropogon saccharoides*, family Gramineae; a coarse, perennial grass with dense, silvery-white, silky panicles, which is rated as good for grazing. Native to the United States. Also called silver bluestem.

Silverleaf Nightshade–*Solanum elaeagnifolium*, family Solanaceae; a perennial, poisonous herb which has silvery-haired stems and leaves and smooth, orange-yellow berries. Poisoning may result if the plant is ingested by livestock. Native to North America. Also called white horse nettle, trompillo. See Nightshade.

Silverleaf Scurfpea–*Psoralea argophylla*, family Leguminosae; a perennial herb which is a poisonous plant. Native to North America.

Silvery Lupine–*Lupinus argenteus*, family Leguminosae; a perennial herb whose pods and seeds (either green or dry) when eaten by livestock (especially by sheep) result in poisoning with symptoms of nausea, frothing at the mouth, nervous trembling, and convulsions. Native to North America. See Lupine.

Silvicide–Any chemical which is capable of killing trees, usually by contact, as by spraying or injection.

Silvics–(1) The natural science which deals with the laws underlying the growth and development of single trees and of the forest as a biological unit. (2) The pure science upon which silviculture is based. See Silviculture.

Silvicultural Rotation–The rotation through which stands a timber should be grown to maintain maximum vigor of growth and reproduction.

Silviculture–(1) The science and art of growing and tending forest crops, based on a knowledge of silvics. See Silvics. (2) More particularly the theory and practice of controlling the establishment, composition, constitution, and growth of forests.

Simple–Not compound, not divided into secondary units, as a simple leaf (not compounded into leaflets), or a simple inflorescence (not branched).

Simple Layering–A method of plant propagation similar to tip layering except that the stem behind the end of the branch is covered with soil and the tip remains above ground.

Simple Leaves–Leaf blades consisting of one unit.

Singe–To burn lightly, as to subject the carcass of a pig, fowl, etc., to a flame in order to remove bristles or hair, or to sear the ends of stems of flowers (which have a copious flow of sticky sap) to prolong freshness.

Single–Designating a blossom having an open center and one whorl each of sepals and petals.

Single Cross–The crossing of one strain, variety, inbred line, or breed of plants or animals with a different strain, variety, inbred line, or breed. The seed, plant, or offspring so produced is designated as a single cross or F_1. See Double Cross, Three-way Cross.

Single Cut–A clover, as mammoth clover, which gives only a single cutting of hay in a season.

Single Fruits–The classification of fleshy or dry fruits which form a single, ripened ovary.

Single-eye Cutting–A stem or scion cut to include one eye or bud, used for rooting or grafting.

Single-grained Structure–A soil in which there is no aggregation of particles, such as sand. See Soil Structure.

Sinker–A log whose specific gravity is greater than that of water and will not float.

Sinuate–Deeply or strongly undulate or wavy.

Sinus–(1) Any of the different cavities in people and some animals, some of which are enclosed by bone, others merely enlargements in a vessel or channel. Among the better known sinuses, because of the frequency with which they become infected in people and birds, are those located in the bones at the sides of the nose or nostrils and above the eyes. (2) A depression between two lobes in a leaf.

Sisal–*Agave sisalana*, family Amaryllidaceae; a tropical plant with a short stem and long, succulent leaves which yield a strong, durable fiber used for making rope, twine, etc. Probably native to America. Also called Bahama hemp, sisal hemp.

Sissoo–*Dalbergia sissoo*, family Leguminosae; a tropical tree grown in frost-free areas for its small, pealike, yellowish-white flowers in short clusters and for its durable lumber, called rosewood, used in railway construction, etc. Native to India.

Site–(1) Location for a specific use. (2) A location for a forest planting or for the natural development of a forest. (3) An area, considered as to its ecological factors with reference to capacity to produce forests or other vegetation; the combination of biotic, climatic, and soil conditions of an area. (4) The situation of a fruit planting regarding elevation, topography, soil, and nearness to large bodies of water.

Site Classes–A classification of forest land in terms of potential growth in cubic feet per year of fully stocked natural stands.

Site Index–A designation of the quality of a forest site based on the height of the dominant stand at an arbitrarily chosen age; e.g., if the average height attained by dominant and codominant trees in a fully stocked stand at the age of fifty years is 75 feet, the site index is 75 feet.

Site Preparation–A general term for removing unwanted vegetation, slash, roots, etc. from a site before reforestation. It is accomplished by using mechanical, chemical, biological means, or fire.

Sitka Spruce–*Picea sitchensis*, family Pinaceae; an evergreen tree grown as an ornamental and as a timber tree. Native to North America.

Six-row Barley–A variety of barley in which all three spikelets at each node are fertile so that the spike has six rows of kernels. It is the most important type grown in the United States. See Barley.

Six-spotted Mite–*Eotetranychus sexmaculatus*, family Tetranychidae; a small, red spider mite that is a serious pest of citrus trees. It works with others in colonies on the underside of leaves, causing damage to foliage.

Sixweeks Gramagrass–*Bouteloua barbata*, family Gramineae; an annual gramagrass of mesas and rocky hills of the southwest range country of New Mexico, Arizona (United States), and northern Mexico. It furnishes forage of value in the early spring.

Skidding Tongs–A hinged, clawlike device used to grasp a log for skidding. A cable or chain is fastened to the tongs and to a skidder or tractor. As the cable or chain is pulled the tongs grip the logs tighter.

Skidway–A prepared place for pulling logs prior to loading which consists of two parallel supports at right angles to the road generally raised at the end nearest the road.

Skin–(1) The flexible integument which forms the external covering of an animal, especially of vertebrates. (2) Anything that resembles skin in nature or use, as the flexible outer covering or peel of fruits, etc. (3) The pelt of a small animal, such as calf, sheep, goat, fox, mink, etc., which is usually dressed, tanned, or intended for such treatment. (4) To strip the skin from. (5) To rub the skin off. (6) To peel.

Skip-row Planting–A custom followed by many cotton producers and to a lesser extent by producers of other commodities. It involves planting in uniform spaces one or more rows to a commodity, then skipping one or more rows. Some of the most common patterns followed are: plant one, skip one; plant two, skip two, skip four; and plant four, skip four.

Skunk Spruce–See White Spruce.

Skyline–In forestry, the cable suspended between the head spar tree and the trial tree on which the trolley travels in cableway logging.

Slab–(1) The exterior portion of a log which is removed in sawing lumber. (2) An unattractive, overripe or broken, dried fruit (prune, apricot, etc.) which is sorted from a drying tray in commercial fruit drying. (3) A piece of unsliced bacon.

Slack Water–(1) The interval between high and low water when the tide is in a state of rest. (2) The apparently motionless water in the upper reaches of a dam. (3) In a log jam, the resulting sluggish flow of water.

Slag–The dross from iron ore furnaces and a by-product of the steel industry which can be converted into forms for agricultural use, as fertilizer and soil amendments. Some prepared material from slag, containing little or no phosphorus, but calcium in the form of silicates, may have value primarily as a liming material and a soil conditioner. Other slags, as basic slag, have some value as fertilizer because of the total phosphorus which they contain.

Slaked Lime–(1) CaO; quicklime or burnt lime, treated with water to produce hydrated lime, which in its pure form has the composition $Ca(OH)_2$. It is marketed in a dry, powdered form and is used as a soil amendment in addition to other uses. Also called calcium hydroxide, hydrated lime. (2) Burnt lime which has been exposed to the open air and thereby lost its causticity.

Slash–The branches, bark, tops, chunks, cull logs, uprooted stumps, and broken or uprooted trees which remain on the ground after logging.

Slash Disposal–Treatment or handling of slash to reduce the fire hazard or for other purposes which includes broadcast burning, strip burning, piling and burning, lopping, scattering, pulling tops away, etc.

Slash Pine–*Pinus elliottii*, family Pinaceae; an evergreen tree grown for its lumber used in general construction and as a source of turpentine and rosin. Native to North America. Also called Cuban pine.

Slashing–Forest area which has been logged off and upon which the slash remains.

Sleepy Seeds–Seeds that are slow or irregular in sprouting, a condition of dormancy.

Sleepygrass–*Stipa robusta*, family Gramineae; a coarse bunchgrass, growing 3 to 6 feet tall, native to dry plains and open woodlands from Colorado and western Texas to southern California (United States), and to Mexico. It is said to produce sleepiness in horses and sheep and to a lesser extent in cattle. Its general forage rating is poor.

Slender Oat–*Avena barbata*, family Gramineae; an annual herb which is a weed pest in the Pacific Coast states of the United States. It provides fair forage, especially when young. Native to Europe and Asia.

Slender Wheatgrass–*Agropyron pauciflorum*, family Gramineae; a widely distributed wheatgrass which makes a choice forage for all kinds of livestock and is used in the reseeding of range lands. Native to western North America.

Sliced Veneer–Very thin sections of wood sliced off by moving a log or bolt of wood against a knife.

Slick Spots–Clayey soils that are black because of humus highly dispersed by sodium and, when plowed with a turning plow, turn over in "rubbery," shiny, slices. Such spots seldom produce a crop because of too little soil oxygen, too little available water, and toxic sodium.

Slick-seeded Cotton–Any variety of cotton having smooth seeds as sea island cotton. Also called smooth-seeded cotton.

Slime–(1) A bacterial rot of lettuce resulting in a wet, slimy decay of the leaves and heads. It is caused by *Erwinia carotovora*, family Enterobacteriaceae; *Pseudomonas viridilivida*, family Pseudomonadaceae; *P. marginalis*, and other species of bacteria. Slime occurs in the field, during warm, muggy weather; in lettuce not refrigerated in transit, and at markets. Also called bacterial rot of lettuce. (2) In the clarification of milk, a collection or deposit of dirt, leucocytes, cells, and other viscous matter thrown to the walls of the clarifier bowl. (3) In the curing or storage of cheese, a condition frequently produced on the surface by yeasts, bacteria, or other contamination, or faulty handling, causing a slimy surface on the cheese. (4) To remove the slimy or viscous coating from animal intestines, as in the preparation of sausage casings.

Slime Flux–A persistent, fermenting exudate of seepage from trunk or branch wounds in deciduous trees. In some cases pathogenic organisms are associated with the flux.

Slimflower Scurfpea–*Psoralea tenuiflora*, family Leguminosae; a bushy-branched herb found on dry prairies in open woods, and on rocky banks of midwestern to southwestern United States. Cattle and horses are reported to have been poisoned by grazing upon its herbage. Native to North America.

Slip–(1) A soft-wood or herbaceous cutting from a plant, used for propagation or grafting. (2) An incompletely castrated male. (3) Curdled milk. (4) To take a cutting or cuttings from a plant. (5) To abort. (6) The downslope movement of a soil mass under wet or saturated conditions; a microlandslide that produces microrelief in soils.

Slips–(1) Cuttings, used to vegetatively propagate plants. (2) Sweet potato sprouts taken from a bed for transplanting.

Slow-release Fertilizers–See Controlled-release Fertilizers.

Sluice–(1) A conduit which carries water at high velocity. (2) An opening in a structure for passing debris. (3) A water gate. (4) A channel which carries or drains off surplus water. (5) To cause water to flow at high velocities for wastage, for excavation, ejecting debris, etc. (6) To move or float logs over or through a dam from the water above into the stream below.

Slurry–(1) A thin mixture of water and any of several fine insoluble materials, as clay, derris or cube powder, and so on. (2) The thick liquid formed by the mixing of dung and urine. (3) A thick suspension of chemicals used in coating seed with a fungicide. (4) A supersaturated, fluid fertilizer.

Small Chestnut Weevil–*Curculio carya*, family Curculionidae; a small weevil which penetrates the burs of chestnuts to lay its eggs. When these weevils are numerous, their larvae may destroy most of the nuts.

Small European Elm Bark Beetle–*Scolytus multistriatus*, family Scolytidae; a beetle which, with its grubs, makes tunnels in the sapwood of elms and is considered the most common vector of Dutch elm disease.

Small Grain–Any of the cereal crops, such as wheat, oats, barley, rye, and rice.

Smartweed–*Polygonum persicaria*; an annual weed pest that is a member of the buckwheat family. Stems can grow to 6 feet tall and can crowd out crops.

Smear–Material smeared on a surface, usually a small piece of glass (slide), and examined under a microscope. Smears are generally stained with dyes before being examined. Materials suitable for examination from a smear include blood, milk, bacteria from a culture, pus, etc.

Smelter Injury–Toxic effects on plant life which are caused by flue gasses or dusts discharged by smelters. It is similar to that caused by smoke in highly industrial areas.

Smilograss–*Oryzopsis miliacea*, family Gramineae; a perennial grass with relatively stout, erect culms, sometimes branching, growing 2 to 4 feet tall, and sometimes used in the reseeding of range lands. Native to North America.

Smoke Chaser–A member of a forest-fire protective unit that goes to reported fires to extinguish them.

Smoldering Fire–A forest fire still smoking, not spreading appreciably, and burning with little or no flame.

Smooth Bromegrass–*Bromus inermis*, family Gramineae; a perennial, awnless, drought-resistant grass which has a moderately spreading growth habit, forms a dense sod, and grows up to 5 feet in height. It is a palatable hay and pasture grass and is used extensively in reseeding pastures and ranges. Native to Europe. Also called Austrian bromegrass, awnless bromegrass, Hungarian bromegrass.

Smooth Paspalum–*Paspalum pubiflorum*, family Gramineae; a good forage grass native to North America. See Paspalum.

Smooth-seeded Pea–Any variety of garden pea having smooth (not wrinkled) seeds, as the Alaska variety.

Smother–(1) To kill by suffocation, as panicky chickens or sheep are smothered when they pile up in a heap. (2) Of plants, to kill by covering thickly with dirt, straw, ice sheet, black sheet plastic, etc. (3) Of plants, to retard in growth or to kill by a dense or more vigorously growing crop. See Smother Crop.

Smother Crop–Any dense-growing crop that suppresses or stops the growth of less competitive plants; a quick-growing crop (buckwheat, rye, etc.) planted especially for the suppression of weeds.

Smothering Disease of Coniferous Seedlings–A disease of coniferous seedlings caused by the smothering fungus *Thelephora laciniata*, family Thelephoraceae.

Smudge–A smoky fire or the dense smoke used to drive insects away.

Smudge Fire–A fire made from damp straw, smoldering manure, or any other fuel which produces a very heavy smoke that hangs low to the ground, erroneously believed to prevent frost injury to crops. See Smudge Pot.

Smudge Pot–A type of open pot or vessel in which an oil fire is maintained to heat the air in orchards and vegetable plots in hope of reducing frost damage.

Smut–(1) The disease which is caused by the true smut fungi of class Basidiomycetes, order Ustilaginales, or the causal fungus. (2) Any disease which is caused by entirely different fungi agreeing with the true smuts only in the production in the diseased host plant of numerous dark, mostly dusty spores, as fig smut.

Smutty Wheat–A United States federal grain-grading designation for wheat in which there is a considerable quantity of smut spores detected by the fishlike odor and/or by the presence of numerous smut balls.

Snag–(1) A sharp point of wood which results from a break. (2) A broken tree trunk more than 20 feet tall.

Snake–(1) Any of the limbless vertebrates of the order Serpentes. Although a few species are poisonous, most snakes are beneficial to humans by the number of rodents they consume. (2) To drag a log or other object across the ground using a rope, chain, or cable.

Snap–(1) A spring fastening or clasp which closes with a clicking sound, used as a fastener on harnesses, gate chains, dog chains, etc. (2) A brief period of cold weather; as a cold snap. (3) To harvest by pulling off with a quick motion, as the ears of corn, the bolls of cotton, etc. (4) In horses, especially in harness or saddle horses, to move briskly with animation; to pick up feet quickly and sharply.

Snap Corn–Ear corn, with all or most of the husks adhering, which is harvested by breaking or snapping from the stalk.

Snapped–Designating an entire mature cotton boll, an unhusked corn ear, asparagus, a garden bean, etc., which has been harvested by breaking from the plant or stalk.

Snip–(1) To remove by clipping, as to snip off a twig, flower, etc. (2) A white marking between the nostrils of a horse.

Snout–(1) The projecting part of an animal's head that contains the nose and jaws, as the snout of a hog. (2) An anterior elongation of the head; a proboscis; a piercing or sucking part of certain insects.

Snout Beetle–Any of the beetles of the family Curculionidae, to which belong the typical weevils which injure fruit and other crops. They are distinguished by the head which is prolonged forward and downward into a cylindrical snout, at the end of which are very small parts which function to chew plant tissues. The long snout is used for making a hole deep into the tissues of fruits or plants into which the beetle lays its eggs.

Snow Mold–A fungal disease of winter cereals in northern Europe and of grasses, grains, and alfalfa in northern United States and Canada, attributed to several fungi such as *Typhula itoana*, *Calonectria graminicola*, and *Pythium* spp. Its outstanding symptom is a conspicuous, white mold growth appearing on the plant as the snow melts.

Snowbreak–(1) An artificial or natural barrier such as a shelterbelt or windbreak to intercept blowing snow and thereby reduce excessive accumulations on roadways or farmsteads. Almost equivalent to a snow fence or a windbreak. (2) The breaking of tree limbs by the accumulation of snow. See Snow Fence.

Soboliferous–Bearing or producing lateral shoots from the ground; clump-forming; usually applied to shrubs or small trees, as species of *Syringa*, *Rhus*, and some Palmae.

Social Insects–Insects which live in a family society, with parents and offspring sharing a common dwelling place and exhibiting some degree of mutual cooperation; e.g., honeybees, ants, termites.

Sod–(1) A tight, closely knit growth of grass or other plant species which may or may not be closely cut at frequent intervals. (2) A surface layer of soil containing grass or other plant species with their matted root systems.

Sod Crop–(1) Any crop or vegetative cover which quickly forms a heavy, close-knit, top growth over the surface of the soil and a root system which binds the soil particles together, thus forming a sod, as white clover, bluegrass, etc. (2) Any forage grasses or legumes.

Sod Culture–A soil management practice of maintaining permanent sod in orchards as opposed to annual tillage.

Sod Grasses–Stoloniferous or rhizomatous grasses that form a sod or turf.

Sod Species–Any grass with rhizomes or stolons, such as Bermudagrass.

Sod Strip–(1) A band or narrow strip of sod used for checking erosion in waterways. (2) A narrow band of grass or other close-rooted crop placed across the channel of a gully to spread and retard the flow of water. (3) In strip cropping, a strip that is in sod.

Sodbound–Designating a condition in which the roots are so restricted and form such a dense mat that growth of the plant is impeded.

Sodding–Removing sod from one area and placing it on a bare soil area in another location.

Sodic Soil–A soil with an exchangeable sodium percentage of 15 percent or greater, a saline content below 4 mmhos, and a pH between 9.5 and 10.0. (Obsolete name: black alkali.)

Sodium–Na; pure metallic sodium is a soft waxy material at ordinary temperatures; it burns in air to form sodium oxide (NaO) which readily takes up water to form sodium hydroxide (NaOH). Combined with chlorine, it forms sodium chloride (NaCl) (common salt), and with nitric acid, it forms sodium nitrate ($NaNO_3$). Sodium can substitute for or replace part of the potassium needed by some plants. It increases the availability of soil phosphorus and reduces fixation of the phosphorus applied in superphosphate. Sodium appears to have some value of its own for such crops as oats and table beets. This, however, may be due to the fact that it can release potassium, calcium, and magnesium by cationic exchange and thus temporarily increase these available cations in the soil.

Sodium Alginate–A sodium salt of the seaweed, algae. It has excellent water-binding properties and is used by some manufacturers to prevent water crystallization or iciness in ice cream.

Sodium Borate–See Borax.

Sodium Fluoracetate–Compound 1080; a powerful rodenticide developed by the United States Fish and Wildlife Service; a fine, white powder readily soluble in water, used to control rats, mice, and other rodents. It is a restricted pesticide.

Sodium Fluoride–NaF; a white, powdered, chemical compound used for combatting chewing lice on animals and poultry, commonly employed in roach and ant powders, as an anthelmintic used to eliminate ascarids (roundworms) in swine, and as a fungicide.

Sodium Fluosilicate–A fluorine compound sometimes substituted for sodium fluoride for the control of lice, roaches, and certain other insects. Also called silicofluoride.

Sodium Hypochlorite–NaOCl; a chlorine compound which is commonly sold in a powdered form or in solutions under various trade names for use in the disinfection of processing equipment and as a household disinfectant and bleach.

Sodium Nitrate–$NaNO_3$; nitrate of soda; Chile saltpeter; a salt recovered from natural deposits in Chile, or produced synthetically by reacting nitric acid with sodium carbonate. Pure sodium nitrate contains 16.48 percent N and 27.05 percent sodium (Na).

Sodium Selenate–A highly poisonous, water-soluble salt sometimes applied in solution to greenhouse soil to control red spider mites and aphids. The compound is taken up from the soil by the plants, and the plant juices become toxic to the insects feeding on them.

Soft Brome–*Bromus mollis*, family Gramineae; a widely distributed, annual grass growing 1 to 1½ feet tall with velvety foliage, which provides fair forage on rangelands. Also called soft chess.

Soft Brown Scale–*Coccus hesperidum*, family Coccidae; a small sucking insect of worldwide distribution which infests a large variety of plants and is found commonly on citrus. In Australia and South Africa, it is considered a major pest of citrus. Also called soft scale.

Soft Cutting–See Slip.

Soft Phosphate with Colloidal Clay–A soft, clayey, low-grade, phosphatic rock, not suitable for use in the manufacture of superphosphate. It is the fine-grained waste which results from the washing of hard rock and pebble phosphate, collected in ponds, subsequently dried and sold as a low-grade fertilizer or used as a filler for mixed fertilizers. It has some value on acid sandy soils. Also called colloidal phosphate, waste-pond phosphate, phosphatic clay.

Soft Pinch–Removing a small part (one-half inch) of the top of a plant to encourage flowering and fruiting. See Hard Pinch.

Soft Red Winter Wheat–A type of wheat that has a long kernel and large germ, the suture is open, very deep, and easily split. These wheats are lower in protein and higher in starch percentage than the hard winter types. They are grown in the more humid climates of the north-central and eastern sections of the United States.

Soft Rot of Vegetables–A rapid, soft, wet rot of carrots, onions, celery, potatoes, and many other vegetables, caused chiefly by bacteria of the genera *Erwinia* and *Pseudomonas*, notably by *Erwinia carotovora*, family Enterobacteriaceae.

Softwood–(1) Any wood of light texture which is easily cut. (2) In forestry, any tree having such wood, particularly a coniferous tree. (3) Immature, succulent wood, such as is used for a softwood cutting. See Hardwood, Softwood Cutting.

Softwood Cutting–Cutting taken from a woody or herbaceous plant before it has matured.

Soil–The mineral and organic surface of the earth capable of supporting upland plants. It has been (and is being) formed by the active factors of climate and biosphere exerting their influence on passive parent material and topography over neutral time.

Soil Acidifier–A material or mixture used, especially in semi-arid western United States to reduce soil alkalinity. Sulfuric acid, phosphoric acid, liquid sulfur dioxide, aluminium sulfate, sulfur, and calcium polysuflide are common soil acidifiers.

Soil Acidity–Excess of hydrogen (H) ions over hydroxyl (OH) ion concentration in the soil, consisting of: (a) active, which is estimated by pH below 7.0, and (b) inactive or reserve, less successfully estimated by base exchange determinations, as opposed to soil alkalinity. See pH, Reaction.

Soil Aeration–Processes by which more air enters the soil such as: plowing, cultivating, disking, burrows made by fauna such as earthworms, soil shrinking upon drying, and root channels resulting from the decay of plant roots.

Soil Aerator–An implement that makes holes in the ground to let air enter the soil. It consists of a roller that has spikes protruding from it.

Soil Aggregate–A single mass consisting of many soil particles held together. See Clod, Ped.

Soil Alkalinity–Excess hydroxyl (OH) ions over hydrogen (H) ion concentration in the soil, as opposed to soil acidity.

Soil Amendment–Any material, such as lime, gypsum, sawdust, compost, animal manures, crop residues, or sewage sludge, that is worked into the soil or applied on the surface to enhance plant growth. Amendments may contain important fertilizer elements, but the term commonly refers to added materials other than those used primarily as fertilizers.

Soil Analysis–Chemical, mechanical, and mineralogical determinations of the separate components of a soil which may be complete or partial depending upon the objectives.

Soil Association–A group of soils with or without common characteristics, geographically associated in an individual pattern. It may constitute a natural land type or a geographical unit in soil mapping. A soil association is described in terms of the taxonomic units included, their relative proportions, and their pattern of association. Sometimes called natural land type. See Land Type, Soil Complex.

Soil Auger–A tool used for boring into the soil and withdrawing a sample for field or laboratory observation. Soil augers are of two general types: those with worm-type bits and those having a hollow cylinder with a cutting edge in place of the worm bit.

Soil Bacteria–Two large groups of bacteria, based on their energy requirements: (a) the heterotrophic bacteria, which obtain their energy and carbon from complex organic substances; (b) the autotrophic bacteria, which can obtain their energy from the oxidation of inorganic elements or compounds, their carbon from carbon dioxide, and their nitrogen and other minerals from inorganic compounds. See Bacteria.

Soil Buffer Compounds—Clay, organic matter, and such compounds as carbonates and phosphates which enable the soil to resist appreciable change in pH value.

Soil Bulk Density—Ratio of dry weight of a soil mass to its volume.

Soil Chemistry—That branch of chemistry which deals with investigations into the chemical nature of soil components and their interrelationships.

Soil Class—See Land Capability Class, Soil Textural Class.

Soil Classification—The systematic grouping of soils into categories from higher to lower. It may be done on a genetic basis, on the basis of the intrinsic properties of different soils, or a combination of the two. The smaller taxonomic units of a scientific system may be regrouped for various purposes, such as agriculture, highway construction, and forestry. See Soil Taxonomy.

Soil Climate—The moisture and temperature conditions existing within the soil.

Soil Colloid—Inorganic and organic matter of very small size having a relatively large surface area per unit of mass. Under certain conditions, soil colloids form a stable suspension in water and are distinguishable from true solutions in that all particles have not dispersed to the molecular state. One micron, or 0.001 millimeter has been arbitrarily set as the upper limit for size of a soil colloid.

Soil Complex—A mapping unit used in detailed soil surveys where two or more defined taxonomic units are so intimately associated geographically that they cannot be separated by boundaries on the scale used. See Soil Association.

Soil Conservation—The efficient use and stability of each area of soil that is needed for use at its optimum level of developed productivity according to the specific patterns of soil and water resources of individual farms, ranches, forests, and other land-managment units. The term includes the positive concept of improvement of soils for use as well as their protection and preservation.

Soil Consistence—The relative neutral attraction of the particles in the whole soil mass or their resistance to separation or deformation, as evidenced in cohesion and plasticity. Soil consistence is described by such general terms as loose or open; slightly, moderately, or very compact; mellow; friable; crumbly; plastic; sticky; soft; firm; hard; and cemented.

Soil Depth—(1) Technically the thickness of the solum (A and B horizons), as distinguished from the C and R horizons of the profile. (2) The thickness of the humus layer or plowed soil. See Soil Profile.

Soil Deterioration—Loss of potential productive capacity of a soil which results from destructive processes accelerated by the activities of people; e.g., through soil erosion, waterlogging, and excessive accumulation of salts.

Soil Development—The work of the natural evolutionary process of soil formation which leads to maturity or complete differentiation into soil horizons. See Soil Profile.

Soil Drainage Classes—Classes of soils grouped according to their ability to hold water. The seven classes are: excessively drained, somewhat excessively drained, well drained, moderately well drained, somewhat poorly drained, poorly drained, and very poorly drained.

Soil Erosion—Removal of soil material from a land surface by wind or water, including normal soil erosion and accelerated soil erosion. (Sometimes used loosely in reference to accelerated erosion only.) See Accelerated Erosion, Erosion, Normal Erosion.

Soil Erosion Service—In the United States, an emergency agency first authorized in 1933 which was placed in the Department of Interior, and subsequently in 1935 transferred to the Department of Agriculture, as the Soil Conservation Service. See Soil Conservation Service.

Soil Family—A group of soils having similar amounts of sand, silt, and clay; similar kinds of minerals and temperatures; and similar other properties important to plant growth. There are about 2,000 soil families recognized.

Soil Fertility—The capability of a soil to supply all essential plant nutrients in balanced and adequate amounts and with proper timeliness. See Soil Productivity.

Soil Formation Factors—The independent variables that define the soil system. Five main groups of soil formation factors are generally recognized by soil scientists: parent material, climate, organisms, topography, and time.

Soil Genesis—The mode of origin of the soil, particularly the processes which are responsible for the development of the solum from the parent material.

Soil Heaving—See Heaving.

Soil Horizon—A natural layer of a vertical section of the soil profile. See Soil Profile.

Soil Injection—Placing pesticide or fertilizer below the surface of the soil.

Soil Loss Equation, Universal—See Universal Soil Loss Equation.

Soil Loss Tolerance—In conservation farming, the maximum average annual soil loss in tons per acre per year that should be permitted on a given soil. The rate of soil formation should equal or exceed the rate of soil erosion loss. See T Factor.

Soil Management—The combination of all tillage operations, cropping practices, fertilizer, lime, and other treatments applied to the soil for the production of plants.

Soil Microbiology—The branch of science which deals with the study of microorganisms in the soil, such as algae, fungi, bacteria, and protozoa.

Soil Mixture—A mixture which consists of soil with varying proportions of one or more of the following: manure, peat moss, leaf mold, and sand used for growing seedlings, pot plants, plants in beds in greenhouses, and as a medium for the rooting of cuttings. Also called potting soil.

Soil Moisture– The water in soil normally expressed as a percentage of the weight of water to the oven-dry weight of soil.

Soil Moisture Capacity– The amount of water expressed in percentage of dry weight that a soil can retain against the pull of gravity. See Field Capacity.

Soil Moisture Tension– The force with which the soil holds onto the soil moisture present. As soil moisture declines, soil moisture tension increases.

Soil Morphology– The kinds and arrangements of horizons in a soil profile.

Soil Mottling– A variation of color in a soil horizon, as mottles, spots, splotches, or veins of one color in a background mass of another; usually gray splotches with a brown to red matrix.

Soil Mulch– (1) The natural cover of dead plant residue on a soil. (2) The corresponding loose matter on the surface of a grassland soil, in contrast to an artificial mulch. See Mulch.

Soil Orders– The classifications of soils. According to the United States Department of Agriculture system of soil taxonomy, the eleven soil orders are: entisols, vertisols, inceptisols, aridisols, mollisols, spodosols, alfisols, ultisols, oxisols, histosols, and andisols.

Soil Organic Matter– The organic fraction of the soil that includes plant and animal residues at various stages of decomposition, cells and tissues of soil organisms, and substances synthesized by the soil population. Commonly determined as the amount of organic carbon contained in a soil sample passed through a 2-millimeter sieve.

Soil Particle– A soil separate in the mechanical analysis of a soil.

Soil Pasteurization– The heat or chemical treatment of soil to destroy all harmful organisms but not necessarily all soil organisms.

Soil Phase– In soil taxonomy and soil mapping, a subdivision of soil series such as surface soil texture, slope, stoniness, saltiness, and erosion. Phase may also be used for a soil order, suborder, great group, subgroup, or family.

Soil Physics– That branch of science which is concerned with the application of the laws, knowledge, and techniques of physics to the study of soils.

Soil Porosity– The degree to which the soil is permeated with pores or cavities. It is expressed in percent of the volume of the soil unoccupied by soil particles.

Soil Probe– A T-shaped metal tube that is forced into the ground for the purpose of obtaining a sample of the soil.

Soil Productivity– The ability of a soil to produce crops under a particular condition of management. See Soil Fertility.

Soil Profile– A vertical section of a soil. The section, or face of an exposure made by a cut, may exhibit with depth a succession of separate layers although these may not be separated by sharp lines of demarcation. See Horizon, Soil.

Soil Reaction– The degree of acidity or alkalinity of the soil mass expressed in pH values. See pH.

Soil Salinity– The amount of soluble salts in a soil.

Soil Science– The science which deals with the scientific study of soil as a natural body and an economic resource. See Pedology.

Soil Separate– Any of a group of soil particles which has definite size limits. The names and average diameters of separates recognized in the United States are: very coarse sand, 2.0 to 1.0 millimeters; coarse sand, 1.0 to 0.05 millimeters; medium sand, 0.05 to 0.25 millimeters; fine sand 0.25 to 0.10 millimeters; very fine sand 0.10 to 0.05 millimeters; silt, 0.05 to 0.002 millimeters; and clay, less than 0.002 millimeters. The class of separates recognized by the International Society of Soil Science are: I, 2.0 to 0.2 millimeters; II, 0.2 to 0.02 millimeters; III, 0.02 to 0.002 millimeters; IV, less than 0.002 millimeters.

Soil Series– A group of soils having horizons similar in characteristics and arrangement in the soil profile, except for the texture of the surface. They are given proper names from place names within the areas where they occur. Thus, Norfolk, Miami, and Houston are names of well-known soil series.

Soil Slaking– The phenomenon expressed by crumbling, flaking, or complete disintegration of a mass of dry soil when immersed in water.

Soil Slip– A downward movement of a mass of soil on hillsides under conditions of complete saturation. The distance covered and the mass of the movement in a soil slip are less than in a landslide or avalanche.

Soil Slope– An incline of a land surface. See Slope.

Soil Solution– (1) The liquid part of soil that surrounds soil particles and contains the elements in solution that are absorbed by plant roots. It is not constant in amount or composition. (2) Technically, the aqueous solution which exists in equilibrium with the soil particles at a particular moisture tension.

Soil Sterilization– Treating soil by fumigation, chemicals, heat, or steam to destroy disease-causing organisms.

Soil Strength– A characteristic of how well soil particles stick together or resist fragmentation. It is used to predict a soil's adaptation to cultivation or to support the foundation of a building.

Soil Structure– See Structure, Soil.

Soil Survey– (1) In the United States, this refers to the National Cooperative Soil Survey conducted by the Soil Conservation Service, United States Department of Agriculture, in cooperation with state agricultural universities and other public and private agencies and organizations. (2) Systematic examination of soils in the field and in laboratories; publishing of descriptions and classifications; mapping of kinds of soils; and interpretation of soils according to their adaptability to various crops such as field crops, fruits, vegetables, and trees.

Soil Survey Report– A written statement accompanying a soil map which describes the geography of the area surveyed, the characteristics and capabilities of the soil-mapping units. It also discusses the principal factors responsible for soil development.

Soil Taxonomy—Classification according to natural relationships. Comparable to plant taxonomy and animal taxonomy. The latest system of soil taxonomy was implemented in the United States in 1965. See Soil Classification, Soil Orders.

Soil Temperature—The temperature in the soil mass which varies according to depth and local climate. It is an important factor in the growth of plants.

Soil Testing—Any of the various laboratory and field examinations of soil samples to determine the amounts, kinds, and availability of plant nutrients. Tests may also be made in relation to plant growth, agriculture, highway and building construction, such as acidity, alkalinity, moisture, porosity, penetrability, etc.

Soil Textural Class—The relative proportion in a soil of the various size groups of individual soil grains. The coarseness or fineness of the soil depends on the predominance of one or the other of these groups, which are silt, clay, and sands. See Soil Separate.

Soil Tilth—Structure or physical condition favorable for germination and development of crop plants. It is usually produced by loosening the soil with plows, harrows, or other agricultural implements, and by other agronomic practices.

Soil Water—(1) Any or all of the various forms of water which may be contained in the soil, as free, capillary, combined. (2) The water below the ground surface in the zone of aeration which may be discharged as evaporation, plant transpiration, and seepage.

Soilage—Freshly cut green fodder fed to confined animals.

Soiling—In livestock management, the cutting and bringing of green forage to livestock in place of allowing the animals to eat the green feed where it grows.

Soilless Gardening—Growing plants in a nutrient solution; water culture. See Hydroponics.

Soilless Mixes—Media mixtures which do not contain soil. The ingredients usually consist of bark, vermiculite, perlite, and similar materials. See Soil Mixture.

Soils, Poorly Drained—Water is not removed because of a compact subsoil or the soil is too low for the water to drain out. The soil is usually a gray color within 6 inches of the surface.

Soils, Well Drained—Water moves readily through the subsoil. They are not discolored but show uniform, bright colors throughout the soil profile.

Solanaceous—Belonging to the nightshade family, of which the Irish potato is a common example.

Solid Manure—Manure in solid form as contrasted to liquid manure.

Solubility—To be most readily available to plants a nutrient must be at least slightly soluble in the soil solution or be held in an exchangeable form on clay and humus particles. See Available.

Soluble—Capable of changing form or changing into a solution.

Solum—The upper part of the soil profile above the parent material, in which the processes of soil formation have taken/are taking place. In mature soils this includes the A and B horizons which may be, and usually are, greatly unlike the parent material beneath. Living roots and life processes are largely confined to the solum. See A Horizon, B Horizon, Parent Material.

Soma—The body, in contrast with the germ or germ plasm.

Somatic—(1) Designating body tissues. (2) Having chromosomes in pairs, one of each pair normally coming from the female parent and one from the male, as contracted with terminal tissue which gives rise to germ cells.

Soot—Very finely divided carbon particles clustered together in long chains. Sometimes used as a low-grade fertilizer.

Sore Shin—A fungal disease of cotton seedlings caused by ***Rhizoctonia solani***, or ***Pellicularia filamentosa***, family Thelephoraceae; characterized by stem canker on the seedling near the soil line. The disease is generally distributed and commonly develops during periods of cold, wet weather. Also called anthracnose seedling blight. See Cotton Anthracnose.

Sorgho—See Sorgo.

Sorghum, Grain—A cereal grass used mainly for feedgrain or silage. Often grown in corn and wheat areas.

Sorghum Molasses—A type of syrup usually farm-produced from sorgo. It is common in the south-central United States.

Sorghum Poisoning—A poisoning of livestock which results from ingestion of sorghum containing prussic acid (hydrocyanic acid), a deadly poison which develops in dangerous amounts under certain conditions of plant stress in many of the sorghums.

Sorghum Silage—Silage made from either sweet or grain sorghums ensiled when the seeds are hard and ripe. It is somewhat inferior to corn silage in feed value.

Sorghum Syrup—A thick syrup made from the juice of sorgo or sweet sorghums by boiling the juice in shallow evaporating pans until it is concentrated.

Sorgo—***Sorghum bicolor*** var. ***saccharatum***, family Gramineae; a tall, sweet herb with a juicy pith high in sugar, whose chief use formerly was for syrup, but now is primarily grown for forage. It is one of the four classes of sorghum commonly grown in the United States. Native to Europe and Asia. Also called saccharine sorghum, sorgho, sweet sorghum, sugar sorghum.

Sour Cherry—***Prunus cerasus***, family Rosaceae; a deciduous tree, grown for its fruit, the red, tart cherry used in cooking, preserves, pies, etc. Important varieties include Montmorency, Early Richmond, and English Morello. Also called red tart cherry. See Cherry.

Sour Orange—***Citrus aurantium***, family Rutaceae; an evergreen tree cultivated in all subtropical regions for its flowers, leaves, and fruit yielding volatile oils (bigarade oils) of characteristic odor much prized for perfume. The fruits are used extensively for the making of marmalade and liqueurs. Native to Asia. Also called Seville orange, bigarade.

Sour Soil—An acid soil.

Southern Armyworm–*Prodenia eridania*, family Noctuidae; an insect whose larvae infest vegetables and several other plants from Florida to South Carolina (United States).

Southern Bacterial Wilt–See Granville Wilt.

Southern Blight–A blight caused by the fungus ***Sclerotium rolfsii***, which attacks a plant through the roots and underground parts of the stem at the soil line, and causes a rapid wilting of the plant. The disease is often serious on peanuts, tomatoes, peppers, and many other crops in the southeastern and Gulf states (United States).

Southern Corn Rootworm–*Diabrotica undecimpunctata howardi*, family Chrysomelidae; an insect whose larvae often cause serious injury by burrowing in the roots, crown, and stems of young corn and many other plants, including peanuts, beans, and cucurbits. The adults eat the foliage of these plants. Also called spotted cucumber beetle.

Southern Corn-leaf Blight–A fungal disease of corn caused by ***Cochliobolus heterostrophus***, which kills the green tissue of the leaves and reduces yield. It occurs throughout the corn areas of the southern United States, where the warm, humid climate favors its development.

Southern Cornstalk Borer–*Diatraea crambidoides*, family Crambidae; a straw-colored moth whose larvae feed on foliage but usually bore inside of stalks of corn and sorghum. It is distributed from Virginia south and west to Louisiana and Missouri, United States.

Southern Grassworm–See Fall Armyworm.

Southern Green Stink Bug–*Nezara viridula*, family Pentatomidae; an insect about two-thirds inch long which sucks sap from soybeans and cowpeas causing pods to drop; forming hardened, knotty areas; or producing stunted or distorted seed. In Florida, United States, these insects often cause severe loss of fruit and kill succulent shoots on young citrus trees.

Southern Leaf Spot–A fungal disease of corn and teosinte caused by ***Helminthosporium maydis***, family Dematiaceae; characterized by numerous, buff or reddish-brown, elongated spots between the veins. It is found worldwide.

Southern Magnolia–*Magnolia grandiflora*, family Magnoliaceae; a magnificent evergreen tree, grown in relatively frost-free areas for its large, fragrant, cup-shaped, white flowers. Its wood is used for furniture, fixtures, Venetian blinds, interior finish, sash, doors, and millwork. Native to North America. Also called evergreen magnolia, bull bay. The state tree of Mississippi.

Southern Pea–Any of several edible cowpeas. See Cowpea.

Southern Pine Beetle–*Dendroctonus frontalis*, family Scolytidae; an insect that infests pine and makes winding egg galleries in the inner bark. It kills healthy trees, as well as attacking weakened or felled specimens. It is a serious pest of pine from Pennsylvania to Florida and west to Texas and Oklahoma, United States.

Southern Wilt–See Granville Wilt.

Southwestern Corn Borer–*Diatraea grandiosella*, family Crambidae; a corn borer, introduced into the United States from Mexico, which very closely resembles the southern cornstalk borer, but differs chiefly in its habit of internally girdling the stalk. It is found from Kansas, Nebraska, and Colorado southward, to Texas, New Mexico, and Arizona.

Sow–(1) A female swine, usually one that shows evidence of having produced pigs or one that is obviously pregnant. (2) To plant seeds by scattering either broadcast or by distribution in a row.

Sow Broadcast–To scatter seeds widely upon a field by hand or by some mechanical device, in contrast to placing such seeds in rows or hills.

Sow Thistle–(1) Any of the several plants of the genus ***Sonchus***, family Compositae. (2) ***S. oleraceus***; an erect, coarse, leafy herb with smooth stems, cut leaves, and yellow, dandelion-like, flowers which is a weed pest. Also called dindle.

Sowbug–*Porcellio laevis*, family Oniscidae; an insectlike creature, commonly found under flower pots or decaying boards, which is a scavenger on rotting plants and manure. In the greenhouse, it may cause some damage by feeding about the roots and on tender portions of plants near the ground.

Soya Bean–See Soybean.

Soybean–*Glycine max*, family Leguminosae; an annual, erect bushy herb, extensively grown for forage, as a cover crop, for green manuring, and particularly for its edible, nutritious, oil-bearing seeds which vary in color from green through yellow and brown to black. Most of the American soybean acreage is grown for the seed, chiefly for the production of soybean oil meal. Native to Asia. Also called soya bean, soja bean. See Soybean Oil, Soybean Oil Meal.

Soybean Anthracnose–There are two distinct fungi concerned with the two types of anthracnose occurring on soybeans: (a) ***Glomerella glycines***, family Gnomoniaceae; of which the perithecia of the perfect stage and the acervuli of the conidial stage (***Colletotrichum glycines***) occur on seedlings and pods; (b) ***Colletotrichum truncatum***, family Melanconiaceae, which is more abundant in the stems. Apparently these fungi attack both soybean (***Glycine max***) and lima bean (***Phaseolus limensis***).

Soybean Oil–An oil obtained from the seeds of the soybean, used extensively in the preparation of many food and industrial products, including vegetable shortening, margarines, salad dressings, paints, soaps, plastics, and linoleum.

Soybean Oil Meal–A livestock feedstuff consisting of ground soybean oil cake or oil chips. If a name descriptive of the process of manufacture, such as expeller-, hydraulic-, or solvent-extracted be used, the product must correspond thereto. Soybean oil meal has an average protein content of about 45 percent and ranks high as a protein supplement for livestock feeding.

Sp. (Spp.)–The abbreviation, singular (plural) for species; e.g., when one species of a genus of plant or animal is referred to, the name may be written ***Canis*** sp. meaning one species of dog; or ***Canis*** spp. meaning more than one species of the genus ***Canis***.

Spanish Clover (Deervetch)–*Lotus americanus*, family Leguminosae; an annual herb, with red or rose-colored flowers, found on dry soils of central United States which is of value as a wildlife plant. Native to the central United States. Also called birdsfoot trefoil, compass plant, prairie lotus. See Birdsfoot Trefoil.

Spaulding Rule–A diagram rule for estimating the lumber content of logs which disregards taper and allows for an 11/32 inch saw kerf, with size of slab varying with size of log. It is the statute rule for California and is widely used elsewhere on the Pacific Coast (United States). Also called Columbia River rule.

Spear–A sprout or shoot of a plant, as a blade of grass, a young stalk of asparagus, etc.

Spearmint–*Mentha spicata*, family Labiatae; an aromatic, perennial herb widely grown for its foliage; used in flavoring and for its aromatic oil, the mint of commerce. Its flowers are a source of nectar for bees. Also called garden mint.

Specialty Crops–Crops with a limited number of producers and limited demand; or crops with high per acre production costs and value.

Specialty Fertilizers–Fertilizers recommended or used principally for growing roses, lawns, home gardens, organic gardens, house plants, or for any purpose other than growing farm crops.

Species–In the naming of plants and animals, Latin is used. Each kind of plant or animal can be identified by genus (plural, genera) and species (both singular and plural); e.g., the generic name (genus) of corn is *Zea* and the species name is *mays*.

Specific–(1) A medicine that cures a particular disease. (2) Pertaining to a species. (3) Produced by a particular microorganism. (4) Restricted by nature to a special animal, thing, etc. (5) Exerting a peculiar influence over any part of the body.

Specimen Plant–A plant with some outstanding quality such as flowers, leaves, fruit, or branching habit and located as a focal point in the landscape.

Speciosus–Showy, good-looking.

Speckle–A small patch or dot of color.

Speckled Alder–*Alnus incana*, family Betulaceae; a deciduous shrub or small tree of several varieties, grown as ornamentals. Native to North America, Europe, and Asia. Also called hoary alder, European alder. See Alder.

Speckled Leaf Blotch–A fungal disease of wheat, rye, and bluegrass caused by *Septoria tritici*, family Sphaerioidaceae, which is characterized by the death of the leaves during the late fall, winter, and early spring. Often confused with winter injury.

Specklepod Loco–*Astragalus lentiginosus*, family Leguminosae; a plant whose foliage, green or dried, is poisonous for livestock. Symptoms of poisoning are dullness, irregularity of gait, dragging of the feet, a solitary habit, loss of flesh, and a shaggy coat. As the animal ceases to eat, it dies. Found from Washington to Nevada, United States. Also called spotted loco. See Loco.

Spelt–*Triticum spelta*, family Gramineae; a cereal which retains its glumes when threshed. Used for livestock feed but is of no value for flour making. Bearded and beardless varieties are grown both in spring and winter, but are of little commercial importance. Native to southern Europe and western Asia.

Spent Bone Black–Bone black or bone charcoal left after use in the refining of sugar. It is valued by fertilizer manufacturers chiefly because of its phosphorus content.

Spent Grains–The nonfermentable solids remaining after fermentation of a grain mash; used as a livestock feed.

Sperm Cell–Male germ cell.

Spermatid–A haploid cell produced from the second division of meiosis in spermatogenesis that has not yet undergone the changes to form a sperm cell.

Spermatium–A spore from a spermogonium (pycnium) in rusts. It acts as a male gamete (pycniospore).

Spermatogenesis–The development of male gametes or sperm cells.

Spermatophore–In spermogonium, the section in which the spores originate.

Spermatophyte–A seed-producing plant, such as a flowering plant or conifer.

Spermiogenesis–That part of the process of spermatogenesis involving the changes that permit spermatids to become spermatozoa.

Sphagnum–Any plant of the genus *Sphagnum*, family Sphagnaceae. Species are mosses found in bogs from the polar region to the equator. The structure of the leaves is such that they have the power to absorb and retain large amounts of water. This gives value to sphagnum in nursery and greenhouse uses. Also called moss peat, sphagnum moss. See Peat.

Sphinx Moth–Any of several moths, family Sphingidae, whose larvae are called hornworms. Several species are important pests of cultivated crops, deciduous trees, and shrubs. See Catalpa Sphinx, Tomato Hornworm.

Spicule–(1) Any of the points on the spore-bearing cells of certain fungi. (2) A small, floral spike.

Spider Mite–A small plant-eating insect. There are many species of spider mites.

Spike–(1) A very large nail or anything shaped like a nail, as the pointed iron teeth of a spike-toothed harrow, etc. (2) A flower cluster or type of inflorescence in which the individual flowers are sessile (stalkless or nearly so) on a central axis, as in the common plantain. (3) The spikelike panicle of wheat, barley, rye, certain grasses, and other plants. (4) A thick, upright stem which carries one or more flowers, as the stems of gladiolus, foxglove, and delphinium.

Spike Crazyweed–*Oxytropis macouni*, family Leguminosae; a poisonous plant whose foliage, green or dried, is poisonous to livestock. Symptoms of poisoning are dullness, irregularity of gait, lack of appetite, dragging of the feet, a solitary habit, loss of flesh, and a shaggy coat. As the animal ceases to eat it dies. Native to North America; found from Montana to Colorado, United States. See Lambert Crazyweed.

Spike Muhly–*Muhlenbergia wrighti*, family Gramineae; one of the important range grasses of the plains and slopes of Oklahoma, Colorado, Utah, New Mexico, Arizona, United States; and northern Mexico. Native to North America.

Spike Trisetum–*Trisetum spicatum*, family Gramineae; a perennial bunchgrass which occurs mostly in scattered stands on alpine slopes and meadows in arctic America and southward to New England, and the western states in the United States, which supplies considerable forage of good quality.

Spikelet–Any small spike, the unit of inflorescence in grasses, which is often provided with a small pedicel. Each spikelet may consist of one, but usually several florets.

Spikesedge–*Eleocharis palustris*, family Cyperaceae; a watergrass growing in marshes and wet places from Labrador to Alaska southward to California and Pennsylvania. It is a weed pest in the rice fields of California, United States; native to North America.

Spinach–*Spinacia olercea*, family Chenopodiaceae; an annual herb whose basal, green leaves are a very important vegetable. Native to Asia.

Spindle–(1) The fine threads of achromatic protoplasm arranged in a fusiform mass within the cell during mitosis. (2) In flower-bearing plants, to produce the stalks on which flowers grow.

Spine–(1) A stiff, sharp-pointed outgrowth on a plant or animal. (2) The vertebral column; the backbone.

Spinose–Covered with, or containing, spines.

Spiny Sow Thistle–*Sonchrus asper*; a serious weed pest of hayfields and pastures. It has yellow flowers in small heads and spines on most of the plant.

Spirability–The ability of cotton fibers to be twisted, as in the manufacture of a yarn or thread. Under the microscope the spiralled cotton fibers give the appearance of long and irregularly twisted ribbons. This properly permits the fibers, when twisted together, to interlock better one with another.

Spiracles–The openings to an insect's internal breathing tubes, the trachea.

Spirit Vinegar–The product made by the acetous fermentation of dilute distilled alcohol which contains not less than 4 grams of acetic acid per 100 cubic centimeters at 20°C. Also called distilled vinegar, grain vinegar.

Spirochete–Any microorganism that is spiral or wavy in form, highly flexible, and capable of contracting, as the organisms that causes leptospirosis and syphilis.

Spittlebug–Any of the small, sucking insects of the family Cercopidae, which are not true bugs but are closely related to leafhoppers. The young nymphs produce masses of whitish, frothy material in which they live. Many spittlebugs are harmless, but several are injurious to plants. Also called froghoppers, because of their leaping habit.

Splash Erosion (Raindrop Erosion)–The spattering of small soil particles caused by the impact of raindrops on bare wet soils. The loosened and spattered particles may or may not be subsequently removed by surface runoff.

Splice Grafting–A simple method of grafting in which both the scion and the stock are cut with a diagonal cut; the two cut surfaces are fitted together, bound, and waxed. It is used for tender wood, which does not split readily, and for small shoots.

Split–(1) A deformed flower. (2) To separate thick hides into layers in the preparation of leather.

Split Application–Breaking up the application of fertilizer into two or more applications throughout the growing season. Split applications are intended to supply nutrients more evenly and at times when the crop can most effectively use them.

Split Crotch–Of a tree, a split or separation at the juncture of two large limbs.

Split Ditch–A lateral ditch used in the sugarcane system of surface land drainage in Louisiana, United States. Such ditches are usually parallel and spaced 100 to 250 feet apart with rows of sugarcane parallel to them.

Split Peas–(1) Peas hulled and dried so that the cotyledons usually separate. (2) A flour made from split peas.

Spodic Horizon–A subsurface soil horizon high in iron, aluminum, and humus but low in clay. See Spodosols.

Spodosols–Soils with illuvial accumulations of amorphous materials in subsurface horizons. The amorphous material is organic matter and compounds of aluminum and usually iron. These soils are formed in acid parent materials and are mainly coarse-textured sands in humid and mostly cool or temperate climates. Also spodosols develop in humid, sandy Florida.

Spoil–(1) To deteriorate by molding or rotting. See Spoilage. (2) Debris or waste material from a coal mine. (3) Dirt or rock that has been removed from its original location, specifically materials that have been dredged from the bottoms of waterways.

Spoil Bank–A small to large ridge of soil excavated from, and piled alongside, a drainage ditch. In good drainage practice, this spoil is leveled, i.e., scattered over a considerable area alongside the ditch.

Spoilage–(1) Hay or forage that has been improperly cured or stored. (2) Any objectionable change which has occurred in a food, feed, or material. (3) Putrefactive changes which occur in canned goods as a result of underprocessing, causing the growth of vegetative cells, spores, or organisms, or as a result of the growth of organisms entering the can after processing.

Spoilage Organisms–Bacteria, yeasts, and molds that cause food to spoil. They live everywhere: in the air, soil, and water and on food, plants, and animals.

Sporadic Disease–A disease which occurs in scattered or isolated instances.

Sporangiophore–A threadlike structure (hypha) which bears a sporangium. See Sporangium.

Sporangium–(Plural, sporangia) A fungus fruiting body that produces nonsexual spores (sporangiospores) within its external walls.

Spore–The one to many-celled unit of a fern, fungus, bacterium, or protozoan that has entered the resting state and is capable of growth and reproduction when conditions become favorable. Spores in lower forms correspond to seeds in higher plants.

Spore Dust–A standardized powder containing spores and used in the biologic control of insect pests. The dust is placed at intervals in the soil, and the spores infect the larvae of certain insects, such as Japanese beetle, causing diseases which destroy them. See Milky Disease.

Sporidium–(Plural, sporidia) A spore which is produced on a fruiting body called a promycelium; the basidiospore of rust and smut fungi.

Sporocyte–A spore mother cell of plants.

Sporogenous–Producing spores; spore bearing.

Sporophyll–A leaflike organ which bears spores, as in *Selaginella*; in angiosperms, the homologous organ is a carpel or stamen.

Sporophyte–The spore-forming generation in the life cycle of plants, normally diploid. The visible seed-bearing plant is the sporophyte. See Gametophyte.

Sporozoite–A small, usually elongated, infective stage of sporozoan parasites, such as *coccidia, plasmodia* (malaria), etc.

Sport–A random mutation.

Sporulaton–(1) In bacteria, the formation of spores within the body of the bacterium. The spores represent the inactive resting, or resistant, forms. (2) In coccidia, a kind of reproduction by which the fertilized cell within the oocyst wall splits up into new individuals, called sporozoites. Sporulation of coccidial oocysts usually occurs after the oocyst has been discharged from the body of the host. (3) In plants and animals, the process of spore formation.

Spot Basis–In marketing in the grain trade, the difference between the present cash price and the futures price of a particular option in a specified market.

Spot Burning–A method of slash disposal in which only the heaviest accumulations of slash are fired, and the fire is not allowed to spread over the entire cut-over area.

Spot Checking–The inspection of an occasional bale of cotton or other commodity to see if the quality of the product is up to the intended grade or specification.

Spot Fire–A fire set in advance of, or away from, the main fire by flying sparks or embers.

Spot Treatment–The application of a pesticide or other material to a restricted or small area of heavier infestation.

Spotted Medic–*Medicago arabica*, family Leguminosae; an annual herb grown extensively for permanent pastures in southern areas (United States) where winters are mild, as a cover crop, and for green manure. Also called southern bur clover, spotted bur clover. See California Bur Clover.

Spotted Waterhemlock–*Cicuta maculata*, family Umbelliferae; a poisonous, fleshy-rooted, perennial herb, which occurs chiefly in wet meadows and pastures, and along ditches and streams, from eastern North America westward to the Great Plains. The plant, especially the roots, contains a violent poison which causes serious or fatal results if eaten by humans or livestock. Native to North America. Also called beaver poison, cowbane, musquash root, water hemlock.

Spray–(1) A small, growing or detached shoot, twig, or branch of a plant with leaves, flowers, berries, etc.; a horizontal arrangement of flowers or greens. (2) A solution or suspension of material, as insecticides, fungicides, etc., which is broken up into fine, liquid particles driven through the air by mechanical means, coating surfaces of plants or other objects that it contacts. (3) A jet of fine liquid particles forced under pressure from a spray nozzle of a spraying device. (4) A device for shooting out a jet or spray. See Sprayer. (5) To scatter a liquid in fine particles. (6) To sprinkle or treat with a spray, as to direct a spray of fine particles of a liquid upon plants or other objects.

Spray Cultivation–The destruction of weeds in growing crops by the use of chemical weed killers (herbicides). See Herbicide.

Spray Drift–The movement of air-borne spray particles from the target plants or target area of application.

Spray Injury–Injury resulting from the application of a chemical compound, as a spray to the foliage or other parts of a plant. The injurious effects may include damage to leaves, blossoms, fruit, twigs, or to the entire plant, sometimes resulting in general necrosis and death.

Spray Irrigation–Irrigation by means of spray from pipes or pipe projections above the soil surface. Also called sprinkler irrigation. See Sprinkler Irrigation.

Spread–(1) A straddle; the difference in price between two delivery months in the same or different markets, or the sale of the one thing against a simultaneous purchase of the other. Straddling between a foreign and the domestic market is often referred to as arbitrage. (2) An extensive tract of land. (3) A ranch, including the buildings and the extent of land grazed by cattle or sheep (western United States). (4) The distribution of a disease. (5) Butter, peanut butter, margarine, or other food mixtures used as a spread on bread to improve its flavor, palatability, and/or nutritive value. (6) To scatter, as to broadcast seed, fertilizer, etc. (7) To disseminate disease.

Spreader-Sticker–A surfactant closely related to wetting agents that facilitates spreading and increases sticking of an herbicide on vegetation. See Surfactant.

Spreading of Water–Spreading excessive run-off water over a large land area to replenish the underground water supply or to provide enough surface water in dry areas to allow grasses to grow at least in a limited area.

Sprig–(1) A shoot, twig, or spray of a plant. (2) A portion of stem and root of grass, used for transplanting. (3) To plant with sprigs.

Sprigging–The planting of a portion of the stem and/or root of grass.

Spring Barley–A variety of barley that is planted in the spring; a very common type grown in the United States.

Spring Crop–Any crop that is planted or starts rapid growth in the early spring and is ready for harvest in the spring season, as turnips, lettuce, and radishes.

Spring Rye–A type of fast-growing rye which may be planted in the early spring and still make a mature crop the same season.

Spring Wheat–Wheat sown early in the spring and harvested in the summer of the same year.

Spring Wood–The less dense, large-celled part of a growth layer formed in woody plants by rapid growth during the spring. See Summer Wood.

Spring-Fall Range–Those grazing areas of the western range which, because of grazing patterns, climatic factors, forage production, quantity of water, etc., are seasonally grazed only in spring and fall. Such grazing may not be necessarily obligatory but it is convenient in a grazing program where summer range at higher elevations is productive and available only during the summer season.

Spring-wheat Region–The region in which spring wheat is commonly grown: North Dakota, South Dakota, Montana, and areas in bordering states to the south and east, United States.

Springer–(1) In the live-poultry industry, a young chicken, larger than a broiler and smaller than a roaster, which commonly weighs from 3¼ to 4¼ pounds; a fryer. (2) In marketing, a pregnant cow or heifer due to calve shortly. (3) A sealed can both ends of which are bulging, either one of which can be easily flattened. (4) A can of spoiled fruits or vegetables.

Springtail–Any one of a large number of minute, wingless insects of the order Collembola which has chewing or piercing mouthparts and which jumps long distances by means of a forked appendage, the furcula, folded forward under the abdomen when at rest. Found in damp places, some species infest seeds, seedlings, mushrooms, and other plants or plant parts close to the ground.

Sprinkle–(1) A light rain. (2) To scatter water, sand, seeds, etc., in drops or particles. (3) To irrigate by a hose and nozzle.

Sprinkler Irrigation–Irrigation by means of above-ground applicators which project water outward through the air making it reach the soil in droplet form approaching rainfall. Applicators commonly used are rotary or fixed sprinklers, oscillating or perforated pipe.

Sprout–A shoot from a dormant bud at the base of a tree or from an exposed root or stump. Also known as root sucker, stump sprout. A stump with one or more sprouts is called a stool.

Spruce–Any tree of the genus *Picea*, family Pinaceae. Species are handsome evergreen trees with dense conical crowns; widely used for timber and pulpwood and valuable for horticultural planting; native to the cooler regions of the Northern Hemisphere. Principal species of spruce are: black spruce, Colorado spruce, Englemann spruce, Norway spruce, red spruce, silver spruce, Sitka spruce, and white spruce.

Spruce Aphid–*Aphis abictina*, family Aphidae, a dull-green insect that is very destructive to Sitka spruce in Oregon and Washington, United States.

Spruce Budworm–*Choristoneura fumiferana*, family Tortricidae; a moth whose larvae infest confiners feeding on the terminal shoots which they have webbed together to form shelters. The tops of infested trees first appear as though scorched by fire, and if heavily infested the trees may die. The budworm is one of the most destructive insects in the forests of northern United States and southern Canada.

Spruce Sawfly–*Neodiprion abietis*, family Diprionidae; a sawfly whose larvae defoliate spruce and pine trees, causing great destruction. In the 1930s, this fly reached epidemic proportions in Canadian forests, before being controlled by spruce sawfly parasites from Europe. Found first in Canada.

Spruce Spider Mite–*Oligonychus sununguis*, family Tetranychidae; a black mite which sucks sap from foliage and spins webbing between needles of evergreens. These mites may kill young spruces the first season and cause older trees to die progressively.

Sprucetop Gramagrass–*Bouteloua chondrosioides*, family Gramineae; an erect, perennial bunchgrass, one to three feet tall, with numerous, slender, flat leaves, and rather naked culms. Largely restricted to mesas and rocky hills of western Texas to southern Arizona, United States, Mexico and Guatemala; it has a fair forage value.

Spud–(1) A narrow, sharp spadelike tool usually fitted with a long handle used for undercutting and lifting out deep-rooted weeds, for removing bark from trees, for tree planting, etc. (2) An Irish potato. (3) In well drilling, to raise and lower the drilling tool.

Spur–(1) A short, stubby shoot, as in some fruit trees where the spurs bear flowers for more than one year or as grapes are pruned to spurs of one to two buds each, etc. (2) A hollow, tubular projection from some part of a flower, very conspicuous in the columbine, usually secreting nectar to induce the visitation of insects. (3) A hornlike protuberance which grows from the inner side of the shank of a fowl. (4) A steel contrivance secured to a rider's heel and used to urge the horse by its pressure. See Spur Rowel.

Spur Lupine–*Lupinus laxiflorus*, family Leguminosae; a perennial, poisonous herb which occurs from Montana and British Columbia southward to Arizona and California. Conditions of poisoning with sheep and other livestock are reported, with conspicuous symptoms of labored breathing, frothing at the mouth, and coma. The poisonous substance is most concentrated in the seed pod. See Lupine.

Spur Terrace–In erosion control, a short terrace which is used to collect and hold or divert runoff.

Squash–The fruit of certain various tender, vinelike, annual plants of the genus *Cucurbita*, family Cucurbitaceae, which is used as a vegetable; *C. maxima; C. maschata*, and certain varieties of *C. pepo*. Also the vine or plant which bears such fruit; e.g., crookneck group, Hubbard group, pattypan group, pumpkin, summer squash, vegetable marrow, winter squash.

Squash Bug–*Anasa tristis*, family Coreidae; an insect whose nymphs feed on cucurbits by sucking the sap. They inject a toxic substance into the vines causing wilting. (It is commonly called stinkbug because a disagreeable odor is given off when the insect is disturbed or crushed.)

Squaw Corn–A strain of corn whose kernels, shaped like flint corn kernels, are composed almost entirely of soft starch. It is grown in all colors, but white and blue are most common. Also called flour corn, soft corn.

Squirrel Corn–*Dicentra canadensis*, family Fumariaceae; a rather weak herb, similar to Dutchman's-breeches. Cattle are reported to have been poisoned by grazing on this plant, the symptoms being sudden trembling, frothing at the mouth, and convulsions. Native to eastern North America.

Squirrel-tail Grass–*Hordeum jubatum*; a grass with long bristles that are a serious weed pest in pastures and hayfields. The bristles stick in the mouths and throats of livestock.

St. Augustine Grass–*Stenotaphrum secundatum*, family Gramineae; a tropical or subtropical, creeping, stoloniferous, perennial, lawn grass which grows in moist, especially mucky, soil mostly near the seashore from South Carolina to Florida and Texas. Native to North America.

St. Lucie Grass–One of the cultural varieties of Bermudagrass. See Bermudagrass.

Stable Manure–The excreta of livestock intermixed with straw or other bedding material which commonly includes stalks and uneaten, coarse stems of forage, corn cobs, etc.; an important organic fertilizer. See Manure.

Stag-headed–A tree dead at the top as a result of injury, disease, deficient moisture, or deficient nutrients.

Staggerbush Lyonia–*Lyonia mariana*, family Ericaceae; a slender, upright shrub with oval or elliptical leaves and white flowers, sometimes grown as an ornamental. Although seldom eaten by livestock, poisoning has been reported with sheep browsing on its herbage with observable symptoms of frothing at the mouth, convulsions and, later, paralysis of the limbs. Native to the United States. Also called wicopy.

Stalacite Frost Structure–Frozen soil which contains tiny icicles or columns of ice separating soil sheets; heaved soil. It is associated with partial thawing and refreezing. See Heaving.

Stalk–A nontechnical term for the more-or-less elongate support of any organ, as a petiole, peduncle, pedicel, filament, or stipe.

Stalk Borer–*Papaipema nebris*, family Noctuidae; an insect that may feed on any plant large enough and soft enough for its operations. It apparently prefers giant ragweed and corn in later stages, but is often a pest in flower and vegetable gardens, where it tunnels into the stalks of dahlia, hollyhocks, aster, rhubarb, pepper, potato, tomato, tobacco, and other plants. It is distributed generally throughout the United States east of the Rocky Mountains.

Stamen–The organ of a flower which bears the pollen (microspores) consisting of the stalk (filament) and the anther. See Flower.

Staminate–Designating a flower that has stamens but no pistil and hence is imperfect. See Pistillate.

Staminodium–A sterile stamen, that is, one that produces no pollen, often a filament only.

Stand–(1) The proper number of uniformly distributed plants per acre. (2) The relative number of plants per area, as a poor, good, or medium stand. In forestry, fully stocked, understocked, pure, mixed, or residual stand. (3) A hive of bees. (4) Density of game per acre, area, etc. (5) A stallion's court. (6) To cease walking or moving; to take or keep a certain position. (7) To rise to the feet. (8) Of a stallion, to be available for breeding purposes.

Stand Improvement–Measures such as thinning, pruning, release cutting, girdling, weeding, or poisoning of unwanted trees aimed at improving growing conditions for the remaining trees.

Standard Length–In softwood lumber, lengths of boards in multiples of even feet; in hardwood lumber, lengths from 3 to 20 feet, including both odd and even number of feet.

Standing Crop–Any crop as it stands in the field before harvest, as corn, wheat, pasture grass, etc.

Standing Water–Any water that collects or stands for a period of time on the surface of soil following a heavy rainfall or fast-melting snow.

Stansbury Cliffrose–*Cowania stansburiana*, family Rosaceae; a leafy, stiff-stemmed shrub or small tree which occurs in New Mexico and southern Colorado to Utah and California in woodland and ponderosa pine forest at elevations of 4,000 to 8,000 feet. Where abundant, it provides good-quality forage for cattle and sheep. Native to North America.

Staple Length–The length of cotton or wool fiber as obtained by measuring the natural staple without stretching.

Stapling–The classing or grading of cotton or wool on the basis of length of fiber or staple.

Starch–A kind of carbohydrate manufactured by plants and stored in the seeds, roots, and fruit as a reserve energy supply; the major component of livestock feeds.

Start–To grow seedling plants, such as cabbage and tomatoes, in specially prepared beds from which they are later transplanted in the field when weather and soil conditions permit.

Starter Fertilizer–Liquid or solid fertilizer, placed near the seeds or the roots of transplants, is commonly considered as a starter fertilizer. For maximum benefit and safety of use, starter fertilizers should have a high concentration of all three primary plant nutrients in readily soluble form. It should have a low salt index, a high proportion of nitrogen in the nitrate form, and a low proportion of sodium and chlorine to reduce the hazard of exosmosis. Also called pop-up fertilizer. See Salt Index.

Starve Out–To hamper the growth of a plant so that it cannot grow and increase in a normal manner. Repeated cuttings, planting other, more vigorous, competitive plants, excessive cultural practices, and the use of black plastic are the most common methods used.

Steapsin–Pancreatic lipase, a lipolytic enzyme secreted by the pancreas which has the power of hydrolizing fats to fatty acids and glycerol.

Stearic Acid–One of the fatty acids occurring in combined form in animal and vegetable oils; one of the major fatty acids in butter and most fats. Commercial stearic acid is used in large quantities, in rubber compounding and in the preparation of soaps, greases, and chemicals.

Stearine–The solid material of any fat which is obtained, in the process of refining, by filtration from an animal or vegetable oil after chilling or freezing.

Steckling–(1) A root such as beet, carrot, turnip, etc., which is stored over winter and replanted the following spring for seed production purposes. Also called stechling. (2) A cutting or a transplanted seedling.

Steershead Bleedingheart–*Dicentra uniflora*, family Fumariaceae; a perennial, poisonous herb occurring from Washington to Utah and California, United States. Native to North America.

Stele–(1) In vascular plants, the central cylinder of the stems and roots. (2) In wood anatomy, that part of the stem which includes the vascular system (xylem and phloem).

Stellate–(1) Star-shaped. (2) A leaf with surface hairs resembling stars.

Stem–Stalk, trunk, branch of a plant. Can be vertical or horizontal.

Stem Cutting–(1) Any part of a stem used for plant propagation, usually of three general types of wood: (a) softwood or green cutting taken from a plant while in leaf, the stem being succulent; (b) hardwood cutting, taken from the leafless, dormant plant; (c) semi-hardwood cutting, intermediate between softwood and hardwood cuttings. (2) In sugarcane propagation, a part of a selected stalk or stem cut into three- or four-eye lengths to be planted, usually in August or September, in field furrows.

Stem Rust of Grasses–Any of numerous species of rusts that infect the stems of grasses. *Puccinia graminis*, family Pucciniaceae, is a common species that attacks many grasses and is better known as a destructive disease of small grain.

Stem Tubers–Short, fleshy underground stems, complete with buds; e.g., Irish potato.

Stem-Foliage Application–An application of an herbicide to both stems and leaves of a plant.

Steno-–A prefix indicating narrowness.

Sterile–(1) In animals, incapable of reproduction; unable to produce normal living young. (2) In soils, unproductive; barren; producing little or nothing, as a sterile soil. (3) In biological products, etc., free from contamination with living bacterial, fungal, or viral organisms; or designating an organism not capable of growing or multiplying.

Sterilization–(1) The destruction of all living organisms. In contrast, disinfection is the destruction of most of the living organism. (2) To make animals infertile.

Sterol–Any of a group of solid cyclic alcohols, such as cholesterol and phytosterol, with wide distribution among animals and plants. The sterols are neutral and comparatively stable substances which occur partly in the free condition and partly esterified with higher fatty acids. They are of great biological importance since irradiation of some sterols leads to the formation of a form of vitamin D.

Sticktight–Any plant of the genus *Bidens*, family Compositae, that bears seeds that stick to animals and to human clothing. Also known as beggarsticks, beggarlice.

Stigma–The receptive surface of the female organ of a flower that receives the pollen.

Stillage–The nonfermentable residue from the fermentation of a mash to produce alcohol.

Stimulus–An activating agent such as heat, moisture, or light.

Sting–(1) The resulting wound or welt caused by the sharp-pointed organ of certain insects, scorpions, etc. (2) The smart or irritation produced by touching certain plants; e.g., the nettle. (3) The puncture or wound in the skin or rind of fruit, etc., made by the ovipositor or mouthparts of an insect.

Stinkgrass–*Eragrostis cilianensis*, family Gramineae; a weedy, annual grass with disagreeable odor when fresh, occurring in cultivated ground, fields and waste places from Maine to Washington, south throughout the United States and Mexico to Argentina. Horses have been reported poisoned by eating stinkgrass in large quantities over a considerable period of time.

Stipe–(1) The stalklike support of a pistil; the leafstalk of a fern frond. (2) In mycology, the stem which supports a cap or pileus, as the stem of a mushroom or toadstool.

Stipule–A leafy appendage attached to the twig at the base of a petiole; usually in pairs, one on each side, often shedding early.

Stock–(1) Plant or plant part upon which a scion is inserted in propagation. (2) Livestock; domesticated farm animals. (3) Material held for future use or distribution. (4) Material destined to be wrought into finished products, as crate stock, barrel stock, cider stock, etc. (5) A plant or plant part that furnishes cuttings for propagation. Also called stock plant. (6) The main stem or trunk of a plant. (7) A rootstock (rhizome). (8) The original type from which a group of plants or animals has been derived. (9) The base or handle of a whip. (10) The stump of a tree (after the tree is felled). (11) (plural) A small enclosure in which an animal is secured in a standing position during shoeing or an operation. (12) To provide or supply with livestock, as to stock a pasture or range with cattle, sheep, etc. (13) To assemble a supply of materials or commodities. (14) Of, or pertaining to, livestock, as stock barn, stock feed, etc.

Stock Solution–A concentrated solution from which a portion is taken and diluted as needed, as a spray, etc.

Stocking–(1) The relative number of livestock per unit area for a specific time. In range management, the relative intensity of animal population, ordinarily expressed as the number of acres of range allowed for each animal for a specific period. (2) In wildlife management, the density of animal population in relation to carrying capacity. (3) In forestry, the density of a stand of trees, such as well-stocked, overstocked, partially stocked. (4) A white leg on an animal.

Stolon–A horizontal stem on the surface of the ground where it propagates vegetatively by forming new roots and shoots at the nodes. See Rhizome, Runner.

Stoma–(Plural, stomata) An opening in the epidermal layer of plant tissues which leads to intercellular spaces. These small openings may open or close, depending on climatic conditions, by means of guard or bullform cells, etc., and are necessary to photosynthesis, transpiration, etc. Also called breathing pore.

Stomach-dirt–An accumulation of dirt in the stomach; a commonly occurring condition in horses and cattle grazing on loosely implanted vegetation.

Stomatal Transpiration–See Transpiration.

Stomate–An opening surrounded by guard cells that opens into an internal air cavity below the epidermis of a leaf. A breathing pore in the epidermis of a leaf.

Stone–(1) In land description, a detached rock fragment on the surface of, or embedded in, the soil. (2) In soil survey mapping, a rock which is greater than 10 inches in diameter or more than 15 inches in length. See Boulder. (3) The hard seed of some pulpy fruits, as peach, plum, etc. See Clingstone, Freestone. (4) A unit of weight, 14 pounds (avoirdupois, British). (5) See Testicles. (6) To take the stone or pit out of stone fruits.

Stone Cell–In plant anatomy, a metamorphosed, parenchyma cell which is short, approximately isodiametric, and characterized by thick, heavily lignified walls and simple, commonly ramiform pits.

Stony Land–Land which contains enough stone, either detached fragments in or on the soil or rock outcrop, to interfere with tillage.

Stool–(1) Fecal material; evacuation from the digestive tract. (2) To develop several stems from the crown of a plant or from a stump.

Storm Resistance–A quality possessed by certain varieties of cotton which have moderately firm locks that cling together well and bolls that hang down so that the bur tends to form a roof over the locks, thus providing drainage and protection from damage by rainwater.

Stover–(1) The stem and leafy parts of corn fodder after the ears have been removed. (2) Sorghum forage.

Straggler–(1) An animal that wanders or strays from a flock or herd. (2) A plant, branch, etc. that grows irregularly.

Strain–(1) A group of plants of common lineage which, although not taxonomically distinct from others of the species or variety, are distinguishable on the basis of productiveness, vigor, resistance to drought, cold, or diseases; or other ecological or physiological characteristics. (2) A group of individuals within a breed which differ in one or more characters from the other members of the breed; e.g., the Milking Shorthorns or Polled Herefords. (3) An organism or group of organisms which differs in origin or in minor respects from other organisms of the same species or variety. (4) A virus entity whose properties and behavior indicate relationship to a type virus and are sufficiently constant to enable the entity to be recognized whenever isolated. (5) A severe muscular effort on the part of a draft or other animal which may result in muscle, ligament, or other damage. (6) The condition of an overworked part operating above optimum load, such as a belt or motor operating under strain. (7) To filter a liquid and free it of impurities by passing it through some medium or fabric which can retain the solid matter and allow the liquid to pass.

Strain Cross–A cross between members of two different strains.

Strand–(1) Each one of the separate units of fibers, threads, wires, etc., which, when twisted or woven together, may make up a rope, cable, fence wire, etc. (2) Any single filament.

Strangleweed–See Dodder.

Strata–(Singular, stratum) (1) In geology, layers or beds of rock which are nearly lithologic units. (2) In ecology, different vertical layers of vegetation, as ground cover, herbaceous growth, shrubs, and trees.

Stratification–(1) The rest period which some seeds must have before they will germinate; generally the seeds must be exposed to a chilling temperature during this period before they will germinate. (2) A method of cross-breeding sheep of different types. (3) Arrangement in layers. The term refers to geologic material. Layers in soils that result from the processes of soil formation are called horizons; those inherited from the parent material are called strata. See Soil Profile.

Straw–(1) The stems and leaves of grain crop plants which remain after threshing. (2) The fallen leaves of pines and similar trees.

Straw Itch Mite–*Pyemotes ventricosus*, family Pyemotidae; a small mite frequently encountered at harvest time in handling wheat and other small grains. On people they cause a hive-like eruption, often over much of the body, accompanied by a very severe itching. Also called harvest mite.

Strawberry–Any of the essentially stemless, perennial herbs of the genus *Fragaria*, family Rosaceae. The strawberry has trifoliate leaves, white flowers, and long slender runners by means of which it naturally propagates itself. In its many horticultural varieties, it is widely grown and much prized for its delicious red fruit which in a strict sense is a much enlarged, juicy, very fleshy receptacle bearing achenes.

Stream–(1) Flowing water in a natural or artificial channel. It may range in volume from a small creek to a major river. (2) A jet of water, as from a nozzle. (3) A continuous flow or succession of anything: air, gas, liquids, light, electricity, persons, animals, and materials.

Stream Runoff–The total outflow of a drainage basin through surface channels; or subsurface channels and surface channels if there are exposed subsurface aquifers.

Streambank Wheatgrass–*Agropyron riparium*, family Gramineae; a forage grass found in dry or moist meadows and hills from North Dakota to Alberta, Canada, and Washington, south to Oregon and Colorado in the United States that is sometimes used in range reseeding. Native to North America.

Strength–(1) The capacity to resist force; solidity or toughness; the quality of bodies by which they endure the application of force without breaking or yielding. (2) Bodily or muscular power; force; vigor. (3) The potency or power of a liquid or other substance; intensity of active properties. (4) The firmness of a market price for a given commodity; a tendency to rise or remain firm in price.

Stress–(1) Abnormal or adverse conditions and factors to which an animal cannot adapt or adjust satisfactorily, resulting in physiological tension and possible disease; the factors may be physical, chemical, and /or psychological. (2) Plants unable to absorb enough water to replace that lost by transpiration. Results may be wilting, halting of growth, or death of the plant.

Striate–With fine grooves, ridges, or lines of color.

Striations–In wood anatomy, fine linear streaks or markings in the cell wall.

Stricture–An abnormal, localized contraction of any passage or duct of the body; a constriction.

Strictus–Erect.

Striga–*Striga asiatica*, family Scrophulariaceae; a weed that has invaded North and South Carolina, where it has caused severe injury to corn. It is an obligate parasite only on grasses, and affects corn, crabgrass, Japanese millet, Johnsongrass, sorghum, Sudangrass, and others. The plants emerge after a period of feeding underground on the roots of the host. They grow 2 to 15 inches tall, have red blossoms with a yellow eye, and produce enormous numbers of sporelike seeds. Native to Asia and South Africa. Also called witchweed.

Strike Root–Of a plant, to throw out lateral roots from the main root, crown, or stem, near the surface or just below the ground, which turn down and take root in the soil, bracing and giving strength as well as added nourishment to the plant. Corn is a typical example of a plant that strikes root.

String–(1) A cord of small diameter. (2) The several fumigating tents which are used by a single crew of people in fumigating orchards. (3) A group of partly broken horses which are assigned to a cowboy or horse trainer for his personal use or for further breaking or training. (4) A group of horses in a pack train. (5) To unroll wire prior to stretching and fastening to posts, etc.

String Out Tile–To lay a number of tiles end to end in the field, prior to placement in the ground; as in a drainage line.

Stringbeans–Any immature beans used in the pod for human food.

Stringless Bean–Any of a number of varieties of kidney bean whose unripe, tender pods are free or practically free from fibrous threads.

Strip–(1) A relatively narrow piece, as a strip of land. (2) Each of the squirts or streams of milk as taken from a cow's teat in milking. (3) To tear off, as the threads from a bolt or nut, the leaves, husks, suckers, from a plant or plant part, or the hide or pelt from a carcass. (4) To take the last of the milk from a cow's udder by hand milking after the machine milker has been removed. (5) A narrow marking extending vertically between the forehead and nostrils of a horse.

Strip Burning–Controlled use of fire to destroy surface material; e.g., an area to be broadcast burned may be successively burned in strips or the method may involve only the burning of inflammable material along roads, firelines, railroads, or highways. See Slash Disposal.

Strip Grazing–See Rotational Grazing.

Strip Rotation–A crop rotation for contour strips within a field, using erosion-resistant crops alternating with cultivated crops.

Strip Sodding–The laying of sod in strips separated by unsodded space.

Stripcropping–The practice of growing crops in a systematic arrangement of strips, or bands. Commonly cultivated crops and sod crops are alternated in strips to protect the soil and vegetation against running water or wind. The alternate strips are laid out approximately on the contour on erodible soils or at approximate right angles to the prevailing direction of the wind where soil blowing is a hazard.

Striped Blister Beetle–An insect which feeds on many vegetable crops and weeds. Its body contains cantharadin, which will blister tender human skin.

Striped Cucumber Beetle–*Acalymma vittata*, family Chrysomelidae; an insect that is a serious pest of cucumber and other cucurbits. The adults devour the leaves and stems and the larvae feed on the underground parts, frequently destroying the root system. Found in central and eastern United States.

Striped Flea Beetle–*Phyllotreta striolata*, family Chrysomelidae; a widely distributed insect pest of cabbage and many other garden plants.

Strobilus–A conelike reproductive structure consisting of an axis bearing sporophylls; i.e., spore-bearing leaves which are either microsporophylls (stamens) or macrosporophylls (carpels), the latter being open in the Gymnospermae, or closed so as to enclose the ovules in the Angiospermae. See Conifer.

Stroma–A mass of fungus hyphae, sometimes including host tissues, which contains or bears spores differing from a sclerotium this respect.

Strong Flour–Flour prepared from hard wheat whose high gluten content makes it suitable for bread making.

Strong Land–Land composed of clay or loam soils. Such soils generally are durable under cultivation and their fertility is less quickly depleted by cropping than that of sandy soils.

Structure, Soil–The arrangement of primary soil particles into compound particles or aggregates that are separated from adjoining aggregates. The principal forms of soil structure are: platy (laminated), prismatic (vertical axis of aggregates longer than horizontal), columnar (prisms with rounded tops), blocky (angular or subangular), and granular. Structureless soils are either single-grained (each grain by itself, as in dune sand) or massive (the particles adhering without any regular cleavage, as in many silty and clayey hardpans).

Strumella Canker of Oak–A destructive, slow-growing, perennial canker, caused by the fungus *Strumella coryneoidea*, family Tuberculariaceae, primarily on oaks, especially the red oak group. Occasionally it affects ornamental plants and other trees, such as hickory, chestnut, beech, and maple. The fungus enters the trunk mainly through dead branch stubs.

Strychnine–An extremely poisonous vegetable alkaloid which has an intensely bitter taste and is used as poison to kill mice, sparrows, pocket gophers, ground squirrels, etc. It is also used in medicines as a stimulant.

Stub–(1) A short blunt projection, as a blunt projecting stem, branch, or root. (2) A tree with a broken stem less than 20 feet. (3) The quill portion of a short feather appearing on the shanks or toes of otherwise clean-shanked birds.

Stubble–(1) The ungrazed bases of grasses and forbs left on the range after grazing. (2) Standing crop residues.

Stubble Crop–(1) A crop developed from the stubble of the previous season. (2) A crop sown on grain stubble after harvest for plowing under, as green manure the following spring.

Stubble Harvesting–The harvesting of a second crop of rice from the root system of plantings which have been previously harvested. See Ratoon.

Stubble Mulch–A protective cover provided by leaving plant residues of any previous crop as a mulch on the soil surface when preparing for and planting the succeeding crop. See Conservation tillage, No-till.

Stubble Mulch Grain Drill–A grain drill used to plant seed in a field that has stubble left from the last crop. The concept is to use minimum tillage methods of production. Also called minimum tillage drill. See Air Drill, Grain Drill, Press Drill.

Stubble Pasture–A field from which a crop of wheat or other grain has been previously harvested and on which animals are placed to consume the crop residues as well as weeds that may follow the grain crop.

Stubborn Disease–A viral disease of citrus characterized by deformed, acorn-shaped fruit, brushlike, stunted growth of twigs and foliage, and irregular blooming.

Stump–The basal part of a tree or plant left after the bole and top have been cut.

Stump Sucker–A shoot arising from a stump and developing into a tree. See Coppice.

Stump Wood–The resinous stump and roots of longleaf or slash pine used for extraction of turpentine and rosin.

Stumpage–The value of standing timber based on cubic volume offered for sale, usually at an agreed-upon log scale.

Stunt–(1) Diseases caused by certain viruses that dwarf a plant and make it unproductive. (2) To check or hinder the growth or development of an animal or plant.

Stupefacient–A drug used to cause birds or other animals to go into a state of stupor so they can be captured and removed.

Style–(1) In the pistil of a flower, the part between the ovary and the stigma; if the style is lacking, the stigma is sessile on the ovary. (2) The manner in which an animal displays itself while at rest or in action. (3) The manifestation of those characteristics which contribute to the general beauty, pleasant appearance, and attractiveness of an animal.

Sub-–A prefix meaning either: (a) nearly, somewhat, slightly; e.g., subcordate, nearly cordate; or (b) below, under; e.g., subaxillary, below the axil.

Subacute–A clinical condition intermediate between acute and chronic.

Subarctic–(1) The region immediately south of the Arctic Circle and those which have a similar climate. (2) That region in which arctic and nonarctic surface waters join. (3) The regions in which mean temperature is not higher than 10°C for more than four months of the year and the mean temperature of the coldest month not more than 0°C.

Suberin–A thin, natural varnishlike layer which seals the moisture of plant cuttings inside the tissues and keeps rot-producing organisms out.

Subglobose–Globe-shaped, but slightly flattened, as the earth.

Subhumid Climate–A climate intermediate between semi-arid and humid which has sufficient precipitation to support a moderate to dense growth of tall and short grasses but in most instances insufficient to support a dense deciduous forest. Some subhumid areas, where the rainfall comes mostly during the growing season, have scattered deciduous trees with grass vegetation in between. The annual rainfall in subhumid climates may be as low as 20 inches in cool regions and as high as 60 inches in hot areas. The precipitation-effectiveness indexes are 32 to 48 for the dry subhumid and 48 to 64 for the moist subhumid.

Subirrigation–Water supplied to the soil from ditches or through underground tile lines, perforated pipe lines, or by natural subsoil moisture, in sufficient amounts to maintain a water table sufficiently close to the soil surface to supply adequate water quantities for crop needs. A low-permeability layer in the soil profile is essential to the maintenance of the watertable. Plants and planted seeds in containers, such as pots, are frequently subirrigated by setting them in a water container.

Sublethal–Less than fatal in effect, as a sublethal dosage or application of a toxic substance, etc.

Sublethal Concentration–A concentration in which an organism can survive, but within which adverse physiological changes may be manifested.

Submarginal Land–Land incapable of sustaining a certain use or ownership status economically. See Marginal Land.

Submergent Plant–A vascular or nonvascular hydrophyte, either rooted or nonrooted, which lies entirely beneath the water surface, except for flowering parts in some species; e.g., wild celery (***Vallisneria americana***) or the stoneworts.

Suborder–A category next below an order and above a family in a classification of animals and plants.

Subsessile–Almost stalkless.

Subsoil–(1) That part of the profile which lies below the usual plow depth without any specific limitation in depth or kind of material; a B horizon, the first change with depth in texture or structure; e.g., a clay layer underlying a sandy surface. (2) To plow deeply into the subsoil. See Soil Profile.

Subsoiling–Tilling a soil below normal plow depth, ordinarily to shatter a hardpan or claypan.

Subspecies–A major subdivision of a species, ranking between species and variety. It has somewhat varying connotations, depending on the user of the term, and often implies a distinct geographic distribution for the taxon.

Substrate–(1) The material on which a plant lives, as soil or rock. (2) The material upon which an enzyme or fermenting agent acts; the material in or upon which a fungus grows or to which it is attached; the matrix.

Substratum–(1) The base or substance upon which an organism is growing or attached. (2) A vague term for the C horizon and/or R horizon of a soil. See Soil Profile.

Subsurface Flow–Water which enters the ground, but which emerges as seepage into streams, ditches, rivers, or springs. See Surface Runoff.

Subterranean Clover–*Trifolium subterraneum*, family Leguminosae; a decumbent, winter annual herb grown mostly for grazing. Also called annual clover.

Succession–(1) The progressive development of vegetation toward its highest ecological expression, the climax. The replacement of one plant community by another. (2) Transfer of operation and of ownership of a farm from one generation to the next, as from father to son.

Succession Planting–The growing of two or more crops per season in the same space; e.g., planting soybeans after wheat is harvested from the land.

Succulence–A condition of plants characterized by juiciness, freshness and tenderness, making them appetizing to humans and animals.

Succulents–A group of plants which have ways of overcoming a deficiency of water as (a) the cacti and other plants that have fleshy leaves or stems or both and (b) yucca, agave, and other plants that have leathery foliage, often reduced in area, so that they are able to survive long droughts.

Succus–Juices or fluids extracted from or secreted by an organism.

Sucker–(1) A secondary shoot which develops from the root, crown, or stem of a plant. (2) A cow that sucks herself. (3) Of a plant, to produce secondary shoots (suckers). (4) To remove suckers or shoots from a plant, as from tobacco, etc. (5) A species of small fish.

Suckering–(1) The act or practice of removing suckers from a plant. (2) Of a plant, currently producing suckers.

Sucking Insects–Any insect which has piercing and sucking mouthparts, as lice, scale insects, chinch bugs, squash beetles, etc.

Sucrose–$C_{12}H_{22}O_{11}$; cane or beet sugar; i.e., common table sugar; a carbohydrate used for sweetening or fermentation reactions.

Sudan Colanut–*Cola acuminata*, family Sterculiaceae; a tree which yields the colanut containing a large amount of caffeine, and is used in the soft drink industry. Native to tropical Africa. Also called kilanut.

Sudangrass–*Sorghum vulgare* var. *sudanense*, family Gramineae; an annual herb grown for summer pasture, hay, soiling, and silage. Sudangrass has been known to produce prussic acid poisoning when grown under stress but is much less likely to do so than other sorghums. Native to Africa.

Suffrutescent–Designating a plant which is partly woody and perennial at the base but with herbaceous, annual growth above, such as the carnation.

Suffruticose–Very low and shrubby and with persistent woody stems, as Epigaea.

Sugar–Any of a group of carbohydrates characterized by being soluble in water, crystalline, and having a relatively sweet taste.

Sugar Beet–*Beta saccharifera*, family Chenopodiaceae; a biennial herb grown as an annual for its large, fleshy roots containing large amounts of sugar, usually from 12 to 18 percent, a source of domestic sugar in the United States. Native to Europe and Asia.

Sugar Bush–A grove of sugar maple trees tapped for sap, which is made into maple syrup or sugar.

Sugar Factory Lime–Lime that has been used in sugar refining. On a dry basis it will usually test about 80 percent calcium carbonate ($CaCO_3$). It has use for liming soils and is usually sold by the cubic yard. Two yards, air dry, have a lime equivalent of about one ton of high-grade ground limestone.

Sugar Maple–*Acer saccharum*, family Aceraceae; a large deciduous tree, chiefly of northern United States and Canada from Quebec to Pennsylvania, westward to Minnesota and Kansas, whose sweet sap may be collected in early spring and used to make syrup and sugar. The tree is sometimes planted for shade and for roadside plantings. Its wood is used in furniture, flooring, etc. Native to North America.

Sugar Pack–Sugar is added directly to fruit and mixed gently to draw juice from fruit before packing into freezer containers.

Sugar Palm–Any palm tree that can yield sugar, especially palms of the genus *Arenga*, family Palmaceae.

Sugar Pine–*Pinus lambertiana*, family Pinaceae; a tall-growing evergreen tree whose easily-worked wood is used for millwork, door, sash, interior and exterior trim, foundry patterns, etc. Native to North America.

Sugar Syrup–(1) A sweet, more-or-less viscid liquid made by concentrating the juices of sugar-producing plants, as sugarcane, sugar maple, and sorgo. (2) A sweet solution made by boiling, to varying degrees of concentration, a mixture of beet or cane sugar and water. Used as as a sweetener in various food and beverage preparations, in the canning of fruit products, etc.

Sugar Weather–Weather (toward the beginning of spring) in which the temperature during the day rises above 32°F and during the night drops below the freezing point causing the sap to start flowing in sugar maple trees and indicating that the time for making maple sugar is at hand.

Sugar-tree Land–A type of agricultural land with a native growth of sugar maple (*Acer saccharum*). It was recognized by early settlers in United States, especially in Indiana and Ohio, as first class land for farming.

Sugarcane–*Saccharum officinarum*, family Gramineae; a tall-growing plant, with big stalks and long, slender, coarse leaves. Chiefly valued for the sweet juices expressed from its stem, the plant is cultivated in tropical regions for sugar and its by-products. Also cultivated in the southern states, United States. Usually propagated from stalk cuttings. Probably a cultigen. See Cultigen.

Sugarcane Borer–*Diatraea saccharalis*, family Crambidae; a moth whose larvae bore into the interior of the stalks of sugarcane, corn, broom corn, sorghums, and Sudangrass, killing the young canes and weakening the larger ones. It is a very close relative of the southern cornstalk borer.

Sugarpod Gardenpea–*Pisum sativum* var. *macrocarpon*, family Leguminosae; an annual herb grown as a vegetable for its edible seed pod. Native to Europe and Asia.

Sul-Po-Mag–See Sulfate of Potash-Magnesia.

Sulfate of Potash-Magnesia–A double sulfate of potash and magnesia; a salt containing not less than 25 percent potash (K_2O, nor less than 25 percent sulfate of magnesia, and not more than 2.5 percent chlorine. Used as a low-grade fertilizer. Also called double manure salt.

Sulfite Yeast–The yeast grown on the waste liquor from the sulfite wood-pulping process. Produced under sanitary conditions and suitable for animal feeding, it contains not less than 40 percent protein.

Sulfiting–The treatment of foods with sulfur dioxide or certain related compounds. The sulfur combines with enzymes in the food and prevents them from causing the quality to deteriorate.

Sulfofication–The oxidation of sulfur, free from organic or inorganic compounds, largely through the action of microorganisms. In the soil, the oxides are quickly changed into sulfites and sulfates.

Sulfur–S; an elementary, yellow mineral, insoluble in water, easily fusible and inflammable. Also called brimstone. One of the secondary but important elements in soil fertility and used in relatively large amounts by most plants, it is an important constituent of both protein and protoplasm. The powder form is an effective insecticide for many insects. The dust is used as a fungicide in the control of mildew, etc. When burned it forms sulfur dioxide, a gas which is highly toxic to insects and has long been used as a fumigant as well as a bleaching agent.

Sulfur Bacteria–*Thiobacillur thiooxidons*; bacteria that obtain their metabolic energy by the oxidation of elemental sulfur.

Sulfur Dioxide–SO_2; a compound produced by burning sulfur; it has a suffocating odor and is used as a fumigant for the control of certain insects, for the prevention of molds on dried fruits, and for bleaching wool, straw goods, etc.

Sulfur Oxides–Pungent, colorless gases formed primarily by the combustion of fossil fuels; considered major air pollutants; sulfur oxides may damage the respiratory tract of animals and people as well as be toxic to vegetation.

Sulfuric Acid–H_2SO_4; a dense, heavy, exceedingly corrosive, oily liquid which can decompose animal and vegetable tissue and has a great affinity for water, giving off heat on combining with it. It is used in the manufacture of superphosphate fertilizers by converting the insoluble rock phosphate to a soluble and available form.

Sulfurous Acid–H_2SO_3; an acid resulting from the combination of sulfur dioxide and water, used as a bleaching agent; e.g., as a preservative in fruit juices.

Sulking–(1) In commercial mushroom culture, a depression of growth and reduction in yield from an excessive accumulation of carbon dioxide and an unsaturated hydrocarbon gas, given off by the growing mushrooms, which are unable to escape due to insufficient ventilation in mushroom houses. (2) In horses, refusal to obey commands promptly.

Sumac–See Swamp Sumac.

Summer Annual–Any plant which grows from seed each spring or summer, produces seed that same season, and does not survive the winter.

Summer Fallow–The plowing of a field prior to or during the summer and cultivating enough to control weeds and to accumulate soil moisture in preparation for a later crop.

Summer Legume–A legume which grows from seed each spring or summer, produces seed that same season, and does not survive the winter; e.g., cowpeas, soybeans, etc.

Summer Squash–Any of several types of squash commonly used as a vegetable soon after harvesting, rather than being kept through late fall and winter, in contrast to the winter squash or Hubbard group.

Summer Wood–The portion of the annual growth ring of a tree formed during the latter part of the yearly growth period. It is usually more dense and stronger mechanically than spring wood. See Ring Porous Wood, Spring Wood.

Sumpweed–Any plant of the genus *Iva*, family Compositae. Its several species are included in a group of plants whose pollen cause hayfever in the Rocky Mountain states (United States) and westward.

Sun Euphorbia–*Euphorbia helioscopia*, family Euphorbiaceae; an annual herb which is a weed in waste places in the northeastern United States and eastern Canada. Contact with the milky sap causes a dermatitis with blisters and inflammation in human beings. Native to Europe.

Sunburn–(1) Injury to the leaves, fruit, or other parts of plants, or to the skin of animals, as a result of exposure to intense sunlight. See Sunscald. (2) Greening of Irish potatoes, onions, and certain root crops due to exposure to sunlight. (3) A superficial inflammation of the skin of hogs (especially white hogs) on rape pasture, when they are wet from the dew on the plants and exposed to bright sunlight.

Sunflower–*Helianthus annuus*, family Compositae; a tall, annual herb growing up to 12 feet or more in height, which has very large ovalish leaves and immense flower heads that are yellow except for the central disk. It is often planted in groups or rows for screening purposes or to provide shade for poultry on summer range. It is also sometimes grown for silage or for its seed which is used as a bird feed. The seed yields an edible oil. Native to North America. Also called Russian sunflower, combflower.

Sunflower Silage–Silage from sunflower plants which is usually made when the heads are just showing bloom; prepared chiefly in regions where the season is too short and cool for corn. Sunflower silage is somewhat less palatable and lower in feeding value than corn silage.

Sunflower-seed Oil Meal–A product obtained by grinding the sunflower-seed oil cake remaining after the extraction of oil from sunflower seed. It is used as a supplement in livestock feeds.

Sunn Hemp–*Crotalaria juncea*, family Leguminosae; a tall shrub with silky stems and bright yellow flowers whose branches yield a fiber, the commercial sunn, used in making ropes, sacking, etc. Native to India.

Sunscald–Injury to plant leaves, stems, fruits, and bark which results from a combination of high temperature, high humidity, and intense sunshine. See Sunburn, Winter Sunscald (of trees).

Super-parasitism–(1) Parasitism upon a parasite. (2) An individual attacked by two or more primary parasites, or by one species more than once. See Autoparasitism.

Superphosphate–A fertilizer product which results from treating rock phosphate with sulfuric acid. The grade shows the available phosphoric acid; e.g., 20 percent superphosphate contains 20 percent phosphoric acid.

Superphosphoric Acid–A phosphoric acid mixture containing a substantial proportion of one or more polyphosphoric acids. It can be made by thermal dehydration of orthophosphoric acid or by absorption of P_2O_5 in water. The most common in fertilizer manufacture is for production of ammonium polyphosphate solution. See Ammonium Polyphosphate. Superphosphoric acid made by dehydration of wet-process phosphoric acid usually contains 68 to 72 percent P_2O_5.

Superspecies–A group of related species that are geographically isolated; without any implication of natural hybridization among them.

Supine–Prostrate.

Supplemental Irrigation–Irrigation carried on in areas where annual rainfall is normally high enough for good yields but is insufficient for some crops during part of a growing season.

Supplemental Pasture–A pasture to augment range forage, particularly during emergency situations. Supplemental pasture may be provided by annual grasses and/or legumes, or by aftermath of meadows, grain fields, etc.

Surface Fire–A fire which runs over the forest floor and burns only the surface litter, the loose debris, and the smaller vegetation. It does not burn tree crowns.

Surface Irrigation–Irrigation distribution of water over the soil surface by flooding or in furrows for storage in the soil for plant use.

Surface Runoff–The portion of runoff which flows to stream channels over the surface of the ground. It is quite distinct from the runoff absorbed by the soil, lost by evaporation, or retained as surface storage. The usual sources of surface runoff are rainfall occurring with intensities exceeding the rate of infiltration and snow upon the ground melting at rates exceeding the rate of infiltration. Surface runoff is the principal source of flood-flow occurring in streams at times of rainstorm or melting snow. See Subsurface Flow.

Surface Sealing–Orientation and packing of dispersed soil particles in the surface layer of soil that makes it almost impermeable to water.

Surface Soil–The soil ordinarily moved in tillage, or its equivalent in uncultivated soil, ranging in depth from 4 to 10 inches (10 to 25 centimeters). Frequently designated as the plow layer, or the A horizon. See Soil Horizon, Soil Profile.

Surface Storage–(1) The storage of water in surface reservoirs, such as constructed reservoirs, ponds, swamps, and lakes to reduce flood flow, increase low-water flow, and retain runoff for irrigation. (2) The retention of water in minute depressions on the soil during a period of runoff. Also called depression storage.

Surface Tension–Tendency of a liquid to contract until its area is the smallest possible for a given volume of liquid. This makes it appear to be covered by an elastic membrane.

Surface Tillage–Cultivating or working the soil to a shallow depth or barely below the surface.

Surface Water–Water on the surface, as lakes and rivers, in contrast to that underground. Surface water properly includes snow and ice.

Surface-active Agent–See Surfactant.

Surfactant–A material that improves the emulsifying, dispersing, spreading, wetting, and other surface-modifying properties of pesticide formulations.

Surge Irrigation–Water applied in a series of regulated pulses or surges into the furrow rather than in the conventional manner of a continuous stream. Surging is done by switching the flow from one furrow to another and back again at regular intervals. Developed at Utah State University, advantages include irrigating more acres in less time, getting more-uniform application without excessive runoff, and making it easier to apply a small amount of water.

Surra–An infectious disease caused by the protozoan *Trypanosoma evansi*, characterized by anemia, fever, edema, marked loss of weight, or the appearance of minor hemorrhages of the mucous and serous membranes. Usually fatal in horses, camels, and mules, it is not so serious in dogs or cattle. Endemic in India, the Philippines, and Africa. Also spelled surrah.

Suspension Fertilizer–Liquid fertilizer in which the solids are held in suspension (prevented from settling) by the use of a suspending agent, usually a swelling type clay. It must be fluid enough to be mixed, pumped, and applied to the soil without major alterations in the application equipment, and yet remain homogeneous during application. See Fluid Fertilizers, Liquid Fertilizers.

Sustained Yield Management–Controlled exploitation of a forest unit in such a way that the annual or periodic yield of timber or other products can be maintained in perpetuity.

Suture–(1) A line along which two things or parts are united. (2) In botany, a line of union or a line of dehiscence. (3) In horticulture, a longitudinal protuberance along one side of drupaceous fruits, as in the peach. (4) To unite or join by sewing.

Swamp–(1) A natural area which has standing water on the surface for all or a considerable period of the year and which has a dense cover of native vegetation. See Bog, Marsh. (2) To clear the ground of underbrush, fallen trees, and other obstructions to facilitate subsequent logging operations, such as skidding.

Swamp Sumac–Poison sumac; ***Rhus vernix***, family Onacardiaceae. A small tree or shrub whose leaves are poisonous to touch.

Sward–A closely grazed or mowed area in which the grass and other plant species are close-growing, making an almost complete ground cover. See Meadow, Pasture.

Sweet Basil–*Ocimum basilicum*, family Labiatae; an annual, aromatic, many-branched herb grown for its leaves used as a flavoring. Native to Europe and Asia.

Sweet Cherry Crinkle Leaf–A nontransmissible disorder of Bing and Black Tartarian varieties of cherry accompanied by unfruitfulness. It has been responsible for great monetary losses to cherry growers in the western United States.

Sweet Corn–*Zea mays* var. *saccharata*, a variety of corn characterized by edible kernels with a high sugar and low starch content which are wrinkled and translucent when dry. It is used as a fresh garden delicacy and a processed food. See Corn.

Sweet Gum–*Liquidamber styraciflua*, a large, fast-growing tree that is native to the southern United States. The trees are grown for shade trees. The lumber is used to build furniture frames.

Sweet Marjoram–*Majorana hortensis*, family Labiatae; a very fragrant perennial herb grown as an annual for its leaves used as a seasoning and garnish and for its essential oil. Also called annual marjoram.

Sweet Orange–*Citrus sinensis*, family Rutaceae; a broad-leaved evergreen tree grown for its edible fruit, the orange of commerce. The plant has become naturalized in many tropical and subtropical regions where new varieties have been developed and commercial culture has become very important. Native to China.

Sweet Pepper–Any variety of pepper which is mild and not hot.

Sweet Pickle Cure–A method of preserving meat in a brine or pickle which consists of salt, sugar, saltpeter, and water.

Sweet Potato–*Ipomoea batatas*, family Convolvulaceae; a tropical and subtropical perennial vine grown for its edible roots, the sweet potato of commerce. A very important food crop of the tropics, it is grown for human food, animal feed, and the manufacture of syrups, pectin, flour, and starch, in warm areas. The vines have antibiotic properties. See Yam.

Sweet Potato Tree–See Cassava.

Sweet Sorghum–Any variety of sorghum which has stems filled with sweet juice and is grown for forage and syrup production, in contrast to the grain or nonsaccharine varieties. See Sorgo.

Sweet Spanish–Any of several varieties of onion of Spanish origin. They are large, and mild in flavor.

Sweet Sudan–A cross between a saccharine sorghum and Sudangrass; a fast-growing sorghum which is more resistant to disease and chinch bugs than sweet sorghum.

Sweetbell Red Pepper–*Capsicum frutescens* var. *grossum*, family Solanaceae; an annual herb grown for its large, bell-shaped, edible, furrowed, variously colored, mild-flavored fruit used in salads and as a flavoring. Native to America. Also called a bell pepper.

Sweetclover–Any herb of the genus ***Melilotus***, family Leguminosae. Species are Asiatic annual or biennial herbs which have been known in the Mediterranean region for 2,000 years and are grown in the United States for pasture, hay, silage, soiling, soil improvement, and as a honey crop; e.g., annual yellow sweetclover, Hubam sweetclover, white sweetclover, yellow sweetclover.

Sweetclover Phytophthora Root Rot–A root rot of sweetclover caused by ***Phytophthora cactorum***, family Pythiaceae. See Phytophthora.

Sweetpea–*Lathyrus odoratus*, family Leguminosae; a very popular, viny, herbaceous, annual grown for its beautiful varicolored blossoms. Native to Sicily.

Sweetpotato Meal–A meal made from dried or dehydrated sweet potatoes which is of some value as a stock feed. It can be included in mash for poultry.

Sweetpotato Weevil–*Cylas formicarius elegantulus*, family Curculionidae; a slender-snout beetle which feeds on the sweet potato and closely related plants of the morning-glory family. The female lays eggs in the roots in the field and honeycombs them with white grubs. Of Asiatic origin, the beetle is now widespread in southern United States.

Sweetpotato Wilt–A stem rot of sweet potato caused by *Fusarium batatatis* and *F. hyperoxysporum*, family Tuberculariaceae. It is a major disease of the sweet potato and is characterized by death of the runners with browning of the vascular tissues.

Swiss Chard–*Beta vulgaris*, family Chenopodiniaceae; a type of beet that has been developed for its tops instead of its roots. The culture of chard is the same as beets, but the plants grow larger and are very productive.

Switchgrass–*Panicum virgatum*, family Gramineae; a tall, perennial grass found growing in patches or bunches, which is widely distributed throughout the eastern United States west to Montana, Nevada, and Arizona. It is palatable as a range species when young. Native to North America.

Syconium–The "fruit" of a fig (*Ficus*), composed of a hollow, globose receptacle open at the apex and thickly beset inside with reduced flowers.

Sylvinite–A potassium salt, a mixture of the minerals sylvite and halite. It is a minor source of potassium chloride used in commercial fertilizers. Also called hardsalt.

Symbiosis–The close association of two dissimilar organisms, each known as symbiont. The associations may have five different characterizations as follows: mutualism: beneficial to both species; commensalism: beneficial to one but with no influence on the other; parasitism:

beneficial to one and harmful to the other; amensalism: no influence on the other; synnecrosis: detrimental to both species of organisms.

Symbiotic Relationship–A relationship between two different types of organisms that is beneficial to both of them.

Sympetalous–The fusion of the flower petals.

Symptom–A perceptible change in any part of the body which indicates disease. A group of symptoms that, considered together, characterize a disease syndrome. A sign, mark, or indication.

Syncarp–A collective or aggregate fruit, such as the mulberry and pineapple.

Syndrome–A group of signs of symptoms that occur together and characterize a disease.

Synecology–The study of plant communities, their taxonomy, life history, and relationship to other biotic communities.

Syneresis–Shrinkage and cracking of a wet gel such as wet clay upon drying.

Syngamy–Union of the gametes in fertilization.

Synsepalous–The fusion of the flower sepals.

Synthetic–Artificially produced; produced by human effort and design rather than naturally occurring; chemically manufactured.

Synthetic Cream–A creamlike material made by the emulsification of nonmilk fats or hardened oils, such as whale oil, ground nut oil, etc., with dried egg, lecithin, soya, glyceryl, monosterate, and other substances.

Synthetic Food/Fiber–A food or fiber produced from a nonagricultural raw material; e.g., a nondairy coffee creamer, a synthetic orange juice, an imitation shoe leather, or a human-made fiber.

Synthetic Manure–(1) Organic material, such as leaves, grass, and straw, to which a mineral fertilizer and lime have been added to hasten decomposition. After decomposition it is spread on, or worked into, the soil in the manner of barnyard manure. Also called artificial manure, compost. (2) Synthetic chemical fertilizers, such as anhydrous ammonia and urea.

Synthetic Organic Chemicals–Calcium cyanamid and urea are produced synthetically for use as fertilizers. They contain organic combinations of elements, but behave in the soil like inorganic fertilizers. The nitrogen in cyanamid and urea are defined as "synthetic nonprotein organic nitrogen."

Synthetics–Artificially produced products that may be similar to natural products.

Syrphid Flies–A group of insects of the family Syrphidae that is distributed throughout the United States. These insects are very beneficial because they pollinate plants and the larvae eat aphids.

Syrup–The thick, sweet liquid made from water, the juice of fruits, etc., boiled with sugar; or the concentrated juice of certain saccharine plants, such as sugarcane, sugar sorghum, sugar maple, etc. Also called sirup. See Corn Syrup.

Syrup Hydrometer–A floating instrument with a special scale for measuring the specific gravity of concentrated sugar solutions. Used to determine the strength of syrups, as in the making of maple syrup.

Systemic–(1) Pesticide material absorbed by plants, making them toxic to feeding insects. Also, pertaining to a disease in which an infection spreads throughout the plant. (2) Pertaining to the body as a whole and not confined to one organ or part of the body, as a systemic infection.

Systemic Fungi–Fungi that grow throughout the body tissues of the host.

Systemic Pesticide–A pesticide that is capable of translocation within the plant.

T Factor–A measure of the amount of erosion in tons per acre per year that a soil can tolerate without losing productivity. For most cropland soils, T values fall in the range of 3 to 5 tons per acre per year.

T-budding–See Shield Budding.

Table Stock Potatoes–Potatoes for human consumption as contrasted to potatoes raised for seed. See Seed Potato.

Tachinid Fly–Any fly of the family Tachinidae, whose larvae are beneficial as insect control, as they are parasitic on many noxious insects.

Tag–(1) A dung-covered lock of wool. Also called dag, daglock. (2) A plastic or metal piece attached to an animal for identification, or a cardboard or cloth label attached to the container of a product, a feed or fertilizer, giving the content analysis, etc. (3) A lock of cotton fiber which adheres to the boll after picking. (4) To place a tag on a product, animal, etc., for identification.

Tail–(1) The posterior part of the vertebral column of animals. It is usually covered with hair, some of which may be quite long. Also called brush. (2) A fanlike row of rather stiff feathers on the posterior part of a bird. (3) The lowest grade of flour. (4) A wisp of hay not properly tucked into a bale. (5) The bottom layer of produce in a container. (6) The weakest and poorest sheep (Australia and New Zealand). (7) The woody part of a plant which has been propagated by tip layerage. (8) To dock an animal's tail (Australia and New Zealand). (9) To keep sheep together in a flock (Australia and New Zealand). (10) To remove the tail from a carcass. (11) To assist an undernourished cow to its feet by pulling on the tail.

Tail Tree–In forestry power skidding, a tree at the end of a run to which the tackle is fastened.

Take–(1) The uniting of a scion with a stock following grafting or budding. (2) To accept a male in coitus. (3) To result in a mild infection after vaccination.

Take-all–A fungal disease of wheat, rye, barley, and oats caused by *Ophiobolus graminis*. It is characterized by stunting, premature ripening, and by black lesions at the base of the culms.

Tall Buttercup–*Ranunculus acris*, family Ranunculaceae; a perennial, fibrous-rooted herb grown for its yellow flowers. It is poisonous to livestock. Native to Europe. Also called blister flower, butter-and-eggs, buttercup, butter rose, field buttercup, tall crowfoot.

Tall Oil–A by-product from the manufacture of chemical wood pulp. Used in making soaps and for various industrial products.

Tallow–(1) The fat extracted from the fat tissue of cattle and sheep. Used in candle making, soap manufacture, etc. See Suet. (2) Designating various plants which yield flammable waxes, have a tallowy taste, form a fattening feed, or yield greasy substances resembling tallow; e.g., Chinese tallow-tree, tallow-weed, wax myrtle, etc.

Tallow Tree–See Chinese Tallow Tree.

Talus–The heap or fragments of rock and soil material which collects at the foot of a cliff or very steep slope chiefly as a result of rock weathering and gravitational force. See Colluvium, Scree.

Tame–(1) Domesticated. (2) Cultivated. (3) Designating an animal which has been made docile or tractable, as a wild horse is tamed or broken. (4) To domesticate. (5) To make docile.

Tame Hay–Hay produced from sown meadows, as contrasted with the forage from wild areas of native forage plants.

Tamed Iodine–Iodine that is combined with an organic material and is used as a disinfectant, mostly as an antiseptic. The iodine is released slowly from the organic compound, thus it is less irritating than tincture of iodine (iodine in alcohol).

Tan Disease–A physiological disorder of plants caused by excess soil moisture, characterized by either the roots or the aerial portions becoming swollen and by the outermost cork layers breaking and peeling off.

Tanbark–The bark of hemlock, chestnut, oak, mimosa (wattle), or tanoak, etc., which is used as a source of tannin. The residue is often used as a ground covering in circus lots, race tracks, livestock pavilions. Also called tanner's bark.

Tangelo–A citrus fruit that is produced by crossing a tangerine with a grapefruit.

Tangerine–A horticultural variety of the mandarin orange (*Citrus nobilis* var *deliciosa*) grown for its deep orange, loose-skinned, edible, sweet fruit.

Tanglefoot–(1) Any sticky substance which can be applied to a tree trunk to catch and hold climbing insects to prevent their damaging the tree. (2) Broom deervetch.

Tangleroot–A physiological abnormality of the pineapple in which the roots mass around the rootstock rather than growing outward.

Tank Agriculture–See Hydroponics.

Tank Gardening–See Hydroponics.

Tanners' Lime–An impure calcium carbonate which is a waste product from the tanning of leather.

Tannic Acid–Tannin; gallic acid, which is obtained from various trees and wood. It is soluble in water and is capable of combining with the proteins in animal skins to convert the skins into leather.

Tannin–See Tannic Acid.

Tansy Ragwort–A yellow flowering plant (*Senecio* spp.) found in the Pacific Northwest that is poisonous to cattle and horses. The whole plant has a strong unpleasant odor when crushed.

Tap–(1) In lumbering, a cut made from the inside of a log. (2) A wooden basket which is used for packing figs. (3) A faucet on a pipe or container containing liquid. (4) To remove a taproot from a plant. (5) To make a cut in a tree for obtaining sap for use in making maple sugar or turpentine. (6) To cut threads in metal that will receive a bolt or screw.

Tap Water–Water which comes from a water faucet, as contrasted to well water or rainwater.

Taper–The gradual reduction in diameter of the trunk of a tree from the base to the top. Also called rise.

Tapioca Plant–See Cassava.

Taproot–The primary descending root, usually conical, of a plant from which lateral branching roots may develop; e.g., as in carrots and alfalfa. See Primary Root, Root.

Tara–See Taro.

Tare–(1) Weed seeds. (2) Deduction made in the weight of packaged products to allow for the weight of the container; gross weight minus the net weight. (3) Deduction made in the weight of sugar beets to allow for earth, crown tops, etc., which might adhere to the beets. (4) Any impurity in seed crops, such as dirt, chaff, weed seeds, broken seeds, etc.

Target Species–A plant or animal species which a pesticide is intended to kill.

Tarnished Plant Bug–*Lygus lineolaris*, family Miridae; an insect that infests many fruits and garden plants, releasing a toxin that sometimes causes a deformity or dwarfing.

Taro–*Colocasia antiquorum*, family Araceae; a tuberous-rooted, perennial herb, grown as a summer bedding plant for its decorative foliage and for its large, edible, starchy root. Native to Southeast Asia. Also called elephant's-ear.

Tarragon–*Artemisia dracunculus*, family Compositae; a perennial herb grown as a source of an aromatic, pungent, flavoring extract used in pickle manufacture and in the production of tarragon vinegar. Native to Europe and Asia. Also called estragon.

Tarragon Vinegar–A white vinegar flavored with tarragon.

Tartaric Acid–An acid occurring widely in plants used in dyeing, photography, medicine, and jelly manufacture.

Tartary Buckwheat–*Fagopyrum tataricum*, family Polygonaceae; an annual herb grown for its seed sometimes used in poultry feeds. Native to India. Also called India wheat, Kangra buckwheat, duck wheat, wildgoose, rye buckwheat, hull-less buckwheat, bloomless buckwheat. See Buckwheat.

Tassel–Male flower of corn.

Tasseling Time–In cornfields, an indefinite period during the growing season between tassel emergence and pollen shedding of corn plants. It generally occurs during July in the Corn Belt of the United States but may vary with season, time of planting, variety, etc., in other places.

Taste–(1) The flavor of a product as determined by placing the substance in the mouth. (2) A small amount or sample. (3) In judging, to place a morsel or a few drops of a product in the mouth to savor, but often not to swallow.

Taungya System–(1) Raising a forest crop in conjunction with an agricultural crop, usually in a modified shifting cultivation system. (2) Intercropping with woody and herbaceous plants.

Tautonomy–Relations that exist if the same word is used for both the generic and specific name in the name of a species.

Taw–(1) To dress and prepare, as the skins of sheep, lambs, goats, and kids, for gloves, etc., with alum, salt, and other softening and bleaching agents. (2) To prepare hemp by beating.

Taxadjuncts–Soils that cannot be classified in a series recognized in the classification system. Such soils are named for a series they strongly resemble and are designated as taxadjuncts to that series because they differ in ways too small to be of consequence in interpreting their use or management.

Taxonomy–(1) The science of classification of organisms and other objects and their arrangement into systematic groups such as species, genus, family, and order. (2) Taxonomy is closely related to classification but it embodies a broader concept. Taxonomy is the science of how to classify and identify. It is the theoretical study of classification including its bases, principles, procedures, and rules. Taxonomy includes classification as well as identification.

Tea–*Camellia sinensis* (*Thea sinensis*), family Theaceae; an evergreen tree or shrub grown on acid soils in the mountains of Asia for its leaves, the tea of commerce.

Tea Garden–A plantation of tea trees.

Teak–*Tectona grandis*, family Verbanaceae; a tall tree grown for its hard, durable, yellowish-brown lumber used in shipbuilding, furniture manufacture, etc. Native to the East Indies.

Teart–(1) A soil or plant that contains a large amount of molybdenum. (2) Molybdenosis caused by a domestic animal ingesting an excess of molybdenum, usually from eating molybdenum-rich forage grown on soils high in molybdenum. The symptoms are diarrhea and general debility (weakness).

Tease–(1) To stimulate an animal to accept coitus. (2) To vex or annoy. (3) Of fibrous materials such as wool or flax, to separate the strands of the fibers, or to prepare the fibers, by combing, for spinning. (4) To form a nap on cloth by stroking the loose fibers in one direction with comblike, natural, or mechanical, teasels.

Technical–(1) Concerned with a particular science, industrial art, profession, sport, etc. (2) Practicing, using, or pertaining to the technique rather than the theory or underlying principles involved in the execution of a project. (3) Designating the grade of a commodity manufactured in the usual commercial manner. (4) Pertaining to or designating a market where prices are controlled by speculation or manipulation.

Tectonics–The study of the broader structural features of the earth and their causes.

Teepees–Pyramidal structures such as tripods or quadripods of lath, lumber, poles, pipe, or bamboo spread at the base and lashed together at the top to provide support for climbing vines.

Teff–*Eragrostis abyssinica*, family Gramineae; a fragrant grass grown for its cereal grain and as an ornamental. Native to North Africa.

Telophase–The phase of cell division between anaphase and the complete separation of the two daughter cells; includes the formation of the nuclear membrane and the return of the chromosomes to long, threadlike and indistinguishable structure.

Temper–(1) The proper relative condition of moisture in grain preparatory to milling. (2) The relative hardness or softness of the metal in implements and tools. (3) The relative mildness or viciousness of an animal. (4) Milk of lime, etc., added to boiling syrup to clarify it.

Temperate Rain Forest–A woodland usually with a dominant species; found in a temperate zone in areas having little frost.

Temperature–(1) The amount of heat or cold measured in degrees on different scales, as Fahrenheit or Centigrade. At sea level, water freezes at 32°F (0°C) and boils at 212°F or 100°C. (2) The degree of heat in a living body. (3) Abnormal heat in a living body. Also called Fever.

Temperature Zero–The temperature below which certain physiological processes of an organism are carried on at a very slow rate.

Tempering–(1) Accustoming planting stock gradually to a material change in temperature. (2) Treatment of tools and implements to impart a degree of hardness or softness to the metal.

Tendril–Slender twining organ found along stems of some plants such as grapes, which helps the vine to climb.

Tensiometer–Any of several types of devices which measure moisture, tension, or condition of water in soil. Used to determine when to irrigate.

Tent Caterpiller–See Eastern Tent Caterpiller.

Tepary Bean–*Phaseolus acutifolius* var. *latifolius*, family Leguminosae; an annual, tropical herb, which is an important vegetable crop grown in arid regions for its edible seeds. Native to North America.

Tequila–An alcoholic beverage made in Mexico, obtained by the crude distillation of the sweet sap of tequila agave and other species of the genus *Agave*, family Amaryllidaceae. The alcohol content averages about 50 percent.

Tequila Agave–*Agave tequilana*, family Amaryllidaceae; a succulent, stemless plant, widely grown as a source of a liquor high in alcohol content. Native to Mexico. Also called tequila mescal.

Terminal–(1) A station for delivery or receipt of produce. (2) Designating growing or being located at the end of a branch or stem.

Terminal Bud–The bud which develops at the end of a branch or stem.

Terminal Cutting–The most commonly used method of plant propagation in which the stem tip of a plant is rooted.

Terminal Moraine–A moraine formed across the course of a glacier at its farthest advance, at or near a relatively stationary edge, or at places marking the termination of important glacial advances.

Terra–(Latin) Soil, earth.

Terra Culta–(Latin) Cultivated land.

Terrace–(1) A long, low embankment or ridge of earth constructed across a slope with a flat or graded channel to control runoff and minimize soil erosion by scouring. Two general types are bench terrace and ridge terrace. Variations of these types are the horizontal bench terrace, sloping bench terrace, broad-base terrace, narrow-base terrace, Nichol's terrace. (2) A raised, level area of earth supported on one side by a wall, bank, etc., as a terraced yard. Often there is one above another on a hillside or slope. (3) A natural land feature which is an elevated, relatively long, flat upland with a short scarp or embankment-like front. The plain may be narrow and the terrace steep or benchlike, or it may be wide and properly designated a terrace plain. Terraces represent former sea, lake, and river levels, and are locally significant in soil and land classification and use. (4) To construct a terrace, especially one across a slope to control erosion and runoff.

Terrace Channel–The depression along the upper side of the terrace ridge in which the water flows to the outlet, or is retained when the channel is not graded so that the water does not have an outlet other than by evaporation and through the soil by infiltration and lateral movement.

Terrace Crown–The highest part of the terrace ridge; the top of the terrace.

Terrace Grade–The slope of the terrace channel in inches or feet per 100 feet length. Terrace grades may be variable or uniform.

Terrace Height–The vertical difference in elevation between the bottom of the terrace channel and the crown of the terrace ridge at adjacent points.

Terrace Interval–Horizontal: the horizontal distance in feet from the center of one terrace line, ridge, or channel to the corresponding point on an adjacent terrace. Vertical: the vertical distance in feet from the center of one terrace line, ridge, or channel to the corresponding point on the adjacent terrace. (The interval of the first terrace is the vertical distance from the top of the hill to the terrace line of the terrace.)

Terrace Outlet–The water channel at the end of the terrace into which the flow from one or more terraces is discharged and conveyed from the field usually through a grassed outlet.

Terrace Outlet Structure–A structure usually installed in or at the end of a terrace outlet to reduce erosion.

Terracing–A practice of constructing a ditch or channel, with a ridge below, across the slope at various vertical intervals to intercept runoff. The channel and ridge are usually constructed in a way that will permit contour operations with farm equipment.

Terrain–(1) Area of ground considered as to its extent and natural features in relation to its use for a particular operation. (2) The tract of ground immediately under observation.

Terrarium–A tightly fitted, glass-enclosed, indoor garden, resembling an aquarium, in which plants are grown for decoration, plant propagation, nature study, etc. Also called Wardian case, bottle garden, crystal garden, glass garden.

Terrestrial–(1) Referring to earth. (2) Designating a plant which lives in soil as contrasted to one which is epiphytic (growing in air) or one growing in water (aquatic or hydrophytic). (3) Designating a ground bird such as a pheasant, partridge, or chicken, as contrasted to an aerial bird.

Terrigenous–Derived from or originating on the land (usually referring to sediments) as opposed to material or sediments produced in the ocean (marine) or as a result of biologic activity (biogenous).

Test Adaptation Trial–In pasture and range research, testing the adaptability of forage species to new edaphic and climatic conditions as to complete their life cycle and/or reseed themselves.

Test Pit–A hole dug in the ground to determine the character of the subsoil or substratum prior to ditching, laying footings, etc.

Testa–The outer coat of a seed.

Testcross–A kind of genetic cross involving one individual expressing a dominant trait and one expressing the recessive trait, the purpose of the cross being to determine whether the individual expressing the dominant trait is heterozygous or homozygous.

Tetrachlorocethane–A chemical compound used as a soil fumigant, nematicide, and herbicide.

Tetraploid–An organism whose cells contain four haploid (monoploid) sets of chromosomes.

Texas Bluegrass–*Poa arachnifera*, family Gramineae; a perennial, vigorous, forage grass, found in the United States from North Carolina to Texas. It is sometimes cultivated for winter pasture. Native to North America.

Texas Croton–*Croton texensis*, family Euphorbiaceae; an annual herb found on dry sandy soils from Alabama to Wyoming, United States, to Mexico. A poisonous plant, it is seldom grazed. Native to North America.

Texas Leaf-cutting Ant–*Atta texans*, family Formicidae; a serous insect pest of garden and field crops which cuts leaves off the plants and invades houses to steal seed and farinaceous (starchy) foods. It is found in eastern and southern Texas and western Louisiana, United States.

Texas Root Rot–See Cotton Root Rot.

Textural Family–Soil groupings according to particle size. Clayey: contains more than 35 percent clay by weight and less than 35 percent rock fragments by volume; fine-silty; less than 15 percent fine sand (0.25 to 0.1 millimeter diameter) or coarser by weight, including fragments up to 7.5 centimeters diameter, 18 to 34 percent clay; fine-loamy: by weight 15 percent or more fine sand (0.25 to 0.1 millimeter) or coarser by weight, including fragments up to 7.5 centimeter diameter, 18 to 34 percent clay; coarse-loamy: same as fine loamy except has less than 18 percent clay; sandy: sand or loamy sand but not loamy very fine sand or very fine sand, rock fragments less than 35 percent.

Texture–(1) The physical properties of a product; that is, the structure and arrangement of the parts which make up the whole. (2) The relative proportions of sand, silt, and clay particles in a mass of soil. The basic textural classes, in order of increasing proportion of fine particles: sand, loamy sand, sandy loam, loam, silt, silt loam, sandy clay loam, clay loam, silty clay loam, sandy clay, silty clay, and clay. The sand, loamy sand, and sandy loam classes may be further divided by specifying coarse, medium, fine, or very fine.

Thallium–A rodenticide, also used to control ants.

Thallophyta–Division of nonvascular plants, those without differentiated roots, stems, or leaves; includes algae and fungi.

Thaw–(1) A period of weather following ice and snow which is warm enough to cause melting. (2) To melt frozen products. (3) To melt ice, snow, and frost, in soil.

Thaw Depression–A hollow formed by the melting of ice in perennially frozen ground. See Permafrost, Thermokarst.

Theobromine–A toxic alkaloid occurring in chocolate, cocoa, tea, and cola nuts that is closely related to caffeine. In practical terms, theobromine is a stimulant and in excess may cause death in dogs and humans.

Therapeutic–Pertaining to the treatment of disease; curative.

Therapy–The sum total of the treatment given to cure disease in plants, animals, and humans.

Thermal Belt–A well-defined area on the sides of valleys, as on the eastern slope of the San Joaquin Valley in California, United States, in which descending cool air is warmed by compression as it nears the lower elevation, while at the lowest elevation cold air accumulates. These belts are of great agricultural importance, for in them the danger of frost is minimized.

Thermal Blanket–A movable cover that is pulled across a greenhouse at night to reduce heat loss.

Thermal Death Point–The amount of heat required to kill a particular organism.

Thermoperiodicity–The response of plants to changes in day and night temperatures.

Thermophyllic–Plants capable of growing at high temperatures.

Therophytes–Annual grasses and forbs that pass through the unfavorable season in the seed stage and complete their life cycle (from seed to seed) in one growing season.

Thicket–A dense growth of shrubs or small trees.

Thief Ant–*Solenopsis molesta*, family Formicidae; a common house ant pest which shows a preference for protein foods. It may also damage germinating grain seeds.

Thin–(1) To reduce the number of plants in a row or area by hoeing, pulling, etc. (2) Designating an animal with little flesh. (3) Designating a pulse which is very feeble.

Thin Land–Soil which is unproductive because of a lack of depth or fertility.

Thin Stand–(1) A relatively smaller number of plants in an area than considered satisfactory. (2) A poor establishment of plants.

Thin Stillage–The water-soluble fraction of a fermented mash plus the mashing water.

Thinning–(1) The removal of plants in a stand to reduce crowding. (2) Removing certain flowers or clusters of flowers or individual fruits after fruit has set and natural dropping has occurred. (3) Removing live branches in a tree crown. (4) In woodland management, the cutting made in immature stands after the sapling stage to increase the rate of growth of the remaining trees.

Thixotropy–The property exhibited by some gels of becoming fluid when shaken. The change is reversible. Some fine clays exhibit thixotropy. See Vertisols.

Thorax–The middle body region of an insect to which the wings and legs are attached.

Thorn–(1) A hard, sharp-pointed, leafless branch of a plant. (2) Any shrub or tree which bears thorns, as the honey locust and rose. (3) Any of the wild species of the genus ***Crataegus***, family Rosaceae.

Thornthwaite Index of Precipitation–A measure of the amount of precipitation corrected for temperature. As the temperature increases, evaporation and transpiration increase. This results in a decrease in the effectiveness of the precipitation. The monthly index of effective precipitation is figured as follows: (P/T-10) 10/9, where P equals mean monthly precipitation in inches and T equals mean monthly temperature in degrees F.

Three-field System–A system of fallowing in which land tracts are divided into three fields, with one field fallowed each year. The practice dates back to the Middle Ages in Europe. See Fallow.

Three-leaved Ivy–See Poison Ivy.

Three-way Cross–The crossing of three different strains or breeds of plants or animals to produce a hybrid. See Double Cross, Single Cross.

Thremmatology–The domestic breeding of plants and animals.

Thresh–To separate grains from the plant, as in removing oat grains from its straw.

Thrips–Very small insects of the order Thysanoptera, for the most part very injurious to plants.

Throatlatch–The strap on a bridle or halter which passes under the horse's throat.

Throughfall–In a forest, all of the precipitation that reaches the soil surface directly plus that which is intercepted by the leaves and drips to the ground. Some authorities include stemflow (down the tree trunk into the soil) in throughfall and some do not.

Throw Up–(1) To hill earth or soil around the crowns of plants. (2) To vomit.

Throwback–(1) In breeding, a reversion to an ancestral type or ancestral characteristic. (2) To revert to an ancestral type or characteristic.

Thyme–(1) Any aromatic, woody, perennial herb of the genus *Thymus*,, family Labiatae. Species are grown as ornamentals and for their aromatic foliage used in seasoning. (2) *T. vulvaris*, an herb cultivated in the garden as a source of an essential oil and for its leaves used in seasoning. Native to southern Europe.

Tideland–(1) The area adjacent to the sea or ocean which is covered during floodtide. (2) The shallow sea bottom (in some instances several miles from shore).

Tier–(1) In grafting and budding, one who ties the bud or scion to the stem or stock. (2) One of the horizontal layers in the vertical stacking of boards in a well-defined pile for seasoning. (3) A layer of lumber, boxes, crates, bags, etc., in storage.

Tierra–(Spanish) Soil, earth.

Tight Soil–A compact clay soil difficult to plow and till and permitting a very slow infiltration of water.

Tiled–Designating an area of land which has been underdrained with tile, as contrasted to open ditches.

Till–(1) A deposit or mixture of earth, sand, gravel, and boulders which has been transported by, and deposited under, glaciers. Till is generally unstratified. Also called boulder clay. (2) To cultivate, plow, fit, and sow land. See Tilth.

Till Plain–A level or undulating land surface covered by glacial till. See Till.

Tillable Acres–That part of the farm land that can be used for cultivated crops without additional drainage, clearing, or irrigation.

Tillage–The mechanical manipulation of soil for any purpose; but in agriculture it is usually restricted to the modifying of soil conditions for crop production.

Tillage Pan–A compacted layer of soil, usually just below the depth of usual tillage. Tillage pans are very serious because they reduce the permeability of the soil to air and water. Crop yields are low on these soils.

Tiller–(1) An erect shoot arising from the crown of a grass. (2) One who tills. (3) An implement for tilling.

Tillering–The production of shoots from the crown of a plant; stooling.

Tilth–The artificial condition of soil structure brought about by tillage (cultivation).

Timber–(1) Forest stands. (2) Sawed lumber 5×5 inches or more. (3) Sawed lumber 4½ × 6 inches or more (United Kingdom). (4) Wood in form for heavy construction.

Timber Danthonia–*Danthonia intermedia*, family Gramineae; a bunchgrass found from Quebec, Canada, to New Mexico, United States,
in the spruce belt, which is of good forage value. Native to North America. Also called timber oatgrass.

Timber Marking–The selection and indication, usually by blaze or paint spot, of trees which are to be cut or retained in a cutting operation.

Timber Poisonvetch–*Astragalus convallarius*, family Leguminosae; a poisonous plant found on mountain slopes from British Columbia, Canada, to Arizona, United States. Characteristics of poisoning in animals are dullness, irregularity in gait, solitary habit, loss of nervous sensibility, loss of flesh, shaggy coat, loss of appetite, and may result in death.

Timber Products–Roundwood products and plant by-products. Timber products output includes roundwood products cut from growing stock on commercial forest land; from other sources, such as cull trees, salvable dead trees, limbs, and saplings; from trees on noncommercial and nonforest lands; and from plant by-products.

Timbo–Any woody plant of the genus *Lonchocarpus*, family Leguminosae, from whose roots is extracted rotenone, the chief and most effective of several insecticides contained in these plants. Native to South America. The crude product is sometimes called derris root. Also called cube. See Rotenone.

Timothy–*Phleum pratense*, family Gramineae; one of the best known hay grasses; easy to establish and easy to grow. The common name *herd's grass* has been applied to timothy in some parts of the United States, especially New England. Native to Europe and Asia.

Tip–(1) The exposed part of the wool fiber in the fleece. (2) The end of a branch, twig, etc. (3) A place where refuse is dumped (England). (4) To remove the ends of branches, stems, twigs, etc.

Tip Cutting–Softwood cutting to produce another plant.

Tip Layering–A method of plant reproduction especially useful for black raspberries. The tips of the branches are covered with soil so that roots may be developed from the nodes at the end of the branch. When the roots have developed, the stem is cut away. Also called arch layering.

Tipburn–Any disease of plants caused by environmental or other factors in which the margins and tips of the leaves burn or turn brown. On potatoes, it is caused by feeding of flea beetles.

Tissue–Groups of cells working together to carry out a common function, such as muscle tissue, connective tissue, and epithelial tissue.

Tissue Culture–The process or technique of making plant or animal tissue grow in a culture medium outside the organism.

Titer–The minimum quantity of a substance required to produce a specific reaction with a given amount of another substance.

Toadstool–The fruiting structure of a fungus of class Basidiomycetes, order Agaricales. The vegetative part of the fungus consists of fine, branching, usually white threads of cells which grow in the soil, in wood (often causing decay in it), in manure, or in other vegetable organic substances. The fruit, produced in the air, is usually fleshy, more rarely tough or even woody consisting of a cap (pileus), the underside of which has radiating plates (gills) on which the spores are produced. More often the cap may sit on a central stalk (stipe) but this may be lateral or wanting in those toadstools that are attached by one edge. Popularly, but not scientifically, the name toadstool is applied to poisonous species, and mushroom to edible ones.

Tobacco–*Nicotiana tobacum*, family Solanaceae; an annual, herbaceous plant which yields the tobacco of commerce. Although the acreage planted is relatively small, it is one of the important money crops of the United States. Probably native to tropical regions in America. Types of tobacco: air-cured: a process that uses natural atmospheric conditions to prepare the crop for use. Artificial heat is sometimes used to control excess humidity during the drying period; fire-cured: a process that uses artificial atmospheric conditions, such as open fires, from which the smoke and fumes of burning wood are partly absorbed by the tobacco; flue-cured: a process of artificial atmospheric conditions, that regulates heat and ventilation, without allowing smoke or fumes from the fuel to come in contact with the tobacco.

Tobacco Barn–A tall building in which tobacco is hung for curing.

Tobacco Budworm–*Heliothis virescens*, family Noctuidae; a caterpillar that infests tobacco, eating into the buds and unfolded leaves. It also infests cotton, geranium, ageratum, and solanaceous plants.

Tobacco Flea Beetle–*Epitrix hirtipennis*, family Chrysomelidae; a serious pest of tobacco. The adult beetles chew small, round holes in the leaves, and the larvae feed on the roots and tunnel into the stalk. It also infests other solanaceous plants.

Tobacco Frenching–A nitrogen-deficiency disease of tobacco in which the foliage is deformed, the leaves narrowed, thickened, and increased in number.

Tobacco Hornworm–*Protoparce sexta*, family Sphingidae; a green and white worm up to 4 inches long which infests tobacco, eating the foliage. It is also a serious pest of tomato and eggplant. See Tomato Hornworm.

Tobacco Mildew–A disease which infects tobacco plants while still in the field, caused by the fungus *Peronospora hyoscyami*.

Tobacco Mosaic–A viral disease of tobacco and numerous other plants outside the family Solanaceae, characterized in young plants by a downward curling and distortion of the youngest leaves along with a chlorosis. Dark green spots appear on the leaves as they grow, developing into irregular, crumpled, blisterlike areas as the rest of the leaf becomes chlorotic. Old leaves may have scattered, necrotic spots, and there is a stunting of the plant. Also called brindle, calico.

Tobacco Sick Soil–Soil infested with the root rot fungus which attacks the tobacco plant. The sickness is commonly ascribed to depletion of available plant nutrients in the soil by continued cultivation of tobacco.

Tobacco Spear–A spearlike device fitted over the end of a lath or stick to aid in piercing the butt ends of tobacco plants and in spacing them on the lath before hanging in the barn for curing.

Tobacco Stems–Waste tobacco products which are ground up to be used as a fertilizer material for nitrogen and potassium. The nicotine may or may not be removed.

Tobacco Stick–A lath on which tobacco leaves or plant are hung for curing.

Tobacco Thrips–*Frankliniella fusca*, family Thripidae; an insect pest of many plants but especially important on cigar-wrapper tobacco. Both adults and young feed along the veins of the leaves, defacing the leaves and reducing their market value.

Tobacco Warehouse–A large building where tobacco is stored and is later auctioned off.

Tobacco Wilt–See Granville Wilt.

Toe–(1) Any one of the digits on the foot. (2) The lowest downstream edge of a dam. (3) The bottom of a seed furrow opened by the furrow opener of a grain drill. (4) The tuberous roots of dahlia, used for propagation. (5) The lower edge or edges of a slope. See Jute.

Toe Slope–The outermost inclined surface at the base of a hill; part of a foot slope.

Tolerance–(1) The ability of a tree to grow in the shade of other trees and in competition with them. (2) The acquired ability of an animal or insect to take poison without ill effects. (3) The ability of a species to develop and survive under unfavorable environmental conditions. (4) A percentage of off-grade or off-sized produce which is allowed in a particular grade, or containers in a lot, that fail to meet grade specifications, especially of fruits and vegetables. (5) The amount of toxic residue permitted by federal law to be present on or in food and forage products that are to be marketed for consumption. (6) The accuracy limits specified in the construction of a machine part as plus or minus 0.005 inch.

Tolerant–Ability of plants to endure a specified pest or an adverse environmental condition, growing and producing despite the disorder.

Tom Thumb–Designating certain dwarf species or varieties.

Tomato–*Lycopersicon esculentum*, family Solanaceae, a tender, annual herb, one of the most important vegetable crops, grown for its delicious, edible fruits eaten raw or cooked. Native to South America. Also called apple of love.

Tomato Anthracnose–The most serious of the tomato fruit rots, caused by the fungus *Colletotrichum villosum*, family Melanconiaceae. It is characterized by cankers on the fruit, rarely on the leaves.

Tomato Hornworm–*Protoparce quinquemaculata*, family Sphingidae; an insect whose larva feeds ravenously on tomato, tobacco, eggplant, pepper, and potato. Found throughout the United States and Canada. See Tobacco Hornworm.

Tomato Pinworm–*Deiferia lycopersicella*, family Gelechiidae; a serious insect pest of tomato, eggplant, and potato in the United States. The young larvae mine the leaves, and the older ones puncture the fruit.

Tomato Root Rot–A root rot of plants caused by *Phytophthora cryptogea*, family Pythiaceae, which attacks potato, tomato, petunia, China aster, and wallflower.

Tomato Wilt–A widespread fungal disease of tomato caused by *Fusarium lycopersici*, family Tuberculariaceae, or by *Verticillium albo-atrum*, or *Pseudomonas solanacearum*, family Bacteriaceae, characterized by a yellowing, upward rolling, and a wilting of the leaves.

Tongue–(1) A part of the mouth, composed chiefly of muscle, on which the taste buds lie. It is one of the organs of taste, and moves the food being chewed around in the mouth and back to the throat for swallowing. (2) A retail cut of beef, veal, mutton, and lamb, consisting of the whole tongue of an animal. (3) The wooden or metal pole which acts as a guide and is fastened to the front end of a vehicle or plow. It separates the two members of a team of draft animals. The eveners for pulling the implement may also be attached to the tongue. (4) To make a slit in the plant's stem in grafting.

Tonka Bean–*Dipteryx odorata (Coumarouna odorata)*, family Leguminosae; a large, tropical tree whose black, almondlike seed, the tonka bean of commerce, is used in flavoring, perfumery, and to scent snuffs. Native to South America.

Tonkabean Wood–*Alyxia buxifolia*, family Apocynaceae; a small, straggling, evergreen shrub sometimes grown for its fragrant, tonkabeanlike odor. Native to Tasmania. Also called scentwood, boxleaf alyxia.

Toonea–(1) *Toona ciliata*, family Meliaceae; a tall, nearly evergreen tree useful for its flowers, which yield a dye, and its soft, reddish wood valuable for making furniture. Native to the Himalaya mountains. Also called Indian mahogany. (2) Panama gumtree.

Toothed Euphorbia–*Euphorbia dentata*, family Euphorbiaceae, a plant which secretes a poisonous, milky juice. The effects of poisoning are swelling of the eyes and mouth, abdominal pains, collapse, excessive scours, and death. Native to North America. Also called toothed spurge.

Top–(1) The aboveground parts of certain plants, especially the leaves, as beet tops, turnip tops. See Crown. (2) Scoured, combed, long wool. (3) The highest price paid for a product within a certain period of time. (4) The upper, branchy portion of a felled tree. It is sold as firewood, charcoal source, etc. (5) To cut off the crown and leaves, as of the sugar beet. (6) To remove the upper portion of the crown of a tree. (7) To place the best articles of produce, as eggs, or fruit, on the top layer, in a container so that the whole container appears to include articles of higher quality than is actually the case. (8) To sort out animals that have reached a certain stage of development or finish.

Top Crop–The bolls of cotton growing at the top of the plant. They are the last to mature and to be picked.

Top Cross–(1) In corn breeding, a cross in which an inbred line is used as the pollen parent and a commercial variety a the seed plant. (2) A cross between purebred males and grade females.

Top Diameter–The diameter at the uppermost end of a log or salable piece of timber.

Top Grade–A market grade of produce which is either the best or the next best after fancy.

Top Grafting–Grafting performed on the branches of a plant, as contrasted to root grafting, crown grafting. Also called top working.

Top Kill–(1) A portion of the top of a tree which is dead, while the remainder of the tree is alive. See Staghead. (2) To kill the top growth of a plant with chemicals.

Top Necrosis–The relatively rapid killing of a bud, branch, or entire top of a plant. Also called acronecrosis.

Top Onion–*Allium cepa* var. *viviparum*, family Liliaceae; a botanical variety of the garden onion, which produces small bulbs in place of some or all of the flowers. The small bulbs are planted for green onions for spring use. Also called Egyptian onion, tree onion, perennial tree onion, winter onion.

Top Quality–Designating generally, but not specifically, produce of the very best sort.

Top Sickness in Tobacco–A boron-deficiency disease of tobacco.

Topato (Pomato)–A cross between a tomato and a potato (of no commercial value).

Topdressing–Applying fertilizer or compost to the soil surface while plants are growing.

Topiary Work–The shaping and training of shrubs into ornamental but unnatural shapes.

Topical Application–The application of a pesticide to the top or upper surface of the plant; thus, applied from above.

Topinambou–See Jerusalem Artichoke, Sunflower.

Toposequence–A sequence or chain of soils whose properties vary with topographic position. Theoretically, the slope is the variable, and other factors of soil formation remain the same.

Topping–(1) In propagation by budding, the removing of the stock above the inserted bud by cutting to cause growth of the bud. (2) In pruning, the removal of the ends of branches, as a hedge is sheared. (3) The breaking off of the top of a plant. (4) The creamy mixture used to decorate ice cream, cakes, molds, specials, etc.

Topsoil–Surface soils and subsurface soils which presumably are fertile soils, rich in organic matter or humus debris. Topsoil is found in the uppermost soil layer called the A horizon. See Soil Horizon, Soil Profile.

Topworking–Grafting one or more varieties to the branches of a tree.

Torula–(1) A group of wild yeasts which are of considerable importance in the fermentation of fruit juices. (2) Lactose fermenting yeasts which produce undesirable changes in dairy products.

Torus–The receptacle of a raspberry or like fruit that stays on the plant when the ripe fruit is picked.

Totipotency–The ability of each cell in a plant to generate or regenerate a whole plant that is an exact copy of the parent.

Tough–(1) Having the quality of flexibility without brittleness; yielding to force without breaking. (2) Designating market grain which contains moisture in excess of 13.5 to 14.5 percent. (3) Designating plant or animal products that are not brittle, crisp, or tender.

Tow–(1) The short fibers of flax and hemp after they have been combed out of the stalk. (2) The parts of flax, jute, and hemp when made ready for spinning. (3) A strong rope.

Tower Silo–A cylindrical tower made of wood, concrete, tile, metal, etc., used for storage of silage. It is the most common type of silo in the United States. Also called upright silo.

Toxaphene–A chlorinated hydrocarbon insecticide.

Toxic–Poisonous; caused by poison.

Toxicant–A substance that injures or kills an organism by physical, chemical, or biological action; e.g., heavy metals, pesticides, and cyanides.

Toxicarol–A rotenoid obtained from the roots of *Tephrosia toxicaria* (fishdeath tephrosia); a poisonous compound

Toxicity–State or degree of being poisonous.

Toxicology–The science which deals with poisons, antidotes, toxins, effects of poisons, and the recognition of poisons.

Toxin–A protein poison produced by some higher plants, certain animals, and pathogenic bacteria. Toxins are differentiated from simple chemical poisons and vegetable alkaloids by their higher molecular weight and antigenicity. See Phytotoxin, Poison.

Toxoid–A toxin which has been chemically altered so that it is no longer toxic but is still capable of uniting with antitoxins and/or stimulating antitoxin formation.

Toy–Designating any small or dwarf variety, as a toy spaniel dog.

Trace–(1) Either of the two leather straps or chains, attached at one end to the hames clamped to the horse collar or breast band, and at the other to tugs on a whippletree. (2) The force to pull the vehicle, etc., as applied to traces. (3) See Trace Element. (4) To follow the course of nutrient elements in plants or animals, as by use of radioisotopes. (5) Designating an artificial flavoring material with little or no true flavoring.

Trace Element–Any of certain chemical elements necessary in minute quantities for optimum growth and development of plants and animals.

Trace Mineral–See Trace Element.

Trace of Precipitation–Precipitation in amounts too small to be measured in a gauge. Less than 0.005 inch for rain and less than 0.05 inch for snow.

Tracer–A radioisotope mixed with a stable substance, by means of which a material can be traced as it undergoes physical and chemical changes. In agricultural research, radioactive isotopes of phosphorus in a chemical fertilizer can be traced through the plant as it is absorbed by the roots.

Tracheid–Any of the water-conducting, tubelike cells in the xylem of a plant. They also act as a support for the plant.

Tract–(1) An area of land of any size, but bigger than a lot. (2) An anatomical structure as the digestive tract, cerebellospinal tract.

Traffic Pan–A compacted soil layer resulting from wheel traffic compressing the soil when wet.

Trafficability–The capability of a terrain to bear traffic. It refers to the extent to which the terrain will permit continued movement of any and/or all types of traffic. Soil type has much to do with the trafficability of the terrain; e.g., vertisols have zero trafficability when saturated, and oxisols can bear some traffic when wet.

Tragacanth–Gum obtained from a shrub; used as an emulsifying agent or as a thickener.

Trailing–(1) The driving of livestock from place to place. (2) The voluntary wandering of livestock about a range, usually in search of forage, water, or salt. (3) Designating a plant which puts forth long, recumbent stems.

Trailing Sumac–See Poison Ivy.

Tramp–(1) To compact soil usually around a transplant by treading on it or by pounding it with a tool. See Tamp. (2) To compact silage in the silo by walking on the material.

Trampling–(1) Treading under feet; the damage to plants or soil brought about by congested movements of livestock, including mechanical injury to tree reproduction and ground cover in woods. (2) Compacting soil in earthen dams and reservoirs by livestock to make the dam or reservoir impervious to water (now replaced by machine compaction).

Transect–A cross section of vegetation; a long, narrow sample area, or a lie used for analyzing vegetation; essentially a cross section of the vegetation.

Transitory Range–Land that is suitable for temporary use for grazing; e.g., on disturbed lands, grass may cover the area for a period of time before being replaced by trees or shrubs.

Translocate–The transfer of the products of metabolism, etc., from one part of a plant to another.

Translocated Herbicide–A herbicide which, when applied to one portion of a plant, travels to another part of the plant.

Translocation–In genetics, change in position of a segment of a chromosome to another part of the same chromosome or to a different chromosome.

Translucent–A term used to describe the light-transmitting property of a material used in making a greenhouse.

Transpiration–The process by which water vapor is released to the atmosphere by the leaves or other parts of a living plant.

Transpiration Pull–A tension within a plant which is generated by transpiration and exerts a pulling force. Transpiration pull is a major factor in the rise of water in plants.

Transpiration Ratio–The ratio of the weight of water transpired by a plant to the dry weight of the plant.

Transplant–(1) A seedling which has been moved one or more times. (2) To transfer seedlings from the seedbed and set them in the ground.

Transported Soil—A soil formed by the consequent or subsequent weathering of materials transported and deposited by some agency, such as water, wind, ice, or gravity.

Trash—(1) Leaves, stalks, husks, etc., left on the ground after harvest. (2) Anything worthless or useless, as trash in seed cotton.

Trash Farming—Stubble mulch farming. See Conservation Tillage, No Till.

Traumatic—Describing a condition resulting from an injury or wound.

Tray—(1) A wide, flat-bottomed, topless, shallow container used for picking, carrying, handling, drying, or storing produce. (2) A short piece of heavy wrapping paper on which seedless grapes are sun dried.

Tray Agriculture—See Hydroponics.

Tray Gardening—See Hydroponics.

Treacle—A heavy dark syrup which is a by-product from the sugar-refining process.

Treat—(1) To care for a sick animal or diseased plant by giving it proper attention and medication. (2) To subject a product to an action or process to improve it in some manner. (3) To subject plants, animals, or soil to various chemicals, practices, etc., in order to learn which are beneficial or harmful.

Treated Area—A place where a pesticide has been applied.

Treble Superphosphate—Double superphosphate.

Tree—Any woody, perennial plant which normally has one well-defined stem and a definitely formed crown. It is usually considered to have a minimum mature height of 15 feet. See Shrub.

Tree Baler—A device used to put a netting around a Christmas tree for shipping.

Tree Caliper—(1) A caliperlike device used to measure diameters of tree trunks and logs. (2) The diameter of a tree trunk; in forestry measured at 4.5 feet above the ground; in horticulture, measured at one-foot height.

Tree Dressing—Paint or paste used to cover and protect wounds of a tree caused by limb breakage or pruning.

Tree Farm—A tract of land on which trees, especially forest species, are grown as a managed crop.

Tree Injection—The introduction of a chemical into the sap stream of a tree (a) to kill it, (b) to prevent disease, (c) to prevent chlorosis, (d) to color the wood prior to sawing, or (e) to destroy insects.

Tree Line—See Timber Line.

Tree Lupine—*Lupinus arboreus*, family Leguminosae; an herbaceous, perennial, shrubby herb grown in the flower border for its fragrant, sulfur-yellow flowers. It is also useful as a sand binder. Native to California, United States. Also called yellow lupine. See Lupine.

Tree of Heaven—*Ailanthus altissima*, family Simaroubaceae; a medium-sized, deciduous tree grown as a street tree for its ability to withstand smoke and city conditions. Native to China. Also called false varnish tree.

Tree Paint—Any one of several types of wound dressings for trees, such as orange shellac, asphalt paints, creosote paints, grafting waxes, house paint, Bordeaux paste, commercial tree paints, etc. Ideally, it should disinfect, prevent fungus entrance, prevent checking, permit callus growth, be toxic to insects, be easily applied, allow excess moisture to evaporate, and should not crack on drying, or injure the tissues.

Tree Planter—A tractor-drawn implement that is used for planting tree seedlings. An operator rides on the planter and places a seedling in the furrow opened up by the implement.

Tree Planting—To grow trees, shrubs, vines, or cuttings to prevent excessive runoff and soil loss. It includes plantings for gully control, post lots, sand dune fixation, bank protection, reinforcement of existing woodlands, windbreaks, etc. Another purpose of tree planting is to obtain wood products or trees to sell for ornament, as Christmas trees, etc.

Tree Planting Bar—A heavy wedge shaped bar of steel attached to a handle that is used for opening the ground to plant tree seedlings.

Tree Processor—A machine that trims and delimbs trees that have been harvested. The trees are fed through the machine and emerge with the limbs removed.

Tree Protector—Any of the several kinds of material for wrapping loosely around the trunk of a tree to protect it from sunburn or injury by rodents.

Tree Ripe—Designating a fruit which has ripened on a tree, as contrasted to one which has been picked and allowed to ripen after harvesting.

Tree Shaker—(1) A machine that grasps a nut tree and shakes the tree causing the nuts to fall to the ground where they can be picked up by a nut sweeper. See Nut Sweeper. (2) A machine that shakes harvested Christmas trees to remove dead needles, debris, etc.

Tree Size Classes—A classification of growing stock trees according to diameter at breast height outside bark, including sawtimber trees, poletimber trees, saplings, and seedlings.

Tree Slash—The debris remaining after logs have been removed during the harvesting of a forest.

Tree Spade—A large hydraulically operated machine that scoops into the ground around a tree and removes the tree from the ground for transplanting.

Tree Species—(1) Commercial species. Tree species currently or prospectively suitable for industrial wood products; excludes so-called weed species such as blackjack oak and blue beech. (2) Hardwoods or dicotyledonous trees, usually broadleaved and deciduous. (3) Softwoods or coniferous trees, usually evergreen, having needles or scale-like leaves.

Tree Surgery—The after-care of trees which includes pruning, repair of injury, spraying, and fertilizing when necessary.

Treetomato—*Cyphomandra betacea*, family Solanaceae; a tender shrub which bears tomatolike, acid fruit, grown as an ornamental in the United States. Native to South America.

Tref–Designating a food which does not meet the dietary laws for Orthodox Jews. See Kosher.

Trefoil, Birdsfoot–See Birdsfoot Trefoil.

Trellis–A latticework frame or wires supported by posts on which vines or flowers are trained. It may be constructed as a bower, arbor, etc.

Trench Layering–A layering method of plant propagation in which a number of new plants may be obtained from a given stock plant of a certain species by burying a part of a live branch.

Trench Silo–A trench excavated in a hillside or on firm ground, usually lined with wood or concrete retaining walls. Commonly about 15 to 25 feet wide, 6 to 8 feet deep and as long as the capacity desired. A trench silo must have good drainage and its use is largely limited to arid or semiarid climates.

Trenching–Deep digging of garden soil and mixing in compost, manure, or some other conditioner.

Tri–A prefix meaning three, as *trilocular*, having three locules.

Tribasic Copper Sulfate–A copper salt used as a fungicide.

Tribasic Phosphate of Soda–Trisodium phosphate, a sodium salt which in a 2.5 percent solution has an appreciable disinfecting value.

Tribe–(1) In botany, a subdivision of a family, roughly equivalent to subfamily. (2) In animal breeding, a group or combination of animals descended through the female line.

Tribulosis–A disease of sheep, also called big head, that results from their feeding on puncervine (*Tribulus terrestris*).

Tricalcium Phosphate–A salt of phosphoric acid, present in phosphate rock and bones. It is the source of almost all phosphorus-containing substances. Reaction of tricalcium phosphate and sulfuric acid yield the widely used fertilizer marketed as superphosphate. Also called bone phosphate of lime, BPL, calcium phosphate, phosphate of lime.

Trichiasis–A turning inward of the eyelashes.

Trichina–See Trichinosis.

Trichinosis–A parasitic disease of animals and people caused by the nematode *Trichinella spiralis*. In severe cases the disease is characterized by diarrhea, anorexia, fever, muscular pains, stiffness, difficulty in breathing and chewing, and loss of weight. Infection in animals occurs by ingestion of tissue (namely muscle) of animals previously infected. In people the disease is contracted by eating raw or improperly cooked pork from pigs harboring the infection. Detection of the disease in swine is not always possible at the time of slaughter, therefore it is assumed that all pork may be infected with trichinae (trichinosis) and should not be consumed until adequately cooked. Also called trichiniasis.

Trichobezoar–See Hair Ball.

Trichomoniasis–An infection with trichomonads. Common types in domestic animals are bovine genital trichomoniasis and trichomoniasis of the upper and lower digestive tracts of poultry.

Trichophytobezoar–Hair and food which an animal has ingested which have gathered in the rumen or the stomach in the form of balls. See Hair Ball.

Trichotomous–Forking regularly and repeatedly into three branches or divisions.

Trickle Irrigation–The application of small quantities of water directly to the root zone through various types of delivery systems on a daily basis. Also known as drip irrigation.

Trifoliate–Having three leaves, as *Trifolium* spp., clovers.

Trigeneric Hybrid–A hybrid which results from crossing plants of three different genera.

Trigo–(Spanish) Wheat.

Trihybrid–An individual that is heterozygous for three pairs of genes.

Trim–To remove unwanted or undesirable portions of a product, plant, etc., as to trim fat from a ham or to trim a tree or hedge.

Trio–(1) A group consisting of three objects or organisms. (2) A male and two female birds of the same variety which are shown as a unit in exhibitions.

Triploidy–A condition where the cells possess three sets of homologous chromosomes rather than two.

Trisomic–Referring to an organism having three chromosomes of one type (chromosome formula 2n/1).

Tristeza–A viral disease of citrus characterized by failure and dying of sweet oranges, tangerines, and grapefruit grown on certain root stock, such as on sour orange stock. It is caused by destruction of food-conducting vessels in bark at the bud union, resulting in starvation and death of roots.

Triticale–A hybrid grain resulting from crossing wheat and rye. The name is derived from the two genera names *Tritticum* and *Secale*. Triticale combines the high yield and protein of wheat with the winter hardiness of rye.

Trockentorf–A relatively undecomposed, peatlike deposit occurring on the surface of well-drained soils under forest cover. It is composed of the remains of leaves and fragments of wood.

Trophic–Of or pertaining to nutrition.

Tropical Zone–(1) The region of the earth bisected by the equator and extending at low elevations to latitude 23° 27" north (Tropic of Cancer) and south (Tropic of Capricorn). (2) Designating a plant, disease, etc., which flourishes in the tropics or where conditions are made to resemble the tropics in temperature, humidity, length of day, etc.

Tropism–A growth reaction of a plant to various external or internal stimuli such as phototropism, the increased growth toward or away from light; geotropism, growth in response to gravity; chemotropism, plant response to chemicals; hydrotropism, plant response to water.

Truck Crop–A vegetable crop usually raised on a relatively large acreage and under intensive methods of farming. See Market Garden.

True–Like the parental type, without change, as a variety which breeds true.

True Leaf—An ordinary leaf, which functions in the production of food by a plant.

True Mountain Mahogany—*Cercocarpus montanus*, family Rosaceae; a tender evergreen shrub grown as an ornamental for its 3 to five inch long, feathery-tailed fruit. It is found on dry ridges at altitudes of 4,000 to 10,000 feet from northern Montana to New Mexico, United States, and furnishes a large amount of high-quality browse.

True-breeding—Designating varieties which conform to the parental type with respect to certain characteristics, such as color, disease resistance, etc.

Truffle—The subterranean fruit of a fungus of the family Tuberaceae, which is highly prized as a food or condiment. The truffles are harvested by the aid of trained pigs or dogs that detect them in the soil by their odor and dig them up. This manner of harvesting is employed in the countries of southern Europe and northern Africa bordering on the Mediterranean Sea.

Truncate—Appearing as if cut off nearly or quite straight across at the end, as the leaf of *Liriodendron*.

Truncated Soil Profile—A soil profile from which one or more of the upper horizons normally present have been removed by accelerated erosion or by other means. The profile may have lost part or all of the A and sometimes the B horizon, leaving as soil only the poor, undeveloped parent material or C horizon. A comparison of eroded soil profiles of the same area, soil series, and slope conditions, indicates the degree of truncation.

Trunk—The main unbranched body, stalk, or stem of a vine or tree.

Tub Gardening—A method of growing certain flowers, fruits, and vegetables in small quantities in a wooden barrel in which 4-inch holes no closer than 12 inches apart are bored in the side. The barrel is filled with rich soil and plants are set in the holes. Strawberries are sometimes so grown.

Tube Well—A small well for obtaining water from shallow strata which consists of a pointed pipe driven into the ground without boring. See Driven Well.

Tuber—Thickened or swollen underground branch or stolon with numerous buds (eyes). Thickening occurs because of the accumulation of reserved food; e.g., Irish potato, Jerusalem artichoke. See Tuberous Roots.

Tuber Indexing—Growing a seed piece of a potato under conditions suitable for virus determination. According to the reaction, the remainder of the tubers produced by that plant are either condemned as unfit to plant or certified as disease-free. Also called hill indexing.

Tuber-unit Method—A system used in the production of certified seed potatoes in which seed pieces from each potato are planted consecutively in the row. The hills are marked off in units of 4 hills. If a diseased hill is found, the entire unit of four is destroyed.

Tubercle—(1) A small tuber. (2) The small nodules ascribed to the action of symbiotic organisms especially on the roots of legumes. (3) A nodule or small eminence in or on soft tissues. (4) A rough, rounded eminence on a bone.

Tuberous—Designating any plant which bears tubers.

Tuberous Roots—Thickened roots, differing from stem tubers in that they lack nodes and internodes, and buds are present only at the crown or stem end; e.g., sweet potato. See Tuber.

Tucked Up—A term used to describe an animal whose belly is under the loin. Also refers to a small-waisted animal.

Tuff—Fine volcanic detritus in various states of stratification and consolidation. *Tufa* applies to similar rocks, but more especially to a kind of porous rock formed as a deposit from springs or streams; usually applied to calcareous deposits as traverteine and calcareous tufa. See Detritus.

Tuft—A small cluster, as of grass, with the roots intertwined.

Tufted Hairgrass—*Deschampsia caespitosa*, family Gramineae; a perennial bunchgrass found from Greenland to Alaska, south to New Jersey and east to California, United States, in bogs and wet places. Its forage rating is good and it is sometimes cut for hay.

Tulip Tree—*Liriodendron tulipifera*, family Magnoliaceae; a deciduous, magnificent, broadly pyramidal tree, up to 150 feet tall, grown for its size, shape, and for its lovely, tuliplike flowers. Its wood has a number of commercial uses including veneer, cabinet making, boxes, etc. Native to North America. Also called yellow poplar (forestry name), white wood, tulip poplar.

Tumble Windmillgrass—*Chloris verticillata*, family Gramineae; a perennial weedy grass, the inflorescence of which breaks away at maturity and rolls as a tumbleweed; however it is sometimes grazed by livestock early in the season. Also called windmill grass.

Tumbleweed—Any weed that breaks away from the ground, usually at or near the surface, at the end of the growing season and because of its form is often blown considerable distances, scattering seeds en route.

Tumbleweed Amaranth—*Amaranthus graecizans*, family Amaranthaceae; a coarse annual weed which is a fairly common weed pest. Also called tumble pigweed.

Tumeric—*Curcuma longa*, family Zingiberaceae; a tropical, aromatic, perennial herb grown for its rootstock, source of commercial tumeric, used as a dye, medicine, and mostly as a condiment, chiefly in India. Native to India.

Tumor—A swelling; a new growth of cells or tissues governed by factors independent of the laws of growth of the host. It may be either benign or malignant.

Tung Oil Tree—*Aleurites fordi*, family Euphorbiaceae; a small tropical tree grown for its nutlike fruit, the seeds of which yield the tung oil used in dyes, varnishes, waterproofing agents, etc. Native to central Asia. Also called China wood oil tree, tung.

Tunnel—(1) A gallery. (2) To burrow, make a passageway, in plant stems, leaves, roots, etc., as by larvae of certain insects.

Turf—(1) A close-growing, well-knit, usually fine-leaved growth of a grass, mixture of grasses, or other plant species, which is best maintained by mowing, fertilizing, and watering so as to present a pleasing appearance. It is useful for lawns, golf courses, horse-racing tracks, athletic fields, etc. (2) A slab of peat used for fuel (British). (3) The surface layer of a peat soil. (4) To produce a turf by the seeding, sodding, or other vegetative propagation of grass or other plant species.

Turf Farm–A farm that produces turf grass for transplanting in lawns, golf courses, etc. The turf is harvested and sold in blocks or plugs.

Turgid–Swollen, or tightly drawn, said of a membrane or covering expanded by pressure from within; e.g., growing plants have turgid cells.

Turgidity–See Turgor.

Turgor–The distension of the cell wall and protoplasmic layer of plants by fluids. It is essential to growth.

Turkey Gnat–*Simulium meridionale*, family Simuliidae; a small, blackish fly that congregates about the eyes, ears, and nostrils of chickens and turkeys, sucking blood from the host. The bite may result in symptoms similar to mastoiditis. It may be a vector of the organism causing onchocerciosis in humans.

Turkeymullein–*Eremocarpus setigerus*, family Euphorbiaceae; an annual weedy plant which yields blackish seeds that are nutritionally fattening for the turkeys. Sheep have been killed from grazing on this plant, for the hairs covering the stems and leaves are indigestible. Native to western United States. Also called dove-weed.

Turn–(1) To change the position of an egg in an incubator or in the nest where the hen is incubating it. (2) To restrain an animal, as a fence turns a cow. (3) To plow, so that the lower part of the soil is brought to the surface and the former surface part is covered. (4) To begin to show signs of ripening, as a fruit. (5) To direct an animal to pasture. (6) To sour, as milk. (7) To change, as leaves in the autumn.

Turning Under–The covering or burying of the surface soil, trash, or a green manure crop, by plowing with a turning plow.

Turnip–*Brassica rapa*, family Cruciferae; a biennial herb grown as an annual for its root and its leaves. Widely used as a vegetable. The leaves of the turnip are used as a potherb and also as a livestock forage. Native to Europe, Asia, and Africa.

Turpentine–(1) Oleoresin. (2) The essential oil derived from the distillation of oleoresin obtained from longleaf and slash pines in the southeastern United States. (3) To tap a tree to obtain oleoresin.

Turpentine Orchard–A stand of coniferous trees from which oleoresin is taken.

Turtle Backing–Preparation of a wide bed, in the form of a low rounded ridge, bordered on each side by furrows; a tillage method used in the planting of sugarcane on flat, wet lands. Known also as drainage-by-beds.

Tussock–A dense, heavy tuft or matted growth of grass or sedge which forms a small hillock.

Twig–A shoot of a woody plant representing the growth of the current season.

Two,Four,Five-T (2,4,5-T)–A herbicide used effectively to kill woody vegetation. Its use is restricted.

Two,Four-D (2,4-D)–A once widely used herbicide that controls most broad-leaf plants. Its use is now restricted or prohibited because of the damage that can occur to plants that are not intended to be killed.

Two-grooved Loco–*Astragalus bisulcataus*, family Leguminosae; a poisonous plant found from Manitoba, Canada, to New Mexico, United States.

Two-story Tree–(1) A pruned fruit tree which develops a water-spout limb strong enough to become a leader and to develop another or second top. Also called sucker top. (2) A fruit tree which has two distinct levels of framework branches.

Typhula Blight–A fungal disease of winter cereals and grasses caused by *Typhula itoana* or *T. idahoensis*, following periods of heavy snow or cloudy, cold weather. It is characterized by a felty, white mycelial mat over the plants and soil. The affected plants fade, wither, and turn brown. The disease disappears as the weather turns warm.

Typical–In appraising, that which most frequently occurs or exists in the particular situation under consideration.

Ubiquitous–Occurring everywhere, as house flies; house sparrows; weeds.

Ultisols–Soils that are low in supply of bases and have subsurface horizons of illuvial clay accumulation. They are usually moist, but during the warm season some are dry part of the time. The balance between liberation of bases by weathering and removal by leaching is normally such that a permanent agriculture is impossible without fertilizers, lime, or shifting cultivation. See Soil Order.

Ultra-low Volume (ULV)–The spraying by air or ground of a pesticide undiluted and in a very concentrated liquid form. The usual dosage is from 2 to 16 fluid ounces per acre but is always less than one-half gallon per acre.

Umbra–(Latin) Shade.

Umbric Epipedon–A thick, dark surface soil horizon that is too acid and/or has too wide carbon to nitrogen ratio to be a mollic epipedon. See Mollic Epipedon.

Unarmed–Designating a plant that is thornless.

Unbroken–(1) In marketing, designating an egg free from cracks or breaks in the shell. (2) Designating an untrained horse. (3) Designating soil that has not been plowed. See Virgin Soil.

Unconfined Groundwater–Groundwater under atmospheric pressure wherein the watertable rises or falls as the volume of stored water changes.

Uncontrolled Burning–See Prescribed Burning, Wildfire.

Under Glass–In a greenhouse.

Under Irrigation–Supplied with water for crop production entirely or in part by means of irrigation. See Subirrigation.

Underbrush–In humid regions, woody plants, usually short, growing in a forest, underneath tree species. In arid and semiarid regions, woody plants, grasses, and forbs.

Undercut–(1) A notch cut in the trunk of a standing tree below the level of the major cut and on the side to which the tree is to fall. It determines the direction of falling. (2) A saw cut made on the underside of a large branch beyond the point of severance, prior to making the actual primary cut, to prevent splitting or tearing. (3) the harvesting of less timber from a stand than that budged. (4) The tenderloin muscle of beef (British).

Underdrain–Drain tiles or drain tubes placed in trenches and covered with soil to such a depth that the surface soil can be cultivated and the plant root profile adequately drained.

Underground Runoff–Water that flows toward stream channels after infiltration into the ground. See Subsurface Flow, Underflow.

Underground Stem–A true subterranean stem that occurs in the form of a rhizome (as in Johnsongrass) or tuber (as in Irish potato).

Underground Water–See Aquifer.

Undergrowth–See Underbrush.

Undershrub–Any low-growing shrub or small tree that will grow when planted beneath a tree or in front of higher shrubs; e.g., dogwood.

Understock–Plants raised from seeds or cuttings and used as a root system for grafted or budded plants of different species.

Understocked Stand–A stand of trees in which the growing space is not effectively occupied by crop trees.

Understory–(1) Grasses and small shrubs growing beneath a tree canopy. (2) Plants growing in the shade of other plants.

Undulate–A leaf margin that forms an irregular, wavy line.

Uni-–A combining form meaning one.

Unicellular–One-celled; refers to an organism the entire body of which consists of a single cell.

Unified Soil Classification System–(Engineering) A classification system based on the identification of soils according to their particle size, gradation, plasticity index, and liquid limit. Indicated for each soil-mapping unit in modern soil survey reports.

Unifoliate–Designating a plant with single leaflets.

Union–(1) The proper, healthy uniting of a stock and scion in a graft. (2) The place on the stem of such a uniting.

Unisexual–Having stamens and pistils in separate flowers.

Unit–(1) A single thing or item of produce. (2) A recognized measure of weight, volume, or distance, such as a bushel, gallon, or mile.

Unit of Plant Nutrient–One percent of a ton, or 20 pounds, of a fertilizer.

Univalent–Designating a chromosome unpaired at meiosis.

Universal Soil Loss Equation–An equation used for the design of water erosion control systems: A = RKLSPC wherein A = average annual soil loss in tons per acre per year; R = rainfall factor; K = soil erodibility factor; L = length of slope; S = percent of slope; P = conservation practice factor; and C = cropping and management factor. (T = soil loss tolerance value that has been assigned each soil series expressed in tons/acre/year.) See T Factor.

Unleached Wood Ashes–Wood ashes that have had no part of their plant nutrients removed and that contain 4 percent or more of water soluble potash. (K_2O).

Unpolished Rice–The whole rice grain that retains its outer natural color.

Unsaturated (Polyunsaturated) Fats–Fats of vegetable origin such as olive oil or cottonseed oil that have more than one double bond in their carbon chain.

Unslaked Lime (CaO)–See Burnt Lime, Calcium Oxide.

Unsweetened Pack–Fruit packed for freezing without any sweetening added. It may be packed dry or covered with water.

Upland–(1) Highland, an elevated plain in association with or in contrast to a valley plain or lowland. (2) Designating crops or crop varieties grown on upland in contrast to lowland areas. See Upland Cotton.

Upland Cotton–*Gossypium hirsutum*, family Malvaceae; a tropical, woody herb grown as an annual for its fiber, the source of much of the commercial cotton in the United States. It is also important for the edible oil derived from the seed, and for the stockfeed residue from the seed after extraction of the oil. It is one of the most important cash crops of the United States. Native to America. Also called American upland cotton.

Upright (Vertical) Silo–A vertical structure used for preserving chopped, green feed silage. Usually it is made of wooden staves, concrete or tile blocks, or metal sheets held together by bands or reinforcements. Diameter and height vary with the capacity desired.

Urea–(1) $CO(NH_2)_2$; a nonprotein, organic compound of nitrogen, made synthetically by a combination of ammonia and carbon dioxide, and used in fertilizers and as a livestock feed supplement. (2) The chief compound of nitrogen in the urine of mammals.

Urea Phosphate–$CO(NH_2)_2H_3PO_4$; a crystalline compound formed in processing mixtures containing urea and phosphoric acid or other phosphatic materials.

Urea-Ammonia Liquors–Any of several solutions of urea in crude aqueous ammonia, used in ammoniating superphosphate fertilizer.

Urea-Ammonium Phosphate–Intimate mixtures containing ammonium orthophosphates and urea. The reaction product of commercial phosphoric acid and ammonia is mixed with molten urea. Heat of reaction dries the granular product. Grades produced include 28-28-0, 36-18-0, and a variety of other grades with total plant nutrient contents up to 60 percent. Potassium can be included in these products. Modified treatment produces urea-ammonium polyphosphate for use as a solid or in liquid fertilizers.

Urea-Ammonium Sulfate–Granulated homogeneous mixtures of urea and ammonium sulfate containing 30 to 40 percent N and 4 to 13 percent S. The products have greater granule strength than urea alone.

Uric Acid–A white, crystalline compound, derived from guano. It contains about 33 percent available nitrogen.

USDA (U.S.D.A.)–United States Department of Agriculture.

Ustoll–A suborder name of a group of soils in semiarid and subhumid climates with complex profiles suggesting several climatic cycles of humidity and aridity.

Vacuolated Cytoplasm–The living substance of cells, exclusive of the nucleus, when filled with or containing bubblelike structures.

Vadose–Water held in soil, or other surficial geological formations, above the level of permanent groundwater.

Vagabond Bee–Any worker bee that lacks an individual foraging area.

Valence–Also called bond or chemical bond; the chemical combining power of an atom. It indicates the number of electrons that can be lost, gained, or shared by an atom in a compound.

Valvate–Opening by valves, as in a capsule or some leaf buds; meeting at the edges without overlapping.

Vanilla–(1) Any one of the several plants of the genus *Vanilla*, family Orchidaceae. (2) The extract derived from the vanilla pods, which is extensively used in confections.

Vanilla Bean–The long, beanlike capsule of the vanilla.

Vanilla Extract–The flavoring extract prepared from vanilla beans, with or without sugar or glycerine.

Vanillin–A sparkling, white, fuzzy crystalline substance that has a characteristic vanillalike odor.

Vapor Drift–The movement of pesticidal vapors from the target area of application.

Var.–The abbreviation for the word variety; as in *Sarcoptes scabiei* var. *bovis*. See Cultivar.

Variable Grade–A terrace channel having a variable slope that decreases as the distance from the outlet increases; e.g., a terrace 1,200 feet long may be built in four 300-feet sections each having different grades, the outlet end having 0.40 feet fall per 100 feet; the next section, 0.30; the third, 0.20, and the fourth (upper) section, 0.10 foot per 100 feet.

Variable Oak Leaf Caterpillar–*Heterocampa manteo*, family Notodontidae; a caterpillar that voraciously feeds on the foliage of the oak, basswood (linden), walnut, birch, elm, hawthorn, and persimmon.

Variance–A statistical measure of the amount of variation that is observed within or among a group of animals or plants.

Variant–A recognized entity different from normal.

Variation–(1) The angle by which the north end of the compass needle (magnetic north) deviates from true north. (2) One of the laws of organic nature; organisms vary in time, from place to place, and also in one locality with time; they vary also in their appearance (morphology).

Variegated–(1) Designating leaves of plants that have a pattern caused by the alternation of scattered areas or stripes on leaf margins of different colors. (2) Designating a flower each petal of which is streaked or striped with a color or colors different from the basic color.

Variegated Cutworm–*Peridroma margaritosa*, family Noctuidae; the larva of a moth, probably the most widely distributed and destructive pests of garden crops, fruit trees, and vines. It eats the foliage, buds, and fruits of many plants.

Varietal Name–The third word in the scientific name of a plant. It is sometimes the name of the person who first described the plant.

Variety–(1) A group of related plants or animals that differs from other similar groups by characters too trivial or inconstant to be recognized as a species; often any category of lower rank than a species. See Cultivar. (2) In domesticated animals, a subdivision of a breed based on some minor character such as color, etc.

Variety Hybrid–A cross between two varieties of the same species.

Variety of Chickens–A subdivision of the breed of chickens, determined by plumage color and the comb.

Variola–See Cow Pox.

Vascular Bundle–In plant anatomy, a unit containing both the phloem (water-conductnig cells) and xylem (food-conducting cells).

Vascular Plant–A plant that has special fibrovascular tissue through which water and the dissolved organic foods are conducted in the xylem and the phloem portions, respectively, of the fibrovascular bundles. These plants consist of the ferns, their allies, and the seed-producing plants, such as the gymnosperms and the anigiosperms.

Vascular Tissues–The fluid-conducting tissues of a plant including both xylem (water-conducting) and phloem (food-conducting) tissues.

Vaseygrass–*Paspalum urvillei*, family Gramineae; a grass grown for hay in the southeastern United States. It has a fair forage rating. Native to South America.

Vection–The passing of a disease from one plant or animal to another.

Vector—Any agent such as an insect or animal that transmits, carries, or spreads disease from one plant or animal to another.

Vega—(1) A fertile, grass-covered tract or extensive plain (southwestern United States). (2) Irrigated land from which a single crop per year is produced.

Vegetable—(1) The edible part of an herbaceous plant. (2) Any plant part that is eaten either cooked or raw during the principal part of a meal rather than as a dessert.

Vegetable Gardening—(1) The art and science of rowing vegetables. (2) The growing of vegetables at home.

Vegetable Mold—See Humus.

Vegetable Weevil—*Listroderes costirostris obliquus*, family Curculionidae; a beetle found in the Gulf States and California (United States) which, along with its larvae, feeds voraciously on beet, cabbage, carrot, cauliflower, lettuce, mustard, onion, radish, potato, turnip, tomato, Swiss chard, and spinach. It defoliates the whole plant with the exception of stems and leaf midribs.

Vegetarian—(1) One who, because of cultural reasons or personal conviction, abstains from eating meat (in the strictest sense, also milk, butter, and eggs). (2) An herbivorous animal or person.

Vegetated Channel—See Grassed Waterway.

Vegetation—Any group or association of plants; the sum of vegetable life; plants in general.

Vegetative Propagation—Increasing the number of plants by such methods as cuttings, grafting, or layering.

Vein—(1) One of the systems of branching tubes, etc., which carry blood back to the heart. (2) One of the fibrovascular bundles that forms the framework of a leaf. (3) In insects, the riblike tubes that strengthen the wings.

Velvet Bean—Any plant of the genus *Stizolobium*, family Leguminosae. Species are grown in warm areas for feed, soil building, green manure, pasturing, and sometimes for ornament. Also called banana bean.

Velvet Bentgrass—*Agrostis canina*, family Gramineae; a perennial herb used as a turfgrass. Native to North America. Also called brown bent, dog bent, dog's grass.

Velvetbean Caterpillar—*Anticarsia gemmatalis*, family Noctuidae; a caterpillar that feeds on soybean, velvet bean, cowpeas, peanuts, kudzu, and black locust.

Velvetpod Mimosa—*Mimosa dysocarpa*, family Leguminosae; a plant with bipinnate foliage, found on the semiarid ranges from Texas to Arizona and Mexico, whose forage rating is fair. Native to the United States and Mexico.

Venation—(1) The arrangement of the veins in a leaf. (2) The arrangement of veins in the wings of insects.

Veneer Grafting—A useful form of grafting, especially for ornamentals, in which a thin, wedge-shaped section, about 1 inch long, is cut from the stock. The lower end of the cut extends about one-third the diameter of the stock. The scion is then cut to match, and the union is tied and waxed.

Venus Flytrap—*Dionaea muscipula*, family Droseraceae; a perennial herb grown for its remarkable habit of catching and digesting insects.

Verdant—Designating fields or forests that are green with growing vegetation.

Verge—(1) In a city or town, the strip of grass lying between the street and the sidewalk. (2) The edge of a flowerbed.

Vermiculite—A mineral, or minerals, classified with the micas, which with treatment at high temperatures, expands into scales and becomes a loose, absorbent mass. It is recognized as a clay-mineral constituent of some clays, but its exact chemical composition is not known. Its chief use is in insulation. Commercial vermiculite is used as a mulch for seedbeds, as a medium for rooting plant cuttings, and in mixing with manure, peat, and soils or potting plants. See Perlite.

Vermiform—Worm-shaped.

Vermifuge—A drug or chemical that expels worms from animals; an anthelmintic. See Vermicide.

Vermin—Any noxious animal; insect, acarid, rodent, etc.

Verminous—Pertaining to or due to worms.

Vernal—Appearing in spring.

Vernalization—Seed treatment in which the germinating seed is held in artificial darkness, at low temperatures, for a fixed period to induce early flowering and fruiting when sown.

Vernis—Of spring.

Vertical Interval—In terrace farming, the vertical distance in feet from the center of one terrace line, ridge, or channel to the corresponding point on an adjacent terrace. The interval of the first terrace is the vertical distance from the top of the hill to the staked channel of the terrace.

Vertical T Budding—A bud graft in which a vertical cut about 1½ inches long is made through the bark of the stock and another cut is made at right angles to the first so as to form a cut similar to the letter T. The bud is placed inside the vertical cut and tied. Also called T budding.

Verticillium Wilt—Wilt produced on many plants by the fungus *Verticillium albo-atrum*, family Moniliaceae, which attacks plant roots and vascular bundles of the stem, discoloring and warping them and often killing the infected plant. Host plants are found mainly in the dicotyledons, in various families, frequently in plants with woody stems. Among the tree species attacked are avocado, ash, apricot, ailanthus, basswood, almond, peach, plum, maple, elm, catalpa, persimmon, black locust. Other plants that are susceptible are: cotton, raspberry, blackberry, sumac, barberry, elderberry, viburnum, rose, currant; and herbaceous plants such as tomato, potato, capsicum, eggplant, muskmelon, cucumber, watermelon, squash, beet, aster, chrysanthemum, peppermint, pumpkin, and a host of others. Also called blue stripe wilt.

Vertisols—Fine clayey soils with high shrink-swell potential that have wide, deep cracks when dry. Most of these soils have distinct wet and dry periods throughout the year. See Soil Orders.

Very Coarse Sand—See Soil Separates.

Very Fine Sand—See Soil Separates.

Vestigial—Imperfectly developed, said of a part or organ that was fully developed and functional in ancestral forms but is now a degenerate relic, usually smaller and less complex than its prototype.

Vetch—(1) Any of the various plants of the genus *Vicia*, family Leguminosae. Species are grown as cover crops, as green manure, and for pasture and forage. See Hairy Vetch. (2) *V. sativa*, an annual or biennial herb grown as a cover crop, as green manure, and for forage in the South and on the Pacific Coast, United States. Native to Europe. Also called fodder vetch, spring vetch, tare.

Viability—(1) Ability to live (immediately after birth or hatching). (2) The capacity of seeds to germinate. (3) The state of being alive. (4) Pertaining to sperm cells in the semen, capable of living and successfully fertilizing the female gamete.

Viable—(1) Living. (2) Specifically in regard to organisms or agents of disease, able to cause infection in animals. (3) Capable of living. (4) Capable of germinating, as seeds.

Vigor—The desirable state of health of any living thing.

Vine—(1) The grape. (2) Any woody or herbaceous plant that trails, climbs, or creeps as contrasted to those that stand without support. (3) A prostrate vegetable garden plant, as a sweet potato vine, tomato vine.

Vine Cuttings—Sections of a plant cut off to be planted for vegetative reproduction.

Vine Fruit—Any fruit grown on woody vines: grapes, muscadines, etc., as contrasted to tree fruits, bush fruits, bramble fruits.

Vine Mesquite—*Panicum obtusum*, family Gramineae; a perennial grass of the southwestern United States whose forage rating is fair. Native to North America.

Vinegar—A condiment made from various sugary and starchy materials with subsequent acetic fermentation, which must contain at least 4 percent acetic acid.

Vinegar Fly—Any fly of the family Drosophilidae, which breeds in fermenting fruits, or in crevices around a vinegar generator. It is a widespread pest of vinegar factories.

Vineless—Designating varieties of the sweet potato to which have short, upright branches.

Vineyard—A plantation, farm, or land on which grapes are grown.

Vintage—(1) The wine production for one year. (2) Designating wines from a particularly good year.

Viral—Having the nature of a virus; pertaining to a virus; like a virus.

Virgin—(1) Any female that has not had coitus, as a virgin queen bee. (2) In turpentining, the face of cut the first year a tree is bled. (3) Undisturbed, unplowed land, uncut forest.

Virgin Forest—A mature or overmature forest that has grown entirely uninfluenced by human activity.

Virgin Sod—Natural sod, especially that of the prairies of the midwestern and western United States, in contrast to that produced by grasses seeded by humans.

Virgin Soil—Soil in its natural state as distinguished from soil or land that has been plowed or otherwise altered by humans for cultivated crops or other uses.

Virginia Sun-cured—A type of tobacco, grown in north-central Virginia, which is a recognized type in manufacturing and for export.

Virginia Tephrosia—*Tephrosia virginiana*, family Leguminosae; a perennial herb grown for its showy, yellowish-purple flowers. Its root is a source of rotenone. Native to North America. Also called goat's-rue, catgut, wild sweet pea, rabbit's pea.

Virginia Wildrye—*Elymus virginicus*, family Gramineae; a tall, coarse perennial grass, which is abundant east of the 100th meridian in the United States, where it is used mainly to stabilize soil. It is also a fairly palatable forage plant. Native to North America. Also called terrel grass.

Viroid—A viruslike disease of potato, tomato, citrus, grape, and other plants.

Virology—The study of viruses and viral diseases.

Virucide—A chemical or physical agent that kills or inactivates viruses; a disinfectant.

Virulence—The disease-producing power of an organism.

Virulent—Highly pathogenic; having great disease-producing capacity; deadly; very poisonous or harmful.

Virus—(Plural, viruses) A self-reproducing agent that is considerably smaller than a bacterium and can multiply only within the living cells of a suitable host. Most viruses are too small to be seen even with the aid of the ordinary microscope, but can be photographed with the aid of the electron microscope. Viruses usually are considered to be living agents of microorganisms but some have characteristics of nonliving matter. They are protein-containing bodies of high molecular weight capable of multiplying and acting like living organisms when in living tissue. They are the cause of many animal, human, and plant diseases, such as smallpox, measles, tobacco mosaic, etc. Recovery from some viral diseases confers lasting immunity.

Virus Interaction—The action of a virus in altering the normal development of other viruses or virus strains that is expressed by partial or complete suppression, by synergistic association, by modification of the type of symptoms in the host plant, or by abnormal increase in concentration of one virus.

Virustatic—A substance that prevents the multiplication of a virus.

Viscosity—The resistance of a fluid to flow; thickness of a fluid.

Viticulture—The art and science of growing grapes.

Vitreous—Resembling glass in hardness or brittleness.

Vitriol—Sulfuric acid or some of its compounds. Blue vitriol is hydrous copper sulfate; green vitriol, copperas; red vitriol is either a sulfate of cobalt or a ferric sulfate; white vitriol is hydrated zinc sulfate.

Vivarium—A glass box resembling an aquarium used to keep or raise animals or plants. See Terrarium.

Vivianite—A hydrous ferrous phosphate, a mineral, and a number of oxidation forms of it, which is found in soils especially those in swamps and bogs. Also called blue ochre, blue iron earth.

Viviparous—(1) The bringing forth of living offspring from the body, as in mammals, as opposed to the laying and hatching of eggs. (2) Said of seeds that germinate or buds that sprout and form plantlets while still on the parent plant. See Oviparous, Ovoviviparous.

Void—(1) To evacuate feces and/or urine. (2) A general term for pore space or other openings in rocks such as vesicles and solution cavities. (3) Pore spaces in soils. (4) The space between kernels in a bulk of a grain that is usually expressed as percent of total volume.

Volcanic Ash—The finely comminuted ejecta of volcanic eruptions which, in thick deposits, are the parent material of soil. The volcanic ash carried by winds is widely distributed in soils.

Volcanic Soil—A soil that is derived from or consists of volcanic rocks, especially ash and other fragmentary ejecta.

Volunteer—Any plant that grows from self-sown seed.

Vulgaris—A species name of several plants, meaning common.

Wainable—Designating that which may be plowed or manured or is tillable.

Wall Tree—An espaliered tree; a tree, shrub, or vine pruned and trained to grow flat against a wall.

Walnut—Any tree of the genus *Juglans*, family Juglandaceae. Species are tall deciduous, valuable nut and timber trees, including butternut, black walnut, Persian walnut.

Walnut Caterpillar—*Datana integerrina*, family Notodonidae; a moth whose larvae feed in colonies on walnut, hickory, and pecans in the United States from Maine to Florida and west to Kansas.

Walnut Husk Fly—*Rhagoletis completa*, family Trypetidae (Tephritidae); a fly infesting walnuts which lays eggs on the husks. The larvae tunnel in the husk resulting in staining of the shell and reducing its value. Found in central and western United States.

Waney—(1) Designating lumber which has bark on a square-edged piece. (2) An imperfect board.

Warm Front—A mass of warm air advancing behind a mass of cool air. Characteristic weather may be several days of heavy overcast and drizzles.

Warm-season Plant—Designating a plant which thrives best when the temperature is regularly quite high; e.g., okra, cotton, grain sorghum. See Cool-season Plant.

Warp—(1) Yarn which runs the long way in a woven fabric. (2) In lumber, any variation of a board from a plane surface often caused by a too rapid loss of moisture in the curing process. (3) To vary from a plane surface, as twisted or buckled board. (4) Sediment deposited by water which acts as a fertilizer.

Wart—(1) A tumor on the skin or mucous membrane of an animal composed of fibrous tissue covered over with epithelial cells similar to those of the part of the body on which they are located. For the most part, they are painless and do not interfere with the function of an animal. (2) An unorganized proliferation of plant cells with an appearance like an animal wart. (3) One of the effects of cucumber mosaic.

Washington Lupine—*Lupinus polyphyllus*, family Leguminosae; a hardy, perennial, herbaceous, poisonous plant which with its many varieties is grown for its white, yellow, or blue flowers. Found wild on the West Coast, United States, it is especially toxic to sheep. Symptoms of poisoning are nervousness, labored breathing, convulsions, and frothing at the mouth. See Lupine.

Washout—(1) The erosion of a portion of a levee, railroad bed, road, etc., by water. (2) Designating a channel made by erosion in the deposit of a certain sediment and filled by fresh material. (3) The flushing out of a pipe, etc.

Wasp—Insects of the family Vespidae. Although sometimes considered a pest because of their painful stings, wasps are truly beneficial insects. They eat large numbers of other insects as adults and some species are parasitic as larvae on other insects.

Waste Lime—Any industrial waste or by-product containing calcium or calcium and magnesium in forms that will neutralize soil acids. It may be designated by a prefix suggesting the name of the industry or process by which it is produced, as gas-house lime, tanner's lime, etc. Also called by-product lime. See Lime.

Watch Tower—In forestry, a tower usually constructed on a hilltop or prominence for a fire warden or ranger to observe the surrounding area and keep a lookout for fires.

Water—H_2O; hydrogen oxide; although the liquid may contain associated molecules. The melting and freezing point is 32°F (0°C) and the boiling point 212°F (100°C). The most valuable natural resource and the most limiting factor in crop production.

Water Application Efficiency—The percentage of irrigation water applied that can be accounted for as moisture increase in the soil occupied by the rooting system of the crop.

Water Application Rate—The rate in inches per hour that irrigation water is applied to fields.

Water Breaker—A device attached on the end of a hose to reduce the velocity of the water.

Water Conservation–The physical control, protection, management, and use of water resources in such a way as to maintain crop, grazing, and forest lands; vegetal cover; wildlife; and wildlife habitat for maximum sustained benefits to people, agriculture, industry, commerce, and other segments of the national economy. See Soil Conservation Service.

Water Content–(1) The water of the soil or habitat; (physiological) the available water supply; (physical) the total amount of soil water. (2) The percentage of water in a material in relation to oven dry weight.

Water Culture–See Hydroponics.

Water Erosion–Erosion by water. See Gully Erosion, Rill Erosion, Sheet Erosion, Splash Erosion.

Water Extract–Whatever can be removed or dissolved out of a substance with water. A substance like sugar is completely soluble in water, whereas when yeast is shaken up with water only a small portion of it goes into solution.

Water Fallow–The maintenance of a rice field under a cover of water for one or two years during which a crop of fish may be produced.

Water Farm–(1) The cultivation of fish in farm ponds. (2) See Hyroponics.

Water Garden–A pool used in landscaping to grow aquatic plants.

Water Hickory–*Carya aquatica*, family Juglandaceae; a deciduous tree grown for its valuable wood used in tool handles, sporting goods, ladders, furniture, implements, woodenware, and for fuel and smoking meat. Native to North America. Also called bitter pecan, swamp hickory.

Water Hyacinth–*Eichhornia crassipes*; a large floating aquatic plant that causes clogging of water areas in the southern United States.

Water Level–(1) The surface of water which is used in measurement of surface or stream water depth. (2) The free water level in the soil. (3) See Water Table.

Water Mark–(1) Any mark, such as stain from suspended sediment, which indicates the level to which surface water has risen. (2) A water-soaked spot on citrus fruit resulting from a moderate freezing temperature.

Water Plant–See Aquatic.

Water Requirement–The quantity of water, regardless of its source, required by a crop in a given period of time for its normal growth under field conditions. It includes surface evaporation and other unavoidable wastes. Usually it is expressed as depth (volume per unit area) for a given time; e.g., acre inches/hour.

Water Retained–Field capacity of a soil. Any surplus water is lost as surface runoff or deep percolation.

Water Spreader–A terrace, dike, or other structure intended to distribute surface water runoff and increase the area of infiltration into the soil for plant use.

Water Spreading–(1) The artificial application of water to lands for the purpose of storing it in the ground for subsequent withdrawal by pumps for crops. (2) Irrigation by surplus waters out of cropping season. (3) The diversion of run-off water from gullies or watercourses and its distribution on adjacent, gently sloping, grazing lands needing additional water. The volume of water flowing down the channels is reduced and the moisture absorbed by the spreading area increases the growth of vegetation.

Water Sprout–Rapid growing succulent shoot which may appear on a trunk or limb of a fruit tree. It frequently appears after a tree has been heavily pruned.

Water Stress–See Wilt, Wilting Coefficient.

Water Suffocation–The killing of a plant by an excessive amount of soil water. See Waterlogged.

Water Table, Apparent–A thick zone of free water in the soil. An apparent water table is indicated by the level at which water stands in an uncased borehole after adequate time is allowed for percolation into the surrounding soil.

Water Table, Artesian–A water table under hydrostatic head, generally beneath an impermeable layer. When this layer is penetrated, the water level rises in an uncased borehole.

Water Table, Perched–A water table standing above an unsaturated zone. In places an upper, or perched, water table is separated from a lower one by a dry zone.

Water Table, True–The upper limit of the soil or underlying rock that is wholly saturated with water.

Water-ground–(1) Designating corn which has been soaked in warm water and passed through crushing mills so regulated that the hulls, endosperms, and embryos are mechanically separated without damage. (2) Designating corn meal that is ground at a water grist mill in contrast to that ground at a conventional mill.

Water-holding Capacity–The ability of a soil and crop system to hold water in the root zone.

Water-tolerant–Designating a plant which will continue to thrive in soil in which there is considerable moisture.

Watercress–*Nasturtium officinale*, family Cruciferae; a hardy, perennial, aquatic herb grown as a salad plant for its edible leaves. Native to Europe.

Waterhemlock–Any plant of the genus *Cicuta*, family Umbelliferae. Some species are among the most poisonous of plants.

Waterhole–(1) A natural depression in which water collects or stands, often in the dry bed of a stream; a spring in the desert. (2) A natural spring or pool on the open range where cattle drink (western United States).

Watering–(1) Furnishing water for the consumption of plants or animals. (2) Designating a device for furnishing water to plants or animals.

Waterlogged–(1) Water saturation of any object. (2) Water saturation of soil to the extent upland plants will not grow normally. The cause of waterlogging may be a naturally poorly drained soil, excess irrigation, too much precipitation, flooding by a stream or river, or seepage from a dam or irrigation canal.

Watermelon–*Citrullus vulgaris*, family Cucurbitaceae; a tropical, tender, annual, herbaceous vine of almost innumerable varieties, grown as an important cash crop for its very popular fruit, the watermelon of commerce. Native to South Africa. Also called anguria.

Wattle–(1) One of the two, thin, leaflike structures suspended from the upper part of the neck of a chicken; they are made of the same kind of tissue as the comb. (2) Fold of skin cut into the dewlap of cattle for identification purposes. (3) A common name for the 800 worldwide species of *Acacia*, family Leguminosae.

Wax–(1) Any substance similar to beeswax in composition and use, as carnauba wax, candelilla wax, bayberry wax, japan wax, sugarcane wax, flax wax, cotton wax, esparto wax, cauassu wax, murumuru sect wax, shellac wax, ceresin wax, montan wax, and paraffin. (2) To apply wax to the surface of a fruit, vegetable, foliage, or flower for preservation.

Wax Bean–A type of either the kidney bean, *Phaseolus vulgaris*, or the bush kidney bean, *P. vulgaris* var. *humilis*, family Leguminosae; important vegetable plants grown for their golden yellow seed pods harvested, before the seeds mature, for immediate consumption or processing. Also called butter bean. See Stringbean.

Wax Myrtle–(1) Any of several plants of the genus *Myrica*, family Myricaceae, which yield fragrant and relatively dripless wax for making candles. Also called tallow bayberry, tallow shrub, wax shrub.

Waxy Corn–An agricultural variety of corn having a waxy kernel and grown as a substitute for tapioca. See Cassava.

Waxy Flour–A flour prepared from certain varieties of rice or corn that contain a type of starch that has waxy adhesive qualities. The flour acts as a stabilizer when it is used as an ingredient in sauces or gravies and binds the mixture together, so there is no separation when the mixture is frozen.

Weak Soil Structure–Soil aggregates (clods or peds) that disrupt easily when it rains.

Weather–(1) The atmospheric conditions prevailing at any specified time and place, or those prevailing during any particular period, as shown by meteorological observations and records of air temperature, barometric pressure, wind velocity, humidity, clouds, and precipitation. (2) Bad weather. (3) To be subjected to the influences of atmospheric conditions, as the abrasive action of rainfall, the disintegration due to frost, the deteriorating influences of oxygen and other gases contained in the atmosphere.

Weathering–(1) The action of all natural forces on a product left exposed to the weather. (2) The deterioration of the unprotected surface layer of wood which has been exposed to the weather. (3) Atmospheric action on rock surfaces producing decomposition, disintegration, or alteration of rocks at or close to the earth's surface. Weathering implies no removal of material other than that in solution in drainage waters.

Webber's Brown Fungus–*Aegerita webberi*, a fungus used in biological control of the citrus whitefly, the cloudy-winged whitefly, the wooly whitefly, and the citrus blackfly.

Weed–(1) A plant out of place. Thus, rye growing in a field of wheat is a weed. (2) More popularly, an herbaceous plant which takes possession of fallow fields or finds its way into lawns or planted fields, crowding the vegetation planted there and robbing it of moisture and nutrients. (3) A plant whose usefulness is not recognized or which is undesirable because of odor, spines, prickles, or poisonous characteristics. (4) A tree of inferior value in a forest, or one growing in a street or lawn, where it is not wanted, like the seedlings of the boxelder. (5) Excessive vegetative growth. (6) Any plant that harbors insects, fungi, or viruses that may spread to nearby crop plants. (7) A horse or other domestic animal that is undersized or a misfit. (8) To remove weeds from a lawn, field, or other places where they are not wanted, usually manually but also by machinery or herbicides. (9) Designating a flavor of dairy products, especially milk, which resembles that of certain vegetation (onion, wild garlic, leeks, etc.).

Weed Control–Any device or system used for the destruction, checking of growth and spreading of weeds by cultivation, by the use of herbicides, cultural practices, etc.

Weed Killer–See Herbicide.

Weed Wiper–A device used to kill weeds that are taller than the crop. Liquid herbicides keep a nylon rope wet. The herbicide-saturated rope is pulled over the field above the level of the crop plants, wiping the taller weeds and killing them.

Weed-free–(1) Designating a standing crop, pasture, lawn, etc., in which there are relatively few weeds. (2) Designating crop seeds from which all weed seeds have been separated in the cleaning process.

Weedy–(1) Designating an off-flavor of milk or its products which suggests the odor or taste of certain weeds. (2) Designating an abundance of weeds in a field, flower bed, lawn, etc. (3) Designating growth of high vigor of a plant, usually associated with fruiting. (4) Designating a thin or undesirable animal.

Weep–(1) To turn syrupy, as a jelly sometimes does. (2) To exude sap, as a cut or broken grapevine. (3) To exude moisture from a plant surface. Also known as guttation.

Weephole–A hole through an abutment or retaining wall to provide for the passage of seepage water.

Weeping–Designating a tree with pendant (drooping) branches; e.g., weeping willow, weeping cherry.

Weevil–Any of the several beetles of the family Curculionidae, which has mouthparts at the end of a more-or-less prolonged snout. They eat plants and seeds.

Weir–(1) Stationary fish trap which acts as a barrier to fish movements and leads fish into pots. (2) Dam or barrier to raise level of water for different purposes.

Weir Head–The depth of water flowing over the weir crest as measured at a point at least 2.5 and preferably 4 times the depth of water flowing over the weir, upstream from the weir. The zero point of the measuring scale should be placed at the exact level of the weir crest.

Weir Notch–The opening in a weir for the passage of water.

Well, Driven–A water well established by pounding a well point into the earth below the permanent water table. This is in contrast to a bored well established in rock-free soil by hand tools, powered augers, or auger buckets. See Well, Bored.

Well Water–Water which is drawn or pumped from a well, in contrast to ditch water, creek water, rain water, spring water, etc.

Well-bred–Designating a plant or animal which has a good pedigree.

Well-drained–Designating a soil (in humid regions) that normally contains only capillary moisture, in contrast to a swampy or water-logged condition which also contains some gravitational water.

Well-graded–Refers to a soil or soil material consisting of particles well-distributed over a wide range in size or diameter. Such a soil normally can be easily increased in density and bearing properties by compaction.

West Indian Locust–*Hymenaea courbaril*, family Leguminosae; a large, hardwood tree, bearing white flowers and woody pods. The brown wood is used for construction, the pods yield an edible pulp, and a resin obtained from the tree has medicinal and wood-finishing value. Native to tropical America. Also called algarroba, courbaril.

West Indies Mahogany–*Swietenia mahogoni*, family Meliaceae; an evergreen tree grown for its hard, red wood, used especially in furniture. Native to tropical America.

Western Azalea–*Rhododendron occidentale*, family Ericaceae; a deciduous shrub grown for its white or pinkish, yellow-blotched flowers and yellow or scarlet, autumnal foliage. It may be poisonous to sheep and goats. Native to western North America.

Western Balsam Bark Beetle–*Dryocoetes confusus*, family Scolytidae; a beetle that infests spruce, fir, and pine in western and northwestern United States. It is capable of killing the trees by constructing a nuptial chamber and radiating egg galleries beneath the bark.

Western Barley–A market classification of barley which includes all white barley grown west of the United States Great Plains, and may not include more than 10 percent barley of other classes.

Western Bracken–*Pteridium aquilinum*, family Polypodiaceae; a fern which is poisonous to livestock. After about two months of feeding on it cattle become sick with very high temperature, difficult breathing, salivation, nosebleed and hemorrhage. Native to western North America. Also called brake fern, hog brake, bracken, bracken fern. See Bracken, Eastern Bracken.

Western Chokecherry–*Prunus virginiana* var. *demissa*, family Rosaceae; a deciduous tree whose leaves are poisonous to livestock. It is found in the United States from the Dakotas west to California on prairies and dry slopes. Symptoms of poisoning are uneasiness, staggering, convulsions, difficulty in breathing, rolling eyes, and tongue hanging out. Then the animal becomes quiet, bloats, and dies within an hour after ingestion.

Western Corn Rootroom–*Diabrotica virgifera*, family Chrysomelidae; a beetle whose larva is sometimes injurious to corn and other grain crops in Colorado, United States. Also called Colorado corn rootworm.

Western Fescuegrass–*Festuca occidentalis*, family Gramineae; a grass found on dry, rocky, wooded slopes from Michigan, United States, to British Columbia, Canada, whose herbage is highly palatable to stock. Its forage rating is high.

Western Fir–See Douglas-fir.

Western Larch–*Larix occidentalis*, family Pinaceae; a deciduous tree, important as a source of wood used as ties, poles, in building construction, planing-mill products, boxes, crates, etc. Native to North America. Also called western tamarack.

Western Pine Beetle–*Dendroctonus brevicomis*, family Scolytidae; a most important insect pest of ponderosa pine and coulter pine along the United States West Coast, causing great losses each year. The adults construct egg galleries between the bark and sapwood.

Western Pitch Pine–Ponderosa pine.

Western Potato Flea Beetle–*Epitrix subcrinita*, family Chrysomelidae; a serious insect pest which infests Irish potato, tomato, and related plants, as well as apple, ash, arbutus, bean, beet, cabbage, carrot, celery, clover, corn, cucumber, holly, honeysuckle, lettuce, maple, muskmelon, pumpkin, phlox, radish, rhubarb, spinach, sunflower, viburnum, violet, and watermelon. The young adult feeds on the foliage riddling the leaves. Especially serious in western United States.

Western Ragweed–*Ambrosia psilostachya*, family Compositae; a perennial herb which is a weed pest and whose pollen often causes hay fever. Native to North America. Also called perennial ragweed.

Western Range–The native livestock grazing areas of the Great Plains, the Rocky Mountains, the Intermountain and the Pacific Coast regions of the United States.

Western Sand Cherry–Bessey cherry; *Prunus besseyi*, family Rosaceae.

Western Snowberry–*Symphoricarpos occidentalis*, family Caprifoliaceae; a deciduous shrub which is a fair browse plant for sheep, goats, and cattle on the western range, and is also grown as an ornamental. Native to North America. Also called buckbrush, wolfberry.

Western Spotted Cucumber Beetle–*Diabrotica undecimpunctata*, family Chrysomelidae; a beetle which feeds on almost every plant except conifers. The larva feeds on roots of corn, sweetpea, and grass, and the adults mine holes in ripening fruit.

Western Spruce–Sitka spruce; *Picea sitchensis*, family Pinaceae.

Western Striped Cucumber Beetle–*Acalymma trivitata*, family Chrysomelidae; a beetle whose adult feeds on foliage of cucumbers, muskmelons, pumpkin, squash, etc., while the larva feeds on their roots. It is found on the West Coast and Arizona, United States.

Western Tamarack–See Western Larch.

Western Tussock Moth–*Hemerocampa vetusta*, family Lymantriidae; a moth whose caterpillar feeds on the leaves and fruit of apple, almond, apricot, blackberry, Brazil peppertree, California buckthorn cherry, hawthorn, oak, pear, prune, plum, walnut, and willow. Sometimes called California tussock moth.

Western Waterhemlock–*Cicuta occidentalis*, family Umbelliferae; an extremely poisonous, perennial herb. Symptoms of poisoning in stock are frothing at the mouth, uneasiness, pain, convulsions, kicking, throwing back of the head, rigid legs, bellowing, and spasmodic contractions of the diaphragm. Death follows ingestion rather rapidly. Native to western North America.

Western Wheatgrass–*Agropyron smithii*, family Gramineae; a tall, coarse, smooth-stemmed grass bearing whitish flowers; widely cultivated as a forage crop. Native to western United States. Also called bluestem, bluestem wheatgrass.

Western White Fir–White fir.

Western White Pine–*Pinus monticola*, family Pinaceae; an evergreen tree important for its wood used for matches, boxes, construction, millwork, fixtures, etc. Native to western Canada and United States.

Western Yarrow–*Achillea lanulosa*, family Compositae; a perennial herb of fair forage value. It is grazed somewhat on the western ranges, United States.

Western Yellow Pine–See Ponderosa Pine.

Wet–(1) To dampen; to sprinkle; to furnish water to a plant. (2) Characterized by rain, as wet weather. (3) Moist, covered, soaked, or saturated with water.

Wet Down–(1) To water a plant so that the soil surrounding its roots is soaked. (2) To bring soil to field capacity by rain, as it wet down six inches.

Wet Feet–(1) A condition in plants which results when the watering is in excess of drainage. (2) A tree or other plant which has its roots in an excessively wet soil.

Wet Pocket–A low area in a field, woods, etc., in which water collects and remains for considerable periods of time.

Wet-weather Blight–A sporadic cotton disease caused by the fungus *Ascochyta gossypii*, family Sphaeropsidaceae, which occurs in serious proportions in cool, moist weather. It is characterized by small, circular, white spots on the cotyledons and leaves. The lesions enlarge, coalesce, turn brown, and become rough. The spots then fall out leaving a ragged appearance and sometimes there is defoliation. Often the lesions girdle the stem and kill the terminal buds, resulting in the loss of the stand. In the summer the lesions are from ½ to 1 inch long, occurring at the branch axils and are ragged at the edges. Also called Ascochyta blight, Ascochyta canker, wet-weather canker, Ascochyta leaf spot.

Wettable Powder (Dust)–Any material manufactured in the form of a powder or dust that can be mixed with or dissolved readily in water.

Wettable Sulfur–That sulfur so treated as to be readily miscible in water. It is widely used in spray mixture, especially for its fungicidal effect.

Wetting Agent–Material included in pesticide solutions that reduces surface tension and helps to completely cover the surface or foliage area of the plant being sprayed. See Spreader.

Weymouth Pine–See Eastern White Pine.

Wheat–(1) Any of the several plant of the genus *Triticum*, family Gramineae. Only a few species, however, are important. (2) *T. aestivum*, an annual herb which is one of the most important crop plants of the world, furnishing grain, one of the principal sources of human food, animal feed, and manufactured products, such as alcohol and glucose. The wheat grain is not commonly fed to livestock in the United States but the by-products of milling are used extensively.

Wheat Beetle–(1) *Silvanus surinamensis*; a small, reddish-brown insect whose larvae feed on cereal crops, especially wheat. (2) Drugstore beetle. (3) Any grain or flour beetle.

Wheat Belt–A region in which wheat is the principal crop. In the United States, it extends from northern Texas and western Oklahoma to North Dakota and the Canadian border. The southern part of the United States Wheat Belt is the winter wheat belt and the northern part is the spring wheat belt.

Wheat Bran–A livestock feedstuff which is the coarse outer covering of the wheat kernel as separated from cleaned and scoured wheat in the usual process of commercial milling. Also called bran.

Wheat Brown Shorts–A livestock feedstuff which consists mostly of the fine particles of wheat bran, wheat germ, and very little of the fibrous material obtained form the tail of the mill. It is obtained in the commercial milling of wheat and contains less than 7.5 percent of crude fiber. Also called brown shorts, wheat red shorts.

Wheat Germ–The embryo of the wheat seed, which is rich in vitamin B_6.

Wheat Germ Oil Cake–A livestock feedstuff which is obtained after oil has been partially expressed from commercial wheat germs. It contains more than 29 percent of protein.

Wheat Gray Shorts–A livestock feedstuff which consists of the fine particles of the outer bran, the inner bran or bee-wing bran, the germ, and the offal or fibrous material obtained from the tail of the mill. It is obtained in the commercial milling of wheat and contains less than 6.0 percent crude fiber. Also called total shorts.

Wheat Jointworm–*Harmolita tritici*, family Eurytomidae; a larva of a fly that is a serious pest of wheat in the United States. The fly lays eggs in the stem tissues of a wheat plant near a joint. The larvae develop and feed inside the stem, weakening it and making it apt to lodge.

Wheat Land–Land ordinarily planted to wheat or which is productive when planted to wheat.

Wheat Midge–*Sitodiplosis mosellana*, family Itonididae; a small insect which lays eggs among the bracts of wheat. The larvae feed on the kernels, causing wilting, discoloration, and lodging. Destructive to wheat in Europe and the United States.

Wheat Poisoning–A condition seen in cattle grazing on lush wheat pasture wherein the animals show a posterior paralysis, tend to drop down, and become semicomatose.

Wheat Red Dog–A livestock feedstuff which is a by-product obtained in commercial flour milling. It consists principally of aleurone with small quantities of wheat flour and fine wheat bran particles and contains less than 4.0 percent crude fiber. Also called red dog flour, wheat red dog flour.

Wheat Shorts–See Wheat Standard Middlings.

Wheat Standard Middlings–A livestock feedstuff which consists mostly of fine particles of wheat bran, wheat germ, and very little of the fibrous material obtained form the tail of the mill. Obtained in the usual commercial process of milling, it contains not more than 9.5 percent crude fiber.

Wheat Stem Maggot–*Meromyza americana*, family Chloropidae; a maggot which infests wheat, rye, barley, oats, bluegrass, and timothy. It tunnels in the stems causing the heads to dry out and whiten while the lower stem and leaves are still green.

Wheat Stem Sawfly–*Cephus cinctus*, family Cephidae; a serious insect pest of wheat which also infests spring rye, barley, spelt, timothy, and some grasses. It tunnels into the stem causing excessive lodging.

Wheat Straw–The dried, mature wheat stem and leaves used for roughage, bedding in barns, and as poultry litter.

Wheat White Shorts–A livestock feedstuff which consists of a small portion of the fine wheat bran particles, the wheat germ, and a large portion of the fibrous offal obtained from the tail of the mill. Obtained in the usual process of flour milling, it contains no more than 3.5 percent crude fiber.

Wheat Wireworm–*Agriotes mancus*, family Elateridae; a beetle whose larva infests wheat, generally feeding on roots of mature plants. Sometimes it eats the seed in the ground or the underground parts of the small plant, causing it to wither and die.

Wheatgrass–Any of the perennial bunchgrasses of the genus *Agropyron*, family Gramineae. In many sections of the northwestern and the Rocky Mountain regions of the United States the wheatgrasses constitute the most important forage species.

Wheel Track Planting–Planting a crop, as corn, on plowed fields, in the tractor wheel tracks. The weight of the wheels crush and firm the soil. The rough surface left reduces the amount of potential soil erosion.

Wheeler Bluegrass–*Poa nervosa*, family Gramineae; a perennial grass of good forage value, found in open woods, dry meadows, and on old burned-over areas at medium altitudes. Native to western North America.

Whie-lined Sphinx–*Celerio lineata*, family Sphingidae; the larva of a moth which feeds on apple, rhododendron, beet, collard, currant, elm, fuchsia, gooseberry, grape, melon, pear, plum, portulaca, prune, tomato, turnip, and other plants. It is common in western United States.

Whip–(1) Any of several kinds of instruments for lashing animals, usually consisting of a handle and a lash. (2) An unbranched shoot of a woody plant, particularly the first year's growth. (3) A tall, slender tree. (4) To lash an animal. (5) To beat cream, egg white, etc., so that it becomes a froth and holds shape.

Whip Graft(ing) (Whip-and-Tongue Graft)–A method of grafting, usually at the collar or on the roots of small seedling trees, as in the propagation of nursery stock, or in joining small branches as in topworking. The stock is cut off with a long slanting cut in which a vertical slit, forming a kind of tongue, is made. The cuts on the lower end of the scion are made to complement those on the stock so that it is possible to insert the tongue of one in the slit of the other. Tying and waxing may or may not be done.

Whiptail–A molybdenum-deficiency disease of cauliflower characterized by malformation of the leaves and unsatisfactory curd formation. The disease is more prevalent on acid soils.

Whiskers–The grayish-white fungous threads (properly, sporangiophores) of the bread-mold fungus, *Rhizopus nigricans*, family Mucoraceae, which grow out of starchy, vegetable matter such as bread, sweet potatoes, spoiling squash, or pumpkin, etc. At their tips, they bear the round, dark sporangia, or spore cases, within which the numerous spores are produced.

White Ash–*Fraxinus americana*, family Oleaceae, a large, deciduous tree grown for its lumber used for handles, furniture, vehicle parts, sporting goods (bats, oars, paddles, etc.), and other articles. It is one of the better street trees in places where the soil is favorable. Native to eastern North America.

White Bud Disease–A zinc-deficiency disease of corn characterized by yellow streaks on old leaves which become necrotic. Young leaves unfolding in the bud may be white or yellow.

White Burley–A type of tobacco for manufacturing and export classification which is thin-bodied, light-colored, and cured.

White Clover–*Trifolium repens*, family Leguminosae; a perennial herb which spreads rapidly by stolons and is seldom unduly injured by mowing or grazing. It is grown as an important pasture plant and is used in lawn mixtures. Native to Europe. Also called Dutch clover, white Dutch clover. See Ladino Clover.

White Corn–(1) Corn with white kernels. See Yellow Corn. (2) A marketing class of corn which should not contain more than 2 percent corn of colors other than white. (3) Wheat, oats, barley (England).

White Falsehellebore–*Veratrum album*, family Liliaceae; a hardy, perennial herb sometimes grown in a shady border or in a wild garden for its greenish flowers which are white inside. Young livestock and chickens may be fatally poisoned by eating the plant. Symptoms of poisoning are salivation, retching, purging, weakness, general paralysis, spasms, rapid threadlike pulse, lowered temperature, shallow respiration, cold skin, blindness, and death. Native to Europe and northern Asia.

White Grubs–White, curled larvae of May beetles, family Scarabaeidae, which are found in the soil. They feed on the roots of various plants, including those of forest trees.

White Ironbark Eucalyptus–*Eucalyptus leucoxylon*, family Myrtaceae; a tall tree, with smooth, pale to dark gray bark, which is of value in California, United States, as a lumber tree, bee tree, and an ornamental. Native to Australia. Also called fat cake, white ironbark.

White Locoweed–See Lambert Crazyweed.

White Lupine–*Lupinus alba*, family Leguminosae; an annual herb grown to some extent for soil improvement. Native to Europe and Asia.

White Mustard–*Brassica hirta*, family Cruciferae; an annual herb grown to some extent for its edible leaves used in salads and as a potherb. Escaped, it can become a weed pest. Native to Europe and Asia.

White Oak–*Quercus alba*, family Fagaceae; a roundheaded, important deciduous, timber tree, which is one of the slowest growing and longest lived trees in the eastern United States. It is a magnificent tree, but it is not grown ordinarily as an ornamental due to difficulty in transplanting. The commercial uses of its wood are extensive.

White Peach Scale–*Pseudaulacaspis pentagona*, family Diaspididae; an armored scale insect found on the United States East Coast, which infests ceanothus, alder, poplar, dogwood, shadbush, and tuliptree. Also called West Indian peach scale.

White Pine–See Eastern White Pine.

White Root Rot of Apples–A fungal disease of apples caused by *Corticium galactinum*, family Thelephoraceae, characterized by a white or cream-colored layer of mycelia on the root surface, and by a knotting or gnarling of the roots.

White Sage–*Salvia apiana*, family Labiatae; a shrubby herb grown for its white-hairy, dense foliage. It is an important bee plant. Native to California, United States. Also called greasewood.

White Snakeroot–*Eupatorium rugosum (E. urticaefolium)*, family Compositae; a perennial herb grown for its branched clusters of white flowers which last from midsummer to frost. A poisonous plant, it is found wild from eastern North America to Minnesota and Texas. Symptoms of poisoning in livestock are trembling of the legs, depression, inactivity, constipation, nausea, rapid breathing, and collapse. The poison may be transmitted to humans by milk from a cow which has eaten the plant. Also called richweed, white sanicle, white snakeweed, Indian sanicle.

White Spruce–*Picea glauca*, family Pinaceae; a hardy, evergreen tree, which is grown as an ornamental. Its wood is used in pulp, cooperage, boxes, general building construction, planing-mill products, furniture, ladders, etc. Native to North America. Also called catpine, skunk spruce.

White Swallowwort–*Cynanchum vincetoxicum*, family Asclepiadaceae; a perennial herb which is a weed in northeastern United States. Cattle, horses, and sheep are reported to have been poisoned from eating it. Native to Europe.

White Sweetclover–*Melilotus alba*, family Leguminosae; a biennial herb, which has a wide range of adaptation and which is grown for pasture, hay, silage, green manure, and as a bee plant. Native to Europe. Also called biennial white sweet clover, Bokhara clover. See Hubam Clover.

White Tip of Rice–A chlorosis of the rice plant attributed to a deficiency of magnesium.

White-fringe Fungus–A fungus, *Fusarium aleyrodis*, used in the biological control of citrus whitefly, cloudy-winged whitefly, purple scale, and perhaps other scale insects.

White-fringed Beetles–*Graphognathus*, family Curculionidae; insect pests which are general feeders, having been observed on over 385 different species of plants. They infest a plant in vast numbers on the lower part of the stem or taproot causing the plant to yellow, wilt, and die.

White-marked Tussock Moth–*Hemerocampa leucostigma*, family Lymantriidae; a moth whose larva feeds on the foliage of deciduous trees: elm, linden, maple, horsechestnut, poplar, sycamore, buckeye, willow, apple, pear, quince, plum, etc., skeletonizing the leaves and scarring the fruit.

White-pine Aphid–*Cinara strobi*, family Aphidae; an aphid which infests the eastern white pine east of the Mississippi River, sometimes killing small trees. A sooty mold develops in the honeydew secreted by the aphids.

White-pine Blister Rust–A destructive, fungal disease of eastern and western white pines, caused by the rust *Cronartium ribicola*, family Melampsoraceae. Hyphal threads of the fungus penetrate the sapwood and inner bark, causing white blisters to appear on the bark. The uredial and telial stages of this rust occur on currants and gooseberries. Occurs in eastern and Pacific United States and Europe. Also called blister rust of pine.

White-pine Cone Beetle–*Conophthorus corniperda*, family Scolytidae; a beetle whose larva is destructive to cones of the eastern white pine, feeding on the scales, seeds, and other parts of the cone. Occurs in Canada and northern United States.

White-pine Sawfly–*Neodiprion pinetum*, family Diprionidae; the larva of an insect that infests the eastern white pine and sometimes other pines in northeastern United States.

White-pine Weevil–*Pissodes strobi*, family Curculionidae; an insect that infests eastern white pine and other pines and spruces in eastern United States. The leader on an infested plant becomes brown in contrast to the rest of the tree as a result of girdling by the weevils.

Whitecedar (Atlantic)–*Chamaecyparis thyoides*; an evergreen, swamp tree whose wood is used for poles, shingles, woodenware, planing-mill products, water tanks, boats, boxes, fences, etc. Native to eastern United States. Also called white cedar, Atlantic white cedar, southern white cedar.

Whitecedar (Northern)–*Thuja occidentalis*; an evergreen swamp tree native to Michigan, southern New York, central Vermont, and New Hampshire. Used for fence posts and poles. The foliage is good winter browse for deer. Also known as aborvitae.

Whitefly–A minute, sucking insect of the family Aleyrodidae, which is, for the most part, tropical.

Whitestem Gooseberry–*Ribes inerme*, family Saxifragaceae; an almost spineless shrub, occurring in the Rocky Mountains and northwestern United States, at 2,000 to 9,500 feet. Sheep, goats, and cattle browse on it when more palatable species are not available. Its food value rating is fair.

Whole Grain–(1) Cereal grain which is not cracked or chopped. (2) Those grains which retain their natural form and all their natural constituents. Whole-kernel corn.

Whole Stillage–The undried "bottoms" from the beer well comprised of nonfermentable solids, distiller's solubles, and the mashing water.

Whorl–(1) A swirl, cowlick, in the hair on an animal's coat. (2) Three or more twigs, flowers, leaves, or floral parts, arranged in a circle at one point on a plant.

Wick Applicator–See Weed Wiper.

Wild Bees–Bees living in hollow trees or other abodes not prepared for them by people, as distinguished from the bees raised in a hive.

Wild Black Cherry–Black cherry, ***Prunus serotina***, family Rosaceae; A tree valued for lumber used in building furniture. The fruit is eaten by many songbirds. The leaves are toxic to some animals.

Wild Blueberry–Any of the several unnamed and uncultivated species of blueberry, genus ***Vaccinium***, family Ericaceae, which grow in acid soils.

Wild Calla–*Calla palustris*, family Araceae; an aquatic herb grown along the edge of a pond in mud or shallow water for its oval or heart-shaped leaves and green flowers. It is poisonous for livestock. Symptoms of poisoning are an intense burning feeling in the mucous membranes of the mouth and throat. Native to the north temperate zone. Also called water arum.

Wild Carrot–*Daucus carota*, family Umbelliferae; an annual or biennial herb, which is a serious weed pest but also a beautiful flower plant. Eaten by cows, it imparts an objectionable flavor to the milk. Native to Europe and Asia. Also called bird's nest, lace-flower, devil's plague, bee's-nest plant, Queen Anne's lace.

Wild Garlic–*Allium vineale*; a perennial weed pest of pastures and hay fields. The plants develop from basal bulbs. The weed gives an off-flavor to milk if eaten by dairy cows. Also known as wild onion.

Wild Hay–Hay made from native or wild, uncultivated grasses and plants.

Wild Leek–*Allium tricoccum*, family Liliaceae; a bulbous, perennial herb, which is a weed pest. Cows feeding on it give milk which has an onion off-flavor. Also called ramps.

Wild Oat–*Avena fatua*, family Gramineae; an annual grass, distributed over most of the United States, and which is a weed pest in small grains. It is grazed on the western ranges and is sometimes used for hay. Its general forage rating is good when pastured or harvested in the immature stages. Native to Europe and Asia.

Wild Onion–*Allium veneale*; a noxious perennial weed pest of lawns, pastures, and grain fields. The plants give milk an off-flavor if eaten by dairy cattle. Also known as wild garlic, field garlic.

Wild Parsnip–*Pastinaca sativa*; a yellow-flowered weed that is frequently confused with poison hemlock and water hemlock. It is somewhat poisonous, but not serious.

Wild Pasture–A pasture of native or volunteer grasses and plants. See Range.

Wild Radish–*Raphanus raphanistrum*, family Cruciferae; an annual herb which is a weed pest in grain fields. Stock feeding on large quantities of the seed mixed with grain may become poisoned. Characteristics of the illness are enteritis, hemorrhagic diarrhea, colic, abortion, nephritis, apathy, and paralysis of heart and respiratory organs.

Wildfire–(1) A bacterial leaf-spot disease of tobacco and soybeans caused by ***Pseudomonas tabaci***, family Pseudomonadaceae, characterized on tobacco by yellow lesions with white spots of dead tissue in the center. In soybeans, it is characterized by light brown to dark necrotic spots each surrounded by a yellow, sometimes indistinct, halo. Severe infestation can cause severe shedding. (2) The promiscuous, indiscriminate, or accidental burning of vegetation. It involves no responsibility for damage to property resulting from escape of the flames.

Wildgoose Plum–*Prunus munsoniana*, family Rosaceae; a variety of plum with fruit yellowish or red with a slight bloom; it is of little commercial importance. Native to south-central and southeastern United States.

Wildling–(1) An escape; a cultivated plant which has become wild and thrives without cultivation. (2) Any wild plant or animal. (3) In forestry, a seedling produced naturally outside of a nursery.

Willow Oak–*Quercus phellos*, family Fagaceae; a deciduous tree grown for shade and ornament in warm areas. Native to North America.

Willow Sawfly–*Nematus ventralis*, family Tenthredinidae; the larva of a sawfly which infests willows and poplars, feeding on the leaves. Trees are sometimes defoliated. Also called yellow-spotted willow slug.

Wilt–(1) Loss of freshness and a drooping of the foliage of a plant due to an inadequate supply of moisture, excessive transpiration by the plant, and as a result of a disease interfering with the utilization of water by the plant. (2) Of a plant, to lose freshness of the foliage and droop, or to become flaccid as a result of inadequate soil moisture, disease, or its being cut. (3) The temporary or transient loss of turgidity in a plant, caused by a rate of transpiration in excess of the rate of absorption of water. See Wilting Coefficient.

Wilt-resistant–Designating a plant variety which does not readily become infected with a wilt disease.

Wilted Silage–Forage that is cut and partially field-cured to reduce moisture content before chopping for silage.

Wilting Coefficient–The moisture, in percentage of dry weight, which remains in the soil within the root-feeding zone as plants reach a condition of permanent wilting. Also know as wilting percentage, permanent wilting point.

Wilts–Plant diseases in which the plant becomes limp and droops, generally because invading organisms obstruct the flow of sap. *Erwinia tracheiphila* is the cause of cucumber wilt, *Xanthomonas campestris* of cabbage wilt, and *Bacterium stewartii* of corn wilt.

Wind Burn–(1) A withered, seemingly blasted, appearance of foliage which results from the action of violent wind on a plant. Fruit may also exhibit mechanical damage from exposure to wind while being shipped, or from dehydration. (2) The bronzed condition of some evergreen foliage caused by an inadequate water supply usually during late winter. It is associated with high transpiration rate and frozen soil or low soil-water content.

Wind Erosion–The removal of the soil by action of the wind. This is a serious problem in arid areas.

Wind Shadow–That portion of a scarp or slope which is protected form the direct action of the wind blowing over it.

Wind Strip Cropping–The production of crops in long, relatively narrow strips which are placed crosswise of the direction of the prevailing winds without regard to the contour of the land.

Wind-borne–Designating diseases, seeds, spores, etc., distributed by air movements. Also called air-borne.

Wind-firm–Designating a tree which withstands a heavy wind.

Wind-pollinated–Designating those plants which have the pollen carried by the wind instead of by insects, as the pines, grasses, etc.

Windbreak–(1) Any object which serves as an obstacle to free movement of surface winds. In forestry, trees serve such a purpose. Tree windbreaks are classified according to their general arrangement as rows and hedgerows, belts and shelterbelts of three or more rows, and groves, or in the most extensive case, forests. They may be of natural or artificial origin. (2) A vegetative barrier used to reduce or check the force of the wind. See Shelterbelt.

Windfall–(1) A tree uprooted or broken off by wind. (2) An area on which the trees have been blown down by wind. (3) A fruit that falls from a tree before the crop is harvested, usually associated with wind. Also called drop. (4) Unexpected income or tax break.

Windmill Grass–The stiff, spreading-radiating, spiked species of the genus *Chloris*, family Gramineae, some of which are weeds but have a fair forage rating.

Winesap–A widely known variety of apple bearing a large, red, round-oblate, good-flavored fruit which keeps well.

Wing–(1) A leaflike, dry or membranous expansion or appendage of a plant part, such as along some stems and petioles and of samaras and some capsules. (2) Either of the lateral petals of a pealike flower. An organ of flight for birds, insects, bats, etc. (3) A piece of bird meat consisting or the organ of flight with the feathers removed. (4) The outside corner of the cutting edge of a plow. (5) Fan or blower blade. (6) An extension on a building. (7) The vane of a windmill.

Winged Yam–*Dioscorea alata*, family Dioscoreaceae; an herbaceous vine widely grown in the Orient for its edible tuber and often in the United States as an ornamental. Native to eastern Asia. Also called white yam.

Winnow–To separate grain, etc., from the husks by means of an air current or by wind.

Winter Annual–An annual plant whose seed germinates in the fall or early winter, overwinters in a vegetative condition, often in a rosette form, completes vegetative growth, matures a crop of seed, and then dies in the late spring or early summer.

Winter Cover Crop–Any of several plants, such as winter rye, wheat, or oats, planted in the early autumn. It makes sufficient growth before winter comes to aid in protecting the soil from wind and water erosion. In the spring it is usually plowed under and used as green manure.

Winter Cress–*Barbarea vulgaris*; a yellow-flowered weed pest in hay fields and pastures.

Winter Fat–*Eurotia lanata*, family Chenopodiaceae; a low shrub found from Texas, United States, and Manitoba, Canada, to the West Coast. One of the outstanding browse plants on the western range, its forage rating is excellent.

Winter Hardiness–The physiological adjustment within a plant that enables it to endure the rigors of winter (snow, freezing, ice) and to retain its vigor the next season.

Winter Injury–Any injury to plants which results from the rigors of winter; the death of plant parts due to drying during cold weather, especially when the soil is frozen deeply.

Winter Irrigation–The application of water to soil in late fall or early spring to store for subsequent use by plants.

Winter Melon–*Cucumis melo* var. *inodorus*, family Cucurbitaceae; a tender, annual herb which is a variety of muskmelon, grown for its sweet edible fruit. It keeps well in storage. Perhaps native to Turkey.

Winter Mulch–Mulch applied in the autumn for the protection of plants or plant parts during the winter months.

Winter Rape–*Brassica napus*, family Cruciferae; an annual herb grown in Europe for its seeds which yield the colza oil or rape oil of commerce, and in the United States as a cover crop and as a temporary pasture crop. It provides excellent forage over a long season, requiring no cultivation. Native to Europe and Asia.

Winter Rye–Rye planted in the late summer or early autumn, which endures winter weather and produces a crop the next spring or early summer.

Winter Sun Scald–Of trees, bark injury on the sun-exposed side of tree trunks resulting from periodic freezing of the tissues. It is characterized by loose, brown, dead bark, sometimes splitting and cracking. Eventually the bark peels away to produce an open wound or dead area on the trunk.

Winter Vegetable–(1) A vegetable produced during the winter, especially one grown in the warm winter sections of the country as contrasted to a hothouse vegetable. (2) A vegetable which stores well and is used during the winter.

Winter Wheat—The type of wheat that is planted in the fall, survives the winter as a young seedling, and matures in the early summer of the following year.

Wintering—The care, feeding, and maintenance of livestock and tender plants through the winter months.

Winterkill—Of plants, to die as a result of the rigors of the winter, as by low temperatures, snow, frost, etc.

Wireworm—The larva of the click beetle, family Elateridae, which eats seeds, underground stems, and small roots, and burrows into potatoes, carrots, etc.

Witch Grass—*Panicum capillare*, family Gramineae; an annual grass widespread throughout the United States, found in waste places, old fields, barnyards, etc. It is a weed pest. Also called old witch grass, tickle grass, fool-hay.

Witches'-broom—An abnormal type of growth of an herbaceous plant or of a branch of a woody plant in which buds normally remaining dormant begin to grow, forming a bushy, broomlike growth, usually with the leaves reduced in size. Usually caused by a fungus or an insect.

Witchweed—A pest under control action by the United States Department of Agriculture. Distribution in United States: North Carolina, South Carolina. Principal hosts: corn, sorghum, sugarcane, and other plants of the grass family.

Withe—Tough flexible branches such as willow or vines.

Wither—Of a plant, to become dehydrated drooped, and limp.

Witloof—See French Endive.

Wolf Plant—A plant that, though of a species generally considered palatable, is not grazed by livestock and becomes rank in growth with much accumulation of stem and leaf material. It competes with more-palatable plants.

Wolf Tree—A tree which occupies more space than its silvicultural value warrants, thus curtailing the space of better species.

Wolftail—(1) *Lycurus phleoides*, family Gramineae; a slender, perennial grass found from west Texas to Arizona, United States, and south to Mexico at elevations of 4,000 to 8,000 feet; its forage rating is good. (2) An old-time imaginary and fictitious disease of cattle. Also called wolf-in-the-tail.

Wood Alcohol—See Methanol.

Wood Preservative—Any of several chemical compounds used to reduce decay in wood by sealing the exposed surfaces or by penetrating the wood; e.g., creosote.

Wood Shavings—Very thin strips of wood resulting from planing boards. Sometimes used as bedding for animals, for a plant mulch, and for kindling.

Wood Sugar—Sugar made from the carbohydrate part of wood by hydrolysis. It is chiefly glucose, containing some xylose and possibly small amounts of galactose, manose, and arabanose.

Wood Technology—The study of wood in all its aspects; the science of wood, including its anatomy, chemistry, physical properties, treatment, and uses.

Wood Turpentine—The essential oil obtained from pine stumps or other resinous wood by destructive or steam distillation.

Wood-wool—A wood product resembling wool, made from waste pulp or fine hairlike shavings. Used primarily as insulating and packing material.

Wooded—Concerning, or designating an area full of trees.

Woodgate Rust—A rust of the Scotch pine caused by the aecial stage of a form of *Cronartium quercuum*, which produces large, round galls on the woody twigs and young branches of the plant. Occurs in New Hampshire, New York, and Michigan, United States.

Woodland Management—The management of existing woodlands and plantations that have passed the establishment stage including all measures designed to improve the quality and quantity of woodland growing stock and to maintain litter and herbaceous ground cover for soil and water conservation. It includes all such measures as planting, improvement-cutting, thinning, pruning, slash disposal, fire protection, and grazing control.

Woodland Pasture—(1) Farm woodlands used for grazing. (2) Wooded areas with grass and other grazing plants growing in open spaces among trees.

Woodlot—A tract of woodland, a part of a farm and usually only a few acres in size, which may have a single or multiple use, as a source of fire wood and logs for lumber, as a woodland pasture, and as a means of providing food and protective cover for wildlife.

Woody—(1) Designating a plant which has some characteristics of wood. (2) Designating a forested or woodland area. (3) Designating a plant with a high content of lignin.

Woody Plant—Any shrub, tree, or certain vines, as distinguished from herbaceous plants, which produces wood and has buds surviving above ground during the winter.

Woolly—Clothed with long, matted hairs.

Woolly Apple Aphid—*Erisoma lanigerum*, family Aphidae; a widely distributed aphid which infests apple, pear, hawthorn, elm, and mountain ash. Infestation is characterized by white cottony masses over masses of purplish aphids in wounds on the trunk and branches. Seriously infested trees may be stunted, retarded, or killed.

Woolly Croton—*Croton capitatus*, family Euphorbiaceae; an annual herb which, as hay, is sometimes poisonous to stock. Native to North America.

Woolly Elm Aphid—*Erisoma americanum*, family Aphidae; an aphid which infests the American elm. Infested trees are characterized by curled or rolled leaves.

Woolly Loco—*Astragalus mollissimus*, family Leguminosae; a poisonous plant, found from South Dakota to Arizona, United States. Both its green and dried foliage are poisonous to livestock. Symptoms of poisoning are dullness, irregularity of gait, lack of appetite, dragging

Woolly Whitefly–Yeast

of the feet, a solitary habit, loss of flesh, and shaggy coat. As the animal ceases to eat, it dies. Also called purple locoweed, Texas loco, woolly crazyweed.

Woolly Whitefly–*Aleurothrix floccosus*, family Aleurodidae; an insect pest of citrus which has been controlled by internal parasites.

Woollyleaf Loco–*Astragalus leucophyllus*, family Leguminosae; a perennial herb which is poisonous to livestock. For symptoms of poisoning see Woolly Loco.

Woolypod Milkweed–*Asclepias eriocarpa*, family Asclepiadaceae; a perennial herb found in California, United States, which is very poisonous to sheep and cattle. Symptoms of poisoning are depression, loss of appetite, diarrhea, low temperature, and rapid pulse. Also called broad-leaved milkweed.

Wooton Loco–*Astragalus wootoni*, family Leguminosae; a poisonous plant, found in southern New Mexico and western Texas, United States, and in northern Mexico. Both its green and dried foliage are poisonous. Symptoms of poisoning are dullness, irregularity of gait, lack of appetite, dragging of the feet, a solitary habit, loss of flesh, and shaggy coat. As the animal ceases to eat, it dies. also called locoweed. See Loco.

Work Over–To bud or graft an individual tree or group of trees in an orchard to other horticultural varieties.

Worm–(1) Any small, soft-bodied, usually limbless animal, such as a larva, grub, maggot, earthworm, silkworm, etc. (2) To rid an animal of internal parasitic worms.

Wormseed Goosefoot–*Chenopodium ambrosioides*, family Chenopodiaceae; an annual herb grown as a source of the chenopodium oil used as medicines and vermifuges. Escaped, it is a weed pest and, if ingested, poisons livestock. Native to tropical America. Also called Mexican tea.

Wort–The liquid remaining from a brewing mash preparation following the filtration of fermentable beer.

Wound–Any violently caused disruption of the continuity of an internal or external tissue.

Wound Parasite–Any parasite which is not able to enter a host except through wounds or injured tissue.

Wrapper Leaf–In marketing, a type of tobacco used for the outer covering in the manufacture of cigars.

Xanthophyll–$C_{40}H_{56}O_2$; a dark brown compound, usually associated with chlorophyll and carotene in plants, which imparts a golden-yellow to ivory color to tissues depending on concentration. Xanthopyhll, or a related compound, is responsible for the color of yellow flower petals and yellow to orange fall foliage.

Xenia–Immediately observable effect in the sperm of seed fertilized with foreign (different) pollen. Common in corn but occurs also in small grain, sorghum, peas, beans, and flax.

Xeric–A soil and climatic environment with a low moisture content or supply. As applied to plants, xeric means those plants that are tolerant of very dry soil or arid conditions. Other words in the same series are hydric and mesic. See Hydric, Mesic.

Xerophyte–Plant which habitually grows where the evaporation stress is high and the water supply is low, as in desert regions, where the soil dries to the wilting coefficient at frequent intervals.

Xerophytic Vegetation–Vegetation which is characteristic of the desert regions, such as thorny brush, cacti, shrubs, and small flowering annual and perennial plants.

Xylem–The "plumbing" system that conducts water and dissolved mineral up the stems from the roots. It is part of the fibrovascular system of a vascular plant that includes the nonliving tracheids and vessels, the immediately associated living parenchyma cells, and the supporting and protective wood fibers. In plants with woody stems, the xylem lies inside the cambium. Xylem is the tissue that makes up most, if not all, of the lumber wood and wood products of commerce. See Phloem.

Yam–The edible root of any of the plants of the genus *Dioscorea*; a sweet potato is sometimes erroneously called a yam.

Yarovization–See Vernalization.

Yarrow–*Achillea millefolium*, family Compositae; a hardy perennial herb, a weed pest in unworked fields, roadsides, etc., generally in North America. It is grazed moderately by stock, and its forage rating is fair. Native to Europe, Asia, and North America. Also called milfoil, thousand-leaf.

Yeast–(1) A yellowish substance composed of microscopic, unicellular fungi of family Saccharomycetaceae, which induces fermentation in juices, worts, doughs, etc. The fungi may reproduce asexually by the division and separation of the cells into two equal parts, as in the fission yeasts, or more frequently, in the budding yeast, by the budding out of small outgrowths which enlarge until they attain full size and then separate form the parent cell. Most yeasts produce enzymes which bring about the fermentation of carbohydrates to produce alcohols and carbon dioxide. The most common of such yeasts is *Saccharomyces cerevisiae* whose fementary process produces the carbon dioxide which leavens bread dough, as well as the alcohol used for beverages (beer, wines, etc.). This is the so-called bread-yeast or brewers'-yeast. In yeast inoculation of a substance, the surface or top yeast ferments it at a faster rate than the sedimentary yeasts at the bottom. (2) The asporogenous yeasts, family Torulopsidaceae, in which the production of internal spores (ascospores) is lacking. Some of them (especially in the genus *Candida*) are parasites of people or animals and may cause various diseases, such as thrush. In recent years, especially in Germany, some yeasts have been cultivated on sugar-containing derivatives of sawdust and have become valuable sources of animal feed. Also called pseudo-yeast.

Yellow Berry of Wheat–A nitrogen-deficiency disease in red wheat characterized by the normally hard and flinty grains becoming partially or entirely starchy in composition and light yellow.

Yellow Birch–*Betula alelghaniensis*, family Betulaceae; a hardy deciduous tree which is important for its timber used in veneer, hardwood distillation, railroad ties, slack-cooperage, furniture, planing-mill products, woodenware, scientific instruments, etc. Native to the Lake States and the northeastern United States. Also known as birch, gray birch, silver birch, and swamp birch.

Yellow Buckeye–*Aesculus octandra*, family Hippocastanaceae; a hardy, deciduous tree grown for shade and in lawns. Its leaves are poisonous to stock and its nutlike seeds have poisoned children. Symptoms of poisoning are inflammation of the mucous membranes, vomiting, depression, weakness, and paralysis. Native to North America. Also called sweet buckeye.

Yellow Corn–(1) Corn which has yellow kernels in contrast to the other common color, white. The yellow corn contains carotenoid pigments which can be converted into vitamin A in the animal body. (2) A marketing class for corn which may not contain more than 5 percent grains of any color other than yellow.

Yellow Cypress–See Baldcypress.

Yellow Dwarf Mosaic of Grasses–A viral disease of barley and certain other cereals and grasses characterized by bright yellow blotches developing from the leaf tip throughout the leaf blade, stunting and producing excessive tillering in the plant, which causes reduced root and spike development, and reduced kernel formation.

Yellow Dwarf of Onion–A viral disease named after the two conspicuous symptoms which it produces: yellowing and dwarfing of the plant. It is transmitted by aphids and leafhoppers, not by seed.

Yellow Dwarf of Potato–A viral disease characterized by dwarfing, yellowing of the foliage, rugosity (roughness) and rolling upward of the leaves, and stem, leaf, and tuber necrosis.

Yellow Fungus of Citrus Whitefly–*Aschersonia goldiana*, family Zythiaceae; a fungus that destroys the citrus whitefly in Florida and the West Indies. Its pure culture in sweet potato strips is used to introduce the fungus into whitefly infested groves.

Yellow Grain Sorghum–A market classification of grain sorghum which includes all varieties with yellow and salmon-pink grains. it may not include more than 10 percent grain sorghums of other colors.

Yellow Indiangrass–*Sorghastrum nutans*, family Gramineae; a tall perennial grass found in the United States from the Atlantic Coast to Montana, and in Mexico, on open prairies, in open woods, and on dry slopes. Its forage rating is excellent.

Yellow Mealworm–*Tenebrio molitor*, family Tenebrionidae; the larva of a beetle which infests stored grain. It may also infest squabs (young pigeons), eating the skin at the neck and vent.

Yellow Monkshood–*Aconitum lutescens*, family Ranunculaceae; a perennial herb found from Idaho to New Mexico, United States, which may be poisonous to livestock feeding on it.

Yellow Mosaic of Bean–Usually a more virulent form of mosaic than the common bean mosaic which differs in not being transmissible in the seed. Also, yellow mosaic occurs on some varieties of bean that are resistant to the common mosaic. Both are transmissible by aphids.

Yellow Nutsedge–*Cyperus eslculentus*; a hard to control, perennial weed grass that grows in the southern coastal plains. The yellow-green plant reproduces by rhizomes, seeds, and tubers.

Yellow Scale–*Aonidiella citrina*, family Diaspididae; a major scale insect pest of citrus.

Yellow Wildindigo–*Baptisia tinctoria*, family Leguminosae; a perennial, tumbleweed-type herb grown for its bright yellow flowers. Stock feeding on it in the wild have been poisoned. Native to North America. Also called clover broom, wild indigo, false indigo.

Yellow Woollybear–*Diacrissia virginica*, family Arctiidae; a woolly, yellowish caterpillar, up to two inches long, which feeds voraciously on field and garden plants in summer and autumn. It is distributed generally over the United States.

Yellow-striped Armyworm–*Prodenia ornithogalli*, family Noctuidae; a cutworm that infests cotton by feeding on young plants, boring into squares and bolls. However, it feeds generally on many different plants. Also called cotton cutworm.

Yellows–(1) Any plant disease in which yellowing (or chlorosis) and stunting are principal symptoms. Also called xanthoses. (2) Jaundice in people, domestic animals, as cattle, horses, sheep, etc.

Yellows of Walnut–A zinc-deficiency disease of Persian walnuts characterized by a yellowish-green mottle between the lateral veins of the leaves; a narrow, small leaf; bunching of the leaves; a tendency of the leaves to stand erect; and death of twigs and small branches.

Yew–Any one of the several evergreen, hardy shrubs or trees of the genus *Taxus*, family Taxaceae. Species are valuable, slow-growing ornamentals and are among the most popular of all evergreens. The wood, bark, leaves, and seed are very poisonous.

Yield–(1) Grade in meat animals, referring to the amount of lean meat produced in a carcass. (2) The quantity of or aggregate products resulting from cultivation or growth, as a yield of 30 bushels of wheat per acre. (3) The percentage of clean wool remaining in a lot after scouring. (4) The ratio of carcass weight to live weight. (5) To produce products as a result of cultivation or growth, as a tree yields fruits.

Yield Goal—In soil testing, the estimated and desired crop yield level that is set for the purpose of recommending lime an d fertilizer to reach that particular level of yield.

Young Soil—Recently deposited or exposed soil material on which the soil profile development has begun but is in the initial stages. See Soil Horizon, Soil Profile.

Zanja—An irrigation ditch (southwestern United States).

Zanjero—An irrigator (southwestern United States).

Zebra Caterpillar—*Ceramica picta*, family Noctuidae (Phalaenidae); a caterpillar that feeds generally in late summer on flowers, gardens, fruit crops, and on shade trees.

Zein—An alcohol-soluble protein, a prolamin, which is found in the corn grain.

Zero Pasture—Cutting green forage and hauling it to stock in corrals or in dairy barns in lieu of pasturing. Sometimes called zero grazing.

Zero Tillage—See Conservation Tillage.

Zinc—Zn; a metallic chemical element, one of the micronutrient elements in soils, essential for both plant and animal growth. Toxic to animals if ingested in too large a quantity. See Micronutrient Elements, Essential.

Zinc Nitrate—$ZnNO_3$; a highly soluble compound used in solution form as a source of zinc in liquid fertilizers.

Zinc Oxide—ZnO; a compound used as a fungicide, a seed fumigant, a source of zinc in livestock rations, and as a fertilizer.

Zinc Oxysulfate—A soluble form of zinc, also known as basic zinc sulfate. It can be added to fertilizers or applied as a spray or dust to correct zinc deficiencies. It contains 52 percent zinc. See Micronutrient Fertilizer.

Zone of Saturation—The zone of groundwater, the upper surface of which is the water table. Water from this zone feeds springs, streams, and wells. Generally, it is beneath the depth of penetration of plant roots.

Zone of Weathering—The surface crust of the earth in which weathering, or subaerial decay of rocks, takes place. Lying above the water table, it includes soil and the parent materials of soil.

Zoospore—An asexual, motile reproductive cell, which, by repeated divisions of the protoplast, develops into a new plant; a means of reproduction of many green algae, lower fungi, and some olive-brown algae.

Zoyzia—*Zoyzia* spp.; any one of five species of warm-weather lawn grasses that reproduce by creeping rhizomes.

Zygote—A fertilized ovum or egg, it is the diploid cell formed from the union of the sperm with an ovum.

Zymogenic Flora—Lower forms of plant life that reproduce in abundance in the soil following the application of large amounts of rapidly biodegradable organic matter such as animal manures, compost, and sewage effluent.

Zymurgy—A branch of applied chemistry dealing with fermentation.